Look for this icon throughout the book. It refers to the following Animated Tutorials and Activities on the Student Website.

E indicates an Experiment Tutorial. * indicates an activity that applies to an entire chapter.

(Continued on inside back cover)

Life The Science of Biology

Seventh Edition

Sinauer Associates, Inc. 

W. H. Freeman and Company 

Life
The Science of Biology
SEVENTH EDITION

William K. Purves *Emeritus, Harvey Mudd College • Claremont, California*

David Sadava *The Claremont Colleges • Claremont, California*

Gordon H. Orians *Emeritus, The University of Washington • Seattle, Washington*

H. Craig Heller *Stanford University • Stanford, California*

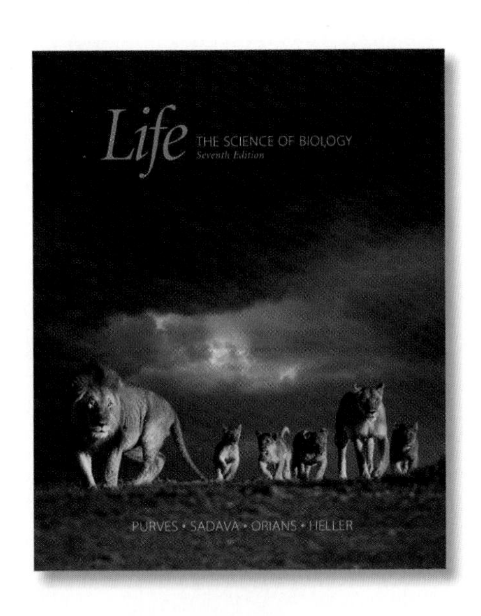

The Cover and Title Photos

The three images of African wildlife on the cover, half-title page, and frontispiece are © Steve Bloom/stevebloom.com.

Life: The Science of Biology, Seventh Edition

Address editorial correspondence to:
Sinauer Associates, Inc., 23 Plumtree Road, Sunderland, MA, 01375 U.S.A.
www.sinauer.com
email: publish@sinauer.com

Address orders to:
VHPS/W.H. Freeman & Co., Order Department, 16365 James Madison Highway, U.S. Route 15, Gordonsville, VA 22942 U.S.A.
www.whfreeman.com
Examination copy information: 1-800-446-8923
Orders: 1-888-330-8477

Library of Congress Cataloging-in-Publication Data

Life, the science of biology / William K. Purves ... [et al.].-- 7th ed.
p. cm.
ISBN 0-7167-9856-5 (hardcover) – ISBN 0-7167-5808-3 (Volume 1)
ISBN 0-7167-5809-1 (Volume 2) – ISBN 0-7167-5810-5 (Volume 3)
ISBN 0-7167-8679-6 (Volume 4)
1. Biology. I. Purves, William K. (William Kirkwood), 1934-

QH308.2 .L565 2003
570--dc22 2003022294

Printed in U.S.A.
First Printing January 2004
Courier Companies Inc.

To our students, especially the 25,000 we have collectively instructed in introductory biology over the years

About the Authors

David Sadava Bill Purves Gordon Orians Craig Heller

Bill Purves is Professor Emeritus of Biology as well as founder and former chair of the Department of Biology at Harvey Mudd College in Claremont, California. He received his Ph.D. from Yale University in 1959 under Arthur Galston. A fellow of the American Association for the Advancement of Science, Professor Purves has served as head of the Life Sciences Group at the University of Connecticut, Storrs, and as chair of the Department of Biological Sciences, University of California, Santa Barbara, where he won the Harold J. Plous Award for teaching excellence. His research interests focus on the chemical and physical regulation of plant growth and flowering. Professor Purves elected early retirement in 1995, after teaching introductory biology for 34 consecutive years, in order to turn his skills to writing and producing multimedia for introductory biology students.

David Sadava is the Pritzker Family Foundation Professor of Biology at the Keck Science Center of Claremont McKenna, Pitzer, and Scripps, three of The Claremont Colleges. Professor Sadava teaches and has taught courses to undergraduates in biotechnology, biochemistry, cell biology, molecular biology, plant biology, introductory biology, and cancer biology. In addition, he has taught courses in cancer to nonacademic staff members. He is a visiting professor in the Division of Biology at Caltech, and a visiting scientist in medical oncology at the City of Hope Medical Center. He is the author or coauthor of five books on cell biology and on plants, genes, and crop biotechnology. His research has resulted in over 50 papers, many coauthored with undergraduates, on topics ranging from plant biochemistry to pharmacology of narcotic analgesics to human genetic diseases. For the past decade, he and his collaborators have investigated multi-drug resistance in human small-cell lung carcinoma cells with a view to understanding and overcoming this clinical challenge.

Gordon Orians is Professor Emeritus of Biology at the University of Washington. He received his Ph.D. from the University of California, Berkeley in 1960 under Frank Pitelka. Professor Orians has been elected to the National Academy of Sciences and the American Academy of Arts and Sciences, and is a Foreign Fellow of the Royal Netherlands Academy of Arts and Sciences. He was President of the Organization for Tropical Studies, 1988–1994, and President of the Ecological Society of America, 1995–1996. He is a recipient of the Distinguished Service Award of the American Institute of Biological Sciences. Professor Orians is a leading authority in ecology, conservation biology, and evolution. His research on behavioral ecology, plant–herbivore interactions, community structure, and environmental policy has taken him to six continents. He now devotes full time to writing and to helping apply scientific information to environmental decision-making.

Craig Heller is the Lorry I. Lokey/Business Wire Professor in Biological Sciences and Human Biology at Stanford University. He earned his Ph.D. from the Department of Biology at Yale University in 1970, and then spent two years as a Postdoctoral Fellow at Scripps Institute of Oceanography studying how the brain regulates body temperature in mammals. Dr. Heller has taught at Stanford since 1972, served as Director of the Program in Human Biology, Chairman of the Biological Sciences Department, and Associate Dean of Research. Dr. Heller is a fellow of the American Association for the Advancement of Science and a recipient of the Walter J. Gores Award for excellence in teaching. His research focus is on the neurobiology of sleep and circadian rhythms, mammalian hibernation, the regulation of body temperature, and the physiology of human performance. Over the years, Dr. Heller has done research on systems ranging from sleeping college students to diving seals to hibernating bears to exercising athletes.

Preface

Like populations, textbooks evolve. Not only must the content change, but our goals must be rethought as well. We have tried, in this Seventh Edition of *Life*, to emphasize those things that will best prepare students for their future careers. The store of biological knowledge increases ever more rapidly. This requires us to seek a careful balance between thoroughness of coverage and appropriate treatment of the process, or processes, of science. We have retained and expanded the emphasis on experiment—on how things were and are learned. The emphasis remains on concepts. However, because different instructors emphasize different topics, and because a key role of the textbook is as a "place to look things up," this book is comprehensive as well. We provide sufficient detail to meet most needs without making the book too voluminous. We have enhanced our emphasis on an evolutionary theme, and have added new material on such important cutting-edge topics as evolutionary developmental biology ("evo–devo") and earth systems.

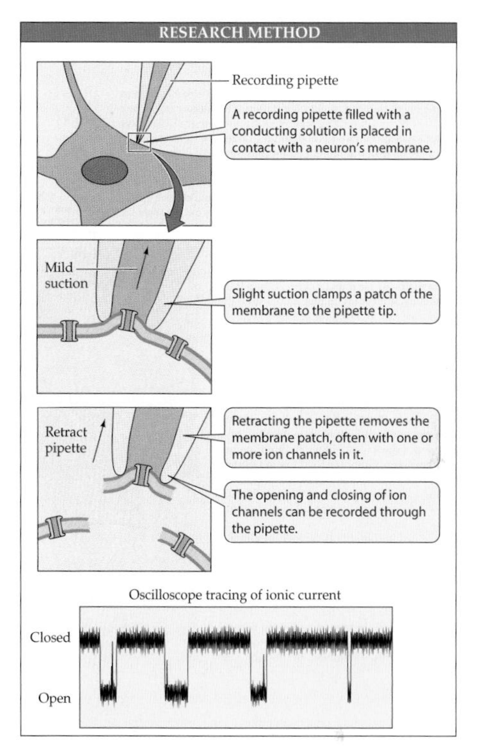

44.11 Patch Clamping The patch clamping technique can record the opening and closing of a single ion channel.

Experimental Focus

Since the First Edition of this book, we have been committed to answering the question, "How do we know?" As the book has evolved, this commitment has steadily deepened.

Obviously, we can't provide the experimental or observational evidence for every fact or theory we discuss. However, we have selected the key experiments underlying some of the most important biological principles. Some are very recent, at the cutting edge of current research; others are classics. To supplement and highlight the text discussions, we have created unique Experiment figures that show how experiments, field observations, and comparative methods help biologists formulate and test hypotheses. Other figures highlight some of the many laboratory and field methods used to do this research. In this edition there are more than 100 Experiment and Research Method figures (examples at right). In addition, we have 20 new Experiment tutorials on www.thelifewire.com, the Website/CD that was created for *Life*.

We hope that, in tandem with the frequent discussions of experimental evidence, these figures and tutorials will help students understand and appreciate the nature of biology as a vital, ongoing experimental science.

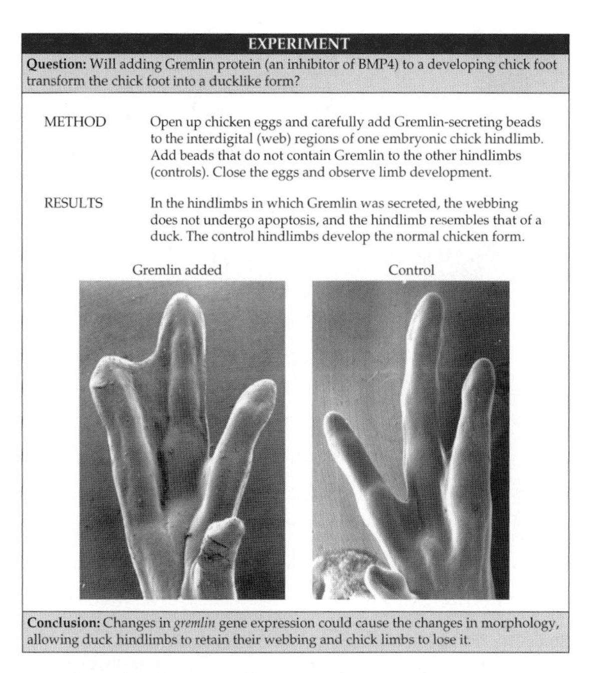

21.7 Changing the Form of an Appendage In this experiment, chick hindlimbs exposed to Gremlin-secreting beads developed ducklike webbed feet.

New Chapters, A New Unit, and New Essays

This edition features two new chapters that reflect current trends in biological research. Chapter 21 ("Development and Evolutionary Change") introduces students to evolutionary developmental biology, a rapidly growing field that deals with how the molecular genetics of the developing organism affects the evolution of complex morphology and biochemistry. In addition, the interaction of environment and embryogenesis on the ultimate form of an organism is covered at length.

Part Seven · THE BIOLOGY OF ANIMALS

41 Physiology, Homeostasis, and Temperature Regulation

The Tour de France, a 3-week, 3,500-km bicycle race, is arguably the most extreme and demanding of all athletic events. Competitors are on their bikes 5 to 7 hours a day, riding at an average speed of over 41 kilometers an hour across terrain that includes the mountains of the French Alps. The Tour can be compared to running 20 marathons at world-class pace in 20 days. In 2003, Lance Armstrong won the Tour for the fifth time.

How can an athlete perform at this level, and what results in a winning performance? A number of factors are involved, including determination, skill, and physiology. It is physiology that is the subject of Part Seven of this book. **Physiology** can be simply defined as the science of how organisms work. Physiological mechanisms span the range from molecular to behavioral.

France on 10,000 Calories a Day   Lance Armstrong is a remarkable athlete largely because of the capacity of his physiological systems.

You learned in earlier parts of this book that cells oxidize glucose to produce ATP, which is then used to do biological work, such as the contraction of muscles. Performance in an event such as the Tour de France is limited ultimately by the maximum sustainable rate at which the athlete's body can convert the chemical energy of food into the mechanical energy of muscles. That rate is determined by more factors than the cellular biochemical reactions you have studied. Oxygen has to be delivered to the blood, and the blood has to be pumped to the muscles and other organs. Food has to be converted to fuel molecules by the digestive system, and those fuel molecules have to be distributed to the mitochondria of the muscle cells. The waste products of cell metabolism have to be carried away and eliminated. The temperature, ion balance, and pH of muscles and other organs have to be maintained at optimal levels. All of these tasks and more are carried out by the physiological systems we will study in this part of the book. How does Lance measure up in terms of some of his physiological characteristics?

One measure of exercise capacity is the maximum rate at which a person can take up and utilize oxygen: the V_{O_2max}. For a healthy man, a typical value is about 40 ml O_2 per kg body mass per minute. Lance's V_{O_2max} is more than twice that value. Whereas a normal, fit man might burn up to 3,500 Calories on a particularly active day, during the Tour, Lance burns about 6,500 Calories a day—10,000 Calories on peak days! Because Lance has an extremely low proportion of body fat—only 4–5 percent (20 percent is normal)—he must eat, and his body must

Chapter 58 ("Earth System Science") introduces students to the new field focusing on Earth as a whole, studying great cycles of materials, inputs of solar energy, and the interactions between living organisms and the physical environment that determine how Earth as a planet functions today.

We have reorganized our treatment of developmental biology to create a new unit, Part Three, Development. Developmental biology is the subdiscipline of biology that draws from a great span of other subdisciplines, from molecular biology to ecology. Thus our new unit begins with updated chapters on "Differential Gene Expression in Development" and "Animal Development: From Genes to Organism," and concludes with the new Chapter 21. This unit leads naturally forward from Part Two, Information and Heredity, and progresses to a transition to Part Four, Evolutionary Processes.

We believe that another mission in the training of new scientists is to make them aware of the links between science and society. New for the Seventh Edition are eight essays, each of which concludes a Part of *Life*. We invited eight eminent humanists and social scientists to address a topic that bridges biology and an important ethical, moral, philosophical, or economic issue. For example, Bonnie Steinbock, SUNY Albany, examines some of the ethical implications arising from today's latest research activities in her essay, "What Are the Moral Issues Surrounding Stem Cell Therapy?" which concludes Part Three, Development.

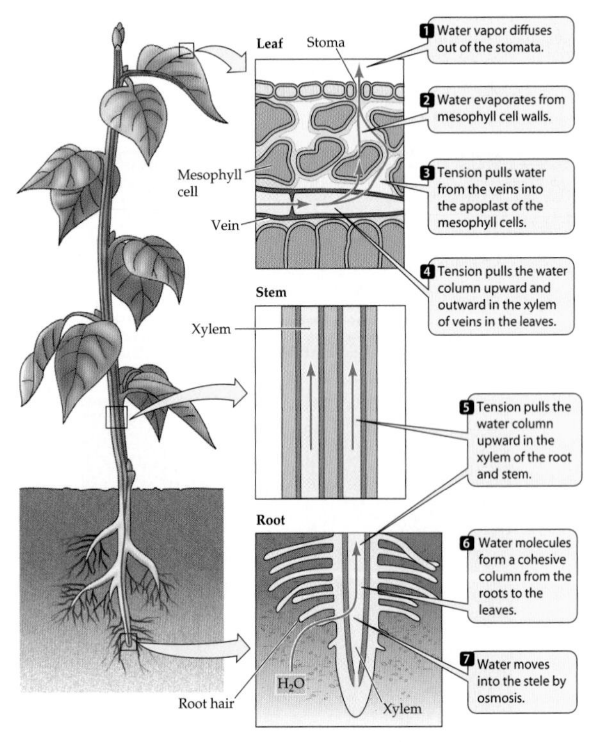

36.8 The Transpiration–Cohesion–Tension Mechanism
Transpiration causes evaporation from mesophyll cell walls, generating tension on the xylem. Cohesion among water molecules in the xylem transmits the tension from the leaf to the root, causing water to move from the soil to the atmosphere.

Pedagogy

In addition to our attention to updates and enhancements of content, we have again set as a major goal making the presentation as clear and helpful as possible for the student. Here are some of the ways in which we intend to make the reader's work easier and more effective:

- Every chapter begins with a short story (example above) to grab the reader's interest and encourage further exploration of the chapter's content.
- We have increased the number of bulleted lists (like this one) that highlight key points.
- All second-level headings (the workhorses) are declarative sentences that describe at a glance the text and figures that follow.
- We have added more interim summaries and bridges between topics to keep the reader on track.
- All chapter summaries include references to key figures and now also provide convenient links to the appropriate tutorials and activities on the Website/CD.
- Responding to adopters' requests, we have put the Self-Quizzes back in the text, since some students prefer to access them directly in the chapters.
- Each chapter concludes with four or five "For Discussion" items that help the reader synthesize the chapter's main concepts.
- The balloon captions (example at left) that we introduced in the Fifth Edition are now further streamlined and positioned for maximum pedagogical effectiveness.

Although our much-praised art program appears very similar to the Sixth Edition's, it has undergone a very significant pedagogical upgrade. Our new artist, Elizabeth Morales, who has been illustrating biology textbooks for more than 20 years, worked with each author to evaluate the effectiveness of every piece of line art in the book. The result is many hundreds of simplifications and improvements in clarity.

Media and New Video Collection

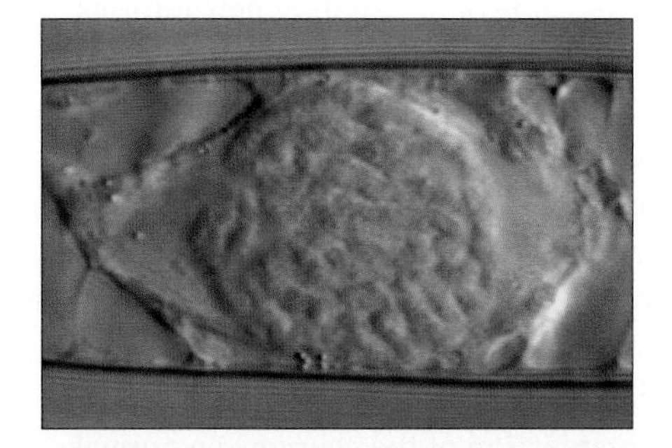

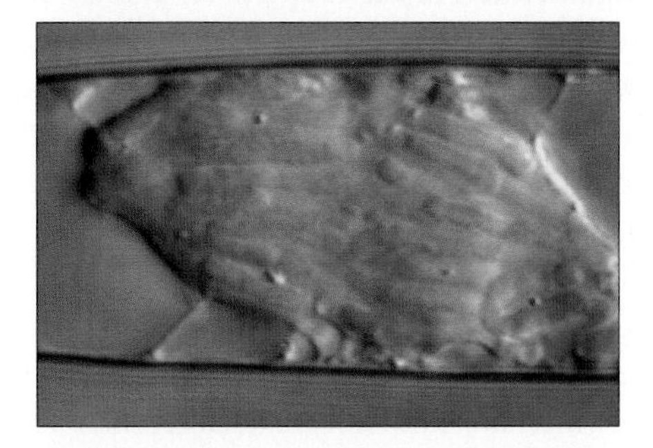

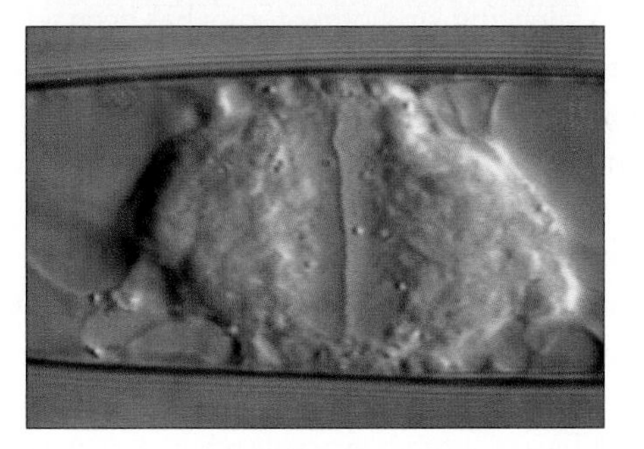

The Seventh Edition media and supplements are built around two main goals: (1) to provide the student with a collection of tools to help digest and truly understand the vast amount of material presented in this textbook; and (2) to provide the instructor with the richest possible collection of resources to aid in effectively teaching the course: preparing, presenting the lecture, providing course materials online, and assessing student comprehension.

Working with a dozen contributing authors and an experienced scientific multimedia studio, we have put together an outstanding package that is built specifically for this textbook. For example, the collection of over 100 in-depth animated tutorials was created using textbook art as the basis for the animations, and the introductions, conclusions, and quizzes were matched in level, terminology, and content to the Seventh Edition of *Life*.

In our continuing effort to provide instructors with outstanding visual resources for the lecture, and in response to many requests from biology professors, we are introducing a new feature for the Seventh Edition: "Seeing Life: Video Sequences in Biology." This collection of approximately 200 outstanding video segments (over two hours of footage that spans the book's coverage; example at right) can help capture the attention and imagination of your students with stunning moving images of biological phenomena. Each video segment is fully narrated.

(For a detailed description of all the media and supplements available for the Seventh Edition, please turn to "*Life*'s Supplements" on page xvi.)

The Eight Parts

Part One, The Cell, leads from basic chemistry to cell structure, membranes, and energetics. Chapters 3 ("Large Molecules") and 4 ("Cells: The Basic Units of Life") now integrate ideas on the origin and evolution of cells. The discussion of thermodynamics in Chapter 6 has been reduced and is now focused on biological applications. The art in the chapters on respiration and photosynthesis has been streamlined for clarity.

Part Two, Information and Heredity, retains the order of principles of genetics and molecular biology in the first chapters followed by applications of them in the later chapters. We have updated all of the material on genomics, and added newly emerging approaches of study, including RNA interference.

Part Three, Development, brings together and integrates topics in developmental biology to build upon the detailed treatment of genetics in Part Two and set the stage for the discussions of evolutionary processes in Part Four. We show how new insights into the ways in which genes and environment interact to yield the forms of adult organisms are providing important new perspectives on the origins of evolutionary novelties.

Part Four, Evolutionary Processes, begins with an overview of the history of life on Earth, followed by a detailed treatment of the evolutionary mechanisms and processes that are being investigated to explain those patterns. Chapter 25 ("Constructing and Using Phylogenies") has been updated to incorporate the most recent methods of inferring evolutionary relationships among organisms. Chapter 26 ("Molecular and Genomic Evolution") describes some of the exciting new information on how the genomes of organisms have evolved and how processes of genomic evolution help us understand the evolution of the diversity of life.

Part Five, The Evolution of Diversity, has been updated to reflect current views on phylogeny. We have retained the strong evolutionary thread, emphasizing lineages over some classically defined groups. The treatment of flowering plants other than monocots and eudicots has been upgraded. We have retained the organization of the chapters on animal diversity to reflect the three great lineages of animals—Lophotrochozoans, Ecdysozoans, and Deuterostomes—while incorporating new information on evolutionary relationships among animals revealed by new fossil finds and improved methods of inferring phylogenetic relationships.

In Part Six, The Biology of Flowering Plants, we have improved the explanations of bulk flow in xylem and phloem by modifying and simplifying the art and focusing the text more directly on key mechanisms. New material on clock genes, auxin carriers, RNA silencing, and other topics has been added.

Part Seven, The Biology of Animals, is about how animals work. Although we give major coverage of human physiology, we try to embed it in the background of comparative animal physiology. We have made an effort to offer a complete and broad coverage of physiology while still introducing the student to new advances. For example, the story of the molecular mechanism of the biological clock has advanced very rapidly since the last edition of this book. The genetic control of sexual behavior in fruit flies is another example where the two ends of the biological spectrum—molecular to behavioral—are coming together. As in previous editions, throughout Part Seven, we bring the student back to issues of control and regulation. These are the most central concepts in all of physiology.

Part Eight, Ecology and Biogeography, has been substantially reorganized and updated. Chapter 53 ("Behavioral Ecology") now emphasizes how the decisions that individual animals make can influence population dynamics and community structure. Chapter 54 ("Population Ecology") is organized around the key questions about populations that ecologists attempt to answer. Chapter 55 ("Communities and Ecosystems") combines and integrates material that was separated into two chapters in the Sixth Edition, providing a more integrated treatment of those topics. Chapter 57 ("Conservation Biology") now gives more emphasis on how science is used in the service of conserving Earth's biological diversity. Finally, the new chapter on "Earth System Science" (Chapter 58) introduces students to this rapidly developing field that looks at Earth as a whole.

Full Book or Paperbacks

We again provide *Life* both as the full book and as a cluster of paperbacks. For the Seventh Edition, the new Part Three is an additional fourth paperback. Thus, instructors who want to use less than the whole book, or who want their students to have more portable units, can choose from these split volumes:

- Volume I, The Cell and Heredity, includes: Part One, The Cell (Chapters 2–8); and Part Two, Information and Heredity (Chapters 9–18).
- Volume II, Evolution, Diversity, and Ecology, includes: Part Four, Evolutionary Processes (Chapters 22–26); Part Five, The Evolution of Diversity (Chapters 27–34); and Part Eight, Ecology and Biogeography (Chapters 53–58).
- Volume III, Plants and Animals, includes: Part Six, The Biology of Flowering Plants (Chapters 35–40); and Part Seven, The Biology of Animals (Chapters 41–52).
- Volume IV, Development, includes Part Three, Development (Chapters 19–21)

Note that Volumes I, II, and III include the book's front matter, Glossary, and Index plus Chapter 1.

There Are Many People To Thank

When we met in Sunderland with the key editorial and marketing people from Sinauer Associates and W. H. Freeman to plan this Seventh Edition, we determined that a central goal would be to involve and seek advice from a greater number of our teaching colleagues. This turned out to be a rich idea. We now have more than twice as many instructors to thank for their help in crafting this edition. We began the process by recruiting adopters of the Sixth Edition to report on what worked and what could be improved. With this input, we created the plan for the Seventh Edition and wrote the first drafts. Then, every chapter was reviewed by at least five introductory biology teachers. In addition to checking for accuracy and clarity, they helped us make decisions on material to cut or add. Many productive e-mail exchanges took place at this stage to the book's benefit.

After the chapters and final art were put into page proofs, we built in another round of reviews to help catch and eliminate lingering errors. This final check also provided suggestions for making the text and figures more precise. Finally, and concurrently with the manuscript and accuracy reviews, we got critical scrutiny of all of the animations, tutorials, and activities for the book's Website. We heartily thank all of the people who contributed these reviews. It's a demanding process but there is no doubt that a better book and supporting media have resulted because of it.

We wish to especially thank Scott Gilbert for providing an excellent draft of the new Chapter 21 on evolution and development.

As mentioned earlier, our new artist, Elizabeth Morales, has also made a very large contribution by assuring that *Life*'s extensive illustration program is as effective as possible. Many of the concepts that are illustrated in the book are complex. It takes an illustrator with Elizabeth's talent to render them both artistically and with maximum clarity.

The exact same team that worked with or within Sinauer Associates on the Sixth Edition on the many facets of editing the book was on board again. James Funston provided forceful and insightful developmental editing. Norma Roche contributed her elegant copy editing. Carol Wigg yet again deftly coordinated the entire editorial process and crafted many of the new captions. Jeff Johnson once again delivered the elegant interior and cover designs and coordinated the layout process. Susan McGlew orchestrated the mammoth reviewing process described above. And David McIntyre produced another dazzling array of new photographs for our selection.

W. H. Freeman's marketing and sales group has again succeeded in bringing *Life* to a wider audience. They are both effective ambassadors and skillful transmitters of information. We depend on their expertise and energy to keep us in touch with how *Life* is perceived by its users.

The constant asset we have had in our efforts to produce a better and better book to help students learn and appreciate the science of *Life* has been Andy Sinauer. For over 34 years Andy has run a company that produces the highest quality books in the biological sciences. His strategy has been to maintain a staff of talented, dedicated people and to give each book his personal attention from recruitment of authors to marketing. We feel that we have had more than our fair share of Andy's attention. He is the constant motivator to find ways to make our book and the teaching of biology more effective. He gently but firmly keeps us on track, and he is always ready to deal with the biggest crisis or the smallest detail. Andy, we are fortunate to work with you and we greatly appreciate all you have done to make *Life* a book of which we are exceedingly proud.

Bill Purves   David Sadava   Gordon Orians   Craig Heller

Reviewers for the Seventh Edition

Between-Editions Reviewers

Annalisa Berta, San Diego State University
Edward L. Braun, University of Florida
John Carlson, Yale University
Susan Cockayne, Brigham Young University
Craig Coleman, Brigham Young University
Karen Curto, University of Pittsburgh
Mark Decker, University of Minnesota
William Eickmeier, Vanderbilt University
William Eldred, Boston University
Ross Feldberg, Tufts University
Steven K. Fisher, University of California, Berkeley
Merrill L. Gassman, University of Illinois, Chicago
Hans-Willi Honegger, Vanderbilt University
Michele Igo, University of California, Davis
Dan Janik, University of Wisconsin, Eau Claire
Jeffrey Jensen, University of Maryland, College Park
Norman Johnson, Ohio State University
Rebecca Kimball, University of Florida
Todd Kostman, University of Wisconsin, Osh Kosh
Mary Lehman, Longwood University
Carl Luciano, Indiana University of Pennsylvania
Paula Mabee, University of South Dakota
David Magrane, Morehead State University
Charles Mallery, University of Miami
Shawn Meagher, Western Illinois University
Ken Mossman, Arizona State University
Darrel Murray, University of Illinois, Chicago
John New, Loyola University Chicago
Lou Pech, Carroll College
Lee Pike, East Tenessee State University
Don Potts, University of California, Santa Cruz
Eric Ribbens, Western Illinois University
Chris Romero, Front Range Community College, Larimer
Al Ruesink, Indiana University
Robert Savage, Williams College
Erik Scully, Towson University
Mark Storey, Texarkana College
James Traniello, Boston University
Nancy Wade, Old Dominion University
Raymond P. White, City College of San Francisco
Elizabeth Willott, University of Arizona

Manuscript Reviewers

Heather Addy, University of Calgary
Sylvester Allred, Northern Arizona University
Robert Angus, University of Alabama, Birmingham
David Armstrong, University of Colorado
Art Ayers, Albertson College of Idaho
Ellen Baker, Santa Monica College
Sharon Balchak, Notre Dame College
Monique Barakat, Stanford University
Ruth Beattie, University of Kentucky
Spencer Benson, University of Maryland, College Park
Andrew Blaustein, Oregon State University
J. Jose Bonner, Indiana University
Thomas Boyle, University of Massachusetts, Amherst
Bryan Brendley, Gannon University
George Brooks, Ohio University, Zanesville
Angela Brown, University of Idaho
James Brown, North Carolina State University
Patrick J. Bryan, Central Washington University
Matthew Buechner, University of Kansas
Art Buikema, Virginia Polytechnic Institute and State University
Warren Burggren, University of North Texas
Nancy T. Burley, University of California, Irvine
Rob Carey, Pima Community College
Clint Carter, Vanderbilt University
Clare Chatot, Ball State University
Helen C. Chuang, Southern Utah University
Elizabeth Conner, University of Massachusetts, Amherst
Ron Cooper, University of California, Los Angeles
Greg Crowther, University of Puget Sound
Sid Das, University of Texas, El Paso
Alan Day, University of Western Ontario
Mark Decker, University of Minnesota
Carmen Domingo, San Francisco State University
Ernest DuBrul, University of Toledo
Stephen Ebbs, Southern Illinois University, Carbondale
Alex Enyedi, Western Michigan University
Gordon L. Fain, University of California, Los Angeles
Paul Ferguson, University of Illinois, Urbana-Champaign
Steven K. Fisher, University of California, Berkeley
Gregory Florant, Colorado State University
Ellen Freund, National Marine Fisheries Service
Javier Gago, Glendale College
Merrill L. Gassman, University of Illinois, Chicago
Daniel Geiger, Natural History Museum of Los Angeles County
Michael Ghedotti, Regis University
Ken Gobalet, California State University, Bakersfield
Deborah Gordon, Stanford University
Dina Gould Halme, Massachusetts Institute of Technology
Dana Haine, Central Piedmont Community College
Leslie Hickok, University of Tennessee, Knoxville

Robert Hinrichsen, Indiana University of Pennsylvania
Christie Howard, University of Nevada, Reno
Andrew Jarosz, Michigan State University
Walter S. Judd, University of Florida
John Kalb, Canisius College
Larry Katz, Rutgers University, Cook College
Laura Katz, Smith College
Steve Kelso, University of Illinois, Chicago
Travis Knowles, Francis Marion University
Allen Kurta, Eastern Michigan University
Andrew Lack, Oxford Brookes University
Ralph Larson, San Francisco State University
Howard Laten, Loyola University Chicago
Carl Luciano, Indiana University of Pennsylvania
Paula Mabee, University of South Dakota
Barbara Mable, University of Guelph
Nancy Magill, Coe College
Charles Mallery, University of Miami
Jim Manser, Harvey Mudd College
Ron Markle, Northern Arizona University
Patrick H. Masson, University of Wisconsin, Madison
Paul Mayes, Muscatine Community College
Kenneth W. McCravy, Western Illinois University
Wayne Merkley, Drake University
Frank Messina, Utah State University
Darrel Murray, University of Illinois, Chicago
Bill Newcomb, Queens University
Gregory Nishiyama, College of the Canyons
Tom Oeltmann, Vanderbilt University
Laura Olsen, University of Michigan
Sanford E. Ostroy, Purdue University
Tom Owens, Cornell University
Julia Thom Oxford, Boise State University
Aparna D.N. Palmer, Mesa State College
M. Theresa Pavlovitch, Pasadena City College
Craig Peebles, University of Pittsburgh
Karen Perkins, Mount St. Mary's College
Jeff Pommerville, Maricopa Community Colleges
Ellen Porzig, Stanford University
Chris Romero, Front Range Community College
Donald Ruch, Ball State University
Al Ruesink, Indiana University
Walter Sakai, Santa Monica College
Stan Schein, University of California, Los Angeles
Daniel Scheirer, Northeastern University
Nicci Schoob, University of Illinois, Urbana-Champaign
Rodney J. Scott, Wheaton College
Neil Shay, University of Notre Dame
Jim Shinkle, Trinity University
Margaret Silliker, DePaul University
Philip Snider, University of Houston, University Park
Mitchell Sogin, Marine Biological Laboratory, Woods Hole
John Stiller, East Carolina University
Mark Sturtevant, University of Michigan, Flint
Cecil Stushnoff, Colorado State University
Kevin Swier, Chicago State University
Iain E.P. Taylor, University of British Columbia
Gerald Thrush, California State University, San Bernardino
Stephen Timme, Pittsburg State University
David Tissue, Texas Tech University
F. Daniel Vogt, Plattsburgh State University of New York
Jonathan Wenger, Concordia University, St. Paul
Lisa Werner, Pima Community College
David Wessner, Davidson College
Dave Westenberg, University of Missouri, Rolla
Mary White, Southeastern Louisiana University
Barny Whitman, University of Georgia
Elizabeth Willott, University of Arizona
Christopher Wills, University of California, San Diego
Michelle Withers, Louisiana State University
Jay Zimmerman, St. John's University, Queens

Accuracy Reviewers

Heather Addy, University of Calgary
Sylvester Allred, Northern Arizona University
Vernon Bauer, Francis Marion University
Wade Bell, Virginia Military Institute
Graeme Berlyn, Yale University
Michael Black, California Polytechnic State University
Andrew Blaustein, Oregon State University
Franklyn Bolander, University of South Carolina
Thomas Boyle, University of Massachusetts, Amherst
Patrick J. Bryan, Central Washington University
Matthew Buechner, University of Kansas
Art Buikema, Virginia Polytechnic Institute and State University
Warren Burggren, University of North Texas
Nancy Burley, University of California, Irvine
Naomi Capuccino, Carleton University
Domenic Castignetti, Loyola University Chicago
David Champlin, University of Southern Maine
Elisabeth Ciletti, Pasadena City College
Keith Clay, Indiana University
William Collins, Stony Brook State University of New York
Ronald Cooper, University of California, Los Angeles
Charles Creutz, University of Toledo
Mike Dalbey, University of California, Santa Cruz
Deborah Dardis, Southeastern Louisiana University
Gerald Deitzer, University of Maryland
Laura DiCaprio, Ohio University
Chuck Duggins, University of South Carolina
Nancy Elwess, Plattsburgh State University of New York
Ray Evert, University of Wisconsin
Marvin Friedman, Hunter College
Eve Gallman, University of Illinois, Urbana-Champaign
Charles Galt, California State University, Long Beach
Michael Ghedotti, Regis University
Diane Gorski, University of Wyoming

Brian K. Hall, Dalhousie University
Susan Han, University of Massachusetts, Amherst
Dennis C. Haney, Furman University
Mike Hart, Dalhousie University
Paul Hasegawa, Purdue University
Albert Herrera, University of Southern California
Hans-Willi Honegger, Vanderbilt University
David Jenkins, University of Alabama, Birmingham
Walter S. Judd, University of Florida
Thomas Kane, University of Cincinnati
Loren Knapp, University of South Carolina
William Kroll, Loyola University Chicago
Josephine Kurdziel, University of Michigan
Sandra F. Larson, Furman University
Howard Laten, Loyola University Chicago
Mary Lehman, Longwood University
Greg Lewis, Furman University
Min-Ken Liao, Furman University
Carol Maillet, Augustana College
Richard Malkin, University of California, Berkeley
Charles Mallery, University of Miami
Richard McCarty, The Johns Hopkins University
Jill Miller, Amherst College
Subhash Minocha, University of New Hampshire
Marty Nemeroff, Rutgers University
Seán O'Connell, Western Carolina University
Peter O'Day, University of Oregon
William H. Outlaw, Jr., Florida State University
Randall Packer, George Washington University
Craig Peebles, University of Pittsburgh
David Polcyn, California State University, San Bernardino
Ron Poole, McGill University
Leo Racich, Southern Illinois University, Edwardsville
Wendy Raymond, Williams College
Robert Reed, Southern Utah University
Eric Ribbens, Western Illinois University
Patricia Rugaber, Coastal Georgia Community College
Andy Schroeder, Harvard University
Stylianos Scordilis, Smith College
Tim Shannon, Francis Marion University
Margaret Silliker, DePaul University
Heidi Sleister, Drake University
Andrew Smith, Arizona State University
Jim Smith, Michigan State University
Philip Snider, University of Houston, University Park
Mitchell Sogin, Marine Biological Laboratory, Woods Hole
Frederick W. Spiegel, University of Arkansas, Fayetteville
Kevin Swier, Chicago State University
Doug Thrower, University of California, Santa Barbara
Briana Timmerman, University of South Carolina
Elizabeth Van Volkenburgh, University of Washington
Benjamin Weeks, Adelphi University
Ted Wilson, Winona State University
Greg Wray, Duke University

Media and Supplements Reviewers

Michael Black, California Polytechnic State University
Steve Brewer, University of Massachusetts, Amherst
Mark Browning, Purdue University
Bob Cabin, Plattsburgh State University of New York
Mark Decker, University of Minnesota
Ernest Dubrul, University of Toledo
William Eldred, Boston University
Joanne Ellzey, University of Texas, El Paso
James Franzen, University of Pittsburgh
Tejendra Gill, University of Houston
Jon Glase, Cornell University
Dominic Lannutti, El Paso Community College
Peter Lortz, North Seattle Community College
Charles Mallery, University of Miami
Philip Meneely, Haverford College
Nancy Raffetto, University of Wisconsin, Madison
Ken Robinson, Purdue University
Robert Schmidt, University of California, San Diego
Allen Shearn, The Johns Hopkins University
Thomas Terry, University of Connecticut
Lisa Werner, Pima Community College

To the Student

There are a few things you can do to help you get the most from this book and from your course. For openers, read the book actively—don't just read passively, but do things that force you to think as you read. If we pose questions, stop and think about them. Ask questions of the text as you go. Do you understand what is being said? Does it relate to something you already know? Is it supported by experimental or other evidence? Does that evidence convince you? How does this passage fit into the chapter as a whole? Annotate the book—write down comments in the margins about things you don't understand, or about how one part relates to another, or even when you find an idea particularly interesting. People remember things they think about much better than they remember things they have read passively. Highlighting is passive; copying is drudge work; questioning and commenting are active and well worthwhile.

"Read" the illustrations actively too. You will find the balloon captions in the illustrations especially useful—they are there to guide you through the complexities of some topics and to highlight the major points.

The chapter summaries will help you quickly review the high points of what you have read. They also identify particular illustrations that you should study to help organize the material in your mind. Add concepts and details to the framework by reviewing the text. Also in the summaries are keyed reminders of the tutorials and activities on the book's website. A way to review the material in slightly more detail after reading the chapter is to go back and look at the boldfaced terms. You can use the boldfaced terms to pose questions—and see if you can answer those questions. The boldfacing will probably be more useful on a second reading than on the first.

The "Self-Quiz" in each chapter is a convenient way to measure your mastery of the material. All answers are at the end of the book. Use the "For Discussion" questions that end each chapter. These questions are usually open-ended and are intended to cause you to reflect on the material.

The glossary and the index can help you a great deal. When you are uncertain of the meaning of a term, check the glossary first—it has more than 1,500 definitions. If you don't find a term in the glossary or you want a more thorough discussion of it, use the index to find where it's discussed.

The Website

The Seventh Edition Student Website (www.thelifewire.com; also available on CD at your instructor's request) is designed to help you learn the vast amount of material we are presenting in this book in a variety of ways. Throughout the book, you will see this icon on headings and figure titles. Wherever you see the icon, you will find a corresponding animated tutorial or activity on the website. They will reinforce your understanding of the key concepts presented in the book. Another important feature of the website is the extensive set of Interactive Quizzes. These quizzes incorporate figures from the book, thorough answer feedback, and links to electronic versions of book pages. There is a second quiz for each chapter, the Online Quiz, the results of which can be emailed to your instructor if he/she requests. Also on the website are key terms flashcards, a full glossary (with audio pronunciations of difficult terms), suggested readings, and two useful documents: Math for Life and Student Survival Skills. The website has been built with you in mind, we hope you find it to be an important resource in your study of introductory biology.

What If the Going Gets Tough?

Most students occasionally have difficulty in courses, including biology courses. If you find that you are slipping behind in the course, or if a particular topic is giving you an unreasonable amount of trouble, here are some useful steps you might take. First, the basics: attend class, take careful lecture notes, and read the textbook assignments. Second, note that one of the most important roles of studying is to discover what you don't know, so that you can do something about it. Use the index, the glossary, the chapter summaries, and the text itself to try to answer any questions you have and to help you organize the material. Make a habit of looking over your lecture notes within 24 hours of when you take them—find out right away what points are unclear, and get them straightened out in your mind. The website can help by providing an additional perspective.

If none of these self-help remedies does the trick, get help! Other students are often a good source of help, because they are dealing with the material at the same level as you are. Study groups can be very useful, as long as the participants are all committed to learning the material. Tutors are almost always helpful, as are faculty members. The main thing is to get help when you need it. It is not a good idea to be strong and silent and drift into a low grade.

But don't make the grade the point of this or any other course. You are in college to learn, to pursue interesting subjects, and to enjoy the subjects you are pursuing. We hope you'll enjoy the pursuit of biology.

Bill Purves   David Sadava   Gordon Orians   Craig Heller

Life's Supplements

For the Student

Student Website: www.thelifewire.com

(Also available as a CD optionally bundled with the book)

The *Life,* Seventh Edition Website offers the student a wealth of in-depth, self-directed review material. With the help of three contributing authors and an experienced scientific multimedia design studio, we've created a collection of resources that take advantage of the flexibility and interactivity of the electronic medium to help the student master the many complex concepts presented in the textbook. Features of the site include:

- **Interactive Summaries:** This is the most convenient way to review the entire chapter. The summary contains links to all the key figures from the chapter as well as all of the relevant animated tutorials and activities.
- **Animated Tutorials:** These 106 in-depth tutorials feature an introduction, a detailed animation, a conclusion, and a quiz on the topic covered. The clear presentation of complex topics makes these tutorials a powerful learning tool. New to the Seventh Edition are 20 **Key Experiment** tutorials that expand upon some of the most important experiments depicted in the book, and 10 additional new tutorials.
- **Activities:** Over 120 interactive activities help the student learn important facts through a wide range of exercises, such as labeling steps in processes or parts of structures, building a phylogenetic tree, and identifying different types of organisms.
- **Flashcards:** For each chapter of the book, there is at least one Flashcard activity that allows students to review and then test themselves on the key terminology from the chapter.
- **Interactive Quizzes:** Every question includes a figure from the book, thorough feedback on answer choices, and links to electronic versions of book pages, where the related material is highlighted for immediate reinforcement of concepts.
- **Online Quizzes:** An additional review tool, these quizzes can be taken online, and, at the instructor's request, students can submit their scores electronically.
- **Plus:** A full **Glossary** with audio pronunciations, **Suggested Readings**, and two useful study aid documents: **Math for Life** and **Student Survival Skills**.

Order ISBN 0-7167-5807-5, Seventh Edition Student CD, or ISBN 0-7167-8851-9, Text with CD

Study Guide

Edward M. Dzialowski, *University of North Texas;* Betty McGuire, *Smith College;* Lindsay Goodloe, *Cornell University;* Nancy Guild, *University of Colorado;* and Paula Mabee, *University of South Dakota*

For each chapter of the text, the Study Guide provides a detailed review of Important Concepts, a Big Picture overview, Common Problem Areas to pay particular attention to, Study Strategies, and a full set of Knowledge and Synthesis questions and Application questions, all with answers and explanations. Order ISBN 0-7167-5811-3

Lecture Notebook

This useful tool consists of all the artwork from the textbook (more than 1,000 images) presented in the order in which they appear in the text, with ample space for note-taking. Since the notebook has already done the drawing, students can focus their attention on the concepts. They will absorb the material more efficiently during class, and their notes will be clearer, more accurate, and more useful when they study from them later. Order ISBN 0-7167-5812-1

Animated Tutorial 11.1: The Meselson–Stahl Experiment

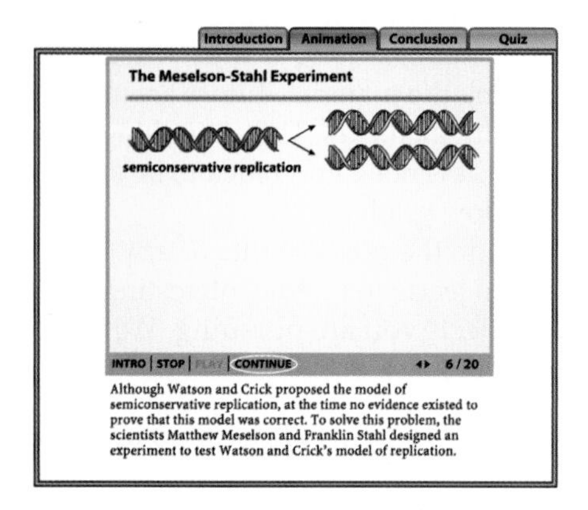

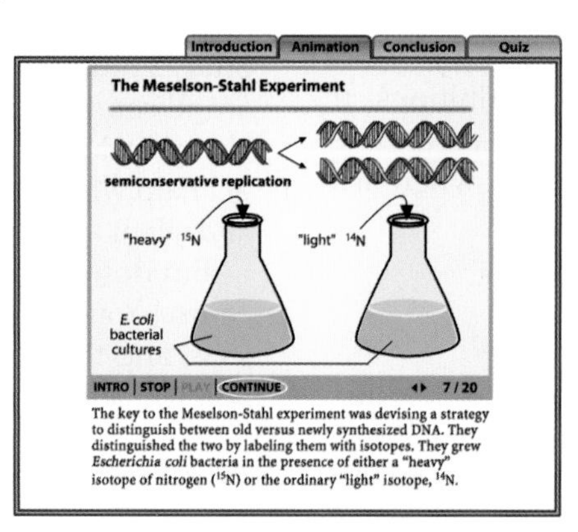

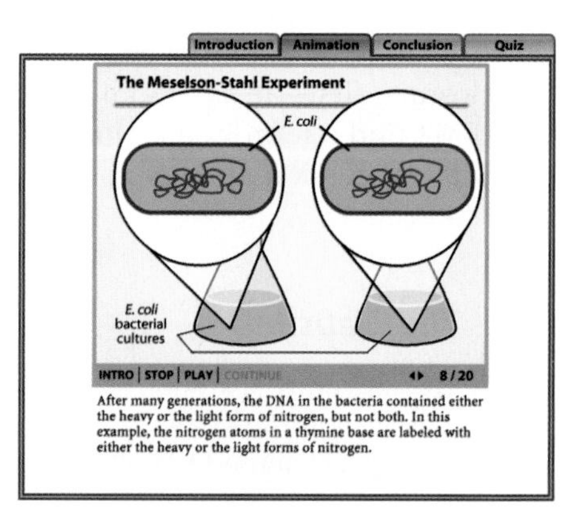

For the Instructor

Instructor's Media Library

As classrooms and introductory biology courses become more and more technically sophisticated, instructors are using an ever wider array of electronic resources in planning, managing, assessing, and in presenting the lecture. In order to give instructors the widest possible range of resources to help them engage students and better communicate the material, we have assembled an unparalleled collection of media resources. The Seventh Edition of *Life* features the most expansive Instructor's Media Library yet, available on a set of CDs and DVDs. The Media Library includes:

- **Textbook Illustrations, Photos, and Tables:** Every image from the textbook is provided in both JPEG (high- and low-res) and PDF formats. These high-quality images have all been optimized for use in PowerPoint® or other presentation software.
- **Unlabeled Figures:** New for the Seventh Edition, and in response to instructor requests, every figure in the textbook is now provided in an additional, unlabeled format, in which all text labels have been removed. These are a useful resource for student quizzing and custom presentation development.
- **Supplemental Photos:** This collection of over 1,500 photos (in addition to those in the text) is a rich resource of visual imagery. All are organized by chapter and include explanatory captions for easy insertion into lectures.
- **Chapter Outlines**, **Lecture Notes**, and the complete **Test File** are all available in Microsoft Word® format for easy use in lecture and exam preparation.
- **An Intuitive Browser Interface** provides a quick and easy way to preview all of the content in the Media Library.
- **Seeing Life: Video Sequences in Biology:** New for the Seventh Edition, this collection of approximately 200 video segments covering topics across the entire textbook helps demonstrate the complexity and beauty of life. The videos are available on both DVD (for stunning image quality) and CD (for use in PowerPoint or other software).

PowerPoint Resources

For each chapter of the textbook, the Media Library offers several different types of PowerPoint presentations. These give instructors the flexibility to build presentations in the manner that best suits their needs.

- **Lecture PowerPoints** are designed to form the basis of a complete lecture. They combine detailed text with selected figures to create a thorough presentation.
- **Complete Art and Photo PowerPoints** include every piece of art, photo, table, and supplemental photo placed onto slides, ready for use in presentations.

Layered PowerPoint 37.7: A Nodule Forms

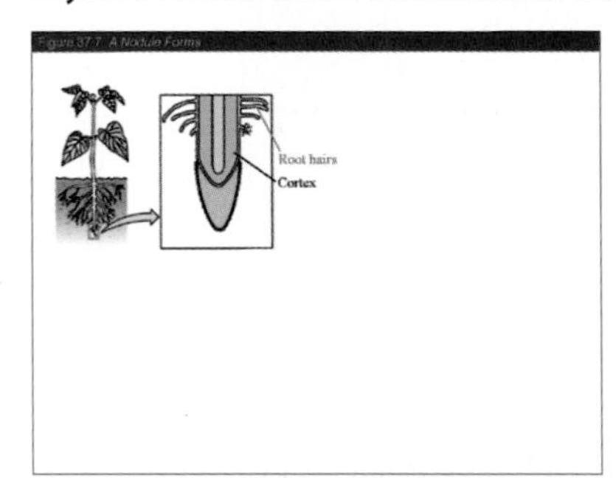

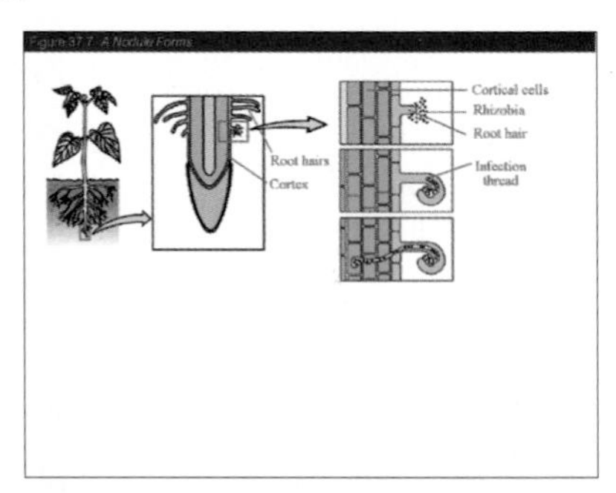

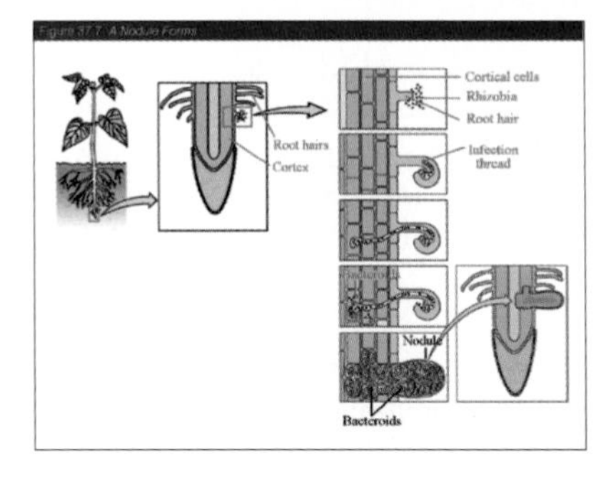

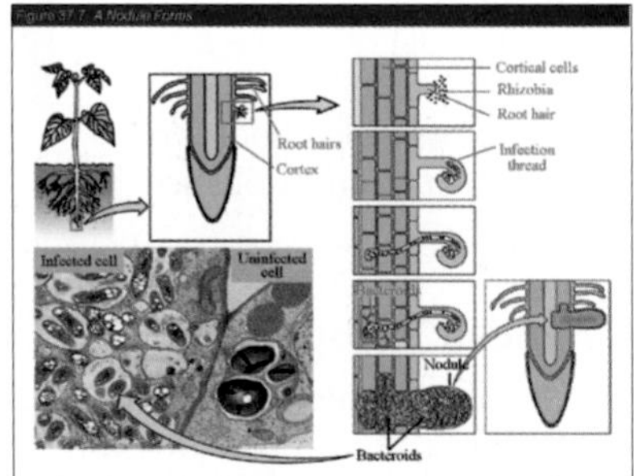

- **Layered Art PowerPoints** offer detailed, customizable presentations of key figures. These have been broken down into their component parts and are presented in a step-by-step, layered, and animated manner, allowing instructors to introduce elements of a figure at their own pace.
- **Video and Animation PowerPoints** include every video and animation in the Instructor's Media Library placed into PowerPoint, allowing instructors to simply copy the slides they want into their own presentations.

Video Sequence: The Venus Flytrap, *Dionaea* sp.

Instructor's Resource Kit

The Resource Kit is the central tool for lecture planning and development using the Seventh Edition of *Life.* The Kit combines several extensive instructor resources into one convenient binder:

- **Instructor's Manual:** The IM includes a chapter Overview, a "What's New" guide to the Seventh Edition, Key Concepts and Information, a Chapter Outline, and the list of Key Terms for each chapter.
- **Lecture Notes:** These detailed bulleted notes cover all the content presented in each chapter and are designed to be used as the basis for a complete lecture.
- **Media Guide:** New for the Seventh Edition, the Media Guide is a visual guide to all the media resources available in the Instructor's Media Library. Thumbnails and descriptions are provided for every animation, activity, video, supplemental photo, and lecture PowerPoint.
- **Custom Labs Information:** This is a list of all the lab separates available for instructors who choose to create a custom-published lab manual.

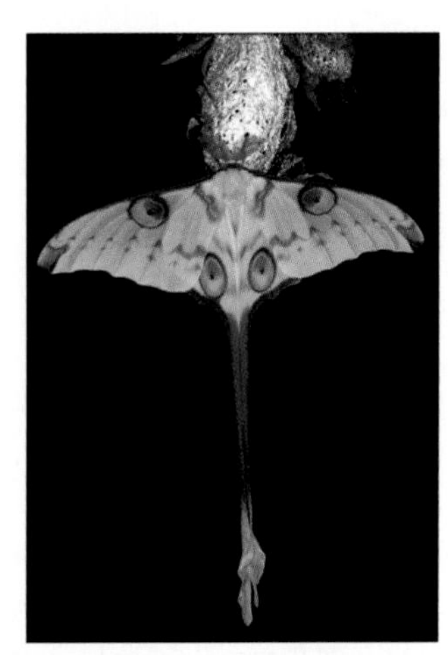

Supplemental Photo 20.33: Comet-tail moth (*Argema mittrei*), adult male

Overhead Transparencies

The set of transparencies includes all the line art figures from the textbook (over 1,000 images), along with a convenient binder. Labels and artwork have been resized and color-corrected for optimal projection.

Printed Test File

Catherine Ueckert, *Northern Arizona University;* Betty McGuire, *Smith College;* Chris Romero, *Front Range Community College;* Paula Mabee, *University of South Dakota;* and Erica Bergquist, *Holyoke Community College*

The Test File offers a bank of thousands of questions from which to create exams. Each chapter includes a set of fill-in-the-blank and multiple-choice questions that cover the full range of content presented in the text. To aid in selecting a range of questions, difficult and conceptual questions are indicated.

Computerized Test Bank CD

The entire printed Test File, plus the textbook end-of-chapter Self-Quizzes, the Student Website Online Quizzes and the Study Guide questions are all included in Brownstone's easy-to-use Diploma® software for an extensive collection of test questions in an easy-to-use interface.

Online Testing

Using the Computerized Test Bank and Diploma Online Testing, instructors can easily create and administer secure exams over a network and over the Internet. For more information, visit the Brownstone Research Group Website: www.brownstone.net.

Instructor's Website (www.thelifewire.com)

An extensive collection of instructor's media, as well as electronic versions of other instructor supplements, is available online for instant access anytime.

Course Management

As a service for adopters using WebCT® or Blackboard® for their courses, full electronic course packs are available. These include all the Student Website contents, plus all the Test File questions and other instructor resources. Contact your sales representative for more information.

Laboratory Manuals

Biology in the Laboratory, Third Edition
Doris R. Helms, Carl W. Helms, Robert J. Kosinski, and John C. Cummings
Order ISBN 0-7167-3146-0

Laboratory Outlines in Biology-VI
Peter Abramoff and Robert G. Thomson
Order ISBN 0-7167-2633-5

Anatomy and Dissection of the Frog, Second Edition
Warren F. Walker, Jr.
Order ISBN 0-7167-2636-X

Anatomy and Dissection of the Rat, Third Edition
Warren F. Walker, Jr. and Dominique Homberger
Order ISBN 0-7167-2635-1

Anatomy and Dissection of the Fetal Pig, Fifth Edition
Warren F. Walker, Jr and Dominique Homberger
Order ISBN 0-7167-2637-8

Custom Publishing for Laboratory Manuals

(available at www.custompub.whfreeman.com)

Instructors can build and order customized lab manuals in just minutes, choosing material from Freeman's acclaimed biology laboratory manuals—lab-tested experiments that have been used successfully by hundreds of thousands of students. Instructors determine the manual's content (including their own material) and a streamlined production process provides a quick turnaround to meet crucial deadlines.

Special Custom Lab Manual Option

The publishers of *Life,* in collaboration with Hayden-McNeil Publishing, offer enhanced customization for instructors adopting 1,000 books or more over two years. This option uses state-of-the-art customization technology to give instructors complete freedom to not only combine Freeman's biology labs with original material, but to also make editorial and page layout changes to the experiments.

Contents in Brief

Contents

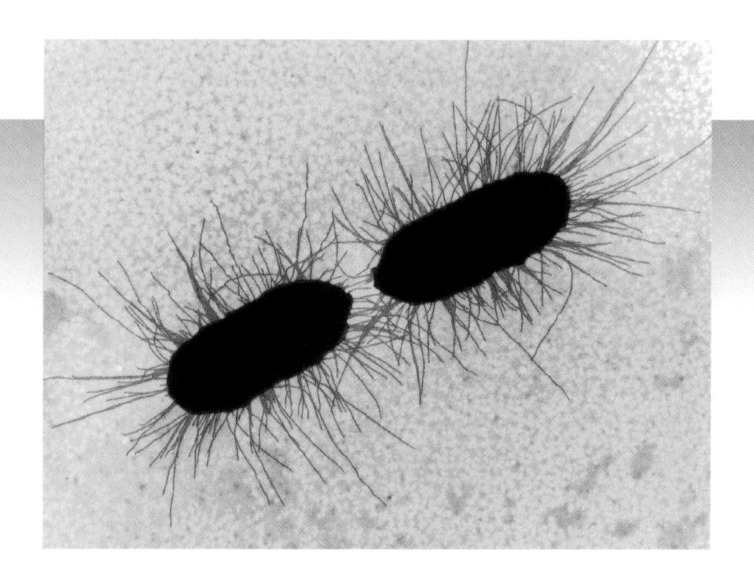

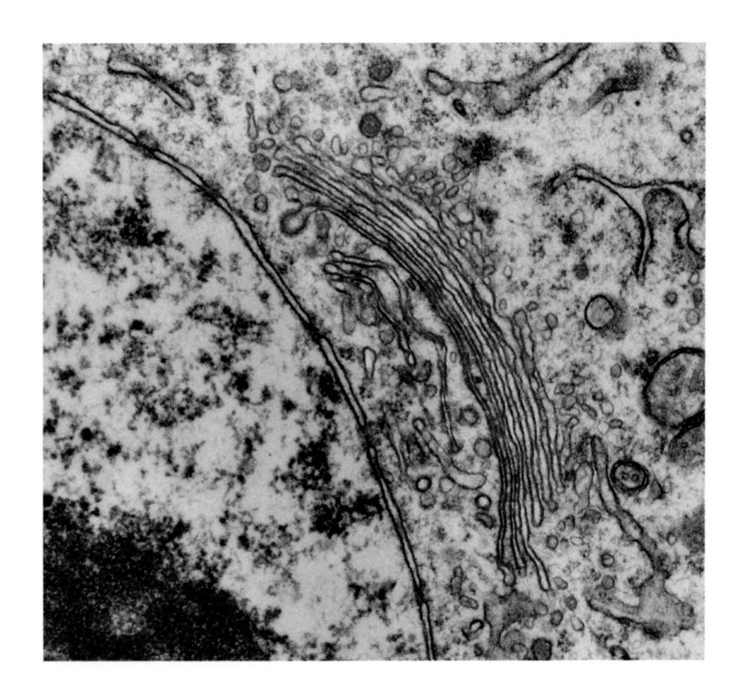

Part Two • INFORMATION AND HEREDITY

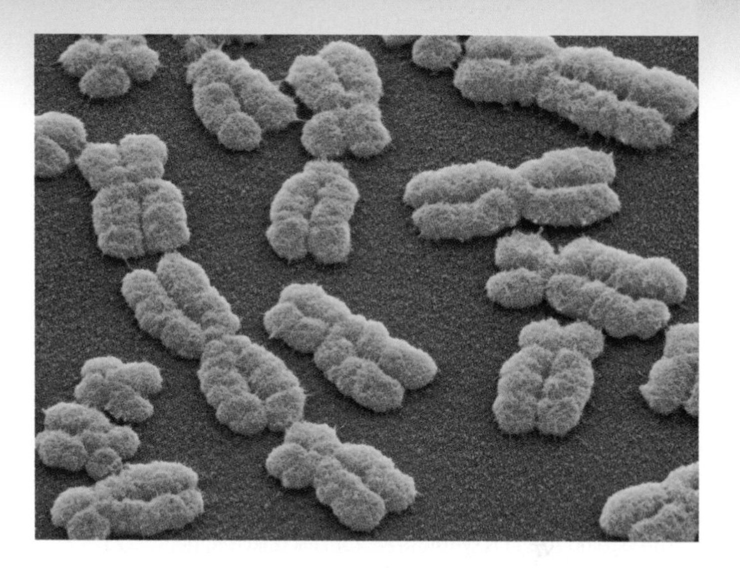

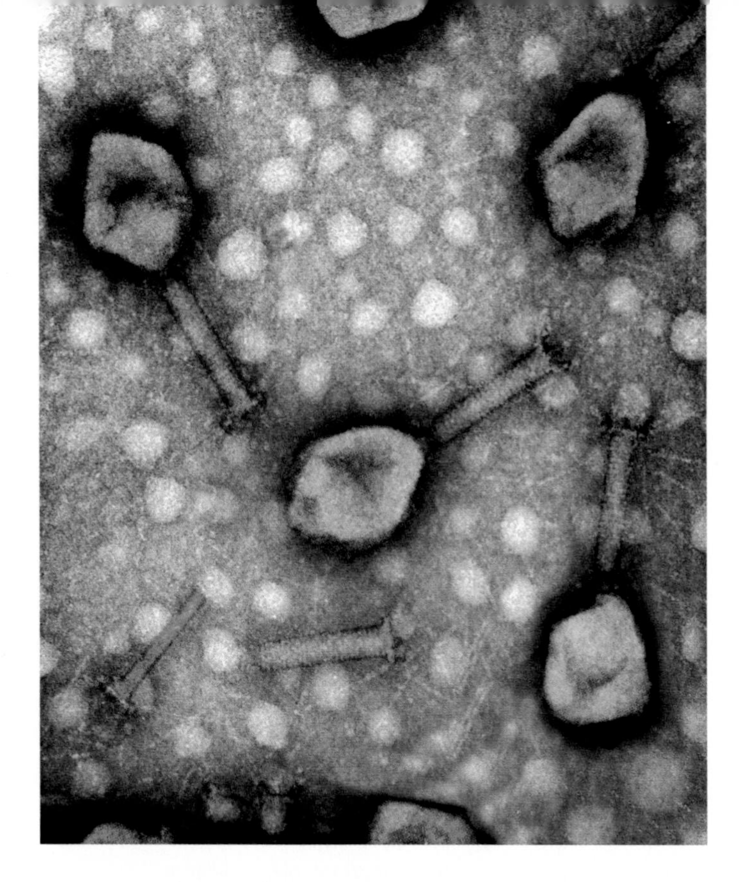

Part Three • DEVELOPMENT

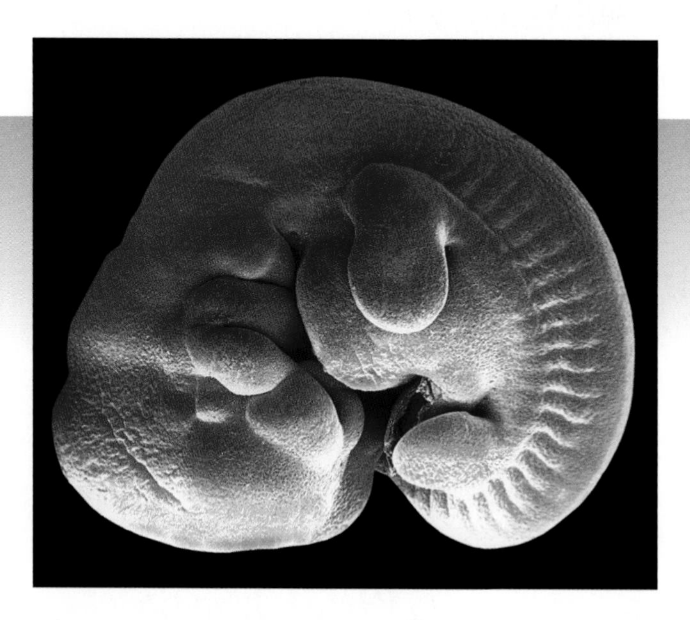

Part Four • EVOLUTIONARY PROCESSES

Part Five • THE EVOLUTION OF DIVERSITY

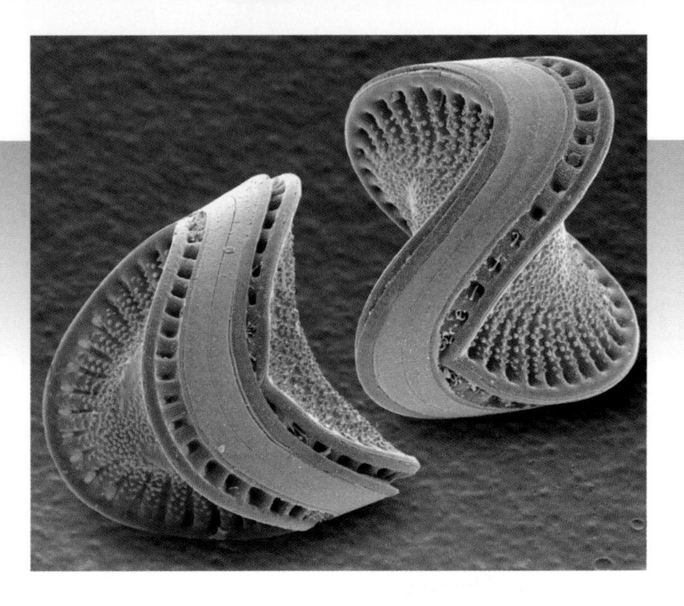

33 *Ecdysozoans: The Molting Animals* 641

34 *Deuterostomate Animals* 655

Part Six • THE BIOLOGY OF FLOWERING PLANTS

35 *The Plant Body* 682

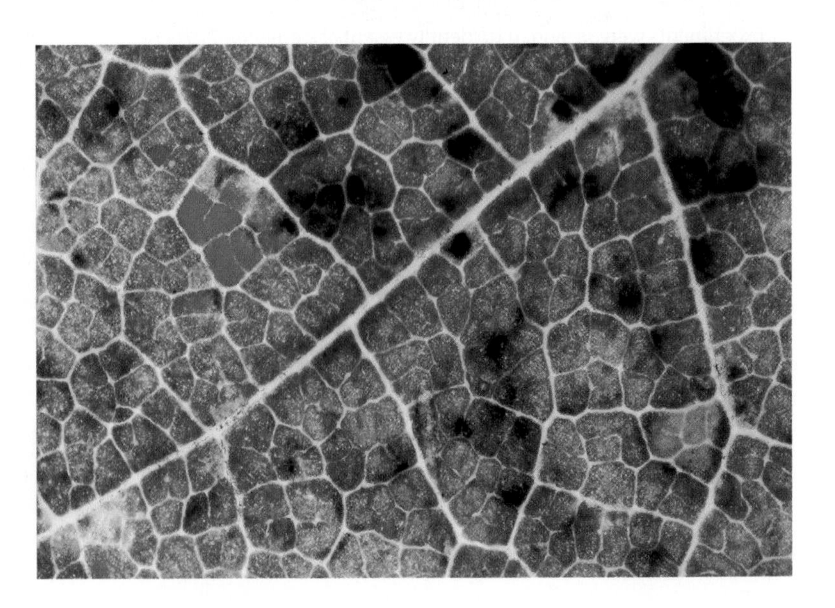

Part Seven • THE BIOLOGY OF ANIMALS

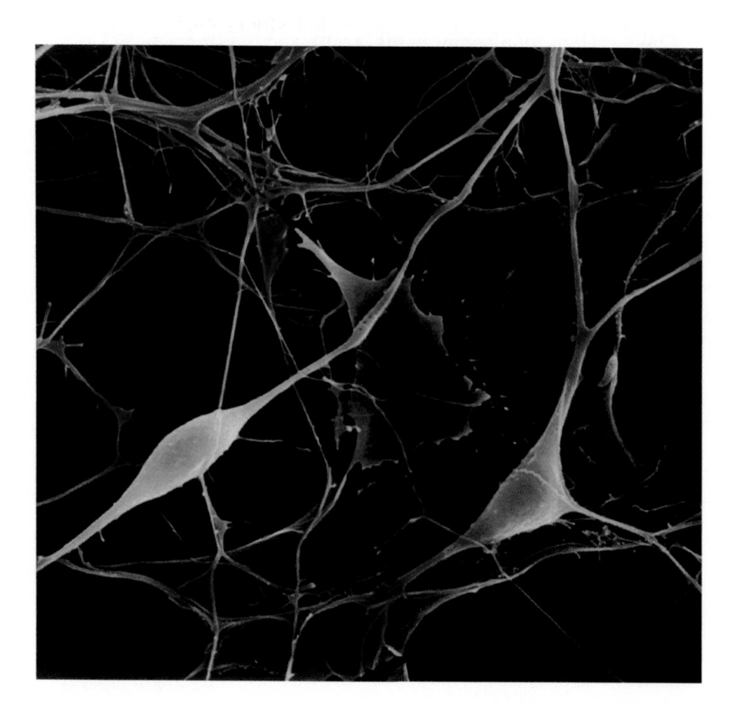

51 Salt and Water Balance and Nitrogen Excretion 985

52 Animal Behavior 1002

Part Eight • ECOLOGY AND BIOGEOGRAPHY

1 An Evolutionary Framework for Biology

Monster frogs—what a great topic for an undergraduate research project! That's what Stanford University sophomore Pieter Johnson thought when he was shown a jar of Pacific tree frogs with extra legs growing out of their bodies. The frogs were collected from a pond on a farm close to the old Almaden mercury mines south of San Jose, California. Scientists from all over the world were reporting alarming declines in populations of many different kinds of frogs, so perhaps these "monster" frogs would hold a clue to why frogs all over the world are in trouble. Possible causes of the deformities could have been agricultural chemicals or heavy metals leaching out of the old mines. Library research, however, suggested other possibilities to Pieter.

Pieter studied 35 ponds in the region where the deformed frogs had been found. He counted frogs in the ponds and measured chemicals in the water. Thirteen of the ponds had Pacific tree frogs, but deformed frogs were found in only four ponds. To Pieter's surprise, analysis of the water samples failed to reveal higher amounts of pesticides, industrial chemicals, or heavy metals in the ponds with deformed frogs. Also surprisingly, when he collected eggs from those ponds and hatched them in the laboratory, he always got normal frogs. The only difference he observed among the ponds he studied was that the ponds with the deformed frogs also contained freshwater snails.

Freshwater snails are hosts for many parasites. Many parasites go through complex life cycles with several stages, each of which requires a specific host animal. Pieter focused on the possibility that some parasite that used freshwater snails as intermediate hosts was infecting the frogs and causing their deformities. Pieter found a candidate with this type of life cycle: a small flatworm called *Ribeiroia*, which was present in the ponds where the deformed frogs were found.

Pieter then did an experiment. He collected frog eggs from regions where there were no records of deformed frogs or of *Ribeiroia*. He hatched the eggs in the laboratory in containers with and without the parasite. When the parasite was present in the contain-

A Monster Phenomenon As a college sophomore, Pieter Johnson studied ponds that were home to Pacific tree frogs (*Hyla regilla*), trying to discover a reason for the presence of so many deformed frogs. What appears in the inset to be a tail is an extra leg.

1.1 The Many Faces of a Life The caterpillar, pupa, and adult are all stages in the life cycle of a monarch butterfly (*Danaeus plexippus*). The caterpillar harvests the matter and energy needed to metabolize the millions of chemical reactions that will result in its growth and transformation, first into a pupa and finally into an adult butterfly specialized for reproduction and dispersal. The transition from one stage to another is triggered by internal chemical signals.

ers, 85 percent of the frogs developed deformities. Further experiments showed why not all the frogs were deformed: The infection had to occur before a tadpole started to grow legs. When tadpoles with already developing legs were infected, they did not become deformed.

Pieter's project started with a question based on an observation in nature. He formulated several possible answers, made observations to narrow down the list of answers, and then did experiments to test what he thought was the most likely answer. His experiments enabled him to reach a conclusion: that these deformities were caused by *Ribeiroia*. Pieter's project is a good example of the application of scientific methods in biology.

Biology is the scientific study of living things. Biologists study processes from the level of molecules to the level of entire ecosystems. They study events that happen in millionths of seconds and events that occur over millions of years. Biologists ask many different kinds of questions and use a wide range of tools, but they all use the same scientific methods. Their goals are to understand how organisms (and assemblages of organisms) function, and to use that knowledge to help solve problems.

In this chapter, we will take a closer look at what biologists do. First, we will describe the characteristics of living things, the major evolutionary events that have occurred during the history of life on Earth, and the evolutionary tree of life. Then we will discuss the methods biologists use to investigate how life functions. At the end of the chapter, we will discuss how scientific knowledge is used to shape public policy.

What Is Life?

Before we probe more deeply into the study of life, we need to agree on what life is. Although we all know a living thing when we see one, it is difficult to define life unambiguously. One concise definition of **life** is: *an organized genetic unit capable of metabolism, reproduction, and evolution*. Much of this book is devoted to describing these characteristics of life and how they work together to enable organisms to survive and reproduce (Figure 1.1). The following brief overview will guide your study of these characteristics.

Metabolism involves conversions of matter and energy

Metabolism, the total chemical activity of a living organism, consists of thousands of individual chemical reactions. Chemical reactions result in the capture of matter and energy and its conversion to different forms, as we will see in Part One of this book. For an organism to function, these reactions, many of which are occurring simultaneously, must be coordinated. Genes provide that control. The nature of the genetic material that controls these lifelong events has been understood only within the last 100 years. Much of Part Two is devoted to the story of its discovery.

The external environment can change rapidly and unpredictably in ways that are beyond an organism's control. An organism can remain healthy only if its internal environment remains within a given range of physical and chemical conditions. Organisms maintain relatively constant internal environments by making metabolic adjustments to conditions such as changes in temperature, the presence or absence of sunlight, or the presence of foreign agents inside their bodies. Maintenance of a relatively stable internal condition, such as a human's constant body temperature, is called **homeostasis**.

The adjustments that organisms make to maintain homeostasis are usually not obvious, because nothing appears to change. However, at some time during their lives, many organisms respond to changing conditions not by maintaining their status, but by undergoing a major reorganization. An early form of such reorganization was the evolution of resting spores, a well protected, inactive form in which organisms survived stressful environments. A striking example that evolved much later is seen in many insects, such as butterflies. In response to internal chemical signals, a caterpillar changes into a pupa and then into an adult butterfly (see Figure 1.1).

Reproduction continues life and provides the basis for evolution

Reproduction with variation is a major characteristic of life. Without reproduction, life would quickly come to an end. The earliest single-celled organisms reproduced by duplicating their genetic material and then dividing in two. The two resulting daughter cells were identical to each other and to the parent cell, except for mutations that occurred during the process of gene duplication. Such errors, although rare, provided the raw material for biological evolution. The combination of reproduction and errors in the duplication of genetic material results in **biological evolution**, a change in the genetic composition of a population of organisms over time.

The diversification of life has been driven in part by variation in the physical environment. There are cold places and warm places, as well as places that are cold during some parts of the year and warm during other parts. Some places (oceans, lakes, rivers) are wet; others (deserts) are usually very dry. No single kind of living thing can perform well in all these environments. In addition, living things generate their own diversity. Once plants evolved, they became a source of food for other living things. Eaters of plants were, in turn, potential food for other organisms. And when living things die, they become food for still other organisms. The differences among living things that enable them to live in different kinds of environments and adopt different lifestyles are called **adaptations**. The great diversity of living things contributes to making biology a fascinating science and Earth a rich and rewarding place to live.

For a long period of time, there was no life on Earth. Then there was an extended period of only unicellular life, followed by a proliferation of multicellular life. In other words, the nature and diversity of life has changed over time. Identification of the processes that result in biological evolution was one of the great scientific advances of the nineteenth century. These processes will be discussed in detail in Part Four of this book. Here we will briefly describe how they were discovered.

Biological Evolution: Changes over Billions of Years

Long before the mechanisms of biological evolution were understood, some people realized that organisms had changed over time and that living organisms had evolved from organisms that were no longer alive on Earth. In the 1760s, the French naturalist Count George-Louis Leclerc de Buffon (1707–1788) wrote his *Natural History of Animals*, which contained a clear statement of the possibility of evolution. Buffon observed that the limb bones of all mammals were remarkably similar in many details (Figure 1.2). He also noticed that the legs of certain mammals, such as pigs, have toes that never touch the ground and appear to be of no use. He found it difficult to explain the presence of these seemingly useless small toes by the commonly held belief that Earth and all its creatures had been divinely created in their current forms relatively recently. To explain these observations, Buffon suggested that the limb bones of mammals might all have been inherited from a common ancestor, and that pigs might have functionless toes because they inherited them from ancestors that had fully formed and functional toes.

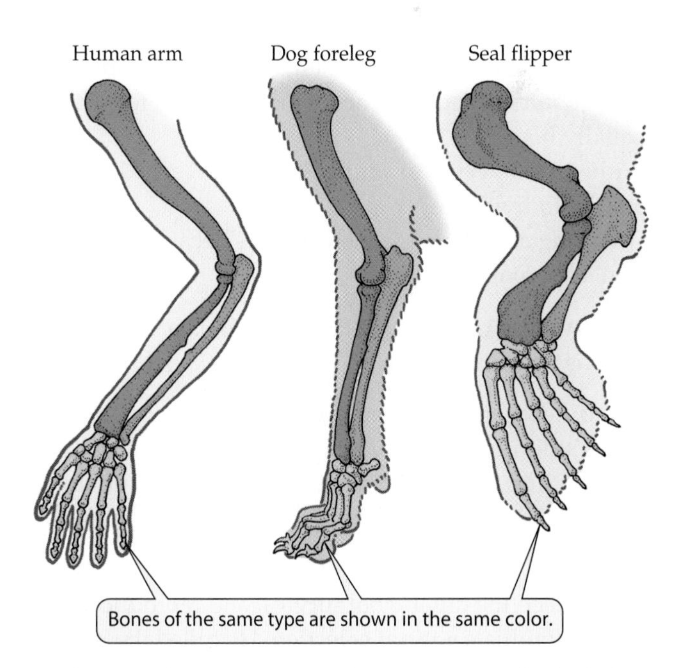

1.2 All Mammals Have Similar Limbs Mammalian forelimbs have different purposes: Humans use theirs for manipulating objects, dogs use theirs for walking, and seals use theirs for swimming. But the numbers and types of their bones are similar, indicating that they have been modified over time from the forelimbs of a common ancestor.

Buffon did not attempt to explain how such changes took place, but his student Jean-Baptiste de Lamarck (1744–1829) proposed a mechanism for such changes. Lamarck suggested that a lineage of organisms could change gradually over many generations as offspring inherited structures that had become larger and more highly developed as a result of continued use or, conversely, had become smaller and less developed as a result of lack of use. Today scientists do not believe that evolutionary changes are produced by this mechanism. But Lamarck had made an important effort to explain how living things change over time.

Darwin provided a mechanistic explanation of biological evolution

By 1858, the climate of opinion (among many biologists, at least) was receptive to a new theory of evolutionary processes proposed independently by Charles Darwin and Alfred Russel Wallace. By that time, geologists had accumulated evidence that Earth had existed and changed over millions of years, not merely a few thousand years, as most people had previously believed.

You will learn more about Darwin's theory of evolution by natural selection in Chapter 23, but its essential features are simple. You will need to be familiar with these ideas to understand the rest of this book. Darwin's theory rests on three observations and one conclusion he drew from them. The three observations are:

- The reproductive rates of all organisms, even slowly reproducing ones, are sufficiently high that populations would quickly become enormous if death rates were not equally high.
- Within each type of organism, there are differences among individuals.
- Offspring are similar to their parents because they inherit their parents' features.

Based on these observations (evidence), Darwin drew the following conclusion:

- The differences among individuals influence how well those individuals survive and reproduce. Any traits that increase the probability that their bearers will survive and reproduce are passed on to their offspring and to their offspring's offspring.

Darwin called the differential survival and reproductive success of individuals **natural selection**. He called the resulting pattern "descent with modification."

Biologists began a major conceptual shift a little more than a century ago with the acceptance of long-term evolutionary change and the gradual recognition that natural selection is the process that adapts organisms to their environments. The shift has taken a long time because it required abandoning many components of an earlier worldview. The pre-Darwinian view held that the world was young, and that organisms had been divinely created in their current forms. In the Darwinian view, the world is ancient, and both Earth and its inhabitants have changed over time. Ancestral forms were very different from the organisms that exist today. Living organisms evolved their particular features because ancestors with those features survived and reproduced more successfully than did ancestors with different features.

Major Events in the History of Life on Earth

The history of life on Earth, depicted on the scale of a 30-day calendar, is outlined in Figure 1.3. The profound changes that have occurred over the 4 billion years of this history are the result of natural processes that can be identified and studied using scientific methods. In this section, we will set the stage for the rest of this book by describing some of the most important of these changes. These six major evolutionary events will provide us with a framework for discussing both life's characteristics and how those characteristics evolved. By recognizing them, you will be able to better appreciate both the unity and diversity of life.

Life arose from nonlife via chemical evolution

The first life must have come from nonlife. All matter, living and nonliving, is made up of chemicals. The smallest chemical units are atoms, which bond together into molecules (the properties of these units are the subject of Chapter 2). The processes of **chemical evolution** that led to the appearance of life began nearly 4 billion years ago, when random inorganic chemical interactions produced molecules that had the remarkable property of acting as templates to form similar molecules. Some of the chemicals involved may have come to Earth from space, but chemical evolution continued on Earth.

The information stored in these simple molecules enabled the synthesis of larger molecules with complex but relatively stable shapes. Because they were both complex and stable, these molecules could participate in increasing numbers and kinds of chemical reactions. Certain types of large molecules are found in all living systems; the properties and functions of these complex molecules are the subject of Chapter 3.

Biological evolution began when cells formed

About 3.8 billion years ago, interacting systems of molecules came to be enclosed in compartments. Within those units—**cells**—control was exerted over the entrance, retention, and exit of molecules, as well as over the chemical reactions taking place. The origin of cells marked the beginning of bio-

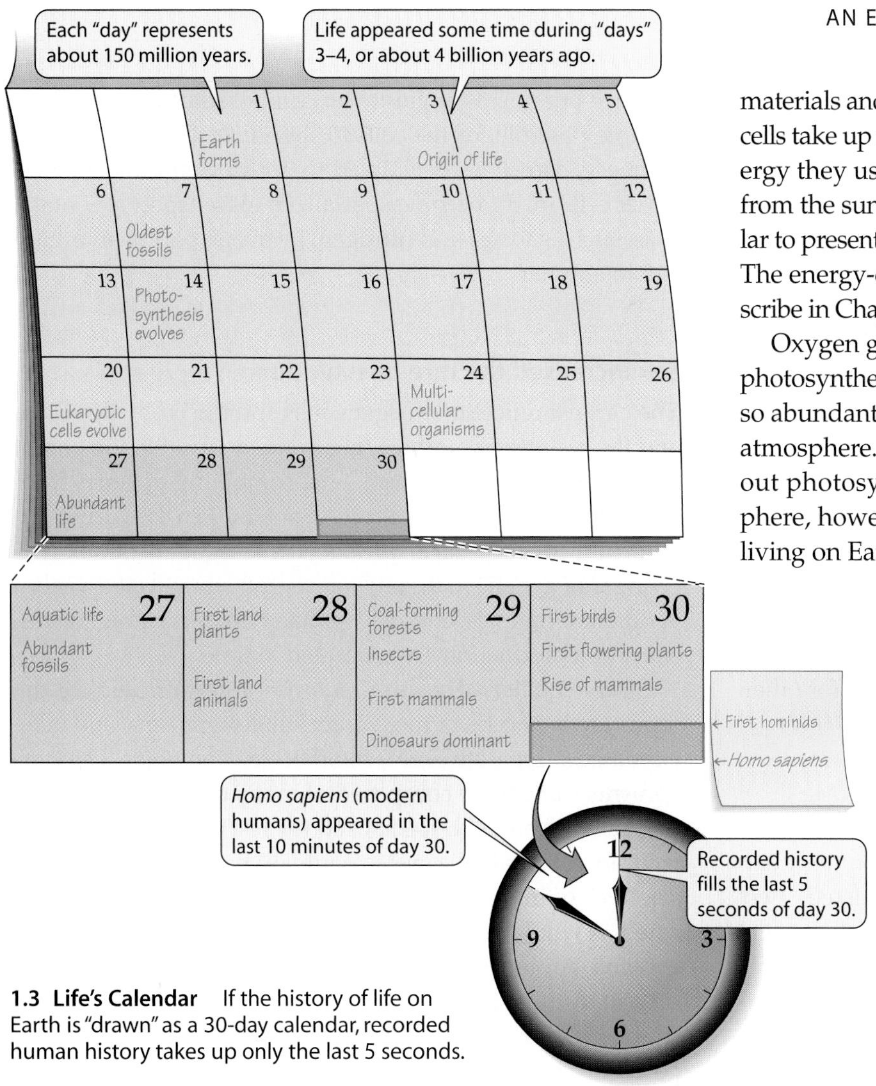

1.3 Life's Calendar If the history of life on Earth is "drawn" as a 30-day calendar, recorded human history takes up only the last 5 seconds.

logical evolution. Cells and the membranes that enclose them are the subjects of Chapters 4 and 5.

Cells are so effective at capturing energy and replicating themselves—two fundamental characteristics of life—that since they evolved, cells have been the unit on which all life is built. Experiments by the French chemist and microbiologist Louis Pasteur and other scientists during the nineteenth century (described in Chapter 3) convinced most scientists that, under present conditions on Earth, cells do not arise from noncellular material, but come only from other cells.

For 2 billion years after cells originated, all organisms were *unicellular* (had only one cell). They were confined to the oceans, where they were shielded from lethal ultraviolet light. These simple cells, called **prokaryotic cells**, had no internal membrane-enclosed compartments.

Photosynthesis changed the course of evolution

A major event that took place about 2.5 billion years ago was the evolution of **photosynthesis**: the ability to use the energy of sunlight to power metabolism. All cells must obtain raw materials and energy to fuel their metabolism. Photosynthetic cells take up raw materials from their environment, but the energy they use to metabolize those chemicals comes directly from the sun. Early photosynthetic cells were probably similar to present-day prokaryotes called *cyanobacteria* (Figure 1.4). The energy-capturing process they used, which we will describe in Chapter 8, is the basis of nearly all life on Earth today.

Oxygen gas (O_2) is a by-product of photosynthesis. Once photosynthesis evolved, photosynthetic prokaryotes became so abundant that they released vast quantities of O_2 into the atmosphere. The O_2 we breathe today would not exist without photosynthesis. When it first appeared in the atmosphere, however, O_2 was poisonous to most organisms then living on Earth. Those prokaryotes that evolved a tolerance to O_2 were able to successfully colonize environments emptied of other organisms and proliferate in great abundance. For those prokaryotes, the presence of oxygen opened up new avenues of evolution. Metabolic reactions that use O_2, called *aerobic metabolism*, are more efficient than the anaerobic (non-oxygen-using) metabolism that earlier prokaryotes had used. Aerobic metabolism allowed cells to grow larger, and it came to be used by most organisms on Earth.

Over a much longer time frame, the vast quantities of oxygen released by photosynthesis had another effect. Formed from O_2, ozone (O_3) began to accumulate in the upper atmosphere. The ozone slowly formed a dense layer that acted as a shield, inter-

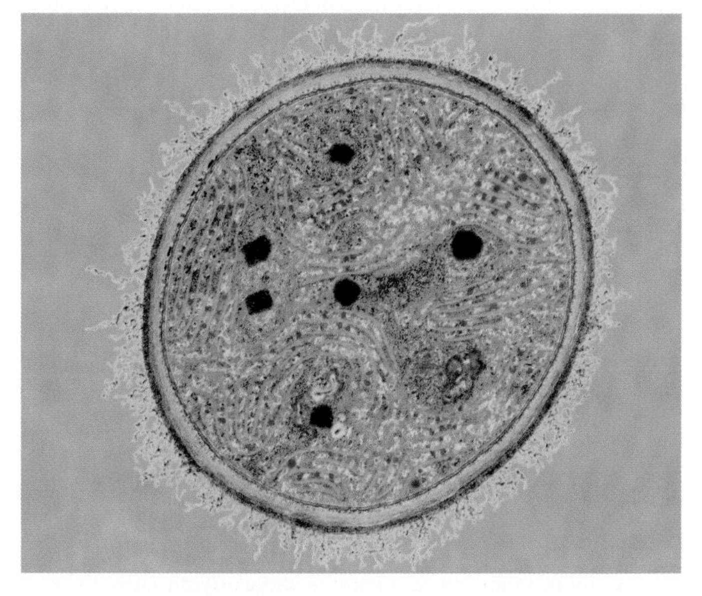

1.4 Oxygen Produced by Prokaryotes Changed Earth's Atmosphere This modern cyanobacterium may be very similar to early photosynthetic prokaryotes.

cepting much of the sun's deadly ultraviolet radiation. Eventually (although only within the last 800 million years of evolution), the presence of this shield allowed organisms to leave the protection of the ocean and establish new lifestyles on Earth's land surfaces.

Cells with complex internal compartments arose

As the ages passed, some prokaryotic cells became large enough to attack, engulf, and digest smaller prokaryotes, becoming the first predators. Usually the smaller cells were destroyed within the predators' cells, but some of these smaller cells survived and became permanently integrated into the operation of their host cells. In this manner, cells with complex internal compartments, called **eukaryotic cells**, arose. The hereditary material of eukaryotic cells is contained within a membrane-enclosed nucleus and is organized into discrete units. Other compartments are specialized for other purposes, such as photosynthesis (Figure 1.5).

Multicellularity arose and cells became specialized

Until slightly more than 1 billion years ago, only unicellular organisms (both prokaryotic and eukaryotic) existed. Two key developments made the evolution of *multicellular* organisms—organisms consisting of more than one cell—possible. One was the ability of a cell to change its structure and functioning to meet the challenges of a changing environment. This was accomplished when prokaryotes evolved the ability to transform themselves from rapidly growing cells into resting spores that could survive harsh environmental conditions. The second development allowed cells to stick together after they divided and to act together in a coordinated manner.

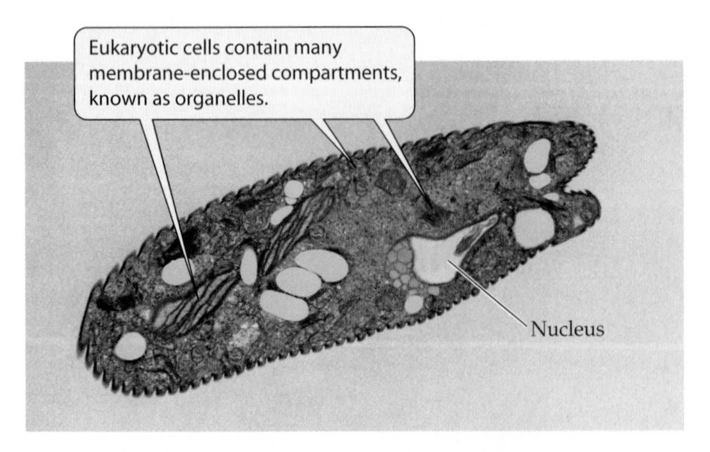
Eukaryotic cells contain many membrane-enclosed compartments, known as organelles.
Nucleus

1.5 Multiple Compartments Characterize Eukaryotic Cells The nucleus and other specialized compartments of eukaryotic cells evolved from small prokaryotes that were ingested by larger prokaryotic cells.

Once organisms began to be composed of many cells, it became possible for the cells to specialize. Certain cells, for example, could be specialized to perform photosynthesis. Other cells might become specialized to transport raw materials, such as water and nitrogen, from one part of an organism to another.

Sex increased the rate of evolution

The earliest unicellular organisms reproduced by dividing, and the resulting daughter cells were identical to the parent cell. But *sexual recombination*—the combining of genes from two different cells in one cell—appeared early during the evolution of life. Early prokaryotes engaged in **sex** (exchanges of genetic material) and reproduction (cell division) at different times. Even today in many unicellular organisms, sex and reproduction are separated in time.

Simple nuclear division—*mitosis*—was sufficient for the reproductive needs of most unicellular organisms, and gene exchange (a separate event) could occur at any time. Once organisms came to be composed of many cells, however, certain cells began to be specialized for sex. Only these specialized sex cells, called *gametes*, could exchange genes, and the sex lives of multicellular organisms became more complicated. A whole new method of nuclear division—*meiosis*—evolved. An intricate and complex process, meiosis opened up a multitude of possibilities for genetic recombination between gametes. Mitosis and meiosis are explained and compared in Chapter 9.

Sex increased the rate of evolution because an organism that exchanges genetic information with another individual produces offspring that are more genetically variable than the offspring of an organism that reproduces by mitotic division of its own cells. Some of these varied offspring are likely to survive and reproduce better than others in a variable and changing environment. It is this genetic variation that natural selection acts on.

Levels of Organization of Life

Biology can be visualized as a hierarchy of units, ordered from the smallest to the largest. These units are molecules, cells, tissues, organs, organisms, populations, communities, and the biosphere (Figure 1.6).

The organism is the central unit of study in biology; Parts Six and Seven of this book discuss organismic biology in detail. But to understand organisms, biologists study life at all its levels of organization. They study molecules, chemical reactions, and cells to understand the functioning of tissues and organs. They study organs and organ systems to determine how organisms maintain homeostasis. At higher levels in the hierarchy, biologists study how organisms interact with one

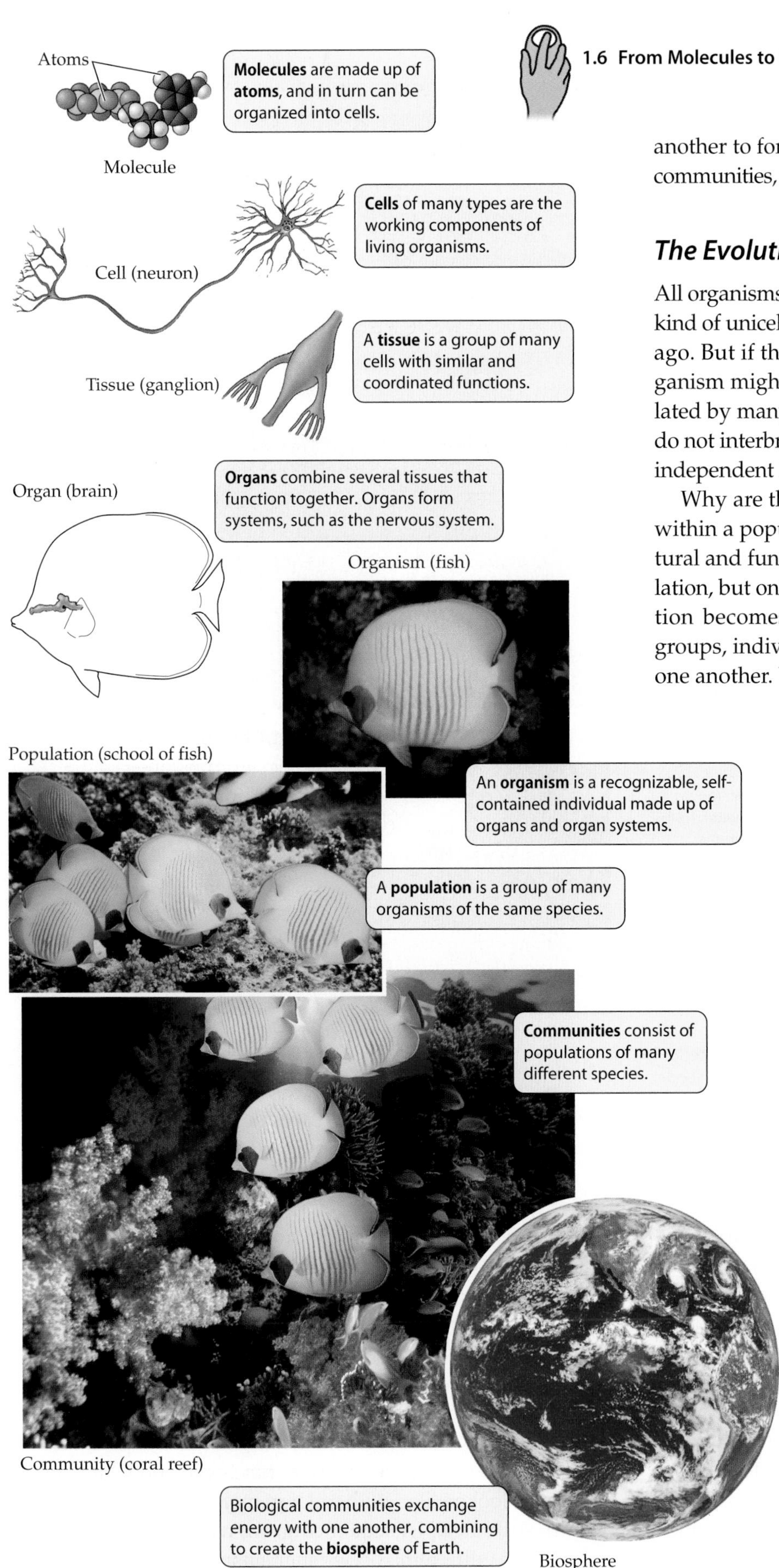

Atoms
Molecule
Molecules are made up of atoms, and in turn can be organized into cells.
Cell (neuron)
Cells of many types are the working components of living organisms.
Tissue (ganglion)
A tissue is a group of many cells with similar and coordinated functions.
Organ (brain)
Organs combine several tissues that function together. Organs form systems, such as the nervous system.
Organism (fish)
An organism is a recognizable, self-contained individual made up of organs and organ systems.
Population (school of fish)
A population is a group of many organisms of the same species.
Communities consist of populations of many different species.
Community (coral reef)
Biological communities exchange energy with one another, combining to create the biosphere of Earth.
Biosphere

1.6 From Molecules to the Biosphere: The Hierarchy of Life

another to form social systems, populations, and ecological communities, which are the subjects of Part Eight of this book.

The Evolutionary Tree of Life

All organisms on Earth today are the descendants of a single kind of unicellular organism that lived almost 4 billion years ago. But if that were the whole story, only one kind of organism might exist on Earth today. Instead, Earth is populated by many millions of different kinds of organisms that do not interbreed with one another. We call these genetically independent kinds **species**.

Why are there so many species? As long as individuals within a population mate at random and reproduce, structural and functional changes may evolve within that population, but only one species will exist. However, if a population becomes separated and isolated into two or more groups, individuals within each group will mate only with one another. When this happens, structural and functional differences between the groups may accumulate over time, and the groups may evolve into different species. The splitting of groups of organisms into separate species has resulted in the great diversity of life found on Earth today. The ways in which species form are explained in Chapter 24.

Sometimes humans refer to a species as "primitive" or "advanced." These and similar terms, such as "lower" and "higher," are best avoided in biology because they imply that some organisms function better than others. In fact, *all* living organisms are successfully adapted to their environments. The shape and strength of a bird's beak, or the form and dispersal mechanisms of a plant's seeds are examples of the rich array of adaptations found among living organisms (Figure 1.7). The abundance and success of prokaryotes—all of which are relatively simple organisms—readily demonstrates that they are highly functional. In this book, we use the terms *simple* and *complex* to refer to the level of complexity of a particular organism. We use the terms *ancestral* and *derived* to distinguish characteristics that appeared earlier from those that appeared later in evolution.

As many as 30 million species of organisms may live on Earth today. Many times that number lived in the past, but are now ex-

(*a*)

The strong, curved beak of the bald eagle is able to tear the flesh from large fish and other sizeable prey.

The curlew uses its long, curved, pointed beak to extract small crustaceans from the surface of mud, sand, and soil.

The roseate spoonbill moves its bill through the water, from which it filters food items.

(*b*)

The coconut seed is covered by a thick husk that protects it as it drifts across thousands of miles of ocean.

Mammals and birds eat blackberries, then disseminate the seeds when they defecate.

The seeds of milkweeds are surrounded by "kites" of fibers that carry them on wind currents.

1.7 Adaptations to the Environment (*a*) Bird beaks are adapted to specific types of food items. (*b*) Plants cannot move, but their seeds have adaptations that allow them to travel varying distances from the parent plant.

tinct. This diversity is the result of millions of splits in populations, known as *speciation events*. The unfolding of these events can be expressed as an evolutionary "tree" showing the order in which populations split and eventually evolved into new species (see Figure 1.8). An evolutionary tree, with its "trunk" and its increasingly finer "branches," traces the descendants coming from ancestors that lived at different times in the past. That is, a tree shows the evolutionary relationships among species and groups of species. The organisms on any one branch share a common ancestor at the base of that branch. The most closely related groups are together on the same branch. More distantly related organisms are on different branches. In this book, we adopt the convention that time flows from left to right, so the tree in Figure 1.8 (and other trees in this book) lies on its side, with its root—the ancestor of all life—at the left.

The U.S. National Science Foundation is sponsoring a major initiative, called Assembling the Tree of Life (ATOL). Its goal is to determine the evolutionary relationships among all species on Earth. Achieving this goal is possible today because, for the first time, biologists have the technology to assemble the complete tree of life, from microbes to mammals. Data for ATOL come from a variety of sources. *Fossils*—the preserved remains of organisms that lived in the past—tell us where and when ancestral organisms lived and what they may have looked like. With modern molecular genetic techniques such as DNA sequencing, we can determine how many genes different species share, and information tech-

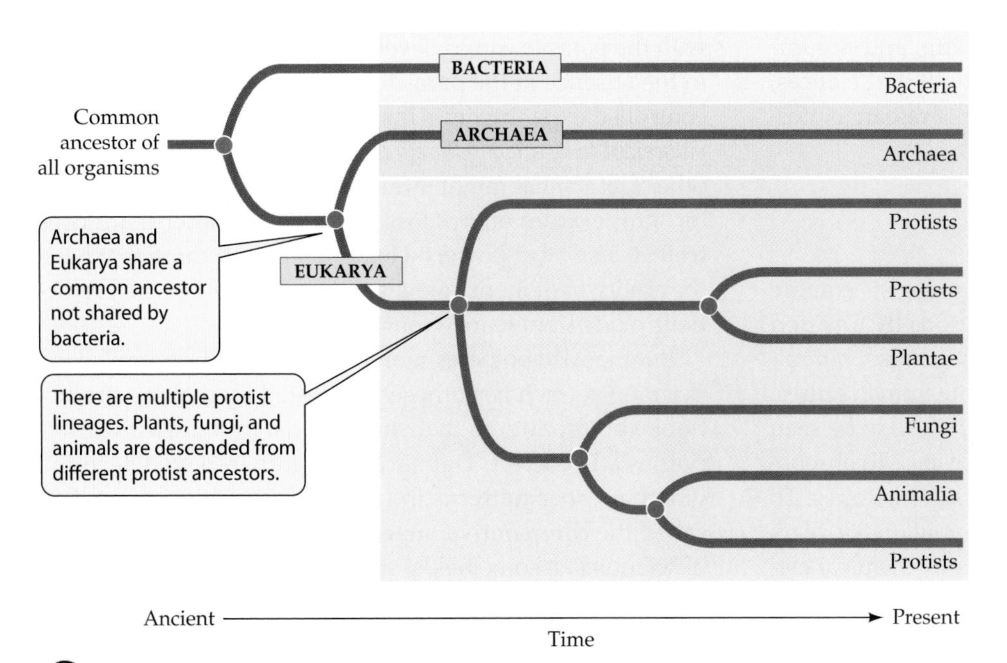

1.8 A Provisional Tree of Life The classification system used in this book divides Earth's organisms into three domains; Bacteria, Archaea, and Eukarya. Protists are descendants of multiple ancestors.

nology enables us to synthesize masses of genetic data. The ATOL initiative, one of the grandest projects of modern biology, is projected to take at least two decades and to involve hundreds of scientists working in a diverse array of fields. The reason it will take so long to complete is that most of Earth' species have not yet been described.

The Tree of Life will be an information framework for biology in much the same way that the periodic table of elements is an information framework for chemistry and physics. Evolution has conducted several billion years of free research and development. Every living thing carries a genetic "package" that has been tested by natural selection. Scientists can now unwrap and study these packages, learning much about the processes that produced them.

Although much remains to be accomplished, biologists know enough to have assembled a provisional tree of life, the broad outlines of which are shown in Figure 1.8. The branching patterns of this tree are based on a rich array of evidence, but no fossils are available to help us determine the earliest divisions in the lineages of life because those simple organisms had no parts that could be preserved as fossils. Therefore, molecular evidence has been used to separate all living organisms into three major **domains**. Organisms belonging to a particular domain have been evolving separately from organisms in the other domains for more than a billion years.

Organisms in the domains **Archaea** and **Bacteria** are prokaryotes. Archaea and Bacteria differ so fundamentally from one another in their metabolic processes that they are believed to have separated into distinct evolutionary lineages very early during the evolution of life. The two prokaryotic domains are described in Chapter 27.

Members of the other domain—**Eukarya**—have eukaryotic cells. The Eukarya are divided into four groups: Protista, Plantae, Fungi, and Animalia. The Protista (protists), the subject of Chapter 28, contains mostly single-celled organisms. The other three groups, referred to as *kingdoms*, are believed to have arisen from ancestral protists.

Some bacteria, some protists, and most members of the kingdom Plantae (plants) convert light energy to chemical energy by photosynthesis. These organisms are called *autotrophs* ("self-feeders"). The biological molecules they produce are the primary food for nearly all other living organisms. The kingdom Plantae is covered in Chapters 29 and 30.

The kingdom Fungi, the subject of Chapter 31, includes molds, mushrooms, yeasts, and other similar organisms, all of which are *heterotrophs* ("other-feeders")—that is, they require a source of energy-rich molecules synthesized by other organisms. Fungi break down food molecules in their environment and then absorb the breakdown products into their cells. They are important as decomposers of the dead materials of other organisms.

Members of the kingdom Animalia (animals) are heterotrophs that ingest their food source, digest the food outside their cells, and then absorb the breakdown products. Animals eat other forms of life to obtain their raw materials and energy. This kingdom is covered in Chapters 32, 33, and 34.

We will discuss the principal levels used in today's classification scheme for living organisms in Chapter 25. But to understand some of the terms we will use in the intervening chapters, you need to know that each species of organism is identified by two Latinized names (a *binomial*). The first name identifies the **genus**—a group of species that share a recent common ancestor—of which the species is a member. The second name is the species name. To avoid confusion, no combination of two names is assigned to more than one species. For example, the scientific name of the human species is *Homo sapiens: Homo* is our genus and *sapiens* is our species. The Pacific tree frogs Pieter Johnson studied are called, in scientific nomenclature, *Hyla regilla.*

Biology is the study of all of Earth's organisms, both those living today and those that lived in the past, so even extinct species are given binomials. These unique and exact names illuminate the tremendous diversity of life, and are important tools for biologists because, as in all the sciences, precise and unambiguous communication of research information is critical.

Biology Is a Science

To study the rich variety of living things, biologists employ many different methods. Direct observations by unaided senses are central to many scientific investigations, but scientists also use many tools that augment the human senses. For example, to study objects that are too small to be seen with the unaided eye, scientists use microscopes. To observe and magnify remote objects, scientists use telescopes. To study events that happened thousands to millions of years ago, scientists "read" radioactive isotopes of chemical elements that decay at specific rates.

Conceptual tools guide scientific research

In addition to such technical tools, scientists use a variety of conceptual tools to help them answer questions about nature. The method that underlies most scientific research is the **hypothesis-prediction (H–P) approach**. The H–P approach allows scientists to modify their conclusions as new information becomes available. The method has five steps:

1. Making *observations*
2. Asking *questions*
3. Forming *hypotheses*, which are tentative answers to the questions
4. Making *predictions* based on the hypotheses
5. *Testing* the predictions by making additional observations or conducting experiments

If the results of the testing support the hypothesis, it is subjected to additional predictions and tests. If they continue to support it, confidence in its correctness increases, and the hypothesis comes to be considered a **theory**. If the results do not support the hypothesis, it is abandoned or modified in accordance with the new information. Then new predictions are made, and more tests are conducted.

Hypotheses are tested in two major ways

Tests of hypotheses are varied, but most are of two types: controlled experiments and the comparative method. When possible, scientists use **controlled experiments** to test predictions from hypotheses. That is what Pieter Johnson was doing when he hatched frog eggs in the laboratory. He predicted that if his hypothesis—that the parasite *Ribeiroia* caused deformities in frogs—was correct, then frogs raised with the parasite would develop deformities and frogs raised in the absence of the parasite would not. The advantage of controlled experiments is that *all factors other than the one hypothesized to be causing the effect can be kept constant*; that is, any other factors that might influence the outcome (such as water temperature and pH in Pieter's experiment) are controlled. The most powerful experiments are those that have the ability to demonstrate that the hypothesis or the predictions made from it are wrong.

But many hypotheses cannot be tested with controlled experiments. Such hypotheses are tested by making predictions about patterns that should exist in nature if the hypothesis is correct. Data are then gathered to determine whether those patterns in fact do exist. This approach is called the **comparative method**. It is the primary approach of scientists in some fields, such as astronomy, in which experiments are rarely possible. Biologists regularly use the comparative method.

A single piece of supporting evidence rarely leads to widespread acceptance of a hypothesis. Similarly, a single contrary result rarely leads to abandonment of a hypothesis. Results that do not support the hypothesis can be obtained for many reasons, only one of which is that the hypothesis is wrong. For example, incorrect predictions can be made from a correct hypothesis. Poor experimental design, or the use of an inappropriate organism, can also lead to erroneous results.

We will now show how the H–P method was used by other researchers to investigate the larger question that concerned Pieter Johnson: Why are amphibian populations declining dramatically in many places on Earth?

STEP 1: MAKING OBSERVATIONS. Amphibian populations, like populations of most organisms, fluctuate over time. Before we decide that the current declines are different from "normal" population fluctuations, we first need to establish that they are unusual. To assess whether the current declines are unusual, an international group of scientists has been gathering worldwide data on amphibian populations. The group's data show that amphibian populations are declining seriously in some parts of the world, especially western North America, Central America, and northeastern Australia, but not others, such as the Amazon Basin. Their data also show that population declines are greater in mountains than in adjacent lowlands. These scientists also discovered that no data on population trends in amphibians are available from Africa or Asia.

STEP 2: ASKING QUESTIONS. Two questions were suggested by these observations: Why are amphibian declines greater at high elevations? Why are amphibians declining in some regions, but not in others?

STEPS 3 AND 4: FORMULATING HYPOTHESES AND MAKING PREDICTIONS. To develop hypotheses about the first question, scientists first identified the environmental factors that change with elevation. Temperatures drop and rainfall increases with elevation worldwide, and in temperate regions, summer levels of ultraviolet-B (UV-B) radiation increase about 18 percent per 1,000 meters of elevation gain. One hypothesis is that declines in the populations of some amphibian species are due to global increases in UV-B radiation resulting from reductions in atmospheric ozone concentrations. If increased levels of UV-B are adversely affecting amphibian populations, we predict that experimentally reducing UV-B over ponds where amphibian eggs are incubating and larvae are developing should improve their survival.

Figure 1.9 describes one of many experiments in which the UV-B hypothesis has been tested. Some other experiments have yielded similar results, while others have shown no effects of UV-B exposure, or have shown a negative effect of UV-B exposure only when it is associated with low pH.

Several hypotheses have also been proposed to account for regional differences in amphibian population declines, including the adverse effects of habitat alteration by humans. Two obvious forms of human habitat alteration are air pollution from areas of urban and industrial growth, and the airborne pesticides used in agriculture.

A straightforward prediction from the habitat alteration hypothesis is that amphibian declines should be more noticeable in areas exposed to higher amounts of human-generated air

STEP 5: TESTING HYPOTHESES. The hypothesis that exposure to increased levels of UV-B might contribute to amphibian population decline was tested by comparing the responses of tadpoles of two species of frogs that live in Australian mountains. One species (*Litoria verreauxii*) had disappeared from high elevations; the other (*Crinia signifera*) had not. Because at higher elevations tadpoles are exposed to higher levels of UV-B radiation, experimenters predicted that *L. verreauxii* would survive less well than *C. signifera* if exposed to UV-B radiation typical of high elevations (Figure 1.9).

As predicted, when exposed to UV-B radiation, individuals of *C. signifera* survived well, but all individuals of *L. verreauxii* died within two weeks. Among control populations raised in tanks covered by filters that blocked UV transmission, individuals of both species survived well. Thus, the results supported the hypothesis.

EXPERIMENT

Hypothesis: Susceptibility to UV-B radiation has contributed to the disappearance of some frogs from high-elevation ponds.

METHOD Establish 3 identical artificial tanks at each of 2 elevations (1,365 meters and 1,600 meters). Set up 6 trays in each tank. Place equal numbers of embryos of one of the two frog species in each tray. In each tank, 2 trays receive unfiltered sunlight; 2 receive sunlight filtered to remove UV-B; and 2 receive filtered sunlight that allowed UV-B transmission. Count the number of surviving individuals 3 times a week for 4 weeks.

RESULTS The probability of dying was much greater for individuals of the species that had disappeared from high elevations (*Litoria verreauxii*) than for individuals of the species surviving there (*Crinia signifera*).

Unfiltered sunlight — Filtered, UV-B blocked — Filtered, UV-B allowed

Elevation 1,365 meters

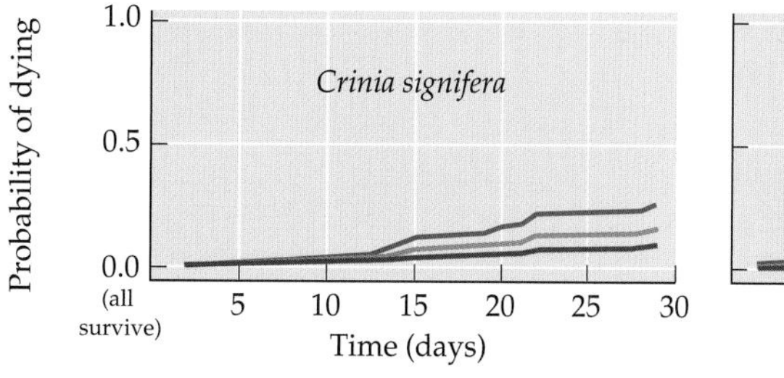

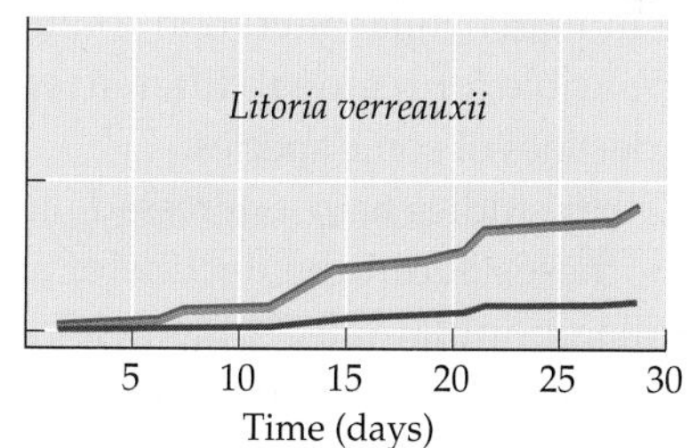

Elevation 1,600 meters

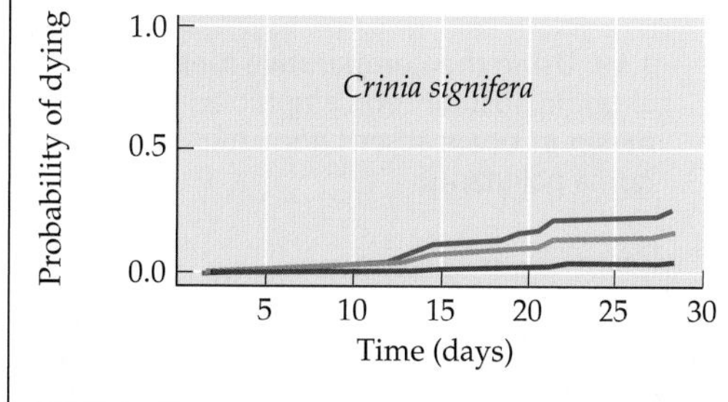

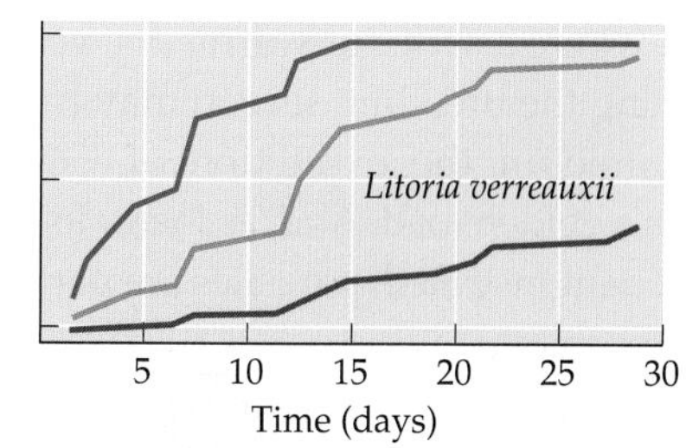

Conclusion: The results support the hypothesis that suceptibility to UV-B radiation has contributed to the disappearance of *Litoria verreauxii* from high elevations.

1.9 A Controlled Experiment Tests the Effects of UV-B The results of this experiment suggest that UV-B susceptibility has contributed to the decline of some amphibian populations. Experimental populations of both species were subjected to different levels of UV radiation; the filtered-light population (no UV-B exposure) acted as a control.

pollutants than in areas with less exposure. This hypothesis has been tested using the comparative method. The extensive tests compared population trends in eight species of amphibians in the state of California. The species studied included four frog species of the genus *Rana*, two species of toads, and a salamander species. The bases of the tests were simple *censuses* (surveys and counts) to determine whether populations of a given species were present or absent at each of the hundreds of study sites across the state. The census results for one of the eight species, the frog *Rana aurora*, are shown in Figure 1.10.

The map in Figure 1.10 shows a significant trend for *R. aurora*: Populations of this amphibian are more likely to be *absent* from sites downwind of large urban and agricultural areas (and thus exposed to heavy airborne pollution), and *present* in sites upwind (not heavily exposed). This type of data is the basis of the comparative method. In this particular study, meticulous tallying and comparison of similar data for all eight species showed that some species exhibited significant declines in exposed areas, but others (including the toads), did not. Therefore, we may conclude that human habitat alteration could be responsible for regional differences in the declines of some species.

Other studies have addressed other hypotheses about the decline of amphibian populations. Some evidence indicates that smoke from extensive fires also is adversely affecting amphibians. Climate change is clearly important in areas such as Central America, where a series of warm, dry years during the breeding season may have resulted in the extinction of Costa Rica's golden toad. And, as Pieter Johnson demonstrated, parasites are part of the problem.

COMPARATIVE METHOD

Hypothesis: Airborne pesticides from agricultural fields and urban air pollutants are contributing to the decline of amphibian populations.

PREDICTION If pesticides and urban air pollutants are factors in amphibian population declines, populations close to and downwind from agricultural and urban areas should have decreased more strikingly than populations upwind and farther away from those sources of air pollutants.

METHOD Census (count) and then compare persistence of populations of species of amphibians at suitable habitat sites that lie upwind and downwind of major agricultural and urban areas.

RESULTS Populations of some species, as illustrated here by *Rana aurora*, persist in areas upwind of or remote from sources of urban and agricultural pollutants, but this amphibian is largely absent from areas close to or downwind of air pollution sources. (Distributions of three other species of *Rana* were similar to that of *R. aurora*.)

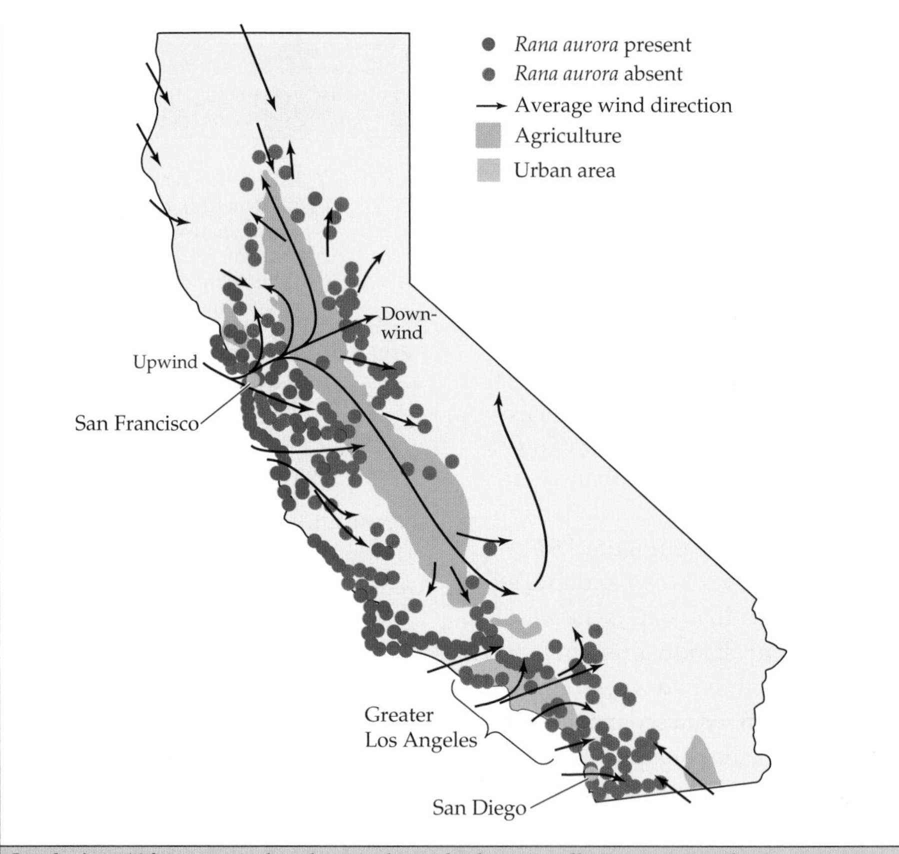

Conclusion: Airborne agricultural pesticides and urban air pollutants are contributing to declines in populations of some amphibian species.

1.10 Using the Comparative Method to Test a Hypothesis The effects of human-generated airborne pollutants on amphibian populations can be assessed by determining whether species persist in, or are absent from, suitable habitats that lie upwind or downwind from sources of airborne pollutants.

Even though much more information needs to be gathered, it is already evident that no single factor is causing amphibian declines. This finding is not surprising, because no two places on Earth are the same, and no two species of amphibians respond in exactly the same way to changes in the environment. In their responses to environmental changes, amphibians are like most living things. They live in complex and changing environments, and they interact with many other species.

Simple explanations that account for everything should not be expected or trusted. Its complexities make biology a difficult science, but they also make it exciting and challenging.

Not all forms of inquiry are scientific

Scientific methods are the most powerful tools that humans have developed to understand how the world works. Their strength is founded on the development of hypotheses that can be tested. The process is self-correcting because if the evidence fails to support a hypothesis, it is either abandoned or modified and subjected to further tests. In addition, because scientists publish detailed descriptions of the methods they use to test hypotheses, other scientists can—and often do—repeat those experiments. Therefore, any error or dishonesty usually is discovered. That is why, in contrast to politicians, scientists around the world usually trust one another's results.

If you understand the methods of science, you can distinguish science from non-science. Art, music, and literature, activities that contribute massively to the quality of human life, are not science. They help us understand what it means to live in a complex world. Religion is not science either. Religious beliefs give us meaning and spiritual guidance, and they form a basis for establishing values. Scientific information helps create the context in which values are discussed and established, but it cannot tell us what those values should be.

Biology has implications for public policy

The study of biology has long had major implications for human life. Agriculture and medicine are two important fields of applied biology. People have been speculating about the causes of diseases and searching for methods of combating them since ancient times. Today, with the deciphering of the genetic code and the ability to manipulate the genetic constitution of organisms, vast new possibilities exist for improvements in the control of human diseases and agricultural productivity. At the same time, these capabilities have raised important ethical and policy issues. How much and in what ways should we tinker with the genetics of people and other species? Does it matter whether organisms are changed by traditional breeding experiments or by gene transfers? How safe are genetically modified organisms in the environment and in human foods?

Another reason for studying biology is to understand the effects of the vastly increased human population on its environment. Our use of renewable and nonrenewable natural resources is putting stress on the ability of the environment to produce the goods and services upon which our society depends. Human activities are changing global climates, causing the extinction of a large number of species, and resulting in the spread of new human diseases and the resurgence of old ones. For example, the rapid spread of SARS and West Nile virus was facilitated by modern modes of transportation. Biological knowledge is vital for determining the causes of these changes, for devising wise policies to deal with them, and for drawing attention to the marvelous diversity of living organisms that provides goods and services for humankind and also enriches our lives aesthetically and spiritually.

Biologists are increasingly called upon to advise governmental agencies concerning the laws, rules, and regulations by means of which society deals with the increasing number of problems and challenges that have at least a partial biological basis. As we discuss these issues in many chapters of this book, you will see that the use of biological information is essential if wise public policies are to be established and implemented.

Throughout this book we will share with you the excitement of studying living things and illustrate the rich array of methods that biologists use to determine why the world of living things looks and functions as it does. The most important motivator of most biologists is curiosity. People are fascinated by the richness and diversity of life and want to learn more about organisms and how they interact with one another.

Humans probably evolved to be curious because individuals who were motivated to learn about their surroundings were likely to have survived and reproduced better, on average, than their less curious relatives. In other words, curiosity is adaptive! There are vast numbers of questions for which we do not yet have answers, and new discoveries usually engender questions no one thought to ask before. Perhaps *your* curiosity will lead to an important new idea.

Chapter Summary

What Is Life?

- Life can be defined as an organized genetic unit capable of metabolism, reproduction, and evolution.
- Metabolism, the total chemical activity of a living organism, is controlled by genes.
- Biological evolution is a change in the genetic composition of a population of organisms over time.

Biological Evolution: Changes over Billions of Years

- Charles Darwin's theory of natural selection rests on three simple observations and one conclusion drawn from them: Any heritable traits that increase the probability that their bearers will survive and reproduce are passed on to their offspring. **Review Figure 1.2**

Major Events in the History of Life on Earth

- Life arose from nonlife about 4 billion years ago by means of chemical evolution. **Review Figure 1.3**

▶ Biological evolution began about 3.8 billion years ago when interacting systems of molecules became enclosed in membranes to form cells.

▶ Photosynthetic prokaryotes released large amounts of oxygen into Earth's atmosphere, making aerobic metabolism possible.

▶ Complex eukaryotic cells evolved by incorporation of smaller cells that survived being ingested.

▶ Multicellular organisms appeared when cells evolved the ability to transform themselves and to stick together and communicate after they divided. The individual cells of multicellular organisms became modified to carry out varied functions within the organism.

▶ The evolution of sex sped up rates of biological evolution.

Levels of Organization of Life

▶ Life is organized hierarchically, from molecules to the biosphere. **Review Figure 1.6.** **See Web/CD Activity 1.1**

The Evolutionary Tree of Life

▶ A major effort called Assembling the Tree of Life (ATOL) is underway to determine the evolutionary relationships among all species on Earth.

▶ The hierarchy of evolutionary relationships can be represented as an evolutionary tree. **Review Figure 1.8.** **See Web/CD Activity 1.2**

▶ Species are grouped into three domains: Archaea, Bacteria, and Eukarya. The domains Archaea and Bacteria consist of prokaryotic cells. The domain Eukarya contains the Protists, Plantae, Fungi, and Animalia.

Biology Is a Science

▶ Biologists use a variety of technical and conceptual tools to study living things.

▶ The hypothesis-prediction (H–P) approach is used in most biological investigations. Hypotheses are tentative answers to questions. Predictions are made on the basis of a hypothesis. The predictions are tested by experiments and comparative observations. **Review Figures 1.9 and 1.10**

▶ Science can tell us how the world works, but it does not form the basis for establishing meaning and values.

▶ Biologists are often called upon to advise governmental agencies on the solution of important problems that have a biological component.

For Discussion

1. The information Darwin used to develop his theory of evolution by natural selection was well known to his contemporaries. Why was it so difficult for people to think of such an obvious mechanism of evolutionary change?
2. According to the theory of evolution by natural selection, a species evolves certain features because they improve the chances that its members will survive and reproduce. There is no evidence, however, that evolutionary mechanisms have foresight or that organisms can anticipate future conditions. What, then, do biologists mean when they say, for example, that wings are "for flying"?
3. The first organisms appeared nearly 4 billion years ago, but multicellular organisms were slow to appear. Why did the evolution of multicellularity take so long?
4. Why is it so important in science that we design and perform tests capable of falsifying a hypothesis?
5. What features characterize questions that can be answered only by using a comparative approach?
6. Experiments show that not all amphibian declines are caused by a single factor. Does this surprise you? What kinds of environmental factors might be capable of affecting amphibian populations everywhere on Earth? What factors are likely to act only locally?

2 Life and Chemistry: Small Molecules

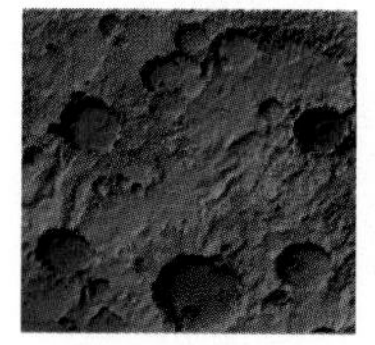

Mars today is a cold, dry place, not suitable for life as we know it. But 3 billion years ago, it was warmer and wetter. An orbiting probe from Earth recently photographed a huge dry lake bed, the size of New Mexico and Texas combined, on the Martian surface. Another probe found evidence of water trapped just below the icy surface of the Martian polar region. These discoveries by geologists have sparked the interest of biologists, for where there is water, there can be life. There is good reason to believe that life as we know it cannot exist without water.

Animals and plants that live on Earth's land masses had to evolve elaborate ways to retain the water that makes up about 70 percent of their bodies. Aquatic organisms living in water do not need these water-retention mechanisms; thus biologists have concluded that the first living things originated in a watery environment. This environment need not have been the lakes, rivers, and oceans with which we are familiar. Living organisms have been found in hot springs at temperatures above the usual boiling point of water, in a lake beneath the frozen Antarctic ice, in water trapped 2 miles below Earth's surface, in water 3 miles below the surface of the sea, in extremely acid and extremely salty water, and even in the water that cools the interiors of nuclear reactors.

With 20 trillion galaxies in the universe, each with 100 billion stars, there are many planets out there, and if our own solar system is typical, some of them have the water needed for life. As biologists contemplate how life could originate from nonliving matter, their attention focuses not just on the presence of water, but on what is dissolved in it.

A major discovery of biology is that living things are composed of the same types of chemical elements as the vast nonliving portion of the universe. This *mechanistic* view—that life is chemically based and obeys universal physicochemical laws—is a relatively recent one in human history. The concept of a "vital force" responsible for life, different from the forces found in physics and chemistry, was common in Western culture until the nineteenth century, and many people still assume such a force exists. However, most scientists adhere to a mechanistic view of life, and it is the cornerstone of medicine and agriculture.

A Grander Canyon on Mars This false color image from the Mars Global Surveyor shows in blue the dry remains of what was once a huge lake on Mars. Just as the Colorado River carved Earth's Grand Canyon, torrents of water from the lake probably carved the mile-deep canyon that is visible as a thin blue line just north of the lake bed.

Before describing how chemical elements are arranged in living creatures, we will examine some fundamental chemical concepts. We will first address the constituents of matter: atoms. We will examine their variety, their properties, and their capacity to combine with other atoms. Then we will consider how matter changes. In addition to changes in state (solid to liquid to gas), substances undergo changes that transform both their composition and their characteristic properties. Then we will describe the structure and properties of water and its relationship to acids and bases. We will close the chapter with a consideration of characteristic groups of atoms that contribute specific properties to larger molecules of which they are part, and which will be the subject of Chapter 3.

Water and the Origin of Life's Chemistry

Astronomers believe our solar system began forming about 4.6 billion years ago when a star exploded and collapsed to form the sun, and 500 or so bodies called planetesimals collided with one another to form the inner planets, including Earth. The first chemical signatures indicating the presence of life here are about 4 billion years old. So it took 600 million years, during a geological time frame called the Hadean, for the chemical conditions on Earth to become just right for life. Key among those conditions was the presence of water.

Ancient Earth probably had a lot of water high in the atmosphere. But the new planet was hot, and this water evaporated into space. As Earth cooled, it became possible for water to remain on its surface, but where did that water come from? One current view is that comets—loose agglomerations of dust and ice that have orbited the sun since the planets formed—struck Earth repeatedly and brought not only water but other chemical components of life, such as nitrogen. As Earth cooled, chemicals from the rocks dissolved in the water and simple chemical reactions took place. Some of these reactions could have led to life, but impacts by large comets and rocky meteorites would have released enough energy to heat the developing oceans almost to boiling, thus destroying any early life. These large impacts eventually subsided, and life gained a foothold about 3.8 to 4 billion years ago. The prebiotic Hadean was over (Figure 2.1). The Archean had begun, and there has been life on Earth ever since.

In Chapter 3 we will return to the question of how the first life could have arisen from inanimate chemicals. But before doing so, we need to understand what the chemistry of life entails. Like the rest of the chemical world, living things are made up of atoms and molecules.

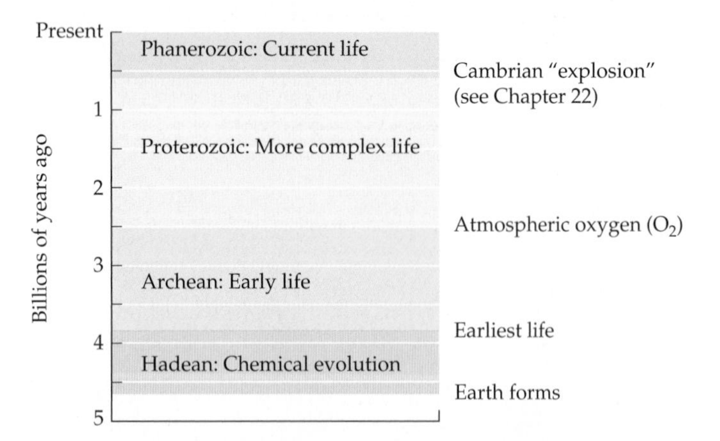

2.1 A Geological Time Scale The Hadean encompasses the time from the formation of Earth (about 4.6 billion years ago) until the earliest life appeared (about 3.8 billion years ago). During the Hadean chemical conditions evolved that were conducive to life, which was able to gain a foothold once the rain of comets and meteorites ended.

Atoms: The Constituents of Matter

More than a trillion (10^{12}) atoms could fit over the period at the end of this sentence. Each atom consists of a dense, positively charged **nucleus**, around which one or more negatively charged **electrons** move (Figure 2.2). The nucleus contains one or more **protons** and may contain one or more **neutrons**. Atoms and their component particles have volume and mass, which are properties of all matter. **Mass** measures the quantity of matter present; the greater the mass, the greater the quantity of matter.

The mass of a proton serves as a standard unit of measure: the atomic mass unit (amu), or *dalton* (named after the English chemist John Dalton). A single proton or neutron has a mass of about 1 dalton (Da), which is 1.7×10^{-24} grams (0.0000000000000000000000017 g). The mass of an electron is 9×10^{-28} g (0.0005 Da). Because the mass of an electron is negligible compare to the mass of a proton or a neutron, the

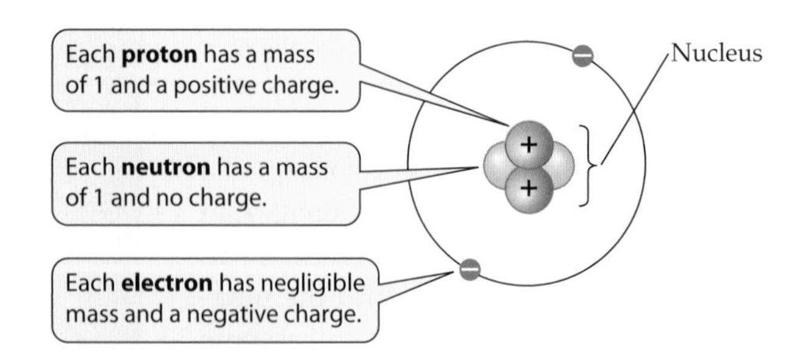

2.2 The Helium Atom This representation of a helium atom is called a Bohr model. It exaggerates the space occupied by the nucleus. In reality, although the nucleus accounts for virtually all of the atomic mass, it occupies only 1/10,000 of the atom's volume.

contribution of electrons to the mass of an atom can usually be ignored when measurements and calculations are made. It is electrons however, that determine how atoms will interact in chemical reactions, and we will discuss them extensively later in this chapter.

Each proton has a positive electric charge, defined as +1 unit of charge. An electron has a negative charge equal and opposite to that of a proton; thus the charge of an electron is –1 unit. The neutron, as its name suggests, is electrically neutral, so its charge is 0 unit. Unlike charges (+/–) attract each other; like charges (+/+ or –/–) repel each other. Atoms are electrically neutral: The number of protons in an atom equals the number of electrons.

An element is made up of only one kind of atom

An **element** is a pure substance that contains only one type of atom. The element hydrogen consists only of hydrogen atoms; the element iron consists only of iron atoms. The atoms of each element have certain characteristics or properties that distinguish them from the atoms of other elements. The more than 100 elements found in the universe are arranged in the periodic table (Figure 2.3). These elements are not found in equal amounts. Stars have abundant hydrogen and helium. Earth's crust, and those of the neighboring planets, are almost half oxygen, 28 percent silicon, 8 percent aluminum, 2–5 percent each of sodium, magnesium, potassium, calcium, and iron, and contain much smaller amounts of the other elements.

About 98 percent of the mass of every living organism (bacterium, turnip, or human) is composed of just six elements: carbon, hydrogen, nitrogen, oxygen, phosphorus, and sulfur. The chemistry of these six elements will be our primary con-

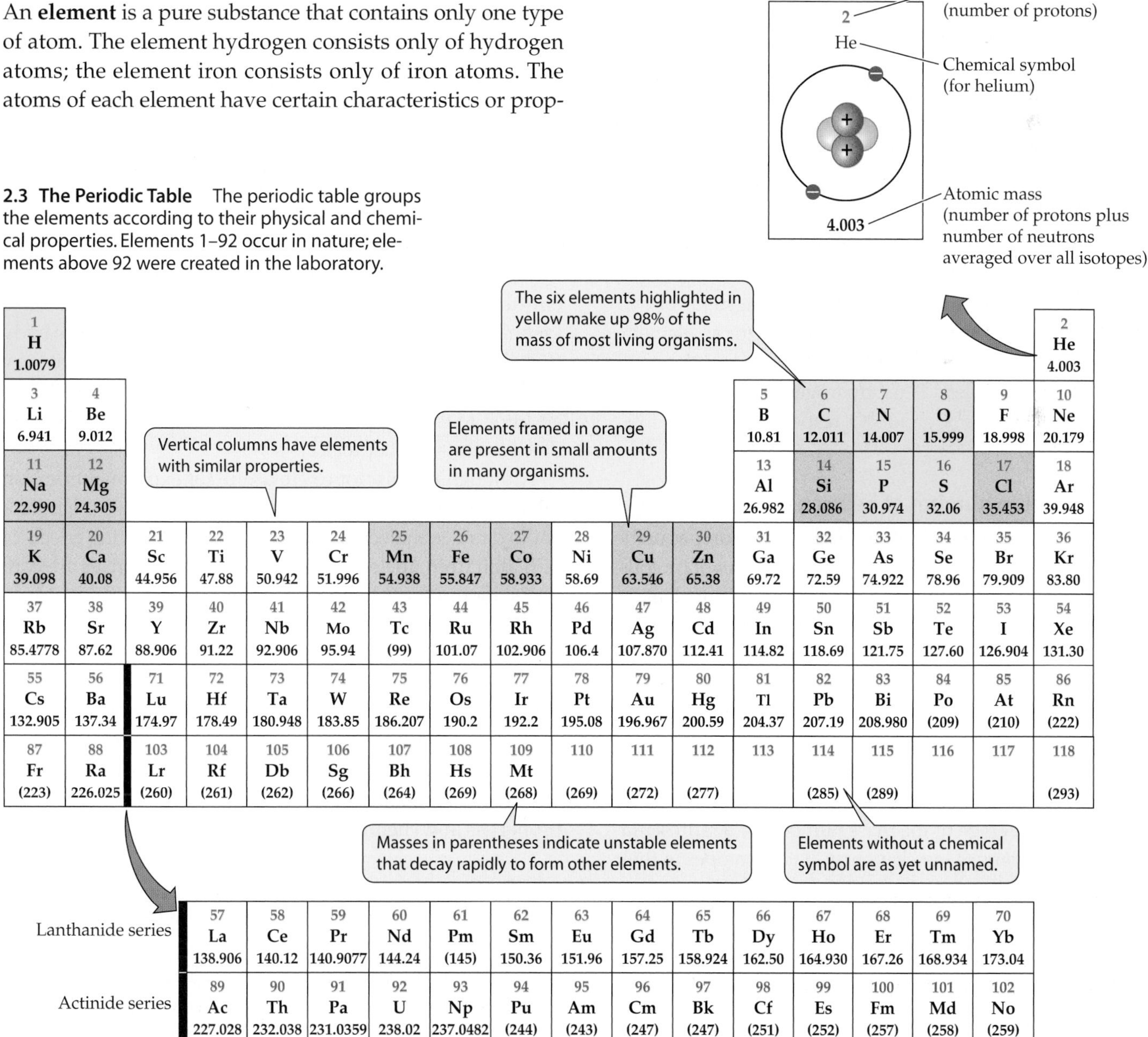

2.3 The Periodic Table The periodic table groups the elements according to their physical and chemical properties. Elements 1–92 occur in nature; elements above 92 were created in the laboratory.

cern here, but the others are not unimportant. Sodium and potassium, for example, are essential for nerves to function; calcium can act as a biological signal; iodine is a component of a vital hormone; and plants need magnesium as part of their green pigment (chlorophyll) and molybdenum in order to incorporate nitrogen into biologically useful substances.

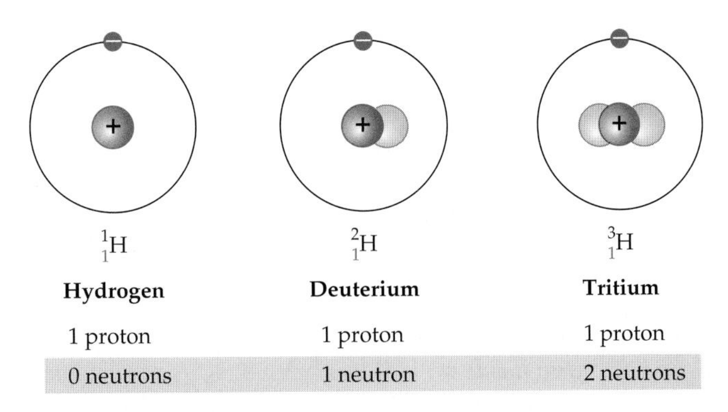

2.4 Isotopes Have Different Numbers of Neutrons The isotopes of hydrogen all have one proton in the nucleus, defining them as that element. Their differing mass numbers are due to different numbers of neutrons.

The number of protons identifies the element

An element is distinguished from other elements by the number of protons in each of its atoms. This number, which does not change, is called the *atomic number*. An atom of helium has 2 protons, and an atom of oxygen has 8 protons; the atomic numbers of these elements are thus 2 and 8, respectively.

Along with a definitive number of protons, every element except hydrogen has one or more neutrons in its nucleus. The *mass number* of an atom is the total number of protons and neutrons in its nucleus. The nucleus of a helium atom contains 2 protons and 2 neutrons; oxygen has 8 protons and 8 neutrons. Therefore, helium has a mass number of 4 and oxygen a mass number of 16. The mass number may be thought of as the mass of the atom in daltons.

Each element has its own one- or two-letter chemical symbol. For example, H stands for hydrogen, He for helium, and O for oxygen. Some symbols come from other languages: Fe (from the Latin, *ferrum*) stands for iron, Na (Latin, *natrium*) for sodium, and W (German, *Wolfram*) for tungsten.

In text, immediately preceding the symbol for an element, the atomic number is written at the lower left and the mass number at the upper left. Thus, hydrogen, carbon, and oxygen are written as ^1_1H, $^{12}_6C$, and $^{16}_8O$, respectively.

Isotopes differ in number of neutrons

Elements can have more than one atomic form. **Isotopes** of the same element all have the same, definitive, number of protons, but differ in the number of neutrons in the atomic nucleus.

In nature, many elements exist as several isotopes. The isotopes of hydrogen shown in Figure 2.4 have special names, but the isotopes of most elements do not have distinct names. For example, the natural isotopes of carbon are ^{12}C, ^{13}C, and ^{14}C (spoken of as carbon-12, carbon-13, and carbon-14). Most carbon atoms are ^{12}C, about 1.1 percent are ^{13}C, and a tiny fraction are ^{14}C. An element's atomic mass, or **atomic weight**,* is the average of the mass numbers of a representative sample of atoms of the element, with all isotopes in their normally occurring proportions. The atomic weight of carbon is thus calculated to be 12.011.

Some isotopes, called *radioisotopes*, are unstable and spontaneously give off energy as α (alpha), β (beta), or γ (gamma) radiation from the atomic nucleus. Such radioactive decay transforms the original atom into another atom, usually of another element. For example, carbon-14 loses a β-particle (actually an electron) to form ^{14}N. Biologists and physicians can incorporate radioisotopes into molecules and use the emitted radiation as a tag to locate those molecules or to identify changes that the molecules undergo inside the body (Figure 2.5). Three radioisotopes commonly used in this way are ^3H (tritium), ^{14}C (carbon-14), and ^{32}P (phosphorus-32). In addition to these applications, radioisotopes can be used to date fossils (see Chapter 22).

Although radioisotopes are useful for experiments and in medicine, even low doses of their radiation have the potential to damage molecules and cells. The devastating effects of radiation from nuclear weapons are well known, as are concerns about possible damage to organisms from isotopes used in nuclear power plants. In medicine, γ-radiation from ^{60}Co (cobalt-60) is used to damage or kill cancer cells.

In discussing isotopes and radioactivity, we have focused on the nucleus of the atom, but the nucleus is not directly involved in the ability of atoms to combine with other atoms. That ability is determined by the number and distribution of electrons. In the following sections, we describe some of the properties and chemical behavior of electrons.

Electron behavior determines chemical bonding

When considering atoms, biologists are concerned primarily with electrons because the behavior of electrons explains how chemical changes occur in living cells. These changes, called *chemical reactions* or just *reactions*, are changes in the atomic

*The concepts of "weight" and "mass" are not identical. Weight is the measure of the Earth's gravitational attraction for mass; on another planet, the same quantity of mass would have a different weight. On Earth, however, the term "weight" is often used as a measure of mass, and in biology one encounters the terms "weight" and "atomic weight" more frequently than "mass" and "atomic mass." Therefore, we will use "weight" for the remainder of this book.

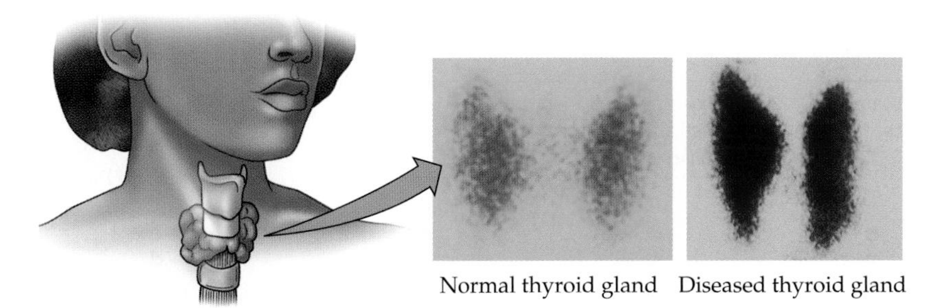
Normal thyroid gland
Diseased thyroid gland

2.5 A Radioisotope Used in Medicine The thyroid gland takes up iodine and uses it to make thyroid hormone. A patient suspected of having thyroid disease can be injected with radioactive iodine, which allows the thyroid gland to be visualized by a scanning device.

composition of substances. The characteristic number of electrons in each atom of an element determines how its atoms will react with other atoms. All chemical reactions involve changes in the relationships of electrons with one another.

The location of a given electron in an atom at any given time is impossible to determine. We can only describe a volume of space within the atom where the electron is likely to be. The region of space where the electron is found at least 90 percent of the time is the electron's *orbital* (Figure 2.6). In an atom, a given orbital can be occupied by at most two electrons. Thus any atom larger than helium (atomic number 2) must have electrons in two or more orbitals. As Figure 2.6 shows, the different orbitals have characteristic forms and orientations in space. The orbitals, in turn, constitute a series of *electron shells*, or energy levels, around the nucleus (Figure 2.7).

- The innermost electron shell consists of only one orbital, called an *s* orbital. Hydrogen ($_1H$) has one electron in its first shell; helium ($_2He$) has two. All other elements have two first-shell electrons, as well as electrons in other shells.
- The second shell is made up of four orbitals (an *s* orbital and three *p* orbitals) and hence can hold up to eight electrons.

The *s* orbitals fill with electrons first, and their electrons have the lowest energy. Subsequent shells have different numbers of orbitals, but the outermost shells usually hold only eight electrons. In any atom, the outermost electron shell determines how the atom combines with other atoms—that is, how the atom behaves chemically. When an outermost shell consisting of four orbitals contains eight electrons, there are no unpaired electrons (see Figure 2.7). Such an atom is *stable* and will not react with other atoms. Examples of chemically stable elements are helium, neon, and argon.

Reactive atoms seek to attain the stable condition of having no unpaired electrons in their outermost shells. They attain this stability by sharing electrons with other atoms, or by gaining or losing one or more electrons. In either case, the atoms are bonded together. Such bonds create stable associations of atoms called molecules.

A **molecule** is two or more atoms linked by chemical bonds. The tendency of atoms in stable molecules to have eight electrons in their outermost shells is known as the octet rule. Many atoms in biologically important molecules—for example, carbon (C) and nitrogen (N)—follow the *octet rule*. However, some biologically important atoms are exceptions to the rule. Hydrogen (H) is the most obvious exception, attaining stability when only two electrons occupy its single shell.

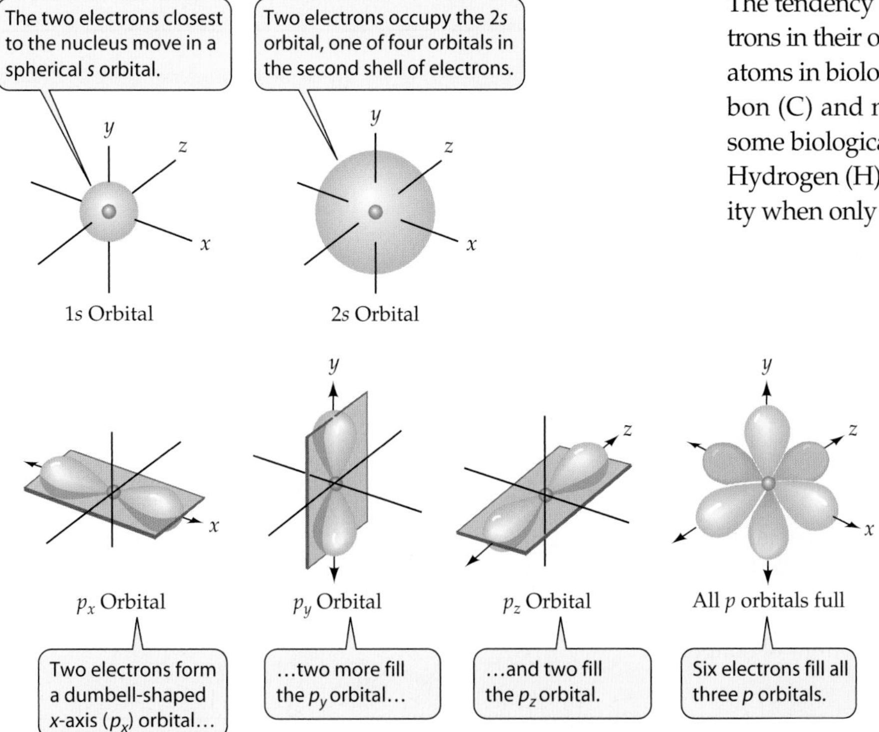
The two electrons closest to the nucleus move in a spherical *s* orbital.
Two electrons occupy the 2*s* orbital, one of four orbitals in the second shell of electrons.
y
z
x
1*s* Orbital
2*s* Orbital
p_x Orbital
p_y Orbital
p_z Orbital
All *p* orbitals full
Two electrons form a dumbell-shaped *x*-axis (p_x) orbital...
...two more fill the p_y orbital...
...and two fill the p_z orbital.
Six electrons fill all three *p* orbitals.

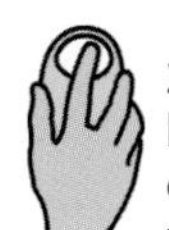

2.6 Electron Orbitals Each orbital holds a maximum of two electrons. The *s* orbitals have a lower energy level and fill with electrons before the *p* orbitals do.

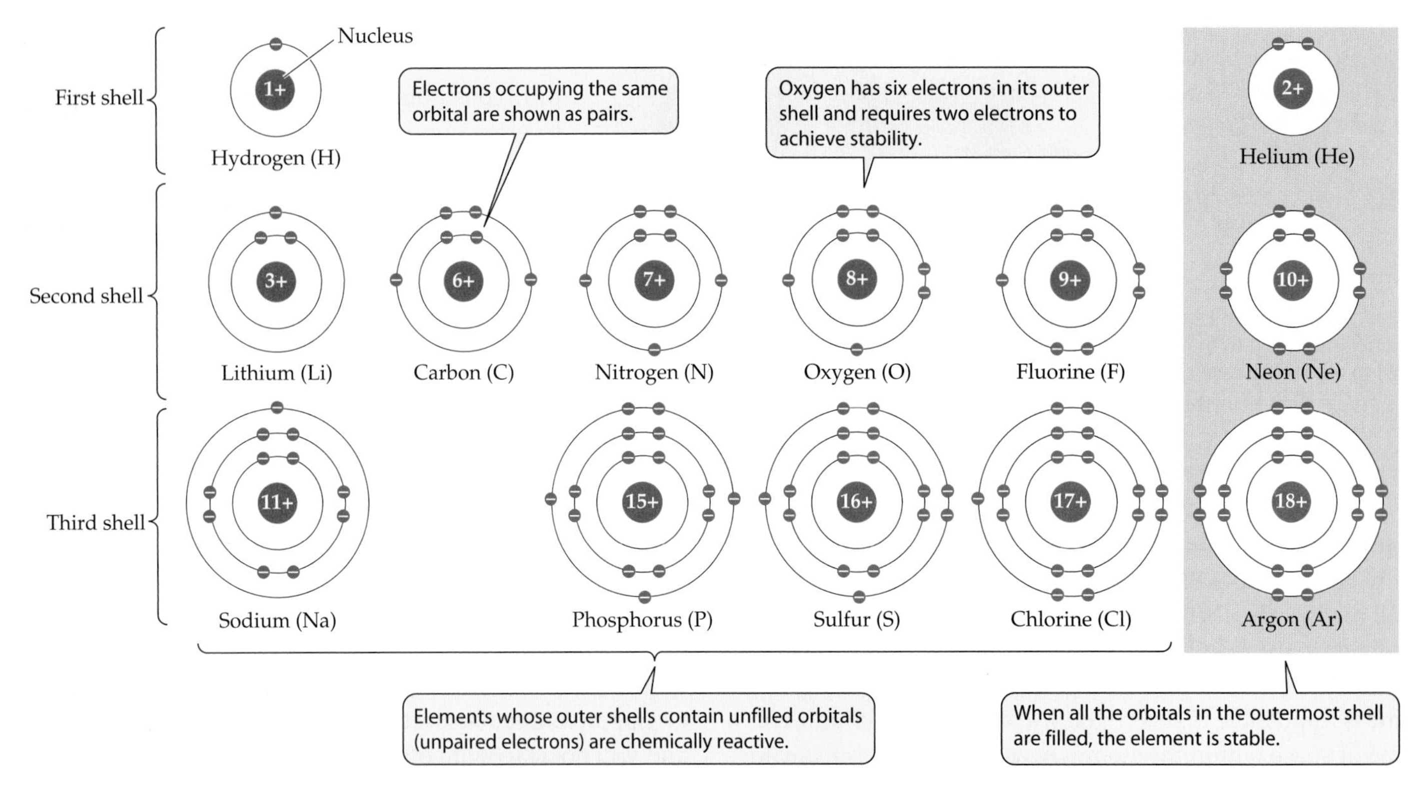

2.7 Electron Shells Determine the Reactivity of Atoms Each orbital holds a maximum of two electrons, and each shell can hold a specific maximum number of electrons. Each shell must be filled before electrons move into the next shell. The energy level of electrons is higher in shells farther from the nucleus. An atom with unpaired electrons in its outermost shell may react (bond) with other atoms.

Chemical Bonds: Linking Atoms Together

A **chemical bond** is an attractive force that links two atoms together to form a molecule. There are several kinds of chemical bonds (Table 2.1). In this section, we will first discuss co-

2.1 *Chemical Bonds and Interactions*

NAME	BASIS OF INTERACTION	STRUCTURE	BOND ENERGY[a] (KCAL/MOL)
Covalent bond	Sharing of electron pairs	—N(H)—C(=O)—	50–110
Hydrogen bond	Sharing of H atom	—N(H)—H$^{\delta+}$ ••••• $^{\delta-}$O=C—	3–7
Ionic bond	Attraction of opposite charges	—N(H)(H)—H ⊕ ⊖ O—C(=O)—	3–7
Hydrophobic interaction	Interaction of nonpolar substances in the presence of polar substances	—C(H)(H)—C(H)(H)—H∘∘∘∘∘H—C(H)(H)—C(H)(H)—	1–2
van der Waals interaction	Interaction of electrons of nonpolar substances	H—H ⋯ CH_4	1

[a]*Bond energy* is the amount of energy needed to separate two bonded or interacting atoms under physiological conditions.

valent bonds, the strong bonds that result from the sharing of electrons. Then we will examine other kinds of interactions, including hydrogen bonds, that are weaker than covalent bonds but enormously important to biology. Finally, we will consider ionic bonding, which is a consequence of the loss or gain of electrons by atoms.

Covalent bonds consist of shared pairs of electrons

When two atoms attain stable electron numbers in their outermost shells by sharing one or more pairs of electrons, a **covalent bond** forms. Consider two hydrogen atoms in close proximity, each with a single unpaired electron in its outer shell. Each positively charged nucleus attracts the other atom's unpaired electron, but this attraction is balanced by each electron's attraction to its own nucleus. Thus the two unpaired electrons become shared by both atoms, filling the outer shells of both of them (Figure 2.8). The two atoms are thus linked by a covalent bond, and a hydrogen gas molecule (H_2) is formed.

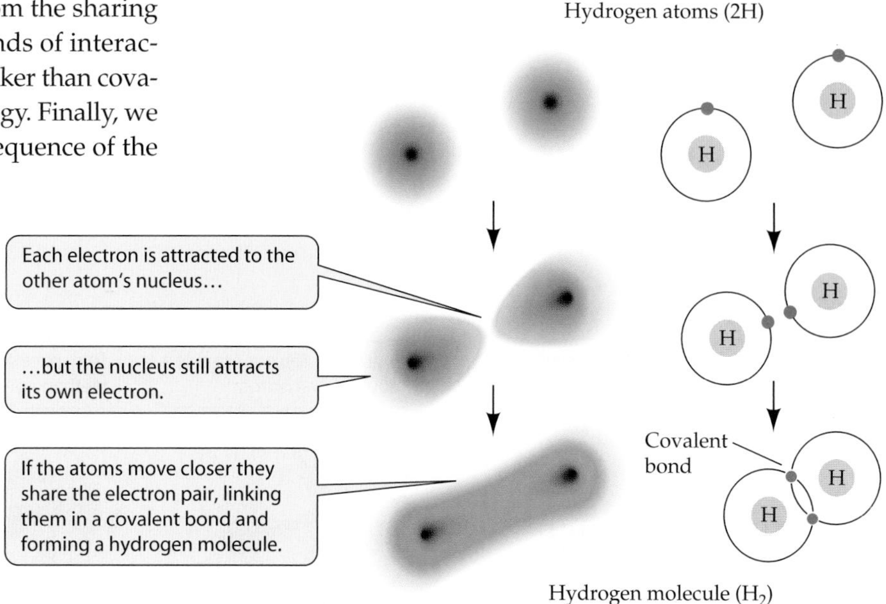
Hydrogen atoms (2H)
H
H
Each electron is attracted to the other atom's nucleus…
…but the nucleus still attracts its own electron.
H
H
If the atoms move closer they share the electron pair, linking them in a covalent bond and forming a hydrogen molecule.
Covalent bond
H
H
Hydrogen molecule (H_2)

2.8 Electrons Are Shared in Covalent Bonds Two hydrogen atoms combine to form a hydrogen molecule. Each electron is attracted to both protons. A covalent bond forms when the electron orbitals of the two atoms overlap.

A molecule made up of more than one type of atom is called a compound. A molecular formula uses chemical symbols to identify the different atoms in a compound and subscript numbers to show how many of each type of atoms are present. Thus, the formula for sucrose—table sugar—is $C_{12}H_{22}O_{11}$. Each compound has a **molecular weight** (molecular mass) that is the sum of the atomic weights of all atoms in the molecule. Looking at the periodic table in Figure 2.3, you can calculate the molecular weight of table sugar to be 342. Molecular weights are usually related to a molecule's size (Figure 2.9).

A carbon atom has a total of six electrons; two electrons fill its inner shell and four are in its outer shell. Because its outer shell can hold up to eight electrons, carbon can share electrons with up to four other atoms—it can form four covalent bonds. When an atom of carbon reacts with four hydrogen atoms, a molecule called methane (CH_4) forms (Figure 2.10*a*). Thanks to electron sharing, the outer shell of methane's carbon atom is filled with eight electrons, and the outer shell of each hydrogen atom is also filled. Four covalent bonds—each consisting of a shared pair of electrons—hold methane together. Table 2.2 shows the covalent bonding capacities of some biologically significant elements.

ORIENTATION OF COVALENT BONDS. Covalent bonds are very strong. The thermal energy that biological molecules ordinarily have at body temperature is less than 1 percent of that needed to break covalent bonds. So biological molecules, most of which are put together with covalent bonds, are quite stable. This means that their three-dimensional structures and the spaces they occupy are also stable. A second property of covalent bonds is that, for a given pair

2.9 Weights and Sizes of Atoms and Molecules The color conventions used here are standard for the atoms. (Yellow is used for sulfur and phosphorus atoms, which are not depicted.)

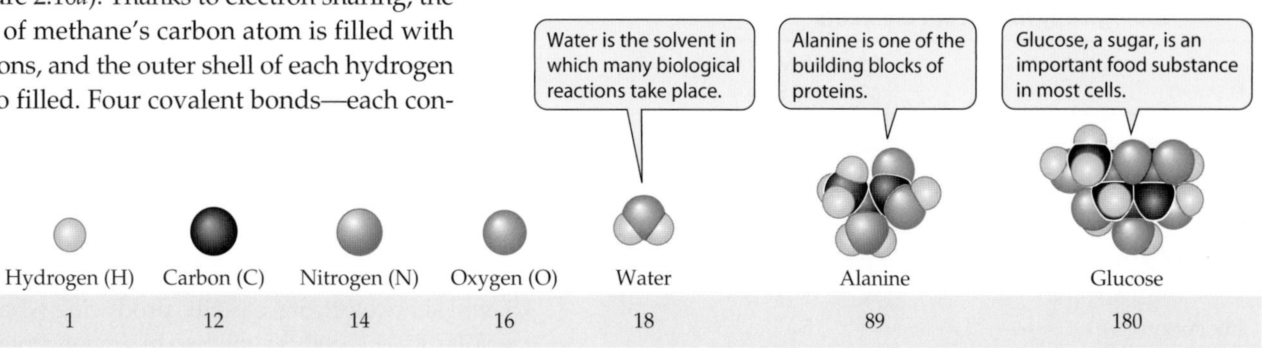
Water is the solvent in which many biological reactions take place.
Alanine is one of the building blocks of proteins.
Glucose, a sugar, is an important food substance in most cells.
Hydrogen (H)
Carbon (C)
Nitrogen (N)
Oxygen (O)
Water
Alanine
Glucose
Molecular weights
1
12
14
16
18
89
180

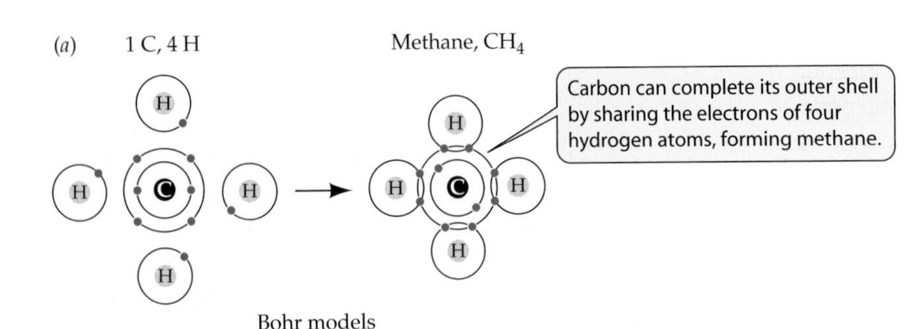
(a)
1 C, 4 H
Methane, CH_4
Carbon can complete its outer shell by sharing the electrons of four hydrogen atoms, forming methane.
Bohr models

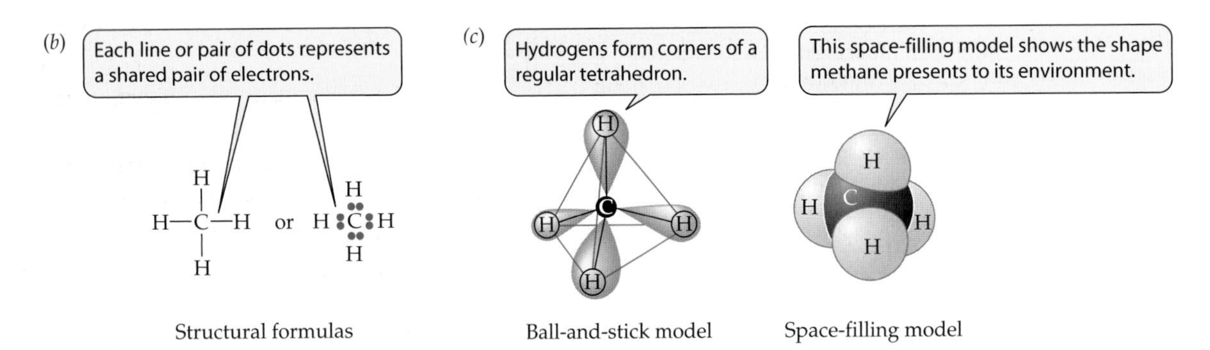
(b)
Each line or pair of dots represents a shared pair of electrons.
Structural formulas
(c)
Hydrogens form corners of a regular tetrahedron.
Ball-and-stick model
This space-filling model shows the shape methane presents to its environment.
Space-filling model

2.10 Covalent Bonding with Carbon Different representations of covalent bond formation in methane, whose molecular formula is CH_4. (*a*) Diagram illustrating the filling and stabilizing of the outer electron shells in carbon and hydrogen atoms. (*b*) Two common structural formulas used to represent bonds. (*c*) Two ways of representing the spatial orientation of bonds.

of atoms, they are the same in length, angle, and direction, regardless of the larger molecule of which the particular bond is a part. The four filled orbitals around the carbon nucleus of methane, for example, distribute themselves in space so that the bonded hydrogens are directed to the corners of a regular tetrahedron with carbon in the center (Figure 2.10*c*). The three-dimensional structure of carbon and hydrogen is the same in a large, complicated protein as it is in the simple methane molecule. This property of covalent bonds makes the prediction of biological structure possible.

Although the orientations of orbitals and the shapes of molecules differ depending on the types of atoms involved and how they are linked together, it is essential to remember that all molecules occupy space and have three-dimensional shapes. The shapes of molecules contribute to their biological functions, as we will see in Chapter 3.

2.2 Covalent Bonding Capabilities of Some Biologically Important Elements

ELEMENT	USUAL NUMBER OF COVALENT BONDS
Hydrogen (H)	1
Oxygen (O)	2
Sulfur (S)	2
Nitrogen (N)	3
Carbon (C)	4
Phosphorus (P)	5

MULTIPLE COVALENT BONDS. A covalent bond is represented by a line between the chemical symbols for the linked atoms:

- A single bond involves the sharing of a single pair of electrons (for example, H—H, C—H).
- A double bond involves the sharing of four electrons (two pairs) (C=C).

Triple bonds (six shared electrons) are rare, but there is one in nitrogen gas (N≡N), the chief component of the air we breathe.

UNEQUAL SHARING OF ELECTRONS. If two atoms of the same element are covalently bonded, there is an equal sharing of the pair(s) of electrons in the outer shell. However, when the two atoms are of different elements, the sharing is not necessarily equal. One nucleus may exert a greater attractive force on the electron pair than the other nucleus, so that the pair tends to be closer to that atom.

The attractive force that an atom exerts on electrons is its **electronegativity**. It depends on how many positive charges a nucleus has (nuclei with more protons are more positive and thus more attractive to electrons) and how far away the electrons are from the nucleus (closer means more electronegativity). The closer two atoms are in electronegativity, the more equal their sharing of electrons will be.

Table 2.3 shows the electronegativities of some elements important in biological systems. Looking at the table, it is obvious that two oxygen atoms, both with electronegativity of 3.5, will share electrons equally, producing what is called a *nonpolar covalent bond*. So will two hydrogen atoms (both 2.1).

2.3 Some Electronegativities

ELEMENT	ELECTRONEGATIVITY
Oxygen (O)	3.5
Chlorine (Cl)	3.1
Nitrogen (N)	3.0
Carbon (C)	2.5
Phosphorus (P)	2.1
Hydrogen (H)	2.1
Sodium (Na)	0.9
Potassium (K)	0.8

But when hydrogen bonds with oxygen to form water, the electrons involved are unequally shared: they tend to be nearer to the oxygen nucleus because it is the more electronegative of the two. The result is called a *polar* covalent bond (Figure 2.11).

Because of this unequal sharing of electrons, the oxygen end of the hydrogen–oxygen bond has a slightly negative charge (symbolized δ^- and spoken as "delta negative," meaning a partial unit of charge), and the hydrogen end is slightly positive (δ^+). The bond is **polar** because these opposite charges are separated at the two ends, or poles, of the bond. The partial charges that result from polar covalent bonds produce polar molecules or polar regions of large molecules. Polar bonds greatly influence the interactions between molecules that contain them.

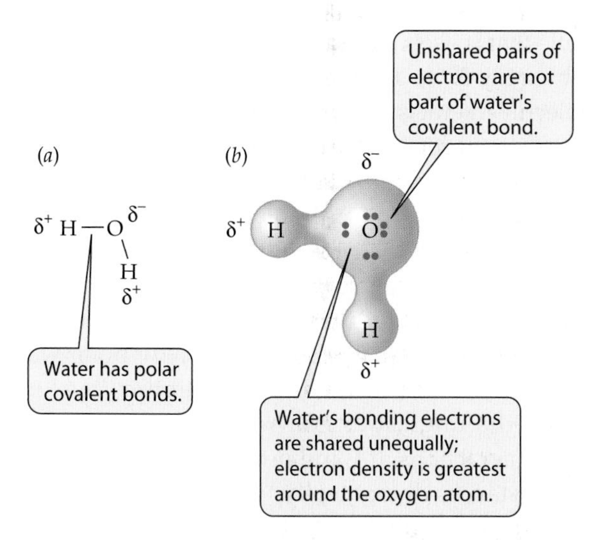

2.11 The Polar Covalent Bond in the Water Molecule (*a*) A covalent bond between atoms with different electronegativities is a polar covalent bond, and has partial (δ) charges at the ends. (*b*) In water, the electrons are displaced toward the oxygen atom and away from the hydrogen atoms.

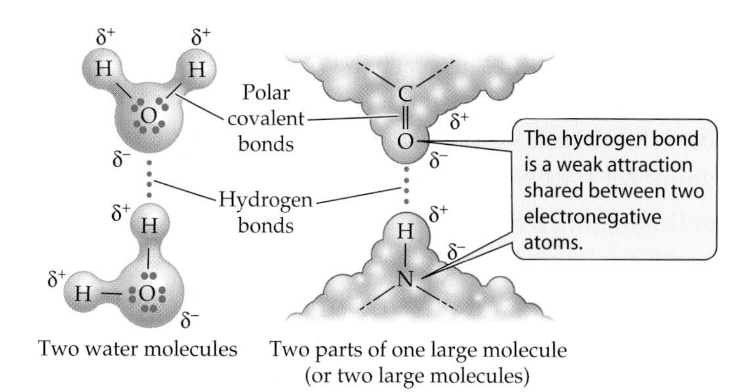

2.12 Hydrogen Bonds Can Form between or within Molecules Hydrogen bonds can form between two molecules or, if a molecule is large, between two different parts of the same molecule. Covalent and polar covalent bonds, on the other hand, are always found within molecules.

Hydrogen bonds may form within or between atoms with polar covalent bonds

In liquid water, the negatively charged oxygen (δ^-) atom of one water molecule is attracted to the positively charged hydrogen (δ^+) atoms of another water molecule. (Remember, negative charges attract positive charges.) The bond resulting from this attraction is called a **hydrogen bond**.

Hydrogen bonds are not restricted to water molecules. They may form between an electronegative atom and a hydrogen atom covalently bonded to a different electronegative atom (Figure 2.12). A hydrogen bond is a weak bond; it has about one-tenth (10%) the strength of a covalent bond between a hydrogen atom and an oxygen atom (see Table 2.1). However, where many hydrogen bonds form, they have considerable strength and greatly influence the structure and properties of substances. Later in this chapter we'll see how hydrogen bonding in water contributes to many of the properties that make water significant for living systems. Hydrogen bonds also play important roles in determining and maintaining the three-dimensional shapes of giant molecules such as DNA and proteins (see Chapter 3).

Ionic bonds form by electrical attraction

When one interacting atom is much more electronegative than the other, a complete transfer of one or more electrons may take place. Consider sodium (electronegativity 0.9) and chlorine (3.1). A sodium atom has only one electron in its outermost shell; this condition is unstable. A chlorine atom has seven electrons in its outer shell—another unstable condition. Since the electronegativities of these elements are so different, any electrons involved in bonding will tend to be much nearer to the chlorine nucleus—so near, in fact, that there is a complete transfer of the electron from one element to the other (Figure 2.13). This reaction between sodium and chlorine makes the resulting atoms more stable. The result is two **ions**. Ions are electrically charged particles that form when atoms gain or lose one or more electrons.

- The sodium ion (Na^+) has a +1 unit charge because it has one less electron than it has protons. The outermost electron shell of the sodium ion is full, with eight electrons, so the ion is stable. Positively charged ions are called **cations**.
- The chloride ion (Cl^-) has a –1 unit charge because it has one more electron than it has protons. This additional electron gives Cl^- a stable outermost shell with eight electrons. Negatively charged ions are called **anions**.

Some elements form ions with multiple charges by losing or gaining more than one electron. Examples are Ca^{2+} (calcium ion, created from a calcium atom that has lost two electrons) and Mg^{2+} (magnesium ion). Two biologically important elements each yield more than one stable ion: Iron yields Fe^{2+} (ferrous ion) and Fe^{3+} (ferric ion), and copper yields Cu^+ (cuprous ion) and Cu^{2+} (cupric ion). Groups of covalently bonded atoms that carry an electric charge are called *complex ions;* examples include NH_4^+ (ammonium ion), SO_4^{2-} (sulfate ion), and PO_4^{3-} (phosphate ion).

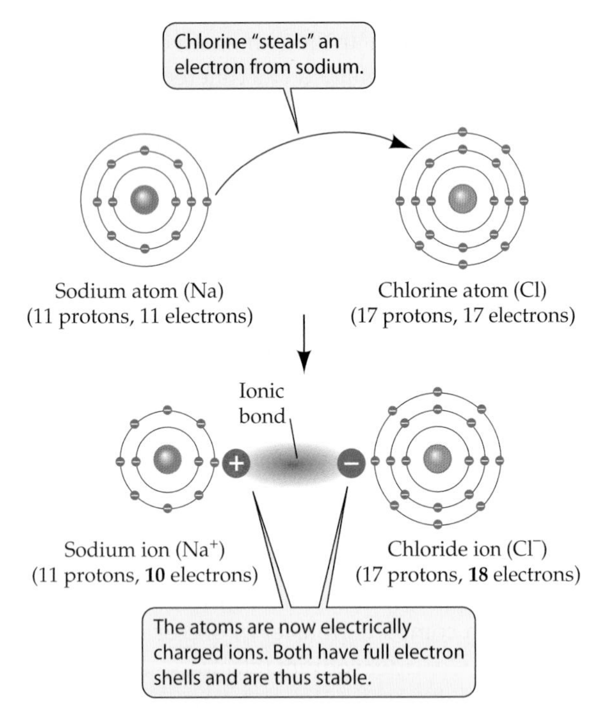

2.13 Formation of Sodium and Chloride Ions When a sodium atom reacts with a chlorine atom, the more electronegative chlorine acquires a more stable, filled outer shell by obtaining an electron from the sodium. In so doing, the chlorine atom becomes a negatively charged chloride ion (Cl^-). The sodium atom, upon losing the electron, becomes a positively charged sodium ion (Na^+).

The charge from an ion radiates from it in all directions. Once formed, ions are usually stable, and no more electrons are lost or gained. Ions can form stable bonds, resulting in stable solid compounds such as sodium chloride (NaCl) and potassium phosphate (K_3PO_4).

Ionic bonds are bonds formed by electrical attraction between ions bearing opposite charges. In sodium chloride—familiar to us as table salt—cations and anions are held together by ionic bonds. In solids, the ionic bonds are strong because the ions are close together. However, when ions are dispersed in water, the distance between them can be large; the strength of their attraction is thus greatly reduced. Under the conditions that exist in the cell, an ionic attraction is less than one-tenth as strong as a covalent bond that shares electrons equally (see Table 2.1).

Not surprisingly, ions with one or more units of charge can interact with polar molecules as well as with other ions. Such interaction results when table salt, or any other ionic solid, dissolves in water: "shells" of water molecules surround the individual ions, separating them (Figure 2.14). The hydrogen bond that we described earlier is a type of ionic bond because it is formed by electrical attraction. However, it is weaker than most ionic bonds because it is formed by partial charges (δ^+ and δ^-) rather than by whole-unit charges (+1 unit, –1 unit).

Polar and nonpolar substances interact best among themselves

"Like attracts like" is an old saying, and nowhere is it more true than in polar and nonpolar molecules, which tend to interact with their own kind. Just as water molecules interact with one another through polarity-induced hydrogen bonds, any molecule that is itself polar will interact with other polar molecules by weak (δ^+ to δ^-) attractions in hydrogen bonds. If a polar molecule interacts with water in this way, it is called **hydrophilic** ("water-loving").

What about nonpolar molecules? For example, carbon (electronegativity 2.5) forms nonpolar bonds with hydrogen (electronegativity 2.1). The resulting *hydrocarbon molecule*—that is, a molecule containing only hydrogen and carbon atoms—is nonpolar, and in water it tends to aggregate with other nonpolar molecules rather than with polar water. Such molecules are known as **hydrophobic** ("water-hating"), and the interactions between them are called hydrophobic interactions. It is important to realize that hydrophobic substances do not really "hate" water; they can form weak interactions with it (recall that the electronegativities of carbon and hydrogen are not exactly the same). But these interactions are far weaker than the hydrogen bonds between the water molecules, and so the nonpolar substances keep to themselves.

2.14 Water Molecules Surround Ions When an ionic solid dissolves in water, polar water molecules cluster around cations or anions, blocking their reassociation into a solid and forming a solution.

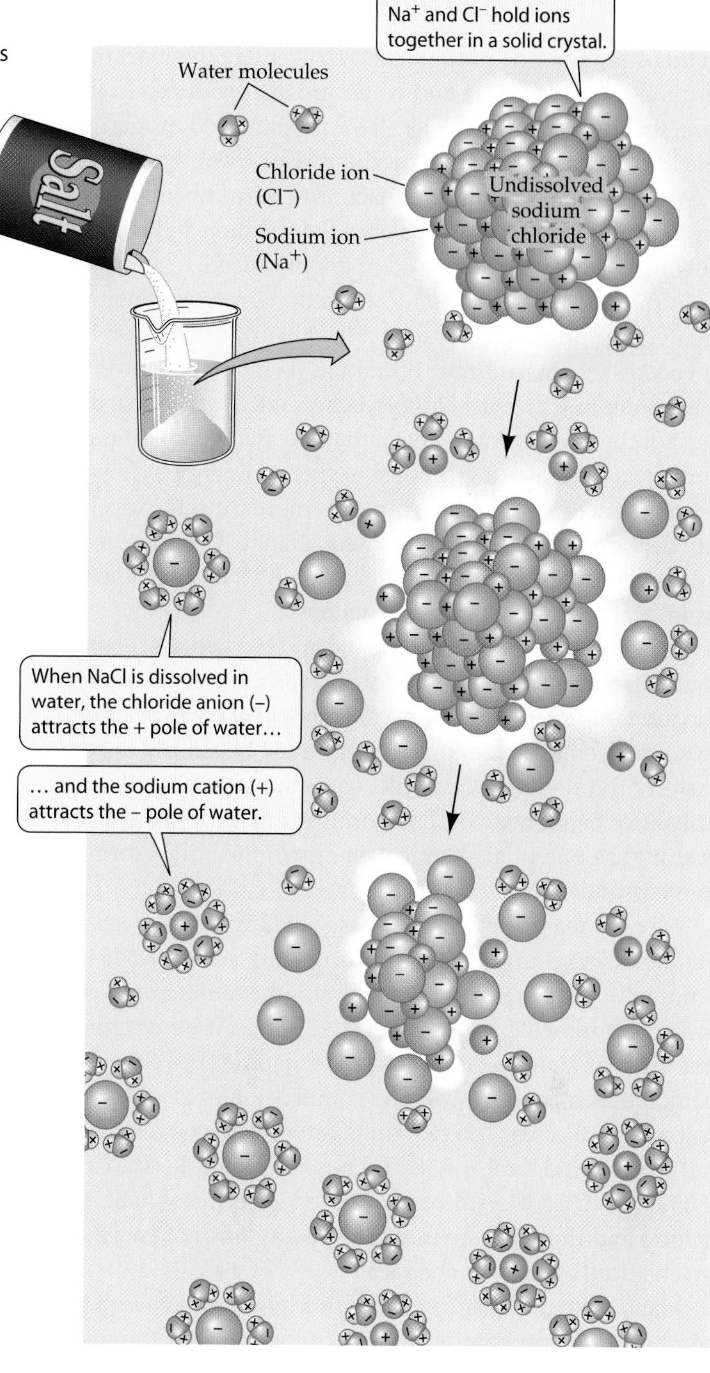
Ionic bonds between Na^+ and Cl^- hold ions together in a solid crystal.
Water molecules
Salt
Chloride ion (Cl^-)
Sodium ion (Na^+)
Undissolved sodium chloride
When NaCl is dissolved in water, the chloride anion (–) attracts the + pole of water…
… and the sodium cation (+) attracts the – pole of water.

These weak interactions between nonpolar substances are enhanced by **van der Waals forces**, which result when two atoms of nonpolar molecules are in close proximity. These brief interactions are a result of random variations in the electron distribution in one molecule, which create an opposite charge distribution in the adjacent molecule. Although a single van der Waals interaction is brief and weak at any given site, the summation of many such interactions over the entire span of a large nonpolar molecule can produce substantial attraction. van der Waals forces are important in maintaining the structures of many biologically important substances.

Chemical Reactions: Atoms Change Partners

A **chemical reaction** occurs when atoms combine or change their bonding partners. Consider the combustion reaction that takes place in the flame of a propane stove. When propane (C_3H_8) reacts with oxygen gas (O_2), the carbon atoms become bonded to oxygen atoms instead of to hydrogen atoms, and the hydrogen atoms become bonded to oxygen instead of carbon (Figure 2.15). As the covalently bonded atoms change partners, the composition of the matter changes, and propane and oxygen gas become carbon dioxide and water. This chemical reaction can be represented by the equation

$$C_3H_8 + 5\ O_2 \rightarrow 3\ CO_2 + 4\ H_2O + \text{energy}$$

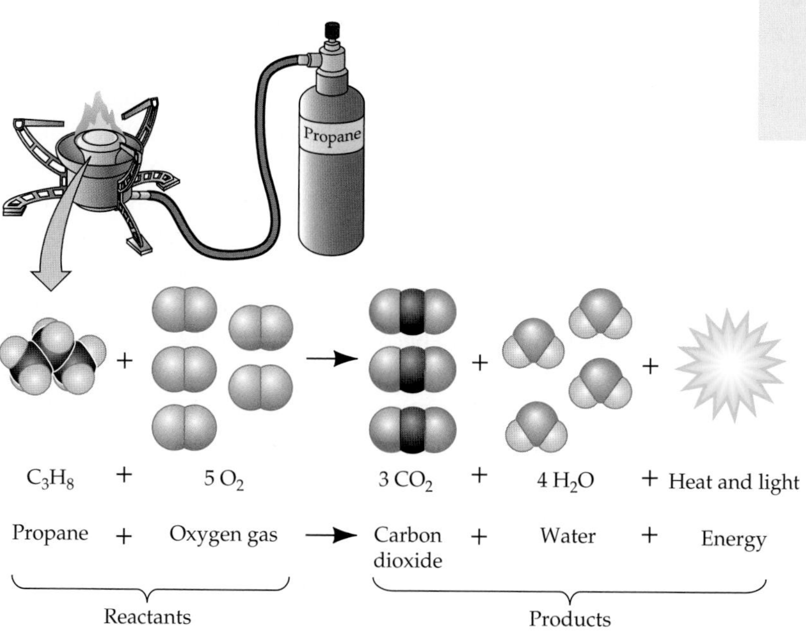
Propane
C_3H_8 + $5\ O_2$ → $3\ CO_2$ + $4\ H_2O$ + Heat and light
Propane + Oxygen gas → Carbon dioxide + Water + Energy
Reactants
Products

2.15 Bonding Partners and Energy May Change in a Chemical Reaction One molecule of propane reacts with five molecules of oxygen gas to give three molecules of carbon dioxide and four molecules of water. This reaction releases energy in the form of heat and light.

In this equation, the propane and oxygen are the **reactants**, and the carbon dioxide and water are the **products**. In this case, the reaction is complete: all the propane and oxygen are used up in forming the two products. The arrow symbolizes the direction of the chemical reaction. The numbers preceding the molecular formulas balance the equation and indicate how many molecules are used or are produced.

In this and all other chemical reactions, matter is neither created nor destroyed. The total number of carbons on the left equals the total number of carbons on the right. However, there is another product of this reaction: energy. The heat and light of the stove's flame reveal that the reaction of propane and oxygen releases a great deal of energy. **Energy** is defined as the capacity to do work, but on a more intuitive level, it can be thought of as the capacity for change. Chemical reactions do not create or destroy energy, but changes in energy usually accompany chemical reactions.

In the reaction between propane and oxygen, the energy that was released as heat and light was already present in the reactants in another form, called *potential chemical energy*. In some chemical reactions, energy must be supplied from the environment (for example, some substances will react only after being heated), and some of this supplied energy is stored as potential chemical energy in the bonds formed in the products.

We can measure the energy associated with chemical reactions using a unit called a **calorie** (cal). A calorie* is the amount of heat energy needed to raise the temperature of 1 gram of pure water from 14.5°C to 15.5°C. Another unit of energy that is increasingly used is the joule (J). When you compare data on energy, always compare joules to joules and calories to calories. The two units can be interconverted: 1 J = 0.239 cal, and 1 cal = 4.184 J. Thus, for example, 486 cal = 2,033 J, or 2.033 kJ. Although defined in terms of heat, the calorie and the joule are measures of any form of energy—mechanical, electric, or chemical.

Many biological reactions have much in common with the combustion of propane. The fuel is different—it is the sugar glucose, rather than propane—and the reactions proceed by many intermediate steps that permit the energy released from the glucose to be harvested and put to use by the cell. But the products are the same: carbon dioxide and water. These reactions were key to the origin of life from simpler molecules.

We will present and discuss energy changes, oxidation–reduction reactions, and several other types of chemical reactions that are prevalent in living systems in the chapters that follow.

*The nutritionist's or dieter's Calorie, with a capital C, is what biologists call a kilocalorie (kcal) and is equal to 1,000 heat-energy calories.

Water: Structure and Properties

Water, like all other matter, can exist in three states: solid (ice), liquid, or gas (vapor). Liquid water is probably the medium in which life originated on Earth, and it is in water that life evolved for its first billion years. In this section, we will explore how the structure and interactions of water molecules make water essential to life.

Water has a unique structure and special properties

The water molecule, H_2O, has unique chemical features. As we learned in the preceding sections, water is a polar molecule that can form hydrogen bonds. In addition, the shape of water is a tetrahedron. The four pairs of electrons in the outer shell of oxygen repel one another, producing a tetrahedral shape.

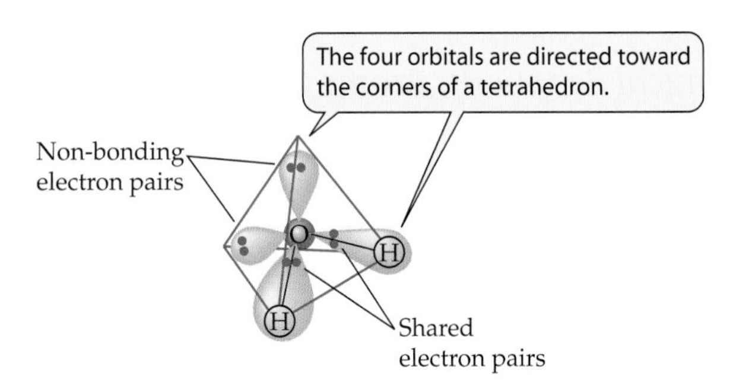

These chemical features explain some of the interesting properties of water, such as the ability of ice to float, the melting and freezing temperatures of water, the ability of water to store heat, and the ability of water droplets to form. These properties are described in detail below.

ICE FLOATS. In water's solid state (ice), individual water molecules are held in place by hydrogen bonds, creating a rigid, crystalline structure in which each water molecule is hydrogen-bonded to four other water molecules (Figure 2.16*a*). Although the molecules are held firmly in place, they are not as tightly packed as they are in liquid water (Figure 2.16*b*). In other words, solid water is less dense than liquid water, which is why ice floats in water.

If ice were to sink in water, as almost all other solids do in their corresponding liquids, ponds and lakes would freeze from the bottom up, becoming solid blocks of ice in winter and killing most of the organisms living in them. Once the whole pond had frozen, its temperature could drop well below the freezing point of water. But, because ice floats, it forms a protective insulating layer on the top of the pond, reducing heat flow to the cold air above. Thus fish, plants, and other organisms in the pond are not subjected to temperatures lower than 0°C, the freezing point of pure water. The recent discovery of liquid water below the polar ice on

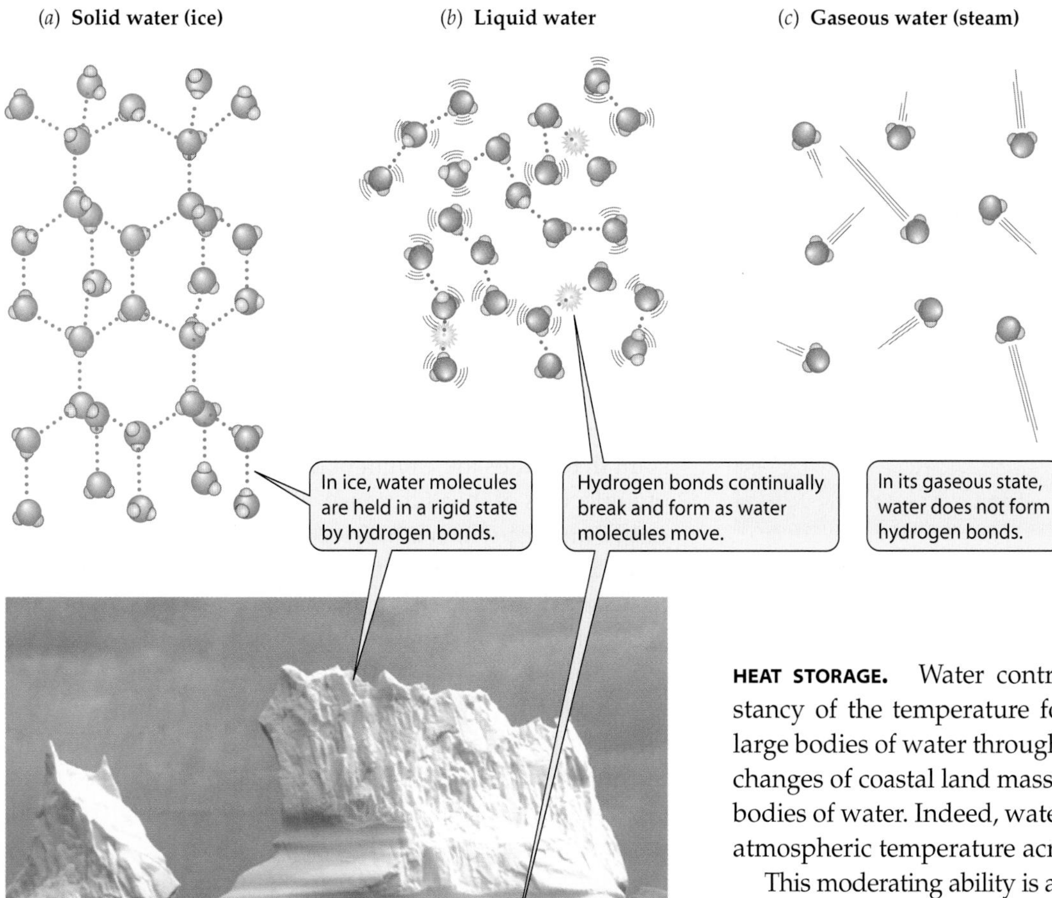
(a) Solid water (ice)
(b) Liquid water
(c) Gaseous water (steam)
In ice, water molecules are held in a rigid state by hydrogen bonds.
Hydrogen bonds continually break and form as water molecules move.
In its gaseous state, water does not form hydrogen bonds.

2.16 Hydrogen Bonds Hold Water Molecules Together Hydrogen bonding exists between the molecules of water in both its liquid and solid states. (*a*) Solid water (ice). (*b*) Liquid water. Although more structured, ice is less dense than liquid water, so it floats. (*c*) Water forms a gas when its hydrogen bonds are broken and molecules move farther apart.

Mars has created speculation that life could exist in that environment.

MELTING AND FREEZING. Water is a moderator of temperature changes. Compared with other nonmetallic substances of the same size, molecular ice requires a great deal of heat energy to melt. Melting 1 mole (6.02×10^{23} molecules, a standard quantity; see page 28) of water requires the addition of 5.9 kJ of energy. This value is high because hydrogen bonds must be broken in order for water to change from solid to liquid. In the opposite process—freezing—a great deal of energy is lost when water is transformed from liquid to solid.

HEAT STORAGE. Water contributes to the surprising constancy of the temperature found in the oceans and other large bodies of water throughout the year. The temperature changes of coastal land masses are also moderated by large bodies of water. Indeed, water helps minimize variations in atmospheric temperature across the planet.

This moderating ability is a result of the high heat capacity of liquid water. The *specific heat* of a substance is the amount of heat energy required to raise the temperature of 1 gram of that substance by 1°C. Raising the temperature of liquid water takes a relatively large amount of heat because much of the heat energy is used to break the hydrogen bonds that hold the liquid together. Compared with other small molecules that are liquids, water has a high specific heat.

EVAPORATION AND COOLING. Water has a high heat of vaporization, which means that a lot of heat is required to change water from its liquid to its gaseous state (the process of *evaporation*). Once again, much of the heat energy is used to break hydrogen bonds. This heat must be absorbed from the environment in contact with the water. Evaporation thus has a cooling effect on the environment—whether a leaf, a forest, or an entire land mass. This effect explains why sweating cools the human body: as sweat evaporates off the skin, it uses up some of the adjacent body heat.

COHESION AND SURFACE TENSION. In liquid water, individual water molecules are free to move about. The hydrogen bonds between the molecules continually form and break. In other words, liquid water has a dynamic structure. On average, every water molecule forms 3.4 hydrogen bonds with other water molecules. This number represents fewer bonds than exist in ice, but it is still a high number. These

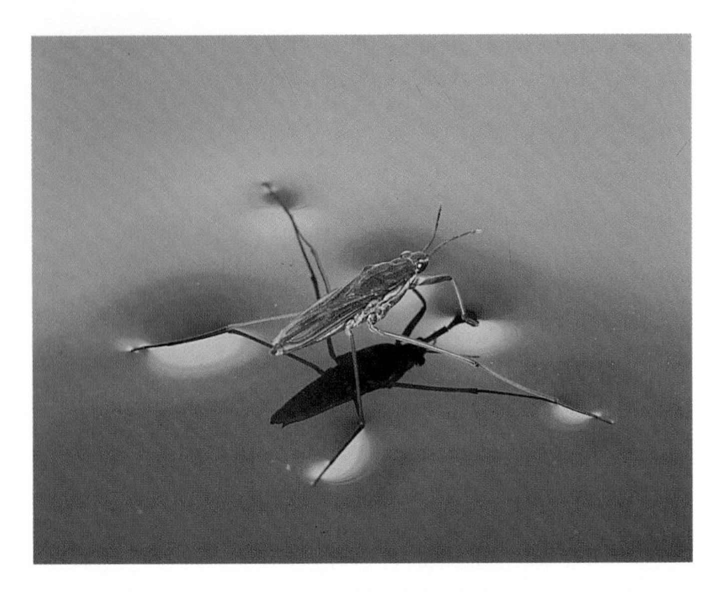

2.17 Surface Tension Water striders "skate" along, supported by the surface tension of the water that is their home.

hydrogen bonds explain the cohesive strength of liquid water. This cohesive strength permits narrow columns of water to stretch from the roots to the leaves of trees more than 100 meters high. When water evaporates from the leaves, the entire column moves upward in response to the pull of the molecules at the top.

Water also has a high surface tension, which means that the surface of liquid water exposed to the air is difficult to puncture. The water molecules in this surface layer are hydrogen-bonded to other water molecules below. The surface tension of water permits a container to be filled slightly above its rim without overflowing, and it permits small animals to walk on the surface of water (Figure 2.17).

Water is the solvent of life

A living organism is over 70 percent water by weight, excluding minerals such as bones. Many substances undergo reactions in this watery environment. Others do not, and thus form biological structures (such as bones).

A *solution* is produced when a substance (the *solute*) is dissolved in a liquid (the *solvent*) such as water (forming an *aqueous solution*). Many of the important molecules in biological systems are polar, and therefore are soluble in water. Reactions that take place in an aqueous solution can be studied in two ways:

- *Qualitative analysis* deals with substances dissolved in water and the chemical reactions that occur there. Qualitative analysis is the subject of much of the next few chapters.
- *Quantitative analysis* measures concentrations, or the amount of a substance in a given amount of solution. What follows is a brief introduction to some of the quantitative chemical terms you will see in this text.

Fundamental to quantitative thinking in chemistry and biology is the mole concept. A **mole** is the amount of an ion or compound (in grams) whose weight is numerically equal to its molecular weight. So a mole of table sugar ($C_{12}H_{22}O_{11}$) weighs 342 grams.

One aim of quantitative analysis is to study the behaviors of precise numbers of molecules in solution. But it is not possible to count molecules directly. Instead, chemists use a constant that relates the weight of any substance to the number of molecules of that substance. This constant is called *Avogadro's number*, which is 6.02×10^{23} molecules per mole. It allows chemists to work with moles of substances (which can be weighed out in the laboratory) instead of actual molecules. The mole concept is analogous to the concept of a dozen: We buy a dozen eggs or a dozen doughnuts, knowing that we will get 12 of whichever we buy. In the same way, when a physician injects a certain molar concentration of a drug into the bloodstream of a patient, a rough calculation can be made of the actual number of drug molecules that will interact with the patient's cells.

In the same way, chemists can dissolve a mole of sugar in water to make 1 liter of solution, knowing that the mole contains 6.02×10^{23} individual sugar molecules. This solution—1 mole of a substance dissolved in water to make 1 liter—is called a 1 molar (1 M) solution.

The many molecules dissolved in water in living tissues are not present at anything close to a 1 molar concentration. Most are in the micromolar (millionths of a mole per liter of solution; mM) to millimolar (thousandths of a mole per liter; μM) range. Some, such as hormone molecules, are even less concentrated than that. While these molarities seem to indicate very low concentrations, remember that even a 1 μM solution has 6.02×10^{17} molecules of the solute per liter.

Acids, Bases, and the pH Scale

When some substances dissolve in water, they release hydrogen ions (H^+), which are actually single, positively charged protons. These tiny bits of charged matter can attach to other molecules, and in doing so, change their properties. For example, the protons in acid rain can damage plants, and you are probably familiar with excess stomach acidity that affects digestion. In this section, we will examine the properties of substances that release H^+ (called **acids**) and substances that attach to H^+ (called **bases**). We will distinguish strong and weak acids and bases and provide a quantitative means for stating the concentration of H^+ in solutions: the pH scale.

Acids donate H^+, bases accept H^+

If hydrochloric acid (HCl) is added to water, it dissolves and ionizes, releasing the ions H^+ and Cl^-:

$$HCl \rightarrow H^+ + Cl^-$$

Because its H^+ concentration has increased, such a solution is acidic. Just like the combustion reaction of propane and oxygen (see Figure 2.15), the dissolution of HCl to form its ions is a complete reaction. HCl is therefore called a *strong acid.*

Acids *release* H^+ ions in solution. HCl is an acid, as is H_2SO_4 (sulfuric acid). One molecule of sulfuric acid may ionize to yield two H^+ and one SO_4^{2-}. Biological compounds that contain —COOH (the carboxyl group; see Figure 2.20) are also acids (such as acetic acid and pyruvic acid), because

$$\text{—COOH} \rightarrow \text{—COO}^- + H^+$$

Not all acids dissolve fully in water. For example, if acetic acid is added to water, at the end of the reaction, there are not just the two ions, but some of the original acid as well. Because the reaction is not complete, acetic acid is a *weak acid.*

Bases *accept* H^+ in solution. Just as with acids, there are strong and weak bases. If NaOH (sodium hydroxide) is added to water, it dissolves and ionizes, releasing OH^- and Na^+ ions:

$$NaOH \rightarrow Na^+ + OH^-$$

Because the concentration of OH^- increases and OH^- absorbs H^+ to form water, such a solution is basic. Because this reaction is complete, NaOH is a strong base.

Weak bases include the bicarbonate ion (HCO_3^-), which can accept a H^+ ion and become carbonic acid (H_2CO_3), and ammonia (NH_3), which can accept a H^+ and become an ammonium ion (NH_4^+). Amino groups (—NH_2) in biological molecules can also accept protons, thus acting as bases:

$$\text{—}NH_2 + H^+ \rightarrow \text{—}NH_3^+$$

The reactions between acids and bases may be reversible

When acetic acid is dissolved in water, two reactions happen. First, the acetic acid forms its ions:

$$CH_3COOH \rightarrow CH_3COO^- + H^+$$

Then, once ions are formed, they re-form acetic acid:

$$CH_3COO^- + H^+ \rightarrow CH_3COOH$$

This pair of reactions is reversible. A **reversible reaction** can proceed in either direction—left to right or right to left—depending on the relative starting concentrations of the reactants and products. The formula for a reversible reaction can be written using a double arrow:

$$CH_3COOH \rightleftharpoons CH_3COO^- + H^+$$

In principle, all chemical reactions are reversible. In terms of acids and bases, there are two types of reactions, depending on the extent of reversibility:

- Ionization of strong acids and bases is virtually irreversible.
- Ionization of weak acids and bases is somewhat reversible.

Many of the acid and base groups on large molecules in biological systems are weak.

Water is a weak acid

The water molecule has a slight but significant tendency to ionize into a hydroxide ion (OH^-) and a hydrogen ion (H^+). Actually, two water molecules participate in this ionization. One of the two molecules "captures" a hydrogen ion from the other, forming a hydroxide ion and a hydronium ion:

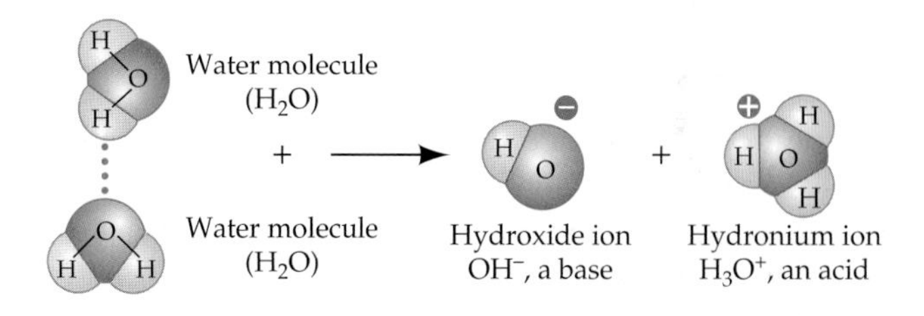

The hydronium ion is in effect a hydrogen ion bound to a water molecule. For simplicity, biochemists tend to use a modified representation of the ionization of water:

$$H_2O \rightarrow H^+ + OH^-$$

The ionization of water is important to all living creatures. This fact may seem surprising, since only about one water molecule in 500 million is ionized at any given time. But we are less surprised if we focus on the abundance of water in living systems and the reactive nature of the H^+ produced by ionization.

pH is the measure of hydrogen ion concentration

The terms "acidic" and "basic" refer only to solutions. How acidic or basic a solution is depends on the relative concentrations of H^+ and OH^- ions in it. The terms "acid" and "base" refer to compounds and ions. A compound or ion that is an acid can donate H^+; one that is a base can accept H^+.

How do we specify how acidic or basic a solution is? First, let's look at the H^+ concentrations of a few contrasting solutions. Remember that these concentrations are expressed in terms of molarity, the number of in moles of a substance in a liter of solution (see page 28). In pure water, the concentration of H^+ is 10^{-7} moles per liter (10^{-7} M). In 1 M hydrochloric acid, the H^+ concentration is 1 M; and in 1 M sodium hydroxide, the H^+ concentration is 10^{-14} M. Because its values range so widely, the H^+ concentration itself is an inconvenient quantity to measure. It is easier to work with the logarithm of the concentration, because logarithms compress this range.

We indicate how acidic or basic a solution is by its pH. The pH value is defined as the negative logarithm of the hydrogen ion concentration in moles per liter (molar concentration). In chemical notation, molar concentration is often indicated by putting square brackets around the symbol for a substance; thus $[H^+]$ stands for the molar concentration of H^+. The equation for pH is

$$pH = -\log_{10}[H^+]$$

Since the H^+ concentration of pure water is 10^{-7} M, its pH is $-\log(10^{-7}) = -(-7)$, or 7. A smaller negative logarithm means a larger number. In practical terms, a lower pH means a higher H^+ concentration, or greater acidity. In 1 M HCl, the H^+ concentration is 1 M, so the pH is the negative logarithm of 1 ($-\log 10^0$), or 0. The pH of 1 M NaOH is the negative logarithm of 10^{-14}, or 14.

A solution with a pH of less than 7 is acidic—it contains more H^+ ions than OH^- ions. A solution with a pH of 7 is neutral, and a solution with a pH value greater than 7 is basic. Figure 2.18 shows the pH values of some common substances.

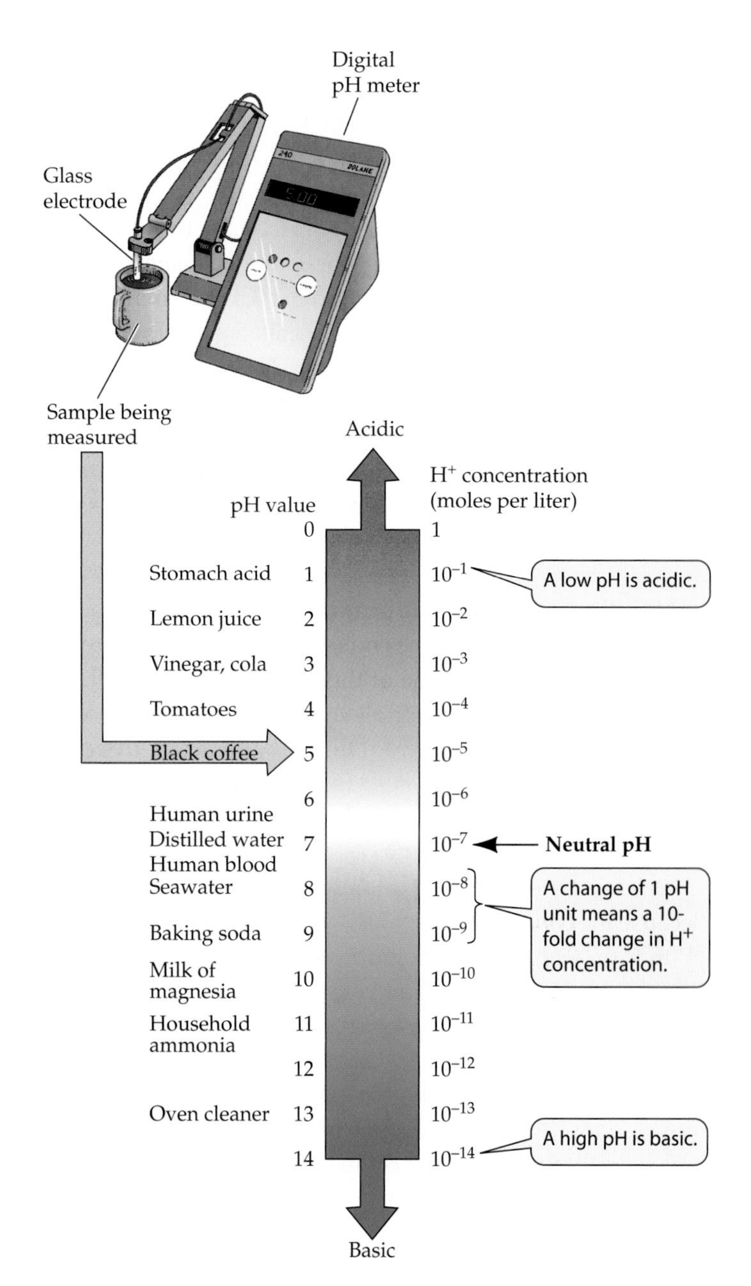

2.18 pH Values of Some Familiar Substances An electronic instrument similar to the one drawn at the top of the figure is used to measure the pH of a solution.

Buffers minimize pH change

Some organisms, probably including the earliest forms of life, live in and have adapted to solutions with extremes of pH. However, most organisms control the pH of the separate compartments within their cells. The normal pH of human red blood cells, for example, is 7.4, and deviations of even a few tenths of a pH unit can be fatal. The control of pH is made possible in part by **buffers**: chemical mixtures that maintain a relatively constant pH even when substantial amounts of an acid or base are added.

A buffer is a mixture of a weak acid and its corresponding base—for example, carbonic acid (H_2CO_3) and bicarbonate ions (HCO_3^-). If an acid is added to a solution containing this buffer, not all the H^+ ions from that acid stay in solution. Instead, many of them combine with the bicarbonate ions to produce more carbonic acid. This reaction uses up some of the H^+ ions in the solution and decreases the acidifying effect of the added acid:

$$HCO_3^- + H^+ \rightleftharpoons H_2CO_3$$

If a base is added, the reaction essentially reverses. Some of the carbonic acid ionizes to produce bicarbonate ions and more H^+, which counteracts some of the added base. In this way, the buffer minimizes the effects of an added acid or base on pH. This is what happens in the blood, where this buffering system is important in preventing significant changes in pH that could disrupt the ability of the blood to function in carrying vital O_2 to tissues. A given amount of acid or base causes a smaller change in pH in a buffered solution than in an unbuffered one (Figure 2.19).

Buffers illustrate an important chemical principle in reversible reactions called the *law of mass action*. Addition of a reactant on one side of a reversible system drives the reaction in the direction that uses up that compound. In this case, addition of an acid drives the reaction in one direction; addition of a base drives it in the other direction.

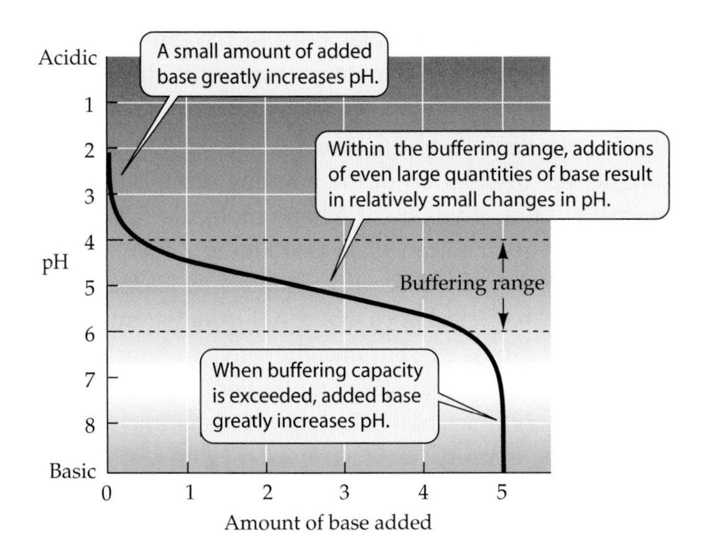

2.19 Buffers Minimize Changes in pH With increasing amounts of added base, the overall slope of a graph of pH is downward. In the buffering range, however, the slope is shallow. At high and low values of pH, where the buffer is ineffective, the slopes are much steeper.

Properties of Molecules

So far, this chapter has discussed many properties of molecules, including size, polarity, solubility, and acid/base properties. Two other important properties that influence the behavior of molecules in a chemical reaction are the presence of recognizable functional groups, and existence of different isomers of molecules with the same chemical formula.

Functional groups give specific properties to molecules

Certain small groups of atoms called **functional groups** are consistently found together in a variety of different molecules, a fact that simplifies our understanding of the reactions that molecules undergo in living cells. Each functional group has specific properties that, when attached to a larger molecule, in turn give the larger molecules specific properties. You will encounter several functional groups in your study of biology, including alcohols, aldehydes, ketones, acids, amines, phosphates, and thiols (Figure 2.20).

An important category of biological molecules containing functional groups is the amino acids, which have both a carboxyl group and an amino group attached to the same carbon atom, called the α carbon. Also attached to the α carbon atom are a hydrogen atom and a side chain, or R group, designated by the letter R:

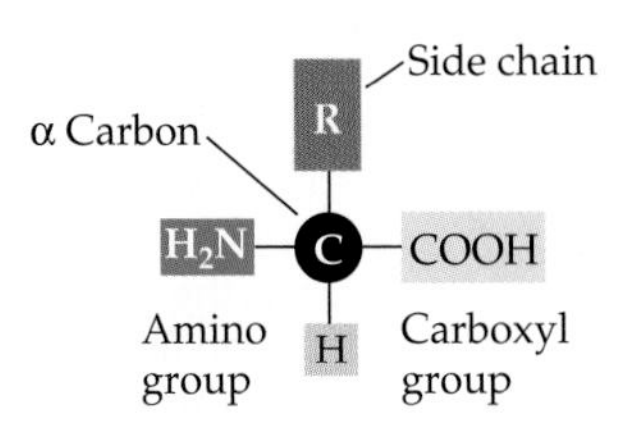

Different side chains have different chemical compositions, structures, and properties. Each of the 20 amino acids found in proteins has a different side chain that gives it its distinctive chemical properties, as we'll see in Chapter 3.

Because they possess both carboxyl and amino groups, amino acids are simultaneously acids and bases. At the pH values commonly found in cells, both the carboxyl and the

This part = R

Functional group	Class of compounds	Structural formula	Example
Hydroxyl —OH	Alcohols	R—OH	CH_3CH_2—OH Ethanol
Aldehyde —CHO	Aldehydes	R—CHO	CH_3—CHO Acetaldehyde
Keto \CO/	Ketones	R—CO—R	CH_3—CO—CH_3 Acetone
Carboxyl —COOH	Carboxylic acids	R—COOH	CH_3—COOH Acetic acid
Amino —NH_2	Amines	R—NH_2	CH_3—NH_2 Methylamine
Phosphate —OPO_3^{2-}	Organic phosphates	R—O—PO_3^{2-}	HOOC—CH(OH)—CH_2—O—PO_3^{2-} 3-Phosphoglyceric acid
Sulfhydryl —SH	Thiols	R—SH	HO—CH_2CH_2—SH Mercaptoethanol

2.20 Some Functional Groups Important to Living Systems These functional groups (highlighted in white boxes) are the most common ones found in biologically important molecules. R represents the "remainder" of the molecule, which may be any of a large number of carbon skeletons or other chemical groups.

amino group are ionized: The carboxyl group has lost a proton, and the amino group has gained one.

Isomers have different arrangements of the same atoms

Isomers are molecules that have the same chemical formula but different arrangements of the atoms. (The prefix *iso-*, meaning "same," is encountered in many biological terms.) Of the different kinds of isomers, we will consider two: structural isomers and optical isomers.

Structural isomers differ in how their atoms are joined together. Consider two simple molecules, each composed of 4 carbon and 10 hydrogen atoms bonded covalently, both with the formula C_4H_{10}. These atoms can be linked in two different ways, resulting in two forms of the molecule:

```
    H  H               CH3
    |  |                |
H3C—C—C—CH3        H3C—C—CH3
    |  |                |
    H  H                H
   Butane           Isobutane
```

The different bonding relationships of butane and isobutane are distinguished in their structural formulas, and the two compounds have different chemical properties.

Optical isomers occur whenever a carbon atom has four different atoms or groups attached to it. This pattern allows two different ways of making the attachments, each the mirror image of the other (Figure 2.21). Such a carbon atom is an asymmetrical carbon, and the pair of compounds are optical isomers of each other. You can imagine your right and left hands as optical isomers. Just as a glove is specific for a particular hand, some biochemical molecules can interact with one optical isomer of a compound, but are unable to "fit" the other.

The α carbon in an amino acid is an asymmetrical carbon because it is bonded to four different functional groups. Therefore, amino acids exist in two isomeric forms, called D-amino acids and L-amino acids. D and L are abbreviations for the Latin terms for right (*dextro*) and left (*levo*), respectively. Only L-amino acids are commonly found in most organisms, and their presence is an important chemical "signature" for life.

Now that we have covered the major properties of all molecules, let's review them in preparation for the next chapter, which focuses on the major molecules of biological systems.

- *Molecules vary in size.* Some are small, such as H_2 and CH_4. Others are larger, such as a molecule of table sugar (sucrose, $C_{12}H_{22}O_{11}$), which has 45 atoms. Still other molecules, especially proteins such as hemoglobin (the oxygen carrier in red blood cells), are gigantic, sometimes containing tens of thousands of atoms. The formation of large molecules from simpler ones in the environment was a key precursor to the emergence of life during the Archean.
- *All molecules have a specific three-dimensional shape.* For example, the orientation of the bonding orbitals around the carbon atom gives the methane molecule (CH_4) the shape of a regular tetrahedron (see Figure 2.10*c*). In carbon dioxide (CO_2), the three atoms are in line. Larger molecules have complex shapes that result from the numbers and kinds of atoms present and the ways in which they are linked together. Some large molecules, such as hemoglobin, have compact, ball-like shapes. Others, such as the protein called keratin that makes up your hair, are long, thin, ropelike structures. Their shapes relate to the roles these molecules play in living cells.
- *Molecules are characterized by certain chemical properties* that determine their biological roles. Chemists use the characteristics of composition, structure (three-dimensional shape), reactivity, and solubility to distinguish a pure sample of one molecule from a sample of a different molecule. The presence of functional groups can impart distinctive chemical properties to a molecules, as does the physical arrangement of atoms into isomers.

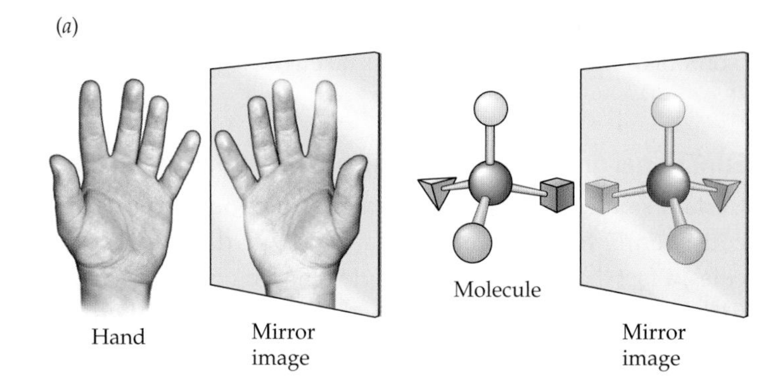

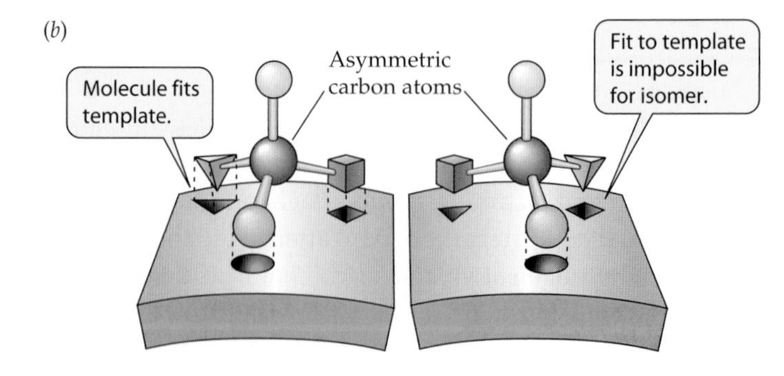

2.21 Optical Isomers (*a*) Optical isomers are mirror images of each other. (*b*) Molecular optical isomers result when four different groups are attached to a single carbon atom. If a template is laid out to match the groups on one carbon atom, the groups on that molecule's mirror-image isomer cannot be rotated to fit the same template.

Between the small molecules discussed in this chapter and the world of the living cell stand the macromolecules. These huge molecules—proteins, lipids, carbohydrates, and nucleic acids—are the subject of the next chapter.

Chapter Summary

Water and the Origin of Life's Chemistry

▶ Current scientific evidence indicates that life as we know it cannot exist without water, and that life on Earth originated in the water of the planet's primordial oceans.

▶ The chemistry of life is ancient. Earth began forming about 4.6 billion years ago, and the first signs of life are 3.8–4 billion years old.

Atoms: The Constituents of Matter

▶ Matter is composed of atoms. Each atom consists of a positively charged nucleus of protons and neutrons, surrounded by electrons bearing negative charges. There are many elements in nature, but only a few of them make up the bulk of living systems. **Review Figures 2.2, 2.3**

▶ Isotopes of an element differ in their numbers of neutrons. Some isotopes are radioactive, emitting radiation as they decay. **Review Figure 2.4**

▶ Electrons are distributed in shells consisting of orbitals. Each orbital contains a maximum of two electrons. **Review Figures 2.6, 2.7. See Web/CD Activity 2.1**

▶ In losing, gaining, or sharing electrons to become more stable, an atom can combine with other atoms to form molecules. **Review Table 2.1**

Chemical Bonds: Linking Atoms Together

▶ Covalent bonds are strong bonds formed when two atomic nuclei share one or more pairs of electrons. Covalent bonds have spatial orientations that give molecules three-dimensional shapes. **Review Figures 2.8, 2.9, 2.10, Table 2.2**

▶ Nonpolar covalent bonds are formed when the electronegativities of two atoms are approximately equal. When atoms with strong electronegativity (such as oxygen) bond to atoms with weaker electronegativity (such as hydrogen), a polar covalent bond is formed, in which one end is δ^+ and the other is δ^-. **Review Figure 2.11, Table 2.3**

▶ Hydrogen bonds are weak electrical attractions that form between a δ^+ hydrogen atom in one molecule and a δ^- nitrogen or oxygen atom in another molecule or in another part of a large molecule. Hydrogen bonds are abundant in water. **Review Figure 2.12**

▶ Ions are electrically charged bodies that form when an atom gains or loses one or more electrons. Ionic bonds are electrical attractions between oppositely charged ions. Ionic bonds are strong in solids, but weaker when the ions are separated from one another in solution. **Review Figures 2.13, 2.14**

▶ Nonpolar molecules interact very little with polar molecules, including water. Nonpolar molecules are attracted to one another by very weak bonds called van der Waals forces.

See Web/CD Tutorial 2.1

Chemical Reactions: Atoms Change Partners

▶ In chemical reactions, substances change their atomic compositions and properties. Energy is released in some reactions, whereas in others energy must be provided. Neither matter nor energy is created or destroyed in a chemical reaction, but both change form. **Review Figure 2.15**

▶ In living cells, chemical reactions take place in multiple steps so that the released energy can be harvested for cellular activities.

Water: Structure and Properties

▶ Water's molecular structure and its capacity to form hydrogen bonds give it unusual properties that are significant for life. Solid water floats in liquid water, and water gains or loses a great deal of heat when it changes its state, a property that moderates environmental temperature changes. **Review Figure 2.16**

▶ Water's high heat of vaporization assures effective cooling when water evaporates. The cohesion of water molecules permits liquid water to rise to great heights in narrow columns and produces a high surface tension.

▶ Solutions are produced when substances dissolve in water. The concentration of a solution is the amount of a given substance in a given amount of solution. Most biological substances are dissolved in water at very low concentrations.

Acids, Bases, and the pH Scale

▶ Acids are substances that donate hydrogen ions. Bases are substances that accept hydrogen ions.

▶ The pH of a solution is the negative logarithm of the hydrogen ion concentration. Values lower than pH 7 indicate an acidic solution; values above pH 7 indicate a basic solution. **Review Figure 2.18**

▶ Buffers are mixtures of weak acids and bases that limit the change in the pH of a solution when acids or bases are added. **Review Figure 2.19**

The Properties of Molecules

▶ Functional groups make up part of a larger molecule and have particular chemical properties. The consistent chemical behavior of functional groups helps us understand the properties of the molecules that contain them. **Review Figure 2.20. See Web/CD Activities 2.2, 2.3**

▶ Structural and optical isomers have the same kinds and numbers of atoms, but differ in their structures and properties. **Review Figure 2.21**

▶ Molecules vary in their size, shape, reactivity, solubility, and other chemical properties.

Self-Quiz

1. The atomic number of an element
   a. equals the number of neutrons in an atom.
   b. equals the number of protons in an atom.
   c. equals the number of protons minus the number of neutrons.
   d. equals the number of neutrons plus the number of protons.
   e. depends on the isotope.
2. The atomic weight (atomic mass) of an element
   a. equals the number of neutrons in an atom.
   b. equals the number of protons in an atom.
   c. equals the number of electrons in an atom.
   d. equals the number of neutrons plus the number of protons.
   e. depends on the relative abundances of its isotopes.

3. Which of the following statements about all the isotopes of an element is *not* true?
   a. They have the same atomic number.
   b. They have the same number of protons.
   c. They have the same number of neutrons.
   d. They have the same number of electrons.
   e. They have identical chemical properties.
4. Which of the following statements about a covalent bond is *not* true?
   a. It is stronger than a hydrogen bond.
   b. One can form between atoms of the same element.
   c. Only a single covalent bond can form between two atoms.
   d. It results from the sharing of electrons by two atoms.
   e. One can form between atoms of different elements.
5. Hydrophobic interactions
   a. are stronger than hydrogen bonds.
   b. are stronger than covalent bonds.
   c. can hold two ions together.
   d. can hold two nonpolar molecules together.
   e. are responsible for the surface tension of water.
6. Which of the following statements about water is *not* true?
   a. It releases a large amount of heat when changing from liquid into vapor.
   b. Its solid form is less dense than its liquid form.
   c. It is the most effective solvent of polar molecules.
   d. It is typically the most abundant substance in an active organism.
   e. It takes part in some important chemical reactions.
7. The following reaction occurs in the human stomach:

   $$HCl \rightarrow H^+ + Cl^-$$

   This reaction is an example of the
   a. cleavage of a covalent bond.
   b. formation of a hydrogen bond.
   c. elevation of the pH of the stomach.
   d. formation of ions by dissolving an acid.
   e. formation of polar covalent bonds.
8. The hydrogen bond between two water molecules arises because water is
   a. polar.
   b. nonpolar.
   c. a liquid.
   d. small.
   e. hydrophobic.
9. Which of the following statements about the carboxyl group is *not* true?
   a. It has the chemical formula —COOH.
   b. It is an acidic group.
   c. It can ionize.
   d. It is found in amino acids.
   e. It has an atomic weight of 75.
10. The three most abundant elements in a human skin cell are
   a. calcium, carbon, and oxygen.
   b. carbon, hydrogen, and oxygen.
   c. carbon, hydrogen, and sodium.
   d. carbon, nitrogen, and oxygen.
   e. nitrogen, hydrogen, and oxygen.

For Discussion

1. Would you expect the elemental composition of Earth's crust to be the same as that of the human body? How could you find out?
2. Some scientists and science fiction writers have envisioned life on other planets based not on carbon, as on Earth, but on silicon (Si). Using the Bohr model (see Figure 2.10*a*), draw the structure of silicon dioxide, showing electrons shared in covalent bonds.
3. Compare a covalent bond between two hydrogen atoms and a hydrogen bond between hydrogen and oxygen atoms with regards to the electrons involved, the role of polarity, and the strength of the bond.
4. The pH of the human stomach is about 2.0, while the pH of the small intestine is about 10.0. What are the hydrogen ion (H^+) concentrations inside these two organs?
5. The side chain of the amino acid glycine is simply a hydrogen atom (—H). Are there two optical isomers of glycine? Explain.

3 Life and Chemistry: Large Molecules

In 1984, a rock was found on the ice in the Allan Hills region of Antarctica. ALH 84001, as it came to be called, was a meteorite that came from Mars. We know this because the composition of the gases trapped within the rock was identical to the Martian atmosphere, which is quite different from Earth's atmosphere. Radioactive dating and mineral analyses determined that ALH 84001 was 4.5 billion years old and had been blasted off the Martian surface 16 million years ago, landing on Earth fairly recently, about 11,000 years ago.

Scientists found water trapped below the Martian meteorite's surface. This discovery was not surprising, considering that surface observations of Mars have indicated that liquid water may once have been abundant there (see Chapter 2). Because water is the *sine qua non* for life, scientists wondered whether the meteorite might contain other signs of life as well. Their analysis revealed two substances related to living systems. First, simple carbon-containing molecules called polycyclic aromatic hydrocarbons were present in small but unmistakable amounts; these substances are formed by decaying organisms, such as microbes. And second, crystals of magnetite, an iron oxide mineral made by many living things on Earth, were isolated from the interior of the rock.

ALH 84001 is not the only visitor from outer space that has been shown to contain the chemistry of life. Fragments of a meteorite that fell around the town of Murchison, Australia in 1969 were found to contain molecules that are unique to life, including purines and pyrimidines (the building blocks of DNA) and amino acids (which link together to form proteins). All of the amino acids showed a "handedness" that is unique to life.

These meteorites suggest that life is not found only on Earth, but they do not answer the question of how or where life arose from nonliving matter. We begin this chapter by presenting two hypotheses for the origin of life on Earth. After discussing these hypotheses, we take a detailed look at the four kinds of large molecules that characterize living organisms: proteins, carbohydrates, lipids (fats), and nucleic acids.

Was Life Once Here? The meteorite ALH 84001, which came from Mars and landed in Antarctica, contains the chemical signatures of life.

Theories of the Origin of Life

Living things are composed of the same elements as the inanimate universe, the 92 elements of the periodic table (see Figure 2.3). But the arrangements of these atoms into molecules in biological systems are unique. You cannot find DNA in rocks unless it came from a once-living organism.

How life began on Earth sometime during the 600 million years of the Hadean is impossible for us to know for certain, given the vast amount of time that has passed. There are two theories of the origin of life: life from extraterrestrial sources, and chemical evolution.

Could life have come from outside Earth?

As we described in Chapter 2, comets probably brought Earth most of its water. The meteorites described at the beginning of this chapter are evidence that molecules characteristic of life may have traveled to Earth from space. Taken together, these two observations suggest that some of life's complex molecules could have come from space. Although the presence of such molecules in rocks may suggest that those rocks once harbored life, it does not prove that there were living things in the rocks when they landed on Earth. Claims that the spherical objects seen in ALH 84001 are the remnants of ancient Martian organisms are far from accepted by all scientists in the field.

Most scientists find it hard to believe that an organism in a meteorite could survive thousands of years traveling through space, followed by intense heat as it passed through Earth's atmosphere. But there is some evidence that the heat inside some meteorites may not have been severe. When weakly magnetized rock is heated, it reorients its magnetic field to align with the magnetic field around it. In the case of ALH 84001, this would have been Earth's powerful magnetic field, which would have affected the meteorite as it approached our planet. Careful measurements indicate that, while reorientation did occur at the surface of the rock, it did not occur in the inside. The scientists who took these measurements, Benjamin Weiss and Joseph Kirschvink at the California Institute of Technology, claim that the inside of ALH 84001 was never heated over 40°C on its trip to Antarctica, making a long interplanetary trip by living organisms more plausible.

Did life originate on Earth?

Both Earth and Mars once had the water and other simple molecules that could, under the right conditions, form the large molecules unique to life. The second theory of the origin of life on Earth, **chemical evolution**, holds that conditions on the primitive Earth led to the emergence of these molecules. Scientists have sought to reconstruct those primitive conditions.

Early in the twentieth century, researchers proposed that there was little oxygen gas (O_2) in Earth's first atmosphere (unlike today, when it constitutes 21 percent of our atmosphere). O_2 is thought to have accumulated in quantity about 2.5 billion years ago as the by-product of the metabolism of single-celled life forms. In the 1950s, Stanley Miller and Harold Urey set up an experimental "primitive" atmosphere, containing hydrogen gas, ammonia, methane gas, and water vapor. Through these gases, they passed a spark to simulate lightning, then cooled the system so the gases would condense and collect in a watery solution, or "ocean" (Figure 3.1). Within days, the system contained numerous complex molecules, including amino acids, purines, and pyrimidines—some of the building blocks of life.

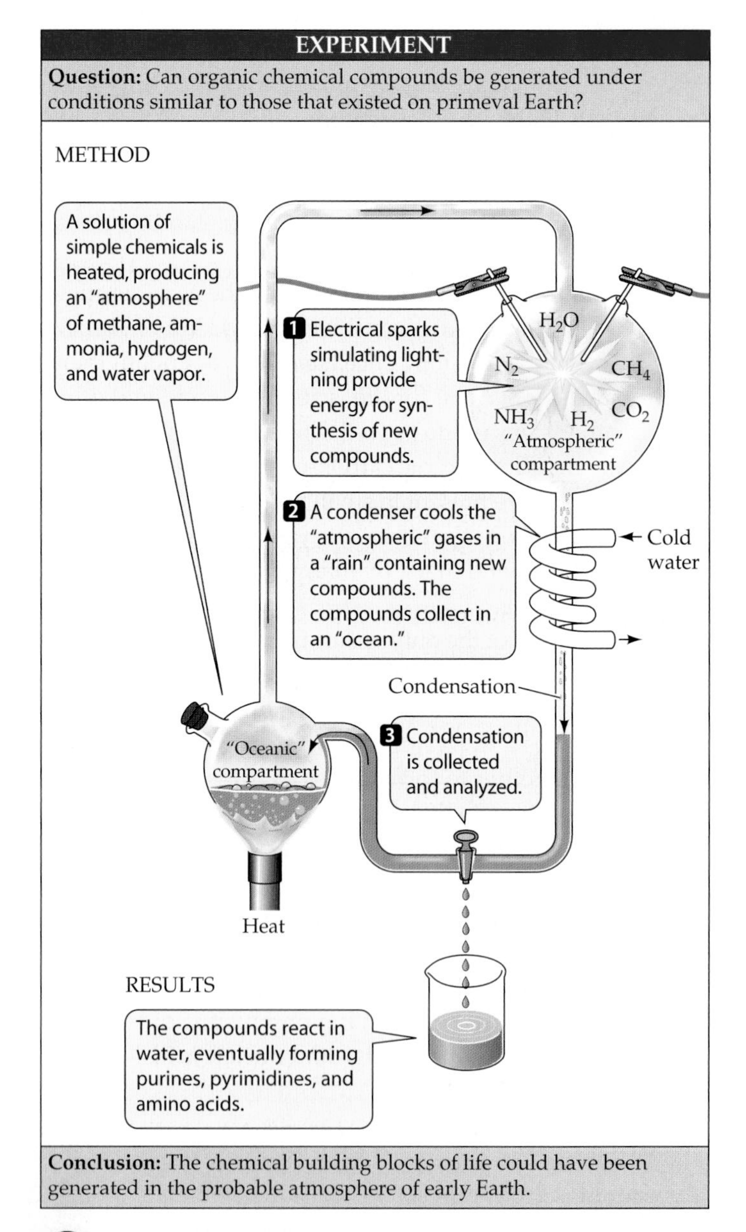

EXPERIMENT
Question: Can organic chemical compounds be generated under conditions similar to those that existed on primeval Earth?
METHOD
A solution of simple chemicals is heated, producing an "atmosphere" of methane, ammonia, hydrogen, and water vapor.
1 Electrical sparks simulating lightning provide energy for synthesis of new compounds.
H_2O
N_2
CH_4
NH_3
H_2
CO_2
"Atmospheric" compartment
2 A condenser cools the "atmospheric" gases in a "rain" containing new compounds. The compounds collect in an "ocean."
Cold water
Condensation
"Oceanic" compartment
3 Condensation is collected and analyzed.
Heat
RESULTS
The compounds react in water, eventually forming purines, pyrimidines, and amino acids.
Conclusion: The chemical building blocks of life could have been generated in the probable atmosphere of early Earth.

3.1 Synthesis of Prebiotic Molecules in an Experimental Atmosphere The Miller-Urey experiment simulated possible atmospheric conditions on primitive Earth to obtain some of the molecular building blocks of biological systems.

In science, an experiment and its results must be constantly reinterpreted and refined as more knowledge accumulates. The results of the Miller-Urey experiments have undergone several such refinements:

- In living organisms, many molecules have a unique three-dimensional "handedness" (see Figure 2.21). The amino acids, for example, are all in the L-configuration. But the amino acids formed in the Miller-Urey experiments were a mixture of the D- and L-forms. Recent experiments show that natural processes could have selected the L-amino acids from the mixture. Some minerals, especially calcite-based rocks, have unique crystal structures that selectively bind to D- or L-amino acids, separating the two. Such rocks were abundant during the Archean.
- Scientists' views of the Earth's original atmosphere have changed since Miller and Urey did their experiment. There is abundant evidence of major volcanic eruptions 4 billion years ago that released carbon dioxide (CO_2), nitrogen (N_2), hydrogen sulfide (H_2S), and sulfur dioxide (SO_2). Prebiotic chemistry experiments using these molecules in addition to the ones in the original "soup" have led to more diverse molecules.
- Long polymers had to be formed from simpler building blocks, called monomers. Scientists have used model systems to try to simulate conditions under which polymers could be made. Solid mineral surfaces, such as finely divided clays, seem to provide the best environment to bind monomers and allow them to polymerize.
- Miller and Urey, as well as others, suggested that life originated in hot pools at the edges of oceans. Because life has been found in many extreme environments on earth, scientists have proposed that such environments—found beneath ice, in deep-sea hydrothermal vents, and within fine clays near the shore—could be the original site of life's emergence.

In whatever way the earliest stages of chemical evolution occurred, they resulted in the emergence of monomers and polymers that have probably remained unchanged in their general structure and function for 3.8 billion years. We now turn our attention to these large molecules.

Macromolecules: Giant Polymers

The four kinds of large molecules are made the same way and they are present in roughly the same proportions in all living organisms (Figure 3.2). A protein that has a certain role in an apple tree probably has a similar role in a human being, because their basic chemistry is the same. One important advantage of this *biochemical unity* is that organisms acquire needed biochemicals by eating other organisms. When you eat an apple, the molecules you take in include carbohydrates, lipids, and proteins that can be refashioned into the special varieties of those molecules used by humans.

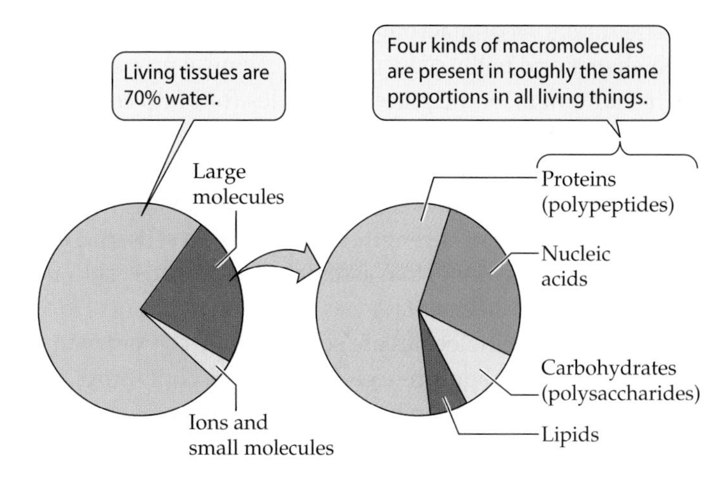

3.2 Substances Found in Living Tissues The substances shown here make up the nonmineral components of living tissue (bone would be an example of a "mineral tissue"). Most tissues are at least 70 percent water.

Macromolecules are giant **polymers** (*poly-*, "many"; *-mer*, "unit") constructed by the covalent linking of smaller molecules called **monomers** (Table 3.1). These monomers may or may not be identical, but they always have similar chemical structures. Molecules with molecular weights exceeding 1,000 are usually considered macromolecules, and the proteins, polysaccharides (large carbohydrates), and nucleic acids of living systems certainly fall into this category.

Each type of macromolecule performs some combination of functions: energy storage, structural support, protection, catalysis, transport, defense, regulation, movement, and information storage. These roles are not necessarily exclusive. For example, both carbohydrates and proteins can play structural roles, supporting and protecting tissues and organisms. However, only nucleic acids specialize in information storage and function as hereditary material, carrying both species and individual traits from generation to generation.

3.1 The Building Blocks of Organisms

MONOMER	SIMPLE POLYMER	COMPLEX POLYMER (MACROMOLECULE)
Amino acid	Peptide or oligopeptide	Polypeptide (protein)
Nucleotide	Oligonucleotide	Nucleic acid
Monosaccharide (sugar)	Oligosaccharide	Polysaccharide (carbohydrate)

The functions of macromolecules are directly related to their shapes and to the sequences and chemical properties of their monomers. Some macromolecules fold into compact spherical forms with surface features that make them water-soluble and capable of intimate interaction with other molecules. Other proteins and carbohydrates form long, fibrous systems that provide strength and rigidity to cells and organisms. Still other long, thin assemblies of proteins can contract and cause movement.

Because macromolecules are so large, they contain many different functional groups (see Figure 2.20). For example, a large protein may contain hydrophobic, polar, and charged functional groups that give specific properties to local sites on the macromolecule. As we will see, this diversity of properties determines the shapes of macromolecules and their interactions with both other macromolecules and smaller molecules.

Condensation and Hydrolysis Reactions

Polymers are constructed from monomers by a series of reactions called **condensation reactions** or dehydration reactions (both terms refer to the loss of water). Condensation reactions result in covalently bonded monomers (Figure 3.3*a*) and release a molecule of water for each bond formed. The condensation reactions that produce the different kinds of polymers differ in detail, but in all cases, polymers form only if energy is added to the system. In living systems, specific energy-rich molecules supply this energy.

The reverse of a condensation reaction is a **hydrolysis reaction** (*hydro*-, "water"; -*lysis*, "break"). Hydrolysis reactions digest polymers and produce monomers. Water reacts with the bonds that link the polymer together, and the products are free monomers. The elements (H and O) of H_2O become part of the products (Figure 3.3*b*).

These two types of reactions are universal in living things, and as we have seen, were an important step in the origin of life in an aqueous environment. We begin our study of biological macromolecules with a very diverse group of polymers, the proteins.

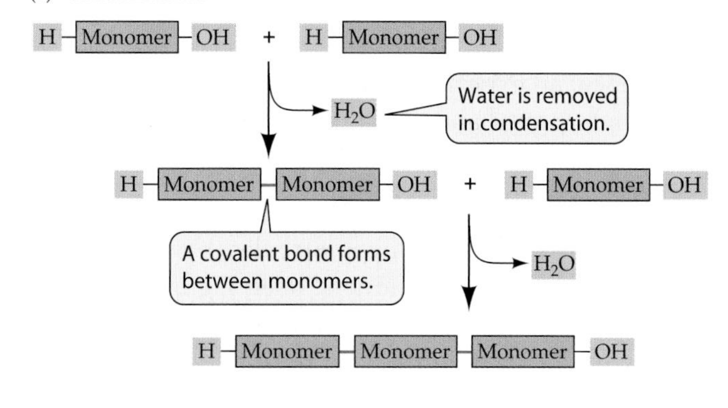

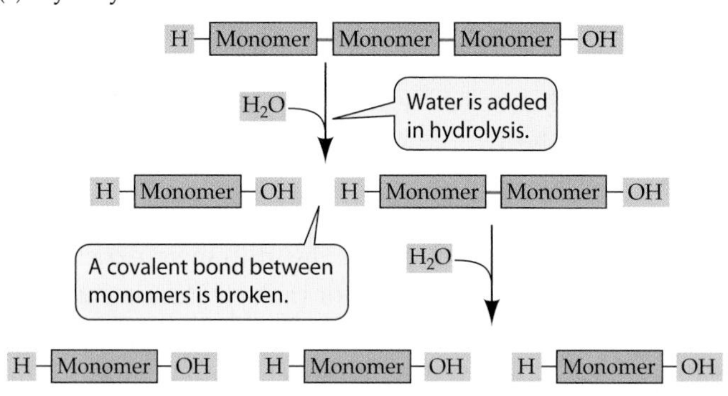

3.3 Condensation and Hydrolysis of Polymers (*a*) A condensation reaction links monomers into polymers. (*b*) A hydrolysis reaction digests polymers into individual monomers.

Proteins: Polymers of Amino Acids

The functions of **proteins** include structural support, protection, transport, catalysis, defense, regulation, and movement. Among the functions of macromolecules listed earlier, only energy storage and information storage are not usually performed by proteins.

Proteins range in size from small ones such as the RNA-digesting enzyme ribonuclease A, which has a molecular weight of 5,733 and 51 amino acid residues, to huge molecules such as the cholesterol transport protein apolipoprotein B, which has a molecular weight of 513,000 and 4,636 amino acid residues. (The word *residue* refers to a monomer when it is part of a polymer.) Each of these proteins consists of a single unbranched polymer of amino acids (a *polypeptide chain*), which is folded into a specific three-dimensional shape.

Many proteins require more than one polypeptide chain to make up the functional unit. For example, the oxygen-carrying protein hemoglobin has four chains that are folded separately and associate together to make the functional protein. As we will see later in this book, numerous functional proteins can associate, forming "multi-protein machines" to carry out complex roles such as DNA synthesis.

The *composition* of a protein refers to the relative amounts of the different amino acids it contains. Not every protein contains all kinds of amino acids, nor an equal number of different ones. The diversity in amino acid content and sequence is the source of the diversity in protein structures and functions.

The next several chapters will describe the many functions of proteins. To understand them, we must first explore protein structure. First we will examine the properties of amino acids and see how they link together to form proteins. Then we will systematically examine protein structure and look at how a linear chain of amino acids is consistently folded into a compact three-dimensional shape. Finally, we will see how this three-dimensional structure provides a specific physical and chemical environment that influences how other molecules can interact with the protein.

Proteins are composed of amino acids

In Chapter 2, we looked at the structure of amino acids and identified four different groups attached to a central (α) carbon atom: a hydrogen atom, an amino group (NH_3^+), a carboxyl group (COO^-), and a unique **side chain**, or **R group**.

The R groups of amino acids are important in determining the three-dimensional structure and function of the protein macromolecule. They are highlighted in white in Table 3.2.

As Table 3.2 shows, amino acids are grouped and distinguished by their side chains. Some side chains are electrically charged (+1, –1), while others are polar (δ^+, δ^-), and still others are nonpolar and hydrophobic.

- The five amino acids that have electrically charged side chains attract water (are hydrophilic) and oppositely charged ions of all sorts.
- The five amino acids that have polar side chains tend to form weak hydrogen bonds with water and with other polar or charged substances. These amino acids are hydrophilic.

3.2 The Twenty Amino Acids Found in Proteins

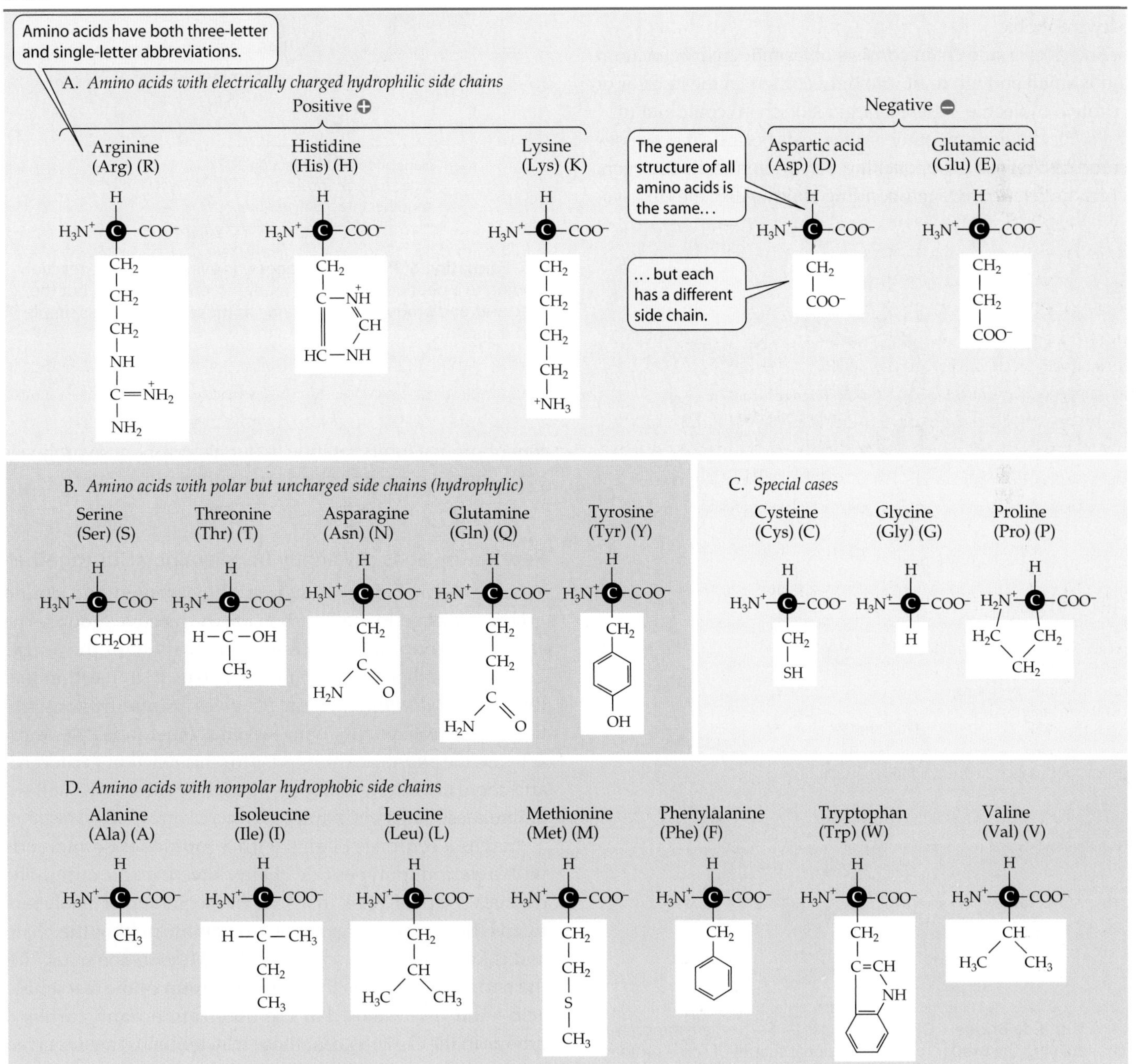

- Seven amino acids have side chains that are nonpolar hydrocarbons or very slightly modified hydrocarbons. In the watery environment of the cell, these hydrophobic side chains may cluster together in the interior of the protein. These amino acids are hydrophobic.
- Three amino acids—cysteine, glycine, and proline—are special cases, although their R groups are generally hydrophobic.

The *cysteine* side chain, which has a terminal —SH group, can react with another cysteine side chain to form a covalent bond called a **disulfide bridge** (—S—S—) (Figure 3.4). Disulfide bridges help determine how a polypeptide chain folds. When cysteine is not part of a disulfide bridge, its side chain is hydrophobic.

The *glycine* side chain consists of a single hydrogen atom and is small enough to fit into tight corners in the interior of a protein molecule, where a larger side chain could not fit.

Proline differs from other amino acids because it possesses a modified amino group lacking a hydrogen on its nitrogen, which limits its hydrogen-bonding ability. Also, the ring system of proline limits rotation about its α carbon, so proline is often found at bends or loops in a protein.

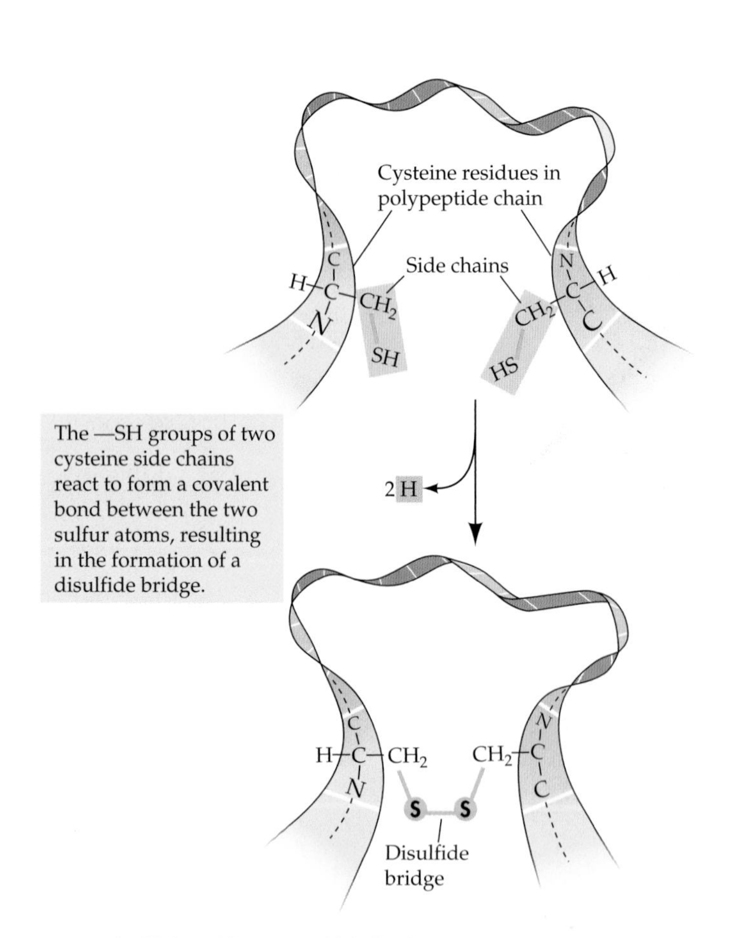

3.4 A Disulfide Bridge Disulfide bridges (—S—S—) are important in maintaining the proper three-dimensional shapes of some protein molecules.

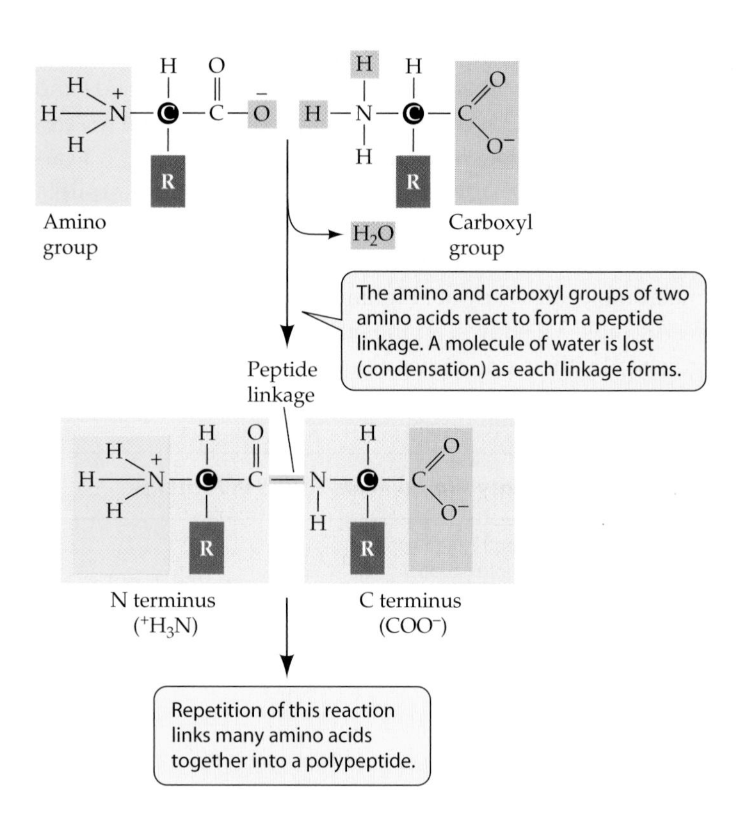

3.5 Formation of Peptide Linkages In living things, the reaction leading to a peptide linkage has many intermediate steps, but the reactants and products are the same as those shown in this simplified diagram.

Peptide linkages covalently bond amino acids together

When amino acids polymerize, the carboxyl and amino groups attached to the α carbon are the reactive groups. The carboxyl group of one amino acid reacts with the amino group of another, undergoing a condensation reaction that forms a *peptide linkage*. Figure 3.5 gives a simplified description of this reaction. (In living systems, other molecules must activate the amino acids in order for this reaction to proceed, and there are intermediate steps in the process. We will examine these steps in Chapter 12.)

Just as a sentence begins with a capital letter and ends with a period, polypeptide chains have a linear order. The chemical "capital letter" marking the beginning of a polypeptide is the amino group of the first amino acid in the chain and is known as the *N terminus*. The "punctuation mark" for the end of the chain is the carboxyl group of the last amino acid—the *C terminus*. All the other amino and carboxyl groups in the chain (except those in side chains) are involved in peptide bond formation, so they do not exist in the chain as "free," intact groups. Biochemists refer to the "N → C," or "amino-to-carboxyl" orientation of polypeptides.

The peptide linkage has two characteristics that are important in the three-dimensional structure of proteins:

- Unlike many single covalent bonds, in which the groups on either side of the bond are free to rotate in space, the C—N peptide linkage is relatively inflexible. The adjacent atoms (the α carbons of the two adjacent amino acids) are not free to rotate because of the partial double-bond character of the peptide bond. This characteristic limits the folding of the polypeptide chain.
- The oxygen bound to the carbon (C—O) in the carboxyl group carries a slight negative charge (δ^-), whereas the hydrogen bound to the nitrogen (N—H) in the amino group is slightly positive (δ^+). This asymmetry of charge favors hydrogen bonding within the protein molecule itself and with other molecules, contributing to both the structure and the function of many proteins. Before we explore the significance of such hydrogen bonds, we need to examine the importance of the order of amino acids.

The primary structure of a protein is its amino acid sequence

There are four levels of protein structure, called primary, secondary, tertiary, and quaternary. The precise sequence of amino acids in a polypeptide constitutes the **primary structure** of a protein (Figure 3.6*a*). The peptide backbone of this primary structure consists of a repeating sequence of three atoms (—N—C—C—): the N from the amino group, the α carbon, and the C from the carboxyl group of each amino acid.

Scientists have deduced the primary structure of many proteins. The single-letter abbreviations for amino acids (see Table 3.2) are used to record the amino acid sequence of a protein. Here, for example, are the first 20 amino acids (out of a total of 124) in the protein ribonuclease from a cow:

KETAAAKFERQHMDSSTSAA

The theoretical number of different proteins is enormous. Since there are 20 different amino acids, there could be $20 \times 20 = 400$ distinct dipeptides (two linked amino acids), and $20 \times 20 \times 20 = 8{,}000$ different tripeptides (three linked amino acids). Imagine this process of multiplying by 20 extended to a protein made up of 100 amino acids (which is considered a small protein). There could be 20^{100} such small proteins, each with its own distinctive primary structure. How large is the number 20^{100}? There aren't that many electrons in the entire universe!

At the higher levels of protein structure, local coiling and folding give the molecule its final functional shape, but all of these levels derive from the primary structure—that is, the precise location of specific amino acids in the polypeptide chain. The properties associated with a precise sequence of amino acids determine how the protein can twist and fold, thus adopting a specific stable structure that distinguishes it from every other protein.

Primary structure is determined by covalent bonds. But the next level of protein structure is determined by weaker hydrogen bonds.

The secondary structure of a protein requires hydrogen bonding

A protein's **secondary structure** consists of regular, repeated patterns in different regions of a polypeptide chain. There are two basic types of secondary structure, both of them determined by hydrogen bonding between the amino acid residues that make up the primary structure.

THE α HELIX. The **α (alpha) helix** is a right-handed coil that is "threaded" in the same direction as a standard wood screw (Figure 3.6*b*). The R groups extend outward from the peptide backbone of the helix. The coiling results from hydrogen bonds that form between the δ^+ hydrogen of the N—H of one amino acid residue and the δ^- oxygen of the C=O of another. When this pattern of hydrogen bonding is established repeatedly over a segment of the protein, it stabilizes the coil, resulting in an α helix. The presence of amino acids with large R groups that distort the coil or otherwise prevent the formation of the necessary hydrogen bonds will keep an α helix from forming.

The α-helical secondary structure is common in the fibrous structural proteins called keratins, which make up hair, hooves, and feathers. Hair can be stretched because stretching requires that only the hydrogen bonds of the α helix, not the covalent bonds, be broken; when the tension on the hair is released, both the hydrogen bonds and the helix re-form.

THE β PLEATED SHEET. A **β (beta) pleated sheet** is formed from two or more polypeptide chains that are almost completely extended and lying next to one another. The sheet is stabilized by hydrogen bonds between the N—H groups on one chain and the C=O groups on the other (Figure 3.6*c*). A β pleated sheet may form between separate polypeptide chains, as in spider silk, or between different regions of the same polypeptide chain that is bent back on itself. Many proteins contain regions of both α helix and β pleated sheet in the same polypeptide chain.

The tertiary structure of a protein is formed by bending and folding

In many proteins, the polypeptide chain is bent at specific sites and then folded back and forth, resulting in the **tertiary**

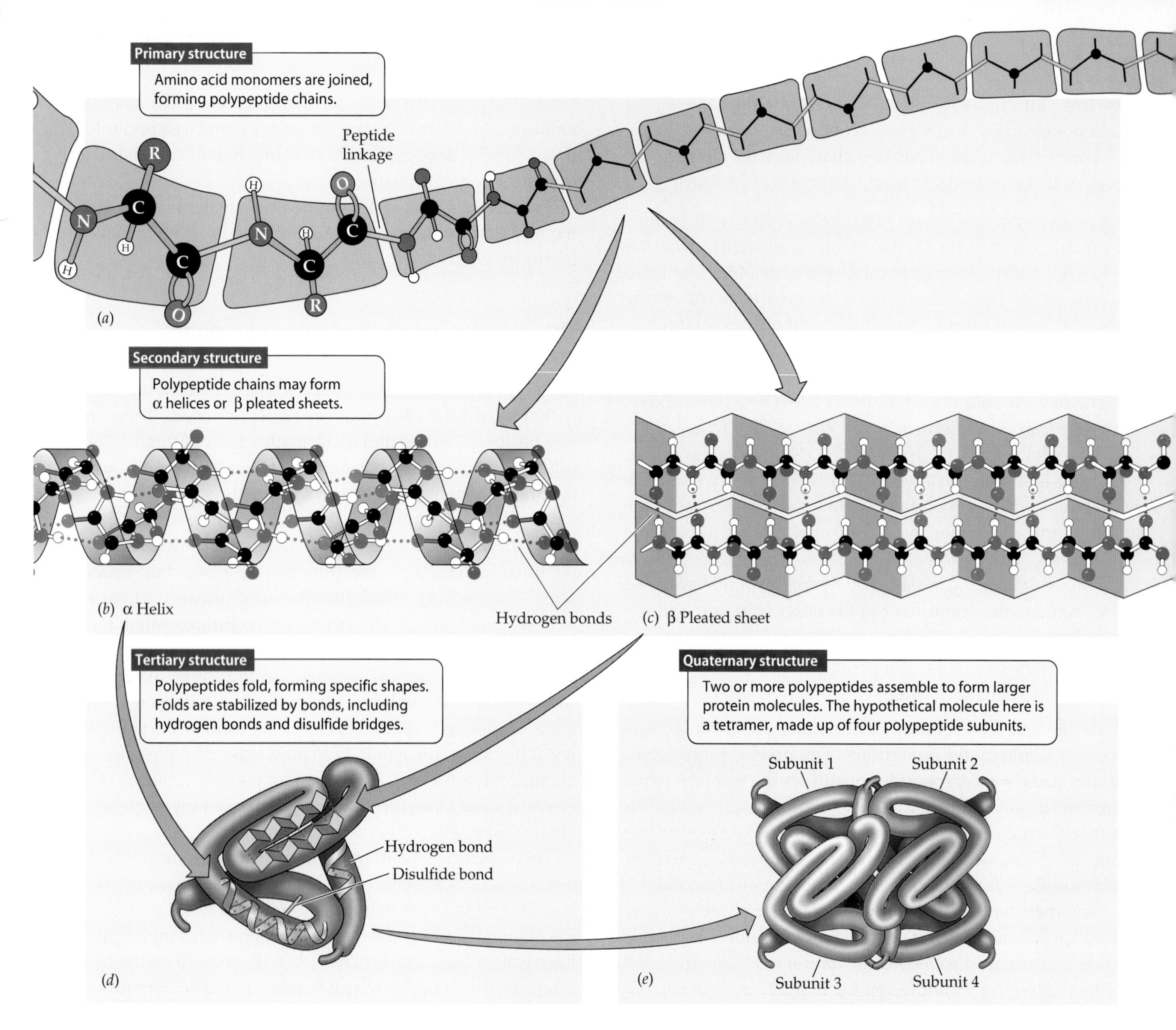
Primary structure
Amino acid monomers are joined, forming polypeptide chains.
Peptide linkage
(a)
Secondary structure
Polypeptide chains may form α helices or β pleated sheets.
(b) α Helix
Hydrogen bonds
(c) β Pleated sheet
Tertiary structure
Polypeptides fold, forming specific shapes. Folds are stabilized by bonds, including hydrogen bonds and disulfide bridges.
Hydrogen bond
Disulfide bond
(d)
Quaternary structure
Two or more polypeptides assemble to form larger protein molecules. The hypothetical molecule here is a tetramer, made up of four polypeptide subunits.
Subunit 1
Subunit 2
Subunit 3
Subunit 4
(e)

3.6 The Four Levels of Protein Structure Secondary, tertiary, and quaternary structure all arise from the primary structure of the protein.

structure of the protein (Figure 3.6*d*). Although the α helices and β pleated sheets contribute to the tertiary structure, only parts of the macromolecule usually have these secondary structures, and large regions consist of structures unique to a particular protein.

While hydrogen bonding between the N—H and C=O groups within and between chains is responsible for secondary structure, the interactions between R groups—the amino acid side chains—determine tertiary structure. We described the various strong and weak interactions between atoms in Chapter 2 (see Table 2.1). Many of these interactions are involved in determining tertiary structure.

- Covalent *disulfide bridges* can form between specific cysteine residues (see Figure 3.4), holding a folded polypeptide in place.
- *Hydrophobic* side chains can aggregate together in the interior of the protein, away from water, folding the polypeptide in the process.
- *van der Waals forces* can stabilize the close interactions between the hydrophobic residues.
- *Ionic bonds* can form between positively and negatively charged side chains buried deep within a protein, away from water, forming a *salt bridge*.

A complete description of a protein's tertiary structure specifies the location of every atom in the molecule in three-dimen-

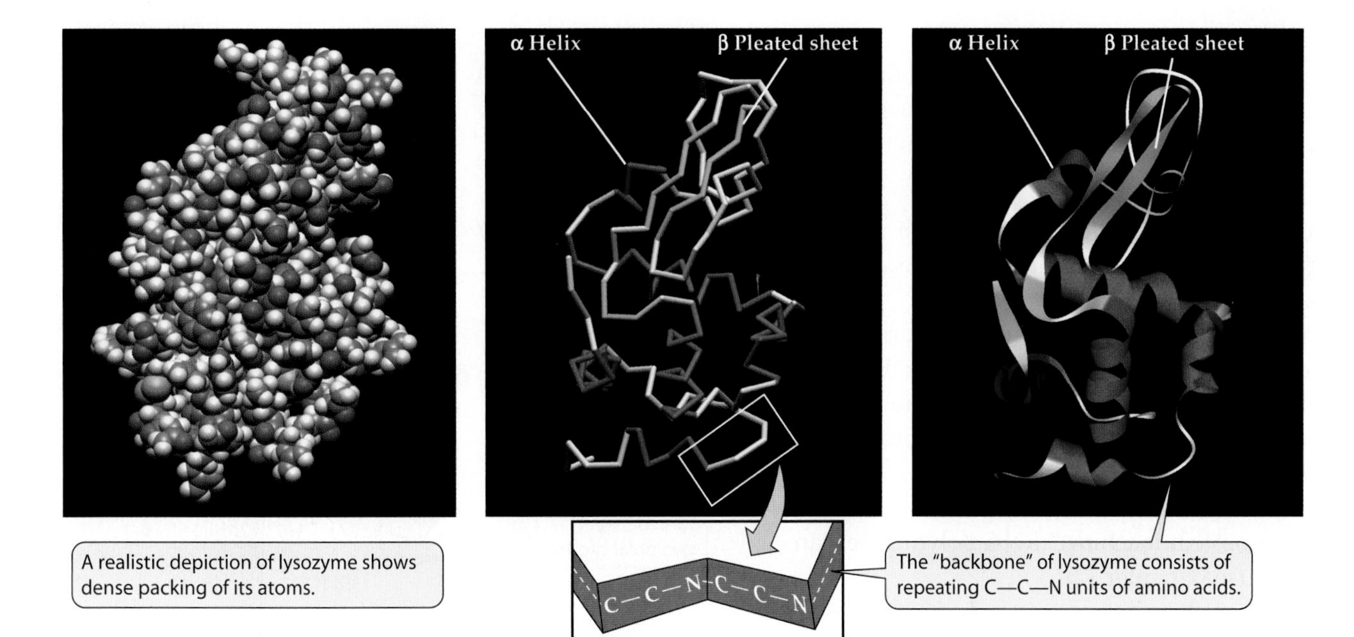

3.7 Three Representations of Lysozyme Different molecular representations of a protein emphasize different aspects of its tertiary structure. These three representations of lysozyme are similarly oriented.

sional space in relation to all the other atoms. Such a description is available for the protein lysozyme (Figure 3.7). The first tertiary structures to be determined took years to figure out, but today, dozens of new structures are published every week. The major advances making this possible have been the ability to produce large quantities of specific proteins by biotechnology and the use of computers to analyze the atomic data.

Bear in mind that both tertiary structure and secondary structure derive from a protein's primary structure. If lysozyme is heated slowly, the heat energy will disrupt only the weak interactions and cause only the tertiary structure to break down. But the protein will return to its normal tertiary structure when it cools, demonstrating that all the information needed to specify the unique shape of a protein is contained in its primary structure.

The quaternary structure of a protein consists of subunits

As mentioned earlier, many functional proteins contain two or more polypeptide chains, called *subunits*, each of them folded into its own unique tertiary structure. The protein's **quaternary structure** results from the ways in which these subunits bind together and interact (see Figure 3.6*e*).

Quaternary structure is illustrated by hemoglobin (Figure 3.8). Hydrophobic interactions, van der Waals forces, hydrogen bonds, and ionic bonds all help hold the four subunits together to form the hemoglobin molecule. The function of hemoglobin is to carry oxygen in red blood cells. As hemoglobin binds one O_2 molecule, the four subunits shift their relative positions slightly, changing the quaternary structure. Ionic bonds are broken, exposing buried side chains that enhance the binding of additional O_2 molecules. The structure changes again when hemoglobin releases its O_2 molecules to the cells of the body.

The surfaces of proteins have specific shapes

Small molecules in a solution are in constant motion. They vibrate, rotate, and move from place to place like corn in a

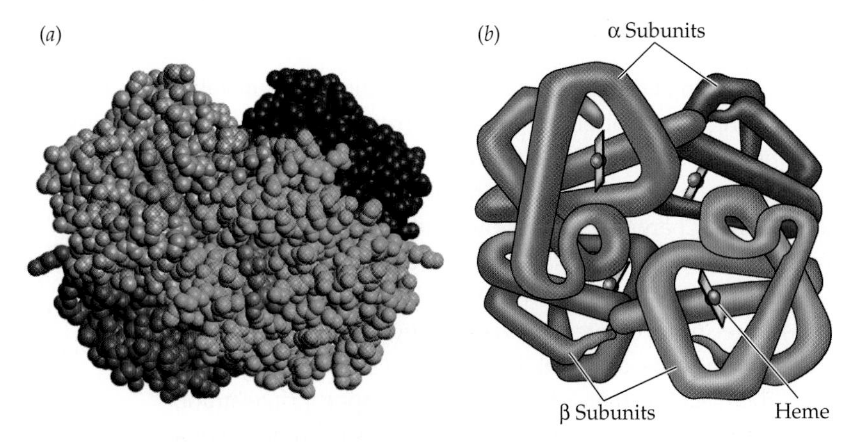

3.8 Quaternary Structure of a Protein Hemoglobin consists of four folded polypeptide subunits that assemble themselves into the quaternary structure shown here. In these two graphic representations, each type of subunit is a different color. The heme groups contain iron and are the oxygen-carrying sites.

popper. If two of them collide in the right circumstances, a chemical reaction can occur. The specific shapes of proteins allow them to bind *noncovalently* to other molecules, which in turn allows other important biological events to occur. Here are just a few examples:

- Two adjacent cells can stick together because proteins protruding from each of the cells interact with each other (see Chapter 5).
- A substance can enter a cell by binding to a carrier protein in the cell surface membrane (see Chapter 5).
- A chemical reaction can be speeded up when an enzyme protein binds to one of the reactants (see Chapter 6).
- A "multi-protein machine," DNA polymerase, can bind to and copy DNA (see Chapter 11).
- Another "multi-protein machine," RNA polymerase, can synthesize RNA (see Chapter 12).
- Chemical signals such as hormones can bind to proteins on a cell's outer surface (see Chapter 15).
- Defensive proteins called antibodies can recognize the shape of a virus coat and bind to it (see Chapter 18).

The biological specificity of protein function depends on two general properties of the protein: its shape and the chemistry of its exposed surface groups.

- *Shape*. When a molecule collides with and binds to a much larger protein, it is like a baseball being caught by a catcher's mitt: The mitt has a shape that binds to the ball and fits around it. A hockey puck or a ping-pong ball would not fit a baseball catcher's mitt. The binding of a molecule to a protein involves a general "fit" between two three-dimensional objects that becomes even more specific after initial binding.
- *Chemistry*. The surface of a protein has certain chemical groups that it presents to a substance attempting to bind to it (Figure 3.9). These groups are the R groups of the exposed amino acids, and are therefore a property of the protein's primary structure.

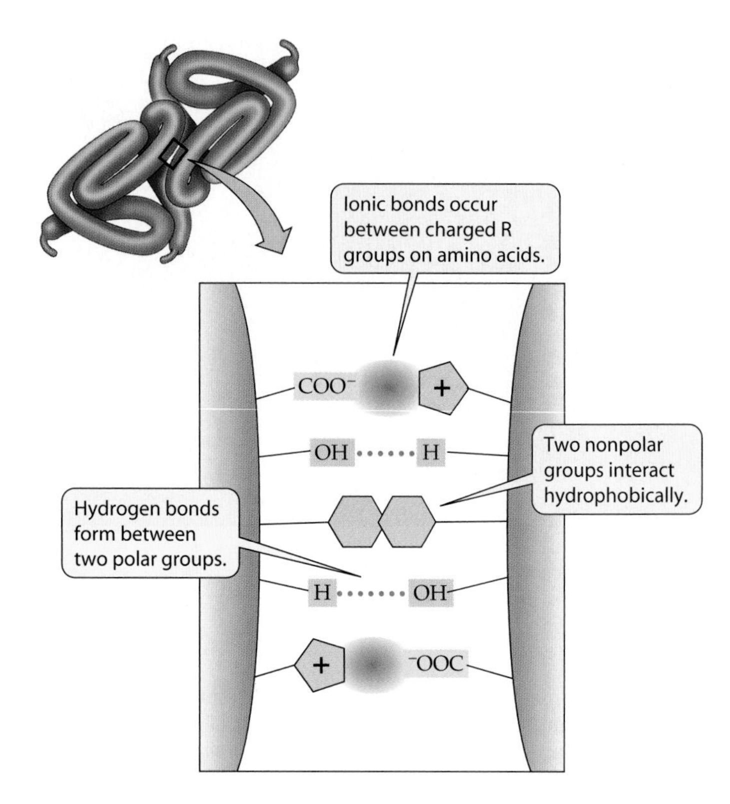

3.9 Noncovalent Interactions between Proteins and Other Molecules Noncovalent interactions allow a protein to bind tightly to another molecule with specific properties, or allow regions within a protein to interact with one another.

Look again at the structures of the 20 amino acids in Table 3.2, noting the properties of the R groups. Exposed hydrophobic groups can bind to similarly nonpolar groups in the substance with which the protein interacts (often called the **ligand**). Charged R groups can bind to oppositely charged groups on the ligand. Polar R groups containing a hydroxyl (—OH) group can form a hydrogen bond with the ligand. These three types of interactions—hydrophobic, ionic, and hydrogen bonding—are weak by themselves, but strong when all of them act together. So the exposure of appropriate amino acid R groups on the protein surface allows the binding of a specific ligand to occur.

Knowing the exact shape of a protein and what can bind to it is important not only in understanding basic biology, but also in applied fields such as medicine. For example, the three-dimensional structure of a protease, a protein essential for the replication of HIV—the virus that causes AIDS—was first determined, then specific proteins were designed to bind to it and block its action. These protease inhibitors have prolonged the lives of countless people living with HIV (Figure 3.10).

Protein shapes are sensitive to the environment

Because it is determined by weak forces, protein shape is sensitive to environmental conditions that would not break covalent bonds, but do upset the weaker noncovalent interactions that determine secondary and tertiary structure.

- Increases in *temperature* cause more rapid molecular movements and thus can break hydrogen bonds and hydrophobic interactions.
- Alterations in *pH* can change the pattern of ionization of carboxyl and amino groups in the R groups of amino acids, thus disrupting the pattern of ionic attractions and repulsions.
- High concentrations of polar substances such as urea can disrupt the hydrogen bonding that is crucial to protein

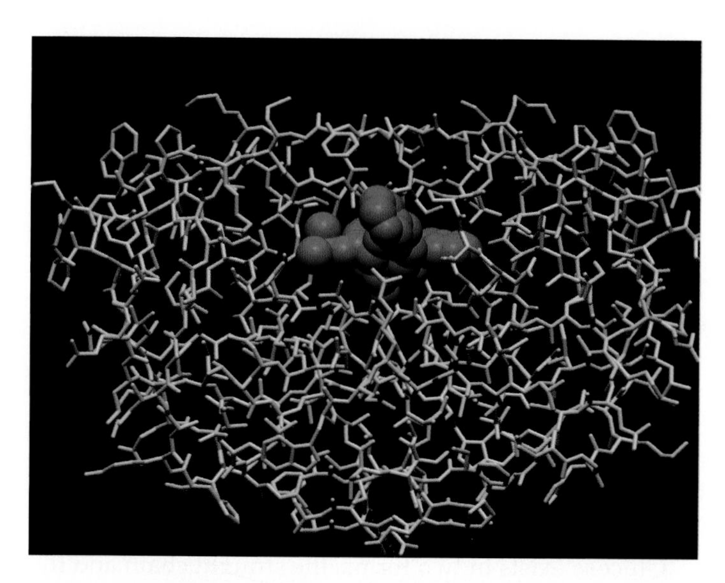

3.10 An HIV Protease Inhibitor After determining the structure of HIV protease (the blue polypeptide chain), a protein essential to the life cycle of HIV, biochemists designed a drug (the red space-filling model) to fit into the protease and block its function. Many people living with HIV and AIDS now take this drug.

structure. Nonpolar solvents may also disrupt normal protein structure.

The loss of a protein's normal three-dimensional structure is called **denaturation**, and it is always accompanied by a loss of the normal biological function of the protein (Figure 3.11).

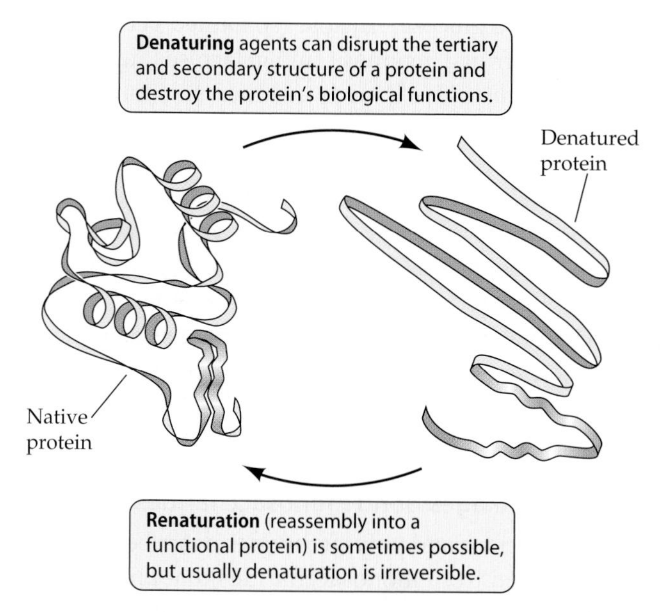
Denaturing agents can disrupt the tertiary and secondary structure of a protein and destroy the protein's biological functions.
Denatured protein
Native protein
Renaturation (reassembly into a functional protein) is sometimes possible, but usually denaturation is irreversible.

3.11 Denaturation Is the Loss of Tertiary Protein Structure and Function Agents that can cause denaturation include high temperatures and certain chemicals.

Denaturation is often irreversible, because amino acids that were buried may now be exposed at the surface, and vice versa, causing a new structure to form or different molecules to bind to the protein. Boiling an egg denatures its proteins and is, as you know, not reversible. However, as we saw earlier in the case of lysozyme, denaturation may be reversible in the laboratory. If the protein is allowed to cool or the denaturing chemicals are removed, the protein may return to its "native" shape and normal function.

Chaperonins help shape proteins

There are two occasions when a polypeptide chain is in danger of binding the wrong ligand. First, following denaturation, hydrophobic R groups, previously on the inside of the protein away from water, become exposed on the surface. Since these groups can interact with similar groups on other molecules, the denatured proteins may aggregate and become insoluble, losing their function. Second, when a protein has just been made and has not yet folded completely, it can present a surface that binds the wrong molecule. In the cell, a protein can sometimes fold incorrectly as it is made. This can have serious consequences: In Alzheimer's disease, misfolded proteins accumulate in the brain and bind to one another, forming fibers in the areas of the brain that control memory, mood, and spatial awareness.

Living systems limit inappropriate protein interactions by making a class of proteins called, appropriately, **chaperonins** (recall the chaperones—usually teachers—at school dances who try to prevent "inappropriate interactions" among the students). Chaperonins were first identified in fruit flies as "heat shock" proteins, which prevented denaturing proteins from clumping together when the flies' temperatures were raised.

Some chaperonins work by trapping proteins in danger of inappropriate binding inside a molecular "cage" (Figure 3.12). This cage is composed of several identical subunits, and is itself a good example of quaternary protein structure. Inside the cage, the targeted protein folds into the right shape, and then is released at the appropriate time and place.

Carbohydrates: Sugars and Sugar Polymers

The second class of biological molecules, the **carbohydrates**, is a diverse group of compounds. Carbohydrates contain primarily carbon atoms flanked by hydrogen atoms and hydroxyl groups (H—C—OH). They have two major biochemical roles:

- They act as a source of energy that can be released in a form usable by body tissues.

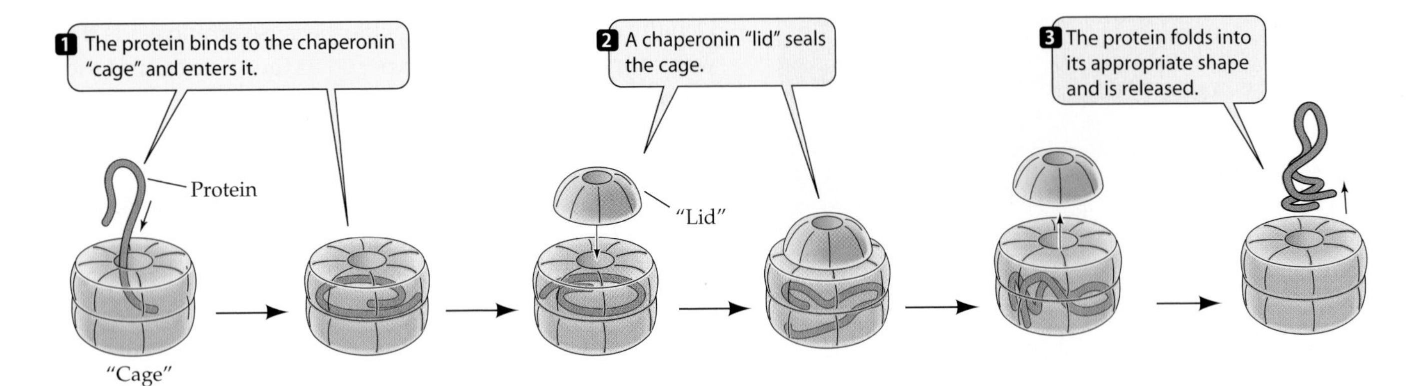

1 The protein binds to the chaperonin "cage" and enters it.
2 A chaperonin "lid" seals the cage.
3 The protein folds into its appropriate shape and is released.
Protein
"Lid"
"Cage"

3.12 Chaperonins Protect Proteins from Inappropriate Binding Chaperonins surround new or denatured proteins and prevent them from binding to the wrong ligand.

- They serve as carbon skeletons that can be rearranged to form other molecules that are essential for biological structures and functions.

Some carbohydrates are relatively small, with molecular weights of less than 100. Others are true macromolecules, with molecular weights in the hundreds of thousands.

There are four categories of biologically important carbohydrates, which we will discuss in turn:

- *Monosaccharides* (*mono-*, "one"; *saccharide*, "sugar"), such as glucose, ribose, and fructose, are simple sugars. They are the monomers out of which the larger carbohydrates are constructed.
- *Disaccharides* (*di-*, "two") consist of two monosaccharides linked together by covalent bonds.
- *Oligosaccharides* (*oligo-*, "several") are made up of several (3 to 20) monosaccharides.
- *Polysaccharides* (*poly-*, "many"), such as starch, glycogen, and cellulose, are large polymers composed of hundreds or thousands of monosaccharides.

The general formula for carbohydrates, CH_2O, gives the relative proportions of carbon, hydrogen, and oxygen in a monosaccharide (i.e., the proportions of these atoms are 1:2:1). In disaccharides, oligosaccharides, and polysaccharides, these proportions differ slightly from the general formula because two hydrogens and an oxygen are lost during each of the condensation reactions that form them.

Monosaccharides are simple sugars

Green plants produce monosaccharides through photosynthesis, and animals acquire them directly or indirectly from plants. All living cells contain the monosaccharide **glucose**. Cells use glucose as an energy source, breaking it down through a series of reactions that release stored energy and produce water and carbon dioxide.

Glucose exists in two forms, the straight chain and the ring. The ring form predominates in more than 99 percent of circumstances because it is more stable under cellular conditions. There are two forms of the ring structure (α-glucose and β-glucose), which differ only in the placement of the —H and —OH attached to carbon 1 (Figure 3.13). The α and β forms interconvert and exist in equilibrium when dissolved in water.

Different monosaccharides contain different numbers of carbons. (The standard convention for numbering carbons in carbohydrates shown in Figure 3.13 is used throughout this book.) Most of the monosaccharides found in living systems belong to the D series of optical isomers (see Chapter 2). But some monosaccharides are structural isomers, which have the same kinds and numbers of atoms, but arranged differently. For example, the *hexoses* (*hex-*, "six"), a group of structural isomers, all have the formula $C_6H_{12}O_6$. Included among the hexoses are glucose, fructose (so named because it was first found in fruits), mannose, and galactose (Figure 3.14).

Pentoses (*pent-*, "five") are five-carbon sugars. Some pentoses are found primarily in the cell walls of plants. Two pentoses are of particular biological importance: Ribose and deoxyribose form part of the backbones of the nucleic acids RNA and DNA, respectively. These two pentoses are not isomers; rather, one oxygen atom is missing from carbon 2 in deoxyribose (*de-*, "absent") (see Figure 3.14). As we will see in Chapter 12, the absence of this oxygen atom has important consequences for the functional distinction of RNA and DNA.

Glycosidic linkages bond monosaccharides together

The disaccharides and polysaccharides described above are all constructed from monosaccharides that are covalently bonded together by condensation reactions that form *glycosidic linkages*. One such linkage between two monosaccharides forms a disaccharide. For example, a molecule of su-

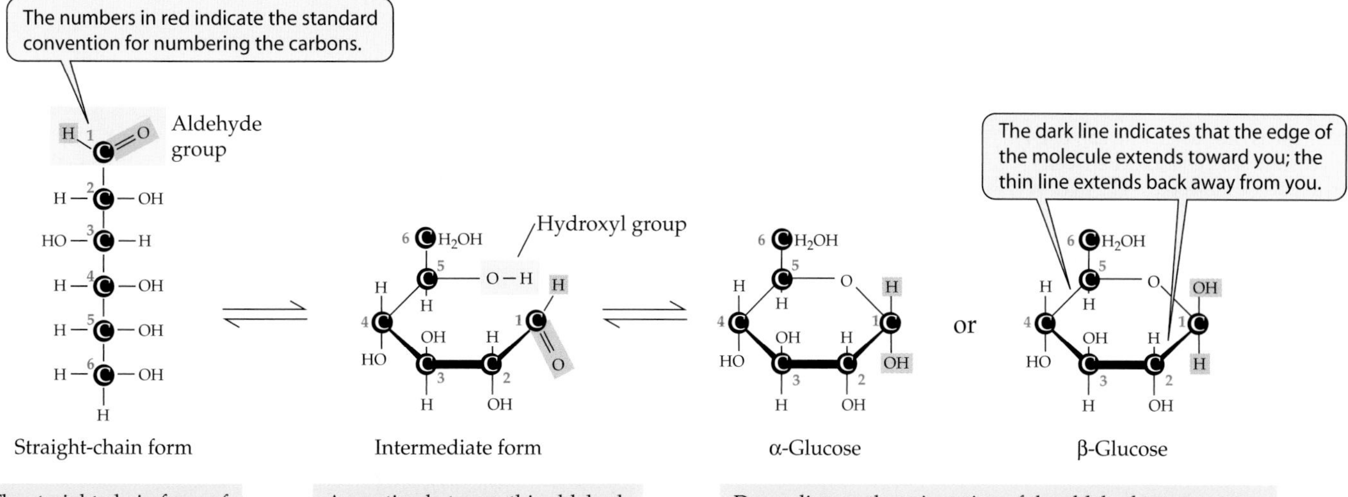
The numbers in red indicate the standard convention for numbering the carbons.
Aldehyde group
Hydroxyl group
or
The dark line indicates that the edge of the molecule extends toward you; the thin line extends back away from you.
Straight-chain form
Intermediate form
α-Glucose
β-Glucose
The straight-chain form of glucose has an aldehyde group at carbon 1.
A reaction between this aldehyde group and the hydroxyl group at carbon 5 gives rise to a ring form.
Depending on the orientation of the aldehyde group when the ring closes, either of two molecules α-glucose or β-glucose forms.

3.13 Glucose: From One Form to the Other All glucose molecules have the formula $C_6H_{12}O_6$, but their structures vary. When dissolved in water, the α and β "ring" forms of glucose interconvert.

crose (table sugar) is formed from a glucose molecule and a fructose molecule, while lactose (milk sugar) contains glucose and galactose.

The disaccharide maltose contains two glucose molecules, but it is not the only disaccharide that can be made from two glucoses. When glucose molecules form a glycosidic linkage, the linkage will be one of two types, α or β, depending on whether the molecule that bonds its carbon 1 is α-glucose or β-glucose (see Figure 3.13). An α linkage with carbon 4 of a second glucose molecule gives maltose, whereas a β linkage gives cellobiose (Figure 3.15).

Maltose and cellobiose are disaccharide isomers, both having the formula $C_{12}H_{22}O_{11}$. However, they are different compounds with different properties. They undergo different chemical reactions and are recognized by different enzymes. For example, maltose can be hydrolyzed to its monosaccharides in the human body, whereas cellobiose cannot. Certain microorganisms have the chemistry needed to break down cellobiose.

Oligosaccharides contain several monosaccharides bound by glycosidic linkages at various sites. Many oligosaccharides have additional functional groups, which give them special properties. Oligosaccharides are often covalently bonded to proteins and lipids on the outer cell surface, where they serve as cell recognition signals. The human blood groups (such as ABO) get their specificity from oligosaccharide chains.

Three-carbon sugar

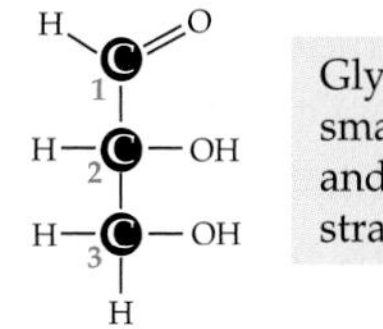
Glyceraldehyde

Glyceraldehyde is the smallest monosaccharide and exists only as the straight-chain form.

Five-carbon sugars (pentoses)

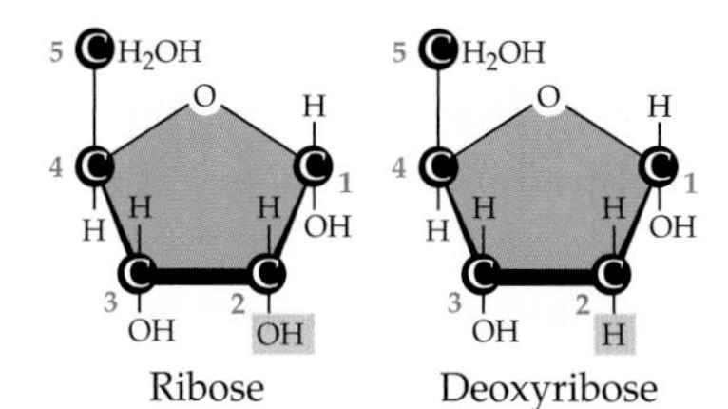
Ribose
Deoxyribose

Ribose and deoxyribose each have five carbons, but very different chemical properties and biological roles.

Six-carbon sugars (hexoses)

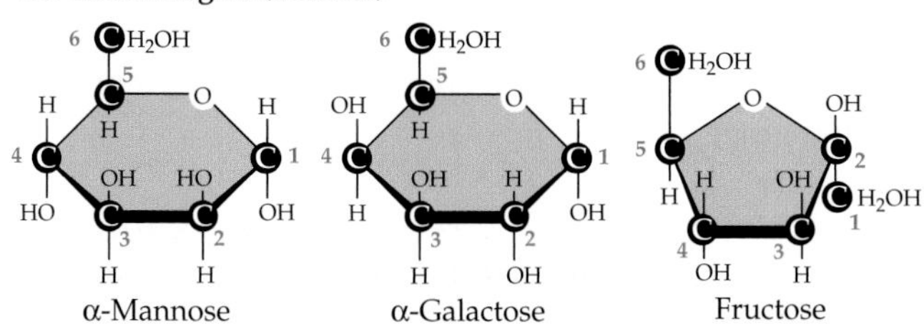
α-Mannose
α-Galactose
Fructose

These hexoses are isomers. All have the formula $C_6H_{12}O_6$, but each has distinct chemical properties and biological roles.

3.14 Monosaccharides Are Simple Sugars Monosaccharides are made up of varying numbers of carbons. Some are structural isomers, which have the same number of carbons, but arranged differently. Fructose, for example, is a hexose, but forms a five-sided ring like the pentoses.

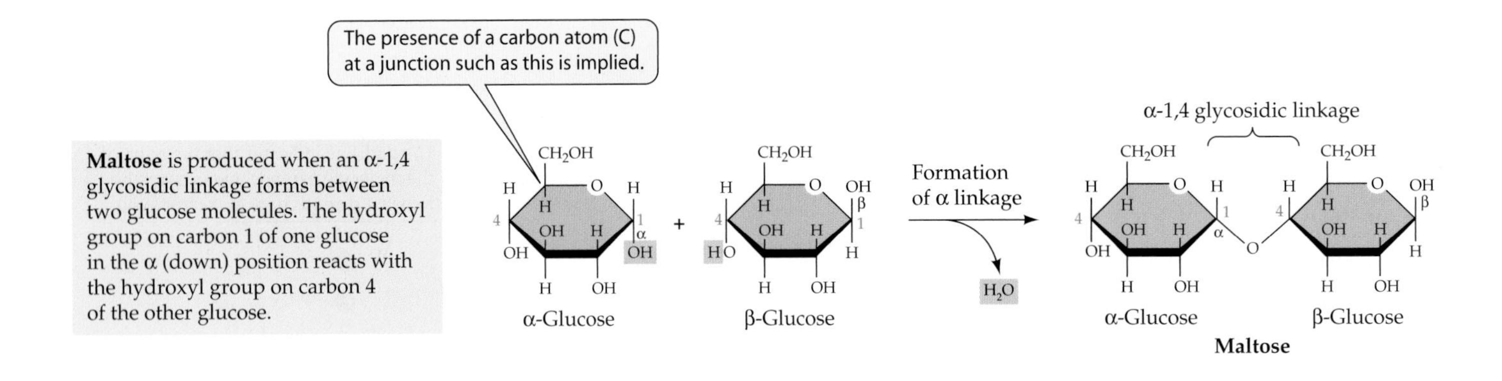

3.15 Disaccharides Are Formed by Glycosidic Linkages
Glycosidic linkages between two monosaccharides create many different disaccharides. Which disaccharide is formed depends on which monosaccharides are linked, and on the site (which carbon atom is linked) and form (α or β) of the linkage.

Polysaccharides serve as energy stores or structural materials

Polysaccharides are giant polymers of monosaccharides connected by glycosidic linkages (Figure 3.16).

- **Starch** is a polysaccharide of glucose with α-glycosidic linkages.
- **Glycogen** is a highly branched polysaccharide of glucose.
- **Cellulose** is also a polysaccharide of glucose, but its individual monosaccharides are connected by β-glycosidic linkages.

Starch actually comprises a large family of giant molecules of broadly similar structure. While all starches are large polymers of glucose with α linkages (Figure 3.16*a*), the different starches can be distinguished by the amount of branching that occurs at carbons 1 and 6 (Figure 3.16*b*). Some plant starches are unbranched, as in plant amylose; others are moderately branched, as in plant amylopectin. Starch readily binds water, and when that water is removed, unbranched starch tends to form hydrogen bonds between the polysaccharide chains, which then aggregate. This is what causes bread to become hard and stale. Adding water and gentle heat separates the chains and the bread becomes softer.

The polysaccharide glycogen stores glucose in animal livers and muscles. Starch and glycogen serve as energy storage compounds for plants and animals, respectively. These polysaccharides are readily hydrolyzed to glucose monomers, which in turn can be further degraded to liberate their stored energy and convert it to forms that can be used for cellular activities. If it is glucose that is actually needed for fuel, why must it be stored as a polymer? The reason is that 1,000 glucose molecules would exert 1,000 times the osmotic pressure (causing water to enter the cells; see Chapter 5) of a single glycogen molecule. If it were not for polysaccharides, many organisms would expend a lot of time and energy expelling excess water.

Cellulose is the predominant component of plant cell walls, and is by far the most abundant **organic** (carbon-containing) compound on Earth. Starch can be easily degraded by the actions of chemicals or enzymes. Cellulose, however, is chemically more stable because of its β-glycosidic linkages (Figure 3.16*a*). Thus starch is a good storage medium that can be easily broken down to supply glucose for energy-producing reactions, while cellulose is an excellent structural material that can withstand harsh environmental conditions without changing.

Chemically modified carbohydrates contain other groups

Some carbohydrates are chemically modified by the addition of functional groups, such as phosphate and amino groups

(*a*) **Molecular structure**

Cellulose

Hydrogen bonding to other cellulose molecules can occur at these points.

Cellulose is an unbranched polymer of glucose with β-1,4 glycosidic linkages that are chemically very stable.

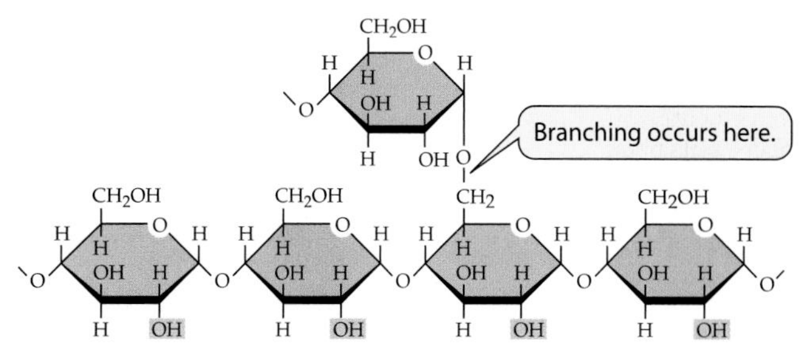

Glycogen and starch are polymers of glucose with α-1,4 glycosidic linkages. α-1,6 glycosidic linkages produce branching at carbon 6.

(*b*) **Macromolecular structure**

Linear (cellulose)

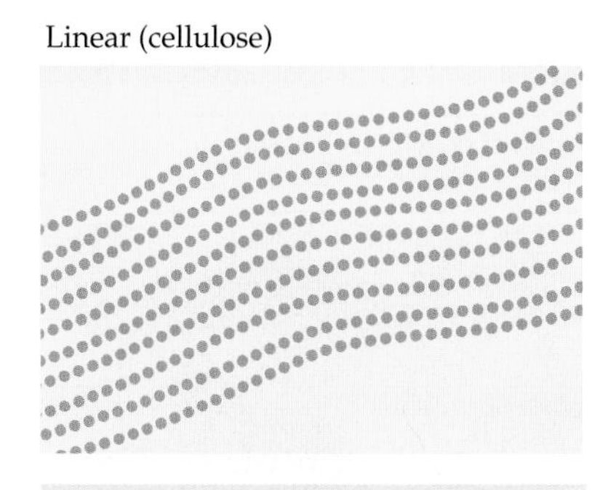

Parallel cellulose molecules from hydrogen-bonds, resulting in thin fibrils.

Branched (starch)

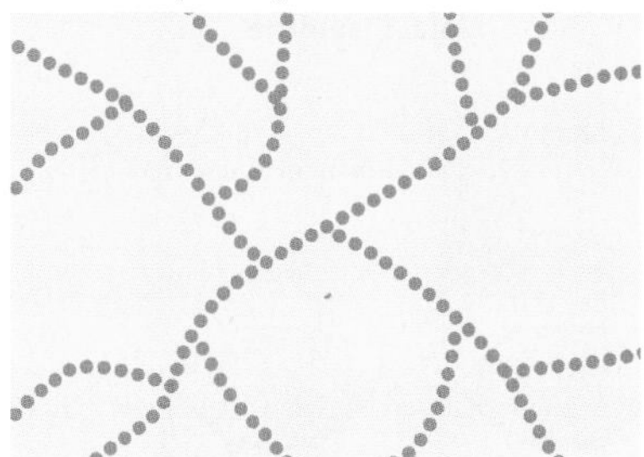

Branching limits the number of hydrogen bonds that can form in starch molecules, making starch less compact than cellulose.

Highly branched (glycogen)

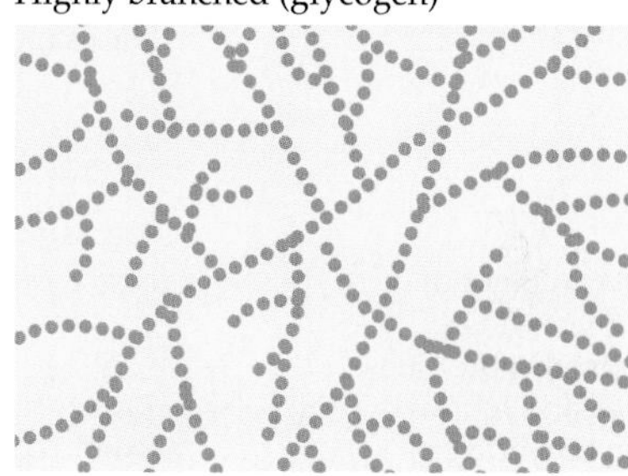

The high amount of branching in glycogen makes its solid deposits more compact than starch.

(*c*) **Polysaccharides in cells**

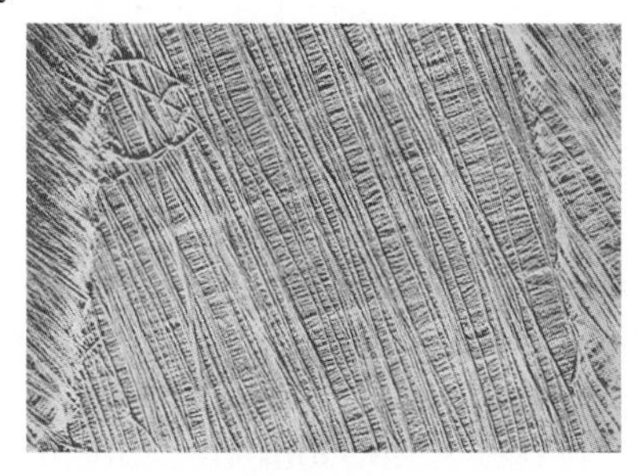

Layers of cellulose fibrils, as seen in this scanning electron micrograph, give plant cell walls great strength.

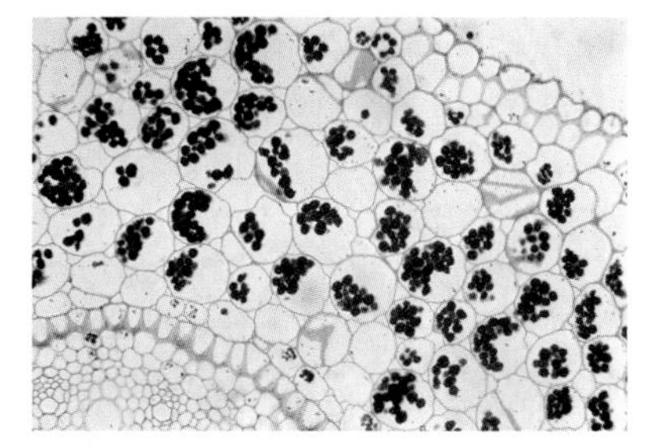

Dyed purple in this micrograph, starch deposits have a large granular shape within cells.

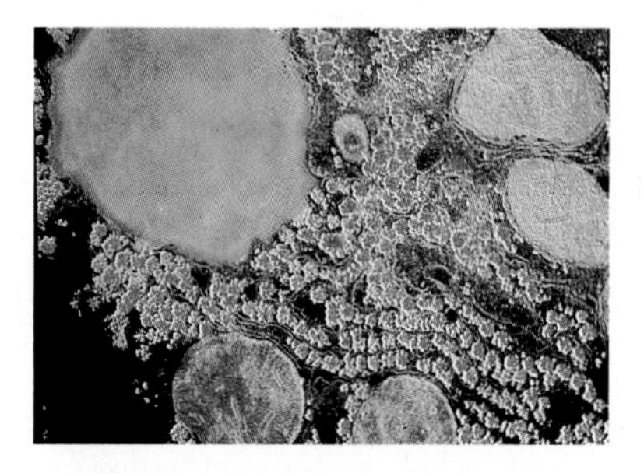

Colored pink in this electron micrograph of human liver cells, glycogen deposits have a small granular shape.

3.16 Representative Polysaccharides Cellulose, starch, and glycogen demonstrate different levels of branching and compaction in polysaccharides.

(Figure 3.17). For example, carbon 6 in glucose may be oxidized from —CH_2OH to a carboxyl group (—COOH), producing glucuronic acid. Or a phosphate group may be added to one or more of the —OH sites. Some of the resulting *sugar phosphates*, such as fructose 1,6-bisphosphate, are important intermediates in cellular energy reactions.

When an amino group is substituted for an —OH group, *amino sugars*, such as glucosamine and galactosamine, are produced. These compounds are important in the extracellular matrix, where they form parts of proteins involved in keeping tissues together. Galactosamine is a major component of cartilage, the material that forms caps on the ends of bones and stiffens the protruding parts of the ears and nose. A derivative of glucosamine produces the polymer *chitin*, which is the principal structural polysaccharide in the skele-

3.17 Chemically Modified Carbohydrates Added functional groups modify the form and properties of a carbohydrate.

(*a*) **Sugar phosphate**

Fructose 1,6 bisphosphate is involved in the reactions that liberate energy from glucose. (The numbers in its name refer to the carbon sites of phosphate bonding; *bis-* indicates that two phosphates are present.)

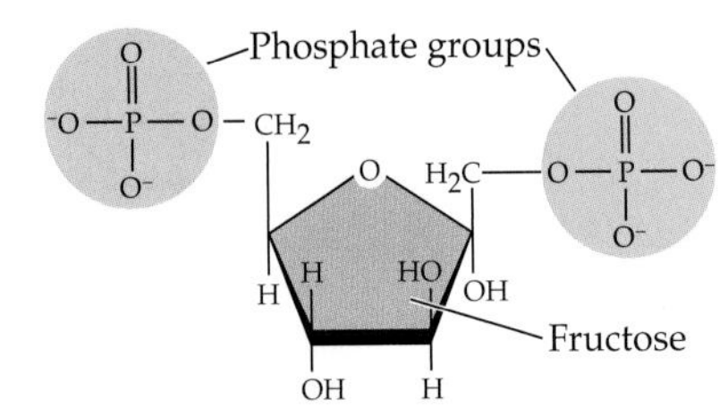

(*b*) **Amino sugars**

The monosaccharides glucosamine and galactosamine are amino sugars with an amino group in place of a hydroxyl group.

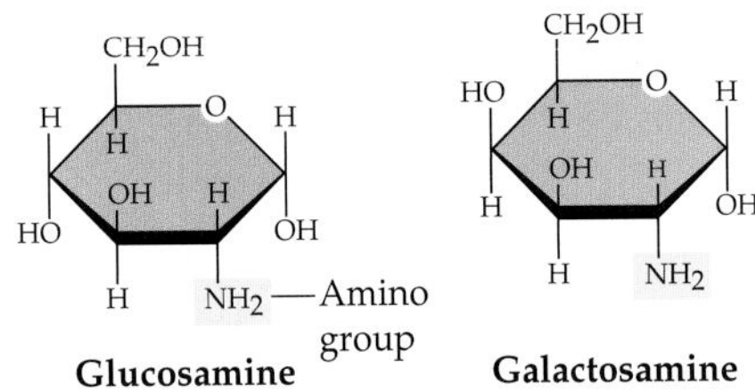

(*c*) **Chitin**

Chitin is a polymer of *N*-acetylglucosamine; *N*-acetyl groups provide additional sites for hydrogen bonding between the polymers.

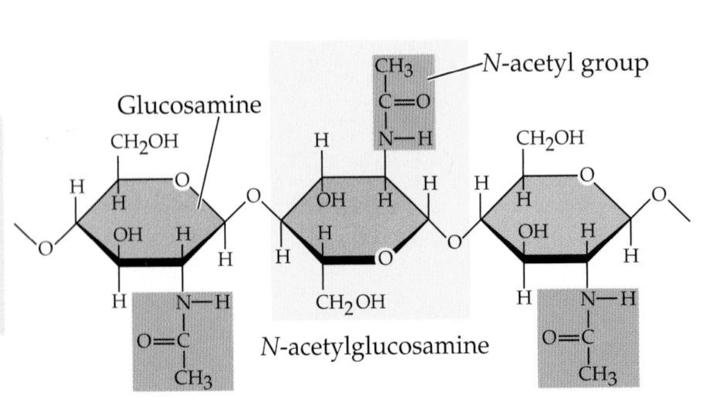

Galactosamine is an important component of cartilage, a connective tissue in vertebrates.

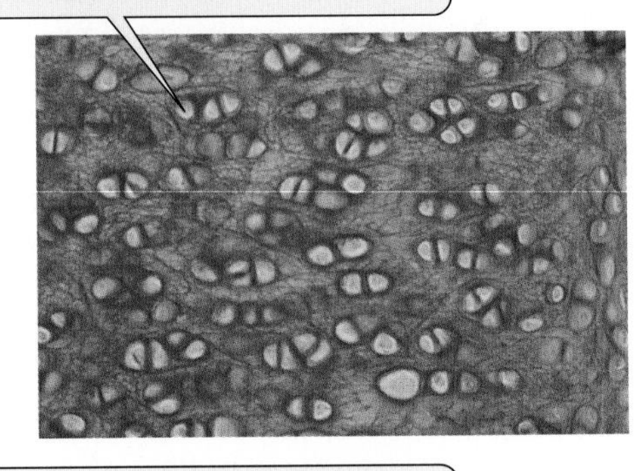

The external skeletons of insects are made up of chitin.

tons of insects, crabs, and lobsters, as well as in the cell walls of fungi. Fungi and insects (and their relatives) constitute more than 80 percent of the species ever described, and so chitin is one of the most abundant substances on Earth.

Lipids: Water-Insoluble Molecules

The **lipids** are a chemically diverse group of hydrocarbons. The property they all share is insolubility in water, which is due to the presence of many nonpolar covalent bonds. As we saw in Chapter 2, nonpolar hydrocarbon molecules are hydrophobic and preferentially aggregate among themselves, away from water, which is polar. When these nonpolar molecules are sufficiently close together, weak but additive van der Waals forces hold them together. These huge macromolecular aggregations are not polymers in a strict chemical sense, since their units (lipid molecules) are not held together by covalent bonds, as are, for example, the amino acids in proteins. But they can be considered polymers of individual lipid units.

In this section, we will describe the different types of lipids. Lipids have a number of roles in living organisms:

- Fats and oils store energy.
- Phospholipids play important structural roles in cell membranes.
- The carotenoids help plants capture light energy.
- Steroids and modified fatty acids play regulatory roles as hormones and vitamins.
- The fat in animal bodies serves as thermal insulation.
- A lipid coating around nerves acts as electrical insulation.
- Oil or wax on the surfaces of skin, fur, and feathers repels water.

Fats and oils store energy

Chemically, fats and oils are **triglycerides**, also known as *simple lipids*. Triglycerides that are solid at room temperature (20°C) are called *fats;* those that are liquid at room temperature are called *oils*. Triglycerides are composed of two types of building blocks: fatty acids and glycerol. *Glycerol* is a small molecule with three hydroxyl (—OH) groups (an alcohol). A *fatty acid* is made up of a long nonpolar hydrocarbon chain and a polar carboxyl group (—COOH). A triglyceride contains three fatty acid molecules and one molecule of glycerol.

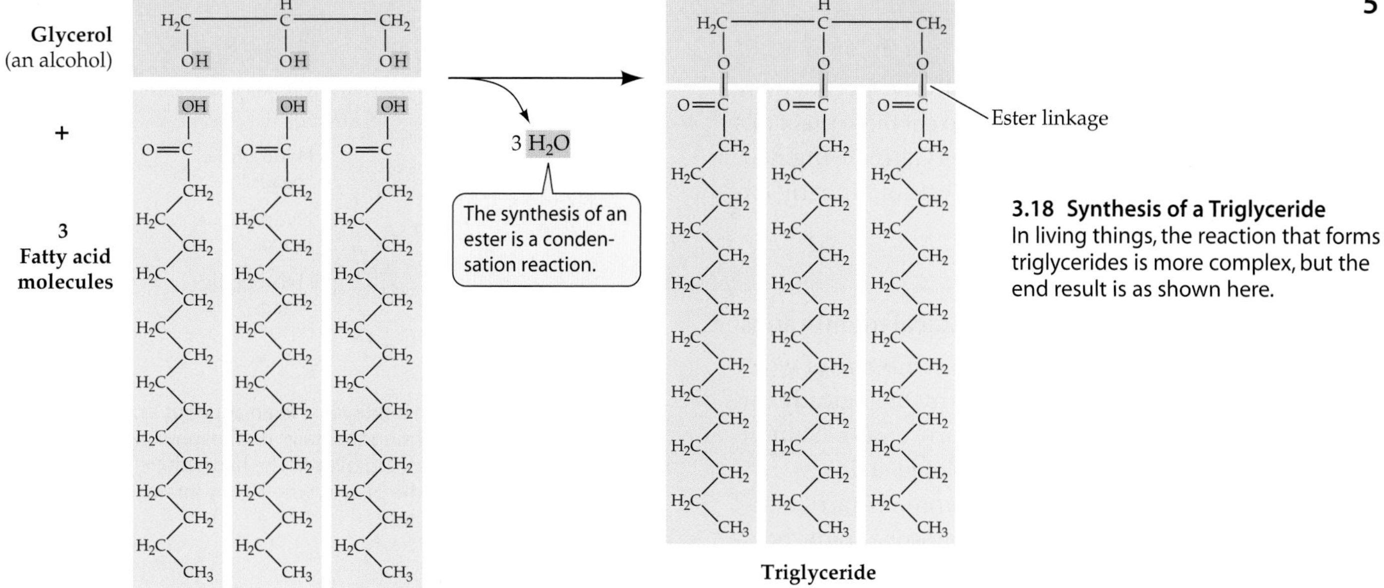
Glycerol
(an alcohol)
+
3
Fatty acid
molecules
3 H_2O
The synthesis of an ester is a condensation reaction.
Ester linkage
Triglyceride

3.18 Synthesis of a Triglyceride In living things, the reaction that forms triglycerides is more complex, but the end result is as shown here.

The carboxyl group of a fatty acid can form a covalent bond with the hydroxyl group of glycerol, resulting a functional group called an *ester* and water (Figure 3.18).

The three fatty acids in a triglyceride molecule need not all have the same hydrocarbon chain length or structure:

- In **saturated** fatty acids, all the bonds between the carbon atoms in the hydrocarbon chain are single bonds—there are no double bonds. That is, all the bonds are saturated with hydrogen atoms (Figure 3.19*a*). These fatty acid molecules are relatively rigid and straight, and they pack together tightly, like pencils in a box.
- In **unsaturated** fatty acids, the hydrocarbon chain contains one or more double bonds. Oleic acid, for example, is a *monounsaturated* fatty acid that has one double bond near the middle of the hydrocarbon chain, which causes a kink in the molecule (Figure 3.19*b*). Some fatty acids have more than one double bond—are *polyunsaturated*—and have multiple kinks. These kinks prevent the molecules from packing together tightly.

The kinks in fatty acid molecules are important in determining the fluidity and melting point of a lipid. The triglycerides of animal fats tend to have many long-chain saturated fatty acids, packed tightly together; these fats are usually solids at room temperature and have a high melting point. The triglycerides of plants, such as corn oil, tend to have short or unsaturated fatty acids. Because of their kinks, these fatty acids pack together poorly and have a low melting point, and these triglycerides are usually liquids at room temperature.

Fats and oils are marvelous storehouses for energy. When they take in excess food, many animal species deposit fat droplets in their cells as a means of storing energy. Some plant species, such as olives, avocados, sesame, castor beans, and all nuts, have substantial amounts of lipids in their seeds

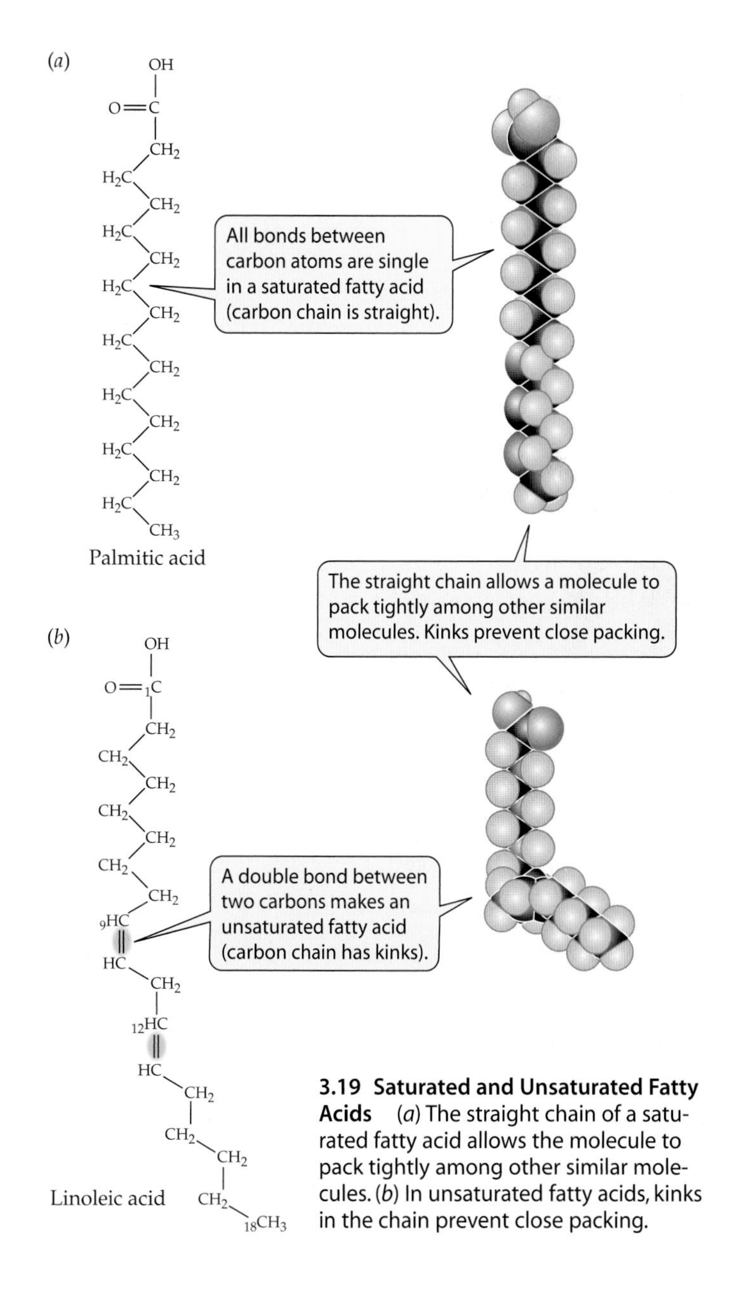
(a)
All bonds between carbon atoms are single in a saturated fatty acid (carbon chain is straight).
Palmitic acid
The straight chain allows a molecule to pack tightly among other similar molecules. Kinks prevent close packing.
(b)
A double bond between two carbons makes an unsaturated fatty acid (carbon chain has kinks).
Linoleic acid

3.19 Saturated and Unsaturated Fatty Acids (*a*) The straight chain of a saturated fatty acid allows the molecule to pack tightly among other similar molecules. (*b*) In unsaturated fatty acids, kinks in the chain prevent close packing.

or fruits that serve as energy reserves for the next generation. This energy can be tapped by people who eat these plant oils or use them for fuel. Indeed, the famous German engineer Rudolf Diesel used peanut oil to power one of his early automobile engines in 1900.

Phospholipids form the core of biological membranes

Because lipids and water do not interact, a mixture of water and lipids forms two distinct layers. Many biologically important substances—such as ions, sugars, and free amino acids—that are soluble in water are insoluble in lipids.

Like triglycerides, **phospholipids** contain fatty acids bound to glycerol by ester linkages. In phospholipids, however, any one of several phosphate-containing compounds replaces one of the fatty acids (Figure 3.20). The phosphate functional group has a negative electric charge, so this portion of the molecule is hydrophilic, attracting polar water molecules. But the two fatty acids are hydrophobic, so they tend to aggregate away from water.

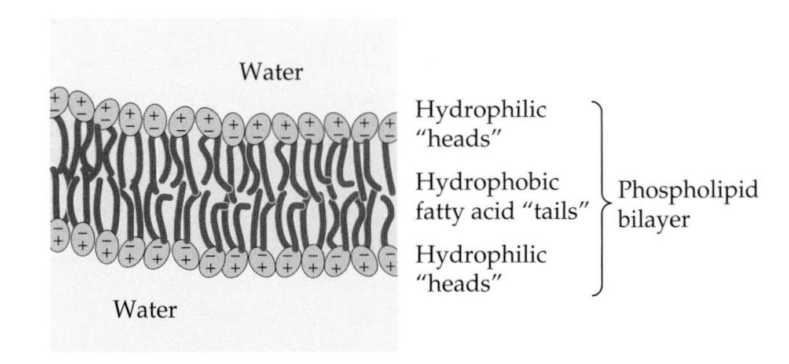

3.21 Phospholipids Form a Bilayer In an aqueous environment, hydrophobic interactions bring the "tails" of phospholipids together in the interior of a phospholipid bilayer. The hydrophilic "heads" face outward on both sides of the bilayer, where they interact with the surrounding water molecules.

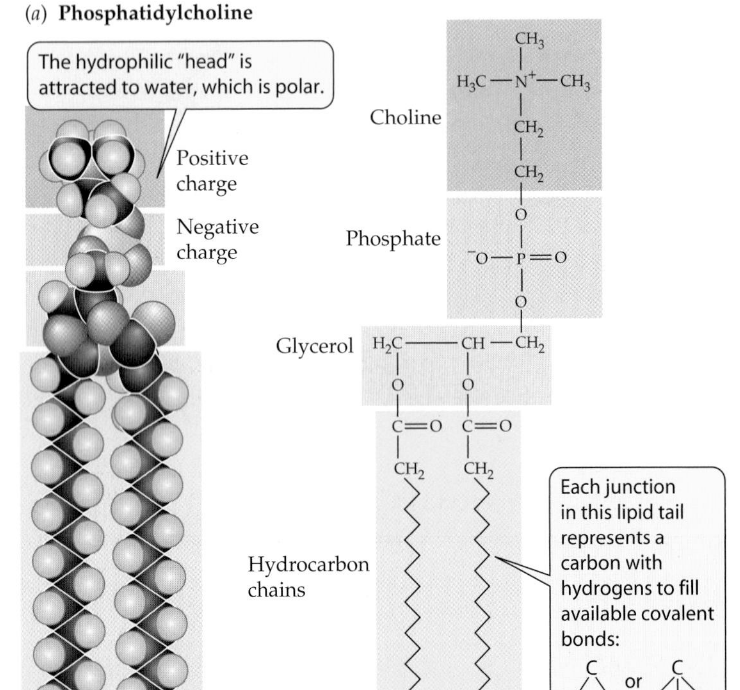

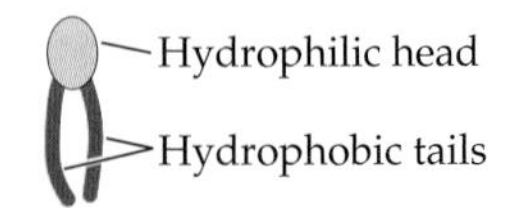

3.20 Phospholipid Structure (*a*) Phosphatidylcholine (lecithin) demonstrates the structure of a phospholipid molecule. In other phospholipids, the amino acid serine, the sugar alcohol inositol, or other compounds replace choline. (*b*) This generalized symbol is used throughout this book to represent a membrane phospholipid.

In an aqueous environment, phospholipids line up in such a way that the nonpolar, hydrophobic "tails" pack tightly together and the phosphate-containing "heads" face outward, where they interact with water. The phospholipids thus form a *bilayer*, a sheet two molecules thick, with water excluded from the core (Figure 3.21). Biological membranes have this kind of phospholipid bilayer structure, and we will devote all of Chapter 5 to their biological functions.

Carotenoids and steroids

The next two lipid classes we'll discuss—the carotenoids and the steroids—have chemical structures very different from those of triglycerides and phospholipids and from each other. Both carotenoids and steroids are synthesized by covalent linking and chemical modification of isoprene to form a series of isoprene units:

$$H_2C{=}C(CH_3){-}CH{=}CH_2$$

CAROTENOIDS TRAP LIGHT ENERGY. The **carotenoids** are a family of light-absorbing pigments found in plants and animals. Beta-carotene (β-carotene) is one of the pigments that traps light energy in leaves during photosynthesis. In humans, a molecule of β-carotene can be broken down into two vitamin A molecules (Figure 3.22), from which we make the pigment rhodopsin, which is required for vision. Carotenoids are responsible for the colors of carrots, tomatoes, pumpkins, egg yolks, and butter.

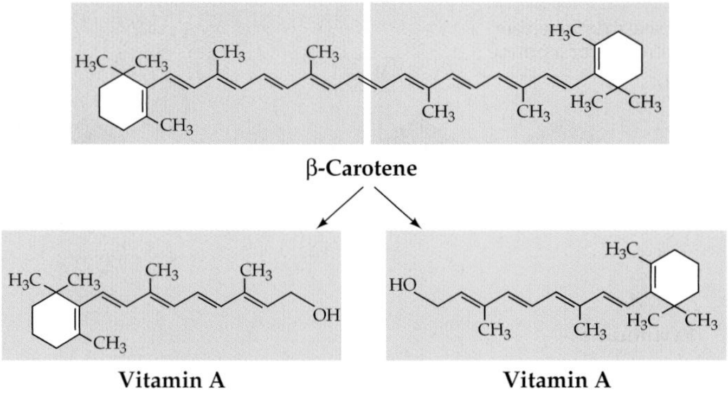

β-Carotene

Vitamin A　　Vitamin A

3.22 β-Carotene is the Source of Vitamin A The carotenoid β-carotene is symmetrical around its central double bond; when split, β-carotene becomes two vitamin A molecules. The simplified structural formula used here is standard chemical shorthand for large organic molecules with many carbon atoms. Structural formulas are simplified by omitting the C (indicating a carbon atom) at the intersections of the lines representing covalent bonds. Hydrogen atoms (H) to fill all the available bonding sites on each C are assumed.

STEROIDS ARE SIGNAL MOLECULES. The **steroids** are a family of organic compounds whose multiple rings share carbons (Figure 3.23). The steroid cholesterol is an important constituent of membranes. Other steroids function as hormones, chemical signals that carry messages from one part of the body to another. Testosterone and the estrogens are steroid hormones that regulate sexual development in vertebrates. Cortisol and related hormones play many regulatory roles in the digestion of carbohydrates and proteins, in the maintenance of salt balance and water balance, and in sexual development.

Cholesterol is synthesized in the liver and is the starting material for making testosterone and other steroid hormones, as well as the bile salts that help break down dietary fats so that they can be digested. Cholesterol is absorbed from foods such as milk, butter, and animal fats.

Some lipids are vitamins

Vitamins are small molecules that are not synthesized by the body, but are necessary for its normal functioning. Vitamins must be acquired from dietary sources.

- *Vitamin A* is formed from the β-carotene found in green and yellow vegetables (see Figure 3.22). In humans, a deficiency of vitamin A leads to dry skin, eyes, and internal body surfaces, retarded growth and development, and night blindness, which is a diagnostic symptom for the deficiency.
- *Vitamin D* regulates the absorption of calcium from the intestines. It is necessary for the proper deposition of calcium in bones; a deficiency of vitamin D can lead to rickets, a bone-softening disease.
- *Vitamin E* seems to protect cells from the damaging effects of oxidation–reduction reactions. For example, it has an important role in preventing unhealthy changes in the double bonds in the unsaturated fatty acids of membrane phospholipids. Commercially, vitamin E is added to some foods to slow spoilage.
- *Vitamin K* is found in green leafy plants and is also synthesized by bacteria normally present in the human intestine. This vitamin is essential to the formation of blood clots.

Inadequate vitamin intake can lead to deficiency diseases.

Wax coatings repel water

The sheen on human hair is not there only for cosmetic purposes. Glands in the skin secrete a waxy coating that repels water and keeps the hair pliable. Birds that live near water have a similar waxy coating on their feathers. The shiny leaves of holly plants, familiar during winter holidays, also have a waxy coating. Finally, bees make their honeycombs out of wax.

All waxes have the same basic structure: They are formed by an ester linkage between a saturated, long-chain fatty acid and a saturated, long-chain alcohol. The result is a very long

3.23 All Steroids Have the Same Ring Structure The steroids shown here, all important in vertebrates, are composed of carbon and hydrogen and are highly hydrophobic. However, small chemical variations, such as the presence or absence of a methyl or hydroxyl group, can produce enormous functional differences.

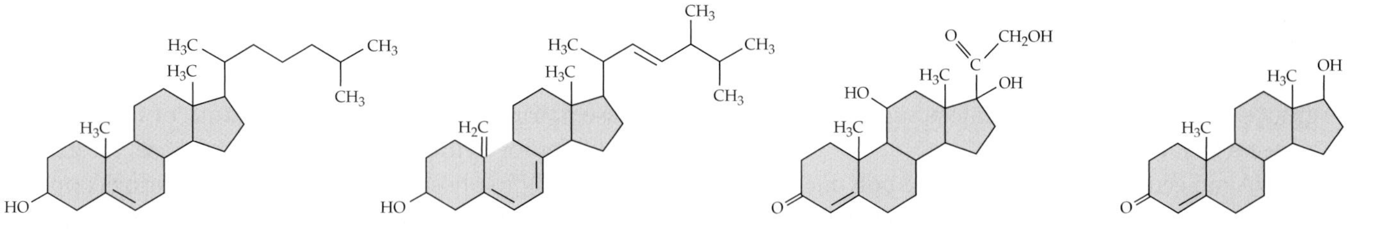

Cholesterol is a constituent of membranes and is the source of steroid hormones.

Vitamin D_2 can be produced in the skin by the action of light on a cholesterol derivative.

Cortisol is a hormone secreted by the adrenal glands.

Testosterone is a male sex hormone.

molecule, with 40–60 CH_2 groups. For example, here is the structure of beeswax:

$$\underbrace{H_3C-(CH_2)_{14}-\overset{\overset{O}{\|}}{C}}_{\textbf{Fatty acid}}-\underbrace{O-CH_2-(CH_2)_{28}-CH_3}_{\textbf{Alcohol}}$$

Ester linkage

This highly nonpolar structure accounts for the impermeability of wax to water.

Nucleic Acids: Informational Macromolecules That Can Be Catalytic

The **nucleic acids** are polymers specialized for the storage, transmission, and use of information. There are two types of nucleic acids: **DNA** (deoxyribonucleic acid) and **RNA** (ribonucleic acid). DNA molecules are giant polymers that encode hereditary information and pass it from generation to generation. Through an RNA intermediate, the information encoded in DNA is also used to specify the amino acid sequence of proteins. Information flows from DNA to DNA in reproduction, but in the nonreproductive activities of the cell, information flows from DNA to RNA to proteins, which ultimately carry out these functions. In addition, certain RNAs act as catalysts for important reactions in cells.

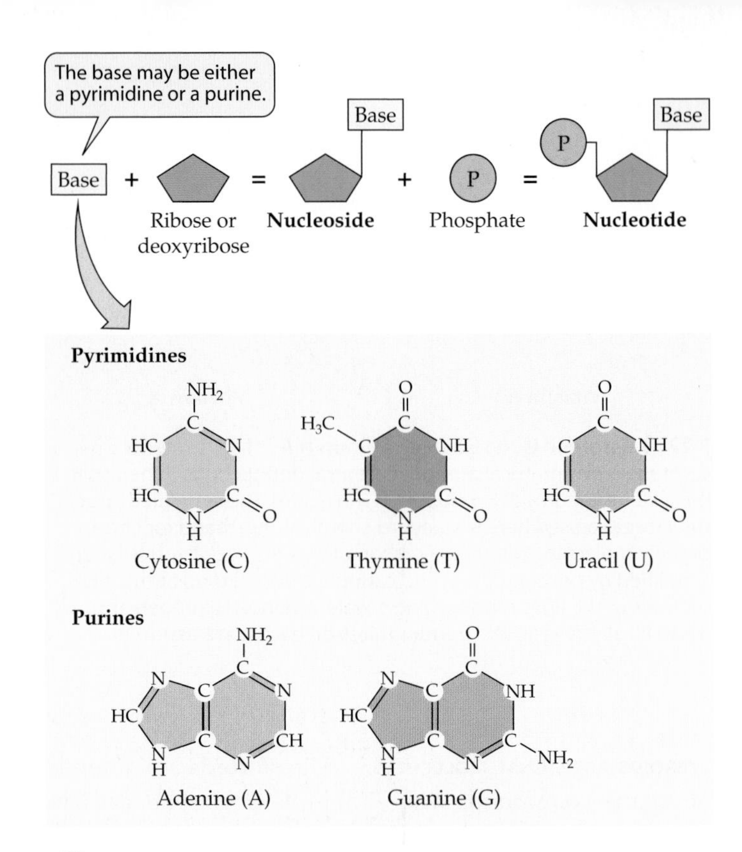

3.24 Nucleotides Have Three Components A nucleotide consists of a phosphate group, a pentose sugar (ribose or deoxyribose), and a nitrogen-containing base, all linked together by covalent bonds. The nitrogenous bases fall into two categories: Purines have two fused rings, and the smaller pyrimidines have a single ring.

The nucleic acids have characteristic chemical properties

Nucleic acids are composed of monomers called **nucleotides**, each of which consists of a pentose sugar, a phosphate group, and a nitrogen-containing **base**—either a pyrimidine or a purine (Figure 3.24). (Molecules consisting of a pentose sugar and a nitrogenous base, but no phosphate group, are called **nucleosides**.) In DNA, the pentose sugar is deoxyribose, which differs from the ribose found in RNA by one oxygen atom (see Figure 3.14).

In both RNA and DNA, the backbone of the macromolecule consists of alternating pentose sugars and phosphates (sugar—phosphate—sugar—phosphate). The bases are attached to the sugars and project from the chain (Figure 3.25). The nucleotides are joined by *phosphodiester linkages* between the sugar of one nucleotide and the phosphate of the next (*-diester* refers to the two covalent bonds formed by —OH groups reacting with acidic phosphate groups). The phosphate groups link carbon 3 in one pentose sugar to carbon 5 in the adjacent sugar.

Most RNA molecules consist of only one polynucleotide chain. DNA, however, is usually double-stranded; it has two polynucleotide strands held together by hydrogen bonding between their nitrogenous bases. The two strands of DNA run in opposite directions. You can see what this means by drawing an arrow through the phosphate group from carbon 5′ to carbon 3′ in the next ribose. If you do this for both strands of the DNA in Figure 3.25, the arrows will point in opposite directions. This antiparallel orientation is necessary for the strands to fit together in three-dimensional space.

The uniqueness of a nucleic acid resides in its nucleotide sequence

Only four nitrogenous bases—and thus only four nucleotides—are found in DNA. The DNA bases and their abbreviations are adenine (A), cytosine (C), guanine (G), and thymine (T). A key to understanding the structure and function of nucleic acids is the principle of **complementary base pairing**. In double-stranded DNA, adenine and thymine always pair (A-T), and cytosine and guanine always pair (C-G).

Base pairing is complementary because of three factors: the sites for for hydrogen bonding on each base, the geometry of the sugar–phosphate backbone, which brings opposite bases near each other, and the molecular sizes of the paired bases. Adenine and guanine are both purines, consisting of two fused rings. Thymine and cytosine are both pyrimidines, consisting of only one ring. The pairing of a large purine with

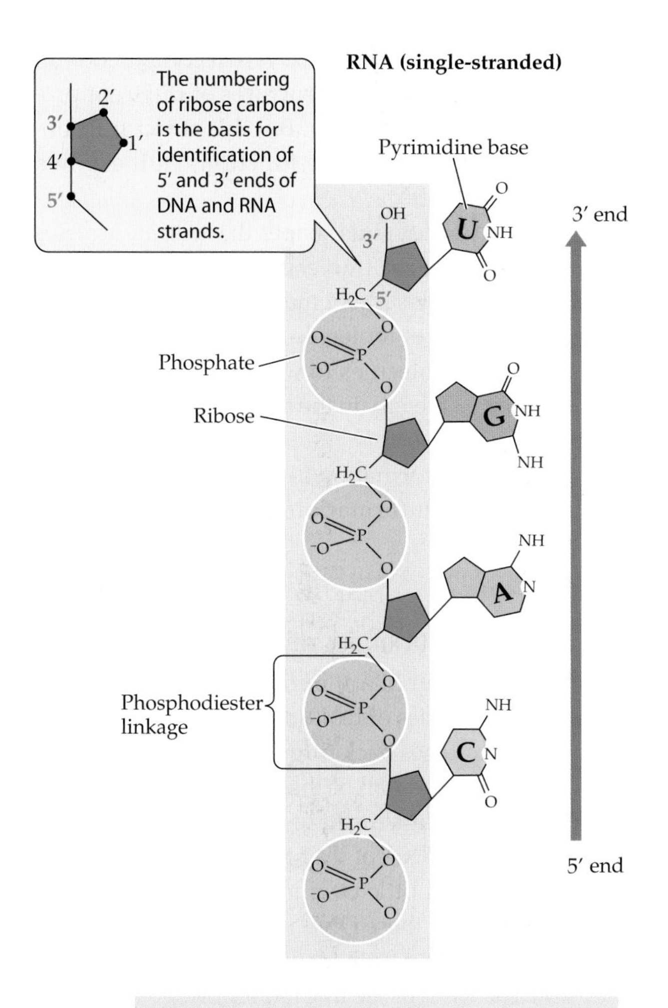

In RNA, the bases are attached to ribose. The bases in RNA are the purines adenine (A) and guanine (G) and the pyrimidines cytosine (C) and uracil (U).

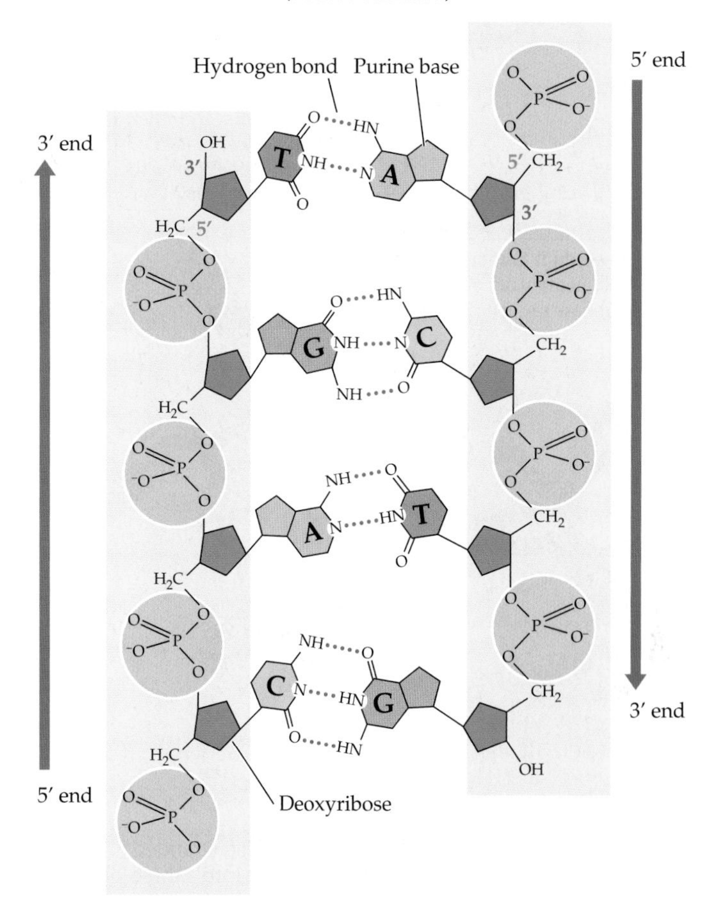

In DNA, the bases are attached to deoxyribose, and the base thymine (T) is found instead of uracil. Hydrogen bonds between purines and pyrimidines hold the two strands of DNA together.

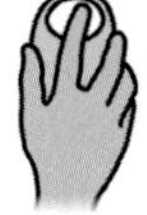

3.25 Distinguishing Characteristics of DNA and RNA
RNA is usually a single strand. DNA usually consists of two strands running in opposite directions.

a small pyrimidine ensures stability and consistency in the double-stranded molecule of DNA.

Ribonucleic acids are also made up of four different monomers, but their nucleotides differ from those of DNA. In RNA the nucleotides are termed *ribonucleotides* (the ones in DNA are *deoxyribonucleotides*). They contain ribose rather than deoxyribose, and instead of the base thymine, RNA uses the base uracil (U) (Table 3.3). The other three bases are the same as in DNA.

Although RNA is generally single-stranded, complementary hydrogen bonding between ribonucleotides can take place. These bonds play important roles in determining the shapes of some RNA molecules and in associations between RNA molecules during protein synthesis (Figure 3.26). When the base sequence of DNA is copied in the synthesis of RNA, complementary base pairing also takes place between ribonucleotides and deoxyribonucleotides. In RNA, guanine and cytosine pair (G-C), as in DNA, but adenine pairs with uracil (A-U). Adenine in an RNA strand can pair either with uracil (in another RNA strand) or with thymine (in a DNA strand).

3.3 Distinguishing RNA from DNA

NUCLEIC ACID	SUGAR	BASES
RNA	Ribose	Adenine Cytosine Guanine Uracil
DNA	Deoxyribose	Adenine Cytosine Guanine Thymine

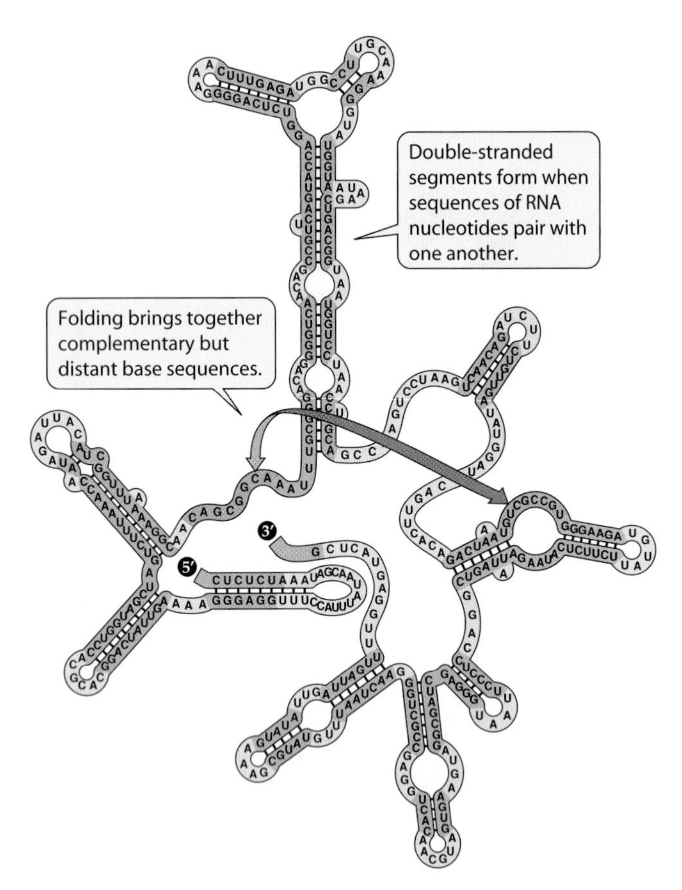
Double-stranded segments form when sequences of RNA nucleotides pair with one another.
Folding brings together complementary but distant base sequences.
3′
5′

3.26 Hydrogen Bonding in RNA When a single-stranded RNA folds in on itself, hydrogen bonds can stabilize it into a three-dimensional shape.

DNA is a purely *informational* molecule. The information in DNA is encoded in the sequence of bases carried in its strands—the information encoded in the sequence TCAG is different from the information in the sequence CCAG. The information can be read easily and reliably, in a specific order.

The three-dimensional appearance of DNA is strikingly uniform. The segment shown in Figure 3.27 could be from any DNA molecule. The variations in DNA—the different sequences of bases—are strictly "internal." Through hydrogen bonding, the two complementary polynucleotide strands pair and twist to form a double helix. When compared with the complex and varied tertiary structures of different proteins, this uniformity is surprising. But this structural contrast makes sense in terms of the functions of these two classes of macromolecules.

It is their different and unique shapes that permit proteins to recognize specific "target" molecules. The unique three-dimensional form of each protein matches at least a portion of the surface of the target molecule. In other words, structural diversity in the molecules to which proteins bind requires corresponding diversity in the structure of the proteins themselves.

In DNA, then, the information is in the sequence of the bases; in proteins, the information is in the shape of the molecule.

DNA is a guide to evolutionary relationships

Because DNA carries hereditary information between generations, a theoretical series of DNA molecules with changes in base sequences stretches back through evolutionary time. Of course, we cannot study all of these DNA molecules, because many of their organisms have become extinct. However, we can study the DNA of living organisms, which are judged to have changed little over millions of years. Comparisons and contrasts of these DNA molecules can be added to evidence from fossils and other sources to reveal the evolutionary record, as we will see in Chapter 24.

Closely related living species should have more similar base sequences than species judged by other criteria to be more distantly related. The examination of base sequences has confirmed many of the evolutionary relationships that have been inferred from the more traditional study of body structures, biochemistry, and physiology. For example, the closest living relative of humans (*Homo sapiens*) is the chimpanzee (genus *Pan*), which shares more than 98 percent of its DNA base sequence with human DNA. This confirmation of well-established evolutionary relationships gives credibility to the use of DNA to elucidate relationships when studies of structure are not possible or are not conclusive. For example,

The yellow phosphorus atoms and their attached red oxygen atoms form the two helical backbones.

The paired bases are stacked in the center of the coil (blue nitrogen atoms and gray carbon atoms).

3.27 The Double Helix of DNA The backbones of the two strands in a DNA molecule are coiled in a double helix. The small white atoms represent hydrogen.

DNA studies revealed a close evolutionary relationship between starlings and mockingbirds that was not expected on the basis of their anatomy or behavior.

DNA studies support the division of the prokaryotes into two domains, Bacteria and Archaea. Each of these two groups of prokaryotes is as distinct from the other as either is from the Eukarya, the third domain into which living things are classified (see Chapter 1). In addition, DNA comparisons support the hypothesis that certain subcellular compartments of eukaryotes (the organelles called mitochondria and chloroplasts) evolved from early bacteria that established a stable and mutually beneficial way of life inside larger cells.

RNA may have been the first biological catalyst

The three-dimensional structure of a folded RNA molecule presents a unique surface to the external environment (see Figure 3.26). These surfaces are every bit as specific as those of proteins. We noted above that an important role of proteins in biology is to act as catalysts, speeding up reactions that would ordinarily take place too slowly to be biologically useful, and that the spatial property of proteins is vital to this role.

As we will see, certain RNA molecules can also act as catalysts, using their three-dimensional shapes and other chemical properties. They can catalyze reactions on their own nucleotides as well as in other cellular substances. These catalytic RNAs are called **ribozymes**. Their discovery had implications for theories of the origin of life.

The Miller-Urey experiment and other such experiments in prebiotic chemistry yielded both amino acids and nucleotides. Organisms can synthesize both RNA and proteins from these monomers. As we noted above, in current organisms on Earth, protein synthesis requires DNA and RNA, and nucleic acid synthesis requires proteins (as enzymes). So the question is, when life originated, which came first, the proteins or the nucleic acids?

The discovery of catalytic RNAs provided a solution to this dilemma and led to the hypothesis that early life was part of an "RNA world." RNA can be informational (in its nucleotide sequence) as well as catalytic. So when RNA was first made, it could have acted as a catalyst for its own replication, as well as for the synthesis of proteins. Then DNA could have eventually evolved by being made from RNA. There is some laboratory evidence supporting this scenario:

- RNAs of different sequences have been put in a test tube and made to replicate on their own. Such self-replicating ribozymes speed up the synthesis of RNA 7 million-fold.
- In living organisms today, the formation of peptide linkages (see Figure 3.5) is catalyzed by a ribozyme.
- In certain viruses called retroviruses, there is an enzyme called reverse transcriptase that catalyzes the synthesis of DNA from RNA.

Nucleotides have other important roles

Nucleotides are more than just the building blocks of nucleic acids. As we will see in later chapters, there are several nucleotides with other functions:

- ATP (adenosine triphosphate) acts as an energy transducer in many biochemical reactions (see Chapter 6).
- GTP (guanosine triphosphate) serves as an energy source, especially in protein synthesis. It also has a role in the transfer of information from the environment to the body tissues (see Chapters 12 and 15).
- cAMP (cyclic adenosine monophosphate), a special nucleotide in which a bond forms between the sugar and phosphate groups within adenosine monophosphate, is essential in many processes, including the actions of hormones and the transmission of information by the nervous system (see Chapter 15).

All Life from Life

The concepts conveyed throughout this chapter—that large molecules obey the mechanistic laws of physics and chemistry, and that life could have arisen from inanimate, self-replicating macromolecules—have come to be generally accepted by the scientific community. So should we expect to see new life forms arise at any time from the biochemical environment?

During the Renaissance (a period from about 1350 to 1700 A.D., marked by the birth of modern science), most people thought that at least some forms of life arose directly from inanimate or decaying matter by *spontaneous generation*. For instance, it was suggested that mice arose from sweaty clothes placed in dim light, frogs came from moist soil, and flies were produced from meat. These ideas were attacked by scientists such as the Italian doctor and poet Francisco Redi using the relatively new idea of using experiments to test an idea. In 1668, Redi proposed that flies arose not by some mysterious transformation of decaying meat, but from other flies, who laid their eggs on the meat. The eggs developed into wormlike maggots (the immature form of flies). Redi set out several jars containing chunks of meat.

- One jar contained meat exposed both to the air and to flies.
- A second jar contained meat in a container wrapped in a fine cloth so that the meat was exposed to the air, but not to flies.
- The meat in the third jar was in a sealed container and thus was not exposed to either air or flies.

As he had hypothesized, Redi found maggots, which then hatched into flies, only in the first container. The idea that a complex organism like a fly could come from a totally different substance was laid to rest.

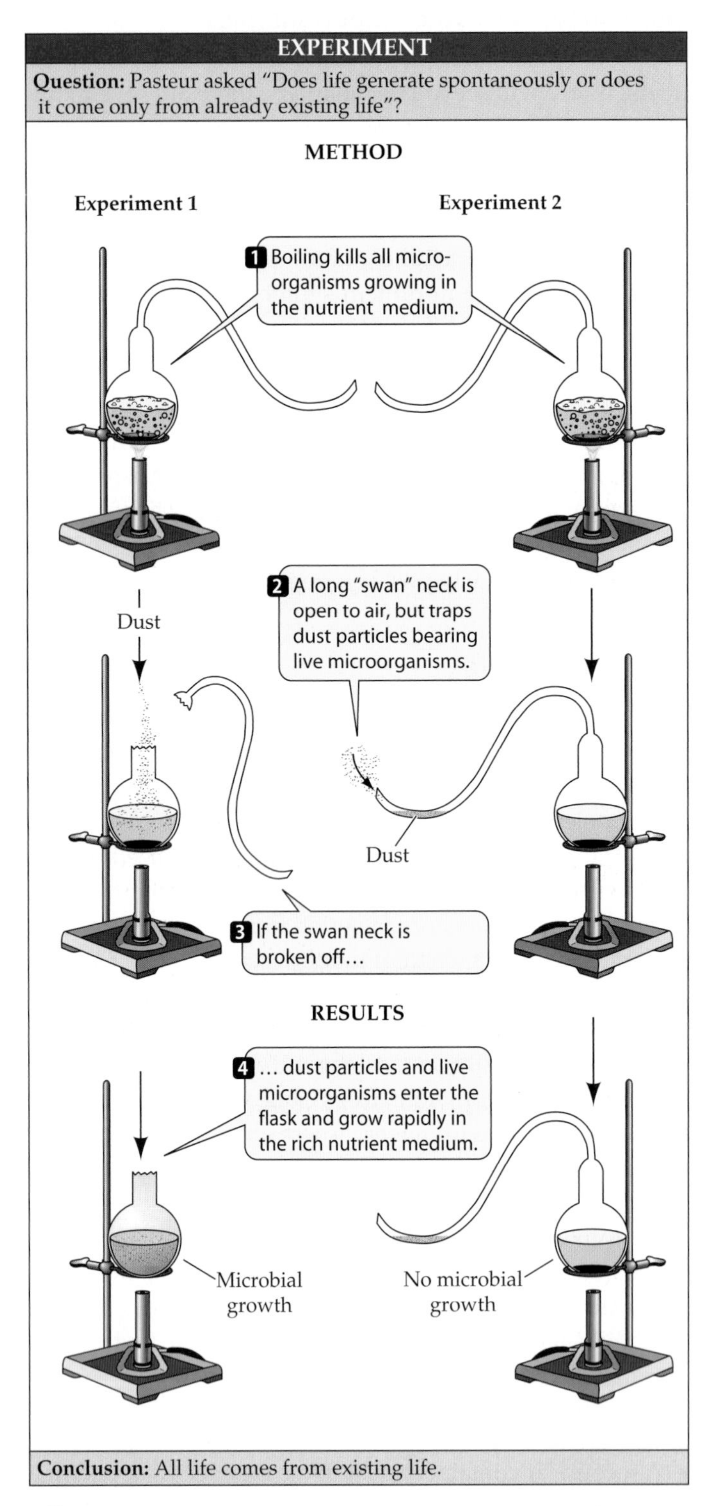

EXPERIMENT
Question: Pasteur asked "Does life generate spontaneously or does it come only from already existing life"?
METHOD
Experiment 1
Experiment 2
1 Boiling kills all microorganisms growing in the nutrient medium.
Dust
2 A long "swan" neck is open to air, but traps dust particles bearing live microorganisms.
Dust
3 If the swan neck is broken off...
RESULTS
4 ... dust particles and live microorganisms enter the flask and grow rapidly in the rich nutrient medium.
Microbial growth
No microbial growth
Conclusion: All life comes from existing life.

3.28 Disproving the Spontaneous Generation of Life Louis Pasteur's classic experiments showed that, under today's conditions, an inanimate solution remains lifeless unless a living organism contaminates it.

With the invention of the microscope in the 1660s, a vast new biological world was unveiled. Under microscopic observation, virtually every environment on Earth was found to be teeming with tiny organisms such as bacteria. Some scientists believed that these organisms arose spontaneously from their rich chemical environment.

The experiments that disproved this idea were done by the great French scientist Louis Pasteur. His experiments showed that microorganisms come only from other microorganisms, and that an environment without life remains lifeless unless contaminated by living creatures (Figure 3.28).

These experiments by Redi, Pasteur, and others provided solid evidence that neither small (bacteria) nor large (flies) organisms come from inanimate matter, but instead come from living parent organisms.

Indeed, life on Earth no longer arises from nonliving materials. This is because the atmospheric and planetary conditions that exist on Earth today are vastly different from those on the prebiotic, anaerobic planet. The oxygen present in today's atmosphere would break down the prebiotic molecules before they could accumulate. In addition, the necessary energy sources—including constant lightning strikes, immense volcanic eruptions, and bombardment by intense ultraviolet light—are no longer present with anything like their primeval force.

Chapter Summary

Theories of the Origin of Life

▶ Life may have come from outside Earth. The evidence for this proposal comes primarily from chemicals contained in meteorites that have landed on Earlth.

▶ The theory of chemical evolution proposes that life on Earth originated on Earth. Experiments using model systems that attempt to duplicate the ancient Earth have shown that chemical evolution could have produced the four types of macromolecules that distinguish living things. **Review Figure 3.1. See Web/CD Tutorial 3.1**

Macromolecules: Giant Polymers

▶ Macromolecules are polymers constructed by the formation of covalent bonds between smaller molecules called monomers. Macromolecules in living organisms include polysaccharides, proteins, and nucleic acids. **Review Figure 3.2 and Table 3.1**

▶ Macromolecules have specific, characteristic three-dimensional shapes that depend on the structure, properties, and sequence of their monomers.

▶ Different functional groups give local sites on macromolecules specific properties that are important for their biological functioning and their interactions with other macromolecules. **See Web/CD Tutorial 3.2**

Condensation and Hydrolysis Reactions

▶ Monomers are joined by condensation reactions, which release a molecule of water for each bond formed. Hydrolysis reactions use water to break polymers into monomers. **Review Figure 3.3**

Proteins: Polymers of Amino Acids

▶ The functions of proteins include support, protection, catalysis, transport, defense, regulation, and movement. Protein function sometimes requires an attached prosthetic group.

▶ There are 20 amino acids found in proteins. Each amino acid consists of an amino group, a carboxyl group, a hydrogen, and a side chain bonded to the α carbon atom. **Review Table 3.2**

▶ The side chains, or R groups, of amino acids may be charged, polar, or hydrophobic; there are also special cases, such as the —SH groups of cysteine, which can form disulfide bridges. The side chains give different properties to each of the amino acids. **Review Table 3.2 and Figure 3.4**

▶ Amino acids are covalently bonded together into polypeptide chains by peptide linkages, which form by condensation reactions between the carboxyl and amino groups. **Review Figure 3.5**

▶ Polypeptide chains are folded into specific three-dimensional shapes to form functional proteins. Four levels of protein structure are possible: primary, secondary, tertiary, and quaternary.

▶ The primary structure of a protein is the sequence of amino acids bonded by peptide linkages. This primary structure determines both the higher levels of structure and protein function. **Review Figure 3.6*a***

▶ The two types of secondary structure—α helices and β pleated sheets—are maintained by hydrogen bonds between atoms of the amino acid residues. **Review Figure 3.6*b,c***

▶ The tertiary structure of a protein is generated by bending and folding of the polypeptide chain. **Review Figures 3.6*d*, 3.7**

▶ The quaternary structure of a protein is the arrangement of two or more polypeptides into a single functional protein consisting of two or more polypeptide subunits. **Review Figures 3.6*e*, 3.8**

▶ Weak chemical interactions are important in the three-dimensional structure of proteins and in their binding to other molecules. **Review Figure 3.9, 3.10**

▶ Proteins denatured by heat, alterations in pH, or certain chemicals lose their tertiary and secondary structure as well as their biological function. Renaturation is not often possible. **Review Figure 3.11**

▶ Chaperonins assist protein folding by preventing binding to inappropriate ligands. **Review Figure 3.12**

Carbohydrates: Sugars and Sugar Polymers

▶ All carbohydrates contain carbon bonded to hydrogen atoms and hydroxyl groups.

▶ Hexoses are monosaccharides that contain six carbon atoms. Examples of hexoses include glucose, galactose, and fructose, which can exist as chains or rings. **Review Figures 3.13, 3.14.** See Web/CD Activity 3.1

▶ The pentoses are five-carbon monosaccharides. Two pentoses, ribose and deoxyribose, are components of the nucleic acids RNA and DNA, respectively. **Review Figure 3.14**

▶ Glycosidic linkages may have either α or β orientation in space. They covalently link monosaccharides into larger units such as disaccharides, oligosaccharides, and polysaccharides. **Review Figure 3.15**

▶ Cellulose, a very stable glucose polymer, is the principal component of the cell walls of plants. It is formed by glucose units linked together by β-glycosidic linkages between carbons 1 and 4. Starches, less dense and less stable than cellulose, store energy in plants. Starches and glycogen are formed by α-glycosidic linkages between carbons 1 and 4 and are distinguished by the amount of branching they exhibit. **Review Figure 3.16**

▶ Chemically modified monosaccharides include the sugar phosphates and amino sugars. A derivative of the amino sugar glucosamine polymerizes to form the polysaccharide chitin, which is found in the cell walls of fungi and the exoskeletons of insects. **Review Figure 3.17**

Lipids: Water-Insoluble Molecules

▶ Although lipids can form gigantic structures, these aggregations are not chemically macromolecules because the individual units are not linked by covalent bonds.

▶ Fats and oils are triglycerides, composed of three fatty acids covalently bonded to a glycerol molecule by ester linkages. **Review Figure 3.18**

▶ Saturated fatty acids have a hydrocarbon chain with no double bonds. The hydrocarbon chains of unsaturated fatty acids have one or more double bonds that bend the chain, making close packing less possible. **Review Figure 3.19**

▶ Phospholipids have a hydrophobic hydrocarbon "tail" and a hydrophilic phosphate "head." **Review Figure 3.20**

▶ In water, the interactions of the hydrophobic tails and hydrophilic heads of phospholipids generate a phospholipid bilayer that is two molecules thick. The head groups are directed outward, where they interact with the surrounding water. The tails are packed together in the interior of the bilayer. **Review Figure 3.21**

▶ Carotenoids trap light energy in green plants. Carotene can be split to form vitamin A, a lipid vitamin. **Review Figure 3.22**

▶ Some steroids, such as testosterone, function as hormones. Cholesterol is synthesized by the liver and has a role in cell membranes, as well as in the digestion of fats. **Review Figure 3.23**

▶ Vitamins are substances that are required for normal functioning, but must be acquired from the diet.

Nucleic Acids: Informational Macromolecules

▶ DNA is the hereditary material. Both DNA and RNA play roles in the formation of proteins. Information flows from DNA to RNA to protein.

▶ Nucleic acids are polymers made up of nucleotides. A nucleotide consists of a phosphate group, a sugar (ribose in RNA and deoxyribose in DNA), and a nitrogen-containing base. In DNA the bases are adenine, guanine, cytosine, and thymine, but in RNA uracil substitutes for thymine. **Review Figure 3.24 and Table 3.3.** See Web/CD Activity 3.2

▶ In the nucleic acids, the bases extend from a sugar–phosphate backbone. The information content of DNA and RNA resides in their base sequences. RNA is single-stranded. DNA is a double-stranded helix in which there is complementary, hydrogen-bonded base pairing between adenine and thymine (A-T) and guanine and cytosine (G-C). The two strands of the DNA double helix run in opposite directions. **Review Figures 3.25, 3.27.** See Web/CD Activity 3.3

▶ Base pairing of single-stranded RNAs can lead to three-dimensional structures, which can be catalytic. This finding has led to the proposal that in the origin of life, RNA preceded protein. **Review Figure 3.26**

▶ Comparing the DNA base sequences of different living species provides information on their evolutionary relationships.

All Life from Life

▶ One of the earliest conclusions from biology as a modern experimental science was that even the tiniest microbe comes from others of the same type—that is, that life begets life. **Review Figure 3.28.** See Web/CD Tutorial 3.3

▶ The conditions on primeval Earth that may have enabled life to arise from inanimate self-replicating chemicals no longer exist. Today all life comes from pre-existing life.

Self-Quiz

1. The most abundant molecule in the cell is
   a. carbohydrate.
   b. lipid.
   c. nucleic acid.
   d. protein.
   e. water.
2. All lipids are
   a. triglycerides.
   b. polar.
   c. hydrophilic.
   d. polymers of fatty acids.
   e. more soluble in nonpolar solvents than in water.
3. All carbohydrates
   a. are polymers.
   b. are simple sugars.
   c. consist of one or more simple sugars.
   d. are found in biological membranes.
   e. are more soluble in nonpolar solvents than in water.
4. Which of the following is *not* a carbohydrate?
   a. Glucose
   b. Starch
   c. Cellulose
   d. Hemoglobin
   e. Deoxyribose
5. All proteins
   a. are enzymes.
   b. consist of one or more polypeptides.
   c. are amino acids.
   d. have quaternary structures.
   e. are more soluble in nonpolar solvents than in water.
6. Which of the following statements about the primary structure of a protein is *not* true?
   a. It may be branched.
   b. It is determined by the structure of the corresponding DNA.
   c. It is unique to that protein.
   d. It determines the tertiary structure of the protein.
   e. It is the sequence of amino acids in the protein.
7. The amino acid leucine (see Table 3.2)
   a. is found in all proteins.
   b. cannot form peptide linkages.
   c. is likely to appear in the part of a membrane protein that lies within the phospholipid bilayer.
   d. is likely to appear in the part of a membrane protein that lies outside the phospholipid bilayer.
   e. is identical to the amino acid lysine.
8. The quaternary structure of a protein
   a. consists of four subunits—hence the name *quaternary*.
   b. is unrelated to the function of the protein.
   c. may be either alpha or beta.
   d. depends on covalent bonding among the subunits.
   e. depends on the primary structures of the subunits.
9. All nucleic acids
   a. are polymers of nucleotides.
   b. are polymers of amino acids.
   c. are double-stranded.
   d. are double-helical.
   e. contain deoxyribose.
10. Which of the following statements about condensation reactions is *not* true?
    a. Protein synthesis results from them.
    b. Polysaccharide synthesis results from them.
    c. Nucleic acid synthesis results from them.
    d. They consume water as a reactant.
    e. Different condensation reactions produce different kinds of macromolecules.

For Discussion

1. Phospholipids make up a major part of most biological membranes; cellulose is the major constituent of the cell walls of plants. How do the chemical structures and physical properties of phospholipids and cellulose relate to their functions in cells?
2. Suppose that, in a given protein, one lysine is replaced by aspartic acid (see Table 3.2). Does this change occur in the primary structure or in the secondary structure? How might it result in a change in tertiary structure? In quaternary structure?
3. If there are 20 different amino acids commonly found in proteins, how many different dipeptides are there? How many different tripeptides? How many different trinucleotides? How many different single-stranded RNAs composed of 200 nucleotides?
4. Contrast the following three structures, emphasizing the surfaces they present to their environment: hemoglobin; a DNA molecule; a protein that spans a biological membrane.
5. Why might RNA have preceded proteins in the evolution of biological macromolecules?

4 Cells: The Basic Units of Life

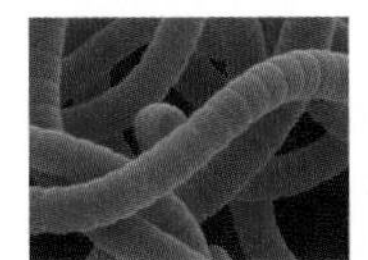

Charles Darwin faced a dilemma. In his great book *On the Origin of Species*, published in 1859, he proposed the theory of natural selection to explain the gradual appearance and disappearance of different forms of animals and plants over long periods of time. But he realized that the fossil record, on which he based his theory, was incomplete, especially for the beginning of life. The oldest fossils that had been found in Darwin's time were complex organisms in rocks dated at about 550 million years ago (the Cambrian period). Where were the missing Precambrian fossils? These would surely provide a link to the origin of life.

As we saw in Chapter 3, conditions on Earth were probably suitable for the emergence of life by 4 billion years ago, about 600 million years after Earth began to form. But until recently there was no evidence for life older than the Cambrian. By the turn of the twentieth century, there was evidence for fossilized clumps of algae (simple aquatic photosynthetic organisms) in rocks at the base of the Grand Canyon that were close to 1 billion years old.

It took nearly another century to push the clock of life back nearer to its origins. In 1993, geologist J. William Schopf found fossilized chains of cylindrical objects, quite similar in size and shape to contemporary cyanobacteria ("blue-green algae"), in rocks in Western Australia that he dated at an astonishing 3.5 billion years old. He then used a chemical analysis method called laser Raman spectroscopy to show that these objects apparently contain carbon deposits that are chemical signatures of life.

Rounded or cylindrical objects in Earth's rocks or in a meteorite from Mars (see Chapter 3) get scientists excited because they realize that life is not just a bunch of macromolecules. Rather, life is macromolecules that can perform unique functions because they are enclosed in a structural compartment that is separate from the external environment. This separation allows living things to maintain a constant internal environment (homeostasis).

The "living compartment" is the cell, the subject of this chapter. The water-insoluble phospholipid structure (see Figure 3.21) that defines and contains cells is called the plasma membrane. It and its functions are so important that we will devote the entire next chapter to membranes. Subsequent chapters will be devoted to the chemical activities that take place inside all cells.

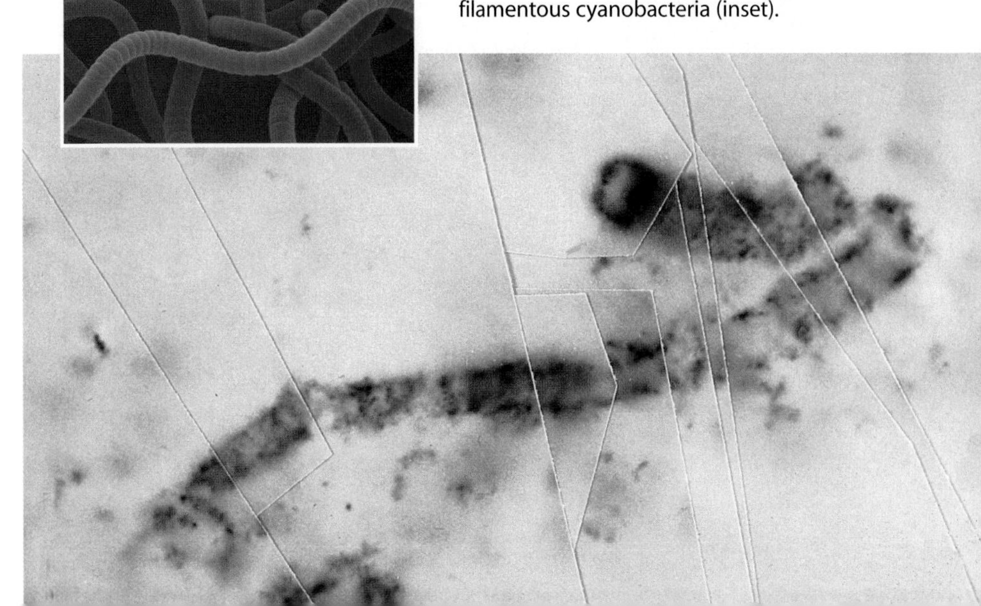

The Earliest Trace of Life? This fossil from Western Australia is 3.5 billion years old and shows carbon traces that indicate life. Its form is similar to that of modern filamentous cyanobacteria (inset).

The Cell: The Basic Unit of Life

Just as atoms are the units of chemistry, cells are the building blocks of life. Three statements constitute the **cell theory**:

- Cells are the fundamental units of life.
- All organisms are composed of cells.
- All cells come from preexisting cells.

Cells are composed of water molecules and the small and large molecules we examined in the previous two chapters. Each cell contains at least 10,000 different types of molecules, most of them present in many copies. Cells use these molecules to transform matter and energy, to respond to their environment, and to reproduce themselves.

The cell theory has three important implications. First, it means that studying cell biology is in some sense the same as studying life. The principles that underlie the functions of the single cell in a bacterium are similar to those governing the 60 trillion cells of your body. Second, it means that life is continuous. All those cells in your body came from a single cell, the fertilized egg, which came from the fusion of two cells, a sperm and an egg from your parents, whose cells came from their fertilized eggs, and so on. Finally, it means that the origin of life on Earth was marked by the origin of the first cells.

Cells may have come from stable bubbles

Isolation from the general environment can be achieved in the laboratory within aggregates produced from molecules made in prebiotic synthesis experiments. Called **protobionts**, these aggregates cannot reproduce, but they can maintain internal chemical environments that differ from their surroundings. Under the microscope, they look a lot like tiny cells (Figure 4.1).

4.1 Protobionts These aggregates, made by agitating a solution of macromolecules, are chemical compartments, can perform some metabolic reactions, and can exchange materials with their environment. They are a model of how cells may have originated.

In the 1920s, Russian scientist Alexander Oparin mixed a large protein and a polysaccharide in solution. When he agitated this mixture, bubbles formed. He could also do this with other polymers. The interiors of these bubbles had much higher concentrations of the macromolecules than their surroundings. Moreover, they catalyzed chemical reactions, and they had some control over what left them and crossed the boundary into the environment. In other words, they were protobionts. Later, other researchers showed that if lipids are mixed in an aqueous environment, they spontaneously arrange themselves into droplets surrounded by a bilayer.

Taken together with the prebiotic chemistry models and RNA world hypothesis described in Chapter 3, these experiments suggest a *bubble theory* for the origin of cells.

Cell size is limited by the surface area-to-volume ratio

Most cells are tiny. The volume of cells ranges from 1 to 1,000 μm^3 (Figure 4.2). The eggs of some birds are enormous exceptions, to be sure, and individual cells of several types of algae and bacteria are large enough to be viewed with the unaided eye. And although neurons (nerve cells) have a volume that is within the "normal" cell range, they often have fine projections that may extend for meters, carrying signals from one part of a large animal to another. But by and large, cells are minuscule. The reason for this relates to the change in the **surface area-to-volume ratio** (SA/V) of any object as it increases in size.

As a cell increases in volume, its surface area also increases, but not to the same extent (Figure 4.3). This phenomenon has great biological significance for two reasons:

- The *volume* of a cell determines the amount of chemical activity it carries out per unit of time.
- The *surface area* of a cell determines the amount of substances the cell can take in from the outside environment and the amount of waste products it can release to the environment.

As a living cell grows larger, its rate of waste production and its need for resources increase faster than its surface area. This explains why large organisms must consist of many small cells: Cells are small in volume in order to maintain a large surface area-to-volume ratio.

In a multicellular organism, the large surface area represented by the multitude of small cells that make up the organism enables it to carry out the multitude of functions required for survival. Special structures transport food, oxygen, and waste materials to and from the small cells that are distant from the external surface of the organism.

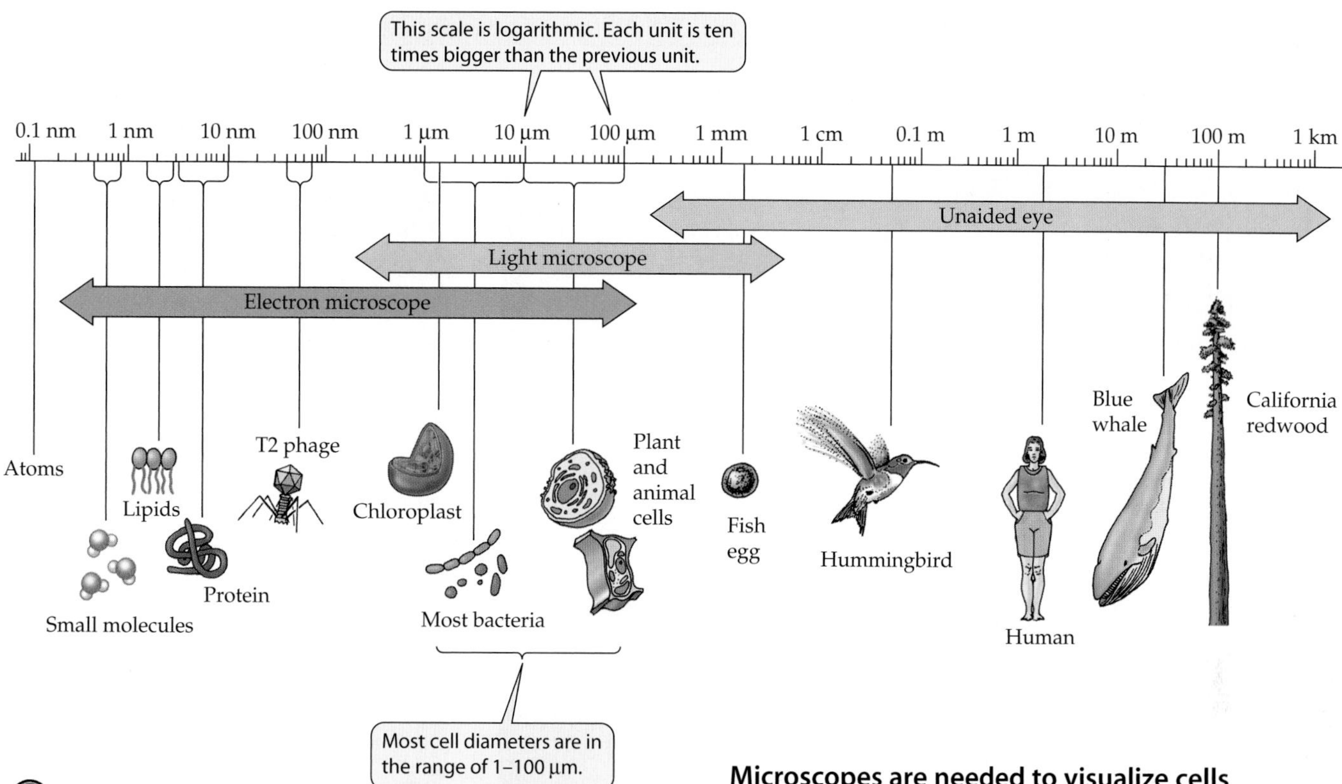

4.2 The Scale of Life This scale shows the relative sizes of molecules, cells, and multicellular organisms.

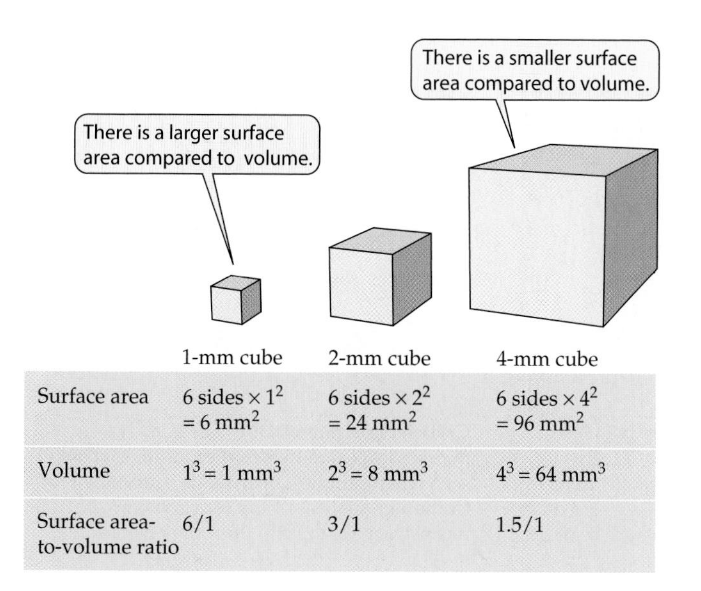

	1-mm cube	2-mm cube	4-mm cube
Surface area	6 sides × 1^2 = 6 mm^2	6 sides × 2^2 = 24 mm^2	6 sides × 4^2 = 96 mm^2
Volume	1^3 = 1 mm^3	2^3 = 8 mm^3	4^3 = 64 mm^3
Surface area-to-volume ratio	6/1	3/1	1.5/1

4.3 Why Cells Are Small As an object grows larger, its volume increases more rapidly than its surface area. Cells must maintain a large surface area-to-volume ratio in order to function, which explains why large organisms must be composed of many small cells rather than a few huge ones.

Microscopes are needed to visualize cells

Most cells are invisible to the human eye. The smallest object a person can typically discern is about 0.2 mm (200 µm) in size. We refer to this measure as *resolution*, the distance apart two objects must be in order for them to be distinguished as separate; if they are closer together, they appear as a single blur. Many cells are much smaller than 200 µm. *Microscopes* are instruments used to improve resolution so that cells and their internal structures can be seen.

There are two basic types of microscopes: light microscopes and electron microscopes. The **light microscope** (LM) uses glass lenses and visible light to form a magnified image of an object. It has a resolving power of about 0.2 µm, which is 1,000 times that of the human eye. It allows visualization of cell sizes and shapes and some internal cell structures. The latter are hard to see under ordinary light, so cells are often killed and stained with various dyes to make certain structures stand out.

An **electron microscope** (EM) uses magnets to focus an electron beam, much as a light microscope uses glass lenses to focus a beam of light. Since we cannot see electrons, the electron microscope directs them at a fluorescent screen or photographic film to create a visible image. The resolving power of electron microscopes is about 0.5 nm, which is 400,000 times that of the human eye. This resolving power permits the details of many subcellular structures to be distinguished.

Many techniques have been developed to enhance the views of cells we see under the light and electron microscopes (Figure 4.4).

RESEARCH METHOD

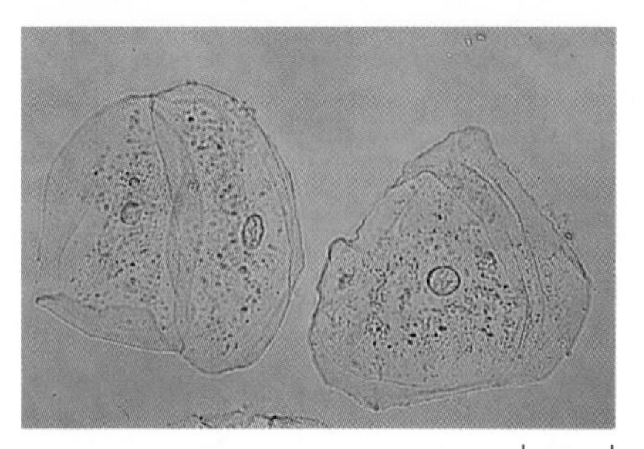

In **bright-field microscopy**, light passes directly through the cells. Unless natural pigments are present, there is little contrast and details are not distinguished.

25 µm

In **phase-contrast microscopy**, contrast in the image is increased by emphasizing differences in refractive index (the capacity to bend light), thereby enhancing light and dark regions in the cell.

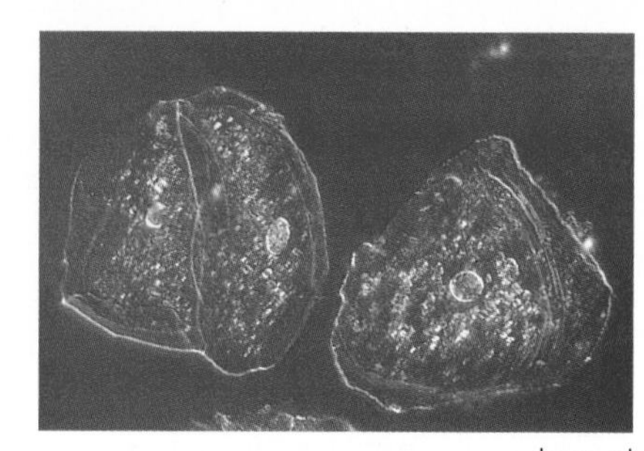

Differential interference-contrast microscopy uses two beams of polarized light. The combined images look as if the cell is casting a shadow on one side.

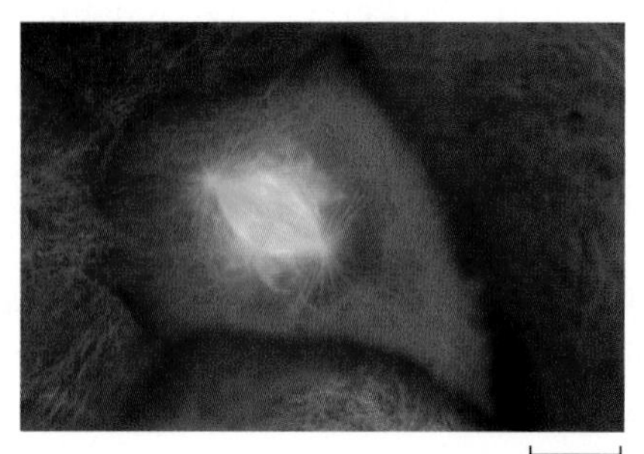

In **fluorescence microscopy**, a natural substance in the cell or a fluorescent dye that binds to a specific cell material is stimulated by a beam of light, and the longer-wavelength fluorescent light is observed coming directly from the dye.

40 µm

Confocal microscopy uses fluorescent materials but adds a system of focusing both the stimulating and emitted light so that a single plane through the cell is seen. The result is a sharper two-dimentional image than with standard fluorescence microscopy.

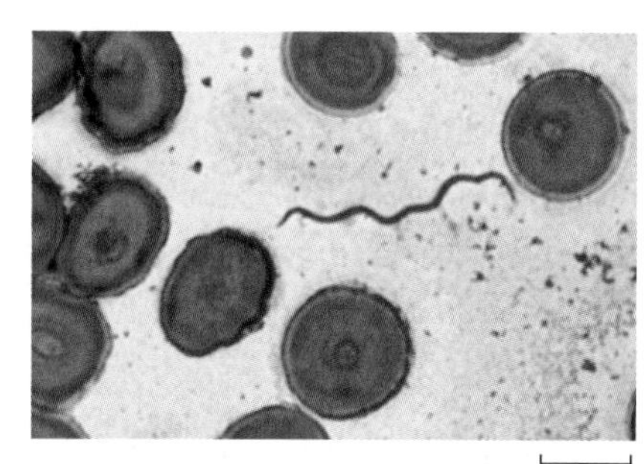

In **stained bright-field microscopy**, a stain added to preserve cells enhances contrast and reveals details not otherwise visible. Stains differ greatly in their chemistry and their capacity to bind to cell materials, so many choices are available.

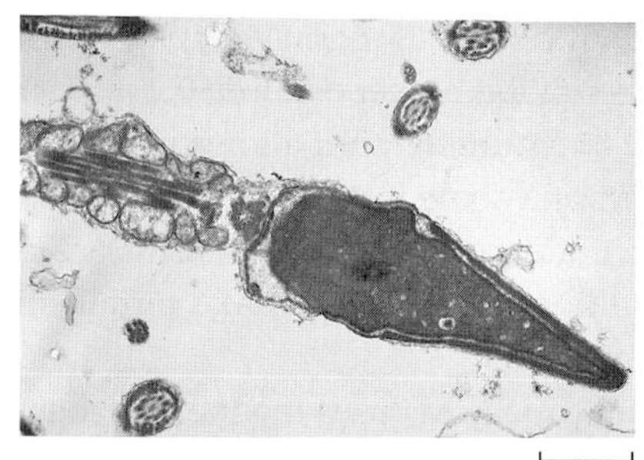

In **transmission electron microscopy** (TEM), a beam of electrons is focused on the object by magnets. Objects appear darker if they absorb the electrons. If the electrons pass through they are detected on a fluorescent screen.

8 µm

Scanning electron microscopy (SEM) directs electrons to the surface of the sample, where they cause other electrons to be emitted. These electrons are viewed on a screen. The three-dimentional surface of the object can be visualized.

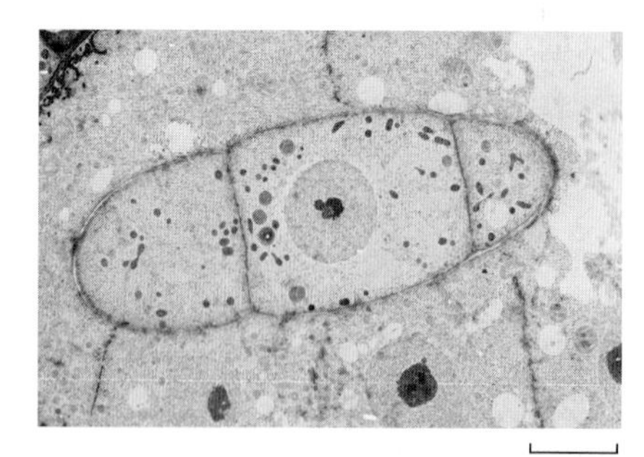

Cryoelectron microscopy uses quickly frozen samples to reduce aberrations that are seen when samples are treated chemically. Computer analysis of thick sections can reconstruct a sample in three dimensions.

4.4 Looking at Cells The top six panels show some techniques used in light microscopy. The lower three images were created using electron microscopes.

Cells are surrounded by a plasma membrane

As we have noted, a **plasma membrane** separates each cell from its environment, creating a segregated (but not isolated) compartment. The plasma membrane is composed of a phospholipid bilayer, with the hydrophilic "heads" of the lipids facing the cell's aqueous interior on one side of the membrane and the extracellular environment on the other (see Figure 3.21). Proteins are embedded in the lipids. In many cases, these proteins protrude into the cytoplasm and into the extracellular environment. We will devote most of Chapter 5 to detailing the structure and functions of the plasma membrane, but summarize its roles here.

- The plasma membrane allows the cell to maintain a more or less constant internal environment. A self-maintaining, *constant internal environment* is a key characteristic of life that will be discussed in detail in Chapter 41.
- The plasma membrane acts as a *selectively permeable barrier*, preventing some substances from crossing while permitting other substances to enter and leave the cell.
- As the cell's boundary with the outside environment, the plasma membrane is important in *communicating with adjacent cells and receiving extracellular signals*. We will describe this function in Chapter 15.
- The plasma membrane often has molecules protruding from it that are responsible for *binding and adhering* to adjacent cells.

Cells show two organizational patterns

Prokaryotic cell organization is characteristic of the domains Bacteria and Archaea. Organisms in these domains are called *prokaryotes*. Their cells do not have membrane-enclosed internal compartments. The first cells ever to form were undoubtedly similar in organization to modern prokaryotes.

Eukaryotic cell organization is found in the domain Eukarya, which includes the protists, plants, fungi, and animals. The genetic material (DNA) of eukaryotic cells is contained in a special membrane-enclosed compartment called the nucleus. Eukaryotic cells also contain other membrane-enclosed compartments in which specific chemical reactions take place. Organisms with this type of cell organization are known as *eukaryotes*.

Both prokaryotes and eukaryotes have prospered for many hundreds of millions of years of evolution, and both are great success stories. Let's look first at prokaryotic cells.

Prokaryotic Cells

Prokaryotes can live off more different and diverse energy sources than any other living creatures, and they inhabit greater environmental extremes, such as very hot springs and very salty water. The vast diversity within the prokaryotic domains is the subject of Chapter 27.

Prokaryotic cells are generally smaller than eukaryotic cells, ranging from 0.25 × 1.2 μm to 1.5 × 4 μm. Each prokaryote is a single cell, but many types of prokaryotes are usually seen in chains, small clusters, or even clusters containing hundreds of individuals. In this section, we will first consider the features that cells in the domains Bacteria and Archaea have in common. Then we will examine structural features that are found in some, but not all, prokaryotes.

Prokaryotic cells share certain features

All prokaryotic cells have the same basic structure:

- The plasma membrane encloses the cell, regulating the traffic of materials into and out of the cell and separating it from its environment.
- A region called the **nucleoid** contains the hereditary material (DNA) of the cell.

The rest of the material enclosed in the plasma membrane is called the **cytoplasm**. The cytoplasm is composed of two parts: the liquid cytosol, and insoluble suspended particles, including ribosomes.

- The **cytosol** consists mostly of water that contains dissolved ions, small molecules, and soluble macromolecules such as proteins.
- **Ribosomes** are granules about 25 nm in diameter that are sites of protein synthesis.

The cytoplasm is not a static region. Rather, the substances in this aqueous environment are in constant motion. For example, a typical protein moves around the entire cell within a minute, and encounters many molecules along the way.

Although structurally less complicated than eukaryotic cells, prokaryotic cells are functionally complex, carrying out thousands of biochemical transformations.

Some prokaryotic cells have specialized features

As they evolved, some prokaryotes developed specialized structures that gave a selective advantage to those cells that had them. These structures include a protective cell wall, an internal membrane for compartmentalization of chemical reactions, and flagella for cell movement through the watery environment. These features are shown in Figures 4.5 and 4.6.

CELL WALLS. Most prokaryotes have a **cell wall** located outside the plasma membrane. The rigidity of the cell wall

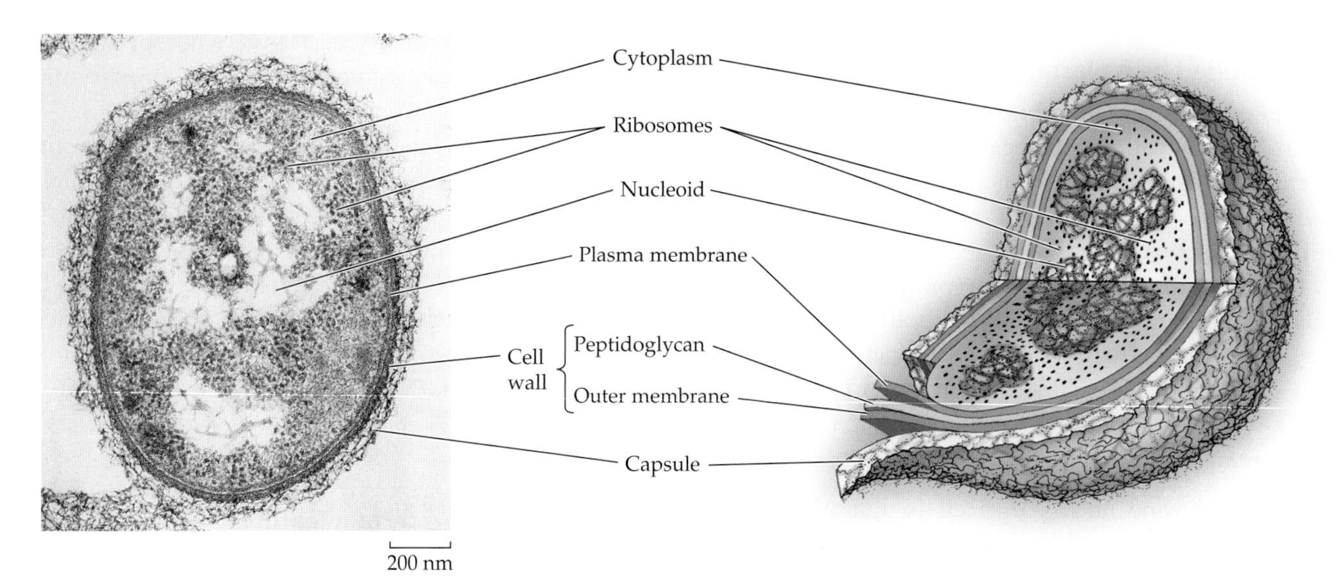

4.5 A Prokaryotic Cell The bacterium *Pseudomonas aeruginosa* illustrates typical prokaryotic cell structures. Note the existence of several protective structures external to the plasma membrane.

supports the cell and determines its shape. The cell walls of most bacteria, but not archaea, contain *peptidoglycan*, a polymer of amino sugars, cross-linked by covalent bonds to form a single giant molecule around the entire cell. In some bacteria, another layer—the outer membrane (a polysaccharide-rich phospholipid membrane)—encloses the peptidoglycan layer. Unlike the plasma membrane, this outer membrane is not a major permeability barrier, and some of its polysaccharides are disease-causing toxins.

Enclosing the cell wall in some bacteria is a layer of slime, composed mostly of polysaccharides and referred to as a *capsule*. The capsules of some bacteria may protect them from attack by white blood cells in the animals they infect. The capsule helps keep the cell from drying out, and sometimes it helps the bacterium attach to other cells. Many prokaryotes produce no capsule, and those that do have capsules can survive even if they lose them, so the capsule is not essential to cell life.

As you will see later in this chapter, eukaryotic plant cells also have a cell wall, but it differs in composition and structure from the cell walls of prokaryotes.

INTERNAL MEMBRANES. Some groups of bacteria—the cyanobacteria and some others—carry on photosynthesis. In these photosynthetic bacteria, the plasma membrane folds into the cytoplasm to form an internal membrane system that contains bacterial chlorophyll and other compounds needed for photosynthesis. The development of photosynthesis, probably by such internal membranes, was an important event in the early evolution of life on Earth. Other prokaryotes have internal membrane folds that remain attached to the plasma membrane. These *mesosomes* may function in cell division or in various energy-releasing reactions.

FLAGELLA AND PILI. Some prokaryotes swim by using appendages called *flagella* (Figure 4.6*a,c*). A single flagellum, made of a protein called *flagellin*, looks at times like a tiny corkscrew. It spins on its axis like a propeller, driving the cell along. Ring structures anchor the flagellum to the plasma membrane and, in some bacteria, to the outer membrane of the cell wall (Figure 4.6*c*). We know that the flagella cause the motion of the cell because if they are removed, the cell cannot move.

Pili project from the surfaces of some groups of bacteria (Figure 4.6*b*). Shorter than flagella, these threadlike structures help bacteria adhere to one another during mating, as well as to animal cells for protection and food.

CYTOSKELETON. Recent evidence suggests that some prokaryotes, especially rod-shaped bacteria, have an internal filamentous helical structure just below the plasma membrane. The proteins that make up this structure are similar in amino acid sequence to actin in eukaryotic cells, and since actin is part of the cytoskeleton in those cells (see below), it has been suggested that the helical filaments in prokaryotes play a role in cell shape.

Eukaryotic Cells

Animals, plants, fungi, and protists have cells that are usually larger and structurally more complex than those of the prokaryotes. To get a sense of the most prominent differ-

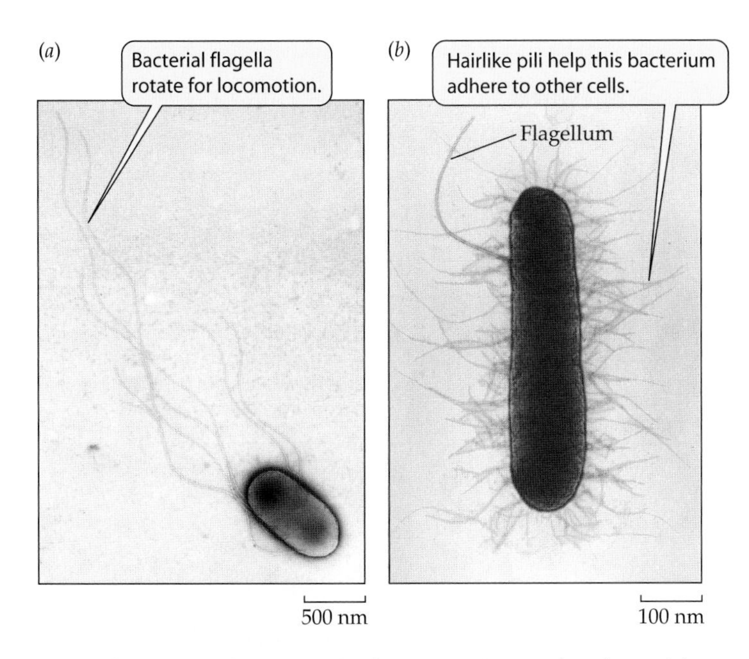

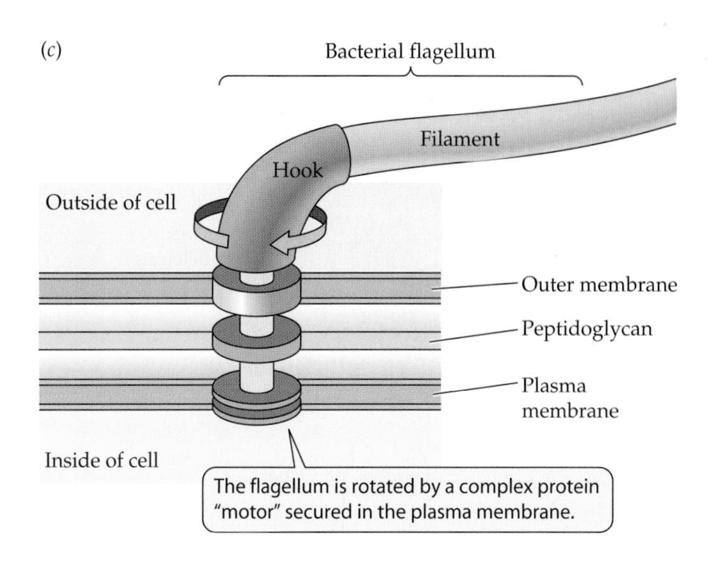

4.6 Prokaryotic Projections Surface projections such as bacterial flagella *(a)* and pili *(b)* contribute to the movement, adhesion, and complexity of prokaryotic cells. *(c)* Complex protein ring structures anchored in the bacterial cell membranes form a motor unit that rotates the flagellum and propels the cell.

ences, compare the eukaryotic plant and animal cells shown in Figure 4.7 with the prokaryotic cell in Figure 4.5.

Eukaryotic cells generally have dimensions ten times greater than those of prokaryotes; for example, the spherical yeast cell has a diameter of 8 μm. Like prokaryotic cells, eukaryotic cells have a plasma membrane, cytoplasm, and ribosomes. But added on to this basic organization are compartments in the cytoplasm whose interiors are separated from the cytosol by a membrane.

Compartmentalization is the key to eukaryotic cell function

Some of the compartments in eukaryotic cells are like little factories that make specific products. Others are like power plants that take in energy in one form and convert it to a more useful form. These membranous compartments, as well as other structures (such as ribosomes) that lack membranes but possess distinctive shapes and functions, are called **organelles** (see Figure 4.7). Each of these organelles has specific roles in its particular cell. These roles are defined by chemical reactions.

- The **nucleus** contains most of the cell's genetic material (DNA). The duplication of the genetic material and the first steps in decoding genetic information take place in the nucleus.
- The **mitochondrion** is a power plant and industrial park, where energy stored in the bonds of carbohydrates is converted to a form more useful to the cell (ATP) and certain essential biochemical conversions of amino acids and fatty acids occur.
- The **endoplasmic reticulum** and **Golgi apparatus** are compartments in which proteins are packaged and sent to appropriate locations in the cell.
- **Lysosomes** and **vacuoles** are cellular digestive systems in which large molecules are hydrolyzed into usable monomers.
- **Chloroplasts** perform photosynthesis.

The membrane surrounding each organelle does two essential things: First, it keeps the organelle's molecules away from other molecules in the cell with which they might react inappropriately. Second, it acts as a traffic regulator, letting important raw materials into the organelle and releasing its products to the cytoplasm. The evolution of compartmentalization was an important development in the ability of eukaryotic cells to specialize, forming the organs and tissues of a complex body.

Organelles can be studied by microscopy or isolated for chemical analysis

Cell organelles were first detected by light and electron microscopy. The use of stains targeted to specific macromolecules has allowed cell biologists to determine the chemical compositions of organelles. (See Figure 4.21, in which a single cell is stained for three different proteins.)

Besides microscopy, another way to look at cells is to take them apart. **Cell fractionation** begins with the destruction of the cell membrane. This allows the cytoplasmic components

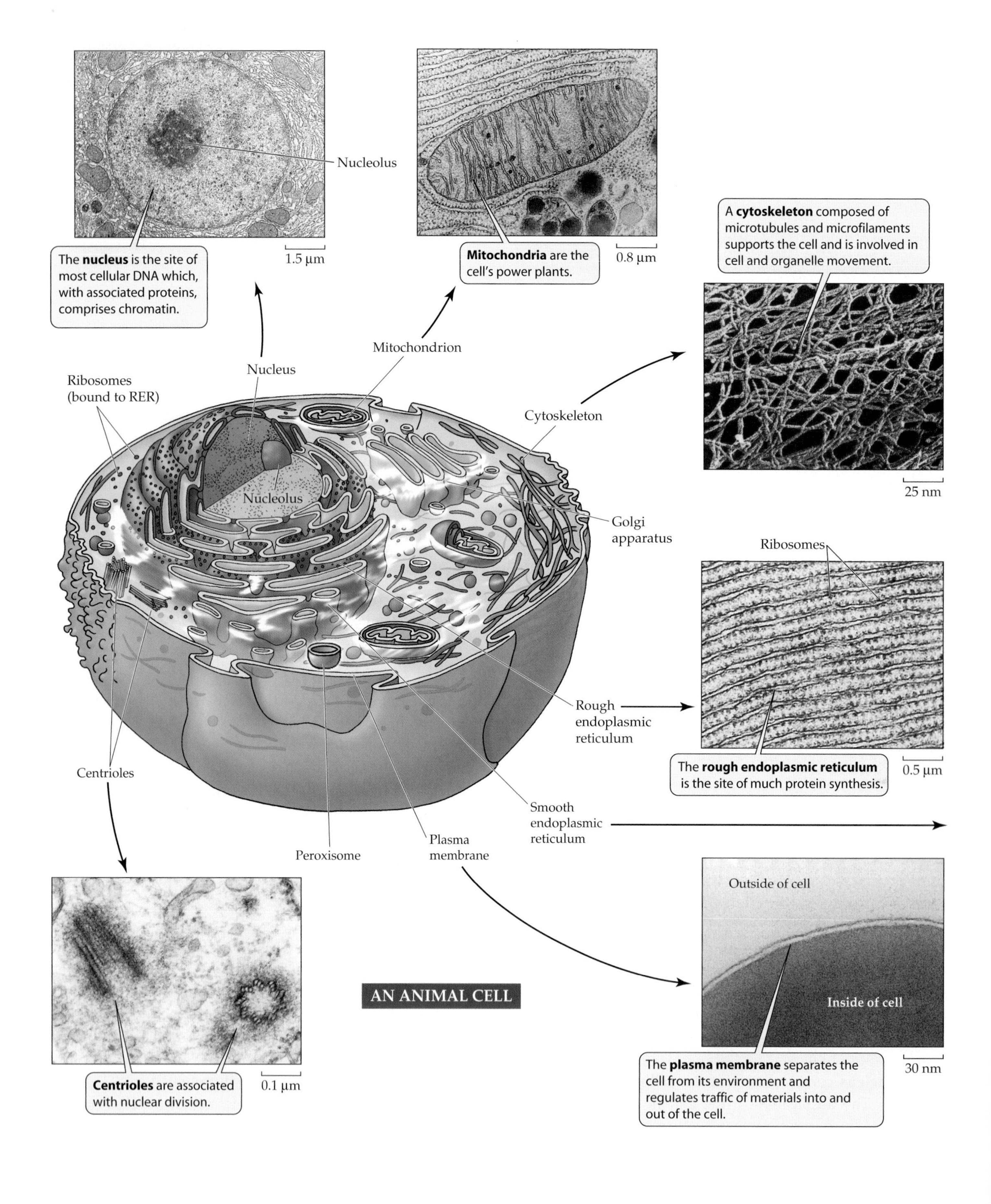

Nucleolus
1.5 µm
The **nucleus** is the site of most cellular DNA which, with associated proteins, comprises chromatin.
Mitochondria are the cell's power plants.
0.8 µm
A **cytoskeleton** composed of microtubules and microfilaments supports the cell and is involved in cell and organelle movement.
25 nm
Mitochondrion
Nucleus
Ribosomes (bound to RER)
Cytoskeleton
Nucleolus
Golgi apparatus
Ribosomes
Rough endoplasmic reticulum
The **rough endoplasmic reticulum** is the site of much protein synthesis.
0.5 µm
Centrioles
Smooth endoplasmic reticulum
Peroxisome
Plasma membrane
Outside of cell
Inside of cell
AN ANIMAL CELL
Centrioles are associated with nuclear division.
0.1 µm
The **plasma membrane** separates the cell from its environment and regulates traffic of materials into and out of the cell.
30 nm

4.7 Eukaryotic Cells In electron micrographs, many plant cell organelles are nearly identical in form to those observed in animal cells. Cellular structures unique to plant cells include the cell wall and the chloroplasts. Animal cells contain centrioles, which are not found in plant cells.

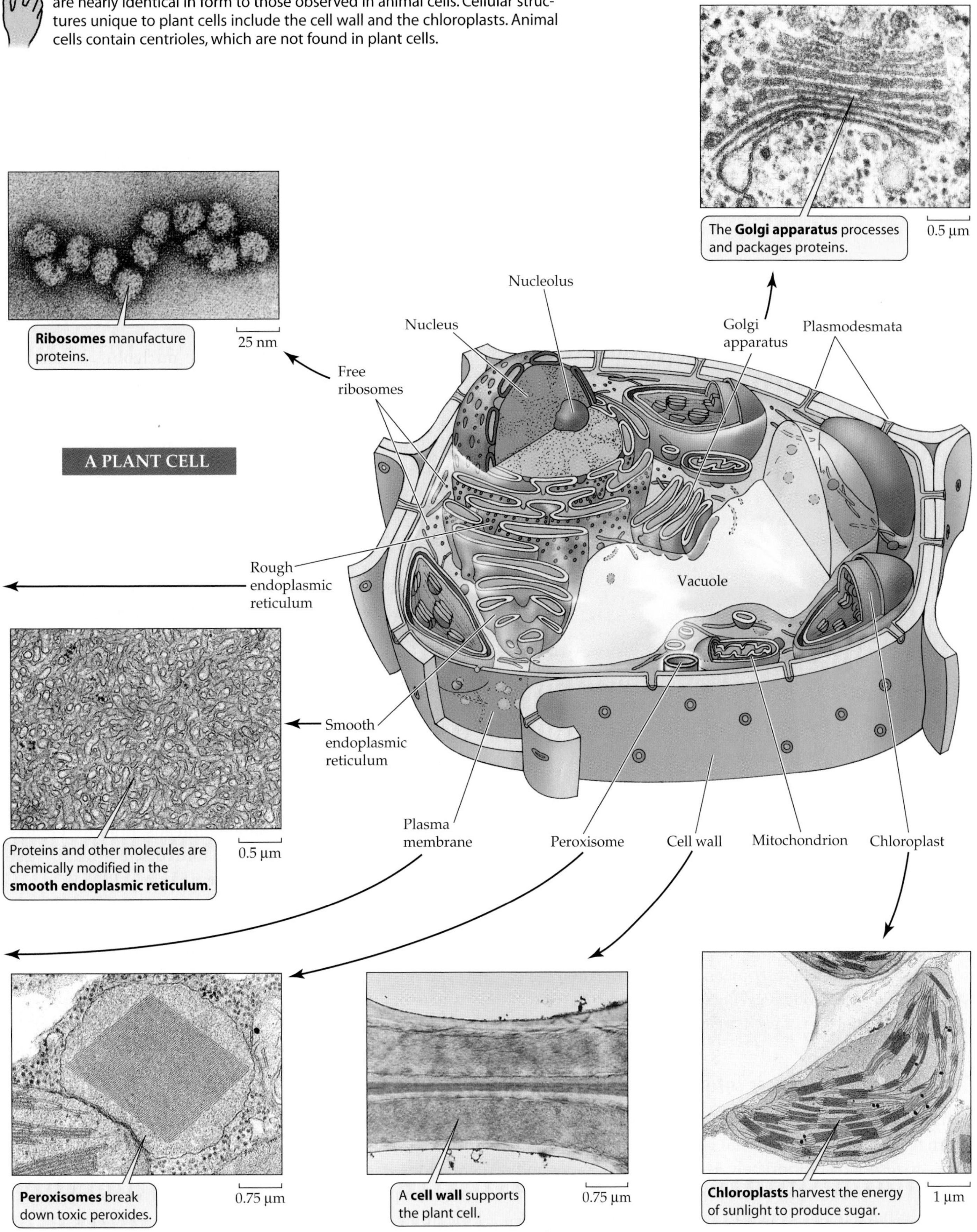

A PLANT CELL

Ribosomes manufacture proteins.

25 nm

The **Golgi apparatus** processes and packages proteins.

0.5 μm

Nucleolus

Nucleus

Free ribosomes

Golgi apparatus

Plasmodesmata

Rough endoplasmic reticulum

Vacuole

Smooth endoplasmic reticulum

Proteins and other molecules are chemically modified in the **smooth endoplasmic reticulum**.

0.5 μm

Plasma membrane

Peroxisome

Cell wall

Mitochondrion

Chloroplast

Peroxisomes break down toxic peroxides.

0.75 μm

A **cell wall** supports the plant cell.

0.75 μm

Chloroplasts harvest the energy of sunlight to produce sugar.

1 μm

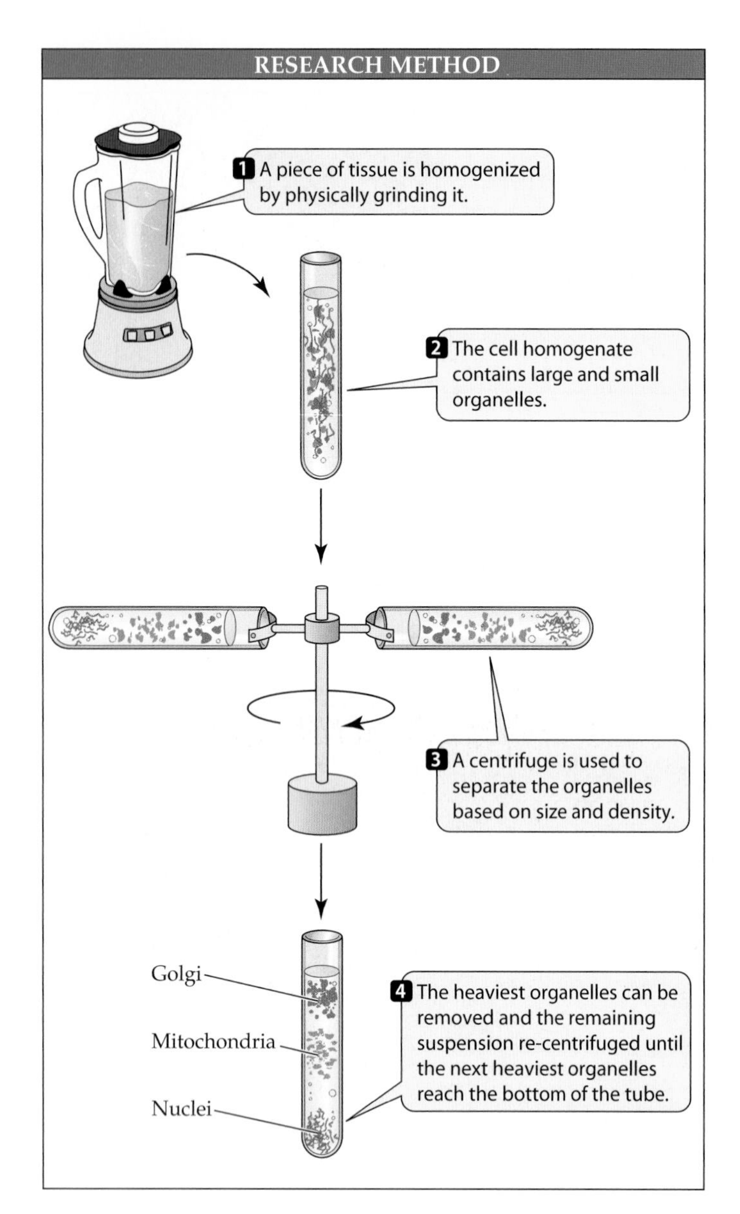
RESEARCH METHOD
1 A piece of tissue is homogenized by physically grinding it.
2 The cell homogenate contains large and small organelles.
3 A centrifuge is used to separate the organelles based on size and density.
4 The heaviest organelles can be removed and the remaining suspension re-centrifuged until the next heaviest organelles reach the bottom of the tube.
Golgi
Mitochondria
Nuclei

4.8 Cell Fractionation The organelles of cells can be separated from one another after cells are broken open and centrifuged.

to flow out into a test tube. The various organelles can then be separated from one another on the basis of size or density (Figure 4.8). Biochemical analyses can then be done on the isolated organelles. Microscopy and cell fractionation have complemented each other, giving a complete picture of the structure and function of each organelle.

Organelles that Process Information

Living things depend on accurate, appropriate information—internal signals, environmental cues, and stored instructions—to respond appropriately to changing conditions and maintain a constant internal environment. In the cell, information is stored in the sequence of nucleotides in DNA molecules. Most of the DNA in eukaryotic cells resides in the nucleus. Information is translated from the language of DNA into the language of proteins at the ribosomes. This process is described in detail in Chapter 12.

The nucleus contains most of the cell's DNA

The single nucleus is usually the largest organelle in a cell (Figure 4.9; see also Figure 4.7). The nucleus of most animal cells is approximately 5 μm in diameter—substantially larger than most entire prokaryotic cells. The nucleus has several roles in the cell:

- The nucleus is the site of DNA duplication.
- The nucleus is the site of genetic control of the cell's activities.
- A region within the nucleus, the **nucleolus**, begins the assembly of ribosomes from specific proteins and RNA.

The nucleus is surrounded by two membranes, which together form the **nuclear envelope**. The two membranes of the nuclear envelope are separated by 10–20 nm and are perforated by **nuclear pores** approximately 9 nm in diameter, which connect the interior of the nucleus with the cytoplasm. At these pores, the outer membrane of the nuclear envelope is continuous with the inner membrane. Each pore is surrounded by a pore complex made up of eight large protein granules arranged in an octagon where the inner and outer membranes merge (see Figure 4.9). RNA and proteins pass through these pores to enter or leave the nucleus.

At certain sites, the outer membrane of the nuclear envelope folds outward into the cytoplasm and is continuous with the membrane of another organelle, the endoplasmic reticulum (discussed later in this chapter).

Inside the nucleus, DNA combines with proteins to form a fibrous complex called **chromatin**. Chromatin consists of exceedingly long, thin, entangled threads. Prior to cell division, the chromatin aggregates to form discrete, readily visible structures called **chromosomes** (Figure 4.10).

Surrounding the chromatin are water and dissolved substances collectively referred to as the **nucleoplasm**. Within the nucleoplasm, a network of apparently structural proteins called the *nuclear matrix* organizes the chromatin. At the periphery of the nucleus, the chromatin is attached to a protein meshwork, called the *nuclear lamina*, which is formed by the polymerization of proteins called *lamins* into filaments. The nuclear lamina maintains the shape of the nucleus by its attachment to both the chromatin and the nuclear envelope.

During most of a cell's life cycle, the nuclear envelope is a stable structure. When the cell divides, however, the nuclear envelope fragments into pieces of membrane with attached pore complexes. The envelope re-forms when distribution of the duplicated DNA to the daughter cells is completed.

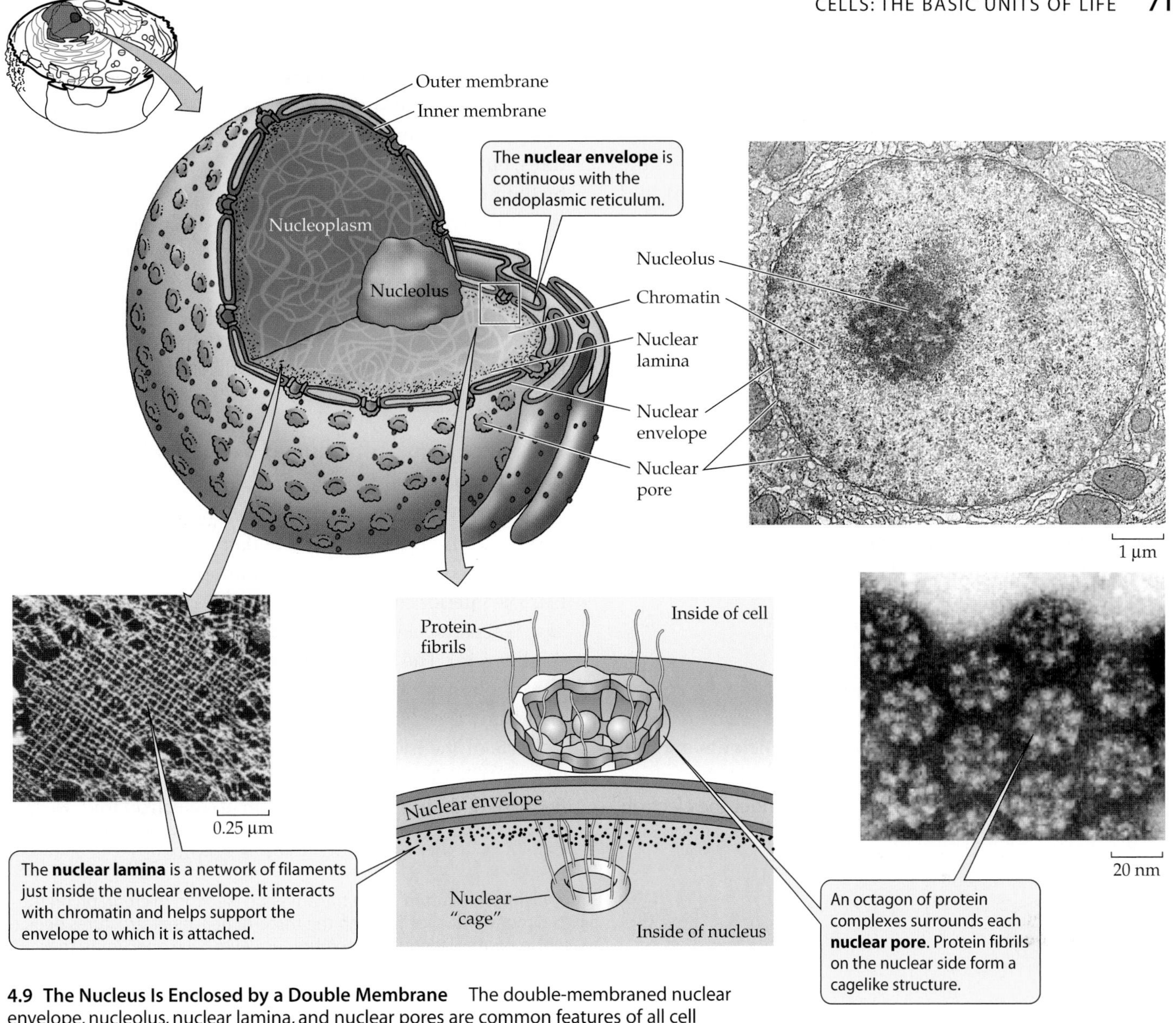

Outer membrane
Inner membrane
The **nuclear envelope** is continuous with the endoplasmic reticulum.
Nucleoplasm
Nucleolus
Nucleolus
Chromatin
Nuclear lamina
Nuclear envelope
Nuclear pore
1 μm
0.25 μm
Protein fibrils
Inside of cell
Nuclear envelope
Nuclear "cage"
Inside of nucleus
20 nm
The **nuclear lamina** is a network of filaments just inside the nuclear envelope. It interacts with chromatin and helps support the envelope to which it is attached.
An octagon of protein complexes surrounds each **nuclear pore**. Protein fibrils on the nuclear side form a cagelike structure.

4.9 The Nucleus Is Enclosed by a Double Membrane The double-membraned nuclear envelope, nucleolus, nuclear lamina, and nuclear pores are common features of all cell nuclei. The pores are the gateways through which proteins from the cytoplasm enter the nucleus and genetic material (mRNA) from the nucleus enters the cytoplasm.

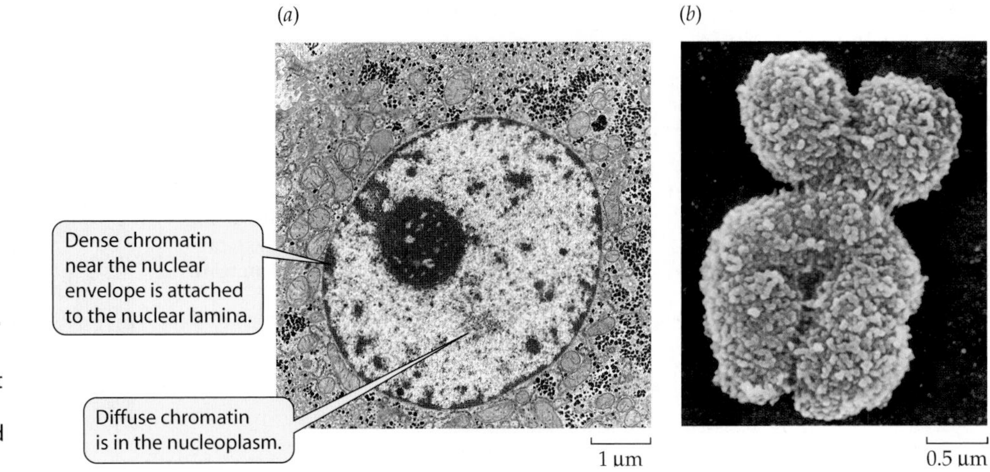

(a)
(b)
Dense chromatin near the nuclear envelope is attached to the nuclear lamina.
Diffuse chromatin is in the nucleoplasm.
1 μm
0.5 μm

4.10 Chromatin and Chromosomes
(*a*) When a cell is not dividing, the nuclear DNA is aggregated with proteins to form chromatin, which is dispersed throughout the nucleus. (*b*) The chromatin in a dividing cell is packed into dense bodies called chromosomes.

Ribosomes are the sites of protein synthesis

In prokaryotic cells, ribosomes float freely in the cytoplasm. In eukaryotic cells they occur in two places: in the cytoplasm, where they may be free or attached to the surface of the endoplasmic reticulum (described in the next section); and inside the mitochondria and chloroplasts, where energy is processed. In each of these locations, the ribosomes are the sites where proteins are synthesized under the direction of nucleic acids. Although they seem small in comparison to the cell in which they are contained, ribosomes are huge machines made up of several dozen kinds of molecules.

The ribosomes of prokaryotes and eukaryotes are similar in that both consist of two different-sized subunits. Eukaryotic ribosomes are somewhat larger, but the structure of prokaryotic ribosomes is better understood. Chemically, ribosomes consist of a special type of RNA, called *ribosomal RNA (rRNA)*, to which more than 50 different protein molecules are noncovalently bound.

The Endomembrane System

Much of the volume of some eukaryotic cells is taken up by an extensive **endomembrane system**. This system includes two main components, the endoplasmic reticulum and the Golgi apparatus. Continuities between the nuclear envelope and the endomembrane system are visible under the electron microscope. Tiny, membrane-surrounded droplets called **vesicles** appear to shuttle between the various components of the endomembrane system. This system has various structures, but all of them are essentially compartments, closed off by their membranes from the cytoplasm.

In this section, we will examine the functional significance of these compartments, and we will see how materials synthesized in one organelle, the endoplasmic reticulum, are transferred to another organelle, the Golgi apparatus, for further processing, storage, or transport. We will also describe the role of the lysosome in cellular digestion.

The endoplasmic reticulum is a complex factory

Electron micrographs reveal a network of interconnected membranes branching throughout the cytoplasm of a eukaryotic cell, forming tubes and flattened sacs. These membranes are collectively called the **endoplasmic reticulum**, or **ER**. The interior compartment of the ER, referred to as the *lumen*, is separate and distinct from the surrounding cytoplasm (Figure 4.11). The ER can enclose up to 10 percent of the interior volume of the cell, and its foldings result in a surface area many times greater than that of the plasma membrane.

Parts of the ER are studded with ribosomes, which are temporarily attached to the outer faces of its flattened sacs. Because of their appearance under the electron microscope,

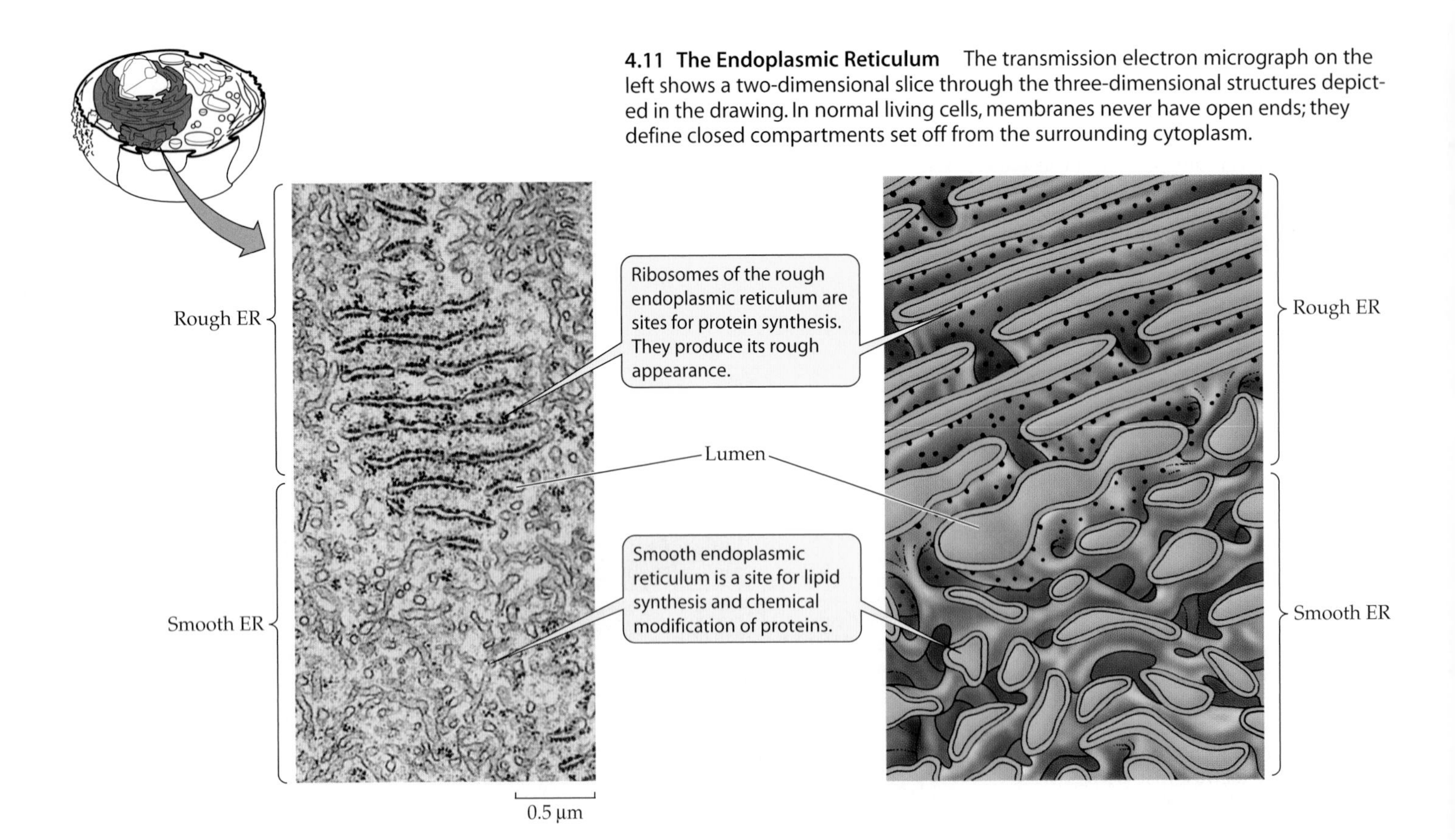

Rough ER
Smooth ER
Ribosomes of the rough endoplasmic reticulum are sites for protein synthesis. They produce its rough appearance.
Lumen
Smooth endoplasmic reticulum is a site for lipid synthesis and chemical modification of proteins.
0.5 µm
Rough ER
Smooth ER

4.11 The Endoplasmic Reticulum The transmission electron micrograph on the left shows a two-dimensional slice through the three-dimensional structures depicted in the drawing. In normal living cells, membranes never have open ends; they define closed compartments set off from the surrounding cytoplasm.

these regions are called **rough endoplasmic reticulum**, or **RER**. RER has two roles:

- As a compartment, it segregates certain newly synthesized proteins away from the cytoplasm and transports them to other locations in the cell.
- While inside the RER, proteins can be chemically modified so as to alter their function and eventual destination.

The attached ribosomes are sites for the synthesis of proteins that function outside the cytosol—that is, proteins that are to be exported from the cell, incorporated into membranes, or moved into the organelles of the endomembrane system. These proteins enter the lumen of the ER as they are synthesized. Once in the lumen of the ER, these proteins undergo several changes, including the formation of disulfide bridges and folding into their tertiary structures (see Figure 3.4).

Proteins gain carbohydrate groups in the RER, thus becoming glycoproteins. In the case of proteins directed to the lysosomes, the carbohydrate groups are part of an "addressing" system that ensures that the right proteins are directed to the organelle.

Smooth endoplasmic reticulum or **SER** is more tubular (less like flattened sacs) and lacks ribosomes (see Figure 4.11). Within the lumen of the SER, proteins that have been synthesized on the RER are chemically modified. In addition, the SER has three other important roles:

- It is responsible for chemically modifying small molecules taken in by the cell. This is especially true for drugs and pesticides.
- It is the site for the hydrolysis of glycogen in animal cells.
- It is the site for the synthesis of lipids and steroids.

Cells that synthesize a lot of protein for export are usually packed with endoplasmic reticulum. Examples include glandular cells that secrete digestive enzymes and plasma cells that secrete antibodies. In contrast, cells that carry out less protein synthesis (such as storage cells) contain less ER. Liver cells, which modify molecules that enter the body from the digestive system, have abundant smooth ER.

The Golgi apparatus stores, modifies, and packages proteins

The exact appearance of the Golgi apparatus (named for its discoverer, Camillo Golgi) varies from species to species, but it always consists of flattened membranous sacs called *cisternae* and small membrane-enclosed vesicles. The cisternae appear to be lying together like a stack of saucers (Figure 4.12). The entire apparatus is about 1 μm long.

The Golgi apparatus has several roles:

- It receives proteins from the ER and may further modify them.
- It concentrates, packages, and sorts proteins before they are sent to their cellular or extracellular destinations.
- It is where some polysaccharides for the plant cell wall are synthesized.

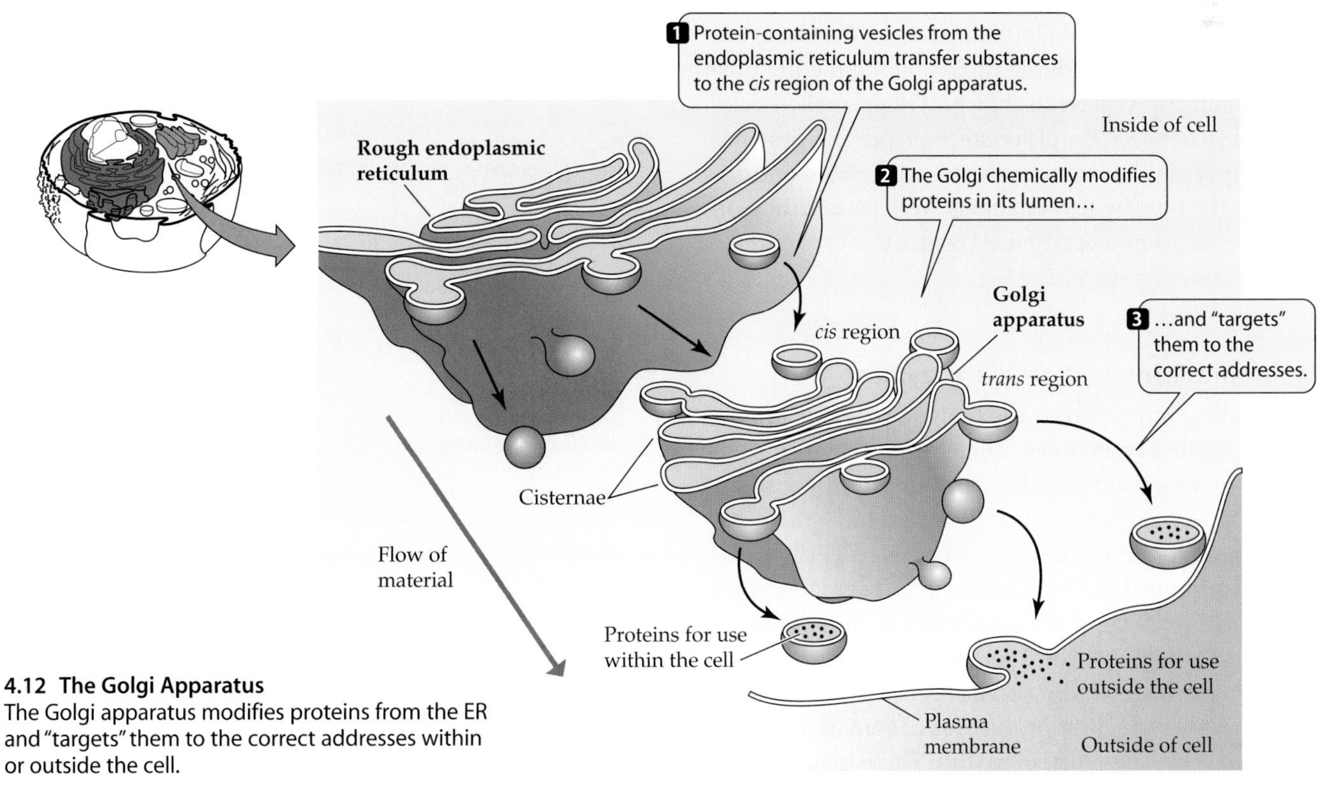
1 Protein-containing vesicles from the endoplasmic reticulum transfer substances to the cis region of the Golgi apparatus.
Inside of cell
Rough endoplasmic reticulum
2 The Golgi chemically modifies proteins in its lumen...
Golgi apparatus
cis region
trans region
3 ...and "targets" them to the correct addresses.
Cisternae
Flow of material
Proteins for use within the cell
Proteins for use outside the cell
Plasma membrane
Outside of cell

4.12 The Golgi Apparatus
The Golgi apparatus modifies proteins from the ER and "targets" them to the correct addresses within or outside the cell.

In the cells of plants, protists, fungi, and many invertebrate animals, the stacks of cisternae are individual units scattered throughout the cytoplasm. In vertebrate cells, a few such stacks usually form a larger, single, more complex Golgi apparatus.

The Golgi apparatus appears to have three functionally distinct parts: a bottom, a middle, and a top. The bottom cisternae, constituting the *cis* region of the Golgi apparatus, lie nearest to the nucleus or a patch of RER (see Figure 4.12). The top cisternae, constituting the *trans* region, lie closest to the surface of the cell. The cisternae in the middle make up the *medial* region of the complex. These three parts of the Golgi apparatus contain different enzymes and perform different functions.

The Golgi apparatus receives proteins from the ER, packages them, and sends them on their way. Since there is often no direct membrane continuity between ER and Golgi apparatus, how does a protein get from one organelle to the other? The protein could simply leave the ER, travel across the cytoplasm, and enter the Golgi apparatus. But that would expose the protein to interactions with other molecules in the cytoplasm. On the other hand, segregation from the cytoplasm could be maintained if a piece of the ER could "bud off," forming a vesicle that contains the protein—and that is exactly what happens. The protein makes the passage from ER to Golgi apparatus safely enclosed in the vesicle. Once it arrives, the vesicle fuses with the membrane of the Golgi apparatus, releasing its cargo.

Vesicles form from the rough ER, move through the cytoplasm, and fuse with the *cis* region of the Golgi apparatus, releasing their contents into the lumen. The vesicles may not have far to go: If living cells are stained specifically for ER and Golgi apparatus, the Golgi apparatus can be seen moving rapidly along the ER, possibly picking up vesicles as they go. Other small vesicles may move between the cisternae, transporting proteins, and it appears that some proteins move from one cisterna to the next by tiny channels. Vesicles budding off from the *trans* region carry their contents away from the complex (see Figure 4.12).

Lysosomes contain digestive enzymes

Originating in part from the Golgi apparatus are organelles called **lysosomes**. They contain digestive enzymes, and they are the sites where macromolecules—proteins, polysaccharides, nucleic acids, and lipids—are hydrolyzed into their monomers (see Figure 3.3). Lysosomes are about 1 μm in diameter, are surrounded by a single membrane, and have a densely staining, featureless interior (Figure 4.13). There may be dozens of lysosomes in a cell, depending on its needs.

Lysosomes are sites for the breakdown of food and foreign objects taken up by the cell. These materials get into the cell by a process called **phagocytosis** (*phago-*, "eating"; *cytosis*, "cellular"), in which a pocket forms in the plasma membrane and eventually deepens and encloses material from outside the cell. This pocket becomes a small vesicle that breaks free of the plasma membrane to move into the cytoplasm as a *phagosome* containing food or other material (see Figure 4.13). The phagosome fuses with a *primary lysosome*, forming a *secondary lysosome* where digestion occurs.

The effect of this fusion is rather like releasing hungry foxes into a chicken coop: The enzymes in the secondary lysosome quickly hydrolyze the food particles. These reactions are enhanced by the mild acidity of the lysosome's interior, where the pH is lower than in the surrounding cytoplasm. The products of digestion exit through the membrane of the lysosome, providing fuel molecules and raw materials for other cell processes. The "used" secondary lysosome, now

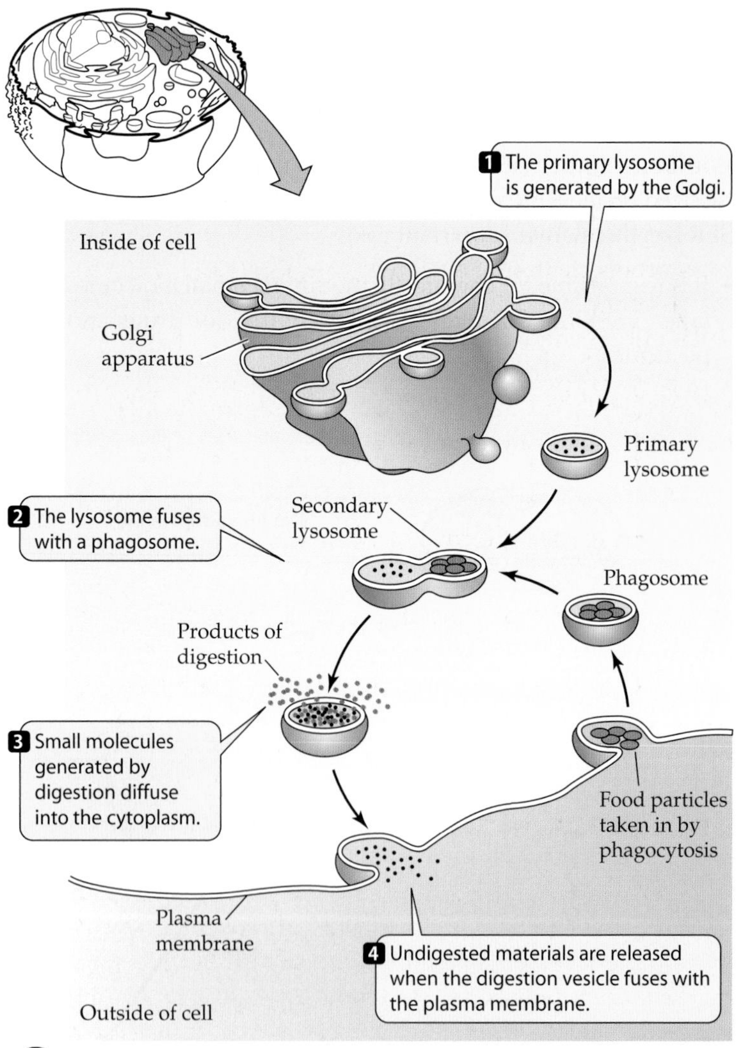
1 The primary lysosome is generated by the Golgi.
Inside of cell
Golgi apparatus
Primary lysosome
2 The lysosome fuses with a phagosome.
Secondary lysosome
Phagosome
Products of digestion
3 Small molecules generated by digestion diffuse into the cytoplasm.
Food particles taken in by phagocytosis
Plasma membrane
4 Undigested materials are released when the digestion vesicle fuses with the plasma membrane.
Outside of cell

4.13 Lysosomes Isolate Digestive Enzymes from the Cytoplasm Lysosomes are sites for the hydrolysis of material taken into the cell by phagocytosis.

containing undigested particles, then moves to the plasma membrane, fuses with it, and releases the undigested contents to the environment.

Lysosomes are also where the cell digests its own material in a process called **autophagy**. Autophagy is an ongoing process in which organelles such as mitochondria are engulfed by lysosomes and hydrolyzed to monomers, which pass out of the lysosome through its membrane into the cytoplasm for reuse.

The importance of lysosome function is indicated by a group of human diseases called *lysosomal storage diseases*. If a cell lacks the ability to hydrolyze one or more macromolecules, these substances pile up in lysosomes, with harmful consequences. An example is Tay-Sachs disease, in which a lipid accumulates in the lysosomes of brain cells, resulting in death in early childhood.

Plant cells do not appear to contain lysosomes, but the central vacuole of a plant cell (which we will describe below) may function in an equivalent capacity because it, like lysosomes, contains many digestive enzymes.

Organelles that Process Energy

A cell uses energy to synthesize cell-specific materials that it can use for activities such as growth, reproduction, and movement. Energy is transformed from one form to another in mitochondria (found in all eukaryotic cells) and in chloroplasts (found in eukaryotic cells that harvest energy from sunlight). In contrast, energy transformations in prokaryotic cells are associated with enzymes attached to the inner surface of the plasma membrane or extensions of the plasma membrane that protrude into the cytoplasm.

Mitochondria are energy transformers

In eukaryotic cells, the breakdown of fuel molecules such as glucose begins in the cytosol. The molecules that result from this partial degradation enter the mitochondria (singular, mitochondrion), whose primary function is to convert the potential chemical energy of those fuel molecules into a form that the cell can use: the energy-rich molecule ATP (adenosine triphosphate). The production of ATP in the mitochondria using fuel molecules and molecular oxygen (O_2) is called **cellular respiration**.

Typical mitochondria are small—somewhat less than 1.5 µm in diameter and 2–8 µm in length—about the size of many bacteria. The number of mitochondria per cell ranges from one contorted giant in some unicellular protists to a few hundred thousand in large egg cells. An average human liver cell contains more than a thousand mitochondria. Cells that require the most chemical energy tend to have the most mitochondria per unit of volume.

Mitochondria have two membranes. The *outer membrane* is smooth and protective, and it offers little resistance to the movement of substances into and out of the mitochondrion. Immediately inside the outer membrane is an *inner membrane*, which folds inward in many places, giving it a much greater surface area than that of the outer membrane (Figure 4.14). These folds tend to be quite regular, giving rise to shelflike structures called *cristae*.

The inner mitochondrial membrane contains many large protein molecules that participate in cellular respiration. The inner membrane exerts much more control over what enters and leaves the mitochondrion than does the outer membrane. The region enclosed by the inner membrane is referred to as the *mitochondrial matrix*. In addition to many proteins, the matrix contains some ribosomes and DNA that are used to make some of the proteins needed for cellular respiration.

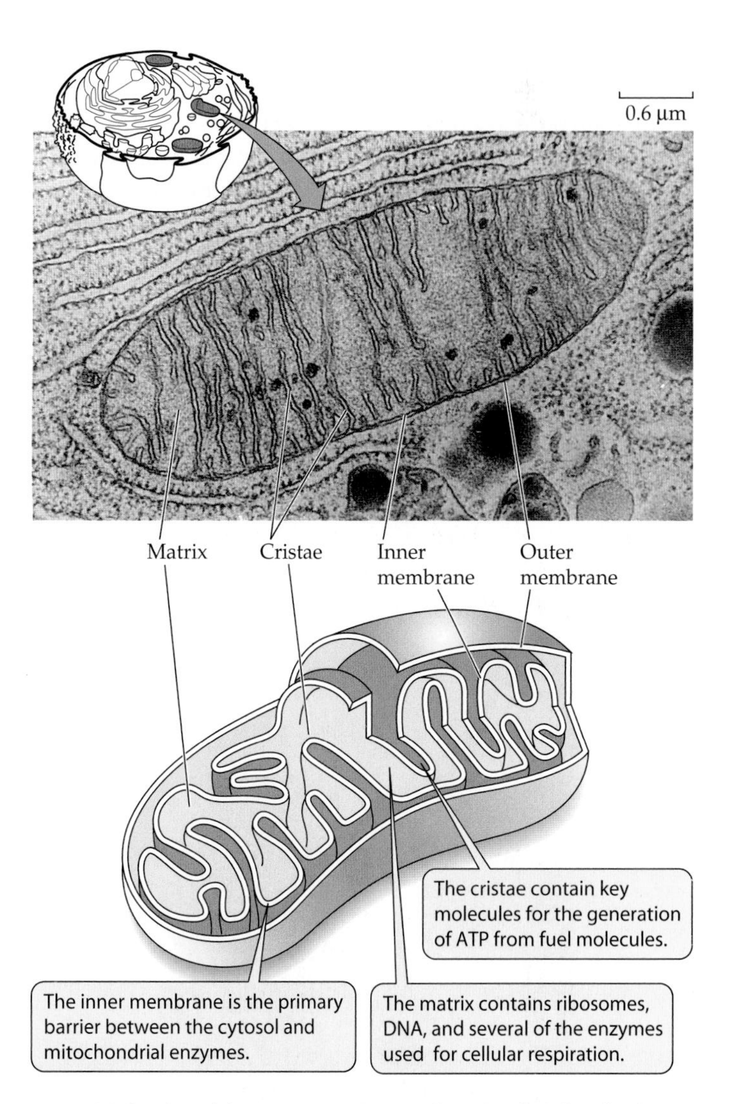

0.6 µm
Matrix
Cristae
Inner membrane
Outer membrane
The cristae contain key molecules for the generation of ATP from fuel molecules.
The inner membrane is the primary barrier between the cytosol and mitochondrial enzymes.
The matrix contains ribosomes, DNA, and several of the enzymes used for cellular respiration.

4.14 A Mitochondrion Converts Energy from Fuel Molecules into ATP The electron micrograph is a two-dimensional slice through a three-dimensional organelle. As the drawing emphasizes, the cristae are extensions of the inner mitochondrial membrane.

In Chapter 7 we will see how the different parts of the mitochondrion work together in cellular respiration.

Plastids photosynthesize or store materials

One class of organelles—the **plastids**—is produced only in plants and certain protists. There are several types of plastids, with different functions.

CHLOROPLASTS. Chloroplasts contain the green pigment **chlorophyll** and are the sites of photosynthesis (Figure 4.15). In **photosynthesis**, light energy is converted into the chemical energy of bonds between atoms. The molecules formed in photosynthesis provide food for the photosynthetic organisms, as well as for other organisms that eat them. Directly or indirectly, photosynthesis is the energy source for most of the living world.

Chloroplasts are variable in size and shape (Figure 4.16). Like a mitochondrion, a chloroplast is surrounded by two membranes. In addition, there is a series of internal membranes whose structure and arrangement vary from one group of photosynthetic organisms to another. Here we concentrate on the chloroplasts of the flowering plants. Even these chloroplasts show some variation, but the pattern shown in Figure 4.15 is typical.

The internal membranes of chloroplasts look like stacks of flat, hollow pita bread. These stacks, called **grana** (singular, granum), consist of a series of flat, closely packed, circular compartments called **thylakoids**. In addition to phospholipids and proteins, the membranes of the thylakoids contain chlorophyll and other pigments that harvest light for photosynthesis. The thylakoids of one granum may be connected to those of other grana, making the interior of the chloroplast a highly developed network of membranes, much like the ER.

The fluid in which the grana are suspended is the **stroma**. Like the mitochondrial matrix, the chloroplast stroma contains ribosomes and DNA, which are used to synthesize some, but not all, of the proteins that make up the chloroplast.

Animal cells do not produce chloroplasts, but some do contain functional chloroplasts. These are either taken up as free chloroplasts derived from the partial digestion of green plants or contained within unicellular algae that live within the animal's tissues. The green color of some corals and sea anemones results from the chloroplasts in algae that live within those animals (Figure 4.16*c*). The animals derive some of their nutrition from the photosynthesis that their chloroplast-containing "guests" carry out. Such an intimate relationship between two different organisms is called **symbiosis**.

OTHER TYPES OF PLASTIDS. Other types of plastids also store pigments or polysaccharides:

- *Chromoplasts* contain red, orange, and/or yellow pigments and give color to plant organs such as flowers (Figure 4.17*a*). The chromoplasts have no known chemical func-

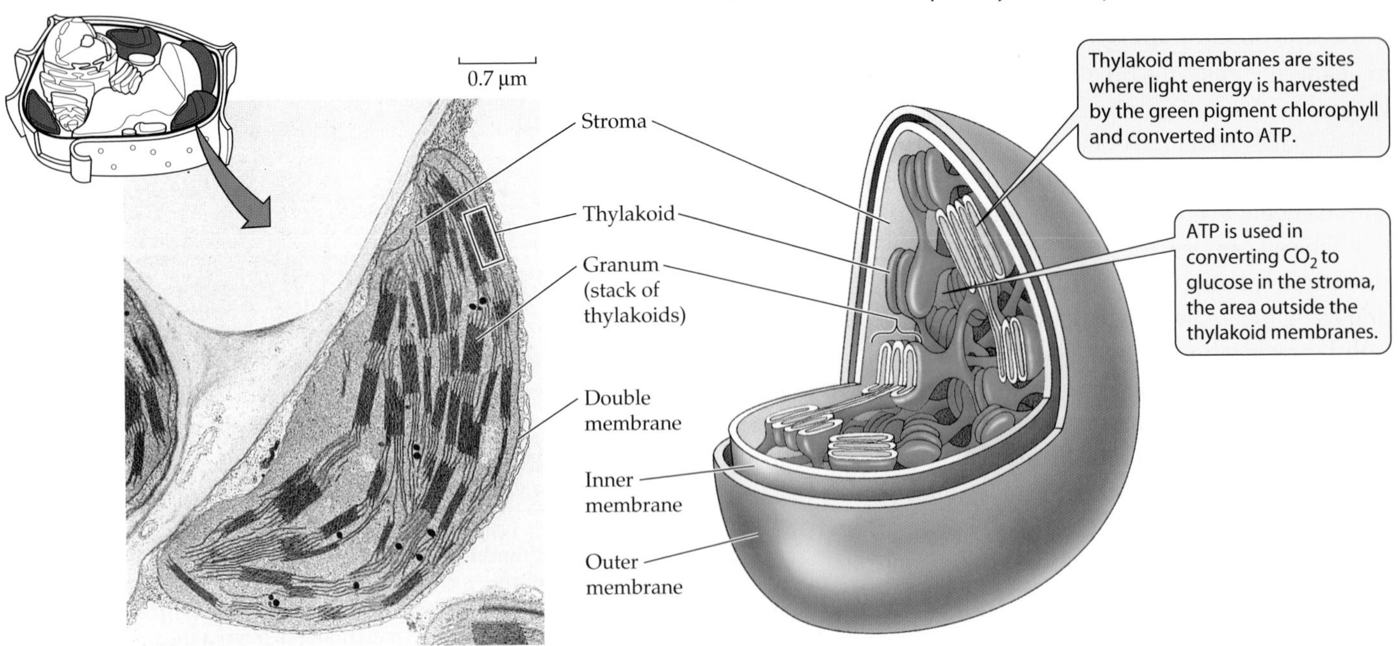
0.7 µm
Stroma
Thylakoid
Granum (stack of thylakoids)
Double membrane
Inner membrane
Outer membrane
Thylakoid membranes are sites where light energy is harvested by the green pigment chlorophyll and converted into ATP.
ATP is used in converting CO_2 to glucose in the stroma, the area outside the thylakoid membranes.

4.15 The Chloroplast: The Organelle That Feeds the World The electron micrograph shows a chloroplast from a leaf of corn. Chloroplasts are large compared with mitochondria and contain an extensive network of photosynthetic thylakoid membranes.

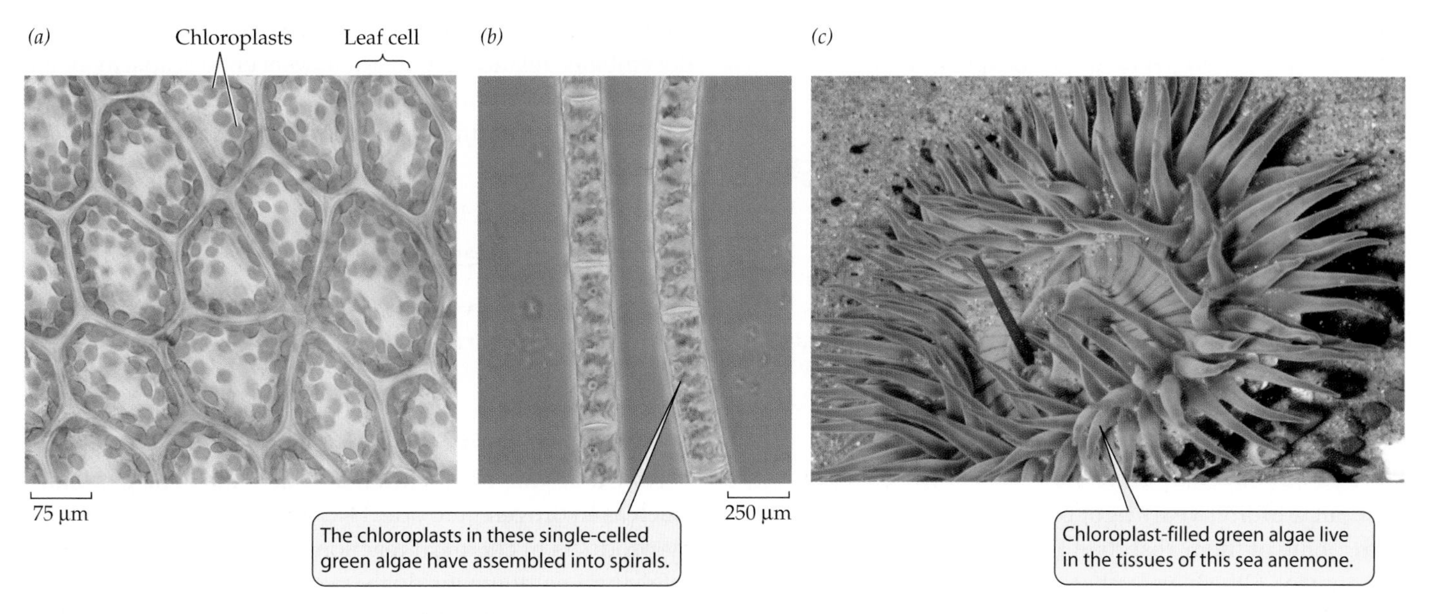

4.16 Being Green (*a*) In green plants, chloroplasts are concentrated in the leaf cells. (*b*) Green algae are photosynthetic and filled with chloroplasts. (*c*) No animal species produces its own chloroplasts, but in this symbiotic arrangement, unicellular green algae nourish a sea anemone.

tion in the cell, but the colors they give to some petals and fruits probably encourage animals to visit flowers and thus aid in pollination, or to eat fruits and thus aid in seed dispersal. (On the other hand, carrot roots gain no apparent advantage from being orange.)

- *Leucoplasts* are storage depots for starch and fats (Figure 4.17*b*).

Endosymbiosis may explain the origin of mitochondria and chloroplasts

Although chloroplasts and mitochondria are about the size of prokaryotic cells and have the genetic material and protein synthesis machinery needed to make some of their own components, they are not independent of control by the nucleus. The vast majority of their proteins are encoded by nuclear DNA, made in the cytoplasm, and imported into the organelle. Observations of these organelles have led to the proposal that they originated by endosymbiosis—that is, that they were once independent prokaryotic organisms.

4.17 Chromoplasts and Leucoplasts (*a*) Colorful pigments stored in the chromoplasts of flowers like this begonia may help attract pollinating insects. (*b*) Leucoplasts in the cells of a potato are filled with white starch grains.

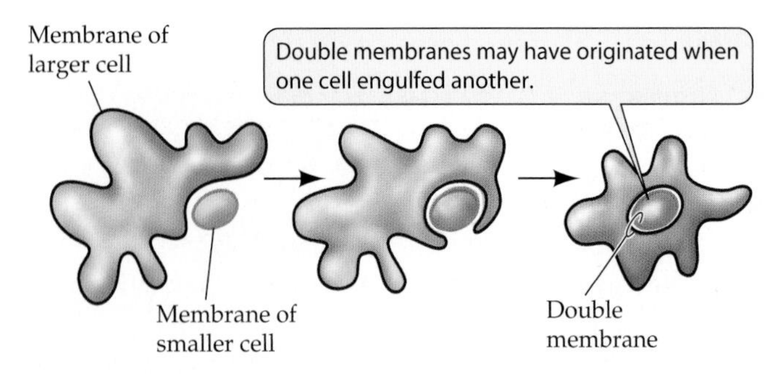

4.18 The Endosymbiosis Theory Chloroplasts and mitochondria may be descended from a small prokaryote that was engulfed by another, larger prokaryote.

About 2 billion years ago, only prokaryotes inhabited Earth. Some of them absorbed their food directly from the environment. Others were photosynthetic. Still others fed on smaller prokaryotes by engulfing them (Figure 4.18).

Suppose that a small, photosynthetic prokaryote was ingested by a larger one, but was not digested. Instead, it somehow survived, trapped within a vesicle in the cytoplasm of the larger cell. The smaller, ingested prokaryote divided at about the same rate as the larger one, so successive generations of the larger cell also contained the offspring of the smaller one. This phenomenon, called **endosymbiosis** (*endo-*, "within"; *symbiosis*, "living together"), exists today, as in the case of the algae that live within sea anemones (see Figure 4.16*c*).

According to this scenario, endosymbiosis provided benefits for both partners: The larger cell obtained the photosynthetic products from the smaller cell, and the smaller cell was protected by the larger one. Over evolutionary time, the smaller cell gradually lost much of its DNA to the nucleus of the larger cell, resulting in the modern chloroplast.

Much circumstantial evidence favors the endosymbiosis theory:

- On an evolutionary time scale of millions of years, there is evidence for DNA moving between organelles in the cell.
- There are many biochemical similarities between chloroplasts and modern photosynthetic bacteria.
- DNA sequencing shows strong similarities between modern chloroplast DNA and that of a photosynthetic prokaryote.
- The double membrane that encloses mitochondria and chloroplasts could have arisen through endosymbiosis. The outer membrane may have come from the engulfing cell's plasma membrane and the inner membrane from the engulfed cell's plasma membrane.

Similar evidence and arguments also support the proposition that mitochondria are the descendants of respiring prokaryotes engulfed by larger prokaryotes. The benefit of this endosymbiotic relationship might have been the capacity of the engulfed prokaryote to detoxify molecular oxygen (O_2), which was increasing in Earth's atmosphere as a result of photosynthesis.

Other Organelles

Eukaryotic cells have several other organelles that are surrounded by a single membrane.

Peroxisomes house specialized chemical reactions

Peroxisomes are organelles that collect the toxic peroxides (such as hydrogen peroxide, H_2O_2) that are the unavoidable by-products of chemical reactions. These peroxides can be safely broken down inside the peroxisomes without mixing with other parts of the cell. Peroxisomes are small organelles, about 0.2 to 1.7 µm in diameter. They have a single membrane and a granular interior containing specialized enzymes (Figure 4.19). Peroxisomes are found at one time or another in at least some of the cells of almost every eukaryotic species.

A structurally similar organelle, the **glyoxysome**, is found only in plants. Glyoxysomes, which are most prominent in young plants, are the sites where stored lipids are converted into carbohydrates for transport to growing cells.

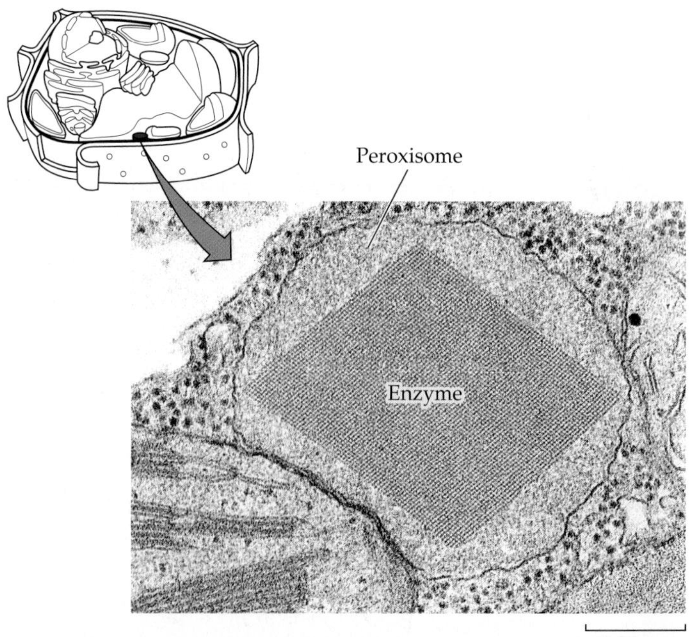

4.19 A Peroxisome A diamond-shaped crystal, composed of an enzyme, almost entirely fills this rounded peroxisome in a leaf cell. The enzyme catalyzes one of the reactions that breaks down toxic peroxides in the peroxisome.

Vacuoles are filled with water and soluble substances

Many eukaryotic cells, but particularly those of plants and protists, contain membrane-enclosed **vacuoles** filled with aqueous solutions containing many dissolved substances (Figure 4.20). Plant vacuoles have several functions:

- *Storage:* Plant cells produce a number of toxic by-products and waste materials, many of which are simply stored within vacuoles. And since they are poisonous or distasteful, these stored materials deter some animals from eating the plants. Thus these stored wastes may contribute to plant survival.
- *Structure:* In many plant cells, enormous vacuoles take up more than 90 percent of the cell volume and grow as the cell grows. The dissolved substances in the vacuole, working together with the vacuolar membrane, provide the *turgor*, or stiffness, of the cell, which in turn provides support for the structure of nonwoody plants. The presence of the dissolved substances causes water to enter the vacuole, making it swell like a balloon. Plant cells have a rigid cell wall, which resists the swelling of the vacuole, providing strength in the process.
- *Reproduction:* Some pigments (especially blue and pink ones) in petals and fruits are contained in vacuoles. These pigments—the *anthocyanins*—are visual cues that help attract the animals that assist in pollination or seed dispersal.
- *Digestion:* In some plants, the vacuoles contain enzymes that hydrolyze seed proteins into monomers that a developing plant embryo can use as food.

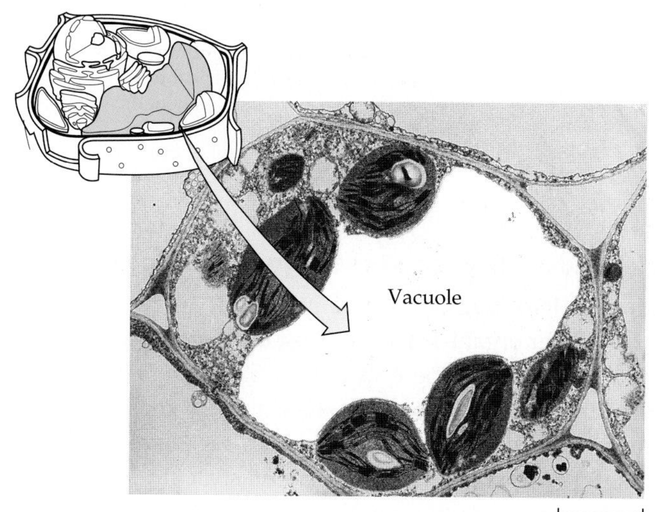

4.20 Vacuoles in Plant Cells Are Usually Large The large central vacuole in this cell is typical of mature plant cells. Smaller vacuoles are visible toward each end of the cell.

Food vacuoles are found in some simple and evolutionarily ancient groups of organisms—single-celled protists and simple multicellular organisms such as sponges, for example. In these organisms, the cells engulf food particles by phagocytosis, generating a food vacuole. Fusion of this vacuole with a lysosome results in digestion, and small molecules leave the vacuole and enter the cytoplasm for use or distribution to other organelles.

Contractile vacuoles are found in many freshwater protists. Their function is to get rid of the excess water that rushes into the cell because of the imbalance in salt concentration between the relatively salty interior of the cell and its freshwater environment. The contractile vacuole enlarges as water enters, then abruptly contracts, forcing the water out of the cell through a special pore structure.

The Cytoskeleton

In addition to its many membrane-enclosed organelles, the eukaryotic cytoplasm contains a set of long, thin fibers called the **cytoskeleton**. The cytoskeleton fills at least three important roles:

- It maintains cell shape and support.
- It provides for various types of cellular movement.
- Some of its fibers act as tracks or supports for motor proteins, which help move things within the cell.

In the discussion that follows, we'll look at three components of the cytoskeleton: microfilaments, intermediate filaments, and microtubules (Figure 4.21).

Microfilaments function in support and movement

Microfilaments can exist as single filaments, in bundles, or in networks. They are about 7 nm in diameter and several micrometers long. They are assembled from **actin**, a protein that exists in several forms and has many functions among members of the animal phyla. The actin found in microfilaments (which are also known as *actin filaments*) is extensively folded and has distinct "head" and "tail" sites. These sites interact with other actin molecules to form long, double helical chains (see Figure 4.21). The polymerization of actin into microfilaments is reversible, and they can disappear from cells, breaking down into units of free actin.

Microfilaments have two major roles:

- They help the entire cell or parts of the cell to move.
- They stabilize cell shape.

In muscle cells, actin fibers are associated with another protein, **myosin**, and the interactions of these two proteins account for the contraction of muscles. In non-muscle cells, actin fibers are associated with localized changes of shape in the cell.

4.21 The Cytoskeleton Three highly visible and important structural components of the cytoskeleton are shown here in detail. These structures maintain and reinforce cell shape and contribute to cell movement.

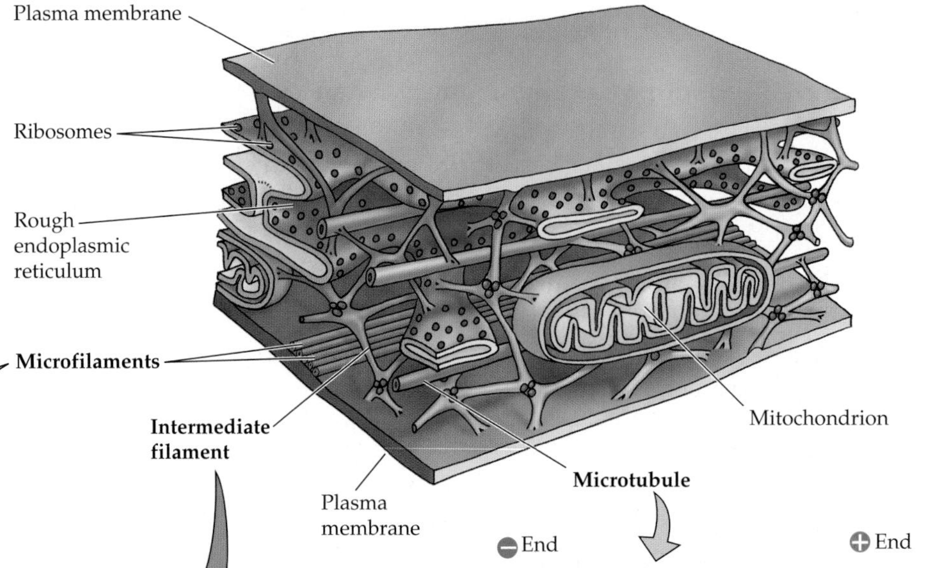

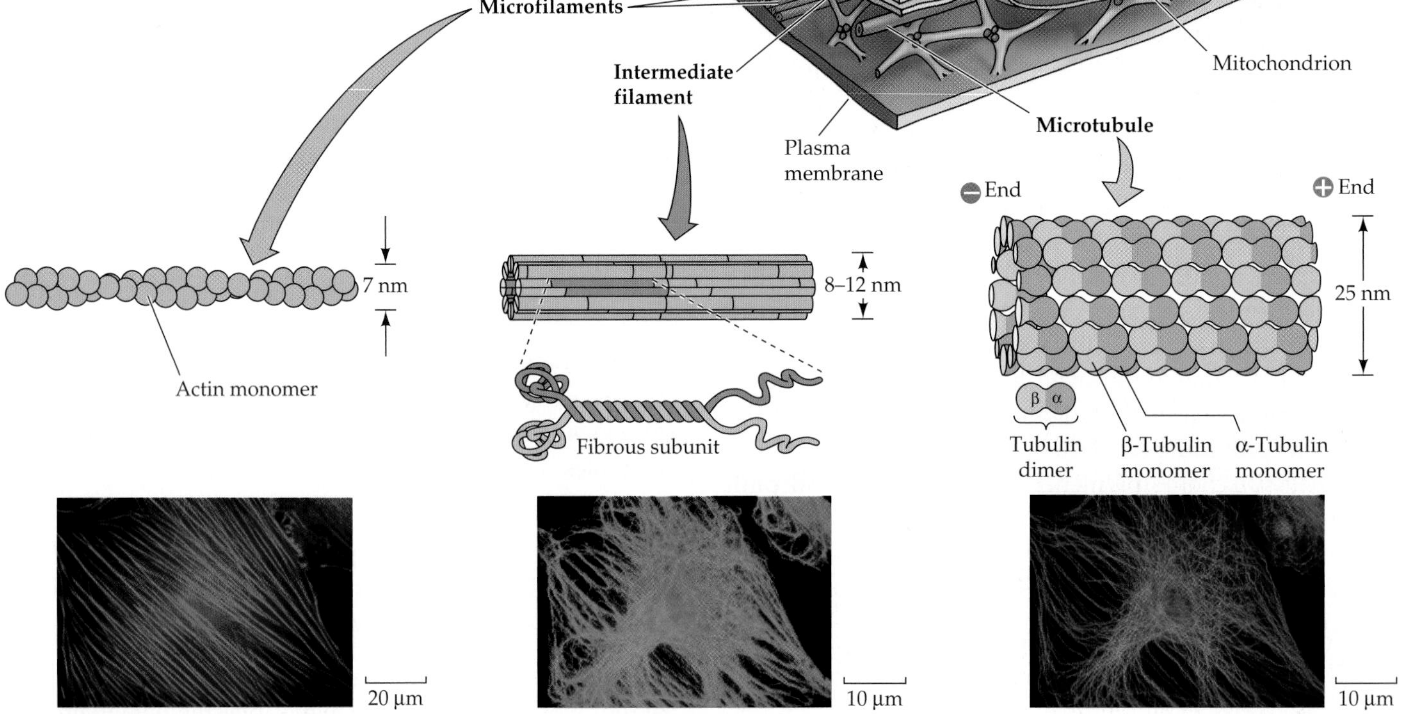

Microfilaments are made up of strands of the protein actin and often interact with strands of other proteins. They change cell shape and drive cellular motion, including contraction, cytoplasmic streaming, and the "pinched" shape changes that occur during cell division. Microfilaments and myosin strands together drive muscle action.

Intermediate filaments are made up of fibrous proteins organized into tough, ropelike assemblages that stabilize a cell's structure and help maintain its shape. Some intermediate filaments help to hold neighboring cells together. Others make up the nuclear lamina.

Microtubules are long, hollow cylinders made up of many molecules of the protein tubulin. Tubulin consists of two subunits, α-tubulin and β-tubulin. Microtubules lengthen or shorten by adding or subtracting tubulin dimers. Microtubule shortening moves chromosomes. Interactions between microtubules drive the movement of cells. Microtubules serve as "tracks" for the movement of vesicles.

For example, microfilaments are involved in a flowing movement of the cytoplasm called *cytoplasmic streaming* and in the "pinching" contractions that divide an animal cell into two daughter cells. Microfilaments are also involved in the formation of cellular extensions called *pseudopodia* (*pseudo-*, "false;" *podia*, "feet") that enable some cells to move.

In some cell types, microfilaments form a meshwork just inside the plasma membrane. Actin-binding proteins then cross-link the microtubules to form a rigid structure that supports the cell. For example, microfilaments support the tiny microvilli that line the intestine, giving it a larger surface area through which to absorb nutrients (Figure 4.22).

Intermediate filaments are tough supporting elements

Intermediate filaments (see Figure 4.21) are found only in multicellular organisms. In contrast to the other components of the cytoskeleton, there are at least 50 different kinds of intermediate filaments, often specific to a few cell types. They generally fall into six molecular classes, based on amino acid sequence, and share the same general structure, being composed of fibrous proteins of the keratin family, similar to the protein that makes up hair and fingernails. In cells, these proteins are organized into tough, ropelike assemblages 8 to 12 nm in diameter.

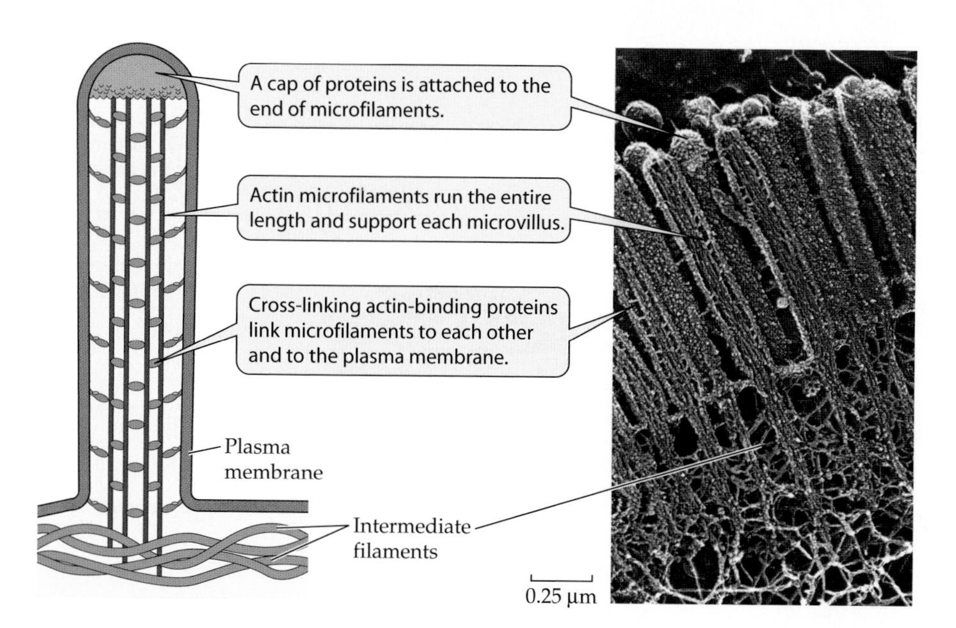
A cap of proteins is attached to the end of microfilaments.
Actin microfilaments run the entire length and support each microvillus.
Cross-linking actin-binding proteins link microfilaments to each other and to the plasma membrane.
Plasma membrane
Intermediate filaments
0.25 µm

4.22 Microfilaments for Support Microfilaments form the backbone of the microvilli that increase the surface area of some cells, such as intestinal cells that absorb nutrients.

Intermediate filaments have two major structural functions:

- They stabilize cell structure.
- They resist tension.

In some cells, intermediate filaments radiate from the nuclear envelope and may maintain the positions of the nucleus and other organelles in the cell. The lamins of the nuclear lamina are intermediate filaments. Other kinds of intermediate filaments help hold a complex apparatus of microfilaments in place in muscle cells. Still other kinds stabilize and help maintain rigidity in surface tissues by connecting "spot welds" called *desmosomes* between adjacent cells (see Figure 5.6).

Microtubules are long and hollow

Microtubules are long, hollow, unbranched cylinders about 25 nm in diameter and up to several micrometers long. Microtubules have two roles in the cell:

- They form a rigid internal skeleton for some cells.
- They act as a framework along which motor proteins can move structures in the cell.

Microtubules are assembled from molecules of the protein **tubulin**. Tubulin is a *dimer*—a molecule made up of two monomers. The polypeptide monomers that make up this protein are known as *α-tubulin* and *β-tubulin*. Thirteen chains of tubulin dimers surround the central cavity of the microtubule (see Figure 4.21). The two ends of a microtubule are different. One end is designated the plus (+) end, the other the minus (–) end.

Tubulin dimers can be added or subtracted, mainly at the plus end, lengthening or shortening the microtubule. This capacity to change length rapidly makes microtubules *dynamic structures*. This dynamic property of microtubules is seen in animal cells, where they are often found in parts of the cell that are changing shape.

Many microtubules radiate from a region of the cell called the *microtubule organizing center*. Tubule polymerization results in rigidity, and tubule depolymerization leads to a collapse of this rigid structure.

In plants, microtubules help control the arrangement of the cellulose fibers of the cell wall. Electron micrographs of plants frequently show microtubules lying just inside the plasma membrane of cells that are forming or extending their cell walls. Experimental alteration of the orientation of these microtubules leads to a similar change in the cell wall and a new shape for the cell.

In many cells, microtubules serve as tracks for **motor proteins**, specialized molecules that use energy to change their shape and move. Motor proteins bond to and move along the microtubules, carrying materials from one part of the cell to another. Microtubules are also essential in distributing chromosomes to daughter cells during cell division. And they are intimately associated with movable cell appendages: the flagella and cilia.

Microtubules power cilia and flagella

Many eukaryotic cells possess flagella* or cilia, or both. These whiplike organelles may push or pull the cell through its aqueous environment, or they may move surrounding liquid over the surface of the cell (Figure 4.23*a*). Cilia and eukaryotic (but not prokaryotic) flagella are both assembled from specialized microtubules and have identical internal structures, but they differ in their relative lengths and their patterns of beating:

- **Flagella** are longer than cilia and are usually found singly or in pairs. Waves of bending propagate from one end of a flagellum to the other in snakelike undulation.
- **Cilia** are shorter than flagella and are usually present in great numbers. They beat stiffly in one direction and recover flexibly in the other direction (like a swimmer's arm), so that the recovery stroke does not undo the work of the power stroke.

*Some prokaryotes have flagella, as we saw earlier, but prokaryotic flagella lack microtubules and dynein. The flagella of prokaryotes are neither structurally nor evolutionarily related to those of eukaryotes. The prokaryotic flagellum is assembled from a protein called flagellin, and it has a much simpler structure and a smaller diameter than a single eukaryotic microtubule. And whereas eukaryotic flagella beat in a wavelike motion, prokaryotic flagella rotate (see Figure 4.6).

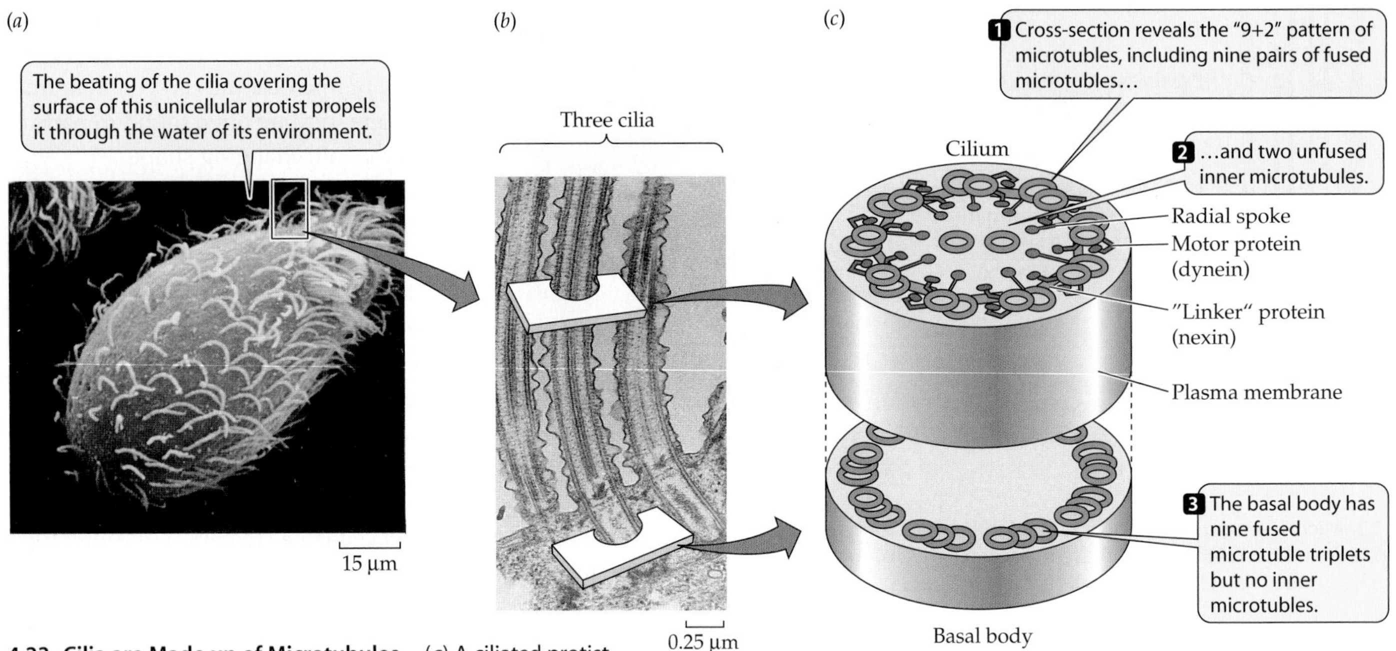
(a)
The beating of the cilia covering the surface of this unicellular protist propels it through the water of its environment.
15 μm
(b)
Three cilia
0.25 μm
(c)
1 Cross-section reveals the "9+2" pattern of microtubules, including nine pairs of fused microtubules...
Cilium
2 ...and two unfused inner microtubules.
Radial spoke
Motor protein (dynein)
"Linker" protein (nexin)
Plasma membrane
3 The basal body has nine fused microtubule triplets but no inner microtubules.
Basal body

4.23 Cilia are Made up of Microtubules (*a*) A ciliated protist. (*b*) Three cilia on a protist cell. (*c*) Cross section of a single cilium.

In cross section, a typical cilium or eukaryotic flagellum is surrounded by the plasma membrane and contains a "9 + 2" array of microtubules. As Figure 4.23*c* shows, nine fused pairs of microtubules—called *doublets*—form an outer cylinder, and one pair of unfused microtubules runs up the center. A spoke radiates from one microtubule of each doublet and connects the doublet to the center of the structure.

In the cytoplasm at the base of every eukaryotic flagellum or cilium is an organelle called a **basal body**. The nine microtubule doublets extend into the basal body. In the basal body, each doublet is accompanied by another microtubule, making nine sets of three microtubules. The central, unfused microtubules do not extend into the basal body.

Centrioles are almost identical to the basal bodies of cilia and flagella. Centrioles are found in all eukaryotes except the flowering plants, pine trees and their relatives, and some protists. Under the light microscope, a centriole looks like a small, featureless particle, but the electron microscope reveals that it is made up of a precise bundle of microtubules arranged as nine sets of three fused microtubules each. Centrioles lie in the microtubule organizing center in cells that are about to divide. As you will see in Chapter 9, they are involved in the formation of the mitotic spindle, to which the chromosomes attach (see Figure 9.8).

Motor proteins move along microtubules

The nine microtubule doublets of cilia and flagella are linked by proteins. The motion of cilia and flagella results from the sliding of the microtubules past each other. This sliding is driven by a motor protein called **dynein**, which can undergo changes in its shape.

All motor proteins work by undergoing reversible shape changes powered by energy from ATP. Dynein molecules attached to one microtubule doublet bind to a neighboring doublet. As the dynein molecules change shape, they move the microtubule past its neighbor (Figure 4.24*a*).

Dynein and another motor protein, **kinesin**, are responsible for carrying protein-laden vesicles from one part of the cell to another. These motor proteins bind to a vesicle or other organelle, then "walk" it along a microtubule by changing their shape. Recall that microtubules have a plus end and a minus end. Dynein moves attached organelles toward the minus end, while kinesin moves them toward the plus end (Figure 4.24*b*).

Extracellular Structures

Although the plasma membrane is the functional barrier between the inside and the outside of a cell, many structures are produced by cells and secreted to the outside of the plasma membrane, where they play essential roles in protecting, supporting, or attaching cells. Because they are outside the plasma membrane, these structures are said to be *extracellular*. The peptidoglycan cell wall of bacteria is such an extracellular structure. In eukaryotes, other extracellular structures—the cell walls of plants and the extracellular matrices found between the cells of multicellular animals—play similar roles. Both of these structures are made up of a prominent fibrous macromolecule embedded in a jellylike medium.

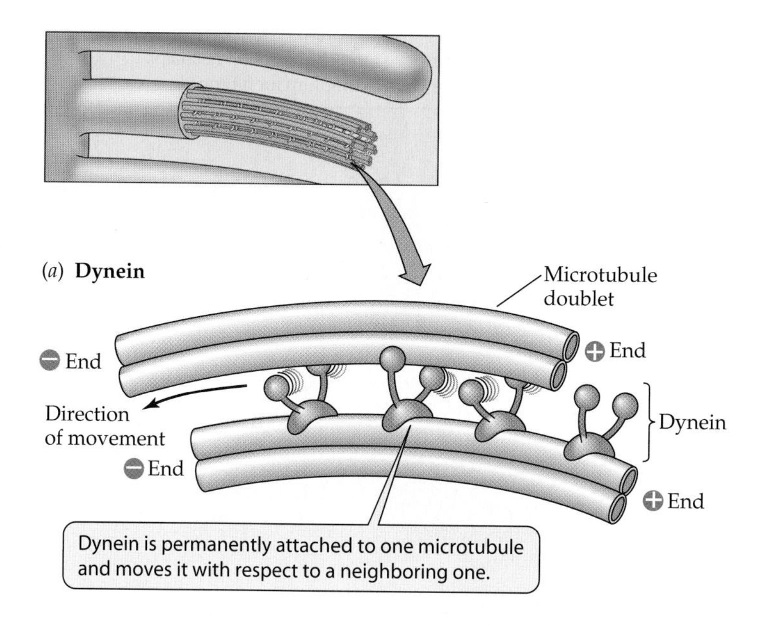

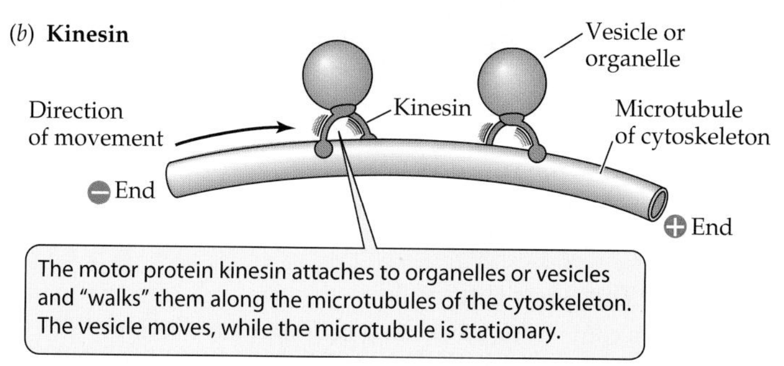

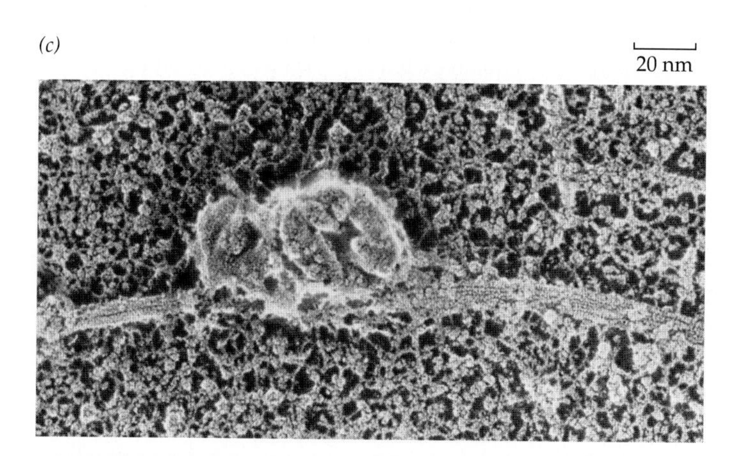

4.24 Motor Proteins Use Energy from ATP to Move Things (*a*) Dynein is responsible for flagellar movement. (*b*) Kinesin, like dynein, delivers vesicles to various parts of the cell. (*c*) This SEM shows a vesicle attached to a microtubule by a motor protein.

The plant cell wall consists largely of cellulose

The cell wall of plant cells is a semirigid structure outside the plasma membrane (Figure 4.25). It consists of cellulose fibers embedded in other complex polysaccharides and proteins. The cell wall has three major roles in plants:

- It provides support for the cell and limits its volume by remaining rigid.
- It acts as a barrier to infections by fungi and other organisms that can cause plant diseases.
- It contributes to plant form by growing as plant cells expand.

Because of their thick cell walls, plant cells viewed under a light microscope appear to be entirely isolated from each other. But electron microscopy reveals that this is not the case. The cytoplasm of adjacent plant cells is connected by numerous plasma membrane-lined channels, called *plasmodesmata*, that are about 20 to 40 nm in diameter and extend through the walls of adjoining cells (see Figure 15.19). Plasmodesmata permit the diffusion of water, ions, small molecules, and RNA and proteins between connected cells. Such diffusion ensures that the cells of a plant have uniform concentrations of these substances.

Animal cells have elaborate extracellular matrices

The cells of multicellular animals lack the semirigid cell wall that is characteristic of plant cells, but many animal cells are surrounded by, or are in contact with, an **extracellular matrix**. This matrix is composed of fibrous proteins such as *collagen* (the most abundant protein in mammals) and glyco-

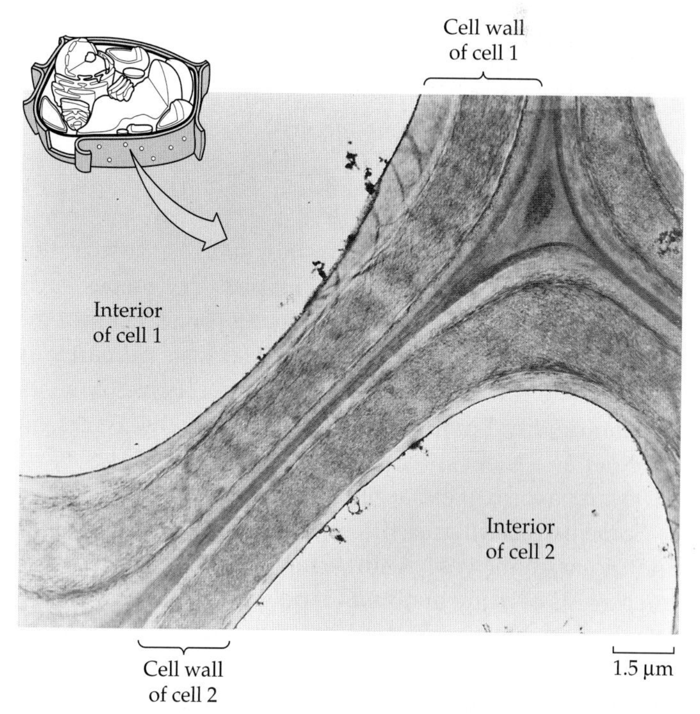

4.25 The Plant Cell Wall The semirigid cell wall provides support for plant cells.

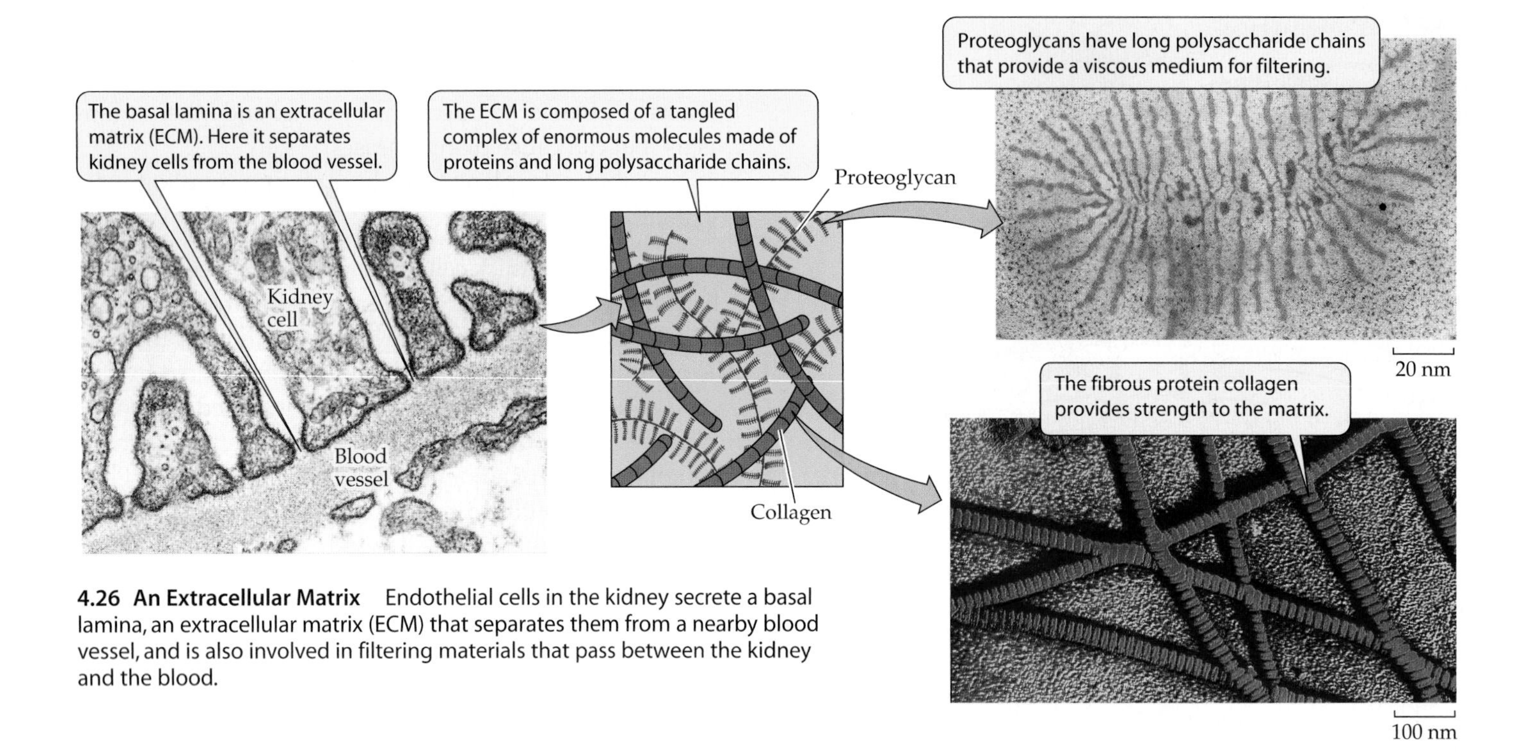

The basal lamina is an extracellular matrix (ECM). Here it separates kidney cells from the blood vessel.
The ECM is composed of a tangled complex of enormous molecules made of proteins and long polysaccharide chains.
Proteoglycans have long polysaccharide chains that provide a viscous medium for filtering.
Proteoglycan
Kidney cell
Blood vessel
Collagen
20 nm
The fibrous protein collagen provides strength to the matrix.
100 nm

4.26 An Extracellular Matrix Endothelial cells in the kidney secrete a basal lamina, an extracellular matrix (ECM) that separates them from a nearby blood vessel, and is also involved in filtering materials that pass between the kidney and the blood.

proteins (Figure 4.26). These proteins, along with other substances that are specific to certain body tissues, are secreted by cells that are present in or near the matrix. The functions of the extracellular matrix are many:

- It holds cells together in tissues.
- It contributes to the physical properties of cartilage, skin, and other tissues.
- It helps filter materials passing between different tissues.
- It helps orient cell movements during embryonic development and during tissue repair.
- It plays a role in chemical signaling from one cell to another.

In the human body, some tissues, such as those in the brain, have very little extracellular matrix; other tissues, such as bone and cartilage, have large amounts. Bone cells are embedded in an extracellular matrix that consists primarily of collagen and calcium phosphate. This matrix gives bone its familiar rigidity. Epithelial cells, which line body cavities, lie together as a sheet spread over a *basal lamina*, or *basement membrane*, a form of extracellular matrix (see Figure 4.26).

Some extracellular matrices are made up, in part, of an enormous *proteoglycan*. A single molecule of this proteoglycan consists of many hundreds of polysaccharides covalently attached to about a hundred proteins, all of which are attached to one enormous polysaccharide. The molecular weight of this proteoglycan can exceed 100 million; the molecule takes up as much space as an entire prokaryotic cell.

Chapter Summary

The Cell: The Basic Unit of Life

- All cells come from preexisting cells and have certain processes, types of molecules, and structures in common.
- The first cells may have arisen from aggregates of macromolecules in bubbles. **Review Figure 4.1**
- To maintain adequate exchanges with its environment, a cell's surface area must be large compared with its volume. **Review Figures 4.2, 4.3. See Web/CD Activity 4.1**
- Cells can be visualized by various methods using microscopes. **Review Figure 4.4. See Web/CD Activity 4.2**
- All cells are surrounded by a plasma membrane.

Prokaryotic Cells

- All prokaryotic cells have a plasma membrane, a nucleoid region with DNA, and a cytoplasm that contains ribosomes, water, and dissolved proteins and small molecules. **Review Figure 4.5**
- Some prokaryotes have additional protective structures: cell wall, outer membrane, and capsule. Some prokaryotes contain photosynthetic membranes or mesosomes, and some have flagella or pili. **Review Figure 4.6**

Eukaryotic Cells

- Like prokaryotic cells, eukaryotic cells have a plasma membrane, cytoplasm, and ribosomes. However, eukaryotic cells are larger and contain many membrane-enclosed organelles. **Review Figure 4.7. See Web/CD Tutorial 4.1**
- The membranes that envelop organelles in the eukaryotic cell are partial barriers, ensuring that the chemical composition of the interior of the organelle differs from that of the surrounding cytoplasm.
- Organelles can be isolated by cell fractionation. **Review Figure 4.8**

Organelles that Process Information

▶ The nucleus is usually the largest organelle in a cell. It is surrounded by a double membrane, the nuclear envelope, which disassembles during cell division. Within the nucleus, the nucleolus is the source of the ribosomes found in the cytoplasm. Nuclear pores have a complex structure. **Review Figure 4.9**

▶ The nucleus contains most of the cell's DNA, which associates with protein to form chromatin. Chromatin is diffuse throughout the nucleus until just before cell division, when it condenses to form chromosomes. **Review Figure 4.10**

The Endomembrane System

▶ The endomembrane system is made up of a series of interrelated compartments enclosed by membranes.

▶ The rough endoplasmic reticulum has attached ribosomes that synthesize proteins. The smooth endoplasmic reticulum lacks ribosomes and is associated with the synthesis of lipids. **Review Figures 4.7, 4.11**

▶ The Golgi apparatus receives materials from the rough ER by means of vesicles that fuse with its *cis* region. Vesicles originating from the *trans* region of the Golgi contain proteins targeted to different cellular locations. Some of these vesicles fuse with the plasma membrane and release their contents outside the cell. **Review Figures 4.7, 4.12. See Web/CD Tutorial 4.2**

▶ Lysosomes contain many digestive enzymes. Lysosomes fuse with the phagosomes produced by phagocytosis to form secondary lysosomes, in which engulfed materials are digested. Undigested materials are secreted from the cell when the secondary lysosome fuses with the plasma membrane. **Review Figure 4.13. See Web/CD Activity 4.3**

Organelles that Process Energy

▶ Mitochondria are enclosed by an outer membrane and an inner membrane that folds inward to form cristae. Mitochondria contain the proteins needed for cellular respiration. **Review Figure 4.14**

▶ The cells of photosynthetic eukaryotes contain chloroplasts. These organelles are enclosed by double membranes and contain an internal system of thylakoids organized as grana. **Review Figures 4.7, 4.15**

▶ Thylakoids within chloroplasts contain the chlorophyll and proteins that harvest light energy for photosynthesis.

▶ Both mitochondria and chloroplasts contain their own DNA and ribosomes and are capable of making some of their own proteins.

▶ The endosymbiosis theory of the evolutionary origin of mitochondria and chloroplasts states that these organelles originated when larger prokaryotes engulfed, but did not digest, smaller prokaryotes. Mutual benefits permitted this symbiotic relationship to be maintained, allowing the smaller cells to evolve into the eukaryotic organelles observed today. **Review Figure 4.18**

Other Organelles

▶ Peroxisomes and glyoxysomes contain special enzymes and carry out specialized chemical reactions inside the cell. **Review Figure 4.19**

▶ Vacuoles are prominent in many plant cells and consist of a membrane-enclosed compartment full of water and dissolved substances. By taking in water, vacuoles enlarge and provide the pressure needed to stretch the cell wall and provide structural support for the plant. **Review Figure 4.20**

The Cytoskeleton

▶ The cytoskeleton within the cytoplasm of eukaryotic cells provides shape, strength, and movement. It consists of three interacting types of protein fibers. **Review Figure 4.21**

▶ Microfilaments consist of two chains of actin units that together form a double helix. Microfilaments strengthen cellular structures and provide the movement in animal cell division, cytoplasmic streaming, and pseudopod extension. Microfilaments may be found as individual fibers, bundles of fibers, or networks of fibers joined by linking proteins. **Review Figures 4.21, 4.22**

▶ Intermediate filaments are formed of keratins and are organized into tough, ropelike structures that hold organelles in place within the celland add strength to cell attachments in multicellular organisms. **Review Figure 4.21**

▶ Microtubules are composed of dimers of the protein tubulin. They can lengthen and shorten by adding and losing tubulin dimers. They are involved in the structure and function of cilia and flagella, both of which have a characteristic "9 + 2" pattern of microtubules. **Review Figures 4.21, 4.23**

▶ The movements of cilia and flagella result from the binding of the motor protein dynein to the microtubules. Dynein and another motor protein, kinesin, also bind to microtubules to move organelles through the cell. **Review Figure 4.24**

▶ Centrioles, made up of triplets of microtubules, are involved in the distribution of chromosomes during cell division.

Extracellular Structures

▶ Materials external to the plasma membrane provide protection, support, and attachment for cells in multicellular systems.

▶ The cell walls of plants consist principally of cellulose. They are pierced by plasmodesmata that join the cytoplasm of adjacent cells. **Review Figure 4.25**

▶ In multicellular animals, the extracellular matrix consists of different kinds of proteins, including proteoglycans. In bone and cartilage, the protein collagen predominates. **Review Figure 4.26**

Self-Quiz

1. Which is present in both prokaryotic cells and in eukaryotic plant cells?
   a. Chloroplasts
   b. Cell walls
   c. Nucleus
   d. Mitochondria
   e. Microtubules
2. The major factor limiting cell size is the
   a. concentration of water in the cytoplasm.
   b. need for energy.
   c. presence of membranous organelles.
   d. ratio of surface area to volume.
   e. composition of the plasma membrane.
3. Which statement about mitochondria is *not* true?
   a. Their inner membrane folds to form cristae.
   b. They are usually 1 μm or smaller in diameter.
   c. They are green because they contain chlorophyll.
   d. Energy-rich substances from the cytosol are oxidized in them.
   e. Much ATP is synthesized in them.
4. Which statement about plastids is *true*?
   a. They are found in prokaryotes.
   b. They are surrounded by a single membrane.
   c. They are the sites of cellular respiration.
   d. They are found in fungi.
   e. They are of several types, with different functions.

5. If all the lysosomes within a cell suddenly ruptured, what would be the most likely result?
   a. The macromolecules in the cytosol would begin to break down.
   b. More proteins would be made.
   c. The DNA within mitochondria would break down.
   d. The mitochondria and chloroplasts would divide.
   e. There would be no change in cell function.
6. The Golgi apparatus
   a. is found only in animals.
   b. is found in prokaryotes.
   c. is the appendage that moves a cell around in its environment.
   d. is a site of rapid ATP production.
   e. packages and modifies proteins.
7. Which organelle is *not* surrounded by one or more membranes?
   a. Ribosome
   b. Chloroplast
   c. Mitochondrion
   d. Peroxisome
   e. Vacuole
8. The cytoskeleton consists of
   a. cilia, flagella, and microfilaments.
   b. cilia, microtubules, and microfilaments.
   c. internal cell walls.
   d. microtubules, intermediate filaments, and microfilaments.
   e. calcified microtubules.
9. Microfilaments
   a. are composed of polysaccharides.
   b. are composed of actin.
   c. provide the motive force for cilia and flagella.
   d. make up the spindle that aids the movement of chromosomes.
   e. maintain the position of the chloroplast in the cell.
10. Which statement about the plant cell wall is *not* true?
   a. Its principal chemical components are polysaccharides.
   b. It lies outside the plasma membrane.
   c. It provides support for the cell.
   d. It completely isolates adjacent cells from one another.
   e. It is semirigid.

For Discussion

1. Which organelles and other structures are found in both plant and animal cells? Which are found in plant but not animal cells? In animal but not plant cells? Discuss these differences in relation to the activities of plants and animals.
2. Through how many membranes would a molecule have to pass in going from the interior of a chloroplast to the interior of a mitochondrion? From the interior of a lysosome to the outside of a cell? From one ribosome to another?
3. How does the possession of double membranes by chloroplasts and mitochondria relate to the endosymbiosis theory of the origins of these organelles? What other evidence supports the theory?
4. Compare the extracellular matrix of the animal cell with the plant cell wall with respect to composition of the fibrous and nonfibrous components, rigidity, and presence of cytoplasmic "bridges."
5. Plastids and mitochondria may have arisen via endosymbiosis. Propose a hypothesis for the origin of the cell nucleus.

5 Cellular Membranes

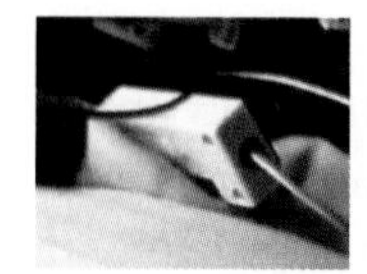

During his years as an undergraduate at Oxford, physics student Stephen Hawking took up rowing. Although he had never been much of an athlete, he was doing a passable job. But he noticed he was getting increasingly clumsy, and by the time he went to Cambridge as a graduate student, he was falling over for no particular reason. After weeks of tests, his physicians told him that he had motor neuron disease, an incurable condition in which the nerve cells that stimulate muscles gradually die and the patient loses all muscular control.

As the years progressed, Hawking made major contributions to theoretical physics, especially to the study of black holes and the origin of the universe. He ascended the ladder of academic success and now holds a professorship at Cambridge once held by Isaac Newton. But his disease has gotten worse, and he has lost almost all muscular control.

A hallmark of living cells is their ability to regulate the substances that enter and leave them. This regulation is a function of the plasma membrane, which is composed of a hydrophobic lipid bilayer with associated proteins. Muscle cells respond to stimulation by nerve cells by opening protein-lined channels in their plasma membranes. Because his nerves cannot stimulate them, the channels of Hawking's muscle cells do not open, and his muscles do not contract. Channels in plasma membranes underlie the biological activities of many organisms, ranging from the beating of an animal's heart to the opening of tiny pores in leaves to let outside air in.

Membranes are dynamic structures whose components move and change. They perform their vital physiological roles by allowing cells to interact with other cells and with molecules in the environment. We describe the structural aspects of those interactions here. Membranes also regulate the traffic of chemicals into and out of the cell. The selective permeability of membranes, which we describe in this chapter, is an important characteristic of life. Later in this book, we will see it in action in such diverse situations as the transduction of light energy into chemical energy in the chloroplast and the retention of water and ions in the mammalian kidney.

Stephen Hawking The effects of motor neuron disease have confined the famous physicist to a wheelchair. A major cellular manifestation of this disease is the lack of ability of the nerve cells to stimulate the opening of channels through muscle cell membranes that would result in normal muscle function.

Membrane Composition and Structure

The physical organization and functioning of all biological membranes depend on their constituents: lipids, proteins, and carbohydrates. The lipids establish the physical integrity of the membrane and create an effective barrier to the rapid passage of hydrophilic materials such as water and ions. In addition, the phospholipid bilayer serves as a lipid "lake" in which a variety of proteins "float" (Figure 5.1). This general design is known as the **fluid mosaic model**.

Proteins embedded in the phospholipid bilayer have a number of functions, including moving materials through the membrane and receiving chemical signals from the cell's external environment. Each membrane has a set of proteins suitable to the specialized function of the cell or organelle it surrounds.

The carbohydrates associated with membranes are attached either to the lipids or to protein molecules. They are located on the outside of the plasma membrane, where they protrude into the environment, away from the cell. Like some of the proteins, carbohydrates are crucial in recognizing specific molecules.

5.1 The Fluid Mosaic Model The general molecular structure of biological membranes is a continuous phospholipid bilayer in which proteins are embedded.

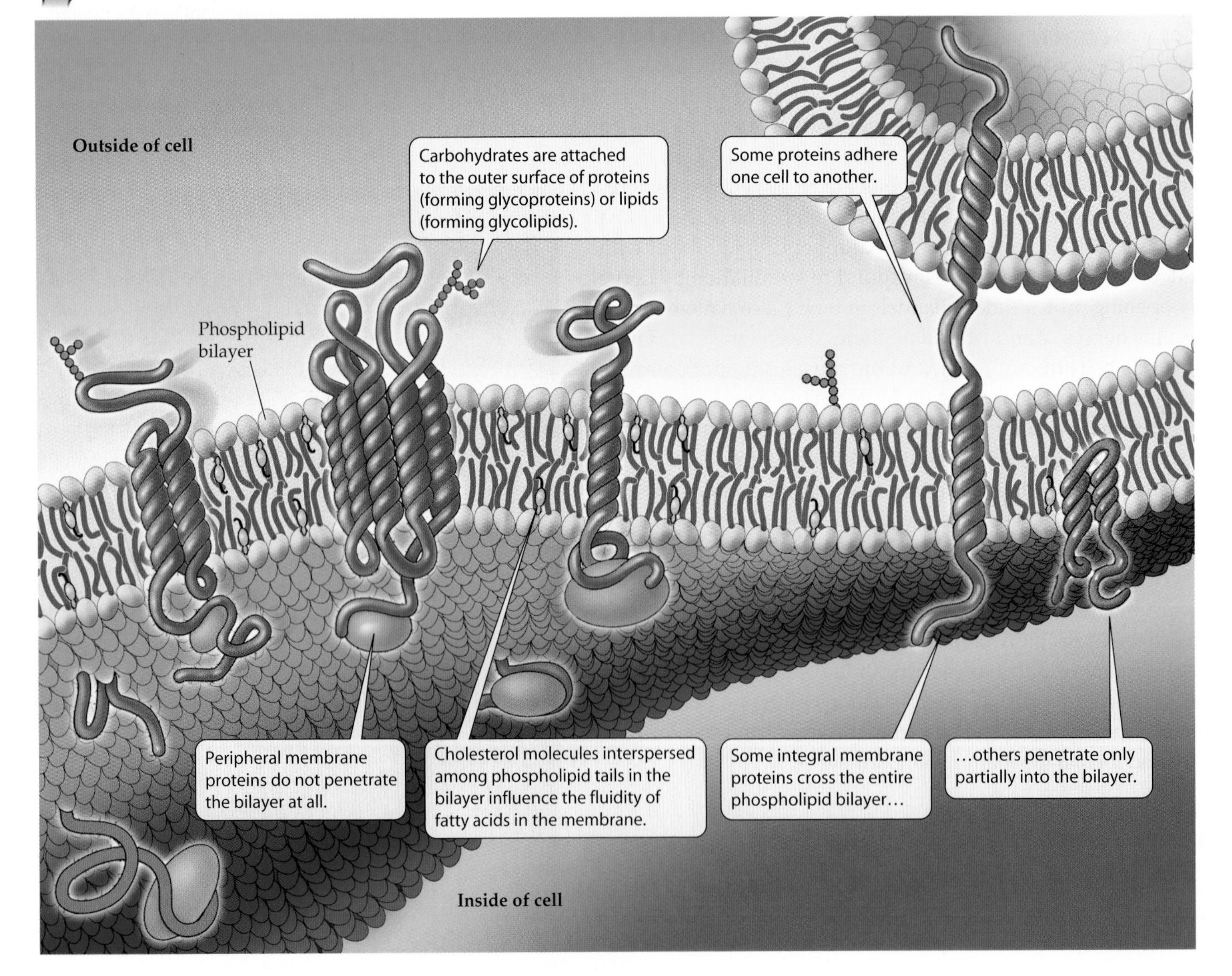
Outside of cell
Carbohydrates are attached to the outer surface of proteins (forming glycoproteins) or lipids (forming glycolipids).
Some proteins adhere one cell to another.
Phospholipid bilayer
Peripheral membrane proteins do not penetrate the bilayer at all.
Cholesterol molecules interspersed among phospholipid tails in the bilayer influence the fluidity of fatty acids in the membrane.
Some integral membrane proteins cross the entire phospholipid bilayer...
...others penetrate only partially into the bilayer.
Inside of cell

Lipids constitute the bulk of a membrane

Most of the lipids in biological membranes are phospholipids. Recall from Chapter 2 that some compounds are hydrophilic ("water-loving") and others are hydrophobic ("water-hating") and from Chapter 3 that phospholipids have both hydrophilic regions and hydrophobic regions:

- *Hydrophilic regions:* The phosphorus-containing "head" of the phospholipid is electrically charged and hence associates with polar water molecules.
- *Hydrophobic regions:* The long, nonpolar fatty acid "tails" of the phospholipid associate with other nonpolar materials, but they do not dissolve in water or associate with hydrophilic substances.

As a consequence of these properties, one way in which phospholipids can coexist with water is to form a *bilayer*, with the fatty acids of the two layers interacting with each other and the polar regions facing the outside aqueous environment (Figure 5.2).

In the laboratory, it is easy to make artificial bilayers with the same organization as natural membranes. In addition, small holes in a bilayer seal themselves spontaneously. This capacity of lipids to associate and maintain a bilayer organization helps biological membranes fuse during vesicle formation, phagocytosis, and related processes.

All biological membranes have a similar structure, but membranes from different cells or organelles may differ greatly in their lipid composition. Phospholipids differ in terms of fatty acid chain length, degree of unsaturation (double bonds) in the fatty acids, and the polar (phosphate-containing) groups present. In addition to phospholipids, membranes may contain cholesterol, a different type of lipid. In some membranes, 25 percent of the lipid is cholesterol (see Chapter 3), but other membranes have no cholesterol at all. When present, cholesterol is important to membrane integrity; most cholesterol in membranes is not hazardous to your health. A molecule of cholesterol is commonly situated next to an unsaturated fatty acid (see Figure 5.1).

The phospholipid bilayer stabilizes the entire membrane structure, but leaves it flexible, not rigid. At the same time, the fatty acids of the phospholipids make the hydrophobic interior of the membrane somewhat *fluid*—about as fluid as lightweight machine oil. This fluidity permits some molecules to move laterally within the plane of the membrane. A given phospholipid molecule in the plasma membrane may travel from one end of the cell to the other in a little more than a second. On the other hand, seldom does a phospholipid molecule in one half of the bilayer flip over to the other side and trade places with another phospholipid molecule. For such a swap to happen, the polar part of each molecule would have to move through the hydrophobic interior of the membrane. Since phospholipid flip-flops are rare, the inner and outer halves of the bilayer may be quite different in the kinds of phospholipids they contain.

The amount of cholesterol present in membranes, along with the degree of saturation of the fatty acids present, can increase or decrease membrane fluidity. Shorter fatty acid chains make for a more fluid membrane, as do unsaturated fatty acids. Adequate membrane fluidity is essential for many membrane functions. Since molecules move more slowly and fluidity decreases at reduced temperatures, membrane functions may decline in organisms that cannot keep their bodies warm. To address this problem, some organisms simply change the lipid composition of their membranes under cold conditions, replacing saturated with unsaturated fatty acids and using fatty acids with shorter tails. Such changes play a part in the survival of plants and hibernating animals and bacteria during the winter.

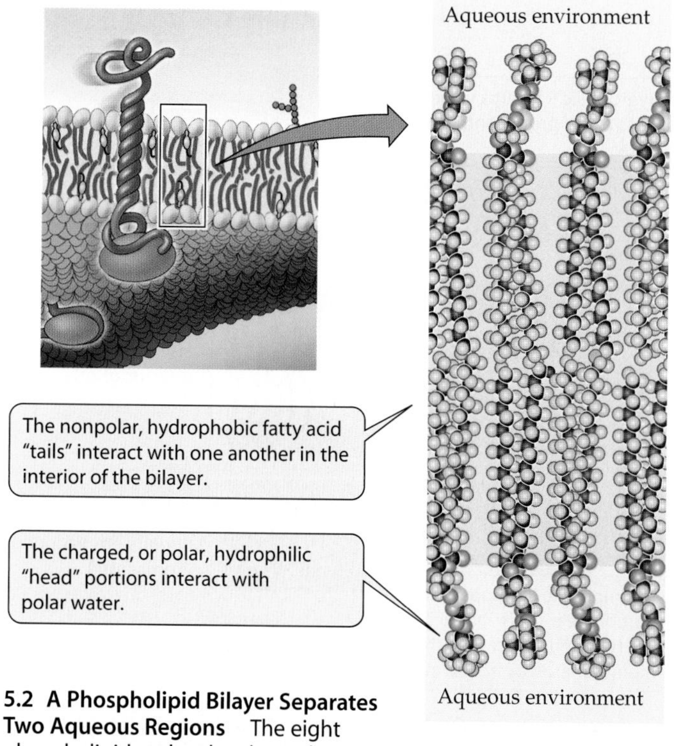

Aqueous environment

The nonpolar, hydrophobic fatty acid "tails" interact with one another in the interior of the bilayer.

The charged, or polar, hydrophilic "head" portions interact with polar water.

Aqueous environment

5.2 A Phospholipid Bilayer Separates Two Aqueous Regions The eight phospholipid molecules shown here represent a small cross section of a membrane bilayer.

Membrane proteins are asymmetrically distributed

All biological membranes contain proteins. Typically, plasma membranes have one protein molecule for every 25 phospholipid molecules. This ratio varies, however, depending on membrane function. In the inner membrane of the mitochondrion, which is specialized for energy processing, there is one protein for every 15 lipids. On the other hand, myelin, a membrane that encloses some nerve cells and uses the properties of lipids to act as an electrical insulator, has only one protein per 70 lipids.

Many membrane proteins are embedded in, or extend across, the lipid bilayer. Like phospholipids, these proteins have both hydrophilic and hydrophobic regions:

- *Hydrophilic regions:* Stretches of amino acids with hydrophilic R groups (side chains; see Table 3.2) give certain regions of the protein a polar character. Those regions, or *domains*, interact with water, sticking out into the aqueous extracellular environment or cytoplasm.
- *Hydrophobic regions:* Stretches of amino acids with hydrophobic R groups give other regions of the protein a nonpolar character. Those domains interact with the fatty acid chains in the interior of the lipid bilayer, away from water.

A special preparation method for electron microscopy, *freeze-fracturing*, reveals proteins embedded in the lipid bilayer of cellular membranes (Figure 5.3). The bumps that can be seen protruding from the interior of these membranes are not observed in pure lipid bilayers.

According to the fluid mosaic model, the proteins and lipids in a membrane are independent of each other and interact only noncovalently. The polar ends of proteins can interact with the polar ends of lipids, and the nonpolar regions of both molecules interact hydrophobically (see Figure 5.1).

There are two general types of membrane proteins:

- **Integral membrane proteins** have hydrophobic regions and penetrate the phospholipid bilayer. Many of these proteins have long hydrophobic α-helical regions that span the hydrophobic core of the bilayer. Their hydrophilic ends protrude into the aqueous environments on either side of the membrane (Figure 5.4).
- **Peripheral membrane proteins** lack hydrophobic regions and are not embedded in the bilayer. Instead, they have polar or charged regions that interact with similar regions on exposed parts of integral membrane proteins or phospholipid molecules (see Figure 5.1).

RESEARCH METHOD
1 Frozen tissue is fractured with a knife.
2 Fracturing causes one half of the membrane to separate from the other.
3 Proteins sticking out of the fractured membrane must have been embedded in the lipid bilayer.

5.3 Membrane Proteins Revealed by the Freeze-Fracture Technique This membrane from a spinach chloroplast was first frozen and then separated so that the membrane bilayer was split open.

Some membrane proteins are covalently attached to fatty acids or other lipid groups. These proteins can be classified as a special type of integral protein, as their hydrophobic lipid component allows them to insert themselves into the lipid bilayer.

Proteins are asymmetrically distributed on the inner and outer surfaces of a membrane. Integral membrane proteins that protrude on both sides of the membrane, known as **transmembrane proteins**, show different "faces" on the two membrane surfaces. Such proteins have certain specific domains on the outer side of the membrane, other domains within the membrane, and still other domains on the inner side of the membrane. Peripheral membrane proteins are localized on one side of the membrane or the other, but not

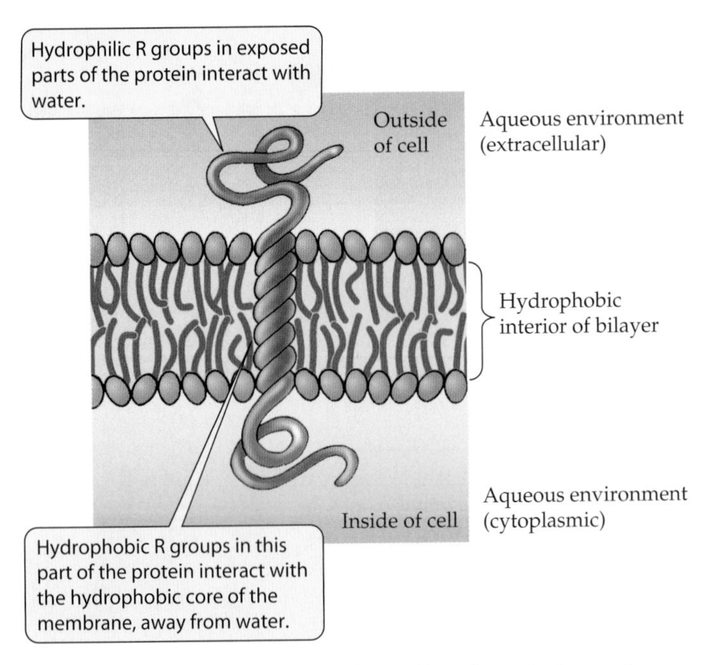
Hydrophilic R groups in exposed parts of the protein interact with water.
Outside of cell
Aqueous environment (extracellular)
Hydrophobic interior of bilayer
Aqueous environment (cytoplasmic)
Inside of cell
Hydrophobic R groups in this part of the protein interact with the hydrophobic core of the membrane, away from water.

5.4 Interactions of Integral Membrane Proteins An integral membrane protein is held in the membrane by the distribution of the hydrophilic and hydrophobic R groups of its amino acids.

both. This arrangement gives the two surfaces of the membrane different properties. As we will soon see, these differences have great functional significance.

Like lipids, many membrane proteins move around relatively freely within the phospholipid bilayer. Experiments using the technique of *cell fusion* illustrate this migration dramatically. When two cells are fused, a single continuous membrane forms and surrounds both cells, and some proteins from each cell distribute themselves uniformly around this membrane.

Although many proteins are free to migrate in the membrane, others are not, but rather appear to be "anchored" to a specific region of the membrane. These membrane regions are like a corral of horses on a farm: The horses are free to move within the fenced area, but cannot get out of it. For instance, the protein in the membrane of a muscle cell that recognizes a molecular signal from nerve cells is normally found only at the site where a nerve cell meets the muscle cell. None of this protein is found elsewhere on the surface of the muscle cell. There are two ways in which the movement of proteins within a membrane can be restricted:

- The *cytoskeleton* may have components just below the inner face of the membrane that are attached to membrane proteins protruding into the cytoplasm.
- *Lipid rafts*, which are groups of lipids in a semisolid (not quite fluid) state, may trap proteins within a region. These lipids have a different composition from the surrounding phospholipids; for example, they may have very long fatty acid chains.

Membrane carbohydrates are recognition sites

In addition to lipids and proteins, many membranes contain significant amounts of carbohydrates. The carbohydrates are located on the outer surface of the membrane and serve as recognition sites for other cells and molecules (see Figure 5.1).

Membrane-associated carbohydrates may be covalently bound to lipids or to proteins:

- *Glycolipids* consist of a carbohydrate covalently bound to a lipid. The carbohydrate units often extend to the outside of the membrane, where they serve as recognition signals for interactions between cells. For example, the carbohydrate of some glycolipids changes when a cell becomes cancerous. This change may allow white blood cells to target cancer cells for destruction.
- *Glycoproteins* consist of a carbohydrate covalently bound to a protein. The bound carbohydrates are oligosaccharide chains, usually not exceeding 15 monosaccharide units in length. Glycoproteins enable a cell to be recognized by other cells and proteins.

An "alphabet" of monosaccharides on membranes can be used to generate a diversity of messages. Recall from Chapter 3 that sugar molecules can be formed from 3–7 carbons attached at different sites to one another, forming linear or branched oligosaccharides with many different three-dimensional shapes. An oligosaccharide of a specific shape from one cell can bind to a mirror-image shape on an adjacent cell. This binding forms the basis of cell-to-cell adhesion.

Cell Recognition and Adhesion

Some organisms, such as bacteria, are *unicellular*; that is, the entire organism is a single cell. Others, such as plants and animals, are *multicellular*—composed of many cells. Often these cells exist in specialized blocks of cells with similar functions, called *tissues*. Your body has about 60 trillion cells, arranged in different kinds of tissues such as muscle, nerve, skin, and so forth. Two processes allow cells to arrange themselves in groups:

- *Cell recognition*, in which one cell specifically binds to another cell of a certain type
- *Cell adhesion*, in which the relationship between the two cells is "cemented"

Both processes involve the plasma membrane. They are most easily studied if the cells in a tissue are separated into individual cells, then allowed to adhere to one another again. Simple organisms provide a good model for the complex tissues of larger species.

A living sponge is a multicellular marine animal with a simple body plan (see Chapter 32). The cells of the sponge are stuck together, but they can be disaggregated mechanically by passing the animal several times through a fine wire screen. What was an animal is now hundreds of individual of cells, suspended in seawater. Remarkably, if the cell suspension is shaken for a few hours, the cells bump into one another and stick together in the same shape as a sponge! The cells recognize and adhere to one another.

There are many different types (species) of sponges. If disaggregated cells from two different species of sponge are placed in the same container, the cells of the two species float around and bump into one another. But the cells of each species stick only to other cells of the same species. Two different sponges form, just like the ones at the start of the experiment.

Such tissue-specific and species-specific cell adhesion is essential in the formation and maintenance of tissues and multicellular organisms. Think of your body. What keeps muscle cells bound to muscle cells and skin to skin? This is so obvious a characteristic of complex organisms that it is easy to overlook. You will see many examples of specific cell adhesion throughout this book; here, we describe its general principles. As you will see, cell recognition and adhesion depend on membrane proteins.

Cell recognition and adhesion involve proteins at the cell surface

The molecule responsible for cell recognition and adhesion in sponges is a huge membrane glycoprotein (80% sugar) that is partly embedded in the plasma membrane, with the recognition part sticking out and exposed to the environment (and to other sponge cells). As we saw in Chapter 3, a macromolecule such as a protein not only has a specific shape, but also has specific chemical groups exposed on its surface where they can interact with other substances, including other proteins. Both of these features allow binding to other specific molecules (Figure 5.5*a*).

In most cases, the binding of cells in a tissue is **homotypic**; that is, the same molecule sticks out of both cells, and the exposed surfaces bind to each other. But **heterotypic** binding (of cells with different proteins) also can occur (Figure 5.5*b*). For example, when the mammalian sperm meets the egg, different proteins on the two types of cells have complementary binding surfaces. Similarly, some algae form similar-appearing male and female reproductive cells (analogous to sperm and eggs) that have flagella to propel them toward each other. Male and female cells can recognize each other by heterotypic proteins on their flagella. In the majority of plant cells, the plasma membrane is covered with a thick cell wall,

(*a*) Homotypic binding
1 Tissue from a red sponge contains similar cells bound to each other.
2 The sponge tissue can be disaggregated into single cells by passing it through a fine mesh screen.
3 Exposed regions of membrane proteins bind to each other…
4 …causing cells to adhere.

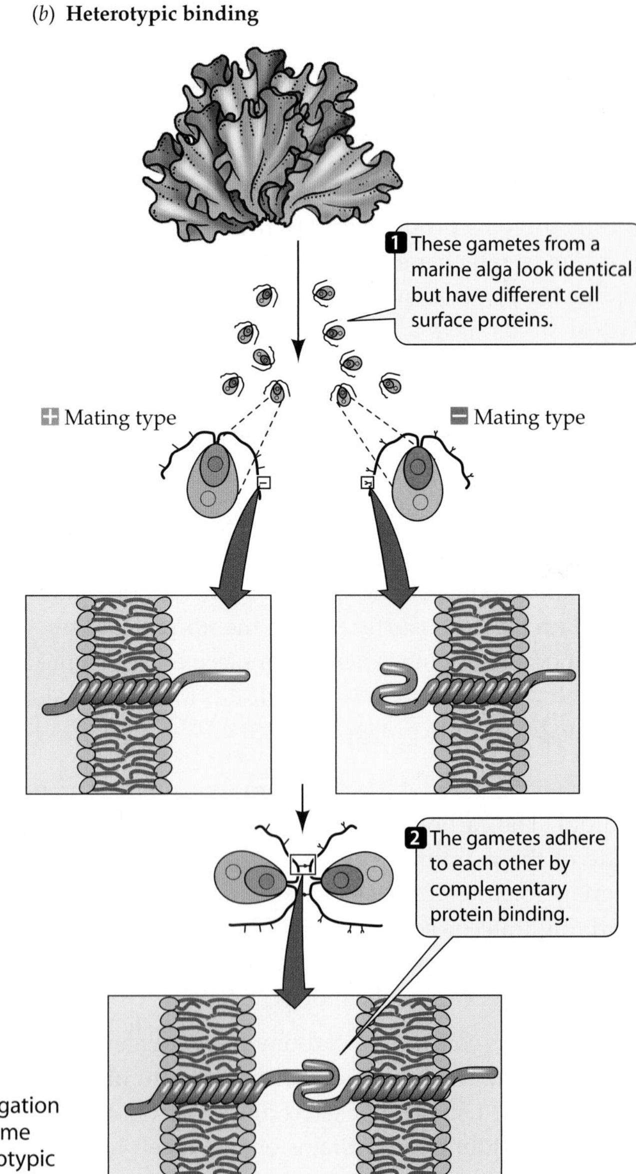
(*b*) Heterotypic binding
1 These gametes from a marine alga look identical but have different cell surface proteins.
+ Mating type
– Mating type
2 The gametes adhere to each other by complementary protein binding.

5.5 Cell Recognition and Adhesion (*a*) In most cases (including the aggregation of animal cells into tissues), protein binding is homotypic: molecules of the same protein occur on the surfaces of two cells and adhere to each other. (*b*) Heterotypic binding occurs between two different but complementary proteins.

but this structure, too, has adhesion proteins that allow cells to bind to one another.

Cell adhesion proteins from many multicellular organisms have been characterized. Some of them do not just bind the two cells together, but initiate the formation of specialized cell junctions. In this case, the functions of cell recognition and cell adhesion reside in different molecules.

Specialized cell junctions

In a complex multicellular organism, cell recognition proteins allow specific kinds of cells to bind to each other. Often, both cells contribute material to additional membrane structures that "cement" their relationship. These specialized structures, called **cell junctions**, are most evident in electron micrographs of *epithelial tissues*, which are layers of cells that line body cavities or cover body surfaces. We will examine three types of cell junctions that enable cells to make direct physical contact and link with one another: tight junctions, desmosomes, and gap junctions.

TIGHT JUNCTIONS SEAL TISSUES AND PREVENT LEAKS. **Tight junctions** are specialized structures at the plasma membrane that link adjacent epithelial cells. They result from the mutual binding of strands of specific membrane proteins, which form a series of joints encircling each epithelial cell (Figure 5.6*a*). They are found in the region surrounding the

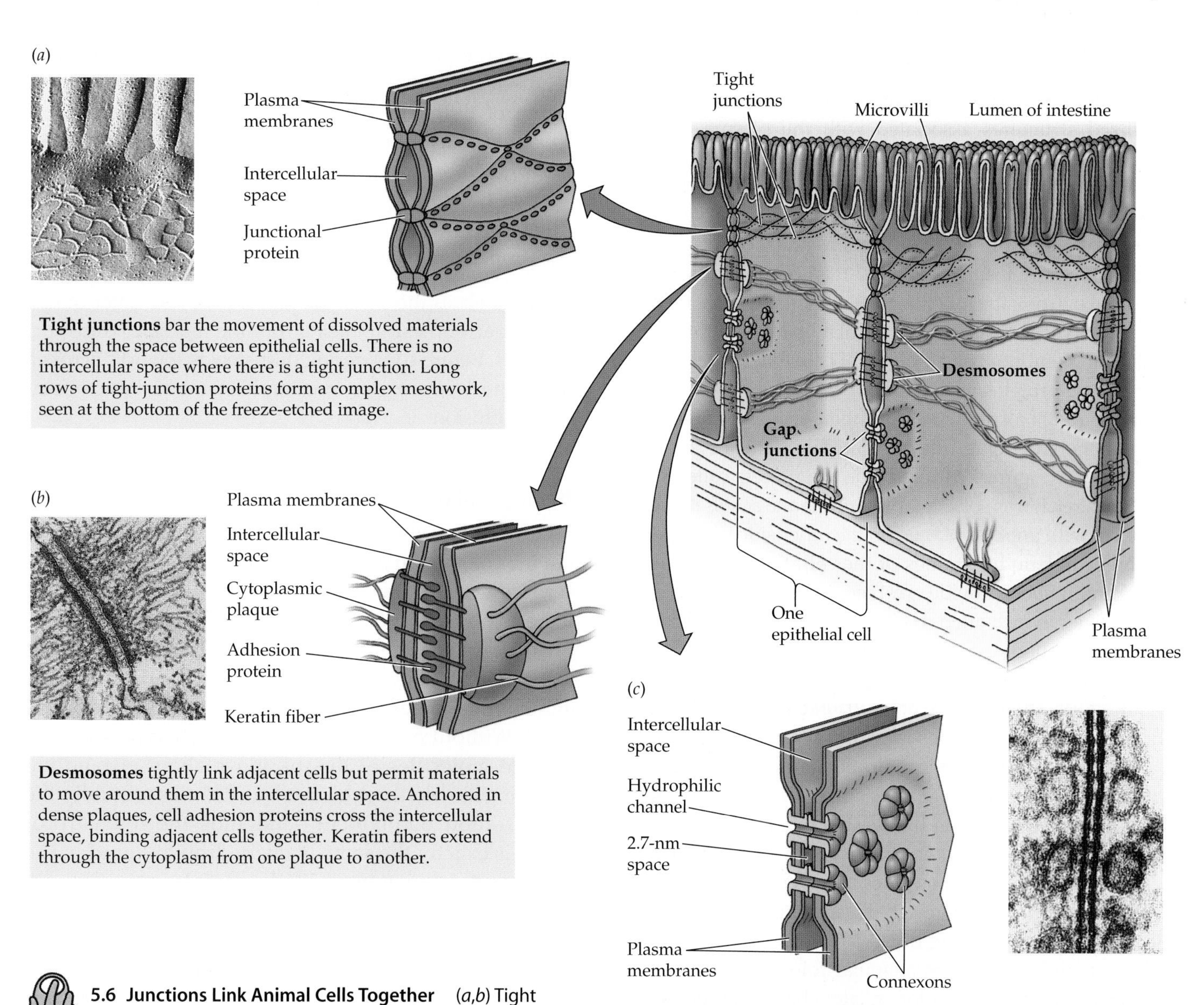

Tight junctions bar the movement of dissolved materials through the space between epithelial cells. There is no intercellular space where there is a tight junction. Long rows of tight-junction proteins form a complex meshwork, seen at the bottom of the freeze-etched image.

Desmosomes tightly link adjacent cells but permit materials to move around them in the intercellular space. Anchored in dense plaques, cell adhesion proteins cross the intercellular space, binding adjacent cells together. Keratin fibers extend through the cytoplasm from one plaque to another.

Gap junctions let adjacent cells communicate. Dissolved molecules and electric signals may pass from one cell to the other through the channel formed by two connexons extending from adjacent cells.

5.6 Junctions Link Animal Cells Together (*a,b*) Tight junctions and desmosomes are abundant in epithelial tissues. (*c*) Gap junctions are also found in some muscle and nerve tissues, in which rapid communication between cells is important.

lumen (cavity) of organs such as the intestine. Tight junctions have two functions:

- They prevent substances from moving through the spaces between cells. Thus, any substance entering the body from the lumen must pass through the epithelial cells.
- They restrict the migration of membrane proteins and phospholipids from one region of the cell to another.

Thus, the proteins and phospholipids in the *apical* (tip) region of the cell facing the lumen can be different from those in the *basolateral* regions facing the sides and bottom of the cell (basolateral: basal = bottom; lateral = side).

By forcing materials to enter certain cells, and by allowing different ends of cells to have different membrane proteins with different functions, tight junctions help ensure the directional movement of materials into the body.

DESMOSOMES HOLD CELLS TOGETHER. **Desmosomes** are also specialized structures associated with the plasma membrane. They hold adjacent cells firmly together, acting like spot welds or rivets (Figure 5.6*b*). Each desmosome has a dense structure called a plaque on the cytoplasmic surface of the plasma membrane. This plaque is attached to special cell adhesion proteins in the plasma membrane. These proteins stretch from the plaque through the plasma membrane of one cell, across the intercellular space, and through the plasma membrane of the adjacent cell, where they bind to the plaque proteins in that cell.

The plaque is also attached to fibers in the cytoplasm. These fibers, which are intermediate filaments of the cytoskeleton (see Figure 4.20), are made of a protein called *keratin*. They stretch from one cytoplasmic plaque across the cell to connect with another plaque on the other side of the cell. Anchored thus on both sides of the cell, these extremely strong keratin fibers provide great mechanical stability to epithelial tissues, which often receive rough wear in protecting the organism's body surface integrity.

GAP JUNCTIONS ARE A MEANS OF COMMUNICATION. Whereas tight junctions and desmosomes have mechanical roles, **gap junctions** facilitate communication between cells. Each gap junction is made up of specialized protein channels, called *connexons*, that span the plasma membranes of two adjacent cells and the intercellular space between them (Figure 5.6*c*). Dissolved molecules and electric signals can pass from cell to cell through these junctions. We will describe their role in more detail, as well as that of plasmodesmata, which perform a similar role in plants, when we discuss cell communication in later chapters, especially in Chapter 15.

Passive Processes of Membrane Transport

We have examined membrane structure and how it is used to perform one major membrane function: the binding of one cell to another. Now we turn to the second major membrane function: the ability to allow some substances, but not others, to pass through the membrane and enter or leave a cell or organelle. This characteristic of membranes is called **selective permeability**.

There are two fundamentally different kinds of processes by which substances cross biological membranes to enter and leave cells or organelles:

- **Passive transport** processes do not require any input of outside energy to drive them. The energy for these processes is in the substances themselves and the difference in their concentration on the two sides of the membrane. Passive transport processes include the different types of diffusion: simple diffusion through the phospholipid bilayer and facilitated diffusion through channel proteins or by means of carrier molecules.
- **Active transport** processes, on the other hand, require the input of chemical energy. They do not use the intrinsic property of a concentration gradient.

We'll discuss the active transport processes later in this chapter, after first focusing on the passive transport processes. Before considering diffusion as it works across a membrane, however, we must understand the basic principles of diffusion.

The physical nature of diffusion

Nothing in this world is ever absolutely at rest. Everything is in motion, although the motions may be very small. An important consequence of all this random jiggling is that all the components of a solution tend eventually to become evenly distributed throughout the system. For example, if a drop of ink is allowed to fall into a container of water, the pigment molecules of the ink are initially very concentrated. Without human intervention such as stirring, the pigment molecules of the ink move about at random, spreading slowly through the water until eventually the concentration of pigment—and thus the intensity of color—is exactly the same in every drop of liquid in the container. A solution in which the particles are uniformly distributed is said to be at *equilibrium*, because there will be no future net change in concentration. Equilibrium does not mean that the particles have stopped moving; it just means that they are moving in such a way that their overall distribution does not change.

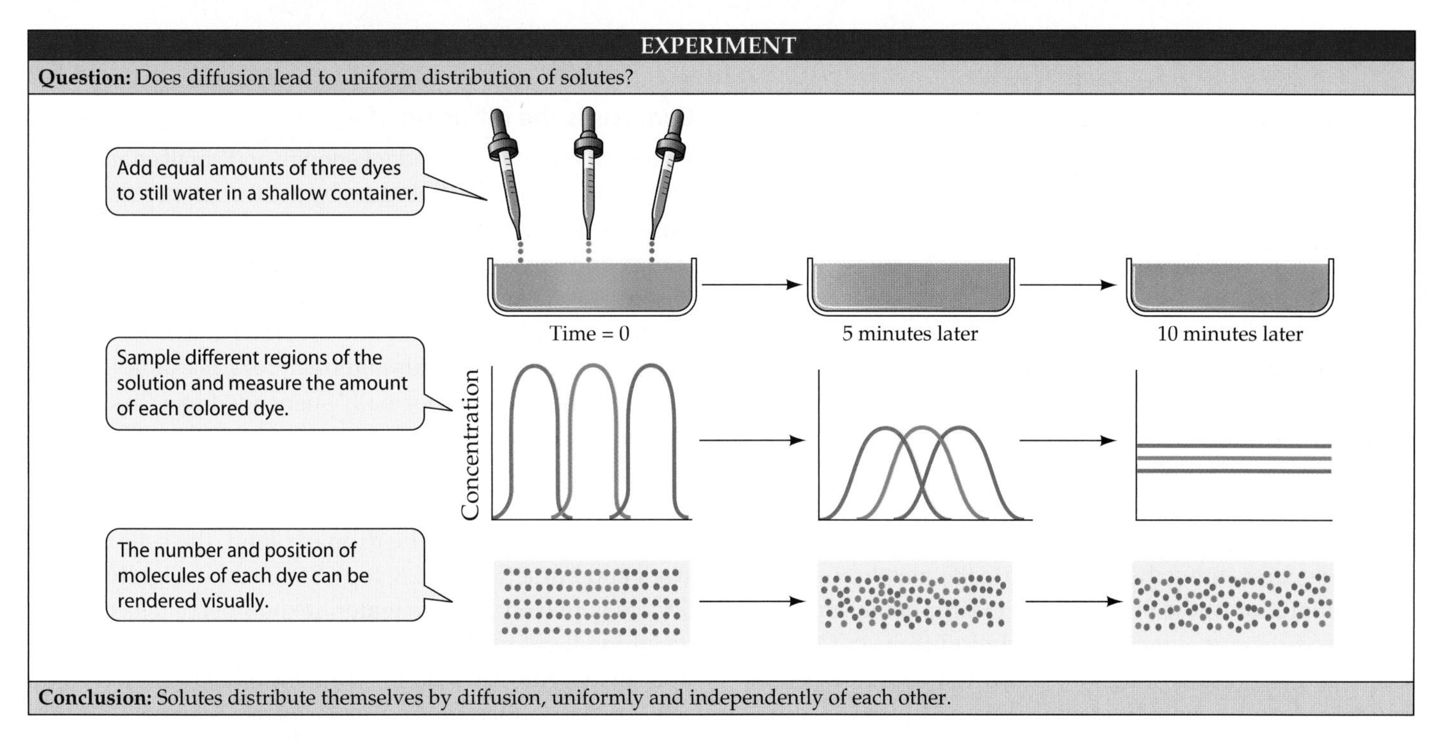

EXPERIMENT
Question: Does diffusion lead to uniform distribution of solutes?
Add equal amounts of three dyes to still water in a shallow container.
Time = 0
5 minutes later
10 minutes later
Sample different regions of the solution and measure the amount of each colored dye.
Concentration
The number and position of molecules of each dye can be rendered visually.
Conclusion: Solutes distribute themselves by diffusion, uniformly and independently of each other.

5.7 Diffusion Leads to Uniform Distribution of Solutes Diffusion is the net movement of a solute from regions of greater concentration to regions of lesser concentration. The speed of diffusion varies with the substances involved, but the process continues until the solution reaches equilibrium.

Diffusion is the process of random movement toward a state of equilibrium. Although the motion of each individual particle is absolutely random, in diffusion the *net* movement of particles is directional until equilibrium is reached. Diffusion is thus net movement from regions of greater concentration to regions of lesser concentration (Figure 5.7).

In a complex solution (one with many different solutes), the diffusion of each substance is independent of that of the others. How fast a substance diffuses depends on four factors:

- the *diameter* of the molecules or ions: smaller molecules diffuse faster
- the *temperature* of the solution: higher temperatures lead to faster diffusion
- the *electric charge*, if any, of the diffusing material: electric charge has a variable effect on diffusion
- the *concentration gradient* in the system—that is, the change in concentration with distance in a given direction. The greater the concentration gradient, the more rapidly a substance diffuses.

DIFFUSION WITHIN CELLS AND TISSUES. Within cells, or wherever distances are very short, solutes distribute themselves rapidly by diffusion. Small molecules and ions may move from one end of an organelle to another in a millisecond (10^{-3} s). On the other hand, the usefulness of diffusion as a transport mechanism declines drastically as distances become greater. In the absence of mechanical stirring, diffusion across more than a centimeter may take an hour or more, and diffusion across meters may take years! Diffusion would not be adequate to distribute materials over the length of the human body, but within our cells or across layers of one or two cells, diffusion is rapid enough to distribute small molecules and ions almost instantaneously.

DIFFUSION ACROSS MEMBRANES. In a solution without barriers, all the solutes diffuse at rates determined by temperature, their physical properties, and the concentration gradient of each solute. If a biological membrane divides the solution into separate compartments, the movement of the different solutes can be affected by the properties of the membrane. The membrane is said to be *permeable* to solutes that can cross it more or less easily, but *impermeable* to substances that cannot move across it.

Molecules to which the membrane is impermeable remain in separate compartments, and their concentrations remain different on the two sides of the membrane. Molecules to which the membrane is permeable diffuse from one compartment to the other until their concentrations are equal on both sides of the membrane. When the concentrations of the diffusing substance are identical on both sides of the permeable membrane, equilibrium is reached. Individual molecules are still passing through the membrane after equilibrium is established, but equal numbers of molecules are moving in each direction, so there is *no net change* in concentration.

Simple diffusion takes place through the membrane bilayer

In **simple diffusion**, small molecules pass through the lipid bilayer of the membrane. The more lipid-soluble the molecule, the more rapidly it diffuses through the bilayer. This statement holds true over a wide range of molecular weights. Only water and the smallest of molecules seem to deviate from this rule, passing through bilayers much more rapidly than their lipid solubilities would predict.

Charged or polar molecules such as amino acids, sugars, and ions do not pass readily through a membrane, for two reasons. First, cells are made up of, and exist in, water, and polar or charged substances form many hydrogen bonds with water, preventing their "escape" to the membrane. Second, the interior of the membrane is hydrophobic, and hydrophilic substances tend to be excluded from it. On the other hand, a molecule that is itself hydrophobic, and hence soluble in lipids, enters the membrane readily and is thus able to pass through it. For example, consider two types of molecules, a small protein of a few amino acids and a steroid of equivalent size. The protein, being polar, will diffuse slowly through the membrane, while the nonpolar steroid will diffuse through it readily.

Osmosis is the diffusion of water across membranes

Water molecules are abundant enough and small enough that they move through membranes by a diffusion process called **osmosis**. This completely passive process uses no metabolic energy and can be understood in terms of the concentrations of solutions. Osmosis depends on the *number* of solute particles present, not on the kinds of particles. We will describe osmosis using red blood cells and plant cells as examples.

Red blood cells are normally suspended in a fluid called plasma, which contains salts, proteins, and other solutes. If a drop of blood is examined under the light microscope, the red cells are seen to have a characteristic doughnut shape. If pure water is added to the drop of blood, the cells quickly swell and burst (Figure 5.8, top). Similarly, if slightly wilted lettuce is placed in pure water, it soon becomes crisp; by weighing it before and after, we can show that it has taken up water (Figure 5.8, bottom). If, on the other hand, red blood cells or crisp lettuce leaves are placed in a relatively concentrated solution of salt or sugar, the leaves become limp (wilt) and the red blood cells pucker and shrink.

From analyses of such observations, we know that the difference in solute concentration between a cell and its surrounding environment determines whether water will move from the environment into the cell or out of the cell into the

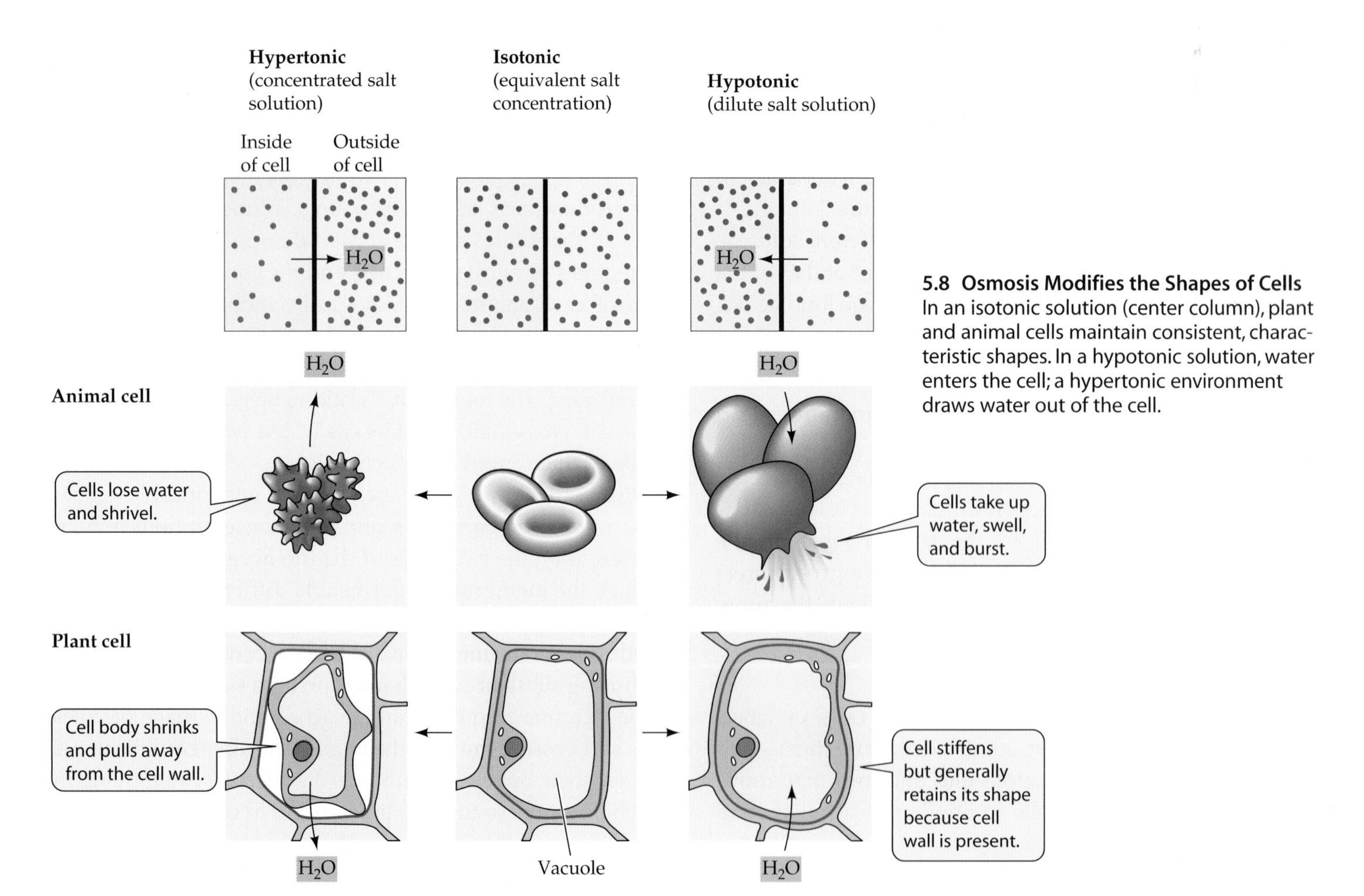

5.8 Osmosis Modifies the Shapes of Cells In an isotonic solution (center column), plant and animal cells maintain consistent, characteristic shapes. In a hypotonic solution, water enters the cell; a hypertonic environment draws water out of the cell.

environment. Other things being equal, if two different solutions are separated by a membrane that allows water, but not solutes, to pass through, water molecules will move across the membrane toward the solution with a higher solute concentration. In other words, *water will diffuse from a region of its higher concentration (lower concentration of solutes) to a region of its lower concentration (higher concentration of solutes).*

Three terms are used to compare the solute concentrations of two solutions separated by a membrane:

- **Isotonic** solutions have equal solute concentrations.
- A **hypertonic** solution has a higher solute concentration than the other solution with which it is being compared.
- A **hypotonic** solution has a lower solute concentration than the other solution with which it is being compared.

Water moves from a hypotonic solution across a membrane to a hypertonic solution (Figure 5.8). When we say that "water moves," bear in mind that we are referring to the net movement of water. Since it is so abundant, water is constantly moving across the plasma membrane into and out of cells. Whether the overall movement is greater in one direction or the other is what concerns us here.

The concentration of solutes in the environment determines the direction of osmosis in all animal cells. A red blood cell takes up water from a solution that is hypotonic to the cell's contents. The cell bursts because its plasma membrane cannot withstand the swelling of the cell (see Figure 5.8). The integrity of red blood cells (and other blood cells) is absolutely dependent on the maintenance of a constant solute concentration in the plasma in which they are suspended: The plasma must be isotonic to the cells if the cells are not to burst or shrink. Regulation of cell volume is an important process for cells without cell walls.

In contrast to animal cells, the cells of plants, archaea, bacteria, fungi, and some protists have cell walls, which limit the volume of the cells and keep them from bursting. Cells with sturdy cell walls take up a limited amount of water and, in so doing, build up internal pressure against the cell wall that prevents further water from entering. This pressure within the cell, called *turgor pressure*, keeps plants upright and is the driving force for the enlargement of plant cells—it is a normal and essential component of plant growth.

Diffusion may be aided by channel proteins

As we saw earlier, polar substances such as amino acids and sugars and charged substances such as ions do not diffuse across membranes. Instead, they cross the hydrophobic lipid barrier in two ways : One way is through integral membrane proteins that form channels through which the polar substance can pass, and the other is by binding to a membrane protein that speeds up its membrane crossing. Both of these processes are forms of **facilitated diffusion**.

Membrane **channel proteins** (Figure 5.9) have polar amino acids and water on the inside of the channel pore (to bind to the polar or charged substance and allow it to pass through) and nonpolar amino acids on the outside of the macromolecule (to allow the channel protein to insert itself into the lipid bilayer). The best-studied channel proteins are the *ion channels*. As you will see in later chapters, the movement of ions into and out of cells is important in many biological processes, ranging from the electrical activity of the nervous system to the opening of the pores in leaves that allow gas exchange with the environment. Hundreds of ion channels have been identified, each of them specific for a particular ion. All of them show the basic structure of a water-lined pore that allows a particular ion to move through it.

Just as the front gate on a fence can be open or closed, most ion channels are *gated:* they can be closed to ion passage or opened. A gated channel opens when something happens to change the shape of the protein. Depending on the channel, this stimulus can range from the binding of a chemical signal to an electrical charge caused by an imbalance of ions. Once the channel opens, millions of ions can rush through it per second. How fast this happens, and in which direction (into or out of the cell), depends on the concentration gradi-

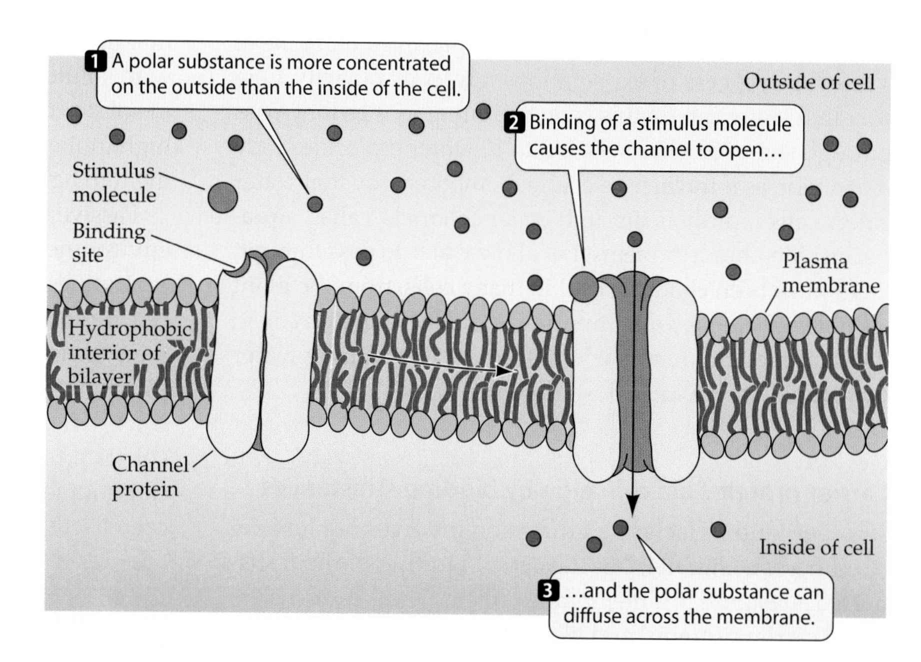
1 A polar substance is more concentrated on the outside than the inside of the cell.
Outside of cell
2 Binding of a stimulus molecule causes the channel to open...
Stimulus molecule
Binding site
Plasma membrane
Hydrophobic interior of bilayer
Channel protein
Inside of cell
3 ...and the polar substance can diffuse across the membrane.

5.9 A Gated Channel Protein Opens in Response to a Stimulus The membrane protein changes its three-dimensional shape when a stimulus molecule binds to it.

ent of the ion between the cytoplasm and the exterior environment of the cell. For example, if the concentration of potassium ion (K^+) is much higher outside of the cell than inside, K^+ will enter the cell through a potassium channel by diffusion; if the concentration is higher inside the cell, K^+ will diffuse out of the cell.

How does an ion channel exert its specificity for one ion and not another? It is not simply a matter of charge or size. For example, sodium ion (Na^+) is 0.095 nm in radius and K^+ is larger, at 0.130 nm; both have the same positive charge. Yet the potassium channel lets only K^+ pass through the membrane, and not the smaller Na^+. How this happens was recently discovered when Roderick MacKinnon determined the structure of the potassium channel from a bacterium (Figure 5.10). The explanation is elegant and provides a good review of cell chemistry.

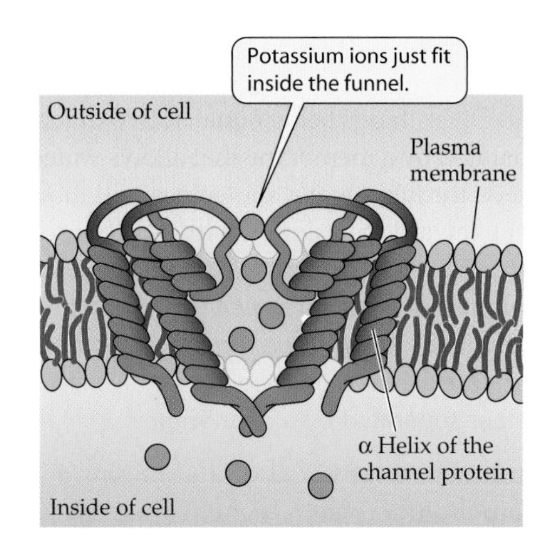

5.10 The K^+ Channel This structure was first worked out for a channel from the bacterium *Streptomyces lividans*.

Being charged, both Na^+ and K^+ are attracted to water molecules. They have water "shells" in solution, held electrostatically by attraction of the positively charged ions to the negatively charged oxygen atom on the polar water molecules (see Figure 2.14). To get through a membrane channel, an ion must let go of its water. The "naked" ion is now attracted to the oxygen atoms on the channel pore protein.

In the potassium channel, oxygen atoms are located at the stem of a funnel-shaped protein region. The K^+ ion just fits the stem, and so can get into a position where it is more strongly attracted to the oxygen atoms there than to those of water. The smaller Na^+ ion, on the other hand, is kept a bit more distant from the oxygen atoms on the stem of the channel, and so prefers to be surrounded by water. So Na^+ does not enter the potassium channel.

As we mentioned, water crosses the plasma membrane at a rate far in excess of expectations, given its polarity. One way that water can do this is by hydrating ions as they pass through some ion channels. Up to 12 water molecules may coat an ion as it traverses a channel. Another way that water enters cells rapidly is through water channels, called **aquaporins**. Membrane proteins that allow water to pass through them have been characterized in many cells, from the plant vacuole, where they are important in maintaining turgor, to the mammalian kidney, where they act in retaining water that would otherwise be lost through urine.

Carrier proteins aid diffusion by binding substances

Another kind of facilitated diffusion involves not just the opening of a channel, but the actual binding of the transported substance to a membrane protein. These proteins are called **carrier proteins**, and like channel proteins, they allow diffusion both into and out of the cell. They are used to transport polar molecules such as sugars and amino acids.

Glucose, for example, is the major energy source for most mammalian cells. The membranes of those cells contain a carrier protein called the *glucose transporter* that facilitates the uptake of glucose (Figure 5.11*a*). Since glucose is rapidly broken down as soon as it gets into a cell, there is almost always a strong concentration gradient favoring glucose entry, with a higher concentration outside the cell than inside.

Transport by carrier proteins is different from simple diffusion. In both processes, the rate of movement depends on the concentration gradient across the membrane. However, in facilitated diffusion, a point is reached at which further increases in the concentration gradient are not accompanied by an increased rate of diffusion. At this point, the facilitated diffusion system is said to be *saturated* (Figure 5.11*b*). Because there are only a limited number of carrier protein molecules per unit of membrane area, the rate of diffusion reaches a maximum when all the carrier molecules are fully loaded with solute molecules. In other words, when the difference in solute concentration across the membrane is sufficiently high, not enough carrier molecules are free at a given moment to handle all the solute molecules.

Passive transport allows substances to enter cells from the environment so that, after equilibrium is reached, the concentrations of a substance inside the cell and just outside the cell are equal. But a hallmark of living things is that they can have a composition quite different from that of their environment. One way that they achieve this is by not relying solely on concentration gradients, but instead by moving substances against their natural tendencies to diffuse. Because it requires an input of chemical energy, this process is called active transport.

Active Transport

In many biological situations, an ion or molecule must be moved across a membrane from a region of lower con-

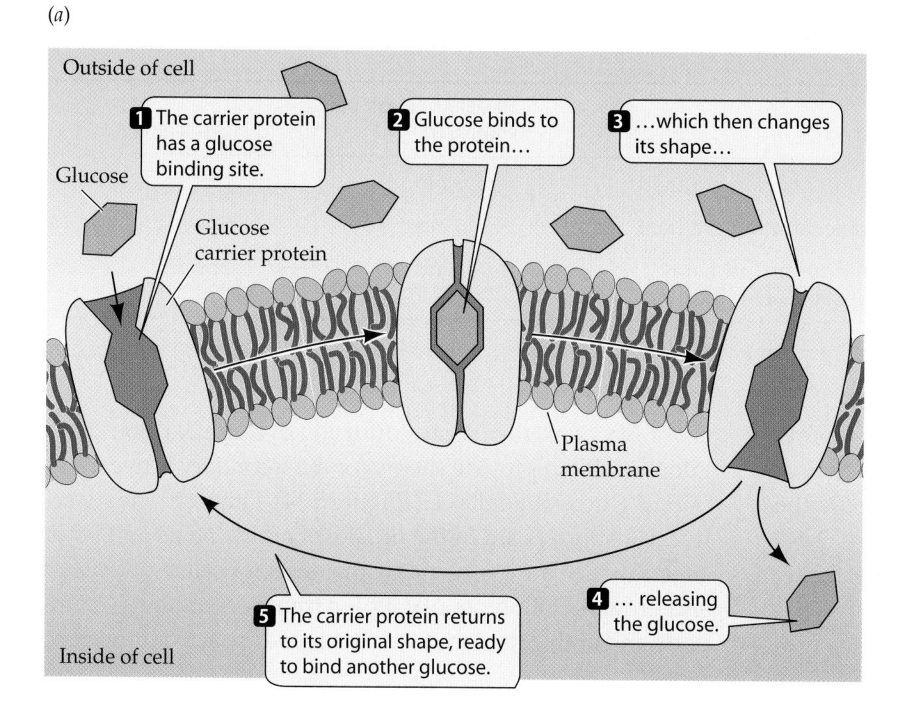
(a)
Outside of cell
1 The carrier protein has a glucose binding site.
2 Glucose binds to the protein...
3 ...which then changes its shape...
Glucose
Glucose carrier protein
Plasma membrane
4 ... releasing the glucose.
5 The carrier protein returns to its original shape, ready to bind another glucose.
Inside of cell

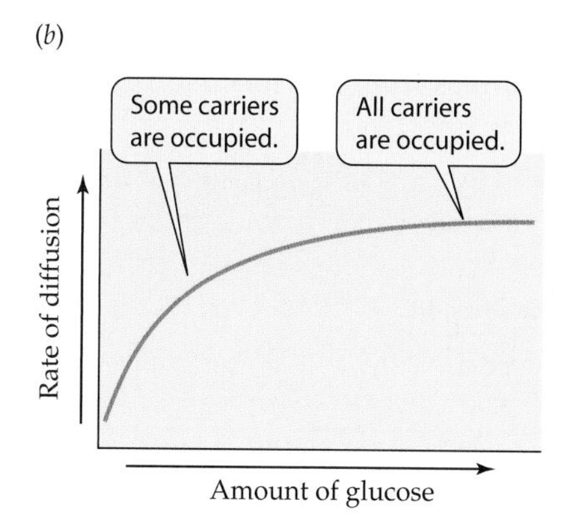
(b)
Some carriers are occupied.
All carriers are occupied.
Rate of diffusion
Amount of glucose

5.11 A Carrier Protein Facilitates Diffusion
(*a*) The carrier protein allows glucose to enter the cell at a faster rate than would be possible by simple diffusion across the membrane barrier. (*b*) At low concentrations of glucose, not all carriers are occupied and so an increase in glucose increases the rate of diffusion. At high concentrations, the carriers are saturated and the rate reaches a plateau.

centration to a region of higher concentration. In these cases, the substance cannot rush into or out of cells by diffusion. The movement of a substance across a biological membrane *against* a concentration gradient—active transport—requires the expenditure of chemical energy. The differences between diffusion and active transport are summarized in Table 5.1.

Active transport is directional

Three types of transporter proteins are involved in active transport (Figure 5.12):

- **Uniports** move a single solute in one direction. For example, a calcium-binding protein found in the plasma membrane and endoplasmic reticulum of many cells actively transports Ca^{2+} to regions of higher concentration either outside the cell or inside the ER.
- **Symports** move two solutes in the same direction. For example, the uptake of amino acids from the intestine into the cells that line it requires the simultaneous binding of Na^+ and an amino acid to the same transporter protein.
- **Antiports** move two solutes in opposite directions, one into the cell and the other out of the cell. For example, many cells have a "sodium–potassium pump" that moves Na^+ out of the cell and K^+ into it.

Symports and antiports are known as *coupled transporters* because they move two solutes at once.

Primary and secondary active transport rely on different energy sources

There are two basic types of active transport processes:

- **Primary active transport** requires the direct participation of the energy-rich molecule ATP.
- **Secondary active transport** does not use ATP directly; rather, its energy is supplied by an ion concentration gradient established by primary active transport.

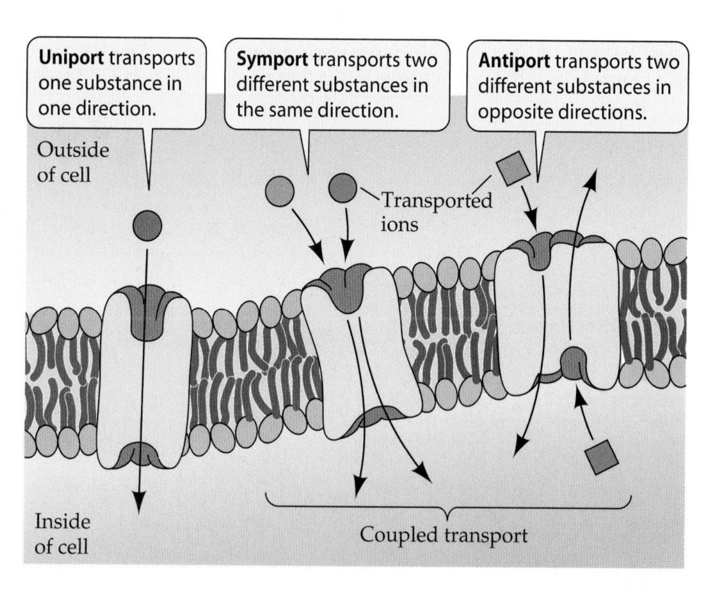
Uniport transports one substance in one direction.
Symport transports two different substances in the same direction.
Antiport transports two different substances in opposite directions.
Outside of cell
Transported ions
Inside of cell
Coupled transport

5.12 Three Types of Proteins for Active Transport Note that in each of the three cases, transport is directional.

5.1 Membrane Transport Mechanisms

TRANSPORT MECHANISM	EXTERNAL ENERGY REQUIRED?	DRIVING FORCE	MEMBRANE PROTEIN REQUIRED?	SPECIFICITY
Simple diffusion	No	With concentration gradient	No	Not specific
Facilitated diffusion	No	With concentration gradient	Yes	Specific
Active transport	Yes	ATP hydrolysis (primary) (against concentration gradient)	Yes	Specific

In primary active transport, energy released by the hydrolysis of ATP drives the movement of specific ions against a concentration gradient. For example, if we compare the concentrations of potassium ions (K^+) and sodium ions (Na^+) inside a nerve cell and in the fluid bathing the nerve, the K^+ concentration is much higher inside the cell, whereas the Na^+ concentration is much higher outside. Nevertheless, a protein in the nerve cell continues to pump Na^+ out and K^+ in against these concentration gradients, ensuring that the gradients are maintained. This *sodium–potassium* (Na^+–K^+) *pump* is found in all animal cells and is an integral membrane glycoprotein. It breaks down a molecule of ATP to ADP and phosphate (P_i) and uses the energy released to bring two K^+ ions into the cell and export three Na^+ ions (Figure 5.13). The Na^+–K^+ pump is thus an antiport.

In secondary active transport, the movement of the solute against its concentration gradient is accomplished using energy "regained" by letting ions move across the membrane *with* their concentration gradient. For example, once the sodium–potassium pump establishes a concentration gradient of Na^+ ions, the passive diffusion of some Na^+ ions back into the cell can provide energy for the secondary active transport of glucose into the cell (Figure 5.14). Other secondary active transporters aid in the uptake of amino acids and other sugars, which are essential raw materials for cell maintenance and growth. Both types of coupled transport proteins—symports and antiports—are used for secondary active transport.

Endocytosis and Exocytosis

Macromolecules such as proteins, polysaccharides, and nucleic acids are simply too large and too charged or polar to pass through membranes. This is a fortunate property. Think of the consequences if these molecules could diffuse out of cells: A red blood cell would not retain its hemoglobin! On the other hand, cells must sometimes take up or secrete intact large molecules. As we saw in Chapter 4, this can be done by means of vesicles that either pinch off from the plasma membrane and enter the cell (endocytosis) or fuse with the plasma membrane and release their contents (exocytosis).

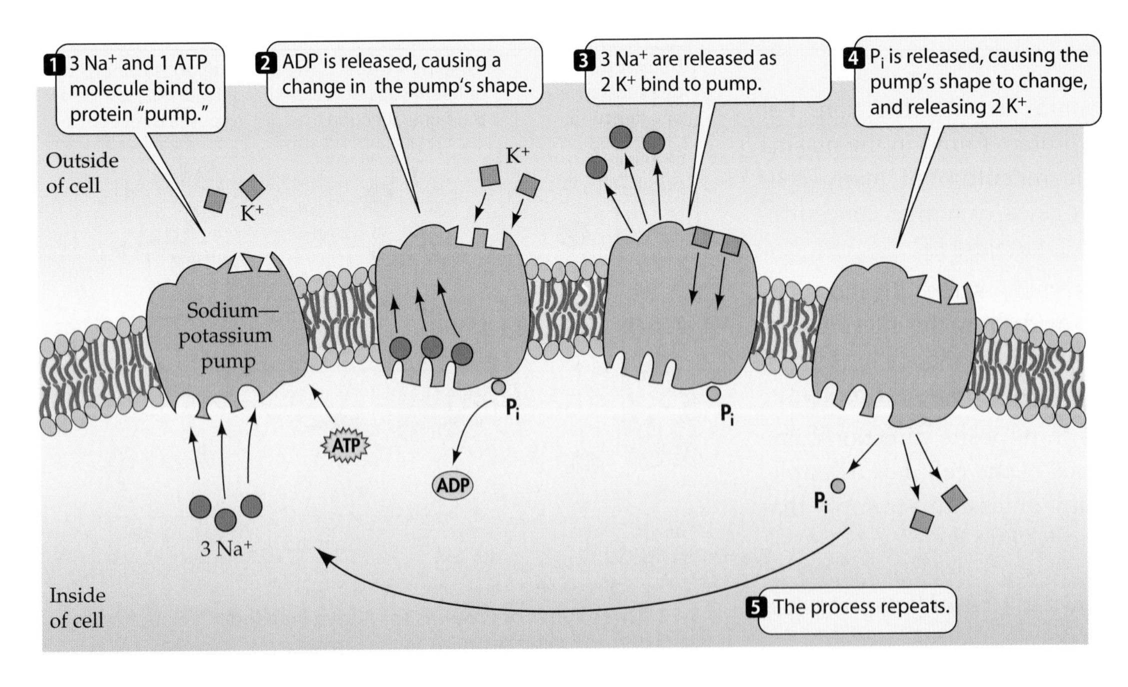
1 3 Na+ and 1 ATP molecule bind to protein "pump."
2 ADP is released, causing a change in the pump's shape.
3 3 Na+ are released as 2 K+ bind to pump.
4 Pi is released, causing the pump's shape to change, and releasing 2 K+.
Outside of cell
K+
K+
Sodium—potassium pump
ATP
ADP
Pi
Pi
Pi
3 Na+
Inside of cell
5 The process repeats.

5.13 Primary Active Transport: The Sodium–Potassium Pump In active transport, energy is used to move a solute against its concentration gradient. Even though the Na^+ concentration is higher outside the cell and the K^+ concentration is higher inside the cell, for each molecule of ATP used, two K^+ are pumped into the cell and three Na^+ are pumped out of the cell.

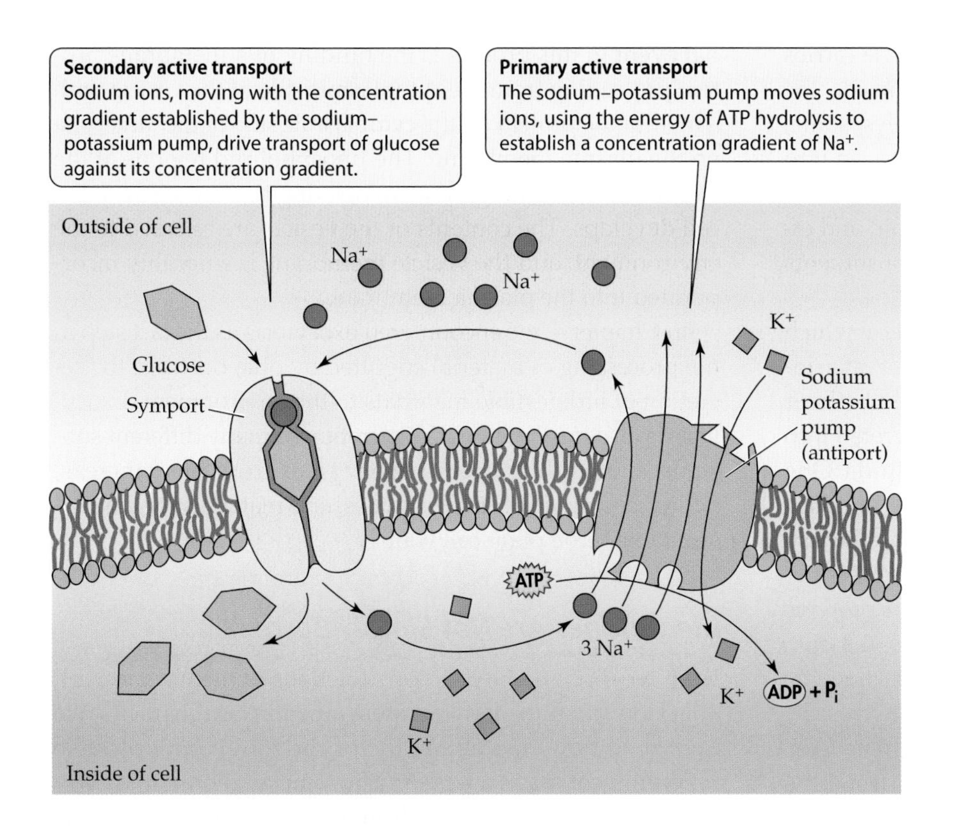

Secondary active transport
Sodium ions, moving with the concentration gradient established by the sodium–potassium pump, drive transport of glucose against its concentration gradient.
Primary active transport
The sodium–potassium pump moves sodium ions, using the energy of ATP hydrolysis to establish a concentration gradient of Na+.
Outside of cell
Na+
Na+
K+
Glucose
Symport
Sodium potassium pump (antiport)
ATP
3 Na+
K+
ADP + Pi
K+
Inside of cell

5.14 Secondary Active Transport The Na^+ concentration gradient established by primary active transport (right) powers the secondary active transport of glucose (left). The movement of glucose across the membrane against its concentration gradient is coupled by a symport protein to the movement of Na^+ into the cell.

Macromolecules and particles enter the cell by endocytosis

Endocytosis is a general term for a group of processes that bring macromolecules, large particles, small molecules, and even small cells into the eukaryotic cell (Figure 5.15*a*). There are three types of endocytosis: phagocytosis, pinocytosis, and receptor-mediated endocytosis. In all three, the plasma membrane invaginates (folds inward) around materials from the environment, forming a small pocket. The pocket deepens, forming a vesicle. This vesicle separates from the plasma membrane and migrates with its contents to the cell's interior.

In **phagocytosis** ("cellular eating"), part of the plasma membrane engulfs large particles or even entire cells. Phagocytosis is used as a cellular feeding process by unicellular protists and by some white blood cells that defend the body by engulfing foreign cells and substances. The food vacuole or phagosome that forms usually fuses with a lysosome, where its contents are digested (see Figure 4.13).

In **pinocytosis** ("cellular drinking"), vesicles also form. However, these vesicles are smaller, and the process operates to bring small dissolved substances or fluids into the cell. Like phagocytosis, pinocytosis is relatively nonspecific as to what it brings into the cell. For example, pinocytosis goes on constantly in the *endothelium,* the single layer of cells that separates a tiny blood capillary from its surrounding tissue, allowing the cells to rapidly acquire fluids from the blood.

In **receptor-mediated endocytosis**, specific reactions at the cell surface trigger the uptake of specific materials. Let's take a closer look at this process.

Receptor-mediated endocytosis is highly specific

Receptor-mediated endocytosis is used by animal cells to capture specific macromolecules from the cell's environment. This process depends on *receptor proteins*, integral membrane proteins that can bind to a specific molecule in the cell's environment. The uptake process is similar to nonspecific endocytosis, as already described. However, in receptor-mediated endocytosis, receptor proteins at particular sites on the extracellular surface of the plasma membrane bind to specific substances. These sites are called **coated pits** because they form a slight depression in the plasma membrane. The cytoplasmic surface of a coated pit is coated by proteins, such as *clathrin*.

When a receptor protein binds to its specific macromolecule outside the cell, its coated pit invaginates and forms a **coated vesicle** around the bound macromolecule. Strength-

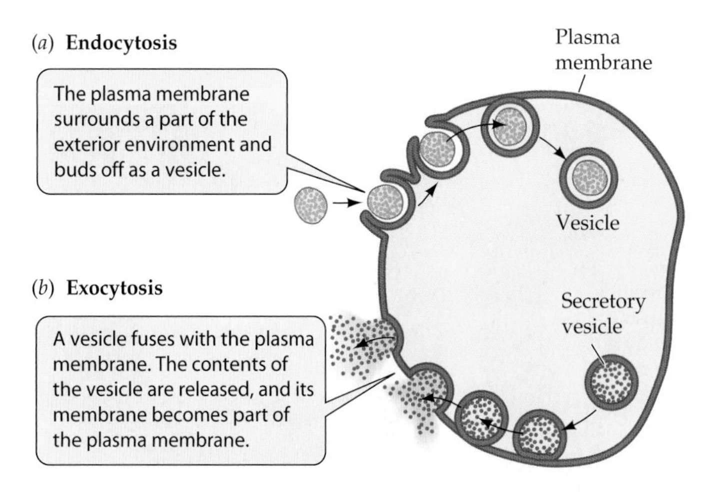

(a) Endocytosis
The plasma membrane surrounds a part of the exterior environment and buds off as a vesicle.
Plasma membrane
Vesicle
(b) Exocytosis
A vesicle fuses with the plasma membrane. The contents of the vesicle are released, and its membrane becomes part of the plasma membrane.
Secretory vesicle

5.15 Endocytosis and Exocytosis Endocytosis and exocytosis are used by all eukaryotic cells to take up substances from and release substances to the outside environment.

ened and stabilized by clathrin molecules, this vesicle carries the macromolecule into the cell (Figure 5.16). Once inside, the vesicle loses its clathrin coat and may fuse with a lysosome, where the engulfed material is processed and released into the cytoplasm. Because of its specificity for particular macromolecules, receptor-mediated endocytosis is a rapid and efficient method of taking up what may be minor constituents of the cell's environment.

Receptor-mediated endocytosis is the method by which cholesterol is taken up by most mammalian cells. Water-insoluble cholesterol and triglycerides are packaged by liver cells into lipoprotein particles, which are then secreted into the bloodstream to provide body tissues with lipids. One type of these particles, called *low-density lipoproteins*, or LDLs, must be taken up by the liver for recycling. This uptake also occurs via receptor-mediated endocytosis. This process begins with the binding of LDLs to specific receptor proteins on the cell surface. Once engulfed by endocytosis, the LDL particle is freed from the receptors. The receptors segregate to a region of the vesicle that buds off to form a new vesicle, which is recycled to the plasma membrane. The freed LDL particle remains in the original vesicle, which fuses with a lysosome in which the LDL is digested and the cholesterol made available for cell use. Persons with the inherited disease *familial hypercholesterolemia* (*-emia*, "blood") have dangerously high levels of cholesterol in their blood because of a deficient receptor for LDL.

Exocytosis moves materials out of the cell

Exocytosis is the process by which materials packaged in vesicles are secreted from a cell when the vesicle membrane fuses with the plasma membrane (see Figure 5.15*b*). The initial event in this process is the binding of a membrane protein protruding from the cytoplasmic side of the vesicle with a membrane protein on the cytoplasmic side of the target site on the plasma membrane. The phospholipid regions of the two membranes merge, and an opening to the outside of the cell develops. The contents of the vesicle are released to the environment, and the vesicle membrane is smoothly incorporated into the plasma membrane.

In Chapter 4, we encountered exocytosis as the last step in the processing of material engulfed by phagocytosis: the secretion of indigestible materials to the environment. Exocytosis is also important in the secretion of many different substances, including digestive enzymes from the pancreas, neurotransmitters from nerve cells, and materials for the construction of the plant cell wall.

Membranes Are Not Simply Barriers

We have discussed two major functions of membranes, cell adhesion and transport, but there are more. In Chapter 4, we described how the membrane of the rough endoplasmic reticulum serves as a site for ribosome attachment. Newly formed proteins are passed from the ribosomes through the membrane and into the interior of the ER for modification and delivery to other parts of the cell. This system sets up a separate compartment that segregates these proteins from the rest of the cell. On the other hand, the plasma membranes of nerve cells, muscle cells, and some eggs are electrically excitable. In nerve cells, the plasma membrane is the conductor of the nerve impulse from one end of the cell to the other.

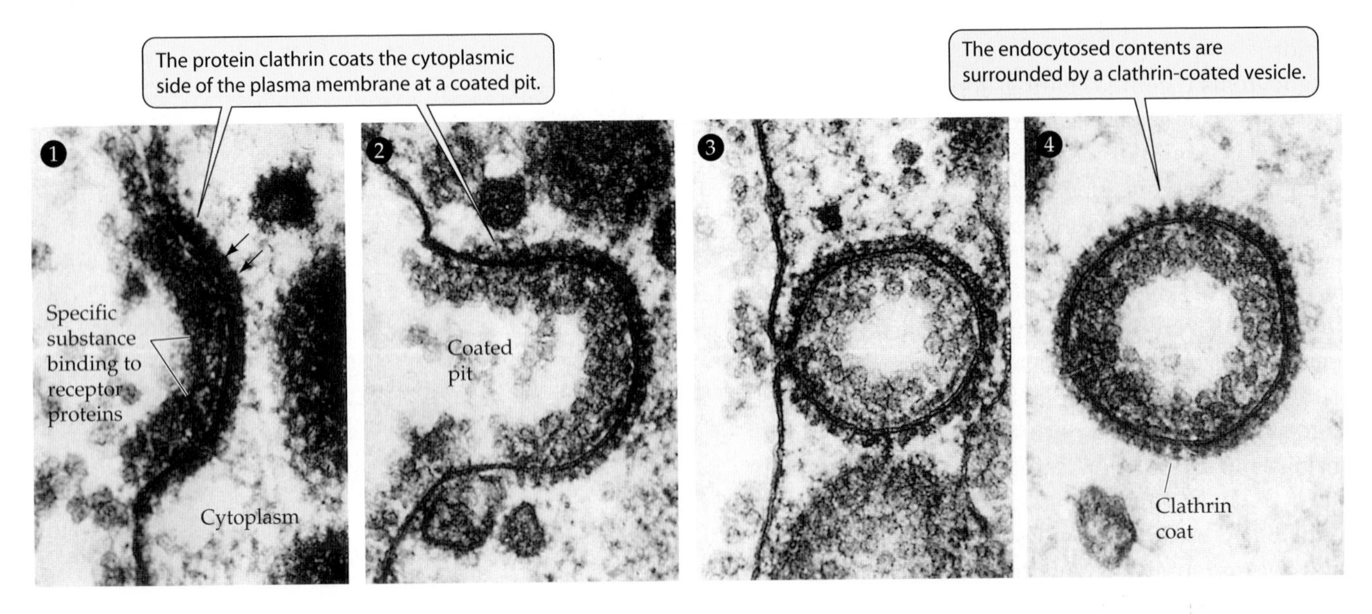

5.16 Formation of a Coated Vesicle In receptor-mediated endocytosis, the receptor proteins in a coated pit bind specific macromolecules, which are then carried into the cell by a coated vesicle.

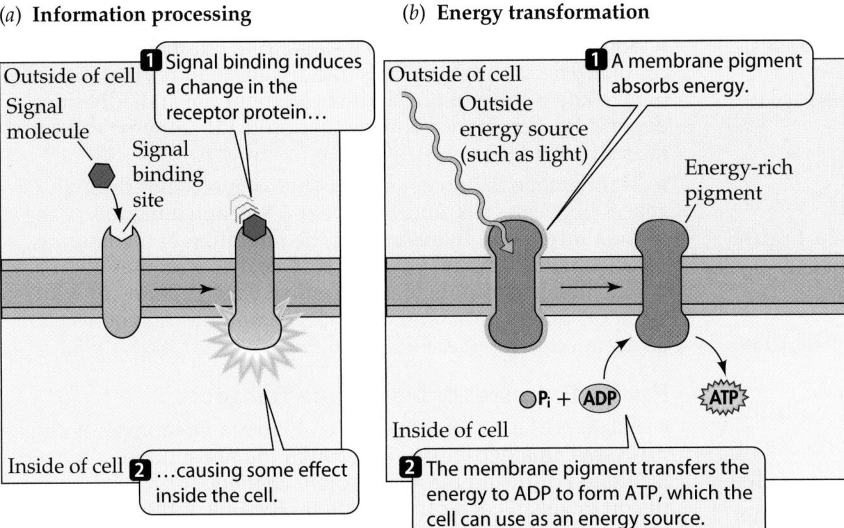
(a) Information processing
Outside of cell
Signal molecule
Signal binding site
1 Signal binding induces a change in the receptor protein...
Inside of cell
2 ...causing some effect inside the cell.
(b) Energy transformation
Outside of cell
Outside energy source (such as light)
1 A membrane pigment absorbs energy.
Energy-rich pigment
Pi + ADP
ATP
Inside of cell
2 The membrane pigment transfers the energy to ADP to form ATP, which the cell can use as an energy source.

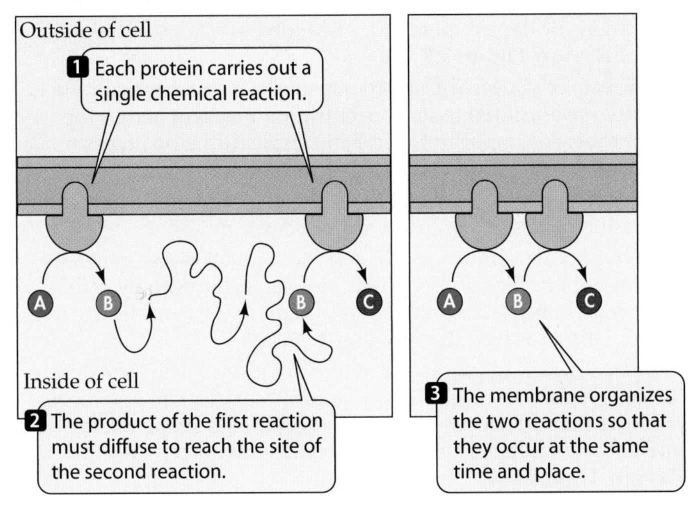
(c) Organizing chemical reactions
Outside of cell
1 Each protein carries out a single chemical reaction.
A
B
C
Inside of cell
2 The product of the first reaction must diffuse to reach the site of the second reaction.
3 The membrane organizes the two reactions so that they occur at the same time and place.

5.17 More Membrane Functions
(*a*) Membrane proteins conduct signals from outside the cell by triggering changes inside the cell. (*b*) The membranes of organelles such as mitochondria and chloroplasts are specialized for the transformation of energy. (*c*) When a series of biochemical reactions must take place in sequence, the membrane can sometimes arrange the needed enzymes in an "assembly line" to ensure that the reactions occur in proximity to each other.

Other biological activities and properties associated with membranes are discussed in the chapters that follow. These activities have been essential to the specialization of cells, tissues, and organisms throughout evolution. We review three of these activities here.

INFORMATION PROCESSING. As we have seen, biological membranes may have protruding integral membrane proteins or attached carbohydrates that can bind to specific substances in the environment. The binding of a specific substance can serve as a signal to initiate, modify, or turn off a cell function (Figure 5.17*a*).

In this type of information processing, specificity in binding is essential. We have already seen the role of a specific receptor protein in the endocytosis of LDL and its cargo of cholesterol. Another example is the binding of a hormone, such as insulin, to specific receptors on a target cell, such as a liver cell, to elicit a response in the cell—in this case, the uptake of glucose. There are many other examples, which we will discuss in Chapter 15.

ENERGY TRANSFORMATION. In a variety of cells, the membranes of organelles are specialized for processing energy (Figure 5.17*b*). For example, the inner mitochondrial membrane helps convert the energy of fuel molecules to the energy of ATP, and the thylakoid membranes of chloroplasts participate in the conversion of light energy to the energy of chemical bonds. These important processes, vital to the life of most eukaryotic organisms, are discussed in detail in Chapters 7 and 8.

ORGANIZING CHEMICAL REACTIONS. Many processes in cells depend on a series of enzyme-catalyzed reactions in which the products of one reaction serve as the reactants for the next. For such a reaction to occur, all the necessary molecules must come together. In a solution, the reactants and enzymes are all randomly distributed, and collisions among them are random. For this reason, a complete series of chemical reactions in solution may occur very slowly. However, if the different enzymes are bound to a membrane in sequential order, the product of one reaction can be released close to the enzyme for the next reaction. With such an "assembly line," reactions proceed more rapidly and efficiently (Figure 5.17*c*).

Membranes Are Dynamic

Membranes are constantly forming, transforming from one type to another, fusing with one another, and breaking down.

- *Phospholipids* in eukaryotes are synthesized on the surface of the smooth endoplasmic reticulum and rapidly distributed to membranes throughout the cell as vesicles form from the ER, move away, and fuse with other organelles.

- *Membrane proteins* are inserted into the rough endoplasmic reticulum as they form on ribosomes.
- *Functioning membranes* also move about within eukaryotic cells. Portions of the rough ER bud away as vesicles and join the *cis* face of the Golgi apparatus (see Figure 4.12). Rapidly—often in less than an hour—these segments of membrane find themselves in the *trans* regions of the Golgi, from which they bud away to join the plasma membrane.
- Membrane from *vesicles* is constantly merging with the plasma membrane by exocytosis, but this process is largely balanced by the removal of membrane in endocytosis, affording a recovery path by which internal membranes are replenished.

Because all membranes appear similar under the electron microscope, and because they interconvert readily, we might expect all subcellular membranes to be chemically identical. However, that is not the case, for there are major chemical differences among the membranes of even a single cell. Membranes are changed chemically when they form parts of certain organelles. In the Golgi apparatus, for example, the membranes of the *cis* face closely resemble those of the endoplasmic reticulum in chemical composition, but the *trans*-face membranes are more similar to the plasma membrane. As a vesicle is formed, the mix of proteins and lipids in its membrane is selected, just as its internal contents are selected, to correspond with the vesicle's target membrane.

In sum, there is a steady flux of membranes within the cell. Ceaselessly moving, functioning, changing their composition and roles, biological membranes are central to life.

Chapter Summary

Membrane Composition and Structure

▶ Biological membranes consist of lipids, proteins, and carbohydrates. The fluid mosaic model of membrane structure describes a phospholipid bilayer in which proteins can move about laterally within the membrane. **Review Figures 5.1, 5.2.** **See Web/CD Activity 5.1**

▶ Integral membrane proteins are at least partially inserted into the phospholipid bilayer. Peripheral membrane proteins are attached to the surface of the bilayer by ionic bonds. **Review Figure 5.1, 5.3, 5.4**

▶ The two surfaces of a membrane may have different properties because of their different phospholipid composition, exposed domains of integral membrane proteins, and peripheral membrane proteins. **Review Figures 5.1, 5.2**

▶ Carbohydrates attached to proteins or phospholipids project from the external surface of the plasma membrane and function as recognition signals for interactions between cells. **Review Figure 5.1**

Cell Recognition and Adhesion

▶ Some organisms consist of a single cell, but many are multicellular. The assembly of cells into tissues requires that they recognize and adhere to one another. Recognition and adhesion depend on membrane proteins that protrude from the cell surface. **Review Figure 5.5**

▶ Tight junctions prevent the passage of molecules through the spaces between cells, and they define functional regions of the plasma membrane by restricting the migration of membrane proteins uniformly over the cell surface. Desmosomes allow cells to adhere strongly to one another. Gap junctions provide channels for chemical and electrical communication between adjacent cells. **Review Figure 5.6.** **See Web/CD Activity 5.2**

Passive Processes of Membrane Transport

▶ Substances can diffuse passively across a membrane by three processes: unaided diffusion through the phospholipid bilayer, facilitated diffusion through protein channels, or facilitated diffusion by means of a carrier protein. **Review Table 5.1**

▶ A solute diffuses across a membrane from a region with a greater concentration of that solute to a region with a lesser concentration of that solute. Equilibrium is reached when the concentrations of the solute are identical on both sides of the membrane. **Review Figure 5.7**

▶ The rate of simple diffusion of a solute across a membrane is directly proportional to its concentration gradient across the membrane. An important factor in simple diffusion across a membrane is the lipid solubility of the solute.

▶ In osmosis, water diffuses from regions of higher water concentration to regions of lower water concentration.

▶ In hypotonic solutions, cells tend to take up water, whereas cells in hypertonic solutions tend to lose water. Animal cells must remain isotonic to the environment to prevent destructive loss or gain of water. **Review Figure 5.8*a,b***

▶ The cell walls of plants and some other organisms prevent the cells from bursting under hypotonic conditions. The turgor pressure that develops under these conditions keeps plants upright and stretches the cell wall during plant cell growth. **Review Figure 5.8*c***

▶ Channel proteins and carrier proteins function in facilitated diffusion. **Review Figures 5.9, 5.10, 5.11*a***

▶ The rate of carrier-mediated facilitated diffusion reaches a maximum when a solute concentration is reached that saturates the carrier proteins so that no increase in rate is observed with further increases in solute concentration. **Review Figure 5.11*b*.** **See Web/CD Tutorial 5.1**

Active Transport

▶ Active transport requires the use of chemical energy to move substances across a membrane against a concentration gradient. **Review Table 5.1**

▶ Active transport proteins may be uniports, symports, or antiports. **Review Figure 5.12**

▶ In primary active transport, energy from the hydrolysis of ATP is used to move ions into or out of cells against their concentration gradients. **Review Figure 5.13**

▶ Secondary active transport couples the passive movement of one solute with its concentration gradient to the movement of another solute against its concentration gradient. Energy from ATP is used indirectly to establish the concentration gradient that results in the movement of the first solute. **Review Figure 5.14.**

See Web/CD Tutorial 5.2

Endocytosis and Exocytosis

▶ Endocytosis transports macromolecules, large particles, and small cells into eukaryotic cells by means of engulfment by and vesicle formation from the plasma membrane. Phagocytosis and pinocytosis are both nonspecific types of endocytosis. **Review Figure 5.15**

▶ In receptor-mediated endocytosis, a specific membrane receptor protein binds to a particular macromolecule. **Review Figure 5.16. See Web/CD Tutorial 5.3**

▶ In exocytosis, materials in vesicles are secreted from the cell when the vesicles fuse with the plasma membrane.

Membranes Are Not Simply Barriers

▶ Membranes function as sites for recognition and initial processing of extracellular signals, for energy transformations, and for organizing chemical reactions. **Review Figure 5.17**

Membranes Are Dynamic

▶ Modifications in membrane composition accompany the conversions of one type of membrane into another type.

Self-Quiz

1. Which statement about membrane phospholipids is *not* true?
   a. They associate to form bilayers.
   b. They have hydrophobic "tails."
   c. They have hydrophilic "heads."
   d. They give the membrane fluidity.
   e. They flop readily from one side of the membrane to the other.
2. Human growth hormone binds to a specific protein on the plasma membrane. This protein is called a
   a. ligand.
   b. clathrin.
   c. receptor.
   d. hydrophobic protein.
   e. cell adhesion molecule.
3. Which statement about membrane proteins is *not* true?
   a. They all extend from one side of the membrane to the other.
   b. Some serve as channels for ions to cross the membrane.
   c. Many are free to migrate laterally within the membrane.
   d. Their position in the membrane is determined by their tertiary structure.
   e. Some play roles in photosynthesis.
4. Which statement about membrane carbohydrates is *not* true?
   a. Most are bound to proteins.
   b. Some are bound to lipids.
   c. They are added to proteins in the Golgi apparatus.
   d. They show little diversity.
   e. They are important in recognition reactions at the cell surface.
5. Which statement about animal cell junctions is *not* true?
   a. Tight junctions are barriers to the passage of molecules between cells.
   b. Desmosomes allow cells to adhere strongly to one another.
   c. Gap junctions block communication between adjacent cells.
   d. Connexons are made of protein.
   e. The fibers associated with desmosomes are made of protein.
6. You are studying how the protein transferrin enters cells. When you examine cells that have taken up transferring, it is inside clathrin-coated vesicles. Therefore, the most likely mechanism for uptake of transferrin is
   a. facilitated diffusion.
   b. proton antiport.
   c. receptor-mediated endocytosis.
   d. gap junctions.
   e. ion channels.
7. Which statement about membrane channels is *not* true?
   a. They are pores in the membrane.
   b. They are proteins.
   c. All ions pass through the same type.
   d. Movement through them is from high concentration to low.
   e. Movement through them is by simple diffusion.
8. Facilitated diffusion and active transport both
   a. require ATP.
   b. require the use of proteins as carriers.
   c. carry solutes in only one direction.
   d. increase without limit as the solute concentration increases.
   e. depend on the solubility of the solute in lipid.
9. Primary and secondary active transport both
   a. generate ATP.
   b. are based on passive movement of sodium ions.
   c. include the passive movement of glucose molecules.
   d. use ATP directly.
   e. can move solutes against their concentration gradients.
10. Which statement about osmosis is *not* true?
   a. It obeys the laws of diffusion.
   b. In animal tissues, water moves into the cell which is hypertonic to the medium.
   c. Red blood cells must be kept in a plasma that is hypoosmotic to the cells.
   d. Two cells with identical osmotic potentials are isosmotic to each other.
   e. Solute concentration is the principal factor in osmosis.

For Discussion

1. In Chapter 47, we will see that the functioning of muscles requires calcium ions to be pumped into a subcellular compartment against a calcium concentration gradient. What types of molecules are required for this to happen?
2. Some algae have complex glassy structures in their cell walls. These structures form within the Golgi apparatus. How do these structures reach the cell wall without having to pass through a membrane?
3. Organisms that live in fresh water are almost always hypertonic to their environment. In what way is this a serious problem? How do some organisms cope with this problem?
4. Contrast nonspecific endocytosis and receptor-mediated endocytosis with respect to mechanism and to performance.
5. The emergence of the phosopholipid membrane was important to the origin of cells. Describe the most important properties of membranes that allowed cells containing them to thrive in comparison with molecular aggregates without membranes.

6 Energy, Enzymes, and Metabolism

Millions of people, including famous athletes such as hockey star Wayne Gretzky, baseball great Hank Aaron, and Olympic decathlete Bruce Jenner, suffer from crippling arthritis. Until recently, physicians prescribed aspirin to calm the swelling that plagues arthritic joints. Aspirin has a long tradition in medicine. For thousands of years, healers in many cultures knew that the bark of a willow tree had anti-inflammatory properties that reduced swelling and pain. In 1829, German chemists isolated the active ingredient in willow bark, and later in that century others modified it chemically to make an even more effective drug—aspirin. While this drug was effective, it had several negative side effects, such as severe stomach irritation and reduced blood clotting.

It was only when biochemists discovered how aspirin works that its beneficial and undesirable effects could be explained. They discovered that aspirin binds to and adds an acetyl group to a particular amino acid (a serine) in a protein called cyclooxygenase, or COX. The normal role of COX in the body is to act as an enzyme: a catalyst to speed up the conversion of a linear fatty acid to a ring structure. The fatty acid enters a channel in the enzyme macromolecule, where it undergoes a specific chemical conversion, then departs from the enzyme. The ring form of the fatty acid stimulates inflammation in joints, repairs damage in the stomach wall, and helps blood clotting. When the serine residue in COX is acetylated by reactions with aspirin, the enzyme no longer speeds the production of the ring structure. Without COX, the formation of the ring structure still occurs, but at an exceedingly slow rate. So when COX is inhibited by aspirin, inflammation of joints is reduced, but stomach damage and a reduction in blood clotting also occur.

The search went on for a "better" anti-inflammatory drug: one that would block COX only in the joints. During the 1990s, biochemists hit paydirt. They found that there are actually *two* COX enzymes, one that acts in the stomach and blood cells (COX-1) and another that acts in the joints (COX-2). When they determined the primary structures of these two enzymes, they found that only one amino acid differs between them: COX-1 has a bulky

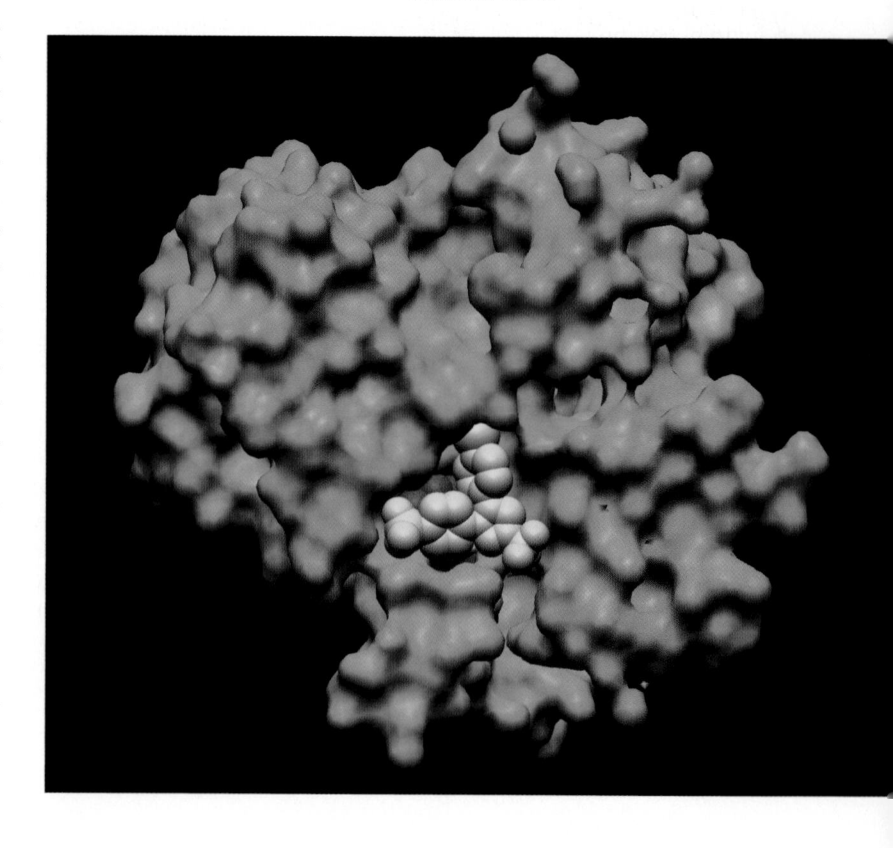

An Inflammatory Enzyme The COX-2 enzyme (green), represented here in a cutaway view, catalyzes the formation of a molecule that stimulates inflammation. Certain drugs (shown in yellow) block the substrate from binding to the active site. Such drugs inhibit the enzyme and depress inflammation. The valine residue that distinguishes COX-2 from the COX-1 enzyme is shown in red.

isoleucine where COX-2 has a smaller valine (see Table 3.2). The effect on tertiary structure was dramatic: The smaller amino acid in COX-2 exposed a side channel in the enzyme macromolecule that was blocked by the larger one in COX-1.

Using their knowledge of protein composition and structure, chemists designed molecules to plug up the COX-2 channel specifically, with no effect on COX-1. These new drugs (celecoxib and rofecoxib) relieve arthritis symptoms without the side effects in the stomach and blood. Their rapid development represents a case study in rational drug design: block a specific chemical transformation in cells by blocking the specific enzyme that catalyzes it.

Thousands of enzyme-catalyzed reactions go on all the time in every organism, each of them catalyzed by a specific protein with a particular three-dimensional structure. Taken together, these reactions make up **metabolism**, which is the total chemical activity of a living organism; at any instant, metabolism consists of thousands of individual chemical reactions. Many metabolic reactions can be classified as either the building up of complexity in the cell, using energy to do so, or the breaking down of complex substances into simpler ones, releasing energy in the process.

This chapter is concerned with energy and enzymes. Without them, neither we nor any other organism would be able to function. Before discovering how enzymes perform their molecular wizardry, we will consider the general principles of energy in biological systems.

Energy and Energy Conversions

Physicists define **energy** as the capacity to do work, which occurs when a force operates on an object over a distance. In biochemistry, it is more useful to consider energy as *the capacity for change*. No cell creates energy—all living things must obtain energy from the environment. Indeed, one of the fundamental physical laws is that energy can neither be created nor destroyed. However, energy can be *transformed* from one type into another, and living cells carry out many such energy transformations. Energy transformations are linked to the chemical transformations that occur in cells—the breaking of chemical bonds, the movement of substances across membranes, and so forth.

Energy changes are related to changes in matter

Energy comes in many forms, such as chemical energy, light energy, and mechanical energy. But all forms of energy can be considered as one of two basic types:

- **Kinetic energy** is the energy of movement. This type of energy does work that alters the state or motion of matter. It can exist in the form of heat, light, electric energy, and mechanical energy, among others.
- **Potential energy** is the energy of state or position—that is, stored energy. It can be stored in chemical bonds, as a concentration gradient, and as electric potential, among other ways.

Water stored behind a dam has potential energy. When the water is released from the dam, some of this potential energy is converted into kinetic energy, which can be harnessed to do work (Figure 6.1). Likewise, fatty acids store chemical energy in their C—H bonds and C—C bonds, and that energy can be released to do biochemical work.

6.1 Energy Conversions and Work The kinetic energy of a flowing river can be converted to potential energy by a dam. Release of water from the dam converts the potential energy back into kinetic energy, which a generator can convert into electric energy.

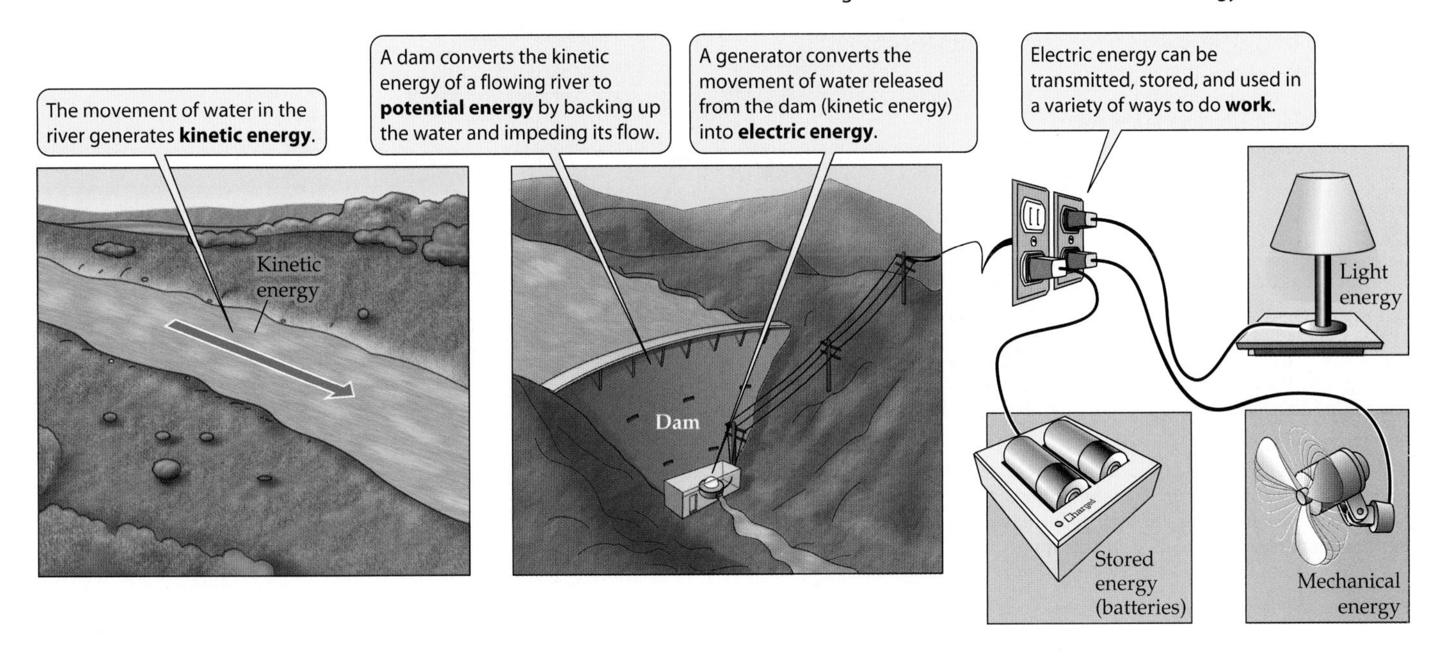

The movement of water in the river generates **kinetic energy**.
A dam converts the kinetic energy of a flowing river to **potential energy** by backing up the water and impeding its flow.
A generator converts the movement of water released from the dam (kinetic energy) into **electric energy**.
Electric energy can be transmitted, stored, and used in a variety of ways to do **work**.
Kinetic energy
Dam
Light energy
Stored energy (batteries)
Mechanical energy

In all cells of all organisms, two types of metabolic reactions occur:

- **Anabolic reactions** (anabolism) link together simple molecules to form more complex molecules. The synthesis of a protein from amino acids is an anabolic reaction. Anabolic reactions require an input of energy and capture it in the chemical bonds that are formed.
- **Catabolic reactions** (catabolism) break down complex molecules into simpler ones and release the energy stored in chemical bonds.

Catabolic and anabolic reactions are often linked. The energy released in catabolic reactions is used to drive anabolic reactions—that is, to do biological work.

Cellular activities such as growth, movement, and active transport of ions across a membrane all require energy, and none of them would proceed without a source of energy. In the discussion that follows, we will discover the physical laws that govern all energy transformations, identify the energy available to do biological work, and consider the direction of energy flow.

The first law: Energy is neither created nor destroyed

Energy can be converted from one form to another. For example, by striking a match, you convert potential chemical energy to light and heat. The **first law of thermodynamics** states that in any such conversion of energy, energy is neither created nor destroyed.

The first law tells us that *in any conversion of energy from one form to another, the total energy before and after the conversion is the same* (Figure 6.2*a*). As you will see in the next two chapters, potential energy in the chemical bonds of carbohydrates and lipids can be converted to potential energy in the form of ATP. This energy can then be used to produce potential energy in the form of concentration gradients established by active transport, and can be converted to kinetic energy and used to do mechanical work, such as muscle contraction.

6.2 The Laws of Thermodynamics (*a*) The first law states that energy cannot be created or destroyed. (*b*) The second law states that during energy transformations, some of the free energy is lost.

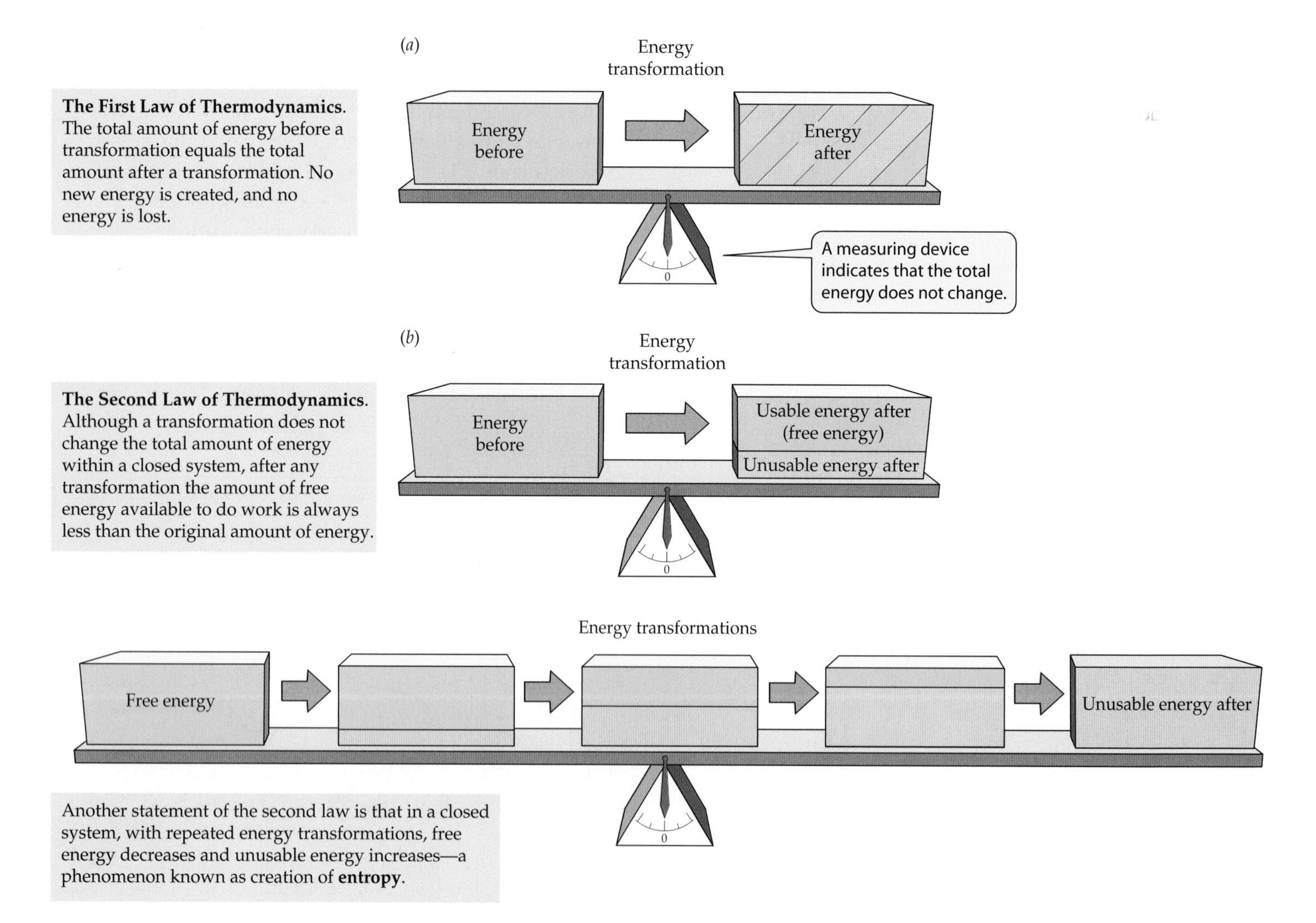

(a)
Energy transformation
The First Law of Thermodynamics. The total amount of energy before a transformation equals the total amount after a transformation. No new energy is created, and no energy is lost.
Energy before
Energy after
0
A measuring device indicates that the total energy does not change.
(b)
Energy transformation
The Second Law of Thermodynamics. Although a transformation does not change the total amount of energy within a closed system, after any transformation the amount of free energy available to do work is always less than the original amount of energy.
Energy before
Usable energy after (free energy)
Unusable energy after
0
Energy transformations
Free energy
Unusable energy after
0
Another statement of the second law is that in a closed system, with repeated energy transformations, free energy decreases and unusable energy increases—a phenomenon known as creation of entropy.

The second law: Not all energy can be used, and disorder tends to increase

The **second law of thermodynamics** states that, although energy cannot be created or destroyed, *when energy is converted from one form to another, some of that energy becomes unavailable to do work* (Figure 6.2*b*). In other words, no physical process or chemical reaction is 100 percent efficient, and not all the energy released can be converted to work. Some energy is lost to a form associated with disorder. The second law applies to all energy transformations, but we will focus here on chemical reactions in living systems.

NOT ALL ENERGY CAN BE USED. In any system, the total energy includes the usable energy that can do work *and* the unusable energy that is lost to disorder:

$$\text{total energy} = \text{usable energy} + \text{unusable energy}$$

In biological systems, the total energy is called **enthalpy** (H). The usable energy that can do work is called **free energy** (G). Free energy is what cells require for all the chemical reactions of cell growth, cell division, and the maintenance of cell health. The unusable energy is represented by **entropy** (S), which is a measure of the disorder of the system, multiplied by the absolute temperature (T). Thus we can rewrite the word equation above more precisely as

$$H = G + TS$$

Because we are interested in usable energy, we rearrange this expression:

$$G = H - TS$$

Although we cannot measure G, H, or S absolutely, we can determine the *change* in each at a constant temperature. Such energy changes are measured in calories (cal) or joules (J) (see Chapter 2). A change in energy is represented by the Greek letter delta (Δ). For example, the change in free energy (ΔG) of any chemical reaction is equal to the difference in free energy between the products and the reactants,

$$\Delta G_{\text{reaction}} = G_{\text{products}} - G_{\text{reactants}}$$

Such a change can be either positive or negative.

At a constant temperature, ΔG is defined in terms of the change in total energy (ΔH) and the change in entropy (ΔS):

$$\Delta G = \Delta H - T\Delta S$$

This equation tells us whether free energy is released or consumed by a chemical reaction:

- If ΔG is negative ($\Delta G < 0$), free energy is released.
- If ΔG is positive ($\Delta G > 0$), free energy is required (consumed).

IF THE NECESSARY FREE ENERGY IS NOT AVAILABLE, THE REACTION DOES NOT OCCUR. The sign and magnitude of ΔG depend on the two factors on the right of the equation:

- ΔH: In a chemical reaction, ΔH is the total amount of energy added to the system ($\Delta H > 0$) or released ($\Delta H < 0$).
- ΔS: Depending on the sign and magnitude of ΔS, the entire term, $T\Delta S$, may be negative or positive, large or small. In other words, in living systems at a constant temperature (no change in T), the magnitude and sign of ΔG can depend a lot on changes in entropy. Large changes in entropy make ΔG more negative in value, as shown by the negative sign in front of the $T\Delta S$ term.

If a chemical reaction increases entropy, its products are more disordered or random than its reactants. If there are more products than reactants, as in the hydrolysis of a protein to its amino acids, the products have considerable freedom to move around. The disorder in a solution of amino acids will be large compared with that in the protein, in which peptide bonds and other forces prevent free movement. So in hydrolysis, the change in entropy (ΔS) will be positive.

If there are fewer products, and they are more restrained in their movements than the reactants, ΔS will be negative. For example, a large protein linked by peptide bonds is less free in its movements than a solution of the hundreds or thousands of amino acids from which it was synthesized.

DISORDER TENDS TO INCREASE. The second law of thermodynamics also predicts that, *as a result of energy conversions, disorder tends to increase*. Chemical changes, physical changes, and biological processes all tend to increase entropy and therefore tend toward disorder or randomness (Figure 6.2*b*). This tendency for disorder to increase gives a directionality to physical processes and chemical reactions. It explains why some reactions proceed in one direction rather than another.

How does the second law apply to organisms? Consider the human body, with its highly complex structures constructed of simpler molecules. This increase in complexity is in apparent disagreement with the second law. But this is not the case! Constructing 1 kg of a human body requires that about 10 kg of biological materials be metabolized and in the process converted to CO_2, H_2O, and other simple molecules, and these conversions require a lot of energy. This metabolism creates far more disorder than the order in 1 kg of flesh. *Life requires a constant input of energy to maintain order*. There is no disagreement with the second law of thermodynamics.

Having seen that the physical laws of energy apply to living things, we'll now turn to a consideration of how these laws apply to biochemical reactions.

Chemical reactions release or take up energy

In cells, anabolic reactions may make a single product, such as a protein (a highly ordered substance), out of many smaller reactants, such as amino acids (less ordered). Such reactions require or consume energy. Catabolic reactions may break down an ordered reactant, such as a glucose molecule, into smaller, more randomly distributed products, such as carbon dioxide and water. Such reactions give off energy. In other words, some reactions release free energy, and others take it up.

The amount of energy released ($-\Delta G$) or taken up ($+\Delta G$) by a reaction is related directly to the tendency of the reaction to run to *completion* (the point at which all the reactants are converted to products):

- Some reactions tend to run toward completion without any input of energy. These reactions, which release free energy, are said to be **exergonic** and have a negative ΔG (Figure 6.3*a*).
- Reactions that proceed toward completion only with the addition of free energy from the environment are **endergonic** and have a positive ΔG (Figure 6.3*b*).

If a reaction runs exergonically in one direction (from reactant A to product B, for example), then the reverse reaction (B to A) requires a steady supply of energy to drive it. If $A \rightarrow B$ is exergonic ($\Delta G < 0$), then $B \rightarrow A$ is endergonic ($\Delta G > 0$).

In principle, chemical reactions can run both forward and backward. For example, if compound A can be converted into compound B ($A \rightarrow B$), then B, in principle, can be converted into A ($B \rightarrow A$), although at given concentrations of A and B, only one of these directions will be favored. Think of the overall reaction as resulting from competition between forward and reverse reactions ($A \rightleftharpoons B$). Increasing the concentration of the reactants (A) speeds up the forward reaction, and increasing the concentration of the products (B) favors the reverse reaction. At some concentration of A and B, the forward and reverse reactions take place at the same rate. At this concentration, no further net change in the system is observable, although individual molecules are still forming and breaking apart. This balance between forward and reverse reactions is known as **chemical equilibrium**. Chemical equilibrium is a static state, a state of no net change, and a state in which $\Delta G = 0$.

Chemical equilibrium and free energy are related

Every chemical reaction proceeds to a certain extent, but not necessarily to completion. In other words, all the reactants present are not necessarily converted to products. Each reaction has a specific equilibrium point, and that equilibrium point is related to the free energy released by the reaction under specified conditions. To understand the principle of equilibrium, consider the following example.

Most cells contain glucose 1-phosphate, which is converted in the cell to glucose 6-phosphate. Imagine that we

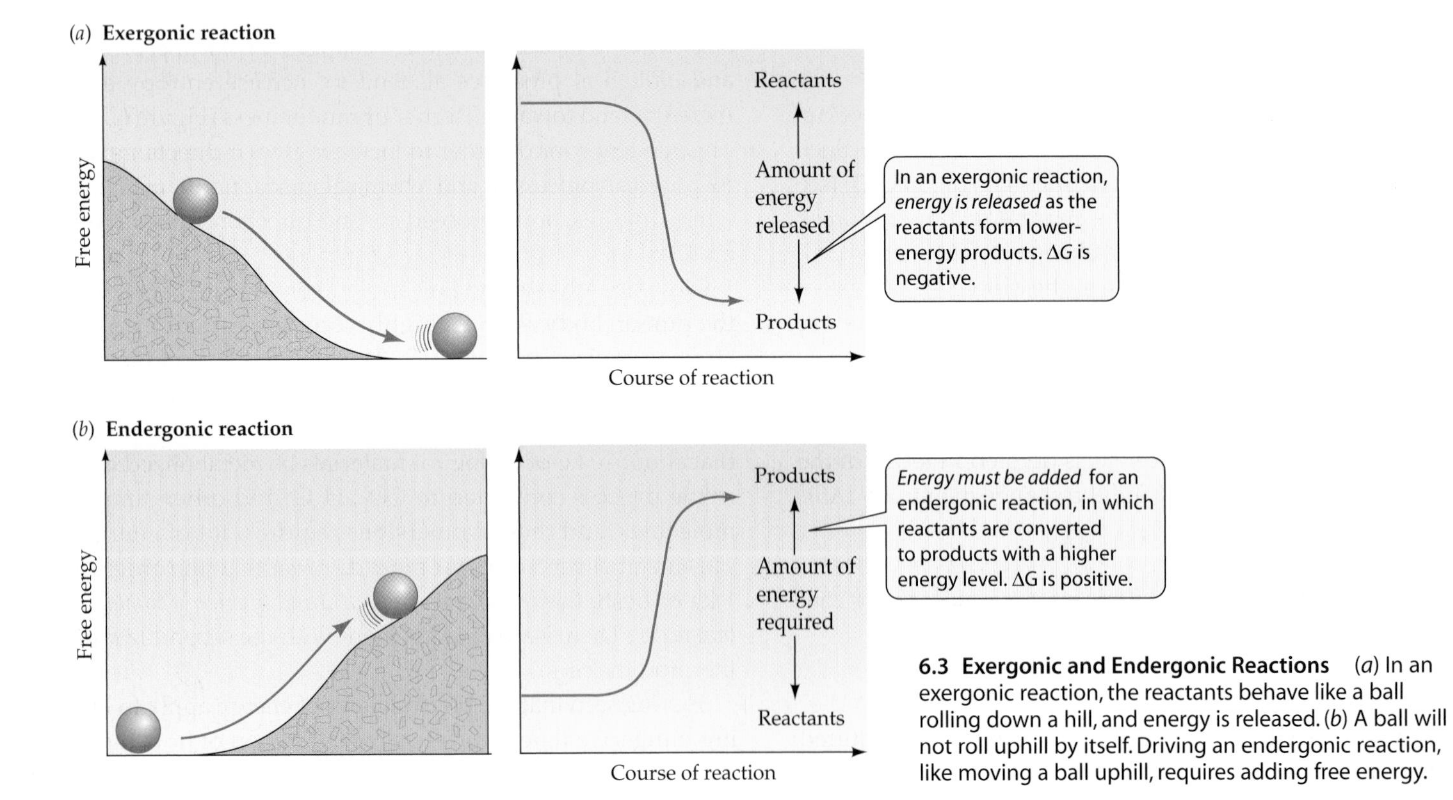
(a) Exergonic reaction
Free energy
Reactants
Amount of energy released
Products
Course of reaction
In an exergonic reaction, energy is released as the reactants form lower-energy products. ΔG is negative.
(b) Endergonic reaction
Free energy
Products
Amount of energy required
Reactants
Course of reaction
Energy must be added for an endergonic reaction, in which reactants are converted to products with a higher energy level. ΔG is positive.

6.3 Exergonic and Endergonic Reactions (*a*) In an exergonic reaction, the reactants behave like a ball rolling down a hill, and energy is released. (*b*) A ball will not roll uphill by itself. Driving an endergonic reaction, like moving a ball uphill, requires adding free energy.

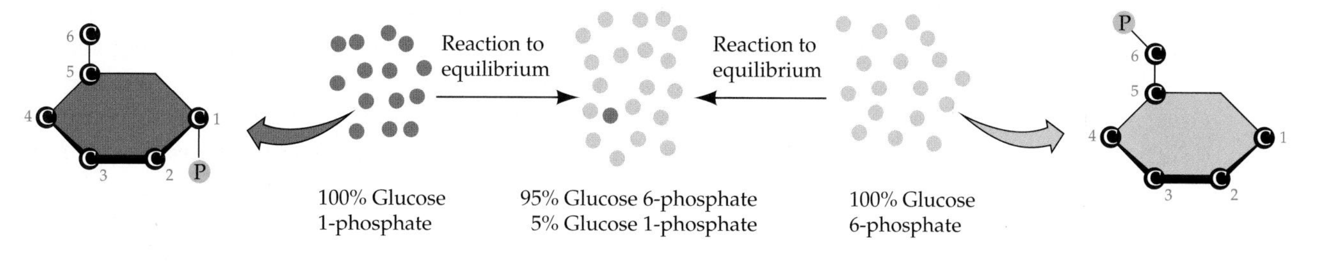

6.4 Concentration at Equilibrium No matter what quantities of glucose 1-phosphate and glucose 6-phosphate are dissolved in water, when equilibrium is attained, there will always be 95% glucose 6-phosphate and 5% glucose 1-phosphate.

start out with an aqueous solution of glucose 1-phosphate that has a concentration of 0.02 M. (M stands for molar concentration; see Chapter 2.) The solution is maintained under constant environmental conditions (25°C and pH 7). As the reaction proceeds slowly to equilibrium, the concentration of the product, glucose 6-phosphate, rises from 0 to 0.019 M, while the concentration of the reactant, glucose 1-phosphate, falls to 0.001 M. At this point, equilibrium is reached (Figure 6.4). From then on, the reverse reaction, from glucose 6-phosphate to glucose 1-phosphate, progresses at the same rate as the forward reaction.

At equilibrium, then, this reaction has a product-to-reactant ratio of 19:1 (0.019/0.001), so the forward reaction has gone 95 percent of the way to completion ("to the right," as written). Therefore, the forward reaction is an exergonic reaction. This result is obtained every time the experiment is run under the same conditions. The reaction is described by the equation

$$\text{glucose 1-phosphate} \rightleftharpoons \text{glucose 6-phosphate}$$

The change in free energy (ΔG) for any reaction is related directly to its point of equilibrium. The further toward completion the point of equilibrium lies, the more free energy is given off. In an exergonic reaction, such as the conversion of glucose 1-phosphate to glucose 6-phosphate, ΔG is a negative number (in this example, $\Delta G = -1.7$ kcal/mol, or –7.1 kJ/mol).

A large, positive ΔG for a reaction means that it proceeds hardly at all to the right ($A \rightarrow B$). But if the product is present, such a reaction runs backward, or "to the left" ($A \leftarrow B$), (nearly all B is converted to A). A ΔG value near zero is characteristic of a readily reversible reaction: reactants and products have almost the same free energies.

The principles of thermodynamics we have been discussing apply to all energy exchanges in the universe, so they are very powerful and useful. Next, we'll apply them to reactions in cells that involve the biological energy currency, ATP.

ATP: Transferring Energy in Cells

All living cells rely on **adenosine triphosphate**, or **ATP**, for the capture and transfer of the free energy needed to do chemical work and maintain the cells. ATP operates as a kind of energy currency. That is, just as you may earn money from a job and then spend it on a meal, some of the free energy released by certain exergonic reactions is captured in ATP, which can then release free energy to drive endergonic reactions.

ATP is produced by cells in a number of ways (which we will describe in the next two chapters), and it is used in many ways. ATP is not an unusual molecule. In fact, it has another important use in the cell: it can be converted into a building block for DNA and RNA. But two things about ATP make it especially useful to cells: it releases a relatively large amount of energy when hydrolyzed, and it can phosphorylate (donate a phosphate group to) many different molecules. We will examine these two properties in the discussion that follows.

ATP hydrolysis releases energy

An ATP molecule consists of the nitrogenous base adenine bonded to ribose (a sugar), which is attached to a sequence of three phosphate groups (Figure 6.5). The hydrolysis of ATP yields **ADP** (**adenosine diphosphate**) and an inorganic phosphate ion (abbreviated P_i, short for HPO_4^{2-}), as well as free energy:

$$ATP + H_2O \rightarrow ADP + P_i + \text{free energy}$$

The important property of this reaction is that it is exergonic, releasing free energy. The change in free energy (ΔG) is about –12 kcal/mol (–50 kJ/mol) at the temperature, pH, and substrate concentrations typical of living cells.*

What characteristics of ATP account for the free energy released by the loss of one of its phosphate groups? First and foremost, the free energy of the P—O bond between phos-

*The "standard" ΔG for ATP hydrolysis is –7.3 kcal/mol or –30 kJ/mol, but that value is valid only at pH 7 and with ATP, ADP, and phosphate present at concentrations of 1 M—concentrations that differ greatly from those found in cells.

(a)

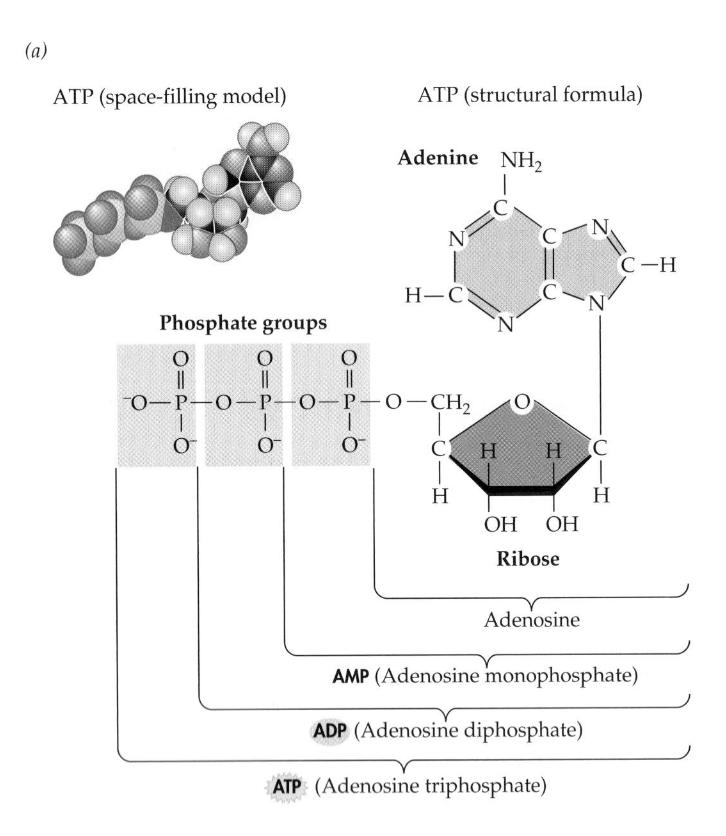

(b)

6.5 ATP *(a)* ATP is richer in energy than its relatives ADP and AMP. The hydrolysis of ATP releases this energy. *(b)* Fireflies use ATP to initiate the oxidation of luciferin. This converts chemical energy into light energy, emitting rhythmic flashes that signal the insect's readiness to mate. Very little of the energy in this conversion is lost as heat.

phate groups is much higher than the energy of the H—O bond that forms after hydrolysis. So some usable energy is released upon hydrolysis. Second, because phosphates are negatively charged and so repel each other, it takes energy to get phosphates near enough to each other to make the covalent bond that links them together (e.g., to add a phosphate to ADP to make ATP).

ATP couples exergonic and endergonic reactions

As we have just seen, the hydrolysis of ATP is exergonic and yields ADP, P_i, and free energy. The reverse reaction, the formation of ATP from ADP and P_i, is endergonic and consumes as much free energy as is released by the breakdown of ATP:

$$ADP + P_i + \text{free energy} \rightarrow ATP + H_2O$$

Many different exergonic reactions in the cell can provide the energy to convert ADP to ATP. In eukaryotes, the most important of these reactions is cellular respiration, in which some of the energy released from fuel molecules is captured in ATP. The formation and hydrolysis of ATP constitute what might be called an "energy-coupling cycle," in which ADP picks up energy from exergonic reactions to become ATP, which donates energy to endergonic reactions.

How does this ATP cycle trap and release energy? An exergonic reaction is coupled to the endergonic reaction that forms ATP from ADP and P_i (Figure 6.6). Coupling of exergonic and endergonic reactions is very common in metabolism. When it forms, ATP captures free energy and retains it like a compressed spring. ATP then diffuses to another site in the cell, where its hydrolysis releases free energy to drive an endergonic reaction.

A specific example of this energy-coupling cycle is shown in Figure 6.7. The formation of the amino acid glutamine has a positive ΔG (is endergonic) and will not proceed without the input of free energy from ATP hydrolysis, which has a negative ΔG (is exergonic). The total ΔG for the coupled reactions is negative (the two ΔGs are added together). Hence the reactions proceed exergonically when they are coupled, and glutamine is synthesized.

An active cell requires millions of molecules of ATP per second to drive its biochemical machinery. An ATP molecule

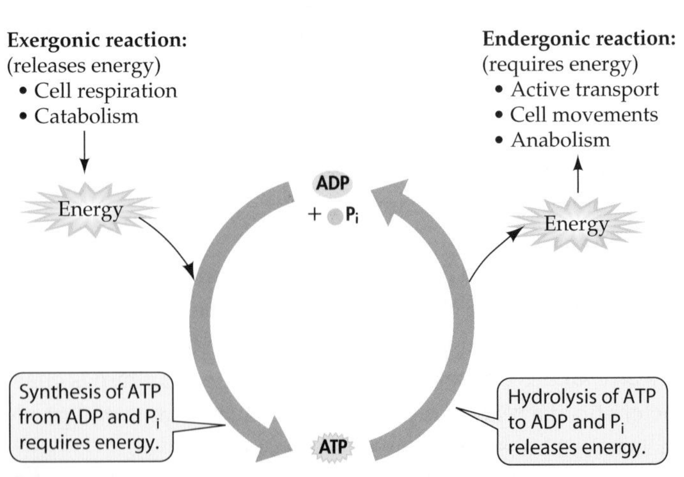

6.6 The Energy-Coupling Cycle of ATP Exergonic cellular processes release the energy needed to make ATP from ADP. The energy released from the conversion of ATP back to ADP can be used to fuel endergonic processes.

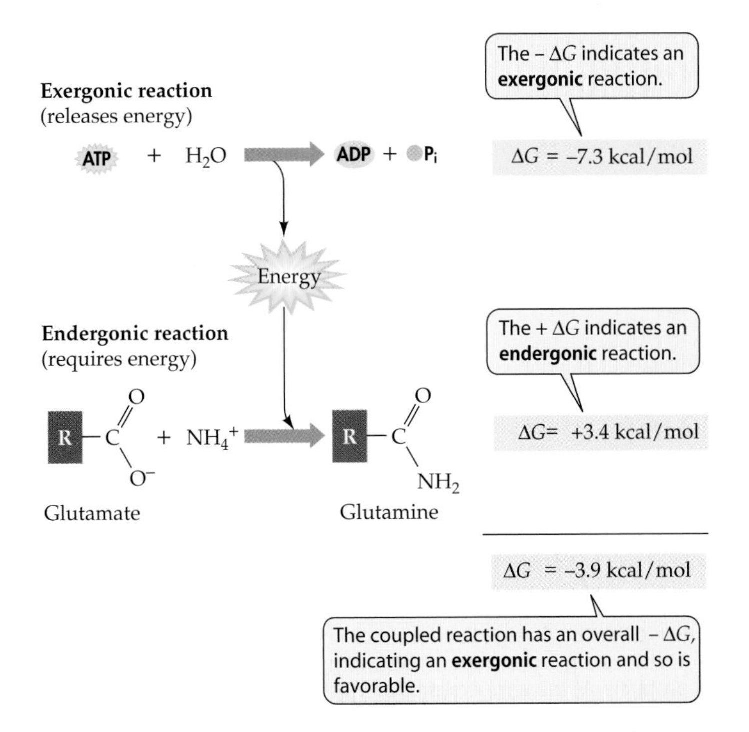

6.7 Coupling ATP Hydrolysis to an Endergonic Reaction The synthesis of the amino acid glutamine from glutamate and an ammonium ion is endergonic and must be coupled with the exergonic hydrolysis of ATP.

is consumed within a second following its formation, on average. At rest, an average person hydrolyzes and produces about 40 kg of ATP per day—as much as some people weigh! This means that each ATP molecule undergoes about 10,000 cycles of synthesis and hydrolysis every day.

Enzymes: Biological Catalysts

When we know the change in free energy (ΔG) of a reaction, we know where the equilibrium point of the reaction lies: The more negative ΔG is, the further the reaction proceeds toward completion. However, ΔG tells us nothing about the *rate* of a reaction—the speed at which it moves toward equilibrium. The reactions that occur in cells are so slow that they could not contribute to life unless the cells did something to speed them up. That is the role of **catalysts**: substances that speed up a reaction without being permanently altered by that reaction. A catalyst does not cause a reaction that would not take place eventually without it, but merely speeds up the rates of both forward and backward reactions, allowing equilibrium to be approached faster.

Most biological catalysts are proteins called **enzymes**. Although we will focus here on proteins, some catalysts—perhaps the earliest ones in the origin of life—are RNA molecules called **ribozymes** (see Chapter 3). A biological catalyst, whether protein or RNA, is a framework or scaffold in which chemical catalysis takes place. It does not matter whether the framework is RNA or protein—indeed, artificial catalysts can be made from DNA. Evolution has selected proteins as catalysts, probably because of their great diversity in three-dimensional structure and variety of chemical functions.

In the discussion that follows, we will identify the energy barrier that controls the rate of reactions. Then we'll focus on the role of enzymes: how they interact with reactants, how they lower the energy barrier, and how they permit reactions to proceed faster. After exploring the nature and significance of enzyme specificity, we'll look at how enzymes contribute to the coupling of reactions.

For a reaction to proceed, an energy barrier must be overcome

An exergonic reaction may release a great deal of free energy, but the reaction may take place very slowly. Some reactions are slow because there is an *energy barrier* between reactants and products. Think about a gas stove. The burning of the natural gas (methane + $O_2 \rightarrow CO_2 + H_2O$) is obviously an exergonic reaction—heat and light are released. Once started, the reaction goes to completion: all of the methane reacts with oxygen to form carbon dioxide and water vapor.

Because burning methane liberates so much energy, you might expect this reaction to proceed rapidly whenever methane is exposed to oxygen. But this does not happen. Simply mixing methane with air produces no reaction. Methane will start burning only if a spark—an input of energy—is provided. (On the stove, this energy is supplied by electricity.) The need for this spark to start the reaction shows that there is an energy barrier between the reactants and the products.

In general, exergonic reactions proceed only after the reactants are pushed over the energy barrier by a small amount of added energy. The energy barrier thus represents the amount of energy needed to start the reaction, known as the **activation energy** (E_a) (Figure 6.8*a*). Recall the ball rolling down the hill in Figure 6.3. The ball has a lot of potential energy at the top of the hill. However, if the ball is stuck in a small depression, it won't roll down the hill, even though that action is exergonic (Figure 6.8*b*). To start the ball rolling, a small amount of energy (activation energy) is needed to get the ball out of the depression (Figure 6.8*c*).

In a chemical reaction, the activation energy is the energy needed to change the reactants into unstable molecular forms called *transition-state species*. Transition-state species have higher free energies than either the reactants or the products. Their bonds may be stretched and hence unstable. Although the amount of activation energy needed for different reactions varies, it is often small compared with the change in free energy of the reaction. The activation energy that starts a reaction is recovered during the ensuing "downhill" phase of the reaction, so it is not a part of the net free energy change, ΔG (see Figure 6.8*a*).

Where does the activation energy come from? In any collection of reactants at room or body temperature, molecules are moving around and could use their kinetic energy of motion to overcome the energy barrier, enter the transition state, and react (Figure 6.9). However, at normal temperatures, only a few molecules have enough energy to do this; most have insufficient kinetic energy for activation, so the reaction takes place slowly. If the system were heated, all the reactant molecules would move faster and have more kinetic energy. Since more of them would have energy exceeding the required activation energy, the reaction would speed up.

However, adding enough heat to increase the average kinetic energy of the molecules won't work in living systems. Such a nonspecific approach would accelerate all reactions, including destructive ones, such as the denaturation of proteins (see Figure 3.11). A more effective way to speed up a reaction in a living system is to lower the energy barrier. In living cells, enzymes accomplish this task.

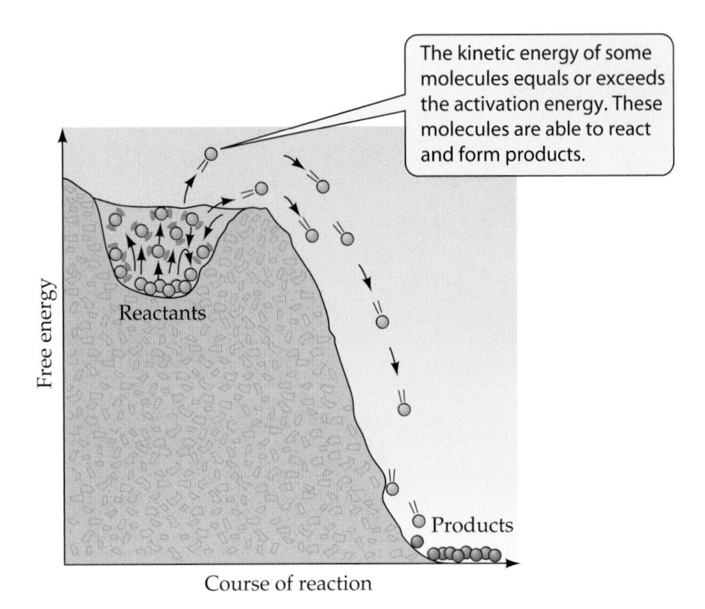

The kinetic energy of some molecules equals or exceeds the activation energy. These molecules are able to react and form products.
Free energy
Reactants
Products
Course of reaction

6.9 Over the Energy Barrier Some molecules have enough kinetic energy to surmount the energy barrier and react, forming products. At the temperatures of most organisms, however, only a small proportion of the molecules have that much kinetic energy.

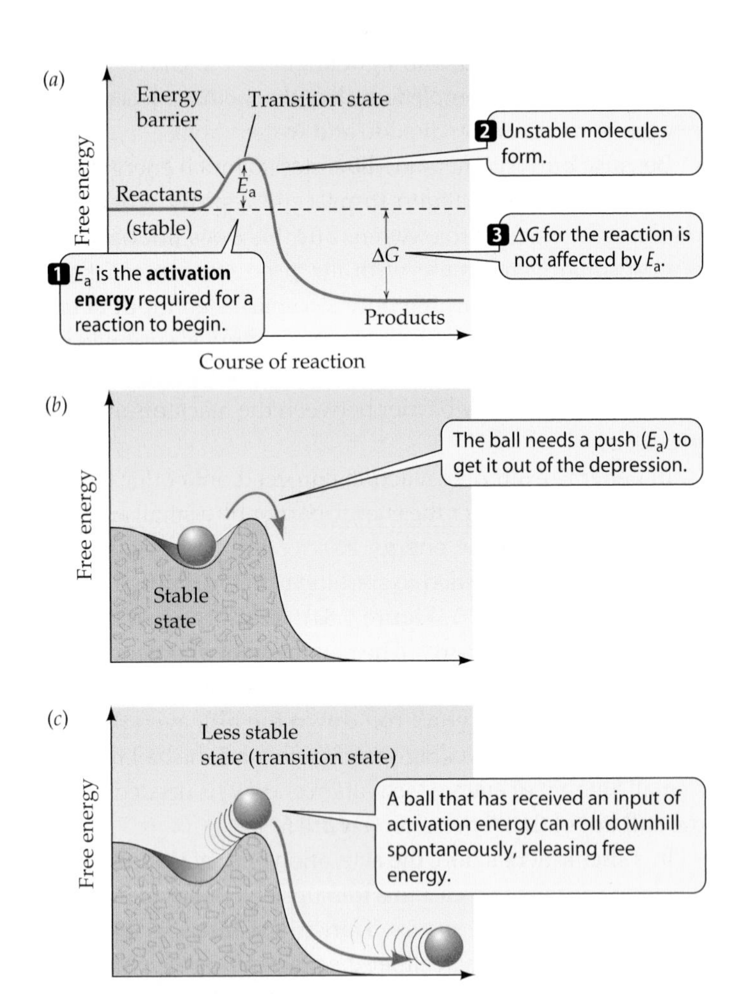

6.8 Activation Energy Initiates Reactions (*a*) In any chemical reaction, an initial stable state must become less stable before change is possible. (*b,c*) A ball on a hillside provides a physical analogy to the biochemical principle graphed in (*a*).

Enzymes bind specific reactant molecules

Catalysts increase the rate of chemical reactions. Most *nonbiological* catalysts are *nonspecific*. For example, powdered platinum catalyzes virtually any reaction in which molecular hydrogen (H_2) is a reactant. In contrast, most *biological* catalysts are *highly specific*. These complex molecules of protein (enzymes) or RNA (ribozymes) catalyze relatively simple chemical reactions. An enzyme or ribozyme usually recognizes and binds to only one or a few closely related reactants, and it catalyzes only a single chemical reaction. In the discussion that follows, we focus on enzymes, but you should note that similar rules of chemical behavior apply to ribozymes as well.

In an enzyme-catalyzed reaction, the reactants are called **substrates**. Substrate molecules bind to a particular site on the enzyme, called the **active site**, where catalysis takes place (Figure 6.10). The specificity of an enzyme results from the exact three-dimensional shape and structure of its active site, into which only a narrow range of substrates can fit. Other molecules—with different shapes, different functional groups, and different properties—cannot properly fit and bind to the active site.

The names of enzymes reflect the specificity of their functions and often end with the suffix "-ase." For example, the

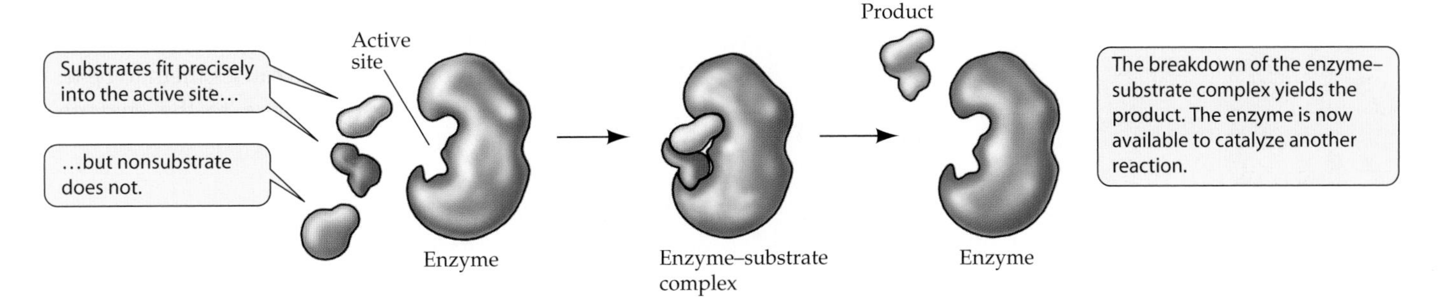

6.10 Enzyme and Substrate An enzyme is a protein catalyst with an active site capable of binding one or more substrate molecules. The enzyme–substrate complex yields product and free enzyme.

enzyme RNA polymerase catalyzes the formation of RNA, but not DNA, and the enzyme hexokinase accelerates the phosphorylation of hexose sugars, but not pentose sugars.

The binding of a substrate to the active site of an enzyme produces an **enzyme–substrate complex** (ES) held together by one or more means, such as hydrogen bonding, ionic attraction, or covalent bonding. The enzyme–substrate complex gives rise to product and free enzyme:

$$E + S \rightarrow ES \rightarrow E + P$$

where E is the enzyme, S is the substrate, P is the product, and ES is the enzyme–substrate complex. The free enzyme (E) is in the same chemical form at the end of the reaction as at the beginning. While bound to the substrate, it may change chemically, but by the end of the reaction it has been restored to its initial form.

Enzymes lower the energy barrier but do not affect equilibrium

When reactants are part of an enzyme–substrate complex, they require less activation energy than the transition-state species of the corresponding uncatalyzed reaction (Figure 6.11). Thus the enzyme lowers the energy barrier for the reaction—it offers the reaction an easier path. When an enzyme lowers the energy barrier, both the forward and the reverse reactions speed up, so the enzyme-catalyzed overall reaction proceeds toward equilibrium more rapidly than the uncatalyzed reaction. The final equilibrium (and ΔG) is the same with or without the enzyme.

Adding an enzyme to a reaction does not change the difference in free energy (ΔG) between the reactants and the products (see Figure 6.11). It does change the activation energy and, consequently, the rate of reaction. For example, if 600 molecules of a protein with arginine as its terminal amino acid just sit in solution, the proteins tend toward disorder, and the terminal peptide bonds break, releasing the arginines (ΔS increases). After 7 years, about half (300) of the proteins will have undergone this reaction. With the enzyme carboxypeptidase A catalyzing the reaction, however, the 300 arginines are released in half a second!

Free energy
E_a
E_a
Uncatalyzed reaction
Reactants
Catalyzed reaction
ΔG
Products
Course of reaction
An uncatalyzed reaction has a greater activation energy than does a catalyzed reaction.
A catalyzed reaction has a lower activation energy.
There is no difference in free energy between catalyzed and uncatalyzed reactions.

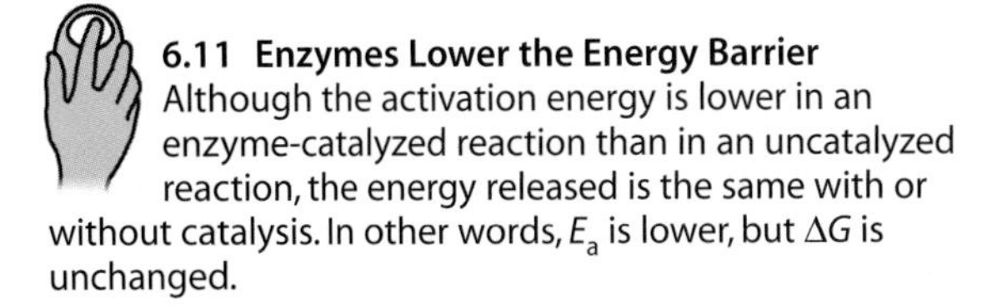
6.11 Enzymes Lower the Energy Barrier Although the activation energy is lower in an enzyme-catalyzed reaction than in an uncatalyzed reaction, the energy released is the same with or without catalysis. In other words, E_a is lower, but ΔG is unchanged.

What are the chemical events at active sites of enzymes?

After formation of the enzyme–substrate complex, chemical interactions occur. These interactions contribute directly to the breaking of old bonds and the formation of new ones (Figure 6.12). In catalyzing a reaction, an enzyme may use one or more of the following mechanisms:

ENZYMES ORIENT SUBSTRATES. While free in solution, substrates are rotating and tumbling around and may not have the proper orientation to interact when they collide. Part of the activation energy needed to start a reaction is used to make the substrates collide with the right atoms for bond formation next to each other. When proteins are synthe-

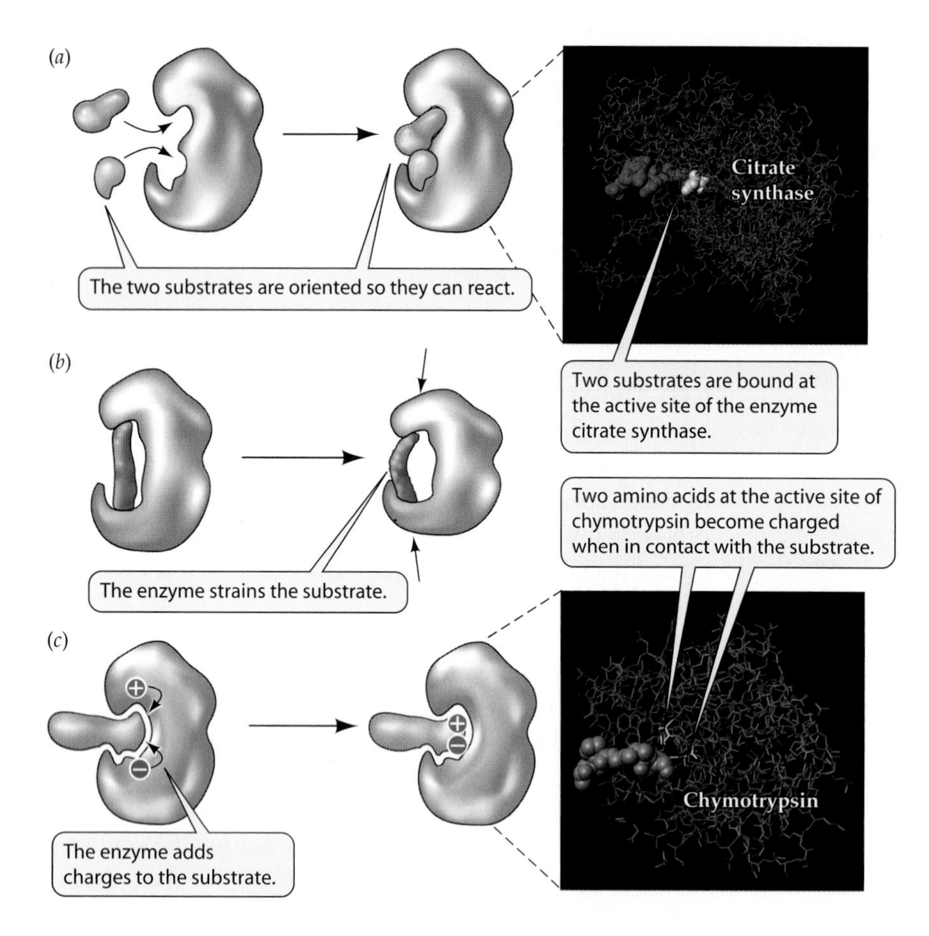
(a)
The two substrates are oriented so they can react.
Citrate synthase
Two substrates are bound at the active site of the enzyme citrate synthase.
(b)
The enzyme strains the substrate.
Two amino acids at the active site of chymotrypsin become charged when in contact with the substrate.
(c)
The enzyme adds charges to the substrate.
Chymotrypsin

6.12 Life at the Active Site Enzymes have several ways of causing their substrates to enter the transition state: (*a*) orientation, (*b*) physical strain, and (*c*) chemical change.

sized, for example, a peptide bond is formed between the carboxyl group of one amino acid and the amino group of the next (see Figure 3.5). If two amino acids are to form a peptide bond, the carboxyl group of one and amino group of the other must be the sites of collision. When the active site of an enzyme binds to one amino acid, however, it is held in the right orientation to react with a second amino acid when that substrate binds to the enzyme.

ENZYMES INDUCE STRAIN IN THE SUBSTRATE. Once a substrate has bound to the active site, the enzyme can cause bonds in the substrate to stretch, putting it in an unstable transition state. For example, the polysaccharide substrate for the enzyme lysozyme enters the active site in a flat-ringed "chair" shape, but the active site quickly causes it to flatten out into a "sofa" (Figure 6.13). The resulting stretching of its bonds causes them to be less stable and more reactive to the enzyme's other substrate, water.

ENZYMES TEMPORARILY ADD CHEMICAL GROUPS TO SUBSTRATES. The side chains (R groups) of an enzyme's amino acids may be direct participants in making its substrates more chemically reactive. For example, in *acid-base catalysis*, the acidic or basic side chains of the amino acids forming the active site may transfer H^+ to or from the substrate, destabilizing a covalent bond in the substrate and permitting it to break. In *covalent catalysis*, a functional group in a side chain forms a temporary covalent bond with a portion of the substrate. In *metal ion catalysis*, metal ions such as copper, zinc, iron, and manganese, which are firmly bound to side chains of the protein, can lose or gain electrons without detaching from the protein (Figure 6.12*c*). This ability makes them important participants in oxidation–reduction reactions, which involve loss or gain of electrons.

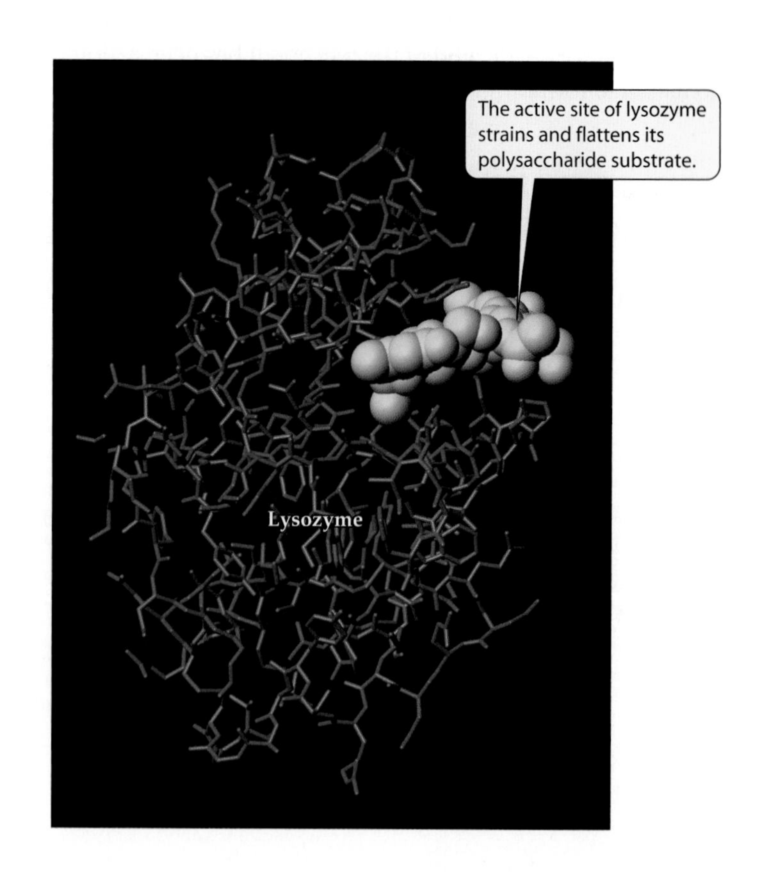
The active site of lysozyme strains and flattens its polysaccharide substrate.
Lysozyme

6.13 Tertiary Structure of Lysozyme Lysozyme is an enzyme that protects the animals that produce it by destroying invading bacteria. To destroy the bacteria, it cleaves certain polysaccharide chains in their cell walls.

Molecular Structure Determines Enzyme Function

Most enzymes (and ribozymes) are much larger than their substrates. An enzyme is typically a protein containing hundreds of amino acids, and may consist of a single folded polypeptide chain or several subunits. Its substrate is generally a small molecule. The active site of the enzyme is usually quite small, not more than 6–12 amino acids. Two questions arise from this observation:

- What is the nature of the active site that allows it to recognize and bind the substrate?
- What is the role of the rest of the huge protein?

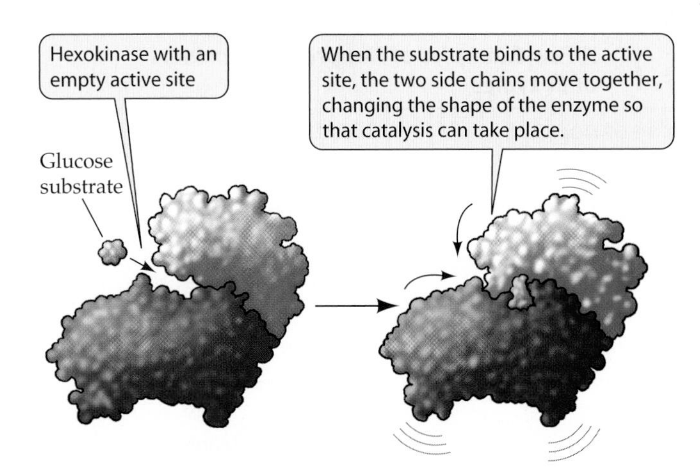

6.14 Some Enzymes Change Shape When Substrate Binds to Them Shape changes result in an induced fit between enzyme and substrate, improving the catalytic ability of the enzyme.

The active site is specific to the substrate

The remarkable ability of an enzyme to select exactly the right substrate depends on a precise interlocking of molecular shapes and interactions of chemical groups at the binding site. The binding of the substrate to the active site depends on the same kinds of forces that maintain the tertiary structure of the enzyme: hydrogen bonds, the attraction and repulsion of electrically charged groups, and hydrophobic interactions.

In 1894, the German chemist Emil Fischer compared the fit between an enzyme and its substrate to that of a lock and key. Fischer's model persisted for more than half a century with only indirect evidence to support it. The first direct evidence came in 1965, when David Phillips and his colleagues at the Royal Institution in London succeeded in crystallizing the enzyme lysozyme and determined its tertiary structure using the techniques of X-ray crystallography (described in Chapter 11). They observed a pocket in lysozyme that neatly fits its substrate (see Figure 6.13).

An enzyme changes shape when it binds a substrate

As proteins, enzymes are not immutable structures. Just as the structure of egg white protein changes when the egg is heated, many enzymes change their structure (albeit less dramatically) when they bind to their substrates. These shape changes expose those regions of the enzyme—the active sites—that actually react with the substrate. Such a change in enzyme shape caused by substrate binding is called **induced fit**.

Induced fit can be observed in the enzyme hexokinase (Figure 6.14) when it is studied with and without one of its substrates, glucose (its other substrate is ATP). It catalyzes the reaction

$$\text{glucose} + \text{ATP} \rightarrow \text{glucose 6-phosphate} + \text{ADP}$$

Induced fit brings reactive side chains from the enzyme's active site into alignment with the substrates, facilitating the catalytic mechanisms described earlier (see Figure 6.12).

Equally important, the folding of hexokinase to fit around the glucose substrate excludes water from the active site. This is essential, because the two molecules binding to the active site are glucose and ATP. If water were present, ATP could be hydrolyzed to ADP and phosphate. But since water is absent, the transfer of a phosphate from ATP to glucose is favored.

Induced fit at least partly explains why enzymes are so large. The rest of the macromolecule may have two roles:

- It provides a framework so that the amino acids of the active site are properly positioned in relation to the substrate.
- It participates in the small but significant changes in protein shape and structure that result in induced fit.

Some enzymes require other molecules in order to operate

As large and complex as enzymes are, many of them require the presence of other, nonprotein molecules in order to function (Table 6.1). Some of these molecular "partners" include:

- **Cofactors**. These are inorganic ions such as copper, zinc, or iron that bind to certain enzymes and are essential to their function.
- **Coenzymes**. These carbon-containing molecules are required for the action of one or more enzymes. Coenzymes are usually relatively small compared with the enzyme to which they temporarily bind (Figure 6.15).

6.1 A Few Examples of Nonprotein "Partners" of Enzymes

TYPE OF MOLECULE	ROLE IN CATALYZED REACTIONS
Cofactors	
Iron (Fe^{2+} or Fe^{3+})	Oxidation/reduction
Copper (Cu^{+} or Cu^{2+})	Oxidation/reduction
Zinc (Zn^{2+})	Helps bind NAD
Coenzymes	
Biotin	Carries $—COO^-$
Coenzyme A	Carries $—CH_2—CH_3$
NAD	Carries electrons
FAD	Carries electrons
ATP	Provides/extracts energy
Prosthetic groups	
Heme	Binds ions, O_2, and electrons; contains iron cofactor
Flavin	Binds electrons
Retinal	Converts light energy

- **Prosthetic groups**. These distinctive molecular groupings are permanently bound to their enzymes. They include the heme groups that are attached to the oxygen-carrying protein hemoglobin (shown in Figure 3.8).

Coenzymes are like substrates in that they are not permanently bound to the enzyme, and must collide with the enzyme and bind to its active site. A coenzyme can be considered a substrate because it changes chemically during the reaction and then separates from the enzyme to participate in other reactions. Coenzymes move from enzyme molecule to enzyme molecule, adding or removing chemical groups from the substrate.

ATP and ADP can be considered coenzymes because they are necessary for some reactions, are changed by reactions, and bind to and detach from the enzyme. In the next chapter, we will encounter coenzymes that function in energy processing by accepting or donating electrons or hydrogen atoms. In animals, some coenzymes are produced from *vitamins* that must be obtained from food—they cannot be synthesized by the body. For example, the B vitamin niacin is used to make the coenzyme NAD.

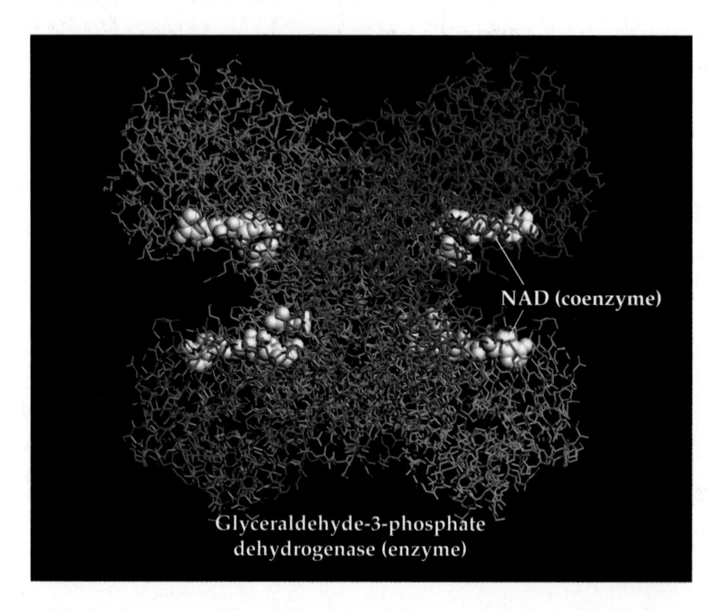

6.15 An Enzyme with a Coenzyme Some enzymes require coenzymes in order to function. This illustration shows the relative sizes of the four subunits (red, orange, green, and purple) of the enzyme glyceraldehyde 3-phosphate dehydrogenase and its coenzyme, NAD (white).

Substrate concentration affects reaction rate

For a reaction of the type A → B, the rate of the uncatalyzed reaction is directly proportional to the concentration of A (Figure 6.16). The higher the concentration of substrate, the more reactions per unit of time. Addition of the appropriate enzyme speeds up the reaction, of course, but it also changes the shape of the plot of rate versus substrate concentration. At first, the rate of the enzyme-catalyzed reaction increases as the substrate concentration increases, but then it levels off. When further increases in the substrate concentration do not significantly increase the reaction rate, the maximum rate is attained.

Since the concentration of an enzyme is usually much lower than that of its substrate, what we are seeing is a *saturation* phenomenon like the one that occurs in facilitated diffusion (see Chapter 5). When all the enzyme molecules are bound to substrate molecules, the enzyme is working as fast as it can—at its maximum rate. Nothing is gained by adding more substrate, because no free enzyme molecules are left to act as catalysts.

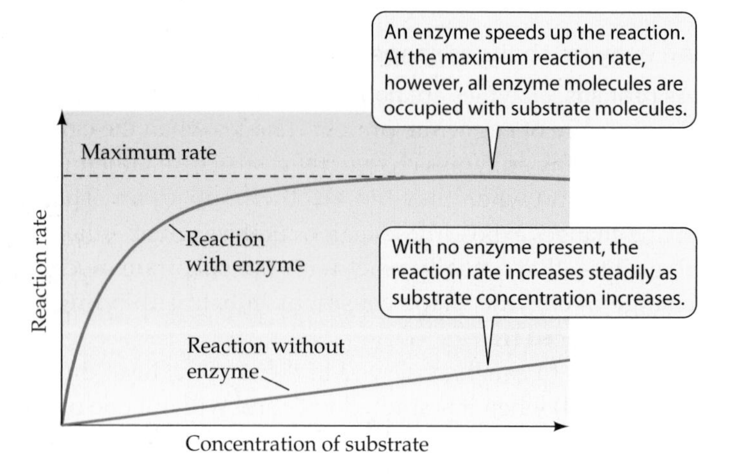

6.16 Catalyzed Reactions Reach a Maximum Rate Because there is usually less enzyme than substrate present, the reaction rate levels off when the enzyme becomes saturated.

The maximum rate of an enzyme reaction can be used to measure how efficient the enzyme can be—that is, how many molecules of substrate are converted to product per unit of time when there is an excess of substrate present. This turnover number ranges from 1 molecule every 2 seconds for lysozyme (see Figure 6.13) to an amazing 40 million molecules per second for the liver enzyme catalase.

Metabolism and the Regulation of Enzymes

A major characteristic of life is *homeostasis*, the maintenance of stable internal conditions. Regulation of the rates at which our thousands of different enzymes operate contributes to metabolic homeostasis. In the remainder of this chapter, we will investigate the role of enzymes in organizing and regulating metabolism. In living cells, the activity of enzymes can be activated or inhibited in various ways, so the presence of an enzyme does not necessarily ensure that it is functioning. There are mechanisms that alter the rate at which some enzymes catalyze reactions, making enzymes the target points at which entire sequences of chemical reactions can be regulated. Finally, we examine how the environment—namely, temperature and pH—affects enzyme activity.

Metabolism is organized into pathways

An organism's **metabolism** is the totality of the biochemical reactions that take place within it. Metabolism transforms raw materials and stored potential energy into forms that can be used by living cells. Metabolism consists of sequences of enzyme-catalyzed chemical reactions called **pathways**. In these sequences, the product of one reaction is the substrate for the next:

$$A \xrightarrow{\text{enzyme 1}} B \xrightarrow{\text{enzyme 2}} C \xrightarrow{\text{enzyme 3}} D$$

Some metabolic pathways are anabolic, synthesizing the important chemical building blocks from which macromolecules are built. Others are catabolic, breaking down molecules for usable free energy, recycling monomers, or inactivating toxic substances. The balance among these anabolic and catabolic pathways may change depending on the cell's (and the organism's) needs. So a cell must regulate all its metabolic pathways constantly.

Enzyme activity is subject to regulation by inhibitors

Various *inhibitors* can bind to enzymes, slowing down the rates of enzyme-catalyzed reactions. Some inhibitors occur naturally in cells; others are artificial. Naturally occurring inhibitors regulate metabolism; artificial ones can be used to treat disease, to kill pests, or in the laboratory to study how enzymes work. Some inhibitors irreversibly inhibit the enzyme by permanently binding to it. Others have reversible effects; that is, they can become unbound from the enzyme. The removal of a natural reversible inhibitor increases an enzyme's rate of catalysis.

IRREVERSIBLE INHIBITION. Some inhibitors covalently bond to certain side chains at the active sites of an enzyme, thereby permanently inactivating the enzyme by destroying its capacity to interact with its normal substrate. At the beginning of this chapter we described aspirin, which adds an acetyl group to a serine residue at the active site of cyclooxygenase, preventing this serine from taking part in chemical catalysis.

Another example of an **irreversible inhibitor** is DIPF (diisopropylphosphorofluoridate), which also reacts with serine (Figure 6.17). DIPF is an irreversible inhibitor of acetylcholinesterase, an enzyme that is essential for the orderly propagation of impulses from one nerve cell to another. Because of their effect on acetylcholinesterase, DIPF and other similar compounds are classified as *nerve gases*. One of them, Sarin, was used in an attack on the Tokyo subway in 1995, resulting in a dozen deaths and hundreds hospitalized. The widely used insecticide malathion is a derivative of DIPF that inhibits only insect acetylcholinesterase, not the mammalian enzyme.

REVERSIBLE INHIBITION. Not all inhibition is irreversible. Some inhibitors are similar enough to a particular enzyme's natural substrate to bind noncovalently to its active site, yet different enough that the enzyme catalyzes no chemical reaction. While such a molecule is bound to the enzyme,

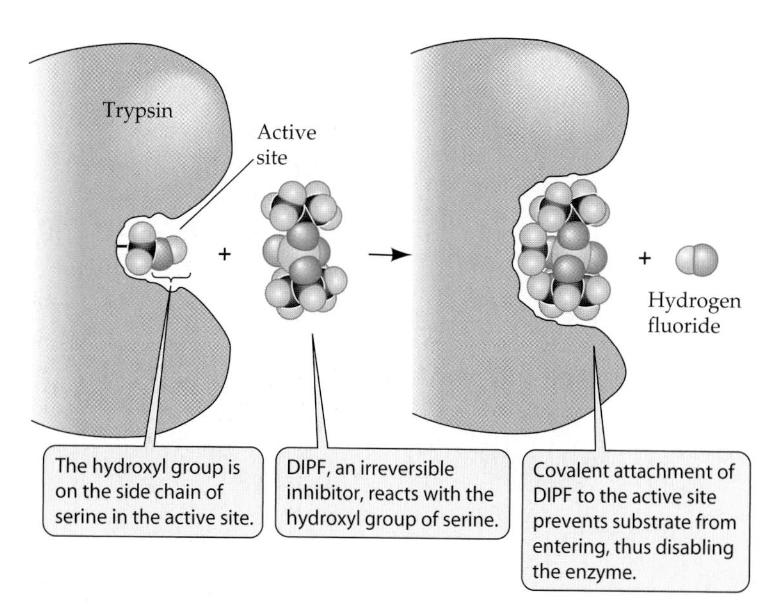
Trypsin
Active site
Hydrogen fluoride
The hydroxyl group is on the side chain of serine in the active site.
DIPF, an irreversible inhibitor, reacts with the hydroxyl group of serine.
Covalent attachment of DIPF to the active site prevents substrate from entering, thus disabling the enzyme.

6.17 Irreversible Inhibition DIPF forms a stable covalent bond with the side chain of the amino acid serine at the active site of the enzyme trypsin.

the natural substrate cannot enter the active site; thus, the inhibitor effectively wastes the enzyme's time, preventing its catalytic action. Such molecules are called **competitive inhibitors** because they compete with the natural substrate for the active site (Figure 6.18*a*). In these cases, the inhibition is reversible. When the concentration of the competitive inhibitor is reduced, it detaches from the active site, and the enzyme is again active.

The enzyme succinate dehydrogenase is subject to competitive inhibition. This enzyme, found in all mitochondria, catalyzes the conversion of the compound succinate to fumarate. A third molecule, oxaloacetate, is similar to succinate and can act as a competitive inhibitor of succinate dehydrogenase by binding to its active site. Once bound to oxaloacetate, the enzyme can do nothing more with it—no reaction occurs. An enzyme molecule cannot bind a succinate molecule until the oxaloacetate molecule has moved out of its active site—which can occur if more substrate (succinate) molecules are added.

Some inhibitors that do not react with the active site are called **noncompetitive inhibitors**. Noncompetitive inhibitors bind to the enzyme at a site distinct from the active site. Their binding can cause a conformational change in the enzyme that alters the active site (Figure 6.18*b*). In this case, the active site may still bind substrate molecules, but the rate of product formation may be reduced. Noncompetitive inhibitors, like competitive inhibitors, can become unbound, so their effects are reversible.

Allosteric enzymes control their activity by changing their shape

The change in enzyme shape due to noncompetitive inhibitor binding is an example of **allostery** (*allo-*, "different"; *-stery*, "shape"). In that case, the binding of the inhibitor *induces* the protein to change its shape. More common are enzymes that

6.18 Reversible Inhibition (*a*) A competitive inhibitor binds temporarily to the active site of an enzyme. Succinate dehydrogenase, for example, is subject to competitive inhibition by oxaloacetate. (*b*) A noncompetitive inhibitor binds temporarily to the enzyme at a site away from the active site, but still blocks enzyme function.

(*a*) **Competitive inhibition**

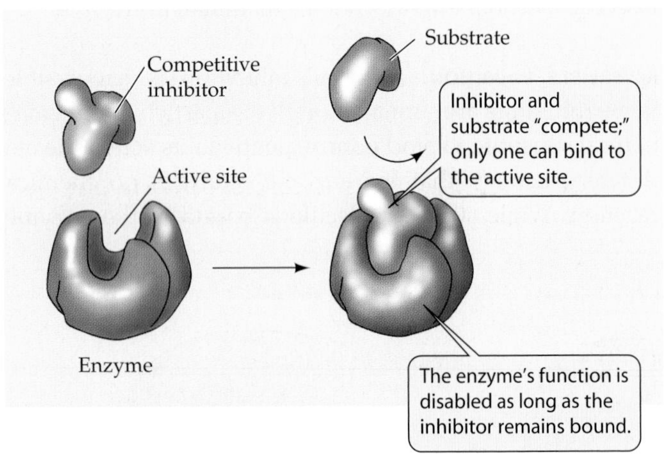

Competitive inhibition of succinate dehydrogenase

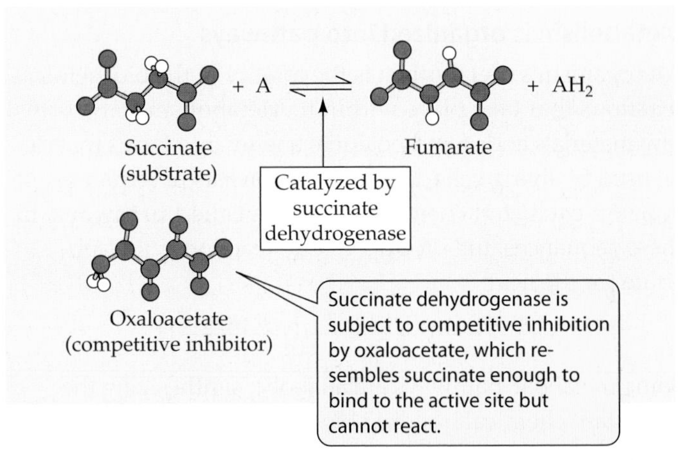

(*b*) **Noncompetitive inhibition**

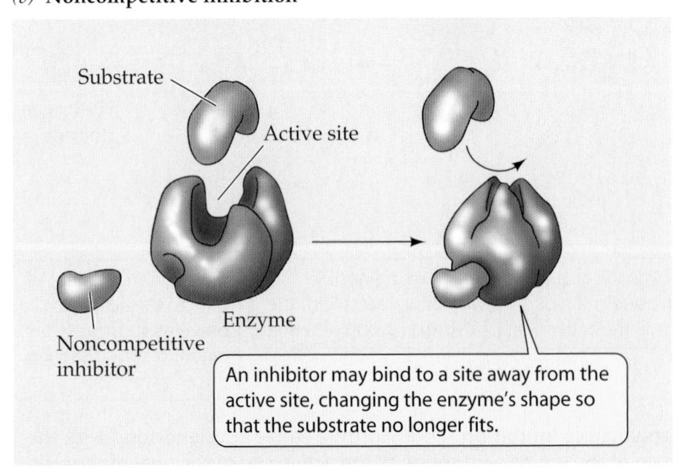

Noncompetitive inhibition of threonine dehydratase

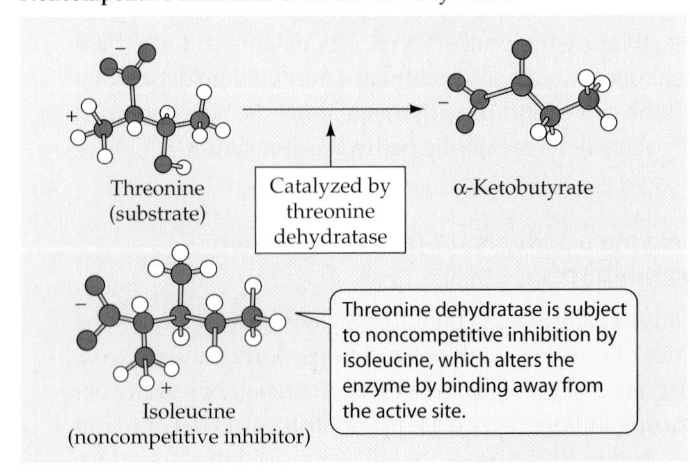

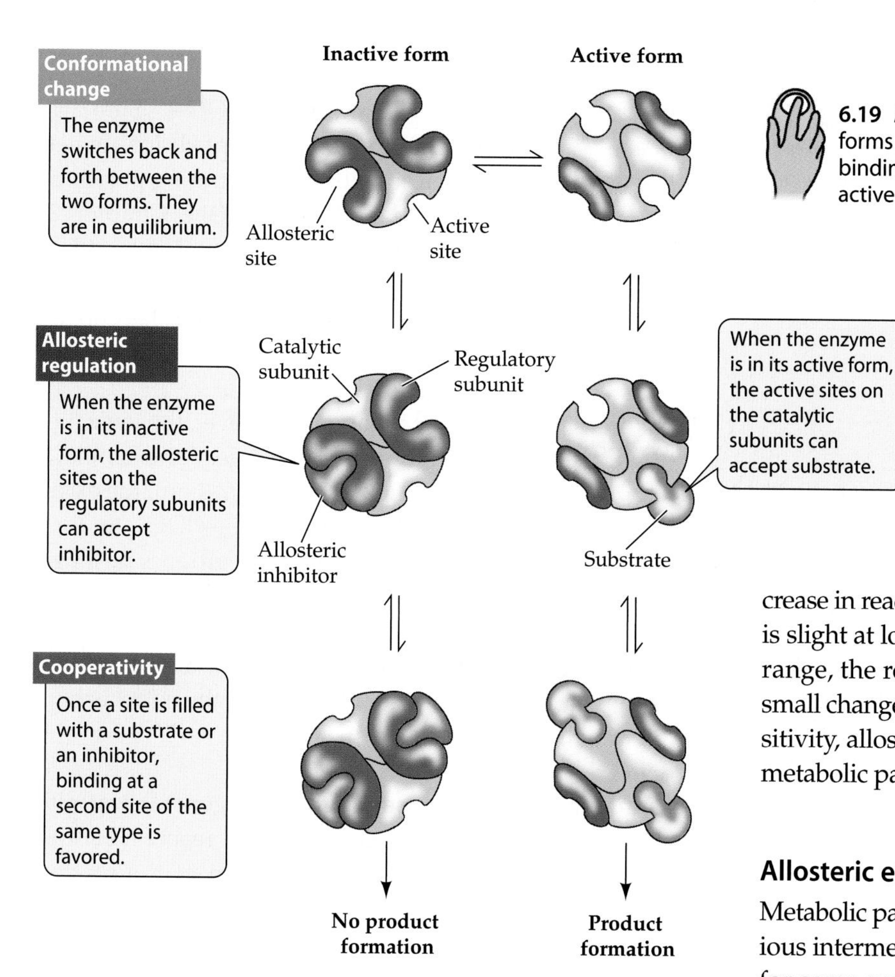

6.19 Allosteric Regulation of Enzymes Active and inactive forms of an enzyme are interconverted, depending on the binding of regulatory molecules at a location distant from the active site.

already exist in the cell in more than one possible shape. The inactive form of the enzyme has a shape that cannot bind the substrate, while the active form has the proper shape at the active site to bind the substrate. These two forms can interconvert, and this process is regulated by the binding of an **allosteric regulator** to a site on the enzyme away from the active site. Regulator binding is just like substrate binding: it is highly specific. So an enzyme may have several sites for binding: one for the substrate(s) and others for regulators.

Allosteric regulators work in two ways:

- *Positive regulators* stabilize the active form of the enzyme.
- *Negative regulators* stabilize the inactive form of the enzyme (Figure 6.19).

Most (but not all) enzymes that are allosterically regulated are proteins with quaternary structure; that is, they are made up of multiple polypeptide subunits. The active site is present on one subunit, called the *catalytic subunit*, while the regulatory site(s) are present on different subunit(s), the *regulatory subunit(s)*.

Allosteric enzymes and nonallosteric enzymes differ greatly in their reaction rates when the substrate concentration is low. Graphs of reaction rate plotted against substrate concentration show this relationship. For an enzyme with a single subunit, the plot looks like that in Figure 6.20*a*. The reaction rate first increases very sharply with increasing substrate concentration, then tapers off to a constant maximum rate as the supply of enzyme becomes saturated with substrate.

The plot for many allosteric enzymes is radically different, having a *sigmoid* (S-shaped) appearance (Figure 6.20*b*). The increase in reaction rate with increasing substrate concentration is slight at low substrate concentrations, but within a certain range, the reaction rate is extremely sensitive to relatively small changes in substrate concentration. Because of this sensitivity, allosteric enzymes are important in regulating entire metabolic pathways.

Allosteric effects regulate metabolism

Metabolic pathways typically involve a starting material, various intermediate products, and an end product that is used for some purpose by the cell. In each pathway, there are a number of reactions, each forming an intermediate product and each catalyzed by a different enzyme. The first step in a pathway is called the **commitment step**, meaning that once this enzyme-catalyzed reaction occurs, the "ball is rolling," and the other reactions happen in sequence, leading to the end product. But what if the cell has no need for that product—for example, if that product is available from its environment in adequate amounts? It would be energetically wasteful for the cell to continue making something it does not need.

One way that cells solve this problem is to shut down the metabolic pathway by having the final product allosterically

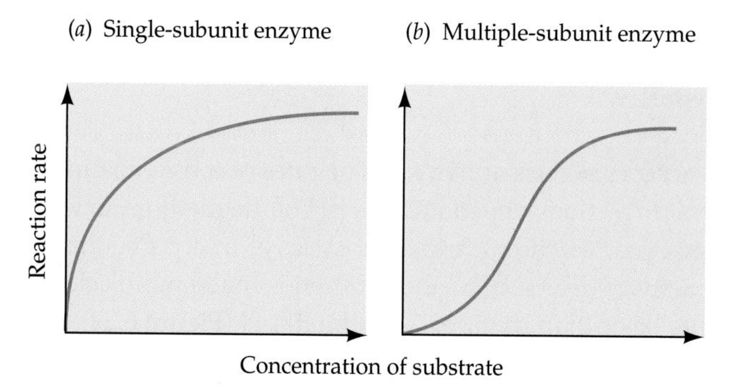

6.20 Allostery and Reaction Rate How the rate of an enzyme-catalyzed reaction changes with increasing substrate concentration depends on whether the enzyme is allosterically regulated.

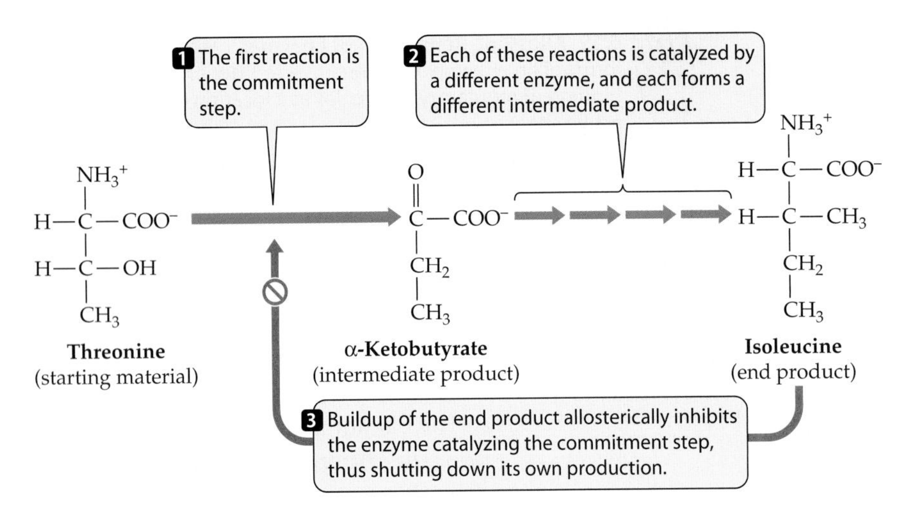
1 The first reaction is the commitment step.
2 Each of these reactions is catalyzed by a different enzyme, and each forms a different intermediate product.
Threonine
(starting material)
α-Ketobutyrate
(intermediate product)
Isoleucine
(end product)
3 Buildup of the end product allosterically inhibits the enzyme catalyzing the commitment step, thus shutting down its own production.

6.21 Inhibition of Metabolic Pathways The commitment step is catalyzed by an allosteric enzyme that can be inhibited by the end product of the pathway. The specific pathway shown here is the synthesis of isoleucine, an amino acid, from threonine in bacteria. This pathway is typical of many enzyme-catalyzed biological reactions.

inhibit the enzyme that catalyzes the commitment step (Figure 6.21). This mechanism is known as **end-product inhibition** or **feedback inhibition**. When the end product is present in a high concentration, some of it binds to an allosteric site on the commitment step enzyme, thereby causing it to become inactive. We will describe many other examples of allosteric interactions in later chapters.

Enzymes are affected by their environment

Enzymes enable cells to perform chemical reactions and carry out complex processes rapidly without using the extremes of temperature and pH employed by chemists in the laboratory. However, because of their three-dimensional structures and the chemistry of the side chains in their active sites, enzymes are highly sensitive to temperature and pH. We described the general effects of these environmental factors on proteins in Chapter 3. Here, we will examine their effects on enzyme function, which, of course, depends on enzyme structure and chemistry.

pH AFFECTS ENZYME ACTIVITY. The rates of most enzyme-catalyzed reactions depend on the pH of the medium in which they occur. Each enzyme is most active at a particular pH; its activity decreases as the solution is made more acidic or more basic than its "ideal" (optimal) pH (Figure 6.22).

Several factors contribute to this effect. One is the ionization of carboxyl, amino, and other groups on either the substrate or the enzyme. In neutral or basic solutions, carboxyl groups (—COOH) release H^+ to become negatively charged carboxylate groups ($—COO^-$). Similarly, amino groups ($—NH_2$) accept H^+ ions in neutral or acidic solutions, becoming positively charged $—NH_3^+$ groups (see Chapter 2). Thus, in a neutral solution, a molecule with an amino group is attracted electrically to another molecule that has a carboxyl group, because both groups are ionized and the two groups have opposite charges.

If the pH changes, however, the ionization of these groups may change. For example, at a low pH (high H^+ concentration), the excess H^+ may react with the $—COO^-$ to form COOH. If this happens, the group is no longer charged and cannot interact with other charged groups in the protein, so the folding of the protein may be altered. If such a change occurs at the active site of an enzyme, the enzyme may no longer have the correct shape to bind to its substrate.

TEMPERATURE AFFECTS ENZYME ACTIVITY. In general, warming increases the rate of an enzyme-catalyzed reaction because at higher temperatures, a greater fraction of the reactant molecules have enough energy to provide the activation energy for the reaction (Figure 6.23). Temperatures that are too high, however, inactivate enzymes, because at high temperatures enzyme molecules vibrate and twist so rapidly that some of their noncovalent bonds break. When heat changes their tertiary structure, enzymes become inactivated, or thermally denatured. Some enzymes denature at temperatures only slightly above that of the human body, but a few are stable even at the boiling or freezing points of water. All enzymes, however, show an optimal temperature for activity.

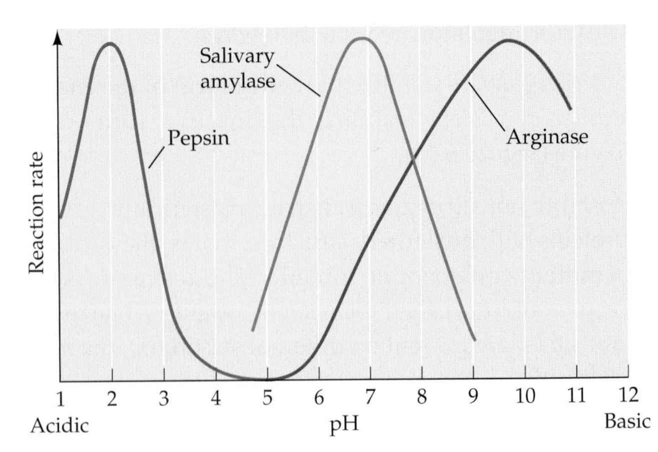
Salivary amylase
Pepsin
Arginase
Reaction rate
1 2 3 4 5 6 7 8 9 10 11 12
Acidic
pH
Basic

6.22 pH Affects Enzyme Activity Each enzyme catalyzes its reaction at a maximum rate at a particular pH. The activity curves peak at the pH where each enzyme is most effective.

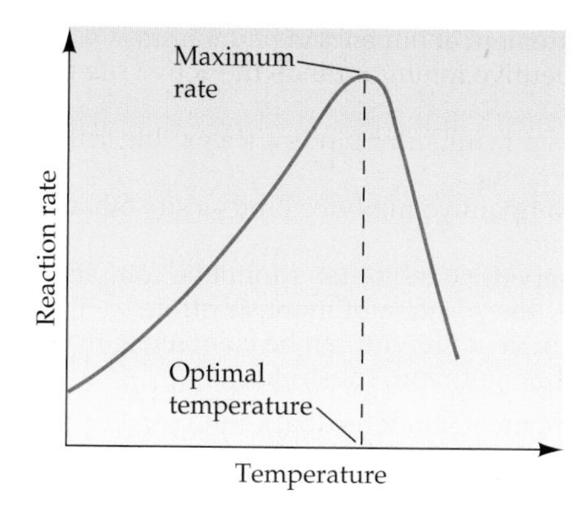

6.23 Temperature Affects Enzyme Activity Each enzyme is most active at a particular optimal temperature. At higher temperatures, denaturation reduces the enzyme's activity.

Individual organisms adapt to changes in the environment in many ways, one of which is based on groups of enzymes, called **isozymes**, that catalyze the same reaction but have different chemical compositions and physical properties. Different isozymes within a given group may have different optimal temperatures. The rainbow trout, for example, has several isozymes of the enzyme acetylcholinesterase, whose operation is essential to the normal transmission of nerve impulses. If a rainbow trout is transferred from warm water to near-freezing water (2°C), the fish produces an isozyme of acetylcholinesterase that is different from the one it produces at the higher temperature. The new isozyme has a lower optimal temperature, allowing the fish to perform normally in the colder water.

In general, enzymes adapted to warm temperatures fail to denature at those temperatures because their tertiary structures are held together largely by covalent bonds, such as disulfide bridges, instead of the more heat-sensitive weak chemical interactions. Most enzymes in humans are more stable at high temperatures than those of the bacteria that infect us, so that a moderate fever tends to denature bacterial enzymes, but not our own.

Chapter Summary

Energy and Energy Conversions

▶ Energy is the capacity to do work. Potential energy is the energy of state or position; it includes the energy stored in chemical bonds. Kinetic energy is the energy of motion (and related forms such as electric energy, light, and heat).

▶ Potential energy can be converted to kinetic energy, which can do work. **Review Figure 6.1**

▶ Living things, like everything else, obey the laws of thermodynamics. The first law of thermodynamics tells us that energy cannot be created or destroyed. The second law of thermodynamics tells us that the quantity of energy available to do work (free energy) decreases and unusable energy (associated with entropy) increases. **Review Figure 6.2**

▶ Changes in free energy, total energy, temperature, and entropy are related by the equation $\Delta G = \Delta H - T\Delta S$.

▶ Exergonic reactions release free energy and have a negative ΔG. Endergonic reactions take up free energy and have a positive ΔG. Endergonic reactions proceed only if free energy is provided. **Review Figure 6.3**

▶ The change in free energy (ΔG) of a reaction determines its point of chemical equilibrium, at which the forward and reverse reactions proceed at the same rate. For exergonic reactions, the equilibrium point lies toward completion (the conversion of all reactants into products). **Review Figure 6.4**

ATP: Transferring Energy in Cells

▶ ATP (adenosine triphosphate) serves as an energy currency in cells. Hydrolysis of ATP releases a relatively large amount of free energy. **Review Figure 6.5**

▶ The ATP cycle couples exergonic and endergonic reactions, transferring free energy from the exergonic to the endergonic reaction. **Review Figures 6.6, 6.7. See Web/CD Activity 6.1**

Enzymes: Biological Catalysts

▶ The rate of a chemical reaction is independent of ΔG, but is determined by the size of the energy barrier. Catalysts speed reactions by lowering the energy barrier. **Review Figures 6.8, 6.9**

▶ Enzymes are biological catalysts, proteins that are highly specific for their substrates. Substrates bind to the active site, where catalysis takes place, forming an enzyme–substrate complex. **Review Figure 6.10**

▶ At the active site, a substrate can be oriented correctly, chemically modified, or strained. As a result, the substrate readily forms its transition state, and the reaction proceeds. **Review Figures 6.11, 6.12. See Web/CD Activity 6.2**

Molecular Structure Determines Enzyme Function

▶ The active site where substrate binds determines the specificity of an enzyme. Upon binding to substrate, some enzymes change shape, facilitating catalysis. **Review Figures 6.13, 6.14**

▶ Some enzymes require cofactors to carry out catalysis. Prosthetic groups are permanently bound to the enzyme. Coenzymes are not usually bound to the enzyme. They can be considered substrates, as they are changed by the reaction and then released from the enzyme. **Review Table 6.1 and Figure 6.15**

▶ Substrate concentration affects the rate of an enzyme-catalyzed reaction. **Review Figure 6.16**

Metabolism and the Regulation of Enzymes

▶ Metabolism is organized into pathways in which the product of one reaction is a reactant for the next reaction. Each reaction in the pathway is catalyzed by an enzyme.

▶ Enzyme activity is subject to regulation. Some inhibitors react irreversibly with enzymes and block their catalytic activity. Others react reversibly with enzymes, inhibiting their action only temporarily. A compound closely similar in structure to an enzyme's normal substrate may competitively inhibit the action of the enzyme. **Review Figures 6.17, 6.18. See Web/CD Tutorial 6.1**

▶ Allosteric regulators bind to a site different from the active site and stabilize the active or inactive form of an enzyme. Many such enzymes have multiple subunits. **Review Figure 6.19. See Web/CD Tutorial 6.2**

▶ For allosteric enzymes, plots of reaction rate versus substrate concentration are sigmoid, in contrast to plots of the same variables for nonallosteric enzymes. **Review Figure 6.20**

▶ The end product of a metabolic pathway may inhibit the allosteric enzyme that catalyzes the commitment step of that pathway. **Review Figure 6.21**

▶ Enzymes are sensitive to their environment. Both pH and temperature affect enzyme activity. **Review Figures 6.22, 6.23**

Self-Quiz

1. Coenzymes differ from enzymes in that coenzymes are
   a. only active outside the cell.
   b. polymers of amino acids.
   c. smaller, such as vitamins.
   d. specific for one reaction.
   e. always carriers of high-energy phosphate.
2. Which statement about thermodynamics is true?
   a. Free energy is used up in an exergonic reaction.
   b. Free energy cannot be used to do work, such as chemical transformations.
   c. The total amount of energy can change after a chemical transformation.
   d. Free energy can be kinetic but not potential energy
   e. Entropy tends always to a maximum.
3. In a chemical reaction,
   a. the rate depends on the value of ΔG.
   b. the rate depends on the activation energy.
   c. the entropy change depends on the activation energy.
   d. the activation energy depends on the value of ΔG.
   e. the change in free energy depends on the activation energy.
4. Which statement about enzymes is *not* true?
   a. They consist of proteins, with or without a nonprotein part.
   b. They change the rate of the catalyzed reaction.
   c. They change the value of ΔG of the reaction.
   d. They are sensitive to heat.
   e. They are sensitive to pH.
5. The active site of an enzyme
   a. never changes shape.
   b. forms no chemical bonds with substrates.
   c. determines, by its structure, the specificity of the enzyme.
   d. looks like a lump projecting from the surface of the enzyme.
   e. changes ΔG of the reaction.
6. The molecule ATP is
   a. a component of most proteins.
   b. high in energy because of the presence of adenine (A).
   c. required for many energy-producing biochemical reactions.
   d. a catalyst.
   e. used in some endergonic reactions to provide energy.
7. In an enzyme-catalyzed reaction,
   a. a substrate does not change.
   b. the rate decreases as substrate concentration increases.
   c. the enzyme can be permanently changed.
   d. strain may be added to a substrate.
   e. the rate is not affected by substrate concentration.
8. Which statement about enzyme inhibitors is *not* true?
   a. A competitive inhibitor binds the active site of the enzyme.
   b. An allosteric inhibitor binds a site on the active form of the enzyme.
   c. A noncompetitive inhibitor binds a site other than the active site.
   d. Noncompetitive inhibition cannot be completely overcome by the addition of more substrate.
   e. Competitive inhibition can be completely overcome by the addition of more substrate.
9. Which statement about feedback inhibition of enzymes is *not* true?
   a. It is exerted through allosteric effects.
   b. It is directed at the enzyme that catalyzes the first committed step in a branch of a pathway.
   c. It affects the rate of reaction, not the concentration of enzyme.
   d. It acts very slowly.
   e. It is an example of negative feedback.
10. Which statement about temperature effects is *not* true?
   a. Raising the temperature may reduce the activity of an enzyme.
   b. Raising the temperature may increase the activity of an enzyme.
   c. Raising the temperature may denature an enzyme.
   d. Some enzymes are stable at the boiling point of water.
   e. All enzymes have the same optimal temperature.

For Discussion

1. How is it possible for endergonic reactions to proceed in organisms?
2. Consider two proteins: One is an enzyme dissolved in the cytosol, the other is an ion channel in a membrane. Contrast the structures of the two proteins, indicating at least two important differences.
3. Plot free energy versus the course of an endergonic reaction and that of an exergonic reaction. Include the activation energy in both plots. Label E_a and ΔG on both graphs.
4. Consider an enzyme that is subject to allosteric regulation. If a competitive inhibitor (not an allosteric inhibitor) is added to a solution of such an enzyme, the ratio of enzyme molecules in the active form to those in the inactive form increases. Explain this observation.
5. If you were presented with a radioactively labeled substance, what experiments would you perform to determine whether it enters cells by simple diffusion or active transport?

7 Cellular Pathways that Harvest Chemical Energy

Agriculture was a key step in the development of human civilizations. The harvesting, planting, and cultivation of seeds began about 10,000 years ago. One of the earliest plants to be domesticated and turned into a reliable crop was barley, and one of the first uses of barley was to brew beer. Living in what is now Iraq, the ancient Sumerians learned that partly germinated and then mashed up barley seeds, stored under the right conditions, could produce a potent and pleasant alcoholic beverage. Beer making spread to Egypt, and was so important in that ancient civilization that the hieroglyphic symbol for food was a pitcher of beer and a loaf of bread.

Fermented beverages such as beer and wine were important to ancient civilizations because pure water, without infectious disease-causing organisms, was hard to obtain. In the nineteenth century, scientists were able to demonstrate that the conversion of seed mash into alcohol is carried out by living cells—in this case, yeast. By the middle of the twentieth century, biochemists had identified the intermediate substances in the metabolic pathway that converts the starch in seeds—a polysaccharide—into alcohol. In addition, they showed that each intermediate step in the pathway is catalyzed by a specific enzyme.

In this chapter, we will describe some aspects of this and related pathways for the breakdown of sugars. The metabolism of sugars is important not only in making alcoholic beverages, but in providing the energy that organisms store in ATP—the energy you use all the time to fuel both conscious actions such as turning the pages of this book and automatic ones such as the beating of your heart.

An Ancient Brewer In the civilizations of ancient Sumeria and Egypt, the important task of brewing beer was usually the domain of women such as the one depicted in this Egyptian statue. The figurine dates from the period known as the "Old Kingdom" and is almost 4,500 years old—about 100 years younger than the Great Pyramid of Giza.

Energy and Electrons from Glucose

We are all familiar with fuels and their uses. Petroleum fuels contain stored energy that is harvested to move cars and heat homes. Wood burning in a stove or campfire releases energy as light and heat. Living organisms also need fuels, which must be obtained from foods. This is true whether we are speaking of organisms that make their own foods through photosynthesis or organisms that obtain foods by eating other organisms. The most common fuel for living cells is the sugar **glucose** ($C_6H_{12}O_6$). Many other compounds serve as foods, but almost all of them are converted to glucose, or to intermediate compounds in the step-by-step metabolism of glucose.

As you will see in this section, cells obtain energy from glucose by the chemical process of oxidation, which is carried out through a series of metabolic pathways. Before we examine that process, let's take a brief look at how metabolic pathways operate in the cell. Several principles govern metabolic pathways:

- Complex chemical transformations in the cell do not occur in a single reaction, but in a number of separate reactions that form a metabolic pathway (see Chapter 6).
- Each reaction in a pathway is catalyzed by a specific enzyme.
- Metabolic pathways are similar in all organisms, from bacteria to humans.
- Many metabolic pathways are compartmentalized in eukaryotes, with certain reactions occurring inside an organelle.
- The operation of each metabolic pathway can be regulated by the activities of key enzymes.

Cells trap free energy while metabolizing glucose

The familiar process of combustion (burning) is very similar to the chemical processes that release energy in cells. If glucose is burned in a flame, it reacts with O_2, rapidly forming carbon dioxide and water and releasing a lot of energy. The balanced equation for this combustion reaction is

$$C_6H_{12}O_6 + 6\ O_2 \rightarrow 6\ CO_2 + 6\ H_2O + \text{energy (heat and light)}$$

The same equation applies to the metabolism of glucose in cells. The metabolism of glucose, however, is a multistep, controlled series of reactions. The multiple steps of the process permit about one-third of the energy released to be captured in ATP. That ATP can be used to do cellular work such as movement or active transport across a membrane, just as energy captured from combustion can be used to do work.

The change in free energy (ΔG) for the complete conversion of glucose and O_2 to CO_2 and water, whether by combustion or by metabolism, is –686 kcal/mol (–2,870 kJ/mol). Thus the overall reaction is highly exergonic and can drive the endergonic formation of a great deal of ATP from ADP and phosphate. It is the capture of this energy in ATP that requires the many steps characteristic of glucose metabolism.

Three metabolic processes play roles in the utilization of glucose for energy: glycolysis, cellular respiration, and fermentation (Figure 7.1). All three involve metabolic pathways made up of many distinct chemical reactions.

- **Glycolysis** begins glucose metabolism in all cells and produces two molecules of the three-carbon product **pyruvate**. A small amount of the energy stored in glucose is captured in usable forms. Glycolysis does not use O_2.
- **Cellular respiration** uses O_2 from the environment and completely converts each pyruvate molecule to three molecules of CO_2 through a set of metabolic pathways. In the process, a great deal of the energy stored in the covalent bonds of pyruvate is released and transferred to ADP and phosphate to form ATP.
- **Fermentation** does not involve O_2. Fermentation converts pyruvate into products such as lactic acid or ethyl alcohol (ethanol), which are still relatively energy-rich molecules. Because the breakdown of glucose is incomplete, much less energy is released by fermentation than by cellular respiration, and no ATP is produced.

Glycolysis and fermentation are **anaerobic** metabolic processes—that is, they do not involve O_2. Cellular respiration is an **aerobic** metabolic process, requiring the direct participation of O_2.

Redox reactions transfer electrons and energy

In Chapter 6, we described the addition of phosphate groups to ADP to make ATP as an endergonic reaction that can extract and store energy from exergonic reactions. Another way of transferring energy is to transfer electrons. A reaction in which one substance transfers one or more electrons to another substance is called an oxidation–reduction reaction, or **redox reaction**.

- **Reduction** is the gain of one or more electrons by an atom, ion, or molecule.
- **Oxidation** is the loss of one or more electrons.

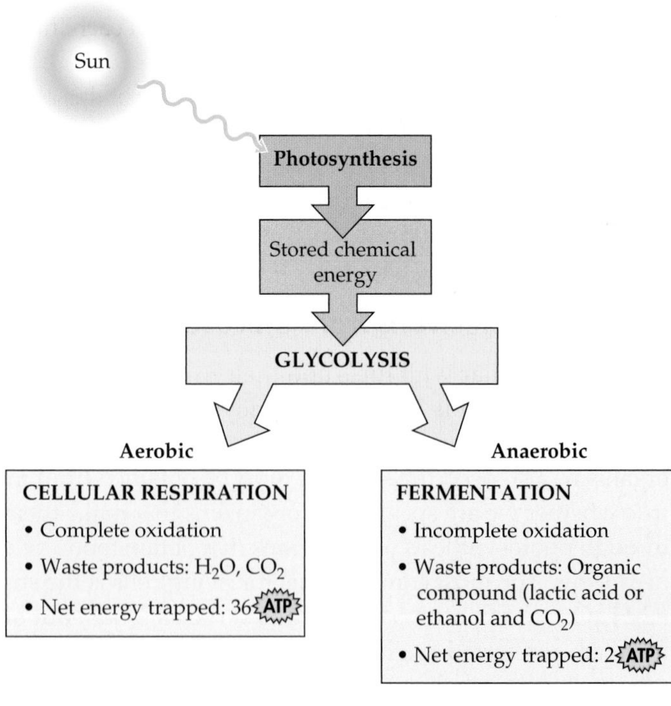

7.1 Energy for Life Both heterotrophic ("other-feeding") and autotrophic ("self-feeding") organisms obtain energy from the food compounds that autotrophs produce by photosynthesis. They convert these compounds to glucose, then metabolize glucose by glycolysis and fermentation or cellular respiration.

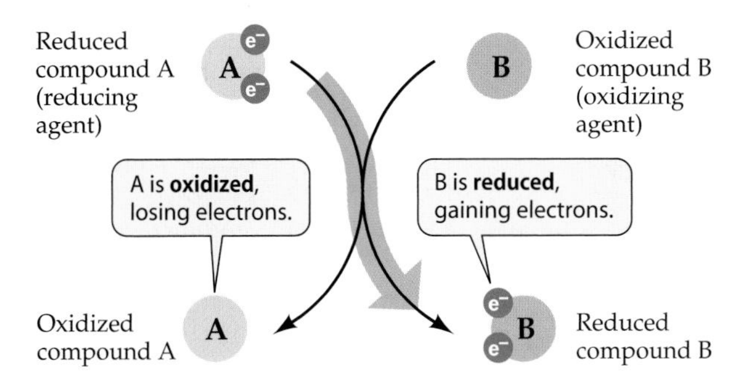
Reduced compound A (reducing agent)
Oxidized compound B (oxidizing agent)
A is **oxidized**, losing electrons.
B is **reduced**, gaining electrons.
Oxidized compound A
Reduced compound B

7.2 Oxidation and Reduction Are Coupled In a redox reaction, reactant A is oxidized and reactant B is reduced. In the process, A loses electrons and B gains electrons. A proton may be transferred along with an electron, so that what is actually transferred is a hydrogen atom.

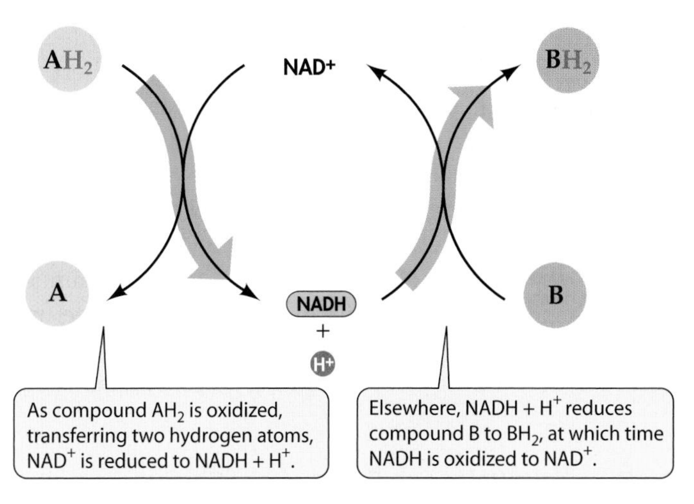
NAD+
NADH
As compound AH2 is oxidized, transferring two hydrogen atoms, NAD+ is reduced to NADH + H+.
Elsewhere, NADH + H+ reduces compound B to BH2, at which time NADH is oxidized to NAD+.

7.3 NAD Is an Energy Carrier Thanks to its ability to carry free energy and electrons, NAD^+ is a major redox carrier and universal energy intermediary in cells.

Although oxidation and reduction are always defined in terms of traffic in electrons, we may also think in these terms when hydrogen atoms (*not* hydrogen ions) are gained or lost, because transfers of hydrogen atoms involve transfers of electrons ($H = H^+ + e^-$). Thus, when a molecule loses hydrogen atoms, it becomes oxidized.

Oxidation and reduction always occur together: As one material is oxidized, the electrons it loses are transferred to another material, reducing that material. In a redox reaction, we call the reactant that becomes reduced an *oxidizing agent* and the one that becomes oxidized a *reducing agent* (Figure 7.2). In both the combustion and the metabolism of glucose, glucose is the reducing agent and oxygen gas is the oxidizing agent.

In a redox reaction, energy is transferred. Much of the energy originally present in the reducing agent becomes associated with the reduced product. (The rest remains in the reducing agent or is lost.) As we will see, some of the key reactions of glycolysis and cellular respiration are highly exergonic redox reactions.

The coenzyme NAD is a key electron carrier in redox reactions

In Chapter 6, we described the role of coenzymes, small molecules that assist in enzyme-catalyzed reactions. ADP acts as a coenzyme when it picks up energy released in an exergonic reaction and uses it to make ATP (an endergonic reaction). In a similar fashion, the coenzyme **NAD** (**nicotinamide adenine dinucleotide**) acts as an energy carrier, in this case in redox reactions (Figure 7.3).

NAD exists in two chemically distinct forms, one oxidized (**NAD**$^+$) and the other reduced (**NADH + H**$^+$) (Figure 7.4). Both forms participate in biological redox reactions. The reduction reaction

$$NAD^+ + 2\,H \rightarrow NADH + H^+$$

is formally equivalent to the transfer of two hydrogen atoms ($2\,H^+ + 2\,e^-$). However, what is actually transferred is a hydride ion (H^-, a proton and two electrons), leaving a free proton (H^+). This notation emphasizes that reduction is accomplished by the addition of electrons.

Oxygen is highly electronegative (see Table 2.3) and readily accepts electrons from NADH. The oxidation of NADH + H^+ by O_2,

$$NADH + H^+ + \tfrac{1}{2}\,O_2 \rightarrow NAD^+ + H_2O$$

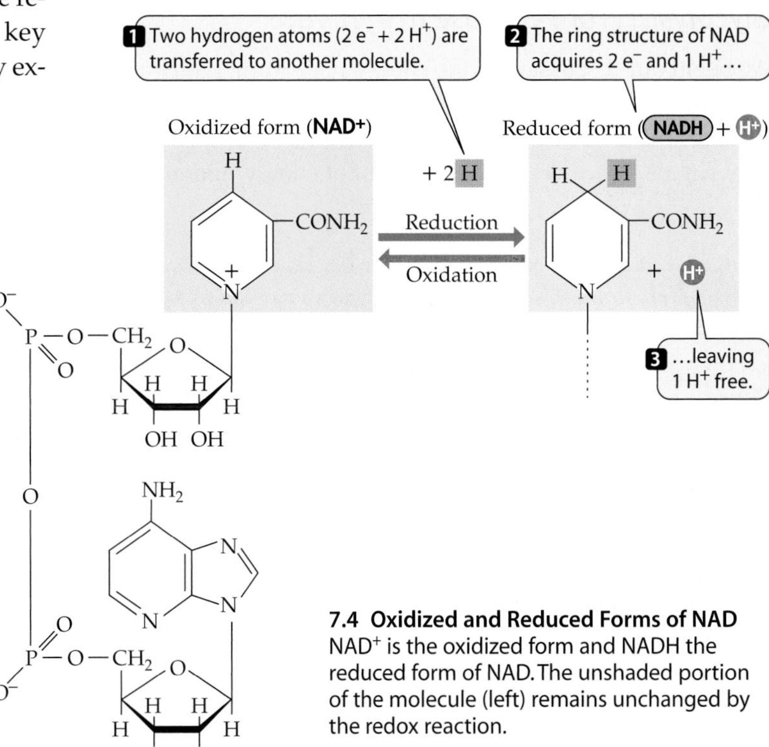
1 Two hydrogen atoms (2 e− + 2 H+) are transferred to another molecule.
2 The ring structure of NAD acquires 2 e− and 1 H+...
Oxidized form (NAD+)
Reduced form (NADH + H+)
+ 2 H
Reduction
Oxidation
3 ...leaving 1 H+ free.

7.4 Oxidized and Reduced Forms of NAD NAD^+ is the oxidized form and NADH the reduced form of NAD. The unshaded portion of the molecule (left) remains unchanged by the redox reaction.

is highly exergonic, with a $\Delta G = -52.4$ kcal/mol (–219 kJ/mol). Note that the oxidizing agent appears here as "½ O_2" instead of "O." This notation emphasizes that it is oxygen gas, O_2, that acts as the oxidizing agent.

Just as ATP can be thought of as packaging free energy in bundles of about 12 kcal/mol (50 kJ/mol), NAD can be thought of as packaging free energy in bundles of approximately 50 kcal/mol (200 kJ/mol). NAD is the most common, but not the only, electron carrier in cells. As you will see, another carrier, **FAD (flavin adenine dinucleotide)**, is also involved in transferring electrons during glucose metabolism.

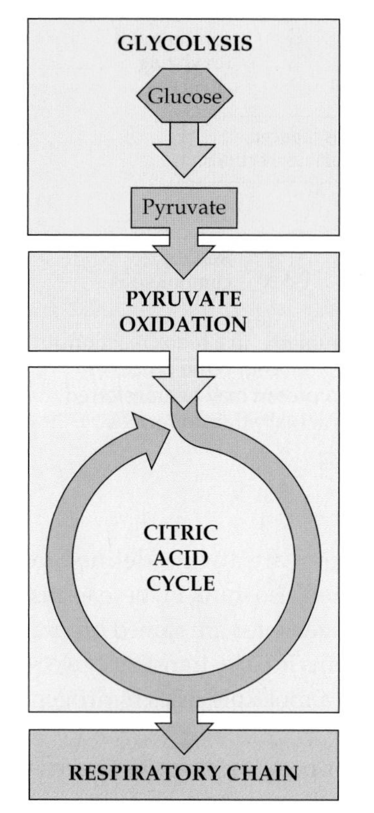

7.5 Energy-Producing Metabolic Pathways Energy-producing reactions can be grouped into five metabolic pathways: glycolysis, pyruvate oxidation, the citric acid cycle, the respiratory chain, and fermentation. The three middle pathways occur only in the presence of O_2 and are collectively referred to as cellular respiration (*a*). When O_2 is unavailable, glycolysis is followed by fermentation (*b*).

An Overview: Releasing Energy from Glucose

Depending on the presence or absence of O_2, the energy-harvesting processes in cells use different combinations of metabolic pathways (Figure 7.5):

- When O_2 is available as the final electron acceptor, four pathways operate. Glycolysis takes place first, and is followed by the three pathways of cellular respiration: **pyruvate oxidation**, the **citric acid cycle**, and the **respiratory chain** (also known as the **electron transport chain**).
- When O_2 is unavailable, pyruvate oxidation, the citric acid cycle, and the respiratory chain do not function, and the pyruvate produced by glycolysis is further metabolized by fermentation.

These five metabolic pathways, which we will consider one at a time, have different locations in the cell (Table 7.1).

Glycolysis: From Glucose to Pyruvate

We begin our discussion of the energy-harvesting pathways with glycolysis, which begins glucose metabolism. Glycolysis takes place in the cytoplasm of cells. It converts glucose to pyruvate, produces a small amount of energy, and generates no CO_2. In glycolysis, a reduced fuel molecule, glucose, gets partially oxidized and in the process releases some of its energy. After ten enzyme-catalyzed reactions, the end products of glycolysis are two molecules of pyruvate (pyruvic acid)* (Figure 7.6). These reactions are accompanied by the net formation of two molecules of ATP and by the reduction

*We tend to use terms such as "pyruvate" and "pyruvic acid" interchangeably. However, at the pH values commonly found in cells, the ionized ("-*ate*") form—pyruvate—is present rather than the acid form—pyruvic acid. Similarly, all carboxylic acids are present as ions at these pH values.

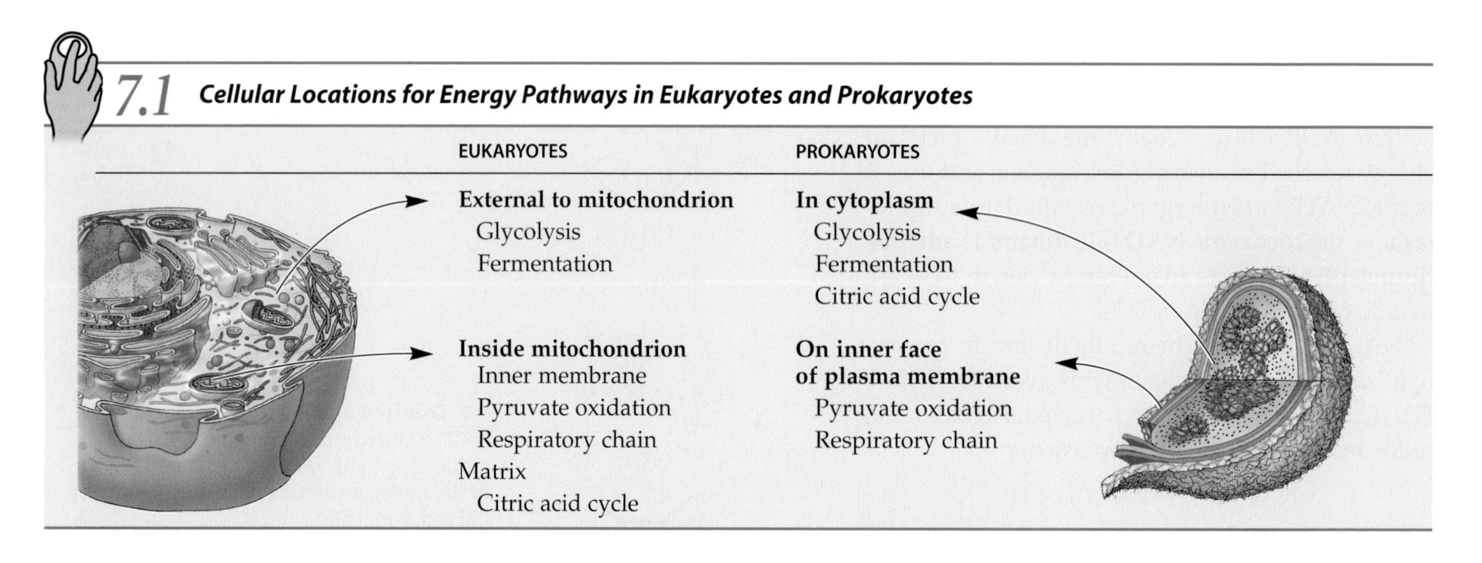

7.1 **Cellular Locations for Energy Pathways in Eukaryotes and Prokaryotes**

EUKARYOTES	PROKARYOTES
External to mitochondrion	**In cytoplasm**
Glycolysis	Glycolysis
Fermentation	Fermentation
	Citric acid cycle
Inside mitochondrion	**On inner face of plasma membrane**
Inner membrane	Pyruvate oxidation
Pyruvate oxidation	Respiratory chain
Respiratory chain	
Matrix	
Citric acid cycle	

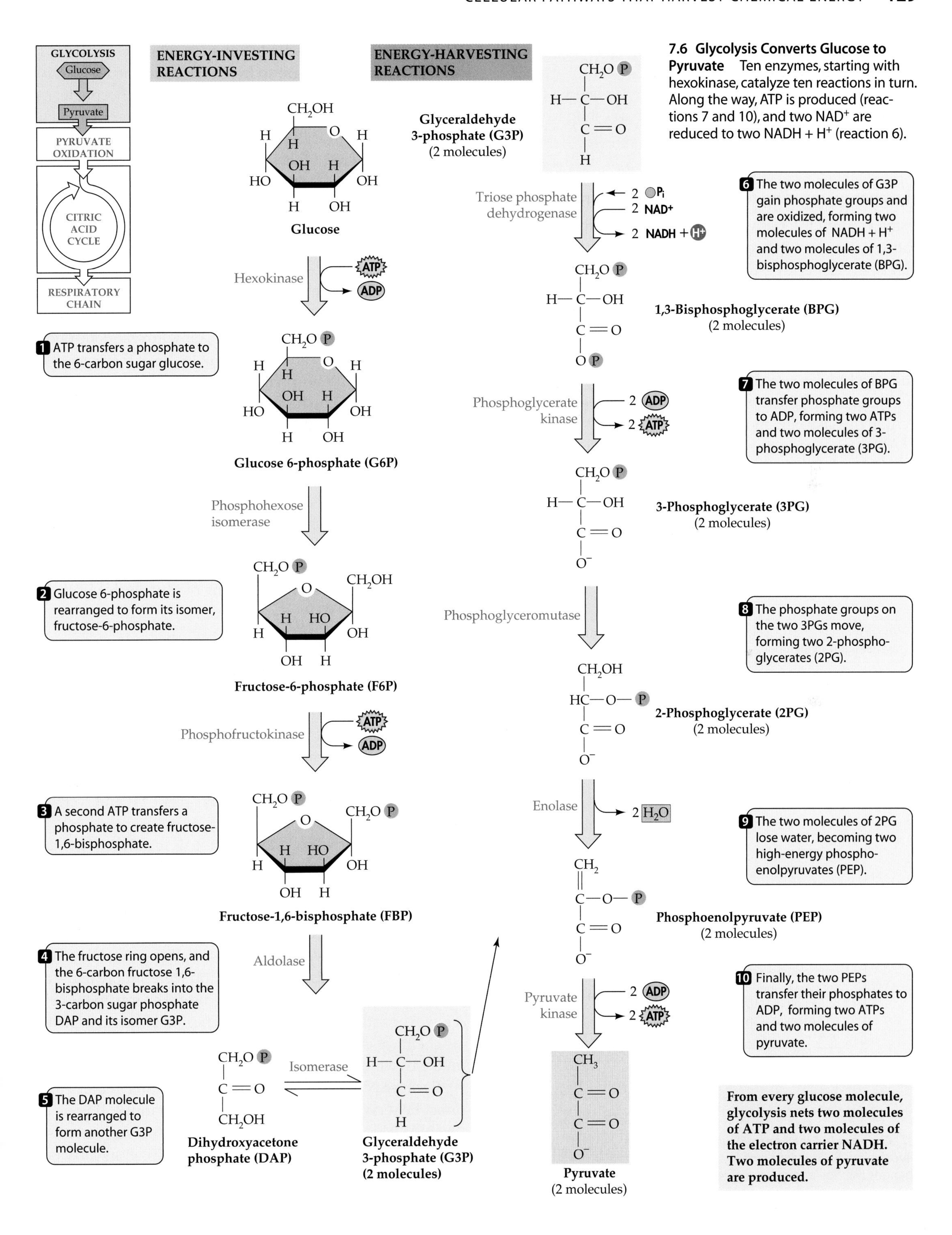

GLYCOLYSIS
Glucose
Pyruvate
PYRUVATE OXIDATION
CITRIC ACID CYCLE
RESPIRATORY CHAIN
ENERGY-INVESTING REACTIONS
ENERGY-HARVESTING REACTIONS
Glucose
Hexokinase
ATP
ADP
1 ATP transfers a phosphate to the 6-carbon sugar glucose.
Glucose 6-phosphate (G6P)
Phosphohexose isomerase
2 Glucose 6-phosphate is rearranged to form its isomer, fructose-6-phosphate.
Fructose-6-phosphate (F6P)
Phosphofructokinase
ATP
ADP
3 A second ATP transfers a phosphate to create fructose-1,6-bisphosphate.
Fructose-1,6-bisphosphate (FBP)
Aldolase
4 The fructose ring opens, and the 6-carbon fructose 1,6-bisphosphate breaks into the 3-carbon sugar phosphate DAP and its isomer G3P.
Isomerase
5 The DAP molecule is rearranged to form another G3P molecule.
Dihydroxyacetone phosphate (DAP)
Glyceraldehyde 3-phosphate (G3P) (2 molecules)
Glyceraldehyde 3-phosphate (G3P) (2 molecules)
Triose phosphate dehydrogenase
2 P_i
2 NAD^+
2 NADH + H^+
6 The two molecules of G3P gain phosphate groups and are oxidized, forming two molecules of NADH + H^+ and two molecules of 1,3-bisphosphoglycerate (BPG).
1,3-Bisphosphoglycerate (BPG) (2 molecules)
Phosphoglycerate kinase
2 ADP
2 ATP
7 The two molecules of BPG transfer phosphate groups to ADP, forming two ATPs and two molecules of 3-phosphoglycerate (3PG).
3-Phosphoglycerate (3PG) (2 molecules)
Phosphoglyceromutase
8 The phosphate groups on the two 3PGs move, forming two 2-phosphoglycerates (2PG).
2-Phosphoglycerate (2PG) (2 molecules)
Enolase
2 H_2O
9 The two molecules of 2PG lose water, becoming two high-energy phosphoenolpyruvates (PEP).
Phosphoenolpyruvate (PEP) (2 molecules)
Pyruvate kinase
2 ADP
2 ATP
10 Finally, the two PEPs transfer their phosphates to ADP, forming two ATPs and two molecules of pyruvate.
Pyruvate (2 molecules)
From every glucose molecule, glycolysis nets two molecules of ATP and two molecules of the electron carrier NADH. Two molecules of pyruvate are produced.

7.6 Glycolysis Converts Glucose to Pyruvate Ten enzymes, starting with hexokinase, catalyze ten reactions in turn. Along the way, ATP is produced (reactions 7 and 10), and two NAD^+ are reduced to two NADH + H^+ (reaction 6).

of two molecules of NAD^+ to two molecules of NADH + H^+ for each molecule of glucose.

Glycolysis can be divided into two stages: energy-investing reactions that use ATP, and energy-harvesting reactions that produce ATP (Figure 7.7).

The energy-investing reactions of glycolysis require ATP

Using Figure 7.6, let us work our way through the glycolytic pathway. The first five reactions of glycolysis are endergonic; that is, the cell is investing free energy in the glucose molecule, rather than releasing energy from it. In two separate reactions (reactions 1 and 3 in Figure 7.6), the energy of two molecules of ATP is invested in attaching two phosphate groups to the glucose molecule to form fructose 1,6-bisphosphate,* which has a free energy substantially higher than that of glucose. Later, these phosphate groups will be transferred to ADP to make new molecules of ATP.

Although both of these first steps of glycolysis use ATP as one of their substrates, each is catalyzed by a different, specific enzyme. The enzyme hexokinase catalyzes **reaction 1**, in which a phosphate group from ATP is attached to the six-carbon glucose molecule, forming glucose 6-phosphate. (A *kinase* is any enzyme that catalyzes the transfer of a phosphate group from ATP to another substrate.) In **reaction 2**, the six-membered glucose ring is rearranged into a five-membered fructose ring. In **reaction 3**, the enzyme phosphofructokinase adds a second phosphate (taken from another ATP) to the fructose ring, forming a six-carbon sugar, fructose 1,6-bisphosphate.

Reaction 4 opens up and cleaves the six-carbon sugar ring to give two different three-carbon sugar phosphates: dihydroxyacetone phosphate and glyceraldehyde 3-phosphate. In **reaction 5**, one of those products, dihydroxyacetone phosphate, is converted into a second molecule of the other one, glyceraldehyde 3-phosphate (G3P).

By this time—the halfway point of the glycolytic pathway—the following things have happened:

- Two molecules of ATP have been invested.
- The six-carbon glucose molecule has been converted into two molecules of a three-carbon sugar phosphate, glyceraldehyde 3-phosphate (G3P, a triose phosphate).

*The root *bis*- means "two." A sugar bisphosphate has two phosphate groups attached to two different carbons. In contrast, the prefix *di*- implies the serial attachment of two phosphate groups to one carbon, as in ADP (adenosine diphosphate).

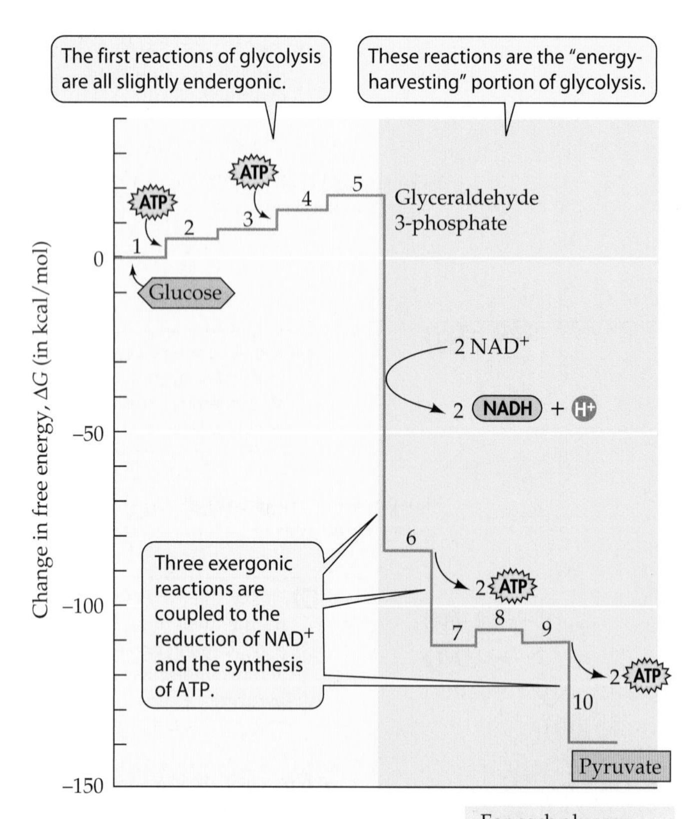

7.7 Changes in Free Energy During Glycolysis Each reaction of glycolysis changes the free energy available.

The energy-harvesting reactions of glycolysis yield NADH + H^+ and ATP

With the investment of two ATPs, the first five reactions of glycolysis have rearranged the six-carbon sugar glucose and split it into two three-carbon sugar phosphates (G3P). In the discussion that follows, remember that *each reaction occurs twice for each glucose molecule* going through glycolysis because each glucose molecule has been split into two molecules of G3P. It is the fate of G3P that now concerns us—its transformation will generate both NADH + H^+ and ATP.

PRODUCING NADH + H^+. **Reaction 6** is catalyzed by the enzyme triose phosphate dehydrogenase, and its end product is a phosphate ester, 1,3-bisphosphoglycerate (BPG). Reaction 6 is an oxidation, and it is accompanied by an enormous drop in free energy—more than 100 kcal of energy per mole of glucose is released in this extremely exergonic reaction. If this big energy drop were simply a loss of heat, glycolysis would not provide useful energy to the cell. However, rather than being lost as heat, this energy is *stored* as chemical energy by reducing two molecules of NAD^+ to make two molecules of NADH + H^+.

Because NAD^+ is present in small amounts in the cell, it must be recycled to allow glycolysis to continue; if none of the NADH is oxidized back to NAD^+, glycolysis comes to a halt. The metabolic pathways that follow glycolysis carry out this oxidation, as we will see.

PRODUCING ATP. In **reactions 7–10**, the two phosphate groups of BPG are transferred one at a time to molecules of ADP, with a rearrangement in between. More than 20 kcal (83.6 kJ/mol) of free energy is stored in ATP for every mole of BPG broken down. Finally, we are left with two moles of pyruvate for every mole of glucose that entered glycolysis.

The enzyme-catalyzed transfer of phosphate groups from donor molecules to ADP molecules (as in reaction 7) is called **substrate-level phosphorylation**. (*Phosphorylation* is the addition of a phosphate group to a molecule. *Substrate-level phosphorylation* is distinguished from the oxidative phosphorylation carried out by the respiratory chain, which we will discuss later in the chapter.)

As an example of substrate-level phosphorylation, when G3P reacts with a phosphate group (P_i) and NAD^+ in reaction 6, a second phosphate is added, an aldehyde is oxidized to a carboxylic acid, NAD^+ is reduced, and BPG is formed. The oxidation provides so much energy that the newly added phosphate group is linked to the rest of the molecule by a covalent bond that has even more energy than the terminal phosphate-to-phosphate bond of ATP.

Another example of substrate-level phosphorylation occurs in reaction 7, where phosphoglycerate kinase catalyzes the transfer of a phosphate group from BPG to ADP, forming ATP. Both reactions 6 and 7 are exergonic, even though a substantial amount of energy is consumed in the formation of ATP.

A review of the glycolytic pathway shows that at the beginning of glycolysis, two molecules of ATP are used per molecule of glucose, but that ultimately four molecules of ATP are produced (two for each of the two BPG molecules)—a net gain of two ATP molecules and two NADH + H^+.

Glycolysis is followed by cellular respiration (if O_2 is present) or fermentation (if no O_2 is present). The first reaction of cellular respiration is the oxidation of pyruvate.

Pyruvate Oxidation

The oxidation of pyruvate to acetate and its subsequent conversion to **acetyl CoA** is the link between glycolysis and all the other reactions of cellular respiration (see Figure 7.8). **Coenzyme A** (CoA), which is attached to the acetyl group to form acetyl CoA, is a complex molecule composed of a nucleotide, the vitamin pantothenic acid, and a sulfur-containing group that is responsible for binding the two-carbon acetate molecule. Acetyl CoA formation is a multi-step reaction catalyzed by the pyruvate dehydrogenase complex, an enormous enzyme complex that is attached to the inner mitochondrial membrane. Pyruvate diffuses into the mitochondrion, where a series of coupled reactions takes place:

1. Pyruvate is oxidized to a two-carbon acetyl group, and CO_2 is released.
2. Part of the energy from the oxidation is captured by the reduction of NAD^+ to NADH + H^+.
3. Some of the remaining energy is stored temporarily by the combining of the acetyl group with CoA, forming acetyl CoA:

$$\text{pyruvate} + NAD^+ + \text{CoA} \rightarrow \text{Acetyl CoA} + \text{NADH} + H^+ + CO_2$$

Acetyl CoA has 7.5 kcal/mol (31.4 kJ/mol) more energy than simple acetate. Acetyl CoA can donate the acetyl group to acceptor molecules, much as ATP can donate phosphate groups to various acceptors. In the next section, we will see that the acetyl CoA donates its acetyl group to the four-carbon compound oxaloacetate to form the six-carbon citrate.

The Citric Acid Cycle

Acetyl CoA is the starting point for the citric acid cycle (also called the Krebs cycle or the tricarboxylic acid cycle) (Figure 7.8). This pathway of eight reactions completely oxidizes the two-carbon acetyl group to two molecules of carbon dioxide. The free energy released from these reactions is captured by ADP and the electron carriers NAD and FAD.

As Figure 7.7 shows, the metabolism of glucose to pyruvate is accompanied by a total drop in free energy of about 140 kcal/mol (585 kJ/mol). About a third of this energy is captured in the formation of ATP and reduced NAD (NADH + H^+). Oxidizing pyruvate to acetate yields much additional free energy. Then, the citric acid cycle takes the acetyl group and essentially breaks it down to two molecules of CO_2, using the hydrogen atoms to reduce electron carriers and passing chemical free energy to those carriers in the process. The reduced carriers are oxidized in the respiratory chain, which transfers an enormous amount of free energy to ATP.

- The *inputs* to the citric acid cycle are acetate (in the form of acetyl CoA), water, and oxidized electron carriers (NAD^+ and FAD).
- The *outputs* are carbon dioxide, reduced electron carriers (NADH + H^+ and $FADH_2$), and a small amount of ATP. Overall, for each acetyl group, the citric acid cycle removes two carbons as CO_2 and uses four pairs of hydrogen atoms to reduce electron carriers.

The citric acid cycle produces two CO_2 molecules and reduced carriers

Acetyl CoA enters the citric acid cycle from pyruvate oxidation, which has released CO_2. At the beginning of the citric acid cycle, acetyl CoA, which has two carbon atoms in its acetyl group, reacts with a four-carbon acid, oxaloacetate, to form the six-carbon compound citrate (citric acid). The remainder of the cycle consists of a series of enzyme-catalyzed

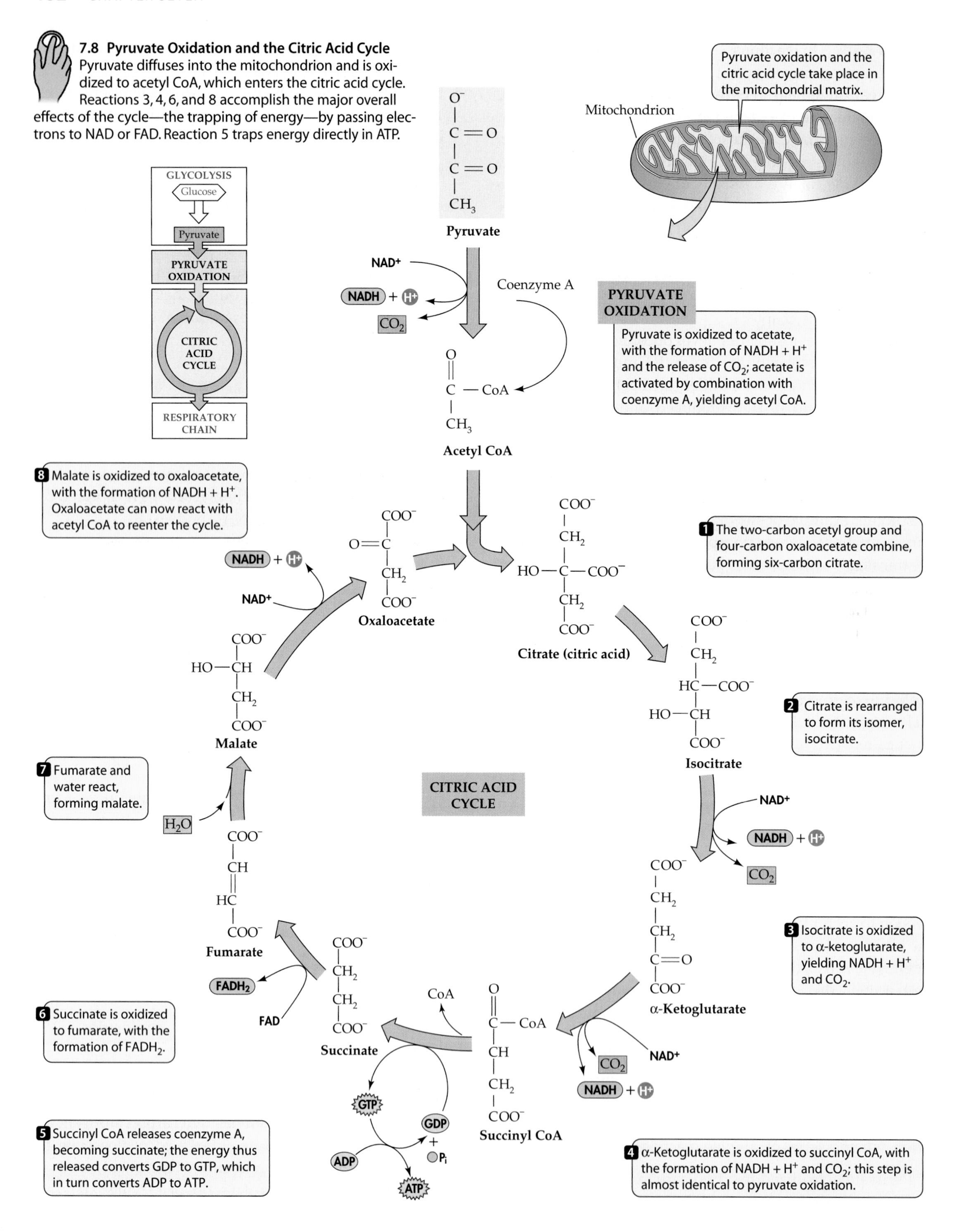
Pyruvate oxidation and the citric acid cycle take place in the mitochondrial matrix.
Mitochondrion
GLYCOLYSIS
Glucose
Pyruvate
PYRUVATE OXIDATION
CITRIC ACID CYCLE
RESPIRATORY CHAIN
Pyruvate
NAD+
NADH + H+
CO2
Coenzyme A
Acetyl CoA
PYRUVATE OXIDATION
Pyruvate is oxidized to acetate, with the formation of NADH + H+ and the release of CO2; acetate is activated by combination with coenzyme A, yielding acetyl CoA.
1 The two-carbon acetyl group and four-carbon oxaloacetate combine, forming six-carbon citrate.
Citrate (citric acid)
2 Citrate is rearranged to form its isomer, isocitrate.
Isocitrate
NAD+
NADH + H+
CO2
3 Isocitrate is oxidized to α-ketoglutarate, yielding NADH + H+ and CO2.
α-Ketoglutarate
NAD+
CO2
NADH + H+
4 α-Ketoglutarate is oxidized to succinyl CoA, with the formation of NADH + H+ and CO2; this step is almost identical to pyruvate oxidation.
Succinyl CoA
CoA
GTP
GDP
+
Pi
ADP
ATP
5 Succinyl CoA releases coenzyme A, becoming succinate; the energy thus released converts GDP to GTP, which in turn converts ADP to ATP.
Succinate
FADH2
FAD
6 Succinate is oxidized to fumarate, with the formation of FADH2.
Fumarate
H2O
7 Fumarate and water react, forming malate.
Malate
NAD+
NADH + H+
8 Malate is oxidized to oxaloacetate, with the formation of NADH + H+. Oxaloacetate can now react with acetyl CoA to reenter the cycle.
Oxaloacetate
CITRIC ACID CYCLE

7.8 Pyruvate Oxidation and the Citric Acid Cycle Pyruvate diffuses into the mitochondrion and is oxidized to acetyl CoA, which enters the citric acid cycle. Reactions 3, 4, 6, and 8 accomplish the major overall effects of the cycle—the trapping of energy—by passing electrons to NAD or FAD. Reaction 5 traps energy directly in ATP.

reactions in which citrate is converted to a new four-carbon molecule of oxaloacetate. This new oxaloacetate can react with a second acetyl CoA, producing a second molecule of citrate and thus enabling the cycle to continue.

The citric acid cycle is maintained in a steady state—that is, although the intermediate compounds in the cycle enter and leave, the concentrations of those intermediates do not change much. Pay close attention to the numbered reactions in Figure 7.8 as you read the next several paragraphs and refer to Figure 7.9, which shows the energetics of the pathway. Also, recall that energy is released upon oxidation and stored in either ATP, $FADH_2$, or NADH + H^+.

The energy temporarily stored in acetyl CoA drives the formation of citrate from oxaloacetate (**reaction 1**). During this reaction, the coenzyme A molecule is removed and can be reused. In **reaction 2**, the citrate molecule is rearranged to form isocitrate. In **reaction 3**, a CO_2 molecule and two hydrogen atoms are removed, converting isocitrate to α-ketoglutarate. This reaction produces a large drop in free energy, some of which is stored in NADH + H^+.

Like the oxidation of pyruvate to acetyl CoA, **reaction 4** of the citric acid cycle is complex. The five-carbon α-ketoglutarate molecule is oxidized to the four-carbon molecule succinate. In the process, CO_2 is given off, some of the oxidation energy is stored in NADH + H^+, and some of the energy is preserved temporarily by combining succinate with CoA to form succinyl CoA. In **reaction 5**, some of the energy in succinyl CoA is harvested to make GTP (guanosine triphosphate) from GDP and P_i, which is another example of substrate-level phosphorylation. GTP is then used to make ATP from ADP.

Free energy is released in **reaction 6**, in which the succinate released from succinyl CoA in reaction 5 is oxidized to fumarate. In the process, two hydrogens are transferred to an enzyme that contains the carrier FAD. After a molecular rearrangement (**reaction 7**), one more NAD^+ reduction occurs, producing oxaloacetate from malate (**reaction 8**). These two reactions illustrate a common biochemical mechanism: Water (H_2O) is added in reaction 7 to form an —OH group, and then the H from that —OH group is removed in reaction 8 to reduce NAD^+ to NADH + H^+. The final product, oxaloacetate, is ready to combine with another acetyl group from acetyl CoA and go around the cycle again. The citric acid cycle operates twice for each glucose molecule that enters glycolysis (once for each pyruvate that enters the mitochondrion).

Although most of the enzymes of the citric acid cycle are located in the mitochondrial matrix, there are two exceptions: succinate dehydrogenase, which catalyzes reaction 6, and α-ketoglutarate dehydrogenase, which catalyzes reaction 4. These enzymes are integral membrane proteins of the inner mitochondrial membrane.

Generations of students have asked the question, "Why did this complicated system evolve to achieve the simple

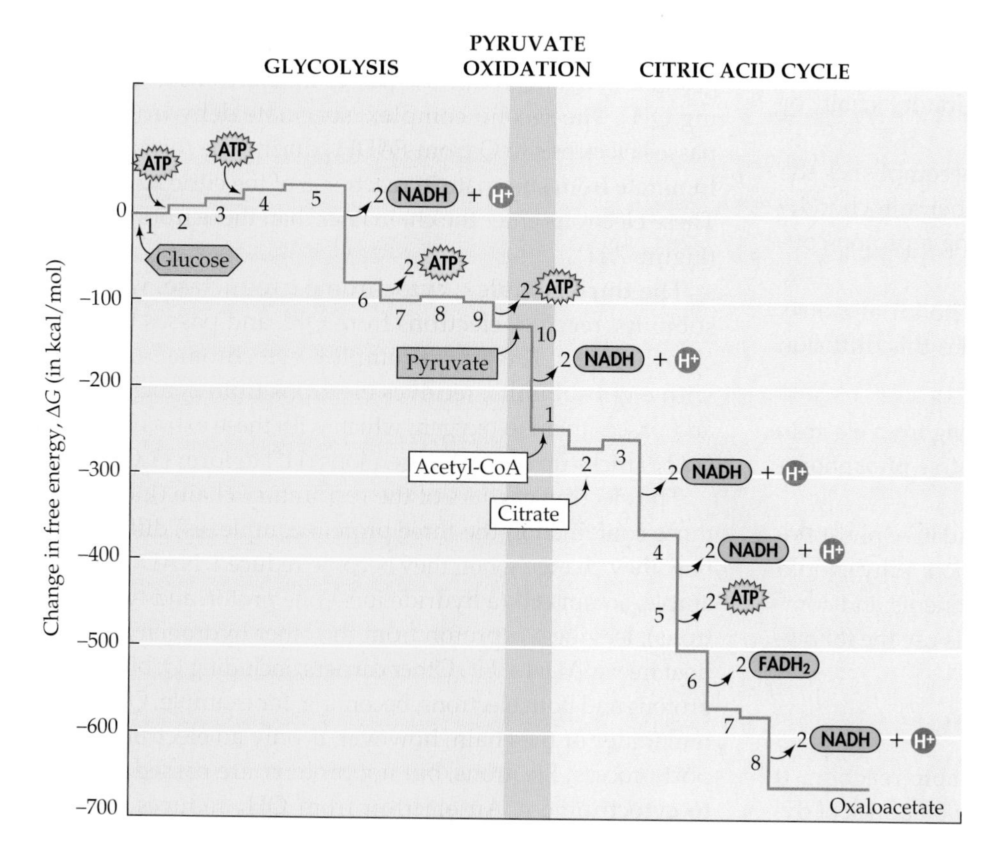

7.9 The Citric Acid Cycle Releases Much More Free Energy Than Glycolysis Does Electron carriers (NAD in glycolysis; NAD and FAD in the citric acid cycle) are reduced and ATP is generated in reactions coupled to other reactions, producing major drops in free energy as metabolism proceeds.

goal of oxidizing two acetyl groups to two molecules of CO_2?" There are three reasons:

- First, the cycle includes molecules that have other roles in the cell. As we will see later in this chapter, the intermediates of the citric acid cycle are themselves catabolic (breakdown) products or anabolic (synthesis) building blocks of other molecules, such as amino acids and nucleotides.
- Second, the citric acid cycle is far more efficient at harvesting energy from acetyl CoA than any single reaction could be.
- Third, evolution is a conservative, add-on process. It rarely operates by inventing an entirely new process.

The Respiratory Chain: Electrons, Protons, and ATP Production

Pyruvate oxidation and the operation of the citric acid cycle generate large amounts of reduced electron carriers containing trapped energy. To liberate this energy and produce ATP, something must happen to these reduced carriers. Furthermore, without NAD^+ and FAD, the oxidative steps of glycolysis, pyruvate oxidation, and the citric acid cycle could not occur. To regenerate NAD^+ and FAD, the reduced forms of these carriers must have some way to get rid of their hydrogens ($H^+ + e^-$). The fate of these protons and electrons is the rest of the story of cellular respiration. The story has three parts:

1. The electrons pass through a series of membrane-associated electron carriers called the respiratory chain or the electron transport chain.
2. The flow of electrons along the chain accomplishes the active transport of protons across the inner mitochondrial membrane, out of the matrix, creating a proton concentration gradient.
3. The protons diffuse back into the mitochondrial matrix through a proton channel, which couples this diffusion to the synthesis of ATP.

The overall process of ATP synthesis resulting from electron transport through the chain is called **oxidative phosphorylation**.

Before we proceed with the details of oxidative phosphorylation, let's reflect on an important question: Why should the respiratory chain have so many components and complex processes? Why, for example, don't cells use the following single step?

$$NADH + H^+ + \frac{1}{2} O_2 \rightarrow NAD^+ + H_2O$$

Fundamentally, this would be an untamable reaction. It would be very exergonic—rather like setting off a stick of dynamite in the cell. There is no biochemical way to harvest that burst of energy efficiently and put it to physiological use (that is, no metabolic reaction that is so endergonic as to consume a significant fraction of that energy in a single step). To control the release of energy during the oxidation of glucose in a cell, evolution has produced the lengthy electron transport chain we observe today: a series of reactions, each releasing a small, manageable amount of energy.

The respiratory chain transports electrons and releases energy

The respiratory chain contains large integral proteins, smaller mobile proteins, and even a smaller lipid molecule:

- Four large protein complexes containing carrier molecules and their associated enzymes are integral proteins of the inner mitochondrial membrane in eukaryotes (see Figure 4.14).
- **Cytochrome** ***c*** is a small peripheral protein that lies in the space between the inner and outer mitochondrial membranes. It is loosely attached to the inner membrane.
- A nonprotein component called **ubiquinone** (abbreviated **Q**) is a small, nonpolar molecule that moves freely within the hydrophobic interior of the phospholipid bilayer of the inner membrane (Figure 7.10).

NADH + H^+ passes electrons to Q by way of the first large protein complex, called NADH-Q reductase, which contains twenty-six polypeptides and attached prosthetic groups. NADH-Q reductase passes the electrons to Q, forming QH_2. The second complex, succinate dehydrogenase, passes electrons to Q from $FADH_2$ during the formation of fumarate from succinate in reaction 6 of the citric acid cycle. These electrons enter the chain later than those from NADH (Figure 7.11).

The third complex, cytochrome *c* reductase, with ten subunits, receives electrons from QH_2 and passes them to cytochrome *c*. The fourth complex, cytochrome *c* oxidase, with eight subunits, receives electrons from cytochrome *c* and passes them to oxygen, which with these extra electrons ($\frac{1}{2} O_2^-$) picks up two hydrogen ions (H^+) to form H_2O.

The electron carriers of the respiratory chain (including those contained in the three protein complexes) differ as to how they change when they become reduced. NAD^+, for example, accepts H^- (a hydride ion—one proton and two electrons), leaving the proton from the other hydrogen atom to float free: NADH + H^+. Other carriers, including Q, bind both protons and both electrons, becoming, for example, QH_2. The remainder of the chain, however, is only an electron transport process. Electrons, but not protons, are passed from Q to cytochrome *c*. An electron from QH_2 reduces a cytochrome's Fe^{3+} to Fe^{2+}. The fate of the protons will be discussed below.

Electron transport within each of the three protein complexes results, as we'll see, in the pumping of protons across the inner mitochondrial membrane, and the return of the protons across the membrane is coupled to the formation of ATP. Thus the energy originally contained in glucose and other foods is finally captured in the cellular energy currency, ATP. For each pair of electrons passed along the chain from NADH + H^+ to oxygen, three molecules of ATP are formed.

If only electrons are carried through the final reactions of the respiratory chain, what happens to the protons? How are proton movements coupled to the production of ATP?

Proton diffusion is coupled to ATP synthesis

As we have seen, all the carriers and enzymes of the respiratory chain except cytochrome *c* are embedded in the inner mitochondrial membrane (see Figure 7.10). The operation of the respiratory chain results in the active transport of protons (H^+), against their concentration gradient, across the inner membrane of the mitochondrion from the mitochondrial matrix to the intermembrane space (the space between the inner and outer mitochondrial membranes). This occurs because the electron carriers contained in the three large protein complexes are arranged such that protons are taken up on one side of the membrane (the mitochondrial matrix) and transported along with electrons to the other side (the intermembrane space) (Figure 7.12). Thus, the protein complexes act as proton pumps. Because of the positive charge on the protons (H^+), this transport causes not only a difference in proton concentration, but also a difference in electric charge, across the membrane, with the inside of the organelle (the matrix) more negative than the intermembrane space.

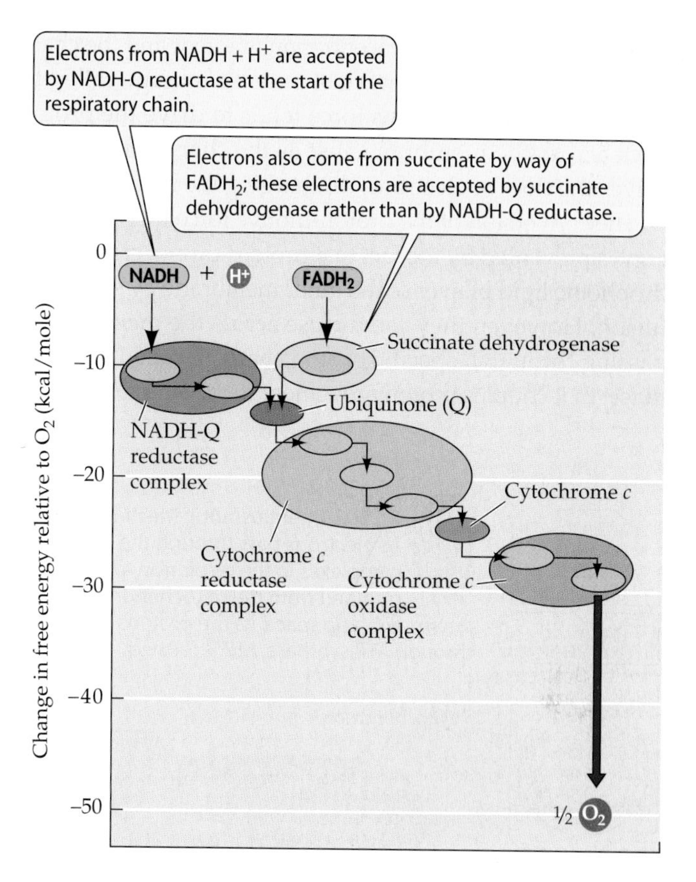

7.11 The Complete Respiratory Chain Electrons enter the chain from two sources, but they follow the same pathway from Q onward.

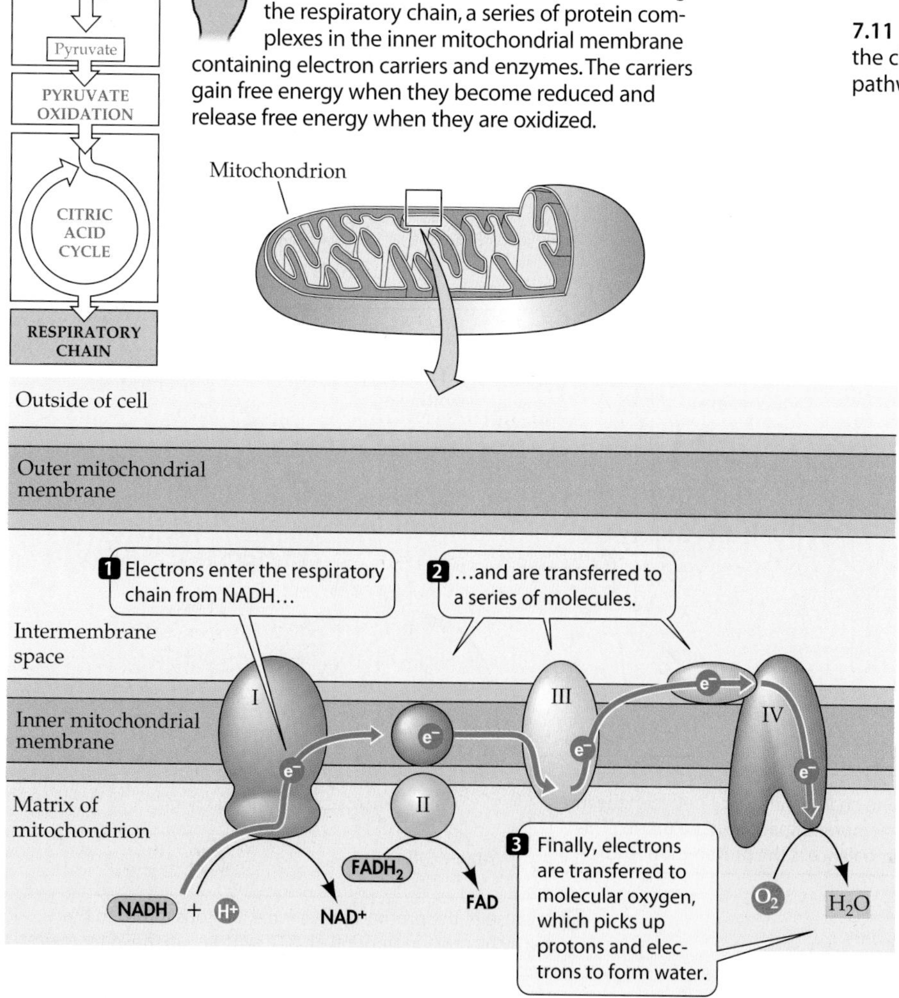

7.10 The Oxidation of NADH + H^+ Electrons from NADH + H^+ are passed through the respiratory chain, a series of protein complexes in the inner mitochondrial membrane containing electron carriers and enzymes. The carriers gain free energy when they become reduced and release free energy when they are oxidized.

Together, the proton concentration gradient and the charge difference constitute a source of potential energy called the **proton-motive force**. This force tends to drive the protons back across the membrane, just as the charge on a battery drives the flow of electrons, discharging the battery.

The conversion of the proton-motive force into kinetic energy is prevented by the fact that protons cannot cross the hydrophobic lipid bilayer of the inner membrane by simple diffusion. However, they can diffuse across the membrane by passing through a specific proton channel, called **ATP synthase**, that couples proton movement to the synthesis of ATP. This coupling of proton-motive force and ATP synthesis is called the chemiosmotic mechanism, or **chemiosmosis**.

THE CHEMIOSMOTIC MECHANISM FOR ATP SYNTHESIS. The chemiosmotic mechanism uses ATP synthase to couple proton diffusion to ATP synthesis. This mechanism has three parts:

1. The flow of electrons from one electron carrier to another in the respiratory chain is a series of exergonic reactions that occurs in the inner mitochondrial membrane.
2. These exergonic reactions drive the endergonic pumping of H^+ out of the mitochondrial matrix and across the

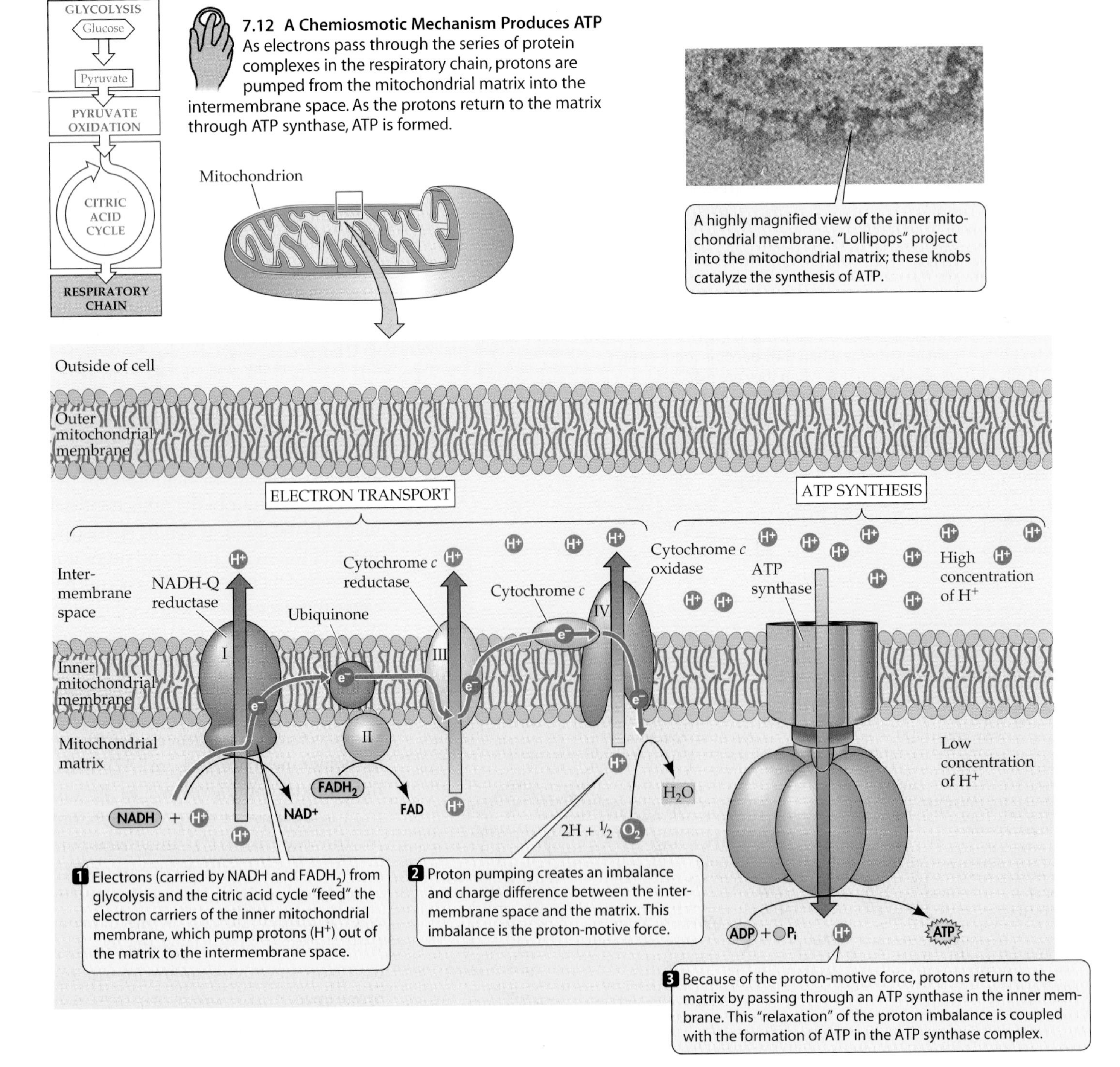

7.12 A Chemiosmotic Mechanism Produces ATP As electrons pass through the series of protein complexes in the respiratory chain, protons are pumped from the mitochondrial matrix into the intermembrane space. As the protons return to the matrix through ATP synthase, ATP is formed.

inner membrane into the intermembrane space. This pumping establishes and maintains a H^+ gradient.

3. The potential energy of the H^+ gradient, or proton-motive force, is harnessed by ATP synthase. This protein has two roles: It acts as a channel allowing the H^+ to diffuse back into the matrix, and it uses the energy of that diffusion to make ATP from ADP and P_i.

ATP synthesis is a reversible reaction, and ATP synthase can also act as an ATPase, hydrolyzing ATP to ADP and P_i:

$$ATP \leftrightarrow ADP + P_i + \text{free energy}$$

If the reaction goes to the right, free energy is released, and that energy is used to pump H^+ out of the mitochondrial matrix. If the reaction goes to the left, it uses free energy from H^+ diffusion into the matrix to make ATP. What makes it prefer ATP synthesis? There are two answers to this question.

- ATP leaves the mitochondrial matrix for use elsewhere in the cell as soon as it is made, keeping the ATP concentration in the matrix low and driving the reaction toward the left. A person hydrolyzes about 10^{25} ATP molecules per day, and clearly the vast majority are recycled using the free energy from the oxidation of glucose.
- The H^+ gradient is maintained by electron transport and proton pumping. (The electrons, you will recall, come from the oxidation of NADH and $FADH_2$, which are themselves reduced by the oxidations of glycolysis and the citric acid cycle. So, one reason you eat is to replenish the H^+ gradient!)

ATP synthase is a large multi-protein machine, containing 16 different polypeptides in mammals. It has two functional components. One of these components is the membrane channel for H^+. The other component sticks out into the mitochondrial matrix like a lollipop (see Figure 7.12) and contains the active site for ATP synthesis. The actual mechanism of transferring energy from the H^+ gradient to the phosphorylation of ADP involves the physical rotation of the core of the enzyme, with this rotational energy transferred to ATP.

EXPERIMENTS DEMONSTRATE CHEMIOSMOSIS. Two key experiments demonstrated (1) that a proton (H^+) gradient across a membrane can drive ATP synthesis; and (2) that the enzyme ATP synthase is the catalyst for this reaction.

Experiment 1 tested the hypothesis that ATP synthesis is driven by the H^+ gradient across an inner mitochondrial membrane (Figure 7.13). In this experiment, mitochondria without a food source were "fooled" into making ATP when researchers raised the H^+ concentration in their environment. A sample of isolated mitochondria that had been exposed to a low H^+ concentration was suddenly put in a medium with a high concentration of H^+. The outer mitochondrial membrane, unlike the inner one, is freely permeable to H^+, so H^+ rapidly diffused into the intermembrane space. This created an artificial gradient across the inner membrane, which the mitochondria used to make ATP from ADP and P_i. This result supported the hypothesis and provided strong evidence for chemiosmosis.

Experiment 2 tested the hypothesis that the enzyme ATPase couples a proton gradient to ATP synthesis. In this experiment, a proton pump isolated from a bacterium was added to artificial membrane vesicles. When an appropriate energy source was provided, H^+ was pumped into the vesicles, creating a gradient. If mammalian ATP synthase was then inserted into the membranes of these vesicles and the energy source removed, the vesicles made ATP even in the absence of the usual electron carriers. Again, the result supported the hypothesis, showing that the enzyme ATP synthase is the coupling factor in the membrane.

UNCOUPLING PROTON DIFFUSION FROM ATP PRODUCTION. For the chemiosmotic mechanism to work, the diffusion of H^+ and the formation of ATP must be tightly coupled; that is, the protons must pass only through the ATP synthase channel in order to move into the mitochondrial matrix. If a second type of H^+ diffusion channel (not ATP synthase) is inserted into the mitochondrial membrane, the energy of the H^+ gradient is released as heat, rather than being coupled to the synthesis of ATP. Such uncoupling molecules are deliberately used by some organisms to generate heat instead of ATP. For example, the natural uncoupling protein thermogenin plays an important role in regulating the temperature of some mammals, especially newborn human infants, who lack the hair to keep warm, and of hibernating animals. We will describe this process in more detail in Chapter 41.

Fermentation: ATP from Glucose, without O_2

Recall that fermentation is the breakdown of the pyruvate produced by glycolysis in the absence of O_2. Because fermentation results in the incomplete oxidation of glucose, it releases much less energy than cellular respiration. Why would such an inefficient process exist?

Suppose the supply of oxygen to a respiring cell is reduced (an anaerobic condition). As a consequence, oxygen is no longer available to pick up electrons at the end of the respiratory chain. As we can deduce from Figure 7.12, the first consequence of an insufficient supply of O_2 is that the cell cannot reoxidize cytochrome *c*, so all of that compound is soon in the reduced form. When this happens, QH_2 cannot be oxidized back to Q, and soon all the Q is in the reduced form. So it goes, until the entire respiratory chain is reduced. Under these circumstances, no NAD^+ and no FAD are regenerated from their reduced forms. Therefore, the oxidative

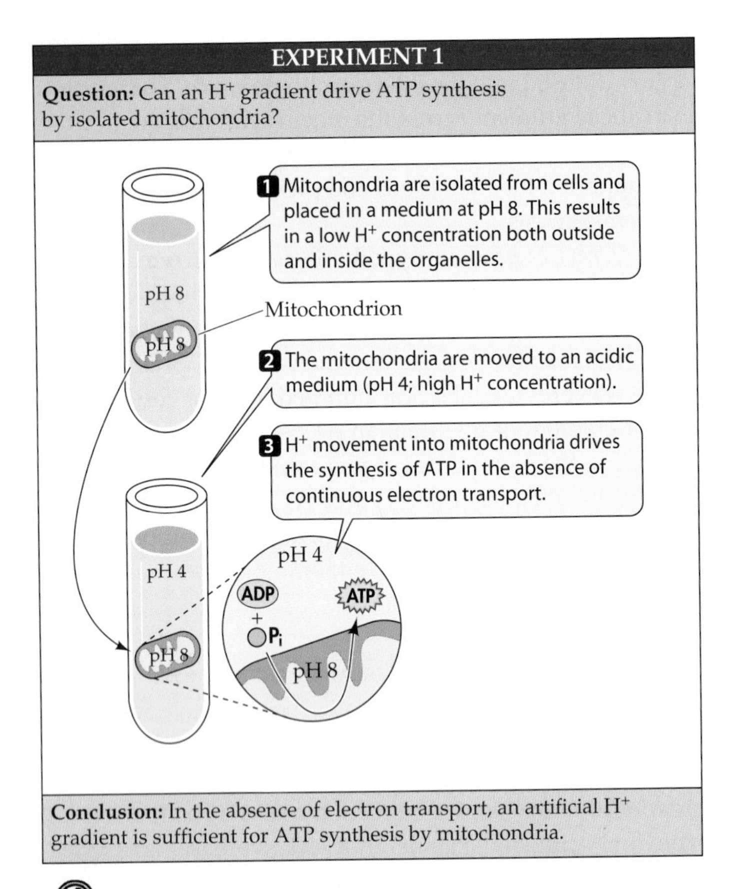

EXPERIMENT 1
Question: Can an H+ gradient drive ATP synthesis by isolated mitochondria?
1 Mitochondria are isolated from cells and placed in a medium at pH 8. This results in a low H+ concentration both outside and inside the organelles.
pH 8
Mitochondrion
pH 8
2 The mitochondria are moved to an acidic medium (pH 4; high H+ concentration).
3 H+ movement into mitochondria drives the synthesis of ATP in the absence of continuous electron transport.
pH 4
pH 4
ADP
+
Pi
ATP
pH 8
pH 8
Conclusion: In the absence of electron transport, an artificial H+ gradient is sufficient for ATP synthesis by mitochondria.

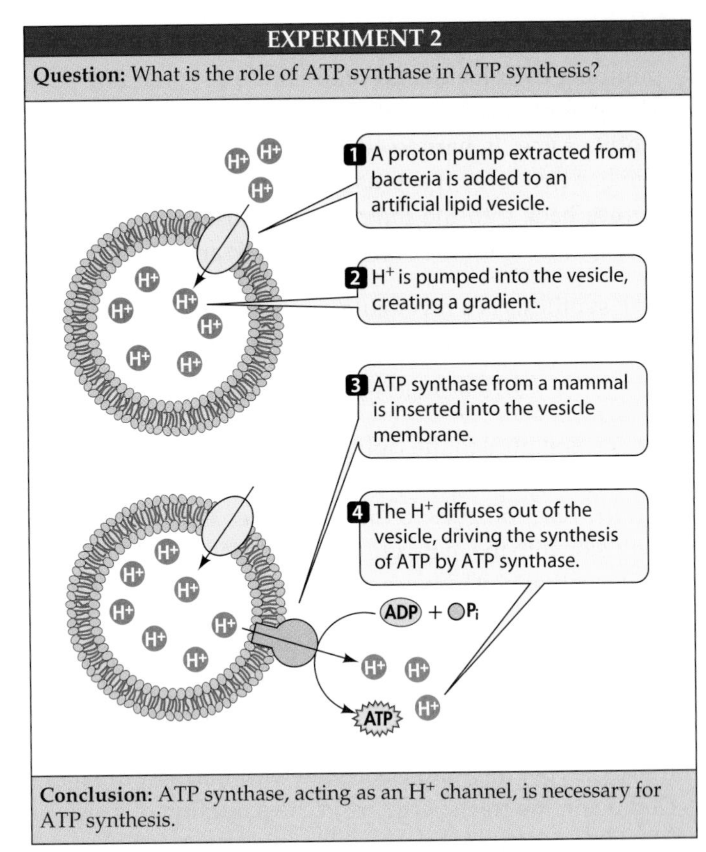

EXPERIMENT 2
Question: What is the role of ATP synthase in ATP synthesis?
1 A proton pump extracted from bacteria is added to an artificial lipid vesicle.
2 H+ is pumped into the vesicle, creating a gradient.
3 ATP synthase from a mammal is inserted into the vesicle membrane.
4 The H+ diffuses out of the vesicle, driving the synthesis of ATP by ATP synthase.
ADP + Pi
ATP
Conclusion: ATP synthase, acting as an H+ channel, is necessary for ATP synthesis.

7.13 Two Experiments Demonstrate the Chemiosmotic Mechanism These two experiments show that an H^+ gradient across a membrane is all that is needed to drive the synthesis of ATP by the enzyme ATP synthase. Whether the H^+ gradient is produced artificially, as in these experiments, or by the respiratory chain found in nature does not matter.

steps in glycolysis, pyruvate oxidation, and the citric acid cycle also stop. If the cell has no other way to obtain energy from its food, it will die. Under anaerobic conditions, many (but not all) cells can produce a small amount of ATP by glycolysis, provided that fermentation metabolizes and regenerates the NAD^+ necessary to keep glycolysis running.

Fermentation, like glycolysis, occurs in the cytoplasm. It has two defining characteristics:

- Fermentation uses NADH + H^+ formed by glycolysis to reduce pyruvate or one of its metabolites, and consequently NAD^+ is regenerated. NAD^+ is required for reaction 6 of glycolysis (see Figure 7.6), so once the cell has replenished its NAD^+ supply in this way, it can carry more glucose through glycolysis.
- Fermentation enables glycolysis to produce a small but sustained amount of ATP. The reactions of fermentation do not themselves produce any ATP. Only as much ATP is produced as can be obtained from substrate-level phosphorylation—not the much greater yield of ATP obtained by cellular respiration using chemiosmosis.

When cells capable of fermentation become anaerobic, the rate of glycolysis speeds up tenfold or even more. Thus a substantial rate of ATP production is maintained, although efficiency in terms of ATP molecules per glucose molecule is greatly reduced compared with cellular respiration under aerobic conditions.

Some organisms are confined to totally anaerobic environments and use only fermentation. Usually, there are two metabolic reasons for this. First, these organisms lack the molecular machinery for oxidative phosphorylation, and second, they lack enzymes to detoxify the toxic by-products of O_2, such as hydrogen peroxide (H_2O_2). An example of such an *obligate anaerobe* is *Clostridium botulinum*, the bacterium that thrives in sealed containers of foods and releases the potentially deadly botulinum toxin. Other bacteria, such as *Mycobacterium tuberculosis*, which causes tuberculosis, cannot carry out fermentation and must grow in aerobic environments. Still others, such as *Escherichia coli*, which grows in the human large intestine, can perform either respiration or fermentation, but prefer the former in an aerobic environment. And several bacteria carry on cellular respiration—not fermentation—without using oxygen gas as an electron acceptor. Instead, to oxidize their cytochromes, these bacteria reduce nitrate ions (NO_3^-) to nitrite ions (NO_2^-). We will return to these organisms when we discuss the nitrogen cycle in Chapter 37.

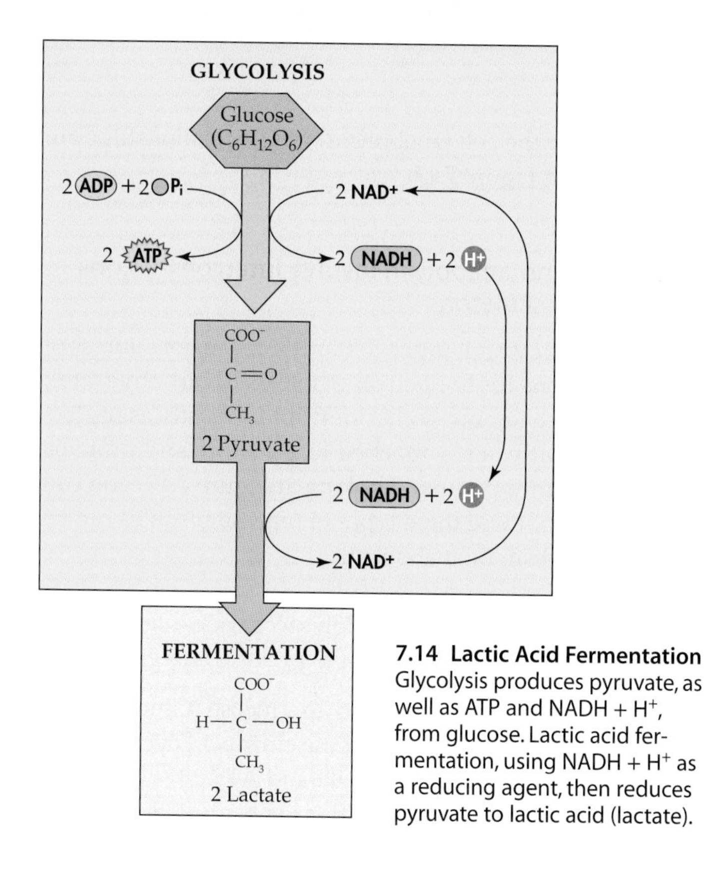

7.14 Lactic Acid Fermentation Glycolysis produces pyruvate, as well as ATP and NADH + H^+, from glucose. Lactic acid fermentation, using NADH + H^+ as a reducing agent, then reduces pyruvate to lactic acid (lactate).

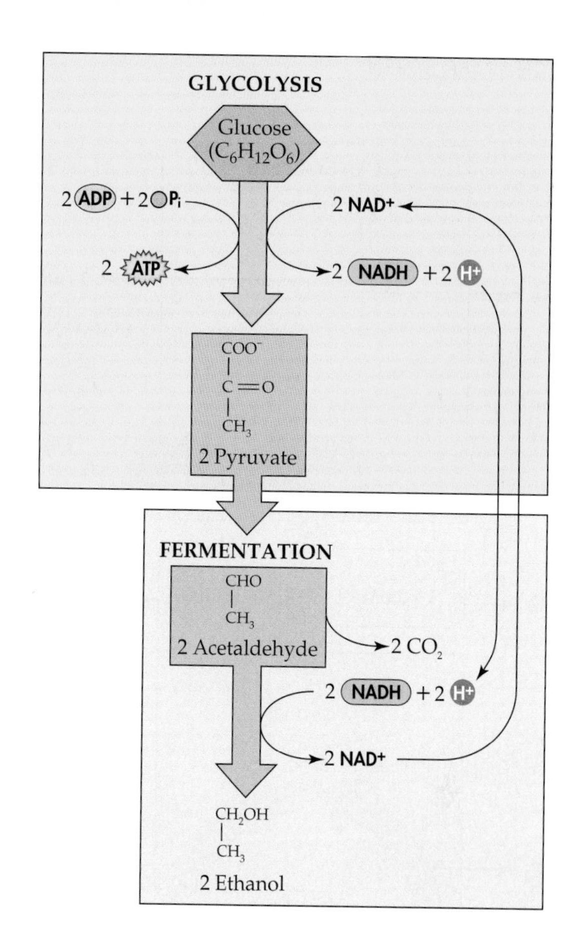

7.15 Alcoholic Fermentation In alcoholic fermentation (the basis for the brewing industry), pyruvate from glycolysis is converted to acetaldehyde and CO_2 is released. The NADH + H^+ from glycolysis acts as a reducing agent, reducing acetaldehyde to ethanol.

Some fermenting cells produce lactic acid and some produce alcohol

Different types of fermentation are carried out by different bacteria and eukaryotic body cells. These different fermentation processes are distinguished by the final product produced. For example, in **lactic acid fermentation**, pyruvate is reduced to lactate (Figure 7.14). Lactic acid fermentation takes place in many microorganisms as well as in our muscle cells. Unlike muscle cells, nerve cells (neurons) are incapable of fermentation because they lack the enzyme that reduces pyruvate to lactate. For that reason, without adequate oxygen the human nervous system (including the brain) is rapidly destroyed; it is the first part of the body to die.

Certain yeasts and some plant cells carry on a process called **alcoholic fermentation** under anaerobic conditions (Figure 7.15). This process requires two enzymes to metabolize pyruvate. First, carbon dioxide is removed from pyruvate, leaving the compound acetaldehyde. Second, the acetaldehyde is reduced by NADH + H^+, producing NAD^+ and ethyl alcohol (ethanol). This is how beer and wine are made.

Contrasting Energy Yields

The total net energy yield from glycolysis using fermentation is two molecules of ATP per molecule of glucose oxidized. In contrast, the maximum yield that can be obtained from a molecule of glucose through glycolysis followed by cellular respiration is much greater—about 36 molecules of ATP (Figure 7.16). (See Figures 7.6, 7.8, and 7.12 to review where these ATP molecules come from.)

Why is so much more ATP produced by cellular respiration? As we have repeatedly stated, glycolysis is only a partial oxidation of glucose, as is fermentation. Much more energy remains in the end products of fermentation, such as lactic acid and ethanol, than in the end product of cellular respiration, CO_2. In cellular respiration, carriers (mostly NAD^+) are reduced in pyruvate oxidation and the citric acid cycle, then oxidized by the respiratory chain, with the accompanying production of ATP (three for each NADH + H^+ and two for each $FADH_2$) by the chemiosmotic mechanism. In an aerobic environment, an organism capable of this type of metabolism will be at an advantage (in terms of energy availability per glucose molecule) over one limited to fermentation.

The total gross yield of ATP from one molecule of glucose processed through glycolysis and cellular respiration is 38. However, we may subtract two from that gross—for a net yield of 36 ATP—because in some animal cells the inner mitochondrial membrane is impermeable to NADH, and a "toll" of one ATP must be paid for each NADH produced in glycolysis that is shuttled into the mitochondrial matrix.

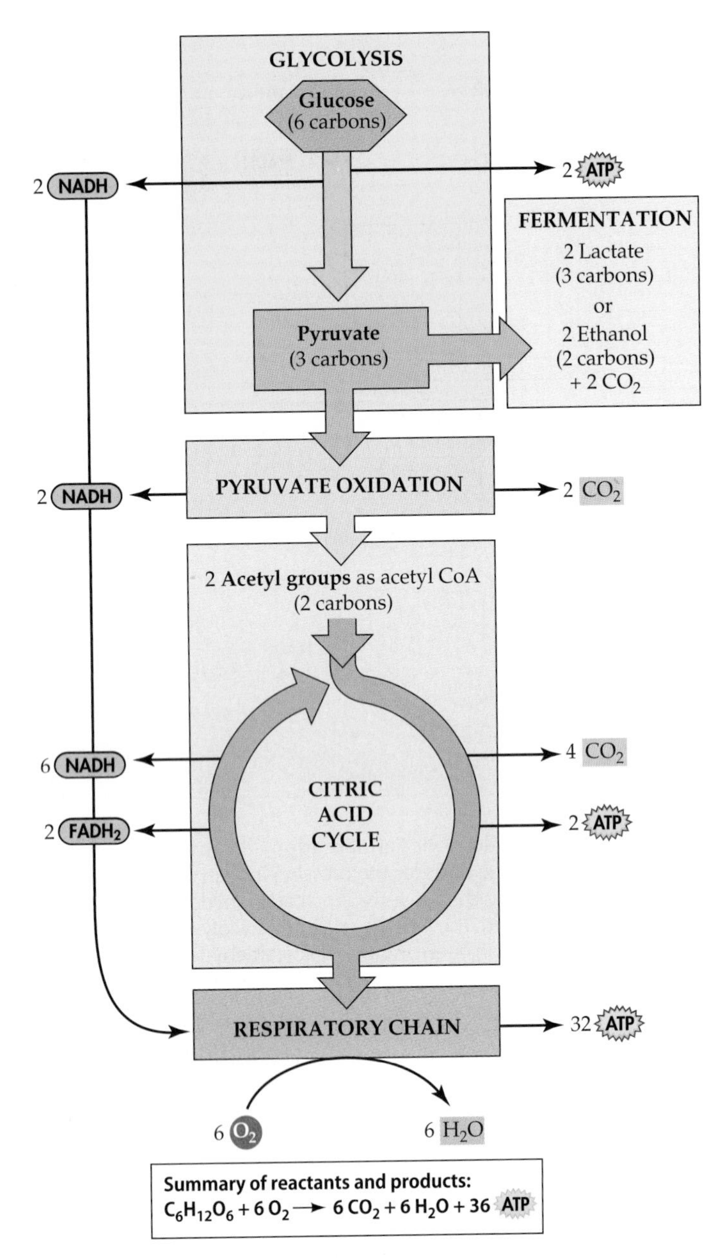

7.16 Cellular Respiration Yields More Energy Than Glycolysis Does Carriers are reduced in pyruvate oxidation and the citric acid cycle, then oxidized by the respiratory chain. These reactions produce ATP via the chemiosmotic mechanism.

Relationships between Metabolic Pathways

Glycolysis and the pathways of cellular respiration do not operate in isolation from the rest of metabolism. Rather, there is an interchange, with biochemical traffic flowing both into these pathways and out of them, to and from the synthesis and breakdown of amino acids, nucleotides, fatty acids, and so forth. Carbon skeletons enter from other molecules that are broken down to release their energy (catabolism), and carbon skeletons leave to form the major macromolecular constituents of the cell (anabolism). These relationships are summarized in Figure 7.17.

Catabolism and anabolism involve interconversions using carbon skeletons

A hamburger or veggiburger contains three major sources of carbon skeletons for the person who eats it: carbohydrates, mostly as starch (a polysaccharide); lipids, mostly as triglycerides (three fatty acids attached to glycerol); and proteins (polymers of amino acids). Looking at Figure 7.17, you can see how each of these three types of macromolecules can be used in catabolism or anabolism.

CATABOLIC INTERCONVERSIONS. Polysaccharides, lipids, and proteins can all be broken down to provide energy:

- *Polysaccharides* are hydrolyzed to glucose. Glucose then passes through glycolysis and the citric acid cycle, where its energy is captured in NADH and ATP.
- *Lipids* are broken down into their substituents, glycerol and fatty acids. Glycerol is converted to dihydroxyacetone phosphate, an intermediate in glycolysis, and fatty acids are converted to acetate and then acetyl CoA in the mitochondria. In both cases, further oxidation to CO_2 and release of energy then occur.
- *Proteins* are hydrolyzed to their amino acid building blocks. The 20 different amino acids feed into glycolysis or the citric acid cycle at different points. A specific example is shown in Figure 7.18, in which an amino acid can be converted to an intermediate in the citric acid cycle.

ANABOLIC INTERCONVERSIONS. Many catabolic pathways can operate in reverse. That is, glycolytic and citric acid cycle intermediates, instead of being oxidized to form CO_2, can be reduced and used to form glucose in a process called **gluconeogenesis** (which means "new formation of glucose"). Likewise, acetyl CoA can be used to form fatty acids. The most common fatty acids have an even number of carbons: 14, 16, or 18. These molecules are formed by adding two-carbon acetyl CoA "units" one at a time until the appropriate chain length is reached. Amino acids can be formed by reversible reactions such as the one shown in Figure 7.18, and can then be polymerized into proteins.

Some intermediates of the citric acid cycle are used in the synthesis of various important cellular constituents. For example, α-ketoglutarate is a starting point for purines and oxaloacetate for pyrimidines, both constituents of the nucleic acids DNA and RNA. α-Ketoglutarate is also a starting point for chlorophyll synthesis. Acetyl CoA is a building block for various pigments, plant growth substances, rubber, and the steroid hormones of animals, among other molecules.

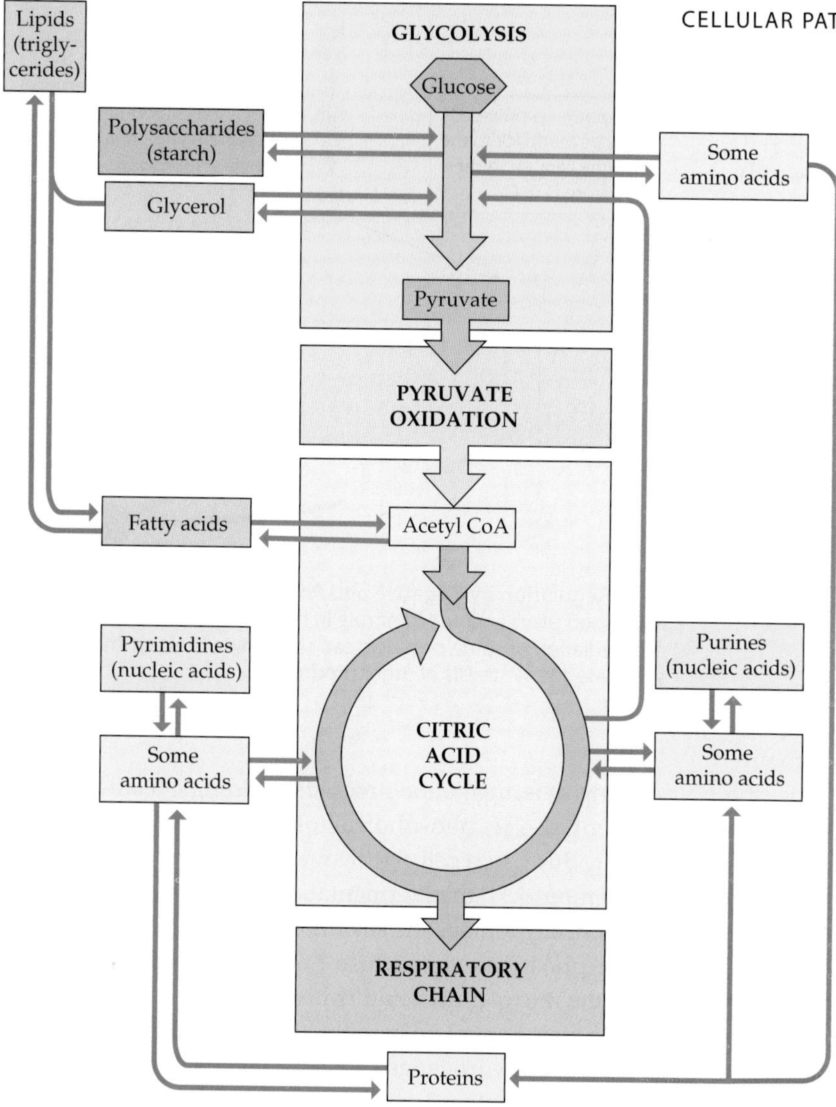

7.17 Relationships Among the Major Metabolic Pathways of the Cell Note the central position of glycolysis and the citric acid cycle in this network of metabolic pathways. Also note that many of the pathways can operate in reverse.

Catabolism and anabolism are integrated

A carbon atom from a protein in your burger can end up in DNA or fat or CO_2, among other fates. How does the cell "decide" which metabolic pathway to follow? With all of these possible interconversions, you might expect that the cellular concentrations of various biochemical molecules would vary widely. For example, the level of oxaloacetate in your cells might depend on what you eat (some food molecules form oxaloacetate) and whether oxaloacetate is used up (in the citric acid cycle or in forming the amino acid aspartate). Remarkably, the levels of these substances in what is called the "metabolic pool"—the sum total of all the small molecules such as metabolic intermediates in a cell—are quite constant. The cell regulates the enzymes of catabolism and anabolism so as to maintain a balance. This *metabolic homeostasis* gets upset only in unusual circumstances. Let's look one such unusual circumstance: undernutrition.

Glucose is an excellent source of energy. From Figure 7.17, you can see that fats and proteins can also serve as energy sources. Any one, or all three, could be used to provide the energy your body needs. In reality, things are not so simple. Proteins, for example, have essential roles in your body as enzymes and structural elements, and using them for energy might deprive you of a catalyst for a vital reaction.

Polysaccharides and fats have no such catalytic roles. But polysaccharides, because they are somewhat polar, can bind a lot of water. Because they are nonpolar, fats do not bind as much water as polysaccharides do. So, in water, fats weigh less than polysaccharides. Also, fats are more reduced than carbohydrates (more C—H bonds as opposed to C—OH) and have more energy stored in their bonds. For these two reasons, fats are a better way for an organism to store energy than polysaccharides. It is not surprising, then, that a typical person has about one day's worth of food energy stored as glycogen, a week's food energy as usable proteins in blood, and over a month's food energy stored as fats.

What happens if a person does not eat enough food to produce sufficient ATP and NADH for anabolism and biological activities? This situation can be the result of a deliberate decision to lose weight, but for too many people, it is forced upon them because not enough food is available. In either case, the first energy stores in the body to be used are the glycogen stores in muscle and liver cells. This doesn't last long, and next come the fats.

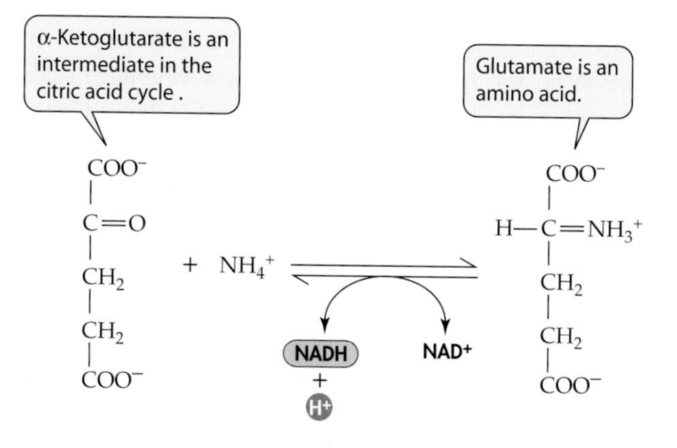

7.18 Coupling Metabolic Pathways This reaction, in which α-ketoglutarate and glutamate and glutamate are interconverted, is catalyzed by the enzyme glutamate dehydrogenase.

The level of acetyl CoA rises as fatty acids are broken down. However, a problem remains: Because fatty acids cannot get from the blood to the brain, the brain can use only glucose as its energy source. With glucose already depleted, the body must convert something else to make glucose for the brain. This gluconeogenesis uses mostly amino acids, largely from the breakdown of proteins. So, without sufficient food intake, both proteins (for glucose) and fats (for energy) are used up. After several weeks of starvation, fat stores become depleted, and the only energy source left is proteins, some of which have already been degraded to supply the brain with glucose. At this point, essential proteins, such as antibodies used to fight off infections and muscle proteins, get broken down, both for energy and for gluconeogenesis. The loss of these proteins can lead to severe illnesses.

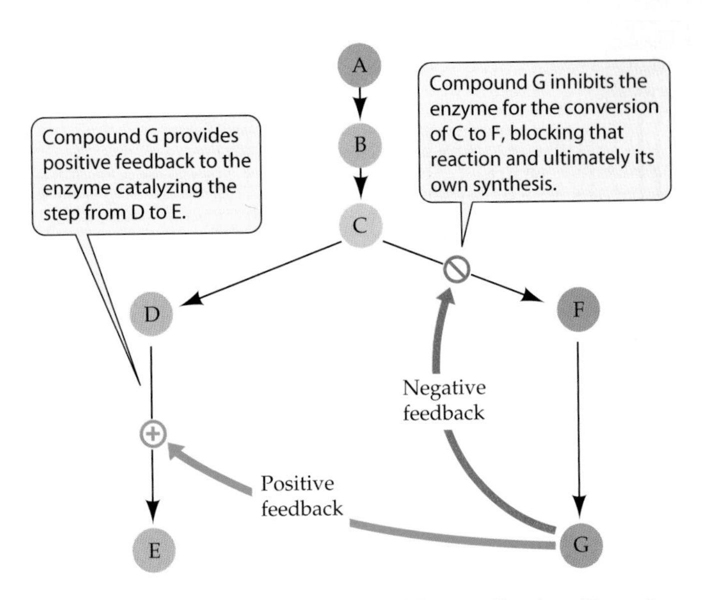

7.19 Regulation by Negative and Positive Feedback Allosteric regulation plays an important role in metabolic pathways. Excess accumulation of some products can shut down their synthesis or stimulate the synthesis of other products.

Regulating Energy Pathways

We have described the relationships between metabolic pathways and noted that these pathways work together to provide homeostasis in the cell and organism. But how does the cell regulate interconversions between these pathways to maintain constant metabolic pools?

Consider what happens to the starch in your burger bun. In the digestive system, starch is hydrolyzed to glucose, which enters the blood for distribution to the rest of the body. Before this happens, however, a "decision" must be made: Is there already enough glucose in the blood to supply the body's needs? If there is, the excess glucose is converted to stored glycogen in the liver. If not enough glucose is supplied by food, liver glycogen is broken down, or other molecules are used to make glucose by gluconeogenesis.

The end result is that the level of glucose in the blood is remarkably constant. We will describe the details of how this happens in Part Seven of this book. For now, it is important to realize that the interconversions of glucose involve many steps, each catalyzed by an enzyme, and it is here that controls often reside.

Glycolysis, the citric acid cycle, and the respiratory chain are regulated by **allosteric control** of the enzymes involved. In metabolic pathways, as we saw in Chapter 6, a high concentration of the products of a later reaction can suppress the action of enzymes that catalyze an earlier reaction. On the other hand, an excess of the product of one branch of a synthetic chain can speed up reactions in another branch, diverting raw materials away from synthesis of the first product (Figure 7.19). These negative and positive feedback control mechanisms are used at many points in the energy-harvesting pathways, which are summarized in Figure 7.20.

The main control point in glycolysis is the enzyme **phosphofructokinase** (reaction 3 in Figure 7.6). This enzyme is allosterically inhibited by ATP and activated by ADP or AMP. As long as fermentation proceeds, yielding a relatively small amount of ATP, phosphofructokinase operates at full efficiency. But when cellular respiration begins producing 18 times more ATP than fermentation does, the abundant ATP allosterically inhibits the enzyme, and the conversion of fructose 6-phosphate to fructose 1,6-bisphosphate declines, as does the rate of glucose utilization.

The main control point in the citric acid cycle is the enzyme **isocitrate dehydrogenase**, which converts isocitrate to α-ketoglutarate (reaction 3 in Figure 7.8). NADH + H^+ and ATP are feedback inhibitors of this reaction; ADP and NAD^+ are activators (Figure 7.20). If too much ATP is accumulating, or if NADH + H^+ is being produced faster than it can be used by the respiratory chain, the conversion of isocitrate is slowed, and the citric acid cycle is essentially shut down. A shutdown of the citric acid cycle would cause large amounts of isocitrate and citrate to accumulate if the conversion of acetyl CoA to citrate were not also slowed by abundant ATP and NADH + H^+. However, a certain excess of citrate does accumulate, and this excess acts as an additional negative feedback inhibitor to slow the fructose 6-phosphate reaction early in glycolysis. Consequently, if the citric acid cycle has been slowed down because of abundant ATP (and not because of a lack of oxygen), glycolysis is shut down as well. Both processes resume when the ATP level falls and they are needed again. Allosteric control keeps these processes in balance.

Another control point involves a method for storing excess acetyl CoA. If too much ATP is being made and the citric acid cycle shuts down, the accumulation of citrate switches acetyl CoA to the synthesis of fatty acids for storage. This is one reason why people who eat too much accumulate fat. These fatty acids may be metabolized later to produce more acetyl CoA.

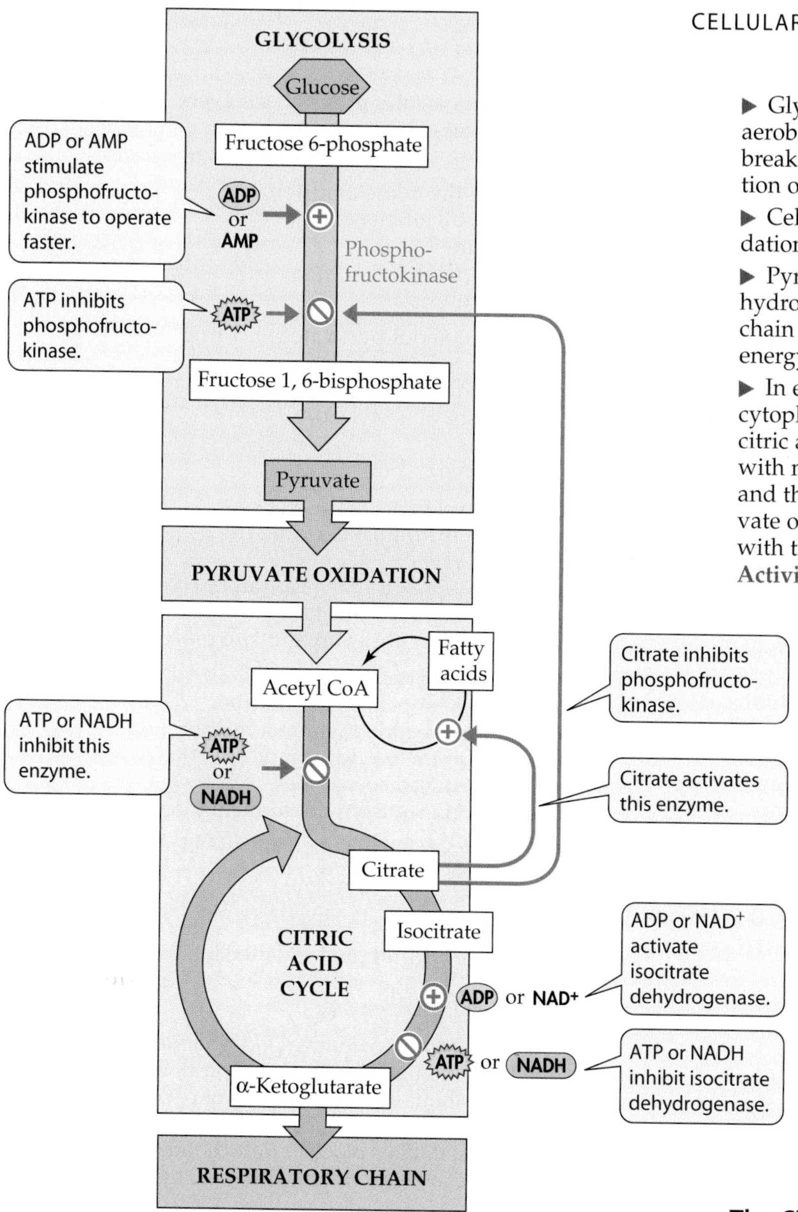

7.20 Feedback Regulation of Glycolysis and the Citric Acid Cycle Feedback controls glycolysis and the citric acid cycle at crucial early steps, increasing their efficiency and preventing the excessive buildup of intermediates.

Chapter Summary

Energy and Electrons from Glucose

▶ Metabolic pathways occur in small steps, each catalyzed by a specific enzyme. They are often compartmentalized.

▶ When glucose burns, energy is released as heat and light. The same equation applies to the metabolism of glucose by cells, but the reaction is accomplished in many separate steps so that the energy can be captured in ATP. **Review Figure 7.1**

▶ Oxidation is the loss of electrons; reduction is the gain of electrons. As a material is oxidized, the electrons it loses are transferred to another material, which is thereby reduced. Such redox reactions transfer large amounts of energy. **Review Figure 7.2**

▶ The coenzyme NAD is a key electron carrier in biological redox reactions. It exists in two forms, one oxidized (NAD^+) and the other reduced ($NADH + H^+$). **Review Figures 7.3, 7.4**

▶ Glycolysis operates in the presence or absence of O_2. Under aerobic conditions, cellular respiration continues the process of breaking down glucose. Under anaerobic conditions, fermentation occurs. **Review Figure 7.5. See Web/CD Activity 7.1**

▶ Cellular respiration consists of three pathways: pyruvate oxidation, the citric acid cycle, and the respiratory chain.

▶ Pyruvate oxidation and the citric acid cycle produce CO_2 and hydrogen atoms carried by NADH and $FADH_2$. The respiratory chain combines these hydrogens with O_2, releasing enough energy for the synthesis of ATP. **Review Figure 7.5**

▶ In eukaryotes, glycolysis and fermentation take place in the cytoplasm outside of the mitochondria; pyruvate oxidation, the citric acid cycle, and the respiratory chain operate in association with mitochondria. In prokaryotes, glycolysis, fermentation, and the citric acid cycle take place in the cytoplasm, and pyruvate oxidation and the respiratory chain operate in association with the plasma membrane. **Review Table 7.1. See Web/CD Activity 7.2**

Glycolysis: From Glucose to Pyruvate

▶ Glycolysis is a pathway of ten enzyme-catalyzed reactions located in the cytoplasm. Glycolysis provides starting materials for both cellular respiration and fermentation. **Review Figure 7.6**

▶ The energy-investing reactions of glycolysis use two ATPs per glucose molecule and eventually yield two G3P molecules. In the energy-harvesting reactions, two NADH molecules are produced, and four ATP molecules are generated by substrate-level phosphorylation. Two pyruvate molecules are produced for each glucose molecule. **Review Figures 7.6, 7.7**

Pyruvate Oxidation

▶ The pyruvate dehydrogenase complex catalyzes three reactions: (1) Pyruvate is oxidized to an acetyl group, releasing one CO_2 molecule and considerable energy. (2) Some of this energy is captured when NAD^+ is reduced to $NADH + H^+$. (3) The remaining energy is captured when the acetyl group is combined with coenzyme A, yielding acetyl CoA.

The Citric Acid Cycle

▶ The energy in acetyl CoA drives the reaction of acetate with oxaloacetate to produce citrate. The citric acid cycle is a series of reactions in which citrate is oxidized and oxaloacetate regenerated (hence a "cycle"). It produces 2 CO_2, 1 $FADH_2$, 3 NADH, and 1 ATP for each acetyl CoA. **Review Figures 7.8, 7.9. See Web/CD Activity 7.3**

The Respiratory Chain: Electrons, Proton Pumping, and ATP Production

▶ NADH and $FADH_2$ from glycolysis, pyruvate oxidation, and the citric acid cycle are oxidized by the respiratory chain, regenerating NAD^+ and FAD. Most of the enzymes and other electron carriers of the chain are part of the inner mitochondrial membrane. Oxygen (O_2) is the final acceptor of electrons and protons, forming water (H_2O). **Review Figures 7.10, 7.11. See Web/CD Activity 7.4**

▶ The chemiosmotic mechanism couples proton transport to oxidative phosphorylation. As the electrons move along the respiratory chain, protons are pumped out of the mitochondrial matrix, establishing a gradient of both proton concentration and electric charge—the proton-motive force. **Review Figure 7.12. See Web/CD Tutorial 7.1**

▶ The proton-motive force causes protons to diffuse back into the mitochondrial matrix through the membrane channel protein

ATP synthase, which couples that diffusion to the production of ATP. Several key experiments demonstrate that chemiosmosis produces ATP. **Review Figure 7.13.** See Web/CD Tutorial 7.2

Fermentation: ATP from Glucose, without O_2

▶ Many organisms and some cells live without O_2, deriving all their energy from glycolysis and fermentation. Together, these pathways partly oxidize glucose and generate energy-containing products such as lactic acid or ethanol. **Review Figures 7.14, 7.15**

Contrasting Energy Yields

▶ For each molecule of glucose used, fermentation yields 2 molecules of ATP. In contrast, glycolysis operating with pyruvate oxidation, the citric acid cycle, and the respiratory chain yields up to 36 molecules of ATP per molecule of glucose. **Review Figure 7.16.** See Web/CD Activity 7.5

Relationships between Metabolic Pathways

▶ Catabolic pathways feed into the energy-harvesting metabolic pathways. Polysaccharides are broken down into glucose, which enters glycolysis. Glycerol from fats also enters glycolysis, and acetyl CoA from fatty acid degradation enters the citric acid cycle. Proteins enter glycolysis and the citric acid cycle via amino acids. **Review Figures 7.17, 7.18**

▶ Anabolic pathways use intermediate components of energy-harvesting pathways to synthesize fats, amino acids, and other essential building blocks. **Review Figures 7.17, 7.18**

Regulating Energy Pathways

▶ The rates of glycolysis and the citric acid cycle are increased or decreased by the actions of ATP, ADP, NAD^+, or NADH + H^+ on allosteric enzymes.

▶ Inhibition of the glycolytic enzyme phosphofructokinase by abundant ATP from cellular respiration slows down glycolysis. ADP activates this enzyme, speeding up glycolysis. The citric acid cycle enzyme isocitrate dehydrogenase is inhibited by ATP and NADH and activated by ADP and NAD^+. **Review Figures 7.19, 7.20.** See Web/CD Activity 7.6

Self-Quiz

1. The role of oxygen gas in our cells is to
   a. catalyze reactions in glycolysis.
   b. produce CO_2.
   c. form ATP.
   d. accept electrons from the electron transport chain.
   e. react with glucose to split water.
2. Oxidation and reduction
   a. entail the gain or loss of proteins.
   b. are defined as the loss of electrons.
   c. are both endergonic reactions.
   d. always occur together.
   e. proceed only under aerobic conditions.
3. NAD^+ is
   a. a type of organelle.
   b. a protein.
   c. present only in mitochondria.
   d. a part of ATP.
   e. formed in the reaction that produces ethanol.
4. Glycolysis
   a. takes place in the mitochondrion.
   b. produces no ATP.
   c. has no connection with the respiratory chain.
   d. is the same thing as fermentation.
   e. reduces two molecules of NAD^+ for every glucose molecule processed.
5. Fermentation
   a. takes place in the mitochondrion.
   b. takes place in all animal cells.
   c. does not require O_2.
   d. requires lactic acid.
   e. prevents glycolysis.
6. Which statement about pyruvate is *not* true?
   a. It is the end product of glycolysis.
   b. It becomes reduced during fermentation.
   c. It is a precursor of acetyl CoA.
   d. It is a protein.
   e. It contains three carbon atoms.
7. The citric acid cycle
   a. takes place in the mitochondrion.
   b. produces no ATP.
   c. has no connection with the respiratory chain.
   d. is the same thing as fermentation.
   e. reduces two NAD^+ for every glucose processed.
8. The electron transport chain
   a. operates in the mitochondrial matrix.
   b. uses proteins embedded within a membrane.
   c. always leads to the production of ATP.
   d. regenerates reduced coenzymes.
   e. operates simultaneously with fermentation.
9. Compared to anaerobic metabolism, aerobic breakdown of glucose produces
   a. more ATP.
   b. pyruvate.
   c. fewer protons for pumping in mitochondria.
   d. less CO_2.
   e. more oxidized coenzymes.
10. Which statement about oxidative phosphorylation is *not* true?
    a. It is the formation of ATP by the respiratory chain.
    b. It is brought about by the chemiosmotic mechanism.
    c. It requires aerobic conditions.
    d. In eukaryotes, it takes place in mitochondria.
    e. Its functions can be served equally well by fermentation.

For Discussion

1. Trace the sequence of chemical changes that occurs in mammalian brain tissue when the oxygen supply is cut off. The first change is that the cytochrome *c* oxidase system becomes totally reduced, because electrons can still flow from cytochrome *c* but there is no oxygen to accept electrons from cytochrome *c* oxidase. What are the remaining steps?
2. Trace the sequence of chemical changes that occurs in mammalian muscle tissue when the oxygen supply is cut off. (The first change is the same as that in Question 1.)
3. Some cells that use the citric acid cycle and the respiratory chain can also thrive by using fermentation under anaerobic conditions. Given the lower yield of ATP (per molecule of glucose) in fermentation, why can these cells function so efficiently under anaerobic conditions?
4. Describe the mechanisms by which the rates of glycolysis and aerobic respiration are kept in balance with one another.
5. You eat a hamburger that has polysaccharides, proteins and lipids. Using your knowledge of the integration of biochemical pathways, explain how the amino acids in the proteins and glucose in the polysaccharides can end up as fats.

8 Photosynthesis: Energy from the Sun

Powered by sunlight, green plants convert CO_2 and water into carbohydrates by a process called photosynthesis. The emergence of this metabolic pathway was a key event in the evolution of life. Photosynthesizing organisms, called *autotrophs* ("self-feeders"), use solar energy to make their own food from simple chemicals in the environment. In this way, they provide an entry point to the biosphere for chemical energy. *Heterotrophs* ("other-feeders") cannot photosynthesize, and they depend on autotrophs (or other heterotrophs) for the raw materials of metabolism, such as glucose.

The "food chain" from autotrophs to heterotrophs requires a lot of photosynthesis. On the African plain, it takes 6 acres of grassland to convert enough CO_2 into plant matter to support the growth of one gazelle that consumes the grass. Globally, more than 10 billion tons of carbon is *fixed*—converted from being part of a simple gas (CO_2) into a more complex molecule (carbohydrate)—by plants every year. This huge amount of photosynthetically-fixed carbon is available for use by all species that need it.

Humans consume a huge amount of Earth's photosynthetic output. Recent calculations of total plant growth in agriculture, pastures, and forests and the products consumed by people indicate that one-third of all the carbon fixed annually is appropriated by humans, leaving two-thirds for the entire remainder of the biosphere. This is by far the greatest proportion of consumption for any single species in known history, Is this situation sustainable? Conferences such as the 2002 United Nations Conference on Sustainability have demonstrated concern for our photosynthetic future.

An important first step in examining ecological sustainability is a thorough understanding of photosynthesis. The process of photosynthesis can be neatly broken down into two steps. The first step is the conversion of energy from light to chemical bonds in reduced electron carriers and ATP. In the second step, these two sources of chemical energy are used to drive the synthesis of carbohydrates from carbon dioxide. In this chapter, we will examine these two processes and show how they are related to each other and to plant growth.

Primary Producers Plants, through photosynthesis, are the basis of life on Earth. The energy stored in these soybeans being cultivated on a farm in Kansas will be harvested and used by human beings.

Identifying Photosynthetic Reactants and Products

By the beginning of the nineteenth century, scientists understood the broad outlines of photosynthesis. It was known to use three principal ingredients—water, carbon dioxide (CO_2), and light—and to produce not only carbohydrates but also oxygen gas (O_2). Scientists had learned several things:

- The water for photosynthesis in land plants comes primarily from the soil and must travel from the roots to the leaves.
- Carbon dioxide is taken in, and water and O_2 are released, through tiny openings in leaves, called **stomata** (singular, stoma) (Figure 8.1).
- Light is absolutely necessary for the production of oxygen and carbohydrates.

By 1804, scientists could summarize photosynthesis as follows:

carbon dioxide + water + light energy → sugar + oxygen

which turns into an equation that is the *reverse* of the overall equation for cellular respiration given in Chapter 7:

$$6\ CO_2 + 6\ H_2O \rightarrow C_6H_{12}O_6 + 6\ O_2$$

Although correct, these statements say nothing about the details of the process, and in fact, in detail, photosynthesis is *not* the reverse of cellular respiration. What are the reactions of photosynthesis? What role does light play in these reactions? How do the carbons become linked to form sugars? And does the oxygen gas come from the CO_2 or from the H_2O?

Almost a century and a half passed before the source of the O_2 released during photosynthesis was determined. One of the first uses of an *isotopic tracer* in biological research resulted in its identification. In these experiments, two groups of green plants were allowed to carry on photosynthesis. Plants in the first group were supplied with water containing the oxygen isotope ^{18}O and with CO_2 containing only the common oxygen isotope ^{16}O; plants in the second group were supplied with CO_2 labeled with ^{18}O and water containing only ^{16}O.

When oxygen gas was collected from each group of plants and analyzed, it was found that O_2 containing ^{18}O was produced in abundance by the plants that had been given ^{18}O-labeled water, but *not* by the plants given ^{18}O-labeled CO_2. These results showed that all the oxygen gas produced during photosynthesis comes from water (Figure 8.2). This discovery is reflected in a revised balanced equation:

$$6\ CO_2 + 12\ H_2O \rightarrow C_6H_{12}O_6 + 6\ O_2 + 6\ H_2O$$

Water appears on both sides of the equation because water is both used as a reactant (the twelve molecules on the left) and released as a product (the six new ones on the right). In this revised equation, there are now sufficient water molecules to account for all the oxygen gas produced.

8.1 The Ingredients for Photosynthesis A typical terrestrial plant uses light from the sun, water from the soil, and carbon dioxide from the atmosphere to form organic compounds by photosynthesis.

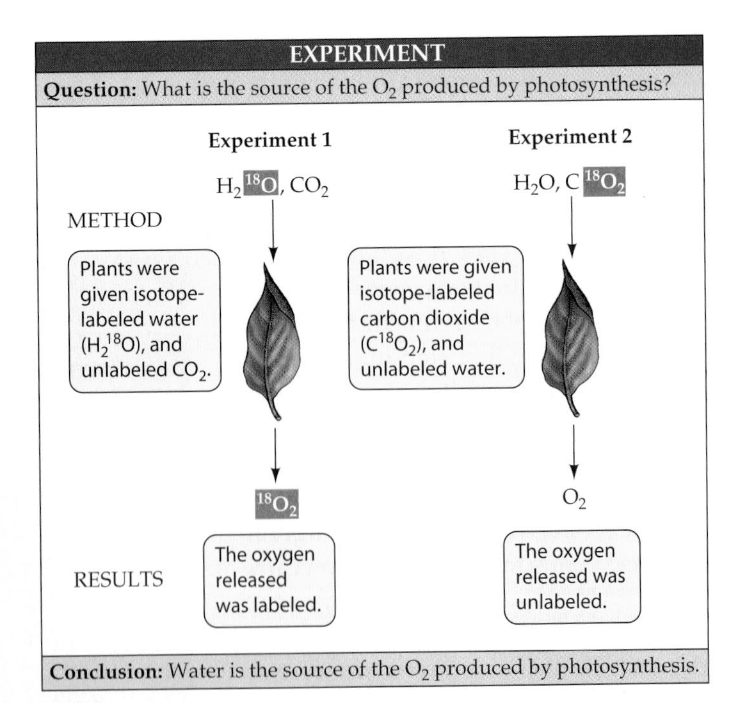

EXPERIMENT
Question: What is the source of the O2 produced by photosynthesis?
Experiment 1
Experiment 2
H2 18O, CO2
H2O, C 18O2
METHOD
Plants were given isotope-labeled water (H2 18O), and unlabeled CO2.
Plants were given isotope-labeled carbon dioxide (C18O2), and unlabeled water.
18O2
O2
RESULTS
The oxygen released was labeled.
The oxygen released was unlabeled.
Conclusion: Water is the source of the O2 produced by photosynthesis.

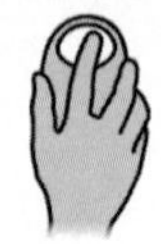

8.2 Water Is the Source of the Oxygen Produced by Photosynthesis Because only plants given isotope-labeled water released labeled O_2, this experiment showed that water is the source of the oxygen released during photosynthesis.

The Two Pathways of Photosynthesis: An Overview

The equation above summarizes the overall process of photosynthesis. However, like glycolysis and the other metabolic pathways that yield energy in cells, photosynthesis consists not of one single reaction but of many reactions. The reactions of photosynthesis can be divided into two pathways:

- The **light reactions** are driven by light energy. This pathway produces ATP and a reduced electron carrier (NADPH + H^+).
- The second pathway, called the **Calvin–Benson cycle**, does not use light directly. It uses ATP, NADPH + H^+ (made by the light reactions), and CO_2 to produce sugars.

The reactions of the Calvin–Benson cycle are sometimes called the *dark reactions* because they do not directly require light energy. However, both pathways stop in the dark because ATP synthesis and $NADP^+$ reduction require light. The reactions of both pathways proceed within the chloroplast, but they reside in different parts of that organelle (Figure 8.3).

The two pathways are linked by the exchange of ATP and ADP, and of $NADP^+$ and NADPH, and the rate of each set of reactions depends on the rate of the other.

We will discuss the light reactions at length, to be followed by the details of the Calvin–Benson cycle. However, because these two photosynthetic pathways are powered by the energy of light, we begin by discussing the physical nature of light and the nature of the specific photosynthetic molecules that capture its energy.

The Interactions of Light and Pigments

Light is a source of both energy and information. In later chapters, we'll examine the many roles of light in the transmission of information. In this chapter, our focus is on light as a source of energy.

Light behaves as both a particle and a wave

Light is a form of electromagnetic radiation. It comes in discrete packets called **photons**. Light also behaves as if it were propagated in waves. The amount of energy contained in a single photon is inversely proportional to its *wavelength:* the shorter the wavelength, the greater the energy of the photons. For example, a photon of red light of wavelength 660 nm has less energy than a photon of blue light at 430 nm.

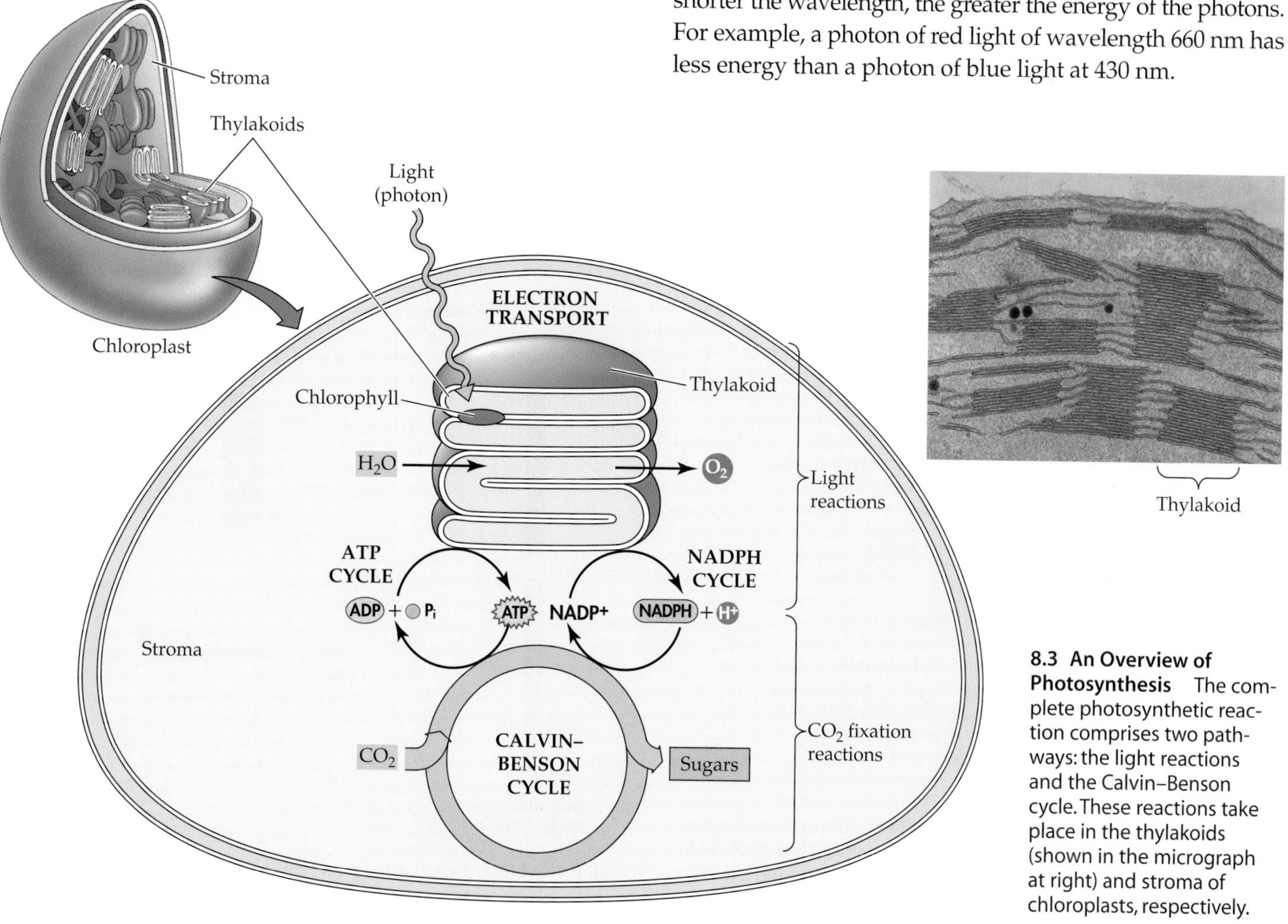

8.3 An Overview of Photosynthesis The complete photosynthetic reaction comprises two pathways: the light reactions and the Calvin–Benson cycle. These reactions take place in the thylakoids (shown in the micrograph at right) and stroma of chloroplasts, respectively.

Two things are required for photons to be active in a biological process:

- Photons must be absorbed by a receptive molecule.
- Photons must have sufficient energy to perform the chemical work required.

Absorbing a photon puts a pigment in an excited state

When a photon meets a molecule, one of three things happens:

- The photon may bounce off the molecule—it may be *scattered*.
- The photon may pass through the molecule—it may be *transmitted*.
- The photon may be *absorbed* by the molecule.

Neither of the first two outcomes causes any change in the molecule. In the third case, the photon disappears. Its energy, however, cannot disappear, because energy is neither created nor destroyed.

When a molecule absorbs a photon, that molecule acquires the energy of the photon. It is thereby raised from a **ground state** (lower energy) to an **excited state** (higher energy) (Figure 8.4*a*). The difference in energy between the molecule's excited state and its ground state is exactly equal to the energy of the absorbed photon. The increase in energy boosts one of the electrons within the molecule into an orbital farther from its nucleus; this electron is now held less firmly (Figure 8.4*b*), making the molecule more chemically reactive, as we will see later in the chapter.

The electromagnetic spectrum (Figure 8.5) shows the wide range of wavelengths (and hence, energy levels) that photons can have. The specific wavelengths absorbed by a particular

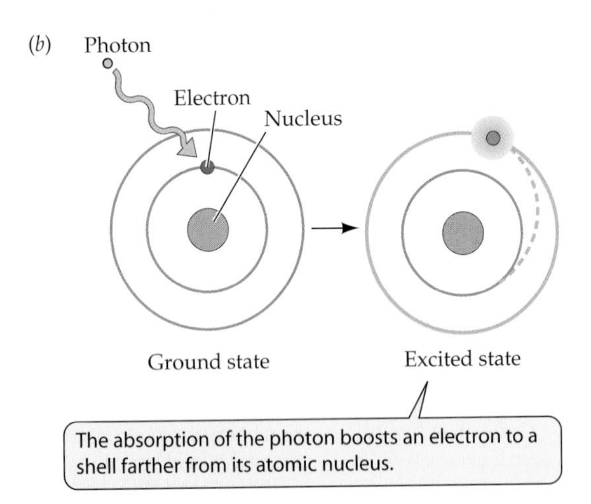

8.4 Exciting a Molecule (*a*) When a molecule absorbs the energy of a photon, it is raised from a ground state to an excited state. (*b*) In the excited state, an electron is boosted to a more distant shell, where it is held less firmly.

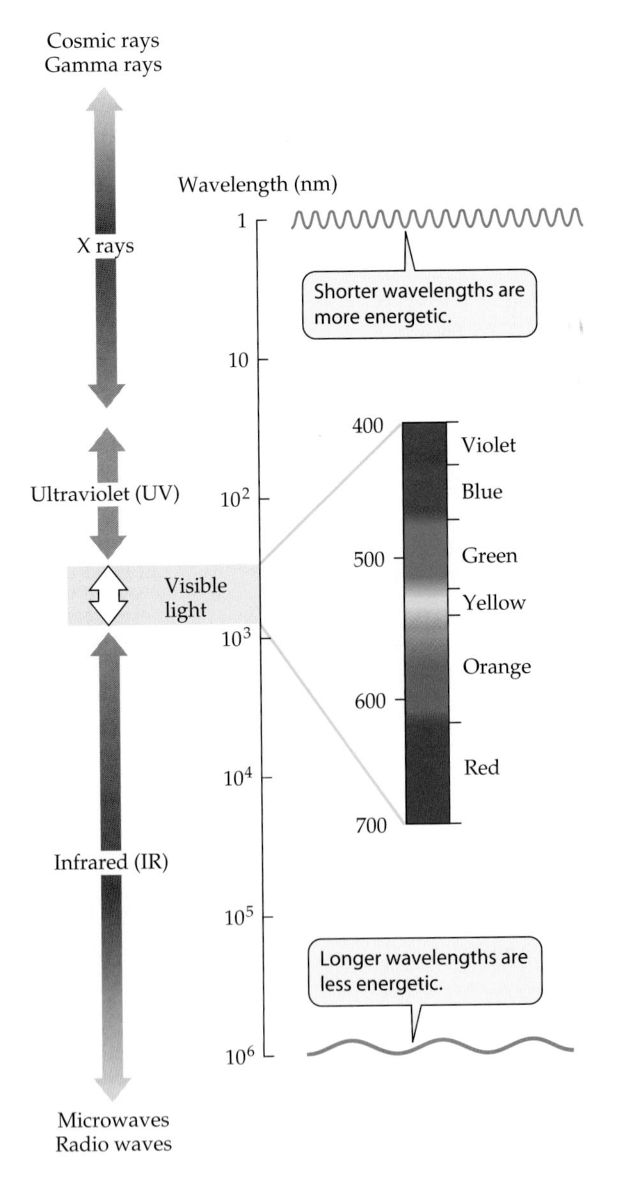

8.5 The Electromagnetic Spectrum The portion of the electromagnetic spectrum that is visible to humans is shown in detail at the right.

molecule are characteristic of that type of molecule. Molecules that absorb wavelengths in the *visible spectrum*—that region of the spectrum that is visible to humans—are called **pigments**.

When a beam of white light (light containing visible light of *all* wavelengths) falls on a pigment, certain wavelengths of the light are absorbed. The remaining wavelengths, which are scattered or transmitted, make the pigment appear to us to be colored. For example, if a pigment absorbs both blue and red light—as chlorophyll does—what we see is the remaining light, which is primarily green.

Absorbed wavelengths correlate with biological activity

If we plot the wavelengths of the light absorbed by a purified molecule, the result is an **absorption spectrum** for that molecule. If we plot the biological activity of a photosynthetic organism as a function of the wavelengths of light to which the organism is exposed, the result is an **action spectrum**. Figure 8.6 shows the absorption spectrum for a pigment, chlorophyll *a*, isolated from the leaves of a plant and the action spectrum for photosynthetic activity for the same plant. A comparison of the two spectra shows that the wavelengths at which photosynthesis is maximal are the same wavelengths at which chlorophyll *a* absorbs light.

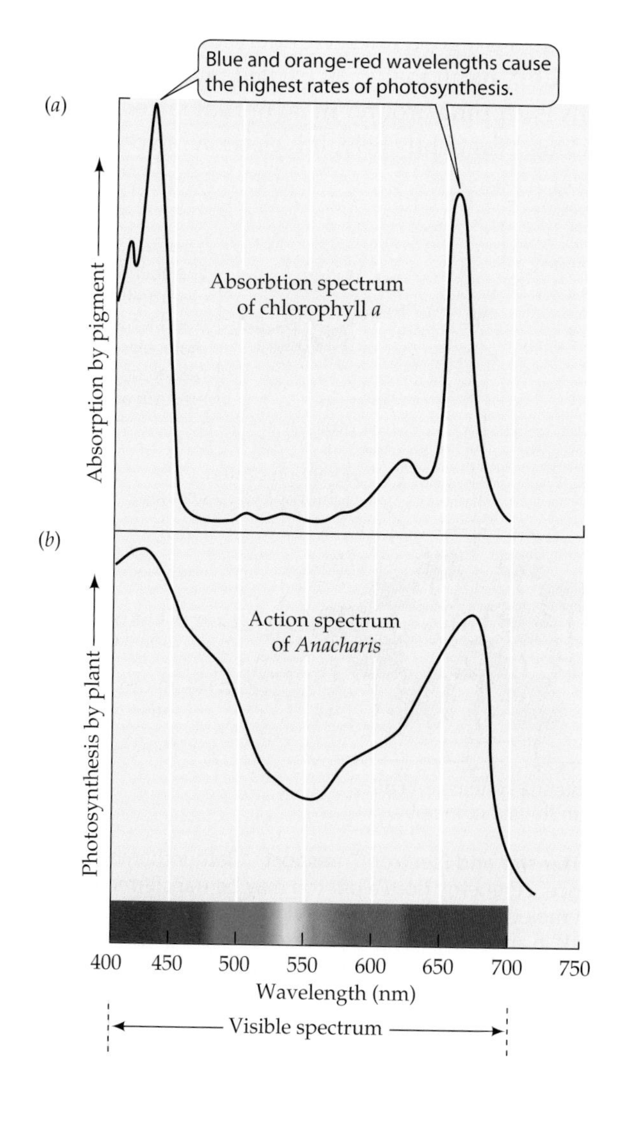

8.6 Absorption and Action Spectra The absorption spectrum *(a)* of the purified pigment chlorophyll *a* from the aquatic plant *Anacharis* is similar to the action spectrum *(b)* obtained when different wavelengths of light are shone on the intact plant and the rate of photosynthesis is measured *(b)*.

Photosynthesis uses energy absorbed by several pigments

The light energy used for photosynthesis is not absorbed by just a single type of pigment. Instead, several different pigments with different absorption spectra absorb the energy that is eventually used for photosynthesis. In photosynthetic organisms of all kinds (plants, protists, and bacteria), these pigments include chlorophylls, carotenoids, and phycobilins.

In plants, two **chlorophylls** predominate: chlorophyll *a* and chlorophyll *b*. These two molecules differ only slightly in their molecular structure. Both have a complex ring structure similar to that of the heme group of hemoglobin. In the center of each chlorophyll ring is a magnesium atom, and attached at a peripheral location on the ring is a long hydrocarbon "tail," which can adhere the chlorophyll molecule to proteins in the hydrophobic portion of the thylakoid membrane (Figure 8.7).

We saw in Figure 8.6 that the chlorophylls absorb blue and red wavelengths, which are near the two ends of the visible spectrum. Thus, if only chlorophyll pigments were active in photosynthesis, much of the visible spectrum would go unused. However, all photosynthetic organisms possess **accessory pigments**, which absorb photons intermediate in energy between the red and the blue wavelengths (for instance, yellow light) and then transfer a portion of that energy to the chlorophylls.

Among these accessory pigments are *carotenoids*, such as β-carotene (see Figure 3.22), which absorb photons in the blue and blue-green wavelengths and appear deep yellow. The *phycobilins*, which are found in red algae and in cyanobacteria, absorb various yellow-green, yellow, and orange wavelengths. Such accessory pigments, in collaboration with the chlorophylls, constitute an energy-absorbing system covering much of the visible spectrum.

Light absorption results in photochemical change

After a pigment molecule absorbs a photon and enters an excited state (see Figure 8.4), that molecule may return to the ground state. When this happens, some of the absorbed en-

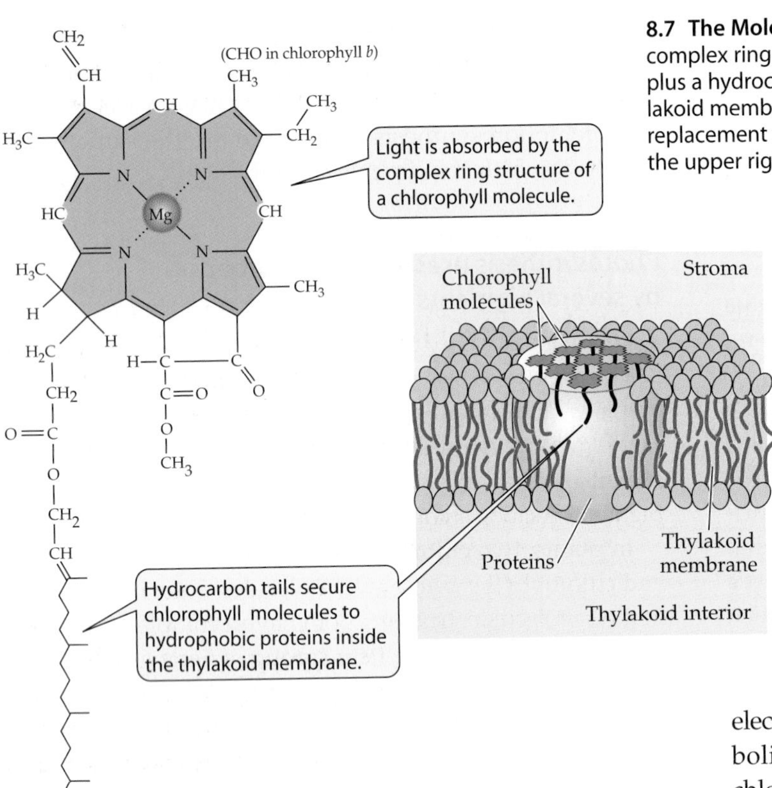

8.7 The Molecular Structure of Chlorophyll Chlorophyll consists of a complex ring structure (shaded area) with a magnesium atom at the center, plus a hydrocarbon "tail." The "tail" anchors chlorophyll molecules to the thylakoid membrane. Chlorophyll *a* and chlorophyll *b* are identical except for the replacement of a methyl group (—CH_3) with an aldehyde group (—CHO) at the upper right.

ergy is given off as heat and the rest is given off as light energy, or *fluorescence*. Because some of the absorbed light energy is lost as heat, the fluorescence has less energy and longer wavelengths than the absorbed light. When there is fluorescence, there are no permanent chemical changes or biological functions—no chemical work is done.

Any pigment molecule can become excited when its absorption spectrum matches the energies of incoming photons. If fluorescence does not occur, that pigment molecule may pass the absorbed energy along to another molecule, provided that the target molecule is very near, has the right orientation, and has the appropriate structure to receive the energy.

The pigments in photosynthetic organisms are arranged into energy-absorbing **antenna systems**. In these systems, the pigments are packed together and attached to thylakoid membrane proteins in such a way that the excitation energy from an absorbed photon can be passed along from one pigment molecule in the system to another (see Figures 8.7 and 8.8). Excitation energy moves from pigments that absorb shorter wavelengths (higher energy) to pigments that absorb longer wavelengths (lower energy). Thus the excitation ends up in the one pigment molecule in the antenna system that absorbs the longest wavelengths; this molecule is in the **reaction center** of the antenna system.

It is the reaction center that converts the light absorbed into chemical energy. It is in the reaction center that a molecule absorbs sufficient energy that it actually gives up its excited electron (is chemically oxidized) and becomes positively charged. In plants, the pigment molecule in the reaction center is always a molecule of chlorophyll *a*. There are many other chlorophyll *a* molecules in the antenna system, but all of them absorb light at shorter wavelengths than does the molecule in the reaction center.

Excited chlorophyll in the reaction center acts as a reducing agent for electron transport

Ultimately, photosynthesis stores chemical energy by using the excited chlorophyll molecule in the reaction center as a reducing agent to reduce a stable electron acceptor (Figure 8.8). Ground-state chlorophyll (symbolized Chl) is not much of a reducing agent, but excited chlorophyll (Chl*) is a good one. To understand the reducing capability of Chl*, recall that in an excited molecule, one of the electrons is zipping around in an orbital farther away from its nucleus. Less tightly held, this electron can be passed

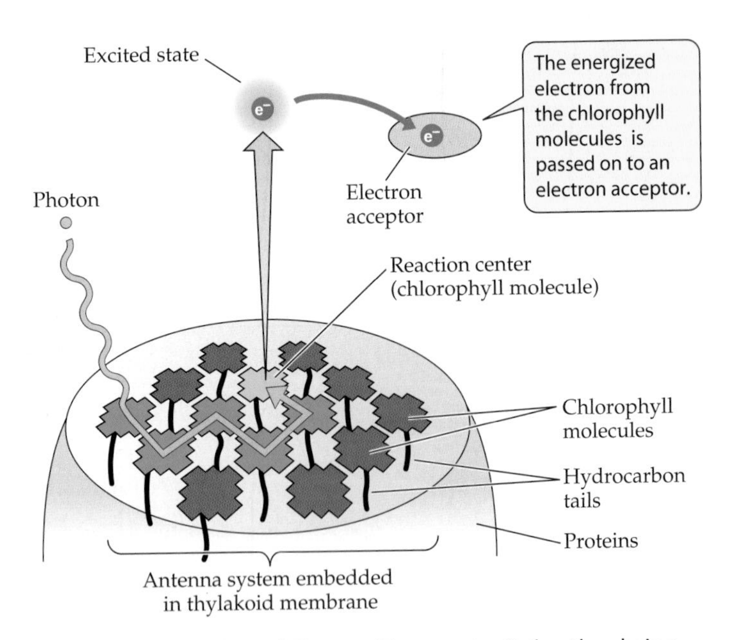

8.8 Energy Transfer and Electron Transport Rather than being lost as fluorescence, energy from a photon may be transferred from one pigment molecule to another, preserving the energy for biochemical work. In an antenna system, an excited pigment molecule can transfer energy through a series of other pigment molecules to a chlorophyll molecule in the reaction center. That molecule may become sufficiently excited that it gives up its excited electron, which can then be passed on to an electron carrier.

on in a redox reaction to an oxidizing agent. Thus Chl* (but not Chl) can react with an oxidizing agent A in a reaction like this:

$$Chl^* + A \rightarrow Chl^+ + A^-$$

This, then, is the first consequence of light absorption by chlorophyll: The chlorophyll becomes a reducing agent and participates in a redox reaction.

As we are about to see, the further adventures of the electrons from chlorophyll reduce the electron carrier $NADP^+$ and generate a proton-motive force that is eventually used to synthesize ATP.

The Light Reactions: Electron Transport, Reductions, and Photophosphorylation

The energized electron that leaves the activated chlorophyll in the reaction center needs somewhere to go. It immediately participates in a series of oxidation-reduction (redox) reactions. The energy-rich electron is passed through a chain of electron carriers in the thylakoid membrane in a process termed **electron transport**. Two energy-rich products of the light reactions, $NADPH + H^+$ and ATP, are the result.

The energy-rich $NADPH + H^+$ is a stable, reduced coenzyme. Its oxidized form is **$NADP^+$ (nicotinamide adenine dinucleotide phosphate)**. Just as NAD^+ couples the metabolic pathways of cellular respiration, $NADP^+$ couples the two photosynthetic pathways. $NADP^+$ is identical to NAD^+ except that the former has an additional phosphate group attached to each ribose (see Figure 7.4). Whereas NAD^+ participates in catabolism, $NADP^+$ is used in anabolic (synthetic) reactions, such as carbohydrate synthesis from CO_2, that require energy from reducing power.

Electron transport in the thylakoid membrane sets up a charge separation, just as electron transport in the inner mitochondrial membrane does (see Chapter 7). This potential energy is captured by the chemiosmotic synthesis of ATP in a process called **photophosphorylation**.

Both $NADPH + H^+$ and ATP are used in the Calvin–Benson cycle as a source of energy for the endergonic synthesis of carbohydrates (see Figure 8.3).

There are two different systems of electron transport in photosynthesis:

- **Noncyclic electron transport** produces $NADPH + H^+$ and ATP.
- **Cyclic electron transport** produces only ATP.

We'll consider these two systems before considering the role of chemiosmosis in phosphorylation—a process that is very similar to oxidative phosphorylation in mitochondria (see Chapter 7).

Noncyclic electron transport produces ATP and NADPH

In noncyclic electron transport, light energy is used to oxidize water, forming O_2, H^+, and electrons. Follow the steps in Figure 8.9 as you read this section.

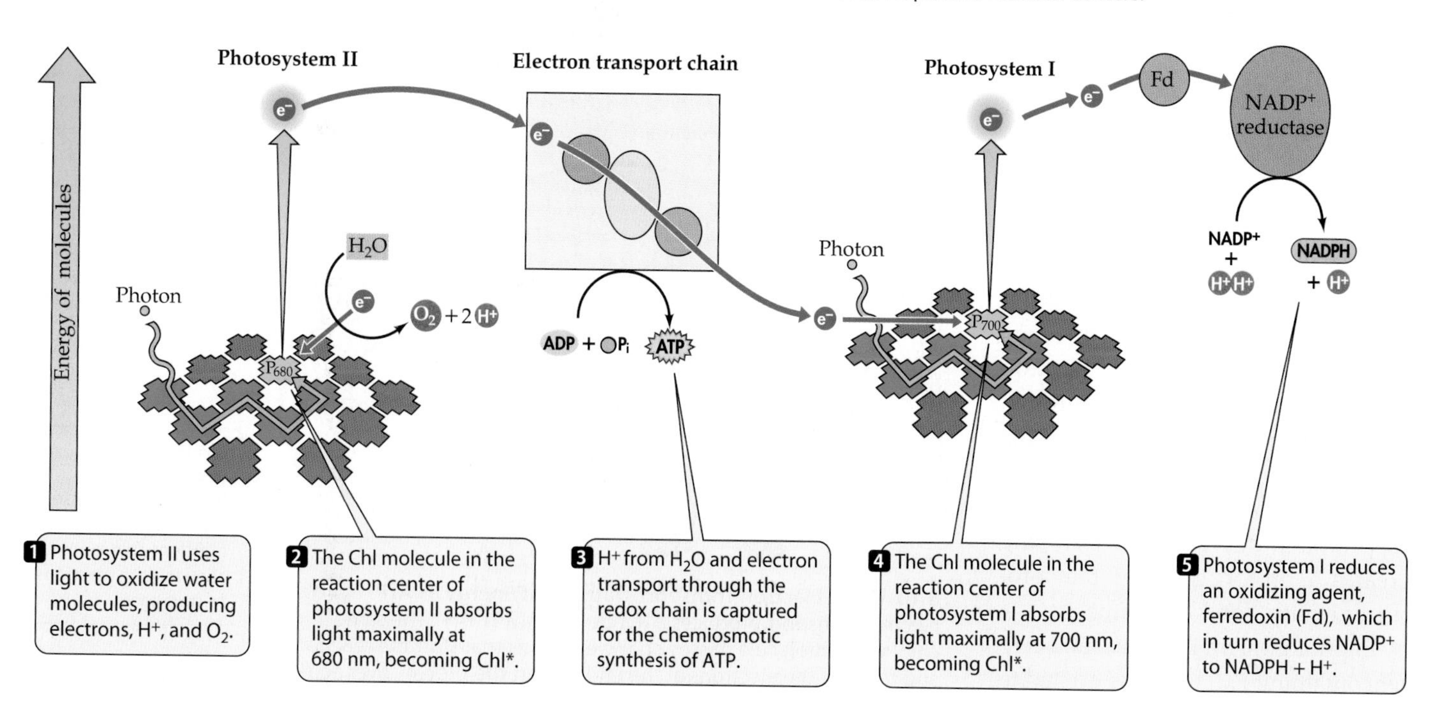

Energy of molecules
Photosystem II
Electron transport chain
Photosystem I
Fd
NADP+ reductase
H2O
O2 + 2H+
Photon
P680
ADP + Pi
ATP
P700
NADP+ + H+H+
NADPH + H+
1 Photosystem II uses light to oxidize water molecules, producing electrons, H+, and O2.
2 The Chl molecule in the reaction center of photosystem II absorbs light maximally at 680 nm, becoming Chl*.
3 H+ from H2O and electron transport through the redox chain is captured for the chemiosmotic synthesis of ATP.
4 The Chl molecule in the reaction center of photosystem I absorbs light maximally at 700 nm, becoming Chl*.
5 Photosystem I reduces an oxidizing agent, ferredoxin (Fd), which in turn reduces NADP+ to NADPH + H+.

8.9 Noncyclic Electron Transport Uses Two Photosystems Photosystems I and II both make use of the excited chlorophyll molecules of their respective reaction centers.

Electrons from water replenish the electrons that chlorophyll molecules lose when they are excited by light. As the electrons are passed from water to chlorophyll, and ultimately to $NADP^+$, they pass through a chain of electron carriers. These redox reactions are exergonic, and some of the free energy released is used ultimately to form ATP by a chemiosmotic mechanism.

TWO PHOTOSYSTEMS ARE REQUIRED. Noncyclic electron transport requires the participation of *two different photosystems*. These photosystems are light-driven molecular units, each of which consists of many chlorophyll molecules and accessory pigments bound to proteins in *separate* energy-absorbing antenna systems.

- **Photosystem I** uses light energy to reduce $NADP^+$ to $NADPH + H^+$.
- **Photosystem II** uses light energy to oxidize water molecules, producing electrons, protons (H^+), and O_2.

The reaction center for photosystem I contains a chlorophyll *a* molecule called P_{700} because it can best absorb light of wavelength 700 nm. The reaction center for photosystem II contains a chlorophyll *a* molecule called P_{680} because it absorbs light maximally at 680 nm. Thus photosystem II requires photons that are somewhat more energetic (i.e., shorter wavelengths) than those required by photosystem I. To keep noncyclic electron transport going, both photosystems I and II must constantly be absorbing light, thereby boosting electrons to higher orbitals from which they may be captured by specific oxidizing agents.

DETAILS OF THE REACTIONS. The reactions of noncyclic electron transport from water to $NADP^+$ are depicted in Figure 8.9. Photosystem II absorbs photons, sending electrons from P_{680} to the primary electron acceptor—the first carrier in the redox chain—and causing P_{680} to become oxidized to P_{680}^+. Electrons from the oxidation of water are passed to P_{680}^+, reducing it once again to P_{680}, which can then absorb more photons. The electron from photosystem II passes through a series of exergonic reactions in the redox chain that are indirectly coupled across the thylakoid membrane to proton pumping. This pumping creates a proton gradient that produces energy for ATP synthesis.

In photosystem I, the reaction center containing P_{700} becomes excited to P_{700}^*, which then leads to the reduction of an oxidizing agent called *ferredoxin* (Fd) and the production of P_{700}^+. Then P_{700}^+ returns to the ground state by accepting electrons passed through the redox chain from photosystem II.

With this accounting for the source of the electrons entering photosystem II, we can now consider the fate of the electrons from photosystem I. These electrons are used in the last step of noncyclic electron transport, in which two electrons and two protons are used to reduce a molecule of $NADP^+$ to $NADPH + H^+$.

In summary:

- Noncyclic electron transport uses a molecule of water, four photons (two each absorbed by photosystems I and II), one molecule each of $NADP^+$ and ADP, and one P_i.
- Noncyclic electron transport produces $NADPH + H^+$ and ATP and half a molecule of oxygen ($\frac{1}{2} O_2$).

Cyclic electron transport produces ATP but no NADPH

Noncyclic electron transport produces ATP and $NADPH + H^+$. However, as we will see, the Calvin–Benson cycle uses more ATP than $NADPH + H^+$. Cyclic electron transport occurs in some organisms when the ratio of $NADPH + H^+$ to $NADP^+$ in the chloroplast is high. This process, which produces only ATP, is called *cyclic* because an electron passed from an excited chlorophyll molecule at the outset cycles *back to the same chlorophyll molecule* at the end of the chain of reactions (Figure 8.10).

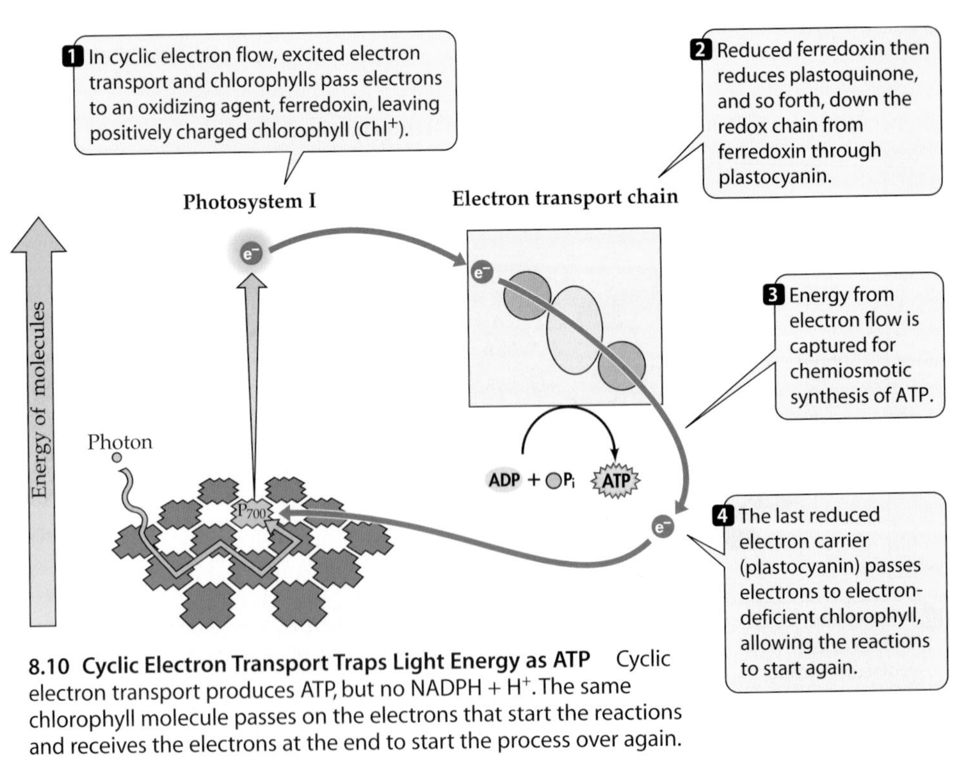

1 In cyclic electron flow, excited electron transport and chlorophylls pass electrons to an oxidizing agent, ferredoxin, leaving positively charged chlorophyll (Chl+).
2 Reduced ferredoxin then reduces plastoquinone, and so forth, down the redox chain from ferredoxin through plastocyanin.
Photosystem I
Electron transport chain
3 Energy from electron flow is captured for chemiosmotic synthesis of ATP.
Energy of molecules
Photon
ADP + Pi
ATP
P700
4 The last reduced electron carrier (plastocyanin) passes electrons to electron-deficient chlorophyll, allowing the reactions to start again.

8.10 Cyclic Electron Transport Traps Light Energy as ATP Cyclic electron transport produces ATP, but no $NADPH + H^+$. The same chlorophyll molecule passes on the electrons that start the reactions and receives the electrons at the end to start the process over again.

Before cyclic electron transport begins, P_{700}, the reaction center chlorophyll of photosystem I, is in the ground state. It absorbs a photon and becomes $P_{700}{}^*$. The $P_{700}{}^*$ then reacts with oxidized ferredoxin (Fd_{ox}) to produce reduced ferredoxin (Fd_{red}). The reaction is exergonic, releasing free energy. Reduced ferredoxin (Fd_{red}) passes its added electron to a different oxidizing agent, *plastoquinone* (PQ, a small organic molecule), which pumps 2 H^+ back across the thylakoid membrane. Thus, Fd_{red} reduces PQ, and PQ_{red} passes the electron to a cytochrome complex (Cyt). The electron continues down the electron transport chain until it completes its cycle by returning to $P_{700}{}^+$, resulting in a restoration of its uncharged form, P_{700}. By the time the electron from $P_{700}{}^*$ travels through the redox chain by way of *plastocyanin* (PC), and comes back to reduce $P_{700}{}^+$, all the energy from the original photon has been released. This cycle is a series of redox reactions, each exergonic, and the released energy is stored in the form of a proton gradient that can be used to produce ATP.

Having seen how a proton gradient is established across the thylakoid membrane, we'll now examine in more detail the role of this gradient in ATP synthesis.

Chemiosmosis is the source of the ATP produced in photophosphorylation

In Chapter 7 we considered the chemiosmotic mechanism for ATP formation in the mitochondrion. The chemiosmotic mechanism also operates in photophosphorylation (Figure 8.11). In chloroplasts, as in mitochondria, electron transport through the redox chain is coupled to the transport of pro-

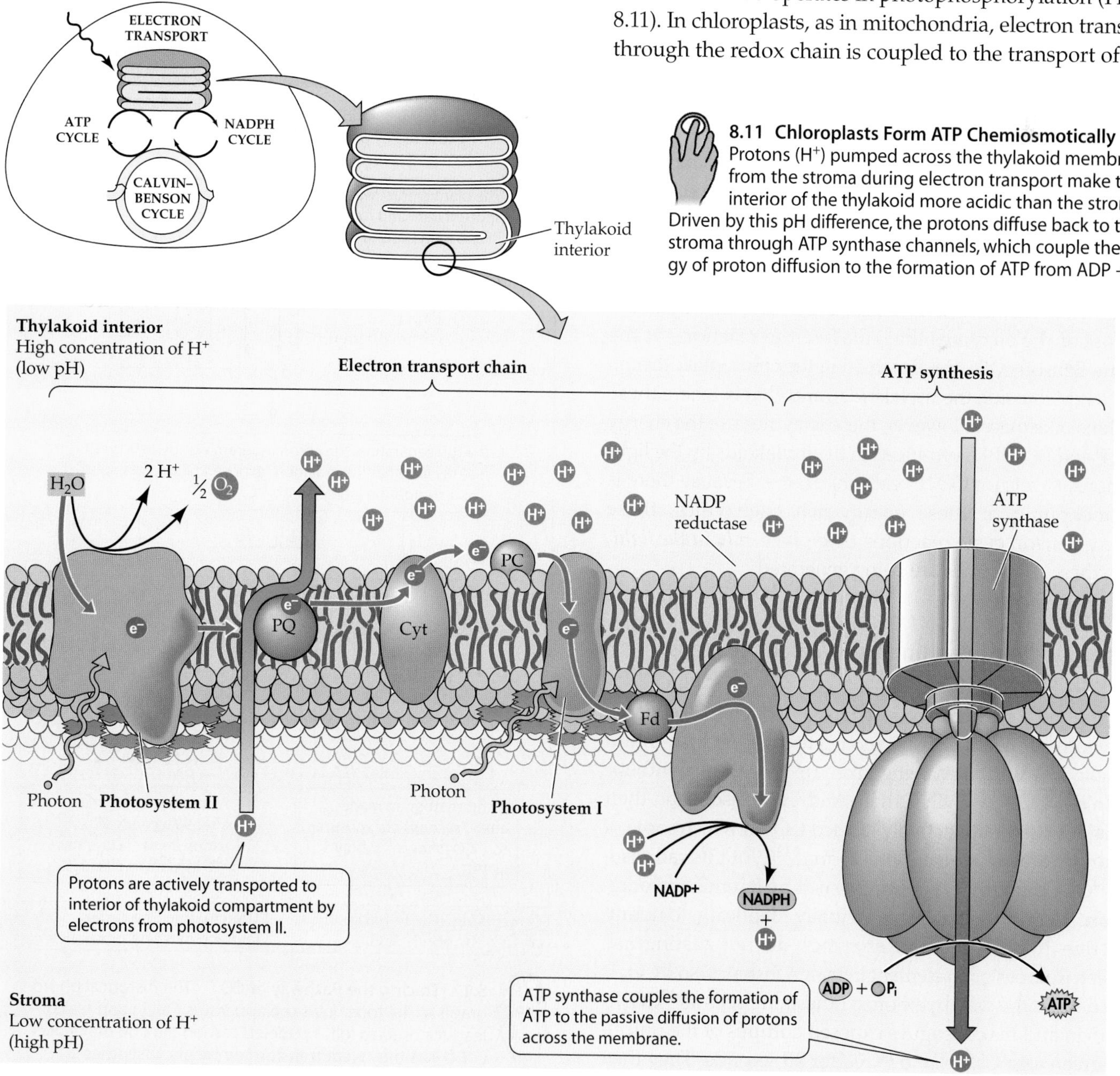

8.11 Chloroplasts Form ATP Chemiosmotically Protons (H^+) pumped across the thylakoid membrane from the stroma during electron transport make the interior of the thylakoid more acidic than the stroma. Driven by this pH difference, the protons diffuse back to the stroma through ATP synthase channels, which couple the energy of proton diffusion to the formation of ATP from ADP + P_i.

tons (H^+) across the thylakoid membrane, which results in a proton gradient across the membrane.

The electron carriers in the thylakoid membranes are oriented so that protons move from the stroma into the interior of the thylakoid. The interior compartment becomes acidic with respect to the stroma. When there is sufficient light, the ratio of H^+ inside versus outside a thylakoid is usually 10,000:1, which is a difference of 4 pH units. This difference leads to the diffusion of H^+ back out of the thylakoid interior through specific protein channels in the thylakoid membrane. These channels are enzymes—ATP synthases—that couple the diffusion of protons to the formation of ATP, just as in mitochondria (see Figure 7.12). In the chloroplast, the ATP is generated in the stroma, where it will be available to provide the energy for the fixation of CO_2 in the production of carbohydrate by the Calvin-Benson cycle.

Making Carbohydrate from CO_2: The Calvin–Benson Cycle

At the start of this chapter we identified two distinct metabolic pathways operating in photosynthesis. We have now discussed the first pathway: the light reactions, which use light energy to produce ATP and NADPH + H^+ in the chloroplasts of green plants. The second pathway, the Calvin–Benson cycle, uses this ATP and NADPH + H^+ to incorporate CO_2 into carbohydrates.

Most of the enzymes that catalyze the reactions of the Calvin–Benson cycle are dissolved in the chloroplast stroma (the "soup" outside the thylakoids), and that is where those reactions take place. However, these enzymes use the energy in ATP and NADPH, produced in the thylakoids by the light reactions, to reduce CO_2 to carbohydrates. Because there is no stockpiling of these energy-rich coenzymes, these Calvin–Benson cycle reactions take place *only in the light*, when these coenzymes are being generated.

Isotope labeling experiments revealed the steps of the Calvin–Benson cycle

To identify the sequence of reactions by which CO_2 ends up in carbohydrates, it was necessary to label CO_2 so that it could be followed after being taken up by a photosynthetic cell. In the 1950s, Melvin Calvin, Andrew Benson, and their colleagues used radioactively labeled CO_2 in which some of the carbon atoms were not the normal ^{12}C, but its radioisotope ^{14}C. Although ^{14}C is distinguished by its emission of radiation, chemically it behaves virtually identically to nonradioactive ^{12}C. In general, enzymes do not distinguish between isotopes of an element in their substrates, so $^{14}CO_2$ is treated the same way by photosynthesizing cells as $^{12}CO_2$.

Calvin and his colleagues exposed cultures of the unicellular green alga *Chlorella* to $^{14}CO_2$ for 30 seconds. They then rapidly killed the cells, extracted their carbohydrates, and separated the different compounds from one another by paper chromatography. Many compounds, including monosaccharides and amino acids, contained ^{14}C (Figure 8.12).

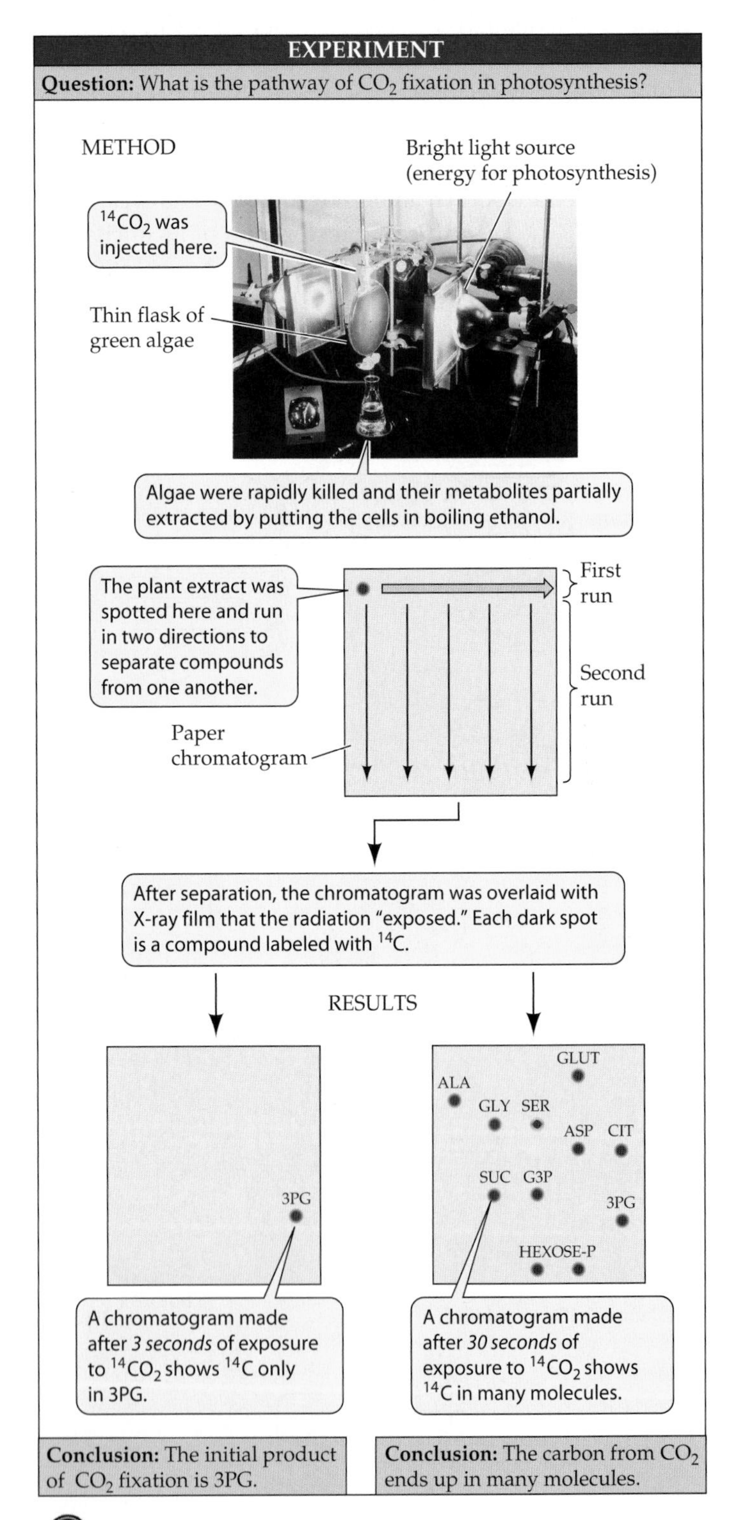

EXPERIMENT
Question: What is the pathway of CO_2 fixation in photosynthesis?
METHOD
Bright light source (energy for photosynthesis)
$^{14}CO_2$ was injected here.
Thin flask of green algae
Algae were rapidly killed and their metabolites partially extracted by putting the cells in boiling ethanol.
The plant extract was spotted here and run in two directions to separate compounds from one another.
First run
Second run
Paper chromatogram
After separation, the chromatogram was overlaid with X-ray film that the radiation "exposed." Each dark spot is a compound labeled with ^{14}C.
RESULTS
3PG
GLUT
ALA
GLY SER
ASP CIT
SUC G3P
3PG
HEXOSE-P
A chromatogram made after *3 seconds* of exposure to $^{14}CO_2$ shows ^{14}C only in 3PG.
A chromatogram made after *30 seconds* of exposure to $^{14}CO_2$ shows ^{14}C in many molecules.
Conclusion: The initial product of CO_2 fixation is 3PG.
Conclusion: The carbon from CO_2 ends up in many molecules.

8.12 Tracing the Pathway of CO_2 The historical photograph at the top shows the apparatus Calvin and his colleagues used to follow labeled carbon dioxide molecules ($^{14}CO_2$) as they were transformed by photosynthesis.

However, if they stopped the exposure after just 3 seconds, only one compound was labeled—a three-carbon sugar phosphate called 3-phosphoglycerate (3PG):

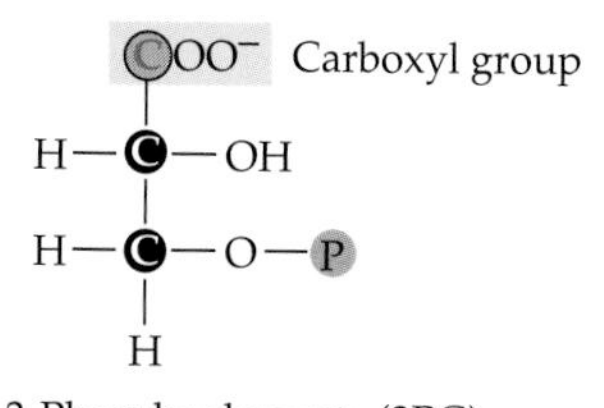

By tracing the steps in this manner, they soon discovered a cycle that "fixes" CO_2 in a larger molecule, produces a carbohydrate, and regenerates the initial CO_2 acceptor. This cycle was appropriately named the **Calvin–Benson cycle** (Figure 8.13).

The initial reaction in the Calvin–Benson cycle adds the one-carbon CO_2 to a receptor, the five-carbon compound **ribulose 1,5-bisphosphate (RuBP)**. The product is an intermediate six-carbon compound, which quickly breaks down and forms two three-carbon molecules of 3PG (as Calvin and colleagues observed; Figure 8.14). The enzyme that catalyzes this fixation reaction, **ribulose bisphosphate carboxylase/oxygenase (rubisco)**, is the most abundant protein in the world, comprising about 20 percent of all the protein in every plant leaf.

The Calvin–Benson cycle is made up of three processes

The Calvin–Benson cycle uses the high-energy coenzymes made in the thylakoids during the light reactions (ATP and NADPH) to reduce CO_2 to a carbohydrate. There are three processes that make up the cycle:

- Fixation of CO_2. As we saw, this reaction is catalyzed by rubisco, and its product is 3PG.
- Reduction of 3PG to form a carbohydrate, glyceraldehyde 3-phosphate (G3P). This series of reactions involves a phosphorylation (using the ATP made in the light reactions) and a reduction (using the NADPH made in the light reactions).

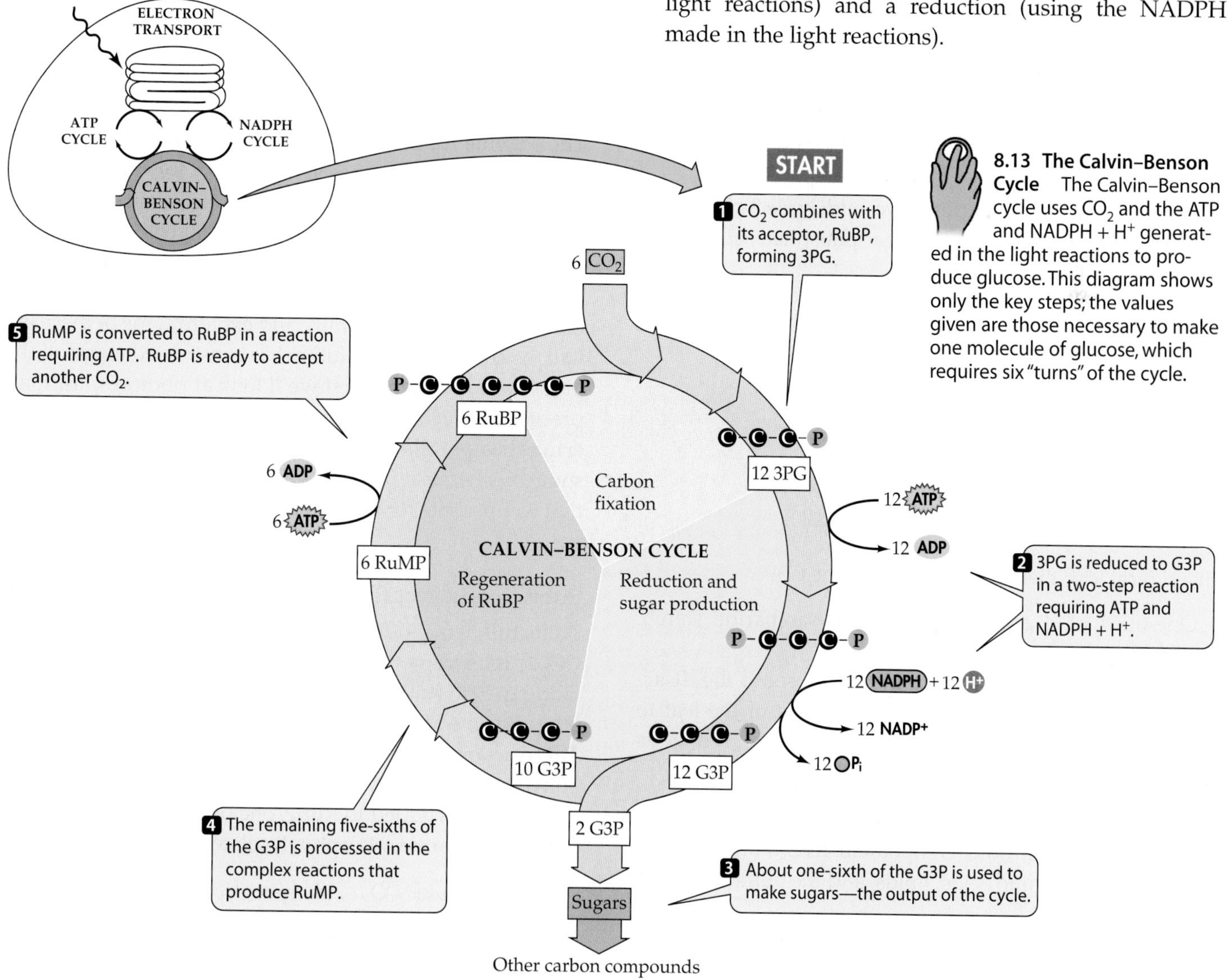

8.13 The Calvin–Benson Cycle The Calvin–Benson cycle uses CO_2 and the ATP and NADPH + H^+ generated in the light reactions to produce glucose. This diagram shows only the key steps; the values given are those necessary to make one molecule of glucose, which requires six "turns" of the cycle.

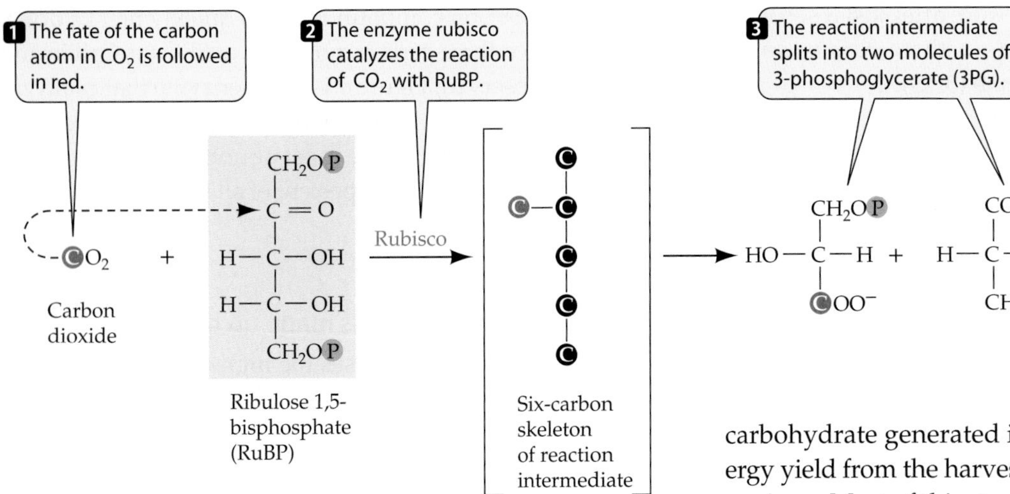
1 The fate of the carbon atom in CO_2 is followed in red.
2 The enzyme rubisco catalyzes the reaction of CO_2 with RuBP.
3 The reaction intermediate splits into two molecules of 3-phosphoglycerate (3PG).
Carbon dioxide
Rubisco
Ribulose 1,5-bisphosphate (RuBP)
Six-carbon skeleton of reaction intermediate

8.14 RuBP Is the Carbon Dioxide Acceptor CO_2 is added to a five-carbon compound, RuBP. The resulting six-carbon compound immediately splits into two molecules of 3PG.

- Regeneration of the CO_2 acceptor, RuBP. Most of the G3P ends up as RuMP (ribulose monophosphate), and ATP is used to convert this compound to RuBP. So for every "turn" of the cycle, with one CO_2 fixed, the CO_2 acceptor is regenerated.

The end product of this cycle is glyceraldehyde 3-phosphate (G3P), which is a three-carbon sugar phosphate, also called triose phosphate:

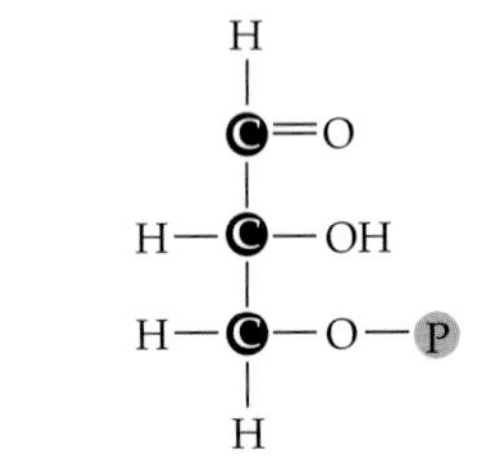

Glyceraldehyde 3-phosphate (G3P)

In a typical leaf, there are two fates for the G3P:

- One-third of it ends up in the polysaccharide starch, which is stored in the chloroplast.
- Two-thirds of it is converted in the cytosol to the disaccharide sucrose, which is transported out of the leaf to other organs in the plant, where it is hydrolyzed to its constituent monosaccharides: glucose and fructose.

The G3P produced in photosynthesis is subsequently used by the plant to make other compounds. Its carbon is thus incorporated into amino acids, lipids, and the building blocks of the nucleic acids.

The products of the Calvin–Benson cycle are of crucial importance to the entire biosphere, for the covalent bonds of the carbohydrate generated in the cycle represent the total energy yield from the harvesting of light by photosynthetic organisms. Most of this stored energy is released by glycolysis and cellular respiration and used to support plant growth, development, and reproduction. Much plant matter ends up being consumed by animals, supplying them with both raw materials and energy sources. Glycolysis and cellular respiration in the animals release free energy from the plant matter for use in the animal cells.

Photorespiration and Its Consequences

The enzyme rubisco, used by the Calvin-Benson cycle to fix CO_2 during photosynthesis, is probably the most abundant enzyme on the planet. Its properties are remarkably identical in all photosynthetic organisms, from bacteria to flowering plants. However, rubisco has properties that severely limit its efficiency under certain conditions. In the discussion that follows, we will identify and explore some of these limitations and see how evolution has constructed metabolic bypasses around them. First we'll look at photorespiration, a process in which rubisco reacts with O_2 instead of CO_2, lowering the overall rate of CO_2 fixation. Then we'll examine some biochemical pathways and features of plant anatomy that compensate for the limitations of rubisco.

Rubisco catalyzes RuBP reaction with O_2 as well as CO_2

As its full name indicates, rubisco is a **carboxylase** (adding CO_2 to the acceptor molecule RuBP) as well as an **oxygenase** (adding O_2 to RuBP). These two reactions compete with each other. So when RuBP reacts with O_2, it cannot react with CO_2. This reaction reduces the overall CO_2 that is converted to carbohydrates, and therefore limits plant growth.

When O_2 is added to RuBP, one of the products is a two-carbon compound, phosphoglycolate:

$$\text{RuBP} + O_2 \rightarrow \text{phosphoglycolate} + \text{3PG}$$

Plants have evolved a metabolic pathway that partially recovers the carbon that has been channeled away from the

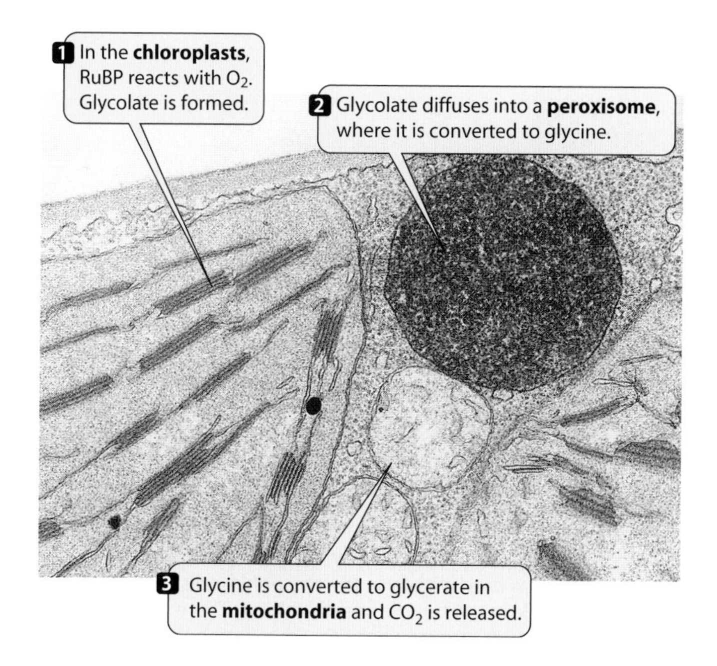

1 In the **chloroplasts**, RuBP reacts with O_2. Glycolate is formed.
2 Glycolate diffuses into a **peroxisome**, where it is converted to glycine.
3 Glycine is converted to glycerate in the **mitochondria** and CO_2 is released.

8.15 Organelles of Photorespiration The reactions of photorespiration take place in the chloroplasts, peroxisomes, and, finally, in the mitochondria.

Calvin–Benson cycle. The phosphoglycolate forms glycerate, which diffuses into membrane-enclosed organelles called **peroxisomes** (Figure 8.15). There, a series of reactions converts it to the amino acid glycine:

$$\text{glycolate} \rightarrow \rightarrow \text{glycine}$$

The glycine then diffuses into a mitochondrion, where two glycine molecules are converted to glycerate (a three-carbon molecule), and CO_2:

$$2\ \text{glycine} \rightarrow \text{glycerate} + CO_2$$

This pathway is called **photorespiration** because it consumes O_2 and releases CO_2. It uses ATP and NADPH produced in the light reactions, just like the Calvin–Benson cycle. The net effect is to take two two-carbon molecules and make one three-carbon molecule. So one carbon of the four is released as CO_2 and three of the carbons (75%) are recovered as fixed carbon. In other words, photorespiration reduces net carbon fixation by 25 percent compared with the Calvin–Benson cycle.

How does rubisco "decide" whether to act as an oxygenase or a carboxylase? First, rubisco has 10 times more affinity for CO_2 than O_2, and so favors CO_2 fixation. Another consideration is the relative concentrations of CO_2 and O_2 in the leaf. If O_2 is relatively abundant, rubisco acts as an oxygenase, and photorespiration ensues. If CO_2 predominates, rubisco fixes it, and the Calvin–Benson cycle occurs. Temperature is also a factor: photorespiration is more likely at high temperatures.

On a hot, dry day, the stomata that allow water to evaporate from the leaf close to prevent water loss (see Figure 8.1). But this also prevents gases from entering and leaving the leaf. The CO_2 concentration in the leaf falls because CO_2 is being used up by the light-driven photosynthetic reactions, and the O_2 concentration rises because of these same reactions. As the ratio of CO_2 to O_2 in the leaf falls, the reaction of rubisco with O_2 is favored, and photorespiration proceeds.

C_4 plants can bypass photorespiration

In plants such as roses, wheat, and rice, the *mesophyll* cells, which lie just below the surface of the leaf, are full of chloroplasts that contain abundant rubisco (Figure 8.16*a*). On a hot day, these leaves close their stomata to conserve water. The level of CO_2 in the air spaces of the leaves falls, and that of O_2 continues to rise, as photosynthesis goes on. As we have

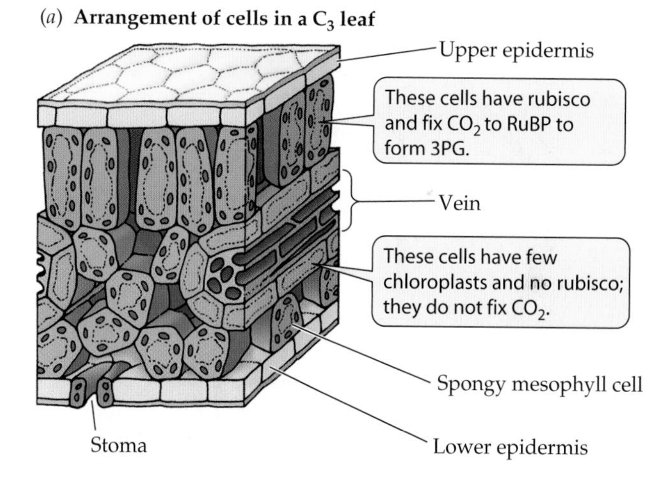

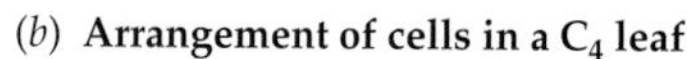

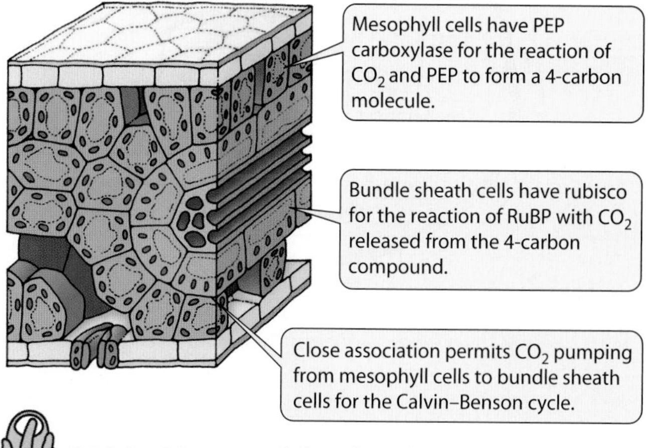

Mesophyll cells have PEP carboxylase for the reaction of CO_2 and PEP to form a 4-carbon molecule.
Bundle sheath cells have rubisco for the reaction of RuBP with CO_2 released from the 4-carbon compound.
Close association permits CO_2 pumping from mesophyll cells to bundle sheath cells for the Calvin–Benson cycle.

8.16 Leaf Anatomy of C_3 and C_4 Plants Carbon dioxide fixation occurs in different organelles and cells of the leaves in (*a*) C_3 and (*b*) C_4 plants.

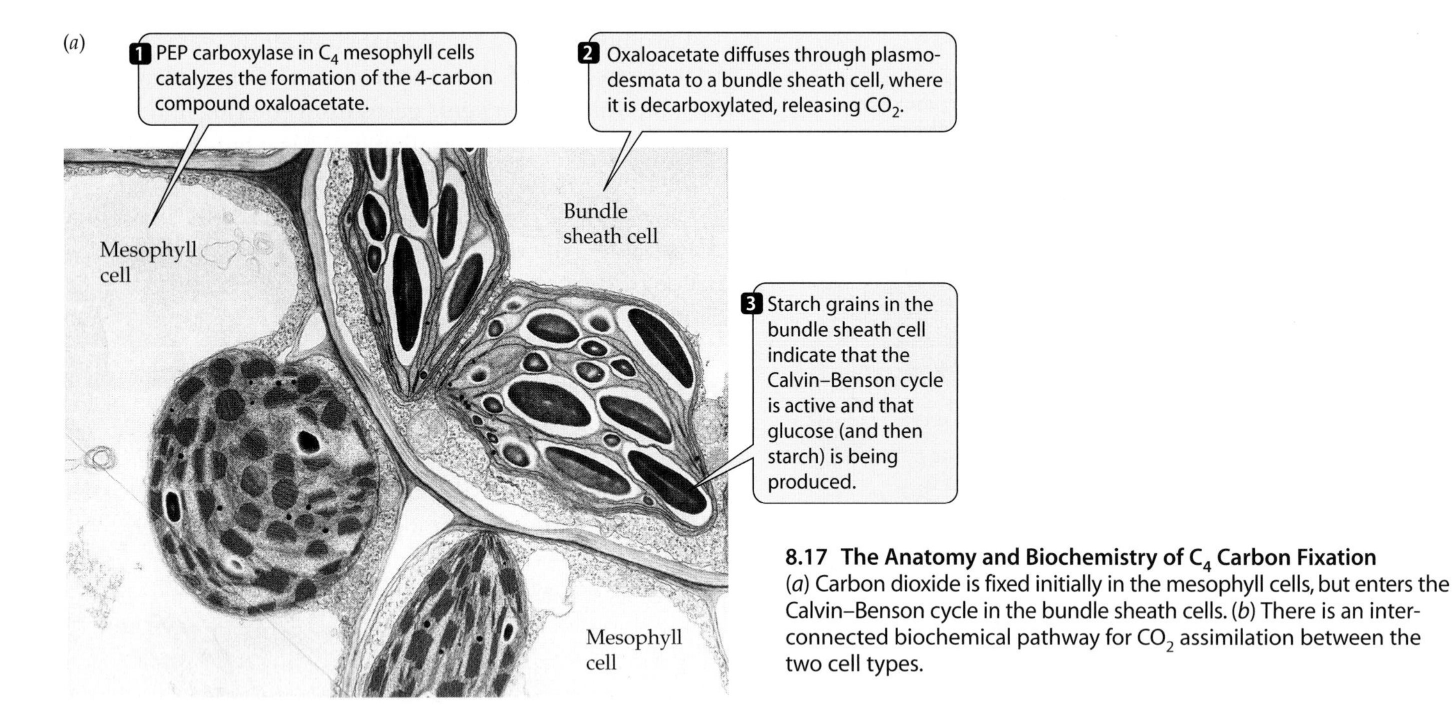

(a)
1 PEP carboxylase in C_4 mesophyll cells catalyzes the formation of the 4-carbon compound oxaloacetate.
2 Oxaloacetate diffuses through plasmodesmata to a bundle sheath cell, where it is decarboxylated, releasing CO_2.
Mesophyll cell
Bundle sheath cell
3 Starch grains in the bundle sheath cell indicate that the Calvin–Benson cycle is active and that glucose (and then starch) is being produced.
Mesophyll cell

8.17 The Anatomy and Biochemistry of C_4 Carbon Fixation
(*a*) Carbon dioxide is fixed initially in the mesophyll cells, but enters the Calvin–Benson cycle in the bundle sheath cells. (*b*) There is an interconnected biochemical pathway for CO_2 assimilation between the two cell types.

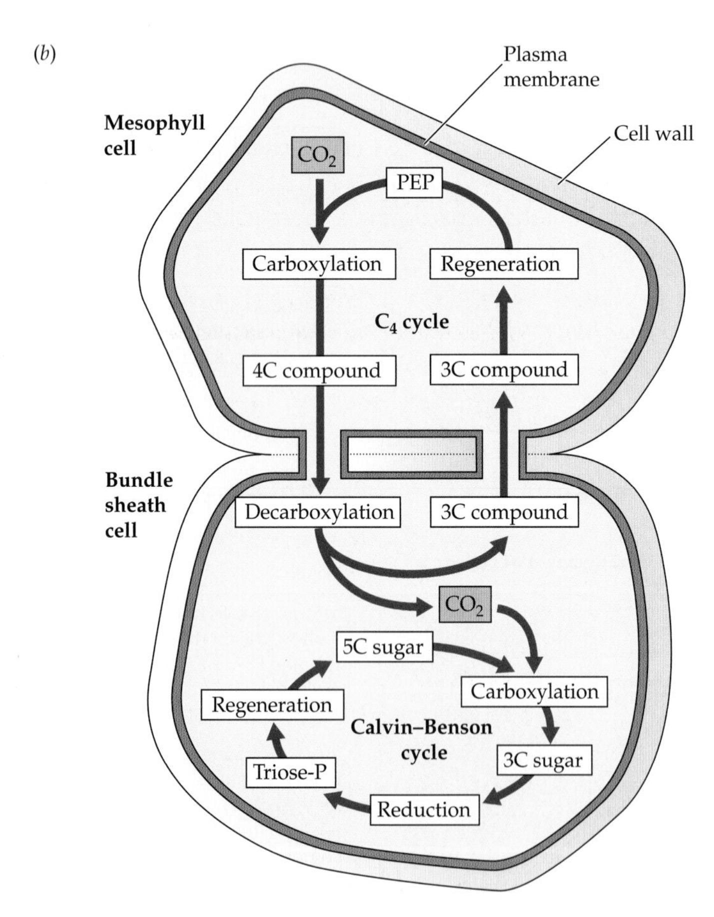

seen, rubisco acts as an oxygenase, and photorespiration occurs, under these conditions. Because the first product of CO_2 fixation in these plants is the three-carbon molecule 3PG, they are called **C_3 plants**.

Corn, sugarcane, and other tropical grasses (Figure 8.16*b*) also close their stomata on a hot day, but their rate of photosynthesis does not fall, nor does photorespiration occur. They keep the ratio of CO_2 to O_2 around rubisco high so that rubisco continues to act as a carboxylase. They do this in part by making a four-carbon compound, *oxaloacetate*, as the first product of CO_2 fixation, and so are called **C_4 plants**.

C_4 plants perform the normal Calvin–Benson cycle, but they have an additional early reaction that fixes CO_2 without losing carbon to photorespiration, greatly increasing the overall photosynthetic yield. Because this initial CO_2 fixation step can function even at low levels of CO_2 and high temperatures, C_4 plants very effectively optimize photosynthesis under conditions that inhibit it in C_3 plants.

C_4 plants have two separate enzymes for CO_2 fixation, located in two different parts of the leaf (Figure 8.17). One enzyme, present in the cytosol of mesophyll cells near the surface of the leaf, fixes CO_2 to a three-carbon acceptor compound, phosphoenolpyruvate (PEP), to produce the four-carbon fixation product, oxaloacetate. This enzyme, **PEP carboxylase**, has two advantages over rubisco:

- It does not have oxygenase activity.
- It fixes CO_2 even at very low CO_2 levels.

8.1 Comparison of Photosynthesis in C_3 and C_4 Plants

VARIABLE	C_3 PLANTS	C_4 PLANTS
Photorespiration	Extensive	Minimal
Perform Calvin–Benson cycle?	Yes	Yes
Primary CO_2 acceptor	RuBP	PEP
CO_2-fixing enzyme	Rubisco (RuBP carboxylase/oxygenase)	PEP carboxylase and rubisco
First product of CO_2 fixation	3PG (3-carbon compound)	Oxaloacetate (4-carbon compound)
Affinity of carboxylase for CO_2	Moderate	High
Photosynthetic cells of leaf	Mesophyll	Mesophyll + bundle sheath
Classes of chloroplasts	One	Two

So even on a hot day when the stomata are closed, the CO_2 concentration in the leaf is low, and the O_2 concentration is high, PEP carboxylase just keeps on fixing CO_2.

Oxaloacetate diffuses out of the mesophyll cells and through plasmodesmata into the *bundle sheath cells,* located in the interior of the leaf. The chloroplasts in bundle sheath cells contain abundant rubisco. There, the four-carbon oxaloacetate loses one carbon, forming CO_2 and regenerating the three-carbon acceptor compound, PEP, in the mesophyll cells. Thus, the role of PEP is to bind CO_2 from the air in the leaf and carry it to the bundle sheath cells, where it is "dropped off" at rubisco. This process essentially pumps up the CO_2 concentration around rubisco, so that it acts as a carboxylase and begins the Calvin–Benson cycle.

Kentucky bluegrass, a C_3 plant, thrives on lawns in April and May. But in the heat of summer, it does not do as well, and crabgrass, a C_4 plant, takes over the lawn. The same is true on a global scale for crops: C_3 plants, such as soybeans, rice, wheat, and barley, have been adapted for human food production in temperate climates, while C_4 plants, such as corn and sugarcane, originated and are grown in the tropics. Table 8.1 compares C_3 and C_4 photosynthesis.

C_3 plants are certainly more ancient than C_4 plants. While C_3 photosynthesis appears to have begun about 3.5 billion years ago, C_4 plants appeared about 12 million years ago. A possible factor in the emergence of the C_4 pathway is the decline in atmospheric CO_2. When dinosaurs ruled Earth 100 million years ago, the concentration of CO_2 in the atmosphere was four times what it is now. As CO_2 levels declined thereafter, the more efficient C_4 plants would have had an advantage over their C_3 counterparts.

CAM plants also use PEP carboxylase

Other plants besides the C_4 species use PEP carboxylase to fix and accumulate CO_2. Such plants include some water-storing plants (called *succulents*) of the family Crassulaceae, many cacti, pineapples, and several other kinds of flowering plants. The CO_2 metabolism of these plants is called **crassulacean acid metabolism**, or **CAM**, after the family of succulents in which it was discovered. CAM is much like the metabolism of C_4 plants in that CO_2 is initially fixed into a four-carbon compound. In CAM plants, however, the processes of initial CO_2 fixation and the Calvin–Benson cycle are separated in time, rather than in space.

- At *night*, when it is cooler and water loss in minimized, the stomata open. CO_2 is fixed in mesophyll cells to form the four-carbon compound oxaloacetate, which is converted to malic acid.
- During the *day*, the accumulated malic acid is shipped to the chloroplasts, where decarboxylation supplies the CO_2 for operation of the Calvin–Benson cycle, and the light reactions supply the necessary ATP and NADPH + H^+.

Metabolic Pathways in Plants

Green plants are autotrophs and can synthesize all the molecules they need from simple starting materials: CO_2, H_2O, phosphate, sulfate, and ammonium ions (NH_4^+). NH_4^+ is needed for amino acids and comes either from the conversion of nitrogen-containing molecules in soil water taken up by the plant's roots, or from the bacterial conversion of N_2 gas from the atmosphere.

The light reactions of photosynthesis generate ATP and NADPH, which are used to synthesize carbohydrates. These compounds can then be used in cellular respiration to provide energy for processes such as active transport and anabolism. Both cellular respiration and fermentation can occur in plants, although the former is far more common. Plant cellular respiration, unlike photosynthesis, takes place both in the light and in the dark. Because glycolysis occurs in the

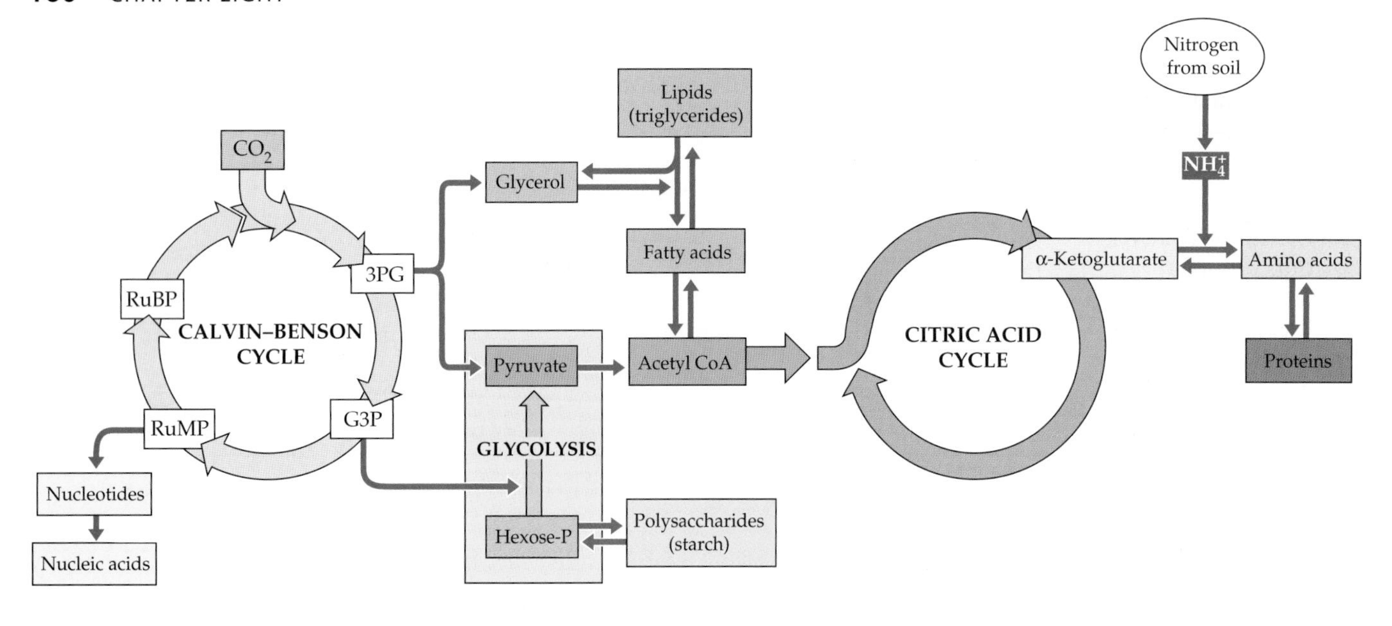

8.18 Metabolic Interactions in a Plant Cell The products of the Calvin–Benson cycle are used in the reactions of cellular respiration (glycolysis and the citric acid cycle).

cytosol, respiration in the mitochondria, and photosynthesis in the chloroplasts, all these processes can proceed simultaneously.

Photosynthesis and respiration are closely linked through the Calvin–Benson cycle (Figure 8.18). The partitioning of G3P is particularly important:

- Some G3P from the Calvin–Benson cycle can be converted to pyruvate, the end product of glycolysis. This pyruvate can be used in cellular respiration for energy, or its carbon skeletons can be used anabolically to make lipids, proteins, and other carbohydrates (see Figure 7.17).
- Some G3P can enter a pathway that is the reverse of glycolysis (the gluconeogenic pathway). In this case, sucrose is formed and transported to the nonphotosynthetic tissues of the plant, such as the root.

Energy flows from sunlight to reduced carbon in photosynthesis to ATP in respiration. Energy can also be stored in the bonds of macromolecules such as polysaccharides, lipids, and proteins. For a plant to grow, energy storage (as body structures) must exceed energy release; that is, overall carbon fixation by photosynthesis must exceed respiration. This principle is the basis of the ecological food chain, as we will see in later chapters.

Chapter Summary

Identifying Photosynthetic Reactants and Products

- Photosynthesizing plants take in CO_2, water, and light energy, producing O_2 and carbohydrates. The overall reaction is $6\ CO_2 + 12\ H_2O + \text{light} \rightarrow C_6H_{12}O_6 + 6\ O_2 + 6\ H_2O$
- The oxygen atoms in the O_2 produced by photosynthesis come from water, not from CO_2. **Review Figures 8.1, 8.2. See Web/CD Tutorial 8.1**

The Two Pathways of Photosynthesis: An Overview

- In plants, photosynthesis takes place in chloroplasts.
- In the light reactions of photosynthesis, electron transport and photophosphorylation produce ATP and reduce $NADP^+$ to $NADPH + H^+$. **Review Figure 8.3**
- ATP and $NADPH + H^+$ are needed for the reactions that fix and reduce CO_2 in the Calvin–Benson cycle, forming carbohydrates. **Review Figure 8.3**

The Interactions of Light and Pigments

- Light energy comes in packets called photons, but it also has wavelike properties.
- Absorption of a photon puts a pigment molecule in an excited state that has more energy than its ground state. **Review Figure 8.4**
- Pigments absorb light in the visible spectrum. **Review Figure 8.5**
- Each compound has a characteristic absorption spectrum. An action spectrum reveals the biological effectiveness of different wavelengths of light. The absorption spectrum of the plant pigment chlorophyll *a* correlates well with the action spectrum for photosynthesis. **Review Figures 8.6**

▶ Chlorophylls and accessory pigments form antenna systems for absorption of light energy. **Review Figure 8.7**

▶ An excited pigment molecule may lose its energy by fluorescence or by transferring it to another pigment molecule. **Review Figure 8.8**

Electron Transport, Reductions, and Photophosphorylation

▶ Noncyclic electron transport uses two photosystems (I and II) and produces ATP, NADPH + H^+, and O_2. Photosystem II uses P_{680} chlorophyll, from which light-excited electrons are passed to a redox chain that drives chemiosmotic ATP production. Light-driven oxidation of water releases O_2 and passes electrons from water to the P_{680} chlorophyll. Photosystem I passes electrons from P_{700} chlorophyll to another redox chain and then to $NADP^+$, forming NADPH + H^+. **Review Figure 8.9**

▶ Cyclic electron transport uses P_{700} chlorophyll and produces only ATP. Its operation maintains the proper balance of ATP and NADPH + H^+ in the chloroplast. **Review Figure 8.10**

▶ Chemiosmosis is the mechanism of ATP production in photophosphorylation. Electron transport pumps protons from the stroma into the thylakoids. Diffusion of the protons back to the stroma via ATP synthase channels drives ATP formation. **Review Figure 8.11. See Web/CD Tutorial 8.2**

Making Carbohydrate from CO_2: The Calvin–Benson Cycle

▶ The Calvin–Benson cycle makes sugar from CO_2. This pathway was elucidated through the use of radioactive tracers. **Review Figure 8.12. See Web/CD Tutorial 8.3**

▶ The Calvin–Benson cycle consists of three phases: fixation of CO_2, reduction and carbohydrate production, and regeneration of RuBP. RuBP is the initial CO_2 acceptor, and 3PG is the first stable product of CO_2 fixation. The enzyme rubisco catalyzes the reaction of CO_2 and RuBP to form 3PG. **Review Figures 8.13, 8.14. See Web/CD Activity 8.1**

Photorespiration and Its Consequences

▶ The enzyme rubisco can catalyze a reaction between O_2 and RuBP in addition to the reaction between CO_2 and RuBP. This reaction with O_2 is called photorespiration and significantly reduces the efficiency of photosynthesis. The reactions that constitute photorespiration are distributed over three organelles: chloroplasts, peroxisomes, and mitochondria. **Review Figure 8.15**

▶ At high temperatures and low CO_2 concentrations, the oxygenase function of rubisco is favored.

▶ C_4 plants bypass photorespiration with special chemical reactions and specialized leaf anatomy. In C_4 plants, PEP carboxylase in chloroplasts of the mesophyll cells initially fixes CO_2 in a four-carbon compound, which then diffuses into bundle sheath cells, where its decarboxylation produces locally high concentrations of CO_2. **Review Figure 8.16, 8.17. See Web/CD Activity 8.2**

▶ CAM plants operate much like C_4 plants, but their initial CO_2 fixation by PEP carboxylase is temporally separated from the Calvin–Benson cycle, rather than spatially separated as in C_4 plants.

Metabolic Pathways in Plants

▶ Photosynthesis and respiration are linked through the Calvin–Benson cycle, the citric acid cycle, and glycolysis. **Review Figure 8.18**

▶ To survive, a plant must photosynthesize more than it respires.

Self-Quiz

1. In noncyclic photosynthetic electron transport, water is used to
   a. excite chlorophyll.
   b. hydrolyze ATP.
   c. reduce chlorophyll.
   d. oxidize NADPH.
   e. synthesize chlorophyll.
2. Which statement about light is true?
   a. An absorption spectrum is a plot of biological effectiveness versus wavelength.
   b. An absorption spectrum may be a good means of identifying a pigment.
   c. Light need not be absorbed to produce a biological effect.
   d. A given kind of molecule can occupy any energy level.
   e. A pigment loses energy as it absorbs a photon.
3. Which statement about chlorophylls is *not* true?
   a. They absorb light near both ends of the visible spectrum.
   b. They can accept energy from other pigments, such as carotenoids.
   c. Excited chlorophyll can either reduce another substance or fluoresce.
   d. Excited chlorophyll may be an oxidizing agent.
   e. They contain magnesium.
4. In cyclic electron transport,
   a. oxygen gas is released.
   b. ATP is formed.
   c. water donates electrons and protons.
   d. NADPH + H^+ forms.
   e. CO_2 reacts with RuBP.
5. Which of the following does *not* happen in noncyclic electron transport?
   a. Oxygen gas is released.
   b. ATP forms.
   c. Water donates electrons and protons.
   d. NADPH + H^+ forms.
   e. CO_2 reacts with RuBP.
6. In the chloroplasts,
   a. light leads to the pumping of protons out of the thylakoids.
   b. ATP forms when protons are pumped into the thylakoids.
   c. light causes the stroma to become more basic than the thylakoids.
   d. protons return passively to the stroma through protein channels.
   e. proton pumping requires ATP.
7. Which statement about the Calvin–Benson cycle is *not* true?
   a. CO_2 reacts with RuBP to form 3PG.
   b. RuBP forms by the metabolism of 3PG.
   c. ATP and NADPH + H^+ form when 3PG is reduced.
   d. The concentration of 3PG rises if the light is switched off.
   e. Rubisco catalyzes the reaction of CO_2 and RuBP.
8. In C_4 photosynthesis,
   a. 3PG is the first product of CO_2 fixation.
   b. rubisco catalyzes the first step in the pathway.
   c. four-carbon acids are formed by PEP carboxylase in bundle sheath cells.
   d. photosynthesis continues at lower CO_2 levels than in C_3 plants.
   e. CO_2 released from RuBP is transferred to PEP.

9. Photosynthesis in green plants occurs only during the day. Respiration in plants occurs
   a. only at night.
   b. only when there is enough ATP.
   c. only during the day.
   d. all the time.
   e. in the chloroplast after photosynthesis.
10. Photorespiration
    a. takes place only in C_4 plants.
    b. includes reactions carried out in peroxisomes.
    c. increases the yield of photosynthesis.
    d. is catalyzed by PEP carboxylase.
    e. is independent of light intensity.

For Discussion

1. Both photosynthetic electron transport and the Calvin–Benson cycle stop in the dark. Which specific reaction stops first? Which stops next? Continue answering the question "Which stops next?" until you have explained why both pathways have stopped.
2. In what principal ways are the reactions of electron transport in photosynthesis similar to the respiratory chain and oxidative phosphorylation discussed in Chapter 7? Differentiate between cyclic and noncyclic electron transport in terms of (1) the products and (2) the source of electrons for the reduction of oxidized chlorophyll.
3. The development of what two experimental techniques made it possible to elucidate the Calvin–Benson cycle? How were these techniques used in the investigation?
4. If water labeled with ^{18}O is added to a suspension of photosynthesizing chloroplasts, which of the following compounds will first become labeled with ^{18}O: ATP, NADPH, O_2, or 3PG? If water labeled with ^{3}H is added to a suspension of photosynthesizing chloroplasts, which of the same compounds will first become radioactive? If CO_2 labeled with ^{14}C is added to a suspension of photosynthesizing chloroplasts, which of those compounds will first become radioactive?
5. The Viking lander was sent to Mars in 1976 to detect signs of life. Explain the rationale behind the following experiments this unmanned probe performed:
   a. A scoop of dirt was inserted into a container and $^{14}CO_2$ was added. After a while during the Martian day, the $^{14}CO_2$ was removed and the dirt was heated to high temperatures. Scientists monitoring the experiment back on Earth looked for the release of $^{14}CO_2$ as a sign of life.
   b. The experiment in *(a)* was performed, except that the dirt was heated to high temperature for 30 minutes and then allowed to cool to Martian temperature right after scooping and *before* the $^{14}CO_2$ was added. If experiment *(a)* released $^{14}CO_2$, then this experiment should not release it, if living things were present.

LIFE ESSAY

What is science?

- By Sal Restivo -

Scientific facts are manufactured out of locally available social, material, and symbolic (interpersonally meaningful) resources. These resources become facts through the social interactions of scientists in a process sometimes described as creating order out of disorder. In the wake of a laboratory experiment, the sequence of writings from laboratory notes to published paper moves statements through different modes, each mode more "objective" than was the previous one. That is, statements describing the experiment progressively erase the subjective, flesh-and-blood human experimenters from the increasingly objective, mechanistic, and technical discussions. Facts attain "universal" status through the international activities of scientists as agents of professions and governments, and as ambassadors for the legitimacy of these facts.

The field of science studies is an alternative to traditional ways of studying and understanding science. According to practitioners of science studies, not only is science a social activity, but scientific activity itself is socially constructed. The idea that science is social is mired in controversy. The controversy has heated up so much that it has spawned the "science wars," pitting physical and natural (P&N) scientists against science studies researchers. The P&N scientists (and some social scientists) are worried that the claim that science is social damages the power of logic and reason to keep irrationalism and superstition at bay. Leading science studies researchers find these attacks and concerns curious because they consider themselves scientists and defenders of science.

We need to understand these issues in the context of the end of modernity, the end of a period in which the nature and value of science was considered to be beyond criticism. "Modern" and "Postmodern" are multifaceted and contentious ideas. Their substance arises from the historical realities of the twentieth century—from crises in logic and mathematics to world wars, atomic bombs, holocausts, and environmental disasters—that challenged uncritical and worshipful attitudes about the value of science, the inevitability of progress, and the transparency and universality of scientific truths. Postmodernism launched social criticisms but also stimulated the development of research and theory on science by social scientists and humanities scholars.

If the practice of science and scientific knowledge are social, critics (including senior P&N scientists and advocates of a purist conception of science coming from all fields of scholarship and research) conclude that must mean they are arbitrary, not objective, not true, and not universal. In fact, however, science studies researchers have not mounted an attack on objectivity and truth. They value science, the methods of science, and the findings of science.

Society and culture are natural phenomena and can be studied scientifically. Some critics claim this leads to confusion, because if science is social, isn't science studies also social? Of course science studies is social. This is only a paradox if you assume that saying science is social is equivalent to saying it is arbitrary and untrue.

Moreover, science studies does not claim jurisdiction over the facts P&N scientists study, but only over the P&N sciences as social phenomena. Science studies researchers do not deny reality, truth, or objectivity. They do, however, claim that we need to revise our understanding of these ideas in the light of what we now know about how society and culture shape science, scientists, and scientific knowledge. In order to appreciate what this means, one must understand the more general idea that self and mind are social phenomena and that humans have social brains shaped by our social lives.

When scientists say that there is a "reality out there," this does not mean that there exists a description of that reality that we approach through closer and closer approximations. Science is at its best when it is not being directly and overtly controlled by powerful interests with the policing power of a state or religious institution behind them. Science is, however, inevitably embedded in a social, cultural, and historical matrix that shapes its methods, theories, and substantive content.

Sal Restivo is Professor of Sociology, Science Studies, and Information Technology at Rensselaer Polytechnic Institute in New York, the Hixon/Riggs Professor of Science, Technology, and Society at Harvey Mudd College in California, and Special Professor of Mathematics Education at Nottingham University in England. He is a social theorist who specializes in the sociology of science, mathematics, and mind. He was one of the pioneers in the ethnography of science and social studies of mathematics.

Discussion Questions

1. For working scientists, what are the implications of the idea that science is socially constructed?
2. If all of our ideas are socially constructed, how is it possible to get a grasp on the concept of "truth"?
3. What is the cultural significance of the "science wars"? What are the roots of this conflict, and what are the possibilities for resolving it?

Web Links

North Carolina State Program on Science, Technology, & Society Information Sources
www.ncsu.edu/chass/mds/stslinks.html

Science Studies Organizations and Societies
www.umkc.edu/scistud/wwg/scilinks/organizations.html

Science Studies References
www.wsulibs.wsu.edu/hist-of-science/

9 Chromosomes, the Cell Cycle, and Cell Division

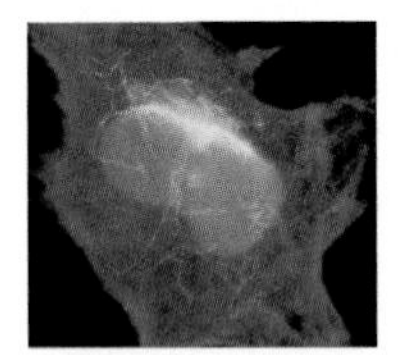

In 1951, 31-year-old Henrietta Lacks entered Johns Hopkins Hospital to be treated for a cancerous tumor. Although she died a few months later, her tumor cells are still alive today. Scientists found that, given adequate nourishment, cancerous cells from the tumor could reproduce themselves indefinitely in a laboratory dish, where they grew as a formless mass. These "HeLa cells" became a test-tube model for many studies of human biology. Over the past half-century, tens of thousands of research articles have been published using information obtained from Henrietta's cells. But are these "immortal" cells really a good model for human biology?

In one sense, they are indeed a good model. Most multicellular organisms come from a single cell: the fertilized egg. This cell reproduces itself to make two cells, which in turn divide to become four cells, and so on until all the cells of a new organism have been produced. An organism is not just a mass of cells like the HeLa culture, however; its cells must form specialized tissues and organs, each with specific roles to perform.

In normal tissues, cell reproduction (cell "births") is offset by cell loss (cell "deaths"). We know that cell death is important from careful studies of a tiny worm, *Caenorhabditis elegans*, in which 1,090 cells are produced from the fertilized egg and exactly 131 of them die before the worm is born. If the cells that are programmed to die do not do so, the worm's organs are severely malformed. Another example occurs in the mammalian brain. Young mice, for instance, lose hundreds of thousands of brain cells each day; if these cells do not die, the mouse's overcrowded brain simply does not work.

A cell's death is often programmed into its genetic instructions: normal cells "sacrifice" themselves for the greater good of the organism. Once an organism reaches its adult size, it stays that way through a combination of cell division and programmed cell death. Unlike most normal cells, but like most cancerous cells, HeLa cells keep growing because they have a genetic imbalance that heavily favors cell reproduction over cell death.

In this chapter, we will first describe how prokaryotic cells produce two new organisms from the original single-celled organism. Then we will describe two types of cell and nuclear division—mitosis and meiosis—and relate them to asexual and sexual reproduction in eukaryotic organisms. Finally, to

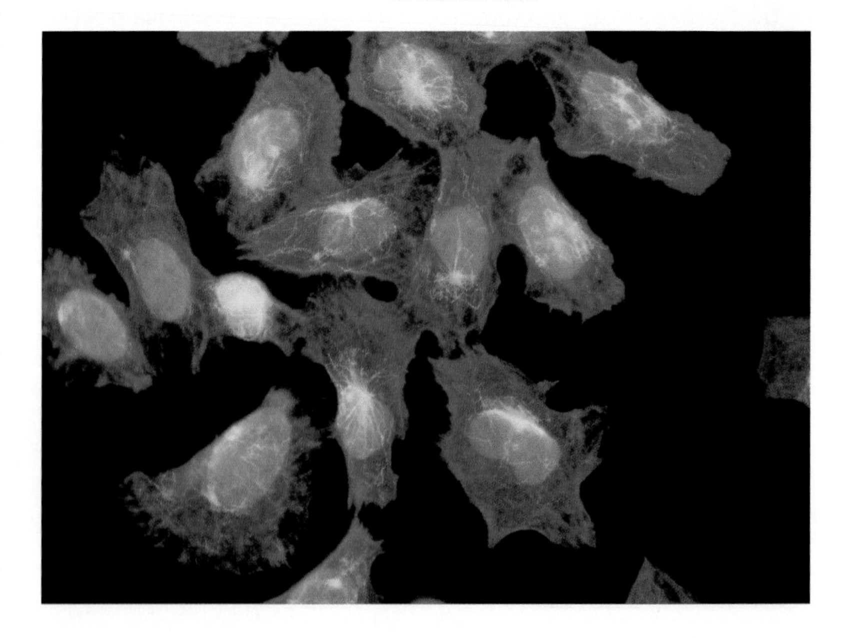

HeLa Cells: More Births Than Deaths These tumor cells grow and reproduce as an unspecialized mass on the surface of a solid medium. They have been cultured in a laboratory since 1951. They are the source of much data relating to the reproduction of human cells.

balance our discussion of cell proliferation through division, we will describe the important process of programmed cell death, also known as apoptosis.

Systems of Cell Reproduction

Unicellular organisms use cell division primarily to reproduce themselves, whereas in multicellular organisms cell division also plays important roles in growth and in the repair of tissues (Figure 9.1). In order for any cell to divide, four events must occur:

- There must be a *reproductive signal*. This signal, which may come either from inside or outside the cell, initiates the cellular reproductive events.
- **Replication** of DNA (the genetic material) and other vital cell components must occur so that each of the two new cells will be identical and have complete cell functions.
- The cell must distribute the replicated DNA to each of the two new cells. This process is called **segregation**.
- New material must be added to the cell membrane (and the cell wall, in organisms that have one) in order to separate the two new cells in a process called **cytokinesis**.

These four events occur in somewhat different ways in prokaryotes and eukaryotes.

Prokaryotes divide by fission

In prokaryotes, cell division results in the reproduction of the entire single-celled organism. The cell grows in size, replicates its DNA, and then essentially divides into two new cells, a process called **fission**.

REPRODUCTIVE SIGNALS. The reproductive rates of many prokaryotes respond to conditions in the environment. The bacterium *Escherichia coli*, a species that is commonly used in genetic studies, is a "cell division machine" that essentially divides continuously. Typically, cell division takes 40 minutes at 37°C. But if there are abundant sources of carbohydrates and salts available, the division cycle speeds up so that cells may divide in 20 minutes. Another bacterium, *Bacillus subtilis*, stops dividing when food supplies are low, then resumes dividing when conditions improve. These observations suggest that external factors, such as materials in the environment, control the initiation of cell division in prokaryotes.

REPLICATION OF DNA. A **chromosome**, as we saw in Chapter 4, is a DNA molecule containing genetic information. When a cell divides, all of its chromosomes must be replicated, and each of the two resulting copies must find its way into one of the two new cells.

Most prokaryotes have only one chromosome, a single long DNA molecule with proteins bound to it. In the bacterium *E. coli*, the DNA is a continuous molecule often referred to as a *circular chromosome*. If the bacterial DNA were actually arranged in a circle, it would be about 1.6 million nm (1.6 mm) in circumference. The bacterium itself is only about 1 μm (1,000 nm) in diameter and about 4 μm long. Thus the bacterial DNA, fully extended, would form a circle over 100 times larger than the cell! To fit it into the cell, the DNA must be packaged. The DNA molecule accomplishes some packaging by folding in on itself, and positively charged (basic) proteins bound to negatively charged (acidic) DNA contribute to this folding. Circular chromosomes appear to be characteristic of all prokaryotes, as well as some viruses, and

(a)
Cell division contributes to the growth of this root tissue.

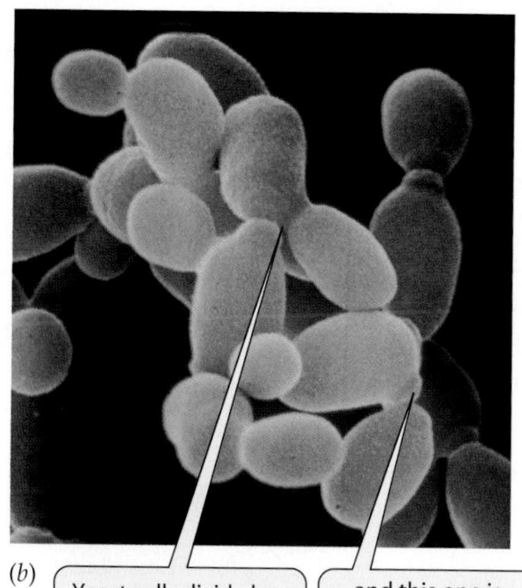

(b)
Yeast cells divide by budding. This one has nearly divided...
...and this one is beginning to bud.

(c)
Cell division contributes to the regeneration of a lizard's tail.

9.1 Important Consequences of Cell Division Cell division is the basis for (*a*) growth, (*b*) reproduction, and (*c*) regeneration.

are also found in the chloroplasts and mitochondria of eukaryotic cells.

Functionally, the prokaryotic chromosome has two regions that are important for cell reproduction:

- The site where replication of the circle starts: the origin of replication, designated *ori*
- The site where replication ends: the terminus of replication, *ter*

The process of chromosome replication occurs as the DNA is threaded through a "replication complex" of proteins at the center of the cell. These proteins include the enzyme DNA polymerase, and their operation will be discussed further in Chapter 11. During the process of prokaryotic DNA replication, the cell grows and provides a mechanism for the ordered distribution of the DNA into the newly formed daughter cells.

SEGREGATION OF DNA. DNA replication actively drives the segregation of the replicated DNA molecules to the two new cells. The first region to be replicated is *ori*, which is attached to the plasma membrane. The two resulting *ori* regions separate as the new chromosome forms and new plasma membrane forms between them as the cell grows longer (Figure 9.2). By the end of replication, there are two chromosomes, one at either end of the lengthened bacterial cell.

CYTOKINESIS. Cell separation, or cytokinesis, begins 20 minutes after chromosome replication is finished. The first event of cytokinesis is a pinching in of the plasma membrane to form a ring similar to a purse string. Fibers composed of a protein similar to eukaryotic tubulin (which makes up microtubules) are major components of this ring. As the membrane pinches in, new cell wall materials are synthesized, which finally separate the two cells.

Eukaryotic cells divide by mitosis or meiosis

Complex eukaryotes, such as humans and flowering plants, originate from a single cell, the fertilized egg. This cell derives from the union of two sex cells, called **gametes**, from the organism's parents—that is, a sperm and egg—and so contains genetic material from both of these parental cells. This means that the fertilized egg contains one set of chromosomes from the male parent and one set from the female parent.

The formation of a multicellular organism from a fertilized egg is called *development*. It involves both cell reproduction and cell specialization. For example, an adult human has several trillion cells, all ultimately deriving from the fertilized egg, and many of them have specialized roles. We will discuss how cells specialize later in this book, in Part Three. For now, we will focus on cell reproduction.

Cell reproduction in eukaryotes, like that in prokaryotes, involves reproductive signals, DNA replication, segregation, and cytokinesis. But, as you might expect, events in eukary-

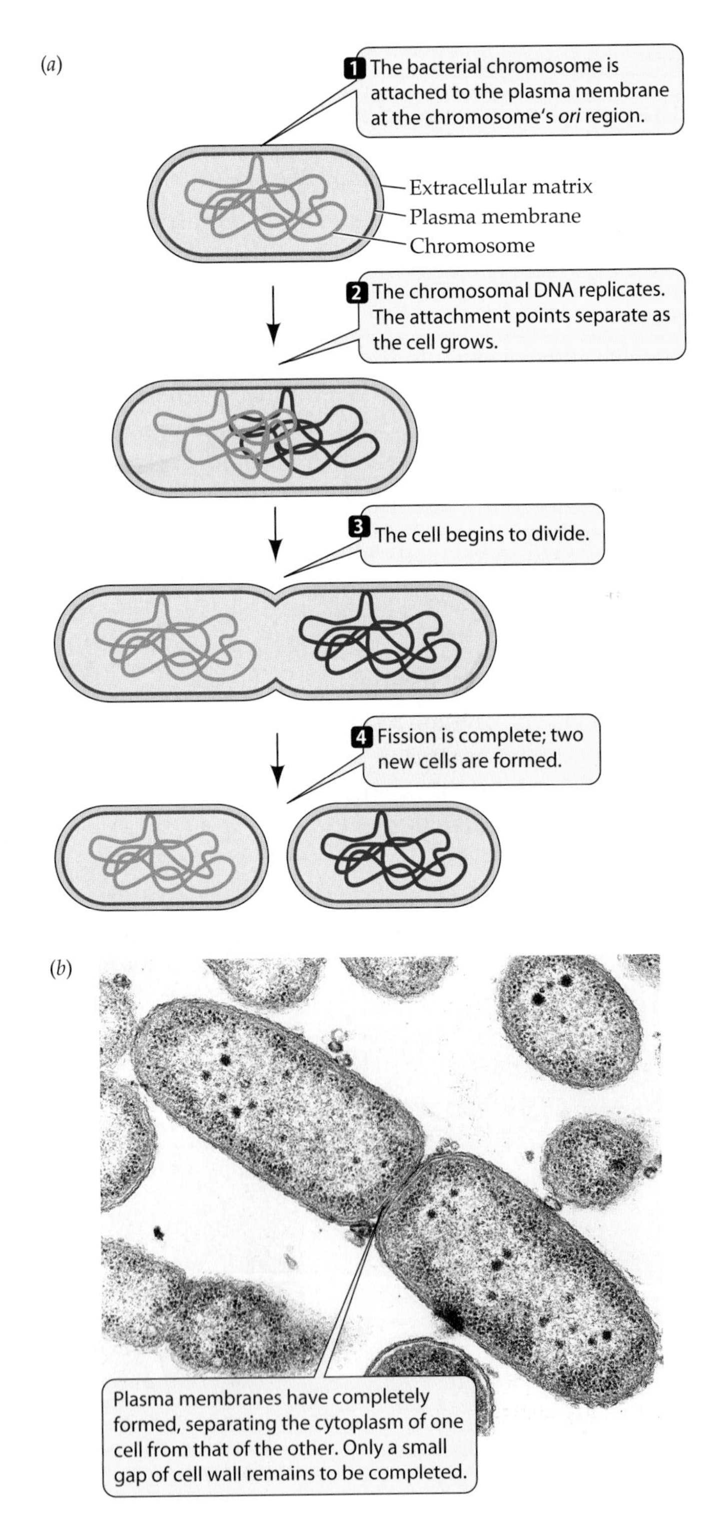
(a)
1 The bacterial chromosome is attached to the plasma membrane at the chromosome's ori region.
Extracellular matrix
Plasma membrane
Chromosome
2 The chromosomal DNA replicates. The attachment points separate as the cell grows.
3 The cell begins to divide.
4 Fission is complete; two new cells are formed.
(b)
Plasma membranes have completely formed, separating the cytoplasm of one cell from that of the other. Only a small gap of cell wall remains to be completed.

9.2 Prokaryotic Cell Division (*a*) The steps of cell division in prokaryotes. (*b*) These two cells of the bacterium *Pseudomonas aeruginosa* have almost completed fission. Each cell contains a complete chromosome, visible as the nucleoid in the center of the cell.

otes are somewhat more complex. First, unlike prokaryotes, eukaryotic cells do not constantly divide whenever environmental conditions are adequate. In fact, eukaryotic cells that are part of a multicellular organism and have become specialized seldom divide. So the signals for cell division are related not to the environment of a single cell, but to the needs of the entire organism. Second, instead of a single chromosome, eukaryotes usually have many (humans have 46), so the processes of replication and segregation, while basically the same as in prokaryotes, are more intricate (see Table 9.1). Third, eukaryotic cells have a distinct nucleus, which has to be replicated and then divided into two new nuclei. Thus, in eukaryotes, cytokinesis is distinct from division of the genetic material. Finally, cytokinesis is different in plant cells (which have a cell wall) than in animal cells (which do not).

The key difference between prokaryotic and eukaryotic cell reproduction is that in the eukaryotes, newly replicated chromosomes remain associated with each other as **sister chromatids**, and a new mechanism, **mitosis**, is used to segregate them into the two new nuclei.

The reproduction of a eukaryotic cell typically consists of three steps:

- The replication of DNA within the nucleus
- The packaging and segregation of the replicated DNA into two new nuclei (nuclear division)
- The division of the cytoplasm (cytokinesis)

A second mechanism of nuclear division, **meiosis**, occurs in germ cells that produce gametes that contribute to the reproduction of a new organism. While the two products of mitosis are genetically identical to the cell that produced them—they both have the same DNA—the products of meiosis are not. As we will see later in the chapter, meiosis generates diversity by shuffling the genetic material, resulting in new gene combinations. It plays a key role in sexual life cycles.

What determines whether a cell will divide? How does mitosis lead to identical cells, and meiosis to diversity? Why do we need both identical copies and diverse cells? Why do most eukaryotic organisms reproduce sexually? In the pages that follow, we will describe the details of interphase, mitosis, and meiosis, as well as their consequences for heredity, development, and evolution.

Interphase and the Control of Cell Division

A cell lives and functions until it divides or dies. Or, if it is a gamete, it lives until it fuses with another gamete. Some types of cells, such as red blood cells, muscle cells, and nerve cells, lose the capacity to divide as they mature. Other cell types, such as cortical cells in plant stems, divide only rarely. Some cells, like the cells in a developing embryo, are specialized for rapid division.

Between divisions—that is, for most of its life—a eukaryotic cell is in a condition called **interphase**. For most types of cells, we may speak of a **cell cycle** that has two phases: mitosis and interphase. In this section, we will describe the cell cycle events that occur during interphase, especially the "decision" to enter mitosis.

A given cell lives for one turn of the cell cycle and then becomes two cells. The cell cycle, when repeated again and again, is a constant source of new cells. However, even in tissues engaged in rapid growth, cells spend most of their time in interphase. Examination of any collection of dividing cells, such as the tip of a root or a slice of liver, will reveal that most of the cells are in interphase most of the time; only a small percentage of the cells will be in mitosis at any given moment.

Interphase consists of three subphases, identified as G1, S, and G2. The cell's DNA replicates during the **S phase** (the S stands for synthesis). The period between the end of mitosis and the onset of the S phase is called **G1**, or Gap 1. Another gap phase—**G2**—separates the end of the S phase and the beginning of mitosis, when nuclear and cytoplasmic division take place and two new cells are formed. Mitosis and cytokinesis are referred to as the **M phase** of the cell cycle (Figure 9.3).

The process of DNA replication, which we will describe in Chapter 11, is completed by the end of S phase. Where there was formerly one chromosome, there are now two, joined

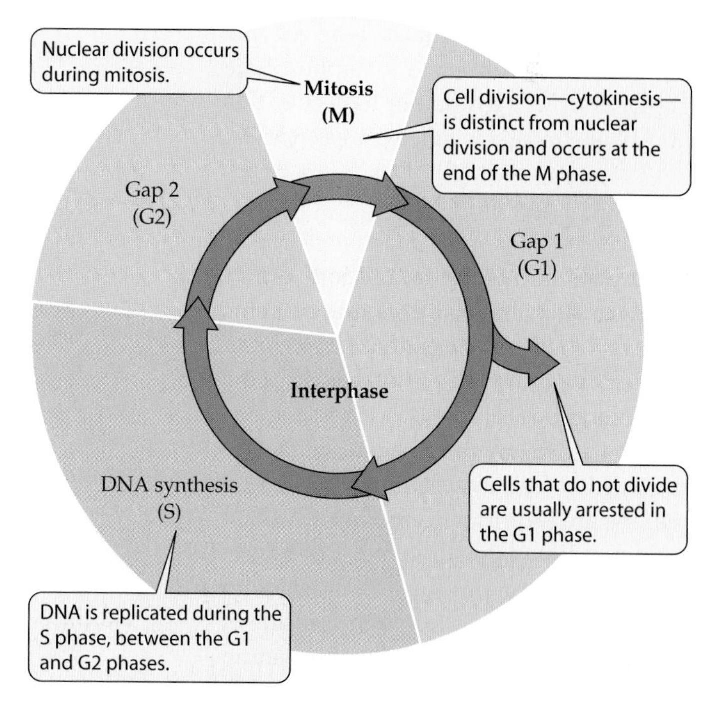

9.3 The Eukaryotic Cell Cycle The cell cycle consists of a mitotic (M) phase, during which first nuclear division (mitosis) and then cell division (cytokinesis) take place. The M phase is followed by a long period of growth known as interphase. Interphase has three subphases (G1, S, and G2) in cells that divide.

together and awaiting segregation into two new cells by mitosis or meiosis.

Although one key event—DNA replication—dominates and defines the S phase, important cell cycle processes take place in the gap phases as well. G1 is quite variable in length in different cell types. Some rapidly dividing embryonic cells dispense with it entirely, while other cells may remain in G1 for weeks or even years. In many cases, these cells enter a resting phase called **G0**. Special internal and external signals are needed to prompt a cell to leave G0 and re-enter the cell cycle at G1.

The biochemical hallmark of a G1 cell is that it is preparing for the S phase, so at this stage each chromosome is a single, unreplicated structure. It is at the G1-to-S transition that the commitment to enter another cell cycle is made.

During G2, the cell makes preparations for mitosis—for example, by synthesizing components of the microtubules that will move the chromosomes to opposite ends of the dividing cell. Because the chromosomes were replicated during the S phase, each chromosome now consists of two identical sister chromatids.

Cyclins and other proteins signal events in the cell cycle

How are appropriate decisions to enter the S or M phases made? These transitions—from G1 to S and from G2 to M—depend on the activation of a type of protein called **cyclin-dependent kinase**, or **Cdk**. Remember that a *kinase* is an enzyme that catalyzes the transfer of a phosphate group from ATP to another molecule; this phosphate transfer is called *phosphorylation*:

$$\text{protein} + \text{ATP} \xrightarrow{\text{kinase}} \text{protein—P} + \text{ADP}$$

What does phosphorylation do to a protein? Recall from Chapter 3 that proteins have both hydrophilic regions (which tend to interact with water on the outside of the macromolecule) and hydrophobic regions (which tend to interact with one another on the inside of the macromolecule). These regions are important in giving a protein its three-dimensional shape. Phosphate groups are charged, so an amino acid with such a group tends to be on the outside of the protein. In this way, phosphorylation changes the shape and function of a protein. By catalyzing the phosphorylation of certain target proteins, Cdk's play important roles in initiating the steps of the cell cycle.

The discovery that Cdk's induce cell division is a beautiful example of research on different organisms and different cell types converging on a single mechanism:

- One group of scientists was studying immature sea urchin eggs, trying to find out how they are stimulated to divide and form mature eggs. A protein called *maturation promoting factor* was purified from maturing eggs, which by itself prodded immature eggs into division.
- Other scientists studying the cell cycle in yeast, a single-celled eukaryote, found a strain that was stalled at the G1–S boundary because it lacked a Cdk. This yeast Cdk was discovered to be very similar to the sea urchin's maturation promoting factor.

Similar Cdk's were soon found to control the G1-to-S transition in many other organisms, including humans.

But Cdk's are not active by themselves. Rather, they must bind to a second type of protein, called **cyclin**. This binding—an example of allosteric regulation—activates the Cdk by altering its shape and exposing its active site (see Figure 6.19). It is the cyclin-Cdk complex that acts as a protein kinase and triggers the transition from G1 to S phase. Then cyclin breaks down, and the Cdk becomes inactive.

Several different cyclin-Cdk combinations act at various stages of the mammalian cell cycle (Figure 9.4):

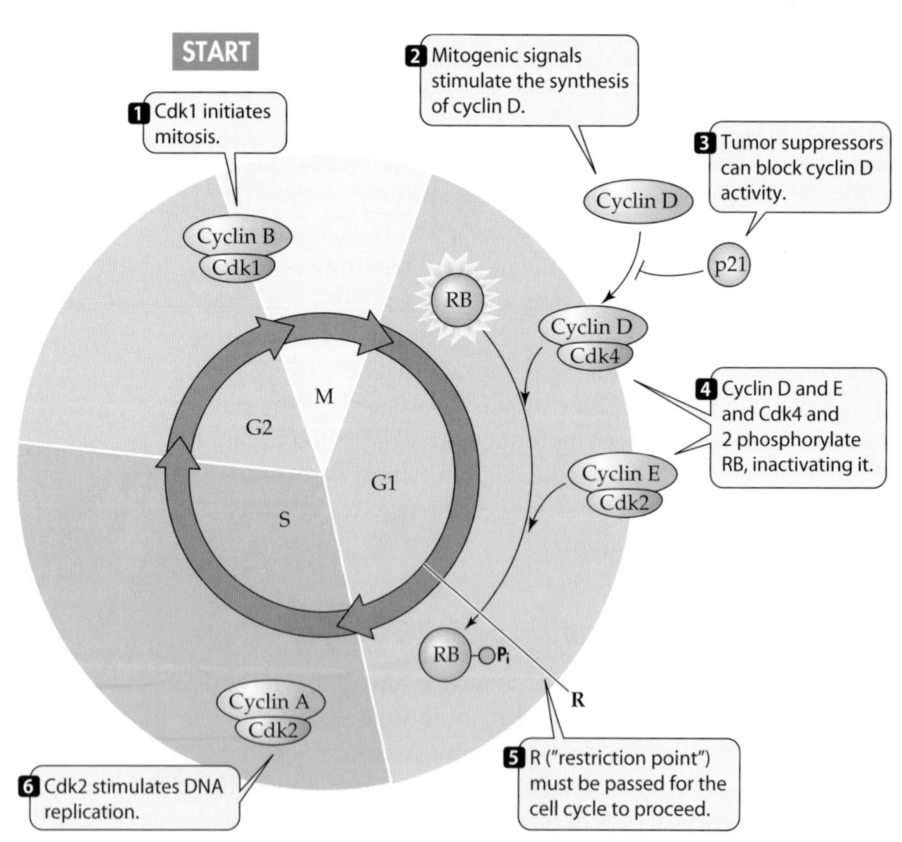

9.4 Cyclin-Dependent Kinases and Cyclins Trigger Transitions in the Cell Cycle There are four cyclin-Cdk controls during the typical cell cycle in humans.

- Cyclin D-Cdk4 acts during the middle of G1. This is the *restriction point* (*R*), a key decision point beyond which the rest of the cell cycle is normally inevitable.
- Cyclin E-Cdk2 also acts in the middle of G1.
- Cyclin A-Cdk2 acts during S, and also stimulates DNA replication.
- Cyclin B-Cdk1 acts at the G2–M boundary, initiating the transition to mitosis.

The key to progress past the restriction point is a protein called *RB* (retinoblastoma protein, named for a childhood cancer in which it was first discovered). RB normally inhibits the cell cycle. But when RB is phosphorylated by a protein kinase, it becomes inactive and no longer blocks the restriction point, and the cell progresses past G1 into S phase (Note the double negative here—a cell function happens because an inhibitor is inhibited! This phenomenon is rather common in the control of cellular metabolism.) The enzymes that catalyze RB phosphorylation are Cdk4 and Cdk2. So what is needed for a cell to pass the restriction point is the synthesis of cyclins D and E, which activate Cdk 4 and 2, which phosphorylate RB, which becomes inactivated.

The cyclin-Cdk complexes act as *checkpoints*, points at which a cell cycle's progress can be monitored to determine whether the next step can be taken. For example, if DNA is damaged by radiation during G1, a protein called p21 is made. (The p stands for "protein" and the 21 stands for its molecular weight—about 21,000 daltons.) The p21 protein then binds to the two G1 Cdk's, preventing their activation by cyclins. So the cell cycle stops while repairs are made to DNA. The p21 protein itself breaks down after the DNA is repaired, allowing cyclins to bind to the Cdk's and the cell cycle to proceed.

Because cancer results from inappropriate cell division, it is not surprising that these cyclin-Cdk controls are disrupted in cancer cells. For example, some fast-growing breast cancers have too much cyclin D, which overstimulates Cdk4 and thus cell division. As we will describe in Chapter 17, a major protein in normal cells that prevents them from dividing is p53, which leads to the synthesis of p21 and therefore inhibition of Cdk's. More than half of all human cancers contain defective p53, resulting in the absence of cell cycle controls.

Growth factors can stimulate cells to divide

Cyclin-Cdk complexes provide an internal control on progress through the cell cycle. But there are tissues in the body in which cells no longer go through the cell cycle, or go through it slowly and divide infrequently. If such cells are to divide, they must be stimulated by external signals (chemical messengers) called **growth factors**. For example, when you cut yourself and bleed, specialized cell fragments called platelets gather at the wound and help to initiate blood clotting. The platelets also produce and release a protein, called *platelet-derived growth factor*, that diffuses to the adjacent cells in the skin and stimulates them to divide and heal the wound.

Other growth factors include *interleukins*, which are made by one type of white blood cell and promote cell division in other cells that are essential for the body's immune system defenses. *Erythropoietin*, made by the kidney, stimulates the division of bone marrow cells and the production of red blood cells. In addition, many hormones promote division in specific cell types.

We will describe the physiological roles of growth factors in later chapters, but all of them act in a similar way. They bind to their target cells via specialized receptor proteins on the target cell surface. This specific binding triggers events within the target cell that initiate the cell cycle. Cancer cells often divide inappropriately because they make their own growth factors, or because they no longer require growth factors to start cycling.

Eukaryotic Chromosomes

Most human cells other than gametes contain two full sets of genetic information, one from the mother and the other from the father. As in prokaryotes, this genetic information consists of DNA molecules. However, unlike prokaryotes, humans and other eukaryotes have more than one chromosome, and during interphase those chromosomes reside within a membrane-enclosed organelle, the nucleus.

The basic unit of the eukaryotic chromosome is a gigantic, linear, double-stranded molecule of DNA complexed with many proteins to form a dense material called **chromatin** (Figure 9.5). Before the S phase, each chromosome contains only one such double-stranded DNA molecule. However,

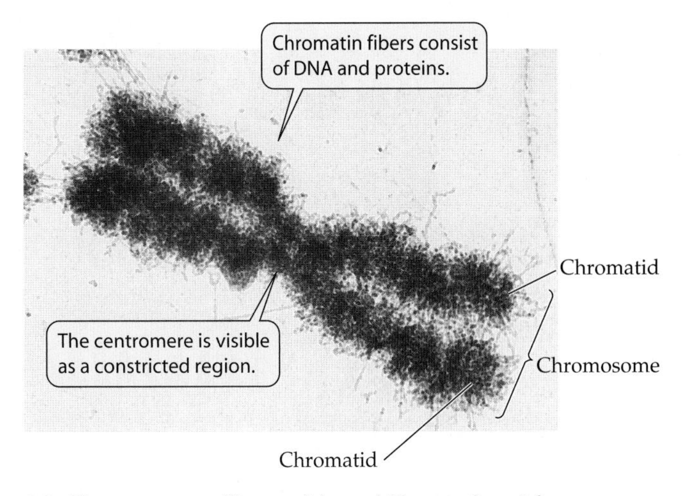

Chromatin fibers consist of DNA and proteins.
Chromatid
The centromere is visible as a constricted region.
Chromosome
Chromatid

9.5 Chromosomes, Chromatids, and Chromatin A human chromosome, shown as the cell prepares to divide.

after the DNA molecule replicates during the S phase, the two resulting DNA molecules, now called chromatids, are held together along most of their length by a protein called **cohesin**. They stay this way until mitosis, when most of the cohesin is removed, except in a region called the **centromere** at which the chromatids are still held together (see Figure 9.9). A second group of proteins called **condensins** coats the DNA molecules at this time and makes them more compact.

The DNA in a typical human cell has a total length of 2 meters. Yet the nucleus is only 5 μm (0.000005 meters) in diameter. So, although the DNA in an interphase nucleus is "unwound," it is still impressively packed! This packing is achieved largely by proteins associated closely with the chromosomal DNA (Figure 9.6).

Chromosomes contain large quantities of proteins called **histones** (from the Greek, "web"). There are five classes of histones. All of them have a positive charge at cellular pH levels because of their high content of the basic amino acids lysine and arginine. These positive charges electrostatically attract the negative phosphate groups on DNA. These interactions, as well as interactions among the histones themselves, form beadlike units called **nucleosomes**. Each nucleosome contains the following components:

- Eight histone molecules, two each of four of the histone classes, united to form a core or spool.
- 146 base pairs of DNA, 1.65 turns of it wound around the histone core.
- Histone H1 (the remaining histone class) on the outside of the DNA, which may clamp it to the histone core.

During interphase, a chromosome is made up of a single DNA molecule running around vast numbers of nucleosomes like beads on a string. Between the nucleosomes stretches a variable amount of non-nucleosomal "linker"

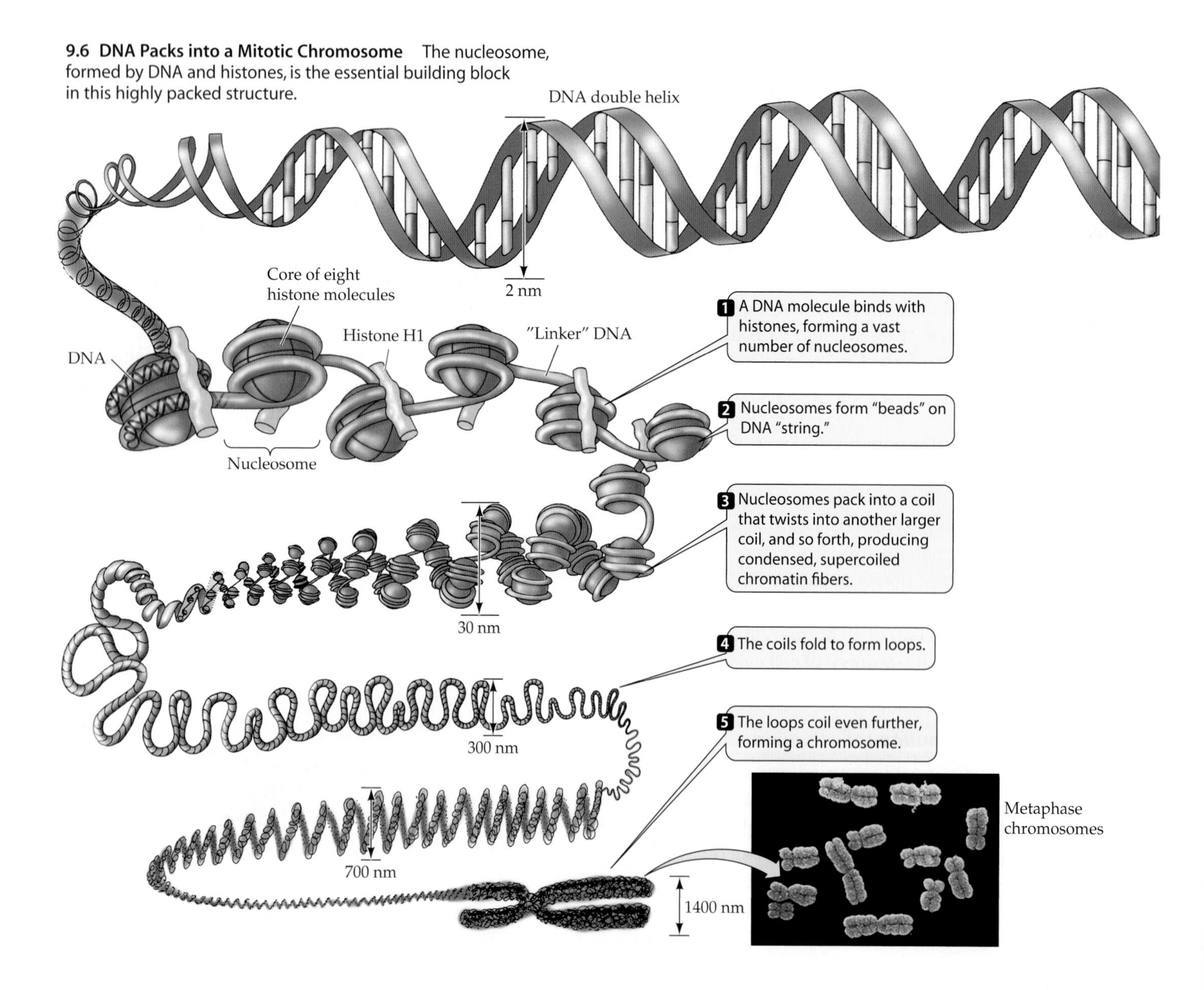
DNA double helix
2 nm
Core of eight histone molecules
Histone H1
"Linker" DNA
DNA
Nucleosome
1 A DNA molecule binds with histones, forming a vast number of nucleosomes.
2 Nucleosomes form "beads" on DNA "string."
3 Nucleosomes pack into a coil that twists into another larger coil, and so forth, producing condensed, supercoiled chromatin fibers.
30 nm
4 The coils fold to form loops.
300 nm
5 The loops coil even further, forming a chromosome.
700 nm
1400 nm
Metaphase chromosomes

9.6 DNA Packs into a Mitotic Chromosome The nucleosome, formed by DNA and histones, is the essential building block in this highly packed structure.

DNA. Since this DNA is exposed to the nuclear environment, it is accessible to proteins involved in its duplication and the regulation of its expression, as we will see in Chapter 14.

During both mitosis and meiosis, the chromatin becomes ever more coiled and condensed as its nucleosomes pack together and coil, with further folding of the chromatin continuing up to the time at which the chromosomes begin to move apart.

Mitosis: Distributing Exact Copies of Genetic Information

In mitosis, a single nucleus gives rise to two nuclei that are genetically identical to each other and to the parent nucleus. This process ensures the accurate distribution of the eukaryotic cell's multiple chromosomes to the daughter nuclei. In reality, mitosis is a continuous process in which each event flows smoothly into the next. For discussion, however, it is convenient to look at mitosis—the M phase of the cell cycle—as a series of separate events: prophase, prometaphase, metaphase, anaphase, and telophase.

The centrosomes determine the plane of cell division

Once the commitment to enter mitosis has been made, the cell enters S phase, and DNA is replicated. At the same time, in the cytoplasm, the **centrosome** ("central body"), an organelle that lies near the nucleus, doubles, forming a pair of centrosomes. In many organisms, each centrosome consists of a pair of **centrioles**, each one a hollow tube lined with nine microtubules. The two tubes are at right angles to each other.

At the G2-to-M transition, the two centrosomes separate from each other, moving to opposite ends of the nuclear envelope. The orientation of the centrosomes determines the plane at which the cell will divide, and therefore the spatial relationship of the two new cells to the parent cell. This relationship may be of little consequence to single free-living cells such as yeasts, but it is important for cells that make up part of a body tissue.

The material around the centrioles initiates the formation of microtubules, which will orchestrate chromosomal movement. Plant cells lack centrosomes, but distinct microtubule organizing centers at either end of the cell serve the same role. The formation of microtubules leads to the formation of the spindle structure that is required for the orderly segregation of the chromosomes in cell division.

Chromatids become visible and the spindle forms during prophase

During interphase, only the nuclear envelope, the nucleoli, and a barely discernible tangle of chromatin are visible under the light microscope. The appearance of the nucleus changes as the cell enters **prophase**—the beginning of mitosis. Most of the cohesin that held the two products of DNA replication together since S phase has been destroyed, so the individual chromatids become visible. They are still held together by a small amount of cohesin at the centromere (see Figure 9.9). Late in prophase, specialized three-layered structures called **kinetochores** develop in the centromere region, one on each chromatid. These structures will be important in chromosome movements.

Each of the two centrosomes serves as a *mitotic center*, or *pole*, toward which the chromosomes will move. Microtubules form between each pole and the chromosomes to make up a **spindle**, which serves both as a structure to which the chromosomes will attach and as a framework keeping the two poles apart. The spindle is actually two half-spindles: Each polar microtubule runs from one pole to the middle of the spindle, where it overlaps with polar microtubules extending from the other half-spindle (Figure 9.7). The polar microtubules are initially unstable, constantly forming and falling apart, until they contact polar microtubules from the other half-spindle and become more stable.

There are two types of microtubules in the spindle:

- *Polar microtubules* have abundant tubulin around the centrioles. Tubulin subunits can aggregate to form long fibers that extend beyond the equatorial plate.
- *Kinetochore microtubules* attach to the kinetochores on the chromosomes.

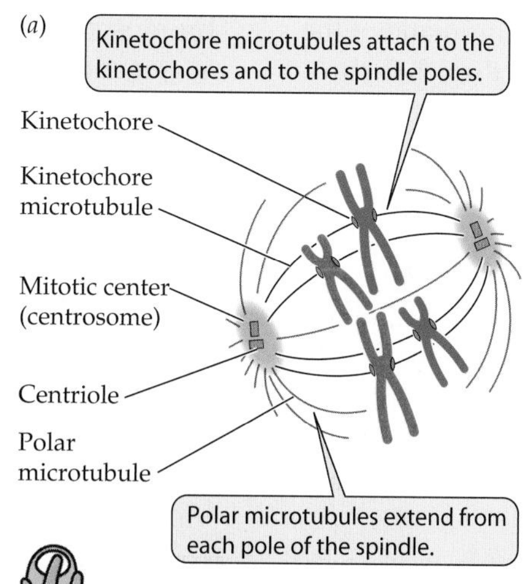

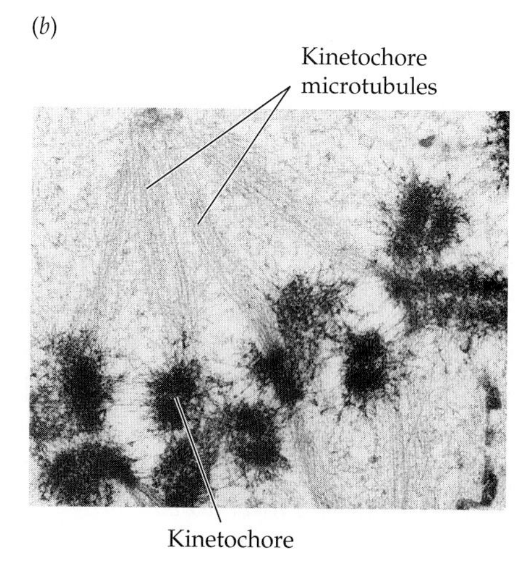

9.7 The Mitotic Spindle Consists of Microtubules (*a*) Diagram of the spindle apparatus in a cell at metaphase. (*b*) An electron micrograph of the stage shown in (*a*).

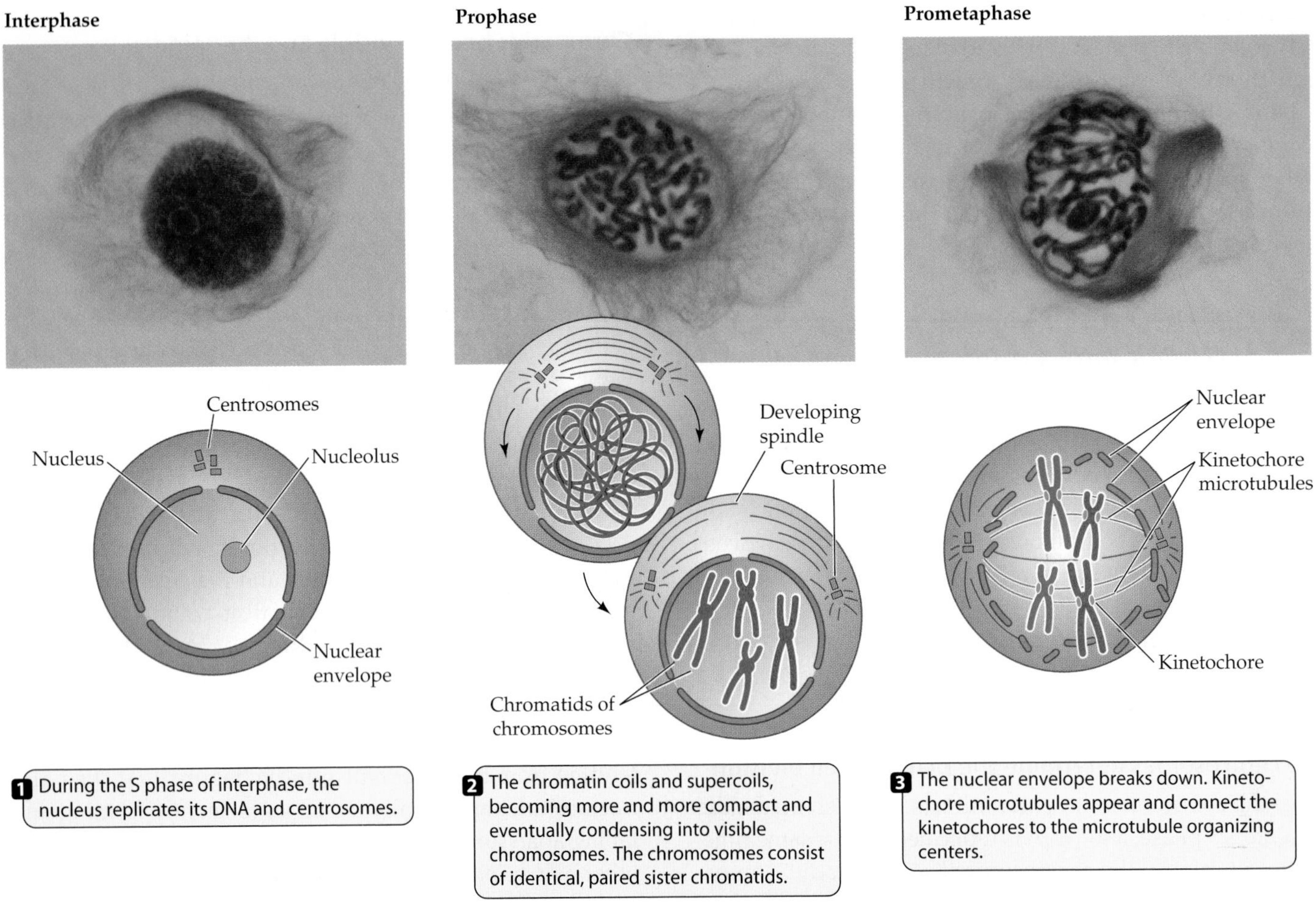

Interphase

1 During the S phase of interphase, the nucleus replicates its DNA and centrosomes.

Prophase

2 The chromatin coils and supercoils, becoming more and more compact and eventually condensing into visible chromosomes. The chromosomes consist of identical, paired sister chromatids.

Prometaphase

3 The nuclear envelope breaks down. Kinetochore microtubules appear and connect the kinetochores to the microtubule organizing centers.

9.8 Mitosis Mitosis results in two new nuclei that are genetically identical to each other and to the nucleus from which they were formed. The photomicrographs are of plant nuclei, which lack centrioles. The diagrams are of corresponding phases in animal cells and introduce the structures not found in plants. In the micrographs, the red dye stains microtubules (and thus the spindle); the blue dye stains the chromosomes. In the diagrams, the chromosomes are stylized to emphasize the fates of the individual chromatids.

The two sister chromatids in each chromosome pair become attached by their kinetochore to microtubules in opposite halves of the spindle. This ensures that one chromatid of the pair will eventually move to one pole and the other chromatid will move to the other pole. Movement of the chromatids is the central feature and accomplishment of mitosis.

Chromosome movements are highly organized

The next three phases of mitosis—prometaphase, metaphase, and anaphase—are the phases during which chromosomes actually move (Figure 9.8). During these phases, the centromeres holding the two chromatids together separate, and the former sister chromatids move away from each other in opposite directions.

PROMETAPHASE. **Prometaphase** is marked by the disappearance of the nuclear envelope. The material of the envelope remains in the cytoplasm, however, to be reassembled when the daughter nuclei re-form. In prometaphase, the chromosomes begin to move toward the poles, but this movement is counteracted by two factors:

- A repulsive force from the poles pushes the chromosomes toward the middle region, or **equatorial plate** (metaphase plate), of the cell.
- The two chromatids are still held together at the centromere by cohesin.

Thus, during prometaphase chromosomes appear to move aimlessly back and forth between the poles and the middle of the spindle. Gradually, the centromeres approach the equatorial plate.

METAPHASE. The cell is said to be in **metaphase** when all the centromeres arrive at the equatorial plate. Metaphase is the best time to see the sizes and shapes of chromosomes

Metaphase

Anaphase

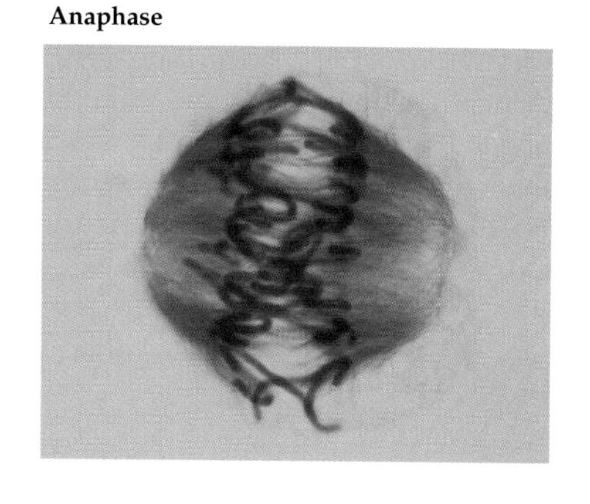

Telophase

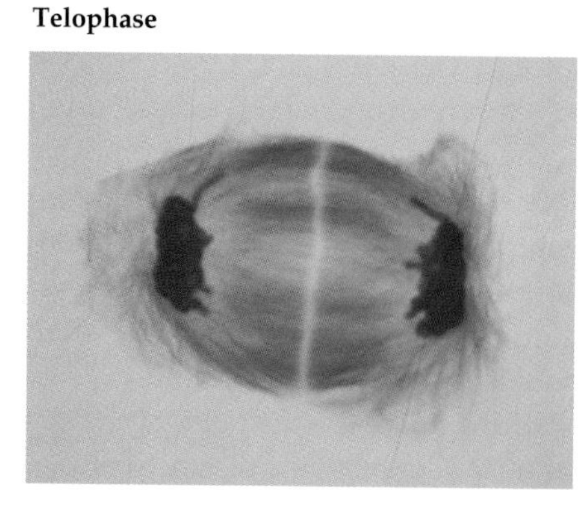

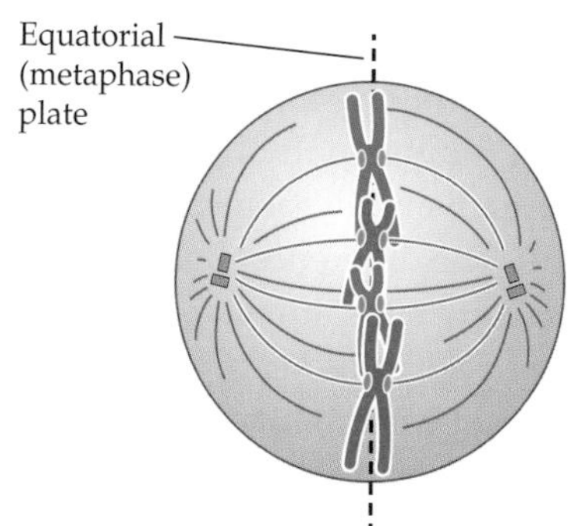

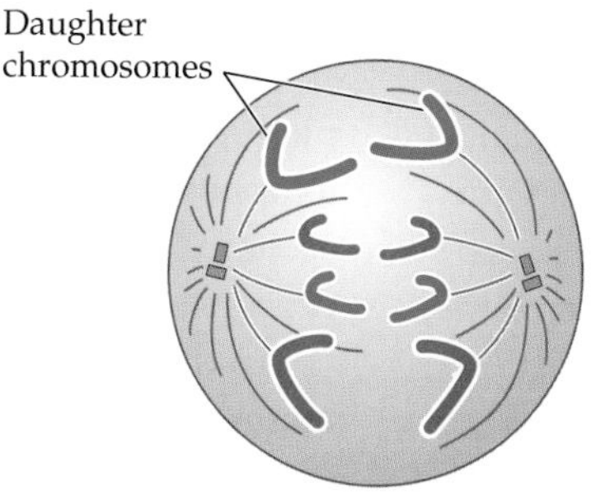

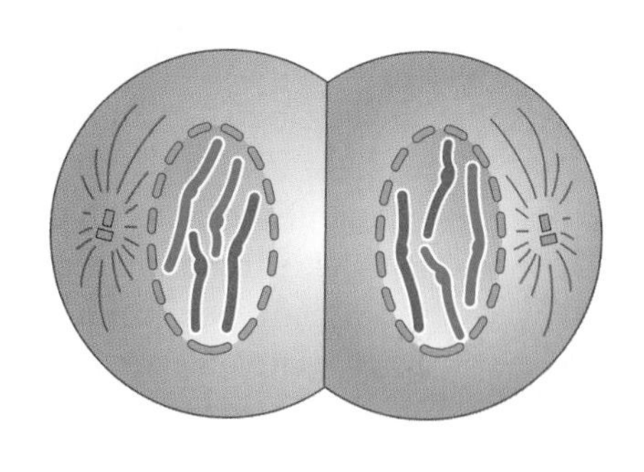

4 The centromeres (regions connecting paired chromatids) become aligned in a plane at the cell's equator.

5 The paired sister chromatids separate, and the new daughter chromosomes begin to move toward the poles.

6 The daughter chromosomes reach the poles. Telophase passes into the next interphase as the nuclear envelopes and nucleoli re-form and the chromatin becomes diffuse.

because they are maximally condensed. The chromatids are now clearly connected to one pole or the other by microtubules. At the end of metaphase, all of the chromatid pairs separate simultaneously.

This separation occurs because the cohesin holding the sister chromatids together is hydrolyzed by a specific protease, appropriately called *separase*. Until this point, separase has been present but inactive, because it has been bound to an inhibitory subunit called *securin*. Once all the chromatids are connected to the spindle, securin is hydrolyzed, allowing separase to catalyze cohesin breakdown (Figure 9.9). In this way, chromosome alignment is connected to chromatid separation. This process, called the *spindle checkpoint*, apparently senses whether there are any kinetochores that are unattached to the spindle. If there are, securin breakdown is blocked, and the sister chromatids stay together.

ANAPHASE. Separation of the chromatids marks the beginning of **anaphase**, during which the two sister chromatids move to opposite ends of the spindle. Each chromatid contains one double-stranded DNA molecule and is now referred to as a **daughter chromosome**.

What propels this highly organized mass migration is not clear. Two things seem to move the chromosomes along.

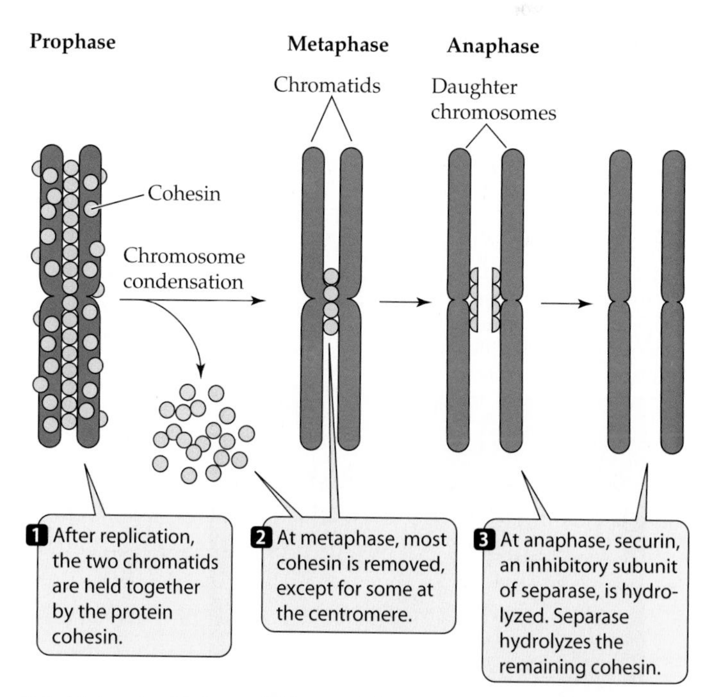

9.9 Molecular Biology of Chromatid Attachment and Separation Cohesin holds sister chromatids together. Separase hydrolyzes cohesin at the onset of anaphase.

First, at the kinetochores are proteins that act as "molecular motors." These proteins, called *cytoplasmic dynein*, have the ability to hydrolyze ATP to ADP and phosphate, thus releasing energy to move the chromosomes along the microtubules toward the poles. These motor proteins account for about 75 percent of the force of motion. Second, the kinetochore microtubules shorten from the poles, drawing the chromosomes toward them. This shortening accounts for about 25 percent of the motion.

During anaphase the poles of the spindle are pushed farther apart, doubling the distance between them. The distance between poles increases because the overlapping polar microtubules extending from opposite ends of the spindle contain motor proteins that cause them to slide past each other, in much the same way that microtubules slide in cilia and flagella (see Figure 4.24*a*). This polar separation further separates one set of daughter chromosomes from the other.

The movements of chromosomes are slow, even in cellular terms. At about 1 μm per minute, it takes about 10–60 minutes for them to complete their journey to the poles. This speed is equivalent to a person taking 9 million years to travel across the United States! This slow speed may ensure that the chromosomes segregate accurately.

Nuclei re-form during telophase

When the chromosomes stop moving at the end of anaphase, the cell enters **telophase**. Two sets of chromosomes (formerly referred to as daughter chromosomes), containing identical DNA and carrying identical sets of hereditary instructions, are now at the opposite ends of the spindle, which begins to break down. The chromosomes begin to uncoil, continuing until they become the diffuse tangle of chromatin that is characteristic of interphase. The nuclear envelopes and nucleoli, which were disaggregated during prophase, coalesce and re-form their respective structures. When these and other changes are complete, telophase—and mitosis—is at an end, and each of the daughter nuclei enters another interphase.

Mitosis is beautifully precise. Its result is two nuclei that are identical to each other and to the parent nucleus in chromosomal makeup, and hence in genetic constitution. Next, the two nuclei must be isolated in separate cells, which requires the division of the cytoplasm.

Cytokinesis: The Division of the Cytoplasm

Mitosis refers only to the division of the nucleus. The division of the cell's cytoplasm, which follows mitosis, is accomplished by cytokinesis, which may actually begin before telophase ends. In different organisms, cytokinesis may be accomplished in different ways. The differences between the process in plants and in animals are substantial.

Animal cells usually divide by a furrowing of the plasma membrane, as if an invisible thread were tightening between the two poles (Figure 9.10*a*). The invisible thread is actually microfilaments of actin and myosin (see Figure 4.21) located in a ring just beneath the plasma membrane. These two proteins interact to produce a contraction, just as they do in muscles, thus pinching the cell in two. These microfilaments assemble rapidly from actin monomers that are present in the interphase cytoskeleton. Their assembly appears to be under the control of Ca^{2+} released from storage sites in the center of the cell.

Plant cell cytoplasm divides differently because plants have cell walls. As the spindle breaks down after mitosis, membranous vesicles derived from the Golgi apparatus appear in the equatorial region roughly midway between the two daughter nuclei. Propelled along microtubules by the

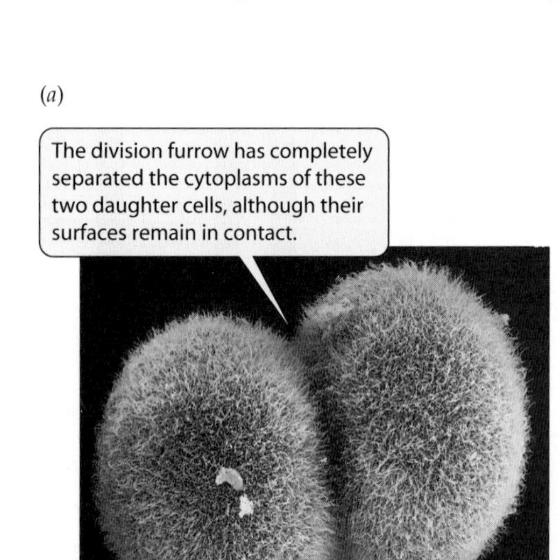

(a)
The division furrow has completely separated the cytoplasms of these two daughter cells, although their surfaces remain in contact.

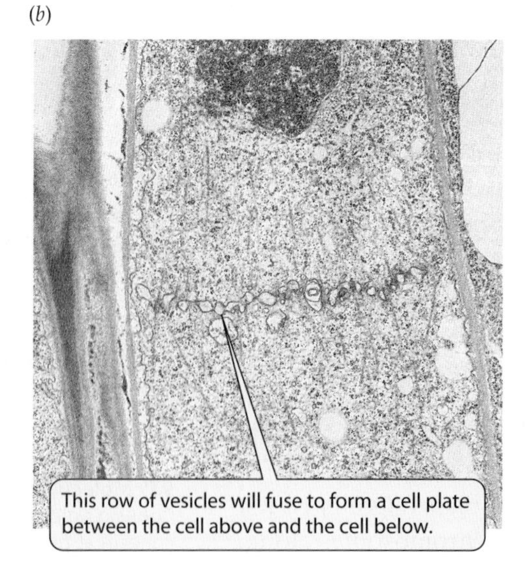

(b)
This row of vesicles will fuse to form a cell plate between the cell above and the cell below.

9.10 Cytokinesis Differs in Animal and Plant Cells Plant cells must divide differently from animal cells because they have cell walls. (*a*) A sea urchin egg that has just completed cytokinesis at the end of the first cell division of its development into an embryo. (*b*) A dividing plant cell in late telophase.

motor protein kinesin, these vesicles fuse to form new plasma membrane and contribute their contents to a *cell plate*, which is the beginning of a new cell wall (Figure 9.10*b*).

Following cytokinesis, both daughter cells contain all the components of a complete cell. A precise distribution of chromosomes is ensured by mitosis. Organelles such as ribosomes, mitochondria, and chloroplasts need not be distributed equally between daughter cells as long as some of each are present in both cells; accordingly, there is no mechanism with a precision comparable to that of mitosis to provide for their equal allocation to daughter cells. As we will see in Part Three of this book, during development the unequal distribution of cytoplasmic components can have functional significance for the two new cells.

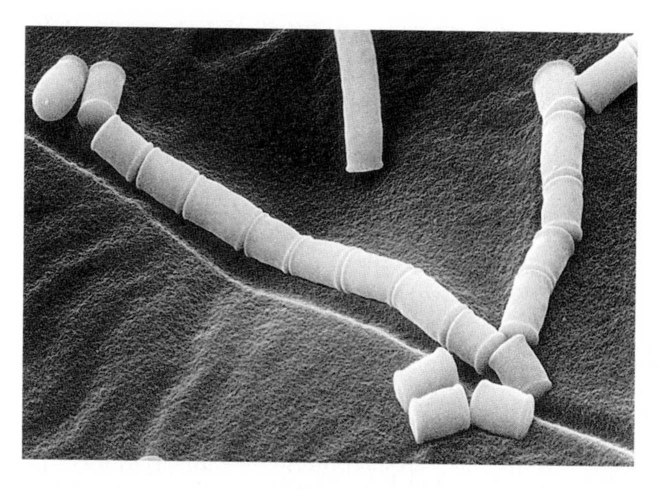

9.11 Asexual Reproduction These spool-shaped cells are asexual spores formed by a fungus. Each spore contains a nucleus produced by a mitotic division. A spore is the same genetically as the parent that fragmented to produce it.

Reproduction: Asexual and Sexual

The mitotic cell cycle repeats itself over and over. By this process, a single cell can give rise to a vast number of other cells. Meiosis, on the other hand, results in only four progeny cells, which usually do not undergo further duplications. These two methods of nuclear and cell division are both involved in reproduction, but they have different reproductive roles.

Reproduction by mitosis results in genetic constancy

A cell undergoing mitosis may be an entire single-celled organism reproducing itself with each cell cycle. Alternatively, it may be a cell produced by a multicellular organism that divides further to produce a new multicellular organism. Some multicellular organisms can reproduce themselves by releasing cells derived from mitosis and cytokinesis or by having a multicellular piece break away and grow on its own (Figure 9.11).

Asexual reproduction, sometimes called *vegetative reproduction*, is based on mitotic division of the nucleus. Accordingly, it produces a *clone* of offspring that are genetically identical to the parent. If there is any variation among the offspring, it is likely to be due to *mutations*, or changes, in the genetic material. Asexual reproduction is a rapid and effective means of making new individuals, and it is common in nature.

Sexual reproduction, which involves meiosis, is very different. In sexual reproduction, two parents each contribute one cell, a gamete, to their offspring. This method produces offspring that differ genetically from each parent as well as from one another. This genetic variation among the offspring means that some of them may be better adapted than others to survive and reproduce in a particular environment. Thus this genetic diversity provides the raw material for natural selection and evolution.

Reproduction by meiosis results in genetic diversity

Sexual reproduction, which combines genetic information from two different cells, generates genetic diversity. All sexual life cycles have certain hallmarks:

- There are two parents, each of which provides chromosomes to the offspring in the form of a gamete produced by meiosis.
- Each gamete contains a single set of chromosomes.
- The two gametes—often identifiable as a female egg and a male sperm—fuse to produce a single cell, the **zygote**, or fertilized egg. The zygote thus contains two sets of chromosomes.

In multicellular organisms, **somatic cells**—those body cells that are *not* specialized for reproduction—each contain two sets of chromosomes, which are found in pairs (see Figure 9.13). One chromosome of each pair comes from each of the organism's two parents. The members of such a **homologous pair** are similar in size and appearance (except for the sex chromosomes found in some species, as we will see in Chapter 10). The two chromosomes (the homologs) of a homologous pair bear corresponding, though generally not identical, genetic information.

Gametes, on the other hand, contain only a single set of chromosomes—that is, one homolog from each pair. The number of chromosomes in such a cell is denoted by n, and the cell is said to be **haploid**. Two haploid gametes fuse to form a new organism in a process called **fertilization**. The resulting zygote thus has two sets of chromosomes, just as somatic cells do. Its chromosome number is denoted by $2n$, and the zygote is said to be **diploid**.

As you can see in Figure 9.12, different kinds of sexual life cycles exhibit different patterns of development after zygote formation:

- In *haplontic* organisms, such as protists and many fungi, the tiny zygote is the only diploid cell in the life cycle; the mature organism is haploid. The zygote undergoes meiosis to produce haploid cells, or **spores**. These spores form the new organism, which may be single-celled or multicellular, by mitosis. The mature haploid organism produces gametes by mitosis, which fuse to form the diploid zygote.
- Most plants and some protists have *alternation of generations*. Here, meiosis does not give rise to gametes but to haploid spores. The spores divide by mitosis to form an alternate, haploid life stage (the *gametophyte*). It is this haploid stage that forms gametes by mitosis. The gametes fuse to form a diploid zygote, which divides by mitosis to become the diploid *sporophyte*.
- In *diplontic* organisms, including animals and some plants, the gametes are the only haploid cells in the life cycle, and the mature organism is diploid. Gametes are formed by meiosis, which fuse to form a diploid zygote. The organism is formed by mitosis of diploid cells.

We will describe all these life cycles in greater detail in Part Five. In this chapter, our focus is on the role of sexual reproduction in generating diversity among individual organisms.

The essence of sexual reproduction is the random selection of half of a parent's diploid chromosome set to make a haploid gamete, followed by the fusion of two such haploid gametes to produce a diploid cell that contains genetic information from both gametes. Both of these steps contribute to a shuffling of genetic information in the population, so that no two individuals have exactly the same genetic constitution. The diversity provided by sexual reproduction opens up enormous opportunities for evolution.

The number, shapes, and sizes of the metaphase chromosomes constitute the karyotype

When nuclei are in metaphase of mitosis, it is often possible to count and characterize the individual chromosomes. This is a relatively simple process in some organisms, thanks to

9.12 Fertilization and Meiosis Alternate in Sexual Reproduction In sexual reproduction, haploid (*n*; peach) cells or organisms alternate with diploid (2*n*; green) cells or organisms.

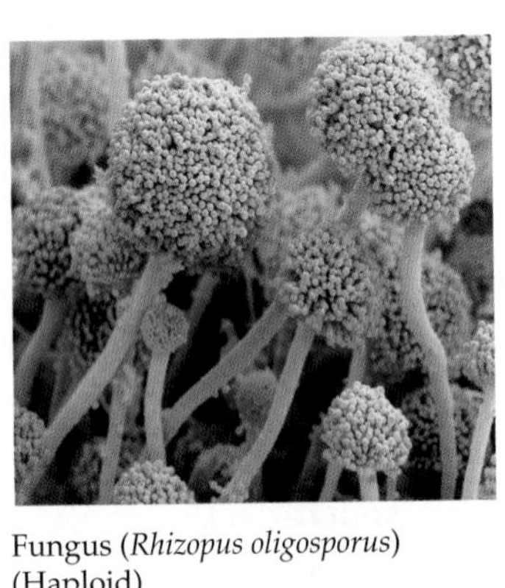

Fungus (*Rhizopus oligosporus*)
(Haploid)

Fern (*Osmunda cinnamomcea*)
(Diploid sporophyte)

Elephant (*Loxodonta africana*)
(Diploid)

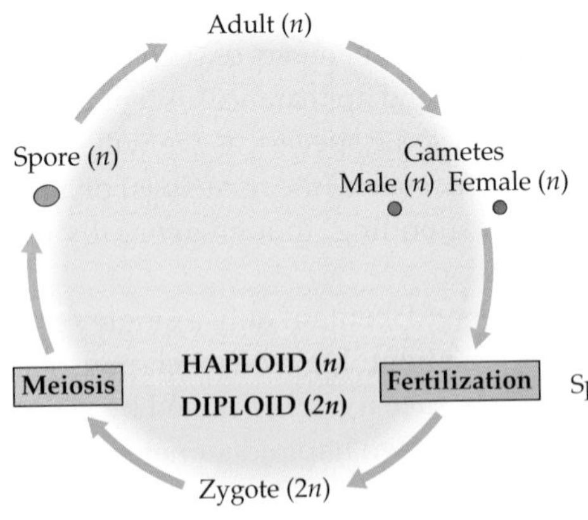

In the **haplontic life cycle**, the organism is haploid and the zygote is the only diploid stage.

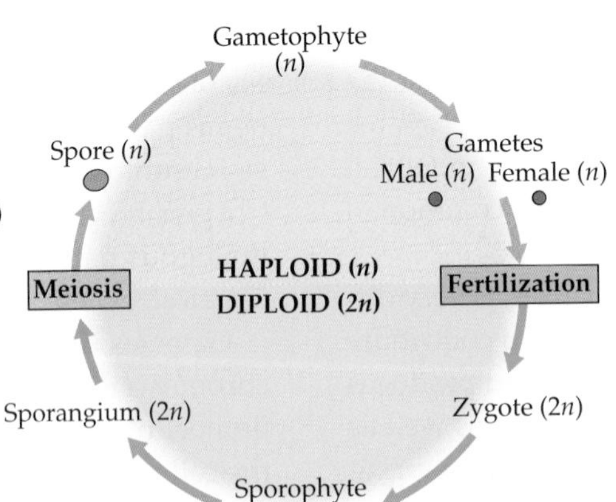

In **alternation of generations**, the organism passes through both haploid and diploid stages.

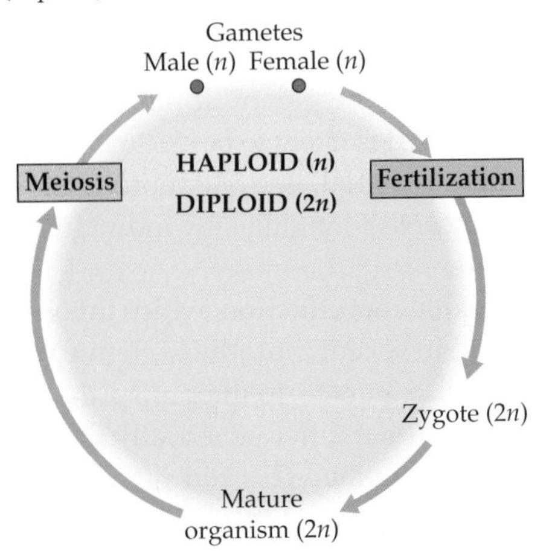

In the **diplontic life cycle**, the organism is diploid and the gametes are the only haploid stage.

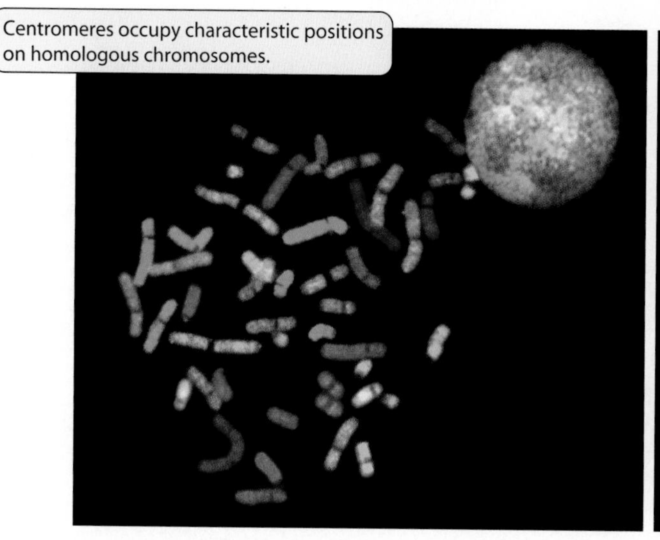
Centromeres occupy characteristic positions on homologous chromosomes.

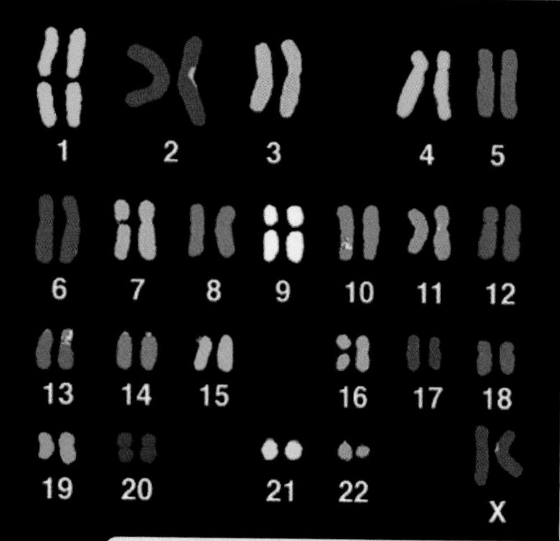
1
2
3
4
5
6
7
8
9
10
11
12
13
14
15
16
17
18
19
20
21
22
X
The karyotype shows 23 pairs of chromosomes, including the sex chromosomes. This female's sex chromosomes are X and X; a male would have X and Y chromosomes.

9.13 Human Cells Have 46 Chromosomes Chromosomes from a human cell are shown in metaphase of mitosis. The DNA of each chromosome has a specific nucleotide sequence that is stained by a specific colored dye, so that each homologous pair shares a distinctive color. Each chromosome at this stage is composed of two chromatids, but these cannot be distinguished by this "chromosome painting" technique. The multicolored globe is an interphase nucleus. The karyotype on the right is produced by computerized analysis of the image on the left.

techniques that can capture cells in metaphase and spread out the chromosomes. A photograph of the entire set of chromosomes can then be made, and the images of the individual chromosomes can be placed in an orderly arrangement. Such a rearranged photograph reveals the number, shapes, and sizes of chromosomes in a cell, which together constitute its **karyotype** (Figure 9.13).

Individual chromosomes can be recognized by their lengths, the positions of their centromeres, and characteristic banding when they are stained and observed at high magnification. When the cell is diploid, the karyotype consists of homologous pairs of chromosomes—23 pairs for a total of 46 chromosomes in humans, and greater or smaller numbers of pairs in other diploid species. There is no simple relationship between the size of an organism and its chromosome number (Table 9.1).

Meiosis: A Pair of Nuclear Divisions

Meiosis consists of two nuclear divisions that reduce the number of chromosomes to the haploid number in preparation for sexual reproduction. Although the *nucleus divides twice* during meiosis, the *DNA is replicated only once*. Unlike the products of mitosis, the products of meiosis are different from one an other and from the parent cell. To understand the process of meiosis and its specific details, it is useful to keep in mind the overall functions of meiosis:

- To reduce the chromosome number from diploid to haploid
- To ensure that each of the haploid products has a complete set of chromosomes
- To promote genetic diversity among the products

The first meiotic division reduces the chromosome number

Two unique features characterize the first of the two meiotic divisions, **meiosis I**. The first is that homologous chromosomes pair along their entire lengths. No such pairing occurs in mitosis. The second is that after metaphase I, the homologous chromosomes separate. The individual chromosomes, each consisting of two sister chromatids, remain intact until the end of metaphase II in the second meiotic division. Throughout this discussion, refer to Figure 9.14 on the two following pages to help you visualize each step.

9.1 Numbers of Pairs of Chromosomes in Some Plant and Animal Species

COMMON NAME	SPECIES	NUMBER OF CHROMOSOME PAIRS
Mosquito	*Culex pipiens*	3
Housefly	*Musca domestica*	6
Toad	*Bufo americanus*	11
Rice	*Oryza sativa*	12
Frog	*Rana pipiens*	13
Alligator	*Alligator mississippiensis*	16
Rhesus monkey	*Macaca mulatta*	21
Wheat	*Triticum aestivum*	21
Human	*Homo sapiens*	23
Potato	*Solanum tuberosum*	24
Donkey	*Equus asinus*	31
Horse	*Equus caballus*	32
Dog	*Canis familiaris*	39
Carp	*Cyprinus carpio*	52

MEIOSIS I

Early Prophase I

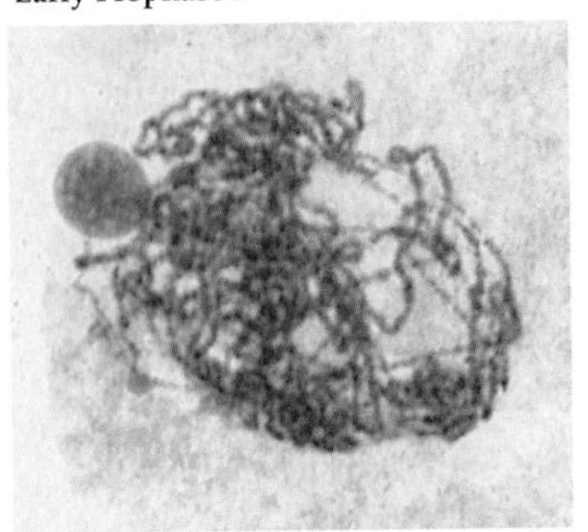

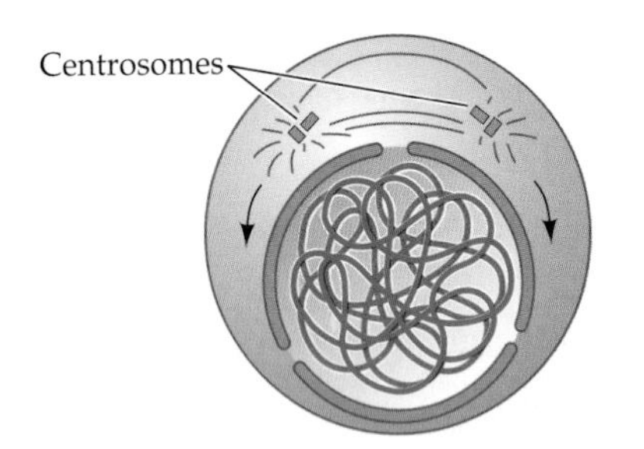

1 The chromatin begins to condense following interphase.

Mid-Prophase I

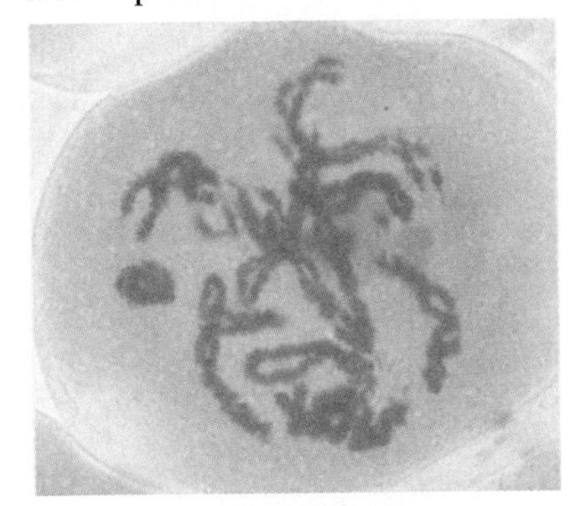

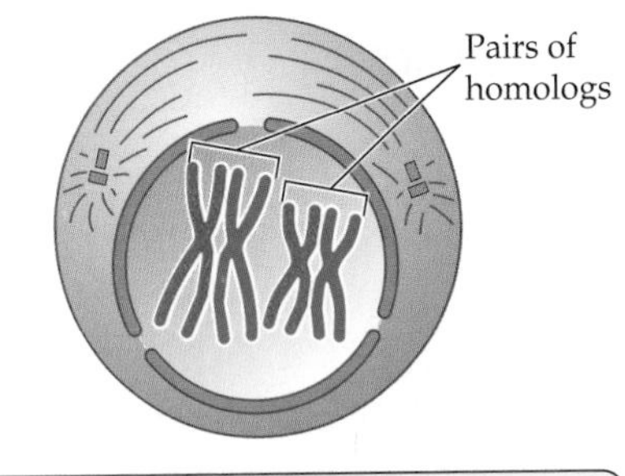

2 Synapsis aligns homologs, and chromosomes condense. Homologs are shown in different colors indicating those coming from each parent. In reality, their differences are very small, usually comprising different alleles of some genes.

Late Prophase I–Prometaphase

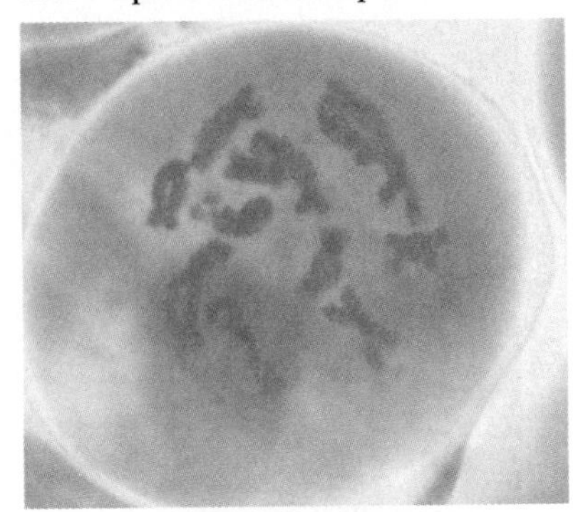

3 The chromosomes continue to coil and shorten. Crossing over results in an exchange of genetic material. In prometaphase the nuclear envelope breaks down.

MEIOSIS II

Prophase II

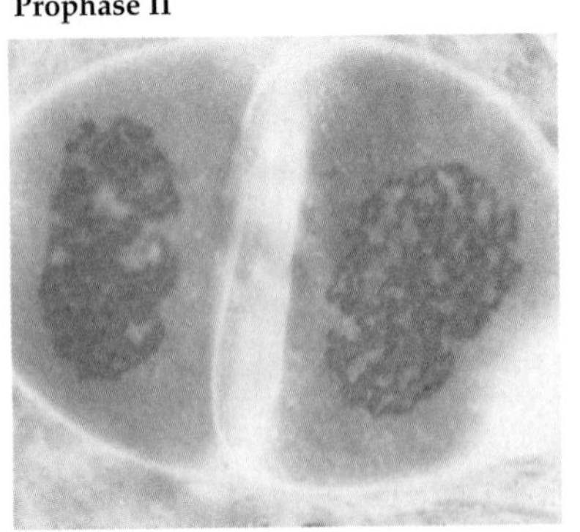

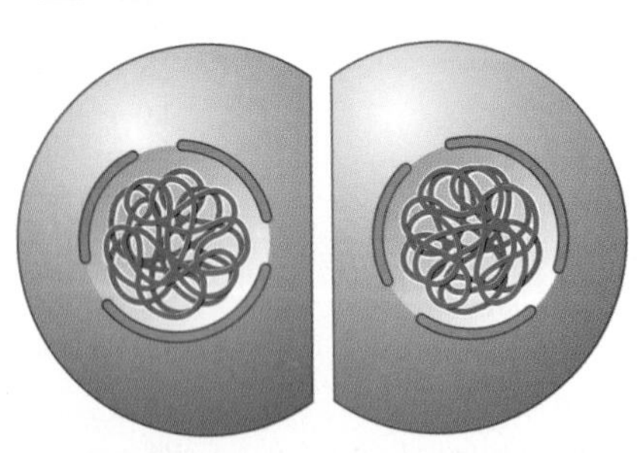

7 The chromosomes condense again, following a brief interphase (interkinesis) in which DNA does not replicate.

Metaphase II

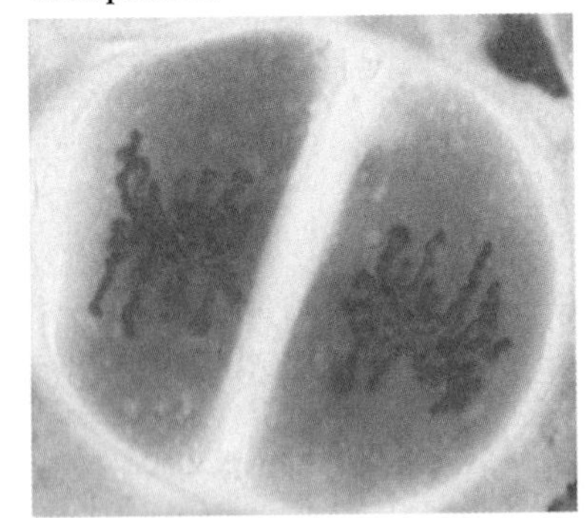

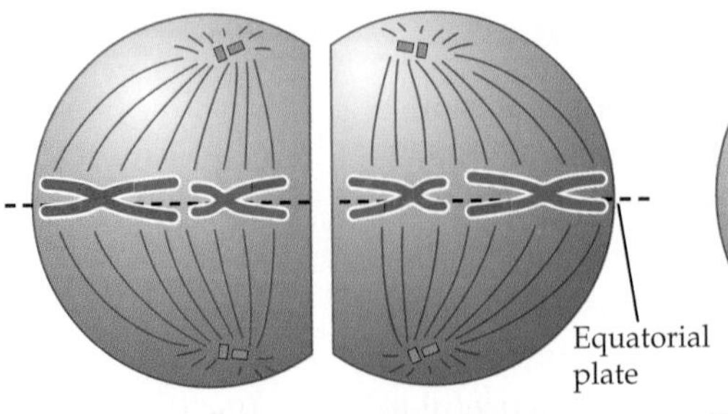

8 Kinetochores of the paired chromatids line up across the equatorial plates of each cell.

Anaphase II

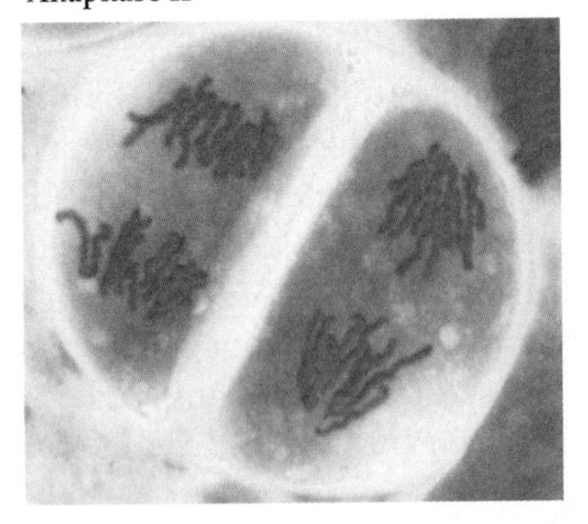

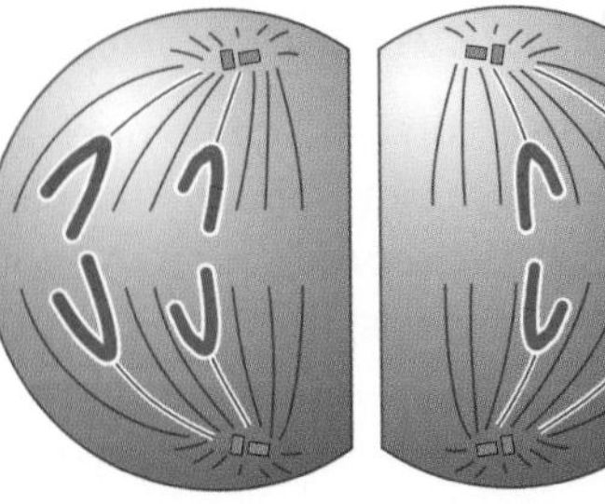

9 The chromatids finally separate, becoming chromosomes in their own right, and are pulled to opposite poles. Because of crossing over in prophase I, each new cell will have a different genetic makeup.

Metaphase I

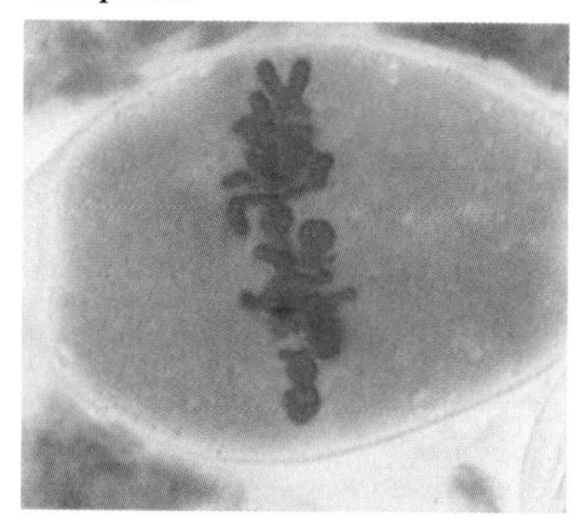

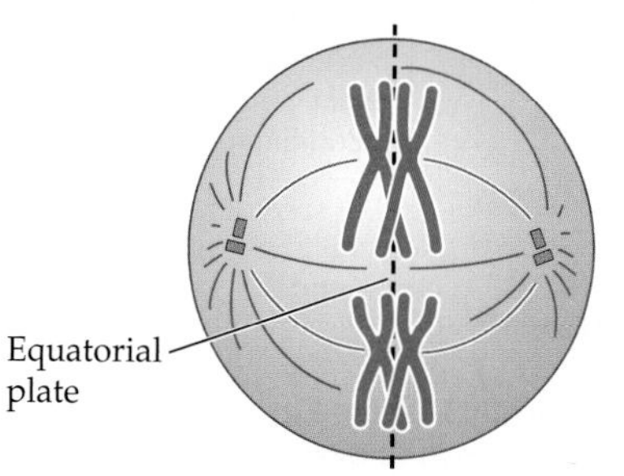

4 The homologous chromosomes line up on the equatorial (metaphase) plate.

Anaphase I

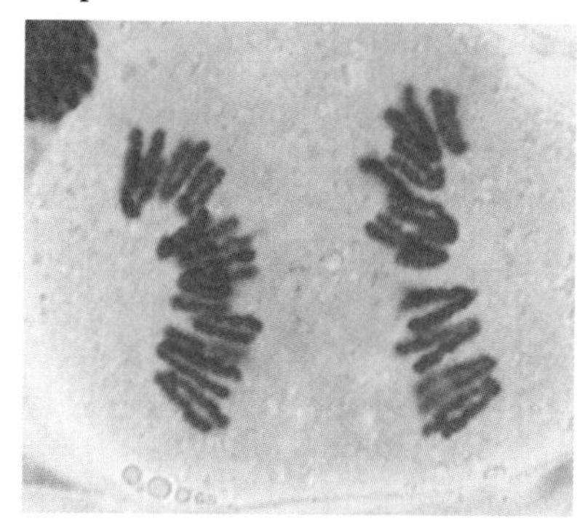

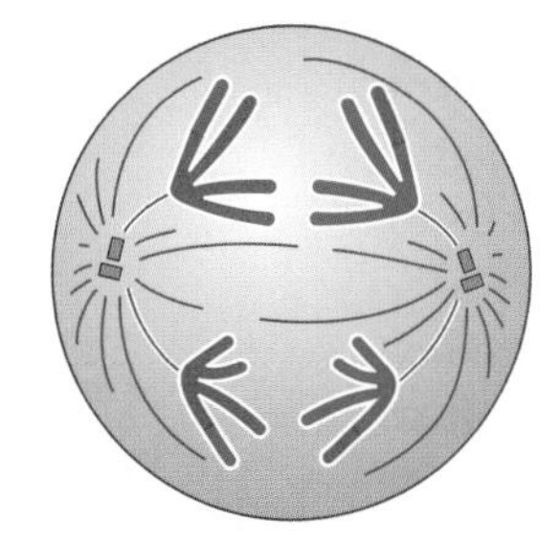

5 The homologous chromosomes (each with two chromatids) move to opposite poles of the cell.

Telophase I

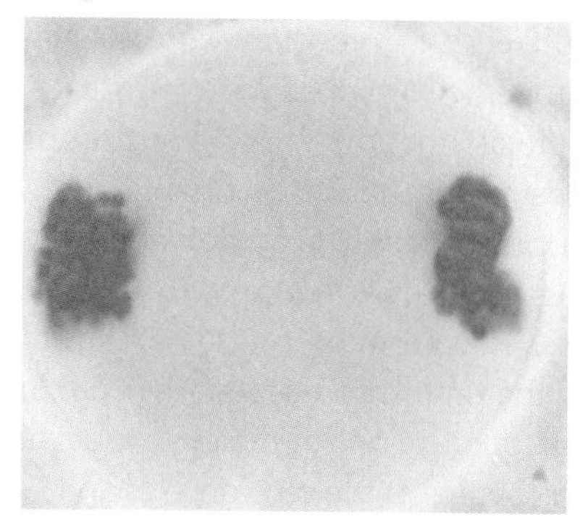

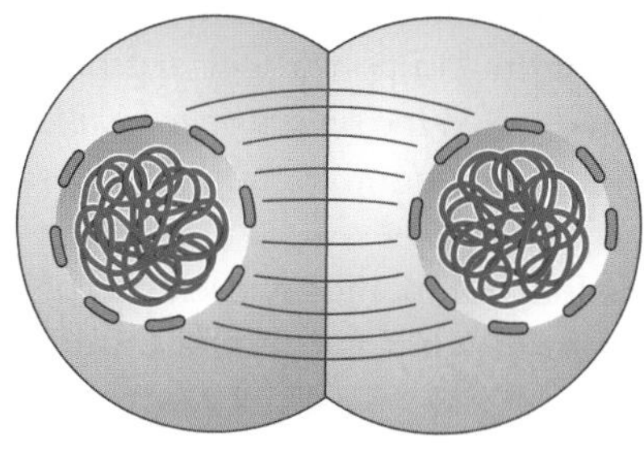

6 The chromosomes gather into nuclei, and the original cell divides.

Telophase II

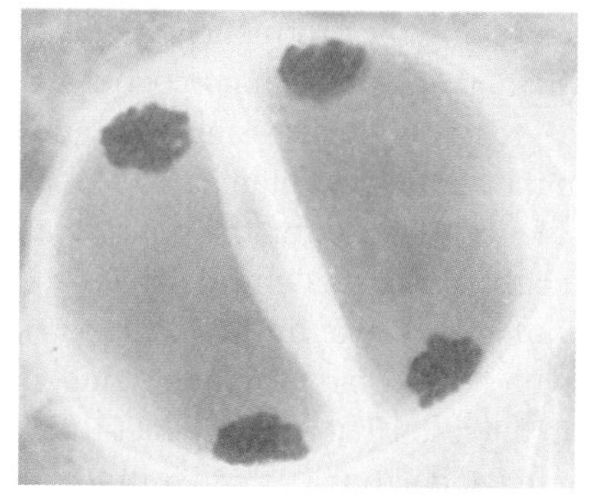

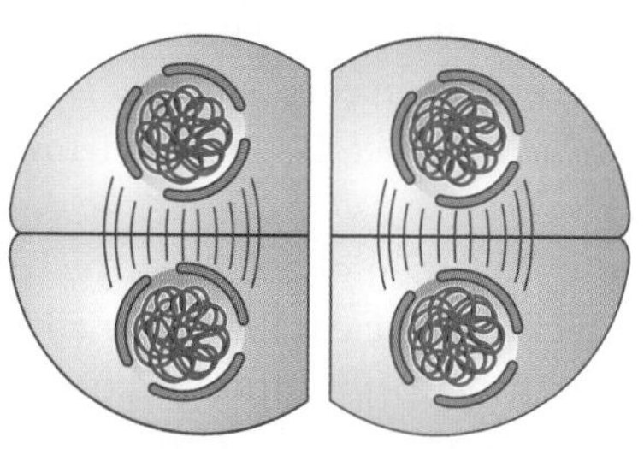

10 The chromosomes gather into nuclei, and the cells divide.

Products

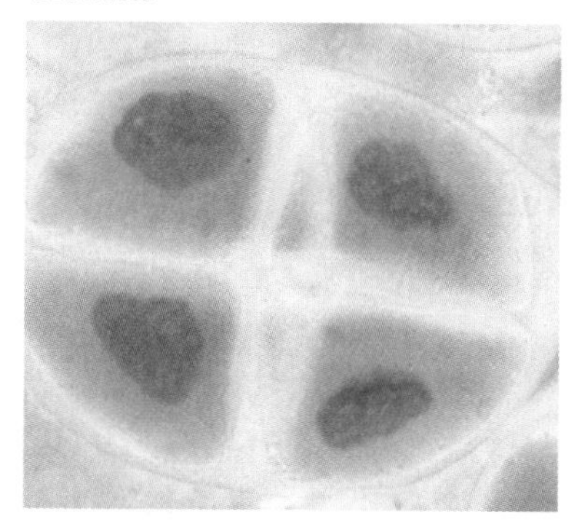

11 Each of the four cells has a nucleus with a haploid number of chromosomes.

9.14 Meiosis In meiosis, two sets of chromosomes are divided among four nuclei, each of which then has half as many chromosomes as the original cell. These four haploid cells are the result of two successive nuclear divisions. The photomicrographs show meiosis in the male reproductive organ of a lily. As in Figure 9.8, the diagrams show corresponding phases in an animal.

Like mitosis, meiosis I is preceded by an interphase with an S phase during which each chromosome is replicated. As a result, each chromosome consists of two sister chromatids, held together by cohesin proteins.

Meiosis I begins with a long prophase I (the first three frames of Figure 9.14), during which the chromosomes change markedly. The homologous chromosomes pair by adhering along their lengths, a process called **synapsis**. This process lasts from prophase I to the end of metaphase I.

By the time chromosomes can be clearly seen under light microscope, the two homologs are already tightly joined. This joining begins at the centromeres and is mediated by a recognition of homologous DNA sequences on homologous chromosomes. In addition, a special group of proteins may form a scaffold called the *synaptonemal complex*, which runs lengthwise along the homologous chromosomes and appears to join them together.

The four chromatids of each pair of homologous chromosomes form what is called a **tetrad**, or *bivalent*. In other words, a tetrad consists of four chromatids, two each from two homologous chromosomes. For example, there are 46 chromosomes in a human diploid cell at the beginning of meiosis, so there are 23 homologous pairs of chromosomes, each with two chromatids (that is, 23 tetrads), for a total of 92 chromatids during prophase I.

Throughout prophase I and metaphase I, the chromatin continues to coil and compact, so that the chromosomes appear ever thicker. At a certain point, the homologous chromosomes seem to repel each other, especially near the centromeres, but they are held together by physical attachments mediated by cohesins. These cohesins are different from the ones holding the two sister chromatids together. Regions having these attachments take on an X-shaped appearance and are called **chiasmata** (from the Greek *chiasma*, "cross"; Figure 9.15).

A chiasma reflects an exchange of genetic material between nonsister chromatids on homologous chromosomes—what geneticists call **crossing over** (Figure 9.16). The chromosomes begin exchanging material shortly after synapsis begins, but chiasmata do not become visible until later, when the homologs are repelling each other. Crossing over increases genetic variation among the products of meiosis by reshuffling genetic information among the homologous pairs. We will have a great deal to say about crossing over and its genetic consequences in the coming chapters.

There seems to be plenty of time for the complicated events of prophase I to occur. Whereas mitotic prophase is usually measured in minutes, and all of mitosis seldom takes more than an hour or two, meiosis can take much longer. In human males, the cells in the testis that undergo meiosis take about a week for prophase I and about a month for the entire meiotic cycle. In the cells that will become eggs, prophase I begins long before a woman's birth, during her early fetal development, and ends as much as decades later, during the monthly ovarian cycle.

Prophase I is followed by prometaphase I (not pictured in Figure 9.14), during which the nuclear envelope and the nucleoli disaggregate. A spindle forms, and microtubules become attached to the kinetochores of the chromosomes. In meiosis I, the kinetochores of both chromatids in each chromosome become attached to the same half-spindle. Thus the entire chromosome, consisting of two chromatids, will migrate to one pole. Which member of a homologous chromosome pair becomes attached to each half-spindle,

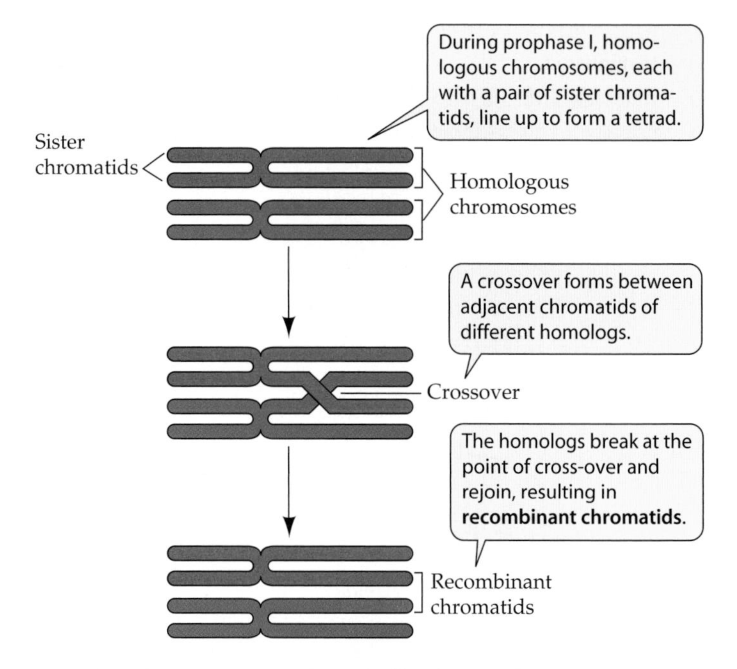

9.16 Crossing Over Forms Genetically Diverse Chromosomes The exchange of genetic material by crossing over may result in new combinations of genetic information on the recombinant chromosomes.

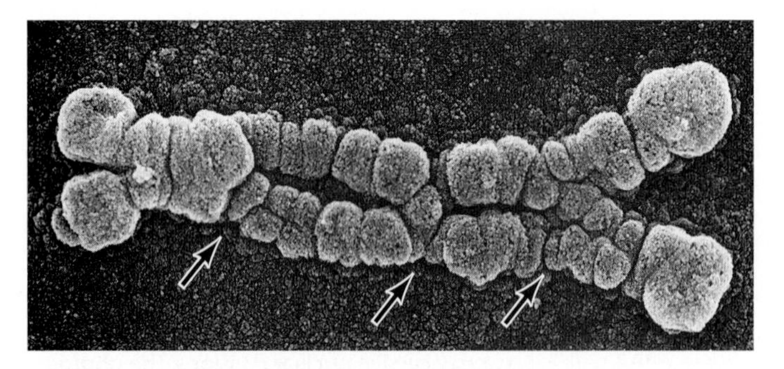

9.15 Chiasmata: Evidence of Exchange between Chromatids Chiasmata are visible near the middle of this scanning electron micrograph of some chromatids from a desert locust, and near the ends of others. Three chiasmata are indicated with arrows.

and thus which member will go to which pole, is random. By metaphase I, all the chromosomes have moved to the equatorial plate. Up to this point, homologous pairs are held together by chiasmata.

The homologous chromosomes separate in anaphase I, when the individual chromosomes, each still consisting of two chromatids, are pulled to the poles, with one homolog of a pair going to one pole and the other homolog going to the opposite pole. (Note that this process differs from the separation of chromatids during mitotic anaphase.) Each of the two daughter nuclei from this division thus contains only one set of chromosomes, not the two sets that were present in the original diploid nucleus. However, because they consist of two chromatids rather than just one, each of these chromosomes has twice the mass that a chromosome at the end of a mitotic division has.

In some organisms, there is a telophase I, with the reappearance of the nuclear envelopes. When there is a telophase I, it is followed by an interphase, called **interkinesis**, similar to the mitotic interphase. During interkinesis the chromatin is partially uncoiled; however, there is no replication of the genetic material, because each chromosome already consists of two chromatids. Furthermore, the sister chromatids in interkinesis are generally not genetically identical, because crossing over in prophase I has reshuffled genetic material between the maternal and paternal chromosomes. In other organisms, the chromosomes move directly into the second meiotic division.

The second meiotic division separates the chromatids

Meiosis II is similar to mitosis in many ways. In each nucleus produced by meiosis I, the chromosomes line up at equatorial plate at metaphase II. The centromeres of the sister chromatids separate because of cohesin breakdown, and the daughter chromosomes move to the poles in anaphase II.

The three major differences between meiosis II and mitosis are:

- DNA replicates before mitosis, but not before meiosis II.
- In mitosis, the sister chromatids that make up a given chromosome are identical. In meiosis II, they may differ over part of their length if they participated in crossing over during prophase I.
- The number of chromosomes on the equatorial plate in meiosis II is half the number in the mitotic nucleus.

The result of meiosis is four nuclei; each nucleus is haploid and has a single set of unreplicated chromosomes that differs from other such sets in its exact genetic composition. The differences among the haploid nuclei result from crossing over during prophase I and from the random segregation of homologous chromosomes during anaphase I.

Meiosis leads to genetic diversity

What are the consequences of the synapsis and segregation of homologous chromosomes during meiosis? In mitosis, each chromosome behaves independently of its homolog; its two chromatids are sent to opposite poles at anaphase. If a mitotic division begins with x chromosomes, we end up with x chromosomes in each daughter nucleus, and each chromosome consists of one chromatid. Each of the two sets of chromosomes (one of paternal and one of maternal origin) is divided equally and distributed equally to each daughter cell. In meiosis, things are very different.

In meiosis, chromosomes of maternal origin pair with their paternal homologs during synapsis. Separation of the homologs during meiotic anaphase I ensures that each pole receives one member of each homologous pair. For example, at the end of meiosis I in humans, each daughter nucleus contains 23 of the original 46 chromosomes. In this way, the chromosome number is decreased from diploid to haploid. Furthermore, meiosis I guarantees that each daughter nucleus gets one full set of chromosomes.

The products of meiosis I are genetically diverse for two reasons:

- Synapsis during prophase I allows the maternal chromosome in each homologous pair to exchange segments with the paternal one by crossing over. The resulting *recombinant* chromatids contain some genetic material from each parent.
- It is a matter of chance which member of a homologous pair goes to which daughter cell at anaphase I. For example, if there are two homologous pairs of chromosomes in the diploid parent nucleus, a particular daughter nucleus could get paternal chromosome 1 and maternal chromosome 2, or paternal 2 and maternal 1, or both maternal, or both paternal. It all depends on the way in which the homologous pairs line up at metaphase I. This phenomenon is termed **independent assortment**.

Note that of the four possible chromosome combinations just described, two produce daughter nuclei that are the same as one of the parental types (except for any material exchanged by crossing over). The greater the number of chromosomes, the less probable that the original parental combinations will be reestablished, and the greater the potential for genetic diversity. Most species of diploid organisms do indeed have more than two pairs of chromosomes. In humans, with 23 chromosome pairs, 2^{23} (8,388,608) different combinations can be produced, just by the mechanism of independent assortment. Taking the extra genetic shuffling afforded by crossing over into account, the number of possible combinations is virtually infinite.

Figure 9.17 compares meiosis with mitosis.

Meiotic Errors

In the complex process of cell division, things occasionally go wrong. A pair of homologous chromosomes may fail to separate during meiosis I, or sister chromatids may fail to separate during meiosis II or during mitosis. This phenomenon is called **nondisjunction**. Conversely, homologous chromosomes may fail to remain together. These problems can result in the production of aneuploid cells. **Aneuploidy** is a condition in which one or more chromosomes are either lacking or present in excess.

Aneuploidy can give rise to genetic abnormalities

One reason for aneuploidy may be a lack of cohesins. Recall that these molecules, formed during prophase I, hold the two homologous chromosomes together into metaphase I. They ensure that when the chromosomes line up at the equatorial plate, one homolog will face one pole and the other homolog will face the other pole. Without this "glue," the two homologs may line up randomly at metaphase I, just like chromosomes during mitosis, and there is a 50 percent chance that both will go to the same pole.

If, for example, during the formation of a human egg, both members of the chromosome 21 pair go to the same pole during anaphase I, the resulting eggs will contain either two of chromosome 21 or none at all. If an egg with two of these chromosomes is fertilized by a normal sperm, the resulting zygote will have three copies of the chromosome: it will be **trisomic** for chromosome 21. A child with an extra chromosome 21 demonstrates the symptoms of Down syndrome: impaired intelligence; characteristic abnormalities of the hands, tongue, and eyelids; and an increased susceptibility to cardiac abnormalities and diseases such as leukemia. If an egg that did not receive chromosome 21 is fertilized by a normal sperm, the zygote will have only one copy: it will be **monosomic** for chromosome 21 (Figure 9.18).

Other abnormal chromosomal events can also occur. In a process called **translocation**, a piece of a chromosome may

MITOSIS

Parent cell (2n)

Prophase
No synapsis of homologous chromosomes

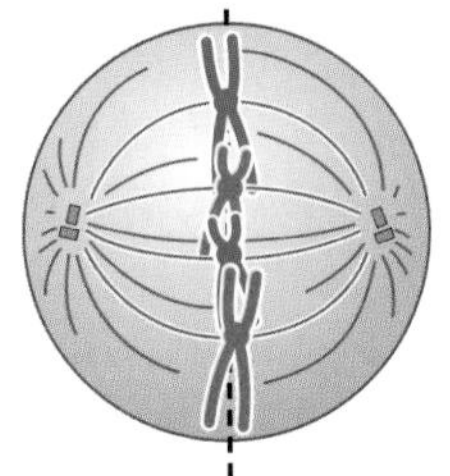
Metaphase
Individual chromosomes align at the equatorial plate.

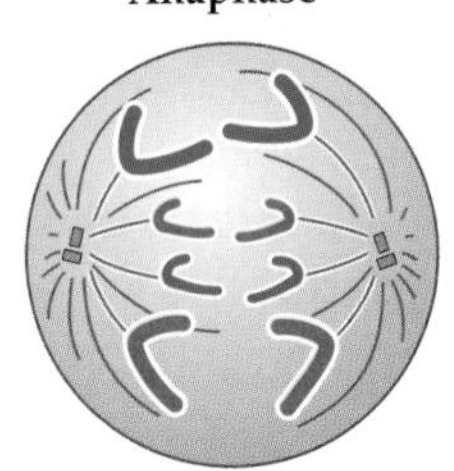
Anaphase
Centromeres separate. Sister chromatids separate during anaphase, becoming daughter chromosomes.

MEIOSIS

Parent cell (2n)

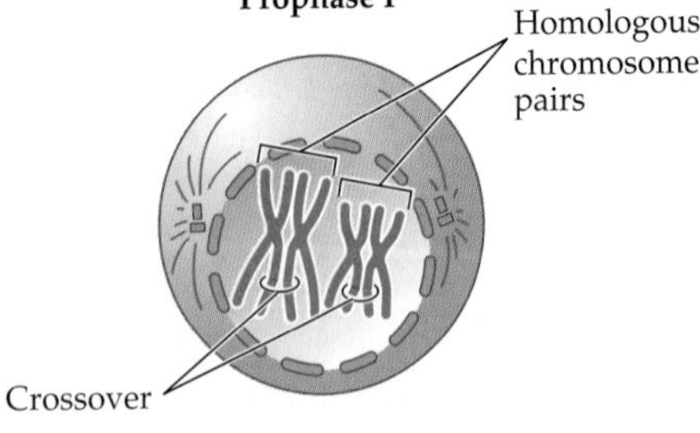
Prophase I
Homologous chromosome pairs
Crossover
Synapsis and crossing over of homologs

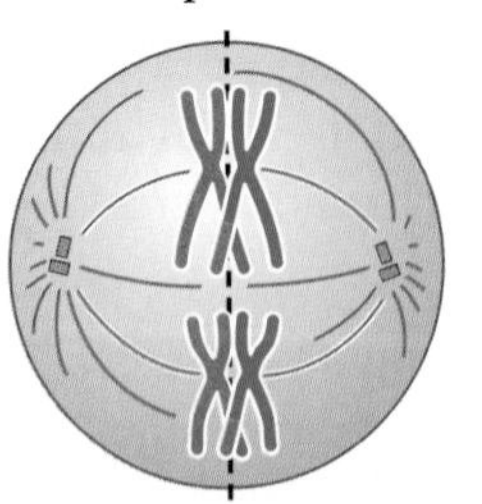
Metaphase I
Homologous pairs align at the equatorial plate.

Anaphase I
Centromeres do not separate; sister chromatids remain together during anaphase; homologs separate; DNA does not replicate before subsequent prophase.

9.17 Mitosis and Meiosis: A Comparison Meiosis differs from mitosis by synapsis and by the failure of the centromeres to separate at the end of metaphase I.

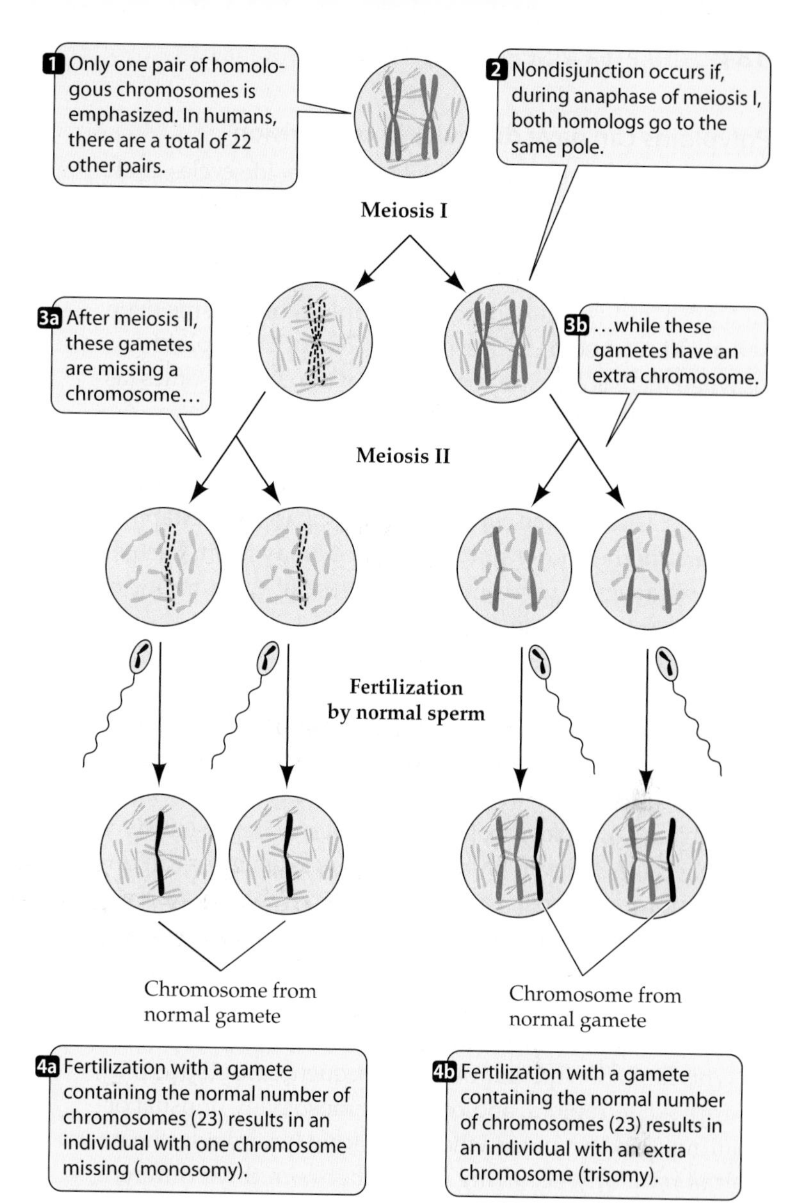

1 Only one pair of homologous chromosomes is emphasized. In humans, there are a total of 22 other pairs.
2 Nondisjunction occurs if, during anaphase of meiosis I, both homologs go to the same pole.
Meiosis I
3a After meiosis II, these gametes are missing a chromosome…
3b …while these gametes have an extra chromosome.
Meiosis II
Fertilization by normal sperm
Chromosome from normal gamete
Chromosome from normal gamete
4a Fertilization with a gamete containing the normal number of chromosomes (23) results in an individual with one chromosome missing (monosomy).
4b Fertilization with a gamete containing the normal number of chromosomes (23) results in an individual with an extra chromosome (trisomy).

9.18 Nondisjunction Leads to Aneuploidy Nondisjunction occurs if homologous chromosomes fail to separate during meiosis I. The result is aneuploidy: One or more chromosomes are either lacking or present in excess.

break away and become attached to another chromosome. For example, a particular large part of one chromosome 21 may be translocated to another chromosome. Individuals who inherit this translocated piece along with two normal chromosomes 21 will have Down syndrome.

Trisomies (and the corresponding monosomies) are surprisingly common in human zygotes, with 10–30 percent of all conceptions showing aneuploidy. But most of the embryos that develop from such zygotes do not survive to birth, and those that do often die before the age of 1 year. Trisomies and monosomies for most chromosomes other than chromosome 21 are lethal to the embryo. At least one-fifth of all recognized pregnancies are spontaneously terminated during the first 2 months, largely because of such trisomies and monosomies. (The actual proportion of spontaneously terminated pregnancies is certainly higher, because the earliest ones often go unrecognized.)

Two daughter cells (each 2*n*)

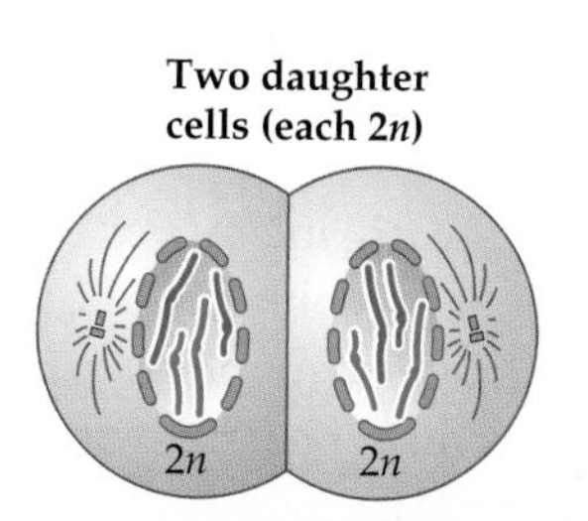

Mitosis is a mechanism for constancy: The parent nucleus produces two identical daughter nuclei.

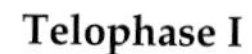

Telophase I

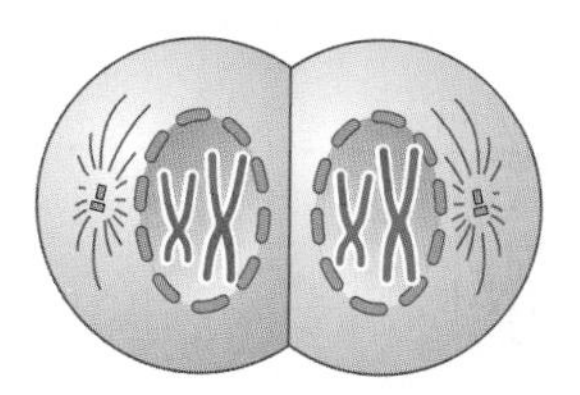

Metaphase II

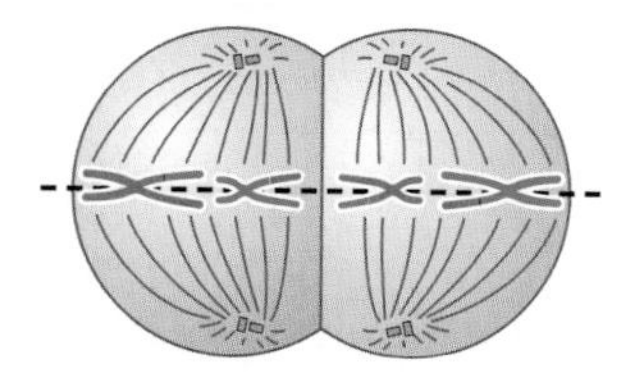

Four daughter cells (each *n*)

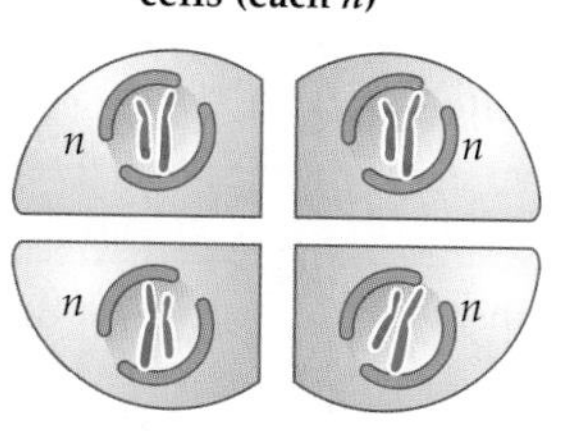

Chromatids separate.

Meiosis is a mechanism for diversity: The parent nucleus produces four different haploid daughter nuclei.

Polyploids can have difficulty in cell division

As we saw earlier in our discussion of sexual life cycles, both diploid and haploid nuclei can divide by mitosis. Multicellular diploid and multicellular haploid individuals both develop from single-celled beginnings by mitotic divisions. Likewise, mitosis may proceed in diploid organisms even when a chromosome is missing from one of the haploid sets or when there is an extra copy of one of the chromosomes (as in people with Down syndrome).

Organisms with complete extra sets of chromosomes may sometimes be produced by artificial breeding or by natural accidents. Under some circumstances, triploid ($3n$), tetraploid ($4n$), and higher-order **polyploid** nuclei may form. Each of these *ploidy levels* represents an increase in the number of complete sets of chromosomes present.

If a nucleus has one or more extra full sets of chromosomes, its abnormally high ploidy in itself does not prevent mitosis. In mitosis, each chromosome behaves independently of the others. In meiosis, by contrast, homologous chromosomes must synapse to begin division. If even one chromosome has no homolog, anaphase I cannot send representatives of that chromosome to both poles. A diploid nucleus can undergo normal meiosis; a haploid one cannot. Similarly, a tetraploid nucleus has an even number of each kind of chromosome, so each chromosome can pair with its homolog. But a triploid nucleus cannot undergo normal meiosis, because one-third of the chromosomes would lack partners.

This limitation has important consequences for the fertility of triploid, tetraploid, and other chromosomally unusual organisms. Modern bread wheat plants are hexaploids, the result of naturally occurring crosses between three different grasses, each having its own diploid set of 14 chromosomes. Over a period of 10,000 years, humans have selected favorable varieties of these hybrids to produce modern wheat strains.

Cell Death

As we mentioned at the start of this chapter, an essential role of cell division in complex eukaryotes is to replace cells that die. In humans, billions of cells die each day, mainly in the blood and in the epithelia lining organs such as the intestine. Cells die in one of two ways. The first, **necrosis**, occurs when cells either are damaged by poisons or are starved of essential nutrients. These cells usually swell up and burst, releasing their contents into the extracellular environment. This process often results in inflammation (see Chapter 18). The scab that forms around a wound is a familiar example of necrotic tissue. More typically, cell death in an organism is due to **apoptosis** (from the Greek, "falling off"). Apoptosis is a genetically programmed series of events that result in cell death. These two modes of cell death are compared in Table 9.2.

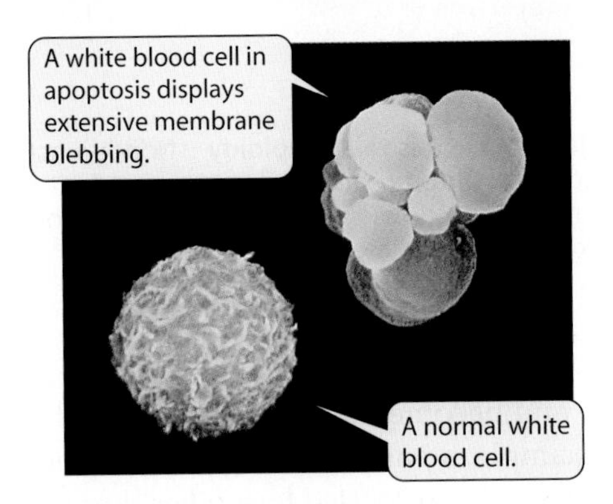

A white blood cell in apoptosis displays extensive membrane blebbing.
A normal white blood cell.

9.19 Apoptosis: Programmed Cell Death Many cells are genetically programmed to "self-destruct" when they are no longer needed, or when they have lived long enough to accumulate a burden of DNA damage that might harm the organism.

Why would a cell initiate apoptosis, which is essentially "cell suicide"? There are two possible reasons:

- The cell is no longer needed by the organism. For example, before birth, a human fetus has weblike hands, with connective tissue between the fingers. As development proceeds, this unneeded tissue disappears as its cells undergo apoptosis (see Figure 19.11).
- The longer cells live, the more prone they are to genetic damage that could lead to cancer. This is especially true of cells in the blood and intestine, which are exposed to high levels of toxic substances. Such cells normally die after only days or weeks.

9.2 ***Two Different Ways for Cells to Die***

	NECROSIS	APOPTOSIS
Stimuli	Low O_2, toxins, ATP depletion, damage	Specific, genetically programmed physiological signals
ATP required	No	Yes
Cellular pattern	Swelling, organelle disruption, tissue death	Chromatin condensation, membrane blebbing, single-cell death
DNA breakdown	Random fragments	Nucleosome-sized fragments
Plasma membrane	Burst	Blebbed (see Figure 9.19)
Fate of dead cells	Ingested by phagocytes	Ingested by neighboring cells
Reaction in tissue	Inflammation	No inflammation

Like the cell division cycle, the cell death cycle is controlled by signals, which may come either from inside or outside the cell. These signals include the lack of a mitotic signal (such as a growth factor), and the recognition of DNA damage. As we will see in Chapter 17, many of the drugs used to treat diseases of excess cell proliferation, such as cancer, work through these signals.

The events of apoptosis are very similar in most organisms. The cell becomes isolated from its neighbors, chops up its chromatin into nucleosome-sized pieces, and then fragments itself (Figure 9.19). In a remarkable example of the economy of nature, the surrounding living cells usually ingest the remains of the dead cell. The genetic signals that lead to apoptosis are also common to many organisms.

Chapter Summary

Systems of Cell Reproduction

▶ Cell division is necessary for the reproduction, growth, and repair of an organism. **Review Figure 9.1**

▶ Cell division must be initiated by a reproductive signal. Cell division consists of three steps: replication of the genetic material (DNA), segregation of the two DNA molecules to separate portions of the cell, and cytokinesis, or division of the cytoplasm.

▶ In prokaryotes, cellular DNA is a single molecule, or chromosome. Prokaryotes reproduce by cell fission. **Review Figure 9.2**

▶ In eukaryotes, cells divide by either mitosis or meiosis.

Interphase and the Control of Cell Division

▶ The mitotic cell cycle has two main phases: interphase (during which cells are not dividing) and mitosis (when cells are dividing).

▶ During most of the cell cycle, the cell is in interphase, which is divided into three subphases: S, G1, and G2. DNA is replicated during the S phase. **Review Figure 9.3**

▶ Cyclin-Cdk complexes regulate the passage of cells through checkpoints in the cell cycle. The most important one is the R point in G1, which determines whether the rest of the cycle will proceed. **Review Figure 9.4**

▶ In addition to the internal cyclin-Cdk complexes, controls external to the cell, such as growth factors and hormones, can also stimulate the cell to begin a division cycle.

Eukaryotic Chromosomes

▶ A eukaryotic chromosome contains a DNA molecule bound to proteins in a complex called chromatin. At mitosis, the replicated chromatids are held together at the centromere. Each chromatid consists of one double-stranded DNA molecule. **Review Figure 9.5**

▶ During interphase, the DNA in chromatin is wound around cores of histones to form nucleosomes. DNA folds over and over again, packing itself within the nucleus. During mitosis or meiosis, it folds even more. **Review Figure 9.6**

Mitosis: Distributing Exact Copies of Genetic Information

▶ After DNA is replicated during the S phase, the first sign of mitosis is the separation of the replicated centrosomes, which initiate microtubule formation for the spindle.

▶ Mitosis can be divided into several phases, called prophase, prometaphase, metaphase, anaphase, and telophase.

▶ During prophase, the chromosomes condense and appear as paired chromatids, and the spindle forms. **Review Figure 9.7.** **See Web/CD Activity 9.1**

▶ During prometaphase, the chromosomes move toward the middle of the spindle. In metaphase, they gather at the middle of the cell with their centromeres on the equatorial plate. At the end of metaphase, the centromeres holding the sister chromatids together separate, and during anaphase, each chromatid, now called the daughter chromosome, migrates to its pole along the microtubule track. **Review Figure 9.8. See Web/CD Activity 9.2**

▶ Cohesin holds sister chromatids together from the time they are formed in DNA replication until the onset of anaphase. Separin hydrolyzes cohesin when an inhibitory subunit, securin, is hydrolyzed. **Review Figure 9.9**

▶ During telophase, the chromosomes become less condensed. The nuclear envelopes and nucleoli re-form, thus producing two nuclei whose chromosomes are identical to each other and to those of the cell that began the cycle.
See Web/CD Tutorial 9.1

Cytokinesis: The Division of the Cytoplasm

▶ Nuclear division is usually followed by cytokinesis. Animal cell cytoplasm usually divides by a furrowing of the plasma membrane, caused by the contraction of cytoplasmic microfilaments. In plant cells, cytokinesis is accomplished by vesicle fusion and the synthesis of new cell wall material. **Review Figure 9.10**

Reproduction: Asexual and Sexual

▶ The cell cycle can repeat itself many times, forming a clone of genetically identical cells.

▶ Asexual reproduction produces a new organism that is genetically identical to the parent. Any genetic variety is the result of mutations.

▶ In sexual reproduction, two haploid gametes—one from each parent—unite in fertilization to form a genetically unique, diploid zygote. **Review Figure 9.12. See Web/CD Activity 9.3**

▶ In sexually reproducing organisms, certain cells in the adult undergo meiosis, a process by which a diploid cell produces haploid gametes. Each gamete contains a random selection of one of each pair of homologous chromosomes from the parent.

▶ The number, shapes, and sizes of the chromosomes constitute the karyotype of an organism. **Review Figure 9.13**

Meiosis: A Pair of Nuclear Divisions

▶ Meiosis reduces the chromosome number from diploid to haploid, ensures that each haploid cell contains one member of each chromosome pair, and results in genetically diverse products. It consists of two nuclear divisions. **Review Figure 9.14. See Web/CD Activity 9.4**

▶ During prophase I of the first meiotic division, homologous chromosomes pair up with each other, and material may be exchanged between the two homologs by crossing over. In metaphase I, the paired homologs line up at the equatorial plate. **Review Figures 9.14, 9.16**

▶ In anaphase I, entire chromosomes, each with two chromatids, migrate to the poles. By the end of meiosis I, there are two nuclei, each with the haploid number of chromosomes. **Review Figures 9.14, 9.17**

▶ In meiosis II, the sister chromatids separate. No DNA replication precedes this division, which in other aspects is similar to mitosis. The result of meiosis is four cells, each with a haploid chromosome content. **Review Figures 9.14, 9.17**

▶ Both crossing over during prophase I and the random selection of which homolog of a pair migrates to which pole during anaphase I ensure that the genetic composition of each haploid gamete is different from that of the parent cell and from that of the other gametes. The more chromosome pairs there are in a diploid cell, the greater the diversity of chromosome combinations generated by meiosis. **Review Figure 9.16, Table 9.1. See Web/CD Tutorial 9.2**

Meiotic Errors

▶ In nondisjunction, one member of a homologous pair of chromosomes fails to separate from the other, and both go to the same pole. Pairs of homologous chromosomes may also fail to stick together when they should. These events may lead to one gamete with an extra chromosome and another lacking that chromosome.

▶ The union of a gamete with an abnormal chromosome number with a normal haploid gamete at fertilization results in aneuploidy and genetic abnormalities that are invariably harmful or lethal to the organism. **Review Figure 9.18**

▶ Polyploid organisms can have difficulty in cell division. Natural and artificially produced polyploids underlie modern agriculture.

Cell Death

▶ Cells may die by necrosis, or they may self-destruct by apoptosis, a genetically programmed series of events that includes the detachment of the cell from its neighbors and the fragmentation of its nuclear DNA. **Review Figure 9.19, Table 9.2**

Self-Quiz

1. Which statement about eukaryotic chromosomes is *not* true?
   a. They sometimes consist of two chromatids.
   b. They sometimes consist of a single chromatid.
   c. They normally possess a single centromere.
   d. They consist of proteins.
   e. They are clearly visible as defined bodies under the light microscope.
2. Nucleosomes
   a. are made of chromosomes.
   b. consist entirely of DNA.
   c. consist of DNA wound around a histone core.
   d. are present only during mitosis.
   e. are present only during prophase.
3. Which statement about the cell cycle is *not* true?
   a. It consists of mitosis and interphase.
   b. The cell's DNA replicates during G1.
   c. A cell can remain in G1 for weeks or much longer.
   d. Proteins are formed throughout all subphases of interphase.
   e. Histones are synthesized primarily during S phase.
4. Which statement about mitosis is *not* true?
   a. A single nucleus gives rise to two identical daughter nuclei.
   b. The daughter nuclei are genetically identical to the parent nucleus.
   c. The centromeres separate at the onset of anaphase.
   d. Homologous chromosomes synapse in prophase.
   e. Mitotic centers organize the microtubules of the spindle fibers.
5. Which statement about cytokinesis is true?
   a. In animals, a cell plate forms.
   b. In plants, it is initiated by furrowing of the membrane.
   c. It generally immediately follows mitosis.
   d. In plant cells, actin and myosin play an important part.
   e. It is the division of the nucleus.
6. Apoptosis
   a. occurs in all cells.
   b. involves the cell membrane dissolving.
   c. does not occur in an embryo.
   d. involves a series of programmed events for cell death.
   e. is not involved with cancer.
7. In meiosis,
   a. meiosis II reduces the chromosome number from diploid to haploid.
   b. DNA replicates between meiosis I and II.
   c. the chromatids that make up a chromosome in meiosis II are identical.
   d. each chromosome in prophase I consists of four chromatids.
   e. homologous chromosomes separate from one another in anaphase I.
8. In meiosis,
   a. a single nucleus gives rise to two daughter nuclei.
   b. the daughter nuclei are genetically identical to the parent nucleus.
   c. the centromeres separate at the onset of anaphase I.
   d. homologous chromosomes synapse in prophase I.
   e. no spindle forms.
9. A plant has a diploid chromosome number of 12. An egg cell of the plant has 5 chromosomes. The most probable explanation of this is
   a. normal mitosis.
   b. normal meiosis.
   c. nondisjunction in meiosis I.
   d. nondisjunction in meiosis I and II.
   e. nondisjunction in mitosis.
10. The number of daughter chromosomes in a human cell in anaphase II of meiosis is
   a. 2.
   b. 23.
   c. 46.
   d. 69.
   e. 92.

For Discussion

1. How does a nucleus in the G2 phase of the cell cycle differ from one in the G1 phase?
2. Compare the roles of cohesins and condensin in mitosis, meiosis I, and meiosis II.
3. Compare and contrast mitosis (and subsequent cytokinesis) in animals and plants.
4. Suggest two ways in which, with the help of a microscope, one might determine the relative duration of the various phases of mitosis.
5. Contrast mitotic prophase and prophase I of meiosis. Contrast mitotic anaphase and anaphase I of meiosis.
6. Compare the sequence of events in the mitotic cell cycle with the sequence in programmed cell death.

10 Genetics: Mendel and Beyond

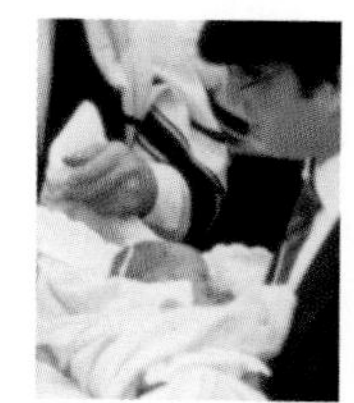

In the Middle Eastern desert 1,800 years ago, the rabbi faced a dilemma. A Jewish woman had given birth to a son. As required by the laws set down by God's commandment to Abraham almost 2,000 years previously and later reiterated by Moses, the mother brought her 8-day-old son to the rabbi for ritual penile circumcision. The rabbi knew that the woman's two previous sons had bled to death when their foreskins were cut. Yet the biblical commandment remained: Unless he was circumcised, the boy could not be counted among those with whom God had made His solemn covenant. After consultation with other rabbis, it was decided to exempt this, the third son.

Almost a thousand years later, in the twelfth century, the physician and biblical commentator Moses Maimonides reviewed this and numerous other cases in the rabbinical literature and stated that in such instances the third son should not be circumcised. Furthermore, the exemption should apply whether the mother's son was "from her first husband or from her second husband." The bleeding disorder, he reasoned, was clearly carried by the mother and passed on to her sons.

Knowing nothing of our modern vision of genetics, these rabbis linked a human disease (which we now know as hemophilia A) to a pattern of inheritance (which we know as sex linkage). Only in the past few decades have the precise biochemical nature of hemophilia A and its genetic determination been worked out.

How do we account for, and predict, such patterns of inheritance? In this chapter, we will discuss how the units of inheritance, called genes, are transmitted from generation to generation, and we will show how many of the rules that govern genetics can be explained by the behavior of chromosomes during meiosis. We will also describe the interactions of genes with one another and with the environment, and we will examine the consequences of the fact that genes occupy specific positions on chromosomes.

An Ancient Ritual
A male infant undergoes ritual circumcision in accordance with Jewish laws. Sons of Jewish mothers who carry the gene for hemophilia may be exempted from the ritual.

The Foundations of Genetics

Much of the early study of biological inheritance was done with plants and animals of economic importance. Records show that people were deliberately crossbreeding date palm trees and horses as early as 5,000 years ago. By the early nineteenth century, plant breeding was widespread, especially with ornamental flowers such as tulips. Half a century later, Gregor Mendel used the existing knowledge of plant reproduction to design and conduct experiments on inheritance. Although his published results were neglected by scientists for more than 30 years, they ultimately became the foundation for the science of genetics.

Plant breeders showed that both parents contribute equally to inheritance

Plants are good experimental subjects for the study of genetics. Many plants are easily grown in large quantities, produce large numbers of offspring (in the form of seeds), and have relatively short generation times. In most plant species, the same individuals have both male and female reproductive organs, permitting each plant to reproduce as a male, as a female, or as both. Best of all, it is often easy to control which individuals mate (Figure 10.1).

Some discoveries that Mendel found useful in his studies had been made in the late eighteenth century by a German botanist, Josef Gottlieb Kölreuter. He had studied the offspring of **reciprocal crosses**, in which plants are crossed (mated with each other) in opposite directions. For example, in one cross, males that have white flowers are mated with females that have red flowers, while in a complementary cross, red-flowered males and white-flowered females are mated. In Kölreuter's studies, such reciprocal crosses always gave identical results, showing that both parents contributed equally to the offspring.

Although the concept of equal parental contributions was an important discovery, the nature of what exactly the parents were contributing to their offspring—the units of inheritance—remained unknown. Laws of inheritance proposed at the time favored the concept of *blending*. If a plant that had one form of a characteristic (say, red flowers) was crossed with one that had a different form of that characteristic (blue flowers), the offspring would be a blended combination of the two parents (purple flowers). According to the blending theory, it was thought that once heritable elements were combined, they could not be separated again (like inks of different colors mixed together). The red and blue genetic determinants were thought to be forever blended into the new purple one. Then, about a century after Kölreuter completed his work, Mendel began his.

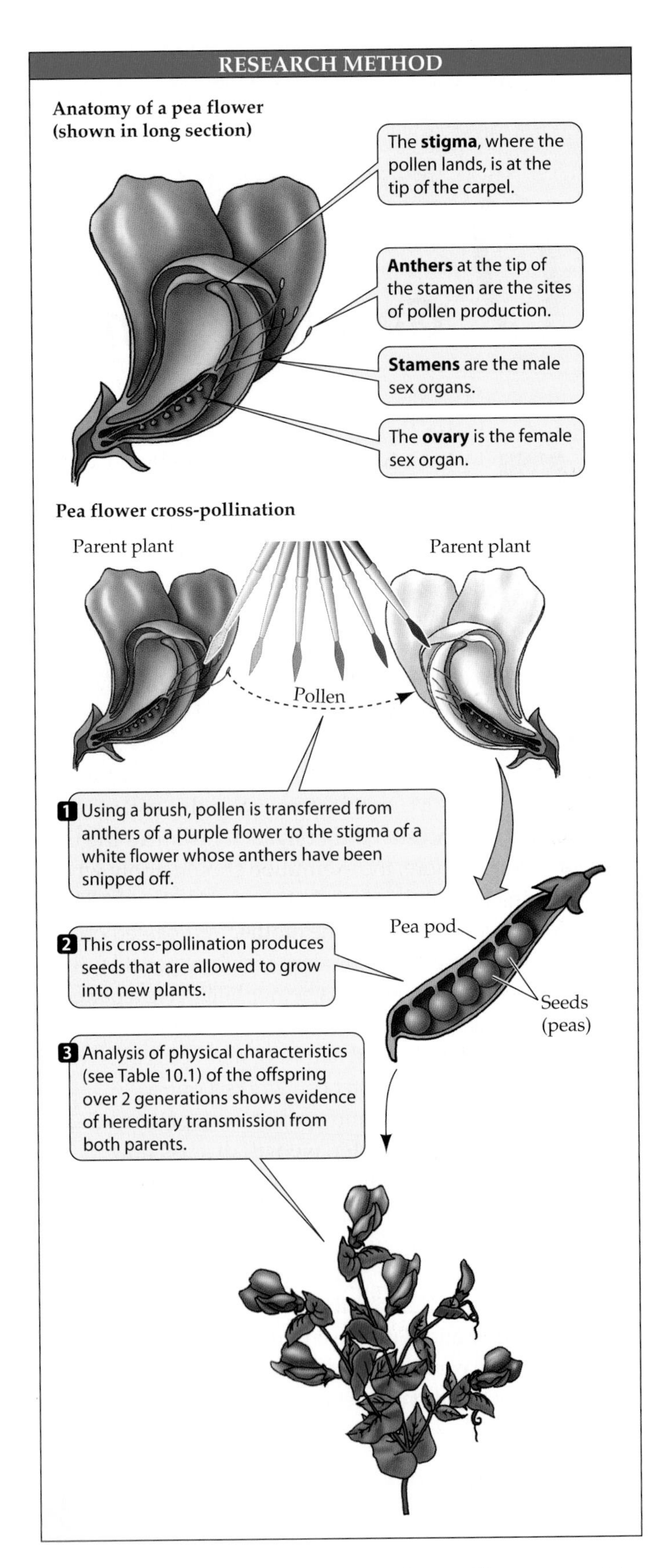

RESEARCH METHOD
Anatomy of a pea flower (shown in long section)
The stigma, where the pollen lands, is at the tip of the carpel.
Anthers at the tip of the stamen are the sites of pollen production.
Stamens are the male sex organs.
The ovary is the female sex organ.
Pea flower cross-pollination
Parent plant
Parent plant
Pollen
1 Using a brush, pollen is transferred from anthers of a purple flower to the stigma of a white flower whose anthers have been snipped off.
2 This cross-pollination produces seeds that are allowed to grow into new plants.
Pea pod
Seeds (peas)
3 Analysis of physical characteristics (see Table 10.1) of the offspring over 2 generations shows evidence of hereditary transmission from both parents.

10.1 A Controlled Cross between Two Plants Plants were widely used in early genetic studies because it is easy to control which individuals mate with which. Mendel used the garden pea (*Pisum sativum*) in many of his experiments.

Mendel brought new methods to experiments on inheritance

Gregor Mendel was an Austrian monk, not an academic scientist, but he was qualified to undertake scientific investigations. Although in 1850 he had failed an examination for a teaching certificate in natural science, he later undertook intensive studies in physics, chemistry, mathematics, and various aspects of biology at the University of Vienna. His work in physics and mathematics probably led him to apply experimental and quantitative methods to the study of heredity, and these methods were the key ingredients in his success.

Mendel worked out the basic principles of inheritance in plants over a period of about 9 years. His work culminated in a public lecture in 1865 and a detailed written publication in 1866. Mendel's paper appeared in a journal that was received by 120 libraries, and he sent reprinted copies (of which he had obtained 40) to several distinguished scholars. However, his theory was not accepted. In fact, it was ignored.

The chief difficulty was that the most prominent biologists of Mendel's time were not in the habit of thinking in mathematical terms, even the simple terms used by Mendel. Even Charles Darwin, whose theory of evolution by natural selection depended on genetic variation among individuals, failed to understand the significance of Mendel's findings. In fact, Darwin performed breeding experiments on snapdragons similar to Mendel's on peas and got data similar to Mendel's, but he missed the point, still relying on the concept of blending. In addition, Mendel had little credibility as a biologist; indeed, his lowest grades were in biology! Whatever the reasons, Mendel's pioneering paper had no discernible influence on the scientific world for more than 30 years.

Then, in 1900, after meiosis had been observed and described, Mendel's discoveries burst into prominence as a result of independent experiments by three plant geneticists, Hugo DeVries, Carl Correns, and Erich von Tschermak. Each carried out crossing experiments and obtained quantitative data about the progeny; each published his principal findings in 1900; each cited Mendel's 1866 paper. They immediately realized that chromosomes and meiosis provided a physical explanation for the theory that Mendel had proposed to explain the data from his crosses. As we go through Mendel's work, we will describe first his experiments and conclusions, and then the chromosomal explanation of his theories.

Mendel's Experiments and the Laws of Inheritance

That Mendel was able to make his discoveries before the discovery of meiosis was due in part to the methods of experimentation he used. Mendel's work is a fine example of preparation, choice of experimental material, execution, and interpretation. Let's see how he approached each of these steps.

Mendel devised a careful research plan

Mendel chose the garden pea for his studies because of its ease of cultivation, the feasibility of controlled pollination (see Figure 10.1), and the availability of varieties with differing traits. He controlled pollination, and thus fertilization, of his parent plants by manually moving pollen from one plant to another. Thus he knew the parentage of the offspring in his experiments. The pea plants Mendel studied produce male and female sex organs and gametes in the same flower. If untouched, they naturally *self-pollinate*—that is, the female organ of each flower receives pollen from the male organs of the same flower. Mendel made use of this natural phenomenon in some of his experiments.

Mendel began by examining different varieties of peas in a search for heritable characters and traits suitable for study:

- A **character** is an observable feature, such as flower color.
- A **trait** is a particular form of a character, such as white flowers.
- A **heritable** character trait is one that is passed from parent to offspring.

Mendel looked for characters that had well-defined, contrasting alternative traits, such as purple flowers versus white flowers. Furthermore, these traits had to be **true-breeding**, meaning that the observed trait was the only form present for many generations. In other words, peas with white flowers, when crossed with one another, would have to give rise only to progeny with white flowers for many generations; tall plants bred to tall plants would have to produce only tall progeny.

Mendel isolated each of his true-breeding strains by repeated inbreeding (done by crossing of sibling plants that were seemingly identical or by allowing individuals to self-pollinate) and selection. In most of his work, Mendel concentrated on the seven pairs of contrasting traits shown in Table 10.1. Before performing any experimental cross, he made sure that each potential parent was from a true-breeding strain—an essential point in his analysis of his experimental results.

Mendel then collected pollen from one parental strain and placed it onto the stigma (female organ) of flowers of the other strain whose anthers were removed. The plants providing and receiving the pollen were the **parental generation**, designated **P**. In due course, seeds formed and were planted. The seeds and the resulting new plants constituted the **first filial generation**, or $\mathbf{F_1}$. Mendel and his assistants examined each F_1 plant to see which traits it bore and then recorded the number of F_1 plants expressing each trait. In some experiments the F_1 plants were allowed to self-pollinate and produce a **second filial generation**, $\mathbf{F_2}$. Again, each F_2 plant was characterized and counted.

In summary, Mendel devised a well-organized plan of research, pursued it faithfully and carefully, recorded great amounts of quantitative data, and analyzed the numbers he recorded to explain the relative proportions of the different kinds of progeny. His results and the conclusions to which they led are the subject of the next several sections.

10.1 Mendel's Results from Monohybrid Crosses

PARENTAL GENERATION PHENOTYPES		F_2 GENERATION PHENOTYPES			
DOMINANT	RECESSIVE	DOMINANT	RECESSIVE	TOTAL	RATIO
Spherical seeds ×	Wrinkled seeds	5,474	1,850	7,324	2.96:1
Yellow seeds ×	Green seeds	6,022	2,001	8,023	3.01:1
Purple flowers ×	White flowers	705	224	929	3.15:1
Inflated pods ×	Constricted pods	882	299	1,181	2.95:1
Green pods ×	Yellow pods	428	152	580	2.82:1
Axial flowers ×	Terminal flowers	651	207	858	3.14:1
Tall stems (1 m) ×	Dwarf stems (0.3 m)	787	277	1,064	2.84:1

Mendel's experiment 1 examined a monohybrid cross

"Experiment 1" in Mendel's paper involved a **monohybrid cross**—one involving offspring of a cross in which each member of the P generation is true-breeding for a different trait. He took pollen from pea plants of a true-breeding strain with wrinkled seeds and placed it on the stigmas of flowers of a true-breeding strain with spherical seeds (Figure 10.2). He also performed the reciprocal cross by placing pollen from the spherical-seeded strain on the stigmas of flowers of the wrinkled-seeded strain.

10.2 Contrasting Traits In experiment 1, Mendel studied the inheritance of seed shape. We know today that wrinkled seeds possess an abnormal form of starch.

In both cases, all the F_1 seeds produced were spherical—it was as if the wrinkled seed trait had disappeared completely. The following spring, Mendel grew 253 F_1 plants from these spherical seeds. Each of these plants was allowed to self-pollinate to produce F_2 seeds. In all, there were 7,324 F_2 seeds, of which 5,474 were spherical and 1,850 wrinkled (Figure 10.3 and Table 10.1).

Mendel observed that the wrinkled seed trait was never expressed in the F_1 generation, even though it reappeared in the F_2 generation. He concluded that the spherical seed trait was **dominant** to the wrinkled seed trait, which he called **recessive**. In each of the other six pairs of traits Mendel studied, one proved to be dominant over the other.

Of most importance, the ratio of the two traits in the F_2 generation was always the same—approximately 3:1. That is, three-fourths of the F_2 generation showed the dominant trait and one-fourth showed the recessive trait (see Table 10.1). In Mendel's experiment 1, the ratio was 5,474:1,850 = 2.96:1. The reciprocal crosses in the parental generation both gave similar outcomes in the F_2; it did not matter which parent contributed the pollen.

By themselves, the results from experiment 1 disproved the widely held belief that inheritance is always a blending

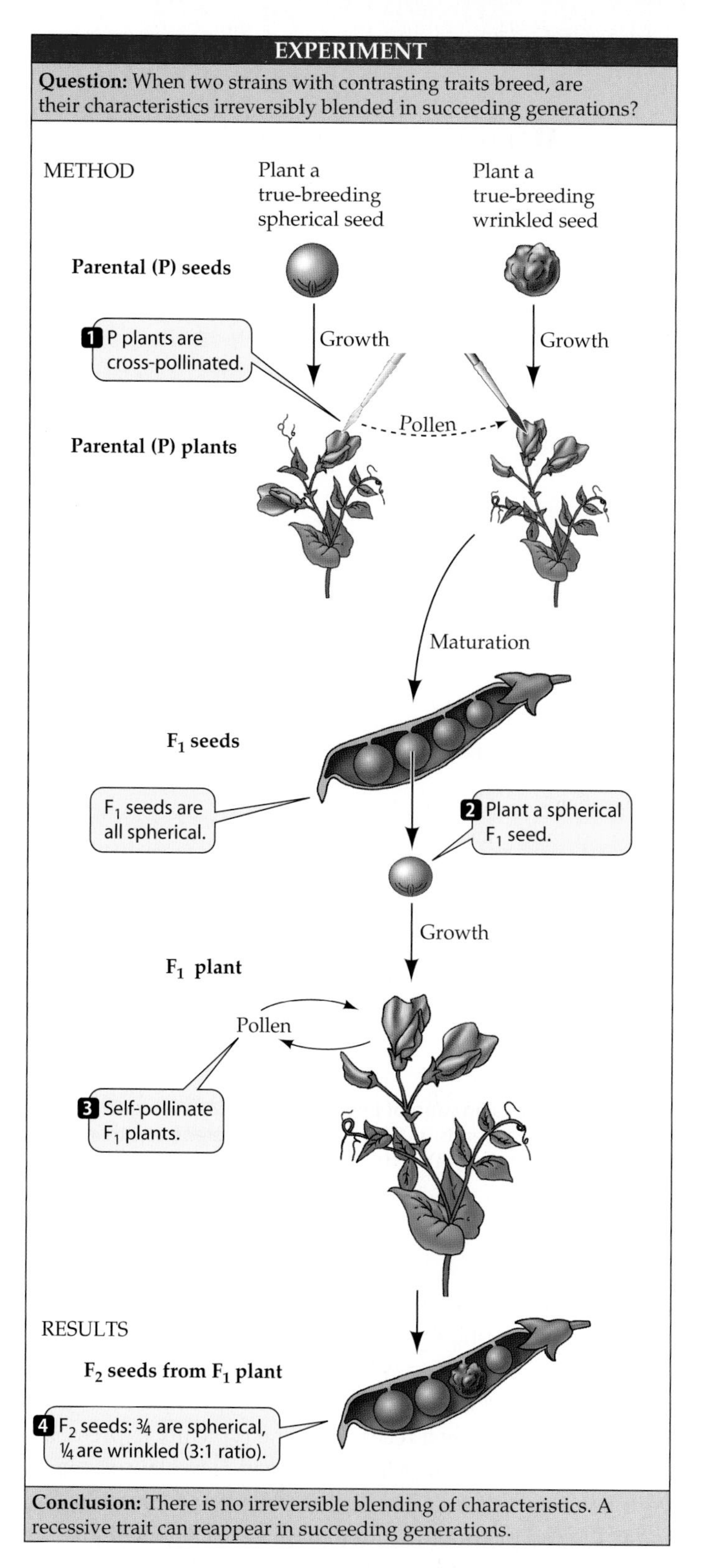
EXPERIMENT
Question: When two strains with contrasting traits breed, are their characteristics irreversibly blended in succeeding generations?
METHOD
Plant a true-breeding spherical seed
Plant a true-breeding wrinkled seed
Parental (P) seeds
1 P plants are cross-pollinated.
Growth
Growth
Pollen
Parental (P) plants
Maturation
F1 seeds
F1 seeds are all spherical.
2 Plant a spherical F1 seed.
Growth
F1 plant
Pollen
3 Self-pollinate F1 plants.
RESULTS
F2 seeds from F1 plant
4 F2 seeds: ¾ are spherical, ¼ are wrinkled (3:1 ratio).
Conclusion: There is no irreversible blending of characteristics. A recessive trait can reappear in succeeding generations.

10.3 Mendel's Experiment 1 The pattern Mendel observed in the F_2 generation—¼ of the seeds wrinkled, ¾ spherical—was the same no matter which strain contributed the pollen in the parental generation.

phenomenon. According to the blending theory, Mendel's F_1 seeds should have had an appearance intermediate between those of the two parents—in other words, they should have been slightly wrinkled. Furthermore, the blending theory offered no explanation for the reappearance of the wrinkled trait in the F_2 seeds after its apparent absence in the F_1 seeds.

Mendel proposed that the units responsible for the inheritance of specific traits are present as discrete particles that occur in pairs and segregate (separate) from one another during the formation of gametes. According to this theory, the units of inheritance retain their integrity in the presence of other units. This **particulate theory** is in sharp contrast to the blending theory, in which the units of inheritance were believed to lose their identities when mixed together.

As he worked mathematically with his data, Mendel reached the tentative conclusion that each pea plant has two units of inheritance for each character, one from each parent. During the production of gametes, only one of these paired units is given to a gamete. Hence each gamete contains one unit, and the resulting zygote contains two, because it is produced by the fusion of two gametes. This conclusion is the core of Mendel's model of inheritance. Mendel's unit of inheritance is now called a **gene**.

Mendel reasoned that in experiment 1, the two true-breeding parent plants had different forms of the gene affecting seed shape. The spherical-seeded parent had two genes of the same form, which we will call *S*, and the parent with wrinkled seeds had two *s* genes. The *SS* parent produced gametes that each contained a single *S* gene, and the *ss* parent produced gametes each with a single *s* gene. Each member of the F_1 generation had an *S* from one parent and an *s* from the other; an F_1 could thus be described as *Ss*. We say that *S* is dominant over *s* because the trait specified by the *s* allele is not evident when both forms of the gene are present.

The different forms of a gene (*S* and *s* in this case) are called **alleles**. Individuals that are true-breeding for a trait contain two copies of the same allele. For example, all the individuals in a population of a strain of true-breeding peas with wrinkled seeds must have the allele pair *ss;* if *S* were present, the plants would produce spherical seeds.

We say that the individuals that produce wrinkled seeds are **homozygous** for the allele *s*, meaning that they have two copies of the same allele (*ss*). Some peas with spherical seeds—the ones with the genotype *SS*—are also homozygous. However, not all plants with spherical seeds have the *SS* genotype. Some spherical-seeded plants, like Mendel's F_1, are **heterozygous**: They have two different alleles of the gene in question (in this case, *Ss*).

To illustrate these terms with a more complex example, one in which there are three gene pairs, an individual with the genotype *AABbcc* is homozygous for the *A* and *C* genes, because it has two *A* alleles and two *c* alleles, but heterozygous for the *B* gene, because it contains the *B* and *b* alleles. An individual that is homozygous for a character is sometimes called a *homozygote;* a *heterozygote* is heterozygous for the character in question.

The physical appearance of an organism is its **phenotype**. Mendel correctly supposed the phenotype to be the result of the **genotype**, or genetic constitution, of the organism showing the phenotype. In experiment 1 we are dealing with two phenotypes (spherical seeds and wrinkled seeds). The F_2 generation contains these two phenotypes, but they are produced by three genotypes. The wrinkled seed phenotype is produced only by the genotype *ss*, whereas the spherical seed phenotype may be produced by the genotypes *SS* or *Ss*.

Mendel's first law says that alleles segregate

How does Mendel's model of inheritance explain the composition of the F_2 generation in experiment 1? Consider first the F_1, which has the spherical seed phenotype and the *Ss* genotype. According to Mendel's model, when any individual produces gametes, the two alleles separate, so that each gamete receives only one member of the pair of alleles. This is Mendel's first law, the **law of segregation**.

In experiment 1, half the gametes produced by the F_1 generation contained the *S* allele and half the *s* allele. In the F_2 generation, since both *SS* and *Ss* plants produce spherical seeds while *ss* produces wrinkled seeds, there are three ways to get a spherical-seeded plant, but only one way to get a wrinkled-seeded plant (*s* from both parents)—predicting a 3:1 ratio remarkably close to the values Mendel found experimentally for all six of the traits he compared (see Table 10.1).

While this simple example is easy to work out in your head, determination of expected allelic combinations for more complicated inheritance patterns can be aided by use of a **Punnett square**, devised in 1905 by the British geneticist Reginald Crundall Punnett. This device reminds us to consider all possible combinations of gametes when calculating expected genotype frequencies. A Punnett square looks like this:

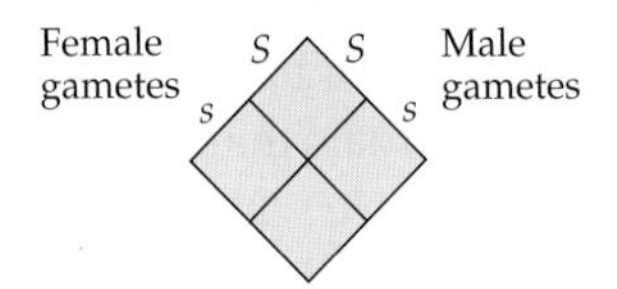

It is a simple grid with all possible male gamete genotypes shown along one side and all possible female gamete genotypes along another side. To complete the grid, we fill in each square with the corresponding pollen genotype and egg genotype, giving the diploid genotype of a member of the F_2 generation. For example, to fill the rightmost square, we put in the *S* from the egg (female gamete) and the *s* from the pollen (male gamete), yielding *Ss* (Figure 10.4).

Mendel did not live to see his theory placed on a sound physical footing based on chromosomes and DNA. Genes are now known to be regions of the DNA molecules in chromosomes. More specifically, a gene is a portion of the DNA that resides at a particular site on a chromosome, called a **locus** (plural, **loci**), and encodes a particular character. Genes are expressed in the phenotype mostly as proteins with particular functions, such as enzymes. So a dominant gene can be thought of as a region of DNA that is expressed as a functional enzyme, while a recessive gene typically expresses a nonfunctional enzyme. Mendel arrived at his law of segregation with no knowledge of chromosomes or meiosis, but today we can picture the different alleles of a gene segregating as chromosomes separate in meiosis I (Figure 10.5).

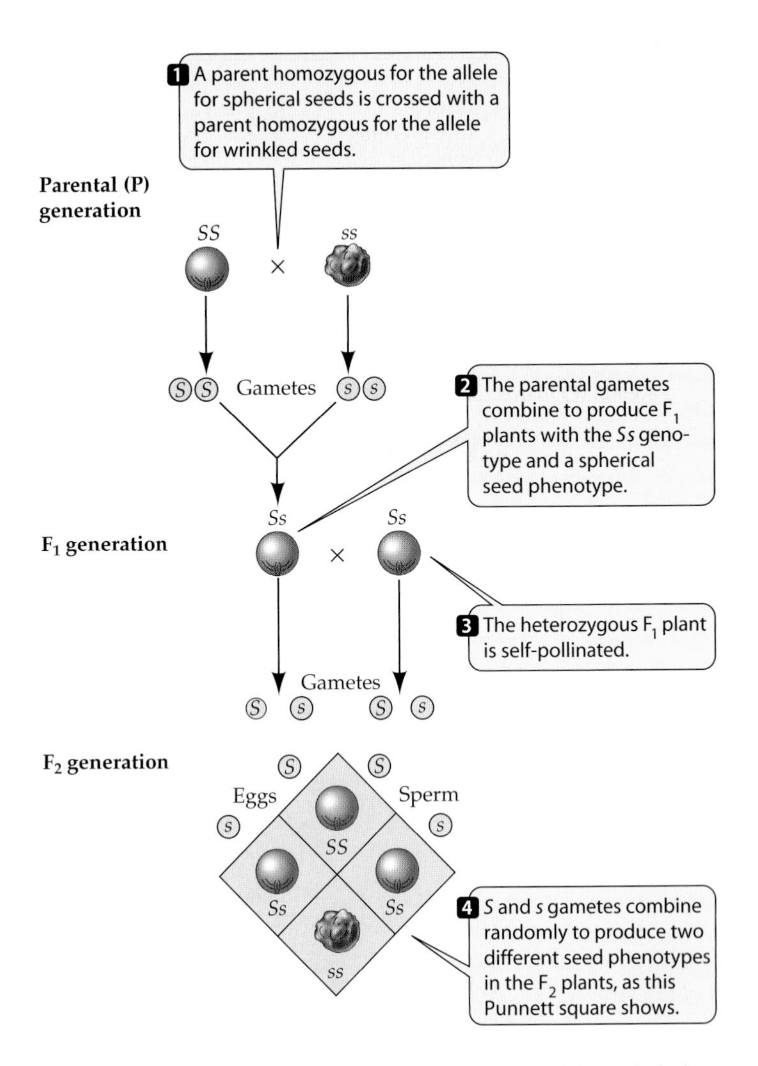

10.4 Mendel's Explanation of Experiment 1 Mendel concluded that inheritance depends on factors from each parent, and that these factors are discrete units that do not blend in the offspring.

Mendel verified his hypothesis by performing a test cross

Mendel set out to test his hypothesis that there were two possible allelic combinations (*SS* and *Ss*) in the spherical-seeded F_1 generation. He did so by performing a **test cross**, which is a way of finding out whether an individual showing a dom-

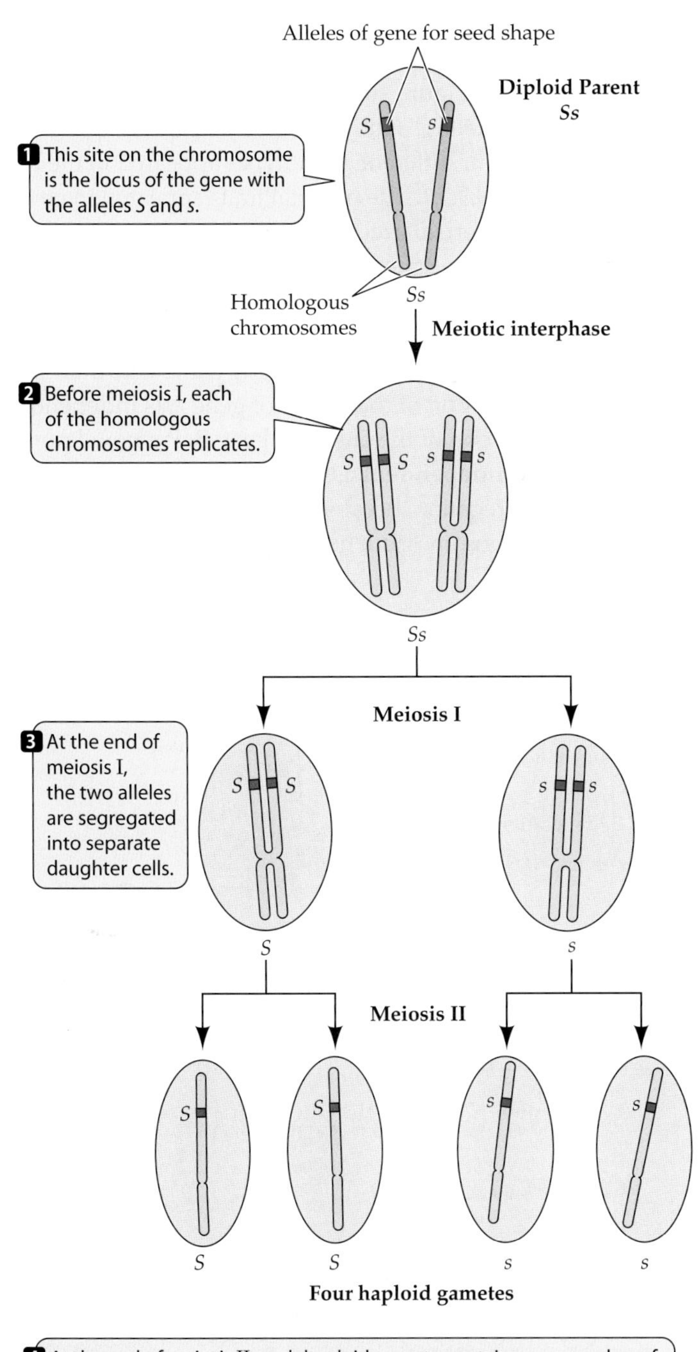

10.5 Meiosis Accounts for the Segregation of Alleles Although Mendel had no knowledge of chromosomes or meiosis, we now know that a pair of alleles resides on homologous chromosomes, and that meiosis segregates those alleles.

inant trait is homozygous or heterozygous. In a test cross, the individual in question is crossed with an individual known to be homozygous for the recessive trait—an easy individual to identify, because in order to have the recessive phenotype, it must be homozygous for the recessive trait.

For the seed shape gene that we have been considering, the recessive homozygote used for the test cross is *ss*. The individual being tested may be described initially as *S*–because we do not yet know the identity of the second allele. We can predict two possible results:

- If the individual being tested is homozygous dominant (*SS*), all offspring of the test cross will be *Ss* and show the dominant trait (spherical seeds).
- If the individual being tested is heterozygous (*Ss*), then approximately half of the offspring of the test cross will be heterozygous and show the dominant trait (*Ss*), but the other half will be homozygous for, and will show, the recessive trait (*ss*) (Figure 10.6).

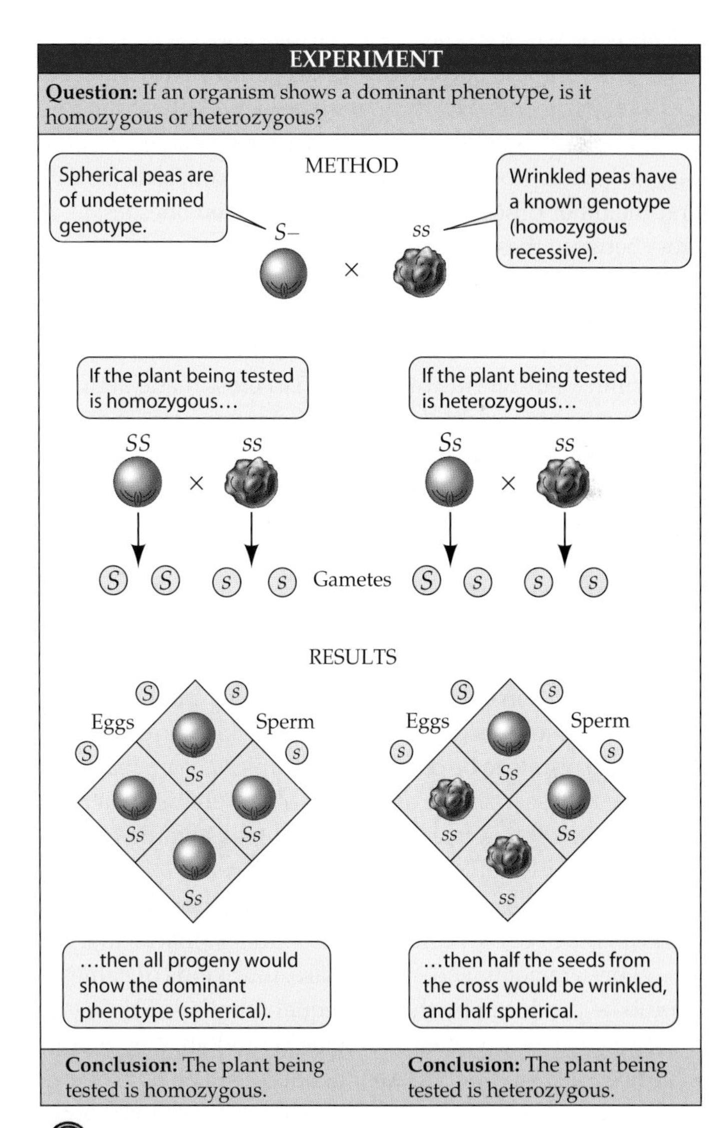

EXPERIMENT
Question: If an organism shows a dominant phenotype, is it homozygous or heterozygous?
METHOD
Spherical peas are of undetermined genotype.
S–
×
ss
Wrinkled peas have a known genotype (homozygous recessive).
If the plant being tested is homozygous...
SS × ss
If the plant being tested is heterozygous...
Ss × ss
S S s s Gametes S s s s
RESULTS
Eggs
Sperm
Ss
Ss
Ss
Ss
Eggs
Sperm
Ss
ss
Ss
ss
...then all progeny would show the dominant phenotype (spherical).
...then half the seeds from the cross would be wrinkled, and half spherical.
Conclusion: The plant being tested is homozygous.
Conclusion: The plant being tested is heterozygous.

10.6 Homozygous or Heterozygous? An individual with a dominant phenotype may be homozygous or heterozygous. Its genotype can be determined by crossing it with a homozygous recessive plant and observing the phenotypes of the progeny produced. This procedure is known as a test cross.

The second prediction closely matches the results that Mendel obtained; thus Mendel's hypothesis accurately predicted the results of his test cross.

With his first hypothesis confirmed, Mendel went on to ask another question: How do different pairs of genes behave in crosses when considered together?

Mendel's second law says that alleles of different genes assort independently

Consider an organism that is heterozygous for two genes (*SsYy*), in which the *S* and *Y* alleles came from its mother and *s* and *y* came from its father. When this organism produces gametes, do the alleles of maternal origin (*S* and *Y*) go together to one gamete and those of paternal origin (*s* and *y*) to another gamete? Or can a single gamete receive one maternal and one paternal allele, *S* and *y* (or *s* and *Y*)?

To answer these questions, Mendel performed another series of experiments. He began with peas that differed in two seed characters: seed shape and seed color. One true-breeding parental strain produced only spherical, yellow seeds (*SSYY*), and the other produced only wrinkled, green ones (*ssyy*). A cross between these two strains produced an F_1 generation in which all the plants were *SsYy*. Because the *S* and *Y* alleles are dominant, the F_1 seeds were all spherical and yellow.

Mendel continued this experiment to the F_2 generation by performing a **dihybrid cross**, which is a cross made between individuals that are identical double heterozygotes. There are two possible ways in which such doubly heterozygous plants might produce gametes, as Mendel saw it. (Remember that he had never heard of chromosomes or meiosis.)

First, if the alleles maintain the associations they had in the parental generation (that is, if they are **linked**), then the F_1 plants should produce two types of gametes (*SY* and *sy*), and the F_2 progeny resulting from self-pollination of the F_1 plants should consist of three times as many plants bearing spherical, yellow seeds as ones with wrinkled, green seeds. Were such results to be obtained, there might be no reason to suppose that seed shape and seed color were regulated by two different genes, because spherical seeds would always be yellow and wrinkled ones always green.

The second possibility is that the segregation of *S* from *s* is independent of the segregation of *Y* from *y* (that is, that the two genes are not linked). In this case, four kinds of gametes should be produced by the F_1 in equal numbers: *SY*, *Sy*, *sY*, and *sy*. When these gametes combine at random, they should produce an F_2 of nine different genotypes. The F_2 progeny could have any of three possible genotypes for shape (*SS*, *Ss*, or *ss*) and any of three possible genotypes for color (*YY*, *Yy*, or *yy*). The combined nine genotypes should produce just four phenotypes (spherical yellow, spherical green, wrinkled yellow, wrinkled green). By using a Punnett square, we can show that these four phenotypes would be expected to occur in a ratio of 9:3:3:1 (Figure 10.7).

The results of Mendel's dihybrid crosses matched the second prediction: Four different phenotypes appeared in the F_2 in a ratio of about 9:3:3:1. The parental traits appeared in new combinations (spherical green and wrinkled yellow). Such new combinations are called **recombinant** phenotypes.

These results led Mendel to the formulation of what is now known as Mendel's second law: Alleles of different genes assort independently of one another during gamete formation. That is, the segregation of the alleles of gene A is independent of the segregation of the alleles of gene B. We now know that this **law of independent assortment** is not as universal as the law of segregation, because it applies to genes located on separate chromosomes, but not necessarily to those located

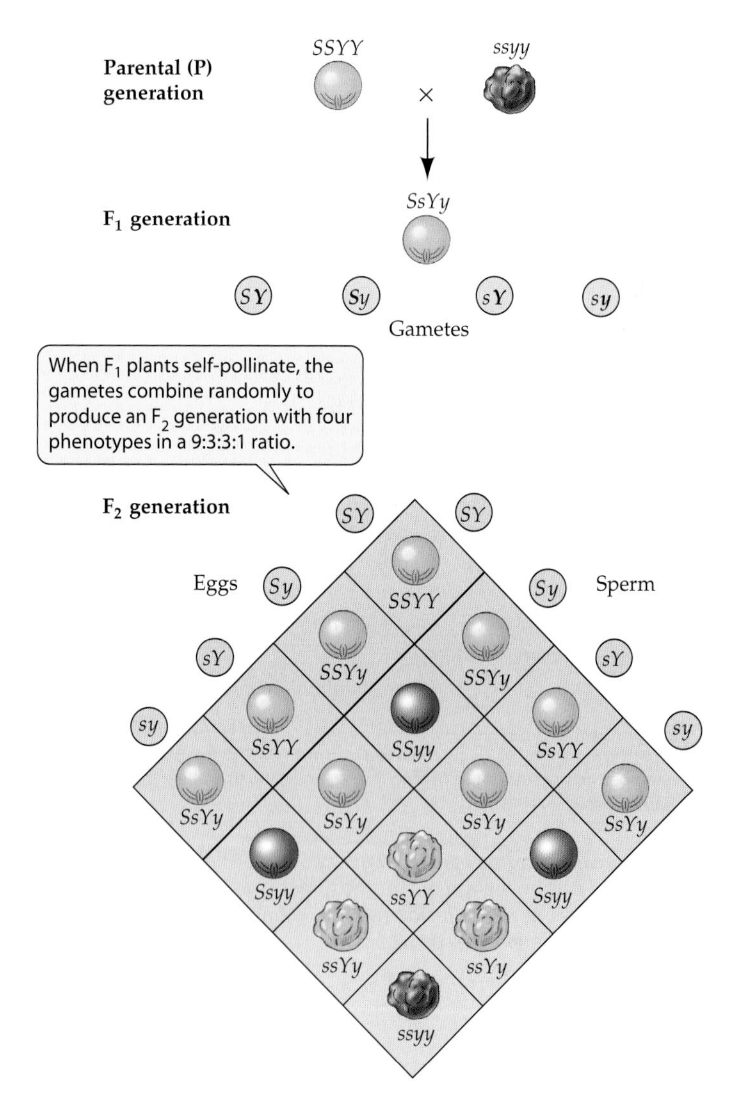

10.7 Independent Assortment The 16 possible combinations of gametes in this dihybrid cross result in 9 different genotypes. Because *S* and *Y* are dominant over *s* and *y*, respectively, the 9 genotypes result in 4 phenotypes in a ratio of 9:3:3:1. These results show that the two genes segregate independently.

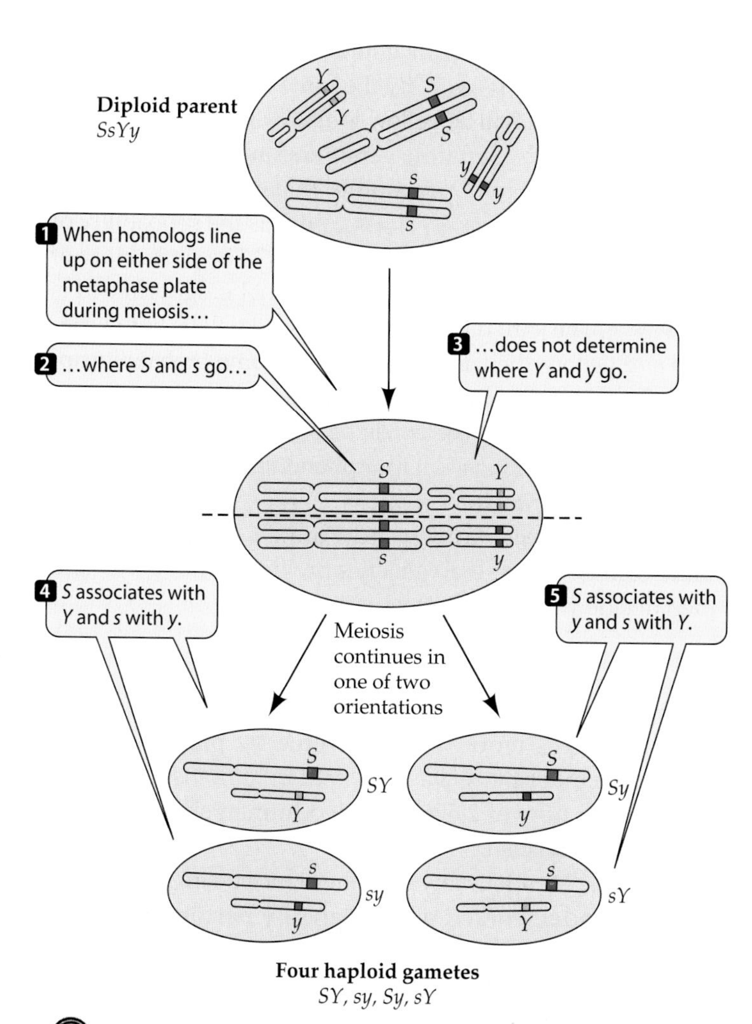
Diploid parent
SsYy
1 When homologs line up on either side of the metaphase plate during meiosis...
2 ...where S and s go...
3 ...does not determine where Y and y go.
4 S associates with Y and s with y.
5 S associates with y and s with Y.
Meiosis continues in one of two orientations
SY
Sy
sy
sY
Four haploid gametes
SY, sy, Sy, sY

10.8 Meiosis Accounts for Independent Assortment of Alleles We now know that alleles of different genes are segregated independently during metaphase I of meiosis. Thus a parent of genotype SsYy can form gametes with four different genotypes.

on the same chromosome, as we will see below. However, it is correct to say that chromosomes segregate independently during the formation of gametes, and so do any two genes on separate homologous chromosome pairs (Figure 10.8).

One of Mendel's major contributions to the science of genetics was his use of the rules of statistics and probability to analyze his masses of data from hundreds of crosses producing thousands of plants. His mathematical analyses led to clear patterns in the data, and then to his hypotheses. Ever since Mendel, geneticists have used simple mathematics in the same ways that Mendel did.

Punnett squares or probability calculations: A choice of methods

Punnett squares provide one way of solving problems in genetics, and probability calculations provide another. Many people find it easiest to use the principles of probability, perhaps because they are so familiar. When we flip a coin, the law of probability states that it has an equal probability of landing "heads" or "tails." For any given toss of a fair coin, the probability of heads is independent of what happened in all the previous tosses. A run of ten straight heads implies nothing about the next toss. No "law of averages" increases the likelihood that the next toss will come up tails, and no "momentum" makes an eleventh occurrence of heads any more likely. On the eleventh toss, the odds of getting heads are still 50/50.

The basic conventions of probability are simple:

- If an event is absolutely certain to happen, its probability is 1.
- If it cannot possibly happen, its probability is 0.
- Otherwise, its probability lies between 0 and 1.

A coin toss results in heads approximately half the time, so the probability of heads is ½—as is the probability of tails.

MULTIPLYING PROBABILITIES. How can we determine the probability of two independent events happening together? If two coins (a penny and a dime, say) are tossed, each acts independently of the other. What, then, is the probability of both coins coming up heads? Half the time, the penny comes up heads; of that fraction, half the time the dime also comes up heads. Therefore, the joint probability of both coins coming up heads is half of one-half, or $\frac{1}{2} \times \frac{1}{2} = \frac{1}{4}$. To find the joint probability of independent events, then, we multiply the probabilities of the individual events (Figure 10.9). How does this method apply to genetics?

THE MONOHYBRID CROSS. To apply the principles of probability to genetics problems, we need only deal with gamete formation and random fertilization instead of coin tosses. A homozygote can produce only one type of gamete, so, for example, the probability of an *SS* individual producing gametes with the genotype *S* is 1. The heterozygote *Ss* produces *S* gametes with a probability of ½, and *s* gametes with a probability of ½.

Consider the F_2 progeny of the cross in Figure 10.4. They are obtained by self-pollination of F_1 plants of genotype *Ss*. The probability that an F_2 plant will have the genotype *SS* must be $\frac{1}{2} \times \frac{1}{2} = \frac{1}{4}$, because there is a 50:50 chance that the sperm will have the genotype *S*, and that chance is independent of the 50:50 chance that the egg will have the genotype *S*. Similarly, the probability of *ss* offspring is $\frac{1}{2} \times \frac{1}{2} = \frac{1}{4}$.

ADDING PROBABILITIES. How are probabilities calculated when an event can happen in different ways? The probability of an F_2 plant getting an *S* allele from the sperm and an *s* allele from the egg is ¼, but remember that the same

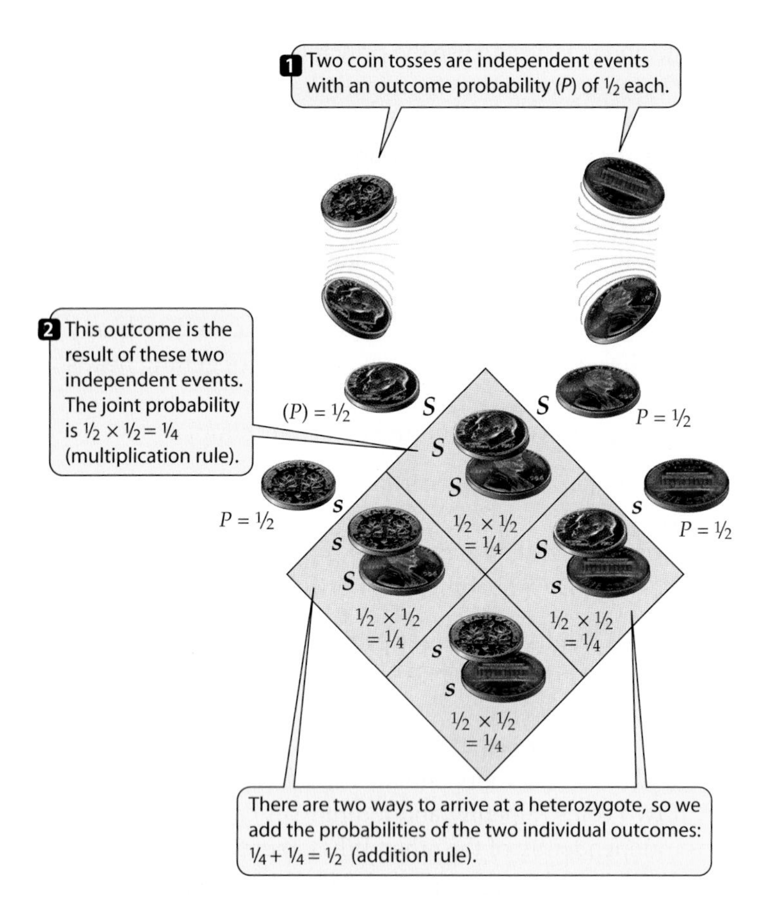

10.9 Using Probability Calculations in Genetics The probability of any given combination of alleles from a sperm and an egg appearing in the offspring of a cross can be obtained by multiplying the probabilities of each event. Since a heterozygote can be formed in two ways, these two probabilities are added together.

genotype can also result from an *s* from the sperm and an *S* from the egg, also with a probability of ¼. The probability of an event that can occur in two or more different ways is the sum of the individual probabilities of those ways. Thus the probability that an F_2 plant will be a heterozygote is equal to the sum of the probabilities of the two ways of forming a heterozygote: ¼ + ¼ = ½ (see Figure 10.9). The three genotypes are therefore expected in the ratio ¼ *SS* : ½ *Ss* : ¼ *ss*—hence the 1:2:1 ratio of genotypes and the 3:1 ratio of phenotypes seen in Figure 10.4.

THE DIHYBRID CROSS. If F_1 plants heterozygous for two independent characters self-pollinate, the resulting F_2 plants express four different phenotypes. The proportions of these phenotypes are easily determined by probability calculations. Let's see how this works for the experiment shown in Figure 10.7.

Using the principle described above, we can calculate that the probability that an F_2 seed will be spherical is ¾: the probability of an *Ss* heterozygote (½) plus the probability of an *SS* homozygote (¼) = ¾. By the same reasoning, the probability that a seed will be yellow is also ¾. The two characters are determined by separate genes and are independent of each other, so the joint probability that a seed will be both spherical and yellow is ¾ × ¾ = 9/16. What is the probability of F_2 seeds being both wrinkled and yellow? The probability of being yellow is again ¾; the probability of being wrinkled is ½ × ½ = ¼. The joint probability that a seed will be both wrinkled and yellow, then, is ¼ × ¾ = 3/16. The same probability applies, for similar reasons, to spherical, green F_2 seeds. Finally, the probability that F_2 seeds will be both wrinkled and green is ¼ × ¼ = 1/16. Looking at all four phenotypes, we see they are expected in the ratio of 9:3:3:1.

Probability calculations and Punnett squares give the same results. Learn to do genetics problems both ways, and then decide which method you prefer.

Mendel's laws can be observed in human pedigrees

After Mendel's work was uncovered by plant breeders, Mendelian inheritance was observed in humans. Currently, the patterns of over 2,500 inherited human characteristics have been described.

How can Mendel's laws of inheritance be applied to humans? Mendel worked out his laws by performing many planned crosses and counting many offspring. Neither of these approaches is possible with humans. So human geneticists rely on **pedigrees**, family trees that show the occurrence of phenotypes (and alleles) in several generations of related individuals.

Because humans have such small numbers of offspring, human pedigrees do not show the clear proportions of offspring phenotypes that Mendel saw in his pea plants (see Table 10.1). For example, when two people who are both heterozygous for a recessive allele (say, *Aa*) marry, there will be, for each of their children, a 25 percent probability that the child will be a recessive homozygote (*aa*). Thus, over many such marriages, one-fourth of all the children will be recessive homozygotes (*aa*). But the offspring of a single marriage are likely to be too few to show the exact one-fourth proportion. In a family with only two children, for example, both could easily be *aa* (or *Aa*, or *AA*).

To deal with this ambiguity, human geneticists assume that any allele that causes an abnormal phenotype is rare in the human population. This means that if some members of a given family have a rare allele, it is highly unlikely that an outsider marrying into that family will have that same rare allele.

Human geneticists may wish to know whether a particular rare allele is dominant or recessive. Figure 10.10 depicts a pedigree showing the pattern of inheritance of a rare domi-

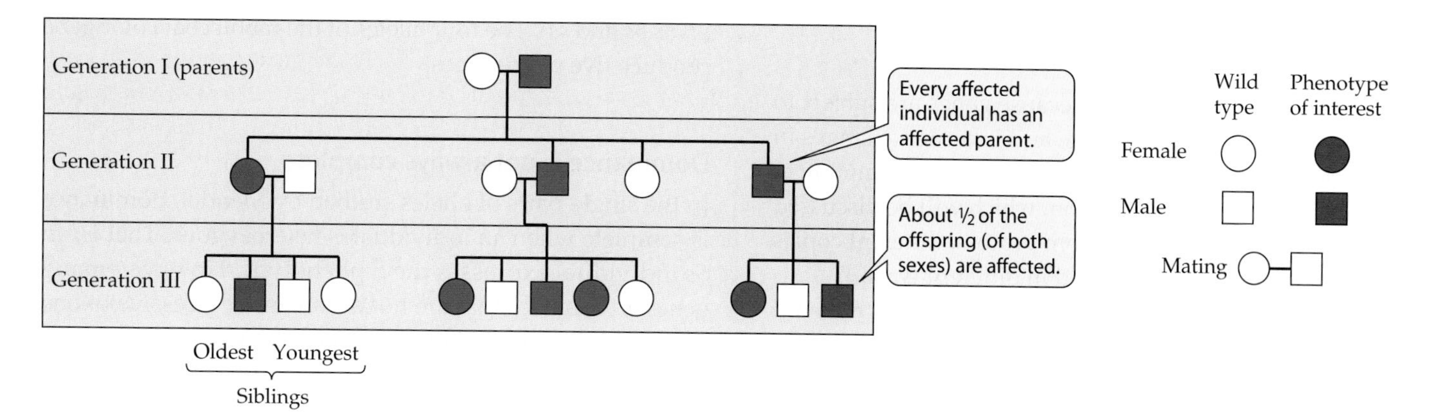

10.10 Pedigree Analysis and Dominant Inheritance This pedigree represents a family affected by Huntington's disease, which results from a rare dominant allele. Everyone who inherits this allele is affected.

nant allele. The following are the key features to look for in such a pedigree:

- Every affected person has an affected parent.
- About half of the offspring of an affected parent are also affected.
- The phenotype occurs equally in both sexes.

Compare this pattern with Figure 10.11, which shows the pattern of inheritance of a rare recessive allele:

- Affected people usually have two parents who are not affected.
- In affected families, about one-fourth of the children of unaffected parents can be affected.
- The phenotype occurs equally in both sexes.

In pedigrees showing inheritance of a recessive phenotype, it is not uncommon to find a marriage of two relatives. This pattern is a result of the rarity of recessive alleles that give rise to abnormal phenotypes. For two phenotypically normal parents to have an affected child (*aa*), the parents must both be heterozygous (*Aa*). If a particular recessive allele is rare in the general population, the chance of two people marrying who are both carrying that allele is quite low. On the other hand, if that allele is present in a family, two cousins might share it (see Figure 10.10). This is why studies on populations isolated either culturally (by religion, as with the Amish in the United States) or geographically (as on islands) have been so valuable to human geneticists. People in these groups tend to have large families, or to marry among themselves, or both.

Because the major use of pedigree analysis is in the clinical evaluation and counseling of patients with inherited abnormalities, a single pair of alleles is usually followed. However, just as pedigree analysis shows the segregation of alleles, it also can show independent assortment if two different allele pairs are considered.

Alleles and Their Interactions

In many cases, alleles do not show the simple relationships between dominance and recessiveness that we have described. In others, a single allele may have multiple phenotypic effects. Existing alleles can give rise to new alleles by mutation, so there can be many alleles for a single character.

Female / Male — Heterozygous for phenotype of interest (inferred)

Mating between relatives

1 One parent is heterozygous...

2 ...and the recessive allele is passed on to one-half of the phenotypically normal offspring.

3 These cousins are heterozygous.

4 Mating of heterozygous recessive parents may produce homozygous recessive (affected) offspring.

Generation I (Parent)

Generation II

Generation III

Generation IV

10.11 Recessive Inheritance This pedigree represents a family that carries the allele for albinism, a recessive trait. Because the trait is recessive, heterozygotes do not have the albino phenotype, but they can pass the allele on to their offspring. Affected persons must inherit the allele from two heterozygous parents or (rarely) from one homozygous and one heterozygous parent. In this family, the heterozygous parents are cousins, but the same result could occur if the parents were unrelated but heterozygous.

New alleles arise by mutation

Different alleles of a gene exist because genes are subject to **mutations**, which are rare, stable, and inherited changes in the genetic material. In other words, an allele can mutate to become a different allele. Mutation, which will be discussed in detail in Chapter 12, is a random process; different copies of the same allele may be changed in different ways.

One particular allele of a gene may be defined as the **wild type**, because it is present in most individuals in nature ("the wild") and gives rise to an expected trait or phenotype. Other alleles of that gene, often called *mutant alleles*, may produce a different phenotype. The wild-type and mutant alleles reside at the same locus and are inherited according to the rules set forth by Mendel. A genetic locus with a wild-type allele that is present less than 99 percent of the time (the rest of the alleles being mutant) is said to be **polymorphic** (from the Greek *poly*, "many," and *morph*, "form").

Many genes have multiple alleles

Because of random mutations, a group of individuals may have more than two alleles of a given gene. (Any one individual has only two alleles, of course—one from its mother and one from its father.) In fact, there are many examples of such multiple alleles.

Coat color in rabbits is determined by one gene with four alleles. There is a dominance hierarchy among these alleles:

$$C > c^{ch} > c^{h} > c$$

Any rabbit with the C allele (paired with any of the four) is dark gray, and a rabbit with *cc* is albino. The intermediate colors result from the different allelic combinations shown in Figure 10.12.

Multiple alleles increase the number of possible phenotypes. In Mendel's monohybrid cross, there was just one pair of alleles (*Ss*) and two possible phenotypes (resulting from *SS* or *Ss* and *ss*). The four alleles of the rabbit coat color gene produce five phenotypes.

Dominance is not always complete

In the single pairs of alleles studied by Mendel, dominance is complete when an individual is heterozygous. That is, an *Ss* individual expresses the *S* phenotype. However, many genes have alleles that are not dominant or recessive to one another. Instead, the heterozygotes show an intermediate phenotype—at first glance, like that predicted by the old blending theory of inheritance. For example, if a true-breeding red snapdragon is crossed with a true-breeding white one, all the F_1 flowers are pink. That this phenomenon can still be explained in terms of Mendelian genetics, rather than blending, is readily demonstrated by a further cross.

The blending theory predicts that if one of the pink F_1 snapdragons is crossed with a true-breeding white one, all the offspring should be a still lighter pink. In fact, approximately ½ of the offspring are white, and ½ are the same shade of pink as the F_1 parent. When the F_1 pink snapdragons are allowed to self-pollinate, the resulting F_2 plants are distributed in a ratio of 1 red:2 pink:1 white (Figure 10.13). Clearly the hereditary particles—the genes—have not blended; they are readily sorted out in the F_2.

We can understand these results in terms of the Mendelian laws of inheritance. All we need to do is recognize that the heterozygotes show a phenotype intermediate between those of the two homozygotes. In such cases, the gene is said to be governed by **incomplete dominance**. Incomplete dominance is common in nature. In fact, Mendel's paper was unusual in that all seven of the examples he described (see Table 10.1) are characterized by complete dominance.

10.12 Inheritance of Coat Color in Rabbits There are four alleles of the gene for coat color in rabbits. Different combinations of two alleles give different coat colors.

Possible genotypes	CC, Cc^{ch}, Cc^{h}, Cc	$c^{ch}c^{ch}$	$c^{ch}c^{h}$, $c^{ch}c$	$c^{h}c^{h}$, $c^{h}c$	cc
Phenotype	Dark gray	Chinchilla	Light gray	Himalayan	Albino

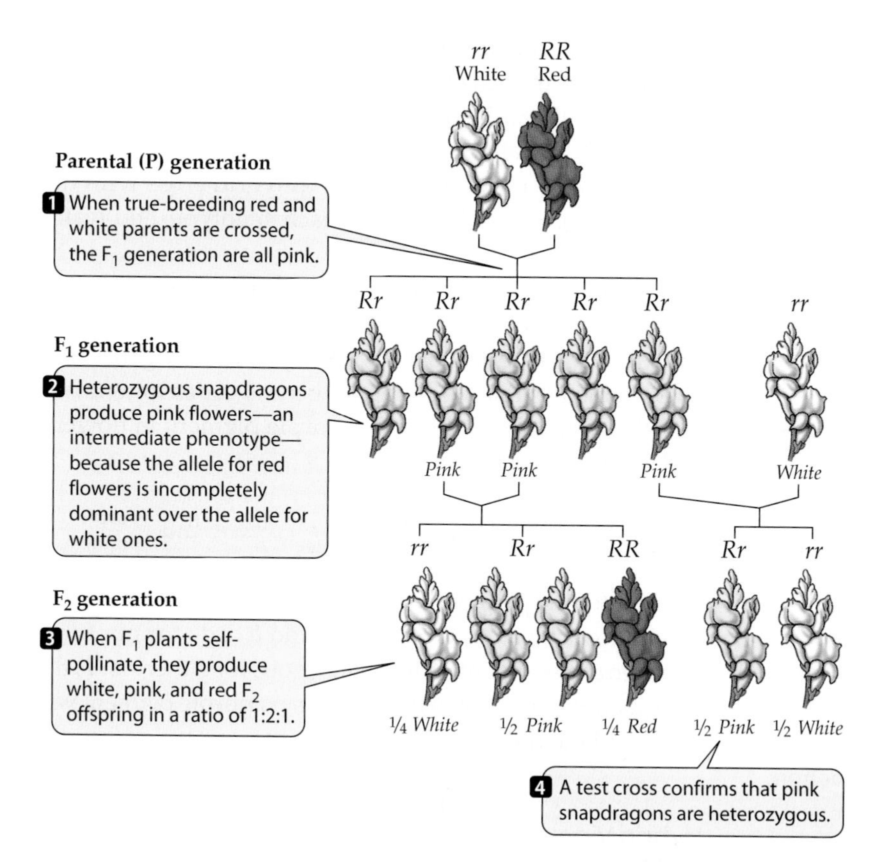

rr
White
RR
Red
Parental (P) generation
1 When true-breeding red and white parents are crossed, the F1 generation are all pink.
Rr
Rr
Rr
Rr
Rr
rr
F1 generation
2 Heterozygous snapdragons produce pink flowers—an intermediate phenotype—because the allele for red flowers is incompletely dominant over the allele for white ones.
Pink
Pink
Pink
White
rr
Rr
RR
Rr
rr
F2 generation
3 When F1 plants self-pollinate, they produce white, pink, and red F2 offspring in a ratio of 1:2:1.
1/4 White
1/2 Pink
1/4 Red
1/2 Pink
1/2 White
4 A test cross confirms that pink snapdragons are heterozygous.

10.13 Incomplete Dominance Follows Mendel's Laws An intermediate phenotype can occur in heterozygotes when neither allele is dominant. The heterozygous phenotype (here, pink flowers) may give the appearance of a blended trait, but the traits of the parental generation reappear in their original forms in succeeding generations, as predicted by Mendel's laws of inheritance.

Blood type of cells	Genotype	Antibodies made by body	Reaction to added antibodies	
			Anti-A	Anti-B
A	I^AI^A or I^Ai^O	Anti-B		
B	I^BI^B or I^Bi^O	Anti-A		
AB	I^AI^B	Neither anti-A nor anti-B		
O	i^Oi^O	Both anti-A and anti-B		

Red blood cells that do not react with antibody remain evenly dispersed.

Red blood cells that react with antibody clump together (speckled appearance).

10.14 ABO Blood Reactions Are Important in Transfusions Red blood cells of types A, B, AB, and O were mixed with serum containing anti-A or anti-B antibodies. As you look down the columns, note that each of the types, when mixed separately with anti-A and with anti-B, gives a unique pair of results; this is the basic method by which blood is typed. People with type O blood are good blood donors because O cells do not react with either anti-A or anti-B antibodies. People with type AB blood are good recipients, since they make neither type of antibody.

In codominance, both alleles are expressed

Sometimes the two alleles at a locus produce two different phenotypes that both appear in heterozygotes. An example of this phenomenon, called **codominance**, is seen in the ABO blood group system in humans.

Early attempts at blood transfusion frequently killed the patient. Around 1900, the Austrian scientist Karl Landsteiner mixed blood cells and serum (blood from which cells have been removed) from different individuals. He found that only certain combinations of blood are compatible. In other combinations, the red blood cells from one individual form clumps in the presence of serum from the other individual. This discovery led to our ability to administer compatible blood transfusions that do not kill the recipient.

Clumps form in incompatible transfusions because specific proteins in the serum, called *antibodies*, react with foreign, or "nonself," cells. The antibodies react with proteins on the surface of nonself cells, called *antigens*. Blood compatibility is determined by a set of three alleles (I^A, I^B, and i^O) at one locus, which determine the antigens on the surface of red blood cells. Different combinations of these alleles in different people produce four different blood types, or phenotypes: A, B, AB, and O (Figure 10.14). The AB phenotype found in individuals of I^AI^B genotype is an example of codominance—these individuals produce cell surface antigens of both the A and B types.

Some alleles have multiple phenotypic effects

Mendel's principles were further extended when it was discovered that a single allele can result in more than one phenotype. When a single allele has more than one distinguishable phenotypic effect, we say that the allele is **pleiotropic**. A familiar example of pleiotropy involves the allele responsible for the coloration pattern (light body, darker extremities) of Siamese cats, discussed later in this chapter. The same allele is also responsible for the characteristic crossed eyes of Siamese cats. Although these effects appear to be unrelated, both result from the same protein produced under the influence of the allele.

Gene Interactions

Thus far we have treated the phenotype of an organism, with respect to a given character, as a simple result of the alleles of a single gene. In many cases, however, several genes interact to determine a phenotype. To complicate things further, the physical environment may interact with the genetic constitution of an individual in determining the phenotype.

Some genes alter the effects of other genes

Epistasis occurs when the phenotypic expression of one gene is affected by another gene. For example, several genes determine coat color in mice. The wild-type color is agouti, a grayish pattern resulting from bands on the individual hairs. The dominant allele *B* determines that the hairs will have bands and thus that the color will be agouti, whereas the homozygous recessive genotype *bb* results in unbanded hairs, and the mouse appears black. A second locus, on another chromosome, affects an early step in the formation of hair pigments. The dominant allele *A* at this locus allows normal color development, but *aa* blocks all pigment production. Thus, *aa* mice are all-white albinos, irrespective of their genotype at the *B* locus (Figure 10.15).

If a mouse with genotype *AABB* (and thus the agouti phenotype) is crossed with an albino of genotype *aabb*, the F_1 mice are *AaBb* and have the agouti phenotype. If the F_1 mice are crossed with each other to produce an F_2 generation, then epistasis will result in an expected phenotypic ratio of 9 agouti:3 black:4 albino. (Can you show why? The underlying ratio is the usual 9:3:3:1 for a dihybrid cross with unlinked genes, but look closely at each genotype, and watch out for epistasis.)

Mice with genotype aa are albino regardless of their genotype for the other locus, because the aa genotype blocks all pigment production.
Mice with bb genotypes are black unless they are also aa (which makes them albino).
Mice that have at least one dominant allele at each locus are agouti.

10.15 Genes May Interact Epistatically Epistasis occurs when one gene alters the phenotypic effect of another gene. In these mice, the presence of the recessive genotype (*aa*) at one locus blocks pigment production, producing an albino mouse no matter what the genotype is at the second locus.

In another form of epistasis, two genes are mutually dependent: The expression of each depends on the alleles of the other. The epistatic action of such **complementary genes** may be explained as follows: Suppose gene *A* codes for enzyme A in the metabolic pathway for purple pigment in flowers, and gene *B* codes for enzyme B:

colorless precursor —enzyme A→ colorless intermediate —enzyme B→ purple pigment

In order for the pigment to be produced, both reactions must take place. The recessive alleles *a* and *b* code for nonfunctional enzymes. If a plant is homozygous for either *a* or *b*, the corresponding reaction will not occur, no purple pigment will form, and the flowers will be white.

Hybrid vigor results from new gene combinations and interactions

If Mendel's paper was the most important event in genetics in the nineteenth century, perhaps an equally important paper in applied genetics was published early in the twentieth century by G. H. Shull, titled "The composition of a field of maize." Farmers growing crops have known for centuries that mating among close relatives (known as *inbreeding*) can result in offspring of lower quality than those from matings between unrelated individuals. The reason for this is that close relatives tend to have the same recessive alleles, some of which may be harmful, as we saw in our discussion of human pedigrees above. In fact, it has long been known that if one crosses two true-breeding, homozygous genetic strains of a plant or animal, the result is offspring that are phenotypically much stronger, larger, and in general more "vigorous" than either of the parents (Figure 10.16).

Shull began his experiment with two of the thousands of existing varieties of corn (maize). Both varieties produced about 20 bushels of corn per acre. But when he crossed them, the yield of their offspring was an astonishing 80 bushels per acre. This phenomenon is known as **heterosis** (short for heterozygosis), or **hybrid vigor**. The cultivation of hybrid corn spread rapidly in the United States and all over the world, quadrupling grain production. The practice of hybridization has spread to many other crops and animals used in agriculture.

The actual mechanism by which heterosis works is not known. A widely accepted hypothesis is *overdominance*, in

10.16 Hybrid Vigor in Corn The heterozygous F_1 offspring is larger and more vigorous than either homozygous parent.

which the heterozygous condition in certain important genes is superior to either homozygote.

The environment affects gene action

The phenotype of an individual does not result from its genotype alone. Genotype and environment interact to determine the phenotype of an organism. Environmental variables such as light, temperature, and nutrition can affect the translation of a genotype into a phenotype.

A familiar example of this phenomenon involves the Siamese cat. This handsome animal normally has darker fur on its ears, nose, paws, and tail than on the rest of its body. These darkened extremities normally have a lower temperature than the rest of the body. A few simple experiments show that the Siamese cat has a genotype that results in dark fur, but only at temperatures below the general body temperature. If some dark fur is removed from the tail and the cat is kept at higher than usual temperatures, the new fur that grows in is light. Conversely, removal of light fur from the back, followed by local chilling of the area, causes the spot to fill in with dark fur.

Two parameters describe the effects of genes and environment on the phenotype:

- **Penetrance** is the proportion of individuals in a group with a given genotype that actually show the expected phenotype.
- **Expressivity** is the degree to which a genotype is expressed in an individual.

For an example of environmental effects on expressivity, consider how Siamese cats kept indoors or outdoors in different climates might look.

Most complex phenotypes are determined by multiple genes and environment

The differences between individual organisms in simple characters, such as those that Mendel studied in peas, are discrete and **qualitative**. For example, the individuals in a population of peas are either short or tall. For most complex characters, however, such as height in humans, the phenotype varies more or less continuously over a range. Some people are short, others are tall, and many are in between the two extremes. Such variation within a population is called **quantitative**, or **continuous**, variation. In most cases, quantitative variation is due to two factors (Figure 10.17): multiple genes, each with multiple alleles, and environmental influences on the expression of these genes.

Geneticists call the genes that together determine a complex character **quantitative trait loci**. Identifying these loci is a major challenge, and an important one. For example, the amount of grain that a variety of rice produces in a growing season is determined by many interacting genetic factors. Crop plant breeders have worked hard to decipher these fac-

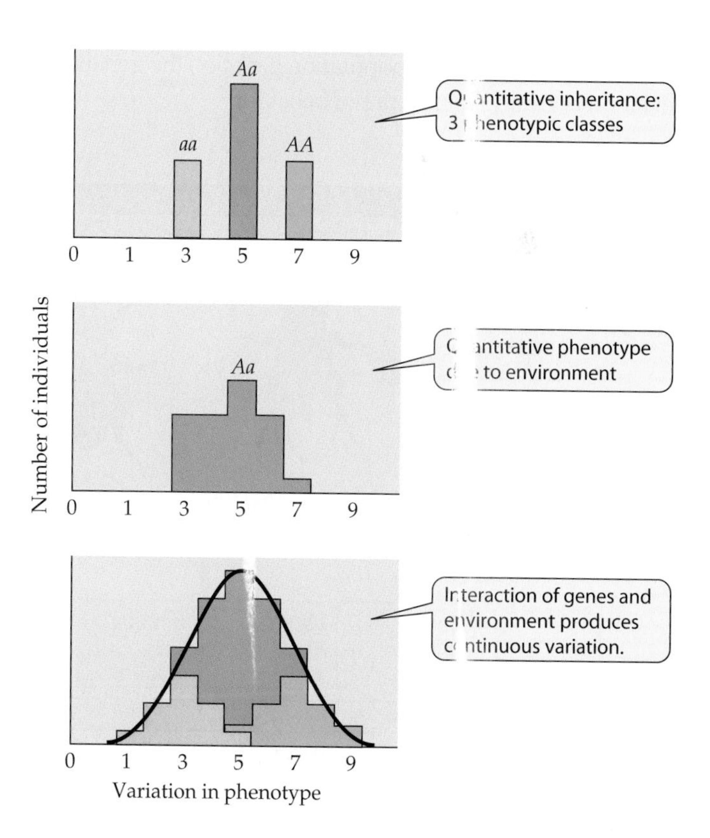

10.17 Quantitative Variation Quantitative variation is produced by the interaction of genes and environment. In this illustration, only a single gene with three alleles is considered. Most complex characters are determined by many genes and alleles, with the environment exerting an influence on each.

tors in order to breed higher-yielding rice strains. In a similar way, human characteristics such as disease susceptibility and behavior are caused in part by quantitative trait loci.

Genes and Chromosomes

The recognition that genes occupy characteristic positions on chromosomes and are segregated by meiosis enabled Mendel's successors to provide a physical explanation for his model of inheritance. It soon became apparent that the association of genes with chromosomes has other genetic consequences as well. We mentioned above that genes located on the same chromosome may not follow Mendel's law of independent assortment. What is the pattern of inheritance of such genes? How do we determine where genes are located on a chromosome, and the distances between them?

The answers to these and many other genetic questions were worked out in studies of the fruit fly *Drosophila melanogaster*. Its small size, its ease of cultivation, and its short generation time made this animal an attractive experimental subject. Beginning in 1909, Thomas Hunt Morgan and his students pioneered the study of *Drosophila* in Columbia University's famous "fly room," where they discovered the phenomena described in this section. *Drosophila* remains extremely important in studies of chromosome structure, population genetics, the genetics of development, and the genetics of behavior.

Genes on the same chromosome are linked

Some of the crosses Morgan performed with fruit flies resulted in phenotypic ratios that were not in accord with those predicted by Mendel's law of independent assortment. Morgan crossed *Drosophila* of two known genotypes, *BbVgvg* × *bbvgvg*, for two different characters, body color and wing shape:

- *B* (wild-type gray body), is dominant over *b* (black body)
- *Vg* (wild-type wing) is dominant over *vg* (vestigial, a very small wing)

(Do you recognize this type of cross? It is a test cross for the two gene pairs; see Figure 10.6.)

Morgan expected to see four phenotypes in a ratio of 1:1:1:1, but that is not what he observed. The body color gene and the wing size gene were not assorting independently; rather, they were for the most part inherited together (Figure 10.18).

These results became understandable to Morgan when he assumed that the two loci are on the same chromosome—that is, that they are linked. After all, since the number of genes in a cell far exceeds the number of chromosomes, each chromosome must contain many genes. The full set of loci on a given chromosome constitutes a *linkage group*. The number of linkage groups in a species equals the number of homologous chromosome pairs.

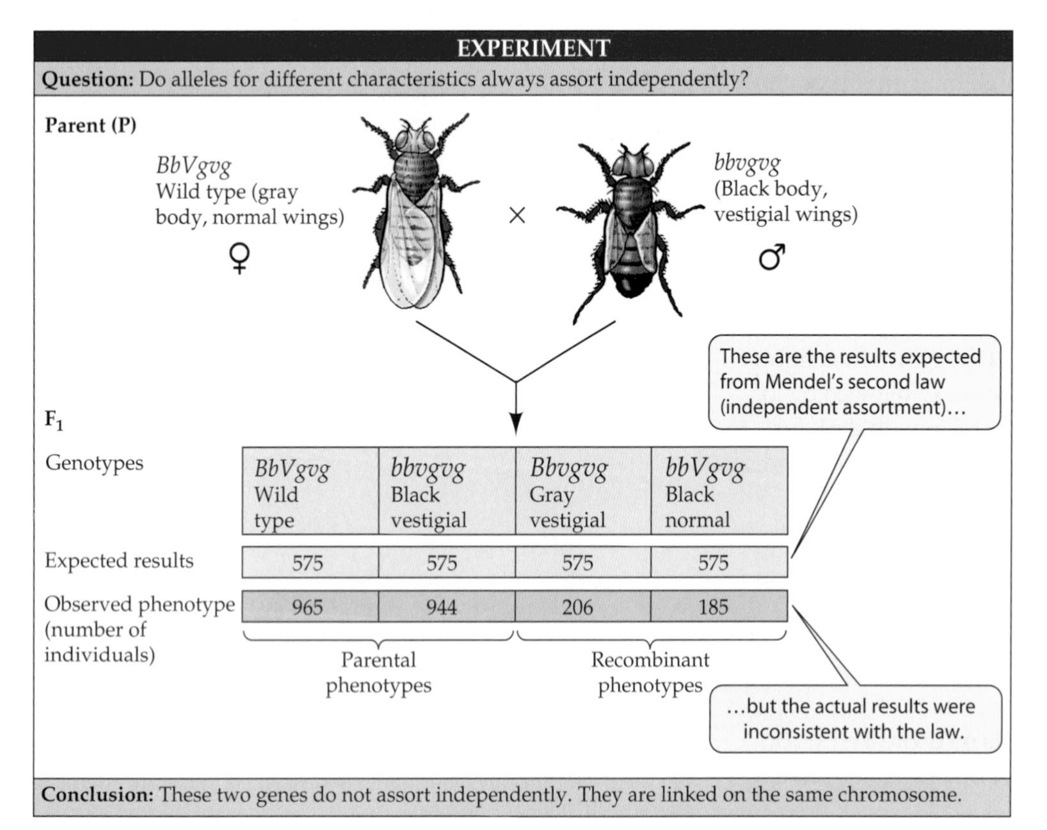

10.18 Some Alleles Do Not Assort Independently Morgan's studies showed that the genes for body color and wing size in *Drosophila* are linked, so their alleles do not assort independently. Linkage accounts for the departure of the phenotype ratios observed from the results predicted by Mendel's law of independent assortment.

Suppose, now, that the *Bb* and *Vgvg* loci are indeed located on the same chromosome. Why, then, didn't all of Morgan's F_1 flies have the parental phenotypes—that is, why did his cross result in anything other than gray flies with normal wings (wild-type) and black flies with vestigial wings? If we assumed that linkage is *absolute*—that is, that chromosomes always remain intact and unchanged—we would expect to see just those two types of progeny. However, this is not always what happens.

Genes can be exchanged between chromatids

Absolute linkage is extremely rare. If linkage were absolute, Mendel's law of independent assortment would apply only to loci on different chromosomes. What actually happens is more complex, and therefore more interesting. Chromosomes are not unbreakable, so recombination of genes can occur. That is, genes at different loci on the same chromosome do sometimes separate from one another during meiosis.

Genes may recombine when two homologous chromosomes physically exchange corresponding segments during prophase I of meiosis—that is, by crossing over (Figure 10.19; see also Figure 9.16). Recall from Chapter 9 that the DNA is replicated during the S phase, so that by prophase I, when homologous chromosome pairs come together to form tetrads, each chromosome consists of two chromatids. The exchange event involves only two of the four chromatids in a tetrad, one from each member of the homologous pair, and can occur at any point along the length of the chromosome. The chromosome segments involved are exchanged reciprocally, so both chromatids involved in crossing over become recombinant (that is, each chromatid ends up with genes from both of the organism's parents). Usually several exchange events occur along the length of each homologous pair.

When crossing over takes place between two linked genes, not all progeny of a cross will have the parental phenotypes. Instead, recombinant offspring appear as well, as they did in Morgan's cross. They appear in proportions called **recombinant frequencies**, which are calculated by dividing the number of recombinant progeny by the total number of progeny (Figure 10.20). Recombinant frequencies will be greater for loci that are farther apart on the chromosome than for loci that are closer together, because an exchange event is more likely to occur between genes that are far apart than between genes that are close together.

Geneticists can make maps of chromosomes

If two loci are very close together on a chromosome, the odds of crossing over between them are small. In contrast, if two loci are far apart, crossing over could occur between them at

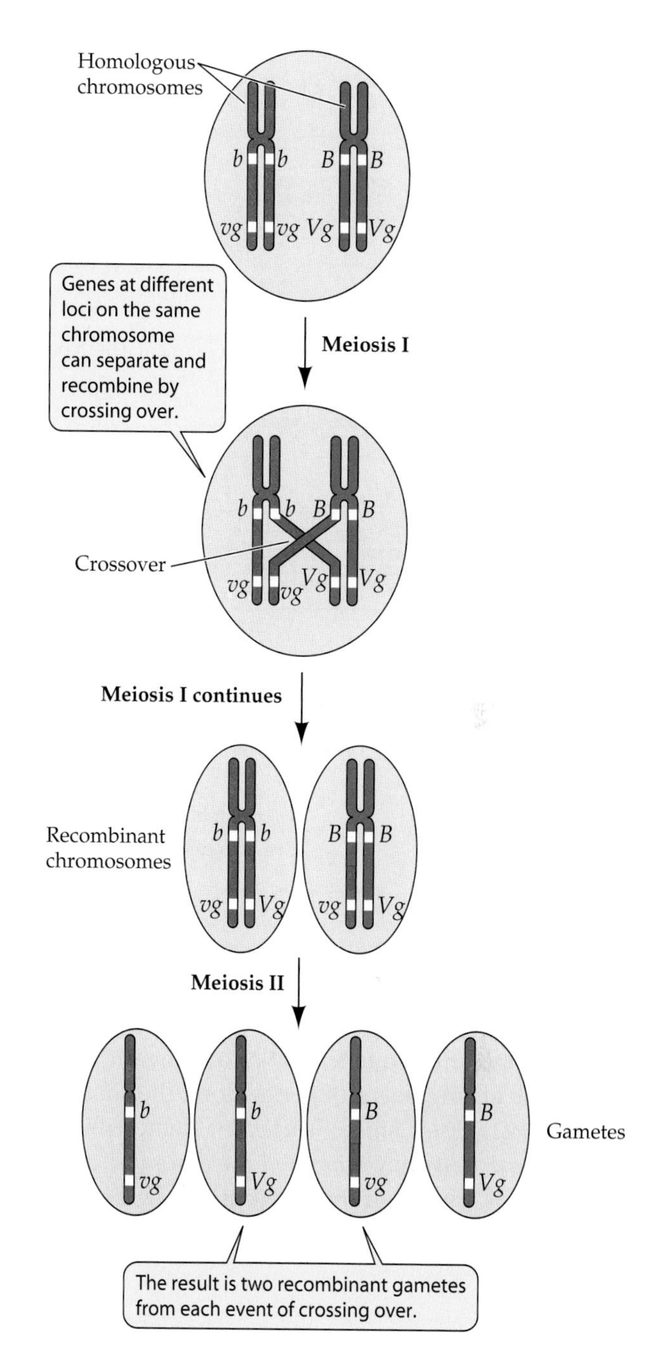

10.19 Crossing Over Results in Genetic Recombination Genes at different loci on the same chromosome can be separated from one another and recombined by crossing over. Such recombination occurs during prophase I of meiosis.

many points. In a population of cells undergoing meiosis, a greater proportion of the cells will undergo recombination between two loci that are far apart than between two loci that are close together. In 1911, Alfred Sturtevant, then an undergraduate student in T. H. Morgan's fly room, realized how that simple insight could be used to show where different genes lie on a chromosome in relation to one another.

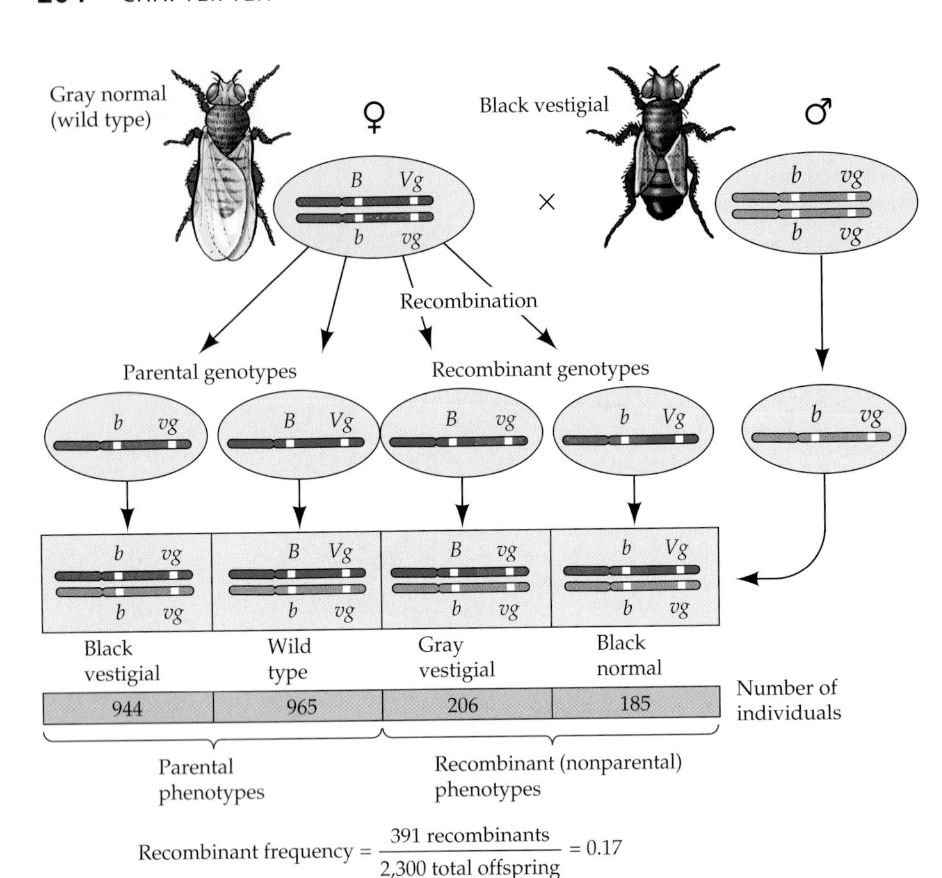

10.20 Recombinant Frequencies The frequency of recombinant offspring (those with a phenotype different from either parent) can be calculated. Recombinant frequencies will be larger for loci that are far apart than for those that are close together on the chromosome.

The Morgan group had determined recombinant frequencies for many pairs of linked genes. Sturtevant used these recombinant frequencies to create **genetic maps** that showed the arrangement of genes along the chromosome (Figure 10.21). Ever since Sturtevant demonstrated this method, geneticists have mapped the chromosomes of eukaryotes, prokaryotes, and viruses, assigning distances between genes in **map units**. A map unit corresponds to a recombinant frequency of 0.01; it is also referred to as a *centimorgan* (cM), in honor of the founder of the fly room. You, too, can work out a genetic map (Figure 10.22).

10.21 Steps toward a Genetic Map Because the chance of a recombinant genotype occurring increases with the distance between two loci on a chromosome, Sturtevant was able to derive this partial map of a *Drosophila* chromosome from the Morgan group's data on the recombinant frequencies of five recessive traits. He used an arbitrary unit of distance—the map unit, or centimorgan (cM)—equivalent to a recombinant frequency of 0.01.

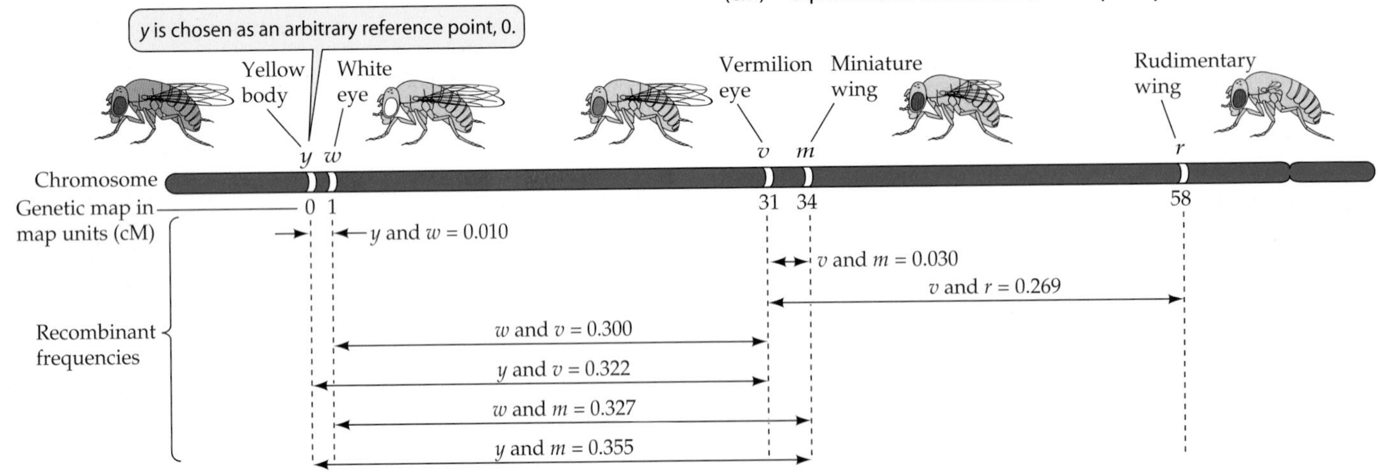

1 At the outset, we have no idea of the individual distances, and there are several possible sequences (*a-b-c*, *a-c-b*, *b-a-c*).

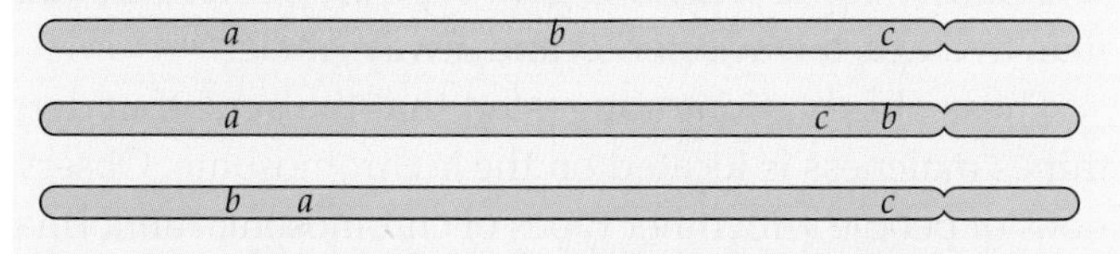

We make a cross *AABB* × *aabb*, and obtain an F_1 generation with a genotype *AaBb*. We test cross these *AaBb* individuals with *aabb*. Here are the genotypes of the first 1,000 progeny:

450 *AaBb*, 450 *aabb*, 50 *Aabb*, and 50 *aaBb*.

2 How far apart are the *a* and *b* genes? Well, what is the recombinant frequency? Which are the recombinant types, and which are the parental types?

Recombinant frequency (*a* to *b*) = (50 + 50)/1,000 = 0.1
So the map distance is
Map distance = 100 × recombinant frequency =
100 × 0.1 = 10 cM

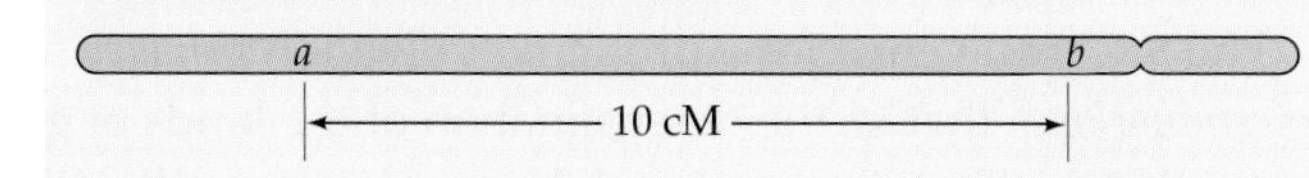

3 Now we make a cross *AACC* × *aacc*, obtain an F_1 generation, and test cross it, obtaining:

460 *AaCc*, 460 *aacc*, 40 *Aacc*, and 40 *aaCc*.

How far apart are the *a* and *c* genes?

Recombinant frequency (*a* to *c*) = (40 + 40)/1,000 = 0.08
Map distance = 100 × recombinant frequency =
100 × 0.08 = 8 cM

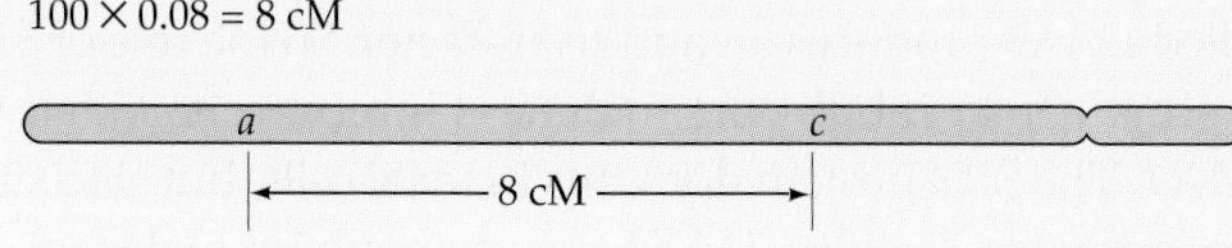

10.22 Map These Genes The object of this exercise is to determine the order of three loci (*a*, *b*, and *c*) on a chromosome, as well as the map distances (in cM) between them.

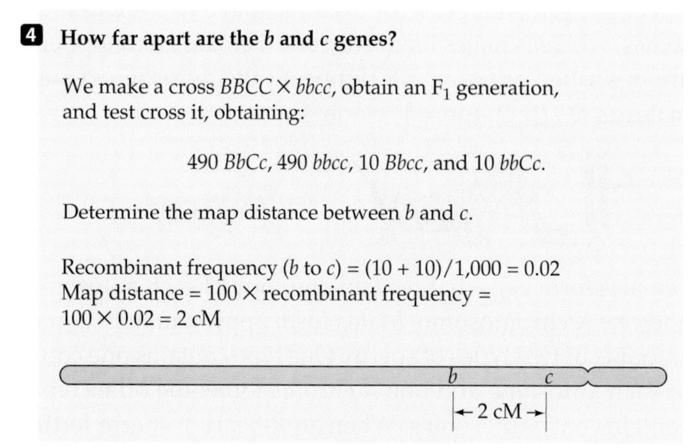

4 How far apart are the *b* and *c* genes?

We make a cross *BBCC* × *bbcc*, obtain an F_1 generation, and test cross it, obtaining:

490 *BbCc*, 490 *bbcc*, 10 *Bbcc*, and 10 *bbCc*.

Determine the map distance between *b* and *c*.

Recombinant frequency (*b* to *c*) = (10 + 10)/1,000 = 0.02
Map distance = 100 × recombinant frequency =
100 × 0.02 = 2 cM

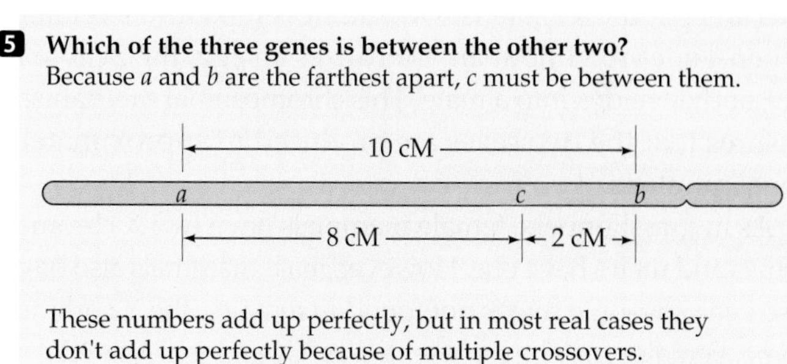

5 Which of the three genes is between the other two?
Because *a* and *b* are the farthest apart, *c* must be between them.

These numbers add up perfectly, but in most real cases they don't add up perfectly because of multiple crossovers.

Sex Determination and Sex-Linked Inheritance

In Mendel's work, reciprocal crosses always gave identical results; it did not matter, in general, whether a dominant allele was contributed by the mother or by the father. But in some cases, the parental origin of a chromosome does matter. For example, as we saw at the beginning of this chapter, human males inherit hemophilia A from their mother, not from their father. To understand the types of inheritance in which the parental origin of an allele is important, we must consider the ways in which sex is determined in different species.

Sex is determined in different ways in different species

In corn, a plant much studied by geneticists, every diploid adult has both male and female reproductive structures. The tissues in these two types of structures are genetically identical, just as roots and leaves are genetically identical. Plants such as corn, in which the same individual produces both male and female gametes, are said to be *monoecious* (from the Greek, "one house"). Other plants, such as date palms and oak trees, and most animals are *dioecious* ("two houses"), meaning that some individuals can produce only male gametes and the others can produce only female gametes. In other words, dioecious organisms have two sexes.

In most dioecious organisms, sex is determined by differences in the chromosomes, but such determination operates in different ways in different groups of organisms. For example, the sex of a honeybee depends on whether it develops from a fertilized or an unfertilized egg. A fertilized egg is diploid and gives rise to a female bee—either a worker or a queen, depending on the diet during larval life (again, note how the environment affects the phenotype). An unfertilized egg is haploid and gives rise to a male drone:

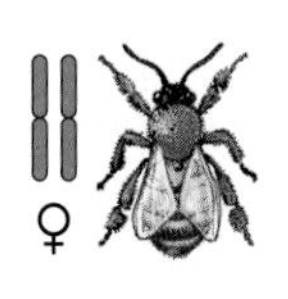

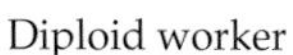
Diploid worker

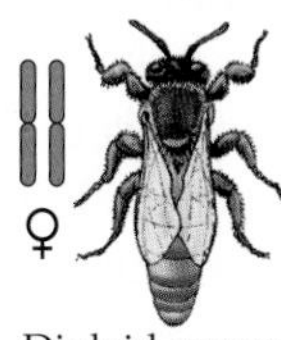

Diploid queen

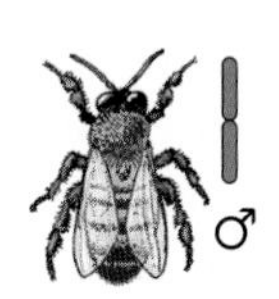

Haploid drone

In many other animals, including humans, sex is determined by a single **sex chromosome**, or by a pair of them. Both males and females have two copies of each of the rest of the chromosomes, which are called **autosomes**.

Female grasshoppers, for example, have two X chromosomes, whereas males have only one. Female grasshoppers are described as being XX (ignoring the autosomes) and males as XO (pronounced "ex-oh"):

Females form eggs that contain one copy of each autosome and one X chromosome. Males form approximately equal amounts of two types of sperm: One type contains one copy of each autosome and one X chromosome; the other type contains only autosomes. When an X-bearing sperm fertilizes an egg, the zygote is XX, and develops into a female. When a sperm without an X fertilizes an egg, the zygote is XO, and develops into a male. This chromosomal mechanism ensures that the two sexes are produced in approximately equal numbers.

As in grasshoppers, female mammals have two X chromosomes and males have one. However, male mammals also have a sex chromosome that is not found in females: the Y chromosome. Females may be represented as XX and males as XY:

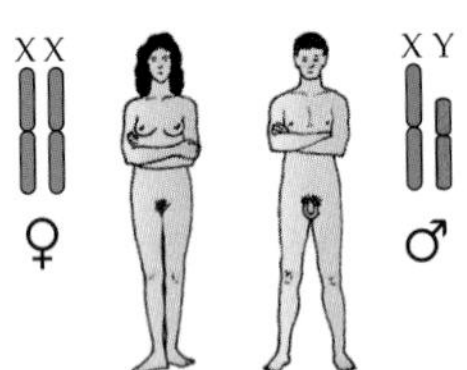

Males produce two kinds of gametes. Each gamete has a complete set of autosomes, but half the gametes carry an X chromosome and the other half carry a Y. When an X-bearing sperm fertilizes an egg, the resulting XX zygote is female; when a Y-bearing sperm fertilizes an egg, the resulting XY zygote is male.

The X and Y chromosomes have different functions

Some subtle but important phenotypic differences show up clearly in mammals with abnormal sex chromosome constitutions. These conditions, which result from nondisjunctions, as described in Chapter 9, tell us something about the functions of the X and Y chromosomes. In humans, XO individuals sometimes appear. Human XO individuals are females who are physically moderately abnormal but mentally normal; usually they are also sterile. The XO condition in humans is called *Turner syndrome*. It is the only known case in which a human can survive with only one member of a chromosome pair (here, the XY pair), although most XO conceptions terminate spontaneously early in development. XXY individuals also occur; this condition is known as *Klinefelter syndrome*. People with this genotype are sometimes taller than average, always sterile, and always male.

These observations suggested that the gene that determines maleness is located on the Y chromosome. Observations of people with other types of chromosomal abnormalities helped researchers to pinpoint the location of that gene:

- Some XY individuals are phenotypically women and lack a small portion of the Y chromosome.
- Some men are genetically XX and have a small piece of the Y chromosome present but attached to another chromosome.

The Y fragment that is missing and present in these two examples, respectively, contains the maleness-determining gene, which was named *SRY* (sex-determining region on the Y chromosome).

The *SRY* gene encodes a protein involved in *primary sex determination*—that is, the determination of the kinds of gametes that will be produced and the organs that will make them. In the presence of functional SRY protein, the embryo develops sperm-producing testes. (Notice that italic type is used for the name of a gene, but roman type is used for the name of a protein.) If the embryo has no Y chromosome, the *SRY* gene is absent, and thus the SRY protein is not made. In the absence of the SRY protein, the embryo develops egg-producing ovaries. In this case, a gene on the X chromosome called *DAX1* produces an anti-testis factor. So the role of *SRY* in a male is to inhibit the maleness inhibitor encoded by *DAX1*. The SRY protein does this in male cells, but since it is not present in females, *DAX1* can act to inhibit maleness.

Primary sex determination is not the same as *secondary sex determination*, which results in the outward manifestations of maleness and femaleness (body type, breast development, body hair, and voice). These outward characteristics are not determined directly by the presence or absence of the Y chromosome. Rather, they are determined by genes scattered on the autosomes and X chromosome that control the actions of hormones, such as testosterone and estrogen.

The Y chromosome functions differently in *Drosophila melanogaster*. Superficially, *Drosophila* follows the same pattern of sex determination as mammals—females are XX and males are XY. However, XO individuals are males (rather than females as in mammals) and almost always are indistinguishable from normal XY males except that they are sterile. XXY *Drosophila* are normal, fertile females:

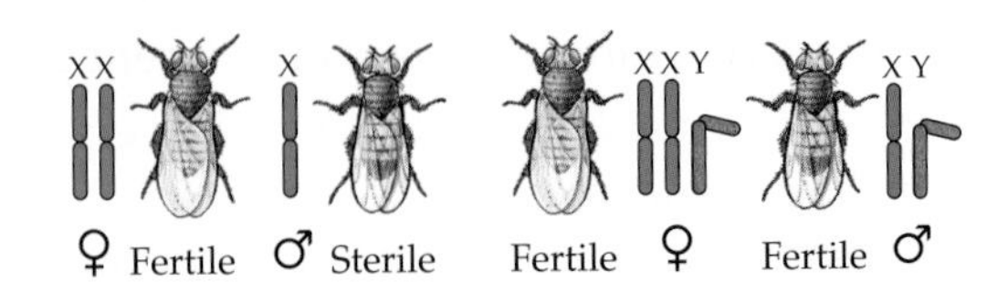

Thus, in *Drosophila*, sex is determined by the ratio of X chromosomes to autosome sets. If there is one X chromosome for each set of autosomes, the individual is a female; if there is only one X chromosome for the two sets of autosomes, the individual is a male. The Y chromosome plays no sex-determining role in *Drosophila*, but it is needed for male fertility.

Caenorhabditis elegans is a favorite model organism for studies of development (see Chapter 19). This tiny worm has two sexes: male and hermaphrodite (self-fertilizing). As in fruit flies, sex is determined by the X:autosome ratio-individuals with a ratio below 0.67 are male.

In birds, moths, and butterflies, males are XX and females are XY. To avoid confusion, these forms are usually expressed as ZZ (male) and ZW (female):

In these organisms, the female produces two types of gametes, carrying Z or W. Whether the egg is Z or W determines the sex of the offspring, in contrast to humans and fruit flies, in which the sperm, carrying either X or Y, determines the sex.

Genes on sex chromosomes are inherited in special ways

Genes on sex chromosomes do not show the Mendelian patterns of inheritance we have described above. In *Drosophila* and in humans, the Y chromosome carries few known genes, but a substantial number of genes affecting a great variety of characters are carried on the X chromosome. Any such gene is present in two copies in females, but in only one copy in males. Therefore, females may be heterozygous for genes that are on the X chromosome, but males will always be **hemizygous** for genes on the X chromosome—they will have only one copy of each, and it will be expressed. Thus, reciprocal crosses do not give identical results for characters whose genes are carried on the sex chromosomes, and these characters do not show the usual Mendelian ratios for the inheritance of genes located on autosomes.

The first and still one of the best examples of inheritance of characters governed by loci on the sex chromosomes (**sex-linked** inheritance) is that of eye color in *Drosophila*. The wild-type eye color of these flies is red. In 1910, Morgan discovered a mutation that causes white eyes. He experimented by crossing flies of the wild-type and mutant phenotypes. His results demonstrated that the eye color locus is on the X chromosome. Study Figure 10.23 as you follow the crosses and results:

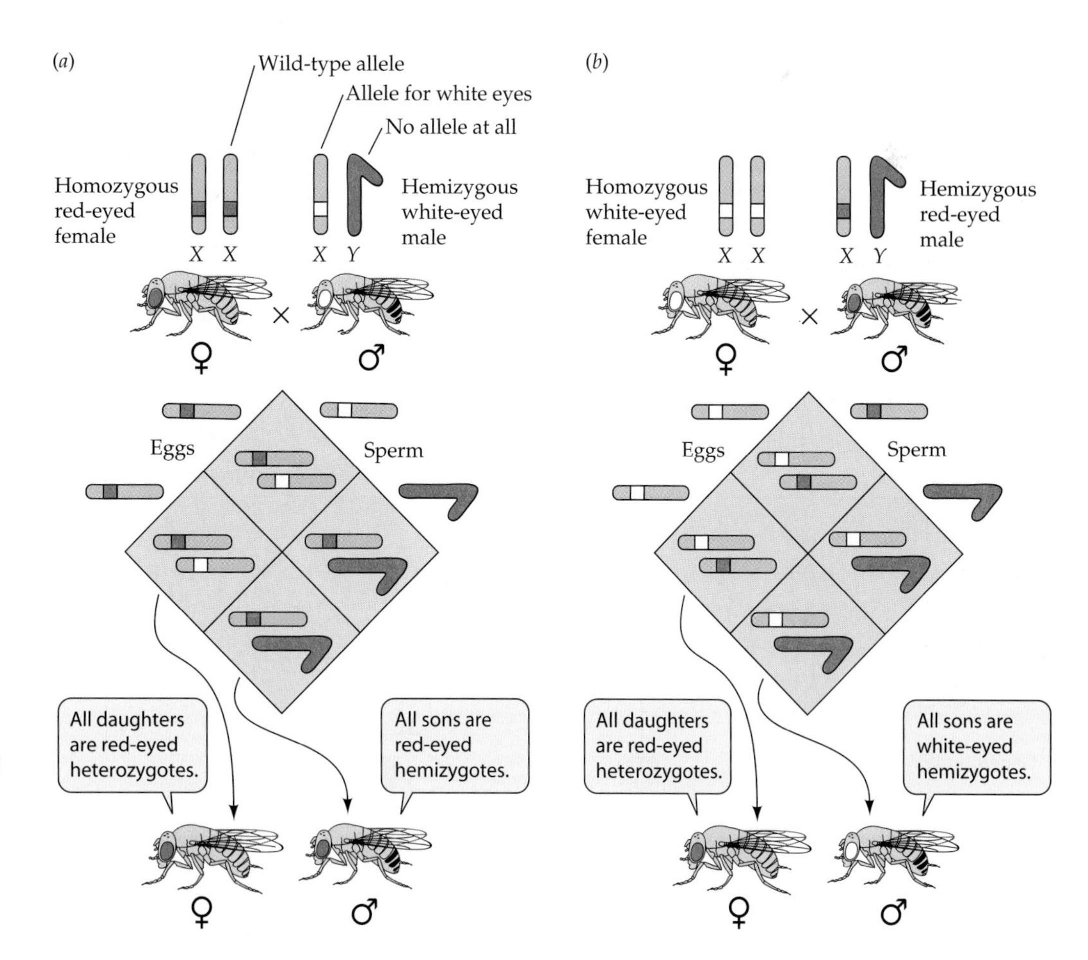

10.23 Eye Color Is a Sex-Linked Trait in *Drosophila* Thomas Hunt Morgan demonstrated that a mutant allele that causes white eyes in *Drosophila* is carried on the X chromosome. Note that in this case, the reciprocal crosses do not have the same results.

- When a homozygous red-eyed female was crossed with a (hemizygous) white-eyed male, all the sons and daughters had red eyes, because red is dominant over white and all the progeny had inherited a wild-type X chromosome from their mothers (Figure 10.23*a*).
- However, in the reciprocal cross, in which a white-eyed female was mated with a red-eyed male, all the sons were white-eyed and all the daughters were red-eyed (Figure 10.23*b*).
- The sons from the reciprocal cross inherited their only X chromosome from their white-eyed mother; the Y chromosome they inherited from their father does not carry the eye color locus (Figure 10.23*b*).
- The daughters, on the other hand, got an X chromosome bearing the white allele from their mother and an X chromosome bearing the red allele from their father; they were therefore red-eyed heterozygotes (Figure 10.23*b*).
- When heterozygous females were mated with red-eyed males, half their sons had white eyes, but all their daughters had red eyes.

Together, these results showed that eye color was carried on the X chromosome and not on the Y.

Humans display many sex-linked characters

The human X chromosome carries about two thousand genes. The alleles at these loci follow the same pattern of inheritance as those for white eyes in *Drosophila*. One human X chromosome gene, for example, has a mutant recessive allele that leads to red-green color blindness, a hereditary disorder. Red-green color blindness appears in individuals who are homozygous or hemizygous for the mutant allele.

Pedigree analysis of X-linked recessive phenotypes (Figure 10.24) reveals the following patterns:

- The phenotype appears much more often in males than in females, because only one copy of the rare allele is needed for its expression in males, while two copies must be present in females.
- A male with the mutation can pass it on only to his daughters; all his sons get his Y chromosome.
- Daughters who receive one mutant X chromosome are heterozygous *carriers*. They are phenotypically normal, but they can pass the mutant X to both sons and daughters (but do so only half of the time, on average, since half of their X chromosomes carry the normal allele).
- The mutant phenotype can skip a generation if the mutation passes from a male to his daughter (who will be phenotypically normal) and thus to her son.

Hemophilia A, which affected the family described at the beginning of this chapter, is an X-linked recessive phenotype, as are several other important human diseases, as we will see in later chapters. Human mutations inherited as X-linked dominant phenotypes are rarer than X-linked recessives because dominant phenotypes appear in every generation, and because people carrying the harmful mutation, even as heterozygotes, often fail to survive and reproduce. (Look at the four points above and try to determine what would happen if the mutation were dominant.)

The small human Y chromosome carries several dozen genes. Among them is the maleness determinant, *SRY*. Interestingly, for some genes on the Y, there are similar, but not identical, genes on the X. For example, one of the proteins

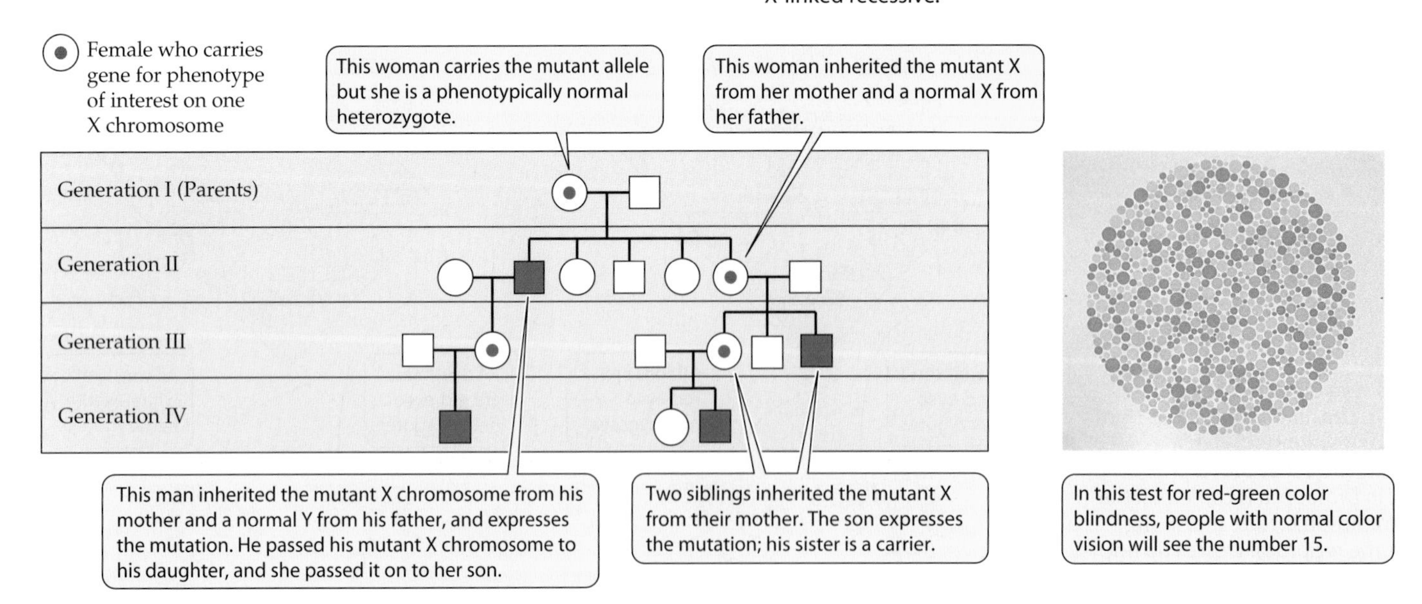
Female who carries gene for phenotype of interest on one X chromosome
This woman carries the mutant allele but she is a phenotypically normal heterozygote.
This woman inherited the mutant X from her mother and a normal X from her father.
Generation I (Parents)
Generation II
Generation III
Generation IV
This man inherited the mutant X chromosome from his mother and a normal Y from his father, and expresses the mutation. He passed his mutant X chromosome to his daughter, and she passed it on to her son.
Two siblings inherited the mutant X from their mother. The son expresses the mutation; his sister is a carrier.
In this test for red-green color blindness, people with normal color vision will see the number 15.

10.24 Red-Green Color Blindness is a Sex-Linked Trait in Humans The mutant allele for red-green color blindness is inherited as an X-linked recessive.

that make up ribosomes has a gene on the Y that is expressed only in male cells, while the X-linked counterpart is expressed in both sexes. This means that there are "male" and "female" ribosomes; the significance of this phenomenon is unknown. Y-linked alleles are passed only from father to son. (You can verify this with a Punnett square.)

Non-Nuclear Inheritance

The nucleus is not the only organelle in a eukaryotic cell that carries genetic material. As we described in Chapter 4, mitochondria and plastids, which may have arisen from prokaryotes that colonized other cells, contain small numbers of genes. For example, in humans, there are about 30,000 genes in the nuclear genome and 37 in the mitochondrial genome. Plastid genomes are about five times larger than those of mitochondria. In any case, several of the genes of cytoplasmic organelles are important for organelle assembly and function, so it is not surprising that mutations of these genes have profound effects on the organism.

The inheritance of organelle genes differs from that of nuclear genes for several reasons:

- In most organisms, mitochondria and plastids are inherited from the mother only. As you will see in later chapters, eggs contain abundant cytoplasm and organelles, but the only part of the sperm that survives to take part in the union of haploid gametes is the nucleus. So you have inherited your mother's mitochondria (with their genes), but not your father's.
- There may be hundreds of mitochondria or plastids in a cell. So a cell is not diploid for organelle genes; rather, it is highly polyploid.
- Organelle genes tend to mutate at much faster rates than nuclear genes, so there are multiple alleles of organelle genes.

The phenotypes of mutations in the DNA of organelles reflect the organelles' roles. For example, in plants and some eukaryotic algae, certain plastid mutations affect the proteins that assemble chlorophyll molecules into photosystems (see Figure 8.9) and result in a phenotype that is essentially white instead of green. Mitochondrial mutations that affect one of the complexes in the electron transport chain result in less ATP production. They have especially noticeable effects in tissues with a high energy requirement, such as the nervous system, muscles, and kidneys. In 1995, Greg Lemond, a professional cyclist who had won the famous Tour de France three times, was forced to retire because of muscle weakness suspected to be caused by a mitochondrial mutation.

Chapter Summary

The Foundations of Genetics

▶ Although it had long been known that both parents contribute to the character traits of their offspring, before Mendel's time it was believed that, once they were brought together, the units of inheritance blended and could never be separated.

▶ Although Gregor Mendel's work was meticulous and well documented, his discoveries, reported in the 1860s, were ignored until decades later.

Mendel's Experiments and the Laws of Inheritance

▶ Mendel used the garden pea for his studies because the plants were easily cultivated and crossed and because they showed numerous characters (such as seed shape) with clearly different traits (spherical or wrinkled). **Review Figure 10.1, Table 10.1**

▶ In a monohybrid cross, the offspring of the first generation (F_1) showed only one of the two parental traits. Mendel proposed that the trait observed in the F_1 was dominant and the other was recessive. **Review Table 10.1**

▶ When the F_1 offspring were self-pollinated, the resulting F_2 generation showed a 3:1 phenotypic ratio, with the recessive phenotype present in one-fourth of the offspring. This reappearance of the recessive phenotype refuted the blending theory. **Review Figure 10.3**

▶ Because some alleles are dominant and some are recessive, the same phenotype can result from different genotypes. Homozygous genotypes have two copies of the same allele; heterozygous genotypes have two different alleles. Heterozygous genotypes yield phenotypes that show the dominant trait.

▶ On the basis of many crosses using different characters, Mendel proposed his first law: that the units of inheritance (now known as genes) are particulate, that there are two alleles of each gene in each parent, and that during gamete formation the two alleles segregate from each other. **Review Figure 10.4**

▶ Geneticists who followed Mendel showed that genes are carried on chromosomes and that alleles are segregated during meiosis I. **Review Figure 10.5**

▶ Using a test cross, Mendel was able to determine whether a plant showing the dominant phenotype was homozygous or heterozygous. The appearance of the recessive phenotype in half of the offspring of such a cross indicates that the parent is heterozygous. **Review Figure 10.6. See Web/CD Activity 10.1**

▶ From studies of the inheritance of two characters using dihybrid crosses, Mendel concluded that alleles of different genes assort independently. **Review Figures 10.7, 10.8. See Web/CD Tutorial 10.1**

▶ We can predict the results of hybrid crosses either by using a Punnett square or by calculating probabilities. To determine the joint probability of independent events, we multiply the individual probabilities. To determine the probability of an event that can occur in two or more different ways, we add the individual probabilities. **Review Figure 10.9**

▶ The analysis of pedigrees can trace Mendelian inheritance patterns in humans. **Review Figures 10.10, 10.11**

Alleles and Their Interactions

▶ New alleles arise by mutation, and many genes have multiple alleles. **Review Figure 10.12**

▶ Dominance is sometimes not complete, since both alleles in a heterozygous organism may be expressed in the phenotype. **Review Figures 10.13, 10.14**

Gene Interactions

▶ In epistasis, the products of different genes interact to produce a phenotype. **Review Figure 10.15**

▶ Environmental variables such as temperature, nutrition, and light affect gene action.

▶ In some cases, the phenotype is the result of the effects of several genes and the environment, and inheritance is quantitative. **Review Figure 10.17**

Genes and Chromosomes

▶ Each chromosome carries many genes. Genes located on the same chromosome are said to be linked, and they are often inherited together. **Review Figure 10.18**

▶ Linked genes can recombine by crossing over in prophase I of meiosis. The result is recombinant gametes, which have new combinations of linked genes because of the exchange. **Review Figures 10.19, 10.20**

▶ The distance between two genes on a chromosome is proportional to the frequency of crossing over between them. Genetic maps are based on recombinant frequencies. **Review Figures 10.21, 10.22.**

See Web/CD Tutorial 10.2

Sex Determination and Sex-Linked Inheritance

▶ Sex chromosomes carry genes that determine whether the organism will produce male or female gametes. The specific functions of X and Y chromosomes differ among species.

▶ In fruit flies and mammals, the X chromosome carries many genes, but the Y chromosome has only a few. Males have only one allele for X-linked genes, so rare alleles show up phenotypically more often in males than in females. **Review Figures 10.23, 10.24**

Non-Nuclear Inheritance

▶ Cytoplasmic organelles such as plastids and mitochondria contain some heritable genes.

▶ Cytoplasmic organelle genes are generally inherited only from the mother because male gametes contribute only their nucleus to the zygote at fertilization.

See Web/CD Activities 10.2 and 10.3 for a concept review of this chapter.

Self-Quiz

1. In a simple Mendelian monohybrid cross, tall plants were crossed with short plants and the F_1 were crossed among themselves. What fraction of the F_2 generation are both tall *and* heterozygous?
   a. 1/8
   b. 1/4
   c. 1/3
   d. 2/3
   e. 1/2
2. The phenotype of an individual
   a. depends at least in part on the genotype.
   b. is either homozygous or heterozygous.
   c. determines the genotype.
   d. is the genetic constitution of the organism.
   e. is either monohybrid or dihybrid.
3. The ABO blood groups in humans are determined by a multiple allelic system where I^A and I^B are codominant and dominant to I^O. A newborn infant is type A. The mother is type O. Possible genotypes of the father are:
   a. A, B or AB
   b. A, B or O
   c. O only
   d. A or AB
   e. A or O
4. Which statement about an individual that is homozygous for an allele is *not* true?
   a. Each of its cells possesses two copies of that allele.
   b. Each of its gametes contains one copy of that allele.
   c. It is true-breeding with respect to that allele.
   d. Its parents were necessarily homozygous for that allele.
   e. It can pass that allele to its offspring.
5. Which statement about a test cross is *not* true?
   a. It tests whether an unknown individual is homozygous or heterozygous.
   b. The test individual is crossed with a homozygous recessive individual.
   c. If the test individual is heterozygous, the progeny will have a 1:1 ratio.
   d. If the test individual is homozygous, the progeny will have a 3:1 ratio.
   e. Test cross results are consistent with Mendel's model of inheritance.
6. Linked genes
   a. must be immediately adjacent to one another on a chromosome.
   b. have alleles that assort independently of one another.
   c. never show crossing over.
   d. are on the same chromosome.
   e. always have multiple alleles.
7. In the F_2 generation of a dihybrid cross
   a. 4 phenotypes appear in the ratio 9:3:3:1 if the loci are linked.
   b. 4 phenotypes appear in the ratio 9:3:3:1 if the loci are unlinked.
   c. 2 phenotypes appear in the ratio 3:1 if the loci are unlinked.
   d. 3 phenotypes appear in the ratio 1:2:1 if the loci are unlinked.
   e. 2 phenotypes appear in the ratio 1:1 whether or not the loci are linked.
8. The sex of a human is determined by
   a. ploidy, the male being haploid.
   b. the Y chromosome.
   c. X and Y chromosomes, the male being XY.
   d. the number of X chromosomes, the male being XO.
   e. Z and W chromosomes, the male being ZZ.
9. In epistasis
   a. nothing changes from generation to generation.
   b. one gene alters the effect of another.
   c. a portion of a chromosome is deleted.
   d. a portion of a chromosome is inverted.
   e. the behavior of two genes is entirely independent.
10. In humans, spotted teeth is caused by a dominant sex-linked gene. A man with spotted teeth whose mother had normal teeth marries a woman with normal teeth. Therefore,
    a. all of their daughters will have normal teeth.
    b. all of their daughters will have spotted teeth.
    c. all of their children will have spotted teeth.
    d. half of their sons will have spotted teeth.
    e. none of their sons will have spotted teeth.

Genetics Problems

1. Using the Punnett squares below, show that for typical dominant and recessive autosomal traits, it does not matter which parent contributes the dominant allele and which the recessive allele. Cross true-breeding tall plants (*TT*) with true-breeding dwarf plants (*tt*).

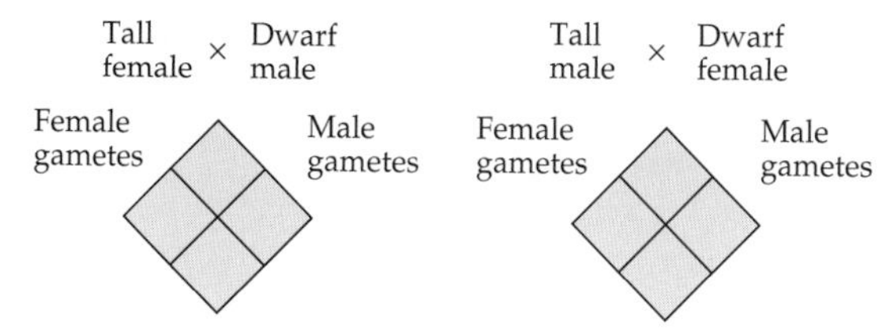

2. The photograph shows the shells of 15 bay scallops, *Argopecten irradians*. These scallops are hermaphroditic; that is, a single individual can reproduce sexually, as did the pea plants of the F_1 generation in Mendel's experiments. Three color schemes are evident: yellow, orange, and black and white. The color-determining gene has three alleles. The top row shows a yellow scallop and a representative sample of its offspring, the middle row shows a black-and-white scallop and its offspring, and the bottom row shows an orange scallop and its offspring. Assign a suitable symbol to each of the three alleles participating in color control; then determine the genotype of each of the three parent individuals and tell what you can about the genotypes of the different offspring. Explain your results carefully.

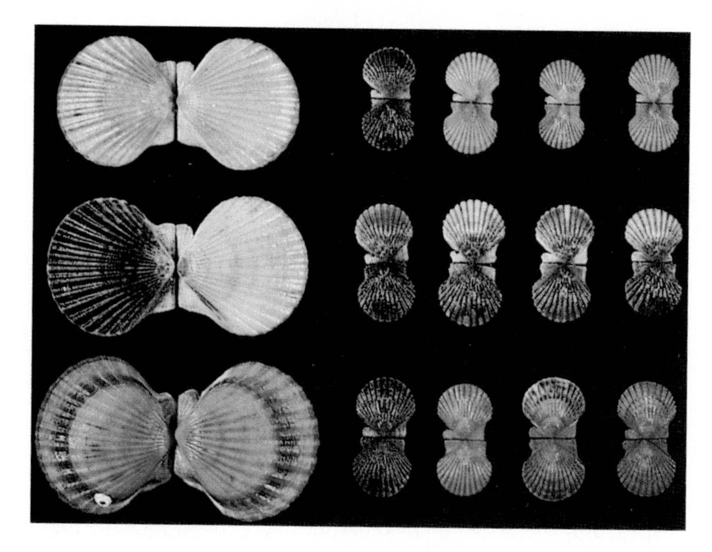

3. Show diagrammatically what occurs when the F_1 offspring of the cross in Question 1 self-pollinate.

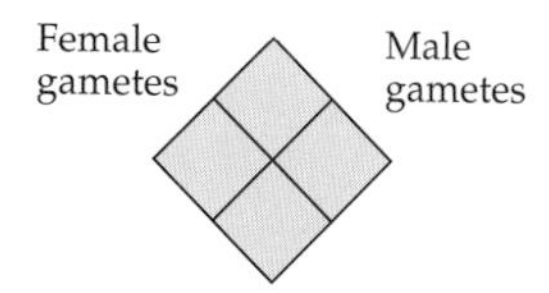

4. A new student of genetics suspects that a particular recessive trait in fruit flies (dumpy wings, which are somewhat smaller and more bell-shaped than the wild-type) is sex-linked. A single mating between a fly having dumpy wings (*dp*; female) and a fly with wild-type wings (*Dp*; male) produces 3 dumpy-winged females and 2 wild-type males. On the basis of these data, is the trait sex-linked or autosomal? What were the genotypes of the parents? Explain how these conclusions can be reached on the basis of so few data.

5. The sex of fishes is determined by the same X-Y system as in humans. An allele of one locus on the Y chromosome of the fish *Lebistes* causes a pigmented spot to appear on the dorsal fin. A male fish that has a spotted dorsal fin is mated with a female fish that has an unspotted fin. Describe the phenotypes of the F_1 and the F_2 generations from this cross.

6. In *Drosophila melanogaster*, the recessive allele *p*, when homozygous, determines pink eyes. *Pp* or *PP* results in wild-type eye color. Another gene, on another chromosome, has a recessive allele, *sw*, that produces short wings when homozygous. Consider a cross between females of genotype *PPSwSw* and males of genotype *ppswsw*. Describe the phenotypes and genotypes of the F_1 generation and of the F_2 generation produced by allowing the F_1 progeny to mate with one another.

7. On the same chromosome of *Drosophila melanogaster* that carries the *p* (pink eyes) locus, there is another locus that affects the wings. Homozygous recessives, *byby*, have blistery wings, while the dominant allele *By* produces wild-type wings. The *P* and *By* loci are very close together on the chromosome; that is, the two loci are tightly linked. In answering these questions, assume that no crossing over occurs.
   a. For the cross *PPByBy* × *ppbyby*, give the phenotypes and genotypes of the F_1 and of the F_2 generations produced by interbreeding of the F_1 progeny.
   b. For the cross *PPbyby* × *ppByBy*, give the phenotypes and genotypes of the F_1 and of the F_2 generations.
   c. For the cross of Question 7*b*, what further phenotype(s) would appear in the F_2 generation if crossing over occurred?
   d. Draw a nucleus undergoing meiosis, at the stage in which the crossing over (Question 7*c*) occurred. In which generation (P, F_1, or F_2) did this crossing over take place?

8. Consider the following cross of *Drosophila melanogaster* (alleles as described in Question 6): Males with genotype *Ppswsw* are crossed with females of genotype *ppSwsw*. Describe the phenotypes and genotypes of the F_1 generation.

9. In the Andalusian fowl, a single pair of alleles controls the color of the feathers. Three colors are observed: blue, black, and splashed white. Crosses among these three types yield the following results:

PARENTS	PROGENY
Black × blue	Blue and black (1:1)
Black × splashed white	Blue
Blue × splashed white	Blue and splashed white (1:1)
Black × black	Black
Splashed white × splashed white	Splashed white

   a. What progeny would result from the cross blue × blue?
   b. If you want to sell eggs, all of which would yield blue fowl, how should you proceed?

10. In *Drosophila melanogaster*, white (w), eosin (w^{e}), and wild-type red (w^{+}) are multiple alleles of a single locus for eye color. This locus is on the X chromosome. A female that has eosin (pale orange) eyes is crossed with a male that has wild-type eyes. All the female progeny are red-eyed; half the male offspring have eosin eyes, and half have white eyes.

    a. What is the order of dominance of these alleles?

    b. What are the genotypes of the parents and progeny?

11. Color blindness is a recessive trait. Two people with normal vision have two sons, one color-blind and one with normal vision. If the couple also has daughters, what proportion of them will have normal vision? Explain.

12. A mouse with an agouti coat is mated with an albino mouse of genotype *aabb*. Half of the offspring are albino, one-fourth are black, and one-fourth are agouti. What are the genotypes of the agouti parents and of the various kinds of offspring? (*Hint*: See the section on epistasis.)

13. The disease Leber's optic neuropathy is caused by a mutation in a gene carried on mitochondrial DNA. What would be the result in their first child if a man with this disease married a woman who did not have the disease? What would be the result if the wife had the disease and the husband did not?

11 DNA and Its Role in Heredity

In the novel and film *Jurassic Park*, fictional scientists were depicted using biotechnology to produce dinosaurs. In the story, the scientists isolated the DNA of dinosaurs from fossilized insects that had sucked the reptiles' blood. The insects, preserved intact in amber (fossilized tree resin), yielded DNA that could be used to produce living individuals of long-extinct organisms such as *Tyrannosaurus rex*. This premise of Michael Crichton's novel was based on an actual scientific paper that claimed to show reptilian DNA sequences in a fossil insect. Unfortunately, the scientific report was not correct; the "preserved" DNA turned out to be a contaminant from modern organisms.

Despite the fact that the preservation of intact DNA over millions of years is highly improbable, the popular success of *Jurassic Park* did bring the idea of DNA as the genetic material to the attention of millions of readers and viewers. Indeed, the double helix of DNA has become a familiar secular icon in the 50 years since that structure was first proposed.

In this and the next several chapters, we will focus on the structure, replication, and function of DNA. As you will see, the structure of DNA determines its function. This chapter will first describe the key experiments that led to the determination that the genetic material is DNA. Then the structure and replication of the molecule will be described. Finally, we will present two practical applications that have arisen from our knowledge of DNA replication: the polymerase chain reaction and DNA sequencing.

Jurassic Park In Michael Crichton's novel, DNA isolated from a fossil is used to produce living dinosaurs. Although such a procedure is strictly fiction, the movie based on the book did bring the role of DNA as the genetic material to the attention of millions of people.

DNA: The Genetic Material

As we saw in Chapter 10, by the early twentieth century geneticists had associated the presence of genes with chromosomes. Research began to focus on exactly which chemical component of chromosomes comprised this genetic material.

By the 1920s, scientists knew that chromosomes were made up of DNA and proteins. At this time a new dye was developed that could bind specifically to DNA and turned red in direct proportion to the amount of DNA present in a cell. This technique provided circumstantial evidence that DNA was the genetic material:

- It was in the right place, since it was an important component of the nucleus and the chromosomes, which were known to carry genes.

- It varied among species. When cells from different species were stained with the dye and their color intensity measured, each species appeared to have its own specific nuclear DNA content.
- It was present in the right amounts. The amount of DNA in somatic cells (body cells not specialized for reproduction) was twice that in the reproductive cells (eggs or sperm)—as might be expected for diploid and haploid cells, respectively.

But circumstantial evidence is *not* a scientific demonstration of cause and effect. After all, proteins are also present in nuclei. The convincing demonstration that DNA is the genetic material came from two lines of experiments, one on bacteria and the other on viruses.

DNA from one type of bacterium genetically transforms another type

The history of biology is filled with incidents in which research on one specific topic has—with or without answering the question originally asked—contributed richly to another, apparently unrelated area. Such a case of *serendipity* is the work of Frederick Griffith, an English physician.

In the 1920s, Griffith was studying the bacterium *Streptococcus pneumoniae*, or pneumococcus, one of the agents that causes pneumonia in humans. He was trying to develop a vaccine against this devastating illness (antibiotics had not yet been discovered). Griffith was working with two strains* of pneumococcus:

- Cells of the S strain produced colonies that looked smooth (S). Covered by a polysaccharide capsule, these cells were protected from attack by a host's immune system. When S cells were injected into mice, they reproduced and caused pneumonia (the strain was *virulent*).
- Cells of the R strain produced colonies that looked rough (R), lacked the protective capsule, and were not virulent.

Griffith inoculated some mice with heat-killed S pneumococci. These heat-killed bacteria did not produce infection. However, when Griffith inoculated other mice with a mixture of living R bacteria and heat-killed S bacteria, to his astonishment, the mice died of pneumonia (Figure 11.1). When he examined blood from the hearts of these mice, he found it full of living bacteria—many of them with characteristics of the virulent S strain! Griffith concluded that, in the presence of the dead S pneumococci, some of the living R pneumococci had been transformed into virulent S-strain organisms.

*A bacterial *strain* is a population of bacterial cells descended from a single parent cell; strains may differ in one or more inherited characteristics.

11.1 Genetic Transformation of Nonvirulent Pneumococci Frederick Griffith's experiments demonstrated that something in the virulent S strain could transform nonvirulent R strain bacteria into a lethal form, even when the S strain bacteria had been killed by high temperatures.

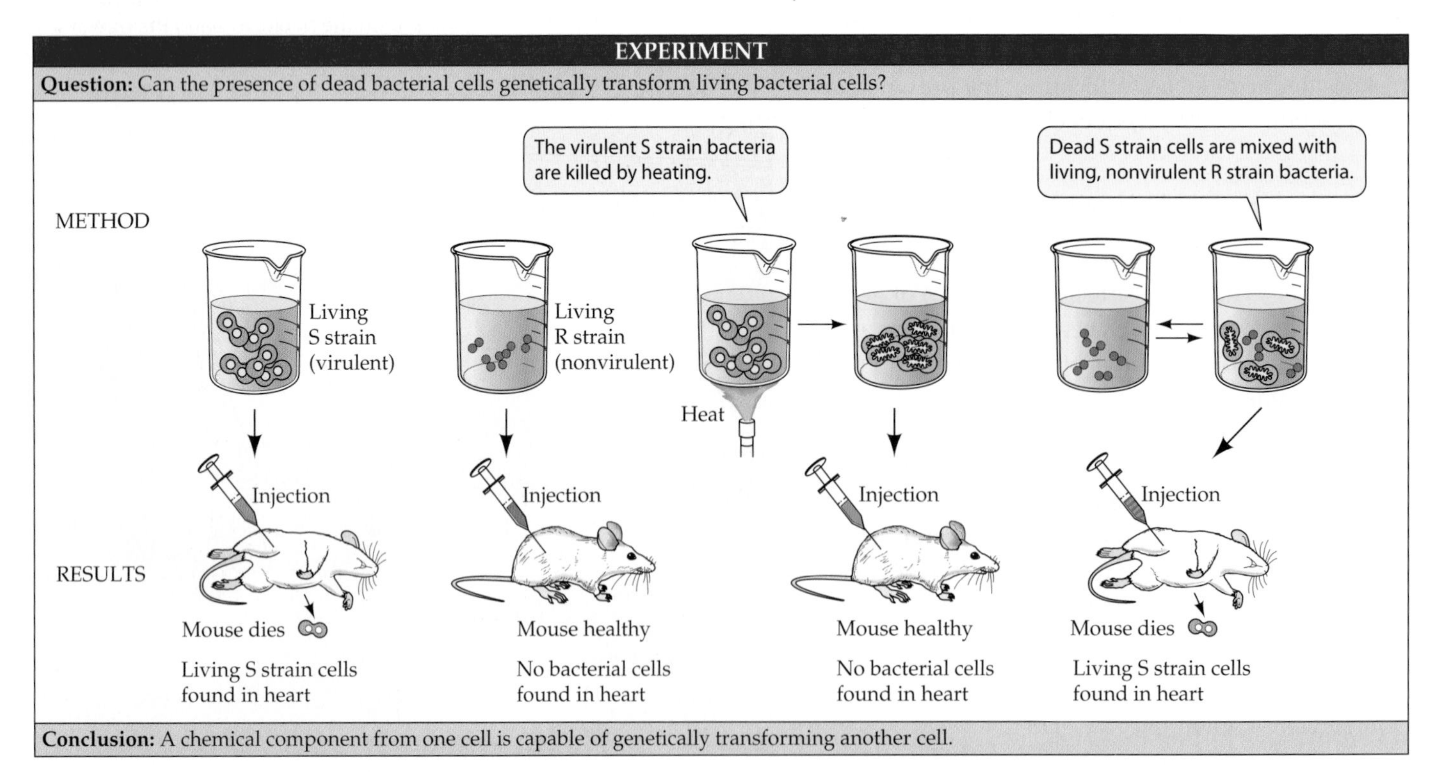
EXPERIMENT
Question: Can the presence of dead bacterial cells genetically transform living bacterial cells?
METHOD
The virulent S strain bacteria are killed by heating.
Dead S strain cells are mixed with living, nonvirulent R strain bacteria.
Living S strain (virulent)
Living R strain (nonvirulent)
Heat
Injection
Injection
Injection
Injection
RESULTS
Mouse dies
Living S strain cells found in heart
Mouse healthy
No bacterial cells found in heart
Mouse healthy
No bacterial cells found in heart
Mouse dies
Living S strain cells found in heart
Conclusion: A chemical component from one cell is capable of genetically transforming another cell.

Did this transformation of the bacteria depend on something that happened in the mouse's body? No. It was shown that simply incubating living R and heat-killed S bacteria together in a test tube yielded the same transformation. Years later, another group of scientists discovered that a cell-free extract of heat-killed S cells also transformed R cells. (A *cell-free extract* contains all the contents of ruptured cells, but no intact cells.) This result demonstrated that some substance—called at the time a chemical **transforming principle**—from the dead S pneumococci could cause a heritable change in the affected R cells. This was an extraordinary discovery: Treatment with a substance permanently changed an inherited characteristic. Now it remained to identify the chemical structure of this substance.

The transforming principle is DNA

Identifying the transforming principle was a crucial step in the history of biology. It was accomplished over a period of years by Oswald Avery and his colleagues at what is now Rockefeller University. They treated samples known to contain the pneumococcal transforming principle in a variety of ways to destroy different types of molecules—proteins, nucleic acids, carbohydrates, and lipids—and tested the treated samples to see if they had retained transforming activity.

The answer was always the same: If the DNA in the sample was destroyed, transforming activity was lost, but there was no loss of activity when proteins, carbohydrates, or lipids were destroyed. As a final step, Avery, with Colin MacLeod and Maclyn McCarty, isolated virtually pure DNA from a sample containing pneumococcal transforming principle and showed that it caused bacterial transformation. We now know that the gene encoding the enzyme that catalyzes the synthesis of the pathogenic polysaccharide capsule was transferred during transformation.

The work of Avery, MacLeod, and McCarty, published in 1944, was a milestone in establishing that DNA is the genetic material in cells. However, it had little impact at the time, for two reasons. First, most scientists did not believe that DNA was chemically complex enough to be the genetic material, especially given the much greater chemical complexity of proteins. Second, and perhaps more important, bacterial genetics was a new field of study—it was not yet clear that bacteria even *had* genes.

Viral replication experiments confirm that DNA is the genetic material

The questions about bacteria were soon resolved as researchers identified genes and mutations in these organisms. Bacteria and viruses seemed to undergo genetic processes similar to those in fruit flies and pea plants. Experiments were designed with these relatively simple systems to discover the nature of the genetic material.

In 1952, Alfred Hershey and Martha Chase of the Carnegie Laboratory of Genetics published a paper that had a much greater immediate impact than Avery's 1944 paper. The Hershey–Chase experiment, which sought to determine whether DNA or protein was the hereditary material, was carried out with a virus that infects bacteria. This virus, called the T2 bacteriophage, consists of little more than a DNA core packed inside a protein coat (Figure 11.2*a*). The virus is thus made of the two materials that were, at the time, the leading candidates for the genetic material.

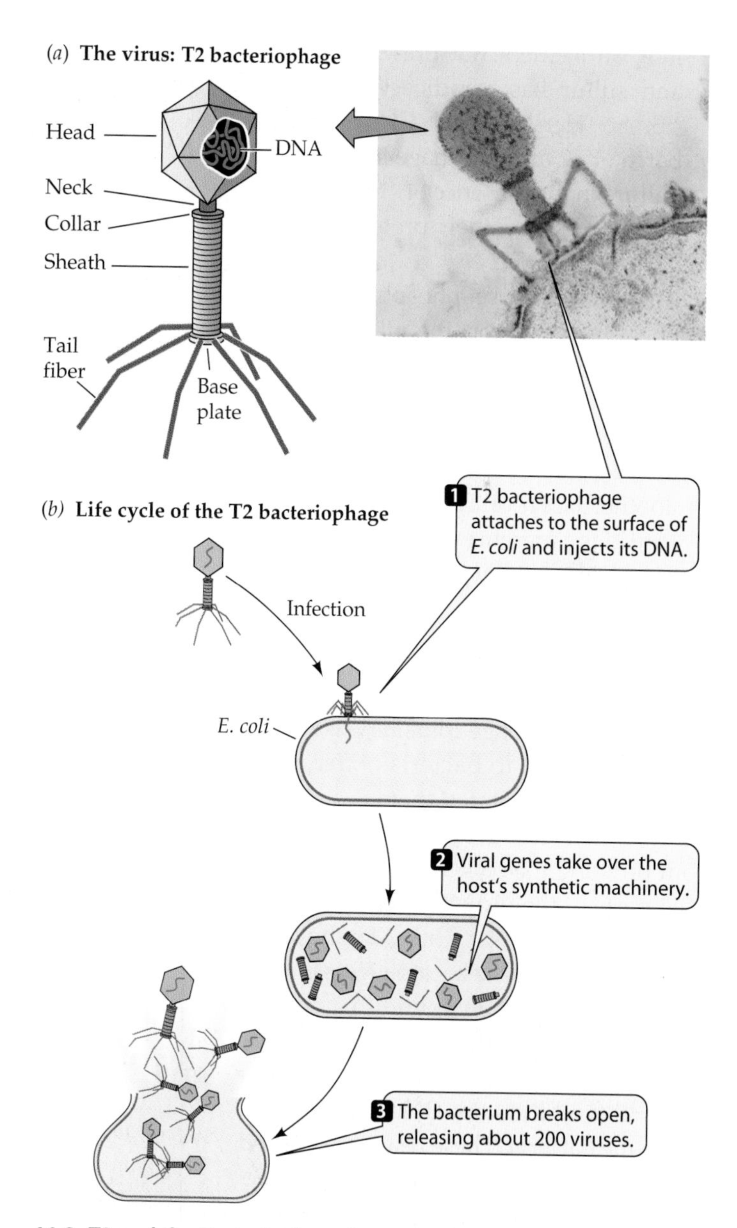
(*a*) The virus: T2 bacteriophage
Head
DNA
Neck
Collar
Sheath
Tail fiber
Base plate
(*b*) Life cycle of the T2 bacteriophage
1 T2 bacteriophage attaches to the surface of *E. coli* and injects its DNA.
Infection
E. coli
2 Viral genes take over the host's synthetic machinery.
3 The bacterium breaks open, releasing about 200 viruses.

11.2 T2 and the Bacteriophage Reproduction Cycle (*a*) The external structures of T2 bacteriophage consist entirely of protein. This cutaway view shows a strand of DNA within the head. (*b*) T2 is parasitic on *E. coli*, depending on the bacterium to produce new viruses.

When a T2 bacteriophage attacks a bacterium, part (but not all) of the virus enters the bacterial cell. About 20 minutes later, the cell bursts, releasing dozens of viruses. The entry of a viral component changes the genetic program of the host bacterial cell: it is converted from a bacterium into a bacteriophage factory. Hershey and Chase set out to determine which part of the virus—protein or DNA—enters the bacterial cell. To trace the two components of the virus over its life cycle (Figure 11.2*b*), Hershey and Chase labeled each with a specific radioactive tracer:

- Viral proteins contain some sulfur (in the amino acids cysteine and methionine), an element not present in DNA, and sulfur has a radioactive isotope, ^{35}S. So Hershey and Chase grew a batch of T2 bacteriophage in a bacterial culture in the presence of ^{35}S; the resulting viruses had their proteins labeled with this isotope.
- The deoxyribose–phosphate "backbone" of DNA, on the other hand, is rich in phosphorus (see Chapter 3), an element that is not present in most proteins, and phosphorus also has a radioactive isotope, ^{32}P. The researchers grew another batch of T2 in a bacterial culture in the presence of ^{32}P, so that all the viral DNA was labeled with ^{32}P.

With these radioactively labeled viruses, Hershey and Chase performed their revealing experiments. They allowed bacteriophage containing either ^{32}P or ^{35}S to attach to bacteria. After a few minutes, they agitated the mixtures vigorously in a kitchen blender, which (without bursting the bacteria) stripped away the parts of the virus that had not penetrated the bacteria. Then Hershey and Chase separated the bacteria from the rest of the material. They found that most of the ^{35}S (and thus the protein) had separated from the bacteria, and that most of the ^{32}P (and thus the DNA) had stayed with the bacteria. These results suggested that the DNA was transferred to the bacteria, whereas the protein remained outside, and thus that it was DNA that redirected the genetic program of the bacterial cell (Figure 11.3).

Hershey and Chase performed other similar but more long-range experiments, allowing a progeny (offspring) generation of viruses to be collected. The resulting viruses contained almost none of the original ^{35}S and none of the parental viral protein. However, they contained about one-

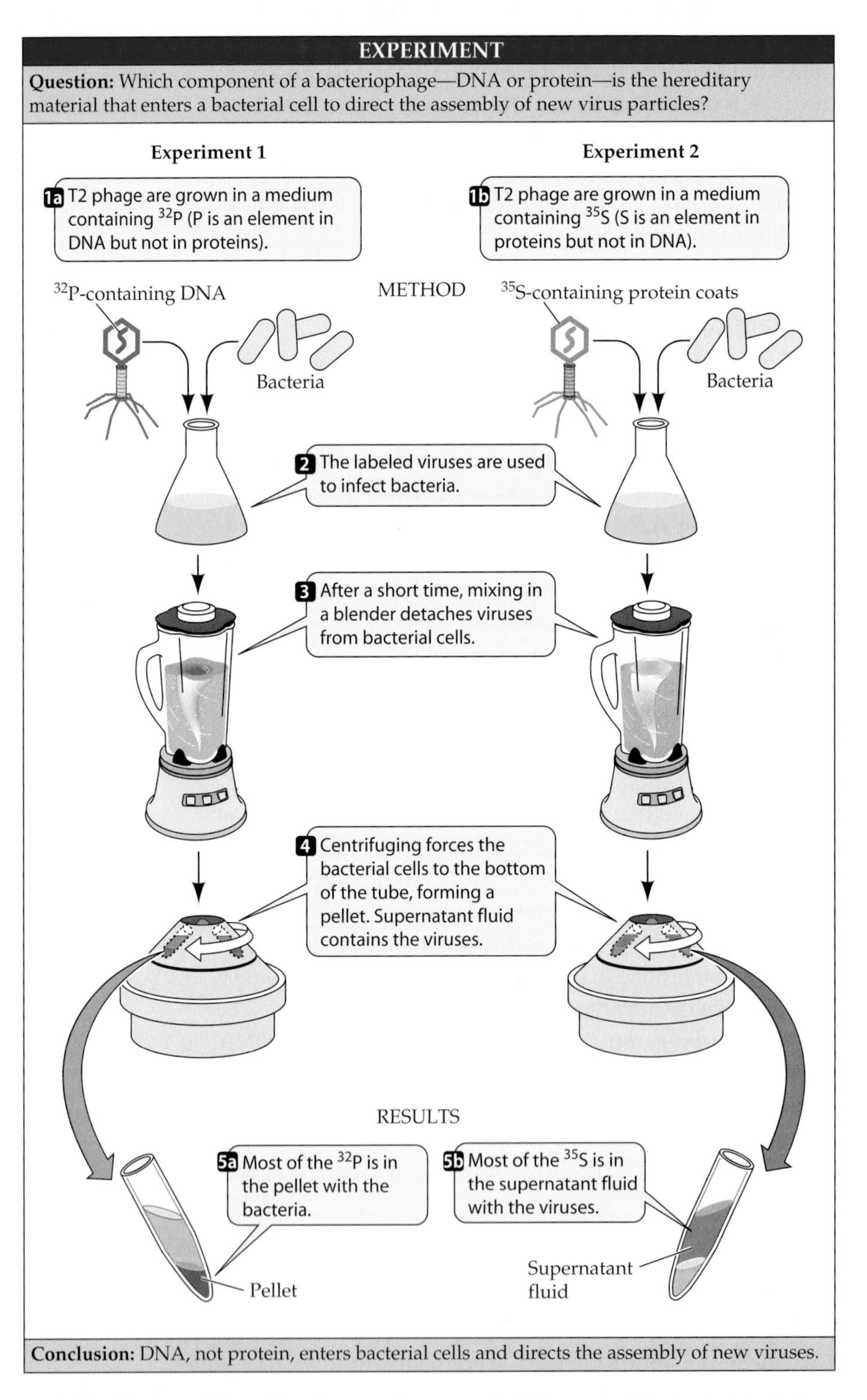
EXPERIMENT
Question: Which component of a bacteriophage—DNA or protein—is the hereditary material that enters a bacterial cell to direct the assembly of new virus particles?
Experiment 1
Experiment 2
1a T2 phage are grown in a medium containing ^{32}P (P is an element in DNA but not in proteins).
1b T2 phage are grown in a medium containing ^{35}S (S is an element in proteins but not in DNA).
^{32}P-containing DNA
METHOD
^{35}S-containing protein coats
Bacteria
Bacteria
2 The labeled viruses are used to infect bacteria.
3 After a short time, mixing in a blender detaches viruses from bacterial cells.
4 Centrifuging forces the bacterial cells to the bottom of the tube, forming a pellet. Supernatant fluid contains the viruses.
RESULTS
5a Most of the ^{32}P is in the pellet with the bacteria.
5b Most of the ^{35}S is in the supernatant fluid with the viruses.
Pellet
Supernatant fluid
Conclusion: DNA, not protein, enters bacterial cells and directs the assembly of new viruses.

11.3 The Hershey–Chase Experiment Because only DNA entered the bacterial cell during infection by labeled bacteriophage, the experiment demonstrated that DNA, not protein, is the hereditary material.

third of the original ^{32}P—and thus, presumably, one-third of the original DNA. Because DNA was carried over in the virus from generation to generation but protein was not, a logical conclusion was that the hereditary information of the virus is contained in the DNA.

The Hershey–Chase experiment convinced most scientists that DNA is the carrier of hereditary information.

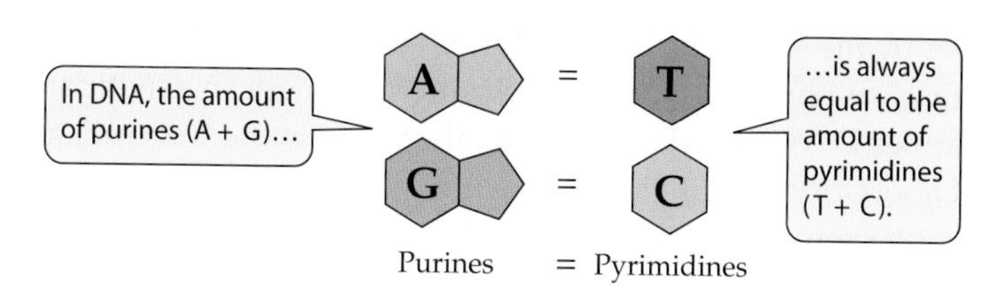

11.5 Chargaff's Rule In DNA, the total abundance of purines is equal to the total abundance of pyrimidines.

The Structure of DNA

As soon as scientists were convinced that the genetic material was DNA, they began efforts to learn its precise, three-dimensional chemical structure. In determining the structure of DNA, scientists hoped to find the answers to two questions: how DNA is replicated between nuclear divisions, and how it causes the synthesis of specific proteins. Both expectations were fulfilled.

X-ray crystallography provided clues to DNA structure

The structure of DNA was deciphered only after many types of experimental evidence and theoretical considerations were combined. The crucial evidence was obtained by *X-ray crystallography* (Figure 11.4). Some chemical substances, when they are isolated and purified, can be made to form crystals. The positions of atoms in a crystalline substance can be inferred from the pattern of diffraction of X-rays passed through it. Even today, however, this is not an easy task when the substance is of enormous molecular weight.

In the early 1950s, even highly talented X-ray crystallographers could (and did) look at the best available images from DNA preparations and fail to see what they meant. Nonetheless, the attempt to characterize DNA would have been impossible without the crystallographs prepared by the English chemist Rosalind Franklin. Franklin's work, in turn, depended on the success of the English biophysicist Maurice Wilkins, who prepared a sample containing very uniformly oriented DNA fibers. These DNA preparations provided samples for diffraction that were far better than previous ones.

The chemical composition of DNA was known

The chemical composition of DNA also provided important clues about its structure. Biochemists knew that DNA was a polymer of nucleotides. Each nucleotide of DNA consists of a molecule of the sugar deoxyribose, a phosphate group, and a nitrogen-containing base (see Figures 3.24 and 3.25). The only differences among the four nucleotides of DNA are their nitrogenous bases: the purines **adenine** (**A**) and **guanine** (**G**), and the pyrimidines **cytosine** (**C**) and **thymine** (**T**).

In 1950, Erwin Chargaff at Columbia University reported some observations of major importance. He and his colleagues found that DNA from many different species—and from different sources within a single organism—exhibits certain regularities. In almost all DNA, the following rule holds: The amount of adenine equals the amount of thymine (A = T), and the amount of guanine equals the amount of cytosine (G = C) (Figure 11.5). As a result, the total abundance of purines (A + G) equals the total abundance of pyrimidines (T + C). The structure of DNA could not have been worked out without this information, now known as *Chargaff's rule*, yet its significance was overlooked for at least three years.

Watson and Crick described the double helix

The solution to the puzzle of the structure of DNA was accelerated by *model building:* the assembly of three-dimensional representations of possible molecular structures using known relative molecular dimensions and known bond angles. This technique, originally exploited in structural stud-

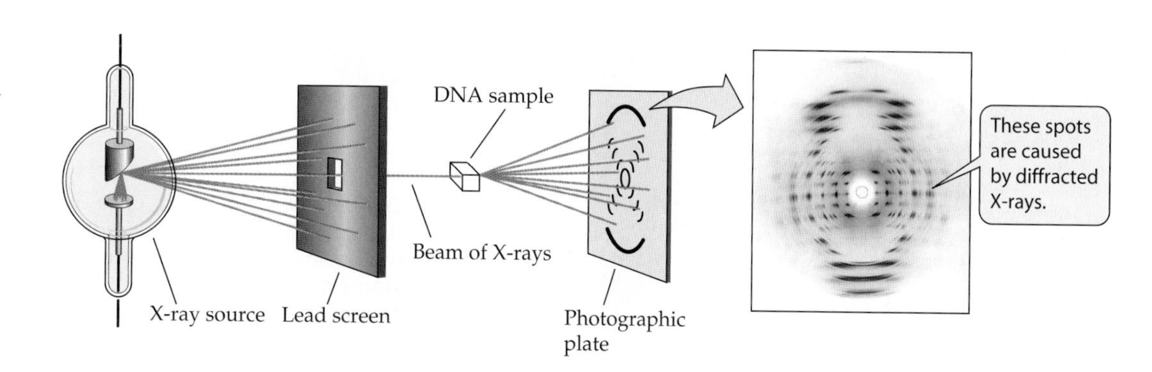

11.4 X-Ray Crystallography Revealed the Basic Helical Structure of the DNA Molecule The positions of atoms in a purified chemical substance can be inferred by the pattern of diffraction of X-rays passed through it, although the task requires tremendous skill.

ies by the American chemist Linus Pauling, was used by the English physicist Francis Crick and the American geneticist James D. Watson (Figure 11.6*a*), then both at the Cavendish Laboratory of Cambridge University.

Watson and Crick attempted to combine all that had been learned so far about DNA structure into a single coherent model. The crystallographers' results (see Figure 11.4) convinced Watson and Crick that the DNA molecule is **helical** (cylindrically spiral) and provided the values of certain distances within the helix. The results of density measurements and previous model building suggested that there are two polynucleotide chains in the molecule. Modeling studies had also led to the conclusion that the two chains in DNA run in opposite directions—that is, that they are **antiparallel**. (We'll clarify this point on the next page.)

Crick and Watson built several large models. Late in February of 1953, they built a model out of tin that established the general structure of DNA. This structure explained all the known chemical properties of DNA, and it opened the door to understanding its biological functions. There have been minor amendments to that first published structure, but its principal features remain unchanged.

Four key features define DNA structure

Four features summarize the molecular architecture of the DNA molecule:

- It is a double-stranded helix.
- It has a uniform diameter.
- It is right-handed (that is, it twists to the right, as do the threads on most screws).
- It is antiparallel (the two strands run in opposite directions).

The sugar–phosphate "backbones" of the polynucleotide chains coil around the outside of the helix, and the nitrogenous bases point toward the center (Figure 11.6*b*).

The two chains are held together by hydrogen bonding between specifically paired bases. Consistent with Chargaff's rule,

- adenine (A) pairs with thymine (T) by forming two hydrogen bonds; and
- guanine (G) pairs with cytosine (C) by forming three hydrogen bonds.

Every base pair consists of one purine (A or G) and one pyrimidine (T or C). This pattern is known as **complementary base pairing** (Figure 11.7).

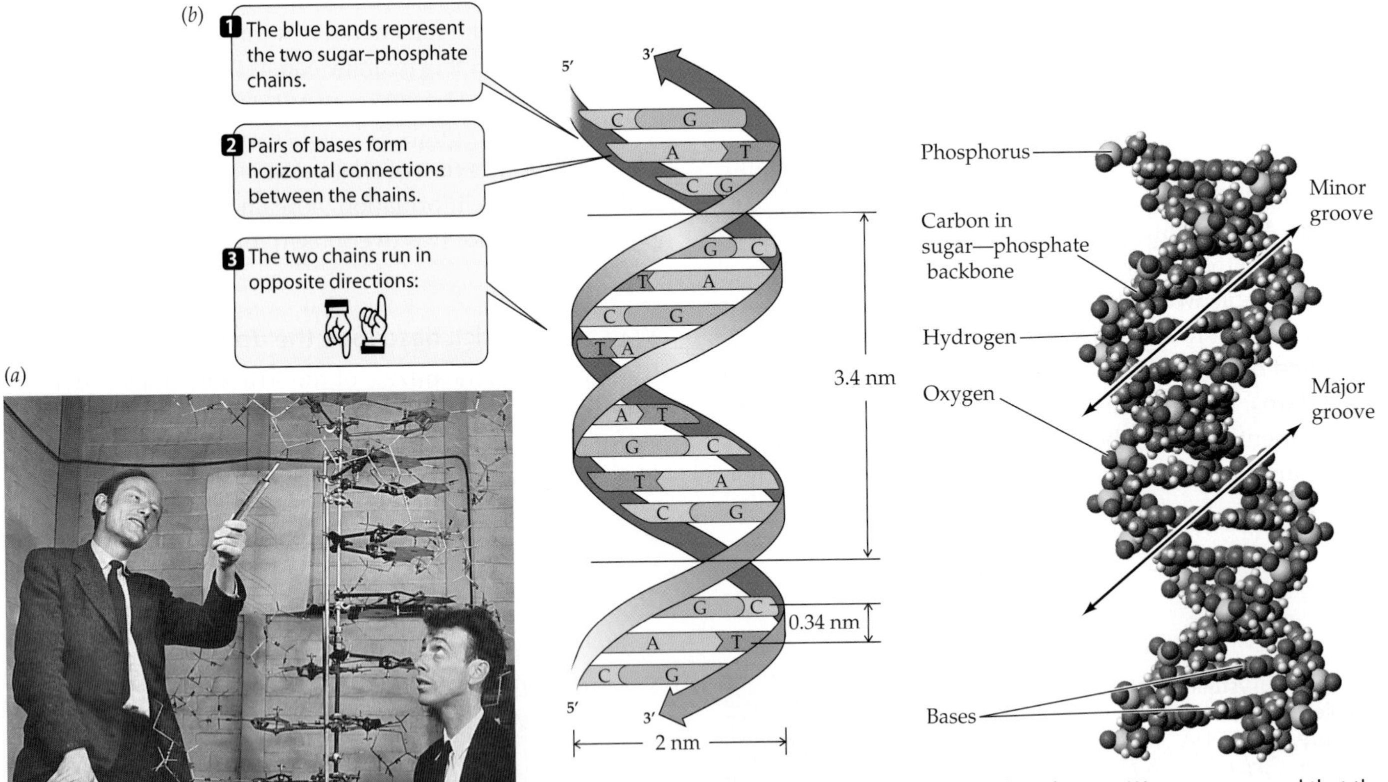

11.6 DNA Is a Double Helix (*a*) Francis Crick and James Watson proposed that the DNA molecule has a double helical structure. (*b*) Biochemists can now pinpoint the position of every atom in a DNA macromolecule. To see that the essential features of the original Watson–Crick model have been verified, follow with your eyes the double helical chains of sugar–phosphate groups and note the horizontal rungs of the bases (see also Figure 3.27).

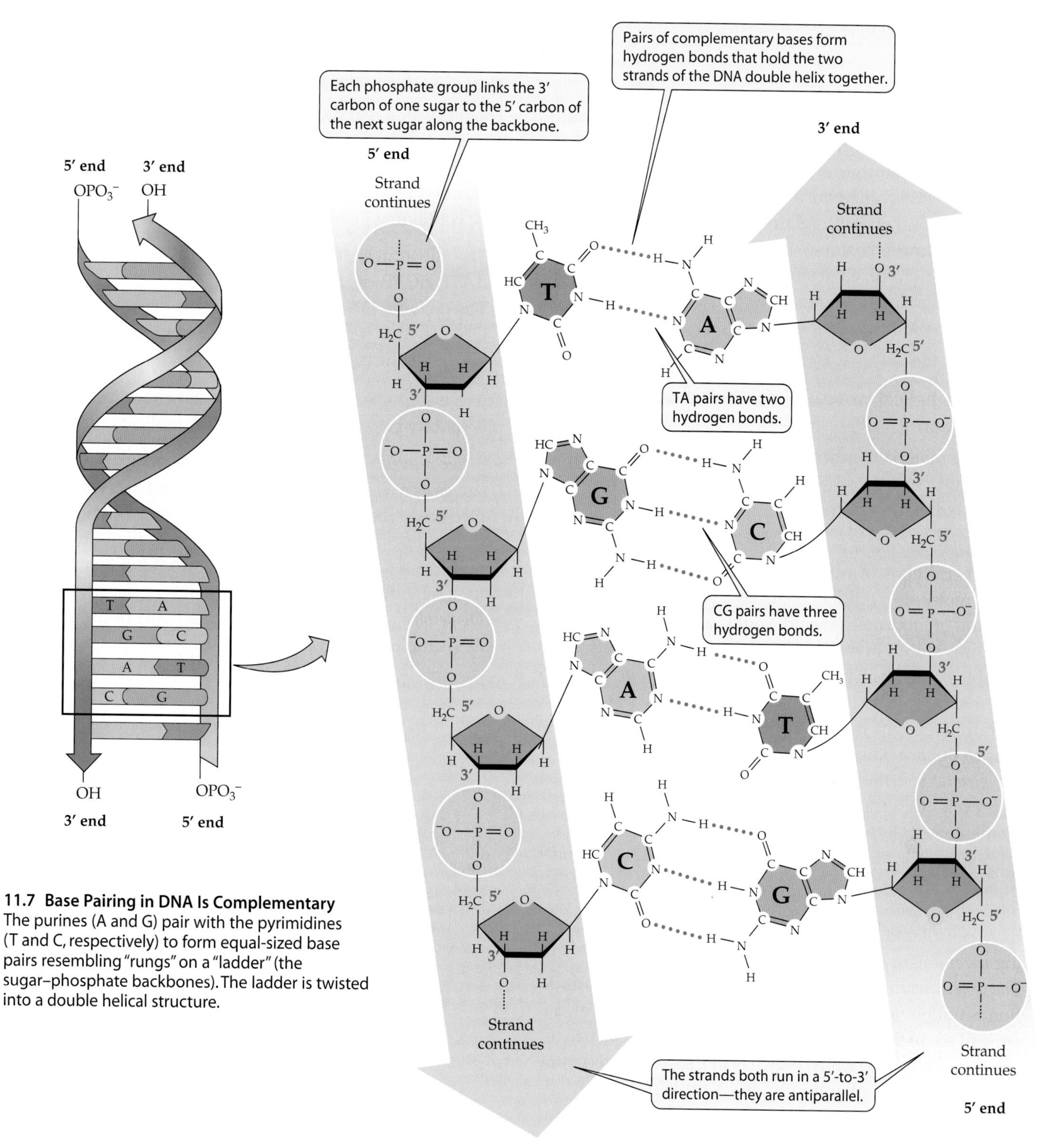

Pairs of complementary bases form hydrogen bonds that hold the two strands of the DNA double helix together.
Each phosphate group links the 3′ carbon of one sugar to the 5′ carbon of the next sugar along the backbone.
5′ end
3′ end
Strand continues
TA pairs have two hydrogen bonds.
CG pairs have three hydrogen bonds.
The strands both run in a 5′-to-3′ direction—they are antiparallel.

11.7 Base Pairing in DNA Is Complementary The purines (A and G) pair with the pyrimidines (T and C, respectively) to form equal-sized base pairs resembling "rungs" on a "ladder" (the sugar–phosphate backbones). The ladder is twisted into a double helical structure.

Because the AT and GC pairs are of equal length and fit identically into the double helix (like rungs on a ladder), the diameter of the helix is uniform. The base pairs are flat (see Figure 11.6), and their stacking in the center of the molecule is stabilized by hydrophobic interactions (see Chapter 2), contributing to the overall stability of the double helix.

What does it mean to say that the two DNA strands are *antiparallel*? The direction of a polynucleotide can be defined by looking at the phosphodiester linkages between adjacent nucleotides (*-diester* refers to the two bonds formed by —OH groups reacting with acidic *phosphate* groups). In the sugar–phosphate backbone of DNA, the phosphate groups connect to the 3′ carbon of one deoxyribose molecule and the 5′ carbon of the next, linking successive sugars together (see Figure 11.7). The number followed by a prime (′) designates the position of a carbon atom in the five-carbon sugar deoxyribose.

Thus the two ends of a polynucleotide chain differ. At one end of a strand is a free (not connected to another nucleotide) 5′ phosphate group (—OPO_3^-); this end is called the 5′ end. At the other end is a free 3′ hydroxyl group (—OH); this end is called the 3′ end. In a DNA double helix, the 5′ end of one strand is paired with the 3′ end of the other strand, and vice versa. In other words, were you to draw an arrow for each strand running from 5′ to 3′, the arrows would point in different directions; thus it is said that the strands are antiparallel.

The double helical structure of DNA is essential to its function

The genetic material performs four important functions, and the DNA structure proposed by Watson and Crick was elegantly suited to three of them.

- *The genetic material stores an organism's genetic information.* With its millions of nucleotides, the base sequence of a DNA molecule could encode and store an enormous amount of information and could account for species and individual differences. DNA fits this role nicely.
- *The genetic material is susceptible to mutation*, or permanent changes in the information it encodes. For DNA, mutations might be simple changes in the linear sequence of base pairs.
- *The genetic material is precisely replicated* in the cell division cycle. Replication could be accomplished by complementary base pairing, A with T and G with C. In the original publication of their findings in the journal *Nature* in 1953, Watson and Crick coyly pointed out, "It has not escaped our notice that the specific pairing we have postulated immediately suggests a possible copying mechanism for the genetic material."
- *The genetic material is expressed as the phenotype.* This function is not obvious in the structure of DNA. However, as we will see in the next chapter, the nucleotide sequence of DNA is copied into RNA, which is in turn converted into a linear sequence of amino acids—a protein. The folded forms of proteins provide much of the phenotype of an organism.

Determining the DNA Replication Mechanism

The mechanism of DNA replication that had suggested itself to Watson and Crick was soon confirmed. First, experiments showed that single strands of DNA could be replicated in a test tube containing simple substrates and an enzyme. Then an elegant experiment showed that each of the two strands of the double helix serves as a template for a new strand of DNA.

Three modes of DNA replication appeared possible

The prediction that the DNA molecule contains the information needed for its own replication was demonstrated by the work of Arthur Kornberg, then at Washington University in St. Louis. He showed that DNA can be synthesized in a test tube containing just three substances:

- The substrates, deoxyribonucleoside triphosphates dATP, dCTP, dGTP, and dTTP
- The enzyme **DNA polymerase**
- DNA, which serves as a **template** to guide the incoming nucleotides

There were three possible patterns that could result in complementary base pairing during DNA replication:

- *Semiconservative replication*, in which each parent strand serves as a template for a new strand, and the two new DNAs each have one old and one new strand (Figure 11.8*a*)
- *Conservative replication*, in which the original double helix serves as a template for, but does not contribute to, a new double helix (Figure 11.8*b*)
- *Dispersive replication*, in which fragments of the original DNA molecule serve as templates for assembling two

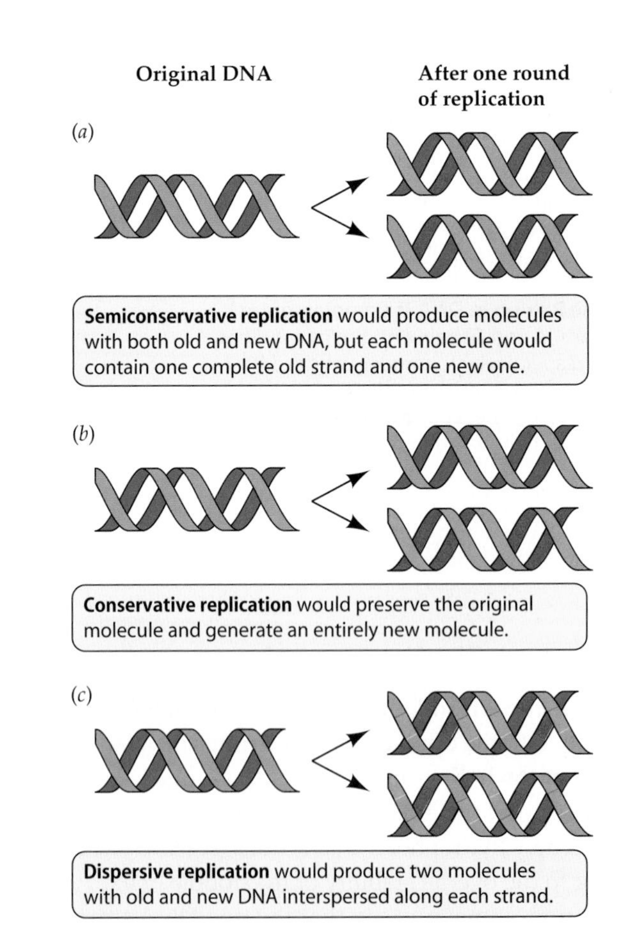

Original DNA
After one round of replication
(a)
Semiconservative replication would produce molecules with both old and new DNA, but each molecule would contain one complete old strand and one new one.
(b)
Conservative replication would preserve the original molecule and generate an entirely new molecule.
(c)
Dispersive replication would produce two molecules with old and new DNA interspersed along each strand.

11.8 Three Models for DNA Replication In each model, original DNA is shown in blue and newly synthesized DNA in red.

new molecules, each containing old and new parts, perhaps at random (Figure 11.8*c*)

Watson and Crick's original paper suggested that DNA replication was semiconservative, but Kornberg's experiment did not provide a basis for choosing among these three models.

Meselson and Stahl demonstrated that DNA replication is semiconservative

A clever experiment conducted by Matthew Meselson and Franklin Stahl convinced the scientific community that **semiconservative replication** is the correct model. Working at the California Institute of Technology in 1957, they devised a simple way to distinguish old strands of DNA from new ones: *density labeling*.

The key to their experiment was the use of a "heavy" isotope of nitrogen. Heavy nitrogen (^{15}N) is a rare, nonradioactive isotope that makes molecules containing it more dense than chemically identical molecules containing the common isotope, ^{14}N. To distinguish DNA of different densities (that is, DNA containing ^{15}N versus DNA containing ^{14}N), Meselson, Stahl, and Jerome Vinograd developed a new procedure using a *centrifuge*. Spinning solutions or suspensions at high speed in a centrifuge causes the solutes or particles to separate and form a gradient according to their density.

Meselson and Stahl grew two cultures of the bacterium *Escherichia coli* for many generations:

- One culture was grown in a medium whose nitrogen source (ammonium chloride, NH_4Cl) was made with ^{15}N instead of ^{14}N. As a result, all the DNA in the bacteria was "heavy."
- Another culture was grown in a medium with ^{14}N, and all the DNA in these bacteria was "light."

When extracts from the two cultures were combined and centrifuged, two separate DNA bands formed, showing that this method could distinguish DNA samples of slightly different densities.

Next, the researchers grew another *E. coli* culture on ^{15}N medium, then transferred it to normal ^{14}N medium and allowed the bacteria to continue growing (Figure 11.9). Under

11.9 The Meselson–Stahl Experiment A centrifuge was used to separate DNAs labeled with isotopes of different densities. This experiment revealed a pattern that supports the semiconservative model of DNA replication.

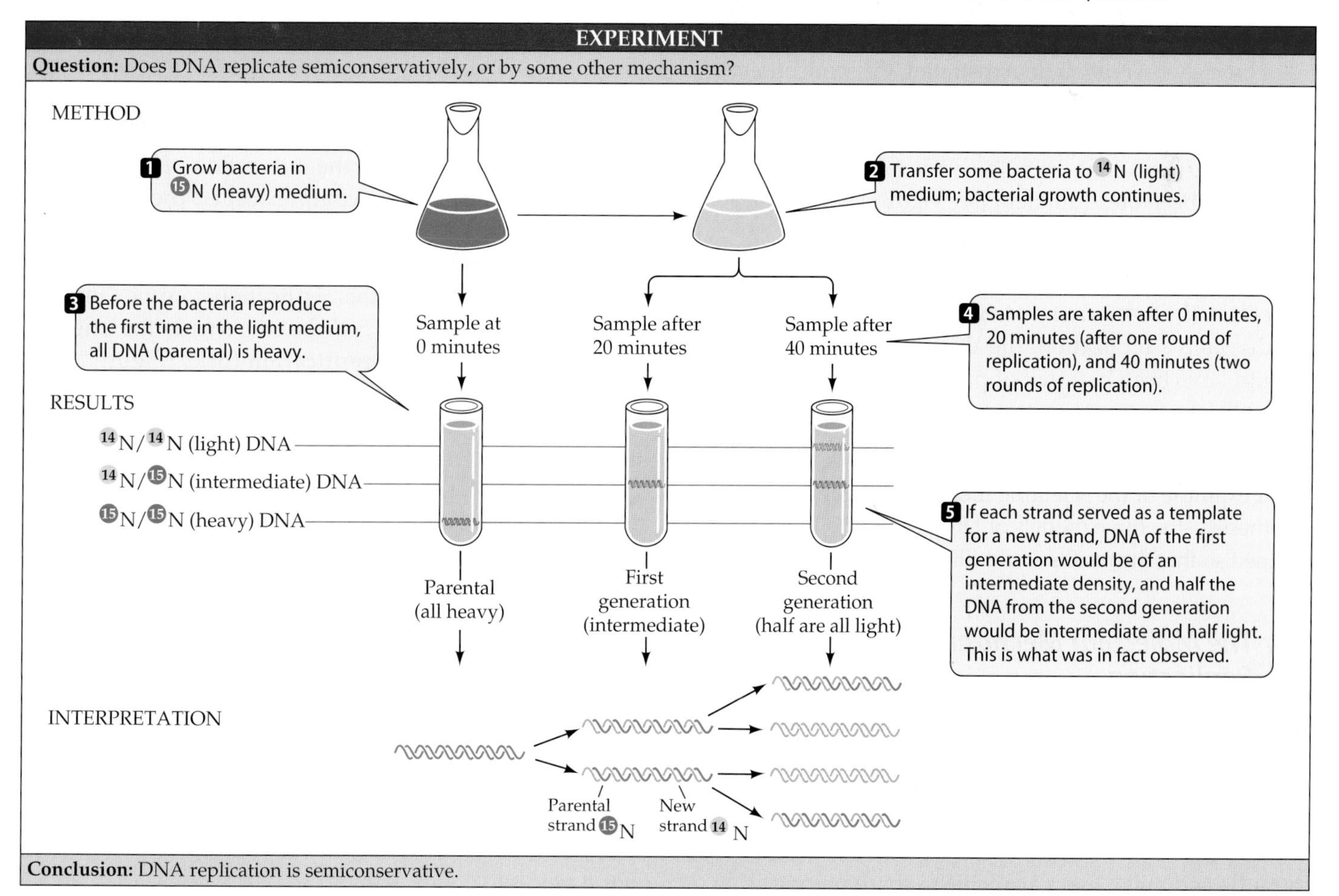
EXPERIMENT
Question: Does DNA replicate semiconservatively, or by some other mechanism?
METHOD
1 Grow bacteria in 15N (heavy) medium.
2 Transfer some bacteria to 14N (light) medium; bacterial growth continues.
3 Before the bacteria reproduce the first time in the light medium, all DNA (parental) is heavy.
Sample at 0 minutes
Sample after 20 minutes
Sample after 40 minutes
4 Samples are taken after 0 minutes, 20 minutes (after one round of replication), and 40 minutes (two rounds of replication).
RESULTS
14N/14N (light) DNA
14N/15N (intermediate) DNA
15N/15N (heavy) DNA
5 If each strand served as a template for a new strand, DNA of the first generation would be of an intermediate density, and half the DNA from the second generation would be intermediate and half light. This is what was in fact observed.
Parental (all heavy)
First generation (intermediate)
Second generation (half are all light)
INTERPRETATION
Parental strand 15N
New strand 14N
Conclusion: DNA replication is semiconservative.

the conditions they used, *E. coli* replicates its DNA every 20 minutes. Meselson and Stahl collected some of the bacteria after each division and extracted DNA from the samples. They found that the DNA banding pattern in the density gradient was different in each bacterial generation:

- At the time of the transfer to the ^{14}N medium, the DNA was uniformly labeled with ^{15}N, and hence was relatively dense.
- After one generation, when the DNA had been duplicated once, all the DNA was of an intermediate density.
- After two generations, there were two equally large DNA bands: one of low density and one of intermediate density.
- In samples from subsequent generations, the proportion of low-density DNA increased steadily.

The results of this experiment can be explained only by the semiconservative model of DNA replication. In the first round of DNA replication, the strands of the double helix—both heavy with ^{15}N—separated. Each strand then acted as the template for a second strand, which contained only ^{14}N and hence was less dense. Each double helix then consisted of one ^{15}N strand and one ^{14}N strand, and was of intermediate density. In the second replication, the ^{14}N-containing strands directed the synthesis of partners with ^{14}N, creating low-density DNA, and the ^{15}N strands formed new ^{14}N partners (see Figure 11.9).

The crucial observation demonstrating the semiconservative model was that intermediate-density DNA (^{15}N–^{14}N) appeared in the first generation and continued to appear in subsequent generations. With the other models, the results would have been quite different (see Figure 11.8):

- In conservative replication, the first generation would have had both high-density DNA (^{15}N–^{15}N) and low-density DNA (^{14}N–^{14}N), but no intermediate-density DNA.
- In dispersive replication, the density of the new DNA would have been half that of parental DNA.

The Meselson-Stahl experiment, called by some scientists among the most elegant ever done by biologists, was an excellent example of the scientific method. It began with three hypotheses—the three models of DNA replication—and was designed so that the results could differentiate between them.

The Molecular Mechanisms of DNA Replication

Semiconservative DNA replication in the cell involves a number of different enzymes and other proteins. It takes place in two steps:

- The DNA is unwound to separate the two template strands and make them available for base pairing.
- New nucleotides are linked by covalent bonding to each growing new strand in a sequence determined by complementary base pairing with the bases on the template strand.

A key observation of virtually all DNA replication is that *nucleotides are always added to the growing strand at the 3′ end*—the end at which the DNA strand has a free hydroxyl (—OH) group on the 3′ carbon of its terminal deoxyribose (Figure 11.10). The three phosphate groups in a deoxyribonucleoside triphosphate are attached to the 5′ position of the sugar (see Figure 11.7). So when a new nucleotide is added to DNA, it can attach only to the 3′ end.

When DNA polymerase brings a deoxyribonucleoside triphosphate with the appropriate base to the 3′ end of a growing chain, the free hydroxyl group on the chain reacts with one of the substrate's phosphate groups. As this happens, the bond linking the terminal two phosphate groups to the rest of the deoxyribonucleoside triphosphate breaks, and stored energy is released as the phosphate groups separate from the molecule. The resulting *pyrophosphate ion*, consisting of the two terminal phosphate groups, also hydrolyzes, forming two separate phosphate ions and in the process releasing additional free energy. The phosphate group still on the nucleotide becomes part of the sugar–phosphate backbone of the growing DNA molecule.

DNA is threaded through a replication complex

DNA is replicated through the interaction of the template DNA with a huge protein complex called the **replication complex**, which catalyzes the reactions involved. All chromosomes have at least one base sequence, called the **origin of replication**, to which this replication complex initially binds. DNA replicates *in both directions* from the origin of replication, forming two **replication forks** (Figure 11.11). Both of the separated strands of the parent molecule act as templates simultaneously, and the formation of the new strands is guided by complementary base pairing.

Until recently, DNA replication was depicted as a locomotive (the replication complex) moving along a railroad track (the DNA) (Figure 11.11*a*). The current view is that this model may not be correct. Instead, the replication complex seems to be stationary, attached to nuclear structures, and it is the DNA that moves, essentially threading through the complex as single strands and emerging as double strands (Figure 11.11*b*). During S phase in eukaryotes, there are about 100 replication complexes, and each of them contains as many as 300 individual replication forks. All replication complexes contain several proteins with different roles in DNA replication; we will describe these proteins as we examine the steps of the process.

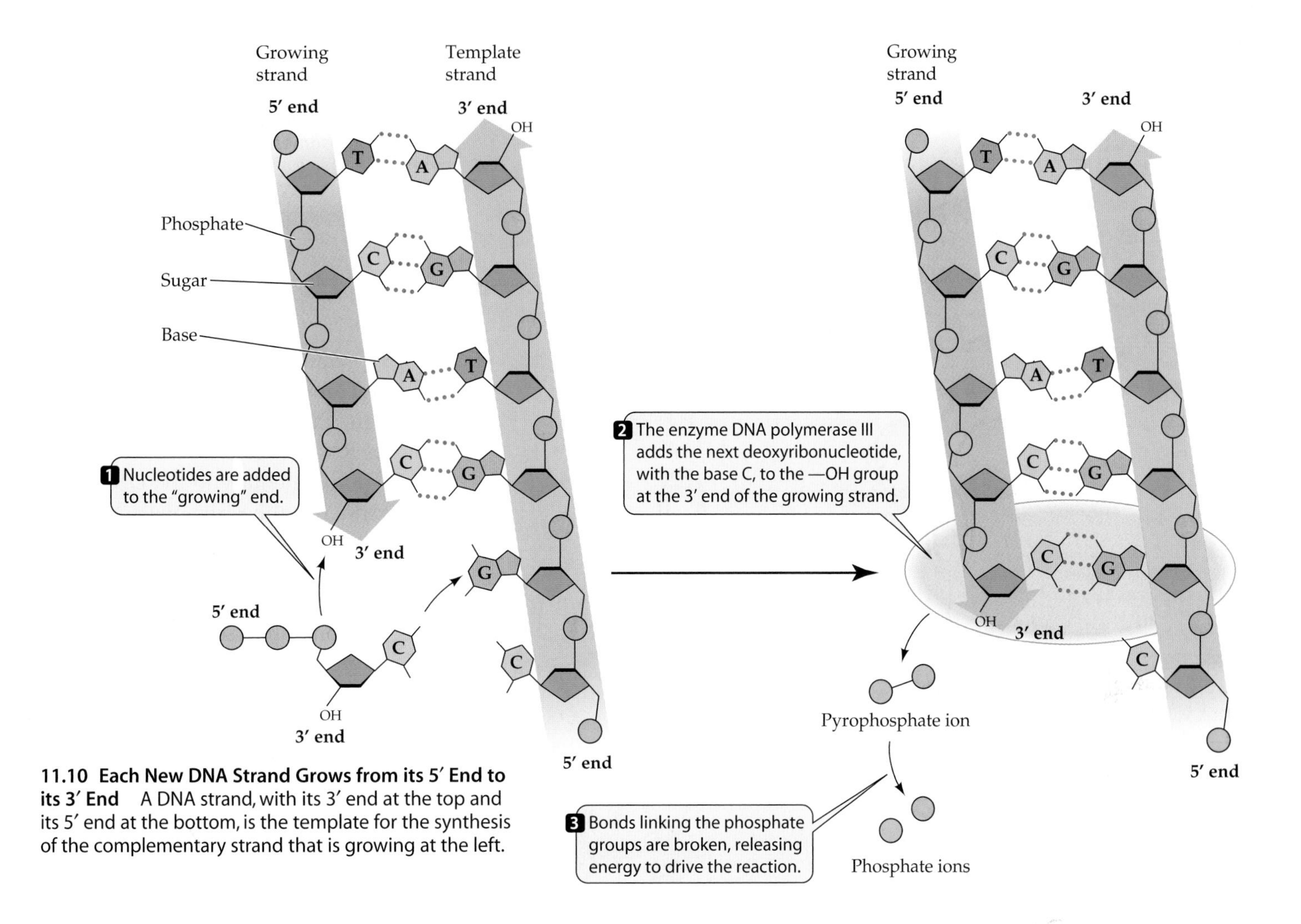
Growing strand
Template strand
5′ end
3′ end
OH
Phosphate
Sugar
Base
T
A
C
G
1 Nucleotides are added to the "growing" end.
OH
3′ end
5′ end
OH
3′ end
5′ end
2 The enzyme DNA polymerase III adds the next deoxyribonucleotide, with the base C, to the —OH group at the 3′ end of the growing strand.
Growing strand
5′ end
3′ end
OH
OH
3′ end
Pyrophosphate ion
5′ end
3 Bonds linking the phosphate groups are broken, releasing energy to drive the reaction.
Phosphate ions

11.10 Each New DNA Strand Grows from its 5′ End to its 3′ End A DNA strand, with its 3′ end at the top and its 5′ end at the bottom, is the template for the synthesis of the complementary strand that is growing at the left.

Small, circular DNAs replicate from a single origin

The first event at the origin of replication is the localized unwinding (denaturation) of DNA. There are several forces that hold the two strands together, including hydrogen bonding and the hydrophobic interactions of bases. An enzyme called **DNA helicase** uses energy from ATP hydrolysis to unwind the DNA, and special proteins called **single-strand binding proteins** bind to the unwound strands to keep them from reassociating into a double helix. This process makes the two template strands available for complementary base pairing.

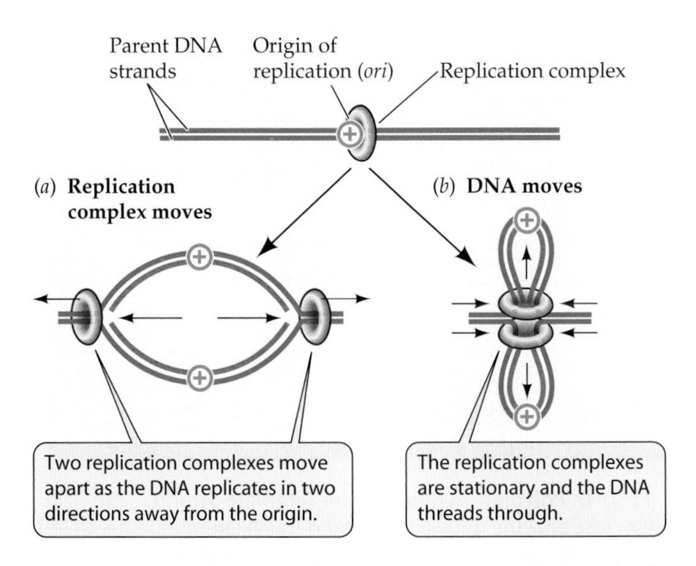
Parent DNA strands
Origin of replication (*ori*)
Replication complex
(*a*) Replication complex moves
(*b*) DNA moves
Two replication complexes move apart as the DNA replicates in two directions away from the origin.
The replication complexes are stationary and the DNA threads through.

11.11 Two Views of DNA Replication (*a*) It was once thought that the replication complex moved along DNA. (*b*) Newer evidence suggests that the DNA is threaded through the stationary complex.

Small circular chromosomes, such as the 3-million-base-pair DNA of bacteria, have a single origin of replication. As the DNA moves through the replication complex, the replication forks grow around the circle (Figure 11.12*a*). Two interlocking circular DNAs are formed, and they are separated by an enzyme called **DNA topoisomerase**.

Large, linear DNAs have many origins

In large linear chromosomes, such as a human chromosome with 80 million base pairs, there are hundreds of origins of replication. Origins of replication that are adjacent to one another along the linear chromosome can be bound by replication complexes at the same time and replicated simultaneously. So there are many replication forks in eukaryotic DNA (Figure 11.12*b*).

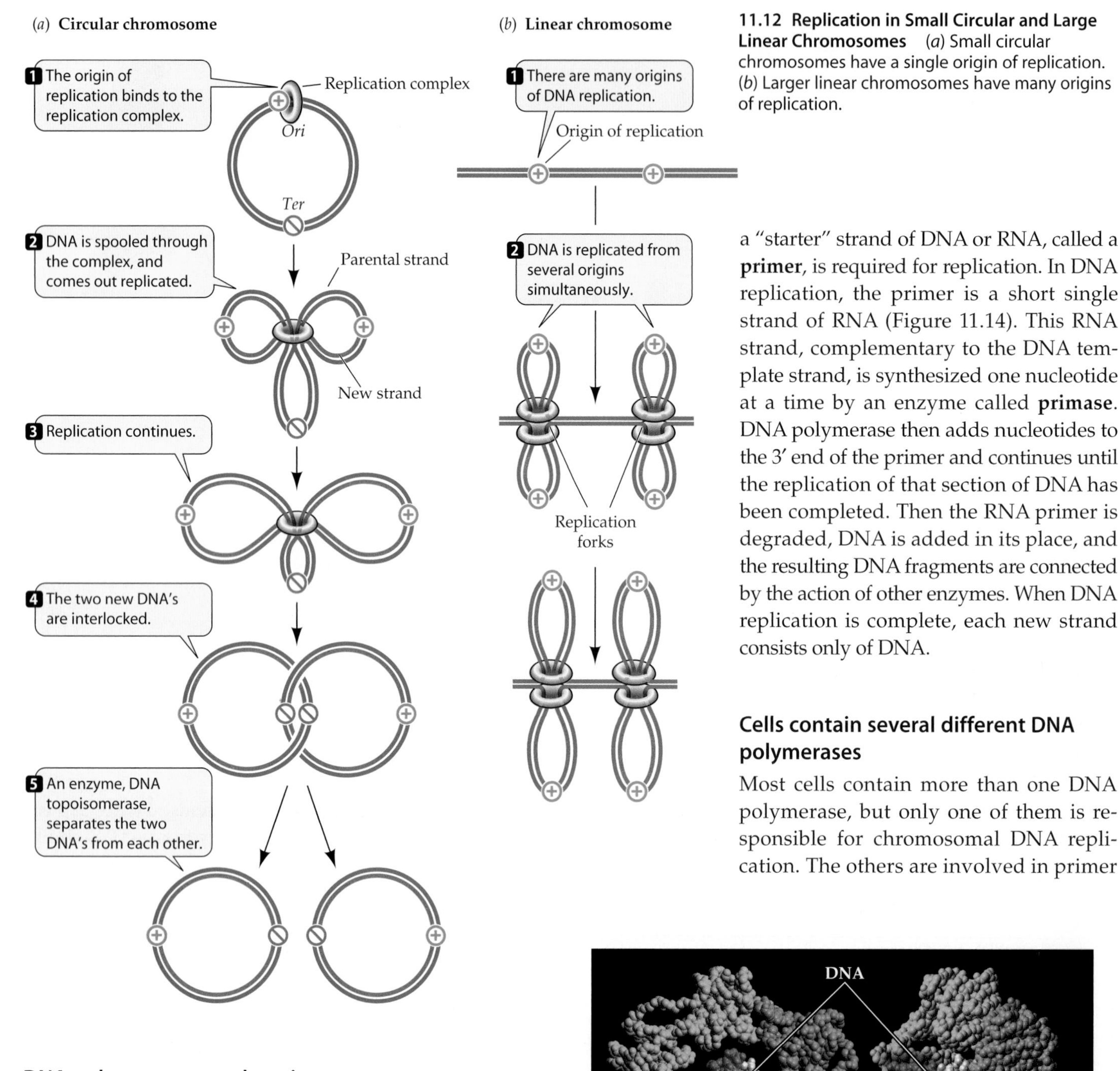

(a) Circular chromosome
1 The origin of replication binds to the replication complex.
Replication complex
Ori
Ter
2 DNA is spooled through the complex, and comes out replicated.
Parental strand
New strand
3 Replication continues.
4 The two new DNA's are interlocked.
5 An enzyme, DNA topoisomerase, separates the two DNA's from each other.
(b) Linear chromosome
1 There are many origins of DNA replication.
Origin of replication
2 DNA is replicated from several origins simultaneously.
Replication forks

11.12 Replication in Small Circular and Large Linear Chromosomes (*a*) Small circular chromosomes have a single origin of replication. (*b*) Larger linear chromosomes have many origins of replication.

DNA polymerases need a primer

DNA polymerases are much larger than their substrates, the deoxyribonucleoside triphosphates, and the template DNA, which is very thin. Molecular models of the enzyme-substrate-template complex from bacteria (Figure 11.13) show that the enzyme is shaped like an open hand with a palm, a thumb, and fingers. The palm holds the active site of the enzyme and brings together the substrate and the template. The finger regions rotate inward and have precise shapes that can recognize the different shapes of the four nucleotide bases.

DNA polymerases can elongate a polynucleotide strand by covalently linking new nucleotides to a previously existing strand, but they cannot start a strand from scratch. Therefore, a "starter" strand of DNA or RNA, called a **primer**, is required for replication. In DNA replication, the primer is a short single strand of RNA (Figure 11.14). This RNA strand, complementary to the DNA template strand, is synthesized one nucleotide at a time by an enzyme called **primase**. DNA polymerase then adds nucleotides to the 3′ end of the primer and continues until the replication of that section of DNA has been completed. Then the RNA primer is degraded, DNA is added in its place, and the resulting DNA fragments are connected by the action of other enzymes. When DNA replication is complete, each new strand consists only of DNA.

Cells contain several different DNA polymerases

Most cells contain more than one DNA polymerase, but only one of them is responsible for chromosomal DNA replication. The others are involved in primer

11.13 DNA Polymerase Binds to the Template Strand The DNA polymerase enzyme (blue and green) is much larger than the DNA molecule. DNA polymerase III is shaped like a hand, and in the side-on view, its "fingers" can be seen curling around the DNA. These "fingers" can recognize the different shapes of the four bases (white; the DNA "backbone" is shown in red).

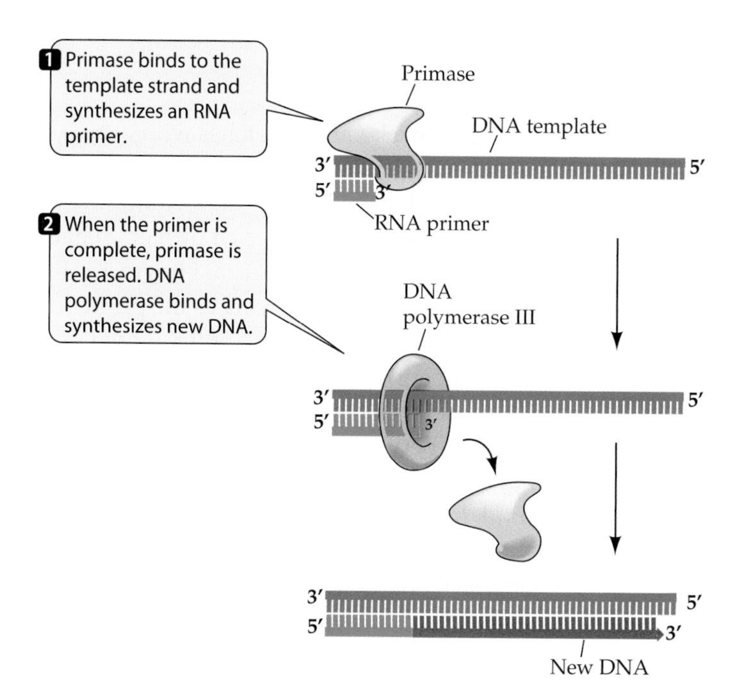

11.14 No DNA Forms without a Primer DNA polymerases require a primer—a "starter" strand of DNA or RNA to which they can add new nucleotides.

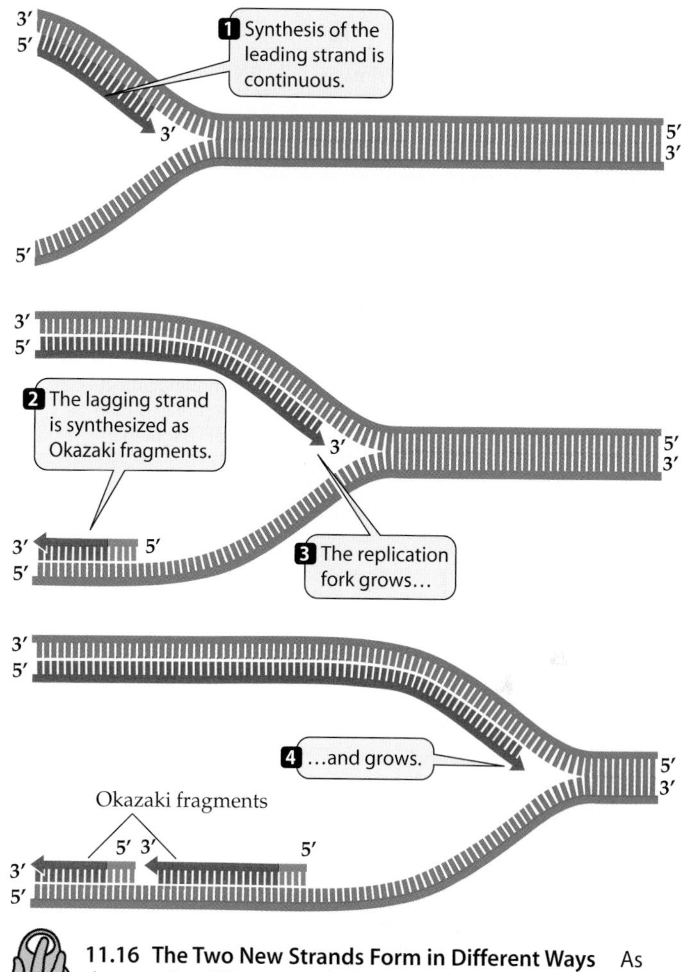

1 Synthesis of the leading strand is continuous.
2 The lagging strand is synthesized as Okazaki fragments.
3 The replication fork grows…
4 …and grows.
Okazaki fragments
3′
5′
5′
3′

11.16 The Two New Strands Form in Different Ways As the template DNA unwinds, both new strands are synthesized in the 5′-to-3′ direction, although their template strands are antiparallel. The leading strand grows continuously forward, but the lagging strand grows in short discontinuous stretches called Okazaki fragments. Eukaryotic Okazaki fragments are hundreds of nucleotides long, with gaps between them.

removal and DNA repair. Fourteen DNA polymerases have been identified in humans; the one catalyzing most replication is DNA polymerase α. In the bacterium *E. coli*, there are three DNA polymerases; the one responsible for replication is DNA polymerase III. Various other proteins play roles in replacing the RNA primer and in other replication tasks; some of these are shown in Figure 11.15.

The DNA at the replication fork opens up like a zipper in one direction. Study Figure 11.16 and try to imagine what is happening over a short period of time. Remember that in DNA the two strands are antiparallel; that is, the 3′ end of one strand is paired with the 5′ end of the other. One newly replicating strand (the **leading strand**) is pointing in the "right" direction to grow continuously at its 3′ end as the fork opens up. But the other strand (the **lagging strand**) is pointing in the "wrong" direction: As the fork opens up further, its exposed 3′ end gets farther and farther away from the fork, and an unreplicated gap is formed, which would get bigger and bigger if there were not a special mechanism to overcome this problem.

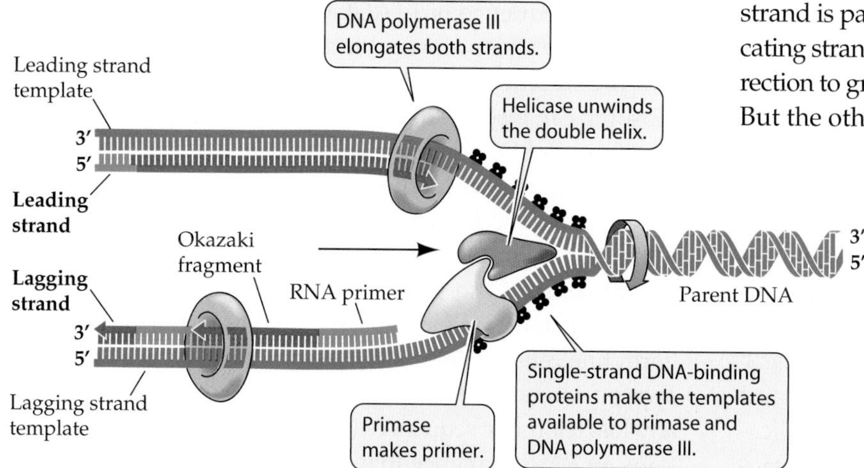

11.15 Many Proteins Collaborate at the Replication Fork Several proteins in addition to DNA polymerase III are involved in DNA replication. The two molecules of DNA polymerase (red) are actually part of the same complex.

The lagging strand is synthesized from Okazaki fragments

Synthesis of the lagging strand requires working in relatively small, discontinuous stretches (100 to 200 nucleotides at a time in eukary-

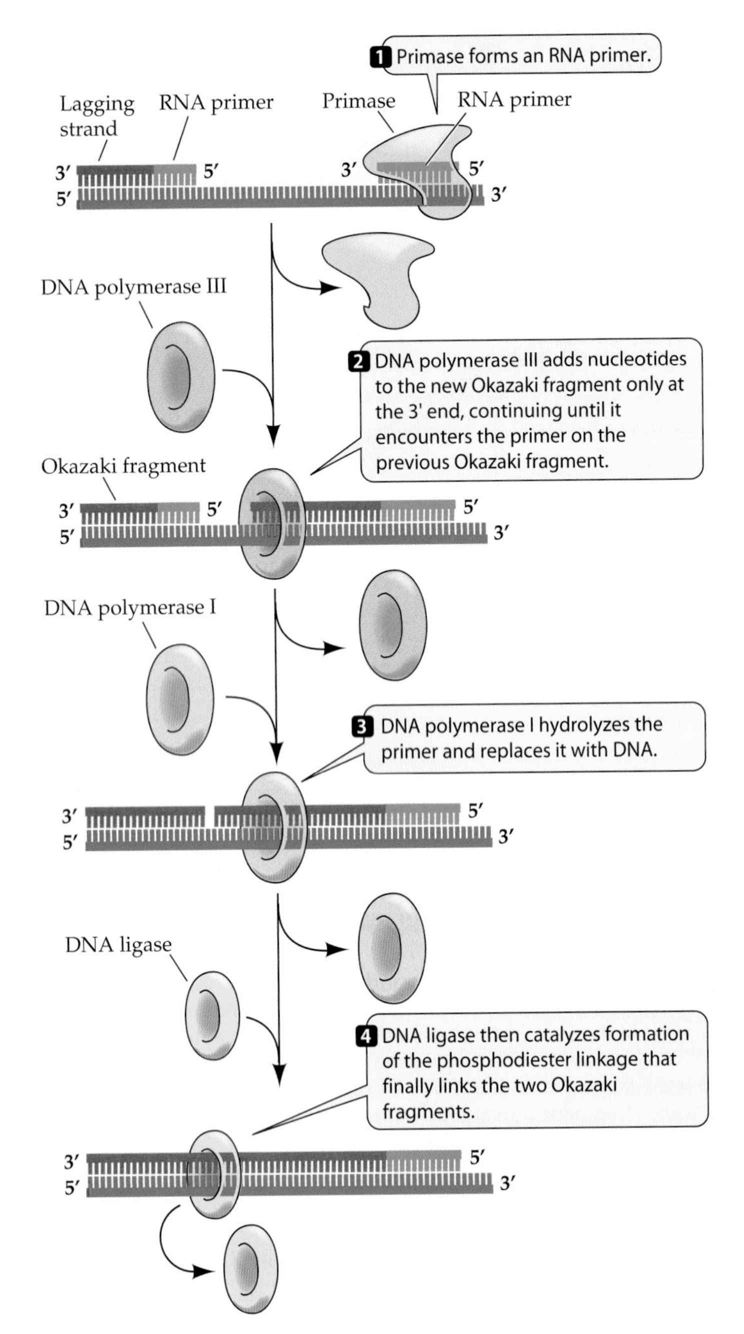

1 Primase forms an RNA primer.
Lagging strand
RNA primer
Primase
RNA primer
3′
5′
3′
5′
5′
3′
DNA polymerase III
2 DNA polymerase III adds nucleotides to the new Okazaki fragment only at the 3' end, continuing until it encounters the primer on the previous Okazaki fragment.
Okazaki fragment
DNA polymerase I
3 DNA polymerase I hydrolyzes the primer and replaces it with DNA.
DNA ligase
4 DNA ligase then catalyzes formation of the phosphodiester linkage that finally links the two Okazaki fragments.

11.17 The Lagging Strand Story In bacteria, DNA polymerase I and DNA ligase cooperate with DNA polymerase III to complete the complex task of synthesizing the lagging strand.

otes; 1,000 to 2,000 at a time in prokaryotes). These discontinuous stretches are synthesized just as the leading strand is, by the addition of new nucleotides one at a time to the 3′ end of the new strand, but the synthesis of this new strand moves in the direction opposite to that in which the replication fork is moving. These stretches of new DNA for the lagging strand are called **Okazaki fragments**, after their discoverer, the Japanese biochemist Reiji Okazaki. While the leading strand grows continuously "forward," the lagging strand grows in shorter, "backward" stretches with gaps between them.

A single primer suffices for synthesis of the leading strand, but each Okazaki fragment requires its own primer. In bacteria, DNA polymerase III synthesizes Okazaki fragments by adding nucleotides to a primer until it reaches the primer of the previous fragment. At this point, DNA polymerase I (the one discovered by Kornberg) removes the old primer and replaces it with DNA. Left behind is a tiny nick—the final phosphodiester linkage between the adjacent Okazaki fragments is missing. The enzyme **DNA ligase** catalyzes the formation of that bond, linking the fragments and making the lagging strand whole (Figure 11.17).

Working together, DNA helicase, the two DNA polymerases, primase, DNA ligase, and the other proteins of the replication complex do the job of DNA synthesis with a speed and accuracy that are almost unimaginable. In *E. coli*, the replication complex makes new DNA at a rate in excess of 1,000 base pairs per second, committing errors in fewer than one base in 10^6, or one in a million.

Telomeres are not fully replicated

As we have just seen, replication of the lagging strand occurs by the addition of Okazaki fragments to RNA primers. Beyond the very end of a linear DNA molecule, however, there is no place for a primer to bind (i.e., there is no complementary DNA strand). So the new chromosome formed after DNA replication has a bit of single-stranded DNA at each end (Figure 11.18*a*). This situation activates mechanisms that cut off the single-stranded region, along with some of the intact double-stranded end. Thus, the chromosome becomes slightly shorter with each cell division.

In many eukaryotes, there are repetitive sequences at the ends of chromosomes called **telomeres**. In humans, the telomere sequence is TTAGGG, and it is repeated about 2,500 times. These repeats bind special proteins that maintain the stability of chromosome ends. Each human chromosome can lose 50–200 base pairs of telomeric DNA after each round of DNA replication and cell division. After 20–30 divisions, the chromosomes are unable to take part in cell division, and the cell dies. This phenomenon explains in part why cells do not last the entire lifetime of the organism: Their telomeres shorten.

Yet constantly dividing cells, such as bone marrow and germ line cells, maintain their telomeric DNA. An enzyme, appropriately called **telomerase**, catalyzes the addition of any lost telomeric sequences (Figure 11.18*b*). Telomerase contains an RNA sequence that acts as a template for the telomeric repeat sequence.

Telomerase is expressed in more than 90 percent of human cancers and may be an important factor in the ability of can-

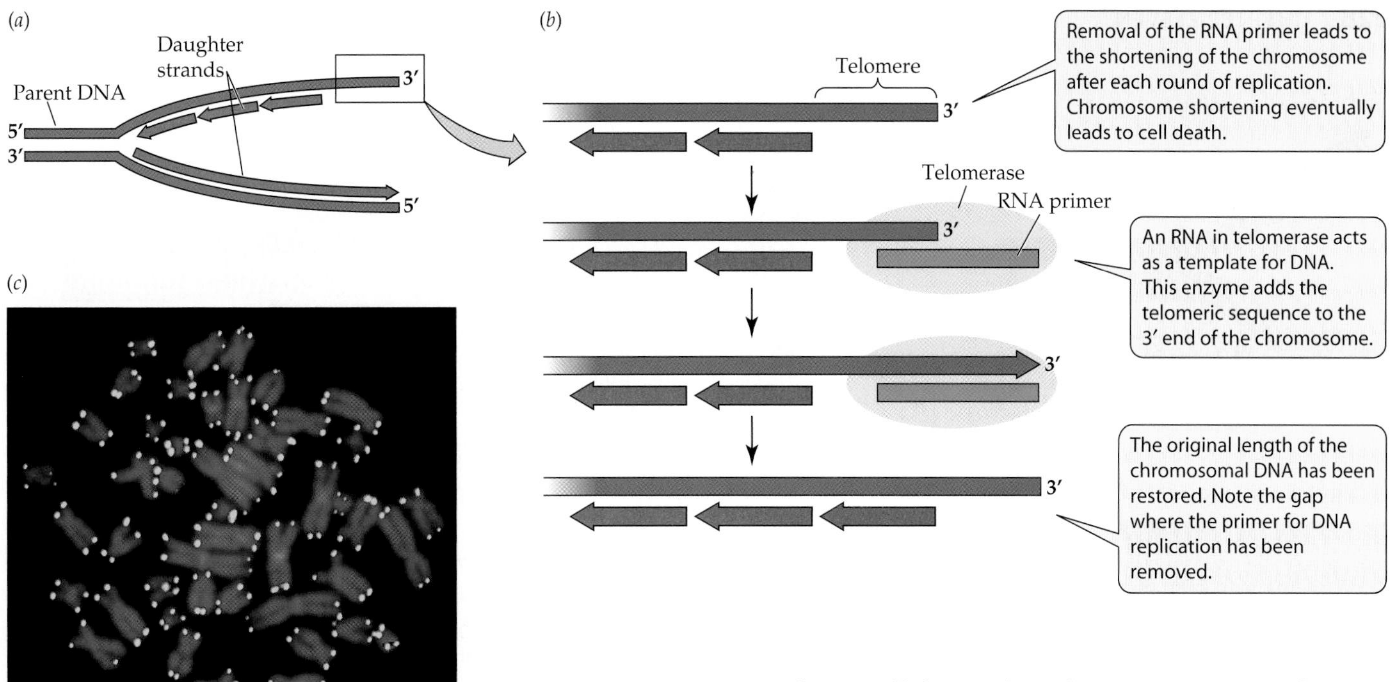
(a)
Daughter strands
Parent DNA
5′
3′
3′
5′
(b)
Telomere
3′
Removal of the RNA primer leads to the shortening of the chromosome after each round of replication. Chromosome shortening eventually leads to cell death.
Telomerase
RNA primer
3′
An RNA in telomerase acts as a template for DNA. This enzyme adds the telomeric sequence to the 3′ end of the chromosome.
3′
3′
The original length of the chromosomal DNA has been restored. Note the gap where the primer for DNA replication has been removed.
(c)

11.18 Telomeres and Telomerase (*a*) Removal of RNA primer at the 3′ end of the lagging strand leaves a region of DNA unreplicated. (*b*) The enzyme telomerase binds to the 3′ end and extends the lagging strand of DNA. An RNA sequence embedded in telomerase provides a template so that, overall, the DNA does not get shorter. (*c*) Bright fluorescent staining marks the telomeric regions on these blue-stained human chromosomes.

cer cells to divide continuously. Since most normal cells do not have this ability, telomerase is an attractive target for drugs designed to attack tumors specifically.

There is also interest in telomerase and aging. When a gene expressing high levels of telomerase is added to human cells in culture, their telomeres do not shorten. Instead of dying after 20–30 cell generations, the cells become immortal. It remains to be seen how this finding relates to the aging of a large organism.

DNA Proofreading and Repair

DNA is accurately replicated and faithfully maintained. The price of failure can be great: the transmission of genetic information is at stake, as is the functioning and even the life of a cell or multicellular organism. Yet the replication of DNA is not perfectly accurate, and the DNA of nondividing cells is subject to damage by environmental agents. In the face of these threats, how has life gone on so long?

The preservers of life are DNA repair mechanisms. DNA polymerases initially make a significant number of mistakes in assembling polynucleotide strands. The observed error rate of one for every 10^6 bases replicated would result in about 1,000 mutations every time a human cell divided. Fortunately, our cells have at least three DNA repair mechanisms at their disposal:

- A **proofreading** mechanism corrects errors in replication as DNA polymerase makes them.
- A **mismatch repair** mechanism scans DNA immediately after it has been replicated and corrects any base-pairing mismatches.
- An **excision repair** mechanism removes abnormal bases that have formed because of chemical damage and replaces them with functional bases.

Proofreading mechanisms ensure that DNA replication is accurate

After introducing a new nucleotide into a growing polynucleotide strand, DNA polymerase performs a proofreading function (Figure 11.19*a*). When a DNA polymerase recognizes a mispairing of bases, it removes the improperly introduced nucleotide and tries again. (Other proteins of the replication complex also play roles in proofreading.) The error rate for this process is only about 1 in 10,000 base pairs, and lowers the overall error rate for replication to about one base in every 10^{10} bases replicated.

Mismatch repair mechanisms correct base-pairing errors

After DNA has been replicated, a second set of proteins surveys the newly replicated molecule and looks for remaining mismatched base pairs (Figure 11.19*b*). For example, this mismatch repair system might detect an AC base pair instead of an AT pair. But how does the repair system "know" whether the AC pair should be repaired by removing the C and replacing it with T or by removing the A and replacing it with G?

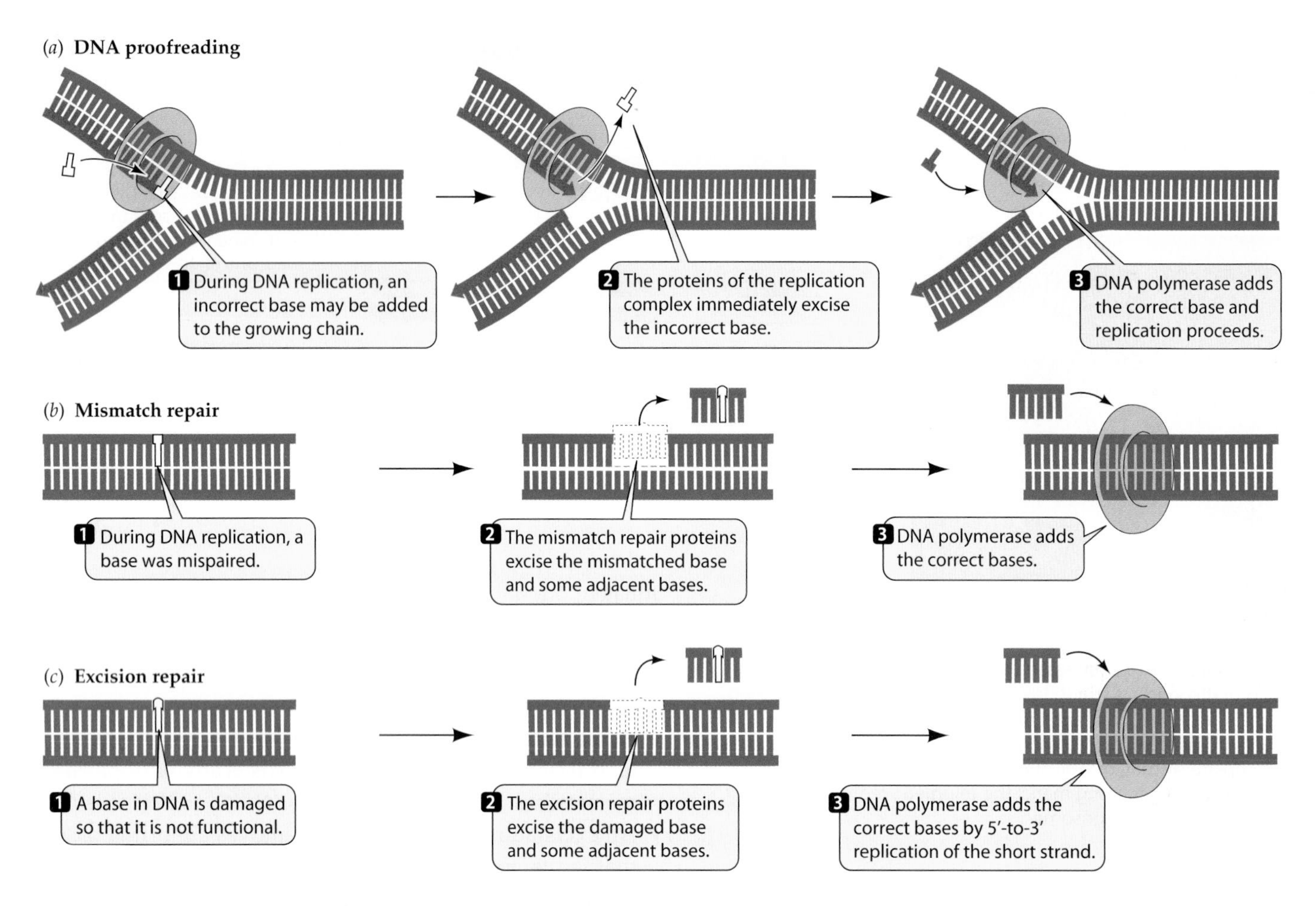

(a) DNA proofreading
During DNA replication, an incorrect base may be added to the growing chain.
The proteins of the replication complex immediately excise the incorrect base.
DNA polymerase adds the correct base and replication proceeds.
(b) Mismatch repair
During DNA replication, a base was mispaired.
The mismatch repair proteins excise the mismatched base and some adjacent bases.
DNA polymerase adds the correct bases.
(c) Excision repair
A base in DNA is damaged so that it is not functional.
The excision repair proteins excise the damaged base and some adjacent bases.
DNA polymerase adds the correct bases by 5′-to-3′ replication of the short strand.

11.19 DNA Repair Mechanisms The proteins of the replication complex also play roles in the life-preserving DNA repair mechanisms, helping to ensure the exact replication of template DNA and repair any damage that occurs.

The repair mechanism can detect the "wrong" base because a DNA strand is chemically modified some time after replication. In prokaryotes, methyl groups (—CH_3) are added to some guanines. Immediately after replication, methylation has not yet occurred, so the newly replicated strand is "marked," by being unmethylated, as the one in which errors should be corrected.

When mismatch repair fails, DNA sequences are altered. One form of colon cancer arises in part from a failure of mismatch repair.

Excision repair mechanisms repair chemical damage

DNA molecules can also be damaged during the life of a cell (e.g., when it is in G1). High-energy radiation, chemicals from the environment, and random spontaneous chemical reactions can all damage DNA.

Certain enzymes constantly "inspect" the cell's DNA (Figure 11.19*c*). When they find mispaired bases, chemically modified bases, or points at which one strand has more bases than the other (with the result that one or more bases of one strand form an unpaired loop), these enzymes cut the defective strand. Another enzyme cuts away the bases adjacent to and including the offending base, and DNA polymerase and DNA ligase synthesize and seal up a new (usually correct) base sequence to replace the excised one.

Our dependence on excision repair is underscored by our susceptibility to diseases that arise from excision repair defects. One example is the skin disease xeroderma pigmentosum. Persons with this disease lack a mechanism that normally repairs the damage caused by ultraviolet radiation in sunlight; they develop skin cancers after even minute exposure to sunlight.

Practical Applications of DNA Replication

The principles underlying DNA replication in cells have been used to develop two laboratory techniques that have been vital in analyzing genes and genomes. The first technique allows researchers to make multiple copies of short DNA sequences, and the second allows them to determine the base sequence of a DNA molecule.

The polymerase chain reaction makes multiple copies of DNA

Since DNA can be replicated in the laboratory, it is possible to make multiple copies of a DNA sequence. The **polymerase chain reaction** (**PCR**) technique essentially automates this process by copying a short region of DNA many times in a test tube.

PCR is a cyclic process in which a sequence of steps is repeated over and over again (Figure 11.20):

- Double-stranded fragments of DNA are separated into single strands by mild heating (denatured).
- A short, artificially synthesized primer is added to the mixture, along with the four deoxyribonucleotide triphosphates (dATP, dGTP, dCTP, and dTTP) and DNA polymerase.
- DNA polymerase catalyzes the production of complementary new strands.

A single cycle takes a few minutes to double the amount of DNA, leaving the new DNA in the double-stranded state. Theoretically, repeating the cycle many times leads to an exponential increase in the number of copies of the DNA sequence.

The PCR technique requires that the base sequences at the 3′ end of each strand of the target DNA sequence be known so that a complementary primer, usually 15–20 bases long, can be made in the laboratory. Because of the uniqueness of DNA sequences, usually only two primers of this length will bind to only one region of DNA in an organism's genome. This specificity in the face of the incredible diversity of target DNA is a key to the power of PCR.

One initial problem with PCR was its temperature requirements. To denature the DNA, it must be heated to more than 90°C—a temperature that destroys most DNA polymerases. The PCR method would not be practical if new polymerase had to be added after denaturation in each cycle.

This problem was solved by nature: In the hot springs at Yellowstone National Park, as well as other locations, lives a bacterium called, appropriately, *Thermus aquaticus*. The means by which this organism survives temperatures up to 95°C was investigated by Thomas Brock and his colleagues. They discovered that *T. aquaticus* has an entire metabolic machinery that is heat-resistant, including DNA polymerase that does not denature at these high temperatures.

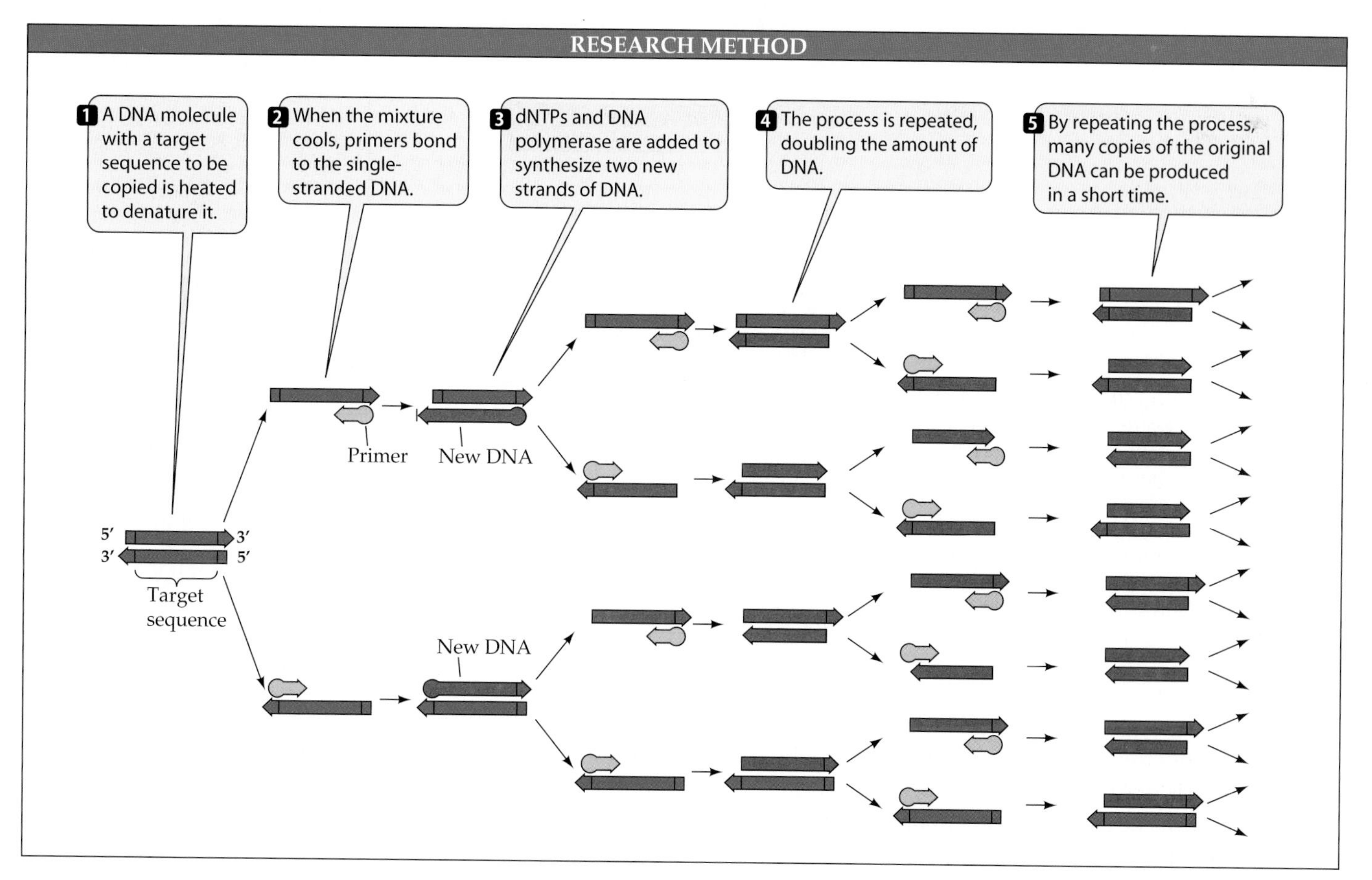

RESEARCH METHOD
1 A DNA molecule with a target sequence to be copied is heated to denature it.
2 When the mixture cools, primers bond to the single-stranded DNA.
3 dNTPs and DNA polymerase are added to synthesize two new strands of DNA.
4 The process is repeated, doubling the amount of DNA.
5 By repeating the process, many copies of the original DNA can be produced in a short time.
5′
3′
3′
5′
Target sequence
Primer
New DNA
New DNA

11.20 The Polymerase Chain Reaction The steps in this cyclic process are repeated many times to produce multiple copies of a DNA fragment.

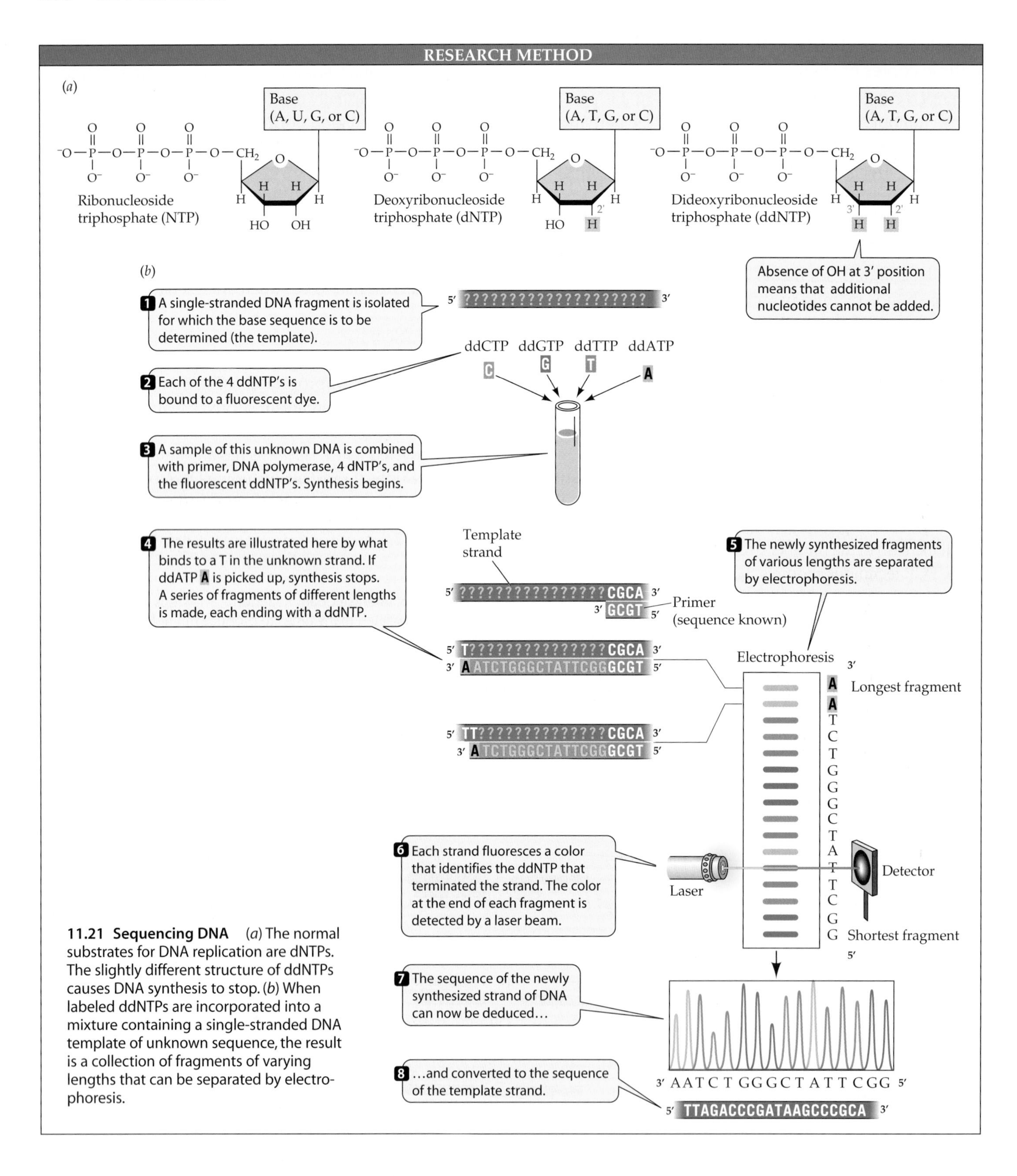

11.21 Sequencing DNA (*a*) The normal substrates for DNA replication are dNTPs. The slightly different structure of ddNTPs causes DNA synthesis to stop. (*b*) When labeled ddNTPs are incorporated into a mixture containing a single-stranded DNA template of unknown sequence, the result is a collection of fragments of varying lengths that can be separated by electrophoresis.

Scientists pondering the problem of copying DNA by PCR read Brock's basic research articles and got a clever idea: Why not use *T. aquaticus* DNA polymerase in the PCR reaction? It could withstand the 90°C temperature and would not have to be added during each cycle. The idea worked, and it earned biochemist Kerry Mullis a Nobel prize. PCR has had an enormous impact on genetic research. Some of its most striking applications will be described in Chapters 13 through 17.

The nucleotide sequence of DNA can be determined

Another important technique allows researchers to determine the base sequence of a DNA molecule. This **DNA sequencing** technique relies on the use of artificially altered nucleosides. As we saw earlier in this chapter, the deoxyribonucleoside triphosphates (dNTPs) that are the normal substrates for DNA replication contain the sugar deoxyribose. If that sugar is replaced with 2,3-dideoxyribose, the resulting dideoxyribonucleoside triphosphate (ddNTP) will still be added by DNA polymerase to a growing DNA chain. However, because ddNTPs lack a hydroxyl group at the 3′ position, the next nucleotide cannot be added (Figure 11.21*a*). Thus, synthesis stops at the position where ddNTP has been incorporated into the growing end of a DNA strand.

To determine the sequence of DNA, a fragment of DNA (usually no more than 700 base pairs long) is denatured. The resulting single strands of DNA are placed in a test tube and mixed with

- DNA polymerase, to synthesize the complementary strand;
- Short, artificially synthesized primers appropriate for that sequence;
- The four dNTPs (dATP, dGTP, dCTP, and dTTP); and
- Small amounts of the four ddNTPs, each bonded to a fluorescent "tag" that emits a different color of light.

DNA replication proceeds and the test tube soon contains a mixture of the template DNA strands and shorter, new complementary strands. The new strands, each ending with a fluorescent ddNTP, are of varying lengths. For example, each time a T is reached on the template strand, DNA polymerase adds either a dATP or ddATP to the growing complementary strand. If dATP is added, the strand continues to grow. If ddATP is added, chain growth stops.

After DNA replication has been allowed to proceed for a while the new DNA fragments are denatured from their templates. The fragments are then subjected to *electrophoresis* (see Figure 16.2). This technique sorts the DNA fragments by length, and can detect differences in fragment length as short as one base. During the electrophoresis run, the fragments pass in order of increasing length through a laser beam that excites the fluorescent tags. The light emitted is then detected, and the resulting information—that is, which color of fluorescence, and therefore which ddNTP, is at the end of a strand of which length—is fed into a computer. The computer processes this information and prints out the DNA sequence of the fragment (Figure 11.21*b*). DNA sequencing has formed the basis of the new science of genomics, as we will describe in Chapters 13, 14, and 17.

Chapter Summary

DNA: The Genetic Material

▶ Circumstantial evidence (its location and quantity in the cell) suggested that DNA might be the genetic material. Two experiments provided a convincing demonstration that this was the case. **Review Figures 11.1, 11.2, 11.3**

The Structure of DNA

▶ X-ray crystallography showed that the DNA molecule is a helix. **Review Figure 11.4**

▶ DNA is composed of nucleotides, each containing one of four bases: adenine, cytosine, thymine, or guanine. Biochemical analysis revealed that the amount of adenine equals the amount of thymine and the amount of guanine equals the amount of cytosine. **Review Figure 11.5**

▶ Putting the accumulated data together, Watson and Crick proposed that DNA is a double-stranded helix in which the strands are antiparallel and the bases are held together by hydrogen bonding. This model accounts for the genetic information, mutation, and replication functions of DNA. **Review Figures 11.6, 11.7**

Determining the DNA Replication Mechanism

▶ An experiment by Meselson and Stahl proved the replication of DNA to be semiconservative. Each parent strand acts as a template for the synthesis of a new strand; thus the two replicated DNA helices each contain one parent strand and one newly synthesized strand. **Review Figures 11.8, 11.9.** See Web/CD Tutorial 11.1

The Mechanisms of DNA Replication

▶ In DNA replication, the enzyme DNA polymerase catalyzes the addition of nucleotides to the 3′ end of each strand. Nucleotides are added by complementary base pairing with the template strand of DNA. The substrates are deoxyribonucleoside triphosphates, which are hydrolyzed as they are added to the growing chain, releasing energy that fuels the synthesis of DNA. **Review Figure 11.10**

▶ The DNA replication complex is attached to nuclear structures, and DNA is threaded through it for replication. **Review Figure 11.11**

▶ Many proteins assist in DNA replication. DNA helicase unwinds the double helix, and the template strands are stabilized by single-strand binding proteins.

▶ Prokaryotes have a single origin of replication; eukaryotes have many. Replication in both cases proceeds in both directions from an origin of replication. **Review Figure 11.12**

▶ An RNA primase catalyzes the synthesis a short RNA primer, to which nucleotides are added. **Review Figure 11.14**

▶ Through the action of DNA polymerase, the leading strand grows continuously in the 5′-to-3′ direction until the replication of that section of DNA has been completed. Then the RNA primer is degraded and DNA is added in its place.

▶ On the lagging strand, DNA is still made in the 5′-to-3′ direction. But synthesis of the lagging strand is discontinuous: The DNA is added as short fragments to primers, then the polymerase skips past the 5′ end to make the next fragment. **Review Figures 11.15, 11.16, 11.17.** See Web/CD Tutorial 11.3

▶ The very ends of linear chromosomes are usually not fully replicated because there is no place for a primer to bind on the lagging strand. This leads to a shortening of the DNA after each

round of replication, and ultimately cell death. Some cells have an enzyme, telomerase, that maintains chromosome length so that the cell can continue to divide. **Review Figure 11.18**
See Web/CD Tutorial 11.2

DNA Proofreading and Repair

▶ The machinery of DNA replication makes about one error in 10^6 nucleotides bases added. DNA is also subject to chemical damage. DNA is repaired by three different mechanisms: proofreading, mismatch repair, and excision repair. **Review Figure 11.19**

Practical Applications of DNA Replication

▶ The polymerase chain reaction technique uses DNA polymerase to repeatedly replicate DNA in the laboratory. **Review Figure 11.20**

▶ The principles of DNA replication can be used to determine the nucleotide sequence of DNA. **Review Figure 11.21**

Self-Quiz

1. Griffith's studies of *Streptococcus pneumoniae*
   a. showed that DNA is the genetic material of bacteria.
   b. showed that DNA is the genetic material of bacteriophages.
   c. demonstrated the phenomenon of bacterial transformation.
   d. proved that prokaryotes reproduce sexually.
   e. proved that protein is not the genetic material.
2. In the Hershey–Chase experiment,
   a. DNA from parent bacteriophages appeared in progeny bacteriophages.
   b. most of the phage DNA never entered the bacteria.
   c. more than three-fourths of the phage protein appeared in progeny phages.
   d. DNA was labeled with radioactive sulfur.
   e. DNA formed the coat of the bacteriophages.
3. Which statement about complementary base pairing is *not* true?
   a. It plays a role in DNA replication.
   b. In DNA, T pairs with A.
   c. Purines pair with purines, and pyrimidines pair with pyrimidines.
   d. In DNA, C pairs with G.
   e. The base pairs are of equal length.
4. In semiconservative replication of DNA,
   a. the original double helix remains intact and a new double helix forms.
   b. the strands of the double helix separate and act as templates for new strands.
   c. polymerization is catalyzed by RNA polymerase.
   d. polymerization is catalyzed by a double helical enzyme.
   e. DNA is synthesized from amino acids.
5. Which of the following does not occur during DNA replication?
   a. Unwinding of the parent double helix
   b. Formation of short pieces that are connected by DNA ligase
   c. Complementary base pairing
   d. Use of a primer
   e. Polymerization in the 3′-to-5′ direction
6. The primer used for DNA replication
   a. is a short strand of RNA added to the 3′ end.
   b. is present only once on the leading strand.
   c. remains on the DNA after replication.
   d. ensures that there will be a free 5′ end to which nucleotides can be added.
   e. is added to only one of the two template strands.
7. One strand of DNA has the sequence 5′–ATTCCG–3′. The complementary strand for this is
   a. 5′–TAAGGC–3′
   b. 5′–ATTCCG–3′
   c. 5′–ACCTTA–3′
   d. 5′–CGGAAT–3′
   e. 5′–GCCTTA–3′
8. The role of DNA ligase in DNA replication is to
   a. add more nucleotides to the growing strand one at a time.
   b. open up the two DNA strands to expose template strands.
   c. ligate base to sugar to phosphate in a nucleotide.
   d. bond Okazaki fragments to one another.
   e. remove incorrectly paired bases.
9. The polymerase chain reaction
   a. is a method for sequencing DNA.
   b. is used to transcribe specific genes.
   c. amplifies specific DNA sequences.
   d. does not require DNA replication primers.
   e. uses a DNA polymerase that denatures at 55°C.
10. The following events occur in excision repair of DNA. What is their proper order?
    (1) Base-paired DNA is made complementary to the template.
    (2) Damaged bases are recognized.
    (3) DNA ligase seals the new strand to existing DNA.
    (4) Part of a single strand is excised.
    a. 1234
    b. 2134
    c. 2413
    d. 3421
    e. 4231

For Discussion

1. Outline a series of experiments using radioactive isotopes to show that bacterial DNA and not protein enters the host cell and is responsible for bacterial transformation.
2. Suppose that Meselson and Stahl had continued their experiment on DNA replication for another ten bacterial generations. Would there still have been any ^{14}N–^{15}N hybrid DNA present? Would it still have appeared in the centrifuge tube? Explain.
3. If DNA replication were conservative rather than semiconservative, what results would Meselson and Stahl have observed? Diagram the results using the conventions of Figure 11.9.
4. Using the following information, calculate the number of origins of DNA replication on a human chromosome: DNA polymerase adds nucleotides at 3,000 base pairs per minute in one direction; replication is bidirectional; S phase lasts 300 minutes; there are 120 million base pairs per chromosome. With a typical chromosome 3 μm long, how many origins are there per μm?
5. The drug dideoxycytidine (used to treat certain viral infections) is a nucleotide made with 2′,3′-dideoxyribose. This sugar lacks —OH groups at both the 2′ and the 3′ positions. Explain why this drug stops the growth of a DNA chain when added to DNA.

12 From DNA to Protein: Genotype to Phenotype

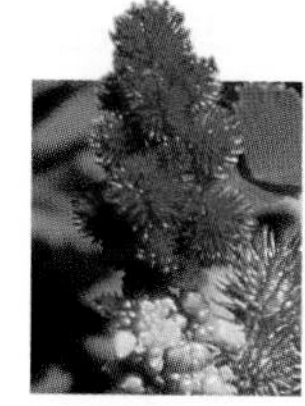

In 1978, Georgi Markov, a journalist who was living in London because he had written articles critical of the then-Communist government of Bulgaria, was standing at a bus stop near Waterloo Station. A man, possibly a Bulgarian secret agent, brushed up against him and, seemingly by accident, poked him with an umbrella. Markov felt a sharp pain, and within a few hours, he started to feel weak. High temperature, vomiting, and more severe symptoms soon followed. Two days later he was dead.

Police investigators found a tiny perforated pellet embedded in Markov's leg, and in that pellet was a small amount of ricin, a highly toxic molecule isolated from the seeds of the tropical castor bean plant, *Ricinus communis*. These seeds have been used for centuries as a source of castor oil, a natural product that used to be commonly given to children to "clean out" the digestive tract and is now used in the plastics industry. The toxin is a protein that is not present in the oil, and people found out the hard way that it is one of the most poisonous substances made by any organism.

Ricin kills cells by blocking protein synthesis. More specifically, it catalyzes the modification and cleavage of one of the large RNA molecules that make up the eukaryotic ribosome, the "workbench" of protein synthesis. Proteins are the major phenotypic expression of the genotype—the genetic information encoded in a cell's DNA. Ricin inhibits the cell's ability to express the genotype as phenotype through protein synthesis, and therefore ricin-poisoned cells cannot survive.

Ricinus communis, the Castor Bean Plant This brightly colored plant, grown in the Tropics as an ornamental, produces ricin, a lethal toxin that inhibits protein synthesis at the ribosome.

This chapter deals with the mechanisms by which genes are expressed as proteins. We will begin with evidence for the relationship between genes and proteins, and then fill in some of the details of the processes of transcription—the copying of the gene sequence of DNA into a sequence of RNA—and translation—the use of the sequence of RNA to make a polypeptide with a defined order of amino acids. Finally, we will define mutations and their phenotypes in specific molecular terms.

One Gene, One Polypeptide

There are many steps between genotype and phenotype. Genes cannot, all by themselves, directly produce a phenotypic result, such as a particular eye color, a specific seed shape, or a cleft chin, any more than a compact disk can play a symphony without the help of a CD player.

The first historical step in relating genes to phenotypes was to define phenotypes in molecular terms. The molecular basis of phenotypes was actually discovered before the discovery that DNA was the genetic material. Scientists had studied the chemical differences between individuals carrying wild-type and mutant alleles in organisms as diverse as humans and bread molds. They found that the major phenotypic differences were the result of differences in specific proteins.

In the 1940s, a series of experiments by George W. Beadle and Edward L. Tatum at Stanford University showed that when an altered gene resulted in an altered phenotype, that altered phenotype was always associated with an altered enzyme. This finding was critically important in defining the phenotype in chemical terms.

The roles of enzymes in biochemistry were being described at this time, and it occurred to Beadle and Tatum that the expression of a gene as phenotype could occur through an enzyme. They experimented with the bread mold *Neurospora crassa*. The nuclei in the body of this mold are haploid (n), as are its reproductive spores. (This fact is important because it means that even recessive mutant alleles are easy to detect in experiments.) Beadle and Tatum grew *Neurospora* on a minimal nutritional medium containing sucrose, minerals, and a vitamin. Using this medium, the enzymes of wild-type *Neurospora* could catalyze the metabolic reactions needed to make all the chemical constituents of their cells, including proteins. These wild-type strains are called *prototrophs* ("original eaters").

Beadle and Tatum treated wild-type *Neurospora* with X rays, which act as a *mutagen* (something known to cause mutations). When they examined the treated molds, they found some mutant strains could no longer grow on the minimal medium, but needed to be supplied with additional nutrients. The scientists hypothesized that these *auxotrophs* ("increased eaters") must have suffered mutations in genes that code for the enzymes used to synthesize the nutrients they now needed to obtain from their environment. For each auxotrophic strain, Beadle and Tatum were able to find a single compound that, when added to the minimal medium, supported the growth of that strain. This result suggested that mutations have simple effects, and that each mutation causes a defect in only one enzyme in a metabolic pathway described as *the one-gene, one-enzyme* hypothesis (Figure 12.1).

One group of auxotrophs, for example, could grow only if the minimal medium was supplemented with the amino acid arginine. (Wild-type *Neurospora* makes its own arginine.) These mutant strains were designated *arg* mutants. Beadle and Tatum found several different *arg* mutant strains. They proposed two alternative hypotheses to explain why these different genetic strains had the same phenotype:

- The different *arg* mutants could have mutations in *the same gene*, as in the case of the different eye color alleles of fruit flies. In this case, the gene might code for an enzyme involved in arginine synthesis.
- The different *arg* mutants could have mutations in *different genes*, each coding for a separate function that leads to arginine production. These independent functions might be different enzymes along the same biochemical pathway.

Some of the *arg* mutant strains fell into each of the two categories. Genetic crosses showed that some of the mutations were at the same chromosomal locus, and so were different alleles of the same gene. Other mutations were at different loci, or on different chromosomes, and so were not alleles of the same gene. Beadle and Tatum concluded that these different genes participated in governing a single biosynthetic pathway—in this case, the pathway leading to arginine synthesis (see the Conclusion in Figure 12.1).

By growing different *arg* mutants in the presence of various compounds suspected to be intermediates in the synthetic metabolic pathway for arginine, Beadle and Tatum were able to classify each mutation as affecting one enzyme or another, and to order the compounds along the pathway. Then they broke open the wild-type and mutant cells and examined them for enzyme activities. The results confirmed their hypothesis: Each mutant strain was indeed missing a single active enzyme in the pathway.

The gene–enzyme connection had been proposed 40 years earlier in 1908 by the Scottish physician Archibald Garrod, who studied the inherited human disease alkaptonuria. He linked the biochemical phenotype of the disease to an abnormal gene and a missing enzyme. Today we know of hundreds of examples of such hereditary diseases, which we will return to in Chapter 17.

The gene–enzyme relationship has undergone several modifications in light of our current knowledge of molecular biology. Many enzymes are composed of more than one polypeptide chain, or subunit (that is, they have a quaternary structure). In this case, each polypeptide chain is specified by its own separate gene. Thus, it is more correct to speak of a *one-gene, one-polypeptide* relationship: The function of a gene is to control the production of a single, specific polypeptide.

Much later, it was discovered that some genes code for forms of RNA that do not become translated into polypeptides, and that still other genes are involved in controlling which other DNA sequences are expressed. While these discoveries have supplanted the idea that all genes code for proteins, they did not invalidate the relationship between genes and polypeptides. But how does this relationship work—that is, how is the information encoded in DNA used to specify a particular polypeptide?

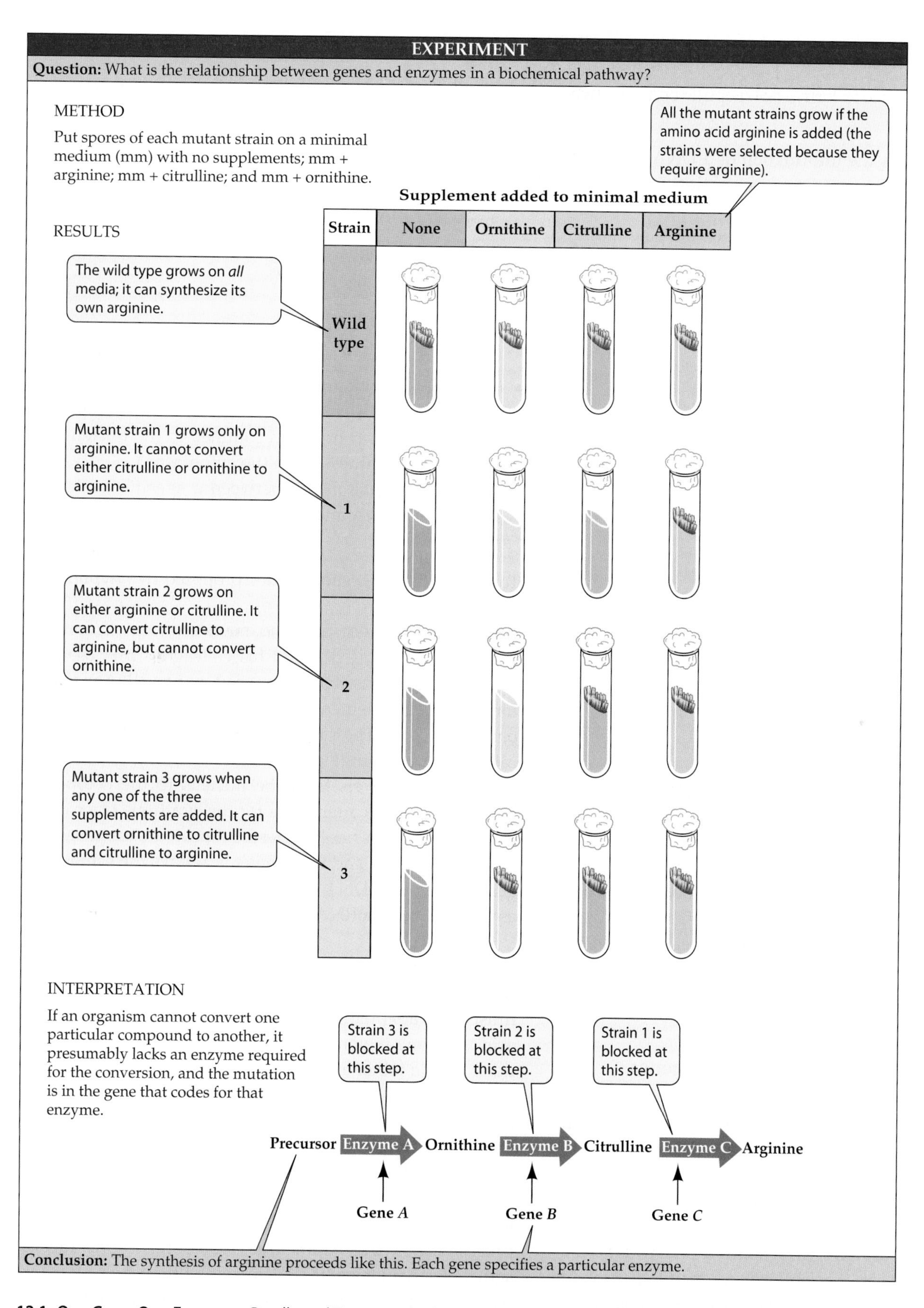

EXPERIMENT
Question: What is the relationship between genes and enzymes in a biochemical pathway?
METHOD
Put spores of each mutant strain on a minimal medium (mm) with no supplements; mm + arginine; mm + citrulline; and mm + ornithine.
All the mutant strains grow if the amino acid arginine is added (the strains were selected because they require arginine).
Supplement added to minimal medium
RESULTS
Strain
None
Ornithine
Citrulline
Arginine
The wild type grows on all media; it can synthesize its own arginine.
Wild type
Mutant strain 1 grows only on arginine. It cannot convert either citrulline or ornithine to arginine.
1
Mutant strain 2 grows on either arginine or citrulline. It can convert citrulline to arginine, but cannot convert ornithine.
2
Mutant strain 3 grows when any one of the three supplements are added. It can convert ornithine to citrulline and citrulline to arginine.
3
INTERPRETATION
If an organism cannot convert one particular compound to another, it presumably lacks an enzyme required for the conversion, and the mutation is in the gene that codes for that enzyme.
Strain 3 is blocked at this step.
Strain 2 is blocked at this step.
Strain 1 is blocked at this step.
Precursor
Enzyme A
Ornithine
Enzyme B
Citrulline
Enzyme C
Arginine
Gene A
Gene B
Gene C
Conclusion: The synthesis of arginine proceeds like this. Each gene specifies a particular enzyme.

12.1 One Gene, One Enzyme Beadle and Tatum studied several *arg* mutants of *Neurospora*. The different *arg* mutant strains required the addition of different compounds in order to synthesize the arginine required for their growth. Step through the figure to follow the reasoning that upheld the "one-gene, one-enzyme" hypothesis.

DNA, RNA, and the Flow of Information

The expression of a gene to form a polypeptide occurs in two major steps:

- *Transcription* copies the information of a DNA sequence (the gene) into corresponding information in an RNA sequence.
- *Translation* converts this RNA sequence into the amino acid sequence of a polypeptide.

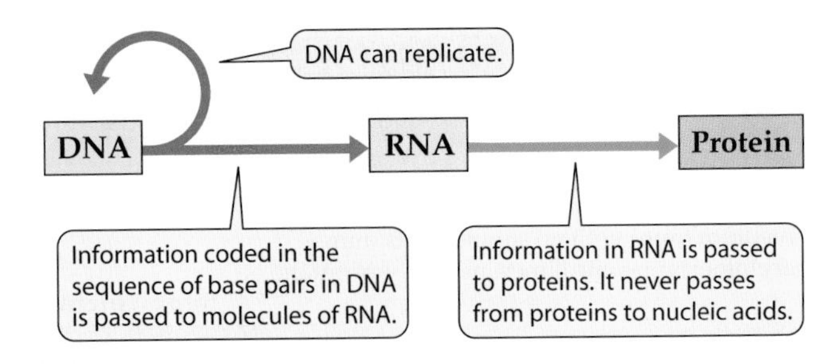

12.2 The Central Dogma Information flows from DNA to RNA to proteins, as indicated by the arrows.

RNA differs from DNA

RNA is a key intermediary between DNA and polypeptide. **RNA (ribonucleic acid)** is a polynucleotide similar to DNA (see Figure 3.25), but it differs from DNA in three ways:

- RNA generally consists of only one polynucleotide strand.
- The sugar molecule found in RNA is ribose, rather than the deoxyribose found in DNA.
- Although three of the nitrogenous bases (adenine, guanine, and cytosine) in RNA are identical to those in DNA, the fourth base in RNA is **uracil (U)**, which is similar to thymine but lacks the methyl (—CH_3) group.

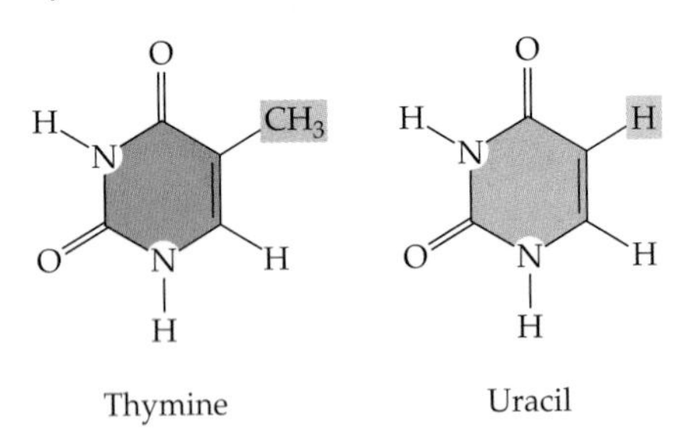

RNA can base-pair with single-stranded DNA. This pairing obeys the same complementary base-pairing rules as in DNA, except that *adenine pairs with uracil* instead of thymine. Single-stranded RNA can fold into complex shapes by internal base pairing, as we will see later in this chapter.

Information flows in one direction when genes are expressed

Soon after he and Watson proposed their three-dimensional structure for DNA, Francis Crick pondered the problem of how DNA is functionally related to proteins. This led him to propose what he called the **central dogma** of molecular biology. The central dogma, simply stated, is that DNA codes for the production of RNA, RNA codes for the production of protein, and protein does not code for the production of protein, RNA, or DNA (Figure 12.2). In Crick's words, "once 'information' has passed into protein it cannot get out again."

The central dogma raised two questions:

- How does genetic information get from the nucleus to the cytoplasm? (As we saw in Chapter 4, most of the DNA of a eukaryotic cell is confined to the nucleus, but proteins are synthesized in the cytoplasm.)
- What is the relationship between a specific nucleotide sequence in DNA and a specific amino acid sequence in a protein?

To answer these questions, Crick proposed two hypotheses.

THE MESSENGER HYPOTHESIS AND TRANSCRIPTION. To answer the first question, Crick and his colleagues proposed the *messenger hypothesis*. They proposed that an RNA molecule forms as a complementary copy of one DNA strand of a particular gene. The process by which this RNA forms is called **transcription** (Figure 12.3). This **messenger RNA**, or **mRNA**, then travels from the nucleus to the cytoplasm, where it serves as a template for the synthesis of proteins. Crick's hypothesis has been tested repeatedly for genes that code for proteins, and the answer is always the same: Each gene sequence in DNA that codes for a protein is expressed as a sequence in mRNA.

THE ADAPTER HYPOTHESIS AND TRANSLATION. To answer the second question, Crick proposed the *adapter hypothesis:* there must be an adapter molecule that can bind a specific amino acid with one region and recognize a sequence of nucleotides with another region. In due course, these adapters, called **transfer RNA**, or **tRNA**, were identified. Because they recognize the genetic message of mRNA and simultaneously carry specific amino acids, tRNAs can translate the language of DNA into the language of proteins. The tRNA adapters line up on the mRNA so that the amino acids are in the proper sequence for a growing polypeptide chain—a process called **translation** (see Figure 12.3). Once again, actual observations of the expression of thousands of genes have confirmed the hypothesis that tRNA acts as the intermediary between the nucleotide sequence information in mRNA and the amino acid sequence in a protein.

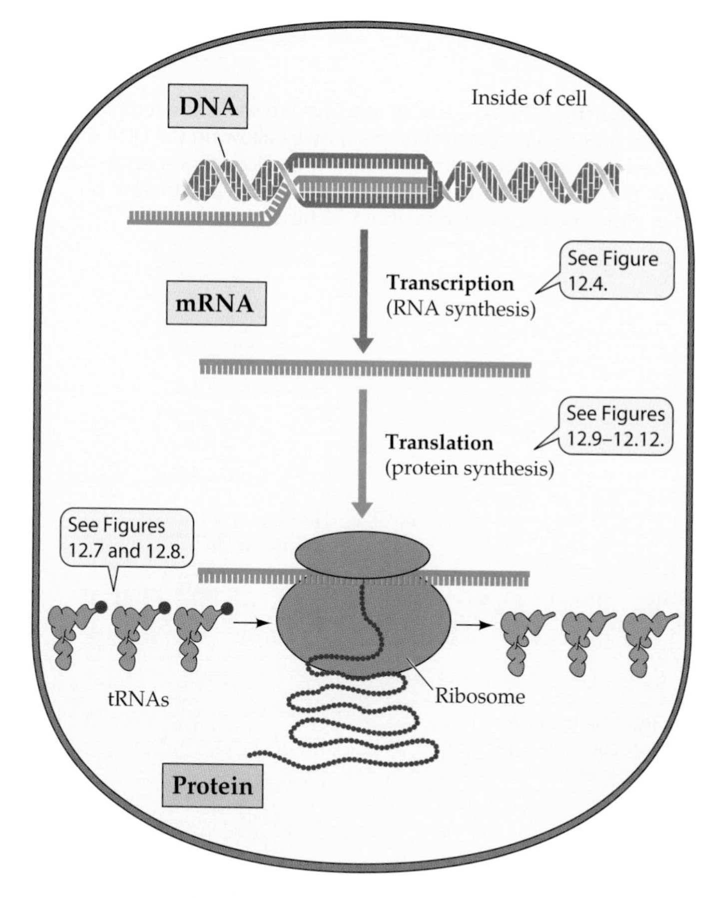

12.3 From Gene to Protein This diagram summarizes the processes of gene expression in prokaryotes. In eukaryotes, the processes are somewhat more complex.

Summarizing the main features of the central dogma, the messenger hypothesis, and the adapter hypothesis, we may say that a given gene is transcribed to produce a messenger RNA (mRNA) complementary to one of the DNA strands, and that transfer RNA (tRNA) molecules translate the sequence of bases in the mRNA into the appropriate sequence of linked amino acids during protein synthesis.

RNA viruses modify the central dogma

Certain viruses are rare exceptions to the central dogma. *Viruses* are infectious particles that reproduce inside cells. Many viruses, such as the tobacco mosaic virus, influenza virus, and poliovirus, have RNA rather than DNA as their genetic material. With its nucleotide sequence, RNA could potentially act as an information carrier and be expressed as proteins. But since RNA is usually single-stranded, its replication is a problem. The viruses generally solve this problem by transcribing from RNA to RNA, making an RNA strand that is complementary to their genome. This "opposite" strand is then used to make multiple copies of the viral genome by transcription:

The human immunodeficiency virus (HIV) and certain rare tumor viruses also have RNA as their genome, but do not replicate it as RNA-to-RNA. Instead, after infecting a host cell, they make a DNA copy of their genome and use it to make more RNA. This RNA is then used both as genomes for more copies of the virus and as mRNA to produce viral proteins.

Synthesis of DNA from RNA is called *reverse transcription*, and not surprisingly, such viruses are called **retroviruses**.

Transcription: DNA-Directed RNA Synthesis

Although the RNA viruses present a modification of the central dogma, the fact remains that in normal prokaryotic and eukaryotic cells, RNA synthesis is directed by DNA. Transcription—the formation of a specific RNA from a specific DNA—requires several components:

- A DNA template for complementary base pairing
- The appropriate ribonucleoside triphosphates (ATP, GTP, CTP, and UTP) to act as substrates
- An enzyme, **RNA polymerase**

Within each gene, only one of the two strands of DNA—the *template strand*—is transcribed. The other, complementary DNA strand, referred to as the *non-template strand*, remains untranscribed. For different genes in the same DNA molecule, different strands may be transcribed. That is, the strand that is the non-template strand in one gene may be the template strand in another.

Not only mRNA is produced by transcription. The same process is responsible for the synthesis of tRNA and ribosomal RNA (rRNA), whose important roles in protein synthesis will be described below. Like polypeptides, these RNAs are encoded by specific genes.

In DNA replication, as we know, the two strands of the parent molecule unwind, and each strand serves as the template for a new strand. In transcription, DNA partly unwinds so that it can serve as a template for RNA synthesis. As the RNA transcript is formed, it peels away, allowing the DNA to be rewound into the double helix (Figure 12.4).

Transcription can be divided into three distinct processes: initiation, elongation, and termination. Let's consider each of these in turn.

Initiation of transcription requires a promoter and RNA polymerase

Initiation begins transcription, and requires a **promoter**, a special sequence of DNA to which RNA polymerase binds very tightly. There is at least one promoter for each gene (or,

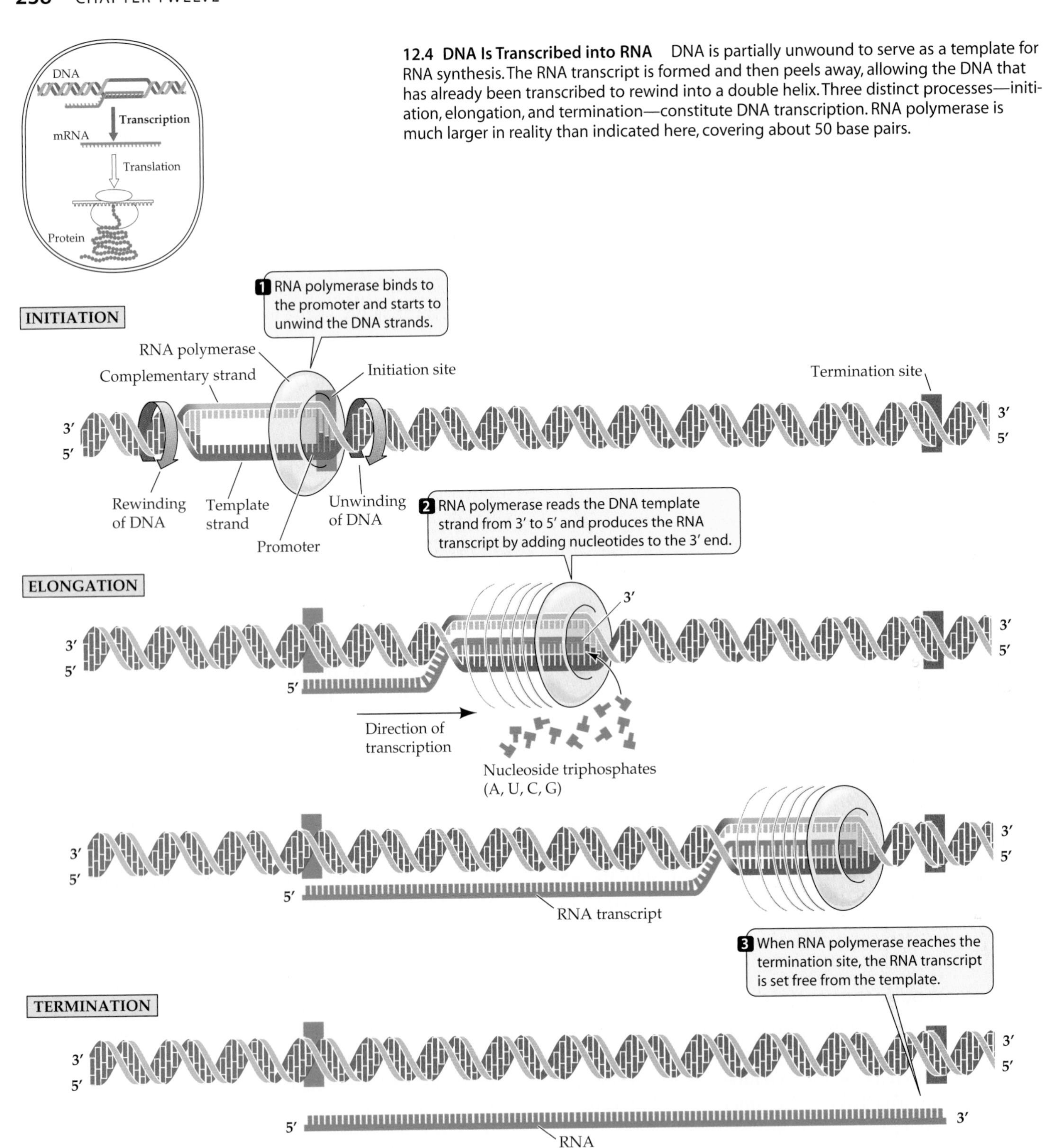

12.4 DNA Is Transcribed into RNA DNA is partially unwound to serve as a template for RNA synthesis. The RNA transcript is formed and then peels away, allowing the DNA that has already been transcribed to rewind into a double helix. Three distinct processes—initiation, elongation, and termination—constitute DNA transcription. RNA polymerase is much larger in reality than indicated here, covering about 50 base pairs.

in prokaryotes, each set of genes). Promoters are important control sequences that "tell" the RNA polymerase three things:

- where to start transcription
- which strand of DNA to read
- the direction to take from the start

A promoter, which is a specific sequence in the DNA that reads in a particular direction, orients the RNA polymerase and thus "aims" it at the correct strand to use as a template. Promoters function somewhat like the punctuation marks that determine how a sequence of words is to be read as a sentence. Part of each promoter is the *initiation site*, where transcription begins. Farther toward the 3′ end of the promoter lie groups of nucleotides that help the RNA polymerase bind. RNA polymerase moves in a 3′-to-5′ direction along the template strand (see Figure 12.4).

Although every gene has a promoter, not all promoters are identical. Some promoters are more effective at transcription initiation than others. Furthermore, there are differences between transcription initiation in prokaryotes and in eukaryotes. These differences will be explored in Chapters 13 and 14.

RNA polymerase elongates the transcript

Once RNA polymerase has bound to the promoter, it begins the process of **elongation**. It unwinds the DNA about 20 base pairs at a time and reads the template strand in the 3′-to-5′ direction (see Figure 12.4). Like DNA polymerase, RNA polymerase adds new nucleotides to the 3′ end of the growing strand, but does not require a primer to get started. The new RNA elongates from the first base that forms its 5′ end to its 3′ end. The RNA transcript is thus antiparallel to the DNA template strand.

Unlike DNA polymerases, RNA polymerases do not inspect and correct their work. Transcription errors occur at a rate of one mistake for every 10^4 to 10^5 bases. Because many copies of RNA are made, and because they often have only a relatively short existence, these errors are not as potentially harmful as mutations in DNA.

Transcription terminates at particular base sequences

What tells RNA polymerase to stop adding nucleotides to a growing RNA transcript? Just as initiation sites specify the start of transcription, particular base sequences in the DNA specify its **termination**. The mechanisms of termination are complex and of more than one kind. For some genes, the newly formed transcript simply falls away from the DNA template and the RNA polymerase. For others, a helper protein pulls the transcript away.

In prokaryotes, in which there is no nuclear envelope and ribosomes can be near the chromosome, the translation of mRNA often begins near the 5′ end of the mRNA before transcription of the mRNA molecule is complete. In eukaryotes, the situation is more complicated. First, there is a spatial separation of transcription (in the nucleus) and translation (in the cytoplasm). Second, the first product of transcription is a pre-mRNA that is longer than the final mRNA and must undergo considerable processing before it can be translated. The advantages of this processing, and its mechanisms, will be discussed in Chapter 14.

The Genetic Code

How do transcription and translation produce specific and functional protein products? These processes require a *genetic code* that relates genes (DNA) to mRNA and mRNA to the amino acids of proteins. The genetic code specifies which amino acids will be used to build a protein. You can think of the genetic information in an mRNA molecule as a series of sequential, nonoverlapping three-letter "words." Each sequence of three nucleotide bases (the three "letters") along the chain specifies a particular amino acid. Each three-letter "word" is called a **codon**. Each codon is complementary to the corresponding triplet in the DNA molecule from which it was transcribed. Thus, the genetic code is the means of relating codons to their specific amino acids.

The complete genetic code is shown in Figure 12.5. Notice that there are many more codons than there are different amino acids in proteins. Combinations of the four available "letters" (the bases) give 64 (4^3) different three-letter codons, yet these codons determine only 20 amino acids. AUG, which codes for methionine, is also the **start codon**, the initiation signal for translation. Three of the codons (UAA, UAG, UGA) are **stop codons**, or termination signals for translation;

First letter	Second letter: U	Second letter: C	Second letter: A	Second letter: G	Third letter
U	UUU, UUC Phenyl-alanine UUA, UUG Leucine	UCU, UCC, UCA, UCG Serine	UAU, UAC Tyrosine UAA Stop codon UAG Stop codon	UGU, UGC Cysteine UGA Stop codon UGG Tryptophan	U C A G
C	CUU, CUC, CUA, CUG Leucine	CCU, CCC, CCA, CCG Proline	CAU, CAC Histidine CAA, CAG Glutamine	CGU, CGC, CGA, CGG Arginine	U C A G
A	AUU, AUC, AUA Isoleucine AUG Methionine; start codon	ACU, ACC, ACA, ACG Threonine	AAU, AAC Asparagine AAA, AAG Lysine	AGU, AGC Serine AGA, AGG Arginine	U C A G
G	GUU, GUC, GUA, GUG Valine	GCU, GCC, GCA, GCG Alanine	GAU, GAC Aspartic acid GAA, GAG Glutamic acid	GGU, GGC, GGA, GGG Glycine	U C A G

12.5 The Universal Genetic Code Genetic information is encoded in mRNA in three-letter units—codons—made up of the bases uracil (U), cytosine (C), adenine (A), and guanine (G). To decode a codon, find its first letter in the left column, then read across the top to its second letter, then read down the right column to its third letter. The amino acid the codon specifies is given in the corresponding row. For example, AUG codes for methionine, and GUA codes for valine.

when the translation machinery reaches one of these codons, translation stops, and the polypeptide is released from the translation complex.

After describing the properties of the genetic code, we will examine some of the scientific thinking and experimentation that went into deciphering it.

The genetic code is redundant but not ambiguous

After the start and stop codons, the remaining 60 codons are far more than enough to code for the other 19 amino acids—and indeed there are repeats. Thus we say that the genetic code is *redundant;* that is, an amino acid may be represented by more than one codon. The redundancy is not evenly divided among the amino acids. For example, methionine and tryptophan are represented by only one codon each, whereas leucine is represented by six different codons (see Figure 12.5).

The term "redundancy" should not be confused with "ambiguity." To say that the code was *ambiguous* would mean that a single codon could specify either of two (or more) different amino acids; there would then be doubt whether to put in, say, leucine or something else. The genetic code is not ambiguous. Redundancy in the code means that there is more than one clear way to say, "Put leucine here." In other words, a given amino acid may be encoded by more than one codon, but a codon can code for only one amino acid. But just as people in different places prefer different ways of saying the same thing—"Good-bye!" "See you!" "Ciao!" and "So long!" have the same meaning—different organisms prefer one or another of the redundant codons.

The genetic code is (nearly) universal

Over 40 years of experiments on thousands of organisms from all the living domains and kingdoms reveal that the genetic code appears to be nearly *universal*, applying to all the species on our planet. Thus the code must be an ancient one that has been maintained intact throughout the evolution of living organisms. Exceptions are known: within mitochondria and chloroplasts, the code differs slightly from that in prokaryotes and in the nuclei of eukaryotic cells; in one group of protists, UAA and UAG code for glutamine rather than functioning as stop codons. The significance of these differences is not yet clear. What is clear is that the exceptions are few and slight.

The common genetic code means that there is also a common language for evolution. As natural selection resulted in one species replacing another, the raw material of genetic variation has remained the same. The common code also has profound implications for genetic engineering, as we will see in Chapter 16, since it means that a human gene is in the same language as a bacterial gene. The differences are more like dialects of a single language than entirely different languages. So the transcription and translation machinery of a bacterium could theoretically utilize genes from a human as well as its own genes.

The codons in Figure 12.5 are *mRNA codons*. The base sequence on the DNA strand that was transcribed to produce the mRNA is complementary and antiparallel to these codons. Thus, for example, 3′-AAA-5′ in the template DNA strand corresponds to phenylalanine (which is encoded by the mRNA codon 5′-UUU-3′), and 3′-ACC-5′ in the template DNA corresponds to tryptophan (which is encoded by the mRNA codon 5′-UGG-3′). How did biologists assign these codons to specific amino acids?

Biologists deciphered the genetic code by using artificial messengers

Molecular biologists broke the genetic code in the early 1960s. The problem was perplexing: How could more than 20 "code words" be written with an "alphabet" consisting of only four "letters"? How, in other words, could four bases (A, U, G, and C) code for 20 different amino acids?

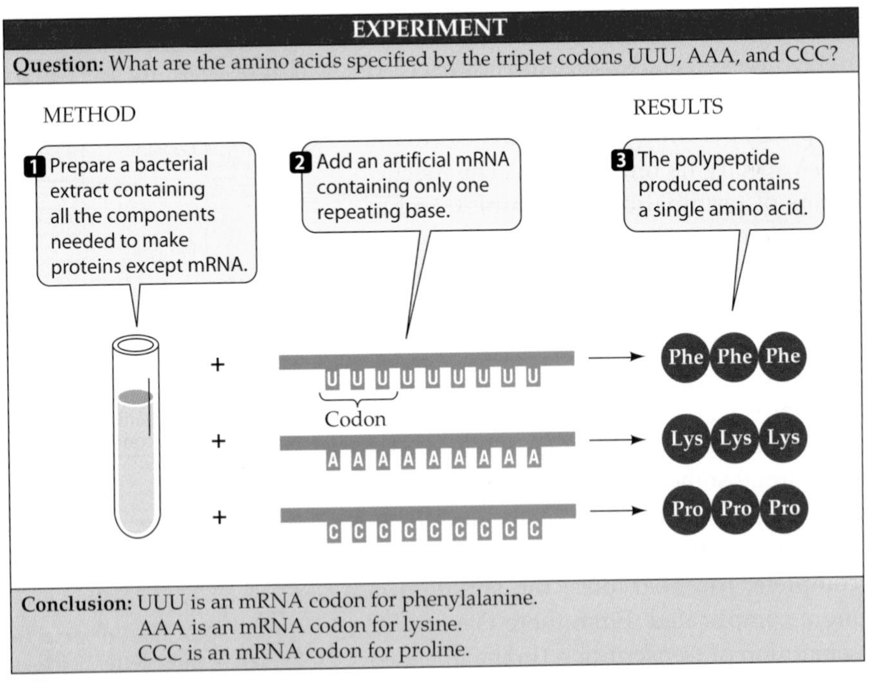

EXPERIMENT
Question: What are the amino acids specified by the triplet codons UUU, AAA, and CCC?
METHOD
RESULTS
1 Prepare a bacterial extract containing all the components needed to make proteins except mRNA.
2 Add an artificial mRNA containing only one repeating base.
3 The polypeptide produced contains a single amino acid.
+
U U U U U U U U U
Codon
A A A A A A A A A
C C C C C C C C C
Phe Phe Phe
Lys Lys Lys
Pro Pro Pro
Conclusion: UUU is an mRNA codon for phenylalanine.
AAA is an mRNA codon for lysine.
CCC is an mRNA codon for proline.

12.6 Deciphering the Genetic Code Nirenberg and Matthaei used a test-tube protein synthesis system to determine the amino acids specified by synthetic mRNAs of known codon composition.

That the code was a triplet code, based on three-letter codons, was considered likely. Since there are only four letters (A, G, C, U), a one-letter code clearly could not unambiguously encode 20 amino acids; it could encode only four of them. A two-letter code could contain only $4 \times 4 = 16$ codons—still not enough. But a triplet code could contain up to $4 \times 4 \times 4 = 64$ codons. This was more than enough to encode the 20 amino acids.

Marshall W. Nirenberg and J. H. Matthaei, at the National Institutes of Health, made the first decoding breakthrough in 1961 when they realized that they could use a simple artificial polynucleotide instead of a complex natural mRNA as a messenger. They could then identify the polypeptide that the artificial messenger encoded.

Scientists prepared an artificial mRNA in which all the bases were uracil (poly U). When poly U was added to a test tube containing all the ingredients necessary for protein synthesis (ribosomes, all the amino acids, activating enzymes, tRNAs, and other factors), a polypeptide formed. This polypeptide contained only one kind of amino acid: phenylalanine (Phe). Poly U coded for poly Phe! Accordingly, UUU appeared to be the mRNA code word—the codon—for phenylalanine. Following up on this success, Nirenberg and Matthaei soon showed that CCC codes for proline and AAA for lysine (Figure 12.6). (Poly G presented some chemical problems and was not tested initially.) UUU, CCC, and AAA were three of the easiest codons; different approaches were required to work out the rest.

Other scientists later found that simple artificial mRNAs only three nucleotides long—each amounting to a codon—could bind to a ribosome, and that the resulting complex could then cause the binding of the corresponding tRNA with its specific amino acid. Thus, for example, simple UUU caused the tRNA carrying phenylalanine to bind to the ribosome. After this discovery, complete deciphering of the genetic code was relatively simple. To find the "translation" of a codon, Nirenberg could use a sample of that codon as an artificial mRNA and see which amino acid became bound to it.

Preparation for Translation: Linking RNAs, Amino Acids, and Ribosomes

As Crick's adapter hypothesis proposed, the translation of mRNA into proteins requires a molecule that links the information contained in mRNA codons with specific amino acids in proteins. That function is performed by tRNA. Two key events must take place to ensure that the protein made is the one specified by mRNA:

- tRNA must read mRNA correctly.
- tRNA must carry the amino acid that is correct for its reading of the mRNA.

Transfer RNAs carry specific amino acids and bind to specific codons

The codon in mRNA and the amino acid in a protein are related by way of an adapter—a specific tRNA with an

12.7 Transfer RNA The tRNA molecule binds to amino acids, associates with mRNA molecules, and interacts with ribosomes. There is at least one specific tRNA molecule for each of the amino acids. When the tRNA is bonded to an amino acid, it is designated as a charged tRNA.

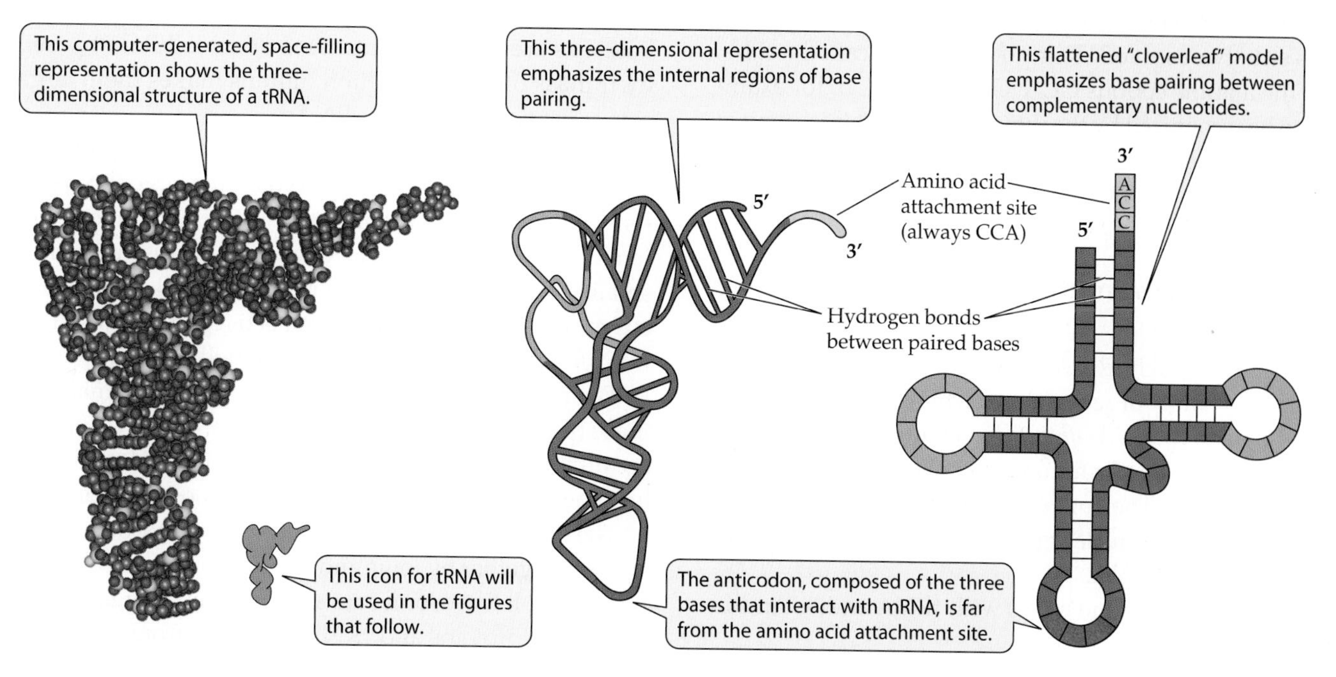

This computer-generated, space-filling representation shows the three-dimensional structure of a tRNA.
This three-dimensional representation emphasizes the internal regions of base pairing.
This flattened "cloverleaf" model emphasizes base pairing between complementary nucleotides.
5′
3′
Amino acid attachment site (always CCA)
3′
A
C
C
5′
Hydrogen bonds between paired bases
This icon for tRNA will be used in the figures that follow.
The anticodon, composed of the three bases that interact with mRNA, is far from the amino acid attachment site.

tached amino acid. For each of the 20 amino acids, there is at least one specific type ("species") of tRNA molecule.

The tRNA molecule has three functions: It carries (is "charged with") an amino acid, it associates with mRNA molecules, and it interacts with ribosomes. Its molecular structure relates clearly to all of these functions. A tRNA molecule has about 75 to 80 nucleotides. It has a *conformation* (a three-dimensional shape) that is maintained by complementary base pairing (hydrogen bonding) within its own sequence (Figure 12.7).

The conformation of a tRNA molecule allows it to combine specifically with binding sites on ribosomes. At the 3′ end of every tRNA molecule is a site to which its specific amino acid binds covalently. At about the midpoint of tRNA is a group of three bases, called the **anticodon**, that constitutes the site of complementary base pairing (hydrogen bonding) with mRNA. Each tRNA species has a unique anticodon, which is complementary to the mRNA codon for that tRNAs amino acid. At contact, the codon and the anticodon are antiparallel to each other. As an example of this process, consider the amino acid arginine:

- The DNA coding region for arginine is 3′-GCC-5′, which is transcribed, by complementary base pairing, to the mRNA codon 5′-CGG-3′.
- That mRNA codon binds by complementary base pairing to a tRNA with the anticodon 3′-GCC-5′, which is charged with arginine.

Recall that 61 different codons encode the 20 amino acids in proteins (see Figure 12.5). Does this mean that the cell must produce 61 different tRNA species, each with a different anticodon? No. The cell gets by with about two-thirds that number of tRNA species, because the specificity for the base at the 3′ end of the codon (and the 5′ end of the anticodon) is not always strictly observed. This phenomenon, called *wobble*, allows the alanine codons GCA, GCC, and GCU, for example, all to be recognized by the same tRNA. Wobble is allowed in some matches, but not in others; of most importance, it does not allow the genetic code to be ambiguous!

Activating enzymes link the right tRNAs and amino acids

The charging of each tRNA with its correct amino acid is achieved by a family of activating enzymes, known more formally as *aminoacyl-tRNA synthetases* (Figure 12.8). Each activating enzyme is specific for one amino acid and for its corresponding tRNA. The enzyme has a three-part active site that recognizes three smaller molecules: a specific amino acid, ATP, and a specific tRNA.

The activating enzyme reacts with tRNA and an amino acid (AA) in two steps:

$$\text{enzyme} + \text{ATP} + \text{AA} \rightarrow \text{enzyme—AMP—AA} + \text{PP}_i$$

$$\text{enzyme—AMP—AA} + \text{tRNA} \rightarrow \text{enzyme} + \text{AMP} + \text{tRNA—AA}$$

The amino acid is attached to the 3′ end of the tRNA (to a free OH group on the ribose) with an energy-rich bond, forming charged tRNA. This bond will provide the energy for the synthesis of the peptide bond that will join adjacent amino acids.

A clever experiment by Seymour Benzer and his colleagues at the California Institute of Technology demonstrated the importance of the specificity of the attachment of tRNA to its amino acid. In their laboratory, the amino acid cysteine, already properly attached to its tRNA, was chemically modified to become a different amino acid, alanine. Which component—the amino acid or the tRNA—would be recognized when this hybrid charged tRNA was put into a protein-synthesizing system? The answer was: the tRNA. Everywhere in the synthesized protein where cysteine was supposed to be, alanine appeared instead. The cysteine-specific tRNA had delivered its cargo (alanine) to every mRNA "address" where cysteine was called for. This experiment showed that the protein synthesis machinery recognizes the anticodon of the charged tRNA, not the amino acid attached to it.

If activating enzymes in nature did what Benzer did in the laboratory and charged tRNAs with the wrong amino acids, those amino acids would be inserted into proteins at inappropriate places, leading to alterations in protein shape and function. The fact that the activating enzymes are highly specific has led to the process of tRNA charging being called the "second genetic code."

The ribosome is the workbench for translation

Ribosomes are required for the translation of the genetic information in mRNA into a polypeptide chain. Although ribosomes are small in contrast to other cellular organelles, their mass of several million daltons makes them large in comparison with charged tRNAs.

Each ribosome consists of two subunits, a large one and a small one (Figure 12.9). In eukaryotes, the large subunit consists of three different molecules of rRNA and about 45 different protein molecules, arranged in a precise pattern. The small subunit consists of one rRNA molecule and 33 different protein molecules. When not active in the translation of mRNA, the ribosomes exist as separated subunits.

The ribosomes of prokaryotes are somewhat smaller than those of eukaryotes, and their ribosomal proteins and RNAs are different. Mitochondria and chloroplasts also contain ribosomes, some of which are similar to those of prokaryotes.

The different proteins and rRNAs in a ribosomal subunit are held together by ionic and hydrophobic forces, not covalent bonds. If these forces are disrupted by detergents, for example, the proteins and rRNAs separate from one another.

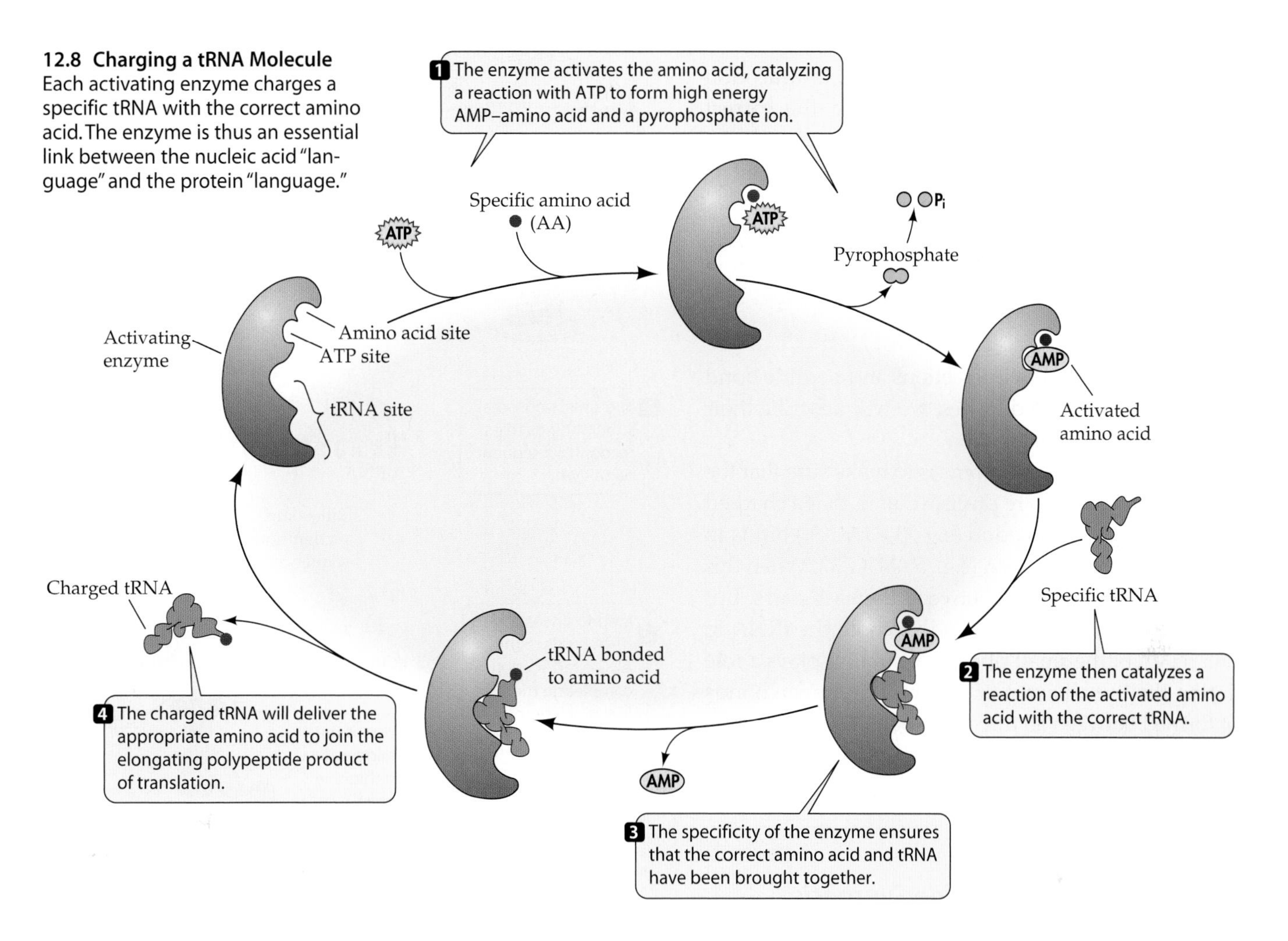

1 The enzyme activates the amino acid, catalyzing a reaction with ATP to form high energy AMP–amino acid and a pyrophosphate ion.
Specific amino acid (AA)
ATP
Pyrophosphate
Pi
Activating enzyme
Amino acid site
ATP site
tRNA site
AMP
Activated amino acid
Specific tRNA
2 The enzyme then catalyzes a reaction of the activated amino acid with the correct tRNA.
Charged tRNA
tRNA bonded to amino acid
4 The charged tRNA will deliver the appropriate amino acid to join the elongating polypeptide product of translation.
3 The specificity of the enzyme ensures that the correct amino acid and tRNA have been brought together.

12.8 Charging a tRNA Molecule Each activating enzyme charges a specific tRNA with the correct amino acid. The enzyme is thus an essential link between the nucleic acid "language" and the protein "language."

When the detergent is removed, the entire complex structure self-assembles. This is like separating the pieces of a jigsaw puzzle and having them fit together again without human hands to guide them!

A given ribosome does not specifically produce just one kind of protein. A ribosome can use any mRNA and all species of charged tRNAs, and thus can be used to make many different polypeptide products. The mRNA, as a linear sequence of codons, specifies the polypeptide sequence to be made; the ribosome is simply the molecular workbench where the task is accomplished. Its structure enables it to hold the mRNA and charged tRNAs in the right positions, thus allowing the growing polypeptide to be assembled efficiently.

On the large subunit of the ribosome are four sites to which tRNA binds (see Figure 12.9). A charged tRNA traverses these four sites in order:

- The *T* (transfer) *site* is where a charged tRNA first lands on the ribosome, accompanied by a special protein "escort" called the *T*, or *transfer, factor*.

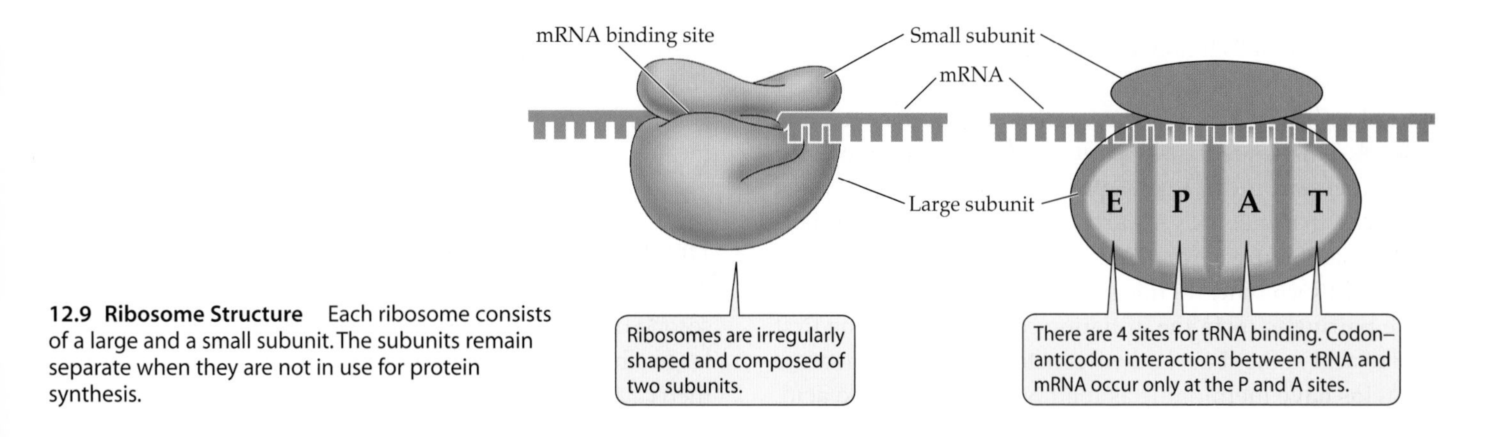

mRNA binding site
Small subunit
mRNA
Large subunit
E
P
A
T
Ribosomes are irregularly shaped and composed of two subunits.
There are 4 sites for tRNA binding. Codon–anticodon interactions between tRNA and mRNA occur only at the P and A sites.

12.9 Ribosome Structure Each ribosome consists of a large and a small subunit. The subunits remain separate when they are not in use for protein synthesis.

- The *A* (amino acid) *site* is where the tRNA anticodon binds to the mRNA codon, thus lining up the correct amino acid to be added to the growing polypeptide chain.
- The *P* (polypeptide) *site* is where the tRNA adds its amino acid to the growing polypeptide chain.
- The *E* (exit) *site* is where the tRNA, having given up its amino acid, resides before leaving the ribosome and going back to the cytosol to pick up another amino acid and begin the process again.

Because codon–anticodon interactions and peptide bond formation occur at the A and P sites, we will describe their function in detail in the next section.

An important role of the ribosome is to make sure that the mRNA–tRNA interactions are precise: that is, that a charged tRNA with the correct anticodon (e.g., 3′-UAC-5′) binds to the appropriate codon in mRNA (e.g., 5′-AUG-3′). When this occurs, hydrogen bonds form between the base pairs. But these hydrogen bonds are not enough to hold the tRNA in place. The rRNA of the small ribosomal subunit plays a role in validating the three-base-pair match. If hydrogen bonds have not formed between all three base pairs, the tRNA must be the wrong one for that mRNA codon, and that tRNA is ejected from the ribosome.

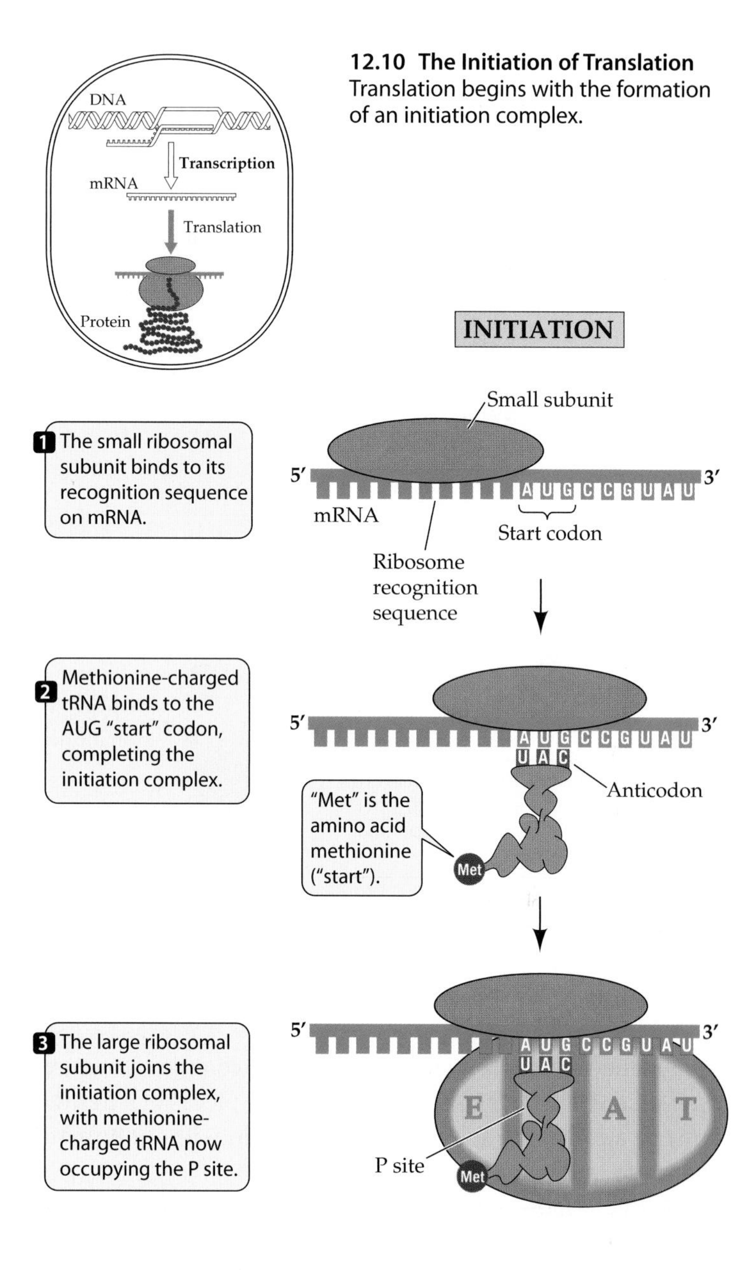

12.10 The Initiation of Translation Translation begins with the formation of an initiation complex.

Translation: RNA-Directed Polypeptide Synthesis

We have been working our way through the steps by which the sequence of bases in the template strand of a DNA molecule specifies the sequence of amino acids in a protein (see Figure 12.3). We are now at the last step: translation, the RNA-directed assembly of a protein. Like transcription, translation occurs in three steps: initiation, elongation, and termination.

Translation begins with an initiation complex

The translation of mRNA begins with the formation of an **initiation complex**, which consists of a charged tRNA bearing what will be the first amino acid of the polypeptide chain and a small ribosomal subunit, both bound to the mRNA (Figure 12.10). The rRNA of the small ribosomal subunit binds to a complementary ribosome recognition sequence on the mRNA. This sequence is "upstream" (toward the 5′ end) of the actual start codon that begins translation.

Recall that the mRNA start codon in the genetic code is AUG (see Figure 12.5). The anticodon of a methionine-charged tRNA binds to this start codon by complementary base pairing to form the initiation complex. Thus the first amino acid in the chain is always methionine. Not all mature proteins have methionine as their N-terminal amino acid, however. In many cases, the initiator methionine is removed by an enzyme after translation.

After the methionine-charged tRNA has bound to the mRNA, the large subunit of the ribosome joins the complex. The methionine-charged tRNA now lies in the P site of the ribosome, and the A site is aligned with the second mRNA codon. These ingredients—mRNA, two ribosomal subunits, and methionine-charged tRNA—are put together properly by a group of proteins called *initiation factors*.

The polypeptide elongates from the N terminus

A charged tRNA whose anticodon is complementary to the second codon on the mRNA now enters the open A site of the large ribosomal subunit. (Figure 12.11). The large subunit then catalyzes two reactions:

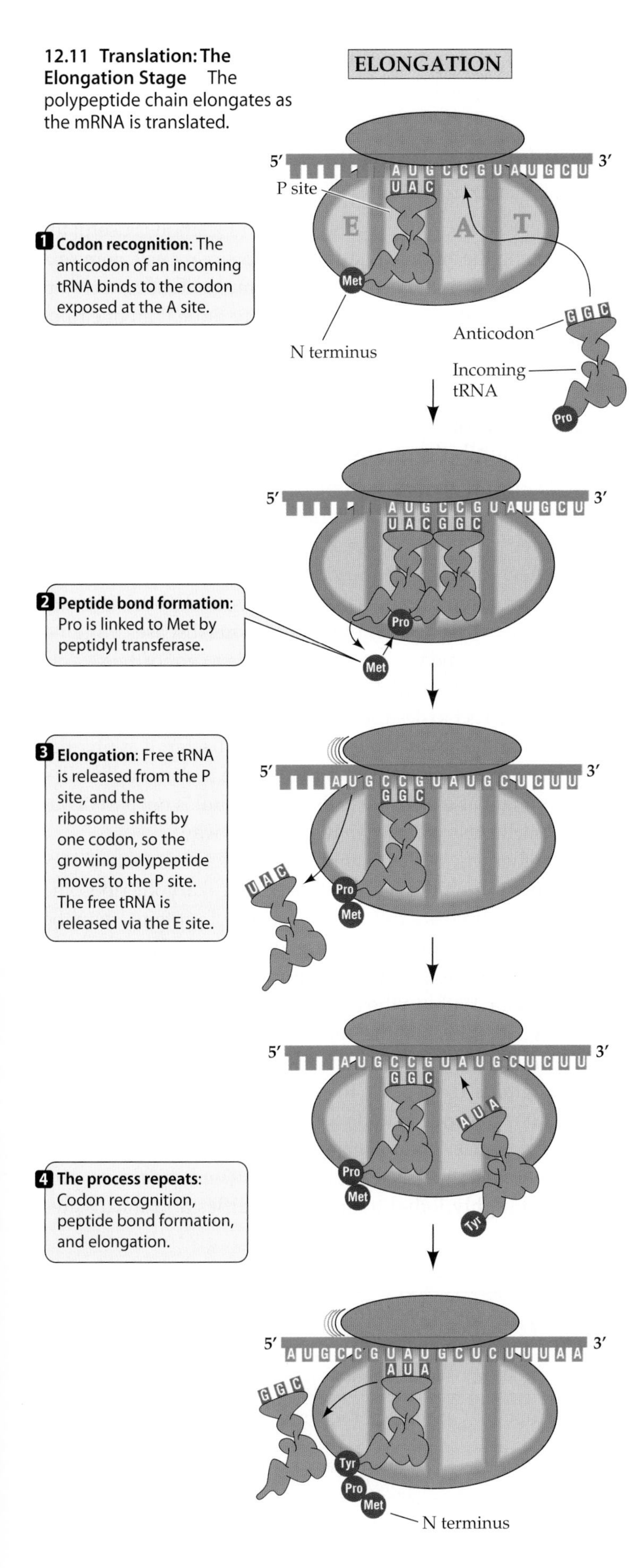

ELONGATION
5′
3′
P site
E
A
T
Met
N terminus
Anticodon
Incoming tRNA
Pro
1 Codon recognition: The anticodon of an incoming tRNA binds to the codon exposed at the A site.
2 Peptide bond formation: Pro is linked to Met by peptidyl transferase.
3 Elongation: Free tRNA is released from the P site, and the ribosome shifts by one codon, so the growing polypeptide moves to the P site. The free tRNA is released via the E site.
4 The process repeats: Codon recognition, peptide bond formation, and elongation.
Tyr
N terminus

12.11 Translation: The Elongation Stage The polypeptide chain elongates as the mRNA is translated.

- It breaks the bond between the tRNA in the P site and its amino acid.
- It catalyzes the formation of a peptide bond between that amino acid and the one attached to the tRNA in the A site.

Because the large subunit performs these two actions, it is said to have *peptidyl transferase activity*. In this way, methionine (the amino acid in the P site) becomes the N terminus of the new protein. The second amino acid is now bound to methionine, but remains attached to its tRNA by its carboxyl group (—COOH) in the A site.

How does the large ribosomal subunit catalyze this binding? In 1992, Harry Noller and his colleagues at the University of California at Santa Cruz found that if they removed almost all the proteins in the large subunit, it still catalyzed peptide bond formation. But if the rRNA was destroyed, so was peptidyl transferase activity. Part of the rRNA in the large subunit interacts with the end of the charged tRNA where the amino acid is attached. Thus rRNA appears to be the catalyst. This situation is very unusual, because proteins are the usual catalysts in biological systems. The recent purification and crystallization of ribosomes has allowed scientists to examine their structure in detail, and the catalytic role of rRNA in peptidyl transferase activity has been confirmed.

Elongation continues and the polypeptide grows

After the first tRNA releases its methionine, it dissociates from the ribosome, returning to the cytosol to become charged with another methionine. The second tRNA, now bearing a *dipeptide*, is shifted to the P site as the ribosome moves one codon along the mRNA in the 5′-to-3′ direction.

The elongation process continues, and the polypeptide chain grows, as the steps are repeated:

- The next charged tRNA enters the open A site.
- Its amino acid forms a peptide bond with the amino acid chain in the P site, so that it picks up the growing polypeptide chain from the tRNA in the P site.
- The tRNA in the P site is released. The ribosome shifts one codon, so that the entire tRNA–polypeptide complex, along with its codon, moves to the newly vacated P site.

All these steps are assisted by proteins called **elongation factors**.

A release factor terminates translation

The elongation cycle ends, and translation is terminated, when a stop codon—UAA, UAG, or UGA—enters the A site (Figure 12.12). These codons encode no amino acids, nor do

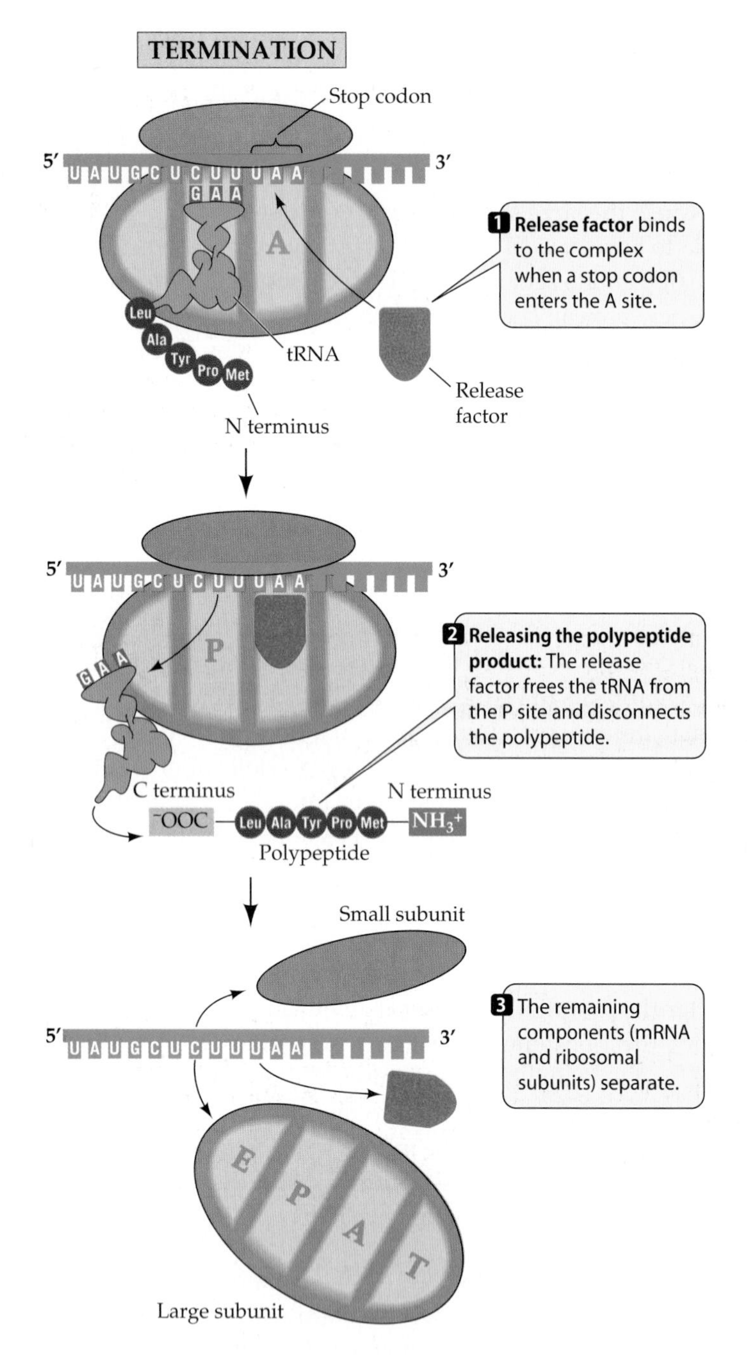
TERMINATION
Stop codon
5′
3′
A
Leu
Ala
Tyr
Pro
Met
tRNA
N terminus
Release factor
Release factor binds to the complex when a stop codon enters the A site.
P
Releasing the polypeptide product: The release factor frees the tRNA from the P site and disconnects the polypeptide.
C terminus
N terminus
Polypeptide
Small subunit
The remaining components (mRNA and ribosomal subunits) separate.
E
P
A
T
Large subunit

12.12 The Termination of Translation Translation terminates when the A site of the ribosome encounters a stop codon on the mRNA.

they bind tRNAs. Rather, they bind a protein *release factor*, which hydrolyzes the bond between the polypeptide and the tRNA in the P site.

The newly completed protein thereupon separates from the ribosome. Its C terminus is the last amino acid to join the chain. Its N terminus, at least initially, is methionine, as a consequence of the AUG start codon. In its amino acid sequence, it contains information specifying its conformation, as well as its ultimate cellular destination.

Table 12.1 summarizes the nucleic acid signals for initiation and termination of transcription and translation.

Regulation of Translation

Like any factory, the machinery of translation can work at varying rates. Variation in the rate of translation is useful for controlling the amount of an active protein in a cell. Some externally applied chemicals, such as some antibiotics, can stop translation. Conversely, the presence of more than one ribosome on an mRNA can speed up protein synthesis.

Some antibiotics and bacterial toxins work by inhibiting translation

Antibiotics are defensive molecules produced by microorganisms such as certain bacteria and fungi. These substances often destroy other microbes, which might compete with the defenders for nutrients. Since the 1940s, scientists have isolated increasing numbers of antibiotics, and physicians use them to treat a great variety of infectious diseases, ranging from bacterial meningitis to pneumonia to gonorrhea.

The key to the medical use of antibiotics is *specificity:* An antibiotic must act to destroy the microbial invader, but not harm the human host. One way in which antibacterial antibiotics achieve this is to block the synthesis of the bacterial cell wall—something that is essential to the microbe but is not part of human biochemistry. Penicillin works in this way.

Another way in which antibiotics work is to inhibit all bacterial protein synthesis. Recall that the prokaryotic ribosome is smaller, and has a different collection of proteins, than the eukaryotic ribosome. Some antibiotics bind only to bacterial ribosomal proteins that are important in protein synthesis (Table 12.2). Without the ability to make proteins, the bacterial invaders die, and the infection is stemmed.

Some bacteria affect their human hosts through mechanisms similar to those we use against them. Diphtheria is an infectious disease of childhood, and before the advent of effective vaccines, it was a major cause of childhood death. The infective agent, the bacterium *Cornybacterium diphtheriae*, produces a highly lethal toxin that modifies and inactivates

12.1 Signals that Start and Stop Transcription and Translation

	TRANSCRIPTION	TRANSLATION
Initiation	Promoter sequence in DNA	AUG start codon in mRNA
Termination	Terminator sequence in DNA	UAA, UAG, or UGA stop codon in mRNA

12.2 Antibiotics that Inhibit Bacterial Protein Synthesis

ANTIBIOTIC	STEP INHIBITED
Chloromycetin	Formation of peptide bonds
Erythromycin	Translocation of mRNA along ribosome
Neomycin	Interactions between tRNA and mRNA
Streptomycin	Initiation of translation
Tetracycline	Binding of tRNA to ribosome
Paromomycin	Validation of mRNA–tRNA match

a protein that is essential for the movement of mRNA and ribosomes during eukaryotic protein synthesis.

Polysome formation increases the rate of protein synthesis

Several ribosomes can work simultaneously at translating a single mRNA molecule, producing multiple molecules of the protein at the same time. As soon as the first ribosome has moved far enough from the initiation point, a second initiation complex can form, then a third, and so on. An assemblage consisting of a thread of mRNA with its beadlike ribosomes and their growing polypeptide chains is called a **polyribosome**, or **polysome** (Figure 12.13). Cells that are actively synthesizing proteins contain large numbers of polysomes and few free ribosomes or ribosomal subunits.

A polysome is like a cafeteria line, in which patrons follow one another, adding items to their trays. At any moment, the person at the start has a little food (a newly initiated protein); the person at the end has a complete meal (a completed protein). However, in the polysome cafeteria, everyone gets the same meal: Many copies of the same protein are made from a single mRNA.

While protein synthesis can be inhibited with antibiotics and speeded up via polysomes, these are not the only ways in which the amount of an active protein in a cell can be controlled. After the protein is synthesized, it may undergo changes that alter its function.

Posttranslational Events

A functional protein is not necessarily the same as the polypeptide chain that is released from the ribosome. Especially in eukaryotic cells, the polypeptide may need to be moved far from the site of synthesis in the cytoplasm, moved into an organelle, or even secreted from the cell. In addition, the polypeptide is often modified by the addition of new chemical groups that have functional significance. In this section, we examine these two *posttranslational* aspects of protein synthesis.

Chemical signals in proteins direct them to their cellular destinations

As a polypeptide chain emerges from the ribosome, it folds into its three-dimensional shape. As described in Chapter 3, this conformation is determined by the sequence of the amino acids that make up the protein, as well as by factors such as the polarity and charge of their R groups. Ultimately, the conformation of the polypeptide allows it to interact with other

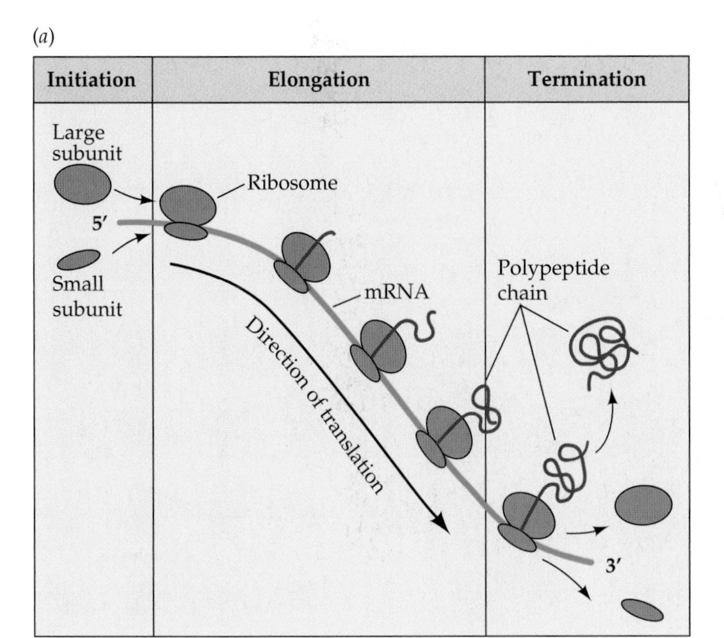

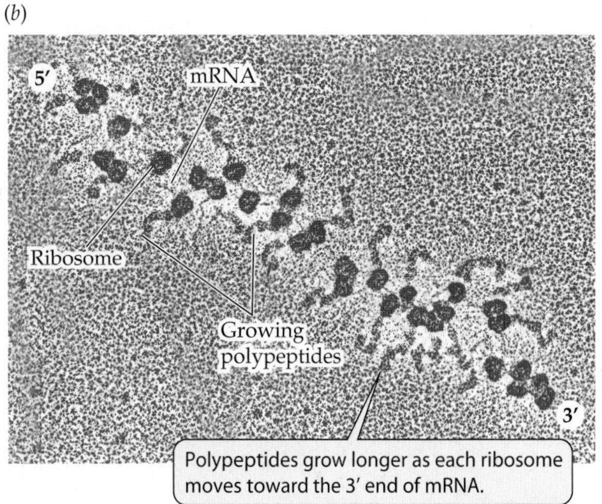

12.13 A Polysome (*a*) A polysome consists of multiple ribosomes and their growing polypeptide chains moving in single file along an mRNA molecule. (*b*) An electron microscopic view of a polysome.

molecules in the cell, such as a substrate or another polypeptide. In addition to this structural information, the amino acid sequence contains an "address label" indicating where in the cell the polypeptide belongs.

All protein synthesis begins on free ribosomes in the cytoplasm. As a polypeptide chain is made, the information contained in its amino acid sequence gives it one of two sets of instructions (Figure 12.14):

- "Finish translation and be released to the cytoplasm." Such proteins are sent to the nucleus, mitochondria, plastids, or peroxisomes, depending on the address in their instructions; or, lacking such specific instructions, they remain in the cytosol.
- "Stop translation, go to the endoplasmic reticulum (ER), and finish synthesis there." After protein synthesis is completed, such proteins may be retained in the ER or sent to lysosomes via the Golgi apparatus. Alternatively, they may be sent to the plasma membrane, or, lacking such specific instructions, they are secreted from the cell via vesicles that emanate from the plasma membrane.

DESTINATION: CYTOPLASM. After translation, some folded polypeptides have a short exposed sequence of amino acids that acts like a postal "zip code," directing them to an organelle. These **signal sequences** are either at the N terminus or in the interior of the amino acid chain. For example, the following sequence directs a protein to the nucleus:

—Pro—Pro—Lys—Lys—Lys—Arg—Lys—Val—

This amino acid sequence occurs, for example, in the histone proteins associated with nuclear DNA, but not in citric acid cycle enzymes, which are addressed to the mitochondria.

The signal sequences have a conformation that allows them to bind to specific receptor proteins, appropriately called *docking proteins*, on the outer membrane of the appropriate organelle. Once the protein has bound to it, the receptor forms a channel in the membrane, allowing the protein to pass through to its organelle destination. (In this process, the protein is usually unfolded by a chaperonin so that it can pass through the channel, then refolds into its normal conformation.)

DESTINATION: ENDOPLASMIC RETICULUM. If a specific hydrophobic sequence of about 25 amino acids occurs at the beginning of a polypeptide chain, the finished product is sent initially to the ER, and then to the lysosomes, the plasma membrane, or out of the cell. In the cytoplasm, before translation is finished, the signal sequence binds to a **signal recognition particle** composed of protein and RNA (Figure 12.15). This binding blocks further protein synthesis until the ribosome can become attached to a specific receptor protein in the membrane of the rough ER. Once again, the receptor protein is converted into a channel, through which the growing polypeptide passes. The elongating polypeptide may be retained in the ER membrane itself, or it may enter the interior space—the lumen—of the ER. In either case, an enzyme in the lumen of the ER

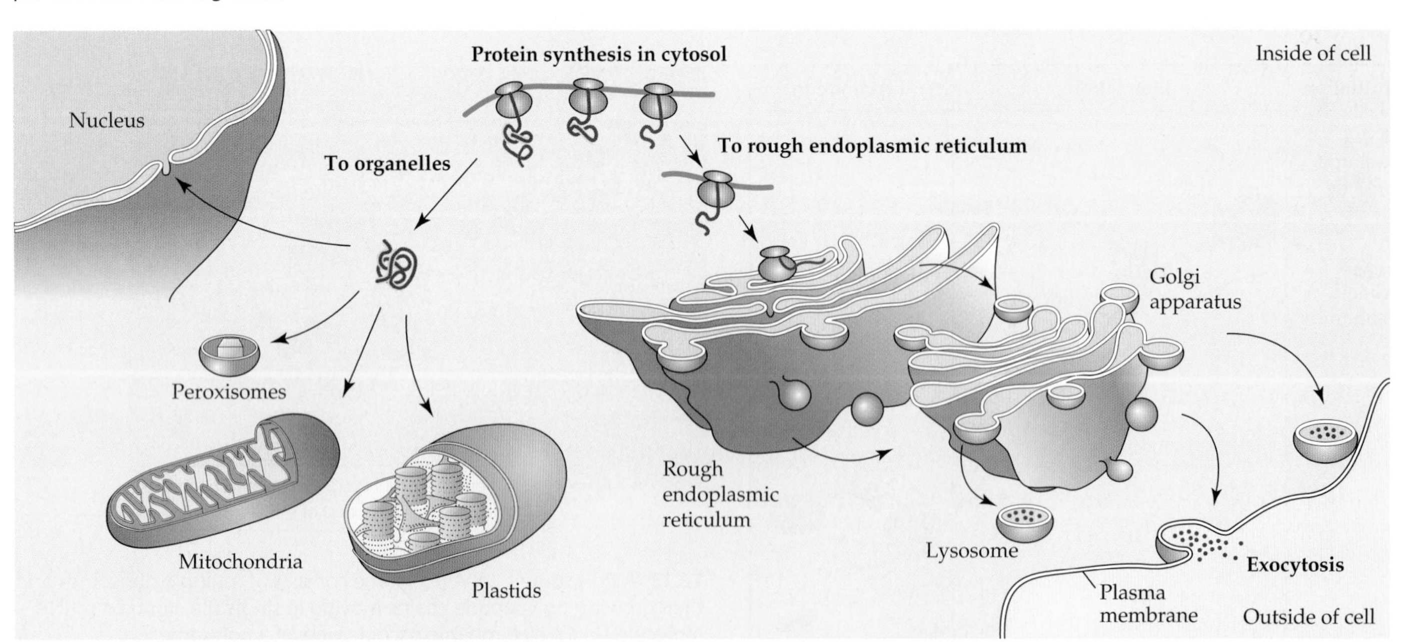

12.14 Destinations for Newly Translated Polypeptides in a Eukaryotic Cell Signal sequences on newly synthesized polypeptides bind to specific receptor proteins on the outer membranes of the organelle to which they are "addressed." Once the protein has bound to it, the receptor forms a channel in the membrane, and the protein enters the organelle.

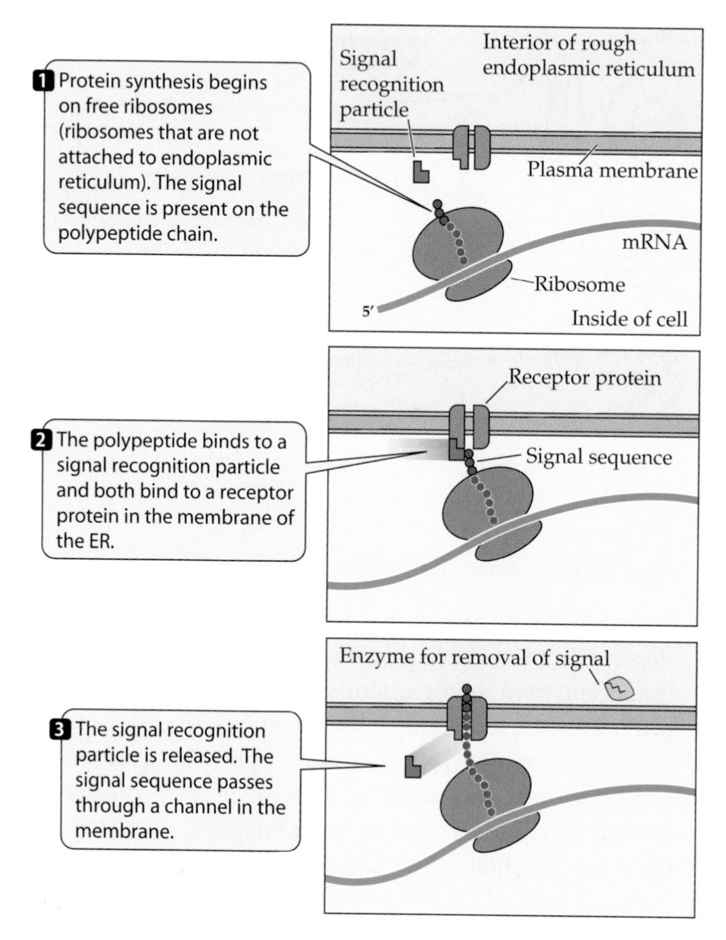
1 Protein synthesis begins on free ribosomes (ribosomes that are not attached to endoplasmic reticulum). The signal sequence is present on the polypeptide chain.
Signal recognition particle
Interior of rough endoplasmic reticulum
Plasma membrane
mRNA
Ribosome
5′
Inside of cell
Receptor protein
2 The polypeptide binds to a signal recognition particle and both bind to a receptor protein in the membrane of the ER.
Signal sequence
Enzyme for removal of signal
3 The signal recognition particle is released. The signal sequence passes through a channel in the membrane.

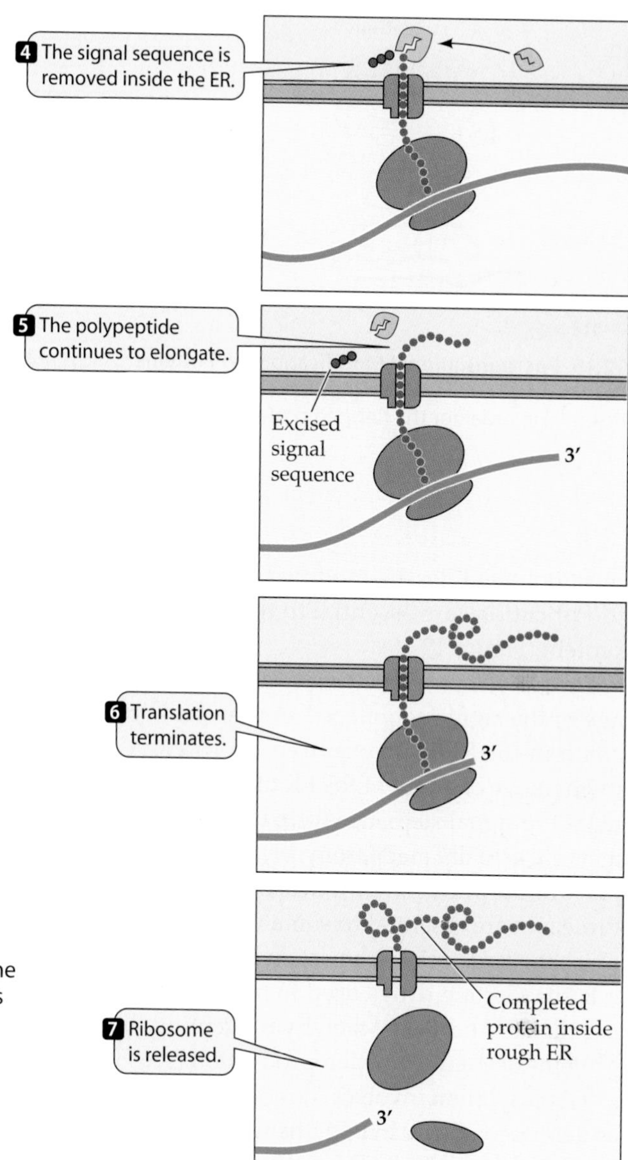
4 The signal sequence is removed inside the ER.
5 The polypeptide continues to elongate.
Excised signal sequence
3′
6 Translation terminates.
3′
7 Ribosome is released.
Completed protein inside rough ER
3′

12.15 A Signal Sequence Moves a Polypeptide into the ER When a signal sequence of amino acids is present at the beginning of the polypeptide chain, the polypeptide will be taken into the endoplasmic reticulum. The finished protein is thus segregated from the cytosol.

removes the signal sequence from the polypeptide chain. At this point, protein synthesis resumes, and the chain grows longer until its sequence is completed. If the finished protein enters the ER lumen, it can be transported to its appropriate location—to other cellular compartments or to the outside of the cell—without mixing with other molecules in the cytoplasm.

Additional signals are needed for sorting the protein further (remember that the signal sequence that sent it to the ER has been removed). These signals are of two kinds:

- Some are sequences of amino acids that allow the protein's retention within the ER.
- Others are sugars added in the Golgi apparatus, to which the protein is transferred in vesicles from the ER. The resulting *glycoproteins* end up either at the plasma membrane or in a lysosome (or plant vacuole), depending on which sugars are added.

Proteins with no additional signals pass from the ER through the Golgi apparatus and are secreted from the cell.

It is important to emphasize that the addressing of a protein to its destination is a property of its amino acid sequence, and so is genetically determined. An example of what can go wrong if a gene for protein targeting is mutated is *mucopolidosis II*, or *I-cell disease*. People with this disease lack an essential enzyme for the formation of the lysosomal targeting signal. Consequently, proteins destined for their lysosomes never get there, but instead either stay in the Golgi (where they form I, or inclusion, bodies) or are secreted from the cell. The lack of normal lysosome functions in a person's cells leads to progressive illness and death in childhood.

Many proteins are modified after translation

Most finished proteins are not identical to the polypeptide chains translated from mRNA on the ribosomes. Instead,

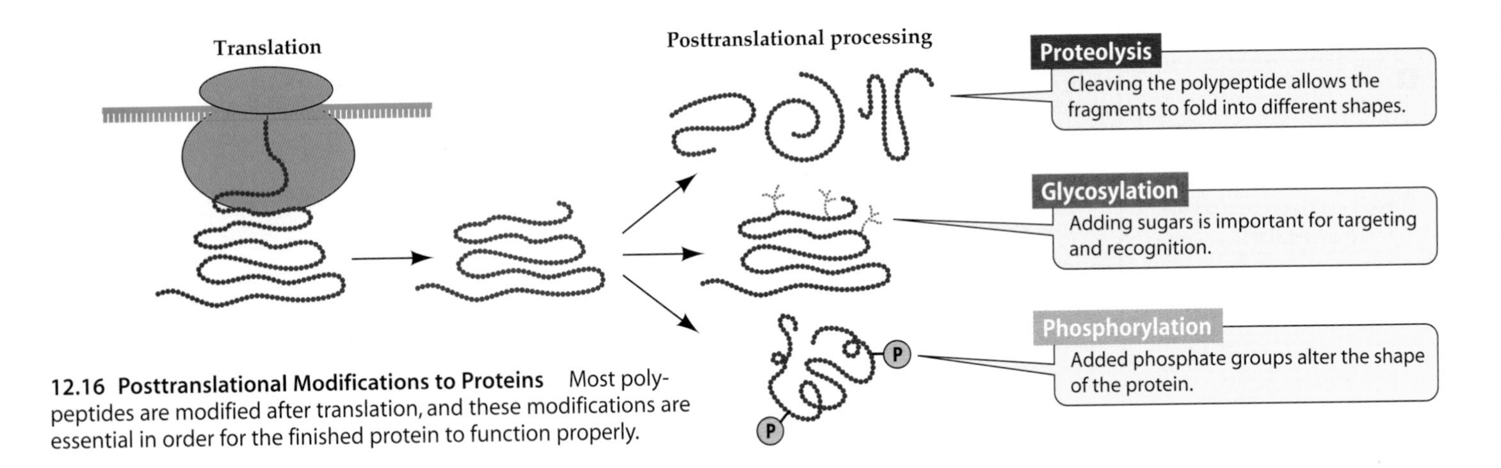

Translation
Posttranslational processing
Proteolysis
Cleaving the polypeptide allows the fragments to fold into different shapes.
Glycosylation
Adding sugars is important for targeting and recognition.
Phosphorylation
Added phosphate groups alter the shape of the protein.
P
P

12.16 Posttranslational Modifications to Proteins Most polypeptides are modified after translation, and these modifications are essential in order for the finished protein to function properly.

most polypeptides are modified after translation, and these modifications are essential to the final functioning of the protein (Figure 12.16).

Proteolysis is the cutting of a polypeptide chain. Cleavage of the signal sequence from the growing polypeptide chain in the ER is an example of proteolysis; the protein might move back out of the ER through the membrane channel if the signal sequence were not cut off. Also, some proteins are actually made from *polyproteins* (long polypeptides) that are cut into final products by enzymes called *proteases*. Proteases are essential to some viruses, including HIV, because the large viral polyprotein cannot fold properly unless it is cut. Certain drugs used to treat AIDS work by inhibiting the HIV protease, thereby preventing the formation of proteins needed for viral reproduction (see Figure 3.10).

Glycosylation involves the addition of sugars to proteins, as described above. In both the ER and the Golgi apparatus, resident enzymes catalyze the addition of various sugar residues or short sugar chains to certain amino acid R groups on proteins as they pass through. One such type of "sugar coating" is essential for addressing proteins to lysosomes discussed in the preceding section. Other types are important in the conformation and the recognition functions of proteins at the cell surface. Still other attached sugar residues help in stabilizing proteins stored in storage vacuoles in plant seeds.

Phosphorylation, the addition of phosphate groups to proteins, is catalyzed by protein kinases. The charged phosphate groups change the conformation of targeted proteins, often exposing an active site of an enzyme or a binding site for another protein—as we will see in Chapter 15.

All of the processes we have just described result in a functional protein only if the amino acid sequence of that protein is correct. If the sequence is not correct, cellular dysfunction and disease may result. Changes in the DNA—mutations—are a major source of errors in amino acid sequences.

Mutations: Heritable Changes in Genes

Accurate DNA replication, transcription, and translation all depend on the reliable pairing of complementary bases. Errors occur, though infrequently, in all three processes—least often in DNA replication. But, the consequences of DNA errors are the most severe because only they are heritable.

Mutations are heritable changes in genetic information. In unicellular organisms, any mutations that occur are passed on to the daughter cells when the cell divides. In multicellular organisms, there are two general types of mutations in terms of inheritance:

- **Somatic mutations** are those that occur in somatic (body) cells. These mutations are passed on to the daughter cells after mitosis, and to the offspring of those cells in turn, but are not passed on to sexually produced offspring. A mutation in a single skin cell, for example, could result in a patch of skin cells, all with the same mutation, but would not be passed on to a person's children.
- **Germ line mutations** are those that occur in the cells of the *germ line*—the specialized cells that give rise to gametes. A gamete with the mutation passes it on to a new organism at fertilization.

Very small changes in the genetic material can lead to easily observable changes in the phenotype. Some effects of mutations in humans are readily detectable—dwarfism, for instance, or the presence of more than five fingers on each hand. A mutant genotype in a microorganism may be obvious if, for example, it results in a change in nutritional requirements, as we described for *Neurospora* earlier (see Figure 12.1).

Other mutations may not be easily observable. In humans, for example, a particular mutation drastically lowers the level of an enzyme called glucose 6-phosphate dehydrogenase that is present in many tissues, including red blood cells. The red blood cells of a person carrying the mutant allele are abnormally sensitive to an antimalarial drug called primaquine;

when such people are treated with this drug, their red blood cells rupture. People with the normal allele have no such problem. Before the drug came into use, no one was aware that such a mutation existed. In bacteria, because of their small sizes and simpler morphologies, distinguishing a mutant from a normal bacterium usually requires sophisticated chemical methods, not just visual inspection.

Some mutations cause their phenotypes only under certain *restrictive* conditions. They are not detectable under other, *permissive* conditions. These phenotypes are known as **conditional mutants**. Many conditional mutants are temperature-sensitive, able to grow normally at a permissive temperature—say, 30°C—but unable to grow at a restrictive temperature—say, 37°C. The mutant allele in such an organism may code for an enzyme with an unstable tertiary structure that is altered at the restrictive temperature.

All mutations are alterations in the nucleotide sequence of DNA. At the molecular level, we can divide mutations into two categories:

- **Point mutations** are mutations of single base pairs and so are limited to single genes: One allele (usually dominant) becomes another allele (usually recessive) because of an alteration (gain/loss or substitution) of a single nucleotide (which, after DNA replication, becomes a mutant base pair).
- **Chromosomal mutations** are more extensive alterations than point mutations. They may change the position or orientation of a DNA segment without actually removing any genetic information, or they may cause a segment of DNA to be irretrievably lost.

Point mutations are changes in single nucleotides

Point mutations result from the addition or subtraction of a nucleotide base, or the substitution of one base for another, in the DNA, and hence in the mRNA. Point mutations can be caused by errors in chromosome replication that are not corrected in proofreading or by environmental mutagens such as chemicals and radiation.

Changes in the mRNA may or may not result in changes in the protein. Silent mutations have no effect on the protein; missense and nonsense mutations will result in changes in the protein, some of them drastic.

SILENT MUTATIONS. Because of the redundancy of the genetic code, some point mutations result in no change in amino acids when the altered mRNA is translated; for this reason, they are called **silent mutations**. For example, there are four mRNA codons that code for proline: CCA, CCC, CCU, and CCG (see Figure 12.5). If the template strand of DNA has the sequence CGG, it will be transcribed to CCG in mRNA, and proline-charged tRNA will bind to it at the ribosome. But if there is a mutation such that the codon in the template DNA now reads AGG, the mRNA codon will be CCU—the tRNA that binds it will still carry proline:

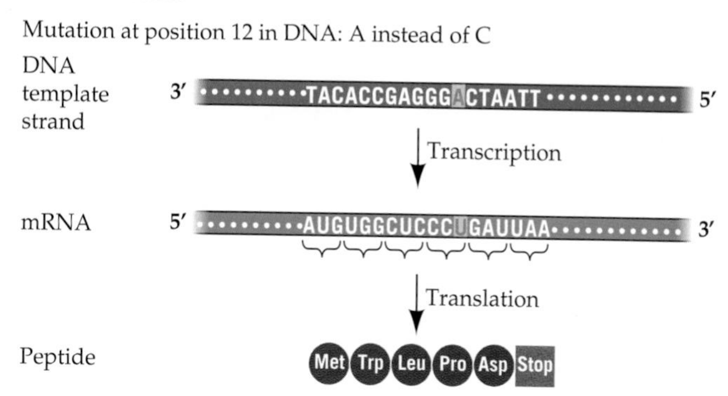

Silent mutations are quite common, and they result in genetic diversity that is not expressed as phenotypic differences.

MISSENSE MUTATIONS. In contrast to silent mutations, some base substitution mutations change the genetic message such that one amino acid substitutes for another in the protein. These changes are called **missense mutations**:

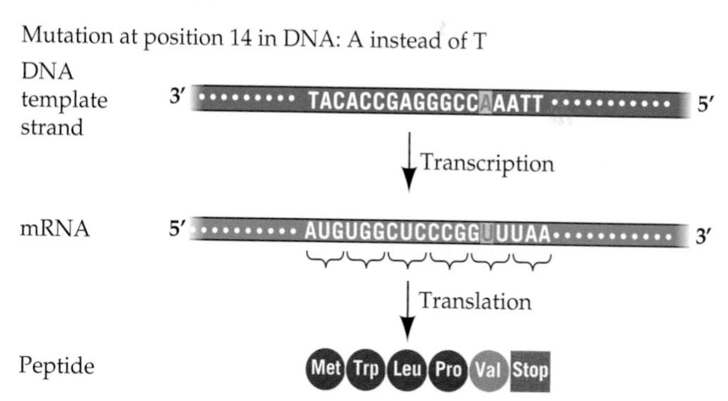

A specific example of a missense mutation is the sickle allele for human β-globin. Sickle-cell disease results from a defect in hemoglobin, a protein in human red blood cells that carries oxygen. The sickle allele of the gene that codes for β-globin (one of the polypeptide subunits in hemoglobin; see Figure 3.8) differs from the normal allele by one amino acid in its coding. Persons who are homozygous for this recessive allele have defective red blood cells. Where oxygen is abundant, as in the lungs, the cells are normal in structure and function. But at the low oxygen levels characteristic of working muscles, the red blood cells collapse into the shape of a sickle (Figure 12.17), causing abnormalities in blood circulation that lead to serious illnesses.

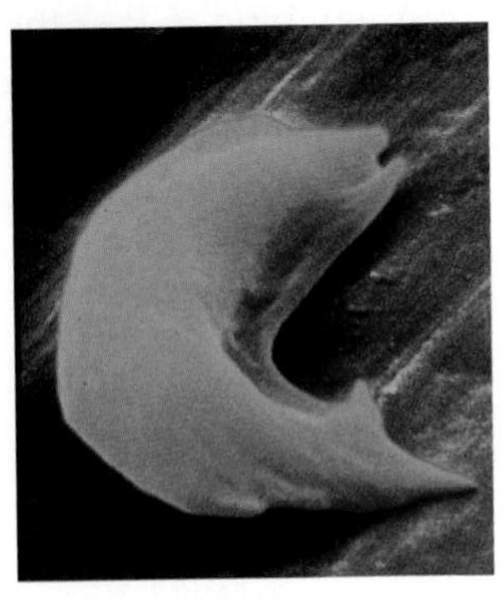

Sickle-cell phenotype

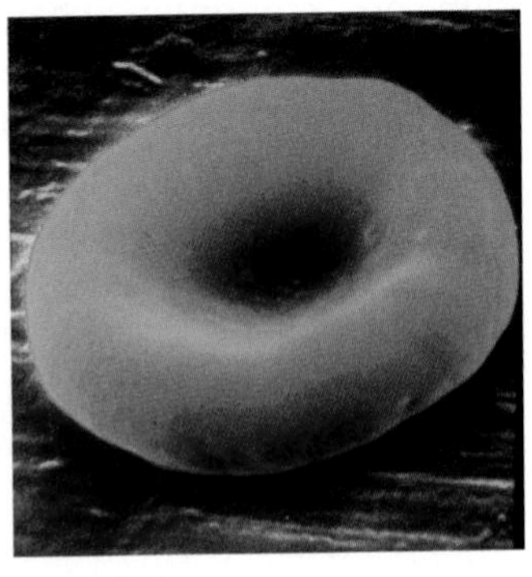

Normal phenotype

12.17 Sickled and Normal Red Blood Cells The misshapen red blood cell on the left is caused by a missense mutation that results in an incorrect amino acid in one of the two polypeptides of hemoglobin.

A missense mutation may cause a protein not to function, but often its effect is only to reduce the functional efficiency of the protein. Therefore, individuals carrying missense mutations may survive, even though the affected protein is essential to life. Through evolution, some missense mutations even improve functional efficiency.

NONSENSE MUTATIONS. **Nonsense mutations**, another type of mutation in which one base is substituted for another, are more often disruptive than missense mutations. In a nonsense mutation, the base substitution causes a stop codon, such as UAG, to form in the mRNA product:

Nonsense mutation

Mutation at position 5 in DNA: T instead of C

DNA template strand 3′ ••••••••• TACATCGAGGGCCTAATT ••••••••••• 5′

Transcription

mRNA 5′ ••••••••• AUGUAGCUCCCGGAUUAA••••••••••• 3′

Translation

Peptide Met Stop

Result: Only one amino acid translated; no protein made

The result is a shortened protein, since translation does not proceed beyond the point where the mutation occurred. Such short proteins are usually not functional.

FRAME-SHIFT MUTATIONS. Not all point mutations are base substitutions. Single base pairs may be inserted into or deleted from DNA. Such mutations are known as **frame-shift mutations** because they interfere with the decoding of the genetic message by throwing it out of register:

Frame-shift mutation

Mutation by insertion of T between bases 6 and 7 in DNA

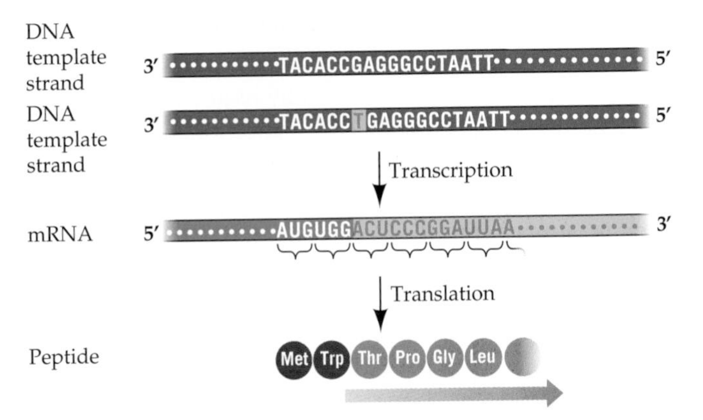

Result: All amino acids changed beyond the insertion

Think again of codons as three-letter words, each corresponding to a particular amino acid. Translation proceeds codon by codon; if a base is added to the message or subtracted from it, translation proceeds perfectly until it comes to the one-base insertion or deletion. From that point on, the three-letter words in the message are one letter out of register. In other words, such mutations shift the "reading frame" of the genetic message. Frame-shift mutations almost always lead to the production of nonfunctional proteins.

Chromosomal mutations are extensive changes in the genetic material

Changes in single nucleotides are not the most dramatic changes that can occur in the genetic material. Whole DNA molecules can break and rejoin, grossly disrupting the sequence of genetic information. There are four types of such chromosomal mutations: deletions, duplications, inversions, and translocations (Figure 12.18). These mutations can be caused by severe damage to chromosomes resulting from mutagens or by drastic errors in chromosome replication.

- **Deletions** remove part of the genetic material (Figure 12.18*a*). Like frame-shift point mutations, their consequences can be severe unless they affect unnecessary genes or are masked by the presence, in the same cell, of normal alleles of the deleted genes. It is easy to imagine one mechanism that could produce deletions: A DNA molecule might break at two points, and the two end pieces might rejoin, leaving out the DNA between the breaks.
- **Duplications** can be produced at the same time as deletions (Figure 12.18*b*). Duplication would arise if homologous chromosomes broke at different positions and then reconnected to the wrong partners. One of the two molecules produced by this mechanism would lack a seg-

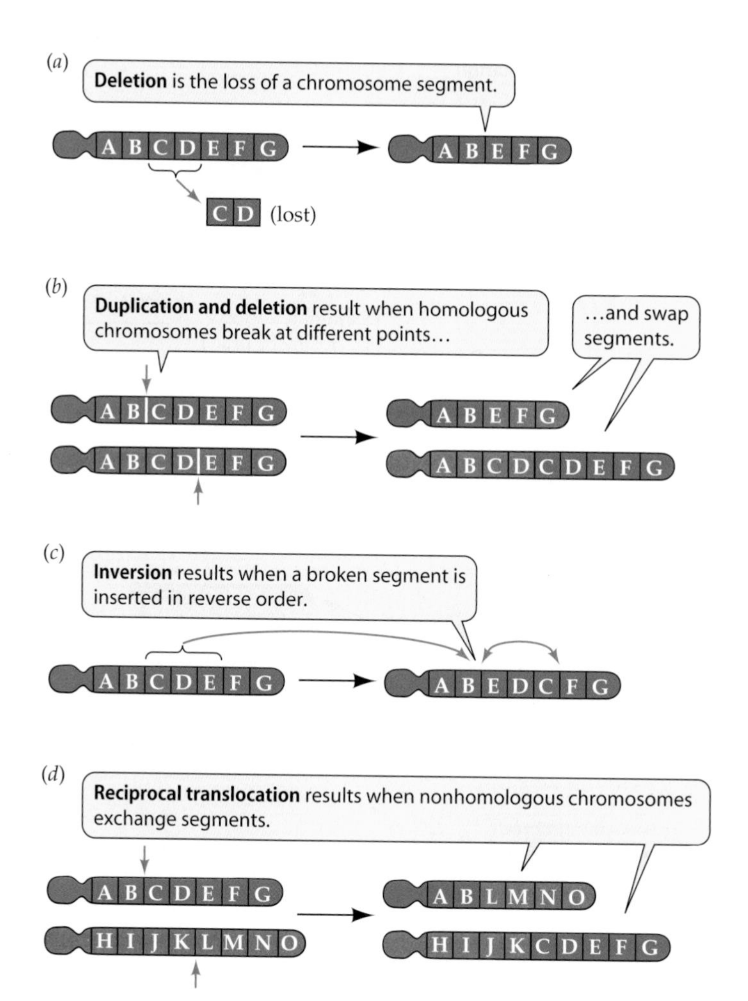

12.18 Chromosomal Mutations Chromosomes may break during replication, and parts of chromosomes may then rejoin incorrectly.

ment of DNA (it would have a deletion), and the other would have two copies (a duplication) of the segment that was deleted from the first.

- **Inversions** also result from breaking and rejoining. A segment of DNA may be removed and reinserted into the same location in the chromosome, but "flipped" end over end so that it runs in the opposite direction (Figure 12.18*c*). If the break site for an inversion includes part of a DNA segment that codes for a protein, the resulting protein will be drastically altered and almost certainly nonfunctional.
- **Translocations** result when a segment of DNA breaks off, moves from its chromosome, and is inserted into a different chromosome. Translocations may be reciprocal, as in Figure 12.18*d*, or nonreciprocal, as the mutation involving duplication and deletion in Figure 12.18*b* illustrates. Translocations often lead to duplications and deletions, and may result in sterility if normal chromosome pairing in meiosis cannot occur.

Mutations can be spontaneous or induced

It is useful to distinguish two types of mutations in terms of their causes. **Spontaneous mutations** are permanent changes in the genome that occur without any outside influence. In other words, they occur simply because the machinery of the cell is imperfect. **Induced mutations** occur when some agent outside the cell—a mutagen—causes a permanent change in DNA.

Spontaneous mutations may occur by several mechanisms:

- *The four nucleotide bases of DNA are somewhat unstable.* They can exist in two different forms (called *tautomers*), one of which is common and one rare. When a base temporarily forms its rare tautomer, it can pair with a different base. For example, C normally pairs with G. But if C is in its rare tautomer at the time of DNA replication, it pairs with (and DNA polymerase will insert) A. The result is a point mutation: G → A (Figure 12.19*a, c*).
- *Bases may change because of a chemical reaction.* For example, loss of an amino group in cytosine (a reaction called *deamination*) forms uracil. When DNA replicates, instead of a G opposite what was C, DNA polymerase adds an A (base-pairs with U).
- *DNA polymerase makes errors in replication* (see Chapter 11); for example, inserting a T opposite a G. Most of these errors are repaired by the proofreading function of the replication complex, but some errors escape and become permanent.
- *Meiosis is not perfect.* Nondisjunction can occur, leading to one too many or one too few chromosomes (aneuploidy; see Figure 9.18). Random chromosome breaks and rejoining can produce deletions, duplications, and inversions, or, when involving nonhomologous chromosomes, translocations.

Mutagens can also alter DNA by several mechanisms:

- *Some chemicals can covalently alter the nucleotide bases.* For example, nitrous acid (HNO_2) and its relatives can turn cytosine in DNA into uracil by deamination: they convert an amino group on cytosine (—NH_2) into a keto group (—C=O). This alteration has the same result as a spontaneous deamination: instead of a G, DNA polymerase inserts an A (base-pairs with U) (Figure 12.19*b,c*).
- *Some chemicals add groups to the bases.* For instance, benzpyrene, a component of cigarette smoke, adds a large chemical group to guanine, making it unavailable for base pairing. When DNA polymerase reaches such a modified guanine, it inserts any of the four bases; of course, three-fourths of the time the inserted base will not be cytosine, and a mutation results.
- *Radiation damages the genetic material in two ways.* Ionizing radiation (X rays) produces highly reactive chemical

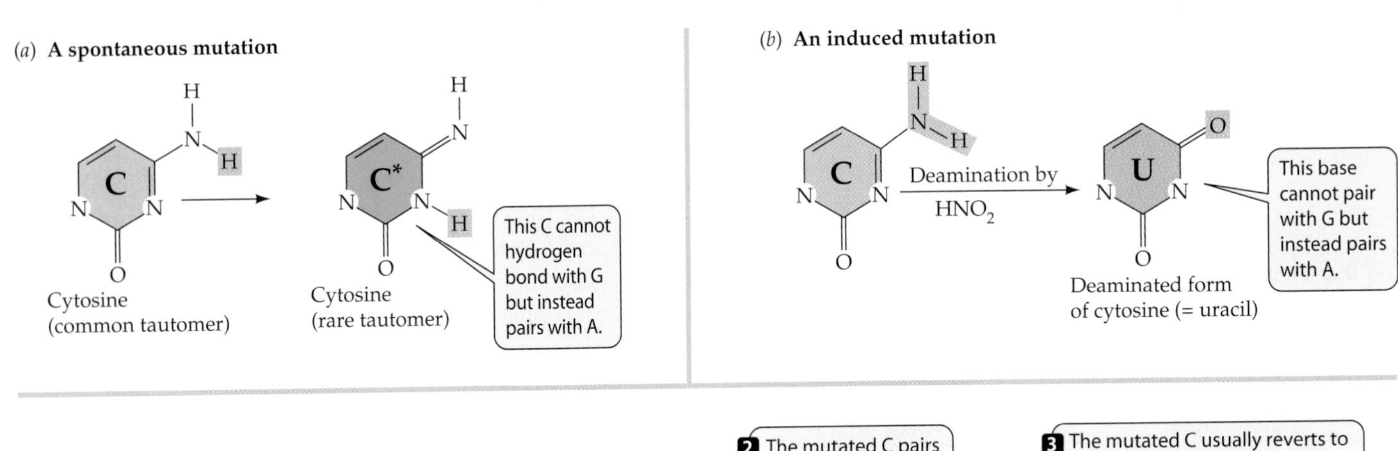

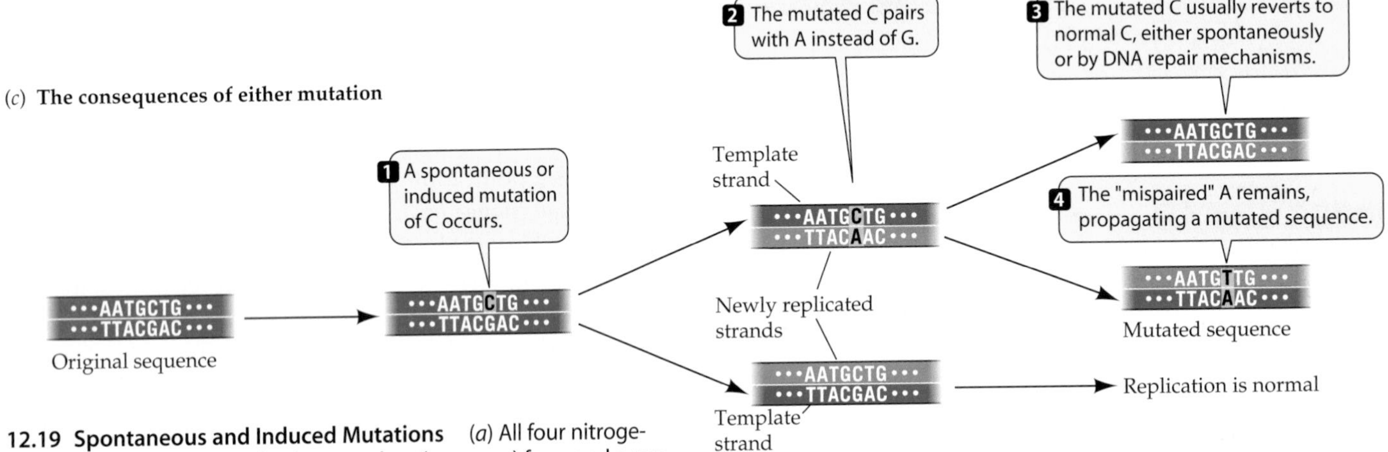

12.19 Spontaneous and Induced Mutations (*a*) All four nitrogenous bases in DNA exist in both a prevalent (common) form and a rare form. When a base spontaneously forms its rare tautomer, it can pair with a different base. (*b*) Mutagenic chemicals such as nitrous acid can induce changes in the bases. (*c*) In both spontaneous and induced mutations, the result is a permanent change in the DNA sequence following replication.

species called *free radicals*, which can change bases in DNA to unrecognizable (by DNA polymerase) forms or break the sugar–phosphate backbone, causing chromosomal abnormalities. Ultraviolet radiation from the sun (or a tanning lamp) is absorbed by thymine in DNA, causing it to form interbase covalent bonds with adjacent nucleotides. This, too, plays havoc with DNA replication.

Mutations have both benefits and costs. Germ line mutations provide genetic diversity for evolution to work on, as we will see below. But they usually produce an organism that does more poorly in its current environment. Somatic mutations do not affect the organism's offspring, but they can lead to cancer. We will return to the effects of germ line and somatic mutations in humans in Chapter 18.

Mutations are the raw material of evolution

Without mutation, there would be no evolution. As we will see in Part 4 of this book, mutation does not drive evolution, but it provides the genetic diversity on which natural selection and other agents of evolution act.

All mutations are rare events, but mutation frequencies vary from organism to organism and from gene to gene within a given organism. The frequency of mutation is usually much lower than one mutation per 10^4 base pairs per DNA duplication, and sometimes as low as one mutation per 10^9 base pairs per duplication. Most mutations are point mutations in which one nucleotide is substituted for another during the synthesis of a new DNA strand.

Mutations can harm the organism that carries them, or they can be neutral (have no effect on the organism's ability to survive or produce offspring). Once in a while, a mutation improves an organism's adaptation to its environment, or it becomes favorable when environmental conditions change.

Most of the complex creatures living on Earth have more DNA, and therefore more genes, than the simpler creatures do. Humans, for example, have 20 times more genes than prokaryotes have. How did these new genes arise? If whole genes were sometimes duplicated by the mechanisms described in the previous section, the bearer of the duplication would have a surplus of genetic information that might be turned to good use. Subsequent mutations in one of the two copies of the gene might not have an adverse effect on survival because the other copy of the gene would continue to produce functional protein. The extra gene might mutate over and over again without ill effect because its original function would be fulfilled by the original copy.

If the random accumulation of mutations in the extra gene led to the production of a useful protein (for example, an enzyme with an altered specificity for the substrates it binds, allowing it to catalyze different—but related—reactions), natural selection would tend to perpetuate the existence of this new gene. New copies of genes may also arise through the activity of transposable elements, which are discussed in Chapters 13 and 14.

Chapter Summary

One Gene, One Polypeptide

▶ Genes are expressed in the phenotype as polypeptides (proteins).

▶ Beadle and Tatum's experiments with the bread mold *Neurospora* resulted in several mutant strains, each lacking a specific enzyme in a biochemical pathway. Their results led to the one-gene, one-polypeptide hypothesis. **Review Figure 12.1**

▶ Certain hereditary diseases in humans had been found to be caused by the absence of certain enzymes. These observations supported the one-gene, one-polypeptide hypothesis.

DNA, RNA, and the Flow of Information

▶ RNA differs from DNA in three ways: It is single-stranded, its sugar molecule is ribose rather than deoxyribose, and its fourth base is uracil rather than thymine.

▶ The central dogma of molecular biology is DNA → RNA → protein. **Review Figure 12.2**

▶ A gene is expressed in two steps: first, DNA is transcribed to RNA; then RNA is translated into protein. **Review Figure 12.3**

▶ Some viruses are exceptions to the central dogma. Some viruses exclude DNA altogether, going directly from RNA to protein. In retroviruses, the central dogma is reversed: RNA → DNA.

Transcription: DNA-Directed RNA Synthesis

▶ RNA is transcribed from a DNA template after the bases of DNA are exposed by unwinding of the double helix.

▶ In a given gene, only one of the two strands of DNA (the template strand) acts as a template for transcription.

▶ RNA polymerase catalyzes transcription from the template strand of DNA.

▶ The initiation of transcription requires that RNA polymerase recognize and bind tightly to a promoter sequence on the DNA.

▶ RNA elongates in a 5′-to-3′ direction, antiparallel to the template DNA. Special sequences and protein helpers terminate transcription. **Review Figure 12.4**

▶ In prokaryotes, translation begins before transcription of the mRNA is completed. In eukaryotes, transcription occurs in the nucleus and translation occurs in the cytoplasm.
See Web/CD Tutorial 12.1

The Genetic Code

▶ The genetic code consists of triplets of nucleotide bases (codons). There are four bases, so there are 64 possible codons.

▶ One mRNA codon indicates the starting point of translation and codes for methionine. Three stop codons indicate the end of translation. The other 60 codons code only for particular amino acids.

▶ Because there are only 20 different amino acids, the genetic code is redundant; that is, there is more than one codon for certain amino acids. But the code is not ambiguous: A single codon does not encode more than one amino acid. **Review Figure 12.5.** See Web/CD Activity 12.1

▶ Test-tube experiments led to the assignment of amino acids to codons. **Review Figure 12.6.** See Web/CD Tutorial 12.2

Preparation for Translation: Linking RNAs, Amino Acids, and Ribosomes

▶ In translation, amino acids are linked in an order specified by the codons in mRNA. This task is achieved by transfer RNAs (tRNAs), which bind to specific amino acids. Each tRNA species has an anticodon complementary to an mRNA codon. **Review Figure 12.7**

▶ A family of activating enzymes attaches specific amino acids to their appropriate tRNAs, forming charged tRNAs. **Review Figure 12.8**

▶ The mRNA meets the charged tRNAs at a ribosome. **Review Figure 12.9**

▶ The small subunit of the ribosome checks to determine whether the tRNA anticodon and mRNA codon have formed hydrogen bonds.

Translation: RNA-Directed Polypeptide Synthesis

▶ An initiation complex consisting of a charged tRNA and a small ribosomal subunit bound to mRNA triggers the beginning of translation. **Review Figure 12.10**

▶ Polypeptides grow from the N terminus toward the C terminus. The ribosome moves along the mRNA one codon at a time in the 5′-to-3′ direction. **Review Figure 12.11**

▶ The presence of a stop codon in the A site of the ribosome terminates translation. **Review Figure 12.12**
See Web/CD Tutorial 12.3

Regulation of Translation

▶ Some antibiotics and bacterial toxins work by blocking events in translation. **Review Table 12.2**

▶ In a polysome, more than one ribosome moves along the mRNA at one time. **Review Figure 12.13**

Posttranslational Events

▶ Signals contained in the amino acid sequences of proteins direct them to their cellular destinations. **Review Figure 12.14**

▶ Protein synthesis begins on free ribosomes in the cytoplasm. Those proteins destined for the nucleus and other organelles are completed there. These proteins have signals that allow them to bind to and enter their destined organelles.

▶ Proteins destined for the ER, Golgi apparatus, lysosomes, and outside the cell complete their synthesis on the surface of the ER. They enter the ER by the interaction of a hydrophobic signal sequence with a channel in the membrane. **Review Figure 12.15**

▶ Modifications of proteins after translation include proteolysis, glycosylation, and phosphorylation. **Review Figure 12.16**

Mutations: Heritable Changes in Genes

▶ Mutations in DNA are often expressed as abnormal proteins. However, the result may not be easily observable phenotypic changes. Some mutations are detectable only under certain conditions.

▶ Point mutations (silent, missense, nonsense, or frame-shift) result from alterations in single base pairs of DNA. **Review pages 251–252**

▶ Chromosomal mutations (deletions, duplications, inversions, or translocations) involve large regions of a chromosome. **Review Figure 12.18**

▶ Mutations can be spontaneous or induced. Spontaneous mutations occur because of instabilities in DNA or chromosomes. Induced mutations occur when a mutagen damages DNA. **Review Figure 12.19**

Self-Quiz

1. Which of the following is *not* a difference between RNA and DNA?
   a. RNA has uracil; DNA has thymine.
   b. RNA has ribose; DNA has deoxyribose.
   c. RNA has five bases; DNA has four.
   d. RNA is a single polynucleotide strand; DNA is a double strand.
   e. RNA is relatively smaller than human chromosomal DNA.
2. Normally, *Neurospora* can synthesize all 20 amino acids. A certain strain of this mold cannot grow in simple growth medium but grows only when the amino acid leucine is added to the medium. This strain is
   a. dependent on leucine for energy.
   b. mutated in the synthesis of all proteins.
   c. mutated in the synthesis of all 20 amino acids
   d. mutated in the synthesis of leucine.
   e. mutated in the syntheses of 19 of the 20 amino acids.
3. An mRNA has the sequence 5′-AUGAAAUCCUAG-3′. What is the template DNA strand for this sequence?
   a. 5′-TACTTTAGGATC-3′
   b. 5′-ATGAAATCCTAG-3′
   c. 5′-GATCCTAAAGTA-3′
   d. 5′-TACAAATCCTAG-3′
   e. 5′-CTAGGATTTCAT-3′
4. The adapters that allow translation of the four-letter nucleic acid language into the 20 letter protein language are called
   a. aminoacyl tRNA synthetases.
   b. transfer RNAs.
   c. ribosomal RNAs.
   d. messenger RNAs.
   e. ribosomes.
5. At a certain location in a gene, the nontemplate strand of DNA has the sequence GAA. A mutation alters the triplet to GAG. This type of mutation is called
   a. silent.
   b. missense.
   c. nonsense.
   d. frame-shift.
   e. translocation.
6. Transcription
   a. produces only mRNA.
   b. requires ribosomes.
   c. requires tRNAs.
   d. produces RNA growing from the 5′ end to the 3′ end.
   e. takes place only in eukaryotes.
7. Which statement about translation is *not* true?
   a. It is RNA-directed polypeptide synthesis.
   b. An mRNA molecule can be translated by only one ribosome at a time.
   c. The same genetic code operates in almost all organisms and organelles.
   d. Any ribosome can be used in the translation of any mRNA.
   e. There are both start and stop codons.
8. Which statement is *not* true?
   a. Transfer RNA functions in translation.
   b. Ribosomal RNA functions in translation.
   c. RNAs are produced in transcription.
   d. Messenger RNAs are produced on ribosomes.
   e. DNA codes for mRNA, tRNA, and rRNA.
9. The genetic code
   a. is different for prokaryotes and eukaryotes.
   b. has changed during the course of recent evolution.
   c. has 64 codons that code for amino acids.
   d. is degenerate.
   e. is ambiguous.
10. A mutation that results in the codon UAG where there had been UGG is
    a. a nonsense mutation.
    b. a missense mutation.
    c. a frame-shift mutation.
    d. a large-scale mutation.
    e. unlikely to have a significant effect.

For Discussion

1. The genetic code is described as degenerate. What does this mean? How is it possible that a point mutation, consisting of the replacement of a single nitrogenous base in DNA by a different base, might not result in an error in protein production?
2. Har Gobind Khorana at the University of Wisconsin synthesized artificial mRNAs such as poly CA (CACA...) and poly CAA (CAACAACAA...). He found that poly CA codes for a polypeptide consisting of threonine (Thr) and histidine (His), in alternation (His–Thr– His–Thr...). There are two possible codons in poly CA, CAC and ACA. One of these must code for histidine and the other for threonine—but which is which? The answer comes from results with poly CAA, which produces three different polypeptides: poly Thr, poly Gln (glutamine), and poly Asn (asparagine). (An artificial messenger can be read, inefficiently, beginning at any point in the chain; there is no specific initiator region.) Thus poly CAA can be read as a polymer of CAA, of ACA, or of AAC. Compare the results of the poly CA and poly CAA experiments, and determine which codon codes for threonine and which for histidine.
3. Look back at Question 2. Using the genetic code (Figure 12.5) as a guide, deduce what results Khorana would have obtained had he used poly UG and poly UGG as artificial messengers. In fact, very few such artificial messengers would have given useful results. For an example of what could happen, consider poly CG and poly CGG. If poly CG were the messenger, a mixed polypeptide of arginine and alanine (Arg–Ala–Ala–Arg . . .) would be obtained; poly CGG would give three polypeptides: poly Arg, poly Ala, and poly Gly (glycine). Can any codons be determined from only these data? Explain.
4. Errors in transcription occur about 100,000 times as often as do errors in DNA replication. Why can this high rate be tolerated in RNA synthesis but not in DNA synthesis?

13 The Genetics of Viruses and Prokaryotes

Robert Stevens felt sicker and sicker, until he finally went to the emergency room of a hospital near his home in Boca Raton, Florida. The medical staff noted his fever, vomiting, and headache and tested his spinal fluid for infectious agents. They saw a few spores of *Bacillus* bacteria, which they might have dismissed as contamination if some of the hospital staff had not just taken a course on identifying possible germ warfare agents at the U.S. Centers for Disease Control and Prevention.

The spores were put in a culture dish with a growth medium, and colonies of bacteria soon appeared. They were identified as *Bacillus anthracis*—the anthrax bacterium. Stevens had respiratory anthrax, a rare form of the disease that he apparently picked up when he inhaled spores deliberately placed in an envelope sent to the newspaper where he worked. The doctors gave him antibiotics to stem the growth of the infection, but it was too late. The rapidly dividing bacteria produced toxins that overwhelmed his body's defenses. Three days later, on October 5, 2001, Robert Stevens died.

When its genome was sequenced, the killer bacterium was found to belong to a strain of *B. anthracis* that had been used in the U.S. government's biological weapons research program until it was disbanded by international agreement in 1969.

Behind the fears of bioterrorism that surrounded the anthrax infections of Stevens and others lie many aspects of prokaryotic genetics and molecular biology. In this chapter, we will describe some of the science behind the headlines, looking at such aspects as bacterial growth and colony formation, exchanges of genetic material, and genome sequencing. Prokaryotes usually reproduce asexually by cell division, but they can acquire new genes in several ways. These mechanisms range from simple recombination in a sexual process to the transport of genes by infective viruses. We will also describe how the expression of prokaryotic genes is regulated and what DNA sequencing has revealed about the prokaryotic genome.

Viruses are not prokaryotes. In fact, they are not even cells, but intracellular parasites that can reproduce only within living cells. We will begin this chapter with a look at the benefits of studying prokaryotes and viruses. Then we will examine the structure, classification, reproduction, and genetics of viruses.

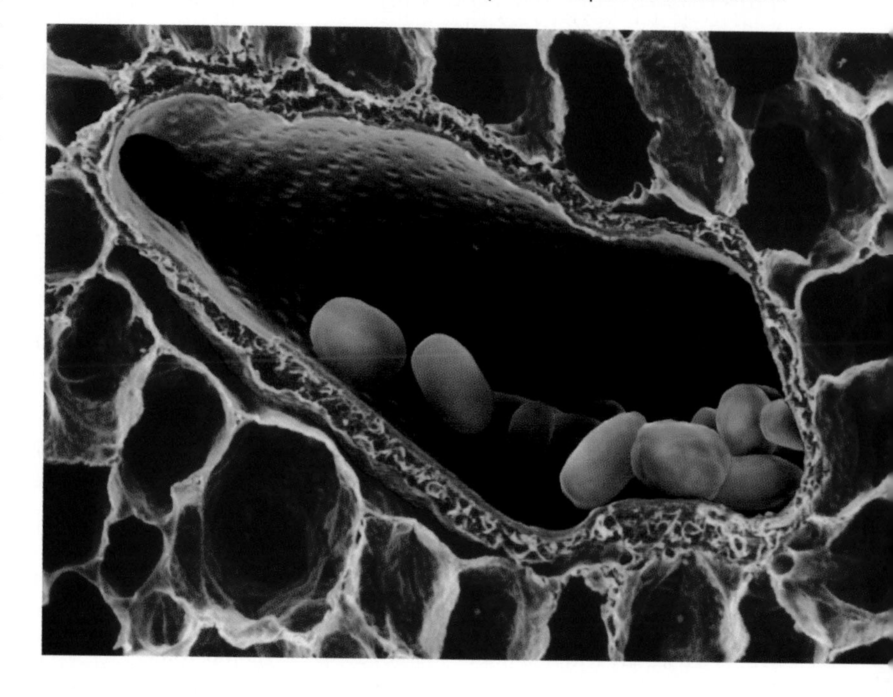

A Prokaryote Weapon This composite micrograph shows spores of *Bacillus anthracis* (yellow) in human lung tissue. *B. anthracis* is the cause of anthrax, a disease that can be fatal to many mammals, including humans. Because anthrax is transmitted by hardy spores that can survive long periods exposed to the environment, it is a prime weapon for bioterrorism.

Probing the Nature of Genes

Prokaryotes and the viruses that infect them have always been important tools for studying the structure, function, and transmission of genes. There are several advantages of working with prokaryotes and viruses:

- Their genomes are small. A typical bacterium contains about a thousandth as much DNA as a single human cell, and a typical bacteriophage contains about a hundredth as much DNA as a bacterium.
- They quickly produce large numbers of individuals. A single milliliter of growth medium can contain more than 10^9 cells of the bacterium *Escherichia coli*, and its numbers can double every 20 minutes.
- Prokaryotes and viruses are usually haploid, which makes genetic analyses easier.

The ease of growing and handling bacteria and their viruses permitted the explosion of genetics and molecular biology that began shortly after the mid-twentieth century. Their relative biological simplicity contributed immeasurably to discoveries about the genetic material, the replication of DNA, and the mechanisms of gene expression. Later they were the first subjects of recombinant DNA technology (see Chapter 16).

Questions of interest to all biologists continue to be studied in prokaryotes, and prokaryotes continue to be important tools for biotechnology and for research on eukaryotes. Prokaryotes also play vital roles in the ecosystem, performing much of the cycling of elements in the soil, atmosphere, and water (see Chapter 58). And, as we saw at the opening of this chapter, prokaryotes and viruses that are **pathogens**—those that cause infectious diseases—continue to challenge humankind.

Viruses: Reproduction and Recombination

Although there are many kinds of viruses, most of them are composed of a nucleic acid and a few proteins. Unlike the organisms that make up the three taxonomic kingdoms of the living world, viruses are *acellular;* that is, they are not cells and do not consist of cells. Viruses do not carry out two of the basic functions of cellular life: they do not regulate the transport of substances into and out of them by membranes, and they do not perform energy metabolism. Furthermore, they can reproduce only in systems that do perform these functions: living cells.

Scientists studied viruses before they could see them

Most viruses are much smaller than even the mycoplasmas—the smallest bacteria (Table 13.1). Viruses have become well

13.1 Relative Sizes of Microorganisms

MICROORGANISM	TYPE	TYPICAL SIZE RANGE (μm^3)
Protists	Eukaryote	5,000–50,000
Photosynthetic bacteria	Prokaryote	5–50
Spirochetes	Prokaryote	0.1–2.0
Mycoplasmas	Prokaryote	0.01–0.1
Poxviruses	Virus	0.01
Influenza virus	Virus	0.0005
Poliovirus	Virus	0.00001

understood only within the last half century, but the first step on this path of discovery was taken by the Russian botanist Dmitri Ivanovsky in 1892. He was trying to find the cause of tobacco mosaic disease, which results in the destruction of photosynthetic tissues in plants and can devastate a tobacco crop. Ivanovsky passed an extract of diseased tobacco leaves through a fine porcelain filter, a technique that had been used previously by physicians and veterinarians to isolate disease-causing bacteria.

To Ivanovsky's surprise, the disease agent in this case was not retained on the filter. It passed through, and the liquid filtrate still caused tobacco mosaic disease. But instead of concluding that the agent was smaller than a bacterium, he assumed that his filter was faulty. Pasteur's recent demonstration that bacteria could cause disease was the dominant idea at the time, and Ivanovsky chose not to challenge it. But, as often happens in science, someone soon came along who did. In 1898, the Dutch microbiologist Martinus Beijerinck repeated Ivanovsky's experiment and also showed that the tobacco mosaic disease agent could diffuse through an agar gel. He called the tiny agent *contagium vivum fluidum*, which later became shortened to *virus*.

Almost 40 years later, the disease agent was crystallized by Wendell Stanley (who won the Nobel prize for his efforts). The crystalline viral preparation became infectious again when it was dissolved. It was soon shown that crystallized viral preparations consist of proteins and nucleic acids. Finally, direct observation of viruses with electron microscopes in the 1950s showed clearly how much they differ from bacteria and other organisms. The simplest infective agents of all are *viroids*, which are made up only of genetic material.

Viruses reproduce only with the help of living cells

Whole viruses never arise directly from preexisting viruses. Viruses are *obligate intracellular parasites;* that is, they develop and reproduce only within the cells of specific hosts. The cells of animals, plants, fungi, protists, and prokaryotes (both bacteria and archaea) can serve as hosts to viruses. Viruses use the host's synthetic machinery to reproduce themselves, usually destroying the host cell in the process. The host cell releases progeny viruses, which then infect new hosts.

Viruses outside of host cells exist as individual particles called **virions**. The virion, the basic unit of a virus, consists of a central core of either DNA or RNA (but not both) surrounded by a **capsid**, or coat, composed of one or more proteins. Because they lack the distinctive cell wall and ribosomal biochemistry of bacteria, viruses are not affected by antibiotics.

There are many kinds of viruses

There are four ways to describe viruses:

- Whether the genome is DNA or RNA
- Whether the nucleic acid is single-stranded or double-stranded
- Whether the shape of the virion is a simple or complex crystal
- Whether the virion is surrounded by a membrane

Some of these variations are shown in Figure 13.1.

Another important descriptor of a virus is the type of organisms it infects. Most viruses have relatively simple means of infecting their host cells. Some can infect a cell but postpone reproduction, remaining inactive in the host cell until conditions are favorable.

(*a*)

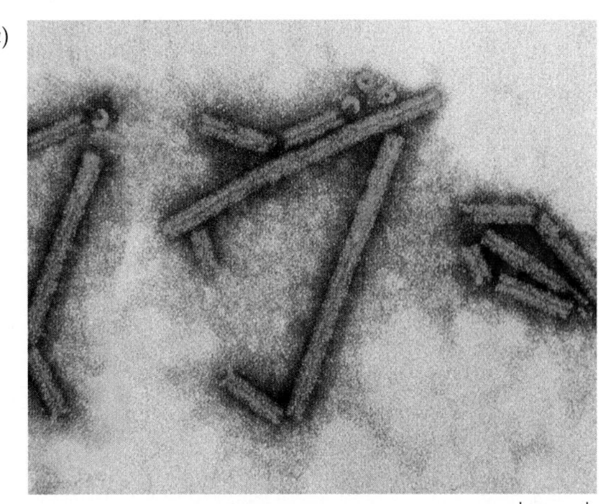

(*b*)

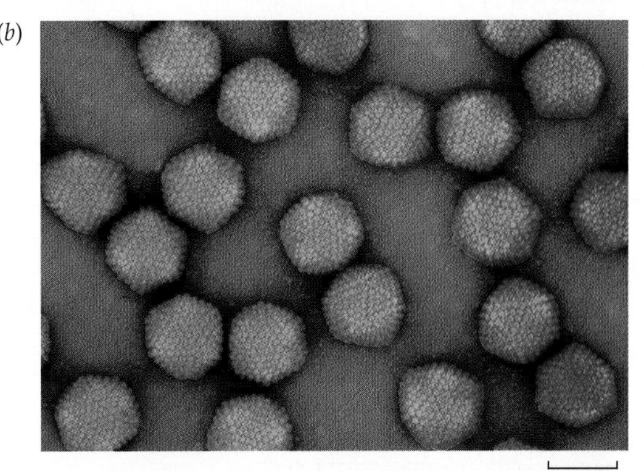

(*c*)

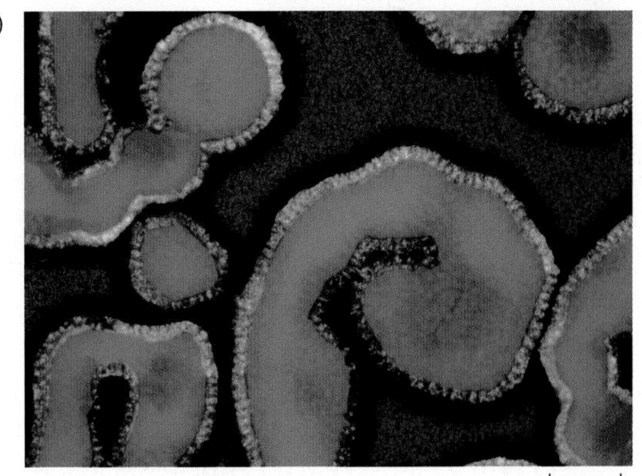

13.1 Virions Come in Various Shapes (*a*) The tobacco mosaic virus (a plant virus) consists of an inner helix of RNA covered with a helical array of protein molecules. (*b*) Many animal viruses, such as this adenovirus, have an icosahedral (20-sided) capsid as an outer shell. Inside the shell is a spherical mass of proteins and DNA. (*c*) Not all virions are regularly shaped. These wormlike virions of the influenza A virus infect humans, causing chills, fever, and sometimes, death.

Bacteriophage reproduce by a lytic cycle or a lysogenic cycle

Viruses that infect bacteria are known as **bacteriophage** or **phage** (Greek *phagos*, "one that eats"). They recognize their hosts by means of proteins in the capsid, which bind to specific receptor proteins or carbohydrates in the host's cell wall. The virions, which must penetrate the cell wall, are often equipped with tail assemblies that inject the phage's nucleic acid through the cell wall into the host bacterium. After the nucleic acid has entered the host, one of two things happens, depending on the kind of phage:

- The virus may reproduce immediately and kill the host cell.
- The virus may postpone reproduction by integrating its nucleic acid into the host cell's genome.

We saw one type of viral reproductive cycle when we studied the Hershey–Chase experiment (see Figure 11.3). That was the **lytic cycle**, so named because the infected bacterium *lyses* (bursts), releasing progeny phage. The alternative fate is the **lysogenic cycle**, in which the infected bacterium does not lyse, but instead harbors the viral nucleic acid for many generations. Some viruses reproduce only by the lytic cycle; others undergo both types of reproductive cycles (Figure 13.2).

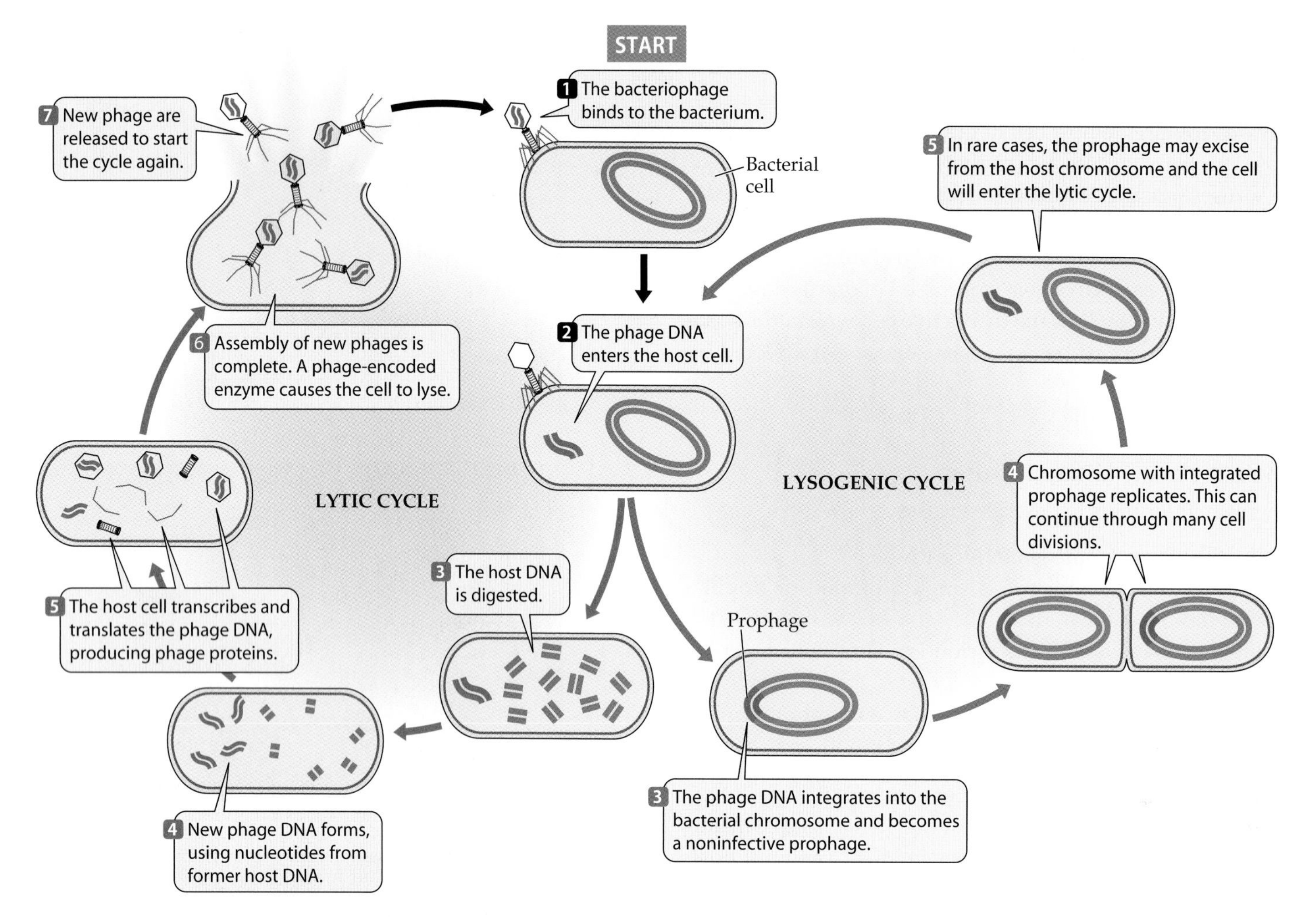

START
1 The bacteriophage binds to the bacterium.
Bacterial cell
2 The phage DNA enters the host cell.
3 The host DNA is digested.
4 New phage DNA forms, using nucleotides from former host DNA.
5 The host cell transcribes and translates the phage DNA, producing phage proteins.
6 Assembly of new phages is complete. A phage-encoded enzyme causes the cell to lyse.
7 New phage are released to start the cycle again.
LYTIC CYCLE
LYSOGENIC CYCLE
Prophage
3 The phage DNA integrates into the bacterial chromosome and becomes a noninfective prophage.
4 Chromosome with integrated prophage replicates. This can continue through many cell divisions.
5 In rare cases, the prophage may excise from the host chromosome and the cell will enter the lytic cycle.

13.2 The Lytic and Lysogenic Cycles of Bacteriophage In the lytic cycle, infection by viral DNA leads directly to the multiplication of the virus and lysis of the host bacterial cell. In the lysogenic cycle, an inactive prophage is replicated as part of the host's chromosome.

THE LYTIC CYCLE. A virus that reproduces only by the lytic cycle is called a **virulent** virus. Once the phage has injected its nucleic acid into the host cell, that nucleic acid takes over the host's synthetic machinery. It does so in two stages (Figure 13.3):

- The viral genome contains a promoter sequence that attracts host RNA polymerase. In the *early stage*, viral genes that lie adjacent to this promoter are transcribed. These *early genes* often code for proteins that shut down host transcription, stimulate viral genome replication, and stimulate late gene transcription. Nuclease enzymes digest the host's chromosome, providing nucleotides for the synthesis of viral genomes.
- In the *late stage*, viral *late genes*, which code for the proteins of the viral capsid and those that lyse the host cell to release the new virions, are transcribed.

This sequence of transcriptional events is carefully controlled: Premature lysis of the host cell before virions are ready for release would stop the infection. The whole process—from binding and infection to lysis of the host cell—takes about half an hour.

Rarely, two viruses infect a cell at the same time. This is an unusual event, as once an infection cycle is under way, there is usually not enough time for an additional infection. In addition, an early protein prevents further infections in some cases. However, the presence of two different viral genomes in the same host cell affords the opportunity for genetic recombination by crossing over (as in prophase I of meiosis in eukaryotes). This phenomenon enables genetically different viruses of the same kind to swap genes and create new strains.

THE LYSOGENIC CYCLE. Phage infection does not always result in lysis of the host cell. Some phage seem to disappear from a bacterial culture, leaving the bacteria immune to further attack by the same strain of phage. In such cultures, however, a few free phage are always present. Bacteria harboring phage that are not lytic are called *lysogenic bacteria*, and the viruses are called **temperate** viruses.

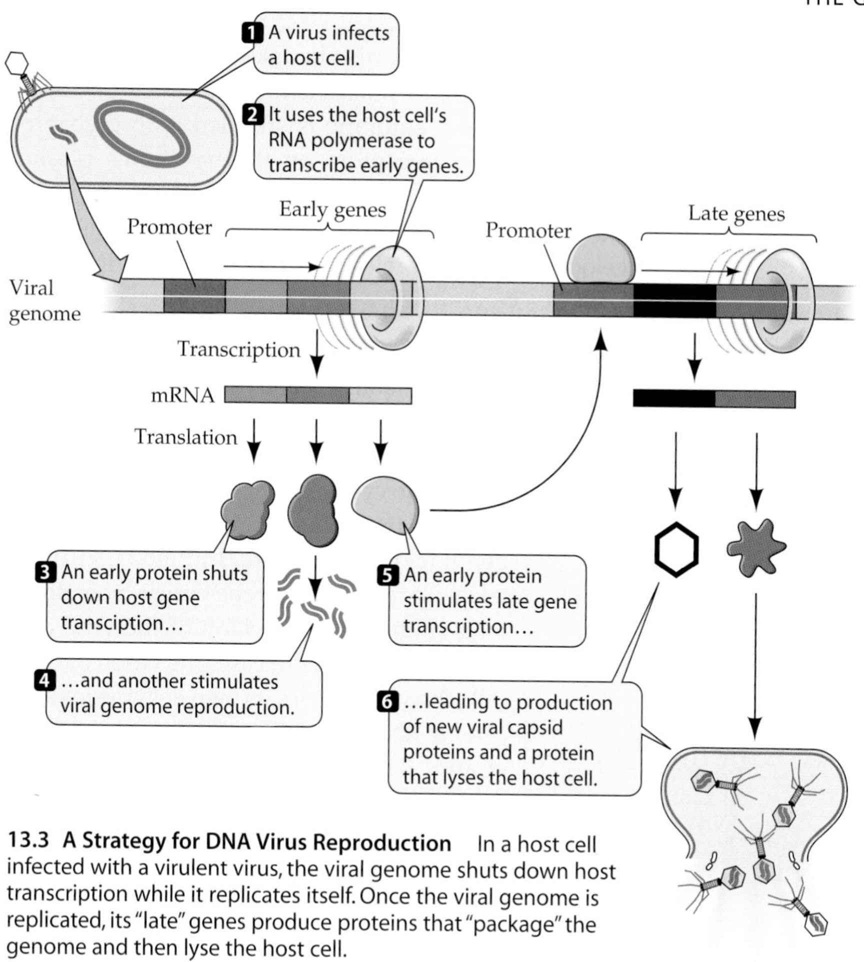

1 A virus infects a host cell.
2 It uses the host cell's RNA polymerase to transcribe early genes.
Early genes
Promoter
Promoter
Late genes
Viral genome
Transcription
mRNA
Translation
3 An early protein shuts down host gene transciption...
4 ...and another stimulates viral genome reproduction.
5 An early protein stimulates late gene transcription...
6 ...leading to production of new viral capsid proteins and a protein that lyses the host cell.

13.3 A Strategy for DNA Virus Reproduction In a host cell infected with a virulent virus, the viral genome shuts down host transcription while it replicates itself. Once the viral genome is replicated, its "late" genes produce proteins that "package" the genome and then lyse the host cell.

Lysogenic bacteria contain a noninfective entity called a **prophage**: a molecule of phage DNA that has been integrated into the bacterial chromosome (see Figure 13.2). The prophage can remain inactive within the bacterial genome through many cell divisions. However, an occasional lysogenic bacterium can be induced to activate its prophage. This activation results in a lytic cycle, in which the prophage leaves the host chromosome and reproduces.

This capacity to switch between the lysogenic and the lytic cycle is very useful to the phage because it enhances the production of the maximum number of progeny viruses. When its host cell is growing rapidly, the phage is lysogenic. When the host is stressed or damaged by mutagens, the prophage is released from its inactive state, and the lytic cycle proceeds. We will see how this switch works later in the chapter when we discuss the regulation of gene expression.

Lytic bacteriophage could be useful in treating bacterial infections

Since lytic bacteriophage destroy their bacterial hosts, they might be useful in treating infectious diseases caused by bacteria. Indeed, one of the early discoverers of phage, the French-Canadian microbiologist Felix D'Herelle, noted in 1917 (before antibiotics were discovered) that when some patients with bacterial dysentery were recovering from the disease, the quantity of phage near the bacteria was much higher than when the disease was at its peak.

D'Herelle tried using phage to control infections of chickens by the bacterium *Salmonella gallinarum*. To do this, he divided chickens into two groups, one that was given phage and another that was not. Then he exposed both groups to the infectious bacteria. The phage-protected group did not get the bacterial disease. Later, he used phage successfully to treat people in Egypt infected with plague-causing bacteria and people in India with infectious cholera.

The emergence of antibiotics and of phage-resistant bacteria reduced interest in phage therapy. However, interest has revived now that bacterial resistance to antibiotics is becoming common. Bacteriophage are even being investigated as a means of treating edible fruits and vegetables to prevent bacterial contamination. In addition to advancing our understanding of fundamental biological processes, the study of bacteriophage opened the door to investigations of viruses that infect eukaryotes.

Animal viruses have diverse reproductive cycles

Almost all vertebrates are susceptible to viral infections, but among invertebrates, such infections are common only in arthropods (the group that includes insects and crustaceans). One group of viruses, called *arboviruses* (short for "arthropod-borne viruses"), is transmitted to a vertebrate through an insect bite. Although they are carried within the arthropod host's cells, arboviruses apparently do not affect that host severely; they affect only the bitten and infected organism. The arthropod acts as a **vector**—an intermediate carrier—by transmitting the disease organism from one host to another.

Animal viruses are very diverse. Some are just particles consisting of proteins surrounding a nucleic acid. Others have a membrane derived from the host cell's plasma membrane and are called *enveloped* viruses. Some animal viruses have DNA as their genetic material; others have RNA. In most cases, the viral genome is small, coding for only a few proteins.

Like that of bacteriophage, the lytic cycle of animal viruses can be divided into early and late stages (see Figure 13.3). Animal viruses enter cells in one of three ways:

- A naked virion (without a membrane) is taken up by endocytosis, which traps it within a membranous vesicle inside the host cell. The membrane of the vesicle breaks down, releasing the virion into the cytoplasm, and the host cell digests the protein capsid, liberating the viral nucleic acid, which takes charge of the host cell.
- Enveloped viruses may also be taken up by endocytosis (see Figure 13.4) and released from a vesicle. In these viruses, the viral membrane is studded with glycoproteins that bind to receptors on the host cell's plasma membrane.
- More commonly, the membranes of the host and the enveloped virus fuse, releasing the rest of the virion into the cell (see Figure 13.5).

Following viral reproduction, enveloped viruses usually escape from the host cell by a budding process in which they acquire a membrane similar to that of the host cell.

The life cycles of influenza virus and HIV illustrate two different styles of infection and genome reproduction. Influenza virus is taken up into a membrane vesicle by endocytosis (Figure 13.4). Fusion of the viral and vesicle membranes releases the virion into the cell. The virus carries its own enzyme to replicate its RNA genome into a complementary strand. The new strand is then used as mRNA to make, by complementary base pairing, more copies of the viral genome.

Retroviruses such as HIV have a more complex reproductive cycle (Figure 13.5). The virus enters a host cell by direct fusion of viral and cellular membranes. A distinctive feature of the retroviral life cycle is the reverse transcription of retroviral RNA. This process produces a DNA **provirus** consisting of **cDNA** (complementary DNA transcribed from the RNA genome), which is the form of the viral genome that gets integrated into the host's DNA. The provirus may reside in the host chromosome permanently, occasionally being expressed to produce new virions. Almost every step in this complex cycle can, in principle, be attacked by therapeutic drugs; this fact is used by researchers in their quest to conquer AIDS, the deadly condition caused by HIV infection in humans, which will be discussed further in Chapter 18.

Animal viruses, including human viruses, take a severe toll on human and animal health. But our well-being is also challenged by plant viruses and the diseases they cause.

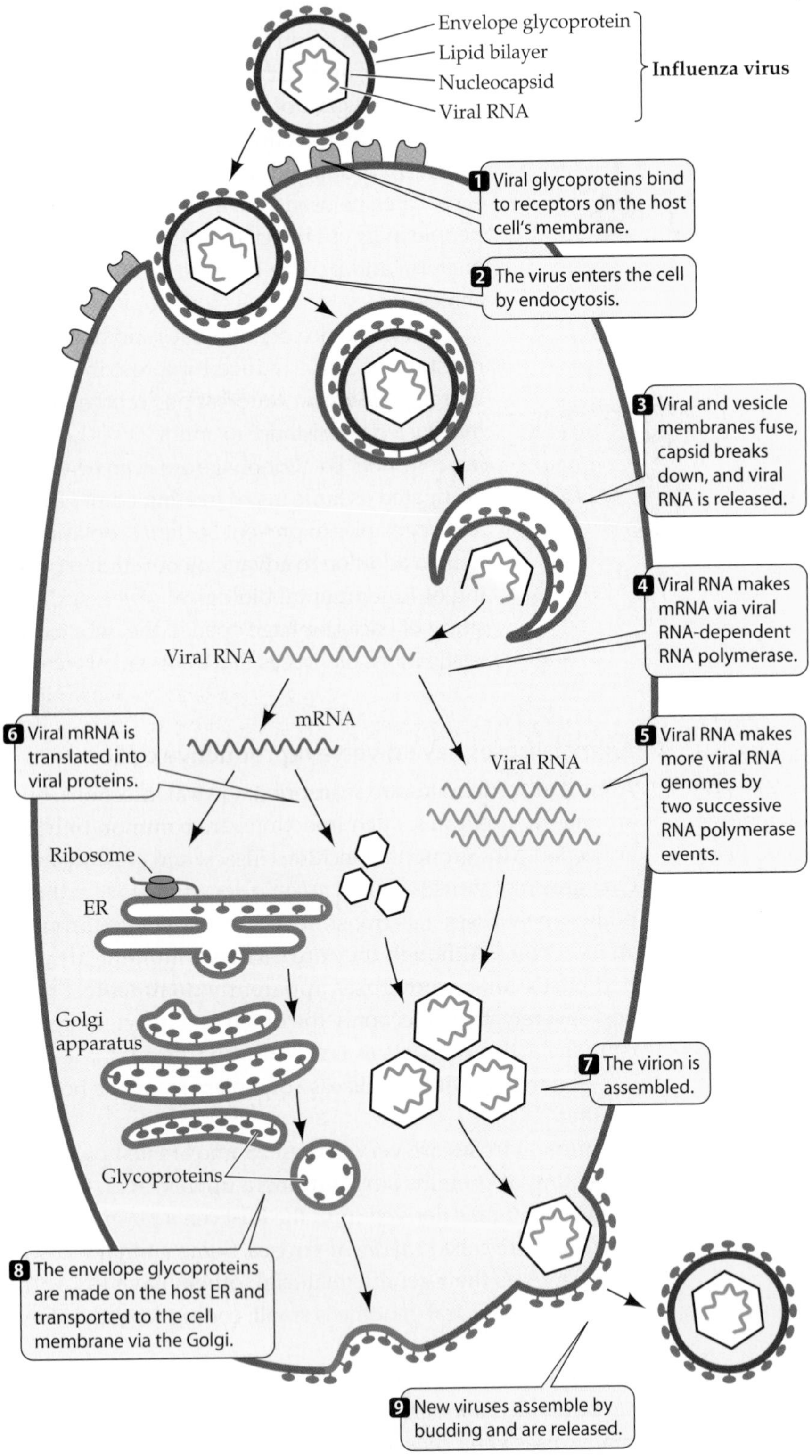

13.4 The Reproductive Cycle of the Influenza Virus The enveloped influenza virus is taken into the host cell by endocytosis. Once inside, fusion of the vesicle and viral membranes releases the RNA genome, which replicates and assembles new virions.

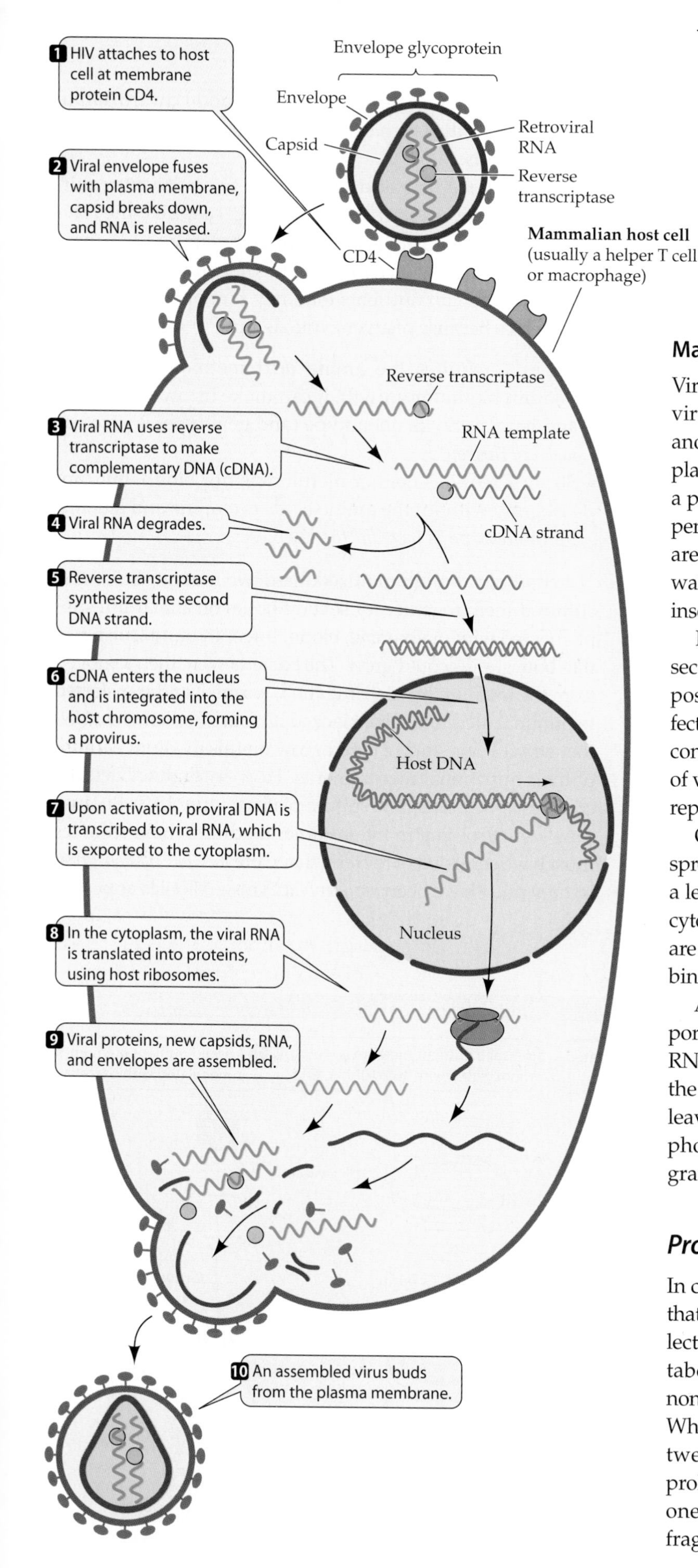

1 HIV attaches to host cell at membrane protein CD4.
Envelope glycoprotein
Envelope
Capsid
Retroviral RNA
Reverse transcriptase
2 Viral envelope fuses with plasma membrane, capsid breaks down, and RNA is released.
CD4
Mammalian host cell (usually a helper T cell or macrophage)
Reverse transcriptase
3 Viral RNA uses reverse transcriptase to make complementary DNA (cDNA).
RNA template
4 Viral RNA degrades.
cDNA strand
5 Reverse transcriptase synthesizes the second DNA strand.
6 cDNA enters the nucleus and is integrated into the host chromosome, forming a provirus.
Host DNA
7 Upon activation, proviral DNA is transcribed to viral RNA, which is exported to the cytoplasm.
Nucleus
8 In the cytoplasm, the viral RNA is translated into proteins, using host ribosomes.
9 Viral proteins, new capsids, RNA, and envelopes are assembled.
10 An assembled virus buds from the plasma membrane.

13.5 The Reproductive Cycle of HIV The retrovirus HIV enters a host cell via fusion of its envelope with the host's plasma membrane. Reverse transcription of retroviral RNA then produces a DNA provirus—a strand of complementary DNA that enters the host nucleus, where it can be transcribed to viral RNA.

Many plant viruses spread with the help of vectors

Viral diseases of flowering plants are very common. Plant viruses can be transmitted *horizontally*, from one plant to another, or *vertically*, from parent to offspring. To infect a plant cell, viruses must pass through a cell wall as well as a plasma membrane. Most plant viruses accomplish this penetration through their association with vectors, which are often insects. When an insect vector penetrates a cell wall with its proboscis (snout), virions can move from the insect into the plant.

Plant viruses can be introduced artificially, without insect vectors, by bruising a leaf or other plant part, then exposing it to a suspension of virions. Horizontal viral infections may also occur in nature if a bruised infected plant contacts an injured uninfected one. Vertical transmission of viral infections may occur through vegetative or sexual reproduction.

Once inside a plant cell, the virus reproduces and spreads to other cells in the plant. Within an organ such as a leaf, the virus spreads through the plasmodesmata, the cytoplasmic connections between cells. Because the viruses are too large to go through these channels, special proteins bind to them to help them squeeze through the pores.

An example of a virus that causes an economically important plant disease is the wheat streak mosaic virus. This RNA virus enters the leaf of a wheat plant via a tiny insect, the mite *Aceria tulipae*. As the infection spreads inside the leaves, they show yellow streaks due to the destruction of photosynthetic tissues. As a result, production of wheat grain can be severely reduced.

Prokaryotes: Reproduction and Recombination

In contrast to viruses, bacteria and archaea are living cells that carry out all the basic cellular functions. They have selectively permeable membranes and perform energy metabolism. Prokaryotes usually reproduce asexually, but nonetheless have several ways of recombining their genes. Whereas in eukaryotes, genetic recombination occurs between the genomes of two parents, recombination in prokaryotes results from the interaction of the genome of one cell with a much smaller sample of genes—a DNA fragment—from another cell.

The reproduction of prokaryotes gives rise to clones

Most prokaryotes reproduce by the division of single cells into two identical offspring (see Figure 9.2). In this way, a single cell gives rise to a **clone**—a population of genetically identical individuals. Prokaryotes reproduce very rapidly. A population of *E. coli*, as we saw above, can double every 20 minutes as long as conditions remain favorable. That is one of the reasons that this bacterium is used so widely in research.

Simple, reliable methods exist for isolating single bacterial cells and rapidly growing them into clones for identification and study. Pure cultures of *E. coli* or other bacteria can be grown in liquid nutrient medium, or on the surface of a solid *minimal medium* that contains a sugar, minerals, a nitrogen source such as ammonium chloride (NH_4Cl), and a solidifying agent such as agar (Figure 13.6). If the number of cells spread on the medium is small, each cell will give rise to a small, rapidly growing *bacterial colony*. If a large number of cells is spread onto the solid medium, their growth will produce one continuous layer—a *bacterial lawn*. Bacteria can also be grown in a liquid nutrient medium. We'll see examples of all these techniques in this chapter.

In recombination, bacteria conjugate

The existence and heritability of mutations in bacteria attracted the attention of geneticists. If there were no form of exchange of genetic information between individuals, bacteria would not be useful for genetic analysis. But can these asexually reproducing organisms exchange genetic information? Luckily, in 1946, Joshua Lederberg and Edward Tatum demonstrated that such exchanges do occur, although they are rare events.

Initially, Lederberg and Tatum grew two nutrient-requiring, or *auxotrophic*, mutant strains of *E. coli*. Like the *Neurospora* studied by Beadle and Tatum (see Figure 12.1), these strains cannot grow on a minimal medium, but require supplementation with nutrients that they cannot synthesize for themselves because of an enzyme defect.

- Strain 1 requires the amino acid methionine and the vitamin biotin for growth; it can make its own threonine and leucine. So its phenotype (and genotype) is given as $met^-bio^-thr^+leu^+$.
- Strain 2 requires neither methionine nor biotin, but cannot grow without the amino acids threonine and leucine. Its phenotype is $met^+bio^+thr^-leu^-$.

Lederberg and Tatum mixed these two mutant strains and cultured them together for several hours on a medium supplemented with methionine, biotin, threonine, and leucine, so that both strains could grow. The bacteria were then removed from the medium by centrifugation, washed, and transferred to minimal medium, which lacked all four supplements. Neither strain 1 nor strain 2 could grow on this medium because of their nutritional requirements. However, a few bacterial colonies appeared on the culture plates (Figure 13.7). Because they were growing in the minimal medium, these colonies must have consisted of bacteria that were $met^+bio^+thr^+leu^+$; that is, they must have been *prototrophic*. These colonies appeared

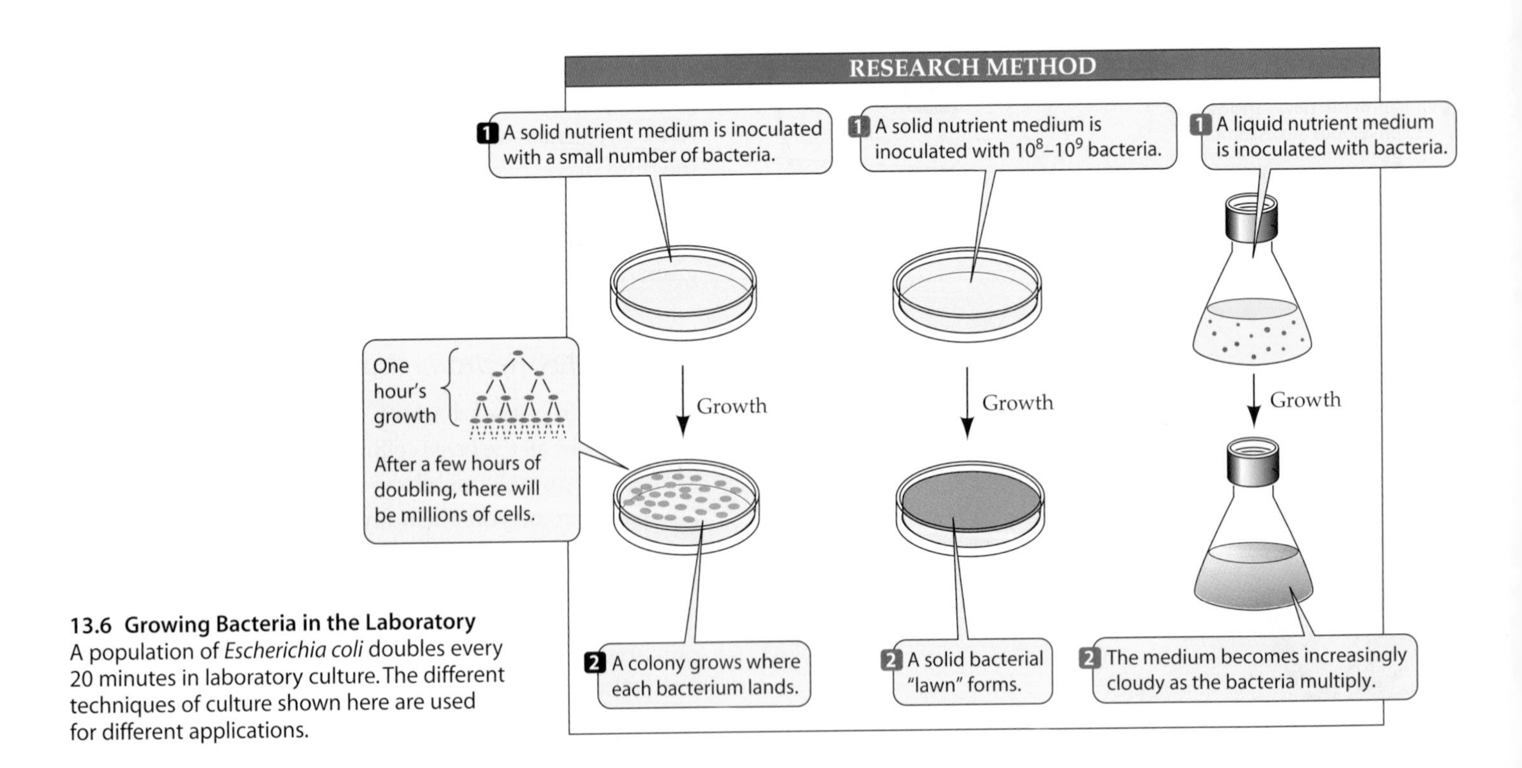

13.6 Growing Bacteria in the Laboratory
A population of *Escherichia coli* doubles every 20 minutes in laboratory culture. The different techniques of culture shown here are used for different applications.

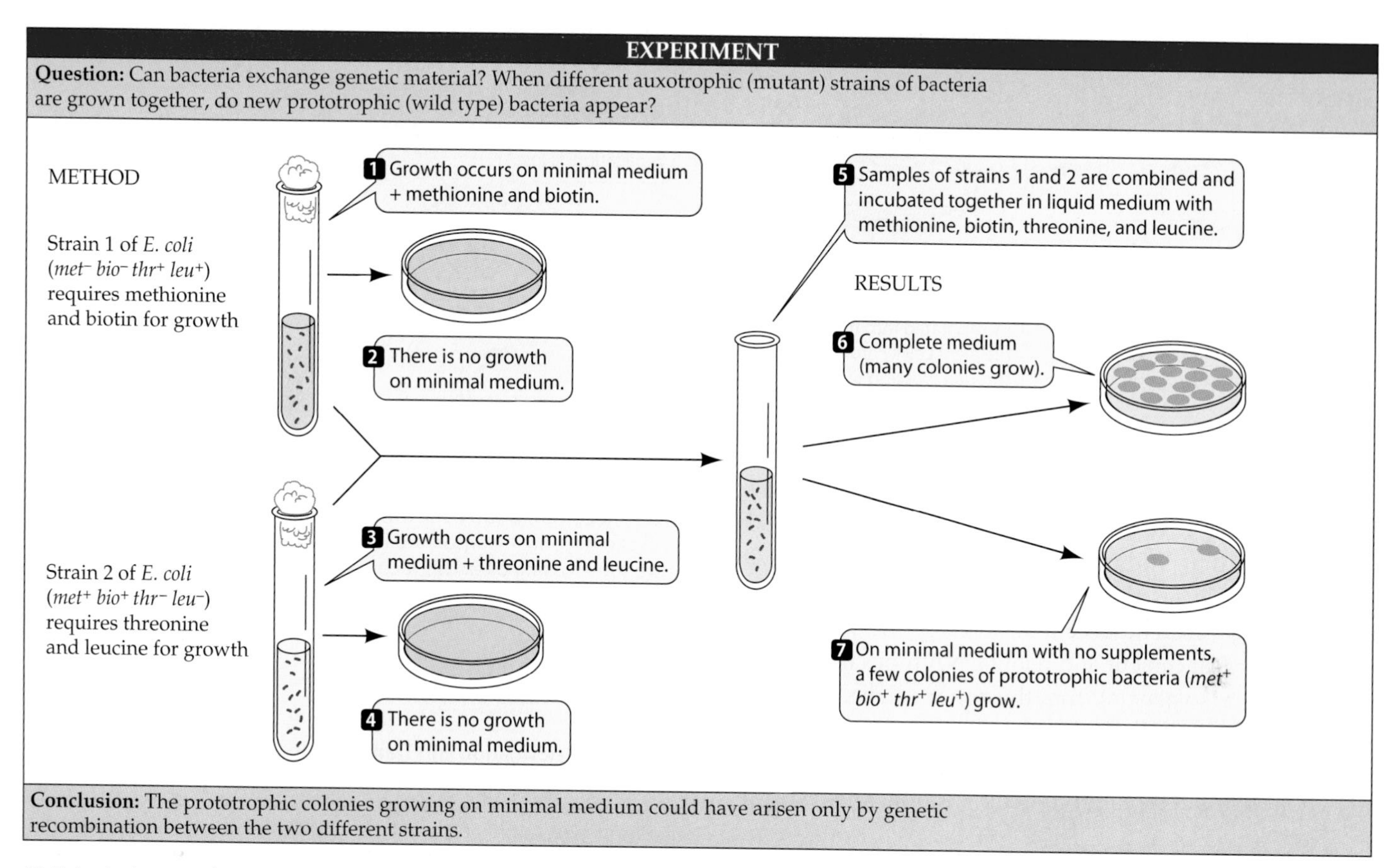

EXPERIMENT
Question: Can bacteria exchange genetic material? When different auxotrophic (mutant) strains of bacteria are grown together, do new prototrophic (wild type) bacteria appear?
METHOD
Strain 1 of E. coli (met⁻ bio⁻ thr⁺ leu⁺) requires methionine and biotin for growth
1 Growth occurs on minimal medium + methionine and biotin.
2 There is no growth on minimal medium.
Strain 2 of E. coli (met⁺ bio⁺ thr⁻ leu⁻) requires threonine and leucine for growth
3 Growth occurs on minimal medium + threonine and leucine.
4 There is no growth on minimal medium.
5 Samples of strains 1 and 2 are combined and incubated together in liquid medium with methionine, biotin, threonine, and leucine.
RESULTS
6 Complete medium (many colonies grow).
7 On minimal medium with no supplements, a few colonies of prototrophic bacteria (met⁺ bio⁺ thr⁺ leu⁺) grow.
Conclusion: The prototrophic colonies growing on minimal medium could have arisen only by genetic recombination between the two different strains.

13.7 Lederberg and Tatum's Experiment After growing together, a mixture of complementary auxotrophic strains of *E. coli* contained a few cells that gave rise to new prototrophic colonies. This experiment proved that genetic recombination takes place in prokaryotes.

at a rate of approximately one for every 10 million cells originally placed on the plates ($1/10^7$).

Where did these prototrophic colonies come from? Lederberg and Tatum were able to rule out mutation, and other investigators ruled out transformation (a process we discussed in Chapter 11 and which we'll look at in more detail below). A third possibility is that the two strains of *E. coli* had exchanged genetic material, producing some cells containing *met*$^+$ and *bio*$^+$ alleles from strain 2 and *thr*$^+$ and *leu*$^+$ alleles from strain 1 (see Figure 13.7). Later experiments showed that such an exchange, called **conjugation**, had indeed occurred. One bacterial cell—the recipient—had received DNA from another cell—the donor—that included the two wild-type ($^+$) alleles that were missing in the recipient. Recombination had then created a genotype with four wild-type alleles.

The physical contact required for conjugation can be observed under the electron microscope (Figure 13.8). It is initiated by a thin projection called a *sex pilus*. Once the sex pili bring the two cells into proximity, the actual transfer of DNA occurs by a thin cytoplasmic bridge called a *conjugation tube*. Since the bacterial chromosome is circular, it must be made linear (cut) so that it can pass through the tube. Contact between the cells is brief—certainly not long enough for the entire donor genome to enter the recipient cell. Therefore, the recipient cell usually receives only a portion of the donor DNA.

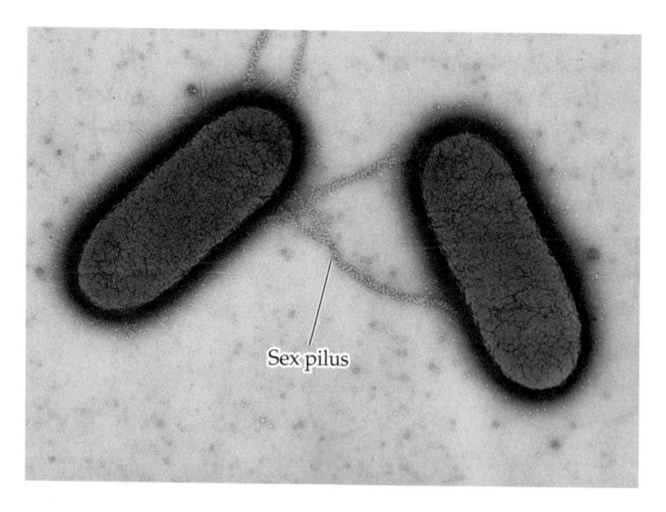

13.8 Bacterial Conjugation Sex pili draw two bacteria into close contact, and a cytoplasmic conjugation tube forms. DNA is transferred from one cell to the other via the conjugation tube.

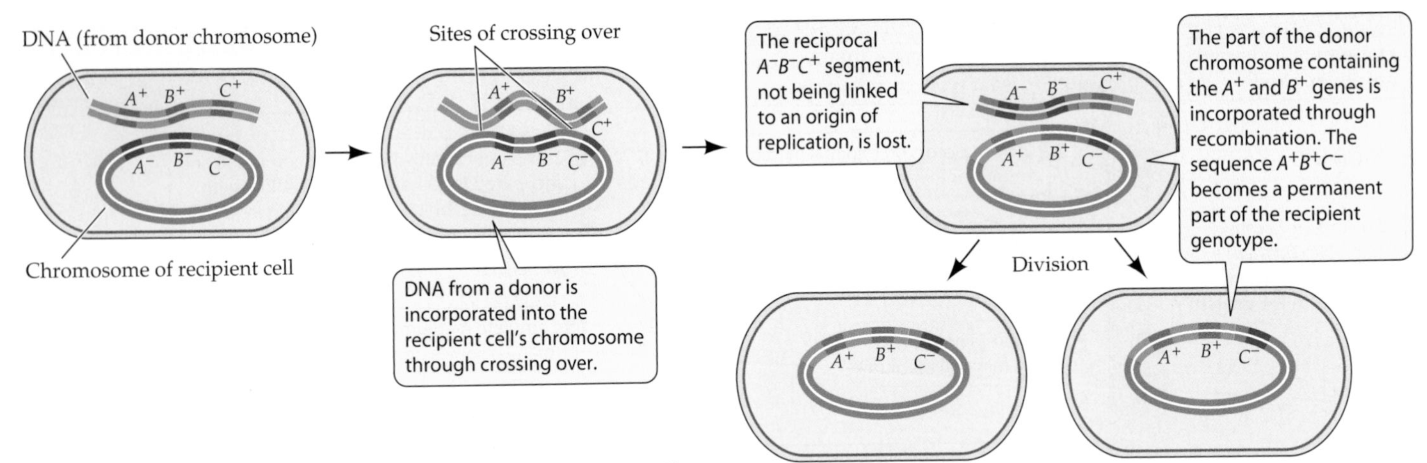

13.9 Recombination Following Conjugation DNA from a donor cell can become incorporated into a recipient cell's chromosome through crossing over. This recombination explains the results of the Lederberg-Tatum experiment shown in Figure 13.7.

Once the donor DNA fragment is inside the recipient cell, it can recombine with the recipient cell's genome. In much the same way that chromosomes pair up, gene for gene, in prophase I of meiosis, the donor DNA can line up beside its homologous genes in the recipient, and crossing over can occur. Enzymes that can cut and rejoin DNA molecules are active in bacteria, so gene(s) from the donor can become integrated into the genome of the recipient, thus changing the recipient's genetic constitution (Figure 13.9), even though only about half the transferred genes become integrated in this way.

In transformation, cells pick up genes from their environment

Frederick Griffith obtained the first evidence for the transfer of prokaryotic genes more than 75 years ago when he discovered the transforming principle (see Figure 11.1). We now know the reason for his results: DNA had leaked from dead cells of virulent pneumococci and was taken up as free DNA by living nonvirulent pneumococci, which became virulent as a result. This phenomenon, called **transformation**, occurs in nature in some species of bacteria when cells die and their DNA leaks out (Figure 13.10*a*). Once transforming DNA is inside a host cell, an event very similar to recombination occurs, and new genes can be incorporated into the host chromosome.

In transduction, viruses carry genes from one cell to another

When bacteriophage undergo a lytic cycle, they package their DNA in capsids. These capsids generally form before the

13.10 Transformation and Transduction After a new DNA fragment enters the host cell, recombination can occur. (*a*) Transforming DNA can leak from dead bacterial cells and be taken up by a living bacterium, which may incorporate the new genes into its chromosome. (*b*) In transduction, viruses carry DNA fragments from one cell to another.

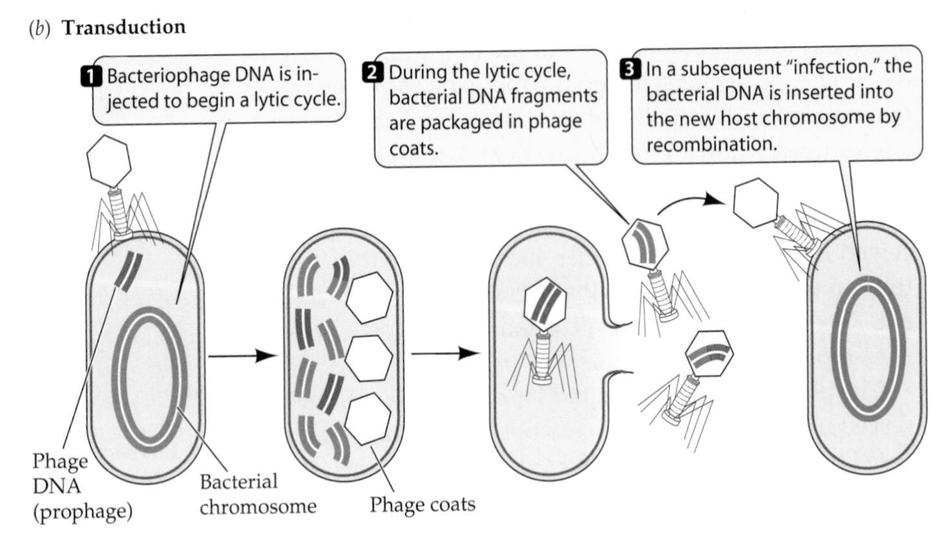

(a) Transformation
1 A lysed bacterium releases DNA fragments...
2 ...which enter a living cell.
3 Recombination occurs between the DNA fragment and host chromosome.
Bacterial cell
Bacterial chromosome
Chromosome of recipient cell
(b) Transduction
1 Bacteriophage DNA is injected to begin a lytic cycle.
2 During the lytic cycle, bacterial DNA fragments are packaged in phage coats.
3 In a subsequent "infection," the bacterial DNA is inserted into the new host chromosome by recombination.
Phage DNA (prophage)
Bacterial chromosome
Phage coats

viral DNA is inserted into them. Sometimes, bacterial DNA fragments are inserted into the empty phage capsids instead of the phage DNA. (Figure 13.10*b*). Recall that the binding of a phage to its host cell and the insertion of phage DNA are carried out by the capsid. So, when a phage capsid carries a piece of bacterial DNA, the latter is injected into the "infected" bacterium. This mechanism of DNA transfer is called **transduction**. Needless to say, it does not result in a productive viral infection. Instead, the incoming DNA fragment can recombine with the host chromosome, resulting in the replacement of host cell genes with bacterial genes from the incoming phage particle.

Plasmids are extra chromosomes in bacteria

In addition to their main chromosome, many bacteria harbor additional smaller, circular chromosomes. These chromosomes, called **plasmids**, usually contain at most a few dozen genes, and, importantly, an origin of replication (the sequence where DNA replication starts), which defines them as chromosomes. Usually plasmids replicate at the same time as the main chromosome, but that is not necessarily the case.

Plasmids are *not* viruses. They do not take over the cell's molecular machinery or make a protein coat to help them move from cell to cell. Instead, they can move between cells during conjugation, thereby adding some new genes to the recipient bacterium (Figure 13.11). Because plasmids exist independently of the main chromosome (the term *episomes* is sometimes used for them), they do not need to recombine with the main chromosome to add their genes to the recipient cell's genome.

There are several types of plasmids, classified according to the kinds of genes they carry. Some code for catabolic enzymes, others enable conjugation, while others code for genes that circumvent antibiotic attack.

SOME PLASMIDS CARRY GENES FOR UNUSUAL METABOLIC FUNCTIONS. Some plasmids, called *metabolic factors*, have genes that allow their recipients to carry out unusual metabolic functions. For example, there are many unusual hydrocarbons in oil spills. Some bacteria can actually thrive on these molecules, using them as a carbon source. The genes for the enzymes involved in breaking down the hydrocarbons are carried on plasmids.

SOME PLASMIDS CARRY GENES FOR CONJUGATION. Other plasmids, called *fertility factors*, or *F factors* for short, encode the genes needed for conjugation. They have approximately 25 genes, including the ones that make both the pilus for attachment and the conjugation tube for DNA transfer. A cell harboring an F factor is referred to as F^+. It can transfer a copy of the F factor to an F^- cell, making the recipient F^+.

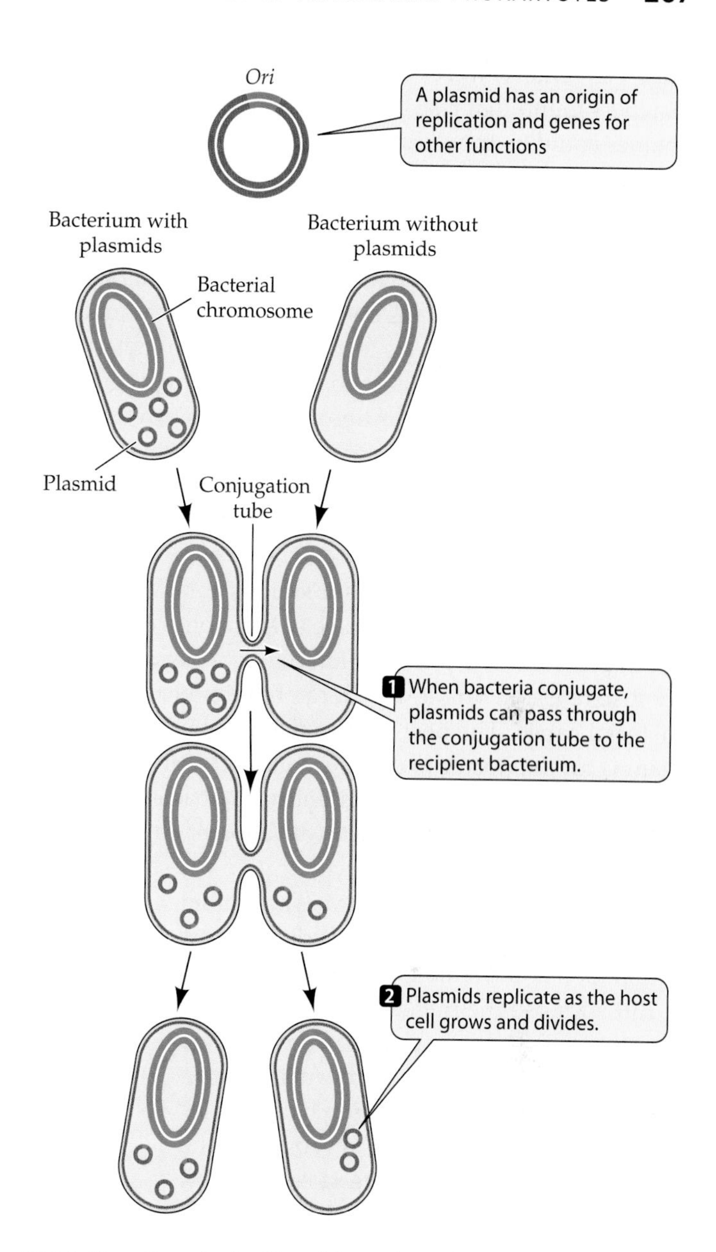

13.11 Gene Transfer by Plasmids When plasmids enter a cell via conjugation, their genes can be expressed in the new cell.

Sometimes the F factor integrates into the main chromosome (at which point it is no longer a plasmid), and when it does, it can bring along other genes from that chromosome when it moves through the conjugation tube from one cell to another.

SOME PLASMIDS ARE RESISTANCE FACTORS. *Resistance factors*, or *R factors*, may carry genes coding for proteins that destroy or modify antibiotics. Other R factors provide resistance to heavy metals that bacteria encounter in their environment.

R factors first came to the attention of biologists in 1957 during an epidemic of dysentery in Japan, when it was discovered that some strains of the *Shigella* bacterium, which causes dysentery, were resistant to several antibiotics. Researchers found that resistance to the entire spectrum of antibiotics could be transferred by conjugation even when no genes on the main chromosome were transferred. Eventually it was shown that the genes for antibiotic resistance are carried on plasmids. Each R factor carries one or more genes conferring resistance to particular antibiotics, as well as genes that code for proteins involved in the transfer of DNA to a recipient bacterium. As far as biologists can determine, R factors providing resistance to naturally occurring antibiotics existed long before antibiotics were discovered and used by humans. However, R factors seem to have become more abundant in modern times, possibly because the heavy use of antibiotics in hospitals selects for bacterial strains bearing them.

Antibiotic resistance poses a serious threat to human health, and the inappropriate use of antibiotics contributes to this problem. You probably have gone to see a physician because of a sore throat, which can have either a viral or a bacterial cause. The best way to determine the causative agent is for the doctor to take a small sample from your inflamed throat, culture it, and identify any bacteria that are present. But perhaps you cannot wait another day for the results. Impatient, you ask the doctor to give you something to make you feel better. She prescribes an antibiotic, which you take. The sore throat gradually gets better, and you think that the antibiotic did the job.

But suppose the infection is viral. In that case, the antibiotic does nothing to combat the disease, which just runs its normal course. However, it may do something harmful: By killing many normal bacteria in your body, the antibiotic may select for bacteria harboring R factors. These bacteria may survive and reproduce in the presence of the antibiotic, and may soon become quite numerous. The next time you get a bacterial infection, there may be a ready supply of resistant bacteria in your body, and antibiotics may be ineffective.

Antibiotic resistance in pathogenic bacteria provides an example of evolution in action. In the years after they were first discovered in the twentieth century, antibiotics were very successful in combating diseases that had plagued humans for millennia, such as cholera, tuberculosis, and leprosy. But as time went on, resistant bacteria appeared. This was, and is, classic natural selection: Genetic variation existed among bacteria, and those that survived the onslaught of antibiotics must have had a genetic constitution that allowed them to do so.

Transposable elements move genes among plasmids and chromosomes

As we have seen, plasmids, viruses, and even phage capsids (in the case of transduction) can transport genes from one bacterial cell to another. There is another type of "gene transport" that occurs within the individual cell. It relies on segments of DNA that can be inserted either at a new location on the same chromosome or into another chromosome. These DNA sequences are called **transposable elements**. Their insertion often produces phenotypic effects by disrupting the genes into which they are inserted (Figure 13.12*a*).

The first transposable elements to be discovered in prokaryotes were large pieces of DNA, typically 1,000 to 2,000 base pairs long, found at many sites on the *E. coli* main chromosome. In one mechanism of transposition, the transposable element replicates independently of the rest of the chromosome. The copy then inserts itself at other, seemingly random sites on the chromosome. The genes encoding the enzymes necessary for this insertion are found within the transposable element itself. Other transposable elements are cut from their original sites and inserted elsewhere without replication. Later, many longer transposable elements were discovered (about 5,000 base pairs). These large elements carry one or more additional genes and are called **transposons** (Figure 13.12*b*).

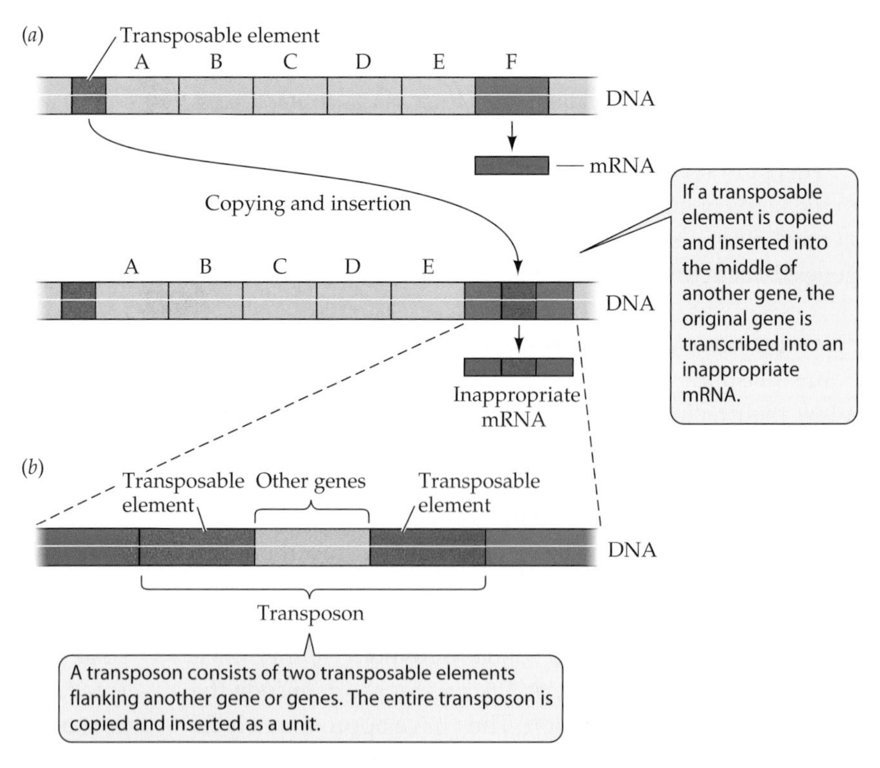
(a)
Transposable element
A B C D E F
DNA
mRNA
Copying and insertion
If a transposable element is copied and inserted into the middle of another gene, the original gene is transcribed into an inappropriate mRNA.
A B C D E
DNA
Inappropriate mRNA
(b)
Transposable element
Other genes
Transposable element
DNA
Transposon
A transposon consists of two transposable elements flanking another gene or genes. The entire transposon is copied and inserted as a unit.

13.12 Transposable Elements and Transposons (*a*) Transposable elements are segments of DNA that can be inserted at new locations, either on the same chromosome or on a different chromosome. (*b*) Transposons consist of transposable elements combined with other genes.

What do transposons and other transposable elements have to do with the genetics of prokaryotes—or with hospitals? Transposable elements have contributed to the evolution of plasmids. R factors probably originally gained their genes for antibiotic resistance through the activity of transposable elements. One piece of evidence for this conclusion is that each resistance gene in an R factor is part of a transposon.

In summary, rapid asexual reproduction can produce enormous clones of prokaryotes. However, these genetically identical cells are all equally vulnerable to some change in the environment. Recombination by means of conjugation, transformation, and transduction, or the acquisition of new genes by means of plasmids and transposable elements, all introduce genetic diversity into bacterial populations, and this diversity allows at least some cells to survive under changing conditions. Prokaryotes can also respond to changes in their environment by regulating the expression of their genes.

Regulation of Gene Expression in Prokaryotes

Prokaryotes can conserve energy and resources by making proteins only when they are needed. The protein content of a bacterium can change rapidly when conditions warrant. There are several ways in which a prokaryotic cell could shut off the supply of an unneeded protein:

- Block the transcription of mRNA for that protein
- Hydrolyze the mRNA after it is made and prior to translation
- Prevent translation of the mRNA at the ribosome
- Hydrolyze the protein after it is made
- Inhibit the function of the protein

These methods would all have to be selective, affecting some genes and proteins and not others. In addition, they would all have to respond to some biochemical signal. Clearly, the earlier the cell intervenes in the process, the less energy it has to expend. Selective inhibition of transcription is far more efficient than transcribing the gene, translating the message, and then degrading or inhibiting the protein. While examples of all five mechanisms for regulating protein levels are found in nature, prokaryotes generally use the most efficient one, transcriptional regulation.

Regulation of transcription conserves energy

As a normal inhabitant of the human intestine, *E. coli* must be able to adjust to sudden changes in its chemical environment. Its host may present it with one foodstuff one hour and another the next. This variation presents the bacterium with a metabolic challenge. Glucose is its preferred energy source, and is the easiest sugar to metabolize, but not all of its host's foods contain an abundant supply of glucose. For example, the bacterium may suddenly be deluged with milk, whose predominant sugar is lactose. Lactose is a β-galactoside—a disaccharide containing galactose β-linked to glucose (see Chapter 3). To be taken up and metabolized by *E. coli*, lactose is acted on by three proteins:

- *β-galactoside permease* is a carrier protein in the bacterial plasma membrane that moves the sugar into the cells.
- *β-galactosidase* is an enzyme that catalyzes the hydrolysis of lactose to glucose and galactose.
- A third protein, the enzyme *β-galactoside transacetylase*, is also required for lactose metabolism, although its role in the process is not yet clear.

When *E. coli* is grown on a medium that does not contain lactose or other β-galactosides, the levels of these three proteins are extremely low—the cell does not waste energy and materials making the unneeded enzymes. If, however, the environment changes such that lactose is the predominant sugar available and very little glucose is present, the bacterium promptly begins making all three enzymes, and they increase rapidly in abundance. For example, there are only two molecules of β-galactosidase present in an *E. coli* cell when glucose is present in the medium. But when glucose is absent, lactose can induce the synthesis of 3,000 molecules of β-galactosidase per cell!

If lactose is removed from *E. coli*'s environment, synthesis of the three enzymes that process it stops almost immediately. The enzyme molecules that have already formed do not disappear; they are merely diluted during subsequent cell divisions until their concentration falls to the original low level within each bacterium.

Compounds that stimulate the synthesis of an enzyme (such as lactose in our example) are called **inducers** (Figure 13.13). The enzymes that are produced are called **inducible**

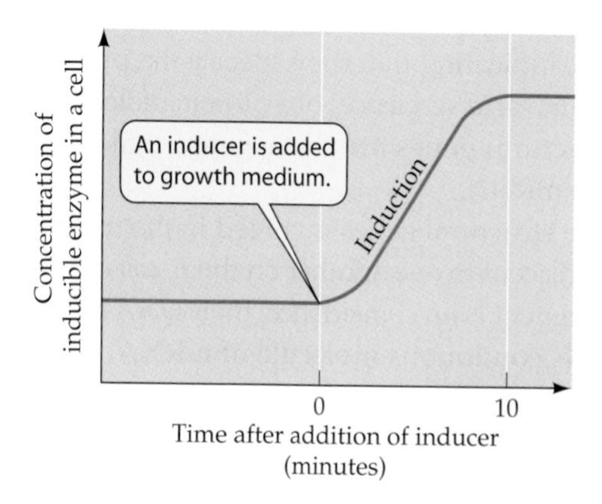

13.13 An Inducer Stimulates the Synthesis of an Enzyme It is most efficient for a cell to produce an enzyme only when it is needed. Some enzymes are induced by the presence of the substance they act upon (for example, β-galactosidase is induced by the presence of lactose).

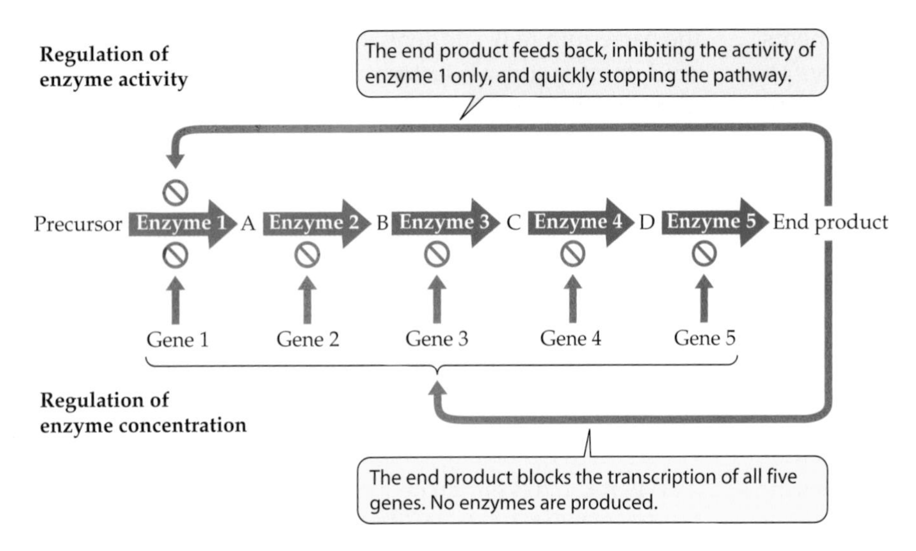

13.14 Two Ways to Regulate a Metabolic Pathway Feedback from the end product can block enzyme activity, or it can stop the transcription of genes that code for the enzymes.

enzymes, whereas enzymes that are made all the time at a constant rate are called **constitutive** enzymes.

We have now seen two basic ways of regulating the rate of a metabolic pathway. Chapter 6 described allosteric regulation of enzyme activity (the rate of enzyme-catalyzed reactions); this mechanism allows rapid fine-tuning of metabolism. Regulation of protein synthesis—that is, regulation of the concentration of enzymes—is slower, but produces a greater savings of energy. Figure 13.14 compares these two modes of regulation.

A single promoter controls the transcription of adjacent genes

The genes that serve as blueprints for the synthesis of the three enzymes that process lactose in *E. coli* are called **structural genes**, indicating that they specify the primary structure (the amino acid sequence) of a protein molecule. In other words, structural genes are those genes that can be transcribed into mRNA.

The three structural genes involved in the metabolism of lactose lie adjacent to one another on the *E. coli* chromosome. This arrangement is no coincidence: their DNA is transcribed into a single, continuous molecule of mRNA. Because this particular messenger governs the synthesis of all three lactose-metabolizing enzymes, either all or none of the enzymes are made, depending on whether their common message—their mRNA—is present in the cell.

The three genes share a single promoter. Recall from Chapter 12 that a *promoter* is a DNA sequence to which RNA polymerase binds to initiate transcription. The promoter for these three structural genes can be very effective, so the maximum rate of mRNA synthesis can be high. However, there is also a mechanism to shut down mRNA synthesis when the enzymes are not needed. That mechanism is the operon, elegantly worked out by François Jacob and Jacques Monod.

Operons are units of transcription in prokaryotes

Prokaryotes shut down transcription by placing an obstacle between the promoter and the structural genes it regulates. A short stretch of DNA called the **operator** lies in this position. It can bind very tightly to a special type of protein molecule, called a **repressor**, to create such an obstacle.

- When the repressor protein is bound to the operator, it blocks the transcription of mRNA (Figure 13.15).
- When the repressor is not attached to the operator, mRNA synthesis proceeds rapidly.

The whole unit, consisting of the closely linked structural genes and the DNA sequences that control their transcription, is called an **operon**. An operon always consists of a promoter, an operator, and two or more structural genes (see Figure 13.16). The promoter and operator are binding sites on DNA and are not transcribed.

E. coli has numerous mechanisms to control the transcription of operons; we will focus on three of them here. Two of these control mechanisms depend on interactions of a repressor protein with the operator, and the third depends on interactions of other proteins with the promoter.

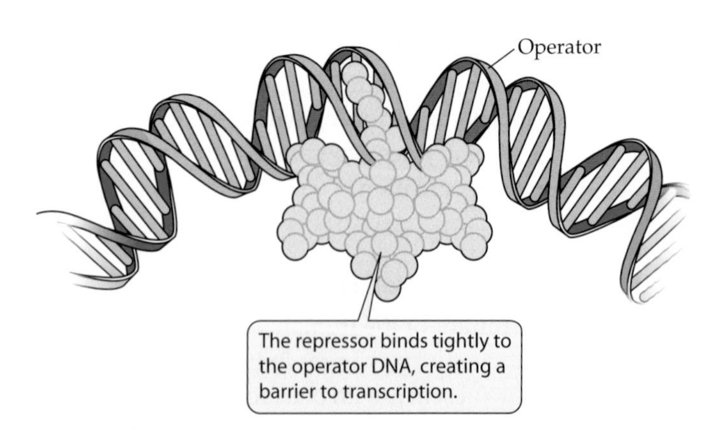

Operator
The repressor binds tightly to the operator DNA, creating a barrier to transcription.

13.15 A Repressor Blocks Transcription An untranscribed DNA sequence called the operator (purple) can control the transcription of a structural gene. When a repressor protein binds to the operator, transcription of the structural gene is blocked.

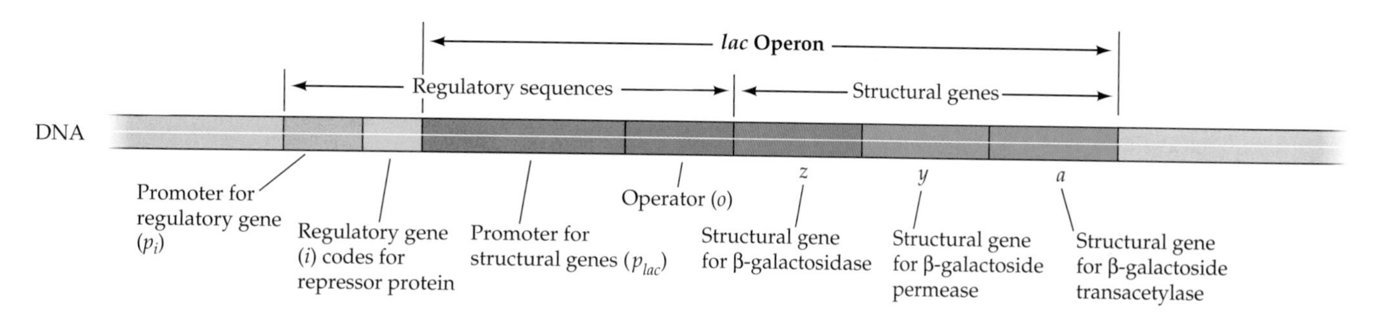
lac Operon
Regulatory sequences
Structural genes
DNA
Promoter for regulatory gene (p_i)
Regulatory gene (i) codes for repressor protein
Promoter for structural genes (p_{lac})
Operator (o)
z
y
a
Structural gene for β-galactosidase
Structural gene for β-galactoside permease
Structural gene for β-galactoside transacetylase

13.16 The *lac* Operon of *E. coli* The *lac* operon of *E. coli* is a segment of DNA that includes a promoter, an operator, and the three structural genes that code for lactose-metabolizing enzymes.

Operator–repressor control that induces transcription: The *lac* operon

The operon containing the genes for the three lactose-metabolizing proteins of *E. coli* is called the *lac operon* (Figure 13.16). As we have just seen, RNA polymerase can bind to the promoter, and a repressor protein can bind to the operator.

The repressor protein has two binding sites: one for the operator and the other for inducers. The inducers of the *lac* operon, as we know, are molecules of lactose and certain other β-galactosides. Binding to an inducer changes the shape of the repressor (by allosteric modification; see Chapter 6). This change in shape prevents the repressor from binding to the operator (Figure 13.17). As a result, RNA polymerase can bind to the promoter and start transcribing the structural genes of the *lac* operon. The mRNA transcribed from these genes is translated on ribosomes to synthesize the three proteins required for metabolizing lactose.

What happens if the concentration of lactose drops? As the lactose concentration decreases, the inducer (lactose) molecules separate from the repressor. Free of lactose molecules, the repressor returns to its original shape and binds to the operator, and transcription of the *lac* operon stops. Translation stops soon thereafter because the mRNA that is already present breaks down quickly. Thus, it is the presence or absence of lactose—the inducer—that regulates the binding of the repressor to the operator, and therefore the synthesis of the proteins needed to metabolize it.

Repressor proteins are encoded by **regulatory genes**. The regulatory gene that codes for the repressor of the *lac* operon is called the *i* (*inducibility*) *gene*. The *i* gene happens to

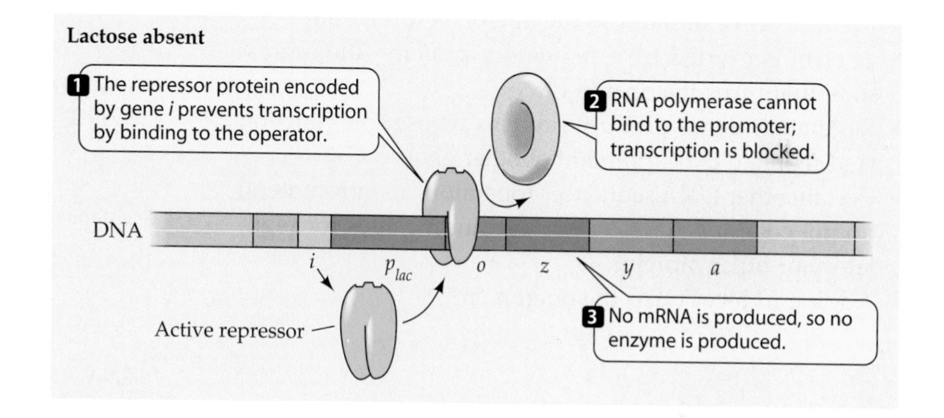
Lactose absent
1 The repressor protein encoded by gene i prevents transcription by binding to the operator.
2 RNA polymerase cannot bind to the promoter; transcription is blocked.
DNA
i
p_{lac}
o
z
y
a
Active repressor
3 No mRNA is produced, so no enzyme is produced.

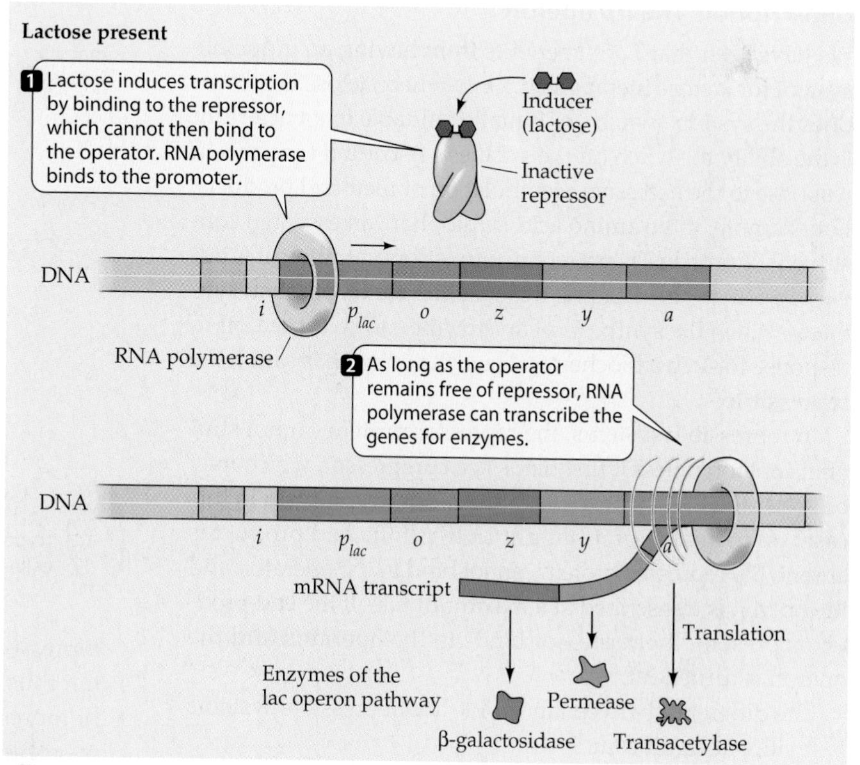
Lactose present
1 Lactose induces transcription by binding to the repressor, which cannot then bind to the operator. RNA polymerase binds to the promoter.
Inducer (lactose)
Inactive repressor
DNA
i
p_{lac}
o
z
y
a
RNA polymerase
2 As long as the operator remains free of repressor, RNA polymerase can transcribe the genes for enzymes.
DNA
mRNA transcript
Translation
Enzymes of the lac operon pathway
Permease
β-galactosidase
Transacetylase

13.17 The *lac* Operon: An Inducible System Lactose (the inducer) leads to enzyme synthesis by preventing the repressor protein (which would have stopped transcription) from binding to the operator.

lie close to the operon that it regulates, but some other regulatory genes are distant from their operons. Like all other genes, the *i* gene itself has a promoter, which can be designated p_i. Because this promoter does not bind RNA polymerase very effectively, only enough mRNA to synthesize about ten molecules of repressor protein per cell per generation is produced. This quantity of the repressor is enough to regulate the operon effectively—to produce more would be a waste of energy. There is no operator between p_i and the *i* gene. Therefore, the repressor of the *lac* operon is a constitutive protein; that is, it is made at a constant rate that is not subject to environmental control.

Let's review the important features of inducible systems such as the *lac* operon:

- In the absence of inducer, the operon is turned off.
- Control is exerted by a regulatory protein—the repressor—that turns the operon off.
- Regulatory genes produce proteins whose sole function is to regulate the expression of other genes.
- Certain other DNA sequences (operators and promoters) do not code for proteins, but are binding sites for regulatory or other proteins.
- Adding inducer turns the operon on.

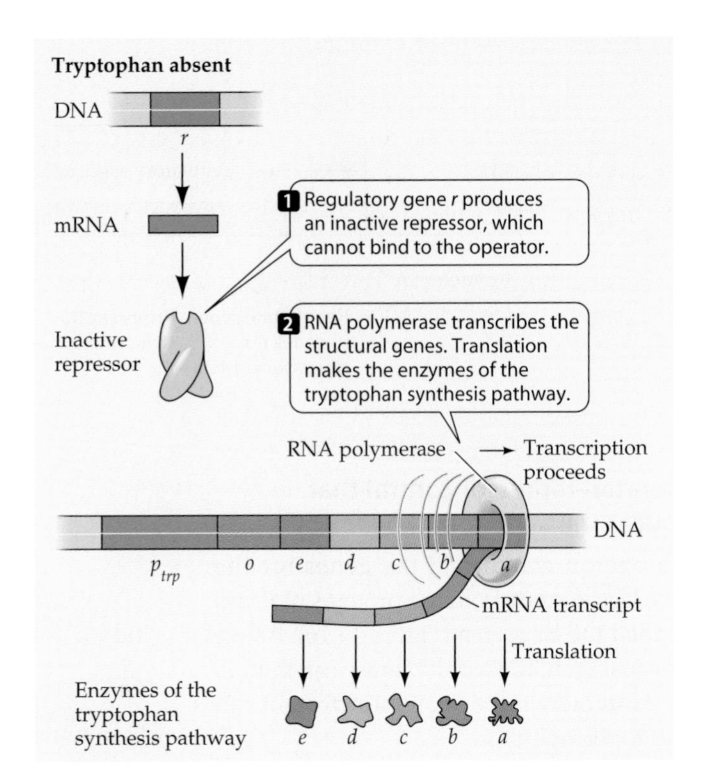

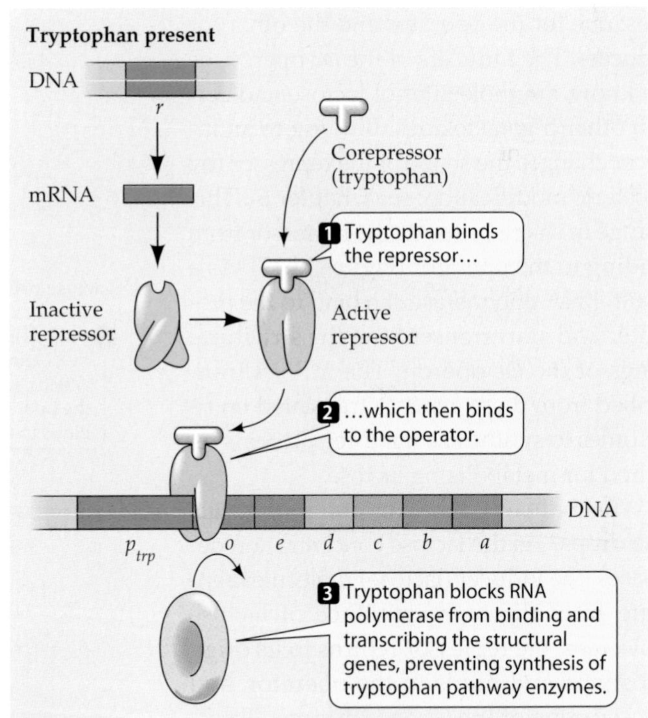

13.18 The *trp* Operon: A Repressible System Because tryptophan activates an otherwise inactive repressor, it is called a corepressor.

Operator–repressor control that represses transcription: The *trp* operon

We have seen that *E. coli* benefits from having an inducible system for lactose metabolism. Only when lactose is present does the system switch on. Equally valuable to a bacterium is the ability to switch off the synthesis of certain enzymes in response to the excessive accumulation of their end products. For example, if the amino acid tryptophan, an essential constituent of proteins, is present in ample concentration, it is advantageous to stop making the enzymes for tryptophan synthesis. When the synthesis of an enzyme can be turned off in response to such a biochemical cue, the enzyme is said to be **repressible**.

In repressible systems, the repressor protein cannot shut off its operon unless it first binds to a **corepressor**, which may be either the metabolic end product itself (tryptophan in this case) or an analog of it (Figure 13.18). If the end product is absent, the repressor protein cannot bind to the operator, and the operon is transcribed at a maximum rate. If the end product is present, the repressor binds to the operator, and the operon is turned off.

The difference between inducible and repressible systems is small, but significant:

- In inducible systems, the substrate of a metabolic pathway (the inducer) interacts with a regulatory protein (the repressor) to render it incapable of binding to the operator, thus allowing transcription.
- In repressible systems, the product of a metabolic pathway (the corepressor) interacts with a regulatory protein to make it capable of binding to the operator, thus blocking transcription.

In general, inducible systems control catabolic pathways (which are turned on only when the substrate is available), whereas repressible systems control biosynthetic pathways (which are turned off until the product becomes unavailable). In both kinds of systems, the regulatory molecule functions by binding to the operator. Next, we will consider an example of control by binding to the promoter.

Protein synthesis can be controlled by increasing promoter efficiency

Suppose an *E. coli* cell lacks a supply of glucose, its preferred energy source, but instead has access to another sugar (such as lactose) that it can break down to obtain energy. Operons encoding enzymes that catabolize such alternative energy sources, such as the *lac* operon, have a mechanism for increasing the transcription of these enzymes by increasing the efficiency of the promoter. In these operons, the promoter binds RNA polymerase in a series of steps (Figure 13.19). First, a protein called CRP (short for *c*AMP *r*eceptor *p*rotein) binds the low-molecular-weight compound adenosine 3′,5′-cyclic monophosphate, better known as cyclic AMP, or cAMP. Next, the CRP–cAMP complex binds to DNA just upstream (5′) of the promoter. This binding results in more efficient binding of RNA polymerase to the promoter, and thus an elevated level of transcription of the structural genes.

When glucose becomes abundant in the medium, the bacterium does not need to break down alternative food molecules, so synthesis of the enzymes that catabolize these molecules diminishes or ceases. The presence of glucose decreases the synthesis of the enzymes by lowering the cellular concentration of cAMP. The lower cAMP concentration leads to less CRP binding to the promoter, less efficient binding of RNA polymerase, and reduced transcription of the structural genes. This mechanism is called **catabolite repression**.

As you will see in later chapters of this book, cAMP is a widely used signaling molecule in eukaryotes, as well as in prokaryotes. The use of this nucleotide in such widely diverse situations as a bacterium sensing glucose levels and a human sensing hunger demonstrates the prevalence of common themes in biochemistry and natural selection.

The inducible *lac* and repressible *trp* systems—the two operator–repressor systems—are examples of **negative control** of transcription because the regulatory molecule (the repressor) in each case prevents transcription. The promoter–catabolite repression system is an example of **positive control** of transcription because the regulatory molecule (the CRP–cAMP complex) enhances transcription. The relationships between these positive and negative control systems are summarized in Table 13.2.

The control of gene expression by regulatory proteins is not unique to prokaryotes. As we will see in the next chapter, it also occurs in eukaryotes and even, as we are about to see, in viruses.

Control of Transcription in Viruses

The mechanisms used by used by viruses within a host cell for the regulation of gene expression are similar to those used by prokaryotes. Even a "simple" biological agent such as a virus is faced with complicated molecular decisions when its genome enters a cell. For example, the viral genome must di-

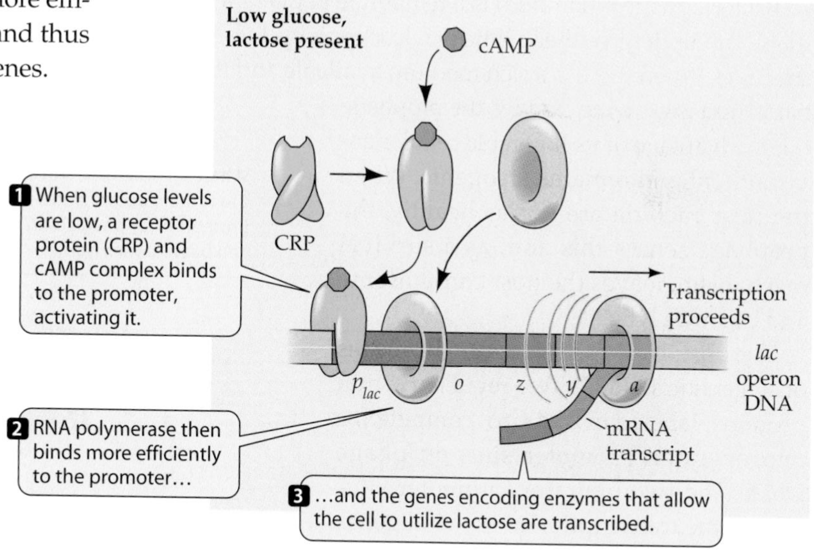

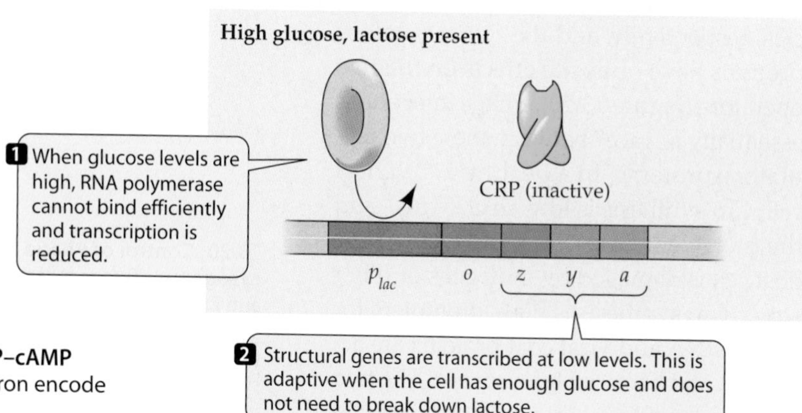

13.19 Transcription Is Enhanced by the Binding of the CRP–cAMP Complex to the Promoter The structural genes of this operon encode enzymes that break down a food source other than glucose.

13.2 Positive and Negative Controls in the *lac* Operon[a]

GLUCOSE	cAMP LEVELS	RNA POLYMERASE BINDING TO PROMOTER	LACTOSE	*LAC* REPRESSOR	TRANSCRIPTION OF *LAC* GENES?	LACTOSE USED BY CELLS?
Present	Low	Absent	Absent	Active and bound to operator	No	No
Present	Low	Present, not efficient	Present	Inactive and not bound to operator	Low level	No
Absent	High	Present, very efficient	Present	Inactive and not bound to operator	High level	Yes
Absent	High	Absent	Absent	Active and bound to operator	No	No

[a]Negative controls are in red type.

rect the shutdown of host transcription and translation, then redirect the host's protein synthesis machinery to virus production and host cell lysis. All the genes involved in this process must be activated in the right order. In temperate viruses, which can insert their genome (or a DNA copy) into the host chromosome, an additional issue arises: When should the provirus leave the host chromosome and undergo a lytic cycle?

Bacteriophage λ (lambda) is a temperate phage, meaning that it can undergo either a lytic or a lysogenic cycle (see Figure 13.2). When there is a rich medium available and its host bacterium is growing rapidly, the prophage takes advantage of its favorable cellular environment and remains lysogenic. When the host bacteria are not as healthy, the prophage senses this and, as a survival mechanism, leaves the host chromosome and becomes lytic.

The phage makes this decision by means of a "genetic switch": Two regulatory viral proteins, labeled cI and Cro, compete for two operator/promoter sites on phage DNA. The two operator/promoter sites control the transcription of the viral genes involved in the lytic and the lysogenic cycles, respectively, and the two regulatory proteins have opposite effects on the two operators (Figure 13.20). Phage infection is essentially a "race" between these two regulatory proteins. In a healthy *E. coli* host cell, Cro synthesis is low, so cI "wins" and the phage enters a lysogenic cycle. If the host cell is damaged by mutagens or other stress, Cro synthesis is high, promoters for phage DNA and viral coat proteins are activated, and bacterial lysis ensues. The two regulatory proteins are made very early in phage infection, and each has a binding site for a specific DNA sequence.

The life cycle of phage λ, which has been greatly simplified here, is a paradigm for viral infections throughout the biological world. The lessons learned from transcriptional controls in this system have been applied again and again to other viruses, including HIV. The control of gene activity in eukaryotic cells is somewhat different, as we will see in the next chapter, but nevertheless usually involves regulatory protein–DNA interactions.

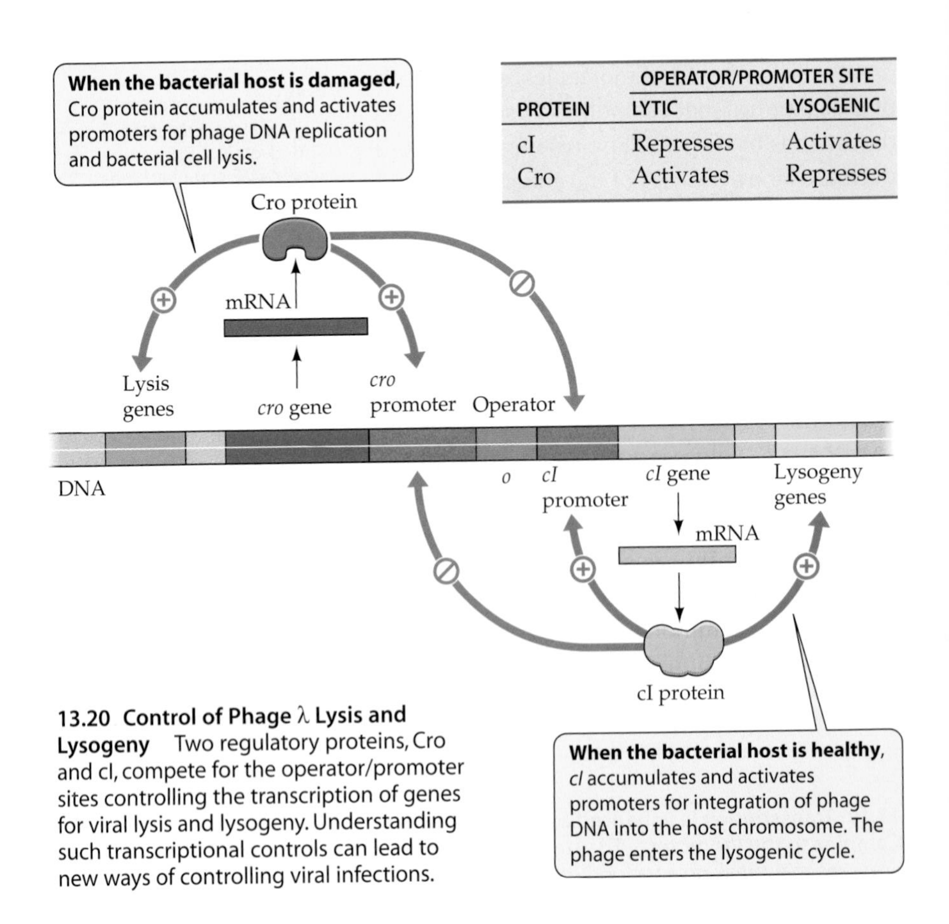
When the bacterial host is damaged, Cro protein accumulates and activates promoters for phage DNA replication and bacterial cell lysis.
OPERATOR/PROMOTER SITE
PROTEIN
LYTIC
LYSOGENIC
cI
Represses
Activates
Cro
Activates
Represses
Cro protein
mRNA
Lysis genes
cro gene
cro promoter
Operator
DNA
o
cI promoter
cI gene
Lysogeny genes
mRNA
cI protein
When the bacterial host is healthy, *cI* accumulates and activates promoters for integration of phage DNA into the host chromosome. The phage enters the lysogenic cycle.

13.20 Control of Phage λ Lysis and Lysogeny Two regulatory proteins, Cro and cI, compete for the operator/promoter sites controlling the transcription of genes for viral lysis and lysogeny. Understanding such transcriptional controls can lead to new ways of controlling viral infections.

Prokaryotic Genomes

When DNA sequencing first became possible in the late 1970s, the first biological agents to be sequenced were the simplest viruses. Soon, over 150 viral genomes, including those of important animal and plant pathogens, had been sequenced. Information on how these virus infect their hosts and reproduce came quickly as a result.

But the manual sequencing techniques used on viruses were not up to the task of elucidating the genomes of prokaryotes and eukaryotes, the smallest of which are a hundred times larger than those of a bacteriophage. In the past decade, however, the automated sequencing techniques described in Chapter 11 have rapidly added many prokaryotic sequences to biologists' store of knowledge.

In 1995, a team led by Craig Venter and Hamilton Smith determined the first sequence of a free-living organism, the bacterium *Haemophilus influenzae*. Many more prokaryotic sequences have followed. These sequences have revealed not only how prokaryotes apportion their genes to perform different cellular roles, but also how their specialized functions are carried out. A beginning has even been made on the provocative question of what the minimal requirements for a living cell might be.

Three types of information can be obtained from a genomic sequence:

- *Open reading frames*, which are the coding regions of genes. For protein-coding genes, these regions can be recognized by the start and stop codons for translation.
- *Amino acid sequences of proteins*. These sequences can be deduced from the DNA sequences of open reading frames by applying the genetic code.
- *Gene control sequences*, such as promoters and terminators for transcription.

Functional genomics relates gene sequences to functions

Functional genomics is the assignment of roles to the products of genes described by genomic sequencing. This field, less than a decade old, is now a major occupation of biologists.

The only host for the bacterium *H. influenzae* is humans. It lives in the upper respiratory tract and can cause ear infections or, more seriously, meningitis in children. Its single circular chromosome has 1,830,137 base pairs (Figure 13.21). In addition to its origin of replication and the genes coding for rRNAs and tRNAs, this bacterial chromosome has 1,743 regions containing amino acid codons as well as the transcriptional (promoter) and translational (start and stop codons) information needed for protein synthesis—that is, regions that are likely to be genes that code for proteins.

When this sequence was first announced, only 1,007 (58%) of the bacterium's genes had amino acid sequences that corresponded to proteins with known functions—in other words, only 58% were genes that the researchers, based on their knowledge of the functions of bacteria, expected to find. The remaining 42% of its genes coded for proteins that were unknown to researchers. The roles of most of the unknown proteins have been identified since that time, a process known as *annotation*.

Of the genes and proteins with known roles, most confirmed a century of biochemical description of bacterial enzymatic pathways. For example, genes for enzymes making up entire pathways of glycolysis, fermentation, and electron transport were found. Some of the remaining gene sequences for unknown proteins may code for membrane proteins, including those involved in active transport. Another important finding was that highly infective strains of *H. influenzae* have genes coding for surface proteins that attach the bacterium to the human respiratory tract, while noninfective strains lack those genes.

Soon after the sequence of *H. influenzae* was announced, smaller (*Mycoplasma genitalium*, 580,070 base pairs) and larger (*E. coli*, 4,639,221 base pairs) prokaryotic sequences were completed. Thus began a new era in biology, the era of **comparative genomics**, in which the genome sequences of different organisms are compared to see what genes one organism has or is missing, in order to relate the results to physiology.

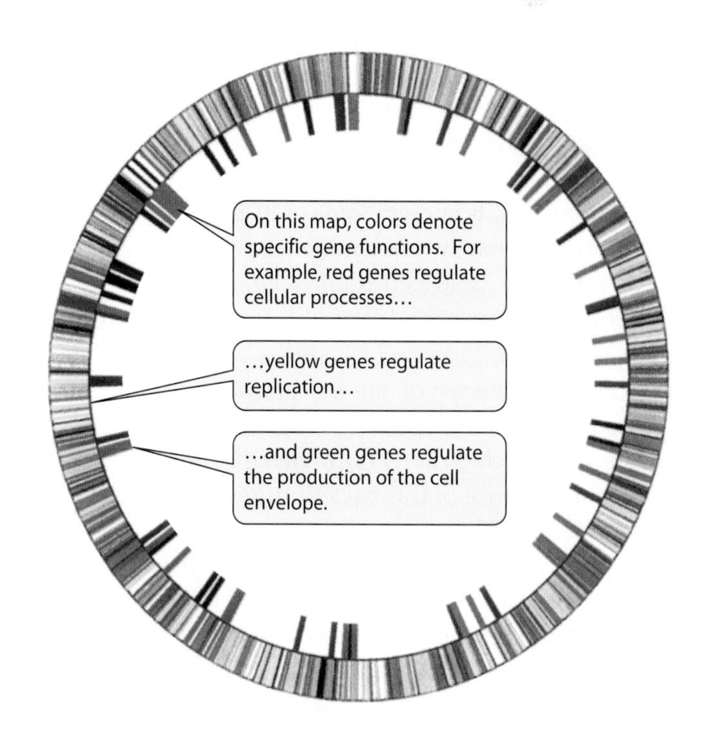

On this map, colors denote specific gene functions. For example, red genes regulate cellular processes...
...yellow genes regulate replication...
...and green genes regulate the production of the cell envelope.

13.21 Functional Organization of the Genome of *H. influenzae* The entire DNA sequence has 1,830,137 base pairs.

M. genitalium, for example, lacks the enzymes needed to synthesize amino acids, which the other two prokaryotes possess. This finding reveals that *M. genitalium* is a parasite, which must obtain all its amino acids from its environment, the human urogenital tract. *E. coli* has 55 regulatory genes coding for transcriptional activators and 58 for repressors; *M. genitalium* has only 3 genes for activators. Comparisons such as these have led to the formulation of specific questions about how an organism lives the way it does. We'll see many more applications of comparative genomics in the next chapter.

The sequencing of prokaryotic genomes has medical applications

Prokaryotic genome sequencing has important ramifications for the study of organisms that cause human diseases, as the previous section suggests. Indeed, most of the early efforts in sequencing have focused on human pathogens.

- *Chlamydia trachomatis* causes the most common sexually transmitted disease in the United States. Because it is an intracellular parasite, it has been very hard to study. Among its 900 genes are several for ATP synthesis—something scientists used to think this bacterium could not do.
- *Rickettsia prowazekii* causes typhus; it infects people bitten by louse vectors. Of its 634 genes, 6 code for proteins that are essential for its virulence. These genes are being used to develop vaccines.
- *Mycobacterium tuberculosis* causes tuberculosis. It has a large (for a prokaryote) genome, coding for 4,000 proteins. Over 250 of these proteins are used to metabolize lipids, so this may be the main way that the bacterium gets its energy. Some of its genes code for previously unidentified cell surface proteins; these genes are targets for potential vaccines.
- *Streptomyces coelicolor* and its close relatives produce two-thirds of all the antibiotics currently in clinical use, including streptomycin, tetracycline, and erythromycin. The genome sequence of this bacterium reveals that there are 22 clusters of genes responsible for antibiotic production, of which only 4 were previously known. This finding may lead to more and better antibiotics to combat resistant pathogens.
- *E. coli* strain O157:H7 in hamburger can cause severe illness when ingested, as happens to at least 70,000 people a year in the United States. Its genome has 5,416 genes, of which 1,387 are different from those in the familiar (and harmless) laboratory strains of this bacterium. Remarkably, many of these unique genes are also present in other pathogenic species, such as *Salmonella* and *Shigella*. This finding suggests that there is extensive genetic exchange between these species, and that "superbugs" are on the horizon.

What genes are required for cellular life?

When the genomes of prokaryotes and eukaryotes are compared, a striking conclusion arises: There are some universal genes that are present in all organisms. There are also some universal gene segments—coding for an ATP binding site, for example—that are present in many genes in many organisms. These findings suggest that there is some ancient, minimal set of DNA sequences that all cells must have. One way to identify these sequences is to look for them (or, more realistically, to have a computer look for them).

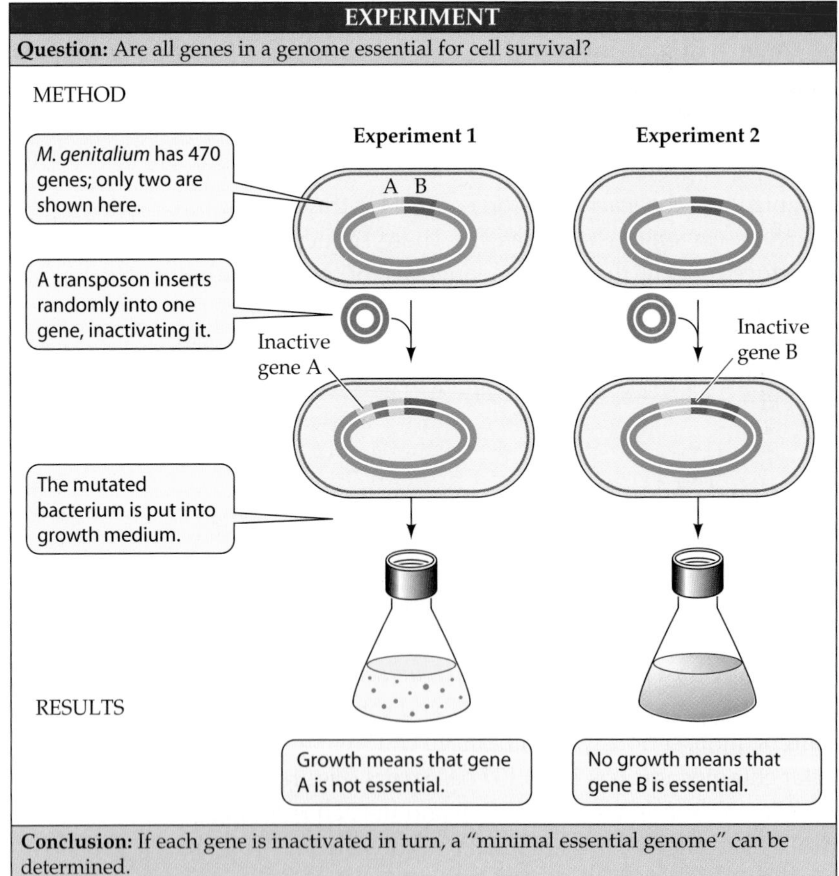

EXPERIMENT
Question: Are all genes in a genome essential for cell survival?
METHOD
Experiment 1
Experiment 2
M. genitalium has 470 genes; only two are shown here.
A B
A transposon inserts randomly into one gene, inactivating it.
Inactive gene A
Inactive gene B
The mutated bacterium is put into growth medium.
RESULTS
Growth means that gene A is not essential.
No growth means that gene B is essential.
Conclusion: If each gene is inactivated in turn, a "minimal essential genome" can be determined.

13.22 Using Transposon Mutagenesis to Determine the Minimal Genome By inactivating genes one by one, scientists can determine which genes are essential for the cell's survival.

Another way to define the minimal genome is to take the organism with the simplest genome, deliberately mutate one gene at a time, and see what happens. *Mycoplasma genitalium* has the smallest known genome—only 470 genes. Even so, some of its genes are dispensable under some circumstances. It has genes for metabolizing both glucose and fructose. In the laboratory, the organism can survive on a medium supplying only one of those sugars, making the genes for metabolizing the other sugar unnecessary. But what about other genes? Experiments using transposons as mutagens have addressed this question. When the bacterium is exposed to transposons, they insert themselves into a gene at random, mutating and inactivating it (Figure 13.22). The mutated cells are sequenced to determine which gene was mutated, and then examined for growth and survival.

The astonishing result of these studies is that *M. genitalium* can survive in the laboratory without the services of 133 of its genes, leaving a minimum genome of 337 genes! This "genomic downsizing" has also been found in other prokaryotes. The bacterium that causes leprosy, *Mycobacterium leprae*, is a cousin of *Mycobacterium tuberculosis*, mentioned above. But *M. leprae* has "discarded" 2,000 of the genes present in its cousin. For example, it lacks genes for the proteins of the electron transport chain (see Chapter 7), and is therefore slow-growing. But it retains the anabolic pathways it needs to survive when external nutrients are scarce.

Chapter Summary

Probing the Nature of Genes

▶ Prokaryotes and viruses are useful for the study of genetics and molecular biology because they contain much less DNA than eukaryotes, grow and reproduce rapidly, and are haploid.

Viruses: Reproduction and Recombination

▶ Viruses were discovered as disease-causing agents small enough to pass through a filter that retains bacteria. The basic viral unit, called a virion, consists of a nucleic acid genome, which codes for a few proteins, and a protein coat called a capsid.

▶ Viruses are obligate intracellular parasites: they need the biochemical machinery of a living cell in order to reproduce.

▶ There are many types of viruses, classified by their size and shape, by their genetic material (RNA or DNA), or by their host organism. **Review Figure 13.1**

▶ Bacteriophage are viruses that infect bacteria. In the lytic cycle, the host cell bursts, releasing new phage particles. Some phage can also undergo a lysogenic cycle, in which their DNA is inserted into the host chromosome, where it replicates for generations. When conditions are appropriate, the phage DNA exits the host chromosome and enters a lytic cycle. **Review Figure 13.2**

▶ Some viruses have promoters for host RNA polymerase, which they use to transcribe their own genes. **Review Figure 13.3**

▶ Most of the many types of RNA and DNA viruses that infect animals cause diseases. Some animal viruses have an envelope derived from the host's plasma membrane.

▶ Retroviruses, such as HIV, have RNA genomes that they reproduce through a complementary DNA intermediate. Other RNA viruses use their RNA to make mRNA to code for enzymes and replicate their genomes without using DNA. **Review Figures 13.4, 13.5**

▶ Many viruses are spread by vectors, such as insects.

Prokaryotes: Reproduction and Recombination

▶ When bacteria divide, they form clones of identical cells that can be observed as colonies when grown on solid media. **Review Figure 13.6**

▶ A bacterium can transfer its genes to another bacterium by conjugation, transformation, or transduction.

▶ In conjugation, a bacterium attaches to another bacterium and passes a fragment of its DNA to the recipient cell. **Review Figures 13.7, 13.8, 13.9**

▶ In transformation, fragments of bacterial DNA are taken up by a cell from the environment. These genetic fragments may recombine with the host chromosome, thereby permanently adding new genes. **Review Figure 13.10*a***

▶ In transduction, phage capsids carry bacterial DNA from one bacterium to another. **Review Figure 13.10*b***

▶ Plasmids are small bacterial chromosomes that are independent of the main chromosome. R factors, which are plasmids that carry genes for antibiotic resistance, are a serious public health threat. **Review Figure 13.11**

▶ Transposable elements are stretches of DNA that can move from one place to another on the bacterial chromosome—either by actually moving or by making a new copy, which is inserted at a new location. **Review Figure 13.12**

Regulation of Gene Expression in Prokaryotes

▶ In prokaryotes, the synthesis of some proteins is regulated so that they are made only when they are needed.

▶ Constitutive enzymes whose products are essential to the cell at all times, are synthesized constantly. A compound that stimulates the synthesis of an enzyme needed to process it is called an inducer, and the enzyme is called an inducible enzyme. **Review Figures 13.13, 13.14**

▶ An operon consists of a promoter, an operator, and two or more structural genes. Promoters and operators do not code for proteins, but serve as binding sites for regulatory proteins. When a repressor protein binds to the operator, transcription of the structural genes is inhibited. **Review Figures 13.15, 13.16**

▶ The mechanisms that regulate the expression of prokaryotic genes include inducible operator–repressor systems, repressible operator–repressor systems, and systems that increase the efficiency of a promoter. **Review Table 13.2**

▶ The *lac* operon is an example of an inducible system. When lactose is absent, a repressor protein binds tightly to the operator. The repressor prevents RNA polymerase from binding to the promoter, turning transcription off. Lactose acts as an inducer by binding to the repressor. This binding changes the repressor's shape so that it can no longer bind to the operator. With the operator unbound, RNA polymerase binds to the promoter, and transcription is turned on. **Review Figure 13.17. See Web/CD Tutorial 13.1**

▶ Repressor proteins are coded by constitutive regulatory genes.

▶ The *trp* operon is an example of a repressible system. The presence of tryptophan, the end product of a metabolic pathway, represses synthesis of the enzymes involved in that pathway. Tryptophan acts as a corepressor by binding to an inactive repressor protein and making it active. When the activated repressor binds to the operator, transcription is turned off. **Review Figure 13.18. See Web/CD Tutorial 13.2**

▶ The efficiency of a promoter can be increased by regulation of the level of cAMP, which binds to a protein called CRP. The CRP–cAMP complex then binds to a site near the promoter, enhancing the effectiveness of RNA polymerase binding and hence transcription. **Review Figure 13.19**

Control of Transcription in Viruses

▶ In bacteriophage that can undergo a lytic or a lysogenic cycle, the decision as to which pathway to take is made by operator–regulatory protein interactions. **Review Figure 13.20**

Prokaryotic Genomes

▶ Functional genomics relates gene sequences to protein functions. **Review Figure 13.21**

▶ By mutating individual genes in a small genome, scientists can determine the minimal genome required for cellular life. **Review Figure 13.22**

See Web/CD Activity 13.1 for a concept review of this chapter.

Self-Quiz

1. Which of the following is *not* true with regard to the *lac* operon?
   a. When lactose binds to the repressor, the latter can no longer bind to the operator.
   b. When lactose binds to the operator, transcription is stimulated.
   c. When the repressor binds to the operator, transcription is inhibited.
   d. When lactose binds to the repressor, the shape of the repressor is changed.
   e. When the repressor is mutated, one possibility is that it does not bind to the operator.
2. Which of the following is *not* a type of virus reproduction?
   a. DNA virus in a lytic cycle
   b. DNA virus in a lysogenic cycle
   c. RNA virus by a double stranded RNA intermediate
   d. RNA virus by reverse transcription to make cDNA
   e. RNA virus by acting as tRNA
3. In the lysogenic cycle of a bacteriophage,
   a. a repressor, cI, blocks the lytic cycle.
   b. a bacteriophage carries DNA between bacterial cells.
   c. both early and late phage genes are transcribed.
   d. the viral genome is made into RNA which stays in the host cell.
   e. many new viruses are made immediately, regardless of host health.
4. An operon is
   a. a molecule that can turn genes on and off.
   b. an inducer bound to a repressor.
   c. regulatory sequences controlling protein-coding genes.
   d. any long sequence of DNA.
   e. a group of linked genes.
5. Which statement about both transformation and transduction is *true*?
   a. DNA is transferred between viruses and bacteria.
   b. Neither occurs in nature.
   c. Small fragments of DNA move from one cell to another.
   d. Recombination between the incoming DNA and host cell DNA does not occur.
   e. A conjugation tube is used to transfer DNA between cells.
6. Plasmids
   a. are circular protein molecules.
   b. are required by bacteria.
   c. are tiny bacteria.
   d. may confer resistance to antibiotics.
   e. are a form of transposable element.
7. The minimal genome can be estimated for a prokaryote
   a. by counting the total number of genes.
   b. by comparative genomics.
   c. as about 5,000 genes.
   d. by transposon mutagenesis, one gene at a time.
   e. does not include any genes coding for tRNA.
8. When tryptophan accumulates in a bacterial cell,
   a. it binds to the operator, preventing transcription of adjacent genes.
   b. it binds to the promoter, allowing transcription of adjacent genes.
   c. it binds to the repressor, causing it to bind to the operator.
   d. it binds to the genes that code for enzymes.
   e. it binds to RNA and initiates a negative feedback loop to reduce transcription.
9. The promoter in the *lac* operon is
   a. the region that binds the repressor.
   b. the region that binds RNA polymerase.
   c. the gene that codes for the repressor.
   d. a structural gene.
   e. an operon.
10. The CRP–cAMP system
    a. produces many catabolites.
    b. requires ribosomes.
    c. operates by an operator–repressor mechanism.
    d. is an example of positive control of transcription.
    e. relies on operators.

For Discussion

1. Viruses sometimes carry DNA from one cell to another by transduction. Sometimes a segment of bacterial DNA is incorporated into a phage protein coat without any phage DNA. These particles can infect a new host. Would the new host become lysogenic if the phage originally came from a lysogenic host? Why or why not?
2. Compare the life cycles of the viruses that cause influenza and AIDS (Figures 13.4 and 13.5) with respect to:
   - How the virus enters the cell
   - How the virion is released in the cell
   - How the viral genome is replicated
   - How new viruses are produced
3. Compare promoters adjacent to "early" and "late" genes in the bacteriophage lytic cycle.
4. In the lactose (*lac*) operon of *E. coli*, repressor molecules are encoded by the regulatory gene. The repressor molecules are made in very small quantities and at a constant rate per cell. Would you surmise that the promoter for these repressor molecules is efficient or inefficient? Is synthesis of the repressor constitutive, or is it under environmental control?
5. A key characteristic of a repressible enzyme system is that the repressor molecule must react with a corepressor (typically, the end product of a pathway) before it can combine with the operator of an operon to shut the operon off. How is this different from an inducible enzyme?

14 The Eukaryotic Genome and Its Expression

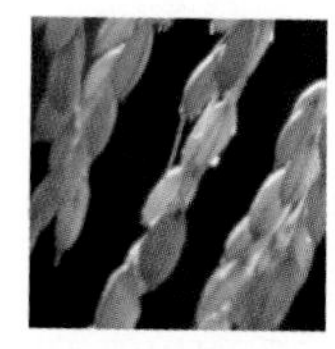

"The most precious things are not jade or pearls, but the five grains." This ancient Chinese saying refers to rice, wheat, maize (corn), sorghum, and millet. Today the saying remains as true as ever, since these crops provide two-thirds of the human diet worldwide. With the recent publication of the genome sequences of the two major cultivated varieties of rice, agricultural scientists are well on the way to dramatic improvements in the nutritional quality and yield of the grain produced in the paddies of Asia. And the rice genome turns out to be a smaller version of the much larger genomes of the other four grains.

Like other eukaryotes, rice has much more DNA than a typical prokaryote—some 430 million base pairs. But unlike the densely packed prokaryotic genome, the rice genome contains many stretches of DNA that do not code for proteins or RNA. Some of these sequences are "spacers," which is another way of saying that they either have no function or that no function has yet been found. Others are repetitive sequences, such as the telomeric DNA at the ends of chromosomes (see Figure 11.18).

In addition to the genes for metabolism that they share with prokaryotes, eukaryotes have genes that mark them as complex organisms: genes for addressing, or targeting, proteins to organelles, and genes for cell–cell interaction and cell differentiation. The transcription and later processing of mRNA is more complicated in eukaryotes than in prokaryotes. Elegant molecular machinery allows the precise regulation of gene expression needed for all the cells of these complex organisms to develop and function.

"The Most Precious Things" The genome of rice (*Oryza sativa*), which directly supplies a third of the overall diet of humanity, was recently sequenced.

The Eukaryotic Genome

As biologists began to unravel the intricacies of gene structure and expression in prokaryotes, they tried to generalize their findings by stating, "What's true for *E. coli* is also true for elephants." Although much of prokaryotic biochemistry does apply to eukaryotes as well, the old saying has its limitations. Table 14.1 lists some of the differences between prokaryotic and eukaryotic genomes.

The eukaryotic genome is larger and more complex than the prokaryotic genome

Comparisons of prokaryotic and eukaryotic genomes reveal several features.

14.1 A Comparison of Prokaryotic and Eukaryotic Genes and Genomes

CHARACTERISTIC	PROKARYOTES	EUKARYOTES
Genome size (base pairs)	10^4–10^7	10^8–10^{11}
Repeated sequences	Few	Many
Noncoding DNA within coding sequences	Rare	Common
Transcription and translation separated in cell	No	Yes
DNA segregated within a nucleus	No	Yes
DNA bound to proteins	Some	Extensive
Promoters	Yes	Yes
Enhancers/silencers	Rare	Common
Capping and tailing of mRNA	No	Yes
RNA splicing required (spliceosomes)	Rare	Common
Number of chromosomes in genome	One	Many

- *Eukaryotic genomes are larger.* The genomes of eukaryotes (in terms of haploid DNA content) are larger than those of prokaryotes. This difference is not surprising, given that in multicellular organisms there are many cell types, many jobs to do, and many proteins—all encoded by DNA—needed to do those jobs. A typical virus contains enough DNA to code for only a few proteins—about 10,000 base pairs (bp). The most thoroughly studied prokaryote, *E. coli*, has sufficient DNA (about 4.5 million bp) to make several thousand different proteins and regulate their synthesis. Humans have considerably more genes and regulators: Nearly 6 billion bp (2 meters of DNA) are crammed into each diploid human cell. However, the idea of a more complex organism needing more DNA seems to break down with some plants. For example, the lily (which produces beautiful flowers each spring, but produces fewer proteins than a human does) has 18 times more DNA than a human.
- *Eukaryotic genomes have more regulatory sequences.* Eukaryotic genomes have many more regulatory sequences—and many more regulatory proteins that bind to them—than prokaryotic genomes do. The great complexity of eukaryotes requires a great deal of regulation, and this fact is evident in the many processes and points of control associated with the expression of the eukaryotic genome.
- *Much of eukaryotic DNA is noncoding.* Interspersed throughout the eukaryotic genome are various kinds of repeated DNA sequences that are not transcribed into proteins. Even the coding regions of genes contain sequences that do not appear in the mRNA that is translated at the ribosome.
- *Eukaryotes have multiple chromosomes.* The genomic encyclopedia of a eukaryote is separated into multiple volumes. This separation requires that each chromosome have, at a minimum, three defining DNA sequences that we have described in previous chapters: an origin of replication recognized by the DNA replication machinery; a centromere region that holds the replicated chromosomes together before mitosis; and a telomeric sequence at each end of the chromosome.
- *In eukaryotes, transcription and translation are physically separated.* The nuclear envelope separates DNA and its transcription (inside the nucleus) from the sites where mRNA is translated into protein (in the cytoplasm). This separation allows for many points of regulation before translation begins: in the synthesis of a pre-mRNA transcript, in its processing into mature mRNA, and in its transport to the cytoplasm for translation (Figure 14.1).

The yeast genome adds some eukaryotic functions to a prokaryotic model

In comparison with *E. coli*, whose genome has about 4,500,000 bp on a single chromosome (one circular DNA molecule), the genome of budding yeast (*Saccharomyces cerevisiae*), a single-celled eukaryote, has 16 linear chromosomes and a haploid content of more than 12,068,000 bp. More than 600 scientists around the world collaborated in mapping and sequencing the yeast genome. When they began, they knew of about 1,000 yeast genes coding for RNAs or proteins. The final sequence revealed 5,900 genes, and sequence analyses have assigned probable roles to about 70 percent of them. Some of these genes are homologous to genes found in prokaryotes, but many are not. The functions of the other 30 percent are being investigated by gene inactivation studies similar to those performed on prokaryotes (see Figure 13.22). This process of discovering the protein product and function of a known gene sequence is called **annotation**. These accomplishments have made yeast an important model for eu-

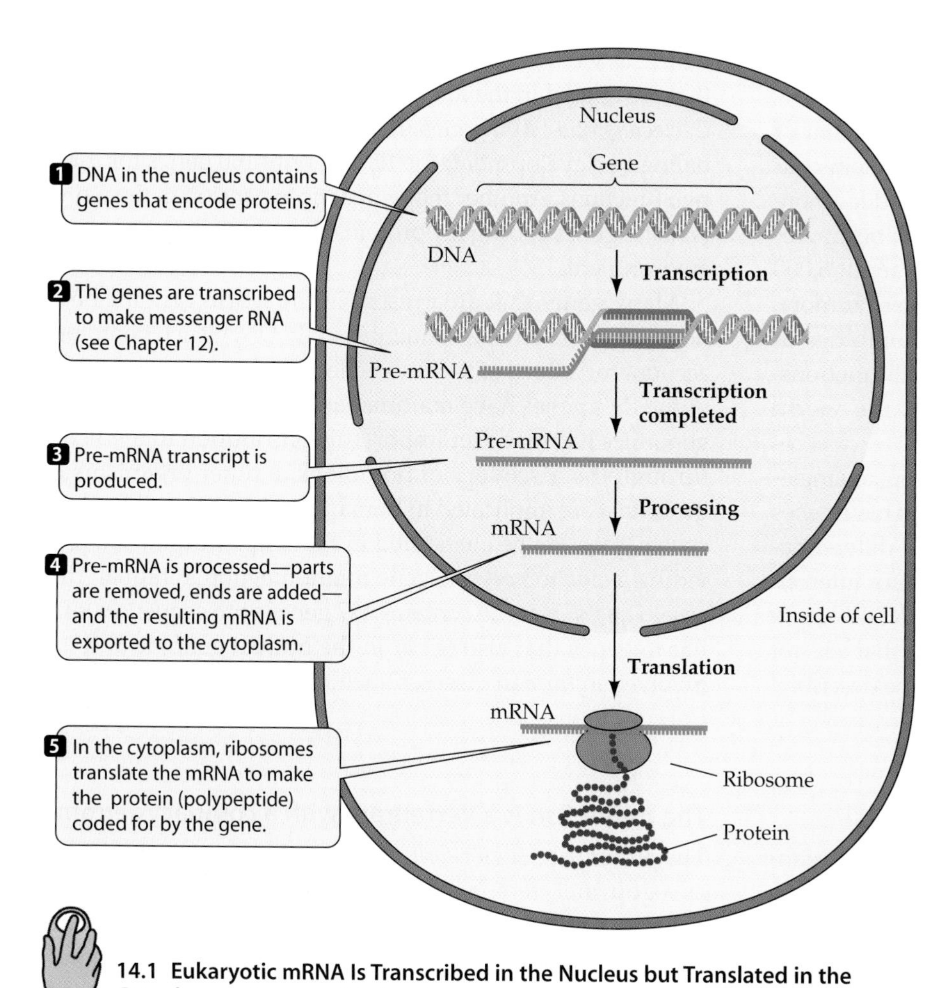

14.1 Eukaryotic mRNA Is Transcribed in the Nucleus but Translated in the Cytoplasm Compare this "road map" to the prokaryotic one shown in Figure 12.3.

karyotic cells, as observations and hypotheses from studies on yeast can be applied to and tested on other eukaryotes.

It is now possible to estimate the proportions of the yeast genome that code for specific metabolic functions. Apparently, 11 percent of yeast proteins function in general metabolism, 3 percent in energy production and storage, 3 percent in DNA replication and repair, 12 percent in protein synthesis, and 6 percent in targeting ("addressing") proteins to organelles and for secretion outside the cell. Many of the other two-thirds of the proteins are involved in cell structure, cell division, and the regulation of gene expression.

The most striking difference between the yeast genome and that of *E. coli* is in the genes for protein targeting (Table 14.2). Both of these single-celled organisms appear to use about the same numbers of genes to perform the basic functions of cell survival. It is the compartmentalization of the eukaryotic yeast cell into organelles that requires it to have so many more genes. This finding is direct, quantitative confirmation of something we have known for a century: The eukaryotic cell is structurally more complex than the prokaryotic cell.

Genes encoding several other types of proteins are present in the yeast and other eukaryotic genomes, but have no homologs in prokaryotes:

- Genes encoding histones that package DNA into nucleosomes
- Genes encoding cytoskeletal and motor proteins such as actin and tubulin
- Genes encoding cyclin-dependent kinases that control cell division
- Genes encoding proteins involved in the processing of RNA

The nematode genome adds developmental complexity

The presence of more than a single cell adds a new level of complexity to the eukaryotic genome. *Caenorhabditis elegans* is a 1-mm-long nematode (roundworm) that normally lives in the soil. But it also lives in the laboratory, where it is a favorite study organism of developmental biologists (see Chapter 19). In fact, the 2002 Nobel prize in physiology and medicine was awarded to researchers who used this worm to study development and the control of cell division. The worm has a transparent body, which scientists can watch over 3 days as a fertilized egg divides and forms an adult worm of nearly 1,000 cells. In spite of its small number of cells, the worm has a nervous system, digests food, reproduces sexually, and ages.

14.2 Comparison of the Genomes of *E. coli* and Yeast

	E. COLI	YEAST
Genome length (base pairs)	4,640,000	12,068,000
Number of proteins	4,300	6,200
Proteins with roles in:		
Metabolism	650	650
Energy production/storage	240	175
Membrane transporters	280	250
DNA replication/repair/recombination	120	175
Transcription	230	400
Translation	180	350
Protein targeting/secretion	35	430
Cell structure	180	250

So it is not surprising that an intense effort was made to sequence the genome of this organism.

The *C. elegans* genome is eight times larger than that of yeast (97 million bp) and has four times as many protein-coding genes (19,099). Once again, sequencing revealed far more genes than expected: When the sequencing effort began, researchers estimated that the worm would have about 6,000 genes and about that many proteins. Clearly, it has far more. About 3,000 genes in the worm have direct homologs in yeast; these genes code for basic eukaryotic cell functions. What do the rest of the genes—the bulk of the worm genome—do?

In addition to surviving, growing, and dividing, as single-celled organisms do, multicellular organisms must have genes for holding cells together to form tissues, for cell differentiation to divide up tasks among those tissues, and for intercellular communication to coordinate their activities (Table 14.3). Many of the genes so far identified in *C. elegans* that are not present in yeast perform these roles, which will be described in the remainder of this chapter and the next one.

The fruit fly genome has surprisingly few genes

The fruit fly *Drosophila melanogaster* is a much larger organism than *C. elegans*, both in size (the fly has 10 times more cells) and complexity. Not surprisingly, the fly's genome is also larger (about 180,000,000 bp). New computerized sequencing technologies made it possible to sequence the entire *Drosophila* genome in about a year.

Even before the complete sequence was announced, decades of genetic studies had identified some 2,500 different genes in the fly. These genes were all found in the complete DNA sequence, along with many other genes whose functions are as yet unidentified. But the big surprise of the *Drosophila* genome sequence was the total number of protein-coding regions. Instead of having more genes than the roundworm, the fly has fewer: only 13,600 genes. One reason for this is that the roundworm has some large gene families, which, as we will see later in this chapter, are groups of genes that are related in their sequence and function. For example, *C. elegans* has 1,100 genes involved in either nerve cell signaling or development; the fly has only 160 genes for these two functions. Another major genetic expansion in the worm is in the genes coding for proteins that sense chemicals in its environment.

Many genes that are present in the worm genome have homologs with similar sequences in fly DNA; such homologs account for a third of the fly's genes. Furthermore, about half of the fly's genes have mammalian homologs. Comparative genomics has made an important contribution to medicine through the discovery of homologs in other organisms of genes that are implicated in human diseases. Often the role of such a gene can be elucidated in the simpler organism, providing a clue to how the gene might function in human disease. The fly genome contains 177 genes with sequences similar to genes that also occur in the human genome and are involved in human diseases, including cancer and neurological conditions.

14.3 *C. elegans* Genes Essential to Multicellularity

FUNCTION	PROTEIN/DOMAIN	NUMBER OF GENES
Transcription control	Zinc finger; homeobox	540
RNA processing	RNA binding domains	100
Nerve impulse transmission	Gated ion channels	80
Tissue formation	Collagens	170
Cell interactions	Extracellular domains; glycotransferases	330
Cell–cell signaling	G protein-linked receptors; protein kinases; protein phosphatases	1,290

The puffer fish is a vertebrate with a compact genome

The puffer fish, *Fugu rubripes*, is prized in the culinary world as a gourmet item from Japan that must be carefully prepared, as it contains a lethal poison called tetrodotoxin that inhibits membrane channels in nerve cells. In the biological world, it is prized for its genome, which is the most compact known among vertebrates. It has 365 million base pairs and about 30,000 genes. The human genome has one-third fewer genes in eight times the amount of DNA.

A comparison of the human and puffer fish genomes showed that many genes in the two organisms are similar, so that, as lead scientist Sydney Brenner (who also led the study of the *C. elegans* genome) put it, "the *Fugu* genome is the '*Reader's Digest* version of the Book of Man.'" A major difference between the two genomes is in repetitive DNA sequences, which make up 40 percent of the human genome but a much smaller proportion in the puffer fish. The significance of this finding is unknown. Of course, humans are obviously much more complex than fish; how we accomplish this with a set of genes that is even smaller in number than the fish's is not known, but certainly points up the fact that it is not genes alone that determine the complexity of an organism.

The rice genome reflects that of a model plant, *Arabidopsis*

About 250,000 species of flowering plants dominate land and fresh water. But in the history of life, the flowering plants are fairly young, having evolved only about 200 million years ago. Given the pace of DNA mutation and other genetic changes, the differences among these plants are likely to be relatively small—at the level of regulation and protein synthesis, rather than in the genes. So, although it is the genomes of the plants used by people as food and fiber that hold the greatest interest for us, it is not surprising that instead of sequencing the huge genomes of wheat (16 billion bp) or corn (3 billion bp), scientists first chose to sequence a simpler flowering plant.

Arabidopsis thaliana, the thale cress, is a member of the mustard family and has long been a favorite model organism for study by plant biologists. It is small (hundreds could grow and reproduce in the space occupied by this page), is easy to manipulate, has only 10 percent repetitive DNA, and has a small (119 million bp) genome. Its DNA sequence reveals about 26,000 protein-coding genes, but remarkably, many of these are duplicates of other genes and have probably originated by chromosomal rearrangements. When these duplicate genes are subtracted from the total, about 15,000 unique genes are left, a number not too dissimilar from the fruit fly and roundworm. Indeed, many of the genes found in these invertebrate animals have homologs in the plant, suggesting that plants and animals have a common ancestor.

But *Arabidopsis* has some genes that distinguish it as a plant (Table 14.4). These genes include those involved in photosynthesis, in the transport of water into the root and throughout the plant, in the assembly of the cell wall, in the uptake and metabolism of inorganic substances from the environment , and in the synthesis of specific molecules used for defense against plant predators.

Justifying its position as a model plant, these "plant" genes in *Arabidopsis* were also found in the genome of rice, the first major crop plant whose sequence has been determined. Rice (*Oryza sativa*) is the world's most important crop; it is the staple diet for 3 billion people, many of them very poor. Actually, two *O. sativa* sequences of have been deciphered: that of *O. sativa indica*, the rice subspecies grown in China and most of tropical Asia, and that of the subspecies *japonica*, which is grown in Japan and other temperate climates (such as the United States). Both genomes are about the same size (430 million bp), yet in this much larger genome is a set of genes remarkably similar to that of *Arabidopsis* (Table 14.5). And many of the genes in rice are also present in the much larger genomes of corn and wheat.

Of course, rice as a whole, and each subspecies, has its own particular set of genes that make it unique. The *indica* subspecies is estimated to have 46,000–55,000 such genes, and *japonica* 32,000–50,000, both numbers higher than *Arabidopsis*. These "extra" genes include genes for characters that are specific to rice, such as a physiology that allows rice to grow for part of the season submerged in water; the nutrient-packed seeds that sustain human lives; and resistance to certain plant diseases, such as viruses and fungi. Analyses of these and other rice genes will no doubt lead to significant improvements in this crop, and to the improvement of the other grain crops as well.

14.4 *Arabidopsis* Genes Unique to Plants

FUNCTION	NUMBER OF GENES
Cell wall and growth	420
Water channels	300
Photosynthesis	139
Defense and metabolism	94

Repetitive Sequences in the Eukaryotic Genome

As you have seen in the genome sequences we have examined so far, the eukaryotic genome contains some base sequences that are repeated many times. Some of these sequences are present in millions of copies in a single genome. In this section, we will examine the organization and possible roles of these repetitive sequences.

Highly repetitive sequences are present in large numbers of copies

Three types of *highly repetitive sequences* are found in eukaryotes:

- *Satellites* are sequences 5–50 base pairs long, repeated side by side up to a million times. Satellites are usually present at the centromeres of chromosomes. They appear to be important in binding the special proteins that make up the centromere.

14.5 Comparison of the Rice and *Arabidopsis* Genomes

FUNCTION	PERCENTAGE OF GENOME	
	RICE	*ARABIDOPSIS*
Cell structure	9	10
Enzymes	21	20
Ligand binding	10	10
DNA binding	10	10
Signal transduction	3	3
Membrane transport	5	5
Cell growth and maintenance	24	22
Other functions	18	20

- *Minisatellites* are 12–100 base pairs long and are repeated several thousand times. Because DNA polymerase tends to make errors in copying these sequences, the number of copies present varies among individuals. For example, one person might have 300 minisatellites and another, 500. This variation provides a set of molecular genetic markers that can be used to identify an individual.
- *Microsatellites* are very short (1–5 bp) sequences, present in small clusters of 10–50 copies. They are scattered all over the genome.

These highly repetitive sequences are not transcribed into RNA. While laboratory scientists have made use of these sequences in genetic studies, their roles in eukaryotes are not clear.

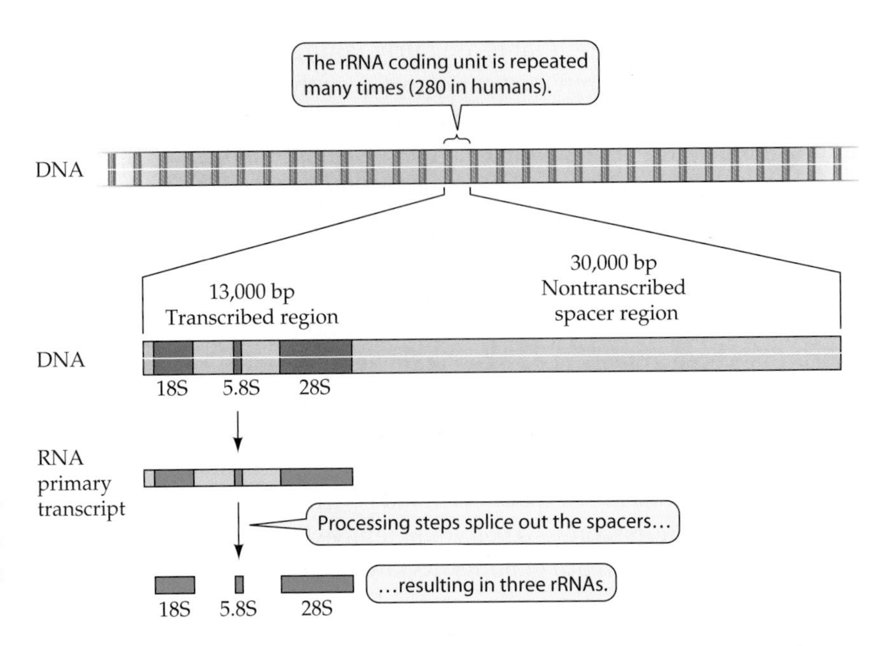

14.2 A Moderately Repetitive Sequence Codes for rRNA This rRNA gene, along with its nontranscribed spacer region, is repeated 280 times in the human genome, with clusters on five chromosomes. Once this gene has been transcribed, posttranscriptional processing removes the spacers within the transcribed region and separates the primary transcript into the three final rRNA products.

Some moderately repetitive sequences are transcribed

In Chapter 11, we described one kind of *moderately repetitive sequence* found at the ends of chromosomes: the telomeres that maintain the length and integrity of the chromosome as it replicates. These sequences are not transcribed into RNA. In contrast, some moderately repetitive DNA sequences are transcribed. These sequences code for tRNAs and rRNAs, which are used in protein synthesis (see Chapter 12).

The cell transcribes tRNAs and rRNAs constantly, but even at the maximum rate of transcription, single copies of the DNA sequences coding for them would be inadequate to supply the large amounts of these molecules needed by most cells; hence, the genome has multiple copies of these sequences. Since these moderately repetitive sequences are transcribed into RNA, they are properly termed "genes," and we can speak of rRNA genes and tRNA genes.

In mammals, there are four different rRNA molecules that make up the ribosome: the 18S, 5.8S, 28S, and 5S rRNAs. (The "S" term describes how a substance behaves in a centrifuge and, in general, is related to the size of a molecule.) The 18S, 5.8S, and 28S rRNAs are transcribed from a repeated DNA sequence as a single precursor RNA molecule, which is twice the size of the three ultimate products (Figure 14.2). Several posttranscriptional steps cut this precursor into the final three rRNA products and discard the nonuseful, or "spacer," RNA. The sequence encoding these RNAs is moderately repetitive in humans: A total of 280 copies of the sequence are located in clusters on five different chromosomes.

These moderately repetitive sequences remain fixed in their locations on the genome. Another class of moderately repetitive sequences, however, can change their location, moving about the genome.

Transposons move about the genome

Most of the remaining scattered moderately repetitive DNA sequences are not stably integrated into the genome. Instead, these sequences can move from place to place in the genome. Such sequences are called **transposons**. They make up about 45 percent of the human genome, far more than the 3–10 percent found in the other sequenced eukaryotes.

There are four main types of transposons in eukaryotes:

- *SINEs* (*s*hort *in*terspersed *e*lements) are up to 500 bp long and are transcribed, but not translated.
- *LINEs* (*l*ong *in*terspersed *e*lements) are up to 7,000 bp long, and some are transcribed and translated into proteins. They constitute about 15 percent of the human genome.

Both of these elements are present in more than 100,000 copies. They move about the genome in a distinctive way: They make an RNA copy of themselves, which acts as a template for new DNA, which then inserts itself at a new location in the genome. In this "copy and paste" mechanism, the original sequence stays where it is and the copy inserts itself at a new location.

- *Retrotransposons* also make an RNA copy of themselves when they move about the genome. They constitute about 17 percent of the human genome. Some of them code for some of the proteins necessary for their own

transposition, and others do not. A single type of retrotransposon, the 300-bp *Alu* element, accounts for 11 percent of the human genome; it is present in a million copies scattered over all the chromosomes.
- *DNA transposons* are similar to their prokaryotic counterparts. They do not use an RNA intermediate, but actually move to a new spot in the genome without replicating (Figure 14.3).

What role do these moving sequences play in the cell? There are few answers to this question. The best answer so far seems to be that transposons are cellular parasites that simply replicate themselves. But these replications can lead to the insertion of a transposon at a new location, which can have important consequences. For example, the insertion of a transposon into the coding region of a gene results in a mutation (see Figure 14.3). This phenomenon has been found in rare forms of several human genetic diseases, including hemophilia and muscular dystrophy. If the insertion of a transposon takes place in the germ line, a gamete with a new mutation results. If the insertion takes place in a somatic cell, cancer may result.

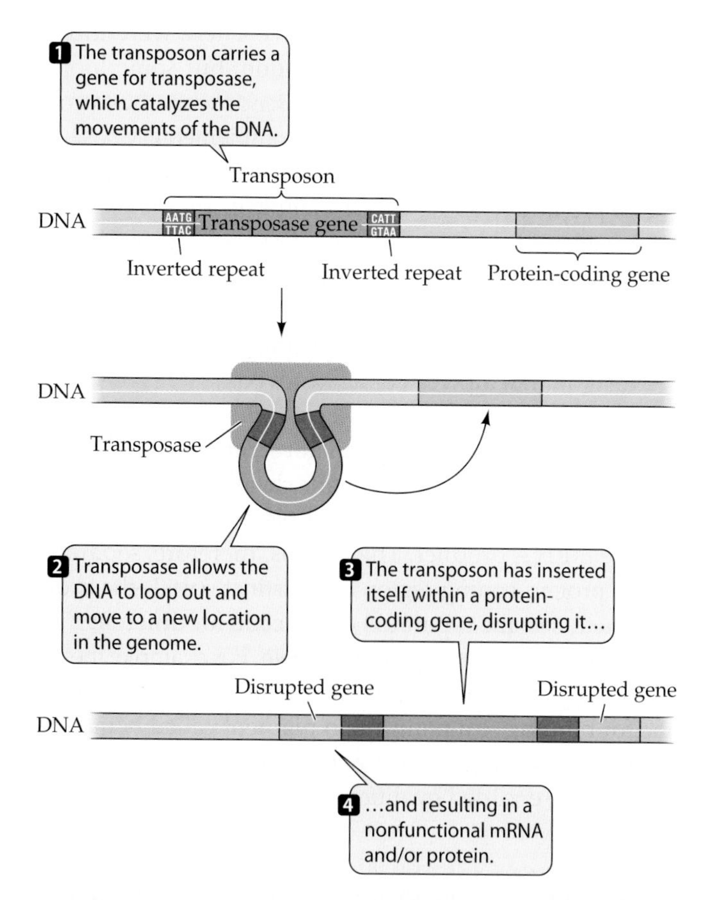

14.3 DNA Transposons and Transposition At the end of each DNA transposon is an inverted repeat sequence that helps in the transposition process.

If a transposon replicates not just itself but also an adjacent gene, the result may be a gene duplication. A transposon can carry a gene, or a part of it, to a new location in the genome, shuffling the genetic material and creating new genes. Clearly, transposition stirs the genetic pot in the eukaryotic genome and thus contributes to genetic variation.

In Chapter 4, we described the theory of endosymbiosis, which proposes that chloroplasts and mitochondria are the descendants of once free-living prokaryotes. Transposons may have played a role in this process. In living eukaryotes, although these organelles contain some DNA, the nucleus contains most of the genes that encode the organelle proteins. If the organelles were once independent, they must originally have contained all of these genes. How did the genes move to the nucleus? The answer may lie in DNA transpositions. Genes in the organelles may have moved to the nucleus by such well-known molecular events, which still occur today. The DNA that remains in the organelles may be the remnants of more complete prokaryotic genomes.

We now turn to the genes that are at the heart of molecular genetics: those that code for proteins.

The Structures of Protein-Coding Genes

Like their prokaryotic counterparts, many protein-coding genes in eukaryotes are single-copy DNA sequences. But eukaryotic genes have two distinctive characteristics that are uncommon among prokaryotes. First, they contain noncoding internal sequences, and second, they form gene families—groups of structurally and functionally related "cousins" in the genome.

Protein-coding genes contain noncoding internal and flanking sequences

Preceding the coding region of a eukaryotic gene is a **promoter**, to which an RNA polymerase binds to begin the transcription process. Unlike the prokaryotic enzyme, however, a eukaryotic RNA polymerase does not recognize the promoter sequence by itself, but requires help from other molecules, as we'll see below. At the other end of the gene, after the coding region, is a DNA sequence appropriately called the **terminator**, which signals the end of transcription when it is synthesized (Figure 14.4).

Eukaryotic protein-coding genes also contain noncoding base sequences, called **introns**. One or more introns are interspersed with the coding regions called **exons**. Transcripts of the introns appear in the primary transcript of RNA, called **pre-mRNA**, but by the time the **mature mRNA**—the mRNA that will be translated—leaves the nucleus, they have been removed. The transcripts of the introns are cut out of the pre-mRNA, and the transcripts of the exons are spliced together.

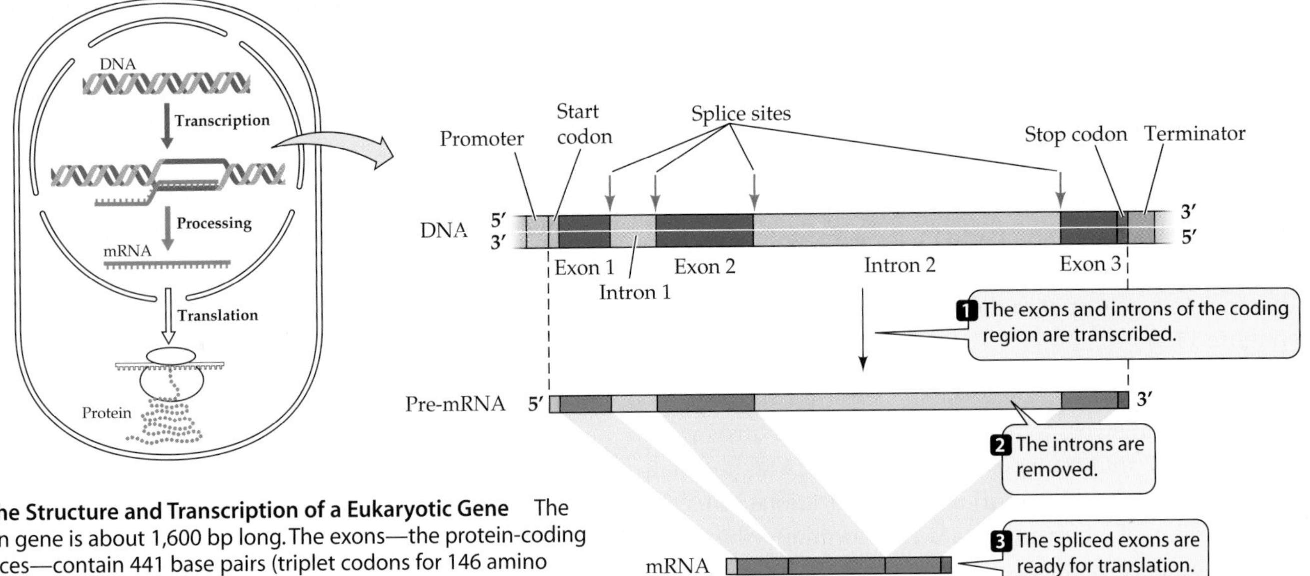

14.4 The Structure and Transcription of a Eukaryotic Gene The β-globin gene is about 1,600 bp long. The exons—the protein-coding sequences—contain 441 base pairs (triplet codons for 146 amino acids plus a triplet stop codon). The introns—noncoding sequences of DNA—between codons 30 and 31 (130 bp long) and 104 and 105 (850 bp long), are initially transcribed, but are spliced out of the initial mRNA transcript.

Where are the introns within a eukaryotic gene? The easiest way to find out is by **nucleic acid hybridization**, the method that originally revealed the existence of introns. This research method, outlined in Figure 14.5, has been crucial for studying the relationship between genes and their transcripts.

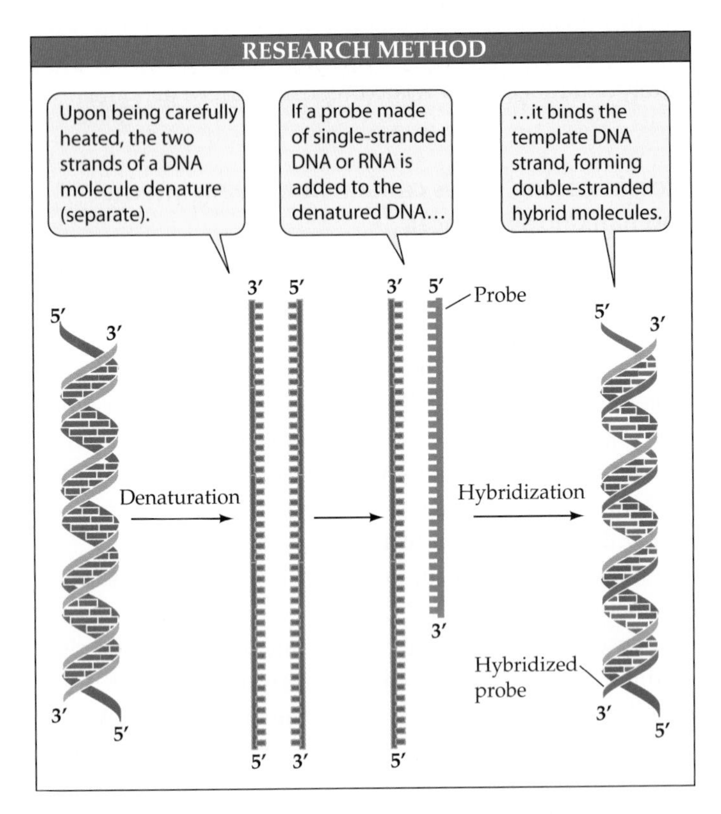

14.5 Nucleic Acid Hybridization Base pairing permits the detection of a sequence complementary to the probe.

Biologists used nucleic acid hybridization to examine the β-globin gene, which encodes one of the globin proteins that make up hemoglobin (Figure 14.6). They first denatured the β-globin DNA by heating it, then added mature β-globin mRNA. As expected, the mRNA bound to the DNA by complementary base pairing. The researchers expected to obtain a linear matchup of the mRNA to the coding DNA. That expectation was met, but only in part: There were indeed stretches of RNA–DNA hybridization, but some looped structures were also visible. These loops were the introns, stretches of DNA that did not have complementary bases on the mature mRNA. Later studies showed that the hybridization of pre-mRNA to DNA was complete, revealing that the introns were indeed transcribed. Somewhere on the path from primary transcript (pre-mRNA) to mature mRNA, the introns had been removed, and the exons had been spliced together. We will examine this splicing process shortly.

Most (but not all) vertebrate genes contain introns, as do many other eukaryotic genes (and even a few prokaryotic ones). Introns interrupt, but do not scramble, the DNA sequence that codes for a polypeptide chain. The base sequence of the exons, taken in order, is complementary to that of the mature mRNA product. The introns, therefore, separate a gene's protein-coding region into distinct parts—the exons. In some cases, the separated exons code for different functional regions, or *domains*, of the protein. For example, the globin proteins that make up hemoglobin have two domains: one for binding to a nonprotein pigment called heme, and another for binding to the other globin subunits. These two domains are encoded by different exons in the globin genes.

Many eukaryotic genes are members of gene families

About half of all eukaryotic protein-coding genes are present in only one copy in the haploid genome. The rest have multiple

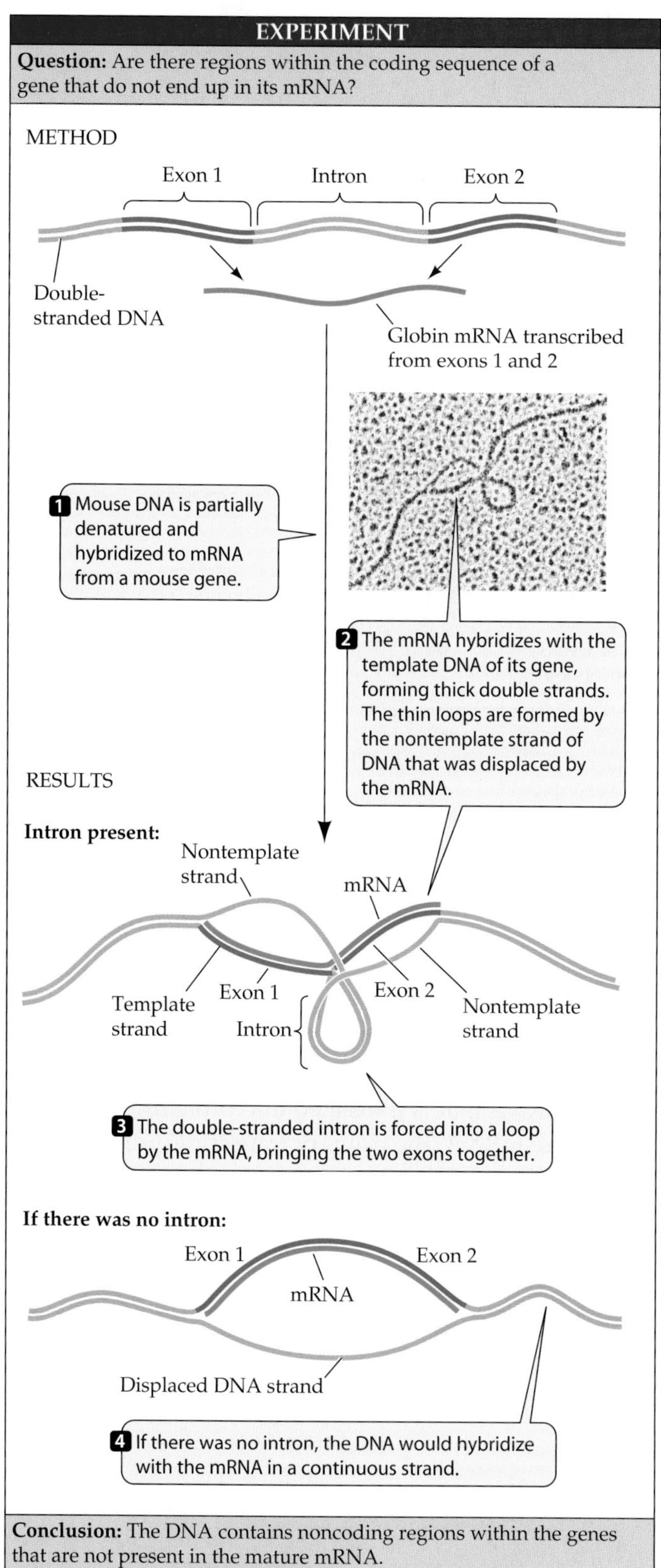

14.6 Nucleic Acid Hybridization Revealed Noncoding DNA When an mRNA transcript of the β-globin gene was experimentally hybridized to the double-stranded DNA of that gene, the introns in the DNA "looped out," demonstrating that the coding region of a eukaryotic gene can contain noncoding DNA that is not present in the mature mRNA transcript.

copies. Often, inexact, nonfunctional copies of a particular gene, called **pseudogenes**, are closely linked to the functional gene. These duplicates may have arisen by an abnormal event in chromosomal crossing over during meiosis or by the action of transposons. In other cases, however, the genome contains slightly altered duplicates of a gene that are functional.

A set of duplicated or related genes is called a **gene family**. Some gene families, such as the genes encoding the globins that are part of hemoglobin, contain only a few members; other families, such as the genes encoding the immunoglobulins that make up antibodies, have hundreds of members. Like the members of any family, the DNA sequences in a gene family are usually different from one another to a certain extent. As long as one member retains the original DNA sequence and thus codes for the proper protein, the other members can mutate slightly, extensively, or not at all. The availability of such "extra" genes is important for "experiments" in evolution: If the mutated gene is useful, it may be selected for in succeeding generations. If the gene is a total loss (a pseudogene), the functional copy is still there to save the day.

The gene family encoding the globins is a good example of the gene families found in vertebrates. These proteins are found in hemoglobin as well as in myoglobin (an oxygen-binding protein present in muscle). The globin genes all arose from a single common ancestor gene long ago. In humans, there are three functional members of the alpha-globin (α-globin) cluster and five in the beta-globin (β-globin) cluster (Figure 14.7). In an adult, each hemoglobin molecule is a tetramer containing four heme pigments (each held inside a globin polypeptide subunit), two identical α-globin subunits, and two identical β-globin subunits (see Figure 3.8).

During human development, different members of the β-globin gene cluster are expressed at different times and in different tissues (Figure 14.8). This differential gene expression has great physiological significance. For example,

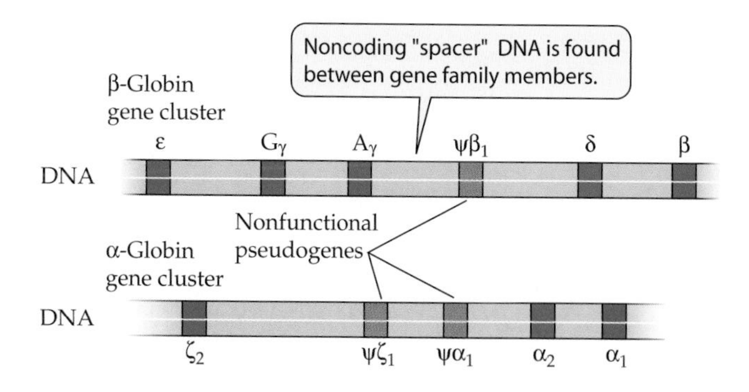

14.7 The Globin Gene Family The α-globin and β-globin clusters of the human globin gene family are located on different chromosomes. The genes of each cluster are separated by noncoding "spacer" DNA. The nonfunctional pseudogenes are indicated by the Greek letter psi (ψ).

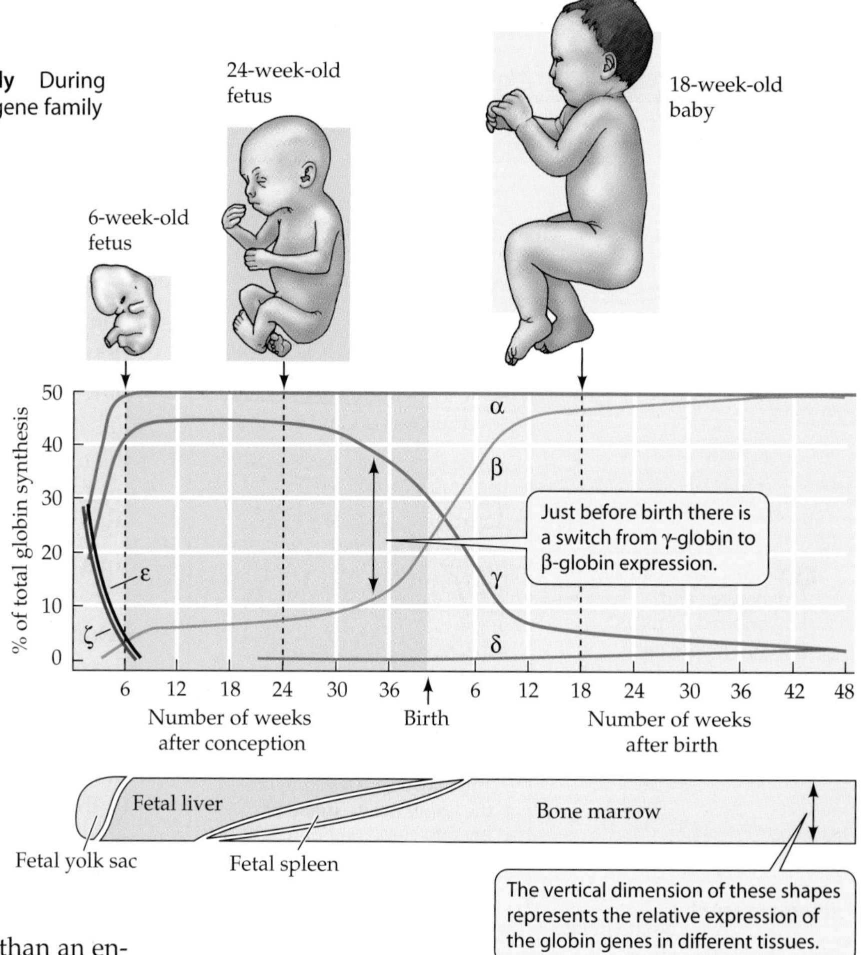

14.8 Differential Expression in the Globin Gene Family During human development, different members of the globin gene family are expressed at different times and in different tissues.

γ-globin, a subunit found in the hemoglobin of the fetus ($\alpha_2\gamma_2$), binds O_2 more tightly than adult hemoglobin ($\alpha_2\beta_2$) does. (Both γ-globin and β-globin are members of the β-globin cluster.) This specialized form of hemoglobin ensures that in the placenta, where the maternal and fetal circulation come close to each other, O_2 will be transferred from the mother's blood to the developing child's blood. Just before birth, the synthesis of fetal hemoglobin in the liver stops, and the bone marrow cells take over, making the adult form. Thus hemoglobins with different binding affinities for O_2 are provided at different stages of human development.

In addition to genes that encode proteins, the globin family includes nonfunctional pseudogenes, designated with the Greek letter psi (ψ). These pseudogenes are the "black sheep" of any gene family: they result from mutations that cause a loss of function, rather than an enhanced or new function. The DNA sequence of a pseudogene may not differ vastly from that of other family members. It may simply lack a promoter, for example, and thus fail to be transcribed. Or it may lack the recognition sites needed for the removal of introns (a process we will describe in the next section) and thus be transcribed into pre-mRNA, but not correctly processed into a useful mature mRNA. In some gene families, pseudogenes outnumber functional genes. Because some members of the family are functional, there appears to be little selective pressure for evolution to eliminate pseudogenes.

RNA Processing

As we saw in the previous section, eukaryotic protein-coding genes contain some sequences that do not appear in the mature mRNA that is translated into proteins. To produce the mature mRNA, the primary transcript (pre-mRNA) is processed in several ways: introns are removed, exons are joined, and bases are added at both ends.

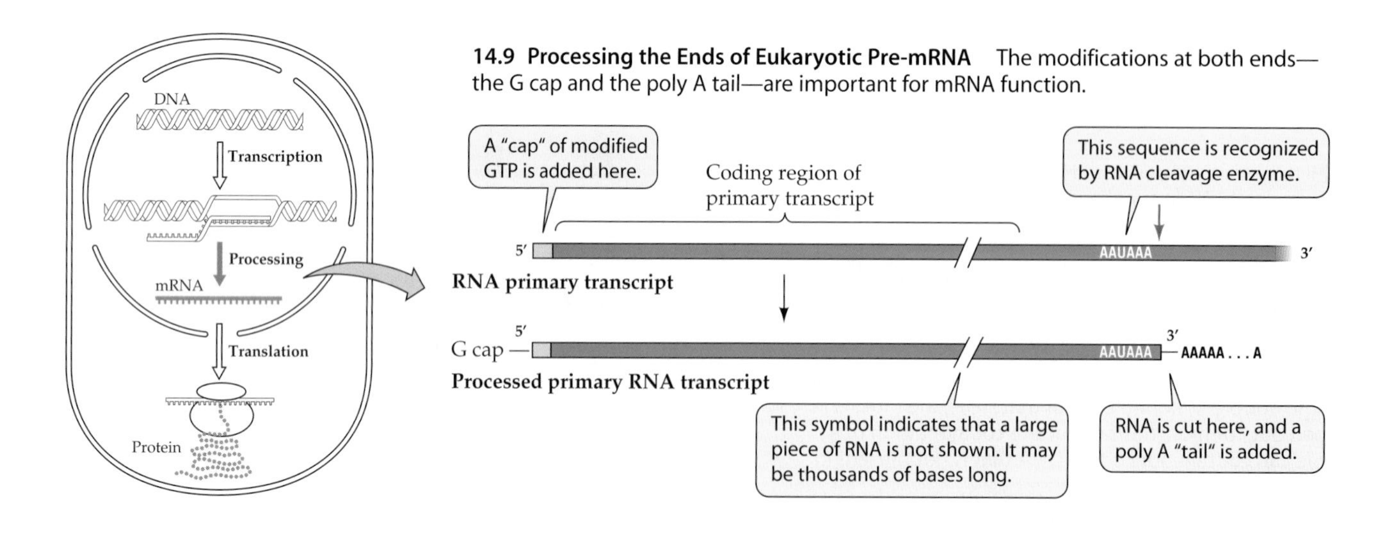

14.9 Processing the Ends of Eukaryotic Pre-mRNA The modifications at both ends—the G cap and the poly A tail—are important for mRNA function.

The primary transcript of a protein-coding gene is modified at both ends

Two early steps in the processing of pre-mRNA take place in the nucleus, one at each end of the molecule (Figure 14.9):

- A **G cap** is added to the 5′ end of the pre-mRNA as it is transcribed. The G cap is a chemically modified molecule of guanosine triphosphate (GTP). It apparently facilitates the binding of mRNA to the ribosome for translation and protects the mRNA from being digested by ribonucleases that break down RNAs.
- A **poly A tail** is added to the 3′ end of pre-mRNA at the end of transcription. Near the 3′ end of pre-mRNA, and after the last codon, is the sequence AAUAAA. This sequence acts as a signal for an enzyme to cut the pre-mRNA. Immediately after this cleavage, another enzyme adds 100 to 300 residues of adenine ("poly A") to the 3′ end of the pre-mRNA. This "tail" may assist in the export of the mRNA from the nucleus and is important for mRNA stability.

Splicing removes introns from the primary transcript

The next step in the processing of eukaryotic pre-mRNA within the nucleus is deletion of the introns. If these RNA regions were not removed, an mRNA producing a very different amino acid sequence, and possibly a nonfunctional protein, would result. A process called **RNA splicing** removes the introns and splices the exons together.

As soon as the pre-mRNA is transcribed, it is quickly bound by several **small nuclear ribonucleoprotein particles** (**snRNPs**, commonly pronounced "snurps"). There are several types of these RNA–protein particles in the nucleus.

At the boundaries between introns and exons are **consensus sequences**—short stretches of DNA that appear, with little variation ("consensus"), in many different genes. The RNA in one of the snRNPs (called U1) has a stretch of bases complementary to the consensus sequence at the 5′ exon–intron boundary, and it binds to the pre-mRNA by complementary base pairing. Another snRNP (U2) binds to the pre-mRNA near the 3′ intron–exon boundary (Figure 14.10).

Next, using energy from ATP, proteins assemble, forming a large RNA–protein complex called a **spliceosome**. This complex cuts the RNA, releases the introns, and joins the ends of the exons together to produce mature mRNA.

Molecular studies of human genetic diseases have been valuable tools in the investigation of consensus sequences and splicing machinery. People with beta thalassemia, for example, make an inadequate amount of the β-globin subunit of hemoglobin. These people suffer from severe anemia because they have an inadequate supply of red blood cells. In some cases, the genetic mutation that causes the disease oc-

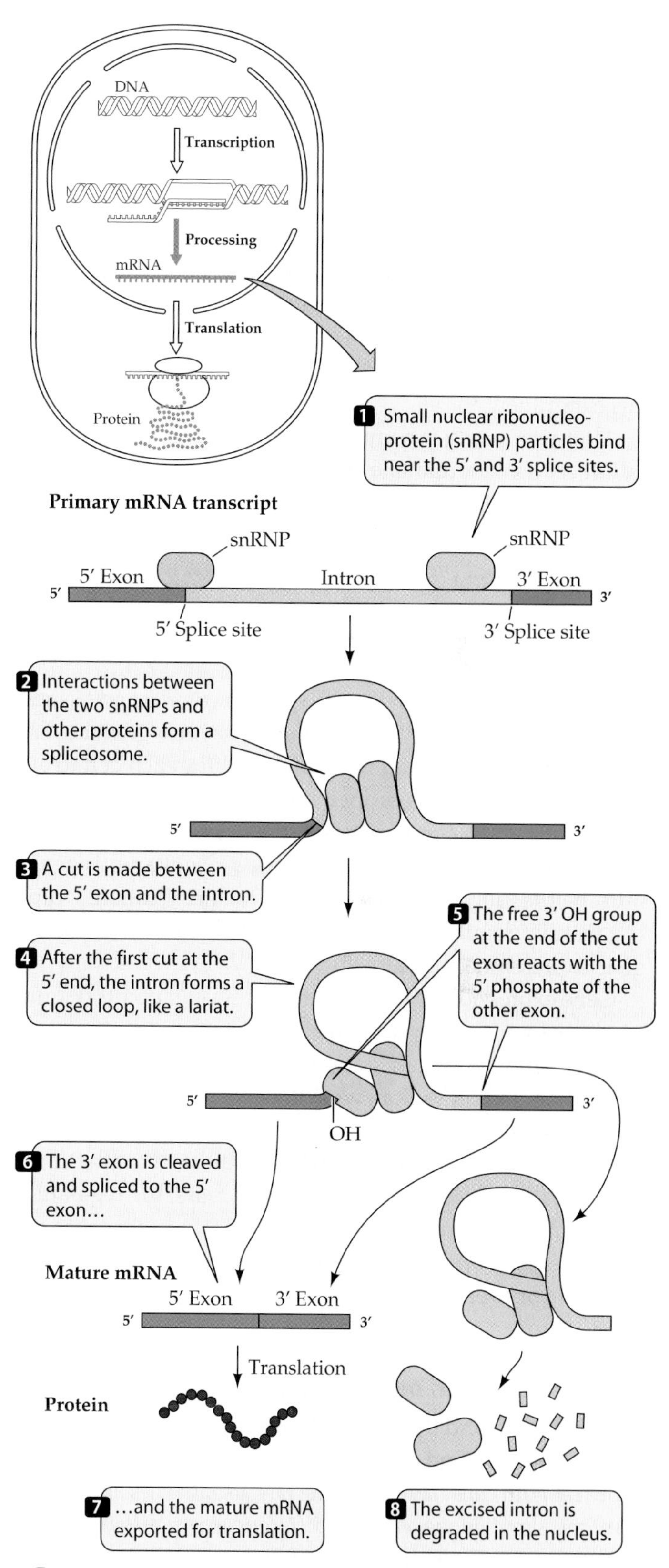

DNA
Transcription
Processing
mRNA
Translation
Protein
1 Small nuclear ribonucleoprotein (snRNP) particles bind near the 5′ and 3′ splice sites.
Primary mRNA transcript
snRNP
snRNP
5′ Exon
Intron
3′ Exon
5′
3′
5′ Splice site
3′ Splice site
2 Interactions between the two snRNPs and other proteins form a spliceosome.
3 A cut is made between the 5′ exon and the intron.
4 After the first cut at the 5′ end, the intron forms a closed loop, like a lariat.
5 The free 3′ OH group at the end of the cut exon reacts with the 5′ phosphate of the other exon.
OH
6 The 3′ exon is cleaved and spliced to the 5′ exon…
Mature mRNA
5′ Exon
3′ Exon
Translation
Protein
7 …and the mature mRNA exported for translation.
8 The excised intron is degraded in the nucleus.

14.10 The Spliceosome, an RNA Splicing Machine The binding of two snRNPs to consensus sequences on the pre-mRNA lines up the splicing machinery. After the snRNPs bind to the pre-mRNA, other proteins join the complex to form a spliceosome. This mechanism determines the exact position of each cut in the primary transcript with great precision.

curs at a consensus sequence in the β-globin gene. Consequently, β-globin pre-mRNA cannot be spliced correctly, and nonfunctional β-globin mRNA is made.

This finding is an excellent example of the use of mutations in determining a cause-and-effect relationship in biology. In the logic of science, merely linking two phenomena (for example, consensus sequences and splicing) does not prove that one is necessary for the other. In an experiment, the scientist alters one phenomenon (for example, the bases of the consensus sequence) to see whether the other (for example, splicing) occurs. In beta thalassemia, nature has done this experiment for us.

After processing is completed in the nucleus, the mature mRNA exits the organelle, apparently through the nuclear pores. A receptor at the nuclear pore recognizes the mature mRNA (or a protein bound to it). Unprocessed or incompletely processed pre-mRNAs remain in the nucleus.

Transcriptional Regulation of Gene Expression

In a multicellular organism with specialized cells and tissues, each cell contains every gene in the organism's genome. For development to proceed normally, and for each cell to acquire and maintain its proper specialized function, certain proteins must be synthesized at just the right times and in just the right cells. Thus, the expression of eukaryotic genes must be precisely regulated. Unlike DNA replication, which is generally regulated in every cell on an all-or-none basis, gene expression is highly selective.

Gene expression can be regulated at several points (Figure 14.11): before transcription, during transcription, after transcription and before translation, during translation, or after translation. In this section, we will describe the mechanisms that result in the selective transcription of specific genes. Some of these mechanisms involve nuclear proteins that alter chromosome function or structure. In other cases, the regulation of transcription involves changes in the DNA itself: genes are selectively replicated to provide more templates for transcription, or even rearranged on the chromosome.

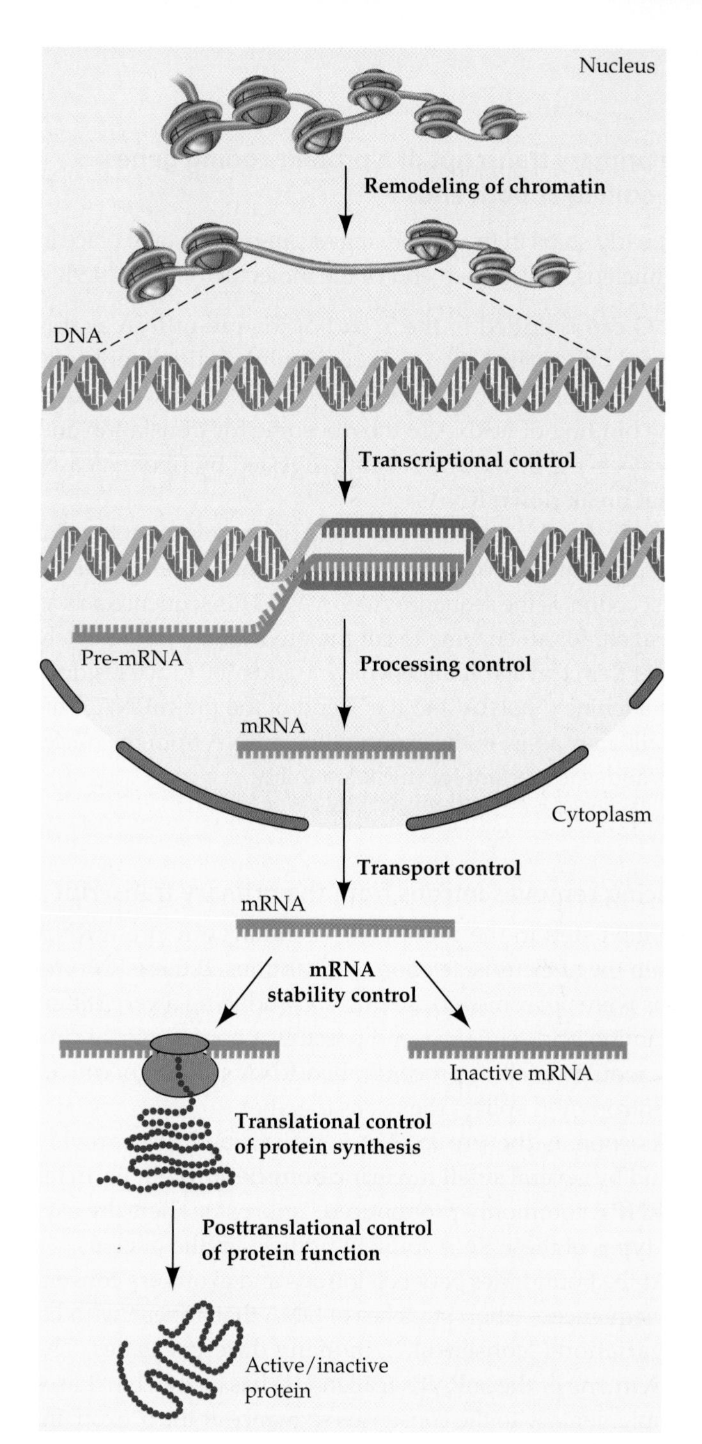

14.11 Potential Points for the Regulation of Gene Expression in Eukaryotes Gene expression can be regulated at four points: at transcription, after transcription (but before translation), at translation, or after translation.

Specific genes can be selectively transcribed

The brain cells and the liver cells of a mouse have some proteins in common and others that are characteristic of each cell type. Yet both cells have the same DNA sequences and, therefore, the same genes. Are the differences in protein content due to differential transcription of the genes? Or is it the case that all the genes are transcribed in both cell types, and some mechanism that acts after transcription is responsible for the differences in proteins?

These two alternatives—**transcriptional regulation** and **posttranscriptional regulation**—can be distinguished by examining the actual RNA sequences made within the nucleus of each cell type. Such analyses indicate that for some proteins, the mechanism of regulation is differential gene transcription. Both brain and liver cells, for example, transcribe "housekeeping" genes—those that encode proteins involved in the basic metabolic processes that occur in every living cell, such as glycolysis enzymes. But liver cells transcribe some genes for liver-specific proteins, and brain cells transcribe some genes for brain-specific proteins. And neither cell type transcribes the genes for proteins that are characteristic of muscle, blood, bone, or the other specialized cell types in the body.

CONTRASTING EUKARYOTES AND PROKARYOTES. Unlike prokaryotes, in which related genes are grouped into operons that are transcribed as a unit, eukaryotes tend to have solitary genes. Thus, the regulation of several genes at once requires common control elements in each of the genes, which allow all of the genes to respond to the same signal.

In contrast to the single RNA polymerase in bacteria, eukaryotes have three different RNA polymerases. Each eukaryotic polymerase catalyzes the transcription of a specific type of gene. Only one (RNA polymerase II) transcribes protein-coding genes. The other two transcribe the DNA that codes for rRNA (polymerase I) and for tRNA and small nuclear RNAs (polymerase III).

The diversity of eukaryotic polymerases is reflected in the diversity of eukaryotic promoters, which tend to be much more varied in their sequences than prokaryotic promoters. Furthermore, most eukaryotic genes have additional sequences that can regulate the rate of their transcription. Whether a eukaryotic gene is transcribed depends on the sum total of the effects of all of these DNA and protein elements; thus there are many points of possible regulation.

Finally, the transcription complex in eukaryotes is very different from that of prokaryotes, in which a single peptide subunit can cause RNA polymerase to recognize the promoter. In eukaryotes, many proteins are involved in initiating transcription. We will confine the following discussion to RNA polymerase II, which catalyzes the transcription of most protein-coding genes, but the mechanisms for the other two polymerases are similar.

TRANSCRIPTION FACTORS. As we saw in Chapter 13, the prokaryotic promoter is a sequence of DNA near the 5′ end of the coding region of a gene or operon where RNA polymerase begins transcription. A prokaryotic promoter has two essential sequences. One is the recognition sequence—the sequence recognized by RNA polymerase. The second, closer to the initiation site, is the **TATA box** (so called because it is rich in AT base pairs), where DNA begins to denature so that the template strand can be exposed.

Things are different in eukaryotes. Eukaryotic RNA polymerase II cannot simply bind to the promoter and initiate transcription. Rather, it does so only after various regulatory proteins, called **transcription factors**, have assembled on the chromosome (Figure 14.12). First, the protein TFIID ("TF" stands for transcription factor) binds to the TATA box. Its binding changes both its own shape and that of the DNA, presenting a new surface that attracts the binding of other

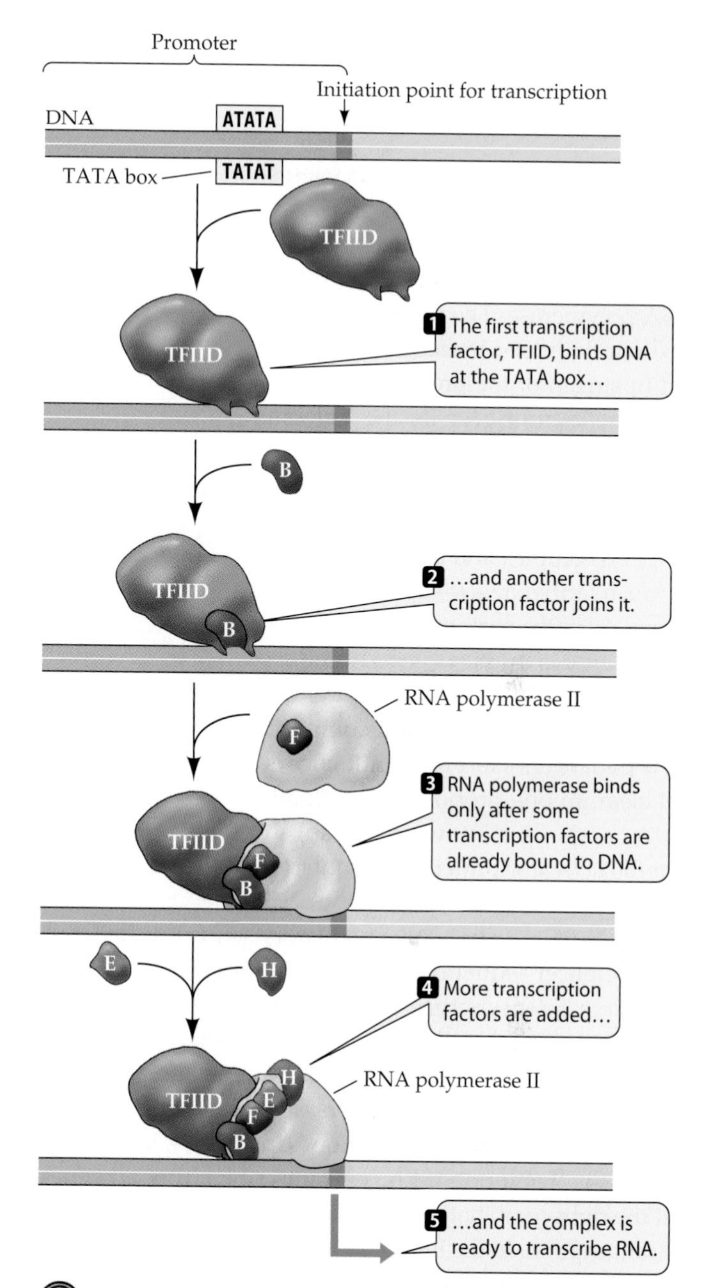
Promoter
Initiation point for transcription
DNA
ATATA
TATA box
TATAT
TFIID
1 The first transcription factor, TFIID, binds DNA at the TATA box...
B
2 ...and another transcription factor joins it.
RNA polymerase II
F
3 RNA polymerase binds only after some transcription factors are already bound to DNA.
E
H
4 More transcription factors are added...
RNA polymerase II
5 ...and the complex is ready to transcribe RNA.

14.12 The Initiation of Transcription in Eukaryotes
Except for TFIID, which binds to the TATA box, each transcription factor in this transcription complex has binding sites only for the other proteins in the complex, and does not bind directly to DNA.

transcription factors to form a **transcription complex**. RNA polymerase II does not bind until several other proteins have bound to this complex.

Some DNA sequences, such as the TATA box, are common to the promoters of many eukaryotic genes and are recognized by transcription factors that are found in all the cells of an organism. Other sequences found in promoters are specific to only a few genes and are recognized by transcription factors found only in certain tissues. These specific transcription factors play an important role in *differentiation*, the specialization of cells during development.

REGULATORS, ENHANCERS, AND SILENCERS IN DNA. In addition to the promoter, two other types of regulatory DNA sequences bind proteins that activate RNA polymerase. The recently discovered **regulator sequences** are clustered just upstream of the promoter. Various *regulator proteins* (seven for the β-globin gene) may bind to these regulator sequences (Figure 14.13). The resulting complexes bind to the adjacent transcription complex and activate it.

Much farther away—up to 20,000 bp away—from the promoter are **enhancer sequences**. Enhancer sequences bind *activator proteins*, and this binding strongly stimulates the transcription complex. How enhancers exert their influence is not clear. In one proposed model, the DNA bends (it is known to do so) so that the activator protein is in contact with the transcription complex.

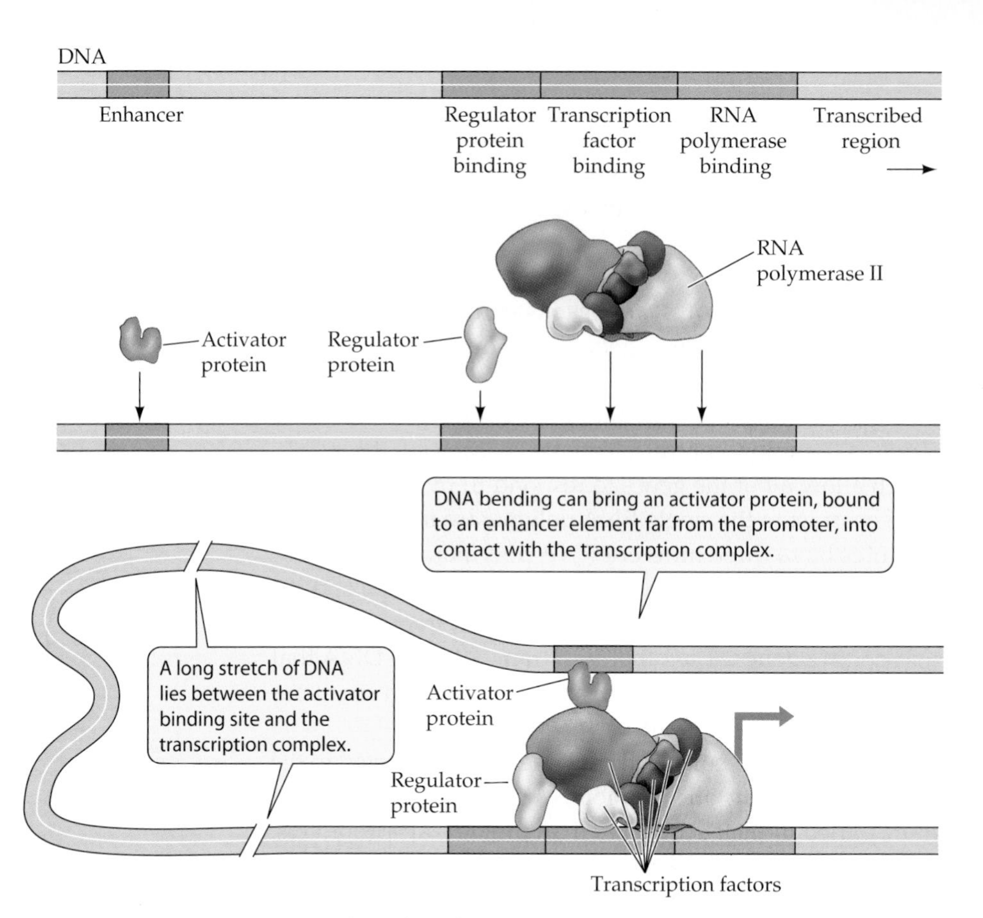

14.13 The Roles of Transcription Factors, Regulators, and Activators The actions of many proteins determine whether and where RNA polymerase II will transcribe DNA.

Finally, there are negative regulatory sequences on DNA, called **silencer sequences**, that have the opposite effect from enhancers. Silencers turn off transcription by binding proteins appropriately called *repressor proteins*.

How do these proteins and DNA sequences—transcription factors, regulators, enhancers, activators, silencers, and repressors—regulate transcription? Apparently, in most tissues, a small amount of RNA is transcribed from all genes. But the right combination of these factors determines the rate of transcription. In the immature red blood cells of bone marrow, for example, which make a large amount of β-globin, transcription of the β-globin gene is stimulated by the binding of seven regulator proteins and six activator proteins. But in white blood cells in the same bone marrow, these thirteen proteins are not made and do not bind to the regulator and enhancer sequences adjacent to the β-globin gene; consequently, the gene is hardly transcribed at all.

COORDINATING THE EXPRESSION OF GENES. How do eukaryotic cells coordinate the regulation of several genes whose transcription must be turned on at the same time? In prokaryotes, in which related genes are linked together in an operon, a single regulatory system can regulate several adjacent genes. But in eukaryotes, the several genes whose regulation requires coordination may be far apart on a chromosome, or even on different chromosomes.

In such a case, regulation can be achieved if the various genes all have the same regulatory sequences, which bind the same regulatory proteins. One of the many examples of this phenomenon is provided by the response of organisms to a stressor—for example, that of plants to drought. Under conditions of drought stress, a plant must synthesize various proteins, but the genes for these proteins are scattered throughout the genome. However, each of these genes has a specific regulatory sequence near its promoter, called the *stress response element* (*SRE*). The binding of a regulator protein to this element stimulates RNA synthesis (Figure 14.14). The proteins made from these genes are involved not only in water conservation, but also in protecting the plant against excess salt in the soil and against freezing. This finding has considerable importance for agriculture, in which crops are often grown under less than optimal conditions.

The regulation and coordination of gene expression requires the binding of many specialized proteins to DNA. Among DNA-binding proteins, there are four common structural themes in the domains that bind to DNA. These themes, called **motifs**, consist of combinations of structures and special components: helix-turn-helix, zinc finger, leucine zipper, and helix-loop-helix (Figure 14.15). DNA-binding proteins with specific motifs are involved in the activation

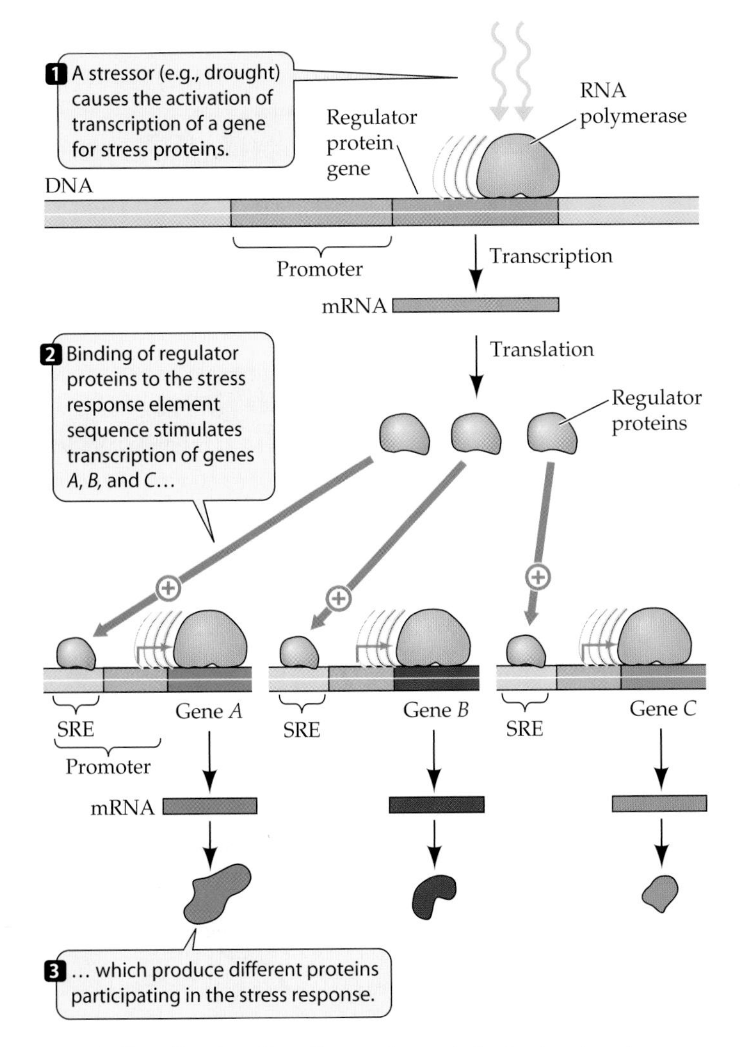

14.14 Coordinating Gene Expression A single signal, such as drought stress, causes the synthesis of a transcriptional regulator for many genes.

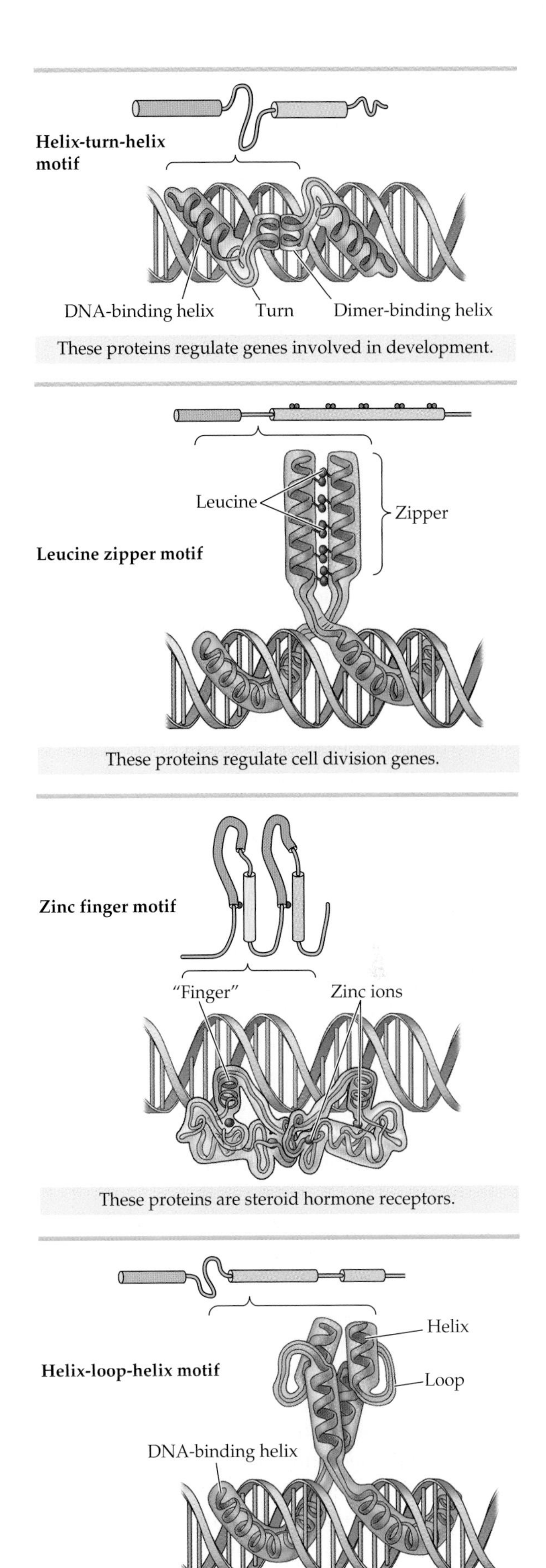

of certain types of genes, both during development and in the adult organism.

Genes can be inactivated by chromatin structure

Other mechanisms that regulate transcription act on the structure of chromatin and chromosomes. As we saw in Chapter 9, chromatin contains a number of proteins as well as DNA. The packaging of DNA into nucleosomes by these nuclear proteins can make DNA physically inaccessible to RNA polymerase and the rest of the transcription apparatus, much as the binding of a repressor to the operator in the prokaryotic *lac* operon prevents transcription. Chromatin structure at both the local and whole-chromosome levels affects transcription.

14.15 Protein-DNA Interactions The DNA-binding domains of most regulatory proteins have one of these four structural motifs.

CHROMATIN REMODELING. Nucleosomes inhibit both the initiation and elongation steps of transcription. To inactivate these blocks, two types of protein complexes bind to chromatin. One of them binds upstream of the initiation site, disaggregating the nucleosomes so that the large transcription complex can bind and begin transcription. The other type of complex binds once transcription is under way, allowing the transcription complex to move through the nucleosomes. These processes are referred to as **chromatin remodeling** (Figure 14.16).

How do the nucleosomes disaggregate to allow transcription (and then reaggregate)? As you will recall, the histone proteins that make up nucleosomes are positively charged, and DNA is negatively charged (owing to its phosphate groups), so the attachment of these two molecules is electrostatic. If the histones are modified to reduce their charge, they will release the DNA. One such modification neutralizes the amino groups of the histones by adding acetyl groups. This *acetylation*, catalyzed by histone acetylase, helps disaggregate nucleosomes. Conversely, *deacetylation*, catalyzed by histone deacetylase, allows nucleosomes to re-form. Thus, acetylation is associated with the activation of genes; deacetylation is associated with gene deactivation.

WHOLE-CHROMOSOME EFFECTS. Other transcriptional regulation mechanisms can act on entire chromosomes. Under a microscope, two kinds of chromatin can be distinguished in the stained interphase nucleus: euchromatin and heterochromatin. *Euchromatin* is diffuse and stains lightly; it contains the DNA that is transcribed into mRNA. *Heterochromatin* stains densely and is generally not transcribed; any genes that it contains are thus inactivated.

Perhaps the most dramatic example of heterochromatin is the inactive X chromosome of mammals. A normal female mammal has two X chromosomes; a normal male has an X and a Y. The Y chromosome has only a few genes that are also present on the X, and it is largely transcriptionally inactive in most cells. So there is a great difference between females and males in the "dosage" of X-linked genes. In other words, each female cell has two copies of the genes on the X chromosome, and therefore has the potential to produce twice as much protein product from these genes as a male cell has. Yet X-linked gene expression is generally the same in males and females. How can this happen?

The answer was found in 1961 independently by Mary Lyon, Liane Russell, and Ernest Beutler. They suggested that one of the X chromosomes in each cell of an XX female is transcriptionally inactivated early in embryonic development. That copy of the X remains inactive in that cell, and in all the cells arising from it. In a given mammalian cell, the "choice" of which X in the pair of X's to inactivate is random. Recall that one X in a female comes from her father and one from her mother. Thus, in one embryonic cell, the paternal X might be the one remaining transcriptionally active, but in a neighboring cell, the maternal X might be active.

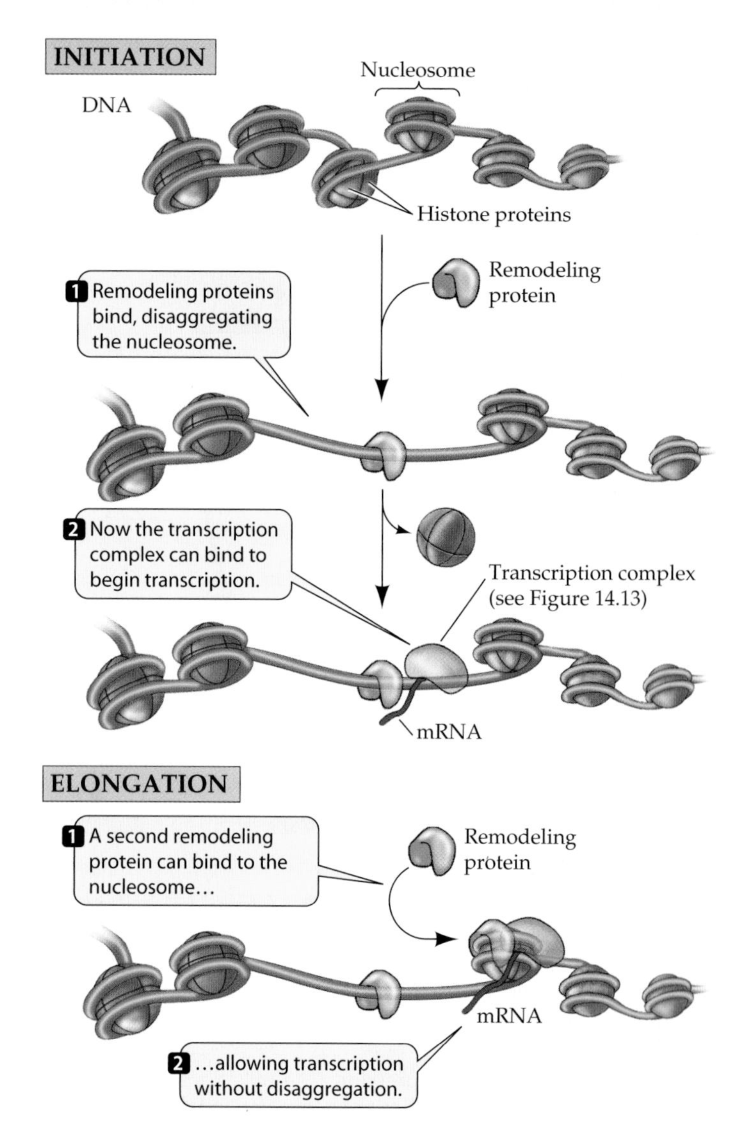

14.16 Local Remodeling of Chromatin for Transcription Initiation of transcription requires that nucleosomes disaggregate. During elongation, however, they can remain intact.

In the interphase cells of human females, a single, stainable nuclear body, called a **Barr body** after its discoverer, Murray Barr, can be seen under the light microscope (Figure 14.17). This clump of heterochromatin, which is not present in males, is the inactivated X chromosome. The number of Barr bodies in a nucleus is equal to the number of X chromosomes minus one (the one represents the X chromosome that remains transcriptionally active). So a female with the normal two X chromosomes will have one Barr body, a rare female with three X's will have two, an XXXX female will have three, and an XXY male will have one. These observations suggest that the interphase cells of each person, male or female, have a single active X chromosome, making the dosage of the expressed X chromosome genes constant across both sexes.

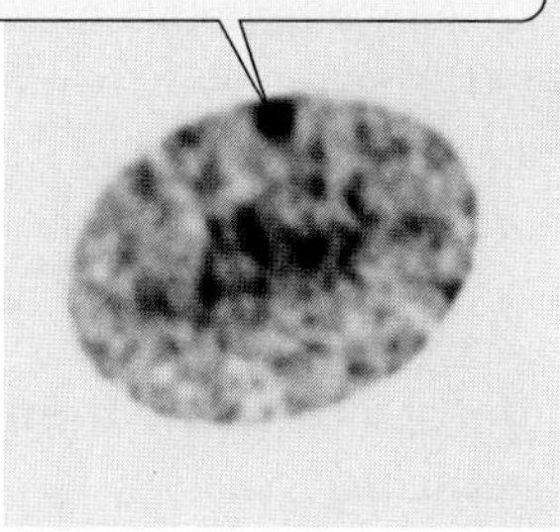

The Barr body is the condensed, inactive member of a pair of X chromosomes in the cell. The other X is not condensed and is active in transcription.

14.17 A Barr Body in the Nucleus of a Female Cell The number of Barr bodies per nucleus is equal to the number of X chromosomes minus one. Thus normal males (XY) have no Barr body, whereas normal females (XX) have one.

Chromosomal condensation in the inactive X chromosome makes its DNA sequences physically unavailable to the transcriptional machinery. One mechanism of inactivation is the addition of a methyl group (—CH_3) to the 5′ position of cytosine on DNA. Such **methylation** seems to be most prevalent in transcriptionally inactive genes. For example, most of the DNA of the inactive X chromosome has many of its cytosines methylated, while few cytosines on the active X are methylated. Methylated DNA appears to bind certain chromosomal proteins that may be responsible for heterochromatin formation.

The otherwise inactive X chromosome has one gene that is only lightly methylated and is transcriptionally active. That gene is called *Xist* (for *X i*nactivation-*s*pecific *t*ranscript), and it is heavily methylated on, and not transcribed from, the other, "active" X chromosome. The RNA transcribed from *Xist* does not leave the nucleus and is not an mRNA. Instead, it appears to bind to the X chromosome from which it is transcribed, and this binding somehow leads to a spreading of inactivation along the chromosome. This RNA transcript is known as **interference RNA** (**RNAi**) (Figure 14.18).

How does the transcriptionally active X overcome the effects of *Xist* RNAi? Apparently, there is an anti-*Xist* gene, appropriately called *Tsix*. This gene codes for an RNAi that binds by complementary base pairing to *Xist* RNA at the active X chromosome.

A DNA sequence can be moved to a new location to activate transcription

In some instances, gene expression is regulated by the movement of a gene to a new location on the chromosome. An example of this mechanism is found in the budding yeast, *Saccharomyces cerevisiae*. This haploid, single-celled fungus exists in two *mating types*, *a* and α. Two cells of different mating types fuse to form a diploid zygote, as we will see in Chapter 31. Although all yeast cells have an allele for each mating type, the allele that is expressed determines the mating type of the cell. In some yeasts, the mating type changes with almost every cell division cycle. How does it change so rapidly?

In the yeast cell, the two mating type alleles (coding for type α and type *a*) have separate, specific locations on the chromosome, away from a third site, called the MAT locus. One allele is transcriptionally silent because a repressor protein is bound to it. However, when a copy of the α or *a* allele is inserted at the MAT locus, the gene for the protein of that mating type is transcribed. A change in mating type requires that one allele be moved out of the MAT locus and the other moved in. This process takes place in three steps:

- First, a new DNA copy of the nonexpressed allele is made (if the cell is now α, a new copy of the *a* allele will be made).
- Second, the current occupant of the MAT locus (in this case, the α allele) is removed by an enzyme.
- Third, the new allele (*a*) is inserted at the MAT locus and transcribed. The *a* proteins are now made, and the mating type of the cell is *a*.

DNA rearrangement is important in producing the highly variable proteins that make up the human repertoire of antibodies. It is also a factor in cancer, in which inactive genes may be moved to positions adjacent to active promoters.

Selective gene amplification results in more templates for transcription

Another way for one cell to make more of a certain gene product than another cell does is to make more copies of the appropriate gene and transcribe them all. The process of cre-

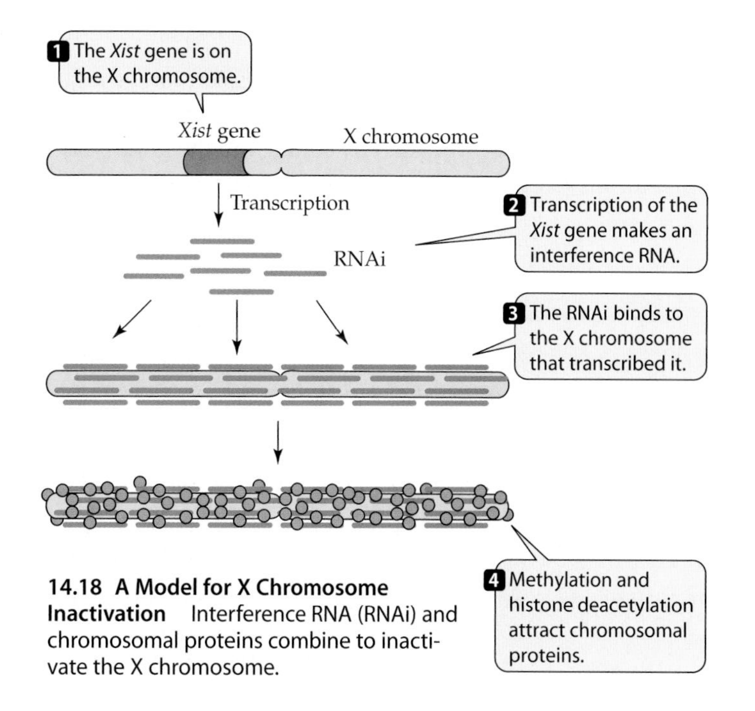

14.18 A Model for X Chromosome Inactivation Interference RNA (RNAi) and chromosomal proteins combine to inactivate the X chromosome.

ating more copies of a gene in order to increase its transcription is called **gene amplification**.

As described earlier, the genes that code for three of the four human ribosomal RNAs are linked together in a unit, and this unit is repeated several hundred times in the genome to provide multiple templates for rRNA synthesis (rRNA is the most abundant kind of RNA in the cell). In some circumstances, however, even this moderate repetition is not enough to satisfy the demands of the cell.

The mature eggs of frogs and fishes, for example, have up to a trillion ribosomes. These ribosomes are used for the massive protein synthesis that follows fertilization. The cell that will differentiate into the egg contains fewer than 1,000 copies of the rRNA gene cluster, and would take 50 years to make a trillion ribosomes if it transcribed those rRNA genes at peak efficiency. How does the egg end up with so many ribosomes (and so much rRNA)?

The egg cell solves this problem by selectively amplifying its rRNA gene clusters until there are more than a million copies. In fact, this gene complex goes from being 0.2 percent of the total genome DNA to 68 percent. These million copies, transcribed at maximum rate (Figure 14.19), are just enough to make the necessary trillion ribosomes in a few days.

The mechanism for selective amplification of a single gene is not clearly understood, but it has important medical implications. In some cancers, a cancer-causing gene called an oncogene becomes amplified (see Chapter 17). Also, when some tumors are treated with a drug that targets a single protein, amplification of the gene for the target protein leads to an excess of that protein, and the cell becomes resistant to the prescribed dose of the drug.

Posttranscriptional Regulation

There are many ways in which gene expression can be regulated even after the gene has been transcribed. As we saw earlier, pre-mRNA is processed by cutting out the introns and splicing the exons together. If exons are selectively deleted from the pre-mRNA by alternative splicing, different proteins can be synthesized. The longevity of mRNA in the cytoplasm can also be regulated. The longer an mRNA exists in the cytoplasm, the more of its protein can be made.

Different mRNAs can be made from the same gene by alternative splicing

Most primary mRNA transcripts contain several introns (see Figure 14.4). We have seen how the splicing mechanism recognizes the boundaries between exons and introns. What would happen if the β-globin pre-mRNA, which has two introns, were spliced from the start of the first intron to the end of the second? Not only the two introns, but also the middle exon, would be spliced out. An entirely new protein (certainly not a β-globin) would be made, and the functions of normal β-globin would be lost.

Alternative splicing can be a deliberate mechanism for generating a family of different proteins from a single gene. For example, a single pre-mRNA for the structural protein tropomyosin is spliced differently in five different tissues to give five different mature mRNAs. These mRNAs are translated into the five different forms of tropomyosin found in these tissues: skeletal muscle, smooth muscle, fibroblast, liver, and brain (Figure 14.20).

Before the sequencing of the human genome began, most scientists estimated that they would find between 100,000 and 150,000 genes. You can imagine their surprise when the actual sequence revealed only 21,000 genes—not many more than *C. elegans* has! In fact, there are many more human mRNAs than there are human genes, and most of this variation comes from alternative splicing. Indeed, recent surveys show that half of all human genes are alternatively spliced. Alternative splicing may be a key to the differences in levels of complexity among organisms.

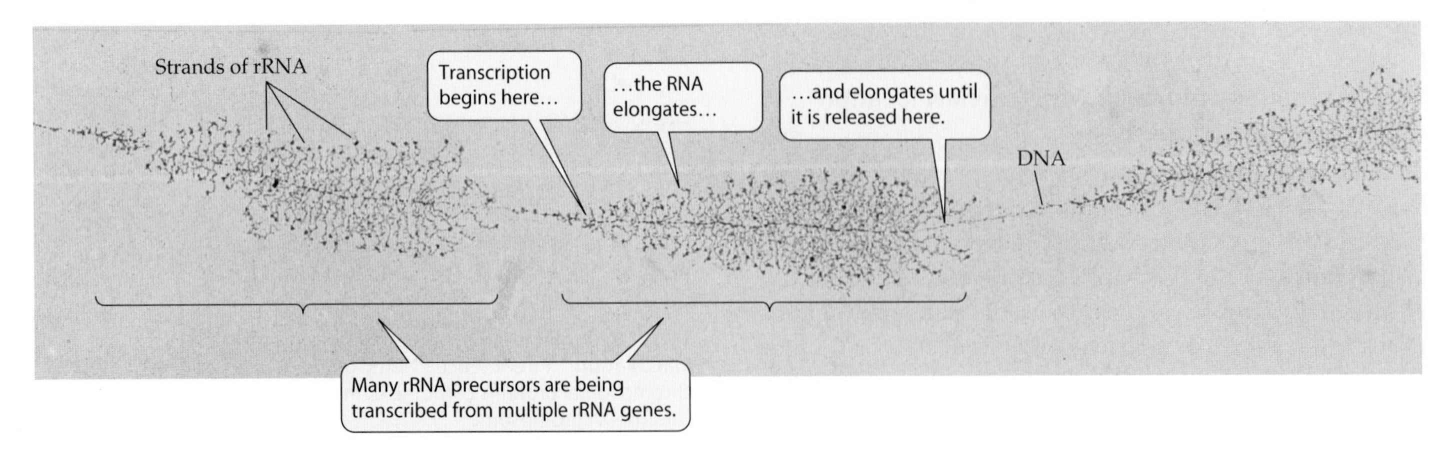

14.19 Transcription from Multiple Genes for rRNA Elongating strands of rRNA transcripts form arrowhead-shaped regions, each centered on a DNA sequence that codes for rRNA.

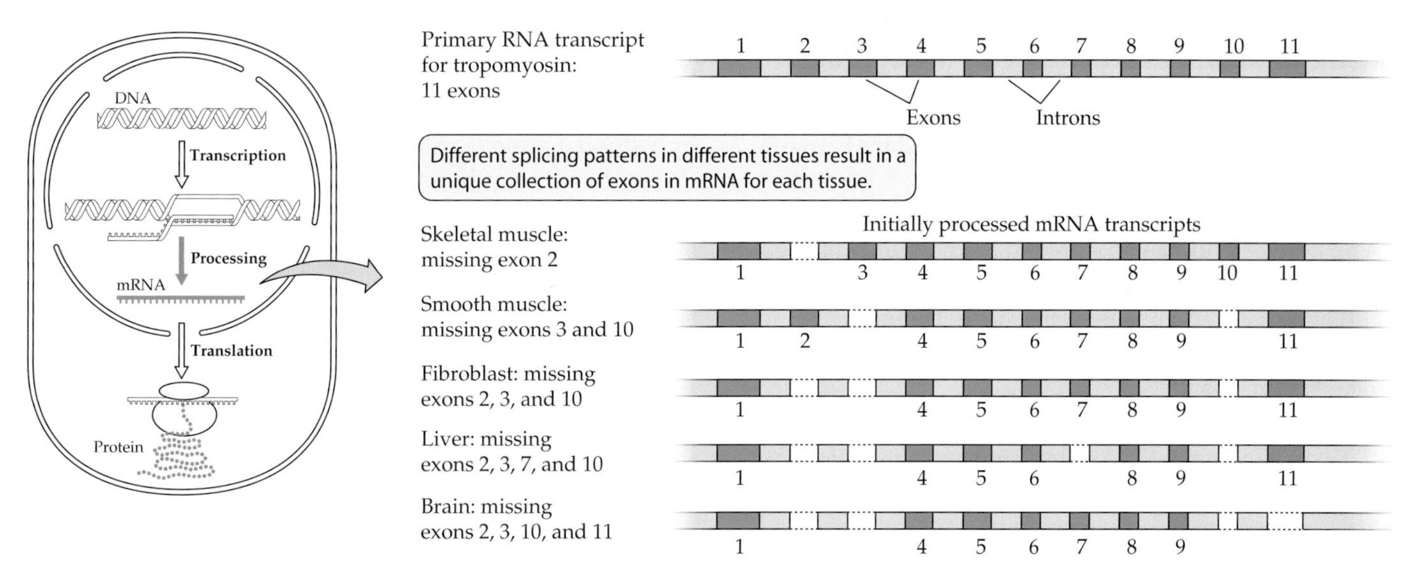

14.20 Alternative Splicing Results in Different mRNAs and Proteins In mammals, the protein tropomyosin is encoded by a gene that has 11 exons. Tropomyosin pre-mRNA is spliced differently in different tissues, resulting in five different forms of the protein.

The stability of mRNA can be regulated

DNA, as the genetic material, must remain stable, and as we have seen, there are elaborate mechanisms for repairing DNA if it becomes damaged. RNA has no such repair mechanism. After it arrives in the cytoplasm, mRNA is subject to breakdown catalyzed by ribonucleases, which exist both in the cytoplasm and in lysosomes. But not all eukaryotic mRNAs have the same life span. Differences in the stabilities of mRNAs provide another mechanism for posttranscriptional regulation of protein synthesis. The less time an mRNA spends in the cytoplasm, the less of its protein can be translated.

Specific AU-rich nucleotide sequences within some mRNAs mark them for rapid breakdown by a ribonuclease complex called the *exosome*. Signaling molecules such as growth factors, for example, are made only when needed, and then break down rapidly. Their mRNAs are highly unstable because they contain an AU-rich sequence.

RNA can be edited to change the encoded protein

The sequence of mRNA can be changed after transcription and splicing by **RNA editing**. This editing can occur in two ways (Figure 14.21):

- *Insertion of nucleotides*. In the parasitic protozoan *Trypanosoma brucei*, certain mRNAs have been found that have a longer base sequence than predicted by the gene coding for them. Stretches of U's are added after transcription, changing the protein that is made.
- *Alteration of nucleotides*. An enzyme can catalyze the deamination of cytosine, forming uracil (see Figure 12.19). This process can affect a membrane channel protein in the mammalian nervous system that normally allows calcium and sodium to pass through. Editing of a certain cytosine in the mRNA for this protein to uracil changes the amino acid at that position in the polypeptide chain from histidine to tyrosine, and the channel protein no longer allows the passage of calcium.

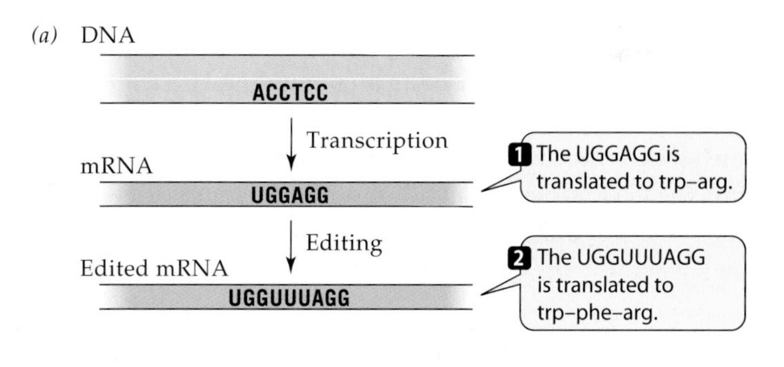

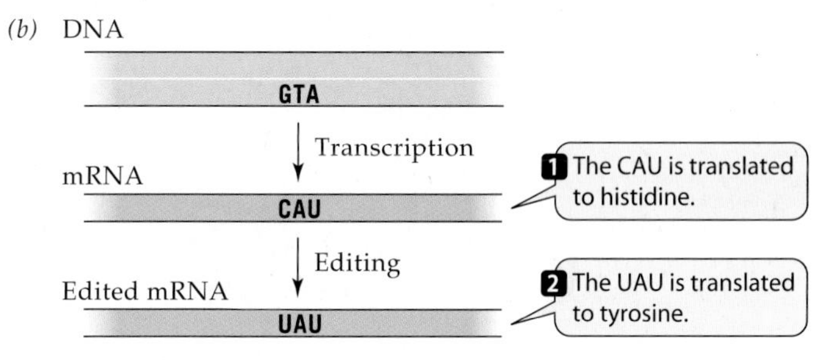

14.21 RNA Editing RNA can be edited in two ways: (*a*) by the insertion of new nucleotides, or (*b*) by the alteration of existing nucleotides.

Translational and Posttranslational Regulation

Is the amount of a protein in a cell determined by the amount of its mRNA? Recently, a survey of the relationships between mRNAs and proteins in yeast cells was made. Dozens of genes were surveyed. For about a third of them, the relationship between mRNA and protein held: more of one led to more of the other. But for two-thirds of the proteins, there was no apparent relationship. The concentrations of these proteins in the cell must be determined by factors acting after the mRNA is made.

Just as certain regulatory proteins can control the synthesis of mRNA by binding to DNA, other proteins can regulate the translation of mRNA by binding to mRNA in the cytoplasm. This mode of control is especially important for long-lived mRNAs. A cell must not continue to make proteins that it does not need. For example, as we saw in Chapter 9, mammalian cells respond to certain stimuli by making cyclins, proteins that stimulate the events of the cell cycle. If the mRNA for a cyclin is still in the cytoplasm and available for translation long after the cyclin is needed, the cyclin will be made and released inappropriately. Its presence might cause a target cell population to divide inappropriately, forming a tumor.

The translation of mRNA can be regulated

Let's look at examples of three general mechanisms by which levels of certain proteins are controlled by regulating the translation of mRNA.

One way to regulate translation is through the G cap on mRNA. As we saw above, mRNA is capped at its 5′ end by a modified guanosine triphosphate molecule (see Figure 14.9). An mRNA that is capped with an unmodified GTP molecule is not translated. For example, stored mRNA in the oocyte of the tobacco hornworm moth has a G cap added to its 5′ end, but the GTP molecule is not modified. Hence, this stored mRNA is not translated. After fertilization, however, the cap is modified, allowing the mRNA to be translated to produce the proteins needed for early embryonic development.

Within mammalian cells, free iron ions (Fe^{2+}) are bound by a storage protein, called ferritin. When iron is present in excess, ferritin synthesis rises dramatically. Yet the amount of ferritin mRNA remains constant. The increase in ferritin synthesis is due to an increased rate of mRNA translation. When the iron level in the cell is low, a translational repressor protein binds to ferritin mRNA and prevents its translation by blocking its attachment to a ribosome. When the iron level rises, the excess iron ions bind to the repressor and alter its three-dimensional structure, causing it to detach from the mRNA, and translation of ferritin proceeds.

Translational regulation also acts on the synthesis of hemoglobin, helping to maintain the balance among its components. As described above, a hemoglobin molecule consists of four globin subunits and four heme pigments. If globin synthesis does not equal heme synthesis, some heme stays free in the cell, waiting for a globin partner. Excess heme increases the rate of translation of globin mRNA by removing a block to the initiation of translation at the ribosome.

The proteasome controls the longevity of proteins after translation

We have considered how gene expression can be regulated by the control of transcription, RNA processing, and translation. However, the story does not end here, because most gene products—proteins—are modified after translation. Some of these changes are permanent, such as the addition of sugars (glycosylation), the addition of phosphate groups, or the removal of a signal sequence after a protein has crossed a membrane (see Figure 12.15).

One way to regulate the action of a protein in a cell is to regulate its lifetime in the cell. Proteins involved in cell division (such as cyclins), for example, are hydrolyzed at just the right moment to time the sequence of events. In many cases,

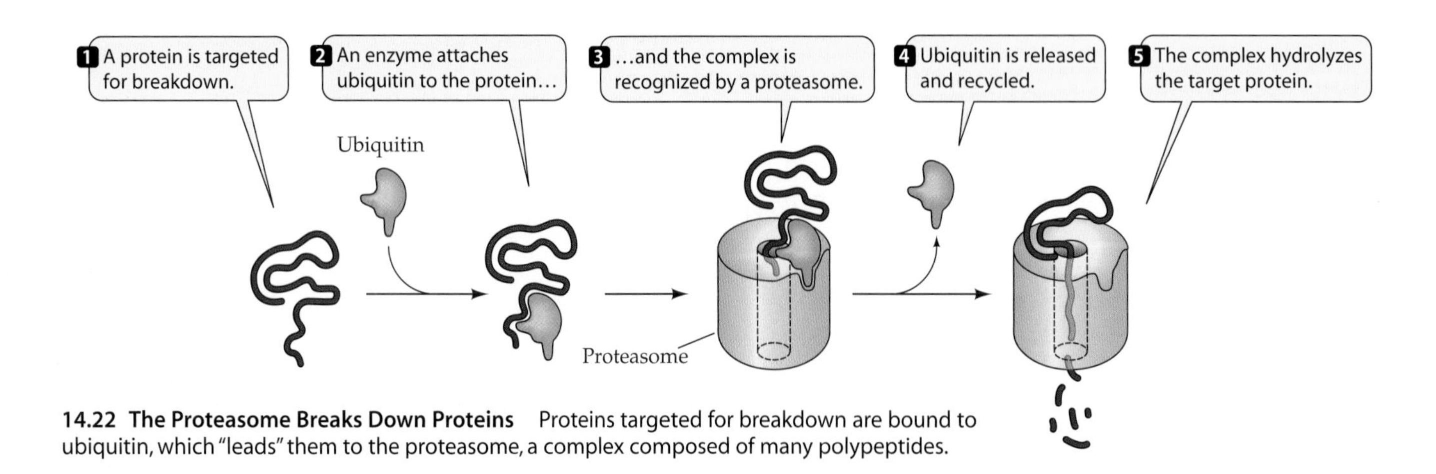

1 A protein is targeted for breakdown.
2 An enzyme attaches ubiquitin to the protein...
3 ...and the complex is recognized by a proteasome.
4 Ubiquitin is released and recycled.
5 The complex hydrolyzes the target protein.
Ubiquitin
Proteasome

14.22 The Proteasome Breaks Down Proteins Proteins targeted for breakdown are bound to ubiquitin, which "leads" them to the proteasome, a complex composed of many polypeptides.

a 76-amino acid protein called **ubiquitin** (so called because it is ubiquitous, or widespread) is covalently linked to a protein targeted for breakdown. The protein–ubiquitin complex then binds to a huge complex of several dozen polypeptide chains called a **proteasome** (Figure 14.22). The entryway to this "molecular chamber of doom" is a hollow cylinder. This part of the complex has ATPase activity, and it uses the released energy to cut off the ubiquitin for recycling and unfold its targeted protein "victim." The protein then passes by three different proteases (thus the name of the complex), which digest it into small peptides and amino acids.

The cellular concentrations of many proteins are determined not by differential transcription of their genes, but by their degradation in proteasomes. Cyclins, for example, are degraded at just the right time during the cell cycle (see Figure 9.4). Transcription regulators are broken down after they are used, lest the affected genes be always "on." Abnormal proteins are often targeted for destruction by a quality control mechanism. Human papillomavirus, which causes cervical cancer, targets the cell division inhibitory protein p53 for proteasomal degradation, so that unregulated cell division—cancer—results.

Chapter Summary

The Eukaryotic Genome

▶ Although eukaryotes have more DNA in their genomes than prokaryotes, there is no apparent relationship between genome size and organism complexity within eukaryotes.

▶ There are many differences between prokaryotic and eukaryotic genomes and their mechanisms of expression. **Review Table 14.1**

▶ Unlike prokaryotic DNA, eukaryotic DNA is contained within a nucleus, so that transcription and translation are physically separated. **Review Figure 14.1. See Web/CD Activity 14.1**

▶ The genome of the single-celled budding yeast contains genes for the same metabolic machinery found in prokaryotes, with the addition of genes for protein targeting in the cell. **Review Table 14.2**

▶ The genome of the multicellular roundworm *Caenorhabditis elegans* contains genes required for intercellular interactions. **Review Table 14.3**

▶ The genome of the fruit fly has fewer genes than that of the roundworm. Many of its genes are homologs of genes found in the roundworm and mammalian genomes.

▶ The puffer fish genome is the most compact vertebrate genome known.

▶ The compact genome of the simple plant *Arabidopsis* is often used in the study of plant genomes. **Review Table 14.4**

▶ The rice genome is similar to that of *Arabidopsis*, and its sequence holds a key to feeding the increasing human population. **Review Table 14.5**

Repetitive Sequences in the Eukaryotic Genome

▶ Highly repetitive DNA is present in up to millions of copies of short sequences. It is not transcribed.

▶ Some moderately repetitive DNA sequences, such as those that code for rRNAs, are transcribed. **Review Figure 14.2**

▶ Some moderately repetitive DNA sequences are transposons, which are able to move about the genome. **Review Figure 14.3**

The Structures of Protein-Coding Genes

▶ A typical eukaryotic protein-coding gene is flanked by promoter and terminator sequences and contains noncoding internal sequences, called introns. **Review Figure 14.4**

▶ Nucleic acid hybridization is an important technique for analyzing eukaryotic genes. **Review Figure 14.5, 14.6**

▶ Some eukaryotic genes exist as families of related genes, which have similar sequences and code for similar proteins. These related proteins may be made at different times and in different tissues. Some sequences in gene families are pseudogenes, which code for nonfunctional mRNAs or proteins. **Review Figure 14.7**

▶ Differential expression of different genes in the β-globin cluster of the globin family ensures important physiological changes during human development. **Review Figure 14.8**

RNA Processing

▶ The transcribed pre-mRNA is altered by the addition of a G cap at the 5′ end and a poly A tail at the 3′ end. **Review Figure 14.9**

▶ The introns are removed from the mRNA precursor by the spliceosome, a complex of snRNPs and proteins. **Review Figure 14.10. See Web/CD Tutorial 14.1**

Transcriptional Regulation of Gene Expression

▶ Eukaryotic gene expression can be regulated at the transcriptional, posttranscriptional, translational, and posttranslational levels. **Review Figure 14.11. See Web/CD Activity 14.2**

▶ The major method of regulation of eukaryotic gene expression is selective transcription, which results from the binding of specific proteins to regulatory sequences on DNA.

▶ A series of transcription factors must bind to one another to form a transcription complex before RNA polymerase can bind. Whether RNA polymerase initiates transcription also depends on the binding of regulator proteins, activator proteins (which bind to enhancers and stimulate transcription), and repressor proteins (which bind to silencers and inhibit transcription). **Review Figures 14.12, 14.13. See Web/CD Tutorial 14.2**

▶ The simultaneous regulation of widely separated genes is possible through common sequences in their promoters, to which the same regulatory proteins bind. **Review Figure 14.14**

▶ The DNA-binding domains of most DNA-binding proteins have one of four structural motifs: helix-turn-helix, zinc finger, leucine zipper, or helix-loop-helix. **Review Figure 14.15**

▶ Chromatin remodeling allows the transcription complex to bind DNA and to move through the nucleosomes. **Review Figure 14.16**

▶ Heterochromatin is a condensed form of DNA that cannot be transcribed. It is found in the inactive X chromosome of female mammals. **Review Figure 14.17**

▶ Interference RNA (RNAi) is important in inhibiting transcription of the inactive X chromosome. **Review Figure 14.18**

▶ The movement of a gene to a new location on a chromosome may alter its ability to be transcribed, as in the change from one mating type to another in budding yeast.

▶ Some genes are selectively amplified in some cells. The extra copies of these genes result in increased transcription of their protein product. **Review Figure 14.19**

Posttranscriptional Regulation

▶ Alternative splicing of pre-mRNA can be used to produce different proteins. The transcripts of over half the genes in the human genome are alternatively spliced, which increases the number of proteins that can be encoded by a single gene. **Review Figure 14.20**

▶ The stability of mRNA in the cytoplasm can be regulated.

▶ Mature mRNA can be edited by the addition of new nucleotides or by the alteration of existing nucleotides. **Review Figure 14.21**

Translational and Posttranslational Regulation

▶ Translational repressors can inhibit the translation of mRNA.

▶ Proteasomes degrade proteins targeted for breakdown by attachment of ubiquitin. **Review Figure 14.22**

Self-Quiz

1. Eukaryotic protein-coding genes differ from their prokaryotic counterparts in that only eukaryotic genes
   a. are double-stranded.
   b. are present in only a single copy.
   c. contain introns.
   d. have a promoter.
   e. transcribe mRNA.
2. Comparison of the genomes of yeast and bacteria shows that only yeast has many genes for
   a. energy metabolism.
   b. cell wall synthesis.
   c. intracellular protein targeting.
   d. DNA binding proteins.
   e. RNA polymerase.
3. The genomes of a fruit fly and nematode work are similar to that of yeast, except that the former have many genes for
   a. intercellular signaling.
   b. synthesis of polysaccharides.
   c. cell cycle regulation.
   d. intracellular protein targeting.
   e. transposable elements.
4. Which of the following does *not* occur after mRNA is transcribed?
   a. binding of RNA polymerase II to the promoter
   b. capping of the 5′ end
   c. addition of a poly A tail to the 3′ end
   d. splicing out of the introns
   e. transport to the cytosol
5. Which statement about RNA splicing is *not* true?
   a. It removes introns.
   b. It is performed by small nuclear ribonucleoprotein particles (snRNPs).
   c. It always removes the same introns.
   d. It is usually directed by consensus sequences.
   e. It shortens the RNA molecule.
6. Eukaryotic transposons
   a. always use RNA for replication.
   b. are approximately 50 bp long.
   c. are made up of either DNA or RNA.
   d. do not contain genes coding for transposition.
   e. make up about half of the human genome.
7. Which statement about selective gene transcription in eukaryotes is *not* true?
   a. Different classes of RNA polymerase transcribe different parts of the genome.
   b. Transcription requires transcription factors.
   c. Genes are transcribed in groups called operons.
   d. Both positive and negative regulation occur.
   e. Many proteins bind at the promoter.
8. Heterochromatin
   a. contains more DNA than does euchromatin.
   b. is transcriptionally inactive.
   c. is responsible for all negative transcriptional control.
   d. clumps the X chromosome in human males.
   e. occurs only during mitosis.
9. Translational control
   a. is not observed in eukaryotes.
   b. is a slower form of regulation than transcriptional control.
   c. can be achieved by only one mechanism.
   d. requires that mRNA be uncapped.
   e. ensures that heme synthesis equals globin synthesis.
10. Control of gene expression in eukaryotes includes all of the following *except*
   a. alternative splicing of RNA transcripts.
   b. binding of proteins to DNA.
   c. transcription factors.
   d. feedback inhibition of enzyme activity by allosteric control.
   e. DNA methylation.

For Discussion

1. In rats, a gene 1,440 bp long codes for an enzyme made up of 192 amino acid units. Discuss this apparent discrepancy. How long would the initial and final mRNA transcripts be?
2. The genomes of rice, wheat, and corn are similar to each other and to the weed, *Arabidopsis*. Discuss how these plants might nevertheless have very different proteins.
3. The activity of the enzyme dihydrofolate reductase (DHFR) is high in some tumor cells. This activity makes the cells resistant to the anticancer drug methotrexate, which targets DHFR. Assuming that you had the complementary DNA for the gene that encodes DHFR, how would you show whether this increased activity was due to increased transcription of the single-copy DHFR gene or to amplification of the gene?
4. Describe the steps in the production of a mature, translatable mRNA from a eukaryotic gene that contains introns. Compare this to the situation in prokaryotes (see Chapter 13).
5. A protein-coding gene has three introns. How many different proteins can be made from alternate splicing of the pre-mRNA transcribed from this gene?

15 Cell Signaling and Communication

It's probably happened to you: it's late at night, you have a paper due, and you've put it off until the last minute. You're exhausted, but you need to stay awake and alert so that you can get your work done. What do you do? You have a cup (or several cups) of coffee. Many people turn to the caffeine in coffee when they need to wake themselves up and give themselves an energy boost.

To understand how caffeine works, we must understand the pathways by which the body's cells respond to signals in their environment. There are three sequential steps involved in the cell's response to any signal. First, the signal binds to a receptor protein in the cell, often on the outside surface of the plasma membrane. Second, the binding of the signal causes a message to be conveyed to the inside of the cell and amplified. Third, the cell changes its activity in response to the signal.

Caffeine acts in different ways in different tissues. A tired person's brain produces adenosine molecules that bind to specific receptor proteins, resulting in decreased brain activity and increased drowsiness. Caffeine's molecular structure is similar to that of adenosine, so it occupies the adenosine receptors without inhibiting brain cell function, and alertness is restored. In heart and liver cells, caffeine indirectly stimulates the same signaling pathway that is normally stimulated by epinephrine, the "fight-or-flight" hormone. In the heart, the result is an increased rate of beating; the liver is stimulated to convert glycogen into glucose and release it into the bloodstream.

We begin this chapter with a discussion of the signals that affect cells. As you will see, these signals include chemicals produced by other cells in the body as well as physical factors in the environment, such as light. Whatever the signal, it affects a cell only if that cell has a receptor protein that binds to that signal. In addition to binding the signal, the receptor must somehow communicate to the rest of the cell that binding has occurred. Finally, this communication process, called signal transduction, must result in a change in the function of the cell. We will describe these three steps of cell signaling—binding, transduction, and the cellular response—in order. We close the chapter with a description of how cells communicate with one another directly via specialized channels in their adjacent plasma membranes.

A Signal to the Body The caffeine in coffee sends signals to cells in the body of this coffee drinker. The effects of these signals help him stay alert.

Signals

Both prokaryotic and eukaryotic cells process information from their environment. This information can be in the form of a physical stimulus, such as the light reaching your eye as you read this book, or chemicals that bathe a cell, such as lactose in the medium surrounding *E. coli*. It may come from outside the organism, such as the scent of a female moth seeking a mate in the dark, or from a neighboring cell within the organism, as in the heart, where thousands of muscle cells contract in unison by transmitting signals to one another.

Of course, the mere presence of a signal does not mean that a cell will respond to it, just as you do not pay close attention to every sound in your environment as you study. To respond, the cell must have a specific receptor protein that can bind to the signal. In the following section, we will describe some of the signals different cells respond to and look at one model signal transduction pathway. After discussing signals, we will consider their receptors.

Cells receive signals from the physical environment and from other cells

The physical environment is full of signals. Our sense organs allow us to respond to light, odors and tastes (chemical signals), temperature, touch, and sound. Bacteria and protists respond to even minute chemical changes in their environment. Plants respond to light as a signal. For example, at sunset, at night, or in the shade, not only the amount , but also the wavelengths of the light reaching Earth's surface differ from that of full sunlight in the daytime. These variations act as signals that affect plant growth and reproduction. Some plants also respond to temperature: when the weather gets cold, they respond either by becoming tolerant to cold or by accelerating flowering. Even magnetism can be a signal: some bacteria and birds orient themselves to Earth's magnetic poles, like a needle on a compass.

A cell inside a large multicellular organism is far away from the exterior environment. Instead, its environment consists of other cells and extracellular fluids. Cells receive their nutrients from, and pass their wastes into, extracellular fluids or gases. Cells also receive signals—mostly chemical signals—from their extracellular fluid environment. Most of these chemical signals come from other cells. Cells also respond to chemical signals coming from the environment via the digestive and respiratory systems. And cells can respond to the concentrations of certain chemicals, such as CO_2 and H^+, whose presence in the extracellular fluids results from the metabolic activities of other cells.

Inside a large multicellular organism, chemical signals reach a target cell by local diffusion or by circulation within the blood. **Autocrine** signals are signals that affect the cells that make them. **Paracrine** signals are signals that diffuse to and affect nearby cells. Signals to distant cells, such as hormones, usually travel through the circulatory system (Figure 15.1).

In all cases, the cell must be able to receive or sense the signal and respond to it. Depending on the cell and the signal,

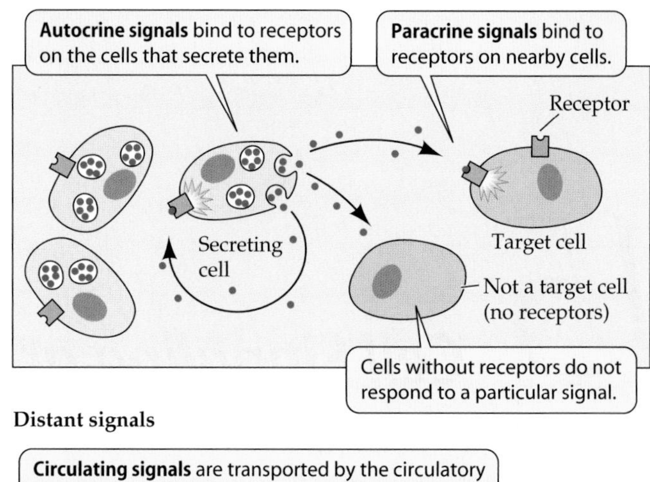
Local signals
Autocrine signals bind to receptors on the cells that secrete them.
Paracrine signals bind to receptors on nearby cells.
Receptor
Secreting cell
Target cell
Not a target cell (no receptors)
Cells without receptors do not respond to a particular signal.

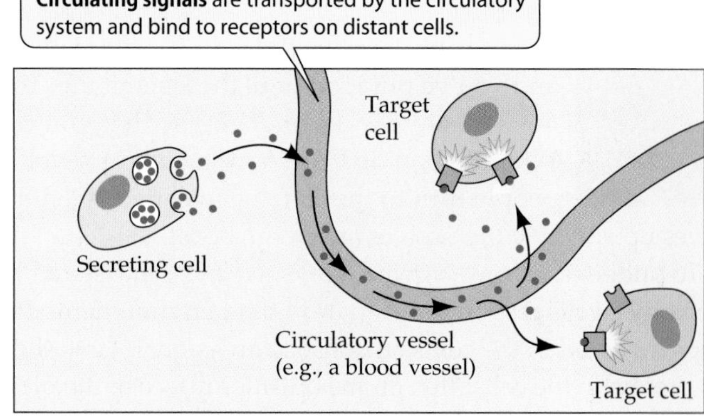
Distant signals
Circulating signals are transported by the circulatory system and bind to receptors on distant cells.
Target cell
Secreting cell
Circulatory vessel (e.g., a blood vessel)
Target cell

15.1 Chemical Signaling Systems A signal molecule can act on the cell that produces it, or on a nearby cell. Many signals act on distant cells, to which they are transported by the organism's circulatory system.

the responses range from entering the cell division cycle to heal a wound, to moving to a new location in the embryo to form a tissue, to releasing enzymes that digest food, to sending messages to the brain about the book you are reading.

A signal transduction pathway involves a signal, a receptor, transduction, and effects

The entire signaling process, from signal detection to final response, is called a **signal transduction pathway**. Let's look at an example of such a pathway in *E. coli* (Figure 15.2). In Chapter 13, we saw that this bacterium responds to changes in the nutrient content of its environment by altering its transcription of certain genes, such as those in the *lac* operon. The bacterium must also be able to sense and respond to other kinds of changes in its environment, such as changes in solute concentration.

In the human intestine, where *E. coli* lives, the solute concentration around the bacterium often rises far above that inside the cell. The principle of diffusion tells us that when this happens, water will diffuse out of the cell and solutes will move into the cell. But the bacterium must maintain

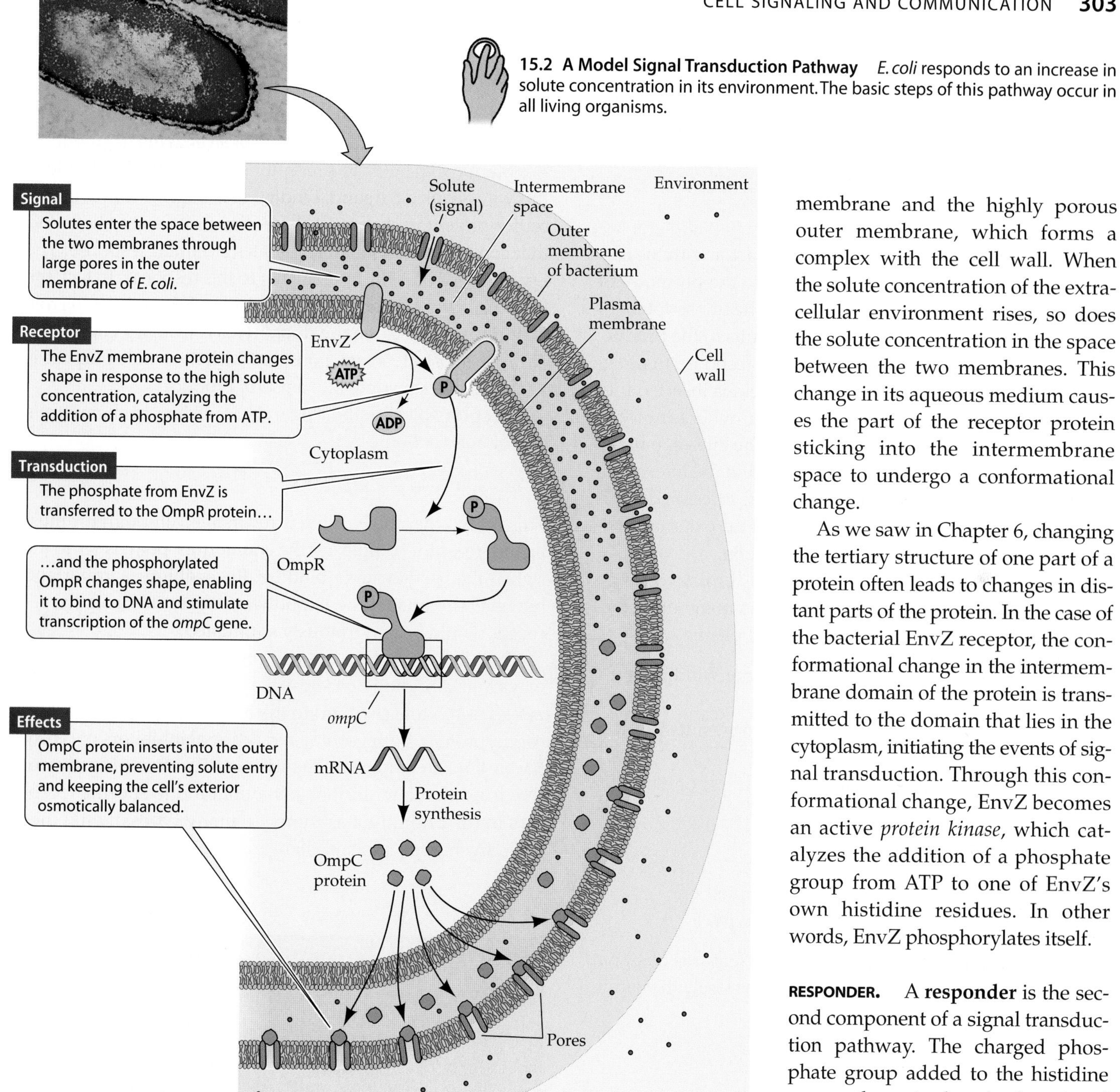

Signal
Solutes enter the space between the two membranes through large pores in the outer membrane of E. coli.
Receptor
The EnvZ membrane protein changes shape in response to the high solute concentration, catalyzing the addition of a phosphate from ATP.
Transduction
The phosphate from EnvZ is transferred to the OmpR protein...
...and the phosphorylated OmpR changes shape, enabling it to bind to DNA and stimulate transcription of the ompC gene.
Effects
OmpC protein inserts into the outer membrane, preventing solute entry and keeping the cell's exterior osmotically balanced.
Solute (signal)
Intermembrane space
Environment
Outer membrane of bacterium
Plasma membrane
Cell wall
EnvZ
ATP
ADP
P
Cytoplasm
OmpR
DNA
ompC
mRNA
Protein synthesis
OmpC protein
Pores

15.2 A Model Signal Transduction Pathway *E. coli* responds to an increase in solute concentration in its environment. The basic steps of this pathway occur in all living organisms.

homeostasis, so it must perceive and respond to this environmental change. The pathway by which *E. coli* does so has much in common with signal transduction pathways in more complex animals and plants. The pathway involves two major components: a receptor and a responder.

RECEPTOR. A **receptor** is the first component of a signal transduction pathway. The receptor protein in *E. coli* for changes in solute concentration is called EnvZ. It is a transmembrane protein that extends through the bacterium's plasma membrane into the space between the plasma membrane and the highly porous outer membrane, which forms a complex with the cell wall. When the solute concentration of the extracellular environment rises, so does the solute concentration in the space between the two membranes. This change in its aqueous medium causes the part of the receptor protein sticking into the intermembrane space to undergo a conformational change.

As we saw in Chapter 6, changing the tertiary structure of one part of a protein often leads to changes in distant parts of the protein. In the case of the bacterial EnvZ receptor, the conformational change in the intermembrane domain of the protein is transmitted to the domain that lies in the cytoplasm, initiating the events of signal transduction. Through this conformational change, EnvZ becomes an active *protein kinase*, which catalyzes the addition of a phosphate group from ATP to one of EnvZ's own histidine residues. In other words, EnvZ phosphorylates itself.

RESPONDER. A **responder** is the second component of a signal transduction pathway. The charged phosphate group added to the histidine causes the cytoplasmic domain of the EnvZ protein to change its shape again. It now binds to a second protein, OmpR, which takes the phosphate group from EnvZ. This phosphorylation changes the shape of OmpR in turn. This change in a responder is a key event in signaling for three reasons:

- The signal on the outside of the cell has now been *transduced* to a protein totally within the cell's cytoplasm.
- The phosphorylated OmpR can *do something*. That "something" is to bind to a promoter on *E. coli* DNA adjacent to the sequence that codes for the protein OmpC. This binding begins the final phase of the signal-

ing pathway: the *effect* of the signal, which is an alteration in cell function.

- The signal has been *amplified*. Because a single enzyme can catalyze the conversion of many substrate molecules, one EnvZ molecule alters the structure of many OmpR molecules.

Phosphorylated OmpR is a transcription factor with the correct three-dimensional structure to bind to the promoter of the *ompC* gene, resulting in an increase in the transcription of that gene. Translation of *ompC* mRNA results in the production of OmpC protein, which leads to the response that regulates osmotic pressure. The OmpC protein is inserted into the outer membrane of the bacterial cell, where it blocks pores and prevents solutes from entering the intermembrane space. As a result, the solute concentration in the intermembrane space is lowered, and osmotic balance is restored. Thus the *E. coli* cell can go on behaving just as if the external environment had a normal osmotic concentration.

Let's highlight the major features of this prokaryotic system, as the same elements will reappear in many other signal transduction pathways in animals and plants:

- A receptor changes its conformation upon binding with a signal.
- A conformational change in the receptor results in protein kinase activity.
- Phosphorylation alters the function of a responder protein.
- The signal is amplified.
- A transcription factor is activated.
- The synthesis of a specific protein is turned on.
- The action of the protein alters cell activity.

Now that we have surveyed the general features of signal transduction pathways, let's consider more closely the nature of the receptors that bind signals.

Receptors

Although a given cell in a multicellular organism is bombarded with many signals, it responds to only a few of them. The reason for this is that any particular cell makes receptors for only some signals. Which cells make which receptors is determined by the regulatory processes we studied in the previous chapter: If a cell transcribes the gene encoding a particular receptor and the resulting mRNA is translated, the cell will have that receptor.

A receptor protein binds to a signal very specifically, in much the same way as an enzyme binds to a substrate or a carrier protein binds to the molecule it is transporting across a membrane. This specificity of binding underlies the specificity of which cells respond to which signals.

Receptors have specific binding sites for their signals

A specific signal molecule fits into a site on its receptor much as a substrate fits into the active site of an enzyme (Figure 15.3). A molecule that binds to a receptor site in another molecule in this way is called a **ligand**. Binding of the ligand causes the receptor protein to change its three-dimensional structure, and that conformational change initiates a cellular response. The ligand does not contribute further to this response. In fact, the ligand usually is not metabolized into useful products. Its role is purely to "knock on the door." This is in sharp contrast to enzyme–substrate interactions, in which the whole purpose is to change the substrate into a useful product.

Receptors bind to their ligands according to chemistry's law of mass action:

$$R + L \rightleftharpoons RL$$

This means that the binding is reversible, although for most ligand–receptor complexes, the equilibrium point is far to the right—that is, they favor binding. Reversibility is important, however, because if the ligand were never released, the receptor would be continuously stimulated.

The binding of a ligand to a receptor is similar in many ways to the binding of a substrate to an enzyme. As with enzymes, inhibitors can bind to the ligand binding site on a receptor protein. Both natural and artificial inhibitors of receptor binding are important in medicine. For example, many of the drugs that alter human behavior bind to specific receptors in the brain. Just as there are many types of enzymes with diverse specificities, there are many kinds of receptors.

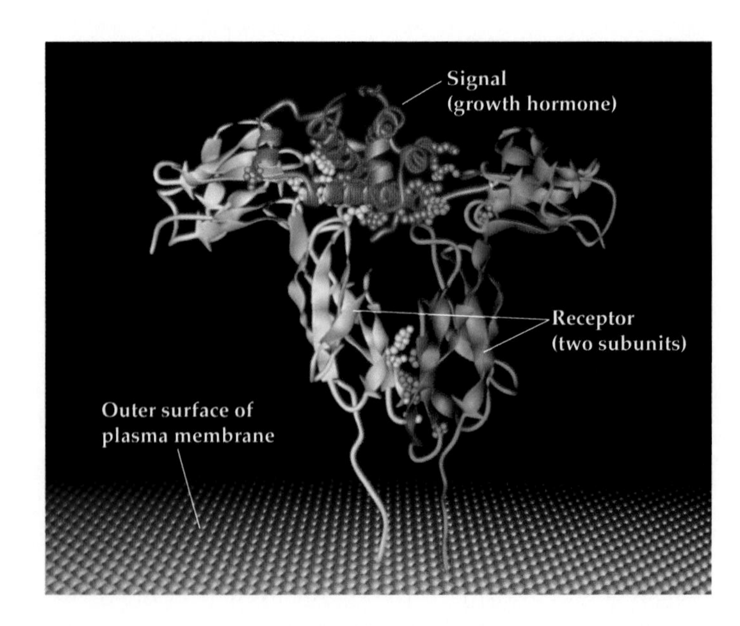

15.3 A Signal Bound to Its Receptor Human growth hormone is shown bound to its receptor, a transmembrane protein. Only the extracellular regions of the receptor are shown.

There are several types of receptors

A major division among receptors is in their cellular location, which largely depends on the nature of their ligands. The chemistry of signal molecules is quite variable, but they can be divided into two classes (Figure 15.4):

- *Ligands with cytoplasmic receptors:* Small and/or nonpolar ligands can diffuse across the lipid bilayer of the plasma membrane and enter the cell. Estrogen, for example, is a lipid-soluble steroid hormone that can easily diffuse across the plasma membrane and enter the cell; it binds to a receptor in the cytoplasm.
- *Ligands with plasma membrane receptors:* Large and/or polar ligands cannot cross the plasma membrane. Insulin, for example, is a protein hormone that cannot diffuse through the plasma membrane; instead, it binds to a receptor that is a transmembrane protein with an extracellular binding domain.

In complex eukaryotes such as mammals, there are three well-studied types of receptors on plasma membranes: ion channels, protein kinases, and G protein-linked receptors.

ION CHANNEL RECEPTORS. In the plasma membranes of many types of cells are channel proteins that can be open or closed. These **ion channels** act as "gates," allowing ions such as Na^+, K^+, Ca^{2+}, or Cl^- to enter or leave the cell. The gate-opening mechanism is an alteration in the three-dimensional structure of the channel protein upon ligand binding. Some ion channels are membrane receptors for signal molecules; others act later in signal transduction pathways. Each type of ion channel receptor has its own signal. These signals include sensory stimuli, such as light and sound, charge differences across the plasma membrane, and chemical ligands such as hormones and neurotransmitters.

The acetylcholine receptor, which is located at the plasma membranes of vertebrate skeletal muscle cells, is an example of a gated ion channel. This receptor protein binds the ligand *acetylcholine*, which is released from nerve cells (Figure 15.5). When two molecules of acetylcholine bind to the receptor, it opens for about a thousandth of a second. That is enough time for Na^+, which is more concentrated outside the cell than inside, to rush into the cell. The change in Na^+ concentration in the cell results in muscle contraction.

PROTEIN KINASES. Like the EnvZ protein of *E. coli*, some eukaryotic receptor proteins become protein kinases when they are activated: that is, they catalyze the transfer of a phosphate group from ATP to a specific protein, referred to as the *target protein*. This phosphorylation can alter the conformation and activity of the target protein.

The receptor for insulin is an example of a protein kinase receptor. Insulin is a protein hormone made by the mammalian pancreas. Its receptor has two copies each of two different polypeptide subunits (Figure 15.6). As with acetylcholine, two molecules of insulin must bind to the receptor. When insulin binds to its extracellular subunits, the recep-

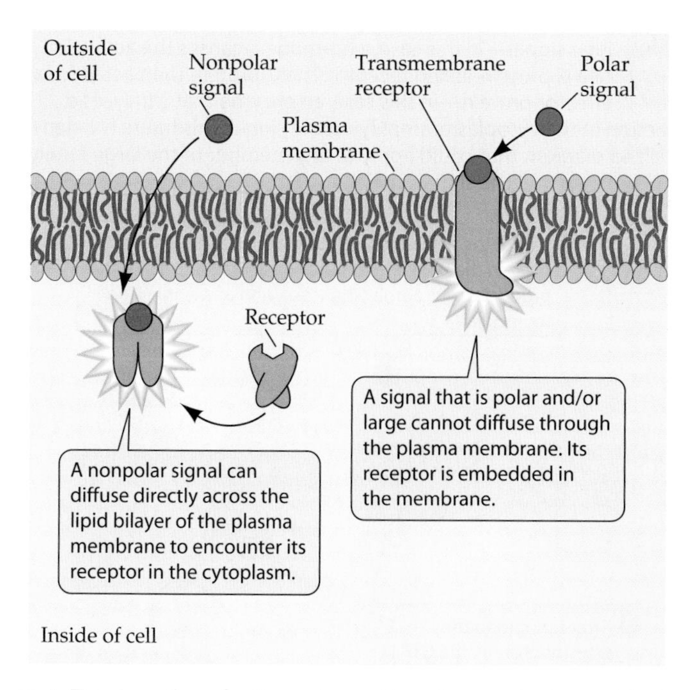

15.4 Two Locations for Receptors Receptors can be located in the plasma membrane or in the interior of the cell.

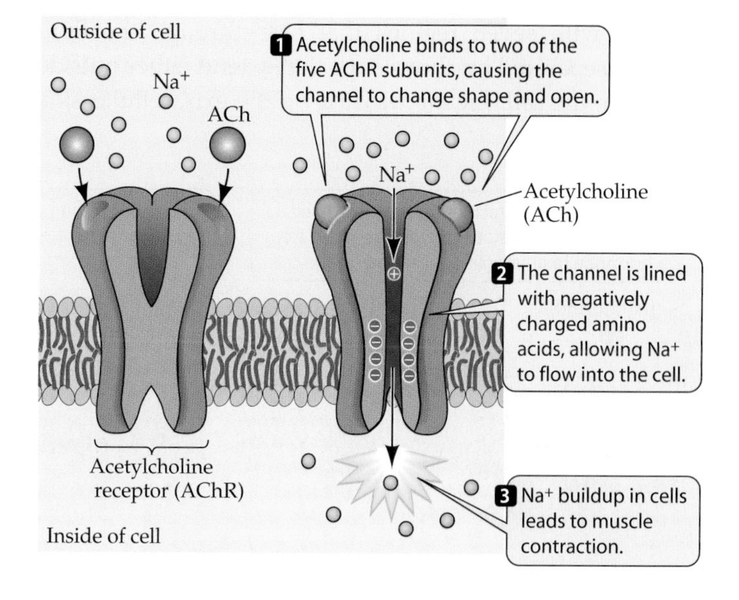

15.5 A Gated Ion Channel The acetylcholine receptor (AChR) is a gated ion channel for sodium ions. It is made up of five polypeptide subunits. When acetylcholine molecules (ACh) bind to two of the subunits, the gate opens and Na^+ flows into the cell.

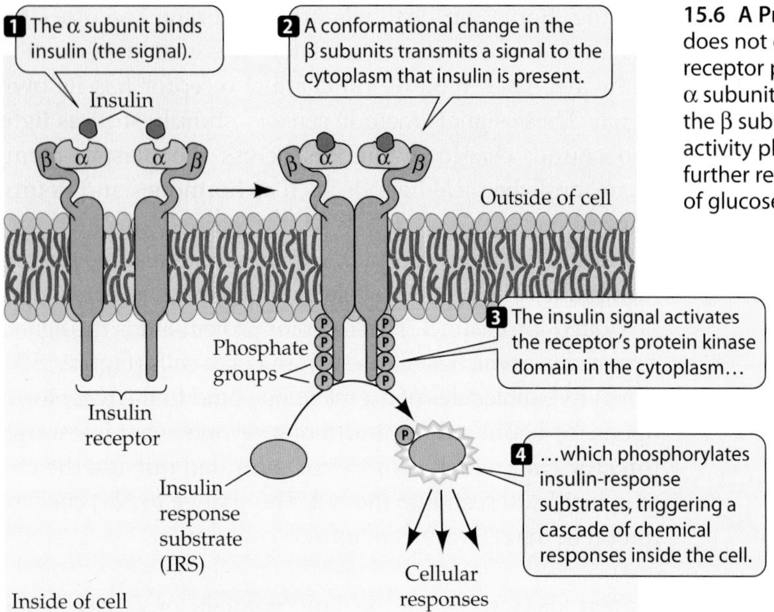
1 The α subunit binds insulin (the signal).
2 A conformational change in the β subunits transmits a signal to the cytoplasm that insulin is present.
Insulin
Outside of cell
3 The insulin signal activates the receptor's protein kinase domain in the cytoplasm...
Phosphate groups
Insulin receptor
4 ...which phosphorylates insulin-response substrates, triggering a cascade of chemical responses inside the cell.
Insulin response substrate (IRS)
Cellular responses
Inside of cell

15.6 A Protein Kinase Receptor The mammalian hormone insulin does not enter the cell, but is bound by the extracellular domain of a receptor protein with four subunits (two α and two β). Binding to the α subunit causes a conformational change in the cytoplasmic domain of the β subunits, exposing a protein kinase active site. This protein kinase activity phosphorylates insulin response substrate proteins, triggering further responses within the cell and eventually resulting in the transport of glucose across the membrane into the cell.

tor changes its shape to expose a cytoplasmic protein kinase active site. Like the EnvZ receptor described above, the insulin receptor autophosphorylates. Then, as a protein kinase, it catalyzes the phosphorylation of certain cytoplasmic proteins, appropriately called insulin response substrates. These proteins then initiate many cellular responses, including the insertion of glucose transporters into the plasma membrane.

G PROTEIN-LINKED RECEPTORS. A third category of eukaryotic plasma membrane receptors is the *seven-spanning G protein-linked receptors*. This long name identifies a fascinating group of receptors, all of which are composed of a single protein with seven regions that pass through the lipid bilayer, separated by short loops that extend either outside or inside the cell. Ligand binding on the extracellular side of the receptor changes the shape of its cytoplasmic region, exposing a binding site for a mobile membrane protein.

This membrane protein, known as a **G protein**, has two important binding sites: one for the G protein-linked receptor and the other for the nucleotide GDP/GTP (Figure 15.7). G proteins have several polypeptide subunits. When the G protein binds to the activated receptor, one of its subunits binds GTP. At the same time, the ligand is released from the extracellular side of the receptor. The GTP-bound subunit of the G protein now separates from the parent G protein, diffusing in the plane of the lipid bilayer until it encounters an effector protein to which it can bind.

An *effector protein* is just what its name implies: It causes an effect in the cell. The binding of the GTP-bearing G protein subunit activates the effector—which may be an enzyme or an ion channel—thereby causing changes in cell function.

After binding to the effector protein, the GTP on the G protein is hydrolyzed to GDP. The now inactive G protein

15.7 A G Protein-Linked Receptor Binding of an extracellular signal—in this case, a hormone—causes the activation of a G protein-linked receptor. The G protein then activates an effector protein—in this case, an enzyme that catalyzes a reaction in the cytoplasm, amplifying the signal. This figure is a generalized diagram that could apply to any member of the large family of G proteins and the signals they react to.

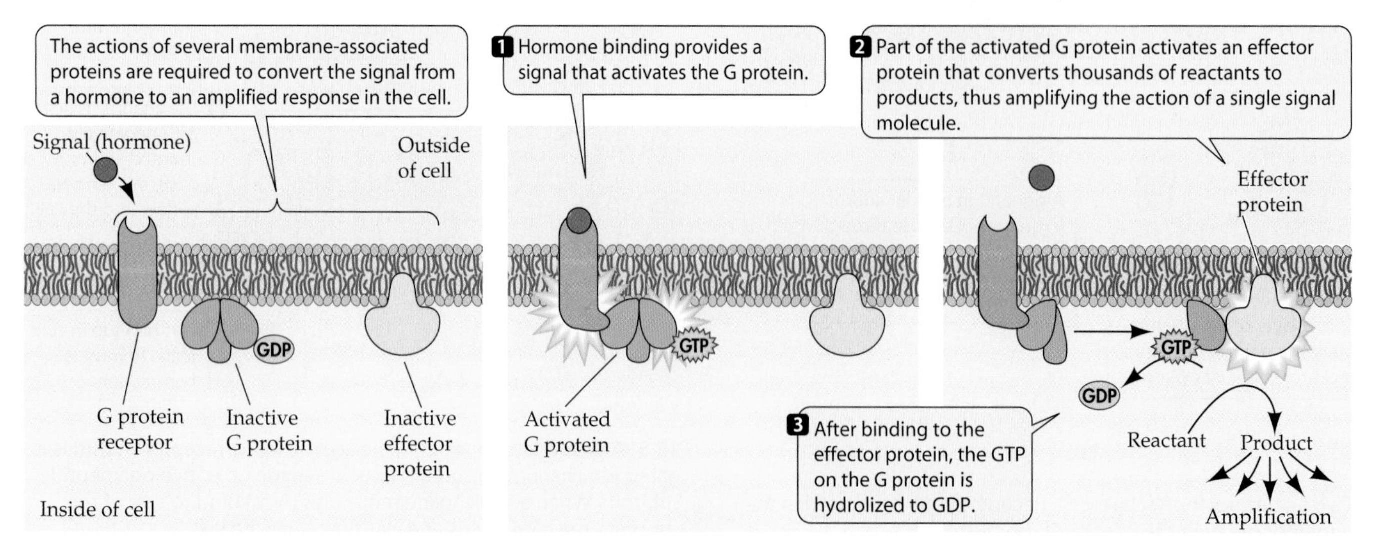
The actions of several membrane-associated proteins are required to convert the signal from a hormone to an amplified response in the cell.
1 Hormone binding provides a signal that activates the G protein.
2 Part of the activated G protein activates an effector protein that converts thousands of reactants to products, thus amplifying the action of a single signal molecule.
Signal (hormone)
Outside of cell
Effector protein
GDP
GTP
G protein receptor
Inactive G protein
Inactive effector protein
Activated G protein
3 After binding to the effector protein, the GTP on the G protein is hydrolized to GDP.
Reactant
Product
Amplification
Inside of cell

subunit separates from the effector protein. The G protein subunit must form a complex with other subunits before binding to yet another activated receptor. When an activated receptor is bound, the G protein exchanges its GDP for GTP, and the cycle begins again.

By means of their diffusing subunits, G proteins can either activate or inhibit an effector. An example of an *activating* response involves the receptor for epinephrine (adrenaline), hormone made by the adrenal gland in response to stress or heavy exercise. In heart muscle, this hormone binds to its G protein-linked receptor, activating a G protein . The GTP-bound subunit then activates a membrane-bound enzyme to produce a small molecule, cyclic AMP (see below), that has many effects on the cell, including glucose mobilization for energy and muscle contraction.

G protein-mediated *inhibition* occurs when the same hormone, epinephrine, binds to its receptor in the smooth muscle cells surrounding blood vessels lining the digestive tract. Again, the epinephrine-bound receptor changes its shape and activates a G protein, and the GTP-bound subunit binds to a target enzyme. But in this case, the enzyme is inhibited instead of being activated. As a result, the muscles relax and the blood vessel diameter increases, allowing more nutrients to be carried away from the digestive system to the rest of the body. Thus the same signal and initial signaling mechanism can have different consequences in different cells, depending on the nature of the responding cell.

CYTOPLASMIC RECEPTORS. Receptors for signals that can diffuse across the plasma membrane are located inside the cell. Binding to the ligand causes the receptor to change its shape so that it can enter the cell nucleus, where it acts as a transcription factor (Figure 15.8). But this general view is somewhat simplified. The receptor for the hormone cortisol, for example, is normally bound to a chaperone protein, which blocks it from entering the nucleus. Binding of the hormone causes the receptor to change its shape so that the chaperone is released. This allows the receptor, which is a transcription factor, to fold into an appropriate conformation for entering the nucleus and initiating transcription.

Having discussed signals and receptors, we now turn our attention to the characteristics of transducers.

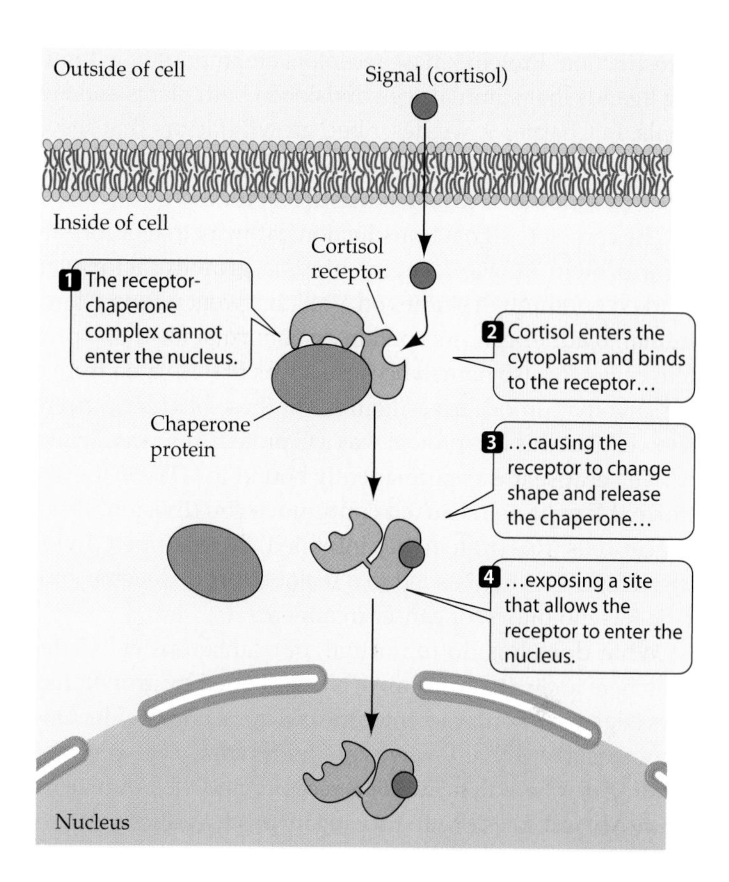

15.8 A Cytoplasmic Receptor The receptor for cortisol is bound to a chaperone protein. Binding of the signal releases the chaperone and allows the receptor protein to enter the cell's nucleus, where it functions as a transcription factor.

Signal Transduction

As we have just seen, the same signal may produce different responses in different tissues. When epinephrine, for example, binds to receptors on heart muscle cells, it stimulates muscle contraction, but when it binds to receptors on smooth muscle cells in the blood vessels of the digestive system, it slows muscle contraction. These different responses to the same signal–receptor complex are mediated by the events of signal transduction. These events, which are critical to the cell's response, may be either direct or indirect.

- *Direct transduction* is a function of the receptor itself and occurs at the plasma membrane.
- In *indirect transduction*, which is more common, another molecule, termed a **second messenger**, mediates the interaction between receptor binding and cellular response.

In neither case is transduction a single event. Rather, the signal initiates a cascade of events, in which proteins interact with other proteins until the final responses are achieved. Through such a cascade, a weak initial signal can be both amplified and distributed to cause several different responses in the target cell.

Protein kinase cascades amplify a response to receptor binding

We have seen that when a signal binds to a protein kinase receptor, the receptor changes its conformation to expose a protein kinase active site, which catalyzes the phosphorylation of target proteins. This process is an example of direct signal transduction. Protein kinase receptors are important in binding ligands that stimulate cell division in both plants and animals. In Chapter 9, we described growth factors that serve as external inducers of the cell cycle. These growth factors work by binding to protein kinase receptors.

The complete signal transduction pathway that occurs after a protein kinase receptor binds a growth factor was worked out through studies on a cell that went wrong. Many human bladder cancers contain an abnormal form of a protein called Ras (so named because it was first isolated from a *rat* sarcoma tumor). Investigations of these bladder cancers showed that this Ras protein was a G protein, but was always active because it was permanently bound to GTP. So the abnormal Ras protein caused continuous cell division. If the cancer cells' Ras protein was inhibited, they stopped dividing. This discovery has led to a major effort to develop specific Ras inhibitors for cancer treatment.

What does Ras do in normal, noncancerous cells? Researchers knew that cells must be stimulated by growth factors (signals) in order to enter the cell cycle and divide. One hypothesis was that Ras was an intermediary between the binding of a growth factor to its receptor and the ultimate response of cell division. To investigate this hypothesis, the researchers treated cells in a culture dish with both a Ras inhibitor and a growth factor. Cell division did not occur, confirming their hypothesis.

After this discovery, the next step was to work out what the activated growth factor receptor did to Ras, and what Ras did to stimulate further events in signal transduction. This signaling pathway has been worked out, and it is an example of a more general phenomenon, called a **protein kinase cascade** (Figure 15.9). Such cascades are key to the external regulation of many cellular activities. Indeed, the eukaryotic genome codes for hundreds, even thousands, of such kinases.

The unbound receptors for growth factors exist in the plasma membrane as separate polypeptide chains (subunits). When the growth factor signal binds to a subunit, it associates with another subunit to form a dimer, which changes its shape to expose a protein kinase active site. The kinase activity sets off a series of events, activating several other pro-

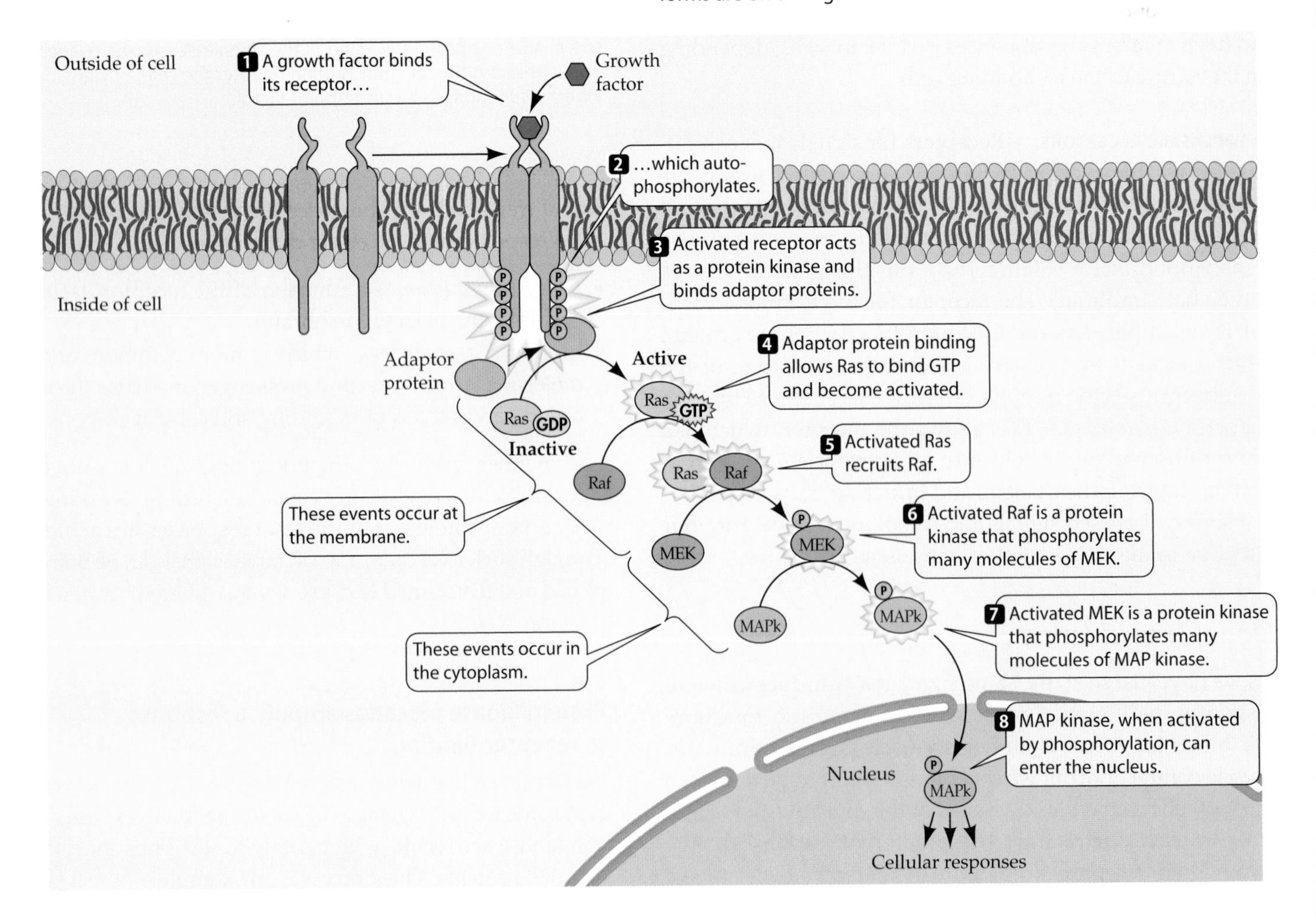

15.9 A Protein Kinase Cascade In a protein kinase cascade, a series of proteins are sequentially activated. In this example, the growth factor receptor protein stimulates the G protein Ras, which mediates a cascading series of reactions. The final product of the cascade, MAP kinase (MAPk), enters the nucleus and causes changes in transcription. Inactive forms of the proteins are on the left, activated forms are on the right.

tein kinases in turn. The final phosphorylated, activated protein—MAP kinase—moves into the nucleus and phosphorylates target proteins that are necessary for cell division.

Protein kinase cascades are useful signal transducers for three reasons:

- At each step in the cascade of events, the signal is amplified, because each newly activated protein kinase is an enzyme, which can catalyze the phosphorylation of many target proteins.
- The information from a signal that originally arrived at the plasma membrane is communicated to the nucleus.
- The multitude of steps provides some specificity to the process. As we have seen with epinephrine, signal binding and receptor activation do not result in the same response in all cells. Different target proteins at each step in the cascade can provide variation in the response.

Cyclic AMP is a common second messenger

As we have just seen, protein kinase receptors initiate the protein kinase cascade right at the plasma membrane. However, the stimulation of events in the cell is more often indirect. In a series of clever experiments, Earl Sutherland, Edwin Krebs, and Edmond Fischer showed that in many cases, there is a small, water-soluble chemical messenger between the membrane receptor and cytoplasmic events. These researchers were investigating the activation of the liver enzyme *phosphorylase* by the hormone epinephrine. Phosphorylase catalyzes the hydrolysis of glycogen stored in the liver so that the resulting glucose molecules can be released to the blood to fuel the fight-or-flight response.

The researchers found that phosphorylase could be activated in liver cells that had been broken open, but only if the entire cell contents, including the plasma membrane fragments, were present. They observed that epinephrine had bound to the plasma membrane, but active phosphorylase was present in the cytoplasm. They hypothesized that there must be some chemical messenger that transmits the message of epinephrine binding (at the membrane) to phosphorylase (in the cytoplasm). To investigate the production of this message, they tried the following steps in sequence:

- First, they incubated plasma membranes of broken liver cells with epinephrine.
- Then they removed the membranes, but kept the solution in which the membranes had been incubated.
- Then they added this solution to the contents of the cytoplasm, which contained inactive phosphorylase.

The phosphorylase became activated, confirming their hypothesis. Hormone binding to the membrane receptor had caused the production of a small, water-soluble molecule that then diffused to the cytoplasm, where it activated the enzyme. This small molecule was identified as **cyclic AMP (cAMP)**, which we encountered in Chapter 13 in the *lac* operon regulatory system in *E. coli*. Here, cAMP was working as a second messenger.

Second messengers are substances released into the cytoplasm after the first messenger—the signal—binds its receptor. In contrast to the specificity of receptor binding, second messengers affect many processes in the cell, and they allow a cell to respond to a single event at the plasma membrane with many events inside the cell. Like the protein kinase cascade, second messengers amplify the signal—a single epinephrine molecule leads to the production of several dozen molecules of cAMP, which then activate many enzyme targets.

Adenylyl cyclase, the enzyme that catalyzes the formation of cAMP from ATP, is located on the cytoplasmic surface of the plasma membrane of target cells (Figure 15.10). Usually, it is activated by the binding of G proteins, themselves activated by receptors.

Second messengers do not have enzymatic activity; rather, they act as cofactors or allosteric regulators of target enzymes. Cyclic AMP has two major target types. In many kinds of sensory cells, cAMP binds to ion channels to open them. Cyclic AMP may also binds to an enzyme in the cytoplasm, such as a protein kinase, whose active site is exposed as a result. A protein kinase cascade ensues, leading to the final effects in the cell.

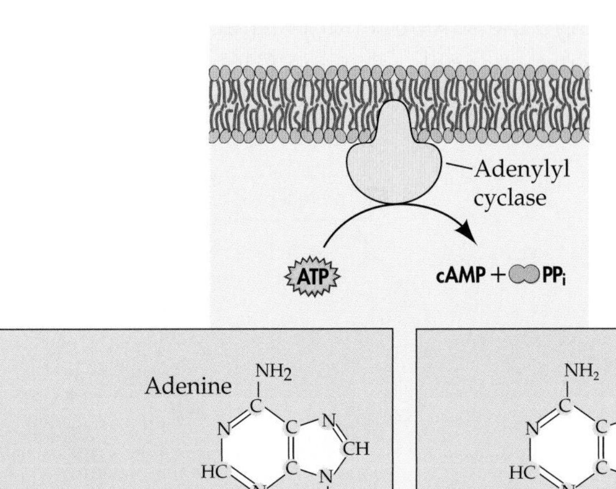

15.10 The Formation of Cyclic AMP The formation of cAMP from ATP is catalyzed by adenylyl cyclase, an enzyme that is activated by G proteins.

Two second messengers are derived from lipids

Phospholipids, in addition to their roles as structural components of the plasma membrane, are involved in signal transduction. When certain phospholipids are hydrolyzed into their component parts (see Figure 3.20) by enzymes called *phospholipases*, second messengers are formed.

The best-studied of these second messengers come from hydrolysis of the phospholipid **phosphatidyl inositol-bisphosphate (PIP2)**, which, like all phospholipids, has a hydrophobic portion (two fatty acid tails attached to a molecule of glycerol, which together form **diacylglycerol**, or **DAG**) embedded in the plasma membrane and a hydrophilic portion (**inositol triphosphate**, or **IP_3**) projecting into the cytoplasm. In mammals, there are over two dozen signals whose actions are mediated by the products of PIP2 hydrolysis. As with cAMP, the receptors involved are often G protein-linked receptors. The activated G protein subunits diffuse within the plasma membrane and activate an enzyme, phospholipase C. This enzyme cleaves off the IP_3 from PIP2, leaving the glycerol and the two attached fatty acids (DAG) in the lipid bilayer:

$$\underset{\text{(in membrane)}}{\text{P1P2}} \xrightarrow{\text{Phospholipase C}} \underset{\substack{\text{released to}\\ \text{cytoplasm}}}{\text{IP}_3} + \underset{\text{(in membrane)}}{\text{DAG}}$$

IP_3 and DAG are both second messengers and have different modes of action that build on each other (Figure 15.11). DAG activates a membrane-bound enzyme, protein kinase C (PKC). PKC is dependent on Ca^{2+} (hence the "C"), and that is where IP_3 plays an essential role. IP_3 diffuses through the cytoplasm to the smooth endoplasmic reticulum, where it opens an ion channel, releasing Ca^{2+} into the cytoplasm. There, in combination with DAG, the Ca^{2+} causes PKC to become active. PKC can then phosphorylate a wide variety of proteins, leading to the ultimate response of the cell.

In this transduction system, DAG and IP_3 function as second messengers, but Ca^{2+} plays a role in the pathway. In some cases, however, Ca^{2+} can itself serve as the second messenger in a signal transduction pathway.

Calcium ions are involved in many signal transduction pathways

Calcium ions are scarce in most cells, with a cytoplasmic concentration of only about 0.1 μM, while the concentrations of Ca^{2+} outside the cell and within the endoplasmic reticulum are usually much higher. This difference is maintained by active transport proteins at the plasma and ER membranes that pump the ion out of the cytoplasm. Unlike cAMP and the lipid second messengers, the level of intracellular Ca^{2+} cannot be increased by making more of it. Instead, the opening and closing of ion channels and the action of membrane pumps regulate levels of the ion in a cellular compartment.

There are many signals that can cause Ca^{2+} channels to open, including IP_3 (as we saw in the previous section) and the entry of a sperm into an egg (Figure 15.12). Whatever the signal, the open channels result in a dramatic increase in cytoplasmic Ca^{2+} concentration, up to a hundredfold within a fraction of a second. As we saw earlier, this increase activates

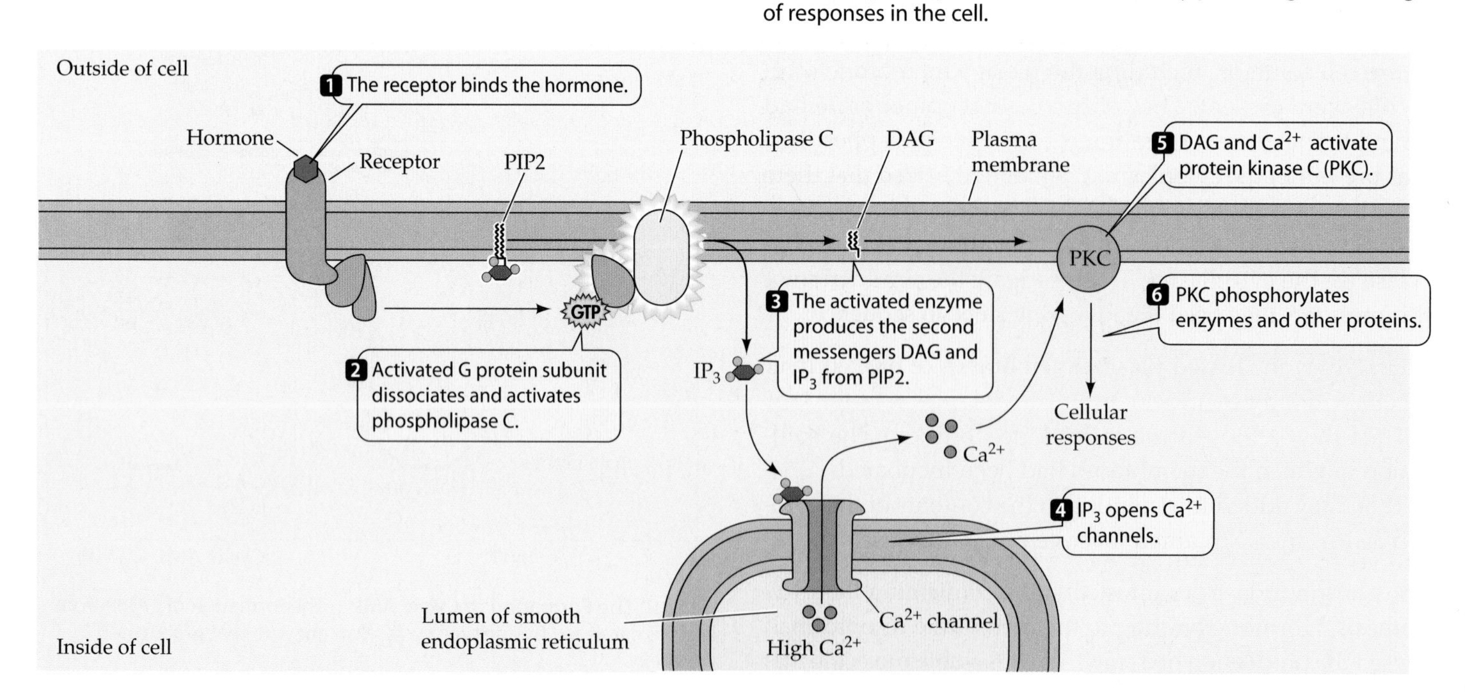

15.11 The IP_3 and DAG Second Messenger System Phospholipase C hydrolyzes the phospholipid PIP2 into its components, IP_3 and DAG, both of which are second messengers. IP_3 and DAG act separately but in concert, ultimately producing a wide range of responses in the cell.

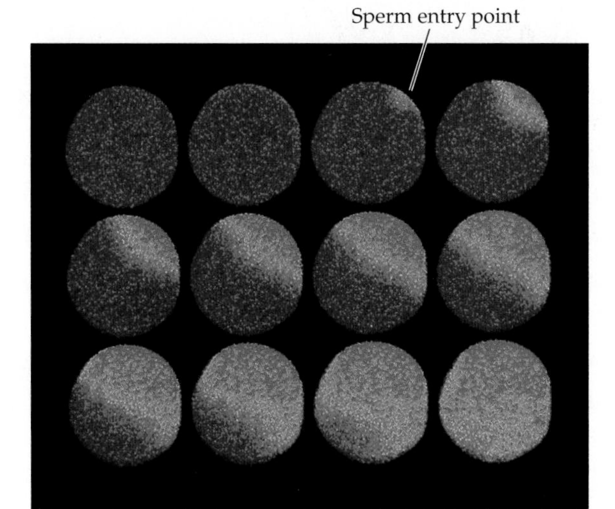

15.12 Calcium Ions as a Second Messenger The concentration of Ca^{2+} can be measured by a dye that fluoresces and turns red when it binds the ion. Here, fertilization causes a wave of Ca^{2+}, photographed at 5-second intervals, to pass through the egg of a sea star (starfish). This signal delivers the message that fertilization is complete and development can begin.

protein kinase C. In addition, Ca^{2+} controls other ion channels and stimulates secretion by exocytosis.

A distinctive aspect of Ca^{2+} signaling is that the ion can stimulate its own release from intracellular stores. For example, in some plant leaf cells, the hormone abscisic acid binds to gated Ca^{2+} channels in the plasma membrane and opens them, causing the ion to rush into the cells. This influx is not enough to trigger the cell's response, however. The ion binds to Ca^{2+} channels in the endoplasmic reticulum and in the membranes of vacuoles, causing those organelles to release their Ca^{2+} stores as well.

In some cases, Ca^{2+} ions act via a calcium-binding protein called **calmodulin**, and it is the Ca^{2+}–calmodulin complex that performs cellular functions by binding to target proteins. Calmodulin, which is present in many cells, has four binding sites for Ca^{2+}. When the cytoplasmic Ca^{2+} concentration is low, calmodulin does not bind enough Ca^{2+} to become activated. But when the cell is stimulated by a signal that causes a rise in the Ca^{2+} level, all four binding sites are filled. The calmodulin then changes its shape and binds to a number of cellular targets, activating them in turn. One such target is a protein kinase in smooth muscle cells that phosphorylates the muscle protein myosin, initiating contraction.

Nitric oxide is a gas that can act as a second messenger

Pharmacologist Robert Furchgott, at the State University of New York in Brooklyn, was investigating how acetylcholine causes the smooth muscles lining blood vessels to relax, thus allowing more blood to flow to certain organs. Acetylcholine appeared to stimulate the IP_3 signal transduction pathway to produce an influx of Ca^{2+}, which led to an increase in the level of another second messenger, **cyclic GMP (cGMP)**. This nucleotide bound to a protein kinase, which then stimulated a kinase cascade leading to muscle relaxation. So far, the pathway seemed straightforward.

But while this pathway seemed to work in intact animals, it did not work on isolated strips of artery tissue. When Furchgott switched to tubular sections of artery, however, signal transduction did occur. There turned out to be a crucial difference between these two tissue preparations: In the strips, the delicate inner layer of cells that lines blood vessels had been lost. Furchgott hypothesized that this layer, the *endothelium*, was making something that diffused into the muscle cells and was needed for their response to acetylcholine. The substance was not easy to isolate. It seemed to break down quickly, with a half-life (the time in which half of it disappeared) of 5 seconds in living tissues. It turned out to be a gas, **nitric oxide (NO)**, that had been thought of only as a toxic air pollutant!

In the body, NO is made from arginine by an enzyme, *NO synthase*. This enzyme is activated by Ca^{2+}, which enters the endothelial cells through a channel opened by PIP2, which is released after acetylcholine binds to its receptor. The NO formed is chemically very unstable, and although it diffuses readily, it does not get far. Conveniently, the endothelial cells are close to the smooth muscle cells, where NO acts as a second messenger. In smooth muscle, NO activates an enzyme called guanylyl cyclase, catalyzing the formation of cGMP, which in turn relaxes the muscle cells (Figure 15.13).

The spectacular discovery of NO as a second messenger explained the action of nitroglycerin, a drug that has been used for over a century to treat angina, the chest pain caused by insufficient blood flow to the heart. Nitroglycerin releases NO, which results in relaxation of the blood vessels and increased blood flow. Penile erection is also caused by the dilation of blood vessels in that organ, and the new drugs that promote erection are NO synthesis activators.

Signal transduction is highly regulated

There are several ways in which cells can regulate the activity of a transducer. The concentration of NO, which breaks down quickly, can be regulated only by how much of it is made. The level of Ca^{2+}, on the other hand, is determined by both membrane pumps and ion channels. For protein kinase cascades, G proteins, and cAMP, there are enzymes that convert the activated transducer back to its inactive precursor:

- *Protein phosphatases* remove the phosphate groups from phosphorylated proteins.
- *GTPases* convert the GTP on an active G protein back to GDP, inactivating the protein.
- *cAMP phosphodiesterase* converts cAMP into its precursor, AMP, which has no second messenger activity.

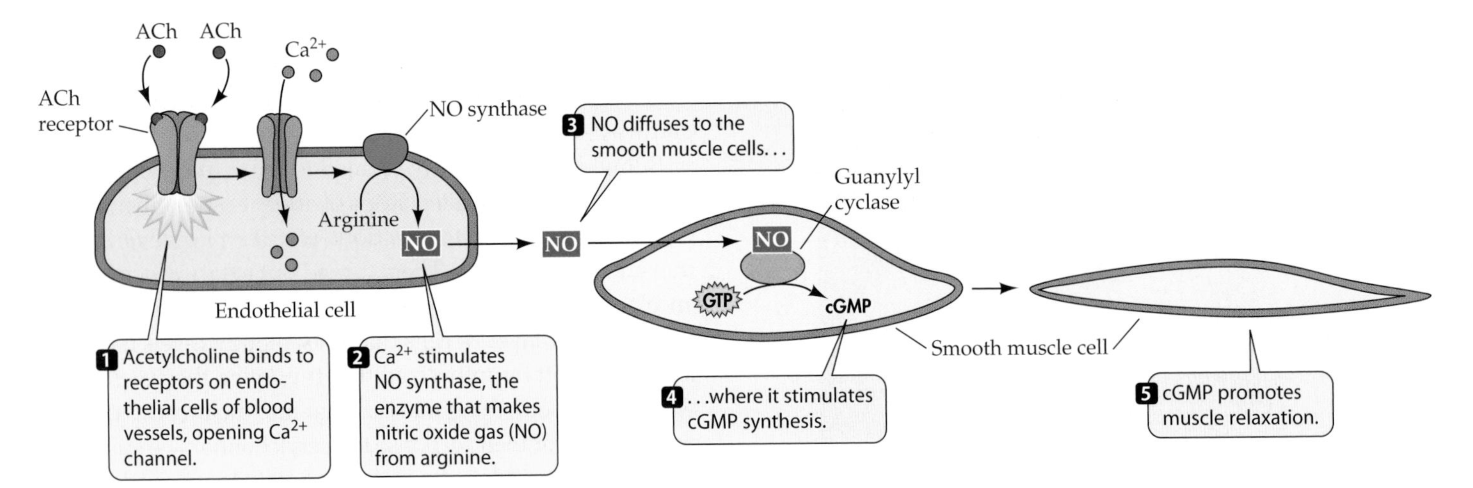

15.13 Nitric Oxide as a Second Messenger Nitric oxide (NO) is an unstable gas, which nevertheless serves as a second messenger between a primary signal, acetylcholine, and its effect, the relaxation of smooth muscles. The endothelial cells that line blood vessels, seen in *(a)*, are crucial intermediaries in this signal transduction pathway *(b)*.

Signal Effects: Changes in Cell Function

We have seen how the binding of an environmental signal to its receptor initiates the response of a cell to the signal, and how the direct or indirect transduction of the signal to the inside of the cell amplifies the signal. In this section, we consider the third and final step in the signal transduction process, the actual effects of the signal on cell function. These effects primarily take the form of the opening of ion channels, changes in the activities of enzymes, or differential gene transcription.

Ion channels are opened

The opening of ion channels is a key step in the response of the nervous system to signals. Sensory nerve cells of the sense organs, for example, become stimulated through the opening of ion channels. We will focus here on one such signal transduction pathway, that for the sense of smell, which responds to gaseous molecules (Figure 15.14).

The sense of smell is well developed in mammals, some of which have an amazing 1,000 genes for odor signal receptors—the largest gene family known. Each of the thousands of nerve cells in the nose expresses one of these receptors. The identification of which chemical signal, or odorant, activates which receptor is just getting under way.

When an odorant molecule binds to its receptor, a G protein becomes activated, which in turns activates adenylyl cyclase, which catalyzes the formation of the second messenger cAMP. This molecule then binds to an ion channel, causing it to open. The resulting influx of Na^+ causes the nerve cell to become stimulated so that it sends a signal to the brain that a particular odor is present.

Enzyme activities are changed

Proteins will change their shape, and their functioning, if they are modified either covalently or noncovalently. We have seen examples of both types of modification in signal transduction. Protein kinases add phosphate groups to a target protein, and this covalent change alters the protein's conformation. Cyclic AMP binds to target proteins allosterically, and this noncovalent interaction changes the protein's conformation. In both cases, previously inaccessible active sites are exposed, and the target protein goes on to perform a cellular role.

The G protein-mediated protein kinase cascade stimulated by epinephrine in liver cells results in the phosphorylation of two key enzymes in glycogen metabolism, with opposite effects (Figure 15.15):

- *Inhibition.* Glycogen synthase, which catalyzes the joining of glucose molecules to synthesize the energy-storing molecule glycogen, is inactivated by phosphorylation. Thus the epinephrine signal prevents glucose from being stored in glycogen.
- *Activation.* Phosphorylase kinase is activated when a phosphate group is added to it. It goes on to stimulate a protein kinase cascade that ultimately leads to the activation by phosphorylation of phosphorylase, the other key enzyme in glucose metabolism. This enzyme liberates glucose molecules from glycogen.

Thus the same signaling pathway inhibits the storage of glucose as glycogen (by inhibiting glycogen synthase) and promotes the release of glucose through glycogen breakdown (by activating glycogen phosphorylase). As we mentioned earlier, the released glucose fuels the ATP-requiring fight-or-flight response to epinephrine.

Different genes are transcribed

Plasma membrane receptors are involved in activating a broad range of gene expression responses. The Ras signaling pathway, for example, ends in the nucleus (see Figure 15.9). The final protein kinase in the cascade, MAPK, enters the nucleus and phosphorylates a leucine zipper protein called AP-1. This activated protein is a transcription factor, and it stimulates the transcription of a number of genes involved in cell proliferation.

As described earlier in this chapter, lipid-soluble hormones can diffuse through the plasma membrane and meet their receptors in the cytoplasm. In this case, binding of the ligand allows the ligand–receptor complex to enter the nucleus, where it binds to hormone-responsive elements at the promoters of a number of genes. In some cases, transcription is stimulated, and in others it is inhibited.

In plants, light acts as a signal to initiate the formation of chloroplasts. Between this signal and response is a transcription-mediated signal transduction pathway. In bright sunlight, red wavelengths are absorbed by a receptor protein called *phytochrome*. We will say more about this important receptor later in the book, but for now it is important to note only that it is activated by red light. The activated phytochrome binds to cytoplasmic regulatory proteins, which enter the nucleus and bind to promoters of genes involved in the synthesis of important chloroplast proteins. Synthesis of these proteins is the key to plant "greening."

15.14 A Signal Transduction Pathway Leads to the Opening of Ion Channels In the signal transduction pathway for the sense of smell, the final effect is the opening of Na^+ channels. The resulting influx of Na^+ stimulates the transmission of a scent message to a specific region of the brain.

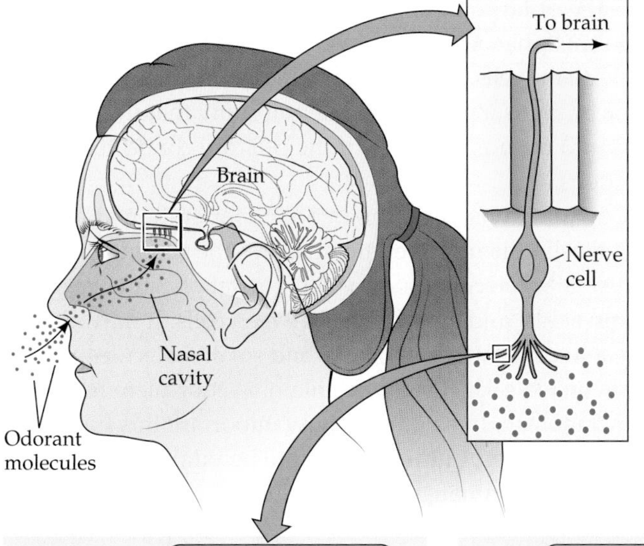

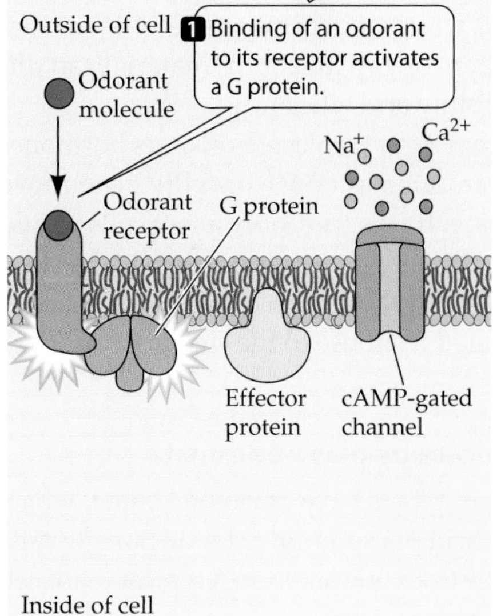

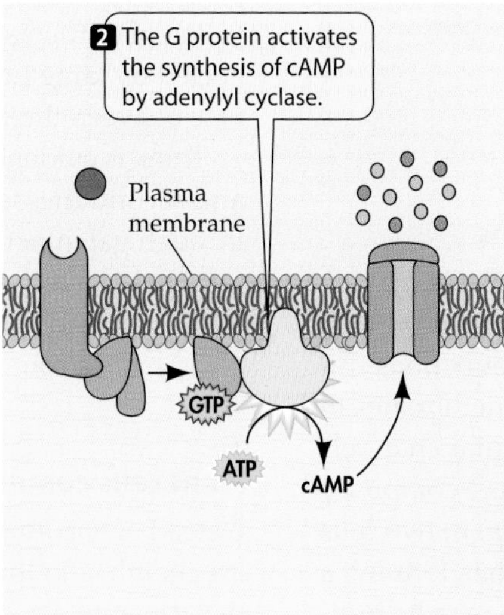

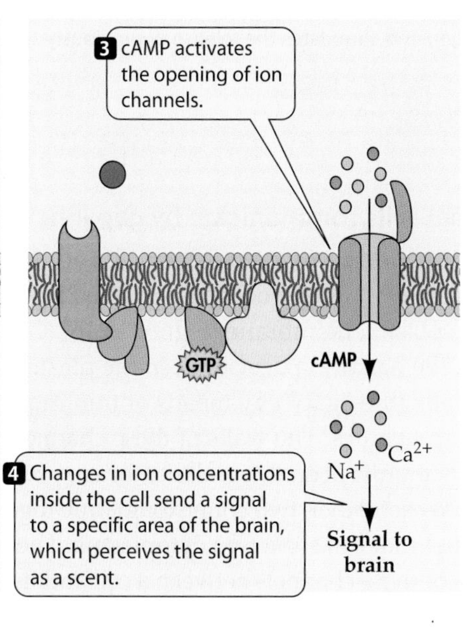

Direct Intercellular Communication

Up to now, we have described how signals from a cell's environment can influence that cell. But the environment of a cell in a multicellular organism is more than the extracellular medium. Most cells are in contact with their neighbors. In Chapter 5, we described how cells adhere to one another by recognition proteins protruding from the cell surface. There are also specialized cell junctions, such as tight junctions and desmosomes, that help "cement" cells together (see Figure 5.6).

However, as we know from our own neighbors (and roommates), just being in proximity does not necessarily mean that there is functional communication. Neither tight junctions nor desmosomes are specialized for intercellular communication. In this section, we look at the specialized junctions between cells that allow them to signal directly one another directly. In animals, these structures are gap junctions; in plants, they are plasmodesmata.

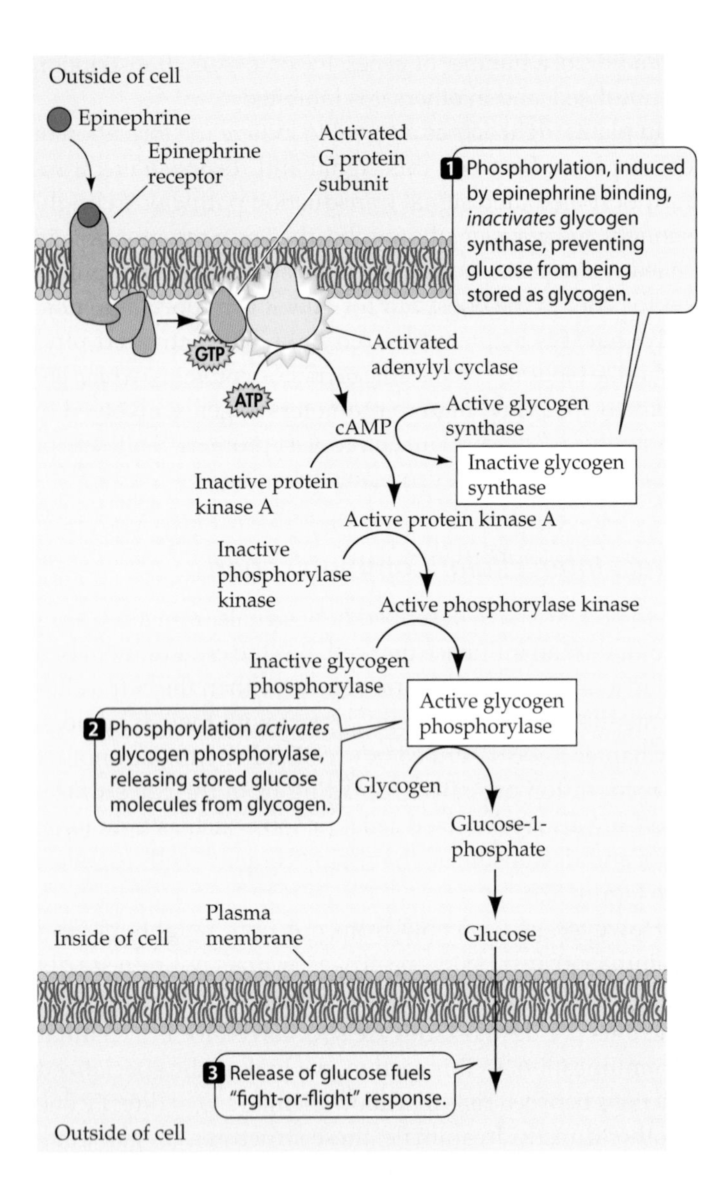

15.15 A Cascade of Reactions Leads to Altered Enzyme Activity Liver cells respond to epinephrine by activating G proteins, which in turn activate cAMP synthesis. The second messenger initiates a protein kinase cascade. The cascade both inhibits the conversion of glucose to glycogen and stimulates the release of previously stored glucose.

Animal cells communicate by gap junctions

Gap junctions are channels between adjacent cells that occur in many animals, occupying up to 25 percent of the area of the plasma membrane (Figure 15.16). Gap junctions traverse the narrow space between the plasma membranes of two cells (the "gap") by means of thin molecular channels called *connexons*. The walls of these channels are composed of six subunits of an integral membrane protein. In two cells close to each other, two connexons come together, forming a channel that links the two cytoplasms. There may be hundreds of these channels between a cell and its neighbors. The

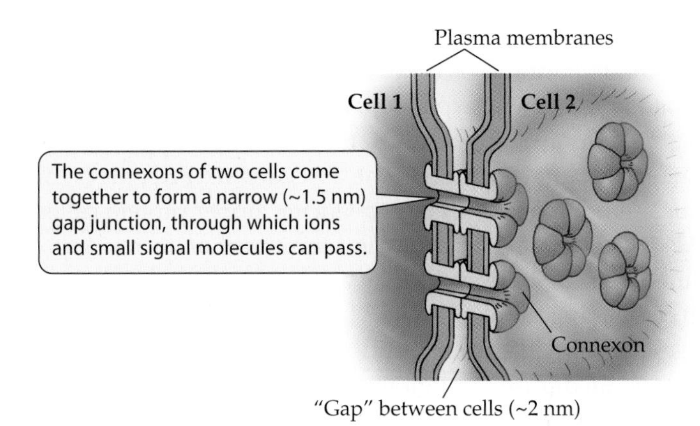

15.16 Gap Junctions Connect Animal Cells An animal cell may contain hundreds of gap junctions connecting it to neighboring cells. Gap junctions are too small for proteins, but small molecules such as ATP, metabolic intermediates, amino acids, and coenzymes can pass through them.

channels about 1.5 nm in diameter—far too narrow for the passage of large molecules such as proteins. But they are wide enough to allow small signal molecules and ions to pass between the cells. Experiments in which a labeled signal molecule or ion is injected into one cell show that it can readily pass into the adjacent cells if the cells are connected by gap junctions.

Gap junctions permit metabolic cooperation among the linked cells. Such cooperation ensures the sharing of important small molecules such as ATP, metabolic intermediates, amino acids, and coenzymes between cells. It may also ensure that concentrations of ions and small molecules are similar in linked cells, thereby maintaining equivalent regulation of metabolism. It is not clear how important this function is in many tissues, but it is known to be vital in some. In the lens of the mammalian eye, for example, only the cells at the periphery are close enough to the blood supply to allow diffusion of nutrients and wastes. But because lens cells are connected by large numbers of gap junctions, material can diffuse between them rapidly and efficiently.

There is evidence that signal molecules such as hormones and second messengers such as cAMP and PIP2 can move through gap junctions. If this is true, only a few cells would need to have receptors binding a signal in order for the stimulus to spread throughout the tissue. In this way, a tissue could have a coordinated response to the signal.

Plant cells communicate by plasmodesmata

Instead of gap junctions, plants have **plasmodesmata**, which are membrane-lined bridges spanning the thick cell walls that separate plant cells from one another. A typical plant cell has several thousand plasmodesmata.

Plasmodesmata differ from gap junctions in one fundamental way: Unlike gap junctions, in which the wall of the channel is made of integral proteins from the adjacent plasma membranes, plasmodesmata are lined by the fused plasma membranes themselves. Plant biologists are so familiar with the notion of a tissue as cells interconnected in this way that they refer to these continuous cytoplasms as a *symplast* (see Chapter 36).

The diameter of a plasmodesma is about 6 nm, far larger than the gap junction channel. But the actual space available for diffusion is about the same—1.5 nm. A look at the interior of the plasmodesma gives the reason for this reduction in pore size: A tubule called the **desmotubule**, apparently derived from the endoplasmic reticulum, fills up most of the opening of the plasmodesma (Figure 15.17). So, typically, only small metabolites and ions move between plant cells. This fact is important physiologically to plants, which lack the tiny circulatory vessels (capillaries) many animals use to bring gases and nutrients to every cell.

Diffusion from cell to cell through plasma membranes is probably inadequate for hormonal responses in plants. Instead, they rely on more rapid diffusion through plasmodesmata to ensure that all cells of a tissue respond to a signal at the same time. In C_4 plants (see Chapter 8), there are abundant plasmodesmata between the mesophyll and bundle sheath cells, which help to rapidly move the carbon fixed in the former cell type to the latter. A similar transport system, found at the junctions of nonvascular tissues and phloem, conducts organic solutes throughout the plant.

Plasmodesmata are not merely passive channels, but can be regulated. Plant viruses may infect cells at one location, then spread rapidly through a plant organ by plasmodesmata until they reach the plant's vascular tissue (circulatory system). These viruses, and even their RNA, would appear to be many times too large to pass through the desmotubules. But they get through, apparently by making "movement proteins" that increase the pore size temporarily while attached to the viral genome. Similar movement proteins made by the plants themselves are involved in transporting mRNAs and even proteins such as transcription factors between plant cells. This finding opens up the possibility of long-distance regulation of transcription and translation.

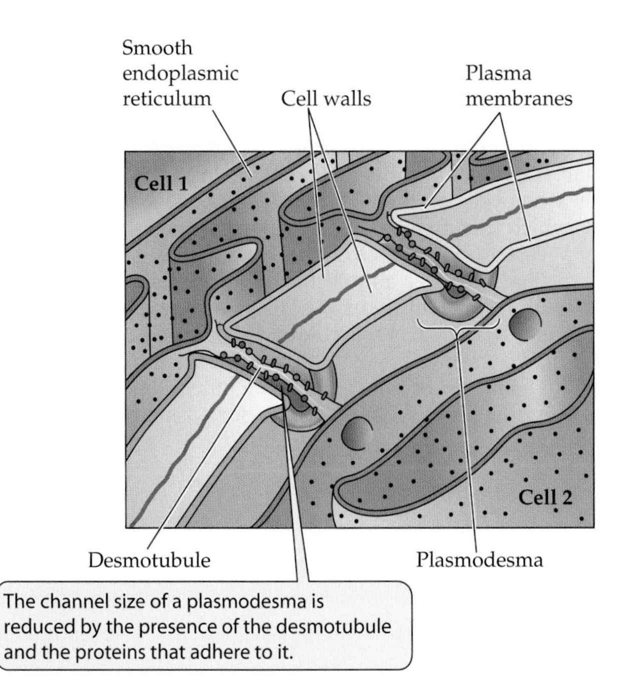

15.17 Plasmodesmata Connect Plant Cells The desmotubule, derived from the smooth endoplasmic reticulum, fills up most of the space inside a plasmodesma, leaving a tiny gap through which small metabolites and ions can pass.

Chapter Summary

Signals

▶ Cells receive many signals from the physical environment and from other cells. **Review Figures 15.1**

▶ A signal transduction pathway involves three steps: the binding of a signal by a receptor, the transduction of the signal within the cell, and the ultimate cellular response. **Review Figure 15.2. See Web/CD Activity 15.1**

Receptors

▶ Cells respond to signals only if they have specific receptor proteins that can bind to those signals. **Review Figure 15.3**

▶ Depending on the nature of its signal, a receptor may be located in the plasma membrane or in the cytoplasm of the target cell. **Review Figure 15.4**

▶ Receptors located in the plasma membrane include ion channels, protein kinases, and G protein-linked receptors. **Review Figures 15.5, 15.6, 15.7. See Web/CD Tutorial 15.1**

▶ When bound by a ligand, cytoplasmic receptors change their shape and enter the cell nucleus. **Review Figure 15.8**

Signal Transduction

▶ The events of signal transduction may be direct, occurring at the plasma membrane, or indirect, involving the formation of a second messenger.

▶ Protein kinase cascades amplify a response to receptor binding. **Review Figure 15.9**

▶ Second messengers include cyclic AMP, the lipid-derived substances inositol triphosphate and diacylglycerol, calcium ions, and the gas nitric oxide. **Review Figures 15.10, 15.11, 15.12, 15.13**

Signal Effects: Changes in Cell Function

▶ The ultimate cell response to a signal may be the opening of ion channels, the alteration of enzyme activities, or changes in gene transcription. **Review Figures 15.14, 15.15**

Direct Intercellular Communication

▶ Most animal cells can communicate with one another directly through small pores in their plasma membranes called gap junctions. Small molecules and ions can pass through these pores. **Review Figure 15.16**

▶ Plant cells are connected by somewhat larger pores called plasmodesmata, which traverse both membranes and cell walls. **Review Figure 15.17**

See Web/CD Activity 15.2 for a concept review of this chapter.

Self Quiz

1. What is the correct order for these events in the interaction of a cell with a signal? (1) alteration of cell function; (2) signal binds to receptor; (3) signal released from source; (4) signal transduction.
   a. 1234
   b. 2314
   c. 3214
   d. 3241
2. Why do some signals ("first messengers") trigger a "second messenger" to activate a target cell?
   a. The first messenger requires activation by ATP.
   b. The first messenger is not water soluble.
   c. The first messenger binds to many types of cells.
   d. The first messenger cannot cross the plasma membrane.
   e. There are no receptors for the first messenger.
3. Steroid hormones act on target cells by
   a. initiating second messenger activity.
   b. binding to membrane proteins.
   c. initiating DNA transcription.
   d. activating enzymes.
   e. binding to membrane lipids.
4. The major difference between a cell that responds to a signal and one that does not is the presence of a
   a. DNA sequence that binds to the signal.
   b. nearby blood vessel.
   c. receptor.
   d. second messenger.
   e. transduction pathway.
5. Which of the following is *not* a consequence of signal binding to a receptor?
   a. Activation of receptor enzyme activity
   b. Diffusion of receptor in the plasma membrane
   c. Change in conformation of the receptor protein
   d. Breakdown of the receptor to amino acids
   e. Release of the signal from the receptor
6. A nonpolar molecule such as a steroid hormone usually binds to a
   a. cytoplasmic receptor.
   b. protein kinase.
   c. ion channel.
   d. phospholipid.
   e. second messenger.
7. Which of the following is *not* a common type of receptor?
   a. Ion channel
   b. Protein kinase
   c. G protein-linked
   d. Transcription factor
   e. Adenylate cyclase
8. Which of the following is *not* true of the protein kinase cascade?
   a. The signal is amplified.
   b. A second messenger is formed.
   c. Target proteins are phosphorylated.
   d. The cascade ends up in the nucleus.
   e. The cascade begins at the plasma membrane.
9. Which of the following is *not* a second messenger for signal transduction?
   a. Calcium ions
   b. Nitric oxide gas
   c. ATP
   d. Cyclic AMP
   e. Diacylglycerol
10. Plasmodesmata and gap junctions
   a. allow small molecules and ions to pass rapidly between cells.
   b. are both membrane-lined channels.
   c. are channels about 1 μm in diameter.
   d. are present only once per cell.
   e. are involved in cell recognition in signaling.

For Discussion

1. Like *ras* itself, the various components of the Ras signaling pathway were discovered when tumors showed mutations in one or another of the components. What might be the biochemical consequences of mutations in the genes for (a) Raf and (b) MAP kinase that result in rapid cell division?
2. Cyclic AMP is a second messenger in many different responses. How can the same messenger act in different ways in different cells?
3. Compare direct communication via plasmodesmata or gap junctions with ligand/receptor-mediated communication between cells. What are the advantages of one method over the other?
4. The tiny invertebrate *Hydra* has an apical region, which has tentacles, and a long, slender body. *Hydra* can reproduce asexually when cells on the body wall differentiate and form a bud, which then breaks off as a new organism. Buds form only at certain distances from the apex, leading to the idea that the apex releases a molecule that diffuses down the body and, at high concentrations (i.e., near the apex), inhibits bud formation. *Hydra* lacks a circulatory system, so the inhibitor must diffuse from cell to cell. If you had an antibody that binds to connexin to plug up the gap junctions, how would you show that *Hydra*'s inhibitory factor passes *through* these junctions?

16 Recombinant DNA and Biotechnology

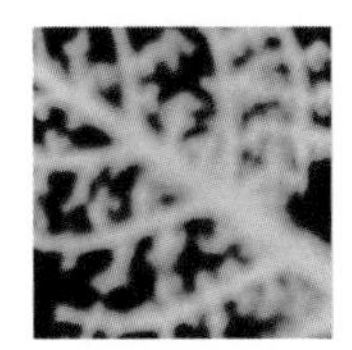

Of the many horrible legacies of the wars of the twentieth century, perhaps none is so lasting as the littering of the countryside with land mines. These inexpensive plastic shells filled with trinitrotoluene (TNT) are built to explode and injure whatever steps on the soil above them. Currently, the most common way of finding land mines is to poke around the soil with a stick—a precarious occupation at best. Since the mines are made of plastic, metal detectors do not work. The agricultural systems of many of the countries with large numbers of land mines, such as Cambodia and Angola, are based on manual labor, so where there are land mines, there cannot be a farm. Clearly, the world needs a sensitive, non-lethal land mine detector.

Enter biotechnology. Neal Stewart, at the University of North Carolina, is developing plants that can detect land mines in a field and show their locations remotely. His method is an excellent example of the application of knowledge of gene transcription and translation as well as of genetic engineering. In Chapter 14, we saw that the control of eukaryotic gene transcription lies at the promoter, a DNA sequence where RNA polymerase and other proteins bind to initiate transcription of the adjacent gene. Certain bacteria have a promoter that is sensitive to TNT, and the binding of a tiny amount of this chemical activates an adjacent gene.

Plants would be ideal biosensors for land mines, as seeds can be spread widely and evenly in a suspect field. But what gene could "announce" the presence of TNT by making a detectable protein? It turns out that certain jellyfish make a protein that fluoresces green when ultraviolet light ("black light") is shone on it. Stewart has introduced this gene into a plant, placing it alongside the TNT-sensitive promoter. When these plants are grown near a land mine, their roots take up TNT, and their leaves will glow in ultraviolet light. Of course, having people plant these seeds or shine a black light on these plants as they walk through the field would bring us back to the old, dangerous way of detecting land mines. To solve these problems, seeding and remote sensing could be done from airplanes or helicopters flying over the field.

This story—from problem to solution—has been repeated many times in the past two decades. The products of recombinant DNA technology range from life-saving drugs that there is no other way to make in adequate amounts to crop plants with improved agricultural characteristics. Although the basic

A Land Mine Detector Plants can be genetically engineered to express green fluorescent protein from a jellyfish gene. When this "glow-in-the-dark" gene is linked so that it is activated by the presence of TNT in the soil, such plants can act as biosensors to detect the presence of explosive land mines.

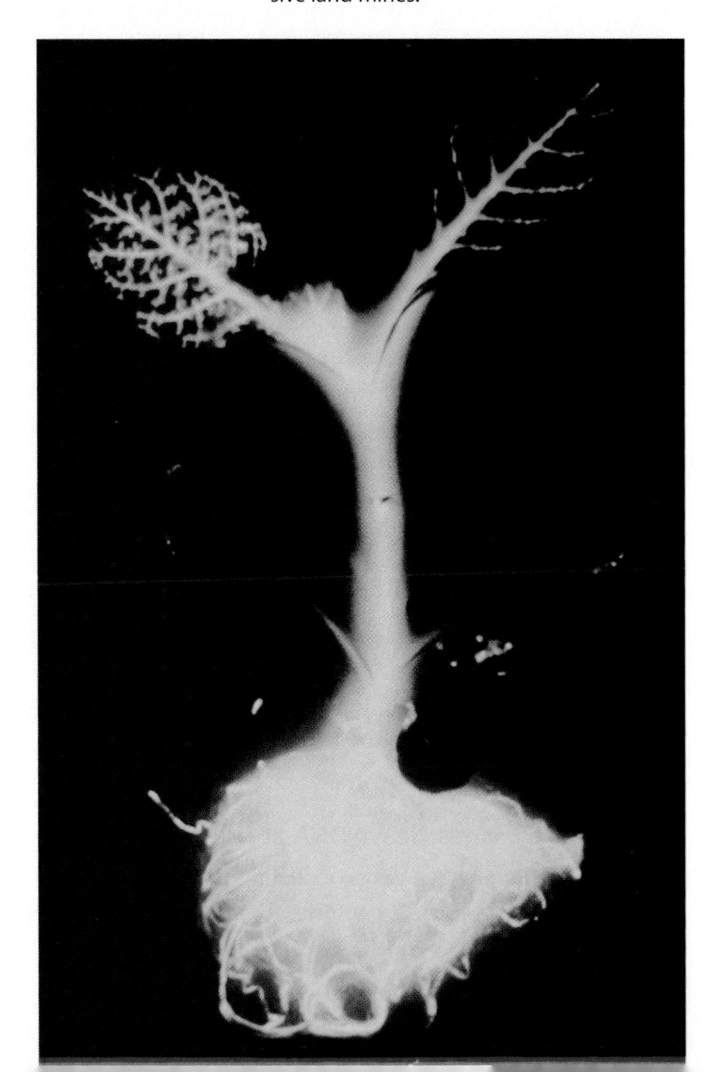

techniques of DNA manipulation have been called revolutionary, most of them come from the knowledge of DNA transcription and translation that we described in earlier chapters. The upshot of this technology is that today we can take a piece of DNA from any source, attach it chemically to any other DNA—making a recombined DNA molecule that has never existed before in the entire evolutionary history of life—and then use this unique DNA for whatever purpose and in whichever target organism we wish. This ability gives humans unprecedented power over life in general.

We begin this chapter with a description of how DNA molecules can be cut into smaller fragments and how fragments from different sources can be covalently linked to create recombinant DNA in a test tube. Recombinant (or any other) DNA can then be introduced into a suitable prokaryotic or eukaryotic host cell. Sometimes, the purpose of adding a new gene to a host cell or organism is to ask an experimental question about the role of that gene that can be answered by placing it in a new environment. In other instances, the purpose is to coax the host cell to make a new gene product.

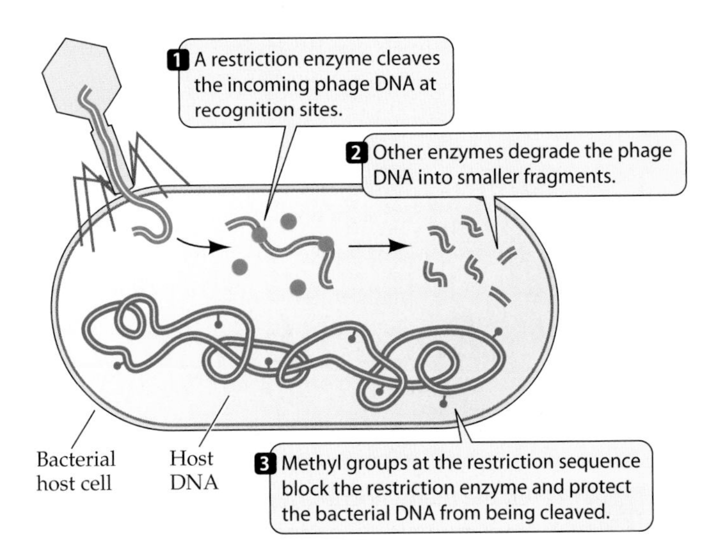

1 A restriction enzyme cleaves the incoming phage DNA at recognition sites.
2 Other enzymes degrade the phage DNA into smaller fragments.
Bacterial host cell
Host DNA
3 Methyl groups at the restriction sequence block the restriction enzyme and protect the bacterial DNA from being cleaved.

16.1 Bacteria Fight Invading Viruses with Restriction Enzymes Bacteria produce restriction enzymes that degrade phage DNA by cleaving it into smaller, double-stranded fragments. Other enzymes protect the bacteria's own DNA from being cleaved.

Cleaving and Rejoining DNA

Scientists have long realized that the chemical reactions used in living cells for one purpose may be applied in the laboratory for other, novel purposes. **Recombinant DNA technology**—the manipulation and combination of DNA molecules from different sources—is based on this realization, and on an understanding of the properties of certain enzymes and of DNA itself.

As we have seen in previous chapters, the complementary pairing of nucleotide bases underlies many fundamental processes of molecular biology. The mechanisms of DNA replication, transcription, and translation all rely on complementary base pairing. Similarly, all the key techniques in recombinant DNA technology—locating, sequencing, rejoining, and amplifying DNA fragments—make use of the complementary base pairing of A with T (or U) and of G with C.

In this section, we will see how some of the numerous naturally occurring enzymes that cleave and repair DNA can be used in recombinant DNA technology. Many of these enzymes have been isolated and purified and are now used in the laboratory to manipulate and recombine DNA. Using these enzymes, fragments of DNA can be separated, covalently linked to other fragments, and employed for many novel and highly useful purposes.

Restriction enzymes cleave DNA at specific sequences

All organisms must have ways of dealing with their enemies. As we saw in Chapter 13, bacteria are attacked by viruses called bacteriophages that inject their genetic material into the host cell. Some bacteria defend themselves against such invasions by producing **restriction enzymes** (also known as **restriction endonucleases**), which catalyze the cleavage of double-stranded DNA molecules—such as those injected by phages—into smaller, noninfectious fragments (Figure 16.1). These enzymes cut the bonds between the 3′ hydroxyl of one nucleotide and the 5′ phosphate of the next one.

There are many such restriction enzymes, each of which cleaves DNA at a specific sequence of bases, called a *recognition sequence*, or **restriction site**. The DNA of the host cell is not cleaved by its own restriction enzymes because specific modifying enzymes called *methylases* add methyl (—CH_3) groups to certain bases at the restriction sites of the host's DNA when it is being replicated. The methylation of the host's bases makes the recognition sequence unrecognizable to the restriction enzyme. But unmethylated phage DNA is efficiently recognized and cleaved.

A specific sequence of bases defines each restriction sequence. For example, the enzyme *Eco*RI (named after its source, a strain of the bacterium *E. coli*) cuts DNA only where it encounters the following paired sequence in the DNA double helix:

5′ . . . G↓AATTC . . . 3′
3′ . . . CTTAA↑G . . . 5′

Notice that this sequence reads the same in the 5′-to-3′ direction on both strands. It is *palindromic*, like the word "mom," in the sense that it is the same in both directions from the 5′ end. The *Eco*RI enzyme has two identical active sites on its two subunits, which cleave the two strands simultaneously between the G and the A of each strand (see Figure 16.4).

The *Eco*RI recognition sequence occurs, on average, about once in every 4,000 base pairs in a typical prokaryotic genome, or about once per four prokaryotic genes. So *Eco*RI can chop a large piece of DNA into smaller pieces containing, on average, just a few genes. Using *Eco*RI in the laboratory to cut small genomes, such as those of viruses that have tens of thousands of base pairs, may result in a few fragments. For a huge eukaryotic chromosome with tens of millions of base pairs, the number of fragments will be very large.

Of course, "on average" does not mean that the enzyme cuts all stretches of DNA at regular intervals. The *Eco*RI recognition sequence does not occur even once in the 40,000 base pairs of the genome of a phage called T7—a fact that is crucial to the survival of this virus, since its host is *E. coli*. Fortunately for *E. coli*, the DNA of other phages does contain the *Eco*RI recognition sequence.

Hundreds of restriction enzymes have been purified from various microorganisms. In the test tube, different restriction enzymes that recognize different restriction sites can be used to cut the same sample of DNA. Thus, restriction enzymes can be used as "knives" for genetic "surgery" to cut a sample of DNA in many different, specific places.

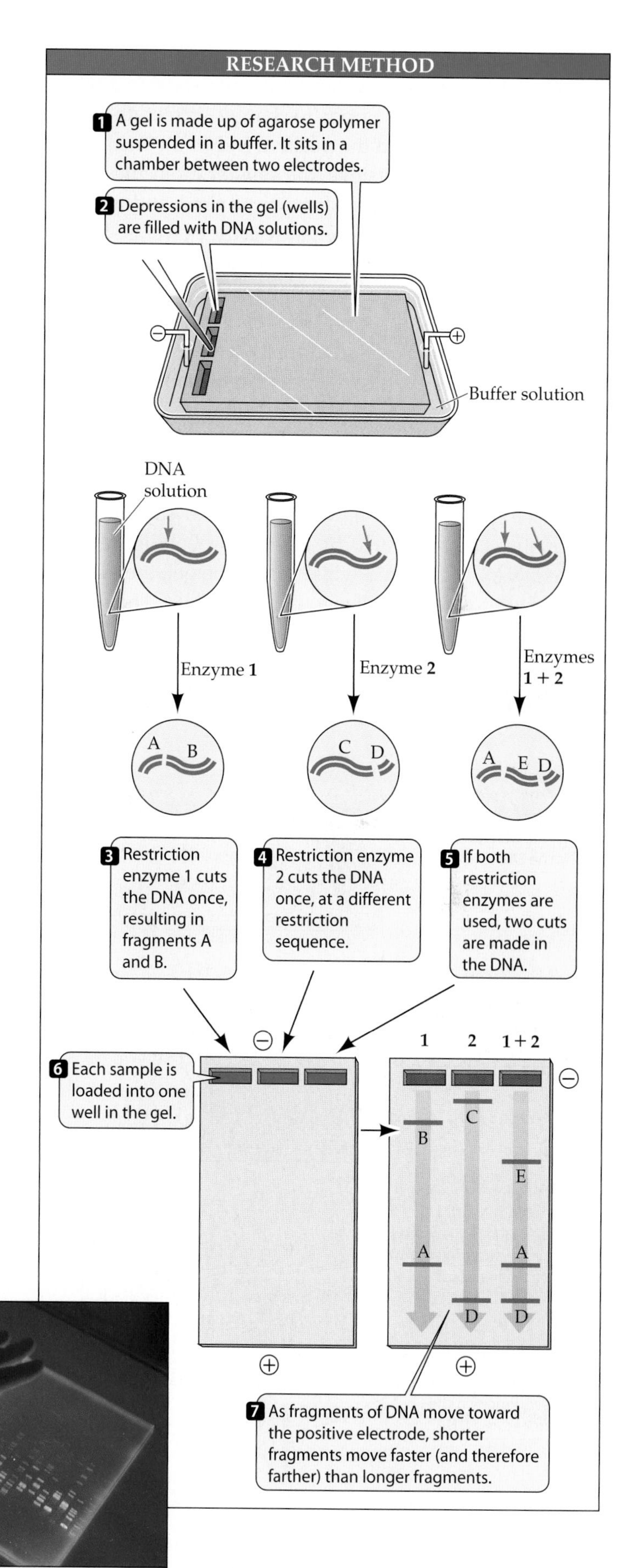
RESEARCH METHOD
1 A gel is made up of agarose polymer suspended in a buffer. It sits in a chamber between two electrodes.
2 Depressions in the gel (wells) are filled with DNA solutions.
Buffer solution
DNA solution
Enzyme 1
Enzyme 2
Enzymes 1 + 2
A
B
C
D
A
E
D
3 Restriction enzyme 1 cuts the DNA once, resulting in fragments A and B.
4 Restriction enzyme 2 cuts the DNA once, at a different restriction sequence.
5 If both restriction enzymes are used, two cuts are made in the DNA.
6 Each sample is loaded into one well in the gel.
1
2
1+2
B
C
E
A
A
D
D
7 As fragments of DNA move toward the positive electrode, shorter fragments move faster (and therefore farther) than longer fragments.

Gel electrophoresis identifies the sizes of DNA fragments

After a laboratory sample of DNA has been cut with a restriction enzyme, the DNA is in fragments, which must be separated. Because the recognition sequence does not occur at regular intervals, the fragments are not all the same size, and this property provides a way to separate them from one another. Separating the fragments is necessary to determine the number and sizes (in base pairs) of fragments produced or to identify and purify an individual fragment of particular interest.

The best way to separate or purify DNA fragments is by **gel electrophoresis** (Figure 16.2). Because of its phosphate groups, DNA is negatively charged at neutral pH. A mixture

16.2 Separating Fragments of DNA by Gel Electrophoresis A mixture of DNA fragments is placed in a gel and an electric field is applied across the gel. The negatively charged DNA moves toward the positive end of the field, with smaller molecules moving faster than larger ones. When the electric power is shut off, the now separated fragments can be analyzed.

of DNA fragments is placed in a well in a porous gel, and an electric field (with positive and negative ends) is applied across the gel. Because opposite charges attract, the DNA fragments move toward the positive end of the field. Since the porous gel acts as a sieve, the smaller molecules move faster than the larger ones. After a fixed time, and while all the fragments are still in the gel, the electric power is shut off. The separated fragments can be visualized by staining them with a dye that fluoresces under ultraviolet light. They can then be seen as bars or spots in the gel (Figure 16.2) and can be examined or removed individually.

Electrophoresis gives two types of information:

- *The sizes of the fragments.* DNA fragments of known molecular size are often placed in a well in the gel next to the sample to provide a size reference.
- *The presence of specific DNA sequences.* A specific DNA sequence can be located by using a probe (Figure 16.3). The DNA is denatured while still in the gel, then the gel is affixed to a nylon filter to make a "blot." The filter is then exposed to a single-stranded DNA probe with a sequence complementary to the one that is being sought. If the sequence of interest in present, the probe will hybridize with it. The probe can be labeled in some way—for example, with radioactivity. After hybridization, spots of radioactivity on the membrane indicates that the probe has hybridized with its target sequence at that location. Unbound probes stay in solution.

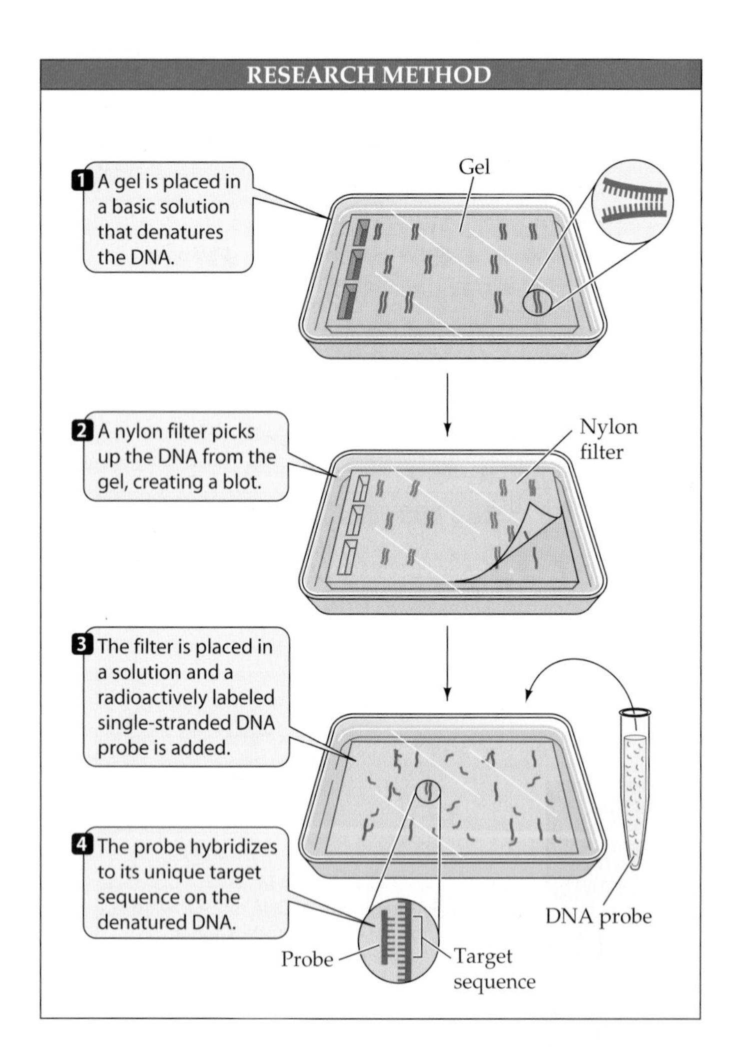

RESEARCH METHOD
1 A gel is placed in a basic solution that denatures the DNA.
Gel
2 A nylon filter picks up the DNA from the gel, creating a blot.
Nylon filter
3 The filter is placed in a solution and a radioactively labeled single-stranded DNA probe is added.
4 The probe hybridizes to its unique target sequence on the denatured DNA.
DNA probe
Probe
Target sequence

16.3 Analyzing DNA Fragments A probe can be used to locate a specific DNA fragment on an electrophoresis gel.

The gel region containing only the desired fragment (in size or sequence) can be cut out as a lump of gel, and the pure DNA fragment can then be removed from the gel by diffusion into a small volume of water.

Recombinant DNA can be made in the test tube

Some restriction enzymes cut the DNA backbone cleanly, cutting both strands exactly opposite one another. Others make staggered cuts, cutting one strand of the double helix several bases away from where they cut the other. Fragments cut in this manner are particularly useful in biotechnology.

*Eco*RI, for example, cuts DNA within its recognition sequence in a staggered manner, as shown at the top of Figure 16.4. After the two cuts in the opposing strands are made, the strands are held together only by the hydrogen bonds between four base pairs. The hydrogen bonds of these few base pairs are too weak to persist at warm temperatures (above room temperature), so the two strands of DNA separate, or *denature*. As a result, there are single-stranded "tails" at the location of each cut. These tails are called **sticky ends** because they have a specific base sequence that can bind by base pairing with complementary sticky ends. If n restriction sites for a given restriction enzyme are present in a linear DNA molecule, then $n + 1$ fragments will be made, all with the same complementary sequences at their sticky ends.

After a DNA molecule has been cut with a restriction enzyme, complementary sticky ends can form hydrogen bonds with one another. The original ends may rejoin, or an end may pair with a complementary end from another fragment. Furthermore, because the ends of all fragments cut by the same restriction enzyme are the same, fragments from one source, such as a human, can be joined to fragments from another source, such as a bacterium.

When the temperature is lowered, the fragments *anneal* (come together by hydrogen bonding) at their sticky ends at random, but these associations are unstable because they are held together by only a few hydrogen bonds. The associated sticky ends can be permanently united by a second enzyme, **DNA ligase**, which forms the one covalent bond needed at each sticky end to "seal" the DNA strands. In the cell, this enzyme unites the Okazaki fragments and mends breaks in DNA, as we saw in Chapter 11.

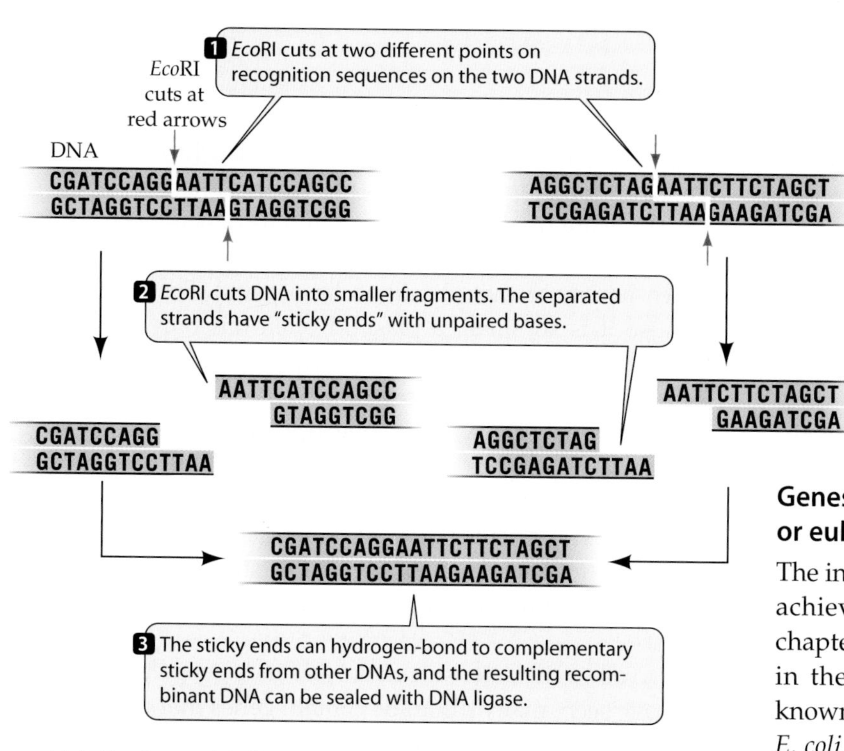

1 EcoRI cuts at two different points on recognition sequences on the two DNA strands.
EcoRI cuts at red arrows
DNA
CGATCCAGGAATTCATCCAGCC
GCTAGGTCCTTAAGTAGGTCGG
AGGCTCTAGAATTCTTCTAGCT
TCCGAGATCTTAAGAAGATCGA
2 EcoRI cuts DNA into smaller fragments. The separated strands have "sticky ends" with unpaired bases.
AATTCATCCAGCC
GTAGGTCGG
CGATCCAGG
GCTAGGTCCTTAA
AATTCTTCTAGCT
GAAGATCGA
AGGCTCTAG
TCCGAGATCTTAA
CGATCCAGGAATTCTTCTAGCT
GCTAGGTCCTTAAGAAGATCGA
3 The sticky ends can hydrogen-bond to complementary sticky ends from other DNAs, and the resulting recombinant DNA can be sealed with DNA ligase.

16.4 Cutting and Splicing DNA Some restriction enzymes (*Eco*RI is shown here) make staggered cuts in DNA. *Eco*RI can be used to cut DNA from two different sources (blue and gold). At warm temperatures, the two DNA strands will separate (denature), leaving sticky ends, exposed bases that can hybridize with complementary fragments. When the temperature is lowered, sticky ends from different DNAs can bind to each other, forming recombinant DNA.

Many restriction enzymes do not produce sticky ends. Instead, they cut both DNA strands at the same base pair within the recognition sequence, making "blunt" ends. DNA ligase can also connect blunt-ended fragments, but it does so with reduced efficiency.

With these two enzyme tools—restriction enzymes and DNA ligase—scientists can cut and rejoin different DNA molecules from any and all sources to form recombinant DNA (see Figure 16.4). These simple techniques have revolutionized biological science in the past 30 years and have given us the power to manipulate genetic material at will.

Getting New Genes into Cells

One goal of recombinant DNA technology is to produce many copies (**clones**) of a particular gene, either for purposes of analysis or to produce its protein product in quantity. If the recombinant DNA is to make its protein, it must be inserted, or **transfected**, into a host cell. Such altered hosts are known as **transgenic** cells or organisms. The choice of a host cell—prokaryotic or eukaryotic—is important in this work.

Once the host species is selected, the recombinant DNA is brought together with a population of host cells and, under specific conditions, enters some of them. Because all the host cells proliferate—not just the few that receive the recombinant DNA—the scientist must be able to determine which cells actually contain the sequence of interest. One common method of identifying cells with recombinant DNA is to tag the inserted sequence with **reporter genes**, whose phenotypes are easily observed. These phenotypes serve as *genetic markers* for the sequence of interest.

Genes can be inserted into prokaryotic or eukaryotic cells

The initial successes of recombinant DNA technology were achieved using bacteria as hosts. As noted in preceding chapters, bacterial cells are easily grown and manipulated in the laboratory. Much of their molecular biology is known, especially for certain well-studied bacteria, such as *E. coli*, and they have numerous genetic markers that can be used to select for cells harboring the recombinant DNA. Bacteria also contain small circular chromosomes called *plasmids*, which, as we will see, can be manipulated to carry recombinant DNA into the cell.

In some important ways, however, bacteria are not ideal organisms for studying and expressing eukaryotic genes. Bacteria lack the splicing machinery to excise introns from the initial RNA transcript of eukaryotic genes. In addition, many eukaryotic proteins are extensively modified after translation by reactions such as glycosylation and phosphorylation. Often these modifications are essential for the protein's activity. Finally, in some instances, the expression of the new gene in a eukaryote is the point of the experiment—that is, the aim is to produce a transgenic organism. In these cases, the host for the new DNA may be a mouse, a wheat plant, a yeast, or a human, to name just a few examples.

Yeasts such as *Saccharomyces* are common eukaryotic hosts for recombinant DNA studies. The advantages of using yeasts include rapid cell division (a life cycle completed in 2–8 hours), ease of growth in the laboratory, and a relatively small genome size (about 20 million base pairs and 6,000 genes). The yeast genome is several times larger than that of *E. coli*, and has one-fourth the number of genes as the human genome. Nevertheless, yeasts have most of the characteristics of other eukaryotes, except for those involved in multicellularity.

Plant cells can also be used as hosts, especially if the desired result is a transgenic plant. The property that makes plant cells good hosts is their *totipotency*—that is, the ability of a differentiated cell to act like a fertilized egg and produce an entire new organism. Isolated plant cells grown in culture can take up recombinant DNA, and by manipulation of the growth medium, these transgenic cells can be induced to form an entire new plant, which can then be reproduced naturally in the field. The transgenic plant will carry and express the gene that is part of recombinant DNA.

Whatever host is chosen, a vehicle for carrying the DNA into the cell is needed. These vehicles are called **vectors**.

Vectors can carry new DNA into host cells

In natural environments, DNA released from one bacterium can sometimes be taken up by another bacterium and genetically transform it (see Chapter 11), but this phenomenon is not common. The challenge of inserting new DNA into a cell lies not just in getting it into the host cell, but in getting it to replicate in the host cell as it divides. DNA polymerase, the enzyme that catalyzes replication, does not bind to just any sequence of DNA to begin replication. Rather, it recognizes a specific sequence, the *origin of replication* (see Chapter 11). If the new DNA is to be replicated, it must become part of a segment of DNA that contains an origin of replication, called a **replicon**, or **replication unit**.

There are two general ways in which the newly introduced DNA can become part of a replication unit. First, it can be inserted into a host chromosome after entering the host cell. Although this insertion is often a random event, it is nevertheless a common method of integrating a new gene into a host cell. Alternatively, the new DNA can enter the host cell as part of a carrier DNA sequence—the vector— that already has the appropriate origin of replication.

A vector should have four characteristics:

- The ability to replicate independently in the host cell
- A recognition sequence for a restriction enzyme, allowing the vector to be cut and combined with the new DNA
- A reporter gene that will announce its presence in the host cell
- A small size in comparison to the host chromosomes for ease of isolation

PLASMIDS AS VECTORS. Plasmids, which, as we saw in Chapter 13, are naturally occurring bacterial chromosomes, have all four of the properties needed for a useful vector. First, they are small (an *E. coli* plasmid has 2,000–6,000 base pairs, as compared with the main *E. coli* chromosome, which has more than 3 million base pairs). Furthermore, because it is so small, a plasmid often has only a single recognition site for a given restriction enzyme (Figure 16.5*a*). This property is essential because it allows for the insertion of new DNA at only one location (see Figure 16.4). When the plasmid is cut with a restriction enzyme, it is transformed into a linear molecule with sticky ends. The sticky ends of another DNA fragment cut with the same restriction enzyme can pair with the sticky ends of the plasmid, resulting in a circular plasmid containing the new DNA.

Two other characteristics make plasmids good vectors. As we have seen, many plasmids contain genes that confer resistance to antibiotics. This property provides a genetic marker for host cells carrying the recombinant plasmid. Finally, plasmids have an origin of replication and can replicate independently of the host chromosome. It is not uncommon for a bacterial cell with a single main chromosome to contain hundreds of copies of a recombinant plasmid.

The plasmids commonly used as vectors in the laboratory have been extensively altered by recombinant DNA technology, and most are combinations of genes and other sequences from several sources. Many of these plasmids have a single marker for antibiotic resistance.

VIRUSES AS VECTORS. Constraints on plasmid replication limit the size of the new DNA that can be inserted into a

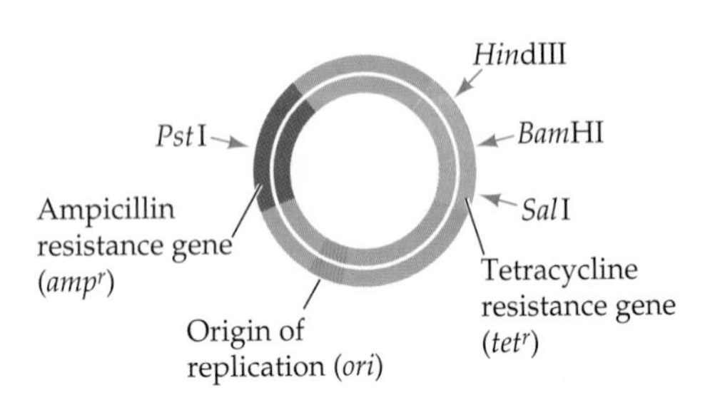

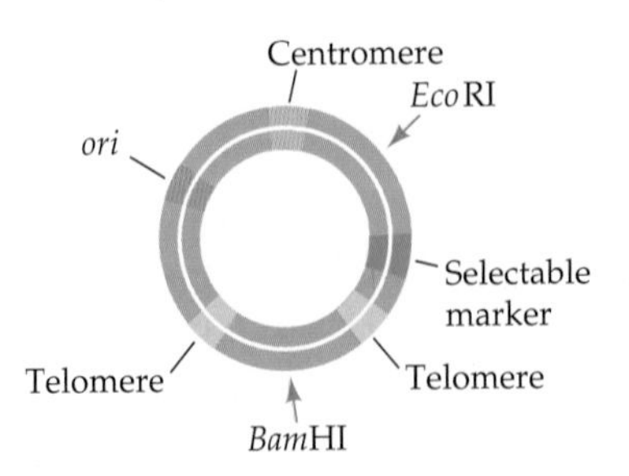

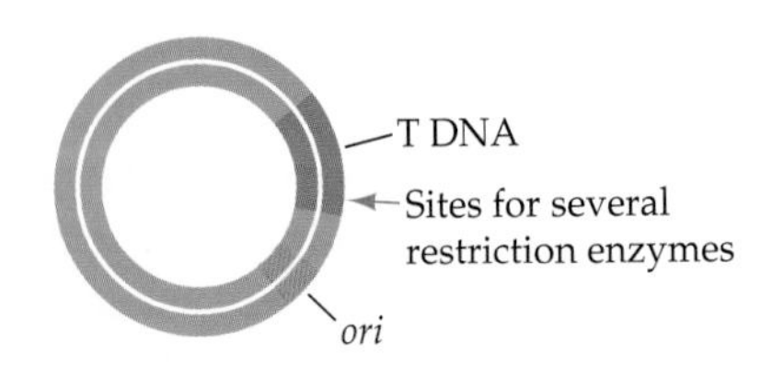

16.5 Vectors for Carrying DNA into Cells (*a*) A plasmid with genes for antibiotic resistance can be incorporated into an *E. coli* cell. (*b*) A DNA molecule synthesized in the laboratory constitutes a chromosome that can carry its inserted DNA into yeasts. (*c*) The Ti plasmid, isolated from the bacterium *Agrobacterium tumefaciens*, is used to insert DNA into many types of plants.

plasmid to about 10,000 base pairs. Although some prokaryotic genes may be this small, 10,000 base pairs is much smaller than most eukaryotic genes with their introns and extensive flanking sequences. A vector that accommodates larger DNA inserts is needed.

Both prokaryotic and eukaryotic viruses are often used as vectors for eukaryotic DNA. Bacteriophage λ, which infects *E. coli*, has a DNA genome of about 45,000 base pairs. If the genes that cause the host cell to die and lyse—about 20,000 base pairs—are eliminated, the virus can still attach to a host cell and inject its DNA. The deleted 20,000 base pairs can be replaced with DNA from another organism, thereby creating usable recombinant viral DNA.

Because viruses infect cells naturally, they offer a great advantage as vectors over plasmids, which often require artificial means to coax them to enter cells. As we will see in Chapter 17, viruses are important vectors in human gene therapy.

ARTIFICIAL CHROMOSOMES AS VECTORS. Bacterial plasmids are not good vectors for yeast hosts because prokaryotic and eukaryotic DNA sequences use different origins of replication. Thus a recombinant bacterial plasmid will not replicate in yeast. To remedy this problem, scientists have created a "minimalist chromosome" called the **yeast artificial chromosome**, or **YAC** (Figure 16.5*b*). This artificial DNA molecule contains not only the yeast origin of replication, but the yeast centromere and telomere sequences as well, making it a true eukaryotic chromosome. YACs also contain artificially synthesized restriction sites and useful reporter genes (for yeast nutritional requirements). YACs are only about 10,000 base pairs in size, but can accommodate 50,000 to 1.5 million base pairs of inserted DNA. These artificial chromosomes carry out eukaryotic DNA replication and gene expression normally in yeast cells.

PLASMID VECTORS FOR PLANTS. An important vector for carrying new DNA into many types of plants is a plasmid that is found in *Agrobacterium tumefaciens*. This bacterium lives in the soil and causes a plant disease called crown gall, which is characterized by the presence of growths, or tumors, in the plant. *A. tumefaciens* contains a plasmid called Ti (for tumor-inducing) (Figure 16.5*c*).

The Ti plasmid contains a transposon, called T DNA, that inserts copies of itself into the chromosomes of infected plant cells. The T DNA contains recognition sequences for restriction enzymes, so that new DNA can be inserted into it. When the T DNA is thus altered, the plasmid no longer produces tumors, but the transposon, with the new DNA, can still be inserted into the host cell's chromosomes. A plant cell containing this DNA can then be grown in culture or induced to form a new, transgenic plant.

Whatever vector is effective, the problem of identifying those host cells that actually contain the recombinant DNA remains.

Reporter genes identify host cells containing recombinant DNA

Even when a population of host cells interacts with an appropriate vector, only a small proportion of the cells actually take up the vector. Also, since the process of cutting the vector and inserting the new DNA to make recombinant DNA is far from perfect, only a few of the vectors that have moved into the host cells will actually contain the DNA sequence of interest. How can we select only the host cells that contain the recombinant DNA?

The procedure we are about to describe illustrates an elegant, commonly used approach to this problem. In this example, we use *E. coli* bacteria as hosts and a plasmid vector (see Figure 16.5*a*) that carries the genes for resistance to the antibiotics ampicillin and tetracycline.

When the pBR322 plasmid is incubated with the restriction enzyme *Bam*HI, the enzyme encounters its recognition sequence, GGATCC, only once, at a site within the gene for tetracycline resistance. If foreign DNA is inserted into this restriction site, the presence of these "extra" base pairs within the tetracycline resistance gene inactivates it. So plasmids containing the inserted DNA will carry an intact gene for ampicillin resistance, but not an intact gene for tetracycline resistance (Figure 16.6). This difference is the key to the selection of the host bacteria that contain the recombinant plasmid.

The cutting and insertion process results in three types of DNA, all of which can be taken up by host bacteria:

- The recombinant plasmid—the one we want—turns out to be the rarest type of DNA. Its uptake confers resistance to ampicillin, but not to tetracycline, on host *E. coli*.
- More common are bacteria that take up plasmids that have sealed their own ends back together. These plasmids retain intact genes for resistance to both ampicillin and tetracycline.
- Even more common are bacteria that take up the foreign DNA sequence alone, without the plasmid; since it is not part of a replication unit, it does not survive as the bacteria divide. These host cells remain susceptible to both antibiotics.

The vast majority (more than 99.9 percent) of host cells take up no DNA at all and remain susceptible to both antibiotics. So the unique drug-resistant phenotype of the cells with recombinant DNA (tetracycline-sensitive and ampicillin-resistant) marks them in a way that can be detected by simply adding ampicillin and/or tetracycline to the medium surrounding the cells.

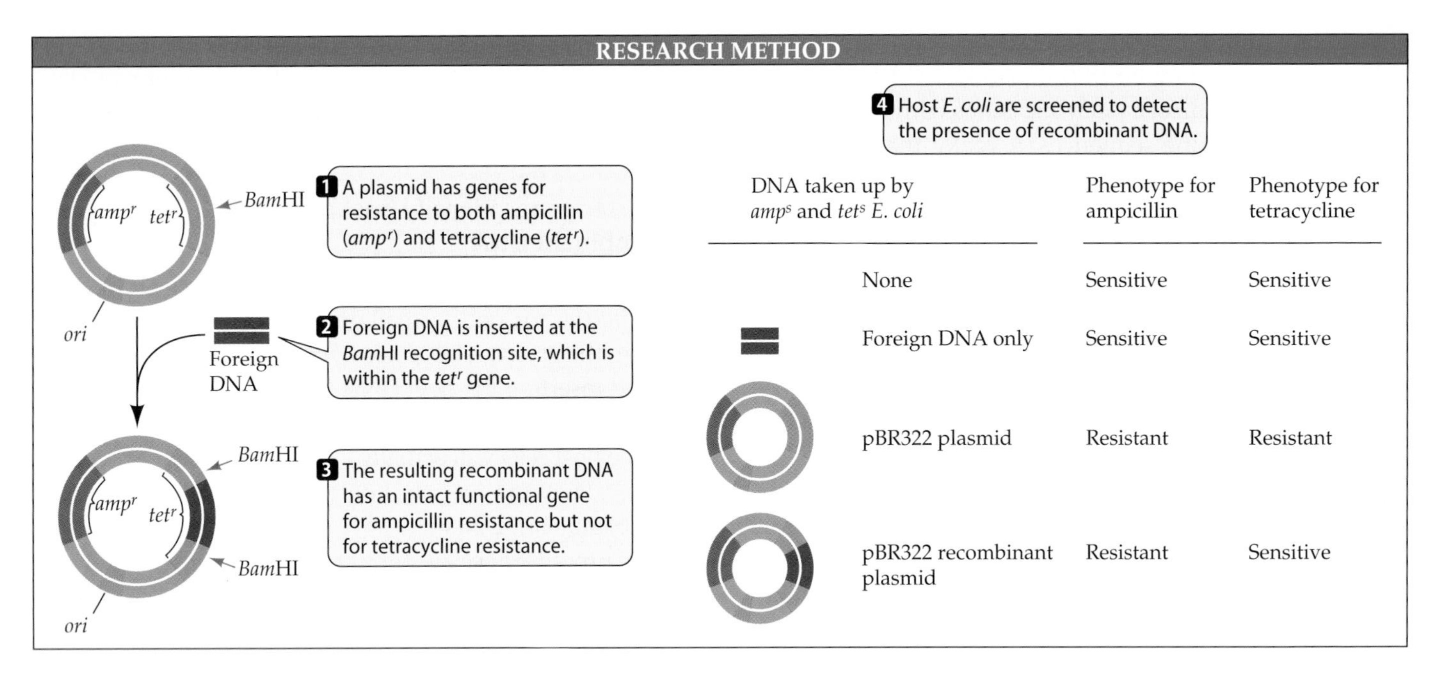
RESEARCH METHOD
ampr
tetr
BamHI
1 A plasmid has genes for resistance to both ampicillin (ampr) and tetracycline (tetr).
ori
Foreign DNA
2 Foreign DNA is inserted at the BamHI recognition site, which is within the tetr gene.
BamHI
3 The resulting recombinant DNA has an intact functional gene for ampicillin resistance but not for tetracycline resistance.
BamHI
ori
4 Host E. coli are screened to detect the presence of recombinant DNA.
DNA taken up by amps and tets E. coli
Phenotype for ampicillin
Phenotype for tetracycline
None
Sensitive
Sensitive
Foreign DNA only
Sensitive
Sensitive
pBR322 plasmid
Resistant
Resistant
pBR322 recombinant plasmid
Resistant
Sensitive

16.6 Marking Recombinant DNA by Inactivating a Gene
Scientists can inactivate reporter genes within plasmids to mark the host cells that have incorporated the recombinant DNA. The host bacteria in this experiment could display any of the three phenotypes indicated in the table.

In addition to genes for antibiotic resistance, several other reporter genes are used to detect recombinant DNA in host cells. Scientists have created several artificial vectors that include restriction sites within the *lac* operon (see Figure 13.17). When the *lac* operon is inactivated by the insertion of foreign DNA, the vector no longer carries its function into the host cell. Other reporter genes that have been used in vectors include the gene for luciferase, the enzyme that causes fireflies to glow in the dark; this enzyme causes host cells to glow when supplied with its substrate. Green fluorescent protein, which normally occurs in the jellyfish *Aequopora victoriana*, does not require a substrate, but emits visible light when exposed to ultraviolet light, and is now widely used as a genetic marker, as described at the beginning of this chapter.

Many commonly used plasmid vectors contain only a single reporter gene for antibiotic resistance, which does not contain a restriction site. In this case, the recombinant plasmid will have the same antibiotic resistance gene that a nonrecombinant plasmid does. The formation of recombinant DNA is favored, however, if there is a high concentration of foreign DNA fragments compared with that of the cut plasmid. So there will be a preponderance of host cells containing recombinant DNA among those that survive in the presence of the antibiotic.

After exposure to the vector, host cells are usually first grown on a solid medium. If the concentration of cells dispersed on the solid medium is low, each cell will divide and grow into a distinct colony (see Chapter 13). The colonies that contain recombinant DNA can be identified by reporter gene expression and removed from the medium, then grown in large amounts in liquid culture. A quick examination of a plasmid can confirm whether the cells of the colony actually have the recombinant DNA. The power of bacterial transfection to amplify a gene is indicated by the fact that a 1-liter culture of bacteria harboring the human β-globin gene in the pBR322 plasmid has as many copies of that gene as the sum total of all the cells in a typical adult human being (10^{14}).

Sources of Genes for Cloning

In the preceding section, we have seen how DNA can be cut, inserted into a vector, and transfected into host cells, and how host cells carrying recombinant DNA can be identified. Now we will pause briefly to consider where the genes or DNA fragments used in these procedures come from. They are obtained from three principal sources: random pieces of chromosomes maintained as gene libraries, complementary DNA obtained by reverse transcription from mRNA, and artificial synthesis or mutation of DNA.

Gene libraries contain pieces of a genome

The 23 pairs of human chromosomes can be thought of as a library that contains the entire genome of our species. Each chromosome, or "volume" in the library, contains, on average, 80 million base pairs of DNA, encoding several thousand genes. Such a huge molecule is not very useful for studying genomic organization or for isolating a specific gene.

To address this problem, researchers can use restriction enzymes to break each chromosome into smaller pieces, then analyze each piece. These smaller DNA fragments still represent a **gene library** (Figure 16.7); however, the information is now in many more than 23 volumes. Each fragment can be inserted into a vector, which can then be taken up by a host cell. When bacteria are used as hosts, proliferation of one cell produces a colony of recombinant cells, each of which harbors many copies of the same fragment of human DNA.

Using plasmids, which are able to insert up to 10,000 base pairs of foreign DNA into a bacterium, about 200,000 separate fragments are required to make a library of the human genome. By using phage λ, which can carry four times as much DNA as a plasmid, the number of volumes can be reduced to about 50,000. Although this seems like a large number, a single growth plate can hold up to 80,000 phage colonies, or plaques, and is easily screened for the presence of a particular DNA sequence by denaturing the phage DNA and applying a particular probe.

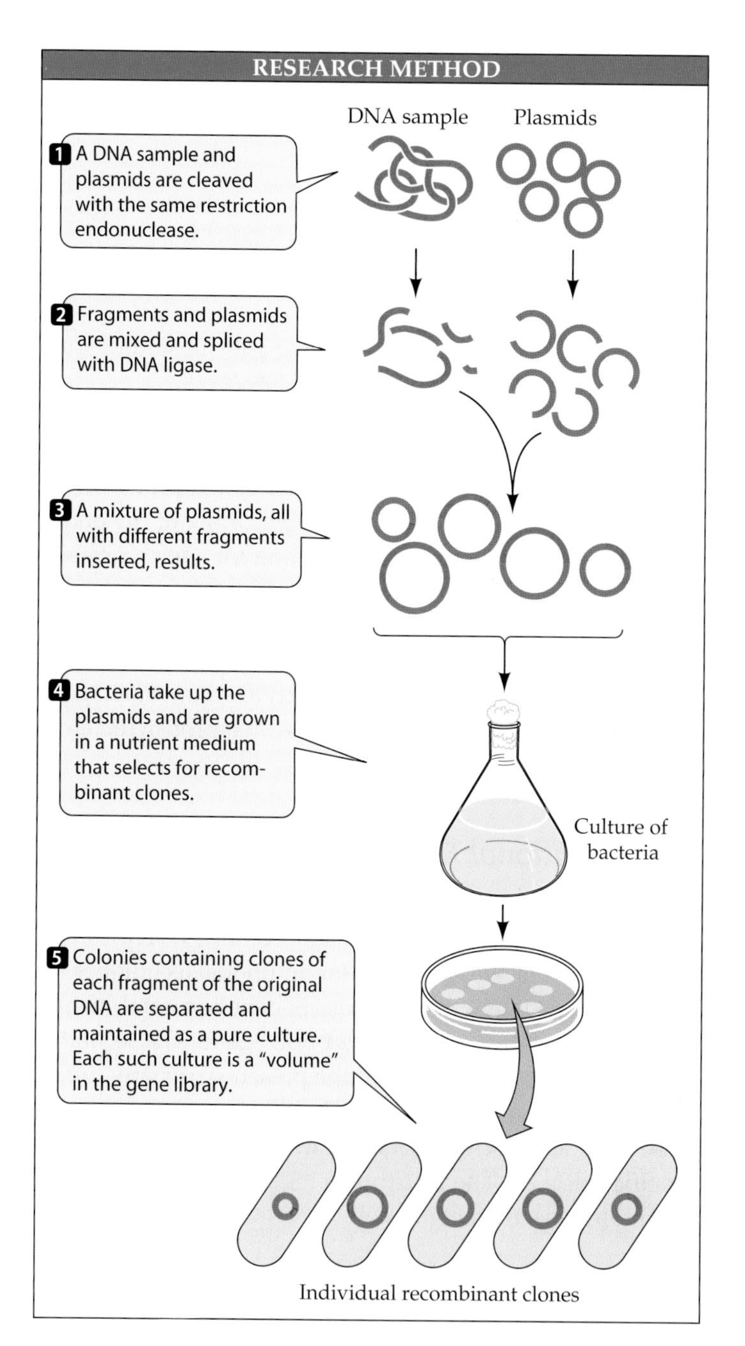

RESEARCH METHOD
DNA sample
Plasmids
1 A DNA sample and plasmids are cleaved with the same restriction endonuclease.
2 Fragments and plasmids are mixed and spliced with DNA ligase.
3 A mixture of plasmids, all with different fragments inserted, results.
4 Bacteria take up the plasmids and are grown in a nutrient medium that selects for recombinant clones.
Culture of bacteria
5 Colonies containing clones of each fragment of the original DNA are separated and maintained as a pure culture. Each such culture is a "volume" in the gene library.
Individual recombinant clones

16.7 Constructing a Gene Library Human chromosomes are broken up into fragments of DNA using restriction enzymes. The fragments are inserted into vectors (plasmids are shown here) and taken up by host bacterial cells, each of which then harbors a single fragment of the human DNA. The information in the resulting bacterial cultures and sets of colonies constitutes a gene library.

A DNA copy of mRNA can be made by reverse transcriptase

A much smaller DNA library—one that includes only the genes transcribed in a particular tissue—can be made from **complementary DNA**, or **cDNA** (Figure 16.8). Recall that most eukaryotic mRNAs have a poly A tail—a string of adenine residues at their 3′ end (see Figure 14.9). The first step in cDNA production is to extract mRNA from a tissue and allow it to hybridize with a molecule called *oligo dT*, which consists of a string of thymine residues (the "d" indicates deoxyribose). The oligo dT hybridizes with the poly A tail of the mRNA. The oligo dT serves as a primer, and the mRNA as a template, for the enzyme reverse transcriptase, which synthesizes DNA from RNA. In this way, a cDNA strand complementary to the mRNA is formed.

A collection of cDNAs from a particular tissue at a particular time in the life cycle of an organism is called a *cDNA library*. Messenger RNAs do not last long in the cytoplasm and are often present in small amounts, so a cDNA library is a "snapshot" that preserves the transcription pattern of the cell. cDNA libraries have been invaluable in comparisons of gene expression in different tissues at different stages of development. Their use has shown, for example, that up to one-third of all the genes of an animal are expressed only during prenatal development. Complementary DNA is also a good starting point for the cloning of eukaryotic genes. It is especially useful for cloning genes expressed at low levels in only a few cell types.

DNA can be synthesized chemically in the laboratory

If we know the amino acid sequence of a protein, we can apply organic chemistry to make the DNA that codes for that protein. Artificial DNA synthesis has even been automated, and at many institutions, a special service laboratory can make short to medium-length sequences overnight for any number of investigators.

How do we design a synthetic gene? Using the genetic code and the known amino acid sequence, we can figure out

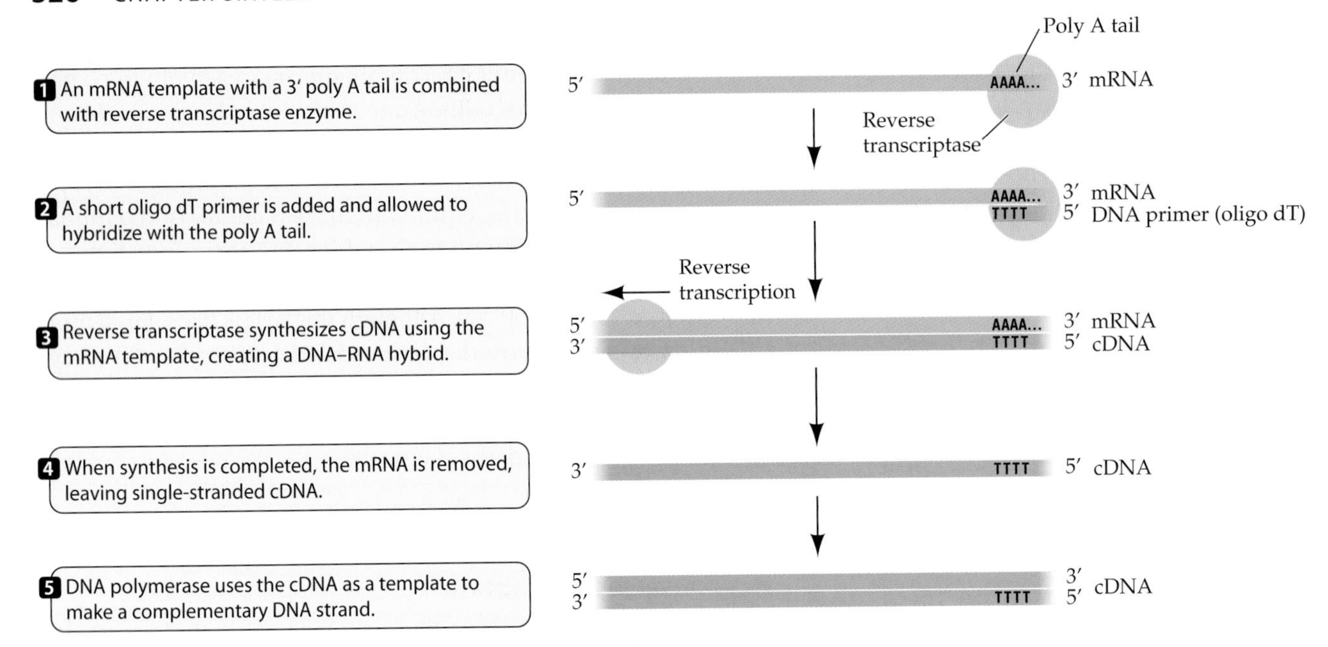

1 An mRNA template with a 3′ poly A tail is combined with reverse transcriptase enzyme.
2 A short oligo dT primer is added and allowed to hybridize with the poly A tail.
3 Reverse transcriptase synthesizes cDNA using the mRNA template, creating a DNA–RNA hybrid.
4 When synthesis is completed, the mRNA is removed, leaving single-stranded cDNA.
5 DNA polymerase uses the cDNA as a template to make a complementary DNA strand.
Poly A tail
5′
AAAA...
3′ mRNA
Reverse transcriptase
5′
AAAA...
TTTT
3′ mRNA
5′ DNA primer (oligo dT)
Reverse transcription
5′
3′
AAAA...
TTTT
3′ mRNA
5′ cDNA
3′
TTTT
5′ cDNA
5′
3′
TTTT
3′
5′
cDNA

16.8 Synthesizing Complementary DNA Gene libraries that include only genes transcribed in a particular tissue at a particular time can be made from complementary DNA. cDNA synthesis is especially useful for identifying mRNAs that are present only in a few copies, and is often a starting point for gene cloning.

an appropriate base sequence for the gene. With this sequence as a starting point, we can add other sequences, such as codons for translation initiation and termination and flanking sequences for transcription initiation, termination, and regulation. Of course, these noncoding DNA sequences must be the ones actually recognized by the host cell if the synthetic gene is to be transcribed. It does no good to have a prokaryotic promoter sequence near a gene if that gene is to be inserted into a yeast cell for expression. Codon usage is also important: as we have seen, many amino acids are encoded by more than one codon, and different organisms stress the use of different synonymous codons.

DNA can be mutated in the laboratory

Mutations that occur in nature have been important in demonstrating cause-and-effect relationships in biology. For example, in Chapter 14, we learned that some people with the disease beta thalassemia have a mutation at a consensus sequence for intron removal in the β-globin gene and so cannot make proper β-globin mRNA. This discovery revealed the importance of the consensus sequence. Recombinant DNA technology has allowed us to ask such "What if?" questions without having to look for mutations in nature. Because synthetic DNA can be made in any sequence desired, we can manipulate DNA to create specific mutations and then see what happens when the mutant DNA expresses itself in a host cell. Additions, deletions, and base-pair substitutions are all possible with isolated or synthetic DNA.

These mutagenesis techniques have led to many cause-and-effect proofs. For example, it was hypothesized that the signal sequence at the beginning of a secreted protein is essential to its passage through the membrane of the endoplasmic reticulum. Thus, a gene coding for such a protein, but with the codons for the signal sequence deleted, was synthesized. Sure enough, when this gene was expressed in yeast cells, the protein did not cross the ER membrane. When the signal sequence codons were added to an unrelated gene encoding a soluble cytoplasmic protein, that protein did cross the ER membrane.

Some Additional Tools for DNA Manipulation

In Chapter 11, we described DNA sequencing and the polymerase chain reaction, two applications of DNA replication techniques. Here, we examine four additional techniques for manipulating DNA. One is the use of genetic recombination to create an inactive, or "knocked-out," gene. The second is the use of DNA chips to detect the presence of many different sequences simultaneously. The third is the use of antisense RNA and RNA interference to block the translation of specific mRNAs. The fourth is a method for determining which proteins interact in a cell.

Genes can be inactivated by homologous recombination

As we have seen, artificial mutations provide an excellent way of asking "What if" questions about the role of a gene in cell function. **Homologous recombination** can be used to ask these questions at the organism level. The aim of this tech-

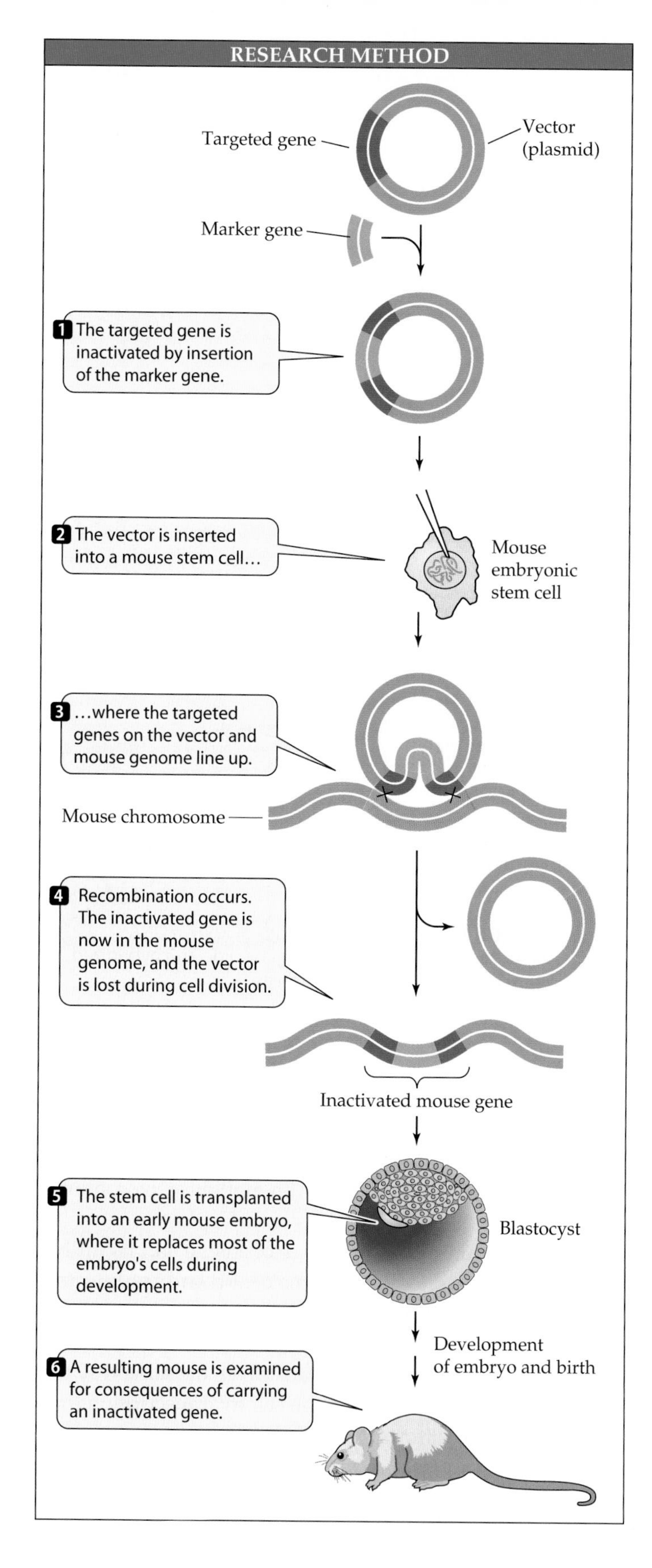
RESEARCH METHOD
Targeted gene
Vector (plasmid)
Marker gene
1 The targeted gene is inactivated by insertion of the marker gene.
2 The vector is inserted into a mouse stem cell...
Mouse embryonic stem cell
3 ...where the targeted genes on the vector and mouse genome line up.
Mouse chromosome
4 Recombination occurs. The inactivated gene is now in the mouse genome, and the vector is lost during cell division.
Inactivated mouse gene
5 The stem cell is transplanted into an early mouse embryo, where it replaces most of the embryo's cells during development.
Blastocyst
Development of embryo and birth
6 A resulting mouse is examined for consequences of carrying an inactivated gene.

16.9 Making a Knockout Mouse Homologous recombination is used to replace a normal mouse gene with an inactivated copy of that gene, thus "knocking out" the gene. Discovering what happens to a mouse with an inactive gene tells us much about the normal role of that gene.

nique is to replace a gene inside a cell with an inactivated form of that gene, then see what happens when the inactive gene is part of an organism. Such a manipulation is called a **knockout** experiment.

Mice are frequently used in knockout experiments (Figure 16.9). The normal allele of the mouse gene to be tested is inserted into a plasmid. Restriction enzymes are then used to insert a fragment containing a genetic marker into the middle of the normal gene. This addition of extra DNA plays havoc with the targeted gene's transcription and translation; a functional mRNA is seldom made from such an interrupted gene. Next, the plasmid is transfected into a stem cell in an early mouse embryo. A *stem cell* is an undifferentiated cell that divides and differentiates into specialized cell(s).

Because much of the targeted gene is still present in the plasmid (although in two separated regions), there is homologous sequence recognition between the inactive allele on the plasmid and the active (normal) allele in the mouse genome. The plasmid lines up with a mouse chromosome, and sometimes, recombination occurs such that the plasmid's inactive allele is swapped for the functional allele in the host cell. Now neither allele can be expressed: the one inserted into the mouse chromosome is still only an interrupted fragment of the normal gene, and the gene inserted into the plasmid usually lacks its promoter.

The reporter gene in the insert is used to identify those stem cells carrying the inactivated gene. A transfected stem cell is now transplanted into an early mouse embryo, and through some clever tricks, a knockout mouse carrying the inactivated gene in homozygous form is produced. The changed phenotype of the mutant mouse gives a clue to the role of the gene in the normal, wild-type animal. The knockout technique has been important in assessing the roles of genes during development.

DNA chips can reveal DNA mutations and RNA expression

The emerging science of genomics must deal with two major quantitative realities. First, there are a large number of genes in eukaryotic genomes. Second, the pattern of gene expression in different tissues at different times is quite distinctive. For example, a skin cancer cell at its early stage may have a unique mRNA "fingerprint" that differs from that of both normal skin cells and the cells of a more advanced skin cancer.

To find these patterns, scientists could isolate the mRNA from a cell and test it by hybridization with each gene in the genome, one gene at a time. But it would be far simpler to do these hybridizations all in one step. To facilitate this, one needs some way to arrange all the genes in a genome in an array on some solid support.

DNA chip technology provides these large arrays of sequences for hybridization. DNA chips were developed by modifying methods that have been used for decades in the semiconductor industry. You may be familiar with the silicon microchip, in which an array of microscopic electric circuits is etched onto a tiny chip. In the same way, DNA chips are glass slides to which a series of DNA sequences are attached in a precise order (Figure 16.10). Typically, the slide is divided into 24 × 24 μM squares, each of which contains about 10 million copies of a particular sequence up to 20 nucleotides long. A computer controls the addition of nucleotides in a predetermined pattern. Each 20-base-long sequence hybridizes to only one genomic DNA (or cDNA) sequence, and thus is a unique identifier of a gene. Up to 60,000 different sequences can be placed on a single chip.

If cellular mRNA is to be analyzed, it is usually incubated with reverse transcriptase (RT) to make cDNA (see Figure 16.8), and the cDNA is amplified by the polymerase chain reaction (PCR) prior to hybridization (see Figure 11.20). This technique is called **RT-PCR**, and it ensures that mRNA sequences naturally present in only a few copies (or in a small sample, such as a cancer biopsy) will be numerous enough to form a signal. The amplified cDNAs are coupled to a fluorescent dye and used to probe the DNA on a chip. Those cDNA sequences that form hybrids can be located by a sensitive scanner. With the number of genes that can be placed on a chip approaching that of the largest genomes, DNA chips will result in an information explosion on mRNA transcription patterns in cells in different physiological states.

Another use for DNA chips is in detecting genetic variants. Suppose one wants to find out if a particular gene, which is 5,500 base pairs long, has any mutations in a particular individual. One way would be to sequence the entire gene, but that would be expensive and time-consuming. On the other hand, DNA chip technology can be used to make 20-nucleotide fragments including the entire gene and all (or nearly all) of its known point mutations and small deletions. Then, hybridization with the individual's DNA might reveal a particular mutation if it hybridized to a mutant sequence on the chip. This rapidly developing technology could be an important step toward individualized diagnosis and therapy for human genetic diseases.

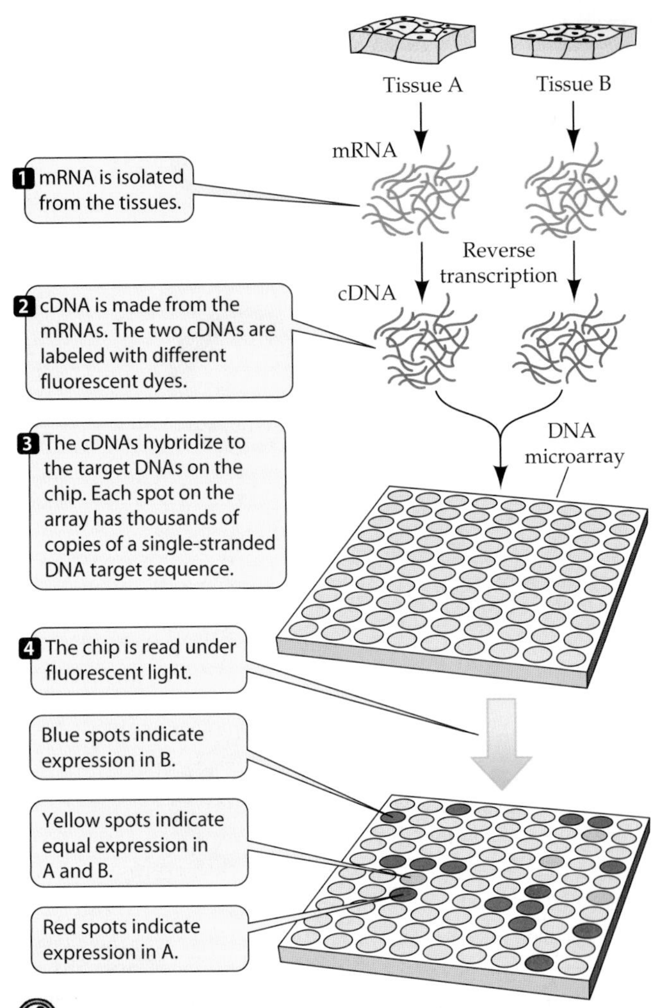

16.10 DNA on a Chip Thousands of known DNA sequences can be attached to a glass slide and hybridized with sequences from tissue samples.

Antisense RNA and RNA interference can prevent the expression of specific genes

The base-pairing rules can be used not only to make artificial genes, but they can also be employed to stop the translation of mRNA. As is often the case, this technique is an example of scientists imitating nature. In normal cells, a rare mechanism for controlling gene expression is the production of an RNA molecule that is complementary to mRNA. This complementary molecule is called **antisense RNA** because it binds by base pairing to the "sense" bases on the mRNA that codes for a protein. The formation of a double-stranded RNA hybrid inhibits translation of the mRNA, and the hybrid tends to be broken down rapidly in the cytoplasm. Although the gene continues to be transcribed, translation does not take place.

After determining the sequence of a gene and its mRNA in the laboratory, scientists can make and add specific antisense RNA to a cell to prevent translation of that gene's mRNA (Figure 16.11, left). The antisense RNA can be added as itself—RNA can be inserted into cells in the same way that DNA is—or it can be made in the cell by transcription from a DNA molecule introduced as a part of a vector.

A related technique takes advantage of **interference RNA (RNAi)**, a rare way of naturally inhibiting mRNA translation, such as occurs in the inactivation of the X chromosome (see Figure 14.18). In this case, a short (about 20 nucleotides) double-stranded RNA is unwound to single strands by a protein complex that guides this RNA to a complementary region on mRNA. The protein complex catalyzes the breakdown of the targeted mRNA.

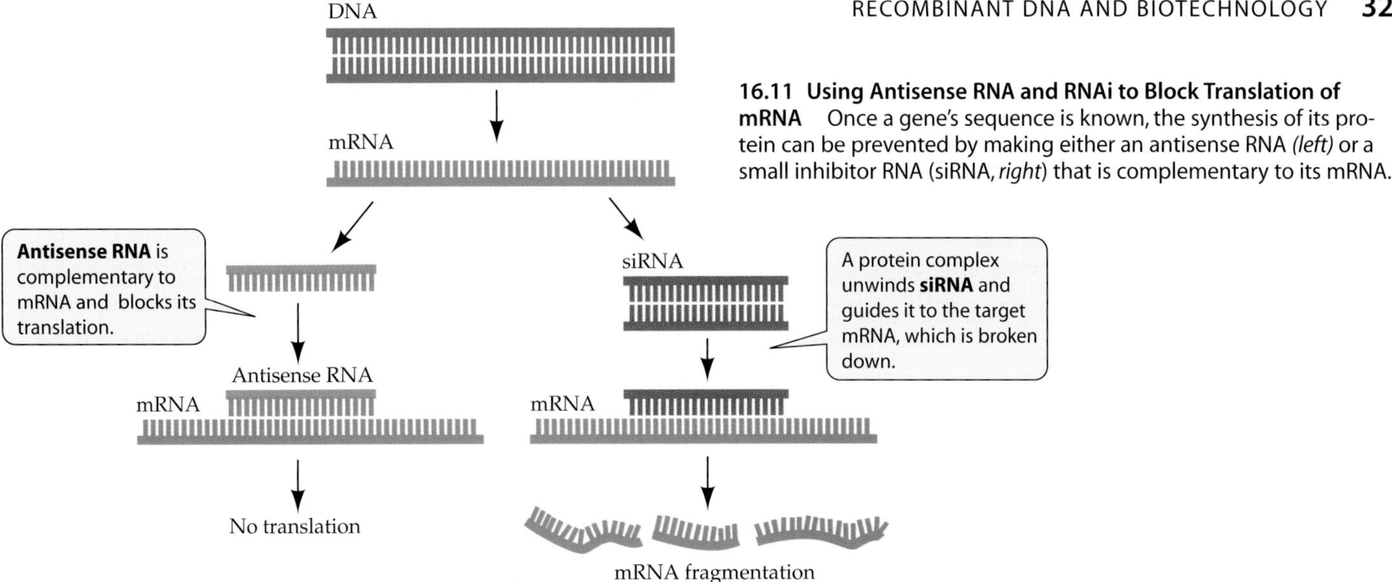

16.11 Using Antisense RNA and RNAi to Block Translation of mRNA Once a gene's sequence is known, the synthesis of its protein can be prevented by making either an antisense RNA *(left)* or a small inhibitor RNA (siRNA, *right*) that is complementary to its mRNA.

Armed with this knowledge, scientists can custom-synthesize a *small interfering RNA* (siRNA) to inhibit the translation of *any* known gene (Figure 16.11, right). Because these double-stranded siRNAs are more stable than antisense RNAs, RNAi is a much easier technique to use than antisense RNA.

Antisense RNA and RNAi have been widely used to test cause-and-effect relationships. For example, when antisense RNA was used to block the synthesis of a protein essential for the growth of cancer cells, the cells reverted to a normal phenotype. Gene silencing offers great potential for the development of drugs to treat diseases that are the result of the inappropriate expression of specific genes.

The two-hybrid system shows which proteins interact in a cell

Proteins often interact with other proteins in a cell, and their interaction leads to an important cellular function. Scientists can determine which proteins bind to which in several ways. One is the test-tube approach, in which one protein acts as a "hook" and the scientist "goes fishing" for proteins that bind to it. A better way is to set up a system that tests for protein interactions in a living cell. That is the purpose of a **two-hybrid system**.

A two-hybrid system uses a transcription factor that activates the transcription of an easily detectable reporter gene. This transcription factor has two domains: one binds to DNA at the promoter, and the other binds to another protein in the transcription complex to activate transcription of the reporter gene.

In the yeast two-hybrid system (Figure 16.12), these two domains of the transcription factor are separated in two different recombinant yeast plasmids. One plasmid makes the DNA-binding domain only, and the gene encoding this domain is fused to a gene that makes the target protein—the one whose binding partner we wish to determine. The resulting hybrid protein will bind to the promoter of the reporter gene, but will not activate it, leaving the target protein exposed. This hybrid protein is the "bait."

The second yeast plasmid has a gene encoding the activating domain of the transcription factor. This gene is fused

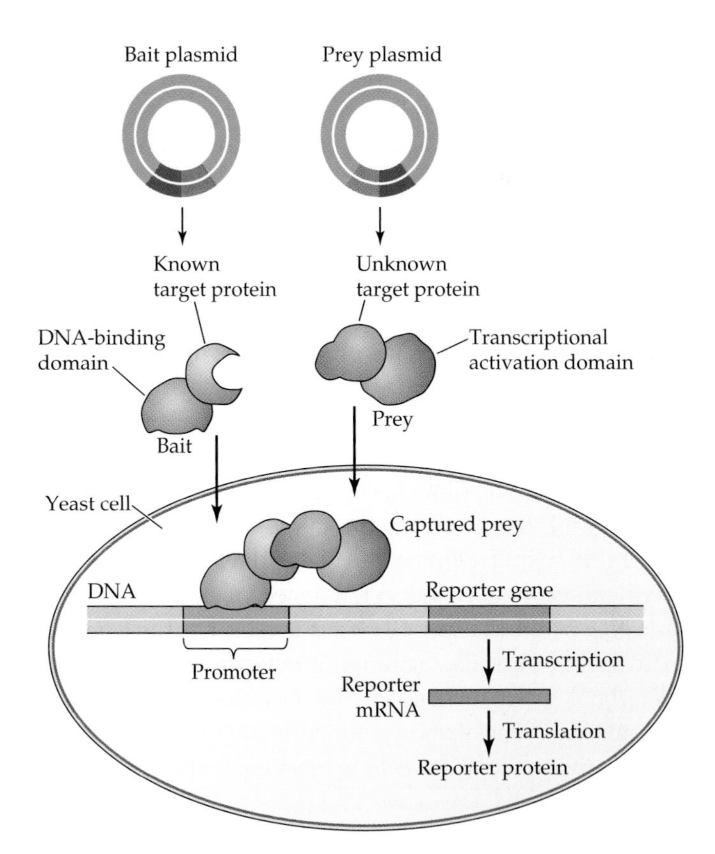

16.12 The Two-Hybrid System The two-hybrid system is a way to determine which proteins interact with one another in a living cell. In this example, the synthesis of the reporter protein is a positive result, revealing that the target protein (the "prey") binds specifically to a possible binding partner (the "bait").

to a gene encoding a protein (the "prey") to be tested for binding to the target protein. When this hybrid gene is introduced into the "bait" strain of yeast cells, the "bait" and "prey" proteins may bind. This binding activates transcription of the reporter gene. If no binding occurs, no reporter protein is made. This method has revealed hundreds of protein-protein interactions in cells.

Biotechnology: Applications of DNA Manipulation

As we have just seen, the development of methods for manipulating DNA has given us the ability to perform experiments that reveal the details of life at the molecular level. But in addition to their use as research tools, these methods are being used for human benefit. **Biotechnology** is the use of living cells to produce materials useful to people, such as foods, medicines, and chemicals. People have been doing this for a very long time. For example, the use of yeasts to brew beer and wine dates back at least 8,000 years, and the use of bacterial cultures to make cheese and yogurt is a technique many centuries old. For a long time, however, people were not aware of the cellular bases of these biochemical transformations.

About 100 years ago, thanks largely to Pasteur's work, it became clear that specific bacteria, yeasts, and other microbes could be used as biological converters to make certain products. Alexander Fleming's discovery that the mold *Penicillium* makes the antibiotic penicillin led to the large-scale commercial culture of microbes to produce antibiotics as well as other useful chemicals. Today, microbes are grown in vast quantities to make much of the industrial-grade alcohol, glycerol, butyric acid, and citric acid that are used by themselves or as starting materials in the manufacture of other products.

In the past, the list of such products was limited to those that were naturally made by microbes. The many products that multicellular eukaryotes make, such as hormones and certain enzymes, had to be extracted from those complex organisms. Yields were low, and purification was difficult and costly. All this has changed with the advent of gene cloning. Our ability to insert almost any gene into bacteria or yeast, along with methods to induce the gene to make its product in large amounts and export it from the cells, has turned these microbes into versatile factories for important products. The key technology for turning cells into factories has been the development of specialized vectors that not only carry genes into cells, but make those cells express them at high levels.

Expression vectors can turn cells into protein factories

If a eukaryotic gene is inserted into a typical plasmid (see Figure 16.5*a*) and transfected into *E. coli*, little, if any, of the product of the gene will be made by the host cell. The reason is that the eukaryotic gene lacks the bacterial promoter for RNA polymerase binding, the terminator for transcription, and a special sequence on mRNA that is necessary for ribosome binding. All of these elements are necessary for the gene to be expressed and its product synthesized in the bacterial cell.

To solve this problem, scientists can make **expression vectors** that have all the characteristics of typical vectors as well as the extra sequences needed for the foreign gene to be expressed in the host cell. For bacterial hosts, these additional sequences include the elements named above (Figure 16.13); for eukaryotes, they include the poly A addition sequence, transcription factor binding sites, and enhancers. Once these sequences are placed at the appropriate location in the vector, a transfected gene can be expressed in almost any kind of host cell.

An expression vector can be modified in various ways. An *inducible promoter*, which responds to a specific signal, can be made part of an expression vector. For example, a specific

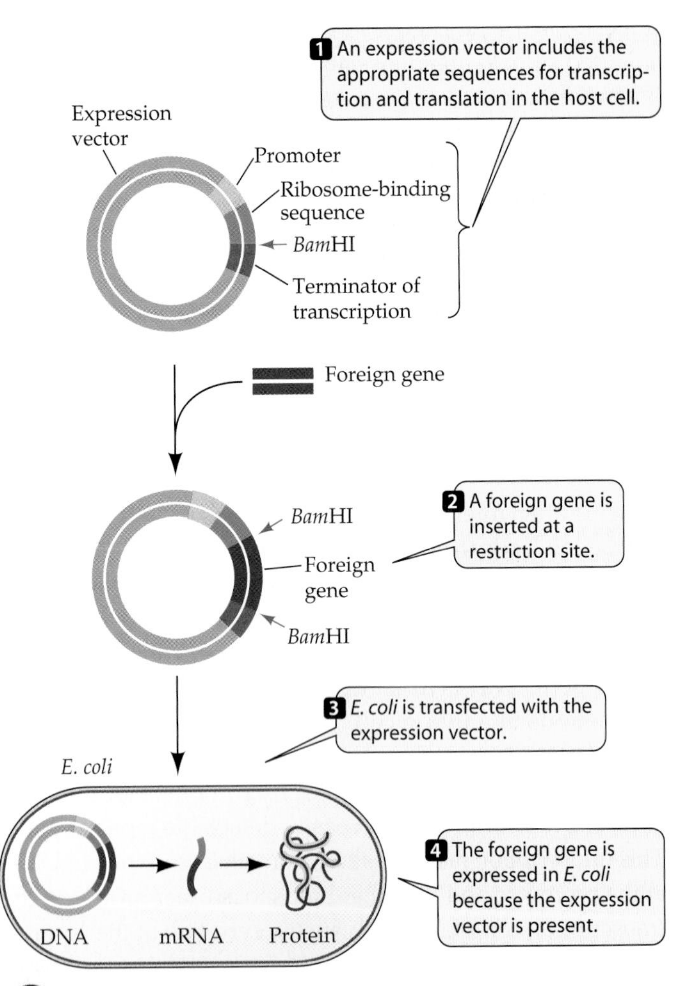

16.13 An Expression Vector Allows a Foreign Gene to Be Expressed in a Host Cell A transfected eukaryotic gene may not be expressed in *E. coli* because it lacks the necessary bacterial sequences for promotion, termination, and ribosome binding. Expression vectors contain these additional sequences, enabling the eukaryotic protein to be synthesized in the prokaryotic cell.

promoter that responds to hormonal stimulation can be used so that the foreign gene can be induced to transcribe its mRNA when the hormone is added. An enhancer that responds to hormonal stimulation can also be added so that transcription and protein synthesis will occur at high rates—a goal of obvious importance in the manufacture of an industrial product.

A *tissue-specific promoter*, which is expressed only in a certain tissue at a certain time, can be used if localized expression is desired. For example, many seed proteins are expressed only in the plant embryo. Coupling a gene to a seed-specific promoter will allow the gene to be expressed only as a seed protein.

Targeting sequences can be added to the expression vector so that the product of the gene is directed to an appropriate destination. For example, when yeast or bacterial cells making a protein are maintained in a large vessel, it is economical to include an export signal for the protein to be secreted into the extracellular medium for easier recovery.

Medically useful proteins can be made by biotechnology

Many medically useful products are being made by biotechnology (Table 16.1), and hundreds more are in various stages of development. The development of one such product, tissue plasminogen activator, illustrates the techniques that have been used.

In most people, when a wound begins bleeding, a blood clot soon forms to stop the flow. Later, as the wound heals, the clot dissolves. How does the blood perform these conflicting functions at the right times? Mammalian blood contains an enzyme called plasmin that catalyzes the dissolution of the clotting proteins. But plasmin is not always active; if it were, a blood clot would dissolve as soon as it formed! Instead, plasmin is "stored" in the blood in an inactive form called plasminogen. The conversion of plasminogen to plasmin is activated by an enzyme appropriately called tissue plasminogen activator (TPA), which is produced by cells lining the blood vessels:

$$\underset{\text{(inactive)}}{\text{plasminogen}} \xrightarrow{\text{TPA}} \underset{\text{(active)}}{\text{plasmin}}$$

Heart attacks and many strokes are caused by blood clots that form in major blood vessels leading to the heart or the brain, respectively. During the 1970s, a bacterial enzyme called streptokinase was found to stimulate the dissolution of clots in some patients. Treating these persons with this enzyme saved lives, but its use had side effects. Streptokinase was a protein foreign to the body, so patients' immune systems reacted against it. More important, the drug sometimes prevented clotting throughout the entire circulatory system, leading to an almost hemophilia-like condition in some patients.

The discovery of TPA and its isolation from human tissues led to the hope that this enzyme would bind specifically to clots, and that it would not provoke an immune reaction. But the amounts of TPA available from human tissues were tiny, certainly not enough to inject at the site of a clot in the emergency room.

Recombinant DNA technology solved this problem. TPA mRNA was isolated and used to make a cDNA copy, which was then inserted into an expression vector and transfected into *E. coli* (Figure 16.14). The transgenic bacteria made the protein in quantity, and it soon became available commercially. This drug has had considerable success in dissolving blood clots in people undergoing heart attacks and, especially, strokes.

16.1 Some Medically Useful Products of Biotechnology

PRODUCT	USE
Colony-stimulating factor	Stimulates production of white blood cells in patients with cancer and AIDS
Erythropoietin	Prevents anemia in patients undergoing kidney dialysis and cancer therapy
Factor VIII	Replaces clotting factor missing in patients with hemophilia A
Growth hormone	Replaces missing hormone in people of short stature
Insulin	Stimulates glucose uptake from blood in people with insulin-dependent (Type I) diabetes
Platelet-derived growth factor	Stimulates wound healing
Tissue plasminogen activator	Dissolves blood clots after heart attacks and strokes
Vaccine proteins: Hepatitis B, herpes, influenza, Lyme disease, meningitis, pertussis, etc.	Prevent and treat infectious diseases

DNA manipulation is changing agriculture

The cultivation of plants and husbanding of animals that constitute *agriculture* give us the world's oldest examples of biotechnology, dating back more than 8,000 years in human history. Over the centuries, people have adapted crops and farm animals to their needs. Through cultivation and selective breeding (artificial selection) of these organisms, desirable characteristics, such as ease of cooking the seeds or fat content of the meat, have been imparted and improved. In addition, people have developed crops with desirable growth char-

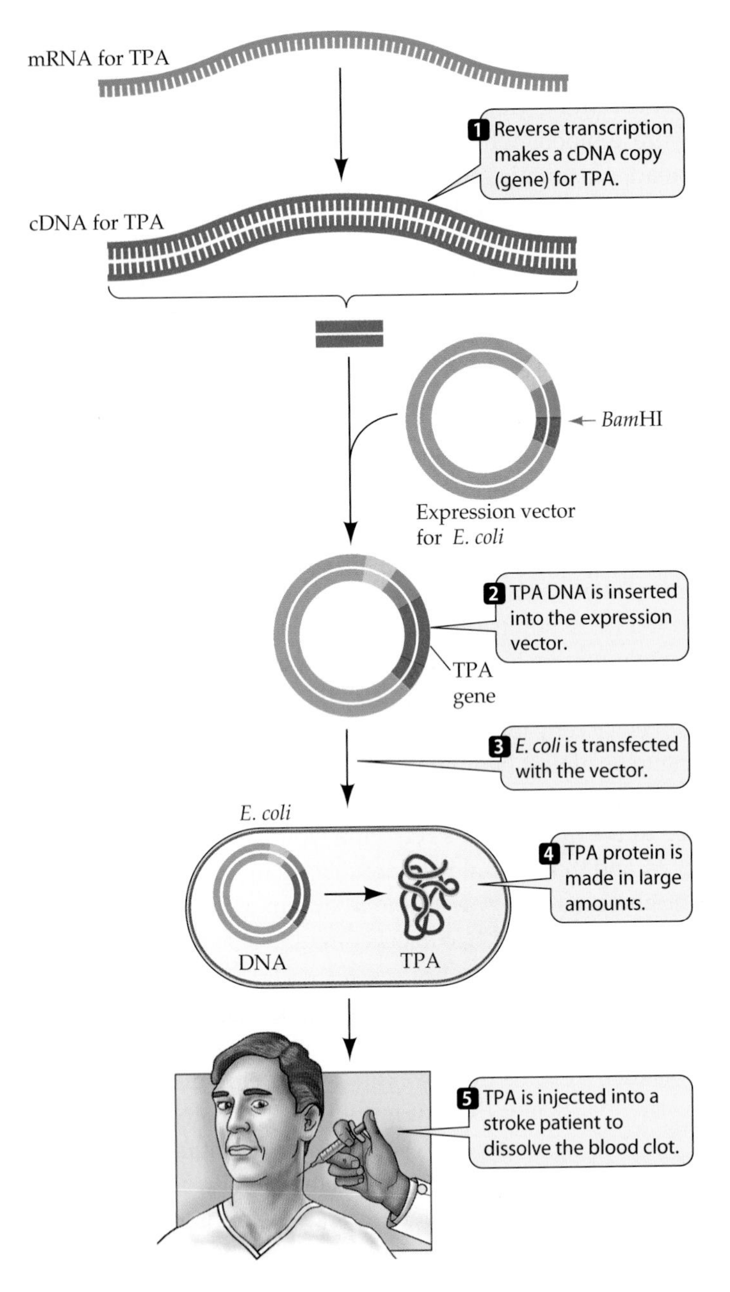
mRNA for TPA
1 Reverse transcription makes a cDNA copy (gene) for TPA.
cDNA for TPA
BamHI
Expression vector for E. coli
2 TPA DNA is inserted into the expression vector.
TPA gene
3 E. coli is transfected with the vector.
E. coli
4 TPA protein is made in large amounts.
DNA
TPA
5 TPA is injected into a stroke patient to dissolve the blood clot.

16.14 Tissue Plasminogen Activator: From Protein to Gene to Drug TPA is a naturally occurring human protein involved in dissolving blood clots. Its isolation and use as a pharmaceutical agent for treating patients suffering from blood clotting in the heart or brain—in other words, heart attacks or strokes—was made possible by recombinant DNA technology.

acteristics, such as high yield, a reliable ripening season, and resistance to diseases.

Until recently, the most common way to improve crop plants and farm animals was to select and breed varieties with desired phenotypes that existed in nature through mutational variation. The advent of genetics a century ago was followed by its application to plant and animal breeding. A crop plant or animal with desirable genes could be identified, and through deliberate crosses, those genes could be introduced into a widely used variety of that crop.

Despite some spectacular successes, such as the breeding of "supercrops" of wheat, rice, and corn, such deliberate crossing remains a hit-or-miss affair. Many desirable characters are complex in their genetics, and it is hard to predict the results of a cross or to maintain a prized combination as a pure-breeding variety year after year. In sexual reproduction, combinations of unlinked genes are quickly separated by genetic recombination. Moreover, traditional crop plant breeding takes a long time: many plants can reproduce only once or twice a year—a far cry from the rapid reproduction of bacteria or fruit flies.

Modern recombinant DNA technology has three advantages over traditional methods of breeding:

- It allows a breeder to choose specific genes, making the process more precise and less likely to fail as a result of the incorporation of unforeseen genes.
- It allows breeders to introduce any gene from any organism into a plant or animal species. This ability, combined with mutagenesis techniques, expands the range of possible new characteristics to an almost limitless horizon.
- The ability to work with cells in the laboratory and then regenerate a whole plant by cloning makes plant breeding much faster than the years needed for traditional breeding.

Biotechnology has found many applications in agriculture (Table 16.2), ranging from improving the nutritional properties of crops to using animals as gene product factories to using edible crops to make oral vaccines. We will describe a few examples here to demonstrate the approaches that have been used.

PLANTS THAT MAKE THEIR OWN INSECTICIDES. Humans are not the only species that consumes crop plants. Plants are subject to infections by viruses, bacteria, and fungi, but probably the most important crop pests are herbivorous insects. From the locusts of biblical (and modern) times to the cotton boll weevil, insects have continually eaten the crops people grow.

The development of insecticides has improved the situation somewhat, but insecticides have their own problems. Most, such as the organophosphates, are relatively nonspecific, killing not only pests in the field but beneficial insects in the ecosystem as well. Some even have toxic effects on other organisms, including people. What's more, insecticides are applied to the surface of crop plants and tend to be blown away to adjacent areas, where they may have unforeseen effects.

Some bacteria have solved their own pest problem by producing proteins that kill insect larvae that eat them. For

16.2 Agricultural Applications of Biotechnology under Development

PROBLEM	TECHNOLOGY/GENES
Improving the environmental adaptations of plants	Genes for drought tolerance, salt tolerance
Improving breeding	Male sterility for hybrid seeds
Improving nutritional traits	High-lysine seeds
Improving crops after harvest	Delay of fruit ripening; sweeter vegetables
Using plants as bioreactors	Plastics, oils, and drugs produced in plants
Controlling crop pests	Herbicide tolerance; resistance to viruses, bacteria, fungi, insects

example, there are dozens of strains of *Bacillus thuringiensis*, each of which produces a protein toxic to the insect larvae that prey on it. The toxicity of this protein is 80,000 times that of the usual commercial insecticides. When a hapless larva eats the bacteria, the toxin becomes activated, binding specifically to the insect's gut to produce holes. The insect starves to death.

Dried preparations of *B. thuringiensis* have been sold for decades as a safe, biodegradable insecticide. But biodegradation is their limitation, because it means that the dried bacteria must be applied repeatedly during the growing season. A more permanent approach would be to have the crop plants make the toxin themselves.

The toxin genes from different strains of *B. thuringiensis* have been isolated and cloned, and they have been extensively modified by the addition of plant promoters and terminators, plant poly A addition sequences, plant codon usage, and plant regulatory elements. These modified genes have been introduced into plant cells in the laboratory using the Ti plasmid vector (see Figure 16.5*c*), and transgenic plants have been grown and tested for insect resistance in the field. So far, transgenic tomato, corn, potato, and cotton crops have been shown to have considerable resistance to their insect predators.

TRANSGENIC ANIMALS EXPRESS USEFUL GENES. A transgene can be inserted into an animal, and if the appropriate promoter is present, the gene can be expressed in a readily available tissue. People with one type of emphysema have lung damage because they lack adequate amounts of a protein called α-1-antitrypsin (α-1AT). This protein inhibits elastase, an enzyme that breaks down connective tissue. Thus, using an inhibitor of elastase could alleviate these symptoms in these patients.

The problem is that only minuscule amounts of α-1AT can be purified from human serum. To overcome this problem, the gene for human α-1AT was introduced into the eggs of sheep, next to the promoter for lactoglobulin, a protein made in large amounts in milk. The resulting transgenic sheep made large amounts of α-1AT in their milk. Since female sheep produce large amounts of milk all year, this natural "bioreactor" produced a large supply of α-1AT, which was easily separated from the other components of the milk.

Goats, sheep, and cows are all being used for what has come to be called **pharming**: the production of medically useful products in milk. These products include blood clotting factors for treating hemophilia and antibodies for treating colon cancer.

CROPS THAT ARE RESISTANT TO HERBICIDES. Herbivorous insects are not the only threat to agriculture. Weeds may grow in fields and compete with crop plants for water and soil nutrients. Glyphosate ("Roundup") is a widely used and effective *herbicide*, or weed killer. It works only on plants, by inhibiting an enzyme system in the chloroplast that is involved in the synthesis of amino acids. Glyphosate is truly a "miracle herbicide," killing 76 of the world's 78 most prevalent weeds. Unfortunately, it also kills crop plants, so great care must be taken with its use. In fact, it is best used to rid a field of weeds before the crop plant starts to grow. But as any gardener knows, when the crop begins to grow, the weeds reappear. If the crop were not affected by the herbicide, the herbicide could be applied to the field at any time, and would kill only the weeds.

Fortunately, some soil bacteria have mutated to develop an enzyme that breaks down glyphosate. Scientists have isolated the gene for this enzyme, cloned it, and added plant sequences for transcription, translation, and targeting to the chloroplast. The gene has been inserted into corn, cotton, and soybean plants, making them resistant to glyphosate. This technology expanded so rapidly in the late 1990s that half of the U.S. crops of these three plants now contain this gene.

GRAINS WITH IMPROVED NUTRITIONAL CHARACTERISTICS. To remain healthy, humans must eat foods (or supplements) containing an adequate amount of β-carotene, which the body converts into vitamin A. About 400 million people worldwide suffer from vitamin A deficiency, which makes them susceptible to infections and blindness. One reason is that rice grains, which do not contain β-carotene, but only a precursor molecule for it, make up a large part of their diet. Other organisms, such as the bacterium *Erwinia* and daffodil plants, have enzymes that can convert the precursor into β-carotene. The genes for this biochemical pathway are present in the bacterial and daffodil genomes, but not in the rice genome.

16.15 Transgenic Rice Is Rich in β-Carotene The grains from a new transgenic strain of rice (left) are yellow because they make the pigment β-carotene, which is converted by humans into vitamin A. Normal rice (right) does not contain β-carotene.

Scientists isolated two of the genes for the β-carotene pathway from the bacterium and the other two from daffodil plants. They added promoter signals for expression in the developing rice grain, and then added each gene to rice plants by using the Ti plasmid vector from *Agrobacterium tumefaciens* (see Figure 16.5*c*). The resulting rice plants produce grains that look yellow because of their high β-carotene content (Figure 16.15). About 300 grams of this cooked rice a day can supply all the β-carotene a person needs. This new transgenic strain is now being crossed with more locally adapted strains, and it is hoped that the diets of millions of people will be improved as a result.

CROPS THAT ADAPT TO THE ENVIRONMENT. Throughout human history, agriculture has involved ecological management—tailoring the environment to the needs of crop plants. A farm field is an unnatural, human-designed system, and when conditions in that field become intolerable, the crops die. The Fertile Crescent, the region between the Tigris and Euphrates rivers in the Middle East where agriculture probably originated 10,000 years ago, is no longer fertile. It is now a desert, largely because the soil has a high salt concentration. Few plants can grow on salty soils, primarily because the environment is hypertonic to the plant roots, and water leaves them, resulting in wilting.

Recently, a gene was discovered in *Arabidopsis thaliana* that allows this tiny weed to thrive in salty soils. The gene codes for a protein that transports sodium ions into the vacuole. When this gene was added to tomato plants, they too grew in soils four times as salty as the normal lethal level (Figure 16.16). This finding raises the prospect of growing useful crops on what were previously unproductive soils.

More important, this example illustrates what could become a fundamental shift in the relationship between crop plants and the environment. Instead of manipulating the environment to suit the plant, biotechnology may allow us to adapt the plant to the environment. As a result, some of the negative effects of agriculture, such as water pollution, could be lessened.

(*a*)

(*b*)

16.16 Salt-Tolerant Tomato Plants Transgenic plants containing a gene for salt tolerance thrive in salty soils (*a*), while plants without the transgene die (*b*).

There is public concern about biotechnology

With the rapid expansion of genetically modified crops, concerns have been raised among the general public. Because of these concerns, some countries have banned foods that come from genetically modified crops. These concerns are centered on three claims:

- Genetic manipulation is an unnatural interference with nature.
- Genetically altered foods are unsafe to eat.
- Genetically altered crop plants are dangerous to the environment.

Advocates of biotechnology tend to agree with the first claim. However, they point out that all major crops are unnatural in the sense that they come from artificially bred plants growing in a manipulated environment (a farmer's field). Recombinant DNA technology just adds another level of sophistication to these techniques.

The concern about safety for humans is countered by the facts that only single genes are added and that these genes are specific for plant function. For example, the *B. thuringiensis* toxin produced by transgenic plants has no effect on people. However, as plant biotechnology moves from adding genes to improve plant growth to adding genes that affect human nutrition, such concerns will become more pressing.

The third concern, about environmental effects, centers on the possible "escape" of transgenes from crops to other species. If the gene for herbicide resistance, for example, were inadvertently transferred from a crop to a nearby weed, that weed could thrive in herbicide-treated areas. Or beneficial insects could eat plant materials containing *B. thuringiensis* toxin and die. Transgenic plants undergo extensive field testing before they are approved for use, but the complexity of the biological world makes it impossible to predict all potential environmental effects of transgenic organisms. Because of the potential benefits of agricultural biotechnology (see Table 16.2), scientists believe that it is wise to "proceed with caution."

DNA fingerprinting is based on the polymerase chain reaction

"Everyone is unique." This old saying certainly applies to the human genome. Mutation and recombination through sexual reproduction ensure that each one of us (unless we have an identical twin) has a unique DNA sequence. The characterization of an individual by his or her DNA base sequence is known as **DNA fingerprinting**.

An ideal way to distinguish an individual from all the other people on Earth would be to describe his or her entire genomic DNA sequence. But the human genome contains more than 3 billion nucleotides, so this idea is clearly not practical. Instead, scientists have looked for genes that are highly *polymorphic*—that is, genes that have multiple alleles (see Chapter 23) in the human population and are therefore most likely to be different in different individuals.

One easily analyzed genetic system consists of short moderately repetitive DNA sequences that occur side by side in the chromosomes. These repeat patterns are inherited. For example, an individual might inherit a chromosome 15 with a short sequence repeated six times from her mother, and the same chromosome with the same sequence repeated two times from her father. These repeats, called **VNTRs** (*v*ariable *n*umber *t*andem *r*epeats), are easily detectable if they lie between two recognition sites for a restriction enzyme. If the DNA from this individual is cut with the restriction enzyme, it will form two different-sized fragments: one larger (the one from the mother) and the other smaller (the one from the father). These patterns are easily seen by the use of gel electrophoresis (Figure 16.17). With several different VNTRs (as many as eight are used, each with numerous alleles), an individual's unique pattern becomes apparent.

DNA fingerprinting methods require 1 μg of DNA, or the DNA content of about 100,000 human cells, but this amount is not always available. The power of the PCR technique (see Figure 11.20) permits the targeted DNA from a single cell to

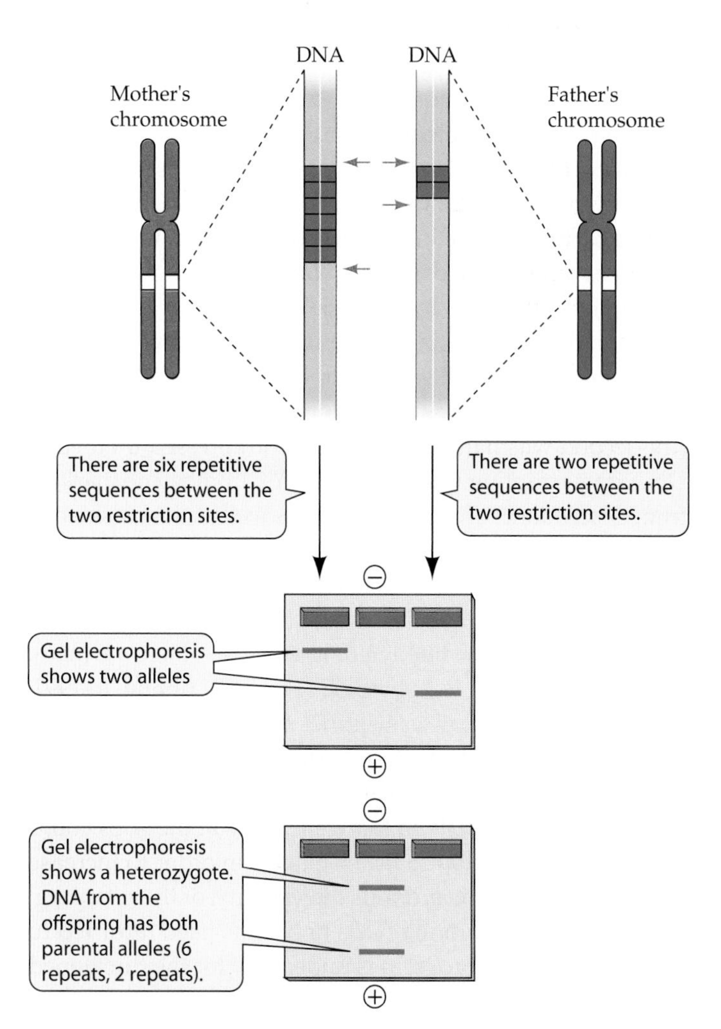

16.17 DNA Fingerprinting The number of VNTRs inherited by an individual can be used to make a DNA fingerprint.

be amplified, producing in a few hours the necessary 1 μg for restriction and electrophoresis.

DNA fingerprints are used in forensics (crime investigation) to help prove the innocence or guilt of a suspect. For example, in a rape case, DNA can be extracted from semen or hair left by the attacker and compared with DNA from a suspect. So far, this method has been used to prove innocence (the DNA patterns are different) more often than guilt (the DNA patterns are the same). It is easy to exclude someone on the basis of these tests, but two people could theoretically have the same patterns, since what is being tested is just a small sample of the genome. Therefore, proof that a suspect is guilty cannot rest on DNA fingerprinting alone.

Two fascinating examples demonstrate the use of DNA fingerprinting in the analysis of historical events. Three hundred years of rule by the Romanov dynasty in Russia ended on July 16, 1918, when Tsar Nicholas II, his wife, and their five children were executed by a firing squad during the Communist revolution. A report that the bodies had been burned to ashes was never questioned until 1991, when a shallow grave with several skeletons was discovered several miles from the presumed execution site. Recent DNA fingerprinting of bone fragments found in this grave indicated that they came from an older man and woman and three female children, who were clearly related to one another (Figure 16.18) and were also related to several living descendants of the Tsar.

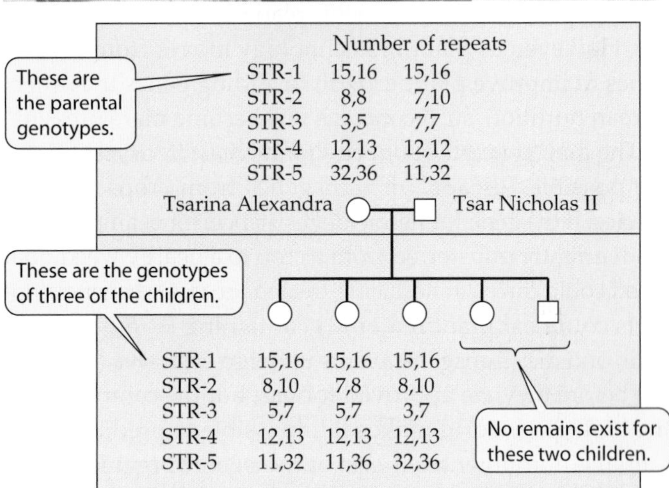

16.18 DNA Fingerprinting the Russian Royal Family The skeletal remains of Tsar Nicholas II, his wife Alexandra, and three of their children were found in 1991 and subjected to DNA fingerprinting. Five VNTRs were tested. The results can be interpreted as follows: Using the VNTR STR-2 as an example, the parents had genotypes 8,8 (homozygous) and 7,10 (heterozygous). The three children all inherited type 8 from the Tsarina and either type 7 or type 10 from the Tsar.

The other example involves Thomas Jefferson, the third president of the United States. In 1802, Jefferson was alleged to have fathered a son by his female slave, Sally Hemmings. Jefferson denied this, and his denial was accepted by many historians because of his vocal opposition to mixed-race relationships. But descendants of Hemmings's two oldest sons (the second was named Eston Jefferson) pressed their case. DNA fingerprinting was done using Y chromosome markers from descendants of these two sons as well as the president's paternal uncle (the president had no acknowledged sons). The results showed that Thomas Jefferson may have been the father of the second son, but was not the father of the first son.

In addition to such highly publicized cases, there are many other applications of PCR-based DNA fingerprinting. In 1992, the California condor was extinct in the wild. There were only 52 California condors on Earth, all cared for by the San Diego and Los Angeles zoos. Scientists made DNA fingerprints of all these birds so that geneticists at the zoos could select unrelated individuals for mating in order to increase the genetic variation, and thus the viability, of the offspring. A number of these young birds have now been returned to the wild. A similar program is under way for the threatened Galápagos tortoises.

Thousands of varieties of crops such as rice, wheat, corn, and grapes have been found in nature or produced by artificial selection. The seeds of many of these varieties are kept in cold storage in "seed banks." Samples of these plants are being fingerprinted to determine which varieties are genetically the same and which are the most diverse—information that will be useful as a guide to future breeding programs. DNA fingerprinting can also be used to identify a product from a crop: the characterization of DNA from grape varieties, for example, will allow wine makers and buyers to tell what they are purchasing. Million-dollar thoroughbred racehorses are also identified by their DNA fingerprints.

A related use of PCR is in the diagnosis of infections. In this case, the test shows whether the DNA of an infectious agent is present in a blood or tissue sample. Two primer strands matching the pathogen's DNA are added to the sample. If the pathogen is present, its DNA will serve as a template for the primer, and will be amplified. Because so little of the target sequence is needed, and because primers can be made to bind only to a specific viral or bacterial genome, this PCR-based test is extremely sensitive. If a pathogen is present in small amounts, PCR testing will detect it.

Finally, the isolation and characterization of genes for various human diseases, such as sickle-cell anemia and cystic fibrosis, has made PCR-based genetic testing a reality. We will discuss this subject in depth in the next chapter.

Chapter Summary

Cleaving and Rejoining DNA

▶ Knowledge of DNA transcription, translation, and replication has been used to create recombinant DNA molecules, made up of sequences from different organisms.

▶ Restriction enzymes, which are made by bacteria as a defense against viruses, bind to DNA at specific recognition sequences and cut it. **Review Figure 16.1**

▶ DNA fragments generated from cleavage by restriction enzymes can be separated by size using gel electrophoresis. The sequences of these fragments can be further identified by hybridization with a probe. **Review Figures 16.2, 16.3. See Web/CD Tutorial 16.1**

▶ Many restriction enzymes make staggered cuts in the two strands of DNA, creating "sticky ends" with unpaired bases. These sticky ends can be used to create recombinant DNA if DNA molecules from different species are cut with the same restriction enzyme. **Review Figure 16.4**

Getting New Genes into Cells

▶ Bacteria, yeasts, and cultured plant cells are commonly used as hosts for recombinant DNA procedures.

▶ Newly introduced DNA must be part of a replication unit if it is to be propagated in host cells. One way to make sure that the transfected DNA is part of such a unit is to insert it into a vector.

▶ There are specialized vectors for transfecting bacteria, yeasts, and plant cells. These vectors must contain an origin of replication, recognition sequences for restriction enzymes, and reporter genes to identify their presence in host cells. **Review Figure 16.5**

▶ Reporter genes conferring nutritional, antibiotic resistance, or fluorescent phenotypes can be used to identify which host cells have taken up the recombinant vector. **Review Figure 16.6**

Sources of Genes for Cloning

▶ The cutting of DNA by a restriction enzyme produces many fragments that can be individually and randomly combined with a vector and inserted into a host to create a gene library. **Review Figure 16.7**

▶ The mRNAs produced in a certain tissue at a certain time can be extracted and used to create complementary DNA (cDNA) by reverse transcription. **Review Figure 16.8**

▶ A third source of DNA is synthetic DNA made by chemists in the laboratory. The methods of organic chemistry can be used to create or mutate DNA sequences.

Some Additional Tools for DNA Manipulation

▶ Homologous recombination can be used to "knock out" a gene in an organism. **Review Figure 16.9**

▶ DNA chip technology permits the screening of thousands of sequences at the same time. **Review Figure 16.10. See Web/CD Tutorial 16.2**

▶ An antisense or interfering RNA complementary to a specific mRNA can prevent translation of the mRNA by hybridizing with it. **Review Figure 16.11**

▶ A two-hybrid system allows scientists to determine which proteins interact in cells. **Review Figure 16.12**

Biotechnology: Applications of DNA Manipulation

▶ Recombinant DNA techniques have made possible many new applications of biotechnology, such as the large-scale production of eukaryotic gene products.

▶ Expression vectors carry sequences such as promoters and transcription terminators that allow a gene of interest to be expressed in a host cell. **Review Figure 16.13. See Web/CD Activity 16.1**

▶ Recombinant DNA techniques have been used to make medically useful proteins that would otherwise have been difficult to obtain in necessary quantities. **Review Figure 16.14, Table 16.1**

▶ Because recombinant DNA technology has several advantages over traditional agricultural biotechnology, it is being extensively applied to agriculture. **Review Table 16.2**

▶ Because plant cells can be cloned to produce adult plants, the introduction of new genes into crop plants has been advancing rapidly. Transgenic crop plants can be adapted to their environment, instead of vice versa.

▶ "Pharming" uses transgenic animals that produce useful products in their milk.

▶ There is public concern about the application of recombinant DNA technology to food production.

▶ Because the DNA of an individual is unique, the polymerase chain reaction can be used to identify an organism from a small sample of its cells—that is, to create a DNA fingerprint. **Review Figures 16.17, 16.18**

Self-Quiz

1. Restriction enzymes
   a. play no role in bacteria.
   b. cleave DNA at highly specific recognition sequences.
   c. are inserted into bacteria by bacteriophages.
   d. are made only by eukaryotic cells.
   e. add methyl groups to specific DNA sequences.
2. When fragments of DNA of different sizes are placed in an electrical field,
   a. the smaller pieces migrate most quickly to the positive pole.
   b. the larger pieces migrate most quickly toward the positive pole.
   c. the smaller pieces migrate most quickly toward the negative pole.
   d. the larger pieces migrate most quickly toward the negative pole.
   e. the smaller and larger pieces migrate at the same rate.
3. From the list below, select the sequence of steps for cloning a piece of foreign DNA into a plasmid vector, introducing the plasmid into bacteria, and verifying that the plasmid and the insert are present:
   (1) Transform competent cells
   (2) Select for the lack of antibiotic resistance gene #1 function
   (3) Select for the plasmid antibiotic resistance gene #2 function

(4) Digest vector and foreign DNA with EcoR1, which inactivates antibiotic resistance gene #1
(5) Ligate the digested DNA together with the foreign DNA
a. 45132
b. 45123
c. 13425
d. 32145
e. 13254

4. Possession of which feature is *not* desirable in a vector for gene cloning?
   a. An origin of DNA replication
   b. Genetic markers for the presence of the vector
   c. Multiple recognition sites for the restriction enzyme to be used
   d. One recognition site each for one to several different restriction enzymes
   e. Genes other than the target for cloning

5. RNA interference (RNAi) inhibits
   a. DNA replication.
   b. RNA synthesis of specific genes.
   c. recognition of the promoter by RNA polymerase.
   d. transcription of all genes.
   e. translation of specific mRNAs.

6. Complementary DNA (cDNA)
   a. is produced from ribonucleoside triphosphates.
   b. is produced by reverse transcription.
   c. is the "other strand" of single-stranded DNA.
   d. requires no template for its synthesis.
   e. cannot be placed into a vector because it has the opposite base sequence of the vector DNA.

7. In a genomic library of frog DNA in *E. coli* bacteria,
   a. all bacterial cells have the same sequences of frog DNA.
   b. all bacterial cells have different sequences of DNA.
   c. each bacterial cell has a random fragment of frog DNA.
   d. each bacterial cell has many fragments of frog DNA.
   e. the frog DNA is transcribed into mRNA in the bacterial cells.

8. An expression vector requires all of the following, except
   a. genes for ribosomal RNA.
   b. a selectable genetic marker.
   c. a promoter of transcription.
   d. an origin of DNA replication.
   e. restriction enzyme recognition sites.

9. "Pharming" is a term that describes
   a. animals used in transgenic research.
   b. plants making genetically altered foods.
   c. synthesis of recombinant drugs by bacteria.
   d. large-scale production of cloned animals.
   e. synthesis of a drug by a transgenic animal in its milk.

10. In DNA fingerprinting,
   a. a positive identification can be made.
   b. a gel blot is all that is required.
   c. multiple restriction digests generate unique fragments.
   d. the polymerase chain reaction amplifies finger DNA.
   e. the variability of repeated sequences between two restriction sites is evaluated.

For Discussion

1. Compare PCR and cloning as methods to amplify a gene. What are the requirements, benefits, and drawbacks of each method?
2. As specifically as you can, outline the steps you would take to (*a*) insert and express the gene for a new, nutritious seed protein in wheat; and (*b*) insert and express a gene for a human enzyme in sheep's milk.
3. The *E. coli* plasmid pSCI carries genes for resistance to the antibiotics tetracycline and kanamycin. The *tet*^{r} gene has a single restriction site for the enzyme *Hin*dIII. Both the plasmid and the gene for glutein protein in corn are cleaved with *Hin*dIII and incubated to create recombinant DNA. The reaction mixture is then incubated with *E. coli* that are sensitive to both antibiotics. What would be the characteristics, with respect to antibiotic sensitivity or resistance, of colonies of *E. coli* containing, in addition to its own genome: (*a*) no new DNA; (*b*) native pSCI DNA; (*c*) recombinant pSCI DNA; and (*d*) corn DNA only? How would you detect these colonies?
4. Compare traditional genetics with molecular methods for producing genetically altered plants. For each case, describe: (*a*) sources of new genes; (*b*) number of genes transferred; and (*c*) how long the process takes.

17 Molecular Biology and Medicine

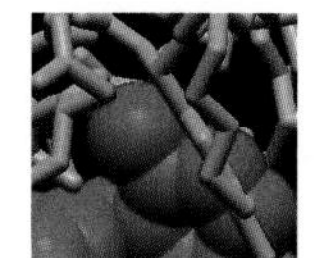

After his fiftieth birthday, Don's wife urged him to get a long-delayed medical checkup. He felt well, but realized that this was a good time to be screened for the various diseases that affect people as they get older. Don's routine blood count showed a surprise: Normally, people have about 5,000 white cells per milliliter of blood; Don had 40 times as many. Within a day, he was diagnosed with chronic myeloid leukemia, a serious cancer where white blood cells from the bone marrow proliferate out of control.

An oncologist (a physician who specializes in treating cancer) put Don on an aggressive regimen of chemotherapy. The three drugs Don took were designed to kill dividing cells—hopefully in the tumor, but such drugs also affect normal cells. One drug blocked microtubules from forming, thus preventing the mitotic spindle from assembling; another inserted into the double helix and damaged DNA replication; the third drug inhibited an enzyme involved in nucleotide synthesis.

Although the side effects were hard on Don, his white cell count gradually got lower. But after 8 months, the decline stalled at 80,000 cells/ml—still dangerously high. Worse, most of them were not mature white blood cells, but were undifferentiated cells from bone marrow. The chemotherapy drugs had killed Don's normal bone marrow cells. Without these specialized mature white blood cells, Don would die within months.

The timely development in the 1990s of a molecular understanding of this leukemia led to a therapy that saved Don's life. Scientists had known for several decades that chronic myelogenous leukemia cells have a particular chromosome translocation between chromosomes 9 and 22. With the advent of DNA sequencing, it became clear that this translocation fuses together parts of two genes: half of a gene called *bcr* on chromosome 22, and half of a gene called *abl* on chromosome 9. The abnormal protein made from this fused gene has strong protein tyrosine kinase activity.

As described in Chapter 15, protein kinases activate proteins by phosphorylation, binding a phosphate group from ATP. During the 1990s, the leukemia protein kinase was purified, crystallized, and the geometry of its binding site for ATP was described in molecular terms. Armed with this knowledge, organic chemists designed a drug that fit into and obstructed the binding site for ATP on the abnormal kinase, inactivating the enzyme. The drug, called Gleevec, was given to Don, and within a few weeks his white cell count was 5,400—normal! The cancer had been virtually wiped out.

The development of Gleevec is an opening chapter in the molecular medicine of the future. Unlike conventional chemotherapy, which uses

A "Smart Drug" The drug Gleevec (shown here as the red molecule) binds to a protein kinase made by certain leukemia cells, preventing the cancer cells' reproduction. Gleevec is the first example of a rationally designed, specifically targeted cancer-fighting drug produced using the knowledge and techniques of molecular medicine.

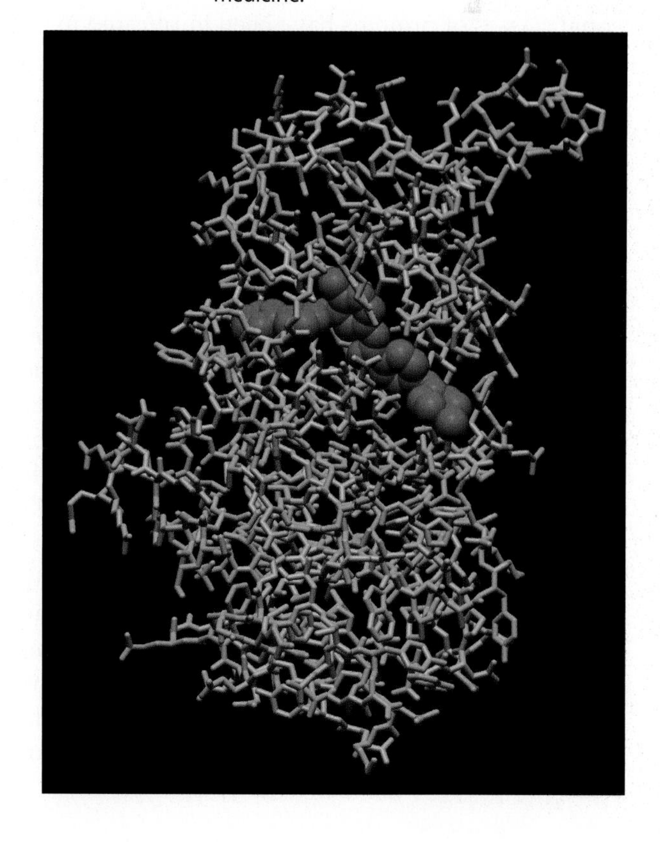

broadly acting nonspecific drugs that stop division in all cells (cancerous or not), Gleevec is highly selective. The new drug is targeted to a specific protein that *only* occurs in a particular cancer cell.

In the first section of this chapter, we identify and discuss the kinds of abnormal proteins that can result from an abnormal allele of a gene, whether the allele is inherited or has its origin from a mutation. Then we will consider the patterns of inheritance of human genetic diseases. Precise descriptions of these genetic abnormalities at the DNA level has come from molecular biology. This knowledge has been extended and applied to the treatment of cancer, among the most dreaded of human afflictions. The rise of molecular medicine is most dramatically shown by undertakings such as gene therapy and the sequencing of the human genome, which are discussed at the end of this chapter.

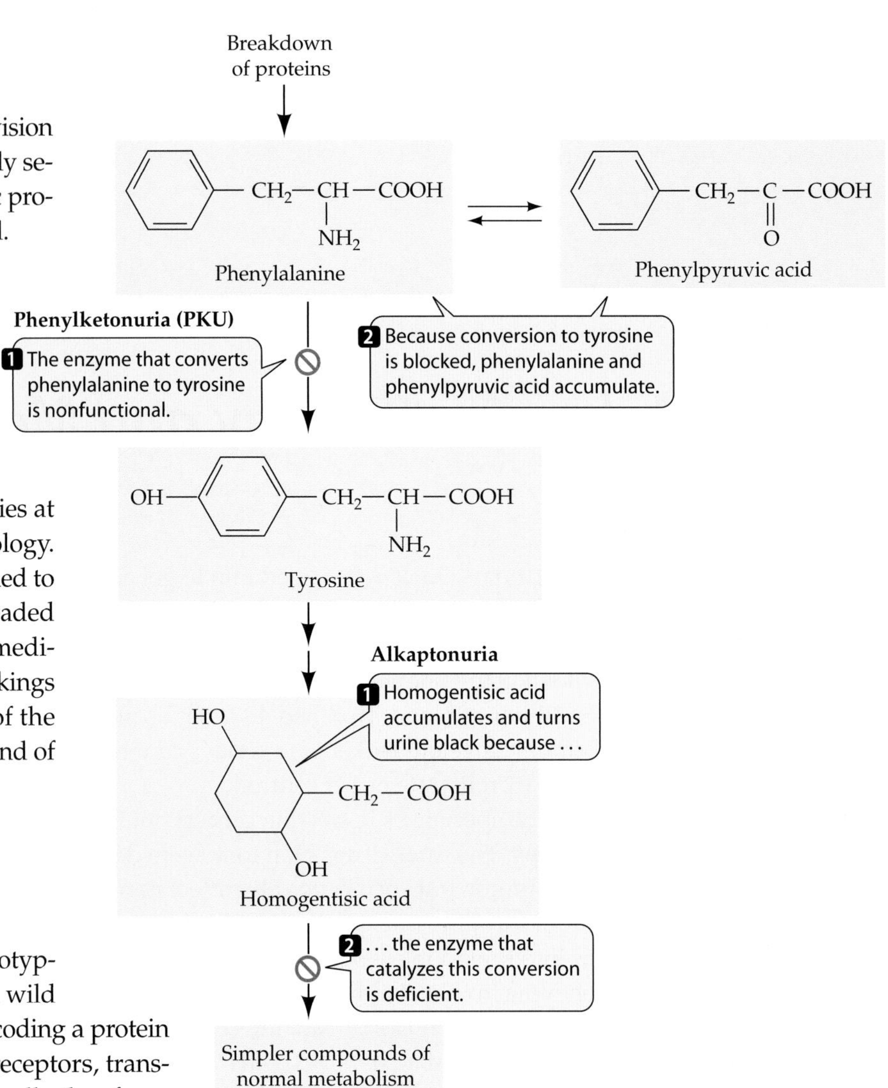

17.1 One Gene, One Enzyme Both phenylketonuria and alkaptonuria are caused by abnormalities in specific enzymes in the metabolic pathway that breaks down the amino acid phenylalanine. Knowing the causes of such single-gene, single-enzyme metabolic diseases can aid in developing screening tests as well as treatments.

Abnormal or Missing Proteins: The Mutant Phenotype

Genetic mutations are often expressed phenotypically as proteins that differ from the normal wild type. In principle, a mutation in any gene encoding a protein could result in a genetic disease. Enzymes, receptors, transport proteins, structural proteins, and nearly all other functional classes of proteins have been implicated in genetic diseases.

Dysfunctional enzymes can cause diseases

In 1934, the urine of two mentally retarded young siblings was found to contain phenylpyruvic acid, an unusual by-product of the metabolism of the amino acid phenylalanine. It was not until two decades later, however, that the complex clinical phenotype of the disease that afflicted these children, called *phenylketonuria* (PKU), was traced back to its molecular phenotype. The disease resulted from an abnormality in a single enzyme, phenylalanine hydroxylase (Figure 17.1). This enzyme normally catalyzes the conversion of dietary phenylalanine to tyrosine, but it was not active in PKU patients' livers. Lack of this conversion led to excess phenylalanine in the blood and explained the accumulation of phenylpyruvic acid. Later, the amino acid sequences of phenylalanine hydroxylase in normal people were compared with those in individuals with PKU. In many cases, the only difference in the 451 amino acids that constitute this long polypeptide chain was that instead of arginine at position 408, many people with PKU had tryptophan.

How does the molecular abnormality in PKU lead to its clinical symptoms? Since the pigment melanin is made from tyrosine, which people with PKU cannot synthesize adequately but must obtain in the diet, these people have light skin and hair color. The exact cause of the mental retardation in PKU remains elusive, but as we will see later in this chapter, it can be prevented.

Hundreds of human genetic diseases that result from enzyme abnormalities have been discovered, many of which lead to mental retardation and premature death. Most of these diseases are rare; PKU, for example, shows up in one newborn out of every 12,000. But these diseases are just the tip of the mutational iceberg. Some mutations result in amino acid changes that have no obvious clinical effects. In fact, at least 30 percent of all proteins whose sequences are known show detectable amino acid differences among individuals. The key point here is that polymorphism does not necessarily mean disease. There can be numerous normal alleles of a gene, each producing normally functioning forms of the protein.

Abnormal hemoglobin is the cause of sickle-cell disease

The first human genetic disease for which an amino acid abnormality was tracked down as the cause was *sickle-cell disease.* This blood disorder most often afflicts people whose ancestors came from the Tropics or from the Mediterranean. About 1 in 655 African-Americans are homozygous for the sickle allele and have the disease. The abnormal allele produces abnormal hemoglobin that leads to sickled red blood cells (see Figure 12.17). These cells tend to block narrow blood capillaries, especially when the oxygen concentration of the blood is low. The result is tissue damage and eventually death by organ failure, such as a heart attack.

Human hemoglobin is a protein with quaternary structure, containing four globin chains—two α chains and two β chains—as well as the pigment heme (see Figure 3.8). In sickle-cell disease, one of the 146 amino acids in the β-globin chain is abnormal: At position 6, the normal glutamic acid has been replaced by valine. This replacement changes the charge of the protein (glutamic acid is negatively charged and valine is neutral), causing it to form long, needle-like aggregates in the red blood cells: The result is *anemia,* a deficiency of normal red blood cells and an impaired ability of the blood to carry oxygen.

Because hemoglobin is easy to isolate and study, its variations in the human population have been extensively documented (Figure 17.2). Hundreds of single amino acid alterations in β-globin have been reported. Some of these polymorphisms are even found at the same amino acid position. For example, at the same position that is mutated in sickle-cell disease, the normal glutamic acid may be replaced by lysine, causing hemoglobin C disease. In this case, the resulting anemia is usually not severe. Many alterations of hemoglobin have no effect on the protein's function, and thus no clinical phenotype. That is fortunate, because about 5 percent of all humans are carriers for one of these variants.

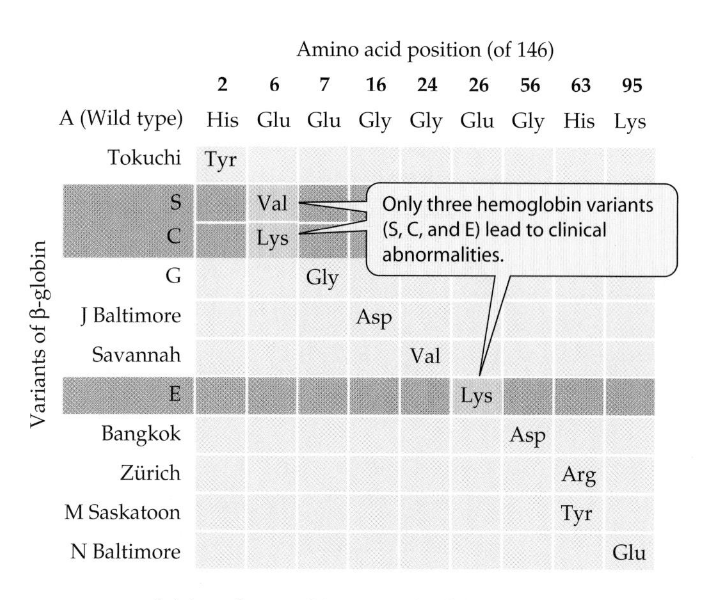

17.2 Hemoglobin Polymorphism Each of these mutant alleles changes a single amino acid in the 146-amino acid chain of β-globin. Only three of the many known variants of β-globin are known to lead to clinical abnormalities.

Altered membrane proteins cause many diseases

Some of the most common human genetic diseases show their primary phenotype as altered membrane receptor and transport proteins. About one person in 500 is born with *familial hypercholesterolemia* (FH), in which levels of cholesterol in the blood are several times higher than normal. The excess cholesterol can accumulate on the inner walls of blood vessels, leading to complete blockage if a blood clot forms. If a clot forms in a major vessel serving the heart, the heart becomes starved of oxygen, and a heart attack results. If a clot forms in the brain, the result is a stroke. People with FH often die of heart attacks before the age of 45.

Unlike PKU, which is characterized by the inability to convert phenylalanine to tyrosine, the problem in FH is not an inability to convert cholesterol to other products. People with FH have all the machinery needed to metabolize cholesterol. The problem is that they are unable to transport cholesterol into the liver and other cells that use it.

Cholesterol travels through the bloodstream in protein-containing particles called *lipoproteins*. One type of lipoprotein, low-density lipoprotein, carries cholesterol to the liver cells (Figure 17.3*a*). After binding to a specific receptor on the plasma membrane of a liver cell, the lipoprotein is taken up by endocytosis and delivers its cholesterol to the interior of the cell. People with FH lack a functional version of the receptor protein. Of the 840 amino acids that make up the receptor, only one may be abnormal, but that is enough to change its structure so that it cannot bind to the lipoprotein.

Among Caucasians, about one baby in 2,500 is born with *cystic fibrosis*. The clinical phenotype of this genetic disease is an unusually thick and dry mucus that lines surface tissues such as the airways of the respiratory system and the ducts of glands. In the respiratory passageways, this thick mucus obstructs the passage of air and also prevents the cilia on the surfaces of the epithelial cells from working efficiently to clear out the bacteria and fungal spores that we take in with every breath. The results are recurrent and serious infections as well as liver, pancreatic, and digestive failures, causing malnutrition and poor growth. People with cystic fibrosis often die in their twenties or thirties.

(*a*) **Hypercholesterolemia**

Normal liver cell: Cholesterol, as part of low-density lipoprotein (LDL), enters the cell after LDL binds to a receptor.

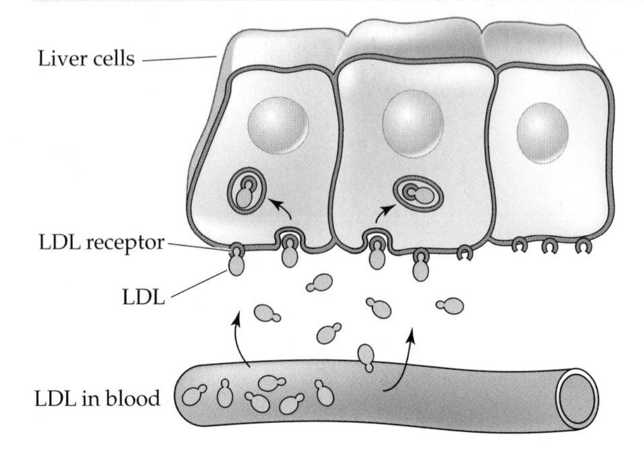

Familial hypercholesterolemia: Absence of a functional LDL receptor prevents cholesterol from entering the cells, and it accumulates in the blood.

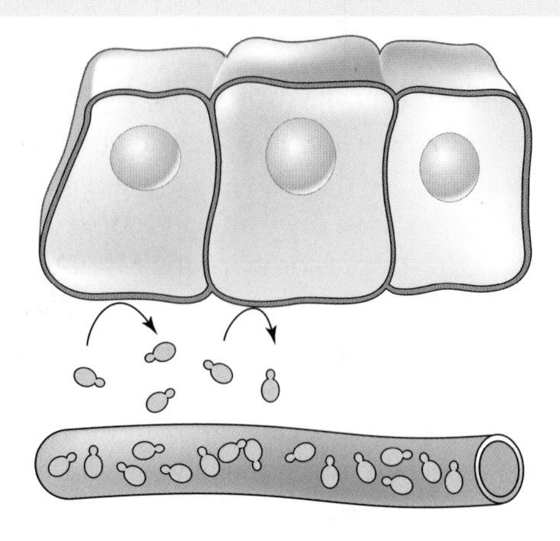

(*b*) **Cystic fibrosis**

Normal cell lining the airway: Cl^- leaves the cell through an ion channel. Water follows by osmosis, and moist thin mucus allows cilia to beat and sweep away foreign particles, including bacteria.

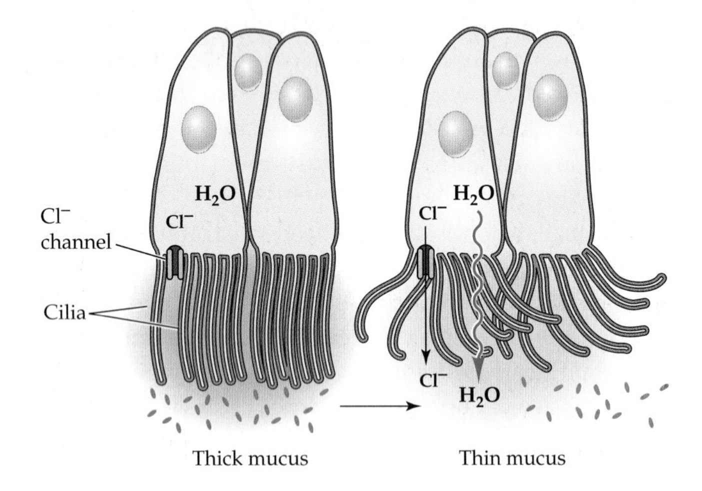

Cystic fibrosis: Lack of a Cl^- channel causes a thick, viscous mucus to form. Cilia cannot beat properly and remove bacteria; infections can easily take hold.

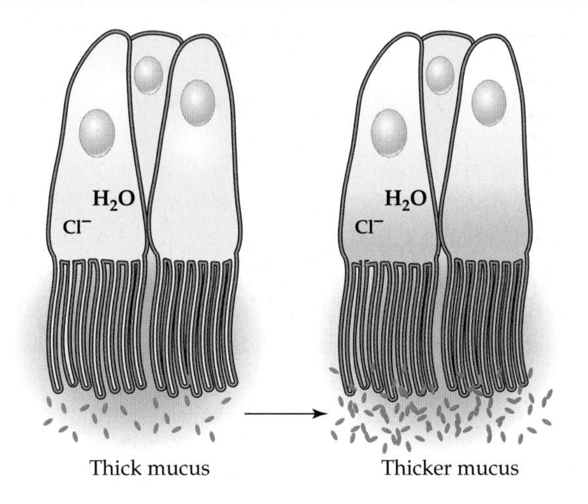

17.3 Genetic Diseases of Membrane Proteins The left two panels illustrate normal cell function, while the two right panels show the abnormalities caused by (*a*) hypercholesterolemia and (*b*) cystic fibrosis.

The reason for the thick mucus is a defective version of a membrane transport protein, the chloride transporter (Figure 17.3*b*). In normal cells, this ion channel opens to release Cl^- to the outside of an epithelial cell. The resulting imbalance of Cl^- ions (because of the channel, there are normally more on the outside of the cell than on the inside) causes water to leave the cell by osmosis, resulting in a moist thin mucus outside the cell. A single amino acid change in the channel protein renders it nonfunctional, leading to thick mucus and the consequent clinical problems.

Altered structural proteins can cause disease

About one boy in 3,000 is born with *Duchenne muscular dystrophy*. In this genetic disease, the problem is not an enzyme or receptor, but a protein involved in biological structure. People with this disease show progressively weaker muscles and are wheelchair-bound by their teenage years. Patients usually die in their twenties, when the muscles that serve their respiratory system fail. Normal people have a protein in their skeletal muscles called *dystrophin*, which connects the actin cytoskeleton of the muscle cells to the extracellular matrix. People with Duchenne muscular dystrophy do not have a working copy of dystrophin, so their muscle cells become structurally disorganized and the muscles stop working.

Coagulation proteins are involved in the clotting of blood at a wound. In normal people, inactive coagulation proteins are always present in the blood and become active only at a wound. People with the genetic disease *hemophilia* lack one of the coagulation proteins. Some people with this disease risk death from even minor cuts, since they cannot stop bleeding.

All of the foregoing protein alterations, and the diseases resulting from them, originate from an altered gene. But some abnormal proteins resulting in diseases may have a very different origin.

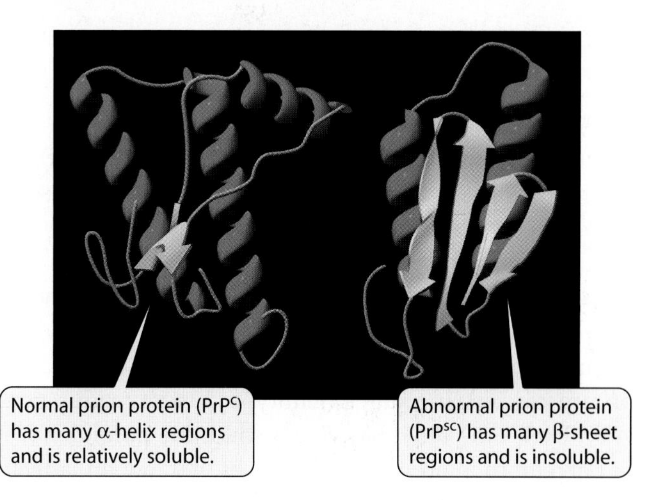

Normal prion protein (PrP^{C}) has many α-helix regions and is relatively soluble.
Abnormal prion protein (PrP^{SC}) has many β-sheet regions and is insoluble.

17.4 Prion Proteins Normal prion proteins (PrP^{c}, left) can be converted to the disease-causing form (PrP^{sc}, right), which has a different three-dimensional structure.

Prion diseases are disorders of protein conformation

Transmissible spongiform encephalopathies (TSEs) are degenerative brain diseases that occur in many mammals, including humans. In these diseases, the brain gradually develops holes, leaving it looking like a sponge. Scrapie, a TSE that causes affected sheep and goats to rub the wool off their bodies, has been known for 250 years. In the 1980s, a TSE that appeared in cows in Britain was traced to the cows having eaten products from sheep that had scrapie. Then, in the 1990s, some people who had eaten beef from cows with this TSE got a human version of the disease (dubbed "mad cow disease" by the media), again suggesting that the causative agent could cross species lines.

Another instance of humans consuming an infective agent and getting a TSE involved kuru, a disease resulting in dementia that occurred among the Fore tribe of New Guinea. In the 1950s, it was discovered that people with kuru had consumed the brains of people who had died of it. When this ritual cannibalism stopped, so did the epidemic of kuru.

Researchers found that TSEs could be transmitted from one animal to another via brain extracts from a diseased animal. At first, a virus was suspected. But when Tikva Alper at Hammersmith Hospital, London, treated infectious extracts with high doses of ultraviolet light to inactivate nucleic acids, they still caused TSEs. She proposed that the causative agent for TSEs was a protein, not a virus. Later, Stanley Prusiner at the University of California purified the protein responsible and showed it to be free of DNA or RNA. He called it a proteinaceous infective particle, or **prion**.

Normal brain cells contain a membrane protein called PrP^{c}. A protein with the same amino acid sequence is present in TSE-affected brain tissues, but that protein, called PrP^{sc}, has an altered shape (Figure 17.4). Thus, TSEs are usually not caused by a mutated gene (the primary structures of the two proteins are the same), but are somehow caused by an alteration in protein conformation. The altered three-dimensional structure of the protein has profound effects on its function in the cell. PrP^{sc} is insoluble, and it piles up as fibers in brain tissue, causing cell death.

How can the exposure of a normal cell to material containing PrP^{sc} result in a TSE? The abnormal PrP^{sc} protein seems to induce a conformational change in the normal PrP^{c} protein so that it too becomes abnormal, just as one rotten apple results in a whole barrel full of rotten apples. Just how the conversion occurs, and how it causes a TSE, are unclear.

Prions appear to represent a highly unusual phenomenon in human disease. The vast majority of infectious and inherited diseases are understood in terms of proteins that are products of functional or dysfunctional genes. But the expression of these genes, like all genes, is influenced by the environment.

Most diseases are caused by both genes and environment

The human diseases for which clinical phenotypes can be traced to a single altered protein and its altered gene may number in the thousands, and in most cases they are dramatic evidence of a one-gene, one-polypeptide relationship. Taken together, these diseases have a frequency of about 1 percent in the total human population.

Far more common, however, are diseases that are **multifactorial**; that is, those that are caused by many genes and proteins interacting with one another and with the environment. Although we tend to call individuals either normal (wild-type) or abnormal (mutant), the sum total of our genes is what determines which of us who eat a high-fat diet will die of a heart attack, or which of us exposed to infectious bacteria will come down with a disease. Estimates suggest that up to 60 percent of all people are affected by diseases that are genetically influenced.

Human genetic diseases have several patterns of inheritance

As in any human genetic system, the alleles that cause genetic diseases are inherited in a dominant or recessive pattern, and are carried on autosomes or on sex chromosomes (see Chapter 10). In addition, some human diseases are caused by more extensive chromosomal abnormalities (see Chapter 9). Different inheritance patterns can be seen when genetic diseases are followed over several human generations.

AUTOSOMAL RECESSIVE PATTERN. PKU, sickle-cell anemia, and cystic fibrosis are all caused by autosomal recessive mutant alleles. Typically, both parents of an affected person are carriers (normal phenotype, heterozygous genotype). Each time they conceive a child, the parents have a 25 percent (one in four) chance of having an affected (homozygous) son or daughter. Because of this low probability and the fact that many families in Western societies now have fewer than four children, it is unusual for more than one child in a family to have an autosomal recessive disease.

In the cells of a person who is homozygous for an autosomal recessive mutant allele, only the nonfunctional, mutant version of the protein it encodes is made. Thus a biochemical pathway or important cell function is disrupted, and disease results. As expected, heterozygotes, with one normal and one mutant allele, often have 50 percent of the normal level of functional protein. For example, people who are heterozygous for the PKU allele have half as many active molecules of phenylalanine hydroxylase in their liver cells as individuals who carry two normal alleles for this enzyme. But by one mechanism or another, this 50 percent suffices for relatively normal cellular function.

AUTOSOMAL DOMINANT PATTERN. Familial hypercholesterolemia is caused by an abnormal autosomal dominant allele. In this case, the presence of only one mutant allele is enough to produce the clinical phenotype. In people who are heterozygous for familial hypercholesterolemia, having half the normal number of functional receptors for low-density lipoprotein on the surface of liver cells is simply not enough to clear cholesterol from the blood. In autosomal dominance, direct transmission from an affected parent to offspring is the rule.

X-LINKED RECESSIVE PATTERN. Hemophilia is an X-linked recessive condition; that is, the gene locus responsible is on the X chromosome. Thus, a son who inherits a mutant allele on the X chromosome from his mother will have the disease, because his Y chromosome does not contain a normal allele. However, a daughter who inherits one mutant allele will be an unaffected heterozygous carrier, since she has two X chromosomes, and hence two alleles. Because, until recently, few males with these diseases lived to reproduce, the most common pattern of inheritance has been from carrier mother to son, and all rare X-linked diseases are much more common in males than in females.

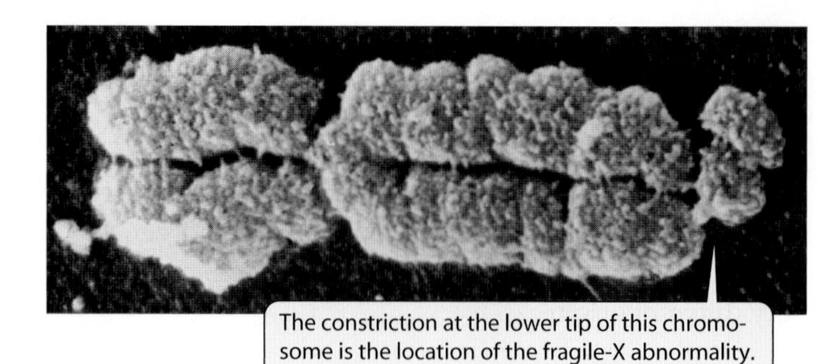

The constriction at the lower tip of this chromosome is the location of the fragile-X abnormality.

17.5 A Fragile-X Chromosome at Metaphase The chromosomal abnormality associated with mental retardation that characterizes fragile-X syndrome shows up under the microscope as a constriction in the chromosome.

CHROMOSOMAL ABNORMALITIES. Chromosomal abnormalities also cause human diseases. Such abnormalities include an excess or loss of one or more chromosomes (aneuploidy), loss of a piece of a chromosome (deletions), and the transfer of a piece of one chromosome to another chromosome (translocations). About one newborn in 200 is born with a chromosomal abnormality. While some of these abnormalities are inherited, many are the result of meiotic events such as nondisjunction (see Chapter 9).

Many zygotes that have chromosomal abnormalities do not survive development and are spontaneously aborted. Of the 20 percent of pregnancies that are spontaneously aborted during the first 3 months of human development, an estimated half of them have chromosomal abnormalities. For example, few human zygotes with only one X chromosome and no Y survive beyond the fourth month of pregnancy.

One common cause of mental retardation is *fragile-X syndrome* (Figure 17.5). About one male in 1,500 and one female in 2,000 are affected. These people have a constriction near the tip of the X chromosome that tends to break during preparation for microscopy, giving the name for this syndrome. Although the basic pattern of inheritance is that of an X-linked recessive trait, there are departures from this pattern. Not all people with the fragile-X chromosomal abnormality are mentally retarded, and we will describe the reason for this variation later in the chapter.

Mutations and Human Diseases

The isolation and description of human mutations has proceeded rapidly since modern molecular biological techniques were developed (see Chapter 16). When the primary phenotype was known, as in the case of abnormal hemoglobins, cloning the gene responsible was straightforward, although time-consuming. In other cases, such as Duchenne muscular dystrophy, a chromosome deletion associated with the disease in a patient pointed the way to the missing gene. In still other cases, such as cystic fibrosis, only a subtle molecular marker was available to lead investigators to the gene. In both of the latter examples, the primary phenotype—the defective protein—was unknown; only when the gene was isolated was the protein found.

In the discussions that follow, we will examine how mRNA, chromosome deletions, and genetic markers can be used to identify both mutant genes and abnormal proteins involved in genetic diseases. We will close this discussion by considering the role of expanding triplet repeats in genetic diseases such as fragile-X syndrome.

One way to identify a gene is to start with its protein

The primary phenotype for sickle-cell anemia was described in the 1950s as a single amino acid change in β-globin. On the basis of the clinical picture of sickled red blood cells, β-globin was certainly the right protein to examine. By the 1970s, researchers were able to isolate β-globin mRNA from immature red blood cells, which transcribe the globins as their major gene product. A cDNA copy of this mRNA was made and used to probe a human DNA library to find the β-globin gene (Figure 17.6*a*). DNA sequencing was then used to compare the normal gene with the gene from people with sickle-cell anemia. As previously described, it was found that a point mutation had changed only one base pair in the entire β-globin gene.

Chromosome deletions can lead to gene and then protein isolation

The inheritance pattern of Duchenne muscular dystrophy is consistent with an X-linked recessive trait. But until the late 1980s, neither the abnormal protein involved nor the gene encoding it had been described. This failure was not from lack of effort: Almost every known muscle protein had been tested without success. Then several boys with the disease were found to have a small deletion in their X chromosome. Comparison of the affected X chromosomes with normal ones made possible the isolation of the gene that was missing in the boys (Figure 17.6*b*).

17.6 Strategies for Isolating Human Genes (*a*) Once the sequence for the normal β-globin gene was established by cloning from the isolated mRNA, it could be compared to the gene sequence in people with sickle-cell anemia. (*b*) When an abnormality is caused by a missing sequence, as in Duchenne muscular dystrophy, researchers can compare the affected chromosome with a normal chromosome and isolate the DNA that is missing, then determine the protein this DNA encodes.

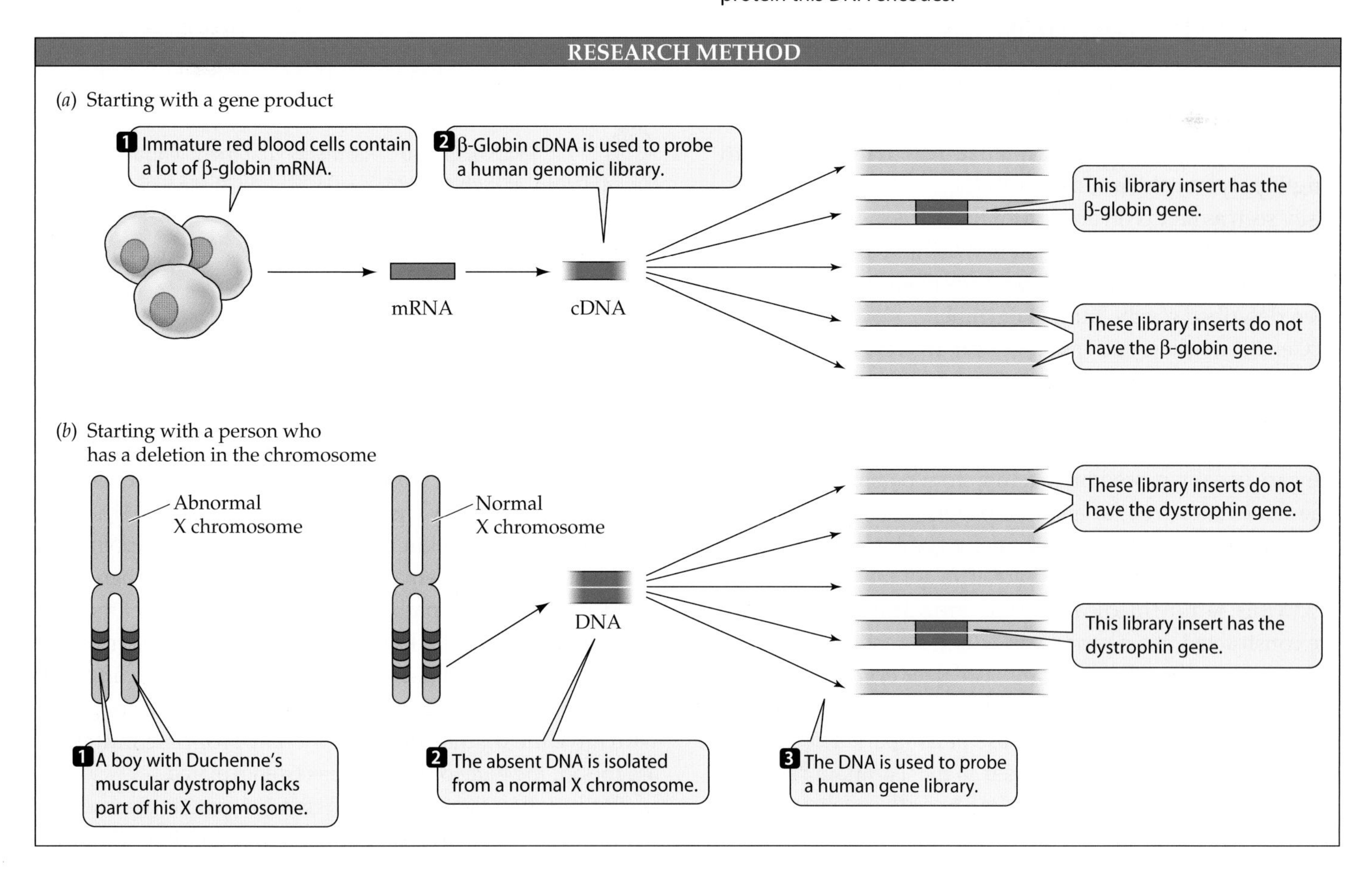
RESEARCH METHOD
(a) Starting with a gene product
1 Immature red blood cells contain a lot of β-globin mRNA.
2 β-Globin cDNA is used to probe a human genomic library.
mRNA
cDNA
This library insert has the β-globin gene.
These library inserts do not have the β-globin gene.
(b) Starting with a person who has a deletion in the chromosome
Abnormal X chromosome
Normal X chromosome
DNA
These library inserts do not have the dystrophin gene.
This library insert has the dystrophin gene.
1 A boy with Duchenne's muscular dystrophy lacks part of his X chromosome.
2 The absent DNA is isolated from a normal X chromosome.
3 The DNA is used to probe a human gene library.

Genetic markers can point the way to important genes

In cases in which no candidate protein nor visible chromosome deletion is available, a technique called **positional cloning** has been invaluable. To understand this method, imagine an astronaut looking down from space, trying to find her son on a park bench on Chicago's North Shore. The astronaut picks out reference points—landmarks that will lead her to the park. She recognizes the shape of North America, then moves to Lake Michigan, the Sears Tower, and so on. Once she has zeroed in on the North Shore park, she can use advanced optical instruments to find her son.

The reference points for positional cloning are genetic markers on the DNA. These markers can be located anywhere in the DNA. The only requirement is that they be polymorphic (have more than one allele).

As we described in Chapter 16, restriction enzymes cut DNA molecules at specific recognition sequences. On a particular human chromosome, a given restriction enzyme may make hundreds of cuts, producing many DNA fragments. The enzyme EcoRI, for example, cuts DNA at

5′... GAATTC ... 3′

Suppose this recognition sequence exists in a certain stretch of human chromosome 7. The restriction enzyme will cut this stretch once and make two fragments of DNA. Now suppose that, in some people, this sequence is mutated as follows:

5′... GAGTTC ... 3′

This sequence will not be recognized by the enzyme; thus it will remain intact and yield one larger fragment of DNA.

Such DNA differences are called **restriction fragment length polymorphisms**, or **RFLPs** (Figure 17.7). They can be easily seen as bands on an electrophoresis gel. An RFLP band pattern is inherited in a Mendelian fashion and can be followed through a pedigree. Thousands of such markers have been described for the human genome.

Genetic markers such as RFLPs can be used as landmarks to find genes of interest if they, too, are polymorphic. The key to this method is the well-known observation that if two genes are located near each other on the same chromosome, they are usually passed on together from parent to offspring. The same holds true for any pair of genetic markers.

To narrow down the location of a gene, a scientist must find a marker and a gene that are always inherited together. To do this, family medical histories are taken and pedigrees are constructed. If a genetic marker and a genetic disease are inherited together, then they must be near each other on the same chromosome. Unfortunately, "near each other" might be as much as several million base pairs apart. The process of locating the gene is thus similar to the astronaut focusing on Chicago: the first landmarks lead to only an approximate location.

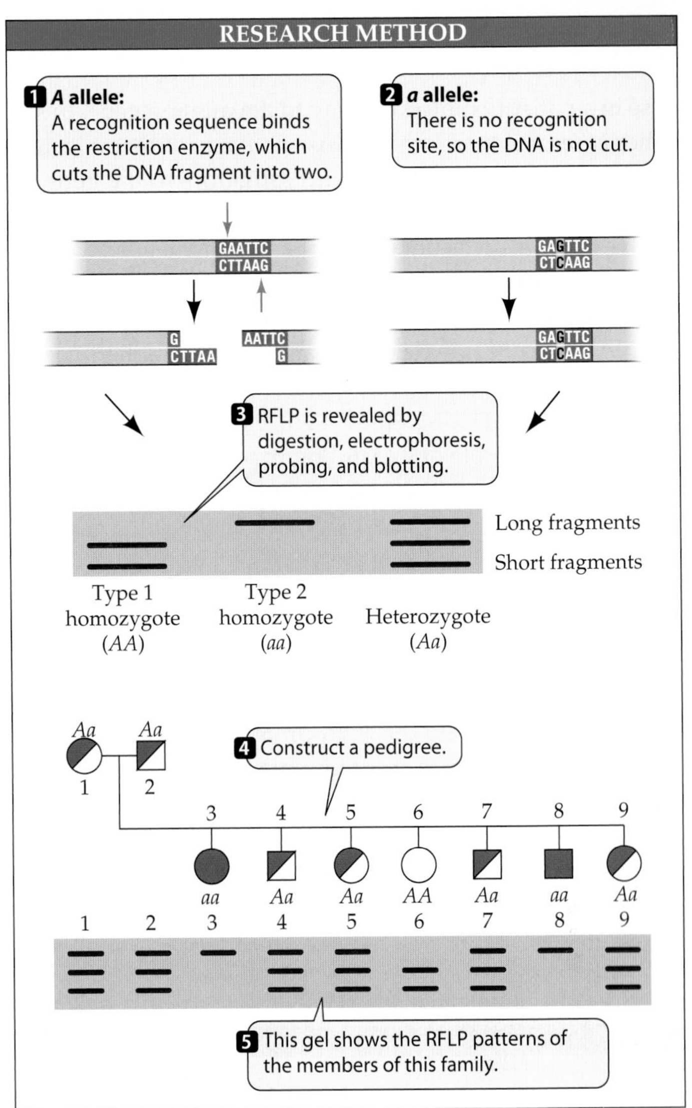

RESEARCH METHOD
1 A allele:
A recognition sequence binds the restriction enzyme, which cuts the DNA fragment into two.
2 a allele:
There is no recognition site, so the DNA is not cut.
GAATTC
CTTAAG
G
CTTAA
AATTC
G
GAGTTC
CTCAAG
3 RFLP is revealed by digestion, electrophoresis, probing, and blotting.
Long fragments
Short fragments
Type 1 homozygote (AA)
Type 2 homozygote (aa)
Heterozygote (Aa)
4 Construct a pedigree.
Aa
Aa
aa
AA
5 This gel shows the RFLP patterns of the members of this family.

17.7 RFLP Mapping Restriction fragment length polymorphisms are differences in DNA sequences that serve as genetic markers. Thousands of such markers have been described for the human genome.

How can the gene be isolated? Many sophisticated methods are available for narrowing the search. For example, the neighborhood around the RFLP can be screened for further RFLPs involving other restriction enzymes. With luck, one of them might be more closely linked to the disease-causing gene. Once a relatively short DNA sequence (several hundred thousand bases) thought to contain the gene is pinpointed, it can be cut into fragments, and those fragments, when denatured, can be used to probe cellular mRNA. If one of the fragments hybridizes with the mRNA, it means that the fragment is part of a gene that is expressed as mRNA. The candidate gene is then sequenced from normal people and from people who have the disease in question. If appropriate mutations are found, the gene of interest has been isolated.

The isolation of genes responsible for genetic diseases has led to spectacular advances in the understanding of human biology. Before the genes, and then the proteins, for Duchenne muscular dystrophy and for cystic fibrosis were isolated, dystrophin and the chloride transporter had never been described. Thus, the identification of mutant genes has opened up new vistas in our understanding of how the human body works.

Human gene mutations come in many sizes

As we saw in Chapter 12, mutations come in many sizes, from changes in a single base pair to changes in entire chromosomes. As we have seen, sickle-cell anemia is caused by a point mutation. Some variants of the β-globin gene cause disease, but most do not (see Figure 17.2). Those point mutations that alter a protein's function usually affect its three-dimensional structure; for example, such a mutation might alter the shape at the active site of an enzyme.

Some mutations lead to a greatly shortened protein chain with total loss of its function. For example, some people with cystic fibrosis have a nonsense mutation such that a codon for an amino acid near the beginning of the long chloride transporter protein chain has been changed to a stop codon. If this happens, only a very short, nonfunctional peptide is made. As we noted in Chapter 12, other point mutations affect RNA processing, leading to nonfunctional mRNA and no protein synthesis.

DNA sequencing has revealed that mutations occur most often at certain base pairs. These "hot spots" are often located where cytosine residues have been methylated to 5-methylcytosine (see Chapter 14). This phenomenon is a result of the natural instability of the bases in DNA. Either spontaneously or with chemical prodding unmethylated cytosine residues can lose their amino group and form uracil (Figure 17.8*a*). But the cell nucleus has a repair system that recognizes this uracil as being inappropriate for DNA (after all, uracil occurs only in RNA). It removes the uracil and replaces with cytosine.

The fate of 5-methylcytosine that loses its amino group is rather different, since the result of that loss is thymine, a natural base for DNA. The uracil repair system ignores this thymine (Figure 17.8*b*). However, since the GC pair is now a mismatched GT pair, a different type of repair system comes in and tries to fix the mismatch. Half the time, the mismatch repair system matches a new C to the G, but the other half of the time, it matches a new A to the T, resulting in a mutation.

Larger mutations may involve many base pairs of DNA. For example, some of the deletions in the X chromosome that result in Duchenne muscular dystrophy cover only part of the dystrophin gene, leading to an incomplete protein and a mild form of the disease. Others cover all of the gene, and thus the protein is missing entirely from muscle, resulting in the severe form of the disease. Still other deletions involve millions of base pairs and cover not only the dystrophin gene but adjacent genes as well; the result may be several diseases simultaneously.

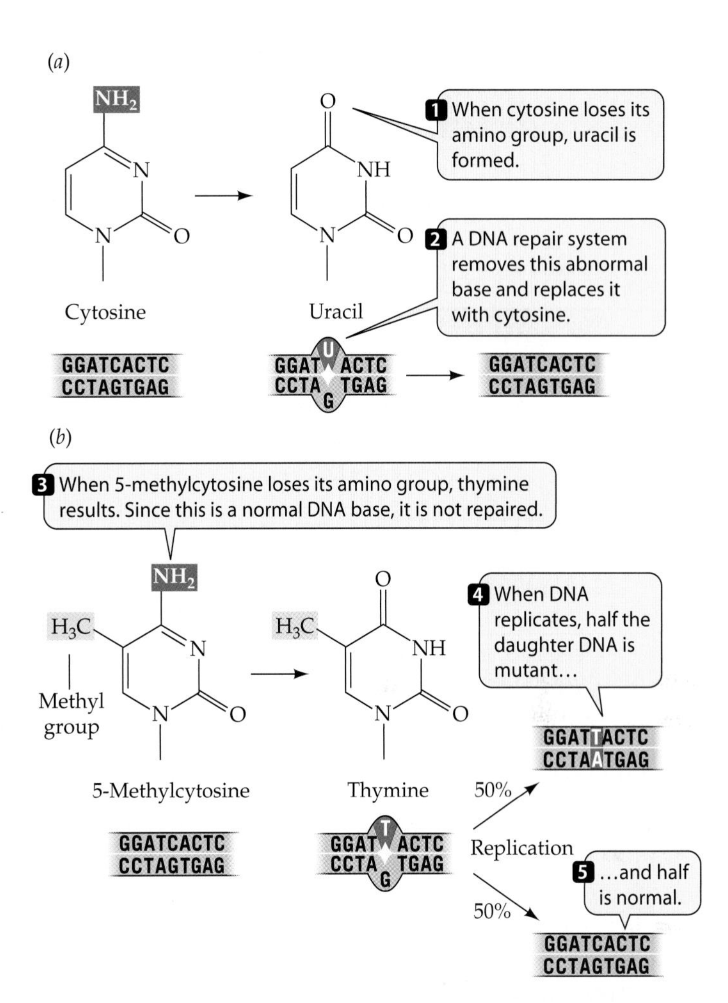

17.8 5-Methylcytosine in DNA Is a "Hot Spot" for Mutagenesis
(*a*) Cytosine can lose an amino group either spontaneously or because of exposure to certain chemical mutagens. Such mutations are usually repaired. (*b*) If the cytosine residue has been methylated to 5-methylcytosine, however, the mutation is unlikely to be repaired.

Expanding triplet repeats demonstrate the fragility of some human genes

About one-fifth of all males that have the fragile-X chromosomal abnormality are phenotypically normal, as are most of their daughters. But many of those daughters' sons are mentally retarded. In a family in which the fragile-X syndrome appears, later generations tend to show earlier onset and more severe symptoms of the disease. It is almost as if the abnormal allele itself is changing—and getting worse. And that's exactly what is happening.

The gene responsible for fragile-X syndrome (*FMR1*) contains a repeated triplet, CGG, at a certain point in the promoter region. In normal people, this triplet is repeated 6 to 54 times (average: 29). In the alleles of mentally retarded peo-

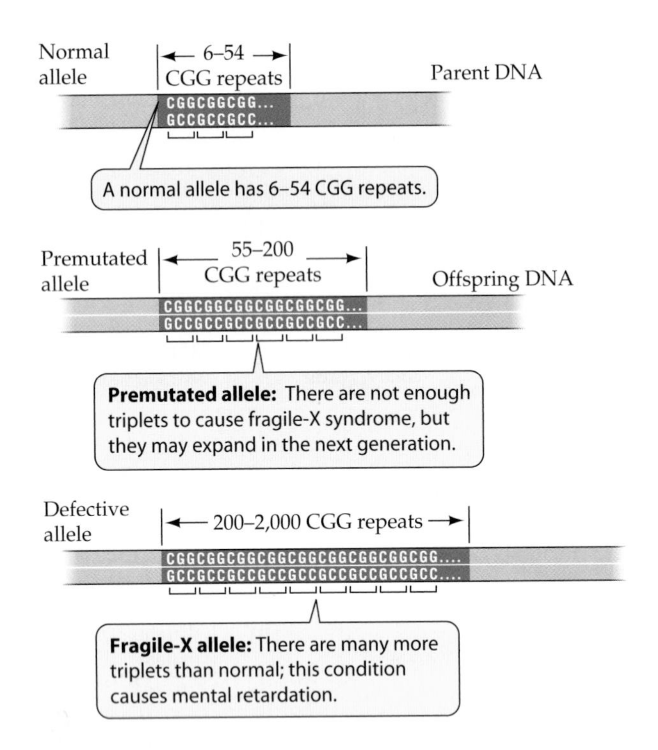

17.9 The CGG Repeat in the Fragile-X Gene Expands with Each Generation The genetic defect in fragile-X syndrome is caused by 200 or more repeats of the CGG triplet.

ple with fragile-X syndrome, the CGG sequence is repeated 200 to 2,000 times.

The "premutated" males that show no symptoms, but have affected descendants, have 52 to 200 repeats. These repeats become more numerous as the daughters of these men pass the chromosome on to their children (Figure 17.9). Expansion to more than 200 repeats leads to increased methylation of the cytosines in the CGG units, accompanied by transcriptional inactivation of the *FMR1* gene. The normal role of the protein product of this gene is to bind to mRNAs involved in nerve cell function and regulate their translation at the ribosome. When the FMR1 protein is not made in adequate amounts, these mRNAs are not properly translated, and nerve cells die. Their loss often results in mental retardation.

Such **expanding triplet repeats** have been found in over a dozen other diseases, such as myotonic dystrophy (involving repeated CTG triplets) and Huntington's disease (in which CAG is repeated). Many benign genes also appear to have these repeats, which may be found within a protein-coding region or outside it. How the repeats expand is not known, but DNA polymerase may slip after copying a repeat and then fall back to copy it again. In all previously known classes of human mutations, the mutation is just as stable as the normal allele, but expanding triplet repeats represent an entirely new class of unstable mutant alleles.

Genomic imprinting shows that mammals need both a mother and a father

Just after fertilization in a mammalian egg, there are two haploid *pronuclei*—one from the sperm and the other from the egg—in the zygote. The two pronuclei can be distinguished from each other, and they can be carefully removed with a micropipette and placed in other eggs. So it is possible to make mouse zygotes in the laboratory with two male or two female pronuclei. These diploid cells should go on to develop into mice—but they don't. Invariably, if the two sets of chromosomes come from only one sex, development begins, but is quickly aborted. The same thing happens in those rare instances when this occurs in humans—for instance, if two sperm enter an empty egg. Again, a fetus never develops.

In addition to showing the obvious need for two sexes, these observations raise the possibility that the male and female genomes are not functionally equivalent. In fact, there are groups of genes that differ in their phenotypic effects depending on which parent they came from. This phenomenon is called **genomic imprinting**.

A dramatic example of genomic imprinting is the inheritance and phenotypic pattern of a certain small deletion on human chromosome 15.

- If the deletion is on the mother's chromosome 15, the result is a thin child with a wide mouth and prominent jaw (*Angelman syndrome*).
- If the deletion is on the father's chromosome 15, the child is short and obese, with small hands and feet (*Prader-Willi syndrome*).

The remaining functional alleles in this region of chromosome 15 must be imprinted in very different ways in the two sexes to result in such different phenotypes. How this happens is not clear.

Detecting Genetic Variations: Screening for Human Diseases

The determination of the precise molecular phenotypes and genotypes of human genetic diseases has given us the ability to diagnose these diseases even before symptoms first appear. **Genetic screening** is the use of a test to identify people who have, are predisposed to, or are carriers of a certain genetic disease. It can be applied at many times of life and used for many purposes.

- *Prenatal* screening can identify an embryo or fetus with a disease so that medical intervention can be applied or decisions about continuing the pregnancy can be made.
- *Newborn* babies can be screened so that proper medical intervention can be initiated quickly for those babies who need it.

- *Asymptomatic* people who have a relative with a genetic disease can be screened to determine whether they are carriers of the disease or are likely to develop the disease themselves.

The existence of genetic screening techniques poses ethical questions concerning the uses of the information they provide, as we will see later in the chapter.

Screening for abnormal phenotypes can make use of protein expression

Screening of newborns for phenylketonuria can free a child from developing mental retardation. Such screening is legally mandatory in many countries, including all of the United States and Canada. Babies who are homozygous for this genetic disease are born with a normal phenotype because excess phenylalanine in their blood before birth diffuses across the placenta to the mother's circulation. Since the mother is almost always heterozygous, and therefore has adequate phenylalanine hydroxylase activity, her body metabolizes the excess phenylalanine from the fetus. Thus, before birth, the baby's blood does not accumulate abnormal levels of phenylalanine.

After birth, however, the situation changes. The baby begins to consume protein-rich food (milk) and to break down some of its own proteins. Phenylalanine enters the baby's blood, and without the mother's phenylalanine hydroxylase to break it down, accumulates there. After a few days, the phenylalanine level in the baby's blood may be ten times higher than normal. Within days, the developing brain is damaged, and untreated children with PKU become severely mentally retarded.

If PKU is detected early, it can be treated with a special diet low in phenylalanine and the brain damage avoided. Thus, early detection is imperative. In 1963, Robert Guthrie described a simple screening test for PKU in newborns that today is used almost universally (Figure 17.10). This elegant method uses auxotrophic bacteria to detect the presence of an amino acid—phenylalanine—in the blood. The test can be automated so that a screening laboratory can process many samples in a day.*

If an infant tests positive for PKU in this screening, he or she must be retested using a more accurate chemical assay for phenylalanine. If that test also shows a high phenylalanine level in the blood, dietary intervention is begun. The whole process must be completed by the end of the second week of life. Since the screening test is inexpensive (about a dollar per test), and since babies with PKU who receive early medical intervention develop practically normally, the benefit of screening is significant.

*Guthrie refused to patent the screening test he developed, or to accept any royalties or payment for it. Its immediate and widespread use was at least in part a result of his generosity in allowing the test to be available to all hospitals at low cost.

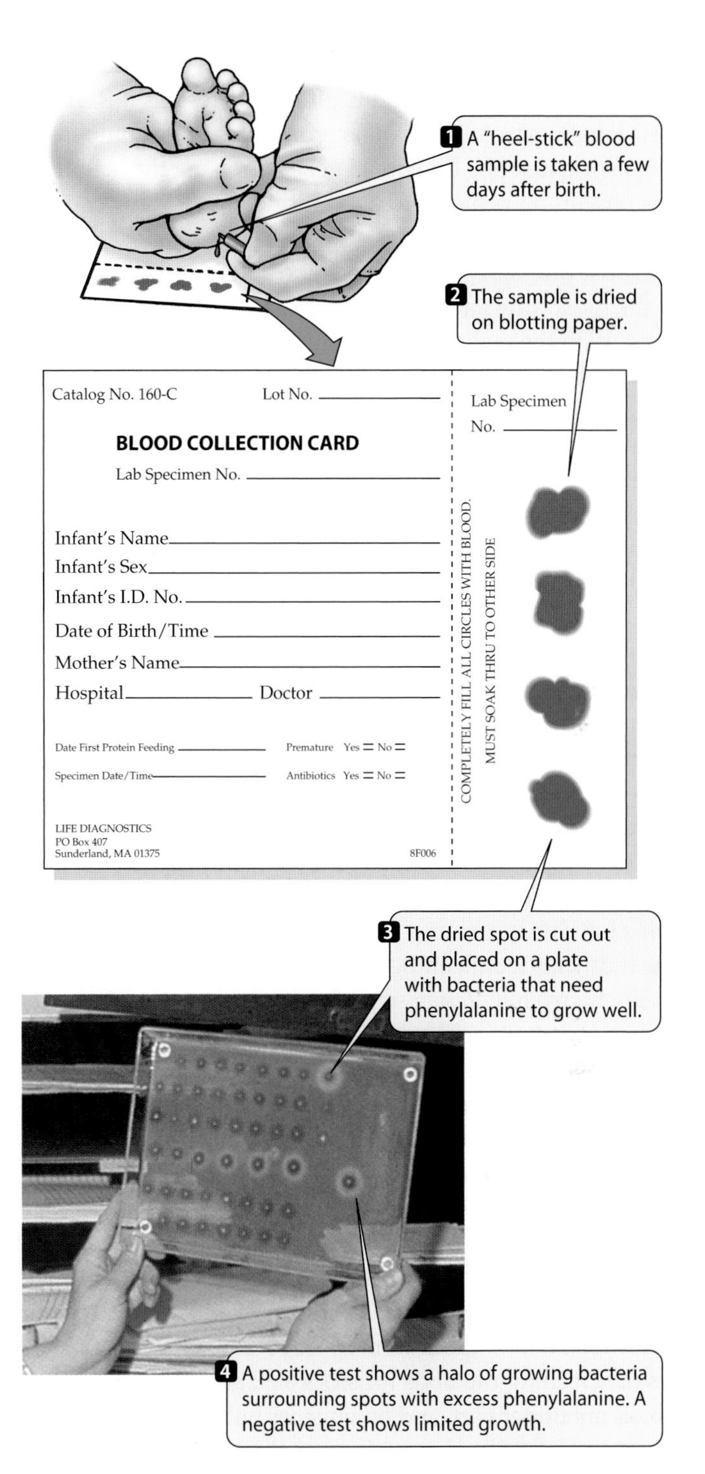
1 A "heel-stick" blood sample is taken a few days after birth.
2 The sample is dried on blotting paper.
Catalog No. 160-C
Lot No.
Lab Specimen No.
BLOOD COLLECTION CARD
Lab Specimen No.
Infant's Name
Infant's Sex
Infant's I.D. No.
Date of Birth/Time
Mother's Name
Hospital
Doctor
Date First Protein Feeding
Premature Yes No
Specimen Date/Time
Antibiotics Yes No
LIFE DIAGNOSTICS
PO Box 407
Sunderland, MA 01375
8F006
COMPLETELY FILL ALL CIRCLES WITH BLOOD.
MUST SOAK THRU TO OTHER SIDE
3 The dried spot is cut out and placed on a plate with bacteria that need phenylalanine to grow well.
4 A positive test shows a halo of growing bacteria surrounding spots with excess phenylalanine. A negative test shows limited growth.

17.10 Genetic Screening of Newborns for Phenylketonuria A simple test devised by Robert Guthrie in 1963 is used today to screen newborns for phenylketonuria. Early detection means that the symptoms of the condition can be prevented by following a therapeutic diet.

Several screening methods can find abnormal genes

The blood level of phenylalanine is an indirect measure of phenylalanine hydroxylase activity in the liver. But how can we screen for genetic diseases that are expressed only in a certain tissue, such as the liver or brain, and are not reflected in the blood? What if blood is difficult to test, as it is in a fetus? Finally, since tissues in heterozygotes often compensate for having just one functional gene by raising the activity of the remaining proteins to near normal levels, how can heterozygotes be identified?

These problems can be overcome by **DNA testing**, which is the most direct and accurate way to test for an abnormal gene. With the molecular description of the genetic mutations responsible for human diseases, it has become possible to examine directly any cell in the body at any time during the life span for mutations. However, these methods work best for diseases caused by only one or a few different mutations.

The polymerase chain reaction (PCR) technique allows testing of the DNA from even a single cell. You will recall from Chapter 11 that PCR amplifies a target sequence of DNA millions of times in the test tube. Consider, for example, two parents who are both heterozygous for the cystic fibrosis allele, have had a child with the disease, and want a normal child. If the mother is treated with the appropriate hormones, she can be induced to "superovulate," releasing several eggs. One of the eggs can be injected with a single sperm from her husband and the resulting zygote allowed to divide to the 8-cell stage. If one of these embryonic cells is removed, it can be tested for the presence of the cystic fibrosis allele(s). The remaining 7-cell embryo can be implanted in the mother's womb and go on to develop normally.

Such *preimplantation screening* is performed only rarely. More typical are analyses of fetal cells after implantation in the womb. Fetal cells can be analyzed at about the tenth week of pregnancy by chorionic villus sampling or during the thirteenth to seventeenth weeks by amniocentesis. These two sampling methods are described in Chapter 43. In either case, only a few fetal cells are required.

Newborns can also be screened for genetic mutations. The blood samples used for screening for PKU and other disorders contain enough of the baby's blood cells to permit extraction of the DNA, its amplification by PCR, and testing. Pilot studies of screening methods for sickle-cell anemia and cystic fibrosis are under way, and other genes will surely follow.

DNA testing is also widely used to test adults for heterozygosity. For example, a sister or female cousin of a boy with Duchenne muscular dystrophy can determine whether she is a carrier of the X chromosome deletion that results in the disease.

Of the numerous methods of DNA testing, two are the most widespread. We will describe their use to detect the mutation in the β-globin gene that results in sickle-cell anemia.

SCREENING FOR ALLELE-SPECIFIC CLEAVAGE DIFFERENCES. There is a difference between the normal and sickle alleles of the β-globin gene with respect to a restriction enzyme recognition sequence. Around codon position 6 in the normal gene is the sequence

5′... CCTGAGGAG... 3′

This sequence is recognized by the restriction enzyme *Mst*II, which will cleave DNA at

5′... CCT*N*AGG... 3′

where *N* is any base.

In the sickle allele, the DNA sequence is changed to

5′... CCTGTGGAG... 3′

The point mutation makes this sequence unrecognizable by *Mst*II. When *Mst*II fails to make the cut in the mutant gene, gel electrophoresis detects a larger DNA fragment (Figure 17.11).

This *allele-specific cleavage* method of DNA testing is similar to the use of RFLPs (see Figure 17.7). It works only if a restriction enzyme exists that can recognize either the sequence at the mutation or the original sequence that is altered by that mutation.

SCREENING BY ALLELE-SPECIFIC OLIGONUCLEOTIDE HYBRIDIZATION. The *allele-specific oligonucleotide hybridization* method uses short DNA strands called oligonucleotides made in the laboratory that will hybridize either with the denatured normal β-globin DNA sequence around position 6 or with the sickle mutant sequence. Usually, an oligonucleotide probe of at least a dozen bases is needed to form a stable double helix with the target DNA. If the probe is labeled with radioactivity or with a colored or fluorescent substrate, hybridization can be readily detected (Figure 17.12). This method is easier and faster than allele-specific cleavage, and will work no matter what the sequence of the normal or mutant allele.

Cancer: A Disease of Genetic Changes

Perhaps no malady affecting people in the industrialized world instills more fear than cancer. One in three Americans will have some form of cancer in their lifetime, and at present, one in four will die of it. With a million new cases and half a million deaths in the United States annually, cancer ranks second only to heart disease as a killer. Cancer was less common a century ago; then, as now in many regions of the world, people died of infectious diseases and did not live long enough to get cancer. Cancer tends to be a disease of the later years of life; children are much less frequently afflicted.

Since the U.S. government declared a "war on cancer" in 1970, a tremendous amount of information on cancer cells—on their growth and spread and on their molecular changes—has been obtained. Perhaps the most remarkable discovery is that cancer is a disease caused primarily by

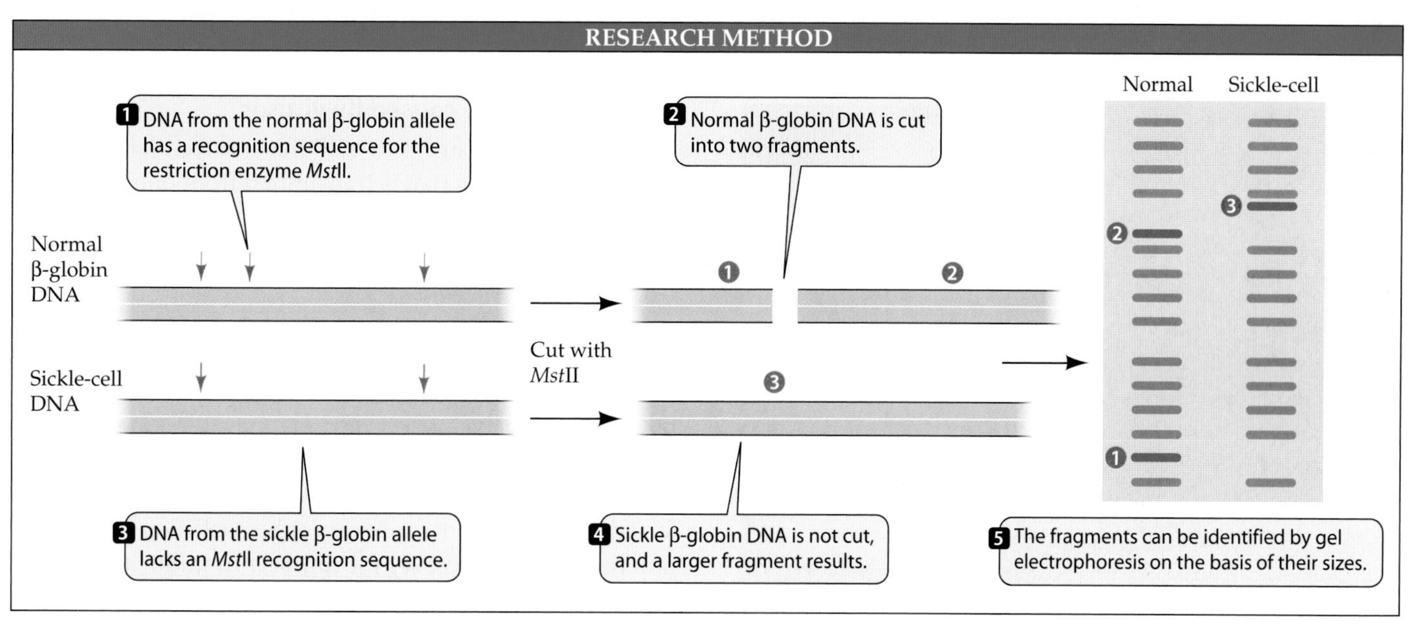
RESEARCH METHOD
1 DNA from the normal β-globin allele has a recognition sequence for the restriction enzyme MstII.
2 Normal β-globin DNA is cut into two fragments.
Normal
Sickle-cell
Normal β-globin DNA
Sickle-cell DNA
Cut with MstII
3 DNA from the sickle β-globin allele lacks an MstII recognition sequence.
4 Sickle β-globin DNA is not cut, and a larger fragment results.
5 The fragments can be identified by gel electrophoresis on the basis of their sizes.

17.11 DNA Testing by Allele-Specific Cleavage Allele-specific cleavage, a technique similar to RFLP analysis, can be used to detect mutations such as the one that causes sickle-cell anemia.

genetic changes. These changes are mostly alterations in the DNA of somatic cells that are propagated by mitosis.

Cancer cells differ from their normal counterparts

Cancer cells differ from the normal cells from which they originate in two major ways.

CANCER CELLS LOSE CONTROL OVER CELL DIVISION. Most cells in the body divide only if they are exposed to extracellular influences, such as growth factors or hormones. Cancer cells do not respond to these controls, and instead divide more or less continuously, ultimately forming **tumors** (large masses of cells). By the time a physician can feel a tumor or see one on an X ray or CAT scan, it already contains millions of cells.

Benign tumors resemble the tissue they came from, grow slowly, and remain localized where they develop. A lipoma, for example, is a benign tumor of fat cells that may arise in the armpit and remain there. Benign tumors are not cancers, but they must be removed if they impinge on an important organ, such as the brain.

Malignant tumors, on the other hand, do not look like their parent tissue at all. A flat, specialized lung epithelial cell in the lung wall may turn into a relatively featureless, round,

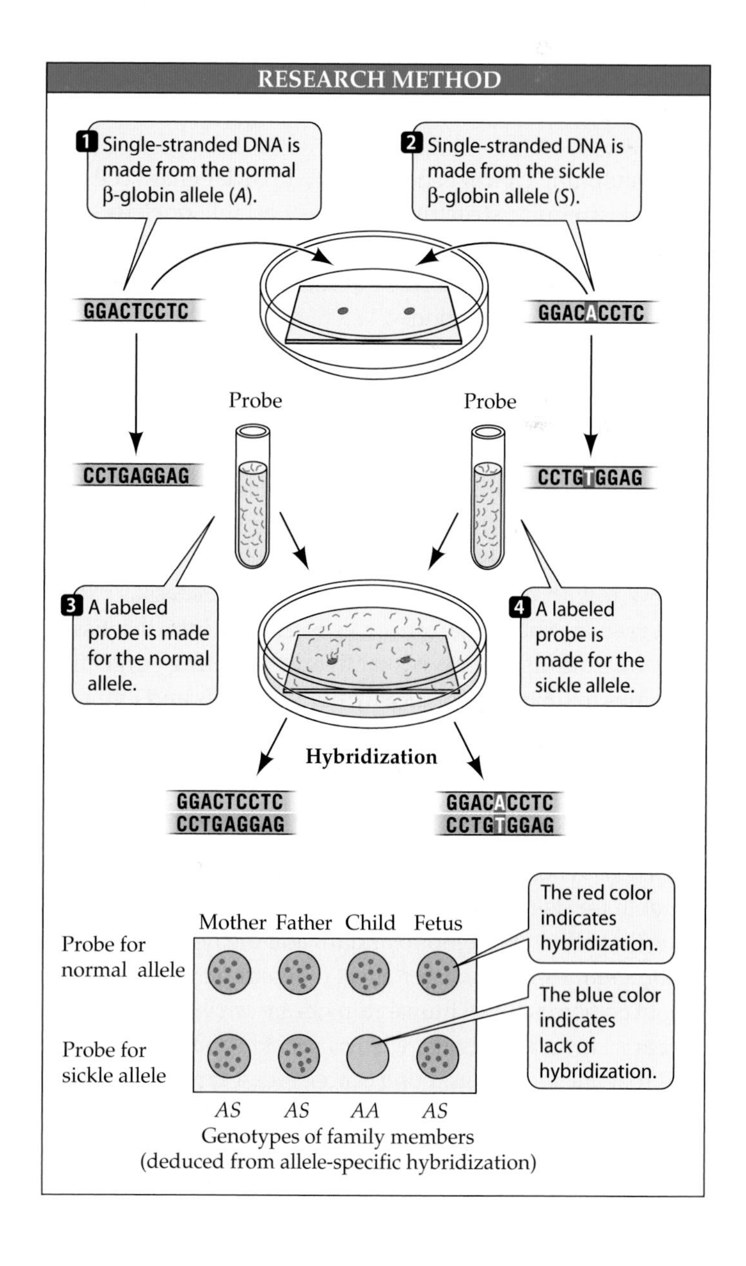
RESEARCH METHOD
1 Single-stranded DNA is made from the normal β-globin allele (A).
2 Single-stranded DNA is made from the sickle β-globin allele (S).
GGACTCCTC
GGACACCTC
Probe
Probe
CCTGAGGAG
CCTGTGGAG
3 A labeled probe is made for the normal allele.
4 A labeled probe is made for the sickle allele.
Hybridization
GGACTCCTC
CCTGAGGAG
GGACACCTC
CCTGTGGAG
Mother Father Child Fetus
Probe for normal allele
Probe for sickle allele
The red color indicates hybridization.
The blue color indicates lack of hybridization.
AS AS AA AS
Genotypes of family members
(deduced from allele-specific hybridization)

17.12 DNA Testing by Allele-Specific Oligonucleotide Hybridization Testing of this family reveals that three of them are heterozygous carriers of the sickle allele. The first child, however, has inherited two normal alleles and is neither affected by the disease nor a carrier.

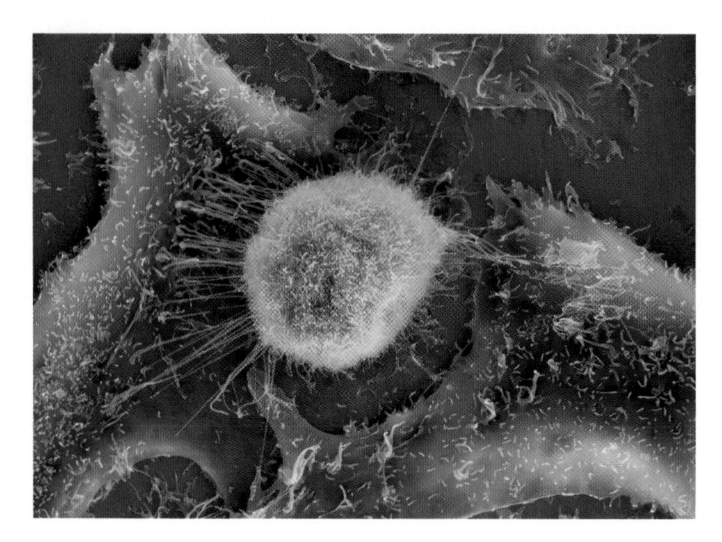

17.13 A Cancer Cell with Its Normal Neighbors This small-cell lung cancer cell (yellow green) is quite different from the surrounding lung epithelial cells from which it came. This particular form of cancer is very lethal, with a 5-year survival rate of less than 10 percent. Most cases are caused by smoking.

malignant lung cancer cell (Figure 17.13). Malignant cells often have irregular structures, such as variable nucleus sizes and shapes. Many malignant cells express the gene for telomerase and thus do not shorten the ends of their chromosomes after each DNA replication.

CANCER CELLS SPREAD TO OTHER TISSUES. The second, and most fearsome, characteristic of cancer cells is their ability to invade surrounding tissues and spread to other parts of the body. This spreading, called **metastasis**, occurs in several stages. First, the malignant tumor secretes chemical signals that cause blood vessels to grow to the tumor and supply it with oxygen and nutrients. This process is called *angiogenesis*. Then, the cancer cells extend into the tissue that surrounds them by actively secreting digestive enzymes to disintegrate the surrounding cells and extracellular materials, working their way toward a blood vessel. Finally, some of the cancer cells enter the bloodstream or the lymphatic system. The journey through these vessels is perilous, and few of the cancer cells survive—perhaps one in 10,000. When by chance a cancer cell arrives at an organ suitable for its further growth, it expresses cell surface proteins that allow it to bind to and invade the new host tissue.

Different forms of cancer affect different parts of the body. About 85 percent of all human tumors are *carcinomas*—cancers that arise in surface tissues such as the skin and the epithelial cells that line the organs. Lung cancer, breast cancer, colon cancer, and liver cancer are all carcinomas. *Sarcomas* are cancers of tissues such as bone, blood vessels, and muscle. *Leukemias* and *lymphomas* affect the cells that give rise to blood cells.

Some cancers are caused by viruses

Peyton Rous's discovery in 1910 that a sarcoma in chickens is caused by a virus that is transmitted from one bird to another spawned an intensive search for cancer-causing viruses in humans. At least five types of human cancer are probably caused by viruses (Table 17.1).

Hepatitis B, a liver disease that affects people all over the world, is caused by the hepatitis B virus, which contaminates blood or is carried from mother to child during birth. The viral infection can be long-lasting and may flare up numerous times. The hepatitis B virus is associated with liver cancer, especially in Asia and Africa, where millions of people are infected. But it does not cause cancer by itself. Some gene mutations that are necessary for tumor formation occur in the infected cells of Asians and Africans, although apparently not in those of Europeans and North Americans.

An important group of virally induced cancers among North Americans and Europeans is the various anogenital cancers caused by papillomaviruses. The genital and anal warts that these viruses cause often develop into tumors. These viruses seem to be able to act on their own, not needing mutations in the host tissue for tumors to arise. Sexual transmission of these papillomaviruses is unfortunately widespread.

Most cancers are caused by genetic mutations

Worldwide, no more than 15 percent of all cancers may be caused by viruses. What causes the other 85 percent? Because most cancers develop in older people, it is reasonable to assume that one must live long enough for a series of events to occur. This assumption turns out to be correct, and the events are genetic mutations.

DNA can be damaged in many ways. As we saw in Chapter 12, spontaneous mutations arise because of chemical changes in the nucleotides. In addition, certain mutagens, called **carcinogens**, can cause mutations that lead to cancer. Familiar carcinogens include the chemicals that are present in tobacco smoke and meat preservatives, ultraviolet light

17.1 *Human Cancers Known To Be Caused by Viruses*

CANCER	ASSOCIATED VIRUS
Liver cancer	Hepatitis B virus
Lymphoma, nasopharyngeal cancer	Epstein–Barr virus
T cell leukemia	Human T cell leukemia virus (HTLV-I)
Anogenital cancers	Papillomavirus
Kaposi's sarcoma	Kaposi's sarcoma herpesvirus

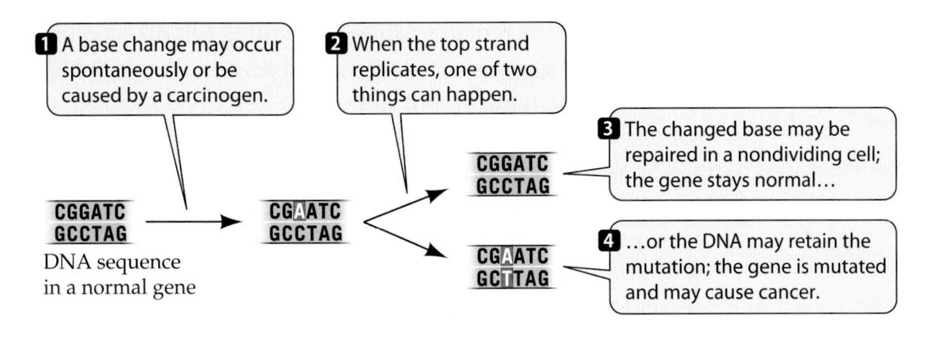

1 A base change may occur spontaneously or be caused by a carcinogen.
2 When the top strand replicates, one of two things can happen.
3 The changed base may be repaired in a nondividing cell; the gene stays normal...
4 ...or the DNA may retain the mutation; the gene is mutated and may cause cancer.
CGGATC
GCCTAG
DNA sequence in a normal gene
CGAATC
GCCTAG
CGGATC
GCCTAG
CGAATC
GCTTAG

17.14 Dividing Cells Are Especially Susceptible to Genetic Damage A base change is more likely to be repaired in a nondividing cell.

from the sun, and ionizing radiation from sources of radioactivity. Less familiar, but just as harmful, are thousands of chemicals that are naturally present in the foods people eat. According to one estimate, these "natural" carcinogens account for well over 80 percent of human exposure to agents that cause cancer.

Both natural and synthetic carcinogens damage DNA, usually by causing changes from one base to another (Figure 17.14). In somatic cells that divide often, such as epithelial and bone marrow cells, there is less time for DNA repair mechanisms to work before replication occurs again. Therefore, such cells are especially susceptible to cancer.

Two kinds of genes are changed in many cancers

The changes in the control of cell division that lie at the heart of cancer can be likened to the controls of an automobile. To make a car move, two things must happen: The gas pedal must be pressed, and the brake must be released. In the human genome, some genes act as **oncogenes**, which "press the gas pedal" to stimulate cell division, and some act as **tumor suppressor genes**, which "put the brake on" to inhibit it.

ONCOGENES. The first hint that oncogenes (from the Greek *onco-*, "mass") were necessary for cells to become cancerous came with the identification of virally induced cancers in animals. In many cases, these viruses bring a new gene into their host cells that stimulates cell division when it is expressed in the viral genome. But few types of human cancers are caused by viruses. It soon became apparent that the viral oncogenes had counterparts in the genomes of host cells that were not usually transcribed. So the search for genes that are damaged by carcinogens quickly zeroed in on these cellular oncogenes. Several dozen such genes were soon found.

Oncogenes are genes that have the capacity to stimulate cell division, but are normally "turned off" in differentiated, nondividing cells. Many of them are involved in the pathways by which growth factors stimulate cell division (Figure 17.15). Some remarkable oncogenes control apoptosis (programmed cell death). Activation of these genes by mutation causes them to prevent apoptosis, allowing cells that normally die to continue dividing.

Some oncogenes can be activated by point mutations, others by chromosome changes such as translocations, and still others by gene amplification. Whatever the mechanism, the result is the same: The oncogene becomes activated, and the "gas pedal" for cell division is pressed.

TUMOR SUPPRESSOR GENES. About 10 percent of all cancer is clearly inherited. Often the inherited form of a cancer is clinically similar to a noninherited form that occurs later in life, called the *sporadic* form. The major differences are that the inherited form strikes much earlier in life and usually shows up as multiple tumors.

In 1971, Alfred Knudson used these observations to predict that for a cancer to occur, a tumor suppressor gene, which normally acts as a "brake" on cell division, must be inactivated. But in contrast to oncogenes, in which one mutated allele is all that is needed for activation, the full inactivation

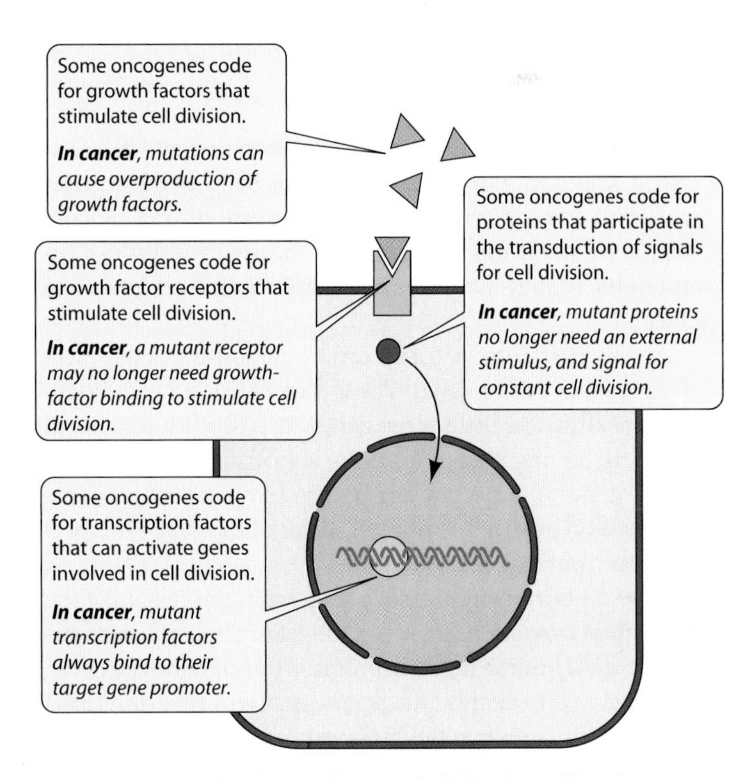

Some oncogenes code for growth factors that stimulate cell division.
In cancer, mutations can cause overproduction of growth factors.
Some oncogenes code for growth factor receptors that stimulate cell division.
In cancer, a mutant receptor may no longer need growth-factor binding to stimulate cell division.
Some oncogenes code for proteins that participate in the transduction of signals for cell division.
In cancer, mutant proteins no longer need an external stimulus, and signal for constant cell division.
Some oncogenes code for transcription factors that can activate genes involved in cell division.
In cancer, mutant transcription factors always bind to their target gene promoter.

17.15 Oncogene Products Stimulate Cell Division Mutations can affect any of the several ways in which oncogenes normally stimulate cell division, thus causing cancer.

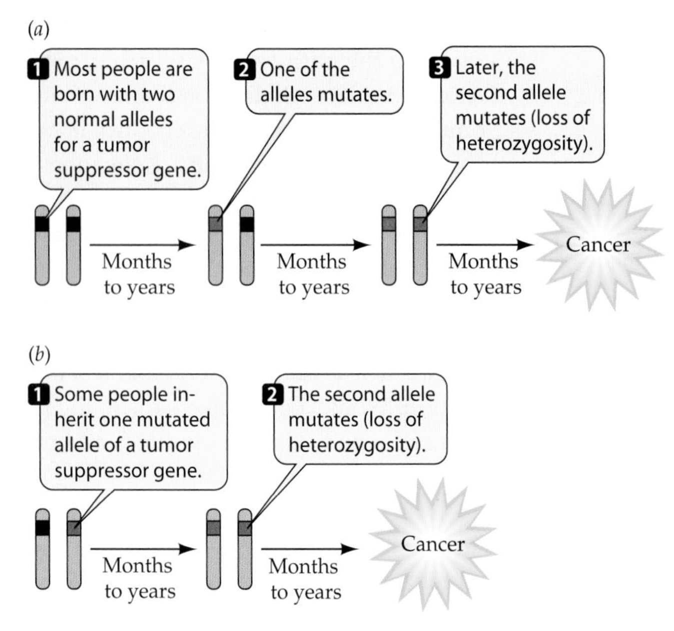

17.16 The "Two-Hit" Hypothesis for Cancer (*a*) Although a single mutation can activate an oncogene, two mutations are needed to inactivate a tumor suppressor gene. (*b*) An inherited predisposition to cancer occurs in people born with one allele already mutated.

of a tumor suppressor gene requires that both alleles be turned off, which requires two mutational events. It takes a long time for both alleles in a single cell to mutate and cause sporadic cancer. But people with inherited cancer are born with one mutant allele for the tumor suppressor gene, and need just one more mutational event for its full inactivation (Figure 17.16).

The isolation of various tumor suppressor genes has confirmed Knudson's "two-hit" hypothesis. Some of these genes are involved in inherited forms of rare childhood cancers such as retinoblastoma (a tumor of the eye) and Wilms' tumor of the kidney as well as in inherited breast and prostate cancers.

An inherited form of breast cancer demonstrates the effect of tumor suppressor genes. The 9 percent of women who inherit one mutated allele of the gene *BRCA1* have a 60 percent chance of having breast cancer by age 50 and an 82 percent chance of developing it by age 70. The comparable figures for women who inherit two normal alleles of the gene are 2 percent and 7 percent, respectively.

How do tumor suppressor genes act in the cell? Like the oncogenes, they are normally involved in vital cell functions (Figure 17.17). Some regulate progress through the cell cycle. The protein encoded by *Rb*, a gene that was first described for its contribution to retinoblastoma, is active during the G1 phase. In its active form, it encodes a protein that binds to and inactivates transcription factors that are necessary for progress to the S phase and the rest of the cell cycle. In nondividing cells, *Rb* remains active, preventing cell division until the proper growth factor signals are present. When the Rb protein is inactivated by mutation, the cell cycle moves forward independently of growth factors.

The protein product of another widespread tumor suppressor gene, *p53*, also stops the cell cycle at G1. It does this by acting as a transcription factor, stimulating the production of (among other things) a protein that blocks the interaction of a cyclin and a protein kinase needed for moving the cell cycle beyond G1. This gene is mutated in many types of cancers, including lung cancer and colon cancer.

The pathway from normal cell to cancerous cell is complex

The "gas pedal" and "brake" analogies we have been using for oncogenes and tumor suppressor genes, respectively, are elegant but simplified. There are many oncogenes and tumor suppressor genes, some of which act only in certain cells at certain times. Therefore, a complex sequence of events must occur before a normal cell becomes malignant.

Because colon cancer progresses to full malignancy slowly, it is possible to describe the oncogene and tumor suppressor gene mutations at each stage in great molecular detail. Figure 17.18 outlines the progress of this form of cancer. At least three tumor suppressor genes and one oncogene must be mutated in sequence for an epithelial cell in the colon to be-

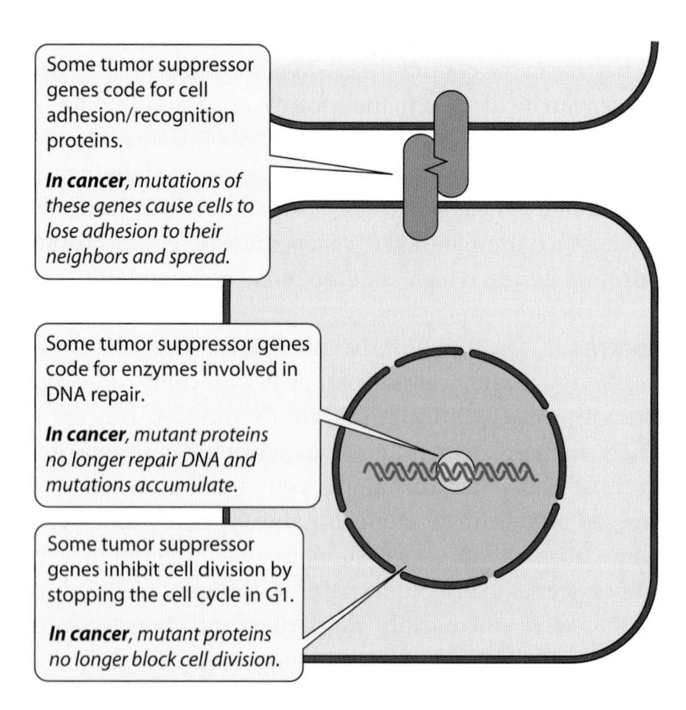

17.17 Tumor Suppressor Gene Products Inhibit Cell Division and Cancer Mutations can affect any of the several ways in which tumor suppressor genes inhibit cell division, allowing cells to divide and form a tumor.

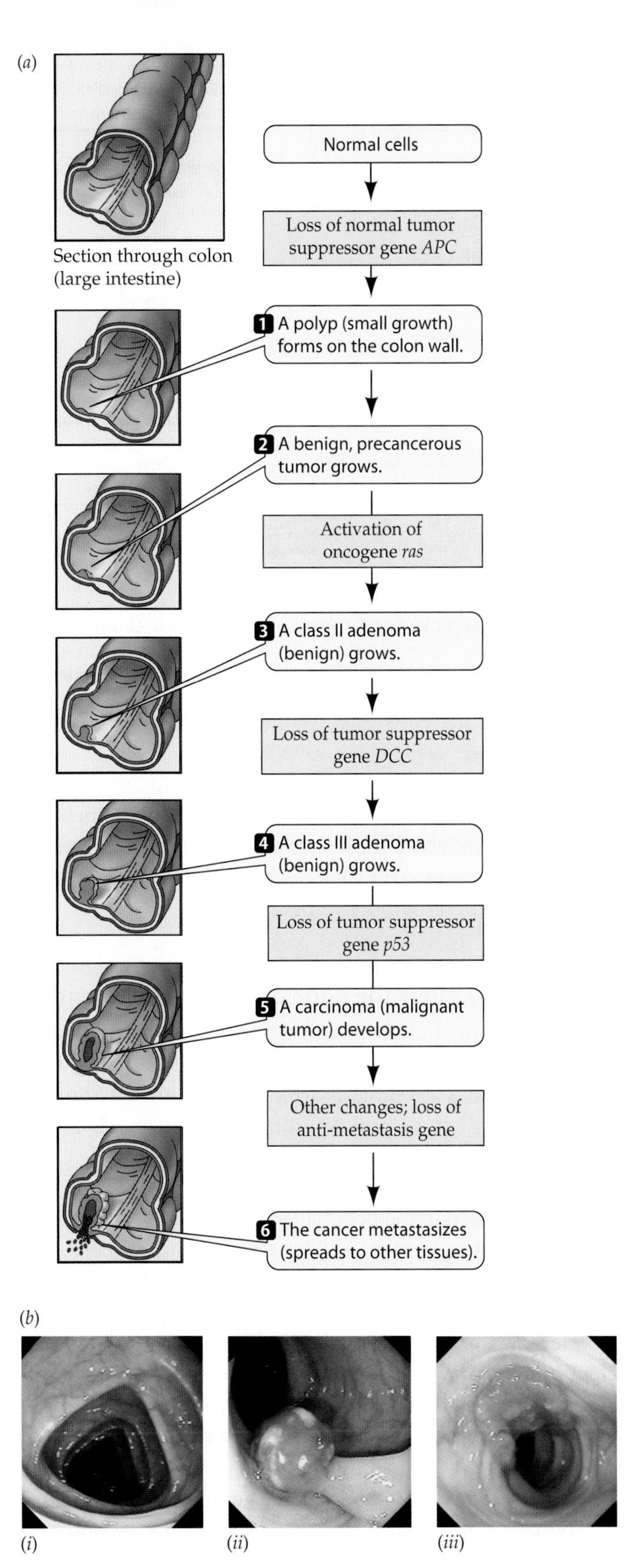
(a)
Section through colon (large intestine)
Normal cells
Loss of normal tumor suppressor gene *APC*
1 A polyp (small growth) forms on the colon wall.
2 A benign, precancerous tumor grows.
Activation of oncogene *ras*
3 A class II adenoma (benign) grows.
Loss of tumor suppressor gene *DCC*
4 A class III adenoma (benign) grows.
Loss of tumor suppressor gene *p53*
5 A carcinoma (malignant tumor) develops.
Other changes; loss of anti-metastasis gene
6 The cancer metastasizes (spreads to other tissues).
(b)
(i)
(ii)
(iii)

17.18 Multiple Mutations Transform a Normal Colon Epithelial Cell into a Cancer Cell *(a)* In colon cancer, at least five genes are mutated in a single cell. *(b)* Colonoscopy is the current standard screening test for colon cancer. These views reveal *(i)* normal colon tissue, *(ii)* a benign adenoma (stalked polyp), and *(iii)* adenocarcinoma (a malignant tumor).

come metastatic. Although the occurrence of all these events in a single cell might appear unlikely, remember that the colon has millions of cells, that the cells giving rise to epithelial cells are constantly dividing, and that these changes take place over many years of exposure to natural and synthetic carcinogens as well as spontaneous mutations.

The characterization of the molecular changes in tumor cells has opened up the possibility of genetic diagnosis and screening for cancer. Many cancers are now commonly diagnosed in part by specific oligonucleotide probes for oncogene or tumor suppressor gene alterations. It is also possible to detect early in life whether an individual has inherited a mutated tumor suppressor gene. A person who inherits mutated copies of the tumor suppressor genes involved in colon cancer, for example, normally would have a high probability of developing this cancer by age 40. Surgical removal of the colon would prevent a metastatic tumor from arising.

Treating Genetic Diseases

Most treatments for genetic diseases simply try to alleviate the symptoms that affect the patient. But to effectively treat diseases caused by genes—whether they affect all cells, as in inherited disorders such as PKU, or only somatic cells, as in cancer—physicians must be able to diagnose the disease accurately, understand how the disease works at the molecular level, and intervene early, before the disease ravages or kills the individual.

Basic research has provided the knowledge needed for accurate diagnostic tests, as well as a preliminary understanding of these diseases at the molecular level. Physicians are now applying this knowledge to develop new treatments for genetic diseases. In this section, we will see that approaches to treatment range from specifically modifying the mutant phenotype to supplying the normal version of a mutant gene.

One approach to treatment is to modify the phenotype

There are three ways of altering the phenotype of a genetic disease so that it no longer harms an individual: restricting the substrate of a deficient enzyme, inhibiting a harmful metabolic reaction, or supplying a missing protein product.

RESTRICTING THE SUBSTRATE. Restricting the substrate of a deficient enzyme is the approach taken when a newborn is diagnosed with PKU. In this case, the deficient enzyme is phenylalanine hydroxylase, and the substrate is phenylalanine. The infant's inability to break down the phenylalanine in food leads to a buildup of the substrate, which causes the clinical symptoms. So the infant is immediately put on a special diet that contains only enough phenylalanine for immediate use. Lofenelac, a milk-based product that is low in phenylalanine, is fed to these infants just like formula. Later, certain fruits, vegetables, cereals, and noodles low in phenylalanine can be added to the diet. Meat, fish, eggs, dairy products, and bread, which contain high amounts of phenylalanine, must be avoided, especially during childhood, when brain development is most rapid. The artificial sweetener aspartame must also be avoided because it is made of two amino acids, one of which is phenylalanine.

People with PKU are generally advised to stay on a low-phenylalanine diet for life. Although maintaining these dietary restrictions may be difficult, it is effective. Numerous follow-up studies since newborn screening was initiated have shown that people with PKU who stay on the diet are no different from the rest of the population in terms of mental ability. This is an impressive achievement in public health, given the extent of mental retardation in untreated patients.

METABOLIC INHIBITORS. As we described earlier, people with familial hypercholesterolemia accumulate dangerous levels of cholesterol in their blood. These people are not only unable to metabolize dietary cholesterol, but also synthesize a lot of it. One effective treatment for people with this disease is the drug mevinolin, which blocks the patient's own cholesterol synthesis. Patients who receive this drug need only worry about cholesterol in their diet, and not about the cholesterol their cells are making.

Metabolic inhibitors also form the basis of chemotherapy for cancer. The strategy is to kill rapidly dividing cells, since rapid cell division is the hallmark of malignancy. But such a strategy is not selective for tumor cells. Many drugs kill dividing cells (Figure 17.19), but most of those drugs also damage other, noncancerous, dividing cells in the body. Therefore, it is not surprising that people undergoing chemotherapy suffer side effects such as loss of hair (due to damage to the skin epithelium), digestive upsets (gut epithelial cells), and anemia (bone marrow stem cells). The effective dose of these highly toxic drugs for treating the cancer is often just below the dose that would kill the patient, so they must be used with utmost care. Often they can control the spread of cancer, but not cure it.

SUPPLYING THE MISSING PROTEIN. An obvious way to treat a disease phenotype in which a functional protein is missing is to supply that protein. This approach is the basis of treatment of hemophilia, in which the missing blood clotting protein is supplied in pure form. The production of human clotting protein by recombinant DNA technology has made it possible for a pure protein to be given instead of crude blood products, which could be contaminated with the AIDS virus or other pathogens.

Unfortunately, the phenotypes of many diseases caused by genetic mutations are very complex. Simple interventions like those we have described do not work for most such diseases. Indeed, a recent survey showed that current therapies for 351 diseases caused by single-gene mutations improved patients' life spans by only 15 percent.

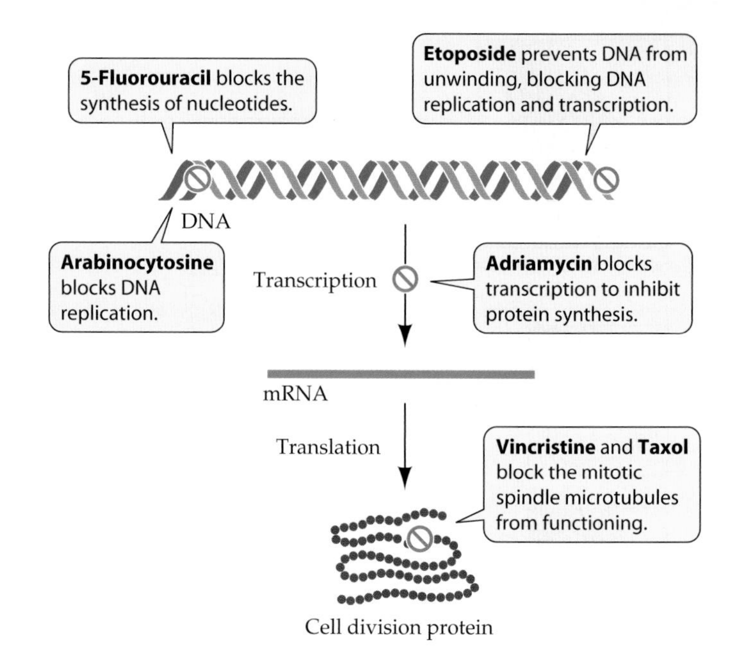

17.19 Strategies for Killing Cancer Cells The medications used in chemotherapy for cancer attack rapidly dividing cancer cells in several ways. Unfortunately, most of them also affect noncancerous dividing cells.

Gene therapy offers the hope of specific treatments

Perhaps the most obvious thing to do when a cell lacks a functional allele is to provide one. Such **gene therapy** approaches to diseases ranging from the rare inherited disorders caused by single-gene mutations to cancer, AIDS, and atherosclerosis are under intensive investigation.

Gene therapy in humans seeks to insert a new gene that will be expressed in the host. Thus, the new DNA is often attached to a promoter that will be active in human cells. The physicians who are developing this "molecular medicine" are confronted by all the challenges of recombinant DNA technology: they must find effective vectors and ensure efficient uptake, precise insertion into the host DNA, appropriate expression and processing of mRNA and protein, and selection within the body for the cells that contain the recombinant DNA.

Which human cells should be the targets of gene therapy? The best approach would be to replace the nonfunctional allele with a functional one in every cell of the body. But vectors to do this are simply not available, and delivery to every cell poses a formidable challenge. Until recently, attempts at gene therapy have used **ex vivo** techniques. That is, physicians have taken cells from the patient's body, added the new gene to those cells in the laboratory, and then returned the

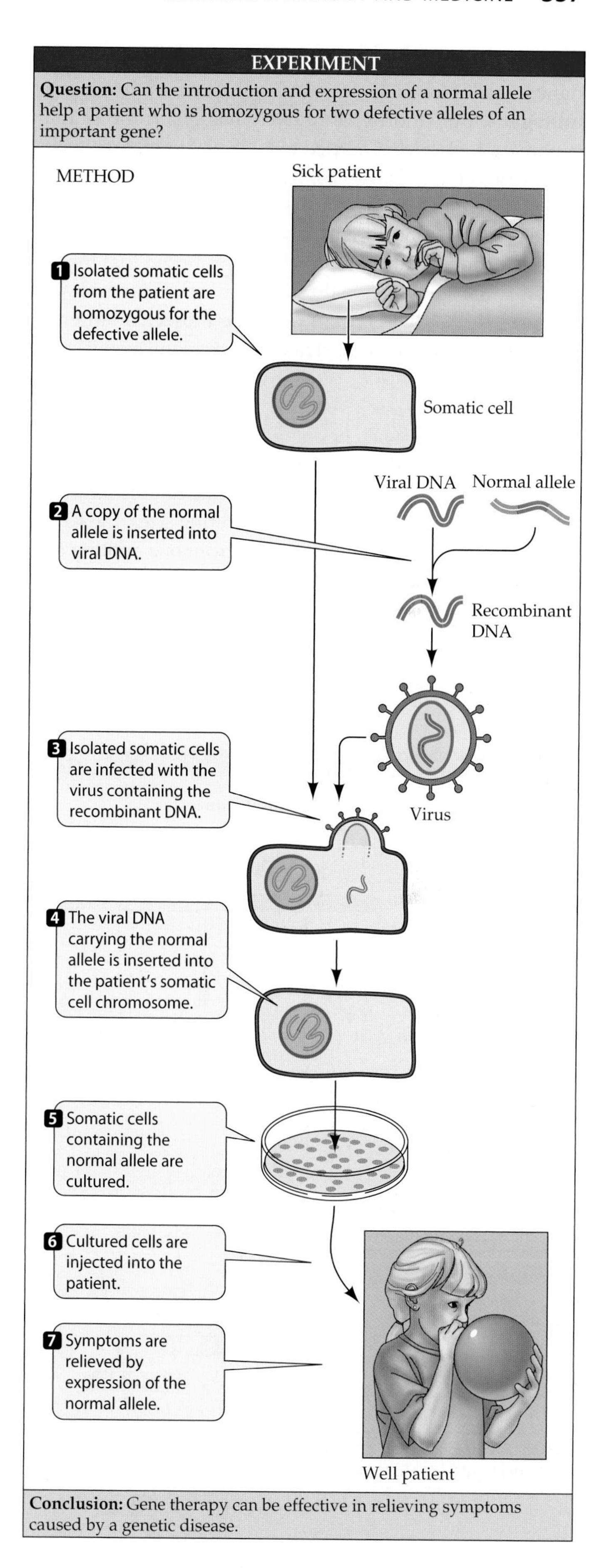

EXPERIMENT
Question: Can the introduction and expression of a normal allele help a patient who is homozygous for two defective alleles of an important gene?
METHOD
Sick patient
1 Isolated somatic cells from the patient are homozygous for the defective allele.
Somatic cell
Viral DNA
Normal allele
2 A copy of the normal allele is inserted into viral DNA.
Recombinant DNA
3 Isolated somatic cells are infected with the virus containing the recombinant DNA.
Virus
4 The viral DNA carrying the normal allele is inserted into the patient's somatic cell chromosome.
5 Somatic cells containing the normal allele are cultured.
6 Cultured cells are injected into the patient.
7 Symptoms are relieved by expression of the normal allele.
Well patient
Conclusion: Gene therapy can be effective in relieving symptoms caused by a genetic disease.

17.20 Gene Therapy: The Ex Vivo Approach New genes are added to somatic cells taken from a patient's body. These transgenic cells are then returned to the body to make the missing gene product.

cells to the patient in the hope that the correct gene product would be made (Figure 17.20). Two examples demonstrate this technique:

- *Adenosine deaminase* is needed for maturation of white blood cells, and people without this enzyme have severe immune system deficiencies. A functional gene for adenosine deaminase was introduced via a viral vector into the white blood cells of a girl with a genetic deficiency of this enzyme. Unfortunately, mature white blood cells were used, and although they survived for a time in the girl and provided some therapeutic benefit, they eventually died, as is the normal fate of such cells. Further clinical trials have used stem cells, the bone marrow cells that constantly divide to produce white blood cells.
- *Hemophilia* is a disease in which patients do not make enough of a blood clotting protein. Some cells from the skin of patients' arms were removed and transfected with a plasmid containing a normal allele of the clotting protein gene. The cells were then reintroduced into the patients' body fat, where they produced adequate protein for normal clotting.

The other approach to gene therapy is to insert the gene directly into cells in the body of the patient. This **in vivo** approach is being attempted for various types of cancer. Lung cancer cells, for example, are accessible to such treatment if the DNA or vector is given as an aerosol through the respiratory system. Vectors carrying functional alleles of the tumor suppressor genes that are mutated in the tumors, as well as vectors expressing antisense RNAs against oncogene mRNAs, have been successfully introduced in this way to patients with lung cancer, with some clinical improvement.

Several thousand patients, over half of them with cancer, have undergone gene therapy. Most of these clinical trials have been at a preliminary level, in which people are given the therapy to see whether it has any toxicity and whether the new gene is actually incorporated into the patient's genome. More ambitious trials are under way, in which a larger number of patients will receive the therapy with the hope that their disease will disappear, or at least improve.

Sequencing the Human Genome

In 1984, the United States government sponsored a conference on the detection of DNA damage in people exposed to low levels of radiation, such as those who had survived the atomic bomb in Japan 39 years earlier. Scientists attending this conference quickly realized that the ability to detect such damage would also be useful in evaluating environmental mutagens. But in order to detect changes in the human genome, scientists first needed to know its normal sequence.

In 1986, Renato Dulbecco, who won the Nobel prize for his pioneering work on cancer-causing viruses, suggested that determining the normal sequence of human DNA could also be a boon to cancer research. He proposed that the scientific community be mobilized for the task. The result was the publicly funded **Human Genome Project**, an international effort. In the 1990s, private industry launched its own sequencing effort.

There are two approaches to genome sequencing

Each human chromosome consists of one double-stranded molecule of DNA. Because of their differing sizes, the 46 human chromosomes can be separated from one another and identified (see Figure 9.13). So it is possible to isolate the DNA of each chromosome for sequencing. The straightforward approach would be to start at one end of a chromosome and simply sequence the entire 50 million base pairs. Unfortunately, this approach is not practical. The DNA of a molecule that is 50 million base pairs long cannot be sequenced all at once; only about 700 base pairs at a time can be sequenced. (See Figure 11.21 to review the DNA sequencing technique.)

To sequence an entire genome, chromosomal DNA is first cut into fragments about 500 base pairs long, then each fragment is sequenced. For the human genome, which has about 3.2 billion base pairs, there are more than 6 million such fragments. The problem then becomes putting these millions of fragments back together, like the pieces of a jigsaw puzzle. This problem can be overcome by breaking up the DNA into "sub-jigsaws" that overlap and aligning the overlapping fragments. There are two ways to do this.

HIERARCHICAL SEQUENCING. The publicly funded sequencing team used a method known as **hierarchical sequencing**. First, they systematically identified short marker sequences along the chromosomes, such that every fragment of DNA to be sequenced would contain a marker (Figure 17.21*a*). This method can be compared to making a road map, showing towns with the mileage separating them. The "towns"

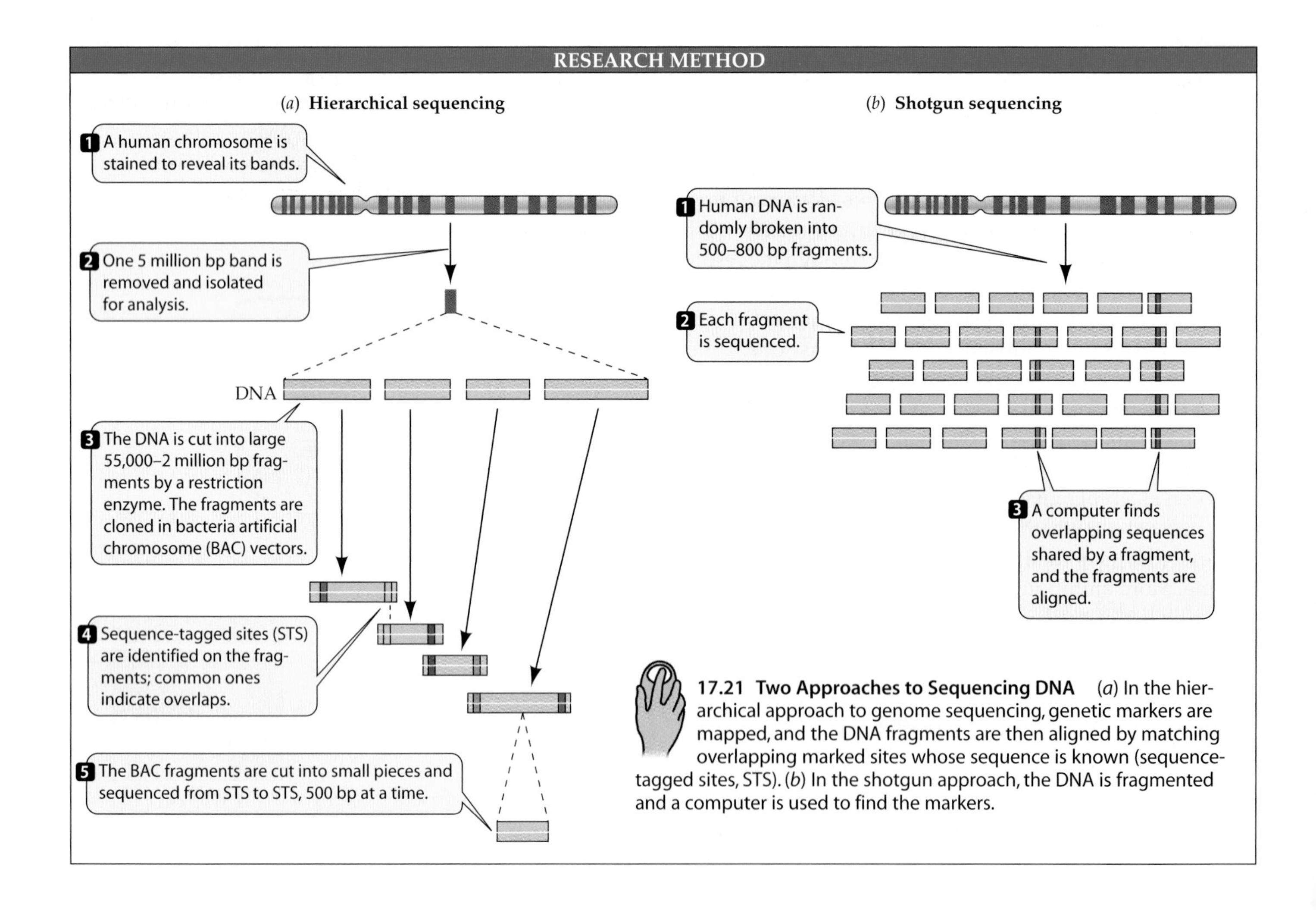

RESEARCH METHOD
(a) Hierarchical sequencing
(b) Shotgun sequencing
1 A human chromosome is stained to reveal its bands.
2 One 5 million bp band is removed and isolated for analysis.
DNA
3 The DNA is cut into large 55,000–2 million bp fragments by a restriction enzyme. The fragments are cloned in bacteria artificial chromosome (BAC) vectors.
4 Sequence-tagged sites (STS) are identified on the fragments; common ones indicate overlaps.
5 The BAC fragments are cut into small pieces and sequenced from STS to STS, 500 bp at a time.
1 Human DNA is randomly broken into 500–800 bp fragments.
2 Each fragment is sequenced.
3 A computer finds overlapping sequences shared by a fragment, and the fragments are aligned.

17.21 Two Approaches to Sequencing DNA (*a*) In the hierarchical approach to genome sequencing, genetic markers are mapped, and the DNA fragments are then aligned by matching overlapping marked sites whose sequence is known (sequence-tagged sites, STS). (*b*) In the shotgun approach, the DNA is fragmented and a computer is used to find the markers.

are the marker sequences, and the "mileage" is in base pairs. The simplest markers are the recognition sequences for restriction enzymes.

Some restriction enzymes recognize 8–12 base pairs in DNA, not just the usual 4–6 base pairs. A DNA molecule with several million base pairs will have relatively few of these larger sites, and thus the enzyme will generate a small number of relatively large fragments. These large fragments can be added to a vector called a *bacterial artificial chromosome* (BAC), which can carry about 250,000 base pairs of inserted DNA, and inserted into bacteria to create a gene library.

The volumes (fragments) in this library can be arranged in the proper order along the chromosome map by using the marker sequences. To arrange the DNA fragments on the map, libraries made with different restriction enzymes are compared. If two large fragments of DNA cut with different enzymes have the same marker, they must overlap. This method works, but is slow.

SHOTGUN SEQUENCING. Instead of finding markers, fragmenting the DNA, and then sequencing it, the "shotgun" approach cuts the DNA at random into small, sequencing-ready fragments and lets powerful computers determine markers that overlap (Figure 17.21*b*). The fragments can then be aligned.

The **shotgun sequencing** method, which has been used by private industry, is much faster than the hierarchical approach because there is no need to make a map. At first there was considerable skepticism about this method. There were concerns that without rigorous prior mapping of marker sites on the chromosomes, the computer might pick out repetitive sequences common to many DNA fragments and line the fragments up incorrectly. But the rapid rate of development of sophisticated computers and software has allowed the shotgun method to be refined to a point at which inaccurate alignment is not a major problem. The entire 180 million-base-pair fruit fly genome (see Chapter 14) was sequenced by the shotgun method in little over a year. This success proved that the shotgun method might work for the much larger human genome, and in fact, it did.

The sequence of the human genome has been determined

The two teams of scientists announced a draft human genome sequence in June 2000 to great fanfare, and published their data simultaneously in February 2001. By the start of 2003, the final sequence was completed, two years ahead of the schedule set over a decade previously and well under budget.

The sequencing of the human genome revealed several interesting characteristics:

- *Of the 3.2 billion base pairs, less than 2 percent are coding regions, containing a total of 21,000 genes.* Before sequencing began, estimates of the number of human genes ranged from 80,000 to 100,000. This lower number of genes, not many more than the fruit fly, means that the observed diversity of proteins, which led to the 100,000 estimate, must be produced posttranscriptionally. An average eukaryotic gene, then, codes for several different proteins.
- *The average gene has 27,000 base pairs.* There is great variation in gene sizes, from 1,000 to 2.4 million base pairs. That is to be expected, as human proteins vary in size (as do RNAs), ranging from 100 to about 5,000 amino acids per polypeptide chain. Virtually all human genes have many introns (Figure 17.22).
- *Over 50 percent of the genome is made up of highly repetitive sequences.* Repetitive sequences near genes are GC-rich, while those farther away from genes are AT-rich.
- *Almost all (99.9%) of the genome is the same in all people.* Even this apparent homogeneity means that there are many individual differences. Scientists have mapped over 2 million **single-nucleotide polymorphisms (SNPs)**—bases that differ in at least 1 percent of people.
- *Genes are not evenly distributed over the genome.* The small chromosome 19 is packed densely with genes, while chromosome 8 has long stretches of "gene desert," with no coding regions. The Y chromosome has the fewest genes (231), while chromosome 1 has the most (2,968).
- *The functions of many genes are not known.* There are 740 genes coding for RNAs that are not translated into proteins. Of these RNAs, several dozen are tRNAs, and a few are rRNAs and splicing RNAs. The roles of the rest are not clear. Nor are the roles of the hundreds of genes encoding protein kinases, although it is a good bet that they are involved in cell signaling.

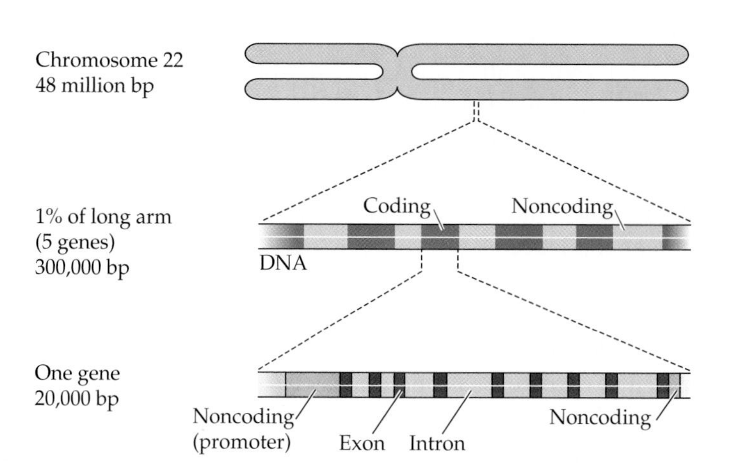

17.22 The Human Genome The genomic anatomy of chromosome 22 is shown here.

The human genome sequence has many applications

Reading the human "book of life" is an achievement that ranks with other recent great events in scientific exploration, such as landing on the moon. But gene sequencing, and the tools developed to carry it out, are changing biology in many other ways as well.

- The sequences of other organisms have provided insights and practical information on both prokaryotic and eukaryotic genomes. Many genes sequenced and identified in "simpler" organisms have homologs in humans, so these findings have facilitated the identification of human genes.
- Mapping technology and SNPs have made the isolation of human genes by positional cloning much easier because of the huge number of genetic markers now available. Disease-related genes have been identified in this way.
- Genetic variation in drug metabolism has been a medical problem for a long time. The emerging field of *pharmacogenomics* is identifying the genes responsible for this variation and developing tests to predict who will react best to which medications.
- DNA chips (see Figure 16.10) are being used to analyze the specific expression of thousands of genes in different cells in different biochemical states. For example, a Cancer Genome Anatomy Project is seeking to make an mRNA "fingerprint" of a tumor at each stage of its development. Finding out which genes are expressed at which stage will be important not only in diagnosis, but also in identifying targets for gene therapy.
- "Genome prospecting" refers to the search for important polymorphisms in specific human populations. For example, the Pima Indians in Arizona have a high frequency of extreme obesity and diabetes. A search of their genomes might reveal genes predisposing them to these conditions.

The end result of all of this knowledge of the human genome may be a new approach to medical care, in which each person's genome will be used to prescribe lifestyle changes and treatments that can maximize that person's genetic potential (Figure 17.23).

How should genetic information be used?

When the genetic defect that causes cystic fibrosis was discovered, many people predicted a "tidal wave" of genetic testing for heterozygous carriers. Everyone, it was thought, would want the test—especially the relatives of people with the disease. But this tidal wave has not developed. To find out why, a team of psychologists, ethicists, and geneticists interviewed 20,000 people in the United States. What the researchers found surprised them. Most people are simply not very interested in their genetic makeup, unless they have a close relative with a genetic disease and are involved in a decision about pregnancy.

There are other people, however, who might be very interested in the results of genetic testing. For example, people who test positive for genetic abnormalities, from hypercholesterolemia to cancer, might be denied employment or health insurance. Consequently, there are laws that prohibit discrimination on the basis of genetic information.

The search for valuable genes in diverse human populations has raised many concerns about exploitation and commercialization of a person's DNA sequence. Is a gene that confers resistance to cancer, for example, the property of an individual, an ethnic group in which it may be frequent, the pharmaceutical company that finds it, or humanity at large?

This issue of ownership is being tested worldwide, perhaps most directly in Iceland, whose 270,000 people trace their ancestry back to the first settlers to arrive on the island 1,100 years ago. Tissues from Iceland's entire population have been sampled and stored for several generations. This tissue bank is a potential gold mine for genetic prospectors. A single company has been set up, with government support, to sell the knowledge that comes from analyzing the genomes of Iceland's people.

The company's approach to mining this genetic lode is illustrated by its search for genes that predispose people to asthma, a respiratory disease:

- The names of Icelanders with asthma were run through a genealogy database.

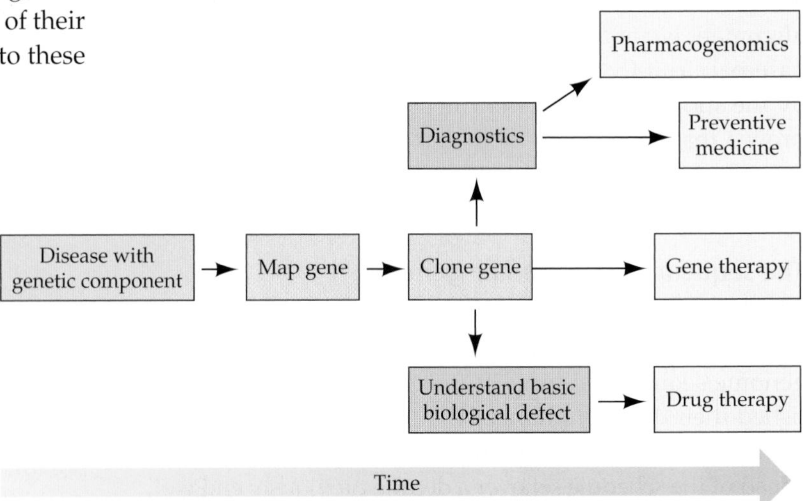

17.23 Is This the Future of Medicine? The sequencing of the human genome may result in a new approach to medicine that is oriented to the genetic and functional individuality of each patient.

- One group of 104 patients were descended from a single ancestor, born in 1710 (11 generations ago). It was considered likely that the same genes for predisposition to asthma were present in all 104 patients.
- Marker genes were sought that would identify alleles that all 104 patients had in common, and a small number of such genes were subsequently identified.

The characterization of these genes will lead to a greater understanding of how a group of genes interacts to produce a complex phenotype.

The proteome is more complex than the genome

Genome sequencing revealed that humans had only about one-third as many genes as had been predicted based on the number of proteins found in human cells. Perhaps this lower number should not have come as a surprise; as we saw in Chapters 12 and 14, many genes encode more than a single protein (Figure 17.24*a*). Alternative splicing leads to different combinations of exons in the mature mRNAs transcribed from a single gene (see Figure 14.20). Posttranslational modifications also add to the forms of a protein that can be made from one gene. Many proteins have peptides clipped off, have sugars added, or are phosphorylated after translation (see Figure 12.16). Therefore, the sum total of the proteins produced by an organism—its **proteome**—is more complex than its genome. The one-gene, one-polypeptide relationship, once a central one in biology, has been laid to rest by genomics.

There are two major ways to analyze the proteome:

- *Two-dimensional gel electrophoresis* attempts to separate all of the individual proteins of a particular cell or tissue into spots that can be analyzed quantitatively and qualitatively (Figure 17.24*b*).
- *Mass spectrometry* employs electromagnets to identify proteins by the masses of their atoms and displays them as peaks on a graph.

The ultimate aim of proteomics is just as ambitious as the aim of genomics. While genomics seeks to describe the genome and its expression, proteomics seeks to describe the phenotypes of the expressed proteins with precision.

An amazing example of proteomics, combined with DNA chip technology, is the recent comparison of brain proteins in chimpanzees and humans. DNA sequencing has shown that humans and chimpanzees differ by no more than 3 percent at the DNA level. Svante Pääbo and his colleagues in Germany, The Netherlands, and the United States examined gene and protein expression in the "thinking" part of the brains (the cortex) of three chimps and three humans who had died of natural causes. Of 12,000 DNA sequences tested for expression as mRNA, only 175 (1.4%) showed differences between the two species, a truly humbling result. But proteomics showed that the *kinds* of proteins expressed by those sequences were 7.4 percent different, probably due to alternative splicing. And the *amounts* of proteins were also quite different (31.4%). So what makes our brain different from a chimpanzee's is more quantitative than qualitative. Thus, the control of gene expression may be the key to human evolution.

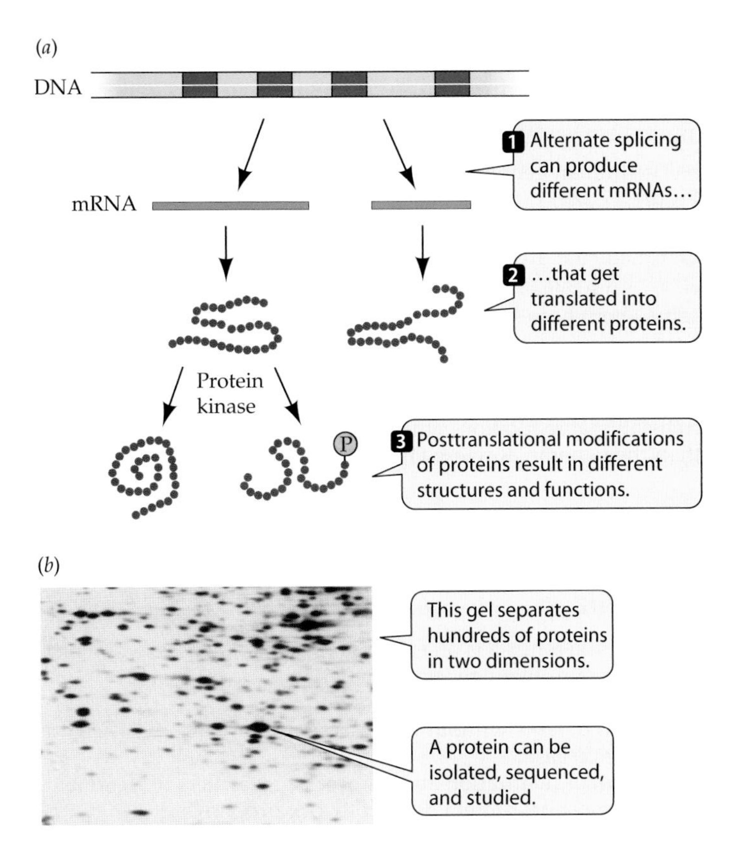
(a)
DNA
mRNA
1 Alternate splicing can produce different mRNAs...
2 ...that get translated into different proteins.
Protein kinase
P
3 Posttranslational modifications of proteins result in different structures and functions.
(b)
This gel separates hundreds of proteins in two dimensions.
A protein can be isolated, sequenced, and studied.

17.24 Proteomics (*a*) A small number of genes can make a large number of proteins. (*b*) A cell's proteins can be separated in two dimensions on the basis of charge and size by gel electrophoresis.

Chapter Summary

Abnormal or Missing Proteins: The Mutant Phenotype

- In some human genetic diseases, a single protein is missing or nonfunctional. **Review Figure 17.1**
- A mutation in a single gene can cause alterations in its protein product that may lead to clinical abnormalities or have no effect. **Review Figure 17.2**
- The genes that code for enzymes, membrane receptors, and membrane transport proteins can be mutated, causing diseases such as phenylketonuria, familial hypercholesterolemia, and cystic fibrosis. **Review Figure 17.3**
- Some diseases are caused by mutations that affect structural proteins; examples include Duchenne muscular dystrophy and hemophilia.

▶ Prions are disease-causing proteins with an altered conformation that can be transmitted from one person to another and alter the same protein in the second person. **Review Figure 17.4**

▶ Relatively few common human diseases are caused by single-gene mutations. Most are caused by the interactions of many genes and proteins with the environment.

▶ Human genetic diseases show different patterns of inheritance. Mutant alleles may be inherited as autosomal recessives, autosomal dominants, or X-linked conditions.

▶ Some human diseases are caused by chromosomal abnormalities.

Mutations and Human Diseases

▶ Molecular biological techniques have made possible the isolation of many genes responsible for human diseases.

▶ One method of identifying the gene responsible for a disease is to isolate the mRNA for the abnormal protein in question and then use that mRNA to locate the gene in a gene library. DNA from a patient with a chromosome deletion can be compared with DNA from a person who does not show this deletion to isolate a missing gene. **Review Figure 17.6**

▶ In positional cloning, genetic markers are used as guides to point the way to a gene. These markers may be restriction fragment length polymorphisms that are linked to a mutant gene. **Review Figure 17.7.** See Web/CD Activity 17.1

▶ Human mutations range from point mutations to large deletions. Some of the most common mutations occur where the modified base 5-methylcytosine is converted to thymine. **Review Figure 17.8**

▶ The effects of the fragile-X chromosome worsen with each generation. This pattern is caused by a triplet repeat that tends to expand with each new generation. **Review Figure 17.9**

▶ Genomic imprinting results in a gene being differentially expressed depending on the sex of the parent it comes from.

Detecting Genetic Variations: Screening for Human Diseases

▶ Genetic screening detects human genetic mutations. Some protein abnormalities can be detected by simple tests, such as tests for the presence of excess substrate or lack of product. **Review Figure 17.10**

▶ The advantage of testing DNA for mutations directly is that any cell can be tested at any time in the life cycle.

▶ There are two predominant methods of DNA testing: allele-specific cleavage and allele-specific oligonucleotide hybridization. **Review Figures 17.11, 17.12.** See Web/CD Tutorial 17.1

Cancer: A Disease of Genetic Changes

▶ Tumors may be benign, growing only to a certain extent and then stopping, or malignant, spreading through organs and to other parts of the body.

▶ At least five types of human cancers are caused by viruses, which account for about 15 percent of all cancers. **Review Table 17.1**

▶ Eighty-five percent of human cancers are caused by genetic mutations of somatic cells. These mutations occur most commonly in dividing cells. **Review Figure 17.14**

▶ Normal cells contain oncogenes, which, when mutated, can become activated and cause cancer by stimulating cell division or preventing cell death. **Review Figure 17.15**

▶ About 10 percent of all cancer is inherited as a mutation of a tumor suppressor gene, which normally acts to slow down the cell cycle. For cancer to develop, both alleles of a tumor suppressor gene must be mutated.

▶ In inherited cancer, an individual inherits one mutant allele of a tumor suppressor gene, and a somatic mutation occurs in the second one. In sporadic cancer, two normal alleles are inherited, so two mutational events must occur in the same somatic cell to produce cancer. **Review Figures 17.16, 17.17**

▶ Mutations must activate several oncogenes and inactivate several tumor suppressor genes for a cell to produce a malignant tumor. **Review Figure 17.18**

Treating Genetic Diseases

▶ Most genetic diseases are treated symptomatically. However, as more knowledge is accumulated, specific treatments are being devised.

▶ One approach to treating genetic diseases is to modify the phenotype—for example, by manipulating the diet to restrict the substrate of a missing enzyme, providing specific metabolic inhibitors to prevent a harmful reaction, or supplying a missing metabolite or protein. **Review Figure 17.19**

▶ In gene therapy, a mutant gene is replaced with a normal gene. The affected cells can be removed, the new gene added, and the cells returned to the body, or the new gene can be inserted via a vector directly into the patient. **Review Figure 17.20**

Sequencing the Human Genome

▶ Sequencing the entire human genome required sequencing many 500-base-pair fragments and then fitting their sequences back together.

▶ In hierarchical gene sequencing, marker sequences are identified and mapped on the chromosome before DNA is fragmented. These markers are then sought in the sequenced fragments and used to align them. In the shotgun approach, the DNA is fragmented and sequenced, and common markers are then identified by computer. **Review Figure 17.21.** See Web/CD Tutorial 17.2

▶ The human genome has only about 21,000 genes. **Review Figure 17.22**

▶ The identification of human genes may lead to a new molecular medicine. **Review Figure 17.23**

▶ As more genes relevant to human health are described, concerns about how such information is used are growing.

▶ Humans make many more proteins than predicted by their number of genes because each gene can encode several different proteins as a result of variation in posttranscriptional and posttranslational regulation. Thus, the proteome is more complex than the genome. **Review Figure 17.24**

See Web/CD Activity 17.2 for a concept review of this chapter.

Self-Quiz

1. Phenylketonuria is an example of a genetic disease in which
   a. a single enzyme is not functional.
   b. inheritance is sex-linked.
   c. two parents without the disease cannot have a child with the disease.
   d. mental retardation always occurs, regardless of treatment.
   e. a transport protein does not work properly.
2. Mutations of the gene for β-globin
   a. are usually lethal.
   b. occur only at amino acid position 6.

c. number in the hundreds.
d. always result in sickling of red blood cells.
e. can always be detected by gel electrophoresis.

3. Multifactorial (complex) diseases
a. are less common than single-gene diseases.
b. involve the interaction of many genes with the environment.
c. affect less than 1 percent of humans.
d. involve the interactions of several mRNAs.
e. are exempified by sickle-cell anemia.

4. In fragile-X syndrome,
a. females are affected more severely than males.
b. a short sequence of DNA is repeated many times to create the fragile site.
c. both the X and Y chromosomes tend to break when prepared for microscopy.
d. all people who carry the gene that causes the syndrome are mentally retarded.
e. the basic pattern of inheritance is autosomal dominant.

5. Most genetic diseases are rare because
a. each person is unlikely to be a carrier for harmful alleles.
b. genetic diseases are usually sex-linked and so uncommon in females.
c. genetic diseases are always dominant.
d. a married couple probably do not carry the same recessive alleles.
e. mutation rates in human are low.

6. Mutational "hot spots" in human DNA
a. always occur in genes that are transcribed.
b. are common at cytosines that have been modified to 5-methylcytosine.
c. involve long stretches of nucleotides.
d. occur where there are long repeats.
e. are very rare in genes that code for proteins.

7. Newborn genetic screening for PKU
a. is very expensive.
b. detects phenylketones in urine.
c. has not led to the prevention of mental retardation resulting from this disorder.
d. must be done during the first day of an infant's life.
e. uses bacterial growth to detect excess phenylalanine in blood.

8. Genetic diagnosis by DNA testing
a. detects only mutant and not normal alleles.
b. can be done only on eggs or sperm.
c. involves hybridization to rRNA.
d. utilizes restriction enzymes and a polymorphic site.
e. cannot be done with PCR.

9. Most human cancers
a. are caused by viruses.
b. are in blood cells or their precursors.
c. involve mutations of somatic cells.
d. spread through solid tissues rather than by the blood or lymphatic system.
e. are inherited.

10. Current treatments for genetic diseases include all of the following *except*
a. restricting a dietary substrate.
b. replacing the mutated gene in all cells.
c. alleviating the patient's symptoms.
d. inhibiting the function of a harmful metabolite.
e. supplying a protein that is missing.

For Discussion

1. How do oncogenes and tumor suppressor genes and their functions change in tumor cells? Propose targets for cancer therapy involving these gene products.
2. In the past, it was common for people with phenylketonuria (PKU) who were placed on a low-phenylalanine diet after birth to be allowed to return to a normal diet during their teenage years. Although the levels of phenylalanine in their blood were high, their brains were thought to be beyond the stage of being harmed. If a woman with PKU becomes pregnant, however, a problem arises. Typically, the fetus is heterozygous but is unable at early stages of development to metabolize the high levels of phenylalanine that arrive from the mother's blood. Why is the fetus heterozygous? What do you think would happen to the fetus during this "maternal PKU" situation? What would be your advice to a woman with PKU who wants to have a child?
3. Cystic fibrosis is an autosomal recessive disease in which thick mucus is produced in the lungs and airways. The gene responsible for this disease codes for a protein composed of 1,480 amino acids. In most patients with cystic fibrosis, the protein has 1,479 amino acids: A phenylalanine is missing at position 508. A baby is born with cystic fibrosis. He has an older brother. How would you test the DNA of the brother to determine if he is a carrier for cystic fibrosis? How would you design a gene therapy protocol to "cure" the cells in the lung and airway?
4. A number of efforts are underway to identify human genetic polymorphisms that correlate with complex diseases such as diabetes, heart disease and cancer. What would be the uses of such information? What concerns do you think are being raised by the people whose DNAs are being analyzed?

18 Natural Defenses against Disease

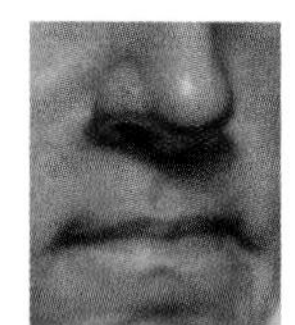

On January 6, 1777, George Washington, commander of the Revolutionary army of the fledgling United States, made a fateful decision. As he wrote to his chief physician, "Finding smallpox to be spreading much, and fearing that no precaution can prevent it from running through the whole of our army, I have determined that the troops shall be inoculated. Should the disease rage with its usual virulence, we should have more to dread from it than the sword of the enemy."

Washington was speaking from experience. He himself had survived the disease in 1751, when he was still a teenager. During 1776 his army lost 1,000 men in battle and 10,000 men to smallpox. This virulent disease, which killed up to 1 of every 4 people exposed to it, had already figured prominently in American history. A century before, it had decimated the native population, making colonization by Europeans easier. Two years previously at Quebec, it had laid waste to an American army that was trying to annex Canada by force.

The death rate due to smallpox in the Revolutionary army plummeted after Washington's order was carried out. How did inoculation, a practice that was learned from the people of the Near East and from African slaves, save the soldiers? And why was Washington himself immune to the disease as it ravaged his army?

The answers to these questions lie in the cells and molecules of the immune system. When Washington caught smallpox in 1751, specialized white blood cells in his body engulfed some of the smallpox viruses by phagocytosis and partly digested them. These cells, called macrophages, displayed fragments of the viruses on their surfaces. Other specialized white blood cells, called T cells, recognized those fragments, which caused them to divide and differentiate. Some descendants of those activated T cells then attacked Washington's virus-infected cells, preventing the lethal spread of the disease. Other descendants of the T cells persisted in his body as "memory cells" and rapidly divided again to defend him when he was exposed to the disease as an adult. Inoculation of Washington's soldiers with powdered scabs from smallpox patients, containing dead smallpox viruses, stimulated the formation of these memory cells in their bodies, once again preventing the virus from spreading following infection. This practice, which had been used for centuries, was finally placed on a more scientific basis by Edward Jenner two decades after Washington's army was inoculated.

George Washington Washington's decision to immunize his army against smallpox saved many lives and probably helped him win the Revolutionary War.

These defensive events in the bodies of Washington and his soldiers required the participation of many kinds of cells and proteins. This chapter begins by introducing the participants in the two types of defense mechanisms found in vertebrate animals. Then we look in greater detail at the nonspecific defense mechanisms. Next we see how the specific defense mechanisms—the immune system—target specific invaders, such as the smallpox virus, for destruction, and how the reshuffling of their genetic material allows them to target an incredible diversity of potential invaders. Finally we look at what happens when this complex system malfunctions.

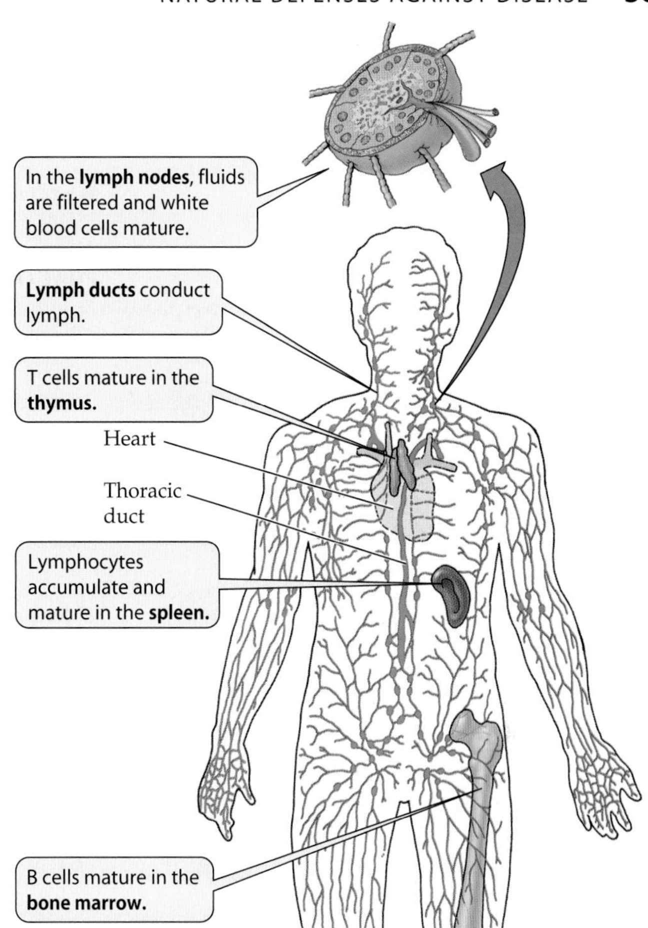

In the **lymph nodes**, fluids are filtered and white blood cells mature.
Lymph ducts conduct lymph.
T cells mature in the **thymus.**
Heart
Thoracic duct
Lymphocytes accumulate and mature in the **spleen.**
B cells mature in the **bone marrow.**

18.1 The Human Lymphatic System A network of ducts and vessels collects lymph from the body's tissues and carries it toward the heart, where it mixes with blood to be pumped back to the tissues. The thymus, spleen, and bone marrow are also essential to the body's defense system.

Animal Defense Systems

Animals have a number of ways of defending themselves against **pathogens**—harmful organisms and viruses that can cause disease. These defense systems are based on the distinction between *self*—the animal's own molecules—and *nonself*, or foreign, molecules. In this section we consider the mechanisms by which animals recognize nonself molecules and combat infection and disease. Many of these mechanisms are based on the principles of genetics and molecular biology that have been discussed in earlier chapters.

In general, there are two types of defense mechanisms:

- **Nonspecific defenses**, or innate defenses, are inherited mechanisms that protect the body from many pathogens. An example is the skin, which acts as a barrier to stop potentially invading viruses from entering the body. Most animals and plants have innate defenses.
- **Specific defenses** are adaptive mechanisms aimed at a specific target. For example, these defense systems can make an antibody protein that will recognize, bind to, and destroy a certain virus if that virus ever enters the bloodstream. Specific defense mechanisms are present in vertebrate animals. DNA rearrangements and mutations play important roles in generating these defenses against a huge variety of targets.

In animals that have both kinds of mechanisms, nonspecific and specific defenses operate together as a coordinated defense system.

Blood and lymph tissues play important roles in defense systems

The components of the mammalian defense system are dispersed throughout the body and interact with almost all of its other tissues and organs. The **lymphoid tissues**, which include the thymus, bone marrow, spleen, and lymph nodes, are essential parts of the defense system (Figure 18.1), but central to their functioning are the blood and lymph.

Blood and lymph are both fluid tissues that consist of water, dissolved solutes, and cells. *Blood plasma* is a yellowish solution containing ions, small molecular solutes, and soluble proteins. Suspended in the plasma are red blood cells, white blood cells, and platelets (cell fragments essential to clotting) (Figure 18.2). While red blood cells are normally confined to the *closed circulatory system* (the heart, arteries, capillaries, and veins), white blood cells and platelets are also found in the lymph.

Lymph is a fluid derived from blood and other tissues that accumulates in intercellular spaces throughout the body. From these spaces, the lymph moves slowly into the vessels of the *lymphatic system*. Tiny lymph capillaries conduct this fluid to larger vessels that eventually join together, forming one large vessel, the thoracic duct, which joins a major vein (the left subclavian vein) near the heart. By this system of vessels, the lymph is eventually returned to the blood and the circulatory system.

At many sites along the lymph vessels are small, roundish structures called **lymph nodes**, which contain a variety of white blood cells. As fluid passes through a node, it is filtered and "inspected" for nonself materials by these defensive cells.

Myeloid progenitor cell

Bone marrow

Pluripotent hematopoietic cell

Lymphoid progenitor cell

TYPE OF CELL	FUNCTION
Red blood cells (erythrocytes)	Transport oxygen and carbon dioxide
Platelets (cell fragments without nuclei)	Initiate blood clotting
White blood cells (leukocytes)	
PHAGOCYTES	
Basophils	Release histamine; may promote the development of T cells
Eosinophils	Kill antibody-coated parasites
Neutrophils	Phagocytose antibody-coated pathogens
Mast cells	Release histamine when damaged
Monocytes	Develop into macrophages
Macrophages	Engulf and digest microorganisms; activate T cells
Dendritic cells	Present antigens to T cells
LYMPHOCYTES	
B cells	Differentiate to form antibody-producing cells and memory cells
Plasma cells	Secrete antibodies
T cells	Kill virus-infected cells; regulate activities of other white blood cells
Natural killer cells	Attack and lyse virus-infected or cancerous body cells

18.2 Blood Cells Pluripotent stem cells in the bone marrow can differentiate into red blood cells, platelets, and the various types of white blood cells.

White blood cells play many defensive roles

One milliliter of blood typically contains about 5 billion red blood cells and 7 million of the larger white blood cells. All of these cells originate from stem cells in the bone marrow (see Figure 18.2). White blood cells (also called *leukocytes*) have nuclei and are colorless, unlike mammalian red blood cells, which lose their nuclei during development. White blood cells can leave the closed circulatory system and enter intercellular spaces where foreign cells or substances are present. The number of white blood cells in the blood and lymph may rise sharply in response to invading pathogens, providing medical professionals with a useful clue for detecting an infection.

Several types of white blood cells are important in the body's defenses. But they are all members of two broad groups, phagocytes and lymphocytes.

- **Phagocytes** engulf and digest nonself materials. Among the most important phagocytes are the **macrophages**. In addition to engulfing nonself materials by phagocytosis, macrophages have the important additional function of presenting partly digested nonself materials to the T cells.
- **Lymphocytes** participate in specific defenses against nonself or altered cells, such as virus-infected cells and tumor cells. There are two types of lymphocytes, **B cells** and **T cells**. Immature T cells migrate from the bone marrow via the blood to the *thymus*, where they mature. The B cells leave the bone marrow and circulate through the blood and lymph vessels. B cells make specialized proteins called antibodies that enter the blood and bind to nonself substances.

Fundamental to the interactions, control, and defensive functioning of these white blood cells are defensive proteins and other signals.

Immune system proteins bind pathogens or signal other cells

The cells that defend mammalian bodies work together like cast members in a drama, interacting with one another and with the cells of invading pathogens. These cell–cell interactions are accomplished by a variety of key proteins, including receptors, other cell surface proteins, signaling molecules, and toxins. These proteins will be discussed later in the chapter, as they appear in the context of our story. However, let's take a brief look at four of the major players here.

- **Antibodies** are proteins that bind specifically to certain substances identified by the immune system as nonself or altered self, thereby denaturing the invading nonself substance. They are secreted by B cells as defensive weapons.
- **T cell receptors** are integral membrane proteins on the surfaces of T cells. They recognize and bind to nonself substances on the surfaces of other cells.
- **Major histocompatibility complex (MHC)** proteins protrude from the surfaces of most cells in the mammalian body. They are important self-identifying labels and play major parts in coordinating interactions among lymphocytes and macrophages.
- **Cytokines** are soluble signal proteins released by T cells, macrophages, and other cells. They bind to and alter the behavior of their target cells. Different cytokines activate or inactivate B cells, macrophages, and T cells. Some cytokines limit tumor growth by killing tumor cells.

Before focusing on the roles of these four major players in specific defenses that constitute the immune system, we will consider the nonspecific defenses, which involve both passive barriers and active roles for molecules and cells.

Nonspecific Defenses

Nonspecific defenses (also called innate defenses) are general protection mechanisms that attempt to stop pathogens from invading the body. Nonspecific defenses in humans include physical barriers as well as cellular and chemical defenses (Table 18.1).

18.1 *Human Nonspecific Defenses*

DEFENSIVE AGENT	FUNCTION
Surface barriers	
Skin	Prevents entry of pathogens and foreign substances
Acid secretions	Inhibit bacterial growth on skin
Mucous membranes	Prevent entry of pathogens
Mucous secretions	Trap bacteria and other pathogens in digestive and respiratory tracts
Nasal hairs	Filter bacteria in nasal passages
Cilia	Move mucus and trapped materials away from respiratory passages
Gastric juice	Concentrated HCl and proteases destroy pathogens in stomach
Acid in vagina	Limits growth of fungi and bacteria in female reproductive tract
Tears, saliva	Lubricate and cleanse; contain lysozyme, which destroys bacteria
Nonspecific cellular, chemical, and coordinated defenses	
Normal flora	Compete with pathogens; may produce substances toxic to pathogens
Fever	Body-wide response inhibits microbial multiplication and speeds body repair processes
Coughing, sneezing	Expels pathogens from upper respiratory passages
Inflammatory response (involves leakage of blood plasma and phagocytes from capillaries)	Limits spread of pathogens to neighboring tissues; concentrates defenses; digests pathogens and dead tissue cells; released chemical mediators attract phagocytes and specific defense lymphocytes to site
Phagocytes (macrophages and neutrophils)	Engulf and destroy pathogens that enter body
Natural killer cells	Attack and lyse virus-infected or cancerous body cells
Antimicrobial proteins	
Interferons	Released by virus-infected cells to protect healthy tissue from viral infection; mobilize specific defenses
Complement proteins	Lyse microorganisms, enhance phagocytosis, and assist in inflammatory response

Barriers and local agents defend the body against invaders

Skin is a primary nonspecific defense against invasion. Fungi, bacteria, and viruses rarely penetrate healthy, unbroken skin. But damage to the skin or to the internal surface tissue greatly increases the risk of infection by pathogens.

The bacteria and fungi that normally live and reproduce in great numbers on our body surfaces without causing disease are referred to as *normal flora*. These natural occupants of our bodies compete with pathogens for space and nutrients and are thus a form of nonspecific defense.

The mucous membranes found at the surfaces of the visual, respiratory, digestive, excretory, and reproductive systems have other defenses against pathogens. Tears, nasal mucus, and saliva contain an enzyme called *lysozyme* that attacks the cell walls of many bacteria. Mucus in the nose traps airborne microorganisms, and most of those that get past this filter end up trapped in mucus deeper in the respiratory tract. Mucus and trapped pathogens are removed by the beating of cilia in the respiratory passageway, which continuously move a sheet of mucus and the debris it contains up toward the nose and mouth. Sneezing is another way to remove microorganisms from the respiratory tract.

Pathogens that reach the digestive tract (stomach, small intestine, and large intestine) are met by other defenses. The gastric juice in the stomach is a deadly environment for many bacteria because of the hydrochloric acid and proteases (protein-digesting enzymes) that are secreted into it. The intact lining of the small intestine is not normally penetrated by bacteria, and some pathogens are killed by bile salts secreted into this part of the digestive tract. The large intestine harbors many bacteria, which multiply freely; however, they are usually removed quickly with the feces. Most of the bacteria in the large intestine are normal flora that provide benefits to their host. We probably add to this beneficial flora when we eat foods such as active-culture yogurt and various cheeses.

All of these barriers and local agents are *nonspecific* defenses because they act on all invading pathogens in the same way. More complex nonspecific defenses await any pathogens that manage to elude this first line of defense.

Nonspecific defenses include chemical and cellular processes

Pathogens that penetrate the body's outer and inner surfaces encounter more complex nonspecific defenses that involve the secretion of various defensive proteins as well phagocytic cells.

COMPLEMENT PROTEINS. Vertebrate blood contains about 20 different antimicrobial proteins that make up the **complement system**. These proteins, in different combinations, provide three types of defenses. In each type, the complement proteins act in a characteristic sequence, or cascade, with each protein activating the next:

- They attach to microbes, which helps phagocytes recognize and destroy them.
- They activate the inflammation response and attract phagocytes to site of infection.
- They lyse (burst) invading cells such as bacteria.

INTERFERONS. When cells are infected by a virus, they produce small amounts of antimicrobial proteins called **interferons** that increase the resistance of neighboring cells to infection by the same *or other* viruses. Interferons have been found in many vertebrates and are one of the body's first lines of nonspecific defense against the internal spread of viral infection.

Interferons differ from species to species, and each vertebrate species produces at least three different interferons. All interferons are glycoproteins (proteins with attached carbohydrate groups) consisting of about 160 amino acids. By binding to receptors in the plasma membranes of uninfected cells, interferons stimulate a signaling pathway that results in the inhibition of viral reproduction inside the infected cells.

PHAGOCYTES AND RELATED CELLULAR DEFENSES. Phagocytes provide another important nonspecific defense against pathogens that penetrate the surface of the host. Some phagocytes travel freely in the circulatory system; others can move out of blood vessels and adhere to certain tissues. Entire pathogenic cells, entire viruses, or fragments of these invaders can become attached to the membrane of a phagocyte (Figure 18.3), which ingests them by phagocytosis. When lysosomes fuse with the phagosome, the pathogens are degraded by lysosomal enzymes (see Figure 4.13*b*).

Several types of phagocytes play roles in nonspecific defenses (see Figure 18.2):

- **Neutrophils** are the most abundant phagocytes, but they are relatively short-lived. They recognize and attack pathogens in infected tissue.
- **Monocytes** mature into macrophages, which live longer than neutrophils and can consume large numbers of pathogens. Some macrophages roam through the body; others reside permanently in lymph nodes, the spleen, and certain other lymphoid tissues, "inspecting" the lymph for pathogens.
- **Eosinophils** are weakly phagocytic. Their primary function is to kill parasites, such as worms, that have been coated with antibodies.
- **Dendritic cells** have highly folded plasma membranes that can capture invading pathogens.

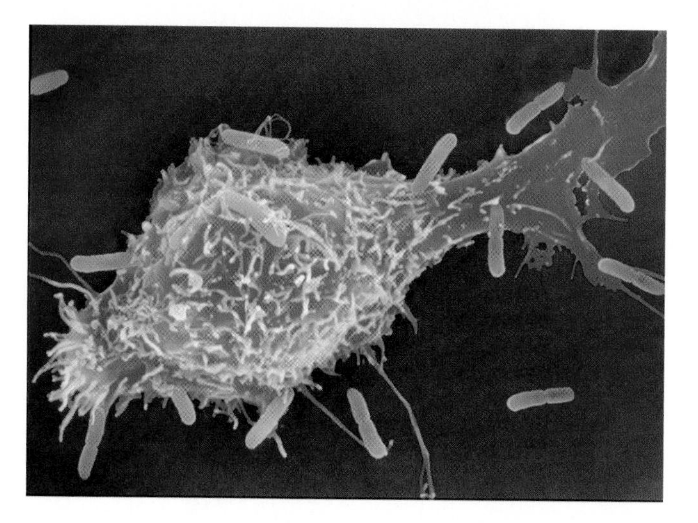

18.3 A Phagocyte and Its Bacterial Prey Some bacteria (which appear yellow in this artificially colored scanning electron micrograph) have become attached to the surface of a phagocyte in the human bloodstream. Many of these bacteria will be engulfed by the phagocyte and destroyed before they can multiply and damage the human host. A single phagocyte can digest many bacteria.

NONPHAGOCYTIC CELLS. A class of nonphagocytic white blood cells, known as **natural killer cells**, can distinguish virus-infected cells and some tumor cells from their normal counterparts and initiate the lysis of these target cells. In addition to this nonspecific action, natural killer cells form part of the specific defenses, as we will describe later in this chapter.

INFLAMMATION. The body employs the **inflammation** response in dealing with infection or with any other process that causes tissue injury, either on the surface of the body or internally. The damaged body cells cause the inflammation by releasing various substances. Cells adhering to the skin and linings of organs, called **mast cells**, release a chemical signal, called **histamine**, when they are damaged, as do white blood cells called **basophils**.

You have no doubt experienced the symptoms of inflammation: redness and swelling, accompanied by heat and pain. The redness and heat of inflammation result from histamine-induced dilation of blood vessels in the infected or injured area (Figure 18.4). Histamine also causes the capillaries (the smallest blood vessels) to become leaky, allowing blood plasma and phagocytes to escape into the tissue, causing the characteristic swelling. The pain of inflammation results from increased pressure (from the swelling) and from the action of leaked enzymes.

In damaged or infected tissue, complement proteins and other chemical signals attract phagocytes—neutrophils first, and then monocytes, which become macrophages. The macrophages, which engulf the invaders and debris, are responsible for most of the healing associated with inflammation. They produce several cytokines, which, among other functions, signal the brain to produce a fever. This rise in

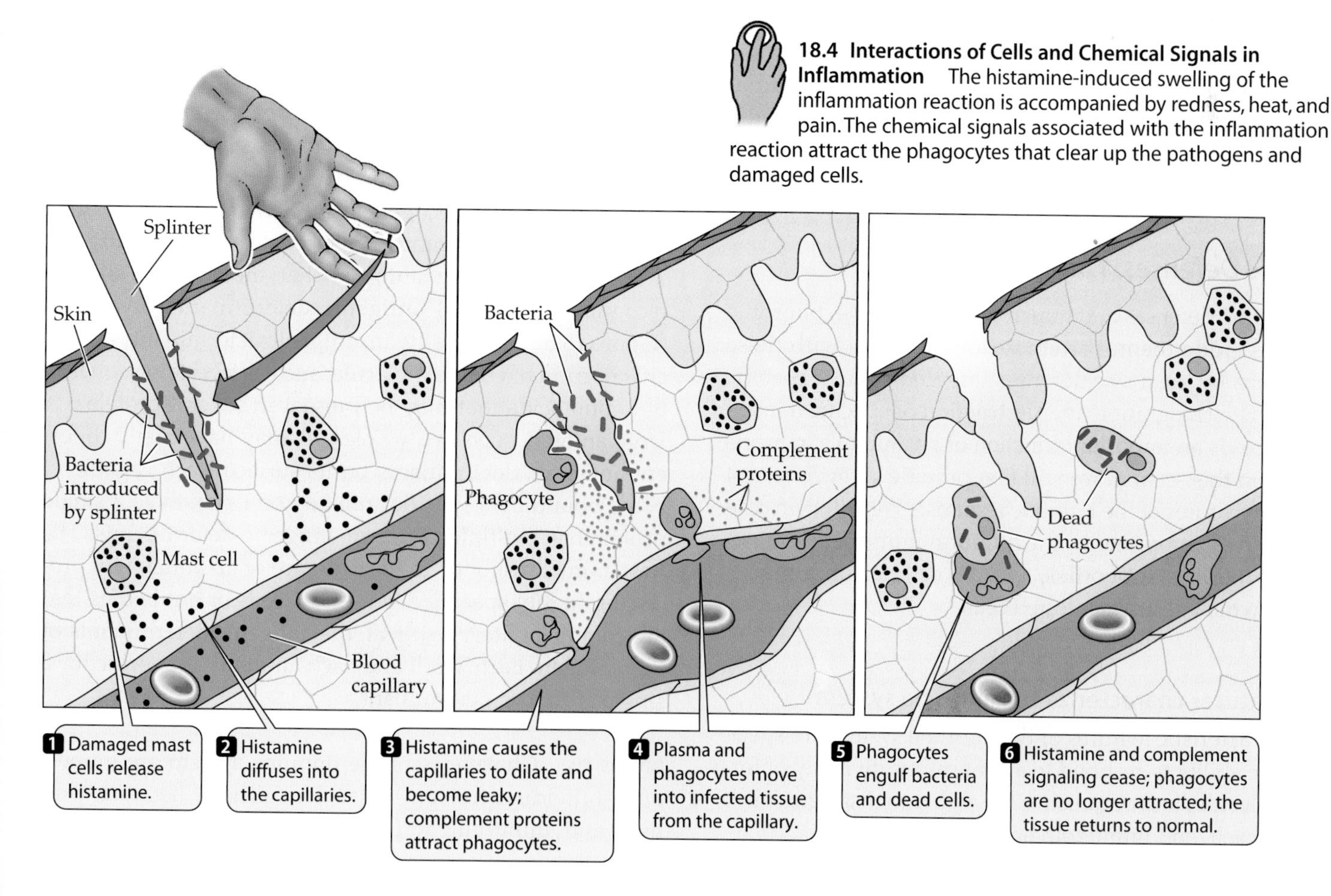
Splinter
Skin
Bacteria introduced by splinter
Mast cell
Blood capillary
Bacteria
Phagocyte
Complement proteins
Dead phagocytes
1 Damaged mast cells release histamine.
2 Histamine diffuses into the capillaries.
3 Histamine causes the capillaries to dilate and become leaky; complement proteins attract phagocytes.
4 Plasma and phagocytes move into infected tissue from the capillary.
5 Phagocytes engulf bacteria and dead cells.
6 Histamine and complement signaling cease; phagocytes are no longer attracted; the tissue returns to normal.

18.4 Interactions of Cells and Chemical Signals in Inflammation The histamine-induced swelling of the inflammation reaction is accompanied by redness, heat, and pain. The chemical signals associated with the inflammation reaction attract the phagocytes that clear up the pathogens and damaged cells.

body temperature inhibits the growth of the invading pathogen. Cytokines may also attract phagocytic cells to the site of injury and initiate a specific response to the pathogen.

Following inflammation, *pus* may accumulate. It is composed of dead cells (neutrophils and the damaged body cells) and leaked fluid. A normal result of inflammation, pus is gradually consumed and digested by macrophages.

A cell signaling pathway stimulates defense

An invading pathogen such as a bacterium can be regarded as a signal. In response to that signal, the body produces molecules such as complement proteins, interferons, and cytokines that regulate phagocytosis and other defense processes. Not surprisingly, the link between signal and response is a signal transduction pathway, similar to the ones we considered in Chapter 15. The receptor in this pathway is a membrane protein called toll. This receptor was originally discovered in fruit flies, in which it plays an essential role in sensing infection by fungi. Comparative genomics has revealed at least ten similar receptors in humans.

Toll is part of a protein kinase cascade that ultimately results in the transcription of at least 40 genes involved in both nonspecific and specific defenses (Figure 18.5). The molecules that set off this pathway are only made by microbes, and include some bacterial and fungal cell wall fragments. Binding of these molecules to toll sets in motion a cascade that results in the phosphorylation of the transcription factor NF-κB. As a result, the transcription factor's conformation changes, allowing it to enter the nucleus, bind to the promoters of genes encoding defensive proteins, and activate their transcription.

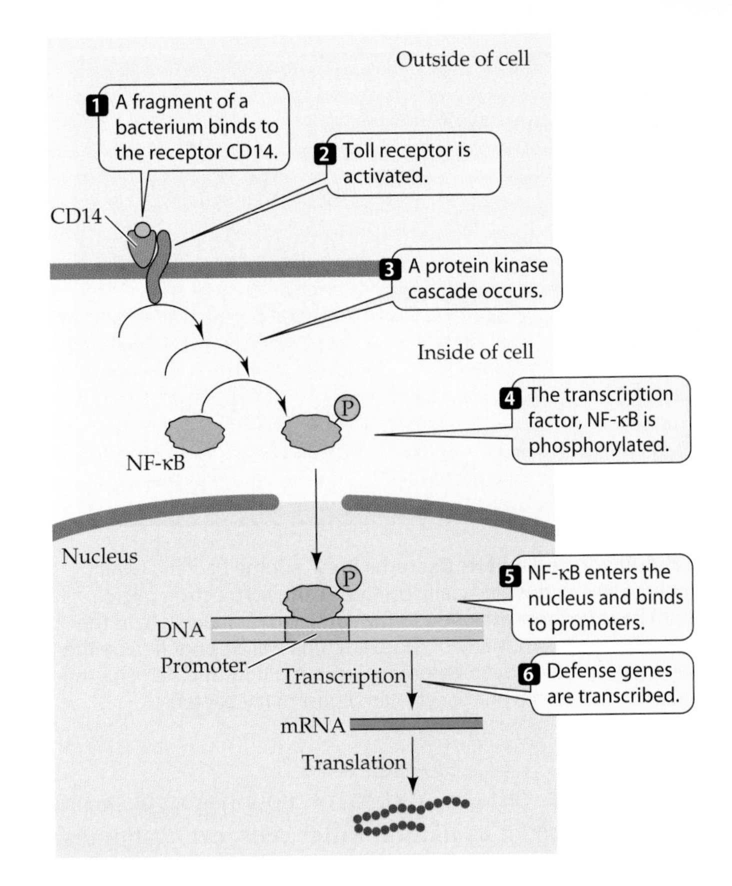

18.5 Cell Signaling and Defense Binding of a molecule from a pathogen to a receptor initiates a signal transduction pathway that results in the transcription of genes whose products are involved in defense against the pathogen.

Specific Defenses: The Immune System

Nonspecific defenses are numerous and effective, but some invaders elude them. Vertebrate animals deal with these pathogens by means of defenses targeted against specific threats. The recognition and destruction of specific nonself substances is an important function of an animal's immune system. In this section, we will first provide an overview of the main features of the immune response. We will then consider its two components: the humoral immune response, which produces antibodies, and the cellular immune response, which destroys infected cells.

Four features characterize the immune system

The characteristic features of the immune system are specificity, the ability to respond to an enormous diversity of foreign molecules and organisms, the ability to distinguish self from nonself, and immunological memory.

SPECIFICITY. As we saw above, the lymphocytes (B cells and T cells) are involved in specific defense mechanisms. T cell receptors and the antibodies produced by B cells recognize and bind to specific nonself substances. Organisms or molecules that are recognized by and interact with these cells to initiate an immune response are called **antigens**. The specific sites on antigens that the immune system recognizes are called **antigenic determinants** or *epitopes* (Figure 18.6). Chemically, an antigenic determinant is a specific portion of a large molecule, such as a certain sequence of amino acids that may be present in several proteins. A large antigen, such as a whole cell, may have many different antigenic determinants on its surface, each capable of being bound by a specific antibody or T cell. Even a single protein has multiple, different antigenic determinants. The host animal responds to the presence of an antigen by producing highly specific defenses—T cells or antibodies that are complementary to, or fit, the antigenic determinants of that antigen. Each T cell and each antibody is specific for a single antigenic determinant.

DIVERSITY. Challenges to the immune system are numerous: individual foreign molecules, viruses, bacteria, protists, and multicellular parasites. Each of these types of

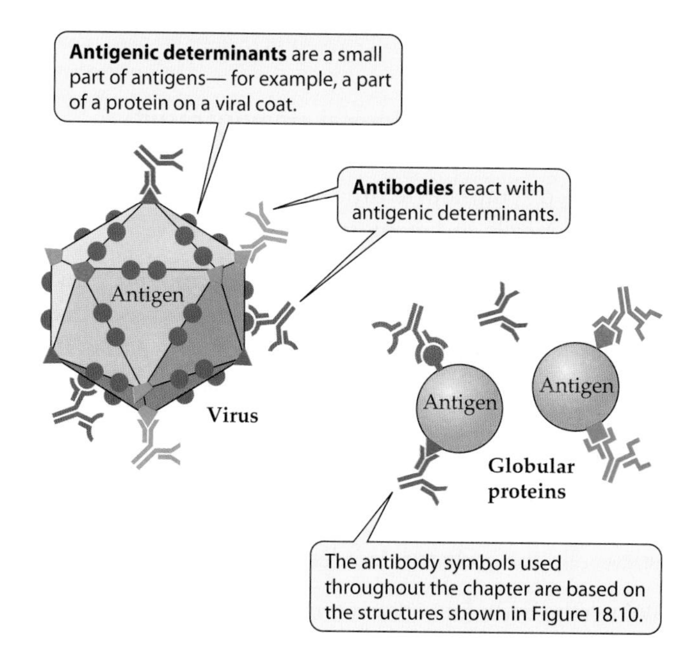

Antigenic determinants are a small part of antigens— for example, a part of a protein on a viral coat.
Antibodies react with antigenic determinants.
Antigen
Virus
Antigen
Antigen
Globular proteins
The antibody symbols used throughout the chapter are based on the structures shown in Figure 18.10.

18.6 Each Antibody Matches an Antigenic Determinant Each antigen has many different antigenic determinants that are recognized by specific antibodies. Each antibody recognizes and binds to its particular antigenic determinant to initiate defensive measures against the antigen.

potential pathogens includes many species, each species includes many subtly differing genetic strains, and each strain possesses multiple surface features. Estimates vary, but a reasonable guess is that humans can respond *specifically* to 10 million different antigenic determinants. Upon recognition of an antigenic determinant, the immune system responds by activating lymphocytes (B cells and T cells) of the appropriate specificity.

DISTINGUISHING SELF FROM NONSELF. The human body contains tens of thousands of different proteins, each with a specific three-dimensional structure capable of generating an immune response. Every cell in the body bears a tremendous number of antigenic determinants. A crucial requirement of an individual's immune system is that it recognize the body's own antigenic determinants and not attack them.

IMMUNOLOGICAL MEMORY. After responding to a particular type of pathogen once, the immune system "remembers" that pathogen and can usually respond more rapidly and powerfully to the same threat in the future. This **immunological memory** usually saves us from repeats of childhood diseases such as chicken pox. Vaccination and inoculation against disease work because the immune system "remembers" the antigenic determinants that are introduced into the body.

These four features of the immune response are seen in both components of the immune system, the humoral response and the cellular response.

There are two interactive immune responses

The immune system has two responses against invaders: the humoral immune response and the cellular immune response. These two responses operate in concert—simultaneously and cooperatively, sharing mechanisms.

HUMORAL IMMUNE RESPONSE. In the **humoral immune response** (from the Latin *humor*, "fluid"), antibodies react with antigenic determinants on pathogens in blood, lymph, and tissue fluids. An animal produces such a diversity of antibodies that between them, they can react with almost any conceivable antigen the animal encounters.

Some antibodies are soluble and travel free in the blood and lymph; others exist as integral membrane proteins on B cells. The first time a specific antigen invades the body, it may be detected and bound by a B cell whose membrane antibody recognizes one of its antigenic determinants. This binding activates the B cell, which makes multiple soluble copies of an antibody with the same specificity as its membrane antibody.

CELLULAR IMMUNE RESPONSE. The **cellular immune response** is directed against an antigen that has become established within a cell of the host animal. It detects and destroys virus-infected or mutated cells.

The cellular immune response is carried out by T cells within the lymph nodes, the bloodstream, and the intercellular spaces. These cells have integral membrane proteins—T cell receptors—that recognize and bind to antigenic determinants while remaining part of the cell's plasma membrane. T cell receptors are rather similar to antibodies in structure and function, each including specific molecular configurations that bind to specific antigenic determinants. Once a T cell is bound to an antigenic determinant, it initiates an immune response that typically results in the total destruction of a nonself or altered self cell.

Genetic processes and clonal selection generate the characteristics of the immune response

Each person possesses an enormous number of different B cells and T cells, apparently capable of dealing with almost any antigenic determinant they are ever likely to encounter. How does this diversity arise? How do lymphocytes specific for certain antigens proliferate? And why don't our antibodies and T cells attack and destroy our own bodies? The diversity of the immune response, the proliferation of specific cells, the ability to distinguish between self and nonself, and immunological memory can all be explained by the process of **clonal selection** and the unique DNA rearrangements upon which it is based.

As we have seen, each individual human contains an enormous variety of different B cells and T cells. This diversity is generated primarily by DNA changes—chromosomal rearrangements and mutations—that occur just after the cells are formed in the bone marrow. Each B cell is able to produce *only one kind of antibody*. Thus there are millions of different B cells, each one producing a particular antibody and displaying it on its cell surface. When an antigen that fits this surface antibody binds to it, the B cell is activated. It divides to form a clone of cells (a genetically identical group derived from a single cell), all of them producing that particular antibody. Thus the antigen "selects" a particular B cell by binding its specific antibody and signaling it to proliferate (Figure 18.7). In the same way, a foreign or abnormal cell "selects" for the proliferation of a T cell expressing a particular T cell receptor on its surface.

Clonal selection accounts nicely for the body's ability to respond rapidly to any of a vast number of different antigens. In the extreme case, even a single B cell might be sufficient for an immunological response, provided that it encounters its antigen and then proliferates into a large clone rapidly enough to combat the invasion.

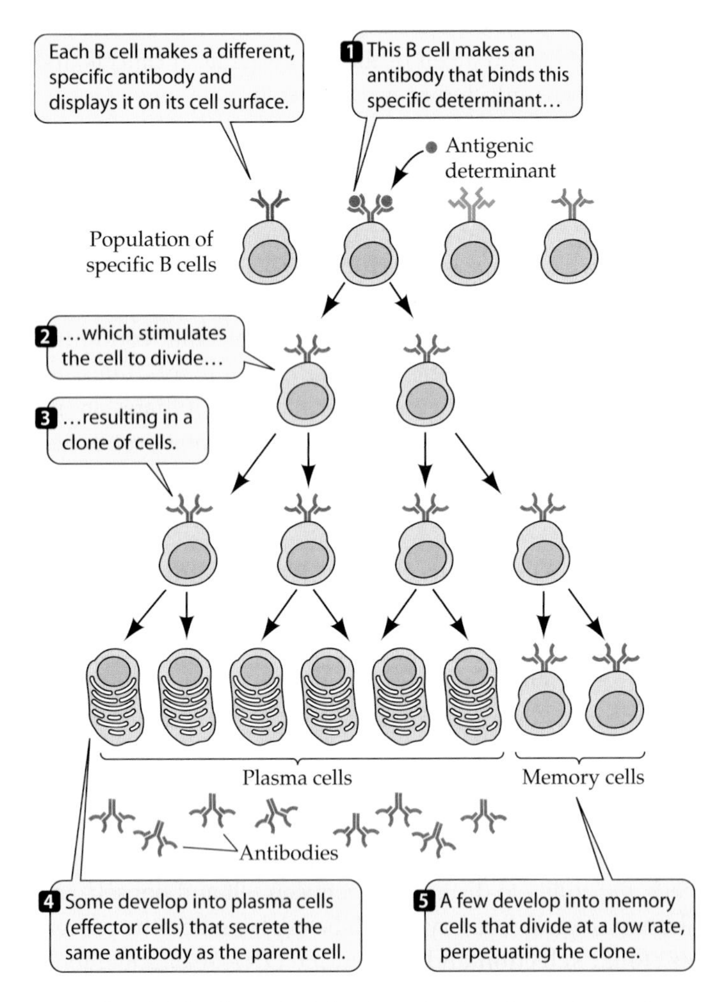

18.7 Clonal Selection in B Cells The binding of an antigenic determinant to a specific antibody on the surface of a B cell stimulates the cell to divide, rapidly producing a clone of cells to fight the invader.

Immunity and immunological memory result from clonal selection

An activated lymphocyte produces two types of daughter cells, effector cells and memory cells.

- **Effector cells** carry out the attack on the antigen. Effector B cells, called **plasma cells**, produce antibodies. Effector T cells release cytokines, which initiate reactions that destroy nonself or altered cells. Effector cells live only a few days.
- **Memory cells** are long-lived cells that retain the ability to start dividing on short notice to produce more effector and more memory cells. Memory B and possibly T cells may survive in the body for decades, dividing at a low rate.

When the body first encounters a particular antigen, a *primary immune response* is activated, in which the lymphocytes that recognize that antigen produce clones of effector and memory cells. The effector cells destroy the invaders at hand and then die, but one or more clones of memory cells have now been added to the immune system and provide immunological memory.

After the body's first immune response to a particular antigen, subsequent encounters with the same antigen will trigger a much more powerful attack. The huge army of plasma and T cells launched by the memory cells at this time is called the *secondary immune response*. The first time a vertebrate animal is exposed to a particular antigen, there is a time lag (usually several days) before the number of antibody molecules and T cells slowly increases (Figure 18.8). But for years afterward—sometimes for life—the immune system "remembers" that particular antigen. The secondary immune response is characterized by a shorter lag time, a greater rate of antibody production, and a larger production of total antibody or T cells than the primary response.

Vaccines are an application of immunological memory

Thanks to immunological memory, recovery from many diseases, such as chicken pox, provides a *natural immunity* to those diseases. However, it is possible to provide *artificial immunity* against many life-threatening diseases by *inoculation*—the introduction of antigenic determinants into the body. **Immunization** is inoculation with antigenic proteins, pathogen fragments, or other molecular antigens. **Vaccination** is inoculation with whole pathogens that have been modified so that they cannot cause disease.

Immunization or vaccination initiates a primary immune response, generating memory cells without making the person ill. Later, if the same or very similar pathogens attack, specific memory cells already exist. They recognize the antigen and quickly overwhelm the invaders with a massive production of lymphocytes and antibodies.

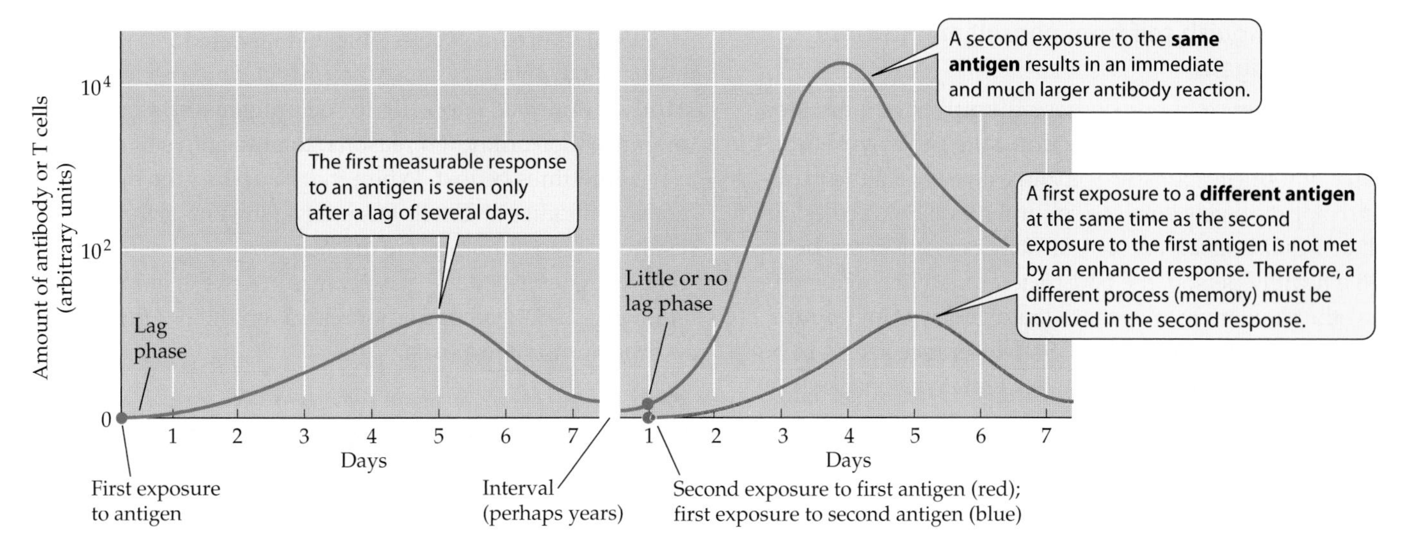

18.8 Immunological Memory The ability of the body to remember an antigen to which it has been exposed is the basis for natural and artificial immunity against a disease.

Because the antigens used for immunization or vaccination are either themselves toxic or are parts of a pathogenic organism, they must be used in a form that is inactive in causing a disease but active in provoking an immune response. There are three principal ways to do this:

- *Attenuation* involves either treating the antigenic molecule or organism with a chemical (for example, formalin) or repeatedly infecting cells with it in the laboratory until its toxicity is reduced.
- *Biotechnology* can be used to produce peptide fragments that activate lymphocytes but do not have the harmful part of a protein toxin.
- *DNA vaccines* are being developed that will introduce a gene encoding an antigen into the body.

For most of the 70 bacteria, viruses, fungi, and parasites that cause serious human diseases, vaccines are already available or will be in the next few years (Table 18.2). Vaccination has almost completely wiped out some deadly diseases, such as diphtheria and polio, in industrialized countries. In fact, smallpox has been eliminated worldwide, thanks to an international effort by the World Health Organization. As far as we know, the only remaining smallpox viruses on Earth are those kept in some laboratories. But there are fears that the smallpox virus and other pathogens, some of which do not have readily available vaccines, may be in the hands of terrorists.

18.2 Some Vaccines against Human Pathogens

INFECTIOUS AGENT	DISEASE	VACCINATED POPULATION
Bacteria		
Bacillus anthracis	Anthrax	Exposed in biological warfare
Bordetella pertussis	Whooping cough	Children and adults
Clostridium tetani	Tetanus	Children
Corynebacterium diphtheriae	Diphtheria	Children
Haemophilus influenzae	Meningitis	Children
Mycobacterium tuberculosis	Tuberculosis	All people
Salmonella typhi	Typhoid fever	Areas exposed to agent
Streptococcus pneumoniae	Pneumonia	Elderly
Vibrio cholerae	Cholera	People in areas exposed to agent
Viruses		
Adenovirus	Respiratory disease	Military personnel
Hepatitis A	Liver disease	Areas exposed to agent
Hepatitis B	Liver disease, cancer	All people
Influenza virus	Flu	All people
Measles virus	Measles	Children and adolescents
Mumps virus	Mumps	Children and adolescents
Poliovirus	Polio	Children
Rabies virus	Rabies	Exposed to agent
Rubella virus	German measles	Children
Vaccinia virus	Smallpox	Laboratory workers, military personnel
Varicella-zoster virus	Chicken pox	Children

Animals distinguish self from nonself and tolerate their own antigens

Given the presence in our bodies of lymphocytes directed against so many antigens, why don't we produce self-destructive immune responses? Sometimes we do. Failure to distinguish appropriately between self and nonself molecules can result in an *autoimmune disease*—an attack on one's own body. But in a healthy person, the body is tolerant of its own molecules—the same molecules that would generate an immune response in another individual. **Self-tolerance** seems to be based on two mechanisms: clonal deletion and clonal anergy.

CLONAL DELETION. **Clonal deletion** physically removes B or T cells from the immune system at some point during their differentiation. Immature B cells in the bone marrow, for example, may encounter self antigens. Any of these cells that shows the potential to mount an immune response against self antigens undergoes programmed cell death (apoptosis) within a short time, and never differentiates enough to make antibodies. Thus, no clones of antiself B cells normally appear in the bloodstream. Clonal deletion eliminates about 90 percent of all the B cells made in the bone marrow. A similar process occurs with T cells in the thymus.

CLONAL ANERGY. **Clonal anergy** is the suppression of the immune response to self antigens. A mature T cell, for example, may encounter and recognize a self antigen on the surface of a body cell. But it does not send out the cytokines that signal the initiation of an immune response. Before it does so, the T cell must encounter not only an antigen, but also a second molecule, CD28, on the cell surface. Most body cells, lacking CD28, will not be attacked by the cellular immune system.

CD28 is a *co-stimulatory signal* that is expressed only on certain *antigen-presenting cells*. Such cells "present" antigens on their surfaces, thus stimulating the cellular immune system. Antigen-presenting cells include the macrophages that wander through the body's fluids, and the dendritic cells that appear among the linings of the respiratory and digestive tracts.

Immunological tolerance was discovered through the observation that some *nonidentical* twin cattle with different blood types contained some of each other's red blood cells. Why didn't these "foreign" blood cells cause immune responses resulting in their elimination? The hypothesis suggested was that the blood cells had passed between the fetal animals in the womb before the lymphocytes had matured. Thus each calf regarded the other's red blood cells as self. This hypothesis was confirmed when it was shown that injecting a foreign antigen into an animal early in its fetal development caused that animal henceforth to recognize that antigen as self.

Self-tolerance must be established repeatedly throughout the life of the animal because lymphocytes are produced constantly. Continued exposure to self antigens helps maintain tolerance. For unknown reasons, tolerance to self antigens may sometimes be lost. When that happens, the body produces antibodies or T cells targeted against its own proteins, resulting in an autoimmune disease.

Having described the general features of the immune system, we will now focus in more detail on the B lymphocytes and the humoral response.

B Cells: The Humoral Immune Response

Every day, billions of B cells survive the test of clonal deletion and are released from the bone marrow into the circulation. B cells are the basis for the humoral immune response.

Some B cells develop into plasma cells

As described above, a B cell is activated by the binding of a specific antigenic determinant to the antibody protein on its surface. Normally, for such a B cell to develop into an antibody-secreting plasma cell, a **helper T cell (T_H)** with the same specificity must also bind to the antigen. Thus, the B cell also functions as an antigen-presenting cell, as we will see below. The division and differentiation of B cells is stimulated by the receipt of chemical signals from the T_H cell. These events lead to the formation of plasma cells (effector B cells) and memory cells (see Figure 18.7).

As plasma cells develop, the number of ribosomes and the amount of endoplasmic reticulum in their cytoplasm increase greatly (Figure 18.9). These increases allow the cells to synthesize and secrete large amounts of antibodies. All the plasma cells arising from a given B cell produce antibodies that are specific for the antigen that originally bound to the parent B cell. Thus antibody specificity is maintained as B cells proliferate.

Different antibodies share a common structure

Antibodies are proteins called **immunoglobulins**. There are several types of immunoglobulins, but all contain a tetramer consisting of four polypeptide chains. In each immunoglobulin molecule, two of these polypeptides are identical *light chains*, and two are identical *heavy chains*. Disulfide bonds hold the chains together.

Each polypeptide chain consists of a constant region and a variable region (Figure 18.10).

- The **constant regions** of both light chains and heavy chains are similar in amino acid sequence among the immunoglobulins. They determine the destination and function—the *class*—of the antibody.

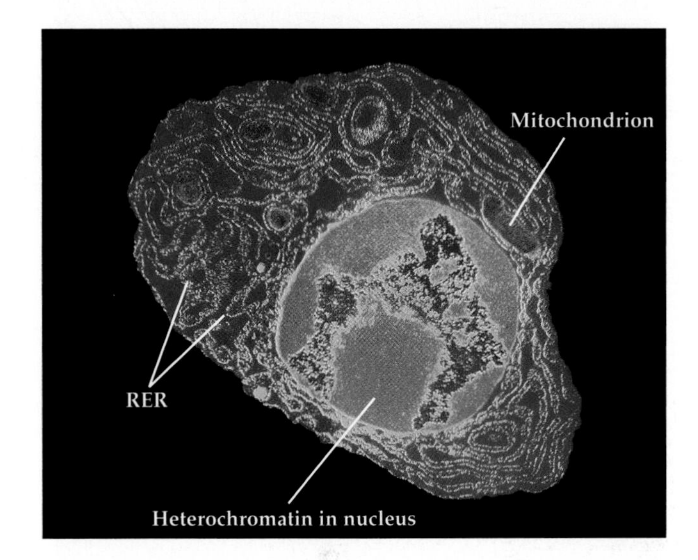

18.9 A Plasma Cell The prominent nucleus with large amounts of heterochromatin (orange) and the cytoplasm (bright blue) crowded with rough endoplasmic reticulum are features of a cell that is actively synthesizing and exporting proteins—in this case, a specific antibody. Whole blocks of genes not needed for this specialized function are kept turned off in the heterochromatin.

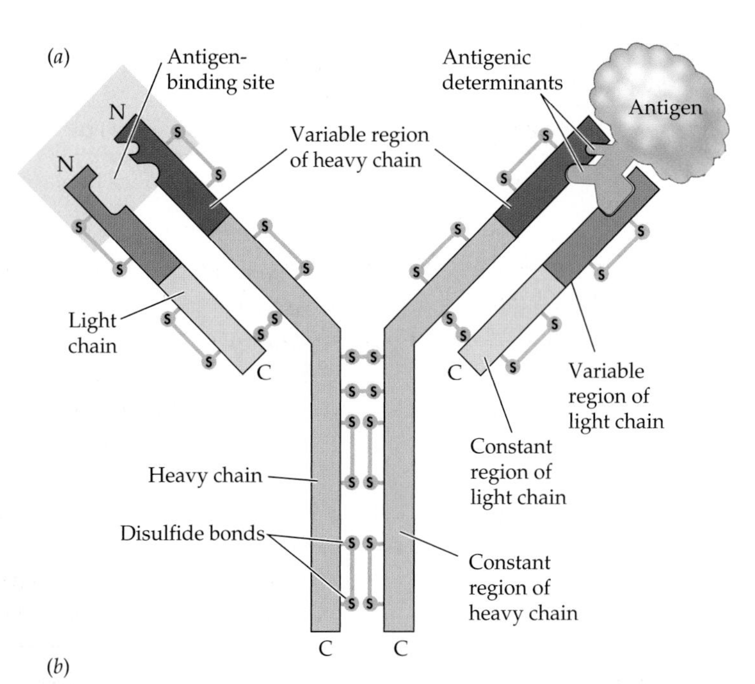

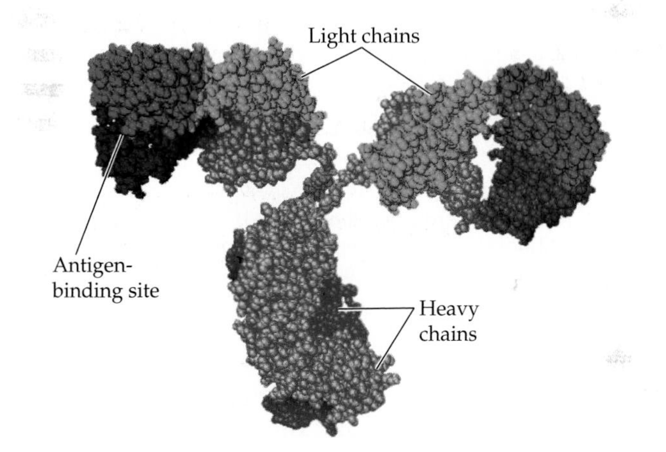

18.10 Structure of Immunoglobulins (*a*) The four polypeptide chains (two light, two heavy) of an immunoglobulin molecule. (*b*) A three-dimensional space-filling model of an antibody molecule in roughly the same orientation as (*a*). In both images, the light chains are shown in green and the heavy chains in blue; the variable regions are shown in a darker color and the constant regions in a lighter color.

- The **variable regions** differ in their amino acid sequences. They contribute directly to the three-dimensional region where the antigen binds—the *antigen-binding site*—and are responsible for the diversity of antibody specificity.

In the initial differentiation of each B cell in the bone marrow, the amino acid sequence of the variable region of both the light and heavy chains is chosen randomly from the DNA information in the genome. This means that the variable region is unique in each of the millions of antigen-specific immunoglobulins. Together, the variable regions of a light and a heavy chain form a highly specific, three-dimensional structure. This part of a particular immunoglobulin molecule is what binds with a particular, unique antigenic determinant. The enormous range of antibody specificities is accomplished by a combination of rearrangements and mutations in the genes that encode the variable regions, as we will see later in this chapter.

The two antigen-binding sites on each immunoglobulin molecule are identical, making the antibody *bivalent* (*bi-*, "two"; *valent*, "binding"). This ability to bind two antigen molecules at once permits the antibody to form a large complex with antigen and other antibody molecules. Such a complex is an easy target for ingestion and breakdown by phagocytic cells.

While the variable regions are responsible for the *specificity* of an immunoglobulin, the constant regions of the heavy chain determine the *class* of the antibody—for example, whether it will be a membrane receptor or a soluble antibody that is secreted into the bloodstream. The five immunoglobulin classes are described in Table 18.3. The most abundant immunoglobulin class is IgG; these soluble antibody proteins make up about 80 percent of the total immunoglobulin content of the bloodstream. They are made in greatest quantity during a secondary immune response. IgG defends the body in several ways. For example, after some IgG molecules bind to antigens, they become attached by their heavy chains to macrophages. This attachment permits the macrophages to destroy the antigens by phagocytosis (Figure 18.11).

Hybridomas produce monoclonal antibodies

The specificity of antibodies suggested to scientists that they might be useful for detecting a specific substance in a fluid.

18.3 Antibody Classes

CLASS	GENERAL STRUCTURE	LOCATION	FUNCTION
IgG	Monomer	Free in plasma; about 80 percent of circulating antibodies	Most abundant antibody in primary and secondary responses; crosses placenta and provides passive immunization to fetus
IgM	Pentamer	Surface of B cell; free in plasma	Antigen receptor on B cell membrane; first class of antibodies released by B cells during primary response
IgD	Monomer	Surface of B cell	Cell surface receptor of mature B cell; important in B cell activation
IgA	Dimer	Monomer found in plasma; polymers in saliva, tears, milk, and other body secretions	Protects mucosal surfaces; prevents attachment of pathogens to epithelial cells
IgE	Monomer	Secreted by plasma cells in skin and tissues lining gastrointestinal and respiratory tracts	Found on mast cells and basophils; when bound to antigens, triggers release of histamine from mast cell or basophil that contributes to inflammation and some allergic responses

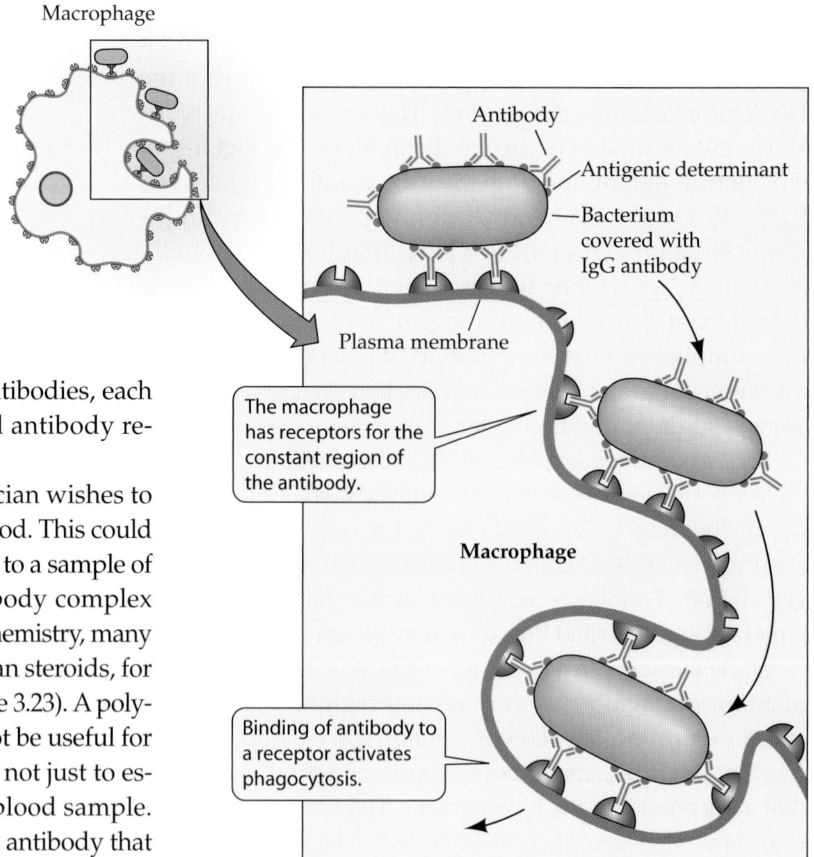
Macrophage
Antibody
Antigenic determinant
Bacterium covered with IgG antibody
Plasma membrane
The macrophage has receptors for the constant region of the antibody.
Macrophage
Binding of antibody to a receptor activates phagocytosis.

18.11 IgG Antibodies Promote Phagocytosis When IgG antibodies cover a bacterium, receptors on a macrophage can recognize, bind to, and engulf it.

An initial challenge to scientists seeking to accomplish this was that the immune response to a complex antigen is not simple. Therefore, they could not simply produce antibodies by injecting an animal with the antigen they wanted to look for. Because most antigens carry many different antigenic determinants, animals injected with a single antigen will produce a complex mixture of antibodies, each made by a different clone of B cells. So the normal antibody response is said to be *polyclonal*.

Suppose that a woman is infertile and her physician wishes to measure the levels of the hormone estrogen in her blood. This could be done adding an antibody directed against estrogen to a sample of her blood and observing how much antigen–antibody complex formed. But, as we have learned in our studies of biochemistry, many molecules share regions of similar structure. All human steroids, for example, have a similar multi-ring structure (see Figure 3.23). A polyclonal group of antibodies against estrogen would not be useful for this test because some of the antibodies would bind not just to estrogen, but to any steroid hormone present in the blood sample. Clearly, a clone of B cells making large amounts of an antibody that binds to only one antigenic determinant—a **monoclonal antibody**—would be needed. How could such a clone be produced?

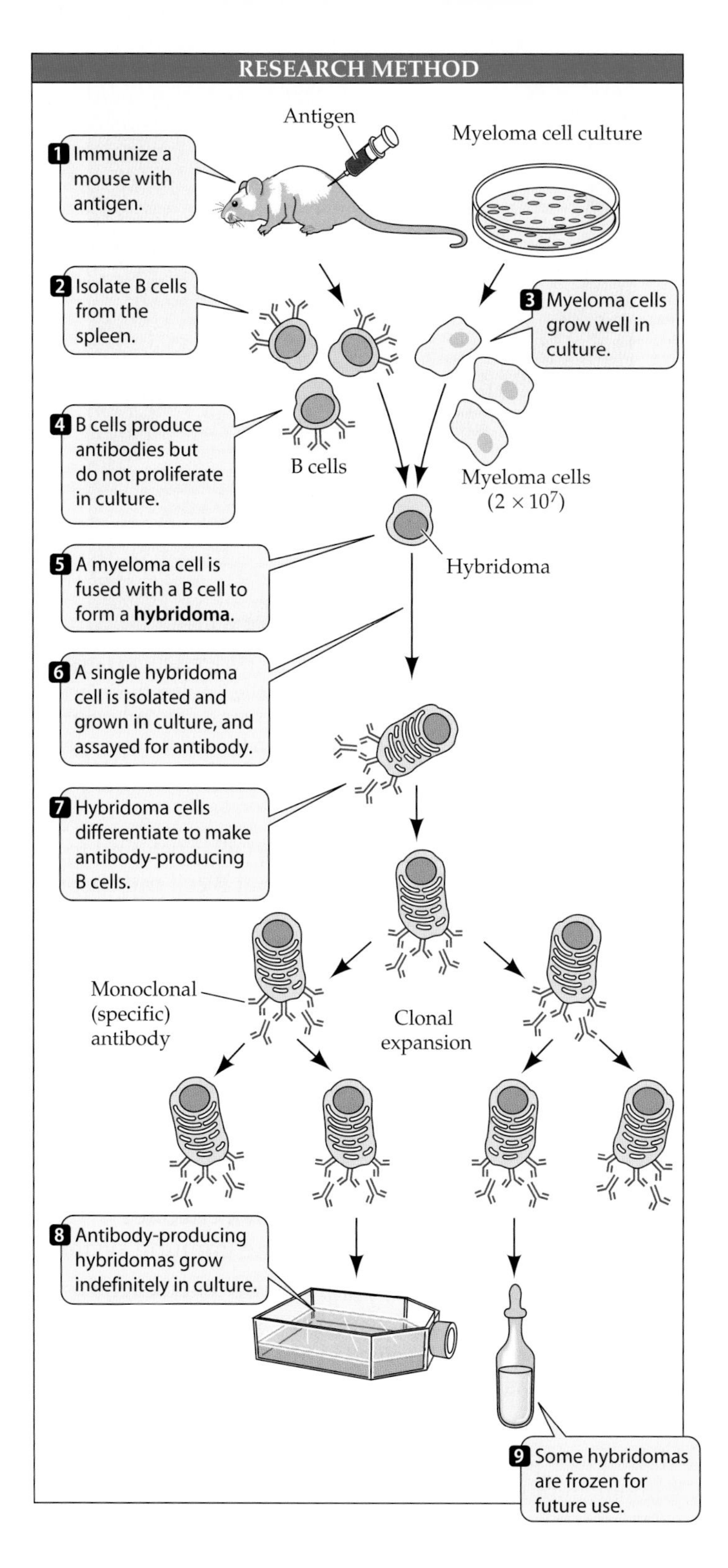
RESEARCH METHOD
Antigen
Myeloma cell culture
1 Immunize a mouse with antigen.
2 Isolate B cells from the spleen.
3 Myeloma cells grow well in culture.
4 B cells produce antibodies but do not proliferate in culture.
B cells
Myeloma cells (2×10^7)
5 A myeloma cell is fused with a B cell to form a **hybridoma**.
Hybridoma
6 A single hybridoma cell is isolated and grown in culture, and assayed for antibody.
7 Hybridoma cells differentiate to make antibody-producing B cells.
Monoclonal (specific) antibody
Clonal expansion
8 Antibody-producing hybridomas grow indefinitely in culture.
9 Some hybridomas are frozen for future use.

18.12 Creating Hybridomas for the Production of Monoclonal Antibodies Cancerous myeloma cells and normal B cells can be hybridized so that the proliferative properties of the myeloma cells are merged with the specificity of the antibody-producing B cells.

A single clone of cells making a single antibody can be made by fusing a B cell (which has a finite lifetime and makes a lot of antibody) with a tumor cell (which has an infinite lifetime). The resulting hybrid cells, called *hybridomas*, each make a specific monoclonal antibody (Figure 18.12).

Monoclonal antibodies have many practical applications:

- *Immunoassays* use the great specificity of the antibodies to detect tiny amounts of molecules in tissues and fluids. This technique is used, for example, to quantify the hormone made by the developing embryo for a pregnancy test.
- *Immunotherapy* uses monoclonal antibodies targeted against antigens on the surfaces of cancer cells. The coupling of a radioactive ligand or toxin to the antibody makes it into a medical "smart bomb." In some cases, binding of the antibody itself is enough to trigger a cellular immune response that destroys the cancer.
- *Passive immunization* is inoculation with an immediately acting, but not long-lasting, specific antibody. This approach is necessary when therapy must be effective quickly (within hours). Examples of such life-threatening situations include the early symptoms of rabies infection, rattlesnake bites, and babies born with hepatitis B virus infection—all cases in which the toxic nature of the infection is so serious that there is not enough time to allow the person's immune system to mount its own defense (several days at least).

A major problem with the clinical use of monoclonal antibodies is that the B cells used to produce them come from inoculated mice, so they are mouse proteins. Since mouse immunoglobulin genes differ somewhat from the human ones, the structure of mouse immunoglobulin proteins will also be different, and so the monoclonal antibody may be antigenic to humans. To circumvent this problem, scientists can use recombinant DNA technology to make immunoglobulin genes containing constant regions from humans and variable regions from the mouse (which are not very antigenic to humans). Such a *humanized antibody* does not provoke an immune response in people.

T Cells: The Cellular Immune Response

Thus far we have been concerned primarily with the humoral immune response, whose effector molecules are the antibodies secreted by plasma cells that develop from activated B cells. T cells, as we have seen, are involved in the humoral immune response, but they are also the effectors of the cellular immune response, which is directed against any factor, such as a virus or mutation, that changes a normal cell into an abnormal cell.

In this section, we will describe two types of effector T cells (helper T cells and cytotoxic T cells). We will also describe the MHC (major histocompatibility complex) proteins, which underlie the immune system's tolerance for the cells of its own body.

T cell receptors are found on two types of T cells

Like B cells, T cells possess specific membrane receptors. T cell receptors are not immunoglobulins, but glycoproteins with molecular weights about half that of an IgG. They are made up of two polypeptide chains, each encoded by a separate gene (Figure 18.13). Thus the two chains are nearly always different in their amino acid sequence, especially in their variable regions.

The genes that code for T cell receptors are similar to those for immunoglobulins, suggesting that both are derived from a single, evolutionarily more ancient group of genes. Like the immunoglobulins, T cell receptors include both variable and constant regions. The variable regions provide the specificity for binding with a single antigenic determinant. There is one major difference between antibodies and T cell receptors: While antibodies bind to an intact antigen, T cell receptors bind to a piece of the antigen displayed on the surface of an antigen-presenting cell.

When a T cell is activated by contact with a specific antigenic determinant, it proliferates and forms a clone. Its descendants differentiate into two sub-clones, giving rise to two types of effector T cells:

- **Cytotoxic T cells**, or T_C cells, recognize virus-infected cells and kill them by inducing lysis (Figure 18.14).
- **Helper T cells**, or T_H cells, assist both the cellular and humoral immune responses.

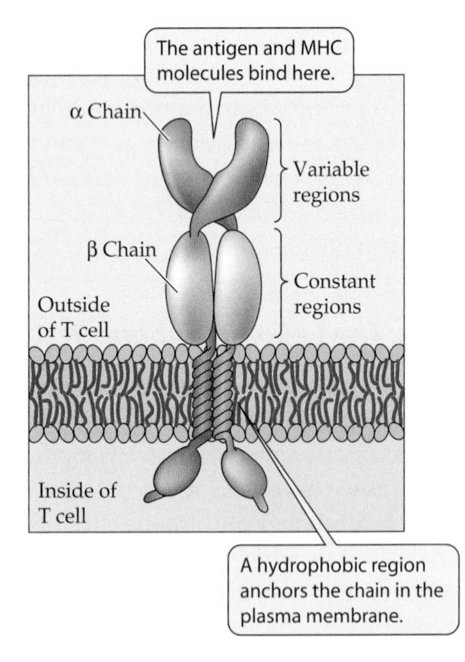

18.13 A T Cell Receptor T cell receptors are glycoproteins, not immunoglobulins, although the structures of the two molecules are similar. In both, each binding site is determined by two polypeptides. T cell receptors are bound more firmly to the plasma membrane of the T cell that produces them than is antibody to B cells.

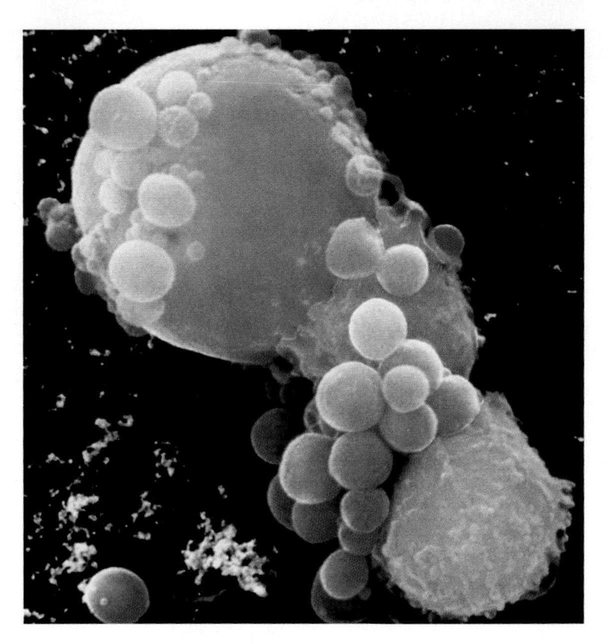

18.14 Cytotoxic T Cells in Action Two cytotoxic T cells (orange) have come into contact with virus-infected cells, causing the infected cells to die. The pink cell at top left has begun cell death, indicated by membrane blisters. The process is complete in the cell in the center.

As mentioned already, a specific T_H cell must bind to an antigen presented on a B cell before that B cell can become activated. The helper cell becomes the "conductor" of the "immunological orchestra" as it sends out chemical signals that not only result in its own proliferation and that of the B cell, but also set in motion the actions of cytotoxic T cells.

Now that we are familiar with the major types of T cells, we can address the question of how T cells meet their antigenic determinants and the role of the MHC proteins in the process.

The major histocompatibility complex encodes proteins that present antigens to the immune system

We have seen that an animal's immune system recognizes its own cells by their surface proteins. Several types of mammalian cell surface proteins are involved in this process, but we will focus here on one very important group, the products of a cluster of genes called the major histocompatibility complex, or MHC. These proteins have important roles in the cellular and humoral immune responses as well as in self-tolerance.

The MHC gene products are plasma membrane glycoproteins. In humans, the MHC proteins are called *human leukocyte antigens* (HLA), while in mice they are called *H-2 proteins*. Their major role is to present antigens on the cell surface to a T cell receptor. There are three classes of MHC proteins:

- *Class I MHC proteins* are present on the surface of every nucleated cell in the animal body. When cellular proteins are degraded into small peptide fragments by a proteasome (see Chapter 14), an MHC I protein may bind to a fragment and travel to the plasma membrane. There, the

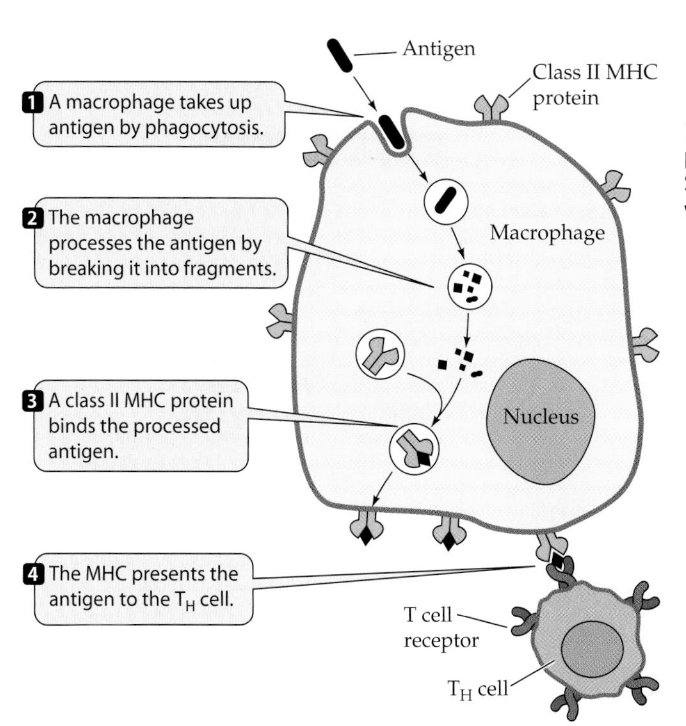
Antigen
Class II MHC protein
1 A macrophage takes up antigen by phagocytosis.
2 The macrophage processes the antigen by breaking it into fragments.
Macrophage
3 A class II MHC protein binds the processed antigen.
Nucleus
4 The MHC presents the antigen to the T_H cell.
T cell receptor
T_H cell

18.15 Macrophages Are Antigen-Presenting Cells A fragment of a processed antigen is displayed by MHC II on the surface of a macrophage. Specific receptors on a helper T cell can then bind to and interact further with the processed antigen/MHC II complex.

MHC I protein "presents" the cellular peptide to T_C cells. The T_C cells have a surface protein called CD8 that recognizes and binds to MHC I.

- *Class II MHC proteins* are found mostly on the surfaces of B cells, macrophages, and other antigen-presenting cells. When an antigen-presenting cell ingests an antigen, such as a virus, the antigen is broken down in a phagosome. An MHC II molecule may bind to one of the fragments and carry it to the cell surface, where it is presented to a T_H cell (Figure 18.15). T_H cells have a surface protein called CD4 that recognizes MHC II.
- *Class III MHC proteins* include some of the proteins of the complement system, which interact with antigen–antibody complexes and result in the lysis of foreign cells.

To accomplish their roles in antigen binding and presentation, both MHC I and MHC II proteins have an antigen-binding site, which can hold a peptide of about 10–20 amino acids (Figure 18.16). The T cell receptor recognizes not just the antigenic fragment, but the fragment *bound to an MHC I or MHC II molecule.* The table in Figure 18.16 summarizes the relationships of T cells and antigen-presenting cells.

In humans, there are three genetic loci for MHC I and three for MHC II; all six loci have as many as 100 different alleles. With so many possible allelic combinations, it is not surprising that different people are very likely to have different MHC genotypes. Similarities in base sequences between the MHC genes and the genes coding for antibodies and T cell receptors suggest that all three may have descended from the same ancestral genes and are part of a gene "superfamily." Major aspects of the immune system in vertebrates seem to be woven together by a common evolutionary thread.

Helper T cells and MHC II proteins contribute to the humoral immune response

When a T_H cell binds to an antigen-presenting macrophage, the T_H cell releases cytokines, which activate the T_H cell to produce a clone of differentiated cells capable of interacting with B cells. The steps to this point constitute the *activation phase* of the humoral immune response, and they occur in the

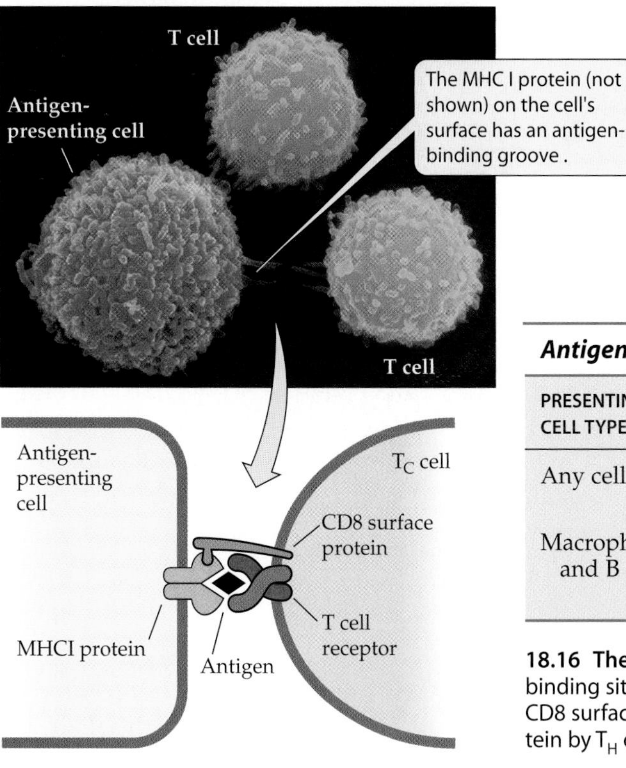
T cell
Antigen-presenting cell
The MHC I protein (not shown) on the cell's surface has an antigen-binding groove .
T cell
Antigen-presenting cell
T_C cell
CD8 surface protein
T cell receptor
MHCI protein
Antigen

Antigen-Presenting and T Cell Types

PRESENTING CELL TYPE	ANTIGEN PRESENTED	MHC CLASS	T CELL TYPE	T CELL SURFACE PROTEIN
Any cell	Intracellular protein fragment	Class I	Cytotoxic T cell (T_C)	CD8
Macrophages and B cells	Fragments from extracellular proteins	Class II	Helper T (T_H)	CD4

18.16 The Interaction between T Cells and Antigen-Presenting Cells An antigen-binding site in the MHC I protein holds an antigen, which it presents to cytotoxic T cells. CD8 surface proteins on the T_C cells ensure binding to MHC I. The binding of MHC II protein by T_H cells works in a similar manner.

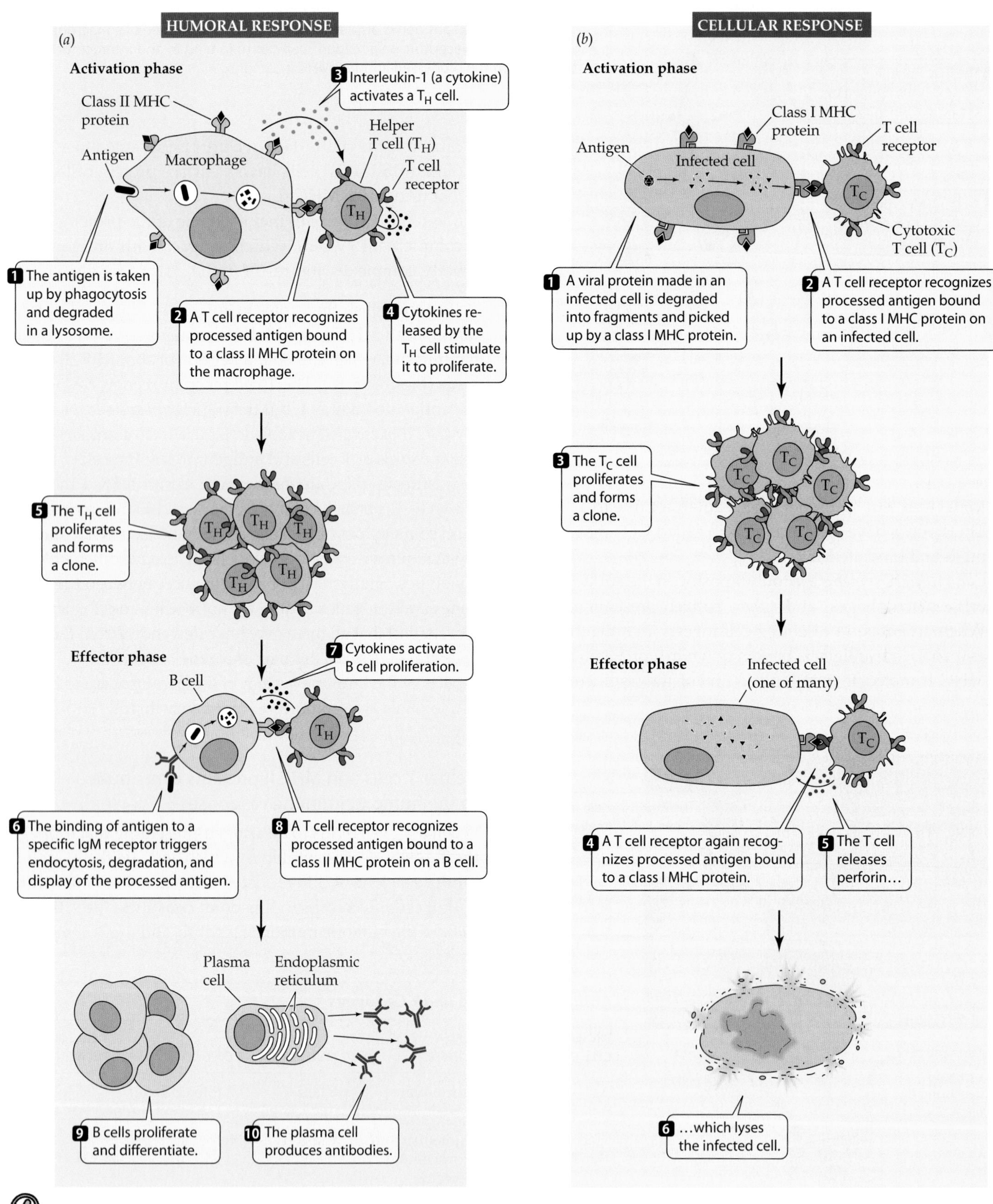

18.17 Phases of the Humoral and Cellular Immune Responses
Both immune responses have activation and effector phases.

lymphoid tissues. Next comes the *effector phase*, in which B cells are activated to produce antibodies (Figure 18.17*a*).

B cells are also antigen-presenting cells. B cells take up antigens bound to their surface immunoglobulin receptors by endocytosis, process them, and display them on class II MHC proteins. When a T_H cell binds to the displayed antigen–MHC II complex, it releases cytokines, which cause the B cell to produce a clone of plasma cells. Finally, the plasma cells secrete antibody, completing the effector phase of the humoral immune response.

Cytotoxic T cells and MHC I proteins contribute to the cellular immune response

Class I MHC proteins play a role in the cellular immune response that is similar to the role played by class II MHC proteins in the humoral immune response. In a virus-infected or mutated cell, foreign or abnormal proteins or peptide fragments combine with MHC I molecules. The resulting complex is displayed on the cell surface and presented to T_C cells. When a T_C cell binds to this complex, it is activated to proliferate (Figure 18.17*b*).

In the effector phase of the cellular immune response, T_C cells recognize and bind to cells bearing the MHC I–antigen complex. These T_C cells produce a substance called perforin, which lyses the target cell. In addition, the T_C cell can bind to a specific receptor (called Fas) on the target cell that initiates apoptosis in that cell. These two mechanisms, cell lysis and programmed cell death, work in concert to eliminate the altered host cell.

Because T cell receptors recognize self MHC proteins complexed with *nonself* antigens, they help rid the body of its own virus-infected cells. Because they also recognize MHC proteins complexed with *altered self* antigens (as a result of mutations), they help eliminate tumor cells, since most tumor cells have been altered by mutations.

In addition to the binding of an antigen–MHC complex to their receptors, T cells must receive a second signal for activation. This co-stimulatory signal occurs after the initial specific binding and involves the interaction of additional proteins on the T cell with the CD28 protein on the antigen-presenting cell, as we saw above. This second binding event leads to T cell activation, including cytokine production and proliferation. It also sets in motion the production of an *inhibitor* of these events, so that the response is appropriately terminated. This inhibitor, a cell surface protein called CTLA4, blocks the activation process, especially for self-antigens.

MHC proteins underlie the tolerance of self

MHC proteins play a key role in establishing self-tolerance, without which an animal would be destroyed by its own immune system. Throughout the animal's life, developing T cells are tested in the thymus. This "test" consists of two "questions":

1. Can this cell recognize the body's MHC proteins? A T cell unable to recognize self MHC proteins would be useless to the animal because it could not participate in any immune reactions. Such a T cell fails the test and dies within about 3 days.
2. Does this cell bind to self MHC protein *and* to one of the body's own antigens? A T cell that satisfied both of these criteria would be harmful or lethal to the animal; it also fails the test and undergoes apoptosis.

T cells that survive this test mature into either T_C cells or T_H cells.

MHC proteins are responsible for transplant rejection

In humans, a consequence of the major histocompatibility complex became important with the development of organ transplant surgery. Because the proteins produced by the MHC are specific to each individual, they act as antigens if transplanted into another individual. An organ or a piece of tissue transplanted from one person to another is recognized as nonself and soon provokes an immune response; the tissue is then killed, or "rejected," by the host's cellular immune system. But if the transplant is performed immediately after birth, or if it comes from a genetically identical person (an identical twin), the material is recognized as self and is not rejected.

The rejection problem can be overcome by treating a patient with drugs, such as cyclosporin, that suppress the immune system. Cyclosporin works by blocking the activation of a transcription factor essential for T cell development. However, this approach compromises the ability of patients to defend themselves against pathogens. These risks are often managed by the use of antibiotics and other drugs.

So far in this chapter, we have only occasionally alluded to the DNA-based events that make the diversity of antibody specificity possible. In the next section, we will address the genetic mechanisms that generate antibody diversity.

The Genetic Basis of Antibody Diversity

A newborn mammal possesses a full set of genetic information for immunoglobulin synthesis. At each of the loci coding for the heavy and light antibody chains, it has one allele from its mother and one from its father. Throughout the animal's life, each of its cells begins with the same full set of immunoglobulin genes. However, as B cells develop, their genomes become modified in such a way that each cell eventually can produce one—and only one—specific type of antibody. In other words, different B cells develop slightly different genomes encoding different antibody specificities. How can a single organism produce millions of different genomes?

One hypothesis was that we simply have millions of antibody genes. However, a simple calculation (the number of base pairs needed per antibody gene multiplied by millions) shows that if this were true, our entire genome would be taken up by antibody genes! More than 30 years ago, an alternative hypothesis was proposed: A relatively small number of genes recombine to produce many unique combinations, and it is this shuffling of the genetic deck, plus the random pairing of light and heavy antibody chains, that produces antibody diversity. This second hypothesis is now the accepted molecular genetic theory.

In this section, we will describe the unusual events that generate the enormous antibody diversity that normally characterizes each individual mammal. Then we will see how similar events produce the five classes of antibodies by producing slightly different constant regions with special properties.

Antibody diversity results from DNA rearrangement and other mutations

Each gene encoding an immunoglobulin is in reality a "supergene" assembled from several clusters of smaller genes scattered along part of a chromosome (Figure 18.18). Every cell in the body has hundreds of genes, located in separate clusters, that are potentially capable of participating in the synthesis of the variable and constant regions of immunoglobulin polypeptide chains. In most body cells and tissues, these genes remain intact and separated from one another. During B cell development, however, these genes are cut out, rearranged, and joined together. Most of the coding and noncoding regions of these genes are deleted, and one gene from each cluster—is chosen randomly for joining (Figures 18.18, 18.19).

In this manner, a unique antibody supergene is assembled from randomly selected "parts." Each B cell precursor in the animal assembles its own two specific antibody supergenes, one for a specific heavy chain and the other, assembled independently, for a specific light chain. This remarkable example of essentially irreversible cell differentiation generates an enormous diversity of antibody specificities from the same starting genome, one for each individual B cell.

In both humans and mice, the gene clusters coding for immunoglobulin heavy chains are on one pair of chromosomes and those for light chains are on others. The variable region of the light chain is encoded by two families of genes; the variable region of the heavy chain is encoded by three families.

Figure 18.18 illustrates the gene families coding for the heavy-chain constant and variable regions in mice. There are multiple genes coding for each of the four kinds of segments in the polypeptide chain: 100 *V*, 30 *D*, 6 *J*, and 8 *C*. Each B cell that becomes committed to making an antibody randomly selects *one* gene for each of these clusters to make the final heavy-chain coding sequence, *VDJC*. So the number of *different* heavy chains that can be made through this random recombination process is quite large.

Now consider that the light chains are similarly constructed, with a similar amount of diversity made possible by random recombination. If we assume that light-chain diversity is the same as heavy-chain diversity (144,000 possible combinations), the number of possible combinations of light and heavy chains is 144,000 different light chains × 144,000 different heavy chains = 21 *billion* possibilities!

Even if this number is an overestimate by severalfold (and it is), the number of different immunoglobulin molecules that B cells can make is huge. But there are other mechanisms that generate even more diversity:

- When the DNA sequences for the *V*, *J*, and *C* regions are rearranged so that they are next to one another, the recombination event is not precise, and errors occur at the junctions. This *imprecise recombination* can create new codons at the junctions, with resulting amino acid changes.
- After the DNA sequences are cut out and before they are joined, an enzyme, *terminal transferase*, often adds some nucleotides to the free ends of the DNAs. These additional bases create *insertion mutations*.
- There is a relatively high *mutation rate* in immunoglobulin genes. Once again, this process creates many new alleles and adds to antibody diversity.

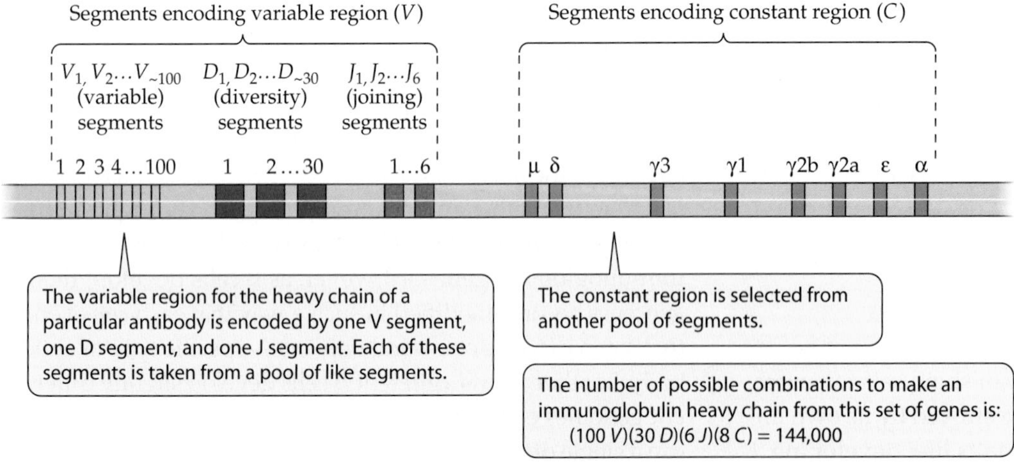
Segments encoding variable region (V)
Segments encoding constant region (C)
$V_1, V_2 \ldots V_{\sim 100}$ (variable) segments
$D_1, D_2 \ldots D_{\sim 30}$ (diversity) segments
$J_1, J_2 \ldots J_6$ (joining) segments
1 2 3 4…100
1 2…30
1…6
μ δ
γ3
γ1
γ2b γ2a
ε
α
The variable region for the heavy chain of a particular antibody is encoded by one V segment, one D segment, and one J segment. Each of these segments is taken from a pool of like segments.
The constant region is selected from another pool of segments.
The number of possible combinations to make an immunoglobulin heavy chain from this set of genes is:
(100 V)(30 D)(6 J)(8 C) = 144,000

18.18 Heavy-Chain Genes Mouse immunoglobulin heavy chains have four domains, each of which is coded for by one of multiple possible genes selected from a cluster of like genes.

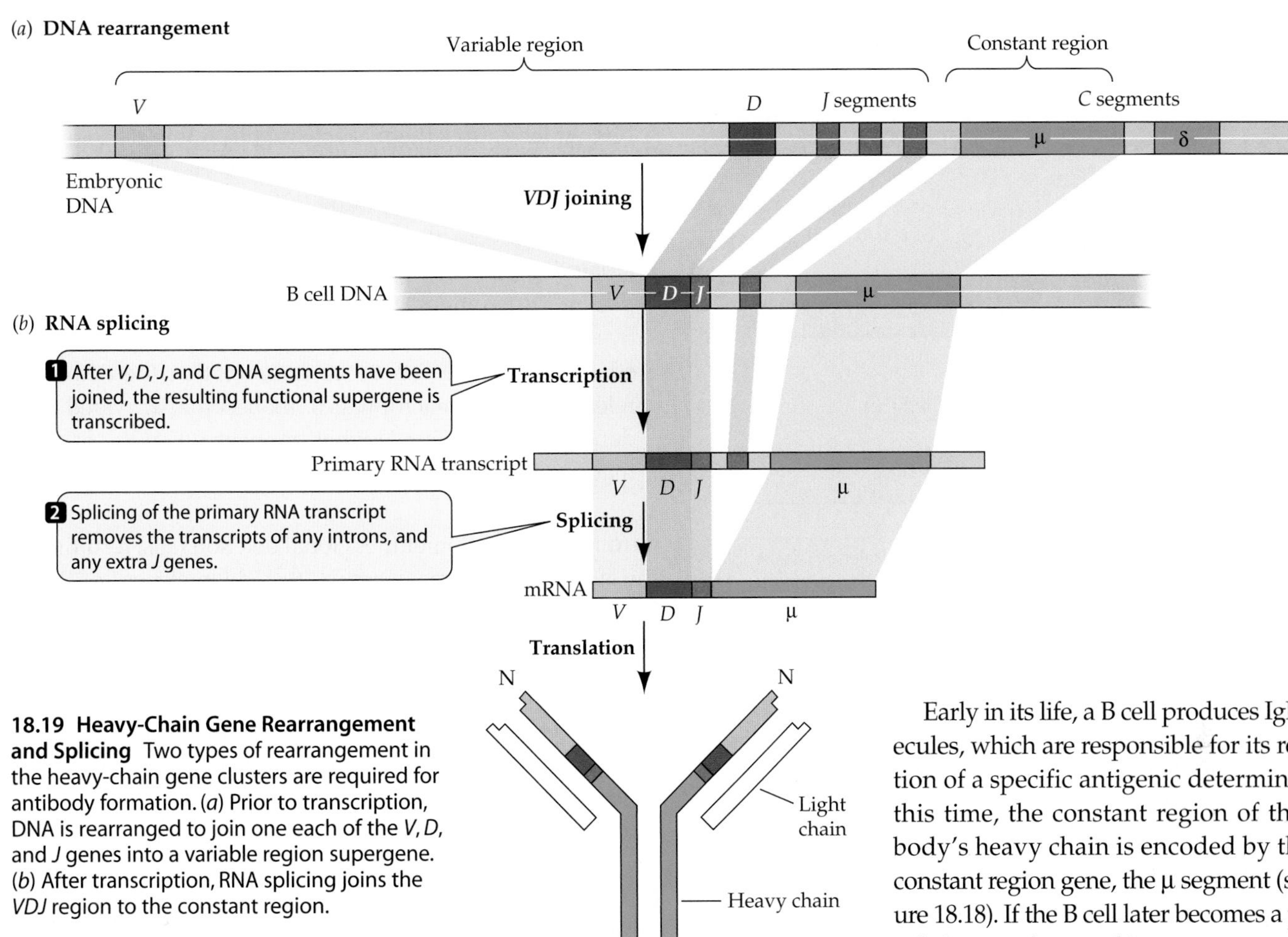

18.19 Heavy-Chain Gene Rearrangement and Splicing Two types of rearrangement in the heavy-chain gene clusters are required for antibody formation. (*a*) Prior to transcription, DNA is rearranged to join one each of the *V*, *D*, and *J* genes into a variable region supergene. (*b*) After transcription, RNA splicing joins the *VDJ* region to the constant region.

When we add these possibilities to the billions of combinations that can be made by random DNA rearrangements, it is not surprising that the immune system can mount a response to almost any natural or artificial substance.

Once this *pretranscriptional* processing in completed, pre-mRNA can be transcribed from each supergene. Posttranscriptional processing removes the remaining introns, so that the mature mRNA contains a continuous coding sequence for an immunoglobulin light chain or heavy chain. Translation then produces the polypeptide chains, which combine to form an active antibody protein.

This genetic system is capable of still other kinds of changes, as seen when a B cell or plasma cell switches the immunoglobulin class it produces, but retains its antibody specificity.

The constant region is involved in class switching

In Table 18.3, we described the different classes of antibodies and their functions. Generally, a B cell makes only one antibody class at a time. But **class switching** can occur, in which a B cell changes which antibody class it synthesizes. For example, a B cell making IgM can switch to making IgG.

Early in its life, a B cell produces IgM molecules, which are responsible for its recognition of a specific antigenic determinant. At this time, the constant region of the antibody's heavy chain is encoded by the first constant region gene, the μ segment (see Figure 18.18). If the B cell later becomes a plasma cell during a humoral immune response, another deletion commonly occurs in the cell's DNA, positioning the heavy-chain variable region gene (consisting of the same *V*, *D*, and *J* segments) next to a constant region gene farther down the original DNA, such as the γ, ε, or α genes (Figure 18.20). Such a DNA deletion results in the production of an antibody with a different constant region of the heavy chain, and therefore a different function. However, the antibody produced has *the same variable regions of the light and heavy chains*, and therefore the same antigen specificity, as before. The new antibody falls into one of the four other immunoglobulin classes (IgA, IgD, IgE, or IgG), depending on which of the constant region genes is placed adjacent to the variable region gene.

After switching classes, the plasma cell cannot go back to making the previous immunoglobulin class, because that part of the DNA has been lost. On the other hand, if additional constant region segments are still present, the cell may switch classes again.

What triggers class switching, and what determines the class to which a given B cell will switch? T_H cells direct the course of an immune response and determine the nature of the attack on the antigen. These T cells induce class switching by sending cytokine signals. The cytokines bind to receptors on the target B cells, generating a signal transduction cascade that results in altered transcription of the immunoglobulin genes.

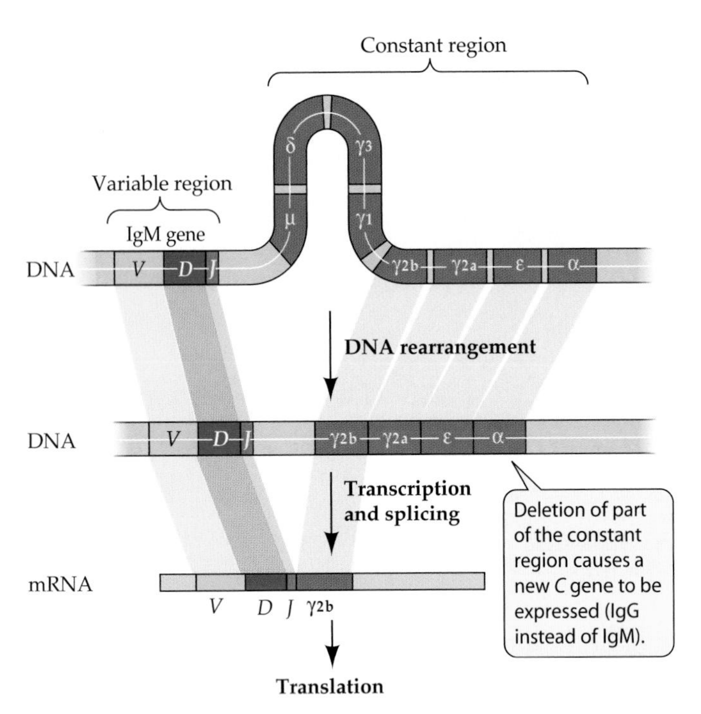

18.20 Class Switching The supergene produced by joining *V*, *D*, *J*, and *C* segments (see Figure 18.19) may later be modified, causing a different C region to be transcribed. This modification, known as class switching, is accomplished by deletion of part of the constant region gene cluster. Shown here is class switching from IgM to IgG.

By now, you can see that the normal functioning of the immune system involves many complex interactions between molecules and cells. We now turn to several situations in which one or more components of this complex system malfunction.

Disorders of the Immune System

Sometimes the immune system fails us in one way or another. It may overreact, as in an allergic reaction; it may attack self antigens, as in an autoimmune disease; or it may function weakly or not at all, as in an immune deficiency disease. After a look at allergies and autoimmune diseases, we will examine the acquired immune deficiency that characterizes AIDS.

HYPERSENSITIVITY. A common type of condition arises when the human immune system overreacts to (is *hypersensitive* to) a dose of antigen. Although the antigen itself may present no danger to the host, the inappropriate immune response may produce inflammation and other symptoms, which can cause serious illness or death. Allergic reactions are the most familiar examples of this phenomenon. There are two types of allergic reactions:

- *Immediate hypersensitivity* occurs when an individual makes large amounts of IgE that react with an antigen in a food, pollen, or the venom of an insect. When this happens, mast cells in tissues and basophils in blood bind the IgE, which causes them to release histamine. The result is symptoms such as dilation of blood vessels, inflammation, and difficulty breathing. If not treated with antihistamines, a severe allergic reaction can lead to death.
- *Delayed hypersensitivity* does not begin until hours after exposure to an antigen. In this case, the antigen is processed by antigen-presenting cells and a T cell response is initiated. The response can be so massive that the cytokines released cause macrophages to become activated and damage tissues. That is what happens when the bacteria that cause tuberculosis colonize the lungs.

AUTOIMMUNITY. Sometimes clonal deletion fails, resulting in the appearance of one or more "forbidden clones" of B and T cells directed against self antigens. This *autoimmunity* does not always result in disease, but in some instances it can.

- People with *systemic lupus erythematosis* (SLE) have antibodies to many cellular components, including DNA and nuclear proteins. These antinuclear antibodies can cause serious damage when they bind to normal tissue antigens to form large circulating antigen–antibody complexes, which become stuck in tissues and provoke inflammation.
- People with *rheumatoid arthritis* have difficulty in shutting down a T cell response. We mentioned earlier that the inhibitor CTLA4 blocks T cells from reacting to self antigens. People with rheumatoid arthritis may have low CTLA4 activity, which results in inflammation of joints due to the infiltration of excess white blood cells.
- *Multiple sclerosis* involves both T cell- and B cell-mediated attack on two major proteins in myelin, the material that coats some nervous tissues. It usually affects young adults, causing progressive damage to the nervous system.
- *Insulin-dependent diabetes mellitus*, or type I diabetes, occurs most often in children. It involves an immune reaction against several proteins in the cells of the pancreas that manufacture the protein hormone insulin. This reaction kills the insulin-producing cells, so people with type I diabetes must take insulin daily in order to survive.

The causes of these autoimmune diseases are not known. Analyses of human pedigrees show that they tend to "run in families," indicating a genetic component. Some alleles of MHC II are strongly linked to certain autoimmune diseases. In some cases, the underlying cause may be molecular mimicry, in which T cells that recognize a nonself antigen also recognize something on the self that has a similar structure.

AIDS is an immune deficiency disorder

People are subject to various *immune deficiency disorders*, such as those in which T or B cells never form and others in which

B cells lose the ability to give rise to plasma cells. In either case, the affected individual is unable to mount an immune response and thus lacks a major line of defense against pathogens.

Because of its essential roles in both the humoral and cellular immune responses, the T_H cell is perhaps the most central of all the components of the immune system—a significant cell to lose to an immune deficiency disorder. This cell is the target of **HIV** (*h*uman *i*mmunodeficiency *v*irus), the retrovirus that eventually results in **AIDS** (*a*cquired *i*mmune *d*eficiency *s*yndrome).

HIV is transmitted from person to person several ways:

- Through blood, such as by a needle contaminated with the virus after being used to inject an infected individual
- Through the exposure of broken skin, an open wound, or mucous membranes to body fluids, such as blood or semen, from an infected individual
- Through the blood of an infected mother to her baby during birth

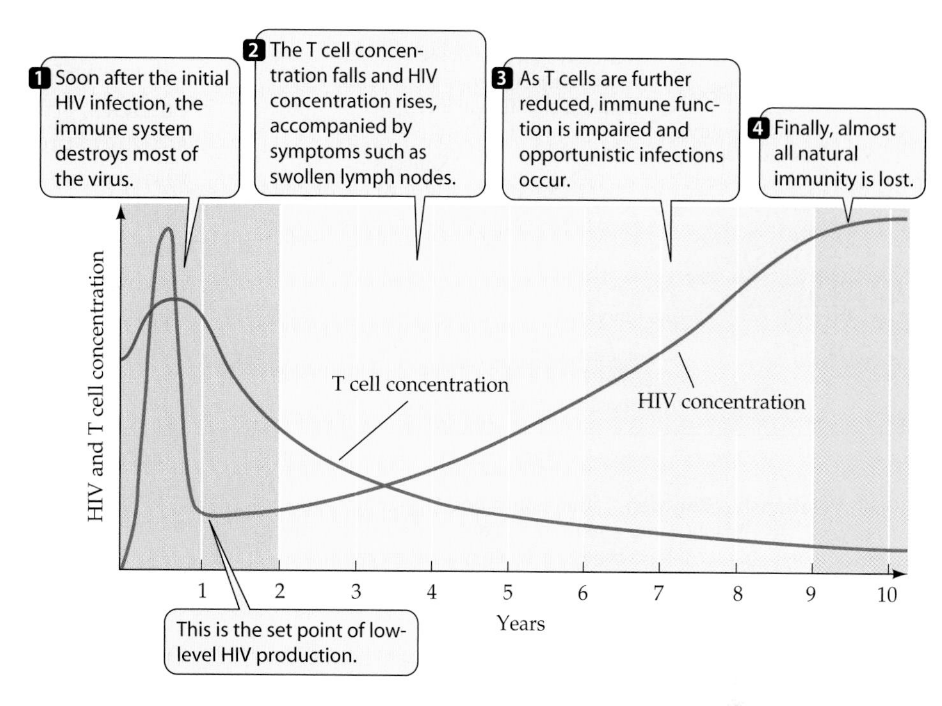

1 Soon after the initial HIV infection, the immune system destroys most of the virus.
2 The T cell concentration falls and HIV concentration rises, accompanied by symptoms such as swollen lymph nodes.
3 As T cells are further reduced, immune function is impaired and opportunistic infections occur.
4 Finally, almost all natural immunity is lost.
HIV and T cell concentration
T cell concentration
HIV concentration
1
2
3
4
5
6
7
8
9
10
Years
This is the set point of low-level HIV production.

18.21 The Course of an HIV Infection HIV infection may be carried, unsuspected, for many years before the onset of symptoms. This long "dormant" period means that the infection is often spread by people who are unaware that they are carrying the virus.

HIV initially infects macrophages, T_H cells, and dendritic cells in blood and tissues. These infected cells carry the virus to the lymph nodes and spleen, where T cells mature and B cells reside.

Normally, the dendritic cells present their captured antigen to T_H cells in the lymph nodes, and this causes the T_H cells to divide and form a clone (see Figure 18.17). But HIV preferentially infects activated, not resting, T_H cells. So the HIV arriving in the lymph nodes proceeds to infect the many activated T_H cells that are already responding to other antigens. These two processes—the transport of the virus to the nodes and the presence in the nodes of cells already receptive to virus infection—combine to ensure that HIV reproduces vigorously. Up to 10 billion viruses are made every day during this initial phase of infection. The numbers of T_H cells quickly drop, and infected people show symptoms similar to mononucleosis, such as enlarged lymph nodes and fever.

These symptoms abate within 3 weeks, however, as T cells recognize infected lymphocytes, an immune response is mounted, and antibodies specific to HIV appear in the blood (Figure 18.21). By this time, the patient has a high level of circulating HIV complexed with antibodies, which is gradually removed by the action of dendritic cells over the next several months. But before they are filtered out, these antibody-complexed viruses can still infect T_H cells that come in contact with them. This secondary infection process reaches a low, steady-state level called the "set point." This point varies among individuals and is a strong predictor of the rate of progression of the disease. For most people, it takes 8–10 years, even without treatment, for the more severe manifestations of AIDS to develop. In some, it can take as little as a year; in others, 20 years. During this dormant period, people carrying HIV generally feel fine, and their T_H cell levels are adequate for them to mount immune responses.

Eventually, however, the virus destroys the T_H cells, and their numbers fall to dangerous levels. At this point, the infected person is considered to have *full-blown AIDS* and is susceptible to infections that the T_H cells would normally eliminate (Figure 18.22). Most notable among these infections are the otherwise rare skin tumor called Kaposi's sarcoma, caused by a herpesvirus; pneumonia, caused by the fungus *Pneumocystis carinii*; and lymphoma tumors, caused by the Epstein-Barr virus. These conditions are called *opportunistic infections* because they take advantage of the crippled immune system of the host. They lead to death within a year or two.

HIV infection and replication occur in T_H cells

As a retrovirus, HIV uses RNA as its genetic material. A central core particle with a protein coat contains two identical copies of the RNA genome as well as the enzymes reverse transcriptase, integrase, and a protease. An envelope, derived from the plasma membrane of the host cell in which the virus

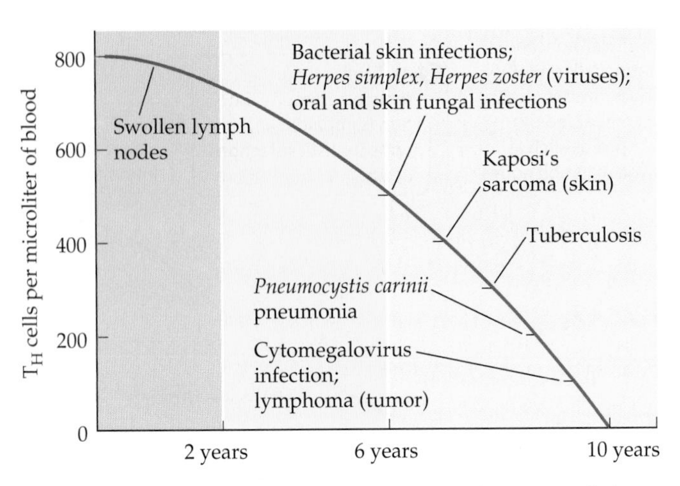

18.22 Relationship Between T_H Cell Count and Opportunistic Infections As HIV kills more and more T_H cells, the immune system is less and less able to defend the body against various pathogens, including many that are not usually infectious to healthy people.

was produced, surrounds the core. The envelope is studded with viral proteins (gp120 and gp41, where "gp" stands for *g*lyco*p*rotein). These virally-coded proteins enable the virus to infect its target cells. Refer to Figure 13.5 for the replication cycle of HIV.

HIV attaches to T_H cells and macrophages via their surface protein CD4, which acts as a receptor for the viral envelope protein gp120. Following binding, the virion enters the cell by membrane fusion. Soon, a cDNA copy of the RNA genome is made via reverse transcriptase. This enzyme lacks the proofreading property of many DNA polymerases, so the errors that inevitably creep into the process are not corrected. Up to 10 incorrect bases out of about 8,000 may end up in each cDNA produced. This is a great advantage to the virus, as genomic mutations allow its proteins to escape the host's immune response; however, the mutations present a challenge to scientists trying to design drugs and vaccines to bind to the constantly changing viral proteins.

The viral core enters the cell nucleus, where integrase catalyzes the insertion of viral cDNA into the host chromosome. The double-helical cDNA thus becomes a permanent part of a T_H cell's DNA, replicating with it at each cell division, and may remain in the T_H cell genome for a decade or more. This provides a molecular camouflage against the defenses of host cells, as well as attempts at therapy.

This latent period ends if the HIV-infected T_H cell becomes activated as it responds naturally to an antigen. The expression of viral genes requires the collaboration of host transcription factors that are made in activated T_H cells and a virally encoded protein called Tat. When the T_H cell is activated, the entire integrated cDNA viral genome is transcribed into RNA, which can either remain as it is or be spliced. Unspliced RNAs become the genomes of new HIV particles; spliced RNAs act as mRNAs to make the viral structural proteins. An important activator of splicing is the viral protein called Rev.

The protease encoded by HIV is needed to complete the formation of individual viral proteins from larger initial products of translation. Packaging domains on viral proteins cause the RNA genomes to fold into them and form core particles. In the meantime, the viral membrane proteins are made on the endoplasmic reticulum of the host cell and transported to the plasma membrane via the Golgi complex. The cytoplasmic tails of the viral gp120 membrane proteins bind to the core particles, and the viruses bud from the infected cell, surrounding themselves with modified plasma membrane from the host.

Treatments for HIV infection rely on knowing its molecular biology

As the AIDS epidemic has grown, so has our knowledge of HIV molecular biology. The general therapeutic strategy is to try to block stages in the viral life cycle without damaging the host cell. Potential therapeutic agents that interfere with the major steps of the life cycle are being tested. Of course, it is crucial to block only steps that are unique to the virus, so that drug therapies do not harm the patient by blocking a step in the patient's own metabolism.

*H*ighly *a*ctive *a*nti*r*etroviral *t*herapy (HAART) was developed in the late 1990s and has had considerable success in delaying the onset of AIDS symptoms in people infected with HIV by 3 years or more, and in prolonging the lives of people with AIDS. The logic of HAART comes from cancer treatment: Employ a combination of drugs acting at different parts of the viral life cycle. Generally, the HAART regimen uses a protease inhibitor and two reverse transcriptase inhibitors.

These drug regimens have had such dramatic effects on patients that they may eliminate HIV entirely in some people, especially in those treated within the first few days after infection, before the virus has arrived in the lymph nodes. Most patients, however, face a lifetime of anti-HIV therapy.

Unfortunately, 80 percent of the patients who take HAART develop mutant strains of HIV that are resistant to this regimen; there is a never-ending race to modify HAART by adding new and/or different drug combinations. There are now 140 different HAART treatments. In short, we seem trapped in an evolutionary struggle: How can we gain a lasting advantage, short of bringing the virus to extinction?

The greatest hope is for the development of a vaccine against HIV. The first major clinical trial of such a vaccine (one directed against the HIV membrane protein, gp120) was not successful, but other vaccines are under development.

What can be done until biomedical science provides the tools to bring the worldwide AIDS epidemic to an end? Above all, people must recognize that they are in danger whenever they have sex with a partner whose total sexual history is not known. The danger rises as the number of sex partners rises, and the danger is much greater if partners participating in sexual intercourse are not protected by a latex condom. The danger that heterosexual intercourse will transmit HIV rises tenfold to a hundredfold if either partner has another sexually transmitted disease.

Chapter Summary

Animal Defense Systems

▶ Animals defend themselves against pathogens by both nonspecific (innate) and specific means.

Defensive Cells and Proteins

▶ Many of our defenses are implemented by cells and proteins carried in the bloodstream and in the lymphatic system. **Review Figure 18.1.** **See Web/CD Activity 18.1**

▶ White blood cells, including lymphocytes (B and T cells) and phagocytes (such as neutrophils and macrophages), play many defensive roles. **Review Figure 18.2.** **See Web/CD Tutorial 18.1**

Nonspecific Defenses

▶ An animal's nonspecific defenses include physical barriers, competing resident microorganisms, and local agents, such as secretions that contain an antibacterial enzyme. **Review Table 18.1**

▶ The inflammation response uses several cells and proteins. Activated mast cells release histamine, which causes blood capillaries to leak and inflame. Complement proteins attract macrophages to the site, where they engulf bacteria and dead cells. **Review Figure 18.4.** **See Web/CD Activity 18.2**

▶ A cell signaling pathway involved the toll receptor stimulates the defense response. **Review Figure 18.5**

Specific Defenses: The Immune Response

▶ Four features characterize the immune response: specificity, the ability to respond to an enormous diversity of antigens, the ability to distinguish self from nonself, and memory.

▶ The immune response is directed against antigens that evade the nonspecific defenses. Each antibody or T cell is directed against a particular antigenic determinant. **Review Figure 18.6**

▶ There are two interactive immune responses: the humoral immune response and the cellular immune response. The humoral immune response employs antibodies secreted by B cells to target antigens in body fluids. The cellular immune response employs T cells to attack body cells that have been altered by viral infection or mutation or to target antigens that have invaded the body's cells.

▶ Clonal selection accounts for the rapidity, specificity, and diversity of the immune response as well as immunological memory and tolerance to self. **Review Figure 18.7**

▶ Immunological memory plays roles in both natural immunity and artificial immunity based on vaccination. **Review Figure 18.8, Table 18.2**

B Cells: The Humoral Immune Response

▶ Activated B cells form plasma cells, which synthesize and secrete specific antibodies.

▶ The basic unit of an antibody, or immunoglobulin, is a tetramer of four polypeptides: two identical light chains and two identical heavy chains, each consisting of a constant and a variable region. **Review Figure 18.10.** **See Web/CD Activity 18.3**

▶ The variable regions of the light and heavy chains collaborate to form the antigen-binding sites of a specific antibody. Each antigen usually has several different antigenic determinants (binding sites for specific antibodies). The variable regions determine each antibody's specificity for a determinant; the constant region determines the destination and function of the antibody.

▶ There are five immunoglobulin classes. IgM, formed first, is a membrane receptor on B cells, as is IgD. IgG is the most abundant antibody class and performs several defensive functions. IgE takes part in inflammation and allergic reactions. IgA is present in various body secretions. **Review Table 18.3**

▶ Monoclonal antibodies consist of identical immunoglobulin molecules directed against a single antigenic determinant. **Review Figure 18.12**

See Web/CD Tutorial 18.2

T Cells: The Cellular Immune Response

▶ The cellular immune response is directed against altered or infected cells of the body. T_C cells attack virus-infected or tumor cells, causing them to lyse. T_H cells activate B cells and influence the development of other T cells and macrophages. **Review Figure 18.13**

▶ T cell receptors in the cellular immune response are analogous to immunoglobulins in the humoral immune response.

▶ The major histocompatibility complex (MHC) encodes many membrane proteins. MHC molecules in macrophages, B cells, or body cells bind processed antigen and present it to T cells. **Review Figures 18.15, 18.16**

▶ In the cellular immune response, class I MHC molecules, T_C cells, CD8, and cytokines collaborate to activate T_C cells with the appropriate specificity. **Review Figure 18.17.** **See Web/CD Tutorial 18.4**

▶ Developing T cells undergo two tests: They must be able to recognize self MHC molecules, and they must *not* bind to both self MHC and any of the body's own antigens. T cells that fail either of these tests die.

▶ The rejection of organ transplants results from the genetic diversity of MHC molecules.

See Web/CD Tutorial 18.3

The Genetic Basis of Antibody Diversity

▶ Immunoglobulin heavy-chain supergenes are constructed from one each of numerous *V*, *D*, *J*, and *C* segments. The *V*, *D*, and *J* segments combine by DNA rearrangement, and transcription yields an RNA molecule that is spliced to form a translatable mRNA. Other gene families give rise to the light chains. **Review Figures 18.18, 18.19**

▶ As a result of these DNA rearrangements, there are millions of possible antibodies as a result of these DNA combinations. Imprecise DNA rearrangements, mutations, and random addition of bases to the ends of the DNAs before they are joined contribute even more diversity.

▶ Class switching after initial immunoglobulin production results in antibodies with the same antigen specificity but a different function. It is accomplished by cutting and rejoining of the genes encoding the constant region. **Review Figure.18.20**

See Web/CD Tutorial 18.5

Disorders of the Immune System

▶ Allergies result from an overreaction of the immune system to an antigen.

▶ Autoimmune diseases result from a failure in the immune recognition of self, with the appearance of antiself B and T cells that attack the body's own cells.

▶ Immune deficiency disorders result from failures of one or another part of the immune system. AIDS is an immune deficiency disorder arising from depletion of the body's T_H cells as a result of infection with HIV. Depletion of the T_H cells weakens and eventually destroys the immune system, leaving the host defenseless against "opportunistic" infections. **Review Figures 18.21, 18.22**

▶ HIV inserts a copy of its genome into a chromosome of a macrophage or T_H cell, where it may lie dormant for years. When the viral genome is transcribed and translated, new viruses form.

▶ Currently the most effective drugs to treat HIV are those directed against reverse transcriptase and protease.

▶ Some treatments may provide a dramatic reduction in HIV levels, but there is as yet no indication that we can prevent infection with HIV, as by vaccination. The only strategy currently available is for people to avoid behaviors that place them at risk.

Self-Quiz

1. Phagocytes kill harmful bacteria by
   a. endocytosis.
   b. producing antibodies.
   c. complement.
   d. T cell stimulation.
   e. inflammation.
2. Which statement about immunoglobulins is *true*?
   a. They help antibodies do their job.
   b. They recognize and bind antigenic determinants.
   c. They encode some of the most important genes in an animal.
   d. They are the chief participants in nonspecific defense mechanisms.
   e. They are a specialized class of white blood cells.
3. Which statement about an antigenic determinant is *not* true?
   a. It is a specific chemical grouping.
   b. It may be part of many different molecules.
   c. It is the part of an antigen to which an antibody binds.
   d. It may be part of a cell.
   e. A single protein has only one on its surface.
4. T cell receptors
   a. are the primary receptors for the humoral immune system.
   b. are carbohydrates.
   c. cannot function unless the animal has previously encountered the antigen.
   d. are produced by plasma cells.
   e. are important in combating viral infections.
5. According to the clonal selection theory,
   a. an antibody changes its shape according to the antigen it meets.
   b. an individual animal contains only one type of B cell.
   c. the animal contains many types of B cells, each producing one kind of antibody.
   d. each B cell produces many types of antibodies.
   e. many clones of antiself lymphocytes appear in the bloodstream.
6. Immunological tolerance
   a. depends on exposure to antigen.
   b. develops late in life and is usually life-threatening.
   c. disappears at birth.
   d. results from the activities of the complement system.
   e. results from DNA splicing.
7. The extraordinary diversity of antibodies results in part from
   a. the action of monoclonal antibodies.
   b. the splicing of protein molecules.
   c. the action of cytotoxic T cells.
   d. the rearrangement of gene segments.
   e. their remarkable nonspecificity.
8. Which of the following play(s) no role in the antibody response?
   a. Helper T cells
   b. Interleukins
   c. Macrophages
   d. Reverse transcriptase
   e. Products of class II MHC gene loci
9. The major histocompatibility complex
   a. codes for specific proteins found on the surface of cells.
   b. plays no role in T cell immunity.
   c. plays no role in antibody responses.
   d. plays no role in skin graft rejection.
   e. is encoded by a single locus with multiple alleles.
10. Which of the following plays no role in HIV reproduction?
   a. Integrase
   b. Reverse transcriptase
   c. gp120
   d. Interleukin-1
   e. Protease

For Discussion

1. Describe the part of an antibody molecule that interacts with an antigenic determinant. How is it similar to the active site of an enzyme? How does it differ from the active site of an enzyme?
2. Contrast immunoglobulins and T cell receptors with respect to their structure and function.
3. Discuss the diversity of antibody specificities in an individual in relation to the diversity of enzymes. Does every cell in an animal contain genetic information for all the organism's enzymes? Does every cell contain genetic information for all the organism's immunoglobulins?
4. The gene family determining MHC on the cell surface in humans is on a single chromosome. A father's MHC type is A1, A3, B5, B7, D9, D11. A mother's phenotype is A2, A4, B6, B7, D11, D12 Their child is A1, A4, B6, B7, D11, D12. What are the parents' haplotypes—that is, which alleles are linked on the diploid chromosomes of each parent? Assuming there is no recombination among the genes determining the MHC type, can these same two parents have a child who is A1, A2, B7, B8, D10, D11?

LIFE ESSAY

What are the ethical issues surrounding genetic modification of nature?

- by Gary Comstock -

Ethical concerns are raised whenever we undertake experiments that may harm sentient individuals. In Chapter 16 we learned that biotechnology can produce genetically modified organisms (GMOs) with the potential to help farmers and consumers. But GMOs also present the possibility of harm to humans and animals. Ethical review of the technology is therefore appropriate.

In 1985, scientists in Maryland inserted a gene responsible for the production of human growth hormone into 19 hog embryos. When the embryos came to term the first nonhuman mammals incorporating a human gene were born. The experiment was conducted to discover ways to grow pigs to market weight more quickly and with less fat.

Had they been realized, these results would have benefited farmers and consumers. However, none of the transgenic hogs grew any faster than the controls, and all were quite sick. Many suffered from renal disease and arthritis, others had decreased immune function. All were sterile.

Are recombinant DNA experiments justified whenever any good might come of them? What if harmful consequences are possible? How can we accurately measure the impact? How much pain and suffering should we allow in research animals in our efforts to produce leaner pork cutlets?

Ethics is the academic discipline that studies such questions. With a history behind them as old as biology, and a set of tools and methods they have developed aggressively in recent decades, ethicists distinguish two types of moral concerns: consequentialist and intrinsic.

Consequentialist questions direct attention to the likely outcomes (positive and negative) of agricultural biotechnology. A standard method for assessing such questions is known as *utilitarianism*—the theory that an action is justified whenever good consequences outweigh bad. But how can we know when good outweighs bad? How can we measure the pain a hog experiences when subjected to genetic engineering against the potential pleasure a consumer might derive from cheaper pork? To answer these questions accurately will require much more detailed empirical information and rigorous moral reflection.

The second type of ethical question focuses on the *intrinsic nature* of an act, asking whether certain types of activities are wrong regardless of any benefits that may follow. Two intrinsic theories are *divine command* and *moral rights*. Divine command theorists assert that God has forbidden certain behavior. For instance, according to Prince Charles of Great Britain, Christian moral theology rules out genetic engineering on intrinsic grounds because it is an act that should be left to God alone. Humans are not wise enough to engineer nature.

But we may ask, Why should God outlaw an activity that may bring relief to suffering humans? And why would God object to biotechnology at the molecular level if God has not objected to biotechnology at the organismic level (that is, traditional selective breeding in agriculture)?

Moral rights theorists, on the other hand, ask whether a proposed action might violate basic human rights. Perhaps we could achieve dramatic decreases in the price of cotton by enslaving people to raise it. But slavery is *intrinsically wrong*. Reducing cotton's price is a good, but not a good we can justify if it entails denying humans basic liberties.

Gary Comstock is Professor of Philosophy and Director of the Ethics Program at North Carolina State University. He wrote *Vexing Nature? On the Ethical Case against Agricultural Biotechnology*, and edited *Life Science Ethics* and *Religious Autobiographies*. A popular speaker, he has lectured across Europe and in Asia, the Middle East, and New Zealand.

Suppose we could guarantee that GMOs will have only positive consequences. Might it be the case, nonetheless, that we should not use GMOs on the grounds that the very act of genetically engineering organisms violates human rights? How would such an argument go? Perhaps something like this: Humans have a moral right to a natural environment free of novel transgenes synthesized by engineers in labs. GMOs inevitably "pollute" the environment and therefore violate human rights. Therefore, we should not pursue GMOs. But is this a good argument?

Modern biology clearly confronts us with profound ethical puzzles. As specialized research advances, so too must democratic discussion and principled deliberation. We stand the best chance of making wise decisions if we insist on two values: an engaged and educated public, and personal integrity. In sum, we must have leaders and societies demonstrating the courage to nurture new technologies that promise to ameliorate human suffering, while avoiding technologies that threaten animals and the environment.

Discussion Questions

1. Scientists have already inserted human genes into pigs. Because genes are largely interchangeable, we also have the ability to insert pig genes into humans. Do you think this is morally permissible? Are your reasons for your answer consequentialist or intrinsic (or both)?
2. Caution is usually a virtue in ethics, and no sound ethical conclusion can rest on faulty science. Is there any additional information you would like to acquire before you answer Question 1?
3. Chapter 16 mentions the benefits of "golden rice"—transgenic rice enriched with β-carotene. Can you think of any possible harmful consequences of this proposed technology?

Web Links

North Carolina State University Ethics Program http://ncsuethics.org

Ethics Updates http://ethics.acusd.edu/

Pew Initiative on Food and Biotechnology http://pewagbiotech.org/

Golden Rice Case Study http://www.biotech.iastate.edu/publications/case_studies/golden_rice/

19 Differential Gene Expression in Development

The banteng is a relative of domestic cattle that lives in the dense jungles of Asia and grazes in adjacent open grasslands. A combination of hunting by humans, loss of its habitat to domestic cattle, and disease has reduced the number of bantengs to fewer than 8,000, mostly in small herds on the island of Java, Indonesia. Restoring this endangered species, in terms of both its numbers and its genetic diversity, is a major challenge.

Working with a large zoo and a biotechnology company, developmental biologists are trying to rescue the banteng using cloning techniques. Over 25 years ago, geneticists at the San Diego Zoo began freezing cells from endangered species, creating a modern-day Noah's Ark in anticipation of the emergence of new knowledge about animal development and its application to reproductive cloning. In 2002, some banteng cells were thawed. Their nuclei were removed and fused into enucleated eggs from their domestic cow relatives. These eggs were then implanted in the uteri of domestic cows. Two of the cows gave birth to banteng clones, one of which has survived. The clone has the genetic characteristics of the nuclear donor.

Meanwhile, in China, scientists are using rabbits as surrogate mothers for cloned pandas, which are only about 6 cm long when born. Behind all of this amazing reproductive technology lies a great deal of basic developmental genetics.

Much of our knowledge of the molecular biology of development has come from studies on certain model organisms such as the fruit fly *Drosophila melanogaster*, the nematode *Caenorhabditis elegans*, frogs, sea urchins, and a flowering plant, the thale cress, *Arabidopsis thaliana*. As we saw in Chapter 14, the genomes of all eukaryotes are surprisingly similar, and the cellular and molecular principles underlying their development also turn out to be similar. Thus, discoveries from one organism aid us in understanding other organisms, including ourselves.

Two major principles have emerged from studies of development, and both principles are vital to the cloning of endangered species. The first is that, in most cases, all types of *somatic cells*—all of an organism's body cells except the gametes—retain all of the genes that were present in the fertilized egg, or zygote. In

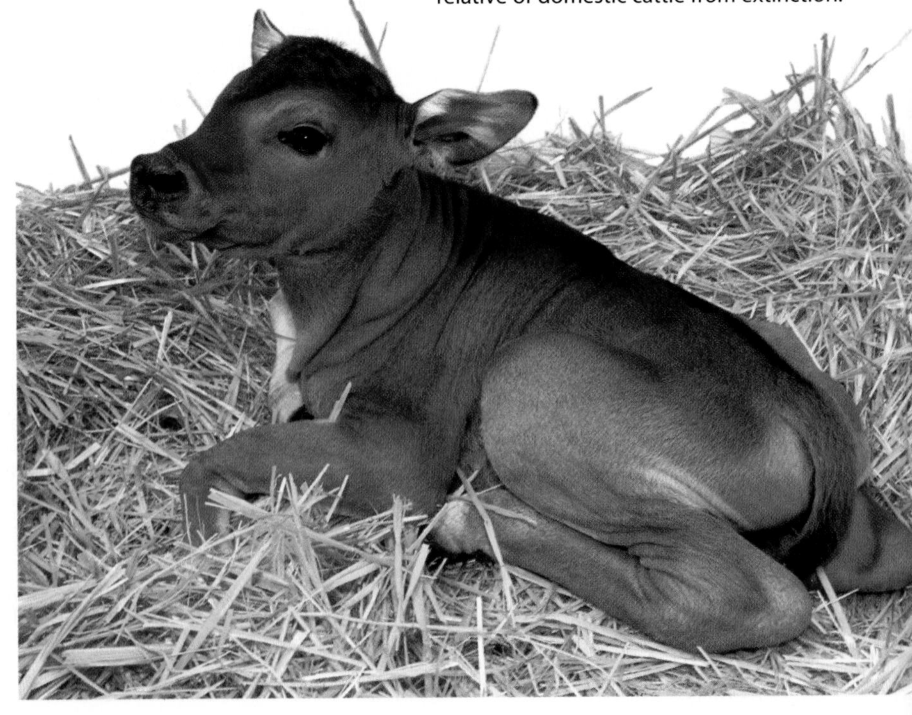

A Newborn Banteng Clone Cloning by nuclear transfer may save this endangered relative of domestic cattle from extinction.

other words, cell differentiation does not usually result from a loss of DNA. The second principle is that cellular changes during development and cell differentiation result from differential expression of genes. During development, the various mechanisms of transcriptional and translational control described in Chapter 14 and the signaling mechanisms described in Chapter 15 work together to produce a complex organism. In this chapter, we will see how these principles apply to normal development as well as to cloning.

The Processes of Development

Development is a process in which an organism undergoes a series of progressive changes, taking on the successive forms that characterize its life cycle (Figure 19.1). In its earliest stages of development, a plant or animal is called an **embryo**. Sometimes the embryo is contained within a protective structure, such as a seed coat, an eggshell, or a uterus. An embryo does not photosynthesize or feed actively; instead, it obtains its food from its mother directly or indirectly (by way of nutrients stored in the seed or egg). A series of embryonic stages may precede the birth of the new, independent organism. Most organisms continue to develop throughout their life cycle; development ceases only with death.

Development consists of growth, differentiation, and morphogenesis

Three processes are responsible for the developmental changes an organism undergoes during its life cycle. **Growth** (increase in size) occurs through cell division and cell expansion. In all multicellular organisms, repeated mitotic divisions generate the multicellular body. In plants, cell expansion begins shortly after the first divisions of the fertilized egg. In animals, on the other hand, cell expansion is often slow to begin: The animal embryo may consist of thousands of cells before it becomes larger than the original fertilized egg. Growth continues throughout the individual's life in some species, but reaches a more or less stable end point in others.

Differentiation is the generation of cellular specializations; that is, differentiation defines the specific structure and function of a cell. Mitosis produces daughter nuclei that are chromosomally and genetically identical to the nucleus that divides to produce them. However, the cells of a multicellular organism are obviously not all identical in structure or

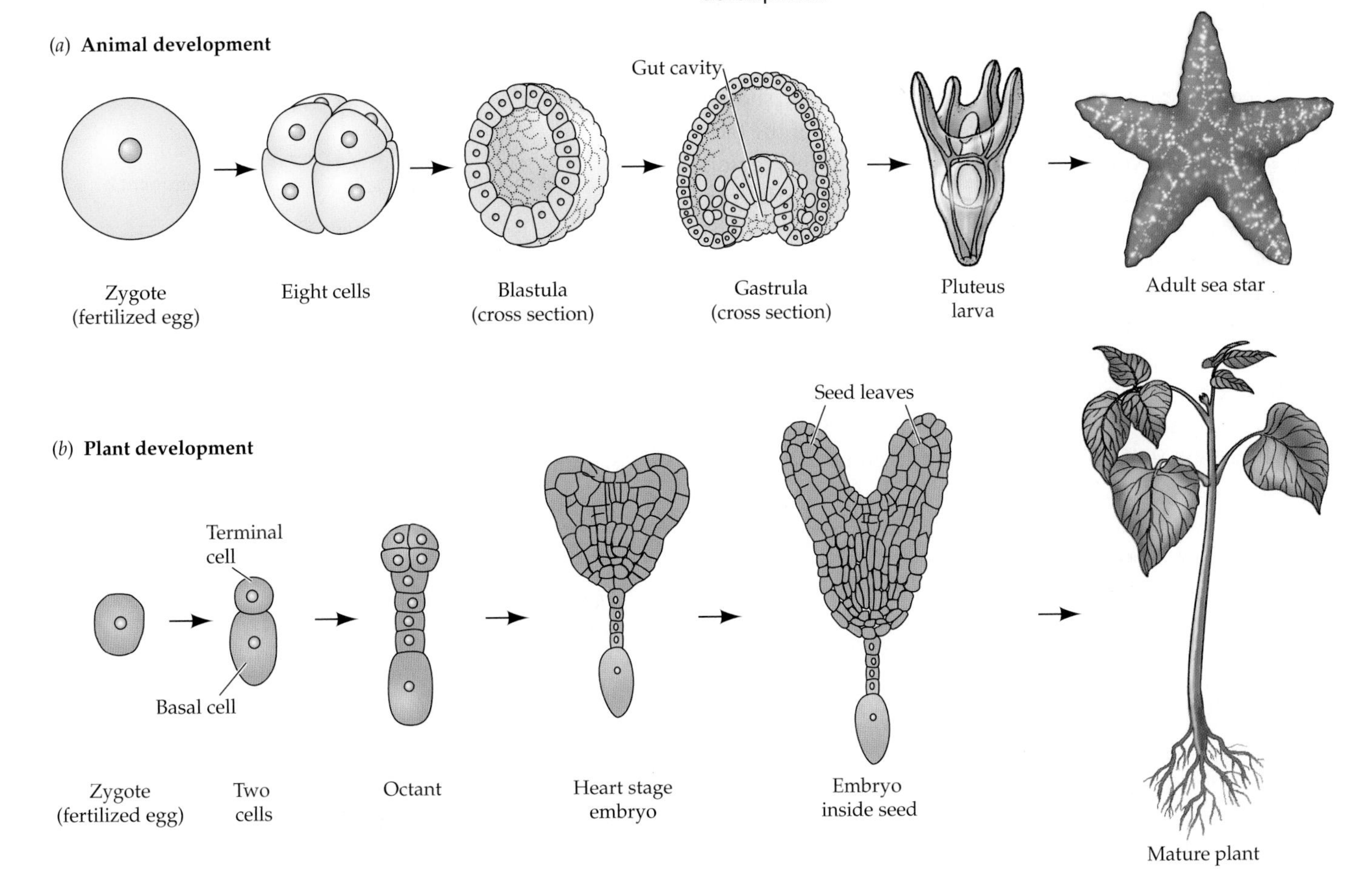

19.1 Stages of Development Stages of development from embryo to adult are shown for a plant and an animal. Growth, differentiation, and morphogenesis are all part of the complex process of development.

function. This apparent contradiction results from regulation of the expression of various parts of the genome. When the embryo consists of only a few cells, each cell has the potential to develop in many different ways. As development proceeds, however, the possibilities available to individual cells gradually narrow, until each cell's fate is fully determined and the cell has differentiated.

Morphogenesis (literally, "creation of form") is the shaping of the multicellular body and its organs. Morphogenesis results from *pattern formation*, the organization of differentiated tissues into specific structures. In plant development, cells are constrained by cell walls and do not move around the body, so organized division and expansion of cells are the major processes that build the plant body. In animals, cell movements are very important in morphogenesis. And in both plants and animals, programmed cell death is essential to orderly development. Like differentiation, morphogenesis results ultimately from the regulated activities of genes and their products, as well as from the interplay of extracellular signals and their transduction in target cells.

As development proceeds, cells become more and more specialized

Experiments in which specific cells of an early embryo were marked with stains have revealed which adult structures are derived from which parts of the embryo. These stained embryos produce what are known as *fate maps*. For instance, we know that the green-shaded area of the frog embryo shown in Figure 19.2 normally becomes part of the skin of the tadpole larva. However, if we cut out a piece from this region and transplant it to another location on another early frog embryo, it does not become skin. The type of tissue it does become is determined by its new environment. The developmental potential of these early embryonic cells—that is, their range of possible fates—is thus greater than their actual fate, which is limited to the cell type that normally develops.

Does embryonic tissue retain its broad developmental potential? Generally speaking, the answer is no. The developmental potential of cells becomes restricted fairly early in normal development. Tissue from a later-stage frog embryo, for example, if taken from a region fated to develop into the brain, becomes brain tissue even if transplanted to a part of an early-stage embryo destined to become another structure.

The cells of the later-stage embryo are thus said to be *determined:* Their fate has been sealed, regardless of their surroundings. By contrast, the cells of the younger tissue transplant in Figure 19.2 have not yet become determined.

Determination, the commitment of a cell to a particular fate, is a process influenced by the action of the extracellular

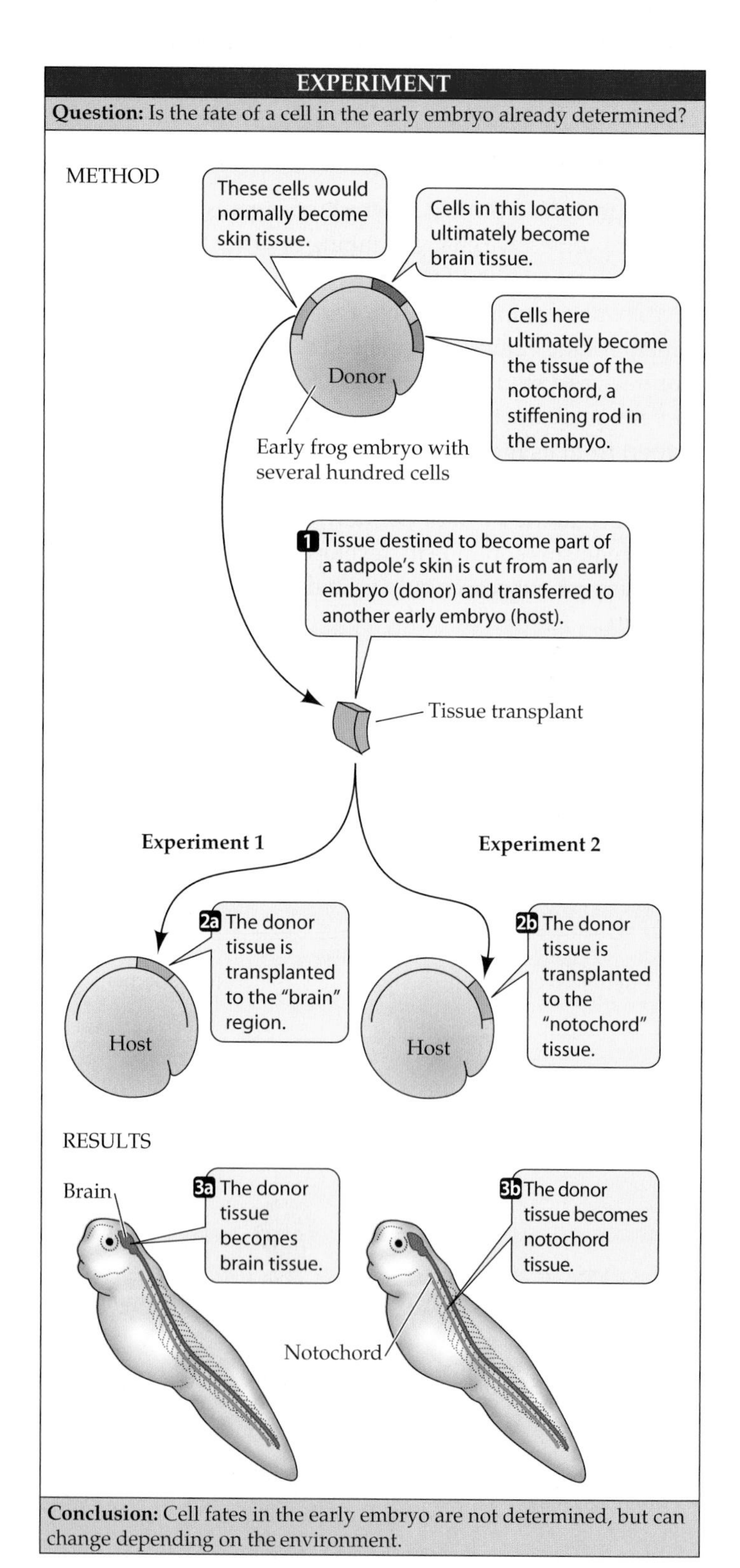
EXPERIMENT
Question: Is the fate of a cell in the early embryo already determined?
METHOD
These cells would normally become skin tissue.
Cells in this location ultimately become brain tissue.
Cells here ultimately become the tissue of the notochord, a stiffening rod in the embryo.
Donor
Early frog embryo with several hundred cells
1 Tissue destined to become part of a tadpole's skin is cut from an early embryo (donor) and transferred to another early embryo (host).
Tissue transplant
Experiment 1
Experiment 2
2a The donor tissue is transplanted to the "brain" region.
2b The donor tissue is transplanted to the "notochord" tissue.
Host
Host
RESULTS
Brain
3a The donor tissue becomes brain tissue.
3b The donor tissue becomes notochord tissue.
Notochord
Conclusion: Cell fates in the early embryo are not determined, but can change depending on the environment.

19.2 Developmental Potential in Early Frog Embryos Cells that would be expected to form one kind of tissue can form completely different tissues when they are experimentally moved to another location. In this experiment, epithelial (skin) tissue from an early-stage frog embryo was transplanted from a donor to a host embryo. The tissue that developed in the host tadpole was not skin, but was consistent with the location to which the "skin graft" was transplanted.

environment and the contents of the cell on the cell's genome. Determination is not something that is visible under the microscope—cells do not change their appearance when they become determined. Determination is followed by *differentiation*, the actual changes in biochemistry, structure, and function that result in cells of different types. Differentiation often involves a change in appearance as well as function. Determination is a commitment; the final realization of that commitment is differentiation.

The Role of Differential Gene Expression in Cell Differentiation

Differentiated cells are recognizably different from one another, sometimes visually as well as in their protein products. For example, certain cells in our hair follicles continuously produce keratin, the protein that makes up hair, nails, feathers, and porcupine quills. Other cell types in the body do not produce keratin. In the hair follicle cells, the keratin-encoding gene is transcribed; in most other cells in the body, that gene is not transcribed. Activation of the keratin-encoding gene is a key step in the differentiation of hair follicle cells.

Generalizing from examples like this one, we may say that *differentiation results from differential gene expression*—that is, from the differential regulation of transcription, posttranscriptional events such as mRNA splicing, and translation in different cell types.

Because the fertilized egg, or **zygote**, has the ability to give rise to every type of cell in the adult body, we say it is **totipotent**. Its genome contains instructions for all of the structures and functions that will arise throughout the life cycle. Later in the development of animals, the cellular descendants of the zygote lose their totipotency and become determined. These determined cells then differentiate into specific types of specialized cells.

With differentiation, there is generally no irreversible change in the genome

Differentiation is irreversible in certain types of cells. Examples include the mammalian red blood cell, which loses its nucleus during development, and the tracheid, a water-conducting cell found in many plants. Tracheid development culminates in the death of the cell, leaving only the pitted cell walls that formed while the cell was alive (see Figure 35.9*e*). In both of these extreme cases, the irreversibility of differentiation can be explained by the absence of a nucleus.

Generalizing about the reversibility of differentiation in mature cells that retain functional nuclei is more difficult. We tend to think of plant differentiation as reversible and of animal differentiation as irreversible, but this is not a hard-and-fast rule. Why is differentiation apparently reversible in some cases, such as a plant cutting, but not in others, such as a mammalian limb? At some stage of development, do changes within the nucleus permanently commit a cell to specialization? For both plants and animals, the answer appears to be no. Under the right environmental circumstances, differentiation is reversible in many cells.

TOTIPOTENCY IN PLANTS. A food storage cell in a carrot root normally faces a dark future. It cannot photosynthesize or give rise to new carrot plants. However, if we isolate that cell from the root, maintain it in a suitable nutrient medium, and provide it with appropriate chemical cues, we can "fool" the cell into acting as if it were a fertilized egg. It can divide and give rise to a mass of undifferentiated cells, called a *callus*, and eventually to a complete plant (Figure 19.3). Since the new plant is genetically identical to the somatic cell from which it came, we call the plant a **clone**.

The ability of scientists to clone an entire carrot plant from a differentiated root cell indicates that the cell contains the entire carrot genome and that it can express the appropriate genes in the right sequence. Many types of cells from other plant species show similar behavior in the laboratory. This ability to generate a whole plant from a single cell has been invaluable in agricultural biotechnology (see Chapter 16).

TOTIPOTENCY IN EARLY EMBRYONIC ANIMAL CELLS. Experiments with plants have established that somatic cells are totipotent. A more direct demonstration that all the genetic material is present in somatic cells has come from nuclear transplantation experiments. Such experiments were first done on frogs by Robert Briggs and Thomas King, who asked whether the nuclei of early frog embryos had lost the ability to do what the totipotent zygote nucleus could do. They first removed the nucleus from an unfertilized egg, thus forming an enucleated egg. Then, with a very fine glass tube, they punctured a cell from an early embryo and drew up part of its contents, including the nucleus, which they injected into the enucleated egg. They stimulated the eggs to divide, and many went on to form embryos, tadpoles, and eventually, frogs. These experiments led to two important conclusions:

- No information is lost from the nuclei of cells as they pass through the early stages of embryonic development. This fundamental principle of developmental biology is known as **genomic equivalence**.
- The cytoplasmic environment around a nucleus can modify its fate.

Similar experiments have been performed on rhesus monkeys, in which a single cell can be removed from an 8-cell embryo and fused with an enucleated egg. This *cell fusion* tech-

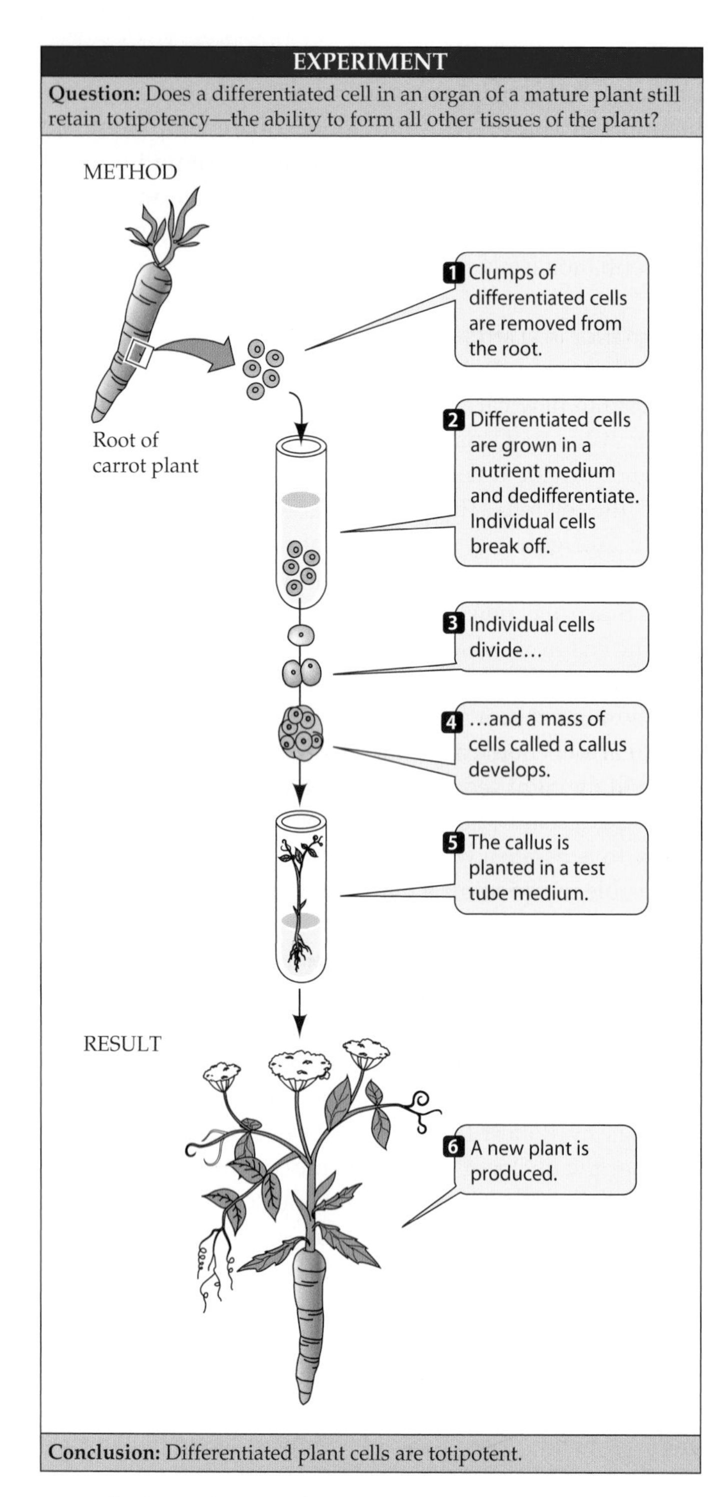

EXPERIMENT
Question: Does a differentiated cell in an organ of a mature plant still retain totipotency—the ability to form all other tissues of the plant?
METHOD
Root of carrot plant
1 Clumps of differentiated cells are removed from the root.
2 Differentiated cells are grown in a nutrient medium and dedifferentiate. Individual cells break off.
3 Individual cells divide...
4 ...and a mass of cells called a callus develops.
5 The callus is planted in a test tube medium.
RESULT
6 A new plant is produced.
Conclusion: Differentiated plant cells are totipotent.

19.3 Cloning a Plant Differentiated, specialized food storage cells from the root of a carrot can be induced by the chemical environment to dedifferentiate. These cells then act like early embryonic cells and form a new plant.

nique causes the nucleus of the embryonic cell to enter the egg cytoplasm. The resulting cell acts like a zygote, forming an embryo, which can be implanted into a foster mother, who ultimately gives birth to a normal monkey. Each of the remaining 7 cells from the original embryo can similarly give rise to offspring by the same cell fusion technique.

In humans, the totipotency of early embryonic cells permits both genetic screening (see Chapter 17) and in vitro fertilization. An 8-cell human embryo can be isolated in the laboratory and a single cell removed and examined to determine whether a harmful genetic condition is present. Each remaining cell, being totipotent, can be stimulated to divide and form an embryo, which can be implanted into the mother's uterus, where it develops into an infant.

TOTIPOTENCY IN ADULT SOMATIC CELLS. Successful cloning of animals was very difficult until the late 1990s, when Ian Wilmut and his colleagues at a biotechnology company in Scotland used the cell fusion procedure to clone sheep (Figure 19.4). Previous attempts to produce mammals by this method had worked, as in the rhesus monkey case, only if the donor nucleus was from an early embryo. Apparently, when mammalian donor cells were in the G2 phase of the cell cycle (see Figure 9.3) and were fused with the cytoplasm of eggs that were also in G2, some extra DNA replication took place that created havoc with the cell cycle in the egg when it attempted to divide.

Wilmut took differentiated cells from a ewe's udder and starved them of nutrients for a week, thus halting the cells in G1 phase of the cell cycle. One of these cells was fused with an enucleated egg from a different breed of ewe. When mitotic inducers in the egg cytoplasm were stimulated, the donor nucleus entered S phase, and the rest of the cell cycle proceeded normally. After several cell divisions, the resulting early embryo was transplanted into the womb of a surrogate mother. Out of 277 successful attempts to fuse adult cells with enucleated eggs, one lamb, named Dolly, survived to be born. DNA analyses confirmed that Dolly was genetically identical to the ewe from whose udder the donor nucleus had been obtained.

A major goal of Wilmut's experiment was to develop a method of cloning sheep that would produce products such as pharmaceuticals in their milk. The cloning procedure could make multiple, identical copies of transgenic sheep that are reliable producers of drugs such as α1-antitrypsin, which is used to treat people with emphysema or cystic fibrosis.

The trick of starving donor cells for cloning has been applied to other mammals. Mice have been cloned using the somatic cells surrounding the egg as a source of donor nuclei (Figure 19.5). Cattle have been cloned to preserve a rare breed in New Zealand. Genetically engineered goats have been cloned to produce several useful proteins in their milk. And, as we described at the beginning of this chapter, cloning is being done to preserve and expand endangered species. A private company has been set up that will clone your pet by nuclear transfer. This flurry of cloning has

19.4 A Clone and Her Offspring In 1996, the experimental procedure described here produced the first cloned mammal, a Dorset sheep named Dolly. Dolly died in 2003 from lung disease, but in her lifetime she mated and gave birth to a "normal" offspring (the lamb on the right in the photo), proving the genetic viability of cloned mammals.

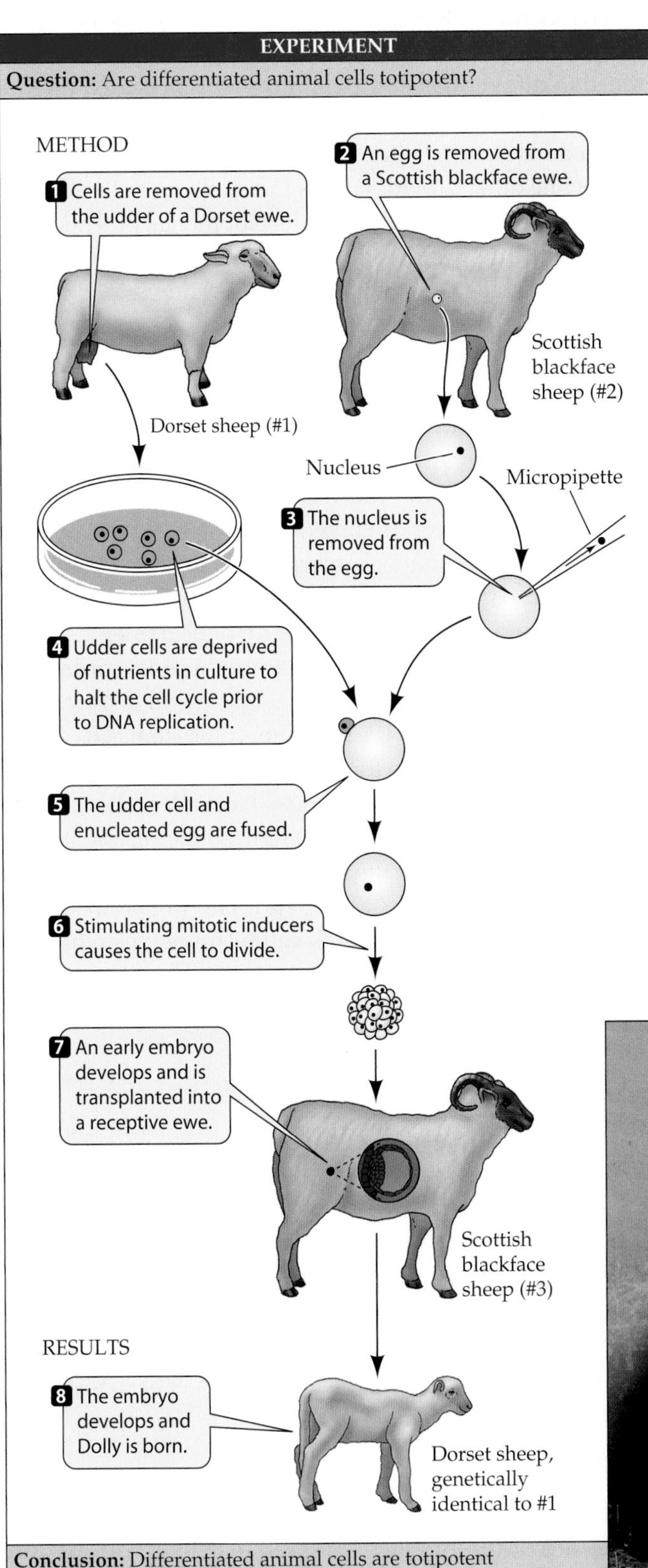

EXPERIMENT
Question: Are differentiated animal cells totipotent?
METHOD
1 Cells are removed from the udder of a Dorset ewe.
2 An egg is removed from a Scottish blackface ewe.
Scottish blackface sheep (#2)
Dorset sheep (#1)
Nucleus
Micropipette
3 The nucleus is removed from the egg.
4 Udder cells are deprived of nutrients in culture to halt the cell cycle prior to DNA replication.
5 The udder cell and enucleated egg are fused.
6 Stimulating mitotic inducers causes the cell to divide.
7 An early embryo develops and is transplanted into a receptive ewe.
Scottish blackface sheep (#3)
RESULTS
8 The embryo develops and Dolly is born.
Dorset sheep, genetically identical to #1
Conclusion: Differentiated animal cells are totipotent in nuclear transplant experiments.

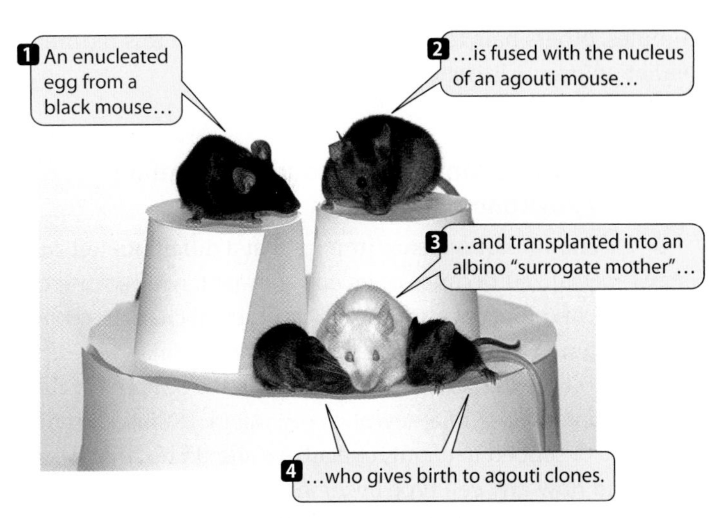

1 An enucleated egg from a black mouse...
2 ...is fused with the nucleus of an agouti mouse...
3 ...and transplanted into an albino "surrogate mother"...
4 ...who gives birth to agouti clones.

19.5 Cloned Mice Because so much is known about mouse genetics and molecular biology, cloned mice may be useful in studies of basic biology.

touched off a flurry of controversy, but cloning is not a new scientific concept. The idea of totipotency was accepted long before Dolly was born, but achieving it is an impressive technical achievement.

An example of nuclear totipotency gone awry occurs in a human tumor called a *teratocarcinoma*. Here, a differentiated cell dedifferentiates to form an unspecialized cell. Then it divides, forming a tumor, as occurs in most cancers. But some cells in the tumor redifferentiate to form specialized tissue arrangements. So the tumor can form a large mass of cells inside the abdomen, with some of the cells forming kidney tubules, others hair, and still others teeth! How this redifferentiation occurs is not clear.

Stem cells can be induced to differentiate by environmental signals

Genomic equivalence implies that a differentiated cell stays specialized because of its environment, not because of its genes, and that appropriate environmental changes could result in a new pattern of differentiation. In normal development, a complex series of signals results in the patterns of differentiation we see in a newborn organism. If these signals could be described in enough detail, we should be able to understand how any cell type becomes any other.

In plants, the growing regions at the tips of the roots and stems contain **meristems**, which are clusters of undifferentiated, rapidly dividing cells. These cells can give rise to the specialized cell types that make up roots and stems, respectively. Plants have many fewer (15–20) cell types than animals (as many as 200). Most plant cell types differ in the structure of their cell walls, whereas most animal cell types have specific cytoplasmic characteristics and many cell-specific proteins.

In mammals, **stem cells** are found in adult tissues that need frequent cell replacement, such as the skin, the inner lining of the intestine, and the blood system. As they divide, stem cells produce cells that differentiate to replace dead cells and maintain tissues. In the body, stem cells have limited abilities to differentiate. The stem cells in bone marrow, for example, produce only the various types of red and white blood cells, while the stem cells in the nervous system produce only the various types of nerve cells.

Can one kind of stem cell be manipulated by its environment to produce cells that differentiate into cells of another tissue type? The answer appears to be yes. For example, when stem cells from the brain were transplanted into the bone marrow of mice whose bone marrow stem cells had been depleted, they proceeded to act like bone marrow stem cells, producing blood cells. In the reverse experiment, bone marrow stem cells were implanted into the brains of mice, where they formed nerve cells. These experiments indicate that some component of the environment—presumably acting through intercellular signals—determines what a stem cell will do.

The stem cell populations that are closest to totipotency are not the ones found in adults, but those of the early embryo. In mice, these embryonic stem cells can be removed from an early embryo (called a *blastocyst*) and then induced to differentiate in some particular way. Normally, these cells are formed a few days after fertilization, and their fate in the developing embryo is soon determined. Before that time, however, they are virtually totipotent. Such cells can be grown indefinitely in the laboratory and, when injected back into a mouse blastocyst, will mix with the resident cells and differentiate to form all the cell types of the mouse. This kind of experiment shows that blastocyst cells do not lose any of their developmental potential while growing in the laboratory.

Embryonic stem cells growing in the laboratory can be induced to differentiate if the right signal is provided (Figure 19.6). For example, treatment of mouse embryonic stem cells with a derivative of vitamin A causes them to form nerve cells, while other growth factors induce them to form blood cells, again demonstrating their developmental potential and the roles of environmental signals. This finding raises the possibility of using stem cell cultures as sources of differentiated cells for clinical medicine. A key advance toward this use has been the ability to grow human embryonic stem cells in the laboratory.

A source for embryonic stem cells could be human embryos made for in vitro fertilization. This medical procedure is used by couples who want a child but cannot conceive naturally. Up to ten eggs are taken from the mother's ovaries and exposed to the father's sperm, with the hope that some early embryos will form. A few of these embryos are then implanted into the mother's uterus for development. Any remaining embryos not used for implantation could be a source of stem cells.

But a problem arises if these embryonic stem cells are induced to differentiate to form a tissue for transplantation—say, pancreatic tissue for a patient with diabetes. The cells and the recipient are genetically different, so the recipient's immune system may reject the transplanted cells . This problem has led to the proposal of **therapeutic cloning**, in which nuclear transplantation and stem cell technologies would be combined. This procedure would require several steps:

- Eggs are removed from a female donor.
- An egg is enucleated.
- A cell is removed from the recipient.
- The entire cell, or its nucleus only, is fused with the enucleated egg.
- The egg is stimulated to divide.
- Embryonic stem cells form; these cells are genetically the recipient's.
- The stem cells are induced to differentiate into the desired tissue for transplantation.

Progress with this ambitious program has been slow but steady. The age of custom-made cells to replace those lost to disease or injury is rapidly approaching.

Genes are differentially expressed in cell differentiation

Nuclear transplantation, cell fusion, and plant cell cloning have demonstrated genomic equivalence in somatic cells of

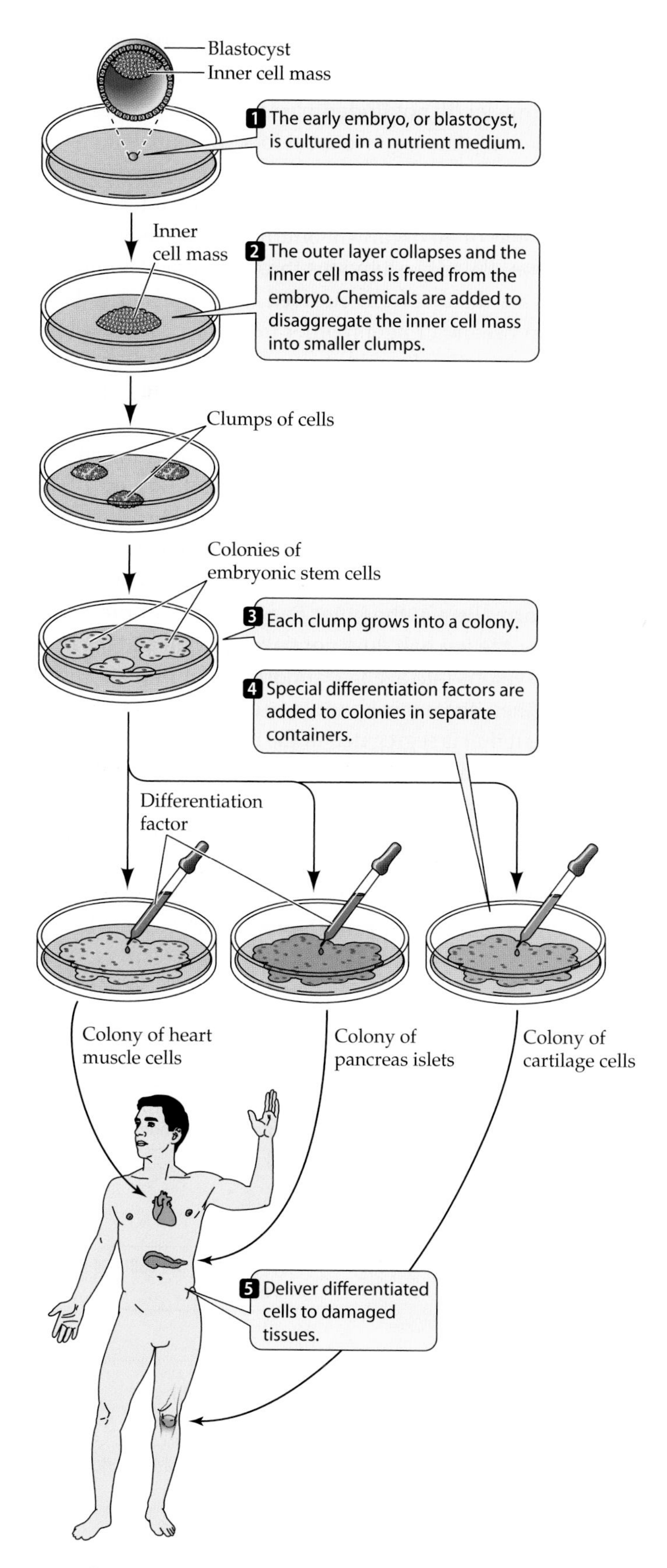
Blastocyst
Inner cell mass
1 The early embryo, or blastocyst, is cultured in a nutrient medium.
Inner cell mass
2 The outer layer collapses and the inner cell mass is freed from the embryo. Chemicals are added to disaggregate the inner cell mass into smaller clumps.
Clumps of cells
Colonies of embryonic stem cells
3 Each clump grows into a colony.
4 Special differentiation factors are added to colonies in separate containers.
Differentiation factor
Colony of heart muscle cells
Colony of pancreas islets
Colony of cartilage cells
5 Deliver differentiated cells to damaged tissues.

19.6 The Potential Use of Embryonic Stem Cells in Medicine Human embryonic stem cells can be cultured in the laboratory and induced to differentiate. Their use as transplants to replace damaged tissue is under intensive investigation.

an organism. Molecular experiments have provided even more convincing evidence. For example, the gene for β-globin, one of the protein components of hemoglobin, is present and expressed in red blood cells as they form in the bone marrow of mammals. Is the same gene also present—but unexpressed—in nerve cells in the brain, which do not make hemoglobin?

Nucleic acid hybridization (see Figure 14.5) can provide an answer. A probe for the β-globin gene can be applied to DNA from both brain cells and immature red blood cells (recall that mature red blood cells lose their nuclei and DNA). In both cases, the probe finds its complement, showing that the β-globin gene is present in both types of cells. On the other hand, if the probe is applied to cellular mRNA, rather than cellular DNA, it finds β-globin mRNA only in the red blood cells, and not in the brain cells. This result shows that the gene is expressed in only one of the two tissues. Many similar experiments have shown convincingly that differentiated cells lose none of the genes that were present in the fertilized egg.

What leads to this differential gene expression? One well-studied example of differentiation is the conversion of undifferentiated muscle precursor cells, called *myoblasts*, into the large, multinucleated *muscle fibers* that make up mammalian skeletal muscles. The key event that starts this conversion is the expression of *MyoD1* (*myo*blast-*d*etermination gene 1). The protein product of this gene is a transcription factor (MyoD1) with a helix-loop-helix domain (see Figure 14.15), which not only binds to the promoters of muscle-determining genes to stimulate their transcription, but also acts on its own promoter to keep its levels high in the myoblasts and in their descendants.

Strong evidence for the controlling role of *MyoD1* in muscle fiber differentiation comes from experiments in which a sequence containing an active promoter adjacent to *MyoD1* is transfected into the precursors of other cell types. For example, if this sequence is added to fat cell precursors, the fat cells are reprogrammed to become muscle cells. Genes such as *MyoD1* that direct fundamental decisions in development, often by regulating genes on other chromosomes, usually encode transcription factors.

The Roles of Cytoplasmic Segregation and Induction in Cell Determination

What initially stimulates the *MyoD1* promoter to begin transcription is not clear, but chemical signals are clearly involved in cell differentiation. In general, two overall mechanisms for producing such signals have been found:

- **Cytoplasmic segregation**. A factor within an egg, zygote, or precursor cell is unequally distributed in the cytoplasm. After cell division, the factor ends up in some daughter cells or regions of cells, but not others.

- **Induction**. A factor is actively produced and secreted by certain cells to induce other cells to differentiate.

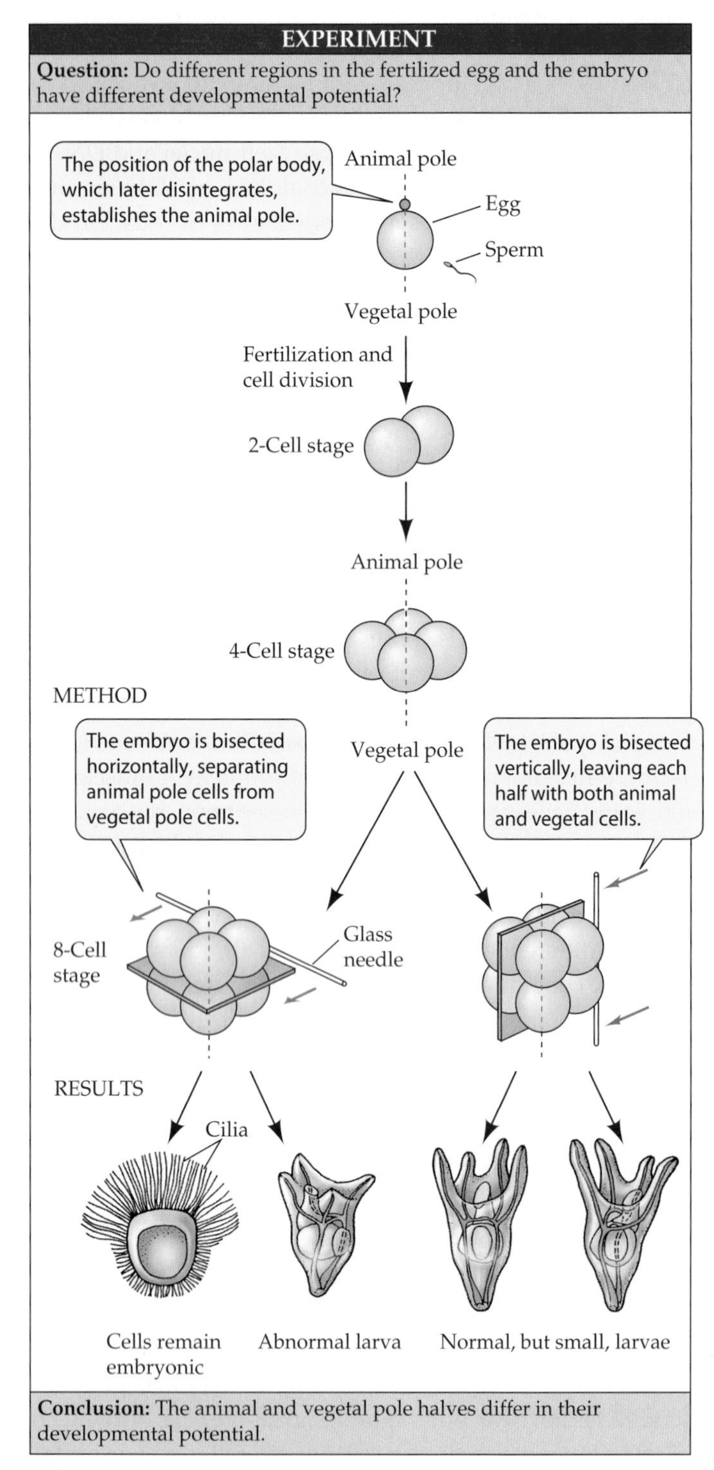

19.7 Asymmetry in the Early Embryo The upper (animal) and lower (vegetal) halves of the sea urchin egg differ in the cytoplasmic determinants they contain. Cells from both halves are necessary to produce a normal larva.

Polarity results from cytoplasmic segregation

As we learned from the cloning experiments described above, cell nuclei do not undergo irreversible changes during early development, so we must look for explanations of some embryological events in the *cytoplasmic* differences between cells. The development of **polarity**—the difference between one end of an organism and the other—is one such phenomenon. Polarity is obvious throughout development. Our heads are distinct from our feet, and the distal ends of our arms (wrists and fingers) differ from the proximal ends (shoulders). An animal's polarity may develop early, even in the egg itself, in which yolk and other factors may be distributed asymmetrically.

An experiment with sea urchins demonstrates the effects of cytoplasmic segregation on development (Figure 19.7). Very early development in this species occurs by equal mitotic divisions of the fertilized egg; there is no increase in size at this stage. If an 8-cell embryo is cut vertically, both halves develop normally. On the other hand, if the embryo is cut horizontally, the top half does not develop at all and the bottom half develops into a small, abnormal embryo.

Clearly, then, there must be at least one factor essential for development that is segregated in the bottom half of the egg, such that the bottom cells of the embryo have it, but the top ones do not. This and many other experiments have established that certain materials, called **cytoplasmic determinants**, are distributed unequally in the egg cytoplasm, and that these materials play a role in directing the embryonic development of many organisms (Figure 19.8).

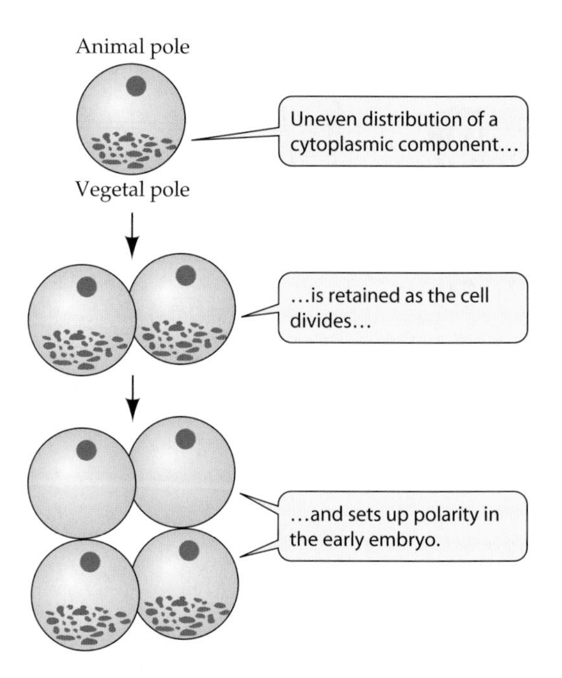

Animal pole
Uneven distribution of a cytoplasmic component...
Vegetal pole
...is retained as the cell divides...
...and sets up polarity in the early embryo.

19.8 The Principle of Cytoplasmic Segregation The distribution of a cytoplasmic substance may determine cell fate.

Tissues direct the development of their neighbors by secreting inducers

Experimental work on developing embryos has clearly established that in many cases, the fates of particular tissues are determined by interactions with other specific tissues in the embryo. In developing animal embryos there are many such instances of induction, in which one tissue causes an adjacent tissue to develop in a particular manner. These effects are mediated by intercellular biochemical communication—that is, by chemical signals and signal transduction mechanisms. We will describe two examples of such induction: one in the developing vertebrate eye, and the other in a developing reproductive structure in the nematode *C. elegans*.

The development of the lens of the vertebrate eye is a classic example of induction. In a frog embryo, the developing forebrain bulges out at both sides to form the *optic vesicles*, which expand until they come into contact with the cells at the surface of the head (Figure 19.9). The surface tissue in the region of contact with the optic vesicles thickens, forming a *lens placode*. The lens placode bends inward, folds over on itself, and ultimately detaches from the surface tissue to produce a structure that will develop into the lens.

If the growing optic vesicle is cut away before it contacts the surface cells, no lens forms. Placing an impermeable barrier between the optic vesicle and the surface cells also prevents the lens from forming. These observations suggest that the surface tissue begins to develop into a lens when it receives a signal—an *embryonic inducer*—from the optic vesicle.

The interaction of tissues in eye development is a two-way street: There is a "dialogue" between the developing optic vesicle and the surface tissue. The optic vesicle induces lens development, and the developing lens determines the size of the *optic cup* that forms from the optic vesicle. If head surface tissue from a frog species with small eyes is grafted over the optic vesicle of one with large eyes, both lens and optic cup will have an intermediate size.

The developing lens also induces the surface tissue over it to develop into a *cornea*, a specialized layer that allows light to pass through and enter the eye. Thus a chain of inductive interactions participates in the development of the parts required to make an eye. Embryonic inducers trigger a sequence of gene expression in the responding cells. Tissues do not induce themselves; rather, different tissues interact and induce one another. We will look at embryonic induction in more detail in Chapter 20.

Single cells can induce changes in their neighbors

The tiny nematode *Caenorhabditis elegans* is used as a model organism in many biological studies, but it is especially useful for studying development. It normally lives in the soil, where it feeds on bacteria, but can also grow in the laboratory if supplied with its food source. The process of development from fertilized egg to larva takes only about 8 hours, and the worm reaches the adult stage in just 3.5 days. The process is easily observed using a low-magnification dissecting microscope because the body covering is transparent (Figure 19.10*a*). For all these reasons, *C. elegans* is a favorite experimental organism. The development of *C. elegans* does not vary, so it has been possible to identify the source of each of the 959 somatic cells of the adult form.

The adult nematode is *hermaphroditic*, containing both male and female reproductive organs. It lays eggs through a pore called the *vulva* on the ventral (belly) surface. During development, a single cell, called the *anchor cell*, induces the vulva to form. If the anchor cell is destroyed by laser surgery, no vulva forms. The eggs develop inside the parent, and a "bag of worms," which eventually consume the parent, results.

The anchor cell controls the fates of six cells on the animal's ventral surface through two molecular switches. Each of these cells has three possible fates: It may become a pri-

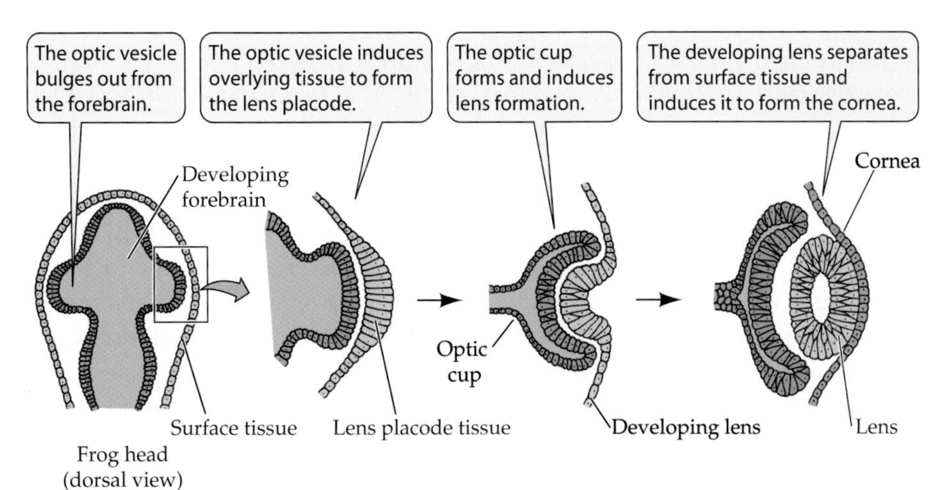

The optic vesicle bulges out from the forebrain.
The optic vesicle induces overlying tissue to form the lens placode.
The optic cup forms and induces lens formation.
The developing lens separates from surface tissue and induces it to form the cornea.
Developing forebrain
Cornea
Optic cup
Surface tissue
Lens placode tissue
Developing lens
Lens
Frog head (dorsal view)

19.9 Embryonic Inducers in the Vertebrate Eye The eye of a frog develops as different tissues take their turns inducing one another.

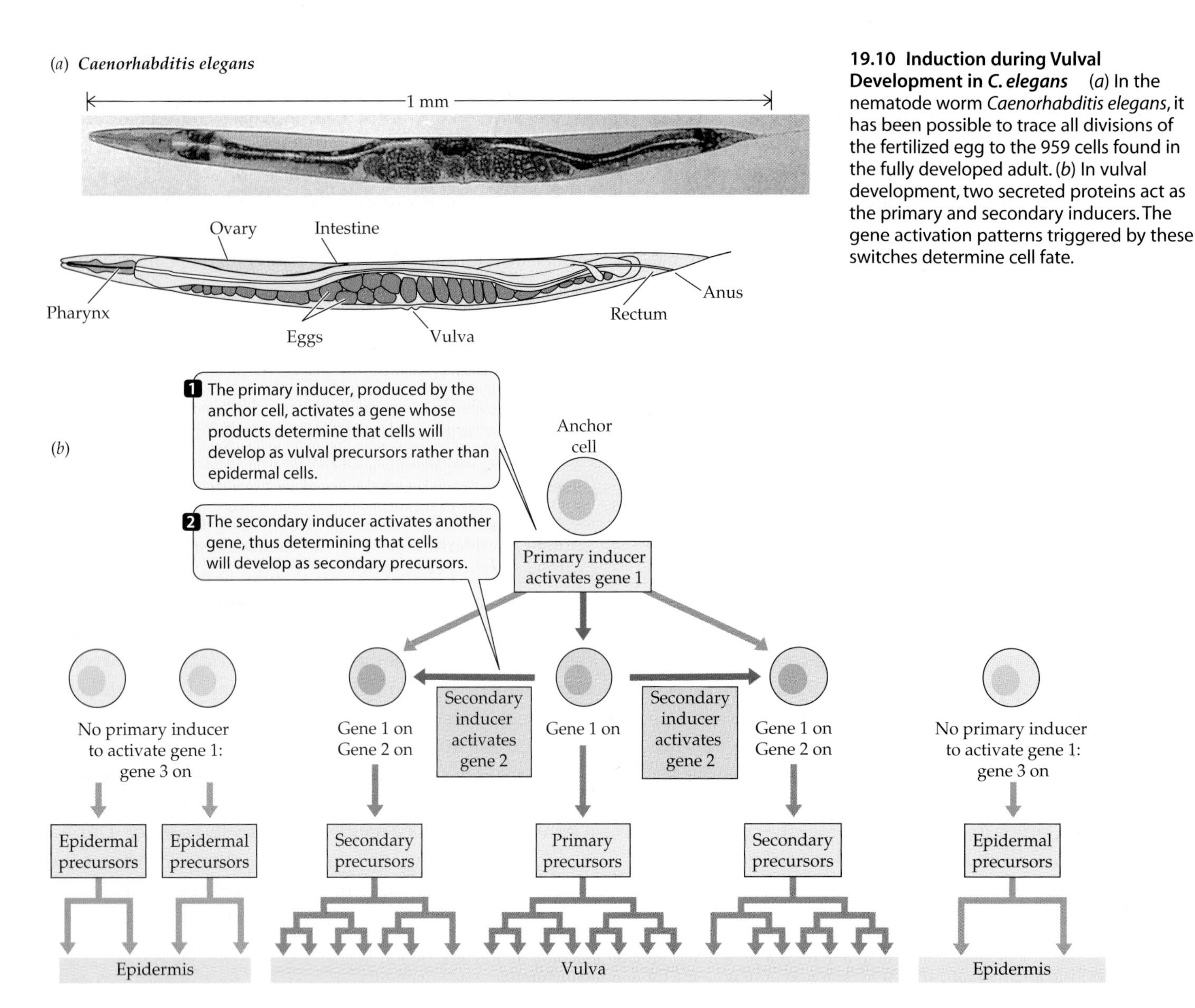

19.10 Induction during Vulval Development in *C. elegans* (*a*) In the nematode worm *Caenorhabditis elegans*, it has been possible to trace all divisions of the fertilized egg to the 959 cells found in the fully developed adult. (*b*) In vulval development, two secreted proteins act as the primary and secondary inducers. The gene activation patterns triggered by these switches determine cell fate.

mary vulval precursor cell, a secondary vulval precursor cell, or simply part of the worm's surface—an epidermal cell (Figure 19.10*b*).

The anchor cell produces an inducer that diffuses out of the cell and interacts with adjacent cells. Cells that receive enough of the inducer become vulval precursor cells; cells slightly farther from the anchor cell become epidermis. The first molecular switch, controlled by the inducer from the anchor cell, determines whether a cell takes the "track" toward becoming part of the vulva or the track toward becoming epidermis.

The cell closest to the anchor cell, having received the most inducer, differentiates into the primary vulval precursor cell. It produces its own inducer, which acts on the two neighboring cells and directs them to become secondary vulval precursor cells. Thus, the primary vulval precursor cell controls a second molecular switch, determining whether a vulval precursor will take the primary track or the secondary track. The two inducers control the activation or inactivation of specific genes in the responding cells.

There is an important lesson to draw from this example: *Much of development is controlled by molecular switches that allow a cell to proceed down one of two alternative tracks*. One challenge for the developmental biologist is to find these molecular switches and determine how they work. The primary inducer released by the *C. elegans* anchor cell appears to be a growth factor homologous to the mammalian epidermal growth factor (EGF). The nematode growth factor, called LIN-3, binds to a receptor on the surface of a potential vulval precursor cell. This binding sets in motion a signal transduction cascade involving the Ras protein and MAP kinases (see Figure 15.9). The end result is increased transcription of the genes involved in the differentiation of vulval cells.

The Role of Pattern Formation in Organ Development

Pattern formation, the spatial organization of a tissue or organism, is inextricably linked to morphogenesis, the appearance of body form. The differentiation of cells is beginning to be understood in terms of molecular events, but how do molecular events contribute to the organization of multitudes of cells into specific body parts, such as a leaf, a flower, a shoulder blade, or a tear duct?

Some cells are programmed to die

Apoptosis is programmed cell death, a series of events caused by the expression of certain genes (see Figure 9.19). Many of these "death genes" have been identified, and related ones have been found in organisms as diverse as nematodes and humans.

Apoptosis is vital to the normal development of all animals. For example, the nematode *C. elegans* produces precisely 1,090 somatic cells as it develops from a fertilized egg to an adult (see Figure 19.10). But 131 of these cells die. The sequential expression of two genes, called *ced-4* and *ced-3* (for *c*ell *d*eath) appears to control this process. In the nervous system, for example, there are 302 nerve cells that come from 405 precursors; thus 103 cells undergo apoptosis. If the protein encoded by either *ced-3* or *ced-4* is nonfunctional, all 405 cells form nerve cells, and disorganization results. A third gene, *ced-9*, codes for an inhibitor of apoptosis; that is, its protein blocks the function of the *ced-4* gene. So, where cell death is required, *ced-3* and *ced-4* are active and *ced-9* is inactive; where cell death does not occur, the reverse is true.

A similar system of cell death genes acts in humans. During early development, human hands and feet look like tiny paddles: The fingers and toes are linked by connective tissue. Between days 41 and 56 of development, the cells between the digits die, freeing the individual fingers and toes (Figure 19.11). The protein—an enzyme called *caspase*—that stimulates this apoptosis is similar in amino acid sequence to the protein encoded by *ced-3*, and a human protein (*bcl-2*) that inhibits apoptosis is similar to *ced-9*. So humans and nematodes, two creatures separated by more than 600 million years of evolutionary time, have similar genes controlling programmed cell death.

Apoptosis plays many other roles in your life. The dead cells that form the outermost layer of your skin and those from the uterine wall that are lost during menstruation have undergone apoptosis. White blood cells live only a few months in the circulation, then undergo apoptosis. In a form of cancer called *follicular large-cell lymphoma*, these white blood cells do not die, but continue to divide. This cancer results from a mutation that causes the overexpression of *bcl-2*, the gene that inhibits cell death.

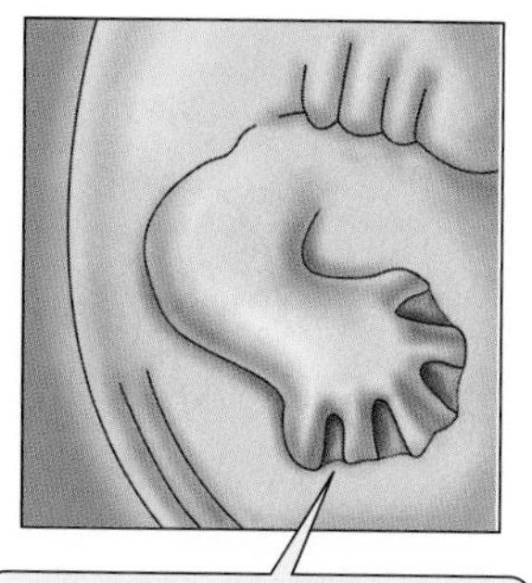

41 days after fertilization: Genes for programmed cell death are expressed in the tissue between the digits.

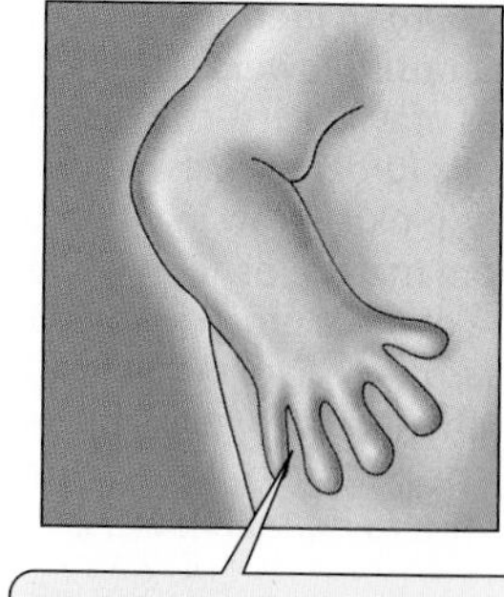

56 days after fertilization: Apoptosis is complete. Cells of the digits have absorbed the remains of the dead cells.

19.11 Apoptosis Removes the Tissue between Fingers Early in the second month of human development, the tissue connecting the fingers is removed by apoptosis, freeing the individual fingers.

Plants have organ identity genes

Like animals, plants have organs—for example, leaves and roots. Many plants form flowers, and many flowers are composed of four types of organs: sepals, petals, stamens, and carpels. These floral organs occur in *whorls*, which are groups of each organ type stacked around a central axis. The whorls develop from meristems in the shape of domes, which develop at growing points on the stem (Figure 19.12*a*). How is the identity of a particular whorl determined? The answer appears to lie in the activities of a group of genes.

These genes have been best described in *Arabidopsis thaliana*. This plant is very useful for studies of development because of its small size (about 25 cm), abundant seed production (over 1,000 seeds per plant), rapid development (from seed to plant to seed in 6 weeks), and small genome (which has been sequenced, as we saw in Chapter 14). Finally, it is easy to produce mutations in this plant by treating the seeds with mutagens.

The development of the flower begins with the meristem, which contains undifferentiated cells. Within this seemingly homogeneous cell population, individual cells "sense" their position and differentiate into the whorls. This happens through the expression of three **organ identity genes**, which code for proteins that act in combination with one another:

- Gene A is expressed in whorls 1 and 2 (which form sepals and petals, respectively).
- Gene B is expressed in whorls 2 and 3 (which form petals and stamens, respectively).
- Gene C is expressed in whorls 3 and 4 (which form stamens and carpels, respectively).

There are two lines of experimental evidence for this model (Figure 19.12*b*):

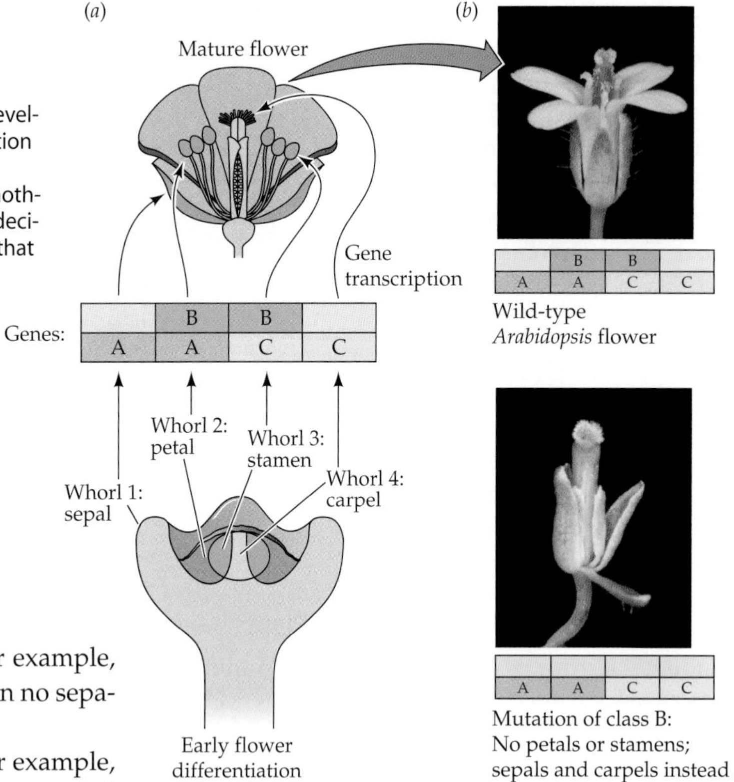

19.12 Organ Identity Genes in *Arabidopsis* Flowers (*a*) The four organs of a flower—carpel (yellow), stamens (green), petals (purple), and sepals (pink)—grow in whorls that develop from meristems. (*b*) When a mutation in one of three organ identity genes occurs, one type of organ replaces another. Such mutations helped scientists decipher the pattern of gene expression that gives rise to normal flowers.

- *Loss of function mutations:* For example, a mutation in gene A results in no sepals or petals.
- *Gain of function mutations:* For example, the promoter for gene C can be coupled to gene A. In this case, A is expressed in all four whorls, resulting in only sepals and petals.

Genes A, B, and C code for subunits of transcription factors, which are active as dimers. Gene regulation in these cases is *combinatorial*—that is, the composition of the dimer determines which other genes will be activated by the transcription factor. For example, a dimer made up only of transcription factor A would activate transcription of the genes that make sepals; a dimer made up of A and B would result in petals, and so forth.

A common feature of the A, B, and C proteins, as well as many other plant transcription factors, is a DNA-binding domain called the **MADS box** (named for homologous regions found in four genes in yeast and in two in plants and humans, that all encode a similar amino acid sequence). These 200-amino acid proteins also have domains for interaction with other proteins.

In addition to being fascinating to biologists, plant organ identity genes have caught the eye of horticultural and agricultural scientists. Flowers filled with petals instead of stamens and carpels often have mutations of the C genes. Many of the foods that make up the human diet, such as the grains of wheat, rice, and corn, come from fruits and seeds. These fruits and seeds form from the carpels (the female reproductive organs) of the flower. Genetically modifying the number of carpels on a particular plant could increase the amount of grain a crop could produce.

A gene called *leafy* codes for a protein that controls the transcription of the ABC genes. Plants with a mutation that causes the underexpression of *leafy* are just that—they make leaves, but no flowers. The protein product of this gene acts as a transcription factor stimulating genes A, B, and C so that they produce flowers (Figure 19.13). This finding, too, has practical applications. It usually takes 6–20 years before a citrus tree produces flowers, and thus the fruits we eat. Scientists have made an orange tree transgenic for *leafy* coupled to a strongly expressed promoter, which flowers and fruits years earlier than a normal tree.

Morphogen gradients provide positional information

During development, cells often need to "know" where they are with respect to the body as a whole, as in the case of the whorls in the flowers just described. This spatial "sense" is called **positional information**. Positional information usually comes in the form of a signal, called a **morphogen**, that diffuses from one group of cells down a body axis, setting up a concentration gradient. There are two requirements for a signal to be considered a morphogen:

- It must directly affect target cells, rather than triggering a secondary signal that affects target cells.
- Different concentrations of the signal must cause different effects.

19.13 A Nonflowering Mutant Mutations in the *leafy* gene of *Arabidopsis* prevent the transcription of the organ identity genes, and the resulting plant does not produce any flowers.

Wild-type

Leafy mutant

The development of the vertebrate limb provides us with an example of a morphogen in action. The limb develops from a round bud. The cells that become the bones and muscles of the limb must receive positional information. If they do not, the limbs will be totally disorganized (imagine fingers growing out of your shoulders). A group of cells at the posterior base of the bud, just where it joins the body wall, makes a morphogen called BMP2, whose gradient determines the anterior–posterior ("thumb to little finger") axis of the developing limb. Cells getting the highest dose of BMP2 make the thumb, and the smallest dose results in the little finger.

The different concentrations of morphogens act through differential regulation of gene expression in their target cells. The model organism often used for studying this process has been the fruit fly.

The Role of Differential Gene Expression in Establishing Body Segmentation

Insects such as the fruit fly *Drosophila melanogaster* develop a highly modular body composed of different types of segments. Complex interactions of different sets of genes underlie the pattern formation of segmented bodies.

Unlike the body segments of segmented worms such as earthworms, which are all essentially alike, the segments of the *Drosophila* body are clearly different from one another. The adult fly has an anterior head (composed of several fused segments), three different thoracic segments, and eight abdominal segments at the posterior end. In the *Drosophila* larva, the thoracic and abdominal segments all appear to be similar, but they have already received their instructions to form these specialized adult segments. Several types of genes are expressed sequentially in the embryo to define these segments. The first step in this process is to establish the polarity of the embryo.

Maternal effect genes encode morphogens that determine polarity

Like those of the sea urchin, *Drosophila* eggs and larvae are characterized by unevenly distributed cytoplasmic determinants (see Figure 19.8). These molecules, which include both mRNAs and proteins, are the products of specific **maternal effect genes**. The maternal effect genes are transcribed in the mother's ovarian cells, which surround and nurture the developing egg and deliver the gene products to specific regions of the egg as it forms. Maternal effect genes exert their effects on the embryo regardless of the genotype of the father. Their products establish the dorsal–ventral (back–belly) and anterior–posterior (head–tail) axes of the embryo.

The fact that these morphogens specify these axes was established by the results of experiments in which cytoplasm was transferred from one egg to another. Females that are homozygous for a particular mutation of the maternal effect gene *bicoid* produce larvae with no head and no thorax. However, if the eggs of these females are inoculated at the anterior end with cytoplasm from the anterior region of a wild-type egg, the treated eggs develop into normal larvae. Conversely, removal of 5 percent or more of the cytoplasm from the anterior end of a wild-type egg results in an abnormal larva that looks like a *bicoid* mutant larva.

Another maternal effect gene, *nanos*, plays a comparable role in the development of the posterior end of the larva. Eggs from homozygous *nanos* mutant females develop into larvae with missing abdominal segments. Injecting cytoplasm from the posterior region of a wild-type egg into a *nanos* mutant egg allows normal development. These findings show that, in wild-type larvae, the overall framework of the anterior–posterior axis is laid down by the activity of these two maternal effect genes (Figure 19.14).

After the axes of the embryo are determined, the next step in pattern formation is the determination of the larval segments.

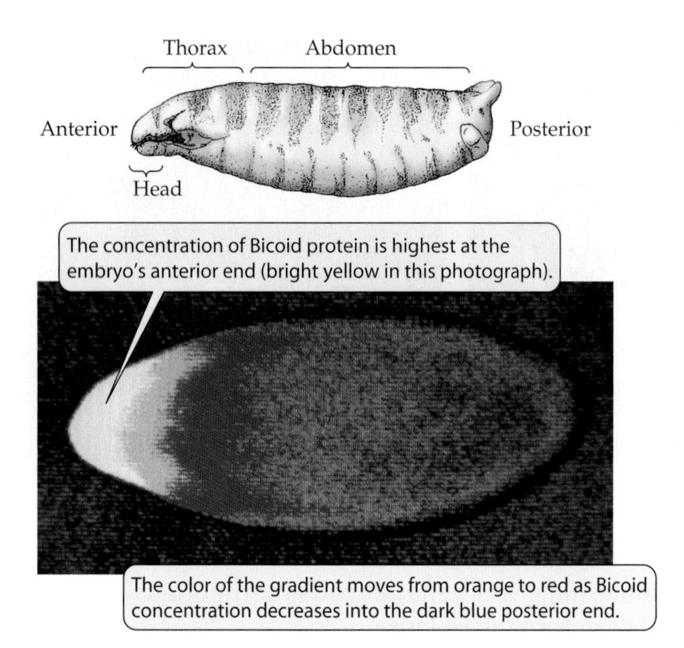

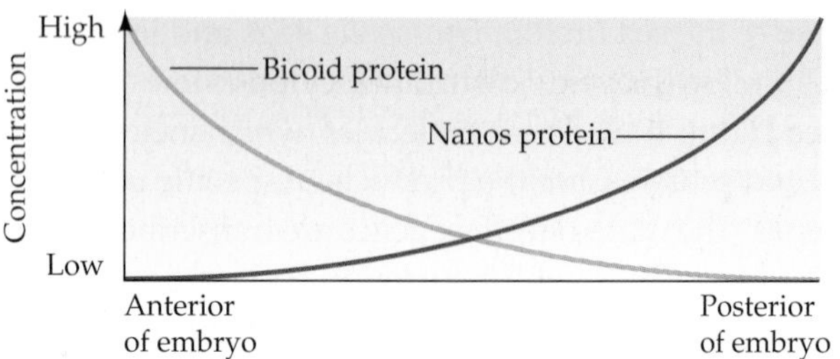

19.14 Bicoid and Nanos Protein Gradients Provide Positional Information The anterior–posterior axis of *Drosophila* arises from morphogens produced by the maternal effect genes *bicoid* and *nanos*. The gradients of these morphogens control the developing body's polarity.

Segmentation and homeotic genes act after the maternal effect genes

The number, boundaries, and polarity of the larval segments are determined by proteins encoded by the **segmentation genes**. These genes are expressed when there are about 6,000 nuclei in the embryo. These nuclei all look the same, but in terms of gene expression, they are not.

The products of the maternal effect genes set the segmentation genes in motion. Three classes of segmentation genes act, one after the other, to regulate finer and finer details of the segmentation pattern (Figure 19.15):

- **Gap genes** organize broad areas along the anterior–posterior axis. Mutations in gap genes result in gaps in the body plan—the omission of several larval segments.
- **Pair rule genes** divide the embryo into units of two segments each. Mutations in pair rule genes result in embryos missing every other segment.
- **Segment polarity genes** determine the boundaries and anterior–posterior organization of the segments. Mutations in segment polarity genes can result in segments in which posterior structures are replaced by reversed (mirror-image) anterior structures.

Finally, after the basic pattern of segmentation has been established by the segmentation genes, differences between the segments are mediated by the activities of **homeotic genes**. These genes are expressed in different combinations along the length of the body and tell each segment what to become. Homeotic genes are analogous to the organ identity genes of plants.

The maternal effect, segmentation, and homeotic genes interact to "build" a *Drosophila* larva step by step, beginning with the unfertilized egg.

Drosophila development results from a transcriptionally controlled gene cascade

One of the most striking and important observations about development in *Drosophila*—and in other animals—is that it results from a sequence of changes, with each change triggering the next. This sequence, or cascade, is largely controlled at the levels of transcription and translation.

Most unfertilized eggs are storehouses of mRNAs, which are supplied by the mother to support protein synthesis dur-

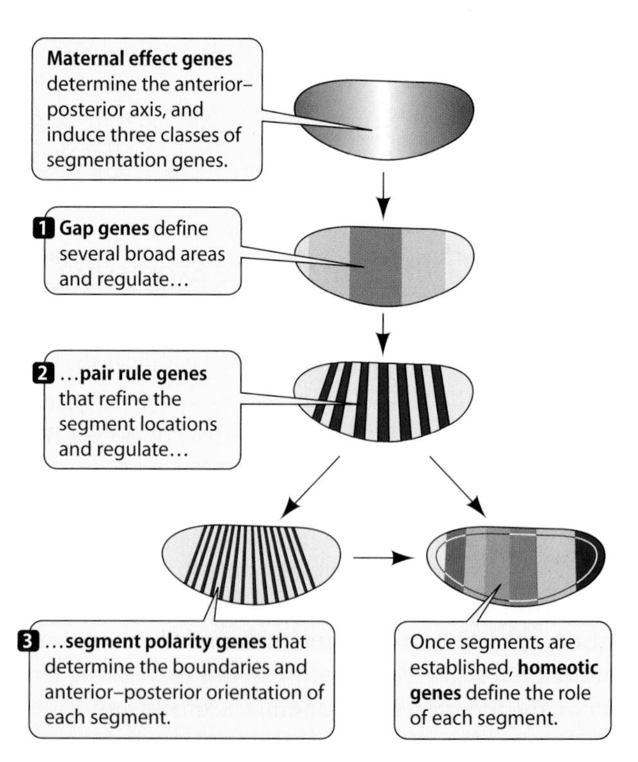

Maternal effect genes determine the anterior–posterior axis, and induce three classes of segmentation genes.
1 Gap genes define several broad areas and regulate...
2 ...pair rule genes that refine the segment locations and regulate...
3 ...segment polarity genes that determine the boundaries and anterior–posterior orientation of each segment.
Once segments are established, homeotic genes define the role of each segment.

19.15 A Gene Cascade Controls Pattern Formation in the *Drosophila* Embryo Gap, pair rule, and segment polarity genes are collectively referred to as the segmentation genes. The shading shows the locations of their gene products in the embryo.

ing the early stages of embryonic development. Indeed, zygotes and early embryos do not carry out transcription. Only after several cell divisions does transcription begin, forming the mRNAs needed for later development.

Cytoplasmic segregation of the prefabricated mRNAs in the egg provides positional information. Before the *Drosophila* egg is fertilized, mRNA for the Bicoid protein is localized at the end that is destined to become the anterior end of the fly. After the egg is fertilized and laid, nuclear divisions begin. (In *Drosophila*, cytokinesis does not begin right away; until the thirteenth nuclear division, the embryo is a single, multinucleated cell called a *syncytium*.) At this early point, *bicoid* mRNA is translated, forming Bicoid protein, which diffuses away from the anterior end, establishing a gradient. At the posterior end, the Nanos protein forms a gradient in the other direction. Thus each nucleus in the developing embryo is exposed to a different concentration ratio of Bicoid and Nanos proteins.

The two morphogens regulate the expression of the gap genes, although in different ways. The Bicoid protein affects their transcription, while the Nanos protein affects their translation. The high concentrations of Bicoid protein in the anterior portion of the egg turn on a gap gene called *hunchback*, while simultaneously turning off another gap gene, *Krüppel*. Nanos at the posterior end reduces the translation of *hunchback*, so a difference in the concentration of these two gap genes' products at the two ends is established.

The proteins encoded by the gap genes control the expression of the pair rule genes. Many pair rule genes, in turn, encode transcription factors that control the expression of the segment polarity genes, giving rise to a complex, striped pattern (see Figure 19.15) of expression that foreshadows the segmented body plan of *Drosophila*.

By this point, each nucleus of the embryo has been exposed to a distinct set of transcription factors. The segmented body pattern of the larva has been established even before any sign of segmentation is visible. When the segments do appear, they are not all identical, because the homeotic genes specify the different structural and functional properties of each segment. Each homeotic gene is expressed over a characteristic portion of the embryo. Let's turn now to the homeotic genes and see how their mutation can alter the course of development.

Homeotic mutations produce changes in segment identity

Two bizarre homeotic mutations in *Drosophila* are the *Antennapedia* mutation, in which legs grow in place of antennae (Figure 19.16), and the *bithorax* mutation, in which an extra pair of wings grows in a thoracic segment (see Figure 21.4*a*). Edward Lewis at the California Institute of Technology found that *Antennapedia* and *bithorax* were mutations not of isolated genes, but of two adjacent clusters of genes that determine the identity of body segments. Moreover, the genes in these clusters were lined up along the chromosome in the same order as the segments they determined. From left to right, genes in the first cluster specified anterior body segments, starting with genes for the different head segments and ending with thoracic segments. The second cluster began with a gene specifying the last thoracic segment, followed by a gene for the anterior abdominal segments, and ended with a gene for the posterior abdominal segments. Lewis hypothesized that all of these genes might have come from the duplication of a single gene in an ancestral, unsegmented organism.

Molecular biologists confirmed Lewis's hypothesis using nucleic acid hybridization. Several scientists found that a probe for a sequence in one of the genes of one cluster bound

(*a*)

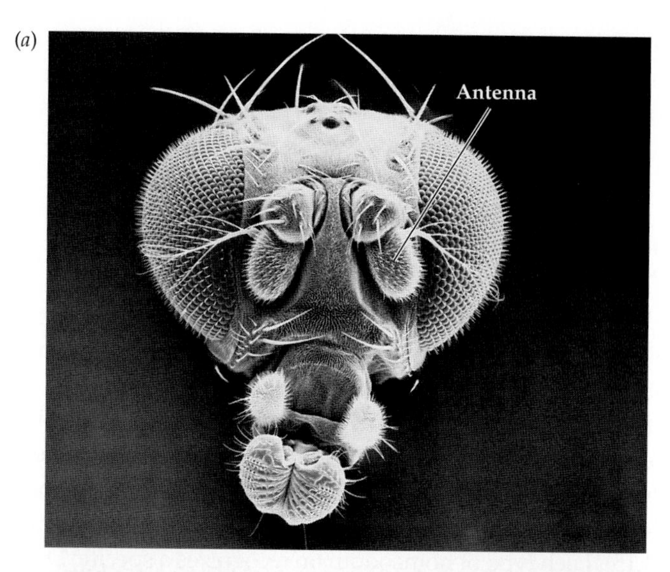

(*b*)

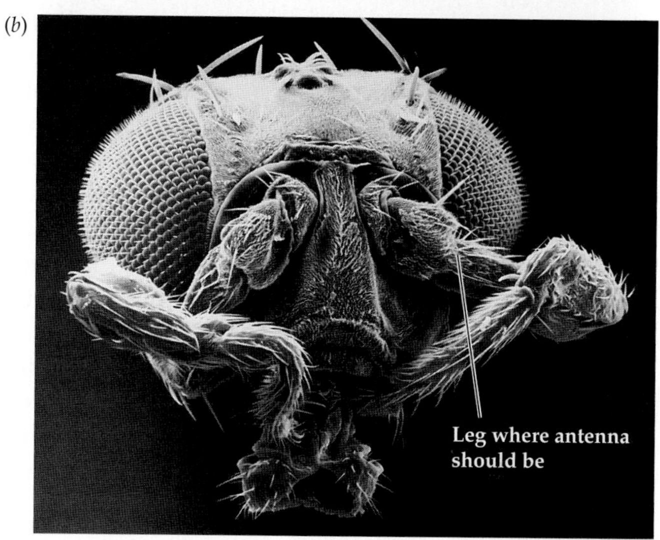

19.16 A Homeotic Mutation in *Drosophila* Mutations of the homeotic genes cause body parts to form on inappropriate segments. (*a*) A wild-type fruit fly. (*b*) An *Antennapedia* mutant fruit fly.

not only to its own gene, but also to adjacent genes in its cluster and to genes in the other homeotic cluster. In other words, this DNA sequence is common to all the homeotic genes in both clusters.

Homeobox-containing genes encode transcription factors

The 180-base-pair DNA sequence that is common to the bithorax and Antennapedia gene clusters is called the **homeobox**. It encodes a 60-amino acid sequence, called the *homeodomain*, that binds to DNA. The homeodomain turns out to be present in other proteins involved in *Drosophila* pattern formation, such as Bicoid. In all cases, the homeodomain portion of the protein has a helix-turn-helix motif (see Figure 14.15). Each type of homeodomain recognizes a specific DNA sequence in the promoter of its target genes. The Bicoid homeodomain, for example, recognizes TCCTAATCCC.

What do homeodomain proteins do when they recognize their target sequence in DNA? Not surprisingly, they are transcription factors. The Bicoid protein, for example, binds to promoters of the gap gene *hunchback*, activating its transcription. The Hunchback protein is also a transcription factor, which binds to enhancers of genes involved in head and thorax formation. In this way, the homeodomain proteins produce the cascade of events that controls *Drosophila* development.

Homeobox genes are found in many animals, including humans. They play a role in development similar to the role the MADS box genes play in plants. The evolutionary significance of these common pathways for development will be discussed in Chapter 21.

Chapter Summary

The Processes of Development

▶ A multicellular organism develops through a series of embryonic stages and eventually into an adult. Development continues until death. **Review Figure 19.1.**

▶ Growth results from a combination of cell division and cell expansion.

▶ Differentiation produces specialized cell types.

▶ Morphogenesis—the creation of the overall form of the multicellular organism—is the result of pattern formation.

▶ In many organisms, the fates of early embryonic cells have not yet been determined. These cells may develop into different tissues if transplanted to a different part of an embryo. **Review Figure 19.2**

▶ As the embryo develops, its cells gradually become determined—committed to developing into particular cell types. Following determination, cells eventually differentiate into their final, often specialized, forms.

The Role of Differential Gene Expression in Cell Differentiation

▶ The zygote is totipotent; it contains the entire genetic constitution of the organism and is capable of forming all adult tissues.

▶ Two lines of evidence show that differentiation does not involve permanent changes in the genome. First, nuclear transplantation and cell fusion experiments show that the nucleus of a differentiated cell retains the ability to act like a zygote nucleus and direct the production of an entire organism. Second, molecular investigations have shown directly that all cells contain all genes for the organism, but that only certain genes are expressed in a given tissue. **Review Figures 19.3, 19.4, 19.5**

▶ Embryonic stem cells are totipotent, and they can be cultured in the laboratory. With suitable environmental stimulation, these cells can be induced to form cells that differentiate into a particular type. **Review Figure 19.6.** See Web/CD Tutorial 19.1

The Role of Cytoplasmic Segregation and Induction in Cell Determination

▶ Unequal distribution of cytoplasmic determinants in the egg, zygote, or embryo can lead to cell determination. Experimentally altering this distribution can alter gene expression and produce abnormal or nonfunctional organisms. **Review Figures 19.7, 19.8.** See Web/CD Tutorial 19.2

▶ Some embryonic animal tissues direct the development of their neighbors by secreting inducers.

▶ Induction is often reciprocal: One tissue induces a neighbor to change, and the neighbor, in turn, induces the first tissue to change, as in eye formation in vertebrate embryos. **Review Figure 19.9**

▶ Induction in the nematode *Caenorhabditis elegans* can be very precise, with individual cells producing specific effects in just two or three neighboring cells. **Review Figure 19.10**

The Role of Pattern Formation in Organ Development

▶ Apoptosis is important in pattern formation. Some genes whose protein products regulate apoptosis have been identified. **Review Figure 19.11**

▶ Plants have organ identity genes that interact to cause the formation of sepals, petals, stamens, and carpels. Mutations of these genes may cause meristem cells to form a different organ. **Review Figure 19.12**

▶ Plant organ identity genes code for transcription factors of the MADS box family.

▶ Both plants and animals use positional information as a basis for pattern formation. Gradients of morphogens provide this information.

The Role of Differential Gene Expression in Establishing Body Segmentation

▶ The fruit fly *Drosophila melanogaster* has provided much information about the development of body segmentation.

▶ The first genes to act in determining *Drosophila* segmentation are maternal effect genes, such as *bicoid* and *nanos*, which encode morphogens that form gradients in the egg. These morphogens act on segmentation genes to define the anterior–posterior organization of the embryo. **Review Figure 19.14**

▶ There are three kinds of segmentation genes. Gap genes organize broad areas along the anterior–posterior axis, pair rule genes divide the axis into pairs of segments, and segment polarity genes define the anterior–posterior axis of each segment. **Review Figure 19.15.** See Web/CD Tutorial 19.3

▶ Segmentation develops as the result of a transcriptionally controlled cascade, with the product of each gene promoting or repressing the expression of the next.

▶ Activation of the segmentation genes leads to the activation of the appropriate homeotic genes in each segment. The homeotic genes define the functional characteristics of the segments.

▶ Mutations of homeotic genes often have bizarre effects, causing structures to form in inappropriate parts of the body.

▶ Homeotic genes contain a sequence called the homeobox, which encodes the homeodomain, an amino acid sequence that is part of many transcription factors.

Self-Quiz

1. Which statement about determination is true?
   a. Differentiation precedes determination.
   b. All cells are determined after two cell divisions in most organisms.
   c. A determined cell will keep its determination no matter where it is placed in an embryo.
   d. A cell changes its appearance when it becomes determined.
   e. A differentiated cell has the same pattern of transcription as a determined cell.
2. The cloning experiments on sheep, frogs, and mice showed that
   a. nuclei of adult cells are totipotent.
   b. nuclei of embryonic cells can be totipotent.
   c. nuclei of differentiated cells have different genes than zygote nuclei have.
   d. differentiation is fully reversible in all cells of a frog.
   e. differentiation involves permanent changes in the genome.
3. The term "embryonic induction" describes a process in which a group of cells
   a. influences the development of another group of cells.
   b. triggers the cell movements in an embryo.
   c. stimulates the transcription of their own genes.
   d. organizes the egg cytoplasm before fertilization.
   e. makes a "fate map" of the embryo.
4. The term "therapeutic cloning" describes
   a. modification of a clone by a transgene.
   b. combining nuclear transplantation and stem cell differentiation.
   c. making clones that produce useful drugs.
   d. producing embryonic stem cells for transplantation.
   e. making many identical copies of an organism.
5. Which statement about cytoplasmic determinants in *Drosophila* is *not* true?
   a. They specify the dorsal–ventral and anterior–posterior axes of the embryo.
   b. Their positions in the embryo are determined by microfilament action.
   c. They are products of specific genes in the mother fruit fly.
   d. They often produce large-scale effects in larvae.
   e. They have been studied by the transfer of cytoplasm from egg to egg.
6. In fruit flies, the following genes are used to determine segment polarity: (k) gap genes; (l) homeotic genes; (m) maternal effect genes; (n) pair rule genes. In what order are these genes expressed during development?
   a. klmn
   b. lknm
   c. mknl
   d. nkml
   e. nmkl
7. Which statement about embryonic induction is *not* true?
   a. One group of cells induces adjacent cells to develop in a certain way.
   b. It triggers a sequence of gene expression in target cells.
   c. Single cells cannot form an inducer.
   d. A tissue may induce itself.
   e. The chemical identification of specific inducers has been difficult.
8. In the process of body segmentation in *Drosophila* larvae,
   a. the first steps are specified by homeotic genes.
   b. mutations in pair rule genes result in embryos missing every other segment.
   c. mutations in gap genes result in the insertion of extra segments.
   d. segment polarity genes determine the dorsal–ventral axes of segments.
   e. segmentation is the same as in earthworms.
9. Homeotic mutations
   a. are often so severe that they can be studied only in larvae.
   b. cause subtle changes in the forms of larvae or adults.
   c. occur only in prokaryotes.
   d. do not affect the animal's DNA.
   e. are confined to the zone of polarizing activity.
10. Which statement about the homeobox is *not* true?
    a. It is transcribed and translated.
    b. It is found only in animals.
    c. All proteins containing the homeodomain bind to DNA.
    d. It is a stretch of DNA shared by many genes.
    e. Its activities often relate to pattern formation.

For Discussion

1. Molecular biologists can insert genes attached to high-level promoters into cells (see Chapter 16). What would happen if the following were inserted and overexpressed? Explain your answers.
   a. ced-9 in embryonic nerve cell precursors in *C. elegans*
   b. MyoD1 in undifferentiated myoblasts
   c. the gene for BMP2 in a chick limb bud
   d. nanos at the anterior end of the *Drosophila* embryo
2. A powerful method to test for the function of a gene in development is to generate a "knockout" organism, in which the gene in question is inactivated (see Chapter 16). What do you think would happen in each of the following cases?
   a. a knocked-out *ced-9* in *C. elegans*
   b. a knocked-out *nanos* in *Drosophila*
3. Look at the chart of organ identity mutations in Figure 19.12. What pattern do you perceive in the results of these mutations, and what might this pattern mean?
4. During development, the potential of a cell becomes ever more limited, until, in the normal course of events, its potential is the same as its original prospective fate. On the basis of what you have learned in this chapter and in Chapter 14, discuss possible mechanisms for the progressive limitation of the cell's potential.
5. How were biologists able to obtain such a complete accounting of all the cells in *Caenorhabditis elegans?* What major conclusions of themes for development came from these studies?

20 Animal Development: From Genes to Organism

The whale blows its nose from the top of its head—as in "thar she blows," the whalers' exclamation. The spout from the blowhole is the whale's exhalation coming out of its nasal passages. It is convenient for a marine mammal to breathe out of the top of its head because not much of its body has to come out of the water, and it can continue moving through the water as it breathes. But in most terrestrial mammals, the nose is on the front of the head. How did the whale's nose get to the top of its head? This is an evolutionary question, but the answer is to be found in development—the processes whereby a fertilized egg becomes an adult organism.

The vertebrate body varies enormously among species in form and function, yet its basic structural design does not. For example, the whale flipper, the bat wing, and the human arm all have the same bones. However, during development, these bones assume different shapes and dimensions to adapt the forelimbs to various functions: swimming, flying, and tool use.

Similarly, all vertebrates have the same bones in their heads, but through development, these bones grow differentially, and therefore the skull takes on different shapes in different species. In both whales and humans, the nasal passages are in the nasal bone, which is just above the bones of the upper jaw. In the human, that places the nasal bone just above the jaw on the front of the face. Things are different in the whale. During development of the whale skull, the bones of the upper jaw grow enormously relative to the other bones of the skull, and project far forward to form the cavernous mouth. As a result of this differential forward growth of the jaw bone, the nasal bone ends up on the top of the skull, rather than on the front. Thus, the answer to why the whale's nose is on the top of its head and how its forelimbs become flippers is found in the processes of development. These processes form and shape the components of the basic vertebrate body plan.

In the previous chapter, we learned that the processes of development include determination, differentiation, growth, and mor-

Thar She Blows! The nasal passages of the whale *Orcinus orca* are on top of its head because of the extreme growth of its jaw bones during development.

phogenesis. In this chapter we will see how these processes are carried out in the early stages of development.

Development begins with the joining of sperm and egg. The fertilized egg goes through an initial rapid series of cell divisions without growth that subdivides the egg cytoplasm into a mass of smaller undifferentiated cells. Although this mass of cells shows no hints of the eventual body plan, the uneven distribution of molecules in the cytoplasm of the fertilized egg provides positional information that will result in the determination of cells and set up the body plan. The body plan then unfolds through orderly movements of cells that create multiple cell layers and set up new cell-to-cell contacts that trigger signal transduction cascades and further steps of determination. These inductive interactions influence the temporal and spatial expression of the genes that control the growth and differentiation of cells, leading to the emergence of the organs of the new individual.

To appreciate both the diversity and the similarity in the development of different animals, we will discuss these early developmental steps in a few model organisms that have been studied extensively by developmental biologists: sea urchins (invertebrates), and frogs, chickens, and humans (all vertebrates).

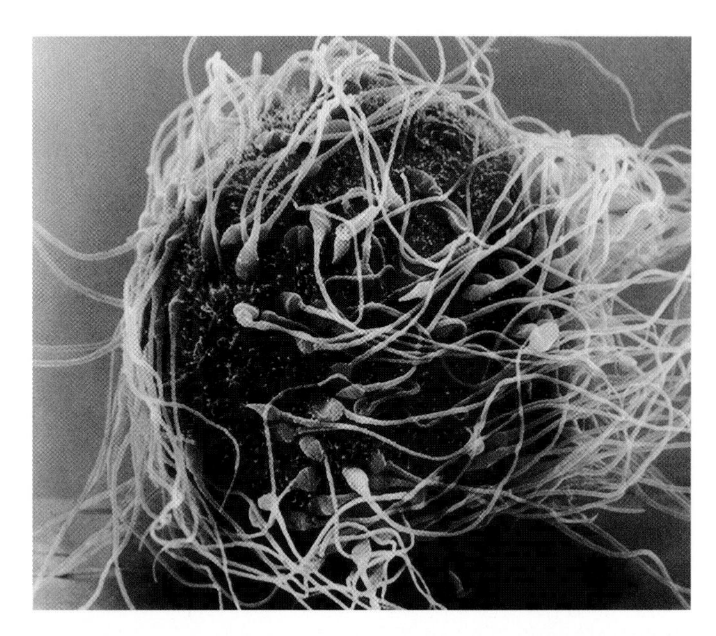

20.1 Sperm and Egg Differ Greatly in Size This artificially colored micrograph of human fertilization illustrates the size difference between the two types of gametes in mammals. The large egg (blue) contributes more cytoplasm to the zygote than the much smaller sperm (yellow).

Development Begins with Fertilization

Fertilization is the union of a haploid sperm and a haploid egg to produce a diploid zygote. Fertilization does more, however, than just restore a full complement of maternal and paternal genes. The entry of a sperm into an egg activates the egg metabolically and initiates the rapid series of cell divisions that produce a multicellular embryo. Also, in many species, the point of entry of the sperm creates an asymmetry in the radially symmetrical egg. This asymmetry is the initiating event that enables a bilateral body plan to emerge from the radial symmetry of the egg. We will describe the mechanisms of fertilization in Chapter 43. Here we take a closer look at the cellular and molecular interactions of sperm and egg that result in the first steps of development.

The sperm and the egg make different contributions to the zygote

Nearly all of the cytoplasm of the zygote comes from the egg (Figure 20.1). Egg cytoplasm is well stocked with nutrients, ribosomes, and a variety of molecules, including mRNAs. Because the sperm's mitochondria degenerate, all of the mitochondria (and therefore all of the mitochondrial DNA) in the zygote come from the mother. In addition to its haploid nucleus, the sperm makes one other important contribution to the zygote in some species: a centriole. This centriole becomes the centrosome of the zygote, which produces the mitotic spindles for subsequent cell divisions.

It had long been assumed that the one thing that sperm and egg contributed equally to the zygote was their haploid nuclei. However, we now know that even though they are equivalent in terms of genetic material, mammalian sperm and eggs are not equivalent in terms of their roles in development. In the laboratory, it is possible to construct zygotes in which both haploid nuclei come from the mother or both come from the father. In neither case does development progress normally. Apparently, in mammals at least, certain genes involved in development are active only if they come from a sperm, and others are active only if they come from an egg. This phenomenon, called *genomic imprinting*, was described in Chapter 17.

Fertilization causes rearrangements of egg cytoplasm

The entry of the sperm into the egg stimulates changes in and rearrangements of the egg cytoplasm that establish the polarity of the embryo. The nutrients and molecules in the cytoplasm of the zygote are not homogeneously distributed, and therefore, they are not divided equally among all daughter cells when cell divisions begin. This unequal distribution of cytoplasmic factors sets the stage for the signal transduction cascades that orchestrate the sequential steps of development: determination, differentiation, and morphogenesis. Let's examine these earliest developmental events in the frog, an organism in which they have been well studied.

The rearrangements of egg cytoplasm in some frog species are easily observed because of pigments in the egg cytoplasm. The nutrient molecules in an unfertilized frog egg are dense, and they are therefore concentrated by gravity in the lower half of the egg, which is called the *vegetal hemisphere*. The haploid nucleus of the egg is located at the opposite end of the egg, in the *animal hemisphere*. The outermost (*cortical*) cytoplasm of the animal hemisphere is heavily pigmented, and the underlying cytoplasm has more diffuse pigmentation. The vegetal hemisphere is not pigmented.

The surface of the frog egg has specific sperm-binding sites located only in the animal hemisphere, so sperm always enter the egg in that hemisphere. When a sperm enters, the cortical cytoplasm rotates toward the site of sperm entry. This rotation reveals a band of diffusely pigmented cytoplasm on the side of the egg opposite the site of sperm entry. This band, called the **gray crescent**, will be the site of important developmental events (Figure 20.2).

The cytoplasmic rearrangements that create the gray crescent bring different regions of cytoplasm into contact on opposite sides of the egg. Therefore, bilateral symmetry is imposed on what was a radially symmetrical egg. In addition to the up–down difference of the animal and vegetal hemispheres, the movement of the cytoplasm sets the stage for the creation of the anterior–posterior and left–right axes. In the frog, the site of sperm entry will become the *ventral* (belly) region of the embryo, and the gray crescent will become the *dorsal* (back) region. Since the gray crescent also marks the posterior end of the embryo, these relationships specify the anterior–posterior and left–right axes as well.

Rearrangements of egg cytoplasm set the stage for determination

The molecular mechanisms underlying the first steps in frog embryo formation are beginning to be understood. The sperm centriole rearranges the microtubules in the vegetal hemisphere cytoplasm into a parallel array that presumably guides the movement of the cortical cytoplasm. Organelles and certain proteins from the vegetal hemisphere move to the gray crescent region even faster than the cortical cytoplasm rotates.

As a result of these movements of cytoplasm, proteins, and organelles, changes in the distribution of critical developmental signals occur. A key transcription factor in early development is β-catenin, which is produced from maternal mRNA and is found throughout the cytoplasm of the egg. Also present throughout the egg cytoplasm is a protein kinase called GSK-3, which phosphorylates and thereby targets β-catenin for degradation. However, an inhibitor of GSK-3 is segregated in the vegetal cortex of the egg. After sperm entry, this inhibitor is moved along microtubules to the gray crescent, where it prevents the degradation of β-catenin. As a result, the concentration of β-catenin is higher on the dorsal side than on the ventral side of the developing embryo (Figure 20.3).

Evidence supports the hypothesis that β-catenin is a key player in the cell–cell signaling cascade that begins the process of cell determination and the formation of the embryo in the region of the gray crescent. But before there can be cell–cell signaling, there must be multiple cells, so let's turn first to the early series of cell divisions that transforms the zygote into a multicellular embryo.

Cleavage: Repackaging the Cytoplasm

The transformation of the diploid zygote into a mass of cells occurs through a rapid series of cell divisions, called **cleavage**. Because the cytoplasm of the zygote is not homogeneous, these first cell divisions result in the differential distribution of nutrients and cytoplasmic determinants among the cells of the early embryo. In most animals, cleavage proceeds with rapid DNA replication and mitosis, but no cell growth and little gene expression. The embryo becomes a solid ball of smaller and smaller cells, called a *morula* (from the Latin word for "mulberry"). Eventually, this ball forms a central fluid-filled cavity called a **blastocoel**, at which point the embryo is called a **blastula**. Its individual cells are called **blastomeres**.

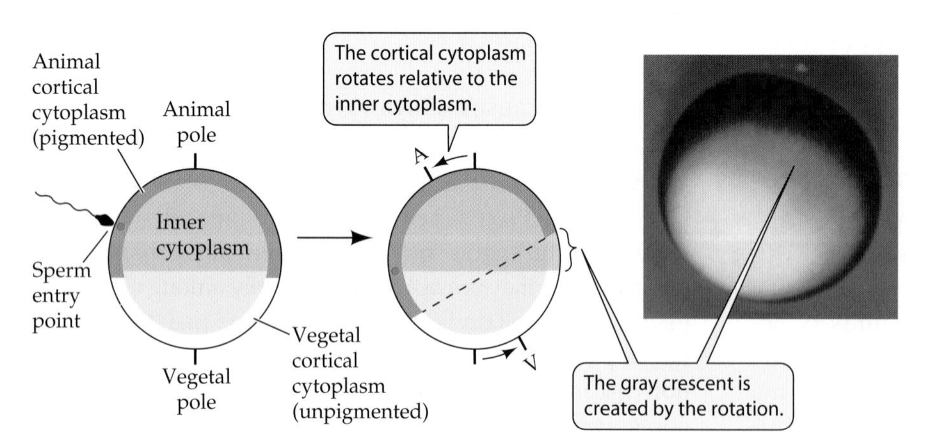

20.2 The Gray Crescent Rearrangements of the cytoplasm of frog eggs after fertilization create the gray crescent.

The pattern of cleavage, and therefore the form of the blastula, is influenced by two major factors. First, the amount of nutrient material, or **yolk**, stored in the egg differs among species. Yolk influences the pattern of cell divisions by impeding the pinching in of the plasma membrane to form a *cleavage furrow* between the daugh-

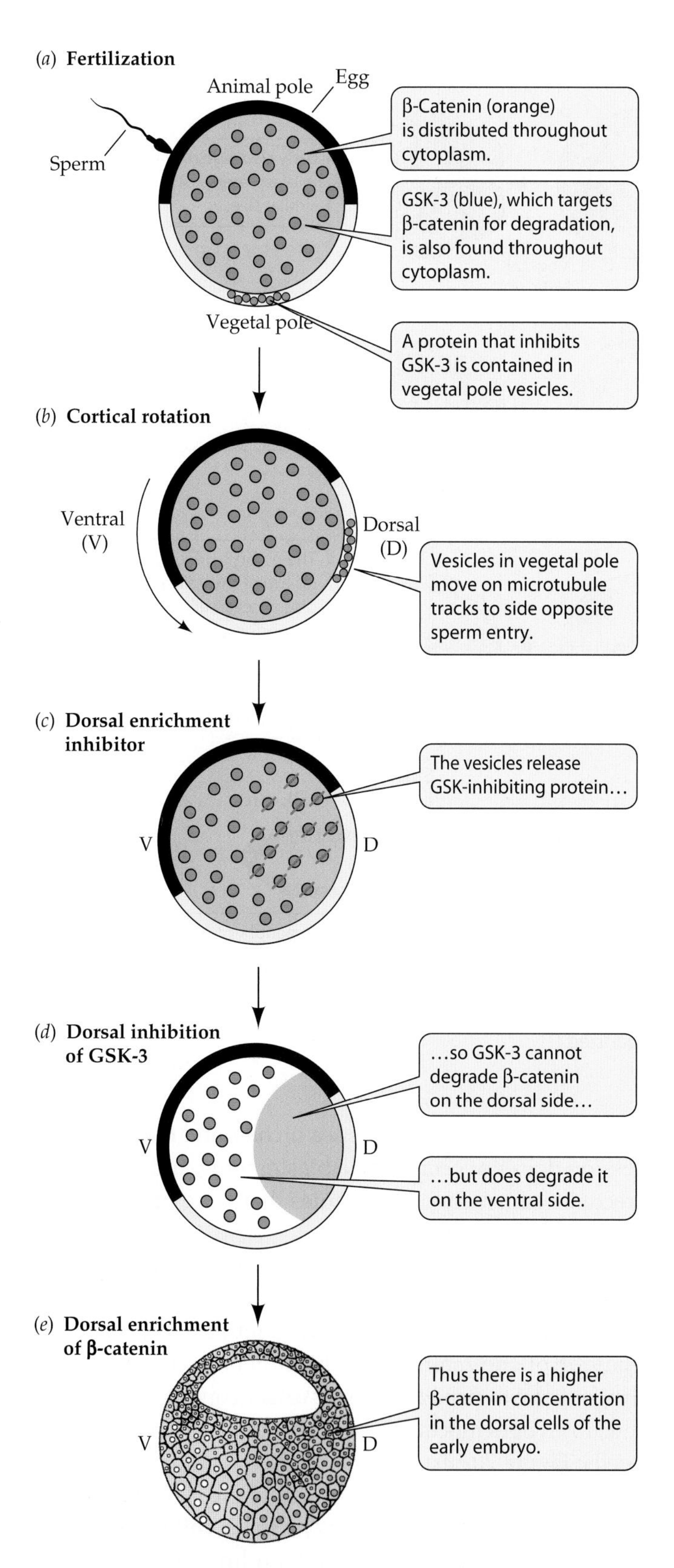

(a) Fertilization
Animal pole
Egg
Sperm
β-Catenin (orange) is distributed throughout cytoplasm.
GSK-3 (blue), which targets β-catenin for degradation, is also found throughout cytoplasm.
Vegetal pole
A protein that inhibits GSK-3 is contained in vegetal pole vesicles.
(b) Cortical rotation
Ventral (V)
Dorsal (D)
Vesicles in vegetal pole move on microtubule tracks to side opposite sperm entry.
(c) Dorsal enrichment inhibitor
The vesicles release GSK-inhibiting protein...
V
D
(d) Dorsal inhibition of GSK-3
...so GSK-3 cannot degrade β-catenin on the dorsal side...
V
D
...but does degrade it on the ventral side.
(e) Dorsal enrichment of β-catenin
Thus there is a higher β-catenin concentration in the dorsal cells of the early embryo.
V
D

20.3 Cytoplasmic Factors Set Up Signaling Cascades
Cytoplasmic movement changes the distributions of critical developmental signals. In the frog zygote, the interaction of the protein kinase GSK-3, its inhibitor, and the protein β-catenin are crucial in specifying the dorsal–ventral (back–belly) axis of the embryo.

ter cells. Second, cytoplasmic determinants stored in the egg by the mother guide the formation of mitotic spindles and the timing of cell divisions.

The amount of yolk influences cleavage

In embryos with little or no yolk, there is little interference with cleavage furrow formation, and all the daughter cells are of similar size; the sea urchin egg provides an example (Figure 20.4*a*). More yolk means more resistance to cleavage furrow formation; therefore, cell divisions progress more rapidly in the animal hemisphere than in the vegetal hemisphere, where the yolk is concentrated. As a result, the cells derived from the vegetal hemisphere are fewer and larger; the frog egg provides an example of this pattern (Figure 20.4*b*).

In spite of this difference between sea urchin and frog eggs, the cleavage furrows completely divide the egg mass in both cases; thus these animals are said to have *complete cleavage*. In contrast, in eggs that contain a lot of yolk, such as the chicken egg, the cleavage furrows do not penetrate the yolk. As a result, cleavage is incomplete, and the embryo forms as a disc of cells, called a **blastodisc**, on top of the yolk mass (Figure 20.4*c*). This type of incomplete cleavage, called *discoidal cleavage*, is common in fishes, reptiles, and birds.

Another type of incomplete cleavage, called *superficial cleavage*, occurs in insects such as the fruit fly (*Drosophila*). In the insect egg, the mass of yolk is centrally located (Figure 20.4*d*). Early in development, cycles of mitosis occur without cytokinesis. Eventually the resulting nuclei migrate to the periphery of the egg, and after several more mitotic cycles, the plasma membrane of the egg grows inward, partitioning the nuclei into individual cells.

The orientation of mitotic spindles influences the pattern of cleavage

The positions of the mitotic spindles during cleavage are not random; rather, they are defined by cytoplasmic determinants that were produced from the maternal genome and stored in the egg. The orientation of the mitotic spindles determines the planes of cleavage and, therefore, the arrangement of the daughter cells.

If the mitotic spindles of successive cell divisions form parallel or perpendicular to the animal–vegetal axis of the zygote, the cleavage pattern is *radial*, as in the sea urchin and the frog. In these organisms, the first two cell divisions are parallel to the animal–vegetal axis and the third is perpendicular to it (Figure 20.4*a,b*). Another cleavage pattern, *spiral cleavage*, results when the mitotic spindles are at oblique angles to the animal–vegetal axis. Mollusks have spiral cleavage, and a visible expression of this is the coiling of snail shells.

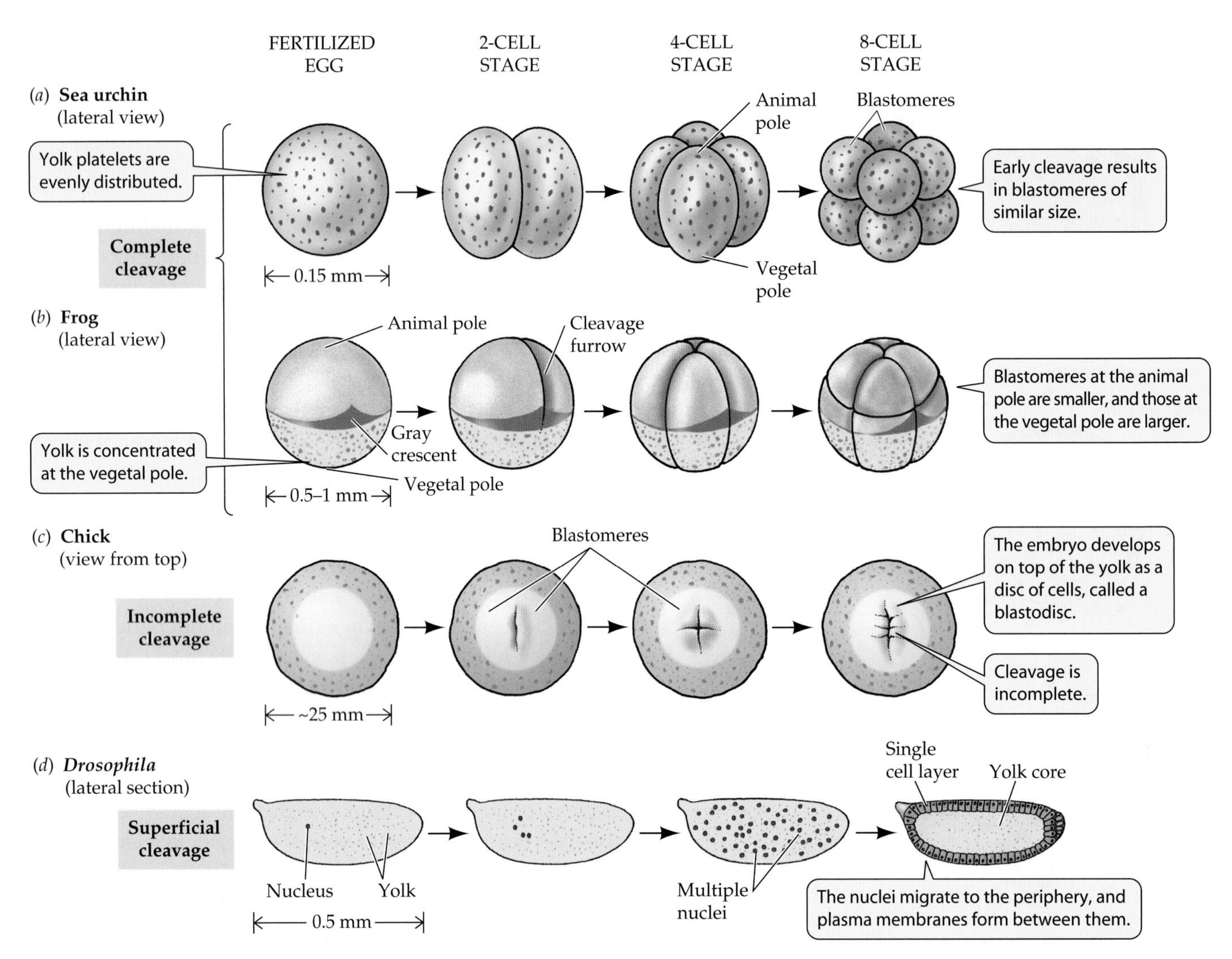

20.4 Patterns of Cleavage in Four Model Organisms Differences in patterns of early embryonic development reflect differences in the way the egg cytoplasm is organized.

Cleavage in mammals is unique

Several features of mammalian cleavage are very different from those seen in other animal groups. First, the pattern of cleavage in mammals is *rotational*: the first cell division is parallel to the animal–vegetal axis, yielding two blastomeres. The second cell division occurs at right angles: one blastomere divides parallel to the animal–vegetal axis, while the other divides perpendicular to it (Figure 20.5*a*).

Cleavage in mammals is very slow; cell divisions are 12–24 hours apart, compared with tens of minutes to a few hours in non-mammalian species. Also, the cell divisions of mammalian blastomeres are not in synchrony with each other. Because the blastomeres do not undergo mitosis at the same time, the number of cells in the embryo does not progress in the regular (2, 4, 8, 16, 32, etc.) progression typical of other species.

Another unique feature of the slow mammalian cleavage is that the products of genes expressed at this time play roles in cleavage. In species such as sea urchins and frogs, gene expression does not occur in the blastomeres, and cleavage is directed exclusively by molecules that were present in the egg prior to fertilization.

As in other animals that have complete cleavage, the early cell divisions in a mammalian zygote produce a loosely associated ball of cells. However, at about the 8-cell stage, the behavior of the mammalian blastomeres changes. They change shape to maximize their surface contact with one another, form tight junctions, and become a very compact mass of cells (Figure 20.5*b*).

At the transition from the 16-cell to the 32-cell stage, the cells separate into two groups. The **inner cell mass** will become the embryo, while the surrounding cells become an encompassing sac called the **trophoblast,** which will become part of the placenta. Trophoblast cells secrete fluid, creating a cavity (blastocoel) with the inner cell mass at one end (see Figure 20.5*b*). At this stage, the mammalian embryo

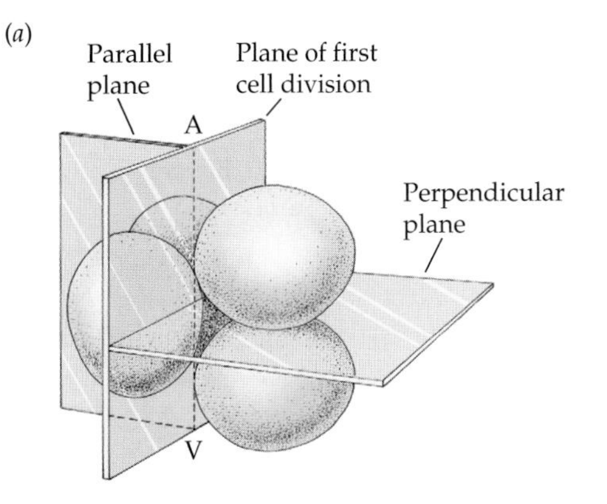
(a)
Parallel plane
Plane of first cell division
A
Perpendicular plane
V

20.5 The Mammalian Zygote Becomes a Blastocyst
(*a*) Mammals have rotational cleavage, in which the plane of the first cleavage is parallel to the animal–vegetal (A, V) axis, but the planes of the second cell division (shown in beige) are at right angles to each other. (*b*) Starting late in the 8-cell stage, the mammalian embryo undergoes compaction of its cells, resulting in a blastocyst—a dense inner cell mass on top of a hollow blastocoel, completely surrounded by trophoblast cells.

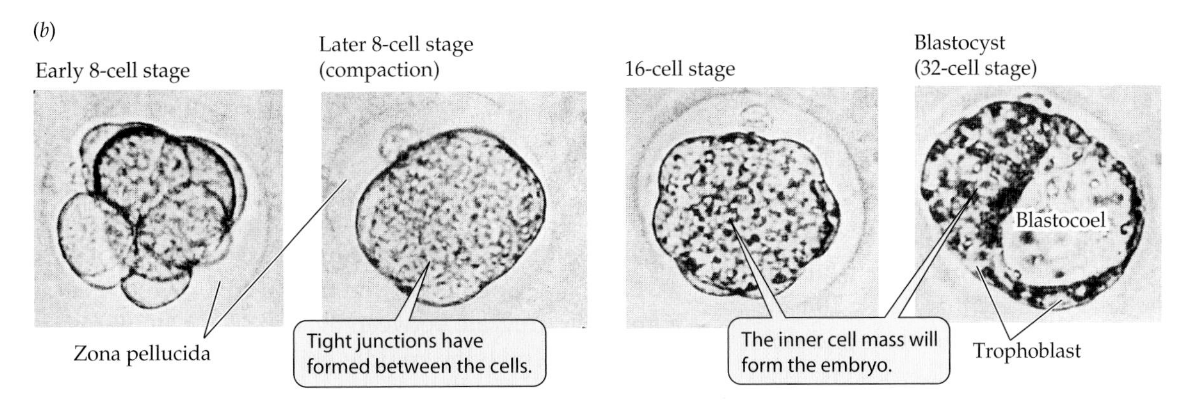
(b)
Early 8-cell stage
Later 8-cell stage (compaction)
16-cell stage
Blastocyst (32-cell stage)
Blastocoel
Zona pellucida
Tight junctions have formed between the cells.
The inner cell mass will form the embryo.
Trophoblast

is called a **blastocyst** to distinguish it from the blastulas of other animals.

Fertilization in mammals occurs in the upper reaches of the mother's oviduct, and cleavage occurs as the zygote travels down the oviduct to the uterus. When the blastocyst arrives in the uterus, the trophoblast adheres to the *endometrium* (the uterine wall). This event begins the process of *implantation* that embeds the embryo in the wall of the uterus (see Figure 20.14). In humans, implantation begins on about the sixth day after fertilization. As the blastocyst moves down the oviduct to the uterus, it must not embed itself in the oviduct wall, or the result will be an ectopic or tubal pregnancy—a very dangerous condition. Early implantation is normally prevented by an external proteinaceous layer called the *zona pellucida*, which surrounds the egg and remains around the cleaving ball of cells. At about the time the blastocyst reaches the uterus, it hatches from the zona pellucida, and implantation can occur.

Specific blastomeres generate specific tissues and organs

In all animal species, cleavage results in a repackaging of the egg cytoplasm into a large number of small cells surrounding a central cavity. Little cell differentiation occurs during cleavage, and in most nonmammalian species, none of the genome of the embryo is expressed. Nevertheless, cells in different regions of the resulting blastula possess different complements of the nutrients and cytoplasmic determinants that were present in the egg.

The blastocoel prevents cells from different regions of the blastula from interacting, but that will soon change. During the next stage of development, the cells of the blastula will move around and come into new associations with one another, communicate instructions to one another, and begin to differentiate. In many animals, these movements of the blastomeres are so regular and well orchestrated that it is possible to label a specific blastomere with a dye and identify the tissues and organs that form from its progeny. Such labeling experiments produce **fate maps** of the blastula (Figure 20.6).

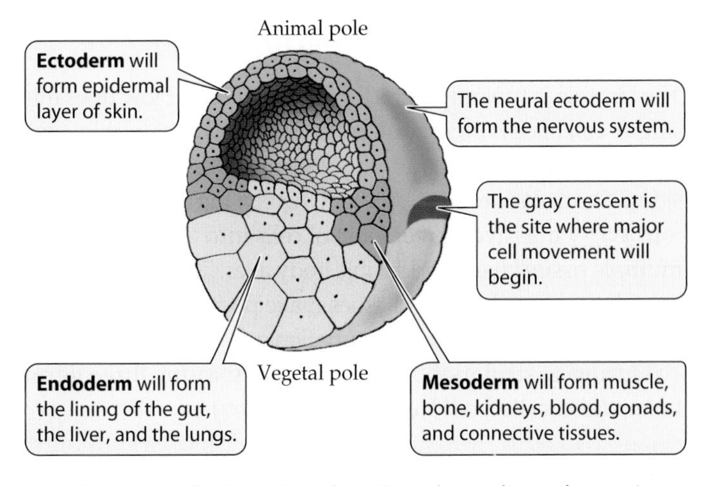
Animal pole
Ectoderm will form epidermal layer of skin.
The neural ectoderm will form the nervous system.
The gray crescent is the site where major cell movement will begin.
Endoderm will form the lining of the gut, the liver, and the lungs.
Vegetal pole
Mesoderm will form muscle, bone, kidneys, blood, gonads, and connective tissues.

20.6 Fate Map of a Frog Blastula The colors indicate the portions of the blastula that will form the three germ layers, and subsequently the frog's tissues and organs.

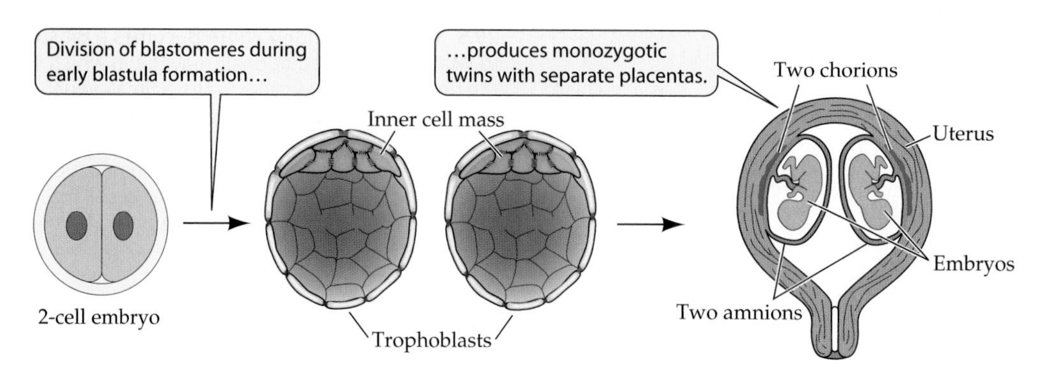

20.7 Twinning in Humans
Because humans have regulative development, remaining cells can compensate when cells are lost in early cleavages. Monozygotic (identical) twins can result when cells in the early blastula become physically separated and each group of cells goes on to produce a separate embryo.

Blastomeres become **determined**—committed to specific fates—at different times in different species. In some species, such as roundworms and clams, blastomeres are determined by the 8-cell stage. If one of these blastomeres is experimentally removed, a particular portion of the embryo will not form. This type of development has been called **mosaic development** because each blastomere seems to contribute a specific set of "tiles" to the final "mosaic" that is the adult animal. In contrast, other species, such as sea urchins and vertebrates, have **regulative development**: The loss of some cells during cleavage does not affect the developing embryo because the remaining cells compensate for the loss.

If some blastomeres can change their fate to compensate for the loss of other cells during cleavage and blastula formation, are those cells capable of forming an entire embryo? To a certain extent, they are. During cleavage or early blastula formation in mammals, for example, if the blastomeres are physically separated into two groups, both groups can produce complete embryos (Figure 20.7). Since the two embryos come from the same zygote, they will be *monozygotic twins*—genetically identical. Non-identical twins occur when two separate eggs are fertilized by two separate sperm. Thus, while identical twins are always of the same sex, non-identical twins have a 50 percent chance of being the same sex.

Gastrulation: Producing the Body Plan

The blastula is typically a fluid-filled ball of cells. How does this simple ball of cells become an embryo, made up of multiple tissue layers, with head and tail ends and dorsal and ventral sides? **Gastrulation** is the process whereby the blastula is transformed by massive movements of cells into an embryo with multiple tissue layers and visible body axes. The resulting spatial relationships between tissues make possible the inductive interactions that trigger differentiation and organ formation.

During gastrulation, the animal body forms three **germ layers** (also called *cell layers* or *tissue layers*):

- Some blastomeres move together as a sheet to the inside of the embryo, creating an inner germ layer called the **endoderm**. The endoderm will give rise to the lining of the digestive tract, respiratory tract, and circulatory system and make up other internal tissues such as the pancreas and liver.
- The cells remaining on the outside of the embryo become the outer germ layer, the **ectoderm**. The ectoderm will give rise to the nervous system, the skin, hair, and nails, sweat glands, oil glands, and milk secretory ducts.
- Other cells migrate between the endoderm and the ectoderm to become the middle germ layer, or **mesoderm**. The mesoderm will contribute tissues to many organs, including blood vessels, muscle, bones, liver, and heart.

Some of the most challenging and interesting questions in animal development have concerned what directs the cell movements of gastrulation and what is responsible for the resulting patterns of cell differentiation and organ formation. In the past 25 years, scientists have answered many of these questions at the molecular level. In the discussion that follows, we'll consider the similarities and differences among gastrulation in sea urchins, frogs, reptiles, birds, and mammals. We'll also review some of the exciting discoveries about the mechanisms underlying these phenomena.

Invagination of the vegetal pole characterizes gastrulation in the sea urchin

The sea urchin blastula is a simple, hollow ball of cells that is only one cell thick. The end of the blastula stage is marked by a dramatic slowing of the rate of mitosis, and the beginning of gastrulation is marked by a flattening of the vegetal hemisphere (Figure 20.8). Some cells at the vegetal pole bulge into the blastocoel, break free, and migrate into the cavity. These cells become *primary mesenchyme* cells—cells of the middle germ layer, the mesoderm. (Mesenchyme cells are unconnected to one another and act as independent units, in contrast to epithelial cells, which are tightly packed into sheets or tubes.)

The flattening at the vegetal pole results from changes in the shape of the individual blastomeres. These cells shift from being rather cuboidal to become wedge-shaped, with constricted outer edges and expanded inner edges. As a result of these shape changes, the vegetal pole bulges inward,

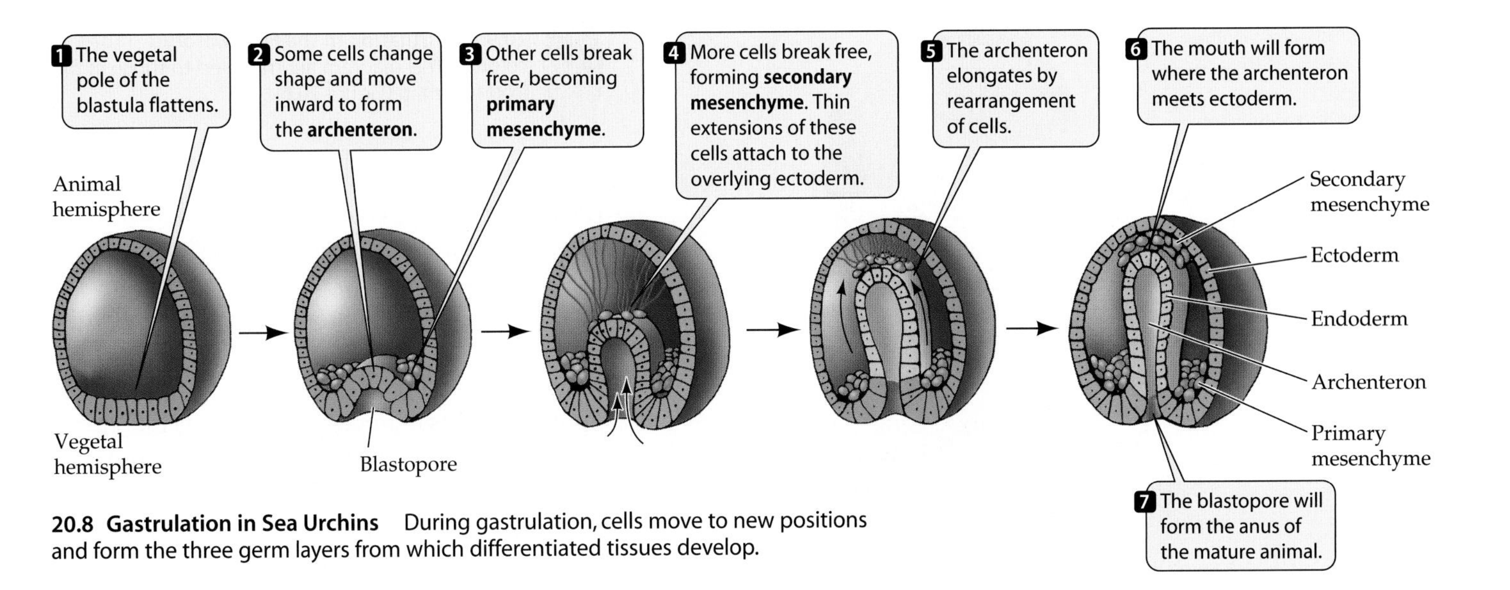

1 The vegetal pole of the blastula flattens.
2 Some cells change shape and move inward to form the archenteron.
3 Other cells break free, becoming primary mesenchyme.
4 More cells break free, forming secondary mesenchyme. Thin extensions of these cells attach to the overlying ectoderm.
5 The archenteron elongates by rearrangement of cells.
6 The mouth will form where the archenteron meets ectoderm.
Animal hemisphere
Vegetal hemisphere
Blastopore
Secondary mesenchyme
Ectoderm
Endoderm
Archenteron
Primary mesenchyme
7 The blastopore will form the anus of the mature animal.

20.8 Gastrulation in Sea Urchins During gastrulation, cells move to new positions and form the three germ layers from which differentiated tissues develop.

or *invaginates*, as if someone were poking a finger into a hollow ball. The cells that invaginate become the endoderm and form the primitive gut, the *archenteron*. At the tip of the archenteron more cells break free, entering the blastocoel to form more mesoderm, the *secondary mesenchyme*.

The early invagination of the archenteron is due to the changes in cell shapes, but eventually it is pulled by the secondary mesenchyme cells. These cells, attached to the tip of the archenteron, send out extensions that adhere to the overlying ectoderm and contract. Where the archenteron eventually makes contact with the ectoderm, the mouth of the animal will form. The opening created by the invagination of the vegetal pole is called the **blastopore**; it will become the anus of the animal.

What mechanisms control the various cell movements of sea urchin gastrulation? The immediate answer is that specific properties of particular blastomeres change. For example, some vegetal cells migrate into the blastocoel to form the primary mesenchyme because they lose their attachments to neighboring cells. Once they bulge into the blastocoel, they move by extending long processes called *filopodia* along an extracellular matrix of proteins that is laid down by the ectodermal cells lining the blastocoel.

A deeper understanding of gastrulation requires that we discover the molecular mechanisms whereby certain blastomeres develop properties different from those of others. Cleavage divides up the cytoplasm of the egg in a very systematic way. The sea urchin blastula at the 64-cell stage is radially symmetrical, but it has polarity. It consists of tiers of cells. As in the frog blastula, the top is the animal pole and the bottom the vegetal pole.

If different tiers of blastula cells are separated, they show different developmental potentials (see Figure 19.7). Only cells from the vegetal pole are capable of initiating the development of a complete larva. It has been proposed that the reason for these differences is an uneven distribution of various transcriptional regulatory proteins in the egg cytoplasm. As cleavage progresses, these proteins end up in different combinations in different groups of cells. Therefore, specific sets of genes are activated in different cells, determining their different developmental capacities. Let's turn now to gastrulation in the frog, in which a number of key signaling molecules have been identified.

Gastrulation in the frog begins at the gray crescent

Amphibian blastulas have considerable yolk and are more than one cell thick; therefore, gastrulation is more complex in amphibians than in sea urchins. Furthermore, there is considerable variation among different species of amphibians. In this brief account, we will mix results from studies done on different species to produce a generalized picture of amphibian development.

Amphibian gastrulation begins when certain cells in the gray crescent change their shape and their cell adhesion properties. The main bodies of these cells bulge inward toward the blastocoel while they remain attached to the outer surface of the blastula by slender necks. Because of their shape, these cells are called *bottle cells*.

The bottle cells mark the spot where the **dorsal lip** of the blastopore will form (Figure 20.9). As the bottle cells move inward, they create this lip, over which successive sheets of cells will move into the blastocoel in a process called *involution*. The first involuting cells are those of the prospective endoderm, and they form the primitive gut, or archenteron. Closely following are the cells that will form the mesoderm. As gastrulation proceeds, cells from the animal hemisphere move toward the site of involution in a process called *epiboly*.

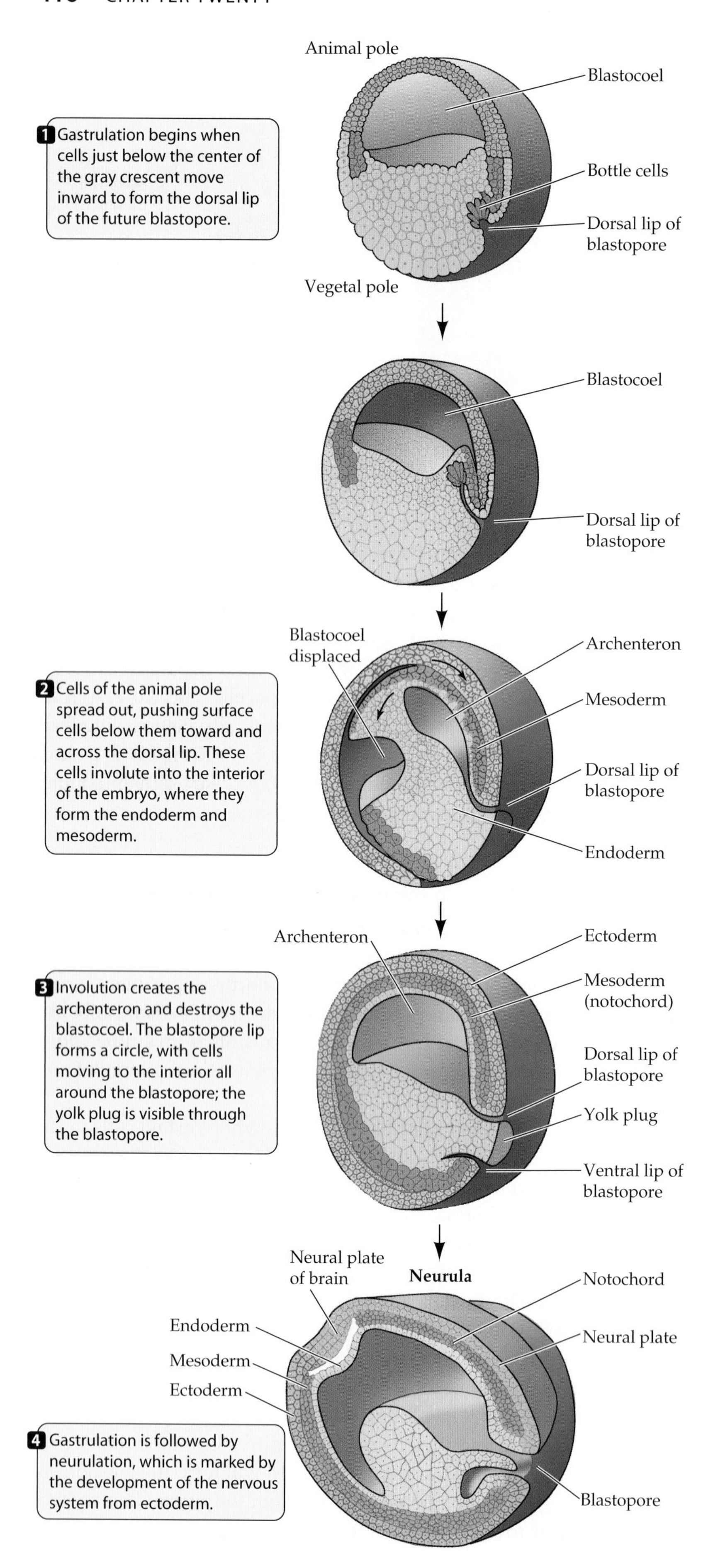
Animal pole
Blastocoel
1 Gastrulation begins when cells just below the center of the gray crescent move inward to form the dorsal lip of the future blastopore.
Bottle cells
Dorsal lip of blastopore
Vegetal pole
Blastocoel
Dorsal lip of blastopore
Blastocoel displaced
Archenteron
Mesoderm
2 Cells of the animal pole spread out, pushing surface cells below them toward and across the dorsal lip. These cells involute into the interior of the embryo, where they form the endoderm and mesoderm.
Dorsal lip of blastopore
Endoderm
Archenteron
Ectoderm
Mesoderm (notochord)
3 Involution creates the archenteron and destroys the blastocoel. The blastopore lip forms a circle, with cells moving to the interior all around the blastopore; the yolk plug is visible through the blastopore.
Dorsal lip of blastopore
Yolk plug
Ventral lip of blastopore
Neural plate of brain
Neurula
Notochord
Endoderm
Neural plate
Mesoderm
Ectoderm
4 Gastrulation is followed by neurulation, which is marked by the development of the nervous system from ectoderm.
Blastopore

20.9 Gastrulation in the Frog Embryo The colors in this diagram are matched to those in the frog fate map (Figure 20.6).

The blastopore lip widens and eventually forms a complete circle surrounding a "plug" of yolk-rich cells. As cells continue to move inward through the blastopore, the archenteron grows, gradually displacing the blastocoel.

As gastrulation comes to an end, the amphibian embryo consists of three germ layers: ectoderm on the outside, endoderm on the inside, and mesoderm in the middle. The embryo also has a dorsal–ventral and anterior–posterior organization. Most importantly, however, the fates of specific regions of the endoderm, mesoderm, and ectoderm have been determined. The discovery of the events whereby determination takes place in the amphibian embryo is one of the most exciting stories in animal development.

The dorsal lip of the blastopore organizes embryo formation

In the 1920s, the German biologist Hans Spemann was studying the development of salamander eggs. He was interested in finding out whether the nuclei of blastomeres remain *totipotent*—capable of directing the development of a complete embryo. With great patience and dexterity, he formed loops from a single human baby hair to constrict fertilized eggs, effectively dividing them in half.

When Spemann's loops bisected the gray crescent, both halves of the zygote gastrulated and developed into complete embryos (Experiment 1 in Figure 20.10). But when the gray crescent was on only one side of the constriction, only that half of the zygote developed into a complete embryo. The half lacking gray crescent material became a clump of undifferentiated cells that Spemann called the "belly piece" (Experiment 2 in Figure 20.10). Spemann thus hypothesized that cytoplasmic determinants in the region of the gray crescent are necessary for gastrulation and thus for the development of a normal organism.

To test his hypothesis, Spemann and his student Hilde Mangold conducted a series of delicate tissue transplantation experiments. They transplanted pieces of early gastrulas to various locations on other gastrulas. Guided by fate maps (see Figure 20.6), they were able to take a piece of ectoderm

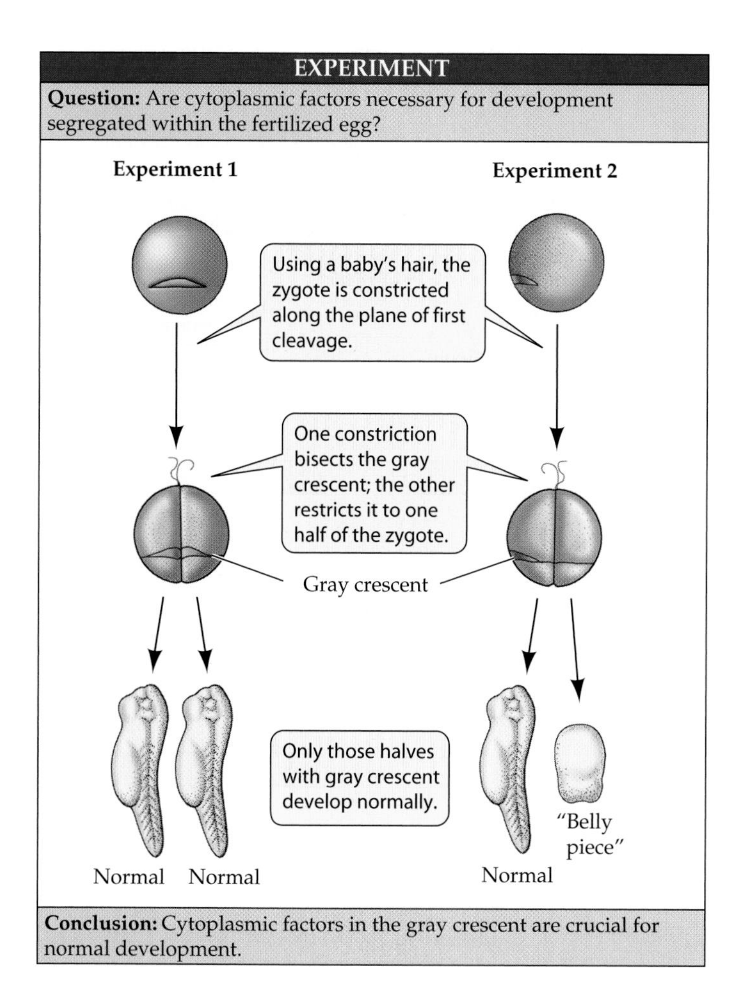

20.10 Spemann's Experiment Spemann's research revealed that gastrulation and subsequent normal development in salamanders depended on cytoplasmic determinants localized in the gray crescent.

20.11 The Dorsal Lip Induces Embryonic Organization In a famous experiment, Spemann and Mangold transplanted the dorsal lip of the blastopore. The transplanted tissue induced a second site of gastrulation and the formation of a second embryo.

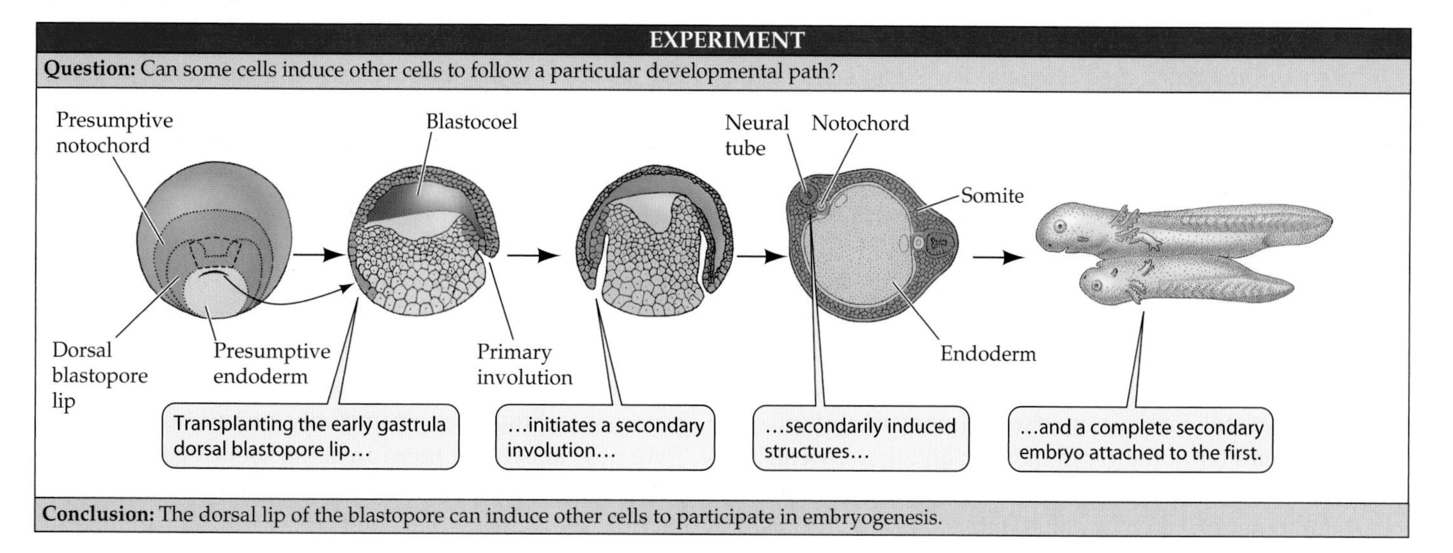

EXPERIMENT
Question: Can some cells induce other cells to follow a particular developmental path?
Presumptive notochord
Blastocoel
Neural tube
Notochord
Somite
Dorsal blastopore lip
Presumptive endoderm
Primary involution
Endoderm
Transplanting the early gastrula dorsal blastopore lip...
...initiates a secondary involution...
...secondarily induced structures...
...and a complete secondary embryo attached to the first.
Conclusion: The dorsal lip of the blastopore can induce other cells to participate in embryogenesis.

they knew would develop into skin and transplant it to a region that normally becomes part of the nervous system, and vice versa.

When they performed these transplants in early gastrulas, the transplanted pieces always developed into tissues that were appropriate for the location where they were placed. Donor presumptive epidermis (that is, cells destined to become skin in their original location) developed into host neural ectoderm (nervous system tissue), and donor presumptive neural ectoderm developed into host skin. Thus, the fates of the transplanted cells had not been determined before the transplantation.

In late gastrulas, however, the same experiment yielded opposite results. Donor presumptive epidermis produced patches of skin cells in the host nervous system, and donor presumptive neural ectoderm produced nervous system tissue in the host skin. Something had occurred during gastrulation to determine the fates of the embryonic cells. In other words, as Spemann and Mangold had hypothesized, the path of differentiation a cell would follow was determined during gastrulation.

Spemann and Mangold next did an experiment that produced momentous results: They transplanted the dorsal lip of the blastopore (Figure 20.11). When this small piece of tissue was transplanted into the presumptive belly area of another gastrula, it stimulated a second site of gastrulation, and second whole embryo formed belly-to-belly with the original embryo. Because the dorsal lip of the blastopore was apparently capable of inducing the formation of an entire embryo, Spemann and Mangold dubbed it the **primary embryonic organizer**, or simply the **organizer**.

MOLECULAR MECHANISMS OF THE ORGANIZER. In recent years, researchers have studied the primary embryonic organizer intensively to discover the molecular mechanisms involved in its action. The distribution of the transcription factor β-catenin in the late blastula corresponds to the location of the organizer in the early gastrula, so β-catenin is a candidate for the initiator of organizer activity. To prove that a protein is an inductive signal, it has to be shown that it is both *necessary* and *sufficient* for the proposed effect. In other words, the effect should not occur if the candidate protein is not present (necessity), and the candidate protein should be capable of inducing the effect where it would otherwise not occur (sufficiency).

The criteria of necessity and sufficiency have indeed been satisfied for the transcription factor β-catenin. If β-catenin mRNA transcripts are depleted by injections of antisense RNA into the egg (see Chapter 16), gastrulation does not occur. If β-catenin is experimentally overexpressed in another region of the blastula, it can induce a second axis of embryo formation, as the transplanted dorsal lip did in the Spemann–Mangold experiments. Thus, β-catenin appears to be both necessary and sufficient for the formation of the primary embryonic organizer—but it is only one component of a complex signaling process.

What follows is a summary of some of the critical early steps in this signaling cascade. This description may contain a confusing amount of detail. However, it is not the arcane names of the genes and gene products involved that are important to remember. Rather, we hope to provide a basic understanding of how these signaling molecules—their interactions and their gradients—can create and convey positional and temporal information.

Studies of early gastrulas revealed that primary embryonic organizer activity is induced by signals emanating from vegetal cells just below the gray crescent. The protein β-catenin appears to play critical roles in generating these signals. One signal critical to stimulating the expression of organizer genes is the transcription factor Goosecoid. Expression of the *goosecoid* gene appears to depend on two signaling pathways, both of which involve β-catenin.

The first of these pathways involves a *goosecoid*-promoting transcription factor called Siamois. The *siamois* gene is normally repressed by a ubiquitous transcription factor called Tcf-3, but in cells where β-catenin is present, an interaction between Tcf-3 and β-catenin induces *siamois* expression (Figure 20.12). But Siamois protein alone is not sufficient for *goosecoid* expression.

Vegetal cells receive mRNA transcripts from the original egg cytoplasm for proteins in the TGF-β (transforming growth factor β) superfamily of cell signaling molecules. One or more of these proteins (candidates include Vg1 and Nodal) interact with Siamois protein by cooperatively binding to the promoter of the *goosecoid* gene and thereby controlling its transcription (see Figure 20.12). Thus it is a particular combination of factors that determine which cells become the primary organizer. Cells that receive other combinations of signaling molecules are induced to become different types of mesoderm.

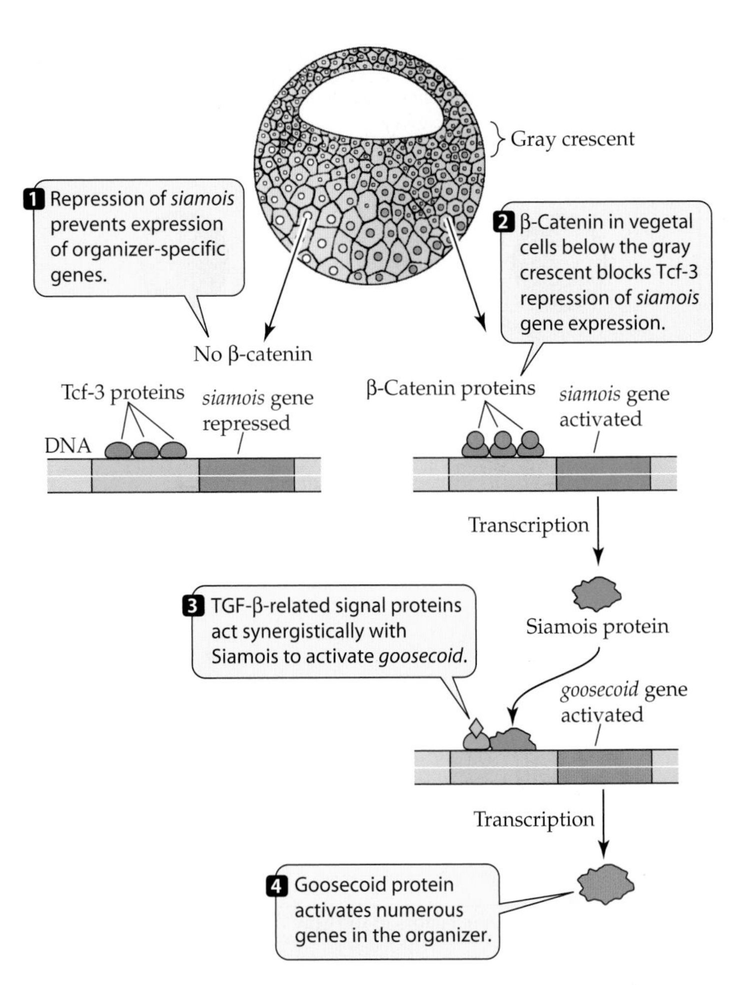

20.12 Molecular Mechanisms of the Primary Embryonic Organizer The organizing potential of the gray crescent depends on the activity of the *goosecoid* gene, which in turn is activated by signaling pathways set up in the vegetal cells below the gray crescent.

MOLECULAR MECHANISMS OF LEFT–RIGHT AXIS FORMATION. We have seen how the distribution of cytoplasmic determinants in the egg can set up a dorsal–ventral axis, and how the site of sperm entry can set up an anterior–posterior axis. What about the left–right body axis? After all, not everything in the animal is bilaterally symmetrical. The internal organs of a vertebrate have many left–right asymmetries: In humans, the heart is tilted to the right side of the body, the aorta comes off of the left side of the heart and the pulmonary artery comes off of the right side of the heart; the spleen is on the left side of the body; and the large intestine goes from right to left, to name only a few.

We now know that there are a number of genes that are necessary for normal left–right organization of the body. If one of these genes is knocked out, it can randomize the left–right organization of the internal organs, with serious, even lethal, consequences. What triggers the asymmetrical expression of these genes?

We do not know the complete answer to this question, but it appears that the mechanism involves a left–right differential distribution of some of the transcription factors that act very early during gastrulation. For example, in frogs, one of the TGF-β proteins involved in organizer determination is also responsible for determining the left–right axis. In mammals, there are cilia that cause a differential flow of fluid in the yolk sac cavity. If these cilia are inactivated, the normal left–right asymmetries of the internal organs become random.

Reptilian and avian gastrulation is an adaptation to yolky eggs

The eggs of reptiles and birds contain a mass of yolk, and therefore the blastulas of these species develop as a disc of cells on top of the yolk (see Figure 20.4*c*). We will use the chicken egg to show how gastrulation proceeds in a flat disc of cells rather than in a ball of cells.

Cleavage in the chick results in a flat, circular layer of cells called a blastodisc. Between the blastodisc and the yolk mass is a fluid-filled space. Some cells from the blastodisc break free and move into this space. Other cells grow into this space from the posterior margin of the blastodisc. These cells come together to form a continuous layer called the **hypoblast**, which will later give rise to extraembryonic membranes that will support and nourish the developing embryo. The overlying cells make up the **epiblast**, which will form the embryo proper. Thus, the avian blastula is a flattened structure consisting of an upper epiblast and a lower hypoblast, which are joined at the margins of the blastodisc. The blastocoel is the fluid-filled space between the epiblast and hypoblast.

Gastrulation begins with a thickening in the posterior region of the epiblast caused by the movement of cells toward the midline and then forward along the midline (Figure 20.13). The result is a midline ridge called the *primitive streak*. A depression called the *primitive groove* forms along the length of the primitive streak. The primitive groove functions as the blastopore, and cells migrate through it into the blastocoel to become endoderm and mesoderm.

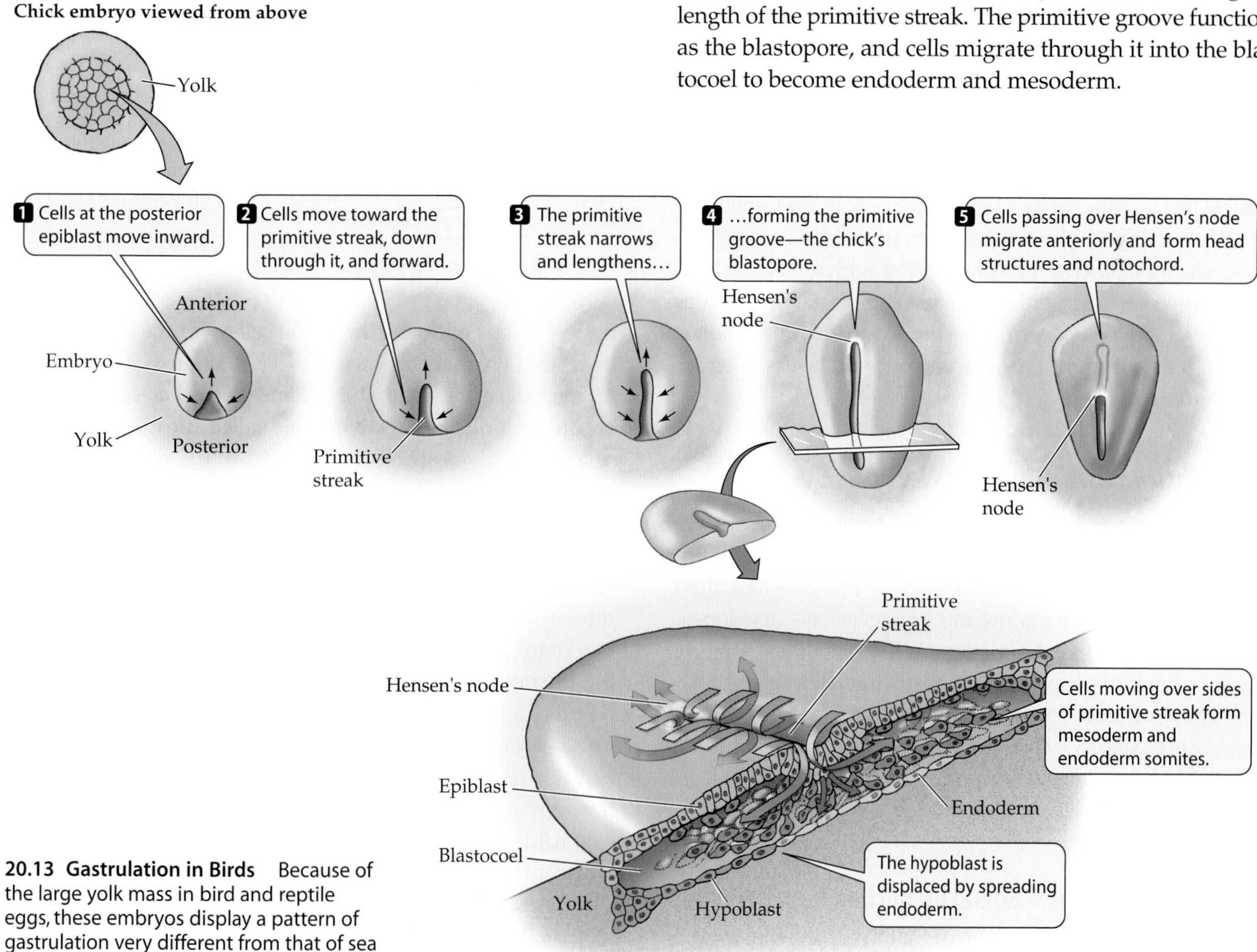
Chick embryo viewed from above
Yolk
1 Cells at the posterior epiblast move inward.
2 Cells move toward the primitive streak, down through it, and forward.
3 The primitive streak narrows and lengthens...
4 ...forming the primitive groove—the chick's blastopore.
5 Cells passing over Hensen's node migrate anteriorly and form head structures and notochord.
Anterior
Embryo
Yolk
Posterior
Primitive streak
Hensen's node
Hensen's node
Primitive streak
Hensen's node
Cells moving over sides of primitive streak form mesoderm and endoderm somites.
Epiblast
Endoderm
Blastocoel
The hypoblast is displaced by spreading endoderm.
Yolk
Hypoblast
Cross section through chick embryo

20.13 Gastrulation in Birds Because of the large yolk mass in bird and reptile eggs, these embryos display a pattern of gastrulation very different from that of sea urchins and amphibians.

In the chick embryo, no archenteron forms, but the endoderm and mesoderm migrate forward to form the gut and other structures. At the anterior end of the primitive groove is a thickening called **Hensen's node**, which is the equivalent of the dorsal lip of the amphibian blastopore. In fact, many signaling molecules that have been identified in the frog organizer are also expressed in Hensen's node. Cells that pass over Hensen's node become determined by the time they reach their final destination, where they differentiate into certain tissues and structures of the head and dorsal midline (but not the nervous system).

Mammals have no yolk, but retain the avian–reptilian gastrulation pattern

Mammals and birds both evolved from reptilian ancestors, so it is not surprising that they share patterns of early development, even though the eggs of mammals have no yolk. Earlier we described the development of the mammalian trophoblast and the inner cell mass, which is the equivalent of the avian epiblast.

As in avian development, the inner cell mass splits into an upper layer called the epiblast and a lower layer called the hypoblast, with a fluid-filled cavity between them. The embryo will form from the epiblast, and the hypoblast will contribute to the extraembryonic membranes (Figure 20.14). The epiblast also contributes to the extraembryonic membranes; specifically, it splits off an upper layer of cells that will form the amnion. The amnion will grow to surround the developing embryo as a membranous sac filled with amniotic fluid. Gastrulation occurs in the mammalian epiblast just as it does in the avian epiblast. A primitive groove forms, and epiblast cells migrate through the groove to become layers of endoderm and mesoderm.

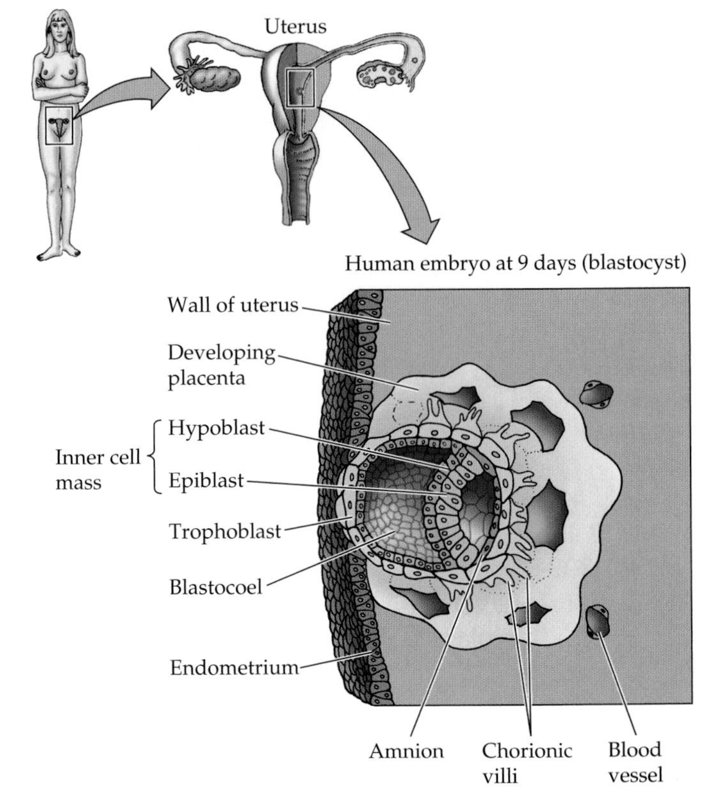

20.14 A Human Blastocyst at Implantation Adehesion molecules and proteolytic enzymes secreted by trophoblast cells allow the blastocyst to burrow into the endometrium. Once implanted within the wall of the uterus, the trophoblast cells send out numerous projections—the chorionic villi—which increase the embyro's area of contact with the mother's bloodstream.

Neurulation: Initiating the Nervous System

Gastrulation produces an embryo with three germ layers that are positioned to influence one another through inductive interactions. During the next phase of development, called **organogenesis**, many organs and organ systems develop simultaneously and in coordination with one another. An early process of organogenesis that is directly related to gastrulation is **neurulation**, the initiation of the nervous system in vertebrates . We will examine this event in the amphibian embryo, but it occurs in a similar fashion in reptiles, birds, and mammals.

The stage is set by the dorsal lip of the blastopore

The first cells to pass over the dorsal lip of the blastopore move anteriorly and become the endodermal lining of the digestive tract. Following these first cells over the dorsal lip are those that will become mesoderm (see Figure 20.9). The dorsal mesoderm closest to the midline (the *chordomesoderm*) will become a rod of connective tissue called the **notochord**. The notochord gives structural support to the developing embryo; it is eventually replaced by the vertebral column. After gastrulation, the chordomesoderm induces the overlying ectoderm to begin forming the nervous system.

Neurulation involves the formation of an internal neural tube from an external sheet of cells. The first signs of neurulation are flattening and thickening of the ectoderm overlying the notochord; this thickened area forms the *neural plate* (Figure 20.15). The edges of the neural plate that run in an anterior–posterior direction continue to thicken to form ridges or folds. Between these neural folds, a groove forms and deepens as the folds roll over it to converge on the midline. The folds fuse, forming both a cylinder, the **neural tube**, and a continuous overlying layer of epidermal ectoderm. The neural tube develops bulges at the anterior end, which become the major divisions of the brain; the rest of the tube becomes the spinal cord.

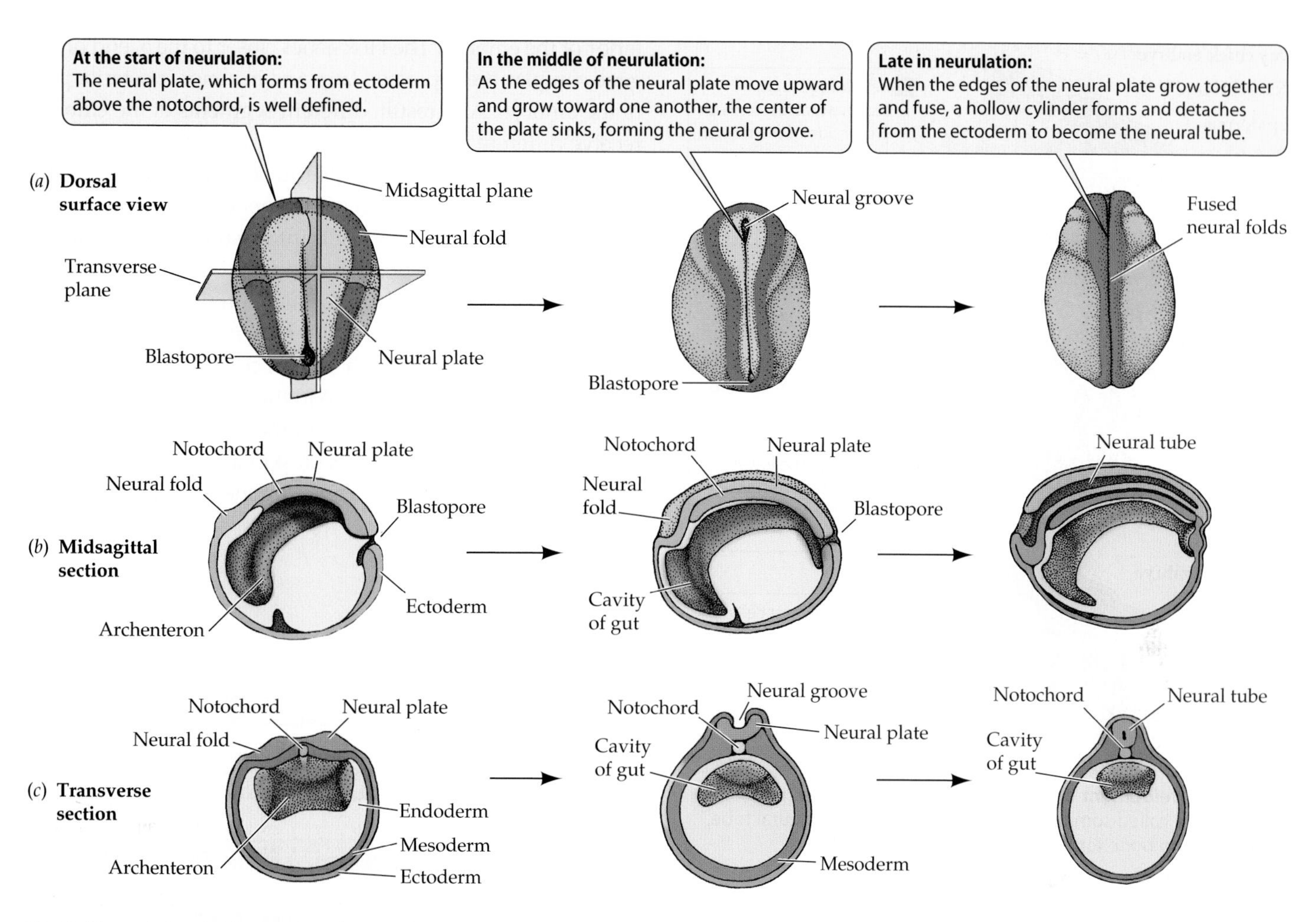

20.15 Neurulation in the Frog Embryo Continuing the sequence from Figures 20.6 and 20.9, these drawings outline the development of the frog's neural tube.

In humans, failure of the neural tube to develop normally can result in serious birth defects. If the neural folds fail to fuse in a posterior region, the result is a condition known as *spina bifida*. If they fail to fuse at the anterior end, an infant can develop without a forebrain—a condition called *anencephaly*. Whereas several genetic factors that can cause neural tube defects have been identified, there are also environmental factors, including dietary ones. The incidence of neural tube defects used to be about 1 in 300 live births, but we now know that this incidence can be cut in half if pregnant women have an adequate amount of folic acid (a B vitamin) in their diets.

Body segmentation develops during neurulation

Like the fruit flies whose development we traced in Chapter 19, vertebrates have a body plan consisting of repeating segments that are modified during development. These segments are most evident as the repeating patterns of vertebrae, ribs, nerves, and muscles along the anterior–posterior axis.

As the neural tube forms, mesodermal tissues gather along the sides of the notochord to form separate blocks of cells called **somites** (Figure 20.16). The somites produce cells that will become the vertebrae, ribs, and muscles of the trunk and limbs.

The nerves that connect the brain and spinal cord with tissues and organs throughout the body are also arranged segmentally. The somites help guide the organization of these peripheral nerves, but the nerves are not of mesodermal origin. When the neural tube fuses, cells adjacent to the line of closure break loose and migrate inward between the epidermis and the somites and under the somites. These cells, called *neural crest cells*, give rise to a number of structures, including the peripheral nerves, which grow out to the body tissues and back into the spinal cord.

As development progresses, the segments of the body become different. Regions of the spinal cord differ, regions of the vertebral column differ in that some vertebrae grow ribs of various sizes and others do not, forelegs arise in the anterior part of the embryo, and hind legs arise in the posterior

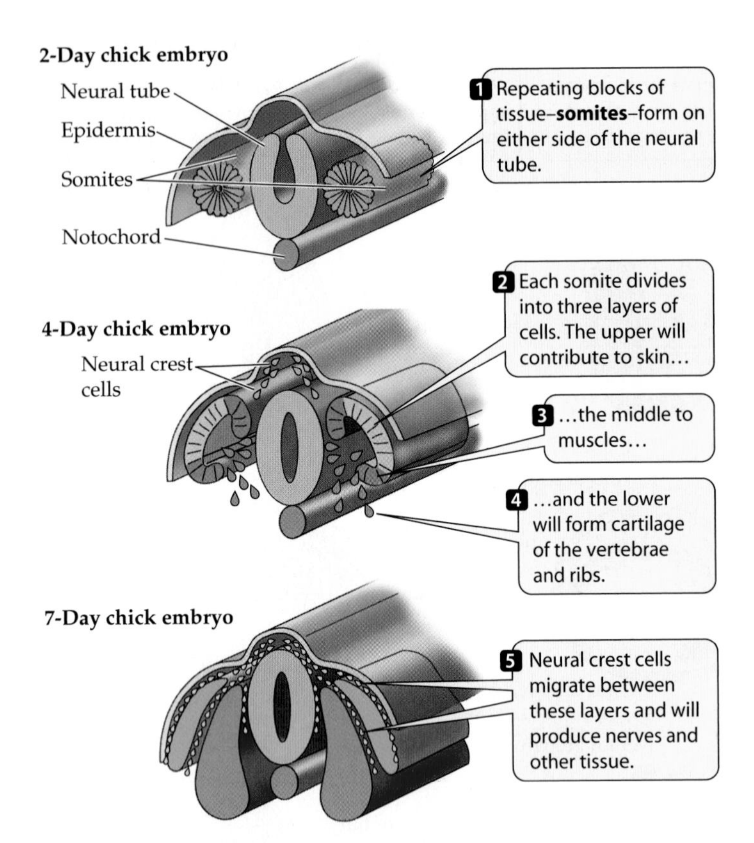

2-Day chick embryo
Neural tube
Epidermis
Somites
Notochord
1 Repeating blocks of tissue–somites–form on either side of the neural tube.
4-Day chick embryo
Neural crest cells
2 Each somite divides into three layers of cells. The upper will contribute to skin…
3 …the middle to muscles…
4 …and the lower will form cartilage of the vertebrae and ribs.
7-Day chick embryo
5 Neural crest cells migrate between these layers and will produce nerves and other tissue.

20.16 The Development of Body Segmentation Repeating blocks of tissue called somites form on either side of the neural tube. Skin, muscle, and bone form from the somites.

region. How is a somite in the anterior part of a mouse embryo programmed to produce forelegs rather than hind legs?

Hox genes control development along the anterior–posterior axis

Homeobox genes are central to the process of anterior–posterior determination and differentiation. In Chapter 19, we saw how homeotic genes control body segmentation in *Drosophila*. In the mouse, four families of homeobox genes, called **Hox genes**, control differentiation along the anterior–posterior body axis.

Each mammalian Hox gene family resides on a different chromosome and consists of about 10 genes. What is remarkable is that the temporal and spatial expression of these genes follows the same pattern as their linear order on their chromosome. That is, the Hox genes closest to the 3′ end of each gene complex are expressed first and are expressed in the anterior of the embryo. The Hox genes closer to the 5′ end of the gene complex are expressed later and in a more posterior part of the embryo. As a result, different segments of the embryo receive different combinations of Hox gene products, which serve as transcription factors (Figure 20.17). What causes the linear, sequential expression of Hox genes is unclear.

Whereas Hox genes give cells information about their position on the anterior–posterior body axis, other genes give cells information about their dorsal–ventral position. Tissues in each segment of the body differentiate according to their dorsal–ventral location. In the spinal cord, for example, sensory nerve connections develop in the dorsal region and motor nerve connections develop in the ventral region. In the somites, dorsal cells develop into skin and muscle and ventral cells develop into cartilage and bone (see Figure 20.16).

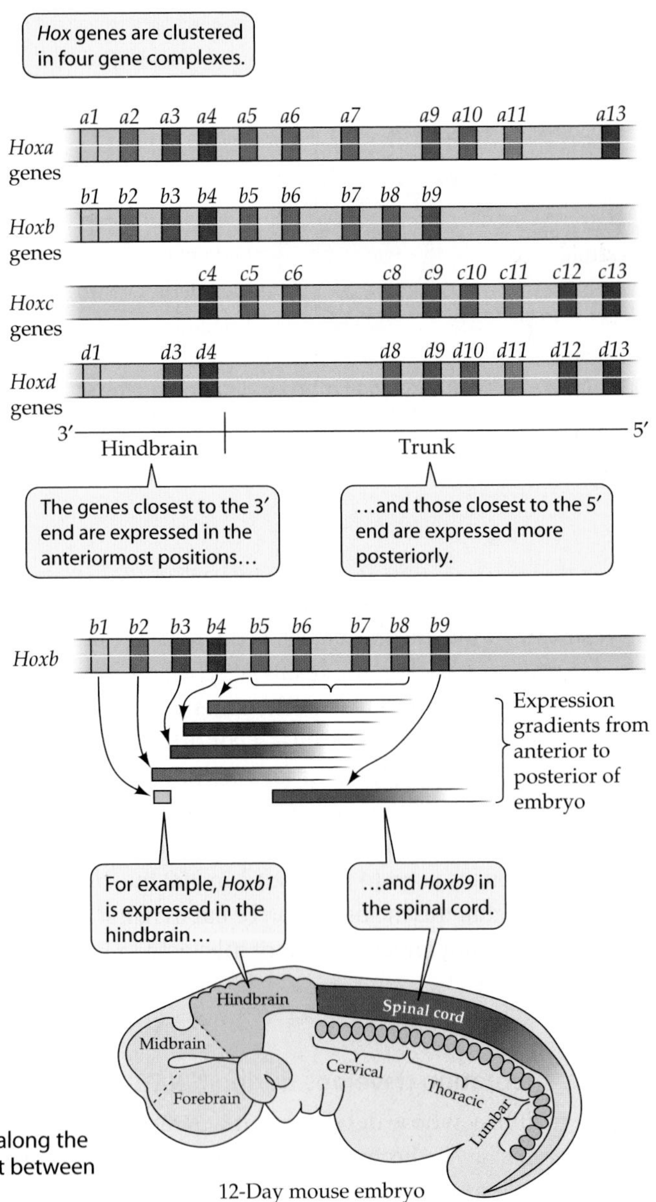

20.17 Hox Genes Control Body Segmentation Hox genes are expressed along the anterior–posterior axis of the embryo in the same order as their arrangement between the 3′ and 5′ ends of the gene complex.

An example of a gene that provides dorsal–ventral information in vertebrates is *sonic hedgehog*, which is expressed in the mammalian notochord and induces cells in the overlying neural tube to have fates characteristic of ventral spinal cord cells. (As with the Hox genes, *sonic hedgehog* is homologous to a *Drosophila* gene, which is known simply as *hedgehog*.)

One family of homeobox genes, the Pax genes, plays many roles in nervous system and somite development. One of these genes, *Pax3*, is expressed in those neural tube cells that will develop into dorsal spinal cord structures. *Sonic hedgehog* represses the expression of *Pax3*, and their interaction is one source of dorsal–ventral information for the differentiation of the spinal cord.

After the development of body segmentation, the formation of organs and organ systems progresses rapidly. The development of an organ involves extensive inductive interactions of the kind we saw in Chapter 19 in the example of the vertebrate eye. These inductive interactions are a current focus of study for developmental biologists.

Extraembryonic Membranes

There is more to a developing reptile, bird, or mammal than the embryo itself. The embryos of these vertebrates are surrounded by several **extraembryonic membranes**, which originate from the embryo but are not part of it. The extraembryonic membranes function in nutrition, gas exchange, and waste removal.

Extraembryonic membranes form with contributions from all germ layers

We will use the chicken to demonstrate how the extraembryonic membranes form from the germ layers created during gastrulation. The **yolk sac** is the first extraembryonic membrane to form, and it does so by extension of the endodermal tissue of the hypoblast layer along with some adjacent mesoderm. The yolk sac grows to encloses the entire body of yolk in the egg (Figure 20.18). It constricts at the top to create a tube that is continuous with the gut of the embryo. However, yolk does not pass through this tube. Yolk is digested by the endodermal cells of the yolk sac, and the nutrients are then transported to the embryo through blood vessels that form from the mesoderm and line the outer surface of the yolk sac. The **allantoic membrane** is also an outgrowth of the extraembryonic endoderm plus adjacent mesoderm. It forms the *allantois*, a sac for storage of metabolic wastes.

Just as the endoderm and mesoderm of the hypoblast grow out from the embryo to form the yolk sac and the allantoic membrane, ectoderm and mesoderm combine and extend beyond the limits of the embryo to form the other extraembryonic membranes. Two layers of cells extend all along the inside of the eggshell, both over the embryo and below the yolk sac. Where they meet, they fuse, forming two membranes, the inner **amnion** and the outer **chorion**. The amnion surrounds the embryo, forming the amniotic cavity. The amnion secretes fluid into the cavity, providing a protective environment for the embryo. The outer membrane, the chorion, forms a continuous membrane just under the eggshell (Figure 20.18). It limits water loss from the egg and also works with the enlarged allantoic membrane to exchange respiratory gases between the embryo and the outside world.

Extraembryonic membranes in mammals form the placenta

In mammals, the first extraembryonic membrane to form is the trophoblast, which is already apparent by the fifth cell

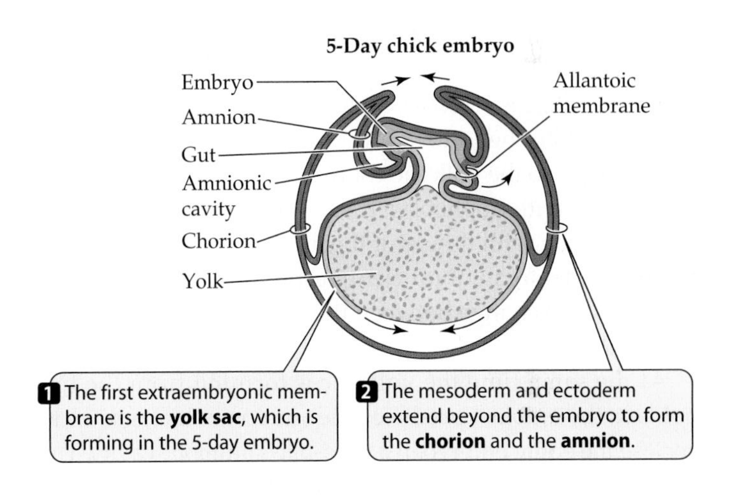
5-Day chick embryo
Embryo
Amnion
Gut
Amnionic cavity
Chorion
Yolk
Allantoic membrane
1 The first extraembryonic membrane is the yolk sac, which is forming in the 5-day embryo.
2 The mesoderm and ectoderm extend beyond the embryo to form the chorion and the amnion.

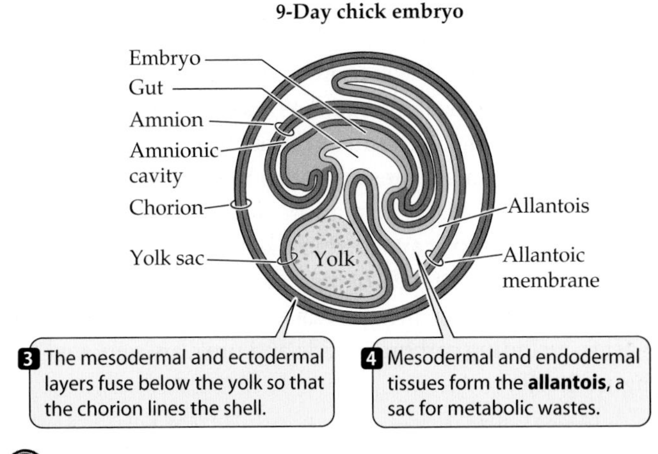
9-Day chick embryo
Embryo
Gut
Amnion
Amnionic cavity
Chorion
Yolk sac
Yolk
Allantois
Allantoic membrane
3 The mesodermal and ectodermal layers fuse below the yolk so that the chorion lines the shell.
4 Mesodermal and endodermal tissues form the allantois, a sac for metabolic wastes.

20.18 The Extraembryonic Membranes In birds, reptiles, and mammals, the embryo constructs four extraembryonic membranes. The yolk sac encloses the yolk, and the amnion and chorion enclose the embryo. Fluids secreted by the amnion fill the amniotic cavity, providing an aqueous environment for the embryo. The chorion, along with the allantois, mediates gas exchange between the embryo and its environment. The allantois stores the embryo's waste products.

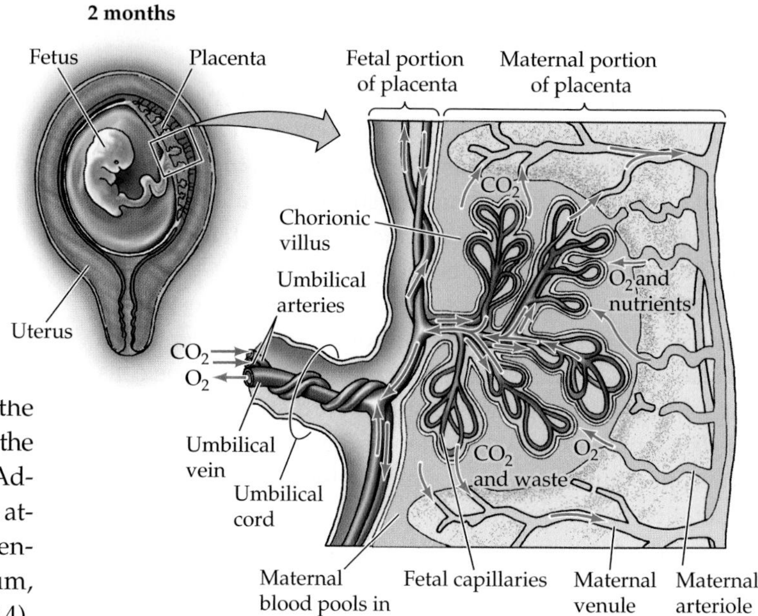

20.19 The Mammalian Placenta In most mammals, nutrients and wastes are exchanged between maternal and fetal blood in the placenta, which forms from the chorion and tissues of the uterine wall. The embryo is attached to the placenta by the umbilical cord. Embryonic blood vessels invade the placental tissue to form fingerlike chorionic villi. Maternal blood flows into the spaces surrounding the villi.

division (see Figure 20.5). When the blastocyst reaches the uterus and hatches from its encapsulating zona pellucida, the trophoblast cells interact directly with the endometrium. Adhesion molecules expressed on the surfaces of these cells attach them to the uterine wall. By excreting proteolytic enzymes, the trophoblast burrows into the endometrium, beginning the process of implantation (see Figure 20.14). Eventually, the entire trophoblast is within the wall of the uterus. The trophoblastic cells then send out numerous projections, or villi, to increase the surface area of contact with maternal blood.

Meanwhile, the hypoblast cells extend to form what in the bird would be the yolk sac. But there is no yolk in mammalian eggs, so the yolk sac contributes mesodermal tissues that interact with trophoblast tissues to form the chorion. The chorion, along with tissues of the uterine wall, produces the **placenta**, the organ of nutrient, respiratory gas, and metabolic waste exchange between the mother and the embryo (Figure 20.19).

At the same time the yolk sac is forming from the hypoblast, the epiblast produces the amnion, which grows to enclose the entire embryo in a fluid-filled amniotic cavity. The rupturing of the amnion and chorion and the loss of the amniotic fluid ("water breaking") herald the onset of labor in humans.

An allantois also develops in mammals, but its importance depends on how well nitrogenous wastes can be transferred across the placenta. In humans the allantois is minor; in pigs it is important. In humans and other mammals, allantoic tissues contribute to the formation of the umbilical cord, by which the embryo is attached to the chorionic placenta. It is through the blood vessels of the umbilical cord that nutrients and oxygen from the mother reach the developing fetus and wastes, including carbon dioxide and urea, are removed (see Figure 20.19).

The extraembryonic membranes provide means of detecting genetic diseases

Cells slough off of the developing human embryo and float in the amniotic fluid that bathes it. Later in development, a small sample of the amniotic fluid may be sampled with a needle as the first step of a process called **amniocentesis**. Cells from the fluid can be cultured and used for biochemical and genetic analyses that can reveal the sex of the fetus, as well as genetic markers for diseases such as cystic fibrosis, Tay-Sachs disease, and Down syndrome.

If amniocentesis is performed, it is usually not until after the fourteenth week of pregnancy, and the tests require two weeks to complete. If abnormalities in the fetus are detected, termination of the pregnancy at that stage would put the mother's health at greater risk than would an earlier abortion. Therefore, a newer technique, called **chorionic villus sampling**, is now in common use. In this test, a small sample of the tissue from the surface of the chorion is taken (Figure 20.20). This test can be done as early as the eighth week of pregnancy, and the results are available in several days.

Human Development

In humans, **gestation**, or pregnancy, lasts about 266 days, or 9 months. In smaller mammals gestation is shorter—for example, 21 days in mice—and in larger mammals it is longer—for example, 330 days in horses and 600 days in elephants. The events of human gestation can be divided into three periods of roughly 3 months each, called *trimesters*.

Intrauterine development can be divided into three trimesters

THE FIRST TRIMESTER. Implantation of the human blastocyst begins on about the sixth day after fertilization. After implantation, gastrulation occurs, the placenta forms, tissues differentiate, and organs begin to develop. The heart begins

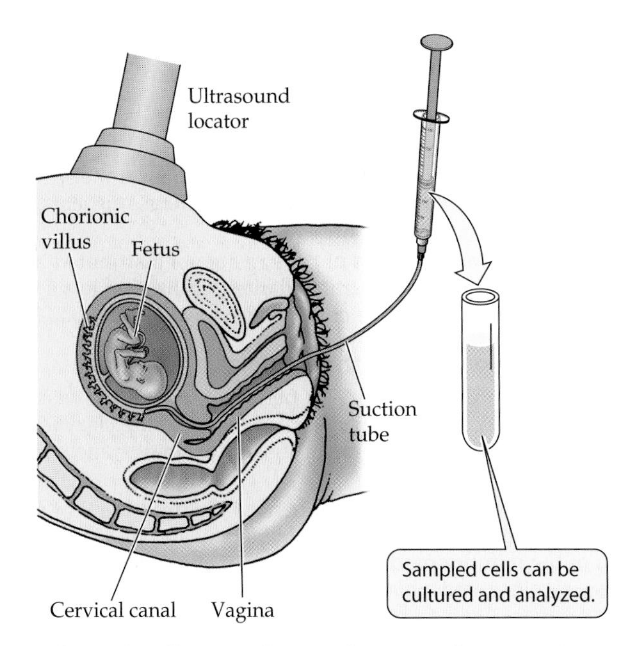

20.20 Chorionic Villus Sampling Information about genetic defects can be obtained from chorionic tissues. The fetus and placenta are imaged by a sonogram to guide a catheter, which samples a chorionic villus.

to beat in week 4, and limbs form by week 8 (Figure 20.21*a*). Most organs have started to form by the end of the first trimester. By that time, the embryo looks like a miniature version of the adult, and is called a **fetus**.

The first trimester is a time of rapid cell division and tissue differentiation. Signal transduction cascades and the resulting branching sequences of developmental processes are in their early stages. Therefore, the first trimester is the period during which the embryo is most sensitive to damage from radiation, drugs, chemicals, and pathogens that can cause birth defects. An embryo can be damaged before the mother even knows she is pregnant. A classic and tragic case is that of thalidomide, a drug widely prescribed in Europe in the late 1950s to treat nausea. Women who took this drug in the fourth and fifth week of pregnancy, when the embryo's limbs are beginning to form, gave birth to children with severely malformed arms and legs.

Hormonal changes cause major and noticeable responses in the mother during the first trimester, even though the fetus at the end of that time is still so small that it would fit into a teaspoon. Soon after the blastocyst implants itself, it begins to secrete human chorionic gonadotropin (hCG). This hormone stimulates the mother's ovary to continue to produce the hormones estrogen and progesterone, which help to maintain the pregnancy. These hormonal changes cause the well-known symptoms of pregnancy: morning sickness, mood swings, changes in the senses of taste and smell, and swelling of the breasts.

THE SECOND TRIMESTER. During the second trimester the fetus grows rapidly to a weight of about 600 g, and the moth-

(*a*)

(*b*)

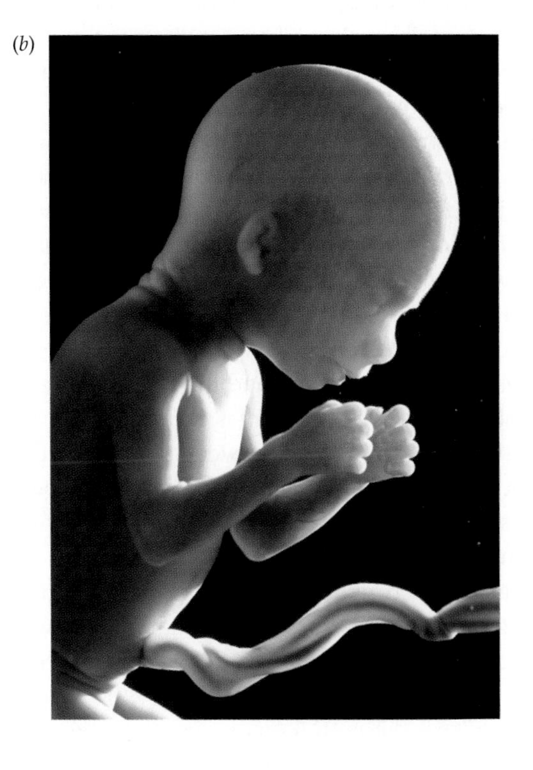

20.21 Stages of Human Development (*a*) The organs and body structures of this 8-week-old embryo are forming rapidly, and it is visibly a human male. The embryo is approximately 4 cm long and weighs less than 10 g. The umbilical cord attaches the embryo to the placenta (upper left). (The dark red structure at the upper right is the remnant of the yolk sac.) (*b*) At 4 months, the fetus is about 14 cm long and weighs about 200 g. It has fully formed limbs and digits (fingers and toes) and moves freely within its protective amniotic cavity.

er's abdomen enlarges considerably. The limbs of the fetus elongate, and the fingers, toes, and facial features become well formed (Figure 20.21*b*). Fetal movements are first felt by the mother early in the second trimester, and they become progressively stronger and more coordinated. By the end of the second trimester, the fetus may suck its thumb.

THE THIRD TRIMESTER. The fetus and the mother continue to grow rapidly during the third trimester. Even though the embryo is most susceptible to adverse effects of drugs, chemicals, and diseases during the first trimester, the potential for serious effects of exposure to many environmental factors continues throughout pregnancy. Severe protein malnutrition, alcohol consumption, and cigarette smoking are examples of factors that can result in low birth weight, mental retardation, and other developmental complications. As the third trimester approaches its end, many internal organs mature. The digestive system begins to function, the liver stores glycogen, the kidneys produce urine, and the brain undergoes cycles of sleep and waking.

Developmental changes continue throughout life

Development does not end with birth. Obviously, growth continues until adult size is reached, and even when growth stops, organs of the body continue to repair and renew themselves through cycles of cell replacement by the progeny of undifferentiated stem cells. In humans, in particular, enormous developmental changes occur in the brain in the years between birth and adolescence. Especially in the early years, there is a great deal of plasticity in the organization of the nervous system as patterns of connection between neurons develop.

For example, if a child is born with its eyes misaligned, a condition known as *strabismus*, he or she will use mostly one eye. The connections to the brain from that eye will become strong, and connections from the other eye will become weak. The child will develop with reduced visual acuity and depth perception. If the eye alignment is corrected in the first 3 years of life, however, the connections between the eyes and the brain will correct themselves, and the child will develop normal vision. If the eye alignment is corrected after 3 years of age, the correct connections between the eyes and the brain will not develop, and the visual impairment will be irreversible. Thus, plasticity in the development of the visual system in humans continues for several years after birth.

A very exciting area of current research is the role of learning in stimulating the production and differentiation of new neurons in the brains of young and even adult animals (see Chapter 46).

Chapter Summary

Development Begins with Fertilization

▶ The sperm and the egg contribute differentially to the zygote. The sperm contributes a haploid nucleus and, in some species, a centriole. The egg contributes a haploid nucleus, nutrients, ribosomes, mitochondria, and mRNAs.

▶ The cytoplasmic contents of the egg are not distributed homogeneously, and they are rearranged after fertilization to set up the major axes of the future embryo. **Review Figures 20.2, 20.3**

Cleavage: Repackaging the Cytoplasm

▶ In most animals, cleavage is a period of rapid cell division without cell expansion or gene expression. During cleavage, the cytoplasm of the zygote is repackaged into smaller and smaller cells.

▶ The pattern of cleavage is influenced by the amount of yolk, which impedes cleavage furrow formation, and by the orientation of the mitotic spindles. The result of cleavage is a ball or mass of cells called a blastula. **Review Figure 20.4**

▶ Cleavage in mammals is unique in that cell divisions are very slow and genes are expressed early in the process. Cleavage results in an inner cell mass that becomes the embryo and an outer cell mass that becomes the trophoblast. The mammalian embryo at this stage is called a blastocyst. **Review Figure 20.5**

▶ A fate map can be created by labeling specific blastomeres and observing what tissues and organs are formed by their progeny. **Review Figure 20.6**

▶ Some species undergo mosaic development, in which the fate of each cell is determined by the 8-cell stage. Other species, including vertebrates, undergo regulative development, in which remaining cells can compensate for cells lost in early cleavages.

Gastrulation: Producing the Body Plan

▶ Gastrulation involves massive cell movements that produce three germ layers and place cells from various regions of the blastula into new associations with one another.

▶ The initial step of sea urchin and amphibian gastrulation is inward movement of certain blastomeres. The site of inward movement becomes the blastopore. Cells that move into the blastula become the endoderm and mesoderm; cells remaining on the outside become the ectoderm. Cytoplasmic factors in the vegetal pole cells are essential to initiate development. **Review Figures 20.8 and 20.9**

▶ Amphibian gastrulation is initiated when cells in the gray crescent move into the blastocoel. This inward migration creates the blastopore.

▶ The dorsal lip of the blastopore is a critical site for cell determination. It has been called the primary embryonic organizer. **Review Figures 20.9, 20.10, 20.11.** **See Web/CD Tutorials 20.1 and 20.2**

▶ The protein β-catenin activates a signaling cascade that induces the primary embryonic organizer and sets up the anterior–posterior body axis. **Review Figure 20.12**

▶ Left–right asymmetries are probably controlled by the asymmetrical distribution of early transcription factors during gastrulation.

▶ Gastrulation in reptiles and birds differs from that in sea urchins and frogs because the large amount of yolk in their eggs causes the blastula to form a flattened disc of cells. **Review Figure 20.13**

▶ Mammals have a pattern of gastrulation similar to that of birds, even though their eggs have no yolk.

Neurulation: Initiating the Nervous System

▶ Gastrulation is followed by organogenesis. Cells that migrate over the dorsal lip of the blastopore are determined to become the notochord. The notochord induces the overlying ectoderm to thicken, form parallel ridges, and fold in on itself to form a neural tube below the epidermal ectoderm. The nervous system develops from this neural tube. **Review Figure 20.15**

▶ The notochord and neural crest cells participate in the segmental organization of tissues called somites along the body axis. Rudimentary organs and organ systems form during these stages. **Review Figure 20.16**

▶ Four families of Hox genes determine the pattern of anterior–posterior differentiation along the body axis in mammals. Other genes, such as *sonic hedgehog*, contribute to dorsal–ventral differentiation. **Review Figure 20.17**

Extraembryonic Membranes

▶ The embryos of reptiles, birds, and mammals are protected and nurtured by four extraembryonic membranes. In birds and reptiles, the yolk sac surrounds the yolk and provides nutrients to the embryo, the chorion lines the eggshell and participates in gas exchange, the amnion surrounds the embryo and encloses it in an aqueous environment, and the allantois stores metabolic wastes. **Review Figure 20.18. See Web/CD Activity 20.1**

▶ In mammals, the chorion and the trophoblast cells interact with the maternal uterus to form a placenta, which provides the embryo with nutrients and gas exchange. The amnion encloses the embryo in an aqueous environment. **Review Figure 20.14, 20.19**

▶ Samples of amniotic fluid or pieces of the chorion can be analyzed for evidence of genetic disease. **Review Figure 20.20**

Human Development

▶ Pregnancy in humans can be divided into three trimesters. The embryo forms in the first trimester; during this time, it is most vulnerable to environmental factors that can lead to birth defects. During the second and third trimesters the embryo grows, the limbs elongate, and the organ systems mature.

▶ Hormonal changes maintain the pregnancy and also cause symptoms of pregnancy in the mother.

▶ Development continues throughout childhood and throughout life.

Self-Quiz

1. Fertilization involves all of the following *except*
   a. joining of mitochondria from sperm and egg.
   b. joining of sperm and egg haploid nuclei.
   c. induction of rearrangements of the egg cytoplasm.
   d. sperm binding to specific sites on the egg surface.
   e. metabolic activation of the egg.
2. Which of the following does *not* occur during cleavage in frogs?
   a. A high rate of mitosis
   b. Reduction in the size of cells
   c. Expression of genes critical for blastula formation
   d. Orientation of cleavage planes at right angles
   e. Unequal division of cytoplasmic determinants
3. How does cleavage in mammals differ from cleavage in frogs?
   a. Slower rate of cell division
   b. Formation of tight junctions
   c. Expression of the embryo's genome
   d. Early separation of cells that will not contribute to the embryo
   e. All of the above
4. Which statement about gastrulation is *true*?
   a. In frogs, gastrulation begins in the vegetal hemisphere.
   b. In sea urchins, gastrulation produces the notochord.
   c. In birds, cells from the epiblast move into the blastocoel through the primitive groove.
   d. In mammals, gastrulation occurs in the hypoblast.
   e. In sea urchins, gastrulation produces only two germ layers.
5. Which of the following was a conclusion from the experiments of Spemann and Mangold?
   a. Cytoplasmic determinants of development are homogeneously distributed in the amphibian zygote.
   b. In the late blastula, certain regions of cells are determined to form skin or nervous tissue.
   c. The dorsal lip of the blastopore can be isolated and will form a complete embryo.
   d. The dorsal lip of the blastopore can initiate gastrulation.
   e. The dorsal lip of the blastopore gives rise to the neural tube.
6. Which of the following is true of human development?
   a. Most organs begin to form during the second trimester.
   b. Gastrulation takes place in the oviducts.
   c. Genetic diseases can be detected by sampling cells from the chorion.
   d. Implantation occurs through interactions of the zona pellucida with the uterine lining.
   e. Exposure to drugs and chemicals is most likely to cause birth defects when it occurs in the third trimester.
7. Which of the following characterizes neurulation?
   a. The notochord forms a neural tube.
   b. The neural tube is formed from ectoderm.
   c. A neural tube forms around the notochord.
   d. The neural tube forms somites.
   e. In birds, the neural tube forms from the primitive groove.
8. Which statement about trophoblast cells is *true*?
   a. They are capable of producing monozygotic twins.
   b. They are derived from the hypoblast of the blastocyst.
   c. They are endodermal cells.
   d. They secrete proteolytic enzymes.
   e. They prevent the zona pellucida from attaching to the oviduct.
9. Which membrane is part of the embryonic contribution to placenta formation?
   a. Amnion
   b. Chorion
   c. Epiblast
   d. Allantois
   e. Zona pellucida
10. A major factor in the determination and differentiation of tissues along the anterior–posterior axis of the mouse is the
    a. differential expression of Hox genes.
    b. concentration gradient of β-catenin.
    c. differential expression of the *sonic hedgehog* gene.
    d. distance of the tissue from the gray crescent.
    e. distribution of GSK-3, which degrades β-catenin.

For Discussion

1. If you found a protein that was localized to a small group of cells in the frog blastula, how would you determine whether that protein played a role in development? Address the issues of sufficiency and necessity.
2. During gastrulation in birds, the *sonic hedgehog* gene is expressed only on the left side of Hensen's node. What might be the significance of this expression pattern?
3. Much of the early work of describing the physiology of animal development was done on sea urchins, amphibians, and chicks. Most recent work on the molecular mechanisms of animal development has been done on nematodes, fruit flies, zebrafish, and mice. Why do you think there has been a shift in the animal models used by developmental biologists?
4. If all the mitochondria and mitochondrial DNA in the embryo come from the egg, what implications does this have for using mitochondrial DNA for molecular evolutionary studies?
5. There is currently much controversy over therapeutic cloning as a way of obtaining embryonic stem cells to treat diseases. Given that human regulative development—in other words, the fact that twinning can occur if an early blastocyst is divided into two cell masses—can you think of a way to guarantee a source of isogenic (i.e., identically matching a person's own body) stem cells for any individual without resorting to therapeutic cloning? Assume isolated cells can be preserved indefinitely in a frozen state.

21 Development and Evolutionary Change

Among many fish species, the sex of other members of the group in which an individual fish lives—its social environment—determines its sex. For example, anemonefish (also known as clownfish) always begin life as males. Anemonefish live in social groups of five or so individuals living within a single sea anemone. The group's leader is the largest fish in the group—and is also its only female. If the female is removed from the group, the largest male in the group changes sex and becomes a female.

Cleaner wrasse, on the other hand, first mature sexually as females. The largest female in a group eventually changes sex to become a male. A female wrasse can be induced to change sex earlier if the dominant male of the group is removed. Within a few hours, the former female adopts male behavior, and within 10 days this new male is producing functional sperm.

The genome has often been believed to provide a "blueprint" for an organism's development. But fish that change sex in response to their social environment demonstrate that an organism's development is not determined entirely by its genes. Genes not only give orders; they also take orders. Thus, an organism's genome is more like a recipe. The final product—the phenotype—may depend on how the ingredients are mixed and the conditions during preparation and cooking.

Genes are molecules that encode other molecules, which in turn affect different parts of an organism. There are no genes "for" complex structures such as eyes, or for behavior patterns. The genome encodes instructions for making enzymes, receptors, signal molecules, structural molecules, and so forth. The phenotype of an adult organism comes into being as a result of interactions between genes, their products, and the environment.

In this chapter we will look at the mechanisms of development from an evolutionary perspective. We will show how mutations in the genes that regulate development can enable new structures to evolve. We will see how the modular nature of organisms can allow structural changes in the phenotype even when the gene sequences themselves have not changed. We will also describe how interactions between the environment and developmental processes help adapt organisms to their environments.

When Size Determines Sex All anemonefish begin life as males. The largest male in the social group will change sex to become the group's only female.

Evolution and Development

Charles Darwin viewed evolution as "descent with modification." He explained *similarities* among organisms as a result of their descent from a common ancestor, and he explained *differences* among them as a result of natural selection, which adapts them to different environments. Darwin's theory of descent with modification led to the recognition that the results of evolution could be visualized as a "tree of life." The tree's root is the ancient common ancestor of all life; its branches represent the divergence of lineages of evolving organisms over time.

Darwin recognized that morphological patterns could be modified by changes in the processes that regulated the growth of an organism from egg to adulthood, but did not explore this theme in great detail. However, in *The Origin of Species*, he showed how similarities among embryos could be used to infer relationships among groups of organisms. For example, he concluded that barnacles are crustaceans (a large group that also includes shrimp and crabs) on the basis of similarities between larval forms, even though adult barnacles look very different from other crustaceans (Figure 21.1)

Early in the twentieth century, embryology and genetics were regarded as a single science. But in the 1920s, geneticists turned their attention almost exclusively to the transmission of inherited characteristics from adult organisms to their offspring. Embryologists turned their attention to the expression of those characteristics during development. That is, geneticists were interested in how genes determine the number of bristles on a fruit fly's back, while embryologists were interested in how the fly forms its back. The two sciences progressed without much intellectual contact until late in the twentieth century, when they began to come together to form the new discipline of **evolutionary developmental biology**.

Evolutionary developmental biologists investigate how the course of evolution has been influenced by heritable changes in the development of organisms. Like geneticists, they are interested in the inheritance of the characteristics of organisms. Developmental biologists now study how changes in the genes that regulate development affect the adult forms of organisms. They also study how those genes have changed during the course of evolution. To understand large evolutionary changes, such as the evolution of eyes, wings, and flowers, biologists study both developing and adult organisms, because the agents of evolution work not only on adults but on the "recipes" for making adults.

Evolutionary developmental biologists look for alterations both in the genes that regulate development and in the genes for their target proteins. Early discoveries showed that many of the genes regulating development are highly **conserved**—that is, the sequences of these genes have changed remarkably little throughout the course of evolution of multicellular organisms. Let's look at these similarities and find out why the genes that govern development have changed so little over evolutionary time.

Gooseneck barnacles (*Lepas* sp.)
Brine shrimp (*Artemia salina*)
Adults
Larvae

21.1 Similarities In Early Developmental Stages Can Be Used to Infer Relationships Adult barnacles do not clearly resemble adults of other crustaceans, such as brine shrimp (right), but Charles Darwin observed the similarities between their larvae (lower photos) and concluded correctly that these animals share a recent common ancestor.

21.2 The Mouse *Pax6* Gene Causes Eye Development in *Drosophila*
The *Pax6* gene for eye development is ancestral to both arthropods and vertebrates. This micrograph shows a compound eye emerging in the leg of a fruit fly in which mouse *Pax6* cDNA was expressed.

Development uses the same sets of genes throughout the animal kingdom

Many of the genes that regulate development in very different animal species are remarkably similar. The compound eyes of fruit flies (*Drosophila*) and the camera-like eyes of house mice (*Mus musculus*) differ greatly in their structure and functioning, but many of the same genes instructs cells to form eyes in both animals. The genetic instructions for eye development in the two species are so similar that fruit fly cells that would normally form part of a leg will form an eye (a *Drosophila* eye) when a mouse *Pax6* gene is expressed in them (Figure 21.2).

Most motile animals (animals that move through their environment as a result of their own exertions) have bilaterally symmetrical bodies with a head (anterior) and a tail (posterior) end, and the bodies of many of them are divided into segments (see Chapter 32). The same sets of homeobox genes provide positional information to cells along the anterior-posterior axis of the body in both insect and human embryos. For example, both the *Drosophila* gap genes *ems*, *tll*, and *otd* and the homologous genes of vertebrates are expressed in the anterior regions of the brain (Figure 21.3).

When certain insect homeotic genes are mutated, the segments differentiate in the wrong way. The *bithorax* mutation causes the developing insect to form two sets of forewings rather than the normal one pair (Figure 21.4*a*), and the *Antennapedia* mutation results in the formation of legs where the antennae should be (see Figure 19.16). In vertebrates, altering the expression patterns of some Hox genes can change lumbar (abdominal) vertebrae into thoracic (ribbed) vertebrae (Figure 21.4*b*). Altering the expression of other genes can replace neck bones with duplications of the ear bones and jaw.

Thus the instructions for forming embryos are provided by homologous genes in vertebrates and invertebrates, even though the structures formed from those instructions are very different. The enormous variation of morphological forms found in the animal kingdom is underlain by a common set of instructions. These instructions have been con-

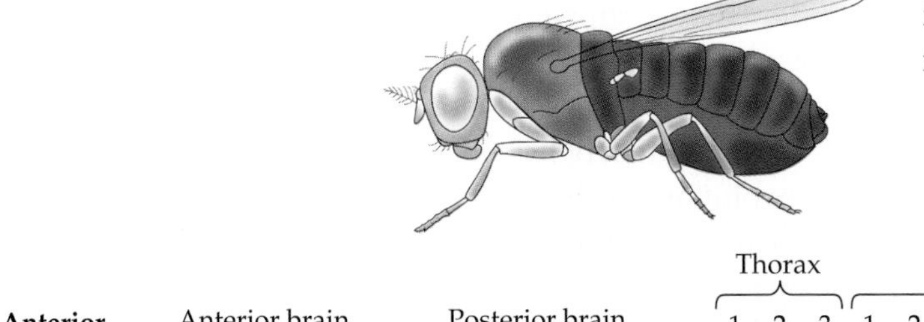

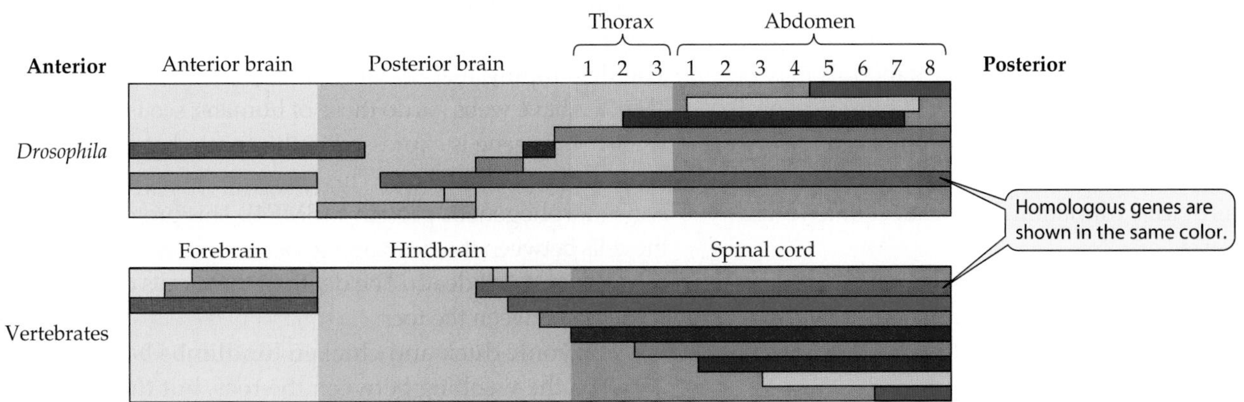

21.3 Genes Show Similar Expression Patterns
Homologous genes for certain transcription factors are expressed in similar patterns along the anterior–posterior axes of both insects and vertebrates.

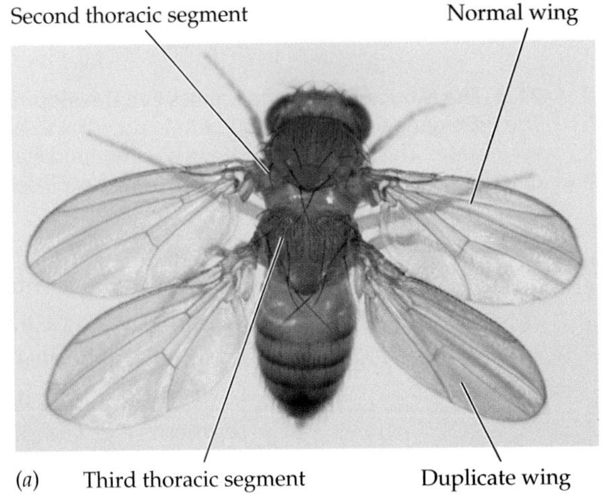

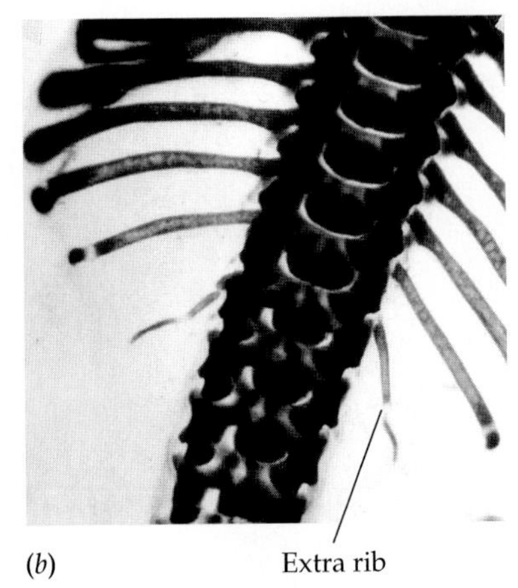

21.4 Altering Homeobox Genes Changes Morphology (*a*) Deletion of the *Ubx* gene in *Drosophila* converts the third thoracic segment, which does not normally bear wings, into a duplication of the second thoracic (forewing-bearing) segment. (*b*) Deletion of the *Hoxc-8* gene in mice transforms a lumbar (abdominal) vertebra into a copy of a thoracic (ribbed) vertebra.

served in thousands of species that display an incredibly vast array of seemingly unrelated morphologies.

However, the vast differences in morphological form that result from such similar genetic instructions means that the instructions cannot be the whole story. The anterior-posterior axes of both human and fruit fly embryos are specified by the same genes, but humans don't develop wings. Humans use *Pax6* to specify eye-forming regions, but the human visual system forms from different precursor structures than the fly system does. So there must be important *differences* as well as important *similarities*. Both similarities and differences are expected under descent with modification, but how can differences in developmental genetics explain structural differences?

Regulatory Genes and Modularity: Modifying Morphology

How can the development pattern of an embryo change without resulting in an adult organism that cannot function well? Such changes are possible because embryos, like adults, are made up of modules. A **module** is a self-contained unit that is part of a larger unit. The form of each module may be changed independently of the other modules in the overall unit. Thus the form of a developing animal's heart can change independently of changes in its limbs because the genes that govern heart formation do not affect limb formation, and vice versa.

Two major ways in which changes in the genes that regulate development can lead to evolutionarily important morphological changes have been elucidated during the past several decades:

- Mutations in genes that regulate developmental processes
- Changes in the time or place of expression of developmental regulatory genes

Both of these pathways of evolution are made easier by the modular nature of most organisms.

Mutations can result in new phenotypes

Insects are highly modular organisms; in Chapter 19 we saw the precision of the segmentation process in insect development. Insects provide examples of the evolution of morphological changes through mutations in the genes encoding transcription factors that regulate segmentation. For example, all arthropods (see Chapter 33) possess the homeotic gene *Ultrabithorax* (*Ubx*), but the insect *Ubx* gene has a mutation not found in the other arthropods (Figure 21.5). The Ubx protein transcribed from this mutated gene represses expression of the *distal-less* gene (*dll*), which is essential for leg formation. The Ubx protein of insects is expressed in the abdomen where it represses *dll*. As a result, insects have only six legs, none of which grow from the abdominal segments. In contrast, the Ubx protein of other arthropods—such as millipedes, centipedes, spiders, mites, and crustaceans—do not repress the expression of *dll*. Consequently, those animals all have abdominal appendages.

The evolution of the webbed feet of ducks is an example of an evolutionary change resulting from an altered spatial expression pattern of a regulatory gene. Ducks have webs that connect their toes, but chickens and most other birds do not. The developing feet of early embryos of both ducks and chickens have webs (as do those of humans; see Figure 19.11). A particular gene is expressed in the spaces between the developing bones of the toes. This gene encodes a protein called bone morphogenetic protein 4 (BMP4). This protein instructs the cells between the developing toes to undergo apoptosis—programmed cell death. The death of these cells destroys the webbing between the toes.

Embryonic duck and chicken hindlimbs both express BMP4 in the webbing between the toes, but they differ in the expression of a *BMP inhibitor* protein, called Gremlin (Figure 21.6). Gremlin expression occurs around the digits in both chick and duck hindlimbs. In ducks, but not in chickens, the *gremlin* gene is also expressed in the webbing cells. The Gremlin protein prevents the BMP4 protein from signaling for cell death in the webbing; the result is a

21.5 A Mutation Changed the Number of Legs in Insects In the insect lineage (blue) of the arthropods, a mutation in the *Ubx* gene resulted in a protein that inhibits a gene that is required for legs to form. Because insects express *Ubx* in their abdominal segments, no legs grow from these segments. Other arthropods, such as centipedes, do grow legs from their abdominal segments.

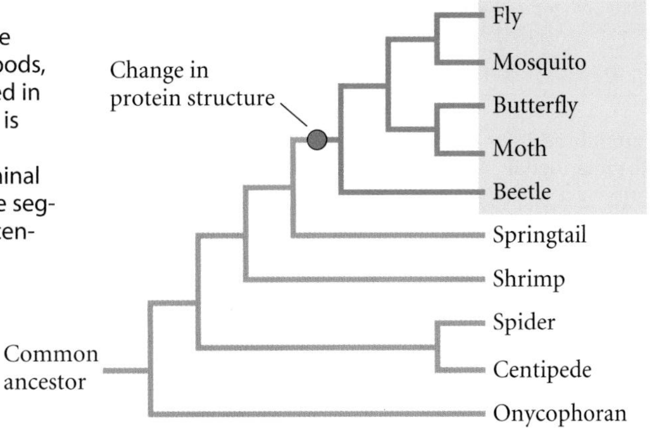
Change in protein structure
Fly
Mosquito
Butterfly
Moth
Beetle
Springtail
Shrimp
Spider
Centipede
Onycophoran
Common ancestor

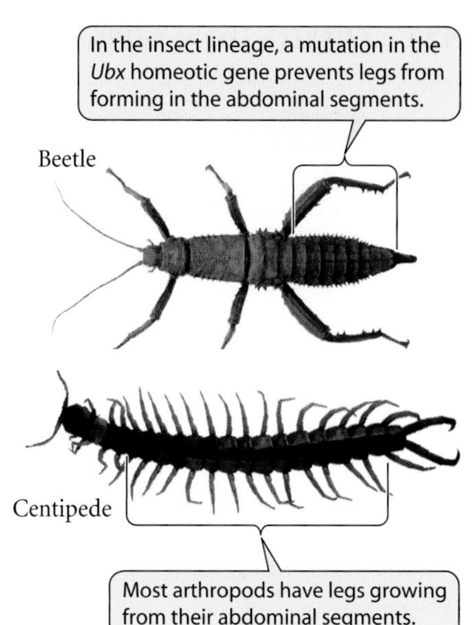
In the insect lineage, a mutation in the *Ubx* homeotic gene prevents legs from forming in the abdominal segments.
Beetle
Centipede
Most arthropods have legs growing from their abdominal segments.

webbed foot. Experimental application of the Gremlin protein to chick hindlimbs converts them into ducklike feet (Figure 21.7).

The timing of a gene's expression can affect morphology

Modularity also allows the relative *timing* of two different developmental processes to shift independently of one another, a process called **heterochrony**. That is, the genes regulating the development of one module (say, the eyes of vertebrates) may be expressed at different times in different species, relative to genes regulating development of other modules.

Heterochrony has been widely studied in salamanders, where extensive examples of this phenomenon are seen. The case of two salamander species of the genus *Bolitoglossa* illustrates how heterochrony can result in new morphology.

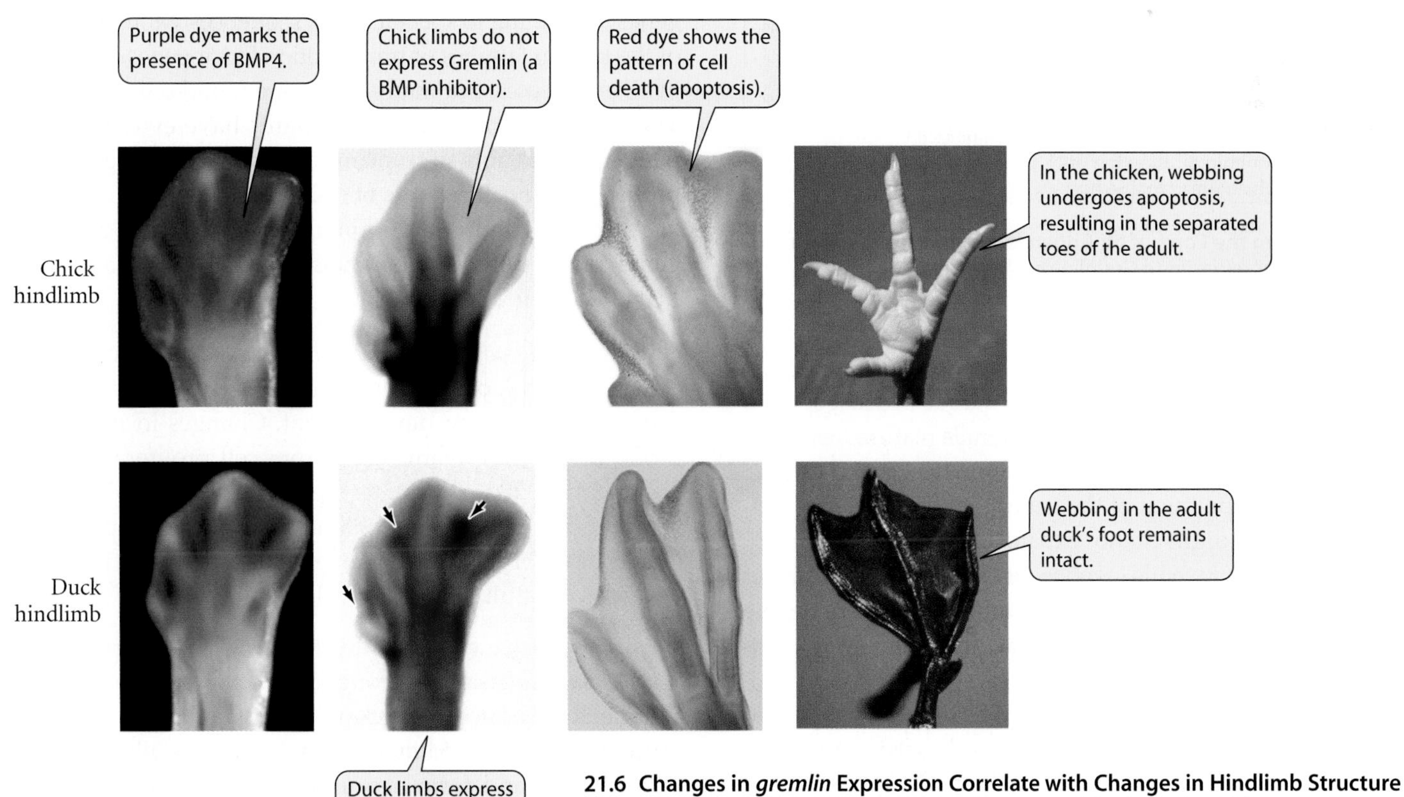
Purple dye marks the presence of BMP4.
Chick limbs do not express Gremlin (a BMP inhibitor).
Red dye shows the pattern of cell death (apoptosis).
In the chicken, webbing undergoes apoptosis, resulting in the separated toes of the adult.
Chick hindlimb
Duck hindlimb
Webbing in the adult duck's foot remains intact.
Duck limbs express Gremlin (arrows).

21.6 Changes in *gremlin* Expression Correlate with Changes in Hindlimb Structure The upper row of photos shows the development of the foot of a chicken; the lower row shows the development of the foot of a duck. Expression of Gremlin protein in the duck foot blocks BMP4 and prevents the embryonic webbing from undergoing apoptosis.

EXPERIMENT

Question: Will adding Gremlin protein (an inhibitor of BMP4) to a developing chick foot transform the chick foot into a ducklike form?

METHOD	Open up chicken eggs and carefully add Gremlin-secreting beads to the interdigital (web) regions of one embryonic chick hindlimb. Add beads that do not contain Gremlin to the other hindlimbs (controls). Close the eggs and observe limb development.
RESULTS	In the hindlimbs in which Gremlin was secreted, the webbing does not undergo apoptosis, and the hindlimb resembles that of a duck. The control hindlimbs develop the normal chicken form.

Gremlin added
Control

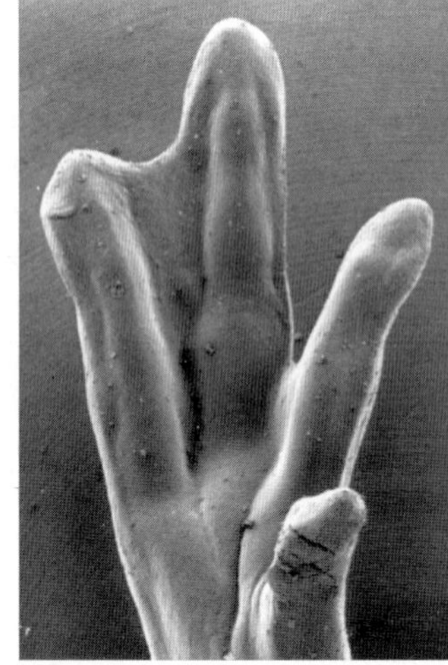

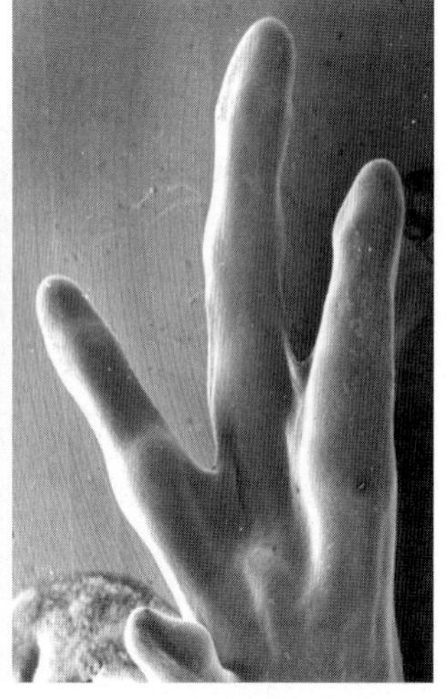

Conclusion: Changes in *gremlin* gene expression could cause the changes in morphology, allowing duck hindlimbs to retain their webbing and chick limbs to lose it.

21.7 Changing the Form of an Appendage In this experiment, chick hindlimbs exposed to Gremlin-secreting beads developed ducklike webbed feet.

The webbing between the feet of the larvae of most species of salamanders disappears as the animals mature, resulting in "toed" feet suited to getting around on the ground (Figure 21.8*a*). But if expression of the genes that dissolve the webbing slows, the digits do not expand as the rest of the body matures. These "juvenile" webbed feet (Figure 21.8*b*) can act like "suction cups," allowing the animal to adhere to tree branches. This ability opens up a new, arboreal way of life to the new species that possesses it.

Modularity also allows structural changes to evolve via gene duplication. When a gene is duplicated, one of the copies can evolve a new function without disrupting the organism, as long as the other copy is still performing the original function. We will discuss this process in detail in Chapter 26.

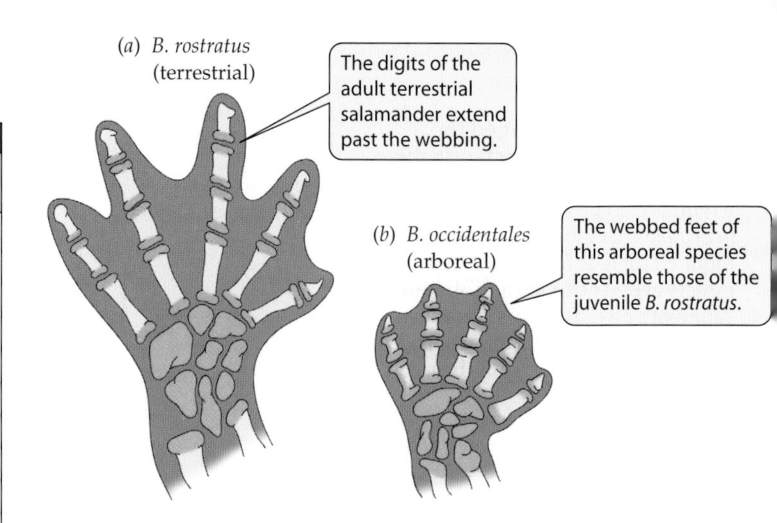
(*a*) *B. rostratus* (terrestrial)
The digits of the adult terrestrial salamander extend past the webbing.
(*b*) *B. occidentales* (arboreal)
The webbed feet of this arboreal species resemble those of the juvenile *B. rostratus*.

21.8 Heterochrony Created an Arboreal Salamander (*a*) The foot of an adult *Bolitoglossa rostratus*, a terrestrial salamander. (*b*) The foot of *B. occidentalis*, a closely related arboreal salamander.

Plant Development and Evolution

Rapid progress has been made during the past decade in identifying the genes that regulate growth and cell differentiation in plants. Much of this information has come from the sequencing of the complete genome of *Arabidopsis thaliana*. About 1,500 of its nearly 26,000 genes encode transcription factors that turn on or repress the expression of genes by binding to specific DNA sequences. More than half of the known families of transcription factor genes are found in all eukaryotes, but many others are found only in plants. The ones that are restricted to plants have evolved independently in plants since plants and animals diverged from protist ancestors hundreds of millions of years ago.

Although plants and animals share many regulatory genes, plants differ from animals in several important ways that affect their development:

- *Plant cells do not move relative to one another*. As we noted in Chapter 19, plant cells do not move around within the plant body during development. Changes in the shape of a developing plant result from cell proliferation and elongation. In contrast, as we saw in Chapter 20, complex movements of cells characterize the development of animal embryos. These movements determine how the shapes of adult animals develop.
- *Future reproductive cells are not set aside early during plant development*. Instead, throughout their lives, plants continue to produce clusters of undifferentiated, actively dividing cells, called **meristems**. Meristems allow a plant to develop and form new organs, such as stems, roots, leaves, and flowers, as long as it grows. In contrast, during animal development, organs are formed according to highly regulated developmental schedules that take

place only once during an individual's lifetime. We humans "make an arm" only once in our life cycle, during our embryonic development.

- *Plants have tremendous developmental plasticity.* If an herbivore eats part of a plant, leaf meristems may grow out and replace the lost part. The new leaves may also produce chemicals that defend them against herbivores. And, as we will see later in this chapter, plant growth responds dramatically to temperature and light. This ability of an organism to change its development in response to environmental conditions is called **developmental plasticity**.

Despite these important differences, members of two families of genes that encode transcription factors—the MADS box genes and the homeobox genes—regulate important developmental processes in both plants and animals (see Chapter 19). However, in plants, these genes govern the development of unique structures—roots, stems, leaves, and flowers—found only in that kingdom.

Much more is known about the genetic control of development in *Arabidopsis* than in any other plant, but enough is known about other plants to suggest that many genes regulating plant development are shared by many species of plants. For example, genes that result in early flowering in *Arabidopsis* have a similar effect in aspen trees.

Why do plants and animals still share so many of the genes that regulate their development, even though they have been evolving separately for such a long time and produce such different tissues and organs? Part of the reason, as we saw above, is that the modular construction of multicellular organisms allows different parts of their bodies to change independently of one another. Another part of the answer involves how the genomes of organisms change, a topic that we will discuss in detail in Chapter 26.

Plants have greater developmental plasticity than animals because developmental plasticity is especially valuable for a sessile organism. Plants cannot move to another place if environmental conditions deteriorate where they are growing, nor can they move to escape from their predators and parasites. Unlike most animals, however, they can be partly eaten but still survive. The combination of repeated production of meristems and developmental plasticity compensates for a plant's lack of mobility.

Environmental Influences on Developmental Patterns

Organisms are adapted to their environments in part because their development has been molded by the agents of evolution. However, the idea that the environment plays an important role in the development of organisms was downplayed until very recently. Part of the reason for this neglect is that, for convenience, developmental biologists studied small organisms that develop rapidly and whose development does not change dramatically in the laboratory under controlled conditions. As we saw in Chapter 1, control of conditions is an important component of the scientific method. But these studies fostered the misleading view that genes had an autonomous existence apart from environmental signals.

Now we know that the development of many organisms is exquisitely sensitive to environmental conditions. Numerous species possess a great deal of developmental plasticity—the ability to express different phenotypes under different environmental conditions. In other words, a single genotype may encode a range of phenotypes, and signals from the environment may determine what phenotype is expressed. But how should organisms respond to signals from the environment in order to develop adaptively?

No single way of responding to signals from the environment results in adaptation because what environmental signals tell an organism varies with the type of signal. We can divide signals from the environment into two major types, based on their significance and how organisms should respond to them:

- *Environmental signals that are accurate predictors of future conditions.* Some of these signals always occur, but organisms may develop without ever encountering others. In either case, we would expect the developmental processes of organisms to respond adaptively to these signals.
- *Environmental signals that are poorly correlated with future conditions.* We would expect organisms to fail to respond to such signals.

Let's look at some of the developmental responses of organisms to these different types of signals.

Organisms respond to signals that accurately predict the future

Seasonal changes in day length occur every year, and these changes are accurate predictors of some future environmental conditions. Increasing day lengths accurately predict the approach of spring and summer; decreasing day lengths accurately predict the coming of fall and winter. Temperature changes also accompany the seasons and signal future environmental conditions. In most tropical regions, wet and dry seasons alternate in a regular pattern during the year. Developing organisms respond to such signals in such a way that the adults they become are adapted to the predicted conditions.

The West African butterfly *Bicyclus anynana* has two color forms. The dry-season form matches the dead brown leaves on the dry-season forest floor, where the butterfly rests much of the time. The more active wet-season form has a white line along the wing and conspicuous ventral hindwing eyespots.

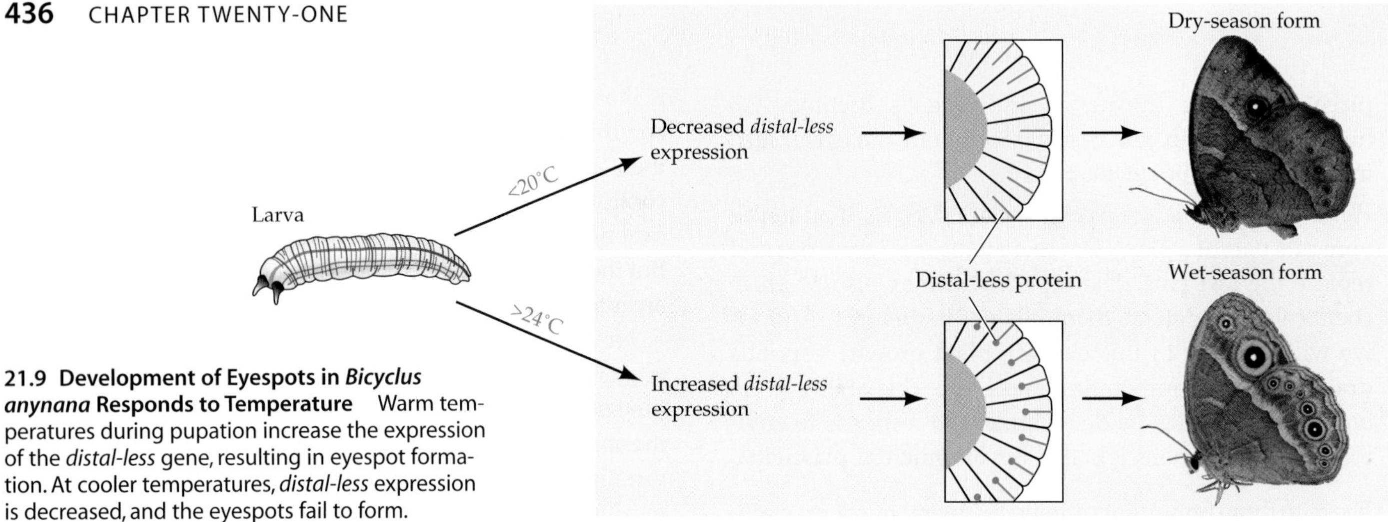

21.9 Development of Eyespots in *Bicyclus anynana* Responds to Temperature Warm temperatures during pupation increase the expression of the *distal-less* gene, resulting in eyespot formation. At cooler temperatures, *distal-less* expression is decreased, and the eyespots fail to form.

These eyespots deceive predatory lizards and birds into attacking the wing, rather than the butterfly's actual eye, increasing the butterfly's chances of escape.

Temperature during pupation determines the color form of the adult butterfly. Pupae developing in the soil experience cooler temperatures during the cooler nights of the dry season. Pupae developing under temperatures less than 20°C produce the dry-season form; temperatures above 24°C produce the wet-season color form. In the late larval stages, transcription of the *distal-less* gene is restricted to several small areas that have the potential to become the centers of eyespots. During pupal development, the area over which *distal-less* is expressed increases with temperature, resulting in eyespots on adults that develop from pupae exposed to warm temperatures (Figure 21.9). Thus, by responding to an environmental signal—temperature—the butterflies develop a form that adapts them to the conditions they will encounter as adults.

Another example of developmental plasticity in response to seasonal changes occurs in the moth *Nemoria arizonaria*, in this case at the larval stage. This moth produces two generations each year. Eggs hatch in spring, and the larvae that hatch from them (caterpillars) feed on oak flowers (catkins). These larvae complete their development, form pupae, and transform themselves into adult moths in summer. These moths then lay their eggs on oak trees. The larvae that hatch from these eggs eat oak leaves, complete their development, and lay their eggs on oak branches. These summer eggs overwinter and hatch the following spring. The spring caterpillars resemble the catkins on which they feed (Figure 21.10a); the summer caterpillars that feed on oak leaves resemble small, year-old oak branches (Figure 21.10b). Thus, both types of caterpillars are well camouflaged in the environment in which they feed. An experimenter was able to convert spring caterpillars into summer caterpillars by feeding them oak leaves. Apparently some chemical in the oak leaves induces them to develop the twiglike summer form.

To complete their development, some organisms need help from another species. Such a requirement could not evolve unless individuals of the other species were reliably present. House mice that are raised in microbe-free environments do not have the bacteria that normally colonize their guts. These mice also do not have normal capillary networks in their intestines. The gut bacteria induce gene expression in the mouse intestine, without which normal capillary development does not occur. In nature, all mice get the bacteria.

21.10 The Spring and Summer Forms of a Caterpillar Differ (*a*) Spring caterpillars of *Nemoria arizonaria* resemble oak catkins. (*b*) Summer caterpillars of the same species resemble oak twigs.

Some conditions that accurately predict the future may not always occur

We have discussed responses to environmental signals that always occur. However, many other changes in an organism's environment are uncertain. Predators may or may not be active in an organism's environment. An individual may live under crowded or uncrowded conditions. The sexes and ages of its associates may change. Nevertheless, if such changes have occurred frequently during the evolution of a species, evolved developmental plasticity may allow individuals to respond to them. For example, as we saw at the beginning of this chapter, individuals of some fish species change sex in response to alterations in their social environment. The social environment, including the sexes of others in the social group, determines which is the most adaptive sex for an individual to have at a particular time.

Similarly, individuals that could sense the presence of predators in their environment and change their development so as to become less likely to be eaten by them would be more likely to survive than individuals whose development did not respond to the presence of predators. Thus, developmental responses to predators have evolved in numerous species. For example, when water fleas (*Daphnia cucullata*) encounter predatory larvae of the fly *Chaeoborus*, the "helmets" on the top of their heads grow to twice their normal size (Figure 21.11). The fly larvae can ingest *Daphnia* with large helmets only with difficulty. Helmet induction also occurs if *Daphnia* are exposed to water in which the fly larvae have been swimming. Moreover, the offspring that are developing in the abdomens of mothers with induced large helmets are born with large helmets. There is a trade-off, however: *Daphnia* with large helmets produce fewer eggs than do *Daphnia* with small helmets. Otherwise, we would expect all individuals to develop large helmets.

Tadpoles of the spadefoot toad (*Scaphiopus couchii*), which breeds in ephemeral ponds in the arid southwestern United States, respond developmentally if their pond begins to dry up while they are growing. At the time she lays her eggs, a mother toad cannot know how long the pond will persist, because that depends on unpredictable future rainfall. If the pond dries up completely before the tadpoles' development has been completed, they will die. Some spadefoot tadpoles respond to crowding in a shrinking pond by developing a wider mouth and powerful jaw muscles. They complete their development rapidly before the pond dries up by eating other tadpoles.

Light exerts a powerful influence on plant development. Low light conditions stimulate the elongation of cells, so that plants growing in the shade become spindly (Figure 21.12). It is obvious why this response is adaptive: A spindly plant is more likely to reach a patch of brighter light than a plant that remains compact. And because they have meristems, plants can continue to respond to light as long as they grow.

21.11 Predator-Induced Developmental Plasticity in *Daphnia* This scanning electron micrograph shows the predator-induced form of *Daphnia* (left), with an enlarged helmet, and the normal form of the crustacean (right). These two individuals are genetically identical from a single asexually produced clone.

21.12 Light Seekers The bean plants on the left were grown experimentally under low-light conditions. The plant's cells have elongated in response to low light, and the overall plant has become spindly. The control plant on the right was grown in normal light conditions.

Organisms do not respond to environmental signals that are poorly correlated with future conditions

We would expect organisms to evolve to ignore environmental signals that are poorly correlated with future conditions, because any responses to those signals would probably be inappropriate. Consider, for example, seed production by plants. The amount of energy a growing plant has available to allocate to seed production depends, among other things, on temperature, rainfall, and the sizes and numbers of its neighbors. But the seeds the plant produces will germinate in future years, with different and unknowable rainfall patterns and densities of neighbors. Plants respond to changing environmental conditions by varying their size, shape, number of flowers, and number of seeds, but they produce seeds of a nearly constant size (Figure 21.13).

A seedling that germinates from a large seed will survive better under conditions of intense competition than a seedling that germinates from a small seed because it can grow larger using the energy in the seed. But, for a given amount of energy, a mature plant cannot produce as many large seeds as small seeds. So, if the next generation of plants grows under more favorable conditions, plants that produced a larger number of smaller seeds in the previous year will have more surviving offspring. Therefore, it is not surprising that plants do not change the sizes of the seeds they produce in response to the conditions under which they grow. Seed size is adjusted to the average conditions encountered by plants over many generations.

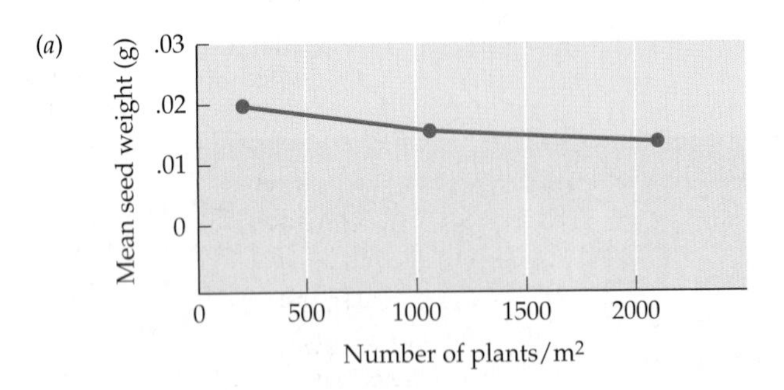

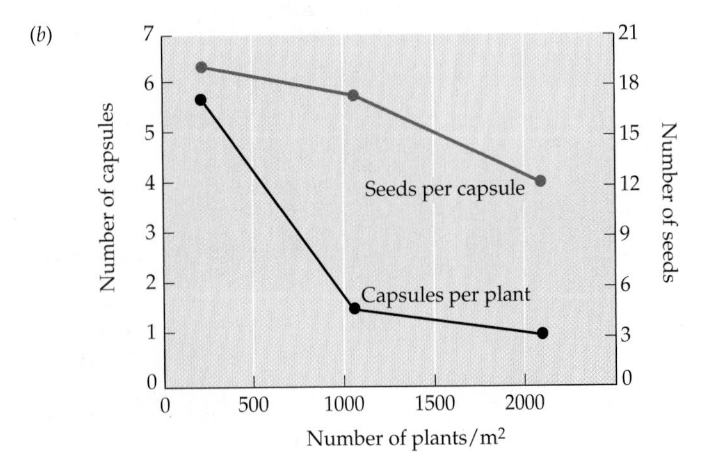

21.13 Seed Production (*a*) No matter how crowded or uncrowded the conditions under which these plants were grown, the seeds they produced were about the same size. (*b*) The number of seeds the plants produced dropped dramatically in more crowded conditions.

Organisms may lack appropriate responses to new environmental signals

As we have seen, organisms can respond adaptively to environmental signals that have occurred frequently during their recent evolutionary histories. But we would not expect organisms to have evolved useful responses to environmental signals that they have not encountered before. The lack of useful developmental responses to new environmental events is an important current problem because modern human societies have changed the environment in so many ways. Humans release thousands of new chemical compounds into the environment, some of which disrupt normal development.

For example, more than 7,000 deformed infants were born to women who took a drug called thalidomide as a mild sedative during pregnancy. Human embryos are especially sensitive to thalidomide for the first 20–36 days after conception, a time when many organs are forming.

And as we saw in Chapter 1, deformed frogs are appearing at high frequencies in some environments. Pieter Johnson's experiments suggested that a parasite causes some of these deformities. Further experiments have shown that tadpoles growing in ponds contaminated by certain pesticides are less able to resist parasite infection than tadpoles growing in pesticide-free ponds. Understanding how and why such chemicals affect development may help us devise substitutes that have fewer or no adverse effects.

Learning: A Modification of Development

A nearly universal modification of development in response to environmental variation is learning. As you know from your struggles to absorb the content of this book, which we authors have struggled just as mightily to compose, learning is costly. Learning takes much effort and time, during which an individual cannot do other useful things. But learning can continue throughout adult life, albeit with continuing costs. Learning also allows an individual to adjust its behavior to the physical, biological, and social environment in which it matures. As we will see in Chapters 46 and 52, learning is especially important in species with complex social systems. Individuals of these species must learn the identities and individual characteristics of many associates and adjust their behavior accordingly. Meanwhile, bear in mind that as difficult as learning may be, ignorance is even more costly.

Evolutionary developmental biologists are generating new insights into the evolution of the diversity of organisms. This rapidly developing field is providing us with new information with which to understand the form of the Tree of Life. In Part Four of this book we will describe the history of life on Earth, and we will show how the agents of evolution act on the survival and reproductive success of adult organisms; that is, we will discuss the approach to evolution that has been developed by population geneticists and ecologists. The insights generated by that approach complement those derived from evolutionary developmental biology that we have just discussed. Together, they constitute a more complete view of the evolution of life on Earth.

Chapter Summary

Evolution and Development

▶ The field of evolutionary developmental biology unites embryology, ecology, and genetics. It is generating new information that helps us understand how the recipes for making organisms interact with environmental signals to produce functional organisms.

▶ Relationships among organisms can often be inferred by similarities in their embryonic or larval features.

▶ Many of the genes regulating development have changed very little throughout animal evolution.

▶ The same sets of genes are involved in specifying the anterior-posterior axis in both vertebrates and invertebrates. **Review Figure 21.3**

Regulatory Genes and Modularity: Modifying Morphology

▶ Major morphological changes can result from mutations in developmental regulatory genes or from alterations in the time or place of expression of these genes. **Review Figures 21.5, 21.6. See Web/CD Tutorial 21.1**

▶ Modularity allows morphological changes to occur without disrupting the entire organism.

▶ Heterochrony, a shift in the relative timing of two different developmental processes, can result in new morphology. **Review Figure 21.8**

Plant Development and Evolution

▶ Plants differ from animals in important ways that influence how they develop. Plant cells do not move relative to one another during development. In addition, the reproductive cells are not set aside early during plant development. Instead, plants continue to produce undifferentiated meristems as long as they grow. And plants have great developmental plasticity. **See Web/CD Activity 21.1**

▶ Despite these differences, plants have MADS box genes and homeobox genes, some of which are shared with animals, although they govern the development of structures unique to plants.

Environmental Influences on Developmental Patterns

▶ A genotype may encode a range of phenotypes.

▶ Developmental plasticity is the ability to express different phenotypes under different environmental conditions.

▶ Some environmental signals always occur and accurately predict future conditions. Others may or may not occur, but are good predictors of future conditions when they do appear. The development of organisms often responds adaptively to such signals. **Review Figure 21.9**

▶ Many organisms have the ability to detect the presence of predators and alter their development to reduce the likelihood that they will be attacked.

▶ Organisms evolve to not respond to environmental signals that are poor predictors of future conditions.

▶ Plants respond to light and other environmental conditions by changing their shapes and the number of flowers or seeds they produce, but not the sizes of their seeds. **Review Figure 21.13**

▶ Knowledge of how developing organisms are influenced by chemical agents that disrupt normal development may help us find alternative materials to use that are less damaging.

Learning: A Modification of Development

▶ Learning is a costly way of modifying development, but it can continue throughout adult life, and it allows an individual to adjust its behavior to the environment in which it matures.

Self-Quiz

1. Genes provide recipes rather than constituting blueprints because
   a. genetic instructions do not specify the form of the adult organism.
   b. an organism's development is not determined entirely by its genes.
   c. genes take orders as well as giving orders.
   d. genes encode other molecules, which in turn affect different parts of an organism.
   e. all of the above
2. Genetics and developmental biology evolved without much intellectual contact during much of the twentieth century because
   a. developmental biologists did not believe that development was genetically controlled.
   b. geneticists underestimated the degree to which the environment could influence development.
   c. geneticists and developmental biologists competed for research funds and therefore emphasized the differences between their disciplines.
   d. geneticists turned their attention to the heritability of traits, whereas embryologists turned their attention toward the unfolding of genetic instructions during the development of individuals.
   e. the genes that control development were of little interest to geneticists because they do not influence the ability of individuals to transmit their genes to offspring.
3. Homeobox genes determine the positions of cells along
   a. the anterior-posterior axis of the bodies of both insect and human embryos.
   b. the anterior-posterior axis of the bodies of insects, but not of vertebrates.
   c. the anterior-posterior axis of the bodies of vertebrates, but not of insects.
   d. the dorsal-ventral axis of the bodies of both insect and human embryos.
   e. the dorsal-ventral axis of the bodies of vertebrates only.

4. Ducks have webbed feet and chickens don't because
   a. ducks need webbed feet to swim, whereas terrestrial chickens do not.
   b. both duck and chicken embryos express BMP4 in the webbing between the toes, but the *gremlin* gene is expressed only in the webbing cells of ducks.
   c. both duck and chicken embryos express BMP4 in the webbing between the toes, but the *gremlin* gene is expressed only in the webbing cells of chickens.
   d. only duck embryos express BMP4 in the webbing between the toes.
   e. only chick embryos express BMP4 in the webbing between the toes.
5. Modularity is important for development because it
   a. guarantees that all units of a developing embryo will change in a coordinated way.
   b. coordinates the establishment of the anterior-posterior axis of the developing embryo.
   c. allows changes in the genes to change one part of the body without affecting other parts.
   d. guarantees that the timing of gene expression is the same in all parts of a developing embryo.
   e. allows organisms to be built up one module at a time.
6. Organisms often respond developmentally to regularly occurring environmental signals that accurately predict future conditions by
   a. stopping development until the signal changes.
   b. altering their development such that the resulting adult is adapted to the future environment.
   c. altering their development such that the resulting adult can produce offspring adapted to the future environment.
   d. producing new mutants.
   e. developing normally because the predicted conditions may not last long.
7. The phenomenon wherein organisms change the relative time of appearance and rate of development of characters is called
   a. heterochrony.
   b. developmental plasticity.
   c. adaptation.
   d. modularity.
   e. mutation.
8. *Daphnia* with large helmets are more difficult for some predators to capture and eat, but not all *Daphnia* produce large helmets because
   a. individuals with large helmets cannot feed efficiently.
   b. individuals with large helmets have trouble mating.
   c. individuals with large helmets produce fewer eggs than individuals with small helmets.
   d. individuals with large helmets become ensnared in vegetation.
   e. some individuals lack the genes that govern helmet formation.
9. Which of the following is *not* an important characteristic of plants that affects their development?
   a. Most plants are sessile.
   b. Plant cells do not move relative to one another during development.
   c. Plants produce meristems regularly as long as they grow.
   d. Plants have great developmental plasticity.
   e. Plants set aside germ cells early during embryonic development.
10. Which of the following plant structures does *not* change in response to the conditions under which a plant grows?
    a. Roots
    b. Seeds
    c. Leaves
    d. Stems
    e. Branches

For Discussion

1. What components of environmental influences on development would likely be missed if investigations were confined to simple organisms?
2. A spadefoot toad tadpole that develops in a rapidly drying pond is likely to eat many of its brothers and sisters. How can eating its siblings, which share half of an individual's genes, be favored by natural selection?
3. If evolutionary novelties can result from rather simple changes in the timing of expression of a few genes, why have such novelties arisen relatively infrequently during evolution?
4. Francois Jacob claimed that evolution was more like tinkering than engineering. Does the observation that developmental genes have changed little over evolutionary time support his assertion? Why?
5. We have learned in this chapter that plants and animals share many of the genes that regulate development. What are the implications of this observation for the ways in which humans can respond to the adverse effects of the many substances we release into the environment that cause developmental abnormalities in plants and animals? What kinds of substances are most likely to have such effects? Why?

LIFE ESSAY

What are the moral issues surrounding stem cell therapy?

- by Bonnie Steinbock -

There are two kind of stem cells: adult and embryonic. All stem cells have potential medical uses, but embryonic stem (ES) cells are thought by scientists to have especially great medical potential because they have the ability to become any kind of cell in the body. Stem cells may one day play an important role in the treatment of such diseases as Parkinson's, Alzheimer's, osteoporosis, macular degeneration, cancer, diabetes, and heart disease, as well as in the treatment of burns and spinal cord injury. While the use of adult stem cells in research does not raise moral questions, ES cell research does because ES cells are derived from early embryos (blastocysts), and thus involves the destruction of an embryo.

ES cells can be derived from embryos created specifically for research or from embryos created originally for infertility treatment but no longer needed ("spare" embryos). Some people think it is not permissible to create embryos for research, but it is morally permissible to use (and destroy) embryos that would be discarded anyway. Others think it is wrong to use either created or spare embryos in research; still others think it is permissible to use embryos in research regardless of their original intended use.

How people think about ES cell research depends largely on their views about the moral status of human embryos. Three views on this have different implications for embryo research:

1. *Embryos are human subjects, entitled to all the protection of any other human subjects.* This view rules out all embryo research that harms or kills embryos.

2. *Embryos, while biologically human, are not human beings in a moral sense.* What makes someone a human being in a moral sense—that is, a possessor of rights, a full member of the moral community; in short, a person—is not the species to which it belongs or the number of chromosomes it has, but rather whether it possesses the morally significant characteristics that human beings typically have—sentience, consciousness, self-consciousness, rationality, and the like. Embryos have none of these characteristics and therefore research on embryos is morally permissible.

3. *Embryos, while not persons, nevertheless deserve special respect and serious moral consideration as a developing form of human life.* This third view, a compromise between the first two views, allows for some embryo research, under carefully controlled conditions.

Critics maintain that the third view is incoherent, that it is impossible to show respect to embryos if one is planning to kill them. The challenge for those who advocate the third view is to give content to the notion of respect for embryos. First, it is important to determine what respect for embryos might mean in contrast to respect for persons. Respect for *persons*, as Kant instructs us, means never treating persons as mere means to our ends, but always treating them as ends in themselves. This means that we must take seriously other people's ends—their plans and goals—and not just our own.

This kind of respect is limited to beings who can have plans and goals, and thus is not appropriate to early embryos. Embryos lack any sort of consciousness or awareness. But even if embryos cannot be treated with the respect due to persons, there may still be moral reasons to treat them in some ways and not others. It is not only persons, or even sentient beings such as animals, that have moral significance. Plants, wilderness areas, entire species, or ecosystems may also deserve respect and protection. The potential of embryos to become human gives us a reason to treat them with more respect than other bodily tissues, even if it is not enough to endow embryos with full moral status and full moral rights.

What, then, does respect for embryos require? The third view does not rule out embryo research any more than respect for the dead precludes autopsies. Respect for embryos rules out frivolous or trivial uses, such as using human embryos in high school science classes, to test the safety of cosmetics, or to create jewelry. These are situations in which there is no pressing need to use human embryos; in such cases their use displays contempt rather than respect for human life. However, respect for human life does not rule out significant medical research that could save many lives or contribute to other worthy medical goals such as improving fertility treatment, reducing pain, or preventing disability.

Bonnie Steinbock received her Ph.D. in philosophy from the University of California, Berkeley. She has taught at the University at Albany since 1977, where she is a Full Professor and currently Chair of the Department of Philosophy. Her areas of specialization within bioethics include reproduction and genetics. She is the author of *Life Before Birth: The Moral and Legal Status of Embryos and Fetuses*, numerous articles, and the editor or co-editor of several anthologies, including the sixth edition of *Ethical Issues in Modern Medicine*.

Discussion Questions

1. Are embryos persons? On what do you base your view?
2. What is the created/spare distinction? Do you think it has moral significance?
3. Is it consistent to allow abortion but ban ES cell research?

Web Links

National Information Resources on Ethics and Human Genetics www.georgetown.edu/research/nrcbl/nirehg/index.html

Yahoo! Biology: Science: Biomedical Ethics Medical Ethics www.yahoo.com/Science/Biology/Biomedical_Ethics/

Center for the Study of Bioethics www.mcw.edu/bioethics/

Bioethics Resources on the Web – National Institute of Health www.nih.gov/sigs/bioethics/index.html

22 The History of Life on Earth

When you want to know what time it is, you probably look at your watch, or at the clock on the wall or on your computer. You could also listen for an announcement of the time on the radio or television. But suppose the electric power system failed and you lost your watch. How would you tell time then? You would use the cue that people have used during most of human history: the cycle of day and night. We are so accustomed to having time-measuring devices all around us that we forget that these devices are recent inventions. When Galileo studied the motion of a ball rolling down an inclined plane about 400 years ago, he used his pulse to mark off equal intervals of time.

The development of the science of biology is intimately linked to changing concepts of time, especially of the age of Earth. Biology as we know it could not and did not develop very far until about 150 years ago, when geologists provided evidence that Earth was ancient. Before 1850, most people believed that Earth was only a few thousand years old. Charles Darwin could not have developed his theory of evolution by natural selection if he had not read the works of Charles Lyell, who was England's leading geologist during Darwin's lifetime. Lyell suggested that existing landforms could be explained by the action, over very long time periods, of the same forces that are still acting on them today. That is, Lyell argued that it is not necessary to postulate sudden catastrophes as the reason for dramatic geological changes. As we pointed out in Chapter 1, Darwin's theory was based on the assumption that Earth was very old and that millions of years were available for life's evolution.

The goals of Part Four of this book are to document the history of life on Earth, to describe patterns of evolutionary change, and to investigate the agents that cause them. We begin this chapter by asking, How do we know that Earth is ancient? What is the evidence that life evolved early during Earth's history and has continued to evolve since then? We will first examine how events in the distant past can be dated. We will review the major changes in physical conditions on Earth during the past 4 billion years, look at how those changes have affected life, and describe some patterns in the evolution of life. In Chapter 23 we will discuss

Sunset at Stonehenge Even the earliest humans felt the need to keep track of time and seasons. The arches of Stonehenge, an astronomical "timepiece" on Salisbury Plain in England, date back to about 2000 B.C., and radioisotope dating has revealed some wooden structures unearthed here to be more than 8,000 years old.

the processes by which life evolves. In subsequent chapters, we will see how biologists determine the evolutionary histories of organisms and how the millions of species that live today (as well as many more that became extinct) were derived from a single common ancestor.

Defining Biological Evolution

Understanding evolution is important to the study of biology because the features of all organisms, including humans, are best understood in light of evolution. Furthermore, evolutionary changes are taking place all around us, some of which have powerful implications for human welfare. For example, our attempts to control populations of species we consider pests and to increase populations of those we consider desirable make humans powerful agents of evolutionary change. In addition to producing the results we desire, these efforts often cause undesirable outcomes, such as the evolution of resistance to antibiotics by pathogens and to pesticides by pests. Medicine and agriculture can respond creatively to the evolutionary changes they are causing only if their practitioners understand how and why those changes happen. But what exactly is biological evolution?

Biological evolution is a change over time in the genetic composition of a population of organisms. Many such changes happen rapidly enough to be studied directly and manipulated experimentally. Plant and animal breeding by agriculturalists and responses of organisms to environmental shifts over decades provide good examples of such short-term evolution. Other changes, such as the appearance of new species and evolutionary lineages, usually take place over much longer time frames. The fossil record is the primary source of direct evidence of those changes.

To understand the long-term patterns of evolutionary change that we will document in this chapter, we must think in time frames spanning many millions of years and imagine events and conditions very different from those we now observe. The Earth of the distant past is, to us, a foreign planet inhabited by strange organisms. The continents were not where they are today, and climates were sometimes dramatically different from those of today. One of the remarkable achievements of twentieth-century science has been the development of sophisticated techniques, using rates of decay of various radioisotopes, changes in Earth's magnetic field, and the presence or absence of certain molecules, for inferring past conditions and dating them accurately.

Determining Earth's Age

It is difficult to age rocks because any type of rock could have been formed at any time during Earth's history. It is easier to determine the ages of rocks relative to one another. The first person to recognize that this could be done was the seventeenth-century Danish physician Nicolaus Steno. Steno realized that in undisturbed sedimentary rock, the oldest layers, or *strata*, lie at the bottom, and successively higher strata are progressively younger (Figure 22.1).

22.1 Young Rocks Lie on Top of Old Rocks The oldest rocks visible in this photo of the Grand Canyon formed about 540 million years ago. The youngest rocks, at the top, are about 500 million years old.

Geologists subsequently combined Steno's insight with their observations of **fossils**—preserved remains of ancient organisms—contained within the rocks. They discovered that fossils of similar organisms were found in widely separated places on Earth, that certain organisms were always found in younger rocks than others, and that organisms in more recent strata were more similar to modern organisms than were those found in lower, more ancient strata. With this information, they learned much about the relative ages of sedimentary rocks and patterns in the evolution of life. But they could not tell how old the rocks were. A method of dating rocks did not become available until the discovery of radioactivity at the beginning of the twentieth century.

Radioactivity provides a way to date rocks

Radioactive isotopes decay in a predictable pattern over long time periods (see Chapter 2). During each successive time interval, an equal fraction of the remaining radioactive material of any radioisotope decays, either changing to another element or becoming the stable isotope of the same element.

For example, in 14.3 days, one-half of any sample of phosphorus-32 (^{32}P) decays to its stable isotope, phosphorus-31 (^{31}P). During the next 14.3 days, one-half of the remaining half decays, leaving one-fourth of the original ^{32}P. After 42.9 days, three *half-lives* have passed, so one-eighth (that is, $1/2 \times 1/2 \times 1/2$) of the original ^{32}P remains.

Each radioisotope has a characteristic half-life. Tritium (^{3}H) has a half-life of 12.3 years, and carbon-14 (^{14}C) has a half-life of about 5,700 years. The half-life of potassium-40 (^{40}K) is 1.3 billion years; that of uranium-238 (^{238}U) is about 4.5 billion years.

To use a radioisotope to date a past event, we must know or estimate the concentration of the isotope at the time of that event. In the case of carbon, we know that the production of new ^{14}C in the upper atmosphere (by the reaction of neutrons with ^{14}N) just balances the natural radioactive decay of ^{14}C. Therefore, the ratio of ^{14}C to its stable isotope, ^{12}C, is relatively constant in living organisms and their environment.

However, as soon as an organism dies, it ceases to exchange carbon compounds with its environment. Its decaying ^{14}C is no longer replenished, and the ratio of ^{14}C to ^{12}C in its remains decreases through time. The ratio of ^{14}C to ^{12}C in fossil organisms can be used to date fossils (and thus the sedimentary rocks that contain those fossils) that are less than 50,000 years old with a fair degree of certainty.

Radioisotope dating methods have been expanded and refined

Dating rocks more ancient than 50,000 years requires estimating isotope concentrations in volcanic (but not in sedimentary) rocks. Sedimentary rocks are formed from materials that existed for varying lengths of time before being transported long distances to the site of their deposition. Therefore, a sedimentary rock does not contain reliable information about the date of its formation. To age sedimentary rocks, geologists search for places where volcanic ash or lava flows have intruded into beds of those rocks. The preliminary estimate of the age of the volcanic rock determines which isotope is used. The decay of potassium-40 to argon-40 has been

22.1 Earth's Geological History

RELATIVE TIME SPAN

Precambrian

ERA	PERIOD	ONSET	MAJOR PHYSICAL CHANGES ON EARTH
Cenozoic	Quaternary	1.8 mya[a]	Cold/dry climate; repeated glaciations
	Tertiary	65 mya	Continents near current positions; climate cools
Mesozoic	Cretaceous	144 mya	Northern continents attached; Gondwana begins to drift apart; meteorite strikes Yucatán Peninsula
	Jurassic	206 mya	Two large continents form: Laurasia (north) and Gondwana (south); climate warm
	Triassic	248 mya	Pangaea slowly begins to drift apart; hot/humid climate
Paleozoic	Permian	290 mya	Continents aggregate into Pangaea; large glaciers form; dry climates form in interior of Pangaea
	Carboniferous	354 mya	Climate cools; marked latitudinal climate gradients
	Devonian	417 mya	Continents collide at end of period; asteroid probably collides with Earth
	Silurian	443 mya	Sea levels rise; two large continents form; hot/humid climate
	Ordovician	490 mya	Gondwana moves over South Pole; massive glaciation, sea level drops 50 m
	Cambrian	543 mya	O_2 levels approach current levels
Precambrian		600 mya	O_2 level at >5% of current level
		1.5 bya[a]	O_2 level at >1% of current level
		3.8 bya	O_2 first appears in atmosphere
		4.5 bya	

[a]mya, million years ago; bya, billion years ago.

used to date most of the ancient events in the evolution of life. Radioisotope dating, combined with fossils, is the most powerful method of determining the ages of rocks.

But there are many places where sedimentary rocks do not contain suitable volcanic intrusions and few fossils are present. In these areas, other dating methods must be used. One method, known as *paleomagnetic dating*, is based on the fact that Earth's magnetic poles move and occasionally reverse themselves. Because both sedimentary and volcanic rocks preserve a record of Earth's magnetic field at the time they were formed, paleomagnetism helps determine the ages of those rocks. Other dating methods, which we will describe in later chapters, use continental drift, sea level changes, and molecular clocks.

Using these methods, geologists have divided the history of life into eras, which in turn are subdivided into periods (Table 22.1). The boundaries between these divisions are based on major differences in the fossil organisms contained in successive layers of rocks. The divisions were established before the ages of the eras and periods were known. The scale at the left of Table 22.1 gives a relative sense of geological time and the vast expanse of the **Precambrian** era, during which early life evolved amid stupendous physical changes on Earth.

MAJOR EVENTS IN THE HISTORY OF LIFE
Humans evolve; many large mammals become extinct
Diversification of birds, mammals, flowering plants, and insects
Dinosaurs continue to diversify; flowering plants and mammals diversify. **Mass Extinction** at end of period (≈76% of species disappear)
Diverse dinosaurs; radiation of ray-finned fishes
Early dinosaurs; first mammals; marine invertebrates diversify; first flowering plants; **Mass Extinction** at end of period (≈65% of species disappear)
Reptiles diversify; amphibians decline; **Mass Extinction** at end of period (≈96% of species disappear)
Extensive "fern" forests; first reptiles; insects diversify
Fishes diversify; first insects and amphibians. **Mass Extinction** at end of period (≈75% of species disappear)
Jawless fishes diversify; first ray-finned fishes; plants and animals colonize land
Mass Extinction at end of period (≈75% of species disappear)
Most animal phyla present; diverse algae
Ediacaran fauna
Eukaryotes evolve; several animal phyla appear
Origin of life; prokaryotes flourish

The Changing Face of Earth

Earth has undergone many physical changes that have influenced the evolution of life. The physical events described in this section, along with the most important milestones in the history of life, are listed in Table 22.1.

The continents have changed their positions

The maps and globes that adorn our walls, shelves, and books give an impression of a static Earth. It would be easy for us to assume that the continents have always been where they are. But we would be wrong. Earth's crust consists of a number of solid *plates* approximately 40 km thick, which float on a fluid *mantle*. The mantle fluid circulates because heat produced by radioactive decay sets up convection patterns in the fluid. The plates move because material from the mantle rises and pushes them aside. Where plates are pushed together, either they move sideways past each other, or one plate moves under the other, pushing up mountain ranges. The movement of the plates and the continents they contain is known as **continental drift**.

At times, the drifting of the plates brought the continents together; at other times, they drifted apart. The positions and sizes of the continents influence ocean circulation patterns, sea levels, and global climate patterns. Mass extinctions of species, particularly marine organisms, have usually accompanied major drops in sea level, which exposed vast areas of the continental shelves, killing the marine organisms that lived in the shallow seas that had covered them (Figure 22.2).

Earth's atmosphere has changed unidirectionally

The continents have moved irregularly over Earth's surface, but some physical changes on Earth have been unidirectional. The atmosphere of early Earth probably contained little or no free oxygen (O_2). Oxygen concentrations in the atmosphere began to increase markedly about 2.5 billion years ago, when certain bacteria evolved the ability to use water as the source of hydrogen ions for photosynthesis. By chemically splitting H_2O ($2\,H_2O \rightarrow 4\,H^+ + O_2 + 4e^-$), these bacteria generated atmospheric O_2 as a waste product; in addition, they made electrons available for reducing CO_2 to form organic compounds.

One lineage of oxygen-generating bacteria evolved into the *cyanobacteria*. These ancient photosynthesizers formed

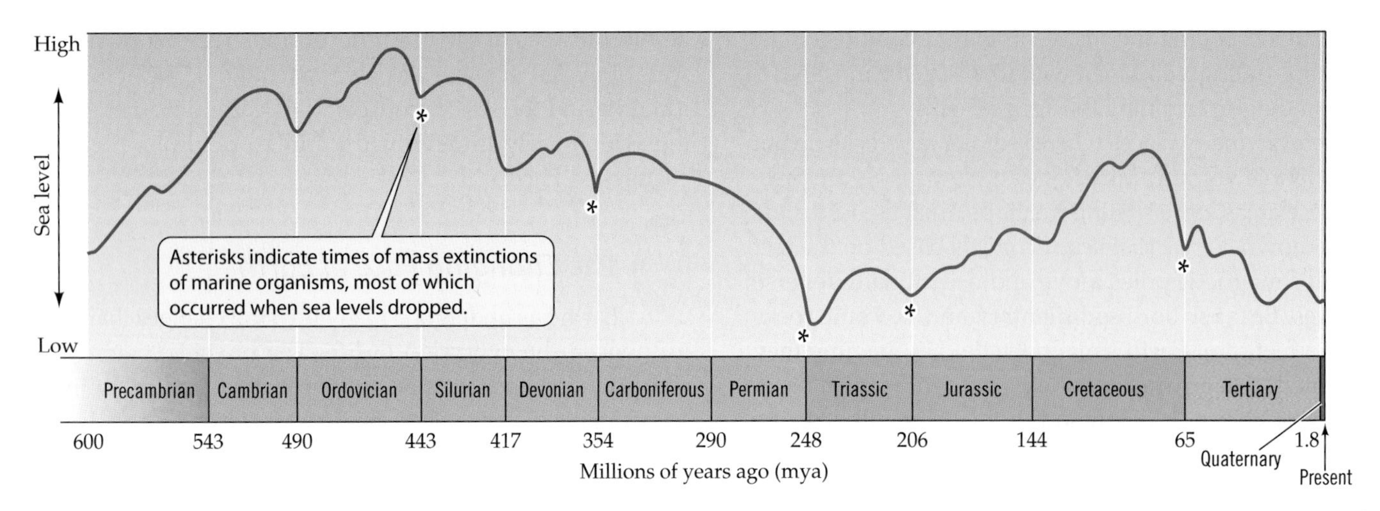

22.2 Sea Levels Have Changed Repeatedly Most mass extinctions (indicated by asterisks) of marine organisms have coincided with periods of low sea levels.

rocklike structures called *stromatolites*, which are abundantly preserved in the fossil record. Cyanobacteria are still forming stromatolites today in a few very salty places on Earth (Figure 22.3). Cyanobacteria liberated enough O_2 to open the way for the evolution of oxidation reactions as the energy source for the synthesis of ATP. Their ability to split water doubtless contributed to their extraordinary success.

The evolution of life irrevocably changed the physical nature of the planet. Living organisms not only added O_2 to Earth's atmosphere, but also removed most of the CO_2 from the atmosphere by taking it up and transferring it to ocean sediments with their remains when they died. When it first appeared, oxygen was poisonous to the anaerobic prokaryotes that inhabited Earth at the time. Those prokaryotes that evolved the ability to metabolize O_2 not only survived, but also gained a number of advantages. Aerobic (oxygen-using) metabolism proceeds at higher rates and can extract more energy from compounds than the anaerobic metabolism prevalent among living things until then (see Chapter 7). Consequently, organisms with aerobic metabolism have replaced anaerobes in most of Earth's environments.

An atmosphere rich in O_2 also made possible larger cells and more complex organisms. Small, unicellular aquatic organisms can obtain enough O_2 by simple diffusion even when O_2 concentrations are very low. Larger unicellular organisms have lower surface area-to-volume ratios (see Figure 4.3). In order to obtain enough O_2 by simple diffusion, they must live in an environment with a relatively high concentration of O_2. Bacteria can thrive on 1 percent of the current atmospheric O_2 levels, but eukaryotic cells require oxygen levels that are at least 2 to 3 percent of current atmospheric concentrations.

About 1,500 million years ago (mya), O_2 concentrations became high enough for large eukaryotic cells to flourish and diversify (Figure 22.4). Further increases in atmospheric O_2

22.3 Stromatolites (*a*) A vertical section through a fossil stromatolite. (*b*) These rocklike structures are living stromatolites that thrive in the very salty waters of Shark Bay, Western Australia. Layers of cyanobacteria are found in the uppermost parts of the structures.

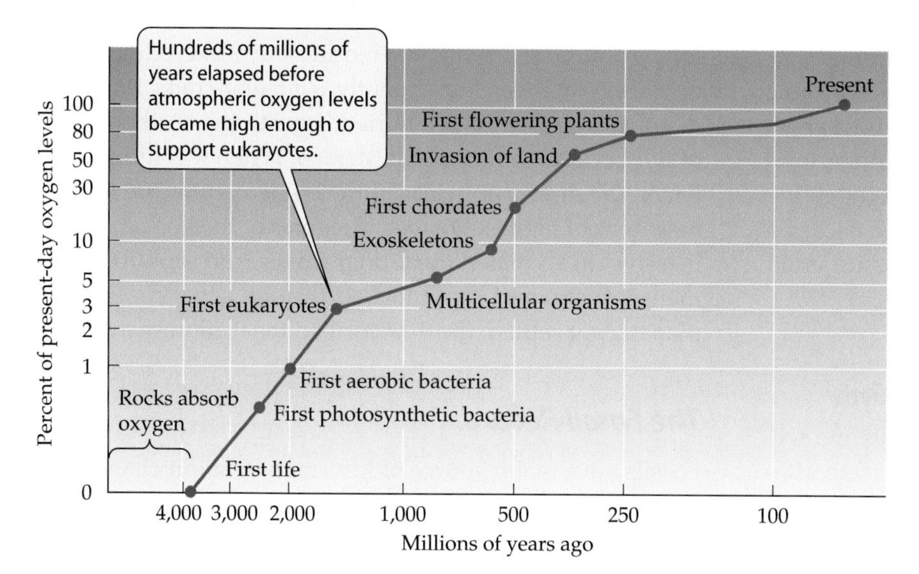

22.4 Larger Cells Need More Oxygen As oxygen concentrations in the atmosphere rose, the complexity of life increased. Although aerobic prokaryotes can flourish with less, larger eukaryotic cells with lower surface area-to-volume ratios require at least 2 to 3 percent of current atmospheric O_2 concentrations. (Both axes of the graph are on logarithmic scales.)

levels 700 to 570 mya enabled multicellular organisms to evolve. The fact that it took many millions of years for Earth to develop an oxygenated atmosphere probably explains why only unicellular prokaryotes lived on Earth for more than a billion years.

In contrast to this largely unidirectional change in atmospheric O_2 concentration, most physical conditions on Earth have oscillated in response to the planet's internal processes, such as volcanic activity and continental drift. External events, such as collisions with meteorites, have also left their mark. In some cases, these events caused **mass extinctions**, exterminating a large proportion of the species living at the time.

Earth's climate shifts between hot/humid and cold/dry conditions

Through much of its history, Earth's climate was considerably warmer than it is today, and temperatures decreased more gradually toward the poles. At other times, however, Earth was colder than it is today. Large areas were covered with glaciers during the end of the Precambrian and during the Carboniferous, Permian, and Quaternary periods, but these cold periods were separated by long periods of milder climates (Figure 22.5). Because we are living in one of the colder periods in the history of Earth, it is difficult for us to imagine the mild climates that were found at high latitudes during much of the history of life.

Weather often changes rapidly; climates usually change slowly. However, major climatic shifts have taken place over periods as short as 5,000 to 10,000 years, primarily as a result of changes in Earth's orbit around the sun. A few climatic shifts appear to have been even more rapid. For example, during one Quaternary interglacial period, the Antarctic Ocean changed from being ice-covered to being nearly ice-free in less than 100 years. Such rapid changes are usually caused by sudden shifts in ocean currents. Climates have sometimes changed rapidly enough that extinctions caused by them appear "instantaneous" in the fossil record.

Volcanoes occasionally changed the history of life

Most volcanic eruptions produce only local or short-lived effects, but a few very large volcanic eruptions have had major consequences for life. The collision of continents during

22.5 Hot/Humid and Cold/Dry Conditions Have Alternated Over Earth's History Throughout Earth's history, periods of cold climates and glaciations (white depressions) have been separated by long periods of milder climates.

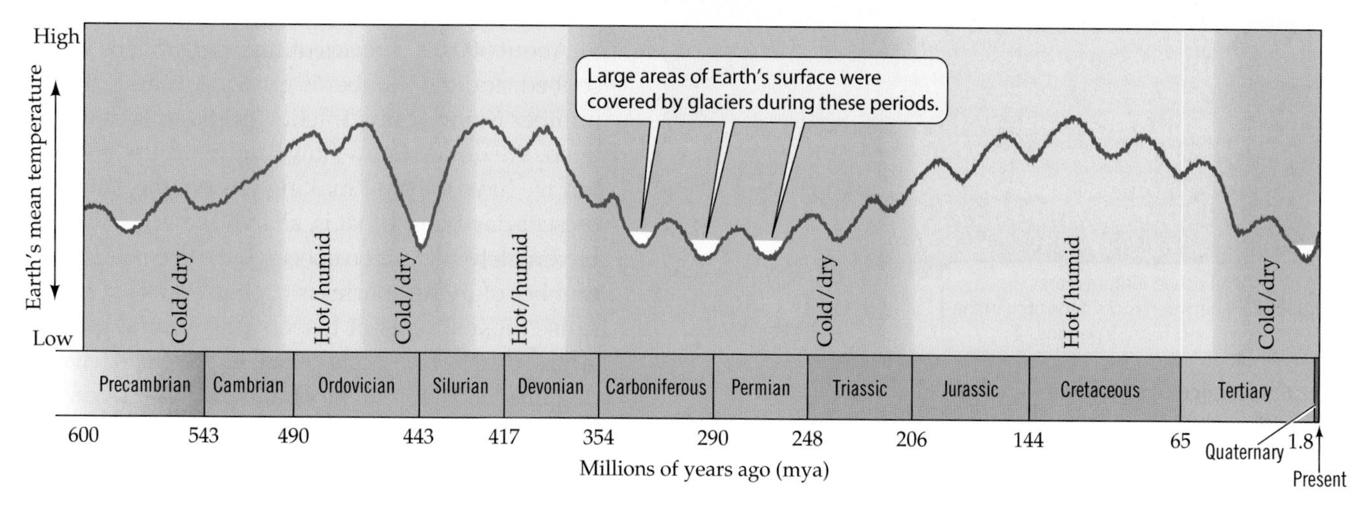

the late Permian period (about 275 mya) to form a single, gigantic land mass, called Pangaea, caused massive volcanic eruptions. The ash the volcanoes ejected into the atmosphere reduced the penetration of sunlight to Earth's surface, lowering temperatures, reducing photosynthesis, and triggering massive glaciation. Massive volcanic eruptions also occurred as the continents drifted apart during the late Triassic period and at the end of the Cretaceous.

External events have triggered changes on Earth

At least 30 meteorites between the sizes of baseballs and soccer balls hit Earth each year. Collisions with large meteorites are rare, but large meteorites have been responsible for several mass extinctions. Evidence for these collisions is found in the craters that resulted from their impact, dramatic disfigurations of rocks (microspherules and shocked quartz crystals), and within giant molecules that contain trapped helium and argon with isotopic ratios characteristic of meteorites, which are very different from the ratios found on Earth. Also, fern fossils are abundant in rocks that formed at the end of the Triassic and Cretaceous periods. Because ferns can more quickly colonize and survive in bare environments than most other plants, their abundance suggests that meteorite impacts had scoured vast areas of Earth's surface.

The first extraterrestrial impact to be documented was that of a meteorite about 10 km in diameter that caused a mass extinction at the end of the Cretaceous period, about 65 mya. The first clue that a meteorite was responsible came from the abnormally high concentrations of the element iridium in a thin layer separating rocks deposited during the Cretaceous from those deposited during the Tertiary (Figure 22.6). Iridium is abundant in some meteorites but is exceedingly rare on Earth's surface. Subsequently, a circular crater 180 km in diameter was discovered buried beneath the northern coast of the Yucatán Peninsula of Mexico. When it collided with Earth, the meteorite released the equivalent of 100 million megatons of high explosives. The force of the impact ignited massive fires, created great tidal waves, and sent up an immense dust cloud that blocked the sun, thus cooling the planet. As it settled, the dust formed the iridium-rich layer.

A thin band rich in iridium marks the boundary between rocks deposited in the Cretaceous and Tertiary periods.

22.6 Evidence of a Meteorite Impact Iridium is a metal common in some meteorites, but rare on Earth. Its high concentration in sediments deposited about 65 million years ago suggests the impact of a large meteorite.

The Fossil Record

Fossils are a major source of information about changes on Earth during the remote past. As we saw above, pre-Darwinian geologists divided geological history into units based on their distinct fossil assemblages of animals (see Table 22.1). These divisions are marked either by mass extinctions or by dramatic increases in the diversity of major groups of organisms (called *evolutionary radiations*).

Life first evolved on Earth about 3.8 billion years ago (bya), and by about 1.5 bya, eukaryotic organisms had evolved. The fossil record of organisms that lived prior to 550 mya is fragmentary, but the available evidence suggests that the major divisions in many animal lineages predate the end of the Precambrian by more than 100 million years. The fossil record is good enough to show that the total number of species and individuals increased dramatically in late Precambrian times.

An organism is most likely to become a fossil if its dead body is deposited in an environment that lacks oxygen. However, most organisms live in aerobic environments; they decompose completely when they die. Thus, many fossil assemblages are collections of organisms that were transported by wind or water to sites that lacked oxygen. Occasionally, however, organisms, or imprints of them, are preserved where they lived. In such cases—especially if the environment in question was a cool, anaerobic swamp, where conditions for preservation were excellent—we can obtain a picture of communities of organisms that lived together.

About 300,000 species of fossil organisms have been described, and the number is growing steadily. However, this number is only a tiny fraction of the species that have ever lived. We do not know how many species lived in the past, but we have ways of making reasonable estimates. Of the present-day **biota**—that is, all living species of all kinds—approximately 1.7 million species have been named. The actual number of living species is probably at least 10 million, because most species of insects (the animal group with the largest number of species; see Chapter 33) have not yet been described. So the number of known fossil species is less than 2 percent of the probable minimum number of living species. Because life has existed on Earth for about 3.8 billion years,

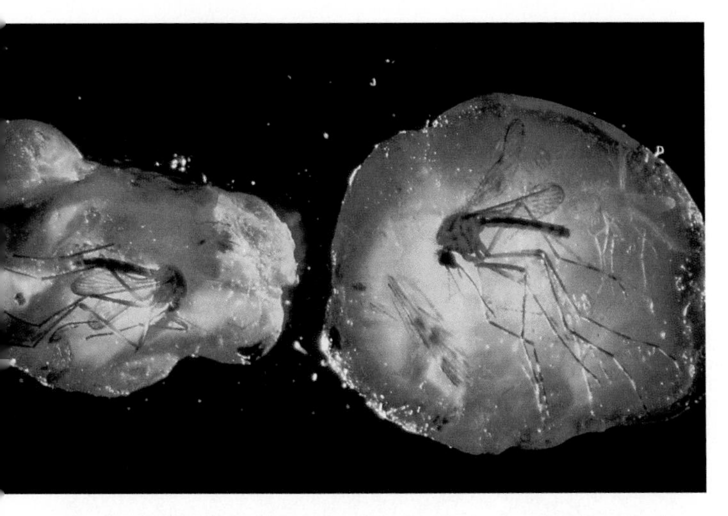

22.7 Insect Fossils These chunks of amber—fossilized tree resin—contain insects that were preserved when they were trapped in the sticky resin some 50 million years ago.

and because species last, on average, less than 10 million years, Earth's biota must have turned over many times during geological history. Thus, the total number of species that lived over evolutionary time must vastly exceed the number living today.

The number of known fossils, although it is a small fraction of the total number of extinct species, is especially large for marine animals that had hard skeletons. Among the nine major animal groups with hard-shelled members, approximately 200,000 species have been described from fossils—roughly twice the number of living marine species in these same groups. *Paleontologists* (scientists who study fossils) lean heavily on these groups in their interpretations of the evolution of life. Insects and spiders are also relatively well represented in the fossil record (Figure 22.7). The fossil record may be incomplete, but it is good enough to demonstrate clearly that organisms of particular types are found in rocks of specific ages and that new organisms appear sequentially in younger rocks. The fossil record also tells us that extinction is the eventual fate of all species.

By combining information about physical changes during Earth's history with evidence from the fossil record, scientists have composed portraits of what Earth and its inhabitants may have looked like at different times. We know in general where the continents were and how life changed over time, but many of the details are poorly known, especially for events in the more remote past. In the next section, we provide an overview of how life changed during its history on Earth. Part Five of this book will look at the evolutionary history of particular groups of organisms in more detail.

Major Patterns in the History of Life on Earth

For much of its history, life was confined to the oceans, and all organisms were small. In the Precambrian era, shallow seas teemed with life. Protists and small multicellular animals fed on floating algae. Small floating organisms, known collectively as *plankton*, were eaten by small animals that filtered them from the water. Other animals ingested sediments on the bottom of the seas and digested the remains of organisms within them. By the late Precambrian, about 650 mya, many kinds of soft-bodied invertebrates had evolved. Some of them were very different from any animals living today. They may be members of animal lineages that have no living descendants (Figure 22.8).

Life expanded rapidly during the Cambrian period

By the early **Cambrian** period (543–490 mya), at the beginning of the Paleozoic era, the O_2 concentration in the atmosphere approached its current level and the continental plates came together to form several land masses. The largest of the land masses was called Gondwana (Figure 22.9*a*). All of the major groups of animals that have species living today appeared during the Cambrian. This rapid diversification of life is referred to as the *Cambrian explosion*.

The most extensive fossil evidence from the Cambrian comes from the unusually well preserved animal fossils recently discovered in northeastern China (Figure 22.9*b*). Arthropods (crabs, shrimps and their relatives) are the most

22.8 Ediacaran Animals These fossils of soft-bodied invertebrates, excavated at Ediacara in southern Australia, formed 600 million years ago. They illustrate the diversity of life that evolved in Precambrian times.

Spriggina floundersi

Mawsonites

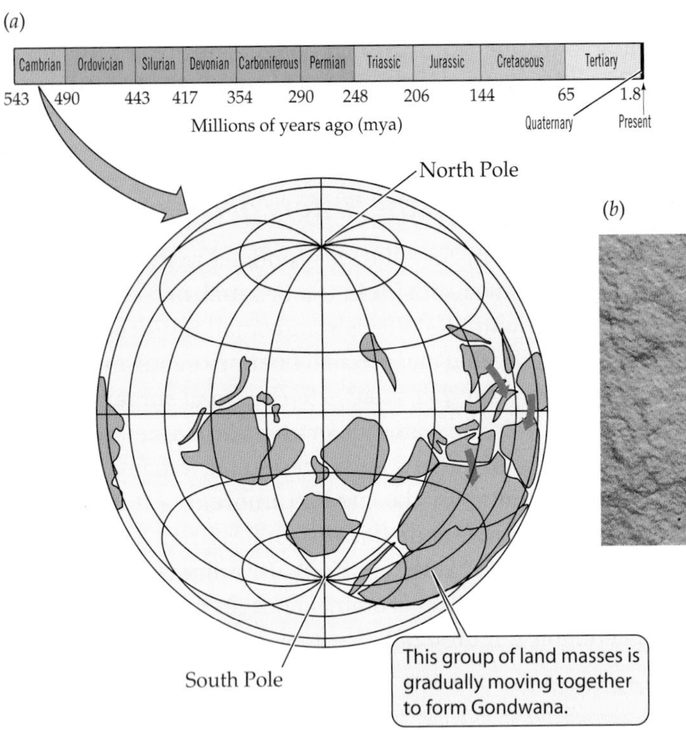

22.9 Cambrian Continents and Animals (*a*) Positions of the continents during mid-Cambrian times (542–490 mya). This view of Earth has been distorted so that you can see both poles. (*b*) Fossil beds in China have yielded well-preserved remains of Cambrian animals such as this one, called *Jianfangia*.

diverse group in this Chinese fauna; some of them were large carnivores. Trilobites, an arthropod group that was abundant and diverse during the Cambrian, suffered a major extinction at the end of the Cambrian, but they recovered and continued to be abundant during subsequent periods.

Major changes continued during the rest of the Paleozoic era

Geologists divide the remainder of the Paleozoic era into five periods: the Ordovician, Silurian, Devonian, Carboniferous, and Permian (see Table 22.1). Each period is characterized by the diversification of specific groups of organisms. Mass extinctions marked the ends of three periods: the Ordovician, Devonian, and Permian.

THE ORDOVICIAN (490–443 MYA). During the **Ordovician** period, the continents, which were located primarily in the Southern Hemisphere, were still devoid of multicellular plants. Evolutionary radiation of marine organisms was spectacular during the early Ordovician, especially among animals that filter small prey from the water, such as brachiopods and mollusks. All animals lived on the seafloor or burrowed in its sediments. At the end of the Ordovician, as massive glaciers formed over Gondwana, sea levels dropped about 50 meters and ocean temperatures dropped. About 75 percent of the animal species became extinct, probably because of these major environmental changes.

THE SILURIAN (443–417 MYA). During the **Silurian** period, the northern continents coalesced, but the general positions of the continents did not change much. Marine life rebounded from the mass extinction at the end of the Ordovician. Animals able to swim and feed above the ocean bottom appeared for the first time, but no new phyla of marine organisms evolved. The tropical sea was uninterrupted by land barriers, and most marine genera were widely distributed. On land, the first known tracheophytes (plants with true vascular tissue; see Chapter 29) appeared late in the Silurian period, about 420 mya. These plants, in the genus *Cooksonia*, were less than 50 cm tall and lacked roots and leaves (Figure 22.10). The first terrestrial arthropods—scorpions and millipedes—appeared at about the same time.

THE DEVONIAN (417–354 MYA). Rates of evolutionary change accelerated in many groups of organisms during the **Devonian** period. Both northern and southern land masses slowly moved northward (Figure 22.11*a*). There were great evolutionary radiations of corals and shelled squidlike cephalopods (Figure 22.11*b*). Fishes diversified as jawed forms replaced jawless ones, and heavy armor gave way to the less rigid outer coverings of modern fishes. All current major groups of fishes were present by the end of the period.

Terrestrial communities also changed dramatically during the Devonian. Club mosses, horsetails, and tree ferns became common toward the end of the Devonian; some attained the size of trees. Their deep roots accelerated the weathering of rocks, resulting in the development of the first forest soils. Distinct floras evolved on the two major land masses toward the end of the period, and the ancestors of gymnosperms, the

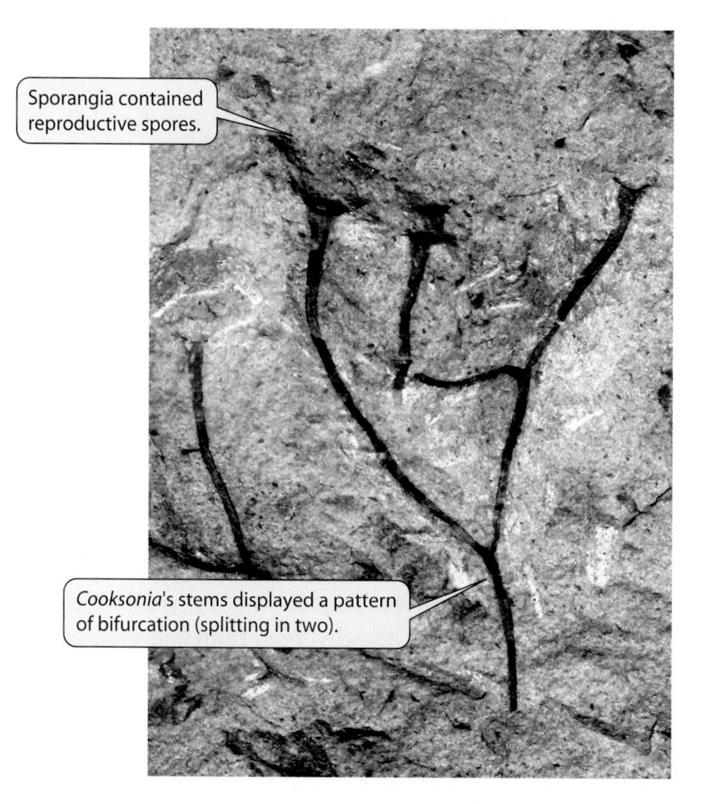

22.10 *Cooksonia*, the Earliest Known Tracheophyte These plants were small and very simple in structure. However, they were true vascular plants (tracheophytes) with internal water-conducting cells (tracheids), well equipped to make the move from the aquatic to the terrestrial environment. This fossil of *Cooksonia pertoni* is from the Silurian (415 mya).

first plants to produce seeds, appeared in the fossil record. The first known fossils of centipedes, spiders, mites, and insects date to this period. Fishlike amphibians began to occupy the land.

An extinction of about 75 percent of all marine species marked the end of the Devonian. Paleontologists are uncertain about the cause of this mass extinction, but two large meteorites that collided with Earth at that time, one in present-day Nevada and the other in Western Australia, may have been responsible.

THE CARBONIFEROUS (354–290 MYA). Large glaciers formed over high-latitude Gondwana during the **Carboniferous** period, but extensive swamp forests grew on the tropical continents. These forests were not made up of the kinds of trees we know today, but were dominated by giant tree ferns and horsetails (see Figure 29.11). Fossilized remains of those trees formed the coal we now mine for energy.

The diversity of terrestrial animals increased greatly. Snails, scorpions, centipedes, and insects were abundant and diverse. Insects evolved wings, becoming the first animals to fly. Flight gave them access to tall plants; plant fossils from this period show evidence of chewing by insects. Amphibians became larger and better adapted to terrestrial existence. From one amphibian stock, the first reptiles evolved late in the period. In the seas, crinoids (sea lilies and feather stars) reached their greatest diversity, forming "meadows" on the seafloor (Figure 22.12)

(*a*)

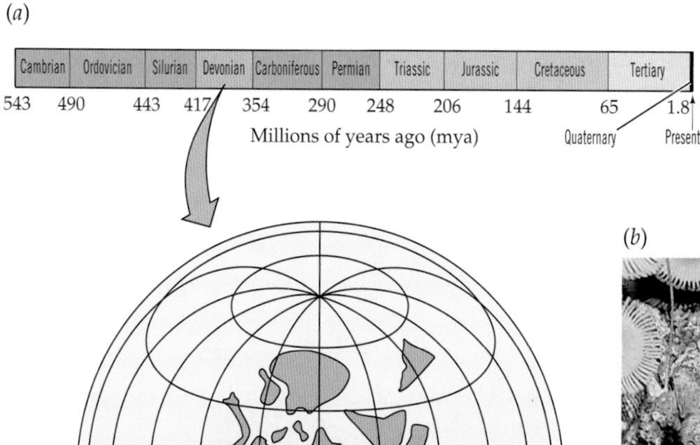

(*b*)

22.11 Devonian Continents and Marine Communities (*a*) Positions of the continents during the Devonian period (417–354 mya). (*b*) This museum reconstruction depicts a Devonian coral reef.

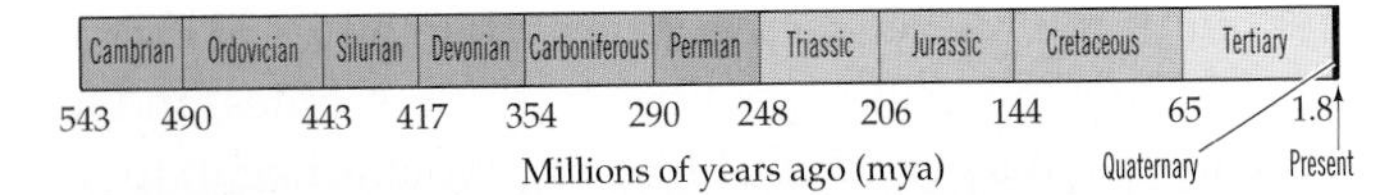

22.12 A Carboniferous "Crinoid Meadow" Crinoids, which were dominant marine animals during the Carboniferous, may have formed communities similar to this one.

THE PERMIAN (290–248 MYA). During the **Permian** period, the continents coalesced into a supercontinent, called Pangaea. Massive volcanic eruptions resulted in outpourings of lava that covered large areas of Earth (Figure 22.13). The ash they produced blocked sunlight and cooled the climate, resulting in the largest glaciers in Earth's history.

Permian deposits contain representatives of most modern groups of insects. By the end of the period, reptiles greatly outnumbered amphibians. Late in the period, the lineage leading to mammals diverged from one reptilian lineage. In fresh waters, the Permian period was a time of extensive radiation of ray-finned fishes.

Toward the end of the Permian period, a large meteorite crashed into northwestern Australia, creating a crater about 190 km in diameter. In addition, a massive outpouring of lava flowed into the oceans, drastically reducing the oxygen content of deep ocean waters. Oceanic turnover then carried these oxygen-depleted waters toward the surface, where they released toxic concentrations of carbon dioxide and hydrogen sulfide into the surface waters and the atmosphere. These gases poisoned most of the species that had survived the impact. All in all, about 96 percent of all species on Earth became extinct at that time.

Geographic differentiation increased during the Mesozoic era

The few organisms that survived the Permian mass extinction found themselves in a relatively empty world at the start of the Mesozoic era (248 mya). As Pangaea slowly separated into individual continents, the climate warmed, the glaciers melted, and the oceans rose and reflooded the continental shelves, forming huge, shallow inland seas. Life again proliferated and diversified, but different lineages came to dominate Earth. The trees that had dominated the Permian forests, for example, were replaced by new plants with seeds.

During the Mesozoic, Earth's biota, which until that time had been relatively homogeneous, became increasingly *provincialized*; that is, distinctive terrestrial floras and faunas evolved on each continent. The biotas of the shallow waters bordering the continents also diverged from one another. The provincialization that began during the Mesozoic continues to influence the geography of life today. By the end of the era, the world and its biota appeared quite modern. The Mesozoic era is divided into three periods—the Triassic, Jurassic, and Cretaceous—the first and third of which were terminated by mass extinctions, probably caused by meteorite impacts.

THE TRIASSIC (248–206 MYA). Pangaea began to break apart during the **Triassic** period. Many invertebrate lineages became more diverse, and many burrowing forms evolved from groups living on the surfaces of bottom sediments. On land, conifers and seed ferns became the dominant trees. The first frogs and turtles appeared. A great radiation of reptiles began, which eventually gave rise to crocodilians, dinosaurs, and birds. The end of the Triassic was marked

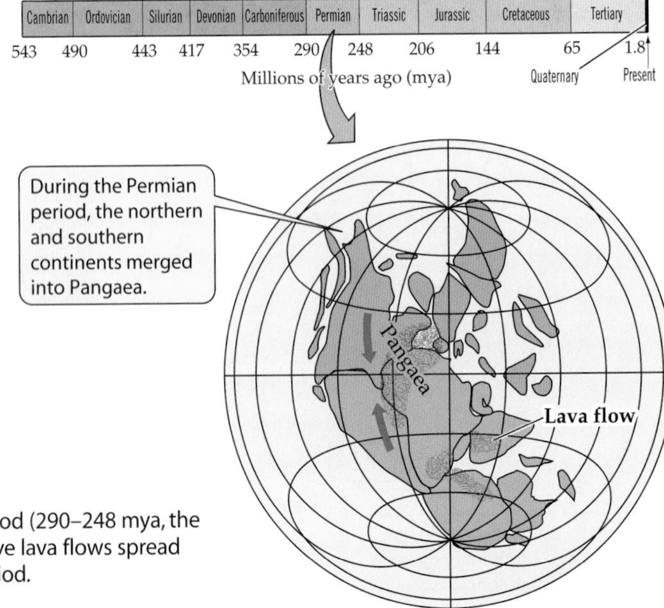

22.13 Pangaea Formed in the Permian Period During the Permian period (290–248 mya, the interior of the "supercontinent" Pangaea experienced harsh climates. Massive lava flows spread over Earth, and the largest glaciers in Earth's history formed during this period.

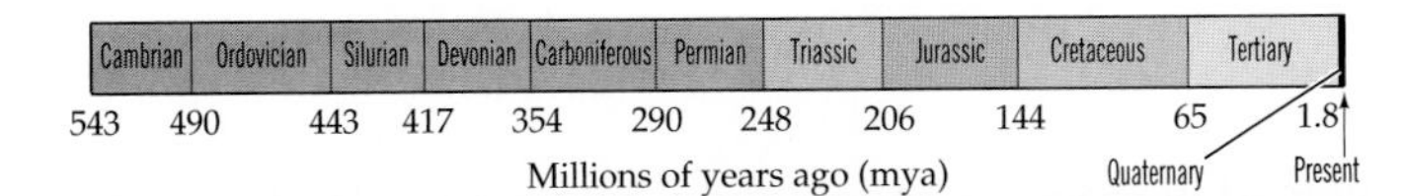

22.14 Mesozoic Dinosaurs The dinosaurs of the Mesozoic era continue to capture our imagination. The horned animals in this artist's depiction are *Triceratops horridus*, a huge (6000 kg) herbivore of the late Cretaceous (70 mya). At the lower right is *Ornithomimus* ("bird mimic"), an ostrich-like dinosaur believed to have been able to run at high speeds.

by a mass extinction that eliminated about 65 percent of the species on Earth. A large meteor that crashed into Quebec may have been responsible.

THE JURASSIC (206–144 MYA). During the **Jurassic** period, two large continents formed—Laurasia in the north, and Gondwana in the south. Diversification of many lineages proceeded. Ray-finned fishes began the great radiation that culminated in their dominance of the oceans. Salamanders and lizards first appeared. Flying reptiles (pterosaurs) evolved, and dinosaur lineages evolved into bipedal predators and large quadrupedal herbivores (Figure 22.14). Several groups of mammals first appeared during this time. Plant evolution continued with the likely emergence of the flowering plant lineage prevalent on Earth today (see Chapter 30).

THE CRETACEOUS (144–65 MYA). By the early **Cretaceous** period, Laurasia was completely separate from Gondwana, which was beginning to break apart. A continuous sea encircled the Tropics (Figure 22.15). Sea levels were high, and Earth was warm and humid. Life proliferated both on land and in the oceans. Marine invertebrates increased in variety and number of species. On land, dinosaurs continued to diversify. The first snakes appeared during the Cretaceous, though their lineages did not radiate until much later. Early in the Cretaceous, flowering plants began the radiation that led to their current dominance on land. Fossils of the earliest known flowering plants (Archaefructaceae), dated at 124 mya, recently were discovered in Liaoning Province in northeastern China (Figure 22.16). By the end of the period, many groups of mammals had evolved, but these mammals were generally small.

Another meteorite-caused mass extinction took place at the end of the Cretaceous period. On land, all vertebrates larger than about 25 kg in body weight, including all of the dinosaurs, apparently became extinct. Many species of insects died out, perhaps because the growth of their food

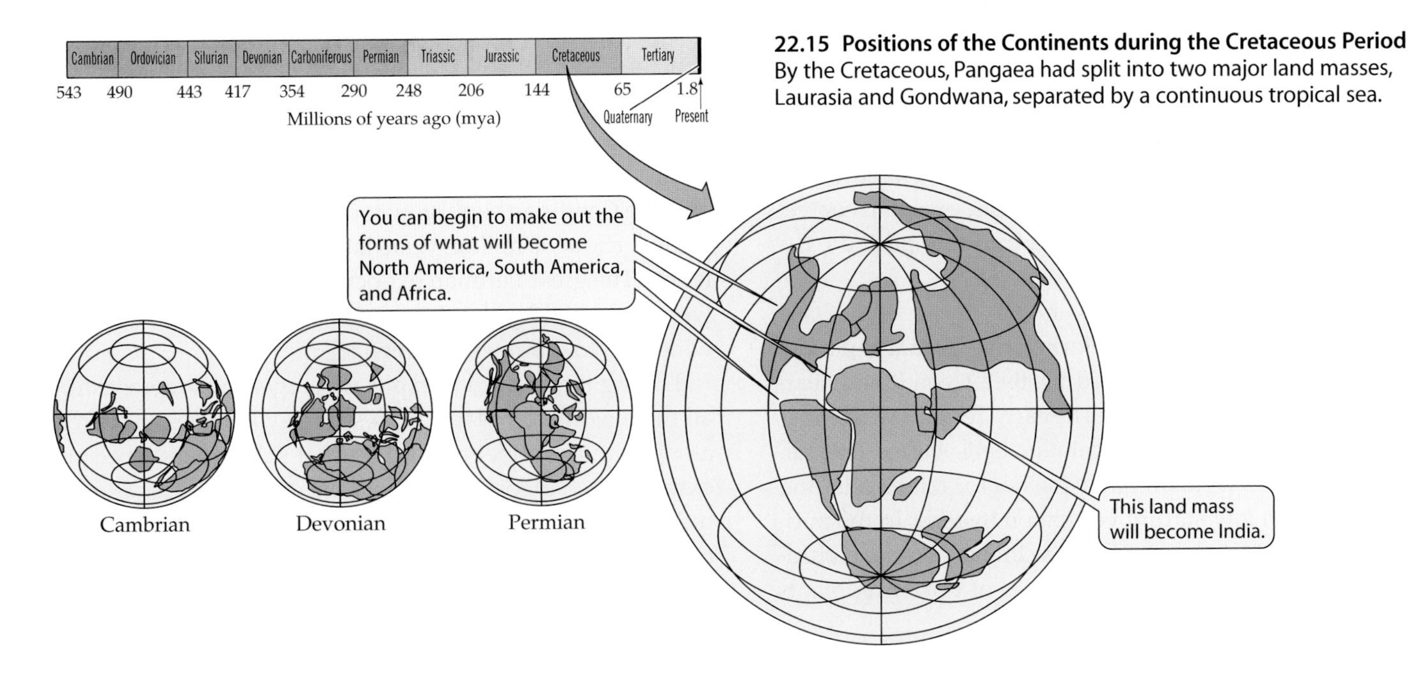

22.15 Positions of the Continents during the Cretaceous Period By the Cretaceous, Pangaea had split into two major land masses, Laurasia and Gondwana, separated by a continuous tropical sea.

The "feather" pattern of its leaves indicates that *Archaefructus* lived in water.

22.16 Flowering Plants of the Cretaceous Fossils of *Archaefructus* found in China are a minimum of 124.6 million years old. These flowering plants are early examples of the type of plants most prevalent on Earth today.

plants was greatly reduced following the impact. In the seas, many planktonic organisms and bottom-dwelling invertebrates became extinct.

The modern biota evolved during the Cenozoic era

By the early Cenozoic era (65 mya), the positions of the continents resembled those of today, but Australia was still attached to Antarctica, and the Atlantic Ocean was much narrower. The Cenozoic era was characterized by an extensive radiation of mammals, but other groups were also undergoing important changes. Flowering plants diversified extensively and came to dominate world forests, except in cool regions. The Cenozoic era is divided into two periods, the Tertiary and the Quaternary.

THE TERTIARY (65–1.8 MYA). During the **Tertiary** period, Australia began its northward drift. By 20 mya it had nearly reached its current position. The early Tertiary was a hot/humid time, during which vegetation belts shifted latitudinally. The Tropics were probably too hot for rainforests, and were clothed in low-stature vegetation instead. In the middle of the Tertiary, however, Earth's climate became considerably drier and cooler. Many lineages of flowering plants evolved herbaceous (nonwoody) forms; grasslands spread over much of Earth.

By the beginning of the Cenozoic era, invertebrate faunas resembled those of today. It is among the vertebrates that evolutionary changes during the Tertiary period were most rapid. Living groups of reptiles, including snakes and lizards, underwent extensive radiations during this period, as did birds and mammals. Three waves of mammals dispersed from Asia to North America about 55 mya. Rodents, marsupials, primates, and hoofed mammals appeared in North America for the first time.

THE QUATERNARY (1.8 MYA TO PRESENT). The current geological period, the **Quaternary**, is subdivided into two *epochs*, the **Pleistocene** and the **Holocene** (also known as the *Recent*). The Pleistocene epoch was a time of drastic cooling and climatic fluctuations. During four major and about 20 minor episodes of glaciation, massive glaciers spread across the continents, and animal and plant populations shifted toward the equator. The last of these glaciers retreated from temperate latitudes less than 15,000 years ago; this retreat marked the beginning of the Holocene epoch. Organisms of the Holocene are still adjusting to these changes. Many high-latitude ecological communities have occupied their current locations for no more than a few thousand years.

Interestingly, few species became extinct during these climatic fluctuations. However, the Pleistocene was the time of hominid evolution and radiation, resulting in the species *Homo sapiens*—modern humans (see Chapter 34). Many large bird and mammal species became extinct in Australia and in the Americas when *H. sapiens* arrived on these continents about 40,000 and 15,000 years ago, respectively. Human hunting may have caused these extinctions, although existing evidence does not convince all paleontologists.

Three major faunas have dominated life on Earth

The fossil record reveals three great radiations that resulted in the evolution of major new faunas (Figure 22.17). The first one, the Cambrian explosion, took place about 540 mya. The second, about 60 million years later, resulted in the Paleozoic fauna. The great Permian extinctions 300 million years later were followed by the third event, the Triassic explosion, which led to our modern fauna.

During the Cambrian explosion, organisms representing all the major body plans of present-day lineages appeared, along with a number of lineages that subsequently became extinct. The Paleozoic and Triassic explosions resulted in many new groups of organisms, but all of them had modifications of body plans that were already present when these great biological diversifications began.

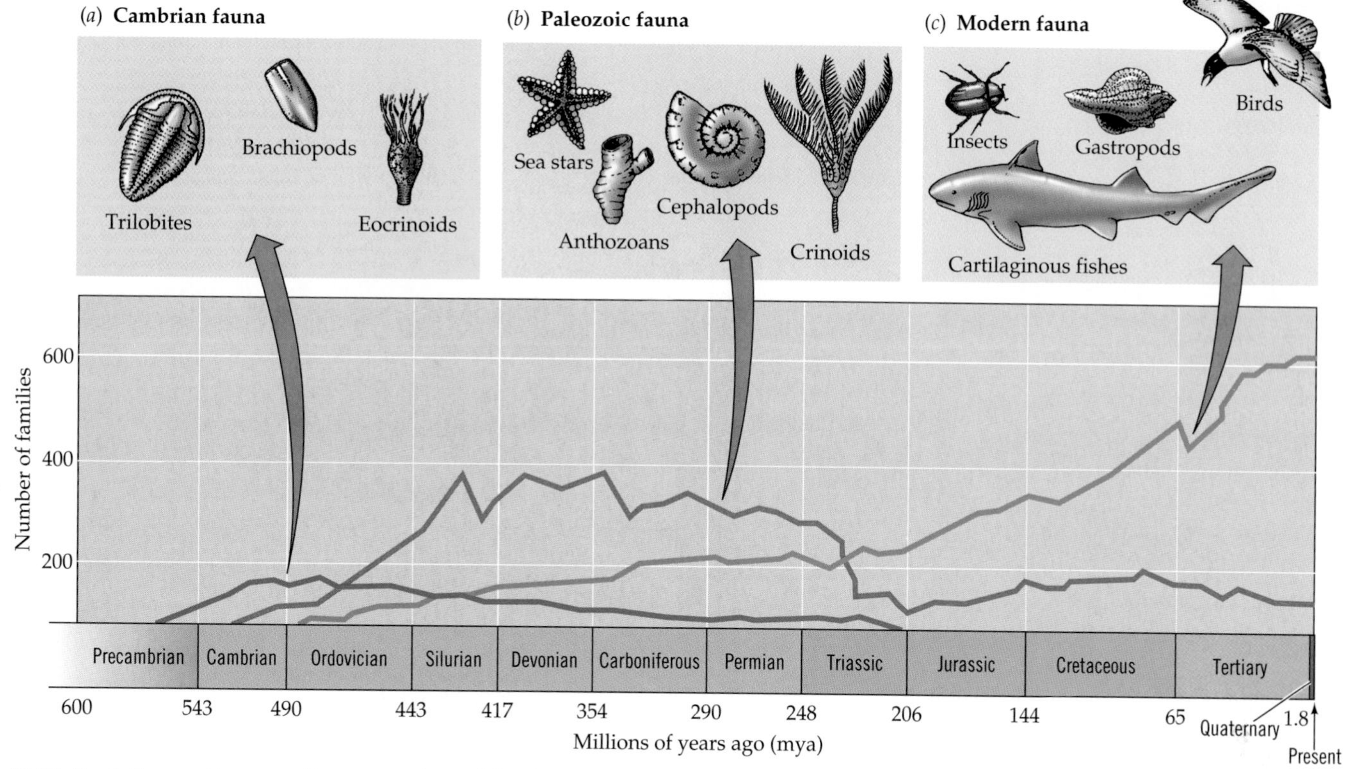

22.17 Evolutionary Faunas Representatives of the three great evolutionary faunas are shown, together with a graph illustrating the number of families in each fauna over time.

Rates of Evolutionary Change within Lineages

In addition to revealing the broad patterns of the evolution of life on Earth, fossils also tell us about rates of change within particular lineages of organisms. The fossil record shows that no single pattern characterizes evolutionary rates. Many species have changed very little over many millions of years. Others have changed gradually over time. Still others have undergone rapid changes over short time periods, followed by long periods of slow change. Change may incorrectly appear to be rapid if the fossil record is very incomplete, but some rapid changes are well documented by excellent temporal series of fossils. Let's look at some examples of these patterns.

Some living species closely resemble ancient ancestors

Species that have changed little over millions of years are known as "living fossils." Fossilized leaves of the genus *Ginkgo* from the Triassic, for example, are very similar to those of living trees (Figure 22.18). Animals in some marine lineages have also evolved slowly. The horseshoe crabs living today are almost identical in appearance to those that lived 300 million years ago. The sandy coastlines where horseshoe crabs spawn (see Figure 33.16*b*) feature extremes in temperature and salt concentration that are lethal to many organisms. These harsh environments have changed relatively little over millennia. The chambered nautiluses of the late Cretaceous are indistinguishable from living species (see Figure 32.26*f*). Chambered nautiluses spend their days in deep, dark ocean waters, ascending to feed in food-rich surface waters only under the protective cover of darkness. Their intricate shells provide little protection against today's visually hunting fish.

(*a*)

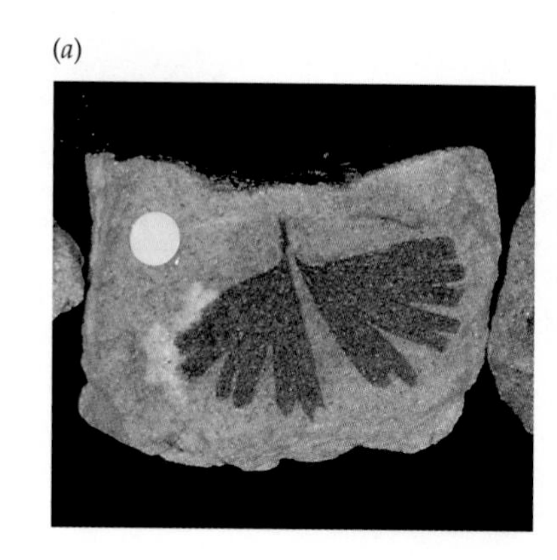

(*b*)

22.18 "Living Fossils" Fossilized *Ginkgo* leaves from the Triassic (*a*) appear very similar to the leaves of living trees (*b*).

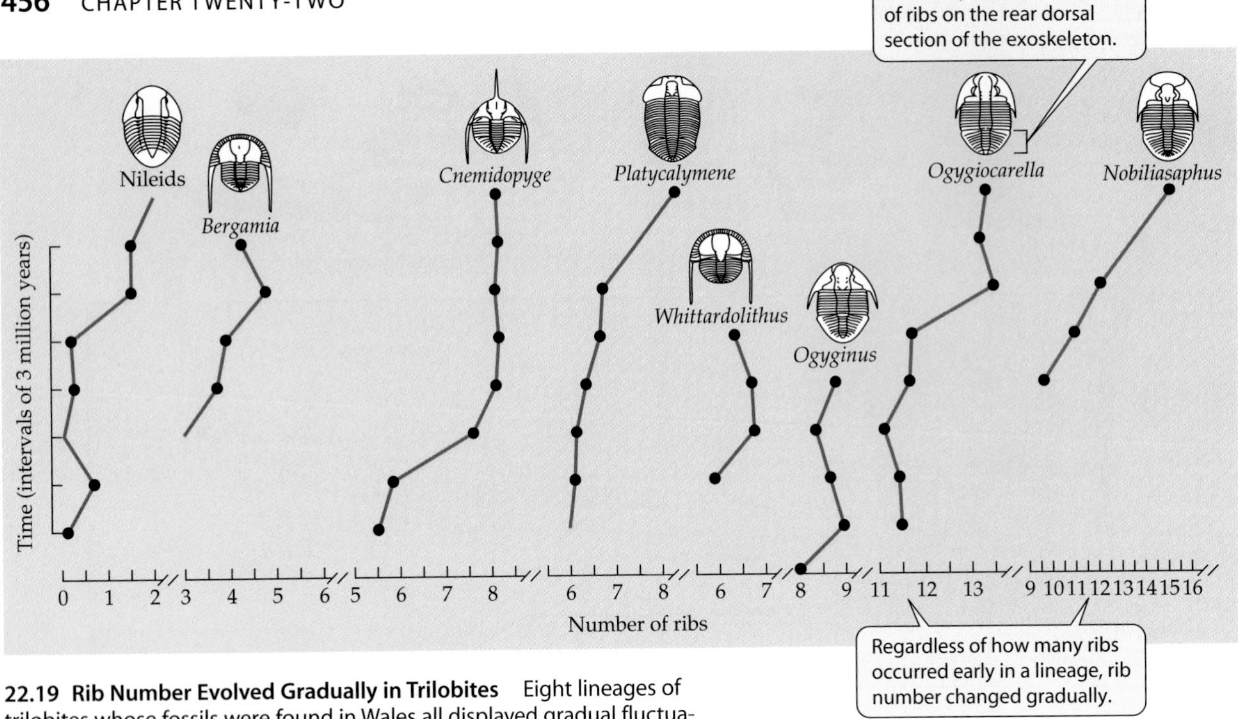
The study measured number of ribs on the rear dorsal section of the exoskeleton.
Nileids
Bergamia
Cnemidopyge
Platycalymene
Whittardolithus
Ogyginus
Ogygiocarella
Nobiliasaphus
Time (intervals of 3 million years)
Number of ribs
Regardless of how many ribs occurred early in a lineage, rib number changed gradually.

22.19 Rib Number Evolved Gradually in Trilobites Eight lineages of trilobites whose fossils were found in Wales all displayed gradual fluctuations in the number of rear dorsal ribs on the exoskeleton.

Evolutionary changes have been gradual in some lineages

The fossil record contains many series of fossils that demonstrate gradual change in lineages of organisms over time. A good example is the series of fossils showing changes in the number of ribs on the exoskeleton in eight lineages of trilobites during the Ordovician (Figure 22.19). Rates of change differed among the lineages, and they did not all change at the same time, but all of the changes were gradual.

Rates of evolutionary change are sometimes rapid

In the histories of some lineages, periods of gradual evolutionary change are broken by periods during which changes, either in the physical or biological environment, create conditions that favor rapid evolution of new traits. Such rapid evolutionary change is illustrated by the three-spined stickleback (*Gasterosteus aculeatus*), a widespread marine fish that

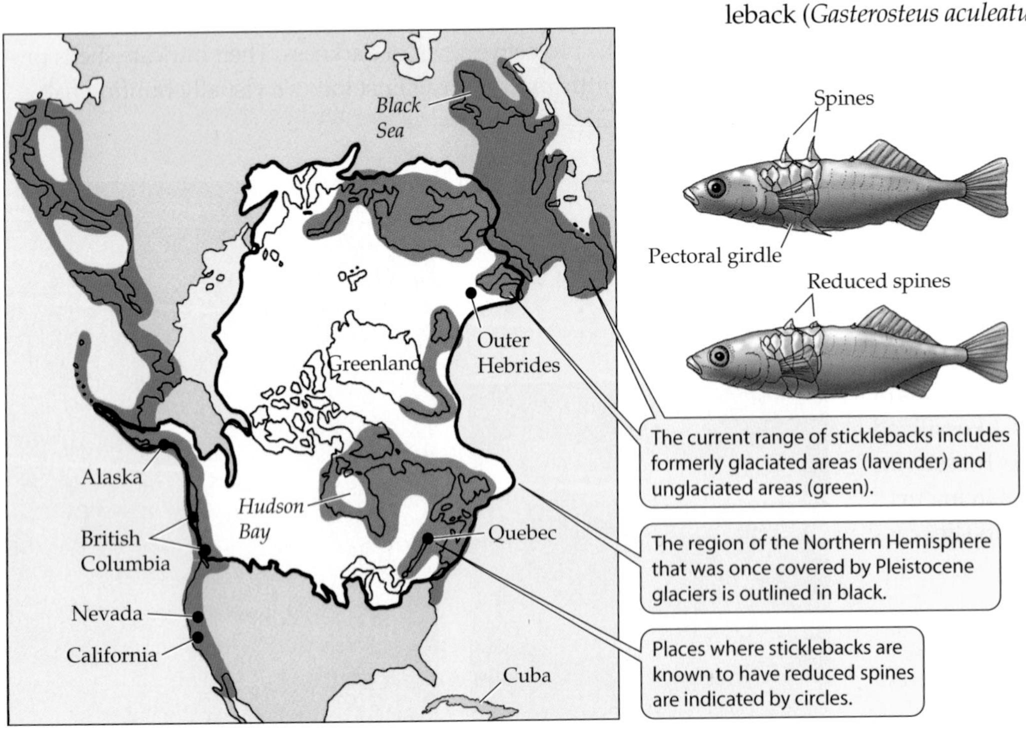
Black Sea
Greenland
Outer Hebrides
Alaska
Hudson Bay
British Columbia
Quebec
Nevada
California
Cuba
Spines
Pectoral girdle
Reduced spines
The current range of sticklebacks includes formerly glaciated areas (lavender) and unglaciated areas (green).
The region of the Northern Hemisphere that was once covered by Pleistocene glaciers is outlined in black.
Places where sticklebacks are known to have reduced spines are indicated by circles.

22.20 Natural Selection Acts on Stickleback Spines Three-spined stickleback populations with reduced spines are found principally in young lakes that were covered by ice during the most recent glacial period. These lakes lack large predatory fish, but contain predatory insects that capture the fish by grasping their spines.

Carpodacus mexicanus (male)

has repeatedly colonized new environments—freshwater lakes—during its evolutionary history.

Sticklebacks are tiny fish, usually less than 10 cm long. All marine and most freshwater populations have well-developed pectoral girdles with prominent spines that make it difficult for other fish to swallow them. However, large predatory insects can readily grasp the stickleback's spines, and they prey selectively on those stickleback individuals with the largest spines. When stickleback populations colonize freshwater habitats where predatory fish are absent but predatory insects are present, they rapidly evolve smaller spines. Populations with reduced spines are found primarily in young lakes that were covered by ice during the most recent glaciation and do not have large predatory fish (Figure 22.20). The extensive fossil record of sticklebacks shows that spine reduction evolved many times in different populations that colonized fresh water. In addition, molecular data show that each freshwater population is most closely related to an adjacent marine population, not to other freshwater populations.

At times, evolutionary change is rapid enough to be measured directly. A good example is provided by the house finch, a bird that as recently as 1939 was confined to the arid and semiarid parts of western North America. That year, some captive finches were released in New York City. Many of them survived to form a small breeding population in the immediate vicinity of the city. During the early 1960s, that population began to grow and increase its range. By the 1990s, the house finch had spread across all of the eastern United States and southern Canada (Figure 22.21). Remarkably, by 2000, birds in finch populations that had been separated for only a few decades were as different in size as birds in finch populations that had been separated for thousands of years.

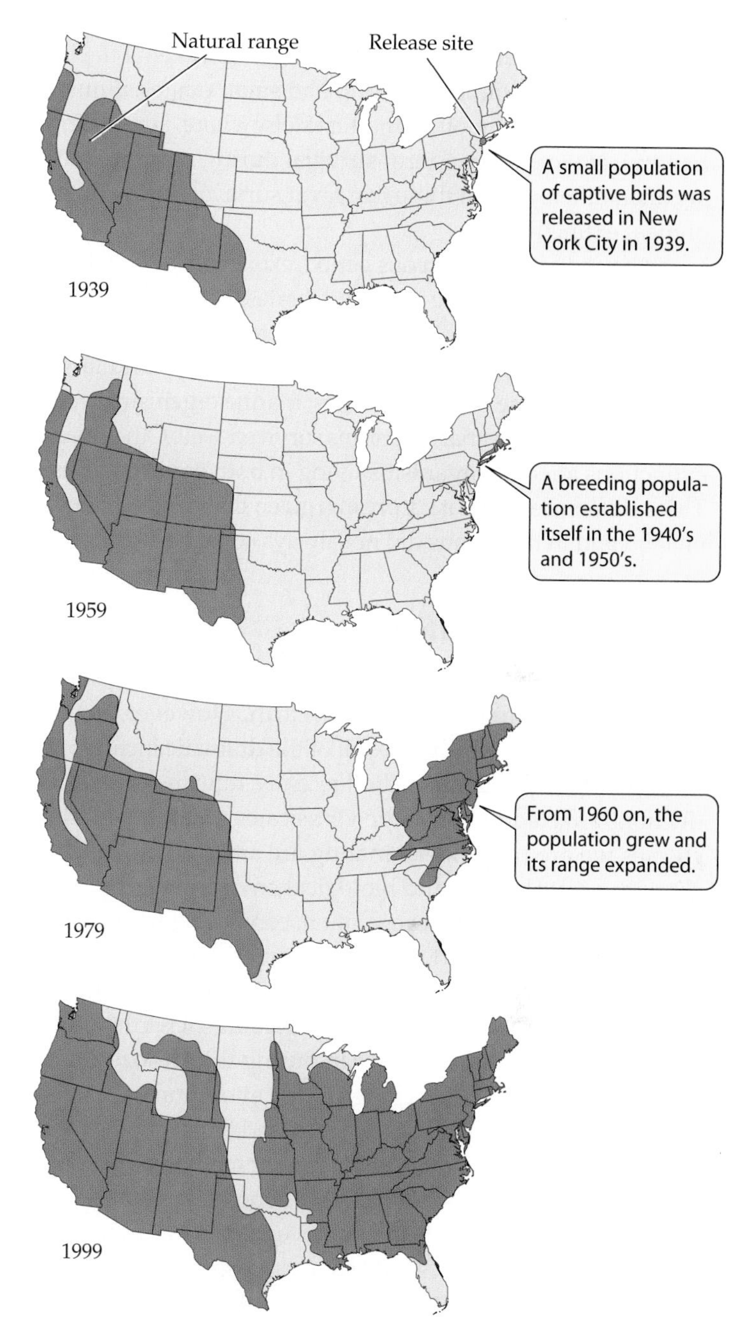

22.21 House Finches Expanded their Range in North America The natural range of house finches in western North America has expanded somewhat during the past few decades, but the most dramatic expansion has been in the East, from a release of a few caged birds in New York City. Differences in body size also evolved rapidly during this short time period, so that northern birds are now larger than southern birds.

Extinction rates vary over time

More than 99 percent of the species that have ever lived are extinct. Species have become extinct throughout the history of life, but extinction rates have fluctuated dramatically. Some groups had high extinction rates while others were proliferating.

A mass extinction changes the biota of the following period by selectively eliminating some types of organisms, thereby allowing others to increase in their relative abundance. For example, among the mollusks of the Atlantic coastal plain of North America, species with broad geographic ranges were less likely to become extinct during normal periods (when no mass extinctions were taking place) than were species with small geographic ranges. During the mass extinction of the late Cretaceous, however, groups of closely related mollusk species with large geographic ranges survived better than groups with small ranges, even if the individual species within the group had small ranges. Similar patterns are found in other mollusks elsewhere, suggesting that traits favoring long-term survival during normal times are often different from those that favor survival during times of mass extinctions.

At the end of the Cretaceous period, extinction rates on land were much higher among large vertebrates than among small ones. The same was true during the Pleistocene, when extinction rates were high only among large mammals and large birds. During some mass extinctions, marine organisms were heavily hit, but terrestrial organisms survived well. Other mass extinctions affected organisms living in both environments. These differences are not surprising, given that major changes on land and in the oceans did not always coincide.

The Future of Evolution

The agents of evolution are operating today just as they have been since life first appeared on Earth. However, major changes are under way as a result of the dramatic increase of Earth's human population. Until recently, human-caused extinctions affected mostly large vertebrates, but many of the species we are now exterminating are small. Humans are changing the physical and biological environment by dramatically altering Earth's vegetation, converting forests and grasslands to crops and pastures. Deliberately or inadvertently, we are moving thousands of species around the globe, reversing the relatively independent evolution of Earth's biota on different continents that began during the Mesozoic era. Humans have also taken charge of the evolution of certain species by means of artificial selection and biotechnology. As we saw in Chapter 16, modern molecular methods enable us to modify species by moving genes among even distantly related species. In short, humans have become a dominant agent of evolution. How we wield our massive influence will powerfully affect the future of life on Earth.

Chapter Summary

Defining Biological Evolution

▶ Biology is intimately linked to concepts of time. The study of biology as we know it could not have been developed until people came to understand the age of Earth.

Determining Earth's Age

▶ The relative ages of rock layers in Earth's crust were determined from their embedded fossils.

▶ Radioisotopes supplied the key for assigning absolute ages to rocks.

▶ Earth's geological history is divided into eras and periods. The boundaries between these divisions are based on differences between their fossil biotas. **Review Table 22.1**

The Changing Face of Earth

▶ Throughout Earth's history continents have drifted about, sometimes separating from one another, at other times colliding. Their collisions typically have led to periods of massive volcanic eruptions, glaciations, and major shifts in sea levels and ocean currents. **Review Figure 22.2**

▶ Earth's early atmosphere lacked free oxygen. Oxygen accumulated after prokaryotes evolved the ability to use water as their source of hydrogen ions for photosynthesis. Increasing concentrations of atmospheric oxygen made possible the evolution of eukaryotes and multicellular organisms. **Review Figure 22.4**

▶ Over Earth's history, hot/humid climatic conditions have alternated with cold/dry conditions. **Review Figure 22.5**

▶ External events, such as collisions with meteorites, also have changed conditions on Earth. Such a collision probably caused the abrupt mass extinction at the end of the Cretaceous period. **See Web/CD Tutorial 22.1**

The Fossil Record

▶ Much of what we know about the history of life on Earth comes from the study of fossils.

▶ The fossil record, although incomplete, reveals broad patterns in the evolution of life. About 300,000 fossil species have been described. The best record is that of hard-shelled animals fossilized in marine sediments.

Major Patterns in the History of Life on Earth

▶ Some lineages that evolved during Precambrian times may not have left living descendants.

▶ The diversity of life exploded during the Cambrian period. Diversification continued throughout the rest of the Paleozoic era. **Review Figures 22.9, 22.11, 22.13**

▶ Geographic differentiation of biotas increased during the Mesozoic era. **Review Figure 22.16**

▶ The modern biota evolved during the Cenozoic era.

▶ After each mass extinction, the diversity of life rebounded, but the groups of organisms that dominated the new biotas differed markedly from those characteristic of earlier biotas. **Review Figure 22.17**

Rates of Evolutionary Change within Lineages

▶ Some species, called "living fossils," closely resemble ancient ancestors.

▶ Evolutionary changes have been gradual in some lineages. **Review Figure 22.19**

▶ Rates of evolutionary change are sometimes rapid because of changes in the physical or biological environment. **Review Figures 22.20, 22.21**

The Future of Evolution

▶ The agents of evolution continue to operate today, but human intervention, both deliberate and inadvertent, now plays an unprecedented role in the history of life.

See Web/CD Activity 22.1 for a concept review of this chapter.

Self-Quiz

1. The number of species of fossil organisms that has been described is about
   a. 50,000.
   b. 100,000.
   c. 200,000.
   d. 300,000.
   e. 500,000.

2. In undisturbed strata of sedimentary rocks,
   a. the oldest rocks lie at the top.
   b. the oldest rocks lie at the bottom.
   c. the oldest rocks are in the middle.
   d. the oldest rocks are distributed among the strata of younger rocks.
   e. None of the above
3. Radioactive carbon can be used to date the ages of fossil organisms because
   a. all organisms contain many carbon compounds.
   b. radioactive carbon has a regular rate of decay to nonradioactive carbon.
   c. the ratio of radioactive to nonradioactive carbon in living organisms is always the same as that in the atmosphere.
   d. the production of new radioactive carbon in the atmosphere just balances the natural radioactive decay of ^{14}C.
   e. All of the above
4. An important unidirectional change in Earth during its history was a
   a. steady increase in volcanic activity.
   b. gradual coming together of the continents.
   c. steady increase in the oxygen content of the atmosphere.
   d. gradual warming of the climate.
   e. steady increase in Earth's precipitation.
5. The total of all species of organisms in a given region is known as the region's
   a. biota.
   b. flora.
   c. fauna.
   d. flora and fauna.
   e. diversity.
6. The coal beds we now mine for energy are the remains of
   a. trees that grew in swamps during the Carboniferous period.
   b. trees that grew in swamps during the Devonian period.
   c. trees that grew in swamps during the Permian period.
   d. small plants that grew in swamps during the Carboniferous period.
   e. None of the above
7. The cause of the mass extinction at the end of the Ordovician was probably
   a. the collision of Earth with a large meteorite.
   b. massive volcanic eruptions.
   c. massive glaciation in Gondwana.
   d. the uniting of all continents to form Pangaea.
   e. changes in Earth's orbit.
8. The cause of the mass extinction at the end of the Mesozoic era probably was
   a. continental drift.
   b. the collision of Earth with a large meteorite.
   c. changes in Earth's orbit.
   d. massive glaciation.
   e. changes in the salt concentration of the oceans.
9. The times during the history of life when many new evolutionary lineages appeared were the
   a. Precambrian, Cambrian, and Triassic.
   b. Precambrian, Cambrian, and Tertiary.
   c. Cambrian, Paleozoic, and Triassic.
   d. Cambrian, Triassic, and Devonian.
   e. Paleozoic, Triassic, and Tertiary.
10. Many scientists believe that the collision of Earth with a large meteorite was a major contributor to the mass extinction at the end of the Cretaceous period because
    a. there is an iridium-rich layer at the boundary of rocks between the Cretaceous and Cenozoic.
    b. a crater that may be the site of the collision has been found off the Yucatán Peninsula.
    c. the mass extinction at the end of the Cretaceous may have been very sudden.
    d. many planktonic organisms and bottom-dwelling invertebrates became extinct.
    e. All of the above
11. We know that organisms can evolve rapidly because
    a. the fossil record reveals periods of rapid evolutionary change.
    b. theoretical models of evolutionary change show that rapid change can be produced by natural selection.
    c. rapid evolutionary changes have been produced under artificial selection.
    d. rapid evolutionary changes have been measured in natural populations of organisms during the past century.
    e. All of the above
12. At which of the following times was there *no* mass extinction?
    a. The end of the Cretaceous period
    b. The end of the Devonian period
    c. The end of the Permian period
    d. The end of the Triassic period
    e. The end of the Silurian period

For Discussion

1. Some lineages of organisms have evolved to contain large numbers of species; other lineages have produced only a few species. Is it meaningful to consider the former more successful than the latter? What does the word "success" mean in evolution? How does your answer influence your thinking about *Homo sapiens*, the only surviving representative of the Hominidae—a family that never had many species in it?
2. Scientists date ancient events using a variety of methods, but nobody was present to witness or record those events. Accepting those dates requires us to believe in the accuracy and appropriateness of indirect measurement techniques. What other basic scientific concepts are also based on the results of indirect measurement techniques?
3. Why is it useful to be able to date past events absolutely as well as relatively?
4. If we are living during one of the cooler periods in Earth's history, why should we be concerned about human activities that are contributing to global climate warming?
5. Large meteors that collided with Earth have caused massive climatic and evolutionary changes. Should we attempt to take steps to prevent future meteorite impacts? What actions might we undertake? What adverse effects might such actions trigger?

23 The Mechanisms of Evolution

Newts and other salamanders can move only slowly, so they are easy prey for garter snakes. But some salamanders have evolved defensive toxic chemicals that make them less desirable as prey. The rough-skinned newt, *Taricha granulosa*, is a salamander that lives on the Pacific Coast of North America. *Taricha* sequesters in its skin a potent neurotoxin called tetrodotoxin (TTX). TTX paralyzes nerves and muscles by blocking sodium channels (see Chapter 5). Most snakes die if they eat a rough-skinned newt, but some populations of the garter snake *Thamnophis sirtalis* have evolved TTX-resistant sodium channels in their nerves and muscles. These snakes are able to eat the newts and survive—but the addition to their diet comes at a price. TTX-resistant snakes can crawl only slowly for several hours after eating a newt, and they never crawl as fast as nonresistant snakes. Thus, TTX-resistant snakes are more vulnerable to their own predators.

Pufferfish, octopuses, tunicates, and some species of frogs also use TTX as a defensive chemical. Many other species use a variety of chemicals to defend themselves against predators, and many predators have evolved resistance to those chemicals. But production of and resistance to defensive chemicals like all other adaptations, has costs as well as benefits. Such adaptations may impose a cost in the form of speed of movement, as they do on garter snakes. They may reduce the ability of the organism to function efficiently, or they may be energetically costly to develop and maintain. That is, to improve its performance in one area, the organism must accept reduced performance in some other area—a trade-off.

An Evolutionary War Rough-skinned newts (below) evolved the ability to secrete a paralyzing neurotoxin in their skin, a trait that deters most of their predators. Some common garter snakes (above) have evolved a resistance that allows them to turn the poisonous newts into a meal.

Biologists try to identify and measure the trade-offs that different adaptations impose because the nature and strength of these trade-offs influences how adaptations evolve. If there were no cost to TTX resistance, then snakes that live in places where toxic newts are rare would probably also be resistant to TTX—which they are not.

Charles Darwin's main contribution to biology was to propose a plausible and testable hypothesis for a mechanism that could result in the adaptation of organisms to their environments. In effect, Darwin offered a mechanistic explanation for the evolution of life on Earth, the last component of the known universe that lacked such an explanation. The mechanism that Darwin proposed can explain the evolution of all forms of life, including humans. It has been difficult for many people to accept that the same processes that determined the evolution-

(a)

(b)

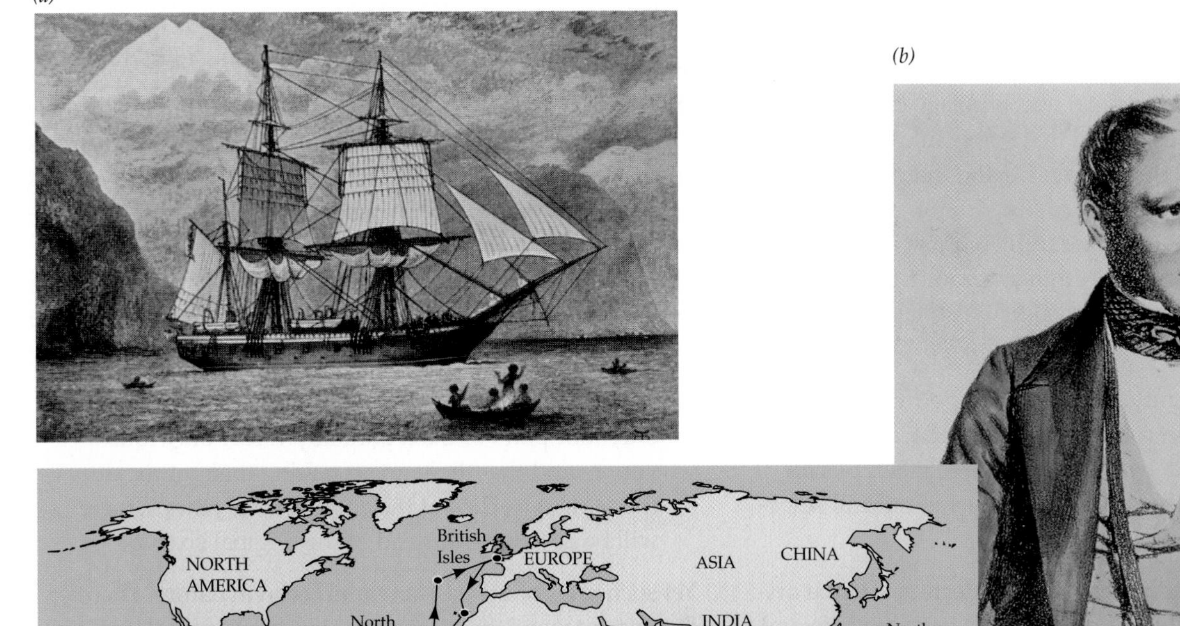

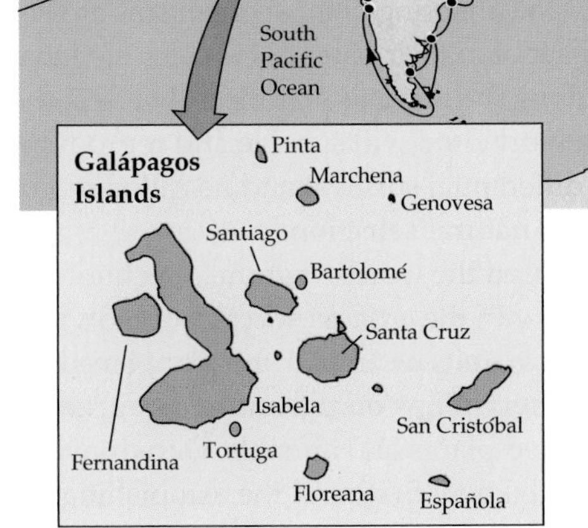

23.1 Darwin and the Voyage of the *Beagle* (*a*) The mission of H.M.S. *Beagle* was to chart the oceans and collect oceanographic and biological information from around the world. The map indicates the ship's path, with emphasis on the Galápagos Islands, whose organisms were an important source of Darwin's ideas on natural selection. (*b*) Charles Darwin at age 24, shortly after the *Beagle* returned to England.

ary pathways of other species also guided human evolution, but as Darwin noted, "there is grandeur in this view of life."

In this chapter we will see how Darwin developed his ideas, and then turn to the advances in our understanding of evolutionary processes since Darwin's time. We will discuss the genetic basis of evolution and show how genetic variation within populations is measured. We will describe the agents of evolution and show how biologists design studies to investigate them. Finally, we will discuss constraints on the pathways evolution can take. When you understand these processes, you will understand the mechanisms of evolution.

Charles Darwin's Theory of Evolution

As a youth, Charles Darwin was passionately interested in *natural history*—the study of how different organisms carry out their lives. He briefly studied medicine at Edinburgh, but he was nauseated by observing surgery conducted without anesthesia. He gave up medicine to study for a career as a clergyman of the Church of England at Cambridge University. However, he was more interested in natural history than theology, and he became a companion of scientists on the faculty, especially the botanist John Henslow. Darwin was given an unprecedented opportunity when in 1831 Henslow recommended him for a position as ship's naturalist on the H.M.S. *Beagle*, which was preparing for a survey voyage around the world (Figure 23.1).

Whenever possible during the 5-year voyage, Darwin (who was often seasick) went ashore to observe and collect specimens of plants and animals. He noticed that the species he saw in South America differed strikingly from those of Europe. He observed that the species of the temperate regions of South America (Argentina, Chile) were more similar to those of tropical South America (Brazil) than they were to European species. When he explored the Galápagos Islands, west of Ecuador, he noted that most of its animal species were found nowhere else, but were similar to those of mainland South America, 1,000 kilometers to the east. Darwin also recognized that the animals of the archipelago differed from island to island. He postulated that some animals had dispersed from mainland South America and then evolved differently on each of the islands.

When he returned to England in 1836, Darwin continued to ponder his observations. Within a decade he had developed the major features of his theory, which had two major components:

- Species are not immutable; they change over time. (In other words, Darwin asserted that evolution is a historical fact that can be demonstrated to have taken place.)
- The agent that produces these changes is natural selection.

Darwin wrote a long essay on natural selection and the origin of species in 1844, but, despite urging from his wife and colleagues, he was reluctant to publish it, preferring to assemble more evidence first.

Darwin's hand was forced in 1858 when he received a letter and manuscript from another traveling naturalist, Alfred Russel Wallace, who was studying plants and animals in the East Indies. Wallace asked Darwin to evaluate the manuscript, in which Wallace proposed a theory of natural selection almost identical to Darwin's. At first Darwin was dismayed, believing that Wallace had preempted his idea. But parts of Darwin's 1844 essay, together with Wallace's manuscript, were presented to the Linnaean Society of London on July 1, 1858, thereby giving credit for the idea to both men. Darwin then worked quickly to finish his own book, *The Origin of Species,* which was published the next year. Although both men conceived of natural selection independently, Darwin developed his ideas first, and *The Origin of Species* provided an enormous amount of evidence from many fields to support both the concept of natural selection and evolution itself, which is why these concepts are more closely associated with the name Darwin than Wallace.

The facts that Darwin used to conceive and develop his theory of evolution by natural selection were familiar to most contemporary biologists. His unique insight was to perceive the significance of relationships among them. On September 28, 1838, Darwin happened to read *An Essay on the Principle of Population* by Thomas Malthus, an economist. Malthus argued that because the rate of human population growth is greater than the rate of increase in food production, unchecked growth inevitably leads to famine. Darwin recognized that populations of all species have the potential for exponential increases in numbers. To illustrate this point, he used the following example:

> Suppose... there are eight pairs of birds, and that only four pairs of them annually... rear only four young, and that these go on rearing their young at the same rate, then at the end of seven years... there will be 2048 birds instead of the original sixteen.

Yet such rates of increase are rarely seen in nature. Therefore, Darwin reasoned that death rates in nature must also be high. Without high death rates, even the most slowly reproducing species would quickly reach enormous population sizes.

Darwin also observed that, although offspring tend to resemble their parents, the offspring of most organisms are not identical to one another or to their parents. He suggested that slight variations among individuals significantly affect the chance that a given individual will survive and reproduce. Darwin called this differential survival and he called reproduction of individuals **natural selection**.

Darwin may have used the words "natural selection" because he was familiar with the *artificial selection* of individuals with certain desirable traits by animal and plant breeders. Many of Darwin's observations on the nature of variation came from domesticated plants and animals. Darwin was a pigeon breeder, and he knew firsthand the astonishing diversity in color, size, form, and behavior that pigeon breeders could achieve (Figure 23.2). He recognized close parallels

23.2 Many Types of Pigeons Have Been Produced by Artificial Selection Charles Darwin raised pigeons as a hobby, and he saw similar forces at work in artificial and natural selection. These are just some of more than 300 varieties of pigeons that have been artificially selected by breeders to display different forms of traits such as color, size, and feather distribution.

between selection by breeders and selection in nature. As he argued in *The Origin of Species*,

> How can it be doubted, from the struggle each individual has to obtain subsistence, that any minute variation in structure, habits or instincts, adapting that individual better to the new conditions, would tell upon its vigour and health? In the struggle it would have a better chance of surviving; and those of its offspring which inherited the variation, be it ever so slight, would have a better chance.

That statement, written almost 150 years ago, still stands as a good expression of the theory of evolution by natural selection.

It is important to remember, as Darwin clearly understood, that *individuals do not evolve; populations do*. A **population** is a group of individuals of a single species that live in a particular geographic area at the same time. A major consequence of the evolution of populations is that their members become adapted to the environments in which they live.

The term **adaptation** has two meanings in evolutionary biology. The first meaning refers to the *processes* by which adaptive traits are acquired—that is, the evolutionary mechanisms that produce them. We will discuss those processes in great detail in this chapter. The second meaning refers to *traits* that enhance the survival and reproductive success of their bearers. For example, wings are adaptations for flight, and a spider's web is an adaptation for capturing flying insects.

Biologists regard an organism as being adapted to a particular environment when they can imagine—or better still, measure the performance of—a slightly different organism that reproduces and survives less well in that environment. To understand adaptation, biologists compare the performance of individuals within or among species that differ in their traits. For example, to investigate the adaptive nature of spiders' webs, we might try to determine the effectiveness of slightly different web structures in capturing insects. We might also measure changes in the webs of the same species in different environments. With these data, we could understand how variations in web structure influenced the survival and reproductive success of their builders.

When Darwin proposed his theory of evolution by natural selection, he had no examples of evolutionary agents operating in nature. Since then many studies of the action of evolutionary agents have been conducted. Similarly, many investigations have documented changes over time in the genetic composition of a population. Darwin understood the importance of heredity for his theory, but he knew nothing of the mechanisms of heredity. He devoted considerable time developing a theory of heredity, but he failed in this effort.

Fortunately, the rediscovery of Gregor Mendel's publications in the early 1900s (see Chapter 10) paved the way for the development of **population genetics**, which provides a major underpinning for Darwin's theory. Population geneticists apply Mendel's laws to entire populations of organisms. They also study variation within and among species to understand the processes that result in evolutionary changes in species through time. The perspective of population genetics given in this chapter, which emphasizes the role of variation in characteristics of adult organisms, complements the perspective of developmental biology we discussed in Chapter 21.

Genetic Variation within Populations

For a population to evolve, its members must possess heritable genetic variation, which is the raw material on which agents of evolution act. In everyday life, we do not directly observe the genetic compositions of organisms. What we do see in nature are *phenotypes*, the physical expressions of organisms' genes. The features of a phenotype are its *characters*—eye color, for example. The specific form of a character, such as brown eyes, is a *trait*. A *heritable trait* is a characteristic of an organism that is at least partly determined by its genes.

The agents of evolution generally act on phenotypes, but for the moment we will concentrate on genetic variation within populations. We will do so because genetic variation is what is passed on to offspring via gametes—eggs and sperm. The genetic constitution that governs a character is called its *genotype*. *A population evolves when individuals with different genotypes survive or reproduce at different rates.*

Recall from Chapter 10 that different forms of a gene, called *alleles*, may exist at a particular locus. A single individual has only some of the alleles found in the population to which it belongs (Figure 23.3). The sum of all copies of alleles found in the population constitutes its **gene pool**. The

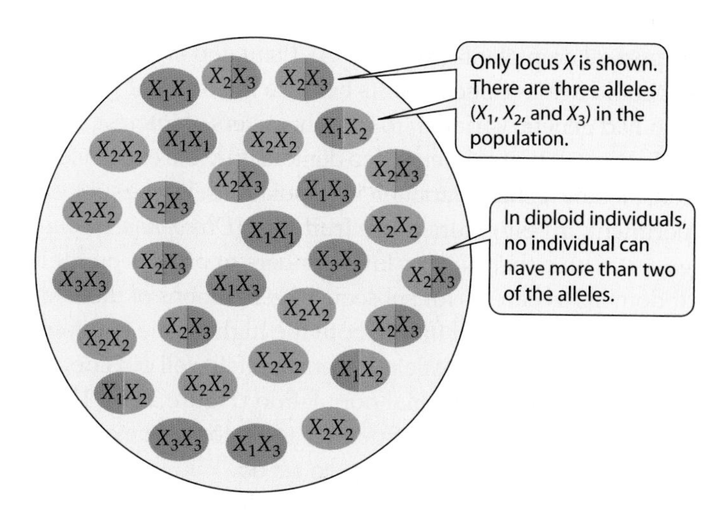

23.3 A Gene Pool A gene pool is the sum of all the alleles found in a population. Each of the colored circles represents an individual. The allele proportions in this gene pool for locus X are 0.20 for X_1, 0.50 for X_2, and 0.30 for X_3.

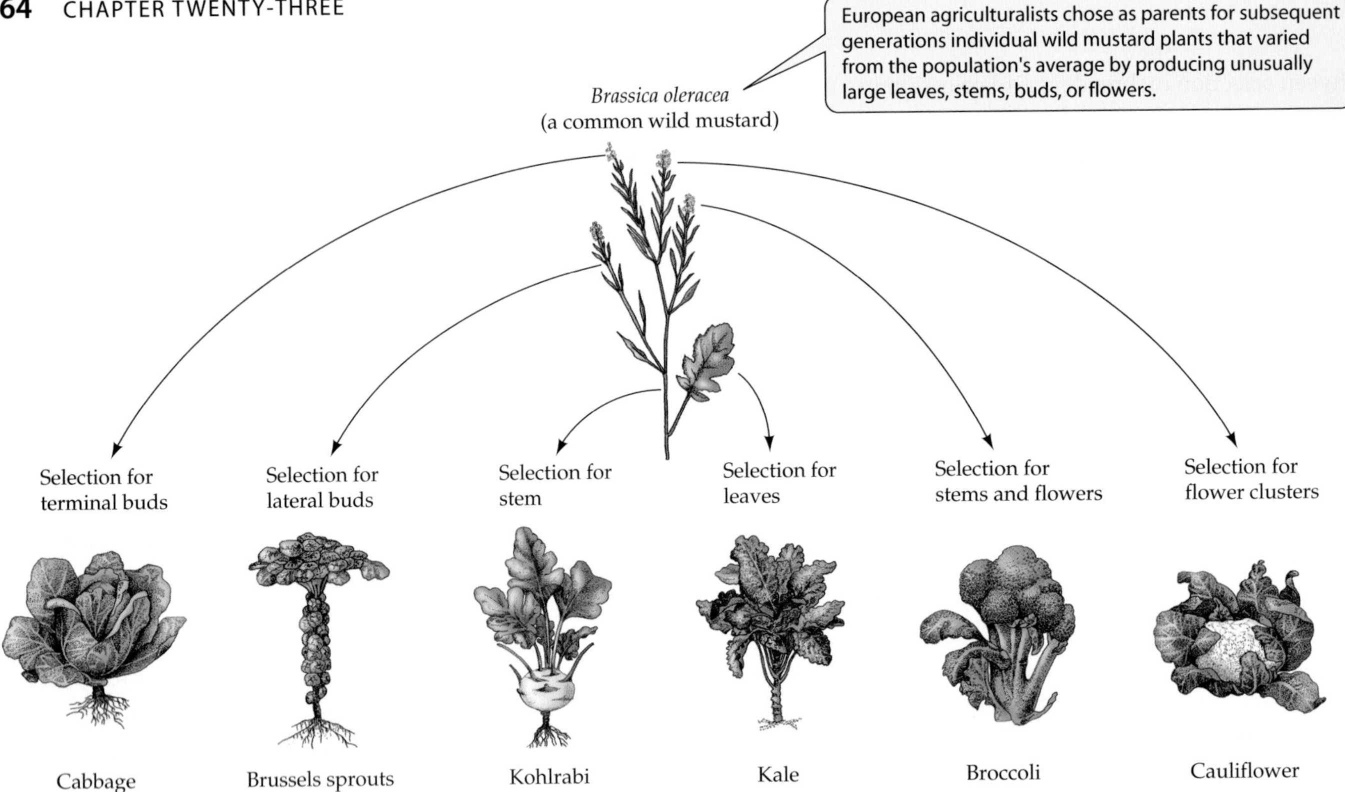

23.4 Many Vegetables from One Species All of these crop plants have been derived from a single wild mustard species. Plant breeders produced these crops by choosing and breeding plants with unusually large buds, stems, leaves, or flowers. The results illustrate the vast amount of variation that can be present in a gene pool.

gene pool contains the variation that produces the phenotypic characters on which agents of evolution act. To understand evolution, we need to know how much genetic variation populations have, the sources of that genetic variation, and how genetic variation is maintained and expressed in populations over space and time.

Most populations are genetically variable

Nearly all populations contain some level of genetic variation for many characters. Artificial selection on different characters in a European species of wild mustard produced many important crop plants (Figure 23.4). Plant and animal breeders could achieve such results because the original population had genetic variation for the characters of interest.

Laboratory experiments also demonstrate the existence of considerable genetic variation in populations. In one such experiment, investigators chose fruit flies (*Drosophila melanogaster*) with either high or low numbers of bristles on their abdomens as parents for subsequent generations of flies. After 35 generations, all flies in both the high-bristle and low-bristle lineages had bristle numbers that fell well outside the range found in the original population (Figure 23.5). Thus, there must have been considerable variation in the original fruit fly population for selection to act on.

The study of the genetic basis of evolution is difficult because genotypes do not uniquely determine phenotypes. With dominance, for example, a particular phenotype can be produced by more than one genotype (e.g., *AA* and *Aa* individuals may be phenotypically identical). Similarly, different phenotypes can be produced by a given genotype, depending on the environment encountered during development. For example, the cells of all the leaves on a tree or shrub are normally genetically identical, yet leaves on the same tree often differ in shape and size. Leaves closer to the top of an oak tree, where they receive more wind and sunlight, may be more deeply lobed than the shaded leaves growing lower down on the same tree. The same differences can be seen between the leaves of individuals growing in sunny and in shady sites.

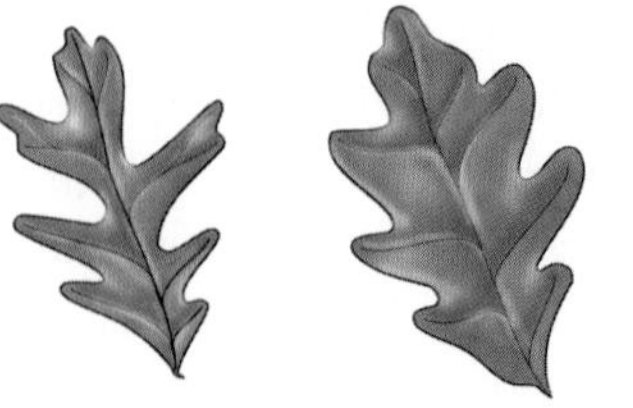

Thus, as we saw in Chapter 21, the phenotype of an organism is the outcome of a complex series of developmental processes that are influenced by both the environment and its genes.

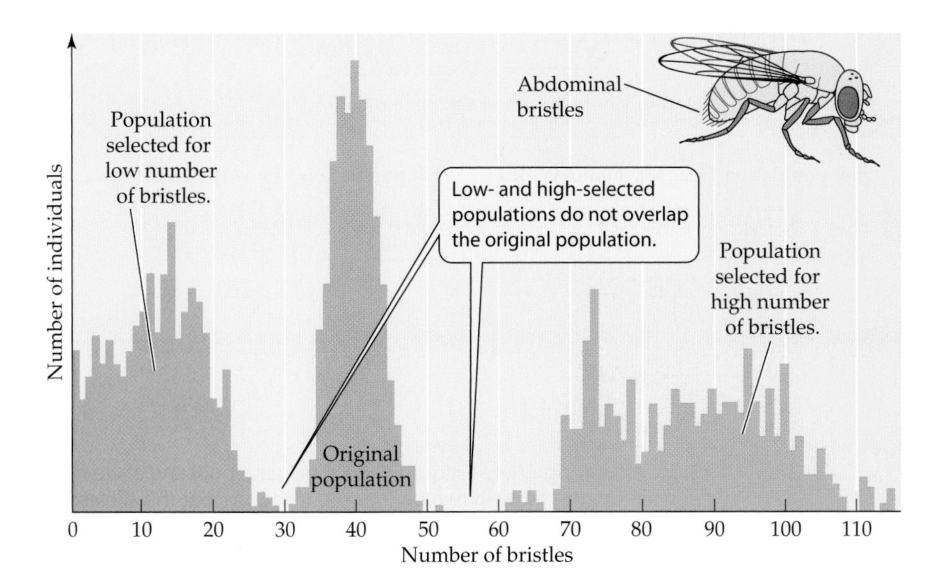

23.5 Artificial Selection Reveals Genetic Variation In artificial selection experiments with *Drosophila melanogaster*, changes in bristle number evolved rapidly. The graphs show the number of flies with different numbers of bristles after 35 generations of artificial selection.

How do we measure genetic variation?

A locally interbreeding group within a geographic population is called a **Mendelian population**. Mendelian populations are often the subjects of evolutionary studies. To measure genetic variation in a Mendelian population precisely, we would need to count every allele at every locus in every individual in it. By doing so, we could determine the relative proportions, or **frequencies**, of all alleles in the population. Fortunately, we do not need to make such complete measurements, because we can reliably estimate *allele frequencies* for a given locus by counting alleles in a sample of individuals from the population. The sum of all allele frequencies at a locus is equal to 1, so measures of allele frequency range from 0 to 1.

An allele's frequency is calculated using the following formula:

$$p = \frac{\text{number of copies of the allele in the population}}{\text{sum of alleles in the population}}$$

If only two alleles (for example, *A* and *a*) for a given locus are found among the members of a diploid population, they may combine to form three different genotypes: *AA*, *Aa*, and *aa*. Using the formula above, we can calculate the relative frequencies of alleles *A* and *a* in a population of *N* individuals as follows:

- Let N_{AA} be the number of individuals that are homozygous for the *A* allele (*AA*).
- Let N_{Aa} be the number that are heterozygous (*Aa*).
- Let N_{aa} be the number that are homozygous for the *a* allele (*aa*).

Note that $N_{AA} + N_{Aa} + N_{aa} = N$, the total number of individuals in the population, and that the total number of copies of both alleles present in the population is $2N$ because each individual is diploid. Each *AA* individual has two copies of the *A* allele, and each *Aa* individual has one copy of the *A* allele. Therefore, the total number of *A* alleles in the population is $2N_{AA} + N_{Aa}$. Similarly, the total number of *a* alleles in the population is $2N_{aa} + N_{Aa}$.

If *p* represents the frequency of *A*, and *q* represents the frequency of *a*, then

$$p = \frac{2N_{AA} + N_{Aa}}{2N}$$

and

$$q = \frac{2N_{aa} + N_{Aa}}{2N}$$

To show how this formula works, Figure 23.6 calculates allele frequencies in two populations, each containing 200 diploid individuals. Population 1 has mostly homozygotes (90 *AA*, 40 *Aa*, and 70 *aa*); population 2 has mostly heterozygotes (45 *AA*, 130 *Aa*, and 25 *aa*).

The calculations in Figure 23.6 demonstrate two important points. First, notice that for each population, $p + q = 1$. If there is only one allele in a population, its frequency is 1. If an allele is missing from a population, its frequency is 0, and the locus in that population is represented by one or more other alleles. Since $p + q = 1$, then $q = 1 - p$. So when there are only two alleles at a given locus in a population, we can calculate the frequency of one allele and then easily obtain the second allele's frequency by subtraction.

The second thing to notice is that both population 1 (consisting mostly of homozygotes) and population 2 (consisting mostly of heterozygotes) have the same allele frequencies for *A* and *a*. Therefore, they have the same gene pool for this locus. However, because the alleles in the gene pool are distributed differently, the *genotype frequencies* of the two populations differ. Genotype frequencies are calculated as the number of individuals that have the genotype divided by the total number of individuals in the population. In population 1 in Figure 23.6, the genotype frequencies are 0.45 *AA*, 0.20 *Aa*, and 0.35 *aa*.

The frequencies of different alleles at each locus and the frequencies of different genotypes in a Mendelian population describe its **genetic structure**. Allele frequencies measure the amount of genetic variation in a population; genotype frequencies show how a population's genetic variation is distributed among its members. With these measurements, it becomes possible to consider how the genetic structure of a population changes or does not change over generations.

In any population:

$$\text{Frequency of allele } A = p = \frac{2N_{AA} + N_{Aa}}{2N} \qquad \text{Frequency of allele } a = q = \frac{2N_{aa} + N_{Aa}}{2N}$$

where N is the total number of individuals in the population.

For population 1 (mostly homozygotes):

$N_{AA} = 90$, $N_{Aa} = 40$, and $N_{aa} = 70$

so

$$p = \frac{180 + 40}{400} = 0.55$$

$$q = \frac{140 + 40}{400} = 0.45$$

For population 2 (mostly heterozygotes):

$N_{AA} = 45$, $N_{Aa} = 130$, and $N_{aa} = 25$

so

$$p = \frac{90 + 130}{400} = 0.55$$

$$q = \frac{50 + 130}{400} = 0.45$$

23.6 Calculating Allele Frequencies The gene pool and allele frequencies are the same in two different populations, but the alleles are distributed differently between heterozygous and homozygous genotypes. In all cases, $p + q$ must equal 1.

The Hardy–Weinberg Equilibrium

If certain conditions are met, the genetic structure of a population may not change over time. The necessary conditions for such an equilibrium were deduced independently in 1908 by the British mathematician Godfrey Hardy and the German physician Wilhelm Weinberg. Hardy wrote his equations in response to a question posed to him by the geneticist Reginald C. Punnett (the inventor of the Punnett square) at the Cambridge University faculty club. Punnett wondered at the fact that even though the allele for short, stubby fingers (a condition called *brachydactyly*) was dominant and the allele for normal-length fingers was recessive, most people in Britain have normal-length fingers. Hardy's equations explain why dominant alleles do not necessarily replace recessive alleles in populations, as well as other features of the genetic structure of populations.

The **Hardy–Weinberg equilibrium** applies to sexually reproducing organisms. The particular example we will illustrate here assumes that the organism in question is diploid, its generations do not overlap, the gene under consideration has two alleles, and allele frequencies are identical in males and females. The Hardy–Weinberg equilibrium also applies if the gene has more than two alleles and generations overlap, but in those cases the mathematics is more complicated.

Several conditions must be met for a population to be at Hardy–Weinberg equilibrium:

- Mating is random
- Population size is very large
- There is no migration between populations
- There is no mutation
- Natural selection does not affect the alleles under consideration

If these conditions hold, two major consequences follow. First, the frequencies of alleles at a locus will remain constant from generation to generation. And second, after one generation of random mating, the genotype frequencies will remain in the following proportions:

Genotype	*AA*	*Aa*	*aa*
Frequency	p^2	$2pq$	q^2

Stated another way, the equation for Hardy–Weinberg equilibrium is

$$p^2 + 2pq + q^2 = 1$$

To see why, consider population 1 in Figure 23.6, in which the frequency of *A* alleles (p) is 0.55. Because we assume that individuals select mates at random, without regard to their genotype, gametes carrying *A* or *a* combine at random—that is, as predicted by the frequencies p and q. The probability that a particular sperm or egg in this example will bear an *A* allele rather than an *a* allele is 0.55. In other words, 55 out of 100 randomly sampled sperm or eggs will bear an *A* allele. Because $q = 1 - p$, the probability that a sperm or egg will bear an *a* allele is $1 - 0.55 = 0.45$.

To obtain the probability of two *A*-bearing gametes coming together at fertilization, we multiply the two independent probabilities of their occurring separately (see the discussion of probability in Chapter 10):

$$p \times p = p^2 = (0.55)^2 = 0.3025$$

Therefore, 0.3025, or 30.25 percent, of the offspring in the next generation will have the *AA* genotype. Similarly, the probability of bringing together two *a*-bearing gametes is

$$q \times q = q^2 = (0.45)^2 = 0.2025$$

Thus, 20.25 percent of the next generation will have the *aa* genotype (Figure 23.7).

Figure 23.7 also shows that there are two ways of producing a heterozygote: An *A* sperm may combine with an *a* egg, the probability of which is $p \times q$; or an *a* sperm may combine with an *A* egg, the probability of which is $q \times p$. Consequently, the overall probability of obtaining a heterozygote is $2pq$.

It is now easy to show that the allele frequencies p and q remain constant for each generation. If the frequency of *A* alleles in a randomly mating population is $p^2 + pq$, this frequency becomes $p^2 + p(1 - p) = p^2 + p - p^2 = p$, the original allele frequencies are unchanged, and the population is at Hardy–Weinberg equilibrium.

If some agent, such as emigration, were to alter the allele frequencies, the genotype frequencies would automatically settle into a predictable new set in the next generation. For instance, if only *AA* and *Aa* individuals left the population, p and q would change, but there would still be *aa* individuals in the population.

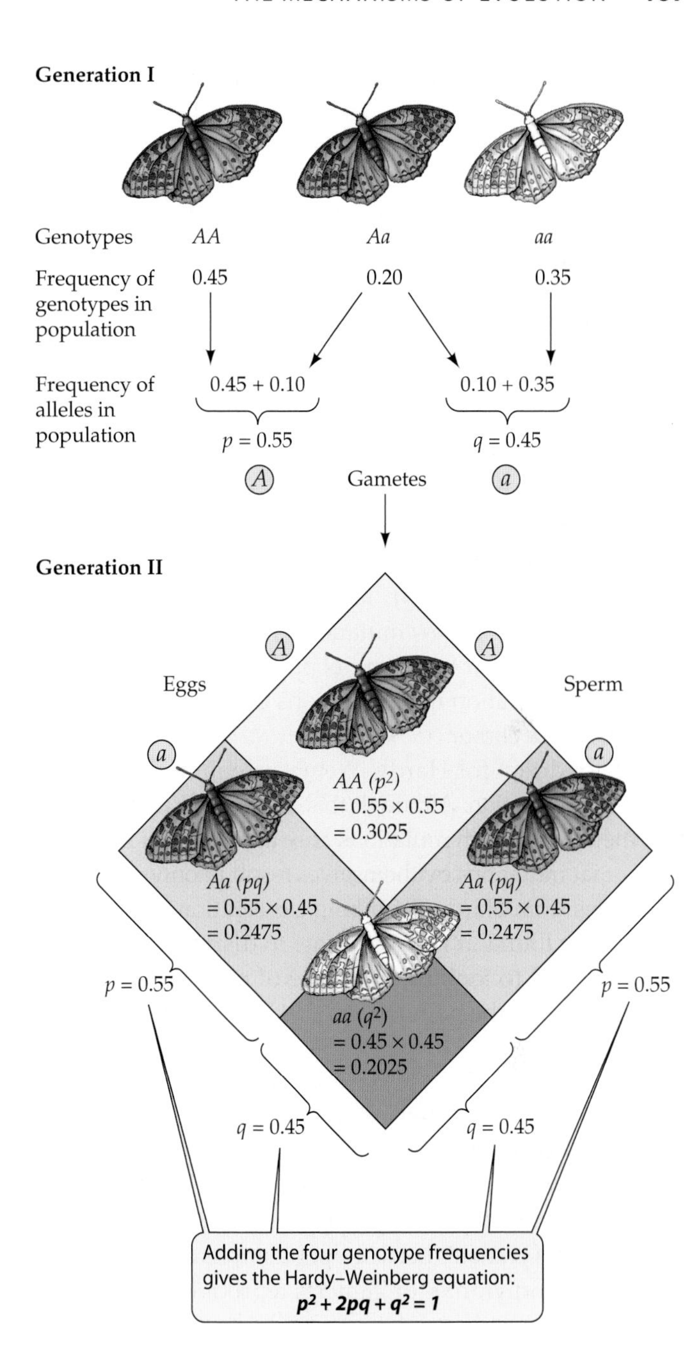

23.7 Calculating Hardy–Weinberg Genotype Frequencies The areas within the squares are proportional to the expected frequencies of possible matings if mating is random with respect to genotype. Because there are two ways of producing a heterozygote, the probability of this event occurring is the sum of the two *Aa* squares.

Why is the Hardy–Weinberg equilibrium important?

You may already have realized that populations in nature rarely meet the stringent conditions necessary to maintain them at Hardy–Weinberg equilibrium. Why, then, is the Hardy-Weinberg equilibrium considered so important for the study of evolution? The answer is that without it, we cannot tell whether or not evolutionary agents are operating. The most important message of the Hardy–Weinberg equilibrium is that *allele frequencies remain the same from generation to generation unless some agent acts to change them.*

In order to ascertain that evolutionary agents are in play, we must estimate the actual allele or genotype frequencies present in a population and then compare them with the frequencies that would be expected at Hardy–Weinberg equilibrium. The pattern of deviation from the Hardy–Weinberg expectations tells us which assumptions are violated. Thus, we can identify the agents of evolutionary change on which we should concentrate our attention.

Evolutionary Agents and Their Effects

Evolutionary agents are forces that change the genetic structure of a population. In other words, they cause deviations from the Hardy–Weinberg equilibrium. The known evolutionary agents are mutation, gene flow, genetic drift, nonrandom mating, and natural selection. Although only natural selection results in adaptation, to understand evolutionary processes we need to discuss all of these evolutionary agents before considering natural selection in detail.

Mutations are changes in the genetic material

The origin of genetic variation is mutation. A mutation, as we saw in Chapter 12, is any change in an organism's DNA. Mutations appear to be random with respect to the adaptive needs of organisms. Most mutations are harmful to their bearers or are neutral, but if environmental conditions change, previously harmful alleles may become advantageous. In addition, mutations can restore to populations alleles that other evolutionary agents remove. Thus mutations both create and help maintain genetic variation within populations.

Mutation rates are very low for most loci that have been studied. Rates as high as one mutation per locus in a thousand zygotes per generation are rare; one in a million is more typical. Nonetheless, these rates are sufficient to create considerable genetic variation because each of a large number of genes may mutate, frame-shift mutations may change many genes simultaneously, and populations often contain large numbers of individuals. For example, if the probability of a point mutation were 10^{-9} per base pair per generation, then in each human gamete, the DNA of which contains 3×10^9 base pairs, there would be an average of three new point mutations ($3 \times 10^9 \times 10^{-9} = 3$). Therefore, each zygote would carry, on average, six new mutations, and the current human population of about 8 billion people would be expected to carry about 48 billion new mutations that were not present one generation earlier.

One condition for Hardy–Weinberg equilibrium is that there be no mutation. Although this condition is never strictly met, the rate at which mutations arise at single loci is usually so low that mutations by themselves result in only very small deviations from Hardy–Weinberg expectations. If large deviations are found, it is appropriate to dismiss mutation as the cause and to look for evidence of other evolutionary agents acting on the population.

Movement of individuals or gametes, followed by reproduction, produces gene flow

Few populations are completely isolated from other populations of the same species. Migrations of individuals and movements of gametes between populations are common. If the arriving individuals or gametes reproduce in their new location, they may add new alleles to the gene pool of the population, or they may change the frequencies of alleles already present if they come from a population with different allele frequencies. For a population to be at Hardy–Weinberg equilibrium, there must be no **gene flow** from populations with different allele frequencies.

Genetic drift may cause large changes in small populations

In very small populations, **genetic drift**—the random loss of individuals and the alleles they possess—may produce large changes in allele frequencies from one generation to the next. Harmful alleles, for example, may increase in frequency because of genetic drift, and rare advantageous alleles may be lost. As we will see later, even in large populations, genetic drift can influence the frequencies of alleles that do not influence the survival and reproductive rates of their bearers.

Populations that are normally large may pass through occasional periods when only a small number of individuals survive. During these **population bottlenecks**, genetic variation can be reduced by genetic drift. How this works is illustrated in Figure 23.8, in which red and yellow beans represent two different alleles. Most of the "surviving" beans in the small sample taken from the bean population are, just by chance, red, so the new population has a much higher frequency of red beans than the previous generation had. In a natural population, the allele frequencies would be said to have "drifted."

Suppose we perform a cross of $Aa \times Aa$ individuals of a species of *Drosophila* to produce an F_1 population in which $p = q = 0.5$ and in which the genotype frequencies are 0.25 *AA*, 0.50 *Aa*, and 0.25 *aa*. If we randomly select 4 individuals (= 8 copies of the gene) from among the offspring to produce the F_2 generation, the allele frequencies in this small sample may differ markedly from $p = q = 0.5$. If, for example, we happen by chance to draw 2 *AA* homozygotes and 2 heterozygotes (*Aa*), the allele frequencies in this "surviving population" will be $p = 0.75$ (6 out of 8) and $q = 0.25$ (2 out of 8). If we replicate this sampling experiment 1,000 times, one of the two alleles

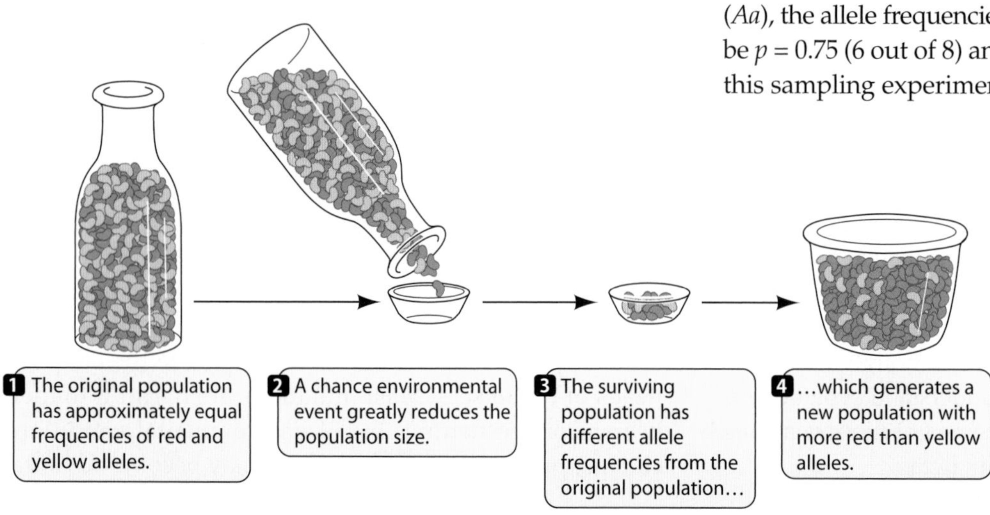

1 The original population has approximately equal frequencies of red and yellow alleles.
2 A chance environmental event greatly reduces the population size.
3 The surviving population has different allele frequencies from the original population…
4 …which generates a new population with more red than yellow alleles.

23.8 A Population Bottleneck Population bottlenecks occur when only a few individuals survive a random event, resulting in a shift in allele frequencies within the population.

Tympanuchus cupido (male)

23.9 A Species with Low Genetic Variation Prairie chickens in Illinois lost most of their genetic variation when the population crashed from millions to fewer than 100 individuals.

will be missing entirely from about 8 of the 1,000 "surviving populations."

These numbers show that, as it passes through a bottleneck, a population may lose much of its genetic variation. This is what happened to greater prairie chickens, millions of which lived in the prairies of North America when Europeans first arrived there. As a result of both hunting and habitat destruction, the Illinois population of prairie chickens plummeted from about 100 million birds in 1900 to fewer than 50 individuals in the 1990s (Figure 23.9). A comparison of DNA from birds collected in Illinois during the middle of the twentieth century with DNA from the surviving population in the 1990s showed that Illinois prairie chickens had lost most of their genetic diversity. As a result, both hatching success and chick survival were low. To increase the genetic diversity of Illinois prairie chickens, birds from Minnesota, Kansas, and Nebraska were introduced to Illinois. They interbred with the Illinois birds, restoring much of the genetic diversity of that population, which is now increasing in size.

When a few pioneering individuals colonize a new region, the resulting population is unlikely to have all the alleles found among members of its source population. The resulting change in genetic variation, called a **founder effect**, is equivalent to that in a large population reduced by a bottleneck. Scientists were given an opportunity to study the genetic composition of a founding population when *Drosophila subobscura*, a well-studied European species of fruit fly, was discovered near Puerto Montt, Chile, in 1978 and at Port Townsend, Washington, in 1982. In both South and North America, populations of the flies grew rapidly and expanded their ranges. Today in North America, *D. subobscura* ranges from British Columbia, Canada, to central California. In Chile it has spread across 23° of latitude, nearly as wide a range as the species has in Europe (Figure 23.10).

The *D. subobscura* founders probably reached Chile and the United States from Europe aboard the same ship, because the two populations are genetically very similar. For example, the North and South American populations have only 20 chromosomal inversions, 19 of which are the same on the two continents, whereas 80 inversions are known from European populations. North and South American populations also have lower allelic diversity at enzyme-producing genes than European populations do. Only alleles that have a frequency higher than 10 percent in European populations are present in the Americas. Thus, as expected for a small founding population, only a small part of the total genetic variation found in Europe reached the Americas. Geneticists estimate that at least ten, but no more than a hundred, flies founded the North and South American populations.

Nonrandom mating changes the frequency of homozygotes

Mating patterns may alter genotype frequencies if individuals in a population choose other individuals of certain genotypes as mates. For example, if they mate preferentially with individuals of the same genotype, then homozygous genotypes will be overrepresented, and heterozygous genotypes underrepresented, in the next generation in comparison with Hardy–Weinberg expectations. Alternatively, individuals

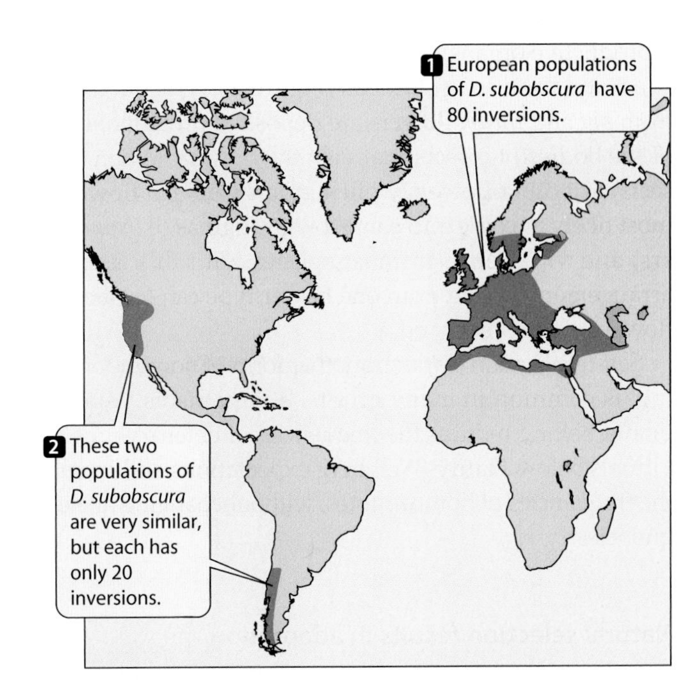

23.10 A Founder Effect Populations of the fruit fly *Drosophila subobscura* in North and South America contain less genetic variation than the European populations from which they came, as measured by the number of chromosome inversions in each population. Within two decades of arriving in the New World, the flies had increased dramatically and spread widely in spite of their reduced genetic variation.

Primula sp. (*pin* type)

An insect visiting a *thrum* flower picks up pollen on its head and body. When it then visits a *pin* flower, pollen is brushed off on the tall stigma.
pin flower
thrum flower
Stigma
Style
Stamen
Anther
Stamen
Stigma
Style
An insect visiting a *pin* flower picks up pollen on its proboscis and head. When it then visits a *thrum* flower, it deposits pollen on the short stigma.

23.11 Flower Structure Fosters Nonrandom Mating The structure of flowers in plant species such as the primroses ensures that pollination usually occurs between individuals of different genotypes.

may mate primarily or exclusively with individuals of different genotypes.

An example of such *nonrandom mating* is provided by plant species, such as primroses (*Primula*), that bear flowers of two different types. One type, known as *pin*, has a long style (female reproductive organ) and short stamens (male reproductive organs). The other type, known as *thrum*, has a short style and long stamens (Figure 23.11). Pollen grains from *pin* and *thrum* flowers are deposited on different parts of the bodies of insects that visit the flowers. When the insects visit other flowers, pollen grains from *pin* flowers are most likely to come into contact with stigmas of *thrum* flowers, and vice versa. In most species with this reciprocal arrangement, pollen from one flower type can fertilize only flowers of the other type.

Self-fertilization (*selfing*), another form of nonrandom mating, is common in many groups of organisms, especially plants. Selfing reduces the frequencies of heterozygous individuals below Hardy–Weinberg expectations and increases the frequencies of homozygotes, without changing allele frequencies.

Natural selection results in adaptation

The evolutionary agents we have just discussed influence the frequencies of alleles and genotypes in populations. As we saw in the previous chapter, major perturbations, such as colliding continents, volcanic eruptions, and meteorite impacts, also have periodically altered the survival and reproductive rates of organisms. All of these agents dramatically affect the course of life's evolution on Earth, but none of them result in adaptations. For adaptation to occur, individuals that differ in heritable traits must survive and reproduce with different degrees of success. When some individuals contribute more offspring to the next generation than others, allele frequencies in the population change in a way that adapts individuals to the environments that influenced their success. This process is known as **natural selection**.

The reproductive contribution of a phenotype to subsequent generations relative to the contributions of other phenotypes is called its **fitness**. The word "relative" is critical: The absolute number of offspring produced by an individual does not influence the genetic structure of a population. Changes in absolute numbers of offspring are responsible for increases and decreases in the *size* of a population, but only the *relative* success of different phenotypes within a population leads to changes in allele frequencies—that is, to evolution. To contribute genes to subsequent generations, individuals must survive to reproductive age and produce offspring. The relative contribution of individuals of a particular phenotype is determined by the probability that those individuals survive multiplied by the average number of offspring they produce over their lifetimes. In other words, the *fitness of a phenotype is determined by the average rates of survival and reproduction of individuals with that phenotype.*

The Results of Natural Selection

To simplify our discussion until now, we have considered only characters influenced by alleles at a single locus. However, as we saw in Chapter 10, most characters are influenced by alleles at more than one locus. Such characters are likely to show quantitative rather than qualitative variation. For ex-

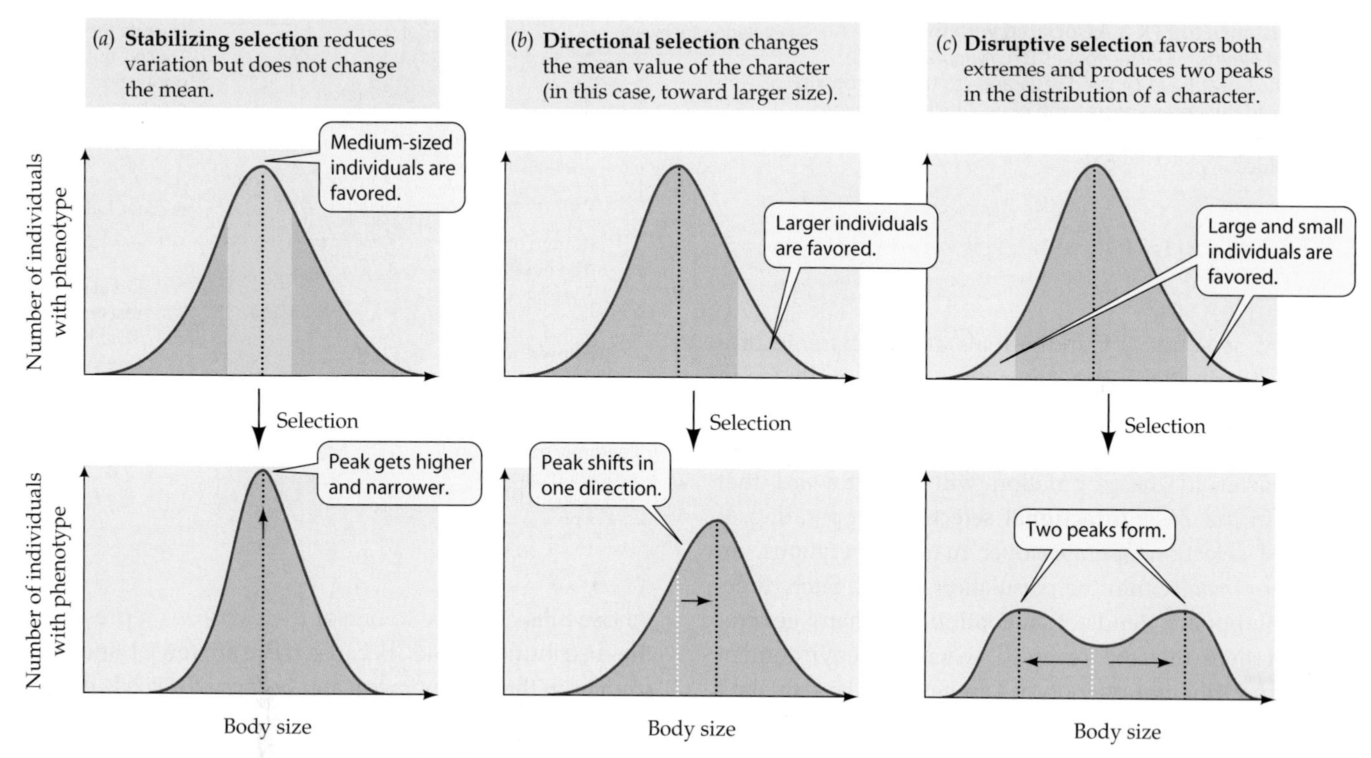

23.12 Natural Selection Can Operate on Quantitative Variation in Several Ways Each curve plots the distribution of body size in a population before selection (top) and after selection (bottom). Natural selection, by favoring the phenotype shown in yellow in the top graphs, changes the shape and position of the original curve (bottom graphs).

ample, the distribution of the sizes of individuals in a population, a character that is influenced by genes at many loci as well as by the environment, is likely to approximate the bell-shaped curves shown in the top row of Figure 23.12.

Natural selection can act on characters with quantitative variation in any one of several different ways, producing quite different results:

- *Stabilizing selection* preserves the average characteristics of a population by favoring average individuals.
- *Directional selection* changes the characteristics of a population by favoring individuals that vary in one direction from the mean of the population.
- *Disruptive selection* changes the characteristics of a population by favoring individuals that vary in both directions from the mean of the population.

STABILIZING SELECTION. If both the smallest and the largest individuals in a population contribute relatively fewer offspring to the next generation than those closer to the average size do, then **stabilizing selection** is operating (Figure 23.12*a*). Stabilizing selection reduces variation, but does not change the mean. Natural selection frequently acts in this way, countering increases in variation brought about by genetic recombination, mutation, or migration. Rates of evolution are typically very slow because natural selection is usually stabilizing. Stabilizing selection operates, for example, on human birth weight. Babies born lighter or heavier than the population mean die at higher rates than babies whose weights are close to the mean (Figure 23.13). This was especially true before modern medical advances.

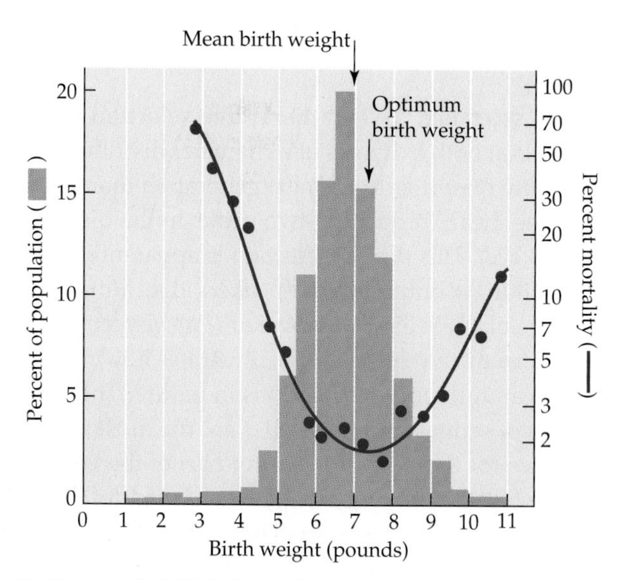

23.13 Human Birth Weight Is Influenced by Stabilizing Selection Babies that weigh more or less than average are more likely to die soon after birth than babies with weights close to the population mean.

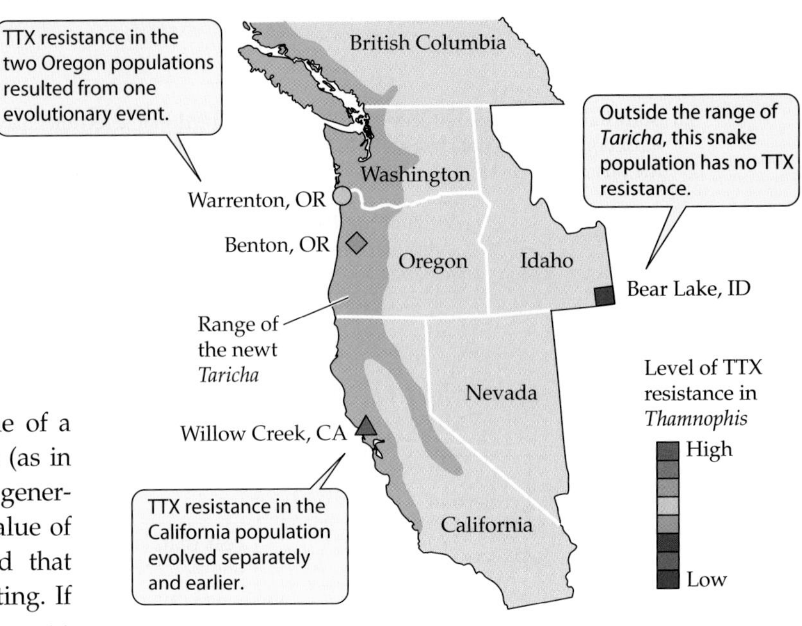

23.14 Resistance to TTX Is Associated with the Presence of Newts Garter snakes (*Thamnophis sirtalis*) of the Pacific Coast have evolved resistance to the neurotoxin TTX produced by a prey species, the newt *Taricha granulosa*. TTX resistance evolved at least twice. The range of the newt is shown in blue.

DIRECTIONAL SELECTION. If individuals at one extreme of a character distribution—the larger ones, for example (as in Figure 23.12*b*)—contribute more offspring to the next generation than other individuals do, then the average value of that character in the population will shift toward that extreme. In this case, **directional selection** is operating. If directional selection operates over many generations, an *evolutionary trend* within the population results. Such directional evolutionary trends often continue for many generations, but they may be reversed when the environment changes and different phenotypes are favored, or they may be halted when an optimum is reached, or when trade-offs oppose further change. The character then falls under stabilizing selection.

Directional selection produced the resistance to tetrodotoxin (TTX) by some garter snakes that we discussed at the beginning of this chapter. The common garter snake, *Thamnophis sirtalis*, is the only predator of the rough-skinned newt, *Taricha granulosa*, known to be resistant to TTX. Resistance to TTX has evolved independently at least twice within *T. sirtalis* populations in western North America, once in California and once in Oregon (Figure 23.14). This resistance is due to genetically based differences in the ability of sodium channels in the snake's nerves and muscles to continue functioning when exposed to variable concentrations of TTX.

DISRUPTIVE SELECTION. When **disruptive selection** operates, individuals at both extremes of a character distribution contribute more offspring to the next generation than do those close to the mean, producing two peaks in the distribution (Figure 23.12*c*). This type of selection is apparently rare.

The strikingly bimodal (two-peaked) distribution of bill sizes in the black-bellied seedcracker (*Pyrenestes ostrinus*), a West African finch (Figure 23.15), illustrates how disruptive selection can influence populations in nature. The seeds of two types of sedges (marsh plants) are the most abundant food source for these finches during part of the year. Birds with large bills can readily crack the hard seeds of the sedge *Scleria verrucosa*. Birds with small bills can crack *S. verrucosa* seeds only with difficulty, but they feed more efficiently on the soft seeds of *S. goossensii* than do birds with larger bills.

Young finches whose bills deviate markedly from the two predominant bill sizes do not survive as well as finches whose bills are close to one of the two sizes represented by the distribution peaks. Because there are few abundant food sources in the environment, and because the seeds of the two sedges do not overlap in hardness, birds with intermediate-

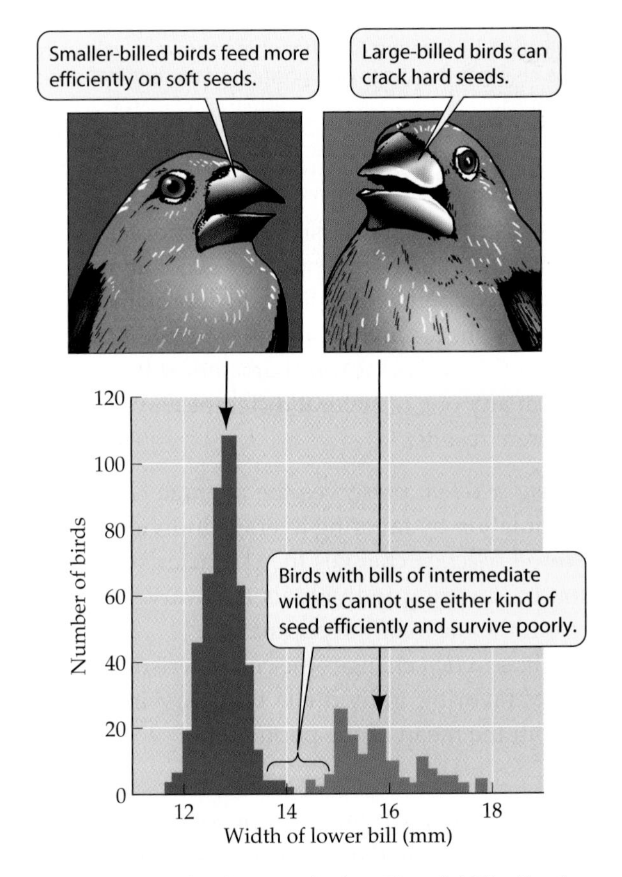

23.15 Disruptive Selection Results in a Bimodal Distribution The bimodal distribution of bill sizes in the black-bellied seedcracker of West Africa is a result of disruptive selection, which favors individuals with larger and smaller bill sizes over individuals with intermediate-sized bills.

sized bills are inefficient in using either one of the principal food sources. Disruptive selection therefore maintains a bimodal bill size distribution.

Sexual selection results in conspicuous traits

In *The Origin of Species*, Darwin devoted a few pages to sexual selection, a topic that he developed at length in another book, *The Descent of Man, and Selection in Relation to Sex*, in 1871. **Sexual selection** was Darwin's explanation for the evolution of apparently useless but conspicuous traits in males of many species, such as bright colors, long tails, horns, antlers, and elaborate courtship displays. He hypothesized that these traits either improved the ability of their bearers to compete for access to members of the other sex (*intrasexual selection*) or made their bearers more attractive to members of the other sex (*intersexual selection*). Darwin argued that female preferences for such features are also the result of sexual selection because "unornamented, or unattractive males would succeed equally in the battle for life and in leaving a numerous progeny, but for the presence of better endowed males." Sexual selection may result in species that are *sexually dimorphic*—that is, species in which males and females differ in size, shape, or color.

The concept of sexual selection was not well received by Darwin's contemporaries. However, many examples of sexual selection have been investigated in the century and a half since he first proposed the idea, and Darwin turned out to be right. For example, sexual selection is responsible for different morphological attributes of male birds that compete with each other for available females. One case in point is the remarkable tails of male African long-tailed widowbirds, which are longer than their heads and bodies combined.

To examine the role of sexual selection in the evolution of widowbird tails, a behavioral ecologist captured some male widowbirds. He shortened the tails of some males by cutting them and lengthened the tails of others by gluing on additional feathers. Male widowbirds normally select, and defend from other males, a site where they perform courtship displays to attract females. Both short-tailed and long-tailed males successfully defended their display sites, indicating that a long tail does not confer an advantage in male–male competition. However, males with artificially elongated tails attracted about four times more females than did males with shortened tails (Figure 23.16).

Why do female widowbirds prefer males with long tails? The ability to grow and maintain a costly feature such as a long tail may indicate that the male bearing it is vigorous and healthy. The hypothesis that having well-developed ornamental traits signals vigor and health has been tested experimentally with captive zebra finches. The bright red bills of male zebra finches are the result of red and yellow carotenoid pigments. Zebra finches (and most other animals) cannot synthesize carotenoids and must obtain them from their food. In addition to influencing bill color, carotenoids are antioxidants and components of the immune system. Males in good health may need to allocate fewer carotenoids to immune function than males in poorer health. If so, then females can use the brightness of his bill to assess the health of a male.

Investigators manipulated blood levels of carotenoids in male zebra finches by means of carotenoid supplements. Experimental males were given drinking water with carotenoids added; control males were given only distilled water. All the males had access to the same food. After one month, the experimental males had higher levels of carotenoids in their blood, had much brighter bills than the control males, and were preferred by female zebra finches (Figure 23.17).

Next, the investigators challenged both groups of males immunologically by injecting phytohemagglutinin (PHA) into the webs of their wings. PHA induces a response by T lymphocytes, resulting in an accumulation of white blood cells and thus a thickening of the skin. Experimental males with enhanced carotenoid levels developed thicker skins

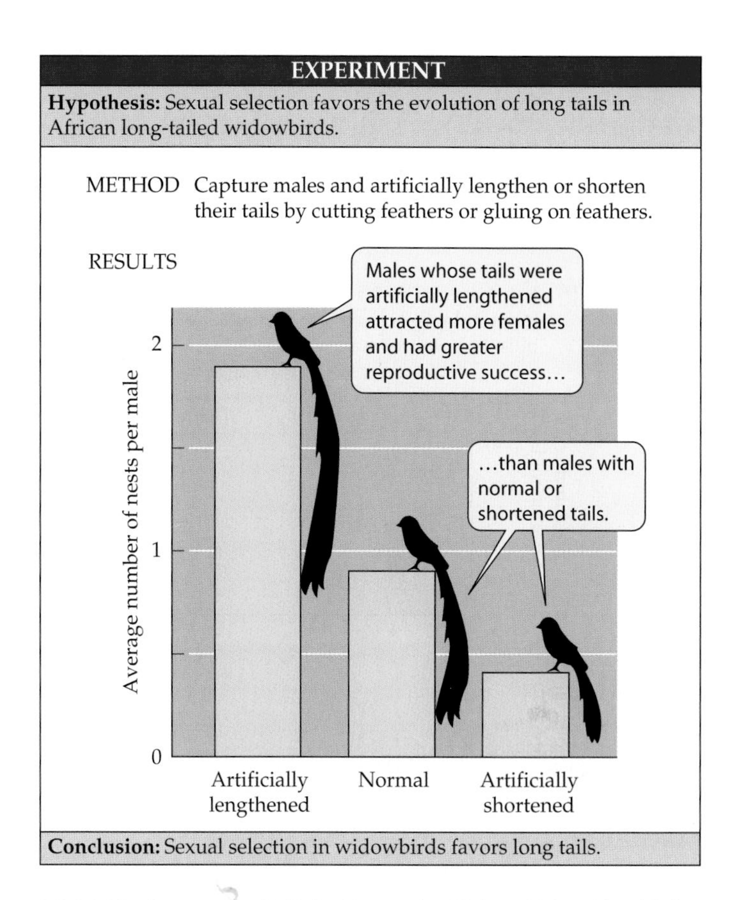

EXPERIMENT
Hypothesis: Sexual selection favors the evolution of long tails in African long-tailed widowbirds.
METHOD Capture males and artificially lengthen or shorten their tails by cutting feathers or gluing on feathers.
RESULTS
Males whose tails were artificially lengthened attracted more females and had greater reproductive success...
...than males with normal or shortened tails.
Average number of nests per male
2
1
0
Artificially lengthened
Normal
Artificially shortened
Conclusion: Sexual selection in widowbirds favors long tails.

23.16 The Longer the Tail, the Better the Male Male widowbirds with shortened tails defended their display sites successfully, but attracted fewer females (and thus fathered fewer nests of eggs) than did males with normal and lengthened tails.

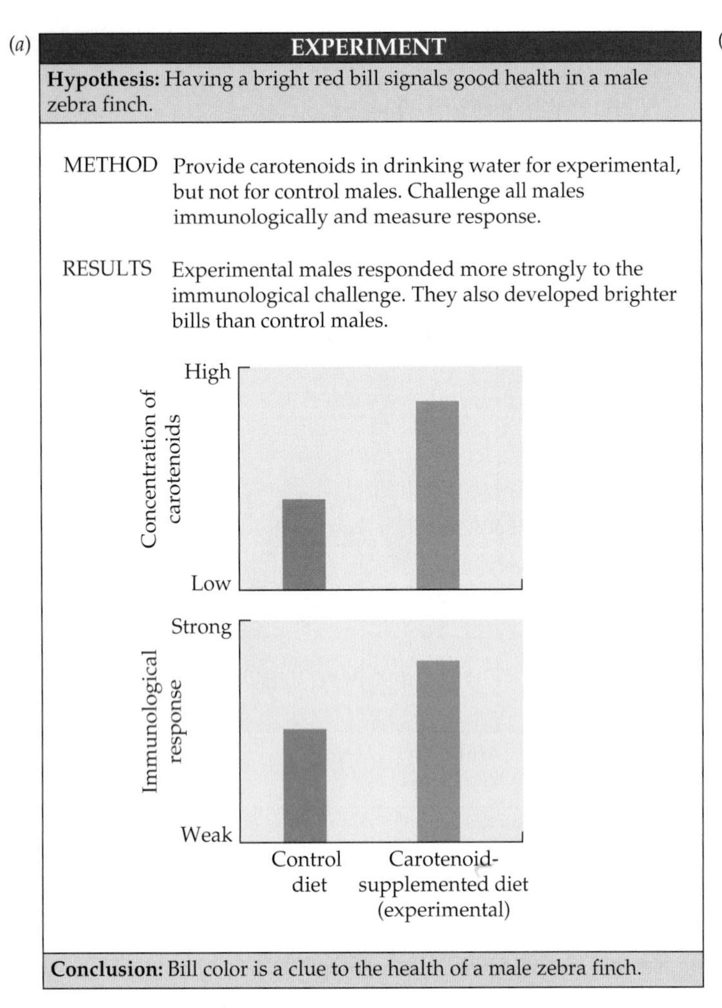

(a)
EXPERIMENT
Hypothesis: Having a bright red bill signals good health in a male zebra finch.
METHOD Provide carotenoids in drinking water for experimental, but not for control males. Challenge all males immunologically and measure response.
RESULTS Experimental males responded more strongly to the immunological challenge. They also developed brighter bills than control males.
High
Concentration of carotenoids
Low
Strong
Immunological response
Weak
Control diet
Carotenoid-supplemented diet (experimental)
Conclusion: Bill color is a clue to the health of a male zebra finch.

Taeniopygia guttata

23.17 Bright Bills Signal Good Health (*a*) This experiment demonstrated that bright bill color in the male zebra finch does indeed indicate a healthy individual. (*b*) Female zebra finches (the bird below) preferentially choose mates with the brightest bill color—thus choosing the healthiest males.

because they responded more strongly to PHA than control males did, indicating a heightened immune system.

This experiment showed that when a female chooses the male with a bright red bill, she probably gets a mate with a healthy immune system. Such males are less likely to become infected with parasites and diseases, so they are less likely to pass on infections to their mates. Healthier males are also better able to assist with parental care than are males with duller bills.

Assessing the Costs of Adaptations

As we mentioned at the beginning of this chapter, adaptations typically impose costs as well as benefits, and the evolution of adaptations depends on the trade-off between those costs and benefits. Garter snakes in some populations, for example, can eat rough-skinned newts without being poisoned, but they pay for this ability by sacrificing crawling speed.

Determining the costs and benefits of a particular adaptation is difficult because individuals differ not only in the degree to which they possess the adaptation, but also in many other ways. How can investigators study individuals that differ only in the genetically based adaptation of interest? Such individuals can be created by recombinant DNA techniques using cloned or highly inbred populations. In plants, for example, plasmids can be used to transfer specific alleles to experimental individuals (see Figure 16.5). Control individuals also receive plasmids, but those plasmids lack the allele of interest.

Plasmid transfer techniques made it possible to measure the cost associated with resistance to the herbicide chlorosulfuron conferred by a single allele in the shale cress, *Arabidopsis thaliana*. The allele, *Csr1-1*, results in the production of an enzyme that is insensitive to chlorosulfuron. However, plants with the *Csr1-1* allele produce 34 percent fewer seeds than nonresistant plants grown under identical conditions in the absence of the herbicide (Figure 23.18).

The reason for the high cost of resistance is not fully understood, but evidence suggests that the resistance allele results in an accumulation of branched-chain amino acids that interfere with metabolism. Agriculturalists wish to alter the genotypes of plants to give them resistance to herbicides so that the herbicides applied to agricultural fields will kill the weeds, but not the crops. This experiment shows that such benefits may impose a trade-off in terms of crop yield.

We saw in the previous section that the possession of certain conspicuous features by males confers reproductive benefits. What kinds of trade-offs do these benefits impose? The cost of long tails was not measured in the experiments with

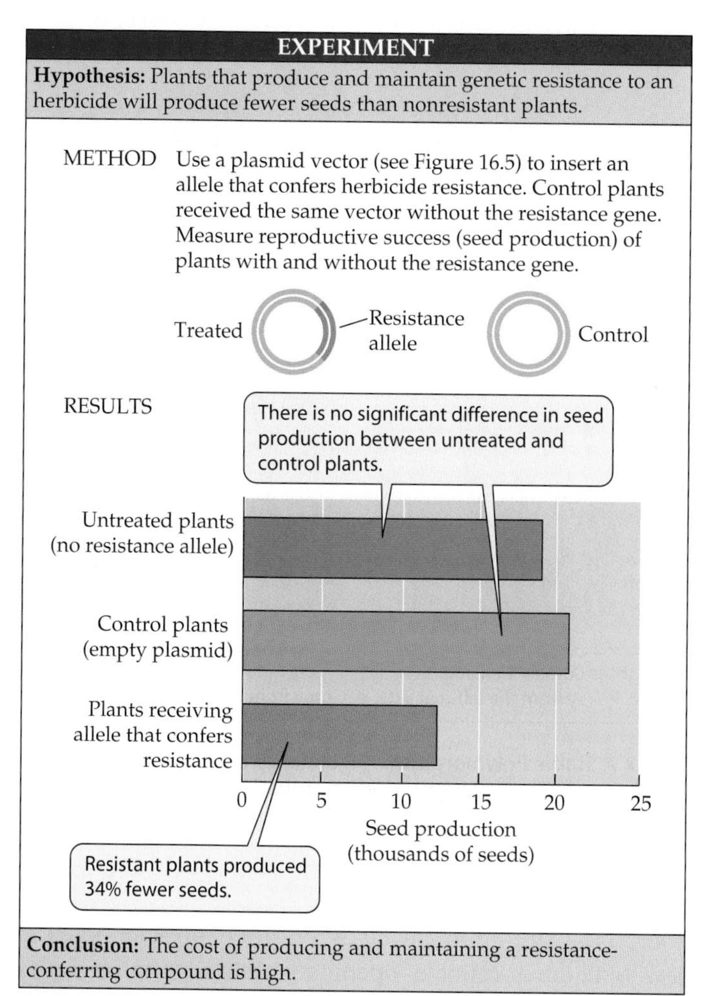

EXPERIMENT
Hypothesis: Plants that produce and maintain genetic resistance to an herbicide will produce fewer seeds than nonresistant plants.
METHOD Use a plasmid vector (see Figure 16.5) to insert an allele that confers herbicide resistance. Control plants received the same vector without the resistance gene. Measure reproductive success (seed production) of plants with and without the resistance gene.
Treated
Resistance allele
Control
RESULTS
There is no significant difference in seed production between untreated and control plants.
Untreated plants (no resistance allele)
Control plants (empty plasmid)
Plants receiving allele that confers resistance
0 5 10 15 20 25
Seed production (thousands of seeds)
Resistant plants produced 34% fewer seeds.
Conclusion: The cost of producing and maintaining a resistance-conferring compound is high.

23.18 Producing and Maintaining Resistance Is Costly Possession of a gene that confers herbicide resistance greatly reduced seed production in *Arabidopsis thaliana*.

widowbirds, but related studies have been done on males of other species.

In some mammalian species, including deer, lions, and baboons, one male controls reproductive access to many females. These *polygynous* species tend to be *sexually dimorphic*—the males appear quite different from the females. Males of these species are significantly larger than females and often bear large weapons (such as horns, antlers, and large canine teeth); size and weaponry are needed to defend a male's multiple mates against other males of the species.

The costs of sexual dimorphism for males of polygynous species were assessed using the comparative method (Figure 23.19). Such males have higher parasite loads and higher mortality rates than females of their own species because maintaining a large size and bearing large weapons makes them more susceptible to parasites. In addition, when compared to parasite loads in males of closely related monogamous species (in which males and females are essentially monomorphic, appearing quite similar), the dimorphic males carried higher parasite loads in almost every case.

Maintaining Genetic Variation

Genetic drift, stabilizing selection, and directional selection all tend to reduce genetic variation within populations. Nevertheless, as we have seen, most populations have considerable genetic variation. What maintains so much genetic variation within populations? To answer this question, we will show how sexual recombination, neutral mutations, and frequency-dependent selection can maintain variation within populations, and how variation may be maintained over geographic space.

Sexual recombination amplifies the number of possible genotypes

In asexually reproducing organisms, the cells resulting from a mitotic division normally contain identical genotypes. Each new individual is genetically identical to its parent, unless there has been a mutation. When organisms exchange genetic

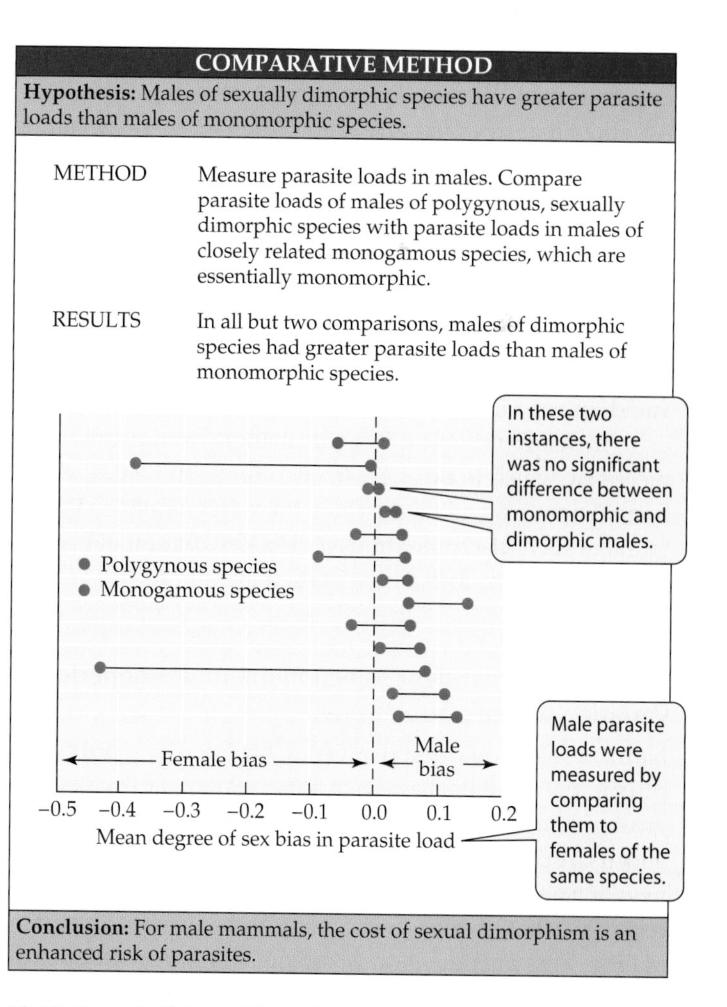

COMPARATIVE METHOD
Hypothesis: Males of sexually dimorphic species have greater parasite loads than males of monomorphic species.
METHOD Measure parasite loads in males. Compare parasite loads of males of polygynous, sexually dimorphic species with parasite loads in males of closely related monogamous species, which are essentially monomorphic.
RESULTS In all but two comparisons, males of dimorphic species had greater parasite loads than males of monomorphic species.
In these two instances, there was no significant difference between monomorphic and dimorphic males.
Polygynous species
Monogamous species
Female bias
Male bias
−0.5 −0.4 −0.3 −0.2 −0.1 0.0 0.1 0.2
Mean degree of sex bias in parasite load
Male parasite loads were measured by comparing them to females of the same species.
Conclusion: For male mammals, the cost of sexual dimorphism is an enhanced risk of parasites.

23.19 Sexually Selected Traits Impose Costs Male mammals of sexually dimorphic, polygynous species have greater parasite loads than do males of closely related species that are not sexually dimorphic.

material during sexual reproduction, however, offspring differ from their parents because chromosomes assort randomly during meiosis, crossing-over occurs, and fertilization brings together material from two different cells (see Chapter 9).

Sexual recombination generates an endless variety of genotypic combinations that increases the evolutionary potential of populations. Because it increases the variation among the offspring produced by an individual, sexual recombination may improve the chance that at least some of those offspring will be successful in the varying and often unpredictable environments they will encounter. Sexual recombination does not influence the frequencies of alleles; rather, *sexual recombination generates new combinations of alleles on which natural selection can act*. It expands variation in a character influenced by alleles at many loci by creating new genotypes. That is why selection for bristle number in *Drosophila* (see Figure 23.5) resulted in flies with more bristles than any flies in the initial population had.

Neutral mutations accumulate within species

As we saw in Chapter 12, some mutations do not affect the functioning of the proteins encoded by the mutated genes. An allele that does not affect the fitness of an organism is called a **neutral allele**. Such alleles, untouched by natural selection, may be lost, or their frequencies may increase with time, purely by genetic drift. Therefore, neutral alleles often accumulate in a population over time, providing it with considerable genetic variation.

Much of the variation in those characters we can observe with our unaided senses is not neutral, but much molecular variation apparently is. Modern molecular techniques enable us to measure variation in neutral alleles and provide the means by which to distinguish adaptive from neutral variation. Chapter 26 will discuss how these techniques enable us to make such discriminations and how variation in neutral traits can be used to estimate rates of evolution.

Frequency-dependent selection maintains genetic variation within populations

Natural selection often preserves variation as a **polymorphism**: the coexistence within a population, at frequencies greater than mutations can produce, of two or more alleles at a locus. A polymorphism may be maintained when the fitness of a genotype (or phenotype) varies with its frequency relative to that of other genotypes (or phenotypes) in a population. This phenomenon is known as **frequency-dependent selection**.

A small fish that lives in Lake Tanganyika, in East Africa, provides an example of frequency-dependent selection. The mouth of this scale-eating fish, *Perissodus microlepis*, opens either to the right or to the left as a result of an asymmetrical jaw joint; the direction of opening is genetically determined (Figure 23.20). *P. microlepis* approaches its prey (another fish) from behind and dashes in to bite off several scales from its flank. "Right-mouthed" individuals always attack from the victim's left; "left-mouthed" individuals always attack from the victim's right. The distorted mouth enlarges the area of teeth in contact with the prey's flank, but only if the scale-eater attacks from the appropriate side.

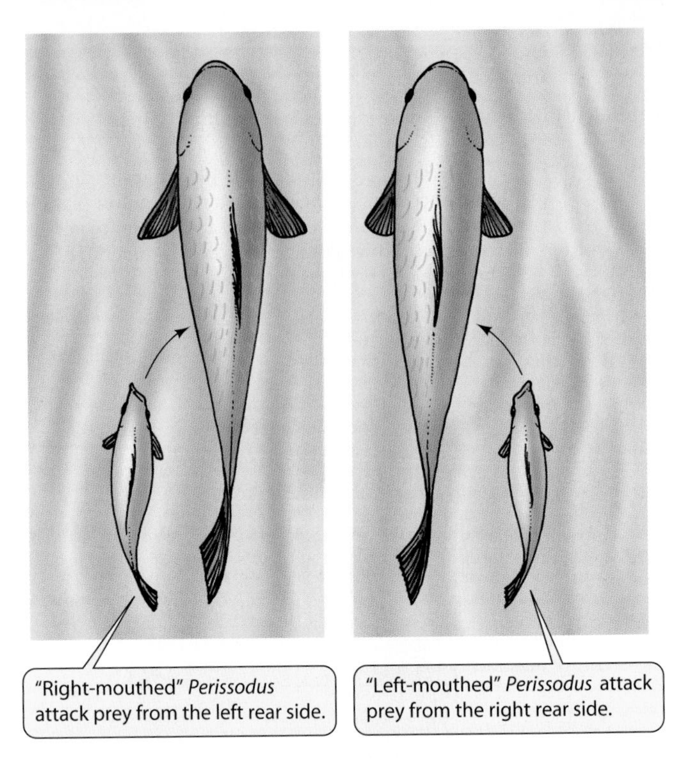

"Right-mouthed" *Perissodus* attack prey from the left rear side.
"Left-mouthed" *Perissodus* attack prey from the right rear side.

23.20 A Stable Polymorphism Frequency-dependent selection maintains equal proportions of left-mouthed and right-mouthed individuals of the scale-eating fish *Perissodus microlepis*.

Prey fish are alert to approaching scale-eaters, so attacks are more likely to be successful if the prey must watch both flanks. Vigilance by the prey favors equal numbers of right-mouthed and left-mouthed scale-eaters, because if one form were more common than the other, prey fish would pay more attention to potential attacks from the corresponding flank. Over an 11-year period in which the scale-eaters in Lake Tanganyika were studied, the polymorphism was found to be stable: The two forms of *P. microlepis* remained at about equal frequencies.

Genetic variation is maintained in geographically distinct subpopulations

Much of the genetic variation in large populations is preserved as differences among members in different places (subpopulations). Subpopulations often vary genetically because they are subjected to different selective pressures in different environments. Plant species, for example, may vary

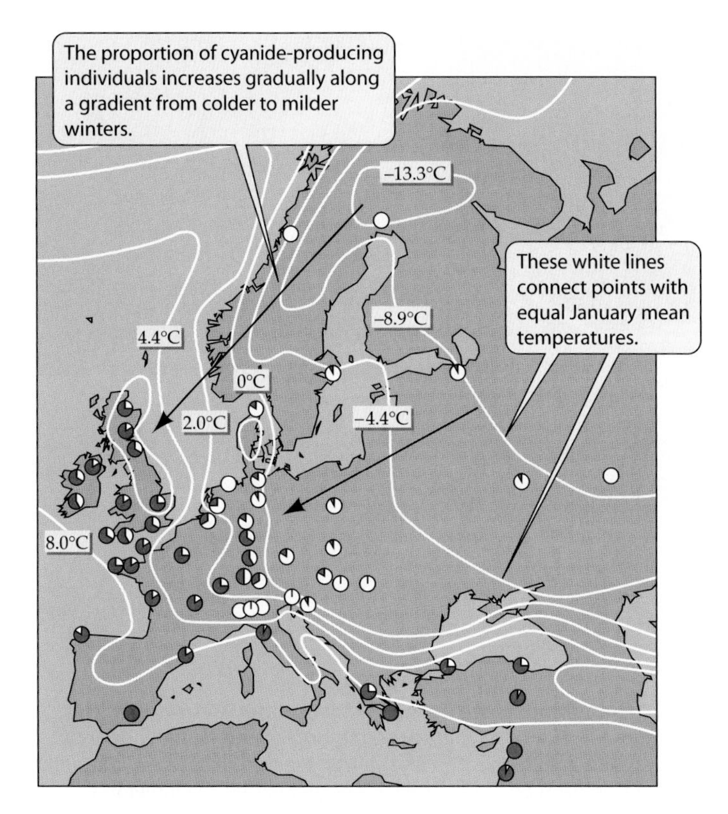
The proportion of cyanide-producing individuals increases gradually along a gradient from colder to milder winters.
These white lines connect points with equal January mean temperatures.
−13.3°C
−8.9°C
4.4°C
0°C
2.0°C
−4.4°C
8.0°C
White indicates proportion of plants not producing cyanide
Red indicates proportion of plants producing cyanide

23.21 Geographic Variation in Poisonous Clovers The frequency of cyanide-producing individuals in each population of white clover (*Trifolium repens*) is represented by the proportion of the circle that is red.

geographically in the chemicals they synthesize to defend themselves against herbivores. Some individuals of the clover *Trifolium repens* produce the poisonous chemical cyanide. Poisonous individuals are less appealing to herbivores—particularly mice and slugs—than are nonpoisonous individuals. However, clover plants that produce cyanide are more likely to be killed by frost, because freezing damages cell membranes and releases the toxic cyanide into the plant's own tissues.

In populations of *Trifolium repens*, the frequency of cyanide-producing individuals increases gradually from north to south and from east to west across Europe (Figure 23.21). Poisonous plants make up a large proportion of clover populations only in areas where winters are mild. Cyanide-producing individuals are rare where winters are cold, even though herbivores graze clovers heavily in those areas.

Constraints on Evolution

The many examples of adaptations that we have just discussed are testimony to the power of natural selection, but evolution is limited by a serious constraint: Evolutionary changes must be based on modifications of previously existing traits, which may come to serve new functions. Engineers are able to design a completely new type of engine (jet) to power an airplane that can replace a previous type (propeller), but evolutionary changes cannot happen that way.

A striking example of such constraints on evolution is provided by the evolution of fish that spend most of their time resting on the sea bottom. One lineage, the bottom-dwelling skates and rays, is beautifully symmetrical. These fishes are descended from sharks, whose bodies were already somewhat flattened; therefore, skates and rays are able to lie on their bellies (Figure 23.22*a*).

Plaice, sole, and flounders, on the other hand, are bottom-dwelling descendants of deep-bodied, laterally flattened ancestors. Unlike sharks, these fishes cannot lie on their bellies; they must flop over on their sides. During development, the eyes of plaice and sole are grotesquely twisted around to bring both eyes to one side of the body (Figure 23.22*b*). Small shifts in the position of one eye probably helped ancestral flatfishes see better, resulting in the form found today.

(*a*) *Taeniura lymma*

(*b*) *Bothus lunatus*

23.22 Two Solutions to a Single Problem (*a*) Stingrays, whose ancestors were dorsally flattened, lie on their bellies. (*b*) Flounders, whose ancestors were laterally flattened, lie on their sides. Their eyes migrate during development so that both eyes are on the same side of the body.

Cultural Evolution

Traits can evolve by natural selection only if they are at least partly heritable. However, individuals may acquire new traits via **cultural evolution**—that is, by learning them from other individuals. Cultural evolution is most highly developed in humans, whose language and remarkable learning abilities enable new innovations to spread and be adopted at rapid rates. But the only requirement for traits to evolve via cultural evolution is that individuals have the ability to learn them. Birds, for example, copy the songs of other individuals, resulting in the evolution of song "dialects."

Many behaviors of the apes (chimpanzees, gorillas, gibbons, and orangutans) are transmitted via learning. In one study, investigators compared the behavior of four orangutan populations on the island of Borneo and two on Sumatra. The investigators identified 24 behaviors that are restricted to a single population. These behaviors are not correlated with any differences in the environments in which the populations live. Ten of the behaviors are specialized feeding techniques (Figure 23.23), including tool use. Six are alternative forms of social signals, such as kiss-squeaks. Thus, orangutan populations develop cultural distinctions as individuals copy the behavior of other individuals.

Pongo pygmaeus

23.23 Orangutans Have Culturally Transmitted Behaviors This orangutan in Indonesia has learned to break open dead twigs and suck out the ants inside. This specialized feeding behavior is culturally transmitted from one individual to another within the individual's social group.

Short-Term versus Long-Term Evolution

The short-term changes in allele frequencies within populations that we have emphasized in this chapter are an important focus of study for evolutionary biologists. These changes can be observed directly, they can be manipulated experimentally, and they show us the actual processes by which evolution occurs. By themselves, however, they do not enable us to predict—or, more properly, "postdict" (because they have already happened)—the long-term evolutionary changes we described in Chapter 22.

The reason is that patterns of evolutionary change can be strongly influenced by events that occur so infrequently or so slowly that they are unlikely to be observed during short-term studies. In addition, the ways in which evolutionary agents act may change with time; even among the descendants of a single ancestral species, different lineages may evolve in different directions. Therefore, additional types of evidence, demonstrating the occurrence of rare and unusual events and trends in the fossil record, must be gathered if we wish to understand the course of evolution over billions of years.

"Postdiction" problems are not unique to evolutionary studies. For example, seismologists know the physical principles that explain how earthquakes occur, and they can pinpoint regions that are prone to earthquakes; but they cannot predict when or where an earthquake will happen.

In subsequent chapters, we will discuss the kinds of information that biologists assemble to study long-term evolutionary changes and infer the processes that led to them.

Chapter Summary

Charles Darwin's Theory of Evolution

- Darwin developed his theory of evolution by natural selection by carefully observing nature, especially during his voyage around the world on the *Beagle*. **Review Figure 23.1**
- Darwin based this theory on well-known facts and some key inferences.
- Modern genetics has discovered the mechanisms of inheritance, which Darwin did not understand.
- Darwin had no examples of the action of natural selection, so he based his arguments on artificial selection.

See Web/CD Tutorial 23.1

Genetic Variation within Populations

- For a population to evolve, its members must possess heritable genetic variation, which is the raw material on which agents of evolution act.
- A single individual has only some of the alleles found in the population of which it is a member. **Review Figure 23.3**
- Considerable genetic variation characterizes most natural populations. **Review Figures 23.4, 23.5**
- Allele frequencies measure the amount of genetic variation in a population. Biologists estimate allele frequencies by measuring a sample of individuals from a population. The sum of all allele frequencies at a locus is equal to 1. **Review Figure 23.6**
- Genotype frequencies show how a population's genetic variation is distributed among its members. Populations that have the same allele frequencies may nonetheless have different genotype frequencies.

The Hardy–Weinberg Equilibrium

▶ Several conditions are required for a population to be at Hardy–Weinberg equilibrium: mating is random, the population is very large, there is no migration, there is no mutation, and natural selection is not acting on the population.

▶ In a population at Hardy–Weinberg equilibrium, allele frequencies remain the same from generation to generation. In addition, genotype frequencies will remain in the proportions $p^2 + 2pq + q^2 = 1$. **Review Figure 23.7**

▶ Biologists can determine whether an agent of evolution is acting on a population by comparing the genotype frequencies of that population with Hardy–Weinberg expectations. **See Web/CD Tutorial 23.2**

Evolutionary Agents and Their Effects

▶ Changes in the genetic structure of populations are caused by several evolutionary agents: mutation, gene flow, genetic drift, nonrandom mating, and natural selection.

▶ The origin of genetic variation is mutation. Most mutations are harmful or neutral to their bearers, but some are advantageous, particularly if the environment changes.

▶ Movement of individuals or gametes from one population to another, followed by reproduction in the new location, produces gene flow. Gene flow may add new alleles to a population or may change the frequencies of alleles already present.

▶ The random loss of alleles, known as genetic drift, produces changes in allele frequencies, which may be especially dramatic in small populations. Organisms that normally have large populations may pass through occasional periods (bottlenecks) when only a small number of individuals survive. **Review Figure 23.8**

▶ New populations established by a few founding immigrants also have gene frequencies that differ from those in the parent population. **Review Figure 23.10**

▶ If individuals mate more often with other individuals of a certain genotype than would be expected on a random basis—that is, when mating is not random—genotype frequencies differ from Hardy–Weinberg expectations. **Review Figure 23.11**

▶ Self-fertilization, an extreme form of nonrandom mating, reduces the frequencies of heterozygous individuals below Hardy–Weinberg expectations without changing allele frequencies.

▶ Natural selection is the only agent of evolution that adapts populations to their environments.

▶ The reproductive contribution of a phenotype to subsequent generations relative to the contributions of other phenotypes is its fitness. The fitness of a phenotype is determined by the average rates of survival and reproduction of individuals with that phenotype.

The Results of Natural Selection

▶ Stabilizing selection reduces variation and preserves the average characteristics of a population. **Review Figures 23.12*a*, 23.13**

▶ Directional selection changes a character by favoring individuals that vary in one direction from the population mean. If directional selection operates over many generations, an evolutionary trend may result. **Review Figures 23.12*b*, 23.14**

▶ Disruptive selection changes a character by favoring individuals that vary in both directions from the population mean. **Review Figures 23.12*c*, 23.15**

▶ Sexually selected traits may evolve because females prefer to mate with males having those traits. **Review Figures 23.16, 23.17**

Assessing the Costs of Adaptations

▶ Possessing resistance to toxic chemicals may involve trade-offs, such as reduced reproductive output. **Review Figure 23.18. See Web/CD Tutorial 23.3**

▶ Sexually selected traits may result in higher parasite loads and mortality rates in males. **Review Figure 23.19**

Maintaining Genetic Variation

▶ Genetic drift, stabilizing selection, and directional selection all tend to reduce genetic variation, but most populations are genetically highly variable.

▶ Sexual recombination increases the evolutionary potential of populations, but it does not influence the frequencies of alleles. Rather, it generates new combinations of genetic material on which natural selection can act.

▶ Genetic variation within a population may be maintained by frequency-dependent selection. **Review Figure 23.20**

▶ Much genetic variation is maintained geographically. **Review Figures 23.21**

Constraints on Evolution

▶ Natural selection acts by modifying what already exists.

Cultural Evolution

▶ Learned traits can spread rapidly via cultural evolution.

Short-Term versus Long-Term Evolution

▶ Patterns of long-term evolutionary change can be strongly influenced by events that occur so infrequently or so slowly that they are unlikely to be observed during short-term evolutionary studies. Additional types of evidence must be gathered to understand why evolution in the long term took the particular course it did.

Self-Quiz

1. The two major components of Darwin's theory of evolution are that
   a. evolution is a fact, and mutations are the agent of evolution.
   b. evolution is a fact, and natural selection is the agent of evolution.
   c. species cannot change into other species, but natural selection can modify them.
   d. species cannot change into other species, but mutations can modify them.
   e. evolution is a hypothesis, and genetic drift is the agent of evolution.
2. To ground his theory, Charles Darwin
   a. developed a comprehensive theory of inheritance.
   b. described several evolutionary changes and identified the agents that caused them.
   c. used patterns of domestication to show how his theory differed from those patterns.
   d. assembled a broad base of supporting information from many fields.
   e. developed a mathematical model of evolutionary change.
3. The phenotype of an organism is
   a. the type specimen of its species in a museum.
   b. its genetic constitution, which governs its traits.
   c. the chronological expression of its genes.
   d. the physical expression of its genotype.
   e. the form it achieves as an adult.

4. The appropriate unit for defining and measuring genetic variation is the
   a. cell.
   b. individual.
   c. population.
   d. community.
   e. ecosystem.
5. Which statement about allele frequencies is *not* true?
   a. The sum of any set of allele frequencies is always 1.
   b. If there are two alleles at a locus and we know the frequency of one of them, we can obtain the frequency of the other by subtraction.
   c. If an allele is missing from a population, its frequency is 0.
   d. If two populations have the same gene pool for a locus, they will have the same proportion of homozygotes at that locus.
   e. If there is only one allele at a locus, its frequency is 1.
6. In a population at Hardy–Weinberg equilibrium in which the frequency of *A* alleles (p) is 0.3, the expected frequency of *Aa* individuals is
   a. 0.21.
   b. 0.42.
   c. 0.63.
   d. 0.18.
   e. 0.36.
7. Natural selection that preserves existing allele frequencies is called
   a. unidirectional selection.
   b. bidirectional selection.
   c. prevalent selection.
   d. stabilizing selection.
   e. preserving selection.
8. The fitness of a genotype is determined by the
   a. average rates of survival and reproduction of individuals with that genotype.
   b. individuals that have the highest rates of both survival and reproduction.
   c. individuals that have the highest rates of survival.
   d. individuals that have the highest rates of reproduction.
   e. average reproductive rate of individuals with that genotype.
9. Laboratory selection experiments with fruit flies have demonstrated that
   a. bristle number is not genetically controlled.
   b. bristle number is not genetically controlled, but changes in bristle number are caused by the environment in which the fly is raised.
   c. bristle number is genetically controlled, but there is little variation on which natural selection can act.
   d. bristle number is genetically controlled, but selection cannot result in flies having more bristles than any individual in the original population had.
   e. bristle number is genetically controlled, and selection can result in flies having more bristles than any individual in the original population had.
10. Disruptive selection maintains a bimodal distribution of bill size in the West African seedcracker because
    a. bills of intermediate shapes are difficult to form.
    b. the two major food sources of the finches differ markedly in size and hardness.
    c. males use their large bills in displays.
    d. migrants introduce different bill sizes into the population each year.
    e. older birds need larger bills than younger birds.
11. A population is said to be polymorphic for a locus if it has at least
    a. three different alleles at that locus.
    b. two different alleles at that locus.
    c. two genotypes for that locus.
    d. three genotypes for that locus.
    e. two alleles for that locus, the rarest of which is more common than expected by mutation alone.

For Discussion

1. During the past 50 years, more than 200 species of insects that attack crop plants have become highly resistant to DDT and other pesticides. Using your recently acquired knowledge of evolutionary processes, explain the rapid and widespread evolution of resistance. Propose ways of using pesticides that would slow down the rate of evolution of resistance. Now that use of DDT has been banned in the United States, what do you expect to happen to levels of resistance to DDT among insect populations? Justify your answer.
2. In what ways does artificial selection by humans differ from natural selection in nature? Was Darwin wise to base so much of his argument for natural selection on the results of artificial selection?
3. In nature, mating among individuals in a population is never truly random, immigration and emigration are common, and natural selection is seldom totally absent. Why then, does it make sense to use the Hardy–Weinberg equilibrium, which is based on assumptions known generally to be false? Can you think of other models in science that are based on false assumptions? How are such models used?
4. As far as we know, natural selection cannot adapt organisms to future events. Yet many organisms appear to respond to natural events before they happen. For example, many mammals go into hibernation while it is still quite warm. Similarly, many birds leave the temperate zone for their southern wintering grounds long before winter has arrived. How can such "anticipatory" behaviors evolve?
5. Populations of most of the thousands of species that have been introduced to areas where they were previously not found, including those that have become pests, began with a few individuals. They should therefore have begun with much less genetic variation than the parent populations have. If genetic variation is advantageous, why have so many of these species been successful in their new environments?
6. The flavors of many crop plants have been enhanced by artificial selection that has removed the bad-tasting chemicals with which they defended themselves in the wild. What problems do growing crop plants with reduced chemical defenses pose for modern agriculture?

24 Species and Their Formation

In May and June, 1993, a previously unknown disease abruptly killed 10 people in the southwestern United States. The victims experienced flu-like symptoms for several days, but then their condition deteriorated rapidly as their lungs filled with fluid. The disease agent was unknown; no cure was available, and initially 70 percent of infected people died.

Researchers from many disciplines focused on the outbreak. Within a few weeks, scientists at the U.S. Centers for Disease Control had identified the agent as a previously undescribed hantavirus—a type of virus known to be transmitted by rodents. Investigators initiated an intensive small mammal field trapping program. They quickly identified the deer mouse as the main host of the virus. The mouse sheds the virus in its feces, urine, and saliva. Humans become infected by inhaling microscopic particles of these substances that are present in the air.

Since the discovery of the hantavirus that caused the 1993 outbreak, about 25 additional hantaviruses have been described in the Western Hemisphere. People have been infected by some of these viruses in Florida, New York, Louisiana, and Texas. Other rodents, in addition to the deer mouse, harbor hantaviruses, and some of these viruses cause human diseases (Figure 24.1). Studies of the evolutionary relationships of hantaviruses and rodents indicate that rodents and hantaviruses have been evolving together for millions of years, and suggest that more kinds of hantaviruses, some of them likely to be human pathogens, have yet to be discovered.

A Source of Disease The deer mouse *Peromyscus maniculatus* harbors a type of hantavirus that infects people.

Efforts to reduce risks to humans and develop effective cures will require an understanding of the distribution of hantavirus types among rodent species and measures of the rate at which the viruses appear to be evolving. The fact that so many new types of hantavirus have been discovered since 1993 suggests that hantaviruses may be evolving rapidly. If so, better knowledge about how new viruses form and how they have coevolved with rodents will serve important human health objectives.

All species, living and extinct, are believed to be descendants of a single ancestral species that lived more than 3 billion years ago. If speciation were a rare event, the biological world would be very different than it is today. How

24.1 Hantaviruses in the New World The map shows the many different hantaviruses found in the Western Hemisphere. Those listed in red are known to be human pathogens.

did these millions of species form? How does one species become two? These questions are the focus of this chapter. We will examine the mechanisms by which a population splits into two or more new species, and we will see how such separations are maintained. We will look at the factors that can make speciation a rapid or a very slow process. Finally, we will look at the conditions that give rise to the great diversifications called evolutionary radiations.

What Are Species?

The word **species** means, literally, "kinds." But what do we mean by "kinds?" Someone who is knowledgeable about a group of organisms, such as orchids or lizards, usually can distinguish the different species found in a particular area simply by examining their visible features. Standard field guides to birds, mammals, insects, and flowers are possible only because most species are cohesive units that change little in appearance over large geographic distances. We can easily recognize male red-winged blackbirds from New York and from California, for example, as members of the same species (Figure 24.2*a*).

But not all members of a species look that much alike. For example, males, females, and young individuals may not resemble one another closely (Figure 24.2*b*). How do we decide whether similar but easily distinguished individuals should be assigned to different species or regarded as members of the same species?

The concept that has guided these decisions for a long time is genetic integration. If individuals of a population mate with one another, but not with individuals of other populations, they constitute a distinct group within which genes recombine; that is, they are independent evolutionary units. These independent evolutionary units are usually called species.

More than 200 years ago, the Swedish biologist Carolus Linnaeus, who originated the system of naming organisms that we use today, described hundreds of species. Because he knew nothing about the mating patterns of the organisms he was naming, Linnaeus classified them on the basis of their appearance; that is, he used a *morphological species* concept. Many of the organisms that he classified as species by their appearance are indeed independent evolutionary units. Their members look alike because they share many of the alleles that code for their body structures. In many groups of organisms for which genetic data are unavailable, species are still recognized by their morphological traits.

In 1940, Ernst Mayr proposed a definition of species that has been used by many biologists since that time. His definition, known as the *biological species concept*, says, "Species are groups of actually or potentially interbreeding natural populations which are reproductively isolated from other such groups." The words "actually or potentially" assert that, even if some members of a species live in different places and hence are unable to mate, they should not be placed in separate species if they would be likely to mate if they were together. The word "natural" is also an important part of Mayr's definition because only in nature does the exchange of genes among individuals from different populations influence evolutionary processes. The interbreeding of such individuals in captivity does not, because the resulting offspring typically spend their lives in captivity without interacting with the wild population. Since genetic integration through interbreeding maintains integrated evolutionary units, the biological species concept, although it does not apply to organisms that reproduce asexually, continues to be used by most evolutionary biologists.

Deciding whether two populations constitute different species may be difficult because speciation is often a gradual process (Figure 24.3). Once a population becomes separated into two or more populations, the daughter populations may evolve independently for a long time before they become re-

(*a*)

Agelaius phoeniceus (male, NY)

Agelaius phoeniceus (male, CA)

(*b*)

Agelaius phoeniceus (female)

24.2 Members of the Same Species Look Alike—or Not (*a*) Both of these male red-winged blackbirds are obviously members of the same species, even though one is from the eastern United States and the other is from California. (*b*) Because red-winged blackbirds are sexually dimorphic (see Chapter 23), the female of the species appears quite different from the male.

productively incompatible. Alternatively, they may become reproductively incompatible before they evolve any noticeable morphological differences. But how do these differences between populations come about?

How Do New Species Arise?

Speciation is the process by which one species splits into two or more daughter species, which thereafter evolve as distinct lineages. Although Charles Darwin titled his book *The Origin of Species*, he did not extensively discuss speciation, a process he called "the mystery of mysteries." He devoted most of his attention to demonstrating that species are

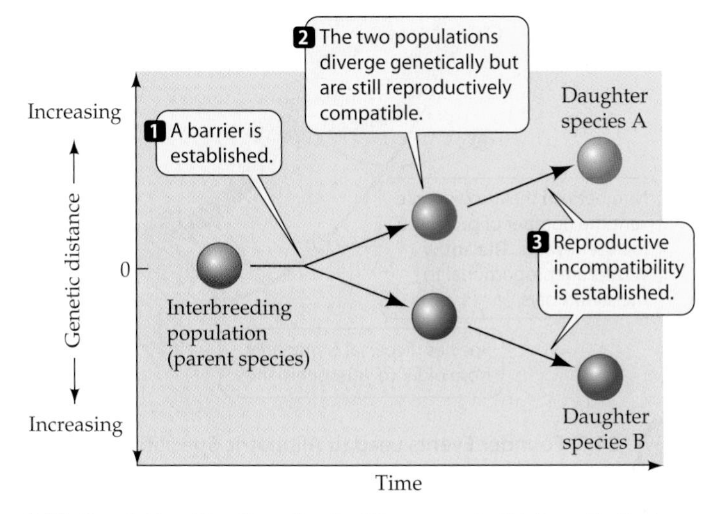

24.3 Speciation May Be a Gradual Process In this hypothetical example, genetic divergence between two separated populations begins before reproductive incompatibility evolves.

altered by natural selection over time. But not all evolutionary changes result in new species: A single lineage may change over time without giving rise to a new species.

The critical event in speciation is the separation of the gene pool of the ancestral species into two or more separate and isolated gene pools. Subsequently, within each isolated gene pool, allele and genotype frequencies may change as a result of the action of evolutionary agents. If two populations are isolated from each other, and sufficient differences in their genetic structure accumulate during the period of isolation, then the two populations may not be able to exchange genes when they come together again. As we will see, the amount of genetic difference that is needed to prevent gene exchange is highly variable. Gene flow among populations may be interrupted in two major ways, each of which characterizes a mode of speciation.

Allopatric speciation requires total genetic isolation

Speciation that results when a population is divided by a physical barrier is known as **allopatric** (*allo-*, "different"; *patris*, "country") or **geographic speciation** (Figure 24.4). Allopatric speciation is thought to be the dominant mode of speciation among most groups of organisms. The physical barrier that divides the range of a species may be a water body for terrestrial organisms, dry land for aquatic organisms, or a mountain range. Barriers can form when continents drift, sea levels rise, glaciers advance and retreat, and climates change. These processes continue to generate physical barriers today. The populations separated by such barriers are often large initially. They evolve differences for a variety of reasons, especially because the environments in which they live are, or become, different.

Allopatric speciation may also result when some members of a population cross an existing barrier and found a new, isolated population. Populations established in this way usually differ genetically from their parent populations because of the *founder effect*: A small group of founding individuals has

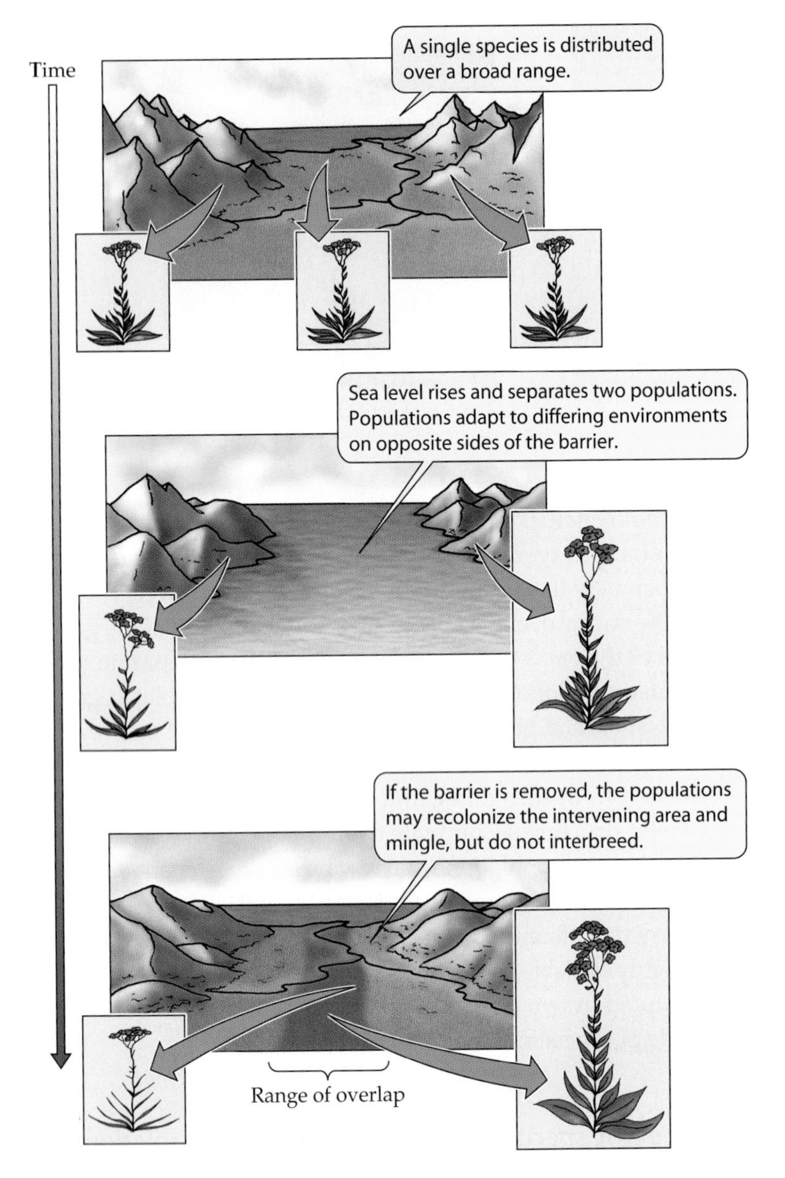

Time
A single species is distributed over a broad range.
Sea level rises and separates two populations. Populations adapt to differing environments on opposite sides of the barrier.
If the barrier is removed, the populations may recolonize the intervening area and mingle, but do not interbreed.
Range of overlap

24.4 Allopatric Speciation Allopatric speciation may result when a population is divided into two separate populations by a physical barrier, such as rising sea levels.

only an incomplete representation of the gene pool of its parent population (see Chapter 23). For example, many of the more than 800 species of the fruit fly genus *Drosophila* in the Hawaiian Islands are restricted to a single island. These species are almost certainly the descendants of new populations founded by individuals dispersing among the islands, because the closest relative of a species on one island is often a species on a neighboring island rather than a species on the same island. Biologists who have studied the chromosomes of picture-winged species of *Drosophila* believe that speciation among this group of flies has resulted from at least 45 such *founder events* (Figure 24.5).

Another example of allopatric speciation is found in the finches of the Galápagos archipelago, 1,000 km off the coast of Ecuador. Darwin's finches (as they are usually called, because Darwin was the first scientist to study them) arose in the Galápagos by speciation from a single South American species that colonized the islands. Today there are 14 species of Darwin's finches, all of which differ strikingly from their closest mainland relative (Figure 24.6). The islands of the Galápagos archipelago are sufficiently isolated from one another that finches seldom disperse between them. Also, environmental conditions differ among the islands. Some are relatively flat and arid; others have forested mountain slopes. Populations of finches on different islands have differentiated enough over millions of years that when occasional immigrants arrive from other islands, they either do not breed with the residents, or, if they do, the resulting offspring usually do not survive as well as those produced by pairs composed of island residents. The genetic distinctness and cohesiveness of the different species is thus maintained.

A physical barrier's effectiveness at preventing gene flow depends on the size and mobility of the species in question. What is an impenetrable barrier to a terrestrial snail may be no barrier at all to a butterfly or a bird. Popu-

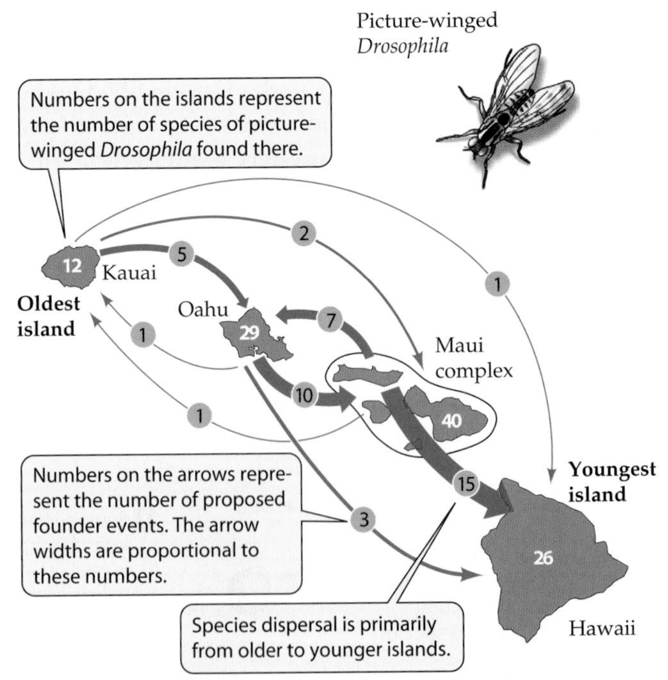

24.5 Founder Events Lead to Allopatric Speciation The large number of species of picture-winged *Drosophila* in the Hawaiian Islands is the result of founder events: new populations founded by individuals dispersing among the islands. The islands, which were formed in sequence as Earth's crust moved over a volcanic "hot spot," vary in age.

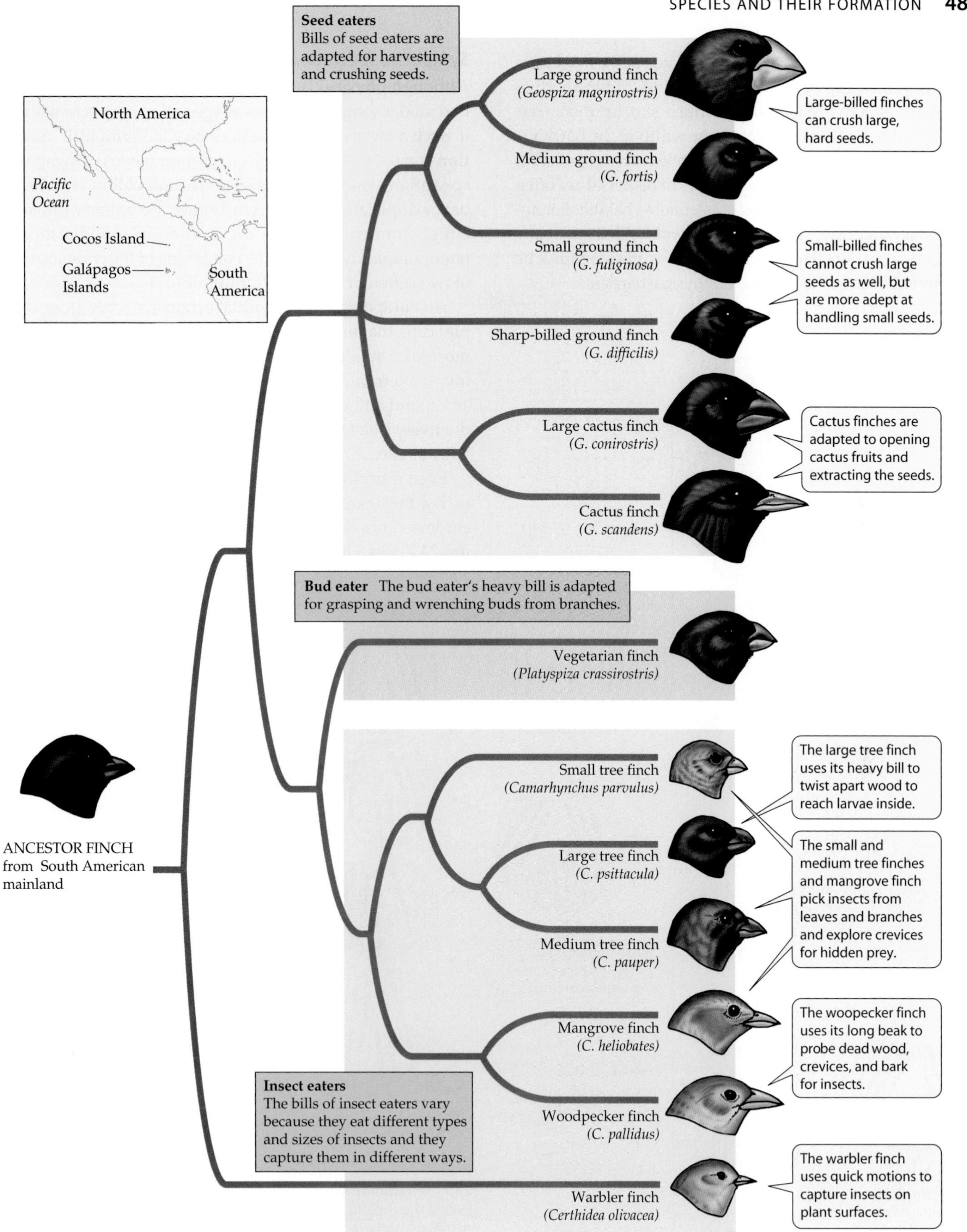

24.6 Allopatric Speciation among Darwin's Finches The descendants of the ancestral finch that colonized the Galápagos archipelago several million years ago evolved into 14 different species whose members are variously adapted to feed on seeds, buds, and insects. (The fourteenth species, not pictured here, lives in Cocos Island, farther north in the Pacific Ocean.)

lations of wind-pollinated plants are isolated at the maximum distance pollen can be blown by the wind, but individual plants are effectively isolated at much shorter distances. Among animal-pollinated plants, the width of the barrier is the distance that animals can travel while carrying pollen or seeds. Even animals with great powers of dispersal are often reluctant to cross narrow strips of unsuitable habitat. For animals that cannot swim or fly, narrow water-filled gaps may be effective barriers. However, gene flow can sometimes be interrupted even in the absence of physical barriers.

Sympatric speciation occurs without physical barriers

Although physical isolation is usually required for speciation, under some circumstances speciation can occur without it. Such a partition of a gene pool is called **sympatric speciation** (*sym-*, "with"). The most common means of sympatric speciation is **polyploidy**, the production within an individual of duplicate sets of chromosomes. Polyploidy can arise either from chromosome duplication in a single species (**autopolyploidy**) or from the combining of the chromosomes of two different species (**allopolyploidy**).

An autopolyploid individual originates when (for example) cells that are normally diploid (with two sets of chromosomes) accidentally duplicate their chromosomes, resulting in a tetraploid (four sets of chromosomes) individual. Tetraploid and diploid plants of the same species are reproductively isolated because their triploid offspring are essentially sterile.

Even if triploid individuals survive to reproductive maturity, they cannot produce viable gametes because their chromosomes do not synapse correctly during meiosis (Figure 24.7). So a tetraploid plant cannot produce viable off-

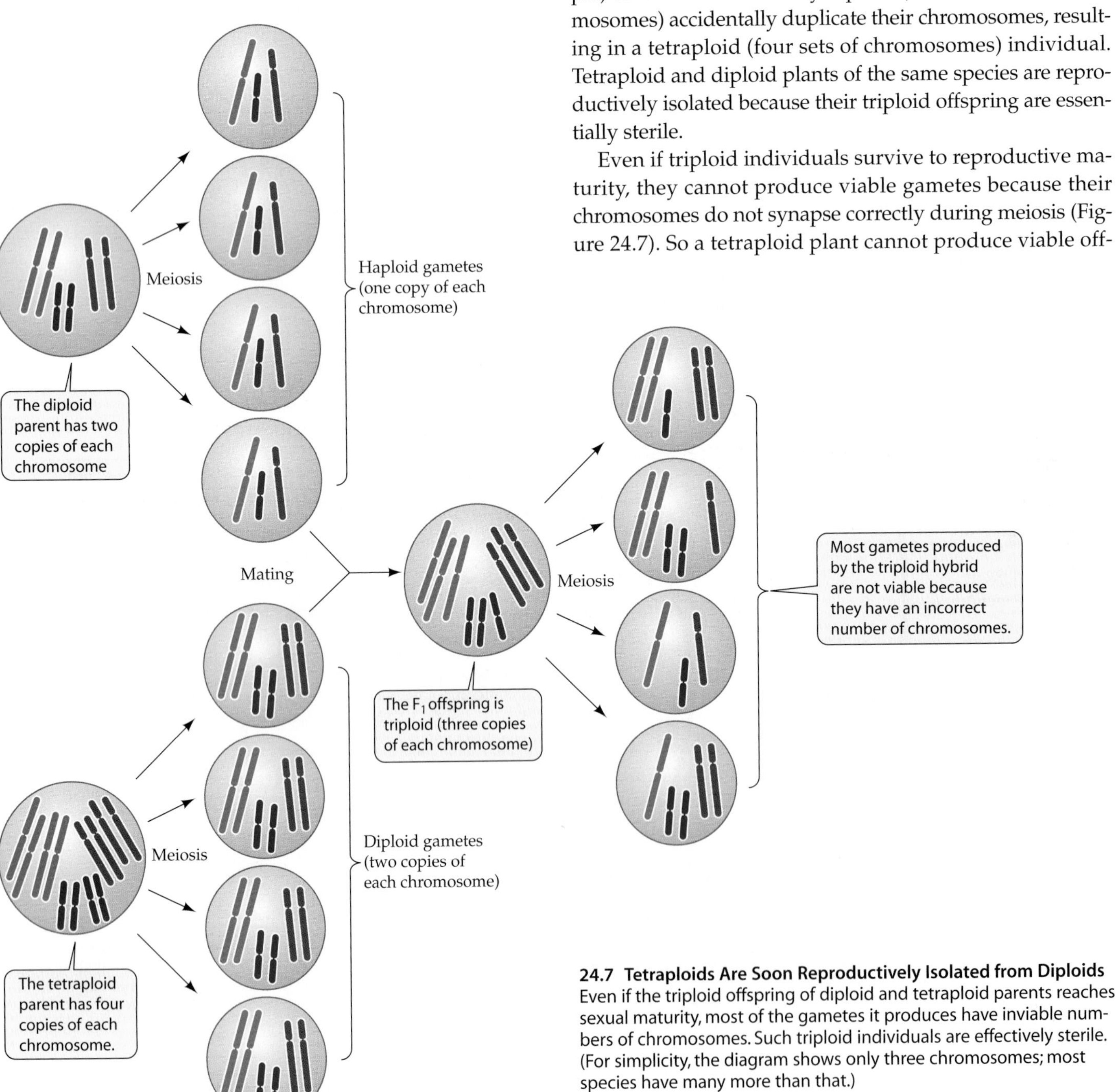
Meiosis
Haploid gametes (one copy of each chromosome)
The diploid parent has two copies of each chromosome
Mating
Meiosis
Most gametes produced by the triploid hybrid are not viable because they have an incorrect number of chromosomes.
The F_1 offspring is triploid (three copies of each chromosome)
Meiosis
Diploid gametes (two copies of each chromosome)
The tetraploid parent has four copies of each chromosome.

24.7 Tetraploids Are Soon Reproductively Isolated from Diploids Even if the triploid offspring of diploid and tetraploid parents reaches sexual maturity, most of the gametes it produces have inviable numbers of chromosomes. Such triploid individuals are effectively sterile. (For simplicity, the diagram shows only three chromosomes; most species have many more than that.)

spring by mating with a diploid individual—but it *can* do so if it self-fertilizes or mates with another tetraploid.

Allopolyploids may be produced when individuals of two different (but closely related) species interbreed, or **hybridize**. Allopolyploids are usually fertile because each of the chromosomes has a nearly identical partner with which to pair during meiosis.

New species arise by means of polyploidy much more easily among plants than among animals because plants of many species can reproduce by self-fertilization. In addition, if polyploidy arises in several offspring of a single parent, the siblings can fertilize one another. Speciation by polyploidy has been very important in the evolution of flowering plants. Botanists estimate that about 70 percent of flowering plant species and 95 percent of fern species are polyploids. Most of these arose as a result of hybridization between two species, followed by self-fertilization.

How easily allopolyploidy can produce new species is illustrated by the salsifies (*Tragopogon*), members of the sunflower family. Salsifies are weedy plants that thrive in disturbed areas around towns. People have inadvertently spread them around the world from their ancestral ranges in Eurasia. Three diploid species of salsify were introduced into North America early in the twentieth century: *T. porrifolius*, *T. pratensis*, and *T. dubius*. Two tetraploid hybrids—*T. mirus* and *T. miscellus*—between the original three diploid species were first discovered in 1950. The hybrids have spread since their discovery and today are more widespread than their diploid parents (Figure 24.8).

Studies of their genetic material have shown that both salsify hybrids have formed more than once. Some populations of *T. miscellus*—a hybrid of *T. pratensis* and *T. dubius*— have the chloroplast genome of *T. pratensis;* other populations have the chloroplast genome of *T. dubius*. Such differences among local populations of *T. miscellus* show that this allopolyploid has formed independently at least 21 times! Scientists seldom know the dates and locations of species formation so well. *T. mirus*, a hybrid of *T. porrifolius* and *T. dubius*, has formed 12 times. *T. porrifolius* prefers wet, shady places; *T. dubius* prefers dry, sunny places. *T. mirus*, however, can grow in partly shaded environments where neither parent does well. The success of these newly formed hybrid species of salsifies illustrates why so many species of flowering plants originated as polyploids.

Polyploidy, as we have just seen, can result in a new species that is completely reproductively isolated from its parent species in one generation. Allopatric speciation proceeds much more slowly, and some populations separated by a physical barrier may never acquire full reproductive isolation. Let's see how reproductive isolation may become established once two populations have been separated from each other.

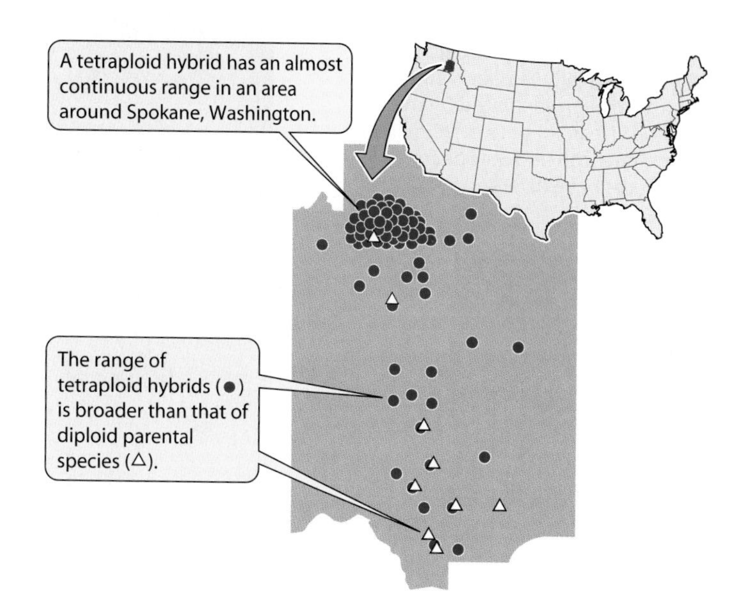
A tetraploid hybrid has an almost continuous range in an area around Spokane, Washington.
The range of tetraploid hybrids (●) is broader than that of diploid parental species (△).

24.8 Polyploids May Outperform Their Parent Species
Tragopogon species (salsifies) are members of the sunflower family. The map shows the distribution of the three diploid parent species and of the two tetraploid hybrid species of *Tragopogon* in eastern Washington and adjacent Idaho.

Completing Speciation: Reproductive Isolating Mechanisms

Once a barrier to gene flow is established, by whatever means, the separated populations may diverge genetically through the action of the evolutionary agents we described in Chapter 23. Over many generations, differences may accumulate that reduce the probability that members of the two populations could mate and produce viable offspring. In this way, reproductive isolation can evolve as an incidental byproduct of genetic changes in allopatric populations.

Geographic isolation does not necessarily lead to reproductive incompatibility. For example, American sycamores and European sycamores (also known as plane trees) have been physically isolated from one another for at least 20 million years. Nevertheless, they are morphologically very similar (Figure 24.9), and they can form fertile hybrids, even though they never have an opportunity to do so in nature.

In other cases, however, genes that result in reproductive isolation between two evolving lineages spread quickly through populations as they diverge. In this section, we will examine the ways in which reproductive isolating mechanisms may arise. In the following section, we will explore what happens when reproductive isolation is incomplete.

24.9 Geographically Separated, Morphologically Similar Although they have been separated by the Atlantic Ocean for at least 20 million years, American and European sycamores have diverged very little in appearance.

(*a*) *Platanus occidentalis* (American sycamore)

(*b*) *Platanus hispanica* (European sycamore)

Prezygotic barriers operate before fertilization

Several processes that operate before fertilization—**prezygotic reproductive barriers**—may prevent individuals of different species from interbreeding:

- *Spatial isolation*. Individuals of different species may select different places in the environment in which to live. As a result, they may never come into contact during their respective mating periods; that is, they are reproductively isolated by location.
- *Temporal isolation*. Many organisms have mating periods that are as short as a few hours or days. If the mating periods of two species do not overlap, they will be reproductively isolated by time.
- *Mechanical isolation*. Differences in the sizes and shapes of reproductive organs may prevent the union of gametes from different species.
- *Gametic isolation*. Sperm of one species may not attach to the eggs of another species because the eggs do not release the appropriate attractive chemicals, or the sperm may be unable to penetrate the egg because the two gametes are chemically incompatible.
- *Behavioral isolation*. Individuals of a species may reject, or fail to recognize, individuals of other species as mating partners.

Two closely related species of crickets from the island of Hawaii, *Laupala paranga* and *Laupala kohalensis*, provide an example of behavioral isolation. These two species live in separate areas and do not hybridize in nature, but they will form interspecific pairs in the laboratory. The males of the two species produce genetically determined songs that differ in the number of pulses per second. The songs of hybrid males have intermediate numbers of pulses (Figure 24.10). Females are much more strongly attracted to the songs of conspecific males (males of their own species) than they are to the songs of males of the other species. We know that this preference is genetically determined because hybrid females are most strongly attracted to the songs of hybrid males.

Sometimes the mate choice of one species is mediated by the behavior of individuals of other species. For example, whether two plant species hybridize may depend on the preferences of their pollinators. Because pollinators visit flowers to gather nutritional rewards, not to pollinate plants, pollinator behavior can be influenced only by changes in floral structures that affect the rewards the pollinators receive. Floral traits can affect reproductive isolation either by influencing pollinator behavior or by altering where pollen is deposited on the bodies of pollinators.

The evolution of floral traits that generate reproductive isolation has been studied in columbines of the genus *Aquilegia*. Columbines have undergone recent and very rapid spe-

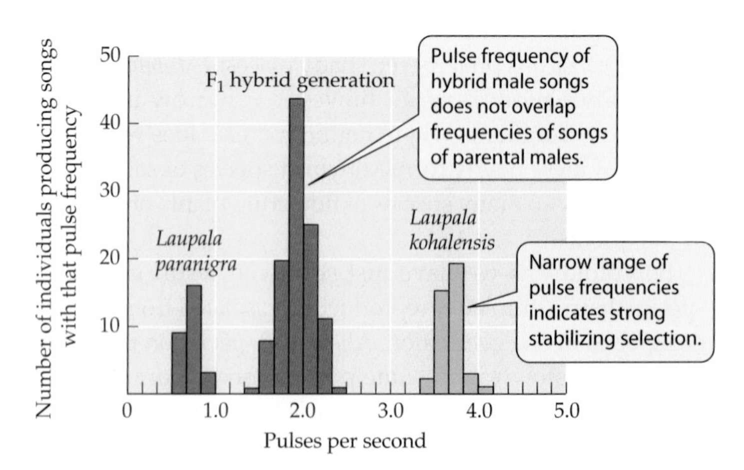

24.10 Songs of Male Crickets are Genetically Determined Hybrid males produce songs with intermediate pulse frequencies.

ciation and, at the same time, have evolved long floral nectar spurs—tubular outgrowths of petals that produce nectar at their tips. Animals pollinate these flowers while probing the spurs to collect nectar. The length of the spurs and the orientation of the flowers influence how efficiently pollinators can extract nectar. Two species, *Aquilegia formosa* and *Aquilegia pubescens*, that grow in the mountains of California can produce fertile hybrids. *A. formosa* has pendant (hanging) flowers and short spurs (Figure 24.11*a*); it is pollinated by hummingbirds. *A. pubescens* has upright flowers and long spurs (Figure 24.11*b*); it is pollinated by hawkmoths.

Investigators tested discrimination among these flowers by hawkmoths by turning *A. formosa* flowers so that they were upright. Hawkmoths still visited mostly *A. pubescens* flowers (Figure 24.11*c*), probably because the flowers of the two species differ strongly in the color of light they reflect. Genetic analyses of hybrids between the two species show that the color differences are caused by a small number of genes. Thus, although these two species are interfertile, hybrids rarely form in nature because the two species attract different pollinators.

24.11 Hawkmoths Favor Flowers of One Columbine Species Flowers of *Aquilegia formosa* are normally pendant (*a*), while those of *A. pubescens* are normally upright (*b*). The plants can interbreed, but the hawkmoths that pollinate *A. pubescens* distinguish between flowers of the two species, even when *A. formosa* flowers are experimentally modified to be upright (*c*).

Postzygotic barriers operate after fertilization

If individuals of two different populations overcome prezygotic reproductive barriers and interbreed, **postzygotic reproductive barriers** may still prevent gene exchange. Genetic differences that accumulated while the populations were isolated from each other may reduce the survival and reproduction of the hybrid offspring in any of several ways:

- *Hybrid zygote abnormality*. Hybrid zygotes may fail to mature normally, either dying during development or developing such severe abnormalities that they cannot mate as adults.
- *Hybrid infertility*. Hybrids may mature normally, but be infertile when they attempt to reproduce. For example, the offspring of matings between horses and donkeys—mules—are strong, but sterile; they produce no descendants.
- *Low hybrid viability*. Hybrid offspring may simply survive less well than offspring resulting from matings within populations.

If hybrid offspring survive poorly, more effective prezygotic barriers may evolve, because individuals that mate with individuals of the other population will leave fewer surviving offspring than individuals that mate only within their own population. This strengthening of prezygotic barriers is known as **reinforcement**. Reinforcement is difficult to detect experimentally, but it can be detected by using the comparative method (see Chapter 1). If reinforcement is occurring, then sympatric pairs of species should evolve prezygotic reproductive barriers more rapidly than allopatric pairs of species do. In a study of related sympatric and allopatric species of *Drosophila*, this was shown to be the case (Figure 24.12).

We can observe speciation in progress

Since speciation is a gradual process, we can find many examples of populations at different stages on the way to complete reproductive isolation. A good example of speciation in progress is found in a picture-winged fruit fly (*Rhagoletis pomenella*) in New York State. Until the mid-1800s, these fruit flies courted, mated, and deposited their eggs only on hawthorn fruits. The larvae recognized the odor of hawthorn

COMPARATIVE METHOD

Hypothesis: If reinforcement has occurred, sympatric species of *Drosophila* should be more strongly isolated prezygotically than allopatric species that have been separated for equal amounts of time.

METHOD Assess length of time *Drosophila* populations have been evolving separately by the amount of genetic distance between them. Compare strength of prezygotic isolation between allopatric and sympatric species pairs.

RESULTS Recently diverged pairs of sympatric species have more prezygotic isolation than allopatric species do.

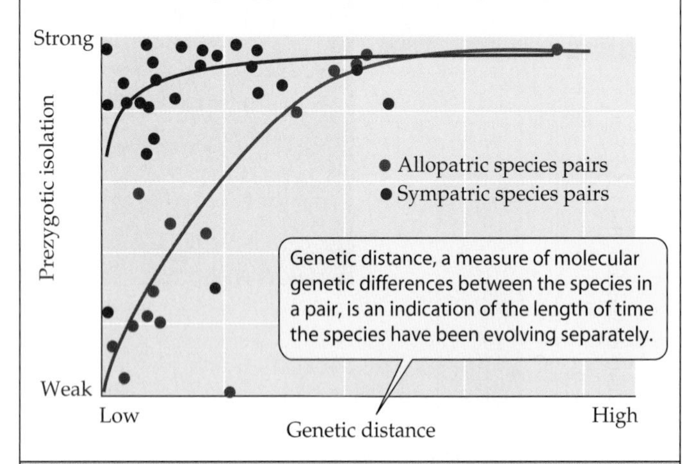
Strong
Weak
Prezygotic isolation
Low
High
Genetic distance
Allopatric species pairs
Sympatric species pairs
Genetic distance, a measure of molecular genetic differences between the species in a pair, is an indication of the length of time the species have been evolving separately.

Conclusion: Reinforcement has resulted in particularly rapid evolution of prezygotic isolation among recently diverged sympatric species of *Drosophila.*

24.12 Prezygotic Barriers Can Evolve Rapidly Reinforcement may occur when individuals who mate outside their own population leave few or no viable offspring. In such a situation, prezygotic isolating mechanisms can evolve extremely rapidly.

as they fed on the fruits, and when they emerged from their pupae, they used this cue to locate other hawthorn plants on which to mate and lay their eggs.

About 150 years ago, large commercial apple orchards were planted in the Hudson River valley. Apple trees are closely related to hawthorns, and a few female *Rhagoletis* laid their eggs on apples, perhaps by mistake. Their larvae did not grow as well as the larvae on hawthorn fruits, but many did survive. These larvae recognized the odor of apples, so when they emerged as adults, they sought out apple trees, where they mated with other flies reared on apples.

Today there are two types of *Rhagoletis* in the Hudson River valley that may be on the way to becoming distinct species. One feeds primarily on hawthorn fruits, the other on apples. The two incipient species are partly reproductively isolated because they mate primarily with individuals raised on the same fruit and because they emerge from their pupae at different times of the year. In addition, the apple-feeding flies have evolved so that they now grow more rapidly on apples than they originally did.

Reproductive isolation does not develop at the same rate in all diverging populations. On the one hand, in some groups, such as Darwin's finches, there has been conspicuous morphological evolution, but many of the 14 species still interbreed and produce fertile hybrid offspring. On the other hand, reproductive isolation may develop rapidly, as it has between diverging sympatric species of *Drosophila*. Generally, reproductive isolation evolves more rapidly in species that have rapid reproductive rates (for example, most plants, fruit flies, and sea urchins) than in species that have slower reproductive rates (for example, birds and mammals).

Many species in nature form hybrids in areas where their ranges overlap, and they may continue to do so for many years. Let's examine what happens in these situations.

Hybrid Zones: Incomplete Reproductive Isolation

If contact is reestablished between formerly isolated populations before complete reproductive isolation has developed, members of the two populations may interbreed. Three outcomes of such interbreeding are possible:

- If hybrid offspring are as successful as those resulting from matings within each population, hybrids may spread through both populations and reproduce with other individuals. The gene pools would then combine, and no new species would result from the period of isolation.
- If hybrid offspring are less successful, complete reproductive isolation may evolve as reinforcement strengthens prezygotic reproductive barriers.
- Even if hybrid offspring are at some disadvantage, a narrow **hybrid zone** may persist if, for one or more reasons, reinforcement does not happen.

Hybrid zones are excellent natural laboratories for the study of speciation. When a hybrid zone first forms, most hybrids are offspring of crosses between purebred individuals of the two species. However, subsequent generations include a variety of individuals with different proportions of their genes derived from the original two populations. Thus, hybrid zones contain recombinant individuals resulting from many generations of hybridization. Detailed genetic studies can tell us much about why hybrid zones may be narrow and stable for long periods of time.

European toads of the genus *Bombina* have been the subject of such studies. The fire-bellied toad (*B. bombina*) lives in eastern Europe. The closely related yellow-bellied toad (*B. variegata*) lives in western and southern Europe. The ranges of the two species meet in a narrow zone stretching 4,800 km

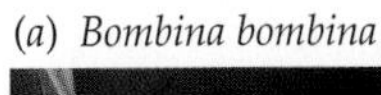

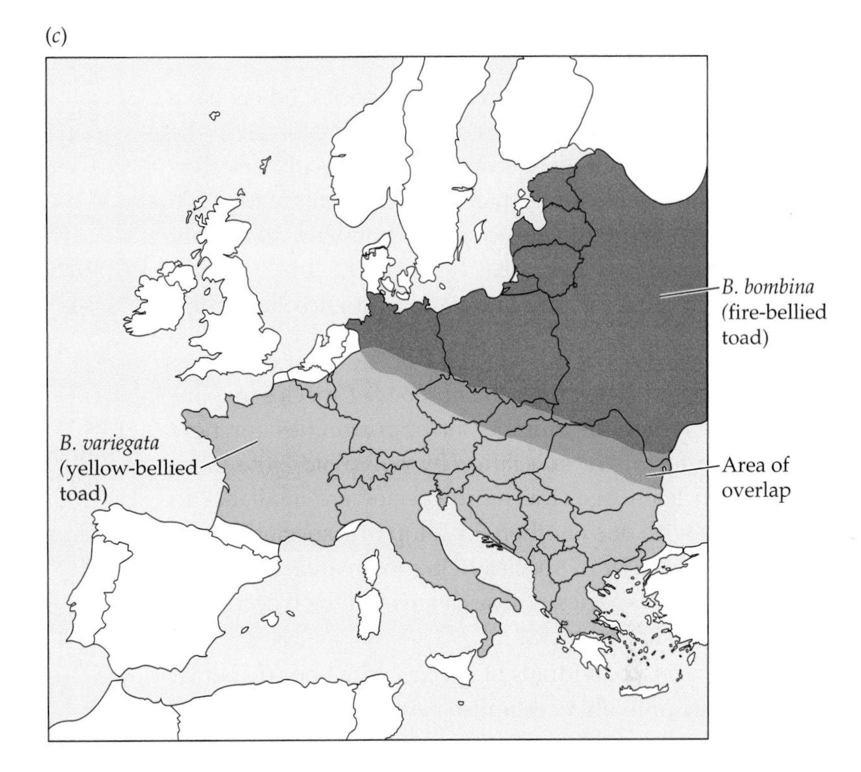

24.13 Hybrid Zones May Be Long and Narrow The narrow zone in Europe where fire-bellied toads (*a*) meet and hybridize with yellow-bellied toads (*b*) stretches across Europe (*c*). This hybrid zone has been stable for hundreds of years, but has never expanded because hybrid toads are much less fit than individuals of the parental species.

from eastern Germany to the Black Sea (Figure 24.13). Hybrids between the two species suffer from a range of defects, many of which are lethal. Those that survive often have skeletal abnormalities, such as misshapen mouths, ribs that are fused to vertebrae, and a reduced number of vertebrae.

By following the fates of thousands of toads from the hybrid zone, investigators have found that a hybrid toad is half as fit as a purebred individual. The hybrid zone is narrow because there is strong selection against hybrids, and because adult toads do not move over long distances. It has persisted for hundreds of years because individuals that move into it have not previously encountered individuals of the other species, so there has been no opportunity for reinforcement to occur.

If two species hybridize, we know that they must be similar genetically, but the absence of interbreeding tells us nothing about how dissimilar two species are. Not until modern molecular genetic techniques were developed could biologists measure genetic differences among species. These techniques show that the genetic differences that separate species are primarily differences among genes involved with reproductive isolation. The extensive data gathered on *Drosophila* indicate that fewer than ten, and often fewer than five, genes are responsible for reproductive isolation. Individuals of different species of Hawaiian *Drosophila* share nearly all of their mitochondrial DNA alleles. All of the hundreds of species of *Drosophila* that have evolved in the Hawaiian Islands during the past 32 million years, even those that have diverged morphologically, are relatively similar genetically (Figure 24.14).

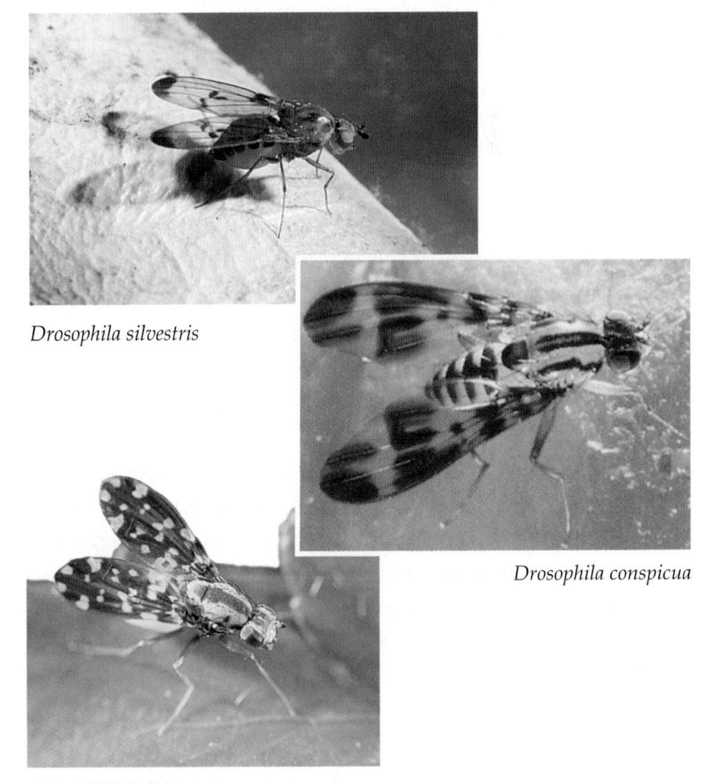

24.14 Morphologically Different, Genetically Similar Although these fruit flies—a small sample of the hundreds of species found only on the Hawaiian Islands—vary greatly in appearance, they are genetically similar.

Variation in Speciation Rates

Some lineages of organisms have many species; others have only a few. Hundreds of species of *Drosophila* evolved in the Hawaiian Islands, but there is only one species of horseshoe crab, even though its lineage has survived for more than 300 million years. Why do rates of speciation vary so widely among lineages?

A number of factors are known to influence speciation rates:

- *Species richness.* The larger the number of species in a lineage, the larger the number of opportunities for new species to form. For speciation by polyploidy, the more species in a lineage, the more species are available to hybridize with one another. For allopatric speciation, the larger the number of species living in an area, the larger the number of species whose ranges will be bisected by a given physical barrier.
- *Dispersal rates.* Individuals of species with poor dispersal abilities are unlikely to establish new populations by dispersing across barriers. Even narrow barriers are effective in dividing species whose members are highly sedentary.
- *Ecological specialization.* Populations of species restricted to habitat types that are patchy in distribution are more likely to diverge than are populations that occupy relatively continuous habitats.
- *Population bottlenecks.* The changes in gene pools that often occur when a population passes through a bottleneck may result in new adaptations.
- *Type of pollination.* Speciation rates in plants are correlated with pollination mode. Animal-pollinated plant families have, on average, 2.4 times as many species as closely related families pollinated by wind. In addition, the switch from animal to wind pollination is strongly associated with a reduction in the rate of speciation in a lineage.
- *Sexual selection.* Animals with complex behavior are likely to form new species at a high rate because they make sophisticated discriminations among potential mating partners. They distinguish members of their own species from members of other species, and they make subtle discriminations among members of their own species on the basis of size, shape, appearance, and behavior (see Figures 23.16 and 23.17). Such discriminations can greatly influence which individuals are most successful in producing offspring and may lead to rapid reinforcement of reproductive isolation between species.
- *Environmental changes.* Oscillations of climates may fragment populations of species that live in formerly continuous habitats.

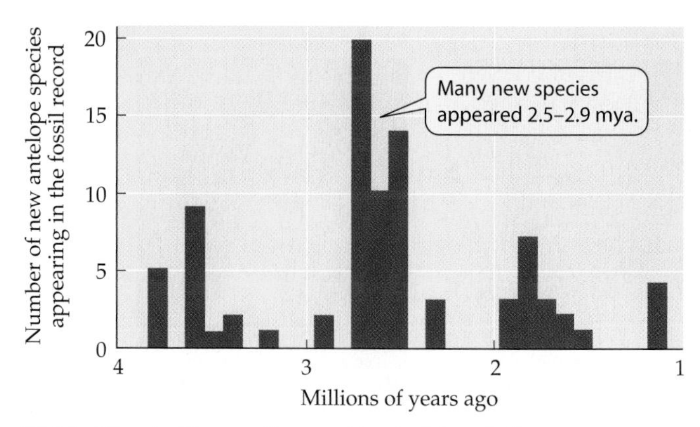

24.15 Climate Change Drove a Burst of Speciation among Antelopes The excellent fossil record of African antelopes reveals that there was a sudden burst of speciation between 2.5 and 2.9 million years ago. At that time, the climate of Africa shifted from being consistently warm and wet to oscillating between warm and wet and cool and dry.

African antelopes provide a striking example of the influence of climate change on speciation rates. These animals experienced a burst of speciation and extinction between 2.5 and 2.9 mya. During that period, the number of known antelope species doubled, and 90 percent of all known species either first appeared or became extinct (Figure 24.15). This burst coincided with a shift in Africa from a warm, wet climate to one that oscillated between warm and wet and cooler but drier conditions. During this period, grassland and savanna environments increased and decreased, repeatedly coalescing and separating over much of Africa as the climate oscillated. The burst of speciation among antelopes resulted in many new species adapted to these environments.

Evolutionary Radiations

The fossil record reveals that, at certain times in certain lineages, speciation rates have been much higher than extinction rates. The result is the proliferation of a large number of daughter species, as happened with finches in the Galápagos archipelago (see Figure 24.6). Such an event is called an **evolutionary radiation**. What conditions cause speciation rates to be much higher than extinction rates?

Evolutionary radiations are likely when a population colonizes a new environment that contains relatively few species. As we saw in Chapter 22, evolutionary radiations occurred in many lineages on continents following mass extinctions. Evolutionary radiations occur on islands because islands lack many plant and animal groups found on the mainland. The ecological opportunities that exist on islands may stimulate rapid evolutionary changes when a new species does reach them. Water barriers also restrict gene flow among the islands in an archipelago, so populations on

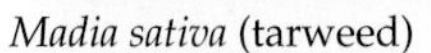

Madia sativa (tarweed)

Argyroxiphium sandwicense

Dubautia menziesii

Wilkesia hobdyi

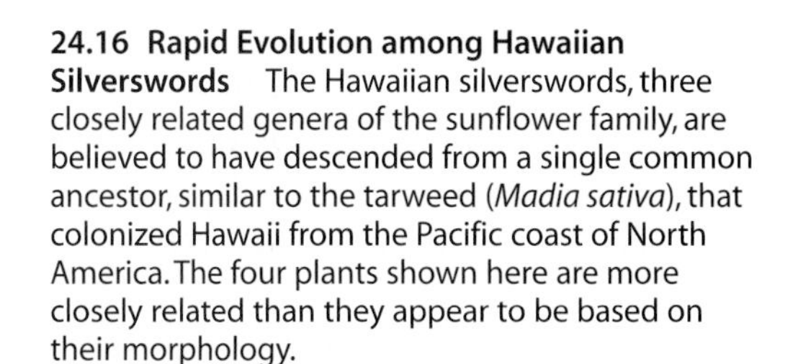

24.16 Rapid Evolution among Hawaiian Silverswords The Hawaiian silverswords, three closely related genera of the sunflower family, are believed to have descended from a single common ancestor, similar to the tarweed (*Madia sativa*), that colonized Hawaii from the Pacific coast of North America. The four plants shown here are more closely related than they appear to be based on their morphology.

different islands may evolve adaptations to their local environments. Together, these two factors make it likely that speciation rates of newly colonizing lineages on island archipelagoes will exceed extinction rates.

Remarkable evolutionary radiations have occurred in the Hawaiian Islands, the most isolated islands in the world. The Hawaiian Islands lie 4,000 km from the nearest major land mass and 1,600 km from the nearest group of islands. The native biota of the Hawaiian Islands includes 1,000 species of flowering plants, 10,000 species of insects, 1,000 land snails, and more than 100 bird species. However, there were no amphibians, no terrestrial reptiles, and only one native mammal—a bat—on the islands until humans introduced additional species. The 10,000 known native species of insects on Hawaii are believed to have evolved from only about 400 immigrant species; only 7 immigrant species are believed to account for all the native Hawaiian land birds.

More than 90 percent of all plant species on the Hawaiian Islands are **endemic**—that is, they are found nowhere else. Several groups of flowering plants have more diverse forms and life histories on the islands, and live in a wider variety of habitats, than do their close relatives on the mainland. An outstanding example is the group of Hawaiian sunflowers called silverswords (the genera *Argyroxiphium*, *Dubautia*, and *Wilkesia*). Chloroplast DNA sequences show that these species share a relatively recent common ancestor with a species of tarweed from the Pacific coast of North America (Figure 24.16). Whereas all mainland tarweeds are small, upright, herbs (that is, nonwoody plants), the silverswords include prostrate and upright herbs, shrubs, trees, and vines. Silverword species occupy nearly all the habitats of the Hawaiian islands, from sea level to above timberline in the mountains. Despite their extraordinary morphological diversification, however, the silverswords have differentiated very little in their chloroplast genes.

The island silverswords are more diverse in size and shape than the mainland tarweeds because the original colonizers arrived on islands that had very few plant species. In particular, there were few trees and shrubs, because such large-seeded plants rarely disperse to oceanic islands. In fact, many island trees and shrubs have evolved from nonwoody ancestors. On the mainland, however, tarweeds live in ecological communities that contain tree and shrub lineages older than their own—that is, where opportunities to exploit the "tree" way of life have already been preempted.

The processes we have discussed in this chapter, operating over billions of years, have produced a world in which life is organized into millions of species, each adapted to live in a particular environment and to use environmental resources in a particular way. How these millions of species are distributed over the surface of Earth and organized into ecological communities will be a major focus of Part Eight of this book.

Chapter Summary

What Are Species?

▶ Species are independent evolutionary units. A commonly accepted definition of species is "groups of actually or potentially interbreeding natural populations which are reproductively isolated from other such groups."

▶ Because speciation is often a gradual process, it may be difficult to recognize boundaries between species. **Review Figure 24.3**

How Do New Species Arise?

▶ Not all evolutionary changes result in new species.

▶ Allopatric (geographic) speciation is the most important mode of speciation among animals and is common in other groups of organisms. **Review Figures 24.4, 24.5, 24.6.**
See Web/CD Tutorial 24.2

▶ Sympatric speciation may occur rapidly by polyploidy because polyploid offspring are sterile in crosses with members of the parent species. Polyploidy is a major factor in plant speciation but is rare among animals. **Review Figures 24.7, 24.8**
See Web/CD Tutorial 24.1

Completing Speciation: Reproductive Isolating Mechanisms

▶ Once two populations have been separated, reproductive isolating mechanisms may prevent the exchange of genes between them.

▶ Prezygotic reproductive barriers operate before fertilization. Some prezygotic barriers affect mate choice; others work by influencing pollinator behavior. **Review Figure 24.10, 24.11**

▶ Postzygotic reproductive barriers operate after fertilization by reducing the survival or fertility of hybrid offspring.

▶ If hybrid offspring survive poorly, more effective prezygotic reproductive barriers may evolve. This process is known as reinforcement. **Review Figure 24.12**

Hybrid Zones: Incomplete Reproductive Isolation

▶ Hybrid zones may develop if barriers to gene exchange fail to develop while diverging species are isolated from each other. **Review Figure 24.13**

▶ Species may differ from one another in very few genes.

Variation in Speciation Rates

▶ Rates of speciation differ greatly among lineages. Speciation rates are influenced by the number of species in a lineage, their dispersal rates, ecological specialization, experience of population bottlenecks, pollinators, and behavior, as well as by climatic changes.

Evolutionary Radiations

▶ Evolutionary radiations occur when speciation rates exceed extinction rates.

▶ High speciation rates often coincide with low extinction rates when species invade islands or other environments that contain few other species.

▶ As a result of speciation, Earth is populated with millions of species, each adapted to live in a particular environment and to use resources in a particular way.

See Web/CD Activity 24.1 for a concept review of this chapter.

Self-Quiz

1. A species is a group of
   a. actually interbreeding natural populations that are reproductively isolated from other such groups.
   b. potentially interbreeding natural populations that are reproductively isolated from other such groups.
   c. actually or potentially interbreeding natural populations that are reproductively isolated from other such groups.
   d. actually or potentially interbreeding natural populations that are reproductively connected to other such groups.
   e. actually interbreeding natural populations that are reproductively connected to other such groups.
2. Allopatric speciation may happen when
   a. continents drift apart and separate previously connected lineages.
   b. a mountain range separates formerly connected populations.
   c. different environments on two sides of a barrier cause populations to diverge.
   d. the range of a species is separated by loss of intermediate habitat.
   e. all of the above
3. Finches speciated in the Galápagos Islands because
   a. the Galápagos Islands are not far from the mainland.
   b. the Galápagos Islands are arid.
   c. the Galápagos Islands are small.
   d. the islands of the Galápagos Archipelago are sufficiently isolated from one another that there is little migration among them.
   e. the islands of the Galápagos Archipelago are close enough to one another that there is considerable migration among them.
4. Which of the following is *not* a potential prezygotic reproductive barrier?
   a. Temporal segregation of breeding seasons
   b. Differences in chemicals that attract mates
   c. Hybrid infertility
   d. Spatial segregation of mating sites
   e. Sperm cannot survive in female reproductive tracts

5. A common means of sympatric speciation is
   a. polyploidy.
   b. hybrid infertility.
   c. temporal segregation of breeding seasons.
   d. spatial segregation of mating sites.
   e. imposition of a geographic barrier.
6. Sympatric species are often similar in appearance because
   a. appearances are often of little evolutionary significance.
   b. the genetic changes accompanying speciation are often small.
   c. the genetic changes accompanying speciation are usually large.
   d. speciation usually requires major reorganization of the genome.
   e. the traits that differ among species are not the same as the traits that differ among individuals within species.
7. Narrow hybrid zones may persist for long times because
   a. hybrids are always at a disadvantage.
   b. hybrids have an advantage only in narrow zones.
   c. hybrid individuals never move far from their birthplaces.
   d. individuals that move into the zone have not previously encountered individuals of the other species, so reinforcement of isolating mechanisms has not occurred.
   e. Narrow hybrid zones are artifacts because biologists generally restrict their studies to contact zones between species.
8. Which statement about speciation is *not* true?
   a. It always takes thousands of years.
   b. It often takes thousands of years, but may happen within a single generation.
   c. Among animals, it usually requires a physical barrier.
   d. Among plants, it often happens as a result of polyploidy.
   e. It has produced the millions of species living today.
9. Speciation is often rapid within lineages in which species have complex behavior because
   a. individuals of such species make fine discriminations among potential mating partners.
   b. such species have short generation times.
   c. such species have high reproductive rates.
   d. such species have complex relationships with their environments.
   e. none of the above
10. Evolutionary radiations
    a. often happen on continents, but rarely on island archipelagos.
    b. characterize birds and plants, but not other taxonomic groups.
    c. have happened on continents as well as on islands.
    d. require major reorganizations of the genome.
    e. never happen in species-poor environments.
11. Speciation is an important component of evolution because it
    a. generates the variation upon which natural selection acts.
    b. generates the variation upon which genetic drift and mutations act.
    c. enabled Charles Darwin to perceive the mechanisms of evolution.
    d. generates the high extinction rates that drive evolutionary change.
    e. has resulted in a world with millions of species, each adapted for a particular way of life.

For Discussion

1. The snow goose of North America has two distinct color forms: blue and white. Matings between the two color forms are common. However, blue individuals pair with blue individuals and white individuals pair with white individuals much more frequently than would be expected by chance. Suppose that 75 percent of all mated pairs consisted of two individuals of the same color. What would you conclude about speciation processes in these geese? If 95 percent of pairs were the same color? If 100 percent of pairs were the same color?
2. Suppose pairs of snow geese of mixed colors were found only in a narrow zone within the broad Arctic breeding range of the geese. Would your answer to Question 1 remain the same? Would your answer change if mixed-color pairs were widely distributed across the breeding range of the geese?
3. Although many butterfly species are divided into local populations among which there is little gene flow, these species often show relatively little morphological variation among populations. Describe the studies you would conduct to determine what maintains this morphological similarity.
4. Evolutionary radiations are common and easily studied on oceanic islands, but in what types of *mainland* situations would you expect to find major evolutionary radiations? Why?
5. Fruit flies of the genus *Drosophila* are distributed worldwide, but most of the species in the genus are found on the Hawaiian Islands. What might account for this distribution pattern?
6. Evolutionary radiations take place when speciation rates exceed extinction rates. What factors can cause extinction rates to exceed speciation rates in a lineage? Name some lineages in which human activities are increasing extinction rates without increasing speciation rates.

25 Reconstructing and Using Phylogenies

Schistosomiasis is a blood infection caused by a parasitic trematode flatworm, *Schistosoma*. More than 200 million people in South America, Africa, China, Japan, and Southeast Asia have this disease. During part of its life cycle, *Schistosoma* inhabits a freshwater snail; people become infected when they come into contact with water where infected snails live. Larval *Schistosoma* swim from a snail through the water and penetrate human skin. The flatworm matures and settles in the abdominal blood vessels. The disease is progressively debilitating, causing a slow death.

Until recently, only one trematode species, *Schistosoma japonicum*, was known to infect humans, and it was believed to be transmitted by a single species of snail in the genus *Oncomelania*. Then, in the 1970s, researchers discovered a different snail species that also transmitted *Schistosoma* to humans. This discovery stimulated anatomical, genetic, and geographic research on the trematode flatworms and snails of Asia.

Investigators found that *S. japonicum* was actually a cluster of at least six different species. They also discovered that the snails that host the various *Schistosoma* parasites are closely related to one another. Of the thirteen species of *Oncomelania* in Southeast Asia, only three can host *Schistosoma*. The other species have a genetic trait that allows them to resist infection by the parasite.

This information on the evolutionary relationships among snails is of great value in efforts to combat schistosomiasis. Scientists can now quickly determine whether or not a snail is likely to be a host for *Schistosoma*, and control efforts can be directed toward only those snails that transmit *Schistosoma*.

How did investigators infer the evolutionary relationships among the *Oncomelania* snails that are hosts of *Schistosoma*? **Systematics**, the scientific study of the diversity of organisms, provides answers to such questions. In this chapter, we will describe the methods systematists use to infer evolutionary relationships among organisms, and we will show how those evolutionary relationships are incorporated into classification systems. **Taxonomy**, which is a subdivision of systematics, is the theory and practice of classifying organisms.

Asian Snails Can Transmit Schistosomiasis Workers in the rice paddies of tropical Asia are at extreme risk of contracting schistosomiasis (known in some parts of the world as bilharzia). The disease is transmitted to humans via freshwater snails that thrive in the standing water of the paddies.

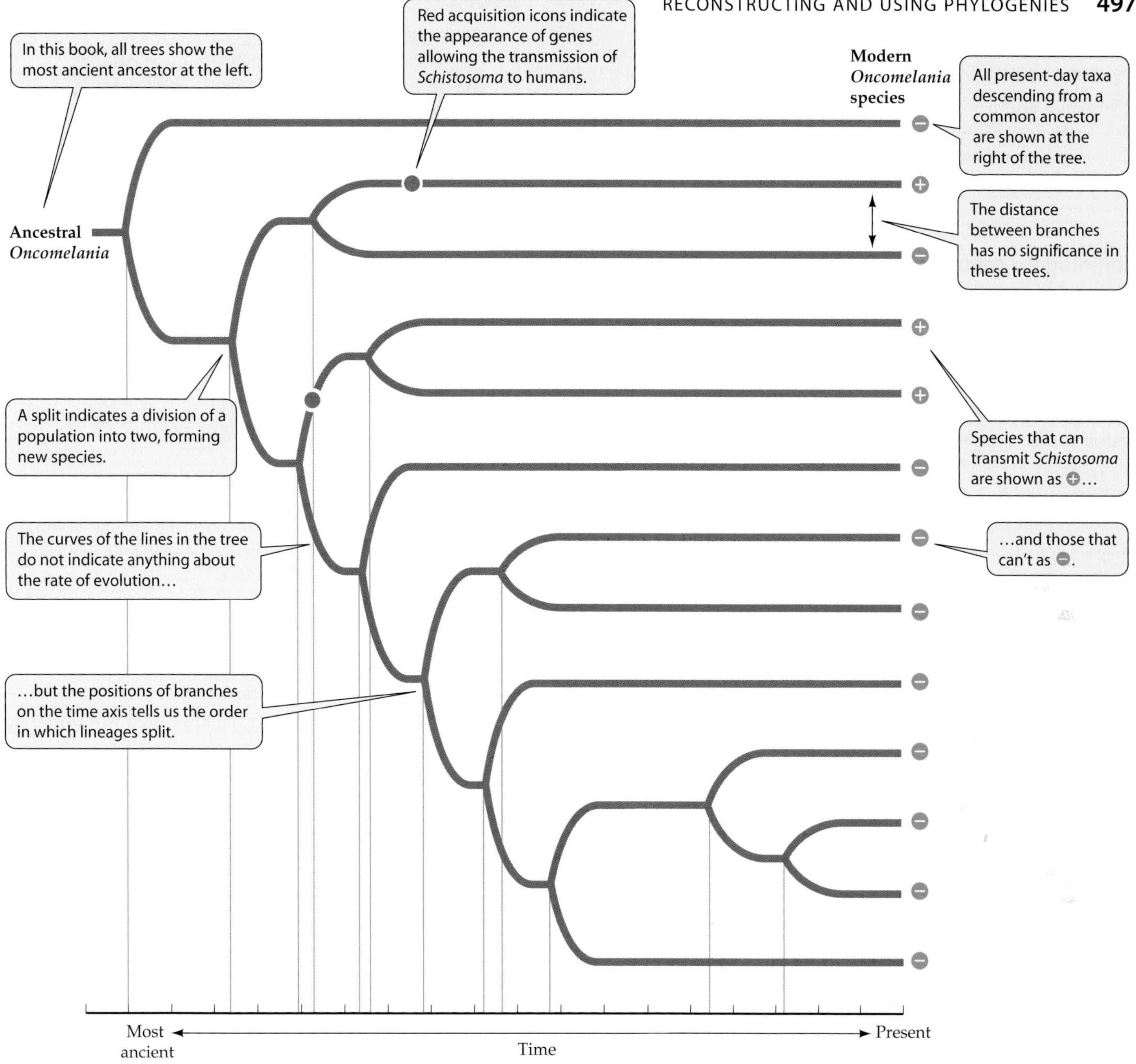

25.1 How to Read a Phylogenetic Tree A phylogenetic tree displays the order in which lineages split. This example shows the phylogeny of *Oncomelania* snails, the intermediate hosts of the human parasite *Schistosoma*. Acquisition of the ability to transmit *Schistosoma* occurred twice during the evolution of this lineage.

Phylogenetic Trees

A **phylogeny** is a hypothesis proposed by a systematist that describes the history of descent of a group of organisms from their common ancestor. A **phylogenetic tree** is a way of portraying that history. In it, a lineage is represented as a branching "tree," in which each node, or split, represents a speciation event. Thus, the tree shows the order in which lineages are hypothesized to have split. To reconstruct a phylogenetic tree, systematists analyze evolutionary changes in the traits of organisms. They are guided by Darwin's fundamental idea of descent with modification, which states that all species are descended from a common ancestor.

A phylogenetic tree may portray the evolution of all life forms; of a major evolutionary lineage such as the insects; or of a small group of organisms, such as the snail genus *Oncomelania* (Figure 25.1). The phylogenetic trees in this book depict time flowing from left (earliest) to right (most recent); it is equally common practice to draw trees with the earliest times at the bottom. Trees show the relative timing of separations between lineages of organisms, but, unless combined with other data, we do not date those separations. In the phylogenetic trees in this book, position on the horizontal axis has meaning, but vertical distance between the branches does not. Vertical distances are adjusted for legibility and clarity of presentation; they do not correlate with the degree of similarity or difference between groups.

Homologous traits are inherited from a common ancestor

The process of descent with modification means that species that share a recent common ancestor are likely to be very similar. In other words, they should share many traits, called **ancestral traits**, which they inherited from their common ancestor. Traits inherited from an ancestor in the distant past are likely to be shared by a large number of species. Traits that first appeared in a more recent ancestor should be shared by fewer species. But in all cases, the sharing of traits by a group of species indicates that they are likely to be descendants of a common ancestor.

Any features shared by two or more species that have been inherited from a common ancestor are said to be **homologous**. These features may be any heritable traits, including anatomical structures, behavior patterns, and DNA sequences. Traits that are shared by most or all of the organisms in a lineage of interest are likely to have been inherited relatively unchanged from an ancestor that lived very long ago. For example, all living vertebrates have a vertebral column, all known fossil vertebrates had a vertebral column, and all vertebrates are descended from the same common ancestor. Therefore, the vertebral column is judged to be homologous in all vertebrates.

A trait that differs from its ancestral form is called a **derived trait**. To determine how traits have changed during evolution, systematists must infer the state of the trait in an ancestor and then determine how it has been modified in the descendants. Doing so is not easy, because real evolutionary patterns are complex. Two processes generate difficulties: convergent evolution and evolutionary reversals.

- Independently evolved traits subjected to similar selective pressures may become superficially similar as a result of **convergent evolution**. For example, although the bones of the wings of bats and birds are homologous, having been inherited from a common ancestor, the wings of bats and birds are not homologous because they evolved independently from the forelimbs of different nonflying ancestors (Figure 25.2).
- A character may revert from a derived state back to an ancestral state—an **evolutionary reversal**. For example, most frogs lack teeth in the lower jaw, but the ancestor of frogs did have such teeth. One frog genus, *Amphignathodon*, has re-evolved teeth in the lower jaw.

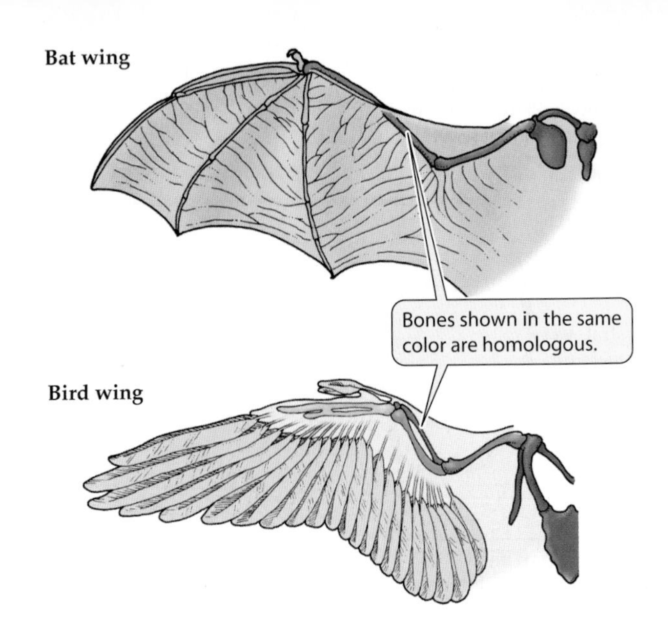

25.2 The Bones Are Homologous, but the Wings Are Not The supporting bone structures of both bat wings and bird wings are derived from a common four-limbed ancestor and are thus homologous. However, the wings themselves—an adaptation for flight—evolved independently in the two groups.

Convergent evolution and evolutionary reversals generate traits that are similar for some reason other than inheritance from a common ancestor. Such traits are called **homoplastic traits** or **homoplasies**.

Depending on the group of interest, a particular trait may be ancestral or derived. For example, rats and mice (both rodents), but not dogs or other mammals, have long, continuously growing incisor teeth. Continuously growing incisors evidently developed in the common ancestor of rats and mice after their lineage separated from the one leading to dogs and other mammals, because no other mammals have that kind of incisors. Thus, if we were reconstructing a phylogeny of a group of rodents, continuously growing incisors would be an ancestral trait because all rodents have them. However, if we were reconstructing a phylogeny of all mammals, continuously growing incisors would be a derived trait unique to the rodents. The distinction between ancestral and derived traits is important in reconstructing phylogenies.

Identifying ancestral traits is sometimes difficult

Distinguishing derived traits from ancestral traits may be difficult because traits often become so dissimilar that ancestral states are unrecognizable. The leaves of plants, for example, have diverged to form many different structures. Several lines of evidence, especially details of their structure and development, indicate that protective spines, tendrils, and brightly colored structures that attract pollinators (Figure 25.3) are all modified leaves; these structures are *homologs* of one another even though they do not resemble one another closely.

The spines of the barrel cactus and the bracts of Heliconia are both modified leaves.

The leaves of the pitcher plant curve to hold water.

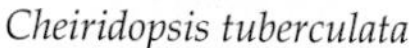

Cheiridopsis tuberculata *Heliconia* sp. *Sarracenia purpurea*

25.3 Homologous Structures Derived from Leaves The leaves of plants have diverged during their evolution to form many different structures, some of which bear very little resemblance to one another.

One method of distinguishing ancestral traits from derived traits is to assume that an ancestral trait should be found not only among the species of the **ingroup** (the lineage of interest), but also in outgroups. An **outgroup** is a lineage that is closely related to the ingroup, but which branches off from the ingroup before its base on the evolutionary tree. Traits found only within the ingroup, on the other hand, are likely to be derived traits.

Steps in Reconstructing Phylogenies

The first step in reconstructing a phylogeny is to select the group of organisms whose phylogeny is to be reconstructed—the ingroup—and an appropriate outgroup. The next step is to choose the characters that will be used in the analysis and to identify the possible forms of those characters (states or traits). A trait may be the presence or absence of a character, or one of the states a particular character may have, such as the number of body segments or number of appendages. The next, and usually the most difficult step, is to determine which traits are ancestral and which are derived. Finally, through phylogenetic analysis, systematists must distinguish homologous from homoplastic traits.

Because organisms differ in many ways, systematists use many characters to reconstruct phylogenies. Some of these characters, such as morphology, are readily preserved in fossils; others, such as behavior and molecular structures, rarely survive fossilization processes. Systematists use physiological, behavioral, molecular, and structural characters that can be assessed in both living and fossil organisms. The more characters that are measured, the more likely it is that the phylogeny will reflect the actual evolutionary pattern.

Morphological and developmental traits are used in reconstructing phylogenies

An important source of information for systematists is **morphology**—that is, the sizes and shapes of body parts. Since living organisms have been studied for centuries, we have a wealth of recorded morphological data, as well as extensive museum and herbarium collections of organisms whose traits can be measured. Technological tools, such as the electron microscope and computer simulations, enable systematists to measure and analyze the structures of organisms at much finer scales, down to the level of molecules, than was formerly possible.

The early developmental stages of many organisms reveal similarities to other organisms, but those similarities may be lost during later development. For example, the larvae of marine creatures called sea squirts have a rod in the back—the *notochord*—that disappears as they develop into adults. All vertebrate animals also have a notochord at some time during their development (Figure 25.4). This shared structure is one of the reasons for believing that sea squirts are more closely related to vertebrates than would be suspected if only adult sea squirts were examined.

Fossils show us where and when organisms lived in the past and give us an idea of what they looked like. Fossils provide important evidence that helps us distinguish ancestral from derived traits. The fossil record also reveals when lineages diverged and began their independent evolutionary histories. However, few or no fossils have been found for some groups whose phylogeny we may wish to determine.

25.4 A Larva Reveals Evolutionary Relationships Sea squirt larvae, but not adults, have a well-developed notochord (orange) that reveals their evolutionary relationship to vertebrates, all of which have a notochord at some time during their life cycle. In adult vertebrates, the vertebral column replaces the notochord as the support structure.

Molecular traits are also useful in reconstructing phylogenies

The molecules that make up organisms are also heritable traits that may diverge among lineages over evolutionary time. Molecular evolution will be discussed in detail in Chapter 26. Here we will briefly mention the molecular traits that are most useful for constructing phylogenies: the primary structures of proteins and nucleic acids (DNA and RNA).

PROTEIN PRIMARY STRUCTURE. Relatively precise information about phylogenies can be obtained by comparison of the primary structures of proteins. We can measure genetic differences between two lineages by obtaining homologous proteins from both of them and determining the number of amino acids that have changed since the lineages diverged from a common ancestor.

DNA BASE SEQUENCES. The base sequences of DNA provide excellent evidence of evolutionary relationships among organisms. The cells of eukaryotes have genes in their mitochondria as well as in their nuclei; plant cells also have genes in their chloroplasts. The chloroplast genome (cpDNA), which is used extensively in phylogenetic studies of plants, has changed little over evolutionarily time. Mitochondrial DNA (mtDNA) has been used extensively for studies of evolutionary relationships among animals (see Figure 25.10).

Relationships among the apes were investigated by sequencing more than 10,000 base pairs making up a segment of DNA that includes a hemoglobin pseudogene (a nonfunctional DNA sequence derived early in primate evolution by duplication of a hemoglobin gene). The outgroups for the analysis were *Ateles*, the spider monkeys of tropical America, and *Macaca*, the Rhesus monkey of Asia. The DNA data strongly indicate that chimpanzees and humans share a more recent ancestor than they do with gorillas, a conclusion supported by other types of molecular data.

Reconstructing a Simple Phylogeny

To show how a phylogeny is constructed, let's consider eight vertebrate animals: the lamprey, perch, pigeon, chimpanzee, salamander, lizard, mouse, and crocodile. We will assume initially that a given derived trait evolved only once during the evolution of these animals, and that no derived traits were lost from any of the descendant groups (Table 25.1). For simplicity, we have selected traits that are either present (+) or absent (–).

As we will see in Chapter 34, a group of jawless fishes called the lampreys is thought to have separated from the lineage leading to the other vertebrates before the jaw arose. Therefore, we will choose the lamprey as the outgroup for our analysis. Derived traits are those that have been acquired by other members of the lineage since they separated from the lamprey.

We begin by noting that the chimpanzee and mouse share two unique traits: mammary glands and fur. Those traits are absent in both the outgroup and the other species of the ingroup. Therefore, we infer that mammary glands and fur are derived traits that evolved in a common ancestor of chimpanzees and mice after that lineage separated from the ones leading to the other vertebrates. In other words, we provisionally assume that mammary glands and fur evolved only once among the animals in our ingroup.

The pigeon has one unique trait: feathers. As before, we provisionally assume that feathers evolved only once, after the lineage leading to birds separated from that leading to the mouse, chimpanzee, and crocodile. By the same reasoning, we assume that keratinous scales evolved only once, after the lineage leading to crocodiles, birds, and lizards separated from the lineage leading to mammals. We assume that claws or nails evolved only once, after the lineage leading to sala-

25.1 Eight Vertebrates Ordered According to Unique Shared Derived Traits

	DERIVED TRAIT[a]							
TAXON	JAWS	LUNGS	CLAWS OR NAILS	GIZZARD	FEATHERS	FUR	MAMMARY GLANDS	KERATINOUS SCALES
Lamprey (outgroup)	–	–	–	–	–	–	–	–
Perch	+	–	–	–	–	–	–	–
Salamander	+	+	–	–	–	–	–	–
Lizard	+	+	+	–	–	–	–	+
Crocodile	+	+	+	+	–	–	–	+
Pigeon	+	+	+	+	+	–	–	+
Mouse	+	+	+	–	–	+	+	–
Chimpanzee	+	+	+	–	–	+	+	–

[a] A plus sign indicates the trait is present, a minus sign that it is absent.

manders separated from the lineage leading to those animals that have claws or nails. We make the same assumption for lungs and jaws, continuing to minimize the number of evolutionary events needed to produce the patterns of shared traits among these eight animals.

Using this information, we can reconstruct a provisional phylogeny (Figure 25.5). We assume that the animals that share unique derived traits have a common ancestor not shared with the animals lacking those traits. We assume, for example, that mice and chimpanzees, the only two animals that share fur and mammary glands, share a more recent common ancestor with each other than they do with birds and crocodiles. Otherwise, we would need to assume that the ancestors of birds and crocodiles also had fur and mammary glands, but that they subsequently lost them—unnecessary additional assumptions.

Figure 25.5 shows a phylogeny for these eight vertebrates, based on the traits we used and the assumption that each derived trait evolved only once. This particular phylogeny was easy to construct because the animals and traits fulfilled the assumptions that derived traits appeared only once in the lineage and that they were never lost after they appeared. If we had included a snake in the group, our second assumption would have been violated, because the lizard ancestors of

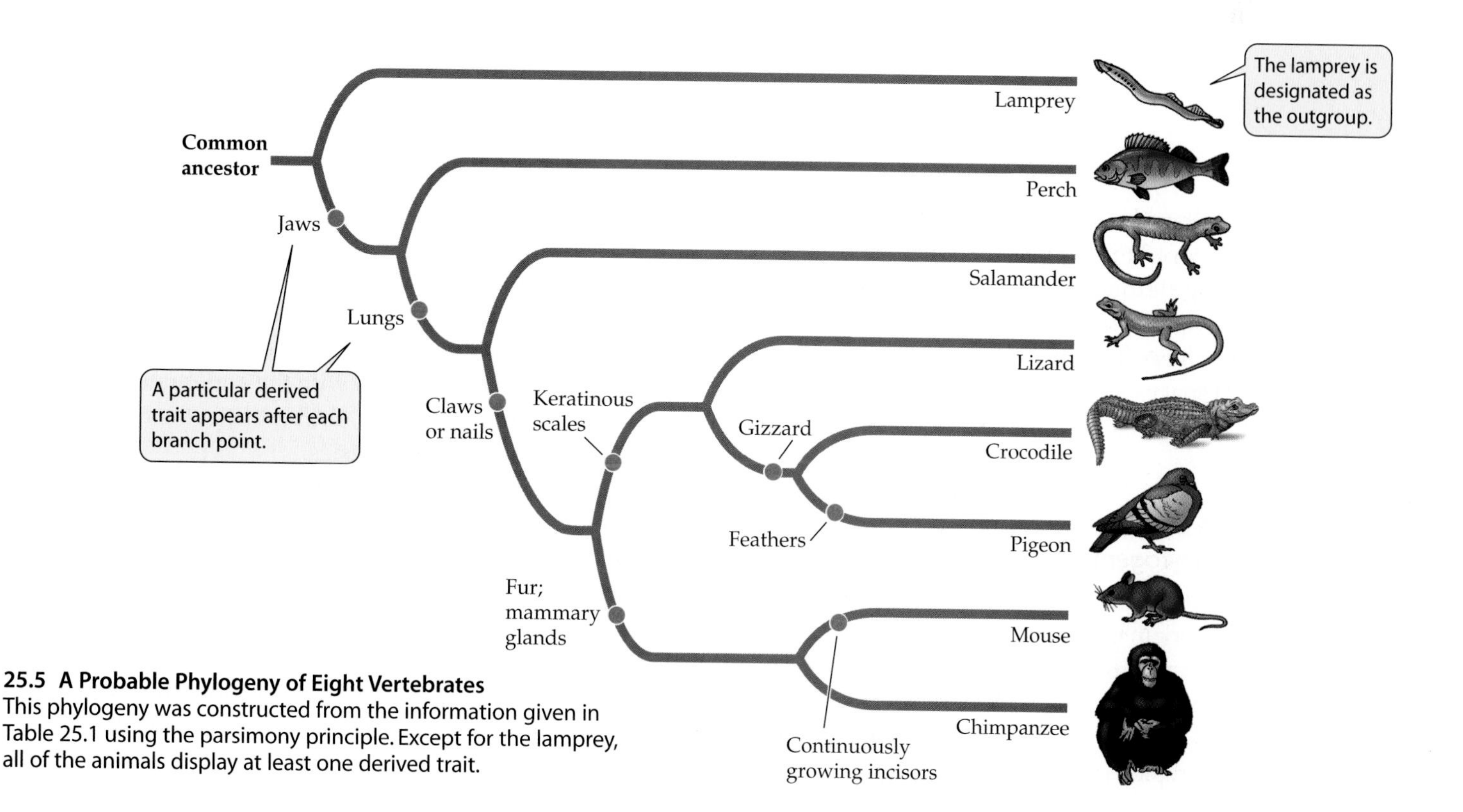

25.5 A Probable Phylogeny of Eight Vertebrates This phylogeny was constructed from the information given in Table 25.1 using the parsimony principle. Except for the lamprey, all of the animals display at least one derived trait.

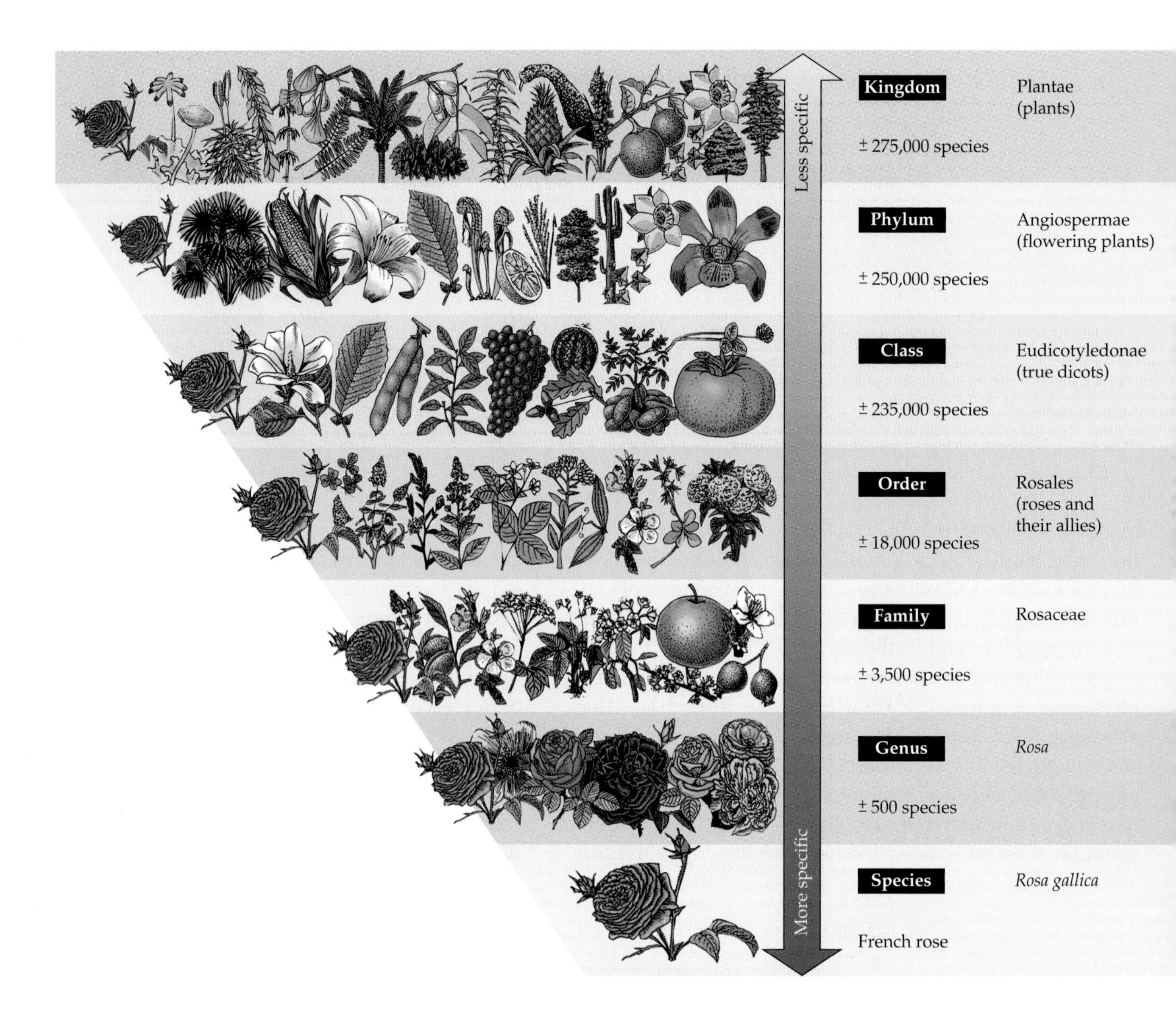

snakes had limbs that were subsequently lost (along with their claws). We would need to examine additional traits to determine that the lineage leading to snakes separated from the one leading to lizards long after the lineage leading to lizards separated from the others. In fact, the analysis of a number of traits shows that snakes evolved from burrowing lizards that became adapted to a subterranean existence.

Systematists use the parsimony principle when reconstructing phylogenies

The simple method we used to reconstruct our vertebrate phylogeny does not work in the vast majority of cases because we know from fossil and other evidence that traits can change more than once or undergo evolutionary reversal. Several methods are used to deal with these complexities. The most widely used ones employ the **parsimony principle**. In its most general form, the parsimony principle states that one should prefer the simplest hypothesis that is capable of explaining the observed data. Its application to the reconstruction of phylogenies means minimizing the number of evolutionary changes that need to be assumed over all characters in all groups in the tree (as we did with the vertebrate example in Figure 25.6). In other words, the best hypothesis is one that requires the fewest homoplasies.

Using the parsimony principle is appropriate, not because all evolutionary changes occurred parsimoniously, but because it is generally wiser to adopt the simplest explanation that can account for the observed data. More complicated explanations are accepted only when the evidence requires them. As we mentioned earlier, phylogenetic trees are hypotheses about evolutionary relationships. They are continually modified as additional traits are measured and as new fossil evidence becomes available.

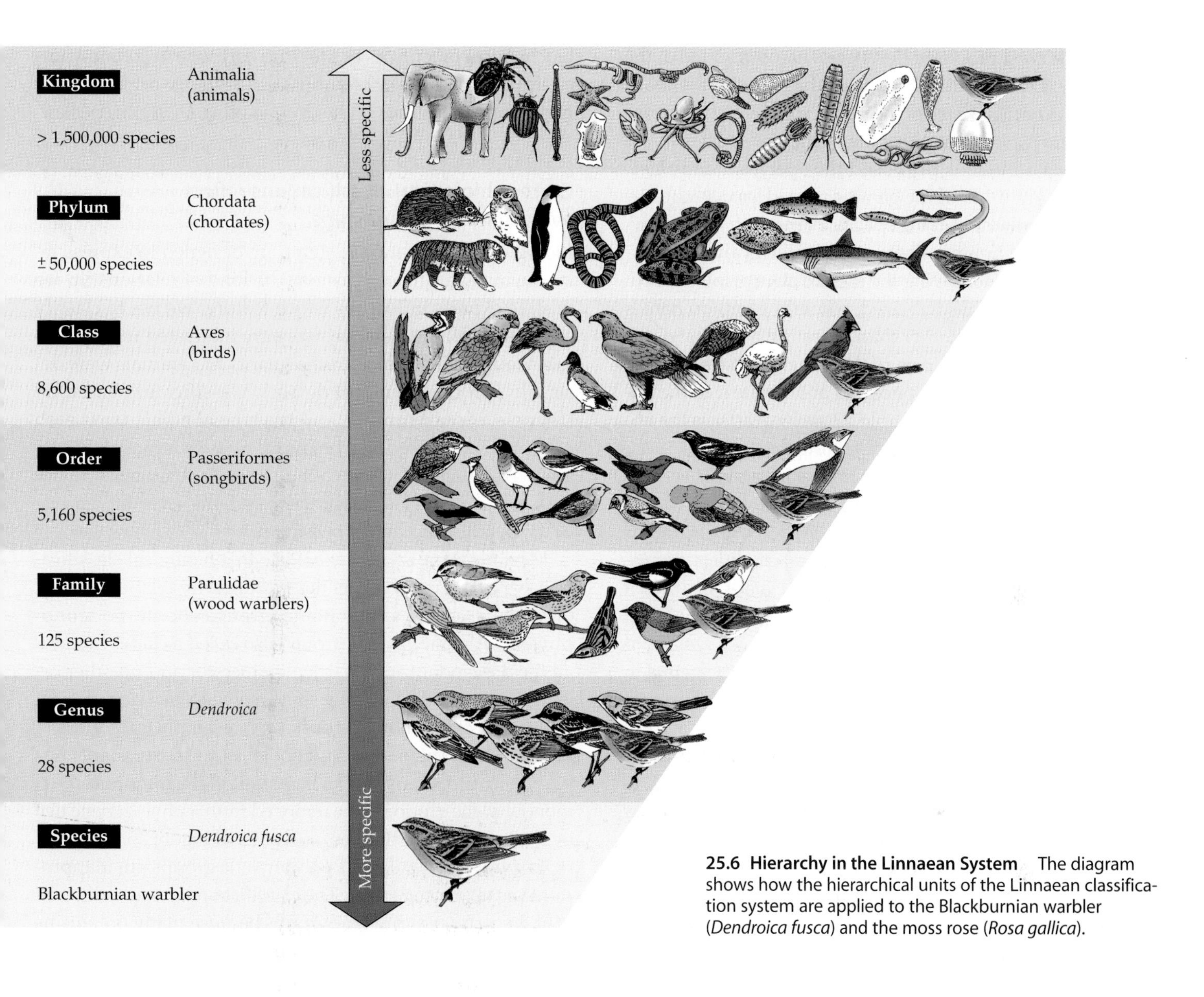

25.6 Hierarchy in the Linnaean System The diagram shows how the hierarchical units of the Linnaean classification system are applied to the Blackburnian warbler (*Dendroica fusca*) and the moss rose (*Rosa gallica*).

Another method of reconstructing phylogenies, called the *maximum likelihood method*, is used primarily with molecular data. The computer programs employed in this method are complex. They are designed to deal with the fact that mutations commonly change nucleotide sequences (see Chapter 26).

Whatever method is employed, reconstructing the phylogeny for any group of organisms is difficult. For example, there are 34,459,425 possible phylogenetic trees for a lineage with only 11 species! Computer programs using the parsimony principle employ various search routines that calculate the shortest phylogenetic tree—the one with the fewest homoplasies—for a given data set and then compare other possible phylogenies with the shortest one. If, as is usually the case, several trees are of equal minimum length, they can be merged into a *consensus tree* that retains only those lineage splits that are found in all the shortest trees. In a consensus tree, groups whose relationships differ among the trees form nodes with more than two branches. These nodes are considered "unresolved" because during a speciation event, a lineage typically splits into only two daughter species.

Biological Classification and Evolutionary Relationships

The biological classification system in use today was developed by the Swedish biologist Carolus Linnaeus and has been used since 1758. Linnaeus's system, referred to as *binomial nomenclature*, allows scientists throughout the world to refer unambiguously to the same organisms by the same names (Figure 25.6).

Linnaeus gave each species two names, one identifying the species itself and the other the genus to which it belongs. A **genus** (plural, genera; adjectival form, generic) is a group of closely related species. In many cases, the name of the taxonomist who first proposed the species name is added at the end. Thus, *Homo sapiens* Linnaeus is the name of the modern human species. *Homo* is the genus to which the species belongs, and *sapiens* identifies the particular species in the genus *Homo*; Linnaeus proposed the species name *Homo sapiens*. You can think of the generic name *Homo* as equivalent to your surname and the specific name *sapiens* as equivalent to your first name. The generic name is always capitalized; the name identifying the species always lowercased. Both names are always italicized, whereas common names of organisms are not. Rather than repeating a generic name when it is used several times in the same discussion, biologists often spell it out only once and abbreviate it to the initial letter thereafter (for example, *D. melanogaster* is the abbreviated form of *Drosophila melanogaster*).

Recognizing and interpreting similarities and differences among organisms is easier if the organisms are classified into groups that are ordered and ranked. Any group of organisms that is treated as a unit in a biological classification system, such as the genus *Oncomelania*, or all snails, is called a **taxon** (plural, taxa). In the Linnaean system, species and genera are further grouped into a hierarchical system of higher taxonomic categories. The taxon above the genus in the Linnaean system is the **family**. The names of animal families end in the suffix "-idae." Thus, Formicidae is the family that contains all ant species, and the family Hominidae contains humans and our recent fossil relatives, as well as chimpanzees and gorillas. Family names are based on the name of a member genus; Formicidae is based on the genus *Formica*, and Hominidae is based on *Homo*. Plant classification follows the same procedures, except that the suffix "-aceae" is used with family names instead of "-idae." Thus, Rosaceae is the family that includes the genus of roses (*Rosa*) and its close relatives.

Families, in turn, are grouped into **orders**, orders into **classes**, and classes into **phyla** (singular, phylum). The phyla of plants, fungi, and animals are grouped into the **kingdoms** Plantae, Fungi, and Animalia.

Biological classification systems and the unique names they provide for organisms are important for several reasons. They improve our ability to infer relationships among organisms. They are also an aid to memory and precise communication. It is impossible to remember the characteristics of many different organisms unless we can group them into categories based on shared characteristics. Classification systems are also useful as predictors. For example, the discovery of biochemical precursors of the drug cortisone in certain yams of the genus *Dioscorea*, stimulated a successful search for higher concentrations of the drug in other *Dioscorea* species.

Current biological classifications reflect evolutionary relationships

Biological classification systems are designed to express relationships among organisms. The kind of relationship we wish to express influences which features we use to classify organisms. If, for instance, we were interested in a system that would help us decide what plants and animals were desirable as food, we might devise a classification based on tastiness, ease of capture, and the type of edible parts each organism possessed. Early Hindu classifications of plants were designed according to these criteria. Biologists do not use such systems today, but those systems served the needs of the people who developed them.

Most taxonomists today believe that biological classification systems should reflect the evolutionary relationships of organisms, and that taxonomic groups should be **monophyletic**. A monophyletic group (also called a **clade**) contains all the descendants of a particular ancestor and no other organisms. A group containing some members that do not share the same common ancestor is said to be **polyphyletic**. A group that contains some, but not all, of the descendants of a particular ancestor is said to be **paraphyletic** (Figure 25.7). A monophyletic group can be removed from a phylogenetic tree by a single "cut" in the tree, as shown in Figure 25.7.

Taxonomists agree that polyphyletic groups are inappropriate as taxonomic units. The classifications used today still contain many polyphyletic groups because many organisms have not been studied well enough to distinguish between characters that are homologies and those that are homoplasies. However, as soon as they detect such homoplasies,

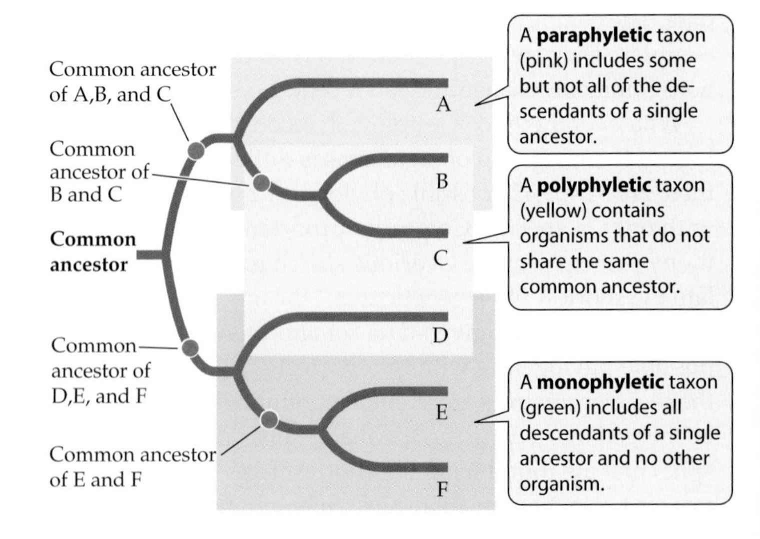

25.7 Monophyletic, Polyphyletic, and Paraphyletic Taxa
Monophyletic groups are preferred by most taxonomists. Polyphyletic groups are considered inappropriate as taxonomic units, but taxonomists sometimes do use paraphyletic taxa.

taxonomists change their classifications to eliminate polyphyletic taxa.

Although most taxonomists prefer strict phylogenetic classifications, some believe that classification systems should also reflect degrees of difference among organisms. According to this view, certain paraphyletic groups that have undergone rapid evolutionary change and diversification and evolved distinctive characters should be retained. Such groups are called **grades**. This perspective can be illustrated using birds, crocodiles, and their relatives.

We have some molecular evidence that birds and crocodilians (a group that includes crocodiles and alligators) share a more recent common ancestor than crocodilians and turtles share with snakes and lizards (Figure 25.8*a*). Traditionally, crocodilians are grouped with snakes, lizards, and turtles in the class Reptilia. Birds are placed in a separate class, Aves (Figure 25.8*b*). This classification came about because crocodilians have evolved more slowly than birds since the two lineages separated. Birds rapidly evolved significant and distinctive adaptations for flight, but crocodilian traits have changed little from those of their ancestors.

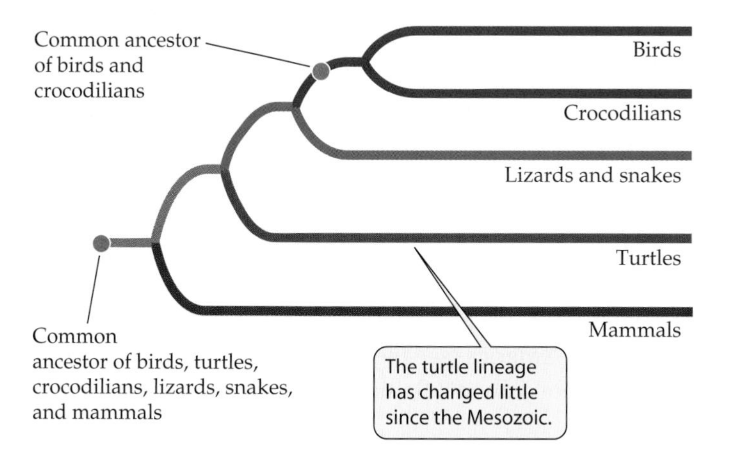

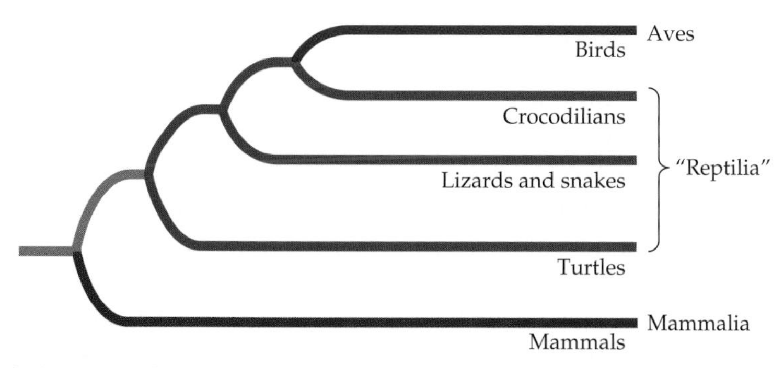

25.8 Phylogeny and Classification (*a*) Evolutionary relationships among mammals, reptiles, and birds. (*b*) The traditional classification unites crocodilians and turtles with lizards and snakes in the paraphyletic taxon "Reptilia," which excludes birds.

As a result, crocodilians are more similar in many ancestral features to snakes and lizards than they are to birds. They look like, and are physiologically similar to, very large lizards.

Thus, the traditional class Reptilia is paraphyletic because it does not include all the descendants of its common ancestor; that is, birds are excluded (Figure 25.8*b*). If only monophyletic taxa were permitted, birds would be grouped with crocodilians in a single taxon separate from snakes and lizards or birds would be included within an expanded class Reptilia. Retaining birds as a separate class (that is, retaining reptiles as a paraphyletic group) emphasizes that birds have evolved unique derived traits since they separated from reptiles, and thus are a distinct grade.

Although the current preference is to change classifications to eliminate paraphyletic groups, some of the most familiar taxonomic categories—gymnosperms and reptiles, for example—are paraphyletic. Because of their familiarity and the extensive literature devoted to them, these categories are likely to remain in use even after their formal taxonomic designations have been changed to better represent their evolutionary relationships.

Phylogenetic Trees Have Many Uses

Phylogenies are usually reconstructed as part of an effort to determine evolutionary relationships among organisms. Information about these relationships is useful to scientists investigating a wide variety of biological questions. Many biological statements are really phylogenetic assertions. Any claim of an association between a trait and a group of organisms is actually a statement about when during the history of the lineage the trait first arose and about the maintenance of the trait since its first appearance. For example, the statement that the cytoskeleton is a trait possessed by all eukaryotes is an assertion that the cytoskeleton is an ancestral trait that has been maintained during the subsequent evolution of all surviving eukaryote lineages. In this section, we will illustrate how phylogenetic trees can be used to determine how many times a particular trait may have arisen during evolution, to assess when lineages split, and to explain how evolutionary radiations came about.

How many times has a trait evolved?

Most flowering plants reproduce by mating with another individual—called *outcrossing*—and have mechanisms to prevent self-fertilization. Individuals of some species, however, are *self-compatible*; that is, they can fertilize themselves with their own pollen. How can we tell how often self-compatibility has evolved in a lineage? We can do so by plotting the outcrossing and selfing species on a phylogenetic tree.

The evolution of fertilization mechanisms was examined in *Linanthus* (a genus in the phlox family), a lineage of plants with a diversity of breeding systems and pollination mechanisms. The outcrossing (self-incompatible) species of *Linanthus* have long petals and are pollinated by long-tongued flies. The self-compatible species all have short petals. The investigators reconstructed a phylogeny for 12 species in the genus using a nuclear ribosomal DNA sequence (Figure 25.9). They determined whether each species was self-compatible by artificially pollinating flowers with their own pollen or pollen from other individuals and observing whether viable seeds formed.

Several lines of evidence suggest that self-incompatibility is the ancestral state in *Linanthus*. First, multiple origins of self-incompatibility have not been found in any other flowering plant family. Second, self-incompatibility depends on physiological mechanisms in both the pollen and the stigma (the female organ on which pollen lands) and requires the presence of at least three different alleles. Therefore, a change from self-incompatibility to self-compatibility would be easier than the reverse change. Third, in all self-incompatible species of *Linanthus*, the site of pollen rejection is the stigma, even though sites of pollen rejection vary greatly among other plant families.

Assuming that self-incompatibility is the ancestral state, the reconstructed phylogeny suggests that self-compatibility has evolved three times within this *Linanthus* lineage (Figure 25.9). The change to self-compatibility has been accompanied by the evolution of reduced petal size. Interestingly, the striking similarity of the flowers in the self-compatible groups led to their being classified as members of a single species. The phylogenetic analysis shows them to be members of three distinct lineages.

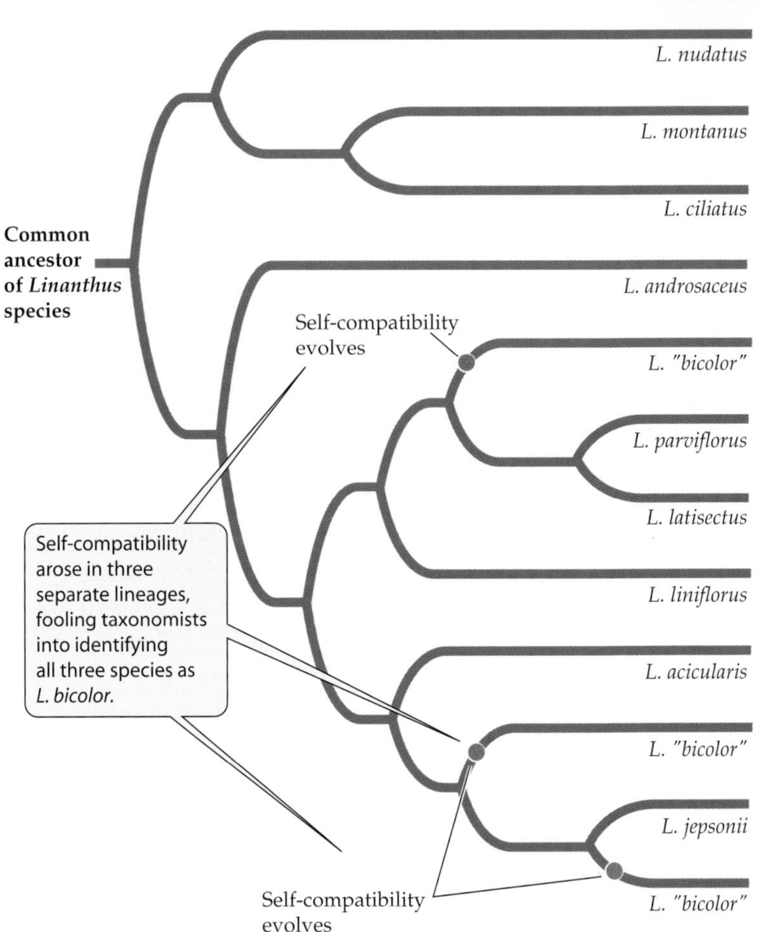

25.9 Phylogeny of a Section of the Phlox Genus *Linanthus* Self-compatibility apparently evolved three times in this lineage. Because the form of the flowers converged in the selfing lineages, taxonomists mistakenly thought that they were all members of a single species.

When did lineages split?

How phylogenetic analyses can help us determine when lineages split is illustrated by studies of characiform fishes. The approximately 1,400 species of these freshwater fishes, which are found in both South America (1,200 species) and Africa (200 species), vary greatly in size, shape, and diet. Analyses of mitochondrial DNA and anatomy suggest that the characiform fishes are monophyletic; a suggested phylogeny of part of the family is shown in Figure 25.10. Analyses using slowly evolving rRNA genes identified three lineages with closely related species on both sides of the Atlantic Ocean. Since all these species live only in fresh water, dispersal across the Atlantic Ocean is unlikely. The genetic differences between the rRNA of the African and the South American species in all three lineages are great enough to be consistent with a split caused by the separation of Africa from South America (see Figure 22.16), which is believed to have happened about 90 million years ago.

How recently did Lake Victoria's cichlid fishes radiate?

The spectacular radiation that produced more than 500 species in one lineage of cichlid fishes in Lake Victoria, in eastern Africa, was initially assumed to have occurred over a period of about 750,000 years, the presumed age of the lake basin. However, recently discovered geological data suggests that Lake Victoria dried up completely between 15,600 and 14,700 years ago! Biologists judged that the hundreds of morphologically diverse cichlids (Figure 25.11*a*) in Lake Victoria could not have evolved in such a short time, so they developed alternative scenarios. One scenario assumed that the lake did not dry up completely. Another postulated that some of the fish species survived in rivers, from which they subsequently recolonized the lake. But no data were avail-

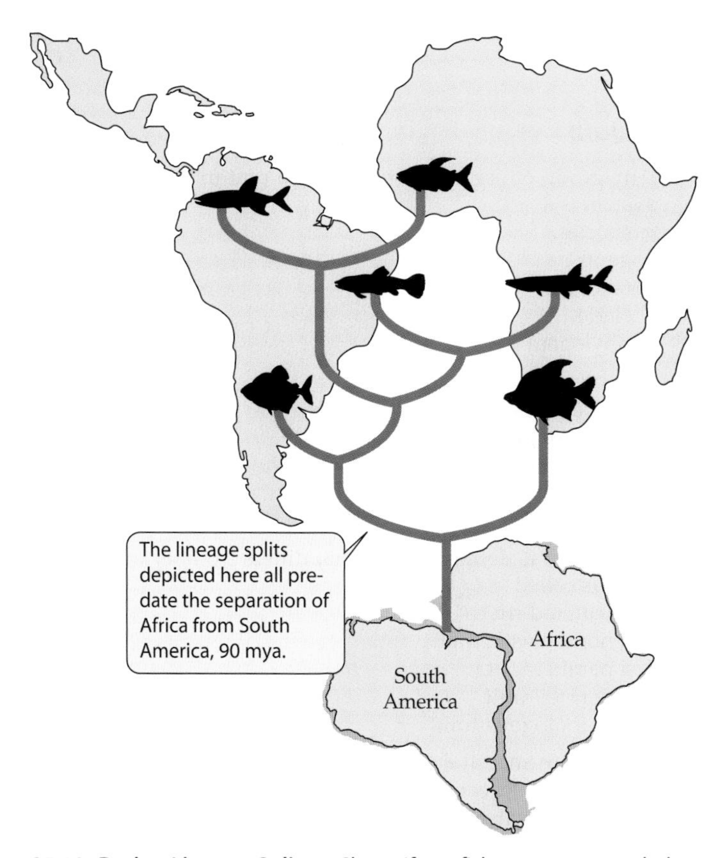

25.10 Dating Lineage Splits Characiform fishes are a monophyletic lineage, and species in several groups on opposite sides of the Atlantic Ocean appear similar. However, rRNA analyses indicate the African and South American lineages have been evolving separately for about 90 million years—the time at which the two continents split apart and the Atlantic Ocean became a barrier to these freshwater fishes.

able to test these possibilities. Phylogenetic analyses based on molecular data helped resolve the problem.

Using 300 mtDNA sequences, investigators reconstructed a phylogeny of the cichlid fishes of Lake Victoria and other lakes in the region. Their phylogeny suggests that the ancestors of Lake Victoria cichlids came from the geologically much older Lake Kivu. Today, Lake Kivu is home to only15 species of cichlids, but the phylogeny suggests that fishes from Lake Kivu colonized Lake Victoria on two different occasions (Figure 25.11*b*). The molecular data also indicate that some of the cichlid lineages that are found only in Lake Victoria, and which therefore probably evolved there, split at least 100,000 years ago. Thus, the reconstructed phylogeny of these fishes strongly suggests that Lake Victoria did not completely dry up about 15,000 years ago, and that many fish species survived in rivers and in pockets of water that remained in the deepest part of the lake throughout the most recent dry period.

These examples illustrate how a plausible phylogeny for a lineage of organisms enables biologists to answer a variety of questions about the history of that group. They also show the value of molecular data in reconstructing phylogenies. We will consider molecules and how they are used to study evolutionary history in greater detail in the next chapter.

25.11 Origins of the Cichlid Fishes of Lake Victoria (*a*) The photographs show only two of the hundreds of cichlid species found in Lake Victoria. (*b*) Phylogenetic analysis suggests that cichlids colonized Lake Victoria by the routes shown on the map.

(*a*)

Harpagochromis sp.

Ptychromis sp.

(*b*)

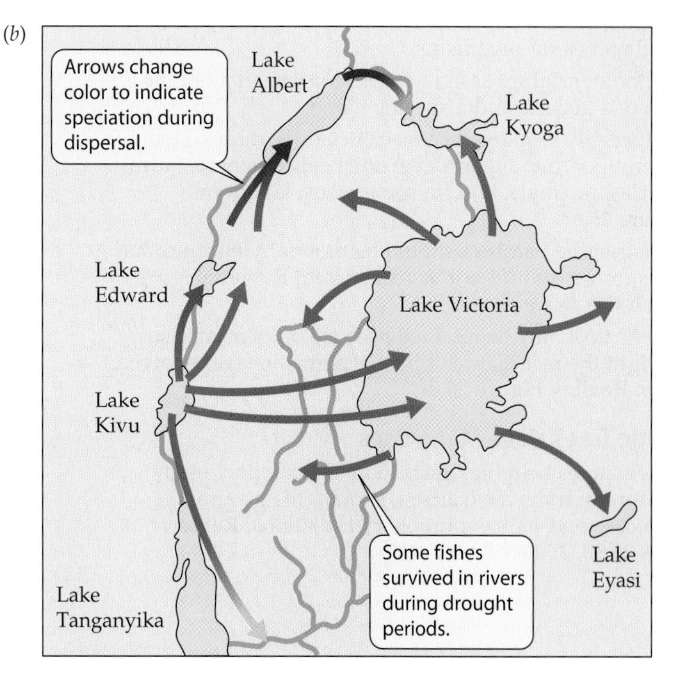

Chapter Summary

▶ Systematics is the scientific study of the diversity of organisms. Taxonomy, a subdivision of systematics, is the theory and method of classifying organisms.

Phylogenetic Trees

▶ A phylogenetic tree displays the order in which lineages split. **Review Figure 25.1**

▶ Traits inherited from a common ancestor, called ancestral traits,—are said to be homologous.

▶ A derived trait is one that differs from its form in the ancestor of a lineage.

▶ Traits that are similar as a result of convergent evolution or evolutionary reversals are said to be homoplastic. **Review Figure 25.2**

Steps in Reconstructing Phylogenies

▶ Systematists use morphological, physiological, behavioral, and molecular characters to reconstruct phylogenies.

▶ Structures in early developmental stages sometimes show evolutionary relationships that are not evident in adults. **Review Figure 25.4**

▶ Protein primary structures and the base sequences of nucleic acids are also important traits that can be used in reconstructing phylogenies.

Reconstructing a Simple Phylogeny

▶ To assess evolutionary relationships, systematists must distinguish between ancestral and derived traits within a lineage. This task is often difficult because traits can change more than once or undergo evolutionary reversal. **Review Figure 25.5**

▶ Systematists use the parsimony principle to reconstruct phylogenetic trees.

See Web/CD Activity 25.1

Biological Classification and Evolutionary Relationships

▶ Classification systems improve our ability to explain relationships among things, aid our memory, and provide unique, universally used names for organisms.

▶ Biological nomenclature assigns to each organism a unique combination of a generic and a specific name.

▶ In the universally employed Linnaean classification system, species are grouped into higher-level units called genera, families, orders, classes, phyla, and (in some cases) kingdoms. **Review Figure 25.6**

▶ Taxonomists agree that taxa should be monophyletic and that polyphyletic groups should not be recognized. **Review Figure 25.7.** See Web/CD Activity 25.2

▶ Paraphyletic taxa may be retained because of their familiarity and to highlight the fact that members of some lineages evolved unique traits. **Review Figure 25.8**

Phylogenetic Trees Have Many Uses

▶ Phylogenetic trees help biologists to determine how many times evolutionary traits have arisen, explain the geographic ranges of species, and date evolutionary radiations. **Review Figures 25.9, 25.10, 25.11**

Self-Quiz

1. Any group of organisms treated as a unit in a classification system is a
   a. species.
   b. genus.
   c. taxon.
   d. clade.
   e. phylogen.
2. A genus is a
   a. group of closely related species.
   b. group of genera.
   c. group of similar genotypes.
   d. taxonomic unit larger than a family.
   e. taxonomic unit smaller than a species.
3. A trait that is defined as one that differs from its ancestral form is called
   a. an altered trait.
   b. a homoplastic trait.
   c. a parallel trait.
   d. a derived trait.
   e. a homologous trait.
4. Identifying ancestral traits is often difficult because
   a. traits often become so dissimilar that ancient states are unrecognizable.
   b. there may be no fossils of appropriate ancestors.
   c. reversals of traits are common during evolution.
   d. traits often evolve rapidly.
   e. All of the above
5. The parsimony principle is typically used when reconstructing phylogenies because
   a. evolution is nearly always parsimonious.
   b. it is better to provisionally adopt the simplest hypothesis capable of explaining the known facts.
   c. it is easier to handle parsimonious data with computers.
   d. parsimony works well for all kinds of traits, both morphological and molecular.
   e. parsimony was used before computers were available and it continues to be used even though new methods are better.
6. Which of the following is a way of identifying ancestral traits?
   a. Determining which traits are found among fossil ancestors
   b. Using an outgroup
   c. Using a lineage that is closely related to the ingroup
   d. Examining the development of the trait
   e. All of the above
7. Traits that evolve very slowly are most useful for determining relationships at the level of
   a. phyla.
   b. genera.
   c. orders.
   d. families.
   e. species.

8. Homologous traits are
   a. similar in function.
   b. similar in structure.
   c. similar in structure but not in function.
   d. derived from a common ancestor.
   e. derived from different ancestral structures and have dissimilar structures.
9. The genes that are most extensively used to determine evolutionary relationships among plants are
   a. nuclear genes.
   b. chloroplast genes.
   c. mitochondrial genes.
   d. genes in flowers.
   e. genes in roots.
10. Which of the following is *not* a way in which phylogenies are used?
    a. To establish evolutionary relationships
    b. To determine how rapidly traits evolve
    c. To determine historical patterns of movement of organisms
    d. To help identify unknown organisms
    e. To infer evolutionary trends
11. Which of the following is *not* a major role of a classification system?
    a. To aid memory
    b. To improve predictive powers
    c. To help explain relationships among things
    d. To provide relatively stable names for things
    e. To design identification keys

For Discussion

1. Why are taxonomists concerned with identifying lineages that share a single common ancestor?
2. How are fossils used to identify ancestral and derived traits of organisms?
3. Taxonomists use the parsimony principle when reconstructing phylogenetic trees. Given that evolutionary processes are not always parsimonious, why is it used as a guiding principle?
4. A student of the evolution of frogs has proposed a strikingly new classification of frogs based on an analysis of a few mitochondrial genes from about 25 percent of frog species. Should frog taxonomists immediately accept the new classification? Why or why not?
5. Linnaeus developed his system of classification before Darwin proposed his theory of evolution by natural selection, and most classifications of organisms initially were proposed by non-evolutionists. Yet, many of these classifications are still used today, with minor modifications, by most evolutionary taxonomists. Why?
6. Classification systems summarize much information about organisms and enable us to remember the traits of many organisms. From our general knowledge, how many traits can you associate with the following names: conifer, fern, bird, mammal?

26 Molecular and Genomic Evolution

There are more species of insects on Earth than of all other animal groups combined. Many of those species transmit diseases to humans. Among the most dangerous are the several genera of mosquitoes that transmit malaria, dengue, lymphatic filariasis, yellow fever, and Japanese encephalitis. Worldwide, malaria is a major killer that causes more than a million deaths among children each year. How can evolutionary biology help us combat this scourge of humankind?

In 1991, a group of scientists met in Tucson, Arizona, to launch a plan to engineer transgenic mosquitoes incapable of carrying the malaria parasite. They would then test possible methods of spreading the new genotypes into wild mosquito populations and develop the basic tools needed for such an effort. In 1999, another group of experts concluded that sequencing the genome of the mosquito *Anopheles gambiae*, the most important vector of malaria in Africa, would help in achieving those goals. A sequencing initiative was launched, and within two years, in October 2002, the complete 278,000-base genome of the mosquito was published. Sequencing has also been completed for *Plasmodium falciparum*, the protist that causes the most deadly form of malaria (see Chapter 28).

Knowing the genomes of *A. gambiae* and *P. falciparum* has provided insights into the genetic makeup of insects and how they have coevolved with the parasites they transmit to humans. But the genomes of the anopheline mosquitoes that carry malaria, such as *A. gambiae,* differ significantly from those of the culicine mosquitoes, which are the major transmitters of the other diseases mentioned above. Comparative investigations of the evolution of insect genomes, as well as the genomes of the parasites they carry, will be needed in our efforts to combat all of these debilitating diseases.

For most of its history, evolutionary biology depended on the study of obvious morphological features of organisms. During his voyage aboard the *Beagle*, Charles Darwin observed morphological differences among species found in different geographic areas. He later synthesized these observations into descriptions of how species change over time. He

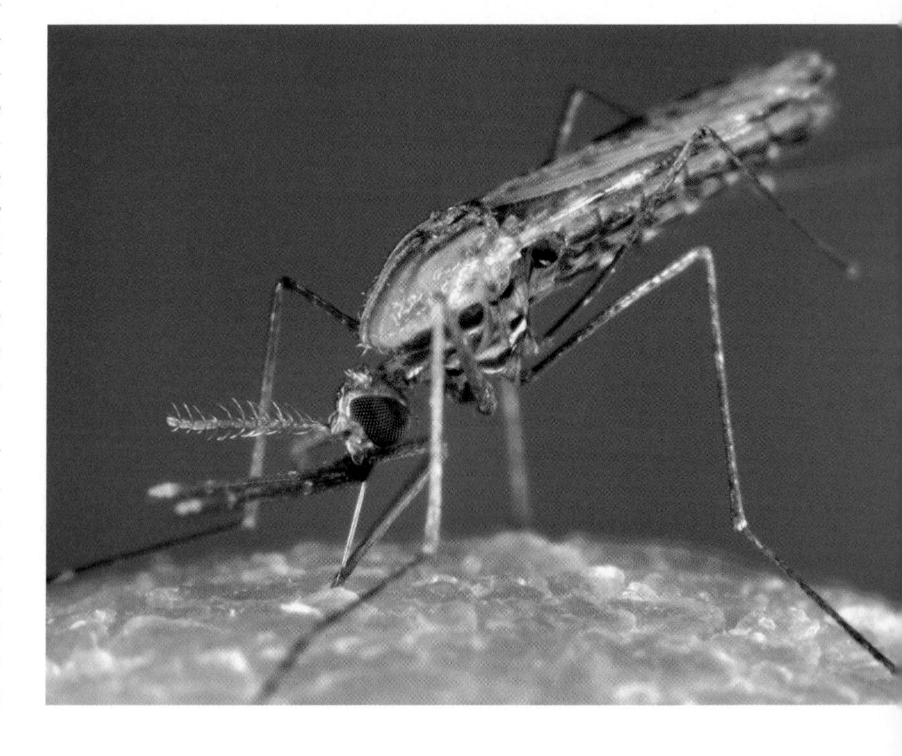

A Deadly Bite The bite of the mosquito *Anopheles gambiae* transmits *Plasmodium falciparum*, a cause of malaria, into the human bloodstream. Malaria is one of the most widespread debilitating (and frequently deadly) human diseases on the planet, affecting over 600 million people.

developed his theory of natural selection to explain *why* many of these morphological changes had happened, but he could not determine *how* they had occurred. Understanding of the mechanisms of evolutionary change had to await new discoveries in biochemistry more than a century later.

In this chapter we will see how molecular biologists determine the structures of nucleic acids and proteins and how they use those structures to infer both the patterns and the causes of molecular evolution. With these insights, we will explore how the functions of molecules change, where new genes come from, and how genomes change in size. Finally, we will see how knowledge of the patterns of molecular evolution can help us to solve other biological problems, such as inferring phylogenetic relationships among organisms, determining the phylogenies of genes, and combating diseases.

Genomes and Their Evolution

An organism's **genome** is the full set of genes that it contains. Most of the genes of eukaryotic organisms are found on chromosomes in the nucleus, but genes are also present in plastids (such as chloroplasts) and mitochondria. In organisms that reproduce sexually, both males and females transmit nuclear genes, but mitochondrial and chloroplast genes usually are transmitted only via the cytoplasm of eggs, as we saw in Chapter 10.

In sexually reproducing organisms, meiosis and fertilization shuffle genes in every generation, and individual gametes transmit partly randomized subsets of parental genes to offspring. For this repeated shuffling to be successful, genes must be able to operate in a wide variety of genetic backgrounds. Closely linked genes, however, are likely to remain together, which is especially important for developmental genes. Many genes that govern developmental processes are closely linked on chromosomes and are inherited together.

A gene will not be passed on to successive generations unless the individual with the gene survives and reproduces. Therefore, the capacity to cooperate with different combinations of other genes is likely to increase a gene's probability of transmission. For this reason, it is useful to view the genes of an individual as interacting members of a group, among which there are divisions of labor, but also strong interdependencies.

Students of genomic evolution look at the genome of an organism as an integrated whole, asking questions such as, How do proteins acquire new functions? Why are the genomes of different organisms so variable in size? How has enlargement of genomes been accomplished? We will return to these questions later in this chapter, but to provide some needed background, we will first describe a related field of study: molecular evolution.

The Evolution of Macromolecules

The field of **molecular evolution** investigates the evolution of macromolecules and uses the findings to reconstruct the evolutionary history of genes and the organisms that carry them. The molecules of special interest to molecular evolutionists are nucleotides, nucleic acids, amino acids, and proteins (see Chapter 3).

Nucleic acids evolve by means of nucleotide base substitutions, which in turn can result in changes in the amino acids they encode. Alterations in the structure and function of proteins result from changes in the ordering of the amino acids of which they are composed. Molecular evolutionists characterize the precise structures of these macromolecules: nucleotide sequences in nucleic acids, and primary structure in proteins. They use that information to determine how rapidly these macromolecules have changed and why they have changed.

Phylogenetic information is essential for determining the order of changes in molecular characters because knowing the order of such changes is usually the first step in inferring their causes. Conversely, knowledge of the pattern and rate of change of a given macromolecule is crucial to attempts to reconstruct the evolutionary history of groups of organisms.

Molecular evolution is driven by changes in nucleotide sequences

As we mentioned in Chapter 23, molecular evolution differs from phenotypic evolution in that random genetic drift and mutation exert important influences on the rates and direction of nucleotide changes. A *mutation*, as you will recall from Chapter 12, is any change in the genetic material. Many mutations do not alter the protein encoded by the mutated genes. The reason is that in the "universal genetic code" (see Figure 12.5), most amino acids are specified by more than one codon. Leucine, for example, is specified by six different codons: UUA, UUG, CUU, CUC, CUA, and CUG. A mutation that does not change the amino acid—UUA to UUG, for example—is known as a **synonymous** or **silent mutation** (Figure 26.1*a*). Synonymous mutations do not affect the functioning of a protein (and hence the organism), and are, therefore, unlikely to be influenced by natural selection.

Because they are unlikely to be influenced by natural selection, synonymous mutations can spread, at rates determined by rates of mutation and genetic drift, so that they completely replace one nucleotide base or longer sequence with another throughout an entire population or species. In this chapter we will call such replacements **substitutions** to distinguish them from a replacement of one nucleotide by another in a single individual (which is also often called a substitution). Molecular evolutionists are primarily interested in population-wide nucleotide substitutions.

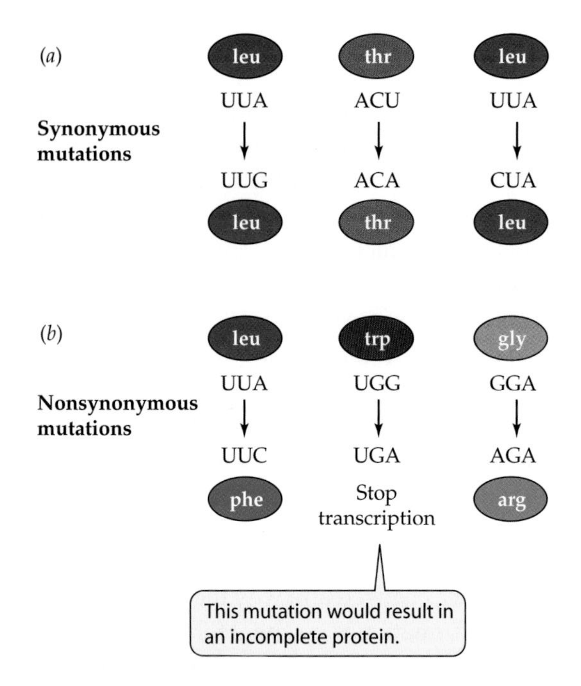

26.1 When One Base Does or Doesn't Make a Difference
(*a*) Synonymous mutations do not change the amino acid specified and do not affect protein function; such mutations are unlikely to be agents of natural selection. (*b*) Nonsynonymous mutations do change the amino acid sequence and are in general likely to have an effect (usually deleterious) on protein function and natural selection.

A mutation that *does* change the amino acid sequence encoded by a gene is known as a **nonsynonymous mutation**. For example, a change from UUA to UUC would result in phenylalanine rather than leucine in the encoded amino acid (Figure 26.1*b*). In general, nonsynonymous mutations are likely to be deleterious to the organism. But not every amino acid change alters a protein's shape (and hence its functional properties). Therefore, some nonsynonymous mutations may be selectively neutral, or nearly so.

As we saw in Chapter 23, most natural populations of organisms harbor much more genetic variation than we would expect if genetic variation were influenced primarily by natural selection. This discovery, combined with the knowledge that many mutations do not change molecular function, stimulated the development of the neutral theory of molecular evolution.

Many mutations may be selectively neutral

In 1968, Motoo Kimura proposed the *neutral theory of molecular evolution*. Kimura suggested that, at the molecular level, the majority of mutations are selectively neutral; that is, they confer neither an advantage nor a disadvantage on their bearers. Thus, the majority of evolutionary changes in macromolecules, and much of the genetic variation within species, could result neither from directional selection of advantageous alleles nor from stabilizing selection, but rather from genetic drift.

To see why this is so, consider a population of size N and a neutral mutation rate of μ per gamete per generation at a locus. The number of new mutations would, on average, be $\mu \times 2N$, because $2N$ gene copies are available to mutate in a diploid organism. According to drift theory, the probability that a mutation will be fixed by drift alone is its frequency, p, which equals $1/(2N)$ for a newly arisen (and hence very rare) mutation. Therefore, the number of neutral mutations that arise per generation that are likely to become fixed is $2N\mu \times 1/(2N) = \mu$—which is the mutation rate!

Thus, the rate of fixation of neutral mutations is theoretically constant and is equal to the mutation rate. So if most mutations of macromolecules do not affect their functioning, macromolecules should diverge from one another at a constant rate. Indeed, in the 1960s, comparative studies of several proteins indicated that, for a particular protein, rates of amino acid substitutions were similar in all evolutionary lineages having that protein. In other words, the rate of evolution of particular proteins might be constant over time, and in effect can be a "molecular clock." We will discuss molecular clocks later in this chapter and show how, with care, they can be used to study several features of molecular evolution.

Determining and Comparing the Structure of Macromolecules

To investigate patterns of molecular evolution, biologists must first determine the precise structure of molecules. The base sequences of nucleic acids provide information about the primary structure of the proteins they encode. The invention of the polymerase chain reaction technique (PCR; see Chapter 11) allowed biologists to determine the sequences of regions of DNA not only from living tissues, but also from fossilized remains, mummified tissues, dried skins in museums, and pressed plants in herbaria, even though these objects contain only tiny amounts of DNA. DNA has been extracted and amplified from human fossils more than 30,000 years old and from plant leaf fossils 40,000 years old.

Once the sequences of molecules from different organisms have been determined, they can be compared. The purpose of comparing them is to identify the locations of deletions, insertions, and substitutions that have occurred in the molecules of interest since the organisms diverged from a common ancestor. A simple hypothetical example illustrates how this is done. In Figure 26.2 we compare two amino acid sequences (1 and 2) from homologous proteins in different organisms. The two sequences differ in the number and identity of their amino acid residues. To compare these sequences, we first observe that, although the sequences appear quite

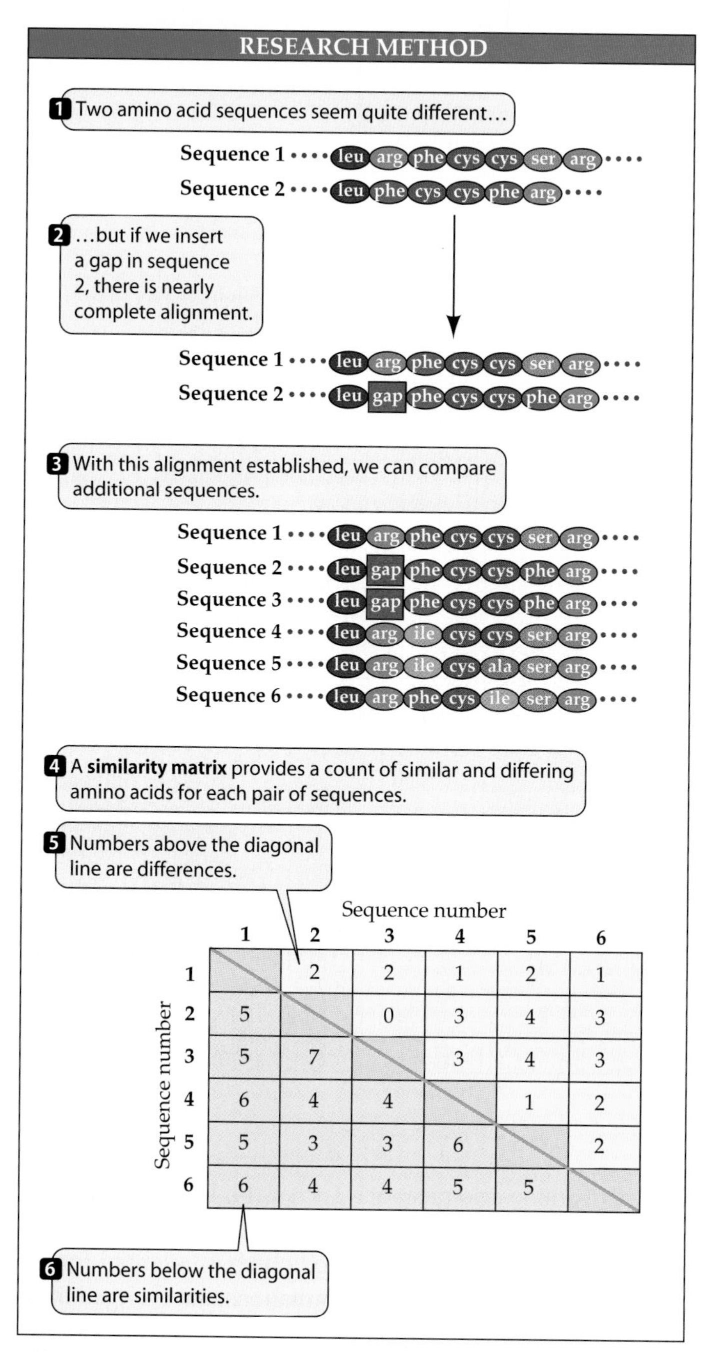

26.2 Amino Acid Sequence Alignment Insertion of a gap allows us to align two homologous amino acid sequences so that we can compare them. Once the alignment is established, sequences from more organisms can be added and compared. A similarity matrix sums similarities and differences between each pair of organisms. The larger the number of similarities, the more recent the presumed common ancestor of the organisms.

different, they would become similar if we were to insert a gap after the first amino acid in sequence 2 (after the leucine residue). In fact, these sequences then differ only by one amino acid at position 6 (serine or phenylalanine). Insertion of a single gap—that is, correcting a deletion—*aligns* these sequences. Longer sequences and those that have diverged more extensively require more elaborate adjustments.

After we have aligned the sequences, we can compare them by counting the number of nucleotides or amino acids that differ between them. If we add more sequences to our original example and sum the number of similar and different amino acids in each pair of sequences, we can construct a *similarity matrix* (Figure 26.2), which gives us a measure of the changes that have occurred during the divergence of the organisms.

The longer molecules have been evolving separately, the more differences they should have. Enough analyses of mammalian genes have been performed to show that the rate of nonsynonymous nucleotide substitution varies from nearly zero to about 3×10^{-9} substitutions per locus per year. Synonymous substitutions in the protein-coding regions of genes have occurred about five times more rapidly than nonsynonymous substitutions. In other words, substitution rates are highest at nucleotide positions that *do not change the amino acid being expressed* (Figure 26.3). The rate of substitution is even higher in **pseudogenes**, duplicate copes of genes that have undergone one or more mutations that eliminate their ability to be expressed. Why are these rates of substitution so dissimilar?

Rates of nucleotide substitution vary because the roles of molecules differ

The observation that rates of substitution are highest at sites and in molecules where they have no functional significance is consistent with the hypothesis that substitution rates at these sites are driven primarily by a combination of mutation and genetic drift. The much slower rates of substitution at sites that *do* affect molecular function is consistent with the

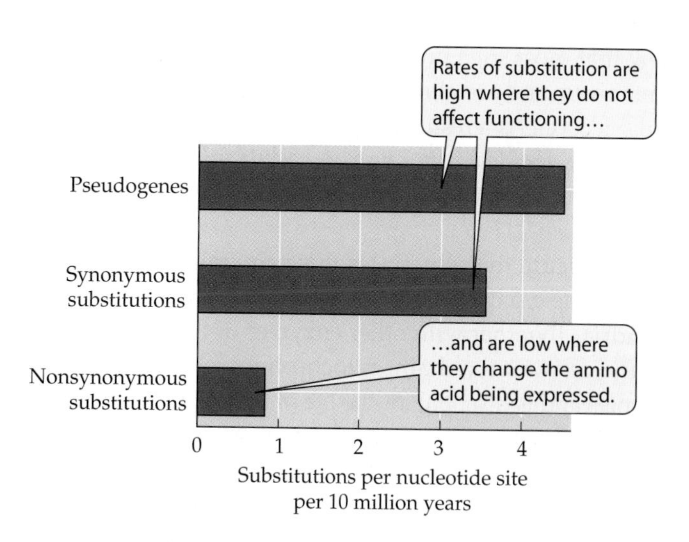

26.3 Rates of Base Substitution Differ Rates of nonsynonymous substitutions are much slower than rates of synonymous substitutions and substitutions in pseudogenes.

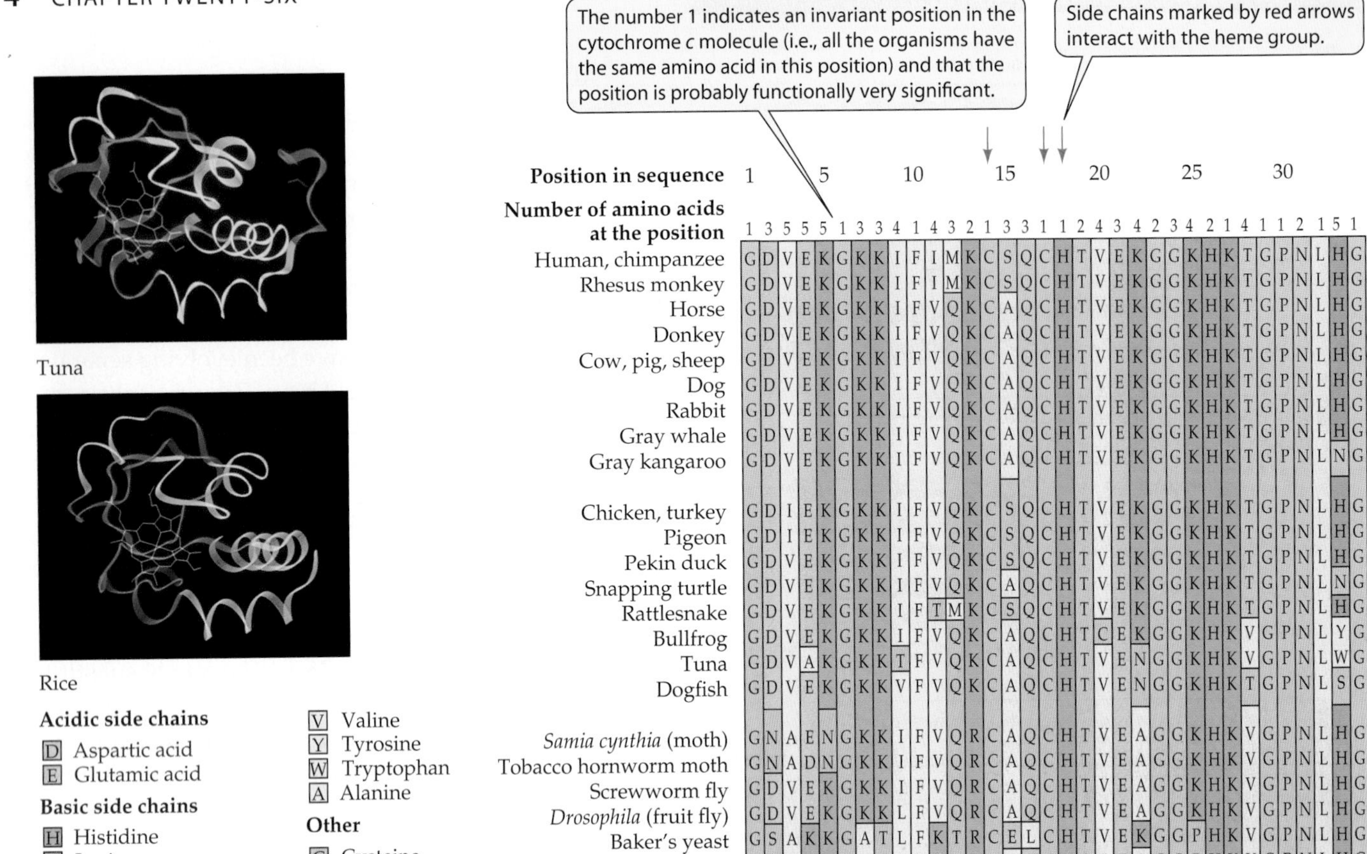

26.4 Amino Acid Sequences of Cytochrome *c* The two computer graphics show how similar the three-dimensional structures of tuna and rice cytochrome *c* are. The amino acid sequences shown in the table were obtained from analyses of cytochromes *c* from 33 species of plants, fungi, and animals.

view that most such nonsynonymous mutations are disadvantageous and are eliminated from the population by natural selection. As a result, *the more essential a molecule is for cell functioning, the slower the rate of its evolution*. These functional constraints provide part of the answer to the question posed above.

A molecule that illustrates this principle is the enzyme cytochrome *c*, a component of the respiratory chain of mitochondria. Together with other enzymes of the citric acid cycle and respiratory chain, cytochrome *c* is found in all eukaryotes and is essential for the life of the eukaryotic cell. The amino acid sequences of cytochrome *c* are known for more than a hundred species of organisms, including protists, plants, fungi, and mammals. Within these cytochromes *c* are regions that have accumulated changes relatively quickly; for example, amino acid positions 44, 89, and 100 differ among many of the organisms compared (Figure 26.4). There are also invariant amino acid positions, such as 14, 17, 18, and 80. This particular set of invariant residues is known to interact with the iron-containing heme group, which is essential for enzyme functioning. Because any mutations that changed these amino acids would have diminished the functioning of cytochrome *c*, they would have been removed by natural selection when they arose.

Changes in macromolecules can serve as molecular clocks

Earlier in this chapter, we mentioned that the rate of evolution of some macromolecules might in effect be a *molecular clock*. To function as a molecular clock, a particular macromolecule would need to evolve at an approximately constant rate in all evolutionary lineages that possess it. But do macromolecules actually behave in this way?

Often they do. For example, if we use the fossil record to determine the time since the divergence of certain organisms, and then plot this time against the number of amino acids by which the nucleotide sequence of the organisms' cytochrome

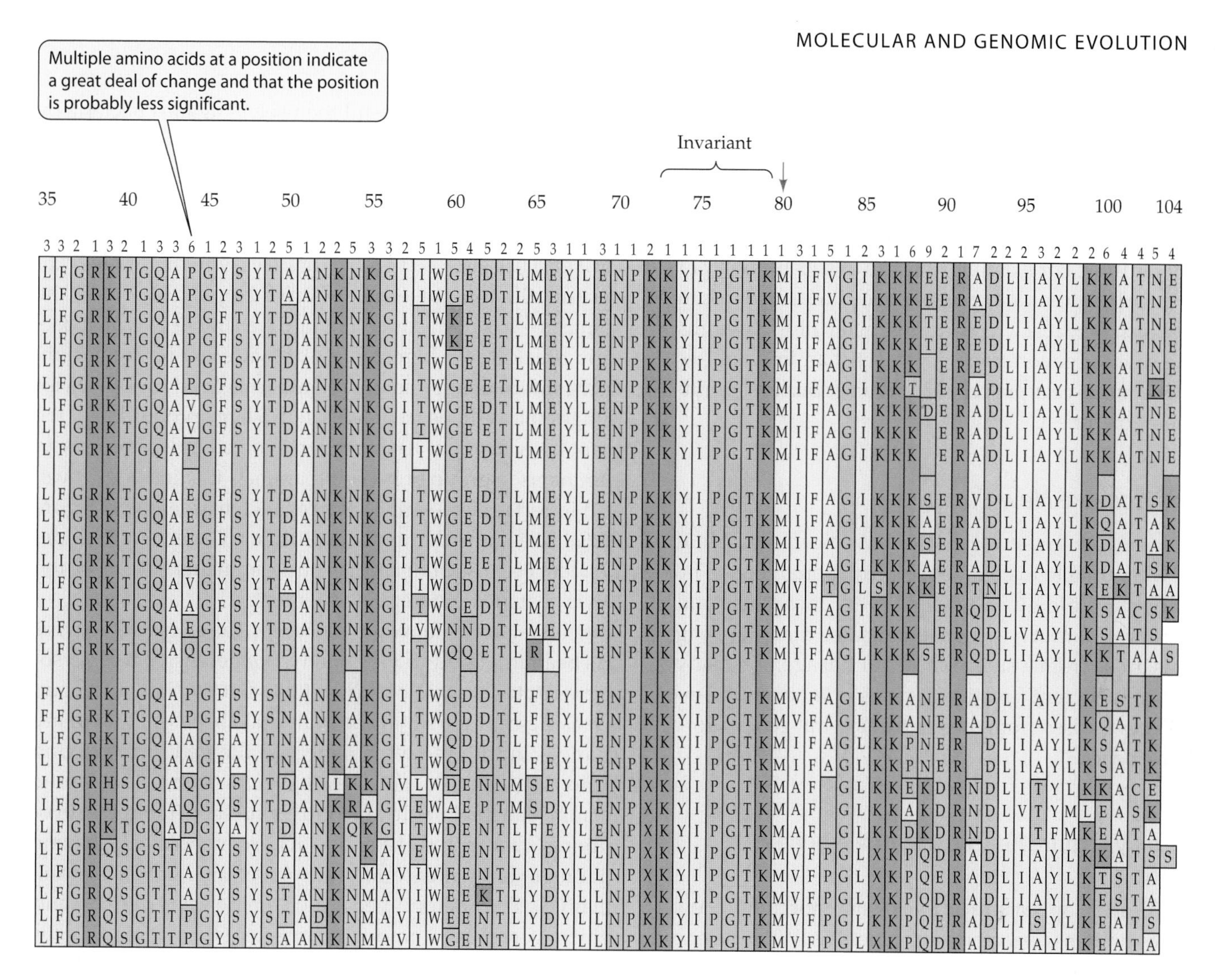

c proteins differ, we find that differences in cytochrome *c* sequences have evolved at a relatively constant rate (Figure 26.5). Many other proteins also show constancy in the rate at which they have accumulated changes over time.

Analyses of patterns of molecular evolution would be relatively easy if the rates of change were the same for all macromolecules. Unfortunately, molecular clocks tick at somewhat different rates over evolutionary time. By comparing the rates of a variety of molecular clocks, insights can be gained into why different molecules have evolved at such different rates. Such differences exist because proteins differ in the nature of the functional constraints on their evolution. For example, the rate of evolution of an enzyme would change drastically if a mutation meant that the enzyme lost its function, or if the population of the species in which the enzyme was found changed dramatically in size.

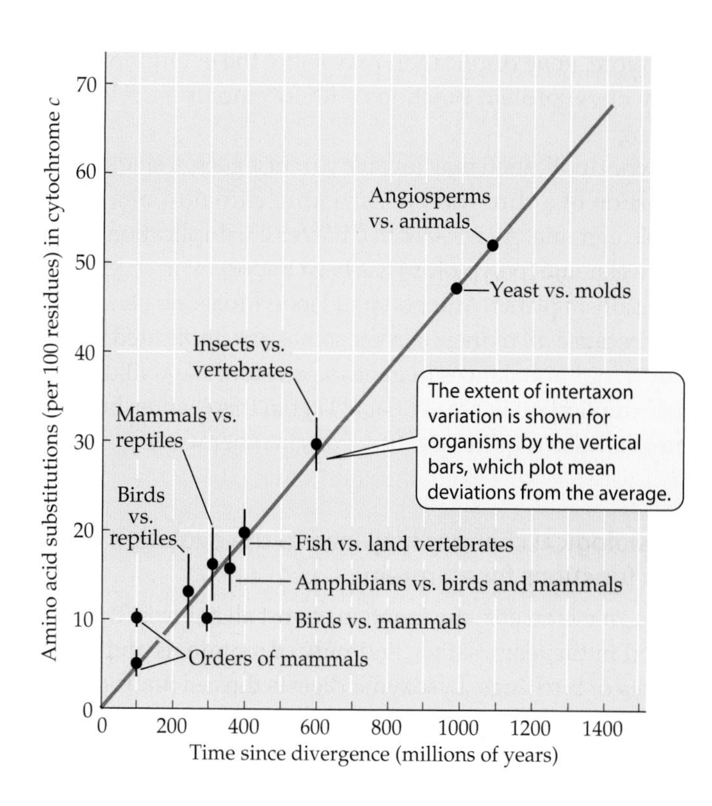

26.5 Cytochrome *c* Has Evolved at a Constant Rate Rates of substitution in cytochrome *c* are constant enough that the evolution of this molecule can be characterized by a molecular clock. The dates in this graph have been inferred from the fossil record.

In addition, rates of molecular evolution are faster in organisms with short generation times than in organisms with long generation times. Short-generation organisms have more rounds of DNA replication, and thus more opportunities for errors in replication, per unit of time than long-generation organisms. For example, the rate of substitution per base per year in introns is from two to four times greater in rodents (which may reach reproductive age within 6 months and have several generations per year) than in primates (which have multi-year pre-reproductive periods and multi-year generation times).

Proteins Acquire New Functions

Evolution as we know it would not have been possible if proteins were unable to change their functional roles. The earliest forms of life must have had very few genes and proteins. Because much evidence indicates that all living organisms arose from a single ancestral lineage, the many thousands of different functional genes in modern organisms must have arisen from those few ancestral genes. How has this happened?

Proteins may acquire new functions via gene duplication

Gene duplication appears to be the most important process that enables proteins to acquire new functions. When a gene is duplicated, one copy of that gene is potentially freed from having to perform its original function; the copy is redundant if the original gene is still producing the original protein. Therefore, gene duplication may allow the evolution of entirely novel protein functions without impairing cell functioning.

Gene duplication may involve part of a gene, a single gene, a portion of a chromosome, an entire chromosome, or the whole genome. As we saw in Chapter 24, duplication of the entire genome (polyploidy) has been important in speciation, especially in plants. Autopolyploid individuals are usually viable because all of their chromosomes are duplicated. Thus, they avoid imbalances in gene expression. We will discuss gene duplication in greater detail later in this chapter, but proteins can also acquire new functions in other ways.

Physiological changes may lead to the evolution of new functions for a protein

Lysozyme is an enzyme found in almost all animals. It is produced in the tears, saliva, and milk of mammals and in the whites of bird eggs. Lysozyme digests the cell walls of bacteria, rupturing and killing them. As a result, lysozyme plays an important role as a first line of defense against invading bacteria. All animals defend themselves against bacteria by digesting them, which is probably why most animals have lysozyme. Some animals, however, also use lysozyme in the digestion of food.

Among mammals, a mode of digestion called *foregut fermentation* has evolved twice. In animals with this mode of digestion, the foregut—the posterior esophagus and/or the stomach—has been converted into a chamber in which bacteria break down ingested plant matter by fermentation. Foregut fermenters can obtain nutrients from the otherwise indigestible cellulose that makes up a large proportion of the plant body. Foregut fermentation evolved independently in ruminants (a group of hoofed mammals that includes cows) and in certain leaf-eating monkeys, such as langurs (Figure 26.6*a*). We know that these evolutionary events were independent because both langurs and ruminants have close relatives that are not foregut fermenters.

In both foregut-fermenting lineages, the enzyme lysozyme has been modified to play a new, nondefensive role. Lysozyme ruptures some of the bacteria that live in the foregut, releasing nutrients that the mammal absorbs. How many changes in the lysozyme molecule allowed it to function amid the digestive enzymes and acidic conditions of the mammalian foregut? To answer this question, molecular evolutionists compared the amino acid sequences of lysozyme in foregut fermenters and in several of their nonfermenting relatives. They determined which amino acids differed and which were shared among the species (Table 26.1). Finally, they compared the pattern of these changes with the phylogenetic relationships among the species that had been established based on fossils and current morphology.

The most striking finding is that amino acid changes in lysozyme, which occurred in the absence of gene duplication, happened about twice as rapidly in the lineage leading to langurs as in any other primate lineage. This high rate of substitution shows that lysozyme went through a period of rapid change in adapting to the stomachs of langurs. The lysozymes of langurs and cows share five amino acid substitutions, all

26.1 Similarity Matrix for Lysozyme in Mammals

SPECIES	LANGUR	BABOON	HUMAN	RAT	COW	HORSE
Langur*		14	18	38	32	65
Baboon	0		14	33	39	65
Human	0	1		37	41	64
Rat	0	1	0		55	64
Cow*	5	0	0	0		71
Horse	0	0	0	0	1	

Shown above the diagonal line is the number of amino acid sequence *differences* between the two species being compared; below the line are the number of sequences uniquely *shared* by the two species. Asterisks (*) indicate foregut-fermenting species.

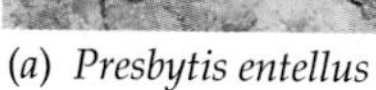

(*a*) *Presbytis entellus*

(*b*) *Opisthocomus hoazin*

26.6 Similar Molecular Evolution Can Take Place in Separate Lineages Foregut-fermenting mammals such as the gray langur (*a*) have been evolving independently from the hoatzin (*b*) for more than 100 million years, but each has evolved similar modifications to the enzyme lysozyme.

of which lie on the surface of the lysozyme molecule, well away from the active site (see Chapter 6). Several of the shared substitutions involve changes from arginine to lysine, which makes the proteins more resistant to attack by the pancreatic enzyme trypsin. By understanding the functional significance of amino acid substitutions, molecular evolutionists can explain the observed changes in amino acid sequences in terms of changes in the functioning of the protein.

A large body of fossil, morphological, and physiological evidence shows that langurs and cows do not share a recent common ancestor. However, langur and ruminant lysozymes share many amino acid residues that neither animal shares with the lysozymes of their own closer relatives. The lysozymes of these two animals have converged on a similar sequence despite having very different ancestry; in other words, they are homoplasies. The amino acid residues they share give these lysozymes the ability to lyse the bacteria that ferment leaves in the foregut.

An even more remarkable story emerges if we look at lysozyme in the crop of the hoatzin, a leaf-eating South American cuckoo, the only known avian foregut fermenter (Figure 26.6*b*). Many birds have an enlarged esophageal chamber called a *crop*. Hoatzins have a crop that contains bacteria and acts as a fermenting chamber. Many of the amino acid changes that occurred in the adaptation of hoatzin crop lysozyme are identical to the changes that evolved in ruminants and langurs. Thus, even though the hoatzin and the foregut-fermenting mammals have been evolving independently from one another for more than 100 million years, they have each evolved a similar molecule that enables them to recover nutrients from their fermenting bacteria in a highly acidic environment. The lysozyme story also illustrates why inferring phylogenies from data on single molecules can be very misleading. Identical molecular changes in this protein are not evidence of common descent.

The Evolution of Genome Size

We can now consider what determines the sizes of the genomes of different organisms. As organisms evolved to become more complex, how did the number of functional genes increase so that the organisms could carry out the greater variety of metabolic activities associated with that complexity?

Complex organisms have more DNA than do simpler ones

As we saw in Chapters 13 and 14, genome size varies tremendously among organisms. The first pattern to be detected was that genome sizes are generally correlated with organisms' complexity. The genome of *Mycoplasma genitalium*, the simplest known prokaryote, has only 470 genes. *Rickettsia prowazekii*, the prokaryote that causes typhus, has 634 genes. *Homo sapiens*, on the other hand, has about 21,000 protein-coding genes. Figure 26.7 shows the relative sizes of several prokaryotic and eukaryotic genomes.

It is not surprising that more complex genetic instructions are needed for building and maintaining a large, complex organism than a small, simple one. What is surprising is that some organisms, such as lungfishes, some salamanders, and lilies, have about 40 times as much DNA as humans do. Clearly, a lungfish or a lily is not 40 times as complex as a human. Why does genome size vary so much?

Some of the apparent variation in genome size disappears when we compare the portion of DNA that actually encodes RNAs or proteins. The size of the coding genome of organisms varies in a way that makes sense. Eukaryotes have more coding DNA than prokaryotes; plants have more coding DNA than single-celled organisms; invertebrates with wings,

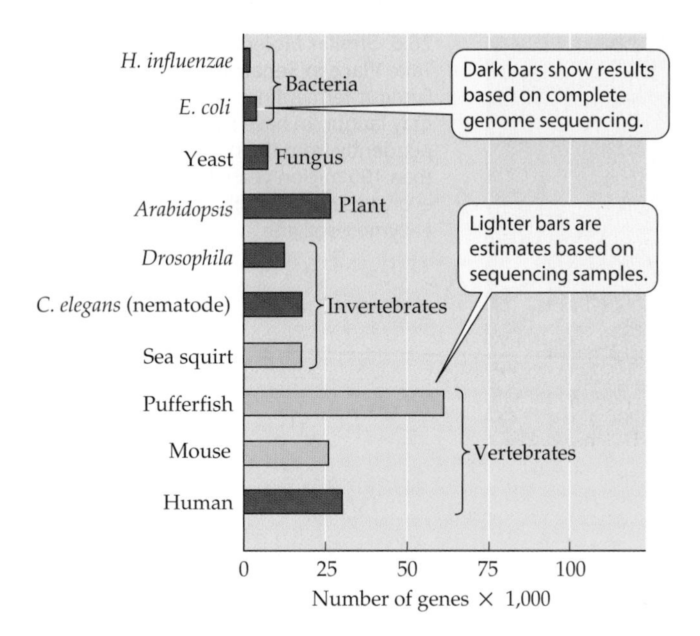

26.7 Complex Organisms Have More Genes than Simpler Organisms Genome sizes have been measured or estimated in a variety of organisms ranging from single-celled prokaryotes to vertebrates.

legs, and eyes have more coding DNA than nematodes; and vertebrates have more coding DNA than invertebrates. The organisms with the largest amount of nuclear DNA (some ferns and flowering plants) have 80,000 times as much as the simplest organisms, but no species has more than 20 times as many protein-coding genes as a bacterium. Therefore, most of the variation in genome size lies not in the number of functional genes, but in the amount of noncoding DNA (Figure 26.8).

What maintains such large quantities of noncoding DNA in the cells of most organisms? Does this noncoding DNA have a function, or is it "junk?" Most of this DNA appears to be nonfunctional. Much of it may consist of pseudogenes that are simply carried in the genome because the cost of doing so is very small. Some of it consists of parasitic transposable elements that spread through populations because they reproduce faster than the host genome. Investigators can use one type of transposable element, retrotransposons, to estimate the rates at which species lose DNA.

Retrotransposons copy themselves with the aid of RNA, as we saw in Chapter 14. The most common type of retrotransposon carries duplicated sequences at each end, called long terminal repeats (LTRs). Occasionally, LTRs join together in the host genome, at which time the DNA between them is excised. When this happens, one of the LTRs is left behind. The number of such "orphaned" LTRs in a genome is a measure of how many retrotransposons have been lost. By comparing the number of LTRs in the genomes of Hawaiian crickets of the genus *Laupala* and those of fruit flies (*Drosophila*), investigators found that *Laupala* loses DNA more than 40 times more slowly than *Drosophila*. As a result, the genome of *Laupala* is 11 times larger than that of *Drosophila*. Why species differ so greatly in the rate at which they lose DNA is not understood.

Gene duplication can increase genome size and complexity

The identical copies of a duplicated gene can have any one of three different fates:

- Both copies of the gene may retain their original function, with the result that the organism produces larger quantities of the gene's RNA or protein product.
- One copy of the gene may be incapacitated by the accumulation of deleterious mutations and become a functionless *pseudogene*.
- One copy of the gene may retain its original function while the second copy accumulates enough mutations that it can perform a different function.

It is the third of the above fates that is most significant for evolution.

How often do gene duplications arise, and which of the three outcomes described above is most likely? These questions can be addressed by counting the number of synonymous nucleotide base changes in the genome of an organism. This number is then compared with the number of base changes that caused protein alterations, to see which number changed faster. Investigators have found that rates of gene duplication are fast enough for a yeast or *Drosophila* population to acquire several hundred duplicate genes over

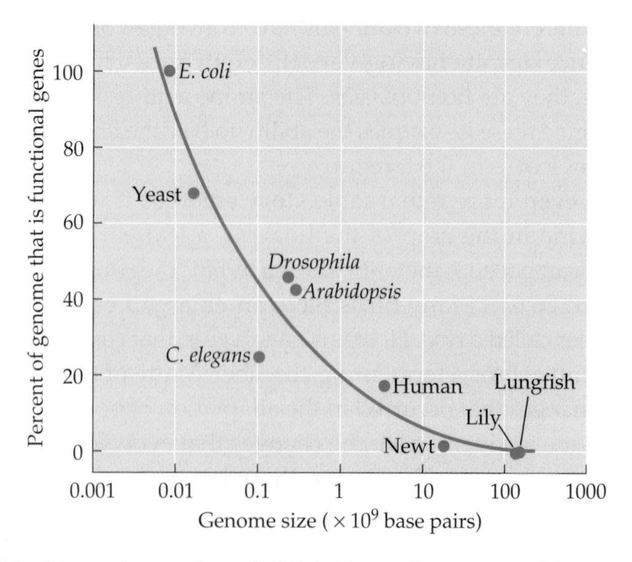

26.8 A Large Proportion of DNA Is Noncoding Most of the DNA of bacteria and yeasts encodes RNAs or proteins, but most of the DNA of more complex organisms is noncoding. Most noncoding DNA is probably nonfunctional.

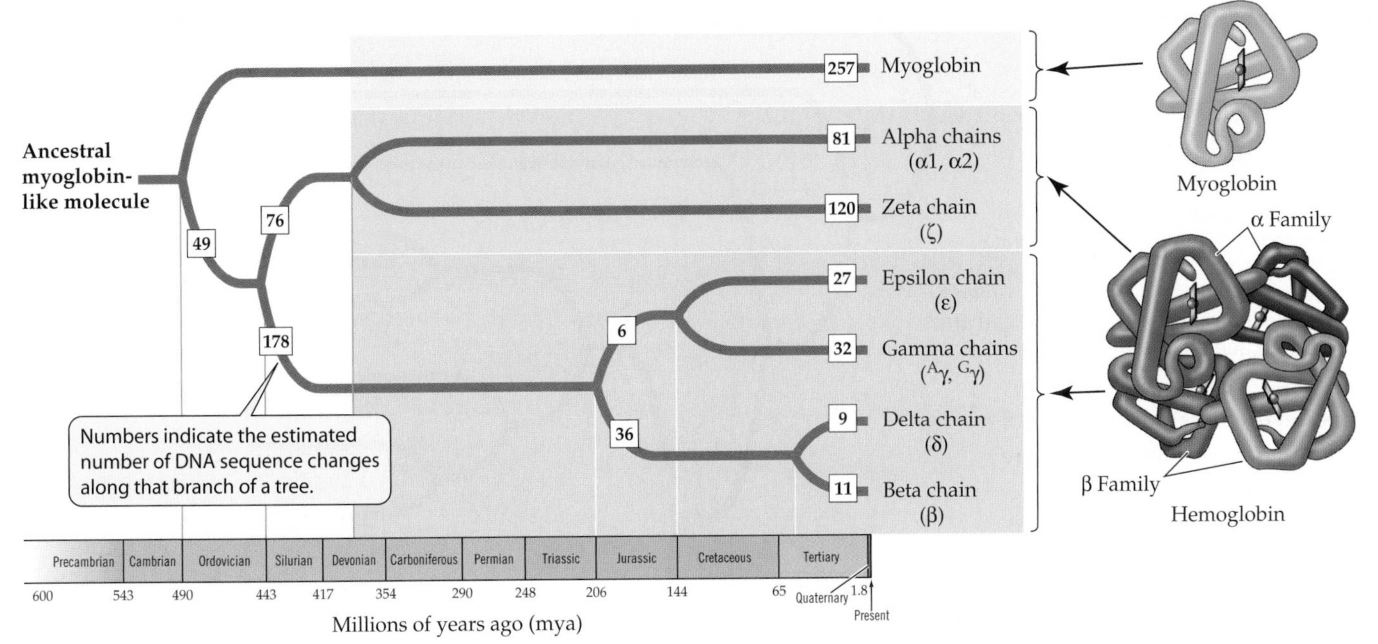

26.9 A Globin Family Gene Tree This gene tree suggests that the α-globin and β-globin gene clusters diverged about 450 mya, at about the time of the origin of the vertebrates.

the course of a million years. They also found that most of the duplicated genes in these organisms are very young. Extra genes typically are lost from a genome within 10 million years (which is rapid on an evolutionary time scale).

Although extra genes usually disappear rapidly, some duplication events lead to the evolution of genes with new functions. Several successive rounds of duplication and mutation may result in a **gene family**, a group of homologous genes with related functions, often arrayed in tandem along a chromosome. An example of this process is provided by the globin gene family (see Figure 14.7). The globins were among the first proteins to be sequenced and compared. Comparisons of their amino acid sequences strongly suggest that the different globins arose via gene duplications. How long the globins have been evolving separately can also be inferred by comparing their amino acid sequences. The greater the number of amino acid differences between two globins, the further back in time was their most recent common ancestor.

Hemoglobin, a tetramer consisting of two α-globin chains and two β-globin chains, carries oxygen in blood. Myoglobin, a monomer, is the primary oxygen storage protein in muscle. Myoglobin's affinity for O_2 is much higher than that of hemoglobin. In contrast, hemoglobin evolved to be more diversified in its role. Hemoglobin binds O_2 in the lungs or gills, where the O_2 concentration is relatively high, transports it to deep body tissues, where the O_2 concentration is low, and releases it in those areas. With its more complex tetrameric structure (see Figure 3.8), hemoglobin is able to carry four molecules of O_2, as well as hydrogen ions and carbon dioxide, in the blood.

To estimate the time of the globin gene duplication that gave rise to the α- and β-globin gene clusters, we can create a **gene tree** based on the estimated number of base substitutions necessary to account for the observed amino acid differences between the globins. Based on this gene tree, and assuming that the rate of amino acid substitution has been relatively constant since then—about 100 substitutions per 500 million years—the two globin gene clusters are estimated to have split about 450 mya (Figure 26.9).

The Uses of Molecular Genomic Information

Information about the genomes of organisms enables biologists to investigate a variety of biological problems in new and powerful ways. Information about the rates at which molecules have evolved is used in reconstructing phylogenetic trees. Molecular genomic information can also be used to determine the phylogenies of genes, which often differ from those of organisms because genes can be transferred from one evolutionary lineage to another, a phenomenon known as *lateral gene transfer*. Recent evidence suggests that such lateral transfers may have occurred repeatedly during the evolution of plants. And, as we discussed at the beginning of this chapter, genomic information has powerful medical applications.

Molecular information is used to reconstruct phylogenies

By comparing the structures of molecules from different species, we both gain insights into how those molecules function and acquire a tool for inferring phylogenies. Molecules that have evolved slowly can be used to reconstruct relationships among organisms that diverged long ago. Molecules

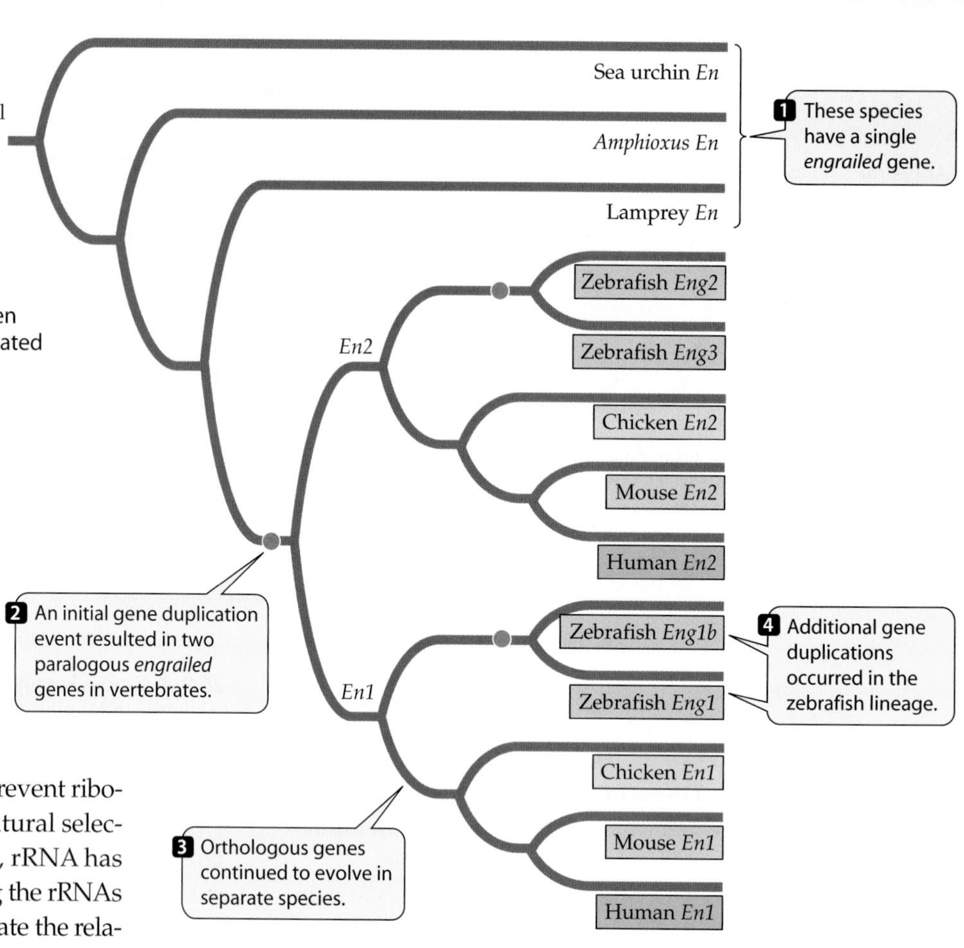

26.10 Phylogeny of the *engrailed* Genes All of the *engrailed* genes are orthologs because they have a common ancestor. Gene duplication events have generated paralogous *engrailed* genes in the vertebrate lineages. Yellow boxes represent orthologous genes that developed after lineage splits. Green and blue boxes indicate paralogous genes created by a gene duplication.

that have evolved rapidly are useful for studying organisms that share more recent common ancestors.

To investigate the evolutionary relationships of all existing organisms, we choose molecules that all organisms possess, such as ribosomal RNA. Equally important to our choice are strong functional constraints, such as those that exist for rRNA. Even minor changes in the rRNA sequence would prevent ribosomes from functioning properly, so natural selection acts to eliminate them. As a result, rRNA has evolved so slowly that differences among the rRNAs of living organisms can be used to estimate the relative timing of lineage splits that may have happened billions of years ago. No fossils exist to document the most ancient splits in the lineages of life on Earth into the three major domains—the Bacteria, the Archaea, and the Eukarya (see Figure 1.8)—so we depend on molecules such as rRNA for insight into these events.

Molecular data are also used in combination with morphological and fossil data to reconstruct phylogenies. Why do we use molecules when morphological data are available? The answer is simple: The more characters we use to reconstruct a phylogeny, the less likely we are to be misled by evolutionary reversals or convergent evolution.

Molecular data are used to determine the phylogenetic histories of genes

A gene tree shows the evolutionary relationships of members of a gene family, just as a phylogenetic tree depicts the evolutionary relationships of members of a lineage. All of the genes of a particular family have similar sequences because they have a common ancestry. Genes found in different organisms that arose from a single gene in their common ancestor are called **orthologs**. Genes that are related through gene duplication events in a single lineage are called **paralogs**.

Figure 26.10 depicts a gene tree for the members of the deuterostome *engrailed* gene family. All of the *engrailed* genes are orthologs because they have a common ancestor. Gene duplication events have generated paralogous *engrailed* genes in some lineages. The homeotic genes of *Drosophila melanogaster* (see Chapter 19) are also paralogs.

Molecular information provides new ways to combat diseases

The effort to determine the genomes of *Anopheles* and *Plasmodium* that we described at the beginning of this chapter has already had medical benefits. Transgenic mosquitoes have been engineered to express an anti-*Plasmodium* molecule that makes them inefficient vectors of malaria in the laboratory. This advance would not have been possible without knowledge of the parasite's genome. In addition to genetic information, success in controlling malaria will require detailed knowledge of the behavior, ecology, and evolution of the many species of mosquitoes that transmit malaria in different parts of the world.

Information provided by the genomic sequence of *Treponema pallidum*, the bacterium that causes syphilis (see Figure 27.12), is being used to help develop a vaccine against this sexually transmitted disease. Syphilis may have been introduced into Europe from the Americas by members of Columbus's expedition, and the disease spread across the world during the sixteenth-century age of exploration. Effective therapies have been available since the discovery of penicillin in the mid-twentieth century, but syphilis remains a serious global health problem.

Treponema pallidum has been a difficult organism to study because it cannot be grown outside a mammalian host. The genome sequence of *T. pallidum* may provide the information necessary for the development of a culture medium. In addition, the sequence revealed that *T. pallidum* contains a family of paralogous genes (*tprA–L*) that encode proteins of the bacterial outer membrane. The identification of this gene family is guiding the search for targets for a vaccine against the disease.

Molecular data cannot solve all disease problems

Molecular data are powerful tools in our struggle with diseases, but the AIDS epidemic shows that they cannot solve all medical problems. As we saw in Chapter 13, the agent that causes AIDS is a retrovirus (HIV) that uses RNA as its genetic material. The central core of the virus contains two identical copies of the RNA genome as well as three enzymes needed to carry out the viral life cycle: reverse transcriptase, integrase, and a protease.

The highly active antiretroviral therapy (HAART) most commonly used against AIDS today employs a combination of protease inhibitors and reverse transcriptase inhibitors, each of which blocks a different stage in the viral life cycle (see Chapter 18). Unfortunately, resistant strains of HIV develop in the blood of most patients that receive HAART. HIV's mutation rate is very high, and since it has no repair enzymes, a wrong base is inserted at about one out of every 8,000 nucleotide positions. The result is that, on average, a new mutant is generated every time HIV replicates its genome—in effect, no two viruses are identical! Although biomedical scientists are now armed with detailed knowledge of the molecules that govern the different stages in the viral life cycle, HIV's extremely high mutation rate has prevented them from coming up with drugs to which the virus cannot rapidly adapt. In the absence of such drugs, reducing both the incidence and severity of HIV will require changes in human sexual behavior.

The development of molecular methods, the sequencing of the genomes of an increasing number of species, and powerful computers have ushered in a new era in the scientific study of the diversity of life. Scientific understanding of the evolutionary patterns of life on Earth and how the agents of evolution have governed these patterns is advancing more rapidly than at any prior time during the study of evolution. The range of data used by systematists is likely to continue to increase because, using modern chemical, biochemical, and microscopic methods, we will be able to measure more traits of organisms than we could previously.

By combining molecular data with information from the fossil record (which is also increasing rapidly), biologists are developing a comprehensive picture of the evolution of life on Earth. In Part Five, we will provide an extensive overview of the remarkable diversity of organisms that is the result of almost 4 billion years of evolution.

Chapter Summary

Genomes and Their Evolution

- A genome is the full set of genes an organism contains.
- The genes in an organism's genome are usefully viewed as interacting members of a group.

The Evolution of Macromolecules

- Molecular evolutionists characterize the structures of macromolecules and use them to determine how rapidly these macromolecules have changed and why they have changed.
- Mutations and genetic drift are important determinants of rates of molecular evolution. **Review Figure 26.1**

Determining and Comparing the Structure of Macromolecules

- Molecules are compared by aligning their sequences and counting the differences between those sequences. **Review Figure 26.2. See Web/CD Activity 26.2**
- Changes evolve slowly in regions of molecules that are functionally significant, but more rapidly in regions where substitutions do not affect the functioning of the molecules. **Review Figure 26.3, 26.4**
- Rates of substitution in some molecules are relatively constant over evolutionary time; that is, these molecules can serve as molecular clocks. **Review Figure 26.5**

See Web/CD Activity 26.1

Proteins Acquire New Functions

- Most new protein functions arise by gene duplication.
- Changes in the functions performed by proteins may also result from changes in the physiological roles of gene products.

The Evolution of Genome Size

- The genome sizes of organisms vary tremendously, but the amount of DNA that actually encodes RNAs or proteins varies much less. **Review Figures 26.7, 26.8**
- Complex organisms have more coding DNA than simpler ones.
- The globin family of proteins evolved via gene duplication. **Review Figure 26.9**

The Uses of Molecular and Genomic Information

- Molecular data can be used to infer phylogenetic relationships among organisms.
- Molecules that have evolved slowly are useful for determining ancient lineage splits. Molecules that have evolved rapidly are useful for determining more recent lineage splits in combination with morphological and fossil data.
- Molecular data are used to determine the phylogenetic histories of genes. **Review Figure 26.10**
- Molecular data are used to find new ways to combat diseases.

See Web/CD Activity 26.3

Self-Quiz

1. Which of the following questions do students of genomic evolution *not* try to answer?
   a. What are the forces that maintain interactions among different genes?
   b. Why are the genomes of organisms so variable in size?
   c. How has enlargement of genomes been accomplished?
   d. Why is DNA the genetic material of most organisms?
   e. How do proteins acquire new functions?
2. Molecular evolution differs from phenotypic evolution in which of the following ways?
   a. It requires changes in molecules if it is to happen.
   b. Random genetic drift and mutations usually exert greater influences on rates and directions of molecular evolution than on rates and directions of phenotypic evolution.
   c. Molecular evolution is not influenced by natural selection.
   d. Rates of molecular evolution are much slower than rates of phenotypic evolution because mutation rates typically are low.
   e. There are no important differences between molecular and phenotypic evolution.
3. Choosing the appropriate molecule for phylogenetic reconstruction does *not* require a consideration of the
   a. question being answered.
   b. rate of evolution of the molecule.
   c. phylogenetic distribution of the molecule.
   d. function of the molecule.
   e. completeness of the fossil record.
4. Ribosomal RNA sequences are useful for addressing the evolutionary relationships of lineages that diverged in ancient times because they
   a. evolve at a rapid rate.
   b. have undergone convergent evolution in many lineages.
   c. are molecules that all organisms have.
   d. consist of mainly neutral characters.
   e. are difficult to align.
5. Mitochondrial DNA sequences are useful in studying the recent evolution of closely related species because
   a. some mitochondrial genes accumulate mutations very rapidly.
   b. they are paternally inherited.
   c. they evolve only in a neutral fashion.
   d. they are highly constrained in function.
   e. they recombine every generation.
6. Issues concerning patterns of molecular evolution include
   a. evolutionary relationships among molecules.
   b. molecular clock.
   c. rate of mutation for neutral characters.
   d. the importance of gene duplication in evolution.
   e. All of the above
7. Molecules are used to reconstruct phylogenies even if a fossil record is available, because
   a. the more characters the better.
   b. molecules are more accurate characters than are fossils.
   c. molecules undergo less homoplasy than do fossil characters.
   d. molecules are less subjective characters than are fossils.
   e. molecules give us the "right" phylogeny.
8. Neutral characters
   a. are not evolving under the influence of natural selection.
   b. have a neutral pH.
   c. are not useful in reconstructing phylogenies.
   d. are subject to strong functional constraints.
   e. are not likely to evolve.
9. The concept of a molecular clock implies that
   a. many proteins show a constancy in rate of change with time.
   b. organisms evolve at a constant rate.
   c. one can date evolutionary events with molecules alone.
   d. all molecules change at the same rate in evolution.
   e. we can predict how rapidly all genes will evolve.
10. Proteins acquire new functions primarily by means of
    a. gene duplication, which frees one copy of a gene from having to perform its original function.
    b. gene duplication, which provides two copies of a gene that, working together, produce a new protein.
    c. deletions, which generate new protein shapes.
    d. deletions, which make proteins nonfunctional, thereby creating new opportunities for other proteins.
    e. None of the above
11. The lysozyme story suggests that
    a. molecules cannot change their function in evolution.
    b. selection does not act at the molecular level.
    c. molecules can help us understand the process of organismic evolution.
    d. all organisms are capable of fermenting bacteria.
    e. lysozyme has a very accurate molecular clock.
12. The actual differences in genome sizes are much less than the apparent differences because
    a. multicellular organisms are really not that much more complicated than eukaryotic protists.
    b. organisms with the largest amounts of nuclear DNA have much more noncoding DNA than organisms with smaller amounts of nuclear DNA.
    c. the sizes of many apparently large genomes have been seriously overestimated.
    d. differences in the sizes of genes account for much of the apparent difference in genome sizes.
    e. species with large genomes preferentially lose DNA by converting it to pseudogenes, which are subsequently lost.

For Discussion

1. If you were interested in reconstructing the phylogeny of a genus of fruit flies using molecular data, what kinds of molecule(s) would you choose to examine? Why? If you wanted to reconstruct the phylogeny of all vertebrates, would you use the same molecule(s)? Why or why not?
2. Discuss the relative importance of molecular characters and morphological characters in reconstructing the phylogeny of a group of organisms.
3. Existing evidence suggests that for some molecules, a molecular clock ticks at a fairly constant rate, but that rates of change differ widely among molecules. How does this variation limit how and in what ways we can use the concept of a molecular clock to help us answer questions about the evolution of both molecules and organisms?
4. One hypothesis proposed to explain the existence of large amounts of noncoding ("junk") DNA is that the cost of maintaining that DNA is so small that natural selection is too weak to reduce it. What other hypotheses might account for the existence of so much noncoding DNA?
5. If fossil evidence and molecular evidence disagree on the date of a major lineage split, which of the two kinds of evidence would you favor? Why?
6. Soon scientists will be able to produce and release into the wild genetically modified mosquitoes unable to harbor and transmit malaria parasites. What ethical issues need to be discussed before such releases are permitted?

LIFE ESSAY

How has Darwin's theory of natural selection transformed our view of humanity's place in the universe?

- by Daniel Dennett -

For as long as our ancestors have been making tools, it has no doubt seemed obvious that an excellent artifact can be created only by something even more excellent: a clever artificer. You never see a shoe creating a cobbler; you never see a house making a carpenter. Darwin overthrew that received wisdom. One of Darwin's earliest critics, Robert Beverley MacKenzie, could not contain his outrage:

> In the theory with which we have to deal, Absolute Ignorance is the artificer; so that we may enunciate as the fundamental principle of the whole system, that, IN ORDER TO MAKE A PERFECT AND BEAUTIFUL MACHINE, IT IS NOT REQUISITE TO KNOW HOW TO MAKE IT. This proposition will be found … to express in a few words all Mr. Darwin's meaning; who, by a strange inversion of reasoning, seems to think Absolute Ignorance fully qualified to take the place of Absolute Wisdom in all the achievements of creative skill.

This is indeed a "strange inversion of reasoning," but once the topsy-turvy perspective it implies has been accepted, most of what we have believed about who we are survives intact. We can still be in awe of the "Wisdom in all the achievements of creative skill" while attributing this wisdom not to a single Creator, but distributing it over billions of years in trillions of lineages of replicators, trying their luck in the great tournament of life, mindlessly discovering and rediscovering the brilliant design principles that constitute the diversity of life. Tradition honors the trickle-down theory of value: what we do and think can be valuable only if it derives its value from something even more valuable—only if we are the servants, in effect, of a greater master. Darwin's "strange inversion" obliges us to rethink what could make something valuable, and then we notice that a bubble-up theory has much to recommend it. There was a time when there was no morality on this planet, and now it has evolved. Just as the air we breathe was created as a by-product of the activities of billions of years of simpler life forms, the very meaning of life on this planet has emerged from the efforts of the life forms that the atmosphere enabled.

We are animals. Are we *just* animals? The ideological tug-of-war over "human exceptionalism" can be damped, if not stopped outright, by emphasizing a few uncontroversial facts. Sight, the capacity to extract huge amounts of relevant information from a relatively safe distance, was an innovation that multiplied the opportunities of intelligent behavior: locomotion, predation, evasion, migration, and so on. Sight and flight have each evolved numerous times, but language has evolved just once, so far as we know—in our genus. (Neanderthals may have been a second talking species for a while.) Language is the key to our huge advantage in knowledge and technology. Other animal species transmit significant amounts of know-how nongenetically from parent to offspring, but without language, the lessons to be learned are rather simple preferences and prohibitions, not elaborate systems of hard-won technique and patiently gathered data.

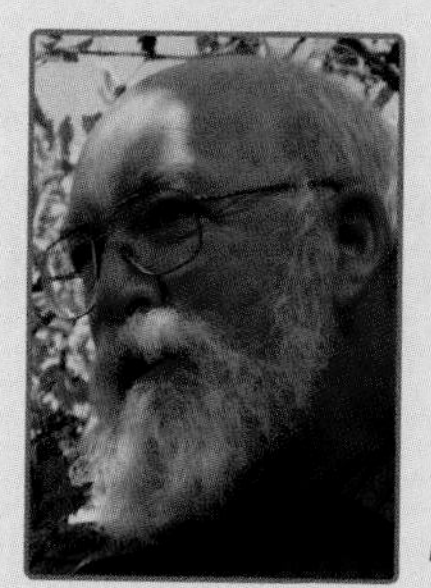

Daniel Dennett is University Professor and Director of the Center for Cognitive Studies at Tufts University, Medford, Massachusetts. He is the author of many books and articles, among them *Consciousness Explained* (1991), *Darwin's Dangerous Idea* (1995), and *Freedom Evolves* (2003).

It has taken our species thousands of years of communication and investigation to begin to find the keys to our own identities. Our newfound capacity for long-distance knowledge gives us powers that dwarf those of all the rest of the life on Earth. It has been estimated that ten thousand years ago, the human population comprised a small fraction of 1 percent of the mass of vertebrate life on land; today, we, together with our livestock and pets, make up about 98 percent of that total. We exploit an ever increasing share of the planet's resources, but we do offer something in return. Now, for the first time in its billions of years of history, our planet is protected by far-seeing sentinels, able to anticipate danger from the distant future—an asteroid on a collision course, or global warming—and devise schemes for doing something about it. The planet has finally grown its own nervous system: us. We are responsible for the future of life on the planet, in a way no other species could ever be.

Discussion Questions

1. What developments would make it possible or likely for language to evolve again, in another species?
2. Are there any good reasons, aside from tradition, for favoring a trickle-down theory of value over a bubble-up theory of value?
3. Beavers build elaborate dams; so do civil engineers. What role does language play in determining the kinds of artifacts a species can make?
4. How would you defend the hypothesis that our ancestors learned to control fire before they mastered language?

Web Links

Center for Cognitive Studies at Tufts University: ase.tufts.edu/cogstud/

Darwin Day, University of Tennessee at Knoxville: fp.bio.utk.edu/darwin/

The C. Warren Irvin, Jr., Collection of Darwiniana: www.sc.edu/library/spcoll/nathist/darwin/darwin.html

Part Five • THE EVOLUTION OF DIVERSITY

27 Bacteria and Archaea: The Prokaryotic Domains

The ancient Phoenicians called it the "river of fire." Today, Spanish astrobiologist Ricardo Amils Pibernat calls Spain's Río Tinto a possible model for the scene of the origin of the life that may have existed on Mars. The river wends its way through a huge deposit of iron pyrite—"fool's gold," or iron disulfide. Prokaryotes in the river and in the damp, acidic soil from which it arises convert the pyrite into sulfuric acid and dissolved iron. The iron gives the river its reddish brown color.

Over a period of at least 300,000 years, these prokaryotes have produced an environment seemingly hostile to life. The Río Tinto has a pH of 2 and exceptionally high concentrations of heavy metals, especially iron. The concentrations of oxygen in the river and in its source soil are extremely low. It is that soil that Amils believes resembles the kind of environment in which life could have begun on Mars. Whatever the truth of that speculation, the Río Tinto represents one of the most unusual habitats for life on Earth.

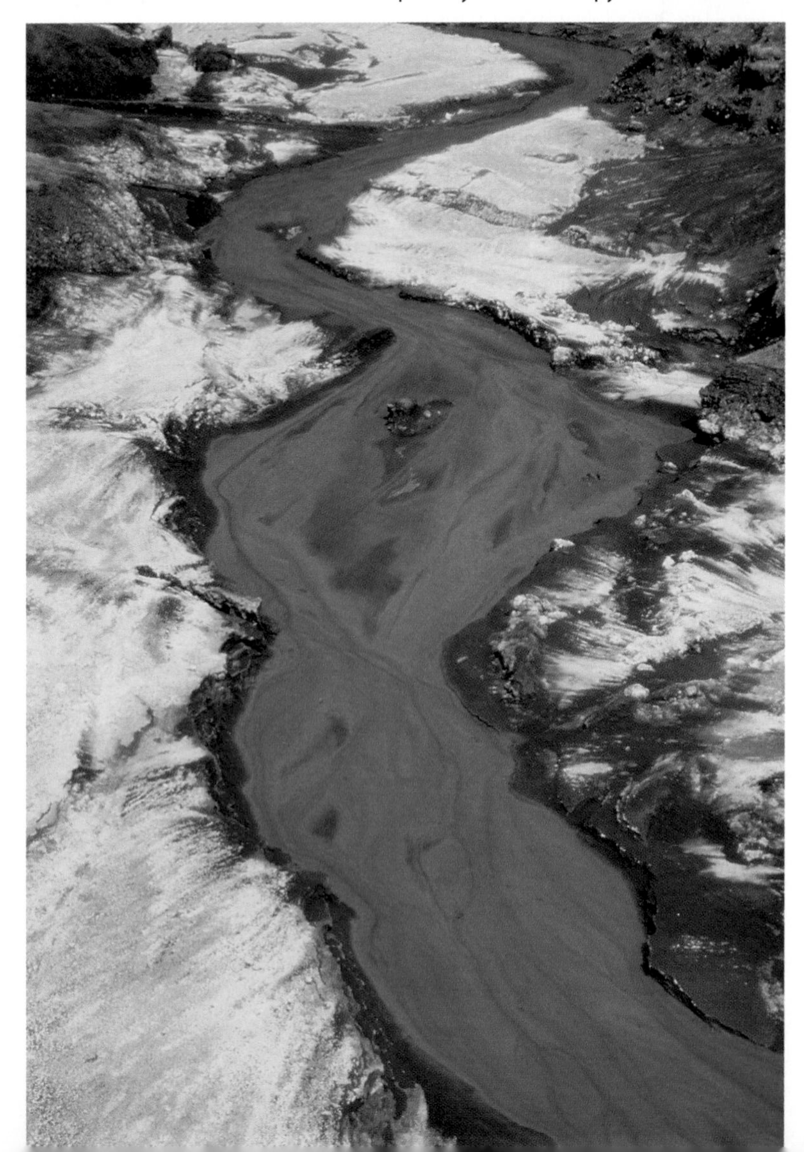

Earth or Ancient Mars? Spain's Río Tinto owes its rusty red color—and its extreme acidity—to the action of prokaryotes on iron pyrite-rich soil.

The organisms most commonly found in such extremely acidic environments belong to the two major groups of prokaryotes: Bacteria and Archaea. The bacteria live in almost every environment on Earth. The archaea are a superficially similar group of microscopic, unicellular prokaryotes. However, both the biochemistry and the genetics of bacteria differ in numerous ways from those of archaea. Not until the 1970s did biologists discover how radically different bacteria and archaea really are. And only with the sequencing of an archaeal genome in 1996 did we realize just how extensively archaea differ from both bacteria and eukaryotes.

Many biologists acknowledge the antiquity of these clades and the importance of their differences by recognizing three domains of living things: Bacteria, Archaea, and Eukarya. The domain Bacteria comprises the "true bacteria." The domain Archaea (Greek *archaios*, "ancient") comprises other prokaryotes once called (inaccurately) "ancient bacteria." The domain Eukarya includes all other living things on Earth.

Dividing the living world in this way, with two prokaryotic domains and a single domain for all the eukaryotes, fits with the current trend toward reflecting evolutionary relationships in classification systems. In the eight chapters of Part Five, we celebrate and describe the diversity of the living world—the products of evolution. This chapter focuses on the two prokaryotic domains. Chapters 28–34 deal with the protists and the kingdoms Plantae, Fungi, and Animalia.

In this chapter, we pay close attention to the ways in which the two domains of prokaryotic organisms resemble each other as well as the ways in which they differ. We will describe the impediments to the resolution of evolutionary relationships among the prokaryotes. Then we will survey the surprising diversity of organisms within each of the two domains, relating the characteristics of the different prokaryotic groups to their roles in the biosphere and in our lives.

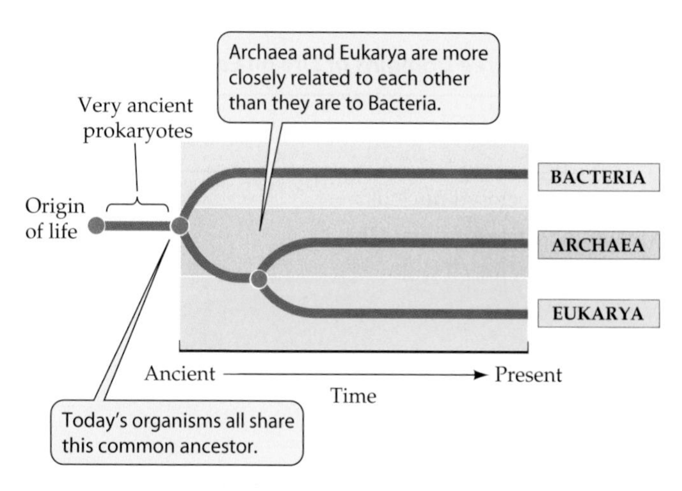

27.2 The Three Domains of the Living World Many biologists believe that the three domains share a common prokaryotic ancestor. The relationships shown here, however, remain controversial.

Why Three Domains?

What does it mean to be *different?* You and the person nearest you look very different—certainly you appear more different than the two cells shown in Figure 27.1. But the two of you are members of the same species, while these two tiny organisms that look so much alike actually are classified in entirely separate domains. Still, all three of you (you in the domain Eukarya and those two prokaryotes in the domains Bacteria and Archaea) have a lot in common. Members of all three domains

- conduct glycolysis;
- replicate their DNA semiconservatively;
- have DNA that encodes polypeptides;
- produce these polypeptides by transcription and translation and use the same genetic code;
- have plasma membranes and ribosomes in abundance.

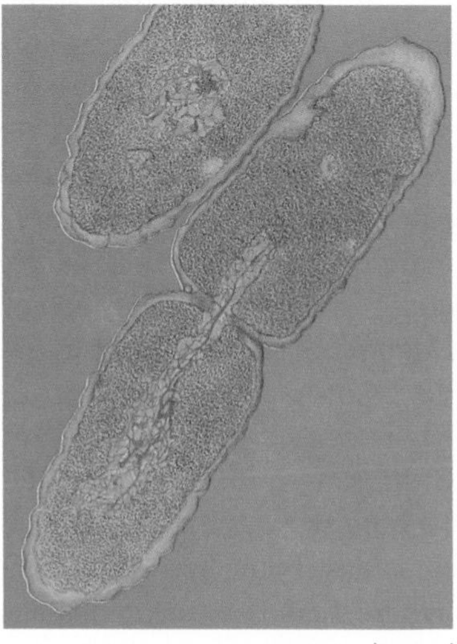

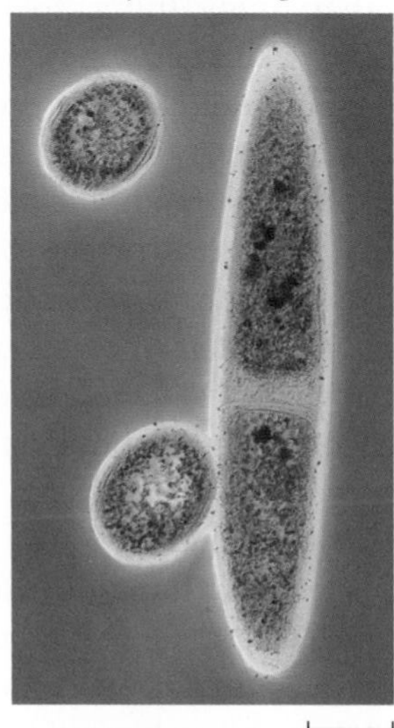

27.1 Very Different Prokaryotes In each image, one of the cells has nearly finished dividing. On the left are bacteria; on the right are archaea, which are more closely related to eukaryotes than they are to the bacteria.

There are also major differences among the domains. Members of the Eukarya have cells with nuclei, membrane-enclosed organelles, and a cytoskeleton—structures that no prokaryote has. And a glance at Table 27.1 will show you that there are also major differences (most of which cannot be seen even under the microscope) between the two prokaryotic domains. In some ways the archaea are more like us; in other ways they are more like bacteria.

Genetic studies have led many biologists to conclude that all three domains had a single common ancestor, and that the present-day archaea share a more recent common ancestor with eukaryotes than they do with bacteria (Figure 27.2). Because of the ancient time at which these three clades diverged, the major differences among the three kinds of organisms, and especially the likelihood that the archaea are more closely related to the eukaryotes than are either of those groups to the bacteria, many biologists agree that it makes sense to treat these three groups as *domains*—a higher taxonomic category than *kingdoms*. To treat all the prokaryotes as a single kingdom within a five-kingdom classification of organisms would result in a kingdom that is paraphyletic. That is, a single kingdom "Prokaryotes" would not include all the descendants of their common ancestor. (See Chapter 25, especially Figure 25.8, for a discussion of paraphyletic groups.) We will use the domain concept in this book, although it is still controversial and may have to be abandoned if new data fail to support it.

The common ancestor of all three domains was prokaryotic. Its genetic material was DNA; its machinery for tran-

27.1 The Three Domains of Life on Earth

CHARACTERISTIC	DOMAIN		
	BACTERIA	ARCHAEA	EUKARYA
Membrane-enclosed nucleus	Absent	Absent	*Present*
Membrane-enclosed organelles	Absent	Absent	*Present*
Peptidoglycan in cell wall	*Present*	Absent	Absent
Membrane lipids	Ester-linked Unbranched	*Ether-linked* *Branched*	Ester-linked Unbranched
Ribosomes[a]	70S	70S	*80S*
Initiator tRNA	*Formylmethionine*	Methionine	Methionine
Operons	Yes	Yes	*No*
Plasmids	Yes	Yes	*Rare*
RNA polymerases	One	One[b]	Three
Ribosomes sensitive to chloramphenicol and streptomycin	*Yes*	No	No
Ribosomes sensitive to diphtheria toxin	*No*	Yes	Yes
Some are methanogens	No	*Yes*	No
Some fix nitrogen	Yes	Yes	*No*
Some conduct chlorophyll-based photosynthesis	Yes	No	Yes

[a] 70S ribosomes are smaller than 80S ribosomes.

[b] Archaeal RNA polymerase is similar to eukaryotic polymerases.

scription and translation produced RNAs and proteins, respectively. It probably had a circular chromosome, and many of its structural genes were grouped into operons (see Chapter 13).

The Archaea, Bacteria, and Eukarya of today are all products of billions of years of natural selection and genetic drift, and they are all well adapted to present-day environments. None are "primitive." The common ancestor of the Archaea and the Eukarya probably lived more than 2 billion years ago, and the common ancestor of the Archaea, the Eukarya, and the Bacteria probably lived more than 3 billion years ago.

The earliest prokaryotic fossils date back at least 3.5 billion years, and these ancient fossils indicate that there was considerable diversity among the prokaryotes even during the earliest days of life. The prokaryotes were alone on Earth for a very long time, adapting to new environments and to changes in existing environments. They have survived to this day—and in massive numbers.

General Biology of the Prokaryotes

There are many, many prokaryotes around us—everywhere. Although most are so small that we cannot see them with the naked eye, the prokaryotes are the most successful of all creatures on Earth, if success is measured by numbers of individuals. The bacteria in one person's intestinal tract, for example, outnumber all the humans who have ever lived, and even the total number of *human* cells in their host's body. Some of these bacteria form a thick lining along the intestinal wall. Bacteria and archaea in the oceans number more than 3×10^{28}. This stunning number is perhaps 100 million times as great as the number of stars in the visible universe.

Although small, prokaryotes play many critical roles in the biosphere, interacting in one way or another with every other living thing. In this section, we'll see that some prokaryotes perform key steps in the cycling of nitrogen, sulfur, and carbon. Other prokaryotes trap energy from the sun or from inorganic chemical sources, and some help animals digest their food. The members of the two prokaryotic domains outdo all other groups in metabolic diversity. Eukaryotes, in contrast, are much more diverse in size and shape, but their metabolism is much less diverse. In fact, much of the energy metabolism of eukaryotes is carried out in organelles—mitochondria and chloroplasts—that are descended from bacteria.

Prokaryotes are found in every conceivable habitat on the planet, from the coldest to the hottest, from the most acidic to the most alkaline, and to the saltiest. Some live where oxygen is abundant and others where there is no oxygen at all. They have established themselves at the bottom of the seas, in rocks more than 2 km into Earth's solid crust, and inside other organisms, large and small. Their effects on our environment are diverse and profound. What do these tiny but widespread organisms look like?

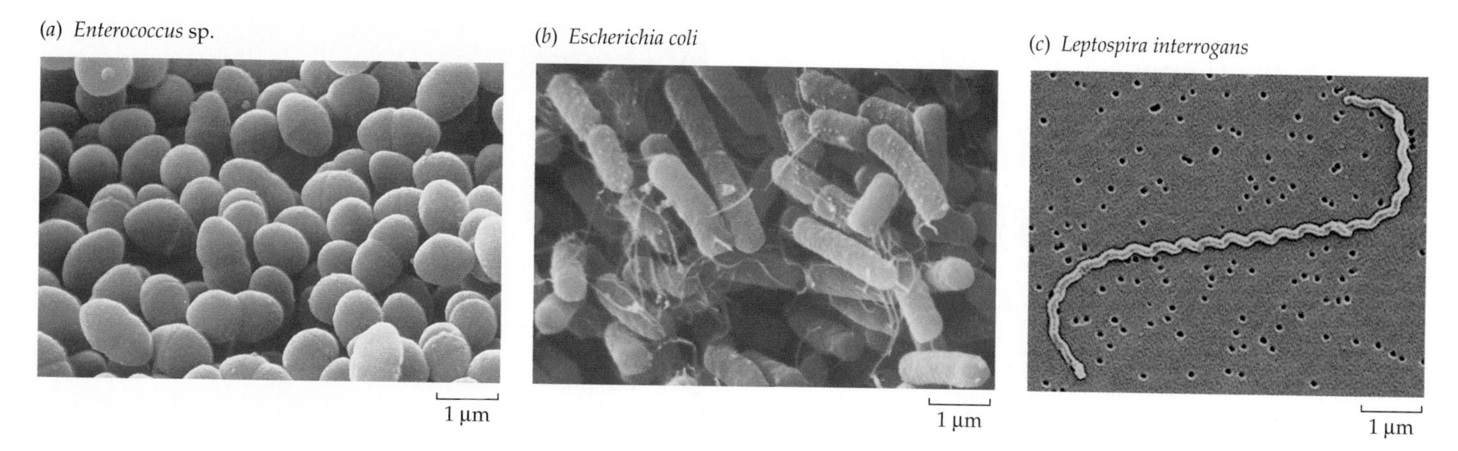

27.3 Shapes of Prokaryotic Cells (*a*) These spherical cocci of an acid-producing bacterium grow in the mammalian gut. (*b*) Rod-shaped *E. coli* are the most thoroughly studied of any bacteria—indeed, of almost any organism on Earth. (*c*) This spiral bacterium belongs to a genus of human pathogens that cause leptospirosis, an infection of the kidney and liver that is spread by contaminated water. The disease has historically been a problem for soldiers in crowded, transient campsites; this particular bacterial strain was isolated in 1915 from the blood of a soldier serving in World War I.

Prokaryotes and their associations take a few characteristic forms

Three shapes are particularly common among the prokaryotes: spheres, rods, and curved or spiral forms (Figure 27.3). A spherical prokaryote is called a **coccus** (plural, cocci). Cocci may live singly or may associate in two- or three-dimensional arrays as chains, plates, blocks, or clusters of cells. A rod-shaped prokaryote is called a **bacillus** (plural, bacilli). Spiral forms are the third main prokaryotic shape. Bacilli and spiral forms may be single or may form chains.

Prokaryotes are almost all unicellular, although some multicellular ones are known. Associations such as chains do not signify multicellularity because each cell is fully viable and independent. These associations arise as cells adhere to one another after reproducing by fission. Associations in the form of chains are called **filaments**. Some filaments become enclosed within delicate tubular sheaths.

Prokaryotes lack nuclei, organelles, and a cytoskeleton

The architectures of prokaryotic and eukaryotic cells were compared in Chapter 4. The basic unit of archaea and bacteria is the prokaryotic cell (see Figure 4.5), which contains a full complement of genetic and protein-synthesizing systems, including DNA, RNA, and all the enzymes needed to transcribe and translate the genetic information into proteins. The prokaryotic cell also contains at least one system for generating the ATP it needs.

In what follows, bear in mind that most of what we know about the structure of prokaryotes comes from studies of bacteria. We still know relatively little about the diversity of archaea, although the pace of research on archaea is accelerating.

The prokaryotic cell differs from the eukaryotic cell in three important ways. First, the organization and replication of the genetic material differs. The DNA of the prokaryotic cell is not organized within a membrane-enclosed nucleus. DNA molecules in prokaryotes (both bacteria and archaea) are usually circular; in the best-studied prokaryotes, there is a single chromosome, but there are often plasmids as well (see Chapter 13).

Second, prokaryotes have none of the membrane-enclosed cytoplasmic organelles that modern eukaryotes have—mitochondria, Golgi apparatus, and others. However, the cytoplasm of a prokaryotic cell may contain a variety of infoldings of the plasma membrane and photosynthetic membrane systems not found in eukaryotes.

Third, prokaryotic cells lack a cytoskeleton, and, without the cytoskeletal proteins, they lack mitosis. Prokaryotic cells divide by their own elaborate method, **fission**, after replicating their DNA.

Prokaryotes have distinctive modes of locomotion

Although many prokaryotes cannot move, others are *motile*. These organisms move by one of several means. Some spiral bacteria, called spirochetes, use a corkscrew-like motion made possible by modified flagella, called *axial filaments*, running along the axis of the cell beneath the outer membrane (Figure 27.4*a*). Many cyanobacteria and a few other bacteria use various poorly understood gliding mechanisms, including rolling. Various aquatic prokaryotes, including some cyanobacteria, can move slowly up and down in the water by adjusting the amount of gas in gas vesicles (Figure 27.4*b*). By far the most common type of locomotion in prokaryotes, however, is that driven by flagella.

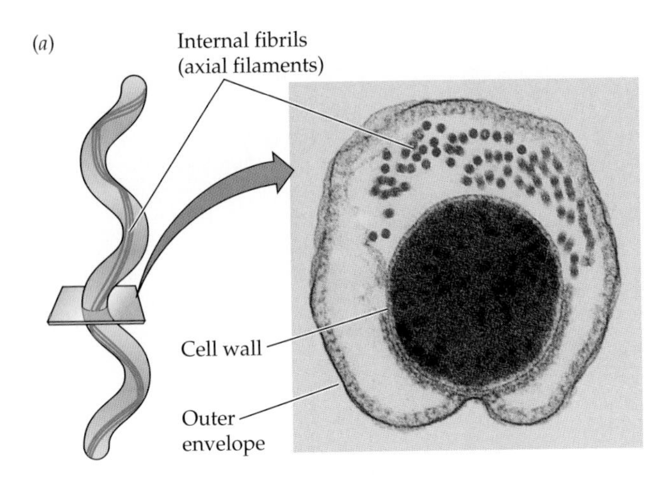

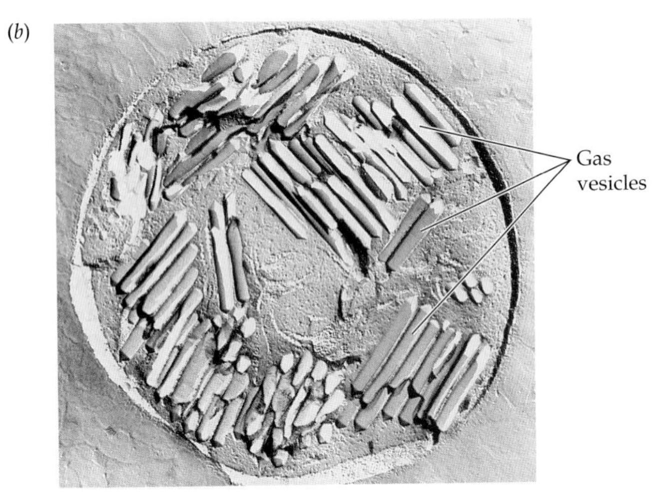

27.4 Structures Associated with Prokaryote Motility (*a*) A spirochete from the gut of a termite, seen in cross section, shows the axial filaments used to produce a corkscrew-like motion. (*b*) Gas vesicles in a cyanobacterium, visualized by the freeze-fracture technique.

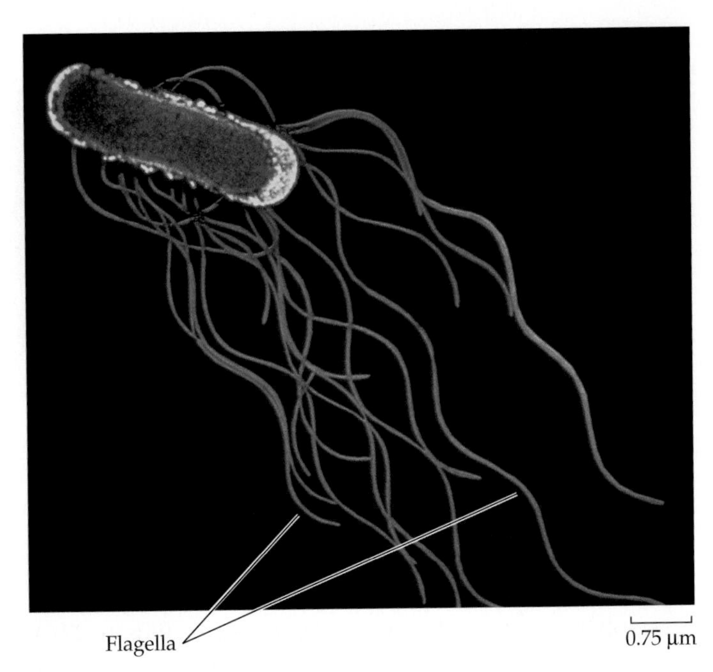

27.5 Some Bacteria Use Flagella for Locomotion Flagella propel this rod-shaped *Salmonella*.

Bacterial flagella are slender filaments that extend singly or in tufts from one or both ends of the cell or are randomly distributed all around it (Figure 27.5). A bacterial flagellum consists of a single fibril made of the protein *flagellin*, projecting from the cell surface, plus a hook and basal body responsible for motion (see Figure 4.6). In contrast, the flagellum of eukaryotes is enclosed by the plasma membrane and usually contains a circle of nine pairs of microtubules surrounding two central microtubules, all containing the protein tubulin, along with many other associated proteins. The prokaryotic flagellum rotates about its base, rather than beating as a eukaryotic flagellum or cilium does.

Prokaryotes have distinctive cell walls

Most prokaryotes have a thick and relatively stiff cell wall. This wall is quite different from the cell walls of plants and algae, which contain cellulose and other polysaccharides, and from those of fungi, which contain chitin. Almost all bacteria have cell walls containing *peptidoglycan* (a polymer of amino sugars). Archaeal cell walls are of differing types, but most contain significant amounts of protein. One group of archaea has pseudopeptidoglycan in its wall; as you have probably already guessed from the prefix *pseudo-*, pseudopeptidoglycan is similar to, but distinct from, the peptidoglycan of bacteria. Peptidoglycan is a substance unique to bacteria; its absence from the walls of archaea is a key difference between the two prokaryotic domains.

In 1884 Hans Christian Gram, a Danish physician, developed a simple staining process that has lasted into our high-technology era as a useful tool for identifying bacteria. The **Gram stain** separates most types of bacteria into two distinct groups, Gram-positive and Gram-negative, on the basis of their staining (Figure 27.6). A smear of cells on a microscope slide is soaked in a violet dye and treated with iodine; it is then washed with alcohol and counterstained with safranine (a red dye). **Gram-positive** bacteria retain the violet dye and appear blue to purple (Figure 27.6*a*). The alcohol washes the violet stain out of **Gram-negative** cells; these cells then pick up the safranine counterstain and appear pink to red (Figure 27.6*b*). Gram-staining characteristics are useful in classifying some kinds of bacteria and are important in determining the identity of bacteria in an unknown sample.

For many bacteria, the Gram-staining results correlate roughly with the structure of the cell wall. Peptidoglycan forms a thick layer outside the plasma membrane of Gram-

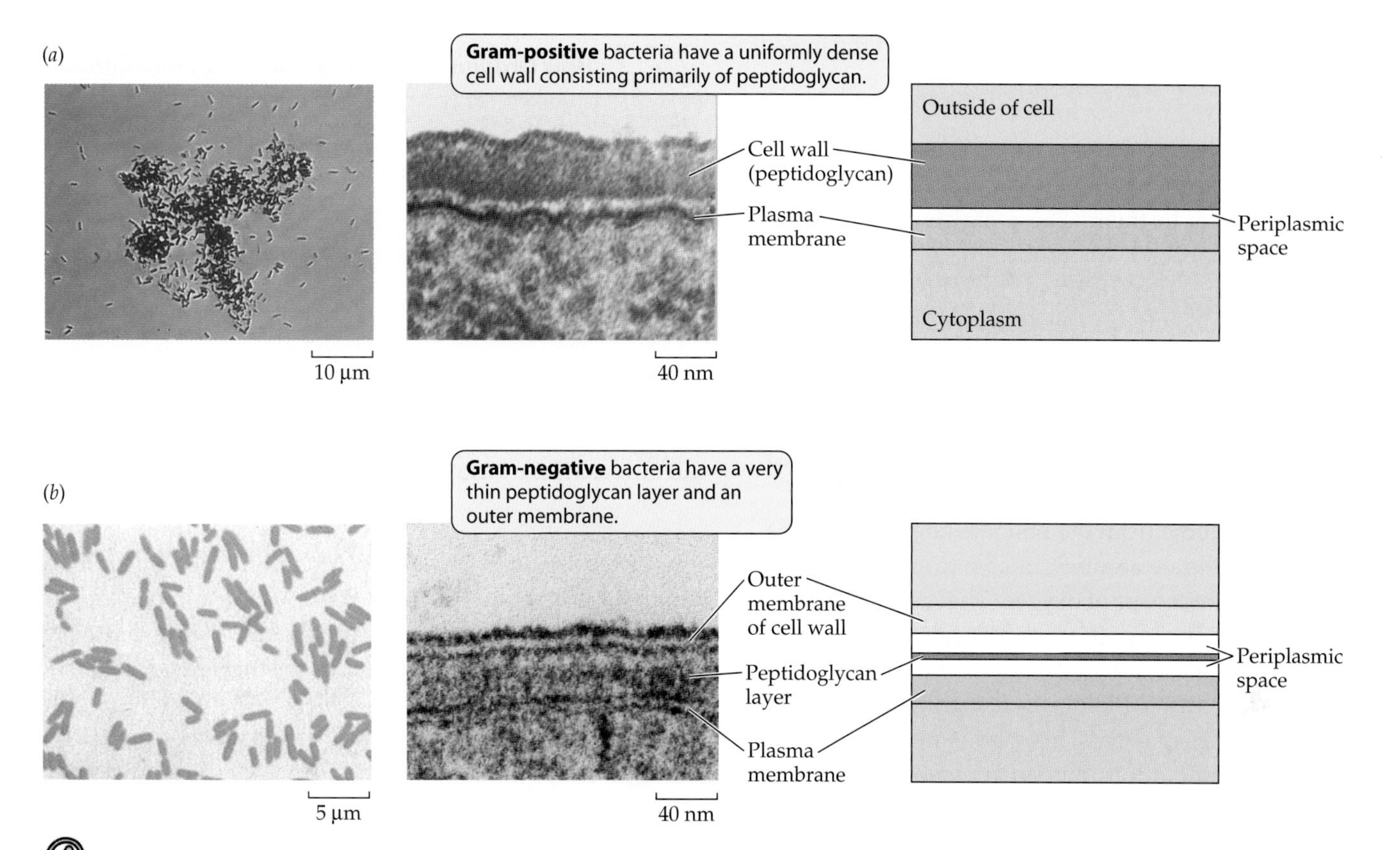

27.6 The Gram Stain and the Bacterial Cell Wall When treated with Gram stain, the cell wall components of different bacteria react in one of two ways. (*a*) Gram-positive bacteria have a thick peptidoglycan cell wall that retains the violet dye and appears deep blue or purple. (*b*) Gram-negative bacteria have a thin peptidoglycan layer that does not retain the violet dye, but picks up the counterstain and appears pink-red.

positive bacteria. The Gram-negative cell wall usually has only one-fifth as much peptidoglycan, and outside the peptidoglycan layer the cell is surrounded by a second, outer membrane quite distinct in chemical makeup from the plasma membrane (see Figure 27.6*b*). Between the inner (plasma) and outer membranes of Gram-negative bacteria is the *periplasmic space*. This space contains enzymes that are important in digesting some materials, transporting others, and detecting chemical gradients in the environment.

The consequences of the different features of prokaryotic cell walls are numerous and relate to the disease-causing characteristics of some prokaryotes. Indeed, the cell wall is a favorite target in medical combat against diseases that are caused by prokaryotes because it has no counterpart in eukaryotic cells. Antibiotics such as penicillin and ampicillin, as well as other agents that specifically interfere with the synthesis of peptidoglycan-containing cell walls, tend to have little, if any, effect on the cells of humans and other eukaryotes.

Prokaryotes reproduce asexually, but genetic recombination does occur

Prokaryotes reproduce by fission, an asexual process. Recall, however, that there are also processes—transformation, conjugation, and transduction—that allow the exchange of genetic information between some prokaryotes quite apart from either sex or reproduction (see Chapter 13).

Some prokaryotes multiply very rapidly. One of the fastest is the bacterium *Escherichia coli*, which under optimal conditions has a generation time of about 20 minutes. The shortest known prokaryote generation times are about 10 minutes. Generation times of 1 to 3 hours are common for others; some extend to days. Bacteria living deep in Earth's crust may suspend their growth for more than a century without dividing and then multiply for a few days before suspending growth again. What kinds of metabolism support such a diversity of growth rates?

Prokaryotes have exploited many metabolic possibilities

The long evolutionary history of the bacteria and archaea, during which they have had time to explore a wide variety of habitats, has led to the extraordinary diversity of their metabolic "lifestyles"—their use or nonuse of oxygen, their energy sources, their sources of carbon atoms, and the materials they release as waste products.

ANAEROBIC VERSUS AEROBIC METABOLISM. Some prokaryotes can live only by anaerobic metabolism because molecular oxygen is poisonous to them. These oxygen-sensitive organisms are called **obligate anaerobes**.

Other prokaryotes can shift their metabolism between anaerobic and aerobic modes (see Chapter 7) and thus are called **facultative anaerobes**. Many facultative anaerobes alternate between anaerobic metabolism (such as fermentation) and cellular respiration as conditions dictate. **Aerotolerant anaerobes** cannot conduct cellular respiration, but are not damaged by oxygen when it is present.

At the other extreme from the obligate anaerobes, some prokaryotes are **obligate aerobes**, unable to survive for extended periods in the *absence* of oxygen. They require oxygen for cellular respiration.

NUTRITIONAL CATEGORIES. Biologists recognize four broad nutritional categories of organisms: photoautotrophs, photoheterotrophs, chemolithotrophs, and chemoheterotrophs. Prokaryotes are represented in all four groups (Table 27.2).

Photoautotrophs perform photosynthesis. They use light as their source of energy and carbon dioxide as their source of carbon. Like the photosynthetic eukaryotes, one group of photoautotrophic bacteria, the cyanobacteria, use chlorophyll *a* as their key photosynthetic pigment and produce oxygen as a by-product of noncyclic electron transport (see Chapter 8).

By contrast, the other photosynthetic bacteria use *bacteriochlorophyll* as their key photosynthetic pigment, and they do not release oxygen gas. Some of these photosynthesizers produce particles of pure sulfur instead because hydrogen sulfide (H_2S), rather than H_2O, is their electron donor for photophosphorylation. Bacteriochlorophyll absorbs light of longer wavelengths than the chlorophyll used by all other photosynthesizing organisms does. As a result, bacteria using this pigment can grow in water beneath fairly dense layers of algae, using light of wavelengths that are not absorbed by the algae (Figure 27.7).

27.2 How Organisms Obtain Their Energy and Carbon

NUTRITIONAL CATEGORY	ENERGY SOURCE	CARBON SOURCE
Photoautotrophs (found in all three domains)	Light	Carbon dioxide
Photoheterotrophs (some bacteria)	Light	Organic compounds
Chemolithotrophs (some bacteria, many archaea)	Inorganic substances	Carbon dioxide
Chemoheterotrophs (found in all three domains)	Organic compounds	Organic compounds

Photoheterotrophs use light as their source of energy, but must obtain their carbon atoms from organic compounds made by other organisms. They use compounds such as carbohydrates, fatty acids, and alcohols as their organic "food." The purple nonsulfur bacteria, among others, are photoheterotrophs.

Chemolithotrophs (chemoautotrophs) obtain their energy by oxidizing inorganic substances, and they use some of that energy to fix carbon dioxide. Some chemolithotrophs use reactions identical to those of the typical photosynthetic cycle (see Figure 8.3), but others use other pathways to fix carbon dioxide. Some bacteria oxidize ammonia or nitrite ions to form nitrate ions. Others oxidize hydrogen gas, hydrogen sulfide, sulfur, and other materials. Many archaea are chemolithotrophs.

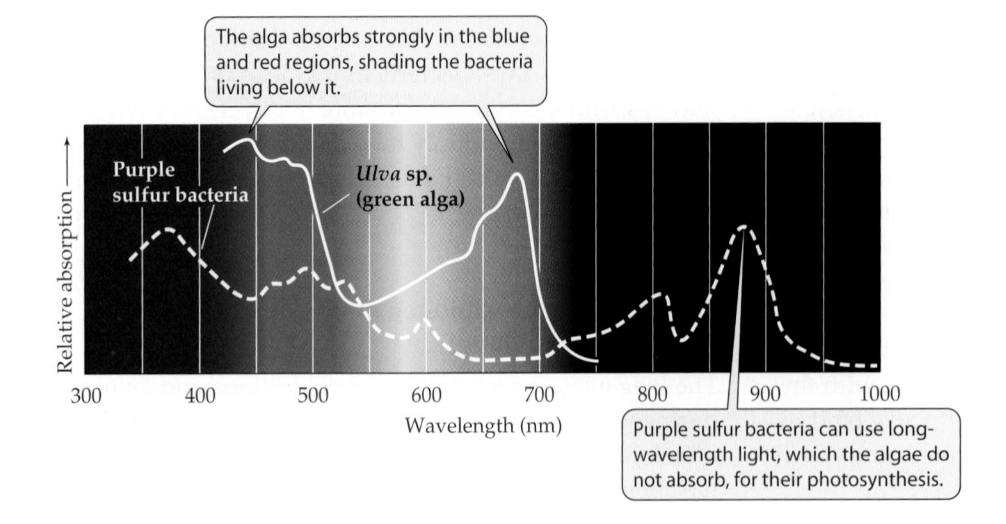
The alga absorbs strongly in the blue and red regions, shading the bacteria living below it.
Purple sulfur bacteria
Ulva sp. (green alga)
Relative absorption
300
400
500
600
700
800
900
1000
Wavelength (nm)
Purple sulfur bacteria can use long-wavelength light, which the algae do not absorb, for their photosynthesis.

27.7 Bacteriochlorophyll Absorbs Long-Wavelength Light The chlorophyll in *Ulva*, a green alga, absorbs no light of wavelengths longer than 750 nm. Purple sulfur bacteria, which contain bacteriochlorophyll, can conduct photosynthesis using the longer wavelengths that pass through the algae.

Deep-sea hydrothermal vent ecosystems are based on chemolithotrophic prokaryotes that are incorporated into large communities of crabs, mollusks, and giant worms, all living at a depth of 2,500 meters, below any hint of light from the sun. These bacteria obtain energy by oxidizing hydrogen sulfide and other substances released in the near-boiling water that flows from volcanic vents in the ocean floor.

Finally, **chemoheterotrophs** obtain both energy and carbon atoms from one or more complex organic compounds. Most known bacteria and archaea are chemoheterotrophs—as are all animals and fungi and many protists.

NITROGEN AND SULFUR METABOLISM. Many prokaryotes base important parts of their metabolism on reactions involving nitrogen or sulfur. For example, some bacteria carry out respiratory electron transport without using oxygen as an electron acceptor. These organisms use oxidized inorganic ions such as nitrate, nitrite, or sulfate as electron acceptors. Examples include the *denitrifiers*, bacteria that release nitrogen to the atmosphere as nitrogen gas (N_2). These normally aerobic bacteria, mostly species of the genera *Bacillus* and *Pseudomonas*, use nitrate (NO_3^-) as an electron acceptor in place of oxygen if they are kept under anaerobic conditions:

$$2\ NO_3^- + 10\ e^- + 12\ H^+ \rightarrow N_2 + 6\ H_2O$$

Nitrogen fixers convert atmospheric nitrogen gas into a chemical form usable by the nitrogen fixers themselves as well as by other organisms. They convert nitrogen gas to ammonia:

$$N_2 + 6\ H \rightarrow 2\ NH_3$$

All organisms require nitrogen for their proteins, nucleic acids, and other important compounds. The vital process of nitrogen fixation is carried out by a wide variety of archaea and bacteria, including cyanobacteria, but by no other organisms. (We'll discuss this process in detail in Chapter 37.)

Ammonia is oxidized to nitrate in the soil and in seawater by chemolithotrophic bacteria called *nitrifiers*. Bacteria of two genera, *Nitrosomonas* and *Nitrosococcus*, convert ammonia to nitrite ions (NO_2^-), and *Nitrobacter* oxidizes nitrite to nitrate (NO_3^-).

What do the nitrifiers get out of these reactions? Their chemosynthesis is powered by the energy released by the oxidation of ammonia or nitrite. For example, by passing the electrons from nitrite through an electron transport chain, *Nitrobacter* can make ATP, and using some of this ATP, it can also make NADH. With this ATP and NADH, the bacterium can convert CO_2 and H_2O to glucose.

Numerous bacteria base their metabolism on the modification of sulfur-containing ions and compounds in their environments. As examples, we have already mentioned the photoautotrophic bacteria and chemolithotrophic archaea that use H_2S as an electron donor in place of H_2O. Such uses of nitrogen and sulfur have environmental implications, as we'll see in the next section.

Prokaryotes in Their Environments

Prokaryotes live in and exploit all sorts of environments and are part of all ecosystems. In the following pages, we'll examine the roles of prokaryotes that live in soils, in water, and even in other living organisms, where they may exist in a neutral, benevolent, or parasitic relationship with their host's tissues.

Prokaryotes are important players in element cycling

Animals depend on photosynthetic plants and microorganisms for their food, directly or indirectly. But plants depend on other organisms—prokaryotes—for their own nutrition. The extent and diversity of life on Earth would not be possible without nitrogen fixation by prokaryotes. Nitrifiers are crucial to the biosphere because they convert the products of nitrogen fixation into nitrate ions, the form of nitrogen most easily used by many plants (see Figure 37.8). Plants, in turn, are the source of nitrogen compounds for animals and fungi. Denitrifiers also play a key role in keeping the nitrogen cycle going. Without denitrifiers, which convert nitrate ions back into nitrogen gas, all forms of nitrogen would leach from the soil and end up in lakes and oceans, making life on land impossible. Other prokaryotes contribute to a similar cycle of sulfur. Prokaryotes, along with fungi, return tremendous quantities of organic carbon to the atmosphere as carbon dioxide.

In the ancient past, the cyanobacteria had an equally dramatic effect on life: Their photosynthesis generated oxygen, converting Earth from an anaerobic to an aerobic environment. The result was the wholesale loss of obligate anaerobic species that could not tolerate the O_2 generated by the cyanobacteria. Only those anaerobes that were able to colonize environments that remained anaerobic survived. However, this transformation to aerobic environments made possible the evolution of cellular respiration and the subsequent explosion of eukaryotic life. What other roles do prokaryotes play in the biosphere?

Archaea help stave off global warming

A time bomb lies deep under the ocean floor. Some ten trillion tons of methane, potentially an overwhelming source of "greenhouse gas," are located there. Will this methane escape to the atmosphere, hastening global warming?

What will prevent such an escape is the presence of legions of archaea, also lying below the bottom of the seas. As methane rises from its deposits, it is metabolized by these archaea, with the result that virtually none of the methane even gets as far as the deepest waters of the ocean. Thus, these archaea play a crucial role in stabilizing the planetary environment.

Prokaryotes live on and in other organisms

Prokaryotes work together with eukaryotes in many ways. In fact, mitochondria and chloroplasts are descended from what were once free-living bacteria. Much later in evolutionary history, some plants became associated with bacteria to form cooperative nitrogen-fixing nodules on their roots (see Figure 37.5).

The tsetse fly, which transmits sleeping sickness by transferring trypanosomes (microscopic protists described in the next chapter) from one person to another, enjoys a profitable association with the bacterium *Wigglesworthia glossinidia*. Biologists who decoded the genome of *W. glossinidia* in 2002 were surprised to learn that the bacterium's tiny genome contains almost nothing but the genes needed for basic metabolism and DNA replication—and 62 genes for making ten B vitamins and other nutritional factors. Without the vitamins provided by the bacterium, the tsetse fly cannot reproduce. The bacteria, living inside the fly's cells, are in effect vitamin pills. Researchers are now trying to determine whether an attack on *W. glossinidia* may succeed in combating sleeping sickness where more obvious direct attacks on tsetse flies or the trypanosomes have failed.

Many animals, including humans, harbor a variety of bacteria and archaea in their digestive tracts. Cows depend on prokaryotes to perform important steps in digestion. Like most animals, cows cannot produce cellulase, the enzyme needed to start the digestion of the cellulose that makes up the bulk of their plant food. However, bacteria living in a special section of the gut, called the rumen, produce enough cellulase to process the cow's daily diet. Humans use some of the metabolic products—especially vitamins B_{12} and K—of bacteria living in the large intestine.

We are heavily populated, inside and out, by bacteria. Although very few of them are agents of disease, popular notions of bacteria as "germs" arouse our curiosity about those few. Let's briefly consider the roles of some bacteria as pathogens.

A small minority of bacteria are pathogens

The late nineteenth century was a productive era in the history of medicine—a time during which bacteriologists, chemists, and physicians proved that many diseases are caused by microbial agents. During this time the German physician Robert Koch laid down a set of four rules for establishing that a particular microorganism causes a particular disease:

1. The microorganism is always found in individuals with the disease.
2. The microorganism can be taken from the host and grown in pure culture.
3. A sample of the culture produces the disease when injected into a new, healthy host.
4. The newly infected host yields a new, pure culture of microorganisms identical to those obtained in the second step.

These rules, called **Koch's postulates**, were very important in a time when it was not widely accepted that microorganisms cause disease. Today medical science makes use of other, more powerful diagnostic tools. However, one important step in establishing that a coronavirus was the causal agent of SARS (Severe Acute Respiratory Syndrome), a disease that first appeared in 2003, was the satisfaction of Koch's postulates.

Only a tiny percentage of all prokaryotes are **pathogens** (disease-producing organisms), and of those that are known, all are in the domain Bacteria. For an organism to be a successful pathogen, it must overcome several hurdles:

- It must arrive at the body surface of a potential host.
- It must enter the host's body.
- It must evade the host's defenses.
- It must multiply inside the host.
- It must damage the host (to meet the definition of a "pathogen").
- It must infect a new host.

Failure to overcome any of these hurdles ends the reproductive career of a pathogenic organism. However, in spite of the many defenses available to potential hosts that we considered in Chapter 18, some bacteria are very successful pathogens.

For the host, the consequences of a bacterial infection depend on several factors. One is the **invasiveness** of the pathogen—its ability to multiply within the body of the host. Another is its **toxigenicity**—its ability to produce chemical substances (*toxins*) that are harmful to the tissues of the host. *Corynebacterium diphtheriae*, the agent that causes diphtheria, has low invasiveness and multiplies only in the throat, but its toxigenicity is so great that the entire body is affected. In contrast, *Bacillus anthracis*, which causes anthrax (a disease primarily of cattle and sheep, but also sometimes fatal in humans, as we saw in Chapter 13), has low toxigenicity but an invasiveness so great that the entire bloodstream ultimately teems with the bacteria.

There are two general types of bacterial toxins: exotoxins and endotoxins. **Endotoxins** are released when certain Gram-negative bacteria grow or lyse (burst). These toxins are lipopolysaccharides (complexes consisting of a polysaccharide and a lipid component) that form part of the outer bacterial membrane (see Figure 27.6). Endotoxins are rarely fatal; they normally cause fever, vomiting, and diarrhea.

Among the endotoxin producers are some strains of *Salmonella* and *Escherichia*.

Exotoxins are usually soluble proteins released by living, multiplying bacteria, and they may travel throughout the host's body. They are highly toxic—often fatal—to the host, but do not produce fevers. Exotoxin-induced human diseases include tetanus (from *Clostridium tetani*), botulism (from *Clostridium botulinum*), cholera (from *Vibrio cholerae*), and plague (from *Yersinia pestis*). Anthrax results from three exotoxins produced by *Bacillus anthracis*.

Remember that in spite of our frequent mention of human pathogens, only a small minority of the known prokaryotic species are pathogenic. Many more species play positive roles in our lives and in the biosphere. We make direct use of many bacteria and a few archaea in such diverse applications as cheese production, sewage treatment, and the industrial production of an amazing variety of antibiotics, vitamins, organic solvents, and other chemicals.

Pathogenic bacteria are often surprisingly difficult to combat, even with today's arsenal of antibiotics. One source of difficulty is the ability of prokaryotes to form resistant films.

Prokaryotes may form biofilms

Many unicellular microorganisms, prokaryotes in particular, tend to form dense films called **biofilms** rather than existing as clouds of individual cells. Upon contacting a solid surface, the cells lay down a gel-like polysaccharide matrix that then traps other bacteria, forming a biofilm. Once a biofilm forms, it is difficult to kill the cells. Pathogenic bacteria are hard for the immune system—and modern medicine—to combat once they form a biofilm. For example, the film may be impermeable to antibiotics. Biofilms often include a mixture of bacterial species.

The biofilm with which you are most likely to be familiar is dental plaque, the coating of bacteria and hard matrix that forms between and on your teeth unless you do a good job of flossing and brushing. Biofilms form on contact lenses, on hip replacements, and on just about any available surface. Other biofilms foul metal pipes and cause corrosion, a major problem in steam-driven electricity generation plants. Biofilms are the object of much current research. For example, some biologists are studying the chemical signals used by bacteria in biofilms to communicate with one another. By blocking the signals that lead to the production of the matrix polysaccharides, they may be able to prevent biofilms from forming.

Prokaryote Phylogeny and Diversity

The prokaryotes comprise a diverse array of microscopic organisms. To explore their diversity, let's first consider how they are classified and some of the difficulties involved in doing so.

The nucleotide sequences of prokaryotes reveal their evolutionary relationships

Why do biologists want to classify bacteria and archaea? There are three primary motivations for classification schemes: to identify unknown organisms, to reveal evolutionary relationships, and to provide universal names (see Chapter 25). Scientists and medical technologists must be able to identify bacteria quickly and accurately—when the bacteria are pathogenic, lives may depend on it.

Until recently, taxonomists based their classification schemes for the prokaryotes on readily observable phenotypic characters such as color, motility, nutritional requirements, antibiotic sensitivity, and reaction to the Gram stain. Although such schemes have facilitated the identification of prokaryotes, they have not provided insights into how these organisms evolved—a question of great interest to microbiologists and to all students of evolution. The prokaryotes and the protists (see Chapter 28) have long presented major challenges to those who attempted phylogenetic classifications. Only recently have systematists had the right tools for tackling this task.

Analyses of the nucleotide sequences of ribosomal RNA have provided us with the first apparently reliable measures of evolutionary distance among taxonomic groups. Ribosomal RNA (rRNA) is particularly useful for evolutionary studies of living organisms for several reasons:

- rRNA is evolutionarily ancient.
- No living organism lacks rRNA.
- rRNA plays the same role in translation in all organisms.
- rRNA has evolved slowly enough that sequence similarities between groups of organisms are easily found.

Let's look at just one approach to the use of rRNA for studying evolutionary relationships.

Comparisons of rRNAs from a great many organisms revealed recognizable short base sequences that are characteristic of particular taxonomic groups. These *signature sequences*, approximately 6 to 14 bases long, appear at the same approximate positions in rRNAs from related groups. For example, the signature sequence AAACUUAAAG occurs about 910 bases from one end of the small subunit of ribosomes in 100 percent of the Archaea and Eukarya tested, but in *none* of the Bacteria tested. Several signature sequences distinguish each of the three domains. Similarly, the major groups within the bacteria and archaea possess unique signature sequences.

These data sound promising, but things aren't as simple as we might wish. When biologists examined other genes and RNAs, contradictions began to appear and new questions arose. Analyses of different nucleotide sequences suggested different phylogenetic patterns. How could such a situation have arisen?

Lateral gene transfer muddied the phylogenetic waters

It is now clear that, from early in evolution to the present day, genes have been moving among prokaryotic species by **lateral gene transfer**. As we have seen, a gene from one species can become incorporated into the genome of another. Mechanisms of lateral gene transfer include transfer by plasmids and viruses and uptake of DNA by transformation. Such transfers are well documented, not just between bacterial species or archaeal species, but also across the boundaries between bacteria and archaea and between prokaryotes and eukaryotes.

A gene that has been transferred will be inherited by the recipient's progeny and in time will be recognized as part of the normal genome of the descendants. Biologists are still assessing the extent of lateral gene transfer among prokaryotes and its implications for phylogeny, especially at the early stages of evolution.

Figure 27.8 is an overview of the major clades in the domains Bacteria and Archaea that we will discuss further in this chapter. This phylogeny is based on the evidence that is currently available, but keep in mind that a new picture is likely to emerge within the next decade, based on new nucleotide sequence data and new information about the currently understudied archaea.

Mutations are a major source of prokaryotic variation

Assuming that the prokaryote groups we are about to describe do indeed represent clades, these groups are amazingly complex. A single lineage of bacteria or archaea may contain the most extraordinarily diverse species; on the other hand, a species in one group may be phenotypically almost indistinguishable from one or many species in another group. What are the sources of these phylogenetic patterns?

Although prokaryotes can acquire new alleles by transformation, transduction, or conjugation, the most important sources of genetic variation in populations of prokaryotes are probably mutation and genetic drift (see Chapter 23). Mutations, especially recessive mutations, are slow to make their presence felt in populations of humans and other diploid organisms. In contrast, a mutation in a prokaryote, which is haploid, has immediate consequences for that organism. If it is not lethal, it will be transmitted to and expressed in the organism's daughter cells—and in their daughter cells, and so on. Thus, a beneficial mutant allele spreads rapidly.

The rapid multiplication of many prokaryotes, coupled with mutation, natural selection, and genetic drift, allows rapid phenotypic changes within their populations. Important changes, such as loss of sensitivity to an antibiotic, can occur over broad geographic areas in just a few years. Think how many significant metabolic changes could have occurred over even modest time spans, let alone over the entire history of life on Earth. When we introduce the proteobacteria, the largest group of bacteria, you will see that its different subgroups have easily and rapidly adopted and abandoned metabolic pathways under selective pressure from their environments.

The Bacteria

The best-studied prokaryotes are the bacteria. We will describe bacterial diversity using a currently popular classification scheme that enjoys considerable support from nucleotide sequence data. More than a dozen clades have been proposed under this scheme; we will describe just of a few of them here. The higher-order relationships among these groups of prokaryotes are not known. Some biologists describe them as kingdoms, some as subkingdoms, and others as phyla; here, we simply call them groups. We'll pay the closest attention to five groups: the proteobacteria, cyanobacteria, spirochetes, chlamydias, and firmicutes (see Figure 27.8). First, however, we'll mention one property that is shared by members of three other groups.

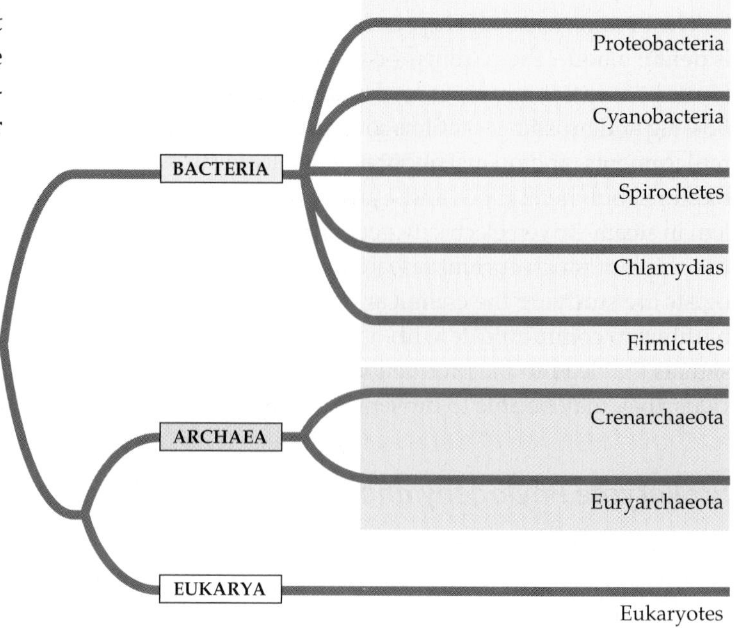

27.8 Two Domains: A Brief Overview This abridged summary classification of the domains Bacteria and Archaea shows their relationships to each other and to the Eukarya. The relationships among the many clades of bacteria, not all of which are listed here, are unresolved at this time.

Some bacteria are heat lovers

Three of the bacterial groups that may have branched out earliest during bacterial evolution are all **thermophiles** (heat lovers), as are the most ancient of the archaea. This observation supports the hypothesis that the first living organisms were thermophiles that appeared in an environment much hotter than those that predominate today.

The Proteobacteria are a large and diverse group

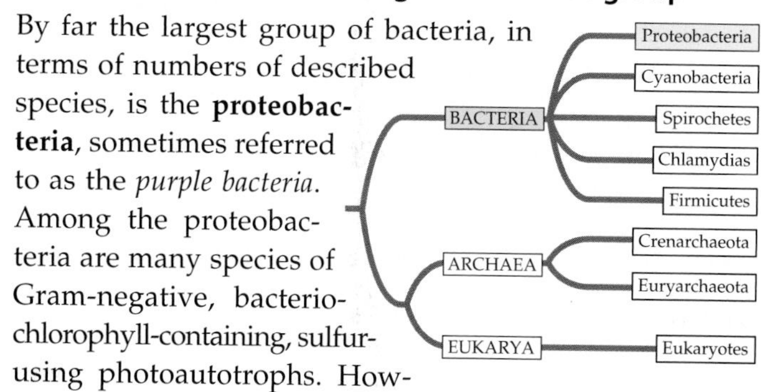

By far the largest group of bacteria, in terms of numbers of described species, is the **proteobacteria**, sometimes referred to as the *purple bacteria*. Among the proteobacteria are many species of Gram-negative, bacteriochlorophyll-containing, sulfur-using photoautotrophs. However, the proteobacteria also include dramatically diverse bacteria that bear no resemblance to those species in phenotype. The mitochondria of eukaryotes were derived from proteobacteria by endosymbiosis.

No characteristic demonstrates the diversity of the proteobacteria more clearly than their metabolic pathways (Figure 27.9). The common ancestor of all the proteobacteria was probably a photoautotroph. Early in evolution, two groups of proteobacteria lost their ability to photosynthesize and have been chemoheterotrophs ever since. The other three groups still have photoautotrophic members, but in *each* group, some evolutionary lines have abandoned photoautotrophy and taken up other modes of nutrition. There are chemolithotrophs and chemoheterotrophs in all three groups. Why? One possibility is that each of the trends in Figure 27.9 was an evolutionary response to selective pressures encountered as these bacteria colonized new habitats that presented new challenges and opportunities.

Among the proteobacteria are some nitrogen-fixing genera, such as *Rhizobium* (see Figure 37.7), and other bacteria that contribute to the global nitrogen and sulfur cycles. *E. coli*, one of the most studied organisms on Earth, is a proteobacterium. So, too, are many of the most famous human pathogens, such as *Yersinia pestis*, *Vibrio cholerae*, and *Salmonella typhimurium*, all mentioned in our discussion of pathogens above.

Fungi cause most plant diseases, and viruses cause others, but about 200 plant diseases are of bacterial origin. *Crown gall*, with its characteristic tumors (Figure 27.10), is one of the most striking. The causal agent of crown gall is *Agrobacterium tumefaciens*, which harbors a plasmid used in recombinant DNA studies as a vehicle for inserting genes into new plant hosts (see Chapter 16).

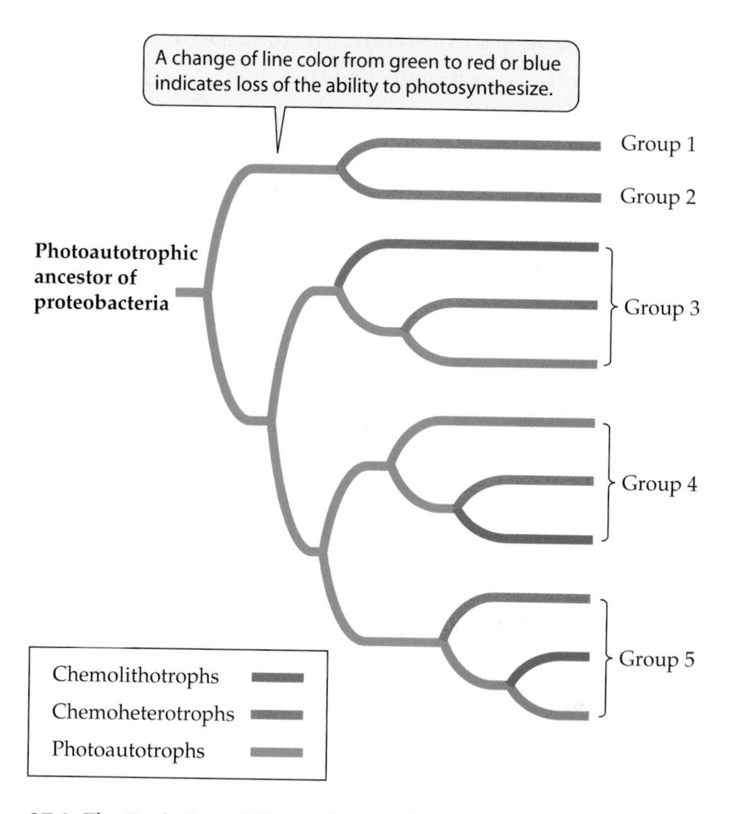

27.9 The Evolution of Metabolism in the Proteobacteria The common ancestor of all proteobacteria was probably a photoautotroph. As they encountered new environments, groups 1 and 2 lost the ability to photosynthesize; in the other three groups, some evolutionary lines became chemolithotrophs or chemoheterotrophs.

27.10 A Crown Gall This colorful tumor growing on the stem of a geranium plant is caused by the Gram-negative bacillus *Agrobacterium tumefaciens*.

Cyanobacteria are important photoautotrophs

Cyanobacteria, sometimes called *blue-green bacteria* because of their pigmentation, are photoautotrophs that require only water, nitrogen gas, oxygen, a few mineral elements, light, and carbon dioxide to survive. They use chlorophyll *a* for photosynthesis and release oxygen gas; many species also fix nitrogen. Their photosynthesis was the basis of the "oxygen revolution" that transformed Earth's atmosphere.

Cyanobacteria carry out the same type of photosynthesis that is characteristic of eukaryotic photosynthesizers. They contain elaborate and highly organized internal membrane systems called *photosynthetic lamellae*, or *thylakoids*. The chloroplasts of photosynthetic eukaryotes are derived from an endosymbiotic cyanobacterium.

Cyanobacteria may live free as single cells or associate in colonies. Depending on the species and on growth conditions, colonies of cyanobacteria may range from flat sheets one cell thick to filaments to spherical balls of cells.

Some filamentous colonies of cyanobacteria differentiate into three cell types: vegetative cells, spores, and heterocysts (Figure 27.11). *Vegetative cells* photosynthesize, *spores* are resting cells that can eventually develop into new filaments, and *heterocysts* are cells specialized for nitrogen fixation. All of the known cyanobacteria with heterocysts fix nitrogen. Heterocysts also have a role in reproduction: When filaments break apart to reproduce, the heterocyst may serve as a breaking point.

Spirochetes look like corkscrews

Spirochetes are Gram-negative, motile, chemoheterotrophic bacteria characterized by unique structures called axial filaments, which are modified flagella running through the periplasm (see Figure 27.4*a*). The cell body is a long cylinder coiled into a spiral (Figure 27.12). The axial filaments begin at either end of the cell and overlap in the middle, and there are typical basal bodies where they are attached to the cell wall. The basal bodies rotate, as they do in other prokaryotic flagella. Many spirochetes live in humans as parasites; a few are pathogens, including those that cause syphilis and Lyme disease. Others live free in mud or water.

27.11 Cyanobacteria (*a*) *Anabaena* is a genus of cyanobacteria that form filamentous colonies containing three cell types. (*b*) A thin neck attaches a heterocyst to each of two vegetative cells in a filament. (*c*) Cyanobacteria appear in enormous numbers in some environments. This California pond has experienced eutrophication; phosphorus and other nutrients generated by human activity have accumulated in the pond, feeding an immense green mat (commonly referred to as "pond scum") that is made up of several species of free-living cyanobacteria.

27.12 A Spirochete This corkscrew-shaped bacterium causes syphilis in humans.

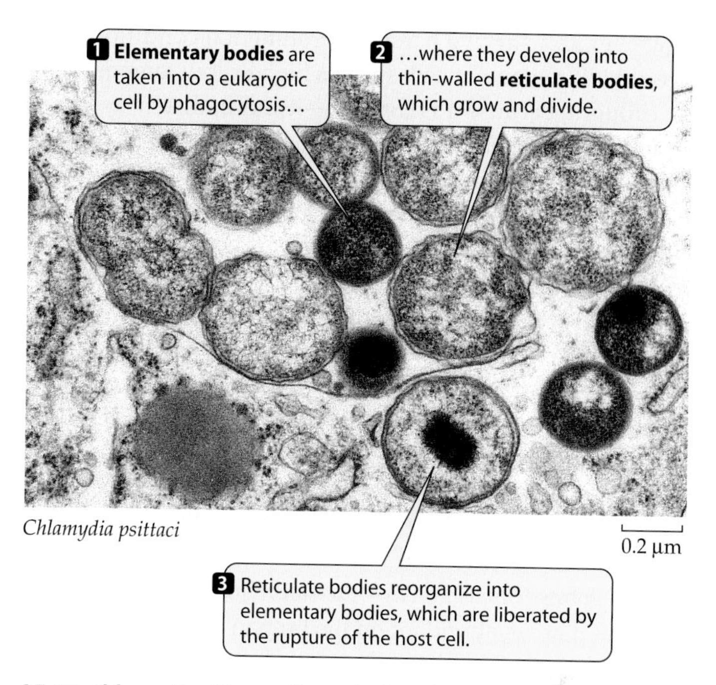

1 **Elementary bodies** are taken into a eukaryotic cell by phagocytosis...
2 ...where they develop into thin-walled **reticulate bodies**, which grow and divide.
Chlamydia psittaci
0.2 µm
3 Reticulate bodies reorganize into elementary bodies, which are liberated by the rupture of the host cell.

27.13 Chlamydias Change Form during Their Life Cycle Elementary bodies and reticulate bodies are the two major phases of the chlamydia life cycle.

Chlamydias are extremely small

Chlamydias are among the smallest bacteria (0.2–1.5 µm in diameter). They can live only as parasites within the cells of other organisms. These tiny Gram-negative cocci are unique prokaryotes because of their complex life cycle, which involves two different forms of cells, *elementary bodies* and *reticulate bodies* (Figure 27.13). In humans, various strains of chlamydias cause eye infections (especially trachoma), sexually transmitted diseases, and some forms of pneumonia.

BACTERIA
Proteobacteria
Cyanobacteria
Spirochetes
Chlamydias
Firmicutes
ARCHAEA
Crenarchaeota
Euryarchaeota
EUKARYA
Eukaryotes

Most firmicutes are Gram-positive

The **firmicutes** are sometimes referred to as the *Gram-positive bacteria*, but some firmicutes are Gram-negative, and some have no cell wall at all. Nonetheless, the firmicutes constitute a clade.

BACTERIA
Proteobacteria
Cyanobacteria
Spirochetes
Chlamydias
Firmicutes
ARCHAEA
Crenarchaeota
Euryarchaeota
EUKARYA
Eukaryotes

Some firmicutes produce **endospores** (Figure 27.14)—heat-resistant resting structures—when a key nutrient such as nitrogen or carbon becomes scarce. The bacterium replicates its DNA and encapsulates one copy, along with some of its cytoplasm, in a tough cell wall heavily thickened with peptidoglycan and surrounded by a spore coat. The parent cell then breaks down, releasing the endospore. Endospore production is not a reproductive process; the endospore merely replaces the parent cell. The endospore,

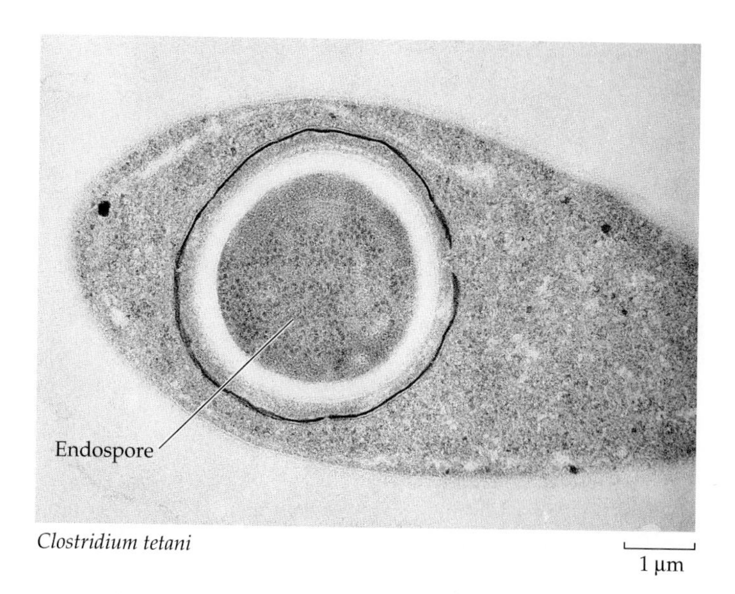

27.14 The Endospore: A Structure for Waiting Out Bad Times This firmicute, which causes tetanus, produces endospores as resistant resting structures.

however, can survive harsh environmental conditions that would kill the parent cell, such as high or low temperatures or drought, because it is *dormant*—its normal activity is suspended. Later, if it encounters favorable conditions, the endospore becomes metabolically active and divides, forming new cells like the parent. Some endospores can be reactivated after more than a thousand years of dormancy. There are credible claims of reactivation of *Bacillus* endospores after millions of years—and even one claim, of uncertain validity, of more than a billion years!

Members of this endospore-forming group of firmicutes include the many species of *Clostridium* and *Bacillus*. The toxins produced by *C. botulinum* are among the most poisonous ever discovered; the lethal dose for humans is about one-millionth of a gram (1 μg). *B. anthracis*, as noted above, is the anthrax pathogen.

The genus *Staphylococcus*—the staphylococci—includes firmicutes that are abundant on the human body surface; they are responsible for boils and many other skin problems (Figure 27.15). *S. aureus* is the best-known human pathogen in this genus; it is found in 20 to 40 percent of normal adults (and in 50 to 70 percent of hospitalized adults). It can cause respiratory, intestinal, and wound infections in addition to skin diseases.

Actinomycetes are firmicutes that develop an elaborately branched system of filaments (Figure 27.16). These bacteria closely resemble the filamentous growth habit of fungi at a reduced scale. Some actinomycetes reproduce by forming chains of spores at the tips of the filaments. In species that do not form spores, the branched, filamentous growth ceases and the structure breaks up into typical cocci or bacilli, which then reproduce by fission.

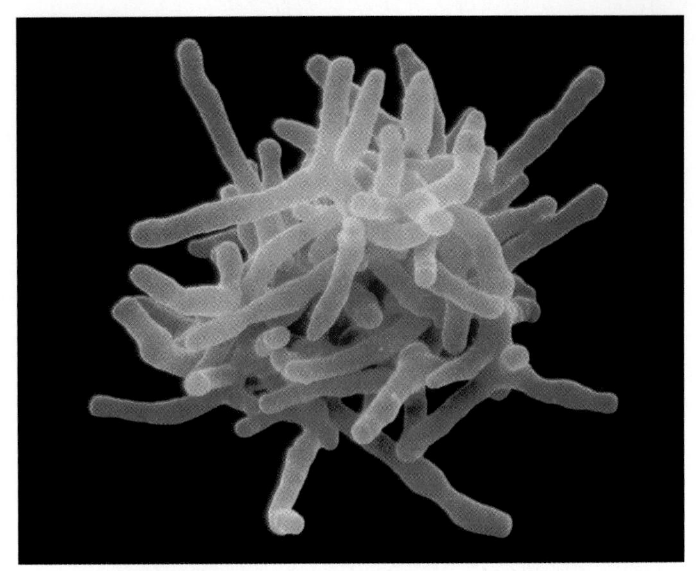

27.16 Filaments of an Actinomycete The branching filaments seen in this scanning electron micrograph are typical of actinomycetes, a medically important bacterial group.

The actinomycetes include several medically important bacteria. *Mycobacterium tuberculosis* causes tuberculosis. *Streptomyces* produces streptomycin as well as hundreds of other antibiotics. We derive most of our antibiotics from members of the actinomycetes.

Another interesting group of firmicutes, the **mycoplasmas**, lack cell walls, although some have a stiffening material outside the plasma membrane. Some of them are the smallest cellular creatures ever discovered—they are even smaller than chlamydias (Figure 27.17). The smallest mycoplasmas capable of multiplication have a diameter of about 0.2 μm. They are small in another crucial sense as well: They

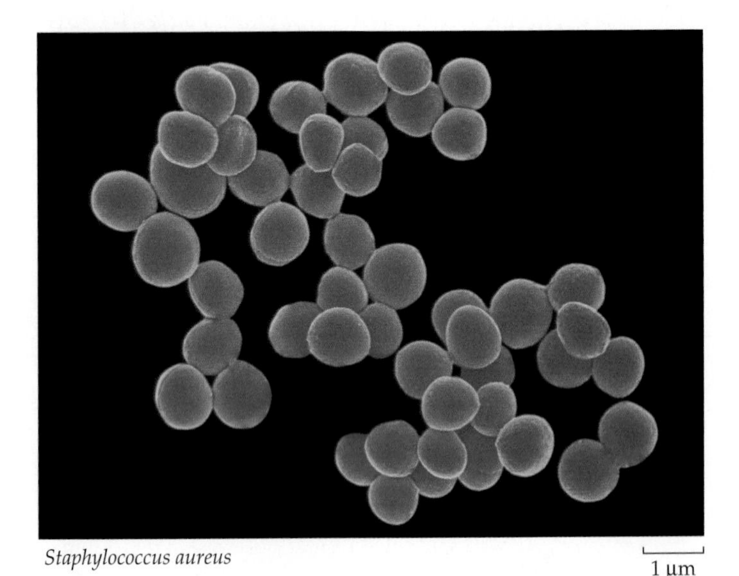

27.15 Gram-Positive Firmicutes "Grape clusters" are the usual arrangement of Gram-positive staphylococci.

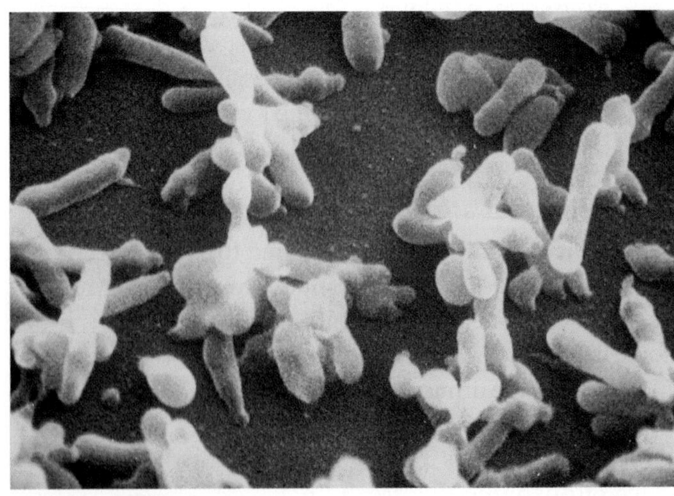

27.17 The Tiniest Living Cells Containing only about one-fifth as much DNA as *E. coli*, mycoplasmas are the smallest known bacteria.

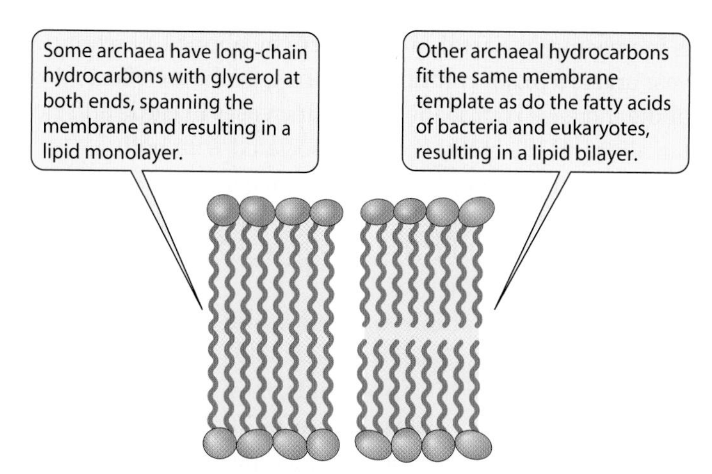

27.18 Membrane Architecture in Archaea The long-chain hydrocarbons of may archaeal membranes are branched, and may have glycerol at both ends. This lipid monolayer structure (on the left) still fits into a biological membrane, however. In fact, all three domains have similar membrane structures.

have less than half as much DNA as do most other prokaryotes—but they still can grow autonomously. It has been speculated that the amount of DNA in a mycoplasma may be the minimum amount required to encode the essential properties of a living cell.

We have discussed five clades of bacteria in some detail, but other bacterial clades are well known, and there may be dozens more waiting to be discovered. This conservative estimate is based on the fact that many bacteria and archaea have never been cultured in the laboratory.

The Archaea

The domain Archaea consists mainly of prokaryotic genera that live in habitats notable for characteristics such as extreme salinity (salt content), low oxygen concentrations, high temperatures, or high or low pH. However, many archaea live in habitats that are not extreme. Perhaps the largest number of archaea live in the ocean depths.

On the face of it, the Archaea do not seem to belong together as a group. One current classification scheme divides the domain into two principal groups, **Euryarchaeota** and **Crenarchaeota**. In fact, we know relatively little about the phylogeny of archaea, in part because the study of archaea is still in its early stages. We do know that archaea share certain characteristics.

The Archaea share some unique characteristics

Two characteristics shared by all archaea are the absence of peptidoglycan in their cell walls and the presence of lipids of distinctive composition in their cell membranes (see Table 27.1). The base sequences of their ribosomal RNAs support a close evolutionary relationship among them. Their separation from the Bacteria and Eukarya was clarified when biologists sequenced the first archaeal genome. It consisted of 1,738 genes, more than half of which were unlike any genes ever found in the other two domains.

The unusual lipids in the membranes of archaea deserve some description. They are found in all archaea, and in no bacteria or eukaryotes. Most bacterial and eukaryotic membrane lipids contain unbranched long-chain fatty acids connected to glycerol by **ester linkages**:

$$-\overset{\overset{\displaystyle O}{\|}}{C}-O-\underset{\underset{\displaystyle H}{|}}{\overset{\overset{\displaystyle H}{|}}{C}}-$$

(Figure 27.18, right; see also Figure 3.18). In contrast, some archaeal membrane lipids contain long-chain hydrocarbons connected to glycerol by **ether linkages**:

$$-\underset{\underset{\displaystyle H}{|}}{\overset{\overset{\displaystyle H}{|}}{C}}-O-\underset{\underset{\displaystyle H}{|}}{\overset{\overset{\displaystyle H}{|}}{C}}-$$

In addition, the long-chain hydrocarbons of the archaea are branched. One class of these lipids, with hydrocarbon chains 40 carbon atoms in length, contains glycerol at *both* ends of the hydrocarbons (Figure 27.18, left). This *lipid monolayer* structure, unique to the domain Archaea, still fits in a biological membrane because the lipids are twice as long as the typical lipids in the bilayers of other membranes (see Figure 27.18). Lipid monolayers and bilayers are both found among the archaea.

In spite of this striking difference in their membrane lipids, all three domains have membranes with similar overall structures, dimensions, and functions.

Most Crenarchaeota live in hot, acidic places

Most known Crenarchaeota are both thermophilic (heat-loving) and *acidophilic* (acid-loving). Members of the genus *Sulfolobus* live in hot sulfur springs at temperatures of 70–75°C. They die of "cold" at 55°C (131°F). Hot sulfur springs are also extremely acidic. *Sulfolobus* grows best in the range from pH 2 to pH 3, but it readily tolerates pH values as low as 0.9. One species of the genus *Ferroplasma* lives at a pH near 0. Some acidophilic hyperthermophiles maintain an internal pH near 7 (neutral) in spite of their acidic environment. These

27.19 Some Would Call It Hell; Archaea Call It Home Masses of heat- and acid-loving archaea form an orange mat inside a volcanic vent on the island of Kyushu, Japan. Sulfurous residue is visible at the edges of the archaeal mat.

and other hyperthermophiles thrive where very few other organisms can even survive (Figure 27.19).

The Euryarchaeota live in many surprising places

Some species of Euryarchaeota share the property of producing methane (CH_4) by reducing carbon dioxide. All of these **methanogens** are obligate anaerobes, and methane production is the key step in their energy metabolism. Comparison of rRNA nucleotide sequences revealed a close evolutionary relationship among all these methanogens, which were previously assigned to several unrelated bacterial groups.

Methanogens release approximately 2 billion tons of methane gas into Earth's atmosphere each year, accounting for 80 to 90 percent of the methane in the atmosphere, including that associated with mammalian belching. Approximately a third of this methane comes from methanogens living in the guts of grazing herbivores such as cows. Methane is increasing in Earth's atmosphere by about 1 percent per year and is a major contributor to the greenhouse effect. Most of the increase is probably due to increases in cattle and rice farming and the methanogens associated with both.

One methanogen, *Methanopyrus*, lives on the ocean bottom near blazing hydrothermal vents. *Methanopyrus* can survive and grow at 110°C. It grows best at 98°C and not at all at temperatures below 84°C.

Another group of Euryarchaeota, the *extreme halophiles* (salt lovers), lives exclusively in very salty environments. Because they contain pink carotenoids, they can be seen easily under some circumstances (Figure 27.20). Halophiles grow in the Dead Sea and in brines of all types: Pickled fish may sometimes show reddish pink spots that are colonies of halophilic archaea. Few other organisms can live in the saltiest of the homes that the extreme halophiles occupy; most would "dry" to death, losing too much water to the hypertonic environment. Extreme halophiles have been found in lakes with pH values as high as 11.5—the most alkaline environment inhabited by living organisms, and almost as alkaline as household ammonia.

Some of the extreme halophiles have a unique system for trapping light energy and using it to form ATP—without using any form of chlorophyll—when oxygen is in short supply. They use the pigment *retinal* (also found in the vertebrate eye) combined with a protein to form a light-absorbing molecule called *bacteriorhodopsin*, and they form ATP by a chemiosmotic mechanism of the sort described in Figure 7.12.

27.20 Extreme Halophiles Commercial seawater evaporating ponds, such as these in San Francisco Bay, are attractive homes for salt-loving archaea, which are easily visible because of their carotenoids.

Another member of the Euryarchaeota, *Thermoplasma*, has no cell wall. It is thermophilic and acidophilic, its metabolism is aerobic, and it lives in coal deposits. It has the smallest genome among the archaea, and perhaps the smallest (along with the mycoplasmas) of any free-living organism—1,100,000 base pairs.

In addition to these archaea that are found in amazing habitats, many Crenarchaeota and Euryarchaeota live in environments that are not extreme.

Chapter Summary

Why Three Domains?

▶ Living organisms can be divided into three domains: Bacteria, Archaea, and Eukarya. Both the Archaea and the Bacteria are prokaryotic; the Eukarya constitute the rest of the living world. The Bacteria and the Archaea are less closely related to each other than are the Archaea and the Eukarya. **Review Figure 27.2, Table 27.1**

▶ The common ancestor of all three domains lived more than 3 billion years ago, and the common ancestor of the Archaea and Eukarya at least 2 billion years ago.

See Web/CD Tutorial 27.1

General Biology of the Prokaryotes

▶ The prokaryotes are the most numerous organisms on Earth, and they occupy an enormous variety of habitats.

▶ Most prokaryotes are cocci, bacilli, or spiral forms. Some link together to form associations, but very few are truly multicellular. **Review Figure 27.3**

▶ Prokaryotes lack nuclei, membrane-enclosed organelles, and cytoskeletons. Their chromosomes are circular. They often contain plasmids. Some prokaryotes contain internal membrane systems.

▶ Many prokaryotes move by means of flagella, gas vesicles, or gliding mechanisms. Prokaryotic flagella rotate rather than beat. **Review Figures 27.4, 27.5**

▶ Prokaryotic cell walls differ from those of eukaryotes. Bacterial cell walls generally contain peptidoglycan. Differences in peptidoglycan content result in different reactions to the Gram stain. **Review Figure 27.6.** See Web/CD Activity 27.1

▶ Prokaryotes reproduce asexually by fission, but also exchange genetic information.

▶ Prokaryotes have diverse metabolic pathways and nutritional modes. They include obligate anaerobes, facultative anaerobes, and obligate aerobes. The major nutritional types are photoautotrophs, photoheterotrophs, chemolithotrophs, and chemoheterotrophs. Some prokaryotes base their energy metabolism on nitrogen- or sulfur-containing ions. **Review Figure 27.7 and Table 27.2**

Prokaryotes in Their Environments

▶ Some prokaryotes play key roles in global nitrogen and sulfur cycles. Important players in the nitrogen cycle are the nitrogen fixers, nitrifiers, and denitrifiers.

▶ Photosynthesis by cyanobacteria generated the oxygen gas that permitted the evolution of aerobic respiration and the appearance of present-day eukaryotes.

▶ Archaea lying beneath the oceans prevent large deposits of methane, a "greenhouse gas," from accumulating in the oceans and the atmosphere.

▶ Many prokaryotes live in or on other organisms, with neutral, beneficial, or harmful effects.

▶ A small minority of bacteria are pathogens. Pathogens vary with respect to their invasiveness and toxigenicity. Some produce endotoxins, which are rarely fatal to their hosts; others produce exotoxins, which tend to be highly toxic.

▶ Prokaryotes and some unicellular eukaryotes form resistant biofilms that present medical and industrial problems.

Prokaryote Phylogeny and Diversity

▶ Phylogenetic classification of prokaryotes is now based on rRNA sequences and other molecular evidence.

▶ Lateral gene transfer among prokaryotes, which has occurred throughout evolutionary history, makes it difficult to infer prokaryote phylogeny.

▶ Evolution, powered by mutation, natural selection, and genetic drift, can proceed rapidly in prokaryotes because they are haploid and can multiply rapidly.

The Bacteria

▶ There are more known bacteria than known archaea. One phylogenetic classification of the domain Bacteria groups them into more than a dozen clades. **Review Figure 27.8**

▶ The three clades that may contain the most ancient bacteria, like the most ancient archaea, are thermophiles, suggesting that life originated in a hot environment.

▶ All four nutritional types occur in the largest bacterial group, the proteobacteria. Metabolism in different groups of proteobacteria has evolved along different lines. **Review Figure 27.9**

▶ Cyanobacteria, unlike other bacteria, photosynthesize using the same pathways plants use. Many cyanobacteria fix nitrogen.

▶ Spirochetes move by means of axial filaments.

▶ Chlamydias are tiny parasites that live within the cells of other organisms.

▶ Firmicutes are diverse; some of them produce endospores as resting structures that resist harsh conditions. Actinomycetes, some of which produce important antibiotics, grow as branching filaments.

▶ Mycoplasmas, the tiniest living things, lack conventional cell walls. They have very small genomes.

The Archaea

▶ Archaea have cell walls lacking peptidoglycan, and their membrane lipids differ from those of bacteria and eukaryotes, containing branched long-chain hydrocarbons connected to glycerol by ether linkages. **Review Figure 27.18**

▶ The domain Archaea can be divided into two principal groups, Crenarchaeota and Euryarchaeota.

▶ Crenarchaeota are mostly heat-loving and often acid-loving archaea.

▶ Methanogens produce methane by reducing carbon dioxide. Some methanogens live in the guts of herbivorous animals; others occupy high-temperature environments on the ocean floor.

▶ Extreme halophiles are salt lovers that often lend a pinkish color to salty environments; some halophiles also grow in extremely alkaline environments.

▶ Archaea of the genus *Thermoplasma* lack cell walls, are thermophilic and acidophilic, and have a tiny genome (1,100,000 base pairs).

▶ Many archaea, including members of both major groups, live in environments that are not extreme.

Self-Quiz

1. Most prokaryotes
   a. are agents of disease.
   b. lack ribosomes.
   c. evolved from the most ancient eukaryotes.
   d. lack a cell wall.
   e. are chemoheterotrophs.
2. The division of the living world into three domains
   a. is strictly arbitrary.
   b. was inspired by the morphological differences between archaea and bacteria.
   c. emphasizes the greater importance of eukaryotes.
   d. was proposed by the early microscopists.
   e. is strongly supported by data on rRNA sequences.
3. Which statement about the archaeal genome is true?
   a. It is much more similar to the bacterial genome than to eukaryotic genomes.
   b. More than half of its genes are genes that are never observed in bacteria or eukaryotes.
   c. It is much smaller than the bacterial genome.
   d. It is housed in the nucleus.
   e. No archaeal genome has yet been sequenced.
4. Which statement about nitrogen metabolism is *not* true?
   a. Certain prokaryotes reduce atmospheric N_2 to ammonia.
   b. Nitrifiers are soil bacteria.
   c. Denitrifiers are strict anaerobes.
   d. Nitrifiers obtain energy by oxidizing ammonia and nitrite.
   e. Without the nitrifiers, terrestrial organisms would lack a nitrogen supply.
5. All photosynthetic bacteria
   a. use chlorophyll *a* as their photosynthetic pigment.
   b. use bacteriochlorophyll as their photosynthetic pigment.
   c. release oxygen gas.
   d. produce particles of sulfur.
   e. are photoautotrophs.
6. Gram-negative bacteria
   a. appear blue to purple following Gram staining.
   b. are the most abundant of the bacterial groups.
   c. are all either bacilli or cocci.
   d. contain no peptidoglycan in their cell walls.
   e. are all photosynthetic.
7. Endospores
   a. are produced by viruses.
   b. are reproductive structures.
   c. are very delicate and easily killed.
   d. are resting structures.
   e. lack cell walls.
8. Actinomycetes
   a. are important producers of antibiotics.
   b. belong to the kingdom Fungi.
   c. are never pathogenic to humans.
   d. are gram-negative.
   e. are the smallest known bacteria.
9. Which statement about mycoplasmas is *not* true?
   a. They lack cell walls.
   b. They are the smallest known cellular organisms.
   c. They contain the same amount of DNA as do other prokaryotes.
   d. They cannot be killed with penicillin.
   e. Some are pathogens.
10. Archaea
    a. have cytoskeletons.
    b. have distinctive lipids in their plasma membranes.
    c. survive only at moderate temperatures and near neutrality.
    d. all produce methane.
    e. have substantial amounts of peptidoglycan in their cell walls.

For Discussion

1. Why do systematic biologists find rRNA sequence data more useful than data on metabolism or cell structure for classifying prokaryotes?
2. Why does lateral gene transfer make it so difficult to arrive at agreement on prokaryote phylogeny?
3. Differentiate among the members of the following sets of related terms:
   a. prokaryotic/eukaryotic
   b. obligate anaerobe/facultative anaerobe/obligate aerobe
   c. photoautotroph/photoheterotroph/chemolithotroph/chemoheterotroph
   d. Gram-positive/Gram-negative
4. Why are the endospores of firmicutes not considered to be reproductive structures?
5. Until fairly recently, the cyanobacteria were called blue-green algae and were not grouped with the bacteria. Suggest several reasons for this (abandoned) tendency to separate the cyanobacteria from the bacteria. Why are the cyanobacteria now grouped with the other bacteria?
6. The actinomycetes are of great commercial interest. Why?
7. Thermophiles are of great interest to molecular biologists and biochemists. Why? What practical concerns might motivate that interest?

28 Protists and the Dawn of the Eukarya

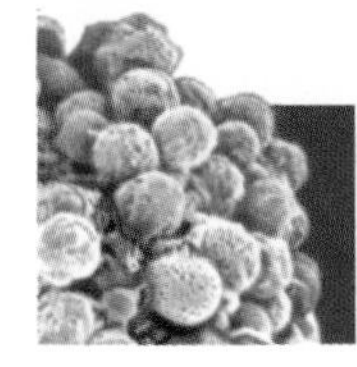

Kwame's illness began with a fever and shaking chills, followed by muscle aches, nausea, and vomiting. The child's kidneys failed, he developed seizures and went into a coma, and finally he died. Every 30 seconds, malaria kills someone somewhere—usually in sub-Saharan Africa, although malaria occurs in more than 100 countries and territories. About *600 million* people have this disease.

Mosquitoes carry the malaria pathogen from person to person. This pathogen is not a bacterium—rather, it is a tiny eukaryote, *Plasmodium falciparum*. Probably the most obvious visible difference between it and the prokaryotes is that *Plasmodium* has numerous compartments—membrane-enclosed organelles that perform specialized functions. This single-celled pathogen has a cytoskeleton, a nucleus enclosed by a nuclear envelope, and several kinds of organelles. As a member of the domain Eukarya, it differs from members of the two prokaryotic domains in other important ways as well.

The flexibility and options that arose once the eukaryotic cell had evolved resulted in a profusion of body forms and myriad specialized functions. Eukaryotic evolution has produced great diversity, especially among the multicellular clades, but even among the unicellular members of the domain. In both multicellular and unicellular forms, however, there are also many cases of convergent evolution; for example, organisms with an amoeba-like body form arose several times. These various amoebas are examples of organisms called protists.

Protists Defined

Many modern members of the Eukarya—trees, mushrooms, and dogs, not to mention ourselves—are familiar to us. We would have no problem recognizing these organisms as members of the kingdoms Plantae, Fungi, and Animalia. However, amoebas and a dazzling assortment of other eukaryotes, mostly microscopic organisms, don't fit into these three kingdoms. We call all those eukaryotes that are neither plants, animals, nor fungi **protists**. *The protists are not a clade; they are a polyphyletic group* (see Figure 1.8). Some protists are more closely related to the animals than they are to other protists. Some protists are motile, while others are stationary; some are photosynthetic, while others are heterotrophic; most are unicellular, while

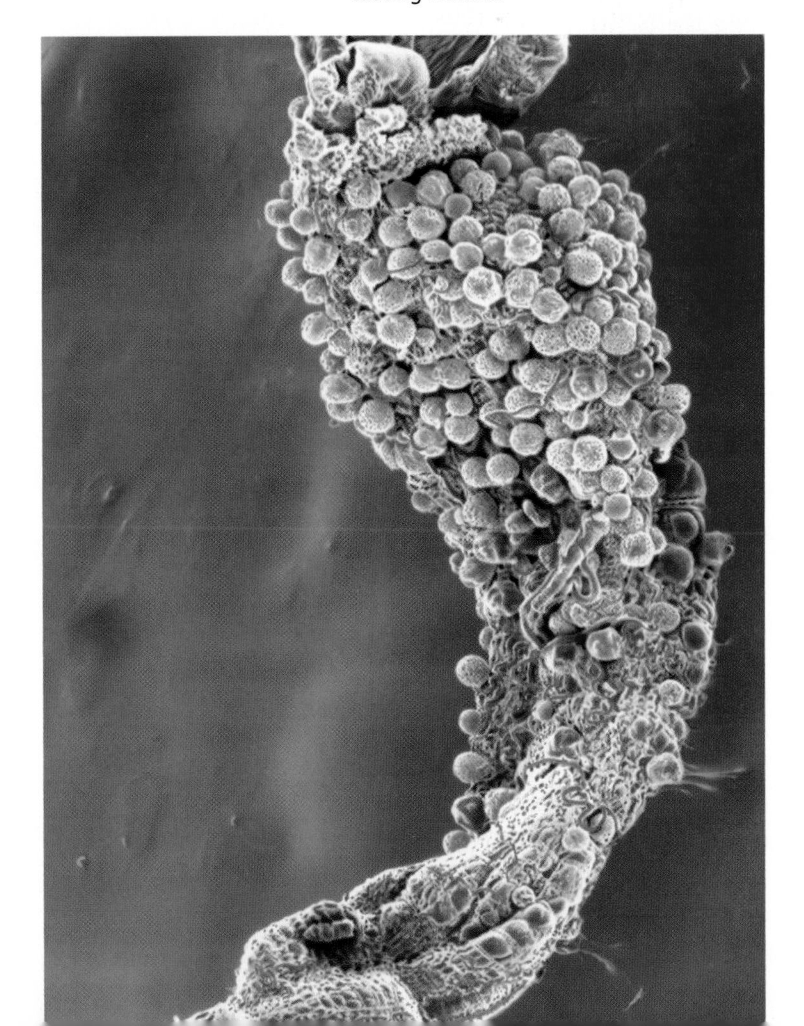

***Plasmodium falciparum*, the Malaria Parasite** This stomach wall of an *Anopheles* mosquito is covered with cells (artificially colored blue) of a particular stage of the *Plasmodium* life cycle. These *Plasmodium* cells will give rise to cells that the mosquito can transmit to humans, causing malaria.

(*c*) *Macrocystis* sp.

(*a*) *Peridinium* sp.

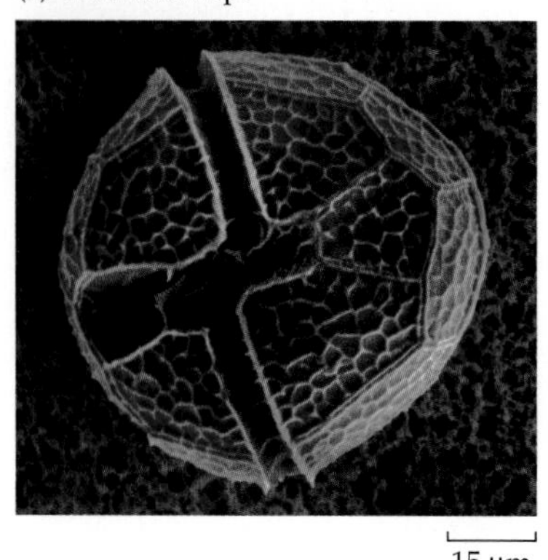

(*b*) *Giardia* sp.

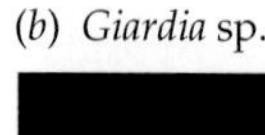

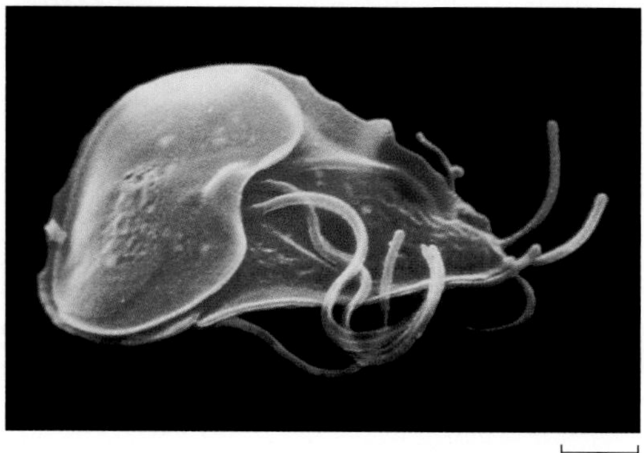

28.1 Three Protists (*a*) Most dinoflagellates are photosynthetic unicellular protists. (*b*) *Giardia* is a unicellular parasite of humans and other mammals. (*c*) Giant kelps are some of the world's longest organisms.

some giant kelps are not only multicellular but also huge, sometimes achieving lengths greater than that of a football field (Figure 28.1).

The protists include some of the most ancient eukaryotic organisms as well as the ancestors of the plants, animals, and fungi. The origin of the eukaryotic cell was one of the pivotal events in evolutionary history. In this chapter, we'll describe the origin and early diversification of the eukaryotes and the complexity achieved by some single cells. Then we'll explore some of the diversity of protist body forms and try to give a sense of developing current views of the evolutionary relationships of some of the protists.

The Origin of the Eukaryotic Cell

The eukaryotic cell differs in many ways from the prokaryotic cell. How did it originate? Given the nature of evolutionary processes, the differences cannot all have arisen simultaneously. We think we can make some reasonable inferences about the most important events, bearing in mind that the global environment underwent an enormous change—from anaerobic to aerobic—during the course of these events. As you read this chapter, keep in mind that the steps we suggest are just that: reasonable inferences. This version of the story is one of a few under current consideration. We present it as a framework for thinking about this challenging problem, not as a set of facts.

The modern eukaryotic cell arose in several steps

The essential steps in the origin of the eukaryotic cell include:

- The origin of a flexible cell surface
- The origin of a cytoskeleton
- The origin of a nuclear envelope
- The appearance of digestive vesicles
- The endosymbiotic acquisition of certain organelles

WHAT A FLEXIBLE CELL SURFACE ALLOWS. Many ancient fossil prokaryotes look like rods, and we presume that they, like most present-day prokaryotic cells, had firm cell walls. The first step toward the eukaryotic condition may have been the loss of the cell wall by an ancestral prokaryotic cell. This may not seem like an obvious first step, but consider the possibilities open to a flexible cell without a wall.

First, think of cell size. As a cell grows larger, its surface area-to-volume ratio decreases (see Figure 4.3). Unless the surface area can be increased, the cell volume will reach an upper limit. If the surface is flexible, it can fold inward and elaborate itself, creating more surface area for gas and nutrient exchange (Figure 28.2). With a surface flexible enough to

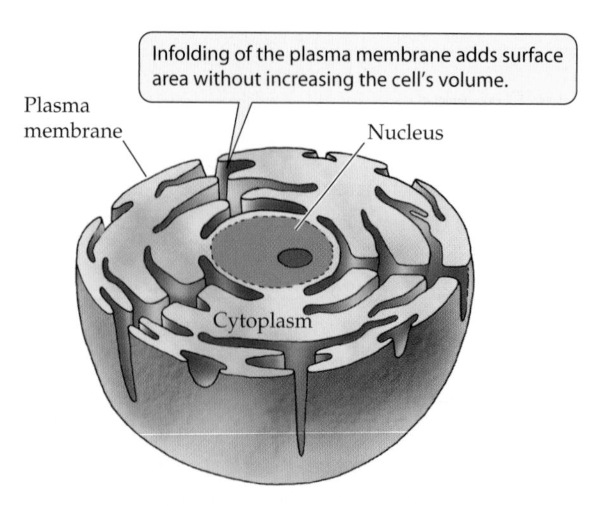

Infolding of the plasma membrane adds surface area without increasing the cell's volume.
Plasma membrane
Nucleus
Cytoplasm

28.2 Membrane Infolding The loss of the rigid prokaryotic cell wall allowed the plasma membrane to fold inward and create more surface area.

allow infolding, the cell can exchange materials with its environment rapidly enough to sustain a larger volume and more rapid metabolism. Further, a flexible surface can pinch off bits of the environment, bringing them into the cell by endocytosis (Figure 28.3).

The chromosome of a bacterial cell is attached to a site on its plasma membrane. If that region of the plasma membrane were to fold into the cell, the first step would be taken toward the evolution of a nucleus, the key feature of the eukaryotic cell. What are some other likely early changes?

CHANGES IN CELL STRUCTURE AND FUNCTION. Other early steps in the evolution of the eukaryotic cell are likely to have included three advances: the formation of ribosome-studded internal membranes, some of which surrounded the DNA (see Figure 28.3); the appearance of a cytoskeleton; and the evolution of digestive vesicles.

A cytoskeleton made up of actin fibers and microtubules would allow the cell to manage changes in shape, to distribute daughter chromosomes, and to move materials from one part of the now much larger cell to other parts. The origin of the cytoskeleton remains a mystery, heightened by the fact that the genes that encode most of the cytoskeleton are present in neither bacteria nor archaea. An intriguing and controversial suggestion is that a fourth domain of life, now long extinct, originated these genes and transferred them laterally to an ancestor of the early eukaryotes.

From an intermediate kind of cell, the next advance was probably to a cell that we could call a *phagocyte*—a motile cell that could prey on other cells by engulfing and digesting them. The first true eukaryote possessed a cytoskeleton and a nuclear envelope. It may have had an associated endoplasmic reticulum and Golgi apparatus, and perhaps one or more flagella of the eukaryotic type.

ENDOSYMBIOSIS AND ORGANELLES. While the processes already outlined were taking place, the cyanobacteria were very busy, generating oxygen gas as a product of photosynthesis. The increasing O_2 levels in the atmosphere had disastrous consequences for most other living things because most organisms of the

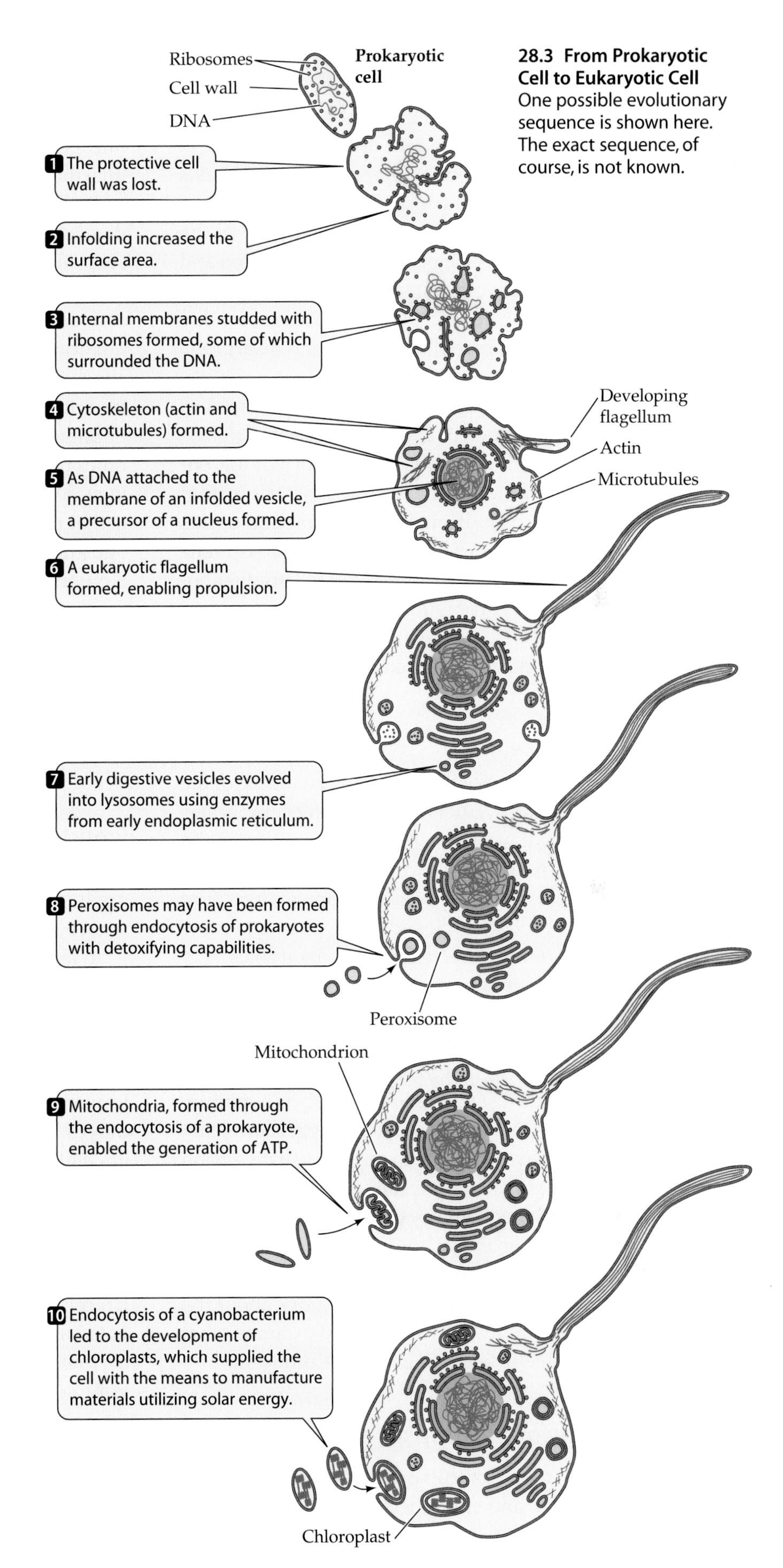

Ribosomes
Cell wall
DNA
Prokaryotic cell
1 The protective cell wall was lost.
2 Infolding increased the surface area.
3 Internal membranes studded with ribosomes formed, some of which surrounded the DNA.
4 Cytoskeleton (actin and microtubules) formed.
Developing flagellum
Actin
Microtubules
5 As DNA attached to the membrane of an infolded vesicle, a precursor of a nucleus formed.
6 A eukaryotic flagellum formed, enabling propulsion.
7 Early digestive vesicles evolved into lysosomes using enzymes from early endoplasmic reticulum.
8 Peroxisomes may have been formed through endocytosis of prokaryotes with detoxifying capabilities.
Peroxisome
Mitochondrion
9 Mitochondria, formed through the endocytosis of a prokaryote, enabled the generation of ATP.
10 Endocytosis of a cyanobacterium led to the development of chloroplasts, which supplied the cell with the means to manufacture materials utilizing solar energy.
Chloroplast

28.3 From Prokaryotic Cell to Eukaryotic Cell One possible evolutionary sequence is shown here. The exact sequence, of course, is not known.

time (archaea and bacteria) were unable to tolerate the newly aerobic, oxidizing environment. But some prokaryotes managed to cope with these changes, and—fortunately for us—so did some of the ancient phagocytes.

In Chapter 4 we introduced the concept of *endosymbiosis* (organisms living together, one inside the other; see Figure 4.18). According to one highly speculative hypothesis, the key to the survival of early phagocytes was the ingestion and incorporation of a prokaryote that took up residence within the phagocyte and evolved into the peroxisomes of today (see Figure 28.3). These organelles were able to disarm the toxic products of oxygen action, such as hydrogen peroxide. This association may have been the first important endosymbiosis in the evolution of the eukaryotic cell.

A crucial endosymbiotic event in the history of the Eukarya was the incorporation of a proteobacterium that evolved into the mitochondrion. Upon completion of this step, the basic modern eukaryotic cell was complete. Some very important eukaryotes are the result of yet another endosymbiotic step, the incorporation of a prokaryote related to today's cyanobacteria, which became the chloroplast. We'll see how this happened later in this chapter.

Many uncertainties remain

Several uncertainties cloud our current understanding of the origins of eukaryotic cells. Lateral gene transfer complicates the study of eukaryotic origins, just as it complicates the study of relationships among the prokaryote clades. At the same time, it may not have been extensive enough to account for the fact that, as genetic studies advance, more and more genes of bacterial origin are being found in eukaryotes.

An endosymbiotic origin of mitochondria and chloroplasts accounts for the presence of bacterial genes encoding enzymes for energy metabolism (respiration and photosynthesis), but it does not explain the presence of many other bacterial genes. The eukaryotic genome clearly is a mixture of genes with two distinct origins. A recent suggestion is that the Eukarya might have arisen from the mutualistic fusion (not endosymbiosis) of a Gram-negative bacterium and an archaean. There are many interesting ideas about eukaryotic origins awaiting additional data and analysis.

We can expect that these and other questions will yield to additional research. Let's leave our speculations about the origin of the eukaryotes for the moment and examine what we do know about them, beginning with the protists.

General Biology of the Protists

Most protists are aquatic. Some live in marine environments, others in fresh water, and still others in the body fluids of other organisms. The slime molds inhabit damp soil and the moist, decaying bark of rotting trees. Many other protists also live in soil water, some of them contributing to the global nitrogen cycle by preying on soil bacteria and recycling their nitrogen compounds into nitrates. Most protists are unicellular, but some are multicellular, and a few are very large.

Protists are strikingly diverse in their structure, but not so diverse in their metabolism as the prokaryotes—which is not surprising, since some of the eukaryotes' most important metabolic pathways were "borrowed" from bacteria through endosymbiosis. However, protists do display a number of nutritional modes. Some are photosynthetic autotrophs, some are heterotrophs, and some switch with ease between the autotrophic and heterotrophic modes of nutrition.

Some protists, formerly classified as animals, are sometimes referred to as **protozoans**, although biologists increasingly regard this term as inappropriate because it lumps together protist groups that are phylogenetically distant from one another. Most protozoans are ingestive heterotrophs. Similarly, there are several kinds of photosynthetic protists that some biologists still refer to as **algae** (singular, alga). Although these two terms are useful in some contexts, they do not correspond with natural phylogeny, and we generally avoid them in this book except as parts of descriptive names such as "brown algae." Let's next consider some of the other ways in which protists differ from one another.

Protists have diverse means of locomotion

Although a few protist groups consist entirely of nonmotile organisms, most groups include cells that move, either by amoeboid motion, by ciliary action, or by means of flagella.

In amoeboid motion, the cell forms **pseudopods** ("false feet") that are extensions of its constantly changing body mass. Cells such as the amoeba in Figure 28.4 simply extend a pseudopod and then flow into it. *Cilia* are tiny, hairlike organelles that beat in a coordinated fashion to move the cell forward or backward (see Figure 4.23). A eukaryotic *flagellum* moves like a whip; some flagella *push* the cell forward, others *pull* the cell forward. Cilia and eukaryotic flagella are identical in cross section; they differ only in length.

Vesicles perform a variety of functions

Unicellular organisms tend to be of microscopic size. As we noted above, an important reason that cells are small is that they need enough membrane surface area in relation to their volume to support the exchange of materials required for their existence. Many relatively large unicellular protists minimize this problem by having membrane-enclosed **vesicles** of various types that increase their effective surface area.

As we saw in Chapter 5, organisms living in fresh water are hypertonic to their environment. Many freshwater pro-

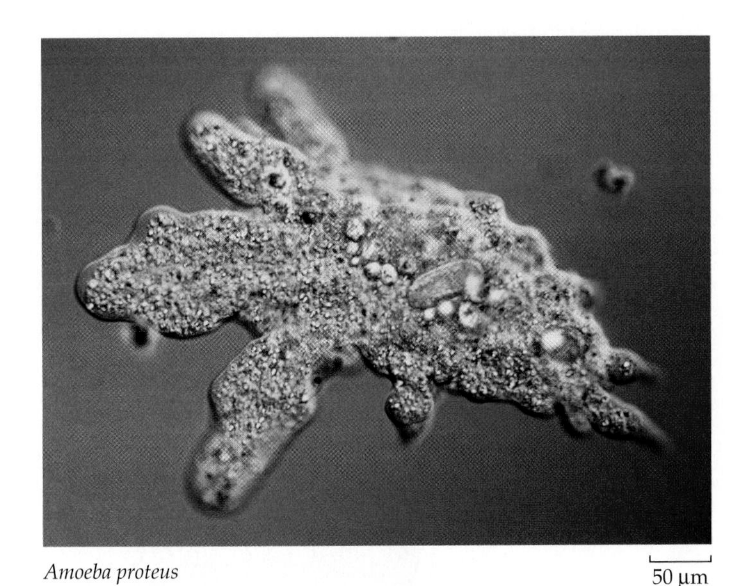

28.4 An Amoeba The flowing pseudopods are constantly changing shape as the amoeba moves and feeds.

tists address this problem by means of specialized vesicles that excrete the excess water they constantly take in by osmosis. Members of several protist groups have such **contractile vacuoles**. The excess water collects in the contractile vacuole, which then expels the water from the cell (Figure 28.5).

A second important type of vesicle found in many protists is the **food vacuole**. Protists such as *Paramecium* engulf solid food by endocytosis, forming a food vacuole within which the food is digested (Figure 28.6). Smaller vesicles containing digested food pinch away from the food vesicle and enter the cytoplasm. These tiny vesicles provide a large surface area across which the products of digestion may be absorbed by the rest of the cell.

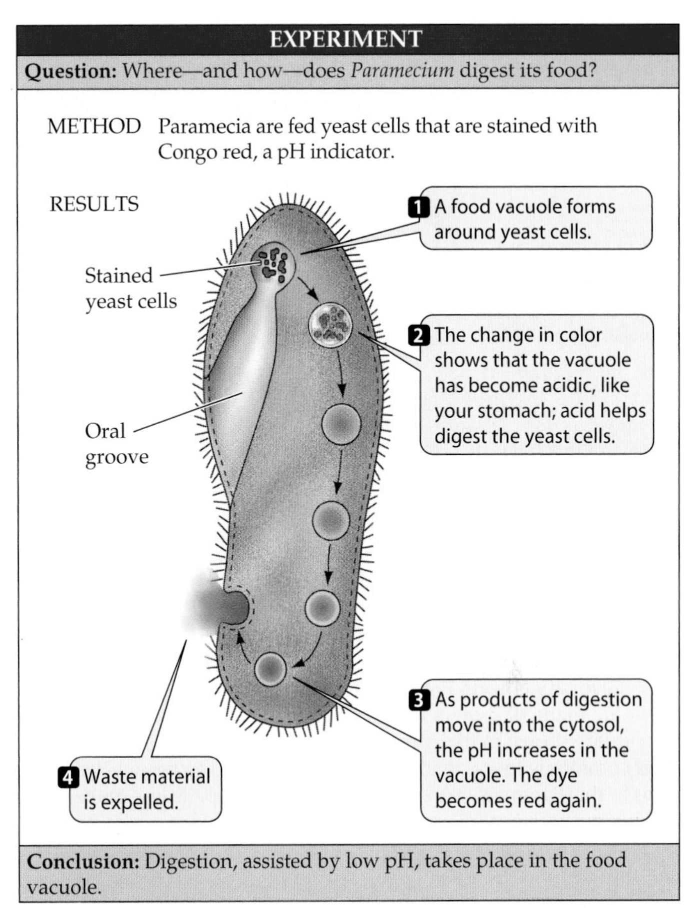

EXPERIMENT
Question: Where—and how—does Paramecium digest its food?
METHOD Paramecia are fed yeast cells that are stained with Congo red, a pH indicator.
RESULTS
Stained yeast cells
Oral groove
1 A food vacuole forms around yeast cells.
2 The change in color shows that the vacuole has become acidic, like your stomach; acid helps digest the yeast cells.
3 As products of digestion move into the cytosol, the pH increases in the vacuole. The dye becomes red again.
4 Waste material is expelled.
Conclusion: Digestion, assisted by low pH, takes place in the food vacuole.

28.6 Food Vacuoles Handle Digestion and Excretion An experiment with *Paramecium* demonstrates the function of food vacuoles. *Paramecium* ingests food by way of the oral groove at the left. The dye Congo red turns green at acidic pH and red at neutral or basic pH.

Contractile vacuole

Inside of cell

Plasma membrane

Outside of cell

1 Water passes from the cytoplasm to radiating canals and to the central vesicle . . .

2 . . .which periodically fuses with the plasma membrane, . . .

3 . . .expels its contents, . . .

4 . . .and detaches from the membrane.

28.5 Contractile Vacuoles Bail Out Excess Water Water constantly enters freshwater protists by osmosis. A pore in the cell surface allows the contractile vacuole to expel the water it accumulates.

The cell surfaces of protists are diverse

A few protists, such as some amoebas, are surrounded by only a plasma membrane, but most have stiffer surfaces that maintain the structural integrity of the cell. Many protists have cell walls, which are often complex in structure. Other protists that lack cell walls have a variety of ways of strengthening their surfaces. Some have internal "shells," which the organism either produces itself, as foraminiferans do, or makes from bits of sand and thickenings immediately beneath the plasma membrane, as some amoebas do (Figure 28.7).

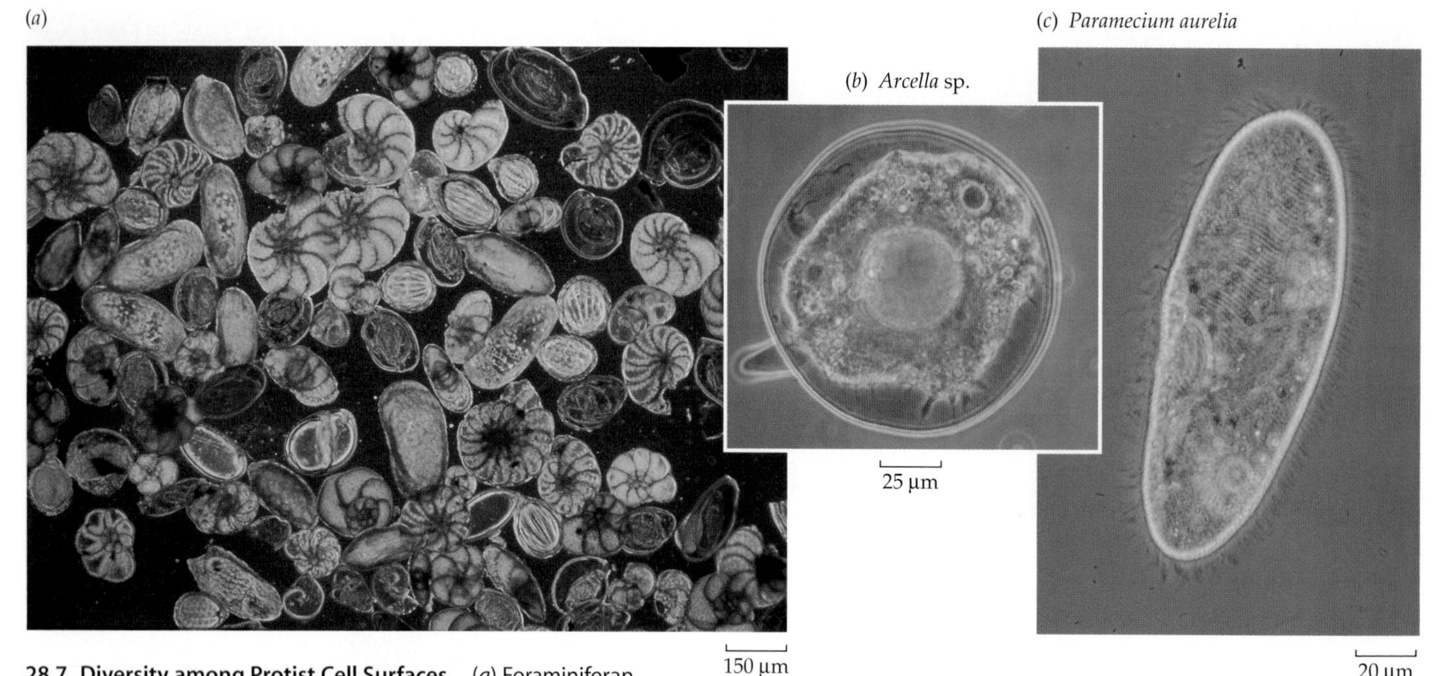

28.7 Diversity among Protist Cell Surfaces (*a*) Foraminiferan shells are made of protein hardened with calcium carbonate. Several species are shown in this photograph. (*b*) This genus of shelled amoeba is commonly found in freshwater ponds and puddles. (*c*) The proteins in this *Paramecium*'s surface—known as its pellicle—make it flexible but resilient.

Many protists contain endosymbionts

Endosymbiosis is very common among the protists, and in some instances both the host and the endosymbiont are protists. Many radiolarians, for example, harbor photosynthetic protists (Figure 28.8). As a result, these radiolarians appear greenish or golden, depending on the type of endosymbiont they contain. This arrangement is beneficial to the radiolarian, for it can make use of the organic nutrients produced by its photosynthetic guest. The guest, in turn, may make use of metabolites made by the host, or it may simply receive physical protection. In other cases, the guest may be a prisoner, exploited for its photosynthetic products while receiving no benefit itself. We will take a more detailed look at the history of endosymbiosis among photosynthetic protists later in this chapter.

Both asexual and sexual reproduction occur among the protists

Although most protists practice both asexual and sexual reproduction, some groups lack sexual reproduction. As we will see, some asexually reproducing protists also engage in genetic recombination that does not directly result in reproduction.

Asexual reproductive processes in the protists include *binary fission* (splitting of the cell, with mitosis followed by cytokinesis), *multiple fission* (splitting into more than two cells), *budding* (the outgrowth of a new cell from the surface of an old one), and the formation of *spores* (cells that are capable of developing into new organisms). Sexual reproduction also takes various forms. In some protists, as in animals, the gametes are the only haploid cells. In some other protists, by contrast, both diploid and haploid cells undergo mitosis, giving rise to alternation of generations, which will be described later in this chapter.

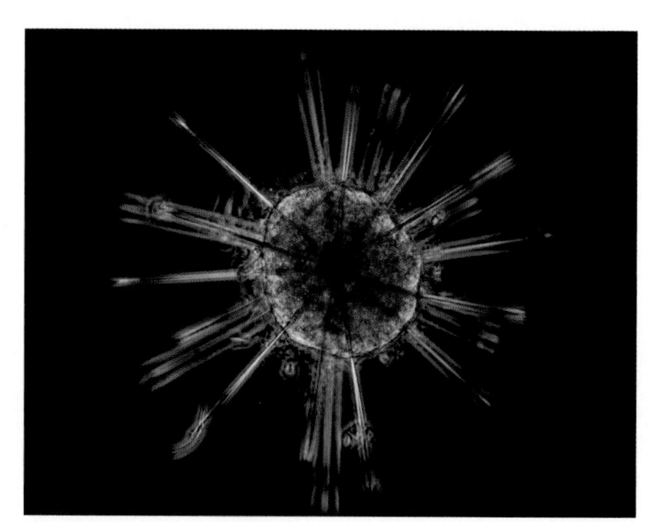

28.8 Protists within Protists Photosynthetic dinoflagellates (see Figure 28.1*a*) are living as endosymbionts within this radiolarian, providing organic nutrients for the radiolarian and imparting the golden-brown pigmentation seen at the center of its glassy skeleton. Both the dinoflagellates and the radiolarian are protists.

28.1 Major Protist Clades

GROUP	ATTRIBUTES	EXAMPLES
Diplomonads	Unicellular, no mitochondria, two nuclei, flagella	*Giardia*
Parabasalids	Unicellular, no mitochondria, flagella and undulating membrane	*Trichomonas*
Euglenozoans	Unicellular, with flagella	
Euglenoids	Mostly photoautotrophic	*Euglena*
Kinetoplastids	Have a single large mitochondrion	*Trypanosoma*
Alveolates	Unicellular; cavities (alveoli) below cell surface	
Dinoflagellates	Pigments give golden-brown color	*Gonyaulax*
Apicomplexans	Apical complex in spores for penetration of host	*Plasmodium*
Ciliates	Cilia; two types of nuclei	*Paramecium*
Stramenopiles	Two unequal flagella, one with hairs	
Diatoms	Unicellular; photoautotrophic; two-part cell walls; no flagellum	*Thalassiosira*
Brown algae	Multicellular; marine; photoautotrophic	*Fucus, Macrocystis*
Oomycetes (water molds, powdery mildews)	Mostly coenocytic; heterotrophic	*Saprolegnia*
Red algae	No flagella; photoautrophic; phycoerythrin and phycocyanin	*Chondrus, Polysiphonia*
Chlorophytes ("Green algae"[a])	Photoautotrophic	*Ulva, Volvox*
Choanoflagellates	Resemble sponge cells; heterotrophic; with flagella	*Codosiga, Choanoeca*

[a]The green algae do not constitute a clade. The chlorophytes are a clade of green algae; a different green algal lineage gave rise to the plant kingdom.

The diversity of form, habitat, metabolism, locomotion, reproduction, and life cycles found among the protists reflects the diversity of avenues pursued during the early evolution of eukaryotes. Many of these avenues led to great success, judging from the abundance and diversity of today's protists and other eukaryotes.

Protist Diversity

The phylogeny of protists is an area of exciting, challenging research. The marvelous diversity of protist body forms and nutritional lifestyles seems reason enough for a fascination with these organisms, but questions about how the multicellular eukaryotic kingdoms originated from the protists stimulate further interest. Fortunately, the tools of molecular biology, such as rRNA sequencing, are making it possible to explore evolutionary relationships among the protists in ever greater detail and with greater confidence (see Chapters 25 and 26).

We will discuss several protist clades in this chapter, as well as a few other groups of more uncertain phylogenetic status. Some biologists refer to many of these clades as kingdoms; others refer to them as subkingdoms, and still others refer to them as phyla. This choice of words is not of immediate concern to us here, so we'll just call them "groups." We'll describe the following groups: diplomonads, parabasalids, euglenozoans, alveolates, stramenopiles, red algae, chlorophytes, and choanoflagellates (Table 28.1; Figure 28.9).

As we shall see, some of these protist clades consist of organisms with very diverse body plans. On the other hand, certain body plans, such as those of amoebas and those of slime molds, have arisen again and again during evolution, in groups only distantly related to one another. We'll begin our tour of protist clades with two of apparently ancient origin.

Diplomonads and Parabasalids

Two clades, the **diplomonads** and the **parabasalids**, appear to represent the earliest surviving branches in today's tree of eukaryotic life. It is likely that other clades diverged even earlier than the diplomonads and parabasalids. However, any such clades either were lost because of massive changes in the environment or remain hidden in rarely studied environments.

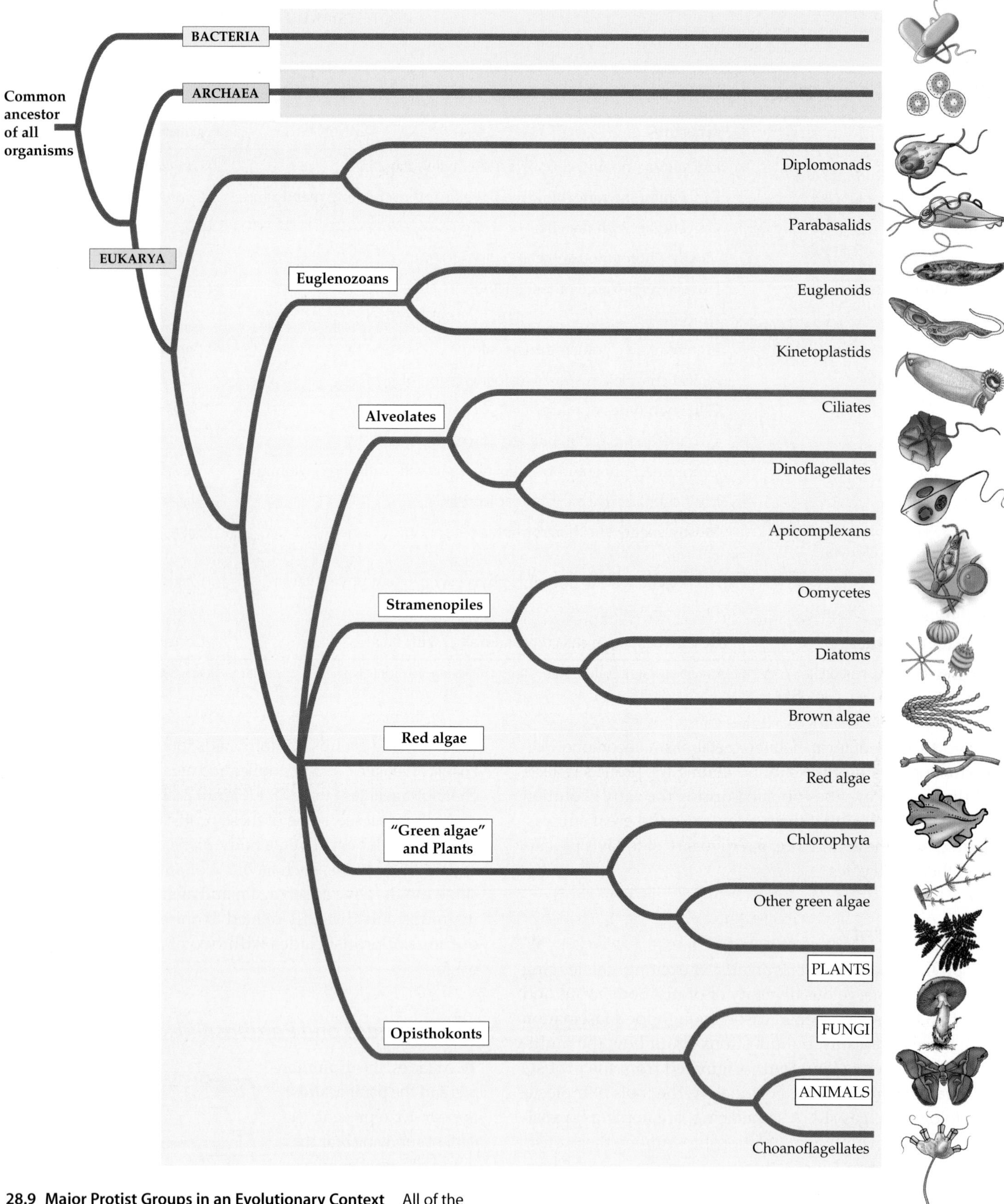

28.9 Major Protist Groups in an Evolutionary Context All of the protist groups shown here, except certain green algae, appear to be clades. The protists themselves do not constitute a clade. We also show the prokaryotic domains and the plant, fungal, and animal kingdoms to provide context. The term "opisthokont" refers to organisms that have or had (ancestrally) a flagellum in a posterior position; this group includes a protist clade as well as the fungi and animals.

Both the diplomonads and the parabasalids are unicellular organisms that lack mitochondria. This absence of mitochondria may be a derived condition: Ancestors of these organisms may have possessed mitochondria that were lost in the course of evolution. The existence of such organisms to-

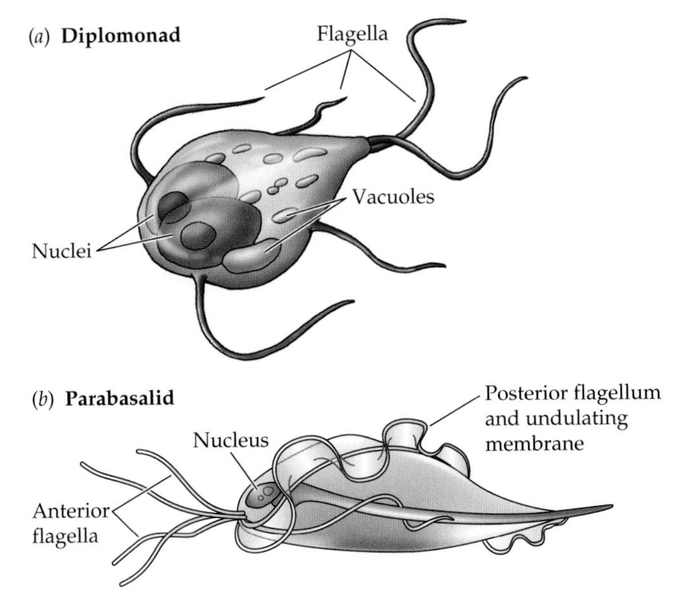

28.10 Two Protist Groups Lack Mitochondria Diplomonads and parabasalids appear to represent the most ancient surviving branches of eukaryotic life. (*a*) *Giardia*, a diplomonad, has flagella and two nuclei (see also Figure 28.1*b*). (*b*) *Trichomonas*, a parabasalid, has flagella and undulating membranes. Neither of these protists possesses mitochondria.

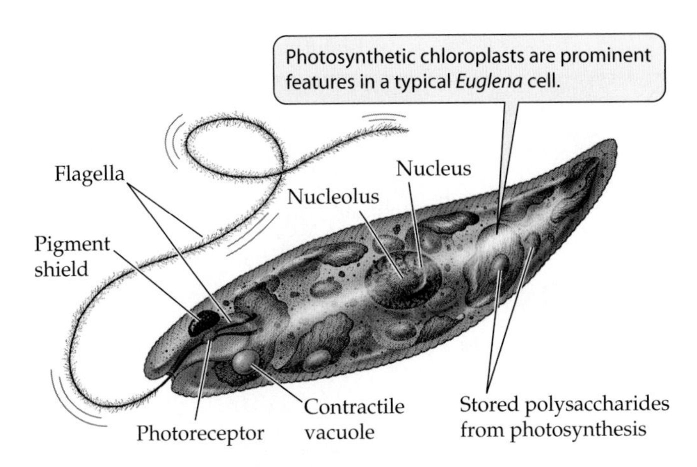

28.11 A Photosynthetic Euglenoid Several *Euglena* species are among the best-known flagellates. In this species, the second flagellum is rudimentary.

day shows that eukaryotic life is feasible without mitochondria, and for that reason, the diplomonads and parabasalids are the focus of much attention.

Giardia lamblia, a diplomonad, is a familiar parasite that contaminates water supplies and causes the intestinal disease giardiasis (Figure 28.10*a*). This tiny organism has no mitochondria, chloroplasts, or other membrane-enclosed organelles, but it contains two nuclei bounded by nuclear envelopes, and it has a cytoskeleton and multiple flagella.

Trichomonas vaginalis is a parabasalid responsible for a sexually transmitted disease in humans (Figure 28.10*b*). Infection of the male urethra, where it may occur without symptoms, is less common than infection of the vagina. In addition to flagella, the parabasalids have undulating membranes that also contribute to the cell's locomotion.

Euglenozoans

BACTERIA
ARCHAEA
EUKARYA
Diplomonads, Parabasalids
Euglenozoans
Alveolates
Stramenopiles
Red algae
"Green algae" and plants
Opisthokonts

The **euglenozoans** are a clade of *flagellates:* unicellular organisms with flagella. They reproduce asexually by binary fission. There are two subgroups of euglenozoans: euglenoids and kinetoplastids.

Euglenoids have anterior flagella

The **euglenoids** possess flagella arising from a pocket at the anterior end of the cell. Many members of the group are photosynthetic. Euglenoids used to be claimed by the zoologists as animals and by the botanists as plants.

Figure 28.11 depicts a cell of the genus *Euglena*. Like most other euglenoids, this common freshwater organism has a complex cell structure. It propels itself through the water with the longer of its two flagella, which may also serve as an anchor to hold the organism in place. The flagellum provides power by means of a wavy motion that spreads from base to tip. The second flagellum is often rudimentary.

Euglena has very flexible nutritional requirements. Many species are always heterotrophic. Other species are fully autotrophic in sunlight, using chloroplasts to synthesize organic compounds through photosynthesis. The chloroplasts of euglenas are surrounded by three membranes (unlike plant chloroplasts, which have only two; we will describe the history of the third membrane later in this chapter). When kept in the dark, these euglenas lose their photosynthetic pigment and begin to feed exclusively on dead organic material floating in the water around them. Such a "bleached" *Euglena* resynthesizes its photosynthetic pigment when it is returned to the light and becomes autotrophic again. But *Euglena* cells treated with certain antibiotics or mutagens lose their photosynthetic pigment completely; neither they nor their descendants are ever autotrophs again. However, those descendants function well as heterotrophs.

Kinetoplastids have mitochondria that edit their own RNA

The **kinetoplastids** are unicellular, parasitic flagellates with a single, large mitochondrion. That mitochondrion contains a *kinetoplast*—a unique structure housing multiple, circular DNA molecules and associated proteins. Some of these DNA molecules encode "guides" that edit RNA within the mitochondrion.

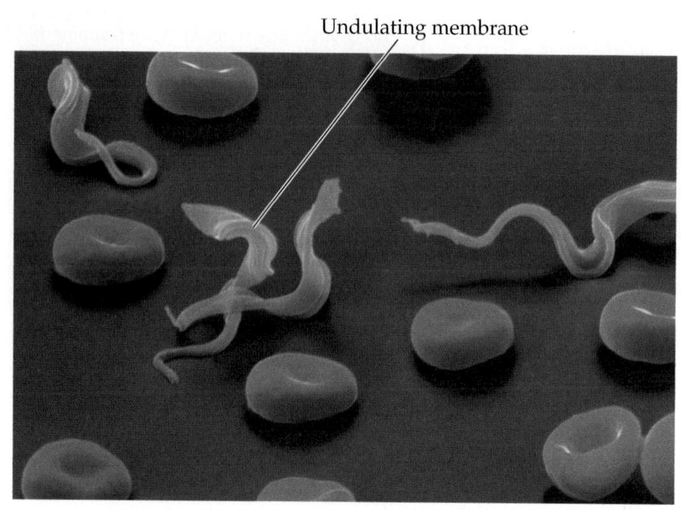

28.12 A Parasitic Kinetoplastid Trypanosomes, shown here among human red blood cells (round), cause sleeping sickness in mammals. A flagellum runs along one edge of the cell as part of a structure called the undulating membrane.

Some kinetoplastids are human pathogens. Sleeping sickness, one of the most dreaded diseases of Africa, is caused by the parasitic kinetoplastid *Trypanosoma* (Figure 28.12). An insect, the tsetse fly, is the *vector* (intermediate host) of *Trypanosoma*. Carrying its deadly cargo, the tsetse fly bites livestock, wild animals, and humans, infecting them with the parasite. *Trypanosoma* then multiplies in the mammalian bloodstream and produces toxins. When these parasites invade the nervous system, the neurological symptoms of sleeping sickness appear and are followed by death. Half a million people now have sleeping sickness, and 80 percent of them will die of the disease. About 3 million head of livestock die from *Trypanosoma* infections each year. Other trypanosomes cause leishmaniasis, Chagas' disease, and East Coast fever; all are major diseases in the Tropics.

Alveolates

The **alveolates** are a clade of unicellular organisms. The shared derived trait (synapomorphy) that characterizes them is the possession of cavities called *alveoli* just below their plasma membranes. They are diverse in body form. The alveolate groups we'll consider here are the dinoflagellates, apicomplexans, and ciliates.

BACTERIA
ARCHAEA
EUKARYA
Diplomonads, Parabasalids
Euglenozoans
Alveolates
Stramenopiles
Red algae
"Green algae" and plants
Opisthokonts

Dinoflagellates are unicellular marine organisms with two flagella

The **dinoflagellates** are all unicellular, and most are marine organisms. A distinctive mixture of photosynthetic and accessory pigments gives their chloroplasts a golden-brown color. The dinoflagellates are of great ecological, evolutionary, and morphological interest. They are among the most important primary photosynthetic producers of organic matter in the oceans.

Many dinoflagellates are endosymbionts living within the cells of other organisms, including various invertebrates and even other marine protists. Dinoflagellates are particularly common endosymbionts in corals, to whose growth they contribute by photosynthesis. As we will see later in this chapter, endosymbiotic events have given rise to dinoflagellates with different numbers of membranes surrounding their chloroplasts. Some dinoflagellates are nonphotosynthetic and live as parasites within other marine organisms.

Dinoflagellates have a distinctive appearance (see Figure 28.1*a*). They generally have two flagella, one in an equatorial groove around the cell, the other starting at the same point as the first and passing down a longitudinal groove before extending into the surrounding medium. Some dinoflagellates, notably *Pfiesteria piscida*, take on different forms, including amoeboid ones, depending on environmental conditions. It has been claimed that *P. piscida* can occur in at least two dozen distinct forms, although this claim is controversial. In any case, this remarkable dinoflagellate is highly toxic to fish and can, when present in great numbers, both stun and feed on them.

Some dinoflagellates reproduce in enormous numbers in warm and somewhat stagnant waters. The result can be a "red tide," so called because of the reddish color of the sea that results from the pigments of the dinoflagellates (Figure 28.13). During a red tide, the concentration of dinoflagellates may reach 60 million cells per liter of ocean water. *Pfiesteria* and certain other red tide species produce a potent nerve toxin that can kill tons of fish. The genus *Gonyaulax* produces a toxin that can accumulate in shellfish in amounts that, although not fatal to the shellfish, may kill a person who eats the shellfish.

Many dinoflagellates are bioluminescent. In complete darkness, cultures of these organisms emit a faint glow. If air is suddenly stirred or bubbled through the culture, the organisms each emit numerous bright flashes. A ship passing through a tropical ocean that contains a rich growth of these species produces a bow wave and wake that glow eerily as billions of these dinoflagellates discharge their light systems.

Apicomplexans are parasites with unusual spores

Exclusively parasitic organisms, the **apicomplexans** derive their name from the *apical complex*, a mass of organelles con-

The pigments of dinoflagellates make the red tide easy to see.

28.13 A Red Tide of Dinoflagellates By reproducing in astronomical numbers, the dinoflagellate *Gonyaulax tamarensis* can cause toxic red tides, such as this one along the coast of Baja California.

tained within the apical end of their spores. These organelles help the apicomplexan spore invade its host's tissues. Unlike many other protists, apicomplexans lack contractile vacuoles.

Apicomplexans generally have an amorphous amoeboid body form. This body form has evolved over and over again in parasitic protists. It appears even among parasitic dinoflagellates, a group of organisms whose nonparasitic relatives, as we have just seen, have highly distinctive, complex body forms.

Like many obligate parasites, apicomplexans have elaborate life cycles featuring asexual and sexual reproduction by a series of very dissimilar life stages. Often these stages are associated with two different types of host organisms.

Toxoplasma, a genus of apicomplexans, causes opportunistic infections in AIDS patients. There are other pathogenic apicomplexans as well. The best-known are the malarial parasites of the genus *Plasmodium*, a highly specialized group of organisms that spend part of their life cycle within human red blood cells (Figure 28.14). Although it has been almost eliminated from the United States, malaria continues to be a serious problem in many tropical countries, as we saw at the beginning of this chapter. In terms of the number of people infected, malaria is one of the world's three most serious diseases, and it kills more than a million people each year.

Female mosquitoes of the genus *Anopheles* transmit *Plasmodium* to humans. The parasite enters the human circula-

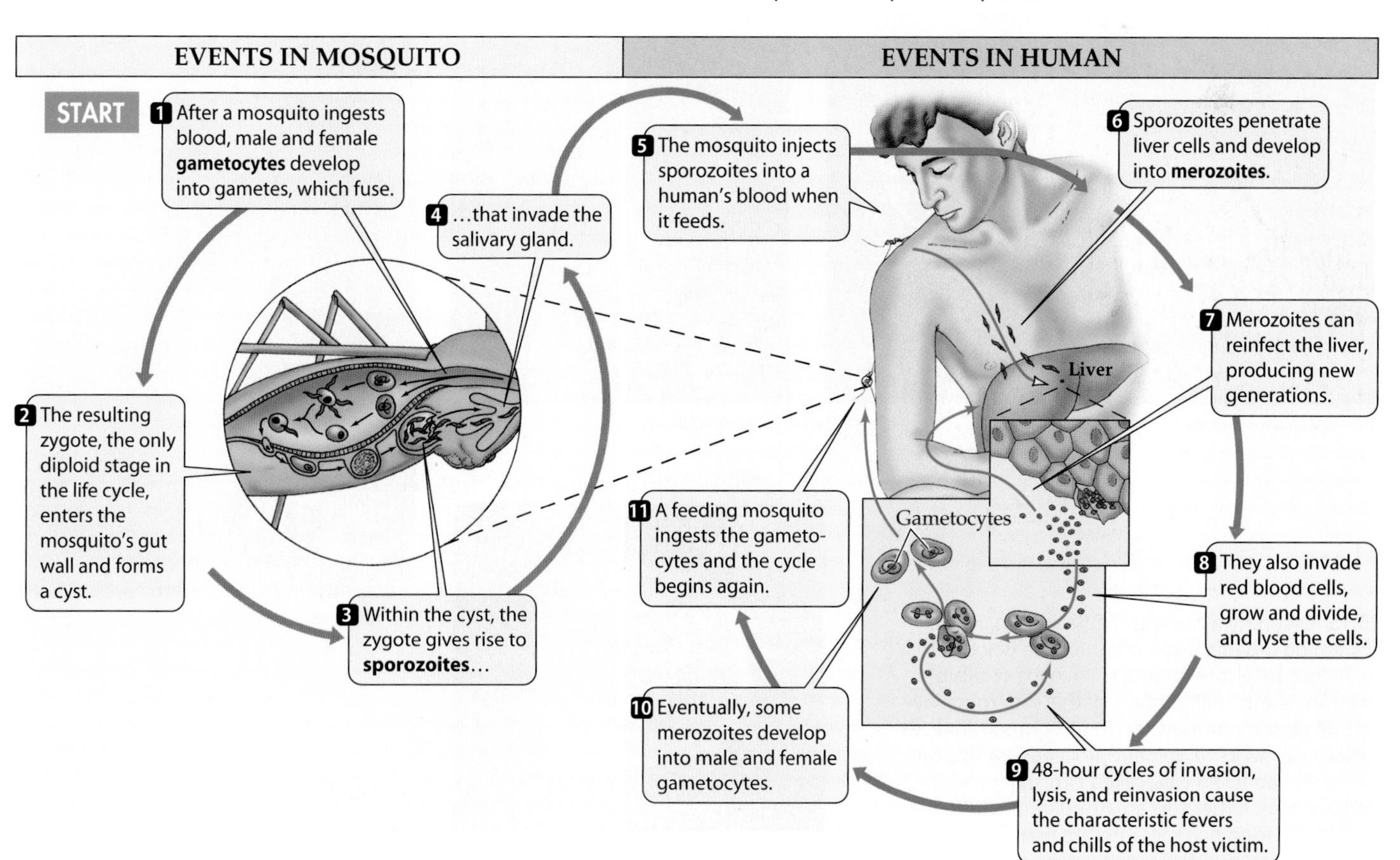

EVENTS IN MOSQUITO
EVENTS IN HUMAN
START
1 After a mosquito ingests blood, male and female gametocytes develop into gametes, which fuse.
2 The resulting zygote, the only diploid stage in the life cycle, enters the mosquito's gut wall and forms a cyst.
3 Within the cyst, the zygote gives rise to sporozoites...
4 ...that invade the salivary gland.
5 The mosquito injects sporozoites into a human's blood when it feeds.
6 Sporozoites penetrate liver cells and develop into merozoites.
7 Merozoites can reinfect the liver, producing new generations.
Liver
Gametocytes
8 They also invade red blood cells, grow and divide, and lyse the cells.
9 48-hour cycles of invasion, lysis, and reinvasion cause the characteristic fevers and chills of the host victim.
10 Eventually, some merozoites develop into male and female gametocytes.
11 A feeding mosquito ingests the gametocytes and the cycle begins again.

28.14 The Life Cycle of an Apicomplexan Malaria-causing *Plasmodium* species spend part of their life cycle in humans and part in mosquitoes. The sporozoite and merozoite forms of the parasite are spores with apical complexes.

tory system when an infected *Anopheles* mosquito penetrates the human skin in search of blood. The parasites find their way to cells in the liver and the lymphatic system, change their form, multiply, and reenter the bloodstream, attacking red blood cells. The apical complex enables *Plasmodium* to enter human liver and red blood cells.

The parasites multiply inside red blood cells, which then burst, releasing new swarms of parasites. If another *Anopheles* bites the victim, the mosquito takes in *Plasmodium* cells along with blood. Some of these cells develop into gametes, which unite to form zygotes that lodge in the mosquito's gut, divide several times, and move into its salivary glands, from which they can be passed on to another human host. Thus, *Plasmodium* is an extracellular parasite in the mosquito vector and an intracellular parasite in the human host.

Plasmodium has proved to be a singularly difficult pathogen to attack. The *Plasmodium* life cycle is best broken by the removal of stagnant water, in which mosquitoes breed. The use of insecticides to reduce the *Anopheles* population can be effective, but their benefits must be weighed against the possible ecological, economic, and health risks posed by the insecticides themselves.

New hope arises from the publication, in 2002, of the genomes of both *Plasmodium falciparum* and one of its vectors, *Anopheles gambiae*, as well as a partial proteome of *P. falciparum*. These advances should lead to a better understanding of the biology of malaria and to the possible development of drugs, vaccines, or other means of dealing with this protist pathogen or its insect vectors.

Ciliates have two types of nuclei

The **ciliates** are so named because they characteristically have hairlike cilia. This group is noteworthy for its diversity and ecological importance (Figure 28.15). Almost all ciliates are heterotrophic (a few contain photosynthetic endosymbionts), and they are much more complex in body form than are most flagellates and other unicellular protists.

The definitive characteristic of ciliates is the possession of two types of nuclei, commonly a single *macronucleus* and, within the same cell, from one to several *micronuclei*. The micronuclei, which are typical eukaryotic nuclei, are essential for genetic recombination. The macronucleus is derived from micronuclei. Each macronucleus contains many copies of the genetic information, packaged in units containing very few genes each. The macronuclear DNA is transcribed and translated to regulate the life of the cell. Although we do not know how this system of macro- and micronuclei came into being,

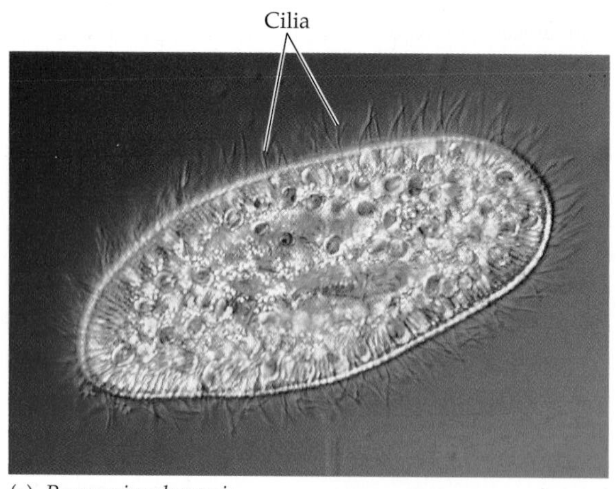

(*a*) *Paramecium bursaria*

(*b*) *Vorticella* sp.

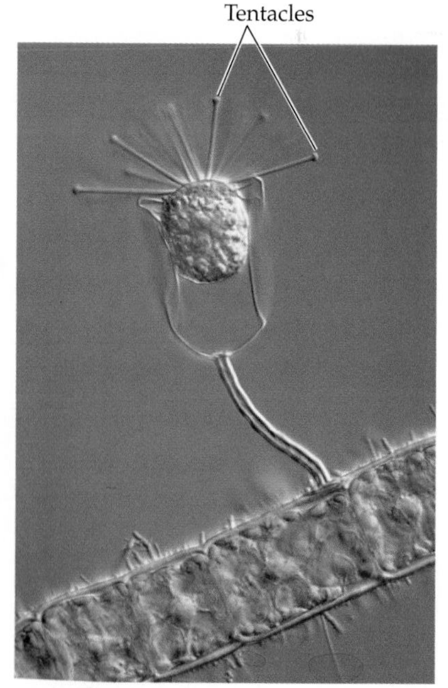

(*c*) *Paracineta* sp.

Cirri

25 μm

(*d*) *Euplotes* sp.

28.15 Diversity among the Ciliates (*a*) A free-swimming organism, this paramecium belongs to a ciliate group whose members have many cilia of uniform length. (*b*) Members of this subgroup have cilia on their mouthparts. (*c*) In this group, tentacles replace cilia as development proceeds. (*d*) This ciliate "walks" on fused cilia, called cirri, that project from its body. Other cilia are fused into flat sheets that sweep food particles into the oral cavity; this individual has ingested green algae.

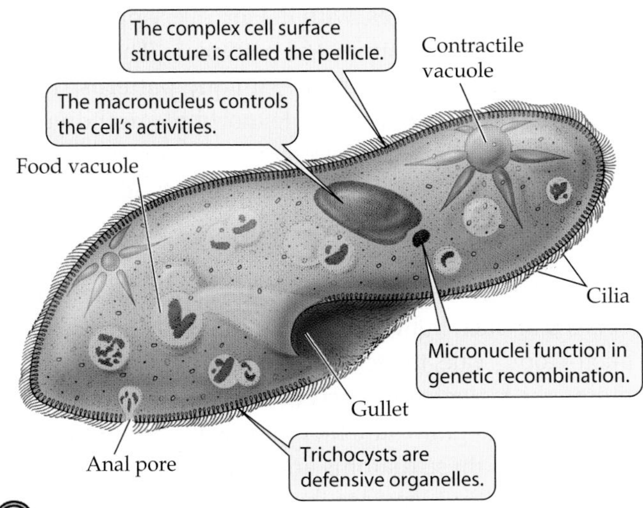

28.16 Anatomy of *Paramecium* This diagram shows the major structures of a typical paramecium.

we do know something about the behavior of these nuclei, which we will discuss after describing the body plan of one important ciliate, *Paramecium*.

A CLOSER LOOK AT ONE CILIATE. *Paramecium*, a frequently studied ciliate genus, exemplifies the complex structure and behavior of ciliates (Figure 28.16). The slipper-shaped cell is covered by an elaborate *pellicle*, a structure composed principally of an outer membrane and an inner layer of closely packed, membrane-enclosed sacs (the alveoli) that surround the bases of the cilia. Defensive organelles called *trichocysts* are also present in the pellicle. In response to a threat, a microscopic explosion expels the trichocysts in a few milliseconds, and they emerge as sharp darts, driven forward at the tip of a long, expanding filament.

The cilia provide a form of locomotion that is generally more precise than locomotion by flagella or pseudopods. A paramecium can direct the beating of its cilia to propel itself either forward or backward in a spiraling manner. It can also back off swiftly when it encounters a barrier or a negative stimulus. The coordination of ciliary beating is probably the result of a differential distribution of ion channels in the plasma membrane near the two ends of the cell.

REPRODUCTION WITHOUT SEX, AND SEX WITHOUT REPRODUCTION. Paramecia reproduce asexually by binary fission. The micronuclei divide mitotically. The macronuclei divide by a still unknown mechanism following a round of DNA replication.

Paramecia also have an elaborate sexual behavior called **conjugation**, in which two paramecia line up tightly against each other and fuse in the oral region of the body. Nuclear material is extensively reorganized and exchanged over the next several hours (Figure 28.17). As a result of this process, each cell ends up with two haploid micronuclei, one of its own and one from the other cell, which fuse to form a new diploid micronucleus. New macronuclei develop from the micronuclei through a series of dramatic chromosomal rearrangements. The exchange of nuclei is fully reciprocal—each of the two paramecia gives and receives an equal amount of DNA. The two organisms then separate and go their own ways, each equipped with new combinations of alleles.

Conjugation in *Paramecium* is a *sexual* process of genetic recombination, but it is not a *reproductive* process. The same two cells that begin the process are there at the end, and no new cells are created. As a rule, each asexual clone of paramecia must periodically conjugate. Experiments have shown that if some species are not permitted to conjugate, the clones can live through no more than approximately 350 cell divisions before they die out.

28.17 Paramecia Achieve Genetic Recombination by Conjugating The exchange of micronuclei by conjugating *Paramecium* individuals permits genetic recombination. After conjugation, the cells separate and continue their lives as two individuals.

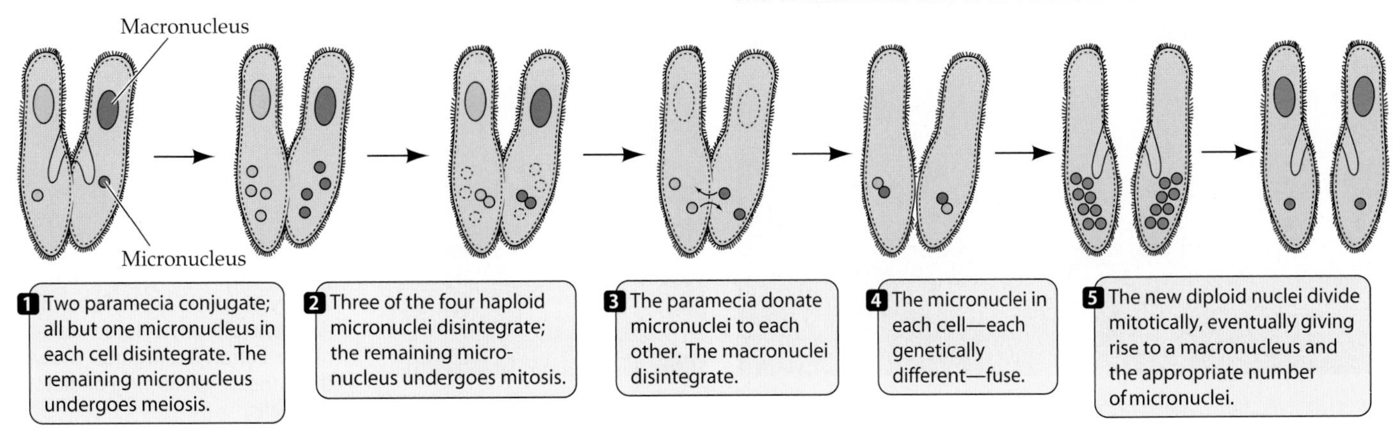

Macronucleus
Micronucleus
1 Two paramecia conjugate; all but one micronucleus in each cell disintegrate. The remaining micronucleus undergoes meiosis.
2 Three of the four haploid micronuclei disintegrate; the remaining micronucleus undergoes mitosis.
3 The paramecia donate micronuclei to each other. The macronuclei disintegrate.
4 The micronuclei in each cell—each genetically different—fuse.
5 The new diploid nuclei divide mitotically, eventually giving rise to a macronucleus and the appropriate number of micronuclei.

Stramenopiles

BACTERIA
ARCHAEA
EUKARYA
Diplomonads, Parabasalids
Euglenozoans
Alveolates
Stramenopiles
Red algae
"Green algae" and plants
Opisthokonts

The shared trait that defines the **stramenopiles** is the possession of two flagella, typically unequal in length. The longer of the two bears rows of tubular hairs. Some stramenopiles lack flagella, but they are presumed to be descended from ancestors that possessed flagella. The stramenopiles include the diatoms and the brown algae, which are photosynthetic, and the oomycetes, which are not. Other stramenopile groups include some lineages that are non-photosynthetic. Most golden algae are photosynthetic, but nearly all of them become heterotrophic when light intensity is limiting or when there is a plentiful food supply; some even feed on diatoms or bacteria. Some botanists refer to the stramenopiles as the "brown plant kingdom."

Diatoms are everywhere in the marine environment

Diatoms are single-celled organisms, although some species associate in filaments. Many have sufficient carotenoids in their chloroplasts to give them a yellow or brownish color. All make *chrysolaminarin* (a carbohydrate) and oils as photosynthetic storage products. Diatoms lack flagella.

Architectural magnificence on a microscopic scale is the hallmark of the diatoms (Figure 28.18*a*). Many diatoms deposit silicon in their cell walls. The cell wall of some species is constructed in two pieces, with the top overlapping the bottom like the top and bottom of a petri plate. The silicon-impregnated walls have intricate, unique patterns (Figure 28.18*b*). Despite their remarkable morphological diversity, however, all diatoms are symmetrical—either bilaterally (with "right" and "left" halves) or radially (with the type of symmetry possessed by a circle).

Why are diatom cell walls so glassy and complex—and why are diatoms so abundant? A 2003 paper by German biologists may shed light on these questions. By measuring, at a microscopic scale, the forces needed to break single, living diatoms, the biologists discovered that their cell walls are exceptionally strong. Evolution of these walls by natural selection has given these diatoms an enhanced defense against predators and, thus, an edge over competitors.

Diatoms reproduce both sexually and asexually. Asexual reproduction is by binary fission and is somewhat constrained by the stiff, silica-containing cell wall. Both the top and the bottom of the "petri plate" become tops of new "plates" without changing appreciably in size; as a result, the new cell made from the former bottom is smaller than the parent cell (Figure 28.19). If this process continued indefinitely, one cell line would simply vanish, but sexual reproduction largely solves this potential problem. Gametes are formed, shed their cell walls, and fuse. The resulting zygote then increases substantially in size before a new cell wall is laid down.

Diatoms are everywhere in the marine environment and are frequently present in great numbers, making them major photosynthetic producers in coastal waters. Diatoms are also common in fresh water. Because the silicon-containing walls of dead diatom cells resist decomposition, certain sedimentary rocks are composed almost entirely of diatom skeletons that sank to the seafloor over time. Diatomaceous earth, which is obtained from such rocks, has many industrial uses, such as insulation, filtration, and metal polishing. It has also been used as an "Earth-friendly" insecticide that clogs the tracheae (breathing structures) of insects.

(*a*)

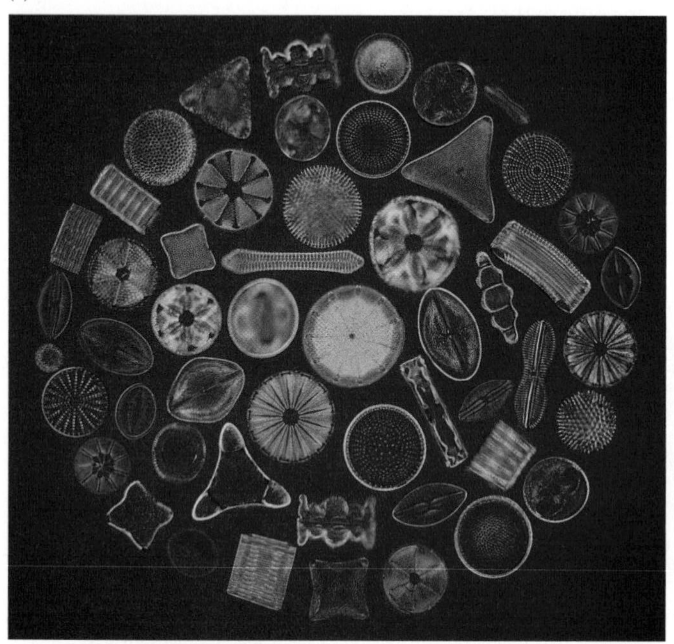

(*b*)

28.18 Diatom Diversity
(*a*) Diatoms exhibit a splendid variety of species-specific forms. (*b*) This artificially colored scanning electron micrograph shows the intricate patterning of diatom cell walls.

28.19 Diatom Reproduction Diatoms reproduce both sexually and asexually. Half of the cells created by asexual reproduction are smaller than the parent cells. Sexual reproduction creates new parent cells with full-sized cell walls.

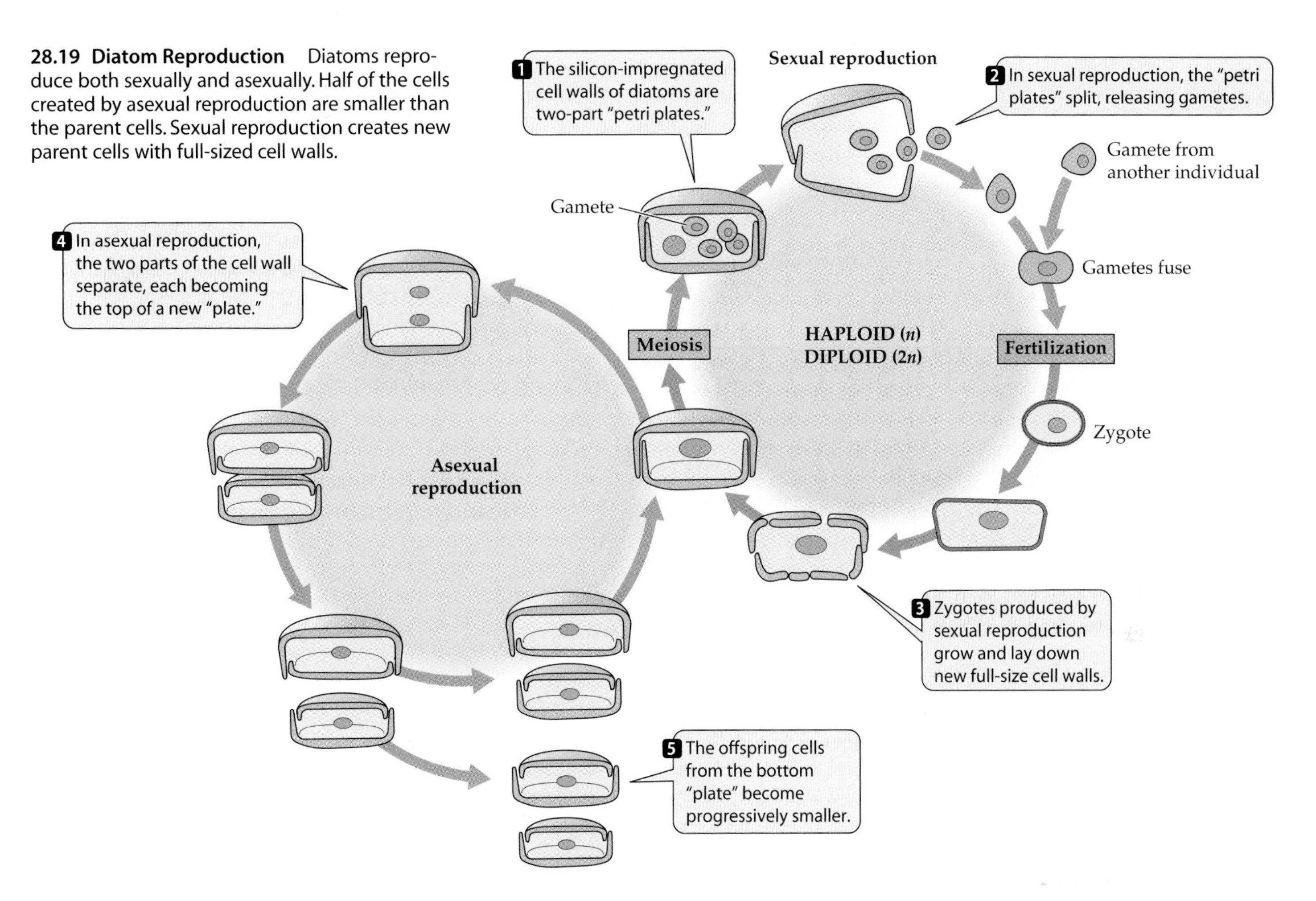

1 The silicon-impregnated cell walls of diatoms are two-part "petri plates."
Sexual reproduction
2 In sexual reproduction, the "petri plates" split, releasing gametes.
Gamete from another individual
Gamete
Gametes fuse
4 In asexual reproduction, the two parts of the cell wall separate, each becoming the top of a new "plate."
Meiosis
HAPLOID (n)
DIPLOID (2n)
Fertilization
Zygote
Asexual reproduction
3 Zygotes produced by sexual reproduction grow and lay down new full-size cell walls.
5 The offspring cells from the bottom "plate" become progressively smaller.

The brown algae include the largest protists

All the **brown algae** are multicellular. They are composed either of branched filaments (Figure 28.20) or of leaflike growths called *thalli* (singular, thallus) (Figure 28.21*a*). The brown algae obtain their namesake color from the carotenoid *fucoxanthin*, which is abundant in their chloroplasts. The combination of this yellow-orange pigment with the green of chlorophylls *a* and *c* yields a brownish tinge.

The brown algae include the largest of the protists. Giant kelps, such as those of the genus *Macrocystis*, may be up to 60 meters long (see Figure 28.1*c*). The brown algae are almost exclusively marine. Some float in the open ocean; the most famous example is the genus *Sargassum*, which forms dense mats in the Sargasso Sea in the mid-Atlantic. Most brown algae, however, are attached to rocks near the shore. A few thrive only

(*a*) *Hormosira banksii*

(*b*) *Ectocarpus* sp.

28.20 Brown Algae (*a*) A filamentous brown alga growing in Australia. This species is sometimes called "Neptune's necklace." (*b*) A filamentous brown alga seen through a light microscope.

(*a*) *Postelsia palmaeformis*

The leaflike structures are the thalli of sea palm.

(*b*)

28.21 Brown Algae in a Turbulent Environment Brown algae growing in the intertidal zone on an exposed rocky shore take a tremendous pounding by the surf. (*a*) Sea palms growing along the California coast. (*b*) The tough, branched holdfast that anchors the sea palm.

where they are regularly exposed to heavy surf; a notable example is the sea palm *Postelsia palmaeformis* of the Pacific coast (Figure 28.21*a*). All of the attached forms develop a specialized structure, called a *holdfast*, that literally glues them to the rocks (Figure 28.21*b*).

Some brown algae differentiate extensively into stemlike stalks and leaflike blades, and some develop gas-filled cavities or bladders. For biochemical reasons that are only poorly understood, these bladders often contain as much as 5 percent carbon monoxide—a concentration high enough to kill a human. In addition to organ differentiation, the larger brown algae also exhibit considerable tissue differentiation. Most of the giant kelps have photosynthetic filaments only in the outermost regions of their stalks and blades. Within these photosynthetic regions lie filaments of long cells that closely resemble the nutrient-conducting tissue of plants. Called *trumpet cells* because they have flaring ends, these tubes rapidly conduct the products of photosynthesis through the body of the organism.

The cell walls of brown algae may contain as much as 25 percent *alginic acid*, a gummy polymer of sugar acids. Alginic acid cements cells and filaments together and provides good holdfast glue. It is used commercially as an emulsifier in ice cream, cosmetics, and other products.

Many protist and all plant life cycles feature alternation of generations

Brown algae, like many other multicellular photosynthetic protists and all plants, exhibit a type of life cycle known as **alternation of generations**, in which a multicellular, diploid, spore-producing organism gives rise to a multicellular, haploid, gamete-producing organism. When two haploid gametes fuse (a process called *fertilization*, or *syngamy*), a diploid organism is formed (Figure 28.22). The haploid organism, the diploid organism, or both may also reproduce asexually.

The two generations (spore-producing and gamete-producing) differ genetically (one has diploid cells and the other has haploid cells), but they may or may not differ morphologically. In **heteromorphic** alternation of generations the two generations differ morphologically; in **isomorphic** alternation of generations they do not, despite their genetic difference. We will see examples of both heteromorphic and isomorphic alternation of generations in some representative brown and green algae. In discussing the life cycles of plants and multicellular photosynthetic protists, we will use the terms **sporophyte** ("spore plant") and **gametophyte** ("gamete plant") to refer to the multicellular diploid and haploid generations, respectively.

Gametes are not produced by meiosis because the gametophyte generation is already haploid. Instead, specialized cells of the diploid sporophyte, called **sporocytes**, divide meiotically to produce four haploid spores. The spores may eventually germinate and divide mitotically to produce multicellular haploid gametophytes, which produce gametes by mitosis and cytokinesis.

Gametes, unlike spores, can produce new organisms only by fusing with other gametes. The fusion of two gametes produces a diploid zygote, which then undergoes mitotic divisions to produce a diploid organism: the sporophyte generation. The sporocytes of the sporophyte generation then undergo meiosis and produce haploid spores, starting the cycle anew.

28.22 Alternation of Generations In many multicellular photosynthetic protists and all plants, a diploid generation that produces spores alternates with a haploid generation that produces gametes.

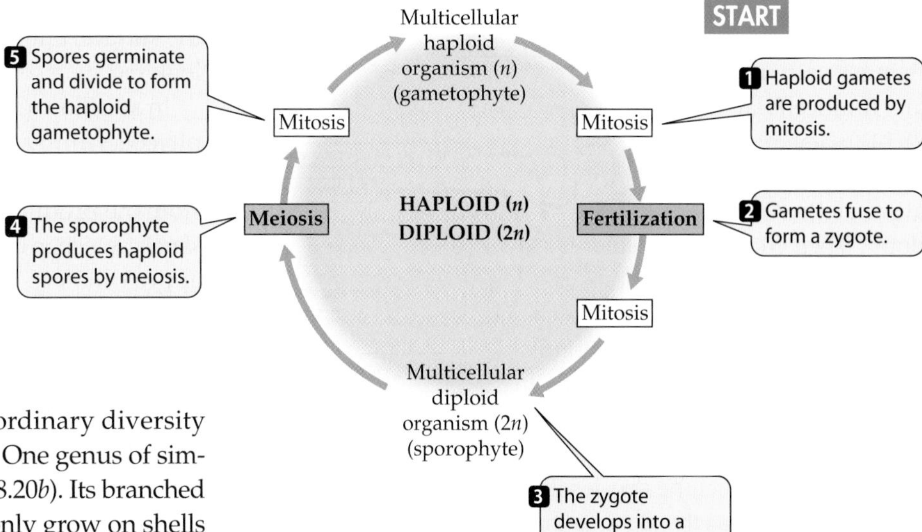
START
Multicellular haploid organism (n) (gametophyte)
1 Haploid gametes are produced by mitosis.
Mitosis
Fertilization
2 Gametes fuse to form a zygote.
Mitosis
HAPLOID (n)
DIPLOID (2n)
Multicellular diploid organism (2n) (sporophyte)
3 The zygote develops into a diploid sporophyte.
Meiosis
4 The sporophyte produces haploid spores by meiosis.
Mitosis
5 Spores germinate and divide to form the haploid gametophyte.

The brown algae exemplify the extraordinary diversity found among the photosynthetic protists. One genus of simple brown algae is *Ectocarpus* (see Figure 28.20*b*). Its branched filaments, a few centimeters long, commonly grow on shells and stones. The gametophyte and sporophyte generations of *Ectocarpus* can be distinguished only by chromosome number or reproductive products (spores or gametes). Thus the generations are isomorphic.

By contrast, some kelps of the genus *Laminaria* and some other brown algae show a more complex heteromorphic alternation of generations. The larger and more obvious generation of these species is the sporophyte. Meiosis in sporocytes located on the leaflike fronds produces haploid **zoospores**—motile spores that are propelled by flagella. These spores germinate to form a tiny, filamentous gametophyte that produces either eggs or sperm. The eggs and sperm of brown algae typically have flagella.

The oomycetes include water molds and their relatives

A nonphotosynthetic stramenopile group called the **oomycetes** consists in large part of the water molds and their terrestrial relatives, such as the downy mildews. Water molds are filamentous and stationary, and they feed by absorption—that is, they secrete enzymes that digest large food molecules into smaller molecules that the water molds can absorb. If you have seen a whitish, cottony mold growing on dead fish or dead insects in water, it was probably a water mold of the common genus *Saprolegnia* (Figure 28.23).

Don't be confused by the "-mycete" in the name. That term means "fungus," and it is there because these organisms were once classified as fungi. However, we now know that the oomycetes are unrelated to the fungi.

The oomycetes are **coenocytes**: They have many nuclei enclosed in a single plasma membrane. Their filaments have no cross-walls to separate the many nuclei into discrete cells. Their cytoplasm is continuous throughout the body of the organism, and there is no single structural unit with a single nucleus, except in certain reproductive stages. A distinguishing feature of the oomycetes is their flagellated reproductive cells. Oomycetes are diploid throughout most of their life cycle and have cellulose in their cell walls.

The water molds, such as *Saprolegnia*, are all aquatic and **saprobic** (they feed on dead organic matter). Some other oomycetes are terrestrial. Although most of the terrestrial oomycetes are harmless or helpful decomposers of dead matter, a few are serious plant parasites that attack crops such as avocados, grapes, and potatoes. The water mold *Phytophthora infestans*, for example, is the causal agent of late blight of potatoes, which brought about the great Irish potato famine of 1845–1847. *P. infestans* destroyed the entire Irish potato crop in a matter of days in 1846. Among the consequences of the famine were a million deaths from starvation and the emigration of about 2 million people, mostly to the United States.

Saprolegnia sp.

28.23 An Oomycete The filaments of a water mold radiate from the carcass of an insect.

Red Algae

Almost all **red algae** are multicellular (Figure 28.24). Some plant biologists now refer to the red algae as the "red plant kingdom." Their characteristic color is a result of the accessory photosynthetic pigment *phycoerythrin*, which is found in relatively large amounts in the chloroplasts of many species. In addition to phycoerythrin, red algae contain phycocyanin, carotenoids, and chlorophyll *a*.

The red algae include species that grow in the shallowest tide pools as well as the algae found deepest in the ocean (as deep as 260 meters if nutrient conditions are right and the water is clear enough to permit the penetration of light). Very few red algae inhabit fresh water. Most grow attached to a substratum by a holdfast.

In a sense, the red algae, like several other groups of algae, are misnamed. They have the capacity to change the relative amounts of their various photosynthetic pigments depending on the light conditions where they are growing. Thus, the leaflike *Chondrus crispus*, a common North Atlantic red alga, may appear bright green when it is growing at or near the surface of the water and deep red when growing at greater depths. The ratio of pigments present depends to a remarkable degree on the intensity of the light that reaches the alga. In deep water, where the light is dimmest, the alga accumulates large amounts of phycoerythrin. The algae in deep water have as much chlorophyll as the green ones near the surface, but the accumulated phycoerythrin makes them look red.

In addition to being the only photosynthetic protists with phycoerythrin and phycocyanin among their pigments, the red algae have two other unique characteristics: First, they store the products of photosynthesis in the form of *floridean starch*, which is composed of very small, branched chains of approximately 15 glucose units. Second, they produce no motile, flagellated cells at any stage in their life cycle. The male gametes lack cell walls and are slightly amoeboid; the female gametes are completely immobile.

Some red algal species enhance the formation of coral reefs. Like coral animals, they possess the biochemical machinery for secreting calcium carbonate, which they deposit both in and around their cell walls. After the death of the corals and algae, the calcium carbonate persists, sometimes forming substantial rocky masses.

Some red algae produce large amounts of mucilaginous polysaccharide substances, which contain the sugar galactose with a sulfate group attached. This material readily forms solid gels and is the source of agar, a substance widely used in the laboratory for making a solid aqueous medium on which tissue cultures and many microorganisms can be grown.

Certain red algae became endosymbionts, long ago, within the cells of other, nonphotosynthetic protists, eventually giving rise to chloroplasts. They are the ancestors of the distinctive chloroplasts of the photosynthetic stramenopiles (the brown algae and the diatoms).

(*a*) *Bossiella orbigniana*

(*b*) *Polysiphonia* sp.

28.24 Red Algae (*a*) A coralline red alga grows along the coast of central Oregon. (*b*) Under the light microscope, both vegetative and reproductive structures can be seen in this red alga.

Chlorophytes

The "green algae" do not form a clade, but they include at least two multicellular clades. One major clade consists of the **chlorophytes**. A sister clade to the chlorophytes contains another green algal clade along with the plant kingdom (see Figure 28.9). The green algal clades share characters that distinguish them from other protists: Like the plants, they contain chlorophylls *a* and *b*, and their reserve of photosynthetic products is stored as starch in plastids.

The chlorophytes are the largest clade of green algae, containing more than 17,000 species. Most are aquatic—some are marine, but more are freshwater forms—but others are terrestrial, living in moist environments. The chlorophytes range in size from microscopic unicellular forms to multicellular forms many centimeters in length.

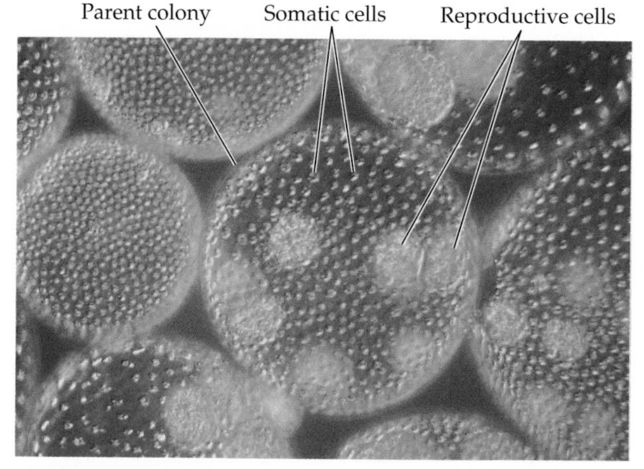

(*a*) *Volvox* sp.

(*b*) *Ulva lactuca*

28.25 Chlorophytes (*a*) *Volvox* colonies are precisely spaced arrangements of cells. Specialized reproductive cells produce daughter colonies, which will eventually release new individuals. (*b*) A stand of sea lettuce submerged in a tidal pool.

Chlorophytes vary in shape and cellular organization

We find among the chlorophytes an incredible variety in shape and construction of the algal body. *Chlamydomonas* is an example of the simplest type: unicellular and flagellated.

Surprisingly large and well-formed colonies of cells are found in such freshwater groups as the genus *Volvox* (Figure 28.25*a*). The cells in these colonies are not differentiated into tissues and organs, as in plants and animals, but the colonies show vividly how the preliminary step of this great evolutionary development might have been taken. In *Volvox*, the origins of cell specialization can be seen in certain cells within the colony that are specialized for reproduction.

While *Volvox* is colonial and spherical, *Oedogonium* is multicellular and filamentous, and each of its cells has only one nucleus. *Cladophora* is multicellular, but each cell is multinucleate. *Bryopsis* is tubular and coenocytic, forming cross-walls only when reproductive structures form. *Acetabularia* is a single, giant uninucleate cell a few centimeters long that becomes multinucleate only at the end of its reproductive stage. *Ulva lactuca* is a thin, membranous sheet a few centimeters across; its unusual appearance justifies its common name: sea lettuce (Figure 28.25*b*).

Chlorophyte life cycles are diverse

The life cycles of chlorophytes show great diversity. Let's examine two chlorophyte life cycles in detail, beginning with that of the sea lettuce *Ulva lactuca* (Figure 28.26). Like many chlorophytes, sea lettuce exhibits alternation of generations. The diploid sporophyte of this common seashore organism is a broad sheet only two cells thick. Some of its cells (sporocytes) differentiate and undergo meiosis and cytokinesis, producing motile haploid spores (zoospores). These swim away, each propelled by four flagella, and some eventually find a suitable place to settle. The spores then lose their flagella and begin to divide mitotically, producing a thin filament that de-

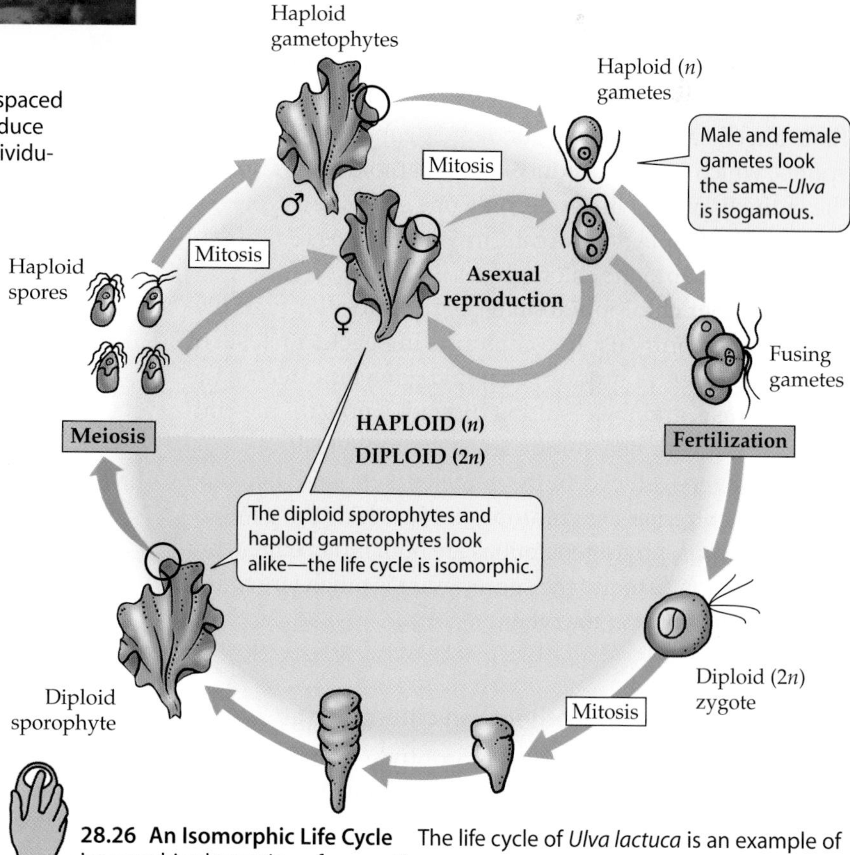

28.26 An Isomorphic Life Cycle The life cycle of *Ulva lactuca* is an example of isomorphic alternation of generations.

velops into a broad sheet only two cells thick. The gametophyte thus produced looks just like the sporophyte—in other words, *Ulva lactuca* has an isomorphic life cycle.

In *Ulva lactuca*, an individual gametophyte can produce only male or female gametes—never both. The gametes arise mitotically within single cells (called *gametangia*), rather than within a specialized multicellular structure, as in plants. Both types of gametes bear two flagella (in contrast to the four flagella of a haploid spore) and hence are motile.

In most species of *Ulva* the female and male gametes are indistinguishable structurally, making those species **isogamous**—having gametes of identical appearance. Other chlorophytes, including some other species of *Ulva*, are **anisogamous**—having female gametes that are distinctly larger than the male gametes.

Female and male gametes come together and unite, losing their flagella as the zygote forms and settles. After resting briefly, the zygote begins mitotic division, producing a multicellular sporophyte. Any gametes that fail to find partners can settle down on a favorable substratum, lose their flagella, undergo mitosis, and produce a new gametophyte directly; in other words, the gametes can also function as zoospores. Few chlorophytes other than *Ulva* have motile gametes that can also function as zoospores.

In contrast to the isomorphic life cycle of *Ulva*, many other chlorophytes have a heteromorphic life cycle, in which sporophyte and gametophyte generations differ in structure. In one variation of the heteromorphic life cycle—the **haplontic** life cycle (Figure 28.27)—a multicellular haploid individual produces gametes that fuse to form a zygote. The zygote functions directly as a sporocyte, undergoing meiosis to produce spores, which in turn produce a new haploid individual. In the entire haplontic life cycle, only one cell—the zygote—is diploid. The filamentous organisms of the genus *Ulothrix* are examples of haplontic chlorophytes.

Some other chlorophytes have a **diplontic** life cycle like that of many animals. In a diplontic life cycle, meiosis of diploid sporocytes produces haploid gametes directly; the gametes fuse, and the resulting diploid zygote divides mitotically to form a new multicellular sporophyte. In such organisms, every cell except the gametes is diploid. Between these two extremes are chlorophytes in which the gametophyte and sporophyte generations are both multicellular, but one generation (usually the sporophyte) is much larger and more prominent than the other.

There are green algae other than chlorophytes

As we mentioned above, the chlorophytes are the largest clade of green algae, but there are other green algal clades as well. Those clades are branches of a clade that also includes the plant kingdom. The green algal clade that is sister to the plant kingdom, containing a group of organisms called *charophytes*, will be described in the next chapter. But now let's consider some close protist relatives of the animals.

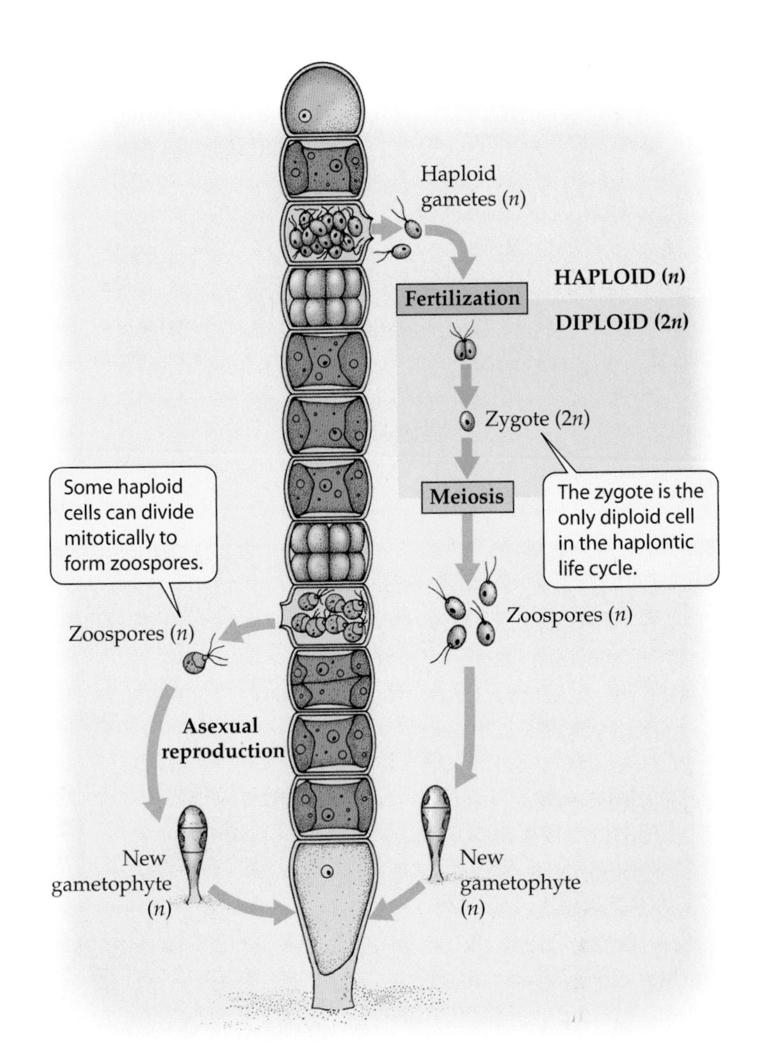

28.27 A Haplontic Life Cycle In the life cycle of *Ulothrix*, a filamentous, multicellular haploid gametophyte generation alternates with a diploid sporophyte generation consisting of a single cell (the zygote). Like *Ulva* gametophytes, *Ulothrix* gametophytes can also reproduce asexually (left side of figure).

Choanoflagellates

One group of heterotrophic protists with flagella, the **choanoflagellates**, is thought to comprise the closest relatives of the animals. The choanoflagellates are sister to the animals, and the animal–choanoflagellate lineage is sister to the fungi (see Figure 28.9). The clade consisting of fungi, animals, and choanoflagellates is called the **opisthokonts**.

BACTERIA
ARCHAEA
EUKARYA
Diplomonads, Parabasalids
Euglenozoans
Alveolates
Stramenopiles
Red algae
"Green algae" and plants
Opisthokonts

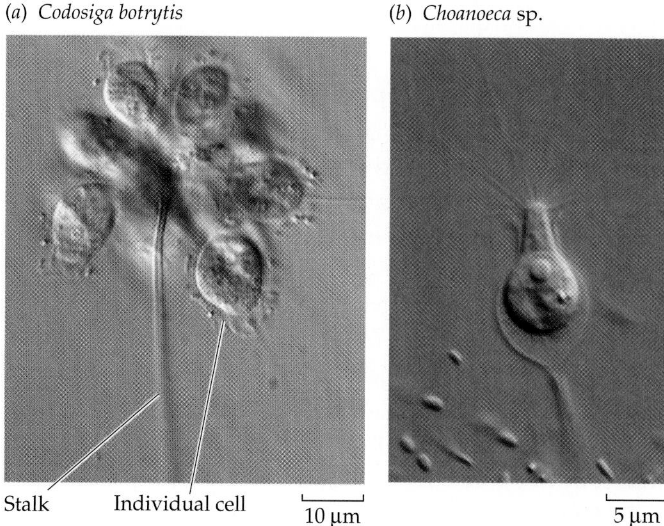

28.28 A Link to the Animal Kingdom Choanoflagellates may be close relatives of the sponges, and thus represent a link between the protists and the kingdom Animalia. (*a*) The formation of colonies by unicellular organisms, as in this species, is one route to the evolution of multicellularity. (*b*) A solitary choanoflagellate illustrates the similarity of this protist group to a cell type present in the multicellular sponges (see Figure 32.4).

Choanoflagellates are colonial and are thought to be closely related to the sponges, the most ancient of the surviving phyla of animals. Choanoflagellates bear a striking resemblance to the most characteristic type of cell found in the sponges (compare Figures 28.28 and 32.4).

Sponges are also colonial rather than truly multicellular, in that they lack organized tissues and their cells can be separated and allowed to reaggregate. We turn now to a topic hinted at several times above: the history of unusual numbers of membranes surrounding the chloroplasts of some photosynthetic protists.

A History of Endosymbiosis

As we have already seen, many protists possess chloroplasts. Groups with chloroplasts appear in several distantly related protist clades. Some of these groups differ in the photosynthetic pigments their chloroplasts contain. And we've seen that not all chloroplasts have a pair of surrounding membranes—in some protists, they are surrounded by three membranes. We now understand these observations in terms of a remarkable series of endosymbioses.

All chloroplasts trace their ancestry back to the engulfment of a cyanobacterium by a larger eukaryotic cell (Figure 28.29). This event is known as *primary endosymbiosis*. The cyanobacterium, a Gram-negative bacterium, had both an inner and an outer membrane. The eukaryote's plasma membrane wrapped around the cyanobacterium as it took it up. The outer membrane and cell wall of the cyanobacterium were eventually lost. Thus, the original chloroplasts had two surrounding membranes—one from the cyanobacterium and one from the eukaryotic host cell.

28.29 A Chloroplast Family Tree One or two primary endosymbioses followed by several secondary and tertiary endosymbioses gave rise to all of today's chloroplasts.

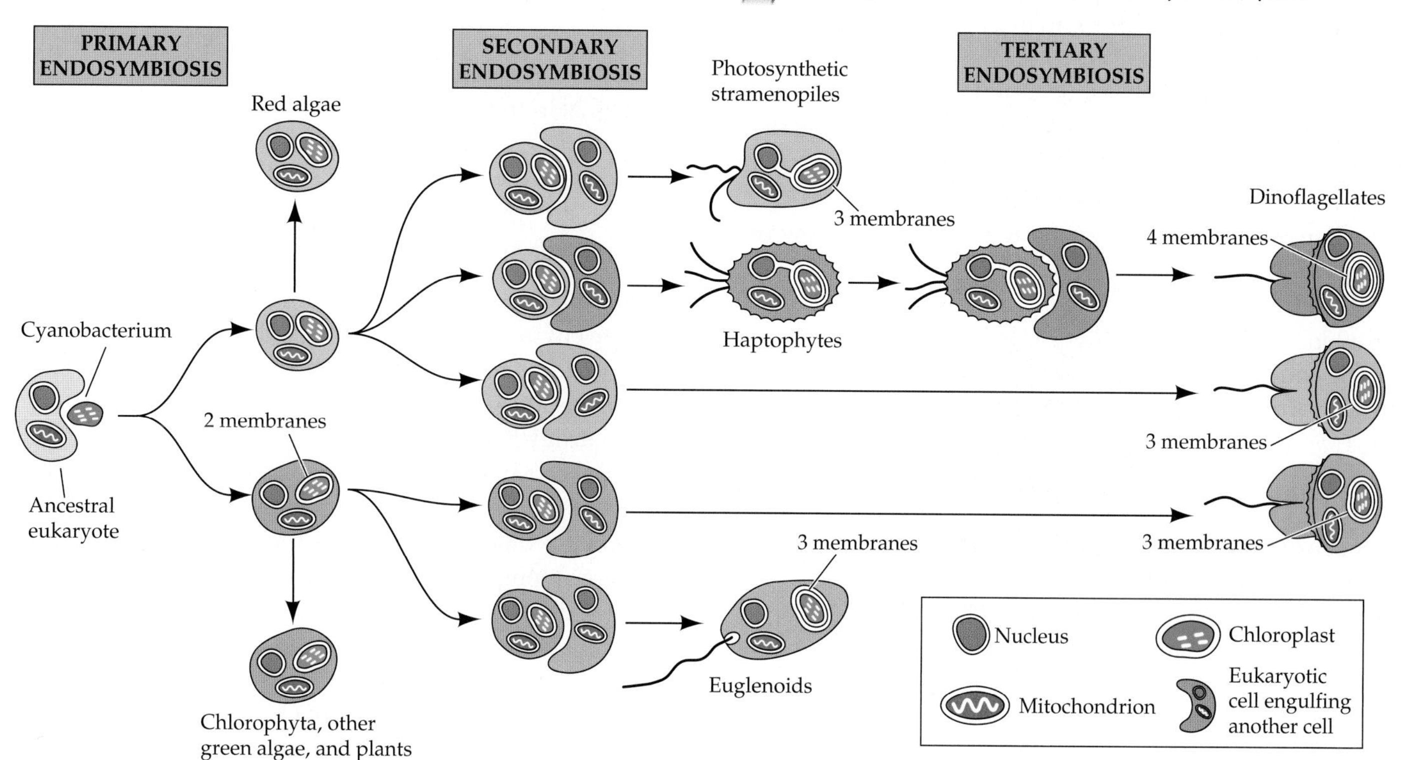

Primary endosymbiosis gave rise to the chloroplasts of the green algae and the red algae. We do not yet know whether both trace back to a single primary endosymbiosis, as is likely, with later divergence, or whether they resulted from independent occurrences of primary endosymbiosis. In either case, each line participated in further endosymbioses.

All remaining photosynthetic eukaryotes are the results of secondary or tertiary endosymbiosis. The photosynthetic euglenoids derived their chloroplasts from *secondary endosymbiosis*. Their ancestor took up a unicellular chlorophyte, retaining the endosymbiont's chloroplast and eventually losing the rest of its constituents. This history accounts for the fact that the photosynthetic euglenoids have the same photosynthetic pigments as the chlorophytes and plants. It also accounts for the third membrane of the euglenoid chloroplast, which is derived from the euglenoid's plasma membrane.

Other photosynthetic protist groups derived their chloroplasts by secondary endosymbiosis with unicellular red algae. Both the green clade and the red clade of chloroplasts appear to have been involved in more than one secondary endosymbiosis. At least one secondary endosymbiosis produced a unicellular protist that became, itself, a partner in a *tertiary endosymbiosis!* In that case, a dinoflagellate lost its plastid and took up a haptophyte protist (itself the result of secondary endosymbiosis). The result is the dinoflagellate *Karenia brevis*.

Although euglenoid chloroplasts are descendants of a chlorophyte and stramenopile chloroplasts are descendants of a red alga, this does not mean that euglenoids themselves are descendants of a chlorophyte, nor are stramenopiles themselves descendants of a red alga. The ancestors that took up green or red algae in secondary endosymbioses had their own evolutionary histories. Thus, the nuclear and chloroplast genomes of these protists have different histories. It has taken much research to piece together the clades as we now understand them.

The clades of protists that we have discussed are summarized in Table 28.1 and Figure 28.9. Now let's consider some of the body forms that have appeared repeatedly in various branches of the eukaryote family tree.

Some Recurrent Body Forms

Amoebas used to be classified together in a single protist group. We now know that the amoebas map to at least two complex assemblages that are difficult to position in phylogenetic trees. Similarly, three kinds of organisms called slime molds, once classified together, may be quite different phylogenetically. In this section, we'll look at some of the variations on these body plans found among the protists.

Amoebas form pseudopods

The pseudopods used by **amoebas** for locomotion are a hallmark of the amoeboid body plan (see Figure 28.4). This body plan has appeared by convergent evolution in various protist groups. The mechanism of amoeboid motion will be discussed in Chapter 47.

Amoebas have often been portrayed in popular writing as blobs—the simplest form of "animal" life imaginable. Superficial examination of a typical amoeba shows how such an impression might have been obtained. An amoeba consists of a single cell. It feeds on small organisms and particles of organic matter by phagocytosis, engulfing them with its pseudopods.

But amoebas are specialized protists. Many are adapted for life on the bottoms of lakes, ponds, and other bodies of water. Their creeping locomotion and their manner of engulfing food particles fit them for life close to a relatively rich supply of sedentary organisms or organic particles. Most amoebas exist as predators, parasites, or scavengers. A few are photosynthetic.

Amoebas of the free-living genus *Naegleria*, some of which can enter humans and cause a fatal disease of the nervous system, have a two-stage life cycle, one stage having amoeboid cells and the other flagellated cells. Some amoebas are shelled, living in casings of sand grains glued together (see Figure 28.7*b*). Others have shells secreted by the organism itself.

ACTINOPODS HAVE THIN, STIFF PSEUDOPODS. The **actinopods** are recognizable by their thin, stiff pseudopods, which are reinforced by microtubules. These pseudopods play at least four roles:

- They greatly increase the surface area of the cell for exchange of materials with the environment.
- They help the cell float in its marine or freshwater environment.
- They provide locomotion in some species.
- They are the cell's feeding organs, trapping smaller organisms and often taking them up by endocytosis.

Radiolarians, a group of actinopods found exclusively in marine environments, are perhaps the most beautiful of all microorganisms (see Figure 28.8). Almost all radiolarian species secrete glassy *endoskeletons* (internal skeletons) from which needlelike pseudopods project. Part of the skeleton is a central capsule within the cytoplasm. The skeletons of the different species are as varied as snowflakes, and many have elaborate geometric designs (Figure 28.30*a*). A few radiolarians are among the largest of the unicellular protists, measuring several millimeters across. Innumerable radiolarian skeletons, some as old as 700 million years, form the sediments under some tropical seas.

Heliozoans are actinopods that lack an endoskeleton (Figure 28.30*b*). Most heliozoans are found in fresh water. They

(*a*) *Podocyrtis mitra*

(*b*) *Actinosphaerium eichhorni*

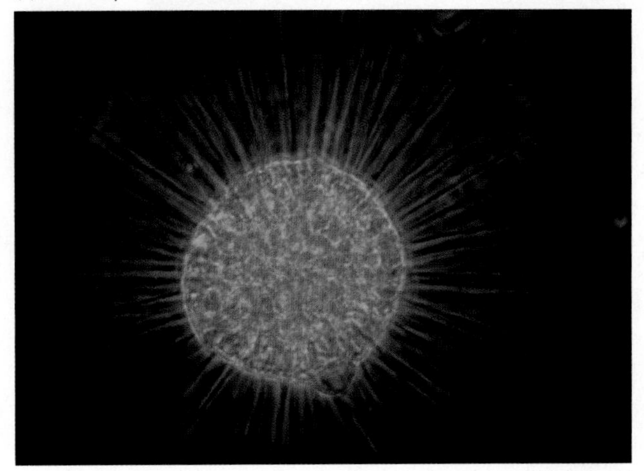

28.30 Two Actinopods (*a*) Radiolarians secrete intricate glassy skeletons such as the one shown here. A living radiolarian is shown in Figure 28.8. (*b*) The radiating pseudopods of a heliozoan (Greek, "sun animalcule") give it an appearance that explains its descriptive name.

roll along the substratum by shortening and elongating their pseudopods.

FORAMINIFERANS HAVE CREATED VAST LIMESTONE DEPOSITS. **Foraminiferans** are marine protists that secrete shells of calcium carbonate (see Figure 28.7*a*). The shells of individual foraminiferan species have distinctive shapes. Some foraminiferans live as **plankton** (free-floating microscopic organisms), and many others live at the bottom of the sea. Their long, threadlike, branched pseudopods reach out through numerous microscopic pores in the shell and interconnect to create a sticky net, which the foraminiferan uses to catch smaller plankton.

After foraminiferans reproduce (by mitosis and cytokinesis), the daughter cells abandon the parent shell and make new shells of their own. The discarded skeletons of ancient foraminiferans make up extensive limestone deposits in various parts of the world, forming a layer hundreds to thousands of meters deep over millions of square kilometers of ocean bottom. Foraminiferan skeletons also make up the sand of some beaches. A single gram of such sand may contain as many as 50,000 foraminiferan shells and shell fragments.

The shells of individual foraminiferans are easily preserved as fossils in marine sediments. Each geological period has distinctive foraminiferan species. For this reason, and because they are so abundant, the remains of foraminiferans are especially valuable as indicators in the classification and dating of sedimentary rocks, as well as in oil prospecting. They also reveal information about temperatures at the time they were alive.

Slime molds release spores from erect fruiting bodies

The three groups of **slime molds** seem so similar at first glance that they were once grouped together in a single phylum. However, the slime molds are actually so different that some biologists now classify them in different *kingdoms*. We will consider two of these groups, called acellular slime molds and cellular slime molds.

The slime molds share only general characteristics. All are motile, all ingest particulate food by endocytosis, and all form spores on erect fruiting bodies. They undergo striking changes in organization during their life cycles, and one stage consists of isolated cells that take up food particles by endocytosis. Some slime molds may cover areas of 1 meter or more in diameter while in their less aggregated stage. Such a large slime mold may weigh more than 50 grams. Slime molds of both types favor cool, moist habitats, primarily in forests. They range from colorless to brilliantly yellow and orange.

ACELLULAR SLIME MOLDS FORM MULTINUCLEATE MASSES. If the nucleus of an amoeba began rapid mitotic division, accompanied by a tremendous increase in cytoplasm and organelles, the resulting organism might resemble the **acellular slime molds** (*myxomycetes*). During its vegetative (feeding) phase, an acellular slime mold is a wall-less mass of cytoplasm with numerous diploid nuclei. This mass streams very slowly over its substratum in a remarkable network of strands called a *plasmodium** (Figure 28.31*a*). The plasmodium of an acellular slime mold is another example of a coenocyte, a body in which many nuclei are enclosed in a single plasma membrane. The outer cytoplasm of the plasmodium (closest to the environment) is normally less fluid

*Do not confuse the plasmodium of an acellular slime mold with the genus *Plasmodium*, the apicomplexan that is the cause of malaria.

(*a*) *Physarum polycephalum*

(*b*) *Physarum* sp.

28.31 Acellular Slime Molds (*a*) Plasmodia of the yellow slime mold *Physarum* cover a rock in Nova Scotia. (*b*) The fruiting structures—sporangiophores (yellow) and sporangia (black)—of *Physarum*.

than the interior cytoplasm and thus provides some structural rigidity.

Acellular slime molds, such as *Physarum*, provide a dramatic example of movement by *cytoplasmic streaming*. The outer cytoplasmic region of the plasmodium becomes more fluid in places, and cytoplasm rushes into those areas, stretching the plasmodium. This streaming somehow reverses its direction every few minutes as cytoplasm rushes into a new area and drains away from an older one, moving the plasmodium over its substratum. Sometimes an entire wave of plasmodium moves across the substratum, leaving strands behind. Actin filaments and a contractile protein called *myxomyosin* interact to produce the streaming movement. As it moves, the plasmodium engulfs food particles by endocytosis—predominantly bacteria, yeasts, spores of fungi, and other small organisms, as well as decaying animal and plant remains.

An acellular slime mold can grow almost indefinitely in its plasmodial stage, as long as the food supply is adequate and other conditions, such as moisture and pH, are favorable. However, one of two things can happen if conditions become unfavorable. First, the plasmodium can form an irregular mass of hardened cell-like components called a *sclerotium*. This resting structure rapidly becomes a plasmodium again when favorable conditions are restored.

Alternatively, the plasmodium can transform itself into spore-bearing fruiting structures (Figure 28.31*b*). These stalked or branched structures, called *sporangiophores*, rise from heaped masses of plasmodium. They derive their rigidity from walls that form and thicken between their nuclei. The nuclei of the plasmodium are diploid, and they divide by meiosis as the sporangiophore develops. One or more knobs, called *sporangia*, develop on the end of the stalk. Within a sporangium, haploid nuclei become surrounded by walls and form spores. Eventually, as the sporangiophore dries, it sheds its spores.

The spores germinate into wall-less, flagellated, haploid cells called *swarm cells*, which can either divide mitotically to produce more haploid swarm cells or function as gametes. Swarm cells can live as separate individual cells, and can become walled and resistant resting cysts when conditions are unfavorable. When conditions improve again, the cysts release flagellated swarm cells. Two swarm cells can also fuse to form a diploid zygote, which divides by mitosis (but without a wall forming between the nuclei) and thus forms a new, coenocytic plasmodium.

CELLS RETAIN THEIR IDENTITY IN THE CELLULAR SLIME MOLDS. Whereas the plasmodium is the basic vegetative unit of the acellular slime molds, an amoeboid cell is the vegetative unit of the **cellular slime molds**. Large numbers of cells called *myxamoebas*, which have single haploid nuclei, engulf bacteria and other food particles by endocytosis and reproduce by mitosis and fission. This simple life cycle stage, consisting of swarms of independent, isolated cells, can persist indefinitely as long as food and moisture are available.

When conditions become unfavorable, however, the cellular slime molds aggregate and form fruiting structures, as do their acellular counterparts. The apparently independent myxamoebas aggregate into a mass called a *slug* or *pseudoplasmodium* (Figure 28.32). Unlike the true plasmodium of the acellular slime molds, this structure is not simply a giant sheet of cytoplasm with many nuclei; the individual myxamoebas retain their plasma membranes and, therefore, their identity.

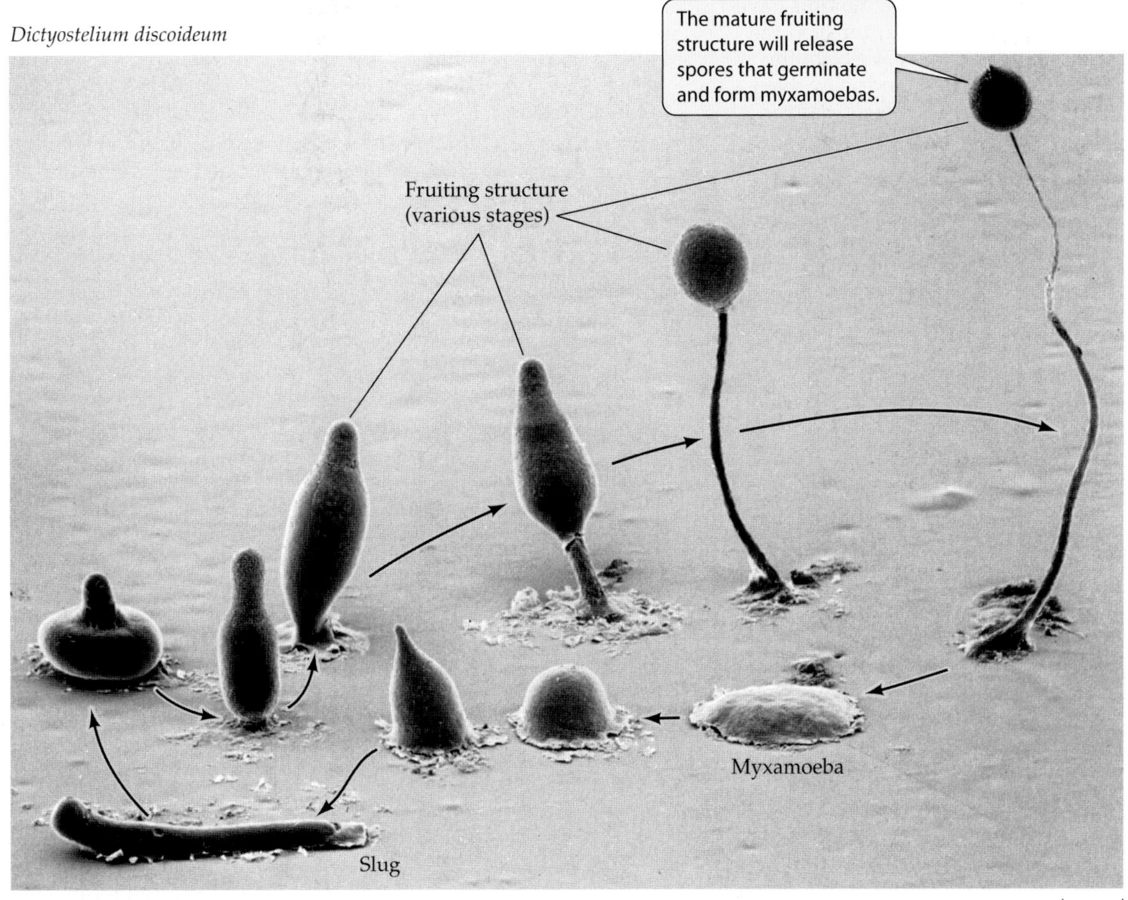

28.32 A Cellular Slime Mold The life cycle of the slime mold *Dictyostelium* is shown here in a composite micrograph.

The chemical signal that causes the myxamoebas of cellular slime molds to aggregate into a slug is 3′,5′-cyclic adenosine monophosphate (cAMP), a compound that plays many important roles as a chemical signal in animals (see Chapter 15).

A slug may migrate over its substratum for several hours before becoming motionless and reorganizing to construct a delicate, stalked fruiting structure (Figure 28.32). Cells at the top of the fruiting structure develop into thick-walled spores, which are eventually released. Later, under favorable conditions, the spores germinate, releasing myxamoebas.

The cycle from myxamoebas through slug and spores to new myxamoebas is asexual. Cellular slime molds also have a sexual cycle, in which two myxamoebas fuse. The product of this fusion develops into a spherical structure that ultimately germinates, releasing new haploid myxamoebas.

This asexual life cycle—individual cells swarming about and then aggregating to form a fruiting body—has evolved, independently, more than once. A bacterial group, the myxobacteria, also has swarming cells that aggregate and form fruiting structures. These bacteria are, of course, prokaryotes and thus have no sexual cycle.

Next we will explore the three classical kingdoms of multicellular eukaryotes. Chapters 29 and 30 deal with the kingdom Plantae (which, combined with the chlorophytes and other green algae, is called by some botanists the "green plant kingdom"). Chapter 31 presents the kingdom Fungi, and Chapters 32–34 describe the kingdom Animalia. All three of these kingdoms arose from protist ancestors.

Chapter Summary

Protists Defined

▶ In this book we define the protists simply as all eukaryotes that are not plants, fungi, or animals. The protists are a paraphyletic group, not a clade.

The Origin of the Eukaryotic Cell

▶ The modern eukaryotic cell arose from an ancestral prokaryote in several steps. Probable steps include the loss of the cell wall and infolding of the plasma membrane. **Review Figure 28.2**

▶ In subsequent steps, an infolded plasma membrane attached to the chromosome may have led to the formation of a nuclear envelope. A primitive cytoskeleton evolved. **Review Figure 28.3**

▶ The first truly eukaryotic cell was larger than its prokaryote ancestor, was probably a phagocyte, and may have possessed one or more flagella of the eukaryotic type.

▶ The incorporation of prokaryotic cells as endosymbionts gave rise to eukaryotic organelles. Peroxisomes, which protected the host cell from an oxygen-rich atmosphere, may have been the first organelles of endosymbiotic origin. Mitochondria evolved from once free-living proteobacteria, and chloroplasts evolved from once free-living cyanobacteria. **Review Figure 28.3**

General Biology of the Protists

▶ Most protists are aquatic; some live within other organisms. The great majority are unicellular and microscopic, but many are multicellular and a few are enormous.

▶ "Protozoan" is an outdated term sometimes applied to protists, mostly ingestive heterotrophs, that were once classified as animals. "Alga" is an outdated term sometimes applied to photosynthetic protists.

▶ Protists vary widely in their modes of nutrition and locomotion. Some protist cells contain contractile vacuoles, and some digest their food in food vacuoles. **Review Figures 28.5, 28.6. See Web/CD Tutorial 28.1**

▶ Protists have a variety of cell surfaces, some of them protective.

▶ Many protists contain endosymbionts. Some protists are endosymbionts in other cells, including other protists. Some endosymbiotic protists perform photosynthesis, to the advantage of their hosts.

▶ Most protists reproduce both asexually and sexually.

Protist Diversity

▶ Molecular and other techniques are enabling biologists to identify many clades of protists. **Review Table 28.1 and Figure 28.9**

Diplomonads and Parabasalids

▶ Diplomonads and parabasalids may have the most ancient roots of today's protists. Both lack mitochondria, having apparently lost them during their evolution.

▶ Diplomonads have two nuclei and multiple flagella. **Review Figure 28.10*a***

▶ Parabasalids have flagella and undulating membranes. **Review Figure 28.10*b***

Euglenozoans

▶ The euglenozoans are a clade of unicellular protists with flagella.

▶ Euglenoids are euglenozoans that are often photosynthetic and have anterior flagella. **Review Figure 28.11**

▶ Kinetoplastids are euglenozoans that have a single, large mitochondrion, in which RNA is edited.

Alveolates

▶ The alveolates are a clade of unicellular organisms with cavities, called alveoli, beneath their plasma membranes.

▶ Dinoflagellates are marine alveolates with a golden-brown color that results from their photosynthetic and accessory pigments. They are major contributors to world photosynthesis. Many are endosymbionts; in that role they are important contributors to coral growth. Dinoflagellates are responsible for toxic "red tides."

▶ Apicomplexans are parasitic alveolates with an amoeboid body form. Their spores, containing a mass of organelles at the apical end, are adapted to the invasion of host tissue. The apicomplexan *Plasmodium*, which causes malaria, uses two alternate hosts (humans and *Anopheles* mosquitoes). **Review Figure 28.14**

▶ Ciliates are alveolates that move rapidly by means of cilia and have two kinds of nuclei. The macronuclei control the cell by means of transcription and translation. The micronuclei are responsible for genetic recombination, accomplished by conjugation, a process that is sexual, but not reproductive. **Review Figures 28.16, 28.17. See Web/CD Activity 28.1**

Stramenopiles

▶ Stramenopiles typically have two flagella of unequal length, the longer bearing rows of tubular hairs. Some stramenopile groups are photosynthetic.

▶ Diatoms are unicellular stramenopiles, many of which have complex, two-part, glassy cell walls. They contribute extensively to world photosynthesis. **Review Figure 28.19**

▶ The brown algae are predominantly multicellular, photosynthetic stramenopiles. They include the largest of all protists, and some show considerable tissue differentiation.

▶ In many multicellular photosynthetic protists and in all plants, both haploid and diploid cells undergo mitosis, leading to an alternation of generations. The diploid sporophyte generation forms spores by meiosis, and the spores develop into haploid organisms. This haploid gametophyte generation forms gametes by mitosis, and their fusion yields zygotes that develop into the next generation of sporophytes. **Review Figure 28.22**

▶ Oomycetes are a group of nonphotosynthetic stramenopiles including water molds and downy mildews. The oomycetes are coenocytic. They are diploid for most of their life cycle.

Red Algae

▶ Red algae are multicellular, photosynthetic protists. They differ from the other photosynthetic protist groups in having a characteristic storage product (floridean starch) and lacking flagellated reproductive cells.

Chlorophytes

▶ The chlorophytes, a clade of green algae, are often multicellular. Like plants, they contain chlorophylls *a* and *b* and use starch as a storage product.

▶ The chlorophytes are sister to a clade that includes other green algae and the plant kingdom.

▶ The chlorophytes have diverse life cycles; among these are the isomorphic alternation of generations of *Ulva* and the haplontic life cycle of *Ulothrix*. **Review Figures 28.26, 28.27. See Web/CD Activities 28.2 and 28.3**

Choanoflagellates

▶ The choanoflagellates are colonial protists with flagella and a body type similar to the most characteristic type of cell found in sponges. The choanoflagellates are sister to the animal kingdom.

A History of Endosymbiosis

▶ Primary endosymbiosis of a cyanobacterium and a eukaryote gave rise to the chloroplasts of green algae, plants, and red algae. **Review Figure 28.29. See Web/CD Tutorial 28.2**

▶ Secondary endosymbioses of eukaryotes with unicellular green or red algae gave rise to the chloroplasts of euglenoids, stramenopiles, and other groups. One of those groups has given rise to another type of chloroplast by tertiary endosymbiosis.

Some Recurrent Body Forms

▶ Some similar body forms are found in several different, unrelated protist groups.

▶ Amoebas, which appear in many protist groups, move by means of pseudopods.

▶ Actinopods have thin, stiff pseudopods that serve various functions, including food capture.

▶ Foraminiferans also use pseudopods for feeding, and they secrete shells of calcium carbonate.

▶ Acellular slime molds and cellular slime molds are superficially similar, moving as slimy masses and producing stalked fruiting structures. However, they differ at the cellular level. Acellular slime molds are coenocytes with diploid nuclei. Cellular slime molds consist of individual haploid cells that aggregate into masses consisting of distinct cells.

Self-Quiz

1. Protists with flagella
   a. appear in several protist clades.
   b. are all algae.
   c. all have pseudopods.
   d. are all colonial.
   e. are never pathogenic.
2. Which statement about amoebas is *not* true?
   a. They are specialized.
   b. They use amoeboid movement.
   c. They include both naked and shelled forms.
   d. They possess pseudopods.
   e. They appeared only once in evolutionary history.
3. Apicomplexans
   a. possess flagella.
   b. possess chloroplasts.
   c. are all parasitic.
   d. are algae.
   e. include the trypanosomes that cause sleeping sickness.
4. The ciliates
   a. move by means of flagella.
   b. use amoeboid movement.
   c. include *Plasmodium*, the agent of malaria.
   d. possess both a macronucleus and micronuclei.
   e. are autotrophic.
5. The acellular slime molds
   a. include the genus *Physarum*.
   b. lack fruiting bodies.
   c. consist of large numbers of myxamoebas.
   d. consist at times of a mass called a pseudoplasmodium.
   e. possess flagella.
6. The cellular slime molds
   a. possess apical complexes.
   b. lack fruiting bodies.
   c. form a plasmodium that is a coenocyte.
   d. use cAMP as a "messenger" to signal aggregation.
   e. possess flagella.
7. The chloroplasts of photosynthetic protists
   a. are structurally identical.
   b. gave rise to mitochondria.
   c. are all descended from a once free-living cyanobacterium.
   d. all have exactly two surrounding membranes.
   e. are all descended from a once free-living red alga.
8. Which statement about the brown algae is *not* true?
   a. They are all multicellular.
   b. They use the same photosynthetic pigments as do plants.
   c. They are almost exclusively marine.
   d. A few are among the largest organisms on Earth.
   e. Some have extensive tissue differentiation.
9. The red algae
   a. are mostly unicellular.
   b. are mostly marine.
   c. owe their red color to a special form of chlorophyll.
   d. have flagella on their gametes.
   e. are all heterotrophic.
10. Which statement about the chlorophytes is *not* true?
   a. They use the same photosynthetic pigments as do plants.
   b. Some are unicellular.
   c. Some are multicellular.
   d. All are microscopic in size.
   e. They display a great diversity of life cycles.

For Discussion

1. For each type of organism below, give a single characteristic that may be used to differentiate it from the other, related organism(s) in parentheses.
   a. Foraminiferans (radiolarians)
   b. *Euglena* (*Volvox*)
   c. *Trypanosoma* (*Giardia*)
   d. Amoeba (flagellate)
   e. *Physarum* (*Dictyostelium*)
2. In what sense are sex and reproduction independent of each other in the ciliates? What does that suggest about the role of sex in biology?
3. Why are dinoflagellates and apicomplexans placed in one group of protists and brown algae and oomycetes in another?
4. Unlike many protists, apicomplexans lack contractile vacuoles. Why don't apicomplexans need a contractile vacuole?
5. Giant seaweeds (mostly brown algae) have "floats" that aid in keeping their fronds suspended at or near the surface of the water. Why is it important that the fronds be suspended in this way?
6. Why are algal pigments so much more diverse than those of plants?
7. Consider the chloroplasts of chlorophytes, euglenozoans, and red algae. For each of these groups, indicate how many membranes surround their chloroplasts, and offer a reasonable explanation in each case. Why do some dinoflagellates have more membranes around their chloroplasts than other dinoflagellates?

29 Plants without Seeds: From Sea to Land

Residents of the coal-producing central Chinese city of Changsha almost never see the sun, because it is hidden behind an atmosphere dense with choking smog. Nine-tenths of the precipitation in Changsha is acid rain. China burns more coal than any other country in the world, and the resulting untreated smoke leads to disastrous conditions such as those in Changsha, the site of a major coal-fired power plant.

Coal is used for 75 percent of China's energy needs—primarily to generate electricity, but also directly for heating, smelting of metals, and other purposes. The United States produces more than half of its electricity by burning coal, and indeed has the largest coal reserves in the world. Extractable coal reserves in the U.S. exceed the total amount of oil available for pumping in all other countries combined. Where did all this coal come from?

Coal comes from the remains of seedless plants that grew in great forests hundreds of millions of years ago. (The two other "fossil fuels"—petroleum and natural gas—come from the remains of plankton that lived in ancient oceans.) Plant parts from those forests sank in swamps that were later covered by soil. Over millions of years, as the buried plant material was subjected to intense pressure and elevated temperatures, coal formed.

At the time those ancient forests flourished, the plant world also included relatives of today's mosses. These "mossy" ancestors were the first plant life on dry land. Today, mosses are among the most abundant plants on Earth, yet they seem at first glance to lack adaptations to life on land. Mosses have no advanced internal "plumbing system" to move water and nutrients within their bodies, and their leafy photosynthetic organs are only one cell thick. They require liquid water in order to reproduce, and indeed, seem at first glance to be highly dependent on external moisture. Mosses and their relatives do have effective adaptations for life in terrestrial environments, however, as is obvious from their wide distribution. Most live in moist habitats, but a few mosses even live in deserts.

An Ingredient of Coal-Based Smog When coal burns, it produces the fly ash shown in this artificially colored image. When too much coal burns where the smoke cannot blow away, the result is disastrous smog.

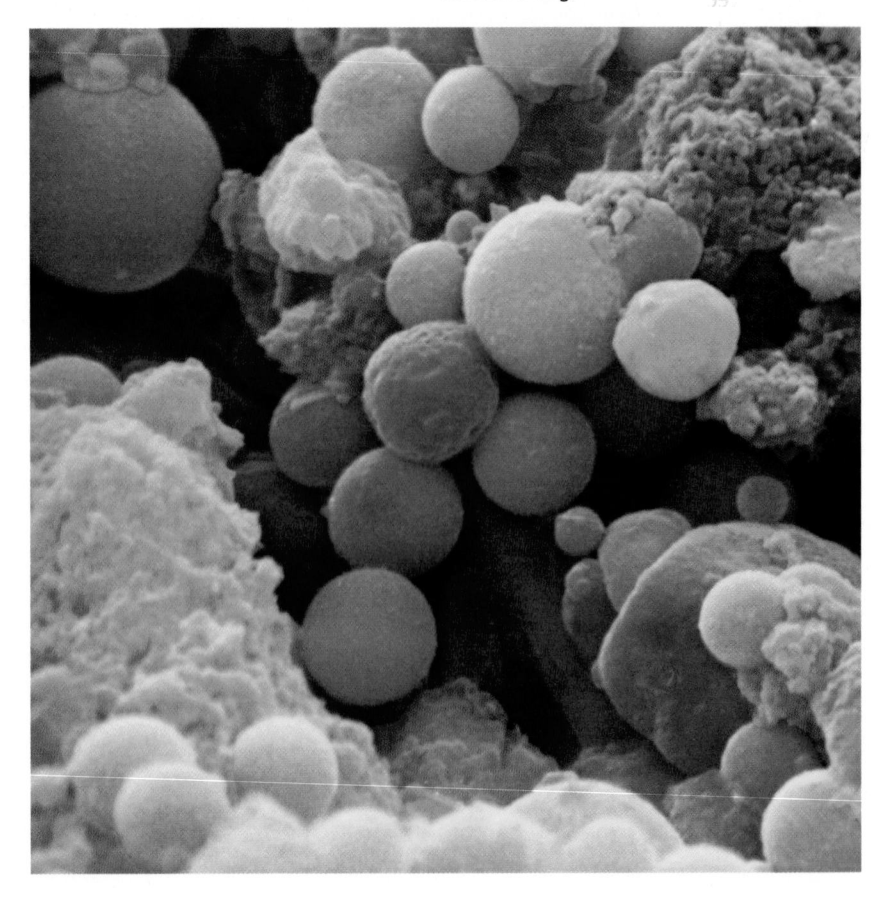

The earliest terrestrial plants invaded the land sometime during the Paleozoic era (see Table 22.1). These plants were tiny, but their metabolic activities helped convert parent rock into soil that could support the needs of their successors. Larger and larger plants evolved rapidly (in geological terms), and by the Carboniferous period (354–290 mya) great forests were widespread. However, few of the trees in those forests were like those we know today. During the tens of millions of years since the Carboniferous, those early trees have been replaced by the modern trees whose adaptations and appearance are familiar to us.

In this chapter, we will see how members of the plant kingdom invaded the land and evolved. Our descriptions here will concentrate on those plants that lack seeds. The next chapter completes our survey of the plant kingdom by considering the seed plants, which dominate the terrestrial scene today.

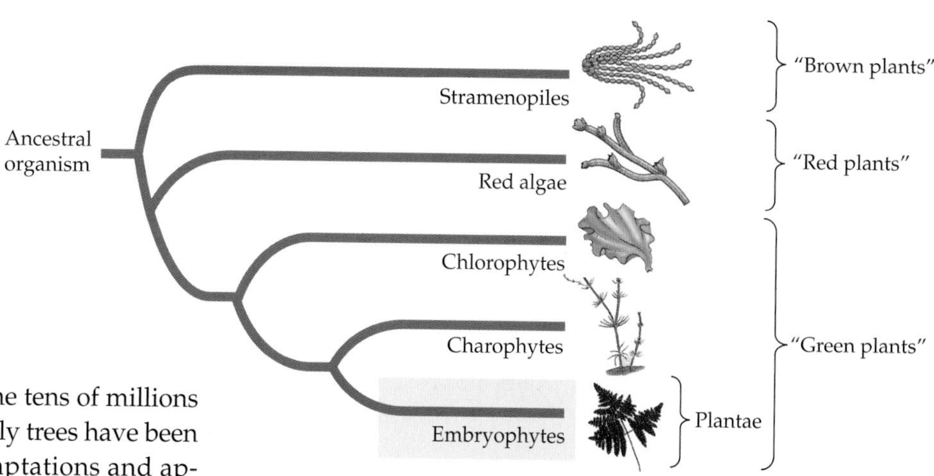

29.1 What Is a Plant? There are three ways to define a plant kingdom, depending on which clade is chosen. In this book, we use the most restrictive definition: plants as embryophytes. Here, the two green algal clades are not considered plants.

The Plant Kingdom

The kingdom Plantae is monophyletic—all plants descend from a single common ancestor and form a branch of the evolutionary tree of life. The shared derived trait, or synapomorphy, of the plant kingdom is development from embryos protected by tissues of the parent plant. For this reason, plants are sometimes referred to as *embryophytes*. Plants retain the derived features that they share with green algae: the use of chlorophylls *a* and *b* and the use of starch as a photosynthetic storage product. Both plants and green algae have cellulose in their cell walls.

There are other ways to define "plant" and "plant kingdom" and still come out with a monophyletic group (clade). For example, combining plants as defined above with a group of green algae called the charophytes results in a monophyletic plant kingdom with several shared derived traits, including the retention of the egg in the parent body. The addition of the chlorophytes (the remainder of the green algae) to the group just described gives another monophyletic group, with synapomorphies including the possession of chlorophyll *b*, that can be called a plant kingdom. There are no hard-and-fast criteria for defining a kingdom (or any other taxonomic rank), so these definitions of the plant kingdom are all valid.

In this book, we choose to use the first definition given above, in which the kingdom Plantae comprises only the embryophytes (Figure 29.1). Some botanists refer to a group consisting of the Plantae plus the green algae as the "green plant kingdom," to the red algae as the "red plant kingdom," and to the stramenopiles as the "brown plant kingdom."

There are ten surviving phyla of plants

The surviving members of the kingdom Plantae fall naturally into ten phyla (Table 29.1). All members of seven of those phyla possess well-developed vascular systems that transport materials throughout the plant body. We call these seven phyla, collectively, the **tracheophytes** because they all possess conducting cells called tracheids. The tracheophytes constitute a clade.

The remaining three phyla (liverworts, hornworts, and mosses), which lack tracheids, were once considered classes of a single larger phylum. In this book we use the term **nontracheophytes** to refer collectively to these three phyla. The nontracheophytes are sometimes collectively called *bryophytes*, but in this text we reserve that term for their most familiar members, the mosses. Collectively, the nontracheophytes are not a monophyletic group. They are the three basal clades of the plant kingdom.

Life cycles of plants feature alternation of generations

A universal feature of the life cycles of plants is the alternation of generations. Recall from Chapter 28 that alternation of generations has two hallmarks:

- The life cycle includes both multicellular diploid individuals and multicellular haploid individuals.
- Gametes are produced by mitosis, not by meiosis. Meiosis produces spores that develop into multicellular haploid individuals.

If we begin looking at the plant life cycle at a single-cell stage—the diploid zygote—then the first phase of the cycle

29.1 Classification of Plants[a]

PHYLUM	COMMON NAME	CHARACTERISTICS
Nontracheophytes		
Hepatophyta	Liverworts	No filamentous stage; gametophyte flat
Anthocerophyta	Hornworts	Embedded archegonia; sporophyte grows basally
Bryophyta	Mosses	Filamentous stage; sporophyte grows apically (from the tip)
Tracheophytes		
Nonseed tracheophytes		
Lycophyta	Club mosses	Microphylls in spirals; sporangia in leaf axils
Pteridophyta	Ferns and allies	Differentiation between main axis and side branches
Seed plants		
Gymnosperms		
Cycadophyta	Cycads	Compound leaves; swimming sperm; seeds on modified leaves
Ginkgophyta	Ginkgo	Deciduous; fan-shaped leaves; swimming sperm
Gnetophyta	Gnetophytes	Vessels in vascular tissue; opposite, simple leaves
Pinophyta	Conifers	Seeds in cones; needlelike or scalelike leaves
Angiosperms		
Angiospermae	Flowering plants	Endosperm; carpels; much reduced gametophytes; seeds in fruit

[a] No extinct groups are included in this classification.

features the formation, by mitosis and cytokinesis, of a multicellular embryo and eventually the mature diploid plant (Figure 29.2). This multicellular, diploid plant is the **sporophyte** ("spore plant").

Cells contained in **sporangia** (singular, sporangium, "spore vessel") on the sporophyte undergo meiosis to produce haploid, unicellular spores. By mitosis and cytokinesis, a spore forms a haploid plant. This multicellular, haploid plant is the **gametophyte** ("gamete plant") that produces haploid gametes. The fusion of two gametes (*syngamy*, or *fertilization*) results in the formation of a diploid cell—the zygote—and the cycle repeats.

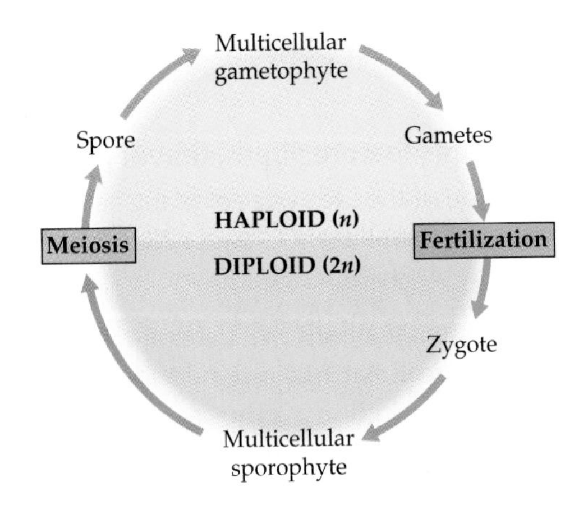

29.2 Alternation of Generations A diploid sporophyte generation that produces spores alternates with a haploid gametophyte generation that produces gametes by mitosis.

The *sporophyte generation* extends from the zygote through the adult, multicellular, diploid plant; the *gametophyte generation* extends from the spore through the adult, multicellular, haploid plant to the gamete. The transitions between the generations are accomplished by fertilization and meiosis. In all plants, the sporophyte and gametophyte differ genetically: The sporophyte has diploid cells, and the gametophyte has haploid cells. In the three basal plant clades, the gametophyte generation is larger and more self-sufficient, while the sporophyte generation is dominant in those groups that appeared later in plant evolution.

Some protist life cycles also feature alternation of generations, suggesting that the plants arose from one of these protist groups. But which one?

The Plantae arose from a green algal clade

Much evidence indicates that the closest living relatives of the plants are members of a clade of green algae called the **charophytes**. The charophytes, along with some other green algae and the plants, form a clade that is sister to the chlorophytes (see Figure 29.1), but we don't yet know which charophyte clade is the true sister group to the plants. Stoneworts of the genus *Chara* are charophytes that resemble plants in terms of their rRNA and DNA sequences, peroxisome con-

29.3 The Closest Relatives of Land Plants The plant kingdom probably evolved from a common ancestor shared with the charophytes, a green algal group. (*a*) Molecular evidence seems to favor stoneworts of the genus *Chara* as sister group to the plants. (*b*) Evidence from morphology indicates that the group including this coleochaete alga may be sister to the land plants.

tents, mechanics of mitosis and cytokinesis, and chloroplast structure (Figure 29.3*a*). On the other hand, strong evidence from morphology-based cladistic analysis suggests that the sister group of the plants is a group of charophytes that includes the genus *Coleochaete* (Figure 29.3*b*). *Coleochaete*-like algae have several features found in plants, such as plasmodesmata and a tendency to protect the young sporophyte.

Whether they were more similar to stoneworts or to *Coleochaete*, the ancestors of the plants lived at the margins of ponds or marshes, ringing them with a green mat. From these marginal habitats, which were sometimes wet and sometimes dry, early plants made the transition onto land.

The Conquest of the Land

Plants, or their immediate ancestors in the green mat, first invaded the terrestrial environment between 400 and 500 million years ago. That environment differs dramatically from the aquatic environment. The most obvious difference is the availability of the water that is essential for life: It is everywhere in the aquatic environment, but hard to find and to retain in the terrestrial environment. Water provides aquatic organisms with support against gravity; a plant on land, however, must either have some other support system or sprawl unsupported on the ground. A land plant must also use different mechanisms for dispersing its gametes and progeny than its aquatic relatives, which can simply release them into the water. How did organisms descended from aquatic ancestors adapt to such a challenging environment?

Adaptations to life on land distinguish plants from green algae

Most of the characteristics that distinguish plants from green algae are evolutionary adaptations to life on land. Several of these features probably evolved in the common ancestor of the plants:

- The *cuticle*, a waxy covering that retards desiccation (drying)
- *Gametangia*, cases that enclose plant gametes and prevent them from drying out
- *Embryos*, which are young sporophytes contained within a protective structure
- Certain *pigments* that afford protection against the mutagenic ultraviolet radiation that bathes the terrestrial environment
- Thick *spore walls* containing a polymer that protects the spores from desiccation and resists decay
- A *mutualistic* association with a fungus* that promotes nutrient uptake from the soil

Further adaptations to the terrestrial environment appeared as plants continued to evolve. One of the most important of these later adaptations was the appearance of vascular tissues.

Most present-day plants have vascular tissues

The first plants were nonvascular, lacking both water-conducting and food-conducting tissue. Although the term "nonvascular plants" is a time-honored name, it is misleading when applied to the entire nontracheophyte group, because some mosses (unlike liverworts and hornworts) do have a limited amount of simple conducting tissue. Thus the more unwieldy name "nontracheophyte" is more descriptive. The first true tracheophytes—possessing specialized conducting cells called tracheids—arose later (Figure 29.4).

The nontracheophytes (the liverworts, hornworts, and mosses) have never been large plants. Except for some of the mosses, they have no water-conducting tissue, yet some are found in dry environments. Many grow in dense masses (see Figure 29.9*a*), through which water can move by capillary action. Nontracheophytes also have leaflike structures that readily catch and hold any water that splashes onto them. These plants are small enough that minerals can be distributed throughout their bodies by diffusion.

*In a mutualistic association, both partners—here, the plant and the fungus—profit.

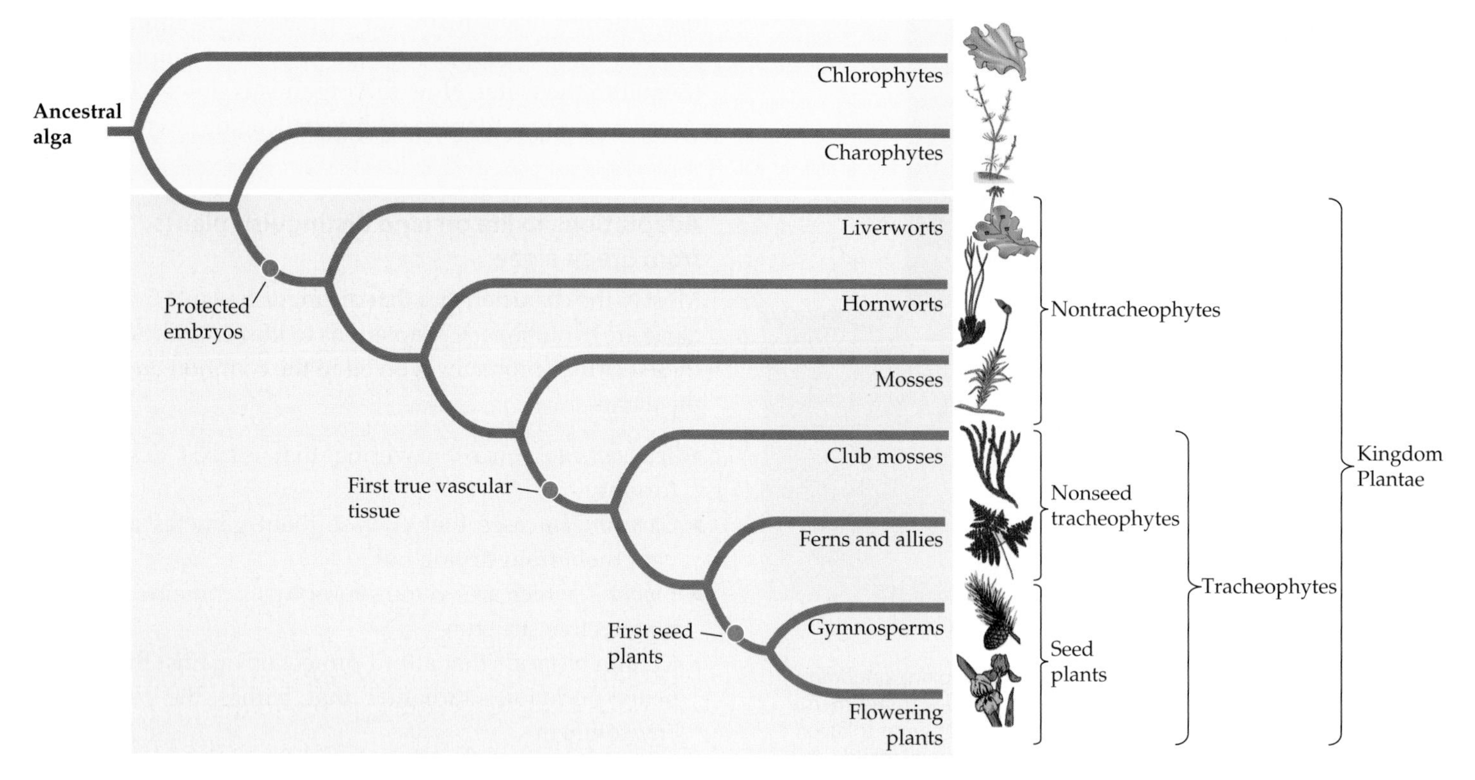

29.4 From Green Algae to Plants Three key characteristics that emerged during plant evolution—protected embryos, vascular tissues, and seeds—are all adaptations to life in a terrestrial environment. Plants with vascular tissue are called tracheophytes.

Familiar tracheophytes include the club mosses, ferns, conifers, and angiosperms (flowering plants). Tracheophytes differ from liverworts, hornworts, and mosses in crucial ways, one of which is the possession of a well-developed **vascular system** consisting of specialized tissues for the transport of materials from one part of the plant to another. One such tissue, the **phloem**, conducts the products of photosynthesis from sites where they are produced or released to sites where they are used or stored. The other vascular tissue, the **xylem**, conducts water and minerals from the soil to aerial parts of the plant; because some of its cell walls are stiffened by a substance called *lignin*, xylem also provides support in the terrestrial environment.

Nontracheophyte plants evolved tens of millions of years before the earliest tracheophytes, even though tracheophytes appear earlier in the fossil record. The oldest tracheophyte fossils date back more than 410 million years, whereas the oldest nontracheophyte fossils are only about 350 million years old, dating from a time when tracheophytes were already widely distributed. This finding simply shows that, given the differences in their structures and the chemical makeup of their cell walls, tracheophytes are more likely to form fossils than nontracheophytes are.

We will examine the adaptations of the tracheophytes later in this chapter, concentrating first on the nontracheophytes.

The Nontracheophytes: Liverworts, Hornworts, and Mosses

Most liverworts, hornworts, and mosses grow in dense mats, usually in moist habitats. The largest of these plants are only about 1 meter tall, and most are only a few centimeters tall or long. Why have the nontracheophytes not evolved to be taller? The probable answer is that they lack an efficient system for conducting water and minerals from the soil to distant parts of the plant body. To limit water loss, layers of maternal tissue protect the embryos of all nontracheophytes. All nontracheophyte clades also have a cuticle, although it is often very thin (or even absent in some species) and thus not highly effective in retarding water loss. Nontracheophytes lack the leaves, stems, and roots that characterize tracheophytes, although they have structures analogous to each.

Most nontracheophytes live on the soil or on other plants, but some grow on bare rock, dead and fallen tree trunks, and even on buildings. Nontracheophytes are widely distributed over six continents and exist very locally on the coast of the

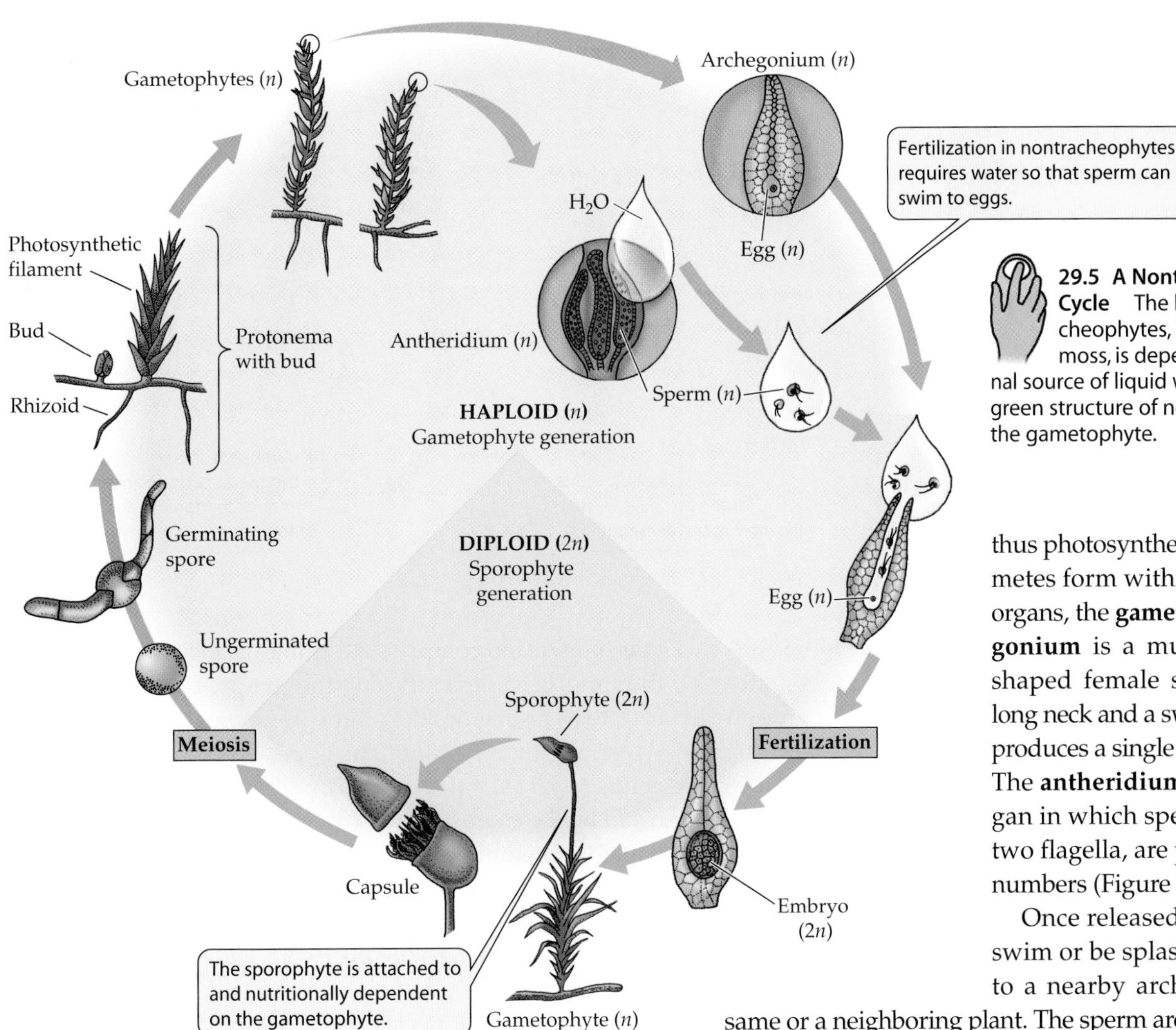

29.5 A Nontracheophyte Life Cycle The life cycle of nontracheophytes, illustrated here by a moss, is dependent on an external source of liquid water. The visible green structure of nontracheophytes is the gametophyte.

seventh (Antarctica). They are successful plants, well adapted to their environments. Most are terrestrial. Some live in wetlands. Although a few nontracheophyte species live in fresh water, these aquatic forms are descended from terrestrial ones. There are no marine nontracheophytes.

Nontracheophyte sporophytes are dependent on gametophytes

In nontracheophytes, the conspicuous green structure visible to the naked eye is the gametophyte (Figure 29.5). In contrast, the familiar forms of tracheophytes, such as ferns and seed plants, are sporophytes. The gametophyte of nontracheophytes is photosynthetic and therefore nutritionally independent, whereas the sporophyte may or may not be photosynthetic, but is always nutritionally dependent on the gametophyte and remains permanently attached to it.

A nontracheophyte sporophyte produces unicellular, haploid spores as products of meiosis within a sporangium, or **capsule**. A spore germinates, giving rise to a multicellular, haploid gametophyte whose cells contain chloroplasts and are thus photosynthetic. Eventually gametes form within specialized sex organs, the **gametangia**. The **archegonium** is a multicellular, flask-shaped female sex organ with a long neck and a swollen base, which produces a single egg (Figure 29.6*a*). The **antheridium** is a male sex organ in which sperm, each bearing two flagella, are produced in large numbers (Figure 29.6*b*).

Once released, the sperm must swim or be splashed by raindrops to a nearby archegonium on the same or a neighboring plant. The sperm are aided in this task by chemical attractants released by the egg or the archegonium. Before sperm can enter the archegonium, certain cells in the neck of the archegonium must break down, leaving a water-filled canal through which the sperm swim to complete their journey. Note that all of these events require liquid water.

On arrival at the egg, the nucleus of a sperm fuses with the egg nucleus to form a zygote. Mitotic divisions of the zygote produce a multicellular, diploid sporophyte embryo. The base of the archegonium grows to protect the embryo during its early development. Eventually, the developing sporophyte elongates sufficiently to break out of the archegonium, but it remains connected to the gametophyte by a "foot" that is embedded in the parent tissue and absorbs water and nutrients from it. The sporophyte remains attached to the gametophyte throughout its life. The sporophyte produces a capsule, within which meiotic divisions produce spores and thus the next gametophyte generation.

The structure and pattern of elongation of the sporophyte differ among the three nontracheophyte phyla—the liverworts (Hepatophyta), hornworts (Anthocerophyta), and mosses (Bryophyta). The probable evolutionary relationships of these three phyla and the tracheophytes can be seen in Figure 29.4.

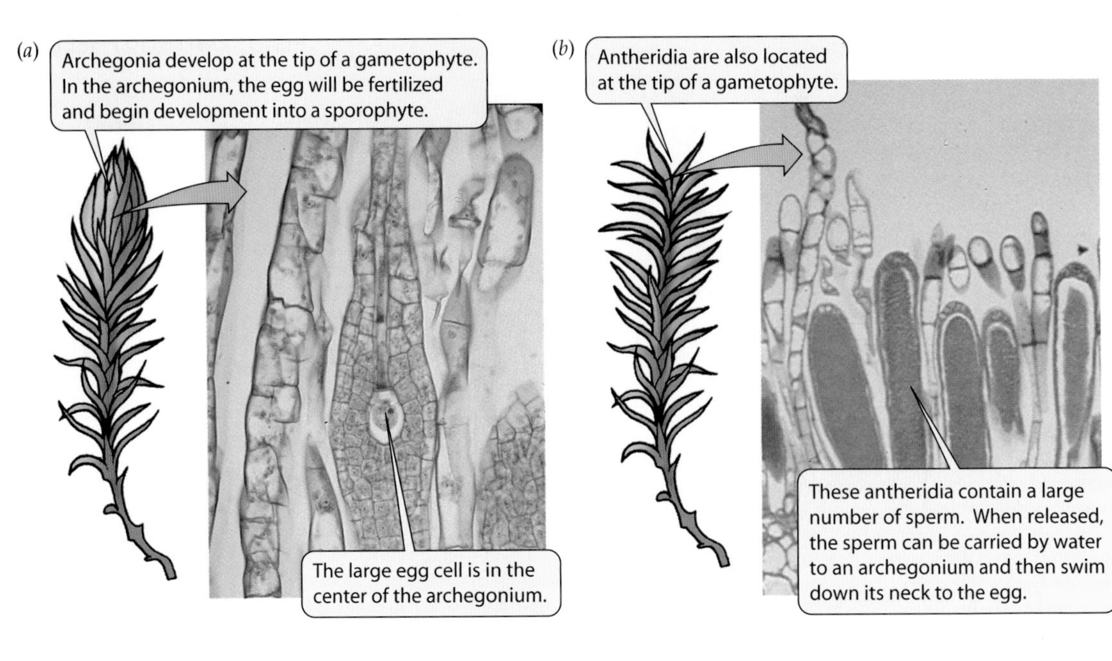

29.6 Sex Organs in Plants (*a*) Archegonia and (*b*) antheridia of the moss *Mnium* (phylum Bryophyta). The gametophytes of all plants have archegonia and antheridia, but they are much reduced in seed plants.

Liverworts may be the most ancient surviving plant clade

The gametophytes of some **liverworts** (phylum **Hepatophyta**) are green, leaflike layers that lie flat on the ground (Figure 29.7*a*). The simplest liverwort gametophytes, however, are flat plates of cells, a centimeter or so long, that produce antheridia or archegonia on their upper surfaces and anchoring and water-absorbing filaments called **rhizoids** on their lower surfaces. Liverwort sporophytes are shorter than those of mosses and hornworts, rarely exceeding a few millimeters.

The liverwort sporophyte has a stalk that connects capsule and foot. In most species, the stalk elongates and thus raises the capsule above ground level, favoring dispersal of spores when they are released. The capsules of liverworts are simple: a globular capsule wall surrounding a mass of spores. In some species of liverworts, spores are not released by the sporophyte until the surrounding capsule wall rots. In other liverworts, however, the spores are thrown from the capsule by structures that shorten and compress a "spring" as they dry out. When the stress becomes sufficient, the compressed spring snaps back to its resting position, throwing spores in all directions.

Among the most familiar liverworts are species of the genus *Marchantia* (Figure 29.7*a*). *Marchantia* is easily recognized by the characteristic structures on which its male and female gametophytes bear their antheridia and archegonia (Figure 29.7*b*). Like most liverworts, *Marchantia* also reproduces asexually by simple fragmentation of the gametophyte. *Marchantia* and some other liverworts and mosses also reproduce asexually by means of *gemmae* (singular, gemma), which are lens-shaped clumps of cells. In a few liverworts, the gemmae are loosely held in structures called *gemmae cups*, which promote dispersal of the gemmae by raindrops (Figure 29.7*c*).

Hornworts evolved stomata as an adaptation to terrestrial life

The phylum **Anthocerophyta** comprises the **hornworts**, so named because their sporophytes look like little horns (Fig-

29.7 Liverwort Structures Members of the phylum Hepatophyta display various characteristic structures. (*a*) Gametophytes. (*b*) Structures bearing antheridia and archegonia. (*c*) Gemmae cups.

(*a*) *Marchantia* sp.

The umbrella-like structures bear archegonia.
The disc-headed structures bear antheridia.

(*b*) *Marchantia* sp.

These cups contain gemmae—small, lens-shaped outgrowths of the plant body, each capable of developing into a new plant.

(*c*) *Lunularia* sp.

The sporophytes of hornworts can reach 20 cm in height.
Gametophytes are flat plates a few cells thick.
Anthoceros sp.

29.8 A Hornwort The sporophytes of hornworts can resemble little horns.

ure 29.8). Hornworts appear at first glance to be liverworts with very simple gametophytes. These gametophytes consist of flat plates of cells a few cells thick.

However, the hornworts, along with the mosses and tracheophytes, share an advance over the liverwort clade in their adaptation to life on land. They have *stomata*—pores that, when open, allow the uptake of CO_2 for photosynthesis and the release of O_2. Stomata may be a shared derived trait (synapomorphy) of hornworts and all other plants except liverworts, although hornwort stomata do not close and may have evolved independently.

Hornworts have two characteristics that distinguish them from both liverworts and mosses. First, the cells of hornworts each contain a single large, platelike chloroplast, whereas the cells of other nontracheophytes contain numerous small, lens-shaped chloroplasts. Second, of all the nontracheophyte sporophytes, those of the hornworts come closest to being capable of indeterminate growth (growth without a set limit). Liverwort and moss sporophytes have a stalk that stops growing as the capsule matures, so elongation of the sporophyte is strictly limited. The hornwort sporophyte, however, has no stalk. Instead, a basal region of the capsule remains capable of indefinite cell division, continuously producing new spore-bearing tissue above. The sporophytes of some hornworts growing in mild and continuously moist conditions can become as tall as 20 centimeters. Eventually the sporophyte's growth is limited by the lack of a transport system.

To support their metabolism, the hornworts need access to nitrogen. Hornworts have internal cavities filled with mucilage; these cavities are often populated by cyanobacteria that convert atmospheric nitrogen gas into a form usable by the host plant.

We have presented the hornworts as sister to the clade consisting of mosses and tracheophytes, but this is only one possible interpretation of the current data. The exact evolutionary status of the hornworts is still unclear, and in some phylogenetic analyses they are placed as the most ancient plant clade.

Water and sugar transport mechanisms emerged in the mosses

The most familiar nontracheophytes are the **mosses** (phylum **Bryophyta**). There are more species of mosses than of liverworts and hornworts combined, and these hardy little plants are found in almost every terrestrial environment. They are often found on damp, cool ground, where they form thick mats (Figure 29.9*a*). The mosses are probably sister to the tracheophytes (see Figure 29.4).

Many mosses contain a type of cell called a *hydroid*, which dies and leaves a tiny channel through which water can

(*a*)
(*b*)
The teeth of this moss capsule help expel the thousands of spores.

29.9 The Mosses (*a*) Dense moss forms hummocks in a valley on New Zealand's South Island. (*b*) The moss capsule, from which spores are dispersed, grows at the tip of the plant.

travel. The hydroid may be the progenitor of the tracheid, the characteristic water-conducting cell of the tracheophytes, but it lacks lignin (a waterproofing substance that also lends structural support) and the cell wall structure found in tracheids. The possession of hydroids and of a limited system for transport of sucrose by some mosses (via cells called *leptoids*) shows that the old term "nonvascular plant" is somewhat misleading when applied to mosses.

In contrast to liverworts and hornworts, the sporophytes of mosses and tracheophytes grow by **apical cell division**, in which a region at the growing tip provides an organized pattern of cell division, elongation, and differentiation. This growth pattern allows extensive and sturdy vertical growth of sporophytes. Apical cell division is a shared derived trait of mosses and tracheophytes.

The moss gametophyte that develops following spore germination is a branched, filamentous structure called a *protonema* (see Figure 29.5). Although the protonema looks a bit like a filamentous green alga, it is unique to the mosses. Some of the filaments contain chloroplasts and are photosynthetic; others, called rhizoids, are nonphotosynthetic and anchor the protonema to the substratum. After a period of linear growth, cells close to the tips of the photosynthetic filaments divide rapidly in three dimensions to form *buds*. The buds eventually differentiate a distinct tip, or apex, and produce the familiar leafy moss shoot with leaflike structures arranged spirally. These leafy shoots produce antheridia or archegonia (see Figure 29.6). The antheridia release sperm that travel through liquid water to the archegonia, where they fertilize the eggs.

Sporophyte development in most mosses follows a precise pattern, resulting ultimately in the formation of an absorptive foot anchored to the gametophyte, a stalk, and, at the tip, a swollen capsule, the sporangium. In contrast to hornworts, whose sporophytes grow from the base, the moss sporophyte stalk grows at its apical end, as tracheophytes do. Cells at the tip of the stalk divide, supporting elongation of the structure and giving rise to the capsule. For a while, the archegonial tissue grows rapidly as the stalk elongates, but eventually the archegonium is outgrown and is torn apart by the expanding sporophyte.

The lid of the capsule is shed after the completion of meiosis and spore development. In most mosses, groups of cells just below the lid form a series of toothlike structures surrounding the opening. Highly responsive to humidity, these structures dig into the mass of spores when the atmosphere is dry; then, when the atmosphere becomes moist, they fling out, scooping out the spores as they go (Figure 29.9*b*). The spores are thus dispersed when the surrounding air is moist—that is, when conditions favor their subsequent germination.

Mosses of the genus *Sphagnum* often grow in swampy places, where the plants begin to decompose in the water after they die. Rapidly growing upper layers compress the deeper-lying, decomposing layers. Partially decomposed plant matter is called *peat*. In some parts of the world, people derive the majority of their fuel from peat bogs. *Sphagnum*-dominated peatlands cover an area approximately half as large as the United States—more than 1 percent of Earth's surface. Long ago, continued compression of peat composed primarily of other nonseed plants gave rise to coal.

With their simple system of internal transport, the mosses are, in a sense, vascular plants. However, they are not tracheophytes because they lack true xylem and phloem.

Introducing the Tracheophytes

Although they are an extraordinarily large and diverse group, the tracheophytes can be said to have been launched by a single evolutionary event. Sometime during the Paleozoic era, probably well before the Silurian period (440 mya), the sporophyte generation of a now long-extinct plant produced a new cell type, the **tracheid** (Figure 29.10). The tracheid is the principal water-conducting element of the xylem in all tracheophytes except the angiosperms, and even in the angiosperms, tracheids persist alongside a more specialized and efficient system of vessels and fibers derived from them.

The evolution of a tissue composed of tracheids had two important consequences. First, it provided a pathway for long-distance transport of water and mineral nutrients from a source of supply to regions of need. Second, its stiff cell walls provided something almost completely lacking—and unnecessary—in the largely aquatic green algae: rigid structural support. Support is important in a terrestrial environment because plants tend to grow upward as they compete for sunlight to power photosynthesis. Thus the tracheid set the stage for the complete and permanent invasion of land by plants.

The tracheophytes feature another evolutionary novelty: a branching, independent sporophyte. A branching sporophyte can produce more spores than an unbranched body, and it can develop in complex ways. The sporophyte of a tracheophyte is nutritionally independent of the gametophyte at maturity. Among the tracheophytes, the sporophyte is the large and obvious plant that one normally notices in nature. This pattern is in contrast to the sporophyte of nontracheophytes such as mosses, which is attached to, dependent on, and usually much smaller than the gametophyte.

The present-day evolutionary descendants of the early tracheophytes belong to seven distinct phyla (see Figure 29.10). The tracheophytes have two types of life cycles, one that involves seeds and another that does not. The nonseed tracheophytes (the two basal phyla) include the club mosses and the ferns and their relatives: horsetails and whisk ferns. We will describe these phyla in detail after taking a closer

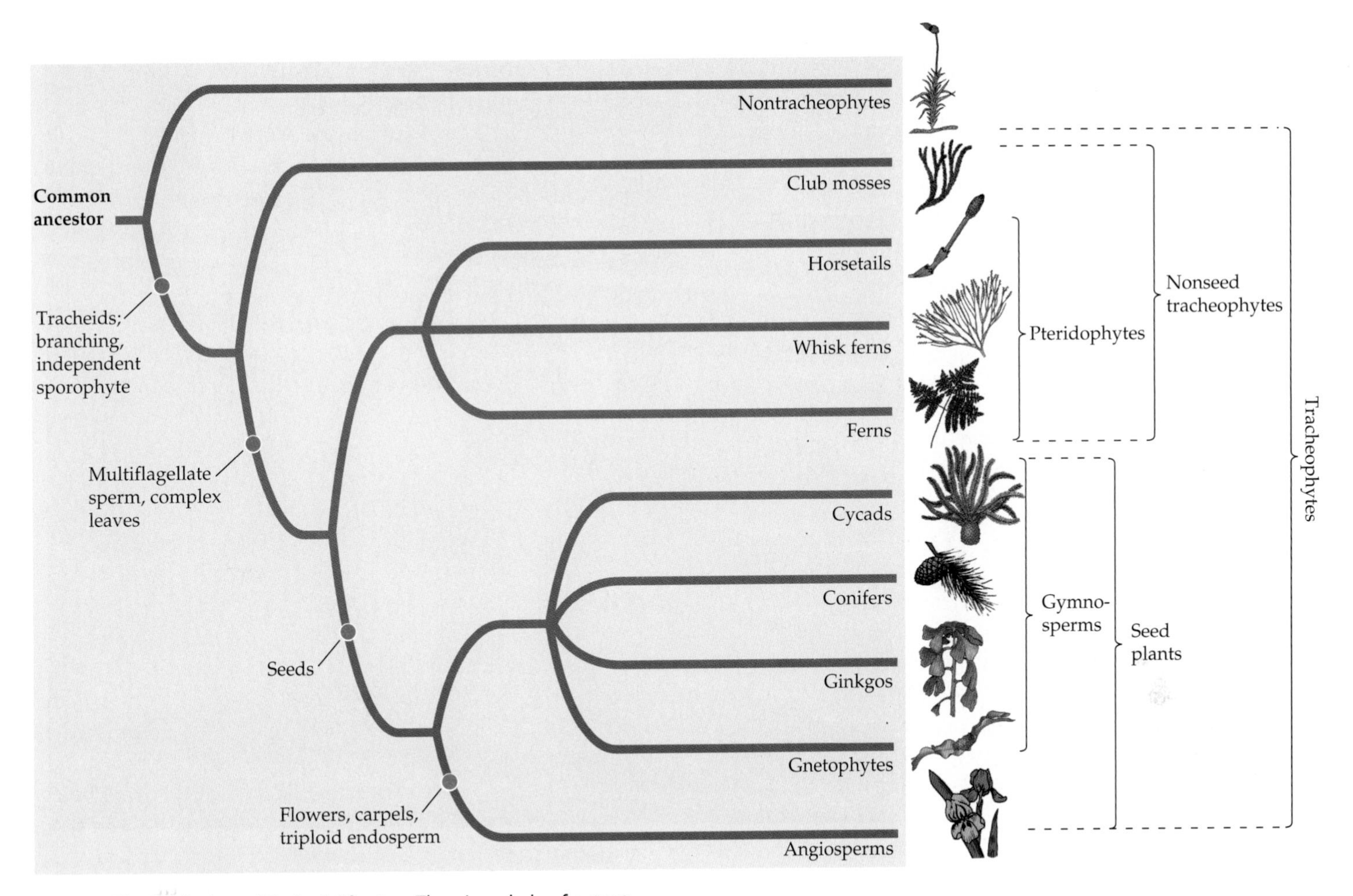

29.10 The Evolution of Today's Plants The nine phyla of extant tracheophytes are divided between those that produce seeds and those that do not.

look at tracheophyte evolution. The five phyla of seed plants will be described in the following chapter.

Tracheophytes have been evolving for almost half a billion years

The evolution of an effective cuticle and of protective layers for the gametangia (archegonia and antheridia) helped make the first tracheophytes successful, as did the initial absence of herbivores (plant-eating animals) on land. By the late Silurian period, tracheophytes were being preserved as fossils that we can study today. Two groups of nonseed tracheophytes that still exist made their first appearances during the Devonian period (409–354 mya): the lycopods (club mosses) and the pteridophytes (including horsetails and ferns). Their proliferation made the terrestrial environment more hospitable to animals. Amphibians and insects arrived soon after the plants became established.

Trees of various kinds appeared in the Devonian period and dominated the landscape of the Carboniferous. Mighty forests of lycopods up to 40 meters tall, horsetails, and tree ferns flourished in the tropical swamps of what would become North America and Europe (Figure 29.11). The remnants of those forests are with us today as huge deposits of coal.

In the subsequent Permian period, the continents came together to form a single gigantic land mass, called Pangaea. The continental interior became warmer and drier, but late in the period glaciation was extensive. The 200-million-year reign of the lycopod–fern forests came to an end as they were replaced by forests of seed plants (gymnosperms), which dominated until other seed plants (angiosperms) became dominant less than 80 million years ago.

The earliest tracheophytes lacked roots and leaves

The earliest known tracheophytes belonged to the now-extinct phylum **Rhyniophyta**. The rhyniophytes were among the only tracheophytes in the Silurian period. The landscape at that time probably consisted of bare ground, with stands of rhyniophytes in low-lying moist areas. Early versions of the structural features of all the other tracheophyte phyla appeared in the rhyniophytes of that time. These shared features strengthen the case for the origin of all tracheophytes from a common nontracheophyte ancestor.

29.11 An Ancient Forest
This reconstruction is of a Carboniferous forest that once thrived in what is now Michigan. The dominant "trees" are lycopods of the genus *Lepidodendron*; ferns are also abundant.

In 1917, the British paleobotanists Robert Kidston and William H. Lang reported their finding of well-preserved fossils of tracheophytes embedded in Devonian rocks near Rhynie, Scotland. The preservation of these plants was remarkable, considering that the rocks were more than 395 million years old. These fossil plants had a simple vascular system of phloem and xylem. Some of the plants had flattened scales on the stems, which lacked vascular tissue and thus were not comparable to the true leaves of any other tracheophytes.

These plants also lacked roots. They were apparently anchored in the soil by horizontal portions of stem, called **rhizomes**, that bore water-absorbing rhizoids. These rhizomes also bore aerial branches, and sporangia—homologous with the nontracheophyte capsule—were found at the tips of these branches. Their branching pattern was dichotomous; that is, the shoot apex divided to produce two equivalent new branches, each pair diverging at approximately the same angle from the original stem (Figure 29.12). Scattered fragments of such plants had been found earlier, but never in such profusion or so well preserved as those discovered by Kidston and Lang.

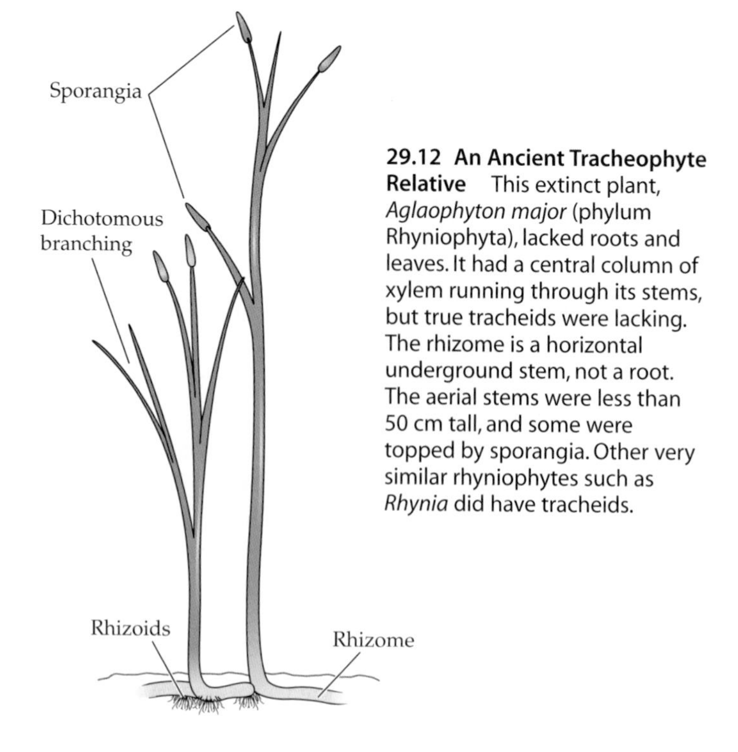

29.12 An Ancient Tracheophyte Relative This extinct plant, *Aglaophyton major* (phylum Rhyniophyta), lacked roots and leaves. It had a central column of xylem running through its stems, but true tracheids were lacking. The rhizome is a horizontal underground stem, not a root. The aerial stems were less than 50 cm tall, and some were topped by sporangia. Other very similar rhyniophytes such as *Rhynia* did have tracheids.

The presence of xylem indicated that these plants were tracheophytes. But were they sporophytes or gametophytes? Close inspection of thin sections of fossil sporangia revealed that the spores were in groups of four. In almost all living nonseed tracheophytes (with no evidence to the contrary from fossil forms), the four products of meiosis and cytokinesis remain attached to one another during their development into spores. The spores separate only when they are mature, and even after separation their walls reveal the exact geometry of how they were attached. Therefore, a group of four closely packed spores is found only immediately after meiosis, and a plant that produces such a group must be a diploid sporophyte—and so the Rhynie fossils must have been sporophytes. Gametophytes of the Rhyniophyta were also found; they, too, were branched, and depressions at the apices of the branches contained archegonia and antheridia.

Although they were apparently ancestral to the other tracheophyte phyla, the rhyniophytes themselves are long gone. None of their fossils appear anywhere after the Devonian period.

Early tracheophytes added new features

A new phylum of tracheophytes—the Lycophyta (club mosses)—also appeared in the Silurian period. Another—the Pteridophyta (ferns and fern allies)—appeared during the Devonian period. These two groups arose from rhyniophyte-like ancestors. These new groups featured specializations not found in the rhyniophytes, including one or more of the following: true roots, true leaves, and a differentiation between two types of spores.

THE ORIGIN OF ROOTS. The rhyniophytes had only rhizoids arising from a rhizome with which to gather water and minerals. How, then, did subsequent groups of tracheophytes come to have the complex roots we see today?

It is probable that roots had their evolutionary origins as a branch, either of a rhizome or of the aboveground portion of a stem. That branch presumably penetrated the soil and branched further. The underground portion could anchor the plant firmly, and even in this primitive condition it could absorb water and minerals. The discovery of fossil plants from the Devonian period, all having horizontal stems (rhizomes) with both underground and aerial branches, supported this hypothesis.

Underground and aboveground branches, growing in sharply different environments, were subjected to very different selection pressures during the succeeding millions of years. Thus the two parts of the plant axis—the aboveground shoot system and the underground root system—diverged in structure and evolved distinct internal and external anatomies. In spite of these differences, scientists believe that the root and shoot systems of tracheophytes are homologous—that they were once part of the same organ.

THE ORIGIN OF TRUE LEAVES. Thus far we have used the term "leaf" rather loosely. We spoke of "leafy" mosses and commented on the absence of "true leaves" in rhyniophytes. In the strictest sense, a **leaf** is a flattened photosynthetic structure emerging laterally from a main axis or stem and possessing true vascular tissue. Using this precise definition as we take a closer look at true leaves in the tracheophytes, we see that there are two different types of leaves, very likely of different evolutionary origins.

The first leaf type, the **microphyll**, is usually small and only rarely has more than a single vascular strand, at least in plants alive today. Plants in the phylum Lycophyta (club mosses), of which only a few genera survive, have such simple leaves. The evolutionary origin of microphylls is thought by some biologists to be sterile sporangia (Figure 29.13*a*). The principal characteristic of this type of leaf is that its vascular

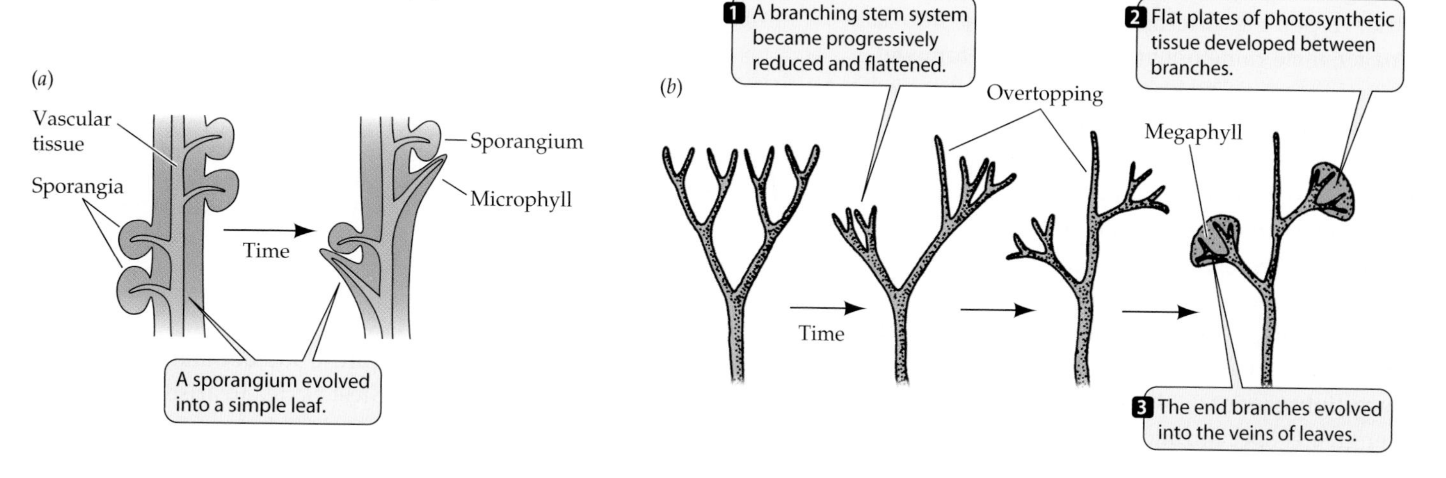
(a)
Vascular tissue
Sporangia
Time
Sporangium
Microphyll
A sporangium evolved into a simple leaf.
(b)
1 A branching stem system became progressively reduced and flattened.
Overtopping
2 Flat plates of photosynthetic tissue developed between branches.
Megaphyll
Time
3 The end branches evolved into the veins of leaves.

29.13 The Evolution of Leaves (*a*) Microphylls are thought to have evolved from sterile sporangia. (*b*) The megaphylls of pteridophytes and seed plants may have arisen as photosynthetic tissue developed between branch pairs that were "left behind" as dominant branches overtopped them.

strand departs from the vascular system of the stem in such a way that the structure of the stem's vascular system is scarcely disturbed. This was true even in the fossil lycopod trees of the Carboniferous period, many of which had leaves many centimeters long.

The other leaf type is found in ferns and seed plants. This larger, more complex leaf is called a **megaphyll**. The megaphyll is thought to have arisen from the flattening of a dichotomously branching stem system and the development of *overtopping* (a pattern in which one branch differentiates from and grows beyond the others). This change was followed by the development of photosynthetic tissue between the members of overtopped groups of branches (Figure 29.13*b*). Megaphylls may have evolved more than once, in different phyla of tracheophytes showing overtopping of branches.

HOMOSPORY AND HETEROSPORY. In the most ancient of the present-day tracheophytes, both the gametophyte and the sporophyte are independent and usually photosynthetic. Spores produced by the sporophytes are of a single type, and they develop into a single type of gametophyte that bears both female and male reproductive organs. The female organ is a multicellular archegonium, typically containing a single egg. The male organ is an antheridium, containing many sperm. Such plants, which bear a single type of spore, are said to be **homosporous** (Figure 29.14*a*).

A different system, with two distinct types of spores, evolved somewhat later. Plants of this type are said to be **heterosporous** (Figure 29.14*b*). One type of spore, the **megaspore**, develops into a larger, specifically female gametophyte (a **megagametophyte**) that produces only eggs. The other type, the **microspore**, develops into a smaller, male gametophyte (a **microgametophyte**) that produces only sperm. The sporophyte produces megaspores in small numbers in **megasporangia** on the sporophyte, and microspores in large numbers in **microsporangia**.

The most ancient tracheophytes were all homosporous, but heterospory evidently evolved independently several times in the early descendants of the rhyniophytes. The fact that heterospory evolved repeatedly suggests that it affords selective advantages. Subsequent evolution in the plant kingdom featured ever greater specialization of the heterosporous condition.

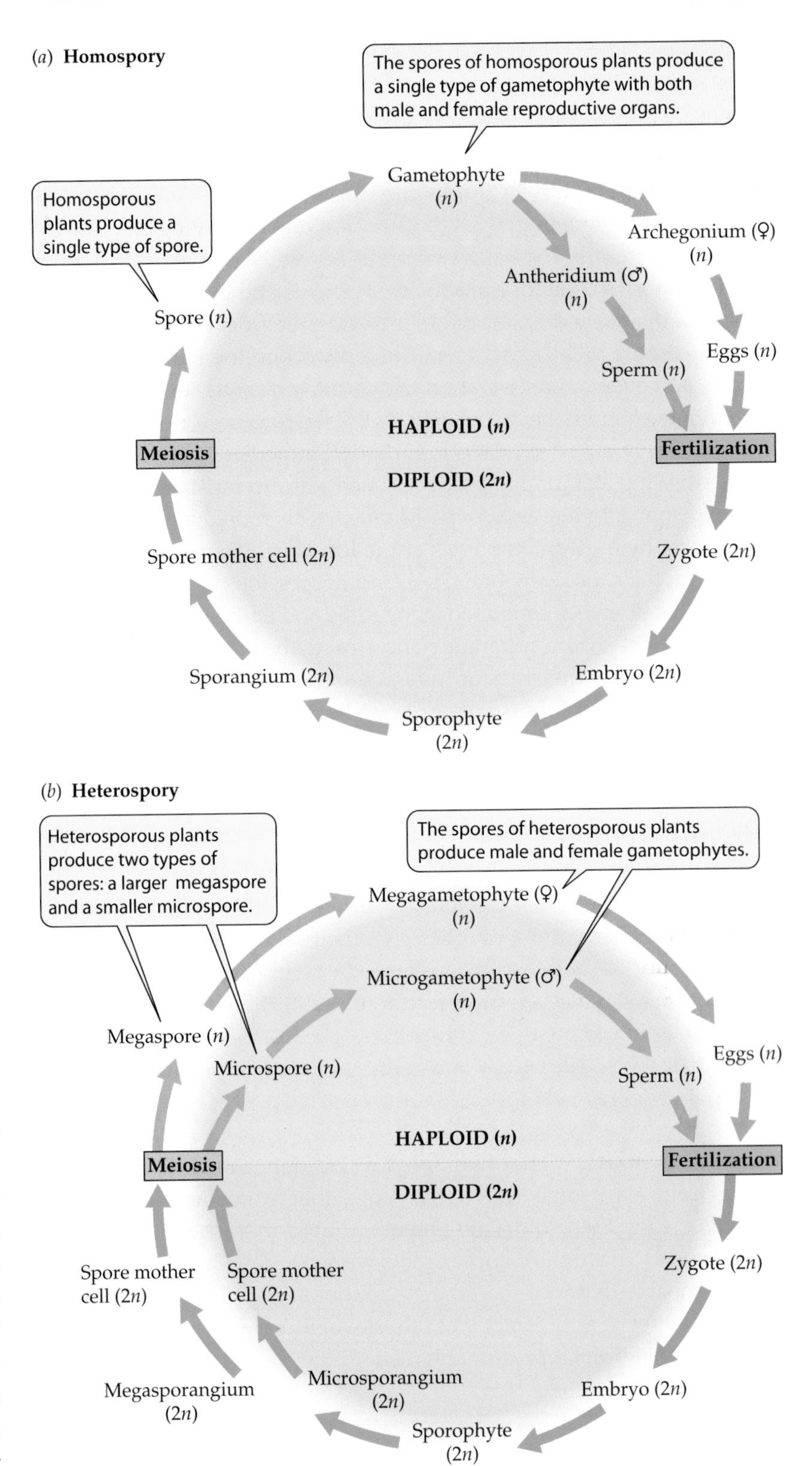

29.14 Homospory and Heterospory (*a*) Homosporous plants bear a single type of spore. Each gametophyte has two types of sex organs, antheridia (male) and archegonia (female). (*b*) Heterosporous plants, which bear two types of spores that develop into distinctly male and female gametophytes, evolved later.

Some tracheophyte clades arose and became extinct in the course of evolution. The earliest clades to arise and survive to this day belong to the nonseed tracheophytes.

The Surviving Nonseed Tracheophytes

The nonseed tracheophytes have a large, independent sporophyte and a small gametophyte that is independent of the sporophyte. The gametophytes of the surviving nonseed tracheophytes are rarely more than 1 or 2 centimeters long and are short-lived, whereas their sporophytes are often highly visible; the sporophyte of a tree fern, for example, may be 15 or 20 meters tall and may live for many years.

The most prominent resting stage in the life cycle of a nonseed tracheophyte is the single-celled spore. This feature makes their life cycle similar to those of the fungi, the green algae, and the nontracheophytes, but not, as we will see in the next chapter, to that of the seed plants. Nonseed tracheophytes must have an aqueous environment for at least one stage of their life cycle because fertilization is accomplished by a motile, flagellated sperm.

The ferns are the most abundant and diverse group of nonseed tracheophytes today, but the club mosses and horsetails were once dominant elements of Earth's vegetation. A fourth group, the whisk ferns, contains only two genera. In this section we'll look at the characteristics of these four groups and at some of the evolutionary advances that appeared in them.

The club mosses are sister to the other tracheophytes

Liverworts
Hornworts
Mosses
Club mosses
Horsetails
Whisk ferns
Ferns
Gymnosperms
Flowering plants

The **club mosses** and their relatives (together called **lycopods**, phylum **Lycophyta**) diverged earlier than all other living tracheophytes—that is, the remaining tracheophytes share an ancestor that was not ancestral to the Lycophyta. There are relatively few surviving species of club mosses.

The lycopods have roots that branch dichotomously. The arrangement of vascular tissue in their stems is simpler than in the other tracheophytes. They bear only microphylls, and these simple leaves are arranged spirally on the stem. Growth in club mosses comes entirely from apical cell division, and branching is dichotomous, by a division of the apical cluster of dividing cells.

The sporangia in many club mosses are contained within conelike structures called *strobili* (singular, strobilus; Figure 29.15). A strobilus is a cluster of spore-bearing leaves inserted on an axis tucked into the upper angle between a specialized leaf and the stem. (Such an angle is called an *axil*.) Other club mosses lack strobili and bear their sporangia in the axil between a photosynthetic leaf and the stem. This placement contrasts with the apical sporangia of the rhyniophytes. There are both homosporous species and heterosporous species of club mosses. Although only a minor element of present-day vegetation, the Lycophyta are one of two phyla that appear to have been the dominant vegetation during the Carboniferous period. One type of coal (cannel coal) is formed almost entirely from fossilized spores of the tree lycopod *Lepidodendron*—which gives us an idea of the abundance of this genus in the forests of that time (see Figure 29.11). The other major elements of Carboniferous vegetation were horsetails and ferns.

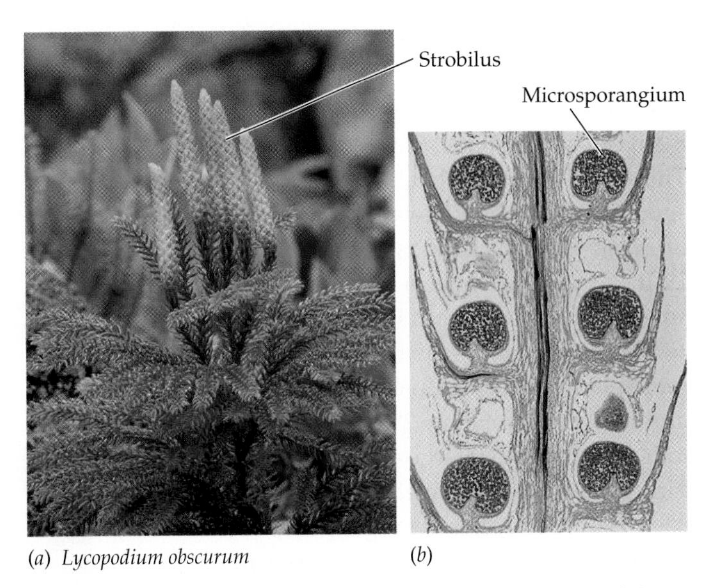

(*a*) *Lycopodium obscurum* (*b*)

29.15 Club Mosses (*a*) Strobili are visible at the tips of this club moss. Club mosses have microphylls arranged spirally on their stems. (*b*) A thin section through a strobilus of a club moss, showing microsporangia.

Horsetails, whisk ferns, and ferns constitute a clade

Liverworts
Hornworts
Mosses
Club mosses
Horsetails
Whisk ferns
Ferns
Gymnosperms
Flowering plants

Once treated as distinct phyla, the horsetails, whisk ferns, and ferns form a clade, the phylum **Pteridophyta** (**pteridophytes**, or "ferns and fern allies"). Within that clade, the whisk ferns and the horsetails are both monophyletic; the ferns are not. However, about 97 percent of all fern species, including those with which you are most likely to be familiar, do belong to a single clade, the *leptosporangiate ferns*. In the pteridophytes—and in all seed plants—there is differentiation (overtopping) between the main axis and side branches.

HORSETAILS GROW AT THE BASES OF STEM SEGMENTS. Like the club mosses, the horsetails are represented by only a few present-day species. All are in a single genus, *Equisetum*. These plants are sometimes called "scouring rushes" because silica deposits found in their cell walls made them useful for cleaning. They have true roots that branch irregularly. Their sporangia curve back toward the stem on the ends of short stalks called *sporangiophores* (Figure 29.16*a*). Horsetails have a large sporophyte and a small gametophyte, both independent.

The small leaves of horsetails are reduced megaphylls and form in distinct whorls (circles) around the stem (Figure 29.16*b*). Growth in horsetails originates to a large extent from discs of dividing cells just above each whorl of leaves, so each segment of the stem grows from its base. Such basal growth is uncommon in plants, although it is found in the grasses, a major group of flowering plants.

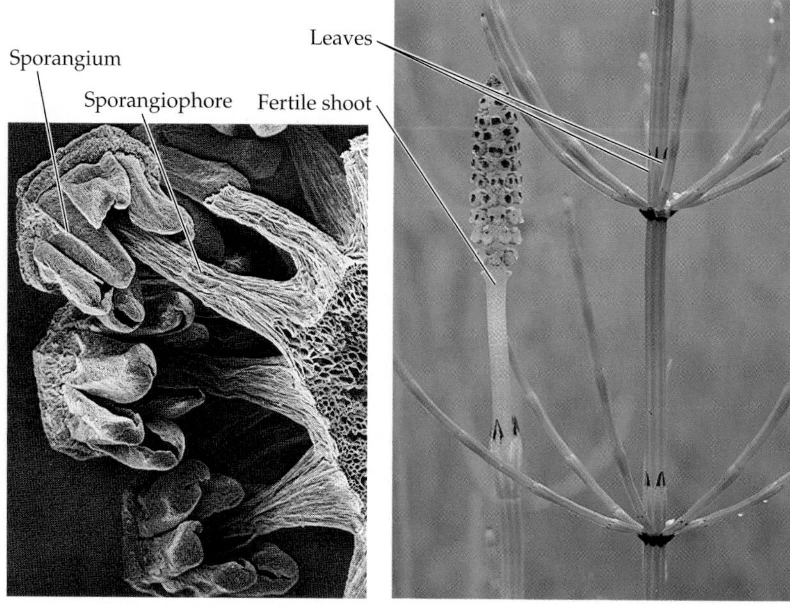

29.16 Horsetails (*a*) Sporangia and sporangiophores of a horsetail. (*b*) Vegetative and fertile shoots of the marsh horsetail. Reduced megaphylls can be seen in whorls on the stem of the vegetative shoot on the right; the fertile shoot on the left is ready to disperse its spores.

PRESENT-DAY WHISK FERNS RESEMBLE THE MOST ANCIENT TRACHEOPHYTES. There once was some disagreement about whether rhyniophytes are entirely extinct. The confusion arose because of the existence today of two genera of rootless, spore-bearing plants, *Psilotum* and *Tmesipteris*, collectively called the whisk ferns. *Psilotum nudum* (Figure 29.17) has only minute scales instead of true leaves, but plants of the genus *Tmesipteris* have flattened photosynthetic organs—reduced megaphylls—with well-developed vascular tissue. Are these two genera the living relics of the rhyniophytes, or do they have more recent origins?

Psilotum and *Tmesipteris* once were thought to be evolutionarily ancient descendants of anatomically simple ancestors. That hypothesis was weakened by an enormous hole in the geological record between the rhyniophytes, which apparently became extinct more than 300 million years ago, and *Psilotum* and *Tmesipteris*, which are modern plants. DNA sequence data finally settled the question in favor of a more modern origin of the whisk ferns from fernlike ancestors. These two genera are a clade of highly specialized plants that evolved fairly recently from anatomically more complex ancestors by loss of complex leaves and true roots. Whisk fern gametophytes live below the surface of the ground and lack chlorophyll. They depend upon fungal partners for their nutrition.

Ferns evolved large, complex leaves

The sporophytes of the ferns, like those of the seed plants, have true roots, stems, and leaves. Their leaves are typically large and have branching vascular strands. Some species have small leaves as a result of evolutionary reduction, but even these small leaves have more than one vascular strand, and are thus megaphylls.

The ferns constitute a group that first appeared during the Devonian period and today consists of about 12,000 species. The ferns are not a monophyletic group, although, as already mentioned, 97 percent of the species—the leptosporangiate ferns—do constitute a monophyletic group. The leptosporangiate ferns differ from the other ferns in having sporangia with walls only one cell thick, borne on a stalk.

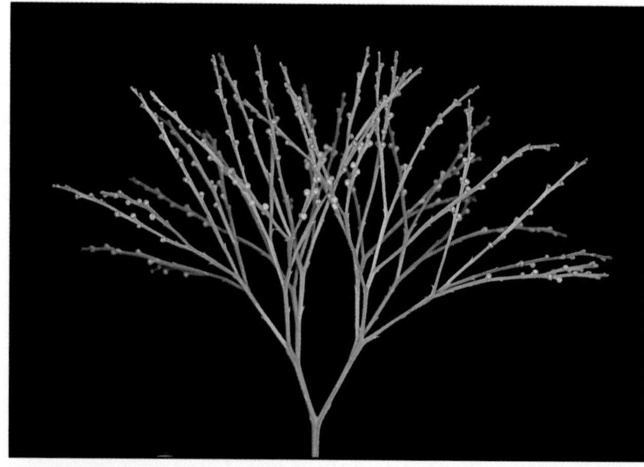

29.17 A Whisk Fern *Psilotum nudum* was once considered by some to be a surviving rhyniophyte and by others to be a fern. It is now included in the phylum Pteridophyta, and it is widespread in the Tropics and Subtropics.

(*a*) *Adiantum pedatum*

(*b*)

(*c*) *Marsilea mutica*

29.18 Fern Fronds Take Many Forms (*a*) The fronds of northern maidenhair fern form a pattern in this photograph. (*b*) The "fiddlehead" (developing frond) of a common forest fern; this structure will unfurl and expand to give rise to a complex adult frond such as those in (*a*). (*c*) The tiny fronds of a water fern.

Ferns are characterized by fronds (large leaves with complex vasculature; Figure 29.18*a*). During its development, the fern frond unfurls from a tightly coiled "fiddlehead" (Figure 29.18*b*). Some fern leaves become climbing organs and may grow to be as much as 30 meters long.

Because they require water for the transport of the male gametes to the female gametes, most ferns inhabit shaded, moist woodlands and swamps. Tree ferns can reach heights of 20 meters. Tree ferns are not as rigid as woody plants, and they have poorly developed root systems. Thus they do not grow in sites exposed directly to strong winds, but rather in ravines or beneath trees in forests. The sporangia of ferns are found on the undersurfaces of the fronds, sometimes covering the whole undersurface and sometimes only at the edges. In most species the sporangia are found in clusters called *sori* (singular, sorus) (Figure 29.19).

Dryopteris intermedia

29.19 Fern Sori Are Clusters of Sporangia Sori, each containing many spore-producing sporangia, have formed on the underside of this frond of the Midwestern fancy fern.

The sporophyte generation dominates the fern life cycle

Inside the sporangia, fern spore mother cells undergo meiosis to form haploid spores. Once shed, the spores travel great distances and eventually germinate to form independent gametophytes. Old World climbing fern, *Lygodium microphyllum*, is currently spreading disastrously through the Florida Everglades, choking off the growth of other plants. This rapid spread is testimony to the effectiveness of windborne spores.

Fern gametophytes have the potential to produce both antheridia and archegonia, although not necessarily at the same time or on the same gametophyte. Sperm swim through water to archegonia—often to those on other gametophytes—where they unite with an egg. The resulting zygote develops into a new sporophyte embryo. The young sporophyte sprouts a root and can thus grow independently of the gametophyte. In the alternating generations of a fern, the gametophyte is small, delicate, and short-lived, but the sporophyte can be very large and can sometimes survive for hundreds of years (Figure 29.20).

Most ferns are homosporous. However, two groups of aquatic ferns, the Marsileaceae and Salviniaceae, are derived from a common ancestor that evolved heterospory. The megaspores and microspores of these plants (which germinate to produce female and male gametophytes, respectively) are produced in different sporangia (megasporangia and microsporangia), and the microspores are always much smaller and greater in number than the megaspores.

29.20 The Life Cycle of a Fern The most conspicuous stage in the fern life cycle is the mature, diploid sporophyte.

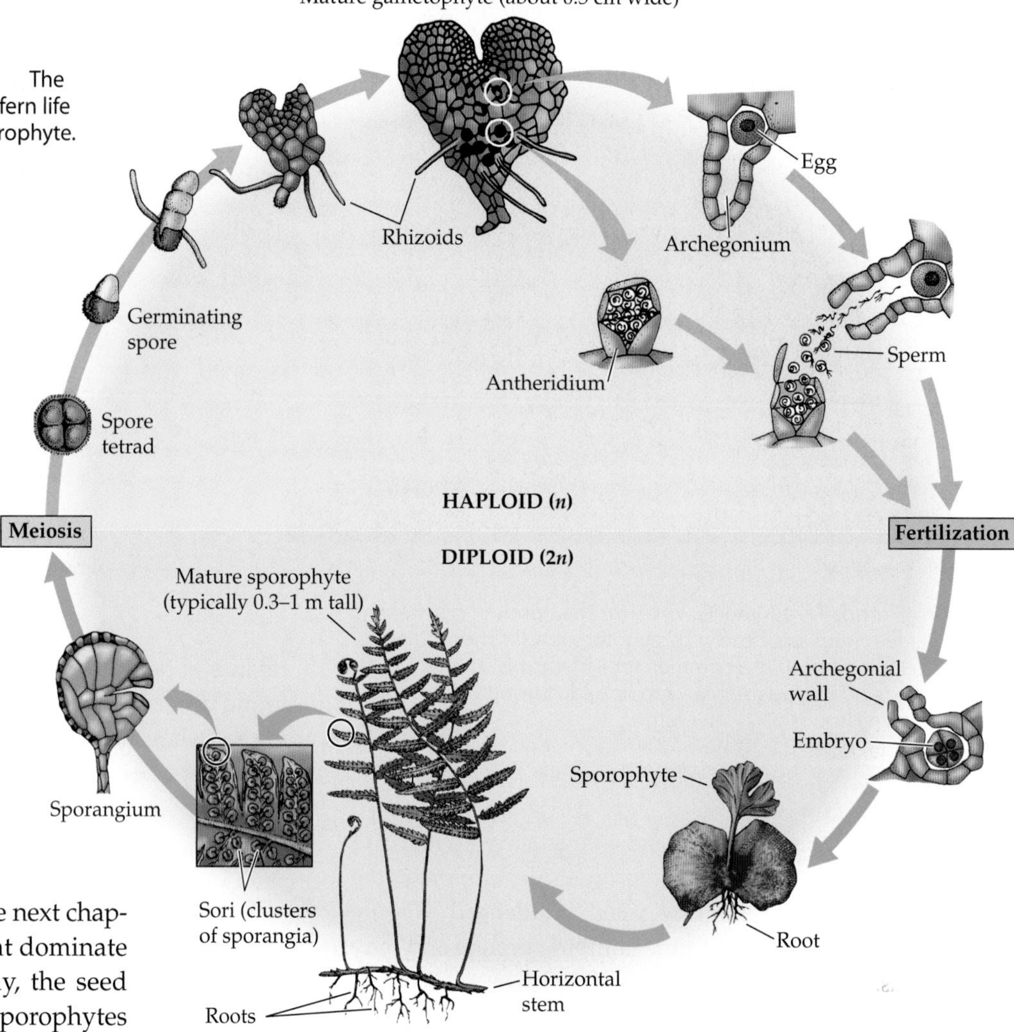

A few genera of ferns produce a tuberous, fleshy gametophyte instead of the characteristic flattened, photosynthetic structure produced by most ferns. Like the gametophytes of whisk ferns, these tuberous gametophytes depend on a mutualistic fungus for nutrition; in some genera, even the sporophyte embryo must become associated with the fungus before extensive development can proceed. In Chapter 31 we will see that there are many other important plant–fungus mutualisms.

All the tracheophytes we have discussed thus far disperse themselves by spores. In the next chapter we will discuss the plants that dominate most of Earth's vegetation today, the seed plants, whose seeds afford new sporophytes protection unavailable to those of the nonseed tracheophytes.

Chapter Summary

The Plant Kingdom

▶ Plants are photosynthetic eukaryotes that develop from embryos protected by parental tissue. Like the green algae, they use chlorophylls *a* and *b* and store carbohydrates as starch. **Review Figure 29.1**

▶ Plant life cycles feature alternation of gametophyte (haploid) and sporophyte (diploid) generations. Both generations include multicellular organisms. **Review Figure 29.2**

▶ There are ten surviving phyla of plants. The three basal phyla are nontracheophytes, and the remaining seven phyla are tracheophytes. **Review Table 29.1**

▶ Plants arose from a common green algal ancestor in the charophyte clade, either a stonewort or a member of the group that includes *Coleochaete*. Descendants of this ancestral charophyte colonized the land.

The Conquest of the Land

▶ The acquisition of a cuticle, gametangia, a protected embryo, protective pigments, thick spore walls with a protective polymer, and a mutualistic association with a fungus are all defining characters of plants, and all are associated with the adaptation of plants to life on land.

▶ Tracheophytes are characterized by possession of a vascular system, consisting of water- and mineral-conducting xylem and nutrient-conducting phloem. Nontracheophytes lack a vascular system. **Review Figure 29.4**

The Nontracheophytes: Liverworts, Hornworts, and Mosses

▶ Nontracheophytes either lack vascular tissues completely or, in the case of certain mosses, have only a rudimentary system of water- and food-conducting cells.

▶ The nontracheophyte sporophyte generation is smaller than the gametophyte generation and depends on the gametophyte for water and nutrition. **Review Figures 29.5, 29.6. See Web/CD Tutorial 29.1**

▶ The nontracheophytes include the liverworts (phylum Hepatophyta), hornworts (phylum Anthocerophyta), and mosses (phylum Bryophyta).

▶ Hornwort sporophytes grow at their basal end.

▶ Hornworts, mosses, and tracheophytes have surface pores (stomata) that allow gas exchange and minimize water loss.

▶ In mosses and tracheophytes, the sporophytes grow by apical cell division.

▶ The hydroids of mosses, through which water may travel, may be ancestral to tracheids, the water-conducting cells of the tracheophytes.

Introducing the Tracheophytes

▶ The tracheophytes have vascular tissue with tracheids and other specialized cells designed to conduct water, minerals, and products of photosynthesis.

▶ Present-day tracheophytes are grouped into seven phyla. The two basal phyla are nonseed tracheophytes, and the rest are seed plants. **Review Figure 29.10**

▶ In tracheophytes, the sporophyte is larger than the gametophyte and independent of the gametophyte generation.

▶ The earliest tracheophytes, known to us only in fossil form, lacked roots and leaves. **Review Figure 29.12**

▶ Roots may have evolved from rhizomes or from branches that penetrated the ground. Microphylls are thought to have evolved from sporangia, and megaphylls may have resulted from the flattening and reduction of an overtopping, branching stem system. **Review Figure 29.13**

▶ Heterospory, the production of distinct female megaspores and male microspores, evolved on several occasions from homosporous ancestors. **Review Figure 29.14**. See Web/CD Activities 29.1 and 29.2

The Surviving Nonseed Tracheophytes

▶ Club mosses (phylum Lycophyta) have microphylls arranged spirally.

▶ Among the pteridophytes (phylum Pteridophyta), horsetails have reduced megaphylls in whorls. Whisk ferns lack roots; one genus has minute scales rather than leaves, and the other has reduced megaphylls with vascular tissue. Leaves with more complex vasculature are characteristic of all other phyla of tracheophytes.

▶ The ferns are not a clade, although 97 percent of fern species do constitute a clade. Ferns have megaphylls with branching vascular strands. **Review Figure 29.20**. See Web/CD Activity 29.3

Self-Quiz

1. Plants differ from photosynthetic protists in that only plants
   a. are photosynthetic.
   b. are multicellular.
   c. possess chloroplasts.
   d. have multicellular embryos protected by the parent.
   e. are eukaryotic.
2. Which statement about alternation of generations in plants is *not* true?
   a. It is heteromorphic.
   b. Meiosis occurs in sporangia.
   c. Gametes are always produced by meiosis.
   d. The zygote is the first cell of the sporophyte generation.
   e. The gametophyte and sporophyte differ genetically.
3. Which statement is *not* evidence for the origin of plants from the green algae?
   a. Some green algae have multicellular sporophytes and multicellular gametophytes.
   b. Both plants and green algae have cellulose in their cell walls.
   c. The two groups have the same photosynthetic and accessory pigments.
   d. Both plants and green algae produce starch as their principal storage carbohydrate.
   e. All green algae produce large, stationary eggs.
4. The nontracheophytes
   a. lack a sporophyte generation.
   b. grow in dense masses, allowing capillary movement of water.
   c. possess xylem and phloem.
   d. possess true leaves.
   e. possess true roots.
5. Which statement is *not* true of the mosses?
   a. The sporophyte is dependent on the gametophyte.
   b. Sperm are produced in archegonia.
   c. There are more species of mosses than of liverworts and hornworts combined.
   d. The sporophyte grows by apical cell division.
   e. Mosses are probably sister to the tracheophytes.
6. Megaphylls
   a. probably evolved only once.
   b. are found in all the tracheophyte phyla.
   c. probably arose from sterile sporangia.
   d. are the characteristic leaves of club mosses.
   e. are the characteristic leaves of horsetails and ferns.
7. The rhyniophytes
   a. possessed vessel elements.
   b. possessed true roots.
   c. possessed sporangia at the tips of stems.
   d. possessed leaves.
   e. lacked branching stems.
8. Club mosses and horsetails
   a. have larger gametophytes than sporophytes.
   b. possess small leaves.
   c. are represented today primarily by trees.
   d. have never been a dominant part of the vegetation.
   e. produce fruits.
9. Which statement about ferns is *not* true?
   a. The sporophyte is larger than the gametophyte.
   b. Most are heterosporous.
   c. The young sporophyte can grow independently of the gametophyte.
   d. The frond is a megaphyll.
   e. The gametophytes produce archegonia and antheridia.
10. The leptosporangiate ferns
    a. are not a monophyletic group.
    b. have sporangia with walls more than one cell thick.
    c. constitute a minority of all ferns.
    d. are pteridophytes.
    e. produce seeds.

For Discussion

1. Mosses and ferns share a common trait that makes water droplets a necessity for sexual reproduction. What is that trait?
2. Are the mosses well adapted to terrestrial life? Justify your answer.
3. Ferns display a dominant sporophyte generation (with large fronds). Describe the major advance in anatomy that enables most ferns to grow much larger than mosses.
4. What features distinguish club mosses from horsetails? What features distinguish these groups from rhyniophytes? From ferns?
5. Why did some botanists once believe that the whisk ferns should be classified together with the rhyniophytes?
6. Contrast microphylls with megaphylls in terms of structure, evolutionary origin, and occurrence among plants.

30 The Evolution of Seed Plants

A violent thunderstorm moves through forested hills and valleys where summer rain has been scarce. A jagged fork of lightning strikes a tree, and it bursts into flame. Soon the flames reach dead and dry underbrush, and fire spreads to the surrounding trees. The fire rages rapidly through the forest, leaving a blackened and smoking landscape behind.

Though devastating, such fires are a natural part of the forest ecosystem. Life returns quickly following a fire in a natural grassland or forest, in part because some plants have adaptations that enable them to live with fire. One example, obvious from its common name, is fireweed. The seeds of fireweed not only survive fires, but are stimulated by high temperatures to break their dormancy and sprout. Another example is the lodgepole pine tree, which covers vast fire-prone areas in the Rocky Mountains and elsewhere. Its cones will not release their seeds unless the heat of a fire causes them to open.

Seeds are remarkable structures. They protect the plant embryo within them from environmental extremes through what may be a very long resting period. This and other properties have contributed to making seed plants the predominant plants on Earth. All of today's forests are dominated by seed plants.

In this chapter we will describe the defining characteristics of the seed plants as a group. We will survey the diversity of seed plants and describe the flowers and fruits that are characteristic of their most dominant group, the flowering plants. Finally, we will consider some of the unsolved problems in seed plant evolution.

A Forest Ablaze Fires like this one in a northern Arizona forest can pose dangers to human life and property. But they play an essential role in the life cycles of many fire-adapted seed plants.

The Seed Plants

The most recent group to appear in the evolution of the tracheophytes is the **seed plants**. The earliest fossil evidence of seed plants is found in Devonian rocks. The earliest seed plants combined characteristics of rhyniophytes and heterosporous ferns, but they had tracheids of the type found in modern seed plants. They also differed from the plants around them by having extensively thickened woody stems, which resulted from the proliferation of xylem. This type of growth in the diameter of stems and roots is called *secondary growth*. By the Carboniferous period, new lines of seed plants had evolved, including various seed ferns, which possessed fernlike foliage but had seeds attached to their leaves.

Two clades of early seed plants are known only as fossils. These clades are basal to the surviving seed plants, which fall into two groups, the **gymnosperms** (such as pines and cycads) and the **angiosperms** (flowering plants). There are four living phyla of gymnosperms and one of angiosperms (Figure 30.1). The phylogenetic relationships among these five clades have not yet been resolved. All living gymnosperms and many angiosperms show secondary growth. The life cycles of all seed plants share major features, as we are about to see.

30.1 The Phyla of Living Seed Plants There are four phyla of gymnosperms and one of angiosperms. Their exact evolutionary relationship is still uncertain.

Seed plants are heterosporous and have tiny gametophytes

In seed plants, the gametophyte generation is reduced even further than it is in the ferns (Figure 30.2). The haploid gametophyte develops partly or entirely while attached to and nutritionally dependent on the diploid sporophyte.

Among the seed plants, only the earliest types of gymnosperms (and their few survivors) had swimming sperm. All other seed plants have evolved other means of bringing eggs and sperm together. The culmination of this striking evolutionary trend was independence from the liquid water that earlier plants needed for sexual reproduction.

Seed plants are heterosporous (see Figure 29.14*b*). They form separate megasporangia and microsporangia on structures that are grouped on short axes, such as the cones and strobili of conifers and the flowers of angiosperms.

As in other plants, the spores of seed plants are produced by meiosis within the sporangia, but in seed plants, the megaspores are not shed. Instead, they develop into female gameto-

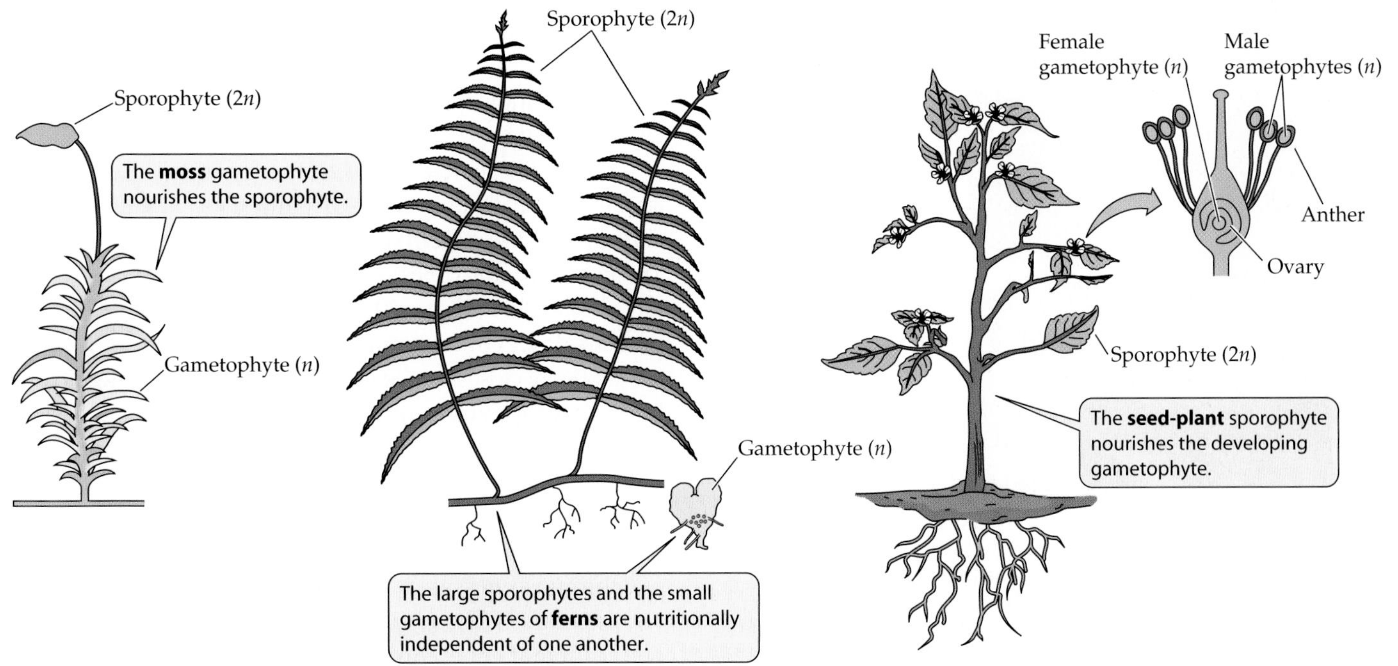

30.2 The Relationship between Sporophyte and Gametophyte Has Evolved In the course of plant evolution, the gametophyte has been reduced and the sporophyte has become more prominent.

phytes within the megasporangia. These megagametophytes are dependent on the sporophyte for food and water.

In most seed plant species, only one of the meiotic products in a megasporangium survives. The surviving haploid nucleus divides mitotically, and the resulting cells divide again to produce a multicellular female gametophyte. This megagametophyte is retained within the megasporangium, where it matures. The megagametophyte, in turn, houses the early development of the next sporophyte generation following fertilization of the egg. The megasporangium is surrounded by sterile sporophytic structures that form a protective **integument**.

Within the microsporangium, the meiotic products are microspores, which divide mitotically within the spore wall one or a few times to form a male gametophyte called a **pollen grain**. Pollen grains are released from the microsporangium to be distributed by wind, an insect, a bird, or a plant breeder (Figure 30.3). A pollen grain that reaches the appropriate surface of a sporophyte of the same species develops further. It produces a slender **pollen tube** that elongates and digests its way through the sporophytic tissue toward the female gametophyte.

When the tip of the pollen tube reaches the female gametophyte, sperm are released from the tube, and fertilization occurs. The resulting diploid zygote divides repeatedly, forming a young sporophyte that develops to an embryonic stage at which growth is temporarily suspended (often referred to as a *dormant* stage). The end product at this stage is a multicellular **seed**.

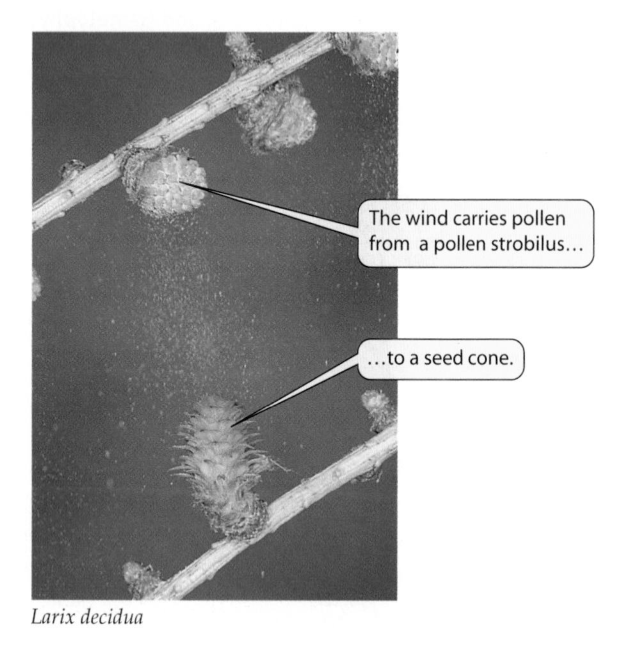

30.3 Pollen Grains Pollen grains are the male gametophytes of seed plants. Conifers have strobili, which produce and release pollen. Their pollen is dispersed by the wind to cones, which contain female gametophytes.

The seed is a complex package

A seed may contain tissues from three generations. The seed coat develops from tissues of the diploid sporophyte parent (the integument). Within the megasporangium is the haploid female gametophytic tissue from the next generation, which contains a supply of nutrients for the developing embryo. (This tissue is fairly extensive in most gymnosperm seeds. In angiosperm seeds its place is taken by a tissue called endosperm, which we will describe below.) In the center of the seed is the third generation, the embryo of the new diploid sporophyte.

The seed of a gymnosperm or an angiosperm is a well-protected resting stage. The seeds of some species may remain *viable* (capable of growth and development) for many years, germinating when conditions are favorable for the growth of the sporophyte. In contrast, the embryos of nonseed plants develop directly into sporophytes, which either survive or die, depending on environmental conditions; there is no dormant stage in the life cycle.

During the dormant stage, the seed coat protects the embryo from excessive drying and may also protect it against potential predators that would otherwise eat the embryo and its nutrient reserves. Many seeds have structural adaptations that promote their dispersal by wind or, more often, by animals. When the young sporophyte resumes growth, it draws on the food reserves in the seed. The possession of seeds is a major reason for the enormous evolutionary success of the seed plants, which are the dominant life forms of Earth's modern terrestrial flora in most areas.

The Gymnosperms: Naked Seeds

The extant gymnosperms are a clade of seed plants that do not form flowers. Although there are probably fewer than 750 species of living gymnosperms, these plants are second only to the angiosperms in their dominance of the terrestrial environment.

Cycads
Ginkgos
Conifers
Gnetophytes
Angiosperms

There are four clades of living gymnosperms today. The **cycads** (phylum **Cycadophyta**) are palmlike plants of the Tropics and Subtropics, growing as tall as 20 meters (Figure 30.4*a*). Of the present-day gymnosperms, the cycads are probably closest to the earliest seed plants. **Ginkgos** (phylum **Ginkgophyta**), which were common during the Mesozoic era, are represented today by a single genus and species, *Ginkgo biloba*, the maidenhair tree (Figure 30.4*b*). There are both male (microsporangiate) and female (megasporangiate) maidenhair trees. The difference is determined by X and Y sex chromosomes, as in humans; few other plants have sex chromosomes. The phylum **Gnetophyta** consists of three very different genera that share certain characteristics with

(*a*) *Cycas* sp.

(*b*) *Ginkgo biloba*

(*c*) *Welwitschia mirabilis*

(*d*) *Sequoiadendron giganteum*

30.4 Diversity among the Gymnosperms (*a*) Many cycads, such as this palmlike tree, have growth forms that resemble both ferns and palms. (*b*) The characteristic fleshy seed coat and broad leaves of the maidenhair tree. (*c*) A gnetophyte growing in the Namib Desert of Africa. Two huge, straplike leaves grow throughout the life of the plant, breaking and splitting as they grow. (*d*) Conifers, like this giant sequoia growing in Sequoia National Park, California, dominate many modern forests.

the angiosperms. One of the gnetophytes is *Welwitschia* (Figure 30.4*c*), a long-lived desert plant with just two straplike leaves that sprawl on the sand and can grow as long as 3 meters. By far the most abundant of the gymnosperms are the **conifers** (phylum **Pinophyta**), cone-bearing plants such as pines and redwoods (Figure 30.4*d*).

All living gymnosperms except the Gnetophyta have only tracheids as water-conducting and support cells in their xylem; they lack the more specialized vessels and fibers found alongside tracheids in the angiosperms. Although this difference may make the gymnosperm water transport and support system seem less efficient than that of the angiosperms, it serves some of the largest trees known. The coast redwoods of California are the tallest gymnosperms; the largest are well over 100 m tall. Secondary xylem—wood—produced by gymnosperms is the principal resource of the timber industry.

During the Permian period, the conifers and cycads flourished. Gymnosperm forests changed over time as the gymnosperm groups evolved. Gymnosperms dominated the

Mesozoic era, during which the continents drifted apart and dinosaurs strode the Earth. They were the principal trees in all forests until less than 100 million years ago, and they still dominate many present-day forests. Let's look at the most abundant gymnosperms, the conifers, in more detail.

Conifers have cones but no motile cells

The great Douglas fir and cedar forests of the northwestern United States and the massive boreal forests of pine, fir, and spruce found in northern regions of Eurasia and North America, as well as on the upper slopes of mountain ranges everywhere, rank among the great vegetation formations of the world. All these trees belong to one phylum of gymnosperms, Pinophyta—the conifers, or cone-bearers. A **cone** is a short axis (a modified *stem*) bearing a tight cluster of scales, which are reduced *branches* specialized for reproduction (Figure 30.5*a*). A **strobilus** is a conelike cluster of scales that are modified *leaves* inserted on an axis (Figure 30.5*b*) . Megaspores are produced in seed cones, and microspores are produced in pollen strobili. Seed cones are much larger than pollen strobili.

We will use the life cycle of a pine to illustrate reproduction in gymnosperms (Figure 30.6). The production of male gametophytes in the form of pollen grains frees the plant completely from its dependence on liquid water for fertilization. Instead of water, wind assists conifer pollen grains in their first stage of travel from the strobilus to the female gametophyte inside the seed cone (see Figure 30.3). The pollen tube provides the sperm with the means for the last stage of travel by elongating and digesting its way through maternal sporophytic tissue. When it reaches the female gametophyte, it releases two sperm, one of which degenerates after the other unites with an egg.

The megasporangium, in which the female gametophyte will form, is enclosed in a layer of sporophytic tissue—the integument—that will eventually develop into the seed coat. The integument, the megasporangium inside it, and the tissue attaching it to the maternal sporophyte constitute the **ovule**. The pollen grain enters through a small opening in the integument at the tip of the ovule, the **micropyle**.

Gymnosperms derive their name (which means "naked-seeded") from the fact that their ovules and seeds are not protected by ovary or fruit tissue. Most conifer ovules (which, upon fertilization, develop into seeds) are borne exposed on the upper surfaces of the modified branches that form the scales of the cone. Each cone scale lies in the angle between a modified leaf and the axis. The only protection of the ovules comes from the scales, which are tightly pressed against each other within the cone. As we have seen, some pines, such as the lodgepole pine, have such tightly closed seed cones that only fire suffices to split them open and release the seeds.

About half of the conifer species have soft, fleshy fruitlike tissues associated with their seeds; examples are the fleshy cones or "berries" of juniper and yew. Animals may eat these tissues and then disperse the seeds in their feces, often carrying them considerable distances from the parent plant. These tissues, however, are not true fruits, which are characteristic of the plant phylum that is dominant today: the angiosperms.

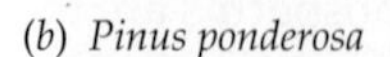

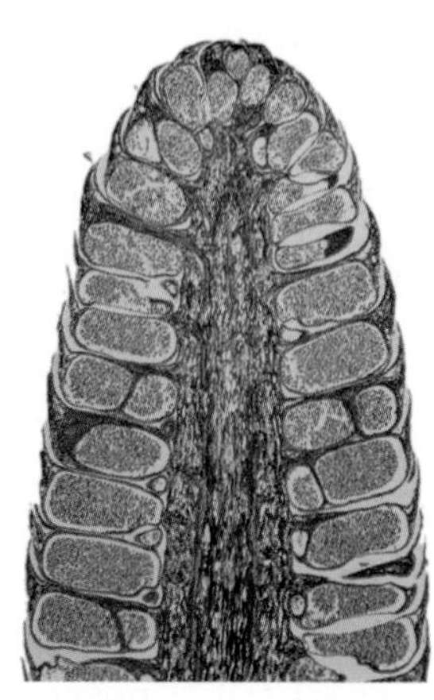

30.5 Cones and Strobili (*a*) The scales of seed cones are modified branches. (*b*) The spore-bearing structures in pollen strobili are modified leaves.

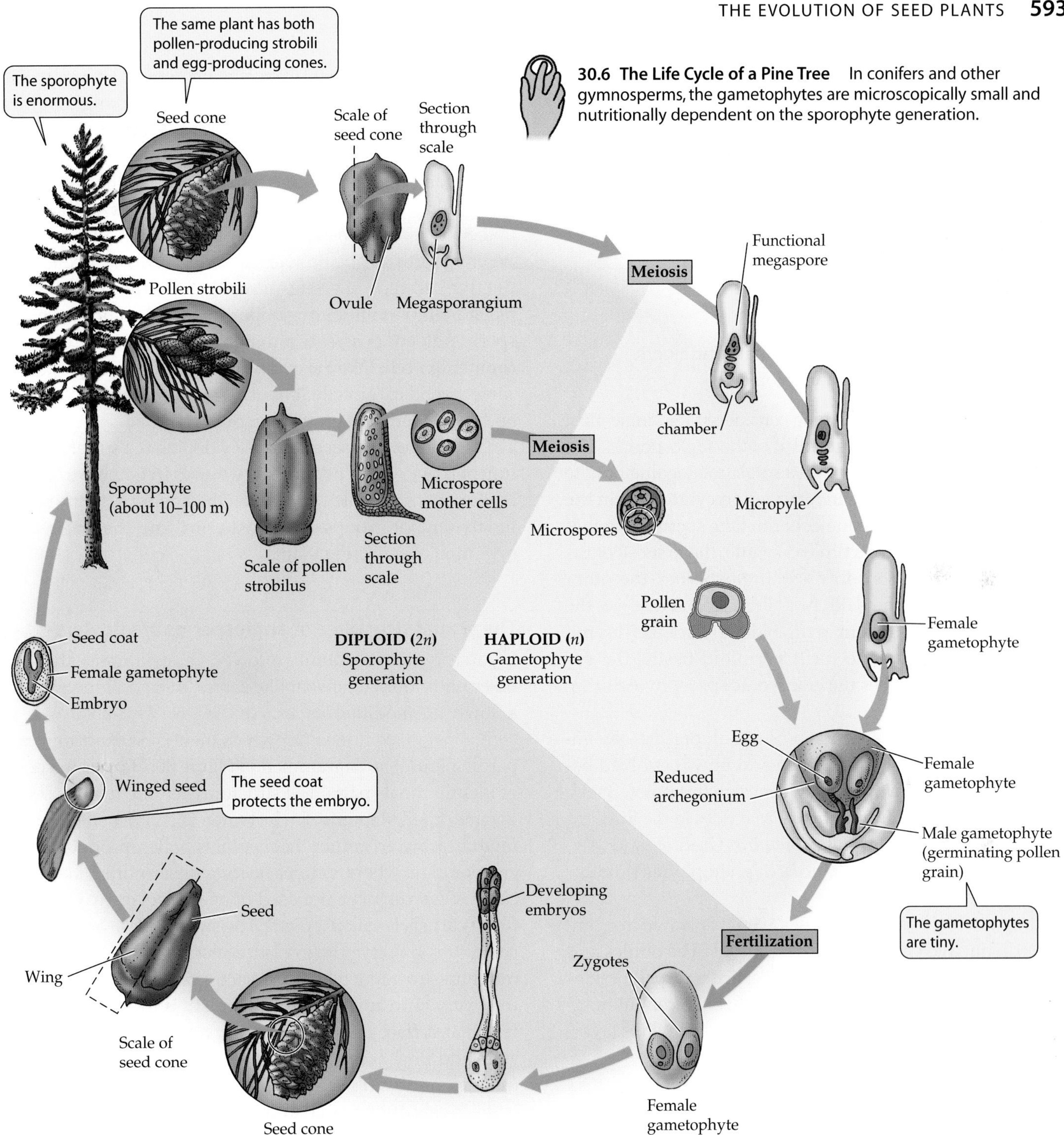

30.6 The Life Cycle of a Pine Tree In conifers and other gymnosperms, the gametophytes are microscopically small and nutritionally dependent on the sporophyte generation.

The Angiosperms: Flowering Plants

The phylum **Angiospermae** consists of the **flowering plants**, also commonly known as the **angiosperms**. This highly diverse phylum includes more than 257,000 species. The oldest evidence of angiosperms dates back to the early Cretaceous period, about 140 million years ago. The angiosperms radiated explosively and, over a period of only about 60 million years, became the dominant plant life of the planet. In later chapters, when we mention "plants," we are generally referring to the angiosperms.

Cycads
Ginkgos
Conifers
Gnetophytes
Angiosperms

The female gametophyte of the angiosperms, consisting of just seven cells, is even more reduced than that of the gymnosperms. Thus, the angiosperms represent the current extreme of an evolutionary trend that runs throughout the tracheophytes: The sporophyte generation becomes larger and more independent of the gametophyte, while the gametophyte generation becomes smaller and more dependent on the sporophyte.

A number of synapomorphies (shared derived traits) characterize the angiosperms:

- They have double fertilization.
- They produce a triploid nutritive tissue called the endosperm.
- Their ovules and seeds are enclosed in a carpel.
- They have flowers.
- They produce fruit.
- Their xylem contains vessel elements and fibers.
- Their phloem contains companion cells.

Double fertilization was long considered the single most reliable distinguishing characteristic of the angiosperms. Two male gametes, contained within a single microgametophyte (pollen grain), participate in fertilization events within the megagametophyte of an angiosperm. One sperm combines with the egg to produce a diploid zygote, the first cell of the sporophyte generation. In most angiosperms, the other sperm nucleus combines with two other haploid nuclei of the female gametophyte to form a triploid ($3n$) nucleus. This nucleus, in turn, divides to form a triploid tissue, the **endosperm**, that nourishes the embryonic sporophyte during its early development.

Double fertilization occurs in nearly all present-day angiosperms. We are not sure when and how it evolved because there is no known fossil evidence on this point. It may have first resulted in two embryos, as it does in the three existing genera of Gnetophyta: *Ephedra*, *Gnetum*, and *Welwitschia*. Both of the fertilizations in gnetophytes produce diploid products.

The name *angiosperm* ("enclosed seed") is drawn from another distinctive character of these plants: The ovules and seeds are enclosed in a modified leaf called a **carpel**. Besides protecting the ovules and seeds, the carpel often interacts with incoming pollen to prevent self-pollination, thus favoring cross-pollination and increasing genetic diversity. Of course, the most evident diagnostic feature of angiosperms is that they have **flowers**. Production of a **fruit** is another of their unique characteristics.

Most angiosperms are also distinguished by the possession of specialized water-transporting cells called **vessel elements** in their xylem, but these cells are also found, in anatomically different form, in gnetophytes and a few ferns. A second distinctive cell type in angiosperm xylem is the **fiber**, which plays an important role in supporting the plant body. Angiosperm phloem possesses another unique cell type, called a **companion cell**. Like the gymnosperms, woody angiosperms show secondary growth, producing secondary xylem and secondary phloem and growing in diameter.

In the following sections we'll examine the structure and function of flowers, evolutionary trends in flower structure, the functions of pollen and fruits, the angiosperm life cycle, the two major groups of angiosperms, and the origin and evolution of flowering plants.

The sexual structures of angiosperms are flowers

If you examine any familiar flower, you will notice that the outer parts look somewhat like leaves. In fact, all the parts of a flower *are* modified leaves.

A generalized flower (for which there is no exact counterpart in nature) is diagrammed in Figure 30.7 for the purpose of identifying its parts. The structures bearing microsporangia are called **stamens**. Each stamen is composed of a **filament** bearing an **anther** that contains pollen-producing microsporangia. The structures bearing megasporangia are the carpels. A structure composed of one carpel or two or more fused carpels is called a **pistil**. The swollen base of the pistil, containing one or more ovules (each containing a megasporangium surrounded by its protective integument), is called the **ovary**. The apical stalk of the pistil is the **style**, and the terminal surface that receives pollen grains is the **stigma**.

In addition, a flower often has several specialized sterile (non-spore-bearing) leaves. The inner ones are called **petals** (collectively, the **corolla**) and the outer ones **sepals** (collectively, the **calyx**). The corolla and calyx, which can be quite showy, often play roles in attracting animal pollinators to the flower. The calyx more commonly protects the immature flower in bud. From base to apex, the sepals, petals, sta-

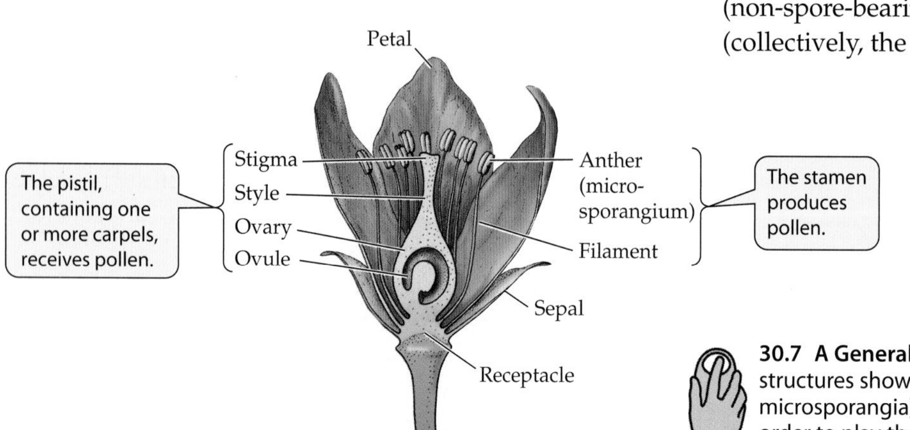

30.7 A Generalized Flower Not all flowers possess all the structures shown here, but they must possess a stamen (bearing microsporangia), a pistil (containing megasporangia), or both in order to play their role in reproduction. Flowers that have both, as this one does, are referred to as perfect.

(*a*) *Daucus carota*

Disk flowers (many)

Ray flowers

(*b*) *Echinacea purpurea*

(*c*) *Pennisetum setaceum*

30.8 Inflorescences (*a*) The inflorescence of Queen Anne's lace is a compound umbel. Each umbel bears flowers on stalks that arise from a common center. (*b*) Coneflowers are members of the aster family; their inflorescence is a head. In a head, each of the long, petal-like structures is a ray flower; the central portion of the head consists of dozens to hundreds of disc flowers. (*c*) Grasses such as this fountain grass have inflorescences called spikes.

mens, and carpels (which are referred to as the floral organs; see Figure 19.12) are usually positioned in circular arrangements or whorls and attached to a central stalk called the **receptacle**.

The generalized flower shown in Figure 30.7 has both megasporangia and microsporangia; such flowers are referred to as **perfect**. Many angiosperms produce two types of flowers, one with only megasporangia and the other with only microsporangia. Consequently, either the stamens or the carpels are nonfunctional or absent in a given flower, and the flower is referred to as **imperfect**.

Species such as corn or birch, in which both megasporangiate and microsporangiate flowers occur on the same plant, are said to be **monoecious** (meaning "one-housed"—but, it must be added, one house with separate rooms). Complete separation is the rule in some other angiosperm species, such as willows and date palms; in these species, a given plant produces either flowers with stamens or flowers with pistils, but never both. Such species are said to be **dioecious** ("two-housed").

Flowers come in an astonishing variety of forms, as you will realize if you think of some of the flowers you recognize. The generalized flower shown in Figure 30.7 has distinct petals and sepals arranged in distinct whorls. In nature, however, petals and sepals sometimes are indistinguishable. Such appendages are called **tepals**. In other flowers, petals, sepals, or tepals are completely absent.

Flowers may be single, or they may be grouped together to form an **inflorescence**. Different families of flowering plants have their own, characteristic types of inflorescences, such as the compound umbels of the carrot family, the heads of the aster family, and the spikes of many grasses (Figure 30.8).

Flower structure has evolved over time

The flowers of the most basal lineages of angiosperms have a large and variable number of tepals (or sepals and petals), carpels, and stamens (Figure 30.9*a*). Evolutionary change within the angiosperms has included some striking modifications of this early condition: reductions in the number of each type of floral organ to a fixed number, differentiation of petals from sepals, and changes in symmetry from radial (as in a lily or magnolia) to bilateral (as in a sweet pea or orchid), often accompanied by an extensive fusion of parts (Figure 30.9*b*).

According to one theory, the first carpels to evolve were modified leaves, folded but incompletely closed, and thus differing from the scales of the gymnosperms. In the groups of angiosperms that evolved later, the carpels fused and became progressively more buried in receptacle tissue (Figure 30.10*a*). In the flowers of the most recent groups, the other flower parts are attached at the very top of the ovary, rather than at the bottom as in Figure 30.7. The stamens of the most ancient flowers may have appeared leaflike (Figure 30.10*b*), little resembling those of the generalized flower in Figure 30.7.

Why do so many flowers have pistils with long styles and anthers with long filaments? Natural selection has favored length in both of these structures, probably because length increases the likelihood of successful pollination. Long filaments may bring the anthers into contact with insect bodies, or they may place the anthers in a better position to catch the wind. Similar arguments apply to long styles.

(a) *Magnolia grandifolia*

(b) *Cypripedium reginae*

30.9 Flower Form and Evolution (*a*) A magnolia flower shows the major features of early flowers: It is radially symmetrical, and the individual tepals, carpels, and stamens are separate, numerous, and attached at their bases. (*b*) Orchids, like this ladyslipper, have a bilaterally symmetrical structure that evolved much later. One of the three petals evolved into the complex lower "lip." Inside, the stamen and pistil are fused. There are two anthers in this species, although most orchids have only a single anther.

(a) **Carpel evolution**

1 According to one theory, the carpel began as a modified leaf with sporangia.
2 In the course of evolution, leaf edges curled inward and finally fused.
3 At the end of the sequence, three carpels have fused to form a three-chambered ovary.
Sporangia
Fused carpel
Modified leaflike structure
Cross section

(b) **Stamen evolution**

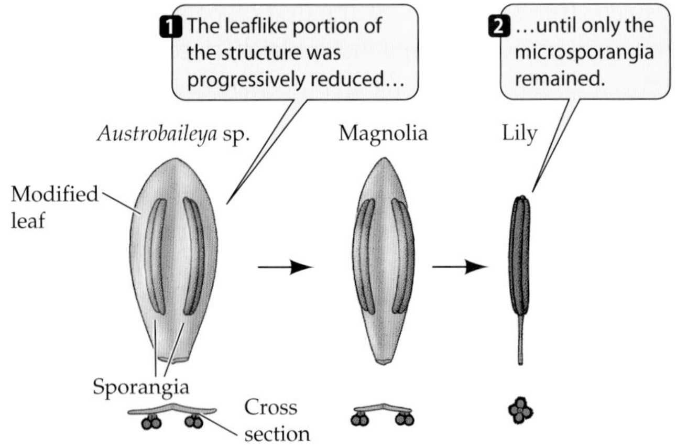

1 The leaflike portion of the structure was progressively reduced...
2 ...until only the microsporangia remained.
Austrobaileya sp.
Magnolia
Lily
Modified leaf
Sporangia
Cross section

30.10 Carpels and Stamens Evolved from Leaflike Structures (*a*) Possible stages in the evolution of a carpel from a more leaflike structure. (*b*) The stamens of three modern plants show the various stages in the evolution of that organ. It is *not* implied that these species evolved one from another; they simply illustrate the structures.

A long style may serve another purpose as well. If several pollen grains land on one stigma, a pollen tube will start growing from each grain down the style toward the ovary. If there are more pollen grains than ovules, there is a "race" to fertilize the ovules. The race down the style can be viewed as "mate selection" by the plant bearing the style.

Angiosperms have coevolved with animals

Pollen has played another crucial role in the evolution of the angiosperms. Whereas many gymnosperms are wind-pollinated, most angiosperms are animal-pollinated. Animals visit flowers to obtain nectar or pollen, and in the process often carry pollen from one flower to another, or from one plant to another. Thus, in its quest for food, the animal contributes to the genetic diversity of the plant population. Insects, especially bees, are among the most important pollinators; birds and some species of bats also play major roles as pollinators.

For more than 130 million years, angiosperms and their animal pollinators have coevolved in the terrestrial environment. The animals have affected the evolution of the plants, and the plants have affected the evolution of the animals. Flower structure has become incredibly diverse under these selection pressures.

Some of the products of coevolution are highly specific; for example, some yucca species are pollinated by only one species of moth. Pollination by just one or a few animal species provides a plant species with a reliable mechanism for transferring pollen from one of its members to another.

Most plant–pollinator interactions are much less specific; that is, many different animal species pollinate the same plant species, and the same animal species pollinate many different plant species. However, even these less specific interactions have developed some specialization. Bird-pollinated flowers are often red and odorless. Many insect-pollinated flowers have characteristic odors, and bee-pollinated flowers may have conspicuous markings, or *nectar guides*, that are evident only in the ultraviolet region of the spectrum, where bees have better vision than in the red region. Coevolution and other aspects of plant–animal interactions are covered in more detail in Chapter 55.

The angiosperm life cycle features double fertilization

The life cycle of the angiosperms is summarized in Figure 30.11. The angiosperm life cycle will be considered in detail in Chapter 39, but let's look at it briefly here and compare it with the conifer life cycle in Figure 30.6.

Like all seed plants, angiosperms are heterosporous. The ovules are contained within carpels, rather than being exposed on the surfaces of scales, as in most gymnosperms. The male gametophytes, as in the gymnosperms, are pollen grains.

The ovule develops into a seed containing the products of the double fertilization that characterizes angiosperms: a diploid zygote and a triploid endosperm. The endosperm

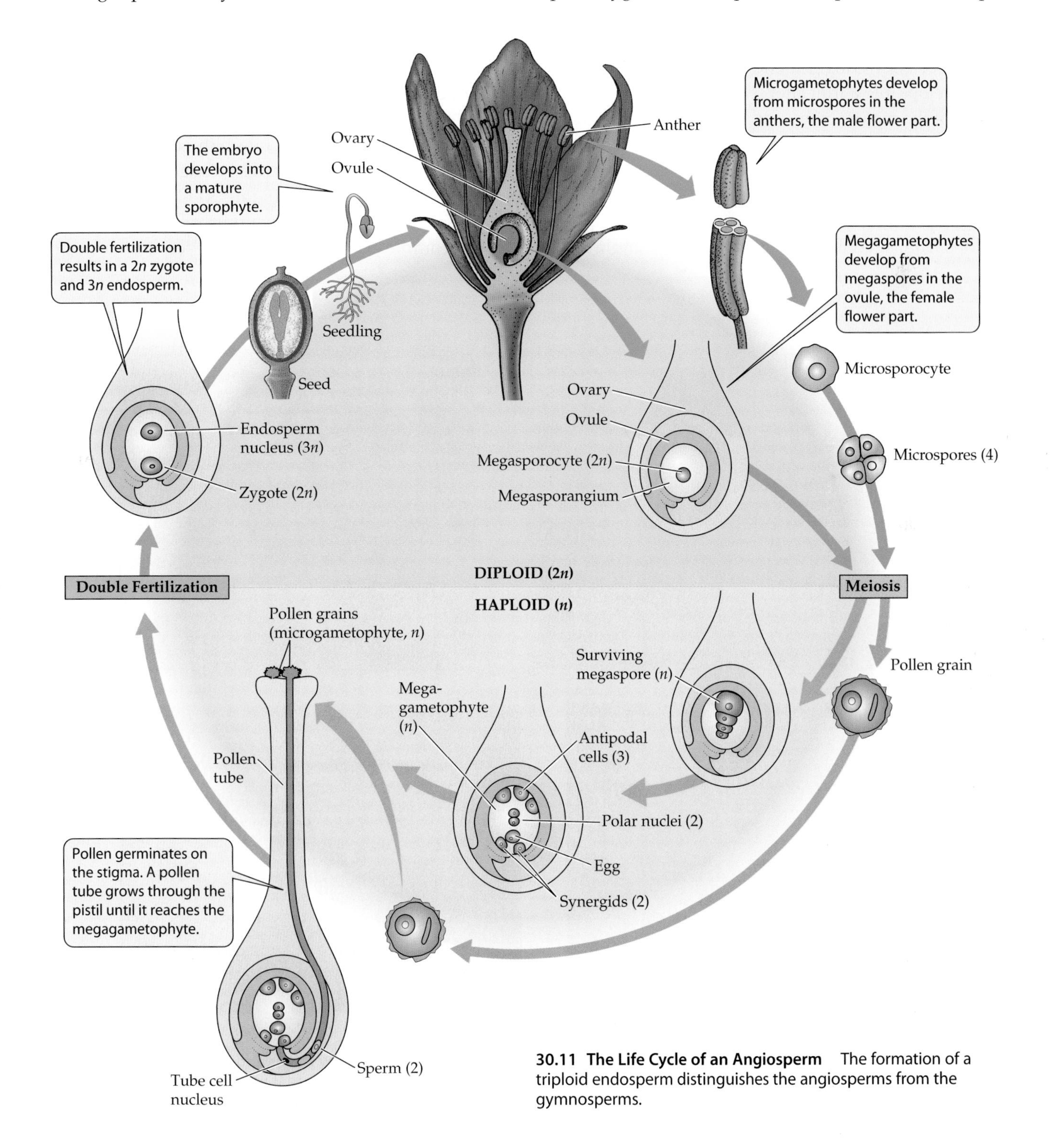

30.11 The Life Cycle of an Angiosperm The formation of a triploid endosperm distinguishes the angiosperms from the gymnosperms.

serves as storage tissue for starch or lipids, proteins, and other substances that will be needed by the developing embryo.

The zygote develops into an embryo, consisting of an embryonic axis and one or two **cotyledons**, or seed leaves. The cotyledons have different fates in different plants. In many, they serve as absorptive organs that take up and digest the endosperm. In others, they enlarge and become photosynthetic when the seed germinates. Often they play both roles.

Angiosperms produce fruits

The ovary of a flowering plant (together with the seeds it contains) develops into a fruit after fertilization. A fruit may consist only of the mature ovary and its seeds, or it may include other parts of the flower or structures associated with it. A *simple fruit*, such as a cherry (Figure 30.12*a*), is one that develops from a single carpel or several united carpels. A raspberry is an example of an *aggregate fruit* (Figure 30.12*b*)—one that develops from several separate carpels of a single flower. Pineapples and figs are examples of *multiple fruits* (Figure 30.12*c*), formed from a cluster of flowers (an inflorescence). Fruits derived from parts in addition to the carpel and seeds are called *accessory fruits* (Figure 30.12*d*); examples are apples, pears, and strawberries. The development, ripening, and dispersal of fruits will be considered in Chapters 38 and 39.

There are several clades of angiosperms

The better-understood relationships among the angiosperm clades are shown in Figure 30.13. Two large clades include the great majority of angiosperm species: the **monocots** and the **eudicots**. The monocots are so called because they have a single embryonic cotyledon; the eudicots have two. We will describe other differences between these groups in Chapter 35.

Some familiar angiosperms belong to clades other than the monocots and eudicots (Figure 30.14). These clades include the water lilies, star anise and its relatives, and the magnoliid complex. The magnoliids are less numerous than the

(*a*)

(*b*)

(*c*)

(*d*)

30.12 Fruits Come in Many Forms and Flavors (*a*) A simple fruit (sour cherry). (*b*) An aggregate fruit (raspberry). (*c*) A multiple fruit (pineapple). (*d*) An accessory fruit (strawberry).

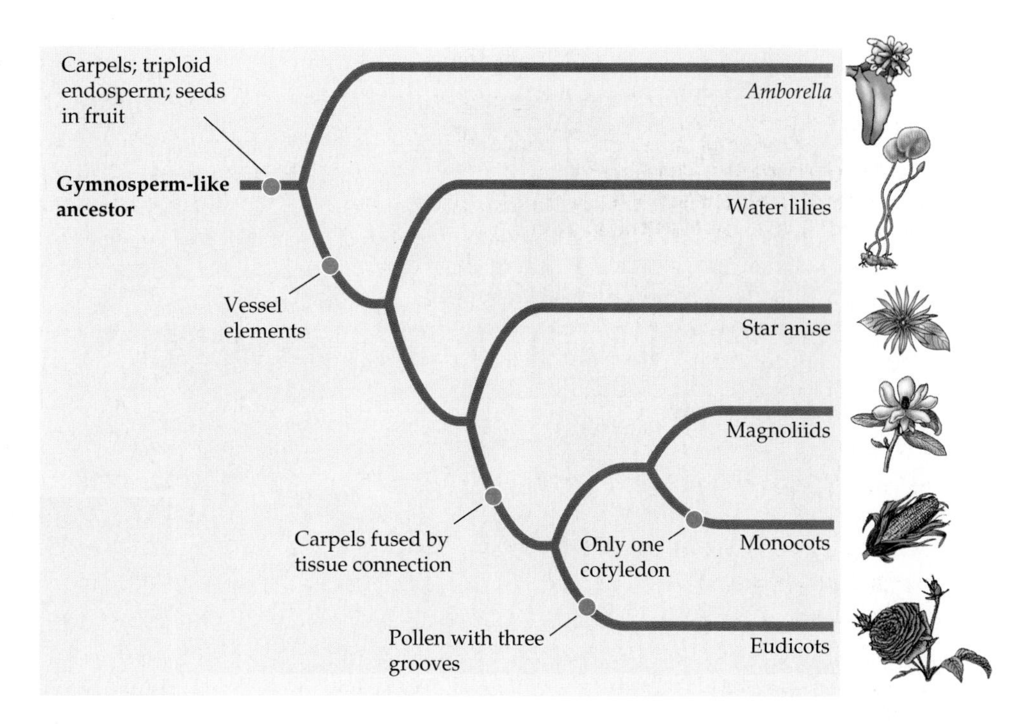

30.13 Evolutionary Relationships among the Angiosperms The monocots and the eudicots are the largest clades among the angiosperms. This diagram is a conservative interpretation of current data on relationships among the clades.

monocots and eudicots, but they include many familiar and often useful plants such as magnolias, avocados, cinnamon, and pepper.

The monocots (Figure 30.15) include grasses, cattails, lilies, orchids, and palms. The eudicots (Figure 30.16) include the vast majority of familiar seed plants, including most herbs, vines, trees, and shrubs. Among them are such diverse plants as oaks, willows, violets, snapdragons, and sunflowers.

(*a*) *Amborella trichopoda*

(*b*) *Nymphaea odorata*

(*c*) *Illicium floridanum*

(*d*) *Piper nigrum*

(*e*) *Aristolochia grandiflora*

(*f*) *Persea* sp.

30.14 Monocots and Eudicots Are Not the Only Surviving Angiosperms (*a*) *Amborella*, a shrub, is the closest living relative of the first angiosperms; its clade is sister to the remaining extant angiosperms. (*b*) The water lily clade is the next most basal clade after *Amborella*'s. (*c*) Star anise and its relatives belong to another basal clade. (*d–f*) The largest clade other than the monocots and eudicots is the magnoliid complex, represented here by (*d*) a black pepper, (*e*) Dutchman's pipe, and (*f*) an avocado tree. The magnolia in Figure 30.9*a* is another magnoliid.

(*a*) *Phoenix dactylifera*

(*b*) *Triticum* sp.

(*c*) *Lilium* sp.

30.15 Monocots (*a*) Palms are among the few monocot trees. Date palms are a major food source in some areas of the world. (*b*) Grasses such as this cultivated wheat and the fountain grass in Figure 30.8*c* are monocots. (*c*) Monocots also include popular garden flowers such as these lilies. Many orchids (Figure 30.9*b*) are highly sought-after monocot flowers.

(*a*) *Borzicactus samaipatanus*

(*b*) *Cornus florida*

(*c*) *Rosa rugosa*

30.16 Eudicots (*a*) The cactus family is a large group of eudicots, with about 1,500 species in the Americas. This cactus bears scarlet flowers for a brief period of the year. (*b*) The flowering dogwood is a small eudicot tree. (*c*) Climbing Cape Cod roses are members of the eudicot family Rosaceae, as are the familiar roses from your local florist.

Determining the oldest angiosperm clade

Which angiosperms were the earliest flowering plants was long a matter of great controversy. Two leading candidates were the magnolia family (see Figure 30.9*a*) and another family, the Chloranthaceae, whose flowers are much simpler than those of the magnolias. At the close of the twentieth century, however, an impressive convergence of evidence led to the conclusion that the most basal living angiosperm belongs to neither of those families, but rather to a clade that today consists of a single species of the genus *Amborella* (see Figure 30.14*a*). This woody shrub, with cream-colored flowers, lives only on New Caledonia, an island in the South Pacific. Its five to eight carpels are in a single whorl, and it has 30 to 100 stamens. The xylem of *Amborella* lacks vessel elements, which appeared later in angiosperm evolution. The characteristics of *Amborella* give us a good sense of what the first angiosperms might have been like. But are there extinct angiosperms that may represent still more ancient clades?

In 2002, Chinese and American botanists examined fossils of two species of a 125-million-year-old aquatic genus, *Archaefructus* (see Figure 22.16). Their studies established an extinct family, Archaefructaceae, that is posited to be the sister taxon of all other angiosperms. The flower of these plants had its ovules enclosed in carpels, as in all angiosperms. The flower had neither petals nor sepals, however, and its carpels and stamens were arranged spirally around elongated shoots. This arrangement of carpels and stamens is seen today in the magnolias.

The origin of the angiosperms remains a mystery

We have learned a lot about evolution within the angiosperm clade. But how did the angiosperms first arise? Are the angiosperms sister to any single gymnosperm phylum? A few years ago, it seemed that we were on the verge of answering these questions. But the puzzle remains as vexing today as it ever was.

Why should this be? Different phylogenetic methods, applied by different investigators, have produced apparently contradictory results. It might seem a simple matter to rectify this situation, but several questions complicate such efforts: What morphological characters should be selected as important, or should they all be treated as equally important? What algorithms should be applied to computerized analysis of data? Are all molecular differences and similarities significant, or are some of them incidental? Which fossils should be chosen for comparisons? What is the likelihood that we can find evidence of double fertilization in ancient fossils? Furthermore, it is possible that the angiosperms have no close relatives at all among living seed plants.

We are left with our original question: Where did the first angiosperm come from? Current progress in methodology gives us reason to hope that our understanding of seed plant evolution will be much improved before the present decade ends. We will see in Chapters 32–34 whether our understanding of animal evolution is any more complete.

Chapter Summary

The Seed Plants

- The seed plants (gymnosperms and angiosperms) are heterosporous and have greatly reduced gametophytes. **Review Figures 30.1, 30.2**
- Modern gymnosperms and many angiosperms have abundant xylem and extensive secondary growth.
- Most modern seed plants have no swimming gametes and do not require liquid water for fertilization. The male gametophyte—the pollen grain—is dispersed by wind or by animals.
- The seed is a well-protected resting stage that often contains nutrients that support the growth of the embryo.

The Gymnosperms: Naked Seeds

- The gymnosperms, once the dominant vegetation on Earth, still dominate forests in the northern parts of the Northern Hemisphere and at high elevations.
- The four surviving gymnosperm phyla are the Cycadophyta (perhaps the most ancient), Ginkgophyta (consisting of a single species, the maidenhair tree), Gnetophyta (which has some characters in common with the angiosperms), and Pinophyta (the familiar cone-bearing trees).
- Conifers have a life cycle in which naked seeds are produced on the scales of cones. Pollen is produced in strobili, which are smaller than cones. Pollen is transferred from strobili to cones by wind. **Review Figures 30.5, 30.6. See Web/CD Tutorial 30.1 and Activity 30.1**

The Angiosperms: Flowering Plants

- Angiosperms (phylum Angiospermae) are distinguished by double fertilization, which results in a triploid nutritive tissue, the endosperm.
- The ovules and seeds of angiosperms are enclosed by a carpel. Angiosperms are also characterized by the production of flowers and fruits.
- The vascular tissues of angiosperms contain three characteristic cell types: vessel elements, fibers, and companion cells. Woody angiosperms show secondary growth.
- Flowers are made up of various combinations of carpels, stamens, petals, and sepals. Perfect flowers have both carpels and stamens. **Review Figure 30.7. See Web/CD Activity 30.2**
- Monoecious plant species have both female and male flowers on the same plant. In dioecious species, female and male flowers are found on separate individuals.
- Carpels and stamens may have evolved from leaflike structures. **Review Figure 30.10**
- Angiosperms and the animals that pollinate them have coevolved.
- The angiosperm seed contains the products of double fertilization: the diploid zygote and the triploid endosperm. **Review Figure 30.11**
- The largest clades of flowering plants, in terms of numbers of species, are the monocots and the eudicots. There are a few other angiosperm clades, notably the water lilies, star anise and its relatives, and the magnoliids. **Review Figure 30.13**

▶ *Amborella*, a tropical shrub, is thought to be the sole living representative of the most ancient living angiosperm clade.

▶ The evolutionary origin of the angiosperms remains a mystery.

Self-Quiz

1. Which of the following statements about seed plants is true?
   a. The phylogenetic relationships among all five phyla have been established.
   b. The sporophyte generation is more reduced than in the ferns.
   c. The gametophytes are independent of the sporophytes.
   d. All seed plant species are heterosporous.
   e. The zygote divides repeatedly to form the gametophyte.
2. The gymnosperms
   a. dominate all land masses today.
   b. have never dominated land masses.
   c. have active secondary growth.
   d. all have vessel elements.
   e. lack sporangia.
3. Conifers
   a. produce ovules in strobili and pollen in cones.
   b. depend on liquid water for fertilization.
   c. have triploid endosperm.
   d. have pollen tubes that release two sperm.
   e. have vessel elements.
4. Angiosperms
   a. have ovules and seeds enclosed in a carpel.
   b. produce triploid endosperm by the union of two eggs and one sperm.
   c. lack secondary growth.
   d. bear two kinds of cones.
   e. all have perfect flowers.
5. Which statement about flowers is *not* true?
   a. Pollen is produced in the anthers.
   b. Pollen is received on the stigma.
   c. An inflorescence is a cluster of flowers.
   d. A species having female and male flowers on the same plant is dioecious.
   e. A flower with both megasporangia and microsporangia is said to be perfect.
6. Which statement about fruits is *not* true?
   a. They develop from ovaries.
   b. They may include other parts of the flower.
   c. A multiple fruit develops from several carpels of a single flower.
   d. They are produced only by angiosperms.
   e. A cherry is a simple fruit.
7. Which statement is *not* true of angiosperm pollen?
   a. It is the male gamete.
   b. It is haploid.
   c. It produces a long tube.
   d. It interacts with the carpel.
   e. It is produced in microsporangia.
8. Which statement is *not* true of carpels?
   a. They are thought to have evolved from leaves.
   b. They bear megasporangia.
   c. They may fuse to form a pistil.
   d. They are floral organs.
   e. They were absent in *Archaefructus*.
9. *Amborella*
   a. was the first flowering plant.
   b. belongs to the first angiosperm clade.
   c. belongs to the oldest angiosperm clade still extant.
   d. is a eudicot.
   e. has vessel elements in its xylem.
10. The eudicots
    a. include many herbs, vines, shrubs, and trees.
    b. and the monocots are the only extant angiosperm clades.
    c. are not a clade.
    d. include the magnolias.
    e. include orchids and palm trees.

For Discussion

1. In most seed plant species, only one of the products of meiosis in the megasporangium survives. How might this be advantageous?
2. Suggest an explanation for the great success of the angiosperms in occupying terrestrial habitats.
3. In many locales, large gymnosperms predominate over large angiosperms. Under what conditions might gymnosperms have the advantage, and why?
4. Not all flowers possess all of the following floral organs: sepals, petals, stamens, and carpels. Which floral organ or organs do you think might be found in the flowers that have the smallest number of floral organ types? Discuss the possibilities, both for a single flower and for a species.
5. The problem of the origin of the angiosperms has long been "an abominable mystery," as Charles Darwin once put it. Scientists still do not know the nearest relatives of the angiosperms. It has often been suggested (correctly or incorrectly) that the gnetophytes are sister to the angiosperms. What pieces of evidence suggested this connection?

31 Fungi: Recyclers, Pathogens, Parasites, and Plant Partners

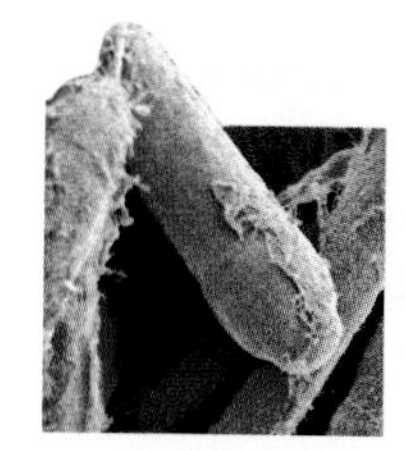

About 300 million Africans in 25 countries are suffering because of the invasion of crops by witchweed (*Striga*), a parasitic flowering plant. This parasite has attacked more than two-thirds of the sorghum, maize, and millet crops in sub-Saharan Africa, doing damage estimated at U.S. $7 billion each year.

In 1991 a team of Canadian scientists began a search for a solution to the *Striga* problem. By 1995 they had begun fieldwork in Mali. What was their strategy? They had isolated a strain of a fungus, the mold *Fusarium oxysporum*, that has two outstanding properties. First, it grows on *Striga*, wiping out a high percentage of the parasites. Second, it is not toxic to humans, nor does it attack the crop plants on which *Striga* is growing. Now farmers apply the fungus to their crops and are rewarded by greatly increased crop yields as *Striga* is held in check.

It may be possible to repeat this story—using a fungus to wipe out a particular type of flowering plant—in a very different context. A different strain of *F. oxysporum* preferentially attacks coca plants (the source of cocaine). There is a controversial proposal to use *F. oxysporum* to wipe out the coca plantations of Andean South America and some countries in other parts of the world.

Some other fungi attack people, not plants. Every breath we take contains large numbers of fungal spores. Some of those spores can be dangerous, and fungal diseases of humans, some of which are as yet incurable, have become a major global threat. However, other fungi are of immense commercial importance to us. Fungi are essential to plants as well. They interact with roots, greatly enhancing the roots' ability to take up water and mineral nutrients. Fungi and plants probably invaded the land together in the Paleozoic era (see Table 22.1).

Earth would be a messy place without the fungi. They are constantly at work in forests, fields, and garbage dumps, breaking down the remains of dead organisms (and even manufactured substances, such as some plastics). For almost a billion years, the ability of fungi to decompose organic substances has been essential for life on Earth, chiefly because by breaking down carbon com-

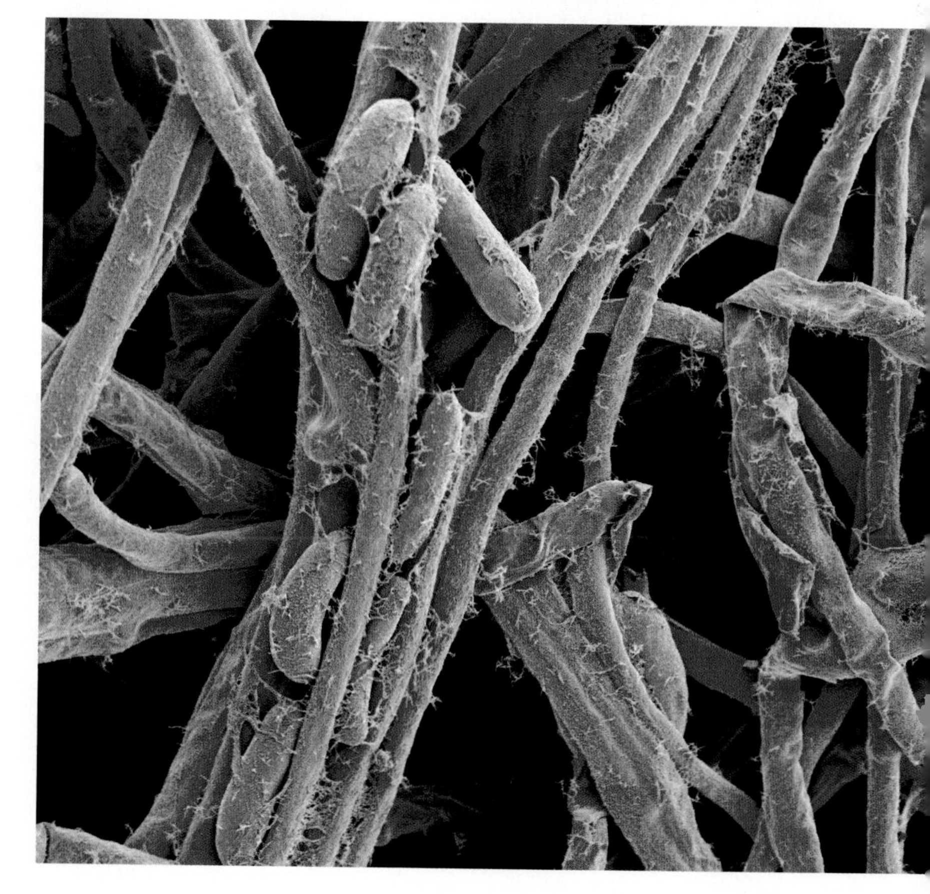

Fungus Trumps Plant The fungus *Fusarium oxysporum* is a potent pathogen of witchweed (*Striga*), a parasitic plant that attacks crops. The fungus spores are shown in blue; the fungal filaments are in tan. Both colors were added to this electron micrograph.

pounds, they return carbon and other elements to the environment, where they can be used again by other organisms.

In this chapter we will examine the general biology of the kingdom Fungi, which differs in interesting ways from the other kingdoms. We will also explore the diversity of body forms, reproductive structures, and life cycles among the four phyla of fungi, as well as the mutually beneficial associations of certain fungi with other organisms. As we begin our study, recall that the fungi and the animals are descended from a common ancestor—molds and mushrooms are more closely related to us than they are to the flowers we admired in the last chapter.

General Biology of the Fungi

The kingdom Fungi encompasses *heterotrophic organisms with absorptive nutrition and with chitin in their cell walls.* The fungi live by **absorptive nutrition**: They secrete digestive enzymes that break down large food molecules in the environment, and then absorb the breakdown products. Many fungi are *saprobes* that absorb nutrients from dead matter, others are *parasites* that absorb nutrients from living hosts (Figure 31.1), and still others are *mutualists* that live in intimate association with other organisms.

The production of **chitin**, a polysaccharide, is a synapomorphy (shared derived trait) for fungi, choanoflagellates, and animals. That is, its presence in fungi is the evidence that all fungi are more closely related to animals than any fungi are to plants. Chitin is used in the cell walls of fungi, but it is used in other ways in animals. The use of chitin in cell walls is a synapomorphy for fungi, and it allows us to distinguish between the fungi and the basal eukaryotes (protists) that resemble them. Some protists that were formerly confused with fungi include the slime molds (see Figures 28.31 and 28.32) and water molds (oomycetes; see Figure 28.23).

The alternation between gametophyte (n) and sporophyte ($2n$) generations that evolved in plants (see Chapter 29) is found in only the most basal group of fungi, the chytrids. The derived condition, which is found in the other three fungal clades, involves a unique state in which two haploid nuclei are present in a single cell, discussed later in this chapter. As one might expect, the chytrids, which are aquatic, possess flagellated gametes (or spores). Flagella have been lost in the terrestrial fungi.

The kingdom Fungi consists of four phyla: Chytridiomycota, Zygomycota, Ascomycota, and Basidiomycota. We distinguish the phyla on the basis of their methods and structures for sexual reproduction and, to a lesser extent, by criteria such as the presence or absence of cross-walls separating their cell-like compartments. This morphologically based phylogeny has proved largely consistent with phylogenies based on DNA sequencing. The term "fungal systematics" has an interesting anagram, "fantastic ugly mess," but we'll see that the situation isn't all that bad.

In the sections that follow, we'll consider some aspects of the general biology of the fungi, including their body structure and its intimate relationship with their environment, their nutrition, and some special aspects of their unusual sexual reproductive cycles.

Some fungi are unicellular

Unicellular forms are found in all of the fungal phyla. Unicellular members of the Zygomycota, Ascomycota, and Basidiomycota are called **yeasts**. Yeasts may reproduce by budding, by fission, or by sexual means (Figure 31.2). Their means of reproduction help us to place them in their appropriate phyla, as we will see below.

The body of a multicellular fungus is composed of hyphae

Most fungi are multicellular. The body of a multicellular fungus is called a **mycelium** (plural, mycelia). It is composed of rapidly growing individual tubular filaments called

31.1 Parasitic Fungi Attack Other Living Organisms (*a*) The gray masses on this ear of corn are the parasitic fungus *Ustilago maydis*, commonly called corn smut. (*b*) The tropical fungus whose fruiting body is growing out of the carcass of this ant has developed from a spore ingested by the ant. The spores of this fungus must be ingested by insects before they will germinate and develop. The growing fungus absorbs organic and inorganic nutrients from the ant's body, eventually killing it, after which the fruiting body produces a new crop of spores.

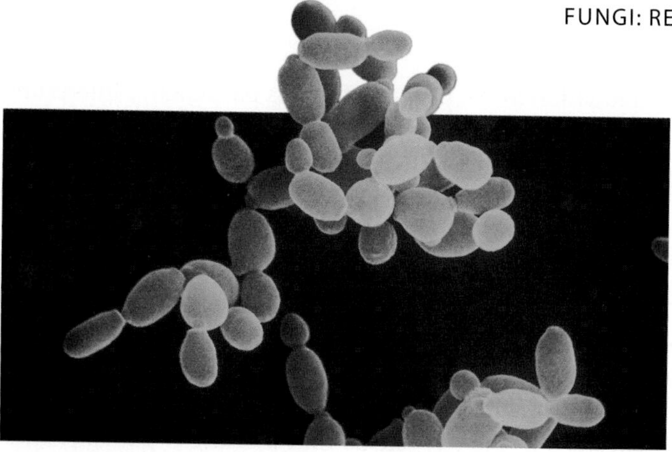

31.2 Yeasts Are Unicellular Fungi Unicellular members of the fungal phyla Zygomycota, Ascomycota, and Basidiomycota are known as yeasts. Many yeasts reproduce by budding—mitosis followed by asymmetrical cell division—as those shown here are doing.

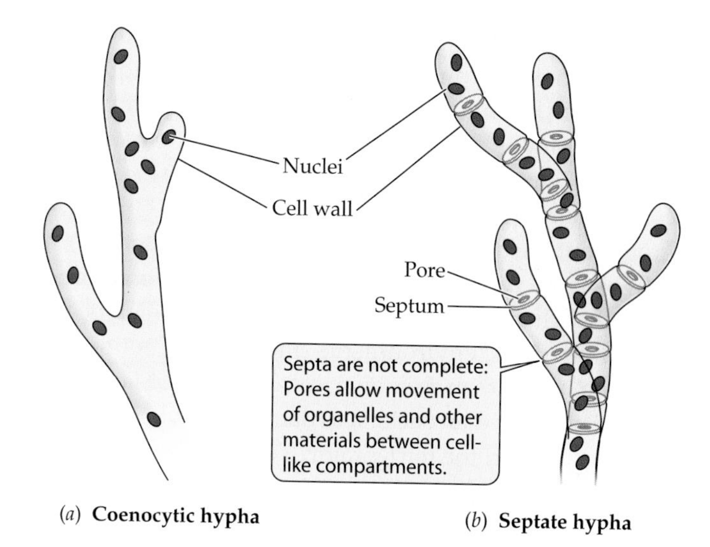

31.3 Most Hyphae Are Incompletely Divided into Separate Cells (*a*) Coenocytic hyphae have no septa between their nuclei. (*b*) Even in septate hyphae, the septa do not block the movement of organelles within the hypha.

hyphae (singular, hypha). Within hyphae of two clades, *incomplete* cross-walls called **septa** (singular, septum) divide the hypha into separate cells. Pores in the septa allow organelles—sometimes even nuclei—to move in a controlled way between cells (Figure 31.3). Other hyphae are **coenocytic** and have no septa.

Certain modified hyphae, called **rhizoids**, anchor chytrids and some other fungi to their substratum (the dead organism or other matter upon which they feed). These rhizoids are not homologous to the rhizoids of plants because they are not specialized to absorb nutrients and water. Parasitic fungi may possess modified hyphae that take up nutrients from their host.

The total hyphal growth of a mycelium (not the growth of an individual hypha) may exceed 1 km per day. The hyphae may be widely dispersed to forage for nutrients over a large area, or they may clump together in a cottony mass to exploit a rich nutrient source. Sometimes, when sexual spores are produced, the mycelium becomes reorganized into a *fruiting* (reproductive) *structure* such as a mushroom.

The way in which a parasitic fungus attacks a plant illustrates the absorptive role of fungal hyphae (Figure 31.4). The hyphae of a fungus invade a leaf through the stomata, through wounds, or in some cases, by direct penetration of epidermal cells. Once inside the leaf, the hyphae form a mycelium. Some hyphae produce **haustoria**, branching projections that push into the living plant cells, absorbing the nutrients within the cells. The haustoria do not break through the plant cell plasma membranes; they simply press into the cells, with the membrane fitting them like a glove. Fruiting structures may form, either within the plant body or on its surface.

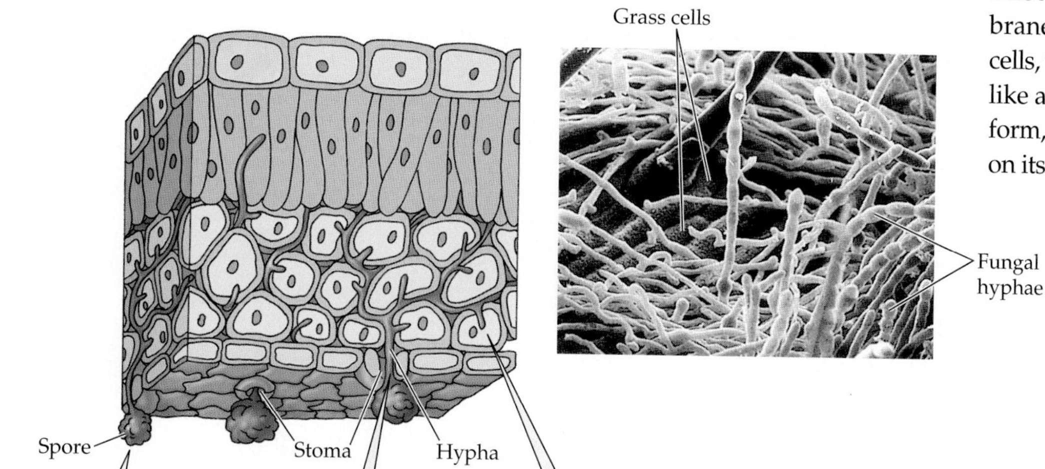

Grass cells
Fungal hyphae
Spore
Stoma
Hypha
Fungal spores germinate on the surface of the leaf.
Elongating hyphae pass through stomata into the interior of the leaf.
Some hyphae penetrate cells within the leaf.

31.4 A Fungus Attacks a Leaf The white structures in the micrograph are hyphae of the fungus *Blumeria graminis*, which is growing on the dark surface of the leaf of a grass.

Fungi are in intimate contact with their environment

The filamentous hyphae of a fungus give it a unique relationship with its physical environment. The fungal mycelium has an enormous surface area-to-volume ratio compared with that of most large multicellular organisms. This large ratio is a marvelous adaptation for absorptive nutrition. Throughout the mycelium (except in fruiting structures), all the hyphae are very close to their environmental food source.

Another characteristic of some fungi is their tolerance for highly hypertonic environments (those with a solute concentration higher than their own; see Chapter 5). Many fungi are more resistant than bacteria to damage in hypertonic surroundings. Jelly in the refrigerator, for example, will not become a growth medium for bacteria because it is too hypertonic to the bacteria, but it may eventually harbor mold colonies. Their presence in the refrigerator illustrates another trait of many fungi: tolerance of temperature extremes. Many fungi tolerate temperatures as low as 5–6°C below freezing, and some tolerate temperatures as high as 50°C or more.

Fungi are absorptive heterotrophs

All fungi are heterotrophs that obtain food by direct absorption from their immediate environment. The majority are saprobes, obtaining their energy, carbon, and nitrogen directly from dead organic matter through the action of enzymes they secrete. However, as we've learned already, some are parasites, and still others form mutualistic associations with other organisms.

Saprobic fungi, along with bacteria, are the major decomposers of the biosphere, contributing to decay and thus to the recycling of the elements used by living things. In the forest, for example, the mycelia of fungi absorb nutrients from fallen trees, thus decomposing their wood. Fungi are the principal decomposers of cellulose and lignin, the main components of plant cell walls (most bacteria cannot break down these materials). Other fungi produce enzymes that decompose keratin and thus break down animal structures such as hair and nails.

Because many saprobic fungi are able to grow on artificial media, we can perform experiments to determine their exact nutritional requirements. Sugars are their favored source of carbon. Most fungi obtain nitrogen from proteins or the products of protein breakdown. Many fungi can use nitrate (NO_3^-) or ammonium (NH_4^+) ions as their sole source of nitrogen. No known fungus can get its nitrogen directly from nitrogen gas, as can some bacteria and plant–bacteria associations (see Chapter 37). Nutritional studies also reveal that most fungi are unable to synthesize their own thiamin (vitamin B_1) or biotin (another B vitamin) and must absorb these vitamins from their environment. On the other hand, fungi can synthesize some vitamins that animals cannot. Like all organisms, fungi also require some mineral elements.

Nutrition in the parasitic fungi is particularly interesting to biologists. *Facultative* parasites can attack living organisms but can also be grown by themselves on artificial media. *Obligate* parasites cannot be grown on any available medium; they can grow only on their specific living hosts, usually plants. Because their growth is limited to living hosts, they must have specialized nutritional requirements.

Some fungi have adaptations that enable them to function as active predators, trapping nearby microscopic protists or animals. The most common strategy is to secrete sticky substances from the hyphae so that passing organisms stick tightly to them. The hyphae then quickly invade the prey, growing and branching within it, spreading through its body, absorbing nutrients, and eventually killing it.

A more dramatic adaptation for predation is the constricting ring formed by some species of *Arthrobotrys*, *Dactylaria*, and *Dactylella* (Figure 31.5). All of these fungi grow in soil. When nematodes (tiny roundworms) are present in the soil, these fungi form three-celled rings with a diameter that just fits a nematode. A nematode crawling through one of these rings stimulates the fungus, causing the cells of the ring to swell and trap the worm. Fungal hyphae quickly invade and digest the unlucky victim.

Two other kinds of relationships between fungi and other organisms have nutritional consequences for the fungal partner. These relationships are highly specific, *symbiotic* (the partners live in close, permanent contact with one another), and *mutualistic* (the relationships benefit both partners). **Lichens** are associations of a fungus with a cyanobacterium, a unicellular photosynthetic protist, or both. **Mycorrhizae** (singular, mycorrhiza) are associations between fungi and the roots of plants. In these associations, the fungus obtains organic com-

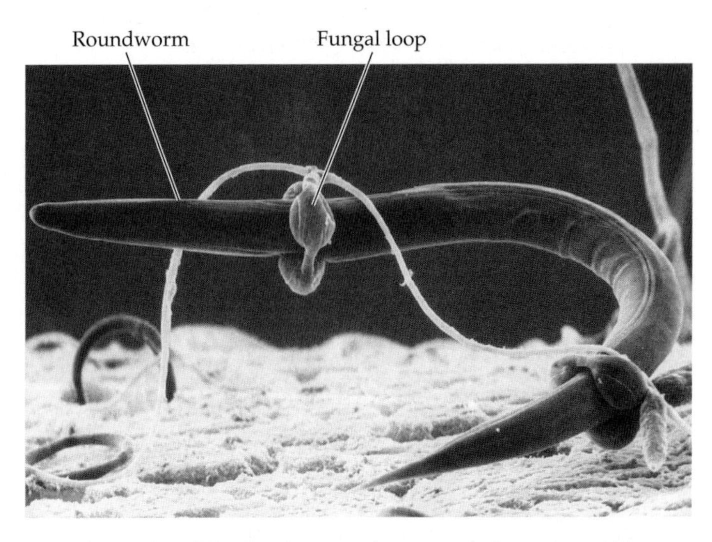

31.5 Some Fungi Are Predators A nematode (roundworm) is trapped in sticky loops of the soil-dwelling fungus *Arthrobotrys anchonia*.

pounds from its photosynthetic partner, but provides it with minerals and water in return, so that the partner's nutrition is also promoted. In fact, many plants could not grow at all without their fungal partners. We will discuss lichens and mycorrhizae more thoroughly later in this chapter.

Most fungi reproduce both asexually and sexually

Both asexual and sexual reproduction are common among the fungi. Asexual reproduction takes several forms:

- The production of (usually) haploid spores within structures called **sporangia**.
- The production of naked spores (not enclosed in sporangia) at the tips of hyphae; such spores are called **conidia** (from the Greek *konis*, "dust").
- Cell division by unicellular fungi—either a relatively equal division (called *fission*) or an asymmetrical division in which a small daughter cell is produced (called *budding*).
- Simple breakage of the mycelium.

Asexual reproduction in fungi can be spectacular in terms of quantity. A 2.5-centimeter colony of *Penicillium* can produce as many as 400 million conidia. The air we breathe contains as many as 10,000 fungal spores per cubic meter.

Sexual reproduction in many fungi features an interesting twist. There is often no morphological distinction between female and male structures, or between female and male individuals. Rather, there is a genetically determined distinction between two *or more* **mating types**. Individuals of the same mating type cannot mate with one another, but they can mate with individuals of another mating type within the same species. This distinction prevents self-fertilization. Individuals of different mating types differ genetically from one another, but are often visually and behaviorally indistinguishable. Many protists also have mating type systems.

Fungi reproduce sexually when hyphae (or, in the chytrids, motile cells) of different mating types meet and fuse. In many fungi, the zygote nuclei formed by sexual reproduction are the only diploid nuclei in the life cycle. These nuclei undergo meiosis, producing haploid nuclei that become incorporated into spores. Haploid fungal spores, whether produced sexually in this manner or asexually, germinate, and their nuclei divide mitotically to produce hyphae. This type of life cycle, called a *haplontic* life cycle, is also characteristic of many protists (see Figure 28.27).

The presence of a dikaryon is a synapomorphy of three phyla

Certain hyphae of some Zygomycota, Ascomycota, and Basidiomycota have a nuclear configuration other than the familiar haploid or diploid states. In these fungi, sexual reproduction begins in an unusual way: The cytoplasms of two individuals of different mating types fuse (*plasmogamy*) long before their nuclei fuse (*karyogamy*), so that *two genetically different haploid nuclei coexist and divide within the same hypha.* Such a hypha is called a **dikaryon** ("two nuclei"). Because the two nuclei differ genetically, such a hypha is also called a **heterokaryon** ("different nuclei").

Eventually, specialized fruiting structures form, within which the pairs of genetically dissimilar nuclei—one from each parent—fuse, giving rise to zygotes long after the original "mating." The diploid zygote nucleus undergoes meiosis, producing four haploid nuclei. The mitotic descendants of those nuclei become spores, which give rise to the next generation of hyphae.

The reproduction of such fungi displays several unusual features. First, there are no gamete *cells*, only gamete *nuclei*. Second, there is never any true diploid tissue, although for a long period the genes of both parents are present in the dikaryon and can be expressed. In effect, the hypha is neither diploid ($2n$) nor haploid (n); rather, it is *dikaryotic* ($n + n$). A harmful recessive mutation in one nucleus may be compensated for by a normal allele on the same chromosome in the other nucleus. Dikaryosis is perhaps the most significant of the genetic peculiarities of the fungi.

Finally, although zygomycetes, ascomycetes, and basidiomycetes grow in moist places, their gamete nuclei are not motile and are not released into the environment. Therefore, liquid water is not required for fertilization.

Some fungi are pathogens

Although most human diseases are caused by bacteria or viruses, fungal pathogens are a major cause of death among people with compromised immune systems. Most people with AIDS die of fungal diseases, such as the pneumonia caused by *Pneumocystis carinii* or the incurable diarrhea caused by some other fungi. *Candida albicans* and certain other yeasts also cause severe diseases in individuals with AIDS and in individuals taking immunosuppressive drugs. Such fungal diseases are a growing international health problem. Our limited understanding of the basic biology of these fungi still hampers our ability to treat the diseases they cause. Various fungi cause other, less threatening human diseases, such as ringworm and athlete's foot.

In plants, the situation is reversed. Fungi are by far the most important plant pathogens, causing crop losses amounting to billions of dollars. Major fungal diseases of crop plants include black stem rust of wheat and other diseases of wheat, corn, and oats. Bacteria and viruses are less important as plant pathogens.

The fungus that causes root and butt rot in pine trees is an important forest pathogen with an interesting, recently dis-

31.1 Classification of Fungi

PHYLUM	COMMON NAME	FEATURES	EXAMPLES
Chytridiomycota	Chytrids	Aquatic; gametes have flagella	*Allomyces*
Zygomycota	Zygote fungi	Zygosporangium; no regularly occurring septa; usually no fleshy fruiting body	*Rhizopus*
Ascomycota	Sac fungi	Ascus; perforated septa	*Neurospora*, baker's yeast
Basidiomycota	Club fungi	Basidium; perforated septa	*Armillariella*, mushrooms

covered property. The virulence (relative ability to cause disease) of some strains of the fungus is controlled by genes in its mitochondria—even though its dikaryotic cells have two different nuclei.

Diversity in the Kingdom Fungi

In this section on fungal diversity, we'll consider four phyla—Chytridiomycota, Zygomycota, Ascomycota, and Basidiomycota (Figure 31.6; Table 31.1). The first two groups are probably not clades, but the Ascomycota and Basidiomycota are clades.

Chytrids probably resemble the ancestral fungi

The earliest-diverging fungal group is the **chytrids** (phylum **Chytridiomycota**). These aquatic microorganisms were formerly classified with the protists. However, morphological (cell walls that consist primarily of chitin) and molecular evidence support their inclusion in the kingdom Fungi as its basal members. In this book, we use the term "chytrid" to refer to the entire phylum, but some mycologists reserve the term to apply to one of the major clades in the phylum.

Like their sister taxon, the animals, the chytrids possess flagellated gametes. The retention of this character reflects the aquatic environment in which fungi first evolved. Chytrids are the only fungi that have flagella at any life cycle stage.

Chytrids are either parasitic (on organisms such as algae, mosquito larvae, and nematodes) or saprobic, obtaining nutrients by breaking down dead organic matter. Chytrids in the compound stomachs of foregut-fermenting animals such as cows may be an exception, living in a mutualistic association with their hosts. Most chytrids live in freshwater habitats or in moist soil, but some are marine. Some chytrids are unicellular; others have mycelia made up of branching, coenocytic hyphae. Chytrids reproduce both sexually and asexually, but they do not have a dikaryon stage.

Allomyces, a well-studied genus of chytrids, displays alternation of generations. A haploid *zoospore* (a spore with flagella) comes to rest on dead plant or animal material in water and germinates to form a small, multicellular haploid mycelium. That mycelium produces female and male *gametangia* (gamete cases) (Figure 31.7). *Mitosis* in the gametangia results in the formation of haploid gametes, each with a single nucleus.

Chytridiomycota
Common ancestor
Zygomycota
Ascomycota
Basidiomycota

31.6 Phylogeny of the Fungi Four phyla are recognized among the fungi.

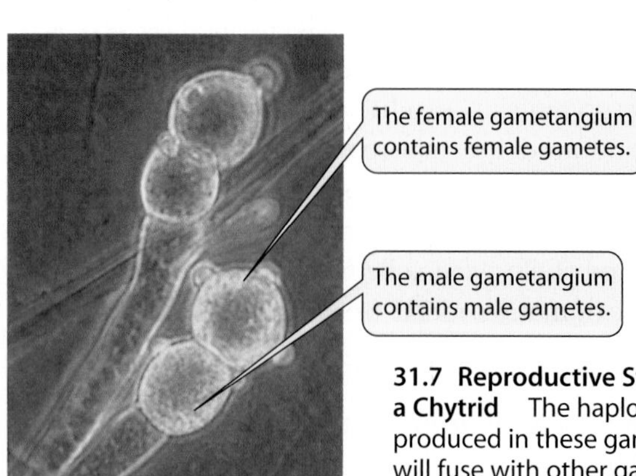

The female gametangium contains female gametes.
The male gametangium contains male gametes.
Allomyces sp.

31.7 Reproductive Structures of a Chytrid The haploid gametes produced in these gametangia will fuse with other gametes to form diploid mycelia. The male gametangia are smaller than the female gametangia and possess a light orange pigment.

Both female and male gametes have flagella. The motile female gamete produces a *pheromone*, a chemical that attracts the swimming male gamete. The two gametes fuse, and then their nuclei fuse to form a diploid zygote. Mitosis and cytokinesis in the zygote gives rise to a small, multicellular diploid organism, which produces numerous diploid flagellate zoospores. These diploid zoospores disperse and germinate to form more diploid organisms. Eventually, the diploid organism produces thick-walled resting sporangia that can survive unfavorable conditions such as dry weather or freezing. Nuclei in the resting sporangia eventually undergo meiosis, giving rise to haploid zoospores that are released into the water and begin the cycle anew.

The presence of flagellated gametes is a distinguishing feature of the chytrids. The loss of flagella is a synapomorphy that unites the remaining three fungal lineages.

Zygomycetes reproduce sexually by fusion of two gametangia

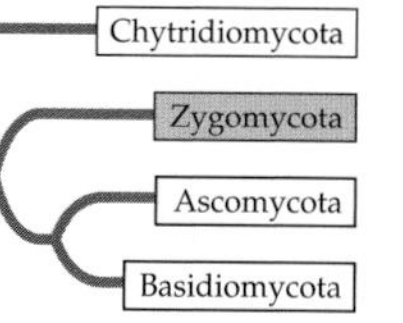

Most **zygomycetes** ("zygote fungi," phylum **Zygomycota**) have coenocytic hyphae. They produce no motile cells, and only one diploid cell—the zygote—appears in the entire life cycle. The mycelium of a zygomycete spreads over its substratum, growing forward by means of vegetative hyphae. Most zygomycetes do not form a fleshy fruiting structure; rather, the hyphae spread in an apparently random fashion, with occasional stalked **sporangiophores** reaching up into the air (Figure 31.8). These reproductive structures may bear one or many sporangia.

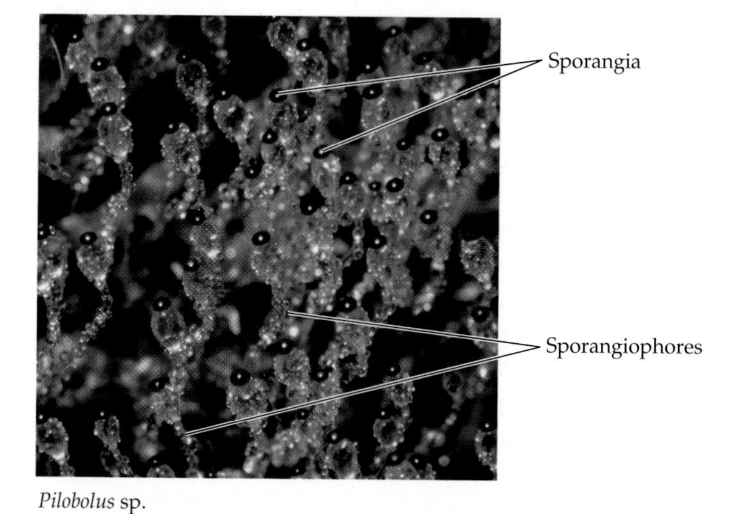

31.8 A Zygomycete This small forest of filamentous structures is made up of sporangiophores. The stalks end in tiny, rounded sporangia.

Almost 900 species of zygomycetes have been described. A very important group of zygomycetes serve as the fungal partners in the most common type of mycorrhizal association with plant roots. A zygomycete that you may be more familiar with is *Rhizopus stolonifer*, the black bread mold. *Rhizopus* reproduces asexually by producing many stalked sporangiophores, each bearing a single sporangium containing hundreds of minute spores (Figure 31.9*a*). As in other filamentous fungi, the spore-forming structure is separated from the rest of the hypha by a wall.

Zygomycetes reproduce sexually when adjacent hyphae of two different mating types release pheromones, which cause them to grow toward each other. These hyphae produce gametangia, which fuse to form a **zygosporangium**. Sometime later, the gamete nuclei now contained within the zygosporangium fuse to form a single multinucleate **zygospore** (Figure 31.9*b*). The zygosporangium develops a thick, multilayered wall that protects the zygospore. The highly resistant zygospore may remain dormant for months before its nuclei undergo meiosis and a sporangiophore sprouts. The sporangium contains the products of meiosis: haploid nuclei that are incorporated into spores. These spores disperse and germinate to form a new generation of haploid hyphae.

The next two fungal lineages that we'll discuss are related groups with many similarities, including a dikaryon stage and hyphae with septa. A key feature distinguishing between them is whether the sexual spores are borne inside a sac (in the ascomycetes) or on a pedestal (in the basidiomycetes).

The sexual reproductive structure of ascomycetes is an ascus

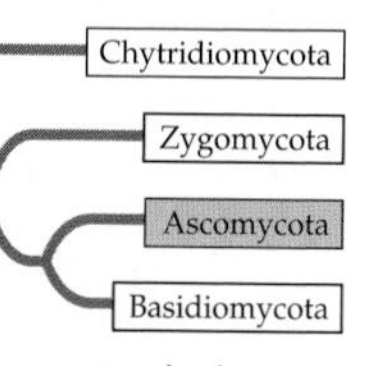

The **ascomycetes** ("sac fungi," phylum **Ascomycota**) are a large and diverse group of fungi distinguished by the production of sacs called **asci** (singular, ascus), which contain sexually produced **ascospores** (Figure 31.10). The ascus is the characteristic sexual reproductive structure of the ascomycetes. Ascomycete hyphae are segmented by more or less regularly spaced septa. A pore in each septum permits extensive movement of cytoplasm and organelles (including the nuclei) from one segment to the next.

The approximately 30,000 known species of ascomycetes can be divided into two broad groups, depending on whether the asci are contained within a specialized fruiting structure. Species that have this fruiting structure, the **ascocarp**, are collectively called **euascomycetes** ("true ascomycetes"); those without ascocarps are called **hemiascomycetes** ("half ascomycetes").

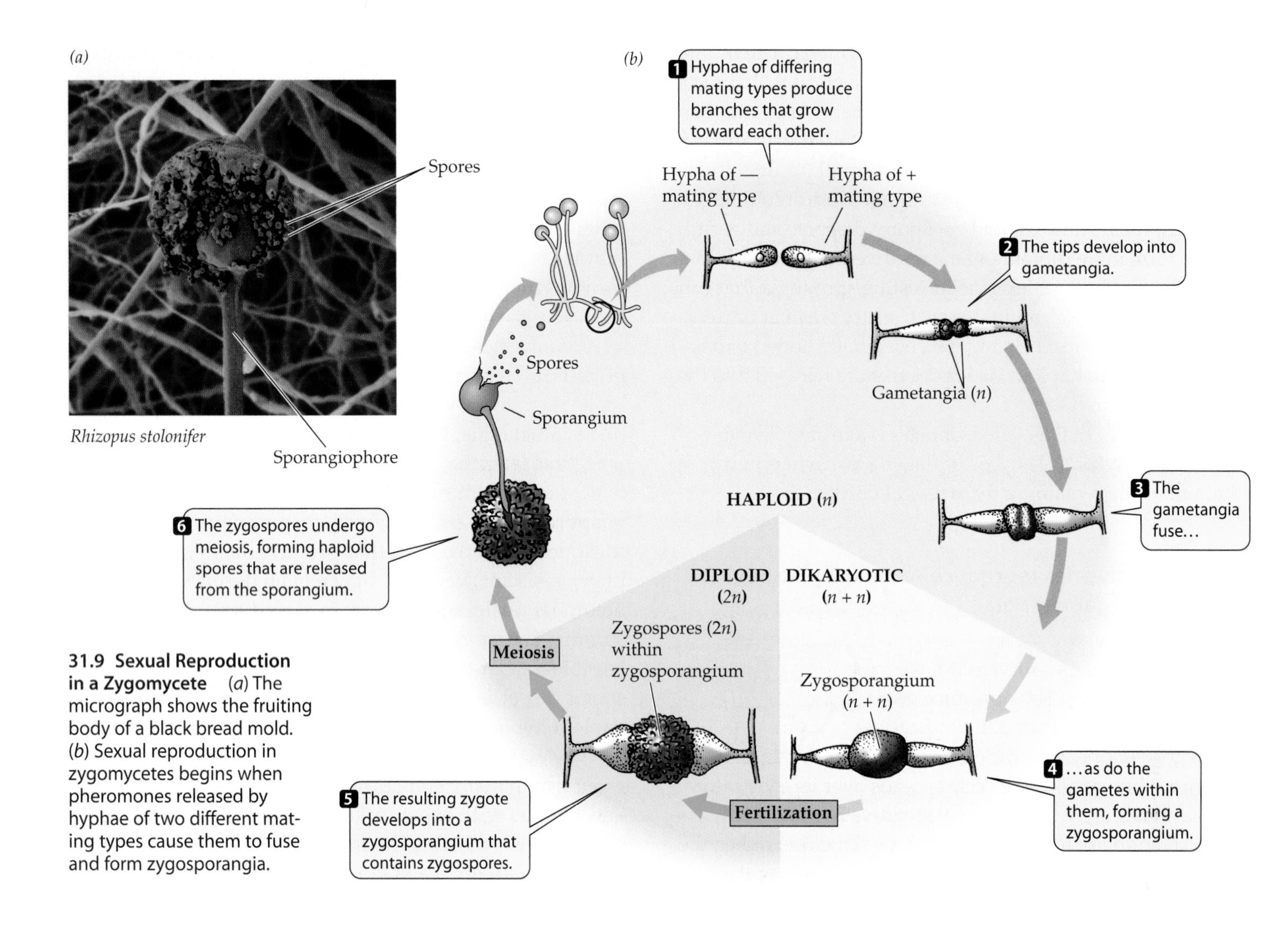

(a)
Spores
Rhizopus stolonifer
Sporangiophore
(b)
1 Hyphae of differing mating types produce branches that grow toward each other.
Hypha of — mating type
Hypha of + mating type
2 The tips develop into gametangia.
Gametangia (n)
3 The gametangia fuse…
4 …as do the gametes within them, forming a zygosporangium.
Zygosporangium (n + n)
Fertilization
Zygospores (2n) within zygosporangium
5 The resulting zygote develops into a zygosporangium that contains zygospores.
Meiosis
6 The zygospores undergo meiosis, forming haploid spores that are released from the sporangium.
Sporangium
Spores
HAPLOID (n)
DIPLOID (2n)
DIKARYOTIC (n + n)

31.9 Sexual Reproduction in a Zygomycete (*a*) The micrograph shows the fruiting body of a black bread mold. (*b*) Sexual reproduction in zygomycetes begins when pheromones released by hyphae of two different mating types cause them to fuse and form zygosporangia.

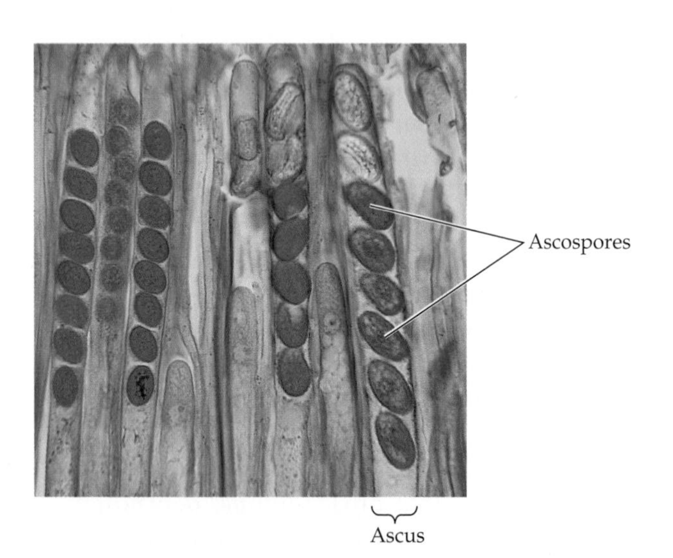

Ascospores
Ascus

31.10 Asci and Ascospores The ascomycetes are characterized by the production of ascospores within sacs called asci. Ascospores are the products of meiosis followed by a single mitotic division. Ascospores and asci do not mature all at once, and they may abort, so not every ascus in this micrograph contains eight mature ascospores.

HEMIASCOMYCETES. Most hemiascomycetes are microscopic, and many species are unicellular. Perhaps the best known are the ascomycete yeasts, especially baker's or brewer's yeast (*Saccharomyces cerevisiae;* see Figure 31.2). These yeasts are among the most important domesticated fungi. *S. cerevisiae* metabolizes glucose obtained from its environment to ethanol and carbon dioxide by fermentation. It forms carbon dioxide bubbles in bread dough and gives baked bread its light texture. Although they are baked away in bread making, the ethanol and carbon dioxide are both retained when yeast ferments grain into beer. Other yeasts live on fruits such as figs and grapes and play an important role in the making of wine.

Hemiascomycete yeasts reproduce asexually either by fission (splitting in half after mitosis) or by budding (an asymmetrical cell division in which a small daughter cell is produced; see Figure 31.2). Sexual reproduction takes place when two adjacent haploid cells of opposite mating types fuse. (We discussed the genetics of yeast mating types in Chapter 14.) In some species, the resulting zygote buds to form a diploid cell population; in others, the zygote nucleus undergoes meiosis immediately. When this diploid nucleus undergoes meiosis, the entire cell becomes an ascus. Depending on whether the products of meiosis then undergo mitosis, a yeast ascus usually contains either eight or four ascospores (see Figure 31.10). The ascospores germinate to become haploid cells. Hemiascomycetes have no dikaryon stage.

Yeasts, especially *Saccharomyces cerevisiae*, are frequently used in molecular biological research. Just as *E. coli* is the best-studied prokaryote, *S. cerevisiae* is the most completely studied eukaryote.

(*a*) *Morchella esculenta*

(*b*) *Sarcoscypha coccinea*

31.11 Two Cup Fungi (*a*) Morels, which have a spongelike ascocarp and a subtle flavor, are considered a delicacy by humans. (*b*) These brilliant red cups are the ascocarps of another cup fungus.

EUASCOMYCETES. The euascomycetes include many of the filamentous fungi known as molds. Among them are several common pink molds, one of which (*Neurospora*) Beadle and Tatum used in their pioneering genetic studies (see Figure 12.1). Many euascomycetes are parasites on flowering plants. Chestnut blight and Dutch elm disease are both caused by euascomycetes. The powdery mildews are euascomycetes that infect cereal grains, lilacs, and roses, among many other plants. They can be a serious problem to grape growers, and a great deal of research has focused on ways to control these agricultural pests.

The euascomycetes also include the cup fungi (Figure 31.11*a*,*b*). In most of these organisms the ascocarps are cup-shaped and can be as large as several centimeters across. The inner surfaces of the cups are covered with a mixture of vegetative hyphae and asci, and they produce huge numbers of spores. Although these fleshy structures appear to be composed of distinct tissue layers, microscopic examination shows that their basic organization is still filamentous—a tightly woven mycelium.

Two particularly delicious euascomycetes ascocarps are morels (Figure 31.11*a*) and truffles. Truffles grow underground in a mutualistic association with the roots of some species of oaks. Europeans traditionally used pigs to find truffles because some truffles secrete a substance that has an odor similar to a pig's sex pheromone. Unfortunately, pigs also eat truffles, so dogs are now the usual truffle hunters.

Penicillium is a genus of green molds, of which some species produce the antibiotic penicillin, presumably for defense against competing bacteria. Two species, *P. camembertii* and *P. roquefortii*, are the organisms responsible for the characteristic flavors of Camembert and Roquefort cheeses, respectively.

Brown molds of the genus *Aspergillus* are important in some human diets. *A. tamarii* acts on soybeans in the production of soy sauce, and *A. oryzae* is used in brewing the Japanese alcoholic beverage sake. Some species of *Aspergillus* that grow on nuts such as peanuts and pecans produce extremely carcinogenic (cancer-inducing) compounds called aflatoxins. In the United States, moldy grain infected with *Aspergillus* is thrown out. In Africa, where food is scarcer, the grain gets eaten, moldy or not, and causes severe health problems.

The euascomycetes reproduce asexually by means of conidia that form at the tips of specialized hyphae (Figure 31.12). Small chains of conidia are produced by the millions and can survive for weeks in nature. The conidia are what give molds their characteristic colors.

The sexual reproductive cycle of euascomycetes includes the formation of a dikaryon. Most euascomycetes form mat-

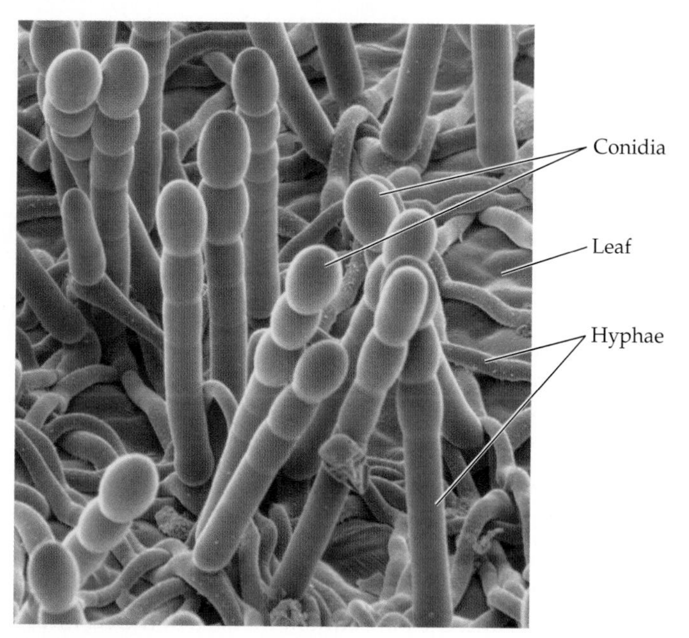

Erysiphe sp.

31.12 Conidia Chains of conidia are developing at the tips of specialized hyphae arising from this powdery mildew growing on a leaf.

31.13 The Life Cycle of a Euascomycete This cup fungus is so named because of its cup-shaped ascocarp.

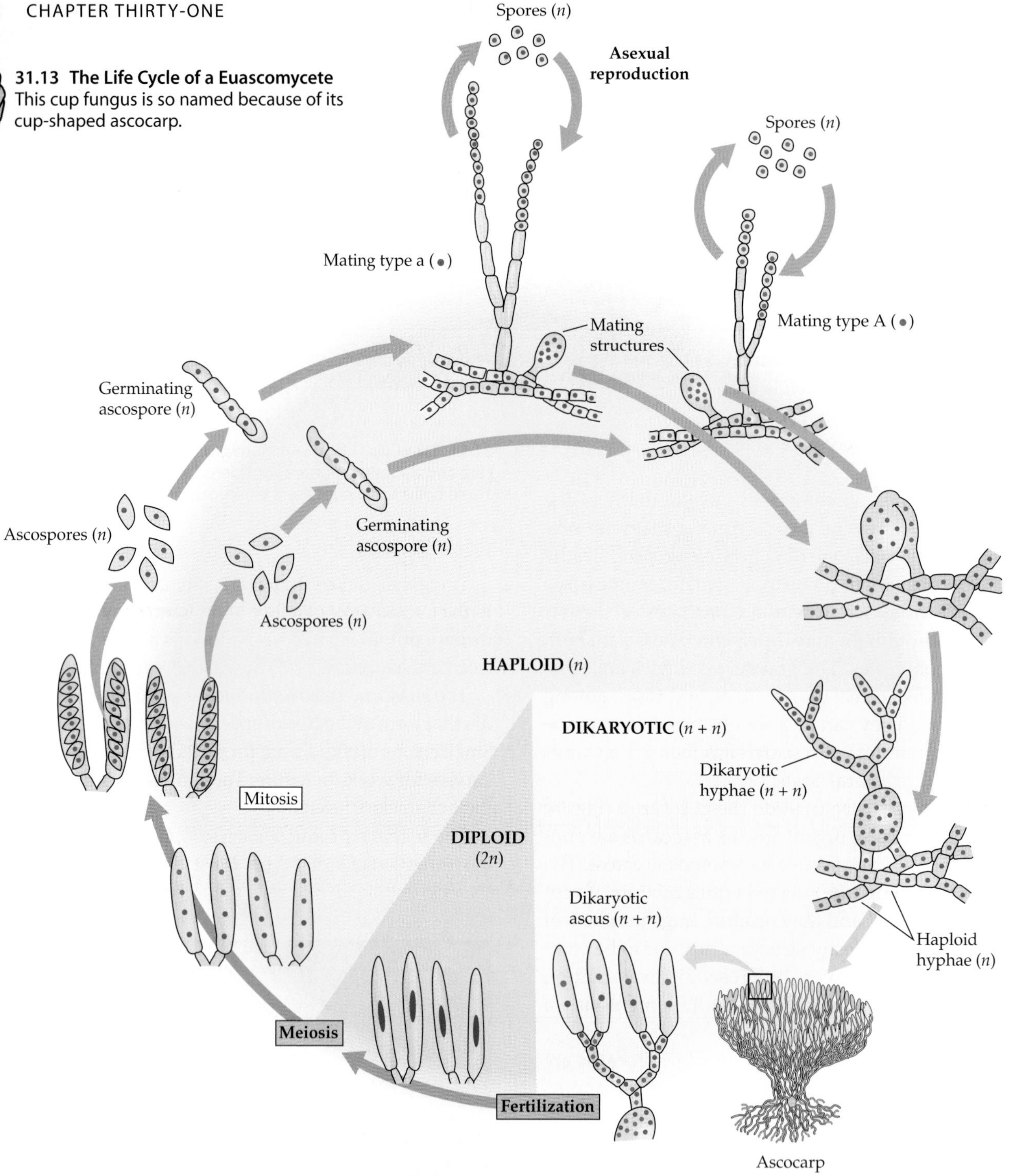

ing structures, some "female" and some "male" (Figure 31.13). Nuclei from a male structure on one hypha enter a female mating structure on a hypha of a compatible mating type. Dikaryotic *ascogenous* (ascus-forming) hyphae develop from the now dikaryotic female mating structure. The introduced nuclei divide simultaneously with the host nuclei. Eventually asci form at the tips of the ascogenous hyphae. Only with the formation of asci do the nuclei finally fuse. Both nuclear fusion and the subsequent meiosis of the resulting diploid nucleus take place within individual asci. The meiotic products are incorporated into ascospores that are ultimately shed by the ascus to begin the new haploid generation.

The sexual reproductive structure of basidiomycetes is a basidium

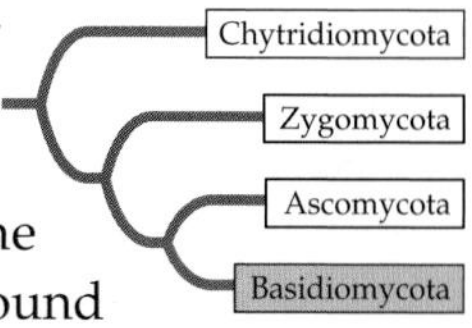

About 25,000 species of **basidiomycetes** ("club fungi," phylum **Basidiomycota**) have been described. Basidiomycetes produce some of the most spectacular fruiting structures found anywhere among the fungi. These fruiting structures, called **basidiocarps**, include puffballs (which may be more than half a meter in diameter), mushrooms of all kinds, and the

(*a*) *Lycoperdon perlatum*

(*b*) *Amanita muscaria*

(*c*) *Laetiporus sulphureus*

31.14 Basidiomycete Fruiting Structures The basidiocarps of the basidiomycetes are probably the most familiar structures produced by fungi. (*a*) When raindrops hit them, these puffballs will release clouds of spores for dispersal. (*b*) These mushrooms were produced by a member of a highly poisonous genus, *Amanita*, that forms mycorrhizal relationships with trees. (*c*) This edible bracket fungus is parasitizing a tree.

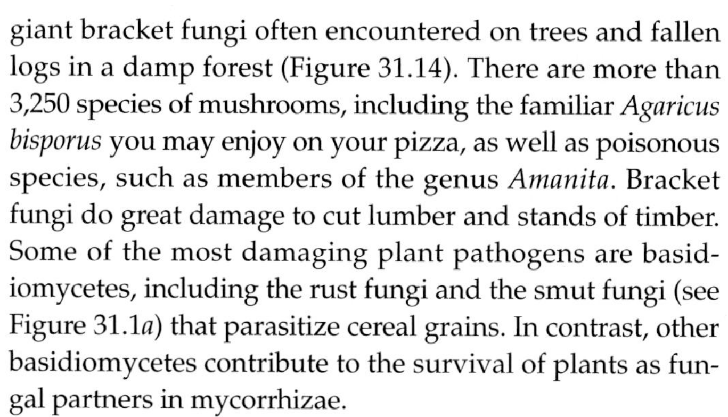

giant bracket fungi often encountered on trees and fallen logs in a damp forest (Figure 31.14). There are more than 3,250 species of mushrooms, including the familiar *Agaricus bisporus* you may enjoy on your pizza, as well as poisonous species, such as members of the genus *Amanita*. Bracket fungi do great damage to cut lumber and stands of timber. Some of the most damaging plant pathogens are basidiomycetes, including the rust fungi and the smut fungi (see Figure 31.1*a*) that parasitize cereal grains. In contrast, other basidiomycetes contribute to the survival of plants as fungal partners in mycorrhizae.

Some of the largest organisms on Earth are basidiomycetes. One such fungus, a member of the genus *Armillariella* growing in Michigan, covers an area of 37 acres. Its effect on plants is evident from the air, but from ground level, it is difficult to realize how large the fungus is. At the surface, only seemingly isolated clumps of mushrooms are visible. The vast body of the fungus, which weighs approximately the same as a blue whale, grows underground and consists almost entirely of microscopic hyphal filaments. Molecular studies indicate that this giant fungus is or was a single individual that arose from a single spore. It is possible that fragmentation over time may have broken it into a few separate—but still gigantic—individuals. Another, larger fungus of the same genus, growing in the state of Washington, occupies parts of three counties.

Basidiomycete hyphae characteristically have septa with small, distinctive pores. The **basidium** (plural, basidia), a swollen cell at the tip of a hypha, is the characteristic sexual reproductive structure of the basidiomycetes. It is the site of nuclear fusion and meiosis. Thus, the basidium plays the same role in the basidiomycetes as the ascus does in the ascomycetes and the zygosporangium does in the zygomycetes.

The life cycle of the basidiomycetes is shown in Figure 31.15. After nuclei fuse in the basidium, the resulting diploid nucleus undergoes meiosis, and the four resulting haploid nuclei are incorporated into haploid **basidiospores**, which form on tiny stalks on the outside of the basidium. These basidiospores typically are forcibly discharged from their basidia and then germinate, giving rise to haploid hyphae. As these hyphae grow, haploid hyphae of different mating types meet and fuse, forming dikaryotic hyphae, each cell of which contains two nuclei, one from each parent hypha. The dikaryotic mycelium grows and eventually, when triggered by rain or another environmental cue, produces a basidiocarp. The dikaryon stage may persist for years—some basidiomycetes live for decades or even centuries. This pattern contrasts with the life cycle of the ascomycetes, in which the dikaryon is found only in the stages leading up to formation of the asci.

The elaborate basidiocarp of some fleshy basidiomycetes, such as the mushroom shown in Figure 31.15, is topped by a

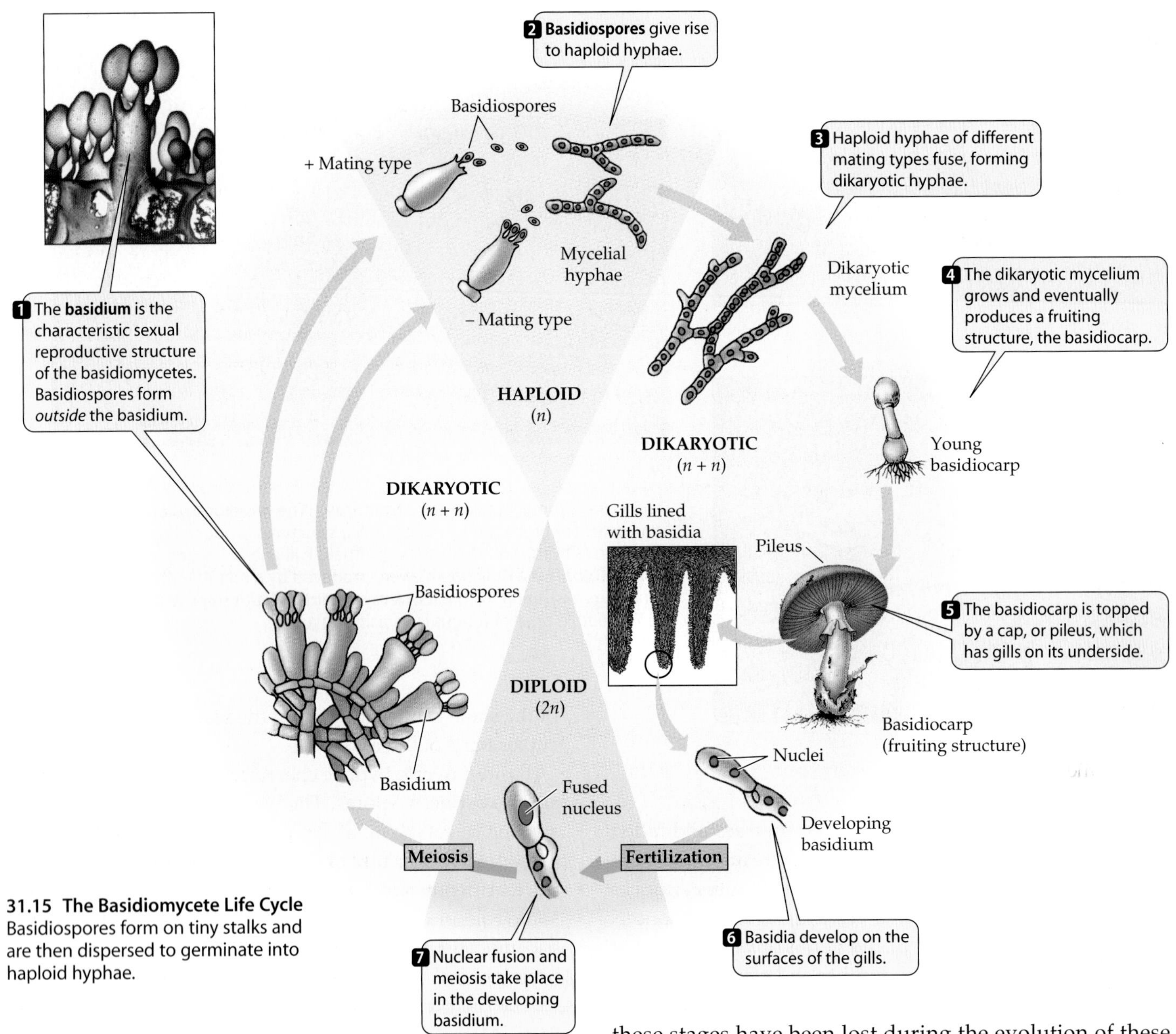

1 The **basidium** is the characteristic sexual reproductive structure of the basidiomycetes. Basidiospores form *outside* the basidium.
2 **Basidiospores** give rise to haploid hyphae.
Basidiospores
+ Mating type
Mycelial hyphae
– Mating type
3 Haploid hyphae of different mating types fuse, forming dikaryotic hyphae.
Dikaryotic mycelium
4 The dikaryotic mycelium grows and eventually produces a fruiting structure, the basidiocarp.
HAPLOID
(n)
DIKARYOTIC
($n + n$)
Young basidiocarp
DIKARYOTIC
($n + n$)
Gills lined with basidia
Pileus
5 The basidiocarp is topped by a cap, or pileus, which has gills on its underside.
Basidiospores
DIPLOID
($2n$)
Basidiocarp (fruiting structure)
Nuclei
Basidium
Fused nucleus
Developing basidium
Meiosis
Fertilization
7 Nuclear fusion and meiosis take place in the developing basidium.
6 Basidia develop on the surfaces of the gills.

31.15 The Basidiomycete Life Cycle Basidiospores form on tiny stalks and are then dispersed to germinate into haploid hyphae.

cap, or *pileus*, which has structures called *gills* on its underside. Enormous numbers of basidia develop on the surfaces of the gills. The basidia discharge their basidiospores into the air spaces between adjacent gills, and the spores sift down into air currents for dispersal and germination as new haploid mycelia. A single basidiocarp of the common bracket fungus *Ganoderma applanatum* can produce as many as 4.5 *trillion* basidiospores in one growing season.

Imperfect fungi lack a sexual stage

As we have just seen, mechanisms of sexual reproduction readily distinguish members of the four phyla of fungi from one another. But many fungi, including both saprobes and parasites, appear to lack sexual stages entirely; presumably these stages have been lost during the evolution of these species or have not yet been observed. Classifying these fungi used to be difficult, but biologists now can assign most such fungi to one of the four phyla on the basis of their DNA sequences.

Fungi that have not yet been placed in any of the existing phyla are pooled together in a polyphyletic group called **deuteromycetes**, informally known as "imperfect fungi." Thus, the deuteromycete group is a holding area for species whose status is yet to be resolved. At present, about 25,000 species are classified as imperfect fungi.

If sexual structures are found on a fungus classified as a deuteromycete, that fungus is reassigned to the appropriate phylum. That happened, for example, to a fungus that produces plant growth hormones called gibberellins (see Chapter 38). Originally classified as the deuteromycete *Fusarium moniliforme*, this fungus was later found to produce asci, whereupon it was renamed *Gibberella fujikuroi* and transferred to the phylum Ascomycota.

Fungal Associations

Earlier in this chapter we mentioned mycorrhizae and lichens, two kinds of symbiotic, mutualistic associations between fungi and other organisms. Now that we have learned a bit about fungal diversity, let's consider mycorrhizae and lichens in greater detail.

Mycorrhizae are essential to many plants

Almost all tracheophytes require a symbiotic association with fungi. Unassisted, the root hairs of such plants do not absorb enough water or minerals to sustain growth. However, their roots usually do become infected with fungi, forming an association called a mycorrhiza.

In *ectomycorrhizae*, the fungus (usually a basidiomycete) wraps around the root, and its mass is often as great as that of the root itself (Figure 31.16*a*). The fungal hyphae do not penetrate the root cells. An extensive web of hyphae penetrates the soil in the area around the root, so that up to 25 percent of the soil volume near the root may be fungal hyphae. The hyphae of the fungi attached to the root increase the surface area for the absorption of water and minerals, and the mass of the mycorrhiza, like a sponge, holds water efficiently in the neighborhood of the root. Infected roots characteristically branch extensively and become swollen and club-shaped, and they lack root hairs.

In *endomycorrhizae*, the fungal (zygomycete) hyphae enter the root and penetrate the root cells, forming tree-like structures inside the cells, which become the primary site of exchange between plant and fungus (Figure 31.16*b*). As with the ectomycorrhizae, the fungus forms a vast web of hyphae leading from the root surface into the surrounding soil.

The mycorrhizal association is important to both partners. The fungus obtains important organic compounds, such as sugars and amino acids, from the plant. In return, the fungus, because of its very high surface area-to-volume ratio and ability to penetrate the fine structure of the soil, greatly increases the plant's ability to absorb water and minerals (especially phosphorus). The fungus may also provide the plant with certain growth hormones and may protect it against attack by microorganisms. Plants that have active endomycorrhizae typically are a deeper green and may resist drought and temperature extremes better than plants of the same species that have little mycorrhizal development. Attempts to introduce some plant species to new areas have failed until a bit of soil from the native area (presumably containing the fungus necessary to establish mycorrhizae) was provided. Trees without ectomycorrhizae normally will not grow at all, so the health of our forests depends on the presence of ectomycorrhizal fungi.

The partnership between plant and fungus results in a plant that is better adapted for life on land. It has been suggested that the evolution of mycorrhizae was the single most important step leading to the colonization of the terrestrial environment by living things. Fossils of mycorrhizal structures more than 300 million years old have been found, and some rocks dating back 460 million years contain structures that appear to be fossilized fungal spores. Some liverworts, which are among the most ancient terrestrial plants (see Chapter 29), form mycorrhizae.

Certain plants that live in nitrogen-poor habitats, such as cranberry bushes and orchids, invariably have mycorrhizae. Orchid seeds will not germinate in nature unless they are already infected by the fungus that will form their mycorrhizae. Plants that lack chlorophyll always have mycorrhizae, which they often share with the

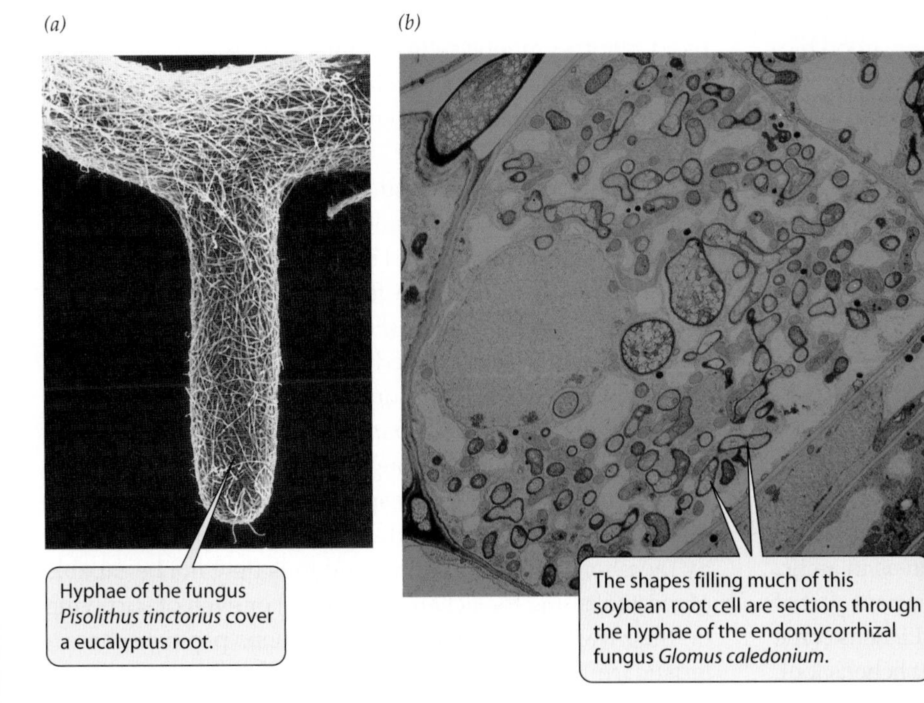
(a)
(b)
Hyphae of the fungus *Pisolithus tinctorius* cover a eucalyptus root.
The shapes filling much of this soybean root cell are sections through the hyphae of the endomycorrhizal fungus *Glomus caledonium*.

31.16 Mycorrhizal Associations (*a*) Ectomycorrhizal fungi wrap themselves around the plant root, increasing the area available for absorption of water and nutrients. (*b*) Endomycorrhizae infect the root internally and penetrate the root cells.

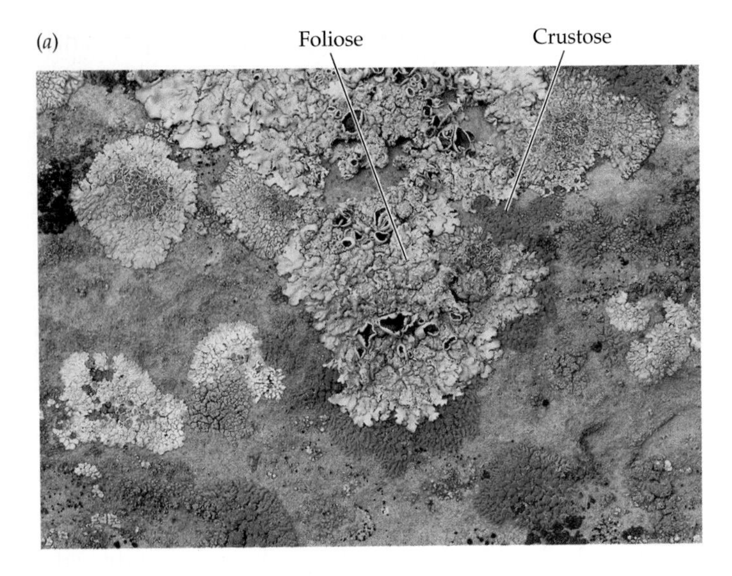

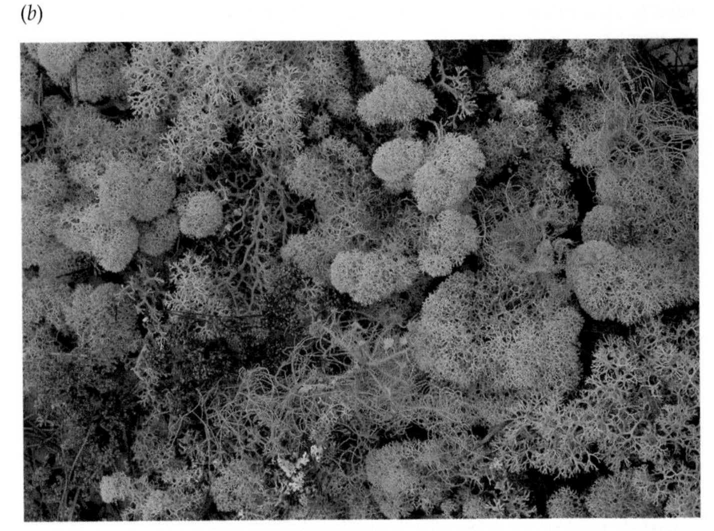

31.17 Lichen Body Forms Lichens fall into three principal classes based on their body form. (*a*) These foliose and crustose lichens are growing on otherwise bare rock. (*b*) A miniature jungle of fruticose lichens.

roots of green, photosynthetic plants. In effect, these plants without chlorophyll are feeding on nearby green plants, using the fungus as a bridge.

Lichens can grow where plants cannot

A lichen is not a single organism, but rather a meshwork of two radically different organisms: a fungus and a photosynthetic microorganism. Together the organisms constituting a lichen can survive some of the harshest environments on Earth. The biota of Antarctica, for example, features more than 100 times as many species of lichens as of plants.

In spite of this hardiness, lichens are very sensitive to air pollution because they are unable to excrete toxic substances that they absorb. Hence they are not common in industrialized cities. Because of their sensitivity, lichens are good biological indicators of air pollution.

The fungal components of most lichens are ascomycetes, but a few are basidiomycetes or imperfect fungi. The photosynthetic component is most often a unicellular green alga but may be a cyanobacterium, or may include both. Relatively little experimental work has focused on lichens, perhaps because they grow so slowly—typically less than 1 centimeter per year.

There are about 13,500 "species" of lichens. Their fungal components may constitute as many as 20 percent of all fungal species, but none of these species are able to grow independently without a photosynthetic partner. Lichens are found in all sorts of exposed habitats: on tree bark, open soil, and bare rock. Reindeer "moss" (actually not a moss at all, but the lichen *Cladonia subtenuis*) covers vast areas in arctic, subarctic, and boreal regions, where it is an important part of the diets of reindeer and other large mammals. Lichens come in various forms and colors. *Crustose* (crustlike) lichens look like colored powder dusted over their substratum (Figure 31.17*a*); *foliose* (leafy) and *fruticose* (shrubby) lichens may have complex forms (Figure 31.17*b*).

The most widely held interpretation of the lichen relationship is that it is a mutually beneficial symbiosis. The hyphae of the fungal mycelium are tightly pressed against the algae or cyanobacteria and sometimes even invade them. The bacterial or algal cells not only survive these indignities, but continue their growth and photosynthesis. In fact, the algal cells in a lichen "leak" photosynthetic products at a greater rate than do similar cells growing on their own. On the other hand, photosynthetic cells from lichens grow more rapidly on their own than when associated with a fungus. On this basis, we could consider lichen fungi as parasitic on their photosynthetic partners.

Lichens can reproduce simply by fragmentation of the vegetative body, which is called the *thallus*, or by means of specialized structures called **soredia** (singular, soredium). Soredia consist of one or a few photosynthetic cells surrounded by fungal hyphae (Figure 31.18*a*). The soredia become detached, are dispersed by air currents, and upon arriving at a favorable location, develop into a new lichen. Alternatively, if the fungal partner is an ascomycete or a basidiomycete, it may go through its sexual cycle, producing either ascospores or basidiospores. When these spores are discharged, however, they disperse alone, unaccompanied by the photosynthetic partner, and thus may not be capable of reestablishing the lichen association, or even of surviving on their own. Nevertheless, many lichens produce characteristic fruiting structures containing asci or basidia.

31.18 Lichen Anatomy (*a*) Soredia of a fruticose lichen. (*b*) Cross section showing the layers of a foliose lichen.

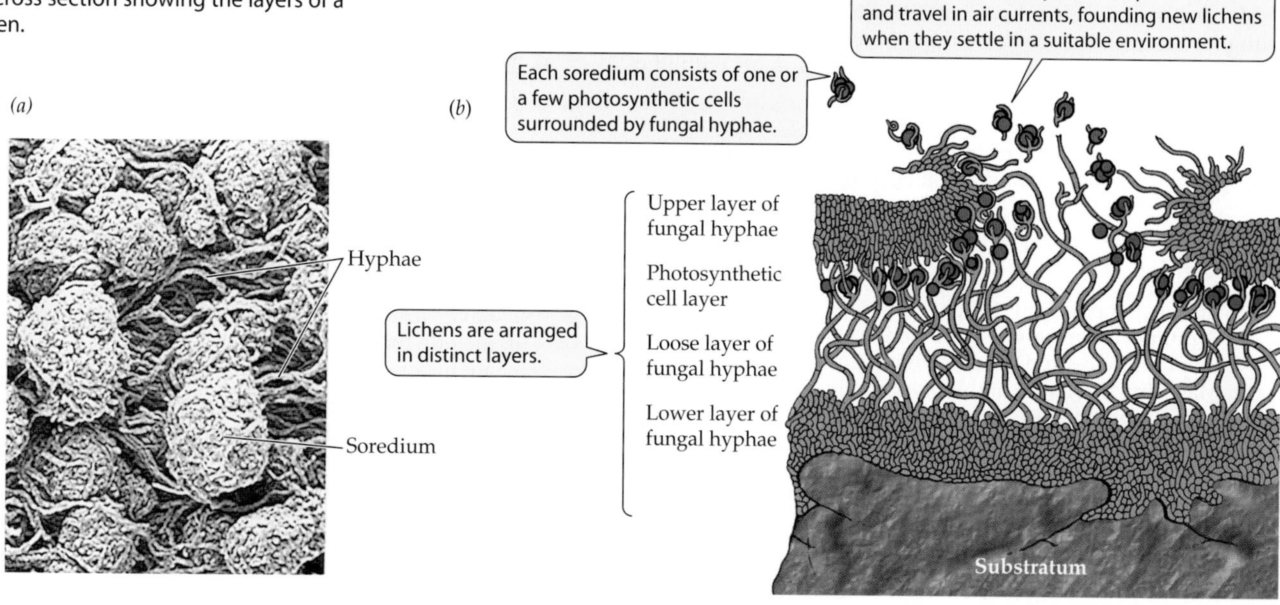
(a)
Hyphae
Soredium
(b)
Each soredium consists of one or a few photosynthetic cells surrounded by fungal hyphae.
Soredia detach readily from the parent lichen and travel in air currents, founding new lichens when they settle in a suitable environment.
Lichens are arranged in distinct layers.
Upper layer of fungal hyphae
Photosynthetic cell layer
Loose layer of fungal hyphae
Lower layer of fungal hyphae
Substratum

Visible in a cross section of a typical foliose lichen are a tight upper region of fungal hyphae, a layer of cyanobacteria or algae, a looser hyphal layer, and finally hyphal rhizoids that attach the whole structure to its substratum (Figure 31.18*b*). The meshwork of fungal hyphae takes up some nutrients needed by the photosynthetic cells and provides a suitably moist environment for them by holding water tenaciously. The fungi derive fixed carbon from the photosynthesis of the algal or cyanobacterial cells.

Lichens are often the first colonists on new areas of bare rock. They satisfy most of their nutritional needs from the air and rainwater, augmented by minerals absorbed from dust. A lichen begins to grow shortly after a rain, as it begins to dry. As it grows, the lichen acidifies its environment slightly, and this acid contributes to the slow breakdown of rocks, an early step in soil formation. After further drying, the lichen's photosynthesis ceases. The water content of the lichen may drop to less than 10 percent of its dry weight, at which point it becomes highly insensitive to extremes of temperature.

Whether living on their own or in symbiotic associations, fungi have spread successfully over much of Earth since their origin from a protist ancestor. That ancestor also gave rise to the choanoflagellates and the animal kingdom, as we will see in Chapter 32.

Chapter Summary

General Biology of the Fungi

▶ Fungi are heterotrophic eukaryotes with absorptive nutrition and with chitin in their cell walls. They may be saprobes, parasites, or mutualists.

▶ These four fungal phyla differ in their reproductive structures, mechanisms of spore formation, and less importantly, the presence and form of septa in their hyphae.

▶ The yeasts are unicellular fungi.

▶ The bodies of multicellular fungi are composed of multinucleate hyphae, often massed to form a mycelium. The hyphae usually have incomplete partitions (septa) that allow the movement of organelles between cells. They give fungi a large surface area-to-volume ratio, enhancing their ability to absorb nutrients. **Review Figures 31.3, 31.4**

▶ Fungi reproduce asexually by means of spores formed within sporangia, by conidia formed at the tips of hyphae, by fission or budding, or by fragmentation.

▶ Fungi reproduce sexually when hyphae of different mating types meet and fuse.

▶ In addition to the haploid and diploid states, many fungi demonstrate a third nuclear condition: the dikaryotic, or $n + n$, state.

Diversity in the Kingdom Fungi

▶ The kingdom Fungi consists of four phyla: Chytridiomycota, Zygomycota, Ascomycota, and Basidiomycota. **Review Figure 31.6, Table 31.1.** **See Web/CD Activity 31.1**

▶ The chytrids, with their flagellated zoospores and gametes, probably resemble the ancestral fungi.

▶ The zygomycetes reproduce sexually by fusion of gametangia. **Review Figure 31.9**

▶ The sexual reproductive structure of ascomycetes is an ascus containing ascospores. The ascomycetes are divided into two groups, euascomycetes and hemiascomycetes, on the basis of whether they have an ascocarp, or fruiting structure. **Review Figure 31.13.** **See Web/CD Activity 31.2**

▶ The sexual reproductive structure of basidiomycetes is a basidium, a swollen cell bearing basidiospores. **Review Figure 31.15**

▶ Imperfect fungi (deuteromycetes) lack sexual structures, but DNA sequencing can sometimes identify the phylum to which they belong.

See Web/CD Tutorial 31.1

Fungal Associations

▶ Mycorrhizae, which are symbiotic associations of a fungus with plant roots, enhance the ability of the roots to absorb water and nutrients. In return, the plant supplies the fungus with photosynthetic products.

▶ Lichens, which are symbiotic associations of a fungus with a green alga or a cyanobacterium, are found in some of the most inhospitable environments on the planet. **Review Figure 31.18**

Self-Quiz

1. Which statement about fungi is *not* true?
   a. A multicellular fungus has a body called a mycelium.
   b. Hyphae are composed of individual mycelia.
   c. Many fungi tolerate highly hypertonic environments.
   d. Many fungi tolerate low temperatures.
   e. Some fungi are anchored to their substrate by rhizoids.
2. The absorptive nutrition of fungi is aided by
   a. dikaryon formation.
   b. spore formation.
   c. the fact that they are all parasites.
   d. their large surface area-to-volume ratio.
   e. their possession of chloroplasts.
3. Which statement about fungal nutrition is *not* true?
   a. Some fungi are active predators.
   b. Some fungi form mutualistic associations with other organisms.
   c. All fungi require mineral nutrients.
   d. Fungi can make some of the compounds that are vitamins for animals.
   e. Facultative parasites can grow only on their specific hosts.
4. Which statement about dikaryosis is *not* true?
   a. The cytoplasm of two cells fuses before their nuclei fuse.
   b. The two haploid nuclei are genetically different.
   c. The two nuclei are of the same mating type.
   d. The dikaryon stage ends when the two nuclei fuse.
   e. Not all fungi have a dikaryon stage.
5. Reproductive structures consisting of one or more photosynthetic cells surrounded by fungal hyphae are called
   a. ascospores.
   b. basidiospores.
   c. conidia.
   d. soredia.
   e. gametes.
6. The zygomycetes
   a. have hyphae without regularly occurring septa.
   b. produce motile gametes.
   c. form fleshy fruiting bodies.
   d. are haploid throughout their life cycle.
   e. have sexual reproductive structures similar to those of the ascomycetes.
7. Which statement about ascomycetes is *not* true?
   a. They include yeasts.
   b. They form reproductive structures called asci.
   c. Their hyphae are segmented by septa.
   d. Many of their species have a dikaryotic state.
   e. All have fruiting structures called ascocarps.
8. The basidiomycetes
   a. often produce fleshy fruiting structures.
   b. have hyphae without septa.
   c. have no sexual stage.
   d. produce basidia within basidiospores.
   e. form diploid basidiospores.
9. The deuteromycetes
   a. have distinctive sexual stages.
   b. are all parasitic.
   c. have "lost" some members to other fungal groups.
   d. include the ascomycetes.
   e. are never components of lichens.
10. Which statement about lichens is *not* true?
   a. They can reproduce by fragmentation of the vegetative body.
   b. They are often the first colonists in a new area.
   c. They render their environment more basic (alkaline).
   d. They contribute to soil formation.
   e. They may contain less than 10 percent water by weight.

For Discussion

1. You are shown an object that looks superficially like a pale green mushroom. Describe at least three criteria (including anatomical and chemical traits) that would enable you to tell whether the object is a piece of a plant or a piece of a fungus.
2. Differentiate among the members of the following pairs of related terms:
   a. hypha/mycelium
   b. euascomycete/hemiascomycete
   c. ascus/basidium
   d. ectomycorrhiza/endomycorrhiza
3. For each type of organism listed below, give a single characteristic that may be used to differentiate it from the other, related organism(s) in parentheses.
   a. Zygomycota (Ascomycota)
   b. Basidiomycota (deuteromycetes)
   c. Ascomycota (Basidiomycota)
   d. baker's yeast (*Neurospora crassa*)
4. Many fungi are dikaryotic during part of their life cycle. Why are dikaryons described as $n + n$ instead of $2n$?
5. If all the fungi on Earth were suddenly to die, how would the surviving organisms be affected? Be thorough and specific in your answer.
6. How might the first mycorrhizae have arisen?
7. What might account for the ability of lichens to withstand the intensely cold environment of Antarctica? Be specific in your answer.

32 Animal Origins and the Evolution of Body Plans

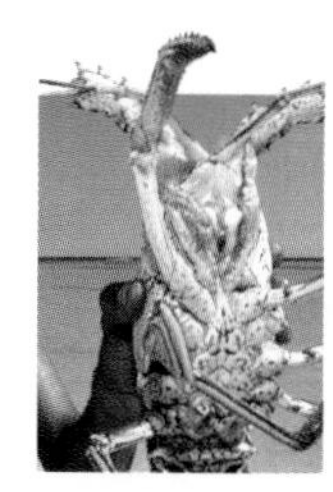

In 1822, nearly forty years before Darwin wrote *The Origin of Species*, a French naturalist, Étienne Geoffroy Saint-Hilaire, was examining a lobster. He noticed that when he turned the lobster upside down and viewed it with its ventral surface up, its central nervous system was located above its digestive tract, which in turn was located above its heart—the same relative positions these systems have in mammals when viewed *dorsally*. His observations led Geoffroy to conclude that the differences between arthropods (such as lobsters) and vertebrates (such as mammals) could be explained if the embryos of one of those groups were inverted during development.

Geoffroy's suggestion was regarded as preposterous at the time and was largely dismissed until recently. However, the discovery of two genes that influence a system of extracellular signals involved in development has lent new support to Geoffroy's seemingly outrageous hypothesis.

A vertebrate gene called *chordin* helps to establish cells on one side of the embryo as dorsal and on the other as ventral. A probably homologous gene in fruit flies, called *sog*, acts in a similar manner, but has the opposite effect. Fly cells where *sog* is active become ventral, whereas vertebrate cells where *chordin* is active become dorsal. However, when *sog* mRNA is injected into an embryo of the frog *Xenopus*, a vertebrate, it causes dorsal development. *Chordin* mRNA injected into fruit flies promotes ventral development. In both cases, injection of the mRNA promotes the development of the portion of the embryo that contains the central nervous system!

Chordin and *sog* are among many genes that regulate similar functions in very different organisms. Such genes are providing evolutionary biologists with information that can help them understand relationships among animal lineages that separated from one another in ancient times. As we saw in Chapter 25, new knowledge about gene functions and gene sequences is increasingly being used to infer evolutionary relationships.

In this chapter, we will apply the methods described in Chapter 25 to infer evolutionary relationships among the animals. First, we will review the defining characteristics of the animal way of life. Then we will describe several lineages of simple animals. Finally, we will describe the lophotro-

Genes that Control Development A human and a lobster carry similar genes that control the development of the body axis, but these genes position their body systems inversely. A lobster's nervous system runs up its ventral (belly) surface, whereas a vertebrate's runs down its dorsal (back) surface.

chozoans, one of the three great evolutionary lineages of animals. In the next two chapters, we will discuss the other two great animal lineages, the ecdysozoans and the deuterostomes.

Animals: Descendants of a Common Ancestor

Biologists have long debated whether animals arose once or several times from protist ancestors, but enough molecular and morphological evidence has now been assembled to indicate that, with the possible exception of sponges (Porifera), the Kingdom Animalia is a monophyletic group—that is, all animals are descendants of a single ancestral lineage. This conclusion is supported by the fact that all animals share several derived traits:

- Similarities in their small-subunit ribosomal RNAs (see Chapter 26)
- Similarities in their Hox genes (see Chapter 20)
- Special types of cell–cell junctions: tight junctions, desmosomes, and gap junctions (see Figure 5.6)
- A common set of extracellular matrix molecules, including collagen (see Figure 4.26)

Animals evolved from colonial flagellated protists as a result of division of labor among their aggregated cells. Within the ancestral colonies of cells—perhaps analogous to those still existing in the chlorophyte *Volvox* or some colonial choanoflagellates (see Figures 28.25*a* and 28.28)—some cells became specialized for movement, others became specialized for nutrition, and still others differentiated into gametes. Once this specialization by function had begun, working groups of cells continued to differentiate while improving their coordination with other groups of cells. Such coordinated groups of cells evolved into the larger and more complex organisms that we now call **animals**.

Animals are multicellular heterotrophs

What traits characterize the animals? In contrast to the Bacteria, Archaea, and most protists, all animals are *multicellular*. Unlike plants, animals must take in pre-formed organic molecules because they cannot synthesize them from inorganic chemicals. They acquire these organic molecules by ingesting other organisms or their products, either living or dead, and digesting them inside their bodies; thus animals are *heterotrophs*. Most animals have *circulatory systems* that take up O_2, get rid of CO_2, and carry nutrients from their guts to other body tissues.

To acquire food, animals must expend energy either to move through the environment and position themselves where food will pass close to them, or to move the environment and the food it contains to them. The foods animals ingest include most other members of the animal kingdom as well as members of all other kingdoms. Much of the diversity of animal sizes and shapes evolved as animals acquired the ability to capture and eat many different kinds of food and to avoid becoming food for other animals. The need to locate food has favored the evolution of sensory structures to provide animals with detailed information about their environment and nervous systems to receive and coordinate that information.

The accounts in this chapter and the following two chapters serve as an orientation to the major groups of animals, their similarities and differences, and the evolutionary pathways that resulted in the current richness of animal lineages and species. But how do biologists infer evolutionary relationships among animals?

Several traits show evolutionary relationships among animals

Biologists use a variety of traits in their efforts to infer animal phylogenies. Clues to these relationships are found in the fossil record, in patterns of embryonic development, in the comparative morphology and physiology of living and fossil animals, and in the structure of animal molecules, especially small subunit rRNAs and mitochondrial genes (see Chapters 25 and 26).

Using this wide variety of comparative data, zoologists concluded that sponges, cnidarians, and ctenophores separated from the remaining animal lineages early in evolutionary history. Biologists have divided the remaining animals into two major lineages: the protostomes and the deuterostomes. Figure 32.1 shows the postulated order of divergence of the major animal groups that we will use in these three chapters.

Several differences in patterns of embryonic development provide clues to animal phylogeny. During the development of an animal from a single-celled zygote to a multicellular adult, distinct layers of cells form. The embryos of **diploblastic** animals have only two of these cell layers: an outer *ectoderm* and an inner *endoderm*. The embryos of **triploblastic** animals have, in addition to ectoderm and endoderm, a third layer, the *mesoderm*, which lies between the ectoderm and the endoderm. The existence of three cell layers distinguishes the protostomes and deuterostomes from those groups of simple animals that diverged from them earlier.

During early development in many animals, a cavity forms in a spherical embryo. The opening of this cavity is called the *blastopore*. Among the **protostomes** (from the Greek, "mouth first"), the mouth arises from the blastopore; the anus forms later. Among the **deuterostomes** ("mouth second"), the blastopore becomes the anus; the mouth forms later.

In the common ancestor of the protostomes and deuterostomes, the pattern of early cleavage of the fertilized egg was

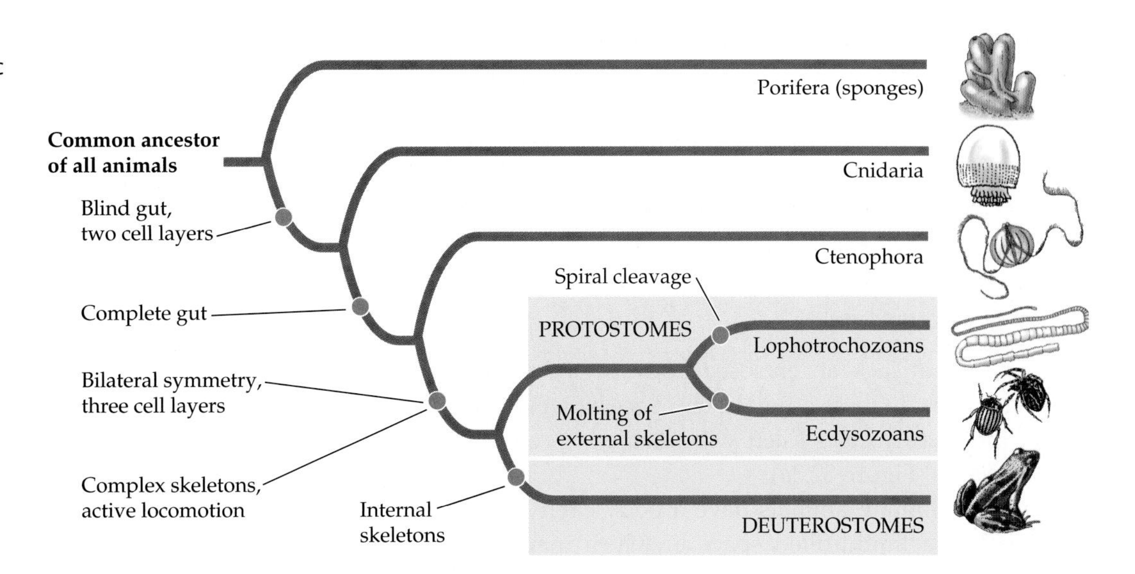

32.1 A Current Phylogeny of the Animals The phylogenetic tree used in this and the following two chapters assumes that the animals are monophyletic. The characters highlighted by red circles on the tree will be explained as we discuss the different phyla.

radial (see Figure 20.4). This cleavage pattern persisted during the evolution of the deuterostomes and in many protostomate lineages, but spiral cleavage evolved in one major protostomate lineage, as we will see.

Body Plans: Basic Structural Designs

The general structure of an animal, its organ systems, and the integrated functioning of its parts are known as its **body plan**. A fundamental aspect of an animal's body plan is its overall shape, described in part by its **symmetry**. A symmetrical animal can be divided along at least one plane into similar halves. Animals that have no plane of symmetry are said to be **asymmetrical**. Many sponges are asymmetrical, but most animals have some kind of symmetry.

The simplest form of symmetry is **spherical symmetry**, in which body parts radiate out from a central point. An infinite number of planes passing through the central point can divide a spherically symmetrical organism into similar halves. Spherical symmetry is widespread among the protists, but most animals possess other forms of symmetry.

An organism with **radial symmetry** has one main axis around which its body parts are arranged. A perfectly radially symmetrical animal can be divided into similar halves by any plane that contains the main axis. Some simple sponges and a few other animals, such as sea anemones (Figure 32.2*a*), have radial symmetry. Most radially symmetrical animals are slightly modified so that fewer planes can divide them into identical halves. Two animal phyla—Cnidaria and Ctenophora—are composed primarily of radially symmetrical animals. These animals move slowly or not at all.

Bilateral symmetry is a common characteristic of animals that move rapidly through their environments. A bilaterally symmetrical animal can be divided into mirror images (left and right sides) by a single plane that passes through the dorsoventral midline of its body from the front (*anterior*) to the back (*posterior*) end (Figure 32.2*b*). A plane at right angles to the first one divides the body into two dissimilar sides; the back side of a bilaterally symmetrical animal is its *dorsal* surface; the belly side is its *ventral* surface.

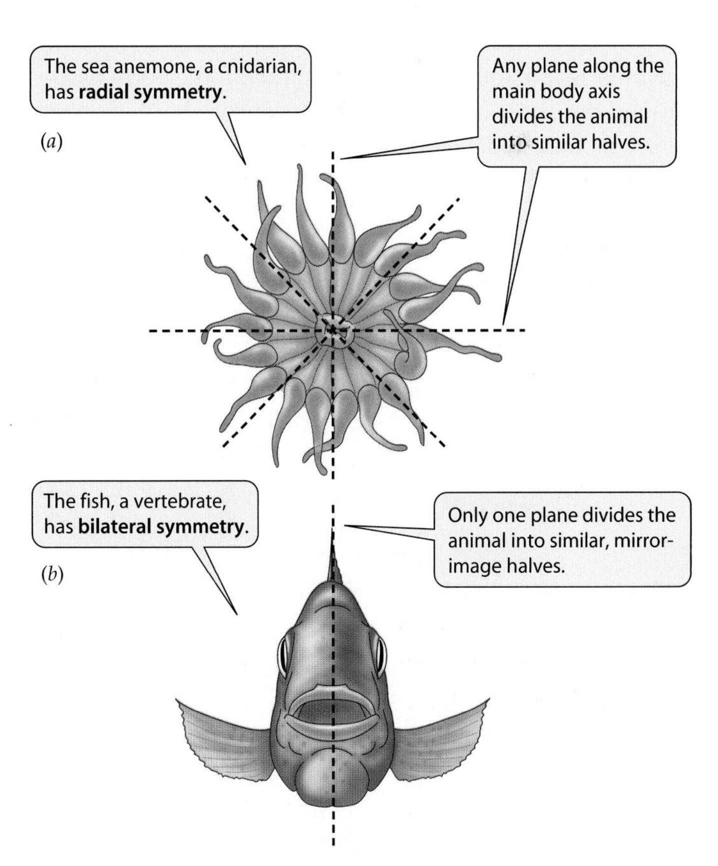

The sea anemone, a cnidarian, has radial symmetry.
(a)
Any plane along the main body axis divides the animal into similar halves.
The fish, a vertebrate, has bilateral symmetry.
(b)
Only one plane divides the animal into similar, mirror-image halves.

32.2 Body Symmetry Most animals are either radially or bilaterally symmetrical.

Bilateral symmetry is strongly correlated with **cephalization**: the concentration of sensory organs and nervous tissues in a head at the anterior end of the animal. Cephalization is favored because the anterior end of a freely moving animal typically encounters new environments first.

Fluid-filled spaces, called **body cavities**, lie between the ectoderm and endoderm of most protostomes and deuterostomes. The type of body cavity an animal has strongly influences the way it moves.

- Animals that lack an enclosed body cavity are called **acoelomates**. In these animals, the space between the gut and the body wall is filled with masses of cells called *mesenchyme* (Figure 32.3*a*).
- **Pseudocoelomate** animals have a body cavity called a *pseudocoel*, a liquid-filled space in which many of the internal organs are suspended. Their control over body shape is crude because the pseudocoel has muscles only on its outside; there is no inner layer of muscle surrounding the organs (Figure 32.3*b*).
- **Coelomate** animals have a *coelom*, a body cavity that develops within the mesoderm. It is lined with a special structure called the *peritoneum* and is enclosed on both the inside and the outside by muscles (Figure 32.3*c*).

The fluid-filled body cavities of simple animals function as **hydrostatic skeletons**. Because fluids are relatively incompressible, they move to another part of the cavity when the muscles surrounding them contract. If the body tissues around the cavity are flexible, fluids squeezed out of one region can cause some other region to expand. The moving fluids can thus move specific body parts. If a temporary attachment can be made to the substratum, the whole animals can move from one place to another.

In animals that have both circular muscles (encircling the body) and longitudinal muscles (running along the length of the body), the action of these antagonistic muscles on the fluid-filled body cavity gives the animal even greater control over its movement. A coelomate animal has better control over the movement of the fluids in its body cavity than does a pseudocoelomate animal, but its control is further improved if the coelom is separated into compartments or segments. Then muscles in each individual segment can change its shape independently of the other segments. Segmentation of the coelom evolved several different times among both protostomes and deuterostomes.

Other forms of skeletons developed in many animal lineages, either as substitutes for, or in combination with, hydrostatic skeletons. Some skeletons are internal (such as vertebrate bones); others are external (such as lobster shells). Some external skeletons consist of a single element (snail shells), others have two elements (clam shells), and still others have many elements (centipedes).

Sponges: Loosely Organized Animals

The lineage leading to modern sponges separated from the lineage leading to all other animals very early during animal evolution. The difference between protist colonies and simple multicellular animals is that the animal cells are differentiated and their activities are coordinated. However, sponge cells do not form true organs.

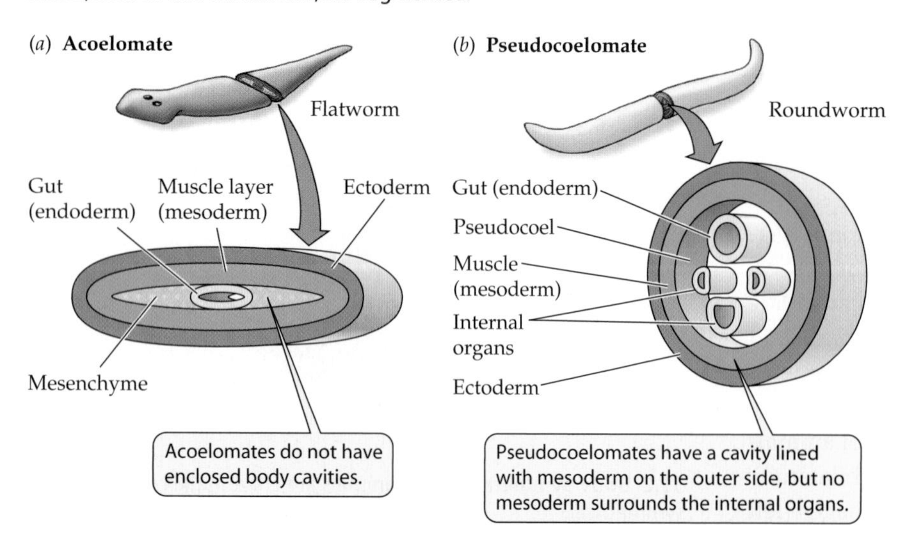

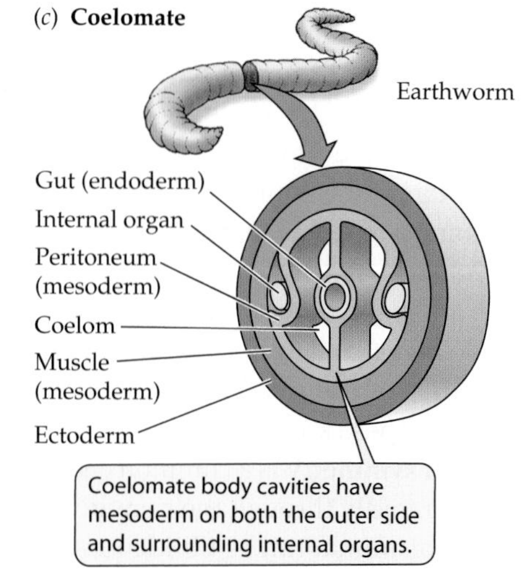

32.3 Animal Body Cavities There are three major types of body cavities among the animals. (*a*) Acoelomates do not have enclosed body cavities. (*b*) Pseudocoelomates have only one layer of muscle, and it lies outside the body cavity. (*c*) Coelomates have a peritoneum surrounding the internal organs. The body cavities of some coelomates, such as this earthworm, are segmented.

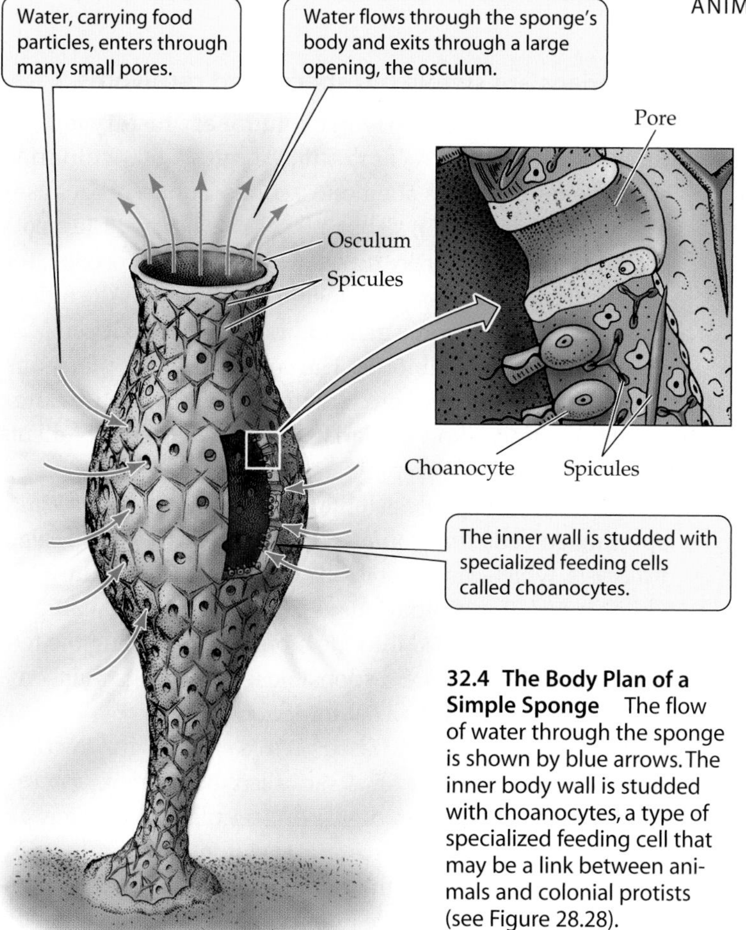
Water, carrying food particles, enters through many small pores.
Water flows through the sponge's body and exits through a large opening, the osculum.
Pore
Osculum
Spicules
Choanocyte
Spicules
The inner wall is studded with specialized feeding cells called choanocytes.

32.4 The Body Plan of a Simple Sponge The flow of water through the sponge is shown by blue arrows. The inner body wall is studded with choanocytes, a type of specialized feeding cell that may be a link between animals and colonial protists (see Figure 28.28).

Sponges (phylum **Porifera**, from the Latin, "pore bearers"), the simplest of animals, are **sessile**: They live attached to the substratum and do not move about. The body plan of all sponges—even large ones, which may reach more than a meter in length—is an aggregation of cells built around a water canal system. Feeding cells called *choanocytes* line the inside of the internal chambers. These cells, with a collar of microscopic villi and a single flagellum, bear a striking resemblance to choanoflagellates (see Figure 28.28). By beating their flagella, choanocytes cause the surrounding water to flow through the animal. The water, along with any food particles it contains, enters by way of small pores and passes into the water canals, where food particles are captured by the choanocytes. Water then exits through one or more larger openings called *oscula* (Figure 32.4).

Between the thin epidermis and the choanocytes is another layer of cells, some of which are similar to amoebas and move about within the body. A supporting skeleton is also present, in the form of simple or branching spines, called *spicules*, and often an elastic, complex, network of fibers. A few species of sponges are carnivores that trap prey on hook-shaped spicules that protrude from the body surface. Sponges also have an extracellular matrix, composed of collagens, adhesive glycoproteins, and other molecules, that holds the cells together. This molecular adhesion system may also be involved in cell–cell signaling.

Thus, sponges are functionally more complex than a superficial look at their morphology might suggest. Nonetheless, sponges are loosely organized. Even if a sponge is completely disassociated by being strained through a filter, its cells can reassemble into a new sponge.

Most of the 5,500 species of sponges are marine animals; only about 50 species live in fresh water. Sponges come in a wide variety of sizes and shapes that are adapted to different movement patterns of water (Figure 32.5). Sponges living in intertidal or shallow subtidal environments, where they are subjected to strong wave action, hug the substratum. Many sponges that live in calm waters are simple, with a single large osculum on top of the body. Most sponges that live in slowly flowing water are flattened and are oriented at right

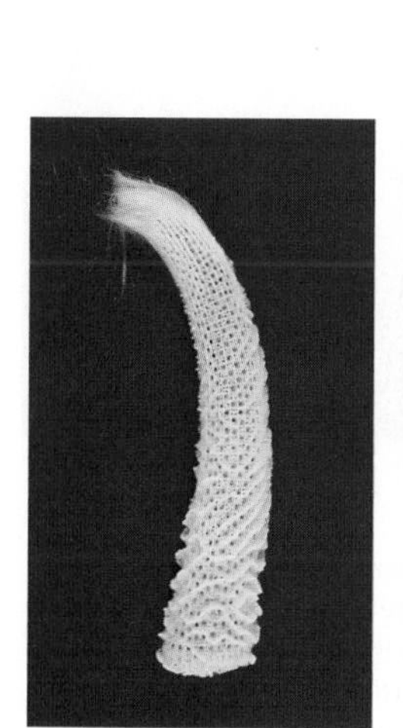
(*a*) *Euplectella aspergillum*

(*b*) *Aplysina lacunosa*

(*c*) *Asbestopluma* sp.

32.5 Sponges Differ in Size and Shape (*a*) Glass sponges are named after their glasslike spicules, which are formed of silicon. (*b*) The purple tube sponge is typical of many simple marine sponges. (*c*) This predatory sponge uses its hook-shaped spicules to capture small prey animals.

angles to the direction of current flow; they intercept water and the prey it contains as it flows past them.

Sponges reproduce both sexually and asexually. In most species, a single individual produces both eggs and sperm, but individuals do not self-fertilize. Water currents carry sperm from one individual to another. Asexual reproduction is by budding and fragmentation.

Cnidarians: Two Cell Layers and Blind Guts

Porifera (sponges)
Cnidaria
Ctenophora
PROTOSTOMES
Lophotrochozoans
Ecdysozoans
DEUTEROSTOMES

Animals in all phyla other than Porifera have distinct cell layers and symmetrical bodies. The next lineage to diverge from the main line of animal evolution after the sponges led to a phylum of animals called the **cnidarians** (phylum **Cnidaria**). These animals are *diploblastic* (have two cell layers) and have a blind gut with only one entrance (the mouth/anus). Despite their relative structural simplicity, cnidarians have structural molecules (such as collagen, actin, and myosin) and homeobox genes.

Cnidarians are simple but specialized carnivores

Cnidarians appeared early in evolutionary history and radiated in the late Precambrian. About 11,000 cnidarian species—jellyfish, sea anemones, corals, and hydrozoans—live today (Figure 32.6), all but a few in the oceans. The smallest cnidarians can hardly be seen without a microscope; the largest known jellyfish is 2.5 meters in diameter. All cnidarians are carnivores; some gain additional nutrition from photosynthetic endosymbionts. The cnidarian body plan combines a low metabolic rate with the ability to capture large prey. These traits allow cnidarians to survive in environments where encounters with prey are infrequent.

All cnidarians possess tentacles covered with *cnidocytes*, specialized cells that contain stinging organelles called *nematocysts*, which can inject toxins into their prey (Figure 32.7). Cnidocytes allow cnidarians to capture large and complex prey, which are carried into the mouth by retracting the tentacles. Nematocysts are responsible for the stings that some jellyfish inflict on human swimmers.

The cnidarian body is based on a "sac plan," in which the mouth is connected to a blind sac called the *gastrovascular cavity*. The sac functions in digestion, circulation, and gas exchange and acts as a hydrostatic skeleton. The single opening serves as both mouth and anus. Cnidarians also have epithelial cells with muscle fibers whose contractions enable the animals to move, as well as simple *nerve nets* that integrate their body activities.

(*a*) *Anthopleura elegantissima*

(*b*) *Ptilosarcus gurneyi*

(*c*) *Pelagia panopyra*

(*d*) *Polyorchis penicillatus*

32.6 Diversity among Cnidarians (*a*) The nematocyst-studded tentacles of this sea anemone from British Columbia are poised to capture large prey carried to the animal by water movement. (*b*) The orange sea pen is a colonial cnidarian that lives in soft bottom sediments and projects polyps above the substratum. (*c*) This purple jellyfish illustrates the complexity of a scyphozoan medusa. (*d*) The internal structure of the medusa of a North Atlantic colonial hydrozoan is visible here.

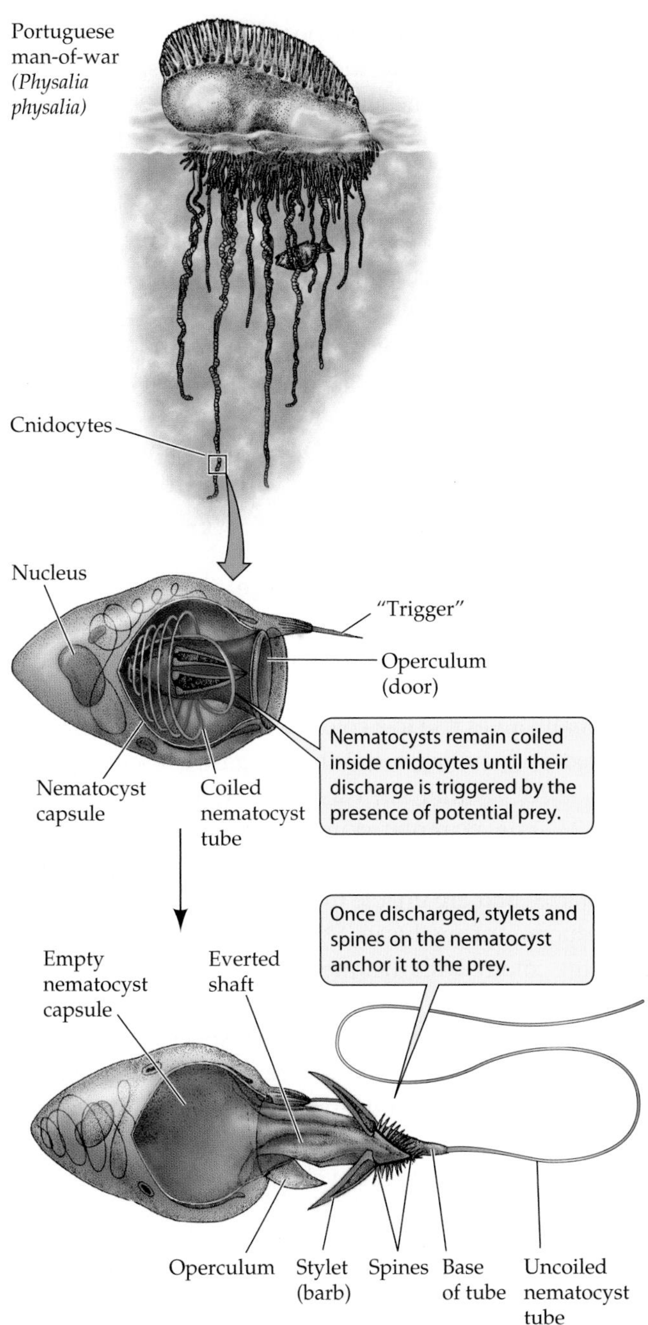

32.7 Nematocysts Are Potent Weapons Cnidarians such as the Portuguese man-of-war, which possesses a large number of nematocysts, can subdue and consume very large prey.

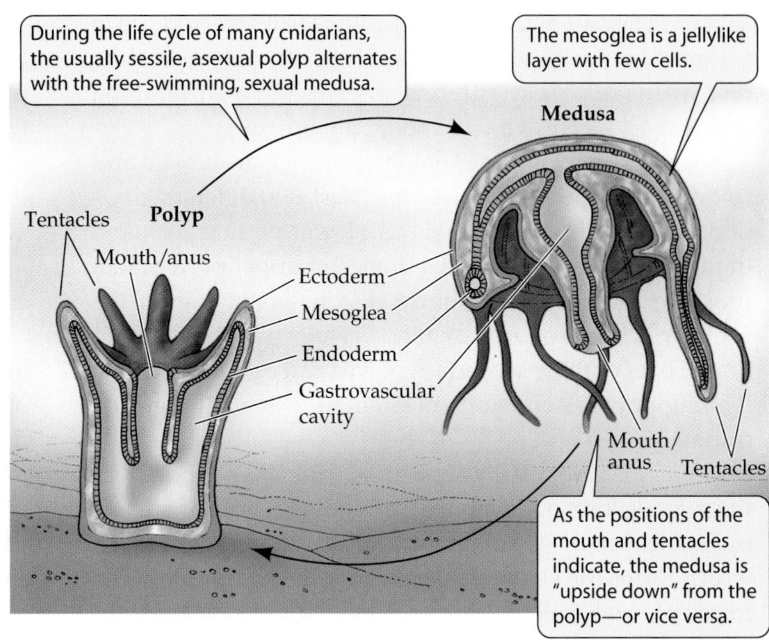

During the life cycle of many cnidarians, the usually sessile, asexual polyp alternates with the free-swimming, sexual medusa.
The mesoglea is a jellylike layer with few cells.
Medusa
Tentacles
Polyp
Mouth/anus
Ectoderm
Mesoglea
Endoderm
Gastrovascular cavity
Mouth/anus
Tentacles
As the positions of the mouth and tentacles indicate, the medusa is "upside down" from the polyp—or vice versa.

32.8 A Generalized Cnidarian Life Cycle Cnidarians typically have two body forms, one asexual (the polyp) and the other sexual (the medusa).

Cnidarian life cycles have two stages

The generalized cnidarian life cycle has two distinct stages (Figure 32.8), although many species lack one of these stages:

- The sessile **polyp** stage has a cylindrical stalk attached to the substratum. Tentacles surround a mouth/anus located at the end opposite from the stalk. Individual polyps may reproduce by budding, thereby forming a colony.
- The **medusa** (plural, medusae) is a free-swimming stage shaped like a bell or an umbrella. It typically floats with its mouth and tentacles facing downward. Medusae of many species produce eggs and sperm and release them into the water. When an egg is fertilized, it develops into a free-swimming, ciliated larva called a **planula**, which eventually settles to the bottom and develops into a polyp.

Although the polyp and medusa stages appear very different, they share a similar body plan. A medusa is essentially a polyp without a stalk. Most of the outward differences between polyps and medusae are due to the *mesoglea*, an internal mass of jellylike material that lies between the two cell layers. The mesoglea contains few cells and has a low metabolic rate. In polyps, the mesoglea is usually thin; in medusae it is very thick, constituting the bulk of the animal.

ANTHOZOANS. All 6,000 species of sea anemones and corals that constitute the **anthozoans** (class **Anthozoa**) are marine animals. Evidence from morphology, rRNA, and mitochondrial genes suggests that the anthozoans, which lack the medusa stage, are sister to the other classes of cnidarians, and that the medusa stage evolved after the anthozoa diverged from those other lineages. In the anthozoans, the polyp produces eggs and sperm, and the fertilized egg develops into a planula that develops directly into another polyp. Many species can also reproduce asexually by budding or fission. Sea anemones (see Figure 32.6*a*) are solitary. They are widespread in both warm and cold ocean waters. Many sea anemones are able to crawl slowly on the discs with which they attach themselves to the substratum. A few species can swim and some can burrow.

Sea pens (see Figure 32.6*b*), by contrast, are sessile and colonial. Each colony consists of at least two different kinds of polyps. The primary polyp has a lower portion anchored in the bottom sediment and a branched upper portion that projects above the substratum. Along the upper portion, the primary polyp produces smaller secondary polyps by budding. Some of these secondary polyps can differentiate into feeding polyps, while others circulate water through the colony.

Corals also are usually sessile and colonial. The polyps of most corals form a skeleton by secreting a matrix of organic molecules upon which they deposit calcium carbonate, which forms the eventual skeleton of the coral colony. The forms of coral skeletons are species-specific and highly diverse. The common names of coral groups—horn corals, brain corals, staghorn corals, and organ pipe corals, among others—describe their appearance (Figure 32.9*a*).

As a coral colony grows, old polyps die, but their calcareous skeletons remain. The living members form a layer on top of a growing bank of skeletal remains, eventually forming chains of islands and reefs (Figure 32.9*b*). The Great Barrier Reef along the northeastern coast of Australia is a system of coral formations more than 2,000 km long and as wide as 150 km. A reef hundreds of kilometers long in the Red Sea has been calculated to contain more material than all the buildings in the major cities of North America combined.

Corals flourish in nutrient-poor, clear, tropical waters. They can grow rapidly in such environments because the photosynthetic dinoflagellates that live symbiotically within their cells provide them with products of photosynthesis and contribute to calcium deposition. In turn, the corals provide the dinoflagellates with a place to live and nutrients. This symbiotic relationship explains why reef-forming corals are restricted to clear surface waters, where light levels are high enough to allow photosynthesis.

Coral reefs throughout the world are being threatened both by global warming, which is raising the temperatures of shallow tropical ocean waters, and by polluted runoff from development on adjacent shorelines. An overabundance of nitrogen in the runoff gives an advantage to algae, which overgrow and eventually smother the corals.

HYDROZOANS. Life cycles are diverse among the **hydrozoans** (class **Hydrozoa**). The polyp typically dominates the life cycle, but some species have only medusae and others only polyps. Most hydrozoans are colonial. A single planula eventually gives rise to a colony of many polyps, all interconnected and sharing a continuous gastrovascular cavity (Figure 32.10). Within such a colony (the man-of-war in Figure 32.7 is an example), some polyps have tentacles with many nematocysts; they capture prey for the colony. Others lack tentacles and are unable to feed, but are specialized for the production of medusae. Still others are fingerlike and defend the colony with their nematocysts.

SCYPHOZOANS. The several hundred species of **scyphozoans** (class **Scyphozoa**) are all marine. The mesoglea of their medusae is thick and firm, giving rise to their common names, jellyfish or sea jellies. The medusa, rather than the polyp, dominates the life cycle of scyphozoans. An indi-

32.9 Corals The South Pacific is home to many spectacular corals. (*a*) This unusually large formation of chalice coral was photographed off the coast of Fiji. (*b*) Many different species of corals and sponges grow together on this reef in Palau.

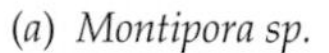
(*a*) *Montipora sp.*

(*b*)

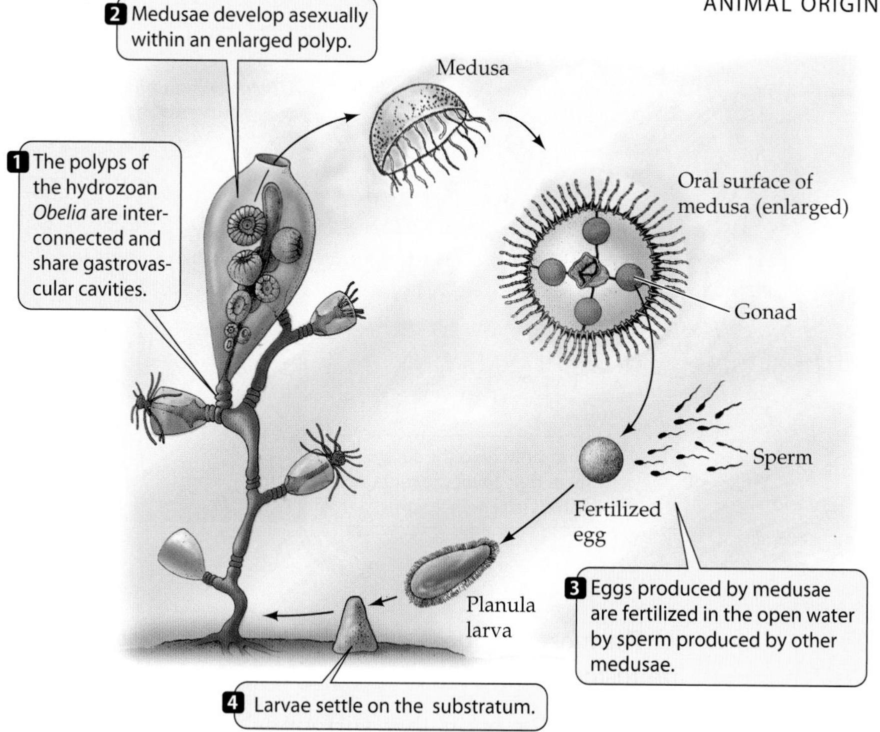

2 Medusae develop asexually within an enlarged polyp.
Medusa
1 The polyps of the hydrozoan *Obelia* are interconnected and share gastrovascular cavities.
Oral surface of medusa (enlarged)
Gonad
Sperm
Fertilized egg
Planula larva
3 Eggs produced by medusae are fertilized in the open water by sperm produced by other medusae.
4 Larvae settle on the substratum.

32.10 Hydrozoans Often Have Colonial Polyps The polyps within a hydrozoan colony may differentiate to perform specialized tasks. In the species whose life cycle is diagrammed here, the medusa is the sexual reproductive stage, producing eggs and sperm in organs called gonads.

vidual medusa is male or female, releasing eggs or sperm into the open sea. The fertilized egg develops into a small planula that quickly settles on a substratum and develops into a small polyp. This polyp feeds and grows and may produce additional polyps by budding. After a period of growth, the polyp begins to bud off small medusae (Figure 32.11). These medusae feed, grow, and transform themselves into adult medusae, which are commonly seen during summer in harbors and bays.

Ctenophores: Complete Guts and Tentacles

Ctenophores (phylum **Ctenophora**) were the next lineage to diverge from the lineage leading to all other animals. Ctenophores, also known as comb jellies, have body plans that are superficially similar to those of cnidarians. Both have two cell layers separated by a thick, gelatinous mesoglea, and both have radial symmetry and feeding tentacles. Like cnidarians, ctenophores have low metabolic rates because they are composed primarily of inert mesoglea. Unlike cnidarians, however, ctenophores have a complete gut. Food enters through a mouth, and wastes are eliminated through two anal pores.

Ctenophores have eight comblike rows of fused plates of cilia, called *ctenes* (Figure 32.12). Ctenophores move by beating these cilia rather than by muscular contractions. Ctenophoran tentacles do not have nematocysts; rather, they are covered with cells that discharge adhesive material when they contact prey. After capturing its prey, a ctenophore retracts its tentacles to bring the food to its mouth. In some species, the entire surface of the body is coated with sticky mucus that captures prey. All of the 100 known species of ctenophores eat small animals. They are common in open seas.

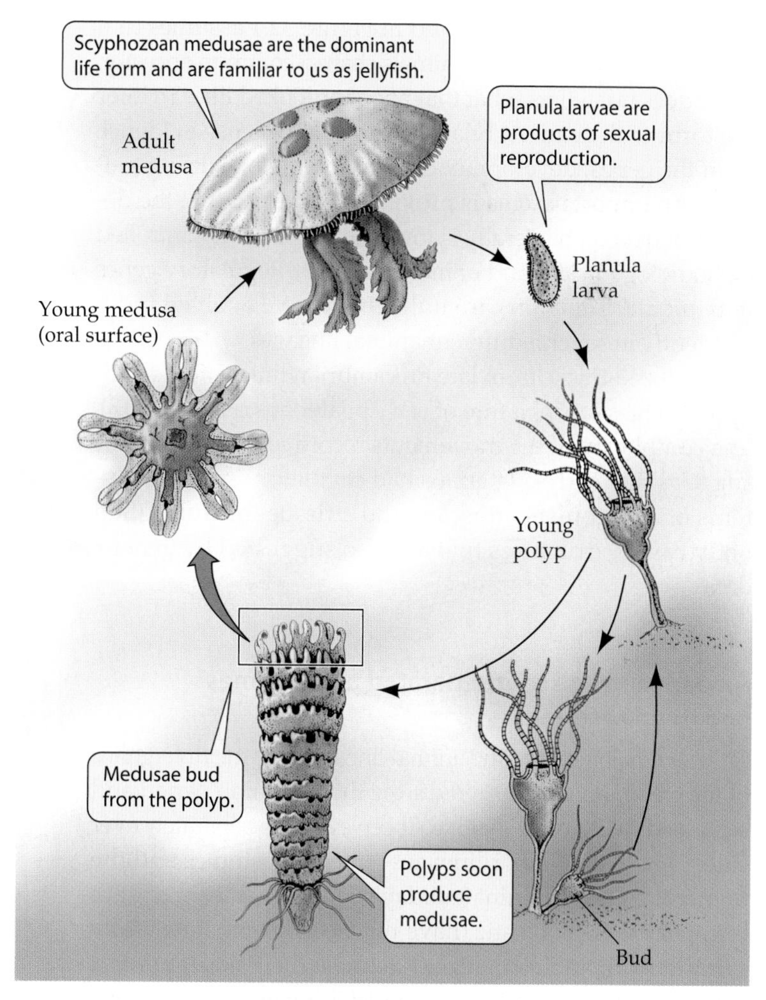

32.11 Medusae Dominate Scyphozoan Life Cycles Scyphozoan medusae are the familiar jellyfish of coastal waters. The small, sessile polyps quickly produce medusae (see Figure 32.6*c*).

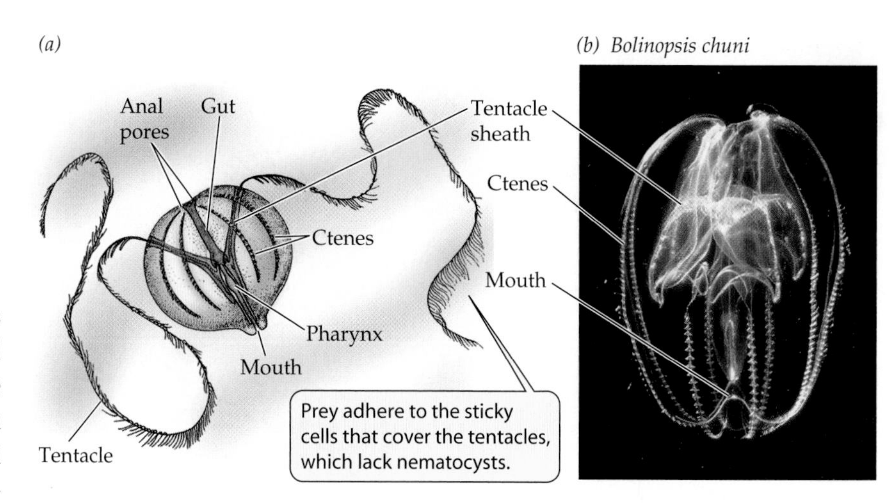

32.12 Comb Jellies Feed with Tentacles (*a*) The body plan of a typical ctenophore. The long, sticky tentacles sweep through the water, efficiently harvesting small prey. (*b*) A comb jelly photographed in Sydney Harbour, Australia, has short tentacles.

Ctenophore life cycles are simple. Gametes are produced in structures called *gonads*, located on the walls of the gastrovascular cavity. The gametes are released into the cavity and then discharged through the mouth or the anal pores. Fertilization takes place in open seawater. In nearly all species, the fertilized egg develops directly into a miniature ctenophore that gradually grows into an adult.

The Evolution of Bilaterally Symmetrical Animals

The phylogenetic tree pictured in Figure 32.1 assumes that all bilaterally symmetrical animals share a common ancestor, but it does not tell us what that ancestor looked like. To infer the form of the earliest bilaterians, zoologists use evidence from the genes, development, and structure of existing animals. An important clue is provided by the fact that the development of all bilaterally symmetrical animals is controlled by homologous *Hox* and homeobox genes. Regulatory genes with similar functions are unlikely to have evolved independently in several different animal lineages.

Fossilized tracks from late Precambrian times provide additional clues to the nature of early bilaterians (Figure 32.13). The complexity of the movements recorded by these tracks suggests that early bilaterians had circulatory systems, systems of antagonistic muscles, and a tissue- or fluid-filled body cavity, structures that are also suggested by genetic data.

An early lineage split separated protostomes and deuterostomes

The next major split in the animal lineage after the divergence of the ctenophores occurred during the Cambrian period and separated two groups that have been evolving separately ever since. These two major lineages—the protostomes and the deuterostomes—dominate today's fauna. Members of both lineages are triploblastic (have three cell layers), bilaterally symmetrical, and cephalized. Because their skeletons and body cavities are more complex than those of the animals we have discussed so far, they are capable of more complex movements.

The most important shared, derived traits that unite the protostomes are

- An anterior brain that surrounds the entrance to the digestive tract
- A ventral nervous system consisting of paired or fused longitudinal nerve cords
- A free-floating larva with a food-collecting system consisting of compound cilia on multiciliate cells
- A blastopore that becomes the mouth
- Spiral cleavage (in some species)

32.13 The Trail of an Early Bilaterian These fossilized tracks indicate that their maker was able to crawl.

The major shared, derived traits that unite the deuterostomes are

- A dorsal nervous system
- A larva, if present, that has a food-collecting system consisting of cells with a single cilium
- A blastopore that becomes the anus
- Radial cleavage

The protostomes split into two lineages

Developmental, structural, and molecular data all suggest that the protostomes soon split into two major lineages that have been evolving independently since ancient times: lophotrochozoans and ecdysozoans. **Lophotrochozoans**, the animals we will discuss in the remainder of this chapter, grow by adding to the size of their skeletal elements. Some of them use cilia for locomotion, and many lineages have a type of free-living larva known as a **trochophore** (see Figure 32.23) The phylogeny of lophotrochozoans we will use in this chapter is shown in Figure 32.14. In contrast, **ecdysozoans**, the animals we will discuss in the next chapter, increase in size by molting their external skeletons. They move by mechanisms other than ciliary action, and they all have a common set of homeobox genes.

32.14 A Current Phylogeny of Lophotrochozoans Three major lineages, including the lophophorate and spiralian phyla, dominate the tree. Some small phyla are not included here.

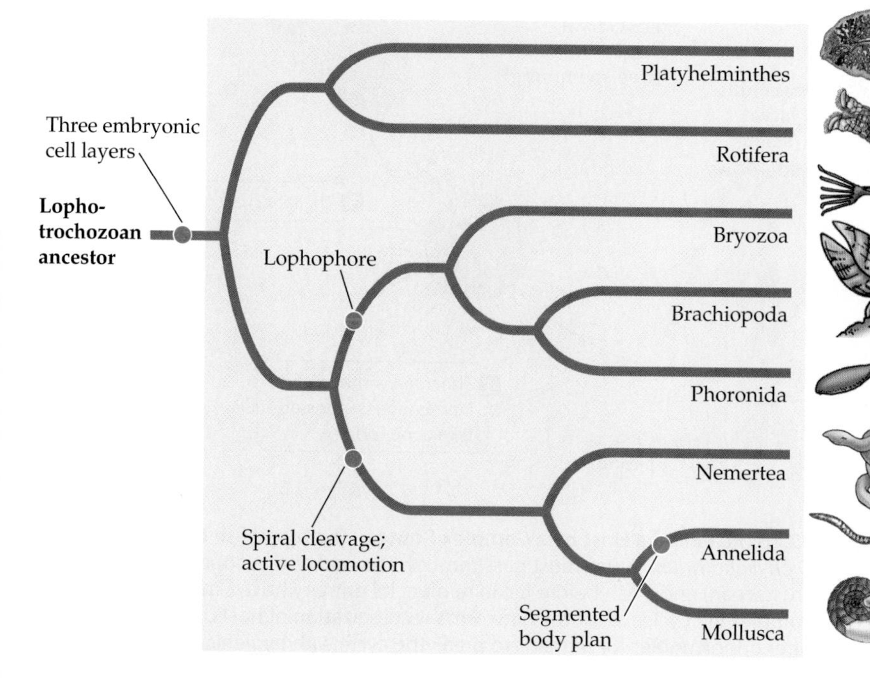

Simple Lophotrochozoans

The simplest lophotrochozoans—flatworms and rotifers—are small aquatic or parasitic animals. They move by rapidly beating cilia, and most have only simple organs.

Flatworms move by beating cilia

Members of the phylum **Platyhelminthes**, or **flatworms**, the simplest lophotrochozoans (Figure 32.15), are bilaterally symmetrical, unsegmented, acoelomate animals. They lack organs for transporting oxygen to internal tissues, and they have only simple cells for excreting metabolic wastes. Their lack of transport systems dictates that each cell must be near a body surface, a requirement met by their dorsoventrally flattened body form.

The digestive tract of a flatworm consists of a mouth opening into a blind sac. However, the sac is often highly branched, forming intricate patterns that increase the surface area available for the absorption of nutrients. Flatworms either feed on animal tissues (living or dead), or absorb nutients from a host's gut. Free-living flatworms glide over surfaces, powered by broad bands of cilia. This form of movement is very slow, but it is sufficient for small, scavenging animals.

The flatworms that are probably most similar to the ancestral bilaterians are the turbellarians (class **Turbellaria**), which are small, free-living marine and freshwater animals (a few live in moist terrestrial habitats). At one end they have a head with chemoreceptor organs, two simple eyes, and a tiny brain composed of anterior thickenings of the longitudinal nerve cords.

Although the earliest flatworms were free-living (Figure 32.15*a*), many species evolved a parasitic existence. A likely evolutionary transition was from feeding on dead organisms to feeding on the body surfaces of dying hosts to invading and consuming parts of living, healthy hosts. Most of the 25,000 species of living flatworms—including the tapeworms (class **Cestoda**) and flukes (class **Trematoda**; Figure 32.15*b*)—are internal parasites. These flatworms absorb digested food from the digestive tracts of their hosts, so many of them lack digestive tracts. They inhabit the bodies of many vertebrates; some cause serious human diseases, such as schistoso-

32.15 Flatworms Live Freely and Parasitically (*a*) Some flatworm species are free-living, like this marine flatworm photographed in the oceans off Sulawesi, Indonesia. (*b*) The flatworm diagrammed here, which lives parasitically in the gut of sea urchins, is representative of parasitic flukes. Because their hosts provide all the nutrition they need, these intestinal parasites do not require elaborate feeding or digestive organs and can devote most of their bodies to reproduction.

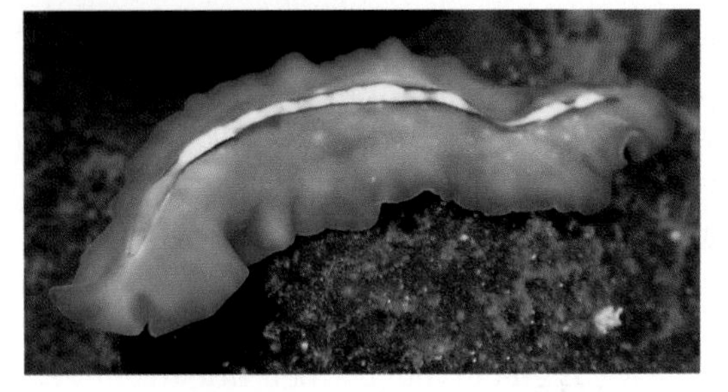

(*a*) *Pseudoceros bifurcus*

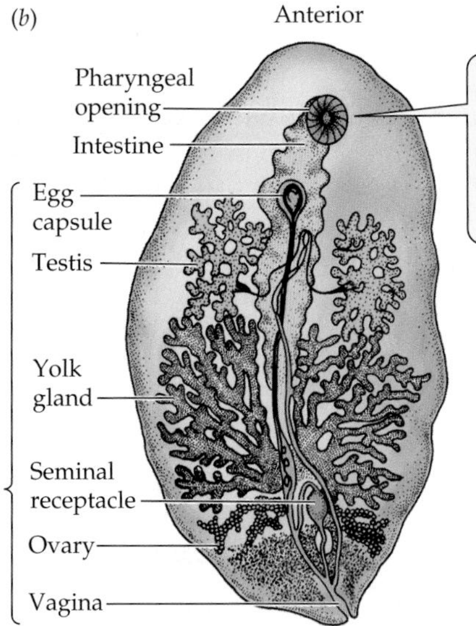
(b)
Anterior
Pharyngeal opening
Intestine
Egg capsule
Testis
Yolk gland
Seminal receptacle
Ovary
Vagina
Posterior
The flatworm gut has a single exterior opening. The pharyngeal opening serves as both "mouth" and "anus."
The flatworm's body is filled primarily with sex organs.

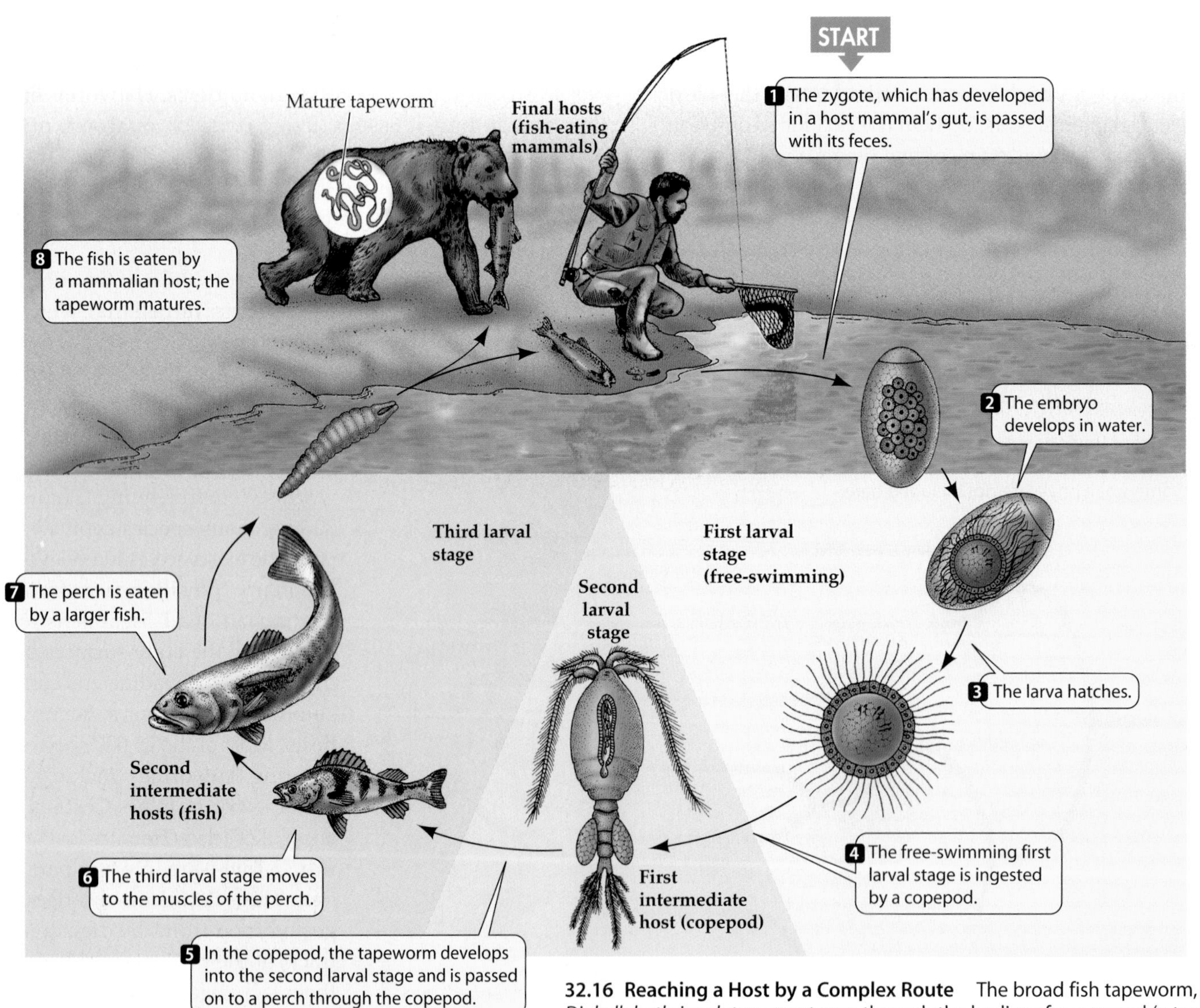
START
1 The zygote, which has developed in a host mammal's gut, is passed with its feces.
Mature tapeworm
Final hosts (fish-eating mammals)
8 The fish is eaten by a mammalian host; the tapeworm matures.
2 The embryo develops in water.
Third larval stage
First larval stage (free-swimming)
Second larval stage
7 The perch is eaten by a larger fish.
3 The larva hatches.
Second intermediate hosts (fish)
6 The third larval stage moves to the muscles of the perch.
4 The free-swimming first larval stage is ingested by a copepod.
First intermediate host (copepod)
5 In the copepod, the tapeworm develops into the second larval stage and is passed on to a perch through the copepod.

32.16 Reaching a Host by a Complex Route The broad fish tapeworm, *Diphyllobothrium latum*, must pass through the bodies of a copepod (a type of crustacean) and a fish before it can reinfect its primary host, a mammal. Such complex life cycles assist the flatworm's recolonization of hosts, but they also offer opportunities for humans to break the cycle with hygienic measures.

miasis. Monogeneans (class **Monogenea**) are external parasites of fishes and other aquatic vertebrates.

Parasites live in nutrient-rich environments where food is delivered to them, but they face other challenges. To complete their life cycle, parasites must overcome the defenses of their host. And because they die when their host dies, they must disperse their offspring to new hosts while their host is still living. The fertilized eggs of some parasitic flatworms are voided with the host's feces and later ingested directly by other host individuals. However, most parasitic species have complex life cycles involving one or more intermediate hosts and several larval stages (Figure 32.16). Such life cycles facilitate the transfer of individual parasites among hosts.

Rotifers are small but structurally complex

Rotifers (phylum **Rotifera**) are bilaterally symmetrical, unsegmented, pseudocoelomate lophotrochozoans. Most rotifers are tiny (50–500 μm long)—smaller than some ciliate protists—but they have highly developed internal organs (Figure 32.17). A complete gut passes from an anterior mouth to a posterior anus; the pseudocoel functions as a hydrostatic skeleton. Most rotifers propel themselves through the water by means of rapidly beating cilia rather than by muscular contraction. This type of movement is effective because rotifers are so small.

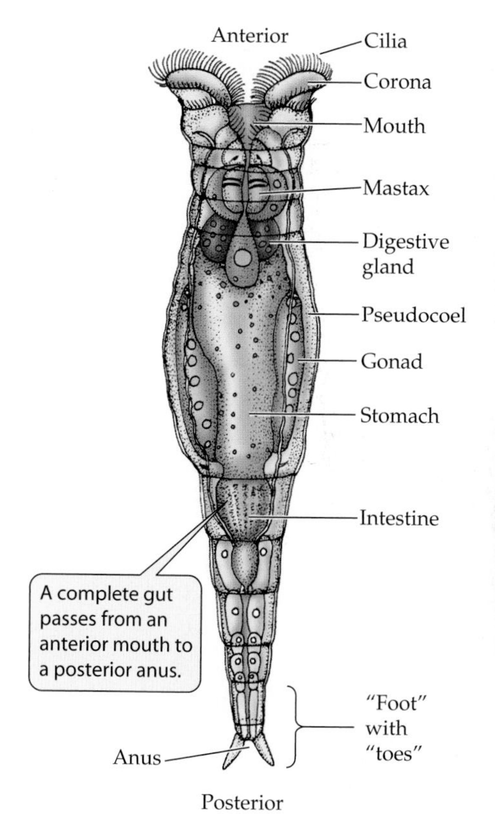

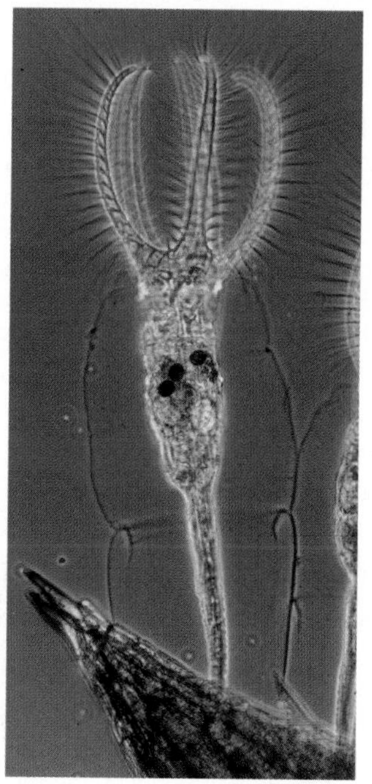

32.17 Rotifers (*a*) The rotifer diagrammed here reflects the general structure of many free-living species in the phylum. (*b*) A micrograph reveals the internal complexity of these living rotifers.

The most distinctive organs of rotifers are those they use to collect and process food. A conspicuous ciliated organ called the *corona* surmounts the head of many species. Coordinated beating of the cilia sweeps particles of organic matter from the water into the animal's mouth and down to a complicated structure called the *mastax*, in which food is ground into small pieces. By contracting the muscles around the pseudocoel, a few rotifer species that prey on protists and small animals can protrude the mastax through the mouth and seize small objects with it. Males and females are found in most species, but some species have only females that produce diploid eggs without being fertilized by a male.

Some rotifers are marine, but most of the 1,800 known species live in fresh water. Members of a few species rest on the surface of mosses or lichens in a desiccated, inactive state until it rains. When rain falls, they absorb water and become mobile, feeding in the films of water that temporarily cover the plants. Most rotifers live no longer than 1 or 2 weeks.

Lophophorates: An Ancient Body Plan

After the platyhelminthes and rotifers diverged from it, the lophotrochozoan lineage divided into two branches. The descendants of those branches became the modern **lophophorates**—the subject of this section—and the **spiralians**, which we will discuss in the following section.

About 4,850 living species of lophophorates are known, but many times that number of species existed during the Paleozoic and Mesozoic eras. Three lophophorate phyla survive today: Phoronida, Brachiopoda, and Ectoprocta. Nearly all members of these phyla are marine; only a few species of ectoprocts live in fresh water.

Lophophorate animals obtain food by filtering it from the surrounding water, a trait they share with many other protostomes. The most conspicuous feature of these animals is the **lophophore**, a circular or U-shaped ridge around the mouth that bears one or two rows of ciliated, hollow tentacles (Figure 32.18). This large and complex structure is an organ for both food collection and gas exchange. Nearly all adult lophophorate animals are sessile, and they use the tentacles and cilia of their lophophore to capture *plankton* (small floating organisms) from the water. Lophophorates also have a U-shaped gut; the anus is located close to the mouth, but outside the tentacles.

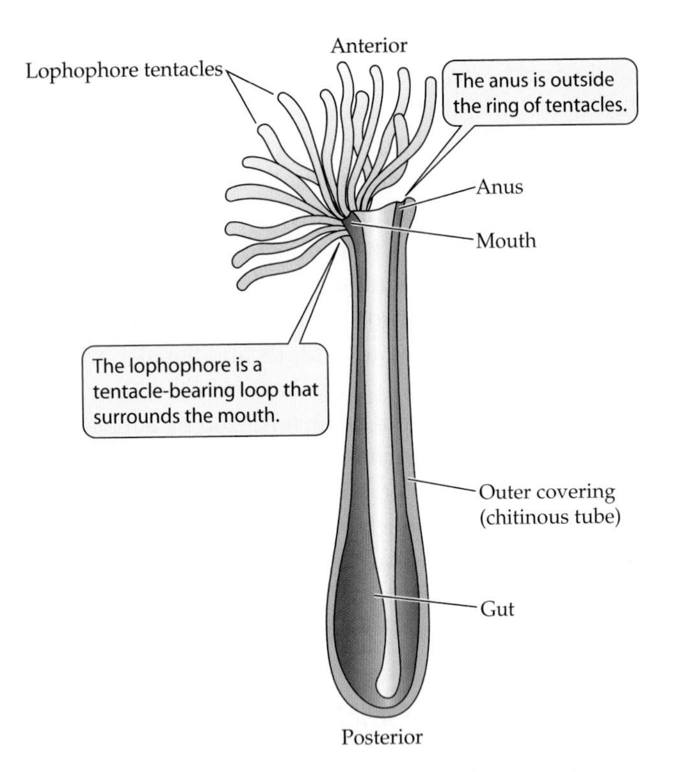

32.18 Lophophore Artistry The lophophore dominates the anatomy of this phoronid. The phoronid gut is U-shaped.

Phoronids are sedentary lophophorates

The 20 known species of **phoronids** (phylum **Phoronida**) are sedentary worms that live in muddy or sandy sediments or attached to a rocky substratum. Phoronids are found in marine waters ranging from intertidal zones to about 400 meters deep. They range in size from 5 to 25 cm in length. They secrete chitinous tubes, in which they live (Figure 32.18). The lophophore is the most conspicuous external feature of the phoronids. Cilia drive water into the top of the lophophore, and water exits through the narrow spaces between the tentacles. Suspended food particles are caught and transported to the mouth by ciliary action. In most species, eggs are released into the water, where they are fertilized, but some species produce large eggs that are fertilized internally, where they are brooded until they hatch.

Ectoprocts are colonial lophophorates

Ectoprocts (phylum **Ectoprocta**) are colonial lophophorates that live in a "house" made of material secreted by the external body wall. A colony consists of many small (1–2 mm) individuals connected by strands of tissue along which materials can be moved (Figure 32.19*a*). Most of the 4,500 species of ectoprocts are marine, but a few live in fresh water. They are able to oscillate and rotate the lophophore to increase contact with prey (Figure 32.19*b*) and can retract it into the tube.

A colony of ectoprocts is created by the asexual reproduction of its founding members. A single colony may contain as many as 2 million individuals. In some species, individual colony members are specialized for feeding, reproduction, defense, or support. Ectoprocts reproduce sexually by releasing sperm into the water, where they are collected by other individuals. Eggs are fertilized internally, and developing embryos are brooded before they exit as larvae to seek suitable sites for attachment to the substratum.

Brachiopods superficially resemble bivalve mollusks

Brachiopods (phylum **Brachiopoda**) are solitary marine lophophorate animals. Their shells are divided into two parts

(*a*) *Lophopus crystallinus*

(*b*)

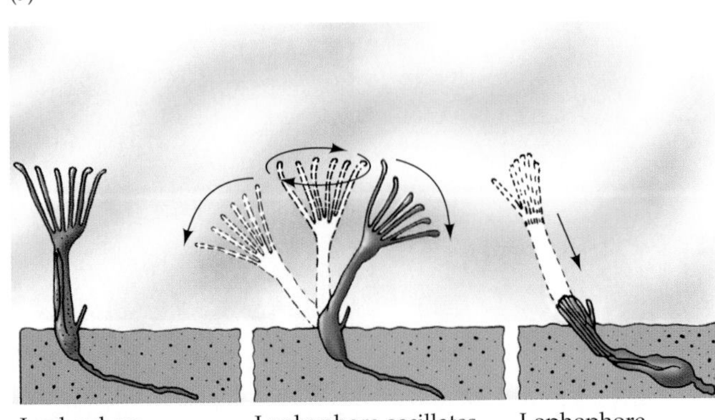

32.19 Ectoprocts (*a*) Branching colonies of ectoprocts may appear plantlike. (*b*) Ectoprocts have greater control over the movement of their lophophores than members of other lophophorate phyla.

32.20 Brachiopods The lophophore of this North Pacific brachiopod can be seen between the valves of its shell.

that are connected by a ligament (Figure 32.20). The two halves can be pulled shut to protect the soft body. Brachiopods superficially resemble bivalve mollusks, but the brachiopod shell differs from that of mollusks in that the two halves are dorsal and ventral rather than lateral. The two-armed lophophore of a brachiopod is located within the shell. The beating of cilia on the lophophore draws water into the slightly opened shell. Food is trapped in the lophophore and directed to a ridge, along which it is transferred to the mouth. Most brachiopods are between 4 and 6 cm long, but some are as long as 9 cm.

Brachiopods live attached to a solid substratum or embedded in soft sediments. Most species are attached by means of a short, flexible stalk that holds the animal above the substratum. Gases are exchanged across body surfaces, especially the tentacles of the lophophore. Most brachiopods release their gametes into the water, where they are fertilized. The larvae remain among the plankton for only a few days before they settle and develop into adults.

Brachiopods reached their peak abundance and diversity in Paleozoic and Mesozoic times. More than 26,000 fossil species have been described. Only about 335 species survive, but they are common in some marine environments.

Spiralians: Spiral Cleavage and Wormlike Body Plans

The spiralian lineage, containing animals that typically have spiral cleavage patterns, gave rise to many phyla. Members of more than a dozen of these phyla are *wormlike*; that is, they are bilaterally symmetrical, legless, soft-bodied, and at least several times longer than they are wide. This body form enables animals to move efficiently through muddy and sandy marine sediments. Most of these phyla have no more than several hundred species. The most species-rich spiralian phylum, the mollusks, shows significant modifications of the wormlike body plan.

Ribbon worms are unsegmented

The carnivorous **ribbon worms** (phylum **Nemertea**) are dorsoventrally flattened. They have nervous and excretory systems similar to those of flatworms, but unlike flatworms, they have a complete digestive tract with a mouth at one end and an anus at the other. Food moves in one direction through the digestive tract and is acted on by a series of digestive enzymes. Small ribbon worms move by beating their cilia. Larger ones employ waves of muscle contraction to move over the surface of sediments or to burrow. Movement by both of these methods is slow.

Within the body of nearly all of the 900 species of ribbon worms is a fluid-filled cavity called the *rhynchocoel*, within which lies a hollow, muscular *proboscis*. The proboscis, which is the feeding organ, may extend much of the length of the worm. Contraction of the muscles surrounding the rhynchocoel causes the proboscis to be everted explosively through an anterior opening (Figure 32.21) without moving the rest of the animal. The proboscis of most ribbon worms

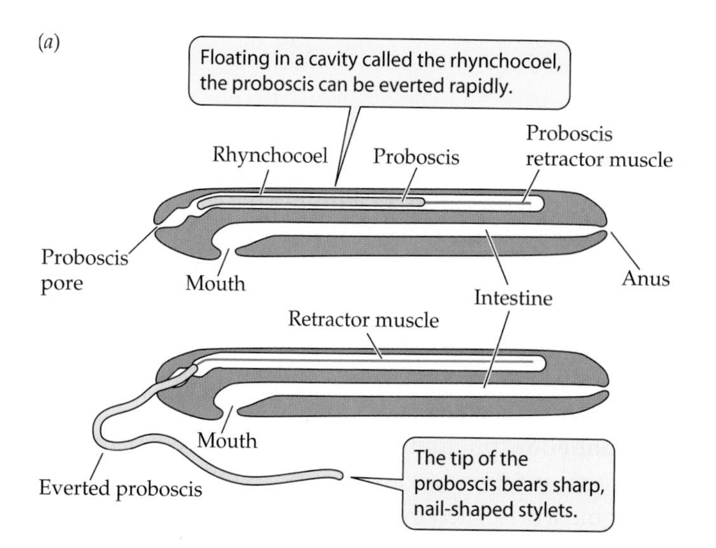

(*b*) *Pelagonemertes* sp.

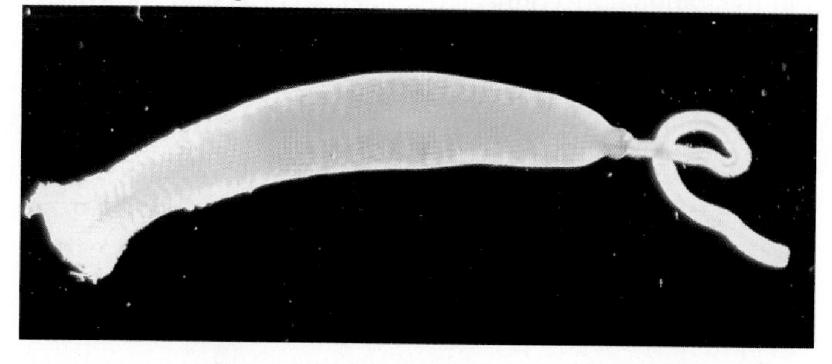

32.21 Ribbon Worms (*a*) The proboscis is the ribbon worm's feeding organ. (*b*) This deep-water nemertean displays an everted proboscis.

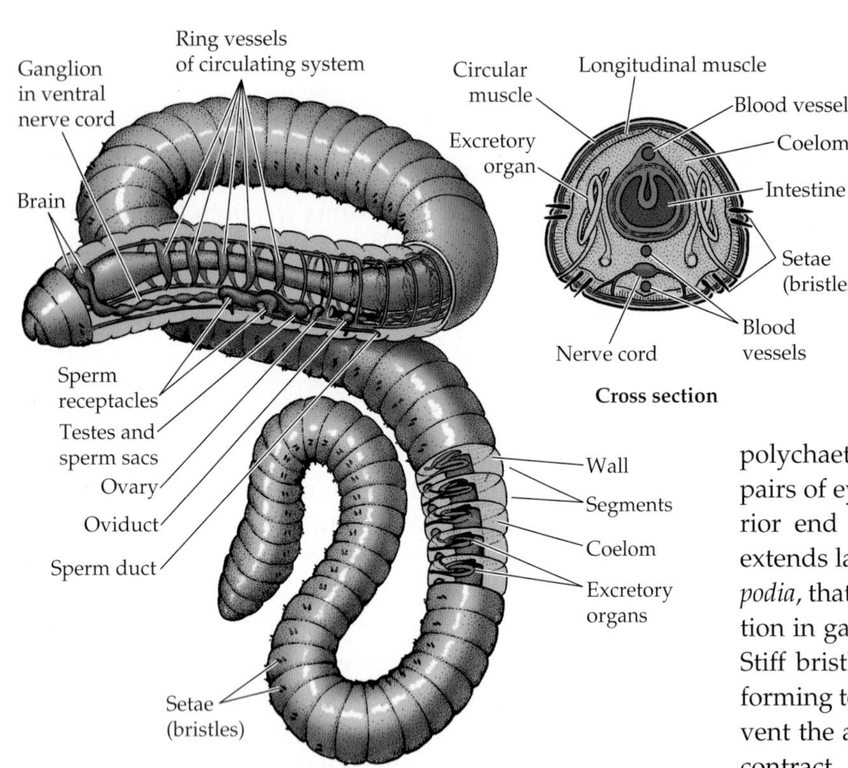

32.22 Annelids Have Many Body Segments The segmented structure of the annelids is apparent both externally and internally. Most organs of this earthworm are repeated serially.

is armed with a sharp stylet that pierces the prey. Paralysis-causing toxins produced by the proboscis are discharged into the wound. Reproduction and development in ribbon worms is highly varied.

Segmentation improved locomotion in the annelids

Segmentation allows an animal to alter the shape of its body in complex ways and to control its movements more precisely. Fossils of segmented worms are known from the middle Cambrian; the earliest forms are thought to have been burrowing marine animals. Segmentation evolved several times among spiralians; we will discuss only one of the phyla with segmented members: the annelids.

The **annelids** (phylum **Annelida**) are a diverse group of segmented spiralian worms (Figure 32.22). The coelom in each segment is isolated from those in other segments. A separate nerve center called a *ganglion* controls each segment, and the ganglia are connected by nerve cords that coordinate their functioning. Most annelids lack a rigid, external protective covering. The body wall serves as a general surface for gas exchange in most species, but this thin, permeable body surface restricts annelids to moist environments; they lose body water rapidly in dry air. The approximately 16,500 described species live in marine, freshwater, and terrestrial environments.

POLYCHAETES. More than half of all annelid species are members of the class **Polychaeta** ("many hairs"). Nearly all polychaetes are marine animals. Most have one or more pairs of eyes and one or more pairs of tentacles at the anterior end of the body. The body wall in most segments extends laterally as a series of thin outgrowths, called *parapodia*, that contain many blood vessels. The parapodia function in gas exchange, and some species use them to move. Stiff bristles called *setae* protrude from each parapodium, forming temporary attachments to the substratum that prevent the animal from slipping backward when its muscles contract.

Typically, males and female polychaetes release gametes into the water, where the eggs are fertilized and develop into trochophore larvae (Figure 32.23). The trochophore is a distinctive larval type found among polychaetes, mollusks, and several other marine lineages with spiral cleavage. The second half of the name "lophotrochozoans" is derived from this larva, which is believed by many researchers to represent an evolutionary link between the annelids and the mollusks.

As the polychaete trochophore develops, it forms body segments at its posterior end; eventually it becomes a small

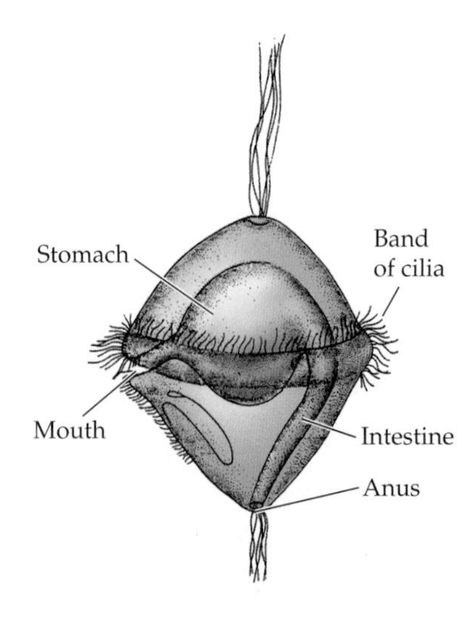

32.23 The Trochophore Larva The trochophore ("wheel-bearer") is a distinctive larval form found in several animal lineages with spiral cleavage, most notably the marine polychaete worms and the mollusks.

(*a*) *Spirobranchus* sp.

(*b*) *Lumbricus* sp.

(*c*) *Microbdella* sp.

(*d*) *Riftia* sp.

32.24 Diversity among the Annelids
(*a*) The feather duster worm is a marine polychaete with striking feeding tentacles. (*b*) Earthworms are hermaphroditic (each individual is simultaneously both male and female). When they copulate, each individual donates and receives sperm. (*c*) This Australian tiger leech is attached to a leaf by its posterior sucker as it waits for a mammalian host. (*d*) Vestimentiferans live around hydrothermal vents deep in the ocean. Their skin secretes chitin and other substances, forming tubes.

adult worm. Many polychaete species live in burrows in soft sediments and filter prey from the surrounding water with elaborate, feathery tentacles (Figure 32.24*a*).

OLIGOCHAETES. More than 90 percent of the approximately 3,000 described species of **oligochaetes** (class **Oligochaeta**) live in freshwater or terrestrial habitats. Oligochaetes ("few hairs") have no parapodia, eyes, or anterior tentacles, and they have relatively few setae. Earthworms—the most familiar oligochaetes (see Figure 32.22)—are scavengers and ingesters of soil, from which they extract food particles.

Unlike polychaetes, all oligochaetes are *hermaphroditic:* that is, each individual is both male and female. Sperm are exchanged simultaneously between two copulating individuals (Figure 32.24*b*). Eggs are laid in a cocoon outside the adult's body. The cocoon is shed, and when development is complete, miniature worms emerge and begin independent life.

LEECHES. Leeches (class **Hirudinea**) probably evolved from oligochaete ancestors. Most species live in freshwater or terrestrial habitats and, like oligochaetes, lack parapodia and tentacles. Like oligochaetes, leeches are hermaphroditic. The coelom of leeches is not divided into compartments; the coelomic space is largely filled with undifferentiated tissue. Groups of segments at each end of the body are modified to form suckers, which serve as temporary anchors that aid the leech in movement (Figure 32.24*c*). With its posterior sucker attached to a substratum, the leech extends its body by contracting its circular muscles. The anterior sucker is then attached, the posterior one detached, and the leech shortens itself by contracting its longitudinal muscles.

Many leeches are external parasites of other animals, but some species also eat snails and other invertebrates. A leech makes an incision in its host, from which blood flows. It can ingest so much blood in a single feeding that its body may enlarge several times. An anticoagulant secreted by the leech into the wound keeps the host's blood flowing. For hundreds of years leeches were widely employed in medicine. Even today they are used to reduce fluid pressure and prevent blood clotting in damaged tissues and to eliminate pools of coagulated blood.

VESTIMENTIFERANS. Members of one lineage of annelids, the **vestimentiferans** (class **Pogonophora**), evolved burrowing forms with a crown of tentacles through which gases are exchanged; they entirely lost their digestive systems (Figure 32.24*d*). Vestimentiferans secrete chitin and other substances to form the tubes in which they live.

A vestimentiferan's coelom consists of an anterior compartment into which the tentacles can be withdrawn, and a long, subdivided cavity that extends much of the length of its body. The posterior end of the body is segmented. Experiments using radioactively labeled molecules have shown that vestimentiferans take up dissolved organic matter at high rates from either the sediments in which they live or the surrounding water.

Vestimentiferans were not discovered until the twentieth century, when deep-sea exploration revealed them living many thousands of meters below the ocean surface. In these deep oceanic sediments, they are abundant, reaching densities of many thousands per square meter. About 145 species have been described. The largest and most remarkable vestimentiferans, which grow to 2 meters in length, live near deep-sea hydrothermal vents—volcanic openings in the sea floor through which hot, sulfide-rich water pours. The tissues of these species harbor endosymbiotic bacteria that fix carbon using energy obtained from oxidation of hydrogen sulfide (H_2S).

Mollusks evolved shells

Mollusks (phylum **Mollusca**) range in size from snails only a millimeter high to giant squids more than 18 meters long—the largest known invertebrates. Mollusks underwent one of the most dramatic of animal evolutionary radiations, based on a unique body plan with three major structural components: a foot, a mantle, and a visceral mass. Animals that appear very different, such as snails, clams, and squids, are all built from these components (Figure 32.25).

The molluscan *foot* is a large, muscular structure that originally was both an organ of locomotion and a support for the internal organs. In the lineage leading to squids and octopuses, the foot was modified to form arms and tentacles borne on a head with complex sense organs. In other groups, such as clams, the foot was transformed into a burrowing organ. In some lineages the foot is greatly reduced.

The *mantle* is a fold of tissue that covers the *visceral mass* of internal organs. In many mollusks, the mantle extends beyond the visceral mass to form a *mantle cavity*. The mantle secretes the hard, calcarous skeleton typical of most mollusks. The *gills*, which are used for gas exchange and, in some

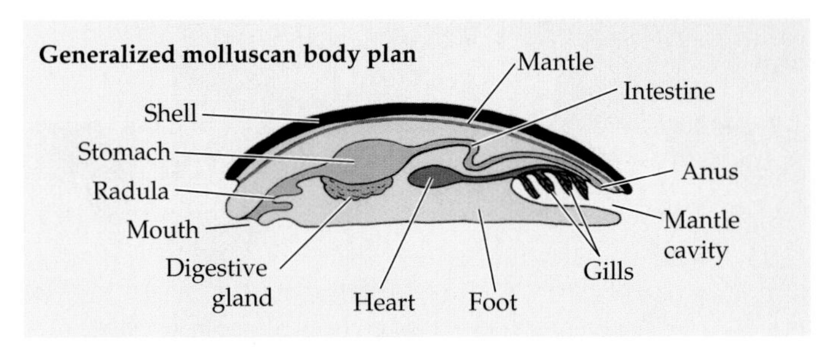

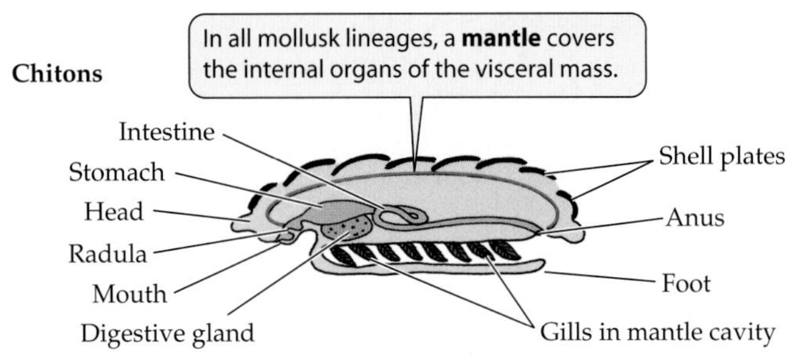

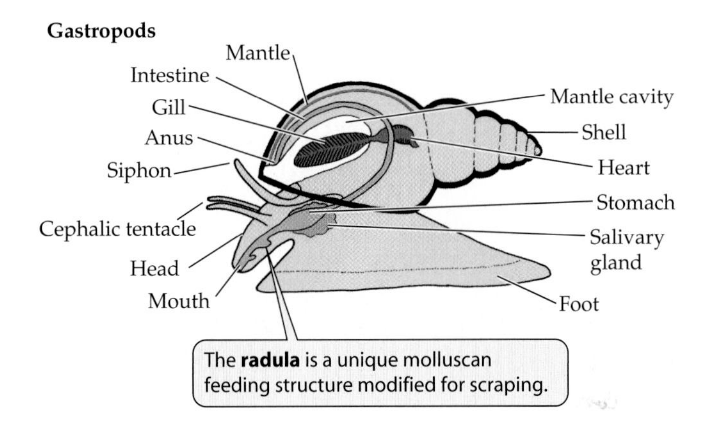

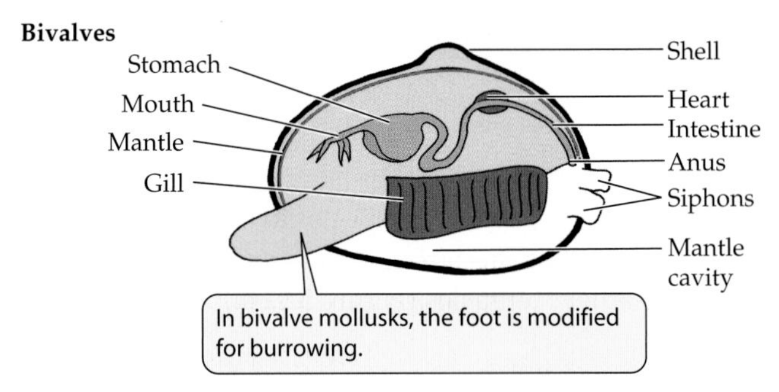

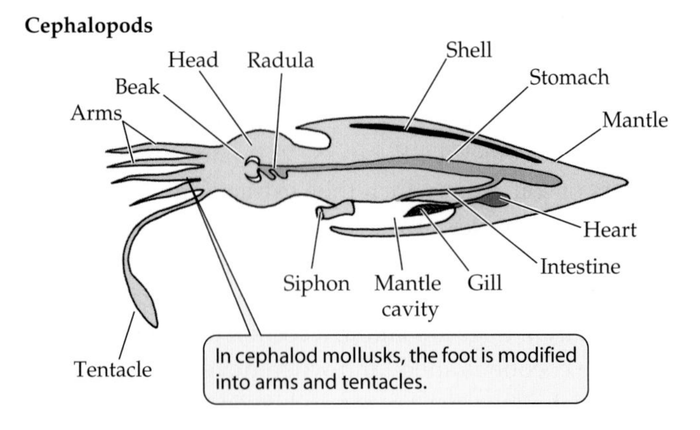

32.25 Molluscan Body Plans The diverse modern mollusks are all variations on a general body plan that includes a foot, a mantle, and a visceral mass of internal organs.

species, for feeding, lie in this cavity. When the cilia on the gills beat, they create a flow of water over the gills. The tissue of the gills, which is highly *vascularized* (contains many blood vessels), takes up O_2 from the water and releases CO_2.

Mollusks have an *open circulatory system* that empties into large fluid-filled cavities, through which fluids move around the animal and deliver O_2 to internal organs. Mollusks also developed a rasping feeding structure known as the *radula.* The radula was originally an organ for scraping algae from rocks, a function it retains in many living mollusks. However, in some mollusks, it has been modified into a drill or poison dart. In others, such as clams, it is absent.

Although individual components have been lost in some lineages, these three unique, shared derived characteristics—the foot, the mantle, and the visceral mass—lead zoologists to believe that all 95,000 species of mollusks have a common ancestor. A small sample of these species is shown in Figure 32.26.

MONOPLACOPHORANS. **Monoplacophorans** (class **Monoplacophora**) were the most abundant mollusks during the Cambrian period, 550 million years ago, but today there are only a few surviving species. Unlike all other living mollusks, the surviving monoplacophorans have respiratory organs, muscles, and excretory pores that are repeated over the length of the body. The respiratory organs are located in a large cavity under the shell, through which oxygen-bearing water circulates.

CHITONS. **Chitons** (class **Polyplacophora**) have multiple gills and shell plates, but the body is not truly segmented (Figure 32.26*a*). The chiton body is bilaterally symmetrical, and its internal organs, particularly the digestive and nervous systems, are relatively simple. The larvae of chitons are almost indistinguishable from those of annelids. Most chitons are marine herbivores that scrape algae from rocks with their sharp radulae. An adult chiton spends most of its life clinging tightly to rock surfaces with its large, muscular, mucus-covered foot. It moves slowly by means of rippling waves of muscular contraction in the foot. Fertilization in most chitons takes place in the water, but in a few species fertilization is internal and embryos are brooded within the body.

BIVALVES. One lineage of early mollusks developed a hinged, two-part shell that extended over the sides of the body as well as the top, giving rise to the **bivalves** (class **Bivalvia**), which include the familiar clams, oysters, scallops, and mussels (Figure 32.26*b*). Bivalves are largely sedentary and have greatly reduced heads. The foot is compressed, and in many clams, it is used for burrowing into mud and sand. Bivalves feed by taking in water through an opening called an *incurrent siphon* and extracting food from the water with their large gills, which are also the main sites of gas exchange. Water and gametes exit through the *excurrent siphon.* Fertilization takes place in open water in most species.

GASTROPODS. Another lineage of early mollusks gave rise to the **gastropods** (class **Gastropoda**), which include snails, whelks, limpets, slugs, abalones, and the often brilliantly ornamented nudibranchs. Gastropods, unlike bivalves, have one-piece shells. Most gastropods are motile, using the large foot to move slowly across the substratum or to burrow through it. Gastropods are the most species-rich and widely distributed of the molluscan classes (Figure 32.26*c,d*). Most species move by gliding on the muscular foot, but in a few species—the sea butterflies and heteropods—the foot is modified into a swimming organ with which the animal moves through open ocean waters. The only mollusks that live in terrestrial environments—land snails and slugs—are gastropods. In these terrestrial species, the mantle tissue is modified into a highly vascularized lung. Fertilization is internal in most species.

CEPHALOPODS. In one lineage of mollusks, the cephalopods (class **Cephalopoda**), the excurrent siphon became modified to allow the animal to control the water content of the mantle cavity. Ultimately, the modification of the mantle into a device for forcibly ejecting water from the cavity enabled these animals to move rapidly through the water. Furthermore, many early cephalopods had chambered shells into which gas could be secreted to adjust buoyancy. Together, these adaptations allow cephalopods to live in open water.

The cephalopods include the squids, octopuses, and nautiluses (Figure 32.26*e, f*). They first appeared about 600 million years ago, near the beginning of the Cambrian period. By the Ordovician period a wide variety of types were present. With their greatly enhanced mobility, some cephalopods, such as squids, became the major predators in the open waters of the Devonian oceans.They remain important marine predators today. Cephalopods capture and subdue their prey with their tentacles; octopuses also use their tentacles to move over the substratum. As is typical of active predators, cephalopods have a head with complex sensory organs, most notably eyes that are comparable to those of vertebrates in their ability to resolve images. The head is closely associated with a large, branched foot that bears the tentacles and a siphon. The large muscular mantle provides a solid external supporting structure. The gills hang in the mantle cavity. As is typical of behaviorally complex animals, many cephalopods have elaborate courtship behavior, which may involve striking color changes.

(*a*) *Mopalia* sp.

(*c*) *Phidiana hiltoni*

(*e*) *Octopus bimaculoides*

32.26 Diversity among the Mollusks (*a*) Chitons are common in the intertidal zones of the temperate zone coasts. (*b*) The giant clam of Indonesia is among the largest of the bivalve mollusks. (*c*) Slugs are gastropods that have lost their shells; this shell-less sea slug is very conspicuously colored. (*d*) Land snails are shelled, terrestrial gastropods. (*e*) Cephalopods such as the octopus are active predators. (*f*) The boundaries of its chambers are clearly visible on the outer surface of this shelled *Nautilus*, another cephalopod.

(*b*) *Tridacna gigas*

(*d*) *Helminthoglypta walkeriana*

(*f*) *Nautilus pompilius*

The earliest cephalopod shells were divided by partitions penetrated by tubes through which liquids could be moved. Nautiloids (genus *Nautilus*) are the only cephalopods with external chambered shells that survive today (Figure 32.26*f*).

Mollusks and brachiopods are among the lophotrochozoans that evolved hard shells that help to protect them from predators and the physical environment. A sturdy outer covering is the main feature of the second protostomate lineage, the ecdysozoans—the subject of the next chapter.

Chapter Summary

Animals: Descendants of a Common Ancestor

- All members of the kingdom Animalia are believed to have a common ancestor, which was a colonial flagellated protist.
- The specialization of cells by function made possible the complex, multicellular body plan of animals.
- Animals are multicellular heterotrophs. They take in complex organic molecules, expending energy to do so.
- Morphological, developmental, and molecular data all support similar animal phylogenies.
- The two major animal lineages—protostomes and deuterostomes—are believed to have diverged early in animal evolution; they differ in several components of their early development. **Review Figure 32.1**

Body Plans Are Basic Structural Designs

- Most animals have either radial or bilateral symmetry. Radially symmetrical animals move slowly or not at all. Bilateral symmetry is strongly correlated with more rapid movement and the concentration of sense organs at the anterior end of the animal. **Review Figure 32.2**
- The body cavity of an animal is strongly correlated with its ability to move. On the basis of their body cavities, animals are classified as acoelomates, pseudocoelomates, or coelomates. **Review Figure 32.3**

Sponges: Loosely Organized Animals

- Sponges (phylum Porifera) are simple animals that lack cell layers and true organs, but have several different cell types.
- Sponges feed by means of choanocytes, feeding cells that draw water through the sponge body and filter out food particles. **Review Figure 32.4**
- Sponges come in a variety of sizes and shapes that are adapted to different movement patterns of water.

Cnidarians: Two Cell Layers and Blind Guts

- Cnidarians (phylum Cnidaria) are radially symmetrical and diploblastic, but with their nematocyst-studded tentacles, they can capture prey larger and more complex than themselves. **Review Figure 32.7**
- Most cnidarian life cycles have a sessile polyp stage and a free-swimming, sexual, medusa stage, but some species lack one of the stages. **Review Figures 32.8, 32.10, 32.11**

See Web/CD Tutorial 32.1

Ctenophores: Complete Guts and Tentacles

- Ctenophores (phylum Ctenophora) are diploblastic marine carnivores with a complete gut and simple life cycles. **Review Figure 32.12**

The Evolution of Bilaterally Symmetrical Animals

- All bilaterally symmetrical animals probably share a common ancestor.
- Protostomes and deuterostomes are each monophyletic lineages that have been evolving separately since the Cambrian period. Their members are structurally more complex than cnidarians and ctenophores.
- Protostomes have a ventral nervous system, paired nerve cords, and larvae with compound cilia.
- Deuterostomes have a dorsal nervous system and larvae with a single cilium per cell.
- The protostomes split into two major groups: lophotrochozoans and ecdysozoans. **Review Figure 32.14**

Simple Lophotrochozoans

- Flatworms (phylum Platyhelminthes) are acoelomate, lack organs for oxygen transport, have only one entrance to the gut, and move by beating their cilia. Many species are parasitic. **Review Figures 32.15, 32.16**
- Although they are no larger than many ciliated protists, rotifers (phylum Rotifera) have highly developed internal organs. **Review Figure 32.17**

Lophophorates: An Ancient Body Plan

- The lophotrochozoan lineage split into two branches, whose descendants became the modern lophophorates and the spiralians.
- The lophophore dominates the anatomy of many lophophorate animals. **Review Figure 32.18**
- Ectoprocts are colonial lophophorates that can move their lophophores. **Review Figure 32.19**
- Brachiopods, which superficially resemble bivalve mollusks, were much more abundant in the past than they are today.

Spiralians: Spiral Cleavage and Wormlike Body Plans

- The spiralian lineage gave rise to many phyla, most of whose members are wormlike.
- Ribbon worms (phylum Nemertea) have a complete digestive tract and capture prey with an eversible proboscis. **Review Figure 32.21**
- Annelids (phylum Annelida) are a diverse group of segmented worms that live in marine, freshwater, and terrestrial environments. **Review Figures 32.22**
- Mollusks (phylum Mollusca) have a body plan with three basic components: foot, mantle, and visceral mass. **Review Figure 32.25**
- The molluscan body plan has been modified to yield a diverse array of animals that superficially appear very different from one another.

See Web/CD Activities 32.1 and 32.2 for a concept review of this chapter

Self-Quiz

1. The body plan of an animal is
   a. its general structure.
   b. the integrated functioning of its parts.
   c. its general structure and the integrated functioning of its parts.
   d. its general structure and its evolutionary history.
   e. the integrated functioning of its parts and its evolutionary history.

2. A bilaterally symmetrical animal can be divided into mirror images by
   a. any plane through the midline of its body.
   b. any plane from its anterior to its posterior end.
   c. any plane from its dorsal to its ventral surface.
   d. any plane through the midline of its body from its anterior to its posterior end.
   e. a single plane through the midline of its body from its dorsal to its ventral surface.
3. Among protostomes, cleavage of the fertilized egg is
   a. delayed while the egg continues to mature.
   b. always radial.
   c. spiral in some species and radial in others.
   d. triploblastic.
   e. diploblastic.
4. The sponge body plan is characterized by
   a. a mouth and digestive cavity but no muscles or nerves.
   b. muscles and nerves but no mouth or digestive cavity.
   c. a mouth, digestive cavity, and spicules.
   d. muscles and spicules but no digestive cavity or nerves.
   e. no mouth, digestive cavity, muscles, or nerves.
5. Which are phyla of diploblastic animals?
   a. Porifera and Cnidaria
   b. Cnidaria and Ctenophora
   c. Cnidaria and Platyhelminthes
   d. Ctenophora and Platyhelminthes
   e. Porifera and Ctenophora
6. Cnidarians have the ability to
   a. live in both salt and fresh water.
   b. move rapidly in the water column.
   c. capture and consume large numbers of small prey.
   d. survive where food is scarce, because of their low metabolic rate.
   e. capture large prey and to move rapidly.
7. Many parasites evolved complex life cycles because
   a. they are too simple to disperse readily.
   b. they are poor at recognizing new hosts.
   c. they were driven to it by host defenses
   d. complex life cycles increase the probability of a parasite's transfer to a new host.
   e. their ancestors had complex life cycles and they simply retained them.
8. Members of which phyla have lophophores?
   a. Phoronida, Brachiopoda, and Nemertea
   b. Phoronida, Brachiopoda, and Ectoprocta
   c. Brachiopoda, Ectoprocta, and Platyhelminthes
   d. Phoronida, Rotifera, and Ectoprocta
   e. Rotifera, Ectoprocta, and Brachiopoda
9. Which of the following is not part of the molluscan body plan?
   a. Mantle
   b. Foot
   c. Radula
   d. Visceral mass
   e. Jointed skeleton
10. Cephalopods control their buoyancy by
    a. adjusting salt concentrations in their blood.
    b. forcibly expelling water from the mantle.
    c. pumping water in and out of internal chambers.
    d. using the complex sense organs in their heads.
    e. swimming rapidly.

For Discussion

1. Differentiate among the members of each of the following sets of related terms:
   a. radial symmetry/bilateral symmetry
   b. protostome/deuterostome
   c. diploblastic/triploblastic
   d. coelomate/pseudocoelomate/acoelomate
2. In this chapter we listed some of the traits shared by all animals that convince most biologists that all animals are descendants of a single common ancestral lineage. In your opinion, which of these traits provides the most compelling evidence that animals are monophyletic?
3. Describe some features that allow animals to capture prey that are larger and more complex than they themselves are.
4. Why is bilateral symmetry strongly associated with cephalization, the concentration of sense organs in an anterior head?
5. Why might mollusks not have evolved segmentation, given that a segmented body enables improved control over locomotion?

33 Ecdysozoans: The Molting Animals

Early in animal evolution, the protostomate lineage split into two branches—the lophotrochozoans and the ecdysozoans—as we saw in the previous chapter. The distinguishing feature of the ecdysozoans is an **exoskeleton**, a nonliving covering that provides an animal with both protection and support. Once formed, however, an exoskeleton cannot grow. How, then, can ecdysozoans increase in size? Their solution is to shed, or **molt**, the exoskeleton and replace it with a new, larger one.

Before the animal molts, a new exoskeleton is already forming underneath the old one. When the old exoskeleton is shed, the new one expands and hardens. But until it has hardened, the animal is very vulnerable to its enemies both because its outer surface is easy to penetrate and because it can move only slowly.

The exoskeleton presented new challenges in other areas besides growth. Ecdysozoans cannot use cilia for locomotion, and most exdysozoans have hard exoskeletons that impede the passage of oxygen into the animal. To cope with these challenges, ecdysozoans evolved new mechanisms of locomotion and respiration.

Despite these constraints, the ecdysozoans—the molting animals—have more species than all other animal lineages combined. An increasingly rich array of molecular and genetic evidence, including a set of homeobox genes shared by all ecdysozoans, suggests that molting may have evolved only once during animal evolution.

In this chapter, we will review the diversity of the ecdysozoans. We will look at the characteristics of animals in the various ecdysozoan phyla and see how having an exoskeleton has influenced their evolution. The phylogeny we will follow is presented in Figure 33.1. In the first part of the chapter, we will look at several small phyla of wormlike ecdysozoans. Then we will detail the characteristics of the arthropods, an incredibly species-rich group of ecdysozoan phyla with hardened exoskeletons. We will close the chapter with an overview of evolutionary themes found in the evolution of the protostomate phyla, including both the lophotrochozoan and ecdysozoan lineages.

Shedding the Exoskeleton This dragonfly has just gone through a molt, a shedding of the outer exoskeleton. Such molts are necessary in order for the insect to grow larger or to change its form.

Cuticles: Flexible, Unsegmented Exoskeletons

Some ecdysozoans have wormlike bodies covered by exoskeletons that are relatively thin and flexible. Such an exoskeleton, called a **cuticle**, offers the animal some protection, but does not provide body support. The action of circular and longitudinal muscles on fluids in

33.1 A Current Phylogeny of the Ecdysozoans Those ecdysozoan phyla with jointed appendages are often placed in a single phylum, Arthropoda. Arthropods are the most numerous animals on the planet.

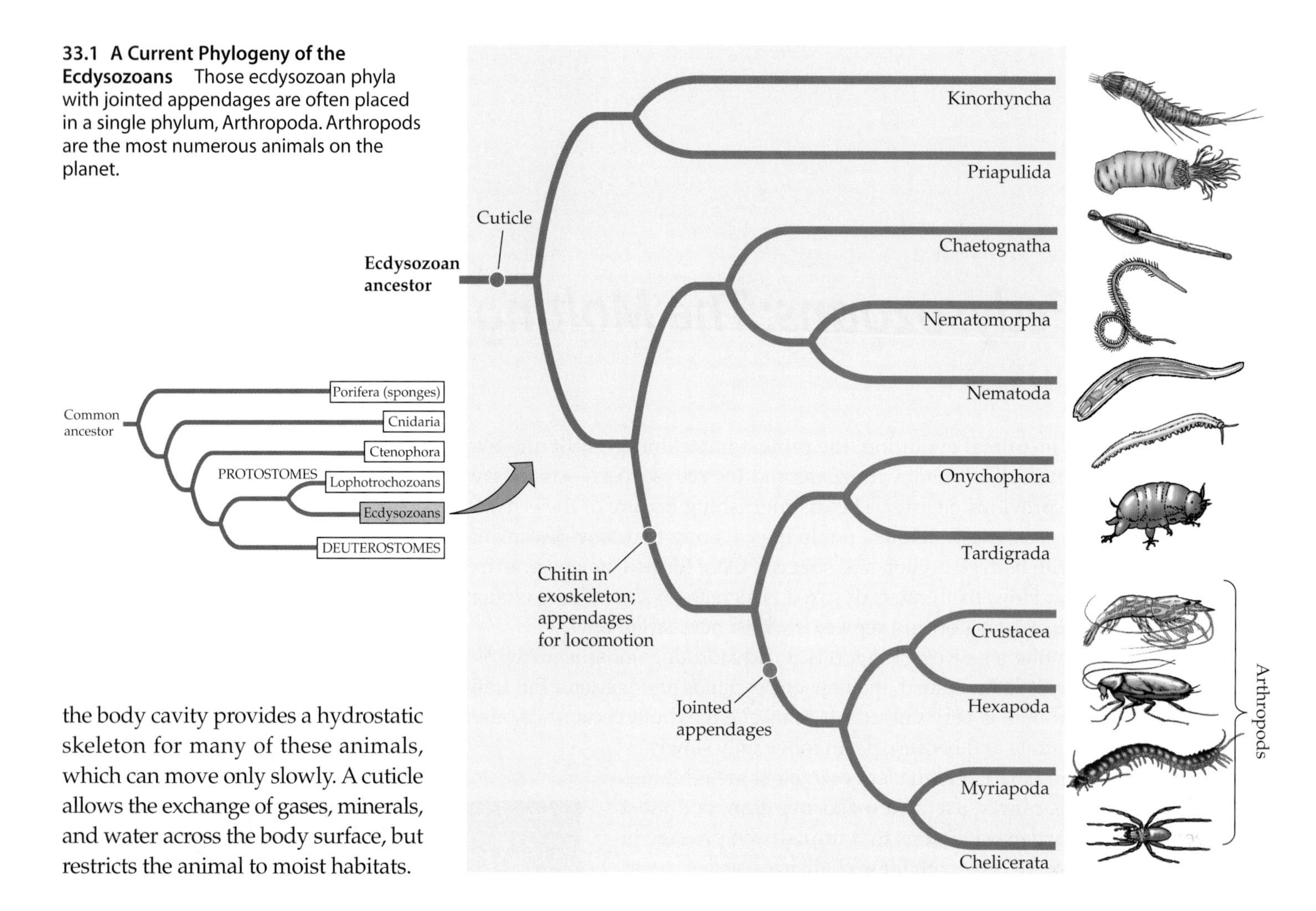

the body cavity provides a hydrostatic skeleton for many of these animals, which can move only slowly. A cuticle allows the exchange of gases, minerals, and water across the body surface, but restricts the animal to moist habitats.

Some marine ecdysozoan phyla have few species

Several phyla of marine wormlike animals branched off early within the ecdysozoan lineage. Each of these phyla contains only a few species. These animals have relatively thin cuticles that are molted periodically as the animals grow to full size.

PRIAPULIDS AND KINORHYNCHS. The 16 species of **priapulids** (phylum **Priapulida**) are cylindrical, unsegmented, wormlike animals that range in size from half a millimeter to 20 centimeters in length (Figure 33.2). They burrow in fine marine sediments and prey on soft-bodied invertebrates, such as polychaete worms. They capture prey with a toothed pharynx, a muscular organ that is everted through the mouth and then withdrawn into the body together with the grasped prey. Fertilization is external, and most species have a larval form that lives in the mud.

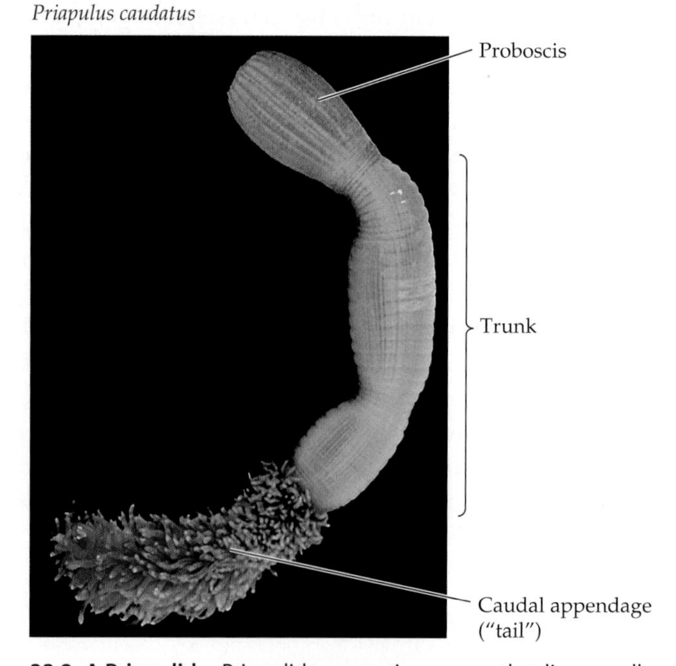

33.2 A Priapulid Priapulids are marine worms that live, usually as burrowers, on the ocean floor. They capture prey with a toothed pharynx that everts through the proboscis. They take their name from Priapus, the Greek god of procreation, who was typically portrayed with an oversize penis.

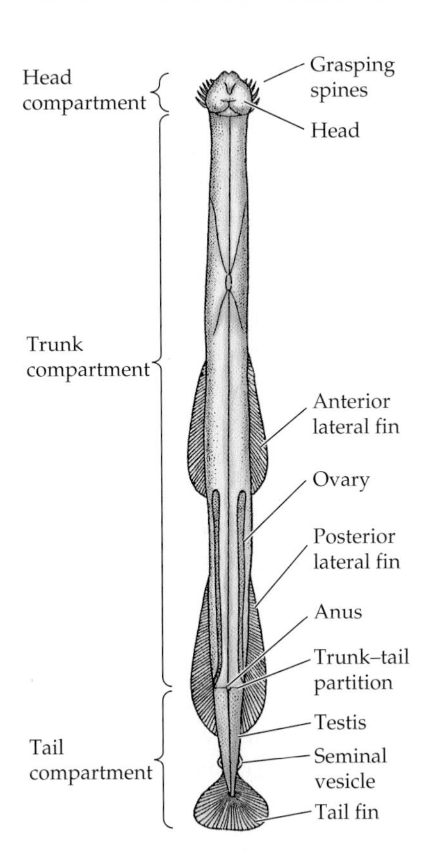

33.3 An Arrow Worm Arrow worms have a three-part body plan. Their fins and grasping spines are adaptations for a predatory lifestyle.

About 150 species of **kinorhynchs** (phylum **Kinorhyncha**) have been described. They are all less than 1 millimeter in length and live in marine sands and muds. Their bodies are divided into 13 segments, each with a separate cuticular plate. These plates are periodically molted during growth. Kinorhynchs feed by ingesting sediments and digesting the organic material found within them, which may include living algae as well as dead matter. Kinorhynchs have no distinct larval stage; fertilized eggs develop directly into juveniles, which emerge from their egg cases with 11 of the 13 body segments already formed.

ARROW WORMS. The phylogeny of the **arrow worms** (phylum **Chaetognatha**) is uncertain. Recent evidence indicates that these animals may in fact belong among the deuterostomes; however, this placement is still in question, and we continue to include them among the ecdysozoans.

The arrow worms body plan is based on a coelom divided into head, trunk, and tail compartments (Figure 33.3). Most arrow worms swim in the open sea, but a few live on the sea floor. Their abundance as fossils indicates that they were common more than 500 million years ago. The 100 or so living species of arrow worms are small enough—less than 12 cm long—that their gas exchange and excretion requirements are met by diffusion through the body surface, and they lack a circulatory system. Wastes and nutrients are moved around the body in the coelomic fluid, which is propelled by cilia that line the coelom. There is no distinct larval stage. Miniature adults hatch directly from eggs that are fertilized internally following elaborate courtship.

Arrow worms are stabilized in the water by means of one or two pairs of lateral fins and a tail fin. They are major predators of small organisms in the open oceans, ranging in size from small protists to young fish as large as the arrow worms themselves. An arrow worm typically lies motionless in the water until water movement signals the approach of prey. The arrow worm then darts forward and grasps the prey with the stiff spines adjacent to its mouth.

Tough cuticles evolved in some unsegmented worms

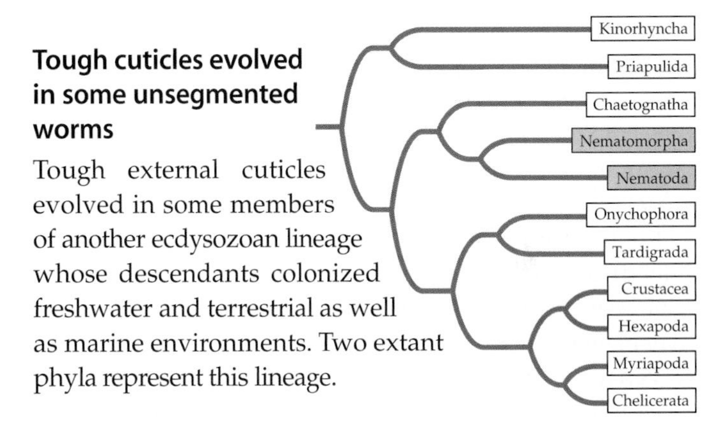

Tough external cuticles evolved in some members of another ecdysozoan lineage whose descendants colonized freshwater and terrestrial as well as marine environments. Two extant phyla represent this lineage.

HORSEHAIR WORMS. About 320 species of horsehair worms (phylum **Nematomorpha**) have been described. As their name implies, horsehair worms are extremely thin, and they range from a few millimeters up to a meter in length (Figure 33.4). Most adult horsehair worms live in fresh water among leaf litter and algal mats near the edges of streams and ponds. The larvae of horsehair worms are internal par-

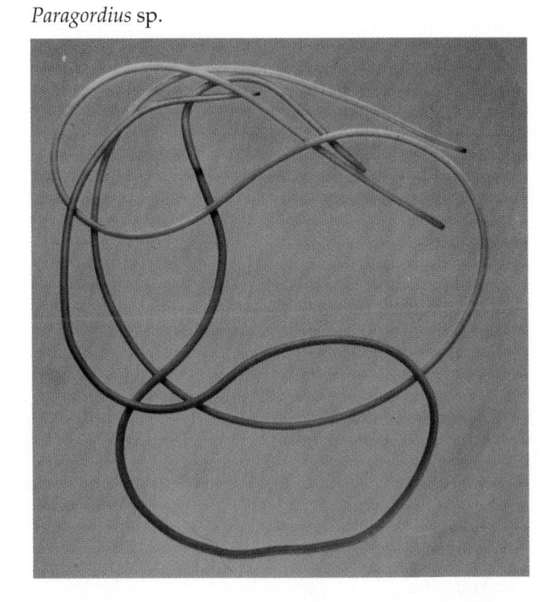

33.4 Horsehair Worms These worms get their name from their hair- or threadlike shape. They can grow to be up to a meter long.

asites of terrestrial and aquatic insects and freshwater crayfish. The horsehair worm's gut is greatly reduced, has no mouth opening, and is probably nonfunctional. These worms may feed only as larvae, absorbing nutrients from their hosts across the body wall, but many continue to grow after they have left their hosts, suggesting that adult worms may also absorb nutrients from their environment.

ROUNDWORMS. Roundworms (phylum **Nematoda**) have a thick, multilayered cuticle secreted by the underlying epidermis that gives their body its shape (Figure 33.5*a*). As a roundworm grows, it sheds its cuticle four times.

Roundworms exchange oxygen and nutrients with their environment through both the cuticle and the intestine, which is only one cell layer thick. Materials are moved through the gut by rhythmic contraction of a highly muscular organ, the *pharynx*, at the worm's anterior end. Roundworms move by contracting their longitudinal muscles.

Roundworms are one of the most abundant and universally distributed of all animal groups. About 25,000 species have been described, but the actual number of living species may be more than a million. Countless roundworms live as scavengers in the upper layers of the soil, on the bottoms of lakes and streams, in marine sediments (Figure 33.5*c*), and as parasites in the bodies of most kinds of plants and animals. The topsoil of rich farmland contains up to 3 billion nematodes per acre.

Many roundworms are predators, preying on protists and other small animals (including other roundworms). Many roundworms live parasitically within their hosts. The largest known roundworm, which reaches a length of 9 meters, is a parasite in the placentas of female sperm whales. The roundworms that are parasites of humans (causing serious tropical diseases such as trichinosis, filariasis, and elephantiasis), domestic animals, and economically important plants have been studied intensively in an effort to find ways of controlling them. One soil-inhabiting nematode, *Caenorhabitis elegans*, is a "model organism" in the laboratories of geneticists and developmental biologists.

The structure of parasitic roundworms is similar to that of free-living species, but the life cycles of many parasitic species have special stages that facilitate the transfer of individuals among hosts. *Trichinella spiralis*, the species that causes the human disease trichinosis, has a relatively simple life cycle. A person may become infected by eating the flesh of an animal (usually a pig) containing larvae of *Trichinella* encysted in its muscles. The larvae are activated in the digestive tract, emerge from their cysts, and attach to the person's intestinal wall, where they feed. Later, they bore through the intestinal wall and are carried in the bloodstream to muscles, where they form new cysts (Figure 33.5*b*). If present in great numbers, these cysts cause severe pain or death.

Arthropods and Their Relatives: Segmented External Skeletons

In Precambrian times, the cuticle of some wormlike ecdysozoan lineages became thickened by the incorporation of layers of protein and a strong, flexible, waterproof polysaccharide called **chitin**. This rigid body covering may originally have had a protective function, but eventually it acquired both support and locomotory functions as well.

A rigid body covering precludes wormlike movement. To move, the animal requires extensions of the body that can be

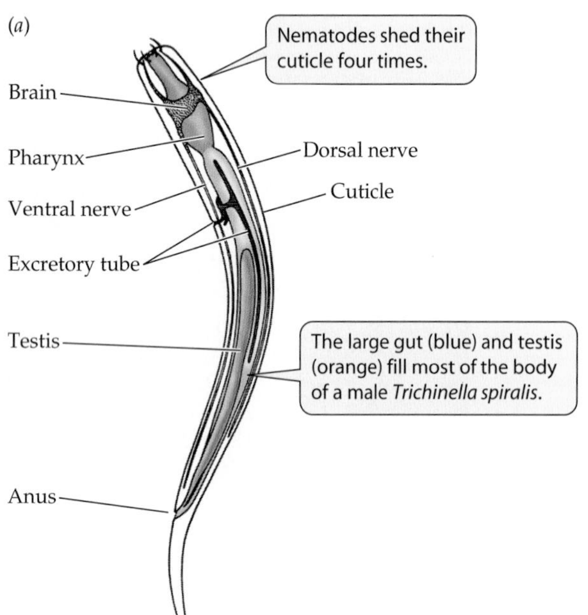

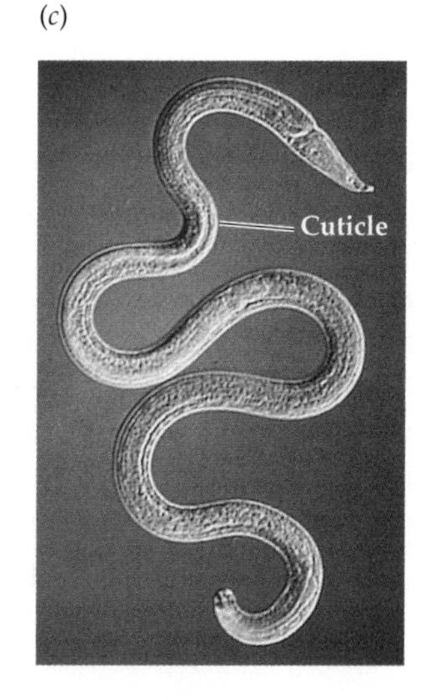

33.5 Roundworms (*a*) The body plan of *Trichinella spiralis*, a roundworm that causes trichinosis. (*b*) A cyst of *Trichinella spiralis* in the muscle tissue of a host. (*c*) This free-living roundworm moves through marine sediments.

manipulated by muscles. Such **appendages** evolved several times in the late Precambrian, leading to the lineages collectively called the **arthropods** ("jointed foot"). Divisions among the arthropod lineages are ancient and have been the subject of much research in the past decade. These phylogenetic relationships are being examined daily in the light of a wealth of new information, much of it concerning gene expression. There is currently no consensus on an exact phylogeny, but most researchers agree that these important animal groups are monophyletic, and some taxonomists consider them as members of a single phylum: **Arthropoda**.

Before presenting one current view of arthropod phylogeny, let's look at some arthropod relatives that have segmented bodies but unjointed legs, and at an early arthropod lineage that disappeared but left an important fossil record.

Some relatives of the arthropods have unjointed legs

Although they were once thought to be closely related to annelid worms, recent molecular evidence links the 110 species of **onychophorans** (phylum **Onychophora**) to the arthropod lineages. Onychophorans have soft bodies that are covered by a thin, flexible cuticle that contains chitin. Onychophorans use their fluid-filled body cavities as hydrostatic skeletons. Their soft, fleshy, unjointed, claw-bearing legs are formed by outgrowths of the body (Figure 33.6*a*). These animals are probably similar in appearance to ancestral arthropods. Fertilization is internal, and the large, yolky eggs are brooded with the body of the female.

Kinorhyncha
Priapulida
Chaetognatha
Nematomorpha
Nematoda
Onychophora
Tardigrada
Crustacea
Hexapoda
Myriapoda
Chelicerata

Like the onychophorans, **water bears** (phylum **Tardigrada**) have fleshy, unjointed legs and use their fluid-filled body cavities as hydrostatic skeletons (Figure 33.6*b*). Water bears are extremely small (0.1–0.5 mm in length), and they lack circulatory systems and gas exchange organs. The 600 extant species of water bears live in marine sands and on temporary water films on plants. When these films dry out, the water bears also lose water and shrink to small, barrel-shaped objects that can survive for at least a decade in a dormant state. They have been found at densities as high as 2 million per square meter of moss.

Jointed legs appeared in the trilobites

The **trilobites** (phylum **Trilobita**) were among the earliest arthropods. They flourished in Cambrian and Ordovician seas, but disappeared in the great Permian extinction at the close of the Paleozoic era (245 mya). Because their heavy exoskeletons provided ideal material for fossilization, they left behind an abundant record of their existence (Figure 33.7).

Trilobites were heavily armored, and their body segmentation and appendages followed a relatively simple, repetitive plan. But their appendages were jointed, and some of them were modified for different functions. This specialization of appendage function became a theme as the evolution of the arthropod lineage continued.

Modern arthropods dominate Earth's fauna

Arthropod appendages have evolved an amazing variety of forms, and they serve many functions, including walking and swimming, gas exchange, food capture and manipulation, copulation, and sensory perception. The pattern of segmentation is similar among most arthropods because their development is governed by a common cascade of regula-

(*a*) *Peripatus* sp.

(*b*) *Echiniscus springer* 50 μm

33.6 Unjointed Legs (*a*) Onychophorans, also called "velvet worms," have unjointed legs and use the body cavity as a hydrostatic skeleton. (*b*) The appendages and general anatomy of water bears superficially resemble those of onychophorans.

Odontochile rugosa

33.7 A Trilobite The relatively simple, repetitive segments of the now-extinct trilobites are illustrated by a fossil trilobite from the shallow seas of the Devonian period, some 400 million years ago.

tory genes (see Figure 19.15), including homeotic genes that determine the kinds of appendages that are borne on each segment.

The bodies of arthropods are divided into segments. Their muscles are attached to the inside of the exoskeleton. Each segment has muscles that operate that segment and the appendages attached to it (Figure 33.8). The arthropod exoskeleton has had a profound influence on the evolution of these animals. Encasement within a rigid body covering provides support for walking on dry land, and the waterproofing provided by chitin keeps the animal from dehydrating in dry air. Aquatic arthropods were, in short, excellent candidates to invade terrestrial environments. As we will see, they did so several times.

There are four major arthropod phyla living today: the crustaceans, hexapods (insects), myriapods, and chelicerates. Collectively, the arthropods (including both terrestrial and marine species) are the dominant animals on Earth, both in numbers of species (about 1.5 million described) and number of individuals (estimated at some 10^{18} individuals, or a billion billion).

Crustaceans: Diverse and Abundant

Crustaceans (phylum **Crustacea**) are the dominant marine arthropods today. The most familiar crustaceans belong to the class Malacostraca, which includes shrimp, lobsters, crayfish, and crabs (decapods; Figure 33.9*a*); and sow bugs (isopods; Figure 33.9*b*). Also included among the crustaceans are a variety of small species, many of which superficially resemble shrimp. The individuals of one group alone, the copepods (class Copepoda; Figure 33.9*c*), are so numerous that they may be the most abundant of all animals.

Barnacles (class Cirripedia) are unusual crustaceans that are sessile as adults (Figure 33.9*d*). With their calcareous shells, they superficially resemble mollusks but, as the zoologist Louis Agassiz remarked more than a century ago, a barnacle is "nothing more than a little shrimp-like animal, standing on its head in a limestone house and kicking food into its mouth."

Most of the 40,000 described species of crustaceans have a body that is divided into three regions: *head*, *thorax*, and *abdomen*. The segments of the head are fused together, and the head bears five pairs of appendages. Each of the multiple thoracic and abdominal segments usually bears one pair of appendages. In some cases, the appendages are branched, with different branches serving different functions. In many species, a fold of the exoskeleton, the *carapace*, extends dorsally and laterally back from the head to cover and protect some of the other segments (Figure 33.10*a*).

The fertilized eggs of most crustacean species are attached to the outside of the female's body, where they remain during their early development. At hatching, the young of some species are released as larvae; those of other species are released as juveniles that are similar in form to the adults. Still other species release eggs into the water or attach them to an object in the environment. The typical crustacean larva, called a **nauplius**, has three pairs of appendages and one central eye (Figure 33.10*b*). In many crustaceans, the nauplius larva develops within the egg before it hatches.

There is a growing recognition among researchers that a crustacean lineage may have been ancestral to all present-day arthropods. Therefore, the phylum Crustacea, as we recognize it here, may be paraphyletic (see Chapter 25). Molecular evidence points especially to a link between the crustaceans and another important lineage, the hexapods.

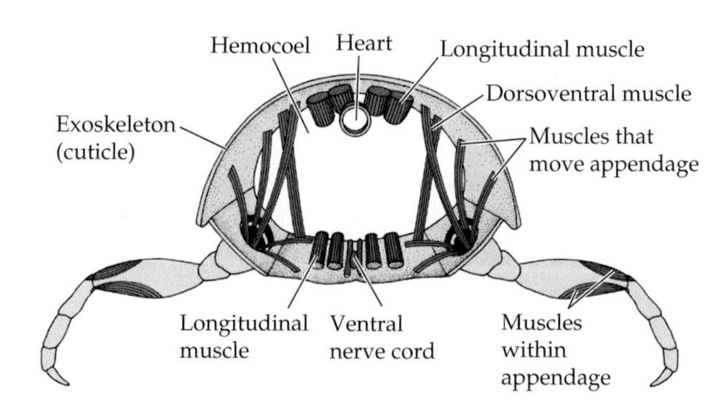

33.8 Arthropod Exoskeletons Are Rigid and Jointed This cross section through a thoracic segment of a generalized arthropod illustrates the arthropod body plan, which is characterized by a rigid exoskeleton with jointed appendages.

(*a*) *Opisthopus transversus*

(*b*) *Ligia occidentalis*

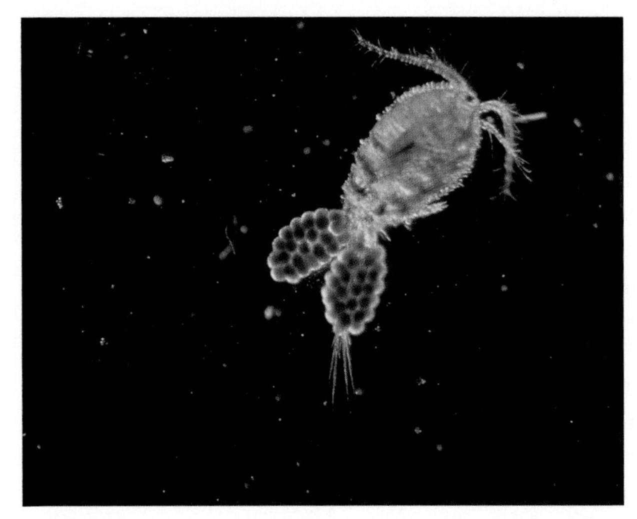

(*c*) *Cyclops* sp.

(*d*) *Lepas pectinata*

33.9 Crustacean Diversity (*a*) This mottled pea crab is a decapod crustacean. Its pigmentation depends on the food it ingests. (*b*) This isopod is found on the beaches of the California coast. (*c*) This microscopic freshwater copepod is only about 30 µm long. (*d*) Gooseneck barnacles attach to a substratum and feed by protruding and retracting feeding appendages from their shells.

Insects: Terrestrial Descendants of Marine Crustaceans

During the Devonian, more than 400 million years ago, arthropods made the leap from the marine environment onto land. Of the several groups who successfully colonized the terrestrial habitat, none is more prominent today than the six-legged individuals of the phylum **Hexapoda**—the insects.

Insects are found in most terrestrial and freshwater habitats, and they utilize nearly all species of plants and many species of animals as food. Some are internal parasites of plants and animals; others suck their host's blood or consume surface body tissues. The 1.4 million species of insects that have been described are believed to be only a small fraction of the total number of species living today.

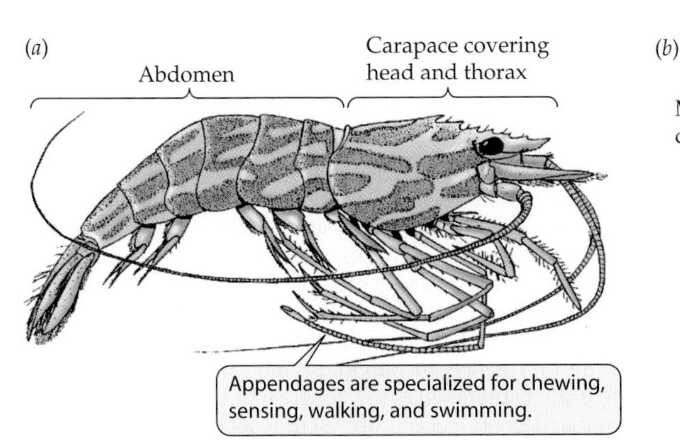

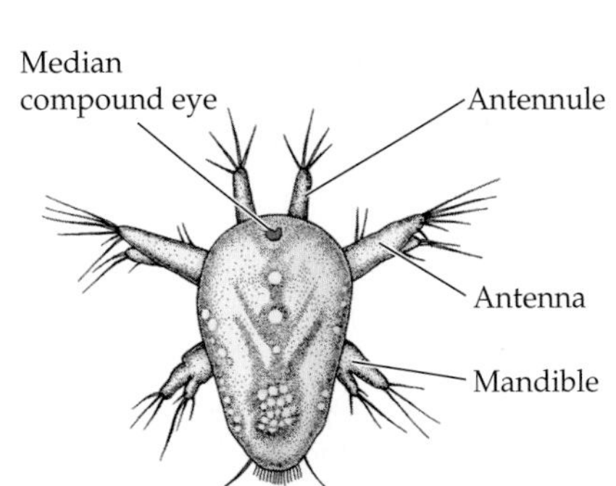

33.10 Crustacean Structure (*a*) The bodies of crustaceans are divided into three regions, each of which bears appendages. (*b*) A nauplius larva has one compound eye and three pairs of appendages.

Very few insect species live in the ocean. In freshwater environments, on the other hand, they are sometimes the dominant animals, burrowing through the substratum, extracting suspended prey from the water, and actively pursuing other animals. Insects were the first animals to achieve the ability to fly, and they are important pollinators of flowering plants.

Insects, like crustaceans, have three basic body regions: head, thorax, and abdomen. They have a single pair of antennae on the head and three pairs of legs attached to the thorax (Figure 33.11). Unlike the other arthropods, insects have no appendages growing from their abdominal segments (see Figure 21.5).

An insect exchanges gases by means of air sacs and tubular channels called *tracheae* (singular, trachea) that extend from external openings inward to tissues throughout the body. The adults of most flying insects have two pairs of stiff, membranous wings attached to the thorax. However, flies have only one pair of wings, and in beetles the forewings form heavy, hardened wing covers.

Wingless insects include springtails and silverfish (Figure 33.12). Of the modern insects, they are probably the most similar in form to insect ancestors. Apterygote insects have a simple life cycle, hatching from eggs as miniature adults.

Development in the winged insects (Figure 33.13) is complex. The hatchlings do not look like adults, and they undergo substantial changes at each molt. The immature stages of insects between molts are called **instars**. A substantial change that occurs between one developmental stage and another is called **metamorphosis**. If the changes between its instars are gradual, an insect is said to have **incomplete metamorphosis**.

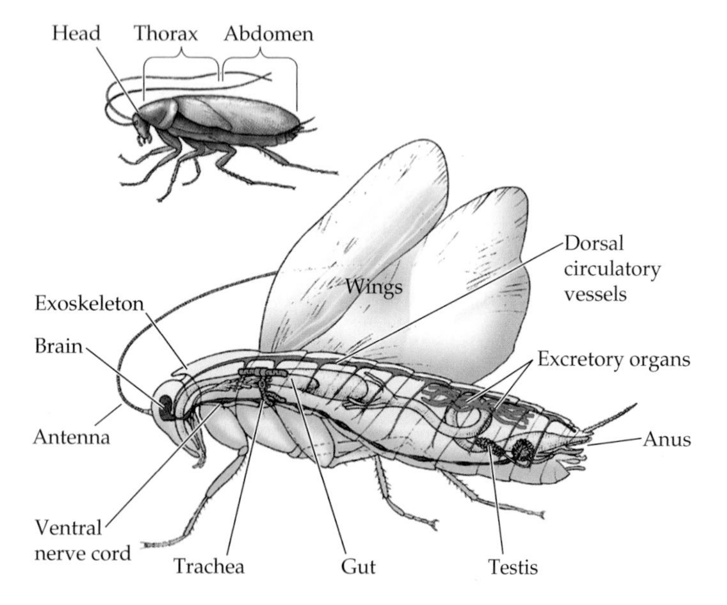

33.11 Structure of an Insect This diagram of a generalized insect illustrates its three-part body plan. The middle region, the thorax, bears three pairs of legs and, in most groups, two pairs of wings.

Hydropodura aquatica

33.12 Wingless Insects The wingless insects have a simple life cycle. They hatch looking like miniature adults, then grow by successive moltings of the cuticles as these springtails are doing.

In some insect groups, the larval and adult forms appear to be completely different animals. The most familiar example of such **complete metamorphosis** occurs in members of the order Lepidoptera, in which the larval caterpillar transforms itself into the adult butterfly (see Figure 1.1). During complete metamorphosis, the wormlike larva transforms itself during a specialized phase, called the **pupa**, in which many larval tissues are broken down and the adult form develops. In many of these groups, the different life stages are specialized for living in different environments and using different food sources. In many species, the larvae are adapted for feeding and growing, and the adults are specialized for reproduction and dispersal.

Entomologists divide the winged insects into about 29 different orders. We can make sense of this bewildering variety by recognizing three major lineages:

- Winged insects that cannot fold their wings against the body
- Winged insects that can fold their wings and that undergo incomplete metamorphosis
- Winged insects that can fold their wings and that undergo complete metamorphosis

Because they can fold their wings over their backs, flying insects belonging to the second and third lineages can tuck their wings out of the way upon landing and crawl into crevices and other tight places.

33.13 The Diversity of Insects (*a*) Unlike most flying insects, ▶ this dragonfly cannot fold its wings over its back. (*b*) The Mexican bush katydid represents the order Orthoptera. (*c*) Harlequin bugs are "true" bugs (order Hemiptera); (*d*) These mating mantophasmatodeans represent a recently discovered Hemipteran lineage found only in the Cape region of South Africa. (*e*) A predatory diving beetle (order Coleoptera). (*f*) The California dogface butterfly is a member of the Lepidoptera. (*g*) The flies, including this Mediterranean fruit fly, comprise the order Diptera. (*h*) Many genera in the order Hymenoptera, such as honeybees, are social insects.

(*a*) *Anax imperator*

(*b*) *Scudderia mexicana*

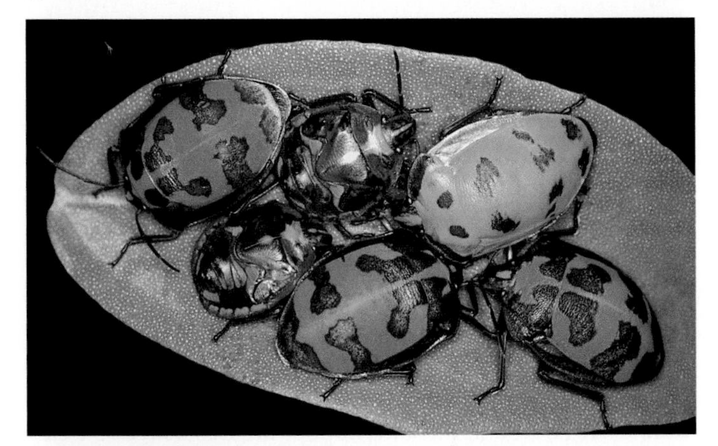

(*c*) *Murgantia histrionica*

(*d*) *Timema* sp.

(*e*) *Dytiscus marginalis*

(*f*) *Colias eurydice*

(*g*) *Ceratitis capitata*

(*h*) *Apis mellifera*

The only surviving members of the lineage whose members cannot fold their wings against the body are the orders Odonata (dragonflies and damselflies, Figure 33.13*a*) and Ephemeroptera (mayflies). All members of these two orders have aquatic larvae that transform themselves into flying adults after they crawl out of the water. Although many of these insects are excellent flyers, they require a great deal of open space in which to maneuver. Dragonflies and damselflies are active predators as adults, but adult mayflies lack functional digestive tracts and live only long enough to mate and lay eggs.

The second lineage, whose members can fold their wings and have incomplete metamorphosis, includes the orders Orthoptera (grasshoppers, crickets, roaches, mantids, and walking sticks; Figure 33.13*b*), Isoptera (termites), Plecoptera (stone flies), Dermaptera (earwigs), Thysanoptera (thrips), Hemiptera (true bugs; Figure 33.13*c*), and Homoptera (aphids, cicadas, and leafhoppers). In these groups, hatchlings are sufficiently similar in form to adults to be recognizable. They acquire adult organ systems, such as wings and compound eyes, gradually through several juvenile instars. Remarkably, a new insect order in this lineage, the Mantophasmatodea, was first described in 2002 (Figure 33.13*d*). These small insects are common in the Cape Region of southern Africa, an area of exceptional species richness and endemism for many animal and plant groups.

Insects belonging to the third lineage undergo complete metamorphosis. About 85 percent of all species of winged insects belong to this lineage. Familiar examples are the orders Neuroptera (lacewings and their relatives), Coleoptera (beetles; Figure 33.13*e*), Trichoptera (caddisflies), Lepidoptera (butterflies and moths; Figure 33.13*f*), Diptera (flies; Figure 33.13*g*), and Hymenoptera (sawflies, bees, wasps, and ants; Figure 33.13*h*).

Members of several orders of winged insects, including the Phthiraptera (lice) and Siphonaptera (fleas), are parasitic. Although descended from flying ancestors, these insects have lost the ability to fly.

Molecular data suggest that the lineage leading to the insects separated from the lineage leading to modern crustaceans about 450 million years ago, about the time of the appearance of the first land plants. These ancestral forms penetrated a terrestrial environment that was ecologically empty, which in part accounts for their remarkable success. But this success of the insects is also due to their wings, which arose only once early during insect evolution. Homologous genes control the development of insect wings and crustacean appendages, suggesting that that the insect wing evolved from a dorsal branch of a crustacean limb (Figure 33.14). The dorsal limb branch of crustaceans is used for respiration and osmoregulation. This finding suggests that the insect wing evolved from a gill-like structure that had a respiratory function.

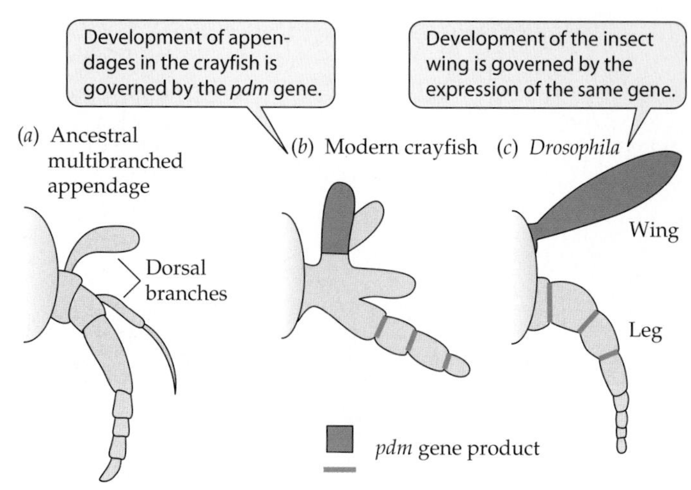

33.14 Origin of Insect Wings The insect wing may have evolved from an ancestral appendage similar to that of modern crustaceans. (*a*) A diagram of the ancestral, multibranched arthropod limb. (*b, c*) The *pdm* gene, a Hox gene, is expressed throughout the dorsal limb branch and walking leg of the thoracic limb of a crayfish (*a*) and in the wings and legs of *Drosophila* (*b*).

Arthropods with Two Body Regions

Insects and most crustaceans have tripartite body plans, with a head, thorax, and abdomen. In two other arthropod lineages, evolution resulted in a body plan with two regions—a head and a trunk.

Myriapods have many legs

Centipedes, millipedes, and the two other groups of animals comprise the phylum **Myriapoda**. Centipedes and millipedes have a well-formed head and a long, flexible, segmented trunk that bears many pairs of legs (Figure 33.15). Centipedes, which have one pair of legs per segment, prey on insects and other small animals. Millipedes, which have two pairs of legs per segment, scavenge and eat plants. More than 3,000 species of centipedes and 10,000 species of millipedes have been described; many more species probably remain unknown. Although most myriapods are less than a few centimeters long, some tropical species are ten times that size.

Most chelicerates have four pairs of walking legs

In the body plan of **chelicerates** (phylum **Chelicerata**), the anterior region (head) bears two pairs of appendages modified to form mouthparts. In addition, many chelicerates have four pairs of walking legs. The 63,000 described chelicerate species are usually placed in three classes: Pycnogonida, Merostomata, and Arachnida; most of them belong to Arachnida.

The **pycnogonids** (class **Pycnogonida**), or sea spiders, are a poorly known group of about 1,000 marine species (Figure 33.16*a*). Most are small, with leg spans less than 1 cm, but some deep-sea species have leg spans up to 60 cm. A few py-

(*a*) *Scolopendra heros*

(*b*) *Harpaphe haydeniana*

33.15 Myriapods (*a*) Centipedes have powerful jaws for capturing active prey. (*b*) Millipedes, which are scavengers and plant eaters, have smaller jaws and legs. They have two pairs of legs per segment, in contrast to the one pair on each segment of centipedes.

cnogonids feed on algae, but most are carnivorous, feeding on a variety of small invertebrates.

The class **Merostomata** contains the horseshoe crabs (order Xiphosura), with five living species, and the extinct giant water scorpions (order Eurypterida). Horseshoe crabs, which have changed very little during their long fossil history, have a large horseshoe-shaped covering over most of the body. They are common in shallow waters along the eastern coasts of North America and Southeast Asia, where they scavenge and prey on bottom-dwelling invertebrates. Periodically they crawl into the intertidal zone in large numbers to mate and lay eggs (Figure 33.16*b*).

Arachnids (class **Arachnida**) are abundant in terrestrial environments. Most arachnids have a simple life cycle in which miniature adults hatch from internally fertilized eggs and begin independent lives almost immediately. Some arachnids retain their eggs during development and give birth to live young.

The most species-rich and abundant arachnids are the spiders, scorpions, harvestmen, mites, and ticks (Figure 33.17). The 30,000 described species of mites and ticks live in soil, leaf litter, mosses, and lichens, under bark, and as parasites of plants, invertebrates, and vertebrates. They are vectors for wheat and rye mosaic viruses, and they cause mange in domestic animals and skin irritation in humans.

Spiders are important terrestrial predators. Some have excellent vision that enables them to chase and seize their prey. Others spin elaborate webs made of protein threads in which they snare prey. The threads are produced by modified abdominal appendages connected to internal glands that secrete the proteins, which dry on contact with air. The webs of different groups of spiders are strikingly varied, and this variation enables the spiders to position their snares in many different environments. Spiders also use protein threads to construct safety lines during climbing and as homes, mating

(*a*) *Decalopoda sp.*

(*b*) *Limulus polyphemus*

33.16 Minor Chelicerate Phyla (*a*) Although they are not spiders, it is easy to see why sea spiders were given their common name. (*b*) This spawning aggregation of horseshoe crabs was photographed on a sandy beach in Delaware.

(*a*) *Phidippus formosus*

(*b*) *Pseudouroctonus minimus*

(*c*) *Hadrobunus maculosus*

(*d*) *Brevipalpus phoenicis*

33.17 Arachnid Diversity (*a*) The black jumping spider's bite produces an inflammatory reaction on mammalian skin. (*b*) Scorpions are nocturnal predators. (*c*) Harvestmen, also called daddy longlegs, are scavengers. (*d*) Mites are blood-sucking, external parasites on vertebrates.

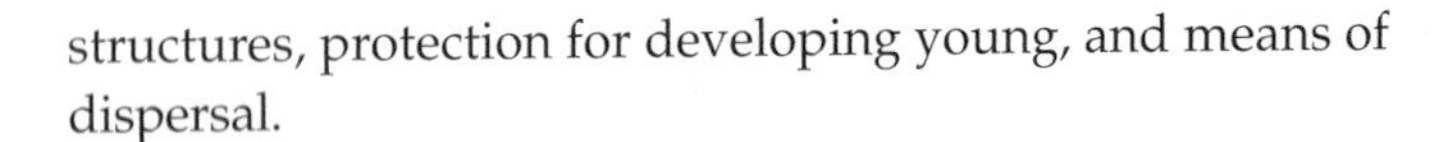

structures, protection for developing young, and means of dispersal.

Themes in the Evolution of Protostomes

We end this chapter by reviewing some of the evolutionary trends we have seen in the animal groups we have discussed so far. Most of protostomate evolution took place in the oceans. Early protostomes used their fluid-filled body cavities as hydrostatic skeletons. Segmentation permitted different parts of the body to be moved independently of one another. Thus species in some protostomate lineages gradually evolved the ability to change their shape in complex ways and to move rapidly over and through the substratum or through the water.

During much of animal evolution, the only food in the water consisted of dissolved organic matter and very small organisms. Consequently, many different lineages of animals, including lophophorates, mollusks, tunicates, and some crustaceans, evolved feeding structures designed to filter small prey from water, as well as structures for moving water through or over their prey-collecting devices. Animals that feed in this manner are abundant and widespread in marine waters today.

Because water flows readily, bringing food with it, sessile lifestyles also evolved repeatedly during lophotrochozoan and ecdysozoan evolution. Most phyla today have at least some sessile members. Being sessile presents certain challenges. For example, sessile animals cannot come together to mate. Some species eject both eggs and sperm into the water; others retain their eggs within their bodies and extrude only their sperm, which are carried by the water to other individuals. Species whose adults are sessile often have motile larvae, many of which have complicated mechanisms for locating suitable sites on which to settle.

A sessile animal gains access to local resources, but forfeits access to more distant resources. Many colonial sessile pro-

33.1 Anatomical Characteristics of the Major Protostomate Phyla

PHYLUM	BODY CAVITY	DIGESTIVE TRACT	CIRCULATORY SYSTEM
Lophotrochozoans			
Platyhelminthes	None	Dead-end sac	None
Rotifera	Pseudocoelom	Complete	None
Bryozoa	Coelom	Complete	None
Brachiopoda	Coelom	Complete in most	Open
Phoronida	Coelom	Complete	Closed
Nemertea	Coelom	Complete	Closed
Annelida	Coelom	Complete	Closed or open
Mollusca	Reduced coelom	Complete	Open except in cephalopods
Ecdysozoans			
Chaetognatha	Coelom	Complete	None
Nematomorpha	Pseudocoelom	Greatly reduced	None
Nematoda	Pseudoceolom	Complete	None
Crustacea	Hemocoel	Complete	Open
Hexapoda	Hemocoel	Complete	Open
Myriapoda	Hemocoel	Complete	Open
Chelicerata	Hemocoel	Complete	Open

Note: All protostomes have bilateral symmetry.

tostomes, however, are able to grow in the direction of better resources or into sites offering better protection. Individual members of colonies, if they are directly connected, can share resources. The ability to share resources enables some individuals to specialize for particular functions, such as reproduction, defense, or feeding. The nonfeeding individuals derive their nutrition from their feeding associates.

Predation may have been the major selective pressure for the development of hard, external body coverings. Such coverings evolved independently in many lophotrochozoan and ecdysozoan lineages. In addition to providing protection, they became key elements in the development of new systems of locomotion. Locomotory abilities permitted prey to escape more readily from predators, but also allowed predators to pursue their prey more effectively. Thus, the evolution of animals has been, and continues to be, a complex "arms race" among predators and prey.

Although we have concentrated on the evolution of greater complexity in animal lineages, many lineages whose members have remained simple have been very successful. Cnidarians are common in the oceans; roundworms are abundant in most aquatic and terrestrial environments. Parasites have lost complex body plans but have evolved complex life cycles.

The characteristics of the major existing phyla of protostomate animals are summarized in Table 33.1. Many major evolutionary trends were shared by protostomes and deuterostomes, the lineage that includes the chordates, the group to which humans belong. We will consider the evolution of diversity among the deuterostomes in the next chapter.

Chapter Summary

- The ecdysozoan lineage is characterized by a nonliving external covering—an exoskeleton, or cuticle. **Review Figure 33.1**
- An animal with an exoskeleton grows by periodically shedding its exoskeleton and replacing it with a larger one, a process called molting.

Cuticles: Flexible, Unsegmented Exoskeletons

- Members of several phyla of marine worms with thin cuticles are descendants of an early split in the ecdysozoan lineage. **Review Figure 33.3**
- Tough cuticles are found in members of two phyla, the horsehair worms and the roundworms.
- Roundworms (phylum Nematoda) are one of the most abundant and universally distributed of all animal groups. Many are parasites. **Review Figure 33.5**

Arthropods and Their Relatives: Segmented External Skeletons

- Animals with rigid exoskeletons lack cilia for locomotion. To move, they have appendages that can be manipulated by muscles. **Review Figure 33.8**
- Although there is currently no consensus on an exact phylogeny, most researchers agree that the arthropod groups are monophyletic.
- Onychophorans and tardigrades have soft, unjointed legs. They are probably similar to ancestral arthropods.
- Trilobites flourished in Cambrian and Ordovician seas, but they became extinct at the close of the Paleozoic era.

Crustaceans: Species-Rich and Abundant

- The segments of the crustacean body are divided among three regions: head, thorax, and abdomen. **Review Figure 33.10**
- The most familiar crustaceans are shrimp, lobsters, crayfish, crabs, sow bugs, and sand fleas. Copepod crustaceans may be the most abundant animals on the planet.
- Recent molecular evidence indicates that the crustacean lineage may be ancestral to all the arthropods.

Insects: Terrestrial Descendants of Marine Crustaceans

- About 1.4 million species of insects (phylum Hexapoda) have been described, but that number is a small fraction of the total number of existing species. Although few species are found in marine environments, they are among the dominant animals in virtually all terrestrial and many freshwater habitats.
- Like crustaceans, insects have three body regions (head, thorax, abdomen). They bear a single pair of antennae on the head and three pairs of legs attached to the thorax. No appendages grow from their abdominal segments. **Review Figure 33.11**
- Wingless insects look like miniature adults when they hatch. Hatchlings of some winged insects resemble adults, but others undergo substantial changes at each molt.

▶ The winged insects can be divided into three major subgroups. Members of one subgroup cannot fold their wings back against the body. Members of the other two subgroups can.

▶ The wings of insects probably evolved from the dorsal branches of multibranched ancestral appendages. **Review Figure 33.14**

Arthropods with Two Body Regions

▶ Individuals of the remaining arthropod phyla generally have segmented bodies with two distinct regions, head and trunk.

▶ Myriapods (centipedes and millipedes) have many segments and many pairs of legs.

▶ Most chelicerates (phylum Chelicerata) have four pairs of legs.

▶ Arachnids—scorpions, harvestmen, spiders, mites, and ticks—are abundant in terrestrial environments.

Themes in the Evolution of Protostomes

▶ Most evolution of protostomes took place in the oceans.

▶ Early animals used fluid-filled body cavities as hydrostatic skeletons. Subdivision of the body cavity into segments allowed better control of movement.

▶ During much of animal evolution, the only food in the water consisted of dissolved organic matter and very small organisms.

▶ Flowing water brings food with it, allowing many aquatic animals to obtain food while being sessile.

▶ Predation may have been the major selective pressure for the development of hard, external body coverings.

See Web/CD Activities 33.1 and 33.2 for a concept review of this chapter.

Self-Quiz

1. The outer covering of ecdysozoans
   a. is always hard and rigid.
   b. is always thin and flexible.
   c. is present at some stage in the life cycle but not always among adults.
   d. ranges from very thin to hard and rigid.
   e. prevents the animals from changing their shapes.
2. The primary support for members of several small phyla of marine worms is
   a. their exoskeletons.
   b. their internal skeletons.
   c. their hydrostatic skeletons.
   d. the surrounding sediments.
   e. the bodes of other animals within which they live.
3. Roundworms are abundant and diverse because
   a. they are both parasitic and free-living and eat a wide variety of foods.
   b. they are able to molt their exoskeletons.
   c. their thick cuticle enables them to move in complex ways.
   d. their body cavity is a pseudocoelom.
   e. their segmented bodies enable them to live in many different places.
4. The arthropod exoskeleton is composed of a
   a. mixture of several kinds of polysaccharides.
   b. mixture of several kinds of proteins.
   c. single complex polysaccharide called chitin.
   d. single complex protein called arthropodin.
   e. mixture of layers of proteins and a polysaccharide called chitin.
5. Which phyla are arthropod relatives with unjointed legs?
   a. Trilobita and Onychophora
   b. Onychophora and Tardigrada
   c. Trilobita and Tardigrada
   d. Onychophora and Chelicerata
   e. Tardigrada and Chelicerata
6. The members of which crustacean group are probably the most abundant of all animals?
   a. Decapoda
   b. Amphipoda
   c. Copepoda
   d. Cirripedia
   e. Isopoda
7. The body plan of insects is composed of which of the three following regions?
   a. Head, abdomen, and trachea
   b. Head, abdomen, and cephalothorax
   c. Cephalothorax, abdomen, and trachea
   d. Head, thorax, and abdomen
   e. Abdomen, trachea, and mantle
8. Insects that hatch from eggs into juveniles that resemble miniature adults are said to have
   a. instars.
   b. neopterous development.
   c. accelerated development.
   d. incomplete metamorphosis.
   e. complete metamorphosis.
9. Which of the following groups of insects cannot fold their wings back against the body?
   a. Beetles
   b. True bugs
   c. Earwigs
   d. Stone flies
   e. Mayflies
10. Factors that may have contributed to the remarkable evolutionary diversification of insects include
    a. the terrestrial environments penetrated by insects lacked any other similar organisms.
    b. insects evolved the ability to fly.
    c. some lineages of insects evolved complete metamorphosis.
    d. insects evolved effective means of delivering oxygen to their internal tissues.
    e. All of the above

For Discussion

1. Segmentation has arisen several times during animal evolution. What advantages does segmentation provide? Given these advantages, why do so many unsegmented animals survive?
2. The British biologist J. B. S. Haldane is reputed to have quipped that "God was unusually fond of beetles." Beetles are, indeed, the most species-rich lineage of organisms. What features of beetles have contributed to the evolution and survival of so many species?
3. In Part Four of this book, we pointed out that major structural novelties have arisen infrequently during the course of evolution. Which of the features of protostomes do you think are major evolutionary novelties? What criteria do you use to judge whether a feature is a major as opposed to a minor novelty?
4. There are more described and named species of insects than of all other animal lineages combined. However, only a very few species of insects live in marine environments, and those species are restricted to the intertidal zone or the ocean surface. What factors may have contributed to the inability of insects to be successful in the oceans?

34 Deuterostomate Animals

Complex social systems, in which individuals associate with one another to breed and care for their offspring, characterize many species of fish, birds, and mammals—the most conspicuous and familiar deuterostomate animals. We tend to think of these social systems as having evolved relatively recently, but some amphibians, members of an ancient deuterostomate group, also have elaborate courtship and parental care behavior. For example, the male of the European midwife toad gathers eggs around his hind legs as the female lays them. He then carries the eggs until they are ready to hatch.

In the Surinam toad, mating and parental care are exquisitely coordinated, as an elaborate mating "dance" results in the female depositing eggs on the male's belly. The male fertilizes the eggs and, as the ritual ends, he presses them against the female's back, where they are carried until they hatch. The female poison dart frog lays clutches of eggs on a leaf or on the ground, which both parents then work to keep moist and protected. When the tadpoles hatch, they wiggle onto the back of one of their parents, who then carries the tadpoles to water.

There are fewer major lineages and many fewer species of deuterostomes than of protostomes (Table 34.1 on page 658), but we have a special interest in the deuterostomes because we are members of that lineage. In this chapter, we will describe and discuss the deuterostomate phyla: Echinodermata, Hemichordata, and Chordata. We close with a brief overview of some major themes in the evolution of animals.

Some Amphibian Parents Nurture Their Young Poison dart frogs (*Dendrobates reticulatus*) of the Amazon basin lay their eggs on land. Both parents protect and nurture the eggs until they hatch, at which time a parent carries the tadpoles to water on its back.

Deuterostome Ancestors

A group of extinct animals known as the yunnanozoans are the likely ancestors of all deuterostomes. Many fossils of these animals have been discovered in China's Yunnan province. These well-preserved fossils show that the animals had a large mouth, six pairs of external gills, and a lightly cuticularized, segmented posterior body section (Figure 34.1). Later in deuterostome evolution, gills became internal and were connected to the exterior via slits in the body wall. These gill slits subsequently were lost in the lineage leading to the modern echinoderms.

Yunnanozoan lividum

34.1 The Ancestral Deuterostomes Had External Gills The extinct Yunnanozoan lineage is probably ancestral to all deuterostomes. This fossil, which dates from the Cambrian, shows the six pairs of external gills and segmented posterior body that characterized these animals.

Modern deuterostomes fall into two major clades (Figure 34.2). One clade, composed of echinoderms and hemichordates, is characterized by a three-part coelom and a bilaterally symmetrical, ciliated larva. The ancestors of the other clade, containing the chordates, had a distinctly different, nonfeeding, tadpole-like larva and a unique dorsal supporting structure.

Echinoderms: Pentaradial Symmetry

During the evolution of one deuterostomate lineage, the **echinoderms** (phylum **Echinodermata**), two major structural features arose. One was a system of calcified internal plates covered by thin layers of skin and some muscles. The calcified plates of early echinoderms later became enlarged and thickened until they fused inside the entire body, giving rise to an internal skeleton.

The other feature was a *water vascular system*, a network of water-filled canals leading to extensions called *tube feet*. This system functions in gas exchange, locomotion, and feeding (Figure 34.3*a*). Seawater enters the system through a perforated *madreporite*. A calcified canal leads from the madreporite to another canal that rings the *esophagus* (the tube leading from the mouth to the stomach). Other canals radiate from this *ring canal*, extending through the arms (in species that have arms) and connecting with the tube feet.

The development of these two structural innovations resulted in a striking evolutionary radiation. About 23 classes of echinoderms, of which only 6 survive today, have been described from fossils. The 13,000 species described from their fossil remains are probably only a small fraction of those that actually lived. Nearly all 7,000 species that survive today live only in marine environments. Some have bilaterally symmetrical, ciliated larvae (Figure 34.3*b*) that feed for some time as planktonic organisms before settling and transforming into adults with *pentaradial symmetry* (symmetry in five or multiples of five).

Living echinoderms are members of two lineages: subphylum **Pelmatozoa** and subphylum **Eleutherozoa**. These two groups differ in the form of their water vascular systems.

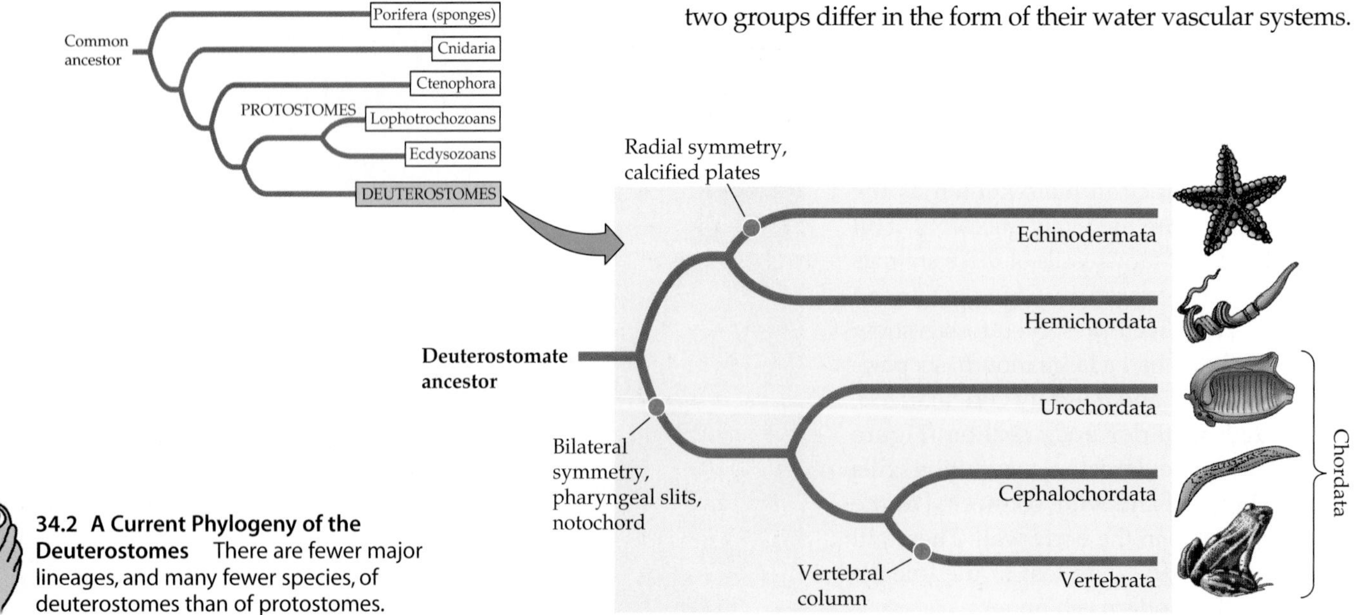

34.2 A Current Phylogeny of the Deuterostomes There are fewer major lineages, and many fewer species, of deuterostomes than of protostomes.

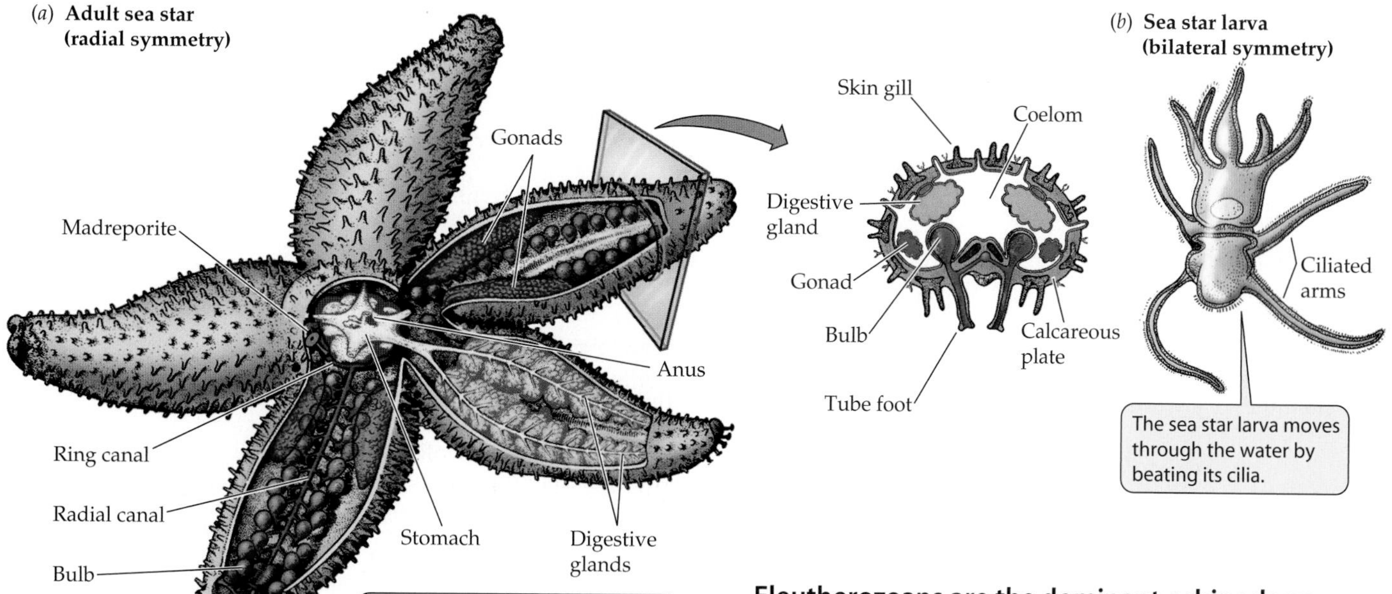

34.3 Echinoderms Display Two Evolutionary Innovations
(*a*) A dorsal view of a sea star displays the canals and tube feet of the echinoderm water vascular system, as well as a calcified internal skeleton. (*b*) The ciliated sea star larva has bilateral symmetry.

Pelmatozoans have jointed arms

Sea lilies and feather stars (class **Crinoidea**) are the only surviving pelmatozoans. Sea lilies were abundant 300–500 million years ago, but only about 80 species survive today. Most sea lilies attach to a substratum by means of a flexible stalk consisting of a stack of calcareous discs. The main body of the animal is a cup-shaped structure that contains a tubular digestive system. Five to several hundred arms, usually in multiples of five, extend outward from the cup. The jointed calcareous plates of the arms enable them to bend.

A sea lily feeds by orienting its arms in passing water currents. Food particles strike and stick to the tube feet, which are covered with mucus-secreting glands. The tube feet transfer these particles to grooves in the arms, where ciliary action carries the food to the mouth. The tube feet of sea lilies are also used for gas exchange and elimination of nitrogenous wastes.

Feather stars are similar to sea lilies, but they have flexible appendages with which they grasp the substratum (Figure 34.4*a*). Feather stars feed in much the same manner as sea lilies. They can walk on the tips of their arms or swim by rhythmically beating their arms. About 600 living species of feather stars have been described.

Eleutherozoans are the dominant echinoderms

Most surviving echinoderms are members of the eleutherozoan lineage. Biochemical data suggest that the eleutherozoa split into two lineages, one leading to sea urchins, sand dollars, and sea cucumbers, and the second leading to sea stars and brittle stars.

Sea urchins and sand dollars (class **Echinoidea**) lack arms, but they share a five-part body plan with all other echinoderms. Sea urchins are hemispherical animals that are covered with spines attached to the underlying skeleton via ball-and-socket joints (Figure 34.4*b*). The spines of sea urchins come in varied sizes and shapes; a few produce toxic substances. Many sea urchins consume algae, which they scrape from rocks with a complex rasping structure. Others feed on small organic debris that they collect with their tube feet or spines. Sand dollars, which are flattened and disc-shaped, feed on algae and fragments of organic matter found on the seafloor or suspended organic material.

The sea cucumbers (class **Holothuroidea**) lack arms, and their bodies are oriented in an atypical manner for an echinoderm. The mouth is anterior and the anus is posterior, not oral and aboral as in other echinoderms. Sea cucumbers use their tube feet primarily for attaching to the substratum rather than for moving. The anterior tube feet are modified into large, feathery, sticky tentacles that can be protruded from the mouth (Figure 34.4*c*). Periodically, the sea cucumber withdraws the tentacles, wipes off the material that has adhered to them, and digests it.

Sea stars (class **Asteroidea**; Figure 34.4*d*) are the most familiar echinoderms. Their digestive organs and gonads are located in the arms. Their tube feet serve as organs of locomotion, gas exchange, and attachment. Each tube foot of a sea star is also an adhesive organ consisting of an internal ampulla connected by a muscular tube to an external suction

34.1 *Summary of Living Members of the Kingdom Animalia*

PHYLUM	NUMBER OF LIVING SPECIES DESCRIBED	MAJOR GROUPS
Porifera: Sponges	10,000	
Cnidaria: Cnidarians	10,000	Hydrozoa: Hydras and hydroids Scyphozoa: Jellyfishes Anthozoa: Corals, sea anemones
Ctenophora: Comb jellies	100	
PROTOSTOMES		
Lophotrochozoans		
Platyhelminthes: Flatworms	20,000	Turbellaria: Free-living flatworms Trematoda: Flukes (all parasitic) Cestoda: Tapeworms (all parasitic) Monogenea (ectoparasites of fishes)
Rotifera: Rotifers	1,800	
Ectoprocta: Bryozoans	4,500	
Brachiopoda: Lamp shells	340	More than 26,000 fossil species described
Phoronida: Phoronids	20	
Nemertea: Ribbon worms	900	
Annelida: Segmented worms	15,000	Polychaeta: Polychaetes (all marine) Oligochaeta: Earthworms, freshwater worms Hirudinea: Leeches
Mollusca: Mollusks	50,000	Monoplacophora: Monoplacophorans Polyplacophora: Chitons Bivalvia: Clams, oysters, mussels Gastropoda: Snails, slugs, limpets Cephalopoda: Squids, octopuses, nautiloids
Ecdysozoans		
Kinorhyncha: Kinorhynchs	150	
Chaetognatha: Arrow worms*	100	
Nematoda: Roundworms	20,000	
Nematomorpha: Horsehair worms	230	
Onychophora: Onychophorans	80	
Tardigrada: Water bears	600	
Chelicerata: Chelicerates	70,000	Merostomata: Horseshoe crabs Arachnida: Scorpions, harvestmen, spiders, mites, ticks
Crustacea	50,000	Crabs, shrimps, lobsters, barnacles, copepods
Hexapoda	1,500,000	Insects
Myriapoda	13,000	Millipedes, centipedes
DEUTEROSTOMES		
Echinodermata: Echinoderms	7,000	Crinoidea: Sea lilies, feather stars Ophiuroidea: Brittle stars Asteroidea: Sea stars Concentricycloidea: Sea daisies Echinoidea: Sea urchins Holothuroidea: Sea cucumbers
Hemichordata: Hemichordates	95	Acorn worms and pterobranchs
Chordata: Chordates	50,000	Urochordata: Sea squirts Cephalochordata: Lancelets Agnatha: Lampreys, hagfishes Chondrichthyes: Cartilaginous fishes Osteichthyes: Bony fishes Amphibia: Amphibians Reptilia: Reptiles Aves: Birds Mammalia: Mammals

* The position of this phylum is uncertain. Many researchers place them in the deuterostomes.

(*a*) *Oxycomanthus bennetti*

(*b*) *Strongylocentrotus purpuratus*

(*c*) *Bohadschia argus*

(*d*) *Henricia leviuscula*

(*e*) *Ophiothrix spiculata*

34.4 Diversity among the Echinoderms (*a*) The flexible arms of this golden feather star are clearly visible. (*b*) Purple sea urchins are important grazers on algae in the intertidal zone of the Pacific Coast of North America. (*c*) This sea cucumber lives on rocky substrata in the seas around Papua New Guinea. (*d*) The blood sea star is typical of many sea stars; some species, however, have more than five arms. (*e*) The arms of the brittle star are composed of hard but flexible plates.

cup. The tube foot is moved by expansion and contraction of the circular and longitudinal muscles of the tube.

Many sea stars prey on polychaetes, gastropods, bivalves, and fish. They are important predators in many marine environments, such as coral reefs and rocky intertidal zones. With hundreds of tube feet acting simultaneously, a sea star can exert an enormous and continuous force. It can grasp a clam in its arms, anchor the arms with its tube feet, and, by steady contraction of the muscles in the arms, gradually exhaust the muscles the clam uses to keep its shell closed. Sea stars that feed on bivalves are able to push the stomach out through the mouth and then through the narrow space between the two halves of the bivalve's shell. The stomach secretes enzymes that digest the prey.

Brittle stars (class **Ophiuroidea**) are similar in structure to sea stars, but their flexible arms are composed of jointed hard plates (Figure 34.4*e*). Brittle stars generally have five arms, but each arm may branch a number of times. Most of the 2,000 species of brittle stars ingest particles from the upper regions of sediments and assimilate the organic material from them, but some species remove suspended food particles from the water; others capture small animals. Brittle stars eject the indigestible particles through their mouths because, unlike most other echinoderms, they have only one opening to the digestive tract.

An additional group, the sea daisies (class **Concentricycloidea**) were discovered only in 1986, and little is known about them. They have tiny disc-shaped bodies with a ring of marginal spines, and two ring canals, but no arms. Sea daisies are found on rotting wood in ocean waters. They apparently eat prokaryotes, which they digest outside their bodies and absorb either through a membrane that covers the oral surface or via a shallow, saclike stomach. Recent molecular data suggest that they are greatly modified sea stars.

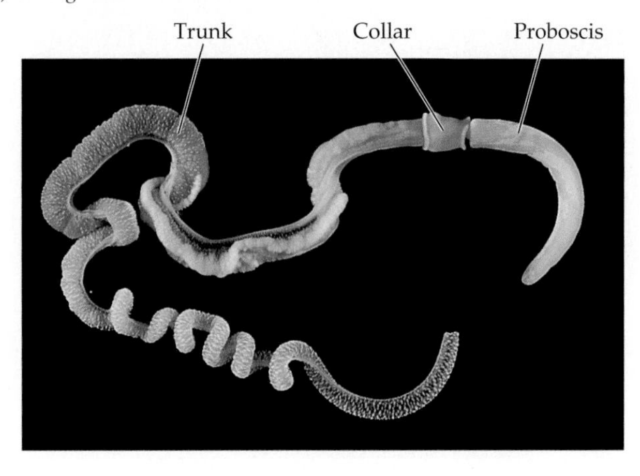

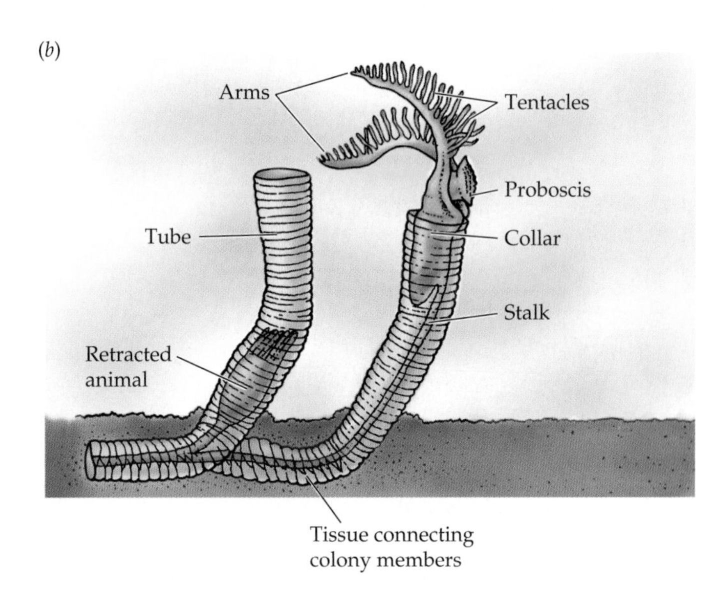

34.5 Hemichordates (*a*) The proboscis of this acorn worm is modified for digging. This individual has been extracted from its burrow. (*b*) Pterobranchs may be colonial or solitary.

Hemichordates: Conservative Evolution

Acorn worms and **pterobranchs** (phylum **Hemichordata**) are probably similar in form to the ancestor they share with the echinoderms. They have a three-part body plan, consisting of a proboscis, a collar, and a trunk.

The 70 species of acorn worms range up to 2 meters in length. They live in burrows in muddy and sandy marine sediments. The large proboscis of an acorn worm is a digging organ (Figure 34.5*a*). It is coated with a sticky mucus that traps small organisms in the sediment. The mucus and its attached prey are conveyed by cilia to the mouth. In the esophagus, the food-laden mucus is compacted into a ropelike mass that is moved through the digestive tract by ciliary action. Behind the mouth is a muscular *pharynx*, a tube that connects the mouth to the intestine. The pharynx opens to the outside through a number of *pharyngeal slits* through which water can exit. Highly vascularized tissue surrounding the pharyngeal slits serves as a gas exchange apparatus. An acorn worm breathes by pumping water into its mouth and out through its pharyngeal slits.

The 10 living species of pterobranchs are sedentary animals up to 12 mm in length that live in a tube secreted by the proboscis. Some species are solitary; others form colonies of individuals joined together. Behind the proboscis is a collar with 1–9 pairs of arms, bearing long tentacles that capture prey and function in gas exchange (Figure 34.5*b*).

Chordates: New Ways of Feeding

Members of the second major lineage of deuterostomes evolved several modifications of the coelom that provided new ways of capturing and handling food. They evolved a strikingly different body plan, characterized by an internal dorsal supporting structure. The pharyngeal slits, which originally functioned as sites for the uptake of O_2 and elimination of CO_2, and for eliminating water, were further enlarged. The result was a phylum (Chordata) of bilaterally

(a) *Rhopalaea crassa*

(b) *Pegea socia*

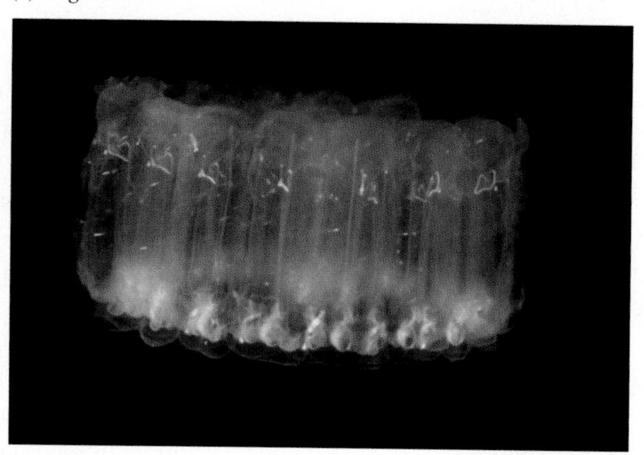

34.6 Urochordates (*a*) The tunic is clearly visible in this transparent sea squirt. (*b*) A chainlike colony of salps floats in tropical waters.

symmetrical animals with body plans characterized by the following shared features at some stage in their development:

- *Pharyngeal slits*
- A dorsal, hollow *nerve cord*
- A ventral *heart*
- A *tail* that extends beyond the anus
- A dorsal supporting rod, the *notochord*

The **notochord** is the distinctive derived trait of the lineage. It is composed of a core of large cells with turgid fluid-filled vacuoles that make it rigid but flexible. In some urochordates, the notochord is lost during metamorphosis to the adult stage. In vertebrates, it is replaced by other skeletal structures that provide support for the body.

The **tunicates** (subphylum **Urochordata**) may be similar to the ancestors of the chordates. All 2,500 species of tunicates are marine animals, most of which are sessile as adults. Their swimming, tadpole-like larvae reveal the close evolutionary relationship between tunicates and chordates (as Darwin realized; see Figure 25.4).

In addition to its pharyngeal slits, a tunicate larva has a dorsal, hollow nerve cord and a notochord that is restricted to the tail region. Bands of muscle surround the notochord, providing support for the body. After a short time swimming in the water, the larvae of most species settle to the seafloor and transform into sessile adults. The tunicate pharynx is enlarged into a *pharyngeal basket*, with which the animal feeds by extracting plankton from the water. Some urochordates are solitary, but others produce colonies by asexual budding from a single founder.

There are three major urochordate groups: ascidians, thaliaceans, and larvaceans. More than 90 percent of the known species of tunicates are *ascidians* (sea squirts). Individual sea squirts range in size from less than 1 mm to 60 cm in length, but colonies may measure several meters across. The baglike body of an adult ascidian is enclosed in a tough tunic that is secreted by epidermal cells. The tunic is composed of proteins and a complex polysaccharide. Much of the body is occupied by a large pharyngeal basket lined with cilia, whose beating moves water through the animal (Figure 34.6*a*).

Thaliaceans (salps and others) float in tropical and subtropical oceans at all depths down to 1,500 meters (Figure 34.6*b*). They live singly or in chainlike colonies up to several meters long. *Larvaceans* are solitary planktonic animals usually less than 5 mm long. They retain their notochords and nerve cords throughout their lives.

The 25 species of **lancelets** (subphylum **Cephalochordata**) are small, fishlike animals that rarely exceed 5 cm in length. Their notochord extends the entire length of the body throughout their lives. Lancelets live partly buried in soft marine sediments. They extract small prey from the water with their pharyngeal baskets (Figure 34.7).

A jointed vertebral column replaced the notochord in vertebrates

In another chordate lineage, the enlarged pharyngeal basket came to be used to extract prey from mud. This lineage gave rise to the **vertebrates** (subphylum **Vertebrata**) (Figure 34.8). Vertebrates take their name from the jointed, dorsal **vertebral column** that replaced the notochord as their primary sup-

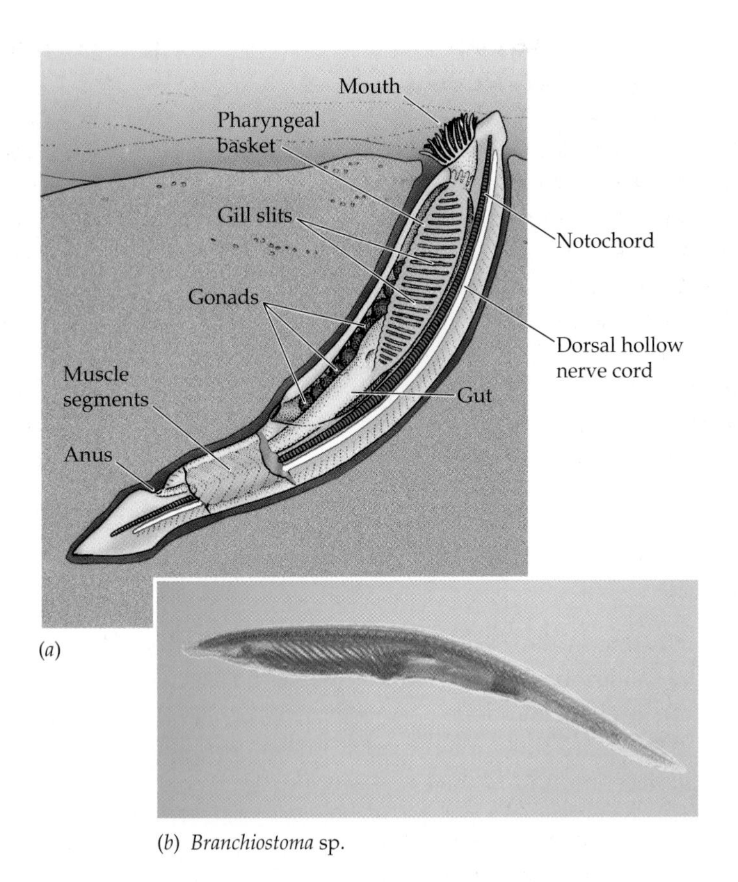

34.7 Lancelets (*a*) The internal structure of a lancelet. Note the large pharyngeal basket with gill slits. (*b*) This lancelet, which is about 6 cm long, has been excavated from the sediment to show its entire body.

port. The vertebrate body plan (Figure 34.9) can be characterized as follows:

- A rigid *internal skeleton*, with the vertebral column as its anchor, that provides support and mobility
- Two pairs of *appendages* attached to the vertebral column
- An anterior *skull* with a large *brain*
- Internal organs suspended in a large *coelom*
- A well-developed *circulatory system*, driven by contractions of a ventral *heart*

The ancestral vertebrates lacked jaws. They probably swam over the bottom, sucking up mud and straining it through the pharyngeal basket to extract microscopic food particles. The vascularized tissues of the basket also served a gas-exchange function. These animals gave rise to the jawless fishes.

One group of jawless fishes, called **ostracoderms** ("shell-skinned"), evolved a bony external armor that protected them from predators. With their heavy armor, these small fish could safely swim slowly above the substratum, which was easier than having to burrow through it, as all previous sediment feeders had done.

Hagfishes
Lampreys
Cartilaginous fishes
Ray-finned fishes
Lobe-finned fishes
Lungfishes
Tetrapods

Jawless fishes could attach to dead organisms and use suction created by the pharynx to pull fluids and partly decomposed tissues into the mouth. Hagfishes and lampreys, the only jawless fishes to survive beyond the Devonian, feed on both dead and living organisms in this way (Figure 34.10). These fishes, often placed in the class **Agnatha**, have tough skins instead of external armor. They lack paired appendages

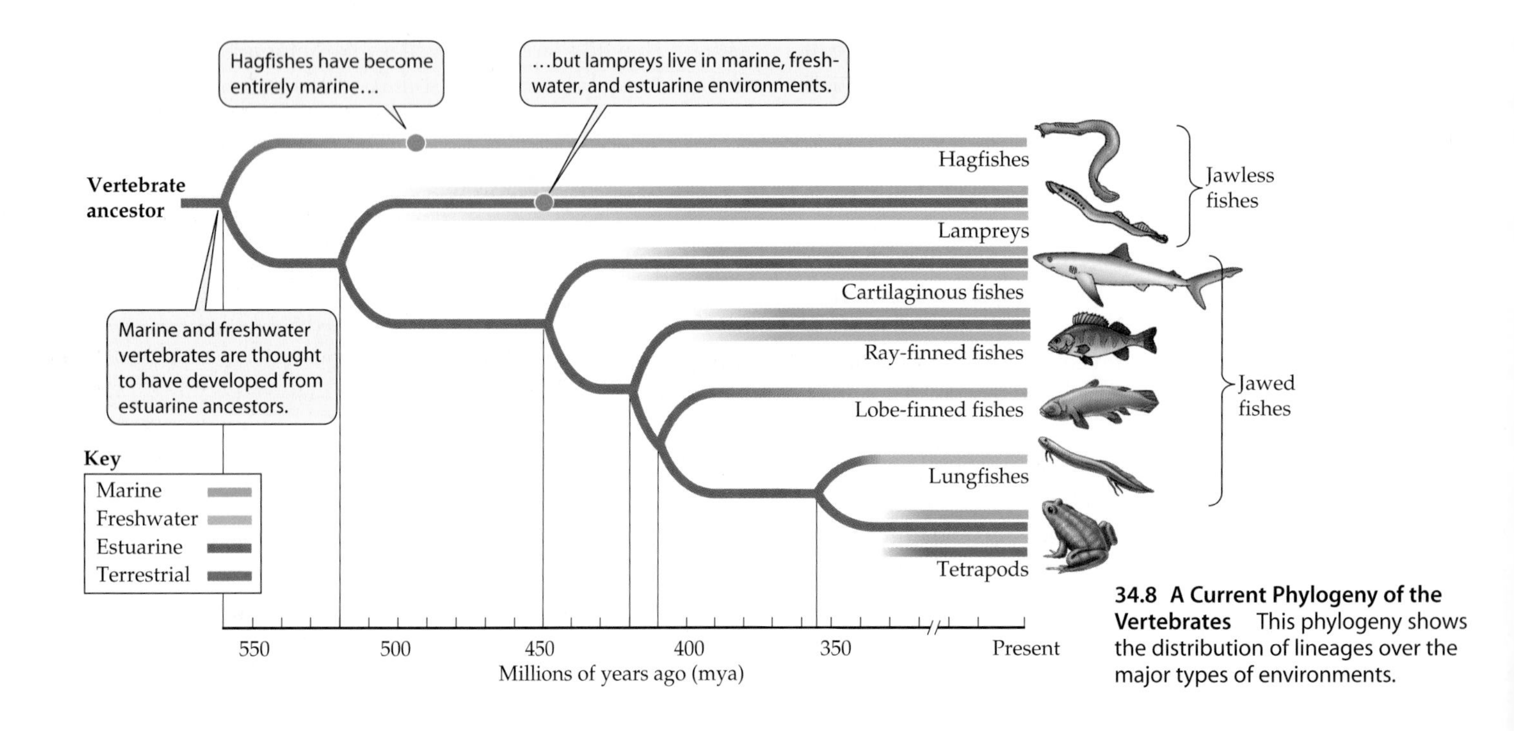

34.8 A Current Phylogeny of the Vertebrates This phylogeny shows the distribution of lineages over the major types of environments.

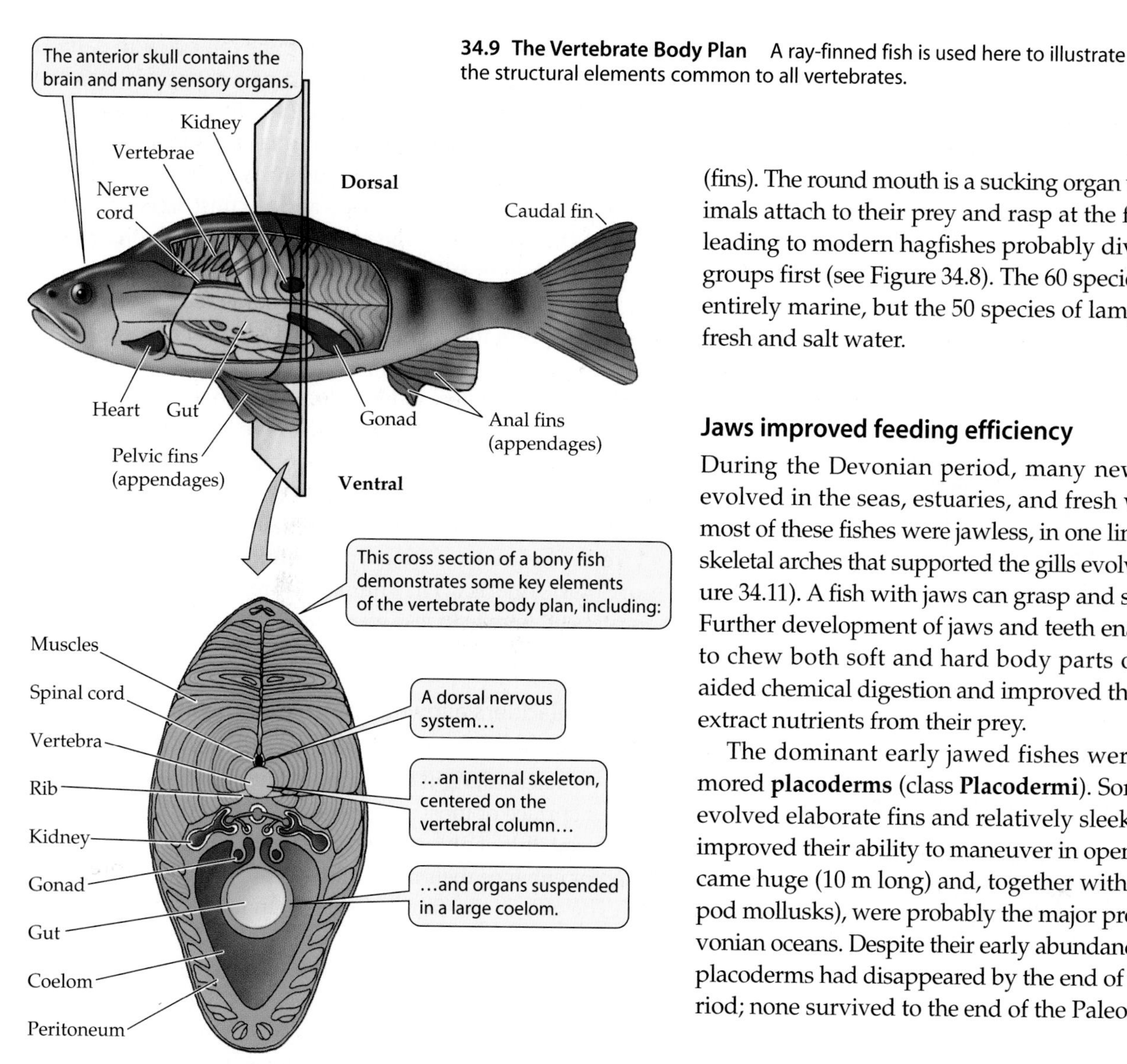
The anterior skull contains the brain and many sensory organs.
Kidney
Vertebrae
Nerve cord
Dorsal
Caudal fin
Heart
Gut
Gonad
Anal fins (appendages)
Pelvic fins (appendages)
Ventral
This cross section of a bony fish demonstrates some key elements of the vertebrate body plan, including:
Muscles
Spinal cord
Vertebra
Rib
Kidney
Gonad
Gut
Coelom
Peritoneum
A dorsal nervous system…
…an internal skeleton, centered on the vertebral column…
…and organs suspended in a large coelom.

34.9 The Vertebrate Body Plan A ray-finned fish is used here to illustrate the structural elements common to all vertebrates.

(fins). The round mouth is a sucking organ with which the animals attach to their prey and rasp at the flesh. The lineages leading to modern hagfishes probably diverged from other groups first (see Figure 34.8). The 60 species of hagfishes are entirely marine, but the 50 species of lampreys live in both fresh and salt water.

Jaws improved feeding efficiency

During the Devonian period, many new kinds of fishes evolved in the seas, estuaries, and fresh waters. Although most of these fishes were jawless, in one lineage, some of the skeletal arches that supported the gills evolved into jaws (Figure 34.11). A fish with jaws can grasp and subdue large prey. Further development of jaws and teeth enabled some fishes to chew both soft and hard body parts of prey. Chewing aided chemical digestion and improved the fishes' ability to extract nutrients from their prey.

The dominant early jawed fishes were the heavily armored **placoderms** (class **Placodermi**). Some of these fishes evolved elaborate fins and relatively sleek body forms that improved their ability to maneuver in open water. A few became huge (10 m long) and, together with squids (cephalopod mollusks), were probably the major predators in the Devonian oceans. Despite their early abundance, however, most placoderms had disappeared by the end of the Devonian period; none survived to the end of the Paleozoic era.

34.10 Modern Jawless Fishes (*a*) The Pacific hagfish. (*b*) Two sea lampreys using their large, jawless mouths to suck blood and flesh from a trout. The sea lamprey can live in either fresh or saltwater.

(*a*) *Eptatretus stouti*

(*b*) *Petromyzon marinus*

Jawless fishes (agnathans)

Extinct and living forms
Skull (cartilage)
Gill slits
Gill arches made of cartilage supported the gills.

Early jawed fishes (placoderms)

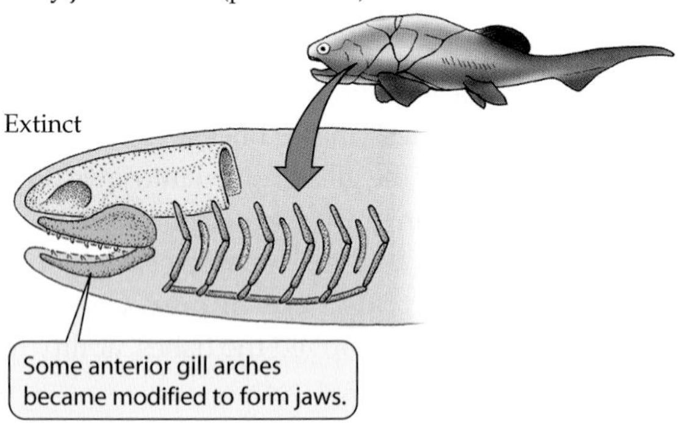

Extinct
Some anterior gill arches became modified to form jaws.

Modern jawed fishes (cartilaginous and bony fishes)

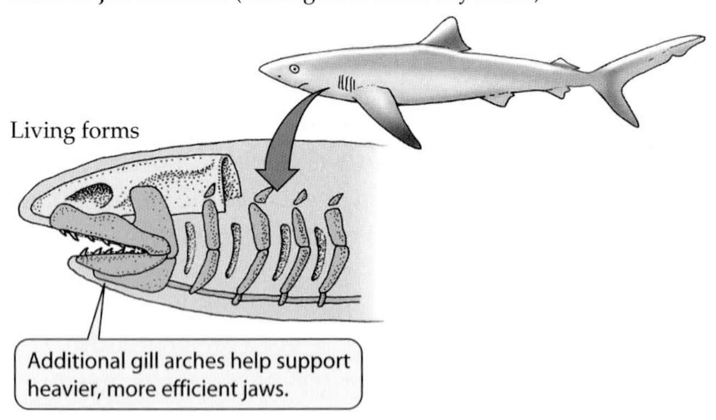

34.11 Jaws from Gill Arches This series of diagrams illustrates one probable scenario for the evolution of jaws from the anterior gill arches of fishes.

Fins improved mobility

Hagfishes
Lampreys
Cartilaginous fishes
Ray-finned fishes
Lobe-finned fishes
Lungfishes
Tetrapods

Several other groups of fishes became abundant during the Devonian period. **Cartilaginous fishes** (class **Chondrichthyes**)—the sharks, skates and rays, and chimaeras (Figure 34.12)—have a skeleton composed entirely of a firm but pliable material called *cartilage*. Their skin is flexible and leathery, sometimes bearing scales that give it the consistency of sandpaper.

Cartilaginous fishes control their movement with pairs of unjointed appendages called *fins:* a pair of pectoral fins just behind the gill slits and a pair of pelvic fins just in front of the anal region (see Figure 34.9). A dorsal median fin stabilizes the fish as it moves. Sharks move forward by means of lateral undulations of their bodies and tail fins. Skates and rays propel themselves by means of vertical undulating movements of their greatly enlarged pectoral fins.

Most sharks are predators, but some feed by straining plankton from the water. The world's largest fish, the whale shark (*Rhincodon typhus*), is a filter feeder. It may grow to more than 12 meters in length and weigh more than 12,000 kilograms. Most skates and rays live on the ocean floor, where they feed on mollusks and other invertebrates buried in the sediments. Nearly all cartilaginous fishes live in the oceans, but a few are estuarine or migrate into lakes and rivers. One group of stingrays is found only in river systems of South America. The chimaeras are found in deep ocean waters and are seen less often than the sharks and rays.

Swim bladders allowed control of buoyancy

Ray-finned fishes (class **Actinopterygii**) have internal skeletons of calcified, rigid bone rather than flexible cartilage. The outer surface of most species of ray-finned fishes is covered with flat, thin, lightweight scales that provide some protection or enhance their movement through the water.

The gills of ray-finned fishes open into a single chamber covered by a hard flap. Movement of the flap improves the flow of water over the gills, where gas exchange takes place. Early ray-finned fishes also evolved gas-filled sacs that supplemented the action of the gills in respiration. These features enabled early ray-finned fishes to live where oxygen was periodically in short supply, as it often is in freshwater environments. The lunglike sacs evolved into *swim bladders*, which function as organs of buoyancy in most ray-finned fishes today. By adjusting the amount of gas in its swim bladder, a fish can control the depth at which it is suspended in the water without expending energy.

Ray-finned fishes radiated during the Tertiary into about 24,000 species, encompassing a remarkable variety of sizes, shapes, and lifestyles (Figure 34.13). The smallest are less than 1 cm long as adults; the largest weigh up to 900 kilograms. Ray-finned fishes exploit nearly all types of aquatic food sources. In the oceans they filter plankton from the water, rasp algae from rocks, eat corals and other colonial invertebrates, dig invertebrates from soft sediments, and prey upon virtually all other fishes. In fresh water they eat plankton, devour insects of all aquatic orders, eat fruits that fall into the water in flooded forests, and prey on other aquatic vertebrates and, occasionally, terrestrial vertebrates.

(*a*) *Triaenodon obesus*

(*b*) *Trygon pastinaca*

(*c*) *Chimaera* sp.

34.12 Cartilaginous Fishes (*a*) Most sharks, such as this whitetip reef shark, are active marine predators. (*b*) Skates and rays, represented here by a stingray, feed on the ocean bottom. Their modified pectoral fins are used for propulsion. (*c*) A chimaera, or ratfish. These deep-ocean fish often possess poisonous dorsal fins.

34.13 Diversity among Ray-Finned Fishes (*a*) The barracuda has the large teeth and powerful jaws of a predator. (*b*) The coral grouper lives on tropical coral reefs. (*c*) Commerson's frogfish can change its color over a range from pale yellow to orange-brown to deep red, thus enhancing its camouflage abilities. (*d*) This weedy sea dragon is difficult to see when it hides in vegetation. It is a larger relative of the more familiar seahorse.

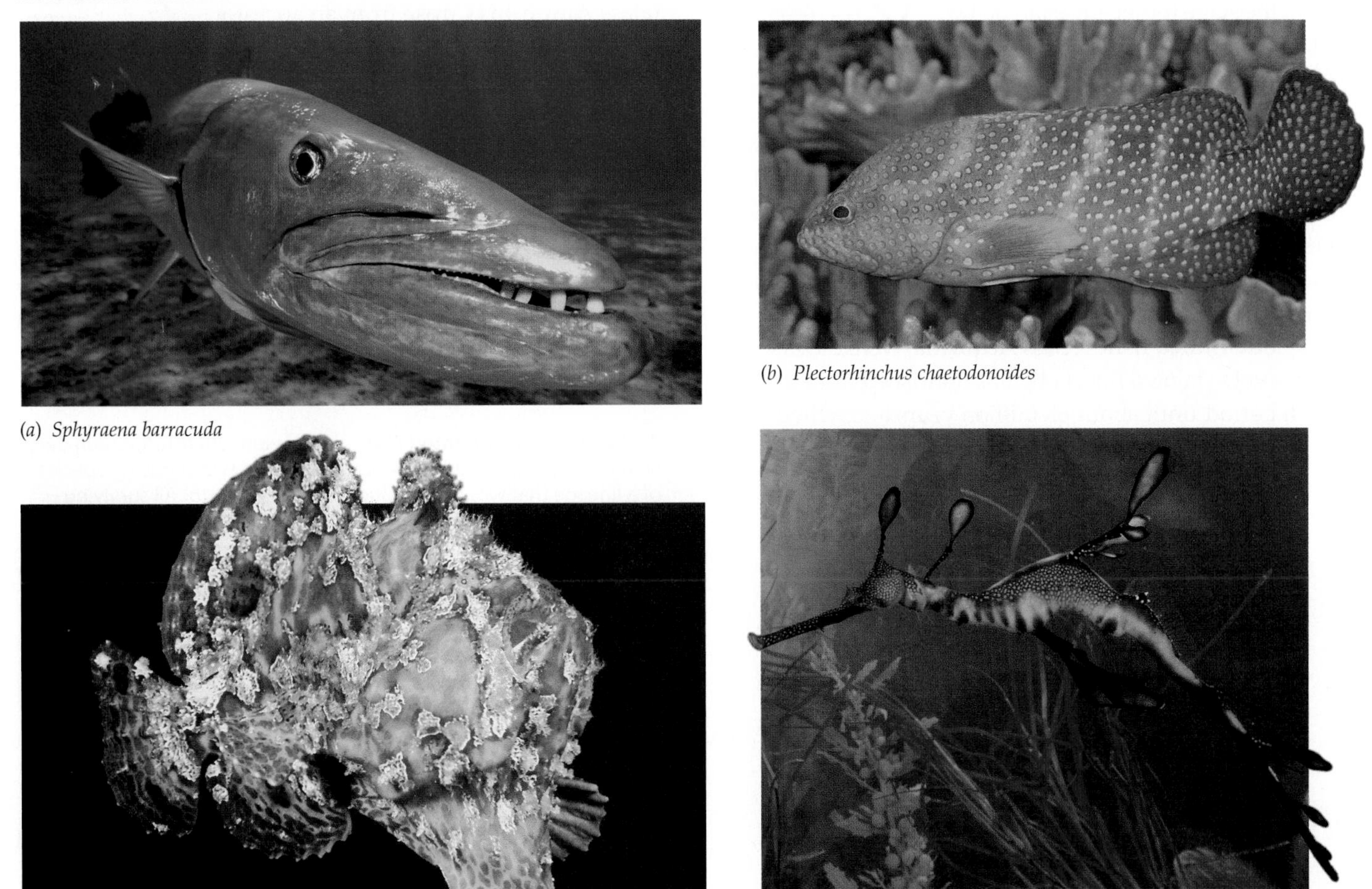

(*a*) *Sphyraena barracuda*

(*b*) *Plectorhinchus chaetodonoides*

(*c*) *Antennarius commersonii*

(*d*) *Phyllopteryx taeniolatus*

Some fishes live buried in soft sediments, capturing passing prey or emerging at night to feed. Many fishes are solitary, but in open water others form large aggregations called *schools*. Many fishes perform complicated behaviors by means of which they maintain schools, build nests, court and choose mates, and care for their young.

Although ray-finned fishes can readily control their positions in open water, their eggs tend to sink. A few species produce small eggs that are buoyant enough to complete their development in the open water. However, most marine fishes move to food-rich shallow waters to lay their eggs, which is why coastal waters and estuaries are so important in the life cycles of many species. Some, such as salmon, actually abandon salt water when they breed, ascending rivers to spawn in freshwater streams and lakes.

Colonizing the Land: Obtaining Oxygen from the Air

The evolution of lunglike sacs in fishes appears to have been a response to the inadequacy of gills for respiration in oxygen-poor waters, but it also set the stage for the invasion of the land. Some early ray-finned fishes probably used their lungs to supplement their gills when oxygen levels in the water were low, as lungfishes do today. This ability would also have allowed them to leave the water temporarily and breathe air when pursued by predators unable to do so. But with their unjointed fins, these fishes could only flop around on land, as most fish out of water do today. Changes in the structure of the fins allowed these fishes to move on land.

Hagfishes
Lampreys
Cartilaginous fishes
Ray-finned fishes
Lobe-finned fishes
Lungfishes
Tetrapods

The **lobe-finned fishes** (class **Actinistia**) were the first lineage to evolve jointed fins. Lobe-fins flourished from the Devonian period until about 65 million years ago, when they were thought to have become extinct. However, in 1938, a living lobe-fin was caught by commercial fishermen off South Africa. Since that time, several dozen specimens of this extraordinary fish, *Latimeria chalumnae*, have been collected. *Latimeria*, a predator on other fish, reaches a length of about 1.8 meters and weighs up to 82 kilograms (Figure 34.14*a*). Its skeleton is mostly composed of cartilage, not bone. A second species, *L. menadoensis*, was discovered in 1998 off the Indonesian island of Sulawesi.

Lungfishes (class **Dipnoi**) were important predators in shallow-water habitats in the Devonian, but most lineages died out. The three surviving species live in stagnant swamps and muddy waters in the Southern Hemisphere, one each in South America, Africa, and Australia (Figure 34.14*b*). Lungfishes have both gills and lungs. When ponds dry up, they can burrow deep into the mud and survive for many months in an inactive state.

It is believed that descendants of some lungfishes began to use terrestrial food sources, became more fully adapted to life on land, and eventually evolved to become the **tetrapods**—the four-legged amphibians, reptiles, birds, and mammals.

Amphibians invaded the land

During the Devonian period, **amphibians** (class **Amphibia**) arose from an ancestor they shared with lungfishes. In this lineage, stubby, jointed fins evolved into walking legs. The basic design of these legs has remained largely unchanged throughout the evolution of terrestrial vertebrates.

Amphibia
"Reptilia"
Aves
Mammalia

The Devonian predecessors of amphibians were probably able to crawl from one pond or stream to another by slowly pulling themselves along on their finlike legs, as do some modern species of catfishes. They gradually evolved the ability to live in swamps and, eventually, on dry land. Modern

34.14 Fishes with Jointed Fins (*a*) This lobe-fin fish, found in deep waters of the Indian Ocean, represents one of two surviving species of a lineage that was once thought to be extinct. (*b*) All surviving lungfish lineages live in the Southern Hemisphere.

(*a*) *Latimeria chalumnae*

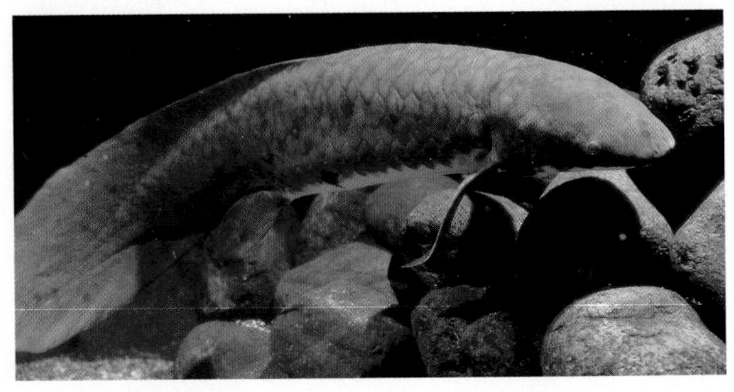

(*b*) *Neoceratodus forsteri*

(a) *Dermophis mexicanus*

(b) *Gyrinophilus porphyriticus*

(c) *Scaphiophryne gottlebei*

34.15 Diversity among the Amphibians (*a*) Burrowing caecilians superficially look more like worms than amphibians. (*b*) A Kentucky spring salamander. (*c*) This rare frog species was discovered in a national park on the island of Madagascar.

amphibians have small lungs, and most species exchange gases through their skins as well. Most terrestrial species are confined to moist environments because they lose water rapidly through their skins when exposed to dry air, and because they require water for reproduction.

About 4,500 species of amphibians live on Earth today, many fewer than the number known only from fossils. Living amphibians belong to three orders (Figure 34.15): the wormlike, limbless, tropical, burrowing caecilians (order Gymnophiona), the frogs and toads (order Anura, which means "tailless"), and the salamanders (order Urodela, which means "tailed"). Most species of frogs and toads live in tropical and warm temperate regions, although a few are found at very high latitudes and altitudes. Some toads have tough skins that enable them to live for long periods of time in dry places. Salamanders are most diverse in temperate regions, but many species are found in cool, moist environments in Central American mountains. Many salamanders that live in rotting logs or moist soil lack lungs. They exchange gases entirely through the skin and mouth lining. Amphibians are the focus of much attention today because populations of many species are declining rapidly (see Chapter 1).

Most species of amphibians live in water at some time in their lives. In the typical amphibian life cycle, part or all of the adult stage is spent on land, but adults return to fresh water to lay their eggs (Figure 34.16). Amphibian eggs can survive only in moist environments because they are enclosed within delicate envelopes that cannot prevent water loss in dry conditions. The fertilized eggs of most species give rise to larvae that live in water until they undergo metamorphosis to become terrestrial adults. Some amphibians, however, are entirely aquatic, never leaving the water at any stage of their lives. Others are entirely terrestrial, laying their eggs in moist places on land and skipping the aquatic larval stage.

Amniotes colonized dry environments

Two morphological changes contributed to the ability of one lineage of tetrapods to control water loss and, therefore, to exploit a wide range of terrestrial habitats:

- Evolution of an egg with a shell that is relatively impermeable to water
- A combination of traits that included a tough skin impermeable to water and kidneys that could excrete concentrated urine

The vertebrates that evolved both of these traits are called **amniotes**. They were the first vertebrates to become widely distributed over the terrestrial surface of Earth.

The amniote egg has a leathery or a brittle, calcium-impregnated shell that retards evaporation of the fluids inside but permits O_2 and CO_2 to pass through. Such an egg does not require a moist environment and can be laid anywhere. Within the shell and surrounding the embryo are membranes that protect the embryo from desiccation and assist its respiration and excretion of waste nitrogen. The egg also stores large quantities of food as *yolk*, permitting the embryo to attain a relatively advanced state of development before it hatches and must feed itself (Figure 34.17).

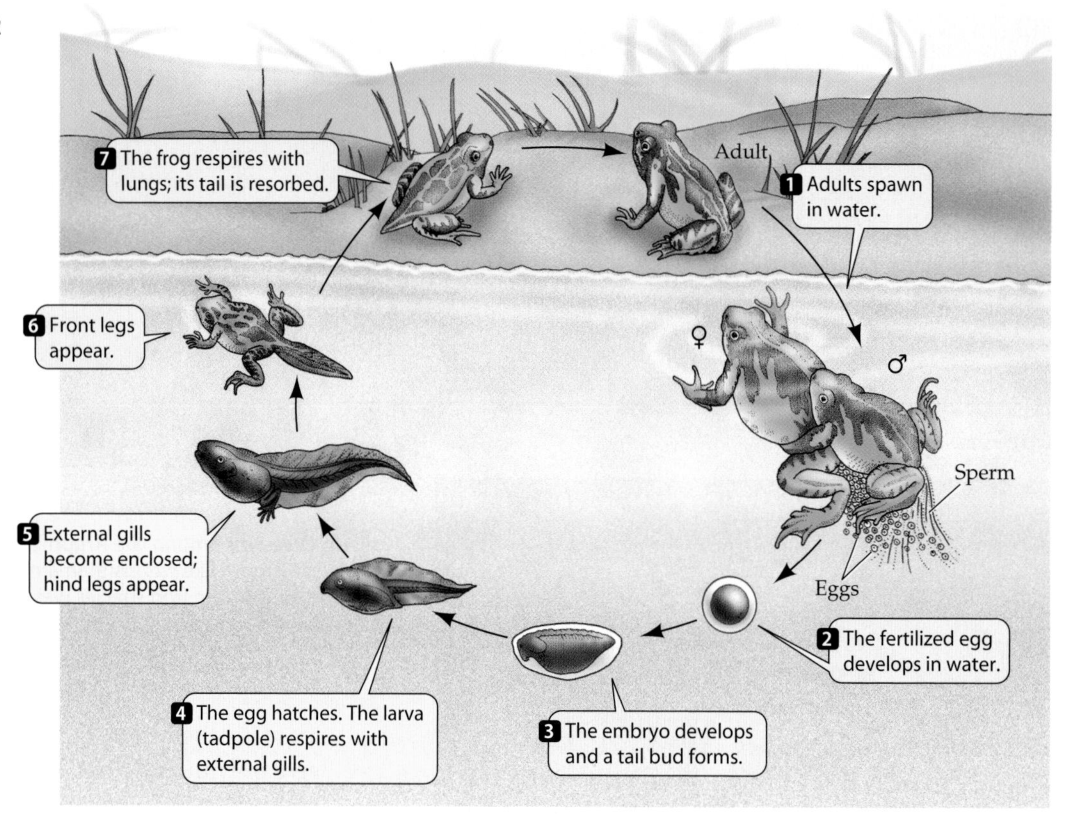

7 The frog respires with lungs; its tail is resorbed.
Adult
1 Adults spawn in water.
♀
♂
Sperm
Eggs
6 Front legs appear.
5 External gills become enclosed; hind legs appear.
2 The fertilized egg develops in water.
4 The egg hatches. The larva (tadpole) respires with external gills.
3 The embryo develops and a tail bud forms.

34.16 In and Out of the Water Most stages in the life cycle of temperate-zone frogs take place in water. The aquatic tadpole is transformed into a terrestrial adult through metamorphosis.

An early amniote lineage, the **reptiles**, arose from a tetrapod ancestor in the Carboniferous period (Figure 34.18). The class "Reptilia," as we use the term here, is a paraphyletic group because some reptiles (crocodilians) are in fact more closely related to the birds than they are to lizards, snakes, and turtles (see Figure 25.8). However, because all members of "Reptilia" are structurally similar, it serves as a convenient group for discussing the characteristics of amniotes. Therefore, we use the traditional classification of "Reptilia" as a basis for our discussion while recognizing that, technically, the birds should be included within it.

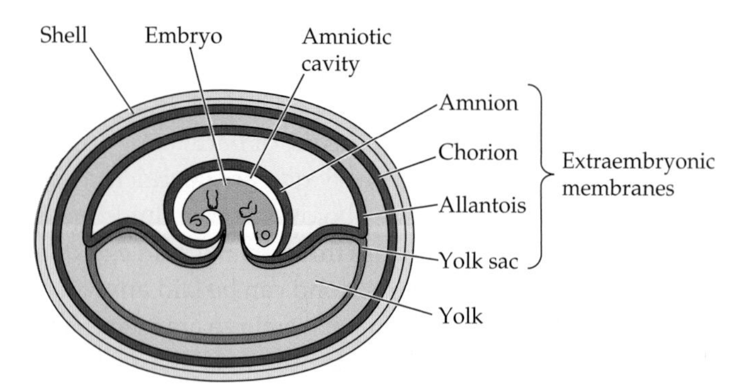

Shell
Embryo
Amniotic cavity
Amnion
Chorion
Allantois
Yolk sac
Yolk
Extraembryonic membranes

34.17 An Egg for Dry Places The evolution of the amniote egg, with its shell, four extraembryonic membranes, and embryo-nourishing yolk, was a major step in the colonization of the terrestrial environment.

About 6,000 species of reptiles live today. Most reptiles do not care for their eggs after laying them. In some species, the eggs do not develop shells, but are retained inside the female's body until they hatch. Some of these species evolved a structure called the *placenta* that nourishes the developing embryos.

The skin of a reptile is covered with horny scales that greatly reduce loss of water from the body surface. These scales, however, make the skin unavailable as an organ of gas exchange. In reptiles, gases are exchanged almost entirely by the lungs, which are proportionally much larger in surface area than those of amphibians. A reptile forces air into and out of its lungs by bellows-like movements of its ribs. The reptilian heart is divided into three and one-half or four chambers that partially separate oxygenated from unoxygenated blood. With this type of heart, reptiles can generate higher blood pressures than amphibians, which have three-chambered hearts, and can sustain higher levels of muscular activity.

Reptilian lineages diverged

The lineages leading to modern reptiles began to diverge about 250 mya. One lineage that has changed very little over the intervening millenia is the turtles (subclass **Testudines**). Turtles have a combination of ancestral traits and highly specialized characteristics that they do not share with any other vertebrate group. For this reason, their phylogenetic relationships are uncertain.

The dorsal and ventral bony plates of modern turtles and tortoises form a shell into which the head and limbs can be

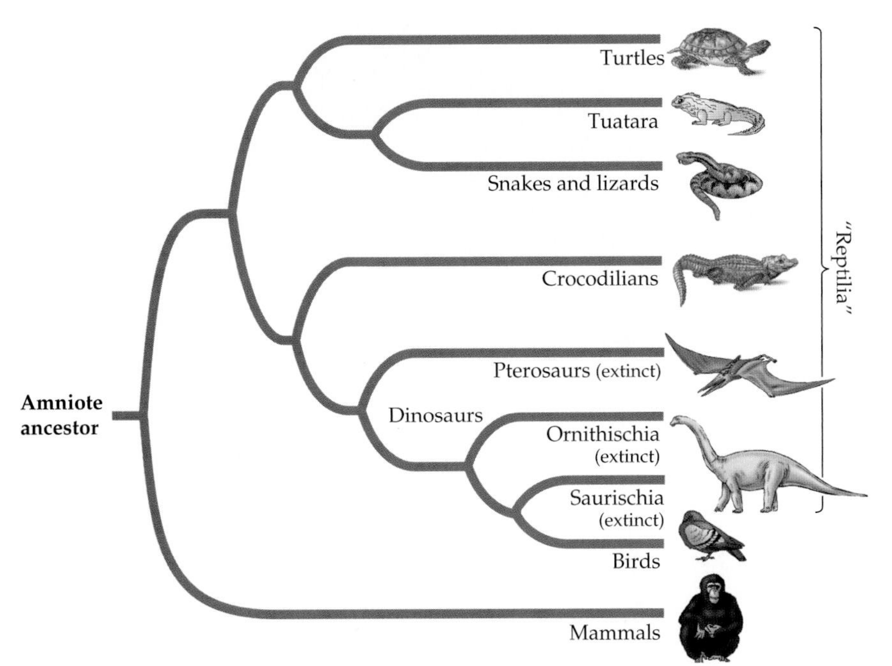

34.18 The Reptiles Form a Paraphyletic Group The traditional classification of the amniotes creates the paraphyletic group "Reptilia." As used here, "Reptilia" does not include the birds (Aves), even though this major lineage split off from a dinosaur lineage relatively recently (in evolutionary terms).

withdrawn (Figure 34.19*a*). Most turtles live in lakes and ponds, but tortoises are terrestrial; some live in deserts. Sea turtles spend their entire lives at sea except when they come ashore to lay eggs. All seven species of sea turtles are endangered. A few species of turtles and tortoises are carnivores, but most species are omnivores that eat a variety of aquatic and terrestrial plants and animals.

The subclass **Squamata** includes lizards and snakes as well as the amphisbaenians (a group of legless, wormlike, burrowing animals with greatly reduced eyes). The tuataras (subclass **Sphenodontida**) are a sister group to the lizards and snakes. Sphenodontids were diverse dur-

(*c*) *Chamaeleo* sp.

(*a*) *Chelonia mydas*

(*b*) *Sphenodon punctatus*

(*d*) *Trimeresurus sumatranus*

(*e*) *Alligator mississippiensis*

34.19 Reptilian Diversity (*a*) The green sea turtle is widely distributed in tropical oceans. (*b*) This tuatara represents one of only two surviving species in a lineage that separated from lizards long ago. (*c*) The African chameleon, a lizard, has large eyes that move independently in their sockets. (*d*) This venomous Sumatran pit viper is coiled to strike. (*e*) Alligators live in warm temperate environments in China and, like this one, in the southeastern United States.

ing the Mesozoic era, but today they are represented only by two species restricted to a few islands off the coast of New Zealand (Figure 34.19*b*). Tuataras superficially resemble lizards, but differ from them in tooth attachment and several internal anatomical features.

Most lizards are insectivores, but some are herbivores; a few prey on other vertebrates. The largest lizards, growing as long as 3 meters, are certain monitor lizards (such as the Komodo dragon) that live in the East Indies. Most lizards walk on four limbs (Figure 34.19*c*), but some are limbless, as are all snakes, which are descendants of burrowing lizards.

All snakes are carnivores; many can swallow objects much larger than themselves. This is the mode of feeding of the largest snakes, the pythons, which can grow to more than 10 meters long. Several snake lineages evolved a combination of venom glands and the ability to inject venom rapidly into their prey (Figure 34.19*d*).

A separate diverging lineage led to the crocodilians (subclass **Crocodylia**) and to the dinosaurs. The crocodilians—crocodiles, caimans, gharials, and alligators—are confined to tropical and warm temperate environments (Figure 34.19*e*). Crocodilians spend much of their time in water, but they build nests on land or on floating piles of vegetation. The eggs are warmed by heat generated by decaying organic matter that the parents place in the nest. Typically the female guards the eggs until they hatch. All crocodilians are carnivorous; they eat vertebrates of all classes, including large mammals.

The **dinosaurs** rose to prominence about 215 mya and dominated terrestrial environments for about 150 million years. During that time, virtually all terrestrial animals more than a meter in length were dinosaurs. Some of the largest dinosaurs weighed up to 100 tons. Many were agile and could run rapidly. The ability to breathe and run simultaneously, which we take for granted, was a major innovation in the evolution of terrestrial vertebrates. Not until the evolution of the lineages leading to the mammals, dinosaurs, and birds did the legs assume vertical positions directly under the body, which reduced the lateral forces on the body during locomotion. Special muscles that enabled the lungs to be filled and emptied while the limbs moved also evolved. We can infer the existence of such muscles in dinosaurs from the structure of the vertebral column in fossils and the capacity of many dinosaurs for bounding, bipedal (two-legged) locomotion.

Several fossil dinosaurs discovered recently in early Cretaceous deposits in Liaoning Province, in northeastern China, clearly show that in some small predatory dinosaurs, the scales had been highly modified to form feathers. One of these dinosaurs, *Microraptor gui*, had feathers on all four limbs, and those feathers were structurally similar to those of modern birds (Figure 34.20*a*).

(*a*)

(*b*)

(*c*)

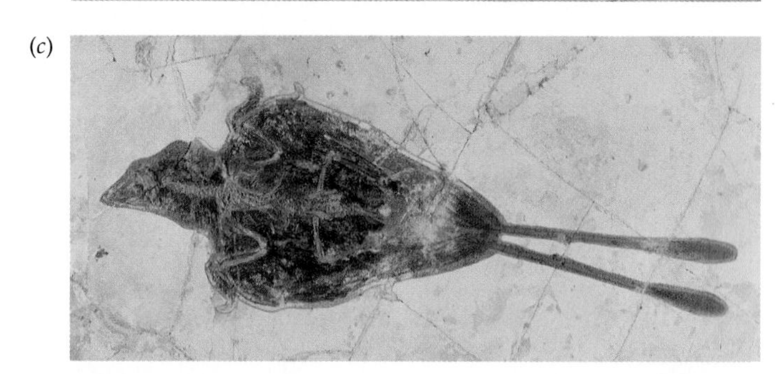

34.20 Mesozoic Birds and Their Ancestors Fossil remains demonstrate the probable evolution of birds from reptilian ancestors. (*a*) *Microraptor gui*, a feathered dinosaur from the early Cretaceous (about 140 mya). (*b*) *Archaeopteryx*, the oldest known bird. (*c*) The elongated tail feathers of a male *Confuciusornis sanctus* ("sacred bird of Confucius") fossil suggest that the males used them in courtship displays.

Birds: More Feathers and Better Flight

Amphibia
"Reptilia"
Aves
Mammalia

During the Mesozoic era, about 175 mya, a dinosaur lineage gave rise to the **birds** (subclass **Aves**). The oldest known avian fossil, *Archaeopteryx*, which lived about 150 mya, had teeth, unlike modern birds, but was covered with feathers that are virtually identical to those of modern birds. It also had well-developed wings, a long tail (Figure 34.20*b*), and a furcula, or "wishbone," to which some of the flight muscles were probably attached. *Archaeopteryx* had clawed fingers on its forelimbs, but it also had typical perching bird claws, suggesting that it lived in trees and shrubs and used the fingers to assist it in clambering over branches. Because the avian lineage separated from other reptiles long before *Archaeopteryx* lived, existing data are insufficient to identify the ancestors of birds with certainty. Most paleontologists believe that birds evolved from feathered terrestrial bipedal dinosaurs that used their forelimbs for capturing prey.

Many remains of other early birds have been discovered in 120–125-million-year-old fossil beds in northeastern China. One of these birds, *Confuciusornis sanctus*, is known from hundreds of complete specimens. The males had greatly elongated tail feathers (Figure 34.20*c*), which they probably used in communal courtship displays. Large numbers of individuals have been found together, as would be expected if many males assembled on communal display grounds, as some birds do today.

Birds range in size from the 2-gram bee hummingbird of the West Indies to the 150-kilogram ostrich (Figure 34.21). Some flightless birds of Madagascar and New Zealand known from fossils were even larger. These birds were exterminated by humans soon after they colonized those islands. There are about 9,600 species of living birds, more than in any other major vertebrate group except ray-finned fishes.

As a group, birds eat almost all types of animal and plant material. A few aquatic species have bills modified for filtering small food particles from water. Insects are the most important dietary items for terrestrial species. Birds are major predators of flying insects during the day, and some species exploit that food source at night. In addition, birds eat fruits and seeds, nectar and pollen, leaves and buds, carrion, and other vertebrates. By eating the fruits and seeds of plants, birds serve as major agents of seed dispersal.

The feathers developed by some dinosaurs may originally have had thermoregulatory or display functions. Birds also use them for flying. Large quills that arise from the skin of the fore-

34.21 Diversity among the Birds (*a*) Penguins such as these gentoos are widespread in the cold waters of the Southern Hemisphere. They are expert swimmers, although they have lost the ability to fly. (*b*) Perching birds, represented here by a male northern cardinal, are the most species-rich of all the bird lineages. (*c*) Parrots are a diverse group of birds, especially in the Tropics of Asia, South America, and Australia. This king parrot is one member of Australia's rich parrot fauna. (*d*) The flightless ostrich is the largest bird species in existence today.

(*a*) *Pygoscelis papua*

(*b*) *Cardinalis cardinalis*

(*c*) *Alisterus scapularis*

(*d*) *Struthio camelus*

limbs create the flying surfaces of wings. Other strong feathers sprout like a fan from the shortened tail and serve as stabilizers during flight. The feathers that cover the body, along with an underlying layer of down feathers, provide insulation.

The bones of birds are modified for flight. They are hollow and have internal struts for strength. The *sternum* (breastbone) forms a large, vertical keel to which the flight muscles are attached. These muscles pull the wings downward during the main propulsive movement in flight. Flight is metabolically expensive. A flying bird consumes energy at a rate about 15–20 times faster than a running lizard of the same weight! Because birds have such high metabolic rates, they generate large amounts of heat. They control the rate of heat loss using their feathers, which may be held close to the body or elevated to alter the amount of insulation they provide.

The brain of a bird is larger in proportion to its body than a lizard or crocodile brain, primarily because the cerebellum, the center of sight and muscular coordination, is enlarged.

Most birds lay their eggs in a nest, where they are warmed by heat from an adult that sits on them. Because birds have such high body temperatures, the eggs of most species hatch within a few weeks. The offspring of many species are *altricial* (hatch at a relatively helpless stage) and are fed for some time by their parents. The young of other bird species, such as chickens, sandpipers, and ducks, are *precocial* (can feed themselves shortly after hatching). Adults of nearly all species attend their offspring for some time, warning them of and protecting them from predators, protecting them from bad weather, leading them to good foraging places, and feeding them.

The Origin and Diversity of Mammals

Mammals (class **Mammalia**) appeared in the early part of the Mesozoic era, about 225 million years ago, branching from a lineage of mammal-like reptiles. Small mammals coexisted with reptiles and dinosaurs for at least 150 million years. After the large reptiles and dinosaurs disappeared during the mass extinction at the close of the Mesozoic era, mammals increased dramatically in numbers, diversity, and size. Today, mammals range in size from tiny shrews and bats weighing only about 2 grams to the endangered blue whale, which measures up to 33 meters long and weighs up to 160,000 kilograms—the largest animal ever to live on Earth.

Amphibia
"Reptilia"
Aves
Mammalia

Skeletal simplification accompanied the evolution of early mammals from their larger reptilian ancestors. During mammalian evolution, some bones from the lower jaw were incorporated into the middle ear, leaving a single bone in the lower jaw. The number of bones in the skull also decreased. The bulk of both the limbs and the bony girdles from which they are suspended was reduced. Mammals have far fewer, but more highly differentiated, teeth than reptiles do. Differences in the number, type, and arrangement of teeth in mammals reflect their varied diets.

Skeletal features are readily preserved as fossils, but the soft parts of animals are seldom fossilized. Therefore, we do not know when mammalian features such as mammary glands, sweat glands, hair, and a four-chambered heart evolved. Mammals are unique among animals in supplying their young with a nutritive fluid (milk) secreted by mammary glands. Mammalian eggs are fertilized within the female's body, and the embryos undergo a period of development, called *gestation*, within a specialized organ, the *uterus*, prior to being born. In many species, the embryos are connected to the uterus and nourished by a placenta. In addition, mammals have a protective and insulating covering of hair, which is luxuriant in some species but has been almost entirely lost in whales, dolphins, and humans. In whales and dolphins, thick layers of insulating fat (blubber) replace hair as a heat-retention mechanism. Clothing assumes the same role for humans. The approximately 4,000 species of living mammals are divided into two major subclasses: Prototheria and Theria. The subclass **Prototheria** contains a single order, the Monotremata, with a total of three species, which are found only in Australia and New Guinea. These mammals, the duck-billed platypus and the spiny anteaters, or echidnas, differ from other mammals in lacking a placenta, laying eggs, and having legs that poke out to the side (Figure 34.22). Monotremes supply milk for their young, but they have no nipples on their mammary glands; rather, the milk simply oozes out and is lapped off the fur by the offspring.

Members of the other subclass, **Theria**, are further divided into two groups. In most species of the first group, the **Marsupialia**, females have a ventral pouch in which they carry and feed their offspring (Figure 34.23*a*). Gestation in marsupials is short; the young are born tiny but with well-developed forelimbs, with which they climb to the pouch. They attach to a nipple, but cannot suck. The mother ejects milk into the tiny offspring until they grow large enough to suckle. Once her offspring have left the uterus, a female marsupial may become sexually receptive again. She can then carry fertilized eggs capable of initiating development and replacing the offspring in her pouch should something happen to them.

There are about 240 living species of marsupials. At one time marsupials were found on all continents, but today the majority of species are restricted to the Australian region, with a modest representation in South America (Figure 34.23*b*). One species, the Virginia opossum, is widely distributed in the United States. Marsupials radiated to become terrestrial herbivores, insectivores, and carnivores, but no marsupial species live in the oceans or can fly, although some are gliders. The largest living marsupial is the red kangaroo of Australia (Figure 34.23*a*), which weighs up to 90 kilo-

(*a*) *Tachyglossus aculeata*

(*b*) *Ornithorhynchus anatinus*

34.22 Monotremes (*a*) The short-beaked echidna is one of the two surviving species of echidnas. (*b*) The duck-billed platypus is the other surviving monotreme species.

grams. Much larger marsupials existed in Australia until they were exterminated by humans soon after they reached the continent (about 50,000 years ago).

Most living mammals belong to the second therian group, the **eutherians**. (Eutherians are sometimes called *placental mammals*, but this name is not accurate because some marsupials also have placentas.) Eutherians are more developed at birth than are marsupials, and no external pouch houses them after birth. The nearly 4,000 species of eutherians are placed into 16 major groups (Figure 34.24), the largest of which is the rodents (order Rodentia) with about 1,700 species. The next largest group, the bats (order Chiroptera), has about 1,000 species, followed by the moles and shrews (order Insectivora) with slightly more than 400 species.

Eutherians are extremely varied in their form and ecology. Several lineages of terrestrial eutherians subsequently colonized marine environments to become whales, dolphins, seals, and sea lions. Eutherian mammals are—or were, until they were greatly reduced in numbers by humans—the most important grazers and browsers in most terrestrial ecosystems. Grazing and browsing have been an evolutionary force intense enough to select for the spines, tough leaves, and difficult-to-eat growth forms found in many plants—a striking example of coevolution.

Primates and the Origin of Humans

A eutherian lineage that has had dramatic effects on ecosystems worldwide is the **primate** lineage, which has undergone extensive recent evolutionary radiation. Primates probably

(*a*) *Macropus rufus*

(*b*) *Caluromys philander*

(*c*) *Sarcophilus harrisii*

34.23 Marsupials (*a*) Australia's red kangaroos are the largest living marsupials. The marsupial radiation also produced (*b*) arboreal species, such as this South American opossum, and (*c*) carnivores, such as the Tasmanian devil.

(*a*) *Citellus parryi*

(*b*) *Carollia perspicillata*

(*d*) *Rangifer tarandus*

(*c*) *Stenella longirostris*

34.24 Diversity among the Eutherians (*a*) The Arctic ground squirrel is one of the many species of small, diurnal rodents found in North America. (*b*) Temperate-zone bats are all insectivores, but many tropical bats, such as this leaf-nosed bat, eat fruit. (*c*) These Hawaiian spinner dolphins represent a eutherian lineage that colonized the marine environment. (*d*) Large hoofed mammals are important herbivores in terrestrial environments. This caribou bull is grazing by himself, although caribou are often seen in huge herds.

Primates and the Origin of Humans

A eutherian lineage that has had dramatic effects on ecosystems worldwide is the **primate** lineage, which has undergone extensive recent evolutionary radiation. Primates probably descended from small *arboreal* (tree-living) insectivorous mammals early in the Cretaceous period. A nearly complete fossil of an early primate species, *Carpolestes*, from Wyoming, dated at 56 mya, had grasping feet with an opposable big toe that had a nail rather than a claw. Such grasping limbs are one of the major adaptations to arboreal life that distinguish primates from other mammals. However, *Carpolestes* did not have eyes positioned on the front of the face to provide good depth perception, as all modern primates do.

Early in its evolutionary history, the primate lineage split into two main branches, the prosimians and the anthropoids (Figure 34.25). **Prosimians**—lemurs, pottos, and lorises—once lived on all continents, but today they are restricted to Africa, Madagascar, and tropical Asia (Figure 34.26). All of the mainland prosimian species are arboreal and nocturnal. However, on the island of Madagascar, the site of a remarkable prosimian radiation, there are also diurnal and terrestrial species.

The **anthropoids**—tarsiers, monkeys, apes, and humans—evolved from an early primate lineage about 55 million years ago in Africa or Asia. New World monkeys diverged from Old World monkeys early enough that they could have reached South America from Africa when those two continents were still close to each other. All New World monkeys are arboreal (Figure 34.27*a*). Many of them have long, *prehensile* (grasping) tails with which they can hold onto branches. Many Old World primates are arboreal as well, but a number of species are terrestrial. Some of these species, such as baboons and macaques, live and travel in large groups (Figure 34.27*b*). No Old World primates have prehensile tails.

About 22 million years ago, the lineage that led to modern **apes** separated from the other Old World primates. Between 22 and 5.5 mya, as many as 100 species of apes ranged over Europe, Asia, and Africa. About 9 mya, members of one

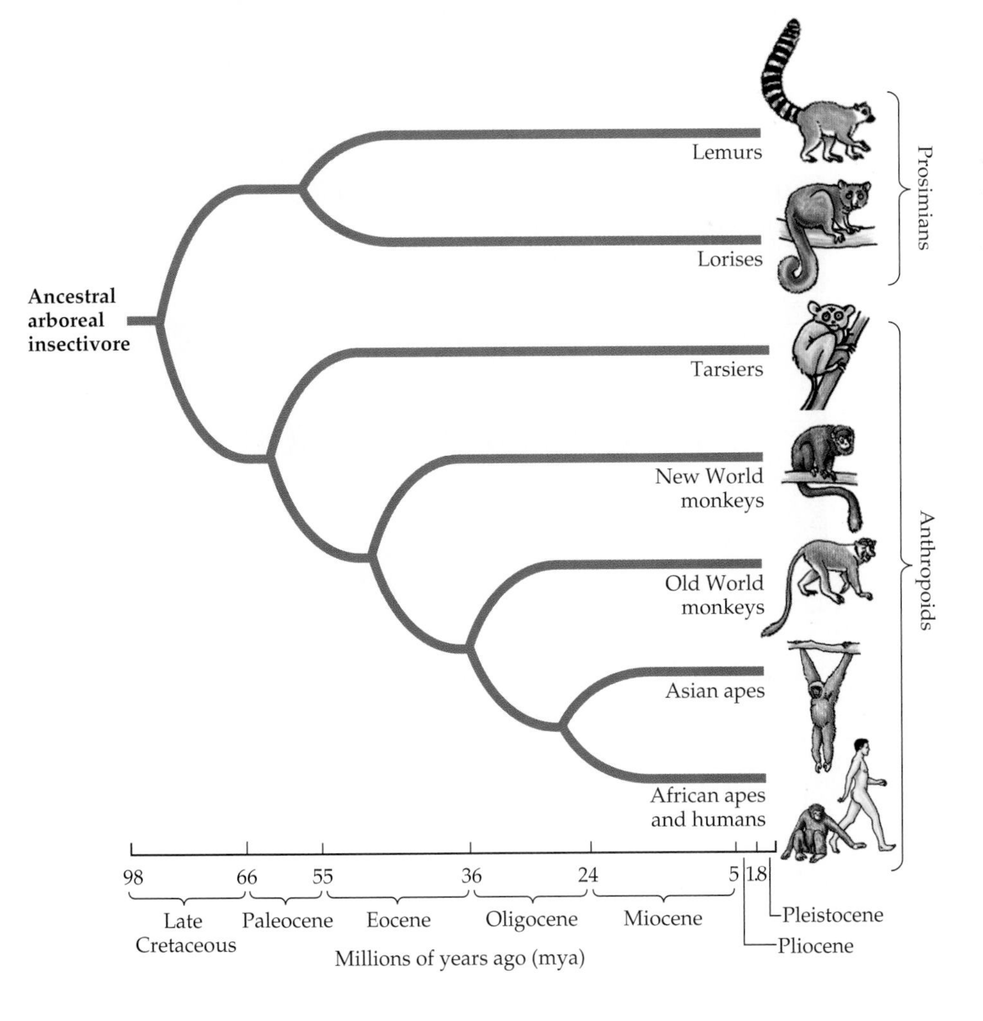

34.25 A Current Phylogeny of the Primates Too few fossil primates have been discovered to reveal with certainty their evolutionary relationships, but this phylogenetic tree is consistent with the existing evidence.

(*a*) *Eulemur fulvus*

(*b*) *Loris tardigradus*

34.26 Prosimians (*a*) The brown lemur is one of the many lemur species found in Madagascar, where they are part of a unique assemblage of plants and animals. (*b*) The slender loris is found in India. Its large eyes tell us that it is nocturnal.

(*a*) *Leontopithecus rosalia*

(*b*) *Macaca sylvanus*

34.27 Monkeys (*a*) Golden lion tamarins are endangered New World monkeys living in coastal Brazilian rainforests. (*b*) Many Old World species, such as these Barbary macaques, live in social groups. Here two members of a group groom each other.

(*a*) *Gorilla gorilla*

(*b*) *Pan troglodytes*

(*c*) *Hylobates lar*

(*d*) *Pongo pygmaeus*

34.28 Apes (*a*) Gorillas, the largest apes, are restricted to humid African forests. This male is a lowland gorilla. (*b*) Chimpanzees, our closest relatives, are found in forested regions of Africa. (*c*) Gibbons are the smallest of the apes. The common gibbon is found in Asia, from India to Borneo. (*d*) Orangutans live in the forests of Indonesia.

Human ancestors evolved bipedal locomotion

The **hominids**—the lineage that led to humans—separated from other ape lineages about 6 mya in Africa. The earliest protohominids, known as **ardipithecines**, had distinct morphological adaptations for **bipedalism**—locomotion in which the body is held erect and moved exclusively by movements of the hind legs. Bipedal locomotion frees the forelimbs to manipulate objects and to carry them while walking. It also elevates the eyes, enabling the animal to see over tall vegetation to spot predators and prey. At walking rates, bipedal movement is also energetically much more economical than quadrupedal (four-legged) locomotion. All three advantages were probably important for the ardipithecines and their descendants, the **australopithecines**.

The first australopithecine skull was found in South Africa in 1924. Since then, australopithecine fossils have been found in many sites in Africa. The most complete fossil skeleton of an australopithecine, approximately 3.5 million years old, was discovered in Ethiopia in 1974. That individual, a young female known to the world as Lucy, was assigned to the species *Australopithecus afarensis*. Fossil remains of more than 100 *A. afarensis* have now been discovered. During the past 5 years, fossils of other australopithecines that lived in Africa 4–5 million years ago have been unearthed.

Experts disagree over how many species are represented by the australopithecine fossils, but it is clear that several million years ago, at least two distinct types lived together over much of eastern Africa. The larger type (about 40 kilograms) is represented by at least two species (*Paranthropus robustus* and *P. boisei*), both of which died out suddenly about 1.5 million years ago.

Humans arose from australopithecine ancestors

Early members of the genus *Homo* lived contemporaneously with australopithecines for perhaps half a million years (Figure 34.29). The oldest fossils of the genus, an extinct species called *H. habilis*, were discovered in the Olduvai Gorge, Tanzania. These fossils are estimated to be 2 million years old. Other fossils of *H. habilis* have been found in Kenya and Ethiopia. Associated with the fossils are tools that these early hominids used to obtain food.

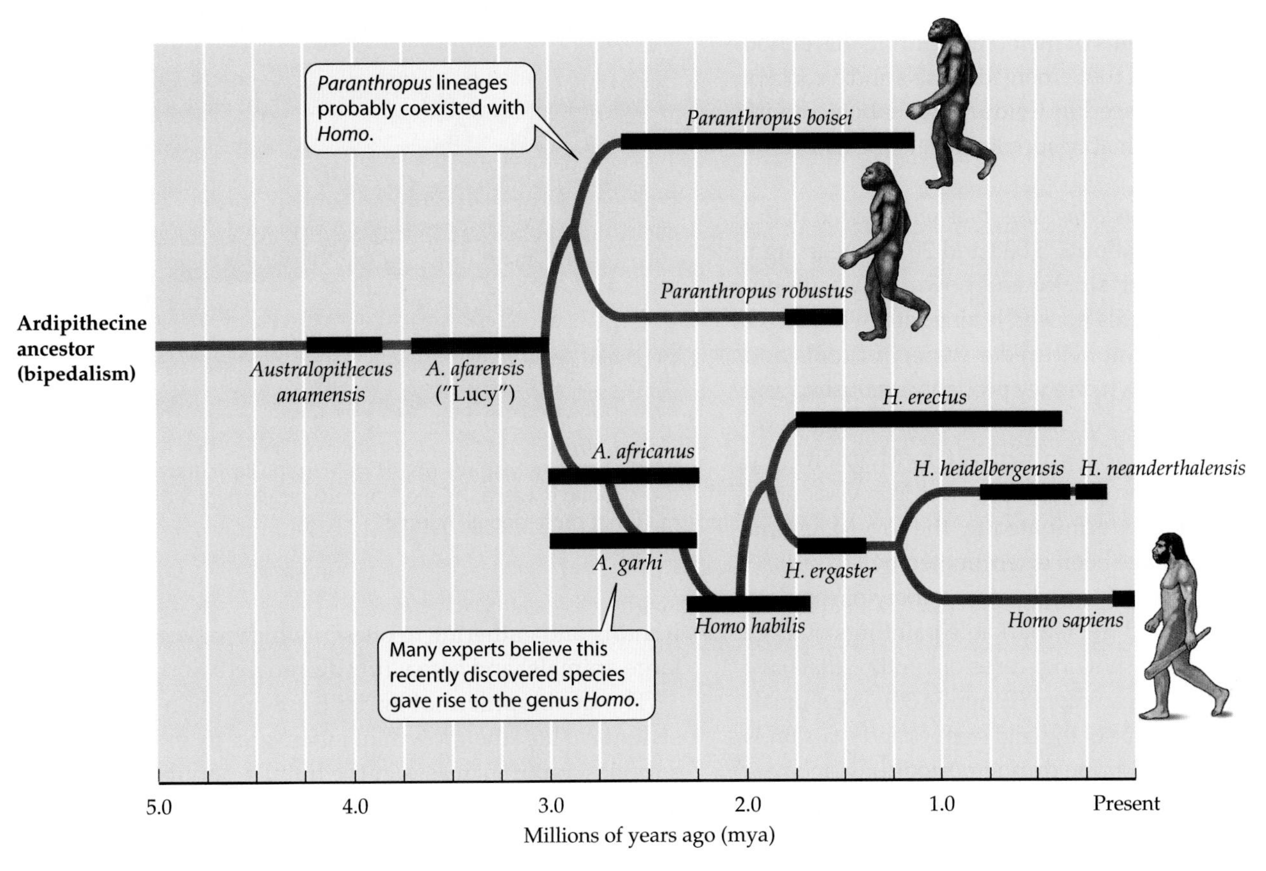

34.29 A Current Phylogeny of *Homo sapiens* At times in the past, more than one species of hominid lived on Earth. The heavy dark blue lines indicate the time frame over which each species lived.

Another extinct species of our genus, *Homo erectus*, evolved in Africa about 1.6 mya. Soon thereafter it had spread as far as eastern Asia. As it expanded its range and increased in abundance, *H. erectus* may have exterminated *H. habilis*. Members of *H. erectus* were as large as modern people, but their bones were considerably heavier. *Homo erectus* used fire for cooking and for hunting large animals, and made characteristic stone tools that have been found in many parts of the Old World. Although *H. erectus* survived in Eurasia until about 250,000 years ago, it was replaced in tropical regions by our species, *Homo sapiens*, about 200,000 years ago.

Human brains became larger

The earliest members of *Homo sapiens* had larger brains than members of the earlier species of *Homo*. Brain size in the lineage increased rapidly, reaching modern size by about 160,000 years ago. This striking change was probably favored by an increasingly complex social life. The ability of group members to communicate with one another would have been valuable in cooperative hunting and gathering and for improving one's status in the complex social interactions that must have characterized early human societies, just as they do ours today. But why did brains become larger only in the human lineage?

A clue to the answer is provided by brain chemistry. The human brain is a fat-rich organ. About 60 percent of its structural material is made up of lipids, most of them long-chain polyunsaturated omega-3 and omega-6 fatty acids. Humans must consume omega fatty acids in their diet because the body cannot synthesize these molecules fast enough from the other fatty acids found in vegetables, nuts, and seeds to supply their brains. Animal brains and livers contain omega fatty acids, but fish and shellfish are by far the best sources.

Therefore, because savannas and open woodlands provide few sources of omega fatty acids, the traditional view that early human evolution took place in those environments is being questioned. In contrast, the shores of Africa's many lakes would have been rich sources of fish and mollusks. Thus, access to fat-rich foods from aquatic environments may have been the key factor that supported the dramatic expansion of the human brain. The archeological record of the past 100,000 years includes hundreds of piles of mollusk shells and fish bones, as well as carved points used for fishing. Chimpanzees remained in the forest and ate fruits and nuts. They may have lacked food sources to support much larger brains.

Several *Homo* species existed during the mid-Pleistocene epoch, from about 1.5 million to about 300,000 years ago. All were skilled hunters of large mammals, but plants continued to be important components of their diets. During this period another distinctly human trait emerged: rituals and a concept of life after death. Deceased individuals were buried with tools and clothing, presumably supplies for their existence in the next world.

One species, *Homo neanderthalensis*, was widespread in Europe and Asia between about 75,000 and 30,000 years ago. Neanderthals were short, stocky, and powerfully built humans whose massive skulls housed brains somewhat larger than our own. They manufactured a variety of tools and hunted large mammals, which they probably ambushed and subdued in close combat. For a short time, their range overlapped that of the *H. sapiens* known as Cro-Magnons, but then the Neanderthals abruptly disappeared. Many scientists believe that they were exterminated by the Cro-Magnons, just as *H. habilis* may have been exterminated by *H. erectus*.

Cro-Magnon people made and used a variety of sophisticated tools. They created the remarkable paintings of large mammals, many of them showing scenes of hunting, that have been discovered in European caves (Figure 34.30). The animals depicted were characteristic of the cold steppes and grasslands that occupied much of Europe during periods of glacial expansion. Cro-Magnon people spread across Asia, reaching North America perhaps as early as 20,000 years ago, although the date of their arrival in the New World is still uncertain. Within a few thousand years, they had spread southward through North America to the southern tip of South America.

34.30 Hunting Inspires Art Cro-Magnon cave drawings such as those found in Lascaux Cave, France, typically depict the large mammals that these people hunted.

Humans evolved language and culture

As our ancestors evolved larger brains, their behavioral capabilities increased, especially the capacity for language. Most animal communication consists of a limited number of signals, which refer mostly to immediate circumstances and are associated with changed emotional states induced by those circumstances. Human language is far richer in its symbolic character than any other animal vocalizations. Our words can refer to past and future times and to distant places. We are capable of learning thousands of words, many of them referring to abstract concepts. We can rearrange words to form sentences with complex meanings.

The expanded mental abilities of humans are largely responsible for the development of **culture**, the process by which knowledge and traditions are passed along from one generation to another by teaching and observation. Culture can change rapidly because genetic changes are not necessary for a cultural trait to spread through a population. A potential disadvantage of culture is that its norms must be taught to each generation.

Cultural learning greatly facilitated the spread of domestic plants and animals and the resultant conversion of most human societies from ones in which food was obtained by hunting and gathering to ones in which *pastoralism* (herding large animals) and *agriculture* dominated.

The development of agriculture led to an increasingly sedentary life, the growth of cities, greatly expanded food supplies, rapid increases in the human population, and the appearance of occupational specializations, such as artisans, shamans, and teachers.

Deuterostomes and Protostomes: Shared Evolutionary Themes

The evolution of deuterostomes paralleled the evolution of protostomes in several important ways. Both lineages exploited the abundant food supplies buried in soft marine substrata, attached to rocks, or suspended in water. Many groups of both lineages developed elaborate structures for moving water and extracting prey from it.

In some lineages of both groups, the body cavity became divided into compartments that allowed better control of shape and movement. Some members of both groups evolved mechanisms for controlling their buoyancy in water using gas-filled internal spaces. Planktonic larval stages evolved in marine members of many protostomate and deuterostomate phyla.

Both protostomes and deuterostomes colonized the land, but the consequences were very different. The jointed external skeletons of arthropods, although they provide excellent support and protection in air, cannot support large animals, as the internal skeletons developed by deuterostomes can.

Terrestrial deuterostomes recolonized aquatic environments a number of times. Suspension feeding evolved once again in several of these lineages. The largest living animals, baleen (toothless) whales, feed upon small prey only a few centimeters long, which they extract from the water with large straining structures in their mouths.

Chapter Summary

Origins of the Deuterostomes

▶ The deuterostomate lineage separated from the protostomate lineage early in animal evolution. The ancestral deuterostome had external gills. **Review Figure 34.1**

▶ There are only two major deuterostomate lineages, and there are fewer species of deuterostomes than protostomes, but as members of the lineage, we have a special interest in its members. **Review Figure 34.2.** See Web/CD Activity 34.1

Echinoderms: Pentaradial Symmetry

▶ Echinoderms have a pentaradially symmetrical body plan, a unique water vascular system, and a calcified internal skeleton. **Review Figure 34.3*a***

▶ Nearly all living species of echinoderms have a bilaterally symmetrical, ciliated larva that feeds as a planktonic organism. **Review Figure 34.3*b***

▶ Six major groups of echinoderms survive today, but 23 other lineages existed in the past. Some groups of echinoderms have arms, but others do not.

Hemichordates: Conservative Evolution

▶ Acorn worms and pterobranchs are similar to ancestral deuterostomes. **Review Figure 34.5**

Chordates: New Ways of Feeding

▶ Members of another deuterostomate lineage evolved enlarged pharyngeal slits used as feeding devices and a dorsal supporting rod, the notochord.

▶ Most urochordates are sessile as adults and filter prey from seawater with large pharyngeal baskets. But some species retain their notochords and nerve cords as planktonic adults.

Evolution of the Chordates

▶ Cephalochordates probably resemble the ancestors of all other chordates. **Review Figure 34.7**

▶ Vertebrates evolved jointed internal skeletons that enabled them to swim rapidly. Early vertebrates used the pharyngeal basket to filter small animals from mud. **Review Figures 34.8, 34.9**

▶ Jaws, which evolved from anterior gill arches, enabled their possessors to grasp and chew their prey. Jawed fishes rapidly became dominant animals in both marine and fresh waters. **Review Figure 34.11**

▶ Fishes evolved two pairs of unjointed fins, with which they control their swimming movements and stabilize themselves in the water, and swim bladders, which help keep them suspended in open water.

▶ Ray-finned fishes come in a wide variety of sizes and shapes. Many species have complex social systems.

Colonizing the Land: Obtaining Oxygen from the Air

▶ Two lineages of fishes—lobe-finned fishes and lungfishes—evolved jointed fins.

▶ Amphibians, the first terrestrial vertebrates, arose from lungfish ancestors.

▶ The 4,500 species of amphibians living today belong to three groups: caecilians, frogs and toads, and salamanders.

▶ Most amphibians live in water at some time in their lives, and their eggs must remain moist. **Review Figure 34.16.** See Web/CD Tutorial 34.1

▶ Amniotes evolved eggs with shells impermeable to water and thus became the first vertebrates to be independent of water for reproduction. **Review Figure 34.17.** See Web/CD Activity 34.2

▶ Modern reptiles are members of four lineages: snakes and lizards, tuataras, turtles and tortoises, and crocodilians. **Review Figure 34.18**

▶ Dinosaurs rose to dominance about 215 mya and dominated terrestrial environments for about 150 million years until they became extinct about 65 mya.

▶ Some dinosaurs evolved feathers and were capable of flight.

Birds: More Feathers and Better Flight

▶ Birds arose about 175 mya from feathered dinosaur ancestors.

▶ The 9,600 species of birds are characterized by feathers, high metabolic rates, and parental care.

The Origin and Diversity of Mammals

▶ Mammals evolved during the Mesozoic era, about 225 mya.

▶ The eggs of mammals are fertilized within the body of the female, and embryos develop for some time within a uterus before being born. Mammals are unique in suckling their young with milk secreted by mammary glands.

▶ The three species of mammals in subclass Prototheria lay eggs, but all other mammals give birth to live young.

▶ Therian mammals are divided into two major groups: the marsupials, which give birth to tiny young that are, in most species, raised in a pouch on the female's belly, and the eutherians, which give birth to relatively well-developed offspring.

Primates and the Origin of Humans

▶ The primates split into two major lineages, one leading to the prosimians (lemurs and lorises) and the other leading to the tarsiers, monkeys, apes, and humans. **Review Figure 34.25**

▶ Hominids evolved in Africa from terrestrial, bipedal ancestors. **Review Figure 34.29**

▶ Early humans evolved large brains, language, and culture. They manufactured and used tools, developed rituals, and domesticated plants and animals. In combination, these traits enabled humans to increase greatly in number and to transform the face of Earth.

Deuterostomes and Protostomes: Shared Evolutionary Themes

▶ Both protostomes and deuterostomes evolved structures to filter prey from the water, mechanisms to control their buoyancy in water, and planktonic larval stages.

See Web/CD Activity 34.3 for a concept review of this chapter.

Self-Quiz

1. Which of the following deuterostomate groups have a three-part body plan?
   a. Acorn worms and tunicates
   b. Acorn worms and pterobranchs
   c. Pterobranchs and tunicates
   d. Pterobranchs and lancelets
   e. Tunicates and lancelets
2. The structure used by adult ascidians to capture food is a
   a. pharyngeal basket.
   b. proboscis.
   c. lophophore.
   d. mucus net.
   e. radula.
3. The pharyngeal gill slits of chordates originally functioned as sites for
   a. uptake of oxygen only.

b. release of carbon dioxide only.
c. both uptake of oxygen and release of carbon dioxide.
d. removal of small prey from the water.
e. forcible expulsion of water to move the animal.

4. The key to the vertebrate body plan is a
   a. pharyngeal basket.
   b. vertebral column to which internal organs are attached.
   c. vertebral column to which two pairs of appendages are attached.
   d. vertebral column to which a pharyngeal basket is attached.
   e. pharyngeal basket and two pairs of appendages.
5. Which of the following fishes do *not* have a cartilaginous skeleton?
   a. Chimaeras
   b. Lungfishes
   c. Sharks
   d. Skates
   e. Rays
6. In most fishes, lunglike sacs evolved into
   a. pharyngeal gill slits.
   b. true lungs.
   c. coelomic cavities.
   d. swim bladders.
   e. none of the above
7. Most amphibians return to water to lay their eggs because
   a. water is isotonic to egg fluids.
   b. adults must be in water while they guard their eggs.
   c. there are fewer predators in water than on land.
   d. amphibians need water to produce their eggs.
   e. amphibian eggs quickly lose water and desiccate if their surroundings are dry.
8. The horny scales that cover the skin of reptiles prevent them from
   a. using their skin as an organ of gas exchange.
   b. sustaining high levels of metabolic activity.
   c. laying their eggs in water.
   d. flying.
   e. crawling into small spaces.
9. Which statement about bird feathers is *not* true?
   a. They are highly modified reptilian scales.
   b. They provide insulation for the body.
   c. They exist in two layers.
   d. They help birds fly.
   e. They are important sites of gas exchange.
10. Monotremes differ from other mammals in that they
    a. do not produce milk.
    b. lack body hair.
    c. lay eggs.
    d. live in Australia.
    e. have a pouch in which the young are raised.
11. Bipedalism is believed to have evolved in the human lineage because bipedal locomotion is
    a. more efficient than quadrupedal locomotion.
    b. more efficient than quadrupedal locomotion, and it frees the forelimbs to manipulate objects.
    c. less efficient than quadrupedal locomotion, but it frees the forelimbs to manipulate objects.
    d. less efficient than quadrupedal locomotion, but bipedal animals can run faster.
    e. less efficient than quadrupedal locomotion, but natural selection does not act to improve efficiency.

For Discussion

1. In what animal phyla has the ability to fly evolved? How do the structures used for flying differ among these animals?
2. Extracting suspended food from the water column is a common mode of foraging among animals. Which groups contain species that extract prey from the air? Why is this mode of obtaining food so much less common than extracting prey from the water?
3. Large size both confers benefits and poses certain risks. What are these risks and benefits?
4. Amphibians have survived and prospered for many millions of years, but today many species are disappearing and populations of others are declining seriously. What features of amphibian life histories might make them especially vulnerable to the kinds of environmental changes now happening on Earth?
5. The body plan of most vertebrates is based on four appendages. Describe the varied forms that these appendages take and how they are used. How do the vertebrates that have kept their four appendages move?
6. Compare the ways that different animal lineages colonized the land. How were those ways influenced by the body plans of animals in the different lineages?

LIFE ESSAY

What is our duty to nature?

- by Holmes Rolston, III -

Environmental ethics seeks appropriate respect for values in and duties regarding nature. This starts with human concerns for a healthy environment. If people have a right to life, they also have a right to a quality environment, needed for human welfare.

Environmental ethics then turns in nonhuman directions. What about the whooping cranes or the sequoia trees, the myriad species with which we co-inhabit Earth? Is there some intrinsic value in their lives we ought to protect? Surely we, *Homo sapiens*, the wise species, the only self-consciously moral species, are less wise than we ought to be if we act only in our collective self-interest.

Western ethics, philosophy, religion, politics, and economics have been dominantly humanistic, or anthropocentric. Contemporary ethics seeks to be inclusive: the poor as well as the rich, women as well as men, indigenous cultures as well as modern ones, future generations beyond the present. Environmental ethics is even more inclusive, concerned about whales slaughtered, whooping cranes and their habitats, ancient forests cut, Earth threatened by global warming.

Science alone does not teach us what we most need to know about nature: how to value it. Still, biology confronts every biologist (researcher and student alike) with an urgent moral concern—caring for life on Earth. Somewhat ironically, just when humans, with their increasing industry and technology, seemed further and further from nature, the natural world has emerged as a focus of ethical concern.

Ought not biologists (above all!) celebrate and cherish Earth's biodiversity? This concern arises, ironically again, despite somewhat uncertain relations between science and ethics, how to move from what *is* (description of biological facts) to what *ought to be* (prescription of duty). It is not simply what a society does to its women, racial minorities, handicapped, children, or future generations, but what it does to its fauna, flora, species, ecosystems, and landscapes that reveals the character of that society.

Animals hunt and howl, care for their young, flee from threats, value their lives. There is "somebody there" behind the fur and feathers. "Man is the measure of things," said Protagoras, an ancient Greek philosopher. But wild animals do not make man the measure at all. Human values may override animal values, but we ought to justify such overriding—especially if we eat animals, exploit them, or experiment on them. Biology teaches that we and they are kin; ethically, their pains and pleasures count morally, too.

Most of the biological world, however, has yet to be taken into account: Plants, lower animals, insects, microbes, all are quite alive with vital interests. Every living organism has a *good-of-its-kind*; it defends its own kind as a *good kind*. Maybe "life" is a better measure of value than "man," or "vertebrate."

Life goes on at multiple levels. An inclusive ethic will be concerned for any ongoing species, for lifelines regenerating. Extinction is a sort of super-killing, a shutdown of life. In threatening Earth's biodiversity, humans are stopping the historical vitality of life.

We reach a "land ethic" (Aldo Leopold) with concern for ecosystems, for living communities, for life processes. Individual animals and plants are what they are not as mere individuals (as though in a zoo or botanical garden), but they flourish in species lines and live in niches in habitats. An organism, a species, is what it is where it is, adapted for living in ongoing ecological and evolutionary systems. The most appropriate unit for moral concern is the whole system, the fundamental unit of development and survival.

Now we can put humans back in the picture. After all, ecology is about living at home (Greek *oikos*, "house"), the inclusive system again. Humans have entwined destinies with the natural world; their richest quality of life requires identifying with these communities.

Environmental ethics becomes Earth ethics. Humans are the only evaluators who can reflect at global scales. When humans do this, they must set up the scales. Animals, plants, insects, species, ecosystems, cannot take part in such inclusive and comprehensive concern for biodiversity on Earth, But they are what is to be measured. Earth (as seen from space) is quite a wonder. We Earthlings ought to care for this home planet.

Holmes Rolston, III, University Distinguished Professor and Professor of Philosophy at Colorado State University, is the author of *Environmental Ethics* (Temple University Press, 1988) and *Conserving Natural Value* (Columbia University Press, 1994). He is past president of the International Society for Environmental Ethics. In 2003 he was awarded the Templeton Prize in Religion, recognizing his work on respect for nature and reverence for life

Discussion Questions

1. Are good biologists always conservationists?
2. Can environmental ethics always be "win–win," people and nature?
3. Does an environmental ethic need to be science-based?
4. In wild nature is there anything bad? Anything ugly? Or that ought not to be respected or conserved?

Web Links

International Society for Environmental Ethics On-Line Bibliography
www.phil.unt.edu/bib/

Environmental Ethics, Systematic Works
www.cep.unt.edu/theo.html

Environmental Ethics, Anthologies
www.cep.unt.edu/anthol.html

Environmental Ethics, Introductory Articles
www.cep.unt.edu/intro.html

35 The Plant Body

On November 1, 2002, John Quigley climbed into the branches of a 70-foot-tall oak tree estimated to be 150 to 400 years old. He stayed perched there until he was removed, 71 days later, to allow a housing developer to cut down the tree. That was a short stay, however, compared with Julia Butterfly Hill's sojourn in a 600-year-old redwood. In the year 2000, Hill created a perch 180 feet above ground and didn't come down to Earth until just over 2 years later, when the Pacific Lumber Company agreed to spare that tree and others in its immediate vicinity.

What prompts some people to tree-sit or protest in other ways against the removal of trees? Clearly, one motivation is their admiration for the sheer longevity of these organisms, which have survived in their environments for decades and even centuries, during which a great deal of human history has taken place. The oldest known individual plant is a bristlecone pine that has lived for more than 4,900 years—almost 50 centuries. In contrast, it is doubtful that any animal has ever lived much longer than 2 centuries.

This longevity is even more impressive when it is understood that plants cannot move from site to site to avoid danger or environmental challenges. Even though plants are not motile, the extreme ages achieved by some trees prove that plants can nevertheless cope successfully with their environment. The plant body creates and maintains an internal environment that differs from the external environment.

Plants accomplish through growth some of the same things that animals achieve through mobility. Growing roots, for example, can reach into new supplies of water and nutrients. By growing, stems and leaves rise out of shaded areas into the sun to obtain energy.

Although plants do not need to obtain complex substances like vitamins from their environments as animals do, they must nevertheless obtain nutrients—not only the raw materials of photosynthesis (carbon dioxide and water), but also mineral elements such as nitrogen, potassium, and calcium. Seed plants—even the tallest trees—transport water and minerals from the soil to their tops, and they transport the products of photosynthesis from the leaves to their roots and other parts.

An Ancient Individual Bristlecone pines (*Pinus longaeva*) can live for centuries. The oldest known living organism is a bristlecone pine that has been alive for almost 5,000 years—long enough to have witnessed all of recorded human history.

Plants also interact with their living and nonliving environments. They respond to environmental cues as they grow and develop. Their responses are mediated by chemical signals that move within cells and throughout the plant body. Among the resulting changes are ones that lead to growth, development, and reproduction.

Because we can understand the functioning of plants only in terms of their underlying structure, this chapter focuses on the structure of the plant body, with a primary emphasis on flowering plants. We'll examine plant structure at the levels of organs, cells, tissues, and tissue systems. Then we'll see how organized groups of dividing cells, called meristems, contribute to the growth of the plant body, both in length and, in woody plants, in width. The chapter concludes with a consideration of how leaf structure supports photosynthesis.

Vegetative Organs of the Flowering Plant Body

You will recall from Chapter 30 that flowering plants (angiosperms) are *tracheophytes* that are characterized by double fertilization, a triploid endosperm, and seeds enclosed in modified leaves called carpels. Their xylem contains cells called vessel elements and fibers, and their phloem contains sieve tube elements and companion cells.

Flowering plants possess three kinds of *vegetative* (nonreproductive) organs: roots, stems, and leaves. Flowers, which are the plant's devices for sexual reproduction, consist of modified leaves and stems; flowers will be considered in detail in a later chapter.

Most flowering plants belong to one of two major lineages. **Monocots** are generally narrow-leaved flowering plants such as grasses, lilies, orchids, and palms. **Eudicots** are broad-leaved flowering plants such as soybeans, roses, sunflowers, and maples. These two lineages account for 97 percent of flowering plant species (Figure 35.1). Most of the remaining species (including water lilies and magnoliids) are structurally similar to the eudicots.*

The basic body plans of a generalized monocot and a generalized eudicot are shown in Figure 35.2. In both lineages, the vegetative plant body consists of two systems: the shoot system and the root system.

The **shoot system** of a plant consists of the stems, leaves, and flowers. Broadly speaking, the **leaves** are the chief organs of photosynthesis. The **stems** hold and display the leaves to the sun and provide connections for the transport of materials between roots and leaves. The locations where leaves attach to a stem are called **nodes**, and the stem regions between successive nodes are **internodes**.

The **root system** anchors the plant in place and provides nutrition. The extreme branching of plant roots and their high surface area-to-volume ratio allow them to absorb water and mineral nutrients from the soil.

Each of the vegetative organs can be understood in terms of its structure. By *structure* we mean both its overall form, called its *morphology*, and its component cells and tissues and their arrangement, called its *anatomy*. Let's first consider the overall forms of roots, stems, and leaves.

Roots anchor the plant and take up water and minerals

Water and minerals usually enter the plant through the root system, which usually lies in the soil, where light does not penetrate. Roots typically lack the capacity for photosynthesis even when removed from the soil and placed in light.

*Botanists traditionally have referred to all flowering plants other than monocots as *dicots*. However, the dicots do not constitute a monophyletic lineage (see Figure 30.13). Because we wish to emphasize lineages, we do not use the term *dicot* here.

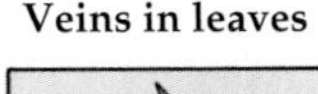

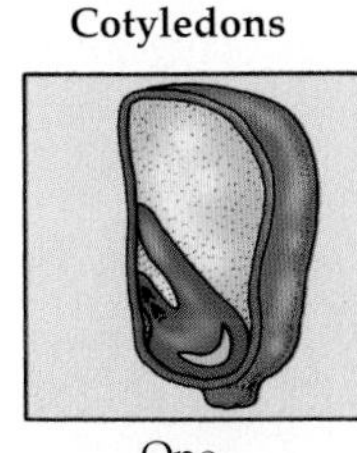

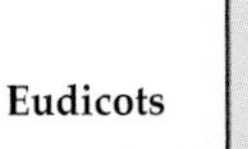

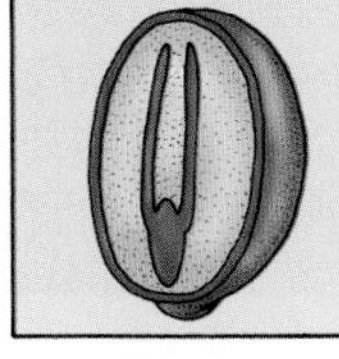

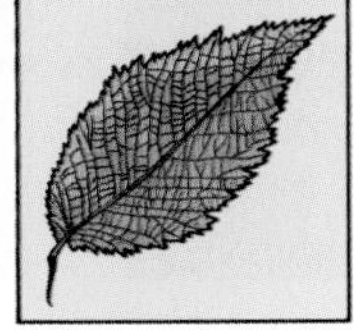

35.1 Monocots versus Eudicots The possession of a single cotyledon clearly distinguishes the monocots from the other angiosperms. Several other anatomical characteristics also differ between the monocots and the eudicots. Most angiosperms that do not belong to either lineage resemble eudicots in the characteristics shown here.

35.2 Vegetative Organs and Systems The basic plant body plan and the principal vegetative organs are similar in monocots and eudicots.

Flowers, made up of specialized leaflike structures, are adapted for sexual reproduction.

The **shoot system** consists of stems and leaves, in which photosynthesis takes place.

The **root system** anchors and provides nutrients for the shoot system.

Monocot

Eudicot

Flower

Apical bud

Node

Internode

Lateral bud

Leaf:
Petiole
Blade

Stem

Roots

There are two principal types of root systems. Many eudicots have a *taproot system:* a single, large, deep-growing primary root accompanied by less prominent lateral roots. The taproot itself often functions as a nutrient storage organ, as in carrots (Figure 35.3*a*).

By contrast, monocots and some eudicots have a *fibrous root system*, which is composed of numerous thin roots that are all roughly equal in diameter (Figure 35.3*b*). Many fibrous root systems have a large surface area for the absorption of water and minerals. A fibrous root system clings to soil very well. Grasses with fibrous root systems, for example, may protect steep hillsides where runoff from rain would otherwise cause erosion.

Some plants have *adventitious roots*. These roots arise above ground from points along the stem; some even arise from the leaves. In many species, adventitious roots can form when a piece of shoot is cut from the plant and placed in water or soil. Adventitious rooting enables the cutting to establish itself in the soil as a new plant. Such a cutting is a form of vegetative reproduction, which we will discuss in a later chapter. Some plants—corn, banyan trees, and some palms, for example—use adventitious roots as props to help support the shoot.

Stems bear buds, leaves, and flowers

Unlike roots, stems bear buds of various types. A *bud* is an embryonic shoot. A stem bears leaves at its nodes, and where each leaf meets the stem there is a **lateral bud** (see Figure 35.2). If it becomes active, the lateral bud can develop into a new *branch*, or extension of the shoot system. The branching patterns of plants are highly variable, depending on the species, environmental conditions, and a gardener's pruning activities.

35.3 Root Systems The taproot system of a carrot (*a*) contrasts with the fibrous root system of a grass (*b*).

At the tip of each stem or branch is an **apical bud**, which produces the cells for the upward and outward growth and development of that shoot. Under appropriate conditions, other buds form that develop into flowers.

Some stems are highly modified. The *tuber* of a potato, for example—the part of the plant eaten by humans—is an underground stem rather than a root. Its "eyes" contain lateral buds; thus, a sprouting potato is just a branching stem (Figure 35.4*a*). The *runners* of strawberry plants and Bermuda grass are horizontal stems from which roots grow at frequent intervals (Figure 35.4*b*). If the links between the rooted portions are broken, independent plants can develop on each side of the break. This phenomenon is a form of vegetative reproduction.

Although stems are usually green and capable of photosynthesis, they usually are not the principal sites of photosynthesis. Most photosynthesis takes place in leaves.

Leaves are the primary sites of photosynthesis

In gymnosperms and most flowering plants, the leaves are responsible for most of the plant's photosynthesis, producing energy-rich organic molecules and releasing oxygen gas. In certain plants, the leaves are highly modified for more specialized functions, as we will see below.

As photosynthetic organs, leaves are marvelously adapted for gathering light. Typically, the **blade** of a leaf is a thin, flat structure attached to the stem by a stalk called a **petiole**. During the daytime, the leaf blade is held by its petiole at an angle almost perpendicular to the rays of the sun. This orientation, with the leaf surface facing the sun, maximizes the amount of light available for photosynthesis. Some leaves track the sun, moving so that they constantly face it.

The leaves at different sites on a single plant may have quite different shapes. These shapes result from a combination of genetic, environmental, and developmental influences. Most species, however, bear similar, if not identical, leaves of a particular broadly defined type. A leaf may be *simple*, consisting of a single blade, or *compound*, with blades, or *leaflets*, arranged along an axis or radiating from a central point (Figure 35.5). In a simple leaf, or in a leaflet of a compound leaf, the veins may be parallel to one another, as in monocots, or in a netlike arrangement, as in eudicots.

The general development of a specific leaf pattern is programmed in the plant's genes and is expressed by differential growth of the leaf veins and of the tissue between the veins. As a result, plant taxonomists have often found leaf forms (outlines, margins, tips, bases, and patterns of arrangement) to be reliable characters for classification and identification. At least some of the forms in Figure 35.5 probably look familiar to you.

During development in some plant species, leaves are highly modified for special functions. For example, modified leaves serve as storage depots for energy-rich molecules, as in the bulbs of onions. In other species, the leaves store water, as in succulents. The spines of cacti are modified leaves (see Figure 35.4*c*). Many plants, such as peas, have modified portions of leaves called *tendrils* that support the plant by wrapping around other structures or plants.

Leaves, like all other plant organs, are composed of cells, tissues, and tis-

35.4 Modified Stems (*a*) A potato is a modified stem called a tuber; the sprouts that grow from its eyes are shoots, not roots. (*b*) The runners of this beach strawberry are horizontal stems that produce roots at intervals. Runners provide a local water supply and allow rooted portions of the plant to live independently if the runner is cut. (*c*) The stem of this barrel cactus is enlarged to store water. Its thorny spines are modified leaves.

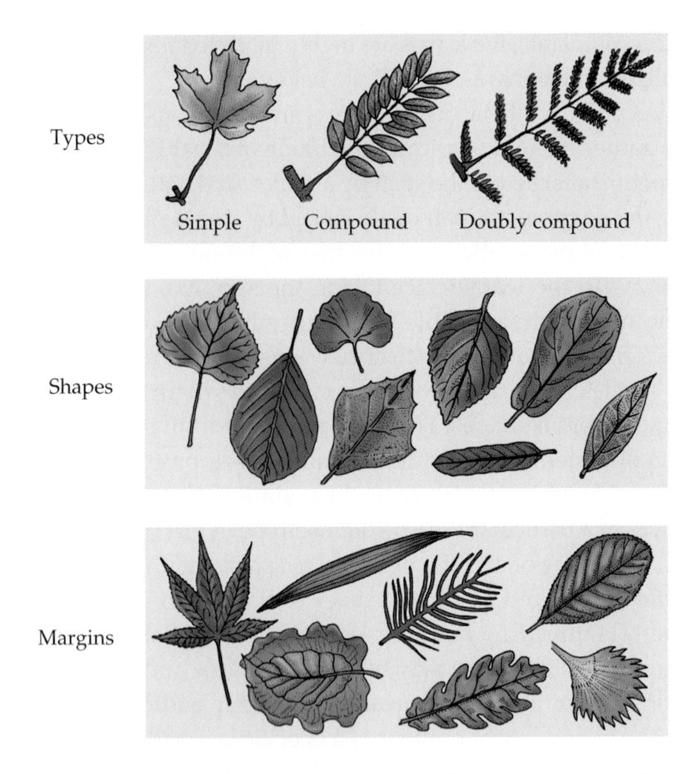

35.5 The Diversity of Leaf Forms Simple leaves are those with a single blade. Some compound leaves consist of leaflets arranged along a central axis. Further division of leaflets results in a doubly compound leaf. Other characters of leaf form can also be used to identify a plant's species.

sue systems. Let's now consider plant cells—the basic structural and functional units of plant organs.

Plant Cells

Plant cells have all the essential organelles common to eukaryotes (see Figure 4.7). In addition, they have certain structures and organelles that distinguish them from many other eukaryotes:

- They contain *chloroplasts* or other plastids.
- They contain *vacuoles*.
- They possess cellulose-containing *cell walls*.

Plant cells are alive when they divide and grow, but certain cells function only after their living parts have died and disintegrated. Other plant cells develop specialized metabolic capabilities; for example, some can perform photosynthesis, and others produce and secrete waterproofing materials. There are several different types of plant cells, which differ dramatically in the composition and structure of their cell walls. The walls of each cell type have a composition and structure that corresponds to its special functions.

Cell walls may be complex in structure

The cytokinesis of a plant cell is completed when the two daughter cells are separated by a cell plate (see Figure 9.10*b*). The daughter cells then deposit a gluelike substance within the cell plate; this substance constitutes the **middle lamella**. Next, each daughter cell secretes cellulose and other polysaccharides to form a **primary wall**. This deposition and secretion continue as the cell expands to its final size (Figure 35.6).

Once cell expansion stops, a plant cell may deposit one or more additional cellulosic layers to form a **secondary wall** internal to the primary wall (Figure 35.6). Secondary walls

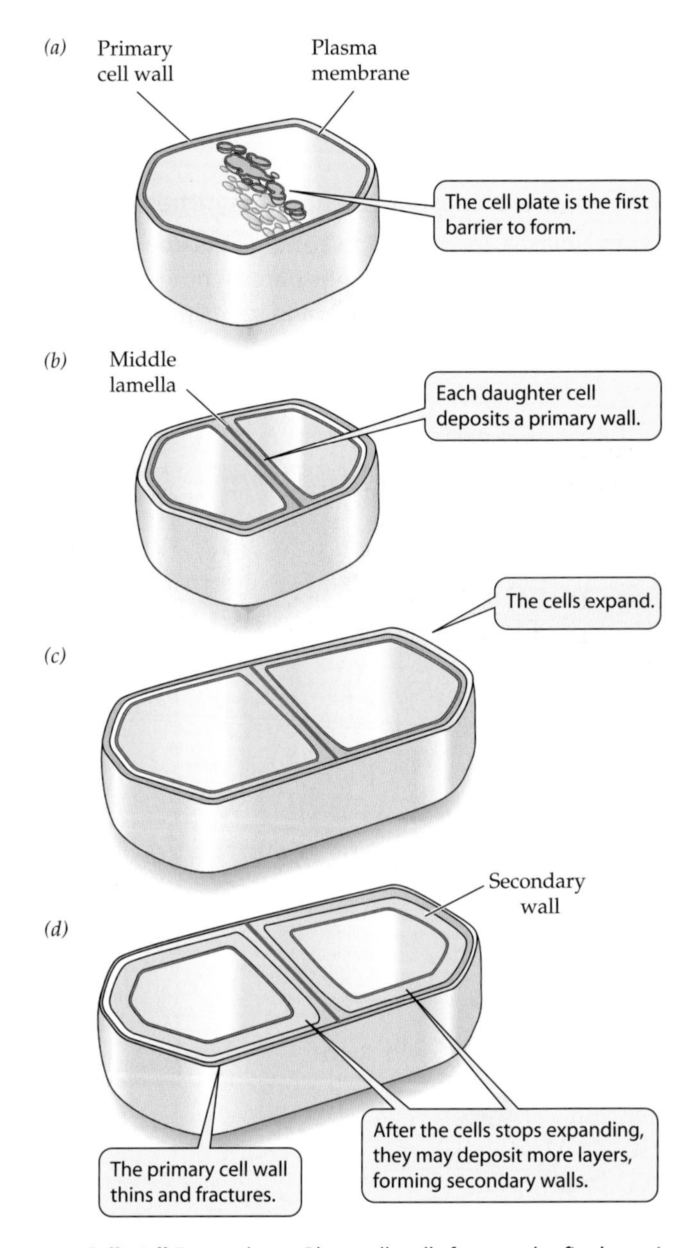

35.6 Cell Wall Formation Plant cell walls form as the final step in cell division.

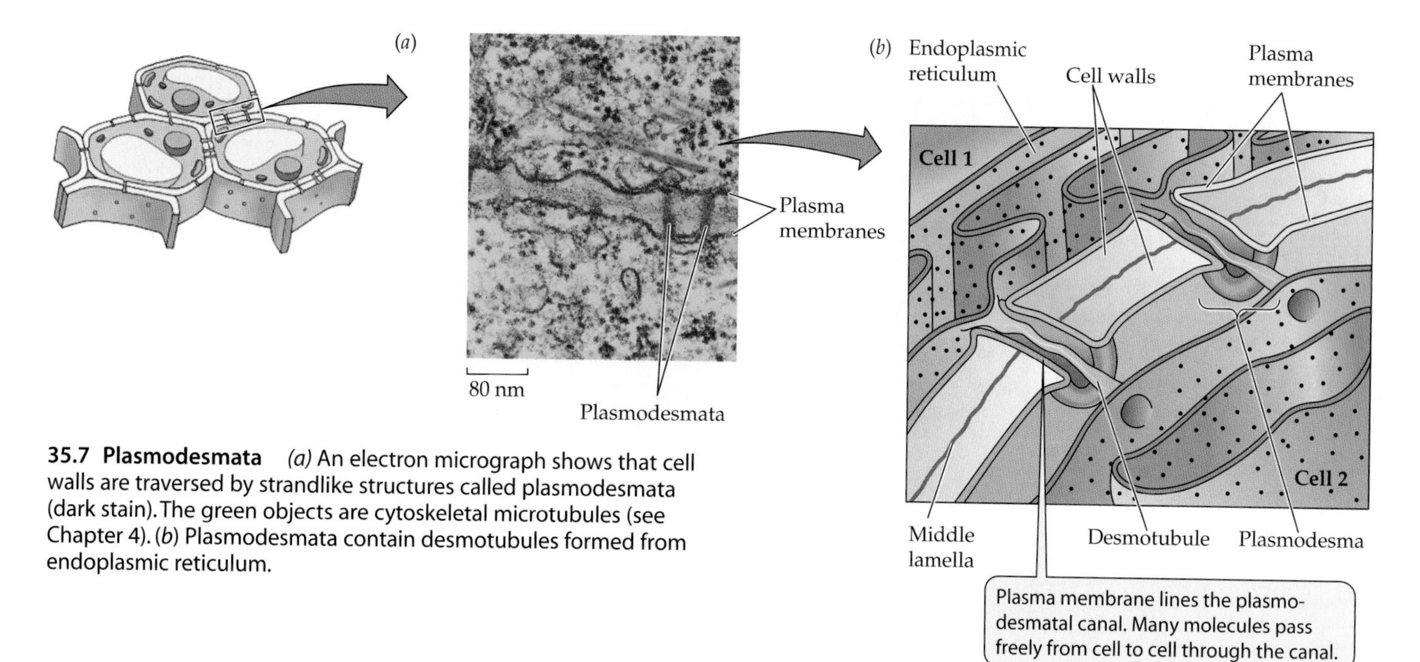

35.7 Plasmodesmata *(a)* An electron micrograph shows that cell walls are traversed by strandlike structures called plasmodesmata (dark stain). The green objects are cytoskeletal microtubules (see Chapter 4). *(b)* Plasmodesmata contain desmotubules formed from endoplasmic reticulum.

are often impregnated with unique substances that give them special properties. Those impregnated with the polymer **lignin** become strong, as in wood cells. Walls to which the complex lipid **suberin** are added become waterproof.

Although it lies outside the plasma membrane, the cell wall is not a chemically inactive region. In addition to cellulose and other polysaccharides, the cell wall contains proteins, some of which are enzymes. Chemical reactions in the wall play important roles in cell expansion and in defense against invading organisms. Cell walls may thicken or be sculpted or perforated as cells differentiate into specialized cell types. Except where the secondary wall is waterproofed, the cell wall is permeable to water, small molecules, and mineral ions.

Localized modifications in the walls of adjacent cells allow water and dissolved materials to move easily from cell to cell. The primary wall usually has regions where it becomes quite thin. In these regions, strands of cytoplasm called **plasmodesmata** (singular, plasmodesma) pass through the primary wall, allowing direct communication between plant cells. A plasmodesma is a plasma-membrane lined canal traversed by a strand of endoplasmic reticulum called the **desmotubule** (Figure 35.7). Under certain circumstances, a plasmodesma can enlarge dramatically, allowing even macromolecules and viruses to pass directly between cells (see Chapter 15). Substances can move from cell to cell through plasmodesmata without having to cross a plasma membrane.

Even in cells with a waterproofed secondary wall, water and dissolved materials can pass from cell to cell by way of structures called **pits**. Pits are interruptions in the secondary wall that leave the thin regions of the primary wall, and thus any plasmodesmata that are present, unobstructed (Figure 35.8).

Parenchyma cells are alive when they perform their functions

The most numerous cell type in young plants is the **parenchyma cell** (Figure 35.9*a*). Parenchyma cells usually have thin walls, consisting only of a primary wall and the shared middle lamella. Many parenchyma cells have shapes with multiple faces. Most have large central vacuoles.

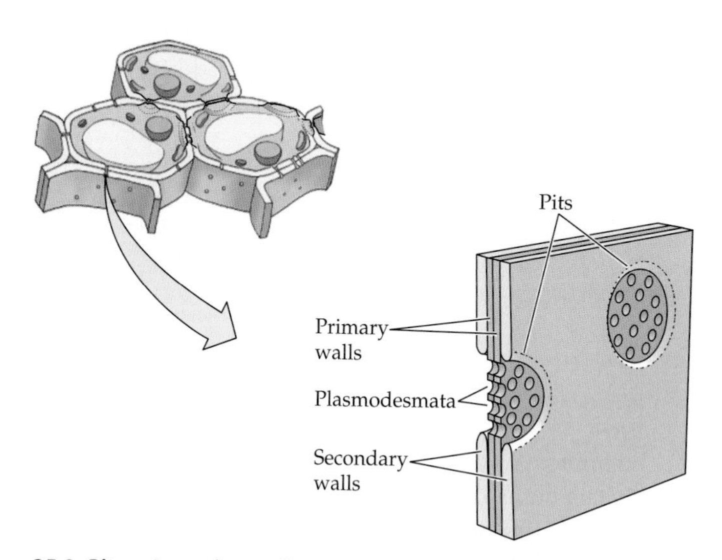

35.8 Pits Secondary walls may be interrupted by pits, which allow the passage of water and other materials between cells.

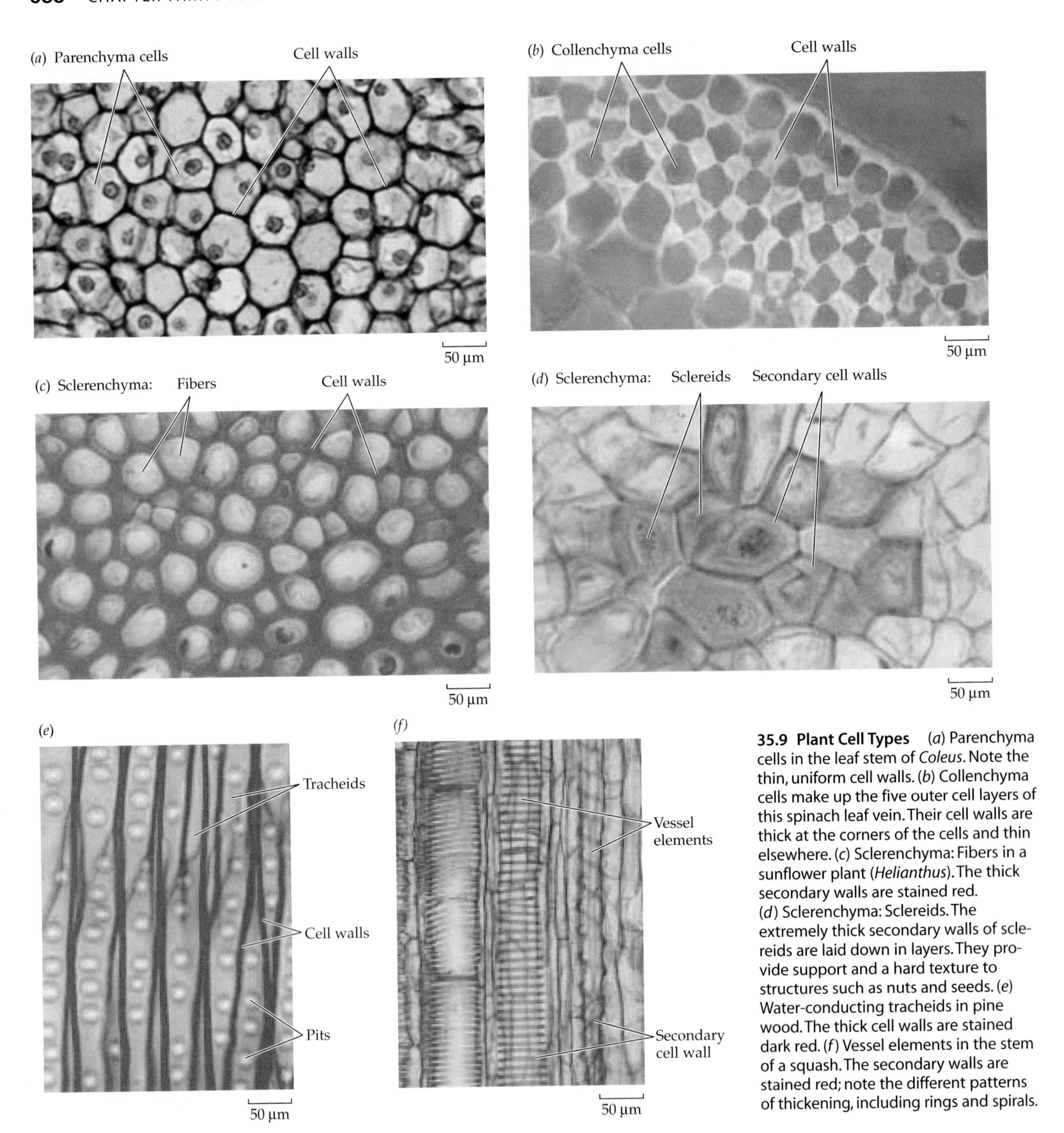

35.9 Plant Cell Types (*a*) Parenchyma cells in the leaf stem of *Coleus*. Note the thin, uniform cell walls. (*b*) Collenchyma cells make up the five outer cell layers of this spinach leaf vein. Their cell walls are thick at the corners of the cells and thin elsewhere. (*c*) Sclerenchyma: Fibers in a sunflower plant (*Helianthus*). The thick secondary walls are stained red. (*d*) Sclerenchyma: Sclereids. The extremely thick secondary walls of sclereids are laid down in layers. They provide support and a hard texture to structures such as nuts and seeds. (*e*) Water-conducting tracheids in pine wood. The thick cell walls are stained dark red. (*f*) Vessel elements in the stem of a squash. The secondary walls are stained red; note the different patterns of thickening, including rings and spirals.

The photosynthetic cells in leaves are parenchyma cells that contain numerous chloroplasts. Some nonphotosynthetic parenchyma cells store substances such as starch or lipids. In the cytoplasm of these cells, starch is often stored in specialized plastids called *leucoplasts* (see Figure 4.17*b*). Lipids may be stored as oil droplets, also in the cytoplasm. Some parenchyma cells appear to serve as "packing material" and play a vital role in supporting the stem. Many retain the capacity to divide and hence may give rise to new cells, as when a wound results in cell proliferation.

Collenchyma cells provide flexible support while alive

Collenchyma cells are supporting cells. Their primary walls are characteristically thick at the corners of the cells (Figure 35.9*b*). Collenchyma cells are generally elongated. In these cells, the primary wall thickens, but no secondary wall forms. Collenchyma provides support to leaf petioles, nonwoody stems, and growing organs. Tissue made of collenchyma cells is flexible, permitting stems and petioles to sway in the wind without snapping. The familiar "strings" in celery consist primarily of collenchyma cells.

Sclerenchyma cells provide rigid support

In contrast to collenchyma cells, **sclerenchyma** cells have a thickened secondary wall that performs their major function: support. Many sclerenchyma cells function when dead. There are two types of sclerenchyma cells: elongated **fibers** and variously shaped **sclereids**. Fibers provide relatively rigid support in wood and other parts of the plant, where they are often organized into bundles (Figure 35.9*c*). The bark of trees owes much of its mechanical strength to long fibers. Sclereids may pack together densely, as in a nut's shell or in some seed coats (Figure 35.9*d*). Isolated clumps of sclereids, called *stone cells*, in pears and some other fruits give them their characteristic gritty texture.

Xylem transports water from roots to stems and leaves

The **xylem** of tracheophytes conducts water from roots to aboveground plant parts. It contains conducting cells called **tracheary elements**, which undergo programmed cell death before they assume their function of transporting water and dissolved minerals. There are two types of tracheary elements. The evolutionarily more ancient tracheary elements, found in gymnosperms and other tracheophytes, are **tracheids**—spindle-shaped cells interconnected by numerous pits in their cell walls (Figure 35.9*e*). When the cell contents—nucleus and cytoplasm—disintegrate upon cell death, water can move with little resistance from one tracheid to its neighbors by way of the pits.

Flowering plants evolved a water-conducting system made up of *vessels*. The individual cells that form vessels, called **vessel elements**, must also die and become empty before they can transport water. These cells secrete lignin into their cell walls, then break down their end walls, and finally die and disintegrate. The result is a hollow tube through which water can flow freely. Vessel elements are generally larger in diameter than tracheids. They are laid down end-to-end, so that each vessel is a continuous hollow tube consisting of many vessel elements, providing an open pipeline for water conduction (Figure 35.9*f*). In the course of angiosperm evolution, vessel elements have become shorter, and their end walls have become less and less obliquely oriented and less obstructed, presumably increasing the efficiency of water transport through them (Figure 35.10). The xylem of many angiosperms also includes tracheids.

Phloem translocates carbohydrates and other nutrients

The transport cells of the **phloem**, unlike those of the mature xylem, are living cells. In flowering plants, the characteristic cells of the phloem are **sieve tube elements** (Figure 35.11). Like vessel elements, these cells meet end-to-end. They form long *sieve tubes*, which transport carbohydrates and many other materials from their sources to tissues that consume or store them. In plants with mature leaves, for example, products of photosynthesis move from leaves to root tissues.

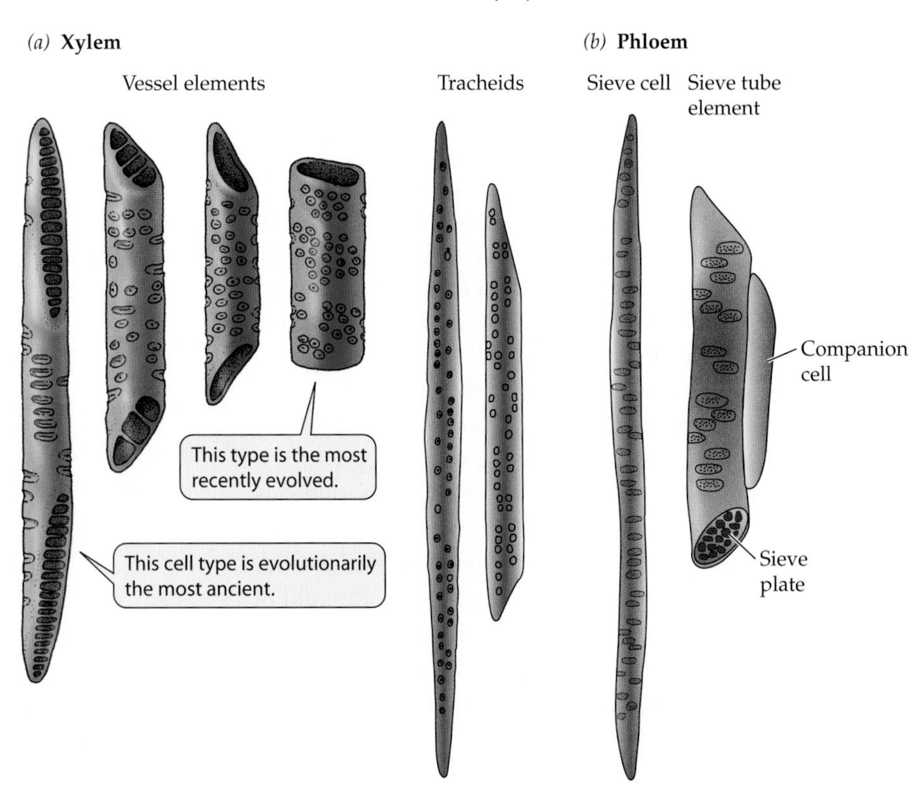

35.10 Evolution of the Conducting Cells of Vascular Systems
The xylem of angiosperms has changed over time. The cells that conduct water and mineral nutrients have become shorter, and the end walls have become more perpendicular to the side walls.

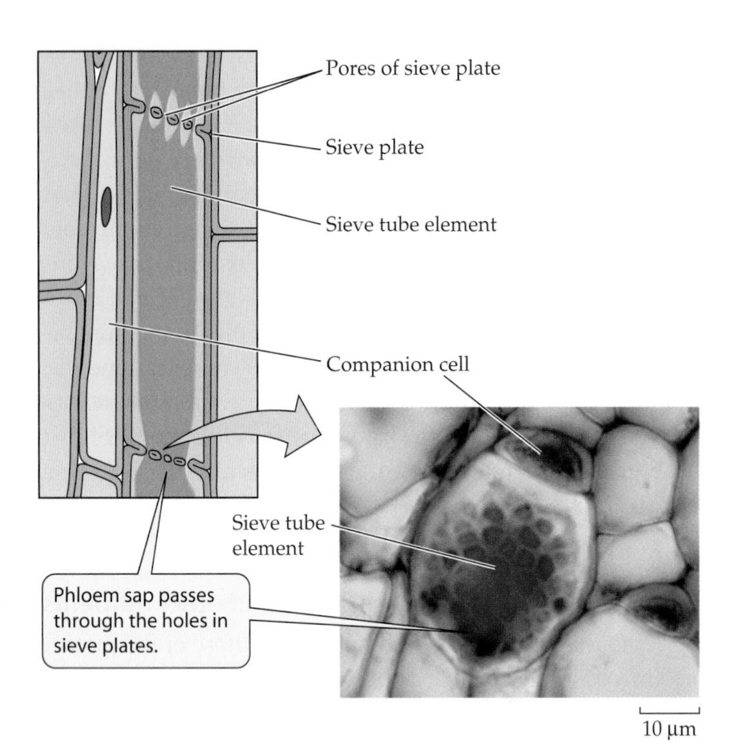

35.11 Sieve Tubes Individual sieve tube elements join together to form long tubes that transport carbohydrates and other nutrient molecules throughout the plant body. Sieve plates form at the ends of each sieve tube element.

As sieve tube elements mature, plasmodesmata in their end walls enlarge to form pores, enhancing the connection between neighboring cells. The result is end walls that look like sieves, called **sieve plates** (see Figure 35.11). As the holes in the sieve plates expand, the membrane that encloses the central vacuole, called the *tonoplast*, disappears. The nucleus and some cytoplasmic components also break down, and thus do not clog the pores of the sieve.

At functional maturity, a sieve tube element is filled with *sieve tube sap*, consisting of water, dissolved sugars, and other solutes. This solution moves from cell to cell along the sieve tube. The moving sap solution is distinct from the layer of cytoplasm at the periphery of a sieve tube element, next to the cell wall. This stationary layer of cytoplasm contains the organelles remaining in the sieve tube element.

Each sieve tube element has one or more **companion cells** (see Figure 35.11), produced as a daughter cell along with the sieve tube element when a parent cell divides. Numerous plasmodesmata link a companion cell with its sieve tube element. Companion cells retain all their organelles and, through the activities of their nuclei, they may be thought of as the "life-support systems" of the sieve tube elements.

All of these types of plant cells play important roles. Next we'll see how they are organized into tissues and tissue systems.

Plant Tissues and Tissue Systems

A *tissue*, as we learned in Chapter 1, is an organized group of cells that have features in common and that work together as a structural and functional unit. Parenchyma cells make up parenchyma tissue, a *simple tissue*—that is, a tissue composed of only one type of cell. Sclerenchyma and collenchyma are other simple tissues, composed, respectively, of sclerenchyma and collenchyma cells.

Different cell types combine to form *complex tissues*. Xylem and phloem are complex tissues, each composed of more than one type of cell. As a result of its cellular complexity, xylem can perform a variety of functions, including transport, support, and storage. The xylem of angiosperms contains vessel elements and tracheids for conduction, thick-walled fibers for support, and parenchyma cells that store nutrients. The phloem of angiosperms includes sieve tube elements, companion cells, fibers, sclereids, and parenchyma cells.

Tissues, in turn, are grouped into *tissue systems* that extend throughout the body of the plant, from organ to organ, in a concentric arrangement. Vascular plants have three tissue systems: vascular, dermal, and ground (Figure 35.12).

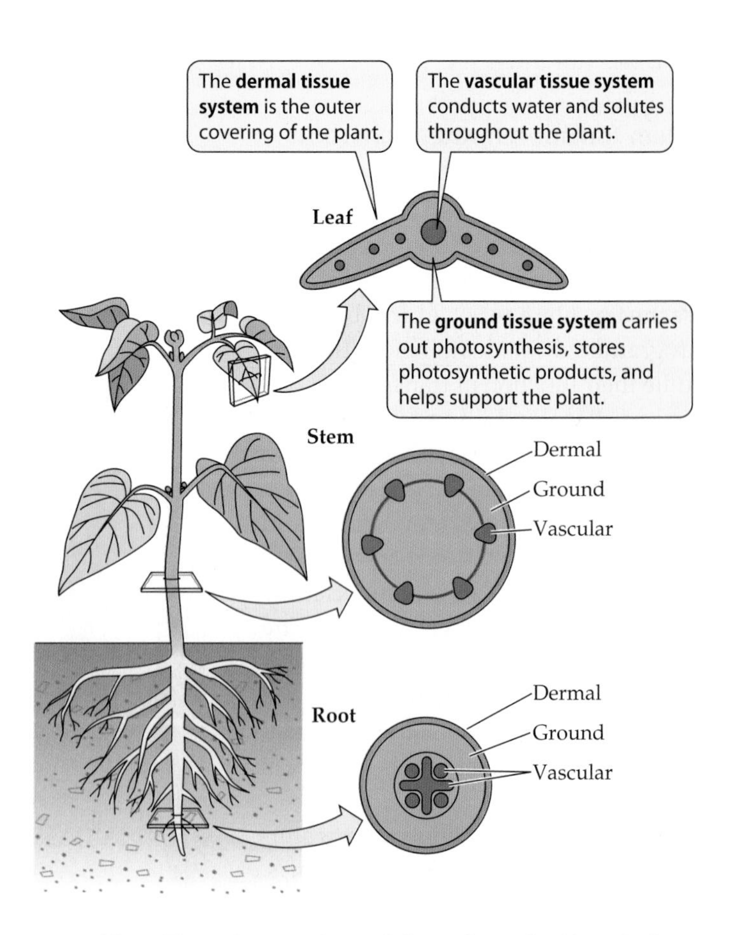

35.12 Three Tissue Systems Extend throughout the Plant Body The arrangement shown here is typical of eudicots.

The **vascular tissue system**, which includes the xylem and phloem, is the plant's plumbing or transport system. All the living cells of the plant body require a source of energy and chemical building blocks. The phloem transports carbohydrates from sites of production (called *sources*, primarily leaves) to sites of utilization or storage (called *sinks*, such as growing tissue, storage tubers, and developing flowers). The xylem distributes water and mineral ions taken up by the roots to all the cells of the stem and leaves.

The **dermal tissue system** is the outer covering of the plant. All parts of the young plant body are covered by an **epidermis**, which may be a single layer of cells or several layers. The epidermis contains *epidermal cells* and may also include specialized cell types, such as the *guard cells* that form stomata (pores) in leaves. The shoot epidermis secretes a layer of wax-covered cutin, the **cuticle**, that helps retard water loss from stems and leaves. The stems and roots of woody plants have a protective covering called the periderm, which will be discussed later in this chapter.

The **ground tissue system** makes up the rest of the plant. It consists primarily of parenchyma tissue, often supplemented by collenchyma or sclerenchyma. Ground tissue functions primarily in storage, support, photosynthesis, and the production of defensive and attractive substances.

In the discussions that follow, we'll examine how the tissue systems are organized in the different organs of a flowering plant. Let's begin by seeing how this organization develops as the plant grows.

Forming the Plant Body

In its early embryonic stages, a plant establishes the basic body plan for its mature form. Two patterns contribute to the plant body plan:

- ▸ The arrangement of cells and tissues along the main axis from root to shoot
- ▸ The concentric arrangement of the tissue systems

Both patterns arise through orderly development and are best understood in developmental terms.

Plants and animals grow differently

As the plant body grows, it may lose parts, and it forms new parts that may grow at different rates. The growing stem consists of **modules** or **units**, laid down one after another. Each module consists of a node with its attached leaf or leaves, the internode below that node, and the lateral bud or buds at the base of that internode (see Figure 35.2). New modules are formed as long as the stem continues to grow.

Each branch of a plant may be thought of as a unit that is in some ways independent of the other branches. A branch of a plant does not bear the same relationship to the remainder of the plant body as a limb does to the remainder of an animal body. Among other things, branches form one after another (unlike limbs, which form simultaneously during embryonic development). Also, branches often differ from one another in their number of leaves and in the degree to which they themselves branch.

Leaves are units of another sort. They are usually short-lived, lasting weeks to a few years. Branches and stems are longer-lived, lasting from years to centuries.

Root systems are also branching structures, and lateral roots are semi-independent units. As the root system grows, penetrating and exploring the soil environment, many roots die and are replaced by new ones.

All parts of the animal body grow as an individual develops from embryo to adult, but in most animals, this growth is *determinate*. That is, the growth of the individual and all its parts ceases when the adult state is reached. Determinate growth is also characteristic of some plant parts, such as leaves, flowers, and fruits. The growth of stems and roots, by contrast, is *indeterminate*, and it is generated from specific regions of active cell division and cell expansion.

The localized regions of cell division in plants are called **meristems**. Meristems are forever young, retaining the ability to produce new cells indefinitely. The cells that perpetuate the meristems, called **initials**, are comparable to the stem cells found in animals (discussed in Chapter 19). When an initial divides, one daughter cell develops into another meristem cell the size of its parent, while the other daughter cell differentiates into a more specialized cell.

A hierarchy of meristems generates a plant's body

There are two types of meristems:

- ▸ **Apical meristems** give rise to the *primary plant body*, which is the entire body of many plants.
- ▸ **Lateral meristems** give rise to the *secondary plant body*. The stems and roots of some plants (most obviously trees) form wood and become thick; it is the lateral meristems that give rise to the tissues responsible for this thickening.

APICAL MERISTEMS. Apical meristems are located at the tips of roots and stems and in buds. They extend the plant body by producing the cells that subsequently expand and differentiate to form all plant organs (Figure 35.13).

- ▸ *Shoot apical meristems* supply the cells that extend stems and branches, allowing more leaves to form and photosynthesize.
- ▸ *Root apical meristems* supply the cells that extend roots, enabling the plant to "forage" for water and minerals.

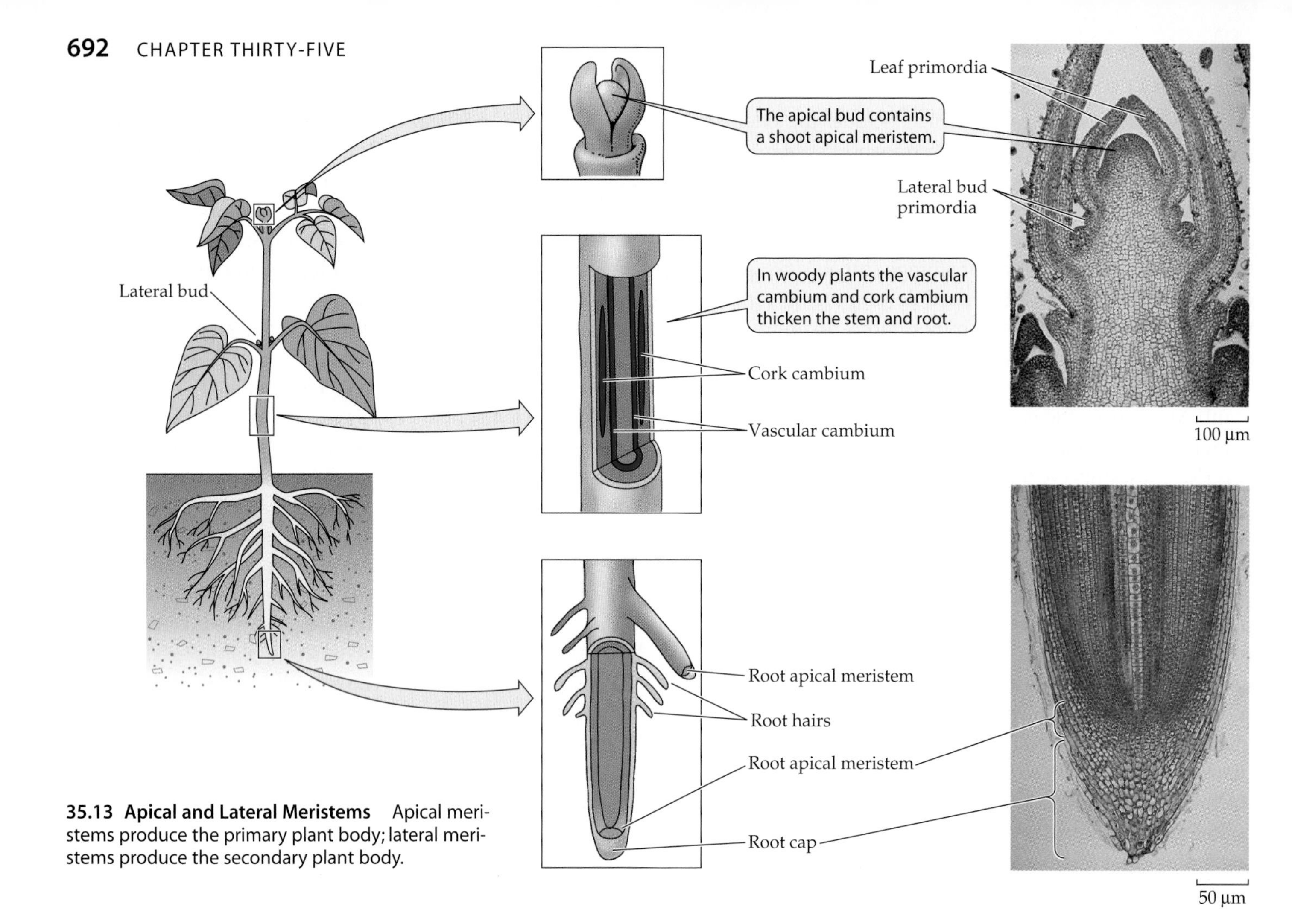

35.13 Apical and Lateral Meristems Apical meristems produce the primary plant body; lateral meristems produce the secondary plant body.

Both root and shoot apical meristems give rise to a set of cylindrical *primary meristems* that produce the primary tissues of the plant body. From the outside to the inside of the root or shoot, which are both cylindrical organs, the primary meristems are the **protoderm**, the **ground meristem**, and the **procambium**. These in turn give rise to the three tissue systems:

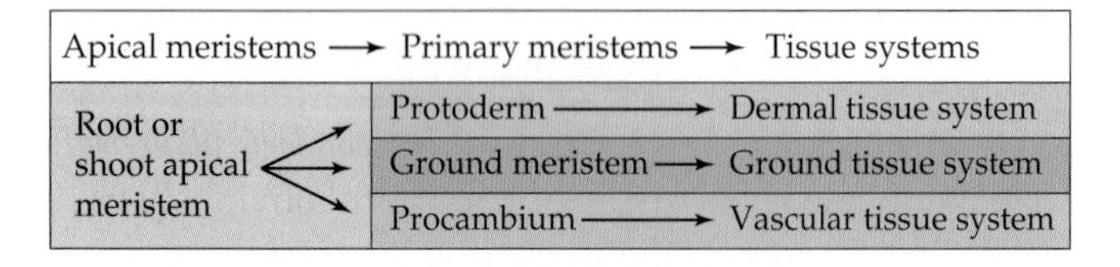

Apical meristems →	Primary meristems →	Tissue systems
Root or shoot apical meristem	Protoderm →	Dermal tissue system
	Ground meristem →	Ground tissue system
	Procambium →	Vascular tissue system

Apical meristems are responsible for **primary growth**, which leads to lengthening of the plant body and organ formation. All plant organs arise ultimately from cell divisions in the apical meristems, followed by cell expansion and differentiation. Primary growth gives rise to the entire body of many plants.

Because meristems can continue to produce new organs throughout the lifetime of the plant, the plant body is much more variable in form than the animal body, whose organs are produced only once.

LATERAL MERISTEMS. Some roots and stems develop a secondary body, the tissues of which we commonly refer to as wood and bark. These complex tissues are derived from two lateral meristems: the vascular cambium and the cork cambium (Figure 35.14).

The **vascular cambium** is a cylindrical tissue consisting predominantly of vertically elongated cells that divide frequently. Toward the inside of the stem or root, the dividing cells form new xylem, the *secondary xylem*, and toward the outside they form new phloem, the *secondary phloem*.

As a tree trunk grows in diameter, the outermost layers of the stem crack and fall off. Without the activity of the **cork cambium**, this sloughing off of tissues, including the epidermis, would expose the tree to potential damage, including excessive water loss or invasion by microorganisms. The cork cambium produces new protective cells, primarily in the outward direction. The walls of these cork cells become impregnated with suberin. The mass of waterproofed cells produced by the cork cambium is called the **periderm**.

Growth in the diameter of stems and roots, produced by the vascular and cork cambia, is called **secondary growth**. It is the source of wood and bark. **Wood** is secondary xylem. **Bark** is everything external to the vascular cambium (periderm plus secondary phloem).

Each year, deciduous trees lose their leaves, leaving bare branches and twigs in winter. These twigs illustrate both pri-

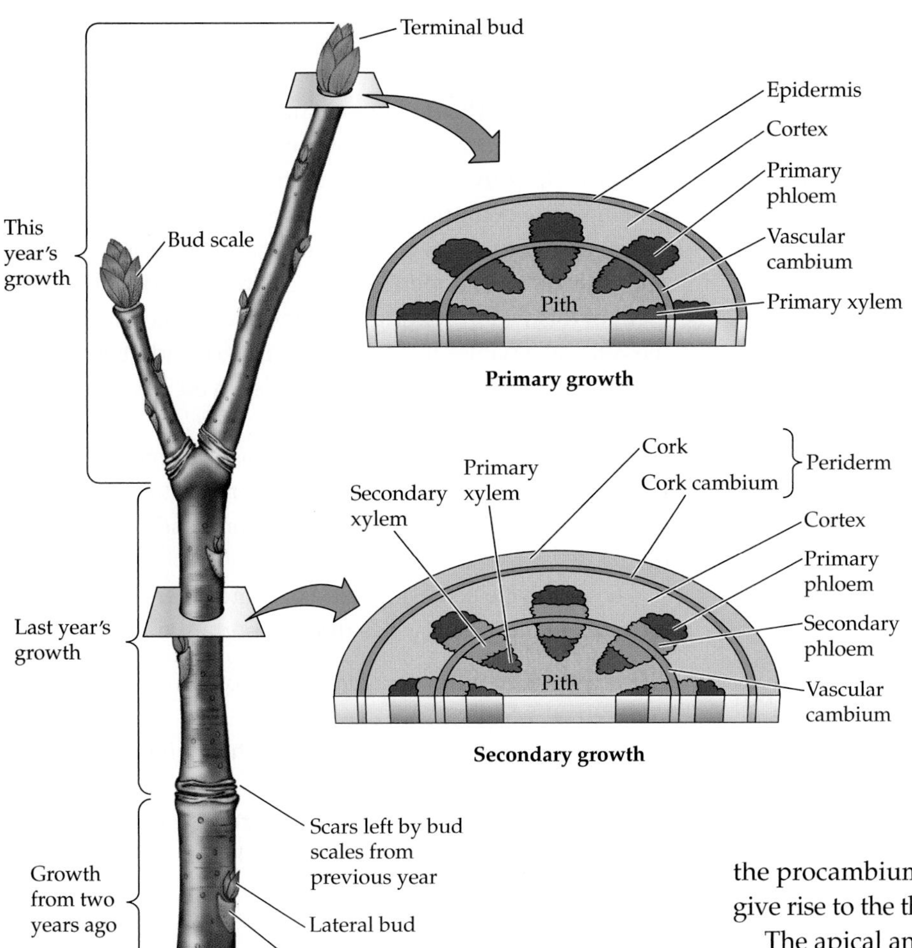

35.14 A Woody Twig Apical meristems produce primary growth. Lateral meristems produce secondary growth.

mary and secondary growth (Figure 35.14). The apical meristems of the twigs and their branches are enclosed in buds protected by bud scales. When the buds begin to grow in the spring, the scales fall away, leaving scars that show us where the bud was and identifying each year's growth. The dormant twig shown in Figure 35.14 is the product of primary and secondary growth. Only the buds consist entirely of primary tissues.

In some plants, meristems may remain active for years—even centuries. The bristlecone pine mentioned at the beginning of this chapter provides a dramatic example. Such plants grow in size, or at least in diameter, throughout their lives. In the sections that follow, we'll examine how the various meristems give rise to the plant body.

The root apical meristem gives rise to the root cap and the primary meristems

The root apical meristem produces all the cells that contribute to growth in the length of the root. Some of the daughter cells from the apical (tip) end of the root apical meristem contribute to a **root cap**, which protects the delicate growing region of the root as it pushes through the soil. The cells of the root cap are often damaged or scraped away and must therefore be replaced constantly. The root cap is also the structure that detects the pull of gravity and thus controls the downward growth of roots.

Part of the root apical meristem nearest the tip of the root forms a **quiescent center**, in which cell divisions are rare. The quiescent center can become more active when needed—following injury, for example.

The daughter cells that are produced at the basal end of the apical meristem (away from the root cap) elongate and lengthen the root. Following elongation, these cells differentiate, giving rise to the various tissues of the mature root. The growing region above the apical meristem comprises the three cylindrical primary meristems: the protoderm, the ground meristem, and the procambium (Figure 35.15). These primary meristems give rise to the three tissue systems of the root.

The apical and primary meristems constitute the *zone of cell division*, the source of all the cells of the root's primary tissues. Just above this zone is the *zone of cell elongation*, where the newly formed cells are elongating and thus causing the root to reach farther into the soil. Above this is the *zone of mat-*

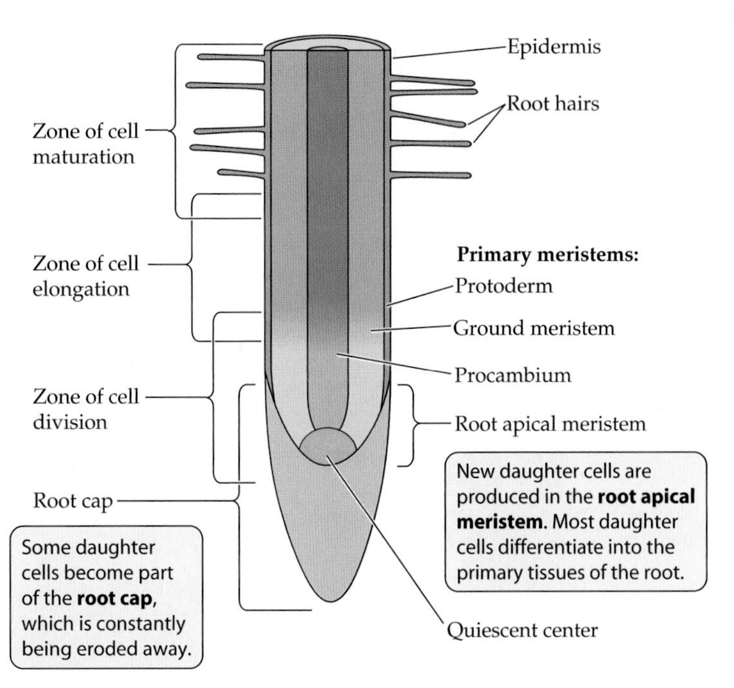

35.15 Tissues and Regions of the Root Tip Extensive cell division creates the complex structure of the root.

uration, where the cells are differentiating, taking on specialized forms and functions such as water transport or mineral uptake. These three zones grade imperceptibly into one another; there is no abrupt line of demarcation.

The products of the root's primary meristems become root tissues

What are the products of the three primary meristems? The protoderm gives rise to the outer layer of cells—the epidermis—which is adapted for protection of the root and for the absorption of mineral ions and water (Figure 35.16). In the zone of maturation, many of the epidermal cells produce amazingly long, delicate **root hairs**, which vastly increase the surface area of the root (Figure 35.16*b*). It has been estimated that the root system of a mature rye plant has a total absorptive surface of more than 600 square meters (almost half again the area of a basketball court). Root hairs grow out among the soil particles, probing nooks and crannies and taking up water and minerals.

Internal to the epidermis, the ground meristem gives rise to a region of ground tissue that is many cells thick, called the **cortex**. The cells of the cortex are relatively unspecialized and often function in nutrient storage.

In the great majority of plants, especially in trees, a fungus is closely associated with the root tips. This association, called a *mycorrhiza*, increases the plant's absorption of minerals and water (see Figure 31.16). Such roots have poorly developed or no root hairs. These plants cannot survive without the mycorrhizae that help them absorb minerals.

Proceeding inward, we come to the **endodermis** of the root, a single cylindrical layer of cells that is the innermost cell layer of the cortex. Unlike those of other cortical cells, the cell walls of the endodermal cells contain suberin. The placement of this waterproofing substance in only certain parts of the cell wall enables the cylindrical ring of endodermal cells to control the access of water and dissolved ions to the vascular tissues.

Moving inward past the endodermis, we enter the vascular cylinder, or **stele**, produced by the procambium. The stele consists of three tissues: pericycle, xylem, and phloem (Figure 35.17).

The **pericycle** consists of one or more layers of relatively undifferentiated cells. It has three important functions:

- It is the tissue within which lateral roots arise (see Figure 35.16*a*).
- It can contribute to secondary growth by giving rise to lateral meristems that thicken the root.
- Its cells contain membrane transport proteins that export nutrient ions into the cells of the xylem.

35.16 Root Anatomy The drawing at the left shows a generalized root structure. (*a*) Cross section through the tip of a lateral root. Cells in the pericycle divide and the products differentiate, forming the tissues of a lateral root. (*b*) Root hairs, seen with a scanning electron microscope. (*c*, *d*) Cross sections showing the primary root tissues of (*c*) a eudicot and (*d*) a monocot. The monocot has a central pith region; the eudicot does not.

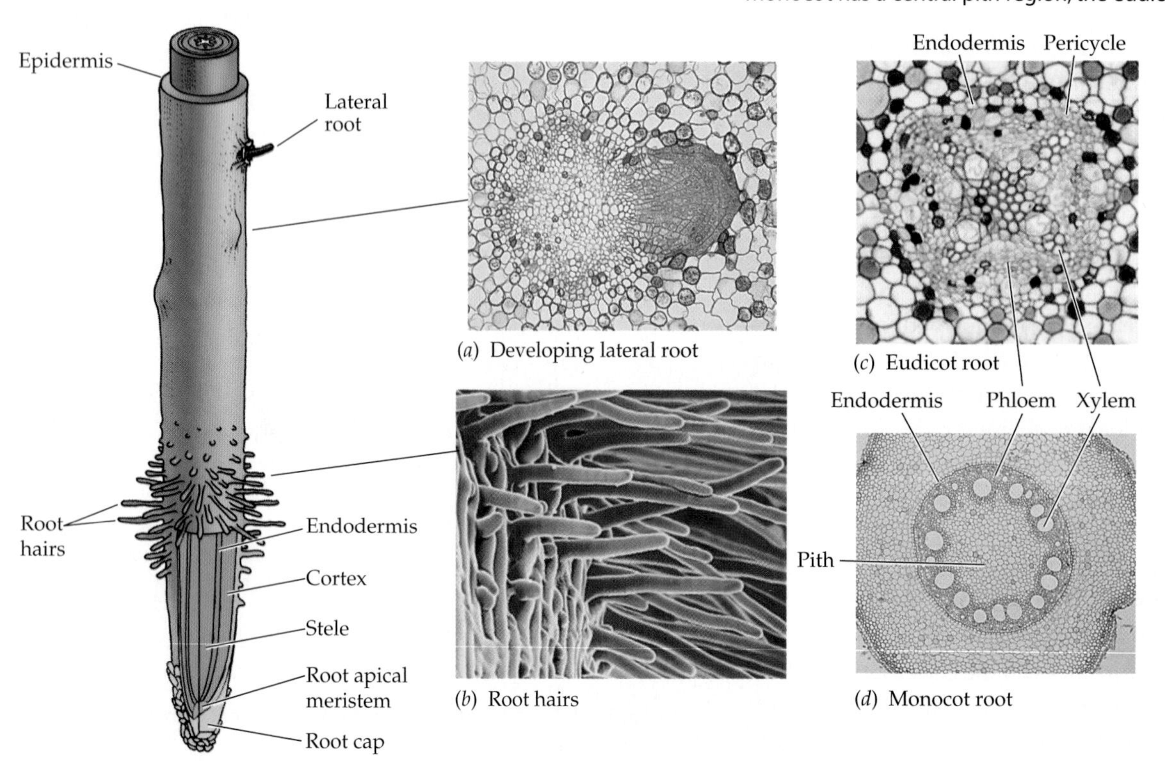

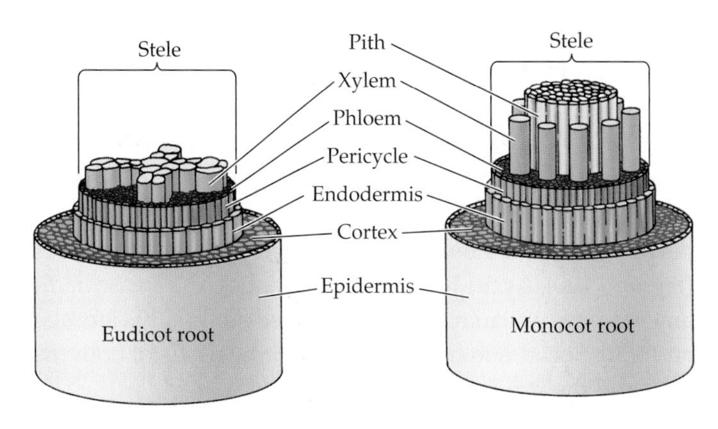

35.17 The Stele The arrangement of tissues in the stele—the region internal to the endodermis—differs in the roots of eudicots and monocots.

At the very center of the root of a eudicot lies the xylem—seen in cross section in the shape of a star with a variable number of points. Between the points are bundles of phloem. In monocots, a region of parenchyma cells, called the **pith**, lies in the center of the root. The pith often stores carbohydrate reserves.

The products of the stem's primary meristems become stem tissues

The shoot apical meristem, like the root apical meristem, forms three primary meristems: the protoderm, the ground meristem, and the procambium. These primary meristems, in turn, give rise to the three tissue systems. The shoot apical meristem also repetitively lays down the beginnings of leaves and lateral buds. Leaves arise from bulges called **leaf primordia**, which form as cells divide on the sides of shoot apical meristems (see Figure 35.13). **Bud primordia** form at the bases of the leaf primordia. The growing stem has no protective structure analogous to the root cap, but the leaf primordia can act as a protective covering for the shoot apical meristem.

The plumbing of angiosperm stems differs from that of roots. In a root, the vascular tissue lies deep in the interior, with the xylem at or near the very center (see Figure 35.16). The vascular tissue of a young stem, however, is divided into discrete **vascular bundles**. Each vascular bundle contains both xylem and phloem. In eudicots, the vascular bundles generally form a cylinder, but in monocots, they are seemingly scattered throughout the stem (Figure 35.18).

In addition to the vascular tissues, the stem contains other important storage and supportive tissues. Internal to the ring of vascular bundles in eudicots is a storage tissue, the pith, and to the outside lies a similar storage tissue, the cortex. The cortex may contain supportive collenchyma cells with thickened walls. The pith, the cortex, and the regions between the vascular bundles in eudicots—called *pith rays*—constitute the ground tissue system of the stem. The outermost cell layer of the young stem is the epidermis, the primary function of which is to minimize the loss of water from the tissues within.

35.18 Vascular Bundles in Stems (*a*) In eudicot stems, the vascular bundles are arranged in a cylinder, with the pith in the center and the cortex outside the cylinder. (*b*) A scattered arrangement of vascular bundles is typical of monocot stems.

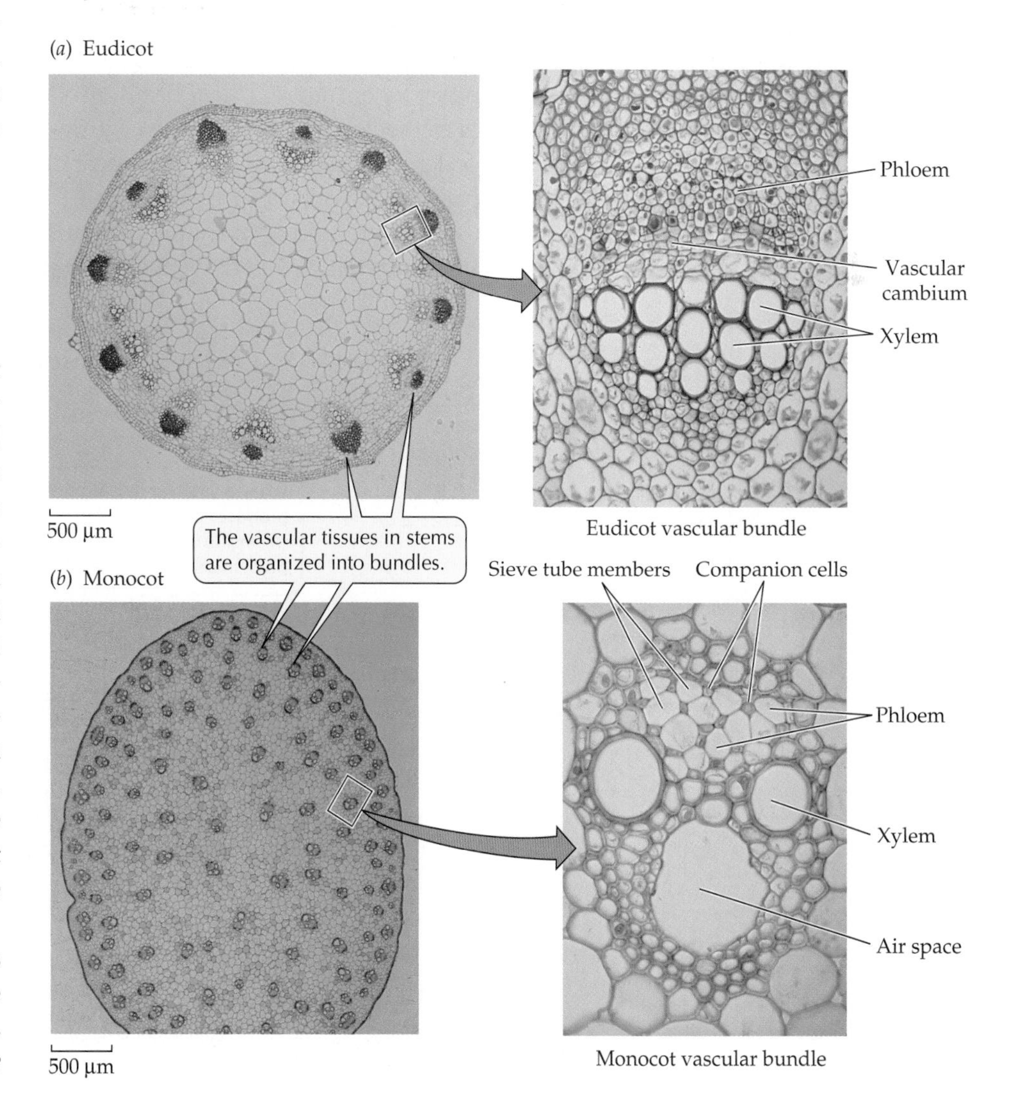

Many stems and roots undergo secondary growth

Some stems and roots remain slender and show little or no secondary growth. However, in many eudicots, secondary growth thickens stems and roots considerably. This process gives rise to wood and bark, and it makes the support of tall trees possible.

Secondary growth results from the activity of the two lateral meristems: vascular cambium and cork cambium (see Figure 35.13). Vascular cambium consists of cells that divide to produce secondary xylem and phloem cells, while cork cambium produces mainly waxy-walled *cork cells*.

Initially, the vascular cambium is a single layer of cells lying between the primary xylem and the primary phloem (see Figure 35.18*a*). The root or stem increases in diameter when the cells of the vascular cambium divide, producing secondary xylem cells toward the inside of the root or stem and producing secondary phloem cells toward the outside (Figure 35.19). In the stems of woody plants, cells in the pith rays between the vascular bundles also divide, forming a continuous cylinder of vascular cambium running the length of the stem. This cylinder, in turn, gives rise to complete cylinders of secondary xylem (wood) and secondary phloem, which contributes to the bark.

As the vascular cambium produces secondary xylem and phloem, its principal cell products are vessel elements, supportive fibers, and parenchyma cells in the xylem and sieve tube elements, companion cells, fibers, and parenchyma cells in the phloem. The parenchyma cells in the xylem and phloem store carbohydrate reserves in the stem and root.

Living tissues such as this storage parenchyma must be connected to the sieve tubes of the phloem, or they will starve to death. These connections are provided by **vascular rays**, which are composed of cells derived from the vascular cambium. These rays, laid down progressively as the cambium divides, are rows of living parenchyma cells that run perpendicular to the xylem vessels and phloem sieve tubes (Figure 35.20). As the root or stem continues to increase in diameter, new vascular rays are initiated so that this storage and transport tissue continues to meet the needs of both the bark and the living cells in the xylem.

The vascular cambium itself increases in circumference with the growth of the root or stem. To do this, some of its cells divide in a plane at right angles to the plane that gives rise to secondary xylem and phloem. The products of each of these divisions lie within the vascular cambium itself and increase its circumference.

Only eudicots and other non-monocot angiosperms have a vascular cambium and a cork cambium and thus undergo secondary growth. The few monocots that form thickened stems—palm trees, for example—do so without using vascular cambium or cork cambium. Palm trees have a very wide apical meristem that produces a wide stem, and dead leaf bases also add to the diameter of the stem. Basically, monocots grow in the same way as do other angiosperms that lack secondary growth.

Wood and bark, consisting of secondary phloem, are unique to plants showing secondary growth. These tissues have their own patterns of organization and development.

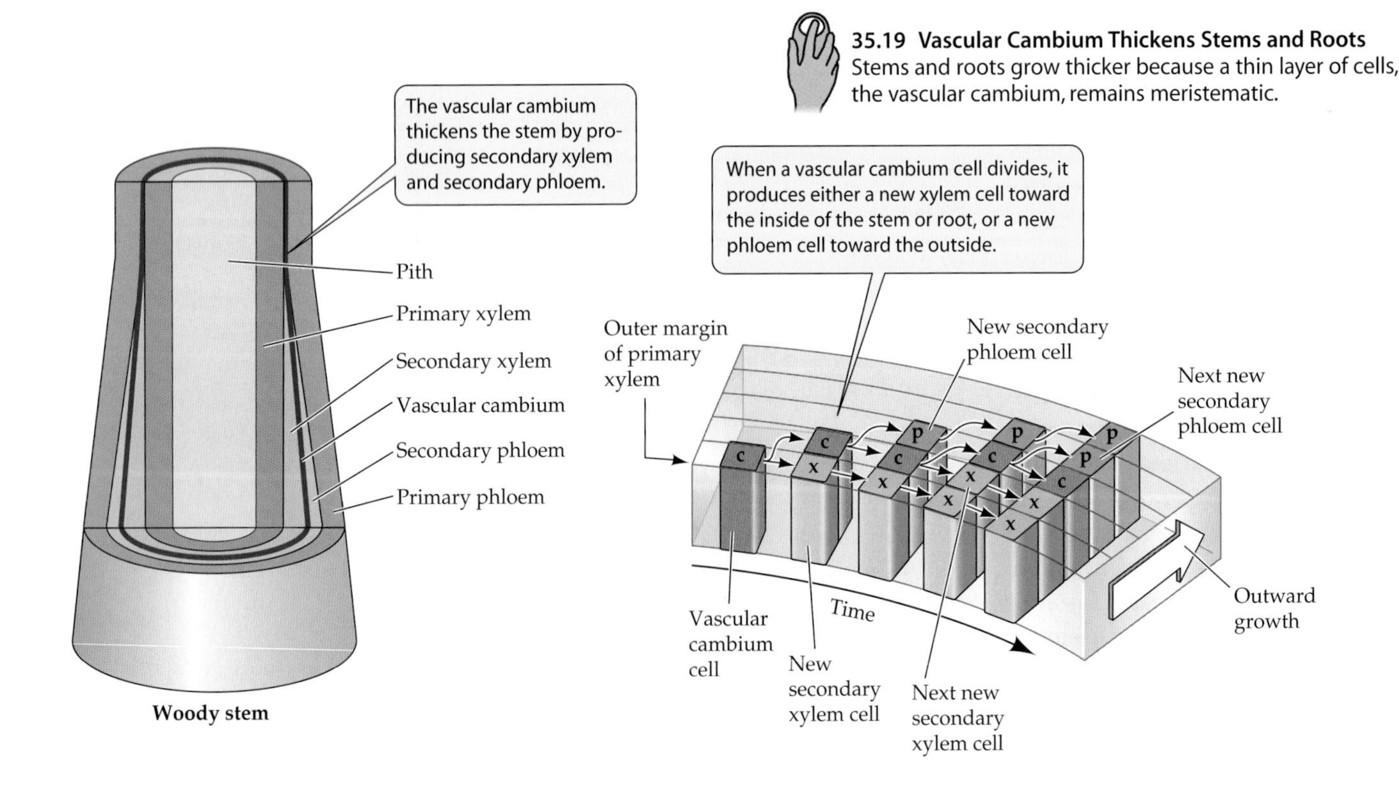

35.19 Vascular Cambium Thickens Stems and Roots Stems and roots grow thicker because a thin layer of cells, the vascular cambium, remains meristematic.

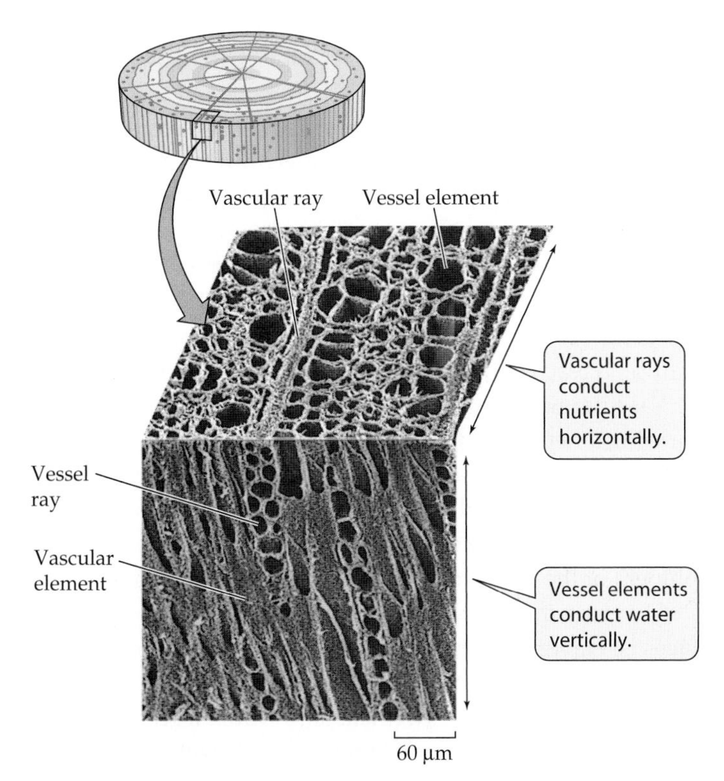

35.20 Vascular Rays and Vessel Elements In this sample of wood from the tulip poplar, the orientation of vascular rays is perpendicular to that of the vessel elements. The vascular rays transport sieve tube sap horizontally from the phloem to storage parenchyma cells.

WOOD. Cross sections of most tree trunks (mature stems) in temperate-zone forests show *annual rings* (Figure 35.21), which result from seasonal environmental conditions. In spring, when water is relatively plentiful, the tracheids or vessel elements produced by the vascular cambium tend to be large in diameter and thin-walled. Such wood is well adapted for transporting water and minerals. As water becomes less available during the summer, narrower cells with thicker walls are produced, making this summer wood darker and perhaps more dense than the wood formed in spring. Thus each growing season is usually recorded in a tree trunk by a clearly visible annual ring. Trees in the moist Tropics do not undergo seasonal growth, so they do not lay down such obvious regular rings. Variations in temperature or water supply can lead to the formation of more than one "annual" ring in a single year.

The difference between old and new regions of wood also contributes to its appearance. As a tree grows in diameter, the xylem toward the center becomes clogged with water-insoluble substances and ceases to conduct water and minerals; this *heartwood* appears darker in color. The portion of the xylem that is actively conducting water and minerals throughout the tree is called *sapwood* and is lighter in color and more porous than heartwood.

The knots that we find attractive in knotty pine but regard as a defect in structural timbers are cross sections of branches. As a trunk grows, the bases of branches become buried in the trunk's new wood and appear as knots when the trunk is cut lengthwise.

BARK. As secondary growth of stems or roots continues, the expanding vascular tissue stretches and breaks the epidermis and cortex, which ultimately flake away. Tissue derived from the secondary phloem then becomes the outermost part of the stem. Before the dermal tissues are broken away, cells lying near the surface of the secondary phloem begin to divide and produce layers of **cork**, a tissue composed of cells with thick walls, waterproofed with suberin. The cork soon becomes the outermost tissue of the stem or root (see Figure 35.14). The dividing cells, derived from the secondary phloem, form a cork cambium. Sometimes the cork cambium produces cells to the inside as well as to the outside; these cells constitute what is known as the *phelloderm*.

Cork, cork cambium, and phelloderm make up the periderm of the secondary plant body. As the vascular cambium continues to produce secondary vascular tissue, the corky layers are in turn lost, but the continuous formation of new cork cambia in the underlying phloem gives rise to new corky layers.

When periderm forms on stems and roots, the underlying tissues still need to release carbon dioxide and take up oxygen. **Lenticels** are spongy regions in the periderm of stems and roots that allow such gas exchange (Figure 35.22).

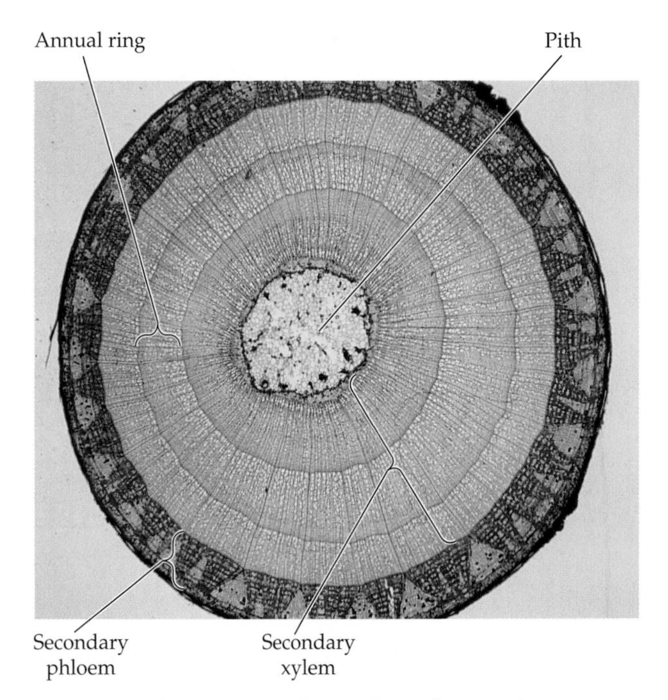

35.21 Annual Rings Rings of secondary xylem are the most noticeable feature of this cross section from a 3-year-old basswood stem.

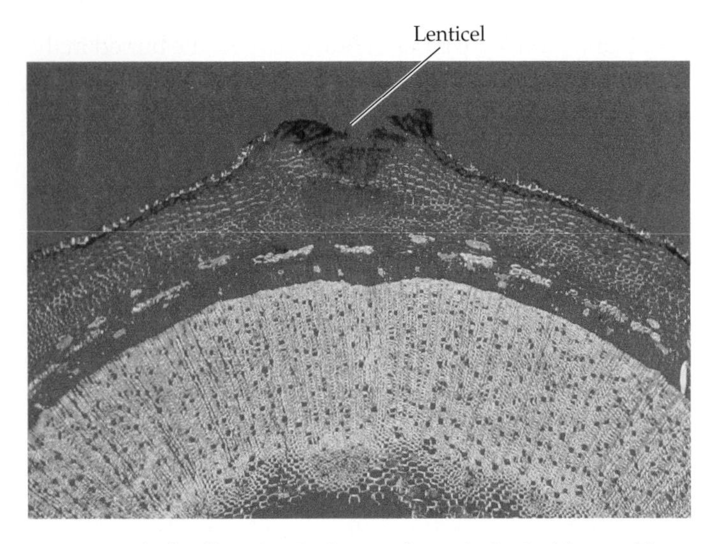

35.22 Lenticels Allow Gas Exchange through the Periderm The region of periderm that appears broken open is a lenticel in a year-old elder twig; note the spongy tissue that constitutes the lenticel.

Leaf Anatomy Supports Photosynthesis

We can think of roots and stems as important supporting actors that sustain the activities of the real stars of the plant body, the leaves—the organs of photosynthesis. Leaf anatomy is beautifully adapted to carry out photosynthesis and to support it by exchanging the gases O_2 and CO_2 with the environment, limiting evaporative water loss, and exporting the products of photosynthesis to the rest of the plant. Figure 35.23*a* shows a typical eudicot leaf in three dimensions.

Most eudicot leaves have two zones of photosynthetic parenchyma tissue referred to as **mesophyll**, which means "middle of the leaf." The upper layer or layers of mesophyll consist of elongated cells; this zone is referred to as *palisade mesophyll*. The lower layer or layers consist of irregularly shaped cells; this zone is called *spongy mesophyll*. Within the mesophyll is a great deal of air space through which carbon dioxide can diffuse to reach and be absorbed by photosynthesizing cells.

Vascular tissue branches extensively throughout the leaf, forming a network of **veins** (Figure 35.23*b*). Veins extend to within a few cell diameters of all the cells of the leaf, ensuring that the mesophyll cells are well supplied with water and minerals. The products of photosynthesis are loaded into the phloem of the veins for export to the rest of the plant.

Covering the entire leaf on both its upper and lower surfaces is a layer of nonphotosynthetic cells, which constitute the epidermis. The epidermal cells have an overlying waxy cuticle that is highly impermeable to water. But this impermeability poses a problem: While keeping water in the leaf, the epidermis also keeps carbon dioxide—the other raw material of photosynthesis—out.

The problem of balancing water retention and carbon dioxide availability is solved by an elegant regulatory system that will be discussed in more detail in the next chapter. *Guard cells* are modified epidermal cells that change their shape, thereby opening or closing pores called *stomata*, which serve as passageways between the environment and the leaf's interior (Figure 35.23*c*). When the stomata are open, carbon dioxide can enter and oxygen can leave, but water vapor can also be lost.

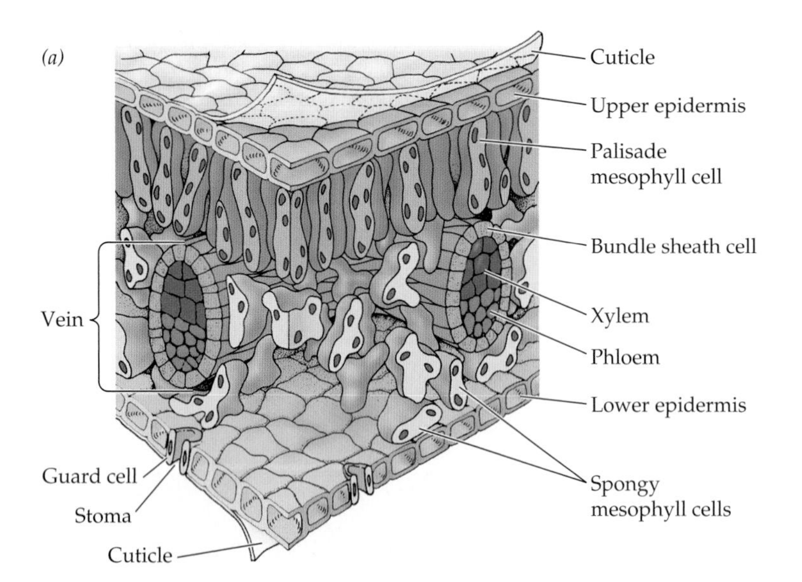

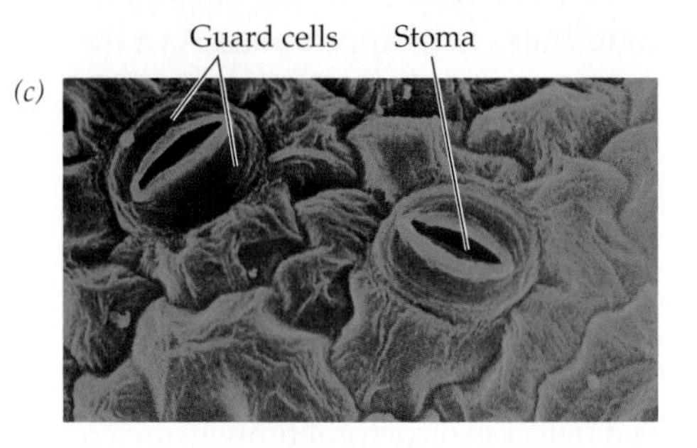

35.23 The Eudicot Leaf (*a*) This three-dimensional diagram shows a eudicot leaf. (*b*) The network of fine veins in this maple leaf carries water to the mesophyll cells and carries photosynthetic products away from them. (*c*) These paired cells on the lower epidermis of a eudicot leaf are guard cells; the gaps between them are stomata, through which carbon dioxide enters the leaf.

In Chapter 8 we described C_4 plants, which can fix carbon dioxide efficiently even when the carbon dioxide supply in the leaf decreases to a level at which the photosynthesis of C_3 plants is inefficient. One adaptation that helps C_4 plants do this is their modified leaf anatomy (see Figure 8.16). The photosynthetic cells in the C_4 leaf are grouped around the veins in concentric layers, forming an outer mesophyll layer and an inner *bundle sheath*. These layers each contain different types of chloroplasts, leading to the biochemical division of labor illustrated in Figure 8.17.

Leaves receive water and mineral nutrients from the roots by way of the stems. In return, the leaves export products of photosynthesis, providing a supply of chemical energy to the rest of the plant body. And, as we have just seen, leaves exchange gases, including water vapor, with the environment by way of the stomata. All three of these processes will be considered in detail in the next chapter.

Chapter Summary

Vegetative Organs of the Flowering Plant Body

▶ Monocots typically have a single cotyledon, narrow leaves with parallel veins, flower parts in threes or multiples of three, and stems with scattered vascular bundles. **Review Figure 35.1**

▶ Eudicots typically have two cotyledons, broad leaves with netlike veins, flower parts in fours or fives, and vascular bundles in a ring. **Review Figure 35.1**

▶ Flowering plants that are neither monocots nor eudicots are generally similar in structure to eudicots.

▶ The vegetative organs of flowering plants are roots, which form a root system, and stems and leaves, which form a shoot system. **Review Figure 35.2**

▶ Roots anchor the plant and take up water and minerals.

▶ Stems bear leaves and buds. Lateral buds form branches. Apical buds produce cells that contribute to the elongation of the stem.

▶ Leaves are responsible for most photosynthesis, for which their flat blades, held perpendicular to the sun's rays, are well adapted. **Review Figure 35.5**

Plant Cells

▶ The walls of plant cells have a structure that often corresponds to the special functions of the cell.

▶ The walls of individual cells are separated by a middle lamella common to two neighboring cells; each cell also has its own primary wall. **Review Figure 35.6**

▶ Some cells produce a thick secondary wall. Adjacent cells are connected by plasmodesmata. **Review Figures 35.7, 35.8**

▶ Parenchyma cells have thin walls. Many parenchyma cells store starch or lipids; some others carry out photosynthesis. **Review Figure 35.9*a***

▶ Collenchyma cells provide flexible support. **Review Figure 35.9*b***

▶ Sclerenchyma cells provide strength and often function when dead. **Review Figure 35.9*c, d***

▶ Tracheids and vessel elements are xylem cells that conduct water and minerals after the cells die. **Review Figures 35.9*e, f*, 35.10**

▶ Sieve tube elements are the conducting cells of the phloem. Their activities are often controlled by companion cells. **Review Figure 35.11**

Plant Tissues and Tissue Systems

▶ Three tissue systems extend throughout the plant body.

▶ The vascular tissue system, consisting of xylem and phloem, conducts water, minerals, and the products of photosynthesis throughout the plant body.

▶ The dermal tissue system protects the body surface.

▶ The ground tissue system produces and stores nutrient materials and performs other functions. **Review Figure 35.12**

Forming the Plant Body

▶ The pattern of cells and tissues along the long axis and the concentric arrangement of the tissue systems are parts of the plant body plan; they arise through orderly development.

▶ The plant body consists of semi-independent modules or units. The growth of stems and roots is indeterminate. Leaves, flowers, and fruits show determinate growth.

▶ Meristems are localized regions of cell division. A hierarchy of meristems generates the plant body.

▶ Apical meristems at the tips of stems and roots produce the primary tissues of those organs. **Review Figure 35.13**

▶ Shoot apical meristems and root apical meristems give rise to primary meristems: the protoderm, the ground meristem, and the procambium. The protoderm produces the dermal tissue system, the ground meristem produces the ground tissue system, and the procambium produces the vascular tissue system.

▶ In some plants, the products of primary growth constitute the entire plant body. Many other plants show secondary growth. Two lateral meristems, the vascular cambium and cork cambium, are responsible for secondary growth. **Review Figure 35.13**

▶ The structure of a winter woody twig reflects both primary and secondary growth. **Review Figure 35.14**

▶ The young root has an apical meristem that gives rise to the root cap and to the three primary meristems, which in turn produce the three tissue systems. Root tips have three overlapping zones: the zone of cell division, the zone of cell elongation, and the zone of maturation. **Review Figure 35.15**

▶ The protoderm gives rise to the epidermis, part of which forms the root hairs that are responsible for absorbing water and minerals. **Review Figure 35.16.** See Web/CD Activities 35.1 and 35.2

▶ The ground tissue system of a young root is the cortex, whose innermost cell layer, the endodermis, controls access to the stele.

▶ The stele, consisting of the pericycle, xylem, and phloem, is the root's vascular tissue system. Lateral roots arise in the pericycle. **Review Figure 35.17.**

▶ The shoot apical meristem also gives rise to three primary meristems, with roles similar to their counterparts in the root. Leaf primordia on the sides of the apical meristem develop into leaves.

▶ The vascular tissue in young stems is divided into vascular bundles, each containing both xylem and phloem. Pith occupies the center of the eudicot stem, and cortex lies outside the ring of vascular bundles, with pith rays lying between the vascular bundles. **Review Figure 35.18.** See Web/CD Activities 35.3 and 35.4

▶ Many eudicot stems and roots show secondary growth in which vascular cambia and cork cambia give rise, respectively, to secondary xylem (wood) and secondary phloem and to cork. **Review Figure 35.19.** See Web/CD Tutorial 35.1

▶ The vascular cambium lays down layers of secondary xylem and phloem. Living cells within these tissues are nourished by vascular rays. **Review Figure 35.20**

▶ The periderm consists of cork, cork cambium, and phelloderm, all pierced at intervals by lenticels that allow gas exchange.

Leaf Anatomy Supports Photosynthesis

▶ The photosynthetic tissue of a leaf is called mesophyll. Veins bring water and minerals to the mesophyll and carry the products of photosynthesis to other parts of the plant body.

▶ A waxy cuticle retards water loss from the leaf and is impermeable to carbon dioxide. Guard cells control the opening of stomata, openings in the leaf that allow CO_2 to enter, but also allow some water to escape. **Review Figure 35.23. See Web/CD Activity 35.5**

Self-Quiz

1. Which of the following is *not* a difference between monocots and eudicots?
   a. Eudicots more frequently have broad leaves.
   b. Monocots commonly have flower parts in multiples of three.
   c. Monocot stems do not generally undergo secondary thickening.
   d. The vascular bundles of monocots are commonly arranged as a cylinder.
   e. Eudicot embryos commonly have two cotyledons.
2. Roots
   a. always form a fibrous root system that holds the soil.
   b. possess a root cap at their tip.
   c. form branches from lateral buds.
   d. are commonly photosynthetic.
   e. do not show secondary growth.
3. The plant cell wall
   a. lies immediately inside the plasma membrane.
   b. is an impermeable barrier between cells.
   c. is always waterproofed with either lignin or suberin.
   d. always consists of a primary wall and a secondary wall, separated by a middle lamella.
   e. contains cellulose and other polysaccharides.
4. Which statement about parenchyma cells is *not* true?
   a. They are alive when they perform their functions.
   b. They typically lack a secondary wall.
   c. They often function as storage depots.
   d. They are the most numerous cells in the primary plant body.
   e. They are found only in stems and roots.
5. Tracheids and vessel elements
   a. die before they become functional.
   b. are important constituents of all plants.
   c. have walls consisting of middle lamella and primary wall.
   d. are always accompanied by companion cells.
   e. are found only in the secondary plant body.
6. Which statement about sieve tube elements is *not* true?
   a. Their end walls are called sieve plates.
   b. They die before they become functional.
   c. They link end-to-end, forming sieve tubes.
   d. They form the system for translocation of organic nutrients.
   e. They lose the membrane that surrounds their central vacuole.
7. The pericycle
   a. separates the stele from the cortex.
   b. is the tissue within which branch roots arise.
   c. consists of highly differentiated cells.
   d. forms a star-shaped structure at the very center of the root.
   e. is waterproofed by Casparian strips.
8. Secondary growth of stems and roots
   a. is brought about by the apical meristems.
   b. is common in both monocots and eudicots.
   c. is brought about by vascular and cork cambia.
   d. produces only xylem and phloem.
   e. is brought about by vascular rays.
9. Periderm
   a. contains lenticels that allow for gas exchange.
   b. is produced during primary growth.
   c. is permanent; it lasts as long as the plant does.
   d. is the innermost part of the plant.
   e. contains vascular bundles.
10. Which statement about leaf anatomy is *not* true?
    a. Stomata are controlled by paired guard cells.
    b. The cuticle is secreted by the epidermis.
    c. The veins contain xylem and phloem.
    d. The cells of the mesophyll are packed together, minimizing air space.
    e. C_3 and C_4 plants differ in leaf anatomy.

For Discussion

1. When a young oak was 5 m tall, a thoughtless person carved his initials in its trunk at a height of 1.5 m above the ground. Today that tree is 10 m tall. How high above the ground are those initials? Explain your answer in terms of the manner of plant growth.
2. Consider a newly formed sieve tube element in the secondary phloem of an oak tree. What kind of cell divided to produce the sieve tube element? What kind of cell divided to produce that parent cell? Keep tracing back until you arrive at a cell in the apical meristem.
3. Distinguish between sclerenchyma cells and collenchyma cells in terms of structure and function.
4. Distinguish between primary and secondary growth. Do all angiosperms undergo secondary growth? Explain.
5. What anatomical features make it possible for a plant to retain water as it grows? Describe the plant tissues and how and when they form.

36 Transport in Plants

From Tarzan of the Apes to George of the Jungle, in legions of comic strips and adventure movies, heroes have swung through the forest canopy on lianas—twining jungle vines. And when these heroes' exertions left them thirsty, they took a machete, chopped open the lianas, and drank the water found in the hollow stems. Lianas are a realistic source of water, in fact as well as fiction. Like all plants, these vines are continually moving water, along with dissolved solutes, from place to place in their bodies.

The water and minerals in a plant's xylem must be transported from the roots to the entire shoot system, all the way to the highest leaves and apical buds. Similarly, carbohydrates produced by photosynthesis in all the leaves, including the highest, must be translocated in the phloem to all the living nonphotosynthetic parts of the plant, such as roots, tubers, and internal stem tissues. Before we consider the mechanisms underlying these processes, we should have some idea about the magnitude of what they accomplish. Let's consider two questions: How much water is transported? And how high can water be transported?

In answer to the first question, consider the following example: A single maple tree 15 meters tall is estimated to have some 177,000 leaves, with a total leaf surface area of 675 square meters—half again the area of a basketball court. During a summer day, that tree loses 220 liters of water *per hour* to the atmosphere by evaporation from the leaves. To prevent wilting, the xylem needs to transport 220 liters of water from the roots to the leaves every hour. (By comparison, a 50-gallon drum holds 189 liters.)

The second question can be rephrased: How tall are the tallest trees? The tallest gymnosperms, the coast redwoods—*Sequoia sempervirens*—exceed 110 meters in height, as do the tallest angiosperms, the Australian *Eucalyptus regnans*. Any successful explanation of water transport in the xylem must account for the transport of water to these great heights.

In this chapter, we will consider the uptake and transport of water and minerals by plants, the control of evaporative water loss from leaves, and the translocation of substances in the phloem.

Hollywood Vines Although the famous swinging-through-the-jungle scenes in the Tarzan movies of the 1940s were pure fiction and the vines the actors used were rope props, the use of lianas—heavy, twining vines of the tropical rain forests—as a source of drinking water is quite plausible.

Uptake and Movement of Water and Solutes

Terrestrial plants must obtain both water and mineral nutrients from the soil, usually by way of their roots. The roots, in turn, obtain carbohydrates and other important materials from the leaves (Figure 36.1). We learned in Chapter 8 that water is one of the ingredients required for carbohydrate production by photosynthesis in the leaves. Water is also essential for transporting solutes both upward and downward, for cooling the plant, and for developing the internal pressure that supports the plant body. Plants lose large quantities of water to evaporation, and this water must be continually replaced.

How do leaves high in a tree obtain water from the soil? What are the mechanisms by which water and mineral ions enter the plant body through the roots and ascend as sap in the xylem? Because neither water nor minerals can move through the plant into the xylem without crossing at least one plasma membrane, we will focus first on osmosis. Then we will examine the uptake of mineral ions and follow the pathway by which both water and minerals move through the root to gain entry to the xylem.

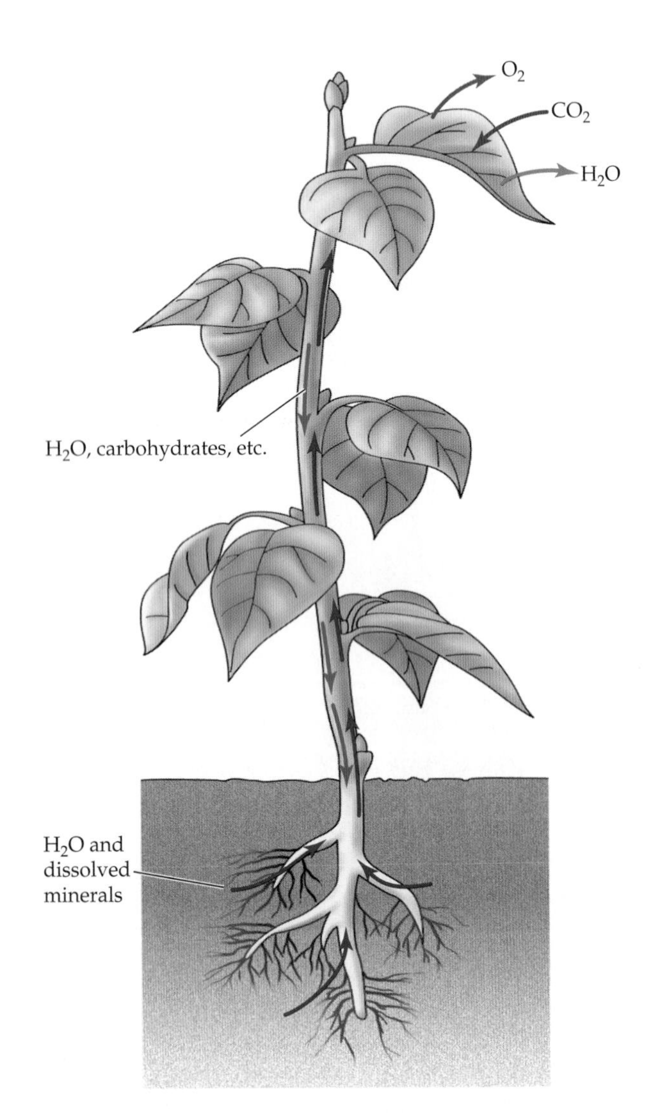

36.1 The Pathways of Water and Solutes in the Plant Water travels from the soil to the atmosphere, and it circulates within the plant, carrying important solutes with it.

Water moves through a membrane by osmosis

Osmosis, the movement of water through a membrane in accordance with the laws of diffusion, was described in Chapter 5 (see Figure 5.8). The **solute potential** (osmotic potential) of a solution is a measure of the effect of dissolved solutes on the osmotic behavior of the solution. The following statement presents an opportunity for confusion, so study it carefully:

- The greater the solute concentration of a solution, the more negative its solute potential, and the greater the tendency of water to move into it from another solution of lower solute concentration (and less negative solute potential).

For osmosis to occur, the two solutions must be separated by a membrane permeable to water but relatively impermeable to the solute. Recall, too, that osmosis is a passive process—energy is not directly required.

Unlike animal cells, plant cells are surrounded by a relatively rigid cell wall. As water enters a plant cell, the entry of more water is increasingly resisted by an opposing **pressure potential** (called *turgor pressure* in plants), owing to the rigidity of the wall. As more and more water enters, the pressure potential becomes greater and greater.

Pressure potential is a hydraulic pressure analogous to the air pressure in an automobile tire; it is a mechanical pressure that can be measured with a pressure gauge. Plant cells do not burst when placed in pure water; instead, water enters by osmosis until the pressure potential exactly balances the solute potential. At this point, the cell is *turgid;* that is, it has a significant pressure potential.

The overall tendency of a solution to take up water from pure water, across a membrane, is called its **water potential**, represented as ψ, the Greek letter psi (pronounced "sigh") (Figure 36.2). The water potential of a solution is simply the sum of its (negative) solute potential (ψ_s) and its (usually positive) pressure potential (ψ_p):

$$\psi = \psi_s + \psi_p$$

For pure water open to the atmosphere and therefore under no applied pressure, all three of these parameters are zero.

We can measure solute potential, pressure potential, and water potential in *megapascals* (MPa), a unit of pressure. (Atmospheric pressure, "one atmosphere," is about 0.1 MPa, or 14.7 pounds per square inch; typical pressure in an automobile tire is about 0.2 MPa.)

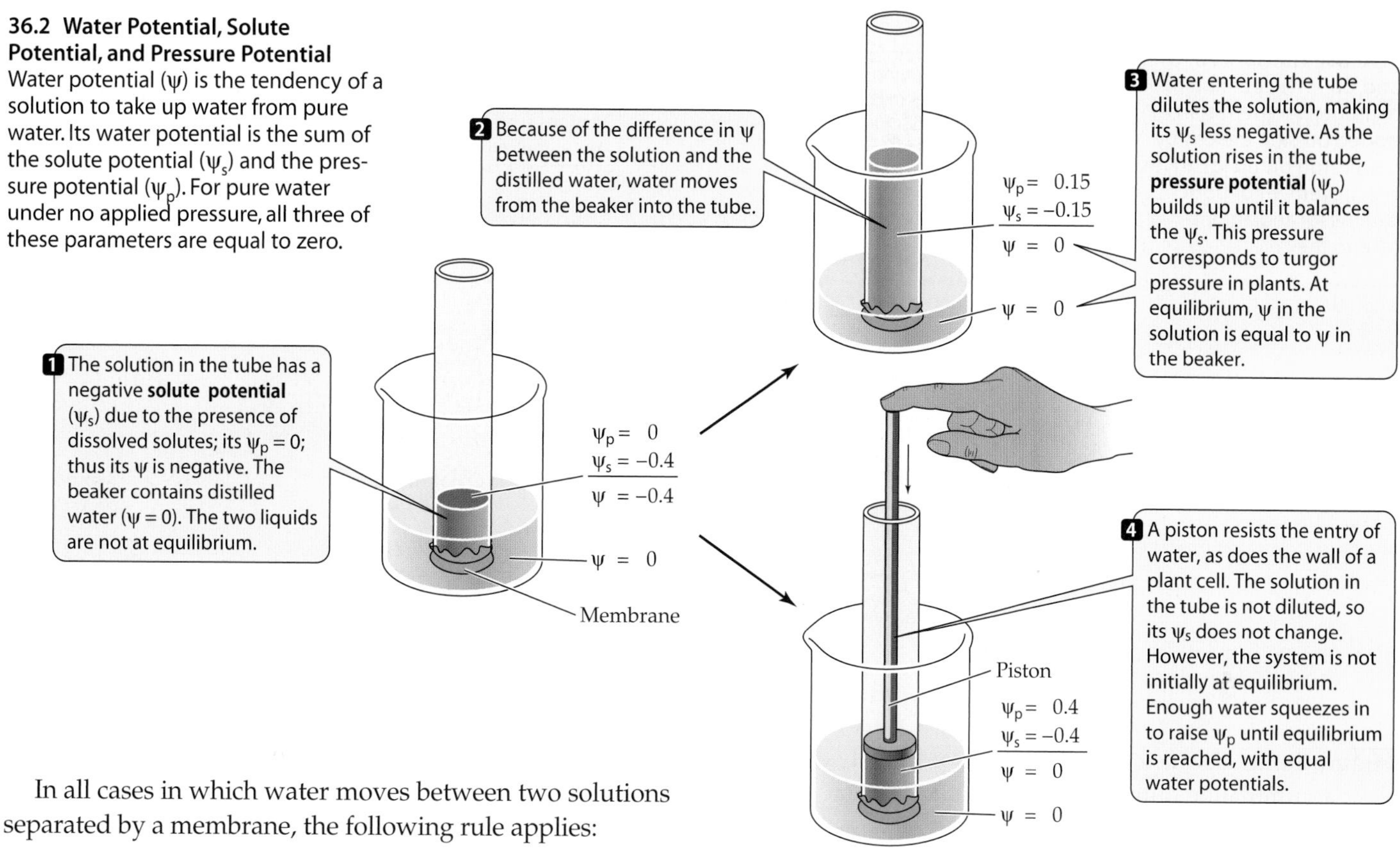

1 The solution in the tube has a negative solute potential (ψ_s) due to the presence of dissolved solutes; its $\psi_p = 0$; thus its ψ is negative. The beaker contains distilled water ($\psi = 0$). The two liquids are not at equilibrium.
$\psi_p = 0$
$\psi_s = -0.4$
$\psi = -0.4$
$\psi = 0$
Membrane
2 Because of the difference in ψ between the solution and the distilled water, water moves from the beaker into the tube.
$\psi_p = 0.15$
$\psi_s = -0.15$
$\psi = 0$
$\psi = 0$
3 Water entering the tube dilutes the solution, making its ψ_s less negative. As the solution rises in the tube, pressure potential (ψ_p) builds up until it balances the ψ_s. This pressure corresponds to turgor pressure in plants. At equilibrium, ψ in the solution is equal to ψ in the beaker.
Piston
$\psi_p = 0.4$
$\psi_s = -0.4$
$\psi = 0$
$\psi = 0$
4 A piston resists the entry of water, as does the wall of a plant cell. The solution in the tube is not diluted, so its ψ_s does not change. However, the system is not initially at equilibrium. Enough water squeezes in to raise ψ_p until equilibrium is reached, with equal water potentials.

36.2 Water Potential, Solute Potential, and Pressure Potential Water potential (ψ) is the tendency of a solution to take up water from pure water. Its water potential is the sum of the solute potential (ψ_s) and the pressure potential (ψ_p). For pure water under no applied pressure, all three of these parameters are equal to zero.

In all cases in which water moves between two solutions separated by a membrane, the following rule applies:

- Water always moves across a selectively permeable membrane toward the region of lower (more negative) water potential.

Osmotic phenomena are of great importance to plants. The structure of many plants is maintained by the pressure potential of their cells; if the pressure potential is lost, the plant *wilts*. Within living tissues, the movement of water from cell to cell follows a gradient of water potential. Over longer distances, in unobstructed tubes such as xylem vessels and phloem sieve tubes, the flow of water and dissolved solutes is driven by a gradient of pressure potential. The movement of a solution due to a difference in pressure potential between two parts of a plant is called **bulk flow**.

Aquaporins facilitate the movement of water across membranes

Aquaporins are membrane channel proteins through which water can move without interacting with the hydrophobic environment of the membrane's phospholipid bilayer. These proteins, important in both plants and animals, allow water to move rapidly from environment to cell and from cell to cell. The permeability of some aquaporins can be regulated, changing the *rate* of osmosis across the membrane. However, water movement through aquaporins is always passive, so the *direction* of water movement is unchanged by alterations in aquaporin permeability.

Uptake of mineral ions requires membrane transport proteins

Mineral ions, which carry electric charges, generally cannot move across a membrane unless they are aided by transport proteins (explained in Chapter 5). When the concentration of these charged ions in the soil is greater than that in the plant, ion channels and carrier proteins can move them into the plant by facilitated diffusion, which is a passive process. The concentrations of most ions in the soil solution, however, are lower than those required inside the plant. Thus the plant must take up these ions against a concentration gradient—a process that requires energy.

Electric charge differences also play a role in the uptake of mineral ions. Movement of a negatively charged ion into a negatively charged region is movement against an electrical gradient and requires energy. The combination of concentration and electrical gradients is called an *electrochemical gradient*. Uptake against an electrochemical gradient is *active transport*, an energy-requiring process, which depends on cellular respiration for a supply of ATP. Active transport, of course, requires specific transport proteins.

Unlike animals, plants do not have a sodium–potassium pump for active transport. Rather, plants have a **proton pump**, which uses energy obtained from ATP to move protons out of the cell against a proton concentration gradient

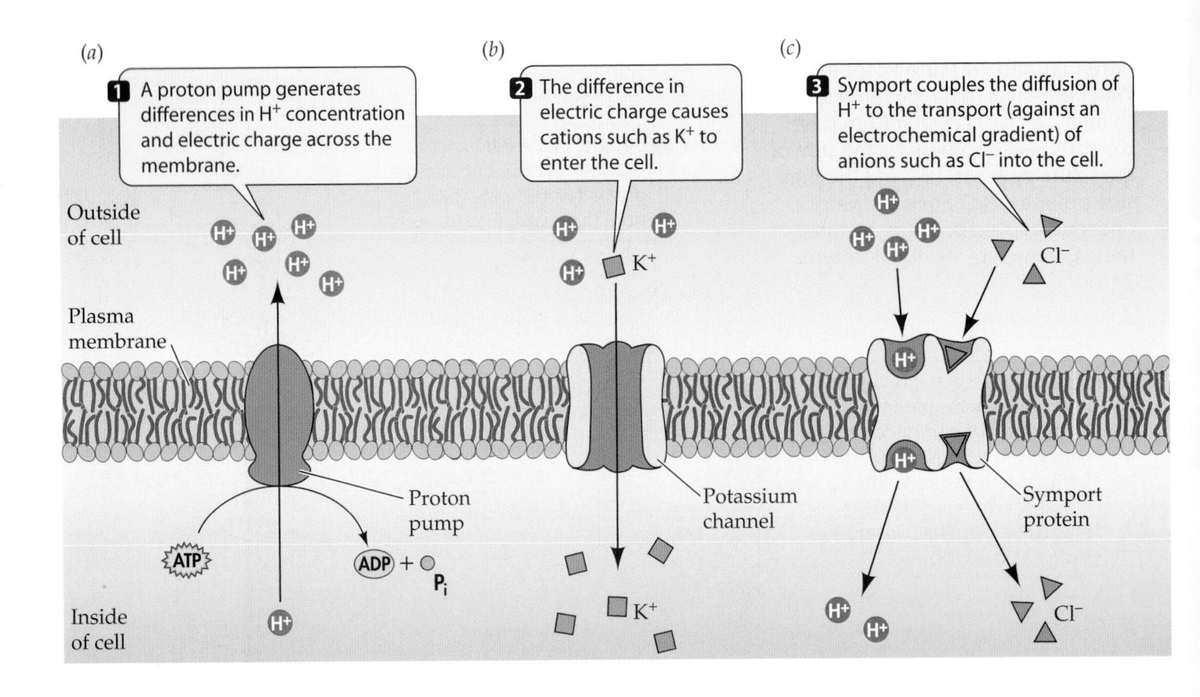

(a)
1 A proton pump generates differences in H+ concentration and electric charge across the membrane.
Outside of cell
Plasma membrane
Proton pump
ATP
ADP + Pi
Inside of cell
(b)
2 The difference in electric charge causes cations such as K+ to enter the cell.
K+
Potassium channel
K+
(c)
3 Symport couples the diffusion of H+ to the transport (against an electrochemical gradient) of anions such as Cl− into the cell.
Cl−
Symport protein
Cl−

36.3 The Proton Pump in Active Transport of K^+ and Cl^- The buildup of hydrogen ions (H^+) transported outside the cell by the proton pump (*a*) drives the movement of both cations (*b*) and anions (*c*) into the cell.

(Figure 36.3*a*). Because protons (H^+) are positively charged, their accumulation outside the cell has two results:

- The region outside the cell becomes positively charged with respect to the region inside.
- A proton concentration gradient develops across the plasma membrane.

Each of these results has consequences for the movement of other ions. Because of the charge difference across the membrane, there is increased movement of cations (positively charged ions), such as potassium (K^+), into the cell through their membrane channels. These ions move into the now more negatively charged interior of the cell by facilitated diffusion (Figure 36.3*b*). In addition, the proton concentration gradient can be harnessed to drive secondary active transport, in which anions (negatively charged ions) such as chloride (Cl^-) are moved into the cell against an electrochemical gradient by a symport protein that couples their movement with that of H^+ (Figure 36.3*c*). In sum, there is a vigorous traffic of ions across plant cell membranes, involving specific membrane transport proteins and both active and passive processes.

The proton pump and the coordinated activities of other membrane transport proteins cause the interior of a plant cell to be very negative with respect to the exterior. Such a difference in charge across a membrane is called a **membrane potential**. Biologists can measure the membrane potential of a plant cell with microelectrodes, just as they can measure similar charge differences in nerve cells and other animal cells (see Chapter 44). Most plant cells maintain a membrane potential of at least –120 millivolts (mV).

Water and ions pass to the xylem by way of the apoplast and symplast

Mineral ions enter and move through plants in various ways. Where water is moving by bulk flow, dissolved minerals are carried along in the stream. Both water and minerals also move by diffusion. At certain sites, where plasma membranes are being crossed, some mineral ions are moved by active transport. One such site is the surface of a root hair, where mineral ions first enter the cells of the plant. Later, within the stele, the ions must cross another plasma membrane before entering the nonliving vessels and tracheids of the xylem.

The movement of ions across membranes can also result in the movement of water. Water moves into a root because the root has a more negative water potential than does the soil solution. Water moves from the cortex of the root into the stele (which is where the vascular tissues are located) because the stele has a more negative water potential than does the cortex.

Water and minerals from the soil may pass through the dermal and ground tissues to the stele via two pathways: the apoplast and the symplast. The **apoplast** (from the Greek, *apo-*, "away from"; *-plast*, "living material") consists of the cell walls, which lie outside the plasma membranes, and the intercellular spaces (spaces between cells) that are common to many tissues. The apoplast is a continuous meshwork through which water and dissolved substances can flow or diffuse without ever having to cross a membrane (Figure 36.4). Movement of materials through the apoplast is thus unregulated—until it reaches the endodermis, as we will soon discuss.

36.4 Apoplast and Symplast Plant cell walls and intercellular spaces constitute the apoplast. The symplast comprises the living cells, which are connected by plasmodesmata. To enter the symplast, water and solutes must pass through a plasma membrane. No such selective barrier limits movement through the apoplast.

The remainder of the plant body is the **symplast** (from the Greek, *sym-*, "together with"). The symplast is the portion of the plant body enclosed by membranes—the continuous cytoplasm of the living cells, connected by plasmodesmata (see Figure 36.4). The selectively permeable plasma membranes of the cells control access to the symplast, so movement of water and dissolved substances into the symplast is tightly regulated.

Water and minerals can pass from the soil solution through the apoplast as far as the endodermis, the innermost layer of the root cortex. The endodermis is distinguished from the rest of the ground tissue by the presence of **Casparian strips**. These waxy, suberin-impregnated regions of the endodermal cell wall form a water-repelling (hydrophobic) belt around each endodermal cell where it is in contact with other endodermal cells. The hydrophobic Casparian strips act as a seal that prevents water and ions from moving between the cells (Figure 36.5).

The Casparian strips of the endodermis thus completely separate the apoplast of the cortex from the apoplast of the stele. However, they do not obstruct the outer or inner faces of the endodermal cells. Accordingly, water and ions can enter the stele only by way of the symplast—that is, by entering and passing through the cytoplasm of the endodermal cells. Thus transport proteins in the plasma membranes of these cells determine which mineral ions pass into the stele, and at what rates.

Once they have passed the endodermal barrier, water and minerals leave the symplast and enter the apoplast of the stele. Parenchyma cells in the pericycle or xylem can aid this

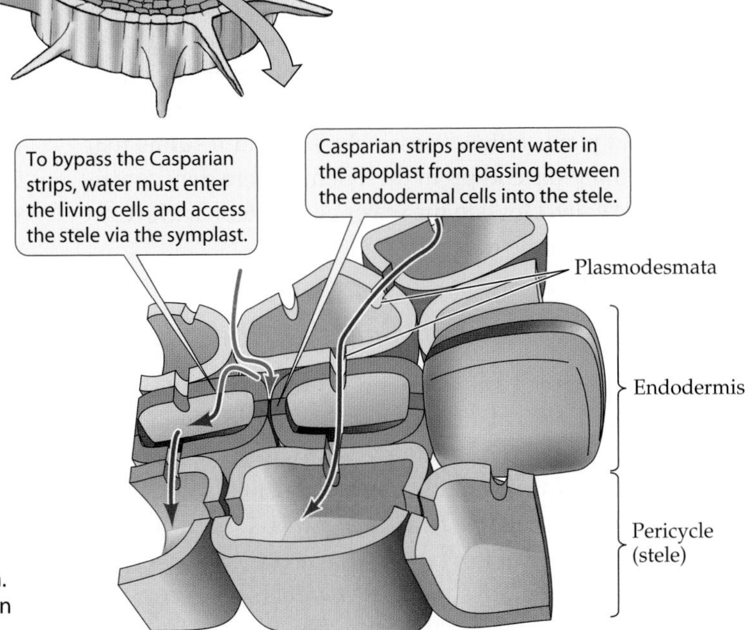

To bypass the Casparian strips, water must enter the living cells and access the stele via the symplast.
Casparian strips prevent water in the apoplast from passing between the endodermal cells into the stele.
Plasmodesmata
Endodermis
Pericycle (stele)

36.5 Casparian Strips Casparian strips in the endodermis of the cortex are impregnated with the water-repelling substance suberin. These strips separate the apoplast in the cortex from the apoplast in the stele.

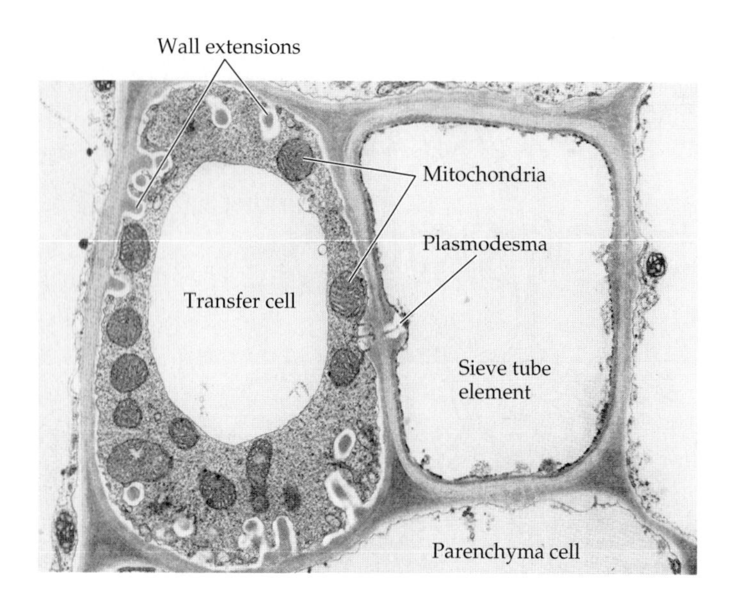

36.6 A Transfer Cell Three walls of this transfer cell in a pea leaf have knobby extensions that face the cells from which the transfer cell imports solutes. A transfer cell exports the solutes to the neighboring sieve tube element.

process. Some of these parenchyma cells, called **transfer cells**, are structurally modified for transporting mineral ions from their cytoplasm (part of the symplast) into their cell walls (part of the apoplast). The cell wall that receives the transported ions has many knobby extensions projecting into the transfer cell, increasing the surface area of the cell's plasma membrane, the number of transport proteins, and thus the rate of transport (Figure 36.6). Transfer cells also have many mitochondria that produce the ATP needed to power the active transport of mineral ions.

As mineral ions move into the solution in the cell walls, the water potential in the apoplast becomes more negative; thus water moves out of the cells and into the apoplast by osmosis. In other words, active transport of ions moves the ions directly, and water follows passively. The end result is that water and minerals end up in the xylem, where they constitute the *xylem sap*. How do the water and materials move on from the xylem of the root system?

Transport of Water and Minerals in the Xylem

So far in this chapter we've described the movement of water and minerals into plant roots and their entry into the root xylem. Now we will consider how xylem sap moves through the remainder of the plant. Let's first consider some early ideas about the ascent of xylem sap and then turn to our current understanding of how it works. We'll describe the experiments that ruled out some early models as well as some evidence in support of the current model.

Experiments ruled out xylem transport by pumping action of living cells

Some of the earliest attempts to explain the rise of sap in the xylem were based on a hypothetical pumping action by living cells in the stem, which might push the sap upward. However, experiments conducted and published in 1893 by the German botanist Eduard Strasburger definitively ruled out such models.

Strasburger worked with trees about 20 meters tall. He sawed through the trunk of each tree at its base and plunged the cut end into a bucket containing a solution of a poison, such as picric acid. The solution rose through the trunk, as was readily evident from the progressive death of the bark higher and higher up. When the solution reached the leaves, the leaves died, too, at which point the movement of the solution stopped (as shown by the liquid level in the bucket, which stopped dropping).

This simple experiment established three important points:

- Living, "pumping" cells were not responsible for the upward movement of the solution, because the solution itself killed all living cells with which it came in contact.
- The leaves played a crucial role in transport. As long as they were alive, the solution continued to move upward; when the leaves died, movement ceased.
- The movement was not caused by the roots, because the trunk had been completely separated from the roots.

Root pressure does not account for xylem transport

In spite of Strasburger's observations, some plant physiologists hypothesized that xylem transport was based on **root pressure**—pressure exerted by the root tissues that would force liquid up the xylem. The basis for root pressure is a higher solute concentration, and accordingly a more negative water potential, in the xylem sap than in the soil solution. This water potential draws water into the stele; once there, the water has nowhere to go but up, so it rises in the vessels and tracheids.

There is good evidence that root pressure exists—for example, the phenomenon of *guttation*, in which liquid water is forced out through openings at the margins of leaves (Figure 36.7). Guttation occurs only under conditions of high atmospheric humidity and plentiful water in the soil, which occur most commonly at night. Root pressure is also the source of the sap that oozes from the cut stumps of some plants, such as *Coleus*, when their tops are cut off.

Root pressure, however, cannot account for the ascent of sap in trees. Root pressure seldom exceeds 0.1–0.2 MPa (1–2 atmospheres). If root pressure were driving sap up the xylem, we would observe a positive pressure potential in the xylem

36.7 Guttation Root pressure is responsible for forcing water through openings in the margins of this strawberry leaf.

at all times. In fact, as we are about to see, the xylem sap in most trees is under *tension*—has a negative pressure potential—when it is ascending. Furthermore, as Strasburger had already shown, materials can be transported upward in the xylem even when the roots have been removed. If the roots are not pushing the xylem sap upward, what causes it to rise?

The transpiration–cohesion–tension mechanism accounts for xylem transport

The obvious alternative to pushing is pulling: The leaves pull the xylem sap upward. The evaporative loss of water from the leaves generates a pulling force (tension) on the water in the apoplast of the leaves. Hydrogen bonding between water molecules makes the sap in the xylem cohesive enough to withstand the tension and rise by bulk flow. Let's see how this process works.

The concentration of water vapor in the atmosphere is lower than that in the leaf. Because of this difference, water vapor diffuses from the intercellular spaces of the leaf, through openings called stomata, to the outside air. This process is called **transpiration** (Figure 36.8). Within the leaf blade, water evaporates from the moist walls of the mesophyll cells and enters the intercellular spaces. The force generated by the evaporation of water from the mesophyll cell walls creates a *tension* that draws more water into the cell walls, replacing that which was lost. The tension in the mesophyll draws water from the xylem of the nearest vein into the apoplast surrounding the mesophyll cells. The removal of water from the veins, in turn, establishes tension on the entire column of water contained within the xylem, so that the column is drawn upward all the way from the roots.

The ability of water to be pulled upward through tiny tubes results from the remarkable *cohesion* of water—the tendency of water molecules to stick to one another through hydrogen bonding. The narrower the tube, the greater the tension the water column can withstand without breaking. The integrity of the column is also maintained by the adhesion of water to the xylem walls. In the tallest trees, such as a 110-meter redwood, the difference in pressure potential between

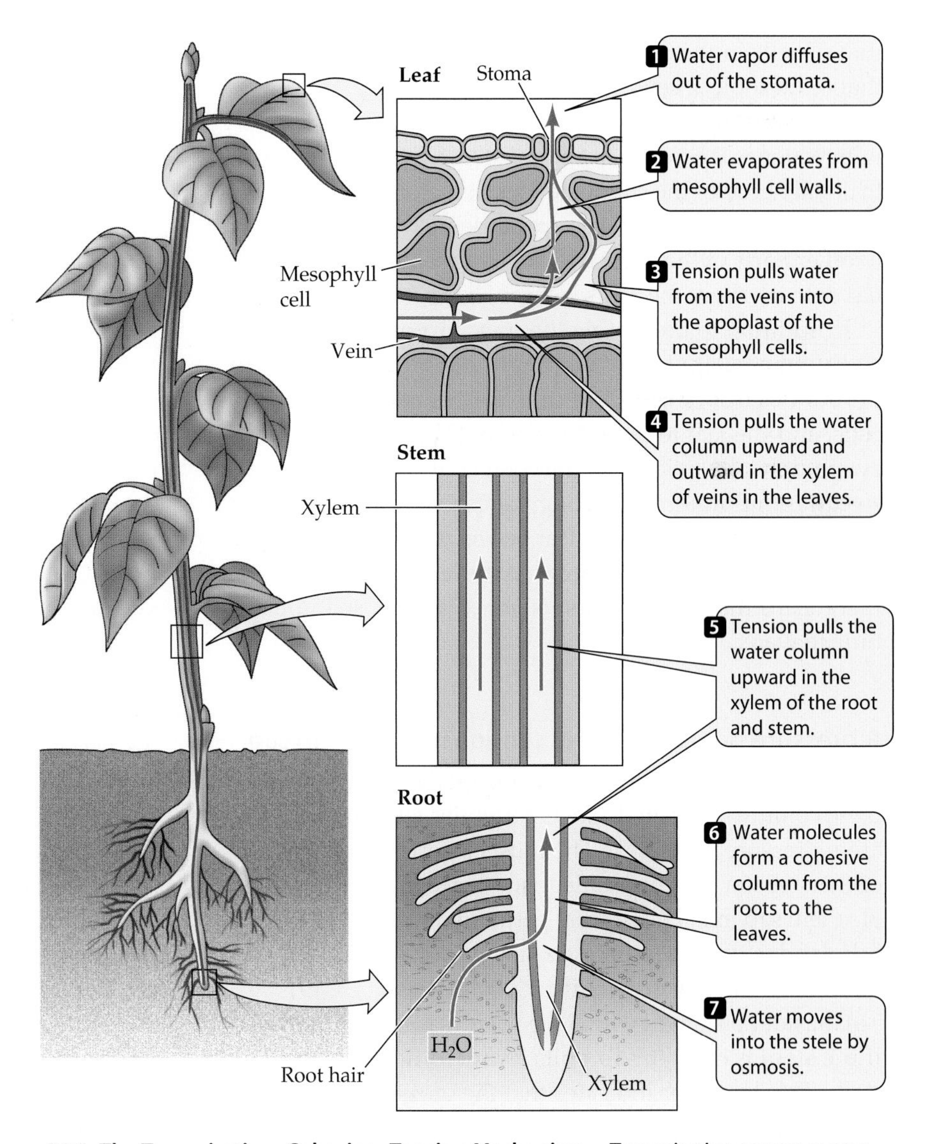

36.8 The Transpiration–Cohesion–Tension Mechanism Transpiration causes evaporation from mesophyll cell walls, generating tension on the xylem. Cohesion among water molecules in the xylem transmits the tension from the leaf to the root, causing water to move from the soil to the atmosphere.

the top and the bottom of the column may be as great as 3 MPa. The cohesion of water in the xylem is great enough to withstand even that great a tension.

In summary, the key elements of water transport in the xylem are

- *Transpiration*, the evaporation of water from the leaves
- *Tension* in the xylem sap resulting from transpiration
- *Cohesion* in the xylem sap from the leaves to the roots

This **transpiration–cohesion–tension mechanism** requires no work (that is, no expenditure of energy) on the part of the plant. At each step between soil and atmosphere, water moves passively toward a region with a more negative water potential. Dry air has the most negative water potential (–95 MPa at 50% relative humidity), and the soil solution has the least negative water potential (between –0.01 and –3 MPa). Xylem sap has a water potential more negative than that of cells in the cortex of the root, but less negative than that of mesophyll cells in the leaf.

Mineral ions contained in the xylem sap rise passively with water as it ascends from root to leaf. In this way the nutritional needs of the shoot are met. Some of the mineral elements brought to the leaves are subsequently redistributed to other parts of the plant by way of the phloem, but the initial delivery from the roots is through the xylem.

In addition to promoting the transport of minerals, transpiration contributes to temperature regulation. As water evaporates from mesophyll cells, heat is taken up from the cells, and the leaf temperature drops. This cooling effect of evaporation (so evident in the cooling of our skin when we sweat) is important in enabling plants to live in hot environments. A farmer can hold a leaf between thumb and forefinger to estimate its temperature; if the leaf doesn't feel cool, that means that transpiration is not occurring, so it must be time to water.

A pressure bomb measures tension in the xylem sap

The transpiration–cohesion–tension model can be true only if the column of sap in the xylem is under tension (has a negative pressure potential). The most elegant demonstrations of this tension, and of its adequacy to account for the ascent of xylem sap in tall trees, were performed by the biologist Per Scholander, who measured tension in stems with an instrument called a **pressure bomb**.

Consider a stem in which the xylem sap is under tension. If the stem is cut, the sap pulls away from the cut, into the stem. This behavior indicates that the pressure in the intact xylem is lower than that of the atmosphere. Now the stem is quickly placed in the pressure bomb, in which the pressure may be raised. The cut surface remains outside the bomb. As gas pressure is applied to the plant parts within the bomb,

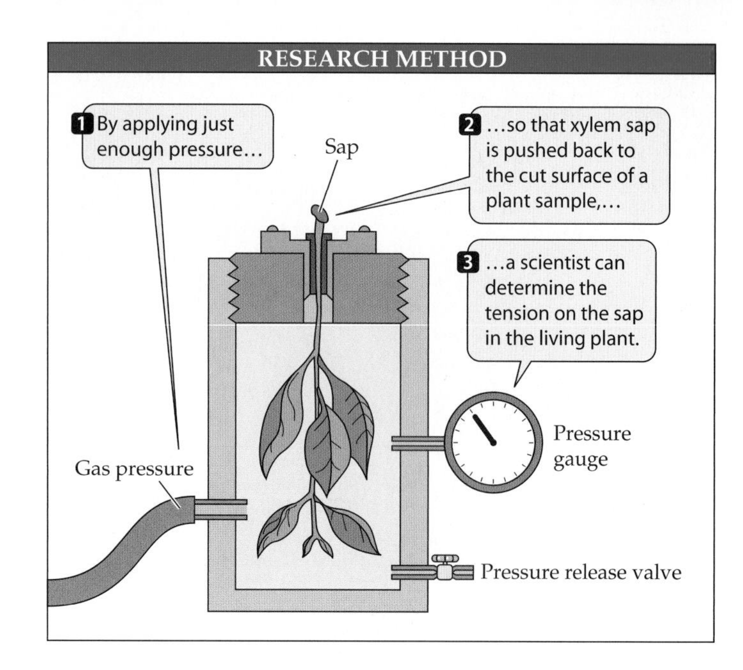
RESEARCH METHOD
1 By applying just enough pressure...
Sap
2 ...so that xylem sap is pushed back to the cut surface of a plant sample,...
3 ...a scientist can determine the tension on the sap in the living plant.
Pressure gauge
Gas pressure
Pressure release valve

36.9 A Pressure Bomb The amount of tension on the sap in different types of plants can be measured with this device.

the xylem sap is pushed back to the cut surface. When the sap first becomes visible again at the cut surface, the pressure in the bomb is recorded. This pressure is equal in magnitude but opposite in sign to the tension (negative pressure potential) originally present in the xylem (Figure 36.9).

Scholander used the pressure bomb to study dozens of plant species, from diverse habitats, growing under a variety of conditions. In all cases in which xylem sap was ascending, it was found to be under tension. The tension disappeared in some of the plants at night, when transpiration ceased. In developing vines, the xylem sap was under no tension until leaves formed. Once leaves developed, transport in the xylem began, and tensions were recorded.

Suppose you wanted to measure tensions in the xylem at various heights in a tall tree to confirm that the tensions are sufficient to account for the rate at which sap is moving up the trunk. How would you obtain stem samples for measurement? Per Scholander used surveying instruments to determine the heights of particular twigs, then had a sharpshooter shoot the identified twigs from the tree with a high-powered rifle. As the twigs fell to the ground, Scholander quickly inserted them in the pressure bomb and recorded their xylem tension. In every case, the differences in tensions at different heights were great enough to keep the xylem sap ascending.

The rate at which the sap ascends is not the same at all times. No flow of xylem sap takes place at night, when there is little or no transpiration. By day, when the sap is ascending, the rate of ascent depends on several factors. One is the concentration of K^+ in the sap: The rate of flow increases as the K^+ concentration increases (Figure 36.10). Other factors, such as

EXPERIMENT

Question: How does K^+ affect xylem flow rate?

METHOD

Two xylem-containing flaps were created on a tobacco plant. One was connected to a source of pure water (the control). The other flap could be connected to either pure water or to a solution containing a known concentration of K^+.

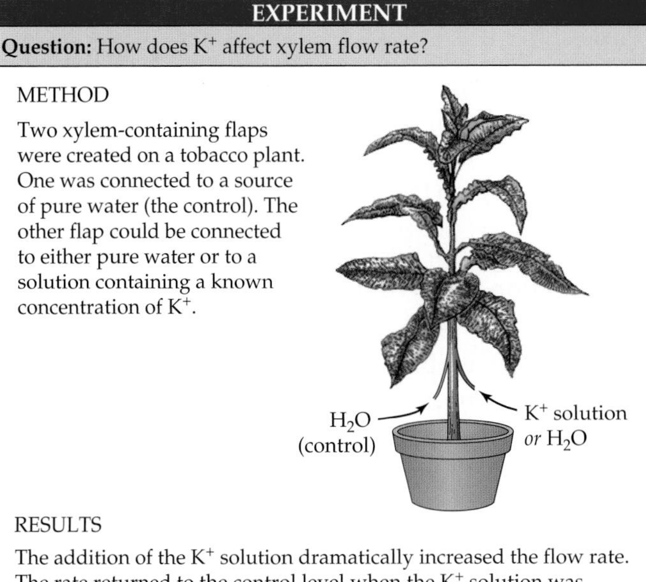

RESULTS

The addition of the K^+ solution dramatically increased the flow rate. The rate returned to the control level when the K^+ solution was replaced by pure water.

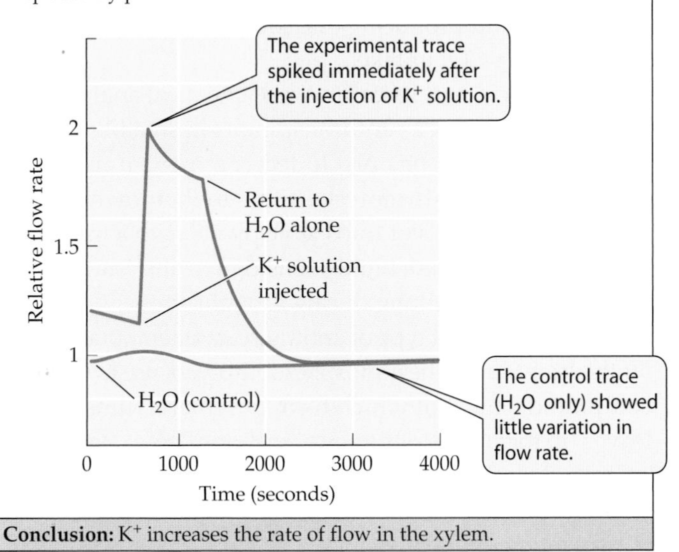

Conclusion: K^+ increases the rate of flow in the xylem.

36.10 Potassium Ions Speed Transport in the Xylem This experiment showed that the rate of fluid ascending through the xylem spiked when a solution with a known concentration of potassium ions was injected. Repeating the experiment with solutions of different concentrations of K^+ showed that the higher the K^+ concentration, the greater the flow rate.

temperature, light intensity, and wind velocity affect the transpiration rate, and hence the rate of sap flow, more directly.

Although transpiration provides the impetus for the transport of water and minerals in the xylem, it also results in the loss of tremendous quantities of water from the plant. How do plants control this loss?

Transpiration and the Stomata

The epidermis of leaves and stems minimizes transpirational water loss by secreting a waxy cuticle, which is impermeable to water. However, the cuticle is also impermeable to carbon dioxide. This poses a problem: How can the leaf balance its need to retain water with its need to obtain CO_2 for photosynthesis?

Plants have evolved an elegant compromise in the form of **stomata** (singular, stoma), or pores, in the epidermis of their leaves. A pair of specialized epidermal cells, called **guard cells**, controls the opening and closing of each stoma (Figure 36.11*a*). When the stomata are open, CO_2 can enter the leaf by diffusion—but water vapor is lost in the same way.

(*a*)

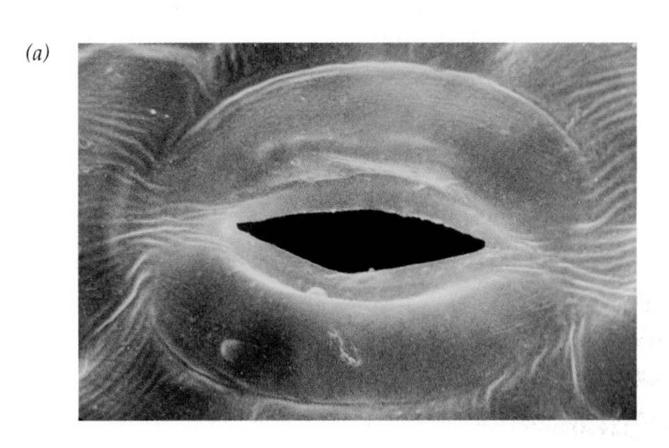

(*b*)

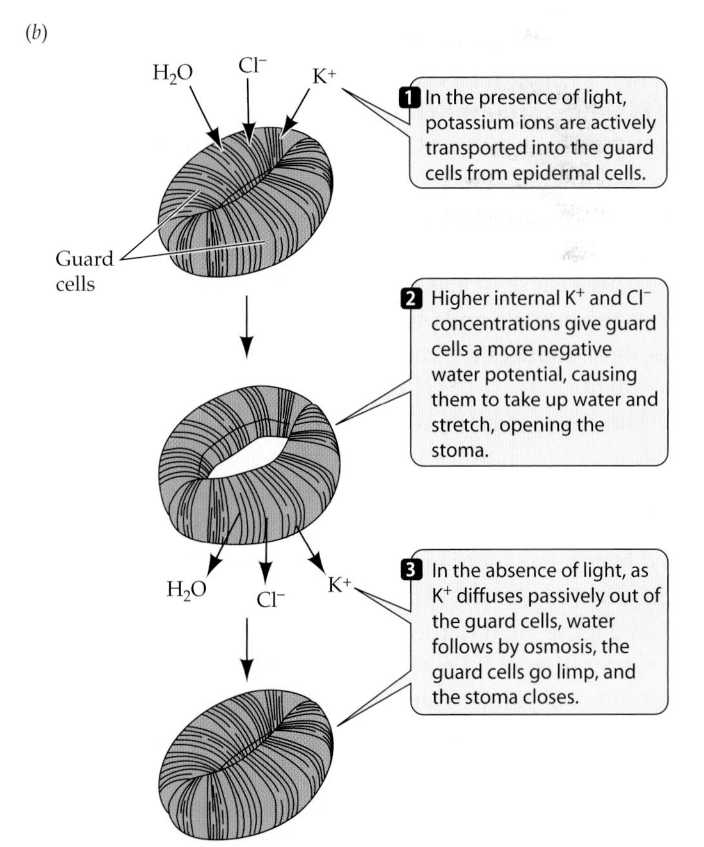

H_2O
Cl^-
K^+
1 In the presence of light, potassium ions are actively transported into the guard cells from epidermal cells.
Guard cells
2 Higher internal K^+ and Cl^- concentrations give guard cells a more negative water potential, causing them to take up water and stretch, opening the stoma.
H_2O
Cl^-
K^+
3 In the absence of light, as K^+ diffuses passively out of the guard cells, water follows by osmosis, the guard cells go limp, and the stoma closes.

36.11 Stomata (*a*) A scanning electron micrograph of an open stoma formed by two sausage-shaped guard cells. (*b*) Potassium ion concentrations affect the water potential of the guard cells, controlling the opening and closing of stomata. Negatively charged ions accompanying K^+ maintain electrical balance and contribute to the changes in solute potential that open and close the stomata.

Closed stomata prevent water loss, but also exclude CO_2 from the leaf.

Most plants open their stomata only when the light intensity is sufficient to maintain a moderate rate of photosynthesis. At night, when darkness precludes photosynthesis, the stomata remain closed; no CO_2 is needed at this time, and water is conserved. Even during the day, the stomata close if water is being lost at too great a rate.

The stoma and guard cells in Figure 36.11*a* are typical of eudicots. Monocots typically have specialized epidermal cells associated with their guard cells. The principle of operation, however, is the same for both monocot and eudicot stomata. In what follows, we describe the regulation and mechanism of stomatal opening and the normal cycle of opening and closing.

The guard cells control the size of the stomatal opening

Light causes the stomata of most plants to open, admitting CO_2 for photosynthesis. Another cue for stomatal opening is the level of CO_2 in the intercellular spaces inside the leaf. A low level favors opening of the stomata, thus allowing the uptake of more CO_2.

Water stress is a common problem for plants, especially on hot, sunny, windy days. Plants have a protective response to these conditions, which uses the water potential of the mesophyll cells as a cue. Even when the CO_2 level is low and the sun is shining, if the mesophyll is too dehydrated—that is, if the water potential of the mesophyll is too negative—the mesophyll cells release a plant hormone called *abscisic acid*. Abscisic acid acts on the guard cells, causing them to close the stomata and prevent further drying of the leaf. This response reduces the rate of photosynthesis, but it protects the plant.

These processes are regulated by control of the K^+ concentration in the guard cells. Blue light, absorbed by a pigment in the guard cell plasma membrane, activates a proton pump, which actively transports protons (H^+) out of the guard cells and into the surrounding epidermis. The resulting proton gradient drives the accumulation of K^+ (Figure 36.11*b*; also review Figure 36.3) in the guard cell. The increasing internal concentration of K^+ makes the water potential of the guard cells more negative. Water enters the guard cells by osmosis, increasing their pressure potential. The arrangement of the cellulose microfibrils in their cell walls causes the guard cells to respond to this increase by changing their shapes so that a gap—the stoma—appears between them.

The stoma closes by the reverse process when active transport ceases in response to the absence of blue light or the presence of abscisic acid. Potassium ions diffuse passively out of the guard cells, water follows by osmosis, the pressure potential decreases, and the guard cells sag together and seal off the stoma. Negatively charged chloride ions and organic ions also move into and out of the guard cells along with the potassium ions, maintaining electrical balance and contributing to the change in the solute potential of the guard cells.

Transpiration from crops can be decreased

Stomata are the "referees" of a compromise between the admission of CO_2 for photosynthesis and the loss of water by transpiration. Farmers would like their crops to transpire less, thus reducing the need for irrigation. Similarly, nurseries and gardeners would like to be able to reduce the amount of water lost by plants that are to be transplanted, because transplanting often damages the roots, causing the plant to wilt or die. What they need is a good *antitranspirant:* a compound that can be applied to plants, reducing water loss from the stomata without producing disastrous side effects by excessively limiting CO_2 uptake.

Abscisic acid and its commercial chemical analogs have been found to work as antitranspirants in small-scale tests, but their high cost has precluded commercial use. What about making plants more sensitive to their own abscisic acid? The guard cells of transgenic plants with a mutant allele of the *era* gene are highly sensitive to abscisic acid and hence resistant to wilting during drought stress.

A totally different type of antitranspirant temporarily seals off the leaves from the atmosphere. Growers use a variety of compounds, most of which form polymeric films around leaves, to form a barrier to evaporation. These compounds cause undesirable side effects, however, and can be used only for short periods of time. Their most common use is in the transplanting of nursery stock.

In the absence of antitranspirants, stomata are normally open during daylight hours, allowing CO_2 to be fixed and converted to the products of photosynthesis. Next we'll see how these products are delivered to other parts of the plant, supporting growth of those parts.

Translocation of Substances in the Phloem

Photosynthesis takes place in the mesophyll cells and, in C_4 plants, in the bundle sheath cells of the leaf (see Figure 8.16). The products of photosynthesis (primarily carbohydrates) diffuse to the nearest small vein, where they are actively transported into sieve tube elements.

Substances in the phloem move from sources to sinks. A **source** is an organ (such as a mature leaf or a storage root) that *produces* (by photosynthesis or by digestion of stored reserves) more sugars than it requires. A **sink** is an organ (such as a root, a flower, a developing fruit or tuber, or an immature leaf) that *consumes* sugars for its own growth and storage needs. Sugars (primarily sucrose), amino acids, some

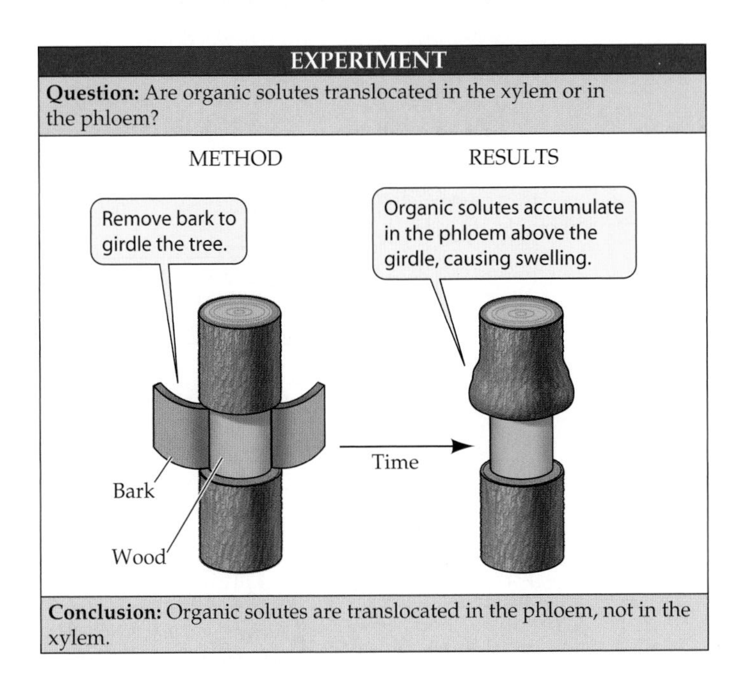

EXPERIMENT
Question: Are organic solutes translocated in the xylem or in the phloem?
METHOD
RESULTS
Remove bark to girdle the tree.
Organic solutes accumulate in the phloem above the girdle, causing swelling.
Time
Bark
Wood
Conclusion: Organic solutes are translocated in the phloem, not in the xylem.

36.12 Girdling Blocks Translocation in the Phloem By removing a ring of bark (containing the phloem), Malpighi blocked the translocation of organic solutes in a tree.

minerals, and a variety of other solutes are translocated between sources and sinks in the phloem.

How do we know that such organic solutes are translocated in the phloem, rather than in the xylem? Just over 300 years ago, the Italian scientist Marcello Malpighi performed a classic experiment in which he removed a ring of bark (containing the phloem) from the trunk of a tree—that is, he *girdled* the tree (Figure 36.12). The bark in the region above the girdle swelled over time. We now know that the swelling resulted from the accumulation of organic solutes that came from higher up the tree and could no longer continue downward because of the disruption of the phloem. Later, the bark below the girdle died because it no longer received sugars from the leaves. Eventually the roots, and then the entire tree, died.

Any explanation of the translocation of organic solutes must account for a few important observations:

- Translocation stops if the phloem tissue is killed by heating or other methods; thus the mechanism must be different from that of transport in the xylem.
- Translocation often proceeds in both directions—up the stem and down the stem—simultaneously.
- Translocation is inhibited by compounds that inhibit respiration and thus limit the ATP supply in the source.

To investigate translocation, plant physiologists needed to obtain samples of pure sieve tube sap from individual sieve tube elements. This difficult task was simplified when it was discovered that a common garden pest, the aphid, feeds on plants by drilling into a sieve tube. An aphid inserts its specialized feeding organ, called a *stylet*, into a stem until the stylet enters a sieve tube (Figure 36.13*a*). The pressure within the sieve tube is greater than that in the surrounding plant tissues or outside the plant, so the nutritious sieve tube sap is forced through the stylet and into the aphid's digestive tract. So great is the pressure that sugary liquid is forced through the insect's body and out the anus (Figure 36.13*b*).

Plant physiologists use aphids to collect sieve tube sap. When liquid appears on the aphid's abdomen, indicating that the insect has connected with a sieve tube, the physiologist quickly freezes the aphid and cuts its body away from the stylet, which remains in the sieve tube element. For hours, sieve tube sap continues to exude from the cut stylet, where it may be collected for analysis. Chemical analysis of sieve tube sap collected in this manner reveals the contents of a single sieve tube element over time. Physiologists can also infer the rates at which different substances are translocated by measuring how long it takes for radioactive tracers administered to a leaf to appear at stylets at different distances from the leaf.

These methods have allowed us to understand how, at times, different substances might move in opposite directions in the phloem of a stem. Experiments with aphid stylets have shown that all the contents of any given sieve tube element move in the same direction. Thus, bidirectional translocation can be understood in terms of different sieve tubes conducting sap in opposite directions. These and other experiments

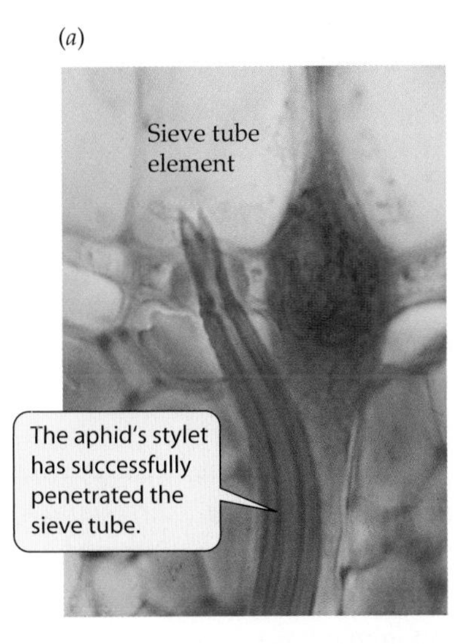

36.13 Aphids Collect Sieve Tube Sap (*a*) Aphids feed on sap drawn from a sieve tube, which they penetrate with a modified feeding organ, the stylet. (*b*) Pressure inside the sieve tube forces sap through the aphid's digestive tract, from which it can be harvested.

led to the general adoption of the pressure flow model as an explanation for translocation in the phloem.

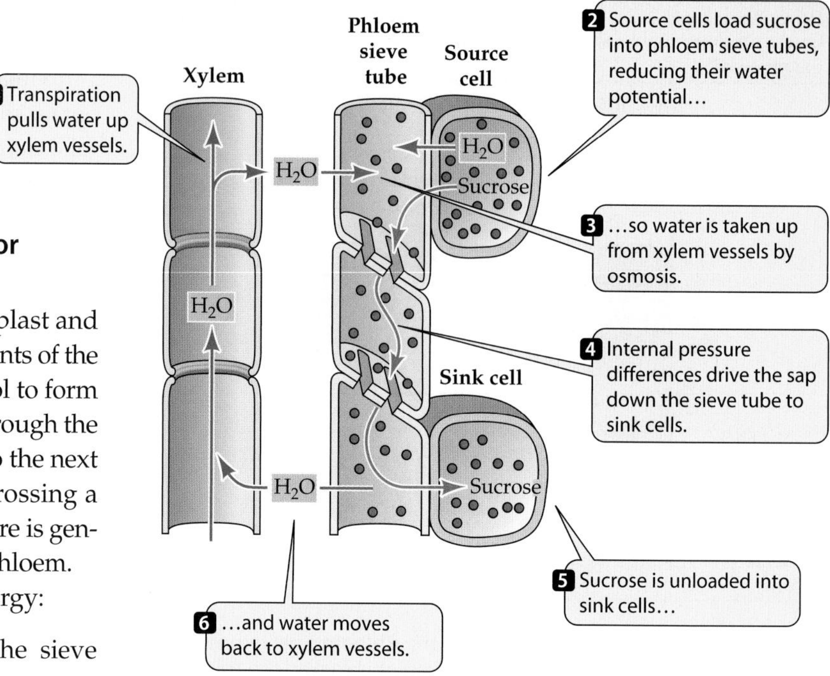

36.14 The Pressure Flow Model Combined pressure potential and water potential differences drive the bulk flow of sieve tube sap from a source to a sink.

The pressure flow model appears to account for translocation in the phloem

During sieve tube element development, the tonoplast and much of the cytosol breaks down, allowing the contents of the central vacuole to combine with much of the cytosol to form the sieve tube sap. The sap flows under pressure through the sieve tubes, moving from one sieve tube element to the next by bulk flow through the sieve plates, without crossing a membrane. We need to understand how this pressure is generated in order to understand translocation in the phloem.

Two steps in translocation require metabolic energy:

- Transport of sucrose and other solutes into the sieve tubes at sources, called *loading*
- Removal of the solutes, called *unloading*, where the sieve tubes enter sinks

According to the **pressure flow model** of translocation in the phloem, sucrose is actively transported into sieve tube elements at a source, giving those cells a greater sucrose concentration than the surrounding cells. Water therefore enters the sieve tube elements by osmosis. The entry of this water causes a greater pressure potential at the source end of the sieve tube, so that the entire fluid content of the sieve tube is pushed toward the sink end of the tube— in other words, the sap moves by bulk flow in response to a pressure gradient (Figure 36.14). In the sink, the sucrose is unloaded by active transport, maintaining the gradients of solute potential and water potential needed for movement.

The pressure flow model of translocation in the phloem is contrasted with the transpiration–cohesion–tension model of xylem transport in Table 36.1.

The pressure flow model has been experimentally tested

The pressure flow model was first proposed more than half a century ago, but some of its features are still being debated. Other mechanisms have been proposed to account for translocation in sieve tubes, but some have been disproved, and none of the rest has as much support as the pressure flow model.

Two essential requirements must be met in order for the pressure flow model to be valid:

- The sieve plates must be unobstructed, so that bulk flow from one sieve tube element to the next is possible.
- There must be an effective method for loading sucrose and other solutes into the phloem in source tissues and removing them in sink tissues.

Let's see whether these requirements are met.

ARE THE SIEVE PLATES CLOGGED OR OPEN? Early electron microscopic studies of phloem samples cut from plants produced results that seemed to contradict the pressure flow model. The pores in the sieve plates always appeared to be plugged

36.1 *Mechanisms of Sap Flow in Plant Vascular Tissues*

	XYLEM	PHLOEM
Driving force for bulk flow	Transpiration from leaves	Active transport of sucrose at source
Site of bulk flow	Non-living vessel elements and tracheids (cohesion)	Living sieve tube elements
Pressure potential in sap	Negative (pull from top; tension)	Positive (push from source; pressure)

with masses of a fibrous protein, suggesting that sieve tube sap could not flow freely. But what is the function of that fibrous protein?

One possibility is that this protein is usually distributed more or less at random throughout the sieve tube elements until the sieve tube is damaged; then the sudden surge of sap toward the cut surface carries the protein into the pores, blocking them and preventing the loss of valuable nutrients. In other words, perhaps the protein does not block the pores unless the phloem is damaged. How might this hypothesis be tested? Could phloem for microscopic observation be obtained without causing the sap to surge to the cut surface?

One way to prevent the surge of the sap is to freeze plant tissue rapidly before cutting it. Another way is to let the tissue wilt so that there is no pressure in the phloem before cutting. When these methods are used, the sieve plates are not clogged by the protein. Thus, the first condition of the pressure flow model is met.

HOW DO NEIGHBORING CELLS LOAD AND UNLOAD THE SIEVE TUBE ELEMENTS? If the pressure flow model is correct, there must be mechanisms for loading sugars and other solutes into the phloem in source regions and for unloading them in sink regions. Such mechanisms exist in all plants.

Sugars and other solutes produced in the mesophyll pass from cell to cell in the leaf and eventually enter the sieve tubes of the phloem. In some plants these substances leave the mesophyll cells and enter the apoplast, sometimes with the help of transfer cells. Then specific sugars and amino acids are actively transported into cells of the phloem, thus reentering the symplast. This passage through the apoplast and back into the symplast allows the selection of substances to be translocated by forcing them to pass through a selectively permeable membrane. In many plants, solutes reenter the symplast at the companion cells, which then transfer the solutes to the adjacent sieve tube elements.

A form of secondary active transport loads sucrose into the companion cells and sieve tubes. Sucrose is carried across the plasma membrane from apoplast to symplast by sucrose–proton symport; thus the entry of sucrose and protons is strictly coupled. For this symport to work, the apoplast must have a high concentration of protons; these protons are supplied by a primary active transport system, the proton pump. The protons then diffuse back into the cell through the symport protein, bringing sucrose with them.

In sink regions, the solutes are actively transported *out* of the sieve tube elements and into the surrounding tissues. This unloading serves two purposes: It helps maintain the gradient of solute potential and hence of pressure potential in the sieve tubes, and it promotes the buildup of sugars and starch to high concentrations in storage regions, such as developing fruits and seeds.

Plasmodesmata and material transfer between cells

Many substances move from cell to cell within the symplast by way of plasmodesmata (see Figure 35.7). Among their other roles, plasmodesmata participate in the loading and unloading of sieve tube elements. The mechanisms vary among plant species, but the story in tobacco plants is a common one. In tobacco, sugars and other solutes in source tissues enter companion cells by active transport from the apoplast and move on to the sieve tube elements through plasmodesmata. In sink tissues, plasmodesmata connect sieve tube elements, companion cells, and the cells that will receive and use the transported compounds.

Plasmodesmata undergo developmental changes as an immature sink leaf matures into a mature source leaf. Plasmodesmata in sink tissues favor rapid unloading: They are more abundant, and they allow the passage of larger molecules. Plasmodesmata in source tissues are few in number.

It was long thought that only substances with molecular weights of less than 1,000 could fit through a plasmodesma. Then biologists discovered that cells infected with tobacco mosaic virus (TMV) could allow molecules with molecular weights of as much as 20,000 to exit. We now know that TMV encodes a "movement protein" that produces this change in the permeability of the plasmodesmata—and that the plants themselves normally produce at least one such movement protein. Even large molecules such as proteins and RNAs, with molecular weights up to at least 50,000, can thus move between living plant cells. We will see some consequences of this movement of macromolecules through plasmodesmata in later chapters. Biologists are exploring possible ways to regulate the permeability, number, and form of plasmodesmata as a means of modifying traffic in the plant. Such modifications might, for example, allow the diversion of more of a grain crop's photosynthetic products into the seeds, increasing the crop yield.

Chapter Summary

Uptake and Movement of Water and Solutes

▶ Plant roots take up water and minerals from the soil. **Review Figure 36.1**

▶ Water moves through biological membranes by osmosis, always moving toward cells with a more negative water potential. The water potential of a cell or solution is the sum of the solute potential and the pressure potential. All three parameters are expressed in megapascals (MPa). **Review Figure 36.2**

▶ Mineral uptake requires transport proteins. Some minerals enter the plant by facilitated diffusion; others enter by active transport. A proton pump facilitates the active transport of many mineral ions across membranes in plants. **Review Figure 36.3**

▶ Water and minerals pass from the soil to the xylem by way of the apoplast and symplast. In the root, water and minerals can move from the cortex into the stele only by way of the symplast because Casparian strips in the endodermis block their movement through the apoplast. **Review Figures 36.4, 36.5. See Web/CD Activity 36.1**

Transport of Water and Minerals in the Xylem

▶ Early experiments established that xylem sap does not move via the pumping action of living cells.

▶ Root pressure is responsible for guttation and for the oozing of s0ap from cut stumps, but it cannot account for the ascent of xylem sap in trees.

▶ Water transport in the xylem is the result of the combined effects of transpiration, cohesion, and tension. Evaporation from the leaf produces tension in the mesophyll cells, which pulls a column of water—held together by cohesion—up through the xylem from the root. Dissolved minerals are carried passively in the water. **Review Figure 36.8**

▶ Evaporation of water cools the leaves, but a plant cannot afford to lose too much water.

▶ Support for the transpiration–cohesion–tension model of water transport comes from studies using a pressure bomb. **Review Figure 36.9**

▶ The role of transport in the xylem depends on several factors, including the K^+ concentration. **Review Figure 36.10**

Transpiration and the Stomata

▶ Transpirational water loss is minimized by the waxy cuticle of the leaves.

▶ Stomata allow a compromise between water retention and carbon dioxide uptake.

▶ A pair of guard cells controls the size of the stomatal opening. A proton pump, activated by blue light, pumps protons from the guard cells to surrounding epidermal cells, setting up a proton gradient that drives the active transport of potassium ions into the cells. Water follows osmotically, swelling the cells and opening the stomata.

▶ Carbon dioxide and water levels in the leaf also affect stomatal opening. **Review Figure 36.11**

Translocation of Substances in the Phloem

▶ Products of photosynthesis, as well as some minerals, are translocated through sieve tubes in the phloem by way of living sieve tube elements. **Review Figure 36.12**

▶ Translocation in the phloem can proceed in both directions in the stem, although in a single sieve tube it goes only one way. Translocation requires a supply of ATP.

▶ Translocation in the phloem is explained by the pressure flow model: The difference in solute concentration between sources and sinks creates a difference in pressure potential along the sieve tubes, resulting in bulk flow. **Review Figure 36.14, Table 36.1. See Web/CD Tutorial 36.1**

▶ The validity of the pressure flow model is supported by the facts that the sieve plates are normally unobstructed, allowing bulk flow, and that the neighboring cells load organic solutes into the sieve tube elements in source regions and unload them in sink regions.

▶ The distribution and properties of plasmodesmata differ between source and sink tissues. It may become possible to regulate plasmodesma permeability in crop plants.

Self-Quiz

1. Osmosis
   a. requires ATP.
   b. results in the bursting of plant cells placed in pure water.
   c. can cause a cell to become turgid.
   d. is independent of solute concentrations.
   e. continues until the pressure potential equals the water potential.
2. Water potential
   a. is the difference between the solute potential and the pressure potential.
   b. is analogous to the air pressure in an automobile tire.
   c. is the movement of water through a membrane.
   d. determines the direction of water movement between cells.
   e. is defined as 1.0 MPa for pure water under no applied pressure.
3. Which statement about aquaporins is *not* true?
   a. They are membrane transport proteins.
   b. Water movement through aquaporins is always active.
   c. The permeability of some aquaporins is subject to regulation.
   d. They are found in both animals and plants.
   e. They enable water to pass through the phospholipid bilayer without encountering a hydrophobic environment.
4. Which statement about proton pumping across the plasma membrane of plants is *not* true?
   a. It requires ATP.
   b. The region inside the membrane becomes positively charged with respect to the region outside.
   c. It enhances the movement of K^+ ions into the cell.
   d. It pushes protons out of the cell against a proton concentration gradient.
   e. It can drive the secondary active transport of negatively charged ions.
5. Which statement is *not* true?
   a. The symplast is a meshwork consisting of the (connected) living cells.
   b. Water can enter the stele without entering the symplast.
   c. The Casparian strips prevent water from moving between endodermal cells.
   d. The endodermis is a cell layer in the cortex.
   e. Water can move freely in the apoplast without entering cells.
6. In the xylem,
   a. the products of photosynthesis travel down the stem.
   b. living, pumping cells push the sap upward.
   c. the motive force is in the roots.
   d. the sap is often under tension.
   e. the sap must pass through sieve plates.
7. Which of the following is *not* part of the transpiration-cohesion-tension mechanism?
   a. Water evaporates from the walls of mesophyll cells.
   b. Removal of water from the xylem exerts a pull on the water column.
   c. Water is remarkably cohesive.
   d. The wider the tube, the greater the tension its water column can withstand.
   e. At each step, water moves to a region with a more strongly negative water potential.
8. Stomata
   a. control the opening of guard cells.
   b. release less water to the environment than do other parts of the epidermis.
   c. are usually most abundant on the upper epidermis of a leaf.
   d. are covered by a waxy cuticle.
   e. close when water is being lost at too great a rate.
9. Which statement about phloem transport is *not* true?
   a. It takes place in sieve tubes.
   b. It depends on mechanisms for loading solutes into the phloem in sources.
   c. It stops if the phloem is killed by heat.
   d. A high pressure potential is maintained in the sieve tubes.
   e. In sinks, solutes are actively transported into sieve tube elements.

10. The fibrous protein in sieve tube elements
    a. may plug leaks when a plant is damaged.
    b. clogs the sieve plates at all times.
    c. never clogs the sieve plates.
    d. serves no known function.
    e. provides the motive force for transport in the phloem.

For Discussion

1. Epidermal cells protect against excess water loss. How do they perform this function?
2. Phloem transports material from sources to sinks. What is meant by "source" and "sink"? Give examples of each.
3. What is the minimum number of plasma membranes a water molecule would have to cross in order to get from the soil solution to the atmosphere by way of the stele? To get from the soil solution to a mesophyll cell in a leaf?
4. Transpiration exerts a powerful pulling force on the water column in the xylem. When would you expect transpiration to proceed most rapidly? Why? Describe the source of the pulling force.

37 Plant Nutrition

Food is an essential commodity that separates prosperous nations from struggling ones. For instance, North Korean agriculture met that entire country's food needs until about a decade ago. The country's farmers were highly efficient and productive. Its food crisis began with the collapse of the Soviet Union, which had provided North Korea with chemicals and petroleum. This loss of support was followed by three years of drought, hailstorms, and floods. Today, North Korea is a starving country with a failed farming system.

Why should a desperate shortage of chemicals and petroleum affect a nation's agriculture? Crop production depends on several factors, but the one that is most commonly limiting is a supply of nitrogen in a form usable by plants. All plants require the element nitrogen, which is an abundant component of proteins and nucleic acids as well as chlorophyll and many other important biochemical compounds. If a plant cannot get enough nitrogen, it cannot synthesize these compounds at a rate adequate to keep itself healthy. To meet their crops' need for nitrogen and other minerals, farmers in all parts of the world apply fertilizers of one kind or another. The industrial production of fertilizers is an energy-intensive process, and the energy needed is most commonly obtained from petroleum. Without petroleum, North Korea cannot begin to provide the fertilizer needed to restore its crop production.

Nitrogen is Essential for Plant Growth In this experimental wheat field in Bangladesh, nitrogen was withheld from the plot on the left. The resulting plants were stunted and unhealthy.

In addition to nitrogen, plants need other materials from their environment. In this chapter, we will explore the differences between the basic strategies of plants and of animals for obtaining nutrition. Then we will look at what nutrients plants require and how they acquire them. Because most nutrients come from the soil, we will discuss the formation of soils and the effects of plants on soils. As any farmer can tell you, nitrogen is the nutrient that most often limits plant growth, so we will devote a section specifically to nitrogen metabolism in plants. The chapter concludes with a look at carnivorous and parasitic plants, which supplement their nutrition in special ways.

The Acquisition of Nutrients

Every living thing must obtain raw materials from its environment. These **nutrients** include the major ingredients of macromolecules: carbon, hydrogen, oxygen, and nitrogen. Carbon and oxygen enter the living world in the form of atmospheric carbon dioxide through the carbon-fixing reactions of photosynthesis. Hydrogen enters living systems through the light reactions of photosynthesis, which split water. For carbon, oxygen, and hydrogen, photosynthesis is the gateway to the living world, and these elements are in plentiful supply.

In the remainder of this chapter, we shall focus our attention on nitrogen, which is in relatively short supply for plants. The movement of nitrogen into organisms begins with processing by some highly specialized bacteria living in the soil. Some of these bacteria act on nitrogen gas, converting it into a form usable by plants. The plants, in turn, provide organic nitrogen (and carbon) to animals, fungi, and many microorganisms.

In addition to nitrogen, other **mineral nutrients** are essential to living organisms. The proteins of organisms contain sulfur (S), and their nucleic acids contain phosphorus (P). There is magnesium (Mg) in chlorophyll, and iron (Fe) in many important compounds, such as the cytochromes. Within the soil, these and other minerals dissolve in water, forming a solution—called the **soil solution**—that contacts the roots of plants. Plants take up most of these mineral nutrients from the soil solution in ionic form.

Autotrophs make their own organic compounds

Plants, some protists, and some bacteria are **autotrophs**; that is, they make their own *organic* (carbon-containing) compounds from simple inorganic nutrients—carbon dioxide, water, nitrogen-containing ions, and a few other soluble mineral nutrients. The plants provide carbon, oxygen, hydrogen, nitrogen, and sulfur to most of the rest of the living world. **Heterotrophs** are organisms that require preformed organic compounds as food. All heterotrophs depend directly or indirectly on autotrophs as their source of nutrition.

Most autotrophs are *photosynthesizers*—that is, they use light as their source of energy for synthesizing organic compounds from inorganic raw materials. Some autotrophs, however, are *chemosynthesizers*, deriving their energy not from light, but from reduced inorganic substances, such as hydrogen sulfide (H_2S), in their environment. All chemosynthesizers are bacteria. As we'll see below, some chemosynthetic bacteria in the soil contribute to the nutrition of plants by increasing the availability of nitrogen and sulfur.

How does a stationary organism find nutrients?

Many heterotrophs can move from place to place to find the nutrients they need. An organism that cannot move, termed a *sessile* organism, must obtain nutrients and energy from sources that are somehow brought to it. Most sessile animals depend primarily on the movement of water to bring them raw materials and energy in the form of food, but a plant's supply of energy arrives at the speed of light from the sun. However, with the exception of carbon and oxygen in CO_2, a plant's supply of nutrients is strictly local, and the plant may use up the water and mineral nutrients in its local environment as it develops. How does a plant cope with the problem of scarce nutrient supplies?

One way is to extend itself by growing in search of new resources. Growth is a plant's version of movement. Among plant organs, the roots obtain most of the mineral nutrients needed for growth. By growing through the soil, they mine the soil for new sources of mineral nutrients and water. The growth of leaves helps a plant secure light and carbon dioxide. A plant may compete with other plants for light by outgrowing and shading them.

As it grows, a plant—or even a single root—must deal with a variable environment. Animal droppings create high local concentrations of nitrogen. A particle of calcium carbonate in the soil may make a tiny area alkaline, while dead organic matter may make a nearby area acidic. Such microenvironments encourage or discourage the proliferation of a root system.

Mineral Nutrients Essential to Plants

As roots grow through the soil, what important mineral nutrients do plants take up from their environment, and what are the roles of those nutrients? Table 37.1 lists the mineral nutrients that have been determined to be essential for plants. Except for nitrogen, they all derive from rock. All of them are usually taken up from the soil solution.

There are three criteria for calling something an **essential element**:

- The element must be *necessary* for normal growth and reproduction.
- The element cannot be *replaceable* by another element.
- The requirement must be *direct*—that is, not the result of an indirect effect, such as the need to relieve toxicity caused by another substance.

In this section, we'll consider the symptoms of particular mineral deficiencies, the roles of some of the mineral nutrients, and the technique by which the essential elements for plants were identified.

There are two categories of essential elements: macronutrients and micronutrients (see Table 37.1).

- Plants need **macronutrients** in concentrations of at least 1 gram per kilogram of their dry matter.*

**Dry matter*, or *dry weight*, is what remains after all the water has been removed from a plant tissue sample.

37.1 Mineral Elements Required by Plants

ELEMENT	ABSORBED FORM	MAJOR FUNCTIONS
Macronutrients		
Nitrogen (N)	NO_3^- and NH_4^+	In proteins, nucleic acids, etc.
Phosphorus (P)	$H_2PO_4^-$ and HPO_4^{2-}	In nucleic acids, ATP, phospholipids, etc.
Potassium (K)	K^+	Enzyme activation; water balance; ion balance; stomatal opening
Sulfur (S)	SO_4^{2-}	In proteins and coenzymes
Calcium (Ca)	Ca^{2+}	Affects the cytoskeleton, membranes, and many enzymes; second messenger
Magnesium (Mg)	Mg^{2+}	In chlorophyll; required by many enzymes; stabilizes ribosomes
Micronutrients		
Iron (Fe)	Fe^{2+}	In active site of many redox enzymes and electron carriers; chlorophyll synthesis
Chlorine (Cl)	Cl^-	Photosynthesis; ion balance
Manganese (Mn)	Mn^{2+}	Activation of many enzymes
Boron (B)	$B(OH)_3$	Possibly carbohydrate transport (poorly understood)
Zinc (Zn)	Zn^{2+}	Enzyme activation; auxin synthesis
Copper (Cu)	Cu^{2+}	In active site of many redox enzymes and electron carriers
Nickel (Ni)	Ni^{2+}	Activation of one enzyme
Molybdenum (Mo)	MoO_4^{2-}	Nitrate reduction

- Plants need **micronutrients** in concentrations of less than 100 milligrams per kilogram of their dry matter.

These two categories differ only with regard to the amounts required by plants. Both the macronutrients and the micronutrients are essential for the plant to complete its life cycle from seed to seed. How do we know if a plant is getting enough of a particular nutrient?

Deficiency symptoms reveal inadequate nutrition

Before a plant that is deficient in an essential element dies, it usually displays characteristic *deficiency symptoms*, such as discoloration or deformation of its leaves. Table 37.2 describes the symptoms of some common mineral deficiencies. Such symptoms help horticulturists diagnose mineral nutrient deficiencies in plants. With proper diagnosis, appropriate treatment can be applied in the form of a **fertilizer** (an added source of mineral nutrients).

Nitrogen deficiency is the most common mineral deficiency in both natural and agricultural environments. Plants in natural environments are almost always deficient in nitrogen, but they seldom display deficiency symptoms. Instead, their growth slows to match the available supply of nitrogen. Crop plants, on the other hand, show deficiency symptoms if a formerly abundant supply of nitrogen runs out. The visible symptoms of nitrogen deficiency include uniform yellowing, or *chlorosis*, of older leaves. Chlorophyll, which is responsible for the green color of leaves, contains nitrogen. Without nitrogen there is no chlorophyll, and without chlorophyll, the yellow carotenoid pigments in the leaves become visible.

Nitrogen deficiency is not the only cause of chlorosis. Inadequate iron in the soil can also cause chlorosis because iron, although it is not contained in the chlorophyll molecule, is required for chlorophyll synthesis. However, iron deficiency commonly causes chlorosis of the *youngest* leaves, with their veins sometimes remaining green. The reason for this difference is that nitrogen is readily translocated in the plant and can be redistributed from older tissues to younger tissues to favor their growth. Iron, on the other hand, cannot

37.2 Some Mineral Deficiencies in Plants

DEFICIENCY	SYMPTOMS
Calcium	Growing points die back; young leaves are yellow and crinkly
Iron	Young leaves are white or yellow with green veins
Magnesium	Older leaves have yellow in stripes between veins
Manganese	Younger leaves are pale with stripes of dead patches
Nitrogen	Oldest leaves turn yellow and die prematurely; plant is stunted
Phosphorus	Plant is dark green with purple veins and is stunted
Potassium	Older leaves have dead edges
Sulfur	Young leaves are yellow to white with yellow veins
Zinc	Young leaves are abnormally small; older leaves have many dead spots

be readily redistributed. Younger tissues that are actively growing and synthesizing compounds needed for their growth show iron deficiency before older leaves, which have already completed their growth.

EXPERIMENT

Question: Is a particular ingredient of a growth medium an essential plant nutrient?

METHOD Grow seedlings in a medium that lacks the element in question (in this case, nitrogen)

Seedling grown in a complete growth medium.

Seedling grown in a medium lacking nitrogen.

RESULTS

Growth is normal.

Growth is abnormal.

Conclusion: Nitrogen is an essential plant nutrient.

37.1 Identifying Essential Elements for Plants This diagram shows the procedure for identifying nutrients essential to plants, using nitrogen as an example.

Several essential elements fulfill multiple roles

Essential elements may play several different roles in plant cells—some structural, others catalytic. Magnesium, as we have mentioned, is a constituent of the chlorophyll molecule and hence is essential to photosynthesis. It is also required as a cofactor by numerous enzymes involved in cellular respiration and other metabolic pathways.

Phosphorus, usually in phosphate groups, is found in many organic compounds, particularly in nucleic acids and in the intermediates of the energy-harvesting pathways of photosynthesis and glycolysis. As we saw in Chapter 7, the transfer of phosphate groups occurs in many energy-storing and energy-releasing reactions, notably those that use or produce ATP. The addition or removal of phosphate groups is also used to activate or inactivate enzymes.

Calcium plays many roles in plants. Its function in the processing of hormonal and environmental cues is a subject of great biological interest, as we'll see in the next chapter. Calcium also affects membranes and cytoskeletal activity, participates in spindle formation for mitosis and meiosis, and is a constituent of the middle lamella of cell walls. Other elements, such as iron and potassium, also play multiple roles in plants.

All of these elements are essential to the life of all plants. How did biologists discover which elements are essential?

Experiments were designed to identify essential elements

An element is considered essential to plants if a plant fails to complete its life cycle, or grows abnormally, when that element is not available, or is not available in sufficient quantities. The essential elements for plants were identified by growing plants *hydroponically*—that is, with their roots suspended in nutrient solutions without soil (Figure 37.1). In the first successful experiments of this type, performed a century and a half ago, plants grew seemingly normally in solutions containing only calcium nitrate, magnesium sulfate, and potassium phosphate. Omission of any of these compounds made the solution incapable of supporting normal growth. Tests with other compounds including these elements soon established the six macronutrients— calcium, nitrogen, magnesium, sulfur, potassium, and phosphorus—as essential elements.

Identifying essential elements by this experimental approach proved to be a more difficult task in the case of the micronutrients. In the nineteenth-century experiments on plant nutrition, some of the chemicals used were so impure that they provided micronutrients that the investigators thought they had excluded. Furthermore, because some micronutrients are required in such tiny amounts, there may be enough in a seed to supply the embryo and the resultant second-generation plant throughout its lifetime and leave enough in the next seed to get the third generation well started. Indeed, simply touching a plant may give it a significant supply of chlorine in the form of chloride ions from sweat. Such difficulties make it necessary to perform nutrition experiments in tightly controlled laboratories with special air filters (to exclude microscopic salt particles in the air) and to use only chemicals that had been purified to the highest degree attainable by modern chemistry. Only rarely are new essential elements reported now. Either the list is nearly complete, or perhaps, we will need more sophisticated techniques to add to it.

Where does the plant find its essential mineral nutrients? How does it absorb them?

Soils and Plants

Most terrestrial plants live their lives anchored to the soil. Of course, soils offer mechanical support for growing plants, but there are many other plant-soil interactions, some of which are much more complex. Plants obtain their mineral nutrients from the soil solution. Water for terrestrial plants also comes from the soil, as does the supply of oxygen for the roots. Soil harbors bacteria, some of which are beneficial to plant life. Soils may also contain organisms harmful to plants.

In this section, we will examine the composition, structure, and formation of soils. We will consider their role in plant nutrition, their care and supplementation in agriculture, and their modification by the plants that grow in them.

Soils are complex in structure

Soils are complex systems made up of living and nonliving components. The living components include plant roots as well as populations of bacteria, fungi, protists, and animals such as earthworms and insects (Figure 37.2). The nonliving portion of the soil includes rock fragments ranging in size from large rocks through *sand* and *silt* and finally to tiny particles called **clay** that are 2 µm or less in diameter. Soil also contains water and dissolved mineral nutrients, air spaces, and dead organic matter. The air spaces are crucial sources of oxygen (in the form of O_2) for plant roots. The characteristics of soils are not static. Soils change constantly through natural phenomena such as rain, temperature extremes, and the activities of plants and animals, as well as human activities—agriculture in particular.

The structure of many soils changes with depth, revealing a *soil profile*. Although soils differ greatly, almost all soils consist of two or more recognizable horizontal layers, called **horizons**, lying on top of one another. Mineral nutrients tend to be **leached** from the upper horizons—dissolved in rain or irrigation water and carried to deeper horizons, where they are unavailable to plant roots.

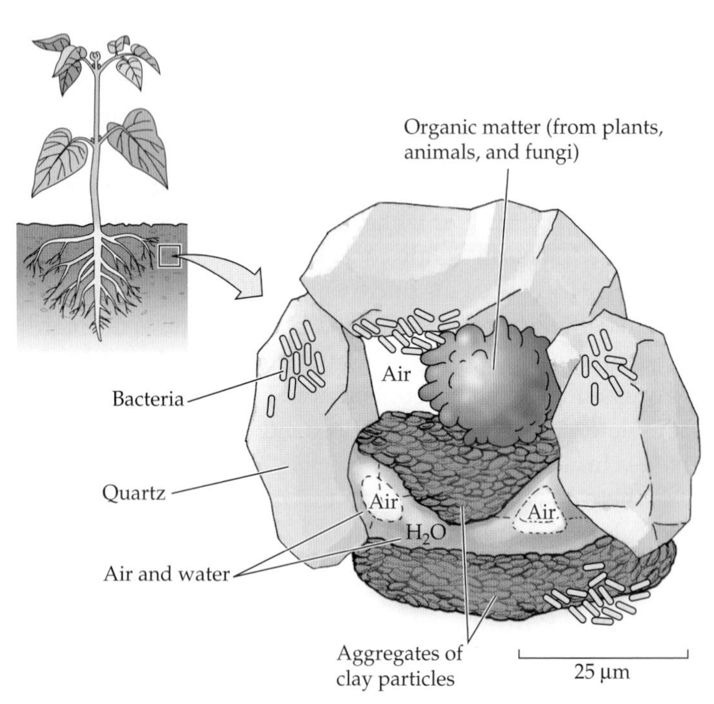

37.2 The Complexity of Soil Even a tiny crumb of soil has both organic and inorganic components.

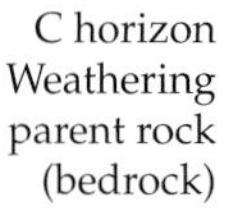

37.3 A Soil Profile The A, B, and C horizons can sometimes be seen in road cuts such as this one in Australia. The dark upper layer (the A horizon) is home to most of the living organisms in the soil.

Soil scientists recognize three major horizons (A, B, and C) in the profile of a typical soil (Figure 37.3). **Topsoil** is the A horizon, from which mineral nutrients may be depleted by leaching. Most of the dead and decaying organic matter in the soil is in the A horizon, as are most plant roots, earthworms, insects, nematodes, and microorganisms. Successful agriculture depends on the presence of a suitable A horizon.

Topsoils are composed of different proportions of sand, silt, and clay. In pure sand there are abundant air spaces between the relatively large particles, but sand is low in water and mineral nutrients. Clay contains many mineral nutrients and more water than sand does, but the tiny clay particles pack tightly together, leaving little space to trap air. A little bit of clay goes a long way in affecting soil properties. A **loam** is a soil that has significant amounts of sand, silt, and clay, and thus has sufficient levels of air, water, and nutrients for plants. Loams also contain organic matter. Most of the best topsoils for agriculture are loams.

Below the A horizon is the B horizon, or *subsoil*, which is the zone of infiltration and accumulation of materials leached from above. Farther down, the C horizon is the *parent rock* that is breaking down to form soil. Some deep-growing roots extend into the B horizon to obtain water and nutrients, but roots rarely enter the C horizon.

Soils form through the weathering of rock

The type of soil in a given area depends on the type of parent rock from which it formed, the climate, the landscape features, the organisms living there, and the length of time that soil-forming processes have been acting (sometimes millions of years). Rocks are broken down into soil in part by *mechanical weathering*, which is the physical breakdown—without any accompanying chemical changes—of materials by wetting, drying, and freezing. The most important parts of soil formation, however, include *chemical weathering*, the chemical alteration of at least some of the materials in the rocks.

Both the physical and chemical properties of soils depend on the amounts and kinds of clay particles they contain. These tiny particles, which bind mineral nutrients and aggregate into larger particles, are extremely important to plant growth. Clay is not produced merely by the mechanical grinding up of rocks. In addition to mechanical weathering, several types of chemical weathering are required:

- Oxidation by atmospheric oxygen makes some essential elements more available to plants.
- Reaction with water (hydrolysis) releases some mineral nutrients from the rock.
- Acids, carbonic acid in particular, free some essential elements from their parent salts.

These reactions leave the surface of clay particles with an abundance of negatively charged chemical groups, to which certain mineral nutrients bind. Let's see how roots take up these mineral nutrients from clay particles.

Soils are the source of plant nutrition

The availability of mineral nutrients to plant roots depends on the presence of clay particles in the soil. The negatively charged clay particles bind the cations of many minerals that are important for plant nutrition, such as potassium (K^+), magnesium (Mg^{2+}), and calcium (Ca^{2+}). To become available to plants, these cations must be detached from the clay particles.

This task is accomplished by reactions with protons (hydrogen ions, H^+). These protons are released into the soil by roots, which also release CO_2 through cellular respiration. The CO_2 dissolves in the soil water and reacts with it to form carbonic acid, which then ionizes to form bicarbonate and free protons ($CO_2 + H_2O \rightleftharpoons H_2CO_3 \rightleftharpoons H^+ + HCO_3^-$). These protons bind more strongly to the clay particles than do the mineral cations, so they trade places with the cations in a process called **ion exchange** (Figure 37.4). Ion exchange puts important cations back into the soil solution, from which they are taken up by the roots. The capacity of a soil to support plant growth, called *soil fertility*, is determined in part by its ability to provide nutrients in this manner.

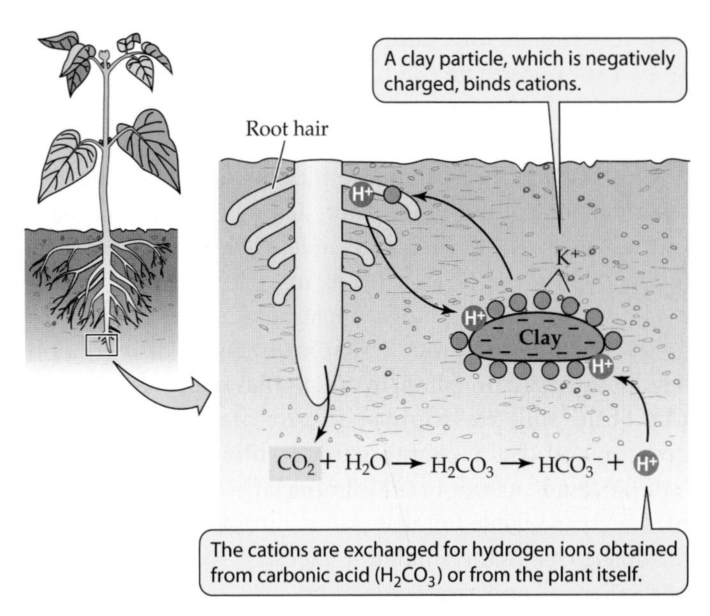
A clay particle, which is negatively charged, binds cations.
Root hair
H+
K+
Clay
H+
H+
$CO_2 + H_2O \rightarrow H_2CO_3 \rightarrow HCO_3^- + H^+$
The cations are exchanged for hydrogen ions obtained from carbonic acid (H_2CO_3) or from the plant itself.

37.4 Ion Exchange Plants obtain mineral nutrients from the soil primarily in the form of positive ions; potassium is the example shown here.

Clay particles effectively hold and exchange cations, and cations tend to be retained in the A horizon. However, there is no comparable mechanism for exchanging anions, the negatively charged ions. As a result, important anions such as nitrate (NO_3^-) and sulfate (SO_4^{2-})—the primary and direct sources of nitrogen and sulfur, respectively—leach rapidly from the A horizon. As a consequence of this leaching, the primary soil reservoir of nitrogen is not in the form of nitrate ions. Most of the nitrogen in the A horizon is found in the organic matter in the soil, which slowly decomposes to release nitrogen in a form that can be absorbed and used by plants.

Fertilizers and lime are used in agriculture

Agricultural soils often require fertilizers because irrigation and rainwater leach mineral nutrients from the soil and because the harvesting of crops removes the nutrients that the plants took up from the soil during their growth. Crop yields decrease if any essential element is depleted. Mineral nutrients may be replaced by organic fertilizers, such as rotted manure, or by inorganic fertilizers of various types.

ORGANIC AND INORGANIC FERTILIZERS. The three elements most commonly added to agricultural soils are nitrogen (N), phosphorus (P), and potassium (K). Commercial fertilizers are characterized by their "N-P-K" percentages. A

5-10-10 fertilizer, for example, contains 5 percent nitrogen, 10 percent phosphate (P_2O_5), and 10 percent potash (K_2O) by weight.* Sulfur, in the form of a sulfate, is also occasionally added to soils.

Either organic or inorganic fertilizers can provide the necessary mineral nutrients for plants. Organic fertilizers release nutrients slowly, which results in less leaching than occurs with a one-time application of an inorganic fertilizer. However, the nutrients from organic fertilizers are not immediately available to plants. Organic fertilizers also contain residues of plant or animal materials that improve the structure of the soil, providing spaces for air movement, root growth, and drainage. Inorganic fertilizers, on the other hand, provide a supply of soil nutrients that is almost immediately available for absorption. Furthermore, inorganic fertilizers can be formulated to meet the requirements of a particular soil and a particular crop.

pH EFFECTS ON NUTRIENTS. The availability of nutrient ions, whether they are naturally present in the soil or added as fertilizer, is altered by changes in soil pH. The optimal soil pH for most crops is about 6.5, but so-called acid-loving crops such as blueberries prefer a pH closer to 4. Rainfall and the decomposition of organic substances lower the pH of the soil. Such acidification can be reversed by **liming**—the application of compounds commonly known as *lime*, such as calcium carbonate, calcium hydroxide, or magnesium carbonate. The addition of these compounds leads to the removal of H^+ ions from the soil. Liming also increases the availability of calcium to plants.

Sometimes, on the other hand, a soil is not acidic enough. In this case, sulfur can be added in the form of elemental sulfur, which soil bacteria convert to sulfuric acid. Iron and some other elements are more available to plants at a slightly acidic pH. Soil pH testing is useful for home gardens and lawns as well as for agriculture. The test results indicate what amendments should be made to the soil.

SPRAY APPLICATION OF NUTRIENTS. Spraying leaves with a nutrient solution is another effective way to deliver some essential elements to growing plants. Plants take up more copper, iron, and manganese when these elements are applied as *foliar* (leaf) sprays than when they are added to the soil. Such foliar application of mineral nutrients is increasingly used in wheat production, but fertilizers are still delivered most commonly by way of the soil.

The relationship between plants and soils is not a one-way affair—soils affect plants, but plants also affect soils.

*The analysis is by weight of the nutrient-containing compound and not as weights of the elements N, P, and K. A 5-10-10 fertilizer actually does contain 5 percent nitrogen, but only 4.3 percent phosphorus and 8.3 percent potassium on an elemental basis.

Plants affect soil fertility and pH

The soil that forms in a particular place depends on the types of plants growing there. Plant litter, such as dead fallen leaves, is the major source of the carbon-rich materials that break down to form **humus**—dark-colored organic material, each particle of which is too small to be recognizable with the naked eye. Soil bacteria and fungi produce humus by breaking down plant litter, animal feces, dead organisms, and other organic material. Humus is rich in mineral nutrients, especially nitrogen that was excreted by animals. In combination with clay, humus favors plant growth by trapping supplies of water and oxygen for absorption by roots. Looking at the big picture, we see that successful plant growth can create conditions that support further plant growth.

Plants also affect the pH of the soil in which they grow. Roots maintain a balance of electric charges. If they absorb more cations than anions, they excrete H^+ ions, thus lowering the soil pH. If they absorb more anions than cations, they excrete OH^- ions or HCO_3^- ions, raising the soil pH.

The mineral nutrient most commonly in short supply, in both natural and agricultural situations, is nitrogen, despite the fact that elemental nitrogen makes up almost four-fifths of Earth's atmosphere. What is the reason for this scarcity? Let's consider how nitrogen is made available to plants.

Nitrogen Fixation

The Earth's atmosphere is a vast reservoir of nitrogen in the form of nitrogen gas (N_2). However, plants cannot use N_2 directly as a nutrient. It is a highly unreactive substance—the triple bond linking the two nitrogen atoms is extremely stable, and a great deal of energy is required to break it. How, then, is nitrogen made available for the synthesis of proteins and nucleic acids?

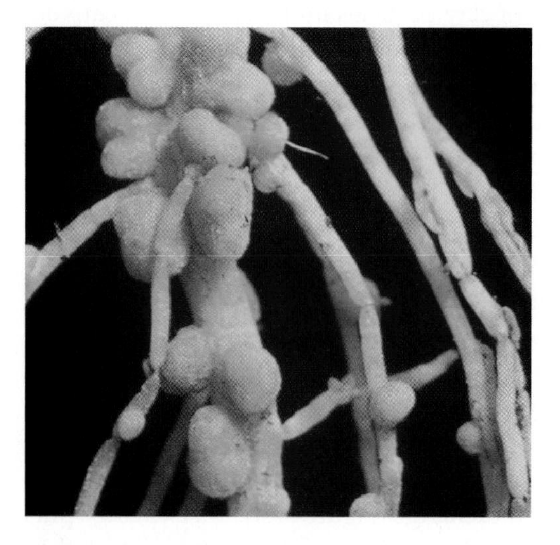

37.5 Root Nodules Large, round nodules are visible in the root system of a pea plant. These nodules house nitrogen-fixing bacteria.

A few species of bacteria have an enzyme that enables them to convert N_2 into a more reactive and biologically useful form by a process called **nitrogen fixation**. These prokaryotic organisms—*nitrogen fixers*—convert N_2 to ammonia (NH_3). There are relatively few species of nitrogen fixers, and their biomass is small relative to the mass of other organisms that depend on them for survival on Earth. This talented group of prokaryotes is just as essential to the biosphere as are the photosynthetic autotrophs.

Nitrogen fixers make all other life possible

By far the greatest share of total world nitrogen fixation is performed biologically by nitrogen-fixing prokaryotes, which fix approximately 170 million Mg (megagrams or metric tons) of nitrogen per year. About 80 million Mg is fixed industrially by humans. A smaller amount of nitrogen is fixed in the atmosphere by nonbiological means such as lightning, volcanic eruptions, and forest fires. Rain brings these atmospherically formed products to the ground.

Several groups of bacteria fix nitrogen. In the oceans, various photosynthetic bacteria, including cyanobacteria, fix nitrogen. In fresh water, cyanobacteria are the principal nitrogen fixers. On land, free-living soil bacteria make some contribution to nitrogen fixation, but they fix only what they need for their own use and release the fixed nitrogen only when they die.

Other nitrogen-fixing bacteria live in close association with plant roots (Figure 37.5). They release up to 90 percent of the nitrogen they fix to the plant and excrete some amino acids into the soil, making nitrogen immediately available to other organisms. The plant obtains fixed nitrogen from the bacterium, and the bacterium obtains the products of photosynthesis from the plant. Such associations are excellent examples of *mutualism*, an interaction between two species in which both species benefit. They are also examples of *symbiosis*, in which two different species live in physical contact for a significant portion of their life cycles.

Bacteria of the genus *Rhizobium* fix nitrogen only in close, mutualistic association with the roots of plants in the legume family. The legumes include peas, soybeans, clover, alfalfa, and many tropical shrubs and trees. The bacteria infect the plant's roots, and the roots develop nodules in response to their presence. The various species of *Rhizobium* show a high specificity for the species of legume they infect. Farmers and gardeners coat legume seeds with *Rhizobium* to make sure the bacteria are present. Some farmers alternate their crops, planting clover or alfalfa occasionally to increase the available nitrogen content of the soil.

The legume–*Rhizobium* association is not the only bacterial association that fixes nitrogen. Some cyanobacteria fix nitrogen in association with fungi in lichens or with ferns, cycads, or nontracheophytes. Rice farmers can increase crop yields by growing the water fern *Azolla*, with its symbiotic nitrogen-fixing cyanobacterium, in the flooded fields where rice is grown. Another group of bacteria, the filamentous actinomycetes, fix nitrogen in association with root nodules on woody species such as alder and mountain lilacs.

How does biological nitrogen fixation work? In the four sections that follow, we'll consider the role of the enzyme nitrogenase, the mutualistic collaboration of plant and bacterial cells in root nodules, the need to supplement biological nitrogen fixation in agriculture, and the contributions of plants and bacteria to the global nitrogen cycle.

Nitrogenase catalyzes nitrogen fixation

Nitrogen fixation is the reduction of nitrogen gas. It proceeds by the stepwise addition of three pairs of hydrogen atoms to N_2 (Figure 37.6). In addition to N_2, these reactions require three things:

37.6 Nitrogenase Fixes Nitrogen Throughout the chemical reactions of nitrogen fixation, the reactants are bound to the enzyme nitrogenase. A reducing agent transfers hydrogen atoms to nitrogen, and eventually the final product—ammonia—is released.

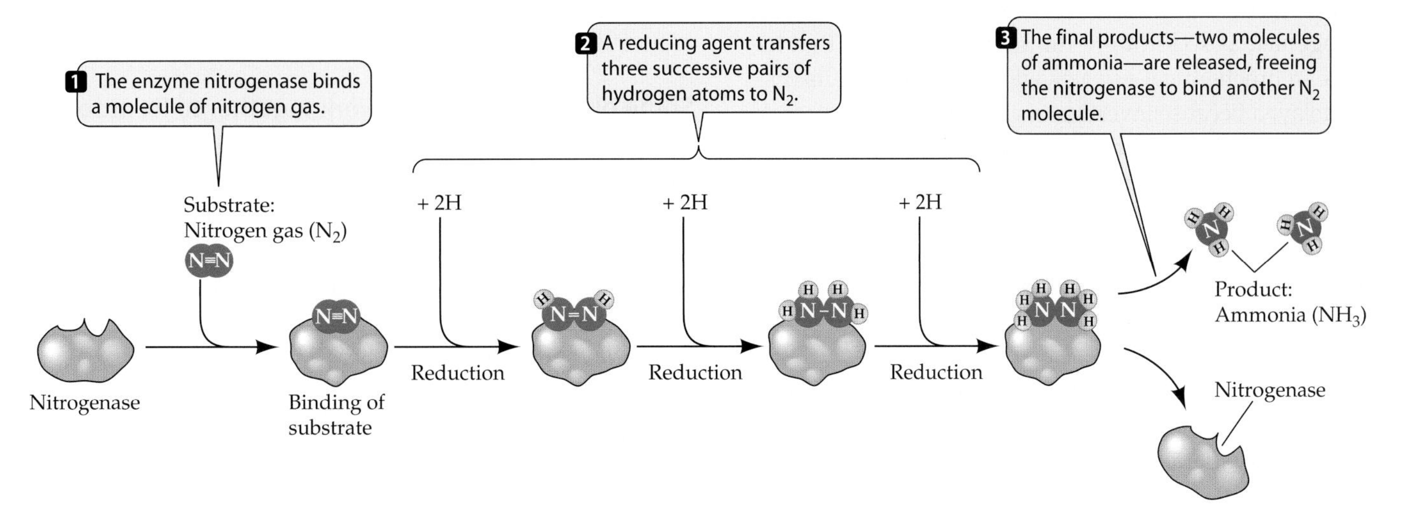
1 The enzyme nitrogenase binds a molecule of nitrogen gas.
2 A reducing agent transfers three successive pairs of hydrogen atoms to N2.
3 The final products—two molecules of ammonia—are released, freeing the nitrogenase to bind another N2 molecule.
Substrate: Nitrogen gas (N2)
+ 2H
+ 2H
+ 2H
Nitrogenase
Binding of substrate
Reduction
Reduction
Reduction
Product: Ammonia (NH3)
Nitrogenase

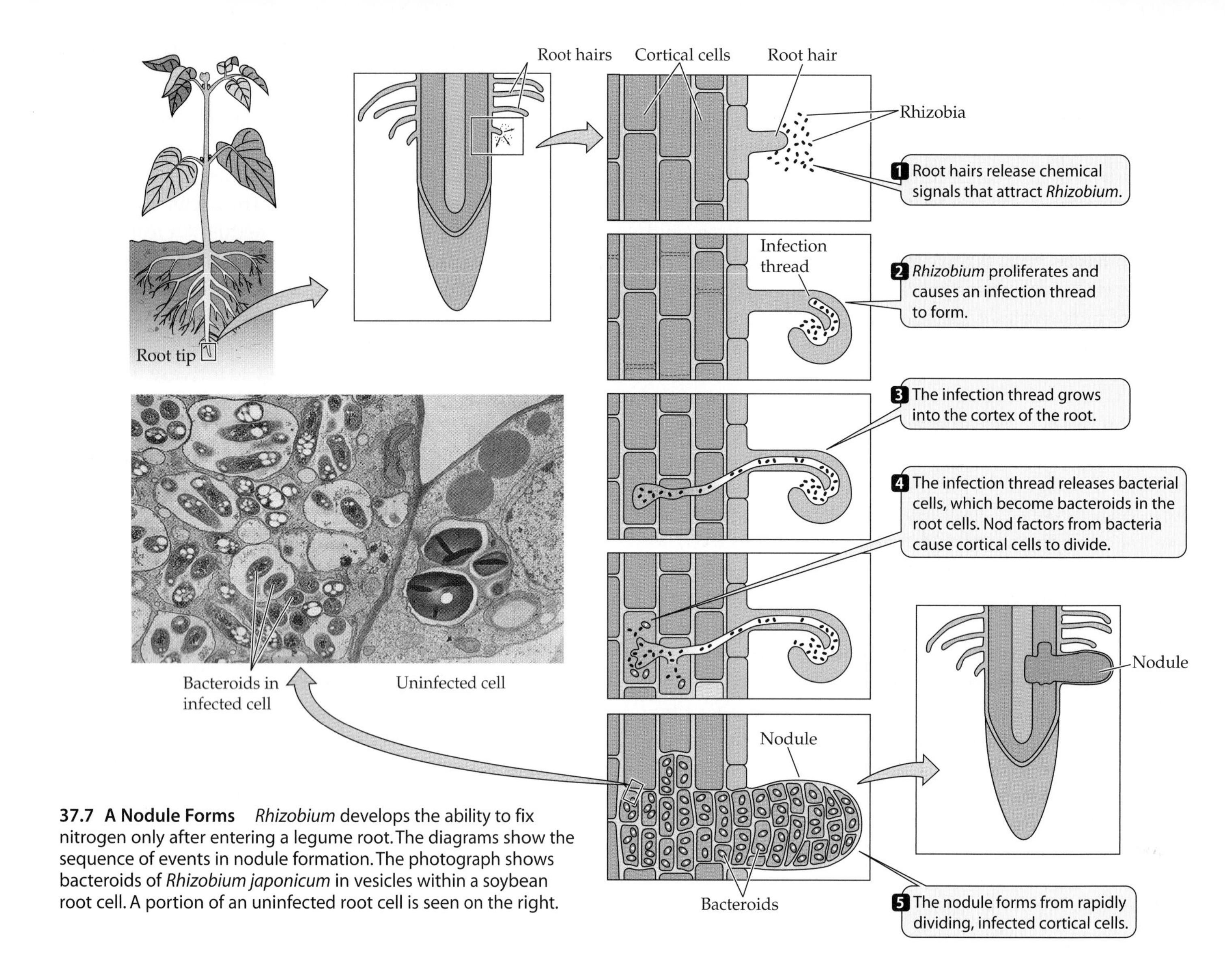

Root hairs
Cortical cells
Root hair
Rhizobia
1 Root hairs release chemical signals that attract *Rhizobium*.
Infection thread
2 *Rhizobium* proliferates and causes an infection thread to form.
Root tip
3 The infection thread grows into the cortex of the root.
4 The infection thread releases bacterial cells, which become bacteroids in the root cells. Nod factors from bacteria cause cortical cells to divide.
Nodule
Bacteroids in infected cell
Uninfected cell
Nodule
Bacteroids
5 The nodule forms from rapidly dividing, infected cortical cells.

37.7 A Nodule Forms *Rhizobium* develops the ability to fix nitrogen only after entering a legume root. The diagrams show the sequence of events in nodule formation. The photograph shows bacteroids of *Rhizobium japonicum* in vesicles within a soybean root cell. A portion of an uninfected root cell is seen on the right.

- a strong reducing agent to transfer hydrogen atoms to N_2 and to the intermediate products of the reaction
- a great deal of energy, which is supplied by ATP
- the enzyme **nitrogenase**, which catalyzes the reaction

(Depending on the species of nitrogen fixer, either respiration or photosynthesis may provide both the necessary reducing agent and ATP.)

Nitrogenase is so strongly inhibited by oxygen that its presence in biochemical extracts was obscured and its discovery delayed because investigators had not thought to seek it under anaerobic conditions. It is therefore not surprising that many nitrogen fixers are anaerobes and live in environments with little or no O_2. Because this crucial enzyme is so inhibited by O_2, it was at first surprising that legumes respire aerobically, as do *Rhizobium*. Investigation of the root nodules where nitrogenase is found revealed how the enzyme could operate there.

Within a root nodule, O_2 is maintained at a low level sufficient to support respiration, but not so high as to inactivate nitrogenase. The plant makes this possible by producing the protein **leghemoglobin** in the cytoplasm of the nodule cells. Leghemoglobin is a close relative of hemoglobin, the oxygen-carrying pigment of animals. Some plant nodules contain enough of it to be bright pink when viewed in cross section. Leghemoglobin, with its iron-containing heme groups, transports enough oxygen to the bacteroids to support their respiration.

Some plants and bacteria work together to fix nitrogen

Neither free-living *Rhizobium* species nor uninfected legumes can fix nitrogen. Only when the two are closely associated in root nodules does the reaction take place. The establishment of this symbiosis between *Rhizobium* and a legume requires a complex series of steps, with active contributions by both the bacteria and the plant root (Figure 37.7). First the root releases flavonoids and other chemical signals that attract soil-living *Rhizobium* to the vicinity of the root. Flavonoids trigger the transcription of bacterial *nod* genes, which encode Nod (nodulation) factors. These factors, secreted by the bacteria, cause cells in the root cortex to divide, leading to the formation of a primary nodule meristem. The meristem gives rise to the plant tissue that constitutes the nodule.

Among the products of the meristem is a layer of cells that excludes O_2 from the interior of the nodule. The function of leghemoglobin is to carry O_2 across this barrier. Within a nodule, the bacteria take the form of **bacteroids** within membranous vesicles. Bacteroids are swollen, deformed bacteria that can fix nitrogen—in effect, nitrogen-fixing organelles.

The partnership between bacterium and plant in nitrogen-fixing nodules is not the only case in which plants depend on other organisms for assistance with their nutrition. Another example is that of *mycorrhizae*, root–fungus associations in which the fungus greatly increases the absorption of water and minerals (especially phosphorus) by the plant (see Figure 31.16). A growing body of evidence suggests that nodule formation depends on some of the same genes and mechanisms that allow mycorrhizae to develop.

Biological nitrogen fixation does not always meet agricultural needs

Bacterial nitrogen fixation is not sufficient to support the needs of agriculture. Traditional farmers used to plant dead fish along with corn so that the decaying fish would release fixed nitrogen that the developing corn could use. Industrial nitrogen fixation is becoming ever more important to world agriculture because of the degradation of soils and the need to feed a rapidly expanding population.

Most industrial nitrogen fixation is done by a chemical process called the *Haber process*, which requires a great deal of energy. An alternative is urgently needed because of the rising cost of energy. At present, in the United States, the manufacture of nitrogen-containing fertilizer takes more energy than does any other aspect of crop production. In biological systems, nitrogen fixation requires a great deal of ATP.

Research on biological nitrogen fixation is being vigorously pursued, with commercial applications very much in mind. One line of investigation centers on recombinant DNA technology as a means of engineering new plant–bacterium associations that produce their own nitrogenase. Currently there are attempts to transfer genes from *Rhizobium* into bacteria that already live in the roots of cereal plants.

Plants and bacteria participate in the global nitrogen cycle

The nitrogen released into the soil by nitrogen fixers is primarily in the form of ammonia (NH_3) and ammonium ions (NH_4^+). Although ammonia is toxic to plants, ammonium ions can be taken up safely at low concentrations. Soil bacteria called *nitrifiers*, which we described in Chapter 27, oxidize ammonia to nitrate ions (NO_3^-)—another form that plants can take up—by the process of **nitrification** (Figure 37.8). Soil pH affects the uptake of nitrogen: Nitrate ions are taken up preferentially under more acidic conditions, ammonium ions under more basic ones.

The steps that we have followed so far are carried out by bacteria: N_2 is *reduced* to ammonia in nitrogen fixation and ammonia is *oxidized* to nitrate in nitrification. The next steps are carried out by plants, which reduce the nitrate they have taken up all the way back to ammonia. All the reactions of **nitrate reduction** are carried out by the plant's own enzymes. The later steps, from nitrite (NO_2^-) to ammonia, take place in the chloroplasts, but this conversion is not part of photosynthesis. The plant uses the ammonia thus formed to

37.8 The Nitrogen Cycle Nitrogen fixation, nitrification, nitrate reduction, and denitrification are the components of an essential chemical cycle that converts atmospheric nitrogen gas into ammonium ions and nitrate ions—forms of nitrogen that can be taken up by plants—and returns N_2 to the atmosphere.

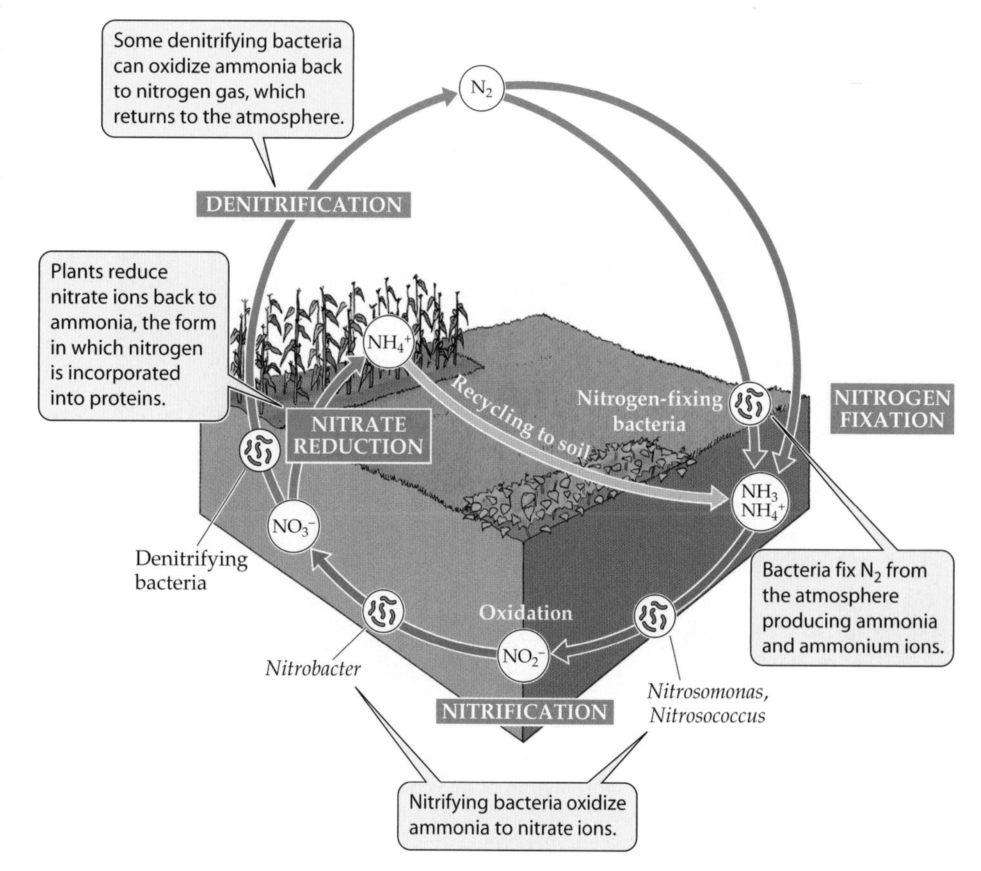

manufacture amino acids, from which the plant's proteins and all its other nitrogen-containing compounds are formed. Animals cannot reduce nitrogen, and they depend on plants to supply them with reduced nitrogenous compounds.

Bacteria called *denitrifiers* return nitrogen from animal wastes and dead organisms to the atmosphere as N_2. This process is called **denitrification** (see Chapter 27). In combination with leaching and the removal of crops, denitrification keeps the level of available nitrogen in soils low.

This *global nitrogen cycle* is complex. It is also essential for life on Earth: Nitrogen-containing compounds constitute 5 to 30 percent of a plant's total dry weight. The nitrogen content of animals is even higher, and all the nitrogen in the animal world arrives there by way of the plant kingdom.

Carnivorous and Heterotrophic Plants

Some plants that are found primarily in nitrogen-deficient soils augment their nitrogen and phosphorus supply by capturing and digesting flies and other insects. There are about 450 of these *carnivorous* species, the best-known of which are Venus flytraps (genus *Dionaea;* Figure 37.9*a*), sundews (genus *Drosera;* Figure 37.9*b*), and pitcher plants (genus *Sarracenia*).

Carnivorous plants are normally found in boggy regions where the soil is acidic. Most decomposing organisms require a less acidic pH to break down the bodies of dead organisms, so relatively little nitrogen is recycled into these acidic soils. Accordingly, the carnivorous plants have adaptations that allow them to augment their supply of nitrogen by capturing animals and digesting their proteins.

(*a*)

Dionaea muscipula

(*b*)

Drosera rotundifolia

37.9 Carnivorous Plants Some plants have adapted to nitrogen-poor environments by becoming carnivorous. *(a)* The Venus flytrap obtains nitrogen and phosphorus from the bodies of insects trapped inside the plant when its hinges snap shut. *(b)* Sundews trap insects on sticky hairs. Secreted enzymes will digest the carcass externally.

The Venus flytraps have specialized leaves with two halves that fold together. When an insect trips trigger hairs on a leaf, its two halves come together, their spiny margins interlocking and imprisoning the insect. The leaf then secretes enzymes that digest its prey. The leaf absorbs the products of digestion, especially amino acids, and uses them as a nutritional supplement.

Pitcher plants produce pitcher-shaped leaves that collect small amounts of rainwater. Insects are attracted into the pitchers either by bright colors or by scent and are prevented from getting out again by stiff, downward-pointing hairs. The insects eventually die and are digested by a combination of enzymes and bacteria in the water. Even rats have been found in large pitcher plants.

Sundews have leaves covered with hairs that secrete a clear, sticky, sugary liquid. An insect touching one of these hairs becomes stuck, and more hairs curve over the insect and stick to it as well. The plant secretes enzymes to digest the insect and eventually absorbs the carbon- and nitrogen-containing products of digestion.

None of the carnivorous plants must feed on insects to survive. They can grow adequately without insects, but in their natural habitats they grow faster and are a darker green when they succeed in capturing insects. They use the additional nitrogen from the insects to make more proteins, chlorophyll, and other nitrogen-containing compounds.

Thus far in this chapter we have considered the mineral nutrition of plants. As you already know, another crucial aspect of plant nutrition is photosynthesis—the principal source of energy and carbon for plants themselves and for the biosphere as a whole. Not all plants, however, are photosynthetic autotrophs. A few, in the course of their evolution, have lost the ability to sustain themselves by photosynthesis. How do these plants get their energy and carbon?

A few plants are heterotrophic parasites that obtain their nutrients directly from the living bodies of other plants. Perhaps the most familiar parasitic plants are the mistletoes and dodders (Figure 37.10). Mistletoes are green and carry on some photosynthesis, but they parasitize other plants for water and mineral nutrients and may derive photosynthetic products from them as well. Mistletoes and dodders extract nutrients from the vascular tissues of their hosts by forming absorptive organs called *haustoria,* which invade the host plant's tissues. Another parasitic plant, the Indian pipe, once was thought to obtain its nutrients from dead organic matter. It is now known to get its nutrients, with the

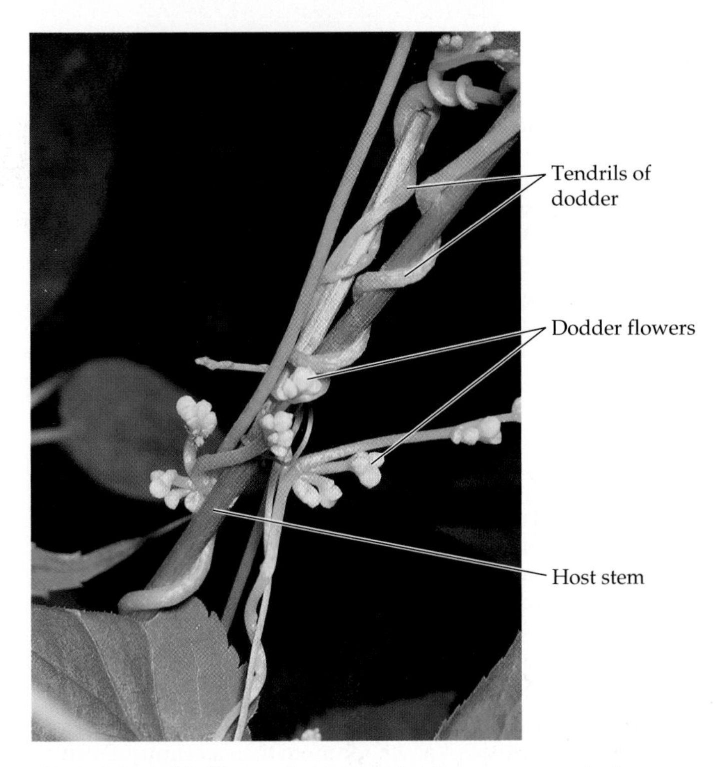

37.10 A Parasitic Plant Tendrils of dodder wrap around other plants. This parasitic plant (genus *Cuscuta*) obtains water, sugars, and other nutrients from its host through tiny, rootlike protuberances that penetrate the surface of the host.

help of fungi, from nearby actively photosynthesizing plants. Hence it, too, is a parasite.

Dwarf mistletoe is a serious parasite in forests of the western United States, destroying more than 3 billion board feet of lumber per year. However, parasitic plants are a much more urgent problem in developing countries. *Striga* (witchweed) imperils more than 300 million sub-Saharan Africans by attacking their cereal and legume crops. In the Middle East and North Africa, *Orobanche* (broomrape) ravages many crops, especially vegetables and sunflowers.

Chapter Summary

The Acquisition of Nutrients

▶ Plants are photosynthetic autotrophs that can produce all the organic compounds they need from carbon dioxide, water, and minerals, including a nitrogen source. They obtain energy from sunlight, carbon dioxide from the atmosphere, and nitrogen-containing ions and mineral nutrients from the soil.

▶ Plants explore their surroundings by growing rather than by movement.

Mineral Nutrients Essential to Plants

▶ Plants require 14 essential mineral elements, all of which come from the soil solution. Several of these essential elements fulfill multiple roles. **Review Table 37.1**

▶ The six mineral nutrients required in substantial amounts are called macronutrients; the eight required in much smaller amounts are called micronutrients. **Review Table 37.1**

▶ Deficiency symptoms suggest what essential element a plant lacks. **Review Table 37.2**

▶ Biologists discovered the requirement for each essential element by growing plants on hydroponic solutions lacking that element. **Review Figure 37.1.** See Web/CD Tutorial 37.1

Soils and Plants

▶ Soils are complex systems with living and nonliving components. They contain water, air, and inorganic and organic substances. They typically consist of two or three horizontal zones called horizons. **Review Figures 37.2, 37.3**

▶ Soils form by mechanical and chemical weathering of rock.

▶ Plants obtain some mineral nutrients through ion exchange between the soil solution and the surface of clay particles. **Review Figure 37.4**

▶ Farmers use fertilizers to make up for deficiencies in soil mineral nutrient content, and they apply lime to raise low soil pH.

▶ Plants affect soils in various ways, such as by adding organic material, removing nutrients (especially in agriculture), and changing pH.

Nitrogen Fixation

▶ A few species of soil bacteria are responsible for almost all nitrogen fixation. Some nitrogen-fixing bacteria live free in the soil; others live symbiotically as bacteroids within the roots of plants.

▶ In nitrogen fixation, nitrogen gas (N_2) is reduced to ammonia (NH_3) or ammonium ions (NH_4^+) in a reaction catalyzed by nitrogenase. **Review Figure 37.6**

▶ Nitrogenase requires anaerobic conditions, but the bacteroids in root nodules require oxygen for their respiration. Leghemoglobin helps maintain the oxygen supply to the bacteroids at the proper level.

▶ The formation of a nodule requires an interaction between the root system of a legume and *Rhizobium* bacteria. **Review Figure 37.7**

▶ Nitrogen-fixing bacteria reduce atmospheric N_2 to ammonia, but most plants take up both ammonium ions and nitrate ions. Nitrifying bacteria oxidize ammonia to nitrate. Plants take up nitrate and reduce it back to ammonia, a feat of which animals are incapable. **Review Figure 37.8.** See Web/CD Activity 37.1

▶ Denitrifying bacteria return N_2 to the atmosphere, completing the global nitrogen cycle. **Review Figure 37.8**

Carnivorous and Heterotrophic Plants

▶ Carnivorous plant species are autotrophs that supplement their nitrogen supply by feeding on insects.

▶ A few heterotrophic plants are parasitic on other plants. Some parasitic plants have major effects on crops, especially in developing countries.

Self-Quiz

1. Macronutrients
   a. are so called because they are more essential than micronutrients.
   b. include manganese, boron, and zinc, among others.
   c. function as catalysts.
   d. are required in concentrations of at least 1 gram per kilogram of plant dry matter.
   e. are obtained by the process of photosynthesis.

2. Which of the following is *not* an essential mineral element for plants?
   a. Potassium
   b. Magnesium
   c. Calcium
   d. Lead
   e. Phosphorus

3. Fertilizers
   a. are often characterized by their N-P-O percentages.
   b. are not required if crops are removed frequently enough.
   c. restore needed mineral nutrients to the soil.
   d. are needed to provide carbon, hydrogen, and oxygen to plants.
   e. are needed to destroy soil pests.

4. In a typical soil,
   a. the topsoil tends to lose mineral nutrients by leaching.
   b. there are four or more horizons.
   c. the C horizon consists primarily of loam.
   d. the dead and decaying organic matter gathers in the B horizon.
   e. more clay means more air space and thus more oxygen for roots.

5. Which of the following is *not* an important step in soil formation?
   a. Removal of bacteria
   b. Mechanical weathering
   c. Chemical weathering
   d. Clay formation
   e. Hydrolysis of soil minerals

6. Nitrogen fixation is
   a. performed only by plants.
   b. the oxidation of nitrogen gas.
   c. catalyzed by the enzyme nitrogenase.
   d. a single-step chemical reaction.
   e. possible because N_2 is a highly reactive substance.

7. Nitrification is
   a. performed only by plants.
   b. the reduction of ammonium ions to nitrate ions.
   c. the reduction of nitrate ions to nitrogen gas.
   d. catalyzed by the enzyme nitrogenase.
   e. performed by certain bacteria in the soil.

8. Nitrate reduction
   a. is performed by plants.
   b. takes place in mitochondria.
   c. is catalyzed by the enzyme nitrogenase.
   d. includes the reduction of nitrite ions to nitrate ions.
   e. is known as the Haber process.

9. Which of the following is a parasite?
   a. Venus flytrap
   b. Pitcher plant
   c. Sundew
   d. Dodder
   e. Tobacco

10. All carnivorous plants
   a. are parasites.
   b. depend on animals as a source of carbon.
   c. are incapable of photosynthesis.
   d. depend on animals as their sole source of phosphorus.
   e. obtain supplemental nitrogen from animals.

For Discussion

1. Methods for determining whether a particular element is essential have been known for more than a century. Since these methods are so well established, why was the essentiality of some elements discovered only recently?
2. If a Venus flytrap were deprived of soil sulfates and hence made unable to synthesize the amino acids cysteine and methionine, would it die from lack of protein? Explain.
3. Soils are dynamic systems. What changes might result when land is subjected to heavy irrigation for agriculture after being relatively dry for many years? What changes in the soil might result when a virgin deciduous forest is cut down and replaced by crops that are harvested each year?
4. We mentioned that important positively charged ions are held in the soil by clay particles, but other, equally important, negatively charged ions are leached deeper into the soil's B horizon. Why doesn't leaching cause an electrical imbalance in the soil? (Hint: Think of the ionization of water.)
5. The biosphere of Earth as we know it depends on the existence of a few species of nitrogen-fixing prokaryotes. What do you think might happen if one of these species were to become extinct? If all of them were to disappear?

38 Regulation of Plant Growth

Plants are creatures of the sun—they require light as their source of energy. Plants regulate the directions and rates of the growth of their shoots in ways that maximize photosynthesis. To obtain enough light to meet their needs, plants display their leaves to the sun. Some follow the sun with their leaves as the day progresses. Many adjust their growth rates to avoid shading by competing plants. Growth regulation thus affects the interactions of a plant with other plants.

Plants monitor their environment with the help of photoreceptor molecules that sense light signals. Chemical signals called hormones carry information throughout the plant body about light and other aspects of the environment inside and outside the plant. These signals regulate the development of the plant throughout its life history, affecting processes as diverse as stem growth, flowering, bud dormancy, and the dropping of leaves in autumn. Recent advances in understanding plant growth and development have come largely from work with *Arabidopsis thaliana*, a little mustard-like weed. This plant is useful to researchers because its shoots and seeds are tiny and because it flowers and forms seeds in a relatively short time after growth begins. In addition, its genome is unusually small for that of a flowering plant. *Arabidopsis* mutants with altered developmental patterns provide evidence for the mechanisms of hormone and photoreceptor action.

Catching Some Rays Most of us have observed the manner in which plants turn their flowers or leaves toward sunlight. Light signals caught by photoreceptor pigments are transmitted by hormones to other parts of the plant in a finely tuned developmental dance.

In this chapter we will give a brief overview of the life of a flowering plant and its developmental stages. We will explore the environmental cues, photoreceptors, and hormones that regulate plant development and consider their multiple roles.

Interacting Factors in Plant Development

The *development* of a plant—the series of progressive changes that take place throughout its life—is regulated in complex ways. Four factors take part in this regulation:

- The plant senses and responds to *environmental cues*.
- In order to sense some environmental cues, the plant uses receptors such as *photoreceptors*, molecules that absorb light.
- Chemical signals, or *hormones*, mediate the effects of the environmental cues including those sensed by receptors.
- The plant's *genome* encodes enzymes that catalyze the biochemical reactions of development.

Several hormones and photoreceptors regulate plant growth

Hormones are regulatory compounds that act at very low concentrations at sites distant from where they are produced. Unlike animals, which produce each hormone in a specific part of the body, plants produce hormones in many of their cells. Each plant hormone plays multiple regulatory roles, affecting several different aspects of plant development (Table 38.1). Interactions among these hormones are often complex.

Like hormones, **photoreceptors** regulate many developmental processes in plants. Unlike plant hormones, which are small molecules, plant photoreceptors are *pigments*: protein molecules that absorb light. Light (an environmental cue) acts directly on photoreceptors, which in turn regulate the processes of development, such as the many changes accompanying the growth of a young seedling emerging from the soil and into the light.

No matter what cues regulate development, ultimately the plant's genome determines the limits within which the plant and its parts will develop. The genome encodes the master plan, but its interpretation depends on conditions in the environment. It is also the target for some hormone actions. For several decades hormones and photoreceptors were the focus of most work on plant development, but recent advances in molecular genetics have allowed us to focus on the underlying processes, such as signal transduction pathways.

Signal transduction pathways mediate hormone and photoreceptor action

We introduced the topic of signal transduction pathways in Chapter 15. Plants, like other organisms, make extensive use of these pathways. Cell signaling in plant development involves three steps: a receptor (for a hormone or for light), a signal transduction pathway, and the ultimate cellular response (see Figure 15.2). Protein kinase cascades amplify responses to signals in plants, as they do in other organisms (see Figure 15.9). The signal transduction pathways of plants differ from those of animals only in the details; for example, protein kinases in plants phosphorylate the amino acids serine or threonine, but not tyrosine.

Before concerning ourselves with molecular details, however, let's look at the general pattern of plant development.

An Overview of Plant Development

A flowering plant goes through many developmental processes during its life history, all of which are regulated by the factors we have just described. As plants develop, these

38.1 *Plant Growth Hormones*

HORMONE	TYPICAL ACTIVITIES
Abscisic acid	Maintains seed dormancy and winter dormancy; closes stomata
Auxins	Promote stem elongation, adventitious root initiation, and fruit growth; inhibit lateral bud outgrowth and leaf abscission
Brassinosteroids	Promote elongation of stems and pollen tubes; promote vascular tissue differentiation
Cytokinins	Inhibit leaf senescence; promote cell division and lateral bud outgrowth; affect root growth
Ethylene	Promotes fruit ripening and leaf abscission; inhibits stem elongation and gravitropism
Gibberellins	Promote seed germination, stem growth, and fruit development; break winter dormancy; mobilize nutrient reserves in grass seeds

factors affect three fundamental processes: cell division, cell expansion, and cell differentiation.

The seed germinates and forms a growing seedling

All developmental activity may be suspended in a seed, even when conditions appear to be suitable for its growth. In other words, a seed may be **dormant**. Cells in dormant seeds do not divide, expand, or differentiate. For the embryo to begin developing, seed dormancy must be broken.

As the seed begins to **germinate**—to develop into a seedling—it first takes up water. The growing embryo must then obtain chemical building blocks—monomers—for its development by digesting the polysaccharides, fats, and proteins stored in the seed. The embryos of some plant species secrete hormones that direct the mobilization of these reserves. Germination is completed when the *radicle* (embryonic root) emerges from the seed coat. The plant is then a **seedling**.

If the seed germinates underground, the new seedling must elongate rapidly and cope with life in darkness or dim light. A photoreceptor controls this stage of development, and ends it when the shoot is exposed to sufficient light to begin photosynthesis.

Early shoot development varies among the flowering plants. Figure 38.1 presents the distinctive shoot development patterns of monocots and eudicots.

Plant growth from seedling to adult is regulated by several hormones. Other hormones are involved in the plant's defenses against herbivores and microorganisms (to be discussed in Chapter 40).

The plant flowers and sets fruit

Flowering—the formation of reproductive organs—may be initiated when the plant reaches an appropriate age or size. Some plant species, however, flower at particular times of the year, meaning that the plant must be capable of distinguishing different times of the year. In these plants, the leaves measure the length of the night (shorter in summer, longer in winter) with great precision. Light absorption by photoreceptors is the first step in this time-measuring process.

Once the leaves have determined that it is time for the plant to flower, that information must be transmitted as a signal to the places where flowers will form. The means by which this signal is transmitted remains a mystery, but it is likely that a "flowering hormone" travels from the leaf to the point of flower formation.

After flowers form, hormones play further roles in reproduction. Hormones and other substances control the growth of the pollen tube that brings sperm and egg together. Following fertilization, a fruit develops and ripens under hormonal control.

38.1 Patterns of Early Shoot Development (*a*) In grasses and some other monocots, growing shoots are protected by a coleoptile until they reach the soil surface. (*b*) In most eudicots, the growing point of the shoot is protected by the cotyledons. (*c*) In some other eudicots, the cotyledons remain in the soil, and the growing point is protected by the first true leaves.

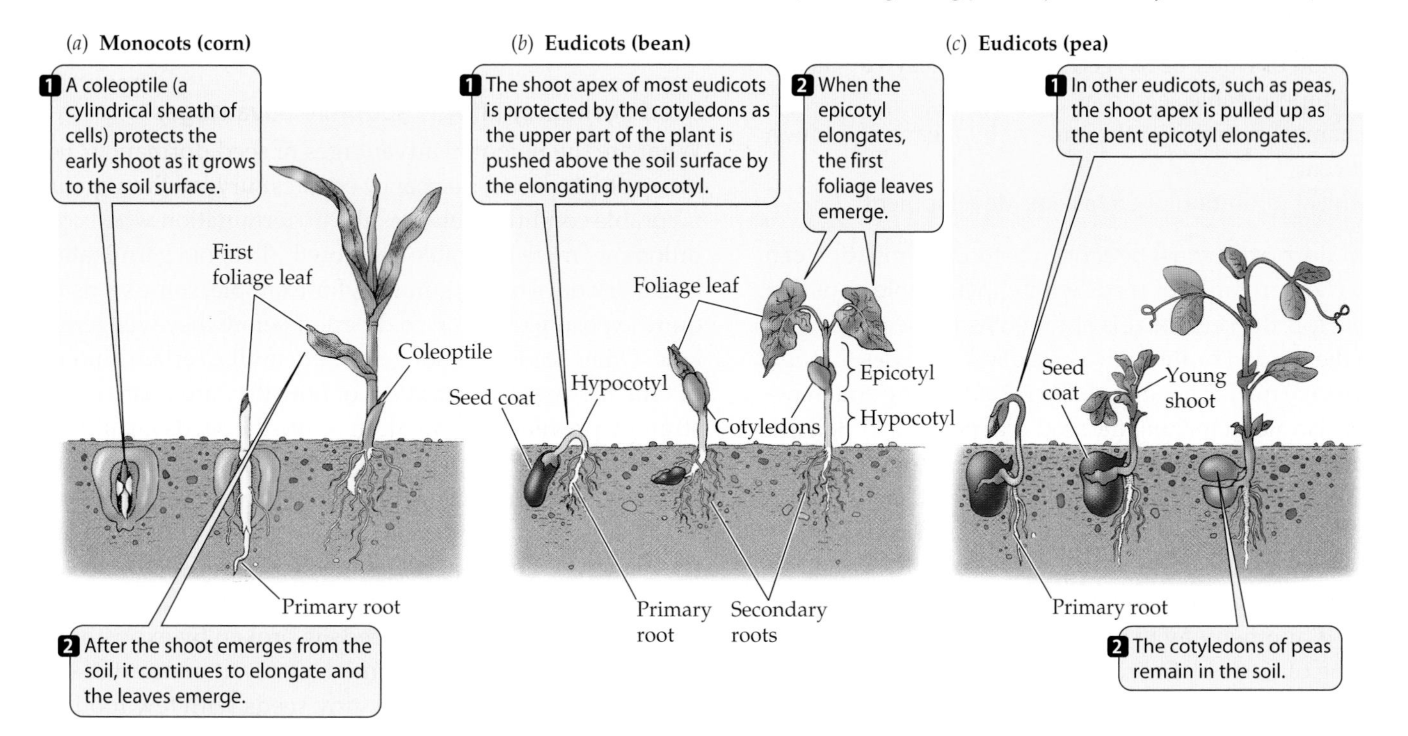
(a) Monocots (corn)
1 A coleoptile (a cylindrical sheath of cells) protects the early shoot as it grows to the soil surface.
First foliage leaf
Coleoptile
Primary root
2 After the shoot emerges from the soil, it continues to elongate and the leaves emerge.
(b) Eudicots (bean)
1 The shoot apex of most eudicots is protected by the cotyledons as the upper part of the plant is pushed above the soil surface by the elongating hypocotyl.
2 When the epicotyl elongates, the first foliage leaves emerge.
Foliage leaf
Seed coat
Hypocotyl
Cotyledons
Epicotyl
Hypocotyl
Primary root
Secondary roots
(c) Eudicots (pea)
1 In other eudicots, such as peas, the shoot apex is pulled up as the bent epicotyl elongates.
Seed coat
Young shoot
Primary root
2 The cotyledons of peas remain in the soil.

The plant senesces and dies

Some plants, known as **perennials**, continue to grow year after year. Many perennials that grow in temperate climates have buds that enter a state of winter dormancy during the cold season. A hormone called abscisic acid helps maintain this dormancy.

In many species, leaves **senesce** (deteriorate because of aging) and fall at the end of the growing season, shortly before the onset of winter. Leaf fall is regulated by an interplay of the hormones ethylene and auxin.

The death of the entire plant, which may be initiated by signals from the environment, follows senescent changes that are controlled by hormones such as ethylene. This life history pattern appears to be an adaptation for producing more offspring by shifting nutrients into the seeds; in so doing, the parent plant essentially starves itself to death, ensuring that sufficient nutrients are available for seed maturation.

We have now reached the end of the plant's life history. Next we'll examine how its various steps are regulated. We'll begin at the start of the life history, with the seed and its germination.

Ending Seed Dormancy and Beginning Germination

The seeds of some plant species are capable of germinating as soon as they have matured. All they need for germination is water. But many other species have seeds that are dormant at maturity. Seed dormancy may last for weeks, months, years, or even centuries. The mechanisms of seed dormancy are numerous and diverse, but three principal strategies dominate:

- Exclusion of water or oxygen from the embryo by means of an impermeable seed coat
- Mechanical restraint of the embryo by means of a tough seed coat
- Chemical inhibition of embryonic development

Seed dormancy must be broken before germination can begin. The dormancy of seeds with impermeable coats can be broken if the seed coat is abraded as the seed tumbles across the ground or through a creek bed or passes through the digestive tract of an animal. Cycles of freezing and thawing can also aid in making the seed coat permeable. Soil microorganisms probably play a major role in softening seed coats. Fire can melt waterproofing wax in seed coats, allowing water to reach the embryo (Figure 38.2). Fire can also release mechanical restraint. *Leaching*—the dissolving out of a water-soluble chemical inhibitor by prolonged exposure to water—is another way in which dormancy can be broken. Scorching of seeds by fire can also break down some chemical inhibitors.

38.2 Fire and Seed Germination This fireweed germinated and flourished after a great fire along the Alaska Highway.

Seed dormancy affords adaptive advantages

What are the potential advantages of seed dormancy? For many plant species, dormancy ensures survival through unfavorable conditions and results in germination when conditions are more favorable for growth. To avoid germination in the dry days of late summer, for example, some seeds require exposure to a long cold period before they will germinate. Other seeds will not germinate until a certain amount of time has passed, regardless of how they are treated. This strategy prevents germination while the seeds are still attached to the parent plant. Seeds that must be scorched by fire in order to germinate avoid competition with other plants by germinating only where an area has been cleared by fire. Dormancy also helps seeds to survive long-distance dispersal, allowing plants to colonize new territory.

The dormancy of some seeds is broken by exposure to light. These seeds, which germinate only at or near the surface of the soil, are generally tiny seeds with few food reserves. Such seeds would be incapable of surviving if germination occurred while they were buried deeply. Conversely, the germination of some other seeds is inhibited by light; these seeds germinate only when buried and thus kept in darkness. Light-inhibited seeds are usually large and well stocked with nutrients.

Seed dormancy helps annual plants (plants that complete their life cycle in a single year) encounter the effects of year-to-year variation in the environment. The seeds of some annuals remain dormant throughout an unfavorable year. The seeds of other plants germinate at specific times during the year, increasing the likelihood that at least some of the seedlings will encounter conditions favorable for their growth.

Dormancy may also increase the likelihood of a seed's germinating in the right place. Some cypress trees, for example, grow in standing water, and their seeds germinate only if inhibitors are leached by water (Figure 38.3).

Seed germination begins with the uptake of water

The breaking of dormancy allows seed germination to begin. The first step in germination is the uptake of water, called **imbibition**. Typically, only 5 to 15 percent of a seed's weight is water, whereas most plant parts contain 80 to 95 percent water. A seed's water potential (see Chapter 36) is very negative, and water can be taken up if the seed coat is permeable. The magnitude of this water potential is demonstrated by the force exerted by seeds expanding in water. Cocklebur seeds that are imbibing can exert a pressure of up to 1,000 atmospheres (about 100 megapascals) against a restraining force.

38.3 Leaching of Germination Inhibitors The seeds of bald cypress, a tree adapted to moist or wet environments, germinate only after being leached by water, which increases the chances that they will germinate in a location suitable for their growth.

As a seed takes up water, it undergoes metabolic changes: Enzymes become activated upon hydration, RNA and then proteins are synthesized, the rate of cellular respiration increases, and other metabolic pathways become activated. In many seeds, there is no DNA synthesis and no cell division during these early stages of germination. Initially, growth results solely from the expansion of small, preformed cells. DNA is synthesized only after the radicle begins to grow and break through the seed coat.

The embryo must mobilize its reserves

To fuel these metabolic activities, the embryo must use the reserves of energy and raw materials stored in the seed. Until the young plant becomes able to photosynthesize, it depends on these reserves, which are stored in the embryonic **cotyledons** (the "seed leaves") or in the **endosperm** (the specialized nutritive tissue) of the seed. The principal reserve of energy and carbon in many seeds is starch. Other seeds store fats or oils. Usually, the endosperm holds amino acid reserves in the form of proteins, rather than as free amino acids.

The giant molecules of starch, lipids, and proteins must be digested by enzymes into monomers that can enter the cells of the embryo. The polymer starch yields glucose for energy metabolism and for the synthesis of cellulose and other cell wall constituents. The digestion of stored proteins provides the amino acids the embryo needs to synthesize its own proteins. The digestion of lipids releases glycerol and fatty acids, both of which can be metabolized for energy. Glycerol and fatty acids can also be converted to glucose, which permits fat-storing plants to make all the building blocks they need for growth.

In germinating barley and other cereal seeds, the embryo secretes **gibberellins**, one of several classes of plant growth hormones. Gibberellins diffuse through the endosperm to a surrounding tissue called the *aleurone layer*, which lies inside the seed coat. The gibberellins trigger a cascade of events in the aleurone layer, causing it to synthesize and secrete enzymes that digest proteins and starch stored in the endosperm (Figure 38.4). Commercially, gibberellins are used in the brewing industry to enhance the "malting" (germination) of barley and the breakdown of its endosperm, producing sugar that is fermented to alcohol.

Gibberellins: Regulators from Germination to Fruit Growth

Gibberellins play a wide variety of roles in plant development in addition to triggering the mobilization of seed reserves. We'll begin our discussion of plant growth hormones by describing the discovery of the gibberellins, as well as their many effects.

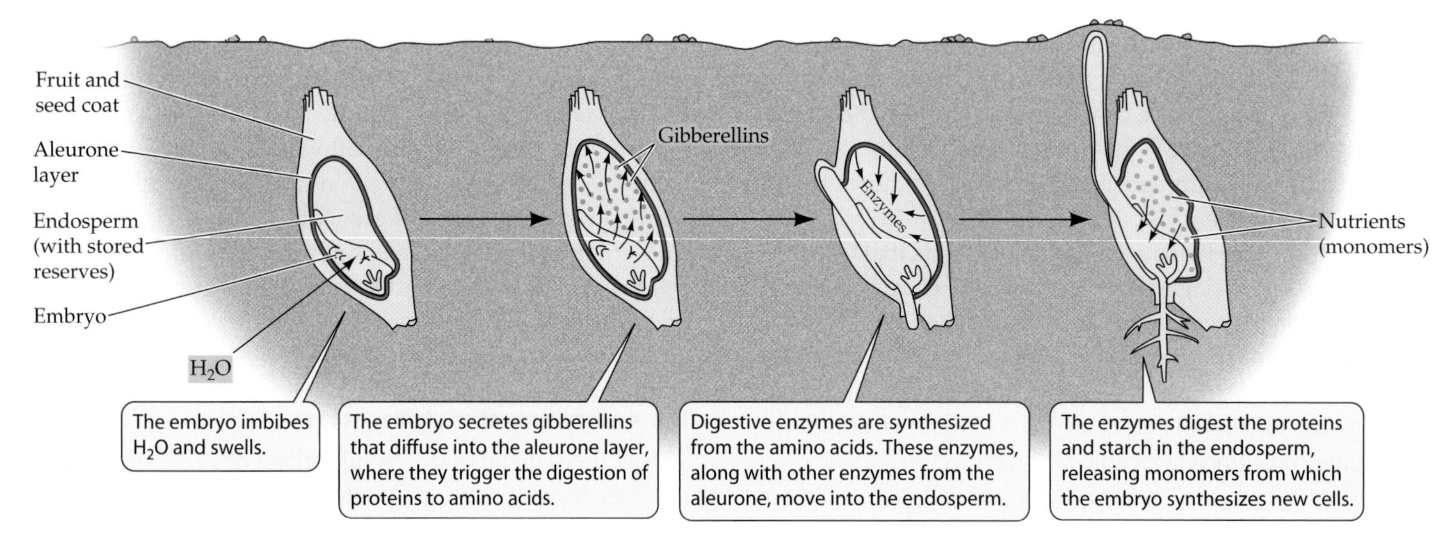

38.4 Embryos Mobilize Their Reserves During seed germination in cereal grasses, gibberellins trigger a cascade of events that results in the conversion of starch and protein reserves into monomers that can be used by the developing embryo.

Foolish seedlings led to the discovery of the gibberellins

The gibberellins are a large family of closely related compounds. Some are found in plants and others in a pathogenic (disease-causing) fungus, where they were first discovered.

Gibberellin A_1 (important in stem growth)

Gibberellin A_3 (commercially available)

In 1809, the study of the gibberellins began indirectly with observations of the *bakanae*, or "foolish seedling," disease of rice. Seedlings affected by this disease grow tall more rapidly than their healthy neighbors, but this rapid growth gives rise to spindly plants that die before producing seeds (the rice grains used for food). The disease has had considerable economic impacts in several parts of the world. It is caused by the ascomycete fungus *Gibberella fujikuroi*.

In 1925, the Japanese biologist Eiichi Kurosawa grew *G. fujikuroi* on a liquid medium, then separated the fungus from the medium by filtration. He heated the filtered medium to kill any remaining fungus, but found that the resulting heat-treated filtrate was still capable of inducing rapid growth in rice seedlings. Medium that had never contained the fungus did not have this effect. This experiment established that *G. fujikuroi* produces a growth-promoting chemical substance, which Kurosawa called a gibberellin.

Were the gibberellins simply exotic products of an obscure fungus, or did they play a more general role in plant growth? Bernard O. Phinney of the University of California, Los Angeles, answered this question in part in 1956, when he reported the spectacular growth-promoting effect of gibberellins on dwarf corn seedlings. He used plants that were known to be genetic dwarfs, in which a particular recessive allele (say, *d1*) was present in the homozygous condition (*d1d1*). Gibberellins applied to nondwarf—wild-type—corn seedlings had almost no effect, whereas dwarf seedlings treated with gibberellins grew as tall as their normal relatives. (A comparable effect of gibberellins applied to a dwarf tomato plant is shown in Figure 38.5.)

22 days after being sprayed with a dilute gibberellin solution, this plant reached the size of a nondwarf plant.
This untreated control plant remained dwarf.

38.5 The Effect of Gibberellins on Dwarf Plants Both of the dwarf tomato plants in this photograph were the same size when the one on the right was treated with gibberellins.

Phinney drew two conclusions from the results of this experiment: first, that gibberellins are normal constituents of corn, and perhaps of all plants, and second, that some dwarf plants are short because they produce insufficient amounts of gibberellins. According to Phinney's hypothesis, nondwarf plants manufacture enough gibberellins to promote their full growth, but dwarf plants do not. Extracts from numerous plant species were found to promote growth in dwarf corn. These findings provided direct evidence that plants that are not genetic dwarfs contain gibberellin-like substances. Phinney's work set the stage for today's use of mutant plants to investigate the control of plant development.

The roots, leaves, and flowers of dwarf corn plants appear normal, but their stems are much shorter than those of wild-type plants. All parts of the dwarf plant contain a much lower concentration of gibberellins than do the organs of a wild-type plant. We can infer, then, that normal stem elongation *requires* gibberellins or the products of gibberellin action. We can further infer that gibberellins play a less essential role in the development of roots, leaves, and flowers.

Although more than 125 gibberellins have been identified, only one, gibberellin A_1, actually controls stem elongation in most plants. The other gibberellins found in stems are simply intermediates in the production of gibberellin A_1. As we will see in the next section, gibberellins affect processes other than stem elongation, but we do not yet know which gibberellin has any other particular effect.

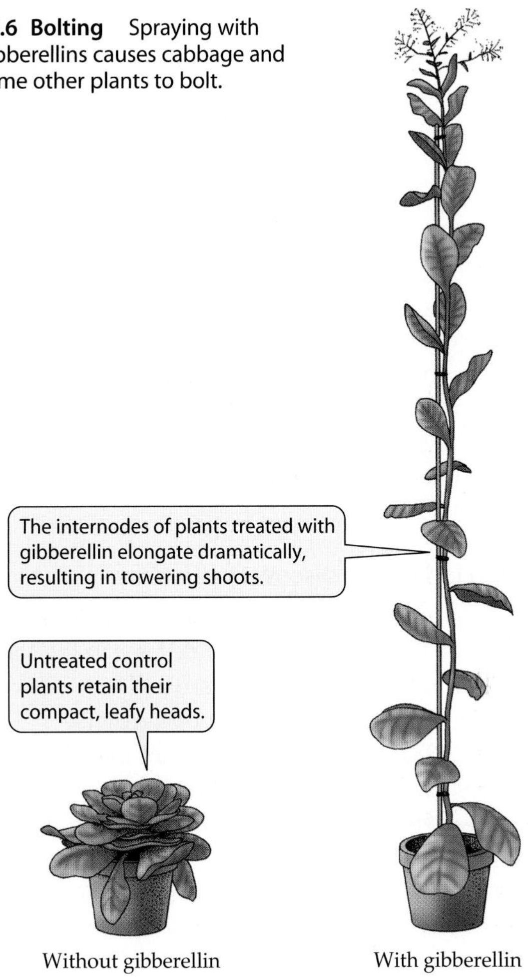

38.6 Bolting Spraying with gibberellins causes cabbage and some other plants to bolt.

The internodes of plants treated with gibberellin elongate dramatically, resulting in towering shoots.
Untreated control plants retain their compact, leafy heads.
Without gibberellin
With gibberellin

The gibberellins have many effects

Gibberellins and other hormones regulate the growth of fruits. It has long been known that grapevines that produce seedless grapes develop smaller fruit than varieties that produce seed-bearing grapes. Experimental removal of seeds from immature seeded grapes prevented normal fruit growth, suggesting that the seeds are sources of a growth regulator. It was then shown that spraying young seedless grapes with a gibberellin solution caused them to grow as large as seeded ones. It is now standard commercial procedure to spray seedless grapes with gibberellins. Biochemical studies showed that the developing seeds produce gibberellins, which diffuse out into the immature fruit tissue.

A different type of gibberellin effect is seen in some **biennials**—plants that grow vegetatively in their first year and then flower in their second year and die. Some biennial plants respond dramatically to an increase in the level of gibberellins. In their second year, the apical meristems of these biennials respond to environmental cues by producing elongated shoots, which eventually bear flowers. This rapid shoot elongation is called **bolting**. When the plant senses the appropriate environmental cue—longer days or a sufficient winter chilling—it produces more gibberellins, raising the gibberellin concentration to a level that causes the shoot to bolt. Some biennial species will bolt when sprayed with a gibberellin solution without exposure to any environmental cue (Figure 38.6).

Gibberellins also cause fruit to grow from unfertilized flowers, promote seed germination in lettuce and several other species, and help bring spring buds out of winter dormancy. Most other plant growth hormones, like the gibberellins, have multiple effects within the plant, and they often interact with one another to regulate developmental processes. In controlling stem elongation, for example, gibberellins interact with another hormone, auxin.

Auxin Affects Plant Growth and Form

If you pinch off the apical bud at the top of a bean plant, inactive lateral buds become active and develop into branches. Similarly, pruning a shrub stimulates the formation of new branches. If you cut off the blade of a leaf but leave its petiole (stalk) attached to the plant, the petiole drops off sooner than it would have if the leaf were intact. If a plant is kept indoors, its shoots will grow toward a window. These diverse responses of shoots are all mediated by plant hormones called **auxins**, of which the most important is *indoleacetic acid* (IAA):

$H_2C-COOH$

N

H

Indoleacetic acid

In this section, we will look at the discovery of auxin, its transport within the plant, and its role as a mediator of the effects of light and gravity on plant growth. We'll discover its many effects on vegetative growth and on fruit development. Then we'll examine its mechanism of action.

Phototropism led to the discovery of auxin

The discovery of auxin and its numerous physiological effects on plants can be traced back to work done in the 1880s by Charles Darwin and his son Francis. The Darwins were interested in plant movements. One type of growth movement they studied was **phototropism**, the growth of plant structures toward light (as in most shoots) or away from it (as in roots). They asked, What part of the plant senses the light?

To answer this question, the Darwins worked with canary grass (*Phalaris canariensis*) seedlings grown in the dark. A young grass seedling has a **coleoptile**—a cylindrical sheath a few cells thick that protects the delicate shoot as it pushes through the soil (see Figure 38.1*a*). When the coleoptile breaks through the soil surface, it soon stops growing, and the shoot emerges unharmed. The coleoptiles of grasses are phototropic—they grow toward the light.

To find the light-receptive region of the coleoptile, the Darwins tried "blindfolding" the coleoptiles of dark-grown canary grass seedlings in various places, then illuminating them from one side (Figure 38.7). The coleoptile grew toward the light whenever its tip was exposed. If the top millimeter or more of the coleoptile was covered, however, there was no phototropic response. Thus, the Darwins were able to conclude that the tip contains the photoreceptor that responds to light. The actual bending toward the light, however, takes place in a growing region a few millimeters below the tip. Therefore, the Darwins reasoned, some type of signal must travel from the tip of the coleoptile to the growing region. Later, others demonstrated that this signal is a chemical substance by showing that it can move through certain permeable materials, such as gelatin, but not through impermeable materials, such as a metal sheet.

Further experiments showed that the tip of the coleoptile produces a hormone that moves down the coleoptile to the growing region. If the tip is removed, the growth of the coleoptile is sharply inhibited. If the tip is carefully replaced, growth resumes, even if the tip and base are separated by a thin layer of gelatin. The hormone moves down from the tip, but it does not move from one side of the coleoptile to the other. If the tip is cut off and replaced so that it covers only one side of the cut end of the coleoptile, the coleoptile curves as the cells on the side below the replaced tip grow more rapidly than those on the other side.

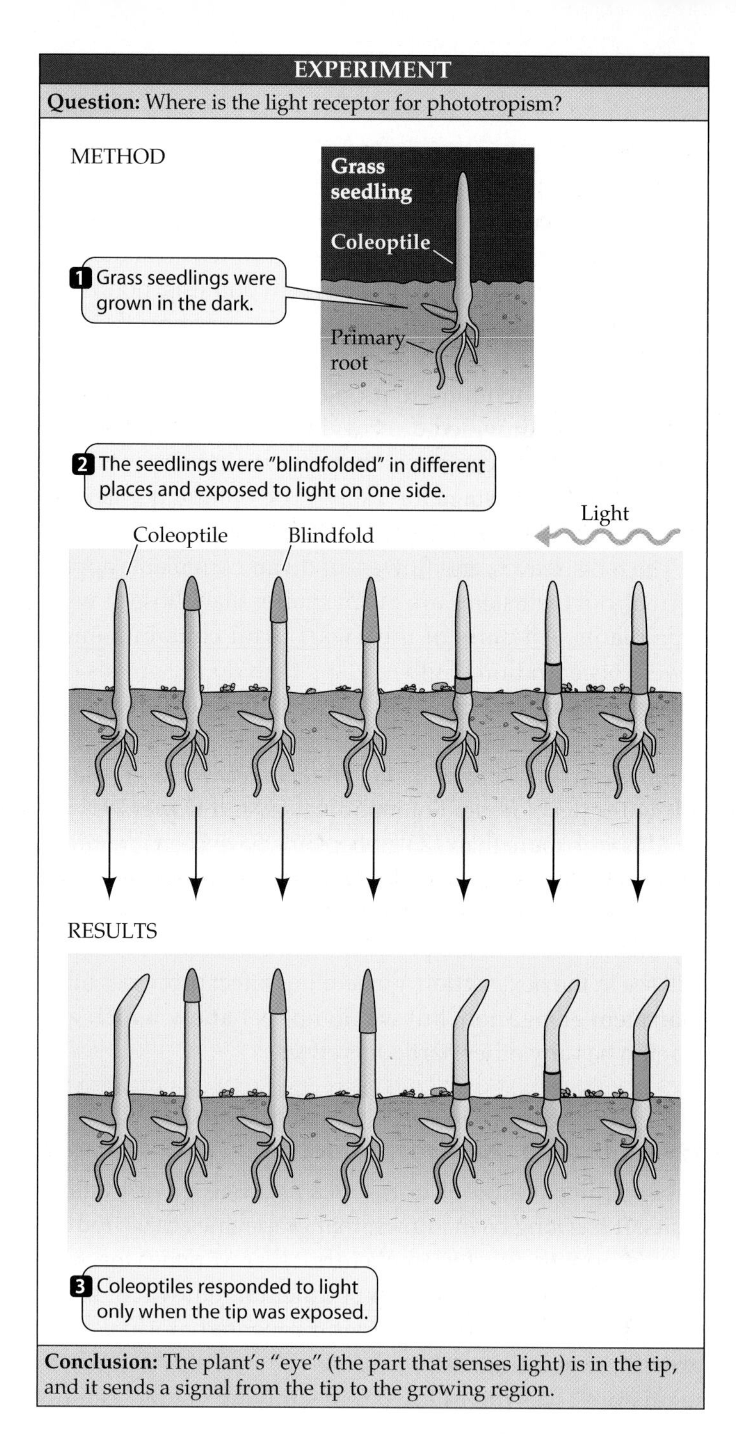

EXPERIMENT
Question: Where is the light receptor for phototropism?
METHOD
Grass seedling
Coleoptile
1 Grass seedlings were grown in the dark.
Primary root
2 The seedlings were "blindfolded" in different places and exposed to light on one side.
Coleoptile
Blindfold
Light
RESULTS
3 Coleoptiles responded to light only when the tip was exposed.
Conclusion: The plant's "eye" (the part that senses light) is in the tip, and it sends a signal from the tip to the growing region.

38.7 The Darwins' Phototropism Experiment The top drawings show some of the ways in which seedlings grown in the dark were "blindfolded"; the lower drawings show the results the Darwins observed in each case. Their observations led them to hypothesize the existence of a growth-promoting signal produced by the coleoptile.

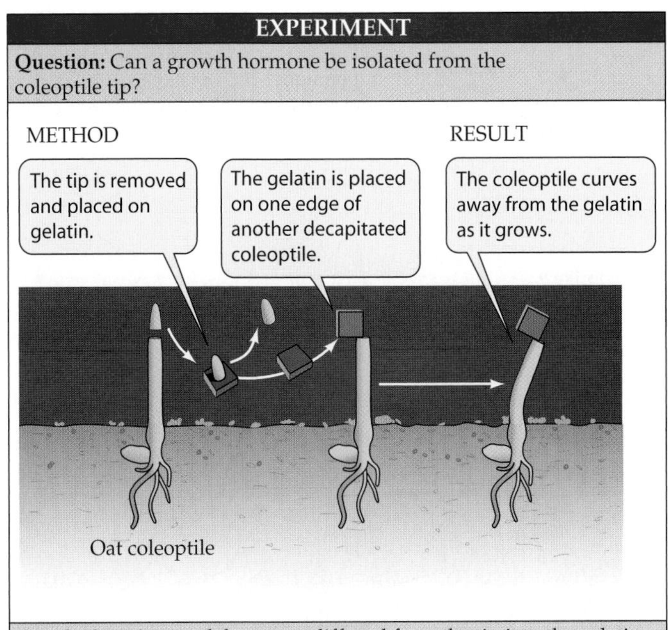

EXPERIMENT
Question: Can a growth hormone be isolated from the coleoptile tip?
METHOD
RESULT
The tip is removed and placed on gelatin.
The gelatin is placed on one edge of another decapitated coleoptile.
The coleoptile curves away from the gelatin as it grows.
Oat coleoptile

Conclusion: A growth hormone diffused from the tip into the gelatin, and from the gelatin into another plant. It had an effect on the growth of the plants similar to that of a coleoptile tip.

38.8 Went's Experiment Went succeeded in isolating the growth-promoting hormone whose existence the Darwins had hypothesized by placing coleoptile tips on a block of gelatin.

In the 1920s, the Dutch botanist Frits W. Went followed up on the Darwins' observations. He removed coleoptile tips and placed their cut surfaces on a block of gelatin. Then he placed pieces of that gelatin on decapitated coleoptiles—positioned to cover only one side, just as coleoptile tips had been placed in earlier experiments (Figure 38.8). As they grew, the coleoptiles curved toward the side away from the gelatin. This curvature demonstrated that a hormone had indeed diffused into the gelatin block from the isolated coleoptile tips. Went had at last isolated a hormone from a plant. Later chemical analysis showed that this hormone, named auxin, was indoleacetic acid.

Auxin transport is polar

Early experiments showed that the movement of auxin through certain plant tissues is strictly *polar*—that is, it is unidirectional along a line from apex to base. By inverting plants and plant parts, scientists determined that the apex-to-base direction of auxin movement has nothing to do with gravity; the polarity of this movement is a totally biological phenomenon.

Auxin transport is completely or partially polar in many plant parts. In most leaf petioles, for example, auxin moves only from the blade end toward the stem end. In roots, however, auxin moves toward the root tip, in the phloem. What regulates these movements of auxin?

Carrier proteins move auxin into and out of cells

Polar transport of auxin depends on the location of *auxin anion efflux carriers*, membrane proteins that are confined to the basal ends of cells. The cytoplasm of plant cells has a pH near neutrality. At this pH, auxin is present as an anion. Auxin anions can leave the cells only by way of the basally located auxin anion efflux carriers.

Proton pumps in the plasma membrane pump hydrogen ions (H^+) out of the cells, rendering the cell walls acidic. At this lower pH, auxin is present both as an anion and as a free acid. Either form of auxin can enter the cell from any direction. About half the auxin entry into cells is by passive diffusion of the free acid and half by means of active transport (symport) of the anions along with H^+. However, once auxin is inside the cell, it takes the ionic form, which can depart only from the base (Figure 38.9). In this way, auxin anion efflux carriers contribute to the establishment of auxin gradients in the plant. Because it forms a gradient, auxin can act as a *mor-*

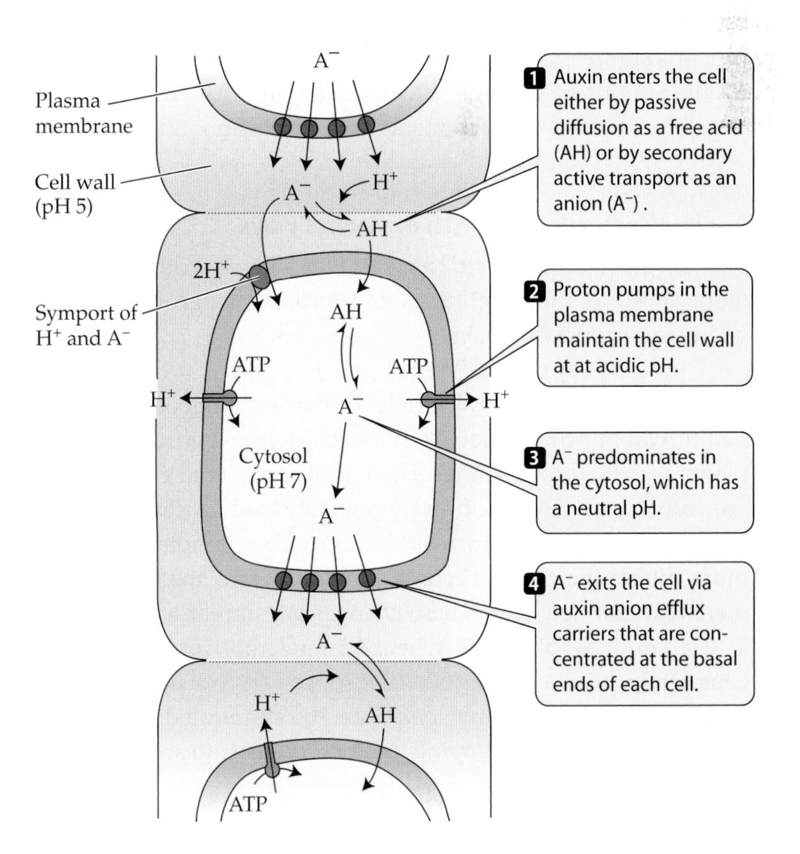

38.9 Polar Transport of Auxin Proton pumps and the basally placed auxin anion efflux carriers lead to a net movement of auxin in one direction.

phogen (see Chapter 19), instructing cells as to their orientation within the plant and determining how they differentiate.

There are other auxin carrier proteins that are specific to certain tissues and cells and participate in specific auxin responses. Such auxin carrier proteins are involved in plant responses to light and gravity.

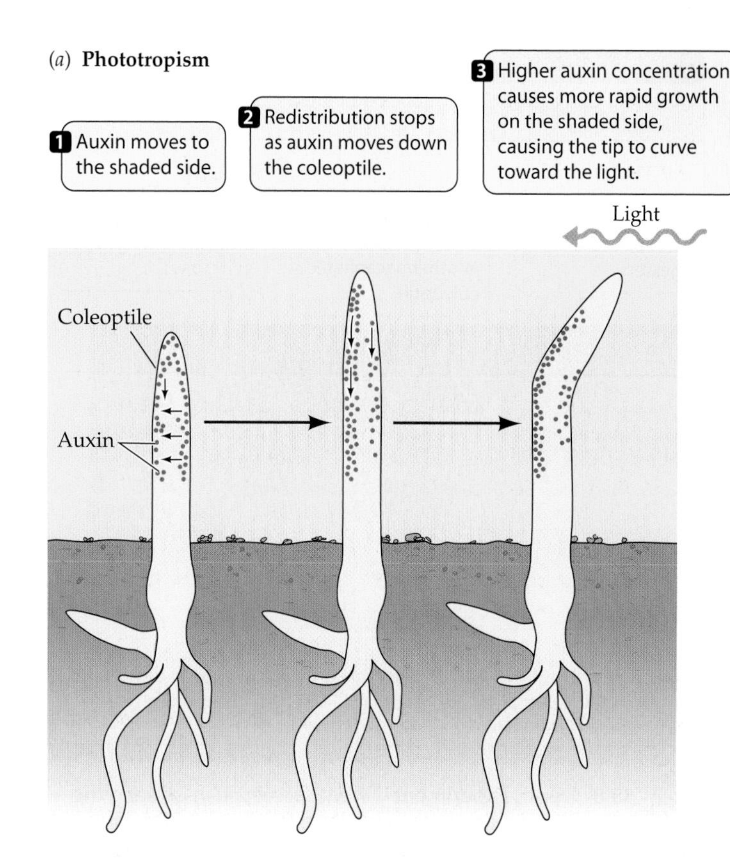
(a) Phototropism
1 Auxin moves to the shaded side.
2 Redistribution stops as auxin moves down the coleoptile.
3 Higher auxin concentration causes more rapid growth on the shaded side, causing the tip to curve toward the light.
Light
Coleoptile
Auxin

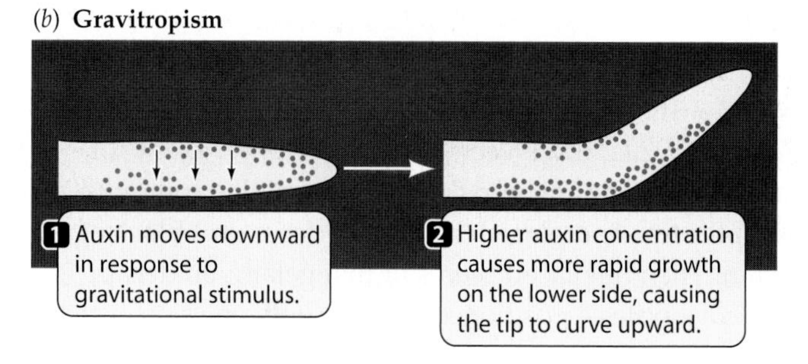
(b) Gravitropism
1 Auxin moves downward in response to gravitational stimulus.
2 Higher auxin concentration causes more rapid growth on the lower side, causing the tip to curve upward.

38.10 Plants Respond to Light and Gravity Phototropism and gravitropism occur in response to a redistribution of auxin.

Light and gravity affect the direction of plant growth

While polar auxin transport establishes the orientation of growth, *lateral* (side-to-side) redistribution of auxin is the mechanism that explains phototropism. This redistribution is carried out by an auxin carrier protein that moves to one side of the cell and thus allows auxin to exit the cell only from that side.

When light strikes a grass coleoptile on one side, auxin at the tip moves laterally toward the shaded side. The imbalance thus established is maintained down the coleoptile, so that in the growing region below, there is more auxin on the shaded side, speeding growth on that side and causing the coleoptile to bend toward the light. This bending toward light is phototropism (Figure 38.10*a*). If you have noticed a house plant bending toward a window, you have seen phototropism.

Even in the dark, auxin moves to the lower side of a shoot that has been tipped over, causing more rapid growth in the lower side and, hence, an upward bending of the shoot. Such growth in a direction determined by gravity is called **gravitropism** (Figure 38.10*b*). The upward gravitropic response of shoots is defined as negative gravitropism; that of roots, which bend downward, is positive gravitropism.

Auxin affects plant growth in several ways

Like the gibberellins, auxin has many roles in plant development. It affects the vegetative and reproductive growth of plants in a number of ways.

INITIATING ROOT GROWTH Cuttings from the shoots of some plants can produce roots and develop into entire new plants. For this to happen, certain undifferentiated cells in the interior of the shoot, originally destined to function only in food storage, must set off on a new mission: They must differentiate and become organized into the apical meristem of a new root. These changes are similar to those in the pericycle of a root when a lateral root forms (see Chapter 35).

Shoot cuttings of many species can be stimulated to develop profuse roots by dipping the cut surfaces into an auxin solution; this observation suggests that the plant's own auxin plays a role in the initiation of lateral roots. Commercial preparations that enhance the rooting of plant cuttings typically contain synthetic auxins.

INHIBITING LEAF ABSCISSION In contrast to its stimulatory effect on root initiation, auxin has an inhibitory effect on the detachment of old leaves from stems. This process, called **abscission**, is the cause of autumn leaf fall. Most leaves consist of a blade and a petiole that attaches the blade to the stem. Abscission results from the breakdown of a specific part of the petiole, the *abscission zone* (Figure 38.11). If the blade of a leaf is cut off, the petiole falls from the plant more rapidly than if the leaf had remained intact. If the cut surface is treated with an auxin solution, however, the petiole remains attached to the plant, often longer than an intact leaf would have. The timing of leaf abscission in nature appears to be determined in part by a decrease in the movement of auxin, produced in the blade, through the petiole.

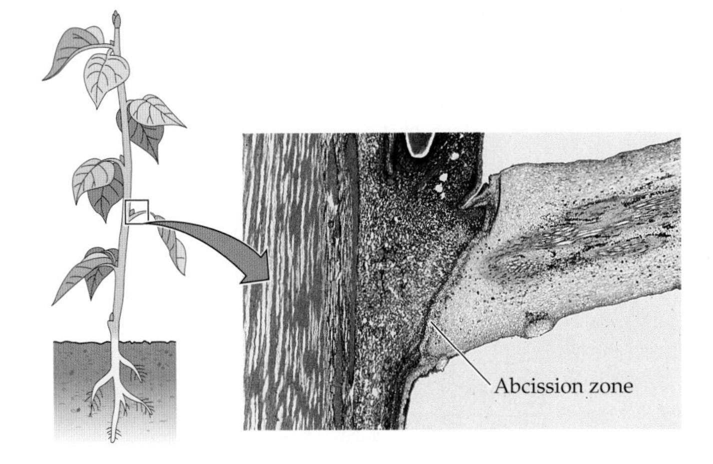

38.11 Changes Occur when a Leaf Is About to Fall The breakdown of the abscission zone of the petiole causes the leaf to fall.

MAINTAINING APICAL DOMINANCE Auxin maintains **apical dominance**, a phenomenon in which apical buds inhibit the growth of lateral buds, resulting in the growth of a single main stem with minimal branching. This phenomenon can be demonstrated by an experiment with young seedlings. If the plant remains intact, the stem elongates, and the lateral buds remain inactive. Removal of the apical bud—the major site of auxin production—results in growth of the lateral buds. If the cut surface of the stem is treated with auxin, however, the lateral buds do not grow (Figure 38.12). The apical buds of branches also exert apical dominance: The lateral buds on the branch are inactive unless the apex of the branch is removed. That is why gardeners prune shrubs to encourage branching.

In the two experiments on leaves and stems that we have just discussed, removal of a particular part of the plant elicits a response—abscission or loss of apical dominance—and that response is prevented by treatment with auxin. These results are consistent with other data showing that the excised part of the leaf or stem is an auxin source and that auxin in the intact plant helps maintain apical dominance and delays the abscission of leaves.

PROMOTING STEM ELONGATION AND INHIBITING ROOT ELONGATION Auxin promotes stem elongation but inhibits the elongation of roots. The question of why different organs respond in opposite ways to the same growth hormone remains unanswered, but is a subject of current research.

CONTROLLING FRUIT DEVELOPMENT Although fruit development normally depends on prior fertilization of the egg, in many species treatment of an unfertilized ovary with auxin or gibberellins causes *parthenocarpy*—fruit formation without fertilization. Parthenocarpic fruits form spontaneously in some cultivated varieties of plants, including seedless grapes, some cucumbers, and bananas.

Auxin analogs as herbicides

All of the preceding activities illustrate the great diversity of important roles that auxin plays in plant growth. Auxin is absolutely essential for plant survival; no mutants lacking auxin

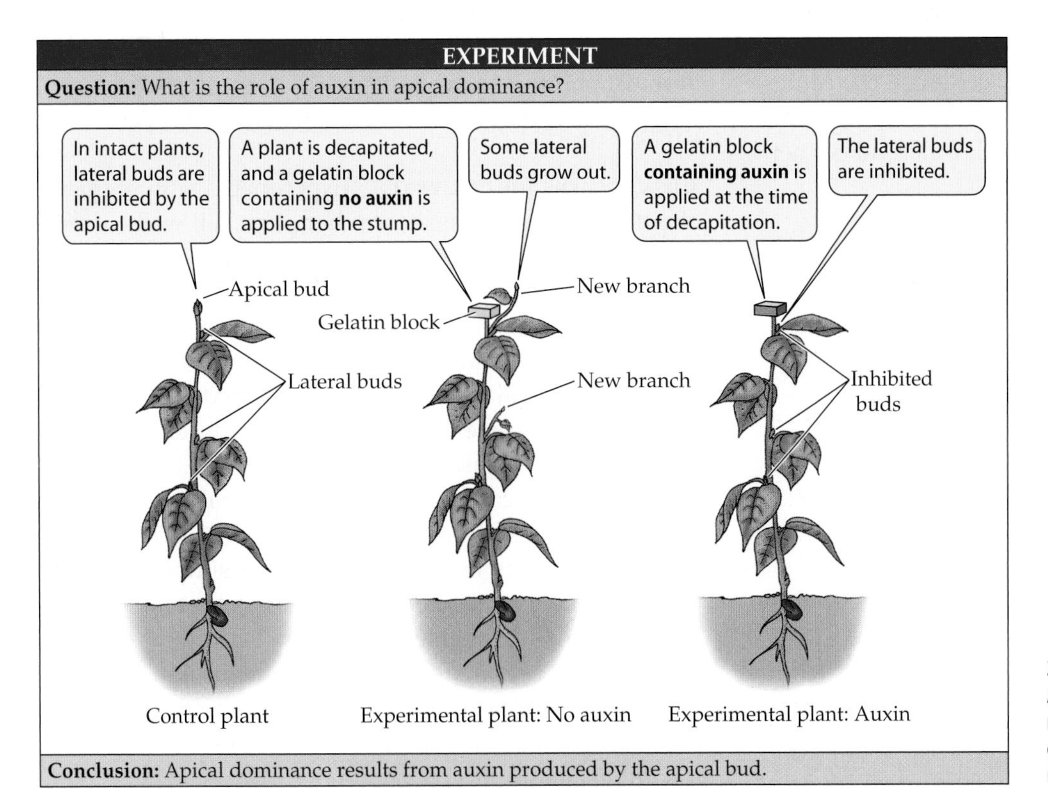

38.12 Auxin and Apical Dominance Auxin produced by the apical bud maintains apical dominance—the growth of a single main stem with minimal branching.

have ever been found. Many synthetic auxins—chemical analogs of indoleacetic acid—have been produced and studied. One of them, 2,4-dichlorophenoxyacetic acid (2,4-D), has the striking property of being lethal to eudicots at concentrations that are harmless to monocots. This property makes 2,4-D an effective *selective herbicide* that can be sprayed on a lawn or a cereal crop to kill those weeds that are eudicots. However, because 2,4-D takes a long time to break down, it pollutes the environment, so scientists have developed herbicides that are less persistent in the environment.

Now let's see *how* auxin plays one of its roles—promoting stem elongation through effects on the cell wall.

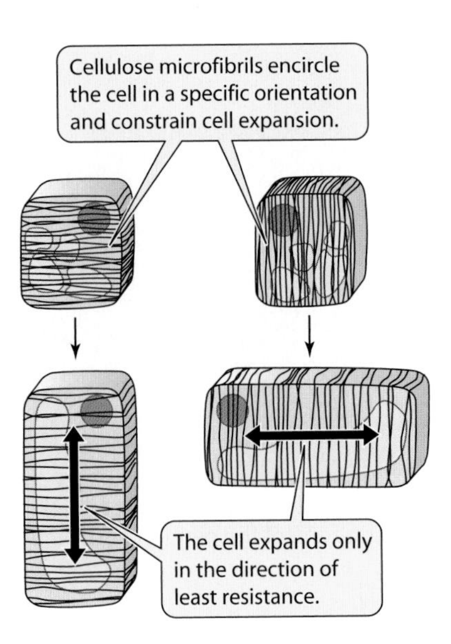

38.14 Plant Cells Expand The orientation of cellulose microfibrils in the plant's cell wall determines the direction of cell expansion.

Auxin promotes growth by acting on cell walls

The expansion of plant cells is what causes plant growth. Thus the cell wall plays key roles in controlling the rate and direction of growth of a plant cell. Auxin acts on cell walls to regulate this process.

CELL WALL ARCHITECTURE DIRECTS CELL EXPANSION. The expansion of a plant cell is driven primarily by the uptake of water, which enters the cytoplasm of the cell and accumulates in its central vacuole. As the vacuole expands, the cell grows rapidly, with the vacuole often making up more than 90 percent of the volume of a mature cell. The vacuole presses the cytoplasm against the cell wall as it expands, and the wall resists this force.

The principal strengthening component of the plant cell wall is *cellulose*, a large polymer of glucose. In the wall, cellulose molecules tend to associate in parallel with one another. Bundles of approximately 250 cellulose molecules make up *microfibrils* that are visible under an electron microscope (Figure 38.13). What makes the cell wall rigid is a network of cellulose microfibrils connected by bridges of other, smaller polysaccharides. The orientation of the majority of cellulose microfibrils determines the direction of cell expansion (Figure 38.14).

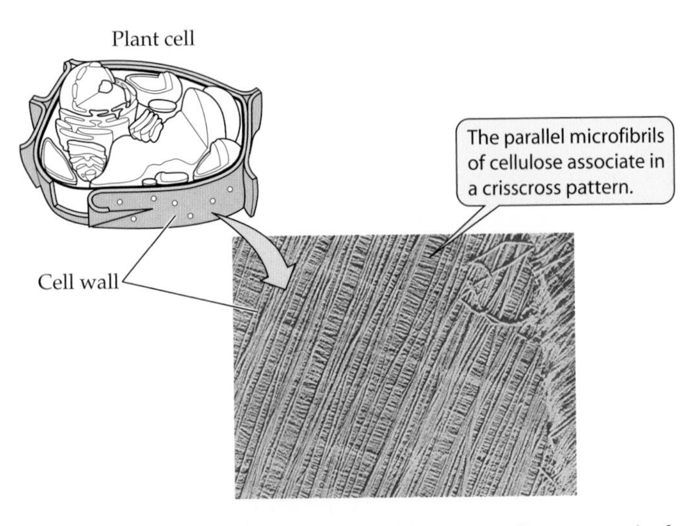

38.13 Cellulose in the Cell Wall The plant cell wall is a network of cellulose microfibrils linked by other polysaccharides.

For the cell to expand, its wall must loosen and be stretched. If the wall were only stretched, however, it would become thinner. Cell expansion involves more than stretching. New polysaccharides are deposited throughout the wall, and new cellulose microfibrils are deposited at the inner surface of the wall, maintaining its thickness. As a consequence of this pattern of cellulose deposition, the microfibrils in the outermost part of the wall are the oldest, and those in the innermost part the youngest. How do these properties of cell walls relate to the action of auxin on plant cell expansion?

AUXIN LOOSENS THE CELL WALL. Experiments with segments of oat coleoptiles have shown that plant cell walls recover incompletely from being stretched (Figure 38.15). Reversible stretching is called *elasticity*, and irreversible stretching is called *plasticity*. Treating the coleoptile segments with auxin before they were stretched significantly increased their plasticity; in other words, it loosened the cell walls. This result suggested that auxin-induced cell expansion might result from just such a loosening effect.

Auxin acts by causing the release of a "wall-loosening factor" from the cytoplasm. Studies in the 1970s indicated that the wall-loosening factor was sometimes simply hydrogen ions (protons, H^+). Acidifying the growth medium (that is, adding H^+) caused segments of stems or coleoptiles to grow as rapidly as segments treated with auxin. Furthermore, treating coleoptile segments with auxin caused acidification of the growth medium. Auxin increases the activity of proton pumps in the plasma membrane, increasing the H^+ con-

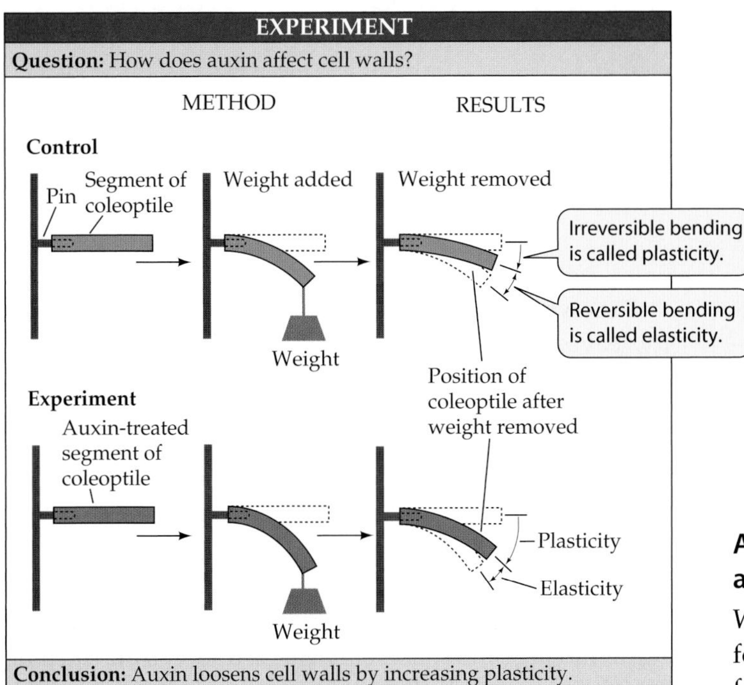

38.15 Auxin Acts on Cell Walls Auxin increases the plasticity, but not the elasticity, of cell walls.

centration in the cell wall. Treatments that block acidification by auxin also block auxin-induced growth. These experimental results led to the hypothesis that hydrogen ions secreted into the cell wall activate one or more cell wall proteins. So, a search began for candidate proteins.

Proteins called *expansins* were isolated and purified from plant cell walls in the 1990s. When the purified proteins were added to isolated cell walls of several plant species, the walls underwent expansion. Expansins are widespread among terrestrial plants. Furthermore, these proteins are activated by hydrogen ions. Expansins modify the pattern of hydrogen bonding between the polysaccharides in the plant cell wall. These modifications may allow polysaccharide macromolecules to slip past each other, so that the wall stretches and the cell expands.

Before auxin can initiate a chain of events such as the one just described, it must first be recognized as a signal by the cell. How does this recognition occur?

Plants contain specific auxin receptor proteins

The initial step in the action of any plant hormone is the binding of that hormone by a specific receptor protein. There are several proteins that can bind various plant hormones, but some of this binding may be nonspecific. To demonstrate that an auxin-binding protein is an auxin receptor, it must be shown that the protein actually mediates the effects of auxin.

Plant molecular biologists showed that the protein ABP1 (*a*uxin-*b*inding *p*rotein *1*) functions as an auxin receptor. They inserted the *ABP1* gene of *Arabidopsis* into other plant species and then induced the expression of the gene in cells that normally show a limited response to auxin. Upon expression of the *ABP1* gene, those cells showed greater responses to both endogenous and applied auxin. Subsequent work has conclusively demonstrated the existence and importance of other auxin receptor proteins. Given the number of processes regulated by auxin, it is hardly surprising that there appear to be multiple receptors and signal transduction pathways for this hormone.

Auxin and other hormones evoke cell differentiation and organ formation

What signals the different types of plant cells and organs to form? Much of the research on this question has been performed using plant tissues grown in culture outside the plant body. One easily grown tissue is pith—the spongy innermost tissue of a stem. Pith cells proliferate rapidly in culture, but show no differentiation. All the cells are similar and unspecialized; they grow into a lump of pith tissue on the surface of the culture medium.

Cutting a notch in the cultured pith tissue and inserting a stem tip into the notch causes the pith cells below the inserted tip to differentiate. Some of them differentiate to form water-conducting xylem cells. Differentiation of pith cells can also be initiated by adding a mixture of auxin and coconut milk (a rich source of plant hormones) to the notch.

A similar effect can be observed in intact plants. If notches are cut in the stems of coleus plants, interrupting some of the strands of vascular tissue, the strands gradually regenerate from the upper side of the cut to the lower (recall that auxin moves from the tip to the base of a stem). If the leaves above the cut are removed, regeneration is slowed. However, when the missing leaves are replaced with an auxin solution, vascular tissue regenerates. These results show that auxin and other plant hormones signal the formation of specific cell types.

Such experiments with cultured plant tissues have helped clarify which hormones control organ formation. Undifferentiated cultures of tobacco pith form roots when treated with an appropriate concentration of auxin. Another group of plant growth hormones—the cytokinins—causes buds and then shoots to form in such cultures. The pattern of organ formation depends on the ratio of auxin to cytokinin in the medium. A high proportion of auxin favors roots, and a high proportion of cytokinin favors buds, but both processes are most active when both hormones are present. These results show that other hormones can modify the effects of auxin, reminding us that plant growth is regulated more by hormone interactions than by a single hormone.

Cytokinins Are Active from Seed to Senescence

The **cytokinins** are derivatives of adenine. In studies of plant cell division, botanists discovered a substance that powerfully stimulated cell division in tissue cultures. This compound, *kinetin*, consists of adenine with an attached group. We now know that kinetin is just one of the family of compounds that are now called cytokinins. Kinetin may be considered a synthetic cytokinin because it has never been isolated from plant tissue. However, two closely related compounds, called *zeatin* and *isopentenyl adenine*, occur naturally in plants.

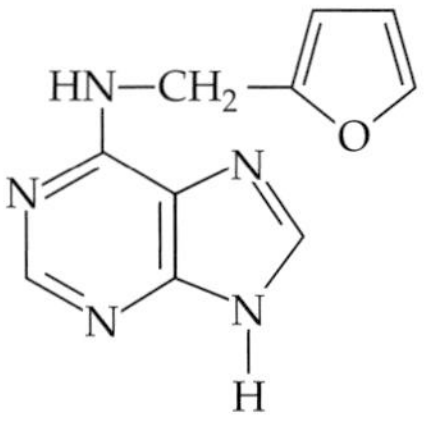

Kinetin

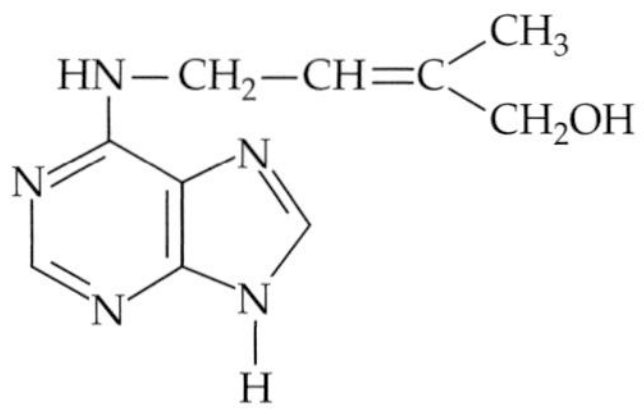

Zeatin

No mutants lacking cytokinins have ever been found. Thus, like auxin, cytokinins seem to be required throughout the life of a plant. Cytokinins form primarily in the roots and move to other parts of the plant. They have a number of different effects:

- Adding an appropriate combination of auxin and cytokinins to a growth medium induces rapid cell proliferation in cultured plant tissues.
- Cytokinins can cause certain light-requiring seeds to germinate even when kept in constant darkness.
- Cytokinins usually inhibit the elongation of stems, but they cause lateral swelling of stems and roots (the fleshy roots of radishes are an extreme example).
- Cytokinins stimulate lateral buds to grow into branches; thus the balance between auxin and cytokinin levels controls the extent of branching (bushiness) of a plant.
- Cytokinins increase the expansion of cut pieces of leaf tissue in culture and may regulate normal leaf expansion.
- Cytokinins delay the senescence of leaves. If leaf blades are detached from a plant and floated on water or a nutrient solution, they quickly turn yellow and show other signs of senescence. If instead they are floated on a solution containing a cytokinin, they remain green and senesce much more slowly.

Ethylene: A Hormone that Hastens Leaf Senescence and Fruit Ripening

Whereas the cytokinins delay senescence, another plant hormone promotes it. That hormone is the gas **ethylene**, which is sometimes called the senescence hormone. Ethylene can be produced by all parts of the plant, and like all plant hormones, it has several effects.

Ethylene (the "senescence hormone")

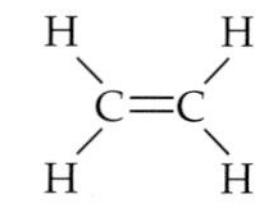

Back when streets were lit by gas rather than by electricity, leaves on trees near street lamps abscised earlier than those on trees farther from the lamps. We now know that ethylene, a combustion product of the illuminating gas, is what caused the early abscission. Auxin delays leaf abscission, but ethylene strongly promotes it; thus a balance of auxin and ethylene controls abscission.

Ethylene hastens the ripening of fruit

By promoting senescence, ethylene also speeds the ripening of fruit. The old saying "one rotten apple spoils the barrel" is true. That rotten apple is a rich source of ethylene, which speeds the ripening and subsequent rotting of the other fruit in a barrel or other confined space. As the fruit ripens, it loses chlorophyll and its cell walls break down; ethylene promotes both of these processes. Ethylene also causes an increase in its own production. Thus, once ripening begins, more and more ethylene forms, and because it is a gas, it diffuses readily throughout the fruit and even to neighboring fruits on the same or other plants.

Farmers in ancient times used to slash developing figs to hasten their ripening. We now know that wounding causes an increase in ethylene production by the fruit and that the raised ethylene level promotes ripening. Today commercial shippers and storers of fruit hasten ripening by adding ethylene to storage chambers. This use of ethylene is the single most important use of a plant hormone in agriculture and commerce. Ripening can also be delayed by the use of "scrubbers" and adsorbents to remove ethylene from the atmosphere in fruit storage chambers.

As flowers senesce, their petals may abscise, to the detriment of the cut-flower industry. Growers and florists often immerse the cut stems of flowers in dilute solutions of silver thiosulfate prior to sale. Silver salts inhibit ethylene action, probably by interacting directly with the ethylene receptor, and thus delay senescence—enabling florists and consumers to keep their cut flowers from senescing prematurely.

Ethylene affects stems in several ways

Although it is associated primarily with senescence, ethylene is active at other stages of plant development as well. The stems of many eudicot seedlings form an *apical hook* that protects the delicate shoot apex while the stem grows through the soil (Figure 38.16). The apical hook is maintained through

38.16 The Apical Hook of a Eudicot Asymmetrical production of ethylene is responsible for the apical hook of this bean seedling, which was grown in the dark.

an asymmetrical production of ethylene gas, which inhibits the elongation of cells on the inner surface of the hook. Once the seedling breaks through the soil surface and is exposed to light, ethylene synthesis stops, and the cells of the inner surface are no longer inhibited. These cells now elongate, and the hook unfolds, raising the shoot apex and the expanding leaves into the sun.

Ethylene also inhibits stem elongation in general, promotes lateral swelling of stems (as do the cytokinins), and causes stems to lose their sensitivity to gravitropic stimulation. Together, these three phenomena constitute the *triple response* observed when normal plants are treated with ethylene.

The ethylene signal transduction pathway is well understood

Analysis of *Arabidopsis* mutants has revealed the steps in the mechanism of ethylene action. Some of these mutants do not respond to applied ethylene, and others act as if they have been exposed to ethylene even though they have not. Studies of the mutant genes and their protein products, coupled with comparisons of their amino acid sequences with those of other known proteins, have revealed some of the details of the signal transduction pathway through which ethylene produces its effects (Figure 38.17). The pathway includes two membrane proteins: The first is an ethylene receptor (ETR1), and the second is a channel (EIN2) that acts through a second messenger to activate a transcription factor (EIN3). The transcription factor turns on the genes that produce ethylene's effects in the cell. Ethylene was the first plant hormone to have its mechanism of action elucidated in this way.

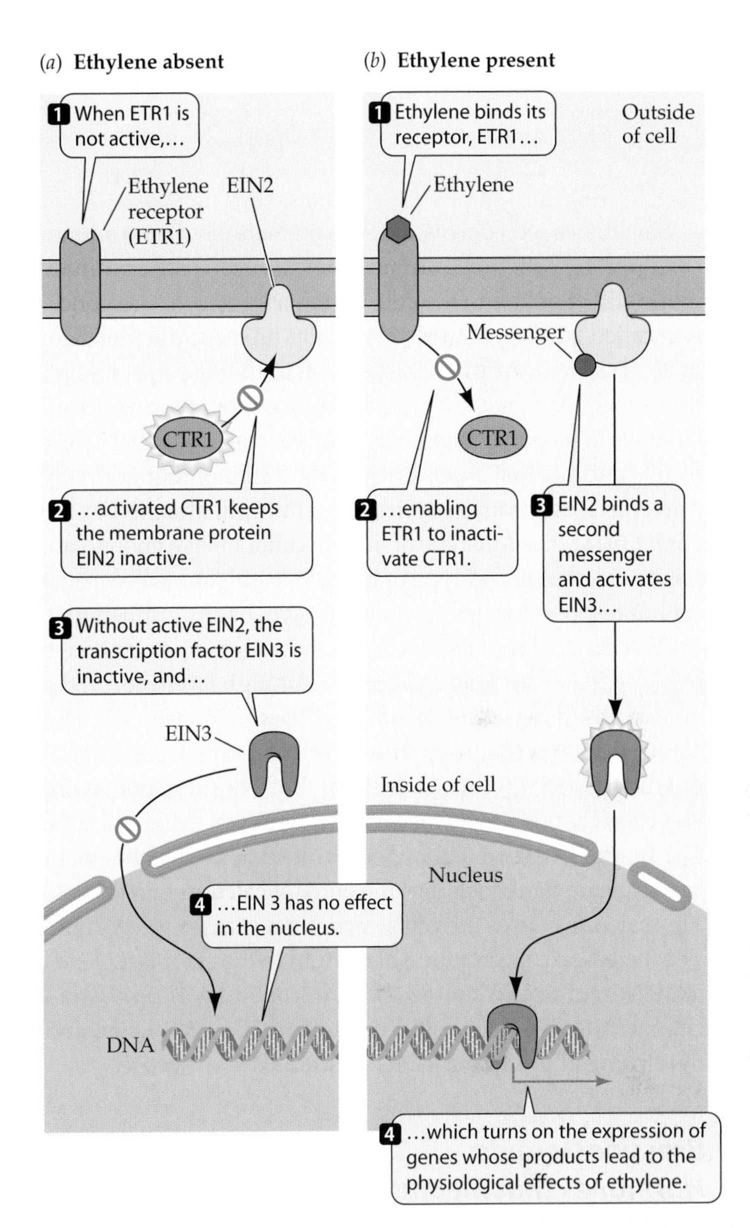

38.17 The Signal Transduction Pathway for Ethylene This slightly simplified diagram shows the roles of four proteins (ETR1, CTR1, EIN2, and EIN3) in the signal transduction pathway through which ethylene exerts its effects.

Abscisic Acid: The Stress Hormone

Abscisic acid is another hormone that has multiple effects in the living plant. During seed formation, abscisic acid promotes the accumulation of storage proteins by allowing the expression of the genes that encode those proteins. It is generally present in high concentrations in dormant buds and some dormant seeds, and it is probably the most common of the chemical inhibitors that initiate and maintain dormancy in mature seeds. Abscisic acid also inhibits stem elongation. It is sometimes referred to as the stress hormone of plants because it accumulates when plants are deprived of water and because of its possible role in maintaining the dormancy of buds in winter.

Abscisic acid
(the "stress hormone")

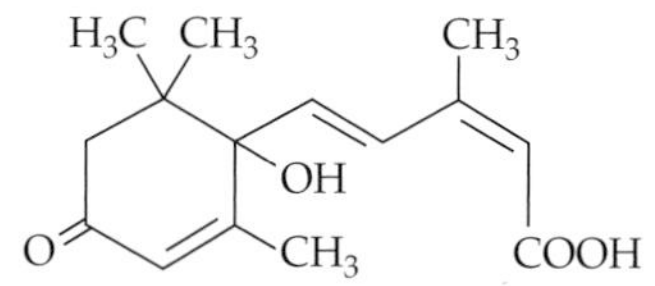

Sometimes seed dormancy ends prematurely. Some mutant corn plants, called *vp* mutants, have seeds that germinate while still attached to the cob, on the parent plant—a condition called *vivipary*. Several *vp* mutants are naturally deficient in abscisic acid. Applying abscisic acid to these mutants reduces their tendency to show vivipary. Another type of *vp* mutant fails to respond in any way to applied abscisic acid. These results indicate that abscisic acid is the inhibitor that normally prevents seeds from germinating while still attached to the parent plant. The first type of mutant cannot make enough abscisic acid; the second type of mutant is viviparous because it cannot respond to abscisic acid—its own, or any applied to it.

Abscisic acid also regulates gas and water vapor exchange between leaves and the atmosphere through its effects on the guard cells of the leaf stomata (see Chapter 36). Abscisic acid causes stomata to close, and it also prevents the stomatal opening normally caused by light. Both of these effects involve ion channels in the plasma membrane of the guard cells. The first response of a guard cell to abscisic acid is the opening of calcium channels and the entry of calcium into the cell. This calcium causes the cell's vacuole to release calcium, too. The increased concentration of calcium in the cytoplasm leads to a chain of events that result in the opening of potassium channels, the loss of K^+ and water from the cytoplasm, and the closing of the stoma as the guard cells sag together.

Brassinosteroids: Hormones that Mediate Effects of Light

More than 20 years ago, biologists isolated an interesting steroid from the pollen of rape, a member of the Brassicaceae, or mustard family. When applied to various plant tissues, this **brassinosteroid** stimulated cell elongation, pollen tube elongation, and vascular tissue differentiation, but it inhibited root elongation. Since then, dozens of chemically related and growth-affecting brassinosteroids have been found in plants. Treatment with as little as a few nanograms of brassinosteroid per plant is enough to promote growth.

Brassinolide
(a brassinosteroid)

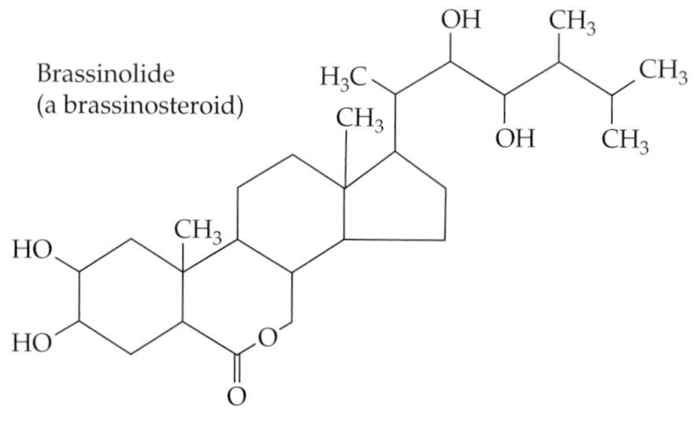

The properties of an *Arabidopsis* mutant called *det2* made it clear that brassinosteroids are naturally occurring plant hormones. When grown in darkness, seedlings homozygous for the *det2* allele differ dramatically from wild-type seedlings: In many respects, they look like wild-type seedlings grown in the light. Treatment of dark-grown *det2* mutant seedlings with brassinosteroids causes them to grow normally—that is, like wild-type plants grown in the dark. These results, supported by chemical analysis, showed that *det2* plants are unable to synthesize their own brassinosteroids, and that lack of the hormone results in abnormal growth.

Some of the effects of light on plant development result from effects on the signal transduction pathway for brassinosteroids. Others may result from alterations in brassinosteroid levels in the plant. Let's now look more closely at how plants sense environmental cues such as light.

Light and Photoreceptors

The length of the night determines the onset of winter dormancy in many plant species. As summer wears on, the days become shorter (that is, the nights become longer). Leaves have a mechanism for measuring the length of the night, as we will see in the next chapter. Measuring night length is an accurate way to determine the season of the year. If a plant determined the season only by the temperature, it might be fooled by a winter warm spell or by unseasonably cold weather in the summer. The length of the night, on the other hand, is determined by Earth's rotation around the sun and does not vary. Plants use the environmental cue of night length to time several aspects of their growth and development.

Night length is one of several environmental cues detected by plants, or by individual organs such as leaves. By its presence or absence, its intensity, its spectral properties (specific wavelengths), and its duration, light provides cues to various environmental conditions. In spite of the reservation just mentioned, temperature, too, provides important environmental cues, both by its value at any particular time and by the distribution of warmer and colder stretches over a period of time. The plant senses these environmental cues and then responds, often by stepping up or decreasing its production of hormones. We'll discuss an example of a temperature cue in the next chapter. In this section, we'll see how certain photoreceptors sense light, its duration, and its wavelength distribution.

Light regulates many aspects of plant development in addition to phototropism. The affected processes range from seed germination to shoot elongation to the initiation of flowering. Several photoreceptors take part in these and other processes. Five **phytochromes** mediate the effects of red and dim blue light. Four or more types of **blue-light receptors**, discovered more recently, mediate the effects of higher-intensity blue light.

Phytochromes mediate the effects of red and far-red light

Some seeds will not germinate in darkness, but do so readily after even a brief exposure to light. Blue and red light are highly effective in promoting germination, whereas green light is not.

Of particular importance to plants is the fact that far-red light *reverses* the effect of a prior exposure to red light. Far-red light is a very deep red, bordering on the limit of human vision and centered on a wavelength of 730 nm; red wavelengths are around 660 nm. If exposed to brief, alternating periods of red and far-red light in close succession, lettuce seeds respond only to the final exposure: If it is red, they germinate; if it is far-red, they remain dormant (Figure 38.18). This reversibility of the effects of red and far-red light regulates many other aspects of plant development, including flowering and seedling growth.

The basis for these effects of red and far-red light resides in certain bluish photoreceptor proteins called phytochromes. They are blue because they absorb red and far-red light and transmit other light. In the cytosol of plants are two interconvertible forms of phytochromes. Light drives the interconversion of the two forms. The form that absorbs principally red light is called P_r. Upon absorption of a photon of red light, a molecule of P_r is converted into P_{fr}. The P_{fr} form absorbs far-red light; when it does so, it is converted to P_r.

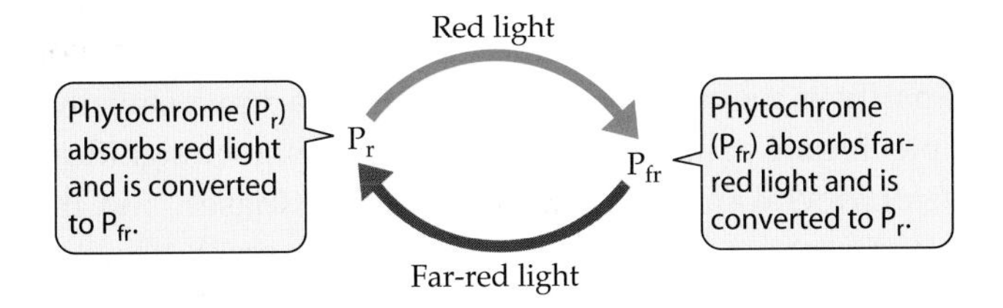

P_{fr} has some important biological effects. As we have just seen, one of them is to initiate germination in certain seeds, such as lettuce.

Phytochromes have many effects on plant growth and development

Phytochromes help to regulate a seedling's early growth. When seeds germinate in the dark below the soil surface, a pale and spindly seedling forms, with undeveloped leaves. Such an **etiolated** seedling cannot carry out photosynthesis. The seedling shoot must reach the soil surface and begin photosynthesis before its nutrient reserves are expended and it starves.

Plants have evolved a variety of ways to cope with the problem of germinating underground. Etiolated flowering plants, for example, do not form chlorophyll. They synthesize chlorophyll and turn green only when exposed to light, thereby conserving the resources needed to make chlorophyll, which would be useless in the dark. An etiolated shoot uses its stored resources to elongate rapidly and hasten its arrival at the soil surface, where photosynthesis quickly begins. To break through the soil yet protect its delicate, underdeveloped leaves, the shoot of an etiolated eudicot seedling forms an apical hook (see Figure 38.16).

All of these etiolation phenomena (lack of chlorophyll, rapid shoot elongation, production of an apical hook, delayed leaf expansion) are regulated by the phytochromes. In a seedling that has never been exposed to light, all the phytochrome is in the red-absorbing (P_r) form. Exposure to light converts P_r to P_{fr} (the far-red-absorbing form). The P_{fr} initiates reversal of the etiolation phenomena: Chlorophyll synthesis begins, shoot elongation slows, the apical hook unfolds, and

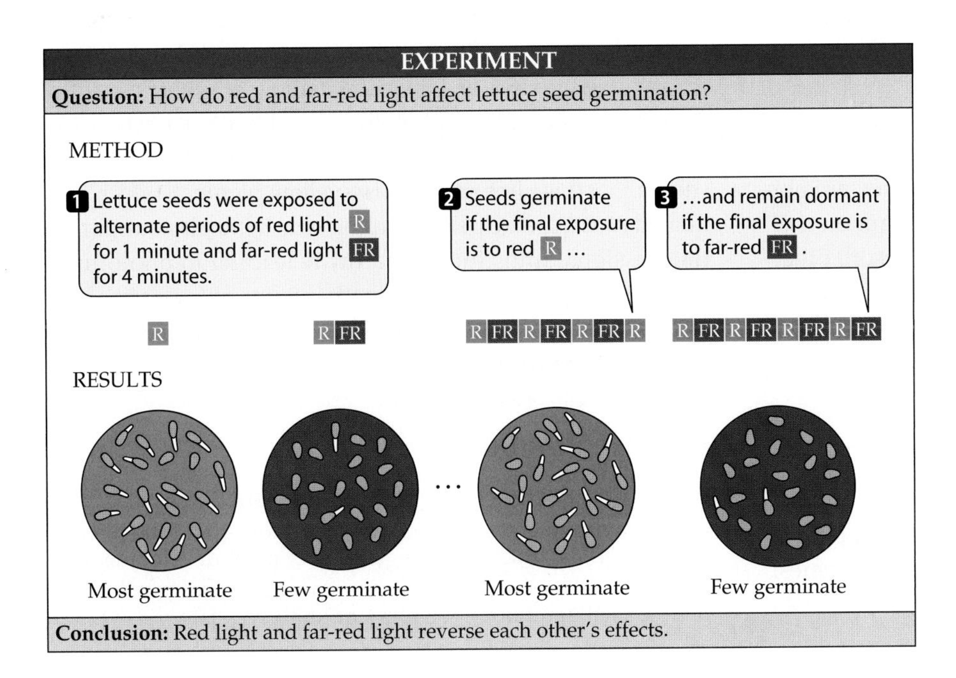

38.18 Sensitivity of Seeds to Red and Far-Red Light In each case, the final exposure reverses the preceding exposure; seeds respond only to the wavelength of the final light exposure.

the leaves begin to expand. These changes constitute *photomorphogenesis*.

There are multiple phytochromes with different developmental roles

For years, plant biologists had difficulty accounting for some aspects of phytochrome action. A solution to these problems may lie in the discovery of multiple forms of phytochromes and other photoreceptors. *Arabidopsis* has five genes that encode different phytochromes, and this diversity has been found throughout the plant kingdom and in algae as well.

The several phytochromes play differing roles during plant development. Some of them even play off each other to fine-tune plant growth during the day. Consider, for example, the light spectrum available to a seedling that is growing in the shade of other plants. Because chlorophyll in the leaves above it absorbs the light first, the shaded seedling receives a spectrum that is relatively rich in far-red light (and poor in red light); the ratio of far-red to red is increased as much as 10 to 20 times in the shade. In some shade-intolerant species, the interplay among signal transduction pathways initiated by the different phytochromes leads to an increased rate of stem elongation that attempts to move the leaves out of the shade.

Phytochromes act on the plant's genome to produce their many effects. In an early step, they activate one or more G proteins. G proteins are membrane proteins that must bind to guanosine triphosphate (GTP) to exert their effects (see Figure 15.7). The phytochrome-activated G proteins may convert GTP into the second messenger cGMP (cyclic guanosine monophosphate) and open channels that admit calcium ions into the cell, where they bind to the protein calmodulin. Both cGMP and the calcium–calmodulin complex can trigger signal transduction pathways leading eventually to the activation of specific genes and the inactivation of others. For example, phytochromes promote the transcription of many genes that encode chloroplast proteins.

Although phytochromes have many effects, they are not the only photoreceptors that participate in the regulation of plant development. Some other photoreceptors absorb other parts of the visible spectrum.

Cryptochromes, phototropins, and zeaxanthin are blue-light receptors

Cryptochromes are yellow photoreceptor pigments that absorb blue and ultraviolet light. They affect some of the same developmental processes, including seedling development and flowering, that phytochromes do. Unlike phytochromes, cryptochromes play important roles in animals as well as plants.

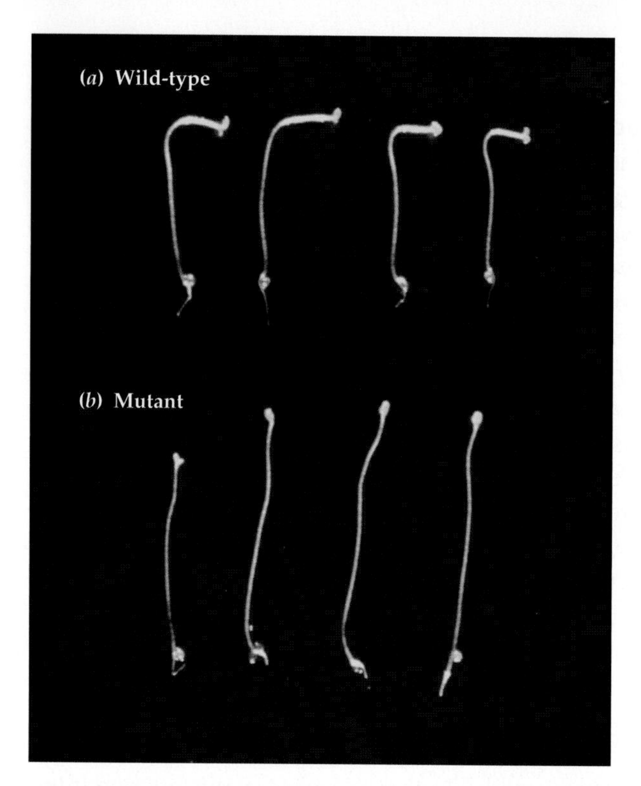

38.19 A Nonphototropic Mutant (*a*) The four etiolated wild-type *Arabidopsis* seedlings in the top row are demonstrating normal phototropism. (*b*) These four mutant seedlings cannot produce phototropin, the photoreceptor that signals the plant to curve toward light.

In contrast to phytochromes, cryptochromes are located primarily in the plant cell nucleus. The exact mechanism of cryptochrome action is not yet known. It may be significant that phytochromes behave like protein kinases, and that cryptochromes can be substrates of such enzymes. It is likely that both classes of photoreceptors participate in protein kinase-based signal transduction pathways (see Chapter 15).

We began this chapter with a photograph of phototropic response. Later we saw that the study of phototropism led to the discovery of auxin. But a question remained: What is the photoreceptor that initiates phototropism? Plant scientists working with phototropic mutants of *Arabidopsis* have recently shown that this photoreceptor is a yellow pigment, which they named *phototropin*. Upon absorbing blue light, phototropin initiates a signal transduction pathway leading to phototropic curvature (Figure 38.19). Still another type of blue-light receptor, the plastid pigment *zeaxanthin*, appears to be responsible for the light-induced opening of stomata.

Chapter Summary

Interacting Factors in Plant Development

▶ The environment, photoreceptors, hormones, and the plant's genome all play roles in the regulation of plant development.

▶ Hormones mediate many developmental phenomena in plants. Each plant hormone plays multiple regulatory roles, affecting several different aspects of development. Interactions among the hormones are often complex. **Review Table 38.1**

▶ Hormones and photoreceptors act through signal transduction pathways.

An Overview of Plant Development

▶ Cell division, cell expansion, and cell differentiation all contribute to plant development.

▶ When seed dormancy ends, the seed germinates and develops into a growing seedling. Photoreceptors and hormones regulate seedling development. **Review Figure 38.1. See Web/CD Activities 38.1 and 38.2**

▶ Eventually the plant flowers and forms fruit. Flowering in some plants is controlled by the length of the night. Hormones, probably including a flowering hormone, play roles in plant reproduction.

▶ Some plant buds exhibit winter dormancy. Eventually, all plants senesce and die. Dormancy and senescence are triggered by environmental cues, mediated by photoreceptors and hormones.

Ending Seed Dormancy and Beginning Germination

▶ Seed dormancy may be caused by exclusion of water or oxygen from the embryo, mechanical restraint of the embryo, or chemical inhibition of embryonic development. In nature, seed dormancy is broken by various mechanisms, including abrasion, fire, leaching, and low temperatures.

▶ Seed dormancy offers adaptive advantages, such as an increased likelihood of germination in a place and at a time favorable for seedling growth.

▶ Seed germination begins with the imbibition of water. Then the embryo mobilizes its reserves to obtain chemical building blocks and energy.

▶ The embryos of cereal seeds secrete gibberellins, which cause the aleurone layer to synthesize and secrete digestive enzymes that break down large molecules stored in the endosperm. **Review Figure 38.4. See Web/CD Activity 38.3**

Gibberellins: Regulators from Germination to Fruit Growth

▶ There are dozens of gibberellins. One, gibberellin A_1, regulates stem growth in most plants.

▶ Mutant plants that cannot produce normal amounts of gibberellins are dwarfs: Their stems are shorter than wild-type stems.

▶ Gibberellins regulate the growth of some fruits and cause bolting in some biennial plants. **Review Figure 38.6**

Auxin Affects Plant Growth and Form

▶ Studies of phototropism led to the discovery and isolation of auxins such as indoleacetic acid. In grass seedlings, the photoreceptor for phototropism is in the tip of the coleoptile. An auxin signal moves from the photoreceptor to the growing region of the coleoptile. **Review Figures 38.7, 38.8. See Web/CD Tutorial 38.2**

▶ Auxin transport is polar. Auxin anion efflux carriers, membrane proteins confined to the basal ends of cells, cause auxin to move from the tip to the base of the shoot. **Review Figure 38.9**

▶ Lateral movement of auxin, mediated by auxin carrier proteins, is responsible for phototropism and gravitropism. **Review Figure 38.10**

▶ Auxin plays roles in root formation, leaf abscission, apical dominance, and parthenocarpic fruit development. Certain synthetic auxins are used as selective herbicides. **Review Figure 38.12**

▶ The arrangement of cellulose microfibrils in the plant cell wall limits the direction of cell expansion. Auxin increases the plasticity of the cell wall, promoting cell expansion. It does so by increasing the pumping of protons from the cytoplasm into the cell wall, where the lowered pH activates proteins called expansins. **Review Figure 38.14, 38.15. See Web/CD Tutorial 38.3**

▶ Like all plant hormones, auxin is bound by receptor proteins.

▶ Auxin and other plant hormones signal cell differentiation and organ formation.

See Web/CD Tutorial 38.1

Cytokinins Are Active from Seed to Senescence

▶ Cytokinins are adenine derivatives. Zeatin and isopentenyl adenine are naturally occurring cytokinins, and kinetin is a synthetic cytokinin.

▶ First studied as promoters of plant cell division, cytokinins also promote seed germination in some species, inhibit stem elongation, promote lateral swelling of stems and roots, stimulate the growth of lateral buds, promote the expansion of leaf tissue, and delay leaf senescence.

Ethylene: A Hormone that Hastens Leaf Senescence and Fruit Ripening

▶ A balance between auxin and ethylene controls leaf abscission.

▶ Ethylene promotes senescence and fruit ripening.

▶ Ethylene causes the formation of a protective apical hook in eudicot seedlings that have not been exposed to light. In stems, it inhibits elongation, promotes lateral swelling, and causes a loss of gravitropic sensitivity.

▶ Ethylene acts through a signal transduction pathway that includes two membrane proteins and leads to the expression of genes. **Review Figure 38.17**

Abscisic Acid: The Stress Hormone

▶ Abscisic acid appears to maintain winter dormancy in buds. It prevents seeds from germinating while still attached to the parent plant, and it inhibits stem elongation. Through its effects on stomatal opening, it also regulates gas and water exchange between leaves and the atmosphere.

Brassinosteroids: Hormones that Mediate Effects of Light

▶ There are dozens of brassinosteroids. These steroid hormones affect cell elongation, pollen tube elongation, vascular tissue differentiation, and root elongation. Some effects of light are mediated by changes in the action and levels of brassinosteroids.

Light and Photoreceptors

▶ Phytochromes are bluish pigments found in the cytosol. Each phytochrome exists in two forms, P_r and P_{fr}, that are interconvertible by light. P_r absorbs red light and is converted to P_{fr}; P_{fr} absorbs far-red light and is converted to P_r. P_{fr} reverts to P_r in complete darkness. **Review Figure 38.18**

▶ Phytochromes have effects on seedling growth, flowering, and etiolation.

▶ The five known phytochromes mediate the effects of red, far-red, and low-energy blue light. They may play different roles in plant development, and their signal transduction pathways may interact to mediate the effects of light environments of differing spectral distribution.

▶ Cryptochromes, yellow pigments that absorb blue and ultraviolet light, interact with phytochromes in controlling seedling development and floral initiation. Cryptochromes mediate the effects of high-energy blue light.

▶ The signaling pathways for phytochromes and cryptochromes are based on protein kinases.

▶ Phototropin, another yellow pigment, is the photoreceptor for phototropism. **Review Figure 38.19**

▶ Zeaxanthin, yet another blue-light receptor, mediates the light-induced opening of stomata.

Self-Quiz

1. Which of the following is *not* an advantage of seed dormancy?
   a. It makes the seed more likely to be digested by birds that disperse it.
   b. It counters the effects of year-to-year variations in the environment.
   c. It increases the likelihood that a seed will germinate in the right place.
   d. It favors dispersal of the seed.
   e. It may result in germination at a favorable time of year.
2. Which of the following does/do *not* participate in seed germination?
   a. Imbibition of water
   b. Metabolic changes
   c. Growth of the radicle
   d. Mobilization of nutrient reserves
   e. Extensive mitotic divisions
3. To mobilize its nutrient reserves, a germinating barley seed
   a. becomes dormant.
   b. undergoes senescence.
   c. secretes gibberellins into its endosperm.
   d. converts glycerol and fatty acids into lipids.
   e. takes up proteins from the endosperm.
4. The gibberellins
   a. are responsible for phototropism and gravitropism.
   b. are gases at room temperature.
   c. are produced only by fungi.
   d. cause bolting in some biennial plants.
   e. inhibit the synthesis of digestive enzymes by barley seeds.
5. In coleoptile tissue, auxin
   a. is transported from base to tip.
   b. is transported from tip to base.
   c. can be transported toward either the tip or the base, depending on the orientation of the coleoptile with respect to gravity.
   d. is transported by simple diffusion, with no preferred direction.
   e. is not transported, because auxin is used where it is made.
6. Which process is *not* directly affected by auxin?
   a. Apical dominance
   b. Leaf abscission
   c. Synthesis of digestive enzymes by barley seeds
   d. Root initiation
   e. Parthenocarpic fruit development
7. Plant cell walls
   a. are strengthened primarily by proteins.
   b. often make up more than 90 percent of the total volume of an expanded cell.
   c. can be loosened by an increase in pH.
   d. become thinner and thinner as the cell grows longer and longer.
   e. are made more plastic by treatment with auxin.
8. Which statement about cytokinins is *not* true?
   a. They promote bud formation in tissue cultures.
   b. They delay the senescence of leaves.
   c. They usually promote the elongation of stems.
   d. They cause certain light-requiring seeds to germinate in the dark.
   e. They stimulate the development of branches from lateral buds.
9. Ethylene
   a. is antagonized by silver salts such as silver thiosulfate.
   b. is liquid at room temperature.
   c. delays the ripening of fruits.
   d. generally promotes stem elongation.
   e. inhibits the swelling of stems, in opposition to cytokinins' effects.
10. Phytochrome
   a. is the only photoreceptor pigment in plants.
   b. exists in two forms interconvertible by light.
   c. is a pigment that is colored red or far-red.
   d. is a blue-light receptor.
   e. is the photoreceptor for phototropism.

For Discussion

1. How may it be advantageous for some species to have seeds whose dormancy is broken by fire?
2. Cocklebur fruits contain two seeds each, and the two seeds are kept dormant by two different mechanisms. How might this use of two mechanisms of dormancy be advantageous to cockleburs?
3. Corn stunt virus causes a great reduction in the growth rate of infected corn plants, so the diseased plants take on a dwarfed form. Since their appearance is reminiscent of the genetically dwarfed corn studied by Phinney, you suspect that the virus may inhibit the synthesis of gibberellins by the corn plants. Describe two experiments you might conduct to test this hypothesis, only one of which should require chemical measurement.
4. Whereas relatively low concentrations of auxin promote the elongation of segments cut from young plant stems, higher concentrations generally inhibit their growth, as shown:

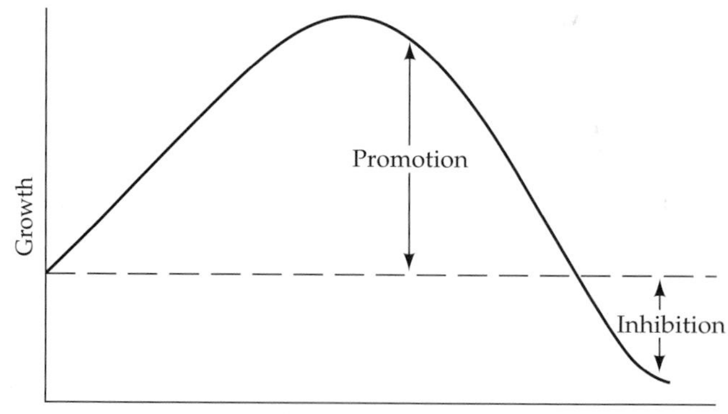

   In some plants, the inhibitory effects of high auxin concentrations appear to be secondary: High auxin concentrations cause the synthesis of ethylene, which is what causes the growth inhibition. Silver thiosulfate inhibits ethylene action. How do you think the addition of silver thiosulfate to the solutions in which the stem segments grew would affect the appearance of the above graph?
5. Some etiolated seedlings develop hairs on their epidermis when exposed to dim light. Describe an experiment to test the hypothesis that a phytochrome is the photoreceptor for this effect.

39 Reproduction in Flowering Plants

For many people, pollen means sneezing and misery because pollen grains of many plant species are potent allergens. However, pollen is not part of nature simply to annoy human beings. What is a pollen grain? It is a tiny, haploid male plant. To the stigma (the pollen "landing pad") of a flower, pollen grains represent an opportunity for mate selection. That is, the stigma may allow some pollen grains to germinate, but not others. If a pollen grain survives the mate selection process and germinates, it may eventually deliver male gametes to a microscopic, haploid female plant embedded in the flower.

Why do angiosperms expend energy and resources to produce flowers and that sometimes obnoxious pollen? The answer is simple: Flowers are sexual reproductive structures, and reproduction is the most important goal in a plant's—or any organism's—life.

In this chapter we will look at several aspects of plant reproduction, including some that are still not well understood. We will contrast sexual and asexual reproduction, and we will consider sexual reproduction in detail. In doing so, we will look at angiosperm gametophytes, pollination, double fertilization, embryonic development, and the roles of fruits in seed dispersal. The transition from the vegetative state to the flowering state is a key event in plant development, and we'll see how changing seasons trigger flowering in some plants—and speculate on the existence of a flowering hormone. We will conclude the chapter with an examination of asexual reproduction in nature and in agriculture.

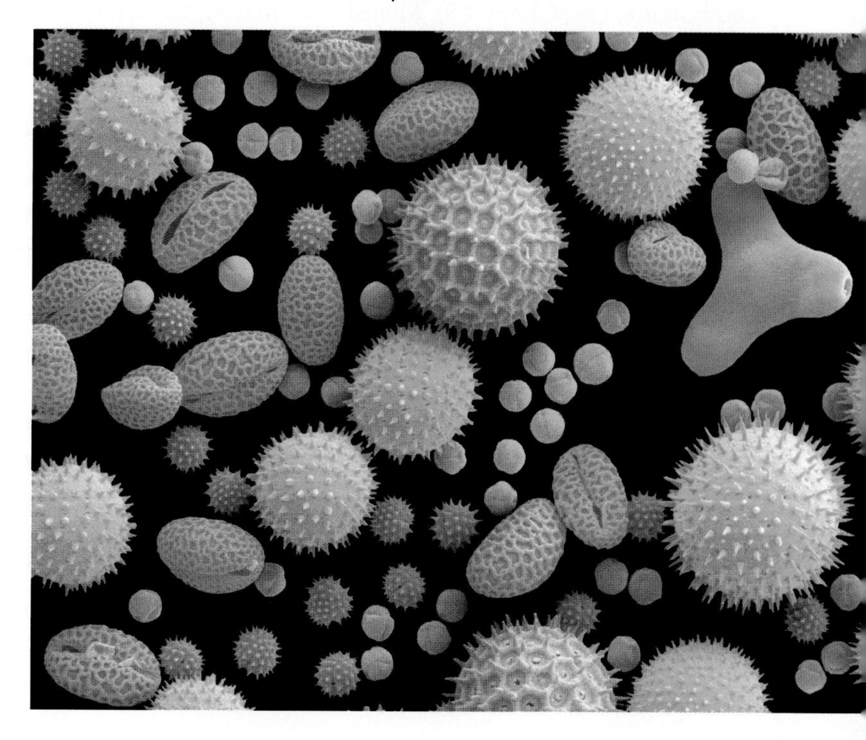

Key Players in Sexual Reproduction Each species' pollen has a characteristic size, shape, and cell wall structure. These structures are the male gametophytes and are essential for sexual reproduction in seed plants.

Many Ways to Reproduce

Plants have many ways of reproducing—and humans have developed even more ways of reproducing them. Flowers contain the sex organs of plants; it is thus no surprise that almost all angiosperms reproduce sexually. But some angiosperms reproduce asexually as well; some even reproduce asexually most of the time. What are the advantages and disadvantages of these two kinds of reproduction? The answers to this question involve genetic recombination. As we have seen, sexual reproduction produces new combinations of genes and diverse phenotypes. Asexual reproduction, in contrast, produces a clone of genetically identical individuals.

Both sexual and asexual reproduction are important in agriculture. Many important annual crops are grown from seeds, which are the products of sexual reproduction. Seed-grown crops include the great grain crops, all of which are grasses—wheat, rice, corn, sorghum, and millet—as well as plants in other families, such as soybeans and safflower. Other crops, such as strawberries, potatoes, and bananas, are produced asexually.

Orange trees, which have been under cultivation for centuries, can be grown from seed—except for the navel orange, which has no seeds. This plant apparently arose only once in history. Early in the nineteenth century, on a plantation on the Brazilian coast, a single orange seed gave rise to one tree that had aberrant flowers. Parts of the flowers aborted, and seedless fruits formed. Asexual reproduction is the only way of propagating this plant, and every navel orange in the world comes from a tree that has been derived asexually from that original Brazilian navel orange tree.

Unlike navel oranges, strawberries are capable of forming seeds and need not be propagated asexually. Nonetheless, asexual propagation of strawberries is common because vast numbers of plants that are genetically and phenotypically identical to a plant humans find particularly desirable can be produced in this way.

We will treat asexual reproduction in greater detail at the end of this chapter. We will begin, however, by considering sexual reproduction.

Sexual Reproduction in Plants

Sexual reproduction provides genetic diversity through recombination (see Chapter 9). Meiosis and mating between different plants shuffle genes into new combinations, giving a population a variety of genotypes in each generation, some of which may be superior to those of their parents. This genetic diversity may serve the population well as the environment changes or as the population expands into new environments. The adaptability resulting from genetic diversity is the major advantage of sexual reproduction over asexual reproduction, although sexual reproduction can also break up well-adapted combinations of alleles through the same process of recombination.

The flower is an angiosperm's device for sexual reproduction

A complete flower consists of four groups of organs that are modified leaves: the carpels, stamens, petals, and sepals (see Figure 30.7). The *carpels* and *stamens* are, respectively, the female and male sex organs. A *pistil* is a structure composed of one or more carpels. The base of the pistil, called the *ovary*, contains one or more *ovules*, each of which contains a megasporangium, within which a female gametophyte may develop. The stalk of the pistil is the *style*, and the end of that stalk is the *stigma*. Each stamen is composed of a *filament* bearing a two-lobed *anther*, which consists of four microsporangia fused together. Male gametophytes begin their development within the microsporangia.

The *petals* and *sepals* of many flowers are arranged in whorls (circles) or spirals around the carpels and stamens. Together, the petals constitute the *corolla*. Below them, the sepals constitute the *calyx*. The petals are often colored, attracting pollinating animals; the sepals are often green and photosynthetic. All the parts of the flower are borne on a stem tip, the *receptacle*. Flower parts are very diverse in form, in contrast to the microscopic gametophytes that develop within them.

Flowering plants have microscopic gametophytes

Before reading this section, you may wish to review the section in Chapter 29 entitled "Life cycles of plants feature alternation of generations" (pages 571–572). Central to understanding plant reproduction is the concept of alternation of generations, in which a multicellular diploid generation alternates with a multicellular haploid generation.

In angiosperms, the diploid sporophyte generation is the larger and more conspicuous one. The sporophyte generation produces flowers. The flowers produce spores, which develop into tiny gametophytes that begin and, in the case of the megagametophyte, end their development enclosed by sporophyte tisue.

The haploid gametophytes—the gamete-producing generation—of flowering plants develop from haploid spores in sporangia within the flower (Figure 39.1):

- Female gametophytes (megagametophytes), which are called **embryo sacs**, develop in megasporangia.
- Male gametophytes (microgametophytes), which are called **pollen grains**, develop in microsporangia.

Within the ovule, a *megasporocyte*—a cell within the megasporangium—divides meiotically to produce four haploid *megaspores*. In most plants, all but one of these megaspores then degenerate. The surviving megaspore usually undergoes three mitotic divisions, producing eight haploid nuclei, all initially contained within a single cell—three nuclei at one end, three at the other, and two in the middle. Subsequent cell wall formation leads to an elliptical, seven-celled megagametophyte with a total of eight nuclei (see Figure 39.1):

- At one end of the elliptical megagametophyte are three tiny cells: the egg and two cells called **synergids**. The egg is the female gamete, and the synergids participate indirectly in fertilization by attracting and accepting the pollen tube.

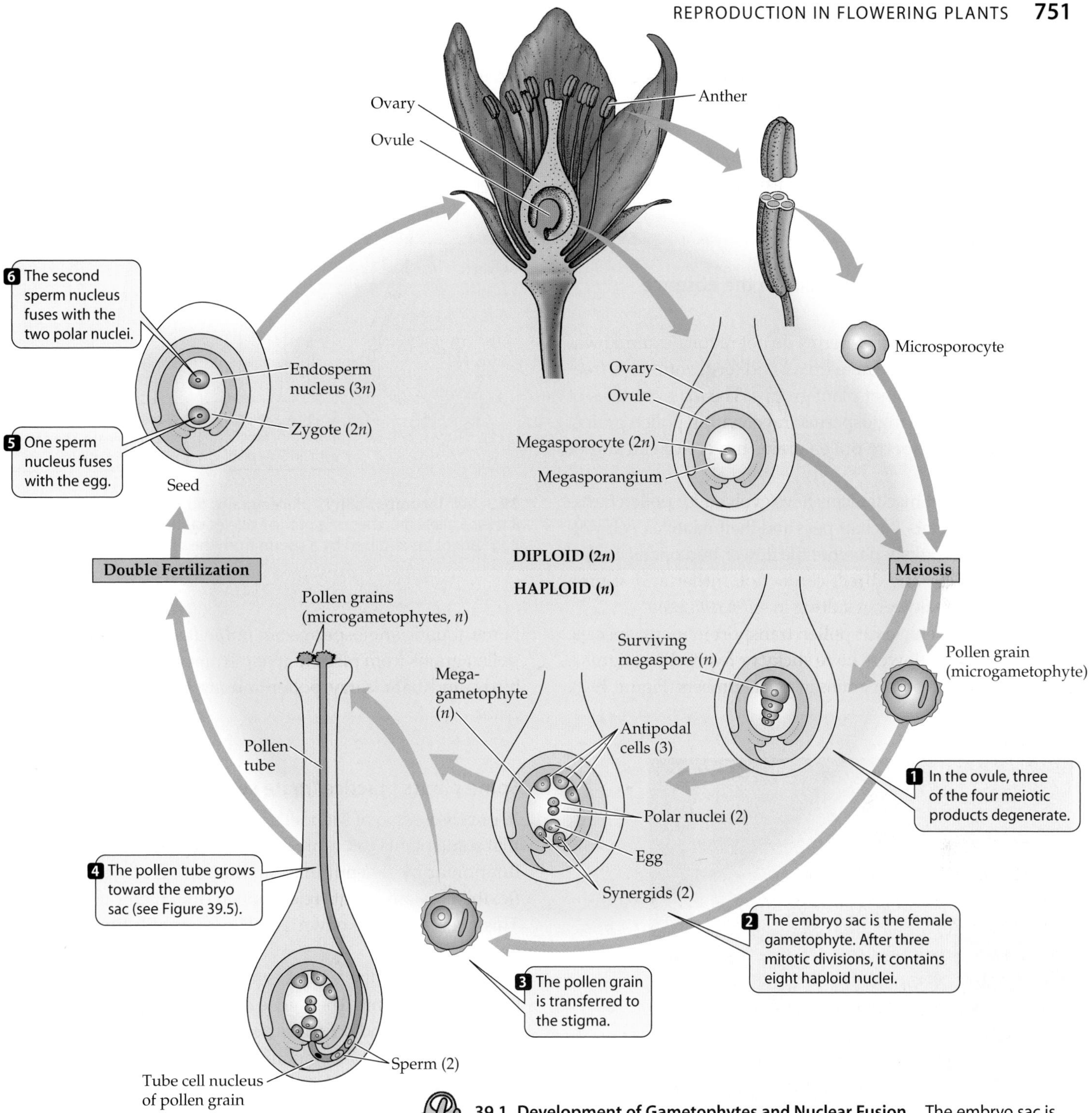

39.1 Development of Gametophytes and Nuclear Fusion The embryo sac is the female gametophyte; the pollen grain is the male gametophyte. The male and female nuclei meet and fuse within the embryo sac. Most angiosperms have double fertilization, in which a zygote and an endosperm nucleus form from separate fusion events—the zygote from one sperm and the egg and the endosperm from the other sperm and two polar nuclei.

- At the opposite end of the megagametophyte are three **antipodal cells**, which eventually degenerate.
- In the large central cell are two **polar nuclei**.

The embryo sac (megagametophyte) is the entire seven-cell, eight-nucleus structure. You can review the development of the embryo sac in Figure 39.1.

The pollen grain (microgametophyte) consists of fewer cells and nuclei than the embryo sac. The development of a pollen grain begins when a microsporocyte within the anther divides meiotically. Each resulting haploid microspore develops a spore wall, within which it normally undergoes one mitotic division before the anthers open and release these two-celled pollen grains. The two cells are the **tube cell** and the **generative cell**. Further development of the pollen grain, which we will describe shortly, is delayed until the pollen arrives at a stigma. In angiosperms, the transfer of pollen from the anther to the stigma is referred to as **pollination**.

Pollination enables fertilization in the absence of liquid water

Gymnosperms and angiosperms do not require external water as a medium for gamete travel and fertilization—a freedom not shared by other plant groups. The male gametes of gymnosperms and angiosperms travel within pollen grains. But how do angiosperm pollen grains travel from an anther to a stigma?

Many different mechanisms have evolved for pollen transport. In some plants, such as peas and their relatives, self-pollination is accomplished before the flower bud opens. Pollen is transferred by the direct contact of anther and stigma within the same flower, resulting in *self-fertilization*.

Wind is the vehicle for pollen transport in many species. Wind-pollinated flowers have sticky or featherlike stigmas, and they produce pollen grains in great numbers (Figure 39.2).

39.2 Wind Pollination The numerous anthers on these inflorescences (groups of flowers) of a hazelnut tree all point away from the stalk and stand free of the plant, promoting dispersal of the pollen by wind.

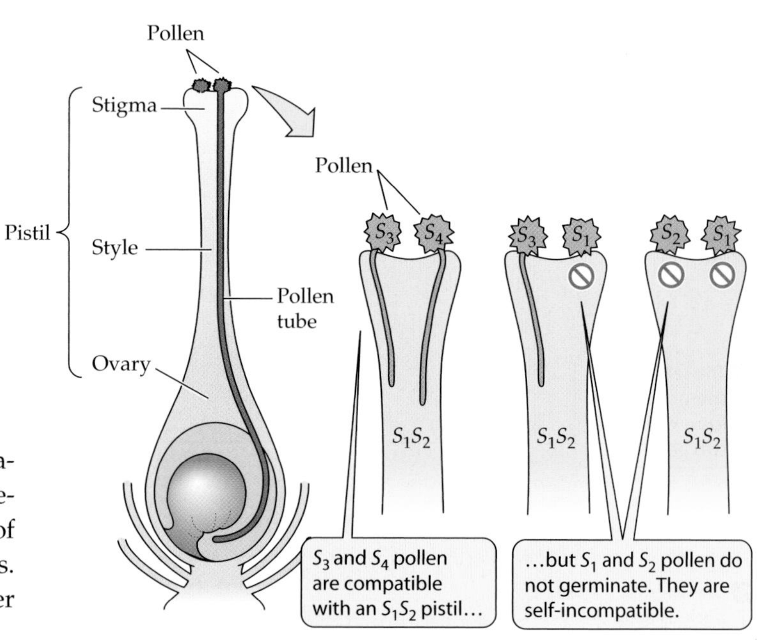

39.3 Self-Incompatibility Pollen grains do not germinate normally if their *S* allele matches one of the *S* alleles of the stigma. Thus, the egg cannot be fertilized by a sperm from the same plant.

Some aquatic angiosperms are pollinated by water carrying pollen grains from plant to plant. Animals, including insects, birds, and bats, carry pollen among the flowers of many plants.

Some plants practice "mate selection"

In our discussion of Mendel's work (see Chapter 10), we saw that some plants can reproduce sexually either by cross-pollination or by self-pollination. But not all plants have this flexibility. Many plants reject pollen from their own flowers. This phenomenon, known as **self-incompatibility**, promotes genetic variation.

A single gene, the *S* gene, is responsible for self-incompatibility in most plants. The *S* gene has dozens of alleles. A pollen grain is haploid and possesses a single *S* allele; the recipient pistil is diploid. In self-incompatible plants, pollen fails to germinate, or the pollen tube fails to traverse the style, if the *S* allele of the pollen matches one of the two *S* alleles in the pistil (Figure 39.3).

The stigma plays an important role in "mate selection" by flowering plants. The stigmas of most plants are exposed to the pollen of many other species as well as their own. Pollen from the same species binds strongly to the stigma due to cell–cell signaling between the stigma and the cell walls of the pollen grains. In contrast, foreign pollen falls off readily or fails to germinate.

A pollen tube delivers male cells to the embryo sac

When a pollen grain lands on the stigma of a compatible pistil, it germinates. Germination, for a pollen grain, is the de-

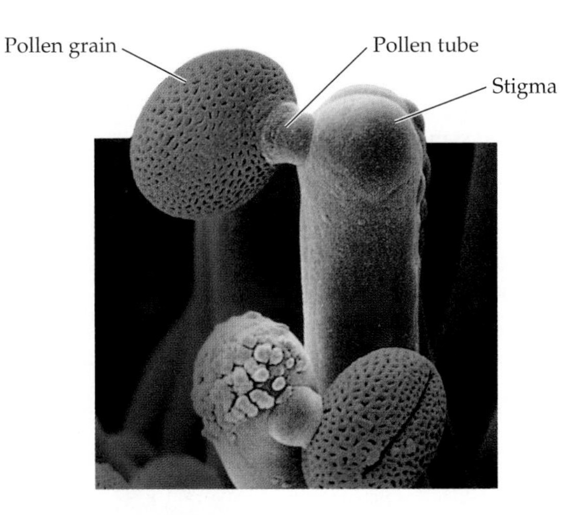

39.4 Pollen Tubes Begin to Grow These pollen grains have landed on hairlike structures on the stigma of an *Arabidopsis* flower, and pollen tubes have penetrated the stigma.

velopment of a **pollen tube** (Figure 39.4). The pollen tube either traverses the spongy tissue of the style or, if the style is hollow, grows downward on the inner surface of this female organ until it reaches an ovule. The pollen tube may grow millimeters or even centimeters in the process.

The rapid growth of the pollen tube requires calcium ions, which are taken up by the growing tip of the tube, as well as cell adhesion proteins. The downward growth of the pollen tube is believed to be guided by a long-distance chemical signal from the synergids within the ovule. If one synergid is destroyed, the ovule still attracts pollen tubes, but destruction of both synergids renders the ovule unable to attract pollen tubes, and fertilization does not occur.

Angiosperms perform double fertilization

In most angiosperm species, the mature pollen grain consists of two cells, the tube cell and the generative cell. The larger tube cell encloses the much smaller generative cell. Guided by the tube cell nucleus, the pollen tube eventually grows through the megasporangial tissue and reaches the embryo sac. The generative cell meanwhile has undergone one mitotic division and cytokinesis to produce two haploid **sperm cells**.

Both of the sperm cells enter the embryo sac, where they are released into the cytoplasm of one of the synergids. This synergid degenerates, releasing the sperm cells (Figure 39.5). Each sperm cell then fuses with a different cell of the embryo sac. One sperm cell fuses with the egg cell, producing the diploid zygote. The other fuses with the central cell, and that sperm cell nucleus and the two polar nuclei unite to form a triploid ($3n$) nucleus. While the zygote nucleus begins mitotic division to form the new sporophyte embryo, the triploid nucleus undergoes rapid mitosis to form a specialized nutritive tissue, the **endosperm**. The endosperm will later be digested by the developing embryo, as we saw in the previous chapter. The antipodal cells and the remaining synergid eventually degenerate, as does the pollen tube nucleus.

This process is known as **double fertilization** because it involves two nuclear fusion events:

- One sperm cell fuses with the egg cell.
- The other sperm cell fuses with the two polar nuclei.

39.5 Sperm Nuclei and Double Fertilization The sperm nuclei contribute to the formation of the diploid zygote and the triploid endosperm. Double fertilization is a characteristic feature of angiosperm reproduction.

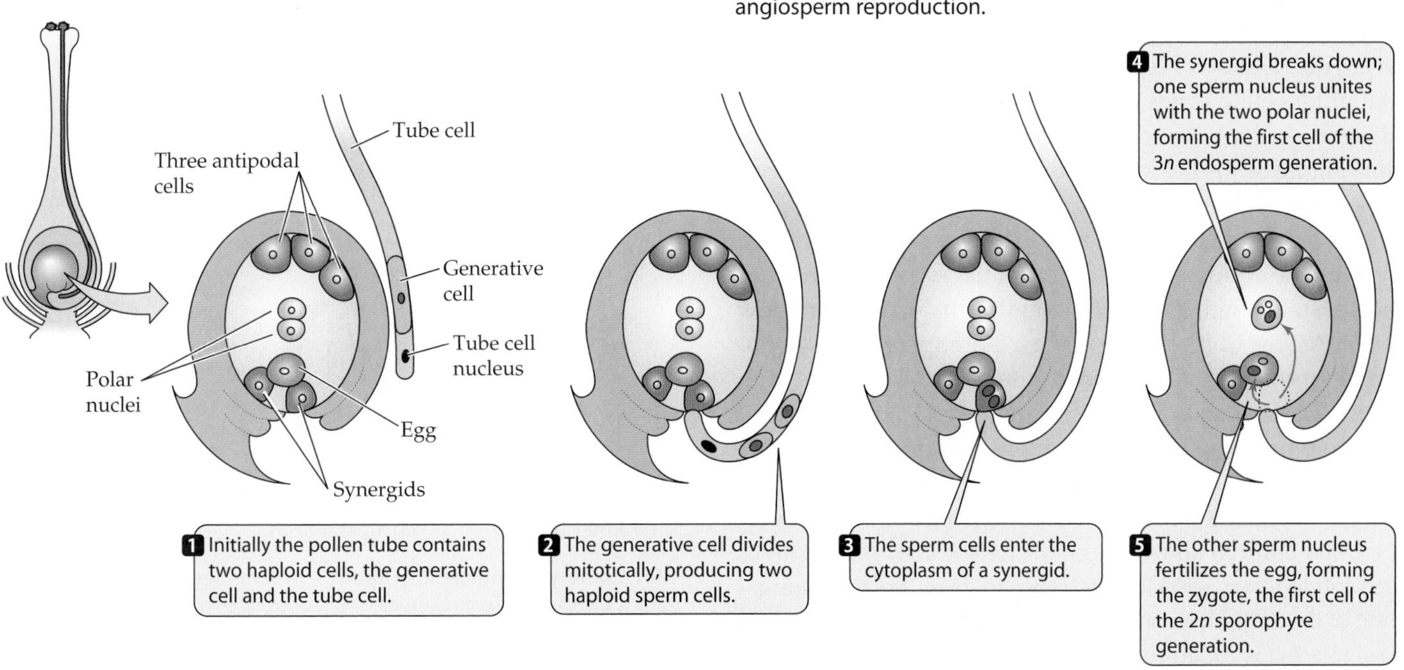

Three antipodal cells
Tube cell
Generative cell
Tube cell nucleus
Polar nuclei
Egg
Synergids
1 Initially the pollen tube contains two haploid cells, the generative cell and the tube cell.
2 The generative cell divides mitotically, producing two haploid sperm cells.
3 The sperm cells enter the cytoplasm of a synergid.
4 The synergid breaks down; one sperm nucleus unites with the two polar nuclei, forming the first cell of the 3n endosperm generation.
5 The other sperm nucleus fertilizes the egg, forming the zygote, the first cell of the 2n sporophyte generation.

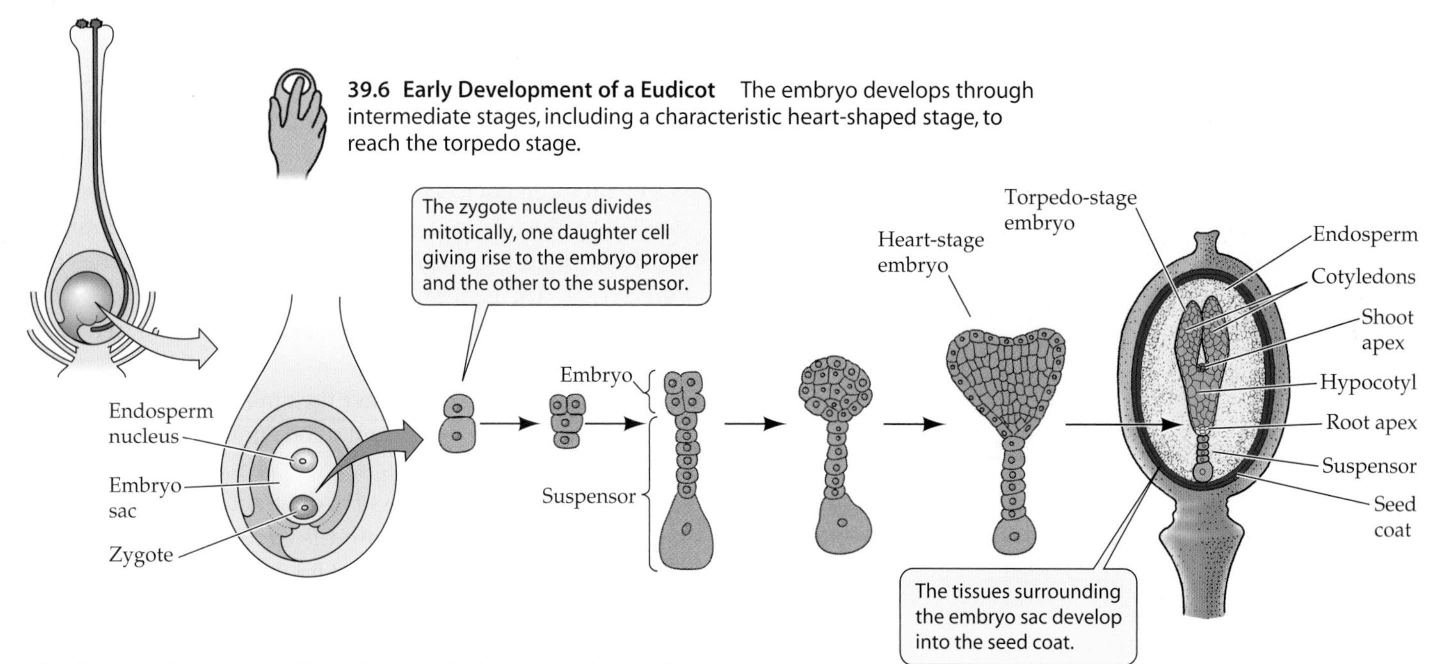

39.6 Early Development of a Eudicot The embryo develops through intermediate stages, including a characteristic heart-shaped stage, to reach the torpedo stage.

The fusion of a sperm cell nucleus with the two polar nuclei to form endosperm takes place only in angiosperms. The fusion of these three nuclei, the possession of flowers, and the formation of fruit are the three most definitive characteristics shared by angiosperms.

Embryos develop within seeds

Shortly after fertilization, highly coordinated growth and development of embryo, endosperm, integuments, and carpel ensues. The integuments—protective tissue layers immediately surrounding the megasporangium—develop into the seed coat, and the carpel ultimately becomes the wall of the fruit that encloses the seed.

The first step in the formation of the embryo is a mitotic division of the zygote that gives rise to two daughter cells. These two cells face different fates. An asymmetrical (uneven) distribution of cytoplasm within the zygote causes one daughter cell to produce the embryo proper and the other daughter cell to produce a supporting structure, the **suspensor** (Figure 39.6). The suspensor pushes the embryo against or into the endosperm and provides one route by which nutrients pass from the endosperm into the embryo.

With the asymmetrical division of the zygote, polarity has been established, as has the longitudinal axis of the new plant. A long, thin suspensor and a more spherical or globular embryo are distinguishable after just four mitotic divisions. The suspensor soon ceases to elongate. However, cell divisions continue, the primary meristems form, and the first organs begin to form within the embryo.

In eudicots (monocots are somewhat different), the initially globular embryo takes on a characteristic *heart stage* form as the cotyledons ("seed leaves") start to grow. Further elongation of the cotyledons and of the main axis of the embryo gives rise to what is called the *torpedo stage*, during which some of the internal tissues begin to differentiate (see Figure 39.6). Between the cotyledons is the shoot apex; at the other end is the root apex. Between the shoot and root apices is the hypocotyl. Each of the apical regions contains an apical meristem whose dividing cells will give rise to the organs of the mature plant.

During seed formation, large amounts of nutrients are moved in from other parts of the plant, and the endosperm accumulates starch, lipids, and proteins. In many species, the cotyledons absorb the nutrient reserves from the surrounding endosperm and grow very large in relation to the rest of the embryo (Figure 39.7*a*). In others, the cotyledons remain thin (Figure 39.7*b*); they draw on the reserves in the endosperm as needed when the seed germinates.

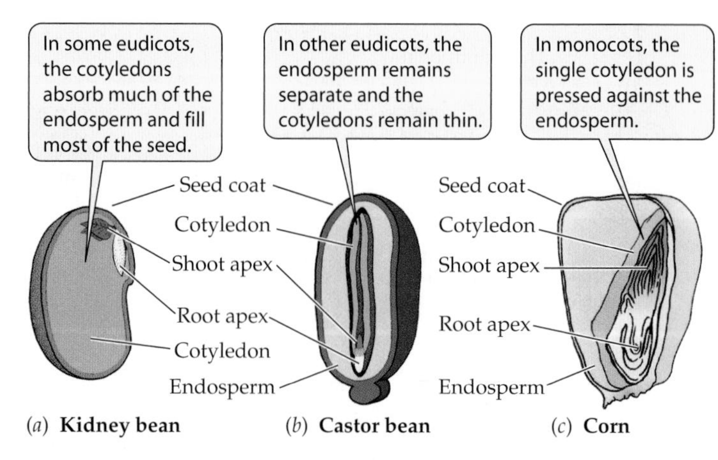

39.7 Variety in Angiosperm Seeds In some seeds, such as kidney beans (*a*), the nutrient reserves of the endosperm are absorbed by the cotyledons. In others, such as castor beans (*b*) and corn (*c*), the reserves in the endosperm will be drawn upon after germination.

In the late stages of embryonic development, the seed loses water—sometimes as much as 95 percent of its original water content. In this desiccated state, the embryo is incapable of further development; it remains quiescent until internal and external conditions are right for germination. (Recall from Chapter 38 that a necessary early step in seed germination is the massive imbibition of water.) In addition to embryo and endosperm development, the structures of the ovary are also undergoing developmental changes to form a seed and fruit.

Some fruits assist in seed dispersal

After fertilization, the ovary wall of a flowering plant—together with its seeds—develops into a fruit. A **fruit** may consist of only the mature ovary and the seeds it contains, or it may include other parts of the flower or structures that are closely related to it. In some species, this process produces fleshy, edible fruits such as peaches and tomatoes, while in other species the fruits are dry or inedible. Some major variations on this theme are illustrated in Figure 30.12, which shows only fleshy, edible fruits. Whatever its form, the fruit serves to assure seed dispersal.

Some fruits help disperse seeds over substantial distances, improving the chances that at least a few of the many seeds produced by a plant will find suitable conditions for germination and growth to sexual maturity. Various trees, including ash, elm, maple, and tree of heaven, produce a dry, winged fruit that may be blown some distance from the parent tree by the wind (Figure 39.8*a*). Water disperses some fruits; coconuts have been spread in this way from island to island in the Pacific Ocean (Figure 39.8*b*). Still other fruits travel by hitching rides with animals—either inside or outside them. Fleshy fruits such as berries provide food for mammals or birds; seeds that are swallowed whole travel safely through the animal's digestive tract and are deposited some distance from the parent plant. In some species, seeds must pass through an animal to break dormancy.

We have now traced the sexual life cycle of angiosperms from the flower to the fruit to the dispersal of seeds. Seed germination and the vegetative development of the seedling were presented in Chapter 38. Now let's complete the cycle by considering the transition from the vegetative to the flowering state, and how this transition is regulated.

39.8 Dispersal of Fruits
(*a*) A samara is a winged fruit characteristic of the maple family. (*b*) A coconut seed germinates where it washed ashore on a beach in the South Pacific.

The Transition to the Flowering State

If we view a plant as something produced by a seed for the purpose of bearing more seeds, then the act of flowering is one of the supreme events in a plant's life. The transition to the flowering state marks the end of vegetative growth for some plants. In other plants, vegetative growth may accompany flowering or resume after flowering is completed. But whatever the specific pattern, flowering always entails major developmental changes.

Apical meristems can become inflorescence meristems

The first visible sign of the transition to the flowering state may be a change in one or more apical meristems in the shoot system. During vegetative growth, an apical meristem continually produces leaves, lateral buds, and internodes (Figure 39.9*a*). This unrestricted growth is *indeterminate* (see Chapter 35).

Flowers may appear singly or in an orderly cluster that constitutes an **inflorescence**. If a vegetative apical meristem becomes an **inflorescence meristem**, it ceases production of leaves, lateral buds, and internodes and produces other structures: smaller leafy structures called *bracts*, as well as new meristems in the angles between the bracts and the internodes (Figure 39.9*b*). These new meristems may also be inflorescence meristems, or they may be **floral meristems**, each of which gives rise to a flower.

Each floral meristem typically produces four consecutive whorls or spirals of organs—the sepals, petals, stamens, and carpels—separated by very short internodes, keeping the flower compact (Figure 39.9*c*). In contrast to vegetative apical meristems and some inflorescence meristems, floral meristems are responsible for *determinate* growth—the limited growth of the flower to a particular size and form.

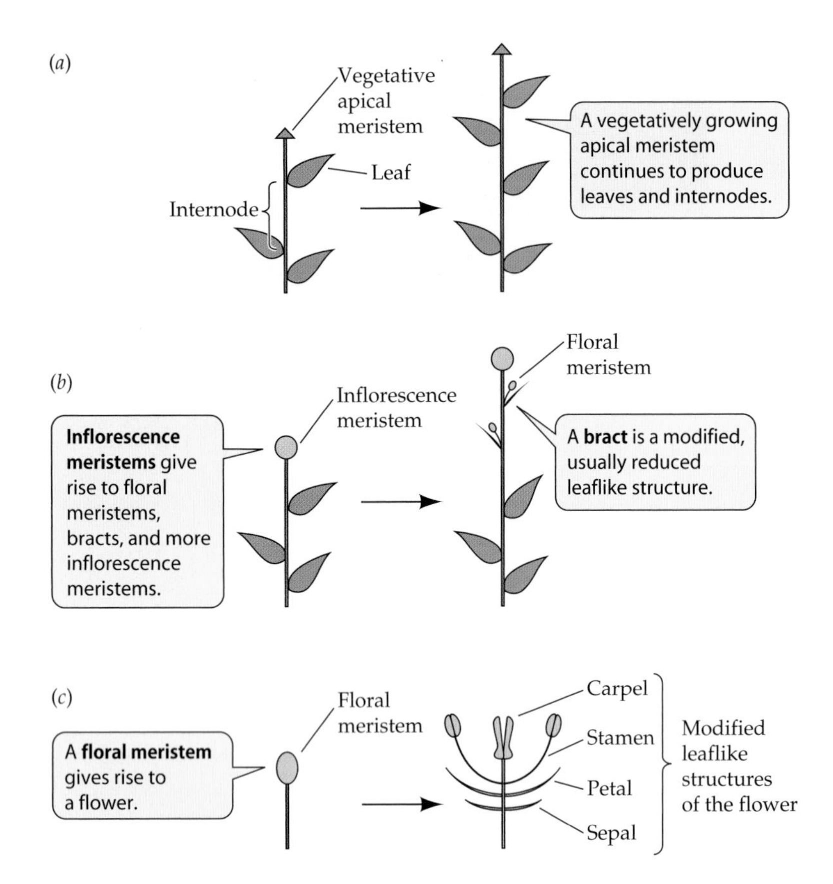

39.9 Flowering and the Apical Meristem A vegetative apical meristem (*a*) grows without producing flowers. Once the transition to the flowering state is made, inflorescence meristems (*b*) give rise to bracts and to floral meristems (*c*), which become the flowers.

A cascade of gene expression leads to flowering

How do apical meristems become inflorescence meristems, and how do inflorescence meristems give rise to floral meristems? How does a floral meristem give rise, in short order, to four different floral organs? How does each flower come to have the correct number of each of the floral organs? Numerous genes collaborate to produce these results. We'll refer here to some of the genes whose actions have been most thoroughly studied in *Arabidopsis* and snapdragons.

- Expression of a group of *meristem identity genes* initiates a cascade of further gene expression.
- This cascade begins with *cadastral genes*, which participate in pattern formation—the spatial organization of the whorls of organs.
- Cadastral genes trigger the expression of *floral organ identity genes*, which work in concert to specify the successive whorls (see Figure 19.12)

Floral organ identity genes are homeotic genes, and their products are transcription factors that mediate the expression of still other genes.

Having seen how flowering occurs, we will now consider how the transition from the vegetative to the flowering state is initiated.

Photoperiodic Control of Flowering

Environmental cues trigger the transition to the flowering state in many cases, but such environmental control is also subject to genetic modification. The life cycles of flowering plants fall into three categories: annual, biennial, and perennial. **Annuals**, such as many food crops, complete their life cycle (seed to flower) in one growing season. **Biennials**, such as carrots and cabbage, grow vegetatively for all or part of one growing season and live on into a second growing season, during which they flower, form seeds, and die. **Perennials**, such as oak trees, live for a few to many growing seasons, during which both vegetative growth and flowering occur. What control systems give rise to these and other differences in flowering behavior?

In 1920, W. W. Garner and H. A. Allard of the U.S. Department of Agriculture studied the behavior of a newly discovered mutant tobacco plant. The mutant, named 'Maryland Mammoth,' had large leaves and exceptional height. When the other plants in the field flowered, 'Maryland Mammoth' plants continued to grow. Garner and Allard took cuttings of 'Maryland Mammoth' into their greenhouse, and the plants that grew from those cuttings finally flowered in December.

Garner and Allard guessed that this flowering pattern had something to do with the mutant's response to some environmental cue. They tested several likely environmental variables, such as temperature, but the key variable proved to be day length. By moving plants between light and dark rooms at different times to vary the day length artificially, they were able to establish a direct link between flowering and day length. We now know that the key variable is the length of the *night*, rather than the day, but Garner and Allard did not make that distinction.

The 'Maryland Mammoth' plants did not flower if the light period they were exposed to was longer than 14 hours per day, but flowering commenced after the days became shorter than 14 hours. Thus, the **critical day length** for 'Maryland Mammoth' tobacco is 14 hours (Figure 39.10). The phenomenon of control by the length of day or night is called **photoperiodism**.

There are short-day, long-day, and day-neutral plants

Plants that flower in response to photoperiodic stimuli fall into several classes. Poinsettias, chrysanthemums, and 'Maryland Mammoth' tobacco are **short-day plants** (SDPs), which flower only when the day is *shorter* than a critical *maximum*.

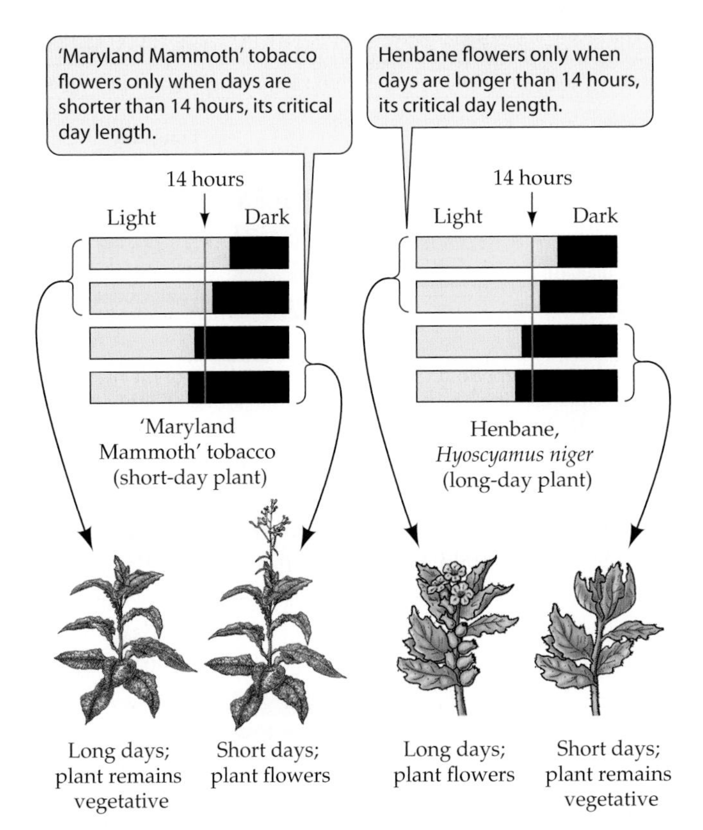

39.10 Day Length and Flowering By artificially varying the length of the day, Garner and Allard showed that the flowering of 'Maryland Mammoth' tobacco is initiated when the days become shorter than a critical length. 'Maryland Mammoth' tobacco is thus called a short-day plant. Henbane, a long-day plant, shows an inverse pattern of flowering.

Spinach and clover are examples of **long-day plants** (LDPs), which flower only when the day is *longer* than a critical *minimum*. Generally, LDPs are triggered to flower in midsummer and SDPs in late summer, fall, or sometimes in the spring. Because short days occur both before and after midsummer, there is a degree of ambiguity in this signal. Could there be a more precise way for plants to regulate flowering?

Some plants require photoperiodic signals that are more complex than just short or long days. One group, the *short-long-day plants*, must experience first short days and then long ones in order to flower. Accordingly, white clover and other short-long-day plants flower during the long days before midsummer. Another group, the *long-short-day plants*, cannot flower until the long days of summer have been followed by shorter ones, so they bloom only in the fall. *Kalanchoe*, seen in Figure 39.16*b*, is a long-short-day plant.

Other processes besides flowering are also under photoperiodic control. We have learned, for example, that short days trigger the onset of winter dormancy in plants. (Animals, too, show a variety of photoperiodic behaviors, as we'll see in Chapter 52.)

The flowering of some angiosperms, such as corn, roses, and tomatoes, is not photoperiodic. In fact, there are more of these **day-neutral plants** than there are short-day and long-day plants. Some plants are photoperiodically sensitive only when young and become day-neutral as they grow older. Others require specific combinations of day length and other factors—especially temperature—to flower.

The length of the night determines whether a photoperiodic plant will flower

The terms "short-day plant" and "long-day plant" became entrenched before scientists learned that photoperiodically sensitive plants actually measure the length of the *night*, or of a period of darkness, rather than the length of the day. This fact was demonstrated by Karl Hamner of the University of California at Los Angeles and James Bonner of the California Institute of Technology (Figure 39.11).

Working with cocklebur, an SDP, Hamner and Bonner ran a series of experiments using two sets of conditions:

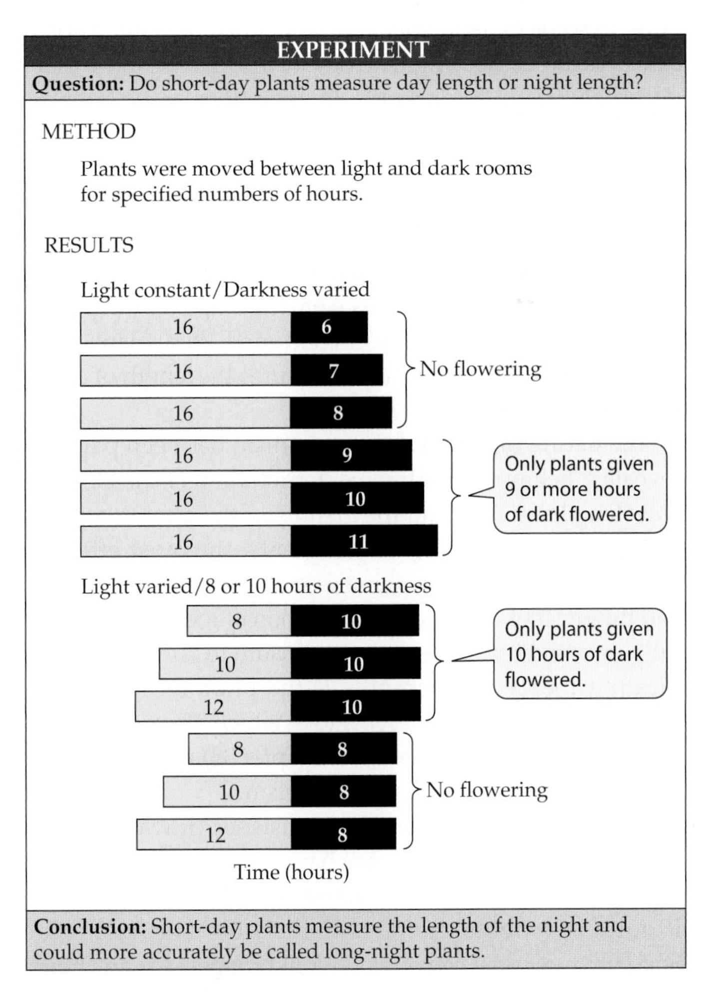

EXPERIMENT
Question: Do short-day plants measure day length or night length?
METHOD
Plants were moved between light and dark rooms for specified numbers of hours.
RESULTS
Light constant/Darkness varied
16 6
16 7
16 8
No flowering
16 9
16 10
16 11
Only plants given 9 or more hours of dark flowered.
Light varied/8 or 10 hours of darkness
8 10
10 10
12 10
Only plants given 10 hours of dark flowered.
8 8
10 8
12 8
No flowering
Time (hours)
Conclusion: Short-day plants measure the length of the night and could more accurately be called long-night plants.

39.11 Night Length and Flowering The length of the dark period, not the length of the light period, determines flowering.

- For one group of plants, the light period was kept constant—either shorter or longer than the critical day length—and the dark period was varied.
- For another group of plants, the dark period was kept constant and the light period was varied.

The plants flowered under all treatments in which the dark period exceeded 9 hours, regardless of the length of the light period. Thus, Hamner and Bonner concluded that it is the length of the *night* that matters; for cocklebur, the *critical night length* is about 9 hours. Thus, it would be more accurate to call cocklebur a "long-night plant" than a short-day plant.

In cocklebur, a single long night is sufficient photoperiodic stimulus to trigger full flowering some days later, even if the intervening nights are short ones. Most plants are less sensitive than cocklebur and require from two to several nights of appropriate length to induce flowering. For some plants, a single shorter night in a series of long ones, even one day before flowering would have commenced, inhibits flowering.

By means of other experiments, Hamner and Bonner gained some insight into how plants measure night length. They grew SDPs and LDPs under a variety of light conditions. In some experiments, the dark period was interrupted by a brief exposure to light; in others, the light period was interrupted briefly by darkness. Interruptions of the light period by darkness had no effect on the flowering of either short-day or long-day plants. Even a brief interruption of the dark period by light, however, completely nullified the effect of a long night (Figure 39.12*a*). An SDP flowered only if the long nights were uninterrupted. An LDP experiencing long nights flowered if those nights were interrupted by exposure to light. Thus, the investigators concluded, these plants must have a timing mechanism that measures the length of a continuous dark period.

The nature of this timing mechanism has been partially revealed, beginning with the determination of the effective wavelengths of light and the identity of the photoreceptors. In the interrupted-night experiments, the most effective wavelengths of light were in the red range (Figure 39.12*b*), and the effect of a red-light interruption of the night could be fully reversed by a subsequent exposure to far-red light, indicating that a phytochrome is the photoreceptor. Phytochromes and blue-light receptors, which affect several aspects of plant development (see Chapter 38), also participate in the photoperiodic timing mechanism.

What might that mechanism consist of? It was once hypothesized that the timing mechanism might simply be the slow conversion of a phytochrome during the night from the P_{fr} form—produced during the light hours—to the P_r form. Such phytochrome conversion would function as an "hourglass," and the effect of a night would depend simply upon whether all the phytochrome had been converted. However,

EXPERIMENT A

Question: How does interrupting a long day or night affect flowering?

Short-day plants	Experimental conditions	Long-day plants
No flowering		Flowering
No flowering		Flowering
Flowering		No flowering
No flowering		Flowering

Conclusion: Photoperiodic plants measure the length of the night, not the day. Interrupting a long night with a brief period of light inhibits flowering in short-day plants. Long-day plants flower when the night is short, but interrupting their long day has no effect.

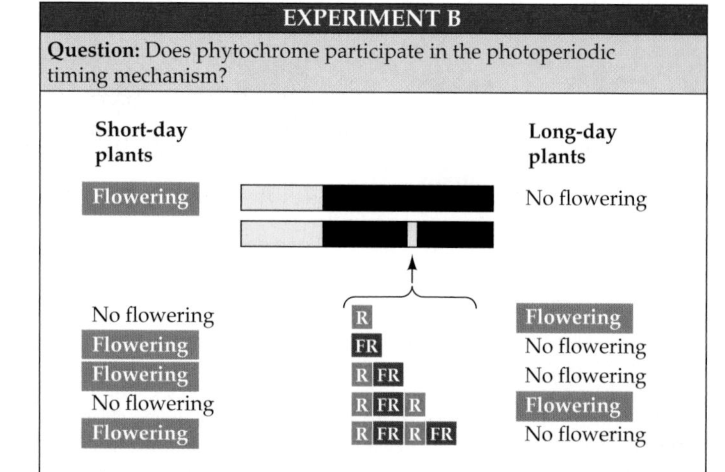

Conclusion: When plants are exposed to red (R) and far-red (FR) light in alternation, the final treatment determines the effect of the light interruption, suggesting that phytochrome participates in photoperiodic responses.

39.12 The Effect of Interrupted Days and Nights
(*a*) Experiments suggest that plants are able to measure the length of a continuous dark period and use this information to trigger flowering. (*b*) Phytochromes seem to be involved in the photoperiodic timing mechanism.

this suggestion is inconsistent with many experimental observations, such as the fact that when a plant is subjected to a dark period several days in duration, the plant's sensitivity to a light flash during the long night varies on a roughly 24-hour cycle. Such data suggest instead that the phytochrome is only a photoreceptor, and that the timekeeping role is played by a biological clock that is linked to the phytochrome (which sets the clock) and also to the production of flowers.

Circadian rhythms are maintained by a biological clock

It is clear that organisms have some way of measuring time, and that they are well adapted to the 24-hour day–night cycle of our planet. A biological clock resides within the cells of all eukaryotes and some prokaryotes. The major outward manifestations of this clock are known as **circadian rhythms** (from the Latin *circa*, "about," and *dies*, "day").

We can characterize circadian rhythms, as well as other regular biological cycles, in two ways: The **period** is the length of one cycle, and the **amplitude** is the magnitude of the change over the course of a cycle (Figure 39.13).

The circadian rhythms of cyanobacteria, protists, animals, fungi, and plants have been found to share some important characteristics:

- The period is remarkably insensitive to *temperature*, although lowering the temperature may drastically reduce the amplitude of the rhythmic effect.
- Circadian rhythms are *highly persistent;* they may continue for days even in an environment in which there are no environmental cues, such as light–dark periods.
- Circadian rhythms can be *entrained*, within limits, by light–dark cycles that differ from 24 hours. That is, the period an organism expresses can be made to coincide with that of the light–dark cycle to which it is exposed.
- A brief exposure to light can shift the peak of the cycle—it can cause a *phase shift*.

Plants provide innumerable examples of circadian rhythms. The leaflets of plants such as clover normally hang down and fold at night and rise and unfold during the day. The flowers of many plants show similar "sleep movements," closing at night and opening during the day. They continue to open and close on an approximately 24-hour cycle even when the light and dark periods are experimentally modified.

The period of circadian rhythms in nature is approximately 24 hours. If a clover plant, for example, were to be placed in light on a day–night cycle totaling exactly 24 hours, it would express a rhythm with a period of exactly 24 hours. However, if an experimenter used a day–night cycle of, say, 22 hours, then over time the rhythm would change—it would be **entrained** to a 22-hour period.

If an organism is maintained under constant darkness, it will express a circadian rhythm with an approximately 24-hour period. However, a brief exposure to light under these circumstances can cause a **phase shift**—that is, it can make the next peak of activity appear either later or earlier than expected, depending on when the exposure is given. Moreover, the organism does not then return to its old schedule if it remains in darkness. If the first peak is delayed by 6 hours, the subsequent peaks are all 6 hours late. Such phase shifts are permanent—until the organism receives more exposures to light.

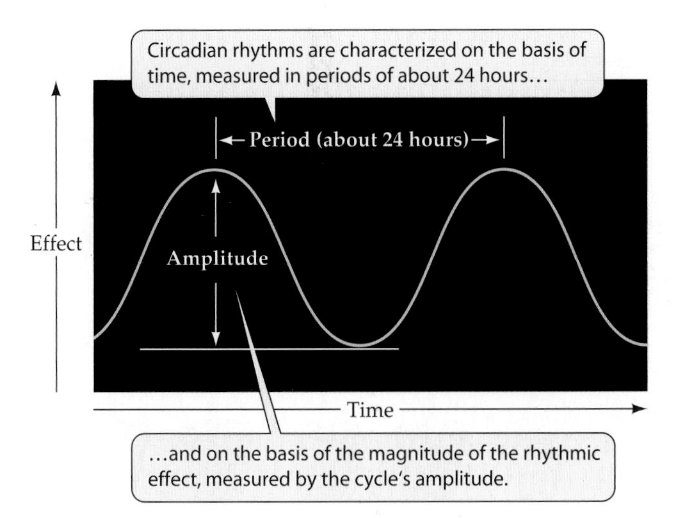

39.13 Features of Circadian Rhythms Circadian rhythms, like all biological rhythms, can be characterized in two ways: by period and by amplitude.

Photoreceptors set the biological clock

Phytochromes and blue-light receptors are known to affect the period of the biological clock, with the different pigments reporting on different wavelengths and intensities of light. This diversity of photoreceptors could be an adaptation to the changes in the light environment that a plant experiences in the course of a day or a season. How do these photoreceptors interact with a plant's biological clock?

The biological clock of *Arabidopsis* is based on the activities of at least three "clock genes." The clock genes encode regulatory proteins that interact to produce a circadian oscillation. How does this oscillating clock interact with photoreceptors and the environment?

Arabidopsis is an LDP. Its clock controls the activity of *CONSTANS* (a gene that is *not* part of the clock mechanism) in such a way that the *CONSTANS* product, CO protein, accumulates in one phase of the clock's cycle—the phase in which night falls. Under long nights (short days), CO protein is found at night. Under short nights (long days), CO is also relatively abundant at dawn and dusk. When CO protein levels are high, light absorbed by phytochrome A and the blue-light receptor cryptochrome 2 leads to flowering (Figure 39.14). Thus, *Arabidopsis* flowering results from the coincidence of light (detected by the two photoreceptors) with a clock-determined phase of the circadian oscillation.

Where is this coincidence-based photoperiodic mechanism located in relation to where flowering occurs? Is the timing device for flowering located in a particular plant part, or are all parts able to sense the length of the night? This question was resolved by "blindfold" experiments, as described next.

Is there a flowering hormone?

It quickly became apparent that each leaf is capable of timing the night. If a cocklebur plant—an SDP—is kept under a

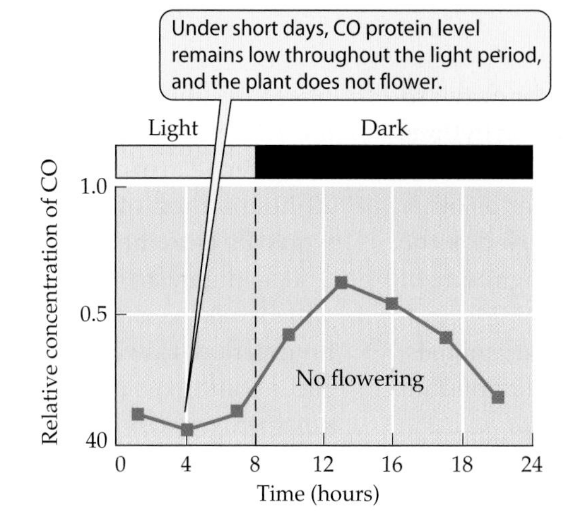

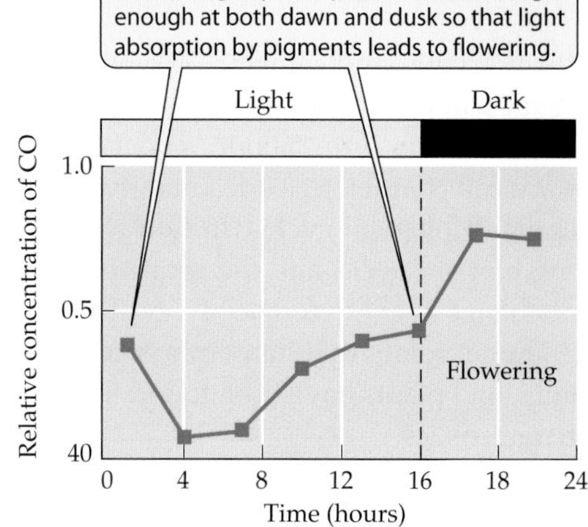

39.14 Photoreceptors and the Biological Clock Interact in Photoperiodic Plants One of the genes regulated by the circadian clock in *Arabidopsis* encodes the CO protein. Flowering depends on enough CO being present when photoreceptors have light available to them.

regime of short nights and long days, but a leaf is covered so as to give it the needed long nights, the plant will flower (experiment A in Figure 39.15). This type of experiment works best if only one leaf is left on the plant. If one leaf is given a photoperiodic treatment conducive to flowering—called an *inductive* treatment—other leaves kept under noninductive conditions will tend to inhibit flowering.

Although it is the leaves that sense an inductive photoperiod, the flowers form elsewhere on the plant. Thus, some kind of signal must be sent from the leaf to the site of flower formation. Three lines of evidence suggest that this signal is a chemical substance—a flowering hormone.

- If a photoperiodically induced leaf is immediately removed from a plant after the inductive dark period, the plant does not flower. If, however, the induced leaf remains attached to the plant for several hours, the plant flowers. This result suggests that something must be synthesized in the leaf in response to the inductive dark period and then move out of the leaf to induce flowering.
- If two cocklebur plants are grafted together, and if one plant is exposed to inductive long nights and its graft partner exposed to noninductive short nights, both plants flower (experiment B in Figure 39.15).
- In at least one species, if an induced leaf from one plant is grafted onto another, noninduced plant, the host plant flowers.

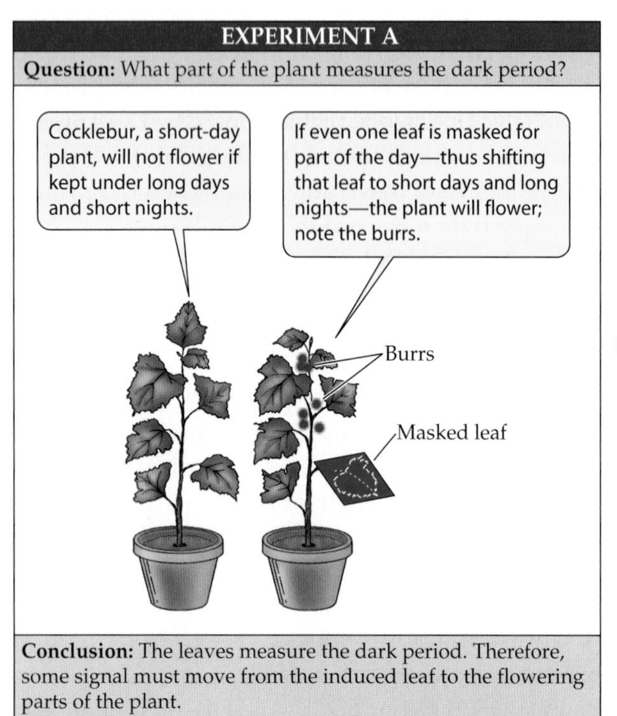

EXPERIMENT A
Question: What part of the plant measures the dark period?
Cocklebur, a short-day plant, will not flower if kept under long days and short nights.
If even one leaf is masked for part of the day—thus shifting that leaf to short days and long nights—the plant will flower; note the burrs.
Burrs
Masked leaf
Conclusion: The leaves measure the dark period. Therefore, some signal must move from the induced leaf to the flowering parts of the plant.

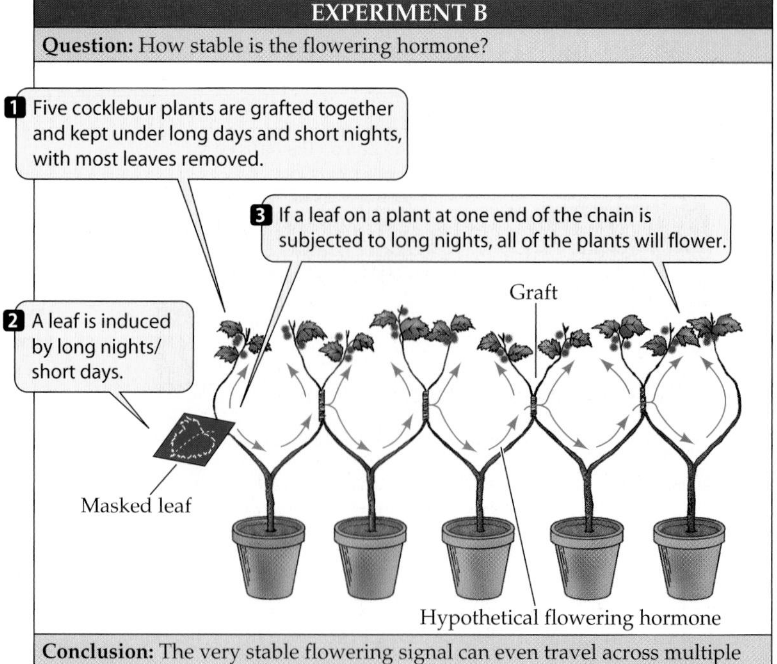

EXPERIMENT B
Question: How stable is the flowering hormone?
1 Five cocklebur plants are grafted together and kept under long days and short nights, with most leaves removed.
2 A leaf is induced by long nights/ short days.
3 If a leaf on a plant at one end of the chain is subjected to long nights, all of the plants will flower.
Graft
Masked leaf
Hypothetical flowering hormone
Conclusion: The very stable flowering signal can even travel across multiple grafts.

39.15 Evidence for a Flowering Hormone If even a single leaf is exposed to inductive conditions, a signal travels to the entire plant (and even to other plants, in grafting experiments), inducing it to flower.

Jan A. D. Zeevaart, a plant physiologist at Michigan State University, performed this last experiment. He exposed a single leaf of the SDP *Perilla* to a short-day/long-night regime, inducing the plant to flower. Then he detached this leaf and grafted it onto another, noninduced, *Perilla* plant—which responded by flowering. The same leaf grafted onto successive hosts caused each of them to flower in turn. As long as 3 months after the leaf was exposed to the short-day/long-night regime, it could still cause plants to flower.

Experiments such as Zeevaart's led to the conclusion that the photoperiodic induction of a leaf causes a more or less permanent change in the leaf, causing it to start and continue producing a flowering hormone that is transported to other parts of the plant, where the hormone initiates the development of reproductive structures. Biologists have named this hypothetical hormone **florigen**, even though, after decades of active searching, it has not been isolated or characterized.

An elegant experiment suggested that the florigen of SDPs is identical to that of LDPs, even though SDPs produce it only under long nights and LDPs only under short nights. An SDP and an LDP were grafted together, and both flowered, as long as the photoperiodic conditions were inductive for one of the partners. Either the SDP or the LDP could be the one induced, but both would always flower. These results suggest that a flowering hormone—the elusive florigen—was being transferred from one plant to the other.

The direct demonstration of florigen activity remains a cherished goal of plant physiologists. For a long time it was thought that florigen could be neither a protein nor an RNA because those molecules were too large to pass from one living plant cell to another. However, we now know that such macromolecules can be transferred by way of plasmodesmata, and biologists are reexamining the possibility that an RNA or a protein is the long-sought florigen.

We have considered the photoperiodic regulation of flowering, from photoreceptors in a leaf to the biological clock to the need for a signal that travels from the induced leaf to the sites of flower formation. However, light is not the only environmental variable that affects flowering. In some plants, low temperatures are an essential cue that eventually triggers flowering.

Vernalization and Flowering

Certain cereal grains serve as classic examples of the control of flowering by temperature. In both wheat and rye, we distinguish two categories of flowering behavior. Spring wheat, for example, is sown in the spring and flowers in the same year. It is an annual plant. Winter wheat is biennial and must be sown in the fall; it flowers in the following summer. If winter wheat is not exposed to cold after its first year, it will not flower normally the next year.

The implications of this finding were of great agricultural interest in Russia because winter wheat is a better producer than spring wheat, but it cannot be grown in some parts of Russia because the winters are too cold for its survival. Several studies performed in Russia during the early 1900s demonstrated that if seeds of winter wheat were premoistened and prechilled, they could be sown in the spring and would develop and flower normally the same year. Thus, high-yielding winter wheat could be grown even in previously hostile regions.

This induction of flowering by low temperatures is called **vernalization**. Vernalization may require as many as 50 days of low temperatures (in the range from about –2° to +12°C). Some plant species require both vernalization and long days to flower. There is a long wait from the cold days of winter to the long days of summer, but because the vernalized state easily lasts at least 200 days, these plants do flower when they experience the appropriate night length.

Asexual Reproduction

Although sexual reproduction takes up most of the space in this chapter, asexual reproduction is responsible for many of the new plant individuals appearing on Earth. This fact suggests that in some circumstances, asexual reproduction must be advantageous.

At the beginning of this chapter, we saw that one of the advantages of sexual reproduction is genetic recombination. Self-fertilization is a form of sexual reproduction, but when a plant self-fertilizes, there are fewer opportunities for genetic recombination than there are with cross-fertilization. A diploid, self-fertilizing plant that is heterozygous for a certain locus can produce both kinds of homozygotes for that locus plus the heterozygote among its progeny, but it cannot produce any progeny that carry alleles that it does not itself possess. Yet many plants continue to be self-compatible, undergo self-fertilization, and produce viable offspring.

Asexual reproduction goes even further than self-fertilization: It eliminates genetic recombination altogether. When a plant reproduces asexually, it produces a clone of progeny that are genetically identical to the parent. If a plant is well adapted to its environment, asexual reproduction may spread its genotype throughout that environment. This ability to exploit a particular environment is an advantage of asexual reproduction.

There are many forms of asexual reproduction

We call stems, leaves, and roots *vegetative organs* to distinguish them from flowers, the reproductive parts of the plant. The modification of a vegetative organ is what makes **vegetative reproduction**—asexual reproduction in plants—possible. In many cases, the stem is the organ that is modified.

Strawberries and some grasses, for example, produce horizontal stems, called *stolons* or *runners*, that grow along the soil surface, form roots at intervals, and establish potentially independent plants (see Figure 35.4*b*). *Tip layers* are upright branches whose tips sag to the ground and develop roots, as in blackberry and forsythia.

Some plants, such as potatoes, form enlarged fleshy tips of underground stems, called *tubers* (see Figure 35.4*a*). *Rhizomes* are horizontal underground stems that can give rise to new shoots. Bamboo is a striking example of a plant that reproduces vegetatively by means of rhizomes. A single bamboo plant can give rise to a stand—even a forest—of plants constituting a single, physically connected entity.

Whereas stolons and rhizomes are horizontal stems, bulbs and corms are short, vertical, underground stems. Lilies and onions form *bulbs* (Figure 39.16*a*), short stems with many fleshy, highly modified leaves that store nutrients. These storage leaves make up most of the bulb. Bulbs are thus large underground buds. They can give rise to new plants by dividing or by producing new bulbs from lateral buds. Crocuses, gladioli, and many other plants produce *corms*, underground stems that function very much as bulbs do. Corms are disclike and consist primarily of stem tissue; they lack the fleshy modified leaves that are characteristic of bulbs.

Not all vegetative organs modified for reproduction are stems. Leaves may also be the source of new plantlets, as in the succulent plants of the genus *Kalanchoe* (Figure 39.16*b*). Many kinds of angiosperms, ranging from grasses to trees such as aspens and poplars, form interconnected, genetically homogeneous populations by means of *suckers*—shoots produced by roots. What appears to be a whole stand of aspen trees, for example, may be a clone derived from a single tree by suckers.

Plants that reproduce vegetatively often grow in physically unstable environments, such as eroding hillsides. Plants with stolons or rhizomes, such as beach grasses, rushes, and sand verbena, are common pioneers on coastal sand dunes. Rapid vegetative reproduction enables these plants, once introduced, not only to multiply but also to survive burial by the shifting sand; in addition, the dunes are stabilized by the extensive network of rhizomes or stolons that develops. Vegetative reproduction is also common in some deserts, where the environment is not often suitable for seed germination and the establishment of seedlings.

Dandelions, citrus trees, and some other plants reproduce by the asexual production of seeds, called **apomixis**. As we have seen, meiosis reduces the number of chromosomes in gametes, and fertilization restores the sporophytic number of chromosomes in the zygote. Some plants can skip over *both* meiosis and fertilization and still produce seeds. Apomixis produces seeds within the ovary without the mingling and segregation of chromosomes and without the union of gametes. The ovule simply develops into a seed, and the ovary wall develops into a fruit. An apomictic embryo has the sporophytic number ($2n$) of chromosomes. The result of apomixis is a fruit with seeds that are genetically identical to the parent plant.

Apomixis sometimes requires pollination. In some apomictic species, a sperm nucleus must combine with the polar nuclei in order for the endosperm to form. In other apomictic species, the pollen provides the signals for embryo and endosperm formation, although neither sperm nucleus participates in fertilization. This observation emphasizes that pollination and fertilization are not the same thing.

Asexual reproduction is important in agriculture

Farmers and gardeners take advantage of some natural forms of vegetative reproduction. They have also developed

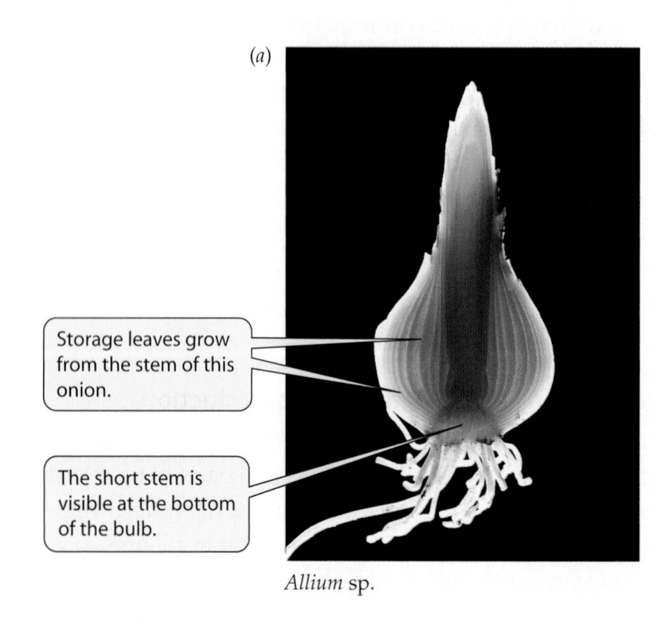

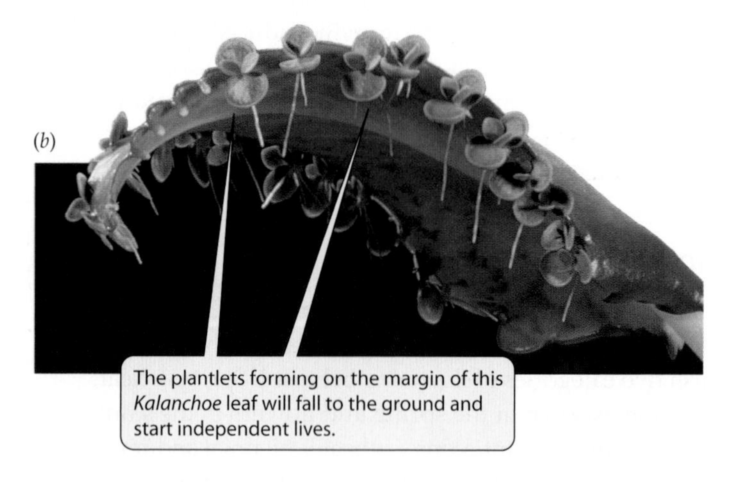

(b)
The plantlets forming on the margin of this *Kalanchoe* leaf will fall to the ground and start independent lives.

39.16 Vegetative Organs Modified for Reproduction (*a*) Bulbs are short stems with large buds that store nutrients and can give rise to new plants. (*b*) In Kalanchoe, new plantlets can form on leaves.

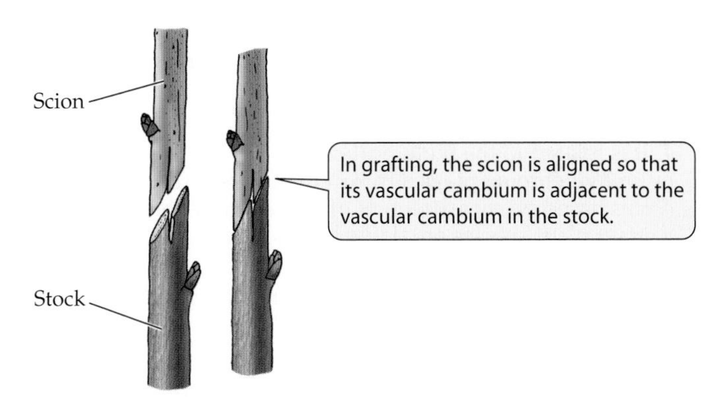

Scion
In grafting, the scion is aligned so that its vascular cambium is adjacent to the vascular cambium in the stock.
Stock

39.17 Grafting Grafting—attaching a piece of a plant to the root or root-bearing stem of another plant—is a common horticultural technique. The "host" root or stem is the stock; the upper grafted piece is the scion.

new types of asexual reproduction by manipulating plants. One of the oldest methods of vegetative reproduction used in agriculture consists of simply making cuttings of stems, inserting them in soil, and waiting for them to form roots and thus become autonomous plants. The cuttings are usually encouraged to root by treatment with a plant hormone, auxin, as described in Chapter 38.

Horticulturists reproduce many woody plants by **grafting**—attaching a bud or a piece of stem from one plant to the root or root-bearing stem of another plant. The part of the resulting plant that comes from the root-bearing "host" is called the *stock*; the part grafted on is the *scion* (Figure 39.17).

In order for a graft to succeed, the vascular cambium of the scion must become associated with that of the stock. By cell division, both cambia form masses of wound tissue. If the two masses meet and fuse, the resulting continuous cambium can produce xylem and phloem, allowing transport of water and minerals to the scion and of photosynthate to the stock. Grafts are most often successful when the stock and scion belong to the same or closely related species. Most fruit grown for market in the United States is produced on grafted trees.

Scientists in universities and industrial laboratories have been developing new ways to produce useful plants via tissue culture. Because many plant cells are totipotent (see Figure 19.3), cultures of undifferentiated tissue can give rise to entire plants, as can small pieces of tissue cut directly from a parent plant. Tissue cultures are used commercially to produce numerous new plants rapidly without resorting to seeds.

Culturing tiny bits of apical meristem can produce plants free of viruses. Because apical meristems lack developed vascular tissues, viruses tend not to enter them. Treatment with hormones causes a single apical meristem to give rise to 20 or more shoots; thus, a single plant can give rise to millions of genetically identical plants within a year by repeated meristem culturing. Using this approach, strawberry and potato producers are able to start each year's crop from virus-free plants.

Recombinant DNA techniques applied to tissue cultures can provide plants with increased resistance to pests or increased nutritive value to humans. There is also interest in making certain valuable, sexually reproducing plants capable of apomixis. By causing cells of different types to fuse, one can obtain plants with exciting new combinations of properties.

Chapter Summary

Many Ways to Reproduce

▶ Almost all flowering plants reproduce sexually, and many also reproduce asexually. Both sexual and asexual reproduction are important in agriculture.

Sexual Reproduction in Plants

▶ Sexual reproduction promotes genetic diversity in a population, which may give the population an advantage under changing environmental conditions or in exploiting new territory.

▶ The flower is an angiosperm's device for sexual reproduction.

▶ Flowering plants have microscopic gametophytes that develop within the flowers of the sporophytes. The megagametophyte is the embryo sac, which typically contains eight nuclei in a total of seven cells. The microgametophyte is the pollen grain, which usually contains two cells. **Review Figure 39.1.** See Web/CD Tutorial 39.1

▶ Pollination enables fertilization in the absence of external water.

▶ In self-incompatible species, the stigma or style rejects pollen from the same plant. **Review Figure 39.3**

▶ The pollen grain delivers sperm cells to the embryo sac by means of a pollen tube.

▶ Most angiosperms perform double fertilization: One sperm nucleus fertilizes the egg, forming a zygote, and the other sperm nucleus unites with the two polar nuclei to form a triploid endosperm. **Review Figure 39.5**

▶ The zygote develops into an embryo (with an attached suspensor), which remains quiescent in the seed until conditions are right for germination. The endosperm supplies the nutritive reserve upon which the embryo depends at germination. **Review Figures 39.6, 39.7.** See Web/CD Activity 39.1

▶ Flowers develop into seed-bearing fruits, which often play important roles in the dispersal of the species.

The Transition to the Flowering State

▶ For a vegetatively growing plant to flower, an apical meristem in the shoot system must become an inflorescence meristem, which gives rise to bracts as well as more meristems. The meristems it produces may become floral meristems or additional inflorescence meristems. **Review Figure 39.9**

▶ Flowering results from a cascade of gene expression. Floral organ identity genes are expressed in floral meristems that give rise to sepals, petals, stamens, and carpels.

Photoperiodic Control of Flowering

▶ Photoperiodic plants regulate their flowering by measuring the length of light and dark periods.

▶ Short-day plants flower when the days are shorter than a species-specific critical day length; long-day plants flower when the days are longer than a critical day length. **Review Figure 39.10**

▶ Some angiosperms have more complex photoperiodic requirements than short-day or long-day plants have, but most are day-neutral.

▶ The length of the night is what actually determines whether a photoperiodic plant will flower. **Review Figure 39.11**

▶ Interruption of the nightly dark period by a brief exposure to light undoes the effect of a long night. **Review Figure 39.12.** **See Web/CD Tutorial 39.2**

▶ The mechanism of photoperiodic control involves phytochromes and a biological clock. **Review Figures 39.13, 39.14**

▶ Evidence suggests that there is a flowering hormone, called florigen, but the substance has yet to be isolated from any plant. **Review Figure 39.15**

Vernalization and Flowering

▶ In some plant species, exposure to low temperatures—vernalization—is required for flowering.

Asexual Reproduction

▶ Asexual reproduction allows rapid multiplication of organisms that are well suited to their environment.

▶ Vegetative reproduction involves the modification of a vegetative organ—usually the stem—for reproduction. Stolons, tip layers, tubers, rhizomes, bulbs, corms, and suckers are means by which plants may reproduce vegetatively.

▶ Some plant species produce seeds asexually by apomixis.

▶ Agriculturalists use natural and artificial techniques of asexual reproduction to reproduce particularly desirable plants.

▶ Horticulturists often graft different plants together to take advantage of favorable properties of both stock and scion. **Review Figure 39.17**

▶ Tissue culture techniques, made possible by the totipotency of many plant cells, are used to propagate plants asexually, to produce virus-free clones of crop plants, and to manipulate plants by recombinant DNA technology.

Self-Quiz

1. Sexual reproduction in angiosperms
   a. is by way of apomixis.
   b. requires the presence of petals.
   c. can be accomplished by grafting.
   d. gives rise to genetically diverse offspring.
   e. cannot result from self-pollination.

2. The typical angiosperm female gametophyte
   a. is called a megaspore.
   b. has eight nuclei.
   c. has eight cells.
   d. is called a pollen grain.
   e. is carried to the male gametophyte by wind or animals.

3. Pollination in angiosperms
   a. never requires external water.
   b. never occurs within a single flower.
   c. always requires help by animal pollinators.
   d. is also called fertilization.
   e. makes most angiosperms independent of external water for reproduction.

4. Which statement about double fertilization is *not* true?
   a. It is found in most angiosperms.
   b. It takees place in the microsporangium.
   c. One of its products is a triploid nucleus.
   d. One sperm nucleus fuses with the egg nucleus.
   e. One sperm nucleus fuses with two polar nuclei.

5. The suspensor
   a. gives rise to the embryo.
   b. is heart-shaped in eudicots.
   c. separates the two cotyledons of eudicots.
   d. ceases to elongate early in embryonic development.
   e. is larger than the embryo.

6. Which statement about photoperiodism is *not* true?
   a. It is related to the biological clock.
   b. A phytochrome plays a role in the timing process.
   c. It is based on measurement of the length of the night.
   d. Most plant species are day-neutral.
   e. It is limited to the plant kingdom.

7. Although florigen has never been isolated, we think it exists because
   a. night length is measured in the leaves, but flowering occurs elsewhere.
   b. it is produced in the roots and transported to the shoot system.
   c. it is produced in the coleoptile tip and transported to the base.
   d. we think that gibberellin and florigen are the same compound.
   e. it may be activated by prolonged (more than a month) chilling.

8. Which statement about vernalization is *not* true?
   a. It may require more than a month of low temperatures.
   b. The vernalized state generally lasts for about a week.
   c. Vernalization makes it possible to have two winter wheat crops each year.
   d. It is accomplished by subjecting moistened seeds to chilling.
   e. It was of interest to Russian scientists because of their native climate.

9. Which of the following does *not* participate in asexual reproduction?
   a. Stolon
   b. Rhizome
   c. Zygote
   d. Tuber
   e. Corm

10. Apomixis involves
   a. sexual reproduction.
   b. meiosis.
   c. fertilization.
   d. a diploid embryo.
   e. no production of a seed.

For Discussion

1. For a crop plant that reproduces both sexually and asexually, which method of reproduction might the farmer prefer? Why?
2. Thompson Seedless grapes are produced by vines that are triploid. Think about the consequences of this chromosomal condition for meiosis in the flowers. Why are these grapes seedless? Describe the role played by the flower in fruit formation when no seeds are being formed. How do you suppose Thompson Seedless grapes are propagated?
3. Poinsettias are popular ornamental plants that typically bloom just before Christmas. Their flowering is photoperiodically controlled. Are they long-day or short-day plants? Explain.
4. You plan to induce the flowering of a crop of long-day plants in the field by using artificial light. Is it necessary to keep the lights on continuously from sundown until the point at which the critical day length is reached? Why, or why not?

40 Plant Responses to Environmental Challenges

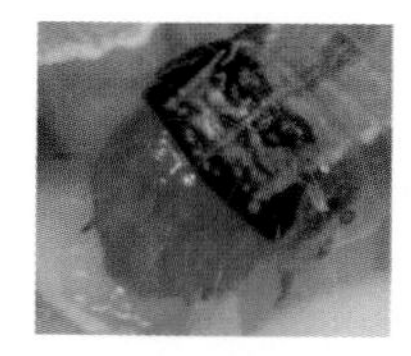

If you are attacked, it makes sense to call for help. Plants do this, too. When caterpillars begin to chew on the leaves of corn, cotton, or some other plant species, the plants synthesize chemical signals and release them into the atmosphere. These substances attract other insects that feed on the caterpillars.

Caterpillars and other herbivores aren't the only challenges plants face, however. The environment teems with organisms that cause plant diseases. We know of more than a hundred diseases that can kill a tomato plant, each of them caused by a different pathogen (including various bacteria, fungi, protists, and viruses). Like animals, plants have a variety of defenses against pathogens. Like the defenses of our own bodies, these mechanisms are not perfect, but they generally keep the plant world in competitive balance with its pathogens.

Environmental challenges to plants are not limited to herbivores and pathogens. Some physical conditions pose substantial problems for plants and thus limit the places where plants can live. The most challenging physical environments include those that are very dry (deserts), that are water-saturated, that are dangerously salty, that contain high concentrations of toxic substances such as heavy metals, and that are very hot or very cold.

This chapter focuses on how plants meet the myriad challenges presented by their biological and physical environments. We will begin by examining interactions between plants and pathogens, then go on to consider interactions between plants and herbivores. Finally, we will discuss some of the adaptations different types of plants have made to their physical environments.

Calling In an Air Strike As caterpillars of the corn earworm moth (*Helicoverpa zea*) munch through the plant's leaves, they may trigger a series of reactions in the plant that can end in the attraction of other insects that will attack the caterpillar.

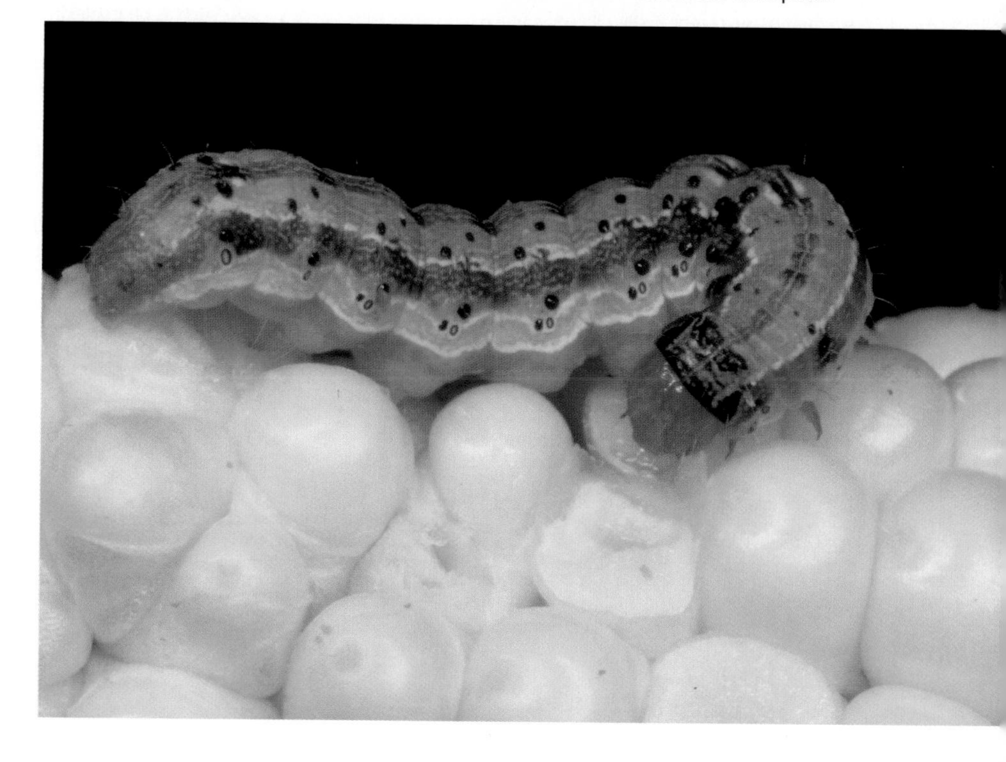

Plant–Pathogen Interactions

Plants and pathogens have evolved together in a continuing "arms race." Pathogens have evolved mechanisms with which to attack plants, and plants have evolved mechanisms for defending themselves against pathogens. Each set of mechanisms uses information from the other. For example, the pathogen's enzymes may break down the plant's cell walls, and the breakdown products may signal to the plant that it is under attack. In turn, the plant's defenses alert the pathogen that it is under attack.

What determines the outcome of a battle between a plant and a pathogen? The key to success for the plant is to respond to the information from the pathogen quickly and massively. Plants use both mechanical and chemical defenses in this effort.

Plants seal off infected parts to limit damage

Tissues such as epidermis or cork protect the outer surfaces of plants, and these tissues are generally covered by cutin, suberin, or waxes. This protection is comparable to the nonspecific defenses of animals. When pathogens pass these barriers, other nonspecific plant defenses are activated.

The defense systems of plants and animals differ. Animals generally repair tissues that have been damaged by pathogens, but plants do not. Instead, they seal off and sacrifice the damaged tissues so that the rest of the plant does not become infected. This approach works because most plants, unlike most animals, are modular and can replace damaged parts by growing new ones.

One of a plant cell's first defensive responses is the rapid deposition of additional polysaccharides on the inside of the cell wall, reinforcing this barrier to invasion by the pathogen (Figure 40.1). These polysaccharides block the plasmodesmata, limiting the ability of viral pathogens to move from cell to cell. They also serve as a base upon which lignin may be laid down. Lignin enhances the mechanical barrier, and the toxicity of lignin precursor chemicals makes the cell inhospitable to some pathogens. These lignin building blocks are only one example of the toxic substances that plants use as chemical defenses.

Plants have potent chemical defenses against pathogens

When infected by certain fungi and bacteria, plants produce a variety of defensive compounds. Two important kinds of defensive compounds are small molecules called phytoalexins and larger proteins called pathogenesis-related proteins (see Figure 40.1).

Phytoalexins are toxic to many fungi and bacteria. (Most are phenolics or terpenes, compounds that are also used to protect plants against herbivores; see Table 40.1.) They are produced by infected cells and their immediate neighbors within hours of the onset of infection. Enzymes from a pathogenic fungus can cause plant cell walls to release signaling molecules called *oligosaccharins*, which trigger phytoalexin production. Because their antimicrobial activity is nonspecific, phytoalexins can destroy many species of fungi and bac-

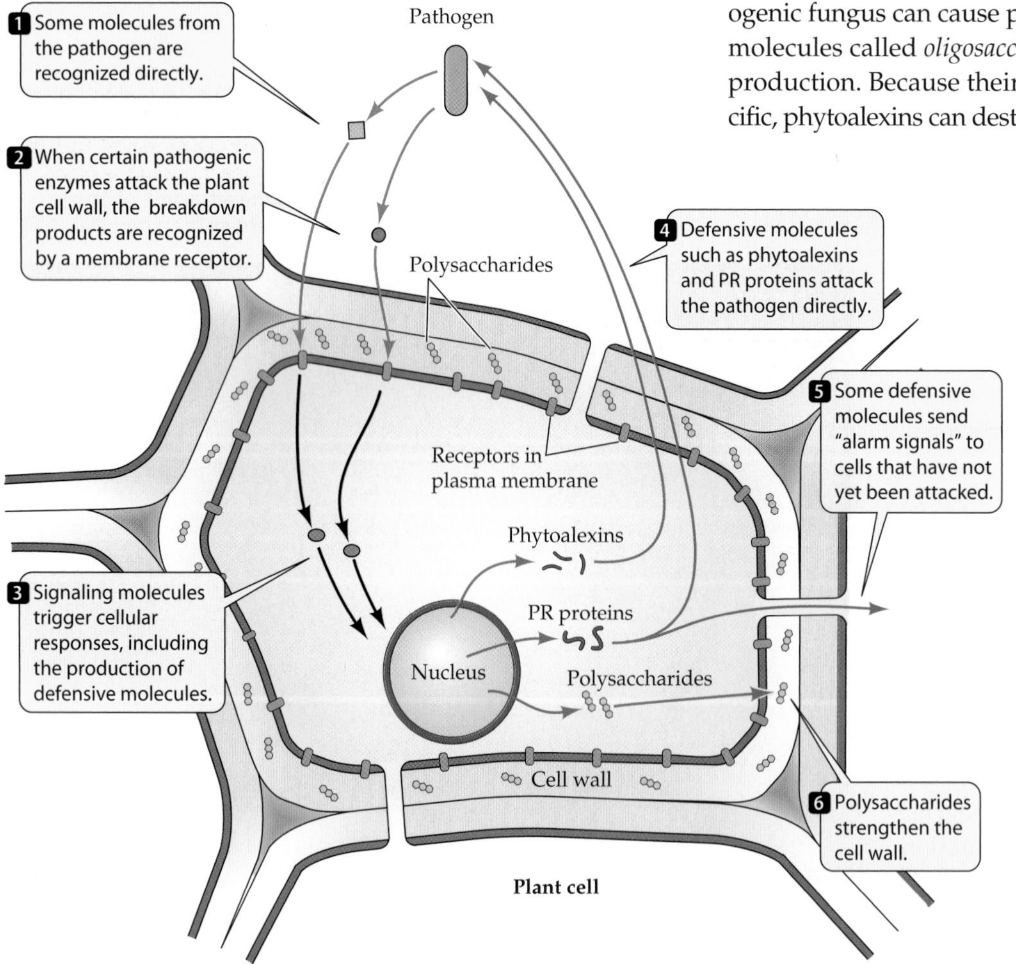

1 Some molecules from the pathogen are recognized directly.
Pathogen
2 When certain pathogenic enzymes attack the plant cell wall, the breakdown products are recognized by a membrane receptor.
Polysaccharides
4 Defensive molecules such as phytoalexins and PR proteins attack the pathogen directly.
5 Some defensive molecules send "alarm signals" to cells that have not yet been attacked.
Receptors in plasma membrane
Phytoalexins
3 Signaling molecules trigger cellular responses, including the production of defensive molecules.
PR proteins
Nucleus
Polysaccharides
Cell wall
6 Polysaccharides strengthen the cell wall.
Plant cell

40.1 Signaling between Plants and Pathogens Chemical interactions between plants and pathogens are highly coevolved. But the presence of a pathogen stimulates the plant to produce defensive molecules that can work in many different ways.

40.1 Secondary Plant Metabolites Used in Defense

CLASS	TYPE	ROLE	EXAMPLE
Nitrogen-containing	Alkaloids	Affect herbivore nervous system	Nicotine in tobacco
	Glycosides	Release cyanide or sulfur compounds	Dhurrin in sorghum
	Nonprotein amino acids	Disrupt herbivore protein structure	Canavanine in jack bean
Phenolics	Flavonoids	Phytoalexins	Capsidol in peppers
	Quinones	Inhibit competing plants	Juglone in walnut
	Tannins	Herbivore and microbe deterrents	Many woods, such as oak
Terpenes	Monoterpenes	Insecticides	Pyrethroids in chrysanthemums
	Sesquiterpenes	Antiherbivores	Gossypol in cotton
	Steroids	Mimic insect hormones and disrupt insect life cycles	α-Ecdysone in ferns
	Polyterpenes	Feeding deterrent?	Rubber in rubber tree

teria in addition to the one that originally triggered their production. Physical injuries, viral infections, and chemical compounds produced in response to damage by herbivores can also induce the production of phytoalexins.

Plants also produce several types of **pathogenesis-related proteins**, or **PR proteins**. Some are enzymes that break down the cell walls of pathogens. These enzymes destroy some of the invading cells, and in some cases the breakdown products of the pathogen's cell walls serve as chemical signals that trigger further defensive responses. Other PR proteins may serve as alarm signals to plant cells that have not yet been attacked. In general, PR proteins appear not to be rapid-response weapons; rather, they act more slowly, perhaps after other mechanisms have blunted the pathogen's attack.

PR proteins and phytoalexins do not act alone. Rather, they are tools used in complex defensive responses, such as the hypersensitive response and systemic acquired resistance.

The hypersensitive response is a localized containment strategy

Plants that are resistant to fungal, bacterial, or viral diseases generally owe this resistance to what is known as the **hypersensitive response**. Cells around the site of infection die, preventing the spread of the pathogen by depriving it of nutrients. Some of the cells produce phytoalexins and other chemicals before they die. The dead tissue, called a *necrotic lesion*, contains and isolates what is left of the microbial invasion (Figure 40.2). The rest of the plant remains free of the infecting microbe.

One of the defensive chemicals produced during the hypersensitive response is a close relative of aspirin. Since ancient times, people in Asia, Europe, and the Americas have used willow (*Salix*) leaves and bark to relieve pain and fever. The active ingredient in willow is *salicylic acid*, the same substance from which aspirin is derived.

Salicylic acid (benzene ring with COOH and OH)

It now appears that all plants contain at least some salicylic acid. This compound often evokes a second complex defensive response, which we will examine next.

Systemic acquired resistance is a form of long-term "immunity"

Systemic acquired resistance is a general increase in the resistance of the entire plant to a wide range of pathogenic species. It is not limited to the pathogen that originally triggered it or to the site of the original infection, and it may have a long-lasting effect.

Systemic acquired resistance is accompanied by the synthesis of PR proteins. Treatment of plants with salicylic acid

40.2 The Aftermath of a Hypersensitive Response The necrotic spots on these strawberry leaves are a response to the fungus that causes strawberry blight.

or aspirin leads to the production of PR proteins and to a resistance to pathogens. Salicylic acid treatment provides substantial protection against tobacco mosaic virus (a well-studied plant pathogen) and some other viruses.

Salicylic acid also serves as a plant hormone. In some cases, microbial infection in one part of a plant leads to the export of salicylic acid to other parts of the plant, where it causes the production of PR proteins before the infection can spread. The PR proteins then limit the extent of the infection. Infected plant parts also produce the closely related compound *methyl salicylate* (also known as oil of wintergreen). This volatile substance travels to other plant parts through the air, and may trigger the production of PR proteins in neighboring plants that have not yet been infected.

How does a plant know when it should activate the hypersensitive response and systemic acquired resistance? An interaction between plant and pathogen initiates these responses.

Some plant genes match up with pathogen genes

Many plants use the hypersensitive response and systemic acquired resistance as nonspecific defenses against various pathogens. However, the triggering of these responses resides in a highly specific mechanism, called **gene-for-gene resistance**. The ability of a plant to defend itself against a specific strain of a pathogen depends on the plant's having a particular allele of a gene that corresponds to a particular allele of a gene in the pathogen (Figure 40.3). Let's see how this matching works.

Plants have a large number of ***R* genes** (resistance genes), and many pathogens have sets of ***Avr* genes** (avirulence genes). Dominant *R* alleles favor resistance, and dominant *Avr* alleles make a pathogen less effective. If a particular plant has the dominant allele of one *R* gene and a pathogen strain infecting it has the dominant allele of the corresponding *Avr* gene, the plant will be resistant to that strain. This is true even when none of the other *R–Avr* pairs features corresponding dominant alleles. (This effect of one *R–Avr* pair overruling the others is an example of *epistasis*, which was discussed in Chapter 10.)

The mechanism of gene-for-gene resistance is not completely understood. There are thousands of specific *R* genes among the plants, and their products have different functions. The *Avr* genes in pathogens are simply the genes that cause the pathogen to produce a substance, often toxic, that elicits a defensive response in the plant. Most gene-for-gene interactions trigger the hypersensitive response.

Before we leave the topic of plant defenses against pathogens, let's consider a recently discovered specific defense mechanism directed against RNA viruses.

Plants develop specific immunity to RNA viruses

Plants respond to attack by RNA viruses by mounting a specific immune response. The plant uses its own enzymes to convert some of the single-stranded RNA of the invading virus into *double-stranded RNA* (dsRNA) and to chop that dsRNA into small pieces called *small interfering RNAs* (siRNAs). Some of the viral RNA is transcribed, forming mRNAs that advance the viral infection. However, the siRNAs interact with another cellular component to degrade those mRNAs, blocking viral replication. This phenomenon is an example of interference RNA (RNAi), or postranscriptional gene silencing (see Figure 16.11). Molecular biologists are exploring applications of RNAi in plant biotechnology.

The immunity conferred by RNAi spreads quickly throughout the entire plant, by mechanisms not yet fully understood. However, the establishment of immunity depends on the extent of the original infection and the speed of the

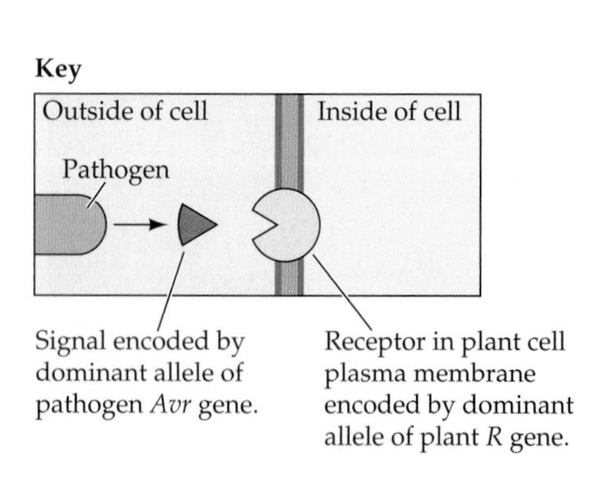
Key
Outside of cell
Inside of cell
Pathogen
Signal encoded by dominant allele of pathogen *Avr* gene.
Receptor in plant cell plasma membrane encoded by dominant allele of plant *R* gene.

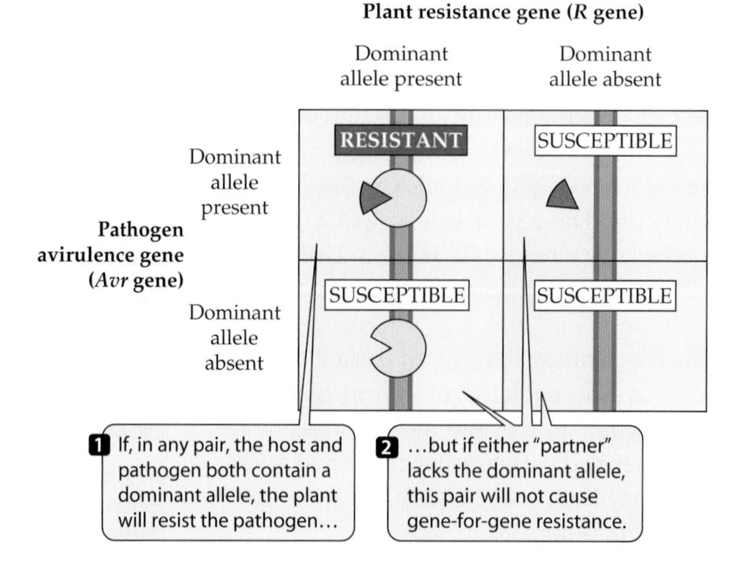
Plant resistance gene (*R* gene)
Dominant allele present
Dominant allele absent
Pathogen avirulence gene (*Avr* gene)
Dominant allele present
Dominant allele absent
RESISTANT
SUSCEPTIBLE
SUSCEPTIBLE
SUSCEPTIBLE
1 If, in any pair, the host and pathogen both contain a dominant allele, the plant will resist the pathogen…
2 …but if either "partner" lacks the dominant allele, this pair will not cause gene-for-gene resistance.

40.3 Gene-for-Gene Resistance A single pair of corresponding dominant alleles promotes resistance even if all the other pairs are mismatches.

plant's response. Also, most plant viruses have mechanisms that sometimes confound RNA silencing—as with other plant–pathogen interactions, natural selection favors both improved attack mechanisms for the pathogens and improved defense mechanisms for the plants.

Not all biological threats to plants come from microorganisms and viruses that cause diseases. Many animals, from inchworms to elephants, *eat* plants.

Plants and Herbivores: Benefits and Losses

Herbivores—animals that eat plants—depend on plants for energy and nutrients, and they often spread disease. Plants have many defense mechanisms that protect them against herbivores, as we will see. First, let's consider how herbivores can have a *positive* effect on some of the plants they eat.

Grazing increases the productivity of some plants

Herbivores are predators that prey on plants, but they often do not kill their prey. In **grazing**, an herbivore eats part of a plant, such as the leaves, without killing the plant, which then has the potential to grow back. What are the consequences of grazing? Is it always detrimental to plants, or are they somehow adapted to their place in the food chain? Certain plants and their predators have evolved together, each acting as the agent of natural selection on the other. Because of this coevolution, grazing actually increases photosynthetic production in some plant species.

Removing some leaves from a plant may increase the rate of photosynthesis in the remaining leaves. This phenomenon probably is the result of several factors. First, nitrogen obtained from the soil by the roots no longer needs to be divided among so many leaves. Second, the export of sugars and other photosynthetic products from the leaves may be enhanced because the demand for those products in the roots is undiminished, while the sources for those products—leaves—have been decreased. The remaining leaves may compensate by photosynthesizing more rapidly.

A third and particularly significant factor increasing photosynthesis, especially in grasses, is an increase in the availability of light to the younger, more active leaves or leaf parts. The removal of older or dead leaves by a grazer decreases the shading of younger leaves. Unlike most other plants, which grow from their shoot and leaf tips, grasses grow from the base of the shoot and leaf, so their growth is not cut short by grazing.

Mule deer and elk graze many plants, including one called scarlet gilia. Although grazing removes about 95 percent of the aboveground plant, the scarlet gilia quickly regrows not one but four replacement stems (Figure 40.4). Grazed plants produce three times as many fruits by the end of the growing season as do ungrazed plants.

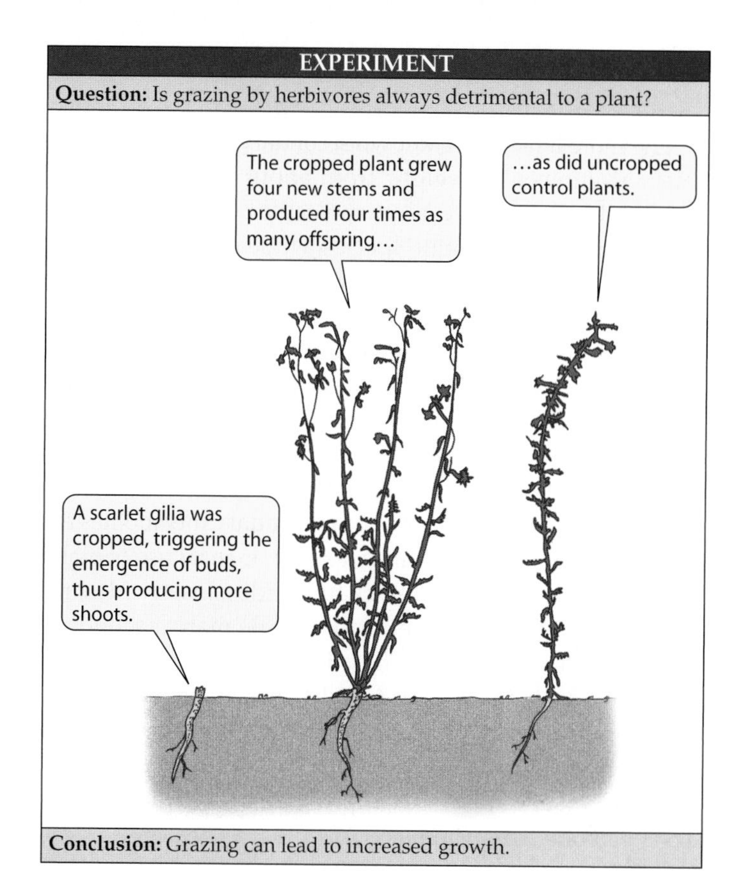

40.4 Overcompensation for Being Eaten Experiments confirm that some plants benefit from the effects of grazing.

Some grazed trees and shrubs continue to grow until much later in the season than do ungrazed but otherwise similar plants. This longer growing season results in part because the removal of apical buds by the grazers stimulates lateral buds to become active, producing a more heavily branched plant. Leaves on ungrazed plants may also die earlier in the growing season than leaves on grazed plants.

A plant may benefit from moderate herbivory by attracting animals that spread its pollen or that eat its fruit and thus disperse its seeds. Nevertheless, resisting attack by herbivores is often to the advantage of a plant.

Some plants produce chemical defenses

Although a plant cannot flee its herbivorous enemies, it may be able to defend itself chemically. Many plants attract, resist, and inhibit other organisms by producing special chemicals known as **secondary metabolites**. *Primary metabolites* are substances, such as proteins, nucleic acids, carbohydrates, and lipids, that are produced and used by all living things. Secondary metabolites are substances that are not used for basic cellular metabolism. Although all organisms use the same kinds of primary metabolites, plants can differ as radically in their secondary metabolites as they do in their external appearance.

The more than 10,000 known secondary plant metabolites range in molecular weight from about 70 to more than 400,000, but most have a low molecular weight. Some are produced by only a single species, while others are characteristic of entire genera or even families. These compounds help plants compensate for being unable to move.

The effects of defensive secondary metabolites on animals are diverse. Some secondary metabolites act on the nervous systems of herbivorous insects, mollusks, or mammals. Others mimic the natural hormones of insects, causing some larvae to fail to develop into adults. Still others damage the digestive tracts of herbivores. Some secondary metabolites are toxic to fungal pests. Humans make commercial use of many secondary plant metabolites as fungicides, insecticides, rodenticides, and pharmaceuticals. While many secondary metabolites have protective functions, others are essential as attractants for pollinators and seed dispersers. Table 40.1 lists the major classes of defensive secondary plant metabolites and their biological roles.

Let's look at a specific example of an insecticidal secondary metabolite, canavanine.

Some secondary metabolites play multiple roles

Canavanine is an amino acid that is not found in proteins, but is closely similar to the amino acid arginine, which is found in almost all proteins. Canavanine has two important roles in those plants that produce it in significant quantities. The first is as a nitrogen-storing compound in seeds. The second, defensive role is based on the similarity of canavanine to arginine:

Arginine	Canavanine
NH_2	NH_2
C=NH	C=NH
N—H	N—H
H—C—H	O
H—C—H	H—C—H
H—C—H	H—C—H
H_2N—C—H	H_2N—C—H
C—OH	C—OH
O	O

A seemingly slight chemical difference... (H—C—H in arginine) ...produces inactive proteins. (O in canavanine)

When an insect larva consumes canavanine-containing plant tissue, the canavanine is incorporated into the insect's proteins in some of the places where the DNA has coded for arginine because the enzyme that charges the tRNA specific for arginine fails to discriminate accurately between the two amino acids. The structure of canavanine is different enough from that of arginine that some of the resulting proteins end up with a modified tertiary structure and hence reduced biological activity. These defects in protein structure and function lead to developmental abnormalities that kill the insect.

A few insect larvae are able to eat canavanine-containing plant tissue and still develop normally. How can this be? In these larvae, the enzyme that charges the arginine tRNA discriminates correctly between arginine and canavanine. The canavanine they ingest is thus not incorporated into the proteins they form, and the larvae are not harmed.

In plants that produce it, canavanine is present whether or not the plant is under attack. Other chemical defenses come into play only when a predator strikes.

Many defenses depend on extensive signaling

Many plant defenses are activated by a series of signals. Insects feeding on tomato leaves damage the cells, leading to a chain of events that includes the formation of hormones and ends with the production of an insecticide. The signaling steps in the production of one defensive compound, shown in Figure 40.5, involve two hormones. **Systemin**, which is

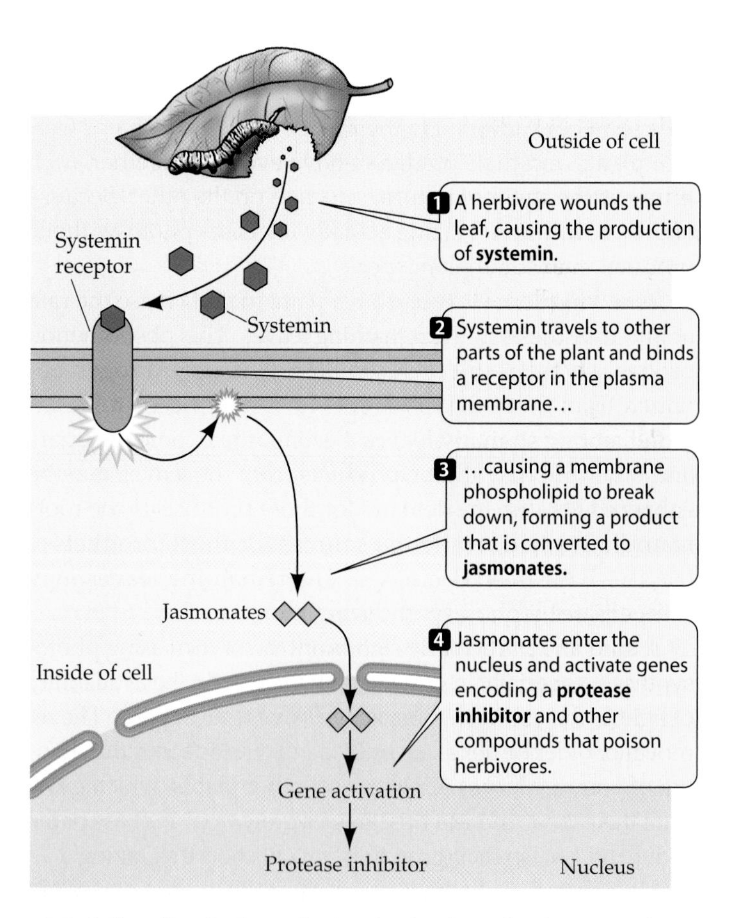

Outside of cell
1 A herbivore wounds the leaf, causing the production of systemin.
Systemin receptor
Systemin
2 Systemin travels to other parts of the plant and binds a receptor in the plasma membrane...
3 ...causing a membrane phospholipid to break down, forming a product that is converted to jasmonates.
Jasmonates
4 Jasmonates enter the nucleus and activate genes encoding a protease inhibitor and other compounds that poison herbivores.
Inside of cell
Gene activation
Protease inhibitor
Nucleus

40.5 A Signaling Pathway for Synthesis of a Defensive Secondary Metabolite The chain of events that is initiated by an insect's attack and leads to the production of a defensive chemical can consist of many steps. These steps may include the synthesis of one or two hormones, binding of receptors, gene activation, and, finally, synthesis of insecticides.

formed in response to an insect attack, is a polypeptide hormone—the first polypeptide hormone to be discovered in plants. **Jasmonates**, whose production is initiated by systemin, are formed from the unsaturated fatty acid linolenic acid. The final step in the chain is the synthesis of a protease inhibitor. The inhibitor, once in an insect's gut, interferes with the digestion of proteins and thus stunts the insect's growth.

Jasmonic acid

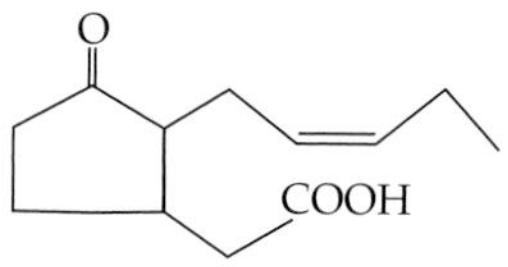

O

COOH

Jasmonates also take part in the "call for help" described at the beginning of this chapter. In that case, a substance released from the leaves by chewing caterpillars is the first signal, leading to the formation of jasmonates by the plant. The jasmonates, in turn, trigger the formation of the volatile compounds that attract the insects that prey on the caterpillars.

Even though plants have so many effective natural defenses, agricultural researchers are attempting to provide crop plants with even more effective ones.

Recombinant DNA technology may confer resistance to insects

Wild and domesticated common beans (*Phaseolus vulgaris*) differ in their resistance to two species of bean weevils. Some wild bean seeds are highly resistant to these insects, but no cultivated bean seeds show such resistance. Scientists discovered that all weevil-resistant bean seeds contain a specific seed protein, *arcelin*. This protein has never been found in cultivated bean seeds. Therefore, the scientists hypothesized that arcelin is responsible for the resistance of some seeds to predation by the weevils.

To rule out other differences between wild and cultivated beans as being responsible for the difference in resistance, the scientists performed two series of experiments. In one series, they crossed cultivated and wild bean plants. All of the progeny seeds of such crosses that contained arcelin showed resistance to weevils. In the other series of experiments, the scientists removed the seed coats of domesticated beans and ground the remainder of the seeds into flour. They added different concentrations of arcelin to different batches and molded the flour into artificial seeds. They then let weevils attack the artificial seeds. The more arcelin the artificial seeds contained, the more resistant they were to weevils.

Next, preliminary tests showed that arcelin in cooked beans was not harmful to rats—a first step toward determining whether arcelin is safe in food for humans. Agricultural scientists must sometimes choose between crop protection and appeal to humans. A plant with sturdy chemical defenses may taste bad, make us sick, or even kill us.

Scientists are now seeking to introduce genes for arcelin and other resistance-conferring proteins into agriculturally important crops such as beans. The development of crop plants that produce their own pesticides is an active area of research in agricultural biotechnology. One of the most widely applied approaches has been the engineering of several crops, such as tomatoes, corn, and cotton, to express the toxin genes from *Bacillus thuringiensis* discussed in Chapter 16.

Why don't plants poison themselves?

Why don't the chemicals that are so toxic to herbivores and microbes kill the plants that produce them? Plants that produce toxic secondary metabolites generally use one of the following measures to protect themselves:

- The toxic material is isolated in a special compartment.
- The toxic substance is produced only after the plant's cells have already been damaged.
- The plant uses modified enzymes or modified receptors that do not recognize the toxic substance.

The first method is the most common. Plants using this method store their poisons in vacuoles if they are water-soluble. If they are hydrophobic, the poisons are stored in **laticifers** (tubes containing a white, rubbery latex) or dissolved in waxes on the epidermal surface. This compartmentalized storage keeps the toxic substance away from the mitochondria, chloroplasts, and other parts of the plant's own metabolic machinery.

Some plants store the precursors of toxic substances in one compartment, such as the epidermis, and store the enzymes that convert those precursors to the active poison in another compartment, such as the mesophyll. These plants produce the toxic substance only after being damaged. When an herbivore chews part of the plant, the cells rupture, and the enzymes come in contact with the precursors, producing the toxic product. The only part of the plant that is damaged by the toxic substance is that which was already damaged by the herbivore. Plants that respond to attack by producing cyanide—a strong inhibitor of cellular respiration in all organisms that respire—are among those that use this protective measure.

The third protective measure is used by the canavanine-producing plants described earlier. These plants produce a tRNA-charging enzyme for arginine that does not bind canavanine. However, as we have seen, some herbivores can evade being poisoned by canavanine in a similar manner, demonstrating that no plant defense is perfect. Like plants and their pathogens, plants and their predators evolve together in a continuing "arms race," and the plant does not always win.

40.6 Disarming a Plant's Defenses This beetle is inactivating a milkweed's defense system by cutting its laticifer supply lines.

The plant doesn't always win

Milkweeds such as *Asclepias syriaca* are latex-producing (laticiferous) plants. When damaged, a milkweed releases copious amounts of toxic latex from its laticifers, which run alongside the veins in its leaves. Latex has long been suspected to deter insects from eating the plant because some insects that feed on neighboring plants of other species do not attack laticiferous plants. This observed behavior is consistent with, but does not prove, the hypothesis that the latex keeps the insects at bay.

Stronger support for this hypothesis was obtained by studying field populations of *Labidomera clivicollis*, a beetle that is one of the few insects that feed on *A. syriaca*. These beetles show a remarkable prefeeding behavior: They cut a few veins in the leaves before settling down to dine (Figure 40.6). Cutting the veins, with their adjacent laticifers, causes massive latex leakage and interrupts the latex supply to a downstream portion of the leaf. The beetles then move to the relatively latex-free portion and eat their fill.

Does this behavior of the beetles negate the adaptive value of latex protection? Not entirely. There are still great numbers of potential insect pests that are effectively deterred by the latex. And evolution proceeds. Over time, milkweed plants producing higher concentrations of toxins may be selected by virtue of their ability to kill beetles that cut their laticifers.

Having discussed how plants defend themselves against other organisms, we now turn our attention to how plants adapt to the nonliving components of their environments.

Water Extremes: Dry Soils and Saturated Soils

Water is often in short supply in the terrestrial environment. Some terrestrial habitats, such as deserts, intensify this challenge. Many plants that inhabit particularly dry areas have one or more adaptations that allow them to conserve water. Plants adapted to dry environments are called **xerophytes**.

40.7 Desert Annuals Evade Drought Seeds of desert plants often lie dormant for long periods awaiting conditions appropriate for germination. When they do germinate, they grow and reproduce rapidly before the short wet season passes. During the long dry spells, only dormant seeds remain alive.

Some plants evade drought

Some desert plants have no special *structural* adaptations for water conservation other than those found in almost all flowering plants. Instead, they have an alternative *strategy*. These desert annuals simply evade the periods of drought. They carry out their entire life cycle—from seed to seed—during a brief period in which rainfall has made the surrounding desert soil sufficiently moist (Figure 40.7).

Some leaves have special adaptations to dry environments

Plants that remain active during dry periods must have structural adaptations that enable them to survive. The secretion of a heavier cuticle over the leaf epidermis to retard water loss is a common adaptation to dry environments. An even more common adaptation is a dense covering of epidermal hairs. Some species have stomata only in sunken cavities below the leaf surface, which reduces the drying effects of air currents; often these stomatal cavities contain hairs as well (Figure 40.8).

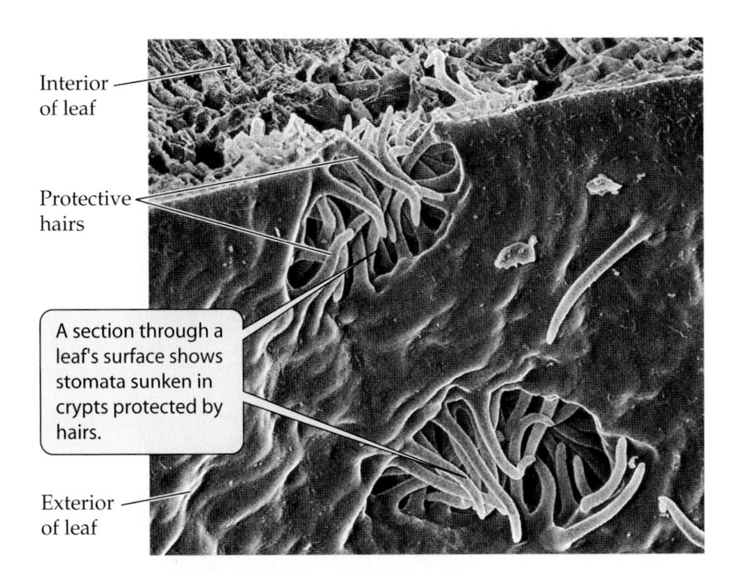

40.8 Stomatal Crypts Stomata in the leaves of some xerophytes are located in sunken pits called stomatal crypts. The hairs covering these crypts trap moist air.

Succulence—the possession of fleshy, water-storing leaves—is an adaptation to dry environments. Ice plants and their relatives have fleshy leaves in which water may be stored. Other xerophytes, such as ocotillo, produce leaves only when water is abundant, shedding them as the soil dries out (Figure 40.9). Cacti and similar plants have spines rather than typical leaves, and photosynthesis is confined to the fleshy stems. The spines may reflect incident radiation, or they may dissipate heat. Corn and some related grasses have leaves that roll up during dry periods, thus reducing the leaf surface area through which water is lost. Some trees, such as eucalyptuses, that grow in arid regions have leaves that hang vertically at all times, thus evading the midday sun.

These xerophytic adaptations of leaves minimize water loss by the plant. However, such adaptations simultaneously minimize the uptake of carbon dioxide and thus limit photosynthesis. In consequence, most xerophytes grow slowly, but they utilize water more efficiently than do other plants—that is, they fix more grams of carbon by photosynthesis per gram of water lost to transpiration than other plants do.

During dry periods, the thorny, leafless stems of an ocotillo appear almost dead.

When water is available, leaves develop rapidly and provide the plant with photosynthetic products.

40.9 Opportune Leaf Production The ocotillo, a xerophyte that lives in the lower deserts of the southwestern United States and northern Mexico, produces leaves only when there is sufficient water for photosynthesis.

Plants have other adaptations to a limited water supply

Roots may also be adapted to dry environments. Mesquite trees (genus *Prosopis;* Figure 40.10) obtain water through taproots that grow to great depths, reaching water supplies far underground, as well as from condensation on their leaves. The Atacama Desert in northern Chile often goes several years without measurable rainfall. The landscape there is almost barren of plant life, save for many surprisingly large mesquite trees.

A more common adaptation of desert plants is a root system that grows rapidly during rainy seasons but dies back during dry periods. Cacti have shallow but extensive fibrous root systems that effectively intercept water at the surface of the soil following even light rains.

Xerophytes and other plants that receive inadequate water may accumulate the amino acid proline or other solutes

40.10 Mining Water with Deep Taproots In Death Valley, California, this mesquite must reach far down into the sand dunes for its water supply.

to substantial concentrations in their vacuoles. As a consequence, the solute potential and water potential of their cells become more negative; thus these plants tend to extract more water from the soil than do plants that lack this adaptation. Plants living in salty environments share this and several other adaptations with xerophytes, as we will see.

In water-saturated soils, oxygen is scarce

For some plants, the environmental challenge is the opposite of that faced by xerophytes: too much water. Some plants live in environments so wet that the diffusion of oxygen to their roots is severely limited. Since most plant roots require oxygen to support respiration and ATP production, most plants cannot tolerate this situation for long.

Some species, however, are adapted to life in a water-saturated habitat. Their roots grow slowly and hence do not penetrate deeply. Because the oxygen level is too low to support aerobic respiration, the roots carry on alcoholic fermentation (see Chapter 7), which provides ATP for the activities of the root system. This adaptation explains why their growth is slow.

The root systems of some plants adapted to swampy environments have **pneumatophores**, which are extensions that grow out of the water and up into the air (Figure 40.11). Pneumatophores have lenticels and contain spongy tissues that allow oxygen to diffuse through them, aerating the submerged parts of the root system. Cypresses and some mangroves are examples of plants with pneumatophores.

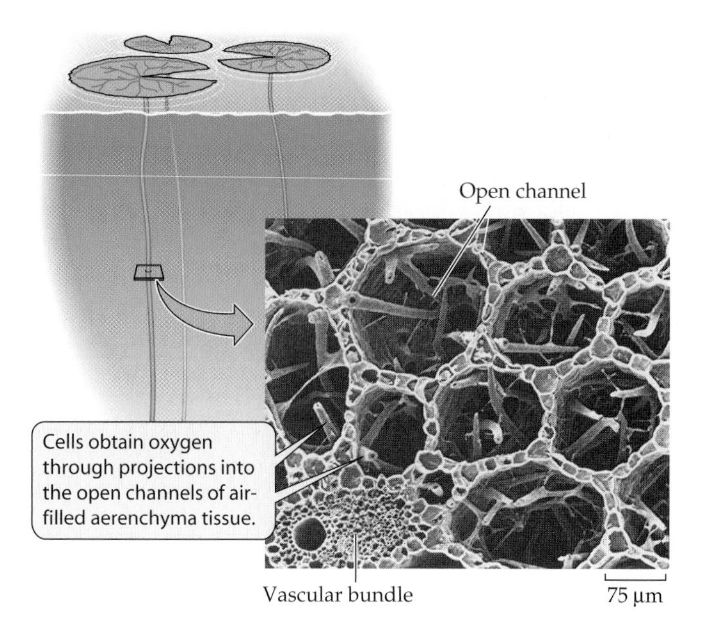

Open channel
Cells obtain oxygen through projections into the open channels of air-filled aerenchyma tissue.
Vascular bundle
75 μm

40.12 Aerenchyma Lets Oxygen Reach Submerged Tissues The scanning electron micrograph, a cross section of a petiole of the yellow water lily, shows the air-filled channels of aerenchyma tissue. The cells that line these channels obtain oxygen by extending projections into these channels.

40.11 Coming Up for Air The roots of the mangroves in this tidal swamp obtain oxygen through pneumatophores.

Submerged or partly submerged aquatic plants often have large air spaces in the leaf parenchyma and in the petioles. Tissue containing such air spaces is called **aerenchyma** (Figure 40.12). Aerenchyma stores oxygen produced by photosynthesis and permits its ready diffusion to parts of the plant where it is needed for cellular respiration. Aerenchyma also imparts buoyancy. Furthermore, because it contains far fewer cells than most other plant tissue, respiratory metabolism in aerenchyma proceeds at a lower rate, and the need for oxygen is much reduced.

Thus far we have considered water supply—either too little or too much—as a factor limiting plant growth. Other substances also can make an environment inhospitable to plant growth. One of these substances is salt.

Too Much Salt: Saline Environments

Worldwide, no toxic substance restricts angiosperm growth more than salt (sodium chloride) does. *Saline*—salty—habitats support, at best, sparse vegetation. Saline habitats themselves are diverse, ranging from hot, dry, salty deserts to moist, cool, salty marshes. Along the seashore are saline environments created by ocean spray. The ocean itself is a saline environment, as are river estuaries, where fresh and salt water meet and mingle. The salinization of agricultural land is an increasing global problem. Even where crops are irrigated with fresh water, sodium ions from the water accumulate in the soil to ever greater concentrations as the water evaporates.

Saline environments pose an osmotic problem for plants. Because of its high salt concentration, a saline environment has an unusually negative water potential. To obtain water from such an environment, a plant must have an even more negative water potential than that of a plant in a nonsaline environment; otherwise, it will lose water, wilt, and die. A second problem is the potential toxicity of high concentrations of certain ions, notably sodium and chloride.

The **halophytes**—plants adapted to saline habitats—belong to a wide variety of flowering plant groups. How can these plants cope with a saline environment?

40.13 Excreting Salt This salty mangrove has special salt glands that excrete salt, which appears here as crystals on the leaves.

Most halophytes accumulate salt

Most halophytes share one adaptation: They accumulate sodium and, usually, chloride ions and transport those ions to the leaves. The accumulated ions are stored in the central vacuoles of leaf cells, away from more sensitive parts of the cells. Nonhalophytes accumulate relatively little sodium, even when placed in a saline environment; of the sodium that is absorbed by their roots, very little is transported to the shoot. The increased salt concentration in the tissues of halophytes makes their water potential more negative, so they can take up water more easily from the saline environment.

Scientists have succeeded in causing the overexpression of a gene in *Arabidopsis* that enables sodium uptake. This gene encodes a Na^+/H^+ antiport protein in the tonoplast (the membrane surrounding the central vacuole). By making the gene produce a greater than normal number of these antiport proteins, the scientists increased sodium transport in *Arabidopsis*, converting this nonhalophyte into a halophyte. Further research along this line may result in a great boost to agriculture in saline environments. Biologists in Israel and elsewhere have had some success in breeding crops that can be watered with seawater or diluted seawater.

Some halophytes have other adaptations to life in saline environments. Some, for example, have **salt glands** in their leaves. These glands excrete salt, which collects on the leaf surface until it is removed by rain or wind (Figure 40.13). This adaptation, which reduces the danger of poisoning by accumulated salt, is found both in some desert plants, such as tamarisk, and in some mangroves growing in seawater in the Tropics.

Salt glands can play multiple roles, as in the desert shrub *Atriplex halimus*. This shrub has glands that secrete salt into small bladders on the leaves, where, by increasing the gradient in water potential, the salt helps the leaves obtain water from the roots. At the same time, by making the water potential of the leaves more negative, the salt reduces the transpirational loss of water to the atmosphere.

The adaptations we have just discussed are specific to halophytes. Several other adaptations are shared by halophytes and xerophytes.

Halophytes and xerophytes have some similar adaptations

Many halophytes, like some xerophytes, accumulate the amino acid proline in their cell vacuoles, making the water potential of their tissues more negative. Unlike sodium, proline is relatively nontoxic.

Succulence is another adaptation that halophytes and xerophytes have in common, as might be expected, since saline environments, like dry ones, make water uptake difficult. Succulence characterizes many halophytes that occupy salt marshes. There the salt concentration in the soil solution may change throughout the day. When the tide is out, evaporation increases the salt concentration. Succulence may offer a reserve of water for the plant during the period of maximum salinity; when the salinity drops as the tide comes in, the leaf's store of water is replenished.

Many succulents—both xerophytes and halophytes—use crassulacean acid metabolism (CAM), which allows them to store CO_2 as carboxyl groups at night and release the CO_2 for use in photosynthesis during the day. They have reversed stomatal cycles that enable them to conserve water by closing their stomata in the daytime (Figure 40.14). Other general adaptations to a saline environment include high root-to-shoot ratios, sunken stomata, reduced leaf areas, and thick cuticles.

Salt is not the only toxic solute found in soils. Some heavy metal ions are more toxic than sodium at equivalent concentrations.

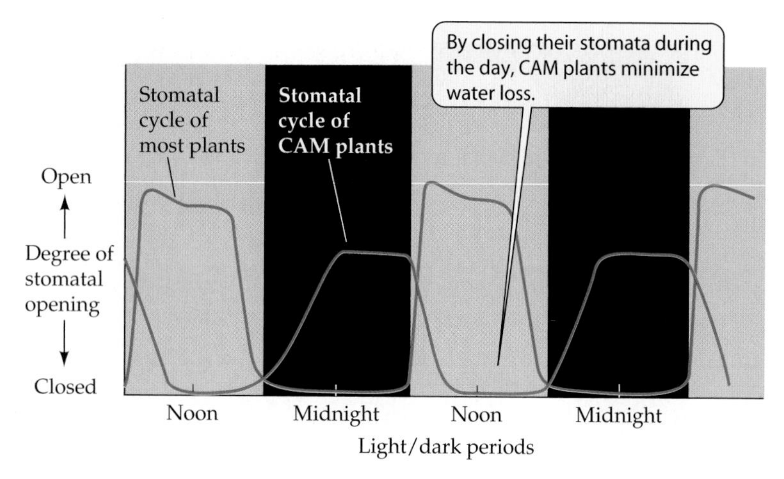

40.14 Stomatal Cycles Most plants open their stomata during the day. CAM plants reverse this stomatal cycle: Their stomata open during the night.

Habitats Laden with Heavy Metals

High concentrations of some heavy metal ions, such as chromium, mercury, lead, and cadmium, poison most plants. Some geographic sites are naturally rich in heavy metals as a result of normal geological processes. In other places, acid rain leads to the release of toxic aluminum ions in the soil. Other human activities, notably the mining of metallic ores, leave localized areas—known as *tailings*—with substantial concentrations of heavy metals and low concentrations of nutrients. Such sites are hostile to most plants, and seeds falling on them generally do not produce adult plants.

Mine tailings rich in heavy metals, however, generally are not completely barren (Figure 40.15). They may support healthy plant populations that differ genetically from populations of the same species on the surrounding normal soils. How can these plants survive?

40.15 Life after Strip Mining Although high concentrations of heavy metals kill most plants, grass is colonizing this eroded strip mine in North Park, Colorado.

Initially, some plants were thought to tolerate heavy metals by excluding them: By not taking up the metal ions, it was believed, the plants avoided being poisoned. However, measurements have shown that tolerant plants growing on mine tailings do take up heavy metals, accumulating them to concentrations that would kill most plants. Thus these plants must have a mechanism for dealing with the heavy metals they take up. Such tolerant plants may be found to be useful agents for *bioremediation*, a decontamination process by which the heavy metal content of some contaminated soils is decreased by living organisms.

We know the mechanism of at least one case of tolerance to a different toxic metal. When the roots of a buckwheat grown in China are exposed to aluminum concentrations high enough to inhibit root growth in other plants, they secrete oxalic acid. Oxalic acid combines with aluminum ions, forming a complex that does not inhibit growth.

From mine to mine, the heavy metals in the soil differ. In Wales and Scotland, bent grass (*Agrostis*) grows near many mines. Samples of bent grass from several such sites were tested for their ability to grow in various solutions, each containing only one heavy metal. In general, the plants tolerated a particular heavy metal—the one most abundant in their habitat—but were sensitive to others. That is, they tolerated only one or two heavy metals, rather than heavy metals as a group.

Tolerant plant populations can evolve and colonize an area surprisingly rapidly. The bent grass population around a particular copper mine in Wales is resistant to copper and is relatively abundant, even though the copper-rich soil dates from mining done only a century ago.

Thus plants are threatened by salt, heavy metals, and other toxic substances in their chemical environment, as well as by pathogens in their biological environment and by shortages or excesses of water in their physical environment. We now examine two more threats in the physical environment: high and low temperatures.

Hot and Cold Environments

Temperatures that are too high or too low can stress plants and even kill them. Plants differ in their sensitivity to heat and cold, but all plants have their limits. Any temperature extreme can damage cellular membranes:

- High temperatures destabilize membranes and denature many proteins, especially some of the enzymes of photosynthesis.

- Low temperatures cause membranes to lose their fluidity and alter their permeabilities to solutes.
- Freezing temperatures may cause ice crystals to form, damaging cellular membranes.

Plants have ways of coping with high temperatures

Transpiration (loss of water through evaporation) can cool a plant, but it also increases the plant's need for water. Therefore, it is not surprising that many plants living in hot environments have adaptations similar to those of xerophytes. These adaptations include epidermal hairs and spines that radiate heat, modified leaf displays that intercept less direct sunlight, and crassulacean acid metabolism.

Plants respond within minutes to high temperatures by producing several kinds of **heat shock proteins**. Among these proteins are chaperonins (see Chapter 3), which help other proteins maintain their structures and avoid denaturation. Threshold temperatures for the production of heat shock proteins vary, but 40°C is sufficient to induce them in most plants. We have much to learn about the dozens of heat shock proteins, but we do know that some other types of stress also induce their formation. Among these stresses are chilling and freezing.

Some plants are adapted to survival at low temperatures

Low temperatures above freezing injure many plants, including important crops such as rice, corn, and cotton. Many plant species can be modified to resist the effects of cold spells by a process called **cold-hardening**, which involves repeated exposure to cool, but not injurious, temperatures. The hardening process is a slow one, requiring many days. A key change that occurs during the hardening process is an increase in the relative fraction of unsaturated fatty acids in membranes. Unsaturated fatty acids solidify at lower temperatures than do saturated ones. Thus, the membranes retain their fluidity and function normally at cooler temperatures.

Low temperatures induce the formation of certain heat shock proteins that protect against chilling damage. There are also cases of "cross-protection" by heat shock proteins that are induced by one type of stress and that protect against other stresses. Tomatoes stressed by 2 days of high temperatures, for example, formed heat shock proteins and became resistant to chilling damage for the next 3 weeks.

If ice crystals form within plant cells, they can kill the cells by puncturing organelles and plasma membranes. Even outside cells, the growth of ice crystals can draw water from the cells and dehydrate them. Freezing-tolerant plants have a variety of adaptations to cope with these problems. A common one is the production of *antifreeze proteins* that inhibit the growth of ice crystals.

Plants have many effective mechanisms for coping with environmental challenges of many kinds. Their success is obvious—just look around you.

Chapter Summary

Plant–Pathogen Interactions

- Plants and pathogens evolve together in a continual "arms race." **Review Figure 40.1. See Web/CD Tutorial 40.1**
- Plants can strengthen their cell walls when attacked.
- Plant chemical defenses include PR proteins and phytoalexins.
- In the hypersensitive response, cells produce phytoalexins and then die, trapping the pathogens in dead tissue.
- The hypersensitive response is often followed by systemic acquired resistance, in which the hormone salicylic acid activates further synthesis of PR proteins and triggers responses in other parts of the plant.
- A specific response, called gene-for-gene resistance, matches up alleles in a plant's resistance genes and a pathogen's avirulence genes. **Review Figure 40.3**
- Plants use short interfering RNAs (siRNAs) to develop immunity to invading RNA viruses.

Plants and Herbivores: Benefits and Losses

- Grazing by herbivores increases the productivity of some plants. **Review Figure 40.4**
- Some plants produce secondary metabolites that function as chemical defenses against herbivores. **Review Table 40.1**
- Various hormones, including systemin and jasmonates, participate in the pathways leading to the production of defensive chemicals. **Review Figure 40.5**
- To avoid poisoning themselves, plants may confine the toxic substances they produce to special compartments, produce those substances only after cells have been damaged, or form enzymes and receptors that are not affected by the substances.

Water Extremes: Dry Soils and Saturated Soils

- Desert annuals evade drought by living only long enough to take advantage of the brief period during which the soil has enough moisture to support them.
- Some leaves have special adaptations to dry environments: a thickened cuticle, epidermal hairs, sunken stomata, fleshy leaves and stems, spines, and altered leaf display angles.
- Other adaptations to dry environments include long taproots and root systems that die back seasonally.
- The submerged roots of some plants form pneumatophores to allow oxygen uptake from the air. Aerenchyma in submerged plant parts stores and permits the diffusion of oxygen. **Review Figure 40.12**

Too Much Salt: Saline Environments

- A saline environment restricts the availability of water to plants. Halophytes are plants that are adapted to such environments.
- Most halophytes accumulate salt. Some have salt glands that excrete the salt to the leaf surface.
- Halophytes and xerophytes have some adaptations in common. **Review Figure 40.14**

Habitats Laden with Heavy Metals

▶ Chromium, mercury, lead, and cadmium are among the heavy metals that are toxic to plants at high concentrations.

▶ Rather than excluding heavy metals, tolerant plants deal with them after taking them up. A given plant's tolerance is limited to only one or two heavy metals.

Hot and Cold Environments

▶ High temperatures destabilize cell membranes and some proteins.

▶ Adaptations to elevated temperatures include the production of heat shock proteins.

▶ Low temperatures cause membranes to lose their fluidity. Plants may respond to cooler temperatures with a change in membrane fatty acid composition.

▶ Ice crystals can puncture organelles and plasma membranes. Plant adaptations to freezing temperatures include the production of antifreeze proteins.

See Web/CD Activity 40.1 for a concept review of this chapter.

Self-Quiz

1. Which of the following is *not* a common defense against bacteria, fungi, and viruses?
   a. Lignin formation
   b. Phytoalexins
   c. A waxy covering
   d. The hypersensitive response
   e. Mycorrhizae
2. Plants sometimes protect themselves from their own toxic secondary metabolites by
   a. producing special enzymes that destroy the toxic substances.
   b. storing precursors of the toxic substances in one compartment and the enzymes that convert those precursors to toxic products in another compartment.
   c. storing the toxic substances in mitochondria or chloroplasts.
   d. distributing the toxic substances to all cells of the plant.
   e. performing crassulacean acid metabolism.
3. Herbivory
   a. is predation by plants on animals.
   b. always reduces plant growth.
   c. usually increases the rate of photosynthesis in the remaining leaves.
   d. reduces the rate of transport of photosynthetic products from the remaining leaves.
   e. is always lethal to the grazed plant.
4. Which statement about secondary plant metabolites is *not* true?
   a. Some attract pollinators.
   b. Some are poisonous to herbivores.
   c. Most are proteins or nucleic acids.
   d. Most are stored in vacuoles.
   e. Some mimic the hormones of animals.
5. Which statement about latex is *not* true?
   a. It is sometimes contained in laticifers.
   b. It is typically white.
   c. It is often toxic to insects.
   d. It is a rubbery solid.
   e. Milkweeds produce it.
6. Which of the following is *not* an adaptation to dry environments?
   a. A less negative solute potential in the vacuoles
   b. Hairy leaves
   c. A heavier cuticle over the leaf epidermis
   d. Sunken stomata
   e. A root system that grows each rainy season and dies back when it is dry
7. Some plants adapted to swampy environments meet the oxygen needs of their roots by means of a specialized tissue called
   a. parenchyma.
   b. aerenchyma.
   c. collenchyma.
   d. sclerenchyma.
   e. chlorenchyma.
8. Halophytes
   a. all accumulate proline in their vacuoles.
   b. have solute potentials that are less negative than those of other plants.
   c. are often succulent.
   d. have low root-to-shoot ratios.
   e. rarely accumulate sodium.
9. Which of the following is *not* a commonly toxic heavy metal?
   a. Chromium
   b. Cadmium
   c. Lead
   d. Potassium
   e. Mercury
10. Plants that tolerate heavy metals commonly
    a. differ genetically from other members of their species.
    b. do not take up the heavy metals.
    c. are tolerant to all heavy metals.
    d. are slow to colonize an area rich in heavy metals.
    e. weigh more than plants that are sensitive to heavy metals.

For Discussion

1. How might plant adaptations affect the evolution of herbivores? How might the adaptations of herbivores affect plant evolution?
2. The stomata of the common oleander, *Nerium oleander*, are located in sunken crypts in its leaves. Whether or not you know what an oleander is, you should be able to describe an important feature of its natural habitat. What is that feature?
3. Explain why halophytes often use the same mechanisms for coping with their challenging environments as xerophytes do for coping with theirs.
4. In ancient times, people used less sophisticated methods for mining than we use today. Thus ancient mines often yield substantial profits to modern-day miners who find and work them. On the basis of the information in this chapter, how might you try to locate the site of an ancient mine?

LIFE ESSAY

How should we manage fire in the forest?

- By David E. Pesonen -

As long as there have been forests there have been forest fires. Before aboriginal inhabitants discovered fire as a game management tool there was lightning. Afternoon thunderstorms have long been a regular phenomenon in the dry western forests of the United States, igniting fires throughout the evolutionary history of western forests and brushlands.

Many tree species would exist only in niche habitats but for their adaptation to fire. Their seed-bearing cones are tightly closed until exposed to heat, when they burst open and spread their seeds. Another evolutionary adaptation to fire occurs among several dominant mature species that are protected from fire by a thick, insulating bark. Periodic ground fires inhibit seedling reproduction and consume woody debris. But in larger trees, these fires cause no harm to the vascular cambium (the tissue most vulnerable to heat). And in arid brushlands, such as the chaparral landscape of Southern California, several dominant species of brush drop their seeds in the soil where they lie dormant for many years until germination is triggered by heat from a passing fire and the brushland is regenerated.

Over the last few decades, fire suppression has become increasingly expensive and less effective. The acreage burned each year has increased, individual fires have burned more acres, and the value of residential and commercial property destroyed has greatly increased. Why?

We now understand that fire is as much a factor in the ecology of our forests and brushlands as precipitation, soil types, and other determinants of a forest community. But, like much human intervention in natural communities, unforeseen consequences are the most predictable result. Two historic kinds of human intervention in the natural forest fire cycles are most significant in limiting efforts to restore natural conditions.

First is our long history of fire suppression. As fire was controlled, undergrowth, which would otherwise have been consumed, flourished. As a result, fires are now not only more difficult to control, but the heat content and the "ladder effect" of young growth threaten stands of older timber that resisted damage from the earlier, faster moving fires burning close to the ground. Everything burns hotter, and the forest is often reduced to its earliest successional stage.

Second, and the most difficult situation to resolve, is the change in rural residential land use. Many people build residences in secluded forest surroundings, on widely separate parcels. When a wildfire gets rolling toward these isolated homes, the traditional approaches to suppressing forest fires—backfires (deliberately burning areas in advance of the fire) and allowing fires to burn to defensible boundaries, such as roads and streams—are not politically acceptable. Homeowners expect fire personnel to protect residences. But if fire-control forces were free of this responsibility, they could manage fires differently to promote restoration of natural, less dangerous conditions.

Public land management agencies have proposed several solutions to resolve the problem created by decades of a philosophy that all fires must be suppressed promptly and the rising demand for homes built in the woods. One approach is to mechanically thin low-value undergrowth and subsidize this work by allowing commercial harvesting of valuable old growth, even though the old growth trees are fire-resistant.

David E. Pesonen attended the University of California at Berkeley, working his way through college as a summer firefighter with the U.S. Forest Service. He graduated with a degree in Forestry in 1960. He later returned to Berkeley as a law student, graduating in 1968. In 1979, he was appointed Director of the California Department of Forestry, one of the nation's largest wildland fire control agencies. He later served as a judge of the Superior Court, then returned to law practice, retiring in 1996.

The best approach is to use fire under controlled conditions to try to recreate natural conditions. Prescribed fire under carefully controlled conditions is cheaper and more effective than mechanical thinning and is increasingly applied in remote areas. But a few such fires have escaped planned controls and caused heavy property damage. As a result, land managers are unwilling to risk the occasional escape of such fires in the areas where this technique would be most useful.

There is no consensus among forest land managers on ways to resolve these issues. Political considerations often drive the search for solutions to the management and control of wildfires, triggered by reaction to losses of life and property, as much as by long-term biological and economic analysis.

Discussion Questions

1. Should public policy discourage residential expansion into rural areas of high fire danger? If not, should the owners of such residential property be required to pay, through taxes or other means, the increased cost of wildfire suppression focused on structures rather than long-term forest health?
2. Should government agencies charged with wildland fire protection subsidize efforts to clear large areas around rural properties to limit the need for fire fighting strategies to concentrate on residential protection?
3. Even if the risk of escape and consequent property damage cannot be eliminated, should public agencies continue the use of prescribed fire as a forest management tool?

Web Links

National Interagency Fire Center
www.nifc.gov/

Fire Ecology Database
www.ttrs.org/feco.html

Western Fire Ecology Center
www.fire-ecology.org/

Biscuit Fire Recovery Information
www.biscuitfire.com

Part Seven • THE BIOLOGY OF ANIMALS

41 Physiology, Homeostasis, and Temperature Regulation

The Tour de France, a 3-week, 3,500-km bicycle race, is arguably the most extreme and demanding of all athletic events. Competitors are on their bikes 5 to 7 hours a day, riding at an average speed of over 41 kilometers an hour across terrain that includes the mountains of the French Alps. The Tour can be compared to running 20 marathons at world-class pace in 20 days. In 2003, Lance Armstrong won the Tour for the fifth time.

How can an athlete perform at this level, and what results in a winning performance? A number of factors are involved, including determination, skill, and physiology. It is physiology that is the subject of Part Seven of this book. **Physiology** can be simply defined as the science of how organisms work. Physiological mechanisms span the range from molecular to behavioral.

France on 10,000 Calories a Day Lance Armstrong is a remarkable athlete largely because of the capacity of his physiological systems.

You learned in earlier parts of this book that cells oxidize glucose to produce ATP, which is then used to do biological work, such as the contraction of muscles. Performance in an event such as the Tour de France is limited ultimately by the maximum sustainable rate at which the athlete's body can convert the chemical energy of food into the mechanical energy of muscles. That rate is determined by more factors than the cellular biochemical reactions you have studied. Oxygen has to be delivered to the blood, and the blood has to be pumped to the muscles and other organs. Food has to be converted to fuel molecules by the digestive system, and those fuel molecules have to be distributed to the mitochondria of the muscle cells. The waste products of cell metabolism have to be carried away and eliminated. The temperature, ion balance, and pH of muscles and other organs have to be maintained at optimal levels. All of these tasks and more are carried out by the physiological systems we will study in this part of the book. How does Lance measure up in terms of some of his physiological characteristics?

One measure of exercise capacity is the maximum rate at which a person can take up and utilize oxygen: the V_{O_2max}. For a healthy man, a typical value is about 40 ml O_2 per kg body mass per minute. Lance's V_{O_2max} is more than twice that value. Whereas a normal, fit man might burn up to 3,500 Calories on a particularly active day, during the Tour, Lance burns about 6,500 Calories a day—10,000 Calories on peak days! Because Lance has an extremely low proportion of body fat—only 4–5 percent (20 percent is normal)—he must eat, and his body must process that number of calories per day. Imagine eating three to four times your normal diet *and* being able to work intensively at the same time.

Oxygen and fuel are delivered to muscles by the blood, and the blood has to be pumped by the heart. When Lance is at rest, his heart has to pump only about half as fast (32 beats per minute) as that of a normal healthy man, but when he is exerting himself, his heart rate can go over 200 beats per minute. Assuming the heart pumps the same amount of blood with each beat, Lance increases his delivery of blood by almost sevenfold. An average healthy person would only be able to achieve a three- to fourfold increase during exercise.

The study of physiology enables us to understand how athletes can achieve high levels of performance and also what ultimately limits their performance. Similarly, the study of physiology introduces us to many fascinating stories of how different kinds of animals have become adapted to exploit unusual or extreme environments.

In this chapter, we will first set the stage for our study of animal and human physiology by presenting an overview of how cells are organized into tissues, tissues into organs, and organs into organ systems with different physiological functions. Then we will see how organ systems are controlled and regulated to achieve constancy in the internal environment. Most of this chapter deals with one feature of the internal environment: temperature. We will see how temperature influences living systems, what adaptations animals have for dealing with temperature challenges, and finally, how mammals regulate their body temperature.

Homeostasis: Maintaining the Internal Environment

Single-celled organisms meet all their needs by direct exchanges with the external environment. Even the cells of some small, simple multicellular animals meet their needs in this way. Such animals are common in the sea. Seawater contains nutrients and salts and provides a relatively unchanging physical environment. Most cells of a sponge or a jellyfish are in direct contact with seawater, or are close enough that they can take up nutrients and eliminate wastes without the aid of specialized organs to transport nutrients and wastes around the animal's body. This lifestyle is quite limiting, however. No part of the body can be more than a few cell layers thick, every cell must be able to take care of all its own needs, and the animal is limited to environments that provide for all of its needs.

The evolution of an *internal environment*, distinct from the external environment, made complex multicellular animals possible. The internal environment consists of *extracellular fluid* that bathes every cell of the body (Figure 41.1). Cells get their nutrients from this extracellular fluid and dump their waste products into it. As long as the conditions in the internal environment are constant and optimal, the cells are protected from changes or harsh conditions in the external environment. Thus, a constant internal environment makes it possible for an animal to occupy a habitat that would kill its cells if they were exposed to it directly. How is the internal environment kept constant?

As multicellular organisms evolved, cells became specialized for maintaining specific aspects of the internal environment. In turn, the development of an internal environment enabled these specializations, since each cell did not have to provide for all of its own needs. Thus, it was possible for some cells to specialize in maintaining salt and water balance, while

41.1 Maintaining Internal Stability while on the Go Organ systems maintain a constant internal environment that provides for the needs of all the cells of the body, making it possible for animals to travel between different and often highly variable external environments. Each organ system controls different aspects of the internal environment.

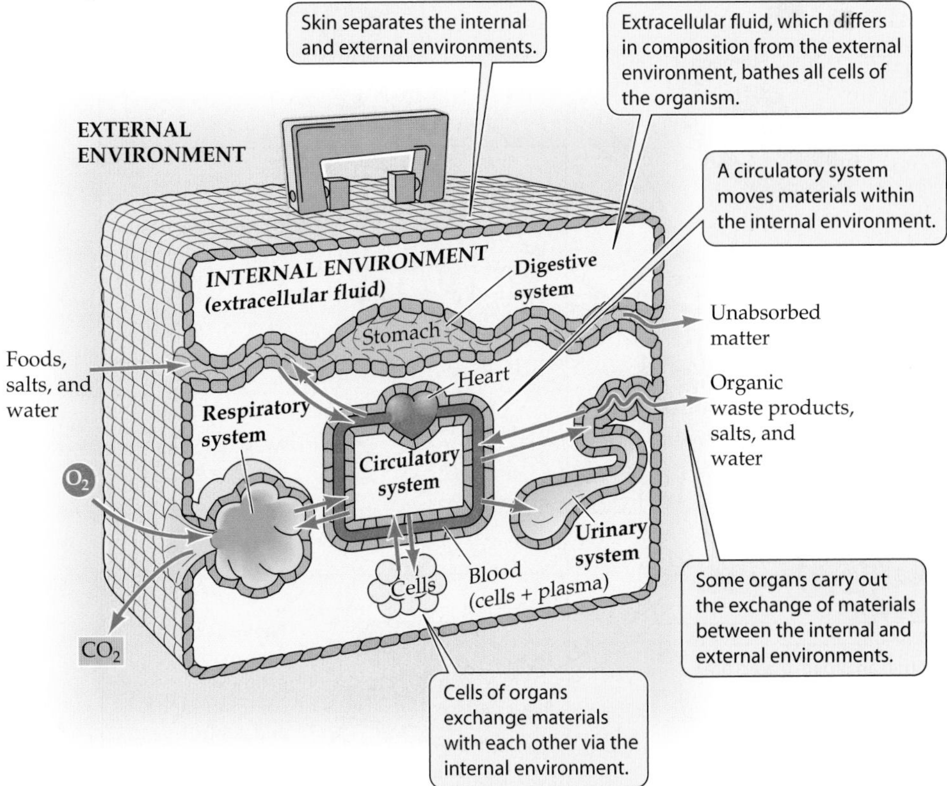

Other cells became specialized to provide nutrients, and still others to maintain appropriate levels of oxygen and carbon dioxide. As multicellular animals evolved to be larger and more complex, specialized cells formed specialized tissues, and different tissues became incorporated into organs and organs became incorporated into organ systems to control various properties of the internal environment (see Figure 41.1).

The composition of the internal environment is constantly being challenged by the external environment and by the activities of cells themselves. For example, during a race, the muscle cells in Lance Armstrong's legs are taking up oxygen and fuel molecules and releasing carbon dioxide and lactic acid. Their metabolism is contributing heat to the internal environment. Lance sweats to dissipate that heat and thereby loses water and salts from his internal environment. For Lance to continue in the race, his cells, tissues, and organs must continuously correct the physical and chemical composition of his internal environment so that his muscle cells can function optimally.

The maintenance of constant conditions in the internal environment is called **homeostasis**. Homeostasis is an essential feature of complex animals. If an organ fails to function properly, homeostasis is compromised, and as a result, cells become damaged and die—not just those of the malfunctioning organ, but the cells of other organs as well. To avoid loss of homeostasis, the activities of organs must be controlled and regulated in response to changes in both the external and internal environments.

Control and regulation require information; hence the organ systems of information—the endocrine and nervous systems—must be included in our discussions of every physiological function. For that reason, we treat the endocrine and nervous systems early in Part Seven. Subsequent chapters deal with the organ systems responsible for controlling various aspects of the internal environment. Although each chapter will focus on different organs, those organs are all made of the same tissue types. What are these tissue types, and what are their general features?

Tissues, Organs, and Organ Systems

Cells are the basic building blocks of multicellular animals. When cells with the same characteristics or specializations are grouped together, they form a **tissue**. There are four basic types of tissues—epithelial, connective, muscle, and nervous—but there are variations on each basic type. An organ is usually made up of several different tissue types (Figure 41.2).

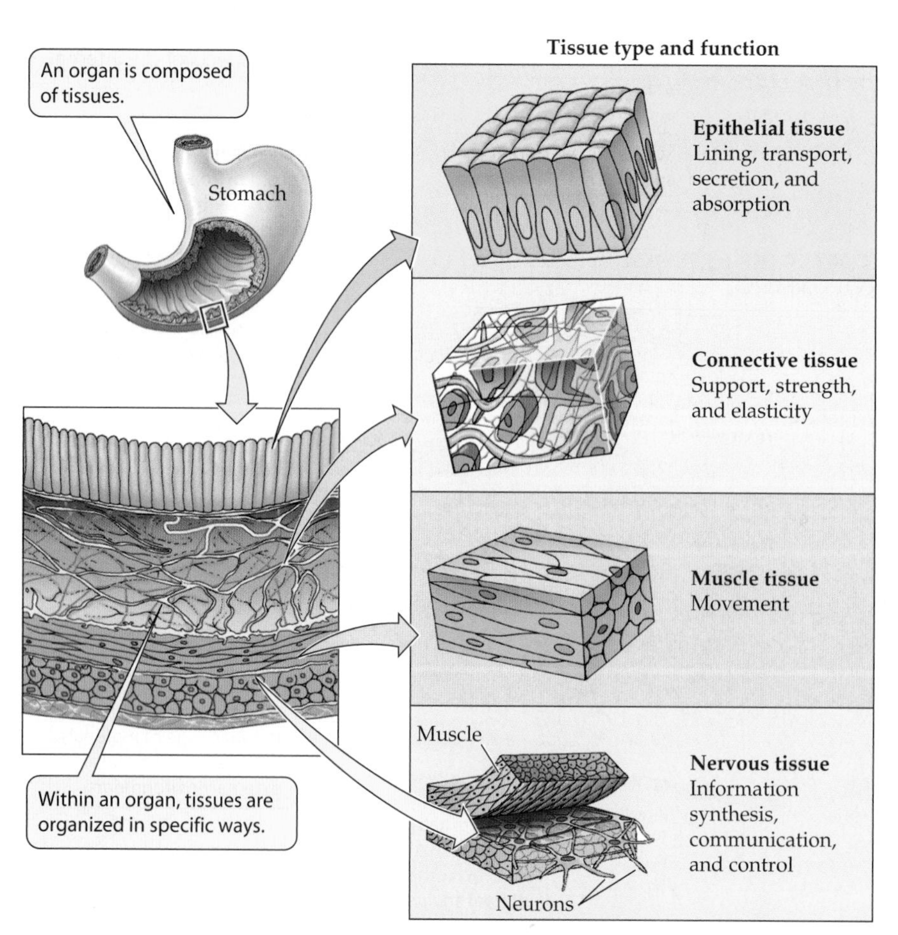

41.2 Four Types of Tissue All cells can be classified into one of four tissue types. An organ such as the stomach is made up of multiple tissue types.

Epithelial tissues cover the body and line organs

Epithelial tissues are sheets of densely packed, tightly connected cells that cover inner and outer body surfaces. They form the skin and line hollow organs of the body, such as the gut (Figure 41.3). Some epithelial cells have secretory functions; examples are the groups of epithelial cells that secrete hormones, milk, mucus, digestive enzymes, or sweat. Other epithelial cells have cilia to help substances move over surfaces or through tubes. Since epithelial cells create boundaries between the inside and the outside of the body and between body compartments, they frequently have protective as well as absorptive and transport functions. Epithelial cells can also form receptors that provide information to the nervous system. Smell and taste receptors, for example, are epithelial cells that detect specific chemicals.

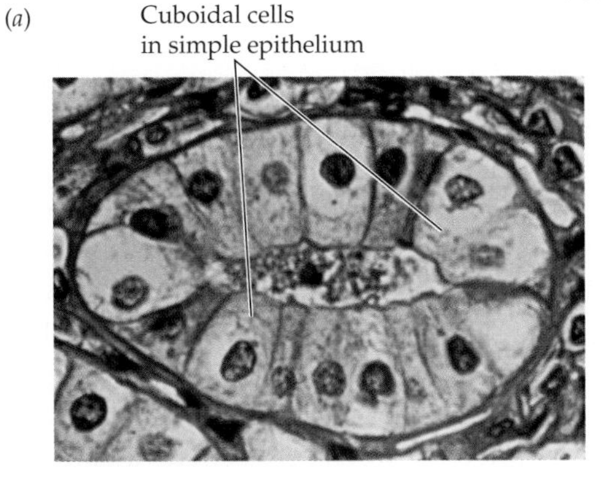

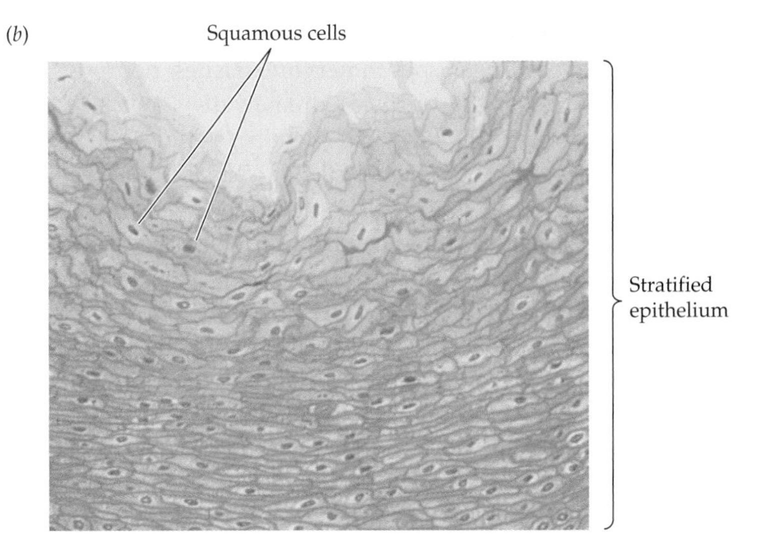

41.3 Epithelial Tissue (*a*) A single layer of cuboidal cells forms a simple epithelium lining the collecting ducts of a human kidney. (*b*) Multiple layers of squamous cells form a stratified epithelium.

Epithelial tissues have distinct inner and outer surfaces. The outer surface faces the air, as in the case of the skin and lungs, or a fluid-filled organ cavity, such as the lumen of the gut. These outer surfaces are the *apical* ends of the epithelial cells, which may have cilia or may be highly folded to increase their surface area. The inner surfaces of an epithelium are the *basal* ends of the epithelial cells, which rest on an extracellular matrix called a *basal lamina* (see Figure 4.26).

The skin and the lining of the gut are examples of epithelial tissues that receive much wear and tear. Accordingly, cells in these tissues have a high rate of cell division to replace cells that die and are shed. Dandruff consists of discarded skin cells.

Connective tissues support and reinforce other tissues

In contrast to densely packed epithelial tissues, **connective tissues** consist of dispersed populations of cells embedded in an *extracellular matrix* that they secrete. The composition and properties of the matrix differ among types of connective tissues.

An important component of the extracellular matrix secreted by connective tissue cells is protein fibers. The dominant protein in the extracellular matrix is *collagen* (see Figure 4.26). Collegen is, in fact, the most abundant protein in the human body, representing 25 percent of total body protein. Collagen fibers are strong. They give the connective tissue of skin, tendons, and ligaments resistance to stretch. Similarly, collagen fibers provide a netlike framework for organs, giving them shape and structural strength. Connective tissue that fills spaces between organs has a low density of collagen fibers.

Another type of protein fiber in the extracellular matrix of connective tissues is the stretchable protein *elastin*. It can be stretched to several times its resting length and then recoil. Fibers composed of elastin are most abundant in tissues that are regularly stretched, such as the walls of the lungs and the large arteries. Gradual loss of elastin fibers with age causes gradual loss of resiliency of the skin.

Cartilage and bone are connective tissues that provide rigid structural support. In *cartilage*, a network of collagen fibers is embedded in a flexible matrix consisting of a protein–carbohydrate complex. Cartilage, which lines the joints of vertebrates, is resistant to compressive forces. Since it is flexible, it provides structural support for flexible structures such as external ears and noses. The extracellular matrix in *bone* also contains many collagen fibers, but it is hardened by the deposition of the mineral calcium phosphate. We will discuss cartilage and bone in greater detail in Chapter 47.

Adipose tissue is a form of loose connective tissue that includes adipose cells, which form and store droplets of lipids. Adipose tissue, or "fat," is a major source of stored energy. It also serves to cushion organs, and layers of adipose tissue under the skin can provide a barrier to heat loss (see Figure 41.15).

Blood is a connective tissue consisting of cells dispersed in an extensive extracellular matrix: the blood *plasma*. The blood plasma is much more liquid than the extracellular matrices of the other connective tissues, but it too contains an abundance of proteins. Many of the proteins and cellular elements of the blood were presented in Chapter 18, and blood will be discussed again in Chapter 49.

Muscle tissues contract

Muscle tissues consist of elongated cells that can contract and cause movement. Muscle tissues are the most abundant tissues in the body, and when animals are active, they use most of the energy produced in the body. We will discuss muscle tissues in detail in Chapter 47.

Nervous tissues process information

There are two basic cell types in **nervous tissues**: neurons and glial cells. *Neurons*, which are extremely diverse in size and form, communicate via electrochemical signals. These nerve impulses can be conducted via long extensions of the neurons to other parts of the body, where they are communicated to other neurons, muscle cells, or secretory cells. Neurons are involved in controlling the activities of most organ systems to achieve homeostasis.

Glial cells do not generate or conduct electrochemical signals, but they provide a variety of supporting functions for neurons. There are more glial cells than neurons in our nervous systems. We will detail and illustrate the properties of nervous tissues in Chapters 44, 45, and 46.

Organs consist of multiple tissues

A discrete structure that carries out a specific function in the body is called an **organ**. Examples are the stomach, the heart, the liver, and the kidney. Most organs include all four tissue types. The wall of the stomach is a good example (see Figure 41.2).

The inner surface of the stomach that contacts food is lined with a sheet of epithelial cells. Some of the epithelial cells secrete mucus, enzymes, or stomach acid. Beneath the epithelial lining is connective tissue. Within this connective tissue are nerves, glands (clusters of secretory epithelial cells), and blood vessels. Concentric layers of smooth muscle tissue enable the stomach to contract to mix food with digestive juices. A network of neurons between the muscle layers controls these movements and also influences the secretions of the stomach. Surrounding the stomach is a sheath of connective tissue.

An individual organ is usually part of an **organ system**—a group of organs that function together. The stomach is part of the digestive system, which also includes the food tube (esophagus), the small and large intestines, the pancreas, which secretes digestive enzymes, and the liver, which secretes bile. The major organ systems of mammals are outlined in Table 41.1.

41.1 The Major Organ Systems of Mammals

SYSTEM	TISSUES AND ORGANS	FUNCTIONS
Nervous system	Brain, spinal cord, sensory organs, peripheral nerves	Receives, integrates, stores information and controls muscles and glands (Chapters 44, 45, 46)
Endocrine system	Glands: pituitary, thyroid, parathyroid, pineal, adrenal, testes, ovaries, pancreas	A system of glands releases chemical messages (hormones) that control and regulate other tissues and organs (Chapter 42)
Muscle system	Skeletal muscle, smooth muscle, cardiac muscle	Produces forces and motion (Chapter 47)
Skeletal system	Bones	Provides structural support for the body (Chapter 47)
Reproductive system	Female: ovaries, oviducts, uterus, vagina, mammary glands Male: testes, sperm ducts, accessory glands, penis	Produces sex cells and hormones necessary to procreate and nurture offspring (Chapter 43)
Digestive system	Mouth, esophagus, stomach, intestines, liver, pancreas, rectum, anus	Acquires and digests food, absorbs and stores nutrients, then makes them available to the cells of the body (Chapter 50)
Respiratory system	Airways, lungs, diaphragm	Exchanges respiratory gases with the environment (Chapter 48)
Circulatory system	Heart and blood vessels	Transports respiratory gases, nutrients, hormones, and heat around the body (Chapter 49)
Lymphatic system	Lymph and lymph vessels, lymph nodes, spleen	Brings extracellular fluids back into the circulatory system; helps the immune system fight invading organisms (Chapters 49 and 18)
Immune system	Many types of white blood cells	Fights invading organisms and infections (Chapter 18)
Skin system	Skin, sweat glands, hair	Protects the body from invading organisms and harsh physical conditions, helps regulate body temperature (Chapter 41)
Excretory system	Kidneys, bladder, ureter, urethra	Regulates the composition of the extracellular fluids; excretes waste products (Chapter 51)

Physiological Regulation and Homeostasis

Homeostasis depends on the ability to regulate the activities of organs and organ systems to keep the internal environment constant. In this section we will discuss the general properties of physiological regulatory systems. In the following sections, we will consider temperature regulation as a specific example.

Generally, the activities of organs and organ systems are controlled—in other words, sped up or slowed down—by actions of the nervous system and the endocrine system. But, to achieve regulation of the internal environment, information is required. Think of it this way. You *control* the speed of your car with the accelerator and the brakes, but you can't use those control mechanisms to *regulate* the speed of your car if you don't know how fast you are going and how fast you want to go. The desired speed is a **set point**, and the reading on your speedometer is **feedback information**. When the set point and the feedback information are compared, any difference between them is an **error signal**. Error signals suggest corrective actions, which you make by using the accelerator or brake (Figure 41.4).

Physiological regulation requires the action of cells, tissues, and organs, which are called *effectors* because they effect changes. Effectors are *controlled systems* because their activities are controlled by commands from regulatory systems. *Regulatory systems* obtain, process, and integrate information, then issue commands to controlled systems.

A fundamental way to analyze a regulatory system is to identify the information it uses. Negative feedback is the most common type of feedback information in regulatory systems. The word "negative" indicates that this feedback information causes the effectors to reduce or reverse the process or counteract the influence that created an error signal. In our car analogy, the recognition that you are going too fast is negative feedback if it causes you to slow down. Negative feedback is a stabilizing influence in physiological regulatory systems. It contributes to homeostasis by stimulating actions that return a variable to its set point.

Is there any such thing as positive feedback in physiology? Although not as common as negative feedback, it does exist. Rather than returning a system to a set point, positive feedback amplifies a response. Examples of regulatory systems that use positive feedback are the responses that empty body cavities, such as urination, defecation, sneezing, and vomiting. Another example is sexual behavior, in which a little stimulation causes more behavior, which causes more stimulation, and so on.

Feedforward information is another feature of regulatory systems. The function of feedforward information is to change the set point. Seeing a deer ahead on the road when you are driving is an example of feedforward information

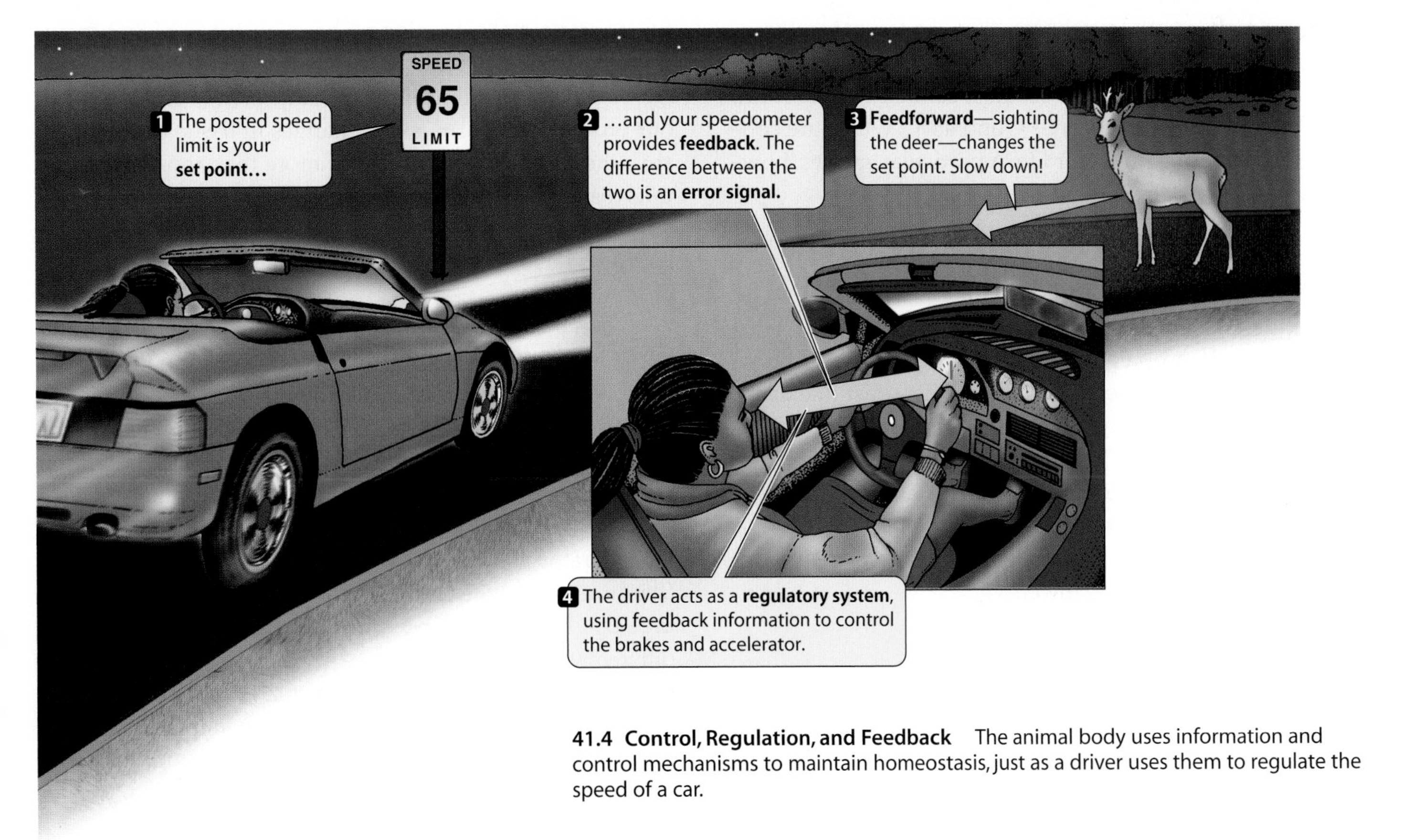

SPEED
65
LIMIT
1 The posted speed limit is your **set point...**
2 ...and your speedometer provides **feedback.** The difference between the two is an **error signal.**
3 **Feedforward**—sighting the deer—changes the set point. Slow down!
4 The driver acts as a **regulatory system,** using feedback information to control the brakes and accelerator.

41.4 Control, Regulation, and Feedback The animal body uses information and control mechanisms to maintain homeostasis, just as a driver uses them to regulate the speed of a car.

(see Figure 41.4); this information takes precedence over the posted speed limit, and you change your set point to a slower speed.

These principles of control and regulation help organize our thinking about physiological systems. Once we understand how an organ or an organ system works, we can then ask how is it regulated. As an example, we will discuss in detail the system that regulates body temperature.

Temperature and Life

Temperatures vary enormously over the face of Earth, from the boiling hot springs of Yellowstone National Park to the interior of Antarctica, where the temperature can fall below –80°C. Because heat always moves from a warmer to a cooler object, any change in the temperature of the environment causes a change in the temperature of an organism in that environment—unless the organism does something to regulate its temperature.

Living cells can function over only a narrow range of temperatures. If cells cool to below 0°C, ice crystals damage their structures. Some animals have adaptations, such as antifreeze molecules in their blood, that help them resist freezing; others have adaptations that enable them to survive freezing. Generally, however, cells must remain above 0°C to stay alive.

The upper temperature limit is less than 45°C for most cells. Some specialized algae can grow in hot springs at 70°C, and some archaea can live at near 100°C, but in general, proteins begin to denature and lose their function as temperatures approach 45°C. Therefore, most cellular functions are limited to the range between 0°C and 45°C, which are considered the thermal limits for life. A particular species, however, generally has much narrower limits.

Q_{10} is a measure of temperature sensitivity

Even within the range from 0° to 45°C, changes in tissue temperature create problems for animals. Most physiological processes, like the biochemical reactions that constitute them, are temperature-sensitive, going faster at higher temperatures (see Figure 6.23). The temperature sensitivity of a reaction or process can be described in terms of Q_{10}, a quotient calculated by dividing the rate of a process or reaction at a certain temperature, R_T, by the rate of that process or reaction at a temperature 10°C lower, R_{T-10}:

$$Q_{10} = \frac{R_T}{R_{T-10}}$$

Q_{10} can be measured for a simple enzymatic reaction or for a complex physiological process, such as rate of oxygen consumption. If a reaction or process is not temperature-sensitive, it has a Q_{10} of 1. Most biological Q_{10} values are between 2 and 3. A Q_{10} of 2 means that the reaction rate doubles as temperature increases by 10°C, and a Q_{10} of 3 indicates a tripling of the rate (Figure 41.5).

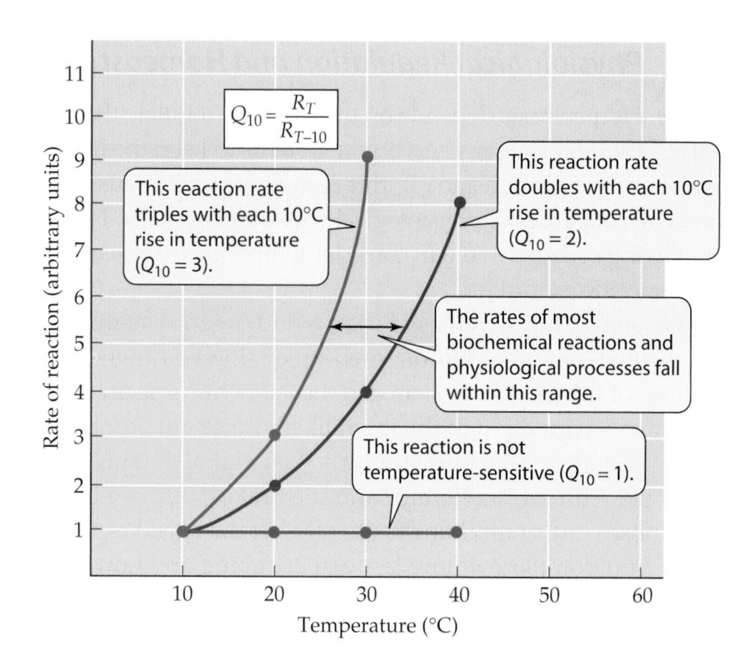

41.5 Q_{10} and Reaction Rate The larger the Q_{10} of a reaction or process, the faster its rate rises in response to an increase in temperature.

Changes in tissue temperature can be particularly disruptive to an animal's functioning because not all of its component reactions have the same Q_{10}. Reactions with different Q_{10} values are linked together in complex networks that carry out physiological processes. Changes in tissue temperature shift the rates of some reactions more than those of others, thus disrupting the balance and integration that the processes require. To maintain homeostasis, organisms must be able to compensate for or prevent changes in temperature.

An animal's sensitivity to temperature can change

The body temperature of some animals is tightly coupled to the environmental temperature. Think of a fish in a temperate-zone pond. As the temperature of the pond changes from 4°C in midwinter to 24°C in midsummer, the body temperature of the fish does the same (Figure 41.6). We can bring the fish into the laboratory in the summer and measure its **metabolic rate** (the sum total of the energy turnover of its cells, often measured by O_2 consumption). If we measure its metabolic rate at different water temperatures, we might plot our data as shown by the red line in Figure 41.6 and calculate a Q_{10} of 2. We might predict from our graph that in winter, when the temperature is 4°C, the fish's metabolic rate will be only one-fourth of what it was in summer. We then return the fish to its pond.

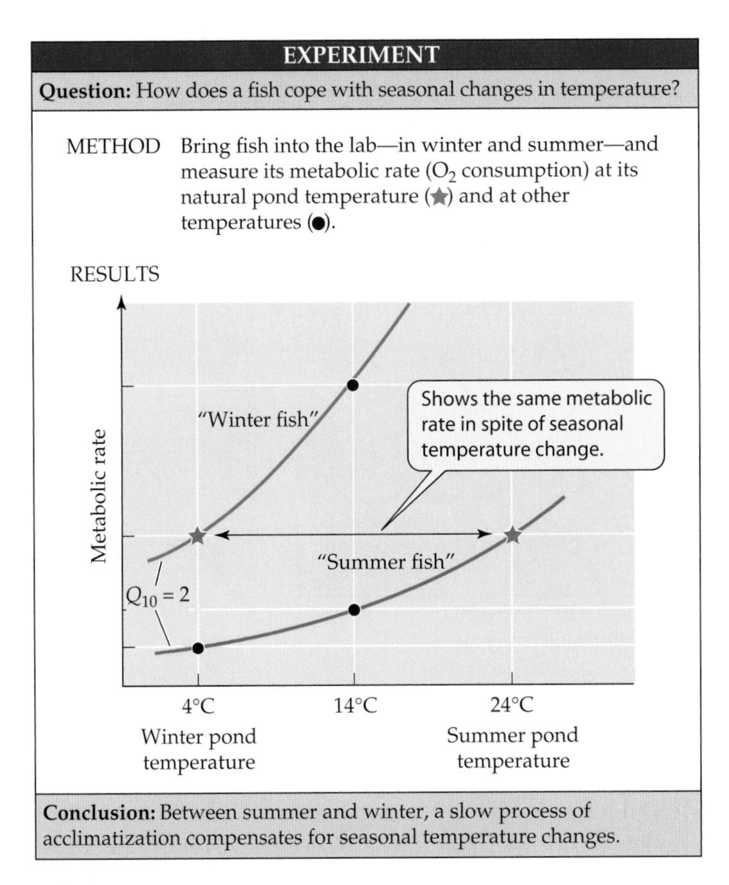

EXPERIMENT
Question: How does a fish cope with seasonal changes in temperature?
METHOD Bring fish into the lab—in winter and summer—and measure its metabolic rate (O_2 consumption) at its natural pond temperature (★) and at other temperatures (●).
RESULTS
"Winter fish"
Shows the same metabolic rate in spite of seasonal temperature change.
"Summer fish"
$Q_{10} = 2$
Metabolic rate
4°C
Winter pond temperature
14°C
24°C
Summer pond temperature
Conclusion: Between summer and winter, a slow process of acclimatization compensates for seasonal temperature changes.

41.6 Metabolic Compensation In its natural environment, a fish's metabolism acclimatizes to compensate for seasonal changes in temperature.

When we bring the fish back into the laboratory in winter and repeat our measurements, we find, as the blue line shows, that its metabolic rate at 4°C is not as low as we predicted; rather, it is almost the same as it was at 24°C in summer. Over a range of temperatures, we find that the fish's metabolic rate is always higher than the rate we predicted from the measurement we took at the same temperature in the summer. This difference is due to **acclimatization**, the process of physiological and biochemical change that an animal undergoes in response to seasonal changes in temperature.

Seasonal acclimatization in the fish has produced **metabolic compensation**—a change in the biochemical machinery that counters the effects of temperature change. What might account for such a change? Recall our discussion of isozymes in Chapter 6. If the fish can express a number of isozymes that operate at different optimal temperatures, it can compensate metabolically, catalyzing reactions with one set of enzymes in summer and another set in winter. The end result is that metabolic functions are much less sensitive to long-term changes in temperature than they are to short-term thermal fluctuations.

Maintaining Optimal Body Temperature

Animals can be classified by their response to environmental temperatures:

- A **homeotherm** is an animal that maintains a constant body temperature over a wide range of environmental temperatures.
- A **poikilotherm** is an animal whose body temperature changes when the temperature of its environment changes.

This system of classification says something about the biology of the animals, but it presents problems. Should a fish in the deep ocean, where the temperature changes very little, be called a homeotherm? Should a hibernating mammal that allows its body temperature to drop to nearly the temperature of its environment be called a poikilotherm? The problem posed by the hibernator has been solved by creating a third category: the **heterotherm**, an animal that maintains a constant body temperature some of the time.

Another set of terms classifies animals on the basis of the sources of heat that determine their body temperatures:

- **Ectotherms** depend largely on external sources of heat, such as solar radiation, to maintain their body temperature above the environmental temperature.
- **Endotherms** can regulate their body temperature by producing heat metabolically or by mobilizing active mechanisms of heat loss.

Mammals and birds are endotherms; most other animals behave as ectotherms most of the time. However, we will see below that ectotherms can occasionally operate as endotherms when they are producing large amounts of metabolic heat.

Ectotherms and endotherms respond differently to changes in environmental temperature

A small lizard can serve as an example of an ectotherm. We can compare it with a mouse, which is an endotherm of the same body size. We can put each animal in a closed chamber and measure its body temperature and metabolic rate as we change the temperature of the chamber from 0°C to 35°C.

The results obtained from the two species differ. The body temperature of the lizard equilibrates with that of the chamber, whereas the body temperature of the mouse remains at 37°C (Figure 41.7*a*). The metabolic rate of the lizard decreases as the temperature decreases (Figure 41.7*b*). In contrast, the mouse's metabolic rate increases as chamber temperature falls below about 27°C (notice that you must read the graph right to left to see this). Based on these observations, we might conclude that the lizard cannot regulate its body tem-

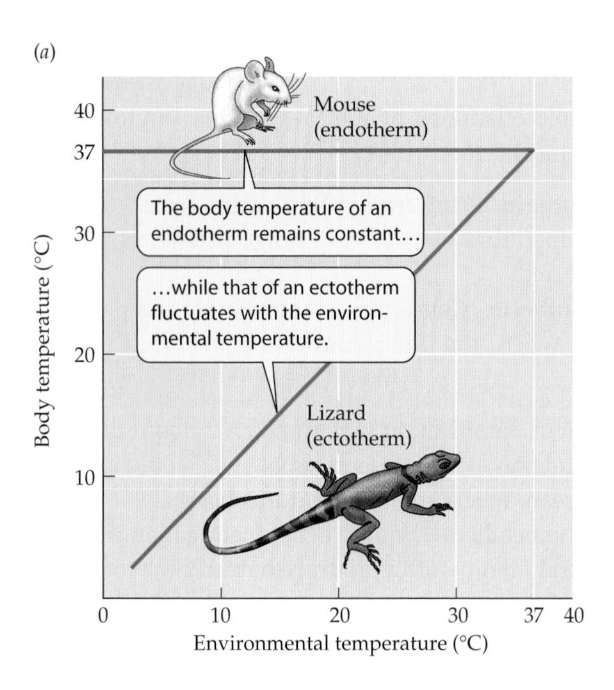

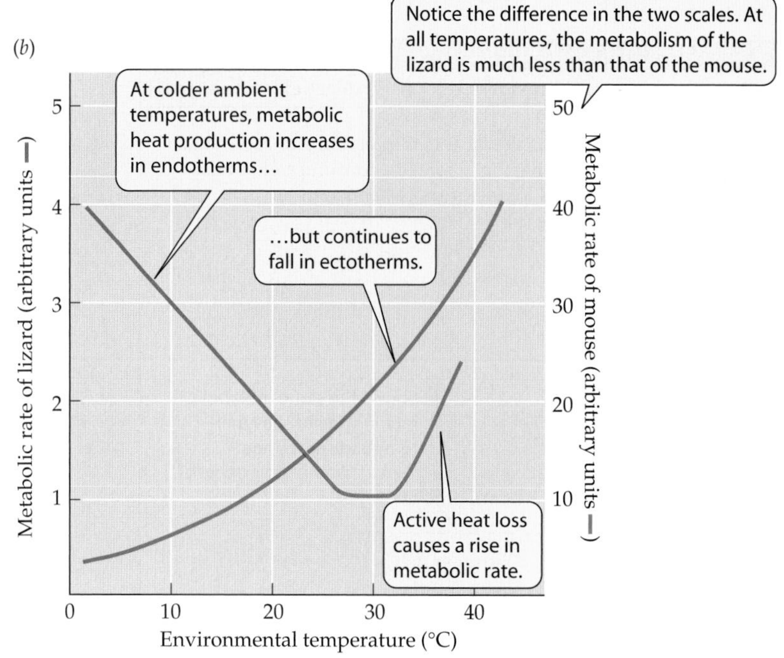

41.7 Ectotherms and Endotherms The body temperatures of a lizard and a mouse of the same body size respond differently to changes in environmental temperature.

perature or metabolism independently of the environmental temperature. The mouse, however, regulates its body temperature by increasing its metabolic rate, which increases its production of body heat. (You may wonder why the mouse curve in Figure 41.7*b* changes shape at higher environmental temperatures. We will come back to that question in the following section.)

Ectotherms and endotherms use behavior to regulate body temperature

We can test our laboratory conclusion that the lizard cannot regulate its body temperature. To do this, we return the lizard to its desert habitat, but we continue to measure its body temperature as it goes about its normal behavior. In this environment, air temperature can change 40°C in a few hours (Figure 41.8).

In contrast to what we observed in the laboratory, the lizard's body temperature is sometimes considerably different from the environmental temperature. The lizard achieves this by using behavior to alter its heat exchange with the environment. Its behavioral strategies include spending time in a burrow, basking in the sun, seeking shade, climbing vegetation, and changing its orientation with respect to the sun.

Our conclusion must be that the lizard can regulate its body temperature quite well, but that it does so by behavioral mechanisms rather than by internal metabolic mechanisms. In our laboratory experiment, the lizard could not use its thermoregulatory behavior, but in its natural environment it could move from place to

41.8 An Ectotherm Uses Behavior to Regulate Its Body Temperature The lizard's body temperature is dependent on environmental heat, but it can regulate its temperature by moving from place to place within its environment.

41.9 Endotherms Use Behavior to Thermoregulate (*a*) Humans must put on many layers of insulating clothing to help their thermoregulatory mechanisms keep pace with the extreme cold of western Siberia. (*b*) When air temperatures on the African savanna soar, an elephant may use a cool shower to thermoregulate.

(*a*) (*b*) *Loxodonta africana*

place to alter the heat exchange between its internal and external environments.

Behavioral thermoregulation is not the exclusive domain of ectotherms (Figure 41.9). Most animals, including endotherms, select comfortable thermal environments whenever possible. They may change their posture, orient to the sun, move between sun and shade, and move between still air and moving air, as demonstrated by the lizard in our field experiment. Examples of more complex thermoregulatory behavior are nest construction and social behavior such as huddling. Humans select appropriate clothing and heat or cool their buildings.

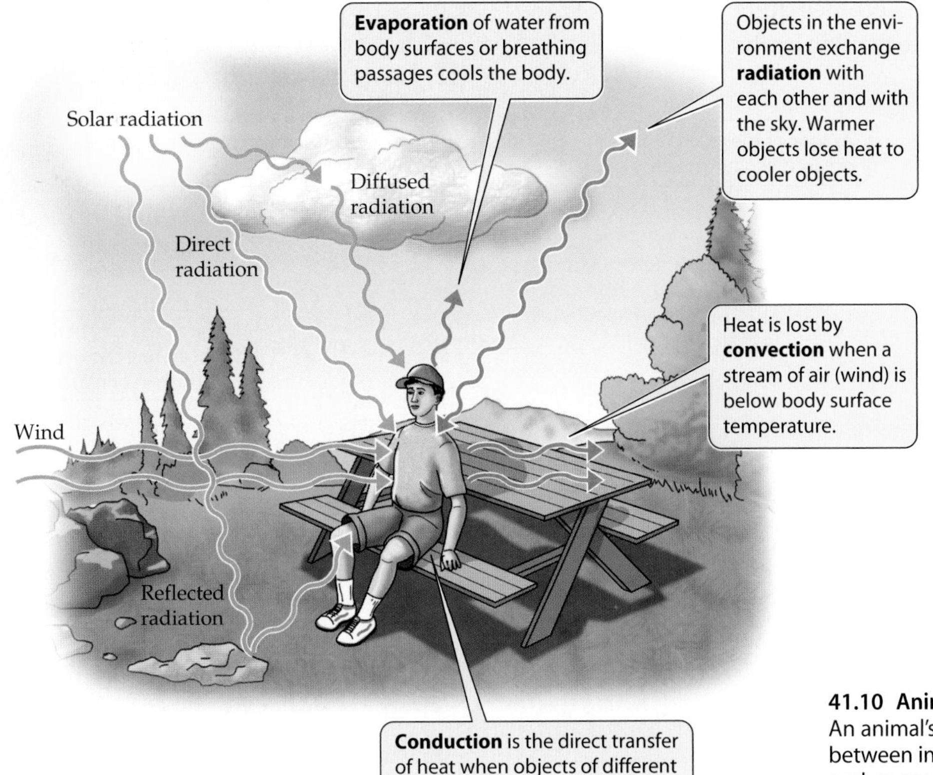

Evaporation of water from body surfaces or breathing passages cools the body.
Objects in the environment exchange radiation with each other and with the sky. Warmer objects lose heat to cooler objects.
Solar radiation
Diffused radiation
Direct radiation
Heat is lost by convection when a stream of air (wind) is below body surface temperature.
Wind
Reflected radiation
Conduction is the direct transfer of heat when objects of different temperatures come into contact.

41.10 Animals Exchange Heat with the Environment An animal's body temperature is determined by the balance between internal heat production and four avenues of heat exchange with the environment: radiation, conduction, convection, and evaporation.

Energy budgets reflect all physiological adaptations for regulating body temperature

Both ectotherms and endotherms can influence their body temperatures by altering four avenues of heat exchange between their bodies and the environment: radiation, conduction, convection, and evaporation (Figure 41.10). The total effects of heat production and heat exchange can be expressed as an **energy budget**, based on the simple fact that if the body temperature of an animal is to remain constant, the heat entering the animal must equal the heat leaving the animal. The heat coming in usually comes from metabolism and solar radiation (R_{abs}, for radiation absorbed). Heat leaves the body via the four mechanisms listed above: radiation emitted (R_{out}), convection, conduction, and evaporative heat loss. The energy budget takes the following form:

$$\underbrace{\text{metabolism} + R_{abs}}_{\text{heat}_{in}} = \underbrace{R_{out} + \text{convection} + \text{conduction} + \text{evaporation}}_{\text{heat}_{out}}$$

Anyone who has experienced a very hot environment will realize that heat can also *enter* the body through convection and conduction; in that case, the sign of those factors will change to negative in the energy budget equation.

The energy budget is a useful concept because any adaptation that influences the ability of an animal to deal with its thermal environment must affect one or more components of the budget. So the energy budget gives us the ability to quantify and compare the thermal adaptations of animals. One interesting observation is that all of the components on the right side of the energy budget equation—the heat loss side—depend on the surface temperature of the animal. That surface temperature can be controlled by altering the flow of blood to the skin.

Both ectotherms and endotherms control blood flow to the skin

Heat exchange between the internal environment and the skin occurs largely through blood flow. For example, when a person's body temperature rises as a result of exercise, blood flow to the skin increases, and the skin surface becomes quite warm. The heat brought from the body core to the skin by the blood is lost to the environment through the four avenues listed above, and this heat loss helps to bring the body temperature back to normal. In contrast, when a person is exposed to cold, the blood vessels supplying the skin constrict, decreasing blood flow and heat transport to the skin and reducing heat loss to the environment.

The control of blood flow to the skin can be an important adaptation for an ectotherm such as the marine iguana of the Galápagos archipelago (Figure 41.11). These volcanic islands lie on the equator, but they are bathed by cold oceanic currents. Marine iguanas are reptiles that bask on black lava rocks on shore and enter the cold ocean water to feed on seaweed. When the iguanas are feeding, they cool to the temperature of the sea. This cooling lowers their metabolism, making them slower and more vulnerable to predators and incapable of efficient digestion. They therefore alternate between feeding in the cold sea and basking in the sun on the hot rocks. It is advantageous for iguanas to retain body heat as long as possible while swimming and to warm up as fast as possible when basking. They accomplish this by changing their heart rate as well as the rate of blood flow to their skin.

Some ectotherms produce heat

Some ectotherms raise their body temperature by producing heat. For example, the powerful flight muscles of many insects must reach 35–41°C before the insects can fly, and they must maintain these high temperatures during flight. Such insects produce the required heat by contracting their flight muscles in a manner analogous to shivering in mammals. The heat-producing ability of these insects can be quite remarkable. Probably the most impressive case is a species of scarab beetle that lives mostly underground in mountains north of Los Angeles, California. To mate, these beetles come

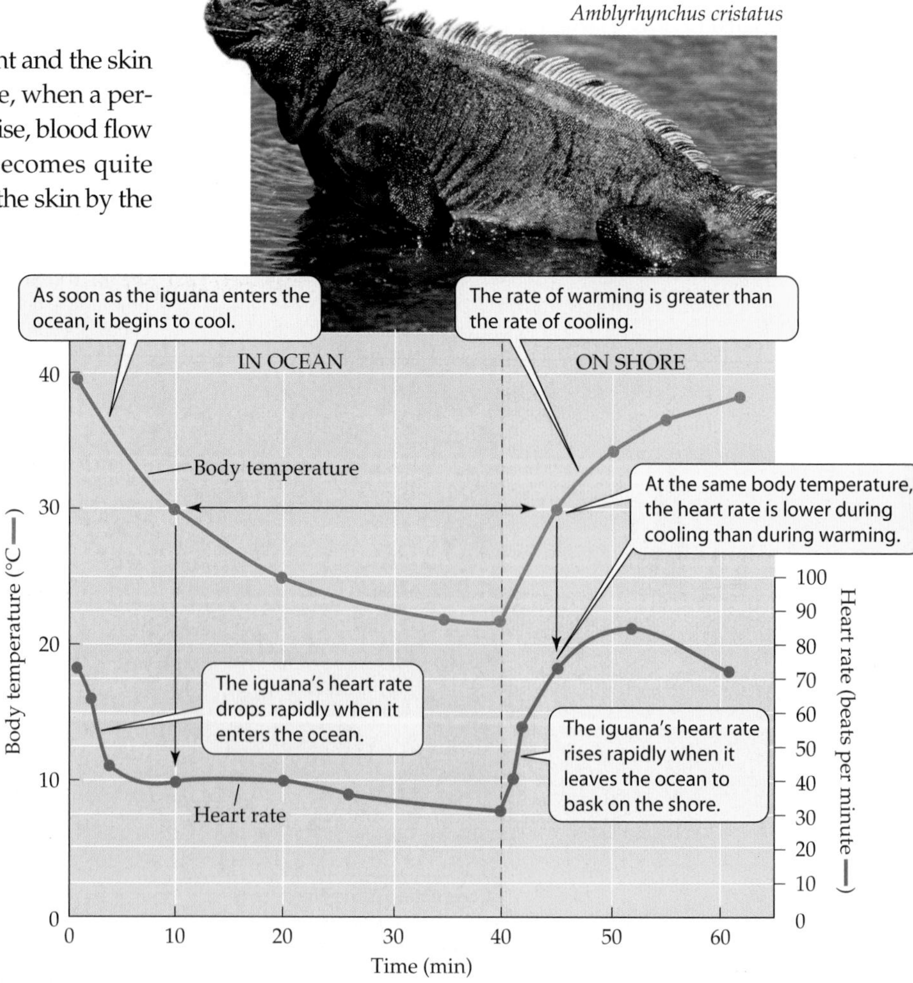

41.11 Some Ectotherms Regulate Blood Flow to the Skin Galápagos marine iguanas control blood flow to the skin to alter their heating and cooling rates.

above ground, and males fly in search of females. They undertake this mating ritual at night, in winter, and only during snowstorms.

Honeybees regulate temperature as a group. They live in large colonies consisting mostly of female worker bees that maintain the hive and rear the larval offspring of the single queen bee. During winter, honeybee workers cluster around the brood of larvae. They adjust their individual metabolic heat production and density of clustering so that the brood temperature remains remarkably constant, at about 34°C, even as the outside air temperature drops below freezing.

Some fish elevate body temperature by conserving metabolic heat

Active fish can produce substantial amounts of metabolic heat, but they have difficulty in retaining any of that heat. Blood pumped from the heart goes directly to the gills, where it comes very close to the surrounding water to exchange respiratory gases. So any heat that the blood picks up from metabolically active muscles is lost to the surrounding water as it flows through the gills. It is thus surprising that some large, rapidly swimming fishes, such as bluefin tuna and great white sharks, can maintain temperature differences as great as 10–15°C between their bodies and the surrounding water. The heat comes from their powerful swimming muscles, and the ability of these "hot" fish to conserve that heat is due to remarkable arrangements of their blood vessels.

In the usual ("cold") fish circulatory system, oxygenated blood from the gills collects in a large dorsal vessel, the aorta, which travels through the center of the fish, distributing blood to all organs and muscles (Figure 41.12*a*). "Hot" fish have a smaller central dorsal aorta. Most of their oxygenated blood is transported in large vessels just under the skin (Figure 41.12*b*). Hence the cold blood from the gills is kept close to the surface of the fish. Smaller vessels transporting this cold blood into the muscle mass run parallel to vessels transporting warm blood from the muscle mass back toward the heart. Since the vessels carrying the cold blood into the muscle are in close contact with the vessels carrying warm blood away, heat flows from the warm to the cold blood and is therefore trapped in the muscle mass.

Because heat is exchanged between blood vessels carrying blood in opposite directions, this adaptation is called a **countercurrent heat exchanger**. It keeps the heat within the muscle mass, enabling the fish to have an internal body temperature considerably above the water temperature. Why is it advantageous for the fish to be warm? Each 10°C rise in muscle temperature increases the fish's sustainable power output almost threefold!

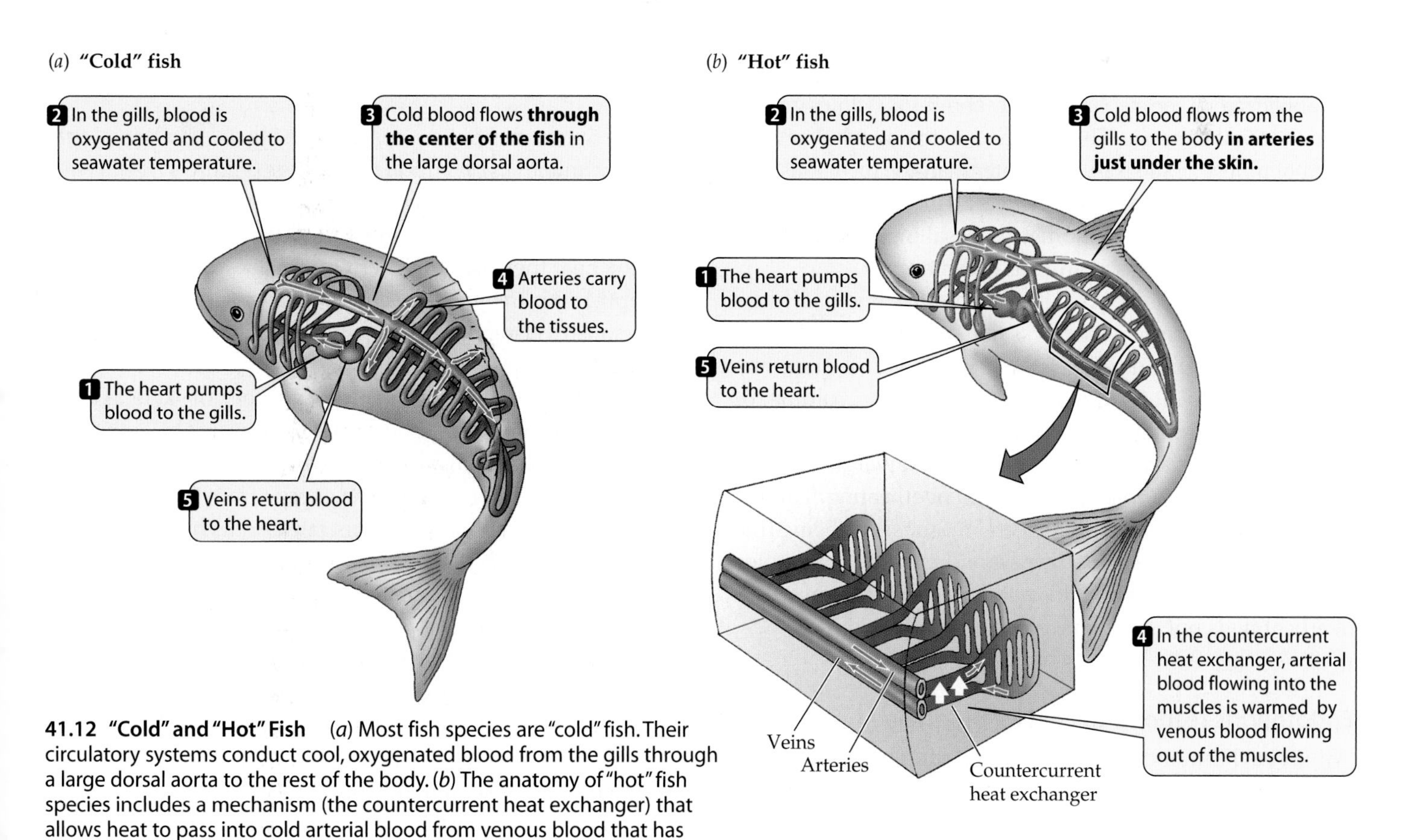

(a) "Cold" fish
2 In the gills, blood is oxygenated and cooled to seawater temperature.
3 Cold blood flows through the center of the fish in the large dorsal aorta.
4 Arteries carry blood to the tissues.
1 The heart pumps blood to the gills.
5 Veins return blood to the heart.
(b) "Hot" fish
2 In the gills, blood is oxygenated and cooled to seawater temperature.
3 Cold blood flows from the gills to the body in arteries just under the skin.
1 The heart pumps blood to the gills.
5 Veins return blood to the heart.
4 In the countercurrent heat exchanger, arterial blood flowing into the muscles is warmed by venous blood flowing out of the muscles.
Veins
Arteries
Countercurrent heat exchanger

41.12 "Cold" and "Hot" Fish (*a*) Most fish species are "cold" fish. Their circulatory systems conduct cool, oxygenated blood from the gills through a large dorsal aorta to the rest of the body. (*b*) The anatomy of "hot" fish species includes a mechanism (the countercurrent heat exchanger) that allows heat to pass into cold arterial blood from venous blood that has been warmed by the metabolism of the muscles.

Thermoregulation in Endotherms

As we saw in Figure 41.7, endotherms respond to changes in environmental temperature by changing their metabolic rate. Within a narrow range of environmental temperatures, called the **thermoneutral zone**, the metabolic rate of endotherms is low and independent of temperature. The metabolic rate of a resting animal at a temperature within the thermoneutral zone is called the **basal metabolic rate.** It is usually measured in animals that are quiet but awake and that are not using energy for digestion, reproduction, or growth. A resting animal consumes energy at the basal metabolic rate just to carry out all of its minimal body functions.

The basal metabolic rate of an endotherm is about six times greater than the metabolic rate of an ectotherm of the same size and at the same body temperature (see Figure 41.7*b*). A gram of mouse tissue consumes energy at a much higher rate than does a gram of lizard tissue when both tissues are at 37°C. This difference results from basic changes in cell metabolism that accompanied the evolution of endotherms from their ectothermic ancestors.

ANIMAL	BODY MASS (KG)	TOTAL O_2 CONSUMPTION (LITERS/HR)	O_2 CONSUMPTION (LITERS/HR) PER KG OF BODY MASS ($l\ O_2\ kg^{-1}\ h^{-1}$)
Shrew	0.005	0.036	7.40
Mouse	0.025	0.041	1.65
Rat	0.29	0.250	0.87
Cat	2.5	1.70	0.68
Dog	11.7	3.87	0.33
Sheep	42.7	9.59	0.22
Human	70	14.76	0.21
Horse	650	71.10	0.11
Elephant	3,833	268.00	0.07

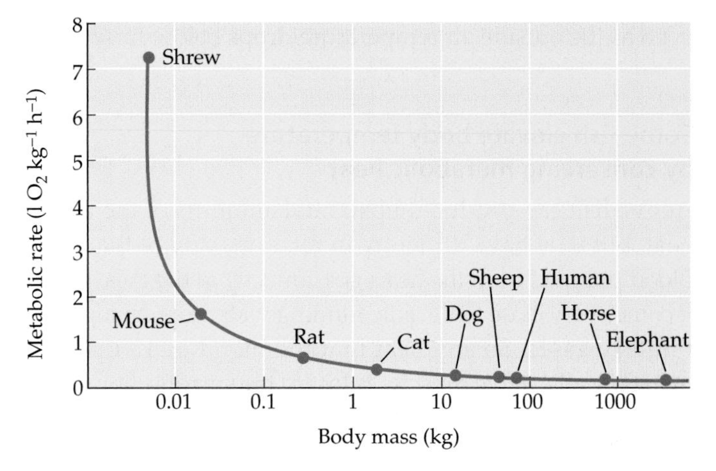

41.13 The Mouse-to-Elephant Curve On a weight-specific basis, the metabolic rate of small endotherms is much greater than that of larger endotherms. Using the data in the table, the graph plots O_2 consumption per kg of body weight (a measure of the metabolic rate) against a logarithmic plot of body weight.

Basal metabolic rates of endotherms are related to body size

Obviously, the total basal metabolic rate of an elephant is greater than that of a mouse. After all, the elephant is more than 100,000 times more massive than the mouse. However, the metabolic rate of the elephant is only about 7,000 times greater than that of the mouse. That means that the metabolism of a gram of mouse tissue is much greater than the metabolism of a gram of elephant tissue—more than 20 times greater (Figure 41.13). Across all of the endotherms, basal metabolic rate per gram of tissue increases as animals get smaller.

Why should this be so? No one actually knows. It was once thought that the reason was that as animals get bigger, they have a smaller ratio of surface area to volume (see Figure 4.3). Since heat production is related to the volume, or mass, of the animal, but its capacity to dissipate heat is related to its surface area, it was proposed that larger animals evolved lower metabolic rates to avoid overheating. This explanation is not sufficient for several reasons, one being that the relationship between body mass and metabolic rate holds for even very small organisms and for ectotherms, in which overheating is not usually a problem.

Endotherms respond to cold by producing heat

Within its thermoneutral zone, an endotherm can generally maintain a constant body temperature by regulating blood flow to the skin. Outside its thermoneutral zone, however, it must expend energy to regulate its body temperature.

The thermoneutral zone is bounded by a *lower critical temperature* and an *upper critical temperature* (Figure 41.14). When the environmental temperature falls below the lower critical temperature, endotherms must produce heat to compensate for the heat they lose to the environment. Mammals can accomplish this *thermogenesis* in two ways: through shivering and nonshivering heat production. Birds use only shivering heat production.

Shivering uses the contractile machinery of skeletal muscles to consume ATP without causing observable behavior. The muscles pull against each other so that little movement other than a tremor results. The energy from the conversion of ATP to ADP in this process is released as heat. Shivering heat production is perhaps too narrow a term, however; increased muscle tone and increased body movements also contribute to increased heat production in cold environments.

Most nonshivering heat production occurs in specialized adipose tissue called *brown fat* (Figure 41.15). This tissue looks brown because of its abundant mitochondria and rich blood supply. In brown fat cells, a protein called *thermogenin* uncouples proton movement from ATP production, allowing protons to leak across the inner mitochondrial membrane rather than having to pass through the ATP synthase protein

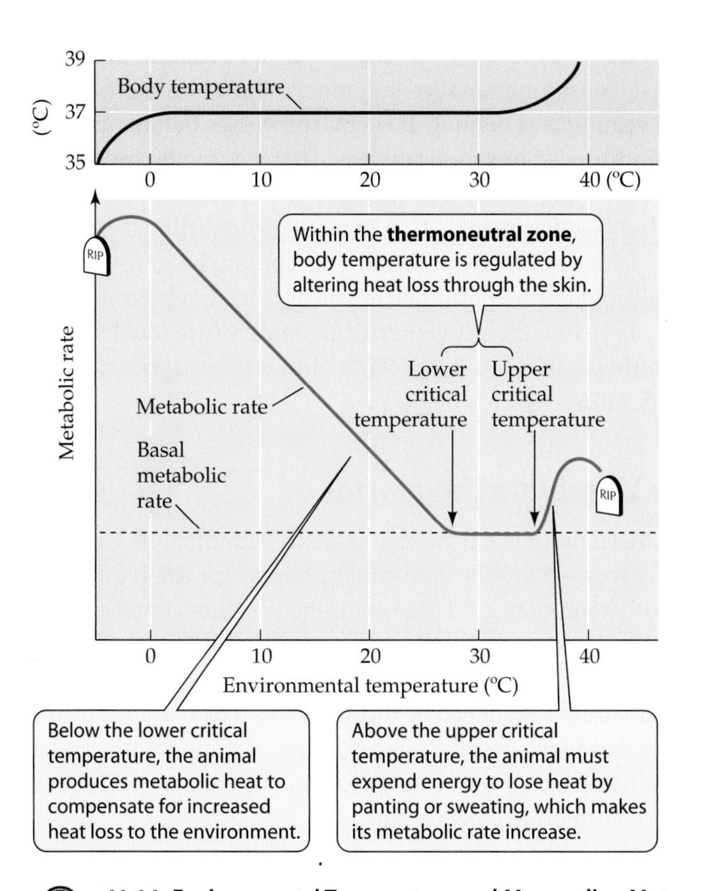
Body temperature
(°C)
39
37
35
0 10 20 30 40 (°C)
Metabolic rate
Within the **thermoneutral zone**, body temperature is regulated by altering heat loss through the skin.
Lower critical temperature
Upper critical temperature
Metabolic rate
Basal metabolic rate
RIP
RIP
0 10 20 30 40
Environmental temperature (°C)
Below the lower critical temperature, the animal produces metabolic heat to compensate for increased heat loss to the environment.
Above the upper critical temperature, the animal must expend energy to lose heat by panting or sweating, which makes its metabolic rate increase.

41.14 Environmental Temperature and Mammalian Metabolic Rates Outside the thermoneutral zone, maintaining a constant body temperature requires the expenditure of energy. Outside extreme limits (0°C and 40°C in this instance), the animal cannot maintain its body temperature and so dies.

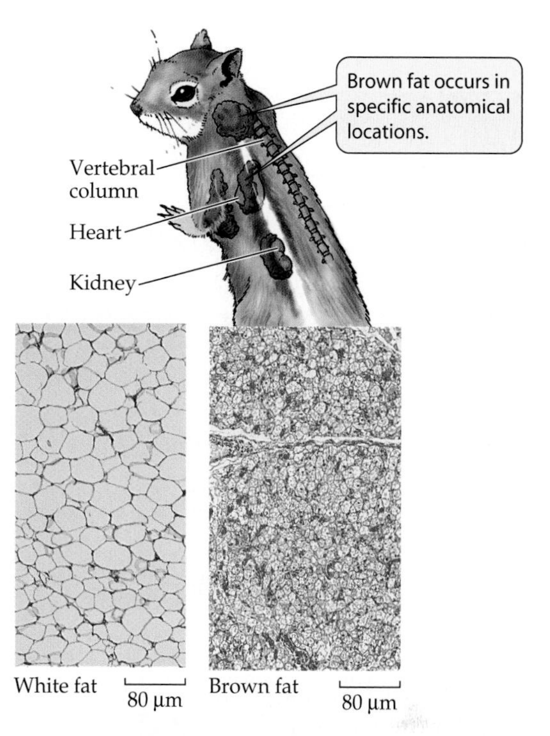
Brown fat occurs in specific anatomical locations.
Vertebral column
Heart
Kidney
White fat
80 μm
Brown fat
80 μm

41.15 Brown Fat In many mammals, specialized brown fat tissue produces heat. When viewed through a microscope at similar magnifications, we see that white fat cells (left) are simple droplets of lipid with few organelles and limited blood supply, while brown fat cells (right) are packed with mitochondria and richly supplied with blood.

and generate ATP (review the discussion of the chemiosmotic mechanism in Chapter 7). As a result, metabolic fuels are consumed without producing ATP, but heat is still released. Brown fat is especially abundant in newborn infants of many mammalian species, including humans, in some adult mammals that are small and acclimatized to cold, and in mammals that hibernate.

Decreasing heat loss is important for life in the cold

The coldest habitats on Earth are in the Arctic, the Antarctic, and at the peaks of high mountains. Many birds and mammals, but almost no reptiles or amphibians, live in these places. What adaptations besides endothermy characterize species that live in the cold?

The most important adaptations of endotherms to cold environments are those that reduce heat loss to the environment. Since most heat is lost from the body surface, many cold-climate species have a smaller surface area than their warm-climate cousins, even when their body masses are the same. Rounder body shapes and shorter appendages reduce the surface area-to-volume ratios of some cold-climate species (Figure 41.16).

Another means of decreasing heat loss is to increase thermal insulation. Animals adapted to cold climates have much thicker layers of fur, feathers, or fat than do their warm-climate relatives. The fur of an arctic fox or a northern sled dog provides such good thermal insulation that those animals don't even begin to shiver until the air temperature drops as low as –20°C to –30°C.

Fur and feathers are good insulators because they trap a layer of still, warm air close to the skin surface. If that air is displaced by water, insulation is drastically reduced. In many species, oil secretions spread through fur or feathers by grooming are critical for resisting wetting and maintaining a high level of insulation.

Decreasing blood flow to the skin is an important thermoregulatory adaptation in the cold. Constriction of blood vessels in the skin, and especially in the appendages, greatly improves the ability of an animal to conserve heat.

Evaporation of water is an effective way to lose heat

When the environmental temperature rises above the upper critical temperature for an endotherm, overheating becomes a problem. For an exercising animal, overheating can become a problem at even low environmental temperatures. Heavily

(*a*) *Otocyon megalotis*

(*b*) *Alopex lagopus*

41.16 Adaptations to Hot and Cold Climates (*a*) The bat-eared fox lives on the dry plains of central and southern Africa. Its large ears serve as heat exchangers, passing heat from the fox's blood to the surrounding air. (*b*) The thick fur of the arctic fox provides insulation in the frigid winter. Its ears and extremities are relatively smaller than those of the desert fox.

insulated arctic species usually have an area on the body surface, such as the abdomen, that has only a thin layer of fur and can act as a window for heat loss. Large mammals, such as elephants, rhinoceroses, and water buffaloes, have little or no fur and seek places where they can wallow in water when the air temperature is too high. Having water in contact with the skin greatly increases heat loss because water has a much greater capacity for absorbing heat than air does.

Evaporation of water from body surfaces can also cool an animal. A gram of water absorbs about 580 calories of heat when it evaporates. Water is heavy, however, so animals do not carry an excess supply of it. Furthermore, hot environments tend to be arid places where water is a scarce resource. Therefore, sweating or panting are usually cooling methods of last resort for animals adapted to hot environments.

Sweating and panting are active processes that require the expenditure of metabolic energy. That is why the metabolic rate increases when the upper critical temperature is exceeded (see Figure 41.14). A sweating or panting animal is producing heat in the process of dissipating heat, which can be a losing battle. Endotherms can survive in environments that are below their lower critical temperature much better than they can in environments above their upper critical temperature.

The Vertebrate Thermostat

The thermoregulatory mechanisms and adaptations we have just discussed are the controlled systems for the regulation of body temperature. These controlled systems must receive commands from a regulatory system that integrates information relevant to the regulation of body temperature. Such a regulatory system can be thought of as a *thermostat*. All animals that thermoregulate, both vertebrate and invertebrate, must have such a regulatory system, but here we will focus on the vertebrate thermostat.

Where is the vertebrate thermostat? Its major integrative center is at the bottom of the brain in a structure called the **hypothalamus**. If you slide your tongue back as far as possible along the roof of your mouth, it will be just a few centimeters below your hypothalamus. The hypothalamus is a part of many regulatory systems, so we will refer to it again in the chapters to come. If the hypothalamus of a mammal's brain is damaged, the animal loses its ability to regulate its body temperature, which then rises in warm environments and falls in cold ones.

The vertebrate thermostat uses feedback information

In many species, the temperature of the hypothalamus itself is the major source of feedback information to the thermostat. Cooling the hypothalamus causes fish and reptiles to seek a warmer environment, and warming the hypothalamus causes them to seek a cooler environment. In mammals, cooling the hypothalamus can stimulate constriction of the blood vessels supplying the skin and increase metabolic heat production. Because it activates these thermoregulatory responses, cooling the hypothalamus causes the body temperature to rise. Conversely, warming the hypothalamus stimulates dilation of the blood vessels supplying the skin and sweating or panting, and the overall body temperature falls (Figure 41.17).

The hypothalamus appears to generate a set point like a setting on the thermostat of a house. When the temperature of the hypothalamus exceeds or drops below that set point,

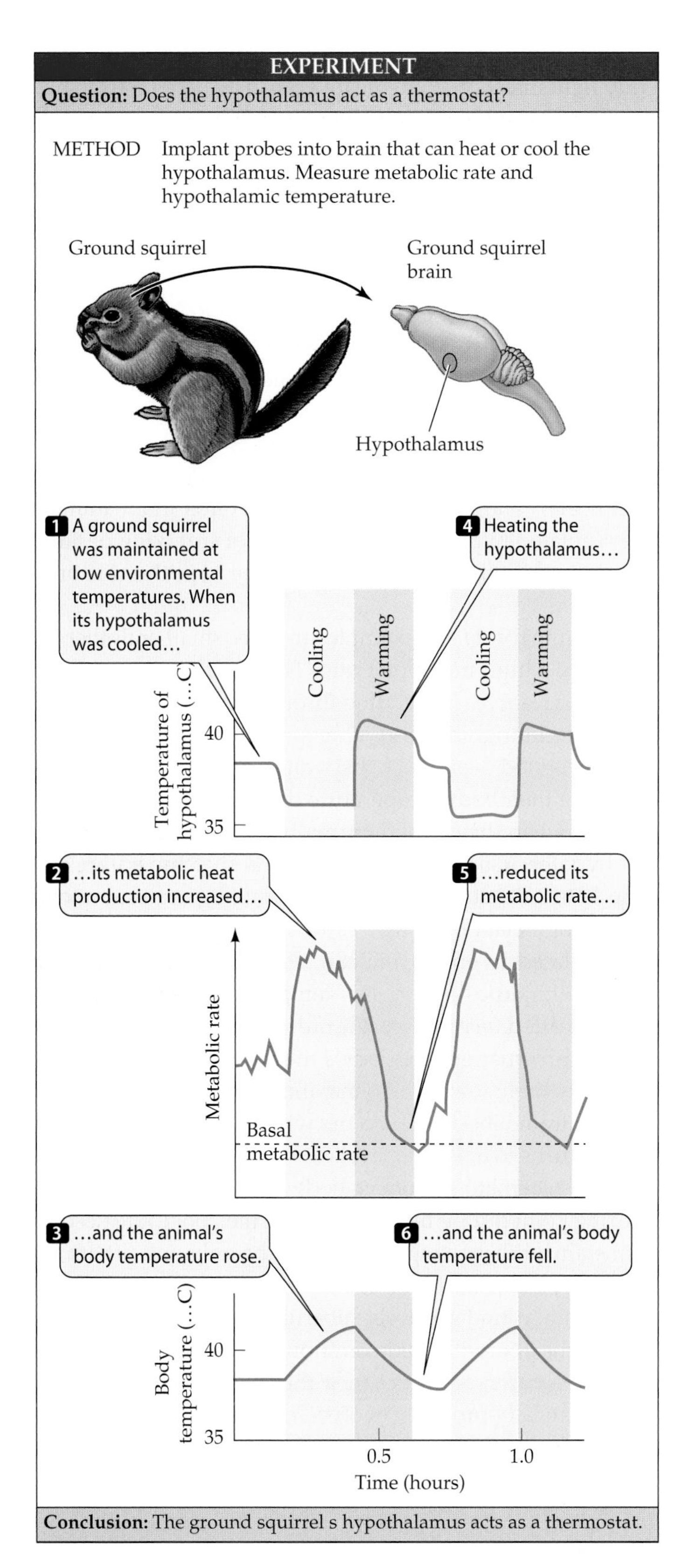
EXPERIMENT
Question: Does the hypothalamus act as a thermostat?
METHOD Implant probes into brain that can heat or cool the hypothalamus. Measure metabolic rate and hypothalamic temperature.
Ground squirrel
Ground squirrel brain
Hypothalamus
1 A ground squirrel was maintained at low environmental temperatures. When its hypothalamus was cooled...
4 Heating the hypothalamus...
Cooling
Warming
Cooling
Warming
Temperature of hypothalamus (...C)
40
35
2 ...its metabolic heat production increased...
5 ...reduced its metabolic rate...
Metabolic rate
Basal metabolic rate
3 ...and the animal's body temperature rose.
6 ...and the animal's body temperature fell.
Body temperature (...C)
40
35
0.5
1.0
Time (hours)
Conclusion: The ground squirrel s hypothalamus acts as a thermostat.

41.17 The Hypothalamus Regulates Body Temperature The observation that damage to the hypothalamus disrupts thermoregulation led to the finding that the hypothalamus acts as a thermostat in the vertebrate body.

thermoregulatory responses (the controlled system) are activated to reverse the direction of temperature change. Hence, hypothalamic temperature is a negative feedback signal.

Experimental warming and cooling of the hypothalamus show that mammals have separate set points for activating different thermoregulatory responses. If the hypothalamus of a mammal is cooled, the vessels supplying blood to the skin constrict at a specific hypothalamic temperature. A slightly lower hypothalamic temperature initiates shivering. If the hypothalamic temperature is then raised, shivering ceases; then blood vessels supplying the skin dilate. At still higher hypothalamic temperatures, panting starts.

We can describe the characteristics of hypothalamic control of each thermoregulatory response. For example, if we measure an animal's metabolic heat production while warming and cooling its hypothalamus, we can describe the results graphically (Figure 41.18). Within a certain range of hypothalamic temperatures, metabolic heat production remains low and constant, but cooling the hypothalamus below a set point stimulates increased metabolic heat production. The increase in heat production is proportional to how much the hypothalamus is cooled below the set point. This regulatory system is much more sophisticated than a simple on–off thermostat like the one in a house.

The vertebrate thermoregulatory system integrates other sources of information in addition to hypothalamic temperature. It uses information about the temperature of the environment as registered by temperature sensors in the skin. Changes in environmental temperature shift the hypothalamic set points for thermoregulatory responses. The set point for the metabolic heat production response is higher when skin is cold and lower when skin is warm (see Figure 41.18).

The temperature of the skin can be considered feedforward information that adjusts the hypothalamic set point. Many other factors also shift hypothalamic set points for thermoregulatory responses. Set points are higher during wakefulness than during sleep, and they are higher during the active part of the daily cycle than during the inactive part, even if the animal is awake at both times.

Fevers help the body fight infections

A **fever** is a rise in body temperature in response to substances called *pyrogens*. Exogenous pyrogens come from foreign substances such as bacteria or viruses that invade the body. Endogenous pyrogens are produced by cells of the immune system when they are challenged. Pyrogens are the reason that we respond to many infectious diseases by getting a fever. Growing evidence suggests that fever is an adaptive response that helps the body fight pathogens.

The presence of a pyrogen in the body causes a rise in the hypothalamic set point for the metabolic heat production re-

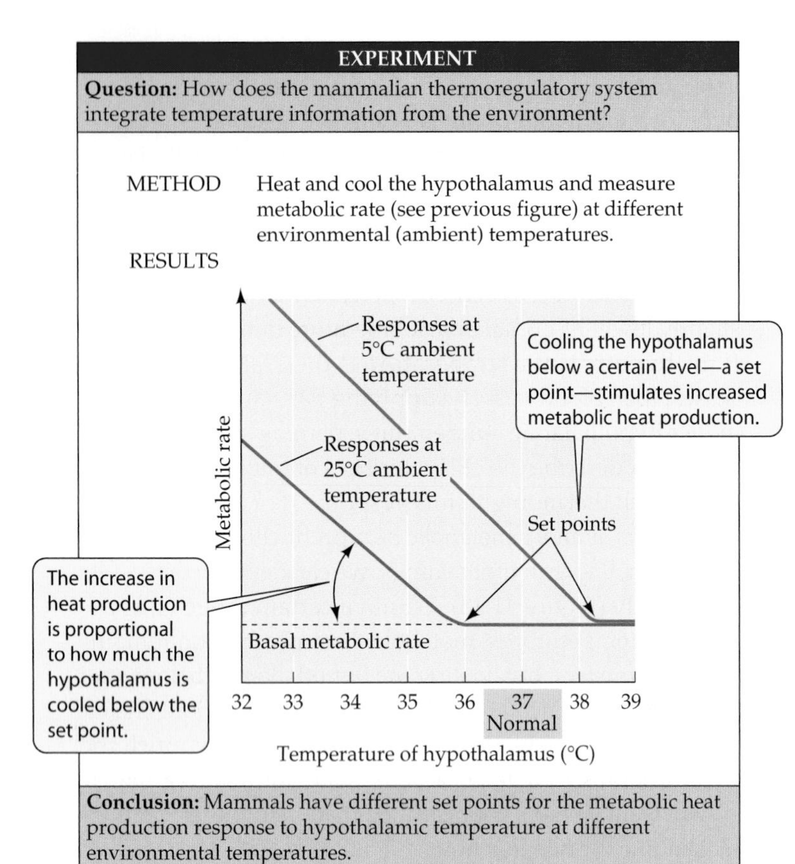
EXPERIMENT
Question: How does the mammalian thermoregulatory system integrate temperature information from the environment?
METHOD
Heat and cool the hypothalamus and measure metabolic rate (see previous figure) at different environmental (ambient) temperatures.
RESULTS
Responses at 5°C ambient temperature
Cooling the hypothalamus below a certain level—a set point—stimulates increased metabolic heat production.
Responses at 25°C ambient temperature
Set points
Metabolic rate
The increase in heat production is proportional to how much the hypothalamus is cooled below the set point.
Basal metabolic rate
32
33
34
35
36
37
38
39
Normal
Temperature of hypothalamus (°C)
Conclusion: Mammals have different set points for the metabolic heat production response to hypothalamic temperature at different environmental temperatures.

41.18 Adjustable Set Points Mammals have different set points for the metabolic heat production response to hypothalamic temperature at different environmental temperatures. Other factors, such as being asleep or awake, the time of day, or the presence of a fever, can also affect the set point.

sponse. As a result, you shiver, put on a sweater, or crawl under a blanket, and your body temperature rises until it matches the new set point. At the higher body temperature you no longer feel cold, and you may not feel hot, but someone touching your forehead will say that you are "burning up." If you take an aspirin, it lowers your set point to normal. Now you feel hot, take off clothes, and even sweat until your elevated body temperature returns to normal.

Why do we take aspirin when we have a fever? The pyrogens entering the body are attacked by cells of the immune system called *macrophages* (see Chapter 18). One of the things the macrophages do is to release chemicals called *interleukins*, which sound the alarm to other cells of the immune system throughout the body and trigger responses that contribute to feeling crummy. The interleukins also raise the hypothalamic set point for metabolic heat production. Among the intracellular signals triggered by interleukins are *prostaglandins*. Aspirin is a potent inhibitor of prostaglandin synthesis, thus explaining how this "miracle drug" reduces fever and makes us feel better.

Clinical evidence suggests that moderate fevers help the body fight infection. Extreme fevers can be dangerous to humans and must be reduced. Even more modest fevers can be dangerous to people who have weakened hearts or those who are seriously ill. A fetus can be endangered when a pregnant woman has a fever. Fever-reducing drugs may be used in such cases.

Turning down the thermostat saves energy

Hypothermia is the condition in which body temperature is below normal. It can result from a natural turning down of the thermostat or from traumatic events such as starvation (lack of metabolic fuel), exposure to cold, serious illness, or treatment by anesthesia. Many species of birds and mammals use regulated hypothermia as a means of surviving periods of cold and food scarcity. Some become hypothermic on a daily basis.

Hummingbirds, for example, are very small endotherms and have a high metabolic rate. They could exhaust their metabolic reserves just getting through a single day without food. Hummingbirds and other small endotherms can extend the period over which they can survive without food by dropping their body temperature during the portion of day or night when they would normally be inactive. This adaptive hypothermia is called *daily torpor*. Body temperature can drop 10 to 20°C during daily torpor, resulting in an enormous savings of metabolic energy.

Regulated hypothermia can also last for days or even weeks, with drops to very low temperatures. This phenomenon is called *hibernation* (Figure 41.19). During the deep sleep of hibernation, the body's thermostat is turned down to an extremely low level to maximize energy conservation. Arousal from hibernation occurs when the hypothalamic set point returns to a normal level.

Many hibernators maintain body temperatures close to the freezing point during hibernation. The metabolic rate needed to sustain a hibernating animal may be only one-fiftieth its basal metabolic rate. Many species of mammals, such as bats, bears, and ground squirrels, hibernate, but only one species of bird (the poorwill) has been shown to hibernate. The ability of hibernators to reduce their thermoregulatory set point so dramatically probably evolved as an extension of the set point decrease that accompanies sleep even in nonhibernating species of mammals and birds.

Chapter Summary

Homeostasis: Maintaining the Internal Environment

▶ The internal environment consists of extracellular fluid. Organs and organ systems are specialized to keep certain aspects of the internal environment in a constant state. **Review Figure 41.1**

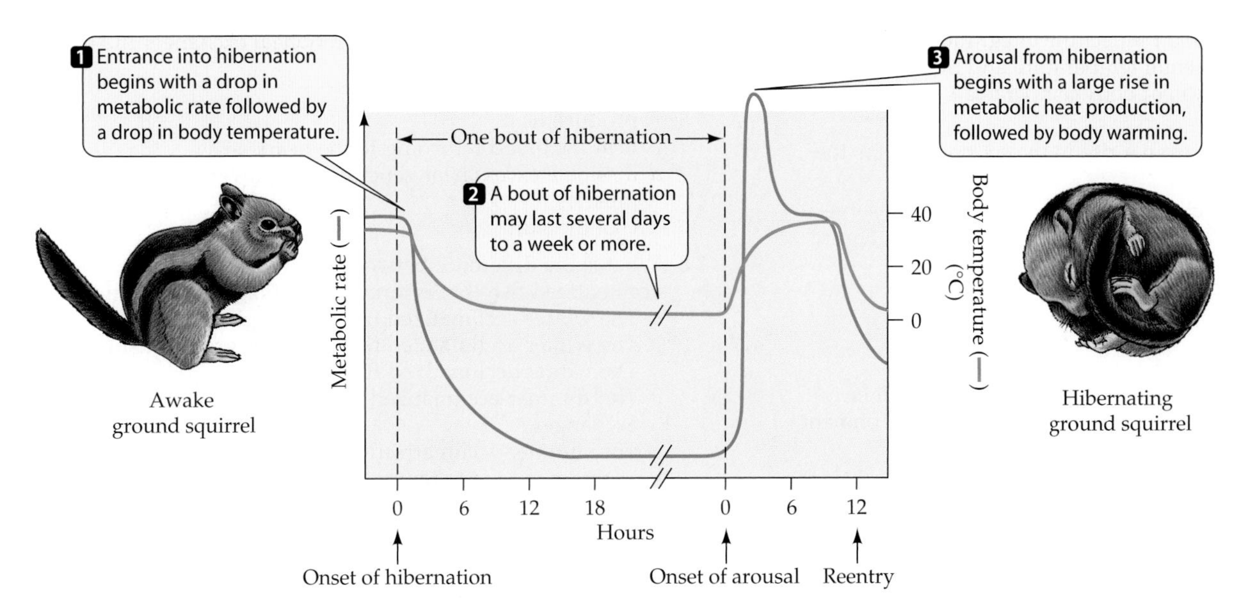

41.19 A Ground Squirrel Enters Repeated Bouts of Hibernation during Winter At the beginning of each bout of hibernation, the ground squirrel's metabolic rate and body temperature fall. Its body temperature may come into equilibrium with the temperature of its nest and stay at that level for days. The bout is ended by a rise in metabolic heat production that returns body temperature to a normal level.

▶ Homeostasis is the maintenance of a constant internal environment.

Tissues, Organs, and Organ Systems

▶ Cells that have a similar structure and function make up a tissue. There are four general types of tissues: epithelial, connective, muscle, and nervous. **Review Figure 41.2**

▶ Organs consist of multiple tissue types and make up organ systems. **Review Table 41.1. See Web/CD Activity 41.1**

Physiological Regulation and Homeostasis

▶ Homeostasis depends on the ability to control and regulate the functions of organs and organ systems.

▶ Regulatory systems have set points and respond to feedback information. Negative feedback corrects deviations from the set point, positive feedback amplifies responses, and feedforward information changes the set point. **Review Figure 41.4. See Web/CD Tutorial 41.1**

Temperature and Life

▶ Living systems require a range of temperatures between the freezing point of water and the temperatures that denature proteins.

▶ Most biological processes and reactions are temperature-sensitive. Q_{10} is a measure of temperature sensitivity. **Review Figure 41.5**

▶ Animals that cannot avoid seasonal changes in body temperature have biochemical adaptations that compensate for those changes. These adaptations enable animals to acclimatize to seasonal changes. **Review Figure 41.6**

Maintaining Optimal Body Temperature

▶ Homeotherms maintain a fairly constant body temperature most of the time; poikilotherms do not.

▶ Endotherms produce metabolic heat to elevate body temperature; ectotherms depend mostly on environmental sources of heat. **Review Figure 41.7**

▶ Ectotherms and endotherms can regulate body temperature through behavior. **Review Figure 41.8**

▶ Heat exchange between a body and the environment occurs via radiation, conduction, convection, and evaporation. **Review Figure 41.10**

▶ Ectotherms and endotherms can control heat exchange with the environment by altering blood flow to the skin. **Review Figure 41.11**

▶ Some ectotherms can produce metabolic heat to raise their body temperatures.

▶ Some fishes have circulatory systems that function as countercurrent heat exchangers to conserve heat produced by muscle metabolism. **Review Figure 41.12**

Thermoregulation in Endotherms

▶ Endotherms have high basal metabolic rates. Over a range of environmental temperatures called the thermoneutral zone, the metabolic rate of resting endotherms remains at basal levels. The basal metabolic rate per gram of tissue decreases as endotherms get bigger. **Review Figures 41.13, 41.14. See Web/CD Activity 41.2**

▶ When the environmental temperature falls below the lower critical temperature, endotherms maintain their body temperatures through shivering and nonshivering metabolic heat production. **Review Figure 41.15**

▶ Endotherms that live in cold climates have adaptations that minimize heat loss, including a reduced surface area-to-volume ratio and increased insulation.

▶ When the environmental temperature rises above an upper critical temperature, metabolic rate increases as a consequence of active evaporative water loss through sweating or panting.

The Vertebrate Thermostat

▶ The vertebrate thermostat is located in the hypothalamus. It has set points for activating thermoregulatory responses.

▶ In mammals, cooling the hypothalamus induces the constriction of blood vessels and increased metabolic heat production. Heating the hypothalamus induces the dilation of blood vessels and panting. **Review Figure 41.17**

▶ Changes in the set point reflect the integration of information, such as environmental temperature and time of day, that is relevant to the regulation of body temperature. **Review Figure 41.18**

▶ Fever, which results from a rise in the set point, helps the body fight infections.

▶ Adaptations in which set points are reduced to conserve energy include daily torpor and hibernation. **Review Figure 41.19**

Self-Quiz

1. If the Q_{10} of the metabolic rate of an animal is 2, then
   a. the animal is better acclimatized to a cold environment than if its Q_{10} is 3.
   b. the animal is an ectotherm.
   c. the animal consumes half as much oxygen per hour at 20°C as it does at 30°C.
   d. the animal's metabolic rate is not at basal levels.
   e. the animal produces twice as much heat at 20°C as it does at 30°C.
2. Which statement about brown fat is true?
   a. It produces heat without producing ATP.
   b. It insulates animals acclimatized to cold.
   c. It is a major source of heat production for birds.
   d. It is found only in hibernators.
   e. It provides fuel for muscle cells.
3. Which of the following is the most important and most general characteristic of endotherms adapted to cold climates in comparison to those adapted to warm climates?
   a. Higher basal metabolic rates
   b. Higher Q_{10} values
   c. Brown fat
   d. Greater insulation
   e. Ability to hibernate
4. Which of the following would cause a decrease in the hypothalamic temperature set point for metabolic heat production?
   a. Entering a cold environment
   b. Taking an aspirin when you have a fever
   c. Arousing from hibernation
   d. Getting an infection that causes a fever
   e. Cooling the hypothalamus
5. Mammalian hibernation
   a. occurs when animals run out of metabolic fuel.
   b. is a regulated decrease in body temperature.
   c. is less common than hibernation in birds.
   d. can occur at any time of year.
   e. lasts for several months, during which body temperature remains close to environmental temperature.
6. Which of the following is an important difference between an ectotherm and an endotherm of similar body size?
   a. The ectotherm has higher Q_{10} values.
   b. Only the ectotherm uses behavioral thermoregulation.
   c. Only the endotherm can constrict and dilate the blood vessels to the skin to alter heat flow.
   d. Only the endotherm can have a fever.
   e. At a body temperature of 37°C, the ectotherm has a lower metabolic rate than the endotherm.
7. The function of the countercurrent heat exchanger in "hot" fish is to
   a. trap heat in the muscles.
   b. produce heat.
   c. heat the blood returning to the heart.
   d. dissipate excess heat generated by powerful swimming muscles.
   e. cool the skin.
8. What is the difference between a winter- and a summer-acclimatized fish that is termed metabolic compensation?
   a. The winter-acclimatized fish has a higher Q_{10}.
   b. The winter-acclimatized fish develops greater insulation.
   c. The winter-acclimatized fish hibernates.
   d. The summer-acclimatized fish has a countercurrent heat exchanger.
   e. The summer-acclimatized fish has a lower metabolic rate at any given water temperature.
9. Which of the following is most characteristic of epithelial cells?
   a. They generate electrochemical signals.
   b. They contract.
   c. They have an extensive extracellular matrix.
   d. They have secretory functions.
   e. They cover the surface of the body and line the body cavities.
10. Negative feedback
   a. works in opposition to positive feedback to achieve homeostasis.
   b. always turns off a process.
   c. reduces an error signal in a regulatory system.
   d. is responsible for metabolic compensation.
   e. is a feature of the thermoregulatory systems of endotherms, but not of ectotherms.

For Discussion

1. In some sheets of epithelial tissue, the cells are joined together with dense membrane proteins that form "tight junctions," which are extremely impermeable (see Chapter 5). In other epithelial sheets, the cells are joined by filamentous extracellular proteins that are strong, but not as impermeable. Where in the body might you expect to find tight junctions? Where might you expect to find epithelial sheets with the leakier connections?
2. If the major adaptation of endotherms to cold climates is their insulation, how would you compare the cold adaptations of a polar bear and a seal?
3. Why is an environment above its upper critical temperature more dangerous to an endotherm than an environment below its lower critical temperature?
4. If the hypothalamic temperature of a mammal is the feedback information for its thermostat, why does the hypothalamic temperature scarcely change when that animal moves between environments hot enough and cold enough to stimulate the animal to pant and to shiver, respectively?
5. There are many places on Earth where cold is a severe challenge to life. Therefore, if global warming occurs, wouldn't it expand the habitats available to many species? What arguments can be advanced to support or rebut this simple proposition?

42 Animal Hormones

In shallow pools around the edge of Lake Tanganyika in east central Africa, brightly colored male cichlid fish stake out territories and vigorously defend them from neighboring males. These dominant males constantly patrol their territories and display their colorful sexual adornments to females, who assemble in groups at the edge of the cichlid colony. The females are hard to see because they are inactive and protectively colored. When a female is impressed by a male's territory and display, she enters his territory and lays her eggs in a spawning pit that the male has prepared.

At any one time, only about 10 percent of the males in the colony are displaying and holding territories. All the other males are small, nondescript, and nonaggressive. Because they look and act like females, they are allowed to feed in the dominant males' territories. If a dominant male is removed by a predator, however, the nondescript males fight over the vacated territory. The winner rapidly assumes the appearance and behavior of a dominant male: brightly colored, big, aggressive, and attractive to females.

What accounts for this dramatic change? Russell Fernald and his students at Stanford University have shown that soon after the nondescript male's victory, certain cells in his brain enlarge and secrete a chemical message. This message triggers cells in the pituitary gland, which is outside of the brain, to secrete chemical messages in turn. Secreted in tiny quantities, these molecules enter the blood and are transported around the body. The responses of cells to these chemical messages produce the characteristics of a dominant male.

This change in the male cichlid is just one example of how chemical messages, or hormones, produce and coordinate anatomical, physiological, and behavioral changes in an animal. We will explore many other examples in this chapter. We will look first at the characteristics of hormones and their receptors, and then examine some of their roles in the control of invertebrate life cycles. Most of this chapter is devoted to vertebrate hormones, examining their functions, control, and molecular mechanisms of action. We will pay particular attention to the extensive interactions between the neural and hormonal information systems. In the process, we will discuss several human diseases involving hormonal dysfunction.

Dominant and Nondescript Male Cichlids
A dominant male cichlid (*Haplochromis burtoni*) displays bright colors that attract females. A non-dominant male (above) is nondescript in comparison.

Hormones and Their Actions

In Chapter 41, we learned that control and regulation require information. In multicellular animals, most of this information is transmitted as electric signals and as chemical signals. The electric signals are nerve impulses, which will be a major focus of later chapters on the nervous system. The chemical signals are **hormones**, which are secreted by cells, diffuse locally in the extracellular fluid, and are picked up by the blood, which distributes them throughout the body (Figure 42.1*a*).

Cells that secrete hormones are called **endocrine cells**. **Target cells** receive the hormonal message if they have appropriate receptors to bind the hormone. The binding of a hormone to its receptor activates mechanisms within the target cell that eventually lead to a response. The response can be developmental, physiological, or behavioral.

The secretion, diffusion, and circulation of hormones is much slower than the transmission of nerve impulses. Therefore, hormones are not useful for controlling rapid actions, such as cichlid fighting behavior. Hormones are most useful for coordinating longer-term developmental or physiological processes, such as the transformation of a nondescript, passive cichlid into a dominant, territorial male.

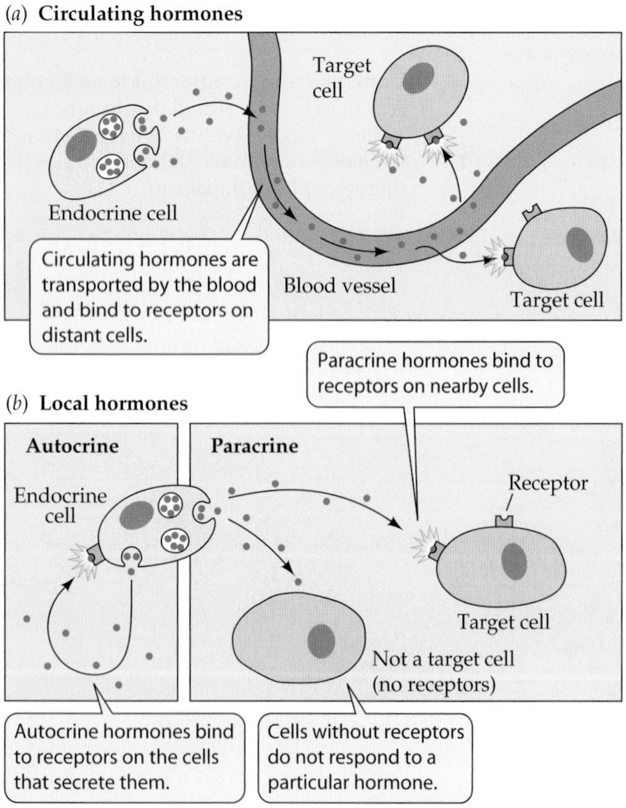

42.1 Chemical Signaling Systems (*a*) Most hormones are distributed throughout the body by the circulatory system. (*b*) An autocrine hormone influences the cell that releases it; a paracrine hormone influences nearby cells.

Hormones can be divided into three chemical groups

There is enormous diversity in the chemical structure of hormones, but most of them can be divided into three groups:

- Peptides or proteins (e.g., insulin)
- Steroid hormones (derivatives of the steroid cholesterol; e.g., testosterone)
- Amines (derivatives of the amino acid tyrosine; e.g., thyroxine)

Most hormones are peptides or proteins. They are water-soluble and therefore easily transported in the blood, but they cannot easily pass through lipid-rich cell membranes. Therefore, peptide and protein hormones are packaged in vesicles in the cells that make them and released by exocytosis.

Steroid hormones are lipid-soluble and can easily dissolve in and pass through cell membranes. Therefore, steroid hormones are not packaged in vesicles; instead, they simply diffuse out of the cells that make them as they are synthesized. Since steroid hormones are not soluble in the blood, they are transported in the blood bound to carrier proteins.

Some amine hormones are water-soluble and others are lipid-soluble; thus, their mode of release differs accordingly.

Hormone receptors are found on the cell surface or in the cell interior

The chemical structure of hormones is related to the location of their receptors. Lipid-soluble hormones can diffuse through plasma membranes, and therefore their receptors are inside the cell, either in the cytoplasm or in the nucleus. In most cases, the complex formed by the lipid-soluble hormone and its receptor acts by altering gene expression in the cell (see Figure 15.8).

Since water-soluble hormones cannot readily pass through plasma membranes, their receptors are on the cell surface. These receptors are large glycoprotein complexes with three domains: a binding domain projecting outside the plasma membrane, a transmembrane domain that anchors the receptor in the membrane, and a cytoplasmic domain that extends into the cytoplasm of the cell. The cytoplasmic domain initiates the target cell's response by activating protein kinases or protein phosphatases (see Figures 15.6 and 15.7).

Some hormones act locally

Hormones can also be classified according to the distance over which their messages are transmitted. Some hormones act only on target cells close to their sites of release; others act on target cells at distant locations in the body. Some chemical messages, called *pheromones*, are secreted to the environment and exert their effects on other individuals. We'll return to pheromones in Chapter 52.

When a hormone binds to receptors on the same cell that is releasing it, the hormone acts as an **autocrine** message (Figure 42.1*b*). Growth factors are examples of **paracrine** messages that can also act as autocrine messages for the purpose of negative feedback. The autocrine response prevents the secretory cell from secreting too much of the hormone.

When the primary effects of a hormone are on cells near its site of release, the hormone is said to have paracrine function (Figure 42.1*b*). Paracrine hormones are released in such tiny quantities, or are so rapidly inactivated by degradative enzymes, or are taken up so efficiently by local cells that they never diffuse into the blood in sufficient amounts to act on distant target cells. An example of a paracrine hormone is histamine, one of the mediators of inflammation (see Figure 18.4).

A major class of paracrine hormones consists of the various **growth factors**, which stimulate the growth and differentiation of cells. Growth factors were first discovered when scientists attempted to culture cells outside the body. Even when given all sorts of nutrients and optimal conditions, the cells did not grow well unless blood plasma or a tissue extract was added to the medium. The components necessary for growth were found to be specific molecules that were present in very small quantities. At present, about 50 specific growth factors are known, along with a complex group of receptors. We discussed growth factors briefly when we discussed the cell cycle in Chapter 9.

Nerve cells, or *neurons*, can also be considered paracrine cells. As we will see in Chapter 44, a neuron communicates with another cell by means of a chemical message called a *neurotransmitter*, which travels over a very small distance to the target cell.

Most hormones are distributed in the blood

Most hormones diffuse through the extracellular fluid and are picked up by the blood, which distributes them throughout the body. Wherever such a hormone encounters a cell with a receptor to which it can bind, it triggers a response. The nature of the response depends on the responding cell. The same hormone can cause different responses in different types of cells.

Consider the hormone epinephrine. If you are walking on a lonely street and a mugger suddenly appears from behind a tree, you jump, your heart starts to thump, and a whole set of protective actions are set in motion. The jump and the initial heart thumping are driven by your nervous system, which reacts very quickly. Simultaneously, your nervous system stimulates endocrine cells just above your kidneys to secrete epinephrine. Within seconds, epinephrine is diffusing into your blood and circulating around your body to activate the many components of the *fight-or-flight response*.

Epinephrine binds to receptors in the heart, blood vessels, liver, and fat cells. Epinephrine sustains the higher heart rate and causes the heart to beat more strongly. Although your heart is now pumping more blood, that blood is needed by your muscles to fuel your escape. Epinephrine causes more of your circulating blood to flow to the muscles by causing blood vessels in your digestive tract to constrict (digestion can wait!). Similarly, it decreases blood flow to the skin and to the kidneys. In the liver, epinephrine stimulates the breakdown of glycogen into glucose for a quick energy supply. In fatty tissue, it stimulates the breakdown of fats to yield another source of energy to the blood. These are just some of the many actions triggered by one hormone. They all contribute to increasing your chances of escaping that dangerous situation.

Endocrine glands secrete hormones

Some endocrine cells exist as single cells within a tissue. Many hormones of the digestive tract, for example, are secreted by isolated endocrine cells in the wall of the stomach and small intestine.

Many hormones, however, are secreted by aggregations of endocrine cells that form secretory organs called **endocrine glands**. The name "endocrine" reflects the fact that these glands do not have ducts that lead to the outside of the body; rather, they secrete their products directly into the extracellular fluid. Vertebrates have nine major endocrine glands, which collectively make up the **endocrine system** (Figure 42.2). A single endocrine gland may secrete several different hormones.

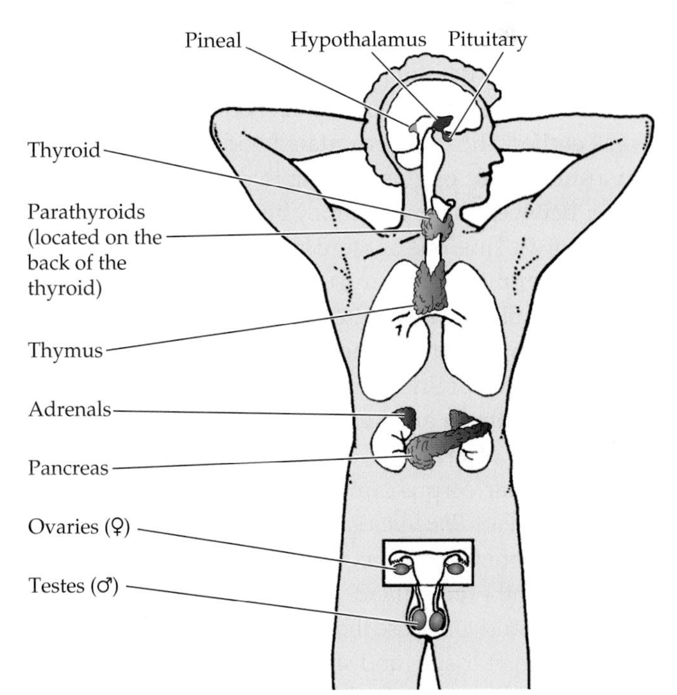

42.2 The Endocrine System of Humans There are nine major glands in the human endocrine system, but hormones are also secreted by endocrine cells that are not part of discrete glands.

Hormonal Control of Molting and Development in Insects

In this chapter we cannot do justice to the diversity of the hormones of the invertebrates, but we'll discuss two important aspects of the lives of many invertebrates that are controlled by hormonal mechanisms: molting and metamorphosis.

Hormones from the head control molting in insects

Because insects have rigid exoskeletons, their growth is episodic, punctuated with *molts* (shedding of the exoskeleton) (see Chapter 33). Each growth stage between two molts is called an *instar*.

The British physiologist Sir Vincent Wigglesworth was a pioneer in the study of the hormonal control of growth and development in insects. Wigglesworth conducted experiments on the blood-sucking bug *Rhodnius*, which undergoes incomplete metamorphosis. Upon hatching, *Rhodnius* looks like a miniature version of an adult, but it lacks some adult features. *Rhodnius* molts five times before developing into a mature adult; a blood meal triggers each episode of molting and growth.

Rhodnius is a hardy experimental animal; it can live a long time even after it is decapitated. If decapitated about an hour after it has had a blood meal, *Rhodnius* may live for up to a year, but it does not molt. If decapitated a week after its blood meal, it does molt (Figure 42.3, experiment 1). These observations led Wigglesworth to the hypothesis that something diffusing slowly from the head controls molting.

Wigglesworth tested his hypothesis with a clever experiment in which he decapitated two *Rhodnius:* one that had just had its blood meal and another that had had its blood meal a week earlier. The two decapitated bodies were connected with a short piece of glass tubing that allowed body fluid transfer between them—and they both molted (Figure 42.3, experiment 2). Thus one or more substances from the bug fed a week earlier must have crossed through the glass tube and stimulated molting in the other bug.

We now know that two hormones working in sequence regulate molting: brain hormone and ecdysone. Cells in the brain produce **brain hormone**, which is transported to and stored in a pair of structures attached to the brain, the *corpora cardiaca* (singular, corpus cardiacum). After appropriate stimulation (which for *Rhodnius* is a blood meal), the corpora cardiaca release brain hormone, which diffuses to an endocrine gland, the prothoracic gland. Brain hormone stimulates the prothoracic gland to release the hormone **ecdysone**. Ecdysone diffuses to target tissues and stimulates molting.

The control of molting by brain hormone and ecdysone is a general mechanism in insects. The nervous system receives various types of information relevant in determining the optimal timing for growth and development (such as the pres-

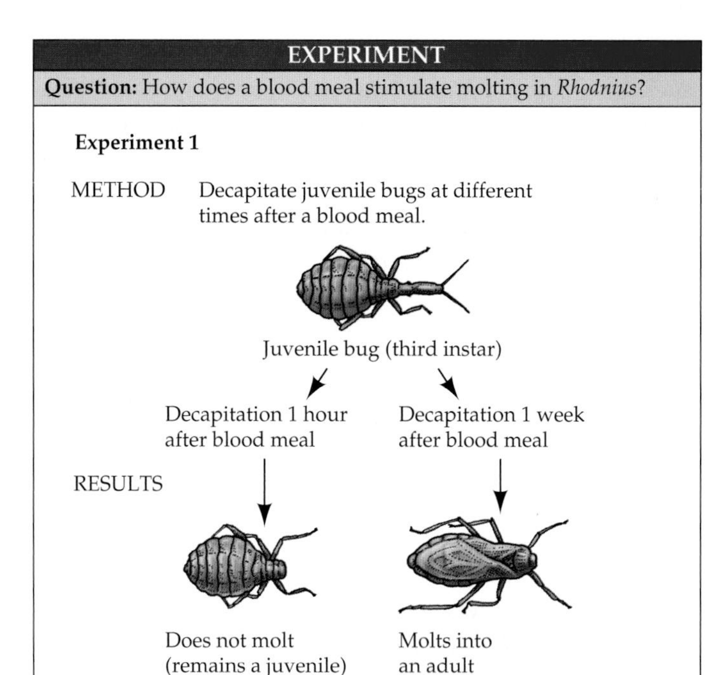
EXPERIMENT
Question: How does a blood meal stimulate molting in *Rhodnius*?
Experiment 1
METHOD Decapitate juvenile bugs at different times after a blood meal.
Juvenile bug (third instar)
Decapitation 1 hour after blood meal
Decapitation 1 week after blood meal
RESULTS
Does not molt (remains a juvenile)
Molts into an adult

Conclusion: Whether a decapitated *Rhodnius* will molt depends on the interval between a blood meal and the decapitation, which supports the idea that a substance must diffuse from head to body.

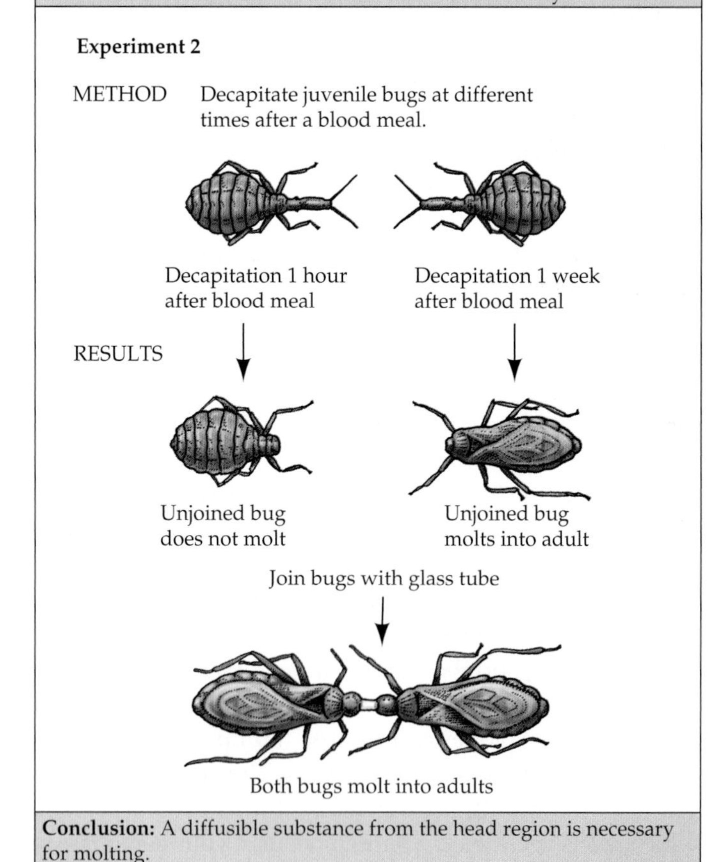
Experiment 2
METHOD Decapitate juvenile bugs at different times after a blood meal.
Decapitation 1 hour after blood meal
Decapitation 1 week after blood meal
RESULTS
Unjoined bug does not molt
Unjoined bug molts into adult
Join bugs with glass tube
Both bugs molt into adults
Conclusion: A diffusible substance from the head region is necessary for molting.

42.3 A Diffusible Substance Triggers Molting The effect of time since the last blood meal on *Rhodnius* molting (experiment 1) led Sir Vincent Wigglesworth to hypothesize that some substance diffusing slowly through the insect's body stimulated molting. Further experiments (experiment 2) showed that molting is indeed controlled by a substance—a hormone—diffusing from the head.

ence of a blood meal in the digestive system). It makes sense, therefore, that the nervous system (the brain) should control the endocrine gland (the prothoracic gland) that produces the hormone that orchestrates all the physiological processes involved in development and molting (ecdysone). Later in this chapter we will see similar links between the nervous system and endocrine glands in vertebrates.

Juvenile hormone controls development in insects

The *Rhodnius* decapitation experiments yielded a curious result: Regardless of the instar used, the decapitated bug always molted directly into an adult form. Additional experiments by Wigglesworth demonstrated that a hormone other than those responsible for molting determines whether a bug molts into another juvenile instar or into an adult.

Because the head of *Rhodnius* is long, it was possible to remove just the front part of the head, which contains the brain and the corpora cardiaca, while leaving the rear part intact. That rear part contains two endocrine structures called the *corpora allata* (singular, corpus allatum). When fourth-instar bugs that had been fed a week earlier were partly decapitated, leaving the corpora allata intact, they molted into fifth instars, not into adults.

This experiment was followed up by more experiments using glass tubes to connect individual bugs. When an unfed, completely decapitated, fifth-instar bug was connected to a fourth-instar bug that had been fed and had had only the front part of its head removed, both bugs molted into juvenile forms. A substance coming from the rear part of the head of the fourth-instar bug prevented the expected result that both bugs would molt into adult forms.

We now know that the substance responsible is **juvenile hormone** and that it comes from the corpora allata. As long as juvenile hormone is present, *Rhodnius* molts into another juvenile instar. The corpora allata normally stop producing juvenile hormone during the fifth instar. When the bug stops producing juvenile hormone, it molts into an adult.

The control of development by juvenile hormone is more complex in insects that, like butterflies, undergo complete metamorphosis. These animals undergo dramatic developmental changes between instars. The fertilized egg hatches into a *larva*, which feeds and molts several times, becoming bigger and bigger. Then it enters an inactive stage called a *pupa*. It undergoes major body reorganization as a pupa, and finally emerges as an adult.

An excellent example of complete metamorphosis is provided by the silkworm moth, *Hyalophora cecropia* (Figure 42.4). As long as juvenile hormone is present in high concentrations, larvae molt into larvae. When the level of juvenile hormone falls, larvae molt into pupae. Because no juvenile hormone is produced in pupae, they molt into adults.

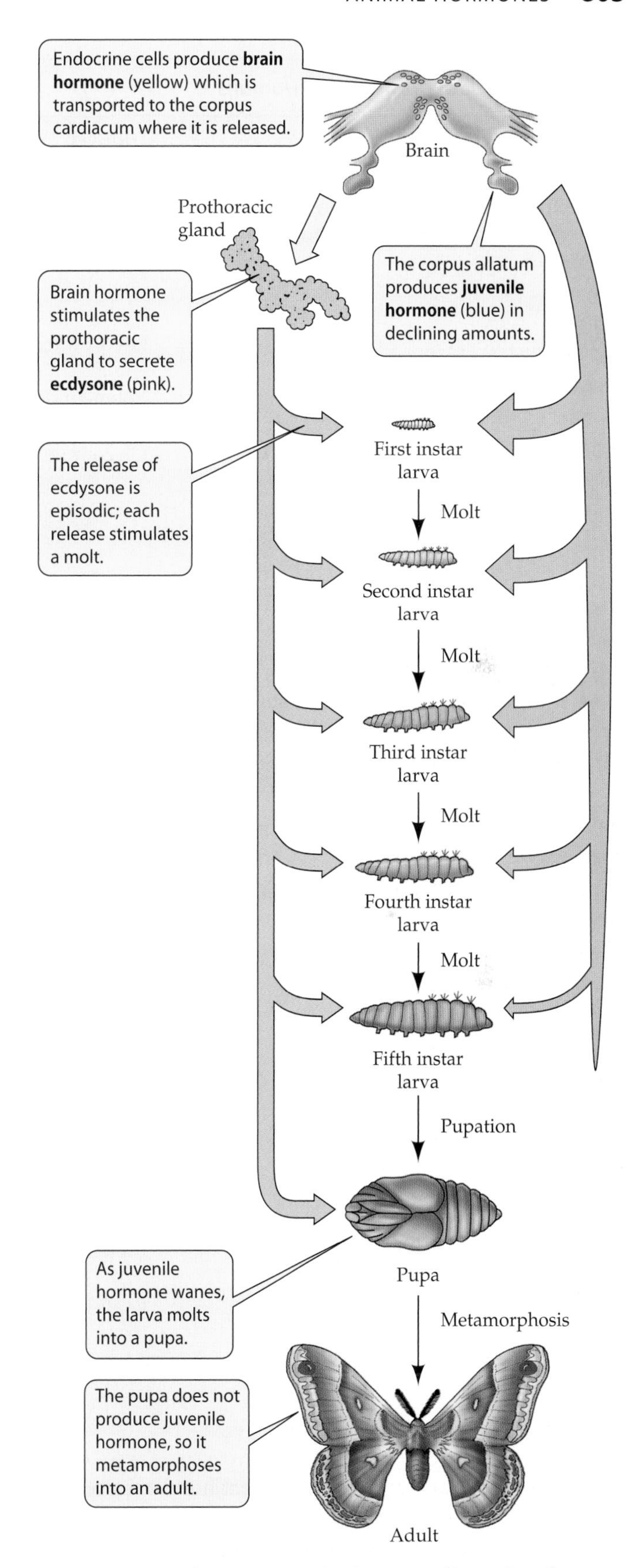
Endocrine cells produce brain hormone (yellow) which is transported to the corpus cardiacum where it is released.
Brain
Prothoracic gland
Brain hormone stimulates the prothoracic gland to secrete ecdysone (pink).
The corpus allatum produces juvenile hormone (blue) in declining amounts.
The release of ecdysone is episodic; each release stimulates a molt.
First instar larva
Molt
Second instar larva
Molt
Third instar larva
Molt
Fourth instar larva
Molt
Fifth instar larva
Pupation
Pupa
As juvenile hormone wanes, the larva molts into a pupa.
Metamorphosis
The pupa does not produce juvenile hormone, so it metamorphoses into an adult.
Adult

42.4 Complete Metamorphosis Butterflies and moths undergo complete metamorphosis, in which the feeding larvae (caterpillars) bear no resemblance to the reproductive adult. Three hormones control molting and metamorphosis in the silkworm moth *Hyalophora cecropia*.

42.1 Principal Hormones of Humans

SECRETING TISSUE OR GLAND	HORMONE	CHEMICAL NATURE	TARGET(S)	IMPORTANT PROPERTIES OR ACTIONS
Hypothalamus	Releasing and release-inhibiting hormones (see Table 42.2)	Peptides	Anterior pituitary	Control secretion of hormones of anterior pituitary
	Oxytocin, antidiuretic hormone	Peptides	(See Posterior pituitary)	Stored and released by posterior pituitary
Anterior pituitary: Tropic hormones	Thyrotropin	Glycoprotein	Thyroid gland	Stimulates synthesis and secretion of thyroxine
	Adrenocorticotropin (ACTH)	Polypeptide	Adrenal cortex	Stimulates release of hormones from adrenal cortex
	Luteinizing hormone (LH)	Glycoprotein	Gonads	Stimulates secretion of sex hormones from ovaries and testes
	Follicle-stimulating hormone (FSH)	Glycoprotein	Gonads	Stimulates growth and maturation of eggs in females; stimulates sperm production in males
Anterior pituitary: Other hormones	Growth hormone (GH)	Protein	Bones, liver, muscles	Stimulates protein synthesis and growth
	Prolactin	Protein	Mammary glands	Stimulates milk production
	Melanocyte-stimulating hormone	Peptide	Melanocytes	Controls skin pigmentation
	Endorphins and enkephalins	Peptides	Spinal cord neurons	Decreases painful sensations
Posterior pituitary	Oxytocin	Peptide	Uterus, breasts	Induces birth by stimulating labor contractions; causes milk flow
	Antidiuretic hormone (ADH) (vasopressin)	Peptide	Kidneys	Stimulates water reabsorption and raises blood pressure
Thyroid	Thyroxine	Iodinated amino acid derivative	Many tissues	Stimulates and maintains metabolism necessary for normal development and growth
	Calcitonin	Peptide	Bones	Stimulates bone formation; lowers blood calcium
Parathyroids	Parathyroid hormone	Protein	Bones	Resorbs bone; raises blood calcium
Thymus	Thymosins	Peptides	White blood cells	Activate immune responses of T cells in the lymphatic system
Pancreas	Insulin	Protein	Muscles, liver, fat, other tissues	Stimulates uptake and metabolism of glucose; increases conversion of glucose to glycogen and fat
	Glucagon	Protein	Liver	Stimulates breakdown of glycogen and raises blood sugar
	Somatostatin	Peptide	Digestive tract; other cells of the pancreas	Inhibits insulin and glucagon release; decreases secretion, motility, and absorption in the digestive tract

The existence and function of insect hormones was experimentally demonstrated many years before the hormones were identified chemically. That is not surprising when you consider the tiny amounts of certain hormones that exist in an organism. In one of the earliest studies of ecdysone, biochemists produced only 250 mg of pure ecdysone (about one-fourth the weight of an apple seed) from 4 tons of silkworms!

Vertebrate Endocrine Systems

The list of known hormones in the bodies of vertebrates is long and growing longer. To make the subject manageable, we will focus mostly on the hormones of mammals—how they function and how they are controlled. Table 42.1 presents an overview of the hormones of mammals. Most of these hormones are shared by other vertebrates as well.

42.1 Principal Hormones of Humans (continued)

SECRETING TISSUE OR GLAND	HORMONE	CHEMICAL NATURE	TARGET(S)	IMPORTANT PROPERTIES OR ACTIONS
Adrenal medulla	Epinephrine, norepinephrine	Modified amino acids	Heart, blood vessels, liver, fat cells	Stimulate fight-or-flight reactions: increase heart rate, redistribute blood to muscles, raise blood sugar
Adrenal cortex	Glucocorticoids (cortisol)	Steroids	Muscles, immune system, other tissues	Mediate response to stress; reduce metabolism of glucose, increase metabolism of proteins and fats; reduce inflammation and immune responses
	Mineralocorticoids (aldosterone)	Steroids	Kidneys	Stimulate excretion of potassium ions and reabsorption of sodium ions
Stomach lining	Gastrin	Peptide	Stomach	Promotes digestion of food by stimulating release of digestive juices; stimulates stomach movements that mix food and digestive juices
Lining of small intestine	Secretin	Peptide	Pancreas	Stimulate secretion of bicarbonate solution by ducts of pancreas
	Cholecystokinin	Peptide	Pancreas, liver, gallbladder	Stimulates secretion of digestive enzymes by pancreas and other digestive juices from liver; stimulates contractions of gallbladder and ducts
	Enterogastrone	Polypeptide	Stomach	Inhibits digestive activities in the stomach
Pineal	Melatonin	Modified amino acid	Hypothalamus	Involved in biological rhythms
Ovaries	Estrogens	Steroids	Breasts, uterus, other tissues	Stimulate development and maintenance of female characteristics and sexual behavior
	Progesterone	Steroid	Uterus	Sustains pregnancy; helps maintain secondary female sexual characteristics
Testes	Androgens	Steroids	Various tissues	Stimulate development and maintenance of male sexual behavior and secondary male sexual characteristics; stimulate sperm production
Many cell types	Prostaglandins	Modified fatty acids	Various tissues	Have many diverse actions
Heart	Atrial natriuretic hormone	Peptide	Kidneys	Increases sodium ion excretion
Skin	Vitamin D (cholecalciferol)	Sterol	Digestive tract, kidneys, bone	Increases blood calcium levels

We will begin this survey with the pituitary gland because it plays a central role in the endocrine system. The pituitary is a link between the nervous system and many other endocrine glands. It secretes some hormones that are actually produced by neurons in the brain, and under the influence of still other brain hormones, it produces a number of its own hormones, which control the activities of various endocrine glands throughout the body.

The pituitary is closely associated with the brain

The **pituitary gland** sits in a depression at the bottom of the skull just over the back of the roof of the mouth (Figure 42.5). It is attached by a stalk to the part of the brain called the *hypothalamus*, which is involved in many homeostatic regulatory systems (see Chapter 41).

The pituitary has two distinct parts that have different functions and separate origins during development. The anterior pituitary originates as an outpocketing of the embryonic mouth cavity, and the posterior pituitary originates as an outpocketing of the developing brain in the region that becomes the hypothalamus.

THE POSTERIOR PITUITARY. The **posterior pituitary** releases two peptide hormones, antidiuretic hormone and oxytocin. Because these hormones are synthesized in neurons in the hypothalamus, they are called **neurohormones**. As antidiuretic hormone and oxytocin are produced, they are packaged in vesicles. These vesicles are then transported down long extensions of the neurons, called *axons*, that run from the hypothalamus through the pituitary stalk and terminate in the posterior pituitary. The vesicles are stored in the

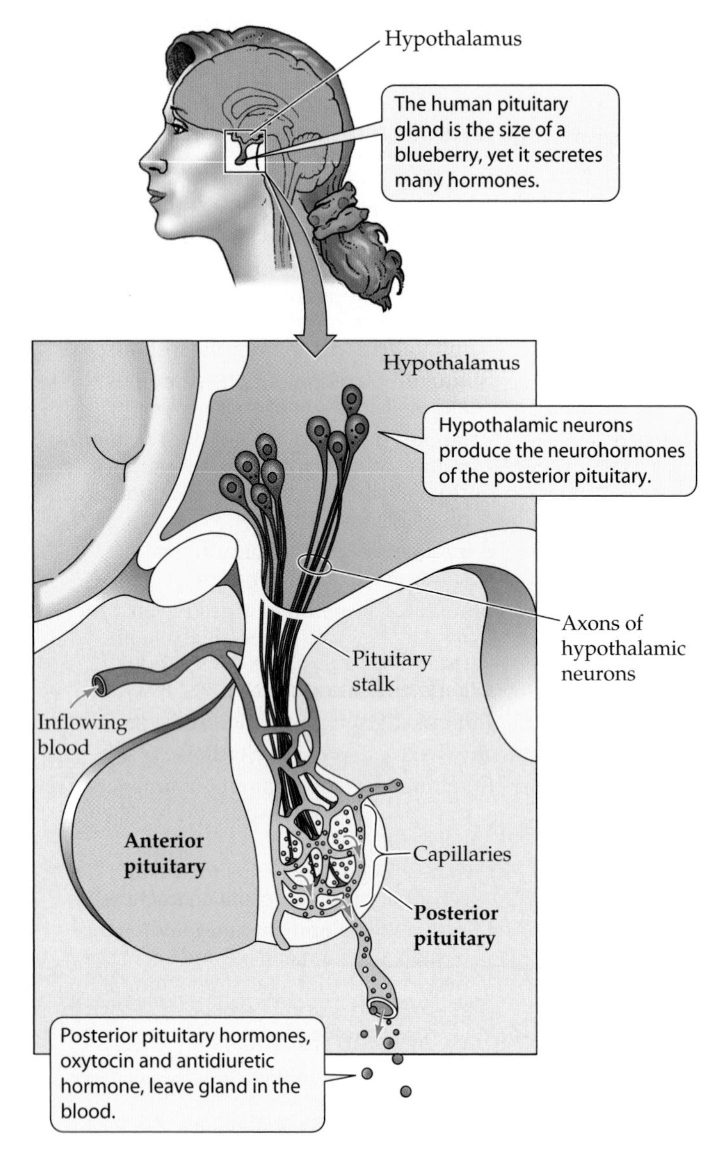

42.5 The Posterior Pituitary Releases Neurohormones The two hormones stored and released by the posterior pituitary are peptide neurohormones produced in the hypothalamus.

axon terminals until a nerve impulse stimulates their release (see Figure 42.5). How do the vesicles move down the axons? Proteins called *kinesins* grab onto the vesicles and, powered by ATP, "walk" step by step down microtubules in the axons.

The main action of **antidiuretic hormone** (**ADH**) in mammals and birds is to increase the amount of water conserved by the kidneys. When ADH secretion is high, the kidneys resorb water and produce only a small volume of highly concentrated urine. When ADH secretion is low, the kidneys produce a large volume of dilute urine. The posterior pituitary increases its release of ADH whenever blood pressure falls or the blood becomes too salty. We will discuss the mechanism of ADH action in Chapter 51. ADH is also known as *vasopressin* because it also causes the constriction of peripheral blood vessels as a means of elevating blood pressure.

When a woman is about to give birth, her posterior pituitary releases **oxytocin**, which stimulates the contractions of the uterus that deliver the baby. Oxytocin also brings about the flow of milk from the mother's breasts. The baby's suckling stimulates neurons in the mother, causing the secretion of oxytocin. Even the sight and sounds of her baby can cause a nursing mother to secrete oxytocin and release milk from her breasts.

THE ANTERIOR PITUITARY. Four peptide and protein hormones released by the **anterior pituitary** (thyrotropin, adrenocorticotropin, luteinizing hormone, and follicle-stimulating hormone) are **tropic hormones**, which control the activities of other endocrine glands (see Figure 42.7). Each tropic hormone is produced by a different type of pituitary cell. We will say more about these tropic hormones when we describe their target glands (thyroid, adrenal cortex, testes, and ovaries) later in this chapter and in the next.

The other peptide–and protein hormones produced by the anterior pituitary influence tissues that are not endocrine glands. These hormones are growth hormone, prolactin, melanocyte-stimulating hormone, endorphins, and enkephalins.

Growth hormone (**GH**) acts on a wide variety of tissues to promote growth directly and indirectly. One of its important direct effects is to stimulate cells to take up amino acids. Growth hormone also promotes growth indirectly by stimulating the liver to produce chemical messages called *somatomedins* or *insulin-like growth factors* (IGFs), which stimulate the growth of bone and cartilage. Thus, in some of its actions, growth hormone can also be considered a tropic hormone in that it stimulates cells to produce and release other hormones.

Overproduction of growth hormone in children causes *gigantism* (individuals may grow to nearly 8 feet tall; Figure 42.6). Underproduction causes *pituitary dwarfism*, in which individuals fail to reach normal adult height.* Beginning in the late 1950s, children diagnosed as having a serious deficiency of growth hormone were treated with growth hormone extracted from human pituitaries from cadavers. The treatment was successful in stimulating substantial growth, but a year's supply the hormone for one individual required up to *50* pituitaries! In the mid-1980s, scientists using genetic

*Pituitary dwarfism, in which the individual is short but normally proportioned, is a distinct condition from the more common *achondroplasia*, in which the trunk and head are normal size but the bones of the arms and legs are foreshortened. Although the result of both conditions is short stature, the underlying causes are completely different.

42.6 **Effects of Excess Growth Hormone** At 7′7″, basketball player turned actor Gheorghe Muresan (shown here in the 1998 film *My Giant*) is the tallest man ever to play in the U.S. National Basketball Association. Mureson was born in Romania to parents who were both under 6 feet tall; his gigantism results from overproduction of pituitary growth hormone during childhood.

engineering technology isolated the gene for human growth hormone and introduced it into bacteria that could be grown in large quantities, making it possible to purify enough of the hormone to make it more widely available.

Prolactin stimulates breast development and the production and secretion of milk in female mammals. In some mammals, prolactin also functions as an important hormone during pregnancy. In human males, prolactin plays a role, along with other pituitary hormones, in controlling the endocrine function of the testes.

Endorphins and **enkephalins** are the body's natural opiates. In the brain, these molecules act as neurotransmitters in pathways that control pain. The significance of their release from the anterior pituitary is unknown. Interestingly, the production of endorphins and enkephalins in the pituitary is governed by the same gene that encodes at least two other pituitary hormones. This gene encodes a large parent molecule called *pro-opiomelanocortin*. This large protein molecule is cleaved to produce several peptides, some of which have hormonal functions. Adrenocorticotropin, melanocyte-stimulating hormone, endorphins, and enkephalins all result from the cleavage of pro-opiomelanocortin.

THE ANTERIOR PITUITARY IS CONTROLLED BY HYPOTHALAMIC NEUROHORMONES. The secretion of hormones by the anterior pituitary is largely under the control of neurohormones from the hypothalamus. The hypothalamus receives information about conditions in the body and in the external environment through both neuronal and hormonal signals. If the connection between the hypothalamus and the pituitary is experimentally cut, pituitary hormones are no longer released in response to changes in the internal or external environment. When pituitary cells were maintained in culture, extracts of hypothalamic tissue stimulated some of those cells to release their hormones into the culture medium. Therefore, scientists hypothesized that secretions of the hypothalamic cells control the activities of anterior pituitary cells.

Although hypothalamic neurons do not extend into the anterior pituitary as they do into the posterior pituitary, a special set of **portal blood vessels** connects the hypothalamus and the anterior pituitary (Figure 42.7). It was thus proposed that secretions from neurons in the hypothalamus enter the blood and are conducted down the portal vessels to

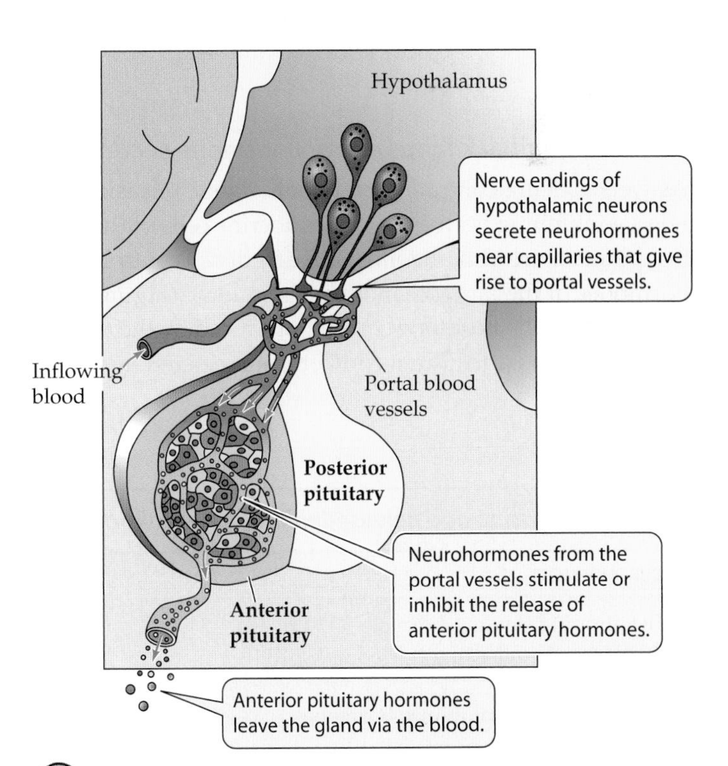

42.7 **Hormones from the Hypothalamus Control the Anterior Pituitary** Neurohormones produced in tiny quantities by cells in the hypothalamus are transported to the anterior pituitary through a system of portal blood vessels. These releasing and release-inhibiting hormones control the activities of endocrine cells in the anterior pituitary.

the anterior pituitary, where they stimulate the release of anterior pituitary hormones.

In the 1960s, two large teams of scientists, led by Roger Guillemin and Andrew Schally, initiated the search for these hypothalamic secretions. Because the amounts of such neurohormones in any individual mammal would be tiny, massive numbers of hypothalami from pigs and sheep were collected from slaughterhouses and shipped to laboratories in refrigerated trucks. One extraction effort began with the hypothalami from 270,000 sheep and yielded only 1 mg of purified **thyrotropin-releasing hormone (TRH)**. Biochemical analysis of this pure sample revealed that TRH is a simple tripeptide consisting of glutamine, histidine, and proline. TRH was the first hypothalamic *releasing* (that is, release-stimulating) *hormone* to be isolated and characterized. It causes certain anterior pituitary cells to release the tropic hormone thyrotropin, which in turn stimulates the activity of the thyroid gland.

Soon after discovering thyrotropin-releasing hormone, Guillemin's and Schally's teams identified **gonadotropin-releasing hormone (GnRH)**, which stimulates certain anterior pituitary cells to release the tropic hormones that control the activity of the gonads (the ovaries and the testes). For these discoveries, Guillemin and Schally received the 1977 Nobel prize in medicine. Many more hypothalamic neurohormones, including both releasing hormones and release-inhibiting hormones, are now known (Table 42.2).

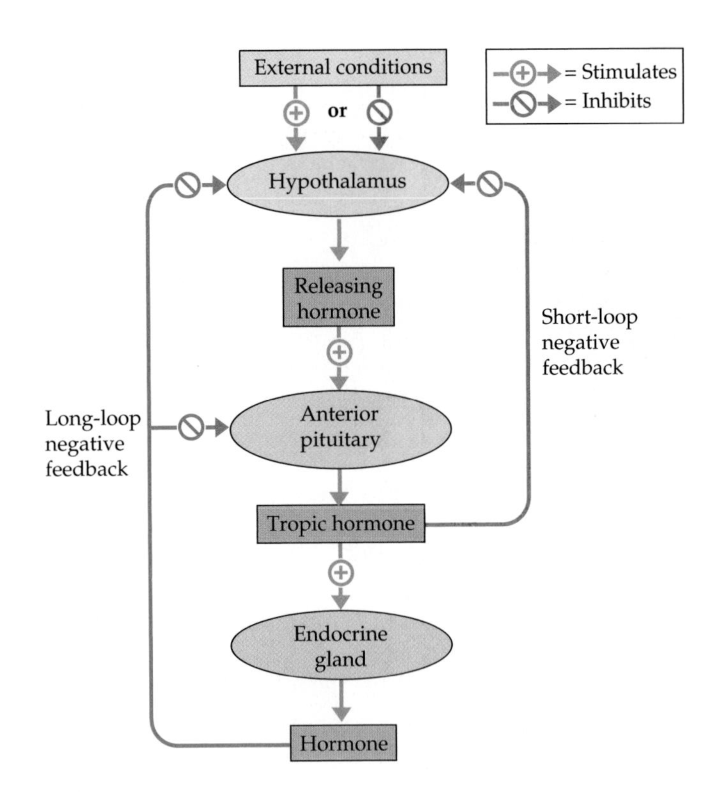

42.8 Multiple Feedback Loops Control Hormone Secretion Multiple feedback loops regulate the chain of command from hypothalamus to anterior pituitary to endocrine glands.

Negative feedback loops control hormone secretion

As well as being controlled by hypothalamic releasing and release-inhibiting hormones, the endocrine cells of the anterior pituitary are also under negative feedback control by the hormones of the target glands they stimulate (Figure 42.8). For example, the hormone cortisol, produced by the adrenal gland in response to adrenocorticotropin secreted by the anterior pituitary, returns to the pituitary in the circulating blood and inhibits further release of that tropic hormone. Cortisol also acts as a negative feedback signal to the hypothalamus, inhibiting the release of adrenocorticotropin-releasing hormone. In some cases a tropic hormone also exerts negative feedback control on the hypothalamic cells producing the corresponding releasing hormone. We will next examine the function of one endocrine gland, the thyroid, and see how it is controlled by negative feedback.

42.2 ***Releasing and Release-Inhibiting Neurohormones of the Hypothalamus***

NEUROHORMONE	ACTION
Thyrotropin-releasing hormone (TRH)	Stimulates thyrotropin release
Gonadotropin-releasing hormone (GnRH)	Stimulates release of follicle-stimulating hormone and luteinizing hormone
Prolactin release-inhibiting hormone	Inhibits prolactin release
Prolactin-releasing hormone	Stimulates prolactin release
Somatostatin (growth hormone release-inhibiting hormone)	Inhibits growth hormone release; interferes with thyrotropin release
Growth hormone-releasing hormone	Stimulates growth hormone release
Adrenocorticotropin-releasing hormone	Stimulates adrenocorticotropin release
Melanocyte-stimulating hormone release-inhibiting hormone	Inhibits release of melanocyte-stimulating hormone

Thyroxine controls cell metabolism

The **thyroid gland** wraps around the front of the windpipe (*trachea*) and expands into a lobe on either side (see Figure 42.2). The thyroid gland produces the hormones thyroxine and calcitonin. It contains many round structures, called *follicles*, that produce, store, and release thyroxine. Cells in the spaces between the follicles produce calcitonin.

Thyroxine is synthesized from two molecules of tyrosine, which then have four atoms of iodine chemically bonded to them. Thus, the thyroxine molecule is also called T_4:

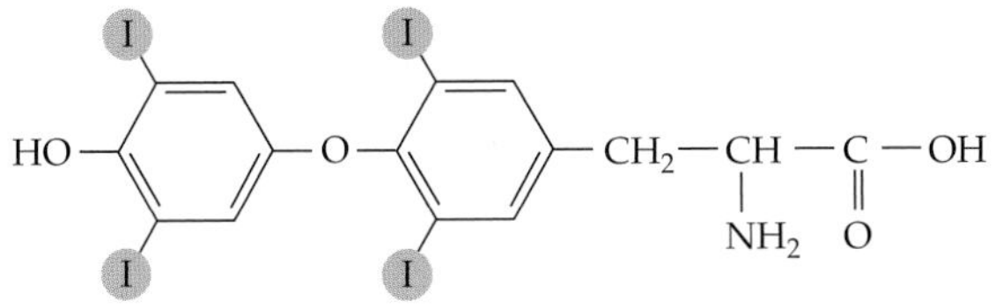

Thyroxine (T_4)

The follicles also produce and release *triiodothyronine*, a version of thyroxine that has only three atoms of iodine and is called T_3:

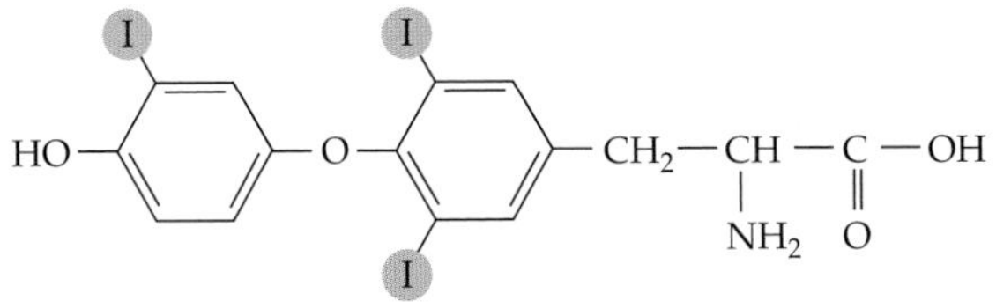

Triiodothyronine (T_3)

The thyroid usually releases about four times as much T_4 as T_3. T_3 is the more active hormone in the cells of the body, but when T_4 is in circulation, it can be converted to T_3 by an enzyme. Therefore, when you read about thyroxine, keep in mind that the actions discussed are primarily those of T_3.

Thyroxine in mammals plays many roles in regulating cell metabolism. It does so by stimulating the transcription of a large number of genes in just about all cells of the body. These genes include those for enzymes in energy pathways, transport proteins, and structural proteins. As a result, thyroxine elevates the metabolic rates of most cells and tissues. It also promotes the use of carbohydrates rather than fats for fuel. Exposure to cold for several days leads to an increased release of thyroxine, an increased conversion of T_4 to T_3, and therefore an increase in basal metabolic rate. Thyroxine is especially crucial during development and growth, as it promotes amino acid uptake and protein synthesis by cells. Insufficient thyroxine in a human fetus or growing child greatly retards physical and mental growth, resulting in a condition known as *cretinism*.

The tropic hormone **thyrotropin,** or **TSH** (*t*hyroid-*s*timulating *h*ormone), produced by the anterior pituitary, activates the follicle cells in the thyroid that produce thyroxine. TRH (thyrotropin-releasing hormone), produced in the hypothalamus and transported to the anterior pituitary through the portal blood vessels, activates the TSH-producing pituitary cells. The hypothalamus uses environmental information, such as temperature or day length, to determine whether to increase or decrease the secretion of TRH. This sequence of steps is also regulated by a negative feedback loop. Circulating thyroxine inhibits the response of pituitary cells to TRH. Therefore, less TSH is released when thyroxine levels are high, and more TSH is released when thyroxine levels are low. To a lesser extent, circulating thyroxine also exerts negative feedback on the production and release of TRH by the hypothalamus.

Thyroid dysfunction causes goiter

A *goiter* is an enlarged thyroid gland, which causes a pronounced bulge on the front and sides of the neck. Goiter can be associated with either **hyperthyroidism** (very high levels of thyroxine) or **hypothyroidism** (very low levels of thyroxine). The negative feedback loop whereby thyroxine controls TSH release helps explain how two very different conditions can result in the same symptom, but it is also necessary to understand how the thyroid makes, stores, and releases thyroxine.

Each thyroid follicle consists of a layer of epithelial cells surrounding a mass of glycoprotein called *thyroglobulin*. The thyroglobulin, which consists of many residues of tyrosine, is made by the epithelial cells . The tyrosine residues are iodinated as the thyroglobulin is secreted into the center of the follicle. When thyroxine is needed, the same epithelial cells that made the thyroglobulin take it back and digest it to release thyroxine molecules. If there was enough iodine available when the thyroglobulin was made, its digestion releases molecules of T_3 and T_4. If there was not enough iodine available when the thyroglobulin was made, many of the residues released will not be T_3 or T_4 and will not bind to receptors on target cells.

Goiter occurs when the production of thyroglobulin is far above normal and the follicles become greatly enlarged. *Hyperthyroid* goiter results when the negative feedback mechanism fails to turn off the follicle cells even though blood levels of thyroxine are high. The most common cause of hyperthyroidism is an autoimmune disease in which an antibody to the TSH receptor is produced. This antibody can bind to the TSH receptor on the follicle cells, causing them to produce and release thyroxine. Even though blood levels of TSH may be quite low because of the negative feedback from high levels of thyroxine, the thyroid remains maximally stimulated, and it grows bigger. Hyperthyroid patients have high metabolic rates, are jumpy and nervous, usually feel hot, and may have a buildup of fat behind the eyeballs, causing their eyes to bulge.

Hypothyroid goiter results when there is not enough circulating thyroxine to turn off TSH production. The most common cause of this condition is a deficiency of dietary iodide, without which the follicle cells cannot make thyroxine. Without sufficient thyroxine, TSH levels remain high, and the thyroid continues to produce large amounts of thyroglobulin. But, because there is insufficient iodine available, few of the tyrosine residues in the thyroglobulin are iodinated. When this iodine-poor thyroglobulin is digested by the follicle cells, it does not yield functional thyroxine. Without functional thyroxine, the TSH levels remain high and stimulate more and more synthesis of thyroglobulin, and the follicles get bigger. The symptoms of hypothyroidism are low metabolism, intolerance of cold, and general physical and mental sluggishness.

Worldwide, goiter affects about 5 percent of the population. The addition of iodide to table salt has greatly reduced the incidence of the condition in industrialized nations, but goiter is still common in the less industrialized countries of the world.

Calcitonin reduces blood calcium

The regulation of calcium levels in the blood is a crucial and difficult task. It is crucial because even small changes in blood calcium levels can cause serious effects. When blood calcium falls more than 30 percent below normal, the nervous system becomes hyperexcitable, resulting in muscle spasms and even seizures. When blood calcium rises above normal, the nervous system becomes depressed and muscles—including the heart—weaken. The reason that regulation of blood calcium is difficult is that only about 0.1 percent of the calcium in the body is located in the extracellular fluids. About 1 percent is within cells, and almost 99 percent is in the bones. Therefore, the body has to regulate a tiny pool of calcium in the blood that can be influenced greatly by relatively small shifts in the much larger pools of calcium in the cells and in the bones.

There are multiple mechanisms for changing blood calcium levels:

- Controlling deposition and absorption of bone
- Controlling excretion of calcium by the kidneys
- Controlling absorption of calcium from the digestive tract

These mechanisms are controlled by the hormones calcitonin, parathyroid hormone, and vitamin D.

Calcitonin acts to lower the concentration of calcium in the blood (Figure 42.9). Bone is continually remodeled through resorption of old bone and laying down of new bone, as we will see in Chapter 47. Cells called *osteoclasts* break down bone and release calcium; *osteoblasts*, on the other hand, take up circulating calcium and deposit new bone. Calcitonin decreases the activity of osteoclasts and thereby shifts the balance of bone turnover to favor deposition of bone and removal of calcium from the blood. Because the turnover of

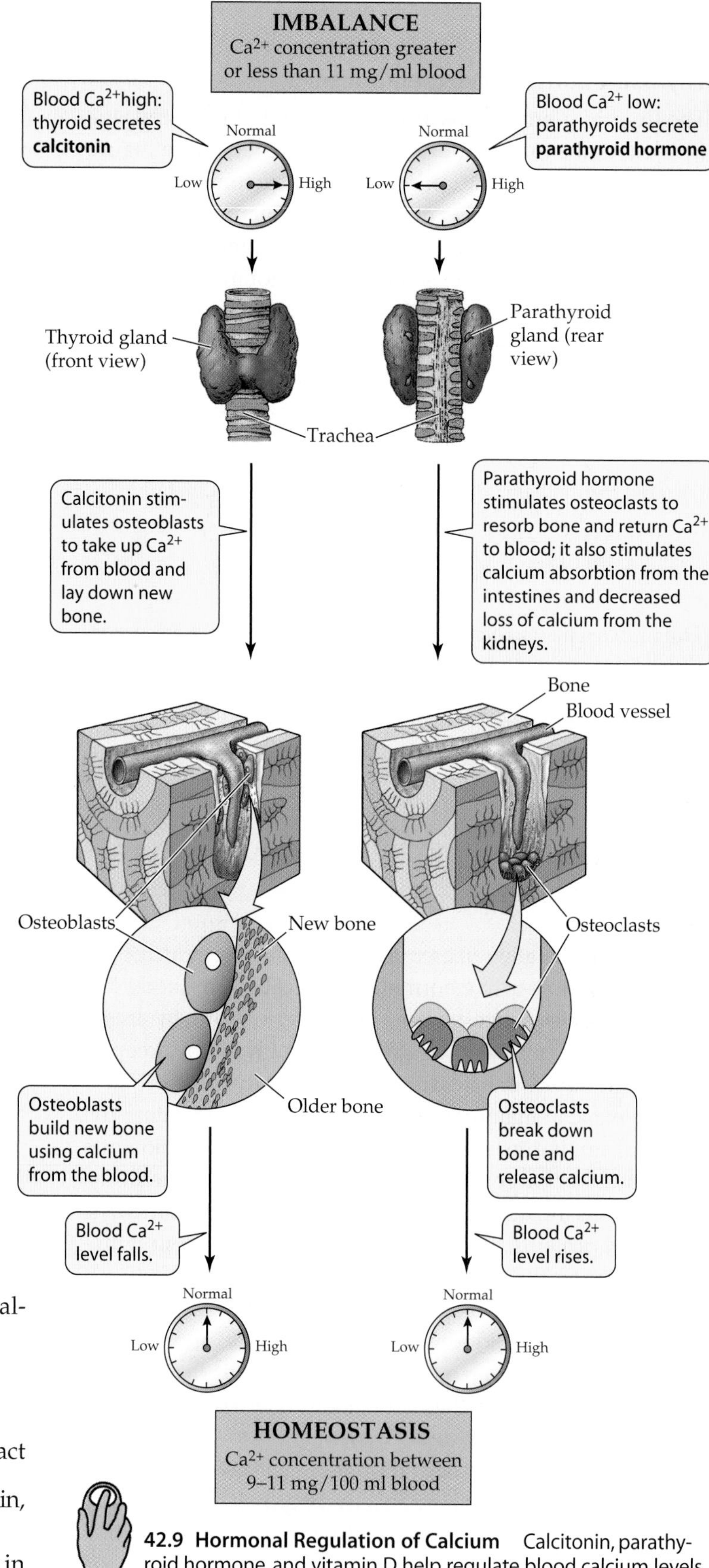
IMBALANCE
Ca2+ concentration greater or less than 11 mg/ml blood
Blood Ca2+ high: thyroid secretes calcitonin
Blood Ca2+ low: parathyroids secrete parathyroid hormone
Normal
Low
High
Thyroid gland (front view)
Parathyroid gland (rear view)
Trachea
Calcitonin stimulates osteoblasts to take up Ca2+ from blood and lay down new bone.
Parathyroid hormone stimulates osteoclasts to resorb bone and return Ca2+ to blood; it also stimulates calcium absorbtion from the intestines and decreased loss of calcium from the kidneys.
Bone
Blood vessel
Osteoblasts
New bone
Osteoclasts
Older bone
Osteoblasts build new bone using calcium from the blood.
Osteoclasts break down bone and release calcium.
Blood Ca2+ level falls.
Blood Ca2+ level rises.
HOMEOSTASIS
Ca2+ concentration between 9–11 mg/100 ml blood

42.9 Hormonal Regulation of Calcium Calcitonin, parathyroid hormone, and vitamin D help regulate blood calcium levels.

bone in adult humans is not very high, calcitonin does not play a major role in calcium homeostasis in adult humans. It is probably more important in young, growing individuals, but overall, calcium levels are more influenced by parathyroid hormone than by calcitonin.

Parathyroid hormone elevates blood calcium

The **parathyroid glands** are four tiny structures embedded in the posterior surface of the thyroid gland. Their single hormone product, **parathyroid hormone** (also called **PTH** or *parathormone*), is the critical hormone in the regulation of blood calcium levels. Levels of calcium in the blood are sensed by receptors in the plasma membrane of the parathyroid cells. When these receptors are activated, they inhibit the synthesis and release of PTH. A fall in blood calcium levels removes this inhibition and triggers the synthesis and release of PTH.

PTH acts in a number of ways to raise blood calcium levels. Its actions provide a good example of the complexity of physiological regulation. PTH does not act directly on osteoclasts, which break down bone and release calcium into the blood. Osteoclasts do not have PTH receptors—but osteoblasts do. What PTH does is to activate bone turnover by activating osteoblasts, which in turn release cytokines that activate the osteoclasts. In this enhanced turnover, there is a net loss of bone, and hence a rise in blood calcium levels (see Figure 42.9). PTH also conserves calcium by stimulating the kidneys to resorb it rather than losing it in the urine. Finally, increased secretion of PTH causes the digestive tract to absorb more calcium from food, but this is an indirect effect dependent on vitamin D. In the kidney, PTH stimulates the activation of vitamin D, and it is vitamin D that acts on the digestive tract to enhance absorption of dietary calcium.

Vitamin D is really a hormone

A *vitamin* is a substance that the body needs in small quantities, but cannot synthesize, and therefore must obtain from the diet. By this definition, **vitamin D** is not a vitamin because the body can and does synthesize it. That synthesis takes place in cells of the skin, where cholesterol is converted into vitamin D (also called calciferol) by ultraviolet light. Long before the biochemistry of vitamin D was known, weak and fragile bones were common among people living at high latitudes if they had diets lacking meat, fish, dairy products, and fresh vegetables. Since the condition could be reversed by improving the diet or giving the person cod-liver oil, it was assumed that a vitamin was involved. However, since vitamin D is produced in skin cells, circulates in the blood, and acts on distant cells, it is actually a hormone.

The vitamin D produced in the skin is not very active, but as it passes through the liver it receives one —OH group, and in the kidneys it receives another —OH group to form (1, 25) dihydroxyvitamin D, which is the most active form. PTH stimulates this final step in the kidneys. Active vitamin D circulates around the body bound to a blood protein. It enters cells, since it is lipid-soluble, and combines with a cytoplasmic receptor, forming a transcription factor. In the digestive tract, this transcription factor acts to increase the synthesis of calcium pumps, calcium channels, and calcium-binding proteins, all of which promote the uptake of calcium.

In the kidneys, vitamin D acts synergistically with PTH to decrease calcium loss in the urine. In bone, vitamin D, like PTH, stimulates bone turnover and liberates calcium—which seems the opposite of what would be expected. However, through all of its actions, vitamin D raises blood calcium levels, and that is essential to promote bone deposition. Vitamin D also acts on parathyroid cells to inhibit the transcription of the PTH gene, thus forming a negative feedback loop for the regulation of PTH.

PTH lowers blood phosphate levels

Bone minerals are a combination of calcium and phosphate. Thus, when PTH stimulates the release of calcium from bone, it also causes the release of phosphate. Increases in blood levels of both calcium and phosphate can be dangerous. The normal levels of calcium and phosphate in the blood approach the concentration at which they would precipitate out of solution as calcium phosphate salts, leading to maladies such as kidney stones and calcium deposits in the arteries (hardening of the arteries). To reduce this problem, PTH acts on the kidneys to increase the elimination of phosphate via the urine.

Insulin and glucagon regulate blood glucose levels

Before the 1920s, *diabetes mellitus** was a fatal disease, characterized by weakness, lethargy, and a dramatic loss of body mass. The disease was known to be connected somehow with the **pancreas**, a gland located just below the stomach (see Figure 42.2), and with abnormal glucose metabolism, but the link was not clear.

Today we know that diabetes mellitus is caused by a lack of the protein hormone **insulin** (type I or juvenile onset diabetes) or by a lack of insulin receptors on the target tissues (type II or adult onset diabetes). For patients in which the hormone is lacking, insulin replacement therapy is an extremely successful treatment. At present, more than 1.5 million people with diabetes in the United States lead almost normal lives through the use of manufactured insulin.

*The name *diabetes* refers to the copious production of urine. *Mellitus* (Greek for "honey") reflects the fact that the urine of an untreated diabetic is sweet.

Insulin binds to a receptor on the plasma membrane of a target cell, and this insulin–receptor complex allows glucose to enter the cell (see Figure 15.6). In the absence of insulin or insulin receptors, glucose fails to enter the cells, and instead accumulates in the blood until it is lost in the urine. High levels of blood glucose cause water to move from cells into the blood by osmosis, and the kidneys increase urine output to remove this excess fluid volume from the blood. Because glucose uptake by most cells is impaired without insulin, those cells must use fat and protein for fuel instead of glucose. As a result, the body of the untreated diabetic wastes away, and critical tissues and organs are damaged.

For centuries the prospects for a person with diabetes were bleak. A change in this outlook came almost overnight in 1921, when medical doctor Frederick Banting and medical student Charles Best of the University of Toronto discovered that they could reduce the symptoms of diabetes by injecting an extract they had prepared from pancreatic tissue. The active component of this extract was found to be a small protein hormone—insulin—consisting of 51 amino acids. For this discovery, Banting was awarded a Nobel prize, which he refused to accept because his student was not also honored.

Insulin is produced in clusters of endocrine cells in the pancreas. These clusters are called **islets of Langerhans** after the German medical student who discovered them. There are several types of cells in the islets:

- Beta (β) cells produce and secrete insulin.
- Alpha (α) cells produce and secrete the hormone glucagon, which has effects opposite from those of insulin.
- Delta (δ) cells produce the hormone somatostatin.

The rest of the pancreas produces enzymes and other secretions that travel through ducts to the intestine, where they play roles in digestion.

After a meal, the concentration of glucose in the blood rises as glucose is absorbed from the food in the gut. This increase stimulates the β cells of the pancreas to release insulin. Insulin stimulates cells to use glucose as fuel and to convert it into storage products, such as glycogen and fat. When the gut contains no more food, the glucose concentration in the blood falls, and the pancreas stops releasing insulin. As a result, most cells of the body shift to using glycogen and fat, rather than glucose, for fuel. If the concentration of glucose in the blood falls substantially below normal, the islet α cells release **glucagon**, which stimulates the liver to convert glycogen back to glucose to resupply the blood. These actions will be discussed in greater detail in Chapter 50.

Somatostatin is a hormone of the brain and the gut

Somatostatin is released from the cells of the pancreas in response to rapid rises of glucose and amino acids in the blood. This hormone has paracrine functions within the islets: It inhibits the release of both insulin and glucagon. Its actions outside the pancreas slow the digestive activities of the gut. Pancreatic somatostatin extends the period of time during which nutrients are absorbed from the gut. Somatostatin is also produced in very small amounts by cells in the hypothalamus. Acting as a neurohormone, hypothalamic somatostatin is transported in the portal vessels to the anterior pituitary, where it inhibits the release of growth hormone and thyrotropin.

The adrenal gland is two glands in one

An **adrenal gland** sits above each kidney, just below the middle of your back. Functionally and anatomically, each adrenal gland consists of a gland within a gland (Figure 42.10). The core, called the **adrenal medulla**, produces the hormone **epinephrine** (also known as *adrenaline*) and, to a lesser degree, **norepinephrine** (or *noradrenaline*), which also acts as a neurotransmitter in the nervous system . Surrounding the medulla is the **adrenal cortex**, which produces other hormones. The medulla develops from nervous tissue and is under the control of the nervous system; the cortex is under hormonal control, largely by **adrenocorticotropin** (**ACTH**) from the anterior pituitary.

THE ADRENAL MEDULLA. The adrenal medulla produces epinephrine and norepinephrine in response to stressful situations, arousing the body to action. As we saw earlier in this chapter, epinephrine increases heart rate and blood pressure and diverts blood flow to active muscles and away from the gut. These fight-or-flight reactions can be stimulated by physically threatening events, such as encounter-

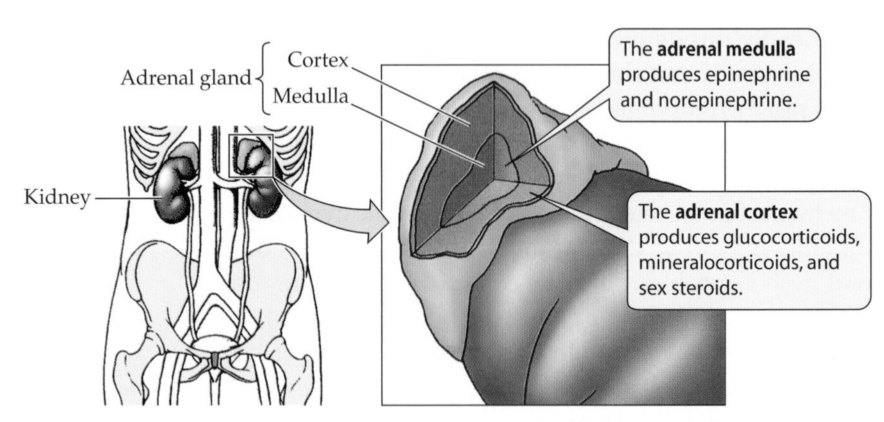

42.10 The Adrenal Gland Has an Outer and an Inner Portion An adrenal gland, consisting of an outer cortex and an inner medulla, sits on top of each kidney. The medulla and the cortex produce different hormones.

ing a mugger, or by events that are mentally stressful, such as giving a public speech or taking a test.

Epinephrine and norepinephrine are both amine hormones derived from the amino acid tyrosine. They are water-soluble, and they both bind to the same receptors on the surfaces of target cells. These receptors can be grouped into two general types, α-adrenergic and β-adrenergic receptors, which stimulate different actions within cells (see Figure 42.14 later in this chapter). Epinephrine acts equally on both types, but norepinephrine acts mostly on α-adrenergic receptors. Therefore, drugs called *beta blockers*, which selectively block β-adrenergic receptors, can reduce the fight-or-flight responses to epinephrine without disrupting the physiological regulatory functions of norepinephrine. Beta blockers are commonly prescribed to reduce anxiety, such as that brought on by public speaking.

THE ADRENAL CORTEX. The cells of the adrenal cortex use cholesterol to produce three classes of steroid hormones, collectively called *corticosteroids*:

- The **glucocorticoids** influence blood glucose concentrations as well as other aspects of fat, protein, and carbohydrate metabolism.
- The **mineralocorticoids** influence the ionic balance of extracellular fluids.
- The **sex steroids** play roles in sexual development, sex drive, and anabolism.

The adult adrenal cortex secretes sex steroids in only negligible amounts. The major producers of sex steroids are the gonads, as we will see in the following section.

Aldosterone, the main mineralocorticoid (Figure 42.11*a*), stimulates the kidneys to conserve sodium and to excrete potassium. If the adrenal glands are removed from an animal, it must have sodium added to its diet, or its sodium will be depleted and it will die. One human patient with a nonfunctional adrenal gland compensated by salting her food heavily and, in addition, ate a 60-pound block of salt in the course of a year.

The main glucocorticoid in humans is **cortisol** (Figure 42.11*b*), which is critical for mediating the body's response to stress. As we have seen, your immediate reaction to a frightening situation is stimulated by your nervous system and by the release of epinephrine. This fight-or-flight response ensures that your muscles will have enough oxygen and glucose to fuel your escape. You have a limited amount of blood glucose, however, and you need to conserve it for your muscles and your brain. Within minutes of the frightening stimulus, your blood cortisol level rises. Cortisol stimulates cells not critical for your escape to decrease their use of blood glucose and shift instead to utilizing fats and proteins for energy. This is not a time to feel sick, have allergic reactions, or heal wounds, so cortisol also blocks immune system reactions. That is why cortisol or pharmacological compounds that mimic cortisol action are useful for reducing inflammation and allergies.

Cortisol release is controlled by ACTH from the anterior pituitary, which in turn is controlled by **adrenocorticotropin-releasing hormone** from the hypothalamus. Because the cortisol response to a stressor has this chain of steps, each involving secretion, diffusion, circulation, and cell activation, it is much slower than the epinephrine response. Also, many of the actions of cortisol involve changes in gene expression, and that takes time.

Turning off the cortisol response is as important as turning it on. A study of stress in rats showed that old rats could turn on their stress responses as effectively as young rats, but that they had lost the ability to turn them off as rapidly. As a result, they suffered from the well-known consequences of stress seen in humans: ulcers, cardiovascular problems, strokes, impaired immune system function, and increased susceptibility to cancers and other diseases. Further research showed that the stress responses are turned off by the negative feedback action of cortisol on cells in the brain, which causes a decrease in the release of adrenocorticotropin-releasing hormone (see Figure 42.8). Repeated activation of this

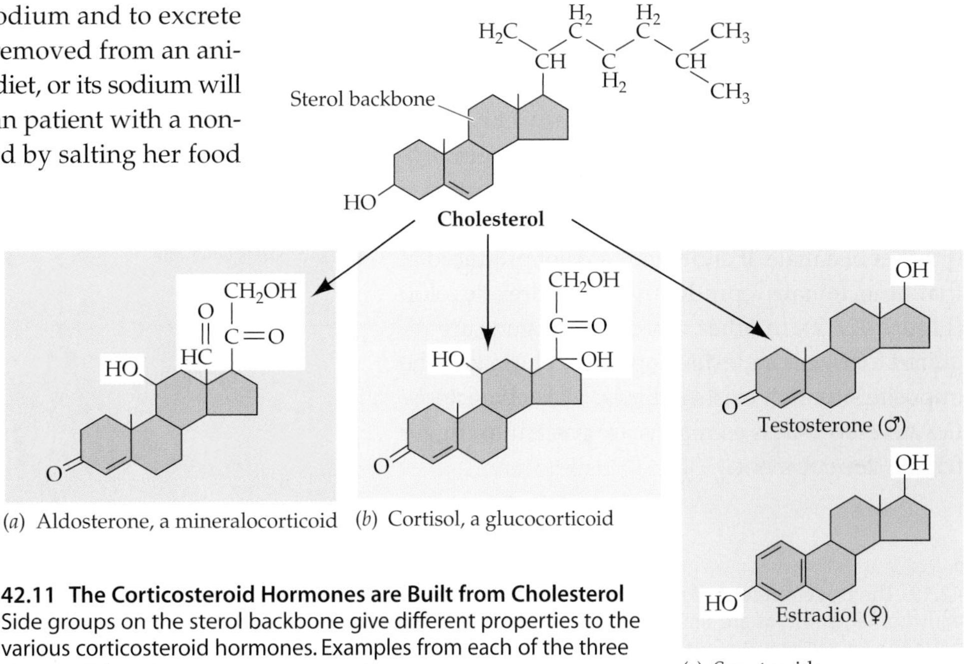

Sterol backbone

Cholesterol

(*a*) Aldosterone, a mineralocorticoid (*b*) Cortisol, a glucocorticoid

Testosterone (♂)

Estradiol (♀)

(*c*) Sex steroids

42.11 The Corticosteroid Hormones are Built from Cholesterol Side groups on the sterol backbone give different properties to the various corticosteroid hormones. Examples from each of the three classes of these hormones are shown here.

negative feedback mechanism, either through repeated stress or through prolonged medical use of cortisol, leads to a gradual loss of cortisol-sensitive cells in the brain, and therefore to a decreased ability to terminate stress responses.

The sex steroids are produced by the gonads

The **gonads**—the testes of the male and the ovaries of the female—produce hormones as well as gametes. Most of the gonadal hormones are steroids synthesized from cholesterol. The male steroids are collectively called **androgens**, and the dominant one is **testosterone**. The female steroids are **estrogens** and **progesterone**. The dominant estrogen is **estradiol,** which is made from testosterone. Thus, males and females both synthesize testosterone, but females have an enzyme that converts testosterone to estradiol (Figure 42.11*c*).

The sex steroids have important developmental effects: They determine whether a fetus develops into a female or a male. (A *fetus* is the latter stage of an embryo; a human embryo is called a fetus from the eighth week of pregnancy to the moment of birth.) After birth, the sex steroids control the maturation of the reproductive organs and the development and maintenance of secondary sexual characteristics, such as breasts and facial hair.

The sex steroids begin to exert their effects in the human embryo in the seventh week of development. Until that time, the embryo has the potential to develop into either sex. In mammals and birds, the instructions for sex determination reside in the genes. In mammals, individuals that receive two X chromosomes normally become females, and individuals that receive an X and a Y chromosome normally become males.

These genetic instructions are carried out through the production and action of the sex steroids. The presence of a Y chromosome normally causes the undifferentiated embryonic gonads to begin producing androgens in the seventh week. In response to the androgens, the reproductive system develops into that of a male. If androgens are not produced at that time, female reproductive structures develop (Figure 42.12). In other words, androgens are required to trigger male development in humans. The opposite situation exists in birds: Male characteristics develop unless estrogens are present to trigger female development.

Changes in control of sex steroid production initiate puberty

Sex steroids have dramatic effects at **puberty**—the time of sexual maturation in humans. Sex steroids are produced at low levels by the juvenile gonads, but their production increases rapidly at the beginning of puberty—around the age of 12 to 13 years. Why does this sudden increase occur?

In the juvenile, as in the adult, the production of sex steroids by the ovaries and testes is controlled by the anterior pituitary tropic hormones **luteinizing hormone** (**LH**) and **follicle-stimulating hormone** (**FSH**), which together are called the **gonadotropins**. The production of these tropic hormones is under the control of the hypothalamic gonadotropin-releasing hormone. Prior to puberty, the gonads are capable of responding to gonadotropins, and the pituitary is capable of responding to GnRH. But prior to puberty, the hypothalamus produces only very low levels of GnRH. Puberty is initiated by a reduction in the sensitivity of hypothalamic GnRH-producing cells to negative feedback from sex steroids and from gonadotropins. As a result,

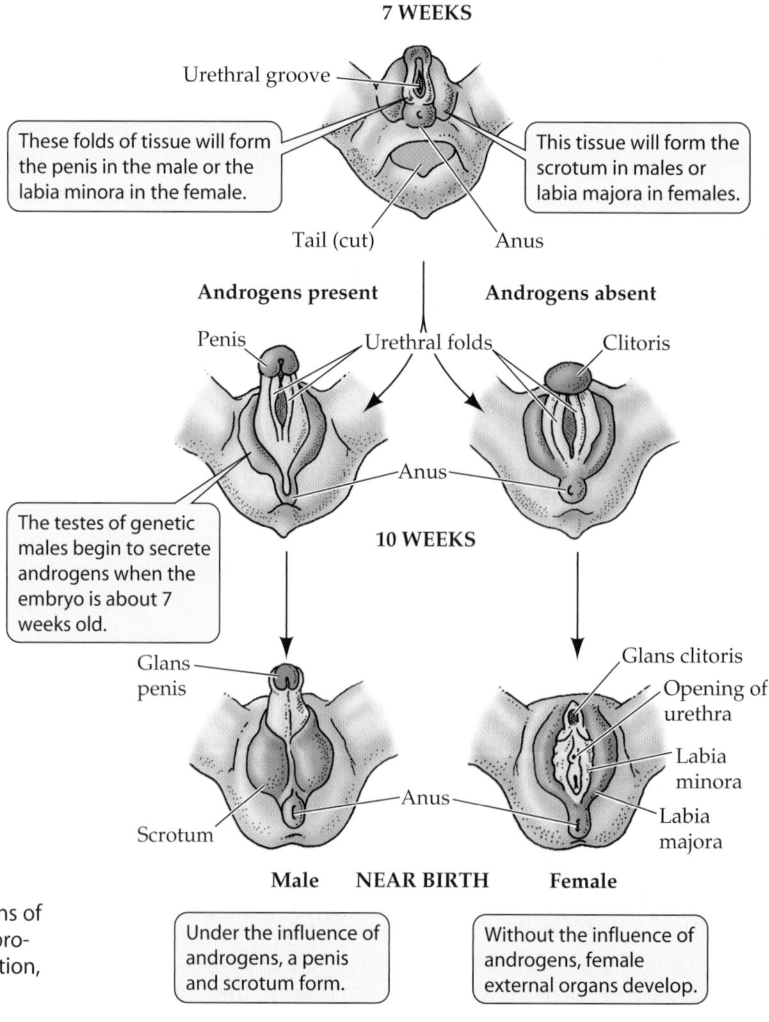
7 WEEKS
Urethral groove
These folds of tissue will form the penis in the male or the labia minora in the female.
This tissue will form the scrotum in males or labia majora in females.
Tail (cut)
Anus
Androgens present
Androgens absent
Penis
Urethral folds
Clitoris
Anus
The testes of genetic males begin to secrete androgens when the embryo is about 7 weeks old.
10 WEEKS
Glans penis
Glans clitoris
Opening of urethra
Labia minora
Labia majora
Anus
Scrotum
Male
NEAR BIRTH
Female
Under the influence of androgens, a penis and scrotum form.
Without the influence of androgens, female external organs develop.

42.12 The Development of Human Sex Organs The sex organs of early human embryos are similar. Male sex steroids (androgens) promote the development of male sex organs. Without androgen action, female sex organs form, even in genetic males.

GnRH release increases, stimulating increased production of gonadotropins and hence increased production of sex steroids.

In females, increasing levels of LH and FSH at puberty stimulate the ovaries to begin producing the female sex hormones. The increased circulating levels of these hormones initiate the development of the traits of a sexually mature woman: enlarged breasts, vagina, and uterus, broad hips, increased subcutaneous fat, pubic hair, and the initiation of the menstrual cycle.

In the male, an increasing level of LH stimulates groups of cells in the testes to synthesize testosterone, which in turn initiate the profound physiological, anatomical, and psychological changes associated with adolescence. The voice deepens, hair begins to grow on the face and body, and the testes and penis grow. Androgens also help bones and skeletal muscles grow, especially when they are exercised regularly.

Natural muscle development can be exaggerated by both men and women who want to increase their maximum strength in athletic competition if they take synthetic androgens—called *anabolic steroids*. However, anabolic steroids have serious negative side effects. In women, their use causes the breasts and uterus to shrink, the clitoris to enlarge, menstruation to become irregular, facial and body hair to grow, and the voice to deepen. In men, the testes shrink, hair loss increases, the breasts enlarge, and sterility can result. You can understand the causes of some of these side effects by considering the negative feedback effects of sex steroids on the production of LH and FSH. Other side effects are even more serious. Continued use of anabolic steroids greatly increases the risk of heart disease, certain cancers, kidney damage, and personality disorders such as depression, mania, psychoses, and extreme aggression. Most official athletic organizations, including the International Olympic Committee, have banned the use of anabolic steroids.

Melatonin is involved in biological rhythms and photoperiodicity

The **pineal gland** is situated between the two hemispheres of the brain and is connected to the brain by a little stalk. It produces the hormone **melatonin** from the amino acid tryptophan. The release of melatonin by the pineal occurs in the dark and therefore marks the length of the night. Exposure to light inhibits the release of melatonin.

In various vertebrates, melatonin is involved in biological rhythms, including **photoperiodicity**—the phenomenon whereby seasonal changes in day length cause physiological changes in animals. Many species, for example, come into reproductive condition when the days begin to get longer (Figure 42.13). Humans are not strongly photoperiodic, but melatonin in humans may play a role in entraining daily biological rhythms to the daily cycle of light and dark. We will learn more about biological rhythms in Chapter 52.

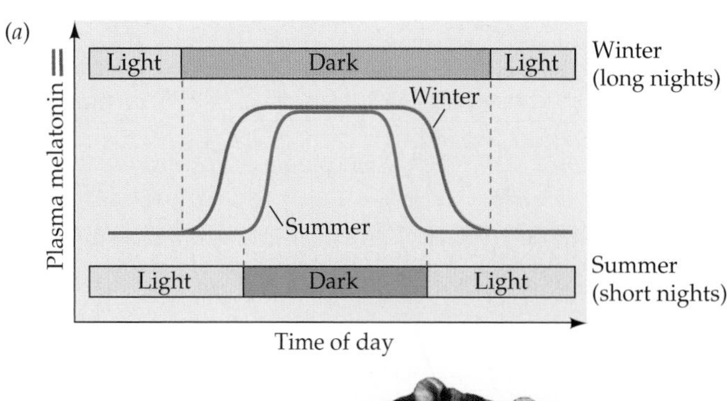

42.13 The Release of Melatonin Regulates Seasonal Changes
(*a*) Melatonin is released in the dark and inhibited by light exposure. The duration of daily melatonin release thus changes as day length (photoperiod) changes, inducing dramatic seasonal physiological changes in some animals. (*b*) In winter, these Siberian hamsters are white and nonreproductive. In summer, they are mottled brown and breed.

The list of hormones is long

We have discussed the major endocrine glands and their hormones in this chapter, but there are many hormones we have not mentioned (see Table 42.1). Even the heart has endocrine functions. When blood volume rises and causes the walls of the heart to stretch, certain cells in those walls release *atrial natriuretic hormone*. This hormone increases the excretion of sodium ions and water by the kidneys, thereby lowering blood volume and blood pressure. As we discuss the organ systems of the body in the chapters that follow, we will frequently mention hormones that their tissues produce, or hormones that control their functions.

Hormone Actions: The Role of Signal Transduction Pathways

Hormones are released in very small quantities, yet they can cause large responses in cells all over the body, as long as those cells express the appropriate hormone receptors. Moreover, different cells can respond to the same hormone in different ways. Testosterone, for example, has many dramatic and diverse effects, as we saw above, yet its concentration in

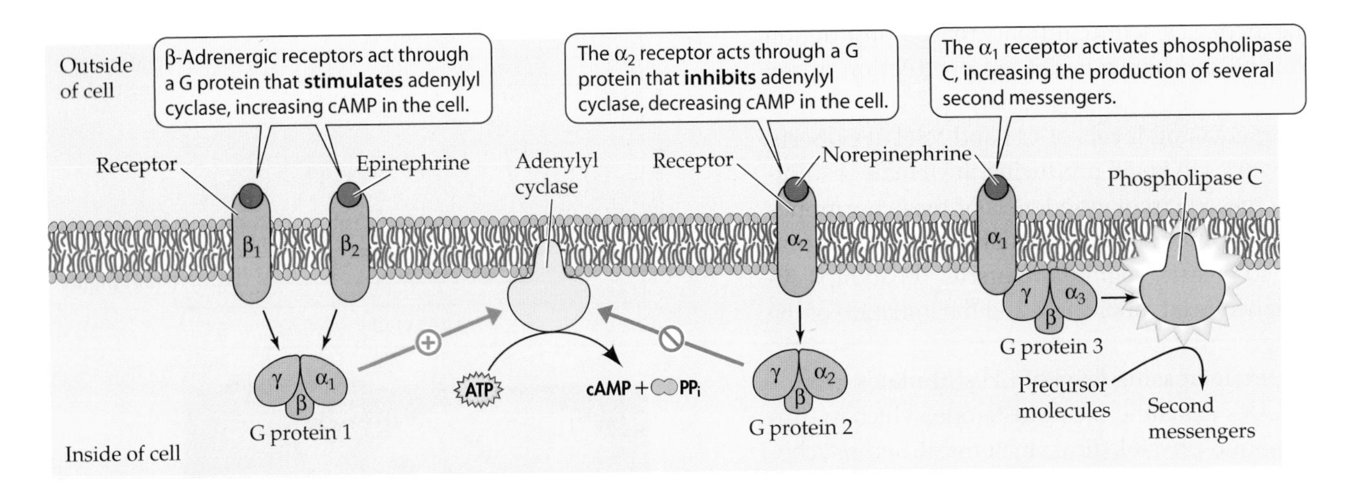

42.14 Some Hormones Can Activate a Variety of Signal Transduction Pathways Epinephrine and norepinephrine bind to G protein-linked adrenergic receptors that act through different signal transduction pathways. Epinephrine acts equally on both α- and β-adrenergic receptors; norepinephrine acts mostly on α-adrenergic receptors.

the blood of adult human males is only about 30 to 100 × 10^{-9} g/ml. How can hormones in such tiny quantities have such strong and selective actions?

The strength of hormone action frequently results from signal transduction cascades that amplify the original signal; an example is the response of liver cells to epinephrine (see Figure 15.15). A single molecule of epinephrine binding to its receptor on a liver cell can result in the release of millions of molecules of glucose into the blood.

The selective action of hormones is explained by the fact that only cells with appropriate receptors respond to a hormone. Also, in different types of cells, the receptors for a particular hormone can be linked to different response mechanisms. The response of a cell to a hormone depends both on the cells receptors and on the signal transduction pathways those receptors activate. For example, the two types of receptors mentioned above for the amine hormones epinephrine and norepinephrine (the α-adrenergic and the β-adrenergic receptors) are cell-surface G protein-linked receptors, but they connect with different signal transduction pathways within cells. (Figure 42.14). Since the cells of different tissues express these various receptors differentially, the same hormones, epinephrine and norepinephrine, stimulate different actions in different cells.

Regulation of hormone receptors controls the sensitivity of cells to hormones

We saw above that the release of hormones can be under feedback control, usually negative feedback control. Similarly, the abundance of receptors for a hormone can be under feedback control. In some cases, continuous high levels of a hormone can decrease the number of its receptors, a process known as **downregulation**. **Upregulation** of receptors, or an increase in their abundance, is a positive feedback mechanism, and is less common.

An example of downregulation is type II diabetes mellitus, which is characterized by elevated levels of circulating insulin, but a loss of insulin receptors. Although genetic factors are likely to be involved, a possible immediate cause of the disease is an overstimulation of pancreatic release of insulin by excessive carbohydrate intake, which leads to downregulation of the insulin receptors.

Responses to hormones can vary greatly

We have discussed two mechanisms for regulating physiological responses to hormones: controlling the amount of hormone released, and controlling the availability of receptors. A technique for measuring hormone and receptor concentrations is therefore necessary for studying hormonal mechanisms. The most common means of quantifying hormones and receptors is an **immunoassay.**

This technique is based on competition for protein-binding sites on antibody. A typical immunoassay begins with a known quantity of an antibody to a specific hormone. A saturating concentration of the hormone is labeled in some manner (often radioactively) and mixed with the antibody until all of the antibody's binding sites are labeled (Figure 42.15). An unlabeled sample of a known concentration of the same hormone is then added to the mixture. This unlabeled hormone will displace some of the labeled hormone from the antibody. The ratio of labeled to unlabeled antibody is a measure of the amount of unlabeled hormone added to the mixture. The process is repeated with different, known concentrations of unlabeled hormone until a standard curve is generated. By comparing a sample of unknown concentra-

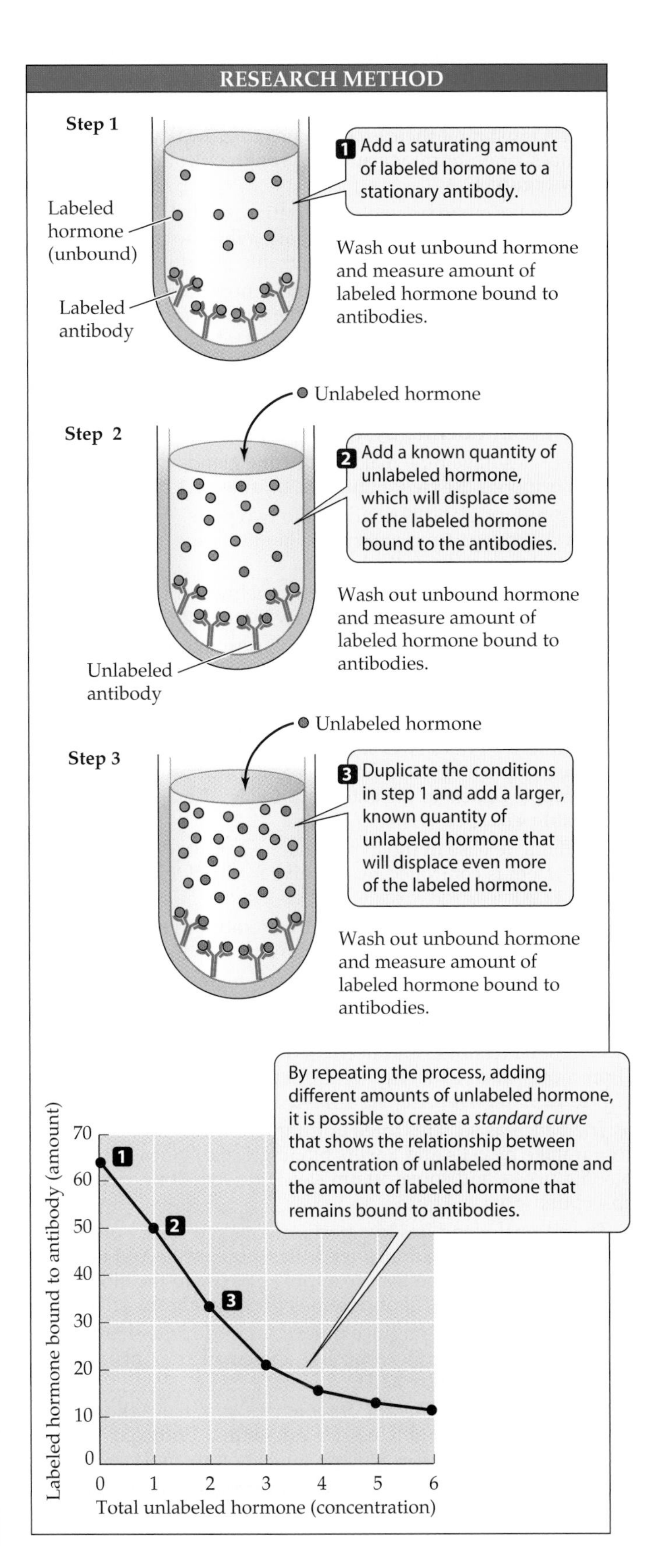

RESEARCH METHOD
Step 1
Labeled hormone (unbound)
Labeled antibody
1 Add a saturating amount of labeled hormone to a stationary antibody.
Wash out unbound hormone and measure amount of labeled hormone bound to antibodies.
Unlabeled hormone
Step 2
2 Add a known quantity of unlabeled hormone, which will displace some of the labeled hormone bound to the antibodies.
Wash out unbound hormone and measure amount of labeled hormone bound to antibodies.
Unlabeled antibody
Unlabeled hormone
Step 3
3 Duplicate the conditions in step 1 and add a larger, known quantity of unlabeled hormone that will displace even more of the labeled hormone.
Wash out unbound hormone and measure amount of labeled hormone bound to antibodies.
By repeating the process, adding different amounts of unlabeled hormone, it is possible to create a standard curve that shows the relationship between concentration of unlabeled hormone and the amount of labeled hormone that remains bound to antibodies.
Labeled hormone bound to antibody (amount)
Total unlabeled hormone (concentration)

42.15 An Immunoassay Measures Hormone Concentration The sample to be measured is added to a mixture that replicates the conditions in step 1. The sample's effect on decreasing the amount of labeled, bound hormone is compared to the standard curve, thus revealing the concentration of unlabeled hormone in the sample of interest.

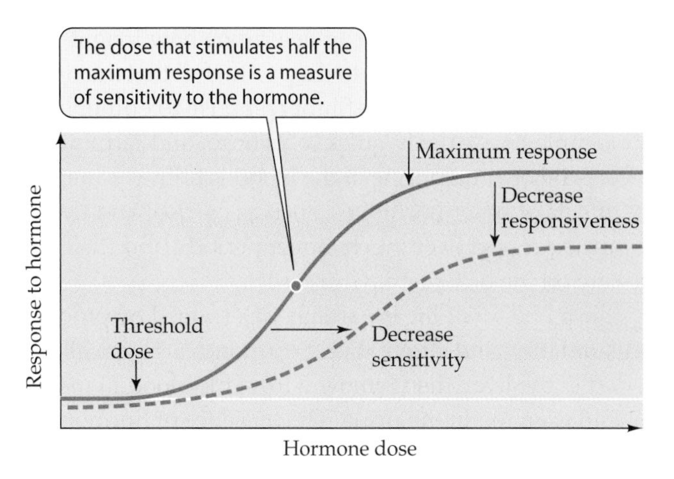

42.16 Dose–Response Curves Quantify Response to a Hormone Between the threshold and maximum values, a dose–response curve frequently has an S shape. Anything that changes the responsiveness of a system—such as a change in the number of receptors in target cells—affects the position of the curve.

tion to that standard curve, the concentration of a hormone can be determined. Similar techniques are used to determine concentrations of receptors.

Factors other than hormone and receptor concentrations can influence physiological responses to hormones; therefore, it is valuable to be able to characterize those responses. One way of doing so is to construct a **dose–response curve**. Using cells and tissues, organs, or even a whole animal, The subjects are experimentally treated with different amounts of a hormone and their response is measured. Responses are then plotted on the y axis of a graph, and the amount of the hormone used is plotted on the x axis (Figure 42.16). The resulting dose–response curve shows the *threshold* dose of the hormone necessary to get a response and the dose of the hormone that produces the *maximum* response. The hormone dose that stimulates half the maximum response indicates the *sensitivity* of the cell, tissue, organ, or animal to the hormone.

Anything that changes the responsiveness of a system to a hormone is reflected in the dose–response curve. A change in the availability of receptors in the responding cells, for example, can result in changes in threshold and in sensitivity. Changes in signaling pathways, rate-limiting enzymes, or the availability of substrates or cofactors can result in changes in the maximum response. Dose–response curves are valuable tools for studying hormone-mediated processes.

Hormones are not simple on–off switches, and an important characteristic of a hormone is the time course over which it acts. This time course can be measured by the hormone's **half-life** in the blood. Soon after endocrine cells are stimulated to secrete their hormone, the hormone reaches its maximum concentration in the blood. By taking subsequent blood samples, researchers can determine how long it takes for the circulating hormone to drop to half of that maximum concentration. The fight-or-flight response to epinephrine, for example, is relatively quick in its onset and termination; the half-life of epinephrine in the blood is only 1–3 minutes. The effects of other hormones, such as cortisol and thyroxine, are expressed over much longer periods, and their half-lives are on the order of days or weeks.

A hormone's half-life is partially determined by processes of degradation and elimination. Hormones are typically degraded in the liver, then removed from the blood in the kidney and excreted in the urine. The presence of hormones or their breakdown products in the urine is the reason urine samples provide important information in clinical tests.

Another factor that influences the half-life of a hormone is its ability to leave the blood. The extent to which a hormone is bound to carrier proteins limits its ability to diffuse out of the blood to reach its target cells, to be degraded in the liver, or to be excreted by the kidney. For example, when the mineralocorticoid aldosterone is released, about 15 percent of it binds to carrier proteins, and its half-life is 25 minutes. In contrast, when thyroxine is released, almost 100 percent of it binds to carrier proteins, and thyroxine has a half-life of 6 days. This variation in the time course of hormone action allows hormone signaling systems to have temporal characteristics appropriate to their functions.

Of course, the nature of the target cell's response to a hormone is also a factor in determining the time course of hormone action. For a hormone that stimulates a developmental effect, the time course of hormone action can be months, years, or even a lifetime. A very good example of a long-term process regulated primarily by hormones is animal reproduction, the topic of the next chapter.

Chapter Summary

Hormones and Their Actions

▶ Endocrine cells secrete chemical messages called hormones, which bind to receptors on or in target cells.

▶ Most hormones are peptides, proteins, steroids, or amines . Peptide and protein hormones and some amines are water-soluble; steroids and some amines are lipid-soluble.

▶ Receptors for water-soluble hormones are on the cell surface. Receptors for lipid-soluble hormones are inside the cell.

▶ Some hormones diffuse to targets near the site of secretion. Autocrine hormones influence the cell that secretes them; paracrine hormones influence nearby cells. **Review Figure 42.1**

▶ Most hormones are distributed throughout the body by the circulatory system.

▶ Hormones cause different responses in different target cells.

▶ Hormones may be secreted by single cells or by cells organized into discrete endocrine glands. **Review Figure 42.2. See Web/CD Activity 42.1**

Hormonal Control of Molting and Development in Insects

▶ Insects must molt their exoskeletons to grow. Two diffusible substances, brain hormone and ecdysone, control molting. **Review Figure 42.3**

▶ Juvenile hormone prevents maturation, so that juvenile instars molt into larger juvenile instars. When an insect stops producing juvenile hormone, it molts into an adult.

▶ Some insects undergo complete metamorphosis. When juvenile hormone drops to a low level, the larval form molts into a pupa. Because no juvenile hormone is secreted during pupation, the pupa molts into an adult. **Review Figure 42.4. See Web/CD Tutorial 42.1**

Vertebrate Endocrine Systems

▶ Vertebrates have nine major endocrine glands that secrete many hormones. **Review Figure 42.2, Table 42.1**

▶ The pituitary gland is divided into two parts. The anterior pituitary develops from embryonic mouth tissue; the posterior pituitary develops from the brain.

▶ The posterior pituitary secretes two neurohormones, antidiuretic hormone and oxytocin. **Review Figure 42.5**

▶ The anterior pituitary secretes tropic hormones (thyrotropin, adrenocorticotropin, luteinizing hormone, and follicle-stimulating hormone), as well as growth hormone, prolactin, melanocyte-stimulating hormone, endorphins, and enkephalins.

▶ The anterior pituitary is controlled by neurohormones produced by cells in the hypothalamus and transported through portal blood vessels to the anterior pituitary. **Review Figure 42.7, Table 42.2. See Web/CD Tutorial 42.2**

▶ Hormone release in the hypothalamus–pituitary–endocrine gland system is controlled by negative feedback loops. **Review Figure 42.8**

▶ The thyroid gland is controlled by thyrotropin and secretes thyroxine, which controls cell metabolism.

▶ The level of calcium in the blood is regulated by three hormones. Calcitonin lowers blood calcium by promoting bone deposition. Parathyroid hormone raises blood calcium by promoting bone turnover and decreased calcium excretion. Vitamin D promotes calcium absorption from the digestive tract. **Review Figure 42.9. See Web/CD Tutorial 42.3**

▶ The pancreas secretes three hormones. Insulin stimulates glucose uptake by cells and lowers blood glucose, glucagon raises blood glucose, and somatostatin slows the rate of nutrient absorption from the gut.

▶ The adrenal gland has two portions, one within the other. The hormones of the adrenal medulla, epinephrine and norepinephrine, stimulate the liver to supply glucose to the blood, as well as other fight-or-flight reactions. **Review Figure 42.10**

▶ The adrenal cortex produce three classes of corticosteroids: glucocorticoids, mineralocorticoids, and small amounts of sex steroids. **Review Figure 42.11**

▶ Aldosterone is a mineralocorticoid that stimulates the kidney to conserve sodium and to excrete potassium. Cortisol is a glucocorticoid that decreases glucose utilization by most cells.

▶ Sex hormones (androgens in males, estrogens and progesterone in females) are produced by the gonads in response to tropic hormones. Sex hormones control sexual development, secondary sexual characteristics, and reproductive functions. **Review Figure 42.12**

▶ The pineal hormone melatonin is involved in controlling biological rhythms and photoperiodism. **Review Figure 42.13**

Hormone Actions: The Role of Signal Transduction Pathways

▶ The response of a cell to a hormone depend on what receptors it has and what signal transduction pathways those receptors activate. **Review Table 42.3 and Figure 42.14**

▶ The sensitivity of a cell to hormones can be altered by up- or downregulation of the receptors in that cell.

▶ Immunoassays are used to measure concentrations of hormones and receptors. **Review Figure 42.15**

▶ Important tools for characterizing hormone action are dose–response curves and measurements of half-life. **Review Figure 42.16**

▶ The time course of a response to a hormone depends on many factors, including binding of the hormone to carrier proteins and elimination of the hormone through degradation and excretion.

See Web/CD Activity 42.2 for a concept review of this chapter.

Self-Quiz

1. Before puberty
   a. the pituitary secretes luteinizing hormone and follicle-stimulating hormone, but the gonads are unresponsive.
   b. the hypothalamus does not secrete much gonadotropin-releasing hormone.
   c. males can stimulate massive muscle development through a vigorous training program.
   d. testosterone plays no role in development of the male sex organs.
   e. genetic females will develop male genitals unless estrogen is present.
2. Both epinephrine and cortisol are secreted in response to stress. Which of the following statements is also true for *both* of these hormones?
   a. They act to increase blood glucose levels.
   b. Their receptors are on the surfaces of target cells.
   c. They are secreted by the adrenal cortex.
   d. Their secretion is stimulated by adrenocorticotropin.
   e. They are secreted into the blood within seconds of the onset of stress.
3. Growth hormone
   a. can cause adults to grow taller.
   b. stimulates protein synthesis.
   c. is released by the hypothalamus.
   d. can be obtained only from cadavers.
   e. is a steroid.
4. PTH
   a. stimulates osteoblasts to lay down new bone.
   b. reduces blood calcium levels.
   c. stimulates calcitonin release.
   d. is produced by the thyroid gland.
   e. is released when blood calcium levels fall.
5. Steroid hormones
   a. are produced only by the adrenal cortex.
   b. have only cell surface receptors.
   c. are water-soluble.
   d. act by altering the activity of proteins in the target cell.
   e. act by altering gene expression in the target cell.
6. The hormone ecdysone
   a. is released from the posterior pituitary.
   b. stimulates molting in insects.
   c. maintains an insect in larval stages unless brain hormone is present.
   d. stimulates the secretion of juvenile hormone from the prothoracic glands.
   e. keeps the insect exoskeleton flexible to permit growth.
7. The posterior pituitary
   a. produces oxytocin.
   b. is under the control of hypothalamic releasing neurohormones.
   c. secretes tropic hormones.
   d. secretes neurohormones.
   e. is under feedback control by thyroxine.
8. Which of the following contributes to a long half-life for a circulating hormone?
   a. The number of receptors on its target cells
   b. The fact that it is water-soluble
   c. The sensitivity to the hormone as expressed by its dose-response curve
   d. Its binding to carrier proteins in the blood
   e. A rapid rate of uptake by liver cells
9. Which of the following is a likely cause of goiter?
   a. The thyroid gland is producing too much PTH.
   b. Circulating levels of thyrotropin are too low.
   c. There is an inadequate supply of functional thyroxine.
   d. There is an oversupply of functional thyroxine.
   e. The diet contains too much iodine.
10. Which statement is true of all hormones?
   a. They are secreted by glands.
   b. They have receptors on cell surfaces.
   c. They may stimulate different responses in different cells.
   d. They target cells that are distant from their site of release.
   e. When the same hormone occurs in different species, it has the same action.

For Discussion

1. Explain how both hyperthyroidism and hypothyroidism can cause goiter. Refer to the roles of the hypothalamus and the pituitary in your answer.
2. There are several apparently enigmatic aspects of the role of PTH in regulating blood calcium. First, PTH raises blood calcium levels, yet osteoclasts do not have PTH receptors. Second, parathyroid cells have calcium-sensing receptors on their plasma membranes, and these receptors are activated by rising levels of blood calcium. Third, bone consists of calcium and phosphate salts, yet PTH causes blood phosphate to decline. Explain these various aspects of PTH actions.
3. Various side effects of anabolic steroid use were mentioned in this chapter. Some of these effects are due to the direct action of the steroid, but others are due to the negative feedback action of the steroid. Discuss an example of each and explain possible mechanisms.
4. Compare the characteristics you would expect of a hormone signaling system that controls a short-term process, such as digestion, with the characteristics you would expect of a hormone signaling system that controls a long-term process, such as development.
5. In the perpetual war between agriculturists and insects, a new weapon has been developed: a chemical similar to juvenile hormone. Explain how this chemical could be used as an insecticide. What factors do you think should be considered before it is released indiscriminately into the environment?

43 Animal Reproduction

Natural selection has created some amazing and bizarre adaptations, but among the most unusual and diverse are the methods that some animals use to reproduce. Just as "unmanned" submersibles are used in deep-sea exploration, some species of polychaete worms use "unwormed" submersibles to reproduce. The adults of these marine worms live in burrows on the seafloor or in coral reefs. Predators make it dangerous for them to leave their burrows to seek a mate, and if they simply released their eggs and sperm at the mouth of the burrow, they would have a poor chance of successful fertilization. So both males and females have evolved the capacity to develop specialized body segments that become stuffed with sperm or eggs. These segments develop sensory organs, but no mouth or gut, since they will not need to feed.

When the time is right—full moon for some species, new moon for others—these "sex-cell transporters" break loose from the main body of the worm, leave the burrow, swim toward the surface, swarm with more of their kind, release their sperm or eggs, and die. Union of sperm and eggs takes place in the open water, and fertilized eggs may drift a long way before they descend to the ocean floor and develop into adult worms. In many places, native people know when and where the sex-cell transporters will swarm, and they harvest them for food.

Most of the examples we will encounter in this chapter will not be as bizarre as this one, but the overall theme of this chapter will be the diverse ways in which animals produce offspring. We will first examine asexual mechanisms of reproduction, in which only a single parent is involved, and then turn to sexual reproduction, which requires two parents. Sexually reproducing organisms produce haploid sex cells—sperm and eggs—through the process of meiosis. An egg and a sperm must then unite through the process of fertilization to create a new diploid individual.

As we will see, much of the diversity in reproductive systems is in mechanisms for getting sperm and eggs together. This chapter, however, focuses mostly on the anatomy, function, and endocrine control of the human reproductive system. This information will allow us to understand the technologies we use both to limit and to enhance fertility. We will end the chapter with a discussion of sexual health and sexually transmitted diseases.

Feasting on Sex Cells During the final quarter of November's moon, the people of Samoa and Fiji harvest the specialized reproductive segments of the polychaete palolo worm (*Eunice viridis*). The worms release such segments into the ocean according to a precise cycle that native people have understood for centuries. The protein-rich segments are prepared by roasting or frying and eaten as a delicacy.

Asexual Reproduction

Sexual reproduction is a nearly universal trait in animals, although many species can reproduce asexually as well. Offspring produced asexually are genetically identical to one another and to their parents. Asexual reproduction is highly efficient because no mating is required. Furthermore, asexual populations can use resources efficiently because all individuals in the population can convert resources into offspring. However, asexual reproduction does not generate genetic diversity, and this can be a disadvantage in changing environments. As we learned in Chapter 9, genetic diversity enables natural selection to shape adaptations in response to environmental change.

A variety of animals, mostly invertebrates, reproduce asexually. They tend to be species that are sessile and cannot search for mates, or species that live in sparse populations and rarely encounter potential mates. Furthermore, asexually reproducing species are likely to be found in relatively constant environments.

There are three common modes of asexual reproduction: *budding, regeneration,* and *parthenogenesis.*

Budding and regeneration produce new individuals by mitosis

Many simple multicellular animals produce offspring by **budding**, in which new individuals form as outgrowths of the bodies of older animals. A bud grows by mitotic cell division, and the cells differentiate before the bud breaks away from the parent (Figure 43.1*a*). The bud is genetically identical to the parent, and it may grow as large as the parent before it becomes independent.

Regeneration is usually thought of as the replacement of damaged tissues or lost limbs, but in some cases pieces of an organism can regenerate complete individuals. Echinoderms, for example, have remarkable abilities to regenerate. If sea stars are cut into pieces, each piece that includes a portion of the central disc grows into a new animal (Figure 43.1*b*).

Regeneration frequently results when an animal is broken by an outside force. A storm, for example, can cause a heavy surf that breaks colonial cnidarians such as corals. Pieces broken off the colony can regenerate into new colonies. In some species, the breakage occurs in the absence of external forces. Some species of segmented marine worms related to the ones we discussed at the beginning of this chapter develop segments with rudimentary heads bearing sensory organs, then break apart. Each fragmented segment forms a new worm.

Parthenogenesis is the development of unfertilized eggs

Not all eggs have to be fertilized to develop. A common mode of asexual reproduction in arthropods is the development of offspring from unfertilized eggs. This phenomenon, called **parthenogenesis**, also occurs in some species of fish, amphibians, and reptiles. Most species that reproduce parthenogenetically also engage in sexual reproduction or sexual behavior.

In some species, parthenogenesis is part of the mechanism that determines sex. For example, in many hymenopterans (ants and most species of bees and wasps), males develop from unfertilized eggs and are haploid. Females develop from fertilized eggs and are diploid. Most females are sterile workers, but a select few become fertile queens. After a queen mates, she has a supply of sperm that she controls, enabling her to produce either fertilized or unfertilized eggs. Thus the queen determines when and how much of the colony resources are expended on males.

Parthenogenetic reproduction in some species requires sexual activity, even though this activity does not fertilize eggs. The eggs of parthenogenetically reproducing ticks and mites, for example, develop only after the animals have mated, even though the eggs remain unfertilized. One case that has been investigated extensively by David Crews and his students at the University of Texas is parthenogenetic reproduction in a species of whiptail lizard. There are no males

(*a*) *Hydra* sp.

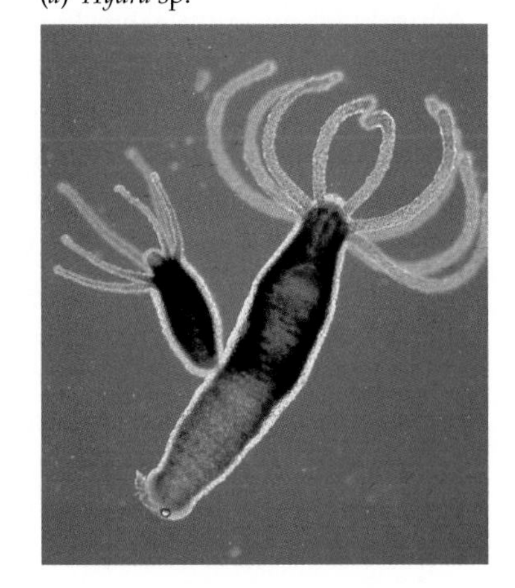

(*b*) *Linckia* sp.

43.1 Asexual Reproduction in Animals (*a*) Budding: A new individual forms as an outgrowth from an adult hydra. (*b*) Regeneration: This five-armed sea star is generating three new arms to replace amputated limbs ones, forming a complete animal.

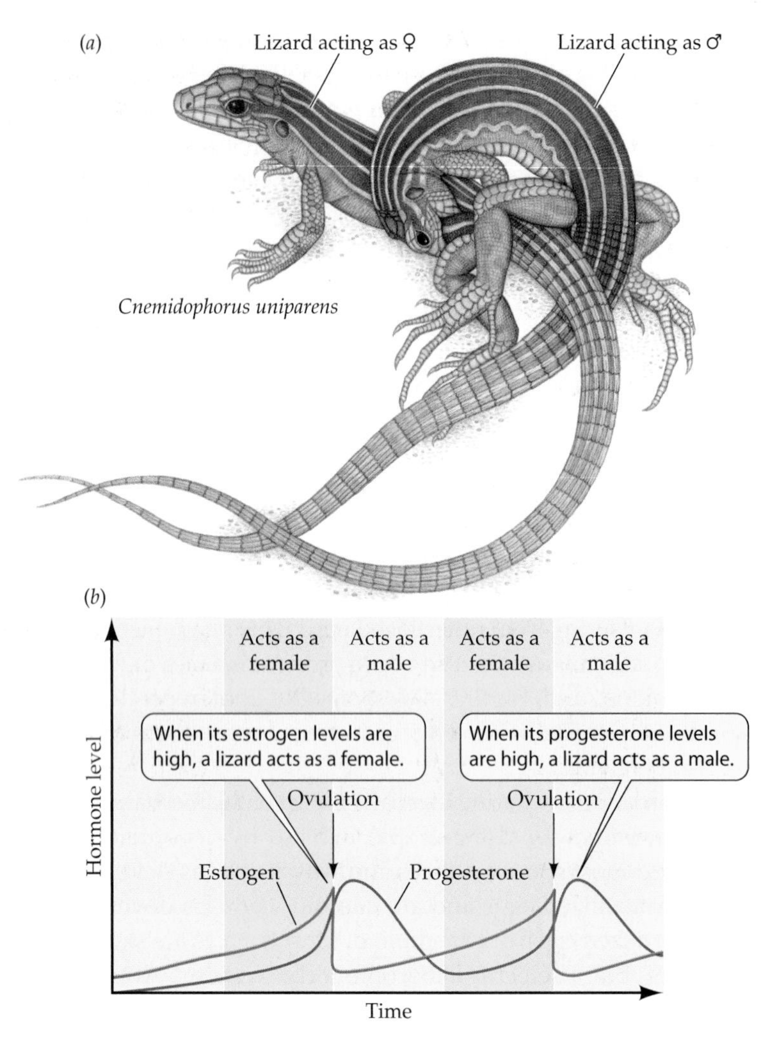

43.2 Sexual Behavior May Be Required for Asexual Reproduction (*a*) Parthenogenetic whiptail lizards are all female, but take turns acting the male role in reproductive behavior. (*b*) The stage of the ovarian cycle determines the role an individual whiptail plays.

in this species, but females act as males, engaging in all aspects of courtship display and mating, even though no sperm are produced or transferred (Figure 43.2). Whether a specific female acts as a female or as a male depends on her hormonal state at the time, but sexual activity is required to stimulate release of eggs from the ovary.

Sexual Reproduction

A large portion of the time and energy budgets of sexually reproducing animals goes into mating, which exposes them to predation, can result in physical damage, and detracts from other useful activities, such as feeding and caring for existing offspring. Furthermore, mating requires that resources be used to maintain a large population of males that do not bear offspring. In spite of all of these disadvantages, there is an overwhelming evolutionary advantage to sexual reproduction: It produces genetic diversity.

Sexual reproduction requires the joining of two haploid sex cells to form a diploid individual. These haploid cells, or *gametes*, are produced through **gametogenesis**, a process that involves meiotic cell divisions. Two events in meiosis contribute to genetic diversity: crossing over between homologous chromosomes and the independent assortment of chromosomes. Both of these genetic phenomena were described in Chapter 10.

Sexual reproduction itself also contributes to genetic diversity. The genetic variation among the gametes of a single individual and the genetic variation between any two parents produce an enormous potential for genetic variation between any two offspring of a sexually reproducing pair of individuals. This genetic diversity is the raw material for natural selection.

Sexual reproduction in animals consists of three fundamental steps:

- *Gametogenesis* (making gametes)
- *Mating* (g etting gametes together)
- *Fertilization* (getting gametes to fuse)

There is not a great deal of diversity in gametogenesis when we compare different groups of animals. Processes of fertilization are also rather similar in widely different species. Therefore, although the discussion of gametogenesis that follows focuses primarily on mammals, and our discussion of fertilization mainly deals with sea urchins, the facts would not be dramatically different if we focused on a different group of animals. Adaptations for mating, on the other hand, show incredible anatomical, physiological, and behavioral diversity.

Eggs and sperm form through gametogenesis

Gametogenesis occurs in the **gonads**, which are **testes** (singular, testis) in males and **ovaries** in females. The tiny gametes of males, called **sperm**, move by beating their flagella. The larger gametes of females, called **eggs** or **ova** (singular, ovum), are nonmotile (see Figure 20.1).

Gametes are produced from **germ cells**, which have their origin in the earliest cell divisions of the embryo and remain distinct from the rest of the body. All other cells of the embryo are called *somatic cells*. Germ cells are sequestered in the body of the embryo until its gonads begin to form. The germ cells then migrate to the gonads, where they take up residence and proliferate by mitosis, producing **oogonia** (singular, oogonium) in females and **spermatogonia** (singular, spermatogonium) in males. Oogonia and spermatogonia, which are diploid, also multiply by mitosis, eventually producing **primary oocytes** and **primary spermatocytes**, which are still diploid cells.

Meiosis, the next step in gametogenesis, reduces the chromosomes to the haploid number, and the resulting haploid cells eventually mature into sperm and ova. (You may want to review the discussion of meiosis in Chapter 9 before reading further.) Although the steps of meiosis are very similar in males and females, there are some significant differences in gametogenesis between the sexes.

SPERMATOGENESIS PRODUCES SPERM. Primary spermatocytes undergo the first meiotic division to form **secondary spermatocytes**, which are haploid. The second meiotic division produces four haploid **spermatids** for each primary spermatocyte that entered meiosis (Figure 43.3*a*). In mammals, these cells remain connected by cross-bridges of cytoplasm after each division.

The reason that mammalian spermatocytes remain in cytoplasmic contact throughout their development probably is the asymmetry of sex chromosomes in males. Half the secondary spermatocytes receive an X chromosome, the other half a Y chromosome. The Y chromosome contains fewer genes than the X chromosome, and apparently some of the products of genes found only on the X chromosome are essential for spermatocyte development. By remaining in cytoplasmic contact, all four spermatocytes can share the gene products of the X chromosomes, even though only half of them have an X chromosome.

A spermatid bears little resemblance to a mature sperm. Through further differentiation, however, it will become compact, streamlined, and motile. We will look at the differentiation of human sperm in more detail below.

OOGENESIS PRODUCES EGGS. Oogonia, like spermatogonia, proliferate through mitosis. The resulting primary oocytes immediately enter prophase of the first meiotic division. In many species, including humans, the development of the oocyte is arrested at this point and may remain so for days,

43.3 Gametogenesis (*a*) Mitosis in diploid spermatogonia produces haploid spermatids, which differentiate into sperm. (*b*) Mitosis in diploid oogonia produces haploid secondary oocytes, which mature into ova.

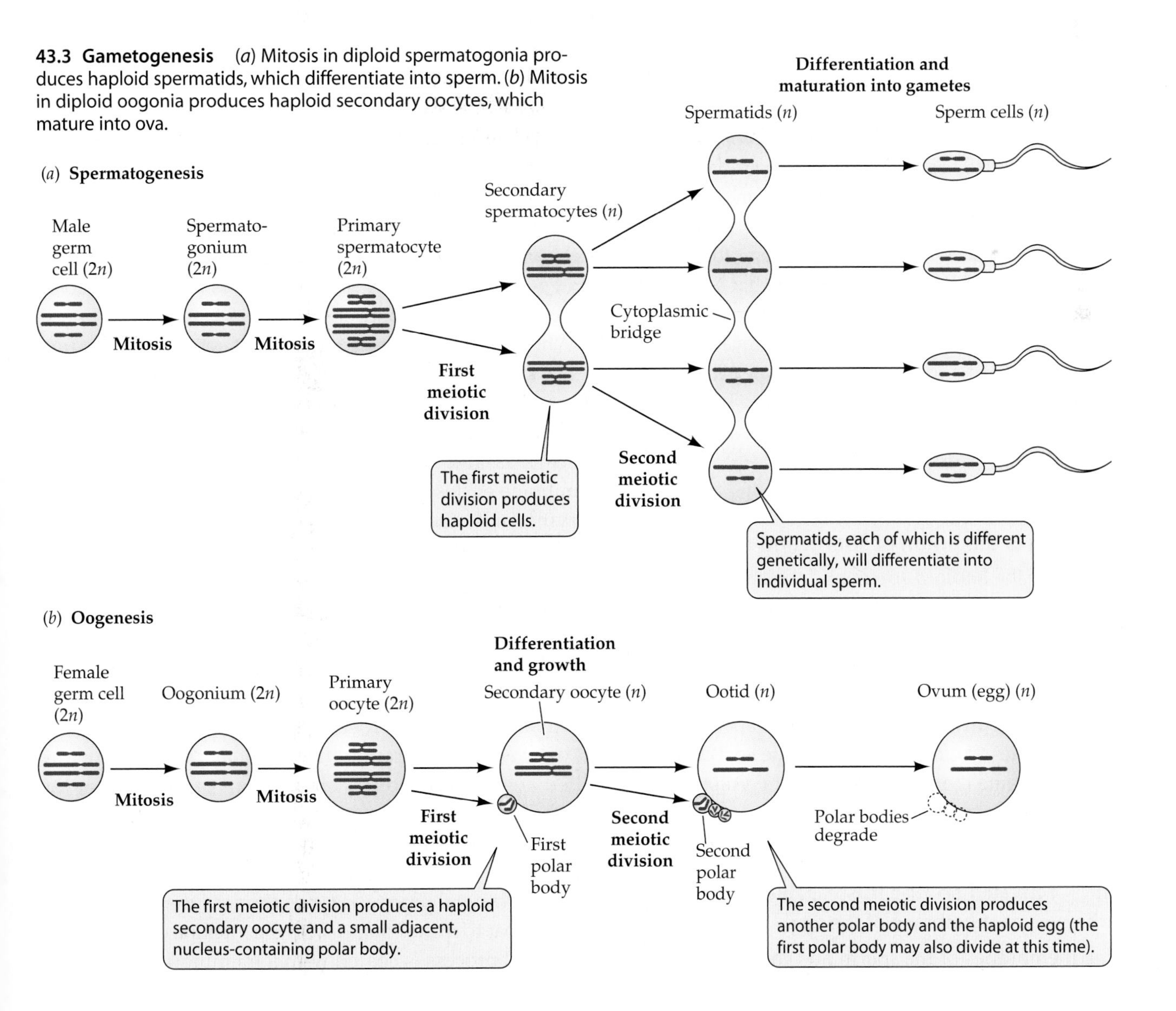

months, or years. In contrast, there is no arrest during male gametogenesis, which goes steadily to completion once the primary spermatocyte has differentiated. In the human female, as we will see, some primary oocytes may remain in arrested prophase I for up to 50 years!

During this prolonged prophase I, or shortly before it ends, the primary oocyte undergoes its major growth phase. It grows larger due to increased production of ribosomes, RNA, cytoplasmic organelles, and energy stores. At this time, the primary oocyte acquires all the energy, raw materials, and RNA that the egg will need to survive its first cell divisions after fertilization. In fact, the nutrients in the egg will have to maintain the embryo until it is either nourished by the maternal system or can feed on its own.

When a primary oocyte resumes meiosis, its nucleus completes the first meiotic division near the surface of the cell. The daughter cells of this division receive grossly unequal shares of cytoplasm. This asymmetry represents another major difference from spermatogenesis, in which cytoplasm is apportioned equally. The daughter cell that receives almost all the cytoplasm becomes the **secondary oocyte**, and the one that receives almost none forms the *first polar body* (Figure 43.3*b*).

The second meiotic division of the large secondary oocyte is also accompanied by an asymmetrical division of the cytoplasm. One daughter cell forms the large, haploid **ootid**, which eventually differentiates into a mature ovum, and the other forms the *second polar body*. Polar bodies degenerate, so the end result of oogenesis is only one mature egg for each primary oocyte that entered meiosis. However, that egg is a very large, well-provisioned cell.

A second period of arrested development occurs after the first meiotic division forms the secondary oocyte. The egg may be expelled from the ovary in this condition. In many species, including humans, the second meiotic division is not completed until the egg is fertilized by a sperm.

Fertilization is the union of sperm and egg

The union of the haploid sperm and the haploid egg creates a single diploid cell, called a **zygote**, which will develop into an embryo. Fertilization does more, however, than just restore the full genetic complement of the animal. The events and processes associated with fertilization help eggs and sperm get together, prevent the union of sperm and eggs of different species, guarantee that only one sperm will enter an egg, and activate the egg metabolically. Fertilization involves a complex series of events:

- The sperm and the egg recognize each other.
- The sperm is activated so that it is capable of gaining access to the plasma membrane of the egg.
- The plasma membranes of the sperm and the egg fuse.
- The egg blocks entry by additional sperm.
- The egg is metabolically activated and stimulated to start development.
- The egg and sperm nuclei fuse to create the diploid nucleus of the zygote.

SPECIFICITY IN SPERM–EGG INTERACTIONS. Specific recognition molecules mediate interactions between sperm and eggs. These molecules ensure that the activities of sperm are directed toward eggs and not other cells, and they help prevent eggs from being fertilized by sperm from the wrong species. The latter function is particularly important in aquatic species that release eggs and sperm into the surrounding water. The sea urchin is such a species, and its mechanisms of fertilization have been well studied.

The eggs of sea urchins and various other marine invertebrates release chemical attractants that increase the motility of sperm and cause them to swim toward the egg. These chemical attractants are species-specific. For example, eggs of one species of sea urchin release a peptide consisting of 14 amino acids. As this peptide diffuses from the egg, it binds to receptors on the sperm the same species. The sperm respond by increasing their mitochondrial respiration and their motility. Before exposure to the peptide, the sperm swim in tight little circles, but after binding the peptide, they swim energetically up the concentration gradient of the peptide until they reach the egg that is releasing it. The peptide released by eggs of one species of sea urchin does not bind to receptors on sperm of other species.

When sperm reach an egg, they must get through two protective layers before they can fuse with the egg plasma membrane. The eggs of sea urchins are covered with a **jelly coat**, which surrounds a proteinaceous **vitelline envelope**. The sperm's assault on these protective layers depends on a membrane-enclosed structure called an **acrosome** containing enzymes and other proteins. The acrosome is located at the front of the sperm head, where it forms a cap over the nucleus (Figure 43.4).

When the sperm makes contact with an egg of its own species, substances in the jelly coat trigger an *acrosomal reaction*, which begins with the breakdown of the plasma membrane covering the sperm head and the underlying acrosomal membrane. The acrosomal enzymes are released, and they digest a hole through the jelly coat. Next, a structure called an *acrosomal process* extends out of the head of the sperm. The acrosomal process forms from globular actin proteins behind the acrosome, which polymerize when the acrosomal membrane breaks down.

The acrosomal process extends through the remainder of the jelly coat to make contact with the vitelline envelope. The acrosomal process is coated with a membrane-bound pro-

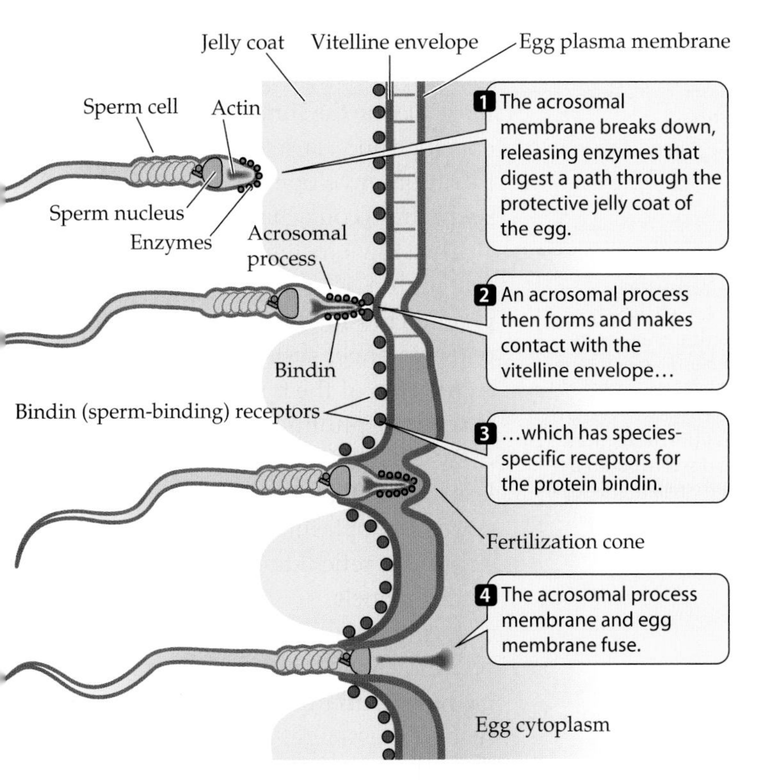
Jelly coat
Vitelline envelope
Egg plasma membrane
Sperm cell
Actin
Sperm nucleus
Enzymes
Acrosomal process
Bindin
Bindin (sperm-binding) receptors
1 The acrosomal membrane breaks down, releasing enzymes that digest a path through the protective jelly coat of the egg.
2 An acrosomal process then forms and makes contact with the vitelline envelope...
3 ...which has species-specific receptors for the protein bindin.
Fertilization cone
4 The acrosomal process membrane and egg membrane fuse.
Egg cytoplasm

43.4 The Acrosomal Reaction The acrosomal reaction allows a sea urchin sperm to recognize an egg of the same species and pass through its protective layers.

tein called *bindin*. Different species have different kinds of bindin molecules. The plasma membrane of the egg has species-specific bindin receptors that extend through the vitelline envelope. The reaction of acrosomal bindin with these receptors stimulates the egg plasma membrane to form a *fertilization cone* that engulfs the sperm head, bringing it into the egg cytoplasm.

In animals that practice internal fertilization, mating behaviors help guarantee species specificity, but egg–sperm recognition mechanisms still exist. The mammalian egg, for example, is surrounded by a thick layer called the **cumulus**, which consists of a loose assemblage of maternal cells in a gelatinous matrix (Figure 43.5). Beneath the cumulus is a glycoprotein envelope called the **zona pellucida**, which is functionally similar to the vitelline envelope of sea urchin eggs. When mammalian sperm are deposited in the female reproductive tract, they become metabolically activated and are made capable of an acrosomal reaction if they should meet an egg. An activated sperm can penetrate the cumulus and interact with the zona pellucida.

Unlike the jelly coat of sea urchin eggs, the cumulus of mammalian eggs does not trigger the acrosomal reaction. When sperm make contact with the zona pellucida, a species-specific glycoprotein in the zona binds to recognition molecules on the head of the sperm. This binding triggers the acrosomal reaction, releasing acrosomal enzymes that digest a path through the zona. When the sperm head reaches the egg plasma membrane, other proteins cause the adhesion of sperm to egg plasma membrane and facilitate fusion of sperm and egg.

The importance of the zona pellucida and its sperm-binding molecules as a species-specific recognition mechanism was revealed in experiments on mammalian eggs and sperm in culture dishes. When the zona was stripped from human eggs and they were exposed to hamster sperm, fertilization took place, resulting in a hamster–human hybrid zygote. The hybrid zygote did not survive its first cell division because of chromosomal incompatibilities, but the experiment demonstrated that the recognition mechanism in mammalian species resides in the zona.

BLOCKS TO POLYSPERMY AND EGG ACTIVATION. The fusion of the sperm and egg plasma membranes and the entry of the sperm into the egg initiate a programmed sequence of events. The first responses to sperm entry are *blocks to polyspermy*—that is, mechanisms that prevent more than one sperm from entering the egg. If more than one sperm enters the egg, the resulting embryo is unlikely to survive.

Blocks to polyspermy have been studied extensively in sea urchin eggs, which can be fertilized in a dish of seawater. Within seconds after a sperm enters a sea urchin egg, there is an influx of sodium ions, which changes the electric charge difference across the egg's plasma membrane. This *fast block to polyspermy* prevents the fusion of other sperm with the egg plasma membrane.

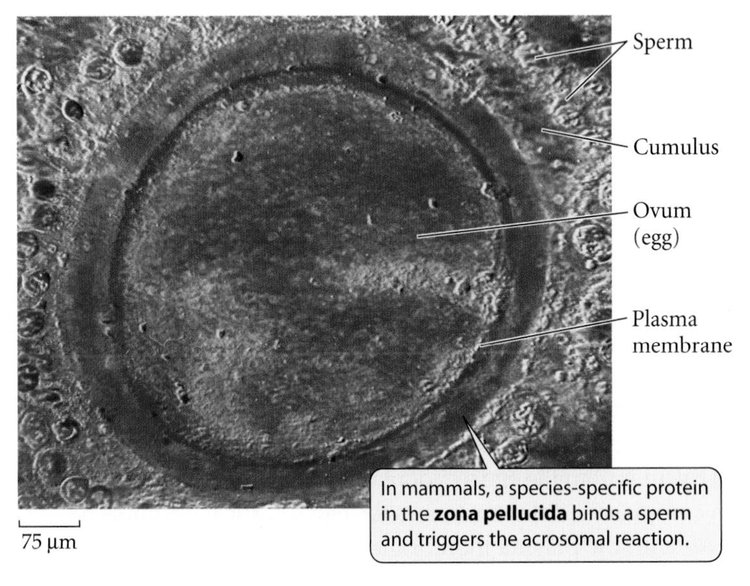
Sperm
Cumulus
Ovum (egg)
Plasma membrane
In mammals, a species-specific protein in the **zona pellucida** binds a sperm and triggers the acrosomal reaction.
75 μm

43.5 A Mammalian Egg Is Surrounded by Barriers to Sperm This human egg is protected by the cumulus and zona pellucida, both of which sperm (shown here in blue) must penetrate to fertilize the egg. Only one sperm will penetrate the zona pellucida.

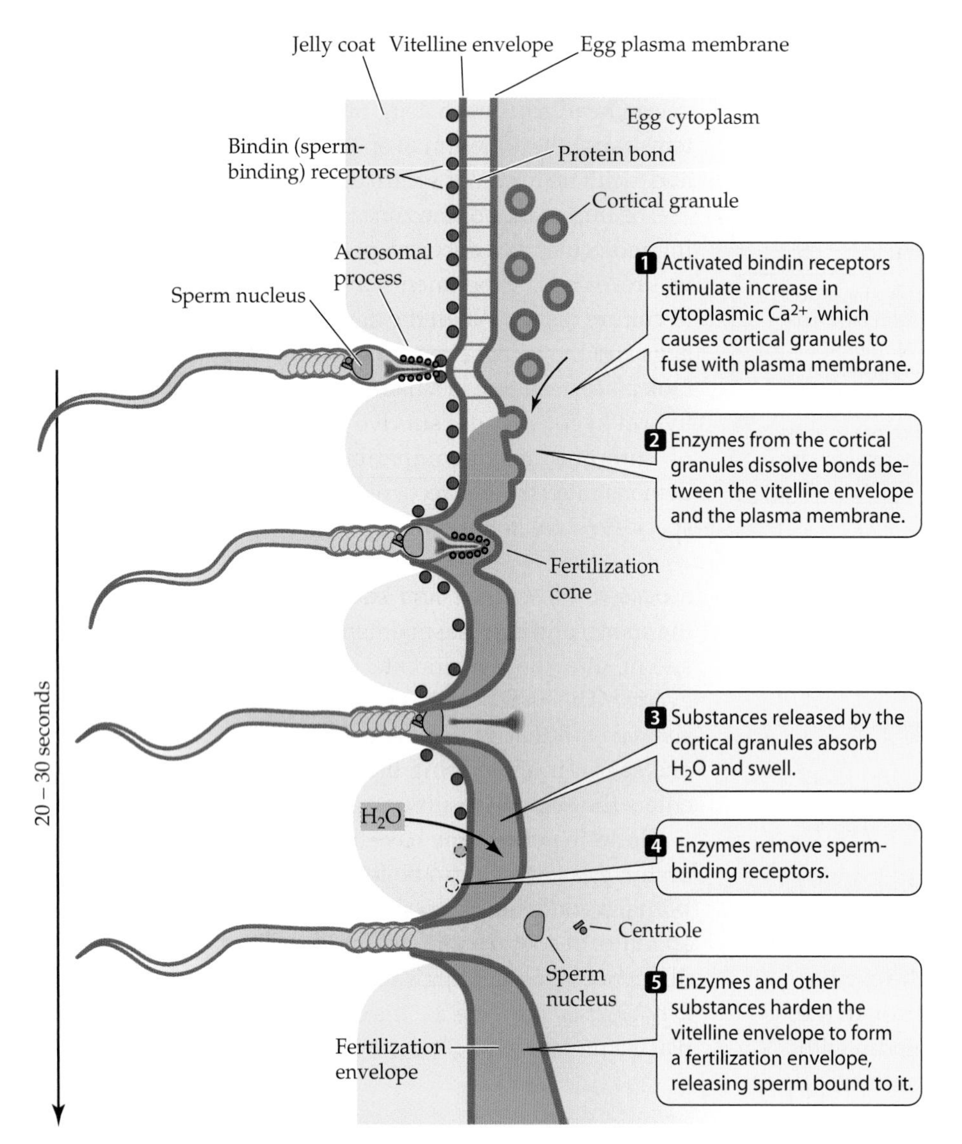
Jelly coat
Vitelline envelope
Egg plasma membrane
Egg cytoplasm
Bindin (sperm-binding) receptors
Protein bond
Cortical granule
Acrosomal process
Sperm nucleus
1 Activated bindin receptors stimulate increase in cytoplasmic Ca2+, which causes cortical granules to fuse with plasma membrane.
2 Enzymes from the cortical granules dissolve bonds between the vitelline envelope and the plasma membrane.
Fertilization cone
3 Substances released by the cortical granules absorb H2O and swell.
H2O
4 Enzymes remove sperm-binding receptors.
Centriole
Sperm nucleus
5 Enzymes and other substances harden the vitelline envelope to form a fertilization envelope, releasing sperm bound to it.
Fertilization envelope
20 – 30 seconds

43.6 The Slow Block to Polyspermy Enzymes from the sea urchin egg's cortical granules trigger the slow block to polyspermy.

The *slow block to polyspermy* takes about a minute and results from the release of calcium (Figure 43.6). Before fertilization, the vitelline envelope is bonded to the egg plasma membrane. Just under the plasma membrane are vesicles called *cortical granules*, which contain enzymes and other proteins. The sea urchin egg, like all animal cells, contains calcium ions that are sequestered in the endoplasmic reticulum.

Sperm binding to the sea urchin egg stimulates the release of calcium from the egg's endoplasmic reticulum. The increase in cytoplasmic calcium causes the egg's cortical granules to fuse with the plasma membrane and release their contents. The cortical granule enzymes break the bonds between the vitelline envelope and the plasma membrane, and other proteins released from the cortical granules attract water into the space between them. As a result, the vitelline envelope rises to form a *fertilization envelope*. Cortical granule enzymes also degrade sperm-binding molecules on the surface of the fertilization envelope and cause it to harden. The fertilization envelope prevents additional sperm from contacting the egg.

The slow block to polyspermy in the sea urchin is mediated by the phosphatidyl inositol-bisphosphate (PIP2) second messenger system (see Figure 15.11). Activation of the bindin receptors activates phospholipase C, which cleaves PIP2 in the egg plasma membrane, releasing inositol triphosphate (IP_3) into the egg cell cytoplasm. IP_3 diffuses to the endoplasmic reticulum, where it opens calcium channels.

In mammals, sperm entry does not seem to cause a rapid change in membrane potential, but it triggers the PIP2 second messenger system, resulting in several events. Calcium is released from the endoplasmic reticulum, and as in the sea urchin, the increased calcium causes the cortical granules to fuse with the egg plasma membrane. A fertilization envelope does not form around the mammalian egg, but the cortical granule enzymes destroy the molecules in the zona pellucida that bind sperm. The rise in cytoplasmic calcium also activates the egg's metabolism and signals it to complete meiosis. The pH of the egg's cytoplasm increases, its oxygen consumption rises, protein synthesis increases, and DNA synthesis is initiated. The stage is set for the first cell division.

Anatomical and behavioral adaptations bring eggs and sperm together

As we have just seen, sexual reproduction requires the production of haploid gametes (gametogenesis) and the joining together of those gametes to form a diploid zygote(fertilization). Mating, the step in between these two processes, gets eggs and sperm close enough together so that fertilization can occur. The simplest distinction in mating systems is whether fertilization occurs externally or internally.

EXTERNAL FERTILIZATION. In an aquatic environment, animals can bring their gametes together by simply releasing them into the water. This practice, called *external fertilization*, is common among simple aquatic animals that are not very mobile. Such animals may produce huge numbers of gametes. A female oyster, for example, may produce 100 million eggs in a year, and the number of sperm produced by a male oyster is astronomical.

But numbers alone do not guarantee that gametes will meet. The reproductive activities of the males and females of a population must be synchronized, since released gametes have a limited life span. Seasonal breeders may use day length, changes in temperature, or changes in weather to time their production and release of gametes. Social stimulation is also important. Sexual activity by one member of a population can stimulate others to engage in it.

Behavior can play an important role in bringing gametes together even when fertilization is external. Many species travel great distances to congregate with potential mates and release their gametes at the same time in a suitable environment. Salmon are an extreme example, traveling hundreds of miles to spawn in the stream where they hatched.

INTERNAL FERTILIZATION. Terrestrial animals cannot simply release their gametes into the environment. Sperm can move only through liquid, and delicate gametes released into air would dry out and die. Terrestrial animals avoid these problems by releasing sperm directly into the female reproductive tract. This practice is called *internal fertilization*.

Animals have evolved an incredible diversity of behavioral and anatomical adaptations for internal fertilization. As we saw above, gametogenesis occurs in the gonads, which are the *primary sex organs*. All of the additional anatomical components of an animal's reproductive system are called *accessory sex organs*. An obvious accessory sex organ in the males of many species is the **penis**, which enables the male to deposit sperm in the female's reproductive tract. Accessory sex organs include a variety of glands, tubules, ducts, and other structures.

Copulation is the physical joining of male and female accessory sex organs. Transfer of sperm in internal fertilization can also be indirect. Males of many invertebrate species (e.g., mites and scorpions) and a few vertebrates (e.g., salamanders) deposit *spermatophores*—packets of sperm—in the environment. When a female mite encounters a spermatophore from a potential mate, she straddles it and opens a pair of plates in her abdomen so that the tip of the spermatophore enters her reproductive tract and allows the sperm to enter.

Male squids and spiders play a more active role in spermatophore transfer. The male spider secretes a drop containing sperm onto a bit of web, then uses a special structure on his foreleg to pick up the sperm-containing web and insert it through the female's genital opening. Male squids use one specialized tentacle to pick up a spermatophore and insert it into the female's genital opening.

Most male insects copulate and transfer sperm to the female's vagina through a penis. The **genitalia**—external sex organs—of insects often have species-specific shapes that match in a lock-and-key fashion. This mechanism ensures a tight, secure fit between the mating pair during the prolonged period of sperm transfer. In some insect species in which females mate with more than one male, the males have elaborate structures on their penises that can scoop sperm deposited by other males out of a female's reproductive tract, replacing it with their own.

A single body can function as both male and female

In most species, gametes are produced by individuals that are either male or female. Species that have separate male and female members are called **dioecious** species (from the Greek for "two houses"). In some species, however, a single individual may produce both sperm and eggs. Such species are called **monoecious** ("one house") or **hermaphroditic** species.

Almost all invertebrate groups contain some hermaphroditic species. An earthworm is an example of a *simultaneous hermaphrodite*, meaning that it is both male and female at the same time. When two earthworms mate, they exchange sperm, and as a result, the eggs of each are fertilized (see Figure 32.24*b*). Some vertebrates, such as the anemonefish described at the beginning of Chapter 21, are *sequential hermaphrodites*, meaning that an individual may function as a male or as a female at different times in its life.

What is the selective advantage of hermaphroditism? Some simultaneous hermaphrodites, such as parasitic tapeworms, have a low probability of meeting a potential mate. Even though a tapeworm may be large and cause lots of trouble for its host, it may be the only tapeworm in the host. Tapeworms can fertilize their own eggs. Most simultaneous hermaphrodites must mate with another individual, but since each member of the population is both male and female, the probability of encountering a possible mate is double what it would be in monoecious species. In some sequential hermaphrodites, all siblings are either male or female at the same time, thus reducing the incidence of inbreeding.

The evolution of vertebrate reproductive systems parallels the move to land

The earliest vertebrates evolved in aquatic environments. The closest living relatives of those earliest vertebrates are modern-day fishes. They remain exclusively aquatic animals, and most practice external fertilization. The most primitive of the fishes, the lampreys and hagfishes, simply release their gametes into the environment. In most fishes, however, mating behaviors bring females and males into close proximity at the time of gamete release. In some sharks and rays, fins have evolved into claspers that hold the male and female together and enable sperm to be transferred directly into the female reproductive tract.

Amphibians were the first vertebrates to live in terrestrial environments. They dealt with the challenge of a dry environment by returning to water to reproduce, as most amphibians still do today.

Reptiles were the first vertebrate group to solve the problem of reproduction in the terrestrial environment. Their solution, the **amniote egg**, is shared with the birds. A good example is the chicken egg, which contains a supply of food (yolk) and water for the developing embryo (see Figure 20.18). A hard shell protects the embryo and impedes water loss while allowing the diffusion of oxygen and carbon dioxide (Figure 43.7*a*). The eggshell creates an obvious problem for fertilization, however: Sperm cannot penetrate the shell, so they have to reach the egg before the shell forms. Hence internal fertilization and the evolution of accessory sex organs were necessary for the evolution of the amniote egg.

Male snakes and lizards have paired *hemipenes*, which can be filled with blood and thereby extruded from the male's body. Only one hemipene is inserted into the female's reproductive tract at a time. It is usually rough or spiny at the end to achieve a secure hold while sperm are transferred down a groove on its surface. Retractor muscles pull the hemipene back into the male's body when mating is completed. Some birds, mostly more primitive species, have erectile penises that channel sperm along a groove into the female's reproductive tract. Bird species with more recent evolutionary origins, however, do not have erectile penises; instead, the male and female simply bring their genital openings close together to transfer sperm. Usually this involves the male standing on the female's back (Figure 43.7*b*).

All mammals practice internal fertilization, but except for the monotremes, they have done away with the shelled egg. They retain the developing embryo in the female reproductive tract, at least through the early stages of development. Mammalian species vary enormously as to the developmental stage of their offspring at the time of birth.

Reproductive systems are distinguished by where the embryo develops

Two patterns of care and nurture of the embryo have evolved in animals: oviparity (egg laying) and viviparity (live bearing). **Oviparous** animals lay eggs in the environment, and their embryos develop outside the mother's body. Oviparity is possible because eggs are stocked with abundant nutrients to supply the needs of the embryo.

Oviparous terrestrial animals, such as insects, reptiles, and birds, protect their eggs from desiccation with tough, waterproof membranes or shells. Some oviparous animals engage in various forms of parental behavior to protect their eggs, but until the eggs hatch, the embryos depend entirely on the nutrients stored in the egg. The only oviparous mammalian

43.7 The Shelled Egg The shelled egg was a major evolutionary step that allowed reptiles and birds to reproduce in the terrestrial environment. (*a*) A female green sea turtle deposits her eggs in the sand. (*b*) Because their environment offers no water to bring sperm and eggs together, terrestrial animals must practice internal fertilization, as these penguins are doing.

(*a*) *Chelonia mydas*

(*b*) *Aptenodytes patagonicus*

species are the monotremes: the echidnas and the duck-billed platypus (see Figure 34.22).

Viviparous animals retain the embryo within the mother's body during its early developmental stages. All mammals (except monotremes) are viviparous. There are examples of viviparity in all other vertebrate groups except the crocodiles, turtles, and birds. Even some sharks retain fertilized eggs in their bodies and give birth to free-living offspring. But there is a big difference between viviparity in mammals and in other species. Mammals have a specialized portion of the female reproductive tract, called the **uterus** or *womb,* that holds the embryo and interacts with it to produce a **placenta**, which enables the exchange of nutrients and wastes between the blood of the mother and that of the embryo. Non-mammalian viviparous animals simply retain fertilized eggs in the mother's body until they hatch. These embryos still receive nutrition from stores in the egg, so this reproductive adaptation is called **ovoviviparity**.

The Human Reproductive System

So far we have seen a small sampling of the fascinating diversity of animal reproductive systems. In this section we will describe the structures and functions of the male and female sex organs in mammals—specifically, in human beings—and discuss hormonal regulation of both male and female systems. Our discussion will include the primary sex organs—testes in males and ovaries in females—that produce gametes and serve endocrine functions. It will also include the accessory sex organs—the ducts through which the gametes pass as well as various glands that empty into those ducts. In humans, the breasts are also considered accessory sex organs. We will also refer on occasion to *secondary sexual characteristics*, which are not directly involved in reproduction, but comprise the external differences between males and females.

Male sex organs produce and deliver semen

Semen is the product of the male reproductive system. Besides sperm, semen contains a complex mixture of fluids and molecules that support the sperm and facilitate fertilization. Sperm make up less than 5 percent of the volume of the semen.

Sperm are produced in the testes, the paired male gonads. In most mammals, the testes are located outside the body cavity in a pouch of skin, called the **scrotum** (Figure 43.8), but there are notable exceptions, including elephants, bats, and marine mammals. Why should the testes be located outside the body cavity? One explanation is that the optimal temperature for spermatogenesis in most mammals is slightly lower than the normal body temperature. The scrotum keeps the testes at this optimal temperature. Muscles in the scrotum contract in a cold environment, bringing the testes closer to the warmth of the body; in a hot environment they relax, suspending the testes farther from the body.

A mammalian testis consists of a great length of tightly coiled **seminiferous tubules**, within which spermatogenesis takes place. Each tubule is lined with a stratified epithelium.

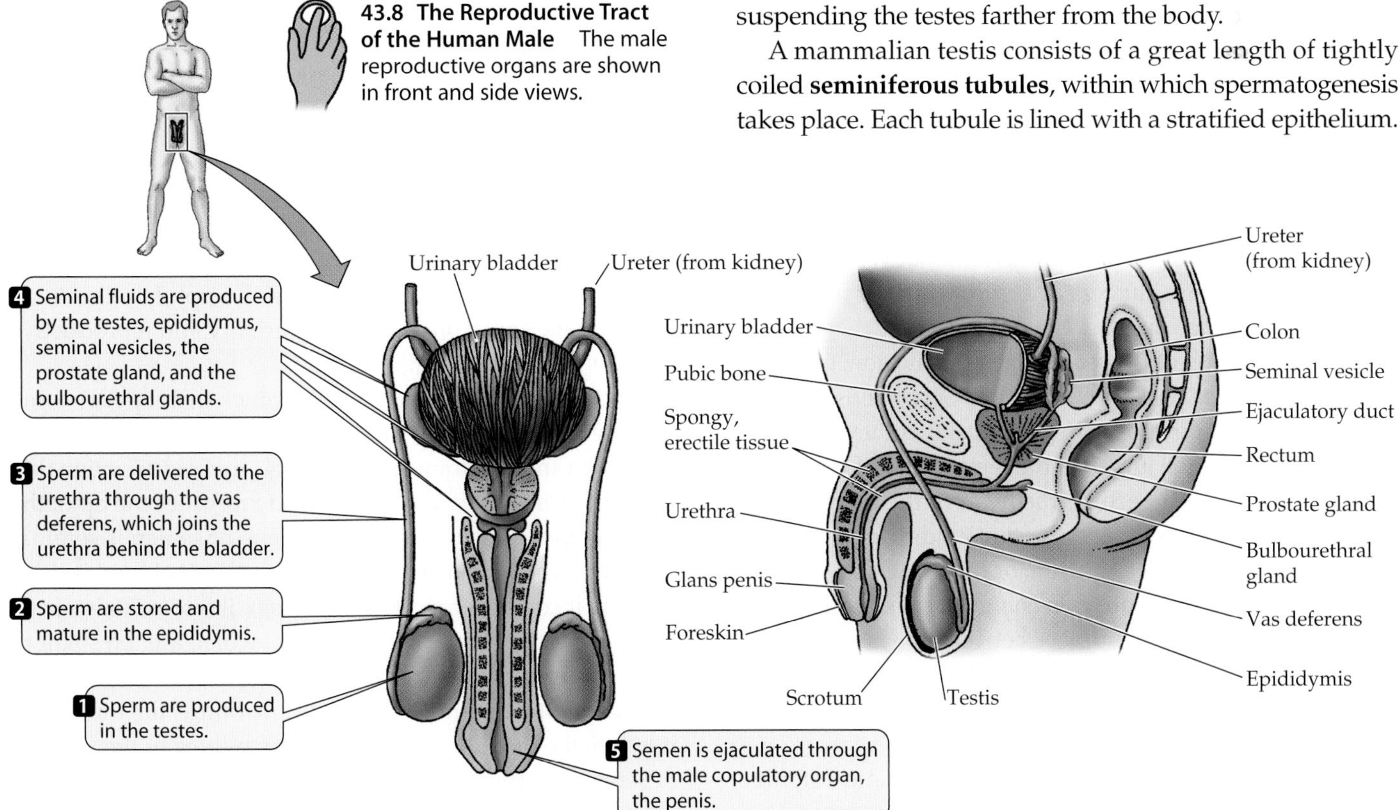

4 Seminal fluids are produced by the testes, epididymus, seminal vesicles, the prostate gland, and the bulbourethral glands.
3 Sperm are delivered to the urethra through the vas deferens, which joins the urethra behind the bladder.
2 Sperm are stored and mature in the epididymis.
1 Sperm are produced in the testes.
Urinary bladder
Ureter (from kidney)
5 Semen is ejaculated through the male copulatory organ, the penis.
Urinary bladder
Pubic bone
Spongy, erectile tissue
Urethra
Glans penis
Foreskin
Scrotum
Testis
Ureter (from kidney)
Colon
Seminal vesicle
Ejaculatory duct
Rectum
Prostate gland
Bulbourethral gland
Vas deferens
Epididymis

43.8 The Reproductive Tract of the Human Male The male reproductive organs are shown in front and side views.

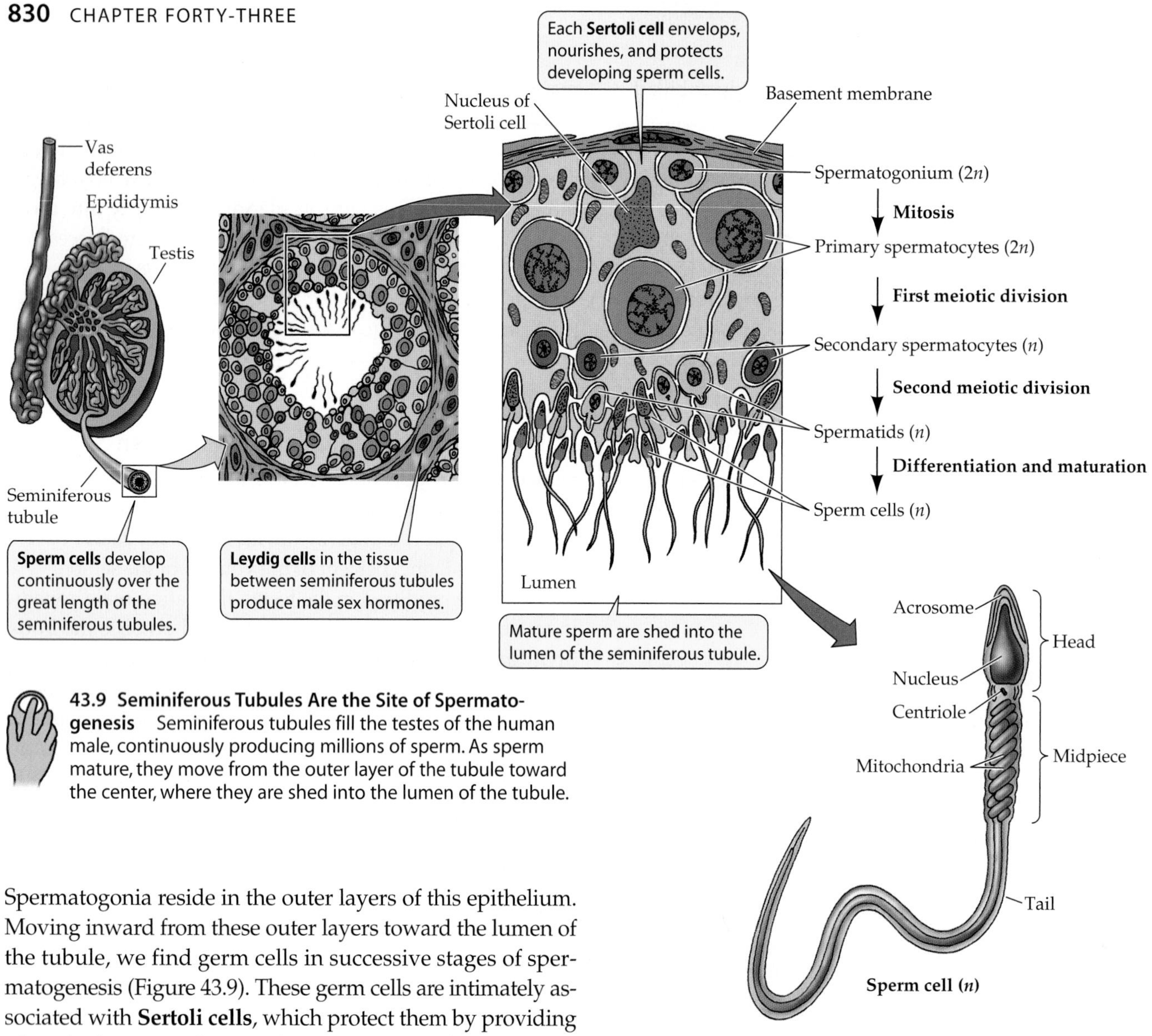

43.9 Seminiferous Tubules Are the Site of Spermatogenesis Seminiferous tubules fill the testes of the human male, continuously producing millions of sperm. As sperm mature, they move from the outer layer of the tubule toward the center, where they are shed into the lumen of the tubule.

Spermatogonia reside in the outer layers of this epithelium. Moving inward from these outer layers toward the lumen of the tubule, we find germ cells in successive stages of spermatogenesis (Figure 43.9). These germ cells are intimately associated with **Sertoli cells**, which protect them by providing a barrier between them and any noxious substances that might be circulating in the blood. The Sertoli cells also provide nutrients for the developing sperm and are involved in the hormonal control of spermatogenesis. Between the seminiferous tubules are clusters of **Leydig cells**, or *interstitial cells*, which produce male sex hormones.

With completion of the second meiotic division, each primary spermatocyte has given rise to four spermatids (see Figure 43.3*a*), which develop into sperm cells as they continue to migrate toward the lumen of the seminiferous tubule. The nucleus in what will become the head of the mammalian sperm becomes compact, and the surrounding cytoplasm is lost (see Figure 43.9). A flagellum, or tail, develops. The mitochondria, which will provide energy for tail motility, become condensed into a midpiece between the head and the tail. An acrosome forms over the nucleus in the head of the sperm. Fully differentiated sperm are shed into the seminiferous tubule.

From the seminiferous tubules, sperm move into a storage structure called the **epididymis**, where they mature and become motile. The epididymis connects to the **urethra** via a tube called the **vas deferens** (plural, vasa deferentia). The urethra originates in the bladder, runs through the penis, and opens to the outside of the body at the tip of the penis. It serves as the common duct for the urinary and reproductive systems (see Figure 43.8).

The components of the semen other than sperm come from several accessory glands. The **bulbourethral glands** produce a small volume of an alkaline, mucoid secretion that neutralizes acidity in the urethra and lubricates the tip of the penis. About two-thirds of the volume of semen is seminal fluid from the paired **seminal vesicles**. Seminal fluid is thick because it contains mucus and protein. It also contains fructose, an energy source for the sperm, which are too small to carry much of their own fuel.

The **prostate gland** completely surrounds the urethra as it leaves the bladder. One-fourth to one-third of the volume of semen is a thin, milky fluid that comes from the prostate gland. Prostate fluid makes the uterine environment more hospitable to sperm. The prostate also secretes a clotting enzyme that works on the protein in seminal fluid to convert semen into a gelatinous mass.

The penis and the scrotum are the male genitalia. The shaft of the penis is covered with normal skin, but the highly sensitive tip, or **glans penis**, is covered with thinner, more sensitive skin that is especially responsive to sexual stimulation. A fold of skin called the *foreskin* covers the glans of the human penis. The cultural practice of *circumcision* removes a portion of the foreskin.

Sexual arousal triggers responses in the nervous system that result in penile **erection**. Nerve endings release a gaseous neurotransmitter, nitric oxide (NO), onto blood vessels leading into the penis, The presence of NO stimulates production of the second messenger cGMP, which causes these vessels to dilate (see Chapter 15). The increased blood flow that results fills and swells shafts of spongy, erectile tissue located along the length of the penis. The enlargement of these blood-filled cavities compresses the vessels that normally carry blood out of the penis. As a result, the erectile tissue becomes more and more engorged with blood. The penis becomes hard and erect, facilitating its insertion into the female's vagina. Many species of mammals, but not humans, have a bone in the penis, but these species still depend on erectile tissue for copulation.

At the climax of copulation, semen is propelled through the vasa deferentia and the urethra in two steps, emission and ejaculation. During **emission**, rhythmic contractions of smooth muscles in the vas deferentia and accessory glands move sperm and the various seminal secretions into the urethra at the base of the penis. **Ejaculation**, which follows emission, is caused by contractions of other muscles at the base of the penis surrounding the urethra. The rigidity of the erect penis allows these contractions to force the gelatinous mass of semen through the urethra and out of the penis. The muscle contractions of ejaculation are accompanied by feelings of intense pleasure known as *orgasm*. They are also accompanied by transient increases in heart rate, blood pressure, breathing, and skeletal muscle contractions throughout the body.

Once ejaculation has been achieved, the autonomic nervous system switches signaling. NO release decreases, and enzymes break down cGMP, causing the blood vessels flowing into the penis to constrict. This constriction causes a decrease in blood pressure in the erectile tissue, thus relieving the compression of the blood vessels leaving the penis, and the erection declines.

Erectile dysfunction, or *impotence*, is the inability to achieve or sustain an erection. *Viagra*, a drug used to treat erectile dysfunction, is one of the largest selling drugs today. Viagra inhibits the breakdown of cGMP and therefore enhances the effect of NO released in the penis, thus improving the ability to achieve and maintain an erection.

Male sexual function is controlled by hormones

Spermatogenesis and maintenance of male secondary sexual characteristics depend on testosterone, which is produced by the Leydig cells of the testes. As we saw in Chapter 42, increased production of testosterone at puberty is due to an increased release of gonadotropin-releasing hormone (GnRH) by the hypothalamus, which stimulates cells in the anterior pituitary to increase their secretion of luteinizing hormone (LH) and follicle-stimulating hormone (FSH) (Figure 43.10). Higher levels of LH stimulate the Leydig cells to increase their production and release of testosterone. Testosterone exerts negative feedback on the anterior pituitary and the hypothalamus. At the time of puberty, the sensitivity of the hypothalamus to negative feedback from testosterone declines, and the level of circulating testosterone increases.

Increased testosterone in the pubertal male causes the development of secondary sexual characteristics and an increased rate of growth. Testosterone also promotes increased muscle mass and maturation of the testes. If a male is castrated (has his testes removed) before puberty, he will not develop a deep voice, typical patterns of body hair, or a muscular build, and his external genitalia will remain childlike. Continued production of testosterone after puberty is essential for the maintenance of secondary sexual characteristics and the production of sperm.

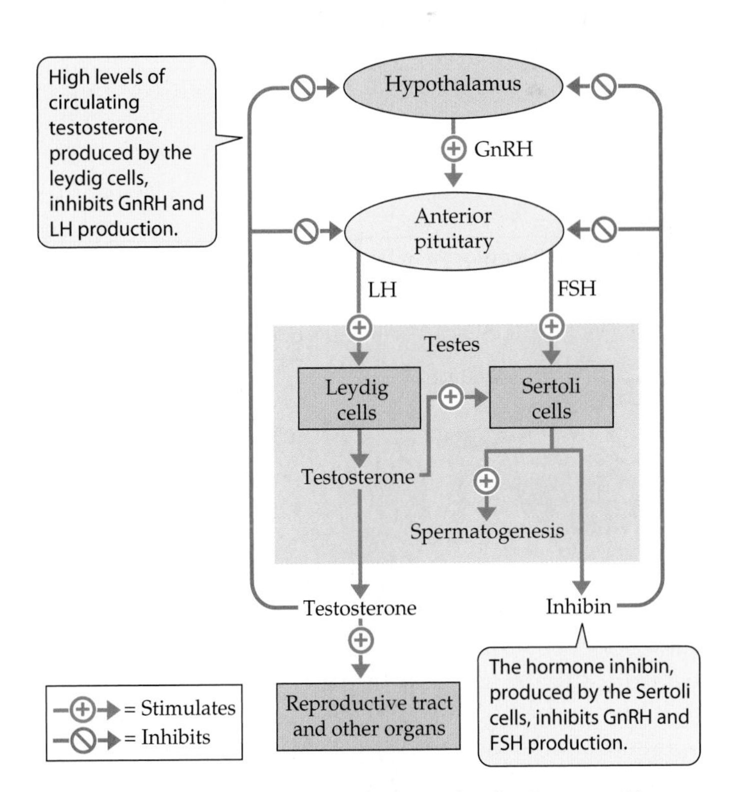

43.10 Hormones Control the Male Reproductive System The male reproductive system is under hormonal control by the hypothalamus and the anterior pituitary.

Spermatogenesis is controlled by the influence of FSH and testosterone on the Sertoli cells in the seminiferous tubules. The Sertoli cells also produce a hormone called *inhibin*, which exerts negative feedback on the production of FSH by cells in the anterior pituitary.

Female sex organs produce eggs, receive sperm, and nurture the embryo

When a mammalian egg matures, it is released from the ovary directly into the body cavity. But the egg does not go far. Each ovary is enveloped by the undulating, fringed opening of an **oviduct** (also known as a *Fallopian tube*), which sweeps the egg into that tube (Figure 43.11). Fertilization takes place in the oviduct. Whether or not the egg is fertilized, cilia lining the oviduct propel it slowly toward the uterus, a muscular, thick-walled cavity shaped in humans like an upside-down pear. The uterus is where the embryo develops if the egg is fertilized. At the bottom, the uterus narrows into a necklike region called the **cervix**, which leads into the **vagina**. Sperm are ejaculated into the vagina during copulation, and the fetus passes through the vagina during birth.

In humans, two sets of skin folds surround the opening of the vagina and the opening of the urethra, through which urine passes. The inner, more delicate folds are the **labia minora** (singular, labium minus); the outer, thicker folds are the **labia majora** (singular, labium majus). At the anterior tip of the labia minora is the **clitoris**, a small bulb of erectile tissue that has the same developmental origins as the penis (see Figure 42.12). The clitoris is highly sensitive and plays an important role in sexual response. The labia minora and the clitoris become engorged with blood in response to sexual stimulation.

The external opening of a female human infant's vagina is partly covered by a thin membrane, the *hymen*. Eventually the hymen becomes ruptured by vigorous physical activity or first sexual intercourse; it can sometimes make first intercourse difficult or painful for the female.

To fertilize an egg, sperm deposited in the vagina swim, and are propelled by contractions of the female reproductive tract, through the cervical opening, the uterus, and most of the oviduct. The egg (actually a secondary oocyte) is fertilized in the upper region of the oviduct. Fertilization stimulates the completion of the second meiotic division, after which the haploid nuclei of the sperm and the egg can fuse to produce a diploid zygote nucleus. Still in the oviduct, the zygote undergoes its first few cell divisions to become a *blastocyst*. The blastocyst moves down the oviduct to the uterus, where it attaches itself to the epithelial lining of the uterus, the **endometrium**.

Once attached to the endometrium, the blastocyst burrows into it—a process called **implantation**—and interacts with it to form the placenta (see Figures 20.14 and 20.19). The placenta exchanges nutrients and waste products between the mother's blood and the embryo's blood. It also produces hormones that help sustain pregnancy.

As the egg matures in the ovary, the endometrium thickens. If a blastocyst does not arrive in the uterus, it regresses or is sloughed off. Thus the female reproductive cycle actu-

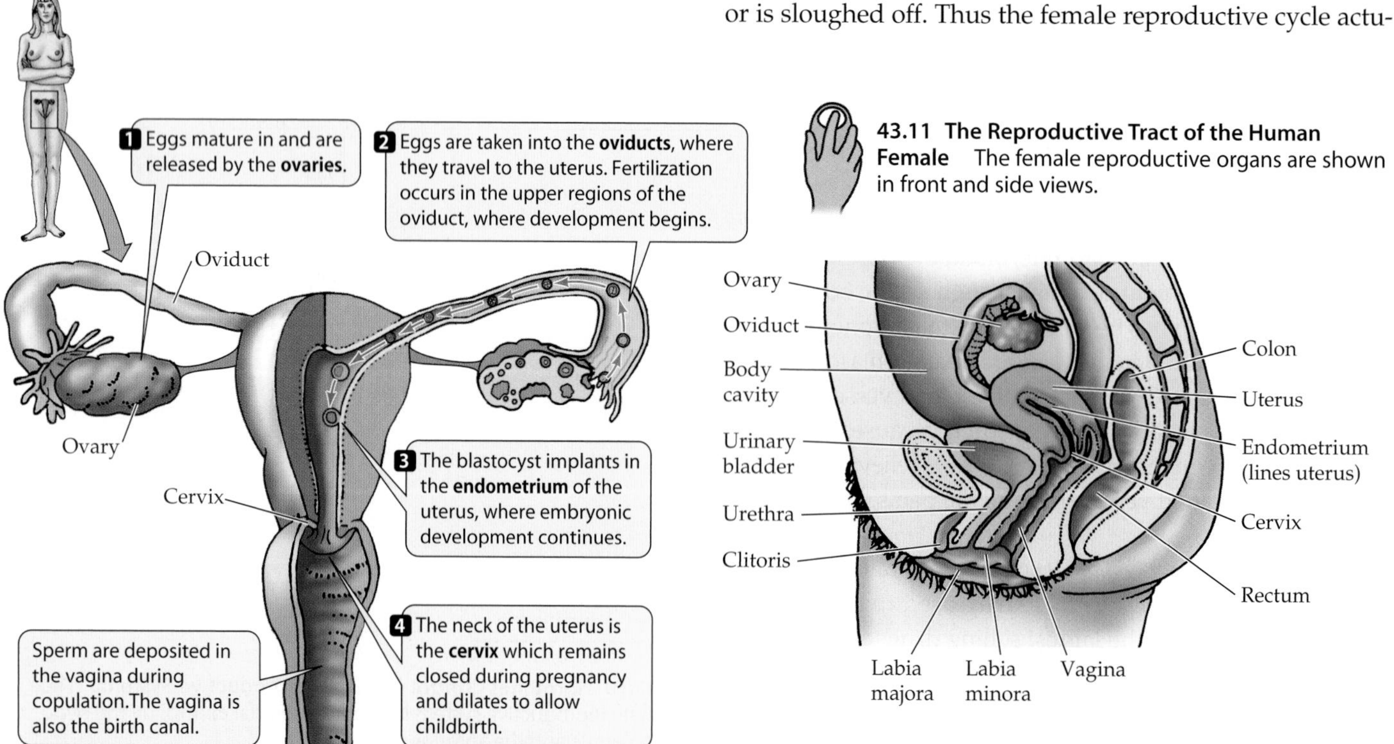

1 Eggs mature in and are released by the ovaries.
2 Eggs are taken into the oviducts, where they travel to the uterus. Fertilization occurs in the upper regions of the oviduct, where development begins.
Oviduct
Ovary
Cervix
3 The blastocyst implants in the endometrium of the uterus, where embryonic development continues.
4 The neck of the uterus is the cervix which remains closed during pregnancy and dilates to allow childbirth.
Sperm are deposited in the vagina during copulation.The vagina is also the birth canal.
Ovary
Oviduct
Body cavity
Urinary bladder
Urethra
Clitoris
Colon
Uterus
Endometrium (lines uterus)
Cervix
Rectum
Labia majora
Labia minora
Vagina

43.11 The Reproductive Tract of the Human Female The female reproductive organs are shown in front and side views.

ally consists of two linked cycles: an ovarian cycle that produces eggs and hormones, and a uterine cycle that prepares the endometrium for the arrival of a blastocyst.

The ovarian cycle produces a mature egg

An **ovarian cycle** is about 28 days long in the human female,* but there is considerable variation among individuals. During the first half of each cycle, at least one primary oocyte matures into a secondary oocyte (egg) and is expelled from the ovary. During the second half of the cycle, cells in the ovary that were associated with the maturing oocyte develop endocrine functions and then regress if the egg is not fertilized. The progression of these events is shown diagrammatically in Figure 43.12.

At birth, a human female has about a million primary oocytes in each ovary. By the time she reaches puberty, she has only about 200,000 in each ovary; the rest have degenerated. During a woman's fertile years, her ovaries will go through about 450 ovarian cycles. During each cycle, a number of oocytes will begin to mature, but usually only one oocyte from one or the other ovary will mature completely and be released; the others will degenerate. At about 50 years of age, she reaches **menopause**, the end of fertility, and may have only a few oocytes left in each ovary. Throughout a woman's life, oocytes are degenerating, and no new ones are produced.

Each primary oocyte in the ovary is surrounded by a layer of ovarian cells. An oocyte and its surrounding cells constitute the functional unit of the ovary, the **follicle**. Between puberty and menopause, six to twelve follicles begin to mature each month. In each follicle, the oocyte enlarges and the surrounding follicular cells proliferate. After about a week, one of these follicles is larger than the rest, and it continues to grow, while the others cease to develop and shrink. In the enlarged follicle, the follicular cells nurture the growing egg, supplying it with nutrients and with the macromolecules and proteins it will use in early stages of development if it is fertilized.

In humans, after 2 weeks of follicular growth, **ovulation** occurs: The follicle ruptures, and the egg is released. Following ovulation, the follicle cells continue to proliferate and form a mass of endocrine tissue about the size of a marble. This structure, which remains in the ovary, is the **corpus luteum** (plural, corpora lutea). It functions as an endocrine gland, producing estrogen and progesterone for about 2 weeks. It then degenerates unless a blastocyst implants in the endometrium.

*Some mammals have ovarian cycles shorter than 28 days; others have longer ones. Rats and mice have ovarian cycles of about 4 days; many seasonally breeding mammals have only one ovarian cycle per year.

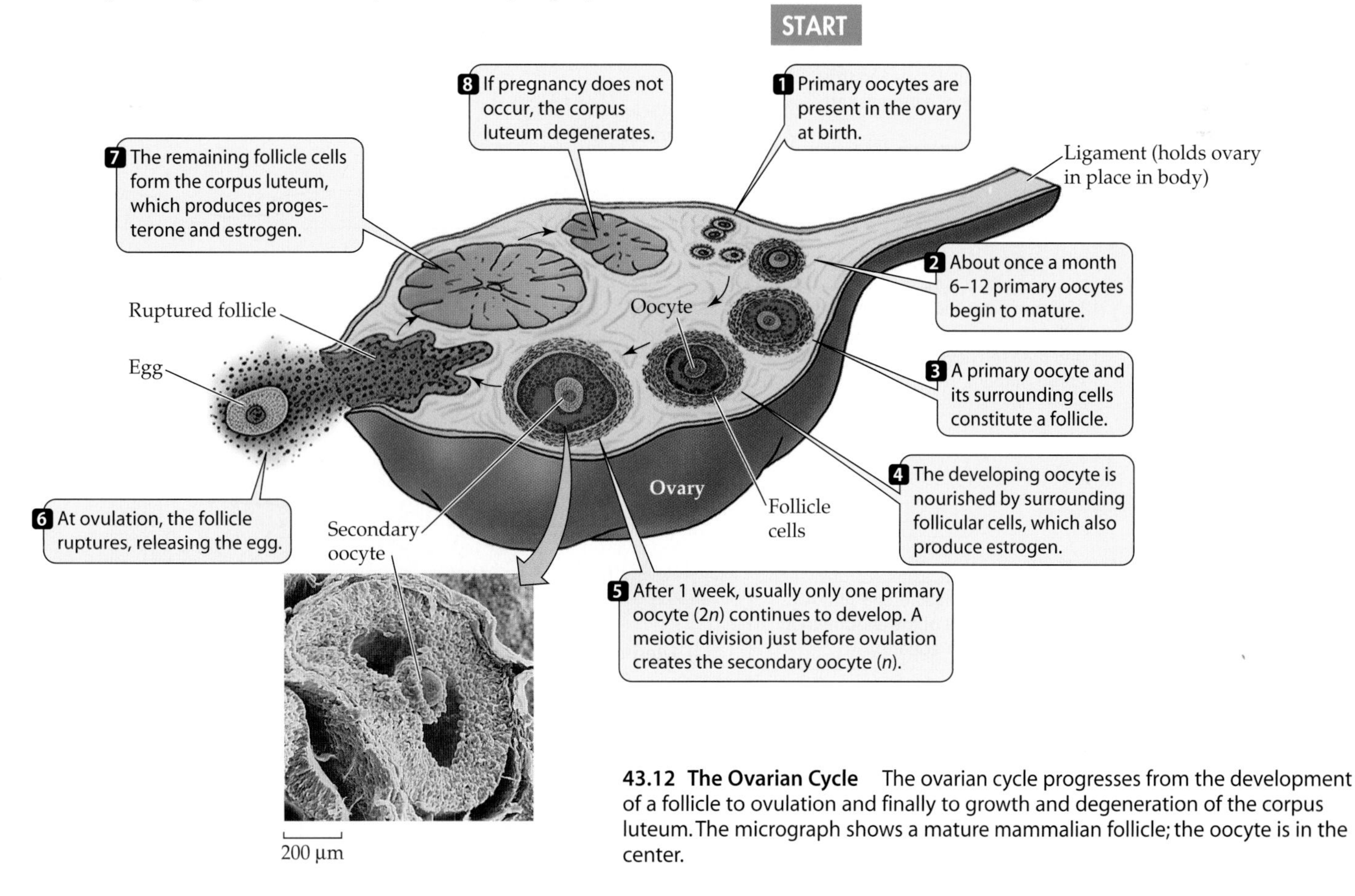

START
1 Primary oocytes are present in the ovary at birth.
2 About once a month 6–12 primary oocytes begin to mature.
3 A primary oocyte and its surrounding cells constitute a follicle.
4 The developing oocyte is nourished by surrounding follicular cells, which also produce estrogen.
5 After 1 week, usually only one primary oocyte (2n) continues to develop. A meiotic division just before ovulation creates the secondary oocyte (n).
6 At ovulation, the follicle ruptures, releasing the egg.
7 The remaining follicle cells form the corpus luteum, which produces progesterone and estrogen.
8 If pregnancy does not occur, the corpus luteum degenerates.
Ligament (holds ovary in place in body)
Oocyte
Ruptured follicle
Egg
Ovary
Follicle cells
Secondary oocyte
200 μm

43.12 The Ovarian Cycle The ovarian cycle progresses from the development of a follicle to ovulation and finally to growth and degeneration of the corpus luteum. The micrograph shows a mature mammalian follicle; the oocyte is in the center.

The uterine cycle prepares an environment for the fertilized egg

The **uterine cycle** of human females, which parallels the ovarian cycle, consists of first a buildup and then a breakdown of the endometrium (Figure 43.13). About 5 days into the ovarian cycle, the endometrium starts to grow in preparation for receiving a blastocyst. The uterus attains its maximum state of preparedness about 5 days after ovulation (about day 19 of the ovarian cycle) and remains in that state for another 9 days. If a blastocyst has not arrived by that time, the endometrium begins to break down, and the sloughed-off tissue, including blood, flows from the body through the vagina—the process of **menstruation** (from *menses*, the Latin word for "months").

The uterine cycles of most mammals other than humans do not include menstruation; instead, the uterine lining typically is resorbed. In these species, the most obvious correlate of the ovarian cycle is a state of sexual receptivity called *estrus* around the time of ovulation. You may be aware of the bloody discharge that occurs in dogs at the time of estrus. This discharge is not the same as menstruation, and in fact is exactly the opposite: Bleeding in dogs occurs during the proliferation of the uterine lining, which occurs just prior to ovulation. When the female mammal comes into estrus, or "heat," she actively solicits male attention and may be aggressive to other females. The human female is unusual among mammals in that she is potentially sexually receptive throughout her ovarian cycle and at all seasons of the year.

Hormones control and coordinate the ovarian and uterine cycles

The ovarian and uterine cycles of human females are coordinated and timed by the same hormones that initiate sexual maturation. Gonadotropins secreted by the anterior pituitary are the central elements of this control. Before puberty (that is, before about 11 years of age), the secretion of gonadotropins is low, and the ovaries are inactive. At puberty, the hypothalamus increases its release of GnRH, thus stimulating the anterior pituitary to secrete FSH and LH.

(*a*) **Gonadotropins (from anterior pituitary)**

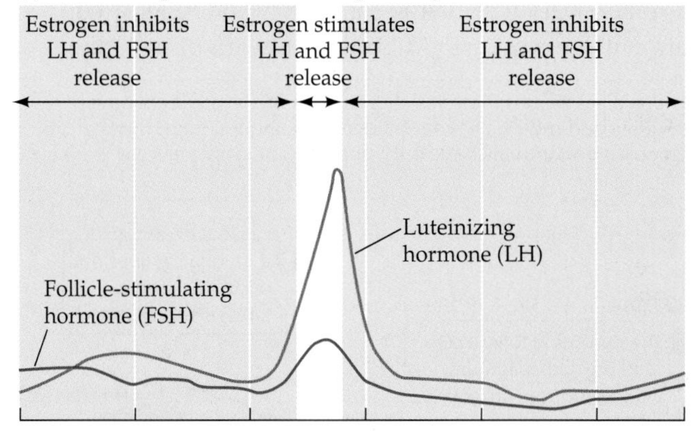

FSH and LH are under control of GnRH from the hypothalamus and the ovarian hormones estrogen and progesterone.

(*b*) **Events in ovary (ovarian cycle)**

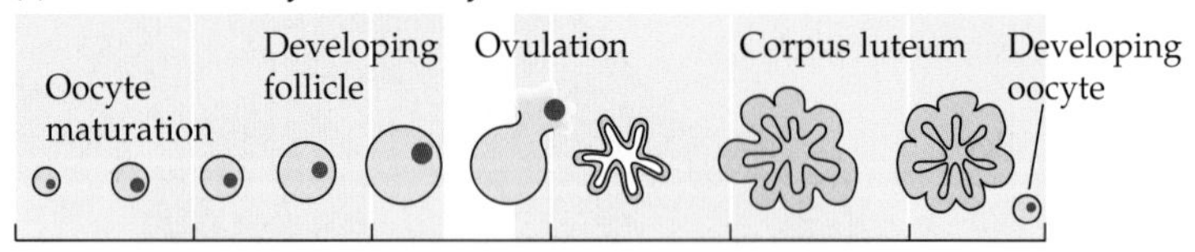

FSH stimulates the development of follicles; the LH surge causes ovulation and then the development of the corpus luteum.

(*c*) **Ovarian hormones and the uterine cycle**

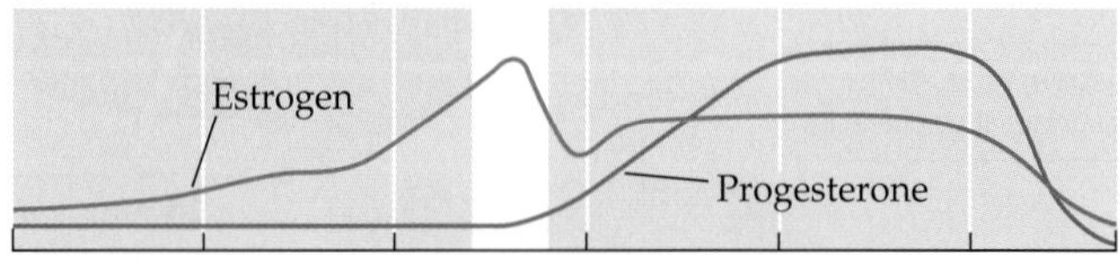

Estrogen and progesterone stimulate the development of the endometrium in preparation for pregnancy.

(*d*) **Uterine lining**

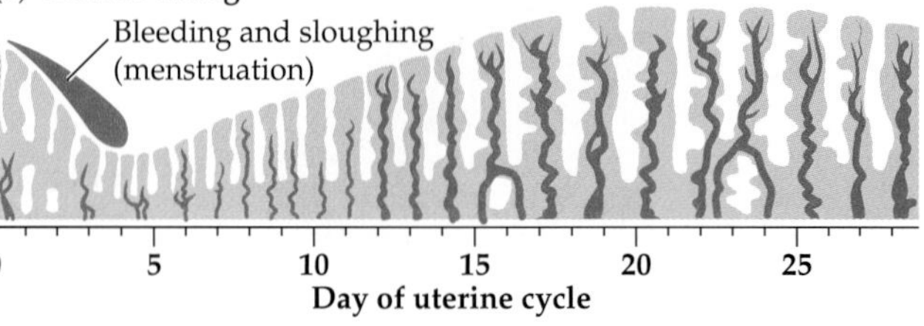

43.13 The Ovarian and Uterine Cycles During a woman's ovarian and uterine cycles, there are coordinated changes in (*a*) gonadotropin release by the anterior pituitary, (*b*) the ovary, (*c*) the release of female sex steroids, and (*d*) the uterus. The cycles begin with the onset of menstruation; ovulation is at midcycle.

In response to FSH and LH, ovarian tissue grows and produces estrogen. The rise in estrogen causes the maturation of the accessory sex organs and the development of female secondary sexual characteristics. Between puberty and menopause, interactions of GnRH, gonadotropins, and sex steroids control the ovarian and uterine cycles.

Menstruation marks the beginning of each uterine and ovarian cycle (see Figure 43.13). A few days before menstruation begins, the anterior pituitary begins to increase its secretion of FSH and LH. In response, some follicles begin to mature in the ovaries, and the follicle cells gradually increase their production of estrogen. After about a week of growth, usually all but one of the follicles wither away. Occasionally more than one follicle continues to develop, making it possible for the woman to bear fraternal (non-identical) twins. The follicle(s) that is still growing secretes increasing amounts of estrogen, stimulating the endometrium to proliferate.

Estrogen exerts negative feedback control on gonadotropin release by the anterior pituitary during the first 12 days of the ovarian cycle. Then, on about day 12, estrogen exerts positive rather than negative feedback control on the pituitary (Figure 43.14). As a result, there is a surge of LH and a lesser surge of FSH (see Figure 43.13*a*). The LH surge triggers the mature follicle to rupture and release its egg, and it stimulates follicular cells to develop into a corpus luteum.

Estrogen and especially progesterone secreted by the corpus luteum following ovulation are crucial to continued growth and maintenance of the endometrium. In addition, these sex steroids exert negative feedback control on the pituitary, inhibiting gonadotropin release and thus preventing new follicles from beginning to mature.

If the egg is not fertilized, the corpus luteum degenerates on about day 26 of the cycle. Without production of progesterone by the corpus luteum, the endometrium sloughs off, and menstruation occurs. The decrease in circulating steroids also releases the hypothalamus and pituitary from negative feedback control, so GnRH, FSH, and LH all begin to increase. The increase in these hormones induces the next round of follicle development, and the ovarian cycle begins again.

In pregnancy, hormones from the extraembryonic membranes take over

If the egg is fertilized, and a blastocyst arrives in the uterus and implants in the endometrium, a new hormone comes into play. A layer of cells covering the blastocyst begins to secrete **human chorionic gonadotropin** (**hCG**). This gonadotropin, a molecule similar to LH, keeps the corpus luteum functional. Because hCG is present only in the blood of pregnant women, the presence of this hormone is the basis for pregnancy testing. Modern pregnancy tests make use of a monoclonal antibody to detect hCG in urine; thus, they take only minutes and can be done at home. These tests are so sensitive that they are 99 percent accurate and in most cases can detect a pregnancy even before the first missed menstrual period.

Tissues derived from the blastocyst also begin to produce estrogen and progesterone, eventually replacing the corpus luteum as the most important source of these sex steroids. Continued high levels of estrogen and progesterone prevent the pituitary from secreting gonadotropins; thus, the ovarian cycle ceases for the duration of pregnancy. The same mechanism is exploited by birth control pills, which contain synthetic hormones resembling estrogen and progesterone that prevent the ovarian cycle by exerting negative feedback control on the hypothalamus and pituitary.

Childbirth is triggered by hormonal and mechanical stimuli

We traced the development of a human blastocyst into an embryo and then a fetus in Chapter 20. Throughout pregnancy, the muscles of the uterine wall periodically undergo slow, weak, rhythmic contractions called *Braxton-Hicks contractions*. These contractions become gradually stronger during the third trimester of pregnancy and are sometimes called *false labor contractions*. True labor contractions usually mark the beginning of childbirth. Both hormonal and mechanical stimuli contribute to the onset of labor.

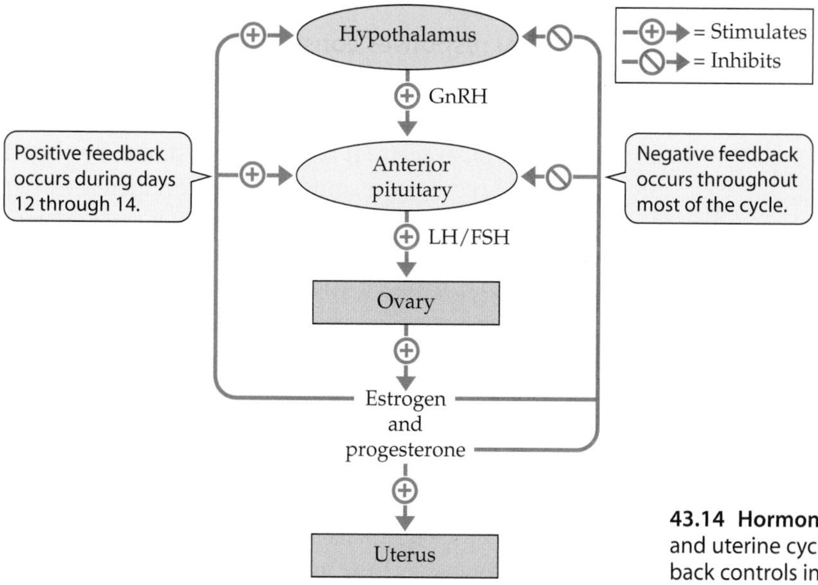

Hypothalamus
GnRH
Anterior pituitary
LH/FSH
Ovary
Estrogen and progesterone
Uterus
= Stimulates
= Inhibits
Positive feedback occurs during days 12 through 14.
Negative feedback occurs throughout most of the cycle.

43.14 Hormones Control the Ovarian and Uterine Cycles The ovarian and uterine cycles are under a complex series of positive and negative feedback controls involving several hormones.

Progesterone inhibits and estrogen stimulates contractions of uterine muscle. Toward the end of the third trimester, the estrogen–progesterone ratio shifts in favor of estrogen. The onset of labor is marked by increased oxytocin secretion by the pituitaries of both mother and fetus. Oxytocin is a powerful stimulant of uterine muscle contraction.

Mechanical stimuli come from the stretching of the uterus by the fully grown fetus and the pressure of the fetal head on the cervix. These mechanical stimuli increase the release of oxytocin by the posterior pituitary, which in turn increases the activity of uterine muscle, which causes even more pressure on the cervix. This positive feedback loop converts the weak, slow, rhythmic Braxton-Hicks contractions into stronger labor contractions (Figure 43.15).

In the early stage of labor, the contractions of the uterus are 15 to 20 minutes apart, and each lasts 45 to 60 seconds. During this time, the contractions pull the cervix open until it is large enough to allow the baby to pass through. Gradually the contractions become more frequent and more intense. This stage of labor lasts an average of 12 to 15 hours in a first pregnancy and 8 hours or less in subsequent ones.

The second stage of labor, called *delivery*, begins when the cervix is fully dilated. The baby's head moves into the vagina and becomes visible from the outside. The usual head-down position of the baby at the time of delivery comes about when the fetus shifts its orientation during the seventh month. If the fetus fails to move into the head-down position, a different part of the fetus enters the vagina first, and the birth is more difficult.

Passage of the fetus through the vagina is assisted by the mother's bearing down with her abdominal and other muscles to help push it along. Once the head and shoulders of the baby clear the cervix, the rest of its body eases out rapidly, but it is still connected to the placenta by the umbilical cord. Delivery may take as little as a minute, or up to half an hour or more in a first pregnancy.

As soon as the baby clears the birth canal, it can start breathing and become independent of its mother's circulation. The umbilical cord may then be clamped and cut. The segment still attached to the baby dries up and sloughs off in a few days, leaving behind its distinctive signature, the belly button—more properly called the *umbilicus*. The detachment and expulsion of the placenta and fetal membranes takes from a few minutes to an hour, and may be accompanied by uterine contractions. If the baby suckles at the breast immediately following birth, its suckling stimulates additional secretion of oxytocin, which augments uterine contractions that reduce the size of the uterus and help stop bleeding.

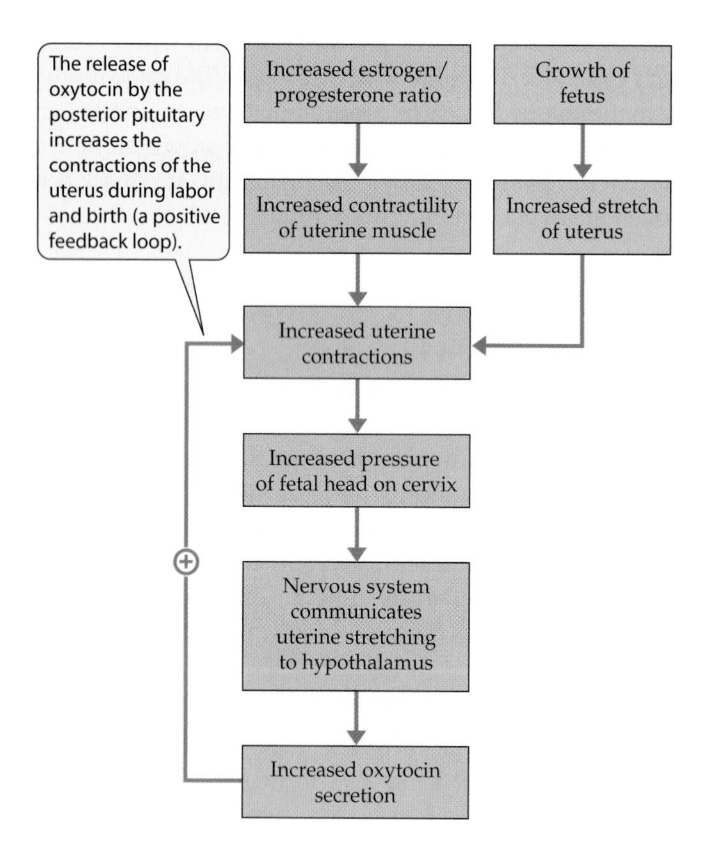

43.15 Control of Uterine Contractions and Childbirth Both mechanical and hormonal signals are involved in the onset of childbirth.

Human Sexual Behavior

Sexual issues and sexual behavior are dominant aspects of our society, and reproductive technologies have had huge impacts on our sexual and reproductive lives. In this section we will discuss the basic human sexual responses, and we will review the technologies available both for contraception (birth control) and for enhancing fertility. The chapter closes with a discussion of sexually transmitted diseases.

Human sexual responses consist of four phases

The responses of both women and men to sexual stimulation consist of four phases: excitement, plateau, orgasm, and resolution. As sexual *excitement* begins in a woman, her heart rate and blood pressure rise, muscular tension increases, her breasts swell, and her nipples become erect. Her external genitals, including the sensitive clitoris, swell as they become filled with blood, and the walls of the vagina secrete lubricating fluid that facilitates copulation.

As a woman's sexual excitement increases, she enters the *plateau* phase. Her blood pressure and heart rate rise further, her breathing becomes rapid, and the clitoris begins to retract—the greater the excitement, the greater the retraction. The sensitivity once focused in the clitoris spreads over the external genitals, and the clitoris itself becomes even more sensitive. *Orgasm* may last as long as a few minutes, and, unlike men, some women can experience several orgasms in rapid succession. During the *resolution* phase, blood drains from the genitals, and body physiology returns to close to normal.

In the male, as in the female, the excitement phase is marked by an increase in blood pressure, heart rate, and muscle tension. The penis fills with blood and becomes hard and erect. In the plateau phase, breathing becomes rapid, the diameter of the glans increases, and a clear lubricating fluid from the bulbourethral gland oozes from the penis. Pressure and friction against the nerve endings in the glans and in the skin along the shaft of the penis eventually trigger orgasm. Massive spasms of the muscles in the genital area and contractions in the accessory reproductive organs result in ejaculation.

Within a few minutes after ejaculation, the penis shrinks to its former size, and body physiology returns to resting conditions. The male sexual response includes a *refractory period* immediately after orgasm. During this period, which may last from minutes to hours, a man cannot achieve a full erection or another orgasm, regardless of the intensity of sexual stimulation.

Humans use a variety of technologies to control fertility

People use many methods to control the number of their children and the time between births. The only sure methods of preventing fertilization and pregnancy are complete abstinence from sexual activity or surgical removal of the gonads. Since those approaches are not acceptable to most people, they turn to a variety of other methods to prevent pregnancy. Many of these methods prevent fertilization or implantation (*conception*) and are therefore referred to as methods of *contraception*.

Some methods of contraception are used by the woman, others by the man. They vary from means of blocking gametogenesis to means of blocking implantation of a blastocyst. Contraceptive methods vary enormously in their effectiveness and in their acceptability to those who use them. Here we will review some of the most common methods and their relative failure rates (Table 43.1). Even with all the contraceptive methods available, a recent study revealed that almost half of the 6.3 million pregnancies that occur in the United States each year are unintended.

NONTECHNOLOGICAL APPROACHES. An approach to contraception that does not involve physical or pharmacological technologies is to separate sperm and egg in time through the **rhythm method.** The couple avoids sex from day 10 to day 20 of the ovarian cycle, when the woman is most likely to be fertile. The cycle can be tracked by use of a calendar, supplemented by the basal body temperature method, which is based on the observation that a woman's body temperature drops on the day of ovulation and rises sharply on the day after. Changes in the stickiness of the cervical mucus also help identify the day of ovulation.

However, sperm deposited in the female reproductive tract may remain viable for up to 6 days. Similarly, the egg remains viable for 12 to 36 hours after ovulation. These facts, added to individual variation in the timing of ovulation, result in an annual failure rate of between 15 and 35 percent for the rhythm method. In other words, 15 to 35 percent of women using only the rhythm method for 1 year will become pregnant during that time.

Another approach is to try to separate sperm and egg in space through **coitus interruptus**—withdrawal of the penis from the vagina before ejaculation. The annual failure rate of this method (mostly due to lack of willpower) may be as high as 40 percent.

BARRIER METHODS. Techniques for placing a physical barrier between egg and sperm have been used for centuries. The **condom** is a sheath made of an impermeable material such as latex that can be fitted over the erect penis. A condom traps semen so that sperm do not enter the vagina. Latex condoms also help prevent the spread of many sexually transmitted diseases. In theory, use of a condom can be highly effective, with a failure rate near zero; in practice, the annual failure rate is about 15 percent due to leakage because of tearing or poor fit (e.g., with the loss of the erec-

43.1 Methods of Contraception

METHOD	MODE OF ACTION	FAILURE RATE[a]
Rhythm method	Abstinence near time of ovulation	15–35
Coitus interruptus	Prevents sperm from reaching egg	10–40
Condom	Prevents sperm from entering vagina	3–20
Diaphragm/jelly	Prevents sperm from entering uterus; kills sperm	3–25
Vaginal jelly or foam	Kills sperm; blocks sperm movement	3–30
Douche	Supposedly flushes sperm from vagina	80
Birth control pills	Prevent ovulation	0–3
Vasectomy	Prevents release of sperm	0.0–0.15
Tubal ligation	Prevents egg from entering uterus	0.0–0.05
Intrauterine device (IUD)	Prevents implantation of fertilized egg	0.5–6
RU-486	Prevents development of fertilized egg	0–15
Unprotected	No form of birth control	85

[a]Number of pregnancies per 100 women per year

tion). A female condom, which creates an impermeable lining of the vagina, is also available.

The **diaphragm** is a dome-shaped piece of rubber with a firm rim that fits over the woman's cervix and thus blocks sperm from entering the uterus. Smaller than the diaphragm is the **cervical cap**, which fits snugly just over the tip of the cervix. Both the diaphragm and the cervical cap are treated first with jelly or cream containing a *spermicide*—a chemical that kills or incapacitates sperm—and then inserted through the vagina before sexual intercourse. Annual failure rates are about 15 percent—the same as for condoms.

Spermicidal foams, jellies, and creams can be used alone by placing them in the vagina with special applicators. Used in this way, they have an annual failure rate of 25 percent or more. *Douching* (flushing the vagina with liquid) after intercourse, in spite of popular belief, is almost useless as a method of birth control. Sperm can reach the upper regions of the oviducts within 10 minutes after ejaculation.

The effectiveness of barrier methods can be greatly improved if different ones are used in combination. For example, if the man uses a condom and the woman a diaphragm, the failure rate is extremely low.

PREVENTING OVULATION. The widely used *oral contraceptives*, or **birth control pills**, work by preventing ovulation. Their mechanisms of action take advantage of the roles of estrogen and progesterone as negative feedback signals to the hypothalamus and the pituitary. The most common pills contain low doses of synthetic estrogens and progesterones (progestins). By keeping the circulating levels of gonadotropins low, these hormones interfere with the maturation of follicles and eggs, suspending the ovarian cycle. The uterine cycle is usually allowed to continue, however, by ceasing the pills every 21 to 23 days.

The negative side effects of oral contraceptives have been the topic of extensive discussion. These side effects include increased risk of blood clot formation, heart attack, stroke, and breast cancer. However, these side effects are associated mostly with pills containing higher hormone concentrations than are used in modern pills. For pills in use today, the risk of these side effects is low, except for women over 35 years old who smoke, for whom the risk is significantly greater. The risk of death from using "the pill" is less than the risk associated with a full-term pregnancy. The pill is the most effective method of contraception other than sterilization or perhaps combined barrier methods. Oral contraceptives have an annual failure rate of less than 1 percent.

The "mini-pill" is an oral contraceptive that contains very low doses of progestins. Although it may interfere with the normal maturation and release of eggs, its principal mode of action is to alter the environment of the female reproductive tract so that it is not hospitable to sperm. Cervical mucus normally becomes watery at the time of ovulation, but low levels of progestins keep the mucus thick and sticky so that it blocks passage of sperm.

Long-lasting injectable or implantable steroids are also used to block ovulation through negative feedback effects. Depo-Provera is an injectable progestin that blocks the release of gonadotropins for several months. Another device, called Norplant, consists of thin, flexible tubes filled with progestin. Several of these tubes are inserted under the skin, where they continue to release progestin slowly for years.

PREVENTING IMPLANTATION. A highly effective method of contraception (with a failure rate varying from 1 percent to about 7 percent) is the **intrauterine device**, or **IUD**. The IUD is a small piece of plastic or copper that is inserted into the uterus. The IUD probably works by causing an inflammatory response that includes the release of prostaglandins, which prevent implantation of the fertilized egg.

Another way of interfering with implantation is through the use of "morning-after pills," which deliver high doses of steroids, primarily estrogens. By acting in several ways on the oviducts and the endometrium, this treatment prevents implantation. Morning-after pills can be effective up to several days after sexual intercourse.

The drug RU-486, developed in France, is not a contraceptive pill, but a *contragestational* pill. It blocks progesterone receptors, thereby interfering with the normal action of progesterone produced by the corpus luteum, which is necessary for the maintenance of the endometrium in early pregnancy. If RU-486 is administered as a "morning-after pill," it prevents implantation. However, RU-486 can be effective even if taken at the time of the first missed menstrual period, after implantation has begun. After a few days of treatment with RU-486, the endometrium regresses and sloughs off, along with the embryo, which is in very early stages of development.

STERILIZATION. One foolproof method of contraception is *sterilization*. Male sterilization by **vasectomy** is a simple operation (cutting and tying of the vasa deferentia) that can be performed under a local anesthetic in a doctor's office (Figure 43.16*a*). After this minor surgery, the semen no longer contains sperm. Sperm production continues, but since the sperm cannot move out of the testes, they are destroyed by macrophages. Vasectomy does not affect a man's hormone levels or his sexual responses, and even the amount of semen he ejaculates is essentially unchanged.

In female sterilization, the aim is to prevent the egg from traveling to the uterus and to block sperm from reaching the egg. The most common method is **tubal ligation**: cutting and tying of the oviducts (Figure 43.16*b*). Alternatively, the oviducts may be burned (cauterized) to seal them off. As in

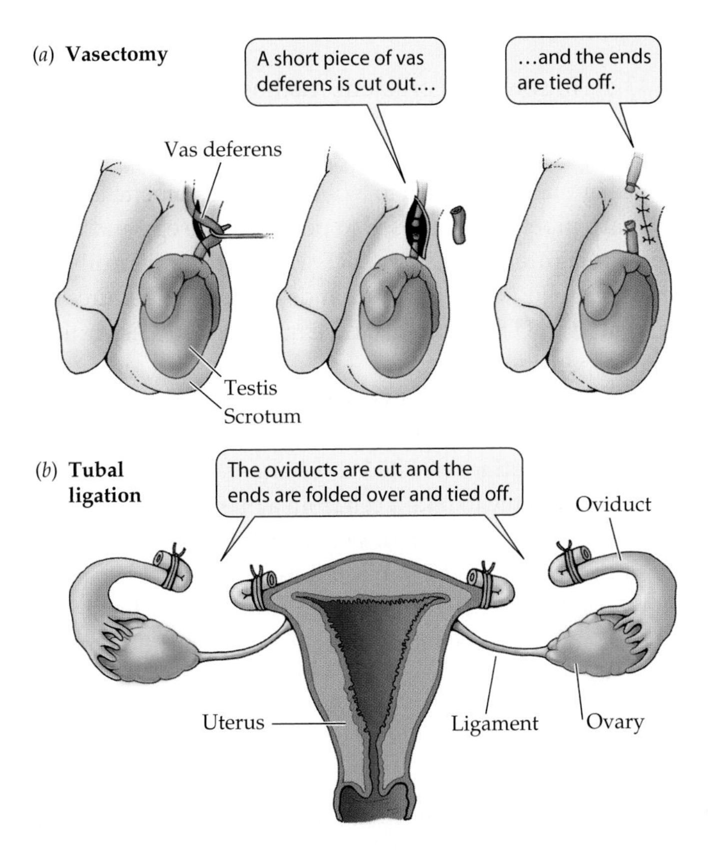

43.16 Sterilization Techniques (*a*) Vasectomy is the technique for male sterilization. (*b*) Tubal ligation is the sterilization procedure most commonly performed on human females.

the male, these procedures do not alter reproductive hormones or sexual responses.

ABORTION. Once a fertilized egg is successfully implanted in the uterus, any termination of the pregnancy is called an **abortion**. A *spontaneous abortion* is the medical term for what most people call a *miscarriage*. Miscarriages are common early in pregnancy (about 10%); most of them occur because of an abnormality in the fetus or in the process of implantation. Abortions that are not spontaneous, but are the result of medical intervention, may be performed either for therapeutic purposes or for fertility control. A therapeutic abortion may be necessary to protect the health of the mother, or it may be performed because prenatal testing reveals that the fetus has a severe defect. Of the approximately 3 million unintended pregnancies in the United States each year, almost half are ended by abortion.

In a medical abortion, the cervix is dilated and the fetus and the endometrium are removed from the uterus by physical means. When performed in the first trimester of a pregnancy, a medical abortion carries less risk of death to the mother than a full-term pregnancy. After the first 12 weeks of pregnancy, the risk rises, but even through the second trimester, it is less than that of a full-term pregnancy.

CONTROLLING MALE FERTILITY. You may ask why all the pharmacological approaches to controlling fertility are applied to females. The control of male fertility is a difficult problem. First, spermatogenesis is a continuous rather than a cyclical event, and it is difficult to block a particular step in a continuous process. The ovarian cycle is more vulnerable to manipulation because certain events must happen at certain times and in a certain sequence for ovulation and implantation to occur. Second, the suppression of spermatogenesis must be total to be effective, since technically it takes only a single sperm to fertilize an egg, and normally millions are produced continuously. Such suppression requires powerful and constant chemical intervention, with associated side effects.

Reproductive technologies help solve problems of infertility

There are many reasons why a man and woman may not be able to have children. The man's rate of sperm production may be low, or his sperm may lack motility. The mucus in the woman's reproductive tract may be thick and not conducive to sperm reaching the oviducts. Structural problems may also exist, such as blockage of the oviducts by scar tissue or by *endometriosis*, a proliferation of endometrial cells outside of the uterus. In some cases, treatment with powerful chemicals to cure cancer damages the ability of the gonads to produce gametes.

Even a couple who are fully fertile and who want children may not want to take the risk of the natural process of fertilization if one or both parents are carriers of a genetic disease. A number of reproductive technologies have been developed to overcome these and other barriers to childbearing.

The oldest and simplest reproductive technology is *artificial insemination*, which involves placing sperm in the appropriate part of the female's reproductive tract. This technique is useful if the male's sperm count is low, if his sperm lack motility, or if problems in the female's reproductive tract prevent the normal movement of sperm up to and through the oviducts. Artificial insemination is used widely in the production of domesticated animals such as cattle.

More recent advances, called **assisted reproductive technologies**, or **ARTs**, involve procedures that remove unfertilized eggs from the ovary, combine them with sperm outside the body, and then place fertilized eggs or egg–sperm mixtures in the appropriate location in the female's reproductive tract for development to take place.

The first successful ART was *in vitro fertilization* (*IVF*). In IVF, the mother is treated with hormones that stimulate many follicles in her ovaries to mature. Eggs are collected from these follicles, and sperm are collected from the father. Eggs and sperm are combined in a culture medium outside the body (*in vitro*, "in glass"), where fertilization takes place. The resulting embryos can be injected into the mother's uterus in the blastocyst stage or kept frozen for implantation later. The first "test-tube baby" resulting from IVF was born in 1978. Since that time, thousands of babies have been produced by this ART. IVF is useful when the woman's oviducts are blocked, but it has a success rate of only 20 to 25 percent.

A technique called *gamete intrafallopian transfer* (*GIFT*) can be used when only the entrance to the oviducts from the ovaries, or the upper segment of the oviducts, is blocked. In this procedure, eggs and sperm are collected and injected directly into the upper regions of the oviducts, where fertilization normally takes place. Then the blastocyst enters the uterus via the normal route. GIFT has a success rate of about 30 percent.

A major cause of failure of IVF and GIFT is failure of sperm to gain access to the egg plasma membrane (see Figure 43.5). To solve this problem, methods have been developed to inject a sperm cell directly into the cytoplasm of an egg. In *intracytoplasmic sperm injection* (*ICSI*), an egg is held in place by suction applied to a polished glass pipette. A slender, sharp pipette is then used to penetrate the egg and inject a sperm (Figure 43.17). This ART was used successfully for the first time in 1992; now thousands of these procedures are performed in U.S. clinics each year, with a success rate of about 25 percent.

IVF, coupled with sensitive techniques of genetic analysis, can eliminate the risk that adults who are carriers of genetic diseases will produce affected children. As we saw in Chapter 20, it is possible to take a cell from a blastocyst at the 4- or 8-cell stage without damaging its developmental potential. The sampled cell can be subjected to molecular analysis to determine whether it carries the harmful gene. This procedure, called *preimplantation genetic diagnosis*, or *PGD*, makes

43.2 Some Sexually Transmitted Diseases

DISEASE	INCIDENCE IN UNITED STATES	SYMPTOMS
Syphilis	80,000 new cases/yr	Primary stage (weeks): skin lesion (chancre) at site of infection Secondary stage (months): skin rash and flu-like symptoms, may be followed by a latent period Tertiary stage (years): deterioration of the cardiovascular and central nervous systems
Gonorrhea	800,000 new cases/yr	Pus-filled discharge from penis or vagina; burning urination. Infection can also start in throat or rectum
Chlamydia	>4,000,000 new cases/yr	Symptoms similar to gonorrhea, although often there are no obvious symptoms. Can result in pelvic inflammatory disease in females (see below)
Genital herpes	500,000 new cases/yr	Small blisters that can cause itching or burning sensations are accompanied by inflammation and by secondary infections
Genital warts	10% of adults infected	Small growths on genital tissues. Increases risk of cervical cancer in women
Hepatitis B	5–20% of population	Fatigue, fever, nausea, loss of appetite, jaundice, abdominal pain, muscle and joint pain. Can lead to destruction of liver or liver cancer
Pelvic inflammatory disease	1,000,000 new cases/yr (females only)	Fever and abdominal pain. Frequently results in sterility
AIDS	Approximately 900,000 cases[a]	Failure of the immune system (see Chapter 18)

[a]AIDS is widespread in other parts of the world, most notably in the southern part of the African continent, where some 9 million people are infected. The virus is also spreading rapidly in India and Southeast Asia.

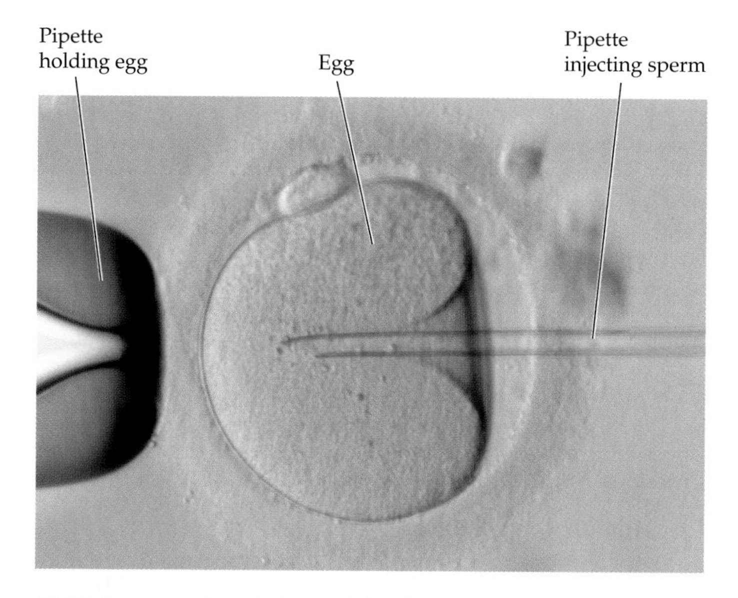

43.17 Intracytoplasmic Sperm Injection In this procedure, sperm are injected directly into a mature egg cell. The fertilized egg can then be placed back in the female reproductive tract.

it possible to determine whether an embryo produced by IVF carries the genetic defect of concern.

Sexual behavior transmits many disease organisms

Disease-causing organisms are parasites and have a very limited ability to survive outside a host organism. Therefore, getting from host to host is a major evolutionary challenge for these organisms. One of the most intimate types of contact that hosts can have is copulation. It is not surprising, then, that many pathogens have evolved to depend on sexual contact between their hosts as their means of transmission. These organisms are the causes of *sexually transmitted diseases* (commonly referred to as *STDs*), and they include viruses, bacteria, yeasts, and protists. A summary of the most common STDs is presented in Table 43.2.

CAUSE	MODE OF TRANSMISSION	CURE/TREATMENT
Spirochete bacterium (*Treponema pallidum*) that penetrates mucosal membranes and abraded skin	Intimate sexual contact (including kissing)	Antibiotics
Bacterium (*Neisseria gonorrheoeae*)	Communicated across mucous membranes	Antibiotics (but antibiotic-resistant strains have arisen)
Bacterium (*Chlamydia trachomatis*)	Communicated across mucous membranes	Antibiotics
Herpes simplex virus	Communicated by contact with infected surfaces, which can be mucous membranes or skin	No cure. Symptoms can be alleviated. Antiviral drugs may lessen subsequent outbreaks
Human papillomavirus	Communicated across mucous membranes through sexual contact	No cure for the virus. Warts can be removed surgically or by burning, freezing, or chemical treatment
Virus	Sexual contact or blood transfusions	No cure. Symptoms can be treated. A vaccine is available that can protect only if given before infection occurs
A variety of bacteria that migrate to the uterus and fallopian tubes	Sexual intercourse	Antibiotics
HIV (see Chapter 13)	The virus enters the bloodstream via cuts or abrasions, including minute ones in the genitalia. Spread primarily by intimate sexual contact, but can also be transmitted via contaminated needles	No cure. Treatments with a variety of medications can slow the course of the infection

STDs have been with humans since ancient times, and they are one of the most serious public health problems today. Over 10 million new cases of STDs occur each year in the United States, and about two-thirds of these cases occur in people between the ages of 15 and 30. About half of U.S. youth will contract an STD before the age of 25. The only contraceptive device that is effective against the transmission of STDs is the condom.

The highly prevalent bacterial diseases *chlamydia* and *gonorrhea* are generally not fatal, but when untreated may develop into painful inflammatory diseases and result in infertility. *Syphilis* is transmitted by a spirochete and is fatal in about half of untreated cases. *AIDS*, a deadly failure of the immune system, is caused by a retrovirus that is most often sexually transmitted (although other forms of transmission are possible). The retrovirus and the disease are described in Chapters 13 and 18, respectively.

Chapter Summary

Asexual Reproduction

▶ Some animals can reproduce asexually, producing offspring that are genetically identical to their parent and to one another. A disadvantage of asexual reproduction is that no genetic diversity is produced.

▶ Means of asexual reproduction include budding, regeneration, and parthenogenesis. **Review Figures 43.1, 43.2**

Sexual Reproduction

▶ Sexual reproduction consists of three basic steps: gametogenesis, mating, and fertilization.

▶ Gametogenesis and fertilization are similar in all animals, but mating includes a great variety of anatomical, physiological, and behavioral adaptations.

▶ In sexually reproducing species, genetic diversity is created by crossing over and independent assortment of chromosomes during gametogenesis. Fertilization also contributes to genetic diversity.

▶ Gametogenesis occurs in testes and ovaries. In spermatogenesis (the production of sperm) and oogenesis (the production of eggs), the germ cells proliferate mitotically, undergo meiosis, and mature into gametes.

▶ Each primary spermatocyte can produce four haploid sperm through the two divisions of meiosis. **Review Figure 43.3*a***

▶ Primary oocytes immediately enter prophase of the first meiotic division, and in many species, including humans, their development is arrested at this point. Each oogonium produces only one egg. **Review Figure 43.3*b***

▶ Fertilization involves sperm activation, species-specific binding of sperm to egg, the acrosomal reaction, digestion of a path through the protective coverings of the egg, and fusion of sperm and egg plasma membranes. **Review Figures 43.4, 43.5.** **See Web/CD Tutorial 43.1**

▶ The fusion of the sperm and egg plasma membranes triggers fast and slow blocks to polyspermy, which prevent additional sperm from entering the egg, and in mammals, signals the egg to complete meiosis and begin development. **Review Figure 43.6**

▶ Fertilization can occur externally, as is common in aquatic species, or internally, as is common in terrestrial species. Internal fertilization usually involves copulation.

▶ Hermaphroditic species have both male and female reproductive systems in the same individual, either sequentially or simultaneously.

▶ Internal fertilization is necessary for terrestrial species. The shelled egg is an important adaptation to the terrestrial environment, but it must be fertilized before the shell forms. All mammals except monotremes retain the embryo internally and have done away with shelled eggs.

▶ Animals can be classified as oviparous or viviparous, depending on whether the early stages of development occur outside or inside the mother's body.

The Human Reproductive System

▶ Males produce semen and deliver it into the female reproductive tract. Semen consists of sperm suspended in a fluid that nourishes them and facilitates fertilization.

▶ Sperm are produced in the seminiferous tubules of the testes, mature in the epididymis, and are delivered to the urethra through the vasa deferentia. Other components of semen are produced in the bulbourethral gland, seminal vesicles, and prostate gland. **Review Figures 43.8, 43.9.** **See Web/CD Activities 43.1 and 43.2**

▶ All components of the semen join in the urethra at the base of the penis and are ejaculated through the erect penis by muscle contractions at the culmination of copulation.

▶ Spermatogenesis depends on testosterone secreted by the Leydig cells of the testes, which are under the control of LH from the anterior pituitary. Spermatogenesis is also controlled by FSH from the pituitary. Hypothalamic GnRH controls pituitary secretion of LH and FSH. The production of these hormones by the hypothalamus and pituitary is controlled by negative feedback from testosterone and another hormone, inhibin, produced by the Sertoli cells of the testes. **Review Figure 43.10**

▶ Eggs mature in the female's ovaries and are released into the oviducts. Sperm deposited in the vagina during copulation move up through the cervix and uterus into the oviducts. **Review Figure 43.11.** **See Web/CD Activity 43.3**

▶ Fertilization occurs in the upper regions of the oviducts. The zygote becomes a blastocyst as it passes down the oviduct. Upon arrival in the uterus, the blastocyst implants in the endometrium and forms a placenta.

▶ The maturation and release of eggs constitute an ovarian cycle. In humans, this cycle takes about 28 days. **Review Figures 43.12**

▶ The uterus also undergoes a cycle that prepares it for receipt of a blastocyst. If no blastocyst is implanted, the lining of the uterus deteriorates and sloughs off in the process of menstruation. **Review Figure 43.13.** **See Web/CD Tutorial 43.2**

▶ Both the ovarian and the uterine cycles are under the control of hypothalamic and pituitary hormones, which in turn are under the feedback control of estrogen and progesterone. **Review Figure 43.14**

▶ Childbirth is initiated by hormonal and mechanical stimuli that increase the contraction of uterine muscle. Oxytocin plays a major role in this positive feedback loop. **Review Figure 43.15**

Human Sexual Behavior

▶ Human sexual responses consist of four phases: excitement, plateau, orgasm, and resolution. In addition, males have a refractory period during which renewed excitement is not possible.

▶ Methods of contraception include abstention from copulation and the use of technologies that decrease the probability of fertilization. **Review Table 43.1**

▶ Barrier methods of contraception, such as condoms, diaphragms, and spermicidal substances, kill sperm or block their passage through the female reproductive tract.

▶ Methods to prevent ovulation, such as birth control pills and other hormonal treatments, interfere with the ovarian cycle so that mature, fertile eggs are not produced and released.

▶ Males and females can be sterilized by surgical blockage of the vasa deferentia (vasectomy) or oviducts (tubal ligation). **Review Figure 43.16**

▶ Methods to prevent implantation include intrauterine devices, excess doses of steroids, and a progesterone receptor blocker. After implantation, the termination of a pregnancy is called an abortion.

▶ Assisted reproductive technologies have been developed to increase fertility. These ARTs include in vitro fertilization and gamete intrafallopian transfer.

▶ Many disease-causing organisms are transmitted through sexual behavior. Many sexually transmitted diseases are curable if treated early, but can have serious long-term consequences if not treated. **Review Table 43.2**

Self-Quiz

1. A species in which the individual possesses both male and female reproductive systems is termed
   a. dioecious.
   b. parthenogenetic.
   c. hermaphroditic.
   d. monoecious.
   e. both *c* and *d.*
2. The major advantage of internal fertilization is that
   a. it ensures paternity.
   b. it permits the fertilization of many gametes.
   c. it reduces the incidence of destructive competitive interactions between the members of a group.
   d. it results in the formation of a stable pair bond.
   e. it gives the developing organism a greater degree of protection during the early phases of development.
3. Which statement about oocytes is true?
   a. At birth, the human female has produced all the oocytes she will ever produce.
   b. At the onset of puberty, ovarian follicles produce new oocytes in response to hormonal stimulation.
   c. At the onset of menopause, the human female stops producing oocytes.
   d. Oocytes are produced by the human female throughout adolescence.
   e. Oocytes produced by the female are stored in the seminiferous tubules.
4. Spermatogenesis and oogenesis differ in that
   a. spermatogenesis produces gametes with greater energy stores than those produced by oogenesis.
   b. spermatogenesis produces four equally functional diploid cells per meiotic event and oogenesis does not.
   c. oogenesis produces four equally functional haploid cells per meiotic event and spermatogenesis does not.
   d. spermatogenesis produces many gametes with meager energy reserves, whereas oogenesis produces relatively few, well-provisioned gametes.
   e. spermatogenesis begins before birth in humans, whereas oogenesis does not start until the onset of puberty.
5. Semen contains all of the following except
   a. fructose.
   b. mucus.
   c. clotting enzymes.
   d. substances to reduce the pH of the uterine environment.
   e. substances to increase the contraction of the uterine muscle.
6. During oogenesis in mammals, the second meiotic division occurs
   a. in the formation of the primary oocyte.
   b. in the formation of the secondary oocyte.
   c. before ovulation.
   d. after fertilization.
   e. after implantation.
7. One of the major differences between the sexual responses of human males and females is
   a. the increase in blood pressure in males.
   b. the increase in heart rate in females.
   c. the presence of a refractory period in females.
   d. the presence of a refractory period in males after orgasm.
   e. the increase in muscle tension in males.
8. Which of the following is *true* of sexually transmitted diseases?
   a. They are always caused by viruses or bacteria.
   b. Using contraception will prevent them.
   c. The organisms that cause them have evolved to depend on intimate physical contact between hosts as their means of transmission.
   d. Their transmission has a high probability of failure.
   e. You cannot catch one from someone you love.
9. Contractions of muscles in the uterine wall and in the breasts are stimulated by
   a. progesterone.
   b. estrogen.
   c. prolactin.
   d. oxytocin.
   e. human chorionic gonadotropin.
10. Which method of contraception is *most* likely to fail?
    a. Rhythm method
    b. Birth control pills
    c. Diaphagms
    d. Vasectomy
    e. Condoms

For Discussion

1. In the very deep ocean, there are species of fish in which the male is very much smaller than the female and actually lives attached to her body. In terms of the selective pressures that operate on sexual and asexual reproduction and in terms of the deep-sea environment, what factors do you think resulted in the evolution of this extreme sexual dimorphism?
2. What are two main differences between the immediate products of the first and second meiotic divisions in spermatogenesis and oogenesis? Why do these differences exist?
3. At the beginning of each ovarian cycle in humans, six to twelve follicles begin to develop in response to rising levels of FSH, but after a week, only one follicle continues to develop, and the others wither away. Given the facts that follicles produce estrogen, estrogen stimulates follicle cells to produce FSH receptors, and estrogen exerts negative feedback on FSH production in the pituitary, can you explain how one follicle gets "selected" to grow?
4. Compare the actions of LH and FSH in the ovaries and testes.
5. Ovarian and uterine events in the month following ovulation differ depending on whether fertilization occurs. Describe the differences and explain their hormonal controls.

44 Neurons and Nervous Systems

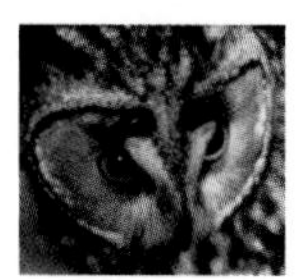

On a dark, moonless winter night a mouse scurries across the icy forest floor. With an almost imperceptible whoosh, an owl swoops down, grabs the mouse in its talons, and rises back into the forest canopy. How did the owl locate the mouse so precisely that it could seize it while flying? It heard the mouse, of course. But place yourself in that same dark forest. If you heard a scratching sound, how accurate do you think you would be at instantly shining your flashlight on the mouse making the sound? Yet your nervous system and the owl's are exposed to the same environmental information. How do nervous systems derive directional information from sound?

Sound is converted into simple electric signals in the inner ear. These signals are produced by nerve cells, or neurons, and carried by long extensions of those neurons to the brain. Different neurons in the ear respond to different pitches or frequencies of sound, and the rate of the electric signals can convey information about the intensity of the sound. Neurons in the owl's brain, and in ours to a lesser degree, are able to detect these tiny differences in the electric signals they receive and integrate that information to decide where the sound is coming from.

The brain and nervous system are capable of many amazing feats such as this. In this chapter and the three that follow, we will learn how nervous systems accomplish these feats. In this chapter, we start by describing the cells of the nervous system, how they generate electric signals, conduct those signals from place to place in the body, and communicate those signals from cell to cell.

Listening in the Dark In total darkness, a long-eared owl (*Asio otus*) catches a mouse by using directional information in the sounds the mouse makes as it moves. Such precise analysis of limited information shows the amazing capabilities of complex nervous systems.

Nervous Systems: Cells and Functions

Nerve cells, or **neurons**, are specialized to receive information, encode it, and transmit it to other cells. Neurons and their specialized supportive cells, called *glial cells*, make up **nervous systems**.

Animals receive various kinds of information from both inside and outside their bodies. This information is received and converted, or *transduced*, by *sensory cells* (also called *receptor cells*) into electric signals that can be transmitted and processed by neurons. To cause behavioral or physiological responses, a nervous system communicates these signals to *effectors*, such as muscles and glands.

Nervous systems process information

Simple animals such as sea anemones can process information with simple networks of neurons that do little more than provide direct lines of communication from sensory cells to effectors (Figure 44.1*a*). The anemone's *nerve net* is most developed around the tentacles and the oral opening, where it facilitates detection of food or danger and causes tentacles to extend or retract. Bilaterally symmetrical animals, such as earthworms, that move more rapidly through their environments need to process and integrate larger amounts of information. This need is met by clusters of neurons called **ganglia** (Figure 44.1*b*). Ganglia serving different functions may be distributed around the body, as in the earthworm or the squid (Figure 44.1*c*). Frequently one pair of ganglia is larger and more central than the others and is therefore given the designation of **brain**.

In vertebrates, most of the cells of the nervous system are found in the brain and the **spinal cord**, the sites of most information processing, storage, and retrieval (Figure 44.1*d*). Therefore, the brain and spinal cord are called the **central nervous system** (**CNS**). Information is transmitted from sensory cells to the CNS and from the CNS to effectors via neurons that extend or reside outside of the brain and the spinal cord; these neurons and their supporting cells are called the **peripheral nervous system**.

Vertebrates differ greatly in their behavioral complexity and in their physiological specializations. Even the smaller and simpler nervous systems of invertebrates can be remarkably complex. Consider the nervous systems of small spiders that have programmed within them the thousands of precise movements necessary to construct a beautiful web without prior experience.

Neurons are the functional units of nervous systems

Although nervous systems vary enormously in structure and function, neurons function similarly in animals as different as squids and humans. Their plasma membranes generate electric signals—called *nerve impulses* or *action potentials*—and conduct these signals from one location on a neuron to the most distant reaches of that cell—a distance that can be more than a meter in a human and many meters in a whale.

Most neurons have four regions—a cell body, dendrites, an axon, and axon terminals (Figure 44.2*a*)—but the variation among different types of neurons is considerable (Figure 44.2*b*). The **cell body** contains the nucleus and most of the cell's organelles. Many projections may sprout from the cell body. Most of these projections are bushlike **dendrites** (from the Greek *dendron*, "tree"), which bring information from other neurons or sensory cells to the cell body. The degree of branching of the dendrites differs among different types of neurons. In most neurons, one projection is much longer than

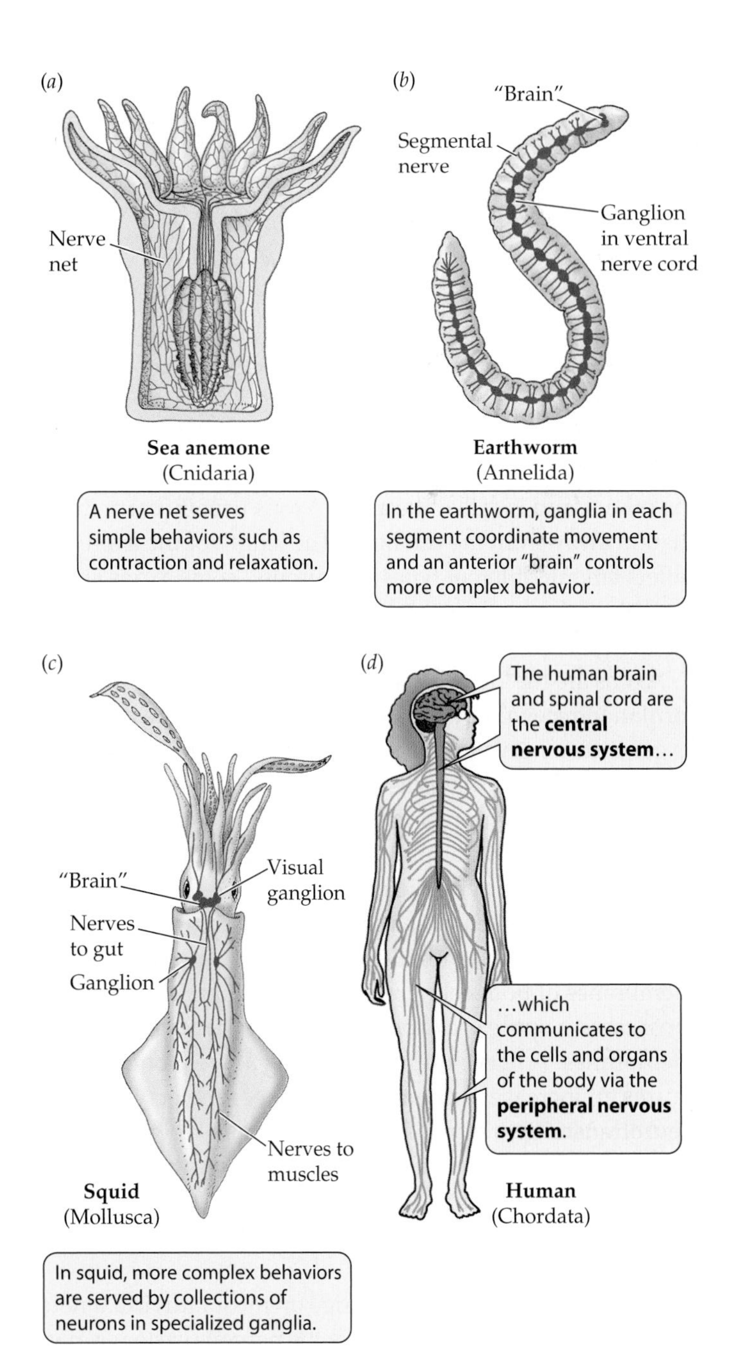

44.1 Nervous Systems Vary in Size and Complexity As we compare animals that have increasingly complex sensory and behavioral abilities, we find that information processing is increasingly centralized in ganglia (collections of neurons) or in a brain.

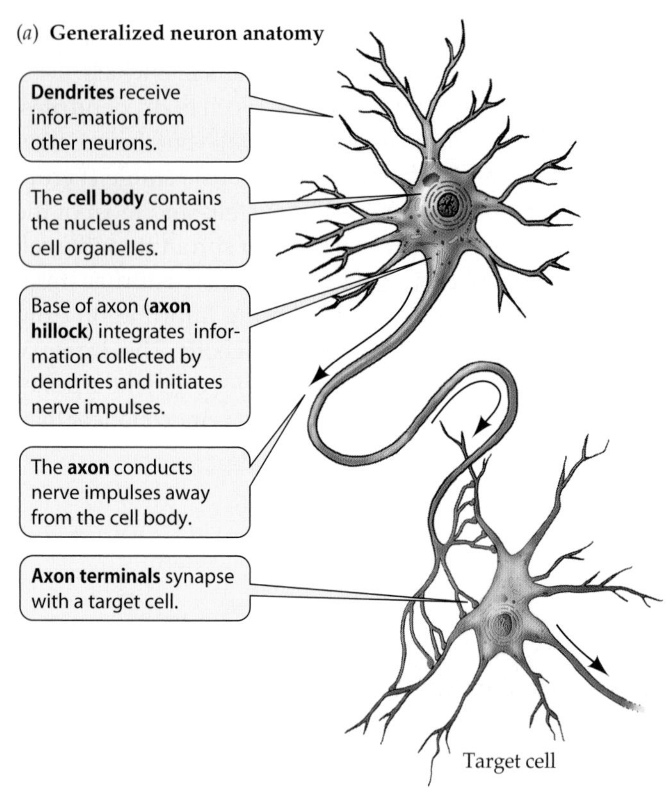
(a) Generalized neuron anatomy
Dendrites receive infor-mation from other neurons.
The cell body contains the nucleus and most cell organelles.
Base of axon (axon hillock) integrates information collected by dendrites and initiates nerve impulses.
The axon conducts nerve impulses away from the cell body.
Axon terminals synapse with a target cell.
Target cell

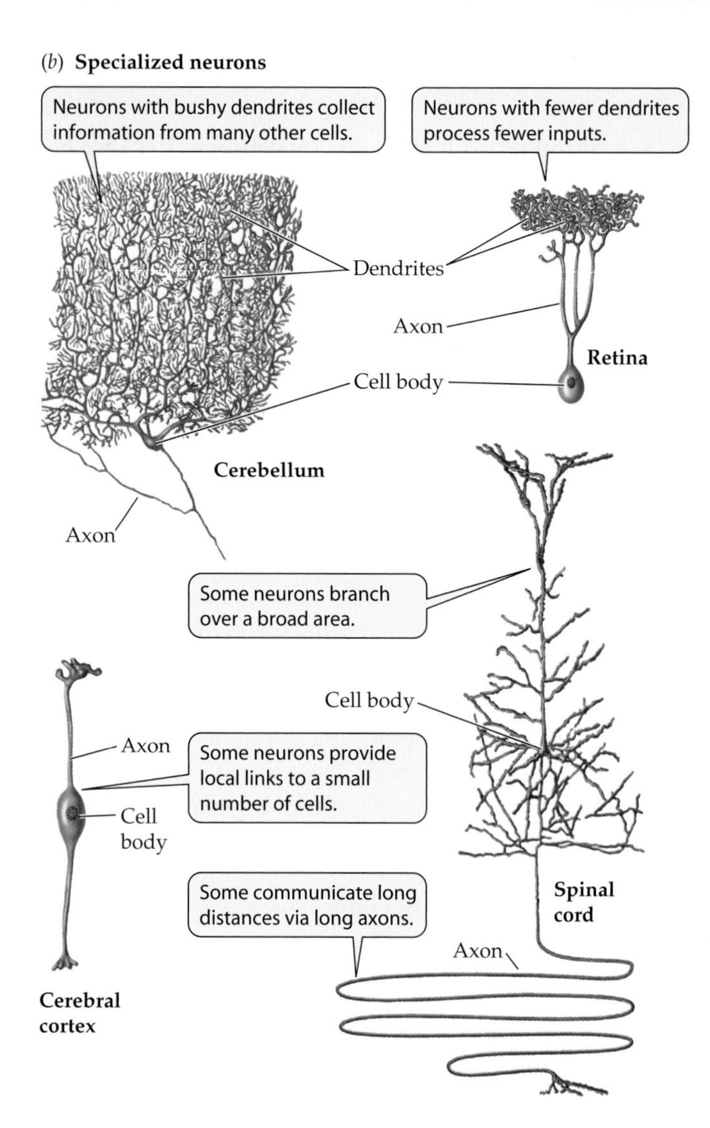
(b) Specialized neurons
Neurons with bushy dendrites collect information from many other cells.
Neurons with fewer dendrites process fewer inputs.
Dendrites
Axon
Cell body
Retina
Cerebellum
Axon
Some neurons branch over a broad area.
Cell body
Axon
Some neurons provide local links to a small number of cells.
Cell body
Some communicate long distances via long axons.
Spinal cord
Axon
Cerebral cortex

44.2 Neurons (*a*) A generalized diagram of a neuron. (*b*) Neurons from different parts of the mammalian nervous system are specifically adapted to their functions.

the others, and is called the **axon**. Axons usually carry information away from the cell body. The length of the axon also differs among different types of neurons—some axons are remarkably long, such as those that run from the spinal cord to the toes.

Axons are the "telephone lines" of the nervous system. Information received by the dendrites can cause the cell body to generate a nerve impulse, which is then conducted along the axon to the cell that is its target. At the target cell—which can be another neuron, a muscle cell, or an endocrine cell—the axon divides into a spray of fine nerve endings. At the tip of each of these tiny nerve endings is a swelling, called an **axon terminal**, that comes very close to the target cell.

Where an axon terminal comes close to another cell, the membranes of both cells are modified to form a **synapse**. In most cases, a space only about 25 nm wide separates the two membranes. A nerve impulse arriving at an axon terminal causes an increase of chemical messenger molecules called **neurotransmitters** stored in the axon terminal to be released. The released neurotransmitters diffuse across the space and bind to receptors on the plasma membrane of the target cell. We will discuss this process of synaptic transmission in more detail later in the chapter.

Thousands of synapses impinge on most individual neurons. Integration of information in the nervous system is possible because a neuron can receive information (synaptic inputs) from many sources before producing nerve impulses that travel down its single axon to target cells.

Glial cells are also important components of nervous systems

Neurons are not the only type of cell in the nervous system. In fact, there are more **glial cells** than neurons in the human brain. Like neurons, glial cells come in several forms and have a diversity of functions. Some glial cells physically support and orient the neurons and help them make the right contacts during embryonic development. Other glial cells insulate axons.

In the peripheral nervous system, **Schwann cells** wrap around the axons of neurons, covering them with concentric layers of insulating plasma membrane (Figure 44.3). Other glial cells called **oligodendrocytes** perform a similar function in the central nervous system. **Myelin** is the covering produced by Schwann cells and oligodendrocytes, and it gives many parts of the nervous system a glistening white appearance. Later in this chapter we will see how the electrical

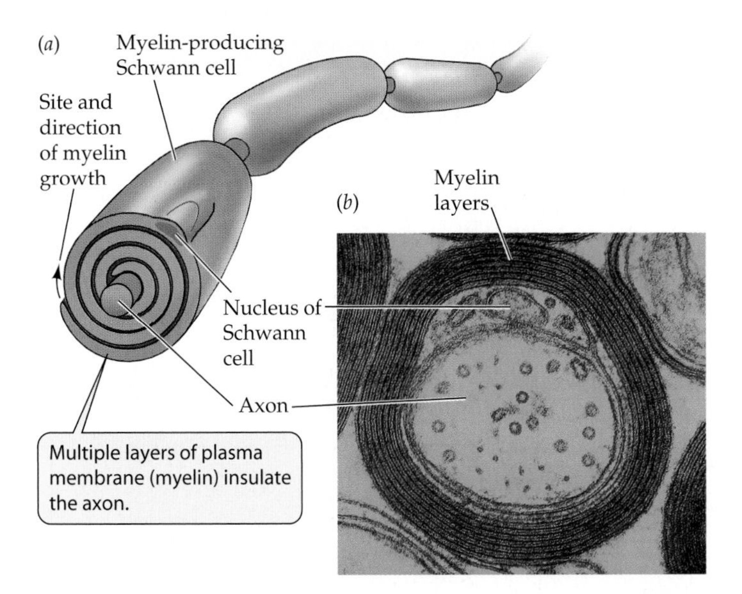

44.3 Wrapping Up an Axon (*a*) Schwann cells wrap axons in the peripheral nervous system with layers of myelin, a type of plasma membrane that provides electrical insulation. (*b*) A myelinated axon, seen in cross section through an electron microscope.

insulation provided by myelin increases the speed with which axons can conduct nerve impulses.

Glial cells are well known for the many supportive roles they play. Some supply neurons with nutrients; others consume foreign particles and cell debris. Glial cells also help maintain the proper ionic environment around neurons. Although they have no axons and do not generate or conduct nerve impulses, some glial cells communicate with one another electrically through *gap junctions*, a special type of connection that enables ions to flow between cells.

Glial cells called **astrocytes** (because they look like stars) contribute to the **blood–brain barrier**, which protects the brain from toxic chemicals in the blood. Blood vessels throughout the body are very permeable to many chemicals, including toxic ones, which would reach the brain if this special barrier did not exist. Astrocytes help form the blood–brain barrier by surrounding the smallest, most permeable blood vessels in the brain. The barrier is not perfect, however. Since it consists of plasma membranes, it is permeable to fat-soluble substances such as anesthetics and alcohol.

Neurons function in networks

As we learn more about the properties of neurons, it is important to keep in mind that nervous systems depend on neurons working together. The simplest neuronal network consists of three cells: a sensory neuron connected to a motor neuron connected to a muscle cell. Most of the neuronal networks that carry out the functions of the human nervous system are much more complex and consist of many more neurons. The human brain contains an estimated 10^{11} neurons, and most of those neurons receive information from a thousand or more synapses; thus, there may be as many as 10^{14} synapses in the human brain. Therein lies the incredible ability of the human brain to process information.

This astronomical number of neurons and synapses is divided into thousands of distinct but interacting networks that function in parallel. But before we can understand how even one of these networks works, we must understand the properties of individual neurons that allow them to generate and conduct nerve impulses.

Neurons: Generating and Conducting Nerve Impulses

The insides of cells are electrically negative in comparison to the outsides. The difference in electric potential, or **voltage**, across the plasma membrane of a cell is called its **membrane potential**. In an *unstimulated* neuron, this voltage difference is called a **resting potential**.

Membrane potentials can be measured with *electrodes*. An electrode can be made from a glass pipette pulled to a very sharp tip and filled with a solution containing ions that conduct electric charges. Using such electrodes, we can record very tiny local electrical events. If a pair of electrodes is placed one on each side of the plasma membrane of a resting axon, they measure a voltage difference of about 60 millivolts (mV) (Figure 44.4).

The resting potential provides a means for neurons to respond to a stimulus. A neuron is sensitive to any chemical or physical factor that causes a change in the resting potential across a portion of its plasma membrane. The most extreme change in membrane potential is an **action potential**, which is a sudden and rapid reversal in the voltage across a portion of the plasma membrane. For 1 or 2 milliseconds, a bioelectric current crosses the membrane and the inside of the cell becomes *more positive* than the outside. **Nerve impulses** are action potentials that move along axons.

Simple electrical concepts underlie neuronal function

Voltage (potential or electric charge difference) is the tendency for electrically charged particles such as electrons or ions to move between two points. Voltage is to the flow of electrically charged particles what pressure is to the flow of water. If the negative and the positive poles of a battery are connected by a copper wire, electrons flow from negative to positive because there is a voltage difference between them. This flow of electrons is an electric current, and it can be used to do work, just as a current of water can be used to do work such as turning a turbine.

Electric current is carried by electrons in wires, but in solutions and across cell membranes, it is carried not by electrons, but by charged ions. The major ions that carry electric

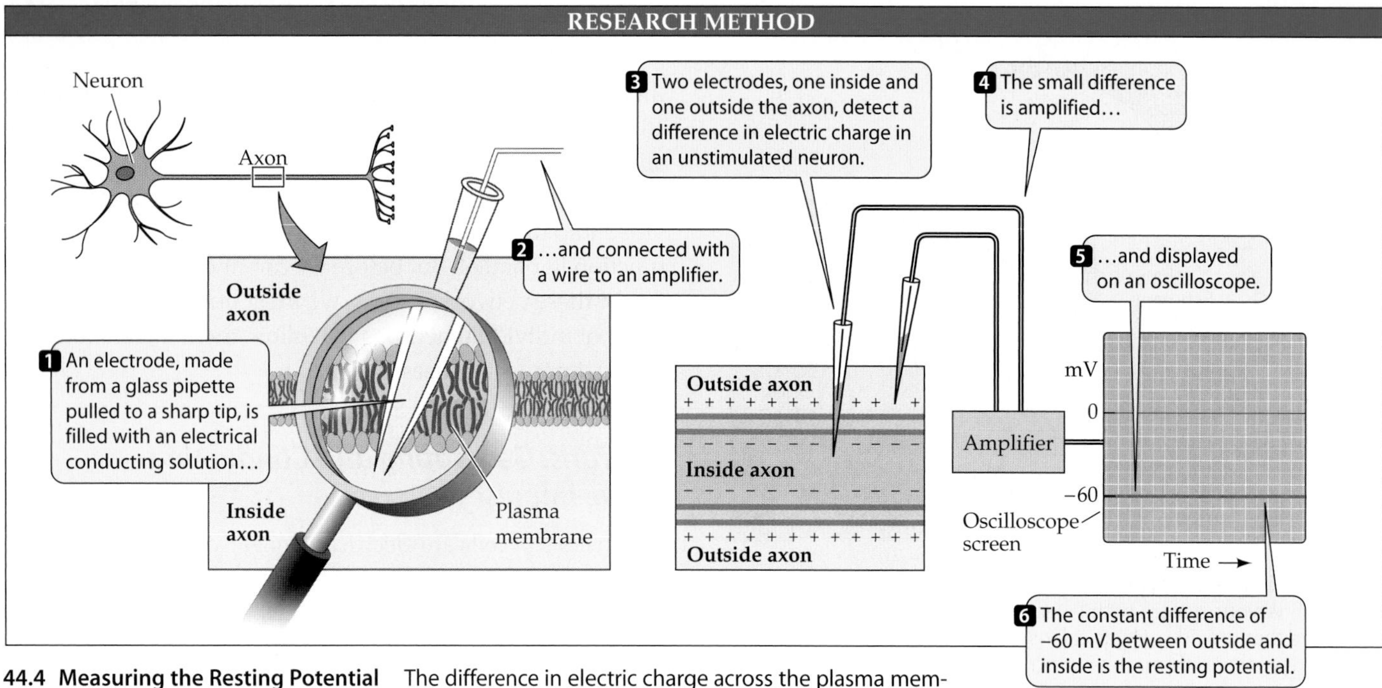
RESEARCH METHOD
Neuron
Axon
Outside axon
Inside axon
Plasma membrane
1 An electrode, made from a glass pipette pulled to a sharp tip, is filled with an electrical conducting solution...
2 ...and connected with a wire to an amplifier.
3 Two electrodes, one inside and one outside the axon, detect a difference in electric charge in an unstimulated neuron.
4 The small difference is amplified...
5 ...and displayed on an oscilloscope.
Outside axon
Inside axon
Outside axon
Amplifier
mV
0
−60
Oscilloscope screen
Time
6 The constant difference of −60 mV between outside and inside is the resting potential.

44.4 Measuring the Resting Potential The difference in electric charge across the plasma membrane of a neuron can be measured using two electrodes, one inside and one outside the cell. In an unstimulated neuron, this difference is constant (about –60 mV), and is known as the resting potential.

charges across the plasma membranes of neurons are sodium (Na^+), chloride (Cl^-), potassium (K^+), and calcium (Ca^{2+}). It is also important to remember that ions with opposite charges attract one another, and those with like charges repel one another. With these basics of bioelectricity in mind, we can ask how the resting potential of the neuronal plasma membrane is created and how the flow of ions through membrane channels is turned on and off to generate action potentials.

Ion pumps and channels generate resting and action potentials

The plasma membranes of neurons, like those of all other cells, are lipid bilayers that are impermeable to ions. However, these impermeable lipid bilayers contain many protein molecules that serve as ion channels and ion pumps (see Chapter 5). Ion pumps and channels are responsible for resting and action potentials.

Ion pumps use energy to move ions or other molecules against their concentration gradients. A major ion pump in the plasma membranes of neurons (and of all other cells) is the **sodium–potassium pump**. The action of this pump expels Na^+ ions from inside the cell, exchanging them for K^+ ions from outside the cell (Figure 44.5*a*; see also Figure 5.13). The sodium–potassium pump keeps the concentration of K^+ inside the cell greater than that of the extracellular fluid, and the concentration of Na^+ inside the cell less than that of the extracellular fluid. The concentration differences established by the pump mean that K^+ would diffuse out of the cell and Na^+ would diffuse in if the ions could cross the lipid bilayer.

Ion channels are pores formed by proteins in the lipid bilayer (see Chapter 5). These water-filled pores allow ions to pass through a membrane, but they are generally *selective*—they allow some types of ions to pass through more easily than others (Figure 44.5*b*). Thus, there are potassium chan-

44.5 Ion Pumps and Channels (*a*) The sodium–potassium pump actively moves K^+ ions to the inside of a neuron and Na^+ ions to the outside. (*b*) Ion channels allow specific ions to diffuse down their concentration gradient; K^+ ions tend to leave neurons when potassium channels are open, and Na^+ ions tend to enter neurons when sodium channels are open.

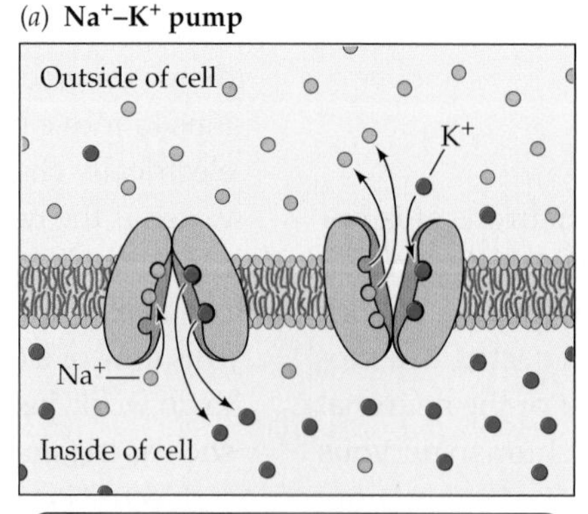
(a) Na+–K+ pump
Outside of cell
K+
Na+
Inside of cell
The Na+–K+ pump moves Na+ and K+ ions against their concentration gradients.

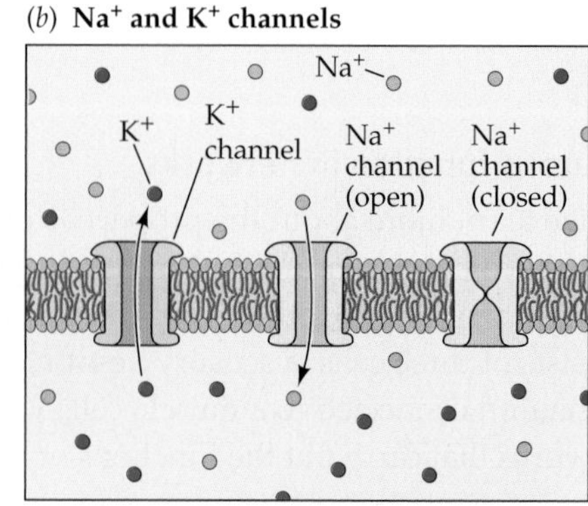
(b) Na+ and K+ channels
Na+
K+
K+ channel
Na+ channel (open)
Na+ channel (closed)
K+ and Na+ ions tend to diffuse down their concentration gradients through ion-specific channels.

nels, sodium channels, chloride channels, and calcium channels, and there are many different kinds of each. Ions move through channels by diffusion, and can move in either direction. The direction and magnitude of net movement of ions through a channel depends on the concentration gradient of that ion type across the plasma membrane, as well as the voltage across that membrane.

Potassium channels are the most common open channels in the plasma membranes of resting (non-stimulated) neurons. As a consequence, resting neurons are more permeable to K^+ than to any other ion. As Figure 44.6 shows, this characteristic explains the resting potential. Because the potassium channels make the plasma membrane permeable to K^+, and because the sodium–potassium pump keeps the concentration of K^+ inside the cell much higher than that outside the cell, K^+ tends to diffuse out of the cell through the channels. As positively charged K^+ ions diffuse out of the cell, they leave behind unbalanced negative charges (mostly Cl^- ions and protein molecules), generating an electric potential across the membrane that tends to pull positively charged K^+ ions back into the cell.

The membrane potential at which the tendency of K^+ ions to diffuse out of the cell is balanced by the negative electric potential pulling them back in is called the *potassium equilibrium potential*. The value of the potassium equilibrium potential can be calculated from the concentrations of K^+ on the two sides of the membrane using an equation called the *Nernst equation*, which is derived from the laws of physical chemistry (Figure 44.7). In general, the resting potential is a bit less negative than this equation predicts because resting neurons are also slightly permeable to other ions, such as Na^+ and Cl^-.

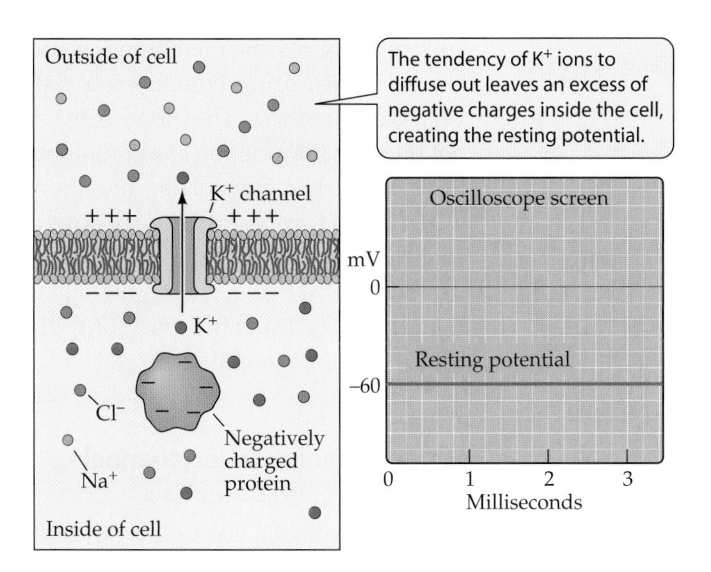

44.6 Open Potassium Channels Create the Resting Potential Open potassium channels allow K^+ ions to diffuse out of the cell, leaving unbalanced negative charges behind (mostly on Cl^- ions and protein molecules).

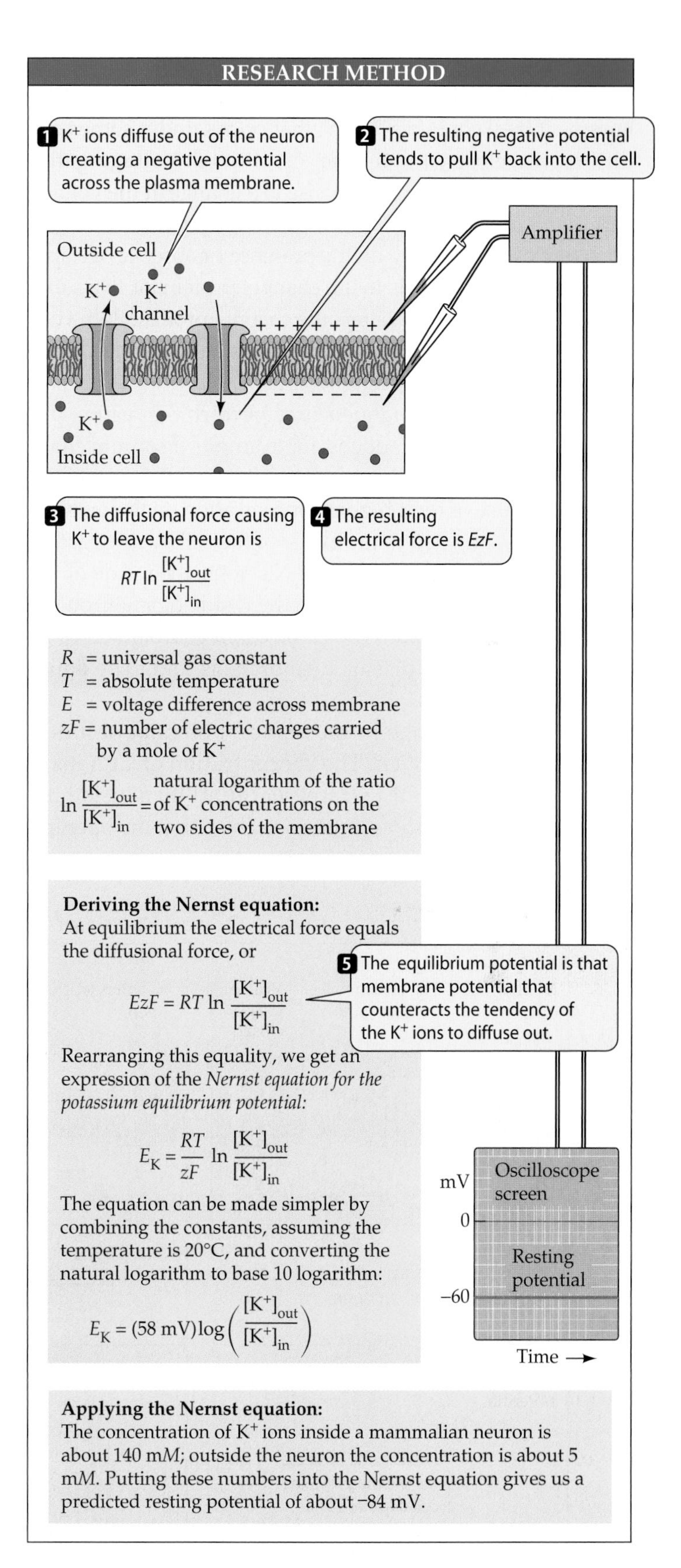

44.7 The Nernst Equation The Nernst equation calculates membrane potential when only one type of ion can cross a membrane that separates solutions with different concentrations of that ion. A resting neuron comes close to that situation because its permeability to K^+ ions is high and its permeability to all other ions is low.

Ion channels can alter membrane potential

Many ion channels in the plasma membranes of neurons behave as if they contain a "gate" that opens under some conditions, but closes under other conditions. **Voltage-gated channels** open or close in response to a change in the voltage across the plasma membrane. **Chemically gated channels** open or close depending on the presence or absence of a specific molecule that binds to the channel protein, or to a separate receptor that in turn alters the channel protein. Both voltage-gated and chemically gated channels play important roles in neuronal function.

Changes in gated channels may perturb the resting potential. Imagine what happens, for example, if sodium channels in the plasma membrane open. Na^+ ions diffuse into the neuron because of their higher concentration on the outside, plus they are attracted to the inside of the cell by its negative charge. As a result of the entry of Na^+ ions, the inside of the cell becomes less negative. When the inside of a neuron becomes less negative (or more positive) in comparison to its resting condition, its plasma membrane is said to be **depolarized** (Figure 44.8*a*).

An opposite change in the resting potential occurs if gated Cl^- channels open. The concentration of Cl^- ions is normally greater in the extracellular fluid than inside the neuron. This difference is large enough so that the opening of Cl^- channels causes Cl^- to enter the cell, even though the membrane potential is negative. The entry of negative charges causes the membrane potential to become even more negative. When the inside of a neuron becomes more negative in comparison to its resting condition, its plasma membrane is said to be **hyperpolarized** (Figure 44.8*b*).

The opening and closing of ion channels, which result in changes in the polarity of the plasma membrane, are the basic mechanisms by which neurons respond to electrical, chemical, or other stimuli, such as touch, sound, and light. How does a neuron use a change in its resting membrane potential to process and transmit information?

A change in resting potential may result from input at a synapse. This input, however, is a very local event that affects only a small patch of plasma membrane. How can that information be passed to other parts of the cell? A local perturbation of the resting potential causes a flow of electrically charged ions, which tends to spread the change in membrane potential to adjacent regions of the membrane. This flow of electrically charged ions is an electric current. For example, if positively charged Na^+ ions enter the cell through open sodium channels at one location, that positively charged area on the inside of the membrane attracts negative charges from surrounding areas, and thus there is a rapid flow of electric current. However, this local flow of electric current does not spread very far before it diminishes and disappears.

The reason why these electric currents do not travel very far is that cell membranes are not completely impermeable to ions. An electric current traveling along a membrane is like water flowing through a leaky hose. The flow of electric current along plasma membranes is useful for transmitting signals over only very short distances. Therefore, axons do not transmit information as a continuous flow of electric current (as telephone wires do). However, the local flow of electric current is an important part of the mechanism that generates the signals that axons do transmit over long distances: action potentials.

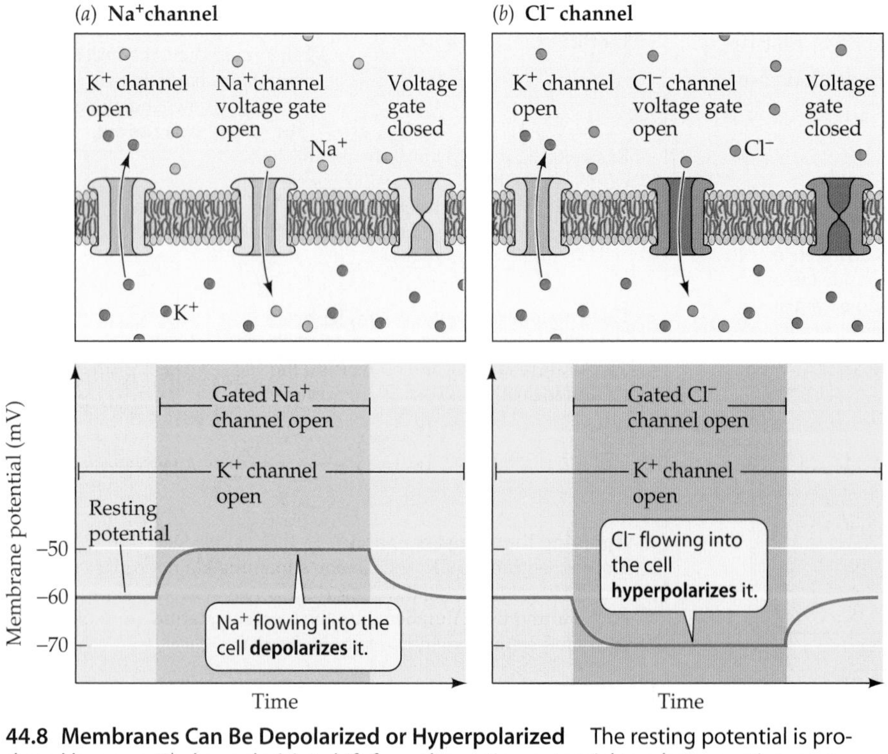

Sudden changes in ion channels generate action potentials

An action potential is a sudden and major change in membrane potential that lasts for only 1 or 2 milliseconds. Action potentials are conducted along the axon of a neuron at speeds of up to 100 meters

44.8 Membranes Can Be Depolarized or Hyperpolarized The resting potential is produced by open K^+ channels. (*a*) A shift from the resting potential to a less negative membrane potential, as occurs when Na^+ enters the cell through a gated sodium channel, is called depolarization. (*b*) Hyperpolarization occurs when the membrane potential becomes more negative, as when Cl^- enters the cell through a gated chloride channel.

per second, which is equivalent to running the length of a football field in a second.

If we place the tips of a pair of electrodes on either side of the plasma membrane of a resting axon and measure the voltage difference, the reading is about –60 mV, as we saw in Figure 44.4. If these electrodes are in place when an action potential travels down the axon, they register a rapid change in membrane potential, from –60 mV to about +50 mV. The membrane potential then rapidly returns to its resting level of –60 mV as the action potential passes (Figure 44.9).

Voltage-gated sodium channels in the plasma membrane of the axon are responsible for action potentials. At the resting potential, most of these channels are closed. They are called voltage-gated channels because depolarization of the membrane causes them to open. For example, if synaptic input to some part of a neuron is sufficiently strong to cause the plasma membrane of its cell body to depolarize, that depolarization can spread by local current flow to the base of the axon (see Figure 44.2), where there are voltage-gated sodium channels. When the plasma membrane in this area is depolarized, the channels open briefly—for less than a millisecond. The Na^+ concentration is much higher outside the

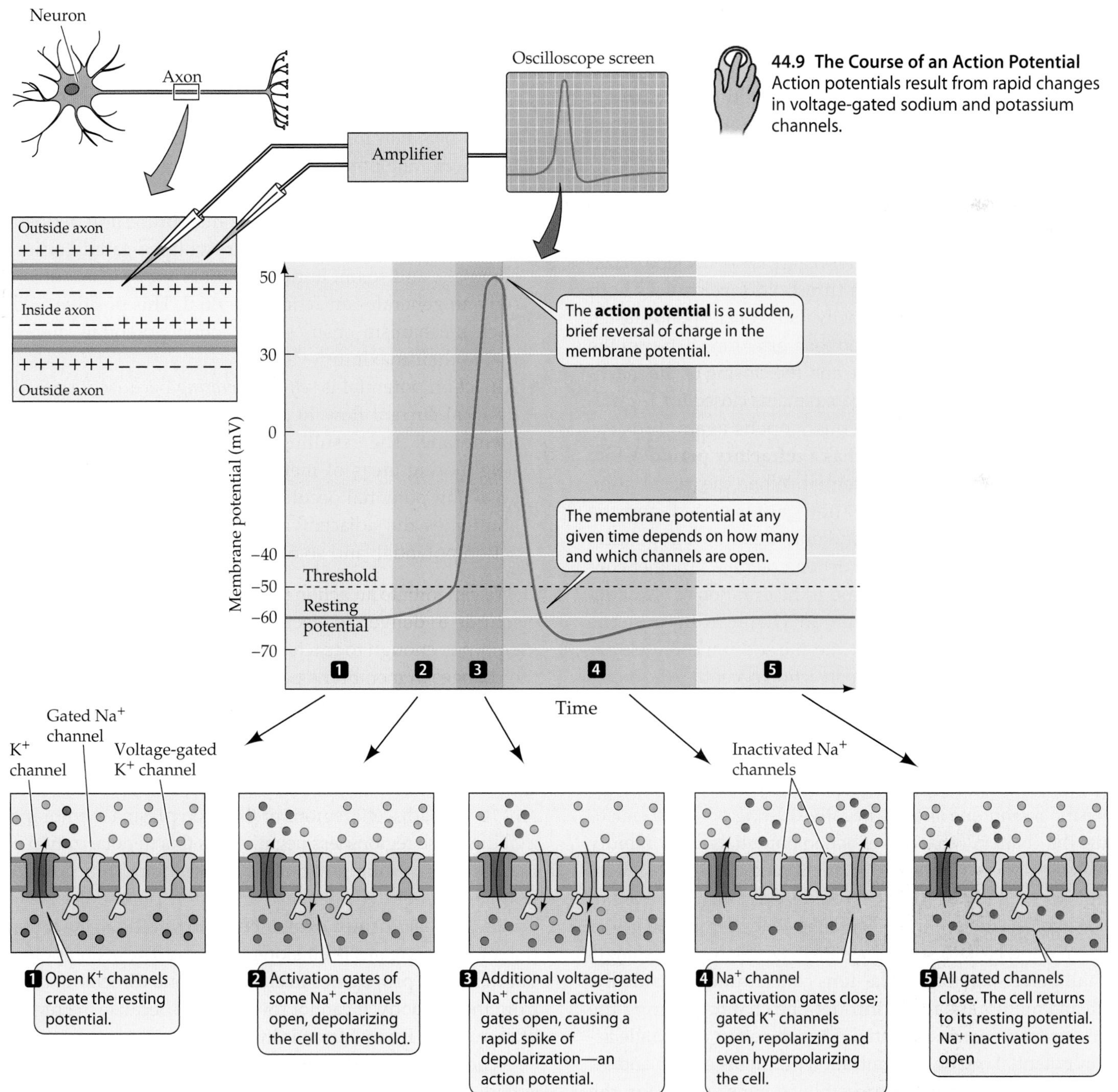
Neuron
Axon
Amplifier
Oscilloscope screen
Outside axon
Inside axon
Outside axon
Membrane potential (mV)
50
30
0
–40
–50
–60
–70
Threshold
Resting potential
Time
1
2
3
4
5
The **action potential** is a sudden, brief reversal of charge in the membrane potential.
The membrane potential at any given time depends on how many and which channels are open.
Gated Na⁺ channel
K⁺ channel
Voltage-gated K⁺ channel
Inactivated Na⁺ channels
1 Open K⁺ channels create the resting potential.
2 Activation gates of some Na⁺ channels open, depolarizing the cell to threshold.
3 Additional voltage-gated Na⁺ channel activation gates open, causing a rapid spike of depolarization—an action potential.
4 Na⁺ channel inactivation gates close; gated K⁺ channels open, repolarizing and even hyperpolarizing the cell.
5 All gated channels close. The cell returns to its resting potential. Na⁺ inactivation gates open

44.9 The Course of an Action Potential Action potentials result from rapid changes in voltage-gated sodium and potassium channels.

axon than inside, so when the channels open, Na^+ ions rush into the axon. The entering Na^+ makes the inside of the plasma membrane electrically positive. When the membrane is depolarized about 5 to 10 mV from the resting potential, a *threshold* is reached, and a large number of sodium channels open, causing a large, sudden depolarization: an action potential (see Figure 44.9).

What causes the depolarized axon to return to resting potential? There are two contributing factors: The voltage-gated sodium channels close, and voltage-gated potassium channels open. The voltage-gated potassium channels open more slowly than the sodium channels and stay open longer, allowing K^+ to carry excess positive charges out of the axon. As a result, the voltage across the membrane returns to its resting level.

Another feature of the voltage-gated sodium channels is that once they open and close, they cannot respond again until after a short delay of 1 to 2 milliseconds. This property can be explained by the assumption that they have two voltage-sensitive gates, an *activation gate* and an *inactivation gate* (see Figure 44.9). Under resting conditions, the activation gate is closed and the inactivation gate is open. Depolarization of the membrane to the threshold level causes both gates to change state, but the activation gate responds faster. As a result, the channel is open for a brief time between the opening of the activation gate and the closing of the inactivation gate. The inactivation gate remains closed for 1–2 milliseconds before it spontaneously opens again, thus explaining why the membrane has a **refractory period** before it can fire another action potential. When the inactivation gate finally opens, the activation gate is closed, and the membrane is poised to respond once again to a depolarizing stimulus by firing another action potential. Another contribution to the refractory period is the duration of the opening of the voltage-gated potassium channels. Because the potassium channels are open and the sodium channels are closed immediately following an action potential, the membrane potential briefly falls below the normal resting potential. This dip in the membrane potential is called the *afterhyperpolarization*.

The difference in the concentration of Na^+ across the plasma membrane and the negative resting potential constitute the "battery" that drives action potentials. How rapidly does the battery run down? It might seem that a substantial number of Na^+ and K^+ ions would have to cross the membrane for the membrane potential to change from –60 mV to +50 mV and back to –60 mV again. In fact, only a vanishingly small percentage of the Na^+ ions concentrated outside the plasma membrane move through the channels during the passage of an action potential. Thus the effect of a single action potential on the concentration gradients of Na^+ and K^+ is very small, and it is possible in most cases for the sodium–potassium pump to keep the "battery" charged, even when the neuron is generating many action potentials every second.

Action potentials are conducted down axons without loss of signal

Action potentials can travel over long distances with no loss of signal. If we place two pairs of electrodes at two different locations along an axon, we can record an action potential at those two locations as it travels down the axon (Figure 44.10*a*). The magnitude of the action potential does not change between the two recording sites. This constancy is possible because an action potential is an all-or-nothing, self-regenerating event.

- An action potential is *all-or-nothing* because of the interaction between the voltage-gated sodium channels and the membrane potential. If the membrane is depolarized slightly, some voltage-gated sodium channels open. Some Na^+ ions cross the plasma membrane and depolarize it even more, opening more voltage-gated sodium channels, and so on, until the membrane reaches threshold and generates an action potential. This positive feedback mechanism ensures that action potentials always rise to their maximum value.
- An action potential is *self-regenerating* because it spreads by local current flow to adjacent regions of the plasma membrane. The resulting depolarization brings those neighboring areas of membrane to threshold. So when an action potential occurs at one location on an axon, it stimulates the adjacent region of axon to generate an action potential, and so on down the length of the axon.

We can initiate an action potential by using a stimulating electrode to deliver an electric current that depolarizes the membrane enough to reach threshold. Now we can observe the changes in membrane potential associated with the passage of that action potential past the recording electrodes (Figure 44.10*b*).

At the site of the action potential, positive ions flood into the neuron. Once inside, those positive ions spread by current flow to adjacent regions of the axon plasma membrane, making those regions less negative. As this depolarization of the adjacent membrane brings it to threshold, an action potential is generated. Because an action potential always brings the adjacent area of membrane to threshold, the action potential propagates itself along the axon. The action potential normally propagates itself in only one direction, away from the cell body. It cannot reverse itself because the region of membrane it came from is in its refractory period.

Action potentials do not travel along all axons at the same speed. They travel faster in large-diameter axons than in

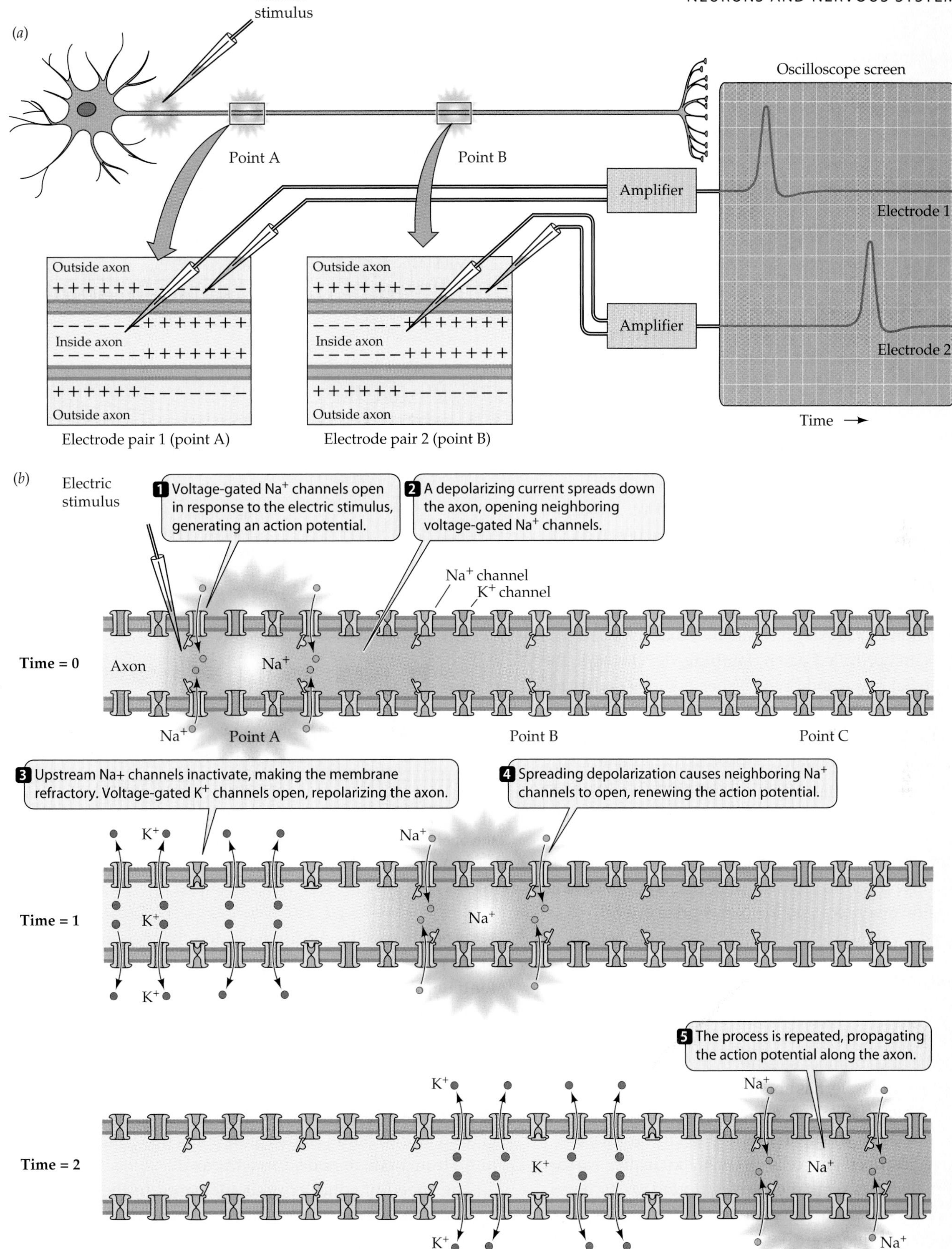

44.10 Action Potentials Travel along Axons (*a*) There is no loss of signal as an action potential travels along an axon. (*b*) When an action potential occurs in one region of membrane, electric current flows to adjacent areas of membrane and depolarizes them. As voltage-gated channels in those areas reach threshold, they generate an action potential. In this way, an action potential continuously regenerates itself along the axon.

small-diameter axons. In invertebrates, the axon diameter determines the rate of conduction, and axons that transmit messages involved in escape behavior are very large. The giant axons that enable squids to escape predators are almost 1 mm in diameter.

Ion channels and their properties can be studied directly

The size of the squid giant axon made it possible for the British neurophysiologists A. L. Hodgkin and A. F. Huxley to study the electrical properties of axonal membranes almost 70 years ago. They used electrodes to measure voltage differences across the plasma membrane of the squid giant axon and to pass electric current into the axon to change its resting potential. They also changed the concentrations of Na^+ and K^+ ions both inside and outside the axon and measured the resulting changes in membrane potential.

On the basis of their many careful experiments, Hodgkin and Huxley developed the story we have discussed so far. However, they could only hypothesize the existence of ion channels and their properties, since they were working long before technology enabled the actual demonstration of their existence. Hodgkin and Huxley received the Nobel prize in 1963.

With a technique called **patch clamping**, developed in the 1980s by Bert Sakmann and Erwin Neher, neurobiologists can record currents caused by the openings and closings of single ion channels. Using a fine pipette and slight suction, patch clamping isolates a small patch of plasma membrane. Using the pipette as an electrode, voltage differences due to movements of ions through channels in the isolated patch can be recorded (Figure 44.11). Frequently, a patch will contain only one or a few ion channels; thus the electrical recording from that patch can show individual channels opening and closing. Sakmann and Neher received the Nobel prize in 1991.

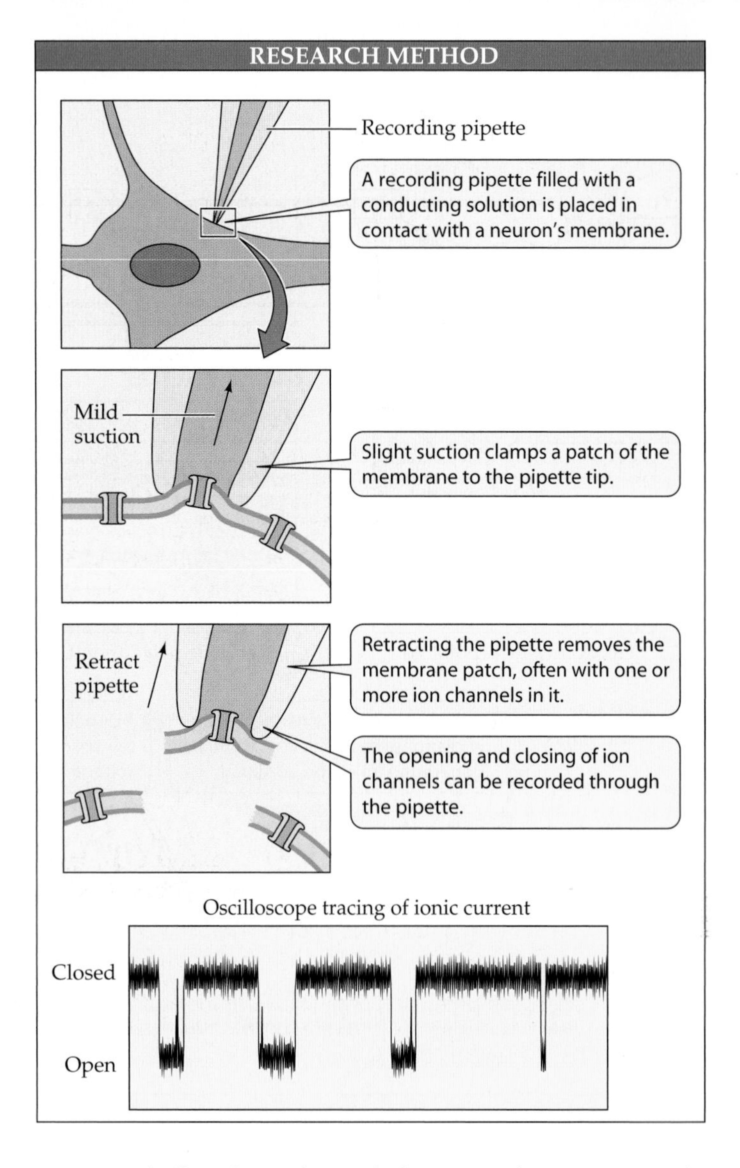

RESEARCH METHOD
Recording pipette
A recording pipette filled with a conducting solution is placed in contact with a neuron's membrane.
Mild suction
Slight suction clamps a patch of the membrane to the pipette tip.
Retract pipette
Retracting the pipette removes the membrane patch, often with one or more ion channels in it.
The opening and closing of ion channels can be recorded through the pipette.
Oscilloscope tracing of ionic current
Closed
Open

44.11 Patch Clamping The patch clamping technique can record the opening and closing of a single ion channel.

Action potentials can jump down axons

In vertebrate nervous systems, increasing the speed of action potentials by increasing the diameter of axons is not feasible because of the huge number of axons in these organisms. Each of our eyes, for example, has about a million axons extending from it. Evolution has increased action potential velocity in vertebrate axons in a way that does not require large size.

When we described glial cells earlier in the chapter, we saw that certain glial cells wrap themselves around axons, covering them with concentric layers of myelin (see Figure 44.3). These myelin wrappings are not continuous along the length of the axon, but have regularly spaced gaps, called **nodes of Ranvier**, where the axon is not covered (Figure 44.12).

Myelin electrically insulates the axon; that is, charged ions cannot cross the regions of the plasma membrane that are wrapped in myelin. Additionally, ion channels are clustered at the nodes of Ranvier. Thus an axon can fire an action potential only at a node, and that action potential cannot be propagated through the adjacent patch of membrane covered with myelin. The positive charges that flow into the axon at the node, however, spread down the inside of the axon. When the spread of current causes the plasma membrane at the next node to depolarize to threshold, an action potential is fired at that node. Action potentials therefore appear to jump from node to node down the axon.

The speed of conduction is increased in these myelin-wrapped axons because electric current flows very fast through the cytoplasm in comparison to the time required for channels to open and close. This form of impulse propagation is called **saltatory** (jumping) **conduction** and is much quicker than continuous propagation of action potentials down an unmyelinated axon.

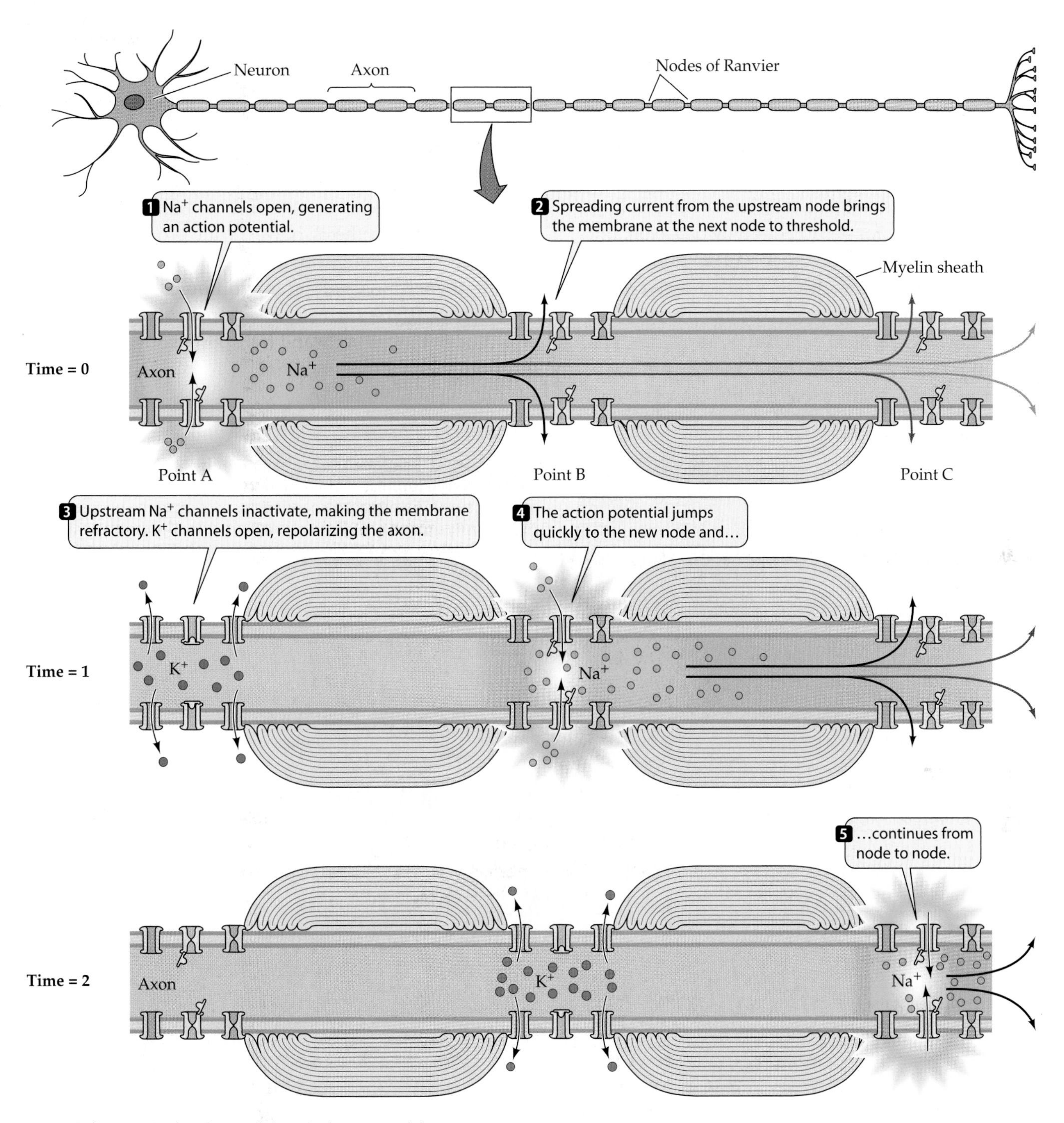
Neuron
Axon
Nodes of Ranvier
1 Na+ channels open, generating an action potential.
2 Spreading current from the upstream node brings the membrane at the next node to threshold.
Myelin sheath
Time = 0
Axon
Na+
Point A
Point B
Point C
3 Upstream Na+ channels inactivate, making the membrane refractory. K+ channels open, repolarizing the axon.
4 The action potential jumps quickly to the new node and...
Time = 1
K+
Na+
5 ...continues from node to node.
Time = 2
Axon
K+
Na+

44.12 Saltatory Action Potentials Action potentials appear to jump from node to node in myelinated axons.

Neurons, Synapses, and Communication

The most remarkable abilities of nervous systems stem from interactions among neurons. It is these interactions that process and integrate information. Our nervous systems can orchestrate complex behaviors, deal with complex concepts, and learn and remember because large numbers of neurons interact with one another. The mechanisms of these interactions depend on synapses between cells. Synapses, as we saw above, are structurally specialized junctions where one cell influences another cell directly through the transfer of an electrical or chemical message. The cell that sends the message is the **presynaptic cell**, and the cell that receives it is the **postsynaptic cell**. The most common type of synapse in the nervous system is the **chemical synapse**—one in which chemical messages released by a presynaptic cell induce changes in a postsynaptic cell.

The neuromuscular junction is a classic chemical synapse

Neuromuscular junctions are synapses between muscle cells and the neurons that innervate them. They are excellent models for how chemical synaptic transmission works. The neurons that innervate muscles cells are called *motor neurons*. Like most other neurons, a motor neuron has only one axon, but that axon can have many branches, each with an axon terminal that forms a neuromuscular junction with a muscle cell. At each axon terminal is an enlarged knob or buttonlike structure that contains many vesicles filled with neurotransmitters. The neurotransmitter used by all vertebrate motor neurons is *acetylcholine*. The portion of the axon terminal plasma membrane that forms a synapse with a muscle cell is called the *presynaptic membrane*. Acetylcholine is released by exocytosis when the membrane of a vesicle fuses with the presynaptic membrane.

Where does the neurotransmitter come from? Some neurotransmitters, like acetylcholine, are synthesized in the axon terminal and packaged in vesicles. The enzymes required for acetylcholine biosynthesis, however, are produced in the cell body of the motor neuron and are transported down the axon to the terminals along microtubules. Other kinds of neurotransmitters, such as peptide neurotransmitters, are produced in the cell body and transported down the axon to the terminals.

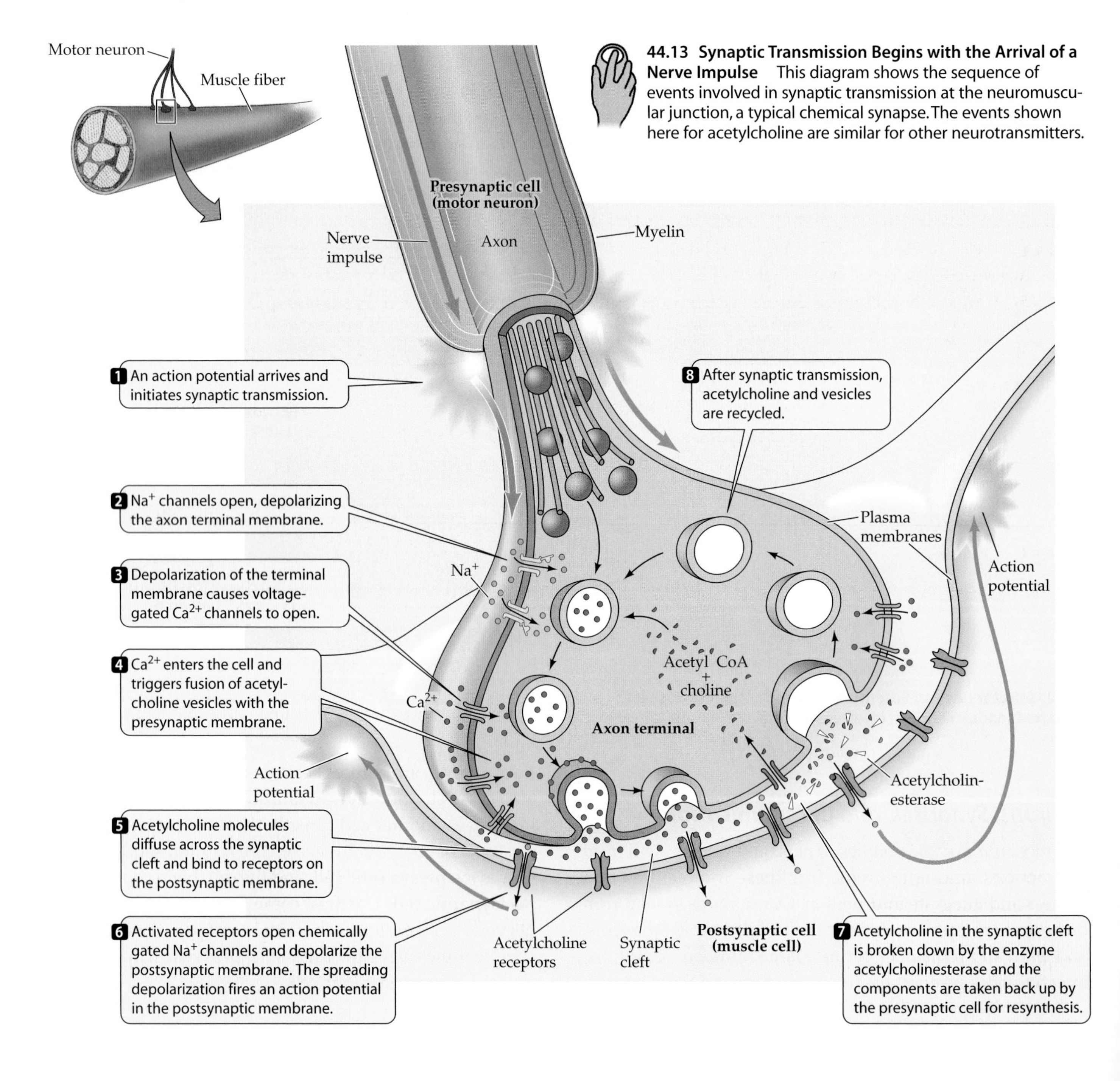

Motor neuron
Muscle fiber
Presynaptic cell (motor neuron)
Nerve impulse
Axon
Myelin
1 An action potential arrives and initiates synaptic transmission.
2 Na$^+$ channels open, depolarizing the axon terminal membrane.
3 Depolarization of the terminal membrane causes voltage-gated Ca^{2+} channels to open.
4 Ca^{2+} enters the cell and triggers fusion of acetylcholine vesicles with the presynaptic membrane.
Action potential
5 Acetylcholine molecules diffuse across the synaptic cleft and bind to receptors on the postsynaptic membrane.
6 Activated receptors open chemically gated Na$^+$ channels and depolarize the postsynaptic membrane. The spreading depolarization fires an action potential in the postsynaptic membrane.
8 After synaptic transmission, acetylcholine and vesicles are recycled.
Na$^+$
Ca^{2+}
Acetyl CoA + choline
Axon terminal
Plasma membranes
Action potential
Acetylcholinesterase
Acetylcholine receptors
Synaptic cleft
Postsynaptic cell (muscle cell)
7 Acetylcholine in the synaptic cleft is broken down by the enzyme acetylcholinesterase and the components are taken back up by the presynaptic cell for resynthesis.

44.13 Synaptic Transmission Begins with the Arrival of a Nerve Impulse This diagram shows the sequence of events involved in synaptic transmission at the neuromuscular junction, a typical chemical synapse. The events shown here for acetylcholine are similar for other neurotransmitters.

The postsynaptic membrane of the neuromuscular junction is a modified part of the muscle cell plasma membrane called a **motor end plate**. The space between the presynaptic membrane and the postsynaptic membrane is called the **synaptic cleft**. In chemical synapses, the synaptic cleft is, on average, about 20–40 nm wide. Neurotransmitter released into the cleft by the presynaptic cell diffuses across to the postsynaptic membrane (Figure 44.13).

The motor end plate contains acetylcholine receptor proteins. These receptors are chemically gated channels that allow both Na^+ and K^+ to pass through. Since the resting membrane of the postsynaptic cell is already permeable to K^+, the major change that occurs when these channels open is the movement of Na^+ into the cell. When a receptor binds acetylcholine, the pore of its channel opens, and Na^+ moves across the membrane, depolarizing the motor end plate (Figure 44.14).

The arrival of a nerve impulse causes the release of neurotransmitter

What causes the presynaptic membrane to release neurotransmitter? Neurotransmitter is released when a nerve impulse arrives at the axon terminal. The presynaptic membrane contains voltage-gated calcium channels. When a nerve impulse depolarizes the axon terminal, it causes these channels to open (see Figure 44.13). Because Ca^{2+} concentration is greater outside the cell than inside the cell, Ca^{2+} enters the axon terminal near the synaptic vesicles.

The increase in Ca^{2+} inside the axon terminal causes the vesicles containing acetylcholine to fuse with the presynaptic membrane and empty their contents into the synaptic cleft. The acetylcholine molecules diffuse within the cleft, and some bind to acetylcholine receptors on the motor end plate.

The postsynaptic membrane integrates synaptic input

The postsynaptic membrane of the neuromuscular junction differs from the presynaptic membrane in an important way. Motor end plates have very few voltage-gated sodium channels; therefore, they do not fire action potentials. This is true not only of motor end plates on muscle cells, but also of most dendrites and most regions of neuronal cell bodies. The binding of acetylcholine to receptors at the motor end plate and the opening of chemically gated ion channels produce a change in the membrane potential of the postsynaptic membrane. This local change in membrane potential spreads to neighboring regions of the plasma membrane of the postsynaptic cell.

The entire plasma membrane of a muscle cell, except for the motor end plates, contains voltage-gated sodium channels. If the axon terminal of a motor neuron releases sufficient amounts of acetylcholine to depolarize a motor end plate enough, that spreading depolarization will reach an area of plasma membrane that contains voltage-gated sodium channels. When that area of membrane is depolarized to threshold, an action potential is fired. This action potential is then conducted throughout the muscle cell's system of membranes, causing the cell to contract. (We'll learn more about muscle membrane action potentials and the contraction of muscle cells in Chapter 47.)

How much neurotransmitter is enough? Neither a single acetylcholine molecule nor the contents of an entire vesicle (about 10,000 acetylcholine molecules) are enough to bring the plasma membrane of a muscle cell to threshold. However, a single action potential in an axon terminal releases the contents of about 100 vesicles, which is enough to fire an action potential in the muscle cell and cause it to contract.

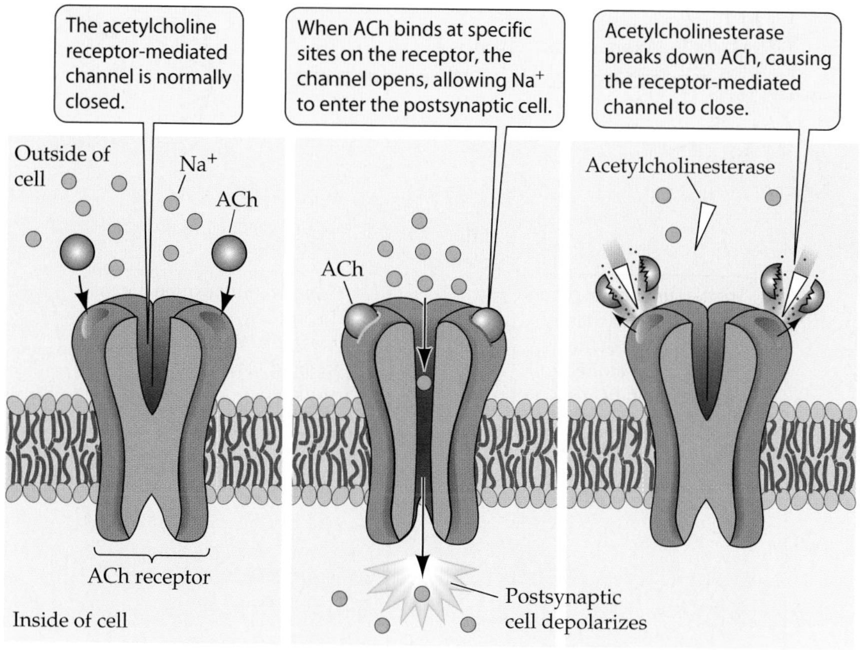

44.14 The Acetylcholine Receptor Is a Chemically Gated Channel The motor end plate contains acetylcholine (ACh) receptors, which are chemically gated ion channels. When one of these receptors binds ACh, its channel pore opens, and Na^+ ions move into the postsynaptic cell, depolarizing its plasma membrane. An enzyme called acetylcholinesterase breaks down ACh in the synapse, closing the channel; the breakdown products are then resynthesized into more ACh.

Synapses between neurons can be excitatory or inhibitory

In vertebrates, the synapses between motor neurons and muscle cells are always **excitatory**; that is, motor end plates always respond to acetylcholine by depolarizing the postsynaptic membrane. Synapses between neurons, however, are not always excitatory.

Recall that a neuron may have many dendrites. Axon terminals from many other neurons may form synapses with those dendrites and with the cell body. The axon terminals of different presynaptic neurons may store and release different neurotransmitters, and the plasma membrane of the dendrites and cell body of a postsynaptic neuron may have receptors for a variety of neurotransmitters. Thus, at any one time, a postsynaptic neuron may receive a variety of different chemical messages. If the postsynaptic neuron's response to a neurotransmitter is depolarization, as at the neuromuscular junction, the synapse is excitatory; if its response is hyperpolarization, the synapse is **inhibitory**.

How do inhibitory synapses work? In vertebrates, the two most common inhibitory neurotransmitters are gamma-aminobutyric acid (GABA) and glycine. The postsynaptic membranes at inhibitory synapses that bind these neurotransmitters have receptors that are chemically gated chloride channels. When these channels bind their neurotransmitter and open, they hyperpolarize the postsynaptic membrane. Thus the release of neurotransmitter at an inhibitory synapse makes the postsynaptic cell *less* likely to fire an action potential.

Neurotransmitters that depolarize the postsynaptic membrane are excitatory; they bring about an *excitatory postsynaptic potential* (EPSP). Neurotransmitters that hyperpolarize the postsynaptic membrane are inhibitory; they bring about an *inhibitory postsynaptic potential* (IPSP).

The postsynaptic cell sums excitatory and inhibitory input

Individual neurons can "decide" whether or not to fire an action potential by summing excitatory and inhibitory postsynaptic potentials. This summation ability is the major mechanism by which the nervous system integrates information. Each neuron may receive a thousand or more synaptic inputs, but it has only one output: an action potential in a single axon. All the information contained in all the inputs a neuron receives is reduced to the rate at which that neuron generates nerve impulses in its axon.

For most neurons, the critical area for "decision making" is the **axon hillock**, the region of the cell body at the base of the axon (see Figure 44.2). The plasma membrane of the axon hillock is not insulated by glial cells and has many voltage-gated channels. Excitatory and inhibitory postsynaptic potentials from synapses anywhere on the dendrites or the cell body spread to the axon hillock by local current flow. If the resulting combined potential depolarizes the axon hillock to threshold, the axon fires an action potential. Because postsynaptic potentials decrease in strength as they spread from the site of the synapse, all postsynaptic potentials do not have equal influences on the axon hillock. A synapse at the tip of a dendrite has less influence than a synapse on the cell body near the axon hillock.

Excitatory and inhibitory postsynaptic potentials can be summed over space or over time. *Spatial summation* adds up the simultaneous influences of synapses at different sites on the postsynaptic cell (Figure 44.15*a*). *Temporal summation* adds up postsynaptic potentials generated at the same site in a rapid sequence (Figure 44.15*b*).

All the neuron-to-neuron synapses that we have discussed up to this point are between the axon terminals of a presynaptic cell and the cell body or dendrites of a postsynaptic cell. Synapses can also form between the axon terminals of one neuron and the axon terminals of another neuron. Such a synapse can modulate how much neurotransmitter the second neuron releases in response to action potentials traveling down

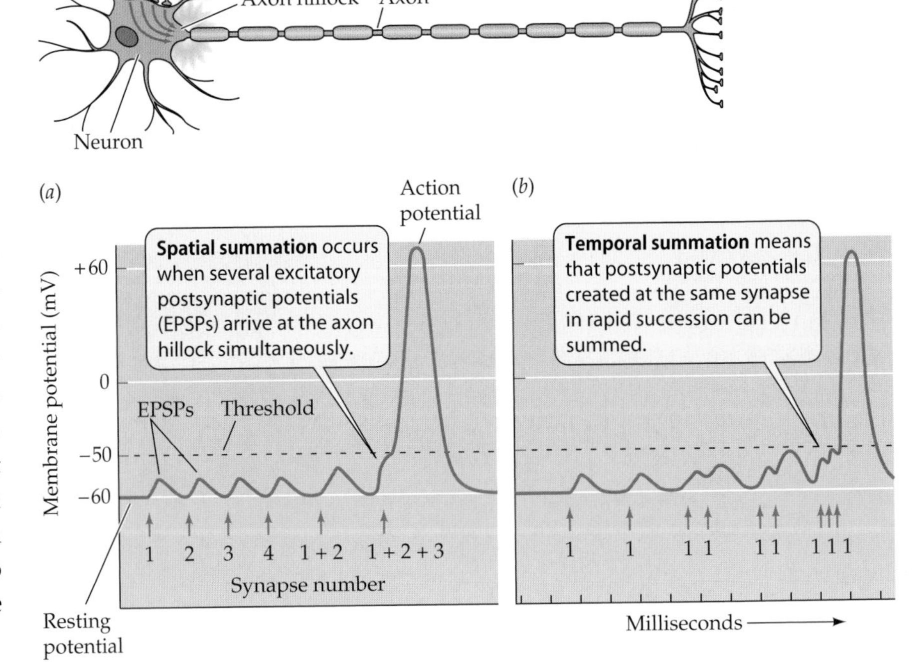

44.15 The Postsynaptic Neuron Sums Information Individual neurons sum excitatory and inhibitory postsynaptic potentials over space (*a*) and time (*b*). When the sum of the potentials depolarizes the axon hillock to threshold, the neuron generates an action potential.

its axon. We refer to this mechanism of regulating synaptic strength as **presynaptic excitation** or **presynaptic inhibition**.

There are two types of neurotransmitter receptors

Most neurotransmitter receptors induce changes in postsynaptic cells by opening or closing ion channels. How they do so is the basis for grouping receptors into two general categories:

- **Ionotropic receptors** are themselves ion channels. Neurotransmitter binding by an ionotropic receptor causes a direct change in ion movement across the plasma membrane of the postsynaptic cell. These proteins are also called *ligand-gated channels*.
- **Metabotropic receptors** are not ion channels, but they induce changes in the postsynaptic cell that can secondarily lead to changes in ion channels.

Postsynaptic cell responses mediated by metabotropic receptors are generally slower and longer-lasting than those induced by ionotropic receptors.

The acetylcholine receptor of the motor end plate is an example of an ionotropic receptor. It consists of five subunits, each of which extends through the plasma membrane (see Figure 44.14). When assembled, the subunits create a central pore that allows ions to pass through. There are several different kinds of subunits, and only one kind has the ability to bind acetylcholine. Each functional receptor has two of the acetylcholine-binding subunits and three other subunits.

Metabotropic receptors are also transmembrane proteins, but instead of acting as ion channels, they initiate an intracellular signaling process that can result in the opening or closing of an ion channel. These receptors have seven transmembrane domains, and they are linked to G proteins (Figure 44.16; see also Figure 15.7). When a neurotransmitter binds to the extracellular domain of a metabotropic receptor, the intracellular domain activates a G protein. In its inactive state, the G protein has three subunits, one of which (the α subunit) is bound to a molecule of GDP. When the receptor binds its neurotransmitter, the GDP is replaced with a GTP molecule, and the α subunit separates from the other two subunits (called β and γ). The α subunit moves laterally in the membrane until it binds to and opens an ion channel or binds to an effector protein that activates a second messenger cascade, which in turn opens an ion channel.

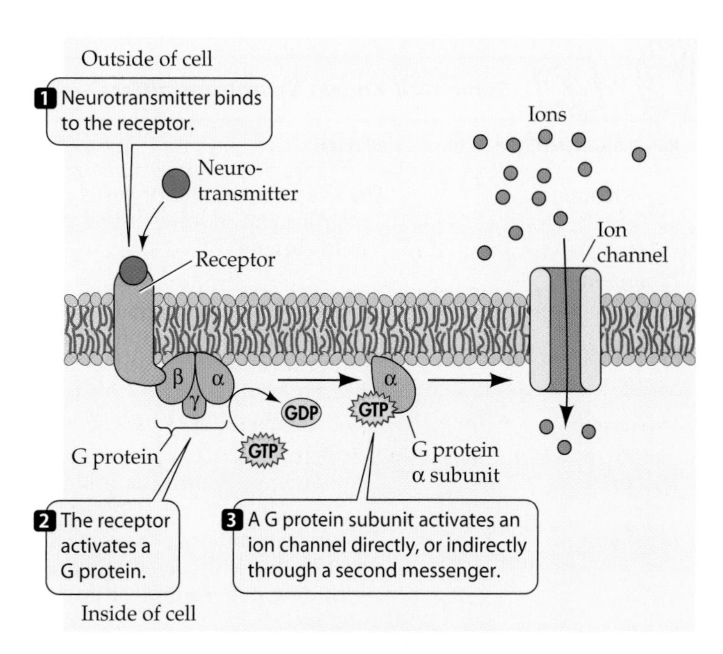

44.16 Metabotropic Receptors Act through G Proteins
Metabotropic receptors activate G proteins, which can influence ion channels directly or through second messengers.

Electrical synapses are fast but do not integrate information well

Electrical synapses are different from chemical synapses because they couple neurons electrically. Electrical synapses are *gap junctions* (see Figures 5.6c and 15.16) At these synapses, the presynaptic and postsynaptic cell membranes are separated by a space of only 2 to 3 nm, and specific membrane proteins called *connexons* link the two neurons by forming molecular tunnels between the two cells. Ions and small molecules can pass directly from cell to cell through the connexons. Transmission at electrical synapses is very fast and can proceed in either direction, whereas transmission at chemical synapses is slower and unidirectional.

Electrical synapses are less common in the nervous systems of vertebrates than are chemical synapses for several reasons. First, electrical continuity between neurons does not allow temporal summation of synaptic inputs. Second, an effective electrical synapse requires a large area of contact between the presynaptic and postsynaptic cells. This condition rules out the possibility of thousands of synaptic inputs to a single neuron—which is the norm in complex nervous systems. Third, electrical synapses cannot be inhibitory. Finally, chemical synapses appear to have considerable plasticity (modifiability), as we will see below, but electrical synapses do not. Thus, electrical synapses are useful for rapid communication, such as that between muscle cells (as we will see in Chapter 47), but they are less useful for processes of integration and learning.

The action of a neurotransmitter depends on the receptor to which it binds

More than 25 neurotransmitters are now recognized, and more will surely be discovered. Table 44.1 describes some of

44.1 Some Well-Known Neurotransmitters

NEUROTRANSMITTER	ACTIONS	COMMENTS
Acetylcholine	The neurotransmitter of vertebrate motor neurons and of some neural pathways in the brain	Broken down in the synapse by acetylcholinesterase; blockers of this enzyme are powerful poisons
Monoamines		
Norepinephrine	Used in certain neural pathways in the brain. Also found in the peripheral nervous system, where it causes gut muscles to relax and the heart to beat faster	Related to epinephrine and acts at some of the same receptors
Dopamine	A neurotransmitter of the central nervous system	Involved in schizophrenia. Loss of dopamine neurons is the cause of Parkinson's disease
Histamine	A minor neurotransmitter in the brain	Thought to be involved in maintaining wakefulness
Serotonin	A neurotransmitter of the central nervous system that is involved in many systems, including pain control, sleep/wake control, and mood	Certain medications that elevate mood and counter anxiety act by inhibiting the reuptake of serotonin
Purines		
ATP	Co-released with many neurotransmitters	Large family of receptors may shape postsynaptic responses to classical neurotranmitters
Adenosine	Transported across cell membranes; not synaptically released	Largely inhibitory effects on postsynaptic cells
Amino acids		
Glutamate	The most common excitatory neurotransmitter in the central nervous system	Some people have reactions to the food additive monosodium glutamate because it can affect the nervous system
Glycine Gamma-aminobutyric acid (GABA)	Common inhibitory neurotransmitters	Drugs called benzodiazepines, used to reduce anxiety and produce sedation, mimic the actions of GABA
Peptides		
Endorphins Enkephalins Substance P	Used by certain sensory nerves, especially in pain pathways	Receptors are activated by narcotic drugs: opium, morphine, heroin, codeine
Gas		
Nitric oxide	Widely distributed in the nervous system	Not a classic neurotransmitter, it diffuses across membranes rather than being released synaptically. A means whereby a postsynaptic cell can influence a presynaptic cell

the best-known neurotransmitters. Acetylcholine, as we have seen, is an important neurotransmitter because it is the means whereby the nervous system commands muscles to contract. Acetylcholine also plays roles in certain synapses between neurons in the central nervous system, but it accounts for only a small percentage of the total neurotransmitter content of the CNS. The workhorse neurotransmitters of the CNS are simple amino acids: glutamate (excitatory) and glycine and GABA (inhibitory). Another important group of neurotransmitters in the CNS is the monoamines, which are derivatives of amino acids. They include dopamine and norepinephrine (derivatives of tyrosine) and serotonin (a derivative of tryptophan). Peptides also function as neurotransmitters. An exciting recent discovery revealed that two gases, carbon monoxide and nitric oxide, are used by neurons as intercellular messengers even though they do not have the characteristics of classic neurotransmitters (that is, they do not have receptors) (see Figure 15.13).

The complexity of neurotransmission is increased by the fact that each neurotransmitter has multiple receptor types. Acetylcholine, for example, has two receptor types: *nicotinic receptors*, which are ionotropic, and *muscarinic receptors*, which are metabotropic. Both types of acetylcholine receptors are found in the CNS, where nicotinic receptors tend to be excitatory and muscarinic receptors tend to be inhibitory. Acetylcholine actions can differ outside of the CNS as well. Acetylcholine acting through nicotinic receptors causes the smooth muscle of the gut to depolarize and therefore increases its motility, but acetylcholine acting through muscarinic receptors causes cardiac muscle to hyperpolarize and therefore decreases the contractility of the heart.

We could give many more examples of neurotransmitters that have different effects in different tissues, but the important thing to remember is that the action of a neurotransmitter depends on the receptor to which it binds.

Glutamate receptors may be involved in learning and memory

Glutamate is a neurotransmitter that can bind to a variety of receptors, including both metabotropic and ionotropic receptors. The glutamate receptors are divided into several classes because they can be differentially activated by other chemicals that mimic the action of glutamate. One class of ionotropic glutamate receptors is the *NMDA receptors*, which can be activated by the chemical *N*-methyl-D-aspartate. Another class of ionotropic glutamate receptors is activated by a different chemical, abbreviated as *AMPA*.

Glutamate is an excitatory neurotransmitter, so activation of glutamate receptors always results in Na^+ entry into the neuron and depolarization. But the *timing* of the response to activation by these different types of receptors differs significantly: The AMPA receptors allow a rapid influx of Na^+ into the postsynaptic cell, while the NMDA receptors allow a slower and longer-lasting influx of Na^+. The NMDA receptors also require that the cell be somewhat depolarized through the action of other receptors before their pores will open and permit Na^+ influx. When they do open, these receptors also allow Ca^{2+} to enter the cell. Ca^{2+} ions act as second messengers in the cell and can trigger a variety of long-term cellular changes.

Figure 44.17 shows how the AMPA and NMDA receptors can work in concert. At resting potential, the NMDA receptor is blocked by a magnesium ion (Mg^{2+}). Strong depolarization of the neuron due to other inputs—such as the activation of AMPA receptors—displaces Mg^{2+} from the NMDA receptors and allows Na^+ and Ca^{2+} to pass through them when they are activated by glutamate. These special properties of the NMDA receptor are probably involved in learning and memory.

Most of the synaptic events we have studied so far happen very quickly. It is therefore a special challenge to understand how the messages carried by action potentials can result in long-term events such as learning and memory. Our understanding of these processes has been greatly affected by a phenomenon called **long-term potentiation**, or **LTP**, that was discovered by neurobiologists working with slices of brain kept alive in dishes of culture medium. Using these brain slice preparations, it is possible to stimulate and record from specific brain regions, or even specific neurons.

In the studies leading to the discovery of LTP, experimenters repeatedly stimulated synaptic inputs to a particular neuron and observed the usual action potential response. When the neuron was stimulated many times in rapid succession, however, they found that the properties of the neuron changed. The magnitude of the postsynaptic response was enhanced, or *potentiated*, and this change lasted for days or weeks.

How does this modification of a synapse occur? The answer in some areas of the brain now seems quite clear. With

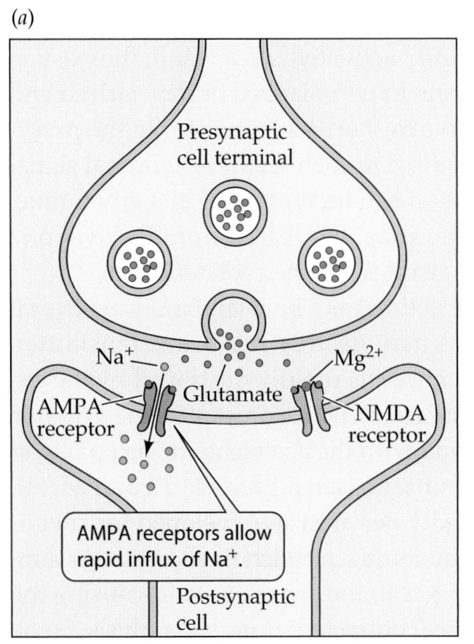
(a)
Presynaptic cell terminal
Na+
Mg2+
Glutamate
AMPA receptor
NMDA receptor
AMPA receptors allow rapid influx of Na+.
Postsynaptic cell

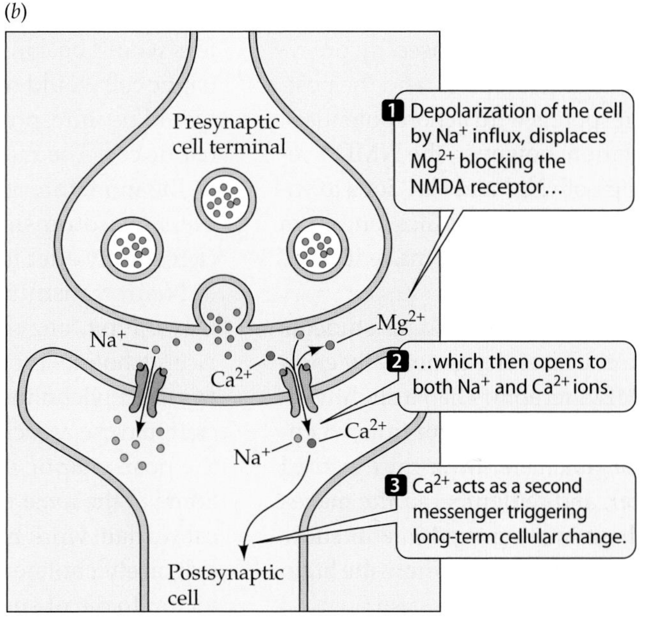
(b)
Presynaptic cell terminal
1 Depolarization of the cell by Na+ influx, displaces Mg2+ blocking the NMDA receptor...
Mg2+
Na+
Ca2+
2 ...which then opens to both Na+ and Ca2+ ions.
Ca2+
Na+
3 Ca2+ acts as a second messenger triggering long-term cellular change.
Postsynaptic cell

44.17 Two Ionotropic Glutamate Receptors
(*a*) AMPA receptors allow rapid influx of Na^+ into the postsynaptic cell. (*b*) NMDA receptors allow both Na^+ and Ca^{2+} to enter the cell, but respond to synaptic input more slowly.

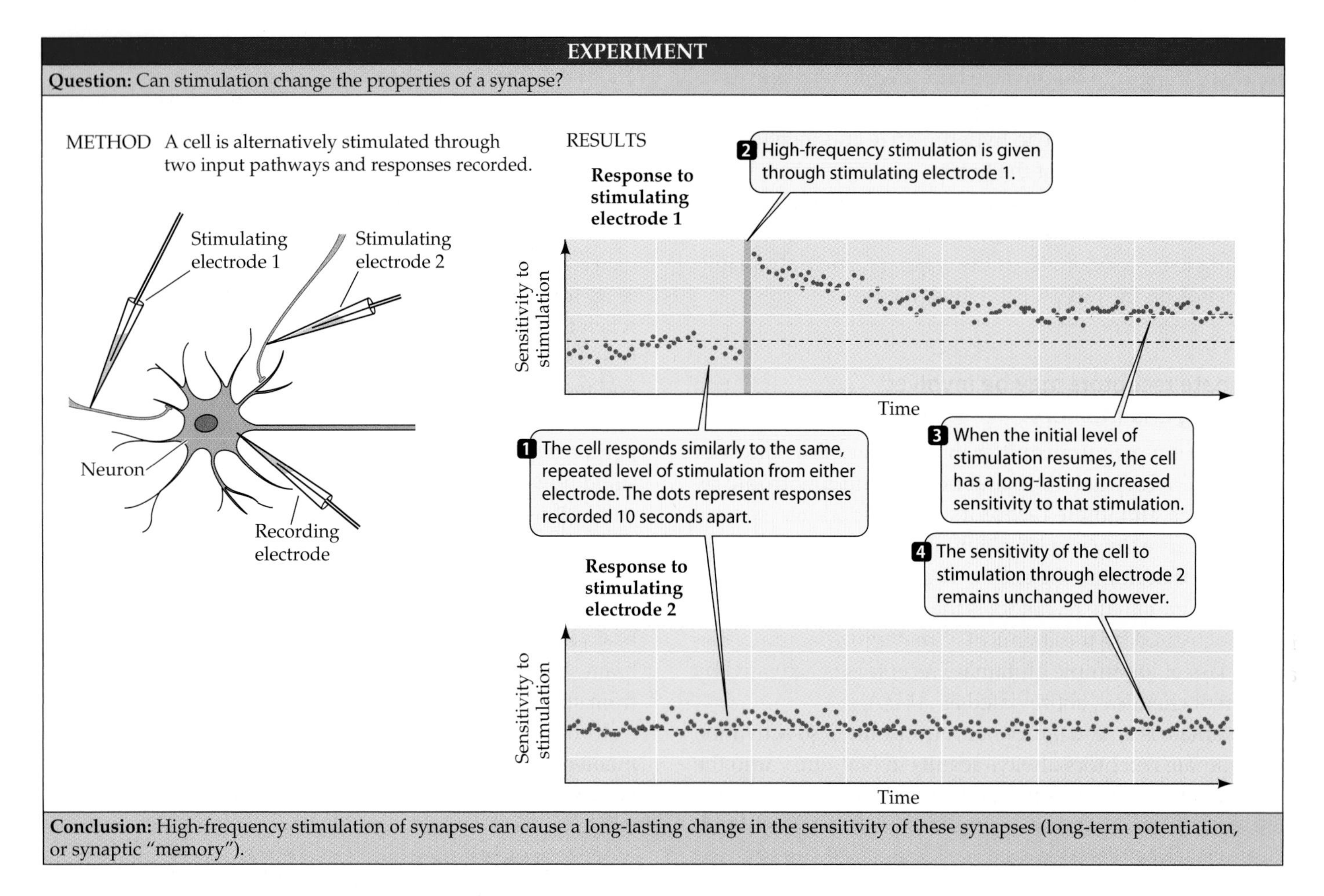

EXPERIMENT
Question: Can stimulation change the properties of a synapse?
METHOD A cell is alternatively stimulated through two input pathways and responses recorded.
Stimulating electrode 1
Stimulating electrode 2
Neuron
Recording electrode
RESULTS
Response to stimulating electrode 1
Sensitivity to stimulation
Time
2 High-frequency stimulation is given through stimulating electrode 1.
1 The cell responds similarly to the same, repeated level of stimulation from either electrode. The dots represent responses recorded 10 seconds apart.
3 When the initial level of stimulation resumes, the cell has a long-lasting increased sensitivity to that stimulation.
4 The sensitivity of the cell to stimulation through electrode 2 remains unchanged however.
Response to stimulating electrode 2
Sensitivity to stimulation
Time
Conclusion: High-frequency stimulation of synapses can cause a long-lasting change in the sensitivity of these synapses (long-term potentiation, or synaptic "memory").

44.18 Repeated Stimulation Can Cause Long-Term Potentiation When a cell receives regular synaptic input, the resulting postsynaptic potential remains constant. If, however, that same synaptic pathway is stimulated briefly at a high frequency, the subsequent sensitivity of the postsynaptic cell to the original level of synaptic input is potentiated for a long time.

low levels of stimulation, the glutamate released by presynaptic cells activates only the AMPA receptors, and the postsynaptic membrane simply responds with action potentials. With higher levels of stimulation, however, the NMDA receptors are activated, allowing both Na^+ and Ca^{2+} ions to enter the postsynaptic neuron. The Ca^{2+} ions induce long-term changes in the postsynaptic membrane that make it more sensitive to synaptic input (Figure 44.18).

Exploiting the LTP system, Dr. Joe Tsien and his students and collaborators at Princeton University genetically engineered mice so that their NMDA receptors had a slightly altered structure and were activated for a longer time whenever they bound a molecule of glutamate. These mice learned tasks better, ran mazes faster, and remembered the mazes longer than normal mice. These exciting experiments show that we are on the right track to understanding how the brain achieves learning and memory.

To turn off responses, synapses must be cleared of neurotransmitter

Turning off the action of neurotransmitters is as important as turning it on. If released neurotransmitter molecules simply remained in the synaptic cleft, the postsynaptic membrane would become saturated with neurotransmitter, and receptors would be constantly activated. As a result, the postsynaptic cell would remain hyperpolarized or depolarized and would be unresponsive to short-term changes in the presynaptic cell. The more discrete each separate neuronal signal is, the more information can be processed in a given time. Thus neurotransmitter must be cleared from the synaptic cleft shortly after it is released by the axon terminal.

Neurotransmitter action may be terminated in several ways. First, enzymes may destroy the neurotransmitter. Acetylcholine, for example, is rapidly destroyed by the enzyme acetylcholinesterase, which is present in the synaptic cleft in close association with the acetylcholine receptors on the postsynaptic membrane (see Figures 44.13 and 44.14). Some of the most deadly nerve gases developed for chemical warfare work by inhibiting acetylcholinesterase. As a result, acetylcholine lingers in the synaptic clefts, causing the victim to die of spastic (contracted) muscle paralysis. Some agricultural insecticides, such as malathion, also inhibit acetylcholinesterase and can poison farm workers if used without safety precautions.

Second, neurotransmitter may simply diffuse away from the cleft. Third, neurotransmitter may be taken up via active transport by nearby cell membranes. Prozac, a drug commonly prescribed to treat depression, slows the reuptake of the neurotransmitter serotonin, thus enhancing its activity at the synapse.

Because neurons can interact in the complex ways we have just discussed, networks of neurons can process and integrate information. Multiple neuronal networks constitute the nervous systems of animals. In the next three chapters, we will see many examples of how neurons work together in networks to accomplish specific tasks. These networks use all of the mechanisms we have discussed in this chapter: excitatory and inhibitory synapses, presynaptic excitation and inhibition, and mechanisms of long-term potentiation. Through these operations, our brains solve puzzles, create inventions, remember experiences, fall in love, and learn about biology. The challenge for the future is to understand how these networks work.

Chapter Summary

Nervous Systems: Cells and Functions

▶ Nervous systems consist of cells called neurons that process and transmit information, along with supporting cells called glial cells.

▶ Sensory cells transduce information from the environment and the body. Neurons receive this information and transmit it to effectors such as muscles or glands.

▶ The nervous systems of different species vary, but all are composed of neurons. **Review Figure 44.1**

▶ In vertebrates, the brain and spinal cord form the central nervous system, which communicates with the rest of the body via the peripheral nervous system.

▶ Neurons generally receive information via their dendrites and transmit information via their axons. **Review Figure 44.2**

▶ Where neurons and their target cells meet, information is transmitted across synapses by the release of neurotransmitters.

▶ Glial cells physically support neurons and perform many housekeeping functions. Schwann cells and oligodendrocytes produce myelin, which insulates neurons. Astrocytes create the blood–brain barrier. **Review Figure 44.3**

▶ Neurons work together in networks.

Neurons: Generating and Conducting Nerve Impulses

▶ Neurons have an electric charge difference across their plasma membranes. This resting potential is created by ion pumps and ion channels. **Review Figure 44.4**

▶ The sodium–potassium pump concentrates K^+ on the inside of a neuron and Na^+ on the outside. Potassium channels allow K^+ to diffuse out of the neuron, leaving behind unbalanced negative charges. **Review Figures 44.5, 44.6.** See Web/CD Tutorial 44.1

▶ A potassium equilibrium potential exists when the tendency of K^+ ions to diffuse out of the neuron is balanced by the negative charges pulling them back in. This potential can be calculated using the Nernst equation. **Review Figure 44.7**

▶ The resting potential is perturbed when ion channels open or close, changing the permeability of the plasma membrane to charged ions. Through this mechanism, the plasma membrane can become depolarized or hyperpolarized. **Review Figure 44.8**

▶ An action potential is a rapid reversal in charge across a portion of the plasma membrane resulting from the sequential opening and closing of voltage-gated sodium and potassium channels. These changes in voltage-gated channels occur when the plasma membrane depolarizes to a threshold level. **Review Figure 44.9.** See Web/CD Tutorial 44.2

▶ Action potentials are all-or-nothing, self-regenerating events. They are conducted down axons because local current flow depolarizes adjacent regions of membrane and brings them to threshold. **Review Figure 44.10**

▶ Patch clamping allows the study of single ion channels. **Review Figure 44.11**

▶ In myelinated axons, action potentials appear to jump between nodes of Ranvier, patches of axonal plasma membrane that are not covered by myelin. **Review Figure 44.12**

Neurons, Synapses, and Communication

▶ Neurons communicate with each other and with other cells at specialized junctions called synapses, where the plasma membranes of two cells come close together.

▶ The classic chemical synapse is the neuromuscular junction, a synapse between a motor neuron and a muscle cell. Its neurotransmitter is acetylcholine, which causes a depolarization of the postsynaptic membrane when it binds to its receptor. **Review Figure 44.13.** See Web/CD Tutorial 44.3

▶ When a nerve impulse reaches an axon terminal, it causes the release of neurotransmitters, which diffuse across the synaptic cleft and bind to receptors on the postsynaptic membrane. **Review Figures 44.13, 44.14**

▶ Synapses between neurons can be either excitatory or inhibitory. A postsynaptic neuron integrates information by summing excitatory and inhibitory postsynaptic potentials in both space and time. **Review Figure 44.15**

▶ Synapses that form between the axon terminals of one neuron and another can influence the release of neurotransmitter by the second cell by presynaptic excitation or presynaptic inhibition.

▶ Ionotropic receptors are ion channels. Metabotropic receptors are G protein-linked receptors that influence the postsynaptic cell through various signal transduction pathways and result in the opening of ion channels. The actions of ionotropic synapses are generally faster than those of metabotropic synapses. **Review Figure 44.16**

▶ Electrical synapses allow electric signals to pass between cells without the use of neurotransmitters.

▶ There are many different neurotransmitters and even more types of receptors. The action of a neurotransmitter depends on the receptor to which it binds. **Review Table 44.1.** See Web/CD Activity 44.1

▶ With repeated stimulation, a neuron can become more sensitive to its inputs. Since this increased sensitivity can last a long time, it is called long-term potentiation, or LTP. The properties of the NMDA glutamate receptor appear to explain LTP. **Review Figure 44.18**

▶ In chemical synapses, the transmitter must be cleared rapidly from the synapse. Some poisons and drugs act by blocking or slowing the clearance of transmitter from the synapse.

Self-Quiz

1. In the human brain, the most abundant cell type is the
   a. motor neuron.
   b. sensory neuron.
   c. parasympathetic neuron.
   d. glial cell.
   e. sympathetic neuron.
2. Within a neuron, information moves from
   a. dendrite to cell body to axon.
   b. axon to cell body to dendrite.
   c. cell body to axon to dendrite.
   d. axon to dendrite to cell body.
   e. dendrite to axon to cell body.
3. The resting potential of a neuron is due mostly to
   a. local current spread.
   b. open Na^+ channels.
   c. synaptic summation.
   d. open K^+ channels.
   e. open Cl^- channels.
4. Which statement about synaptic transmission is *not* true?
   a. The synapses between neurons and muscle cells use acetylcholine as their neurotransmitter.
   b. A single vesicle of neurotransmitter cannot cause a muscle cell to contract.
   c. The release of neurotransmitter at the neuromuscular junction causes the motor end plate to fire action potentials.
   d. In vertebrates, the synapses between motor neurons and muscle fibers are always excitatory.
   e. Inhibitory synapses cause the resting potential of the postsynaptic membrane to become more negative.
5. Which statement accurately describes an action potential?
   a. Its magnitude increases along the axon.
   b. Its magnitude decreases along the axon.
   c. All action potentials in a single neuron are of the same magnitude.
   d. During an action potential the membrane potential of a neuron remains constant.
   e. An action potential permanently shifts a neuron's membrane potential away from its resting value.
6. A neuron that has just fired an action potential cannot be immediately restimulated to fire a second action potential. The short interval of time during which restimulation is not possible is called
   a. hyperpolarization.
   b. the resting potential.
   c. depolarization.
   d. repolarization.
   e. the refractory period.
7. The rate of propagation of an action potential depends on
   a. whether or not the axon is myelinated.
   b. the axon's diameter.
   c. whether or not the axon is insulated by glial cells.
   d. the cross-sectional area of the axon.
   e. All of the above
8. The binding of neurotransmitter to the postsynaptic receptors in an inhibitory synapse results in
   a. depolarization of the membrane.
   b. generation of an action potential.
   c. hyperpolarization of the membrane.
   d. increased permeability of the membrane to sodium ions.
   e. increased permeability of the membrane to calcium ions.
9. Whether a synapse is excitatory or inhibitory depends on the
   a. type of neurotransmitter.
   b. presynaptic axon terminal.
   c. size of the synapse.
   d. nature of the postsynaptic receptors.
   e. concentration of neurotransmitter in the synaptic space.
10. Which of the following is a likely mechanism for long-term potentiation?
    a. When glutamate binds to postsynaptic AMPA receptors, it activates G proteins that trigger intracellular changes.
    b. When glutamate binds to NMDA receptors, it allows Mg^{2+} ions to enter the cell, which initiate intracellular changes.
    c. When sufficient glutamate is released by the presynaptic cell, it causes an increase in the number of AMPA receptors on the postsynaptic cell.
    d. When sufficient glutamate is released, both AMPA and NMDA receptors are activated, and NMDA receptors allow Ca^{2+} as well as Na^+ to enter the cell, thus initiating intracellular changes.
    e. When both glutamate and acetylcholine are released together, they create a long-lasting depolarization of the postsynaptic cell.

For Discussion

1. The language of the nervous system consists of one "word," the action potential. How can this single message convey a diversity of information, how can that information be quantitative, and how can it be integrated?
2. If you stimulate an axon in the middle, action potentials are conducted in both directions. Yet when an action potential is generated at the axon hillock, it goes only toward the axon terminals and does not backtrack. Explain why action potentials are bidirectional in the first example and unidirectional in the second.
3. The nature of synapses presents various opportunities for plasticity in the nervous system. Discuss at least four synaptic mechanisms that could be altered to change the response of a neuron to a specific input.
4. If Dr. Tsien had genetically engineered the AMPA receptor to remain open longer when activated, would it have made his mice learn faster? Why or why not?
5. Benzodiazepines are drugs that act through GABA receptors and open chloride channels. What effects would you expect these drugs to have?

45 Sensory Systems

Animals perceive the world through their senses. Different species look through different sensory windows, so their views of the world are not the same. Dogs, for example, do not see color well, but they have far keener senses of hearing and smell than humans do. Thus while you watch a beautiful sunset, your dog is probably sniffing around the bushes and listening for the sounds of small animals in the underbrush.

Human hunters have exploited the remarkable sensory abilities of dogs for thousands of years. Most recently the hunt has extended to illicit drugs, smuggled contraband, bombs, and firearms. Dogs can be trained to detect the signature odors of such items, so they are used by police, customs agents, and other investigators to identify those odors wherever suspicious activities are likely to occur.

A black Labrador named Charlie (badge K9-001) was the first dog trained by the U.S. Treasury Department's Bureau of Alcohol, Tobacco, and Firearms to sniff out firearms and explosives. Charlie has sniffed out more than 200 illegal guns and 500 pounds of hidden explosives. With a nose that outperforms electronic sensors, Charlie helped solve a terrorist bombing case by discovering a tiny fragment of the bomb hundreds of yards from the site of the explosion. Charlie's nose is never off duty; on a recreational visit to a Civil War battlefield, it smelled out cannonball fragments that had been buried for 130 years. ATF dogs receive expert training, but their careers are based on their remarkable sense of smell.

In this chapter, we will look at the general properties of sensory cells and see how they convert environmental stimuli to neuronal information. We will examine in detail the cells responsible for our senses and see how they are incorporated into sensory systems that provide the central nervous system with information about the world around and within us. In the course of our study of sensory systems, we will learn about the unusual sensory abilities of many other animals.

Special Agent K9-001 Charlie's remarkable sense of smell enables him and his partner to discover illicit firearms and explosives.

Sensory Cells and Transduction of Stimuli

Sensory cells *transduce* (convert) physical or chemical stimuli into signals that are transmitted to other parts of the nervous system for processing and interpretation. Sensory cells are generally called *receptors*, which

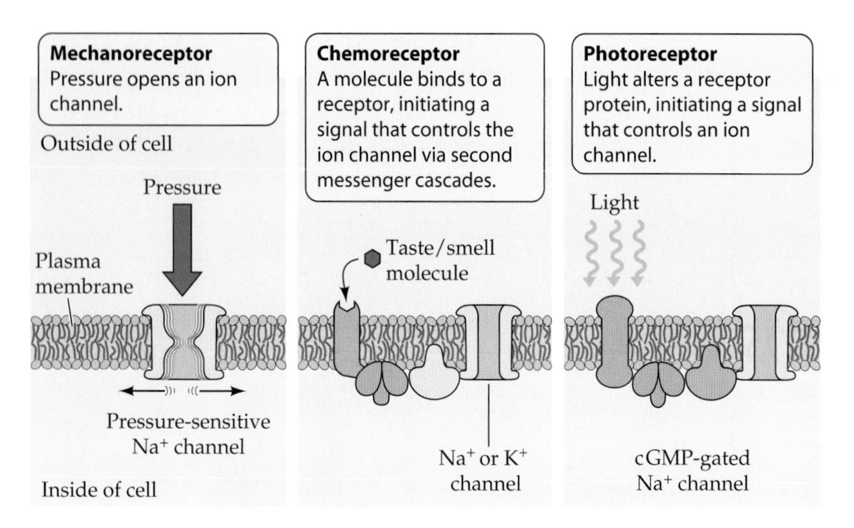
Mechanoreceptor
Pressure opens an ion channel.
Chemoreceptor
A molecule binds to a receptor, initiating a signal that controls the ion channel via second messenger cascades.
Photoreceptor
Light alters a receptor protein, initiating a signal that controls an ion channel.
Outside of cell
Pressure
Plasma membrane
Taste/smell molecule
Light
Pressure-sensitive Na^+ channel
Na^+ or K^+ channel
cGMP-gated Na^+ channel
Inside of cell

45.1 Sensory Cell Membrane Receptor Proteins Respond to Stimuli The receptor proteins in mechanoreceptors are ion channels. The activated receptor proteins of chemoreceptors and photoreceptors initiate signal transduction cascades that eventually open or close ion channels.

creates some confusion with the *receptor proteins* that bind signaling molecules. To avoid this confusion, we will use the terms *sensory cells* or *receptor cells* in this chapter. Most sensory cells are modified neurons, but some are other types of cells closely associated with neurons. Sensory cells are specialized for detecting specific kinds of stimuli, such as pressure, heat, or light.

Most sensory cells possess membrane receptor proteins that detect a stimulus and respond by altering the flow of ions across the cell's plasma membrane (Figure 45.1). The resulting change in membrane potential causes the sensory cell either to fire action potentials itself or to change its secretion of a neurotransmitter onto an associated neuron that fires action potentials. The intensity of the stimulus is encoded in the frequency of the action potentials.

Sensation depends on which neurons receive action potentials from sensory cells

If the messages derived from all sensory cells are the same—they are all action potentials—how do we perceive different sensations? Sensations such as heat, pressure, pain, light, smell, and sound differ because the messages from different kinds of sensory cells arrive at different places in the central nervous system (CNS). Action potentials arriving in the visual cortex of the brain are interpreted as light, in the auditory cortex as sound, in the olfactory bulb as smell, and so forth.

A small patch of skin on your arm contains sensory cells that increase their firing rates when the skin is warmed and others that increase their activity when the skin is cooled. Other types of cells in the same patch of skin respond to touch, movement of hairs, irritants such as mosquito bites, and pain from cuts or burns. These sensory cells transmit their messages through axons that enter the CNS at the spinal cord. The synapses made by those axons in the spinal cord and the subsequent pathways of transmission determine whether the stimulation of the patch of skin on your arm is perceived as warmth, cold, touch, tickle, itch, or pain.

Some sensory cells transmit information about internal conditions in the body, but we may not be consciously aware of that information. The brain receives continuous information about body temperature, blood carbon dioxide and oxygen concentrations, arterial pressure, muscle tension, and the positions of the limbs. All this information is important for the maintenance of homeostasis. All sensory cells produce information that the nervous system can use, but that information does not always result in conscious sensation.

Some sensory cells are assembled with other types of cells into *sensory organs*, such as eyes, ears, and noses, that enhance the ability of the sensory cells to collect, filter, and amplify stimuli. We therefore refer to *sensory systems*, which include the sensory cells, the associated structures, and the neuronal networks that process the information.

Sensory transduction involves changes in membrane potentials

In this chapter we will examine several sensory systems. In each case, we can ask the same general question: How do sensory cells transduce energy from a stimulus into a change in membrane potential? The details differ for different sensory cells, but those details all fit into a general pattern.

In most cases, the first step of sensory transduction is the activation of a receptor protein in the plasma membrane of the sensory cell by a specific stimulus (see Figure 45.1). The activated receptor protein opens or closes ion channels in the membrane. Like neurotransmitter receptors, receptor proteins in sensory cells may do this by one of several mechanisms. The receptor protein may itself be part of an ion channel and, by changing its conformation, may directly open or close the channel pore. This mechanism is called *ionotropic* sensory detection. Alternatively, the receptor protein may be linked to a G protein that activates a cascade of intracellular events that eventually open or close ion channels (see Figure 15.14). This mechanism is called *metabotropic* sensory detection. For some sensory receptor cells (such as electroreceptors) there is no channel; instead the stimulus alters the membrane potential of the receptor cell directly.

A change in the resting membrane potential of a sensory cell in response to a stimulus is called a **receptor potential**. Receptor potentials can spread by local current flow over short distances, but to travel long distances in the nervous system, they must be converted into action potentials. Receptor potentials produce action potentials in two ways: by generating action potentials within the sensory cell itself, or by releasing a neurotransmitter that induces an associated neuron to generate action potentials.

A good example of a sensory cell that can generate action potentials (a *primary sensory cell*) is the stretch receptor of a crayfish (Figure 45.2). By placing an electrode in the cell body of a crayfish stretch receptor cell, we can record the receptor potentials that result from stretching of the muscle to which the dendrites of the cell are attached. These receptor potentials spread to the base of the sensory cell's axon (the axon hillock), where there are voltage-gated sodium channels. Action potentials generated here travel down the axon to the CNS. The rate at which action potentials are fired by the axon depends on the magnitude of the receptor potential; that, in turn, depends on how much the muscle is stretched.

In a sensory cell that does not fire action potentials (a *secondary sensory cell*), the spreading receptor potential reaches a presynaptic patch of plasma membrane and induces the release of a neurotransmitter. Whether or not the sensory cell itself fires action potentials, ultimately the stimulus is transduced into action potentials, and the intensity of the stimulus is encoded by the frequency of action potentials.

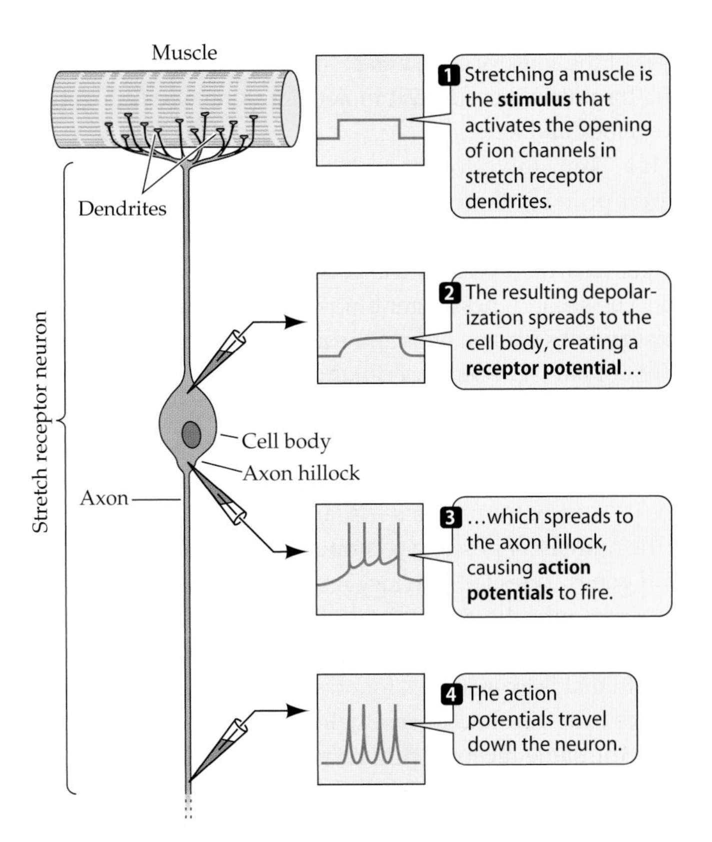

45.2 Stimulating a Sensory Cell Produces a Receptor Potential The stretch receptor of a crayfish produces a receptor potential when the muscle is stretched. When the receptor potential spreads to the axon hillock, it causes the firing of one or more action potentials that travel down the axon.

Many receptors adapt to repeated stimulation

Some sensory cells give gradually diminishing responses to maintained or repeated stimulation. This phenomenon is known as **adaptation** (desensitization), and it enables an animal to ignore background or unchanging conditions while remaining sensitive to changes or to new information. (Note that this use of the term "adaptation" is different from its application in an evolutionary context.) When you dress, you feel each item of clothing touch your skin, but the sensation of clothes touching your skin is not constantly on your mind throughout the day. You are immediately aware, however, when a seam rips, your shoe comes untied, or someone lightly touches your back.

The ability of animals to discriminate between continuous and changing stimuli comes partly from the fact that some sensory cells adapt; it is also a result of information processing by the CNS. Some sensory cells adapt very little or very slowly; examples are some types of pain receptors and mechanoreceptors for balance.

In the rest of this chapter we will learn how sensory systems gather and filter stimuli, transduce specific stimuli into action potentials, and transmit action potentials to the CNS.

Chemoreceptors: Responding to Specific Molecules

A colony of corals responds to a small amount of meat extract in the seawater around it by extending bodies and tentacles and searching for food. A solution of a single amino acid can stimulate this response. Humans have similar reactions to chemical stimuli. When we smell freshly baked bread, we salivate and feel hungry, but we gag and retch when we smell diamines from rotting meat. Animals receive information about chemical stimuli through **chemoreceptors**. Chemoreceptors are responsible for smell, taste, and the monitoring of aspects of the internal environment such as the level of carbon dioxide in the blood. Chemoreception is universal among animals and is even found in bacteria. Information from chemoreceptors can cause powerful behavioral and physiological responses.

Arthropods provide good examples for studying chemoreception

Arthropods use chemical signals to attract mates. These signals, called *pheromones*, demonstrate the sensitivity of

The female moth releases a pheromone from a gland at the tip of her abdomen.

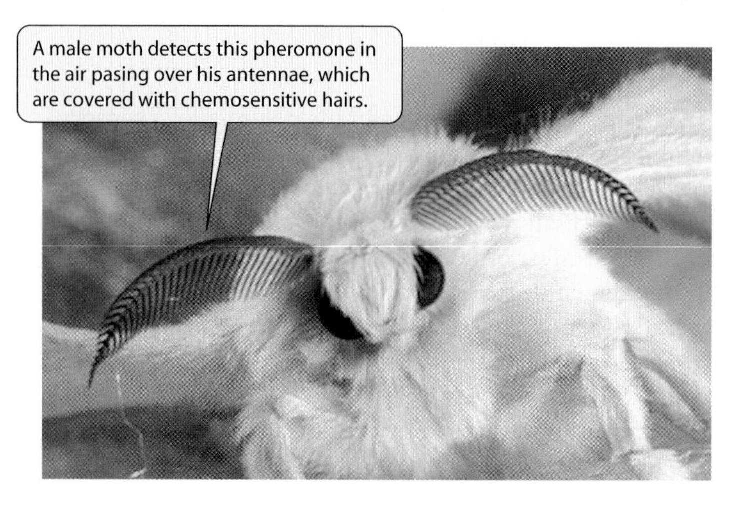
A male moth detects this pheromone in the air pasing over his antennae, which are covered with chemosensitive hairs.

45.3 Some Scents Travel Great Distances Mating in silkworm moths of the genus *Bombyx* is coordinated by a pheromone called bombykol.

chemosensory systems. One of the best-studied examples of this phenomenon is the silkworm moth.

To attract a mate, the female silkworm moth releases a pheromone called bombykol from a gland at the tip of her abdomen. The male silkworm moth has receptors for this molecule on his antennae (Figure 45.3). Each feathery antenna carries about 10,000 bombykol-sensitive hairs. A single molecule of bombykol may be sufficient to generate action potentials in the antennal nerve that transmits the signal to the CNS. Because of the male's high degree of sensitivity, the sexual message of a female moth is likely to reach any male that happens to be within a downwind area stretching over several kilometers. When approximately 200 hairs per second are activated, the male flies upwind in search of the female. Because the rate of firing in the male's sensory nerves is proportional to the bombykol concentration in the air, he can follow the airborne concentration gradient and home in on the signaling female.

Many arthropods have chemosensory hairs, each containing one or more specific types of receptors. Crabs and flies, for example, have chemoreceptor hairs on their feet; these hairs can respond to sugars, amino acids, salts, and even distilled water. A fly tastes a potential food by stepping in it.

Olfaction is the sense of smell

The sense of smell, known as **olfaction**, also depends on chemoreceptors. In vertebrates, the olfactory sensors are neurons embedded in a layer of epithelial tissue at the top of the nasal cavity. These neurons project their axons to the olfactory bulb of the brain, while their dendrites end in olfactory hairs on the surface of the nasal epithelium. A protective layer of mucus covers the epithelium. Molecules from the environment must diffuse through this mucus to reach the receptor proteins on the olfactory hairs. When you have a cold, the amount of mucus in your nose increases, and the epithelium swells. With this in mind, study Figure 45.4, and you will easily understand why respiratory infections can cause you to lose your sense of smell.

A dog has up to 40 million nerve endings per square centimeter of nasal epithelium, many more than we do. Humans have a fairly sensitive olfactory system, but we are unusual among mammals in that we depend more on vision than on olfaction (for example, we tend to join bird-watching societies more often than mammal-smelling societies).

How does an olfactory sensory cell transduce the structure of a chemical in the environment into action potentials? An *odorant* is a molecule that activates an olfactory receptor protein. Odorants bind to receptor proteins on the olfactory hairs of the sensory cells. Olfactory receptor proteins are specific for particular odorant molecules—the two fit together like a lock and key.

If a "key" (an odorant molecule) fits the "lock" (the receptor protein), then a G protein is activated, which in turn activates an enzyme that causes an increase of a second messenger (cAMP in vertebrates) in the cytoplasm of the sensory cell. The second messenger binds to sodium channels in the sensory cell's plasma membrane and opens them, causing an influx of Na^+. The sensory cell thus depolarizes to threshold and fires action potentials.

The olfactory world has an enormous number of "keys"—molecules that produce distinct smells. The number of "locks"—receptor proteins—is large, but not nearly as large as the number of possible odorants. A family of about a thousand genes codes for olfactory receptor proteins. In humans, however, only about one-third of these genes are expressed and produce functional proteins. Each functional receptor protein that is expressed is found in a limited number of sensory cells in the olfactory epithelium. All of the cells that express the same receptor protein project to the same regions in the olfactory bulb. A given odorant molecule may bind to one or to more than one receptor protein. Therefore, each odorant molecule can excite a unique combination of cells in the olfactory bulb, so an olfactory system with even hundreds of different receptor proteins can discriminate a large number of smells.

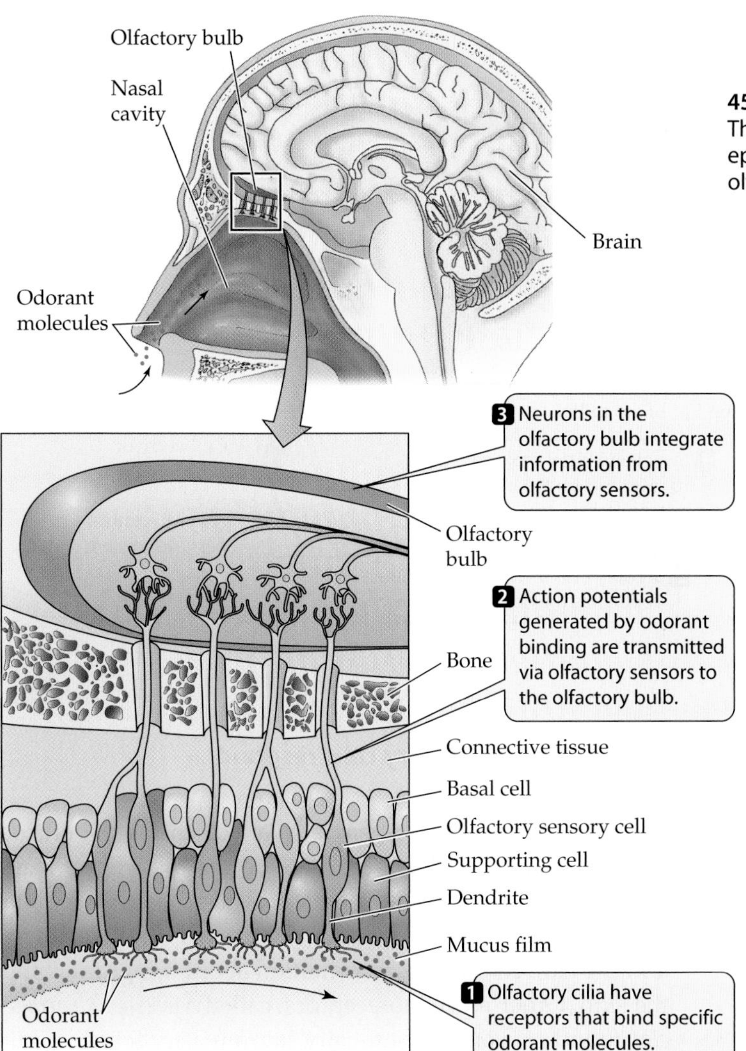
Olfactory bulb
Nasal cavity
Brain
Odorant molecules
3 Neurons in the olfactory bulb integrate information from olfactory sensors.
Olfactory bulb
2 Action potentials generated by odorant binding are transmitted via olfactory sensors to the olfactory bulb.
Bone
Connective tissue
Basal cell
Olfactory sensory cell
Supporting cell
Dendrite
Mucus film
1 Olfactory cilia have receptors that bind specific odorant molecules.
Odorant molecules

45.4 Olfactory Receptors Communicate Directly with the Brain The receptor cells of the human olfactory system are embedded in epithelial tissues lining the nasal cavity and send their axons to the olfactory bulb of the brain.

How does the sensory cell signal the intensity of a smell? The more odorant molecules that bind to receptors, the more action potentials are generated, and the greater the intensity of the perceived smell.

The vomeronasal organ senses pheromones

The **vomeronasal organ** (**VNO**) is a small, paired tubular structure embedded in the nasal epithelium. In mammals it is located anteriorly on the septum dividing the two nostrils. The VNO has a pore opening into the nasal cavity. When the animal sniffs, the VNO pulsates and draws a sample of nasal fluid over the chemoreceptors embedded in its walls. The information from these chemoreceptors goes to an accessory olfactory bulb in the brain, and information from there goes to brain regions involved in sexual and other instinctive behaviors.

In some exciting recent studies at Duke University, Lawrence Katz and colleagues succeeded in recording from neurons in the accessory olfactory bulbs of conscious, behaving mice. These neurons were activated when another mouse was placed in the same cage and was sniffed by the mouse with the implanted electrodes. What confirmed the hypothesis that the VNO detects pheromones, however, was the fact that neurons fired differentially depending on whether the strange mouse being sniffed was a male or a female and whether it was of the same or a different strain.

In snakes, the VNO opens into the roof of the mouth cavity. Each time the snake's forked tongue darts in and out, the forks fit into the VNO openings and present to the chemoreceptors a sample of molecules from the surrounding air. Thus the snake is really using its tongue to smell its environment, not to taste it. Why doesn't the snake simply use the flow of air to and from its lungs, as we do, to smell the environment? In reptiles, air flows to and from the lungs slowly (and can even stop entirely for long periods of time), but the tongue can dart in and out many times in a second. It is a quick source of olfactory information.

Gustation is the sense of taste

The sense of taste, or **gustation**, in humans and other vertebrates depends on clusters of sensory cells called **taste buds**. The taste buds of terrestrial vertebrates are confined to the mouth cavity, but some fishes have taste buds in the skin that enhance their ability to sense their environment. Some fishes living in murky water are very sensitive to small amounts of amino acids in the water around them and can find food without the use of vision. The duck-billed platypus, a monotreme mammal (see Figure 34.22*b*), has similar talents as a result of taste buds on the sensitive skin of its bill.

A human tongue has approximately 10,000 taste buds. The taste buds are embedded in the epithelium of the tongue, and many are found on the raised papillae of the tongue. (Look at your tongue in a mirror—the papillae make it look fuzzy.) Each papilla has many taste buds. The outer surface of a taste bud has a pore that exposes the tips of the sensory cells. Microvilli (tiny hairlike projections) increase the surface area of the sensory cells where their tips converge at the pore (Figure 45.5). These sensory cells, unlike olfactory receptors, are not neurons. At their bases, they form synapses with dendrites of sensory neurons.

The tongue does a lot of hard work, so its epithelium, along with cells of its taste buds, are shed and replaced at a rapid rate. Individual taste bud cells last only a few days before they are replaced, but the sensory neurons associated with them live on, always forming new synapses as new taste buds form.

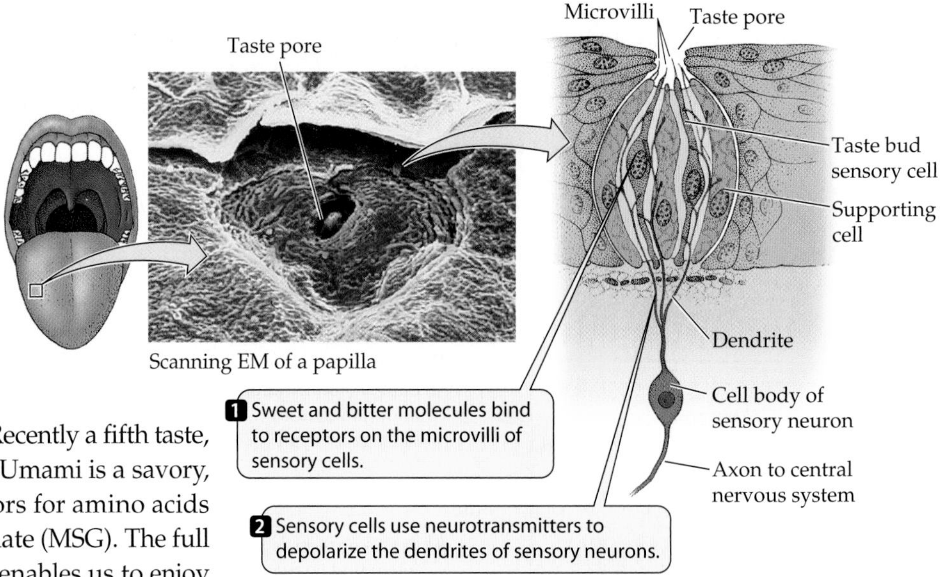

45.5 Taste Buds Are Clusters of Sensory Cells Each taste bud contains a number of sensory cells that are not neurons.

You have probably heard that humans can perceive only four tastes: sweet, salty, sour, and bitter. Although controversial, there is some evidence that taste buds can distinguish among a variety of sweet-tasting molecules and a variety of bitter-tasting molecules. Recently a fifth taste, called *umami*, has been added to the list. Umami is a savory, meaty taste that originates from receptors for amino acids and is enhanced by monosodium glutamate (MSG). The full complexity of the chemosensitivity that enables us to enjoy the subtle flavors of food comes from the combined activation of gustatory and olfactory receptors; that is the reason you lose some of your sense of flavor when you have a cold.

Gustation begins with receptor proteins in the membranes of the microvilli. The nature of these proteins and the mechanisms by which they depolarize the sensory cell differ for the different basic tastes. Saltiness is simply due to Na^+ ions diffusing through open Na^+ channels and depolarizing the sensory cell. Sourness is similarly due to H^+ ions diffusing in through open channels and depolarizing sensory cells, but the mechanism is unknown. Bitterness is due to a large family of receptor proteins that bind specific molecules, much like the olfactory receptor proteins. The bitter taste probably evolved as a protective mechanism enabling animals to avoid toxic plant compounds. Since plants have evolved a variety of such molecules to make them distasteful to predators, a variety of receptors is essential. The mechanism for sweet taste is not fully understood, but it probably involves receptor proteins that bind sugar molecules. In all cases of taste sensation, however, changes in the membrane potential of the sensory cells cause them to release neurotransmitters onto the dendrites of the sensory neurons. The sensory neurons fire action potentials that are conducted to the CNS, where the information is interpreted as specific taste sensations.

Mechanoreceptors: Detecting Stimuli that Distort Membranes

Mechanoreceptors are cells that are sensitive to mechanical forces. Physical distortion of a mechanoreceptor's plasma membrane causes ion channels to open, altering the membrane potential of the cell, which in turn leads to the generation of action potentials. The rate of action potentials tells the CNS the strength of the stimulus exciting the mechanoreceptor. Mechanoreceptor cells are involved in many sensory systems, ranging from skin sensations to sensing blood pressure.

Many different sensory cells respond to touch and pressure

Objects touching the skin generate varied sensations because skin is packed with diverse mechanoreceptors (Figure 45.6). The most important tactile receptors found in both hairy and non-hairy skin are *Merkel's discs*, which adapt rather slowly and provide continuous information about things touching the skin. Other mechanoreceptors, called *Meissner's corpuscles*, found primarily in non-hairy skin, are very sensitive, but adapt rapidly, so they provide information about changes in things touching the skin. The rapid adaptation of these tactile sensors is why you roll a small object between your fingers, rather than holding it still, to discern its shape and texture. As you roll it, you continue to stimulate Meissner's corpuscles anew.

Two other kinds of mechanoreceptors are found deeper in the skin. *Ruffini endings*, which are rather slowly adapting, are good at providing information about vibrating stimuli of low frequencies. *Pacinian corpuscles*, which are rapidly adapting, are good at providing information about vibrating stimuli of higher frequencies. Even deeper in the skin, dendrites of sensory neurons wrap around hair follicles. When the hairs are displaced, those neurons are stimulated.

The density of tactile mechanoreceptors varies across the surface of the body. A two-point spatial discrimination test demonstrates this fact. If you lightly touch someone's skin with two toothpicks, you can determine how far apart the two stimuli have to be before the person can distinguish whether he or she was touched with one or two toothpicks. On the back, the stimuli have to be rather far apart. The same test applied to the person's lips or fingertips reveals finer spatial discrimination; that is, the person can identify as separate two stimuli that are close together.

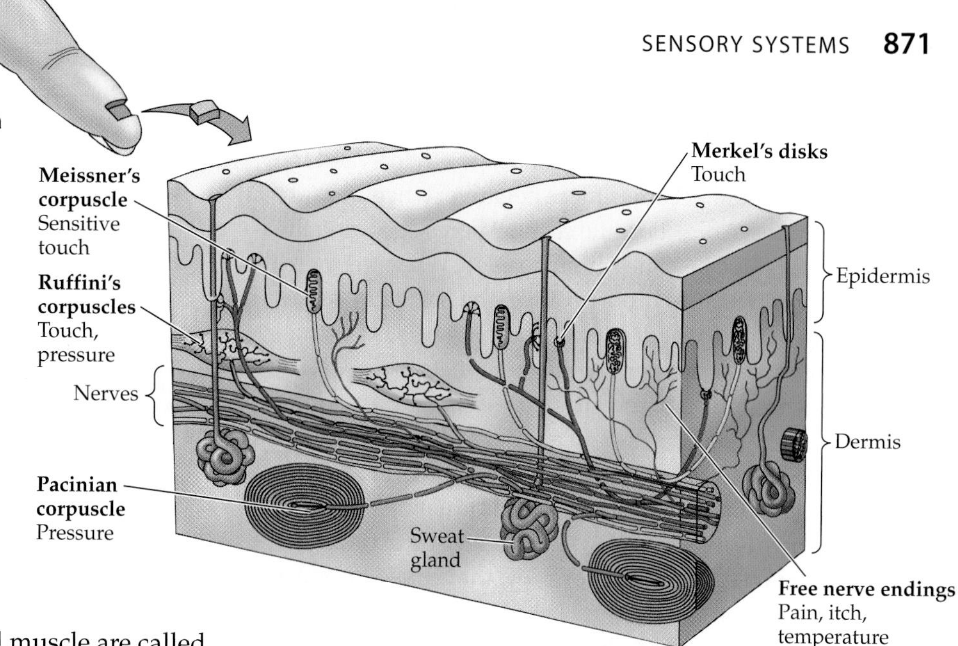

45.6 The Skin Feels Many Sensations Even a very small patch of skin contains a variety of sensory cells.

Stretch receptors are found in muscles, tendons, and ligaments

An animal receives information from **stretch receptors** about the position of its limbs and the stresses on its muscles and joints. These mechanoreceptors are activated by being stretched. The information they feed continuously to the CNS is essential for the coordination of movements.

The stretch receptors found in skeletal muscle are called **muscle spindles**. These receptors, which are embedded in connective tissue within muscles, consist of modified muscle cells that are innervated in the center by extensions of sensory neurons. Whenever the muscle stretches, muscle spindles are also stretched, and the neurons transmit action potentials to the CNS (Figure 45.7*a*). The CNS uses this information to maintain *muscle tone*, keeping muscles taut and ready for action. Earlier in this chapter, we saw how crayfish stretch receptors transduce physical force into action potentials (see Figure 45.2). The actions of muscle spindles are similar.

Another type of stretch receptor, the **Golgi tendon organ**, is found in tendons and ligaments. It provides information about the force generated by a contracting muscle. When a contraction becomes too forceful, the information from the Golgi tendon organ feeds into the spinal cord, inhibits the motor neurons innervating the muscle, and causes the contracting muscle to relax, thus protecting the muscle from tearing (Figure 45.7*b*).

Hair cells provide information about balance, orientation in space, and motion

Hair cells are also mechanoreceptors. Projecting from the surface of each hair cell is a set of *stereocilia*, which looks like a set of organ pipes. When these stereocilia (which are really microvilli) are bent, they alter ionotropic receptor proteins in the hair cell's plasma membrane. These receptors are gated by the

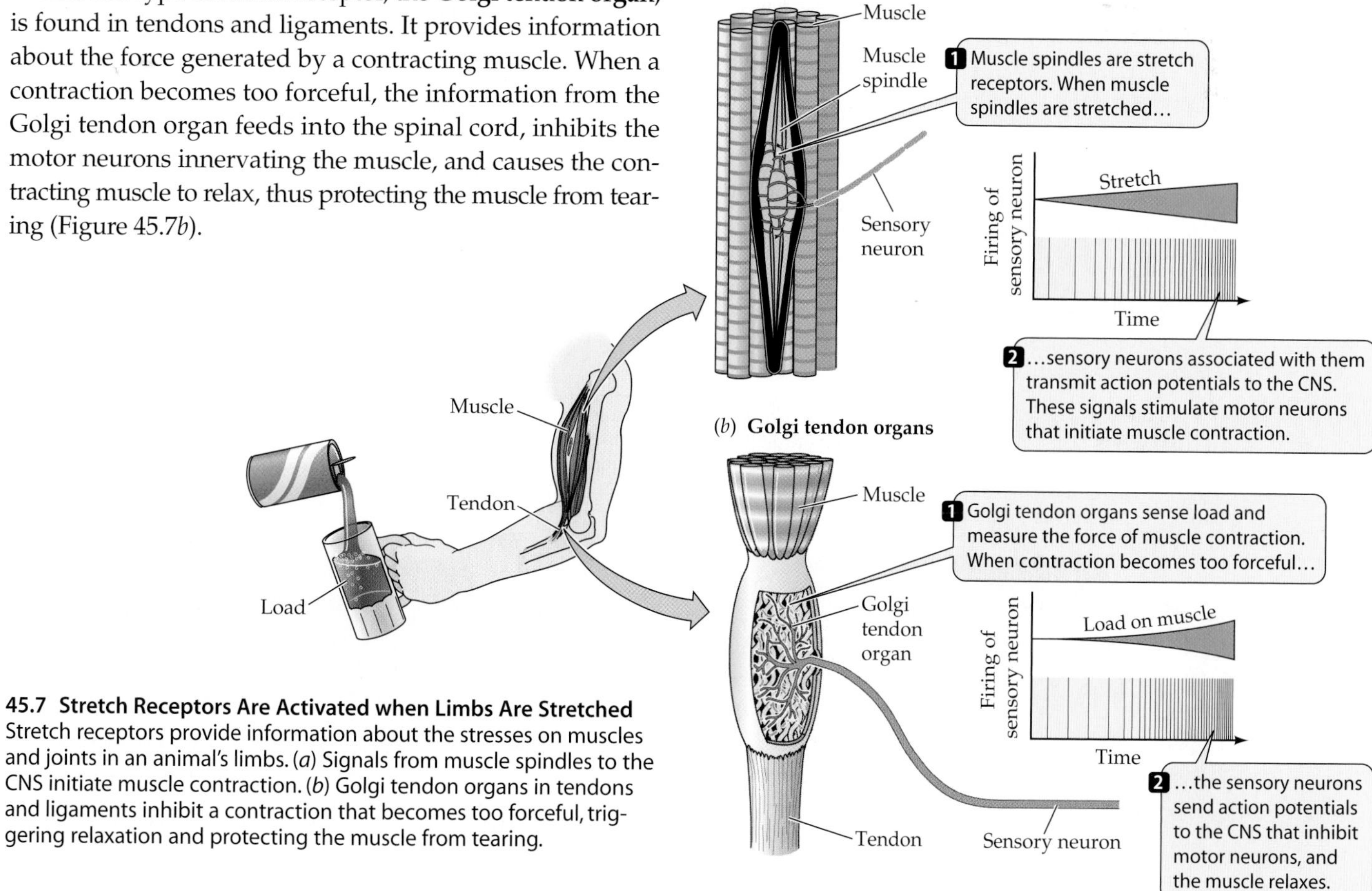

45.7 Stretch Receptors Are Activated when Limbs Are Stretched Stretch receptors provide information about the stresses on muscles and joints in an animal's limbs. (*a*) Signals from muscle spindles to the CNS initiate muscle contraction. (*b*) Golgi tendon organs in tendons and ligaments inhibit a contraction that becomes too forceful, triggering relaxation and protecting the muscle from tearing.

movement of the stereocilia. When the stereocilia of some hair cells are bent in one direction, the channel pores close, and the membrane potential becomes more negative; when they are bent in the opposite direction, the channel pores open, and the membrane potential becomes more positive. When the membrane potential becomes more positive, the hair cell releases a neurotransmitter to the sensory neuron associated with it, and the sensory neuron sends action potentials to the CNS.

Hair cells are found in the **lateral line** sensory system of fishes. The lateral line consists of a canal just under the surface of the skin that runs down each side of the fish (Figure 45.8). The lateral line system provides information about the fish's movements through the water, as well as about moving objects, such as predators or prey, that cause pressure waves in the water.

Vertebrate organs of equilibrium use hair cells to detect the position of the body with respect to gravity. Within the mammalian inner ear, three **semicircular canals** at right angles to one another sense the position and orientation of the head. The **vestibular apparatus** has two chambers that sense the static position of the head as well as linear acceleration produced by movement. The structure and function of these organs are described in Figure 45.9.

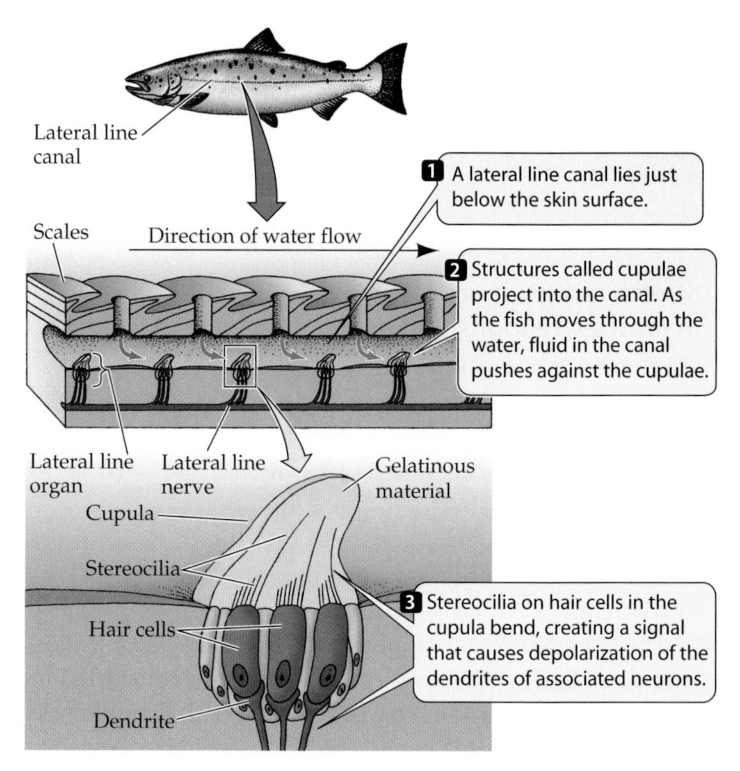

Lateral line canal
1 A lateral line canal lies just below the skin surface.
Scales
Direction of water flow
2 Structures called cupulae project into the canal. As the fish moves through the water, fluid in the canal pushes against the cupulae.
Lateral line organ
Lateral line nerve
Gelatinous material
Cupula
Stereocilia
3 Stereocilia on hair cells in the cupula bend, creating a signal that causes depolarization of the dendrites of associated neurons.
Hair cells
Dendrite

45.8 The Lateral Line System Contains Mechanoreceptors Hair cells in the lateral line of a fish detect movement of the water around the animal, giving the fish information about its own movements and the movements of objects nearby.

45.9 Organs in the Inner Ear of Mammals Provide the Sense of Equilibrium The bony inner ear includes organs of equilibrium—three semicircular canals and two vestibular organs—as well as the snail-shaped cochlea, which is part of the auditory system.

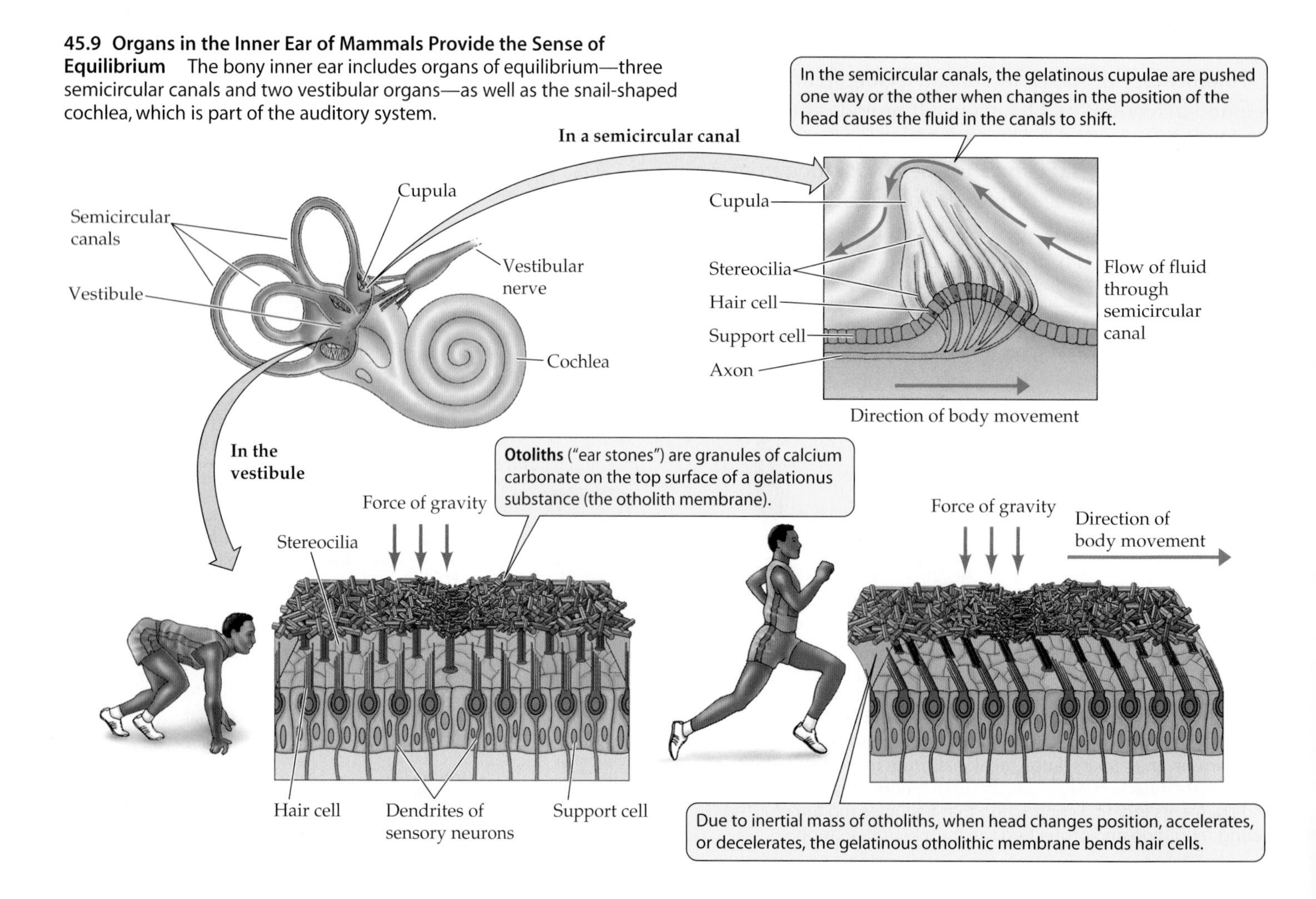

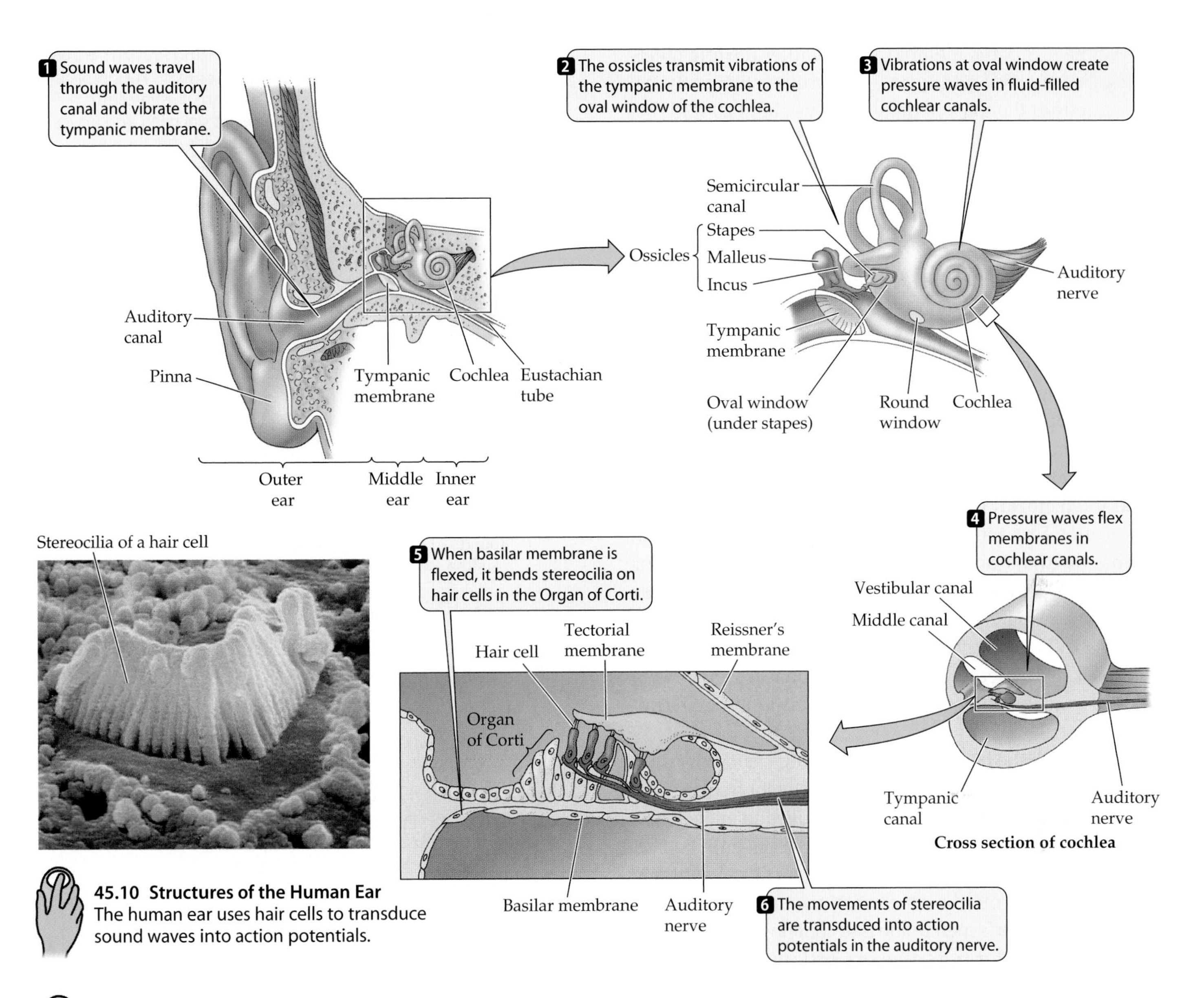
1 Sound waves travel through the auditory canal and vibrate the tympanic membrane.
2 The ossicles transmit vibrations of the tympanic membrane to the oval window of the cochlea.
3 Vibrations at oval window create pressure waves in fluid-filled cochlear canals.
Auditory canal
Pinna
Tympanic membrane
Cochlea
Eustachian tube
Outer ear
Middle ear
Inner ear
Semicircular canal
Ossicles
Stapes
Malleus
Incus
Tympanic membrane
Oval window (under stapes)
Round window
Cochlea
Auditory nerve
4 Pressure waves flex membranes in cochlear canals.
Vestibular canal
Middle canal
Tympanic canal
Auditory nerve
Cross section of cochlea
Stereocilia of a hair cell
5 When basilar membrane is flexed, it bends stereocilia on hair cells in the Organ of Corti.
Hair cell
Tectorial membrane
Reissner's membrane
Organ of Corti
Basilar membrane
Auditory nerve
6 The movements of stereocilia are transduced into action potentials in the auditory nerve.

45.10 Structures of the Human Ear
The human ear uses hair cells to transduce sound waves into action potentials.

Auditory systems use hair cells to sense sound waves

The stimuli that animals perceive as sounds are pressure waves. **Auditory systems** use mechanoreceptors to convert pressure waves into receptor potentials. Auditory systems include special structures that gather sound waves, direct them to the sensory organ, and amplify their effect on the mechanoreceptors.

Human hearing provides a good example of an auditory system. The organs of hearing are the ears. The two prominent structures on the sides of our heads usually thought of as ears are the *pinnae*. The pinna of an ear collects sound waves and directs them into the *auditory canal*, which leads to the actual hearing apparatus in the middle ear and the inner ear (Figure 45.10). If you have ever watched a rabbit, a horse, or a cat change the orientation of its ear pinnae to focus on a particular sound, then you have witnessed the role of pinnae in hearing.

The eardrum, or **tympanic membrane**, covers the end of the auditory canal. The tympanic membrane vibrates in response to pressure waves traveling down the auditory canal. The *middle ear*, an air-filled cavity, lies on the other side of the tympanic membrane.

The middle ear is open to the throat at the back of the mouth through the *eustachian tube*. Because the eustachian tube is also filled with air, pressure equilibrates between the middle ear and the outside world. When you have a cold or allergy, the tube can become blocked by mucus or by tissue swelling, so you have difficulty "clearing your ears," or equilibrating the pressure in the middle ear with the outside air pressure. As a result, the flexible tympanic membrane bulges in or out, dampening your hearing and sometimes causing earaches.

The middle ear contains three delicate bones called the **ossicles**, individually named the *malleus* (hammer), *incus* (anvil), and *stapes* (stirrup). The ossicles transmit the vibrations of the tympanic membrane to another flexible membrane called the **oval window**. The leverlike action of the ossicles amplifies the force of the vibrations about 20-fold. Behind the oval window lies the fluid-filled inner ear. Movements of the oval window result in pressure changes in the inner ear. These pressure waves are transduced into action potentials.

The *inner ear* is a long, tapered, coiled chamber called the **cochlea** (from Latin and Greek words for "snail" or "shell"). A cross section of this chamber reveals that it is composed of three parallel canals separated by two membranes: **Reissner's membrane** and the **basilar membrane** (see Figure 45.10). Sitting on the basilar membrane is the **organ of Corti**, the apparatus that transduces pressure waves into action potentials. The organ of Corti contains hair cells whose stereocilia are in contact with an overhanging, rigid shelf called the **tectorial membrane**. Hair cells are not neurons and do not fire action potentials. However, they form synapses with associated sensory neurons, whose axons make up the auditory nerve. When the basilar membrane flexes, the tectorial membrane bends the hair cell stereocilia, changing the rate at which they release a neurotransmitter onto the sensory neurons. As a result, there are changes in the rates of action potentials traveling to the brain in the auditory nerve.

What causes the basilar membrane to flex, and how does this mechanism distinguish sounds of different frequencies? In Figure 45.11, the cochlea is shown uncoiled to make it easier to understand its structure and function. To simplify matters, we have left out Reissner's membrane, thus combining the upper and the middle canals into one upper canal. (The purpose of Reissner's membrane is to contain a specific aqueous environment for the organ of Corti separate from the aqueous environment in the rest of the cochlea.)

The simplified diagram of the cochlea shown in Figure 45.11 reveals two additional features that are important to its function. First, the upper and lower canals separated by the basilar membrane are joined at the distal end of the cochlea (the end farthest from the oval window), making one continuous canal that turns back on itself. Second, just as the oval window is a flexible membrane at the beginning of the cochlea, the **round window** is a flexible membrane at the end of the long cochlear canal.

Air is highly compressible, but fluids are not. Therefore, a pressure wave can travel through air without much displacement of the air, but a pressure wave in fluid causes displacement of the fluid. When the stapes pushes the oval window in, the fluid in the upper canal of the cochlea is displaced. The cochlear fluid pressure wave travels down the upper canal, around the bend, and back through the lower canal. At the end of the lower canal, the displacement pressure is dissipated by the outward bulging of the round window.

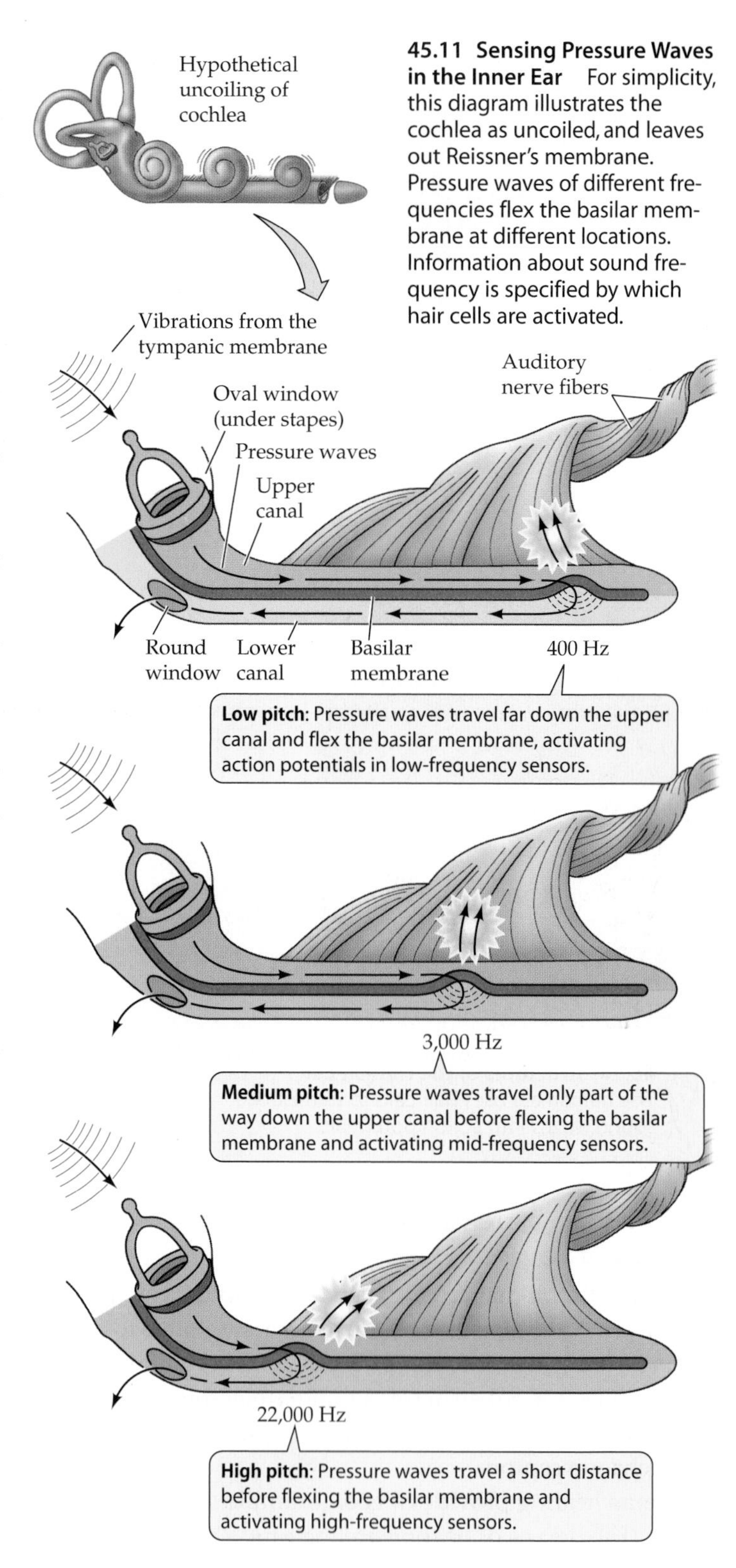
Hypothetical uncoiling of cochlea
Vibrations from the tympanic membrane
Oval window (under stapes)
Pressure waves
Upper canal
Auditory nerve fibers
Round window
Lower canal
Basilar membrane
400 Hz
Low pitch: Pressure waves travel far down the upper canal and flex the basilar membrane, activating action potentials in low-frequency sensors.
3,000 Hz
Medium pitch: Pressure waves travel only part of the way down the upper canal before flexing the basilar membrane and activating mid-frequency sensors.
22,000 Hz
High pitch: Pressure waves travel a short distance before flexing the basilar membrane and activating high-frequency sensors.

45.11 Sensing Pressure Waves in the Inner Ear For simplicity, this diagram illustrates the cochlea as uncoiled, and leaves out Reissner's membrane. Pressure waves of different frequencies flex the basilar membrane at different locations. Information about sound frequency is specified by which hair cells are activated.

If the oval window vibrates in and out rapidly, the waves of fluid pressure do not have enough time to travel all the way to the end of the upper canal and back through the lower canal. Instead, they take a shortcut by crossing the basilar membrane, causing it to flex. The more rapid the vibration, the closer to the oval and round windows the pressure wave will flex the basilar membrane. Thus, different pitches of sound flex the basilar membrane at different locations and activate different sets of hair cells.

The ability of the basilar membrane to respond to vibrations of different frequencies is enhanced by its structure. Near the oval and round windows, at the proximal end, the basilar membrane is narrow and stiff, but it gradually becomes wider and more flexible toward the opposite (distal) end. So it is easier for the proximal basilar membrane to resonate with high frequencies and for the distal basilar membrane to resonate with lower frequencies. A complex sound made up of many frequencies distorts the basilar membrane at many places simultaneously and activates a unique subset of hair cells. Action potentials stimulated by the mechanoreceptors at different positions along the organ of Corti travel to the brain stem along the auditory nerve.

Deafness, the loss of the sense of hearing, has two general causes. *Conduction deafness* is caused by the loss of function of the tympanic membrane and the ossicles of the middle ear. Repeated infections of the middle ear can cause scarring of the tympanic membrane and stiffening of the connections between the ossicles. The consequence is less efficient conduction of sound waves from the tympanic membrane to the oval window. With increasing age, the ossicles progressively stiffen, resulting in a gradual loss of the ability to hear high-frequency sounds. *Nerve deafness* is caused by damage to the inner ear or the auditory pathways. A common cause of nerve deafness is damage to the hair cells of the delicate organ of Corti by exposure to loud sounds such as jet engines, pneumatic drills, or highly amplified music. This damage is cumulative and permanent.

Photoreceptors and Visual Systems: Responding to Light

Sensitivity to light—**photosensitivity**—confers on the simplest animals the ability to orient to the sun and sky and gives more complex animals rapid and extremely detailed information about objects in their environment. It is not surprising that both simple and complex animals can sense and respond to light. What is remarkable is that across the entire range of animal species, evolution has conserved the same basis for photosensitivity: a family of pigments called **rhodopsins**.

In this section we will learn how rhodopsin molecules respond when stimulated by light energy and how that response is transduced into neuronal signals. We will also examine the structures of eyes, the organs that gather light energy and focus it onto photoreceptor cells.

Rhodopsins are responsible for photosensitivity

Photosensitivity depends on the ability of rhodopsins to absorb photons of light and to undergo a change in conformation. A rhodopsin molecule consists of a protein, **opsin** (which alone is not photosensitive), and a light-absorbing prosthetic group, **11-*cis*-retinal**. The light-absorbing group is cradled in the center of the opsin and is bound covalently to it. The entire rhodopsin molecule sits within the plasma membrane of a photoreceptor cell (Figure 45.12).

When the 11-*cis*-retinal absorbs a photon of light energy, it changes into a different isomer of retinal, called all-*trans*-retinal. This change puts a strain on the bonds between retinal and opsin, changing the conformation of opsin. This change signals the detection of light. In vertebrate eyes, the retinal and the opsin eventually separate from each other—a process called *bleaching*, which causes the molecule to lose its photosensitivity. A series of enzymatic reactions is then required to return the all-*trans* retinal to the 11-*cis* isomer, which then recombines with opsin so that it once again becomes the photosensitive pigment rhodopsin.

How does the conformational change of rhodopsin transduce light into a cellular response? After retinal is converted from the 11-*cis* into the all-*trans* form, its interactions with

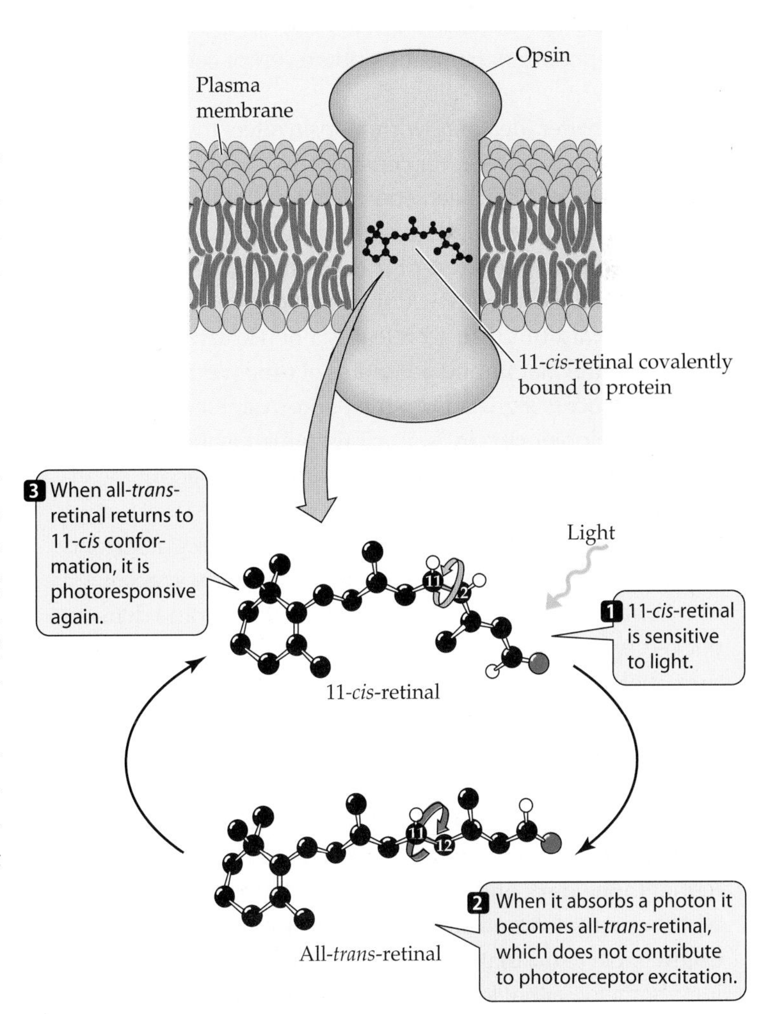
Opsin
Plasma membrane
11-*cis*-retinal covalently bound to protein
3 When all-*trans*-retinal returns to 11-*cis* conformation, it is photoresponsive again.
Light
1 11-*cis*-retinal is sensitive to light.
11-*cis*-retinal
11
12
All-*trans*-retinal
2 When it absorbs a photon it becomes all-*trans*-retinal, which does not contribute to photoreceptor excitation.

45.12 Rhodopsin: A Photosensitive Molecule Rhodopsin changes its conformation when it absorbs light.

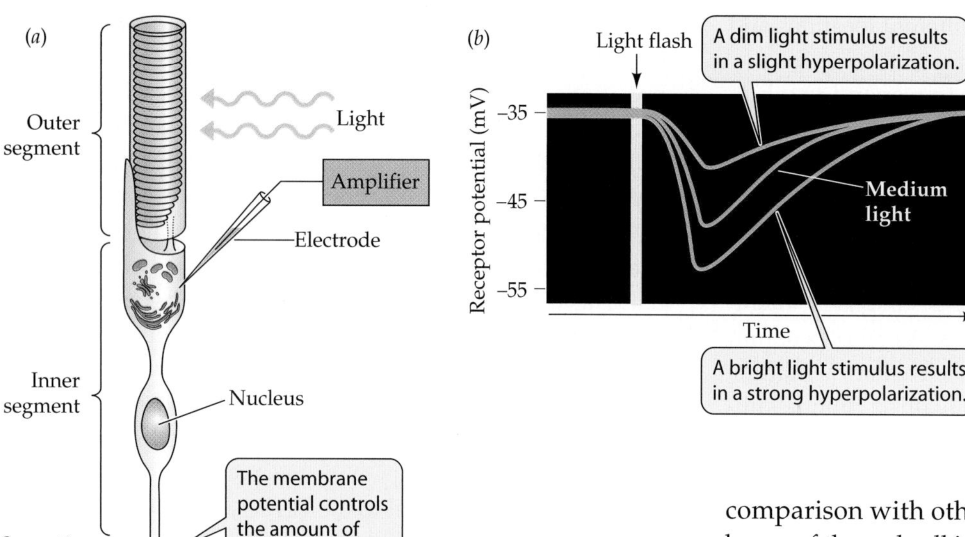

45.13 A Rod Cell Responds to Light (*a*) The vertebrate rod cell is a neuron modified for photosensitivity. The membranes of a rod cell's discs are densely packed with rhodopsin. (*b*) The plasma membrane of a rod cell hyperpolarizes—becomes more negative—in response to a flash of light.

opsin pass through several unstable intermediate stages. One of these stages is known as *photoexcited rhodopsin* because it triggers a cascade of reactions that result in the alteration of membrane potential that is the photoreceptor cell's response to light.

To get a better idea of how rhodopsin alters the membrane potential of a photoreceptor cell and how that photoreceptor cell signals that it has been stimulated by light, let's look at one type of vertebrate photoreceptor cell, the **rod cell**. Like other vertebrate photoreceptor cells, the rod cell is a modified neuron (Figure 45.13*a*). However, it does not produce action potentials, but instead releases a neurotransmitter that influences the membrane potentials of other neurons. Those neurons process signals from the photoreceptor cells to produce action potentials in their axons, which make up the optic nerve. The optic nerve transmits information about the visual world to the brain.

Each rod cell has an outer segment, an inner segment, and a synaptic terminal. The outer segment is highly specialized and contains a stack of discs of plasma membrane densely packed with rhodopsin. The function of the discs is to capture photons of light passing through the rod cell. The inner segment contains the cell nucleus and abundant mitochondria. The synaptic terminal is where the rod cell communicates with other neurons.

To see how a rod cell responds to light, we can penetrate a single rod cell with an electrode and record its membrane potential in the dark and in the light (see Figure 45.13*a*). From what we have learned about other types of sensory cells, we might expect stimulation of the rod cell by light to make its membrane potential less negative. But photoreceptor cells are atypical, and the opposite is true. When a rod cell is kept in the dark, it has a relatively depolarized resting potential in comparison with other neurons. In fact, the plasma membrane of the rod cell is almost as permeable to Na^+ ions as to K^+ ions, and Na^+ ions are continually entering the outer segment of the cell.

When a light is flashed on the dark-adapted rod cell, its membrane potential becomes more negative—it hyperpolarizes (Figure 45.13*b*). The rod cell changes its rate of neurotransmitter release as its membrane potential changes (since the rod cell hyperpolarizes, neurotransmitter release decreases).

How does the absorption of light by rhodopsin hyperpolarize the rod cell? When rhodopsin is excited by light, it initiates a cascade of events. The photoexcited rhodopsin combines with and activates another protein, a G protein called *transducin*. Activated transducin in turn activates a phosphodiesterase, which converts cyclic GMP (cGMP) to GMP. This reaction plays a central role in phototransduction. In the dark, the cGMP in the outer segment binds to both cation channels, keeping them open and allowing Na^+ and Ca^{2+} to enter the outer segment. As cGMP is converted to GMP, the sodium channels close, and the cell hyperpolarizes.

This mechanism may seem like a roundabout way of doing business, but its advantage is its enormous amplification ability. Each molecule of photoexcited rhodopsin can activate several hundred transducin molecules, thus activating a large number of phosphodiesterase molecules. The catalytic capacity of a molecule of phosphodiesterase is great: It can hydrolyze more than 4,000 molecules of cGMP per second. The bottom line is that a single photon of light can cause a huge number of sodium channels to close (Figure 45.14).

Invertebrates have a variety of visual systems

Photoreceptors are incorporated into a variety of visual systems, from simple to complex. Flatworms obtain directional information about light from photoreceptor cells that are organized into *eye cups*. The eye cups are paired bilateral structures, each partly shielded from light by a layer of pigmented

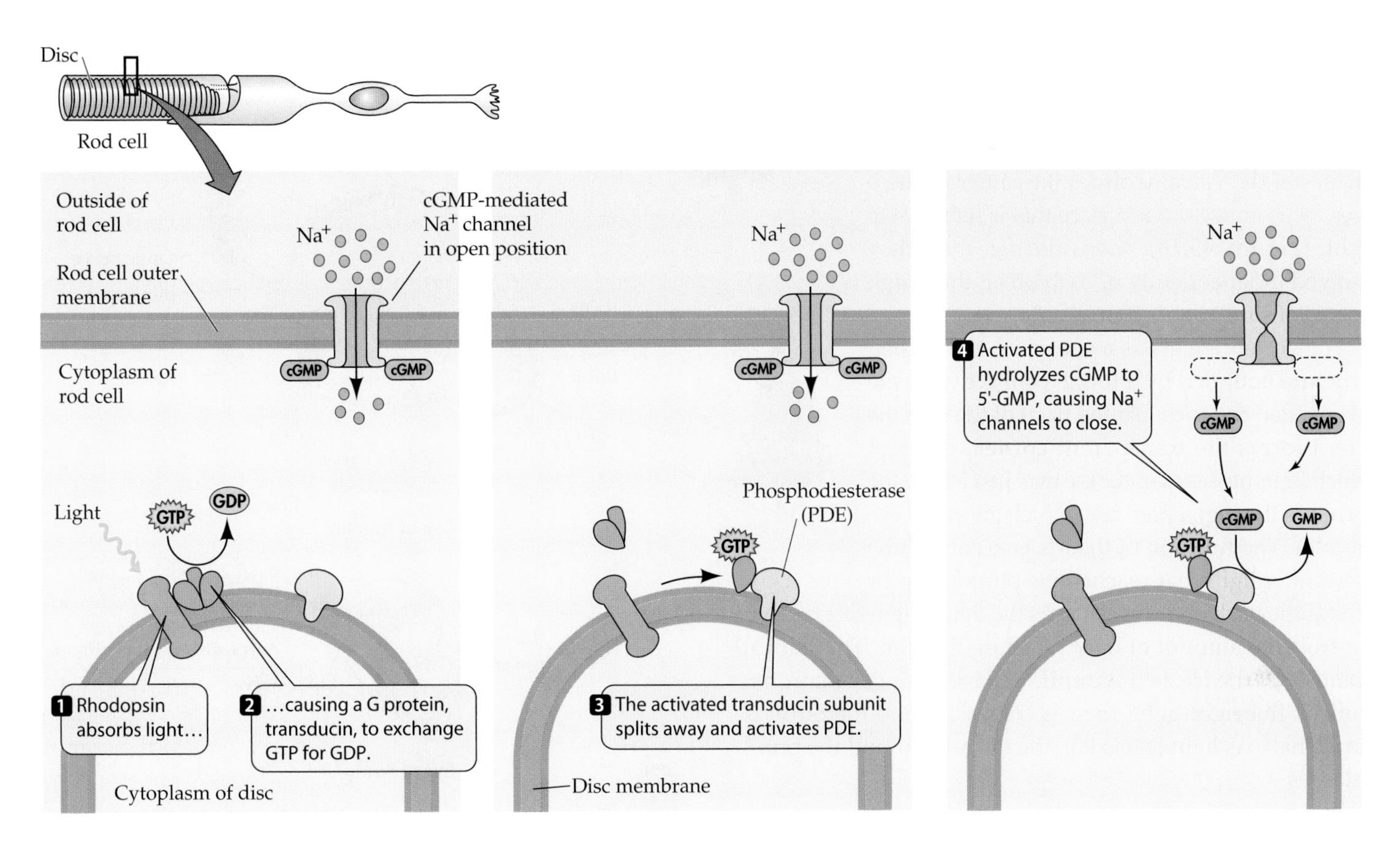

45.14 Light Absorption Closes Sodium Channels The absorption of light by rhodopsin initiates a cascade of events resulting in the hyperpolarization of the rod cell.

cells lining the cup. The photoreceptors on the two sides of the animal are unequally stimulated unless the animal is facing directly toward or away from a light source. The flatworm generally uses directional information from the eye cups to move away from light.

Arthropods have evolved **compound eyes** that provide them with information about patterns or images in the environment. Each compound eye consists of many optical units called **ommatidia** (singular, ommatidium) (Figure 45.15). The number of ommatidia in a compound eye varies from only a few in some ants, to 800 in fruit flies, to 10,000 in some dragonflies.

Each ommatidium has a lens structure that directs light onto photoreceptors. Flies, for example, have eight elongated photoreceptors in each ommatidium. The inner borders of the photoreceptors are covered with microvilli that contain rhodopsin and trap light. Axons from the photoreceptors communicate with the nervous system. Since each ommatidium of a compound eye is directed at a slightly different part of the visual world, only a crude, or perhaps a broken-up, image can be communicated from the compound eye to the CNS.

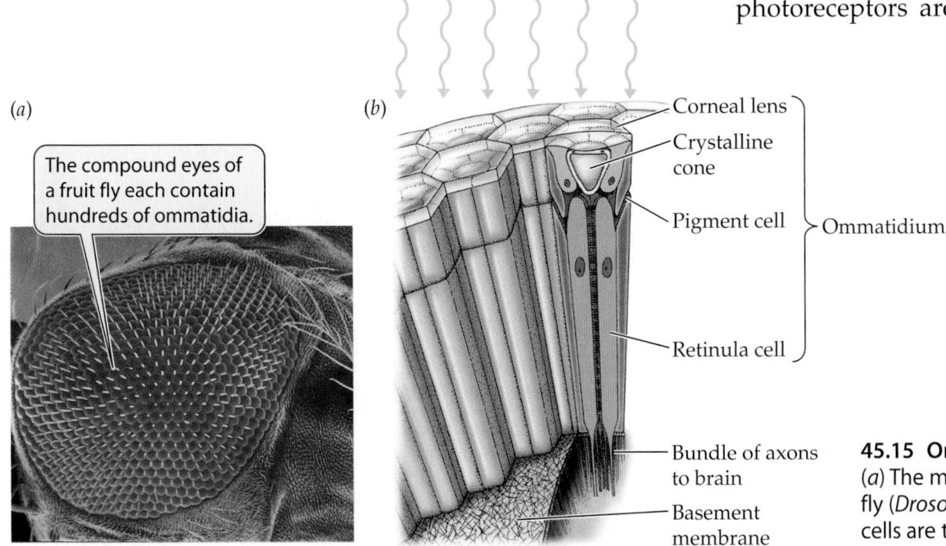

45.15 Ommatidia: The Functional Units of Insect Eyes (*a*) The micrograph shows the compound eye of a fruit fly (*Drosophila*). (*b*) The rhodopsin-containing retinula cells are the photoreceptors in ommatidia.

Image-forming eyes evolved independently in vertebrates and cephalopods

Both vertebrates and cephalopod mollusks have evolved eyes with exceptional abilities to form images of the visual world. Like cameras, these eyes focus images on a surface that is sensitive to light (Figure 45.16). Considering that they evolved independently of each other, their high degree of similarity is remarkable.

The vertebrate eye is a spherical, fluid-filled structure bounded by a tough connective tissue layer called the *sclera*. At the front of the eye, the sclera forms the transparent **cornea**, through which light passes to enter the eye. Just inside the cornea is the pigmented **iris**, which gives the eye its color. The function of the iris is to control the amount of light that reaches the photoreceptor cells at the back of the eye, just as the diaphragm of a camera controls the amount of light reaching the film. The central opening of the iris is the **pupil**. The iris is under neuronal control. In bright light, the iris constricts, and the pupil is very small. As light levels fall, the iris relaxes, and the pupil enlarges.

Behind the iris is the crystalline protein **lens**, which makes fine adjustments in the focus of images on the photosensitive layer, the **retina**, at the back of the eye. The most sensitive area of the retina is called the **fovea**. The cornea and the fluids within the eye are mostly responsible for focusing light on the retina, but the lens allows the eye to *accommodate*—that is, to focus on objects at various locations in the near visual field. To focus a camera on objects close at hand, you adjust the distance between the lens and the film. Fishes, amphibians, and reptiles accommodate in a similar manner, moving the lenses of their eyes closer to or farther from their retinas. Mammals and birds use a different method: They alter the shape of the lens.

The lens is contained in a connective tissue sheath that tends to keep it in a spherical shape, but it is attached to suspensory ligaments that pull it into a flatter shape. Circular muscles called the *ciliary muscles* counteract the pull of the suspensory ligaments and permit the lens to round up. When the ciliary muscles are at rest, the flatter lens has the correct optical properties to focus distant images on the retina, but not close images. Contracting the ciliary muscles rounds up the lens, changing its light-bending properties to bring close images into focus (Figure 45.17). As we age, our lenses become less elastic, and we lose the ability to focus on objects close at hand without the help of corrective lenses. As a consequence, most adults over the age of 45 need the assistance of bifocal lenses or reading glasses to compensate for their lost ability to accommodate.

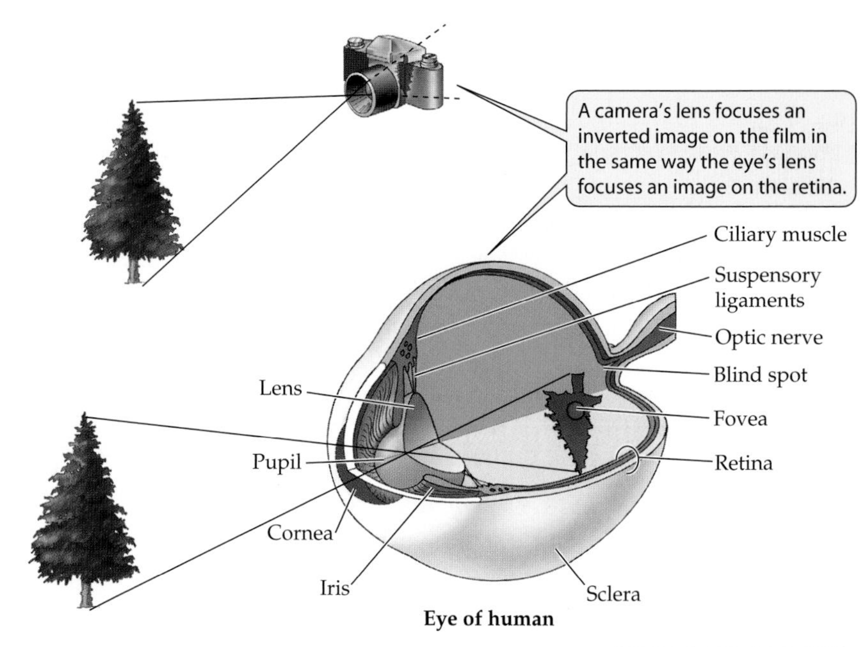

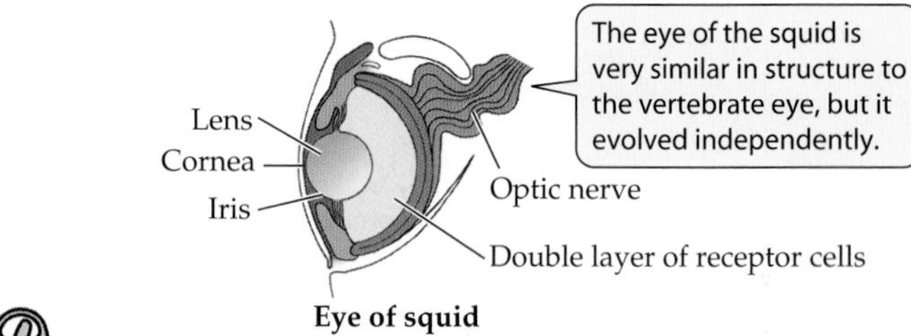

45.16 Eyes Like Cameras The lenses of cephalopod and vertebrate eyes focus images on layers of photoreceptor cells, just as a camera's lens focuses images on film.

The vertebrate retina receives and processes visual information

During embryonic development, neuronal tissue grows out from the brain to form the retina. In addition to a layer of photoreceptor cells, the retina includes four layers of cells that process visual information from the photoreceptors and produce an output signal that is transmitted to the brain via

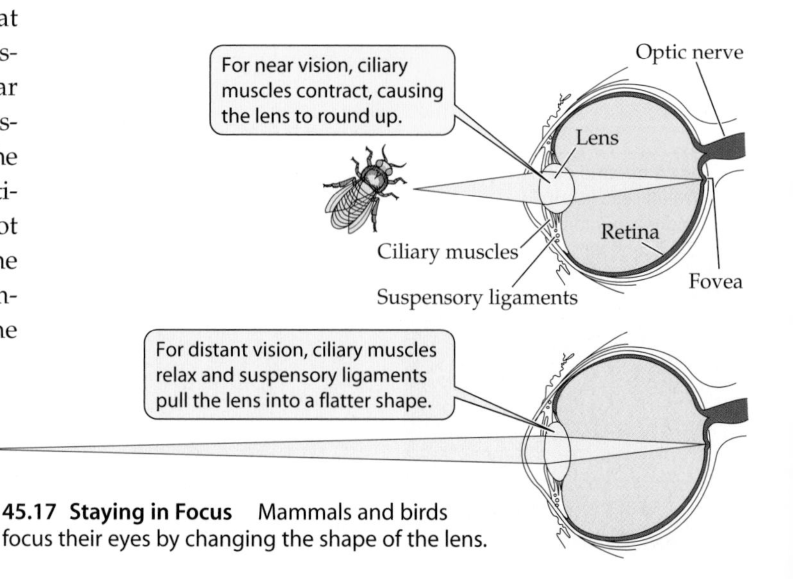

For near vision, ciliary muscles contract, causing the lens to round up.
Optic nerve
Lens
Retina
Ciliary muscles
Suspensory ligaments
Fovea
For distant vision, ciliary muscles relax and suspensory ligaments pull the lens into a flatter shape.

45.17 Staying in Focus Mammals and birds focus their eyes by changing the shape of the lens.

the optic nerve. The light-absorbing outer segments of the photoreceptor cells are all the way at the back of the retina. Light must pass through all the layers of retinal cells before being captured by rhodopsin. The outer segments are partly buried in a layer of pigmented epithelium that absorbs photons not captured by rhodopsin and prevents any backscattering of light that might decrease visual sharpness.

THE PHOTORECEPTORS OF THE RETINA. Until now we have referred to only one kind of photoreceptor cell, the rod cell. But there are two major kinds of vertebrate photoreceptors, both named for their shapes: rod cells and cone cells (Figure 45.18). A human retina has about 5 million cones and about 100 million rods. Rod cells are highly sensitive to light, so they are well suited for vision under dim light, are saturated by daylight conditions, and do not contribute to color vision. **Cone cells** are less sensitive to light, so they are better suited for daylight and color vision. Cones are also responsible for our sharpest vision. Even though there are many more rods than cones in human retinas, our foveas contain only cones.

Because cones have low sensitivity to light, they are of no use in dim light. At night our vision is not very sharp, and we see mostly in shades of gray. You may have trouble seeing a small object such as a keyhole at night when you are looking straight at it—that is, when its image is falling on your fovea. If you look a little to the side, so that the image falls on a rod-rich area of your retina, you can see the object better. Astronomers looking for faint objects in the sky learned this trick a long time ago. Animals that are nocturnal (such as flying squirrels) have retinas containing a high percentage of rods and may have poor color vision. By contrast, some animals that are active only during the day (such as chipmunks) have mostly cones in their retinas.

The human retina has three kinds of cone cells, each containing slightly different isomers of opsin. These opsin molecules differ in the wavelengths of light they absorb best. Although the same 11-*cis*-retinal group is the light absorber in all three kinds of cones (see Figure 45.12), its molecular interactions with opsin determine the spectral sensitivity of the rhodopsin molecule as a whole. One isomer of opsin causes the retinal group to absorb short-wavelength light (e.g., violet and blue) most efficiently; the others result in absorption of middle wavelengths (e.g., green) long wavelengths (e.g., yellow and red) (Figure 45.19).

The density of rods and cones is not the same across the entire retina. In humans, light coming from the center of the visual field falls on the fovea, where the density of cone cells is highest. The human fovea has about 160,000 photoreceptors per square millimeter. A hawk has about a million photoreceptors per square millimeter, making its vision sharper than ours. In addition, the hawk has two foveas in each eye: One receives light from straight ahead, while the other receives light from below. Thus, while the hawk is flying, it sees both its projected flight path and the ground below, where it might detect a mouse scurrying in the grass.

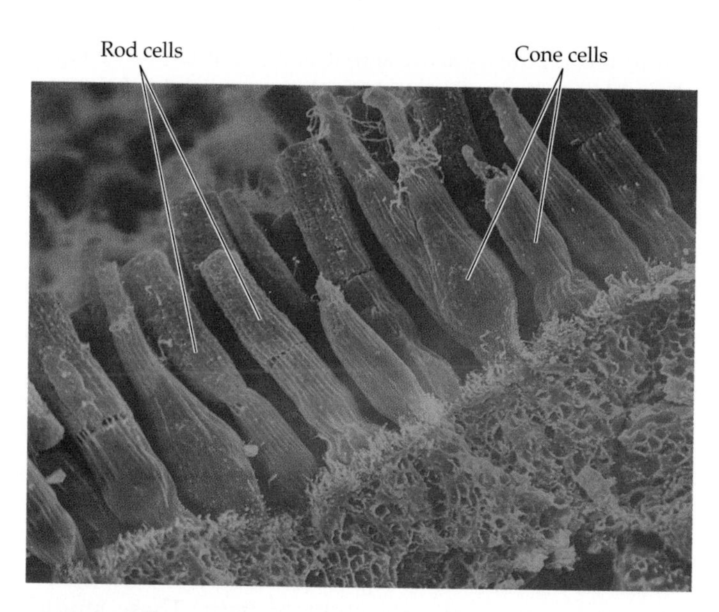

45.18 Rods and Cones This scanning electron micrograph of photoreceptors in the retina of a mud puppy (an amphibian) shows cylindrical rods and tapered cones.

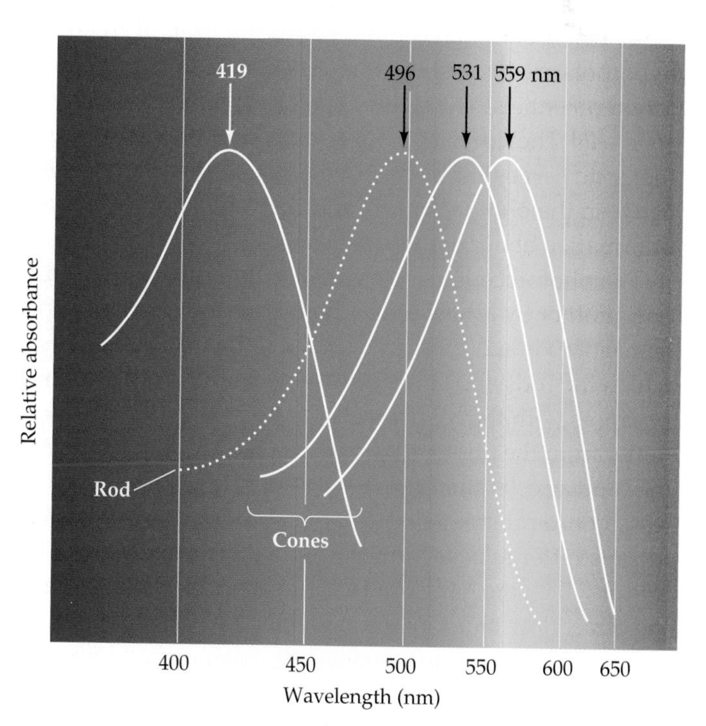

45.19 Absorption Spectra of Cone Cells The three kinds of cone cells contain slightly different opsin molecules, which absorb different wavelengths of light.

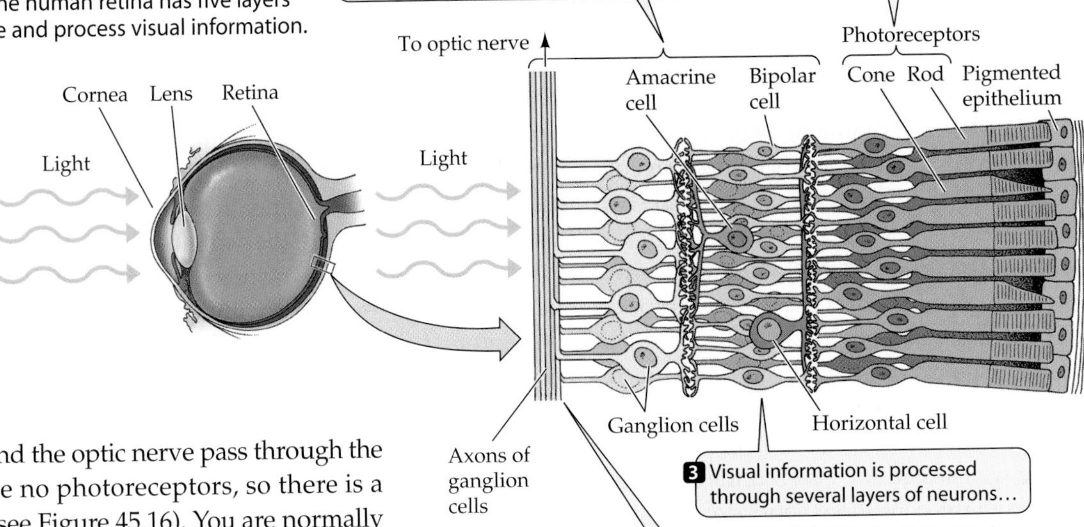
1 Light travels through layers of transparent neurons—ganglion, amacrine, bipolar, and horizontal cells…
2 … and is absorbed by the rods and cones (the photoreceptive layer) at the back of the retina.
To optic nerve
Photoreceptors
Cornea
Lens
Retina
Amacrine cell
Bipolar cell
Cone
Rod
Pigmented epithelium
Light
Light
Ganglion cells
Horizontal cell
Axons of ganglion cells
3 Visual information is processed through several layers of neurons…
4 …and finally converges on ganglion cells, which send their axons to the brain.

45.20 The Retina The human retina has five layers of neurons that receive and process visual information.

Where blood vessels and the optic nerve pass through the back of the eye, there are no photoreceptors, so there is a blind spot on the retina (see Figure 45.16). You are normally not aware of your blind spot, but you can find it. Stare straight ahead, holding a pencil in your outstretched hand so that the eraser is in the center of your field of vision. While continuing to stare straight ahead, slowly move the pencil to the side until the eraser disappears. When this happens, the light from the eraser is focused directly on your blind spot.

INFORMATION FLOW IN THE RETINA. The human retina is organized into five layers of neurons that receive visual information and process it before sending it to the brain (Figure 45.20). A first step in understanding how the retina tells the brain what it is seeing is to study how these layers are interconnected and how they influence one another. As we know, the photoreceptor cells at the back of the retina hyperpolarize in response to light and do not generate action potentials. The cells at the front of the retina are *ganglion cells*. They fire action potentials, and their axons form the optic nerve that travels to the brain. The layers of cells between the photoreceptors and the ganglion cells process information about the visual field.

The photoreceptors and ganglion cells are connected to one another by *bipolar cells*. Changes in the membrane potential of rods and cones in response to light alter the rates at which the rods and cones release neurotransmitter at their synapses with the bipolar cells. In response to neurotransmitter from the photoreceptors, the membrane potentials of the bipolar cells change, altering the rate at which they release neurotransmitter onto ganglion cells. The rate of neurotransmitter release from the bipolar cells determines the rate at which the ganglion cells fire action potentials. Thus, the direct flow of information in the retina is from photoreceptor to bipolar cell to ganglion cell. The ganglion cells send the information to the brain.

The other two cell layers, the horizontal cells and the amacrine cells, communicate laterally across the retina. *Horizontal cells* form synapses with neighboring photoreceptors. Thus, light falling on one photoreceptor can influence the sensitivity of its neighbors to light. This lateral flow of information enables the retina to sharpen the perception of contrast between light and dark patterns.

Amacrine cells form local interconnections between bipolar cells and ganglion cells. The role of amacrine cells is still not entirely understood. Some amacrine cell types are highly sensitive to changing illumination or to motion. Others assist in adjusting the sensitivity of the eyes according to the overall level of light falling on the retina. When background light levels change, amacrine cell connections to the ganglion cells help the ganglion cells remain sensitive to temporal changes in stimulation. Thus, even with large changes in background illumination, the eyes are sensitive to smaller, more rapid changes in the pattern of light falling on the retina.

INFORMATION PROCESSING IN THE RETINA. Knowing the path of information flow through the retina still does not tell us how that information is processed. What does the eye tell the brain in response to a pattern of light falling on the retina? One aspect of information processing in the retina is *convergence of information*. There are more than 100 million photoreceptors in each retina, but only about 1 million ganglion cells sending messages to the brain. How is the information from all those photoreceptors integrated by the ganglion cells?

This question was addressed in some elegant, classic experiments in which electrodes were used to record the activity of single ganglion cells in living animals while their retinas were stimulated with spots of light. These studies revealed that each ganglion cell has a well-defined **receptive field** that consists of a specific group of photoreceptor cells.

Stimulating these photoreceptors with light activates the ganglion cell (Figure 45.21). Information from many photoreceptor cells is integrated in this way to produce a single message to the brain.

The receptive fields of many ganglion cells are circular, but the way a spot of light influences the activity of the ganglion cell depends on where in the receptive field it falls. The receptive field of a ganglion cell can be divided into two concentric areas, called the *center* and the *surround*. There are two kinds of receptive fields, *on-center* and *off-center*. Stimulating the center of an on-center receptive field excites the ganglion cell, and stimulating the surround inhibits it. Stimulating the center of an off-center receptive field inhibits the ganglion cell, and stimulating the surround excites it. Center effects are always stronger than surround effects.

45.21 What Does the Eye Tell the Brain? When the retina is stimulated with dots and rings of light, individual ganglion cells show different responses.

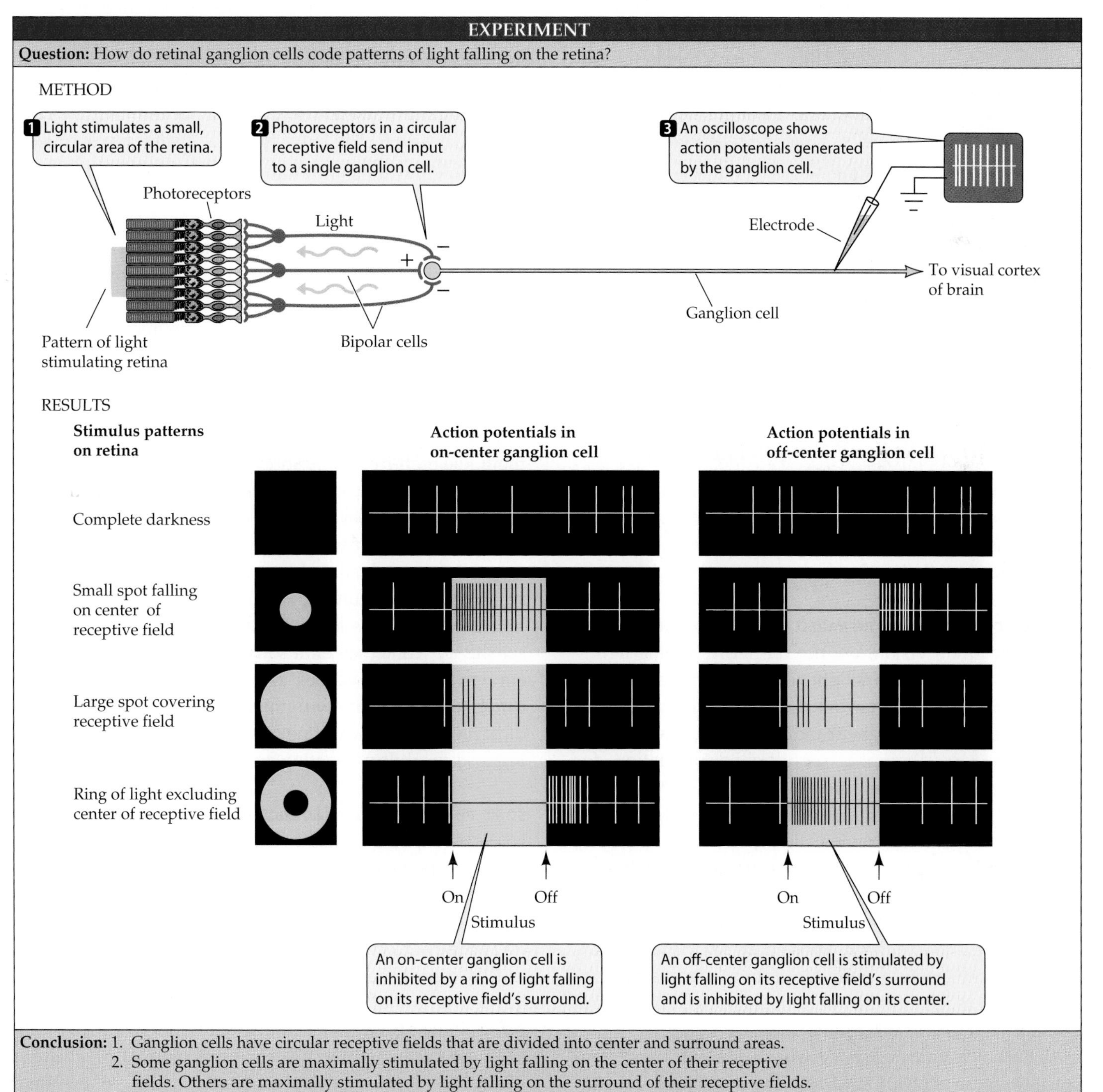
EXPERIMENT
Question: How do retinal ganglion cells code patterns of light falling on the retina?
METHOD
1 Light stimulates a small, circular area of the retina.
2 Photoreceptors in a circular receptive field send input to a single ganglion cell.
3 An oscilloscope shows action potentials generated by the ganglion cell.
Photoreceptors
Light
Electrode
To visual cortex of brain
Ganglion cell
Pattern of light stimulating retina
Bipolar cells
RESULTS
Stimulus patterns on retina
Action potentials in on-center ganglion cell
Action potentials in off-center ganglion cell
Complete darkness
Small spot falling on center of receptive field
Large spot covering receptive field
Ring of light excluding center of receptive field
On
Off
Stimulus
On
Off
Stimulus
An on-center ganglion cell is inhibited by a ring of light falling on its receptive field's surround.
An off-center ganglion cell is stimulated by light falling on its receptive field's surround and is inhibited by light falling on its center.
Conclusion: 1. Ganglion cells have circular receptive fields that are divided into center and surround areas.
2. Some ganglion cells are maximally stimulated by light falling on the center of their receptive fields. Others are maximally stimulated by light falling on the surround of their receptive fields.
3. Ganglion cells encode patterns of contrast between light and dark.

The response of a ganglion cell to stimulation of the center of its receptive field depends on how much of the surround is also stimulated. A small dot of light directly on the center has the maximal effect. A bar of light hitting the center and parts of the surround has less of an effect, and a large, uniform patch of light falling equally on center and surround has very little effect. Ganglion cells thus communicate information about contrasts between light and dark that fall on different regions of their receptive fields.

How are receptive fields related to the connections among the neurons of the retina? The photoreceptors in the center of the receptive field of a ganglion cell are connected to that ganglion cell by bipolar cells. The photoreceptors in the surround modify the communication between the center photoreceptors and their bipolar cells through the lateral connections of horizontal cells. Thus the receptive field of a ganglion cell consists of a pattern of synapses among photoreceptors, horizontal cells, bipolar cells, and ganglion cells.

The receptive fields of neighboring ganglion cells can overlap greatly; a given photoreceptor can be effectively connected to several ganglion cells. Thus the ganglion cells send simple messages to the brain about the pattern of light intensities falling on small, circular patches of retina. In Chapter 47 we will see how the brain reassembles that information into our view of the world.

Sensory Worlds Beyond Human Experience

Humans make use of only a subset of the information available to us in the environment. Other animals have sensory systems that enable them to use different subsets and different types of information.

SOME SPECIES CAN SEE INFRARED AND ULTRAVIOLET LIGHT. When discussing vision, we use the term "visible light," but what we really mean is light visible to humans. Our visible spectrum is a very narrow region of the entire, continuous range of electromagnetic radiation in the environment (see Figure 8.5). We cannot see ultraviolet radiation, for example, but many other animals can.

One of the eight photoreceptors in each ommatidium of a fruit fly is sensitive to ultraviolet light. The visual sensitivity of many pollinating insects includes the ultraviolet part of the spectrum. Some flowers have patterns that are invisible to us, but show up if we photograph them with film that is sensitive to ultraviolet light. Those patterns provide information to prospective pollinators, but humans are not equipped to receive that information (though some other mammalian species, such as mice, are).

At the other end of the spectrum is infrared radiation, which we sense as heat. Other animals extract much more information from infrared radiation—especially that emitted by potential prey. Pit vipers such as rattlesnakes have *pit organs*, one just in front of each eye, that use highly sensitive heat detectors and a simple pinhole camera arrangement to sense and locate infrared radiation. In total darkness, these snakes can locate a mouse, orient to it, and strike it with great accuracy.

ELEPHANTS CAN COMMUNICATE WITH INFRASOUND. Just as there are wavelengths of light beyond our visual capabilities, there are frequencies of sound that we cannot hear. A researcher observing elephants had the sensation that there was a throbbing in the air around her. When she obtained microphones and recorders capable of picking up very low frequencies of sound, she discovered that the elephants were communicating with sounds below the range of human hearing. The advantage of using low-frequency sound is that it carries over very long distances, so elephants in the wild can communicate even when very far apart.

ECHOLOCATION IS SENSING THE WORLD THROUGH REFLECTED SOUND. Some species emit intense sounds that are above the range of human hearing, and they use the echoes from those sounds to create images of their environments. Bats, porpoises, and dolphins are all accomplished *echolocators*. Some species of bats have elaborate modifications of their noses to direct the sounds they emit, as well as impressive ear pinnae to collect the returning echoes. Although we cannot hear them, the high-frequency sounds they emit as pulses (about 20 to 80 per second) are extremely loud in contrast to the resulting faint echoes bouncing off small insects. An echolocating bat is similar to a construction worker who is trying to overhear a whispered conversation on a street corner while using a pneumatic drill. To avoid deafening themselves, bats use muscles in their middle ears to dampen their sensitivity while they emit sounds, then relax them quickly enough to hear the echoes. The ability of bats to use echolocation to sense their environment is so good that in a totally dark room strung with fine wires, they can capture tiny flying insects while navigating around the wires.

SOME FISH CAN SENSE ELECTRIC FIELDS. The lateral lines of some fish species, such as sharks and rays, contain electroreceptors as well as mechanoreceptors. These sensory cells enable the fish to detect weak electric fields, which can help them locate prey.

The use of electroreceptors is quite sophisticated in species called electric fishes. These fishes have evolved electric organs in their tails that generate a continuous series of electric pulses, creating a weak electric field around their bodies. Any objects in the environment, such as rocks, plants, or other fish, disrupt the electric field, and the electroreceptors of the lateral line detect those disruptions. Such electroreceptors allow electric fishes to find food, avoid obstacles, and locate mates in extremely murky waters, such as those of the Amazon River. Electroreception can also play a role in social interactions. In some electric fish species, each individual in a group emits its electric pulses at a different frequency. If a new fish is added to the group, they all readjust their frequencies.

Chapter Summary

Sensory Cells and Transduction of Stimuli

▶ Sensory cells transduce information about an animal's external and internal environment into action potentials.

▶ The interpretation of action potentials as particular sensations depends on which neurons in the CNS receive them.

▶ Sensory cells have membrane receptor proteins that cause ion channels to open or close, generating receptor potentials. **Review Figures 45.1, 45.14**

▶ Receptor potentials can spread to regions of the sensory cell plasma membrane that generate action potentials, or they can influence the release of neurotransmitters from the sensory cell. **Review Figure 45.2**

▶ Adaptation enables the nervous system to ignore irrelevant or continuous stimuli while remaining responsive to relevant or new stimuli.

Chemoreceptors: Responding to Specific Molecules

▶ Chemoreceptors are responsible for smell, taste, and the sensing of pheromones.

▶ Olfactory sensory cells contain receptor proteins that can bind a specific molecule or ion. The binding of an odorant molecule to a receptor protein causes the production of a second messenger, which opens ion channels and creates an action potential in the sensory cell. **Review Figure 45.4**

▶ Taste buds in the mouth cavities of vertebrates are responsible for the sense of gustation. **Review Figure 45.5**

Mechanoreceptors: Detecting Stimuli that Distort Membranes

▶ The skin contains a variety of mechanoreceptors that respond to touch and pressure. The density of mechanoreceptors in any skin area determines the sensitivity of that area. **Review Figure 45.6**

▶ Stretch receptors in muscles, tendons, and ligaments inform the CNS of the positions of and loads on parts of the body. **Review Figure 45.7**

▶ Hair cells are also mechanoreceptors. The bending of their stereocilia alters receptor proteins and therefore their membrane potentials. Hair cells are found in the auditory organs and organs of equilibrium such as the lateral line system of fishes and the semicircular canals and vestibular apparatus of mammals. **Review Figures 45.8, 45.9**

▶ In mammalian auditory systems, ear pinnae collect and direct sound waves to the tympanic membrane, which vibrates in response to sound waves. The movements of the tympanic membrane are amplified through a chain of ossicles that conduct the vibrations to the oval window. Movements of the oval window create pressure waves in the fluid-filled cochlea. **Review Figure 45.10.** See Web/CD Activity 45.1 and Tutorial 45.1

▶ The basilar membrane running down the center of the cochlea is distorted by sound waves at specific locations that depend on their frequency. These distortions cause the bending of hair cells in the organ of Corti, which rests on the basilar membrane. Receptor potentials in hair cells cause them to release neurotransmitter, which creates action potentials in the auditory nerve, which conducts the information to the CNS. **Review Figure 45.11**

Photoreceptors and Visual Systems: Responding to Light

▶ Photosensitivity depends on the absorption of photons of light by rhodopsin, a photoreceptor molecule that consists of a protein called opsin and a light-absorbing prosthetic group called retinal. Absorption of light by retinal is the first step in a cascade of intracellular events leading to a change in the membrane potential of the photoreceptor cell. **Review Figure 45.12**

▶ When excited by light, vertebrate photoreceptor cells hyperpolarize and release less neurotransmitter onto the neurons with which they form synapses. They do not fire action potentials. **Review Figures 45.13, 45.14**

▶ Visual systems vary from the simple eye cups of flatworms, which enable the animal to sense the direction of a light source, to the compound eyes of arthropods, which enable the animal to detect shapes and patterns, to the image-forming eyes of cephalopods and vertebrates. **Review Figures 45.15, 45.16.** See Web/CD Activity 45.2

▶ The image-forming eyes of vertebrates focus detailed images of the visual field onto dense arrays of photoreceptors that transduce the visual image into neuronal signals. **Review Figure 45.17**

▶ Vertebrates have two types of photoreceptors, rod cells and cone cells. In humans, the fovea contains almost exclusively cone cells, which are responsible for color vision but are not very sensitive in dim light.

▶ Color vision is based on the fact that different cone cells contain different isomers of opsin, which give them different spectral absorption properties. **Review Figure 45.19**

▶ The vertebrate retina consists of five layers of neurons lining the back of the eye. The light-absorbing photoreceptor cells are at the back of the retina. **Review Figure 45.20.** See Web/CD Activity 45.3

▶ The innermost layer of the retina consists of the ganglion cells, which send their axons in the optic nerve to the brain. Between the photoreceptors and the ganglion cells are neurons that process information from the photoreceptors.

▶ Each ganglion cell is stimulated by light falling on a small circular patch of photoreceptors called a receptive field. Receptive fields have a center and a surround, which have opposing effects on the ganglion cell. If the center is excitatory, the surround is inhibitory, and vice versa. **Review Figure 45.21** See Web/CD Tutorial 45.2

Sensory Worlds beyond Human Experience

▶ Many animals have sensory abilities that humans do not share. Insects see ultraviolet radiation, pit vipers "see" infrared radiation, elephants communicate with low-frequency sound, bats echolocate, and some fishes sense electric fields.

Self-Quiz

1. Which statement about sensory systems is *not* true?
   a. Sensory transduction involves the conversion (direct or indirect) of a physical or chemical stimulus into changes in membrane potentials.

b. In general, a stimulus causes a change in the flow of ions across the plasma membrane of a sensory cell.
c. The term "adaptation" refers to the process by which a sensory system becomes insensitive to a continuing source of stimulation.
d. The more intense a stimulus, the greater the magnitude of each action potential fired by a sensory neuron.
e. Sensory adaptation plays a role in the ability of organisms to discriminate between important and unimportant information.

2. The female silkworm moth releases a chemical called bombykol from a gland at the tip of her abdomen. Bombykol is
a. a sex hormone.
b. detected by the male only when present in large quantities.
c. not species-specific.
d. detected by hairs on the antennae of male silkworm moths.
e. a chemical basic to the taste process in arthropods.

3. Which statement about olfaction is *not* true?
a. Dogs are unusual among mammals in that they depend more on olfaction than on vision as their dominant sensory modality.
b. Olfactory stimuli are recognized by the interaction between an odorant molecule and a specific receptor protein on olfactory hairs.
c. The more odorant molecules that bind to receptors, the more action potentials are generated.
d. The greater the number of action potentials generated by an olfactory receptor, the greater the intensity of the perceived smell.
e. The perception of different smells results from the activation of different combinations of olfactory receptors.

4. The touch receptors located very close to the skin surface
a. are relatively insensitive to light touch.
b. adapt very quickly to stimuli.
c. are uniformly distributed throughout the surface of the body.
d. are called Pacinian corpuscles.
e. adapt slowly and only partially to stimuli.

5. The membrane that gives us the ability to discriminate different pitches of sound is the
a. round window.
b. oval window.
c. tympanic membrane.
d. tectorial membrane.
e. basilar membrane.

6. Which statement is *not* true?
a. The transmembrane potential of a rod cell becomes more negative when the rod cell is exposed to light after a period of darkness.
b. A photoreceptor releases the most neurotransmitter when in total darkness.
c. Whereas in vision the intensity of a stimulus is encoded by the degree of hyperpolarization of photoreceptors, in hearing the intensity of a stimulus is encoded by changes in firing rates of sensory cells.
d. Stiffening of the ossicles in the middle ear can lead to deafness.
e. The interaction among hammer (malleus), anvil (incus), and stirrup (stapes) conducts sound waves across the fluid-filled middle ear.

7. In humans, the region of the retina where the central part of the visual field falls is called the
a. central ganglion cell.
b. fovea.
c. optic nerve.
d. cornea.
e. pupil.

8. The region of the vertebrate eye where the optic nerve passes out of the retina is called the
a. fovea.
b. iris.
c. blind spot.
d. pupil.
e. visual cortex.

9. Which statement about the cone cells in a human eye is *not* true?
a. They are responsible for our sharpest vision.
b. They are responsible for color vision.
c. They are more sensitive to light than rods are.
d. They are fewer in number than rods.
e. They exist in high numbers at the fovea.

10. The color in color vision results from the
a. ability of each cone cell to absorb all wavelengths of light equally.
b. lens of the eye acting like a prism and separating the different wavelengths of light.
c. different absorption of wavelengths of light by different kinds of rod cells.
d. three different isomers of opsin in cone cells.
e. absorption of different wavelengths of light by amacrine and horizontal cells.

For Discussion

1. Compare and contrast the functioning of olfactory receptors and photoreceptors. How do these sensory cells enable the CNS to discriminate between an apple and an orange?

2. Amplification of signal is an important feature of sensory systems. Compare mechanisms of amplification in olfactory, visual, and auditory systems.

3. If you were blindfolded and placed in a wheelchair, how would you know if you were being pushed forward or backward?

4. Describe and contrast two sensory systems that enable animals to "see" in the dark. What problems or limitations are inherent in these systems in comparison with vision?

5. Animals can use visual, olfactory, tactile, and auditory signals to communicate. From what you know about these sensory systems, discuss the relative advantages and disadvantages of these systems for communication.

46 The Mammalian Nervous System: Structure and Higher Functions

A "Far Side" cartoon by Gary Larson shows a classroom with pupils at their desks. One pupil with a noticeably small head and his hand raised says, "Teacher, may I be excused? My head is full." This amusing cartoon suggests a deep question: What is the capacity of the human brain? Is it limited by size, number of neurons, number of synapses?

Since the earliest studies of the cellular structure of the brain, the dogma has been that we are born with a certain number of neurons, we steadily lose neurons throughout life, and we do not get any new neurons. The evidence: In adult brains, no neurons with mitotic structures, or neurons at different stages of maturation, are visible. The explanation for our lifelong ability to learn has been that we have enormous potential to form and re-form synapses. Synaptic plasticity is certainly an important property of the nervous system; however, one of the most exciting and revolutionary recent discoveries in neurobiology is that new neurons are formed in adult avian and mammalian brains, and that the formation of these new neurons seems to be stimulated by experience and learning.

The birth of new cells can be revealed by giving an animal an injection of radioactively labeled thymidine, which becomes incorporated into new DNA. The surprise discovery was the appearance of some labeled neurons when adult rats were injected. Lots of reasons were generated as to why they could not really be new neurons, and few scientists gave up the old dogma. This debate received much more attention when Fernando Nottebohm and his colleagues at Rockefeller University showed that new neurons are formed in those parts of the bird brain responsible for song at the time of the year when birds come into reproductive condition. Furthermore, they showed that sex hormones and hearing song stimulated the birth of new neurons. (We will return to this story in Chapter 52.)

The exciting results from bird brains gave new impetus to mammalian studies. Two parts of the adult mammalian brain have now been shown to acquire new neurons. One is the olfactory bulb, which is not surprising, since the olfactory bulb neurons extend into the nasal epithelium, which is regularly shed. The other area is the hippocampus, which, as we will learn in this chapter, is a region involved in the formation of long-term memories. As in birds, experience and learning stimulate neurogenesis in the mammalian hippocampus. For those who still doubt, Fred Gage and his colleagues at the Salk Institute have recorded action potentials from new hippocampal neurons and have

New Neurons Are Born in an Adult Mouse An adult mouse was injected with viral particles that carried a gene for green fluorescent protein (GFP). The genetic material from the virus was taken up and subsequently expressed by dividing cells. When the mouse brain was examined two weeks later, it contained neurons labeled with GFP—neurons that must have arisen *after* the viral injection. Mice that were allowed to run in wheels or mazes produced even greater numbers of labeled (new) neurons.

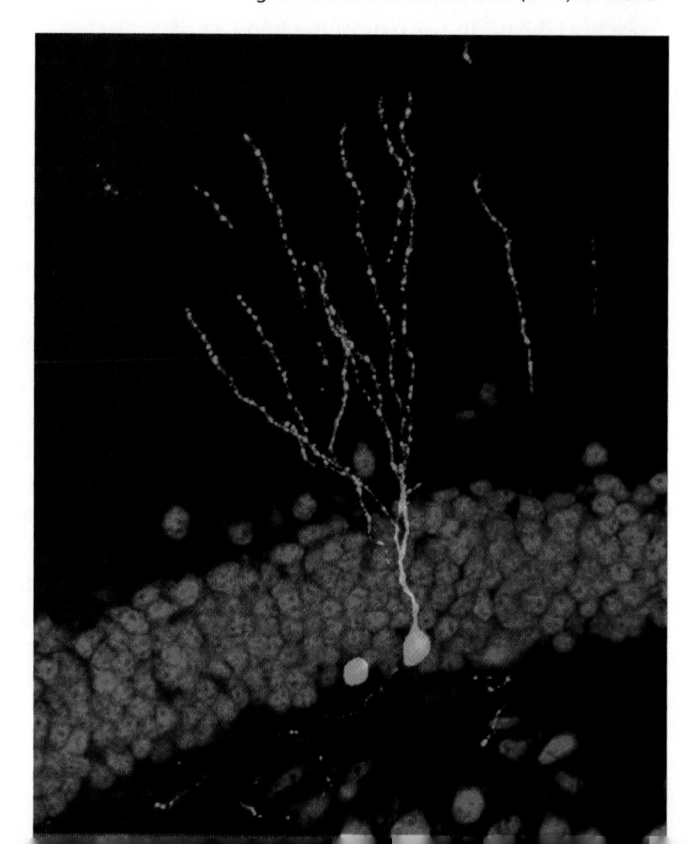

shown that they mature into functional neurons with properties identical to those of older neighboring cells.

Many new questions arise from these studies. Could neurogenesis be stimulated in other parts of the nervous system to repair damage and to counter the effects of aging? Since brains don't continue to get bigger, is there regular loss and replacement of neurons? Does loss of neurons contribute to forgetting? Do new neurons facilitate learning? Will you generate new neurons by studying this chapter?

The unit of function of the brain is the neuron. The human brain consists of about 100 billion neurons, which account for its ability to handle vast amounts of information. In the previous two chapters we learned about the cellular properties of neurons. In this chapter we take on the challenge of understanding some functions of the human nervous system in terms of these cellular mechanisms.

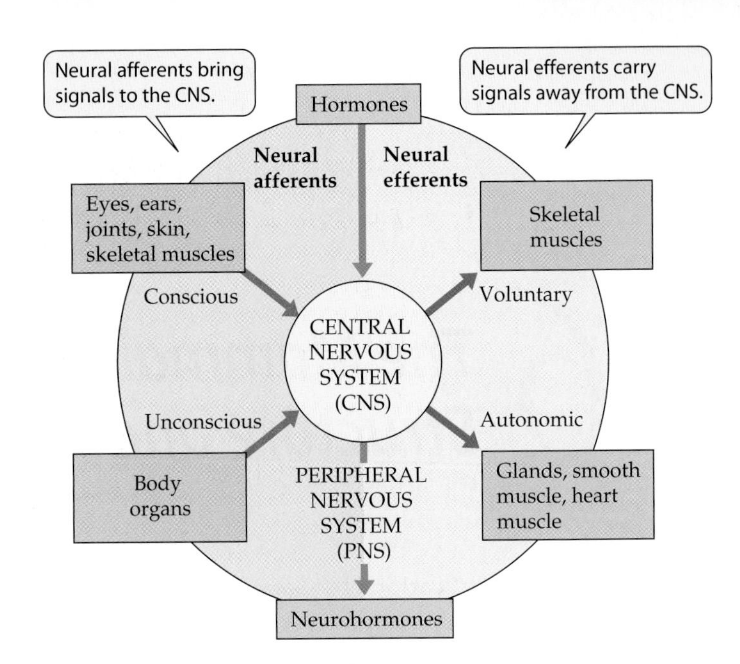

46.1 Organization of the Nervous System The peripheral nervous system (indicated by pink and blue) carries information both to and from the central nervous system. The CNS also receives hormonal inputs and produces hormonal outputs.

The Nervous System: Structure, Function, and Information Flow

The human nervous system consists of three major components. The brain and spinal cord together constitute the **central nervous system** (**CNS**). Information is transmitted to and from the CNS by means of an enormous network of nerves that make up the **peripheral nervous system** (**PNS**). The PNS reaches every tissue of the body. It connects to the CNS via *spinal nerves* and *cranial nerves*.

A **nerve** is a bundle of axons that carries information about many things simultaneously. It is important to distinguish between the axon of a single neuron and a nerve. Some axons in a nerve may be carrying information to the CNS, while other axons in the same nerve are carrying information from the CNS to the organs of the body.

A conceptual diagram of the nervous system traces information flow

The major avenues of information flow through the nervous system are illustrated in Figure 46.1. The **afferent** portion of the peripheral nervous system carries information to the CNS. We are consciously aware of much of the information that moves through these afferent pathways (for example, vision, hearing, temperature, pain, the position of limbs). We are not consciously aware of other afferent information that is important for physiological regulation (for example, blood pressure, deep body temperature, blood oxygen supply).

The **efferent** portion of the peripheral nervous system carries information from the CNS to the muscles and glands of the body. Efferent pathways can be divided into a *voluntary* division, which executes our conscious movements, and an *involuntary*, or **autonomic**, division, which controls physiological functions.

In addition to neuronal information, the CNS receives chemical information in the form of hormones circulating in the blood. *Neurohormones* released by neurons into the extracellular fluids of the brain can send chemical information to other neurons in the brain or can leave the brain and enter the circulation. In Chapter 42 we learned of the important role of neurohormones (such as GnRH) in the control of the anterior pituitary, and we saw that other neurohormones (such as oxytocin) are released from the posterior pituitary into the circulation.

The vertebrate CNS develops from the embryonic neural tube

Early in the development of all vertebrate embryos, a hollow tube of neural tissue forms. This *neural tube* runs the length of the embryo on its dorsal side. At the anterior end of the embryo, the neural tube forms three swellings that become the basic divisions of the brain: the **hindbrain**, the **midbrain**, and the **forebrain**. The rest of the neural tube becomes the spinal cord (Figure 46.2). The cranial and spinal nerves, which make up the peripheral nervous system, sprout from the neural tube and grow throughout the embryo.

Each of the three regions of the embryonic brain develops into several structures in the adult brain. From the hindbrain come the **medulla**, the **pons**, and the **cerebellum**. The medulla is continuous with the spinal cord. The pons is in front of the medulla, and the cerebellum is a dorsal outgrowth of the pons. The medulla and pons contain distinct groups of neurons that are involved in the control of physi-

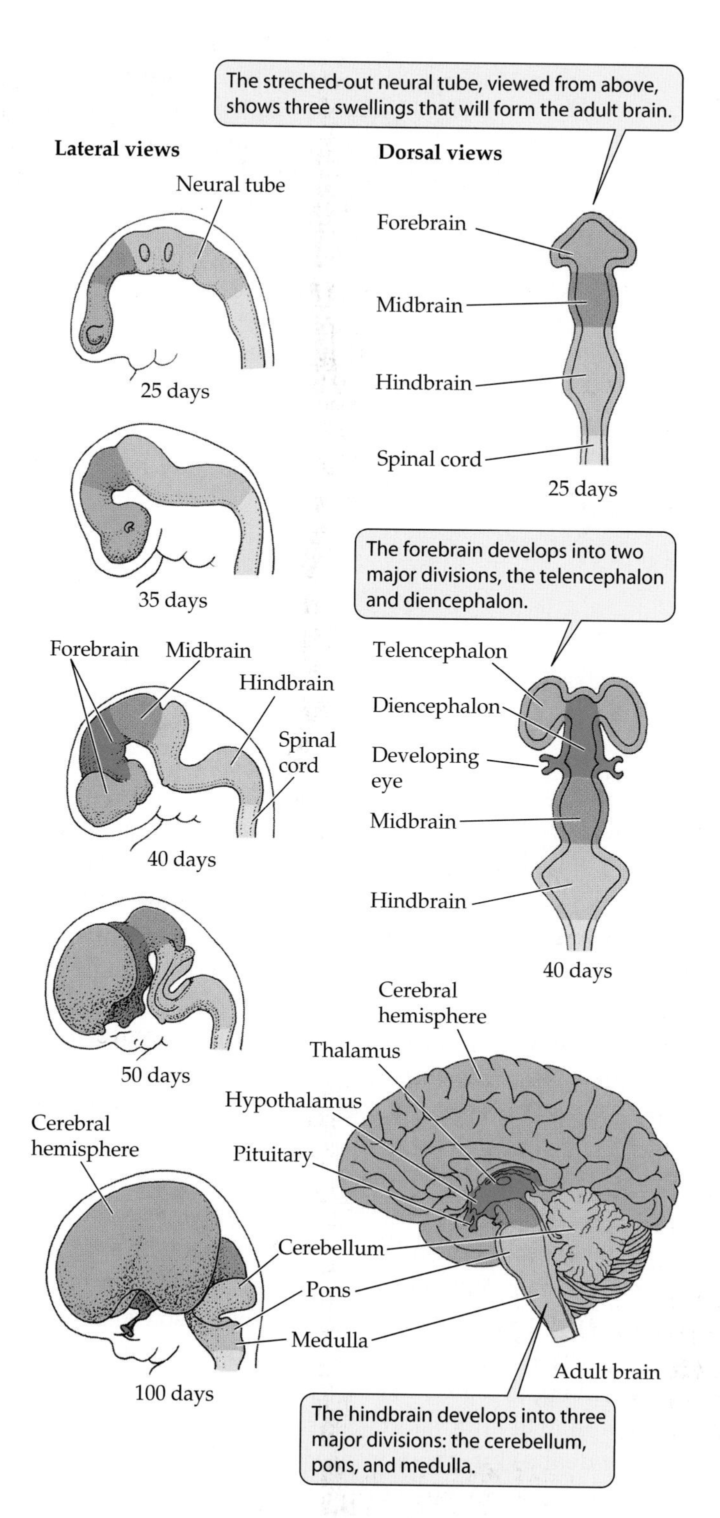

46.2 Development of the Human Nervous System Three swellings at the anterior end of the hollow neural tube in the early vertebrate embryo develop into the parts of the adult brain. The final view is an adult human brain cut in half through the midline.

ological functions such as breathing and circulation or basic motor patterns such as swallowing and vomiting. All information traveling between the spinal cord and higher brain areas must pass through the pons and the medulla.

The cerebellum is like the conductor of an orchestra; it receives "copies" of the commands going to the muscles from higher brain areas, and it receives information coming up the spinal cord from the joints and muscles. Thus it can compare the motor "score" with the actual behavior of the muscles and refine the motor commands.

From the embryonic midbrain come structures that process aspects of visual and auditory information. In addition, all information traveling between higher brain areas and the spinal cord must pass through the midbrain. The structures that develop from the hindbrain and the midbrain are collectively known as the **brain stem**.

The embryonic forebrain develops a central region called the **diencephalon** and a surrounding structure called the **telencephalon**. The diencephalon is the core of the forebrain and consists of an upper structure called the **thalamus** and a lower structure called the **hypothalamus**. The thalamus is the final relay station for sensory information going to the telencephalon, and the hypothalamus is responsible for the regulation of many physiological functions and biological drives.

The telencephalon consists of two **cerebral hemispheres**, left and right (and is also referred to as the **cerebrum**). In humans, the telencephalon is by far the largest part of the brain and plays major roles in sensory perception, learning, memory, and conscious behavior.

As we go up the vertebrate phylogenetic scale from fishes to mammals, the telencephalon increases in size, complexity, and importance. The forebrain dominates the nervous systems of mammals, and damage to this region results in severe impairment of sensory, motor, or cognitive functions, and even coma. In contrast, a shark with its telencephalon removed can swim almost normally.

Functional Subsystems of the Nervous System

We have just surveyed the development of the nervous system in terms of anatomically distinct structures. At any one time, these various structures are engaged in many simultaneous tasks—a property known as *parallel processing* of information. Specific tasks are carried out by subsystems that may involve several different anatomical regions or structures of the nervous system. We will now examine several of these functional subsystems.

The spinal cord receives and processes information from the body

The spinal cord conducts information in both directions between the brain and the organs of the body. It also integrates a great deal of the information coming from the peripheral nervous system, and it responds to that information by issuing motor commands.

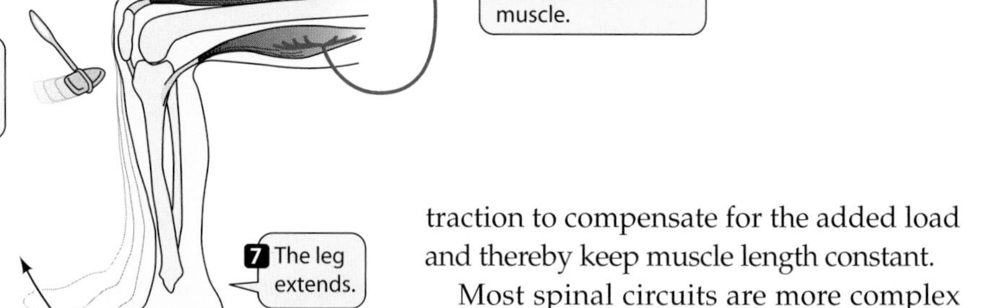

Gray matter
White matter
Dorsal root (afferent nerves)
Dorsal horn
Ventral horn
Ventral root (efferent nerves)
Motor neurons
1 A hammer tap stretches the tendon in the knee, stretching a receptor in the extensor muscle.
2 A stretch receptor fires an action potential.
3 In a monosynaptic pathway, the sensory neuron synapses with a motor neuron in the ventral horn of the spinal cord.
4 The motor neuron conducts an action potential to the extensor muscle, causing contraction.
5 In this polysynaptic pathway, an action potential travels from the sensory neuron via a spinal interneuron…
6 …that inhibits the motor neuron of the antagonistic flexor muscle.
7 The leg extends.

46.3 The Spinal Cord Processes Information Sensory (afferent) information enters the spinal cord through the dorsal horns (red pathway), and motor (efferent) output leaves it via the ventral horns (blue pathways). The extensor component of the knee-jerk response is a monosynaptic reflex circuit, but the flexor inhibition component involves a spinal interneuron (green).

A cross section of the spinal cord reveals a central area of gray matter in the shape of a butterfly, surrounded by an area of white matter (Figure 46.3). In the nervous system, **gray matter** is tissue rich in neuronal cell bodies, and **white matter** contains axons. The gray matter of the spinal cord contains the cell bodies of the spinal neurons; the white matter contains the axons that conduct information up and down the spinal cord. Spinal nerves extend from the spinal cord at regular intervals on each side. Each spinal nerve has two roots, one connecting with the *dorsal horn* of the gray matter, and the other connecting with the *ventral horn*. Each spinal nerve carries both afferent and efferent information. The afferent axons enter the spinal cord through the *dorsal root*, and the efferent axons leave the spinal cord through the *ventral root*.

The conversion of afferent to efferent information in the spinal cord without participation of the brain is called a **spinal reflex**. The simplest type of spinal reflex involves only two neurons and one synapse and is therefore called a **monosynaptic reflex**. An example is the knee-jerk reflex, which your physician checks with a mallet tap just below your knee. We can diagram the wiring of a monosynaptic reflex by following the flow of information through the spinal cord.

In the case of the knee-jerk reflex, sensory information comes from stretch receptors in the leg muscle that is suddenly stretched when the mallet strikes the tendon that runs over the knee. Each stretch receptor initiates action potentials that are conducted by the axon of a sensory neuron through the dorsal horn of the spinal cord and all the way to the ventral horn. In the ventral horn, the sensory neuron synapses with motor neurons, causing them to fire action potentials that are then conducted back to the leg extensor muscle, causing it to contract. The function of this simple circuit is to sense an increased load on the limb and to increase the strength of muscle contraction to compensate for the added load and thereby keep muscle length constant.

Most spinal circuits are more complex than this monosynaptic reflex, as we can demonstrate by building on the circuit we have just traced. Limb movement is controlled by *antagonistic* sets of muscles—muscles that work against each other. When one member of an antagonistic set of muscles contracts, it bends, or flexes, the limb; it is therefore called a *flexor*. The antagonist to this muscle straightens, or extends, the limb, and is called an *extensor*. For a limb to move, one muscle of the pair must relax while the other contracts. Thus, sensory input that activates the motor neuron of one muscle also inhibits its antagonist. This coordination is achieved by an **interneuron**, which makes an inhibitory synapse onto the motor neuron of the antagonistic muscle (see Figure 46.3). Thus the reciprocal inhibition of antagonistic muscles involves an interneuron between the sensory cell and the motor neuron of the inhibited muscle, and therefore at least two synapses.

Information entering the dorsal horn is also transmitted by axons up the spinal cord to the brain. We are aware of the mallet hitting the knee, but the reflex response actually begins before that information registers in our consciousness. A great deal of information processing takes place in the spinal cord without any input from the brain. Spinal circuits can even generate repetitive motor patterns, such as the swimming movements of the shark that had its telencephalon removed.

The reticular system alerts the forebrain

Sensory information ascending the spinal cord to final destinations in the forebrain passes through the brain stem. Many sensory fibers give off collateral branches that form synapses with a network of brain stem neurons called the **reticular sys-**

tem. The reticular system is a highly complex network of axons and dendrites. Within the reticular system are many discrete groups of neurons. Such an anatomically distinct group of neurons in the CNS is called a **nucleus** (not to be confused with the nucleus of a single cell).

The reticular system is distributed through the core of the medulla, pons, and midbrain. Afferent information passes through the reticular system, where many connections are made to neurons involved in controlling many functions of the body. Information from joints and muscles, for example, is directed to nuclei in the pons and cerebellum that are involved in balance and coordination, whereas information from pain receptors is directed to nuclei that control sensitivity to pain. This information continues upward to the forebrain, where it results in conscious sensations that can be localized to the specific sites in the body where the information originated.

The information routed through the reticular system also influences the level of arousal of the nervous system. Nuclei in the reticular system are involved in the control of sleep and waking. High levels of activity in the reticular system influence these nuclei to maintain the brain in a waking condition; low levels of activity enable sleep. Because of the alerting function of the reticular core of the brain stem, it has been called the *reticular activating system*.

If the brain of a person is damaged at midbrain or higher levels and the alerting action of the reticular system cannot reach the forebrain, the person loses the ability to be in a conscious, waking state and becomes comatose. Damage to the brain stem or the spinal cord below the reticular system does not interfere with the ascending alerting actions of the reticular system and leaves the person with normal patterns of sleep and waking. However, such damage can cause loss of sensation and loss of motor function.

The limbic system supports basic functions of the forebrain

The telencephalon of fishes, amphibians, and reptiles consists of only a few structures surrounding the diencephalon. In birds and mammals, these primitive forebrain structures are completely covered by the evolutionarily more recent elaborations of the telencephalon called the *neocortex*, but these primitive forebrain structures still have important functions. These structures are collectively referred to as the **limbic system** (Figure 46.4).

The limbic system is responsible for basic physiological drives, instincts, and emotions. Within the limbic system are areas that, when stimulated with small electric currents, can cause intense sensations of pleasure, pain, or rage. If a rat is given the opportunity to stimulate its own pleasure centers by pressing a switch, it will ignore food, water, and even sex, pushing the switch until it is exhausted. Pleasure and pain

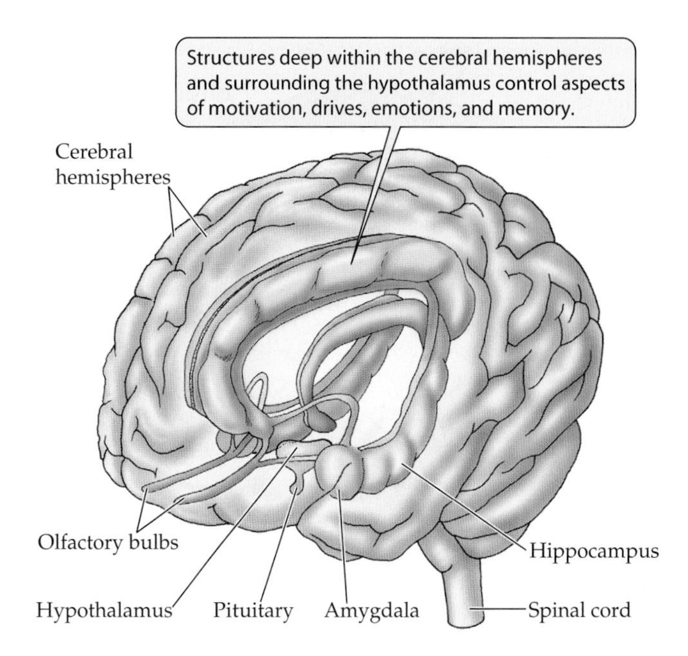
Structures deep within the cerebral hemispheres and surrounding the hypothalamus control aspects of motivation, drives, emotions, and memory.
Cerebral hemispheres
Olfactory bulbs
Hippocampus
Hypothalamus
Pituitary
Amygdala
Spinal cord

46.4 The Limbic System The evolutionarily primitive parts of the telencephalon (blue) are referred to as the limbic system.

centers in the limbic system are believed to play roles in learning and in physiological drives.

A component of the limbic system, the **amygdala**, is involved in fear and fear memory. If a certain portion of the amygdala is damaged or chemically blocked, an animal cannot learn to be afraid of a stimulus or a situation that would normally induce a strong fear reaction. Moreover, blocking protein synthesis in this part of the limbic system blocks the formation of fear memory.

Another part of the limbic system, the **hippocampus**, is necessary in humans for the transfer of short-term memory to long-term memory. If you are told a new telephone number, you may be able to hold it in short-term memory for a few minutes, but within half an hour it is forgotten unless you make a real effort to remember it. The phenomenon of remembering something for more than a few minutes requires its transfer to long-term memory.

Regions of the cerebrum interact to produce consciousness and control behavior

The cerebral hemispheres are the dominant structures in the mammalian brain. In humans, they are so large that they cover all other parts of the brain except the cerebellum (Figure 46.5*a*). A sheet of gray matter called the **cerebral cortex** covers each cerebral hemisphere. It is about 4 mm thick and covers a total surface area over both hemispheres of 1 square meter. The cerebral cortex is convoluted, or folded, into ridges called *gyri* (singular, *gyrus*) and valleys called *sulci* (singular, *sulcus*). These convolutions allow it to fit into the skull.

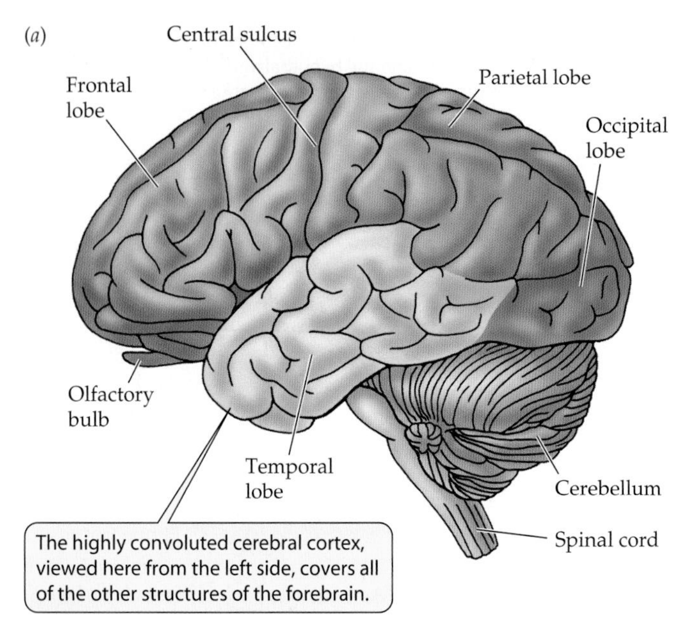

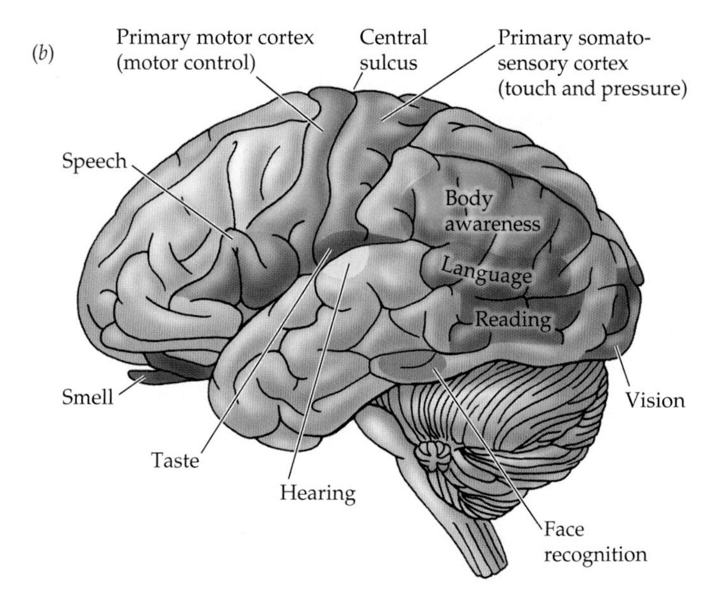

46.5 The Human Cerebrum (*a*) Each cerebral hemisphere is divided into four lobes. (*b*) Different functions are localized in particular areas of the cerebral lobes.

Under the cerebral cortex is white matter, made up of the axons that connect the cell bodies in the cortex with one another and with other areas of the brain.

A curious feature of our nervous systems is that the left side of the body is served (in both sensory and motor aspects) mostly by the right side of the brain, and the right side of the body is served mostly by the left side of the brain. Thus, sensory input from the right hand goes to the left cerebral hemisphere, and sensory input from the left hand goes to the right cerebral hemisphere. The two hemispheres, however, are not exactly symmetrical. Language abilities, for example, reside predominantly in the left hemisphere, as we will see below.

Different regions of the cerebral cortex have specific functions (Figure 46.5*b*). Some of those functions are easily defined, such as receiving and processing sensory information, but most of the cortex is involved in higher-order information processing that is less easy to define. These latter areas are given the general name of **association cortex**.

To understand the cerebral cortex, it helps to have an anatomical road map. As viewed from the left side, a left cerebral hemisphere looks like a boxing glove for the right hand with the fingers pointing forward, the thumb pointing out, and the wrist at the rear (see Figure 46.5*a*). The "thumb" area is the **temporal lobe**, the fingers the **frontal lobe**, the back of the hand the **parietal lobe**, and the wrist the **occipital lobe**. A mirror image of this arrangement characterizes the right cerebral hemisphere. Let's look at each lobe of the cerebrum separately.

THE TEMPORAL LOBE. The upper region of the temporal lobe receives and processes auditory information. The association areas of the temporal lobe are involved in the recognition, identification, and naming of objects. Damage to the temporal lobe results in disorders called *agnosias* in which the individual is aware of a stimulus, but cannot identify it.

Damage to one area of the temporal lobe results in the inability to recognize faces. Even old acquaintances cannot be identified by facial features, although they may be identified by other attributes such as voice, body features, and characteristic style of walking. Using monkeys, it has been possible to record the activity of neurons in this region that respond selectively to faces in general (Figure 46.6). These neurons do not respond to other stimuli in the visual field, and their responsiveness decreases if some of the features of the face are missing or appear in inappropriate locations. Damage to other association areas of the temporal lobe causes deficits in understanding spoken language, even though speaking, reading, and writing abilities may be intact.

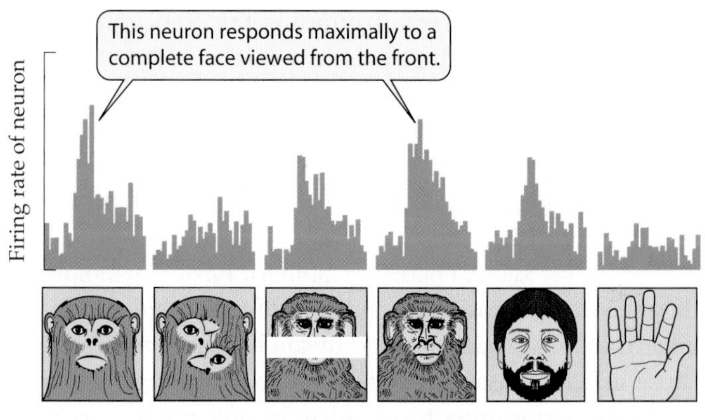

46.6 Neurons in One Region of the Temporal Lobe Respond to Faces The traces represent the firing rate of a neuron in the temporal lobe of a monkey in response to the pictures shown below them.

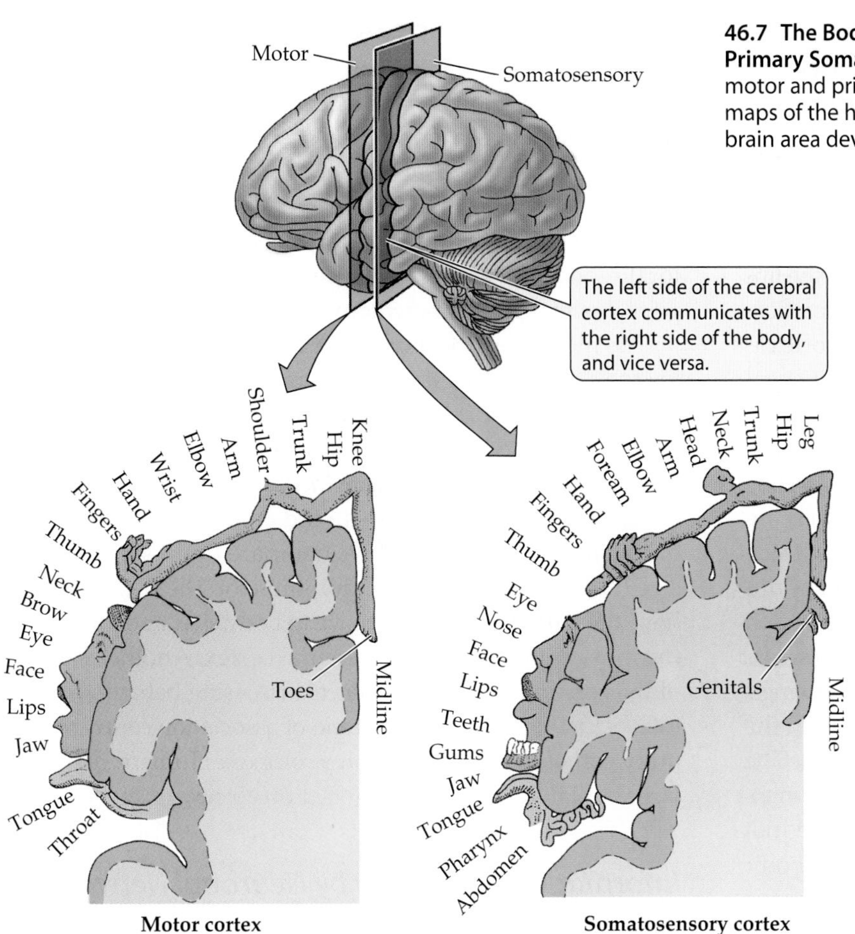

46.7 The Body Is Represented in the Primary Motor Cortex and the Primary Somatosensory Cortex Cross sections through the primary motor and primary somatosensory cortexes can be represented as maps of the human body. Body parts are shown in proportion to the brain area devoted to them.

THE FRONTAL LOBE. The frontal and parietal lobes are separated by a deep valley called the *central sulcus*. A strip of the frontal lobe cortex just in front of the central sulcus is called the **primary motor cortex** (see Figure 46.5*b*). The neurons in this region control muscles in specific parts of the body. The parts of the body can be mapped onto the primary motor cortex, from the head region on the lower side to the lower part of the body at the top. Areas with fine motor control, such as the face and hands, have the greatest representation (Figure 46.7). If a neuron in the primary motor cortex is electrically stimulated, the response is the twitch of a muscle, but not a coordinated, complex behavior.

The association functions of the frontal lobe are diverse. They are best described as having to do with planning, and they contribute very significantly to personality. People with frontal lobe damage have drastic alterations of personality because they cannot create an accurate view of themselves in the context of the world around them and cannot plan for future events. A dramatic case of frontal lobe damage is the story of Phineas Gage, who was an industrious, responsible, considerate young railroad construction foreman in 1848. Then a blasting accident shot a meter-long, 3-cm-wide iron tamping rod through his brain. The tamping iron entered Gage's head below his left eye, passed through his frontal lobe, and exited the top of his head (Figure 46.8).

Remarkably, Gage survived this terrible accident, but he was a completely different person. He was quarrelsome, bad-tempered, lazy, and irresponsible. He was impatient and obstinate, and he used profane language, which he had never done before. He spent the rest of his days as a drifter, earning money by telling his story, exhibiting his scars and the tamping iron. If you are in Cambridge, Massachusetts, you can pay him a visit. His skull, death mask, and the tamping iron are on display in the Museum of the Medical College of Harvard University.*

*The careful reader may have noted that we have mentioned two Gages in this chapter: Phineas and Fred. Fred Gage, the neuroscientist, is the great-grand-nephew of Phineas Gage.

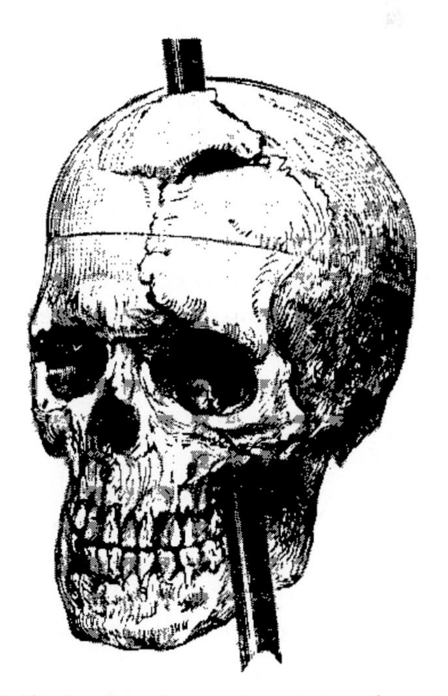

46.8 A Mind-Altering Experience In a nineteenth-century railroad construction accident, an explosion blew a tamping iron through the brain of Phineas Gage. Unbelievably, Gage survived, but his personality was radically changed. This drawing of Gage's skull was made at the time of his death.

THE PARIETAL LOBE. The strip of parietal lobe cortex just behind the central sulcus is the **primary somatosensory cortex** (see Figure 46.5*b*). This area receives touch and pressure information through the thalamus.

The whole body surface can be mapped onto the primary somatosensory cortex (see Figure 46.7). Areas of the body that have a high density of tactile mechanoreceptors and are capable of making fine discriminations in touch (such as the lips and the fingers) have disproportionately large representation. If a very small area of the primary somatosensory cortex is stimulated electrically, the subject reports feeling specific sensations, such as touch, in a very localized part of the body.

A major association function of the parietal lobe is attending to complex stimuli. Damage to the right parietal lobe causes a condition called *contralateral neglect syndrome*, in which the individual tends to ignore stimuli from the left side of the body or the left visual field. Such individuals have difficulty performing complex tasks, such as dressing the left side of the body; an afflicted man may not be able to shave the left side of his face. When asked to copy simple drawings, a person who exhibits this syndrome can do well with the right side of the drawing, but not the left (Figure 46.9). The parietal cortex is not symmetrical with respect to its role in attention, however. Damage to the left parietal cortex does not cause the same degree of neglect of the right side of the body. We will see similar asymmetries in cortical function when we discuss language.

THE OCCIPITAL LOBE. The occipital lobe receives and processes visual information; we'll learn more about the details of that process later in this chapter. The association areas of the occipital cortex are essential for making sense of the visual world and translating visual experience into language. Some deficits resulting from damage to these areas are specific. In one case, a woman with limited damage was unable to see motion. Her vision was intact, but she could see a waterfall only as a still image, and a car approaching only as a series of scenes of a stationary object at different distances.

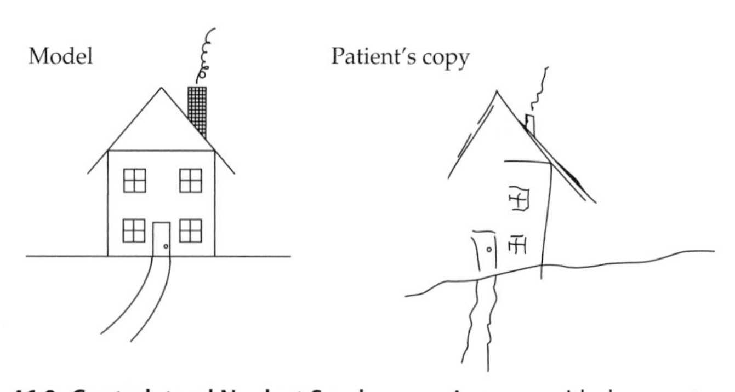

46.9 Contralateral Neglect Syndrome A person with damage to the right parietal association cortex will neglect the left side of a drawing when asked to copy a model.

The cerebrum has increased in size and complexity

As mentioned earlier, the size of the telencephalon relative to the rest of the brain increases substantially as we move up the phylogenetic scale from fishes to amphibians, to reptiles, to birds and mammals. Even when we consider only mammals, the cerebral cortex increases in size and complexity when we compare animals such as rodents, whose behavioral repertoires are relatively simple, with animals such as primates that have much more complex behavior.

The most dramatic increase in the size of the cerebral cortex took place during the last several million years of human evolution. The incredible intellectual capacities of *Homo sapiens* are associated with this enlargement of the cerebral cortex. Humans do not have the largest brains in the animal kingdom; elephants, whales, and porpoises have larger brains in terms of mass. If we compare brain size to body size, however, humans and dolphins top the list. Humans have the largest ratio of brain size to body size, and they have the most highly developed cerebral cortex. Another feature of the cerebral cortex that reflects increasing behavioral and intellectual capabilities is the ratio of association cortex to primary somatosensory and motor cortexes. Humans have the largest relative amount of association cortex.

Information Processing by Neuronal Networks

The functions of the nervous system can be understood in terms of neuronal networks. In this section we will use two subsystems of the nervous system to demonstrate the functioning of such networks. The first example, the autonomic nervous system, consists of efferent pathways; the second, the visual system, consists of afferent and integrative pathways. Techniques that have allowed neurobiologists to trace neuronal connections, identify neurotransmitters at synapses, and record action potentials in single cells and groups of cells have advanced our understanding of how certain subsystems of the nervous system work.

The autonomic nervous system controls the physiological functions of organs and organ systems

The autonomic nervous system is divided into two parts: the **sympathetic** and **parasympathetic** divisions. These two divisions work in opposition to each other in their effects on most organs, one causing an increase in activity and the other causing a decrease. The two divisions of the autonomic nervous system are easily distinguished from each other by their anatomy, their neurotransmitters, and their actions (Figure 46.10).

The best-known functions of the autonomic nervous system are those of the sympathetic division that produce the "fight-or-flight" response, increasing heart rate, blood pres-

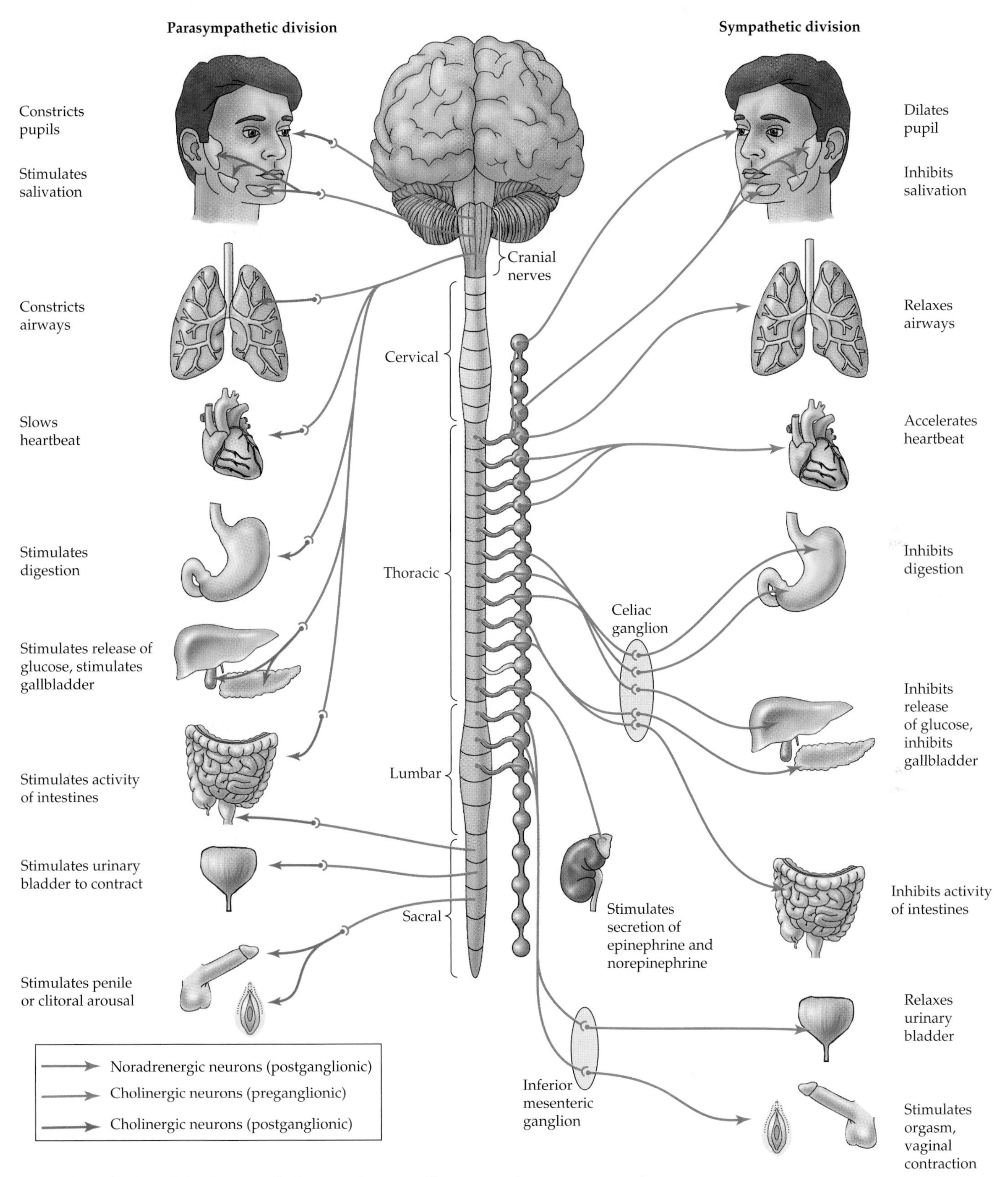

46.10 Organization of the Autonomic Nervous System The autonomic nervous system is divided into the sympathetic and parasympathetic divisions, which work in opposition to each other in their effects on most organs (one causing an increase and the other a decrease in activity).

sure, and cardiac output and preparing the body for emergencies (see Chapter 42). In contrast, the parasympathetic division slows the heart and lowers blood pressure. It is tempting to think of the sympathetic division as the one that speeds things up and the parasympathetic division as the one that slows things down, but that is not always a correct distinction. The sympathetic division slows the digestive system, and the parasympathetic division accelerates it.

Both divisions of the autonomic nervous system are efferent pathways. Each autonomic efferent pathway begins with a *cholinergic* neuron (one that uses acetylcholine as its neurotransmitter) that has its cell body in the brain stem or spinal cord. These cells are called *preganglionic neurons* because the second neuron in the pathway with which they synapse resides in a **ganglion** (a collection of neuronal cell bodies that is outside of the CNS). The second neuron is called a *postganglionic neuron* because its axon extends out from the ganglion. The axon of the postganglionic neuron synapses with cells in the target organs.

The postganglionic neurons of the sympathetic division are *noradrenergic* (use norepinephrine as their neurotransmitter), while the postganglionic neurons of the parasympathetic division are cholinergic. In organs that receive both sympathetic and parasympathetic input, the target cells respond in opposite ways to norepinephrine and to acetylcholine. A region of the heart called the *pacemaker*, which generates the heartbeat, is an example. Stimulating the sympathetic nerve to the heart or dripping norepinephrine onto the pacemaker region depolarizes the pacemaker cells, increases their firing rate, and causes the heart to beat faster. Stimulating the parasympathetic nerve to the heart or dripping acetylcholine onto the pacemaker region hyperpolarizes the pacemaker cells, decreases their firing rate, and causes the heart to beat more slowly. In contrast, in the digestive tract, norepinephrine hyperpolarizes muscle cells, which slows digestion, and acetylcholine depolarizes muscle cells, which accelerates digestion.

The sympathetic and parasympathetic divisions of the autonomic nervous system can also be distinguished by anatomy (see Figure 46.10). The preganglionic neurons of the parasympathetic division come from the brain stem and the last segment of the spinal cord (the *sacral* region). The preganglionic neurons of the sympathetic division come from the upper regions of the spinal cord below the neck (the *thoracic* and *lumbar* regions). Most of the ganglia of the sympathetic division are lined up in two chains, one on either side of the spinal cord. The parasympathetic ganglia are close to—sometimes sitting on—the target organs.

The autonomic nervous system is an important link between the CNS and many physiological functions of the body. Its control of diverse organs and tissues is crucial to homeostasis. In spite of its complexity, work by neurobiologists and physiologists over many decades has made it possible to understand its functions in terms of neuronal properties and circuits. In Chapter 49, for example, we will see how information from pressure receptors in the blood vessels is transmitted to the CNS, where it produces autonomic signals that control the rate of the heartbeat.

Neurons and circuits in the occipital cortex integrate visual information

In Chapter 45, we learned that the information conveyed to the brain in the optic nerve consists of action potentials that are stimulated by light falling on small circular areas of the retina called *receptive fields*. A receptive field contains many photoreceptor cells connected together in a circuit in such a way that the signals they produce are integrated and transmitted to the brain by a single retinal ganglion cell. The axon of each ganglion cell travels to the brain in the optic nerve. How does the brain construct visual images from this information about circular patches of light falling on the retina?

Information from the retina is transmitted through the optic nerve to a relay station in the thalamus, and then to the brain's visual processing area, in the occipital cortex at the back of the cerebral hemispheres (see Figure 46.5*b*). David Hubel and Torsten Wiesel of Harvard University studied the activity of neurons in this *visual cortex*. They recorded the activities of single cells in the brains of living animals while they stimulated the animals' retinas with spots and bars of light. They found that cells in the visual cortex, like retinal ganglion cells, have receptive fields—specific areas of the retina that, when stimulated by light, influence the rate at which the cells fire action potentials.

Cells in the visual cortex, however, have receptive fields that differ from the simple circular receptive fields of retinal ganglion cells. Cortical cells called *simple cells* are maximally stimulated by bars of light that have specific orientations. Simple cells probably receive input from several ganglion cells whose circular receptive fields are lined up in a row.

Complex cells in the visual cortex are also maximally stimulated by a bar of light with a particular orientation, but the bar may fall anywhere on a large area of retina described as that cell's receptive field. The receptive field of a complex cell appears to be built from the receptive fields of several simple cells that share a certain stimulus orientation, but have receptive fields in different places on the retina (Figure 46.11). Some complex cells respond most strongly when the bar of light moves in a particular direction.

The concept that emerges from these experiments is that the brain assembles a mental image of the visual world by analyzing edges in patterns of light falling on the retina. This

46.11 Receptive Fields of Cells in the Visual Cortex Cells in the visual cortex respond to specific patterns of light falling on the retina. Ganglion cells that transmit information about circular receptive fields converge on simple cells in the cortex in such a way that the simple cells have linear receptive fields. Simple cells transmit information to complex cells in such a way that the complex cells can respond to linear stimuli falling on different areas of the retina. ▶

analysis is conducted in a massively parallel fashion. Each retina sends a million axons to the brain, but there are hundreds of millions of neurons in the visual cortex. Each bit of information from a retinal ganglion cell is received by hundreds of cortical cells, each responsive to a different combination of orientation, position, and even movement of contrasting lines in the pattern of light falling on the retina.

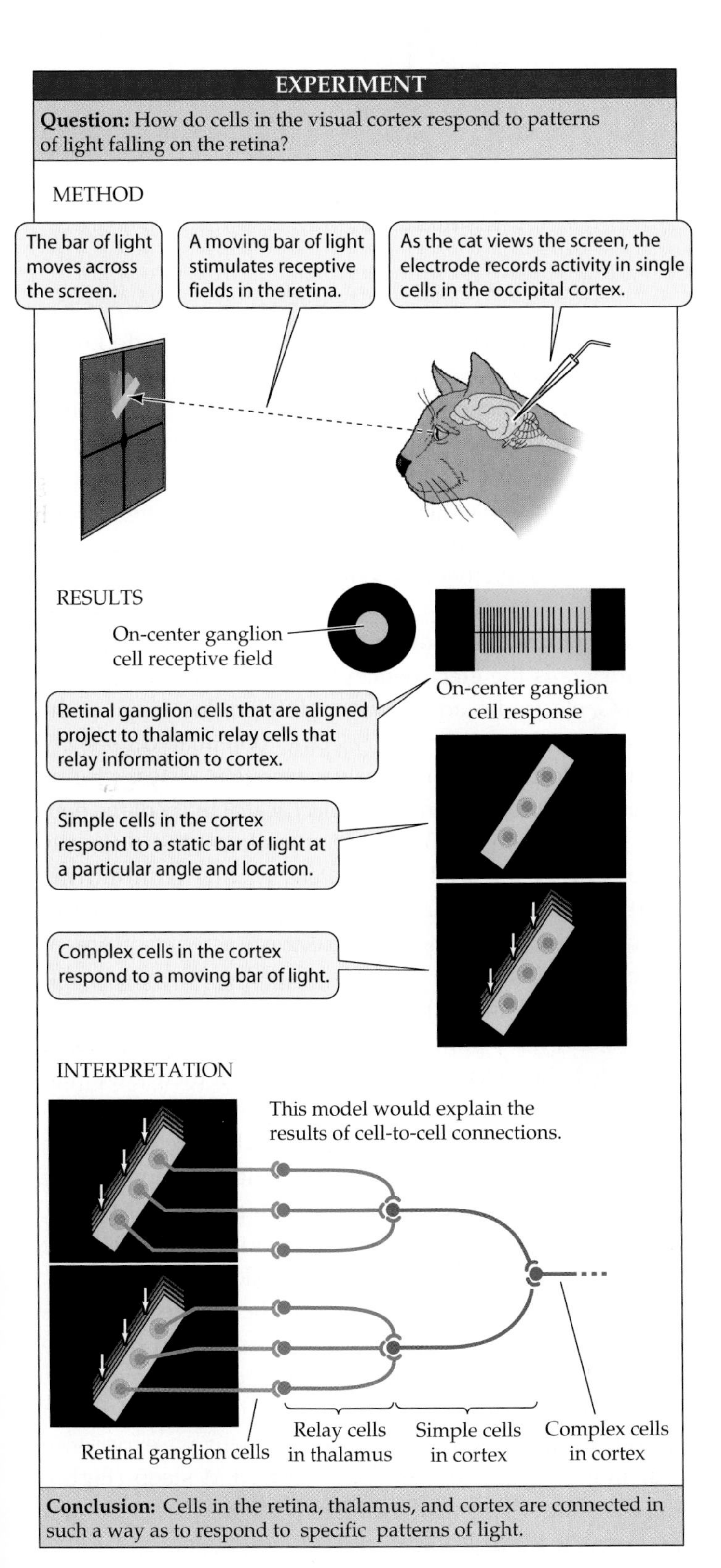

EXPERIMENT
Question: How do cells in the visual cortex respond to patterns of light falling on the retina?
METHOD
The bar of light moves across the screen.
A moving bar of light stimulates receptive fields in the retina.
As the cat views the screen, the electrode records activity in single cells in the occipital cortex.
RESULTS
On-center ganglion cell receptive field
On-center ganglion cell response
Retinal ganglion cells that are aligned project to thalamic relay cells that relay information to cortex.
Simple cells in the cortex respond to a static bar of light at a particular angle and location.
Complex cells in the cortex respond to a moving bar of light.
INTERPRETATION
This model would explain the results of cell-to-cell connections.
Retinal ganglion cells
Relay cells in thalamus
Simple cells in cortex
Complex cells in cortex
Conclusion: Cells in the retina, thalamus, and cortex are connected in such a way as to respond to specific patterns of light.

Cortical cells receive input from both eyes

How do we see objects in three dimensions? The quick answer is that our two eyes, located at the front of the head, see overlapping, yet slightly different, visual fields; that is, we have **binocular vision**. A person who is blind in one eye has great difficulty discriminating distances. Animals whose eyes are on the sides of the head have minimal overlap in their fields of vision and, as a result, poor depth vision, but they can see predators creeping up from all sides.

The story of how the brain integrates information from two eyes begins with the paths of the optic nerves. If you look at the underside of the brain, the optic nerves from the two eyes appear to join together just under the hypothalamus and then separate again. The place where they join is called the **optic chiasm** (Figure 46.12). Axons from the half of each retina closest to your nose cross in the optic chiasm and go to the opposite side of your brain. The axons from the other half of each retina go to the same side of the brain.

The result of this division of axons in the optic chiasm is that all visual information from your left visual field (everything left of straight ahead) goes to the right side of your brain, as shown in red in Figure 46.12. All visual information from your right visual field goes to the left side of your brain, as indicated in green in the figure. Both eyes transmit information about a specific spot in your right visual field, for example, to the same place in the left visual cortex.

Cells in the visual cortex are organized in columns. These columns alternate according to the source of their input: left eye, right eye, left eye, right eye, and so on. Cells closest to the border between two columns receive input from both eyes and are therefore called *binocular cells*. Binocular cells interpret distance by measuring the disparity between where the same stimulus falls on the two retinas.

What is disparity? Hold your finger out in front of you and look at it, closing one eye and then the other. Your finger appears to jump back and forth because its image falls on a different position on each retina. Repeat the exercise with an object at a distance. It doesn't appear to jump back and forth as much because there is less disparity in the positions of the image on the two retinas. Certain binocular cells respond optimally to a stimulus falling on both retinas with a particular disparity. Which set of binocular cells is stimulated depends on how far away the stimulus is.

When we look at something, we can detect its shape, color, depth, and movement. Where does all this information come together? Is there a single cell that fires only when a red sports car drives by? Probably not. A specific visual experience comes from simultaneous activity in a large collection

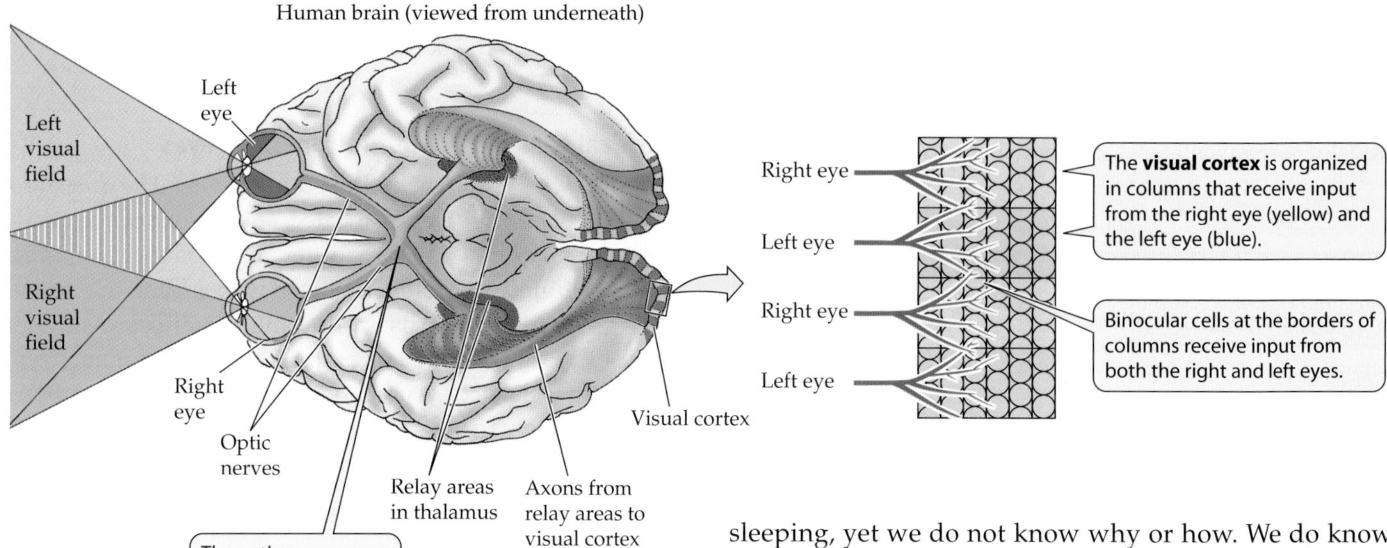

46.12 The Anatomy of Binocular Vision Each eye transmits information to both sides of the brain; however, the right side of the brain processes all information from the left visual field, and the left side of the brain processes all information from the right visual field. The visual cortex sorts visual field information according to whether it comes from the right eye or the left eye.

of cells. In addition, most visual experiences are enhanced by information from the other senses and from memory as well. This realization helps explain why about 75 percent of the cerebral cortex is association cortex.

Understanding Higher Brain Functions in Cellular Terms

Very few functions of the nervous system have been worked out to the point of identifying the underlying neuronal networks. The processes responsible for the higher brain functions discussed in the remaining pages of this chapter are undeniably complex. Nevertheless, neurobiologists, using a wide range of techniques, are making considerable progress in understanding some of the cellular and molecular mechanisms involved in those processes. The following discussion presents several complex aspects of brain and behavior that present challenges to neurobiologists: sleep and dreaming, learning and memory, language use, and consciousness.

Sleep and dreaming produce electrical patterns in the cerebrum

A dominant feature of human behavior is the daily cycle of sleep and waking. All birds and mammals, and probably all other vertebrates, sleep. We spend one-third of our lives sleeping, yet we do not know why or how. We do know, however, that we need to sleep. Loss of sleep impairs alertness and performance. Most people in our society—certainly most college students—are chronically sleep-deprived. Large numbers of accidents and serious mistakes that endanger lives can be attributed to impaired alertness due to sleep loss. Yet insomnia (difficulty in falling or staying asleep) is one of the most common medical complaints.

THE ELECTROENCEPHALOGRAM. A common tool of sleep researchers is the **electroencephalogram** (**EEG**). To record an EEG, electrodes are placed at different locations on the scalp, and changes in the electric potential differences between electrodes are recorded over time. These electric potential differences reflect the electrical activity of the neurons in the brain regions under the electrodes, primarily regions of the cerebral cortex. Pens writing on a moving chart are used to record the patterns of these differences (Figure 46.13*a,b*). Usually, the electrical activity of one or more skeletal muscles is also recorded on the chart; this record is called an *electromyogram* (EMG).

EEG and EMG patterns reveal the transition from being awake to being asleep. They also reveal that there are different states of sleep. In mammals other than humans, two major sleep states are easily distinguished: They are **slow-wave sleep** and **rapid-eye-movement** (**REM**) **sleep**. In humans, we characterize sleep states as **non-REM sleep** and **REM sleep**. Human non-REM sleep is divided into four stages. Only the two deepest stages are considered true slow-wave sleep.

When a person falls asleep at night, the first sleep state entered is non-REM sleep, which progresses from stage 1 to stage 4. Stages 3 and 4 are deep, restorative, slow-wave sleep. This first episode of non-REM sleep is followed by an episode of REM sleep. Throughout the night, we experience four or five cycles of non-REM and REM sleep (Figure 46.13*c*). About 80 percent of our sleep is non-REM sleep, and 20 percent is REM sleep.

(a)

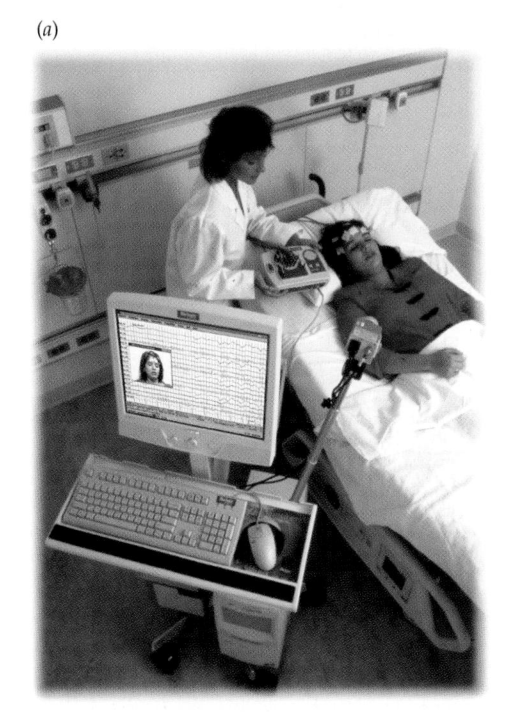

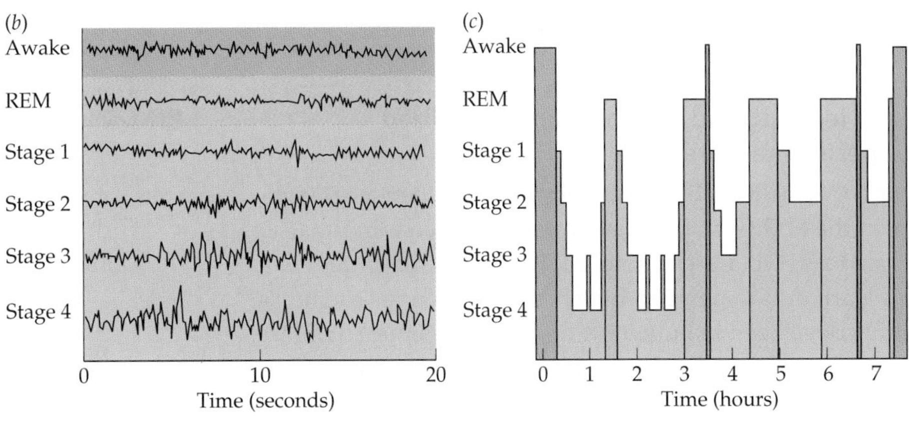

46.13 Patterns of Electrical Activity in the Cerebral Cortex Characterize Stages of Sleep (*a*) Electrical activity in the cerebral cortex is detected by electrodes placed on the scalp and recorded on moving chart paper by a polygraph. (*b*) The resulting record is an electroencephalogram (EEG). (*c*) During a night, humans cycle through different stages of sleep.

We have vivid dreams and nightmares during REM sleep, which gets its name from the jerky movements of the eyeballs that occur during this state. The most remarkable feature of REM sleep is that inhibitory commands from the brain almost completely paralyze the skeletal muscles. Occasional muscle twitches break through the paralysis, as can be seen in a dog that appears to be trying to run in its sleep. If you look closely at a sleeping dog when its legs and paws are twitching, you will be able to see the rapid eye movements as well. Probably the function of muscle paralysis during REM sleep is to prevent the acting out of dreams. Sleepwalking occurs during non-REM sleep.

CELLULAR CHANGES DURING SLEEP. There are striking neurophysiological differences between non-REM and REM sleep. Non-REM sleep is characterized by a decrease in the responsiveness of neurons in the thalamus and cerebral cortex. Remember that neurons have a negative resting membrane potential and a threshold for firing action potentials. Usually the resting potential is below the threshold potential, so the neuron is not firing. When synaptic input causes the membrane potential to become less negative (depolarized), the cell can reach threshold and fire action potentials.

During waking, several nuclei in the brain stem are continuously active. Many axons from these nuclei extend to the thalamus and the cerebral cortex, where the neurotransmitters (acetylcholine, norepinephrine, and serotonin) they release are generally depolarizing. Therefore, these broadly distributed neurotransmitters keep the resting potential of the neurons of the thalamus and cortex close to threshold and sensitive to synaptic inputs, thereby maintaining waking.

With the onset of sleep, activity in these brain stem nuclei decreases, and their axon terminals release less neurotransmitter. With the withdrawal of the depolarizing neurotransmitters, the resting potentials of the cells of the thalamus and cortex become more negative (hyperpolarized), and the cells are less sensitive to synaptic input. Their processing of information is inhibited, and consciousness is lost.

An interesting neuronal event happens as a result of this hyperpolarization: The cells begin to fire action potentials in bursts. The synchronization of these bursts over broad areas of cerebral cortex results in the EEG slow-wave pattern that characterizes non-REM sleep. Studies of neurons of the thalamus and the cortex have shown that their hyperpolarization during non-REM sleep is due to increased opening of K^+ channels, and that the bursting is due to Ca^{2+} channels whose inactivation gates close rapidly and require hyperpolarization to be reopened. We can therefore explain the EEG pattern of non-REM sleep in terms of the properties of neurons and ion channels.

At the transition from non-REM to REM sleep, dramatic changes occur. Some of the brain stem nuclei that were inactive during non-REM sleep become active again, causing a general depolarization of cortical neurons. Thus the synchronized bursts of firing cease, and the EEG resembles that of the waking brain. Because the resting potentials of the neurons return to near threshold levels, the cortex can process information, and vivid dreams occur. However, the brain inhibits both afferent and efferent pathways; therefore, the activity in the cortex is unconstrained by its usual sources of information. One example of the effect of this loss of motor output and sensory input is the frequently reported dream in which a person is trying to run but cannot move.

We do not know the function of REM sleep, but since a wide variety of mammals have about the same percentage of total sleep time that is REM sleep, it is probably a rather basic, cellular function. A prominent hypothesis about the functions of sleep is that it is essential for the maintenance and repair of neuronal connections and for the neuronal changes that are involved in learning and memory. However, evidence for such functions is still meager.

Some learning and memory can be localized to specific brain areas

Learning is the modification of behavior by experience. *Memory* is the ability of the nervous system to retain what is learned and what is experienced. Even very simple animals can learn and remember, but these two abilities are most highly developed in humans. Consider the amount of information associated with learning a language. The capacity of memory and the rate at which items can be retrieved are remarkable features of the human nervous system.

LEARNING. Learning that leads to long-term memory and modification of behavior must involve long-lasting synaptic changes. Synaptic changes that last for weeks have been observed (see Figure 44.18). High-frequency electrical stimulation of certain identifiable circuits of the mammalian hippocampus makes them more sensitive to subsequent stimulation. This phenomenon is called **long-term potentiation** (**LTP**). In contrast, continuous, repetitive, low-level stimulation of these hippocampal circuits reduces their responsiveness, a phenomenon that has been called **long-term depression** (**LTD**). LTP and LTD have been demonstrated in circuits other than hippocampal circuits, and they may be fundamental cellular or molecular mechanisms involved in learning and memory.

A form of learning that is widespread among animal species is *associative learning*, in which two unrelated stimuli become linked to the same response. The simplest example of associative learning is the *conditioned reflex*, discovered by the Russian physiologist Ivan Pavlov. Pavlov was studying the control of digestive functions in dogs and observed that a dog salivates at the sight or smell of food—a simple autonomic reflex. He discovered that if he rang a bell just before food was presented to the dog, after a few trials the dog would salivate at the sound of the bell, even if no food followed. The salivation reflex was conditioned to be associated with the sound of a bell, a stimulus that normally is unrelated to feeding and digestion.

This simple form of learning has been studied extensively in efforts to understand its underlying neuronal mechanisms. In a series of studies led by Richard Thompson, the eye-blink reflex of a rabbit in response to a puff of air directed at its eyes was conditioned to be associated with a tone stimulus. After conditioning, the rabbit blinked when it heard the tone (Figure 46.14). A small and specific area of the cerebellum

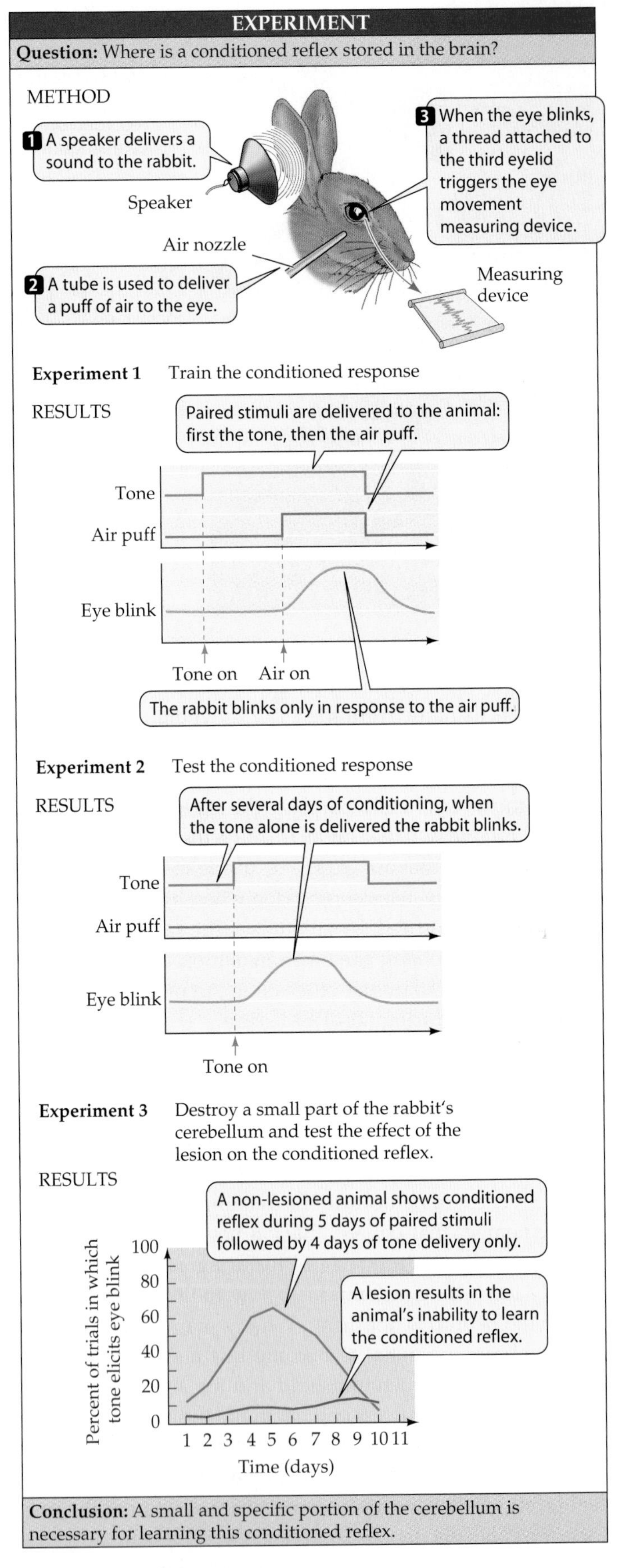

46.14 The Conditioned Eye-Blink Reflex Depends on a Cerebellar Circuit A small and specific area of the cerebellum is necessary for a rabbit to form a conditioned reflex.

was discovered to be necessary for this conditioned reflex. Thus, it was possible to localize learning to an identifiable set of synapses in the mammalian brain.

MEMORY. Attempts to treat human neurological diseases have led to the localization of areas of the brain involved in the formation and recall of memories. Epilepsy is a disorder characterized by uncontrollable increases in neuronal activity in specific parts of the brain. The resulting *seizures*, or "epileptic fits," can endanger the afflicted individual. In the past, serious cases of epilepsy were sometimes treated by destroying the part of the brain from which the surge of activity originated.

To find the right area, the surgery was done under local anesthesia, and different regions of the brain were electrically stimulated with electrodes while the patient reported on the resulting sensations. When some regions of association cortex were stimulated, patients reported vivid memories. Such observations were the first evidence that memories have anatomical locations in the brain and exist as properties of neurons and networks of neurons. Yet the destruction of a small area of the brain does not completely erase a memory, so it is postulated that memory is a function distributed over many brain regions and that a memory may be stimulated via many different routes.

You can recognize several forms of memory from your own experience. There is *immediate memory* for events that are happening now. Immediate memory is almost perfectly photographic, but it lasts only seconds. *Short-term memory* contains less information, but it lasts longer—on the order of 10 to 15 minutes. If you are introduced to a group of new people, you may remember most of their names for 5 or 10 minutes, but you will have forgotten them in an hour or so if you have not repeated them, written them down, or used them in a conversation. Repetition, use, or reinforcement by something that gets your attention (such as the title President) facilitates the transfer of short-term memory to *long-term memory*, which can last for days, months, or years.

Knowledge about neuronal mechanisms for the transfer of short-term memory to long-term memory has come from observations of persons who have lost parts of the limbic system, notably the hippocampus. A famous case is that of a man identified as H.M., whose hippocampus on both sides of the brain was removed in an effort to control severe epilepsy. Since that surgery, H.M. has not been able to transfer information to long-term memory. If someone is introduced to him, has a conversation with him, and then leaves the room, when that person returns, he or she is unknown to H.M.—it is as if the previous conversation had never taken place. H.M. retains memories of events that happened before his surgery, but he remembers postsurgery events for only 10 or 15 minutes.

Memory of people, places, events, and things is called *declarative memory* because you can consciously recall and describe them. Another type of memory, called *procedural memory*, cannot be consciously recalled and described: It is the memory of how to perform a motor task. When you learn to ride a bicycle, ski, or use a computer keyboard, you form procedural memories. Although H.M. is incapable of forming declarative memories, he is capable of forming procedural memories. When taught a motor task day after day, he cannot recall the lessons of the previous day, yet his performance steadily improves. Thus procedural learning and memory must involve mechanisms different from those used in declarative learning and memory.

Our understanding of learning and memory in cellular terms is very rudimentary. New techniques that enable functional imaging of the brain in ways that reveal changes in the metabolic activity of specific regions and structures are greatly enhancing progress in this area.

Language abilities are localized in the left cerebral hemisphere

No aspect of brain function is as integrally related to human consciousness and intellect as is language. Therefore, studies of the brain mechanisms that underlie the acquisition and use of language are extremely interesting to neuroscientists. A curious observation about language abilities is that they are usually located in only one cerebral hemisphere—which in 97 percent of people is the left hemisphere. This phenomenon is referred to as the *lateralization* of language functions.

Some of the most fascinating research on this subject was conducted by Roger Sperry and his colleagues at the California Institute of Technology; Sperry received a Nobel prize for this work. The two cerebral hemispheres are connected by a tract of white matter called the *corpus callosum*. In one severe form of epilepsy, bursts of action potentials travel from hemisphere to hemisphere across the corpus callosum. Cutting the tract eliminates the problem, and patients function nearly normally following the surgery. But experiments revealed interesting deficits in the language abilities of these "split-brain" persons. Without the connection between the two hemispheres, the knowledge or experience of the right hemisphere could no longer be expressed in language, nor could language be used to communicate with the right hemisphere.

The mechanisms of language in the left hemisphere have been the focus of much research. The experimental subjects are persons who have suffered damage to the left hemisphere and are left with one of many forms of *aphasia*, a deficit in the ability to use or understand words. These studies have identified several language areas in the left hemisphere (Figure 46.15).

Broca's area, located in the frontal lobe just in front of the primary motor cortex, is essential for speech. Damage to

46.15 Language Areas of the Cortex Different regions of the left cerebral cortex participate in the processes of (*a*) repeating a word that is heard and (*b*) speaking a written word.

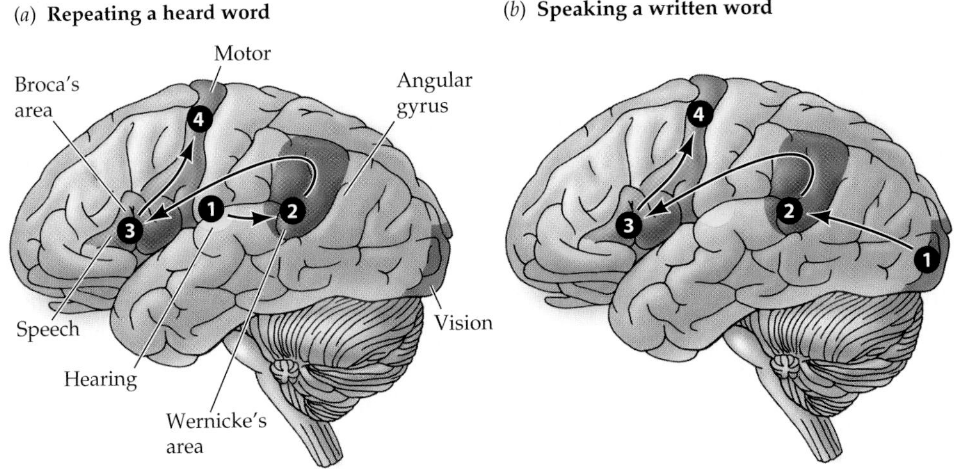

Broca's area results in halting, slow, poorly articulated speech or even complete loss of speech, but the patient can still read and understand language. In the temporal lobe, close to its border with the occipital lobe, is *Wernicke's area*, which is more involved with sensory than with motor aspects of language. Damage to Wernicke's area can cause a person to lose the ability to speak sensibly while retaining the abilities to form the sounds of normal speech and to imitate its cadence. Moreover, such a patient cannot understand spoken or written language. Near Wernicke's area is the *angular gyrus*, which is believed to be essential for integrating spoken and written language.

Normal language ability depends on the flow of information among various areas of the left cerebral cortex. Input from spoken language travels from the auditory cortex to Wernicke's area (Figure 46.15*a*). Input from written language travels from the visual cortex to the angular gyrus to Wernicke's area (Figure 46.15*b*). Commands to speak are formulated in Wernicke's area and travel to Broca's area and from there to the primary motor cortex. Damage to any one of those areas or the pathways between them can result in aphasia. Using modern methods of functional brain imaging, it is possible to see the metabolic activity in different brain areas when the brain is using language (Figure 46.16).

What is consciousness?

This chapter has only scratched the surface of our knowledge about the organization and functions of the human brain, but it may give you some idea of the incredible challenge that neurobiologists face in trying to understand their own brains. Progress is being aided by powerful new technologies such as patch clamping (see Figure 44.11), functional imaging, and neurochemical and molecular methods. However, even these sophisticated new research tools may not allow us to answer the question "What is consciousness?"

If you see a black dog running across a field, you are conscious of the fact that it is a dog, it is black, and it is a Labrador retriever. You may remember that the dog's name is Sarina, that he belongs to your friend Meera, and that he is 6 years old. From what you have learned in this chapter, imagine how many neurons would be active during this experience: neurons in the visual system, the language areas, and in different regions of association cortex. But is being conscious of the black dog simply a result of the fact that all of these neurons are firing at the same time? Your brain is simultaneously processing many other sensory inputs, but you are not necessarily conscious of those inputs. What makes you conscious of the black dog and associated memories and not of other information the brain is processing at the same time?

If we could describe all the neurons and all the synapses involved in the conscious experience of seeing and naming a black dog, and then build a computer with devices that modeled all these neurons and connections, would that computer be conscious? It has been said that the question of consciousness resolves into two types of problems: "easy" and "hard." The easy problems deal with all the cells and circuits that process the information that is involved in conscious ex-

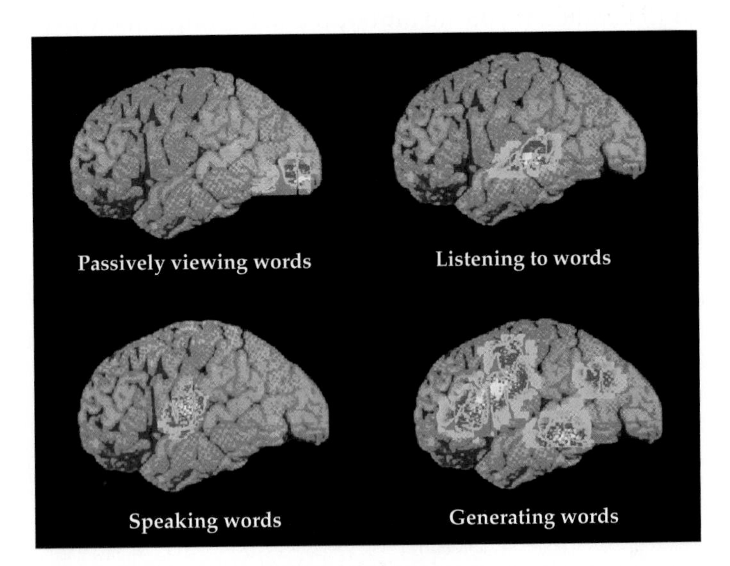

46.16 Imaging Techniques Reveal Active Parts of the Brain Positron emission tomography (PET) scanning reveals the brain regions that are activated by different aspects of language use. Radioactively labeled glucose is given to the subject. Brain areas take up radioactivity in proportion to their metabolic use of glucose. The PET scan visualizes levels of radioactivity in specific brain regions when a particular activity is performed.

perience. The implication of "easy" is that we seem to have the tools to solve these kinds of problems, as complex as they may be. The hard problems involve explaining how properties of cells and networks result in consciousness, and we seem to lack the proper tools or concepts even to begin to solve these problems.

Chapter Summary

The Nervous System: Structure, Function, and Information Flow

▶ The brain and spinal cord make up the central nervous system; the cranial and spinal nerves make up the peripheral nervous system. A nerve is a bundle of many axons carrying information to and from the central nervous system.

▶ The nervous system can be modeled conceptually in terms of the direction of information flow and whether or not we are conscious of the information. **Review Figure 46.1**

▶ The vertebrate nervous system develops from a hollow dorsal neural tube. The brain forms from three swellings at the anterior end of the neural tube, which become the hindbrain, the midbrain, and the forebrain. **Review Figure 46.2**

▶ The forebrain develops into the cerebral hemispheres (the telencephalon) and the underlying thalamus and hypothalamus (the diencephalon). The midbrain and hindbrain develop into the brain stem.

Functional Subsystems of the Nervous System

▶ The nervous system is composed of many subsystems that function simultaneously. Some important subsystems are the spinal cord, the reticular system, the limbic system, and the cerebrum.

▶ The spinal cord communicates information between the brain and the rest of the body. It also processes and integrates much information, and can issue some commands to the body without input from the brain. **Review Figure 46.3. See Web/CD Tutorial 46.1**

▶ The reticular system is a complex network that directs incoming information to appropriate brain stem nuclei that control autonomic functions, as well as transmitting the information to the forebrain that results in conscious sensation. The reticular system controls the level of arousal of the nervous system.

▶ The limbic system is an evolutionarily primitive part of the telencephalon that is involved in emotions, physiological drives, instincts, and memory. **Review Figure 46.4**

▶ The cerebral hemispheres are the dominant structures of the human brain. Their surfaces consist of a layer of neurons called the cerebral cortex.

▶ Most of the cerebral cortex is involved in higher-order information processing; these areas are generally called association cortex.

▶ The cerebral hemispheres can be divided into temporal, frontal, parietal, and occipital lobes. Many motor functions are localized in parts of the frontal lobe. Information from many sensory receptors around the body projects to a region of the parietal lobe. Visual information projects to the occipital lobe, and auditory information projects to a region of the temporal lobe. **Review Figures 46.5, 46.6, 46.7. See Web/CD Activity 46.1**

Information Processing by Neuronal Networks

▶ The functions of the nervous system are beginning to be understood in terms of the properties of cells organized in neuronal networks.

▶ The autonomic nervous system consists of efferent pathways that control the physiological function of organs and organ systems. Its sympathetic and parasympathetic divisions normally work in opposition to each other. These divisions are characterized by their anatomy, neurotransmitters, and effects on target tissues. **Review Figure 46.10**

▶ Neuronal circuits in the occipital cortex integrate visual information. Information from the receptive fields of retinal ganglion cells is communicated to the brain in the optic nerves. This information is transmitted to the visual cortex in such a way as to create receptive fields for cortical cells.

▶ A simple cell in the visual cortex is stimulated by a bar of light with a specific orientation falling at a specific location on the retina. A complex cell is maximally stimulated by such a stimulus moving across the retina. The visual cortex seems to assemble a mental image of the visual world by analyzing edges of patterns of light. **Review Figure 46.11**

▶ Binocular vision results from circuits that communicate information from both eyes to binocular cells in the visual cortex. These cells interpret distance by measuring the disparity between where the same stimulus falls on the two retinas. **Review Figure 46.12**

Understanding Higher Brain Functions in Cellular Terms

▶ Humans have a daily cycle of sleep and waking. Sleep can be divided into slow-wave (non-REM) sleep and rapid-eye-movement (REM) sleep. Human non-REM sleep is divided into four stages of increasing depth. **Review Figure 46.13**

▶ Some learning and memory processes have been localized to specific brain areas. Repeated activation of identified circuits in the hippocampus has revealed long-lasting changes in synaptic properties referred to as long-term potentiation and long-term depression, which may be involved in learning and memory. **Review Figure 46.14**

▶ Complex memories can be elicited by stimulating small regions of association cortex. Damage to the hippocampus can destroy the ability to form long-term declarative memories, but not procedural memories.

▶ Language abilities are localized mostly in the left cerebral hemisphere, a phenomenon known as lateralization.

▶ Different areas of the left hemisphere—including Broca's area, Wernicke's area, and the angular gyrus—are responsible for different aspects of language. **Review Figure 46.15. See Web/CD Activity 46.2**

See Web/CD Activity 46.3 for a concept review of this chapter.

Self-Quiz

1. Which of the following describes the route of sensory information from the foot to the brain?
   a. Ventral horn, spinal cord, medulla, cerebellum, midbrain, thalamus, parietal cortex
   b. Dorsal horn, spinal cord, medulla, pons, midbrain, hypothalamus, frontal cortex
   c. Dorsal horn, spinal cord, medulla, pons, midbrain, thalamus, parietal cortex
   d. Ventral horn, spinal cord, pons, cerebellum, midbrain, thalamus, parietal cortex
   e. Dorsal horn, spinal cord, medulla, pons, midbrain, thalamus, frontal cortex
2. Which statement about the reticular system is *not* true?
   a. Increased activity in the reticular system induces sleep.
   b. The reticular system is located in the brain stem.
   c. Damage to the reticular system in the midbrain can result in coma.

d. Information from the spinal cord is routed to different nuclei in the reticular system and to the forebrain.
e. There are groups of neurons called nuclei in the reticular system.

3. Which statement about afferent and efferent pathways is *not* true?
 a. Sensory afferent pathways carry information of which we are consciously aware.
 b. Visceral afferents carry information about physiological functions of which we are not consciously aware.
 c. The voluntary division of the efferent portion of the peripheral nervous system executes conscious movements.
 d. The cranial nerves and spinal nerves are part of the peripheral nervous system.
 e. Afferent and efferent axons never travel in the same nerve.

4. Which statement about the limbic system is *not* true?
 a. In the spinal cord, the white matter contains the axons conducting information up and down the spinal cord.
 b. The limbic system is involved in basic physiological drives, instincts, and emotions.
 c. The limbic system consists of primitive forebrain structures.
 d. In humans, the limbic system is the largest part of the brain.
 e. In humans, a part of the limbic system is necessary for the transfer of short-term memory to long-term memory.

5. Which of the following represents the largest portion of the human cerebral cortex?
 a. The frontal lobes
 b. The primary somatosensory cortex
 c. The temporal cortex
 d. The association cortex
 e. The occipital cortex

6. Which statement about the autonomic nervous system is true?
 a. The sympathetic division is afferent, and the parasympathetic division is efferent.
 b. The transmitter norepinephrine is always excitatory, and acetylcholine is always inhibitory.
 c. Each pathway in the autonomic nervous system includes two neurons, and the neurotransmitter of the first neuron is acetylcholine.
 d. The cell bodies of many sympathetic preganglionic neurons are in the brain stem.
 e. The cell bodies of most parasympathetic postganglionic neurons are in or near the thoracic and lumbar spinal cord.

7. Which statement about cells in the visual cortex is *not* true?
 a. Many cortical cells receive inputs directly from single retinal ganglion cells.
 b. Many cortical cells respond most strongly to bars of light falling at a specific location on the retina.
 c. Some cortical cells respond most strongly to bars of light falling anywhere over large areas of the retina.
 d. Some cortical cells receive inputs from both eyes.
 e. Some cortical cells respond most strongly to an object when it is a certain distance from the eyes.

8. Which of the following characterizes non-REM sleep?
 a. Dreaming
 b. Circadian rhythms
 c. EEG slow waves
 d. Rapid and jerky eye movements
 e. It makes up about 20 percent of total sleep time

9. Which conclusion was supported by experiments on split-brain patients?
 a. Language abilities are localized mostly in the left cerebral hemisphere.
 b. Language abilities require both Wernicke's area and Broca's area.
 c. The ability to speak depends on Broca's area.
 d. The ability to read depends on Wernicke's area.
 e. The left hand is served by the left cerebral hemisphere.

10. In the knee-jerk reflex,
 a. spinal interneurons inhibit the motor neuron of the antagonistic muscle.
 b. activity in the stretch receptor neuron causes contraction of the leg flexor muscles.
 c. the cell body of the motor neuron is in the dorsal horn of the spinal cord.
 d. action potentials in the sensory neuron release inhibitory neurotransmitter onto the motor neurons.
 e. the sensory neuron forms a monosynaptic loop with the motor neuron to the antagonistic muscle.

For Discussion

1. A person receives a stab wound to the left side of his neck. Miraculously, blood vessels are spared. However, following this trauma, his left pupil remains more constricted than his right pupil, and he drools out of the left side of his mouth. How can you explain these symptoms?
2. The stretch receptors in muscles are modified muscle cells, and they have their own motor neurons. What is the function of those motor neurons? To think about this question, remember that the function of the monosynaptic reflex is to adjust muscle tension to a change in load so that the position of the limb does not change.
3. A patient is unable to speak coherently. He can read and write, and he has no obvious loss of muscle function. Where would you expect to find an abnormality if you did brain scans of this patient?
4. We described the organization of the visual cortex as columns of cells that alternately receive input from the left eye and the right eye. If a young kitten is allowed to see light out of only one eye for a day, more synapses are maintained in the cortical columns receiving input from that eye, while synapses decrease in the intervening columns. This redistribution of synapses does not occur, however, if the kitten is not allowed to sleep. What hypotheses could you propose on the basis of these results?
5. As a result of a car accident, a woman has her right arm amputated just below the shoulder. Following her recovery, she continues to experience sensations and even severe pain in her nonexistent right hand and forearm. Explain the basis for this phantom limb experience.

47 Effectors: Making Animals Move

An animal's central nervous system is more than a processor and storage medium for information. It also allows the animal to respond to information. **Effectors** are tissues and organs commanded by the CNS to carry out these responses, most of which involve movement.

A fascinating array of adaptations enable animals to move. Consider the act of jumping. When you jump, neuronal signals from the motor cortex of your brain are routed through spinal circuits that tell specific leg muscles to contract, extending your legs. Highly skilled and trained athletes can actually outleap their own body height.

But as "record jumpers" go, many other animals—cats, spiders, kangaroos, and fleas, to name just a few—far surpass even the Olympian feats of humans. A flea, for example, can jump more than 200 times its body length. Unlike that of a human jumper, the flea's jumping mechanism does not involve muscles, but works like a slingshot. The flea is so small, and its initial acceleration is so great, that no muscle could contract fast enough to cause such a movement. Instead, at the base of its jumping legs is an elastic material that is compressed by muscles while the flea is resting. When a trigger mechanism is released, the elastic material recoils and "fires" the flea up and over to its target (or away from an enemy).

Many Signals to Many Muscles Practitioners of the martial arts can achieve amazingly coordinated high jumps, with the aim of focusing the force of the jump. For a human, such a movement requires years of training and practice.

Jumping is just one adaptation an animal can use to respond to information received by its sensory receptors. Effectors also include the internal organs and organ systems that the animal uses to control its internal environment. We will begin this chapter by looking at the mechanisms that power the movement of cilia, flagella, and single cells. The focus of the chapter is on muscles and skeletons—the mechanisms that create mechanical forces and use those forces to change shape and move, and which are the basis for most animal behavior. At the end of the chapter, we will consider a few effectors other than those that create movement.

Microtubules, Microfilaments, and Cell Movement

Two components of the cytoskeleton—microtubules and microfilaments—generate cell movement (see Figure 4.21). Both of these structures consist of long protein molecules that can change their length or shape.

Microtubules are components of the cytoskeleton

Microtubules are important intracellular effectors for changing cell shape, moving organelles, and enabling cells to respond to their environment. Microtubules generate forces by polymerizing and depolymerizing the protein *tubulin*. The spindle that moves chromosomes to the mitotic poles at anaphase is made up of microtubules (see Figure 9.8). Another example of microtubule involvement in cell movement is the growth of the axons of neurons in the developing nervous system. Neurons find and make their appropriate connections by sending out long extensions that search for the correct target cells. If polymerization of tubulin is chemically inhibited, these neurons do not extend.

Microtubules also generate the small-scale movements of cilia and flagella (see Figures 4.23 and 4.24). We have seen a number of the functions of these structures in the previous chapters of this book. Whereas many protists and small invertebrates use cilia for locomotion, larger multicellular animals typically use ciliated cells to move liquids and particles over cell surfaces. Many mollusks, for example, use cilia to circulate a current of water across their gas exchange and feeding surfaces. In humans, the cilia continuously sweep a layer of mucus from deep down in the lungs, up through the windpipe, and into the throat. The mucus carries particles of dirt and dead cells. We can then either swallow or spit out the mucus, and with it, the trapped detritus. Ciliated cells lining the female reproductive tract create currents that sweep eggs from the ovaries into the oviducts and all the way down to the uterus. Flagellated cells maintain a flow of water through the bodies of sponges, bringing in food and oxygen and removing carbon dioxide and wastes. Flagella power the movement of the sperm of most species.

Microfilaments change cell shape and cause cell movements

Microfilaments are proteins that change conformation as a means of generating forces. The dominant microfilament in animal cells is the protein **actin**. Bundles of cross-linked actin strands form important structural components of cells. The microvilli that increase the absorptive surface area of the cells lining the gut are stiffened by actin microfilaments (see Figure 4.22), as are the stereocilia of the sensory hair cells in the mammalian ear. Actin microfilaments can change the shape of a cell by polymerizing and depolymerizing. Microfilaments reach their highest level of organization in muscle cells, which generate large-scale movements.

Together with the protein **myosin**, actin microfilaments generate the contractile forces responsible for many aspects of cell movement and changes in cell shape. The contractile ring that divides an animal cell undergoing mitosis into two daughter cells is composed of actin microfilaments in association with myosin. The mechanisms that many cells employ to engulf materials (endocytosis; see Figure 5.15) also rely on nets of actin and myosin beneath the plasma membrane.

Certain cells in multicellular animals travel within the body by **amoeboid motion**, which is generated by the activity of actin microfilaments and myosin. During development, many cells migrate by amoeboid motion. Throughout an animal's life, phagocytic cells circulate in the blood, squeeze through the walls of the blood vessels, and wander through the tissues by amoeboid motion. The mechanisms of amoeboid motion have been studied extensively in the protist for which this type of movement was named—the amoeba (see Figure 28.4).

Amoeboid motion is accomplished by the cell extending a lobe-shaped projection called a *pseudopod* and then seemingly squeezing itself into that pseudopod. The cytoplasm in the core of the cell is relatively liquid and is called *plasmasol*, but just beneath the plasma membrane the cytoplasm is much thicker, and is called *plasmagel*. To form a pseudopod, the thick plasmagel in one area of the cell thins, allowing a bulge to form. Just under the cell surface, in the plasmagel, is a network of actin microfilaments that interacts with myosin to squeeze plasmasol into the bulge. As the microfilament network continues to contract, cytoplasm streams in the direction of the pseudopod. When the cytoplasm at the leading edge of the pseudopod converts to plasmagel, the pseudopod stops forming. Thus the basis for amoeboid motion is the ability of the cytoplasm to cycle through sol and gel states and the ability of the microfilament network under the plasma membrane to contract and cause the cytoplasmic streaming that pushes out a pseudopod.

Muscle Contraction

Most behavioral and many physiological responses depend on muscle cells. Muscle cells are specialized for contraction and have high densities of actin and myosin. Muscle cells are found throughout the animal kingdom. Wherever whole tissues contract in animals, muscle cells are responsible.

In muscle cells, actin and myosin molecules are organized into microfilaments consisting of two or more molecules. Actin filaments consist of a twisted chain of actin molecules. Myosin filaments are bundles of many myosin molecules. The actin and myosin filaments lie parallel to each other. When contraction is triggered, the actin and myosin filaments slide past each other in a telescoping fashion. This sliding is the mechanism by which muscle cells contract. Like those of neurons, muscle cell plasma membranes can generate action potentials, and it is these action potentials that trigger the contractile machinery.

There are three types of vertebrate muscle: smooth muscle, cardiac (heart) muscle, and skeletal muscle (Figure 47.1). Although they all use the same contractile mechanism, these three muscle types have important differences that adapt them to their particular functions.

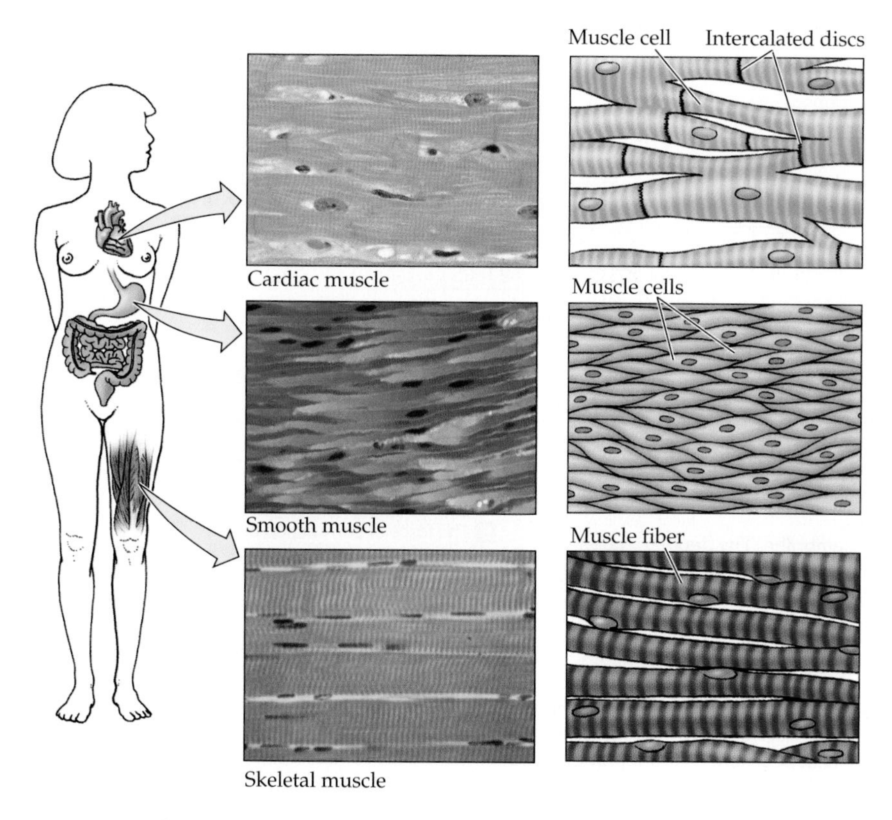

47.1 Types of Vertebrate Muscle Tissue The cells of cardiac, or heart, muscle (*top*), branch and create a meshwork that resists tearing or breaking. Intercalated discs provide strong mechanical adhesion between the cells. In smooth muscle (*center*), the cells are usually arranged in sheets. Skeletal muscle (*bottom*) appears striped, or striated. The individual cells, called muscle fibers, are very large and are multinucleated.

Smooth muscle causes slow contractions of many internal organs

Smooth muscle provides the contractile force for most of our internal organs, which are under the control of the autonomic nervous system. Smooth muscle moves food through the digestive tract, controls the flow of blood through blood vessels, and empties the urinary bladder. Structurally, smooth muscle cells are the simplest muscle cells. They are usually long and spindle-shaped, and each cell has a single nucleus. They are called "smooth" because of their microscopic appearance. The actin and myosin filaments in the other types of muscle cells have such a regular arrangement that they give the cells a striped, or *striated*, appearance. Such regular arrangement of the filaments does not occur in smooth muscle, hence they appear smooth, rather than striated (Figure 47.1, center).

Smooth muscle tissue, such as that from the wall of the digestive tract, has interesting properties. The cells are arranged in sheets, and individual cells in a sheet are in electrical contact with one another through gap junctions. As a result, an action potential generated in the membrane of one smooth muscle cell can spread to all the cells in the sheet of tissue. Thus the cells in the sheet can contract in a coordinated fashion.

Another interesting property of smooth muscle cells is that their plasma membranes are sensitive to being stretched. If the wall of the digestive tract is stretched in one location (as by receiving a swallowed mouthful of food), the membranes of the stretched cells depolarize, reach threshold, and fire action potentials, which cause the cells to contract. Thus, smooth muscle contracts after being stretched, and the harder it is stretched, the stronger the contraction.

Other factors that alter the membrane potential of smooth muscle cells are the neurotransmitters of the autonomic nervous system (see Figure 46.10). In the case of the digestive tract, acetylcholine causes smooth muscle cells to depolarize and thus makes them more likely to fire action potentials and contract. Norepinephrine causes these muscle cells to hyperpolarize and therefore makes them less likely to fire action potentials and contract (Figure 47.2).

Cardiac muscle causes the heart to beat

Cardiac muscle looks different from smooth muscle or skeletal muscle when viewed under the microscope (Figure 47.1, top). Cardiac muscle cells are striated because of the regular arrangement of their actin and myosin filaments. Cardiac muscle cells also branch, and the branches of adjoining cells interdigitate into a meshwork that allows cardiac muscle to resist tearing. As a result, the heart walls can withstand high pressures while pumping blood without the danger of developing leaks. Adding to the strength of cardiac muscle are *intercalated discs* that provide strong mechanical adhesions between adjacent cells.

Like smooth muscle cells, the individual cells in a sheet of cardiac muscle are in electrical contact with one another. Gap junctions in the intercalated discs offer low resistance to ions or electric currents. Therefore, an action potential initiated at one point in the heart spreads rapidly through a large mass of cardiac muscle.

Certain specialized cardiac muscle cells, called **pacemaker cells**, initiate the rhythmic contractions of the heart. We'll learn about the molecular basis for this pacemaking function in Chapter 49. Because of these pacemaker cells, the heartbeat is *myogenic*—generated by the heart muscle itself. The autonomic nervous system modifies the rate of the pacemaker cells, but is not essential for their continued rhythmic function. A heart removed from an animal continues to beat with no input from the nervous system. The myogenic nature of the heartbeat is a major factor in making heart transplants possible.

Question: What stimulates contraction of smooth muscle?

METHOD Incubate a strip of smooth (intestinal) muscle in a saline bath. Measure action potentials and force of contraction.

Experiment 1 Stretch intestinal muscle and analyze response.

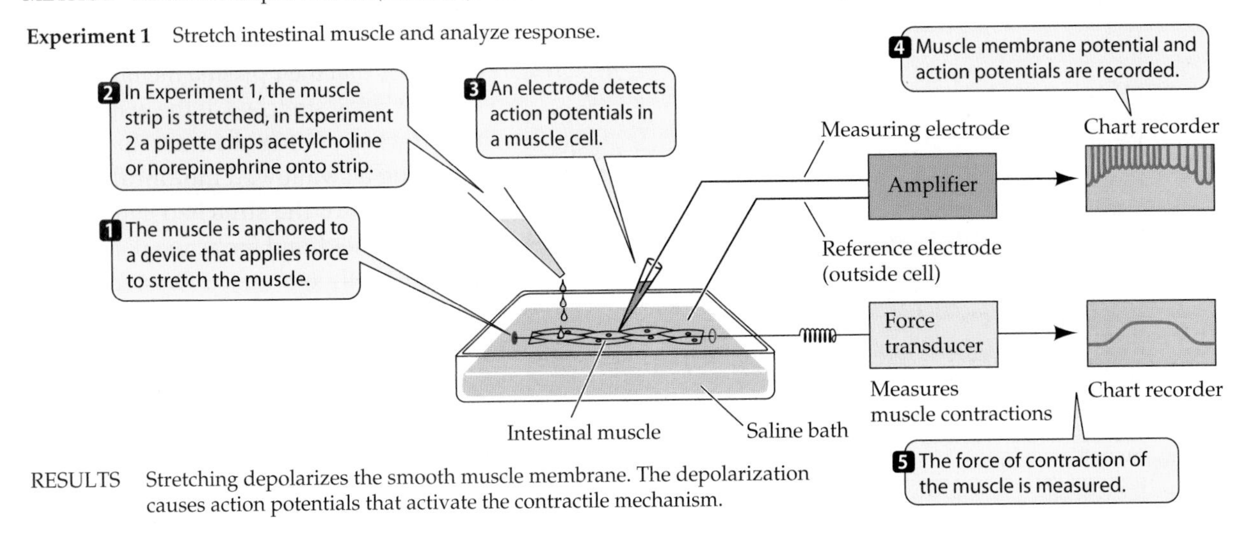

1 The muscle is anchored to a device that applies force to stretch the muscle.
2 In Experiment 1, the muscle strip is stretched, in Experiment 2 a pipette drips acetylcholine or norepinephrine onto strip.
3 An electrode detects action potentials in a muscle cell.
4 Muscle membrane potential and action potentials are recorded.
5 The force of contraction of the muscle is measured.
Measuring electrode
Amplifier
Chart recorder
Reference electrode (outside cell)
Force transducer
Measures muscle contractions
Chart recorder
Intestinal muscle
Saline bath

RESULTS Stretching depolarizes the smooth muscle membrane. The depolarization causes action potentials that activate the contractile mechanism.

Experiment 2 Response of muscle strip to neurotransmitters of the autonomic nervous system.

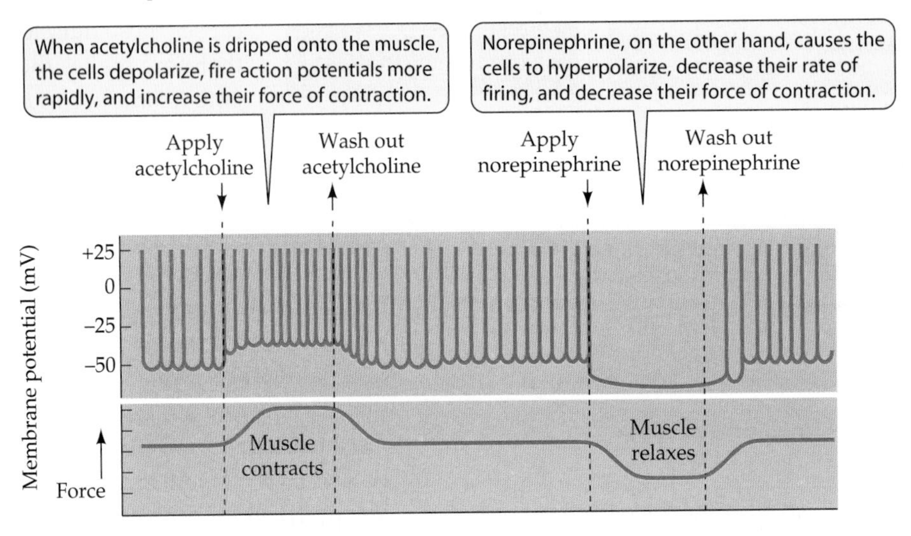

RESULTS Autonomic neurotransmitters alter membrane resting potential and thereby determine the rate that smooth muscle cells fire action potentials.

Conclusion: Smooth muscle contraction is stimulated by stretch and by the parasympathetic neurotransmitter acetylcholine.

47.2 Mechanisms of Smooth Muscle Activation
Stretching depolarizes the membrane of smooth muscle cells, and this depolarization causes action potentials that activate the contractile mechanism. The neurotransmitters acetylcholine and norepinephrine also alter the membrane potential of smooth muscle, making it more or less likely to contract.

Sliding filaments cause skeletal muscle to contract

Skeletal muscle carries out, or *effects*, all voluntary movements, such as running or playing a piano, and generates the movements of breathing. Skeletal muscle is also called *striated muscle* because of its striped appearance (Figure 47.1, bottom). Skeletal muscle cells, called **muscle fibers**, are large. Unlike smooth muscle and cardiac muscle cells, each of which has a single nucleus, skeletal muscle fibers have many nuclei because they develop through the fusion of many individual cells. A muscle such as your biceps (which bends your arm) is composed of many muscle fibers bundled together by connective tissue.

What is the relation between a skeletal muscle fiber and the actin and myosin filaments responsible for its contraction? Each muscle fiber is packed with **myofibrils**—bundles of contractile filaments made up of actin and myosin (Figure 47.3). Within each myofibril are thin actin filaments and thick myosin filaments. If we cut across the myofibril at certain locations, we see only thick filaments; if we cut at other loca-

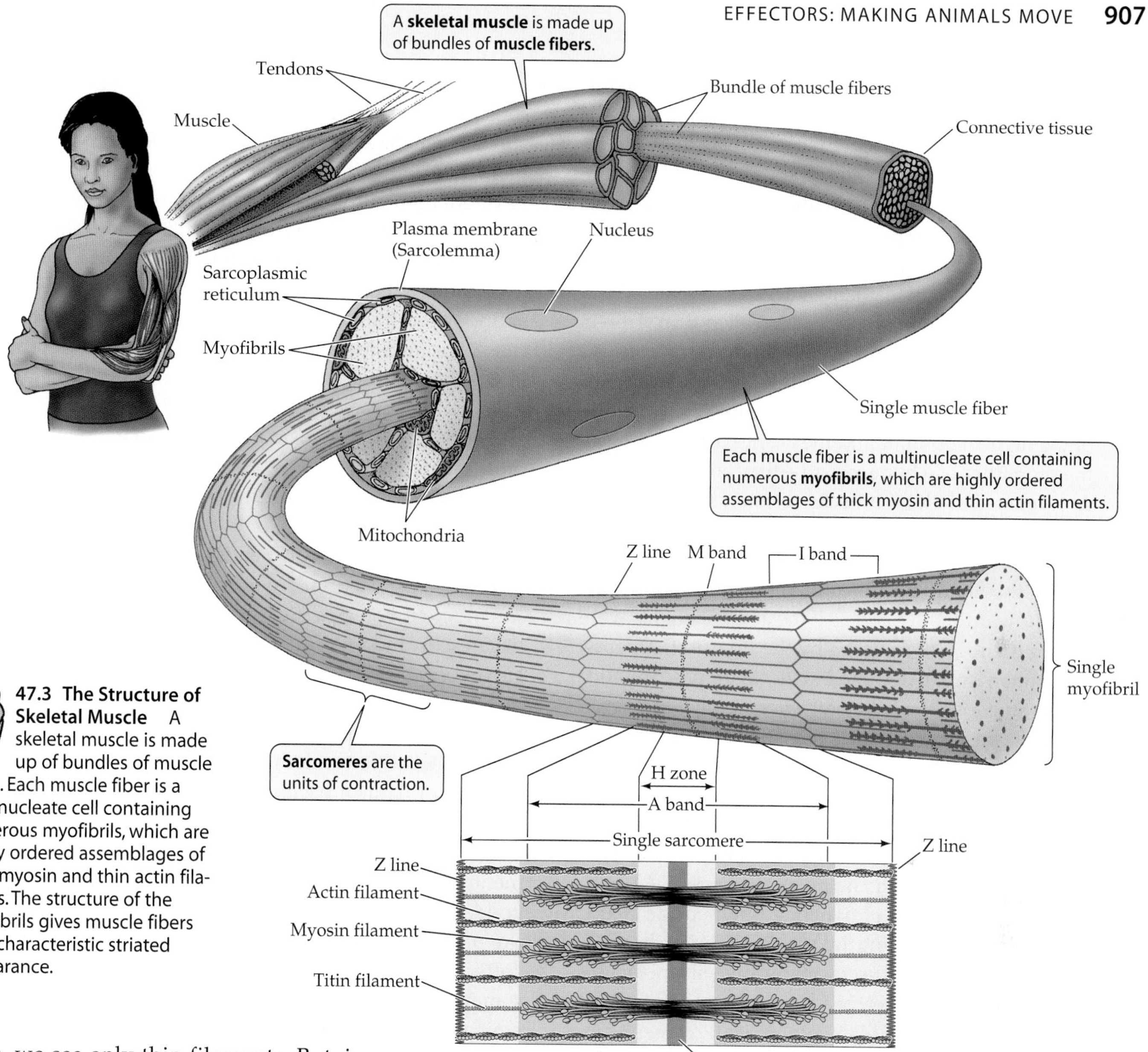

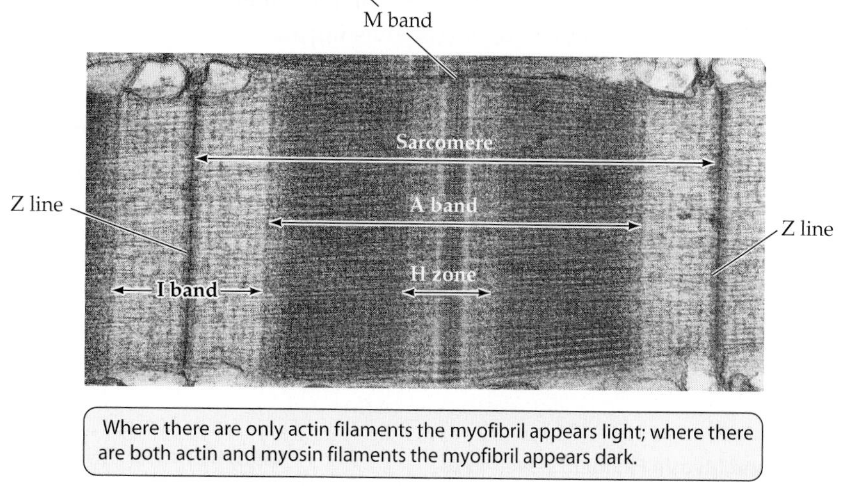

A **skeletal muscle** is made up of bundles of **muscle fibers.**

Each muscle fiber is a multinucleate cell containing numerous **myofibrils**, which are highly ordered assemblages of thick myosin and thin actin filaments.

Sarcomeres are the units of contraction.

Where there are only actin filaments the myofibril appears light; where there are both actin and myosin filaments the myofibril appears dark.

47.3 The Structure of Skeletal Muscle A skeletal muscle is made up of bundles of muscle fibers. Each muscle fiber is a multinucleate cell containing numerous myofibrils, which are highly ordered assemblages of thick myosin and thin actin filaments. The structure of the myofibrils gives muscle fibers their characteristic striated appearance.

tions, we see only thin filaments. But, in most regions of the myofibril, each thick myosin filament is surrounded by six thin actin filaments, and conversely, each thin actin filament sits within a triangle of three thick myosin filaments.

A longitudinal view of a myofibril reveals the reason for the striated appearance of skeletal muscle (and cardiac muscle). The myofibril consists of repeating units, called **sarcomeres**, which are the units of contraction. Each sarcomere is made of overlapping filaments of actin and myosin, which create a distinct band pattern. As the muscle contracts, the sarcomeres shorten, and the appearance of the band pattern changes.

The observation that the widths of the bands in the sarcomeres change when a muscle contracts led two British biologists, Hugh Huxley and Andrew Huxley, to propose a molecular mechanism of muscle contraction. Let's look at the band pattern of a sarcomere in detail (see the micrograph in Figure 47.3). Each sarcomere is bounded by *Z lines*, which are structures that anchor the thin actin filaments. Centered in the sarcomere is the *A band*, which contains all the myosin filaments. The *H zone* and the *I band*, which appear light, are regions where actin and myosin filaments do not overlap in the relaxed muscle. The dark stripe within the H zone is called the *M band;* it contains proteins that help hold the myosin filaments in their regular arrangement.

The bundles of myosin filaments are held in a centered position within the sarcomere by a protein called **titin**. Titin is probably the longest polypeptide in the body; it runs the full length of the sarcomere from Z line to Z line, and each titin molecule runs right through a myosin bundle. Between the ends of the myosin bundles and the Z lines, titin molecules have the properties of a bungee cord—they are very stretchable. In a relaxed skeletal muscle, resistance to stretch is mostly due to the elasticity of the titin molecules.

When the muscle contracts, the sarcomere shortens. The H zone and the I band become much narrower, and the Z lines move toward the A band as if the actin filaments were sliding into the region occupied by the myosin filaments. This observation led Huxley and Huxley to propose the **sliding filament theory** of muscle contraction: Actin and myosin filaments slide past each other as the muscle contracts.

Actin–myosin interactions cause filaments to slide

To understand what makes the filaments slide, we must examine the structures of actin and myosin (Figure 47.4). Each myosin molecule consists of two long polypeptide chains coiled together, each ending in a large globular head. A myosin filament is made up of many myosin molecules arranged in parallel, with their heads projecting laterally from one or the other end of the filament. An actin filament consists of a helical arrangement of two chains of actin monomers twisted together like two strands of pearls. Twisting around the actin chains is another protein, tropomyosin, and attached to it at intervals are molecules of troponin. We'll discuss the roles of these last two proteins in the following section.

The myosin heads have sites that can bind to actin and thereby form cross-bridges between the myosin and the actin filaments. The myosin heads also have ATPase activity; that is, they bind and hydrolyze ATP. The energy released when this happens changes the conformation, and therefore the orientation, of the myosin head.

Together, these details explain the cycle of events that cause the actin and myosin filaments to slide past each other and shorten the sarcomere. A myosin head binds to an actin filament (see Figure 47.6). Upon binding, the head changes its orientation with respect to the myosin filament, thus exerting a force that causes the actin filament to slide about 5 to 10 nm relative to the myosin filament. Next, the myosin head binds a molecule of ATP, which causes it to release the actin. When the ATP is hydrolyzed, the energy released causes the myosin head to return to its original conformation, in which it can bind again to actin. It is as if the energy from ATP hydrolysis is being used to cock the hammer of a pistol, and contact of the myosin head with an actin binding site pulls the trigger.

We have been discussing the cycle of contraction in terms of a single myosin head. Don't forget that each myosin filament has many myosin heads at both ends and is surrounded by six actin filaments; thus the contraction of the sarcomere involves a great many cycles of interaction between actin and myosin molecules. That is why when a single myosin head breaks its contact with actin, the actin filaments do not slip backward.

An interesting aspect of this contractile mechanism is that ATP is needed to break the actin–myosin bonds, but not to form them. Thus muscles require ATP to *stop* contracting. This fact explains why muscles stiffen soon after animals die, a condition known as *rigor mortis*. Death stops the replenishment of the ATP stores of muscle cells, so the actin–myosin bonds cannot be broken, and the muscles stiffen. Eventually the proteins begin to lose their integrity, and the muscles soften. These events have regular time courses that differ somewhat for different regions of the body; therefore, an examination of the stiffness of the muscles of a corpse can help a coroner estimate the time of death.

Actin–myosin interactions are controlled by calcium ions

Muscle contractions are initiated by action potentials from motor neurons arriving at the neuromuscular junction (see

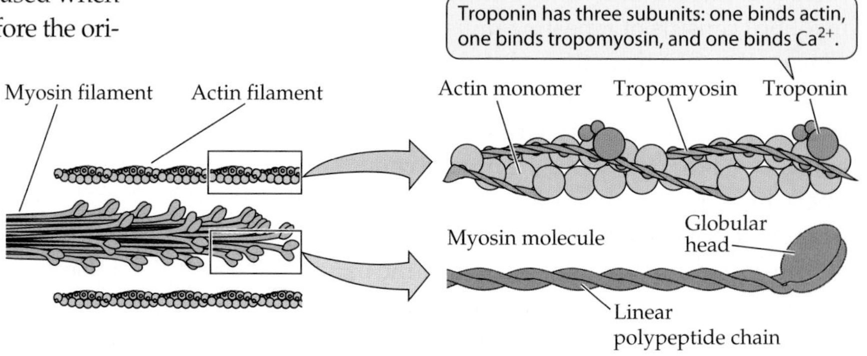

47.4 Actin and Myosin Filaments Overlap to Form Myofibrils Myosin filaments are bundles of molecules with globular heads and polypeptide tails. Actin filaments consist of two chains of actin monomers twisted together. They are wrapped by chains of the polypeptide tropomyosin and studded at intervals with another protein, troponin.

Figure 44.13). The axons of motor neurons are generally highly branched and can synapse with up to a hundred muscle fibers each. All the fibers activated by a single motor neuron constitute a **motor unit** and contract simultaneously in response to action potentials fired by that motor neuron.

Like neurons, muscle cells are *excitable;* that is, their plasma membranes can generate and conduct action potentials. In the case of skeletal muscle fibers (but not smooth or cardiac muscle fibers), all action potentials are initiated by motor neurons. When an action potential arrives at the neuromuscular junction, the neurotransmitter acetylcholine is released from the motor neuron, diffuses across the synaptic cleft, binds to receptors in the postsynaptic membrane, and causes ion channels in the motor end plate to open. Most of the ions that flow through these channels are Na^+, and therefore the motor end plate is depolarized. The depolarization spreads to the surrounding plasma membrane of the muscle fiber, which contains voltage-gated sodium channels. When threshold is reached, the plasma membrane fires an action potential that is conducted rapidly to all points on the surface of the muscle fiber.

An action potential in a muscle fiber also travels deep within the cell. The plasma membrane is continuous with a system of tubules that descends into and branches throughout the cytoplasm of the muscle fiber (also called the **sarcoplasm**) (Figure 47.5). The action potential that spreads over the plasma membrane also spreads through this system of transverse tubules, or **T tubules**.

The T tubules come very close to a network of intracellular membranes called the **sarcoplasmic reticulum**. The sarcoplasmic reticulum forms a membrane-enclosed compartment that surrounds every myofibril. Calcium pumps in the sarcoplasmic reticulum cause it to take up Ca^{2+} ions from the sarcoplasm. Therefore, when the muscle fiber is at rest, there is a high concentration of Ca^{2+} in the sarcoplasmic reticulum and a low concentration of Ca^{2+} in the sarcoplasm.

Spanning the space between the membranes of the T tubules and the membranes of the sarcoplasmic reticulum are two proteins. One protein, which is located in the T tubule membrane, is voltage-sensitive and changes its conformation when an action potential reaches it. The other protein is located in the sarcoplasmic reticulum membrane and is a Ca^{2+} channel. When it is activated by an action potential, the voltage-sensitive protein opens the Ca^{2+} channel, and Ca^{2+} ions diffuse out of the sarcoplasmic reticulum and into the sarcoplasm surrounding the actin and myosin filaments. It is these Ca^{2+} ions that trigger the interaction of actin and myosin and the sliding of the filaments. How do the Ca^{2+} ions do this?

An actin filament, as we have seen, is a helical arrangement of two strands of actin monomers. Lying in the grooves between the two actin strands is the two-stranded protein **tropomyosin** (see Figure 47.4). At regular intervals, the filament also includes a globular protein, **troponin**. The troponin molecule has three subunits: One binds actin, one binds tropomyosin, and one binds Ca^{2+}.

When Ca^{2+} is sequestered in the sarcoplasmic reticulum, the tropomyosin strands block the sites on the actin filament where myosin heads can bind. When the T tubule system depolarizes, Ca^{2+} is released into the sarcoplasm, where it binds to troponin, changing its conformation. Because the troponin is bound to the tropomyosin, this conformational change of the troponin twists the tropomyosin enough to expose the actin–myosin binding sites. Thus the cycle of making and breaking actin–myosin bonds is initiated, the filaments are pulled past each other, and the muscle fiber contracts. When the T tubule system repolarizes, the calcium pumps remove the Ca^{2+} ions from the sarcoplasm, causing the tropomyosin to return to the position in which it blocks the binding of myosin heads to actin, and the muscle fiber returns to its resting condition. Figure 47.6 summarizes this cycle.

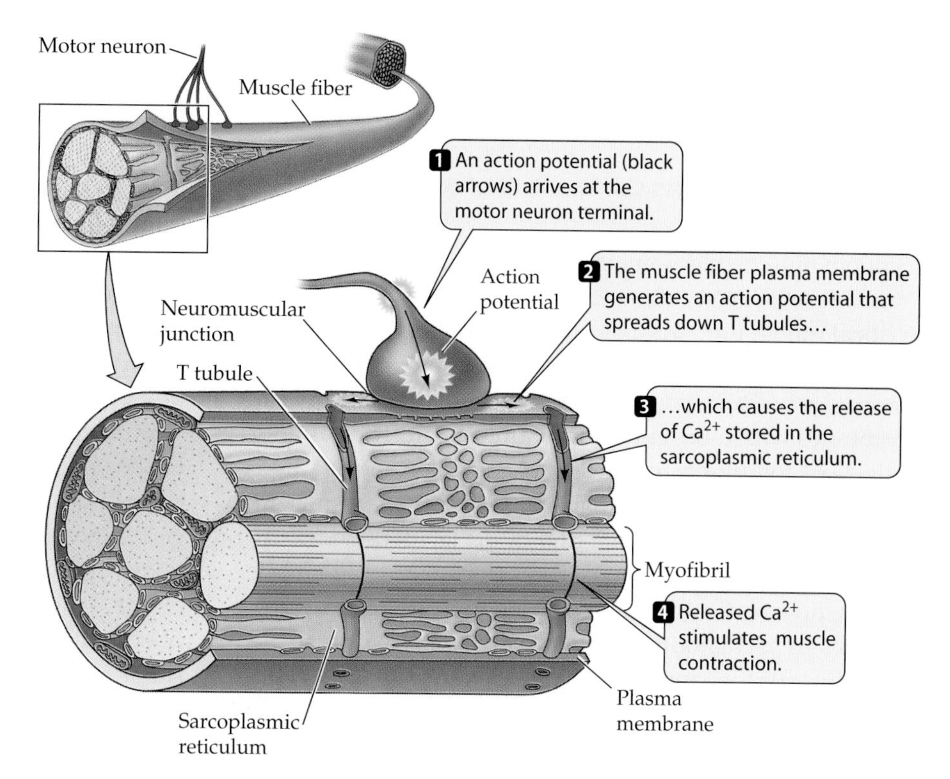
Motor neuron
Muscle fiber
1 An action potential (black arrows) arrives at the motor neuron terminal.
Action potential
2 The muscle fiber plasma membrane generates an action potential that spreads down T tubules...
Neuromuscular junction
T tubule
3 ...which causes the release of Ca^{2+} stored in the sarcoplasmic reticulum.
Myofibril
4 Released Ca^{2+} stimulates muscle contraction.
Plasma membrane
Sarcoplasmic reticulum

47.5 T Tubules in Action An action potential at the neuromuscular junction spreads throughout the muscle fiber via a network of T tubules, triggering the release of Ca^{2+} from the sarcoplasmic reticulum.

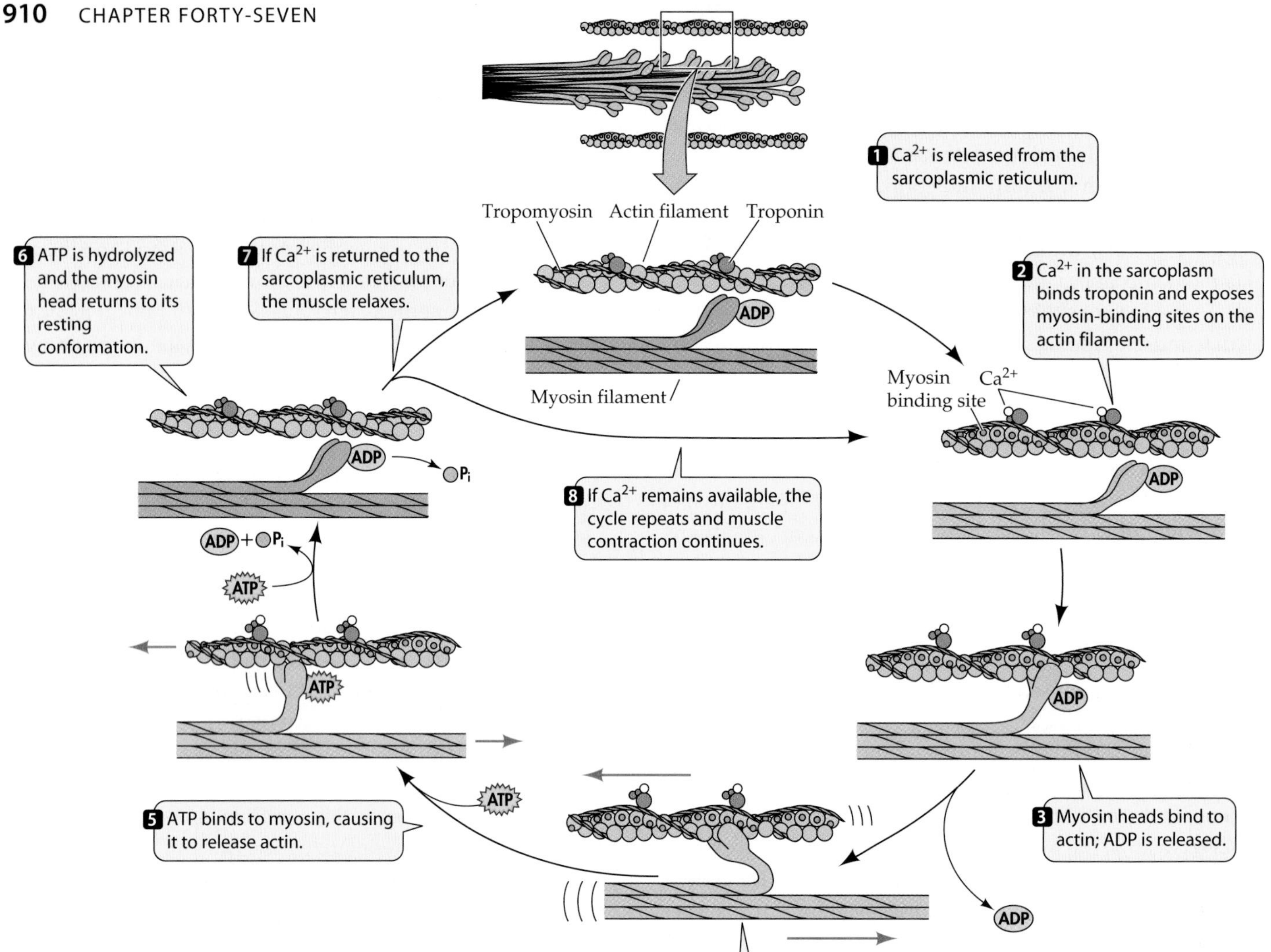

47.6 The Release of Ca^{2+} from the Sarcoplasmic Reticulum Triggers Muscle Contraction When Ca^{2+} binds to troponin, it exposes actin–myosin binding sites. As long as binding sites and ATP are available, the cycle of actin and myosin interactions continues, and the filaments slide past each other.

Calmodulin mediates Ca^{2+} control of contraction in smooth muscle

Smooth muscle cells do not have the troponin–tropomyosin mechanism for controlling contraction, but Ca^{2+} still plays a critical role. A Ca^{2+} influx into the sarcoplasm of a smooth muscle cell can be stimulated by action potentials, by hormones, or by stretching. The Ca^{2+} that enters the sarcoplasm combines with a protein called **calmodulin**. The calmodulin–Ca^{2+} complex activates an enzyme called *myosin kinase*, which can phosphorylate myosin heads. When the myosin heads in smooth muscle are phosphorylated, they can undergo cycles of binding and releasing actin, causing muscle contraction. As Ca^{2+} is removed from the sarcoplasm, it dissociates from calmodulin, and the activity of myosin kinase falls. In addition, another enzyme, *myosin phosphatase*, dephosphorylates the myosin and helps stop the actin–myosin interactions.

Single skeletal muscle twitches are summed into graded contractions

In skeletal muscle, the arrival of an action potential at a neuromuscular junction causes an action potential in a muscle fiber. The spread of that action potential through the T tubule system of the muscle fiber causes a minimum unit of contraction, called a **twitch**. A twitch can be measured in terms of the *tension*, or force, it generates (Figure 47.7*a*). A single action potential stimulates a single twitch, but the ultimate force generated by a muscle can vary enormously depending on how many muscle fibers are in its motor units. In muscles responsible for fine movements, such as those of the fingers, a motor neuron may innervate only one or a few muscle fibers, but in a muscle that produces large forces, such as the biceps, a motor neuron innervates a large number of muscle fibers. Still, however, at the level of the single muscle fiber, a single action potential stimulates a single twitch.

If action potentials reaching the muscle fiber are adequately separated in time, each twitch is a discrete, all-or-

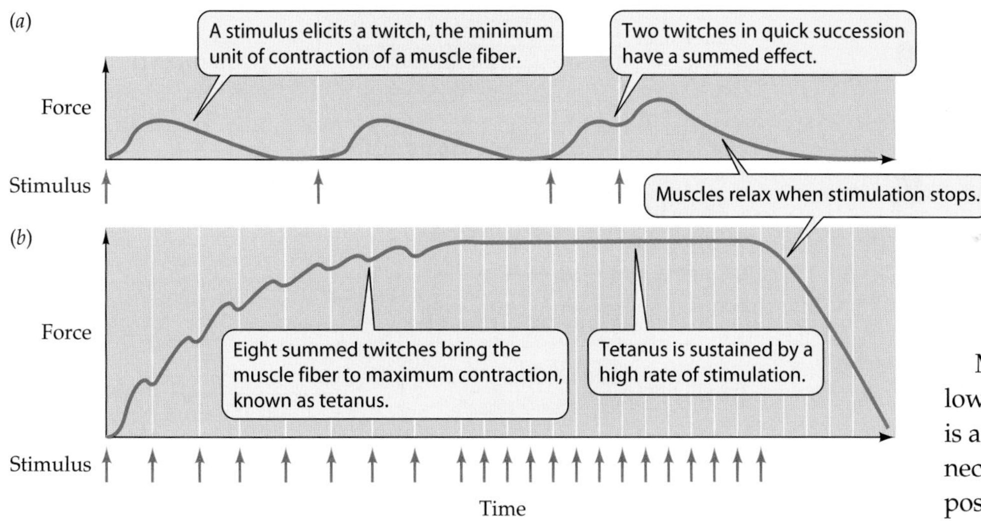
(a)
A stimulus elicits a twitch, the minimum unit of contraction of a muscle fiber.
Two twitches in quick succession have a summed effect.
Force
Stimulus
Muscles relax when stimulation stops.
(b)
Force
Eight summed twitches bring the muscle fiber to maximum contraction, known as tetanus.
Tetanus is sustained by a high rate of stimulation.
Stimulus
Time

47.7 Twitches and Tetanus
(*a*) Action potentials from a motor neuron cause a muscle fiber to twitch. Twitches in quick succession can be summed. (*b*) Summation of many twitches can bring the muscle fiber to the maximum level of contraction, known as tetanus.

none phenomenon. If action potentials are fired more rapidly, however, new twitches are triggered before the myofibrils have had a chance to return to their resting condition. As a result, the twitches sum, and the tension generated by the fiber increases and becomes more sustained. Thus an individual muscle fiber can show a graded response to increased levels of stimulation by its motor neuron.

At high levels of stimulation, the calcium pumps in the sarcoplasmic reticulum can no longer remove Ca^{2+} ions from the sarcoplasm between action potentials, and the contractile machinery generates maximum tension—a condition known as **tetanus** (Figure 47.7*b*). (Do not confuse this condition with the disease *tetanus*, which is caused by a bacterial toxin and is characterized by spastic contractions of skeletal muscles.)

How long a muscle fiber can maintain a tetanic contraction depends on its supply of ATP. Eventually the fiber will become fatigued. It may seem paradoxical that the *lack* of ATP causes fatigue, since the action of ATP is to break actin–myosin bonds. But remember that the energy released from the hydrolysis of ATP "re-cocks" the myosin heads, allowing them to cycle through another power stroke. When a muscle is contracting against a load, the cycle of making and breaking actin–myosin bonds must continue to prevent the load from stretching the muscle. The situation is like rowing a boat upstream: You cannot maintain your position relative to the stream bank by just holding the oars out against the current; you have to keep rowing. Likewise, actin–myosin bonds have to keep cycling to maintain tension in the muscle.

The level of tension generated by a muscle depends on how many motor units in that muscle are activated. Whether a muscle contraction is strong or weak depends both on how many of the motor neurons that synapse with that muscle are firing and on the rate at which those neurons are firing. These two factors can be thought of as *spatial summation* and *temporal summation*, respectively.

Many muscles of the body maintain a low level of tension even when the body is at rest. For example, the muscles of the neck, trunk, and limbs that maintain our posture against the pull of gravity are always working, even when we are standing or sitting still. **Muscle tone** comes from the activity of a small but changing number of motor units in a muscle; at any one time, some of the muscle's fibers are contracting and others are relaxed. Muscle tone is constantly being readjusted by the nervous system.

Muscle Strength and Performance

Muscle fiber types determine endurance and strength

Not all skeletal muscle fibers are alike, and a single muscle contains more than one type of fiber. The two major types of skeletal muscle fibers differ in the properties of their myosin molecules, and these myosin variants have different rates of ATPase activity. Those with high ATPase activity can recycle their actin–myosin cross-bridges rapidly and are therefore called fast-twitch fibers. Slow-twitch fibers have lower ATPase activity, so they can develop tension more slowly and spread maintain it over a longer period of time.

Slow-twitch fibers are also called *red muscle* because they contain lots of the oxygen-binding molecule **myoglobin**, they have many mitochondria, and they are well supplied with blood vessels. These characteristics increase their capacity for oxidative metabolism. The maximum tension a slow-twitch fiber produces is low and develops slowly, but it is highly resistant to fatigue. Because slow-twitch fibers have substantial reserves of fuel (glycogen and fat), their abundant mitochondria can maintain steady, prolonged production of ATP as long as oxygen is available. Muscles with high proportions of slow-twitch fibers are good for long-term *aerobic* work (that is, work that requires lots of oxygen). Champion long-distance runners, cross-country skiers, swimmers, bicyclists, and other athletes whose activities require endurance have leg and arm muscles consisting mostly of slow-twitch fibers (Figure 47.8).

Fast-twitch fibers are also called *white muscle* because, in comparison to slow-twitch fibers, they have fewer mito-

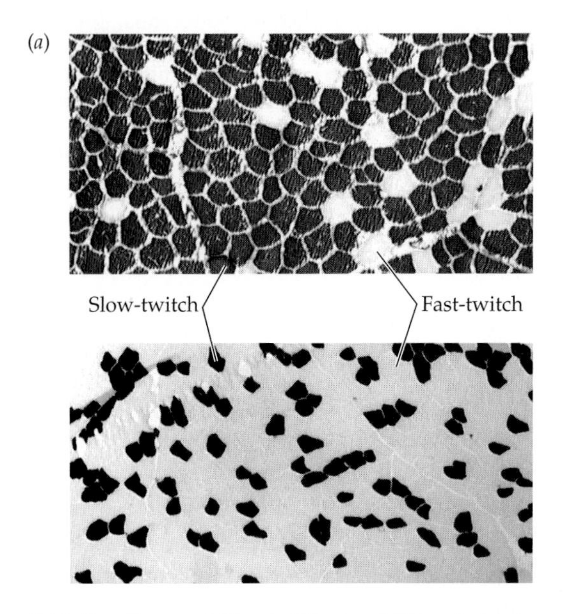

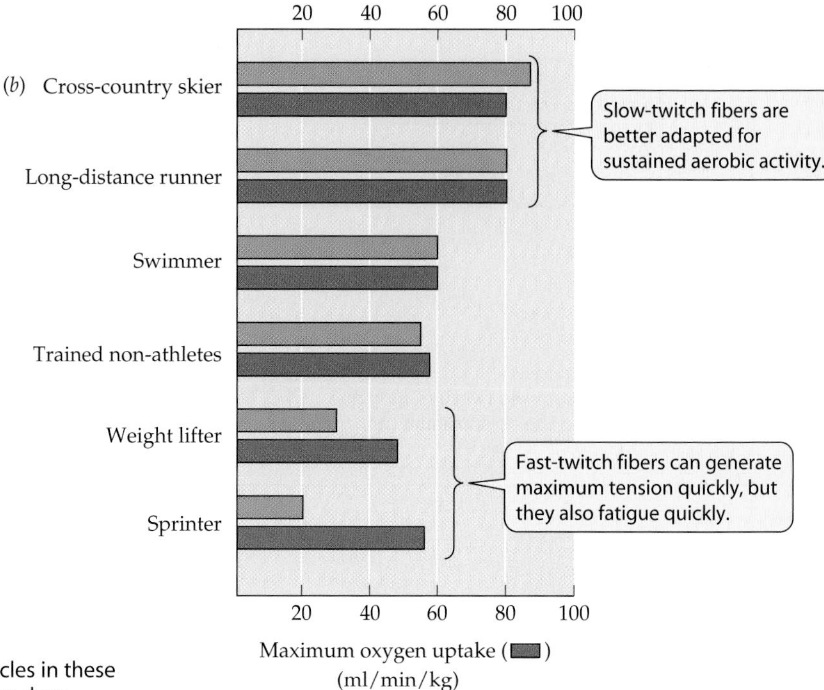

47.8 Two Types of Muscle Fibers (*a*) The skeletal muscles in these micrographs have been stained with a reagent that shows slow-twitch fibers as dark. The upper photo shows muscle from a professional cyclist. The lower photo shows muscle from a nonathlete who has about 75 percent fast-twitch fibers; this person would probably perform better as a sprinter than as a distance runner. (*b*) Athletes in different sports have different distributions of muscle fiber types.

chondria, little or no myoglobin, and fewer blood vessels. The white meat of domestic chickens has a high percentage of fast-twitch fibers. Fast-twitch fibers can develop maximum tension more rapidly than slow-twitch fibers can, and that maximum tension is greater, but fast-twitch fibers fatigue rapidly. The myosin of fast-twitch fibers can put the energy of ATP to work very rapidly, but the fibers cannot replenish ATP quickly enough to sustain contraction for a long time. Fast-twitch fibers are especially good for short-term work that requires maximum strength. Champion weight lifters and sprinters have leg and arm muscles with high proportions of fast-twitch fibers.

What determines the proportion of fast- and slow-twitch fibers in your skeletal muscles? The most important factor is genetic heritage, so there is some truth to the statement that champions are born, not made. To a certain extent, you can alter the properties of your muscle fibers through training. But a person born with a high proportion of fast-twitch fibers will never become a champion marathon runner, and one born with a high proportion of slow-twitch fibers will never become a champion sprinter.

The strength of a muscle fiber is related to its length

Have you ever done a pull-up? If you have, you know that two parts of this exercise are especially difficult. When you are hanging from the bar with your arms fully extended, it is hard to get the pull-up started; and when your chin is just about to the bar, pulling yourself up the last small distance is difficult. These experiences are explained by the structure of the sarcomere.

When a muscle is stretched and the sarcomeres are lengthened, there is less overlap between the actin and myosin filaments; therefore, fewer cross-bridges can form, and less force can be produced. In fact, if the sarcomeres are stretched too much, there is no overlap between the actin and myosin, and no force can be produced. How would a muscle recover from such a difficult situation? The bungee cord-like titin molecules create enough elastic recoil to pull the actin and myosin fibrils back into an overlapping arrangement.

When the muscle is fully contracted, the actin and myosin filaments overlap so much that the myosin bundles are pressed up against the Z lines. Because they have no place to go, additional shortening is difficult. You can see the relationship between the length of a muscle fiber and its ability to develop tension in Figure 47.9.

Exercise increases muscle strength and endurance

Different types of exercise produce different physical conditioning responses. In general, anaerobic activities, such as weight lifting, increase strength, and aerobic activities, such as jogging, increase endurance. What is the physiological basis for these differences? Strength is quite simply a function of the cross-sectional area of muscles: the more actin and myosin filaments in a muscle or a muscle fiber, the more tension it can produce. When athletes undertake strength train-

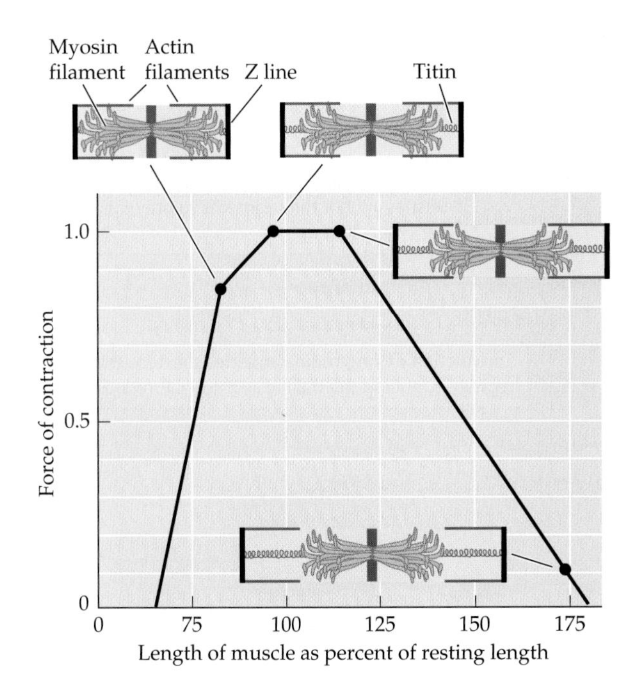

47.9 Strength and Length The amount of force a sarcomere can generate depends on its resting length. When a muscle is stretched, the sarcomeres lengthen, there is less overlap between the actin and myosin filaments, and less force is produced. Overstretched sarcomeres produce no force because there is no overlap between the actin and myosin.

ing, they use weights or exercises such as push-ups to repeatedly contract specific muscles under heavy loads. Repetitions are usually done until the muscle is completely fatigued. Such stress on a muscle probably does minor tissue damage—hence the soreness the day after a hard workout—but it also induces the formation of new actin and myosin filaments in existing muscle fibers. The muscle fibers, and hence the muscles, get bigger and stronger. In extreme cases, and after serious muscle damage, new muscle fibers can also be produced from stem cells called *satellite cells* in the muscle. In general, however, the major effect of strength training is to produce bigger, rather than more, muscle fibers.

Aerobic exercise has a completely different effect on muscles: it enhances their oxidative capacity. This effect comes from increases in the number of mitochondria, increases in enzymes involved in energy utilization, and an increase in the density of capillaries that deliver oxygen to the muscle. There is also an increase in myoglobin, which facilitates the diffusion of oxygen throughout the muscle fibers and provides a store of oxygen for use when oxygen delivery by the blood is insufficient. In this way, aerobic training can stimulate many fast-twitch fibers to increase their oxidative capacity.

Muscle fuel supply limits performance

Muscles have three systems for obtaining the ATP they need for contraction:

- The *immediate system:* preformed ATP and creatine phosphate
- The *glycolytic system:* metabolizing carbohydrates to lactate and pyruvate
- The *oxidative system:* metabolizing carbohydrates or fats all the way to H_2O and CO_2

The capacity of these three systems and the rates at which they can produce ATP determine both work capacity and endurance (Figure 47.10).

ATP is stored in muscles in very small amounts. However, muscle fibers also contain a storage compound called *creatine phosphate* (CP). This molecule stores energy in a phosphate bond, which it can transfer to ADP. The total energy available in all the muscles of your body in the form of ATP and CP—the immediate energy system—is only about 10 Calories.

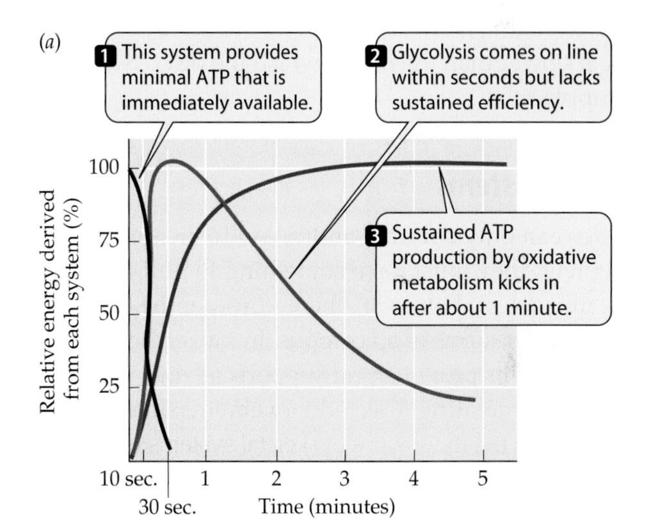

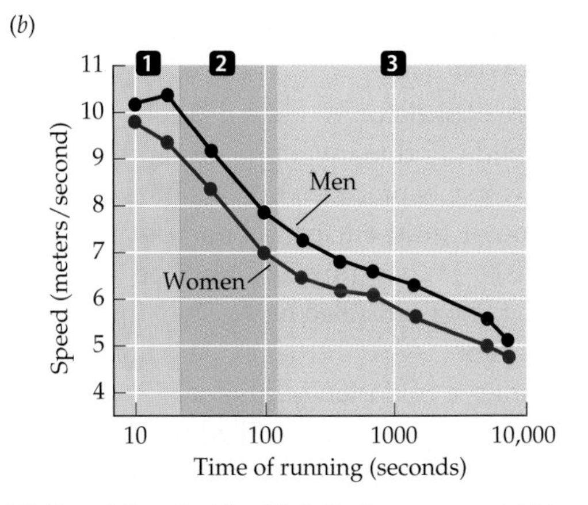

47.10 Supplying Fuel for High Performance (*a*) Muscles have three systems for obtaining the ATP they need for contraction during exertion (such as running). (*b*) Plotting the time course of world records for running events, you can see that the three segments of performance in world-class athletes correspond to the time courses of the three energy systems.

However, it is available immediately, and it enables fast-twitch fibers to generate a lot of force quickly. The immediate system is exhausted in only a few seconds.

The glycolytic system is able to come on line within a few seconds to replace the ATP depleted at the onset of muscle activity. The glycolytic enzymes are located in the cytoplasm of the muscle fiber, and therefore the ATP they generate is rapidly available to the myosin filaments. However, as we saw in Chapter 7, glycolysis alone is an inefficient way to produce ATP, and it rapidly leads to the accumulation of lactic acid, which slows the process. Thus, the glycolytic system and the immediate system together can provide most of the energy for active muscles for less than a minute (see Figure 47.10).

Oxidative metabolism can come on line fully in about a minute, producing relatively huge amounts of ATP due to its ability to completely metabolize carbohydrates and fats. However, it requires many reactions (see Chapter 7), and it takes place in the mitochondria, so the ATP has to diffuse to the myosin filaments in the muscle. Therefore, the rate at which oxidative metabolism can make ATP available to do work is slower than the rate at which the other two systems can supply ATP.

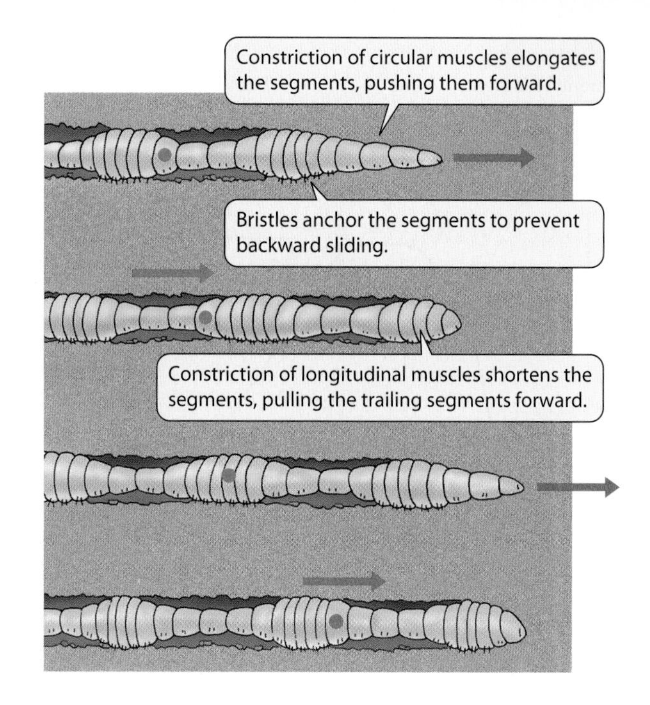

47.11 A Hydrostatic Skeleton Alternating waves of muscle contraction move the earthworm through the soil. The red dot enables you to follow the changes in one segment as the worm moves to the right.

Skeletal Systems

Muscles can only contract and relax. To create significant movement, they must have something to pull on. In some cases, muscles pull on each other—consider the trunk of the elephant or the arms of an octopus. In most cases, however, skeletal systems provide rigid supports against which muscles can pull, creating directed movements. In this section, we'll examine the three types of skeletal systems found in animals: hydrostatic skeletons, exoskeletons, and endoskeletons.

A hydrostatic skeleton consists of fluid in a muscular cavity

The simplest type of skeleton is the **hydrostatic skeleton** of cnidarians, annelids, and many other soft-bodied invertebrates. As we saw in Chapter 32, a hydrostatic skeleton consists of a volume of fluid enclosed in a body cavity surrounded by muscle. When muscles oriented in a certain direction contract, the fluid-filled body cavity bulges out in the opposite direction.

The sea anemone, a cnidarian (see Figure 32.6*a*), has a hydrostatic skeleton. Its body cavity is filled with seawater. To extend its body and its tentacles, the anemone closes its mouth and constricts muscle fibers that are arranged in circles around its body. Contraction of these circular muscles puts pressure on the water in the body cavity, and that pressure forces the body and tentacles to extend. The anemone retracts its tentacles and body by contracting muscle fibers that are arranged longitudinally (lengthwise) in the body wall and along the tentacles.

An earthworm uses its hydrostatic skeleton to crawl. The earthworm's body cavity is divided into many separate segments, each of which contains a compartment filled with extracellular fluid. The body wall surrounding each segment has two muscle layers: a circular layer and a longitudinal layer. If the circular muscles in a segment contract, the compartment in that segment narrows and elongates. If the longitudinal muscles in a segment contract, the compartment shortens and bulges outward. Alternating contractions of the earthworm's circular and longitudinal muscles create waves of narrowing and widening, lengthening and shortening, that travel down the body. Bulging, shortened segments serve as anchors as long, narrow segments project forward and longitudinal contractions pull other segments forward. Bristles help the widest parts of the body to hold firm against the substratum (Figure 47.11).

Another type of locomotion made possible by hydrostatic skeletons is the jet propulsion used by squids and octopuses. Muscles surrounding a water-filled cavity in these cephalopods contract, forcefully expelling water from the animal's body. As the water shoots out under pressure, the animal is propelled in the opposite direction.

Exoskeletons are rigid outer structures

An **exoskeleton** is a hardened outer surface to which muscles can be attached. Contractions of the muscles cause jointed segments of the exoskeleton to move relative to each other. The simplest example of an exoskeleton is the shell of a mollusk. Some marine mollusks, such as clams and snails, have shells composed of protein strengthened by crystals of calcium carbonate (a rock-hard material). These shells can be massive, affording significant protection against predators. The shells of land snails generally lack the hard mineral component and are much lighter. Molluscan shells can grow as the animal grows, and growth rings are usually apparent on the shells.

The most complex exoskeletons are found among the arthropods. An exoskeleton, or *cuticle*, covers all the outer surfaces of the arthropod's body and all its appendages. It is made up of plates secreted by a layer of cells just below the exoskeleton. The cuticle contains stiffening materials everywhere except at the joints, where flexibility must be retained. Muscles attached to the inner surfaces of the arthropod exoskeleton move its parts around the joints (see Figure 33.8).

The layers of the cuticle include an outer, thin, waxy *epicuticle* that protects the body from drying out, and a thicker, inner *endocuticle* that forms most of the structure. The endocuticle is a tough, pliable material found only in arthropods. It consists of a complex of protein and *chitin*, a nitrogen-containing polysaccharide. In marine crustaceans, the endocuticle is further toughened by insoluble calcium salts. The thickness of the cuticle varies among species, but it can be thick enough to form a protective armor.

An exoskeleton protects all the soft tissues of the animal, but is itself subject to damage by abrasion and crushing. The greatest drawback of the arthropod exoskeleton is that it cannot grow. Therefore, if the animal is to become larger, it must *molt*, shedding its exoskeleton and forming a new, larger one. A molting animal is vulnerable because the new exoskeleton takes time to harden. The animal's body is temporarily unprotected, and without a firm exoskeleton against which its muscles can exert maximum tension, it is unable to move rapidly. Soft-shelled crabs, a gourmet delicacy, are crabs caught when they are molting.

Vertebrate endoskeletons provide supports for muscles

The **endoskeleton** of vertebrates is an internal scaffolding. Muscles are attached to it and pull against it. Endoskeletons are composed of rodlike, platelike, and tubelike bones connected to one another at a variety of joints that allow a wide range of movements. An advantage of endoskeletons over exoskeletons is that they can grow. Because bones are inside the body, the body can enlarge without shedding its skeleton.

The human skeleton consists of 206 bones, some of which are shown in Figure 47.12. It can be divided into an *axial skeleton*, which includes the skull, vertebral column, and ribs, and an *appendicular skeleton*, which includes the pectoral girdle, the pelvic girdle, and the bones of the arms, legs, hands, and feet.

Two kinds of connective tissue cells produce large amounts of extracellular matrix material to create the vertebrate endoskeleton. The matrix material produced by *cartilage cells* is a rubbery mixture of polysaccharides and proteins—mainly fibrous collagen. Collagen fibers run in all directions like reinforcing cords through the gel-like matrix and give it the well-known strength and resiliency of "gristle." This matrix, called **cartilage**, is found in parts of the endoskeleton where both stiffness and resiliency are required,

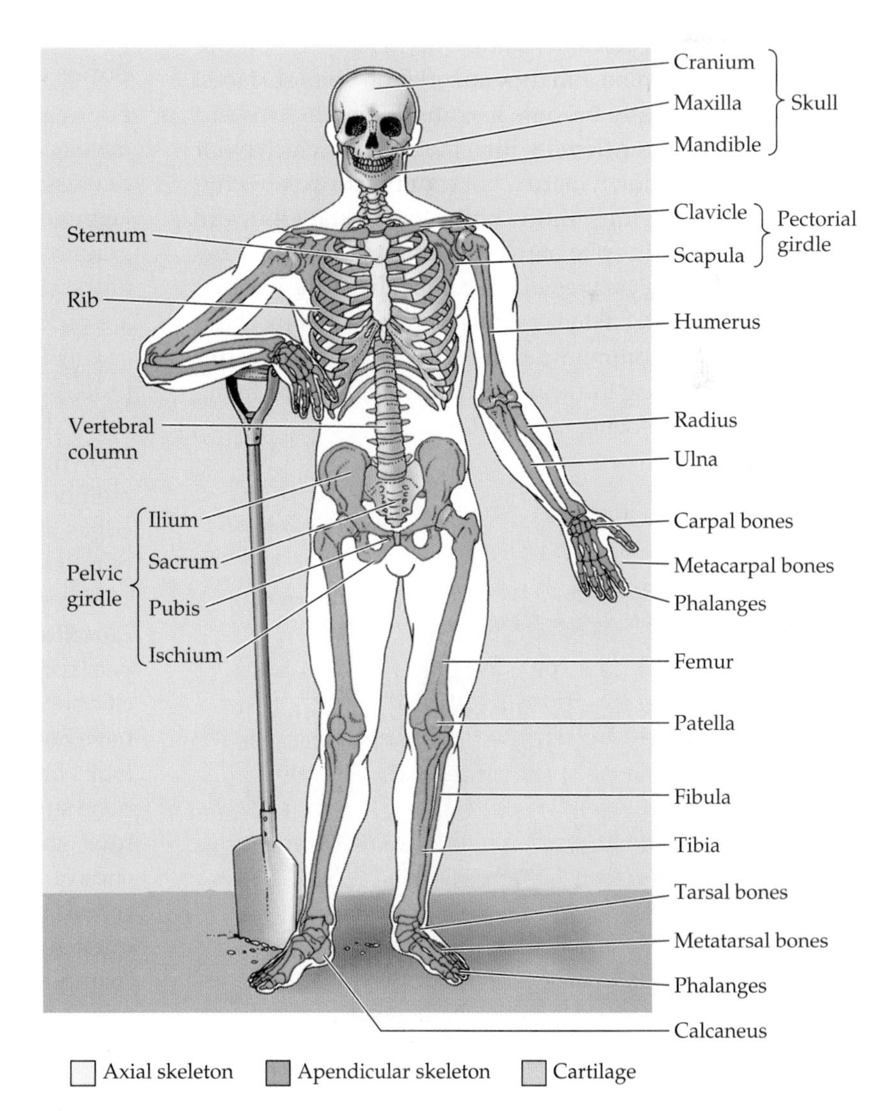

47.12 The Human Endoskeleton Cartilage and bone make up the internal skeleton of a human being.

such as on the surfaces of joints, where bones move against one another. Cartilage is also the supportive tissue in stiff but flexible structures such as the larynx (voice box), the nose, and the ear pinnae. Sharks and rays are called *cartilaginous fishes* because their skeletons are composed entirely of cartilage. In all other vertebrates, cartilage is the principal component of the embryonic skeleton, but during development most of it is gradually replaced by bone.

Bone consists mostly of extracellular matrix material that contains crystals of insoluble calcium phosphate, which give bone its rigidity and hardness, as well as collagen fibers. The skeleton serves as a reservoir of calcium for the rest of the body and is in dynamic equilibrium with soluble calcium in the extracellular fluids of the body. This equilibrium is under the control of calcitonin and parathyroid hormone (see Figure 42.9). If too much calcium is taken from the skeleton, the bones are seriously weakened.

The living cells of bone—called osteoblasts, osteocytes, and osteoclasts—are responsible for the dynamic remodeling of bone that is constantly under way (Figure 47.13). **Osteoblasts** lay down new matrix material on bone surfaces. These cells gradually become surrounded by matrix and eventually become enclosed within the bone, at which point they cease laying down matrix, but continue to exist within small lacunae (cavities) in the bone. In this state they are called **osteocytes**. In spite of the vast amounts of matrix between them, osteocytes remain in contact with one another through long cellular extensions that run through tiny channels in the bone. Communication between osteocytes is important in controlling the activities of the cells that are laying down or removing bone.

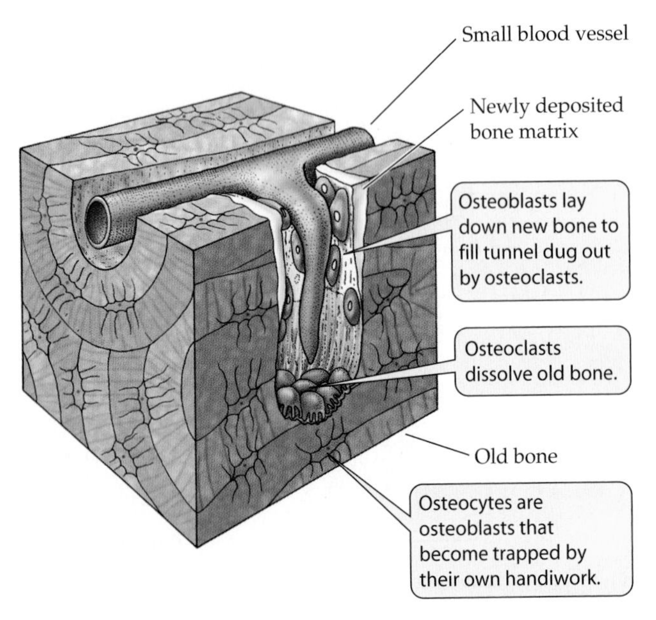
Small blood vessel
Newly deposited bone matrix
Osteoblasts lay down new bone to fill tunnel dug out by osteoclasts.
Osteoclasts dissolve old bone.
Old bone
Osteocytes are osteoblasts that become trapped by their own handiwork.

47.13 Renovating Bone Bones are constantly being remodeled by osteoblasts, which lay down bone, and osteoclasts, which resorb bone.

The cells resorb bone are the **osteoclasts**. They are derived from the same cell lineage that produces the white blood cells. Osteoclasts erode bone, forming cavities and tunnels. Osteoblasts follow osteoclasts, depositing new bone. Thus the interplay of osteoblasts and osteoclasts constantly replaces and remodels the bones.

How the activities of the bone cells are coordinated is not understood, but stress placed on bones somehow provides them with information. A remarkable finding in studies of astronauts who spent long periods in zero gravity was that their bones decalcified. Conversely, certain bones of athletes thicken during training. Both thickening and thinning of bones are experienced by anyone who has had a leg in a cast for a long time. The bones of the uninjured leg carry the person's weight and thicken, while the bones of the inactive leg in the cast thin. The jawbones of people who lose their teeth experience less compressional force during chewing and become considerably reduced.

Bones develop from connective tissues

Bones are divided into two types on the basis of how they develop. **Membranous bone** forms on a scaffold of connective tissue membrane. **Cartilage bone** forms first as a cartilaginous structure resembling the future mature bone, then gradually hardens (*ossifies*) to become bone. The outer bones of the skull are membranous bones; the bones of the limbs are cartilage bones.

Cartilage bones can grow throughout the ossification process. The long bones of the legs and arms, for example, ossify first at the centers and later at each end (Figure 47.14). Growth can continue until these areas of ossification join. The membranous bones forming the skull cap grow until their edges meet. The soft spot on the top of a baby's head is the point at which the skull bones have not yet joined.

The structure of bone may be **compact** (solid and hard) or **cancellous** (having numerous internal cavities that make it appear spongy, even though it is rigid). The architecture of a specific bone depends on its position and function, but most bones have both compact and cancellous regions. The shafts of the long bones of the limbs, for example, are cylinders of compact bone surrounding central cavities that contain the bone marrow, where the cellular elements of the blood are made. The ends of the long bones are cancellous (see Figure 47.14). Cancellous bone is lightweight because of its numerous cavities, but it is also strong because its internal meshwork constitutes a support system. It can withstand considerable forces of compression. The rigid, tubelike shaft of compact bone can withstand compression and bending forces. Architects and nature alike use hollow tubes as lightweight structural elements.

Most of the compact bone in mammals is called *Haversian bone* because it is composed of structural units called **Haver-**

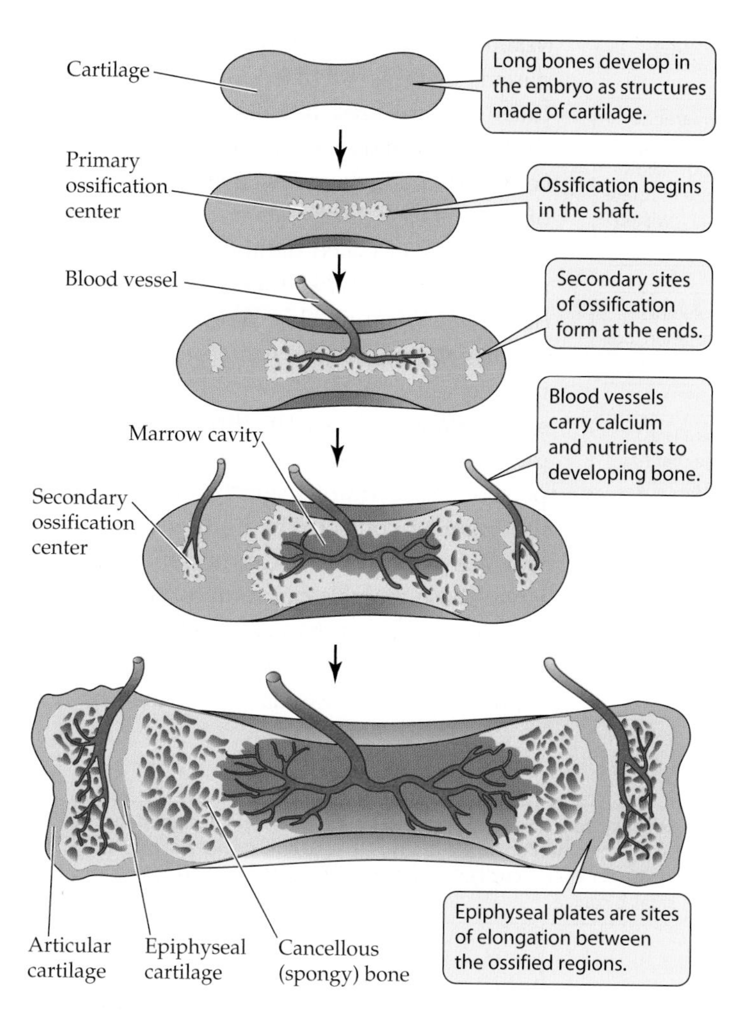

Cartilage
Long bones develop in the embryo as structures made of cartilage.
Primary ossification center
Ossification begins in the shaft.
Blood vessel
Secondary sites of ossification form at the ends.
Marrow cavity
Blood vessels carry calcium and nutrients to developing bone.
Secondary ossification center
Articular cartilage
Epiphyseal cartilage
Cancellous (spongy) bone
Epiphyseal plates are sites of elongation between the ossified regions.

47.14 The Growth of Long Bones In the long bones of human limbs, ossification occurs first at the centers and later at each end.

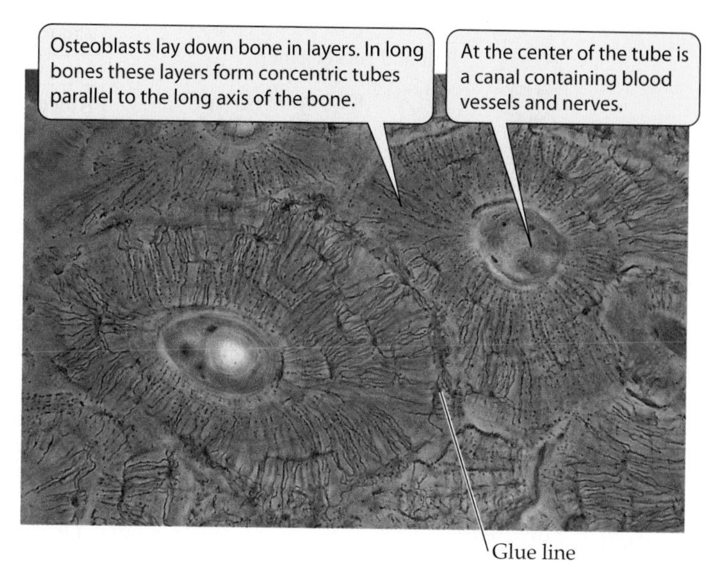

Osteoblasts lay down bone in layers. In long bones these layers form concentric tubes parallel to the long axis of the bone.
At the center of the tube is a canal containing blood vessels and nerves.
Glue line

47.15 Most Compact Bone Is Composed of Haversian Systems A micrograph of a section of a long bone shows Haversian systems with their central canals. Glue lines separate Haversian systems.

sian systems (Figure 47.15). Each Haversian system is a set of thin, concentric bony cylinders, between which are the osteocytes in their lacunae. Through the center of each Haversian system runs a narrow canal containing blood vessels and nerves. Adjacent Haversian systems are separated by boundaries called *glue lines*. Haversian bone is resistant to fracturing because cracks tend to stop at glue lines.

Bones that have a common joint can work as a lever

Muscles and bones work together around **joints**, where two or more bones come together. Since muscles can only contract and relax, they create movement around joints by working in antagonistic pairs: When one contracts, the other relaxes. When both contract, the joint becomes rigid. With respect to a particular joint, such as the knee, we can refer to the muscle that bends, or flexes, the joint as the **flexor** and the muscle that straightens, or extends, the joint as the **extensor**. The bones that meet at the joint are held together by **ligaments**, which are flexible bands of connective tissue. Other straps of connective tissue, called **tendons**, attach the muscles to the bones (Figure 47.16). In many kinds of joints, only the tendon spans the joint, sometimes moving over the surfaces of the bones like a rope over a pulley. The tendon of the quadriceps muscle traveling over the knee joint is what is tapped to elicit the knee-jerk reflex (see Figure 46.3). The human skeleton has a wide variety of joints with different ranges of movement (Figure 47.17).

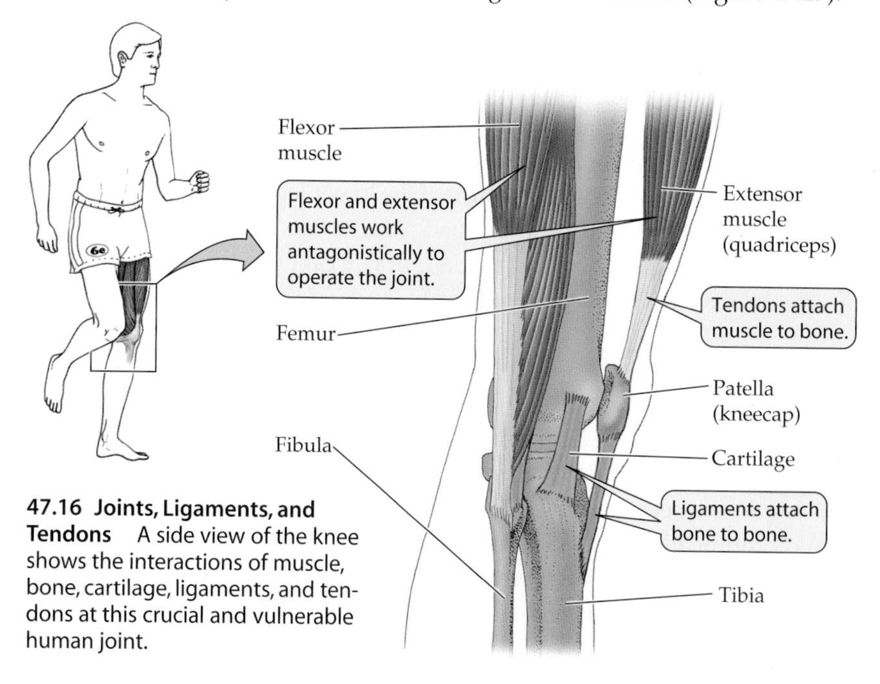

47.16 Joints, Ligaments, and Tendons A side view of the knee shows the interactions of muscle, bone, cartilage, ligaments, and tendons at this crucial and vulnerable human joint.

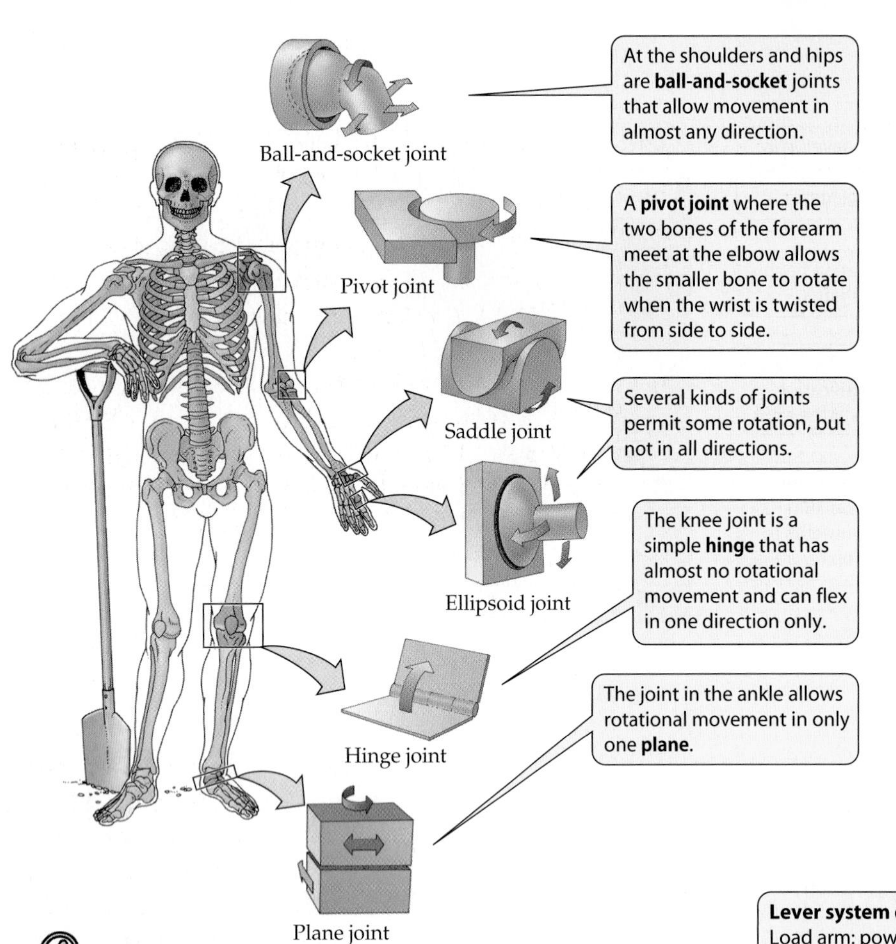

Ball-and-socket joint
At the shoulders and hips are **ball-and-socket** joints that allow movement in almost any direction.
Pivot joint
A **pivot joint** where the two bones of the forearm meet at the elbow allows the smaller bone to rotate when the wrist is twisted from side to side.
Saddle joint
Several kinds of joints permit some rotation, but not in all directions.
Ellipsoid joint
The knee joint is a simple **hinge** that has almost no rotational movement and can flex in one direction only.
Hinge joint
The joint in the ankle allows rotational movement in only one **plane**.
Plane joint

47.17 Types of Joints The designs of joints are similar to mechanical counterparts and enable a variety of movements.

Bones can be thought of as a system of levers that are moved around joints by the muscles. A lever has a *power arm* and a *load arm* that work around a *fulcrum* (pivot). The length ratio of the two arms determines whether a particular lever can exert a lot of force over a short distance or is better at translating force into large or fast movements. Compare the jaw joint and the knee joint, for example (Figure 47.18). The power arm of the jaw is long relative to the load arm, allowing the jaw to apply great force over a small distance, as when you crack a nut with your teeth. The power arm of the lower leg, on the other hand, is short relative to the load arm, so you can run fast, jump high, and deliver swift kicks, but you cannot apply nearly the force with a leg that you can with your jaws.

Other Effectors

Muscles are universal in animals, but many effectors are more specialized and are shared by only a few animal species. Some specialized effectors are used for defense, some for communication, and some for capturing prey or avoiding predators. In this section we mention only a few of these specialized effectors to give a sampling of their evolutionary diversity.

NEMATOCYSTS. **Nematocysts**, found in cnidarians such as jellyfishes, are cellular structures that are fired like miniature missiles to capture prey and repel predators. They are concentrated in huge numbers on the outer surface of the tentacles. Each nematocyst consists of a slender thread coiled tightly within a capsule, armed with a spinelike trigger projecting to the outside (see Figure 32.7). When potential prey brushes the trigger, the nematocyst fires, turning the thread inside out and exposing little spines along its base. The thread either entangles or penetrates the body of the victim, and a poison may be simultaneously released around the point of contact. The Portuguese man-of-war has tentacles that can be several meters long. These animals can capture, subdue, and devour full-grown mackerel, and the poison of their nematocysts is so potent that it can kill a human who becomes entangled in the tentacles.

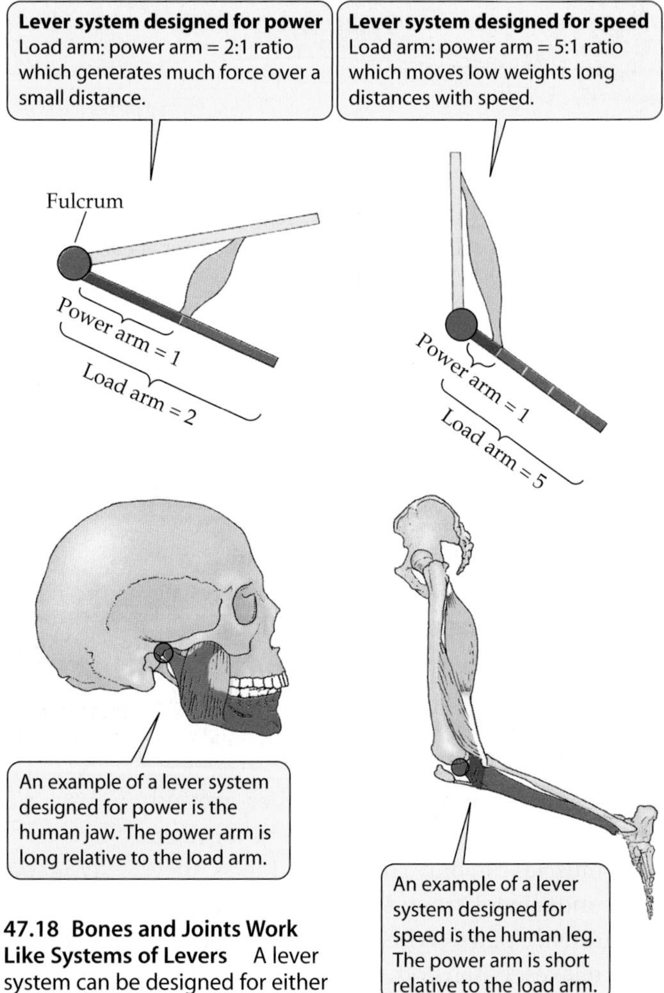

47.18 Bones and Joints Work Like Systems of Levers A lever system can be designed for either power or speed.

CHROMATOPHORES. A change in body color is a response that some animals use to camouflage themselves in a particular environment or to communicate with other animals. **Chromatophores** are pigment-containing cells in the skin that can change the color and pattern of the animal. Chromatophores are under neuronal or hormonal control, or both; in most cases, they can effect a change within minutes or even seconds.

Chromatophores enable squids, sole, and flounder, all of which spend much time on the seafloor, as well as the famous chameleons (a group of African lizards; see Figure 34.19) and a few other animals, to blend in with the background on which they are resting and thus escape discovery by predators. Chromatophores with different pigments enable animals to assume different hues or to become mottled to match the background more precisely. In other mollusks, fishes, and lizards, a color change sends a signal to potential mates and territorial rivals of the same species.

There are three principal types of chromatophore cells. The most common type has fixed cell boundaries, within which pigmented granules may be moved about by microfilaments. When the pigment is concentrated in the center of each chromatophore, the animal is pale; the animal turns darker when the pigment is dispersed throughout the cell. Another type of chromatophore is capable of amoeboid motion. These cells can mold themselves into shapes with a minimal surface area, leaving the tissue relatively pale, or they can flatten out to make the tissue appear darker.

The third type of chromatophore changes shape as a result of the action of muscle fibers radiating outward from the cell (Figure 47.19*a*). When the muscle fibers are relaxed, the chromatophores are small and compact, and the animal is pale. To darken the animal, the muscle fibers contract and spread the chromatophores over more of the body surface. These chromatophores can change so rapidly that they are used in some species for communication during courtship and aggressive interactions. For example, the cuttlefish, a cephalopod, can signal courtship intentions to a potential mate on one side of its body while signaling aggressive threats to a rival on the other side (Figure 47.19*b*).

GLANDS. **Glands** are effector organs that produce and release chemicals. Endocrine glands, as we saw in Chapter 42, produce hormones for internal signaling. Other glands secrete substances into the gut or onto the body surface. Some of these secretions are used defensively or to capture prey. Others are *pheromones*, chemical signals released into the environment for communication with other individuals.

Certain snakes, frogs, salamanders, spiders, mollusks, and fishes have poison glands, which are used for capturing prey or defending against predators. Many of the poisons produced by these glands are extremely specific in their modes of action. For example, the poison dendrotoxin, which certain tribes of the Amazonian rainforest use on the tips of their arrows for hunting, comes from the skin of a frog and blocks certain potassium channels. The snake venom bungarotoxin inactivates the acetylcholine receptors at the neuromuscular junction. The puffer fish poison tetrodotoxin blocks voltage-gated sodium channels. A poison from a mollusk, conotoxin, blocks calcium channels. Not all defensive secretions are poisonous, however. A well-known example is mercaptan, the odoriferous chemical sprayed by skunks.

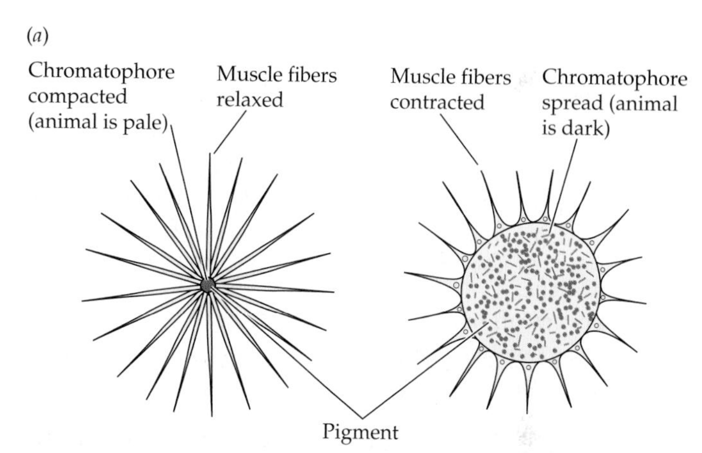

(*b*) *Sepia latimanus*

47.19 Chromatophores Help Animals Camouflage Themselves or Communicate (*a*) Muscle fibers around chromatophores cause the chromatophores to contract. (*b*) Cuttlefish are cephalopod mollusks that can change color patterns so fast that these changes can be used for rapid communication.

ELECTRIC ORGANS. Various fishes can generate electricity, as we saw in Chapter 45. These species include the electric eel, the knife fish, the torpedo (a type of ray), and the electric catfish. The electric fields they generate are used for sensing the environment, for communication, and also for stunning potential predators or prey. The electric organs of these animals evolved from muscles, and they produce electric potentials in the same general way as nerves and muscles do.

Electric organs consist of very large, disc-shaped cells arranged in long rows like stacks of batteries. When these cells discharge simultaneously, the electric organ can generate far more voltage and current than can nerve or muscle tissue. Electric eels, for example, can produce up to 600 volts with an output of approximately 100 watts—enough to light a row of light bulbs or to temporarily stun a person.

Chapter Summary

▶ Effectors enable animals to respond to information from their internal and external environments. Most effectors generate mechanical forces and cause movement.

Microtubules, Microfilaments, and Cell Movement

▶ Cell movement is generated by two components of the cytoplasmic skeleton, microtubules and microfilaments, both of which consist of long protein molecules that can change their length or shape.

▶ The movements of cilia and flagella depend on microtubules.

▶ Microfilaments allow animal cells to change their shape and move.

Muscle Contraction

▶ The three types of vertebrate muscle are smooth, cardiac, and skeletal (striated). **Review Figure 47.1**

▶ Smooth muscle provides contractile force for internal organs. Smooth muscle cells are electrically connected by gap junctions, so action potentials can spread rapidly throughout smooth muscle tissue. Smooth muscle cells are sensitive to stretching and to neurotransmitters from the autonomic nervous system. **Review Figure 47.2. See Web/CD Tutorial 47.2**

▶ The walls of the heart consist of sheets of branching cardiac muscle cells. The cells are electrically connected by gap junctions, so that action potentials spread rapidly throughout sheets of cardiac muscle and cause coordinated contractions. Some cardiac muscle cells are pacemaker cells that generate the heartbeat.

▶ Skeletal, or striated, muscle consists of bundles of muscle fibers. Each muscle fiber is a huge cell containing multiple nuclei.

▶ Skeletal muscles contain numerous myofibrils, which are bundles of actin and myosin filaments. The regular, overlapping arrangement of the actin and myosin filaments into sarcomeres gives skeletal muscle its striated appearance. During contraction, the actin and myosin filaments slide past each other in a telescoping fashion. **Review Figure 47.3. See Web/CD Activity 47.1**

▶ The molecular mechanism of muscle contraction involves the binding of the globular heads of myosin molecules to actin. Upon binding, the myosin head changes its conformation, causing the two filaments to slide past each other. Release of the myosin heads from actin and their return to their original conformation requires ATP. **Review Figure 47.4**

▶ The plasma membrane of the muscle fiber is continuous with a system of T tubules that extends deep into the sarcoplasm. **Review Figure 47.5. See Web/CD Activity 47.2**

▶ When an action potential spreads across the plasma membrane and through the T tubules, it causes Ca^{2+} ions to be released from the sarcoplasmic reticulum. The Ca^{2+} ions bind to troponin and change its conformation, pulling the tropomyosin strands away from the myosin binding sites on the actin filament. Cycles of actin–myosin binding and release occur, and the muscle fiber contracts until the Ca^{2+} is returned to the sarcoplasmic reticulum. **Review Figure 47.6**

▶ In skeletal muscle, a single action potential causes a minimum unit of contraction called a twitch. Twitches occurring in rapid succession can be summed, thus increasing the strength of contraction. **Review Figure 47.7**

See Web/CD Tutorial 47.1

Muscle Strength and Performance

▶ Slow-twitch muscle fibers are adapted for extended, aerobic work; fast-twitch fibers are adapted for generating maximum forces for short periods of time. The ratio of slow-twitch to fast-twitch fibers in the muscles of an individual is genetically determined. **Review Figure 47.8**

▶ The force that a muscle fiber can produce depends on its initial state of extension or contraction. **Review Figure 47.9**

▶ Anaerobic exercise stimulates the enlargement of muscle fibers through production of new microfilaments. Through aerobic conditioning, muscle fibers can acquire greater oxidative capacity.

▶ Muscle performance depends on fuel supply. Available ATP and creatinine phosphate can fuel maximum tension immediately, but are exhausted in seconds. Glycolysis can regenerate ATP rapidly, but is rapidly slowed by accumulation of lactic acid. Oxidative metabolism delivers ATP more slowly, but can continue to do so for a long time. **Review Figure 47.10**

Skeletal Systems

▶ Skeletal systems provide rigid supports against which muscles can pull.

▶ Hydrostatic skeletons are fluid-filled body cavities that can be squeezed by muscles. **Review Figure 47.11**

▶ Exoskeletons are hardened outer surfaces to which internal muscles are attached.

▶ Endoskeletons are internal systems of rigid rodlike, platelike, and tubelike supports, consisting of bone and cartilage, to which muscles are attached. **Review Figure 47.12**

▶ Bone is continually being remodeled by osteoblasts, which lay down new bone, and osteoclasts, which erode bone. **Review Figure 47.13**

▶ Bones develop from connective tissue membranes (membranous bone) or from cartilage through ossification (cartilage bone). Cartilage bone can grow until centers of ossification meet. **Review Figure 47.14**

▶ Bone can be solid and hard (compact bone), or it can contain numerous internal spaces (cancellous bone). Most of the compact bone of mammals is composed of Haversian systems. **Review Figure 47.15**

▶ Tendons connect muscles to bones; ligaments connect bones to one another. **Review Figure 47.16**

▶ Muscles and bones work together around joints as systems of levers. **Review Figures 47.17, 47.18. See Web/CD Activity 47.3**

Other Effectors

▶ Effector organs other than muscles include nematocysts, chromatophores, glands, and electric organs. **Review Figure 47.19**

Self-Quiz

1. Smooth muscle differs from both cardiac and skeletal muscle in that
   a. it can act as a pacemaker for rhythmic contractions.
   b. contractions of smooth muscle are not due to interactions between neighboring microfilaments.
   c. neighboring cells are electrically connected by gap junctions.
   d. neighboring cells are tightly coupled by intercalated discs.
   e. the membranes of smooth muscle cells are depolarized by stretching.
2. Fast-twitch fibers differ from slow-twitch fibers in that
   a. they are more common in the leg muscles of champion sprinters.
   b. they have more mitochondria.
   c. they fatigue less rapidly.
   d. their abundance is more a product of training than of genetics.
   e. they are more common in the leg muscles of champion cross-country skiers.
3. The role of Ca^{2+} in the control of muscle contraction is to
   a. cause depolarization of the T tubule system.
   b. change the conformation of troponin, thus exposing myosin binding sites.
   c. change the conformation of myosin heads, thus causing microfilaments to slide past each other.
   d. bind to tropomyosin and break actin–myosin cross-bridges.
   e. block the ATP binding site on myosin heads, enabling muscles to relax.
4. Fifteen minutes into a 10-km run, what is the major energy source of the leg muscules?
   a. Preformed ATP
   b. Glycolysis
   c. Oxidative metabolism
   d. Pyruvate and lactate
   e. High-protein drink consumed right before the race
5. Which statement about muscle contractions is *not* true?
   a. A single action potential at the neuromuscular junction is sufficient to cause a muscle to twitch.
   b. Once maximum muscle tension is achieved, no ATP is required to maintain that level of tension.
   c. An action potential in the muscle cell activates contraction by releasing Ca^{2+} into the sarcoplasm.
   d. Summation of twitches leads to a graded increase in the tension that can be generated by a single muscle fiber.
   e. The tension generated by a muscle can be varied by controlling how many of its motor units are active.
6. Which statement about the structure of skeletal muscle is *true*?
   a. The light bands of the sarcomere are the regions where actin and myosin filaments overlap.
   b. When a muscle contracts, the A bands of the sarcomere lengthen.
   c. The myosin filaments are anchored in the Z lines.
   d. When a muscle contracts, the H zone of the sarcomere shortens.
   e. The sarcoplasm of the muscle cell is contained within the sarcoplasmic reticulum.
7. The long bones of our arms and legs are strong and can resist both compressional and bending forces because
   a. they are solid rods of compact bone.
   b. their extracellular matrix contains crystals of calcium carbonate.
   c. their extracellular matrix consists mostly of collagen and polysaccharides.
   d. they have a very high density of osteoclasts.
   e. they consist of lightweight cancellous bone with an internal meshwork of supporting elements.
8. If we compare the jaw joint with the knee joint as lever systems,
   a. the jaw joint can apply greater compressional forces.
   b. their ratios of power arm to load arm are about the same.
   c. the knee joint has greater rotational abilities.
   d. the knee joint has a greater ratio of power arm to load arm.
   e. only the jaw is a hinged joint.
9. Which statement about skeletons is *true*?
   a. They can consist of mostly cartilage.
   b. Hydrostatic skeletons can be used only for amoeboid motion.
   c. An advantage of exoskeletons is that they can continue to grow throughout the life of the animal.
   d. External skeletons must remain flexible, so they never include calcium carbonate crystals, as bones do.
   e. Internal skeletons consist of four different types of bone: compact, cancellous, membranous, and Haversian.
10. Which of the following effectors *is not* used both for avoiding predators and for communication?
   a. Chromatophores
   b. Nematocysts
   c. Electric organs
   d. Skeletal muscle
   e. Pheromones

For Discussion

1. You can see from the structure of a sarcomere that it can shorten only by a certain percentage of its resting length. Yet, muscles can cause a wide variety of ranges of movement—compare the range of movement of a toe and a leg. What are two adaptive design features of muscles and skeletons that can maximize the ability of a muscle to cause a greater range of movement of an appendage?
2. Exercising to exhaustion depletes muscle glycogen. The rate at which that glycogen is replenished depends on diet. On a carbohydrate diet it may take more than a day, but on a high fat and protein diet it may take several days. If athletes train up until a day before competition and then take a day of rest, performance in which types of events will be most affected by what they eat during the rest day?
3. Wombats are powerful digging animals, and kangaroos are powerful jumping animals. How do you think the structures of their legs would compare in terms of their designs as lever systems?
4. Why are ducks better long-distance fliers than chickens?
5. If an adolescent breaks a leg bone close to the ankle joint, after the break heals, that leg may not grow as long as the other one. Why?

48 Gas Exchange in Animals

And so back up the gradual slopes, the wind behind me. A much greater effort this, stopping every few yards with a slight anxiety lest I should not make the distance. As I approached the tents, I was astonished to see a bird, a chough, strutting about on the stones near me. … During this day, too, Charles Evans saw what must have been a migration of small grey birds. … Neither of us had thought to find any signs of life as high as this.

—*Sir John Hunt, Ascent of Everest, 1953*

In his book about the first expedition to successfully reach the top of Mount Everest, Sir John Hunt related the above episode, which occurred at the last camp before the summit attempt—at 8,000 meters (more than 5 miles) high. At that altitude, climbers are incapacitated if they do not breathe supplemental oxygen from pressurized bottles. Prior to the moment described, Hunt had gone a short distance downhill from his tent without supplemental oxygen.

Birds do indeed exist at such extreme altitudes, and some can fly over the highest mountains. A bird in flight can consume oxygen at a rate that a well-trained athlete can sustain for only minutes. Birds sustain such high rates of oxygen consumption during very long flights and during flights at high altitude. How do they do it?

Another remarkable metabolic capability is shown by fish that can swim much faster, farther, and longer than the best human swimmer. Yet these fish are breathing water that has less than 5 percent of the oxygen content of the air breathed by the human. How do they do it? The abilities of these animals to maintain high rates of metabolism depend in part on the capacities of their respiratory gas exchange systems.

Both water breathers and air breathers have respiratory systems with adaptations that facilitate exchanges of oxygen and carbon dioxide with the environment. In this chapter we explore those adaptations. We begin by discussing the physical factors that determine respiratory gas exchange, and we identify those factors that natural selection has been able to optimize. We will then examine the respiratory gas exchange organs of a variety of

"I Was Astonished To See a Bird" Many birds can sustain the high metabolic costs of flight even at very high altitudes, where oxygen is scarce.

species, including some highly efficient ones, such as fish gills and bird lungs, and less efficient ones, such as our own. We will also look at the adaptations of the blood for transporting respiratory gases. Finally, we see how respiratory gas exchange systems are controlled and regulated.

Physical Processes of Respiratory Gas Exchange

The **respiratory gases** that animals must exchange are oxygen (O_2) and carbon dioxide (CO_2). Cells need to obtain O_2 from the environment to produce an adequate supply of ATP by cellular respiration (see Chapter 7). CO_2 is an end product of cellular respiration, and it must be removed from the body to prevent toxic effects.

Diffusion is the only means by which respiratory gases are exchanged between the internal body fluids of an animal and the outside medium (air or water). There are no active transport mechanisms to move respiratory gases across biological membranes. Because diffusion is a physical process, knowing the physical factors that influence rates of diffusion helps us understand the diverse adaptations of gas exchange systems. You might want to review what you learned about the physical nature of diffusion in Chapter 5. Here, we will discuss environmental factors that influence diffusion rates, then describe the adaptations of respiratory systems for facilitating the diffusion of respiratory gases.

Air is a better respiratory medium than water

Oxygen can be obtained more easily from air than from water, for several reasons:

- The O_2 content of air is much higher than the O_2 content of an equal volume of water. The maximum O_2 content of a rapidly flowing stream is less than 10 ml of O_2 per liter of water. The O_2 content of the air over the stream is about 200 ml of O_2 per liter of air.
- Oxygen diffuses about 8,000 times more rapidly in air than in water. That is why the O_2 content of a stagnant pond can be zero only a few millimeters below the surface.
- When an animal breathes, it does work to move water or air over its specialized gas exchange surfaces. More energy is required to move water than to move air because water is 800 times more dense than air and about 50 times more viscous.

The slow diffusion of O_2 molecules in water affects air-breathing animals as well as water-breathing ones. Eukaryotic cells carry out cellular respiration in their mitochondria, which are located in the cytoplasm—an aqueous medium. Cells are bathed in extracellular fluid—also an aqueous medium. The slow rate of O_2 diffusion in water limits the efficiency of O_2 distribution from gas exchange surfaces to the sites of cellular respiration.

Diffusion of O_2 in water is so slow that even animal cells with low rates of metabolism can be no more than a couple of millimeters away from a good source of environmental O_2. Therefore, there are severe size and shape limits for the many species of invertebrates that lack internal systems for transporting O_2. Most of these species are very small, but some have grown larger by having a flat, leaflike body (Figure 48.1*a*). Still others have very thin bodies that are built around a central cavity through which water circulates (Figure 48.1*b*). A critical factor enabling larger, more complex animal bodies has been the evolution of specialized respiratory systems with a large surface area for enhancing respiratory gas exchange (Figure 48.1*c*).

48.1 Keeping in Touch with the Medium (*a*) No cell in the leaflike body of this marine flatworm is more than a millimeter away from seawater. (*b*) Sponges have body walls perforated by many channels, which communicate with the outside world and with a central cavity. No cell in the sponge is more than a millimeter away from seawater. (*c*) A feathery fringe of gills on this larval salamander provides a large surface area for gas exchange. Blood circulating through the gills comes into close contact with the respiratory medium.

High temperatures create respiratory problems for aquatic animals

Animals that breathe water are in a double bind when environmental temperatures rise. Most water breathers are *ectothermic;* that is, their body temperatures are closely tied to the temperature of the water around them. As the temperature of the water rises, so do their body temperature and metabolic rate (see Figure 41.6). Thus, with rising temperatures, water breathers need more O_2. But warm water holds less dissolved gas than cold water does (just think of the gases that escape when you open a warm bottle of soda). In addition, if the animal performs work to move water across its gas exchange surfaces (as fish do), the energy the animal must expend to breathe increases as water temperature rises. Therefore, as water temperature goes up, the water breather must extract more and more O_2 from an environment that is increasingly O_2 deficient, and a lower percentage of that O_2 is available to support activities other than breathing (Figure 48.2).

O_2 availability decreases with altitude

Just as a rise in temperature reduces the supply of O_2 available to water breathers, an increase in altitude reduces the O_2 supply for air breathers. The amount of O_2 in the atmosphere decreases with altitude.

One way biologists express the amounts of different gases in air and in water is by the **partial pressures** of those gases. At sea level, the pressure exerted by the atmosphere is equivalent to the pressure produced by a column of mercury 760 mm high. Therefore, *barometric pressure* (atmospheric pressure) at sea level is 760 mm of mercury (Hg). Because dry air is 20.9 percent O_2, the *partial pressure of oxygen* (P_{O_2}) at sea level is 20.9 percent of 760 mm Hg, or about 159 mm Hg. That is, the contribution of O_2 alone to the total air pressure is about 159 mm Hg.

At higher elevations, where there is less air, barometric pressure declines. For example, at an altitude of 5,800 m, barometric pressure is only half as much as it is at sea level, so the P_{O_2} at that altitude is only about 80 mm Hg. At the summit of Mount Everest (8,848 m), the P_{O_2} is only about 50 mm Hg— roughly a third what it is at sea level. Since the movement of O_2 across respiratory gas exchange surfaces and into the body depends on diffusion, its rate of movement depends on the P_{O_2} difference between the air and the body fluids. Therefore, the drastically reduced P_{O_2} in the air at high altitudes constrains O_2 uptake. Because of these constraints, mountain climbers who venture to the heights of Mount Everest usually breathe O_2 from pressurized bottles.

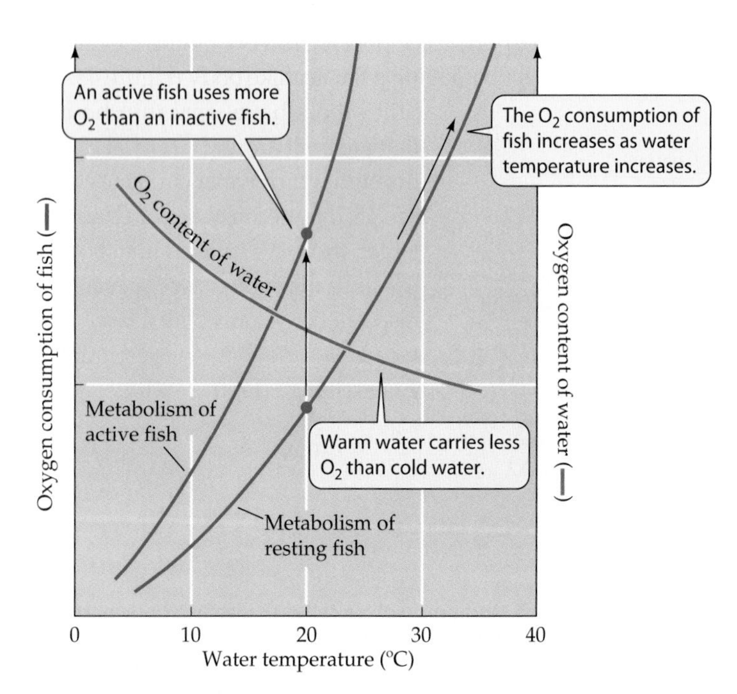

48.2 The Double Bind of Water Breathers Fish need *more* O_2 when the water is warmer, but warm water carries *less* O_2 than cold water.

Carbon dioxide is lost by diffusion

Respiratory gas exchange is a two-way process: CO_2 diffuses out of the body as O_2 diffuses in. The direction and rate of gas diffusion depends on the partial pressure gradient of the gas. The partial pressure gradients of O_2 and CO_2 across gas exchange surfaces are not the same. The amount of CO_2 in the atmosphere is extremely low (0.03 percent), so regardless of altitude, there is always a steep partial pressure gradient favoring loss of CO_2 by air-breathing animals.

Water-breathing animals are much more likely than air breathers to experience high partial pressures of CO_2 in their environments, especially if the water is stagnant, not well aerated, and contains a lot of decomposing organic material. In such an environment, however, lack of O_2 will probably be a more serious problem. In general, getting rid of CO_2 is not a problem for water-breathing animals.

Fick's law applies to all systems of gas exchange

All adaptations that maximize respiratory gas exchange influence one or more components of a simple equation called **Fick's law of diffusion**, which shows how various physical factors influence the rate of diffusion. Fick's law is written

$$Q = DA\frac{P_1 - P_2}{L}$$

where

- Q is the rate at which a gas such as O_2 diffuses between two locations
- D is the *diffusion coefficient*, which is a characteristic of the diffusing substance, the medium, and the temperature

(for example, perfume has a higher D than motor oil vapor, and all substances diffuse faster in air than in water)
- A is the cross-sectional area over which the gas is diffusing
- P_1 and P_2 are the partial pressures of the gas at the two locations
- L is the path length, or distance, between the two locations

Therefore, $(P_1 - P_2)/L$ is a partial pressure gradient.

Animals can maximize D for respiratory gases by using air rather than water as their gas exchange medium whenever possible; doing so greatly increases Q. All other adaptations for maximizing respiratory gas exchange must influence the surface area (A) for gas exchange or the partial pressure gradient $[(P_1 - P_2)/L]$ across that surface area.

Adaptations for Respiratory Gas Exchange

Now that we know the physical factors that influence the rates of diffusion of respiratory gases between an animal and its environment, let's take a look at some of the adaptations animals have evolved for maximizing their respiratory gas exchange. They include adaptations for increasing the surface area over which diffusion of gases can occur, maximizing partial pressure gradients, and minimizing the diffusion path length through an aqueous medium.

Respiratory organs have large surface areas

Many anatomical adaptations maximize the specialized body surface area (A) over which respiratory gases can diffuse. **External gills** are highly branched and folded extensions of the body surface that provide a large surface area for gas exchange with water (Figure 48.3*a*). External gills are found in larval amphibians and in many insect species. Because they consist of thin, delicate tissues, they minimize the length of the path (L) traversed by diffusing molecules of O_2 and CO_2. Because external gills are vulnerable to damage and are tempting morsels for carnivorous organisms, protective body cavities for gills have evolved. Many mollusks, arthropods, and fish have **internal gills** in such cavities (Figure 48.3*b*).

Air-breathing vertebrates also have large surface areas for gas exchange. **Lungs** are internal cavities for respiratory gas exchange with air. Their structure is quite different from that of gills (Figure 48.3*c*). Lungs have a large surface area because they are highly divided, and they are elastic so that they can be inflated and deflated with air.

The most abundant air-breathing invertebrates are insects, which have a respiratory gas exchange system consisting of a highly branched network of air-filled tubes called **tracheae** that branch through all tissues of the insect's body (Figure 48.3*d*). The terminal branches of these tubes are so numerous that they have an enormous surface area.

Transporting gases to and from the exchange surfaces optimizes partial pressure gradients

Fick's law of diffusion points to other possible adaptations besides increasing the surface area for gas exchange. Animals can maximize the partial pressure gradients ($P_1 - P_2/L$) that drive the diffusion of respiratory gases across their gas exchange surfaces in several ways:

- Very thin tissues in gills and lungs reduce the diffusion path length (L).
- Breathing actions move the respiratory medium past the environmental side of the exchange surfaces. This process, called **ventilation**, supplies the exchange surfaces with fresh respiratory medium that has maximum O_2 and minimum CO_2 concentrations.
- Circulatory systems transport respiratory gases to and from the internal side of the exchange surfaces. This process, called **perfusion**, helps maintain the lowest possible O_2 concentration and the highest possible CO_2 concentration on the inside of the exchange surfaces.

An animal's **gas exchange system** is made up of its gas exchange surfaces and the mechanisms it uses to ventilate and perfuse those surfaces. The following sections describe four

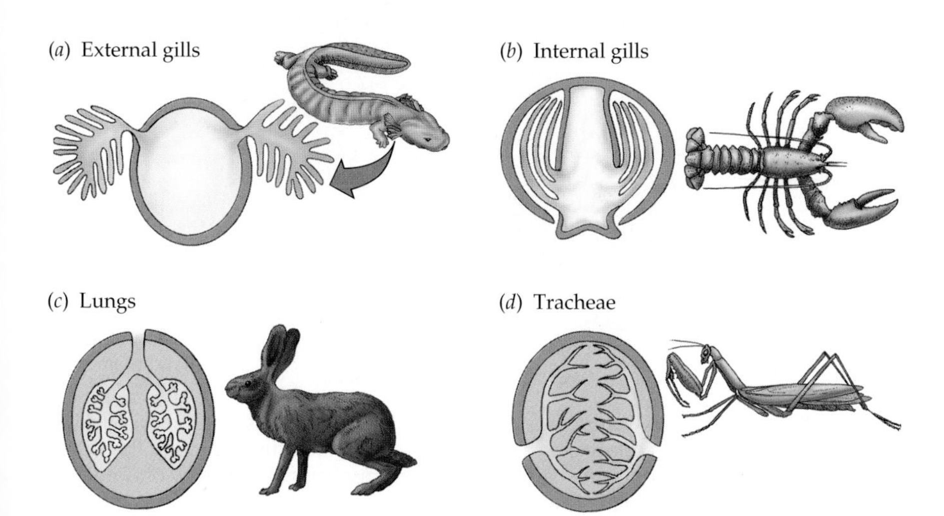

48.3 Gas Exchange Systems Large surface areas (blue in these diagrams) for the diffusion of respiratory gases are common features of animals. Both external (*a*) and internal (*b*) gills are adaptations for gas exchange with water. Lungs (*c*) and tracheae (*d*) are organs for gas exchange with air.

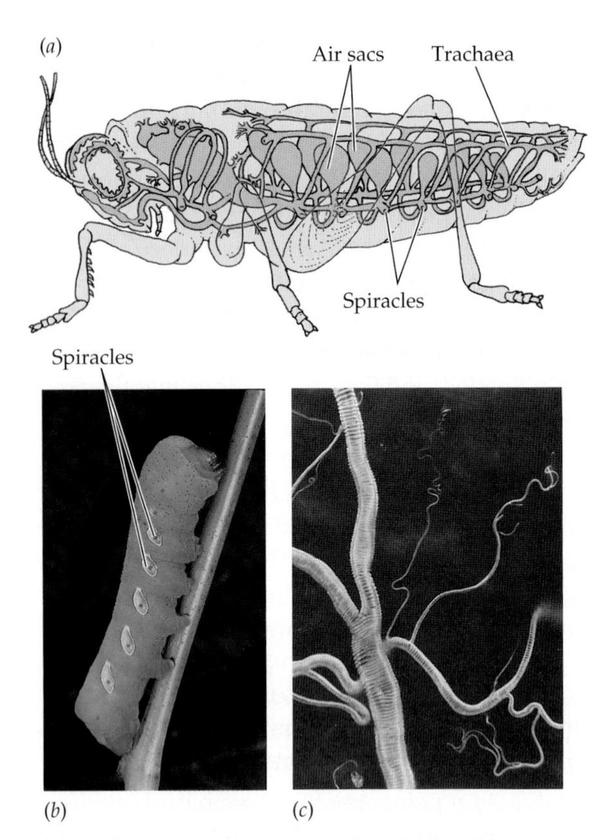
(a)
Air sacs
Trachaea
Spiracles
Spiracles
(b)
(c)

48.4 The Tracheal Gas Exchange System of Insects (*a*) In insects, respiratory gases diffuse through a system of air tubes (tracheae) that open to the external environment through holes called spiracles. (*b*) The spiricles of a sphinx moth larva run down its sides and are visually obvious. (*c*) A scanning electron micrograph shows an insect trachea dividing into smaller tracheoles and still finer air capillaries.

gas exchange systems. First we'll look at the gas exchange system of insects. Then we'll describe two remarkably efficient systems: fish gills and bird lungs. Finally, we'll discuss human lungs.

INSECT TRACHEAE. Respiratory gases can diffuse through air most of the way to and from every cell of an insect's body. This diffusion is achieved through a system of air tubes, or tracheae, that communicate with the outside environment through gated openings called *spiracles* in the sides of the abdomen (Figure 48.4*a*,*b*). The spiracles can open to allow gas exchange, and then close to decrease water loss. The tracheae branch into even finer tubes, or *tracheoles*, until they end in tiny *air capillaries* (Figure 48.4*c*). In the insect's flight muscles and other highly active tissues, no mitochondrion is more than a few micrometers away from an air capillary.

Some species of insects that dive and stay under water for long periods make use of an interesting variation on diffusion. These insects carry with them a bubble of air. A small bubble may not seem like a very large reservoir of O_2, yet these insects can stay under water for a long time with their small air supplies. The secret has to do

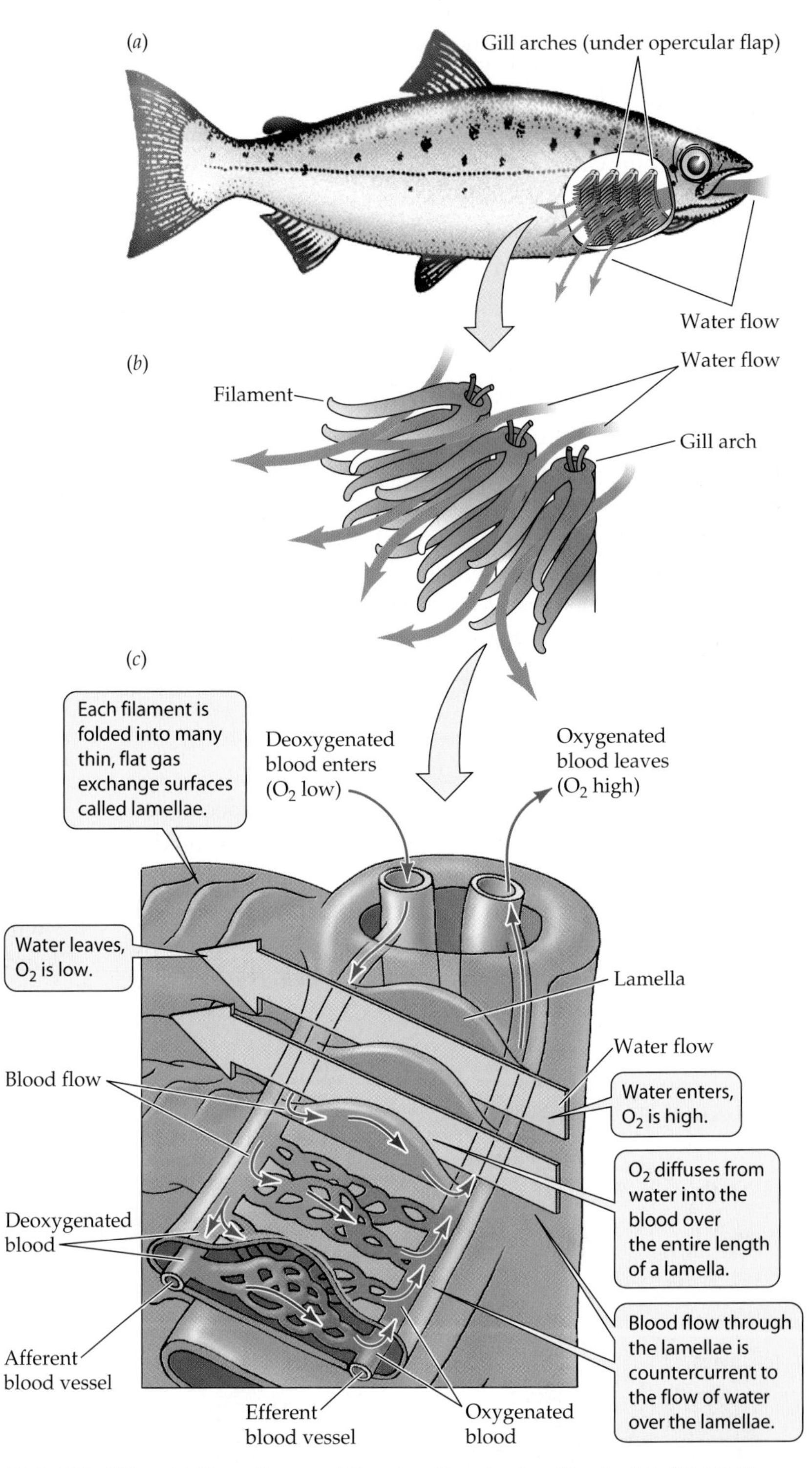
(a)
Gill arches (under opercular flap)
Water flow
(b)
Water flow
Filament
Gill arch
(c)
Each filament is folded into many thin, flat gas exchange surfaces called lamellae.
Deoxygenated blood enters (O_2 low)
Oxygenated blood leaves (O_2 high)
Water leaves, O_2 is low.
Lamella
Water flow
Water enters, O_2 is high.
Blood flow
O_2 diffuses from water into the blood over the entire length of a lamella.
Deoxygenated blood
Blood flow through the lamellae is countercurrent to the flow of water over the lamellae.
Afferent blood vessel
Efferent blood vessel
Oxygenated blood

48.5 Fish Gills (*a*) Water flows unidirectionally over the gills of a fish. (*b*) Gill filaments have a large surface area and thin tissues. (*c*) Blood flows through the lamellae in the direction opposite (left to right, in this depiction) to the flow of water (right to left) over the lamellae.

with the P_{O_2} in the bubble. When the insect dives, the air bubble contains about 80 percent nitrogen and 20 percent O_2. As the insect consumes the O_2 in its bubble, the bubble shrinks a little. The bubble doesn't disappear, however, because it consists mostly of nitrogen, which the insect does not consume. When the P_{O_2} in the bubble falls below the P_{O_2} in the surrounding water, O_2 diffuses from the water into the bubble. The bubble acts as an auxiliary lung, and for these small animals, the rate of O_2 diffusion into the bubble is enough to meet their O_2 demand while they are under water.

FISH GILLS. The internal gills of fish are supported by *gill arches* that lie between the mouth cavity and the protective *opercular flaps* on the sides of the fish just behind the eyes (Figure 48.5*a*). Water flows unidirectionally into the fish's mouth, over the gills, and out from under the opercular flaps, so that the gills are continuously bathed with fresh water. This constant, one-way flow of water moving over the gills maximizes the P_{O_2} on the external surfaces. On the internal side, the circulation of blood minimizes the P_{O_2} by sweeping the O_2 away as rapidly as it diffuses across.

The gills have an enormous surface area for gas exchange because they are so highly divided. Each gill consists of hundreds of leaf-shaped *gill filaments* (Figure 48.5*b*). The upper and lower flat surfaces of each gill filament are covered with rows of evenly spaced folds, or *lamellae*. The lamellae are the actual gas exchange surfaces. Their delicate structure minimizes the path length (L) for diffusion of gases between blood and water. The surfaces of the lamellae consist of highly flattened epithelial cells, so the water and the fish's red blood cells are separated by little more than 1 or 2 μm.

The flow of blood perfusing the inner surfaces of the lamellae, like the flow of water over the gills, is unidirectional. *Afferent* blood vessels bring blood to the gills, while *efferent* blood vessels take blood away from the gills (Figure 48.5*c*). Blood flows through the lamellae in the direction opposite to the flow of water over the lamellae. This **countercurrent flow** optimizes the P_{O_2} gradient between water and blood, making gas exchange more efficient than it would be in a system using concurrent (parallel) flow (Figure 48.6).

Some fish, including anchovies, tuna, and certain species of sharks, ventilate their gills by swimming almost constantly with their mouths open. Most fish, however, ventilate their gills by means of a two-pump mechanism. The closing and contracting of the mouth cavity pushes water over the gills, and the expansion of the opercular cavity prior to opening of the opercular flaps pulls water over the gills.

These adaptations allow fish to extract an adequate supply of O_2 from meager environmental sources by maximizing the surface area (A) for diffusion, minimizing the path length (L) for diffusion, and maximizing the P_{O_2} gradient by means of constant, unidirectional, countercurrent flow of blood and water over the opposite sides of their gas exchange surfaces.

BIRD LUNGS. As we saw at the beginning of this chapter, birds can sustain extremely high levels of activity much longer than mammals can—even at very high altitudes, where mammals cannot even survive. Yet the lungs of a bird are smaller than the lungs of a similar-sized mammal. How can this be? Another unusual feature of bird lungs is that they expand and contract less during a breathing cycle than mammalian lungs do. To make things even more puzzling, bird lungs contract during inhalation and expand during exhalation!

The structure of bird lungs allows air to flow unidirectionally through the lungs, rather than having to flow in and

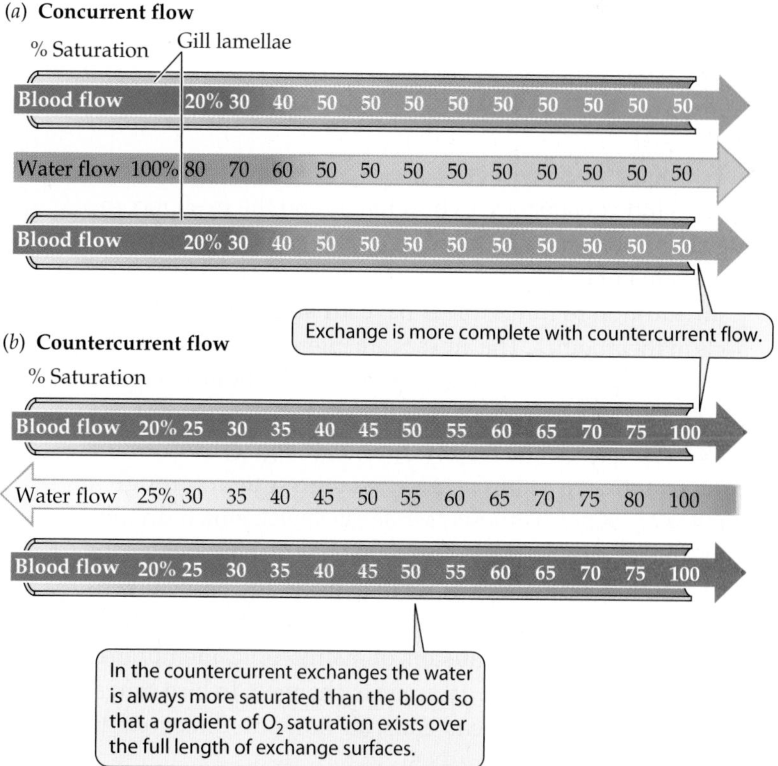

48.6 Countercurrent Exchange Is More Efficient than Concurrent Exchange In these models of concurrent and countercurrent gas exchange, the numbers represent the O_2 saturation of blood and water. (*a*) In a concurrent exchanger, the percentages of saturation of blood and water would reach equilibrium even before the water had flowed halfway across the exchange surface. (*b*) A countercurrent exchanger, such as that found in fish gills, allows more complete exchange because a gradient of O_2 saturation is always maintained.

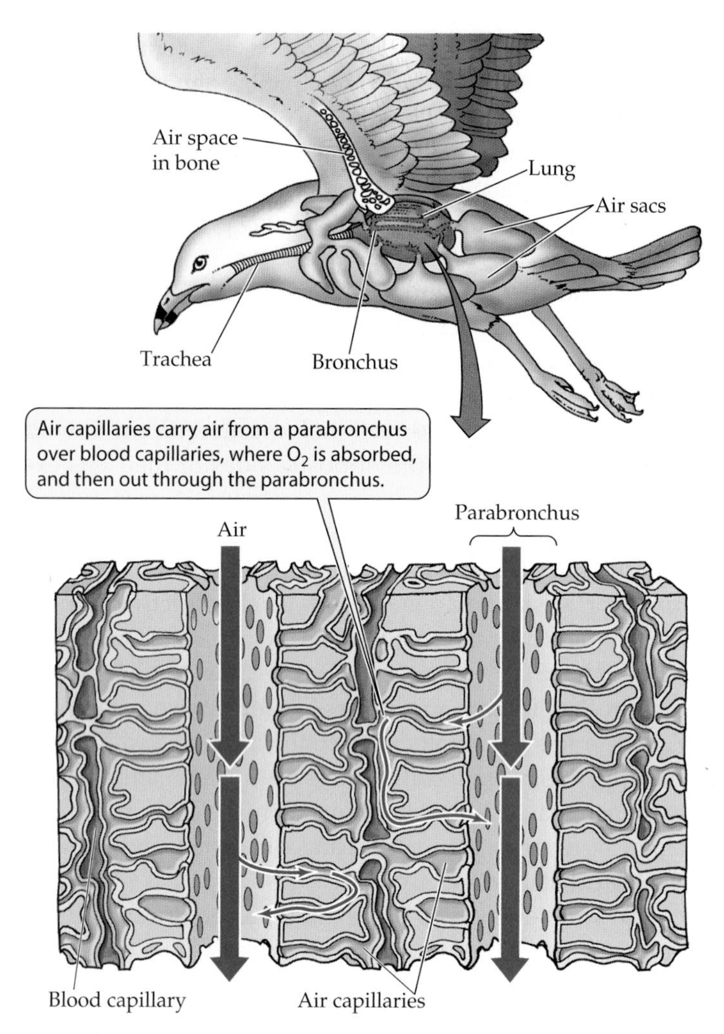

48.7 The Respiratory System of a Bird (*a*) The air sacs and air spaces in the bones are unique to birds. (*b*) Air flows through bird lungs unidirectionally in parabronchi. Air capillaries, the site of gas exchange, branch off the parabronchi.

out through the same airways, as it does in mammals. Thus there is little dead space in bird lungs, and the fresh incoming air is not mixed with stale air. In this way, a high P_{O_2} gradient is maintained.

In addition to lungs, birds have **air sacs** at several locations in their bodies. The air sacs are interconnected with the lungs and with air spaces in some of the bones (Figure 48.7*a*). The air sacs receive inhaled air, but they are not gas exchange surfaces. As in other air-breathing vertebrates, air enters and leaves a bird's gas exchange system through a **trachea** (commonly known as the *windpipe*), which divides into smaller airways called **bronchi** (singular, bronchus). In air-breathing vertebrates other than birds, the bronchi generate trees of branching airways that become finer and finer until they dead-end in clusters of microscopic, membrane-enclosed air sacs, where gases are exchanged. In bird lungs, however, there are no dead ends; air flows unidirectionally through the lungs (Figure 48.8).

In bird lungs, the bronchi divide into tubelike **parabronchi** that run parallel to one another through the lungs (Figure 48.7*b*). Branching off the parabronchi are numerous tiny airways called *air capillaries*. Air flows through the lungs in the parabronchi and diffuses into the air capillaries, which are

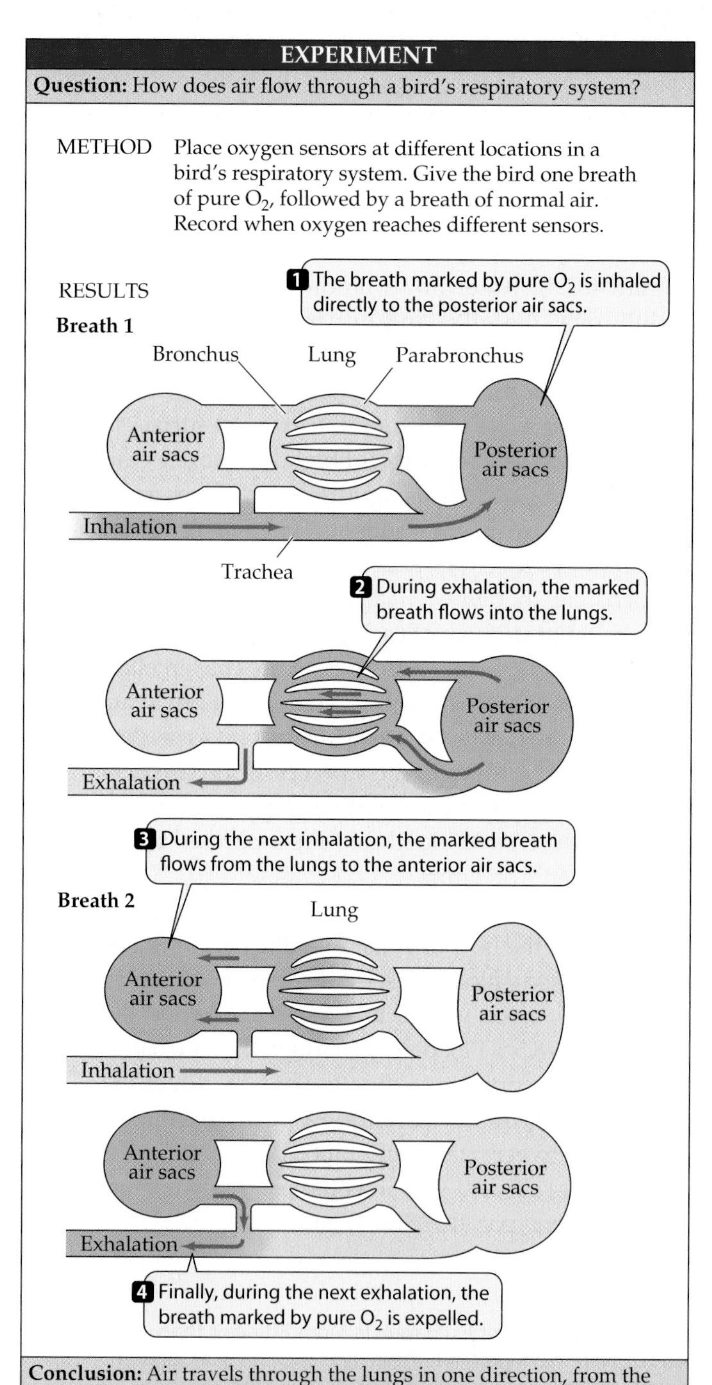

EXPERIMENT
Question: How does air flow through a bird's respiratory system?
METHOD Place oxygen sensors at different locations in a bird's respiratory system. Give the bird one breath of pure O_2, followed by a breath of normal air. Record when oxygen reaches different sensors.
RESULTS
1 The breath marked by pure O_2 is inhaled directly to the posterior air sacs.
Breath 1
Bronchus
Lung
Parabronchus
Anterior air sacs
Posterior air sacs
Inhalation
Trachea
2 During exhalation, the marked breath flows into the lungs.
Anterior air sacs
Posterior air sacs
Exhalation
3 During the next inhalation, the marked breath flows from the lungs to the anterior air sacs.
Breath 2
Lung
Anterior air sacs
Posterior air sacs
Inhalation
Anterior air sacs
Posterior air sacs
Exhalation
4 Finally, during the next exhalation, the breath marked by pure O_2 is expelled.
Conclusion: Air travels through the lungs in one direction, from the posterior to the anterior air sacs. Two cycles of inhalation and exhalation are required for the air to travel through the bird's respiratory tract.

48.8 The Path of Air Flow through Bird Lungs The air a bird takes in with one breath (blue) travels through the lungs in one direction, from the posterior to the anterior air sacs.

the gas exchange surfaces. They are so numerous that they provide an enormous surface area for gas exchange.

The puzzle of how birds breathe was solved by an experiment that placed small oxygen sensors at different locations in the air sacs and airways of birds. The bird could then be exposed to pure oxygen for just a single breath, and the progress of that single breath through the bird's gas exchange system could be followed by the oxygen sensors. This experiment showed that a single breath remains in the bird's gas exchange system for two cycles of inhalation and exhalation, and that the air sacs work as bellows, maintaining a continuous and unidirectional flow of fresh air through the lungs (see Figure 48.8).

The advantages of the bird gas exchange system are similar to those of fish gills. The air sacs keep fresh air flowing unidirectionally and continuously over the gas exchange surfaces. Thus, the bird can supply its gas exchange surfaces with a continuous flow of fresh air that has a P_{O_2} close to that of the ambient air. Even when the P_{O_2} of the ambient air is only slightly above the P_{O_2} of the blood, O_2 can diffuse from air to blood.

TIDAL BREATHING IN HUMANS. Lungs evolved in the first "air-gulping" vertebrates as outpocketings of the digestive tract. Although their structure has evolved considerably, lungs remain dead-end sacs in all air-breathing vertebrates except birds. Because lungs are dead-end sacs, ventilation cannot be constant and unidirectional, but must be **tidal**: Air flows in and exhaled gases flow out by the same route.

A *spirometer* is an old-fashioned device that has been replaced by electronic flow meters, but we can use a virtual spirometer to visualize how we use our lung capacity in breathing (Figure 48.9). When we are at rest, the amount of air that moves in and out per breath is called the *tidal volume* (about 500 ml for an average human adult). We can breathe much more deeply and inhale more air than our resting tidal volume; the additional volume of air we can take in above normal tidal volume is our *inspiratory reserve volume*. Conversely, we can forcefully exhale more air than we normally do during a resting exhalation. This additional amount of air is the *expiratory reserve volume*. The combined tidal volume, inspiratory reserve volume, and expiratory reserve volume is the *vital capacity*. The vital capacity of an athlete is generally greater than that of a nonathlete, and vital capacity decreases with age.

Vital capacity is not the entire lung volume. Even after the most extreme exhalation possible, some air remains in the lungs. The lungs and airways cannot be collapsed completely; they always contain a *residual volume*. The *total lung capacity* is the sum of the residual volume and the vital capacity.

Tidal breathing severely limits the partial pressure gradient available to drive the diffusion of O_2 from air into the blood. Fresh air is not moving into the lungs during part of the breathing cycle; therefore, the average P_{O_2} of air in the lungs is considerably less than it is in the air outside the

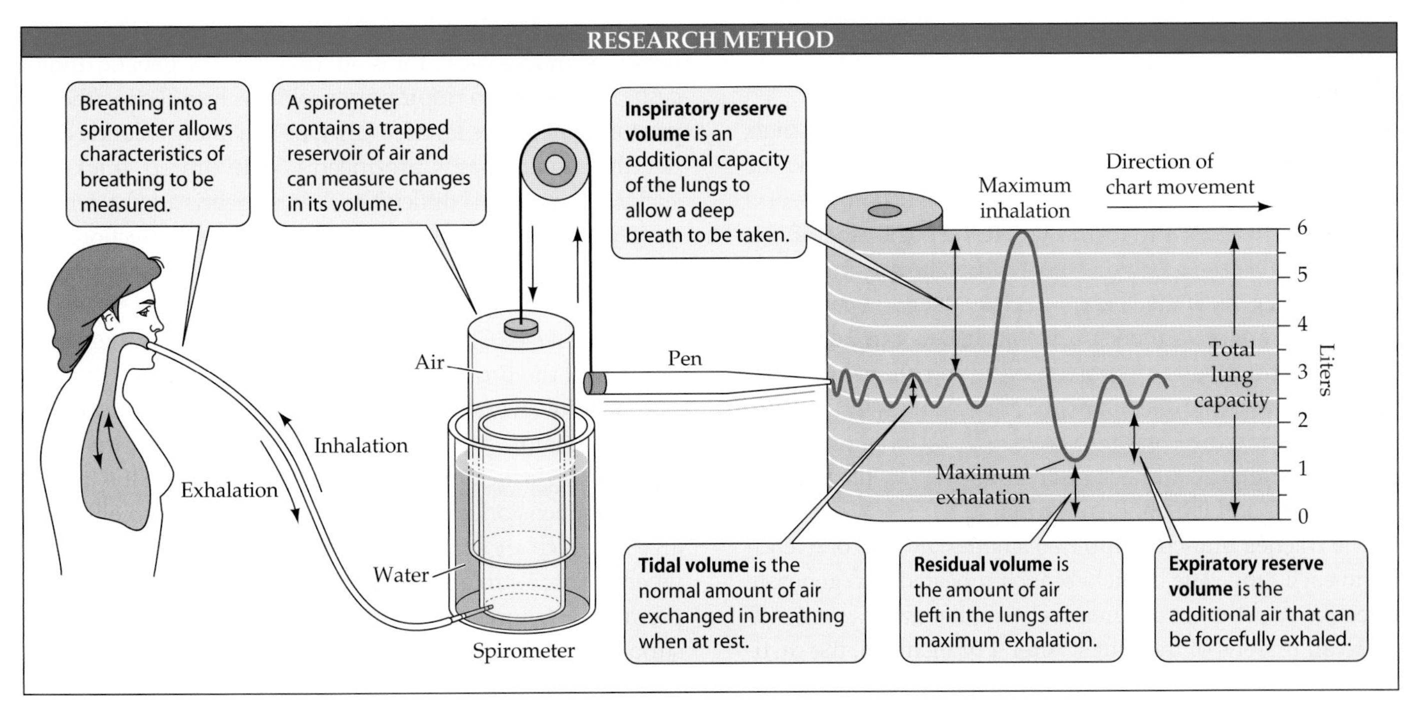
RESEARCH METHOD
Breathing into a spirometer allows characteristics of breathing to be measured.
A spirometer contains a trapped reservoir of air and can measure changes in its volume.
Inspiratory reserve volume is an additional capacity of the lungs to allow a deep breath to be taken.
Maximum inhalation
Direction of chart movement
Air
Pen
Inhalation
Exhalation
Water
Spirometer
Total lung capacity
Liters
6
5
4
3
2
1
0
Maximum exhalation
Tidal volume is the normal amount of air exchanged in breathing when at rest.
Residual volume is the amount of air left in the lungs after maximum exhalation.
Expiratory reserve volume is the additional air that can be forcefully exhaled.

48.9 Measuring Lung Ventilation The spirometer has been superceded by more modern instruments, but it provides a simple demonstration of the characteristics of mammalian tidal breathing. The subject breathes from a closed reservoir of air; the spirometers simply measures the changes in the volume of the reservoir.

lungs. Furthermore, the incoming fresh air mixes with the stale air that was not expelled by the previous exhalation. The volume of this stale air is the sum of the residual volume and, depending on how deeply one is breathing, some or all of the expiratory reserve volume.

The scale in Figure 48.9 shows a typical resting breathing pattern in which a tidal volume of 500 ml of fresh air mixes with over 2,000 ml of stale air before reaching the gas exchange surfaces in our lungs. In this situation, even though the P_{O_2} in the ambient air is 150 mm Hg, the P_{O_2} of the air that reaches the gas exchange surfaces is only about 100 mm Hg.

About 150 ml of the residual volume exists in airways, where gas exchange cannot occur; this volume is called the *anatomical dead space*. Because the anatomical dead space has to be filled with a portion of each incoming breath, shallow breathing can greatly reduce the amount of O_2 that is available at the gas exchange surfaces. For this reason, patients recovering from surgery are encouraged to breathe deeply even though it hurts. Some lung diseases, such as emphysema, increase the anatomical dead space of the lungs and make this problem worse.

In addition to reducing the P_{O_2} gradient, tidal breathing eliminates the possibility of a significant adaptation we noted in fish gills: countercurrent gas exchange. Because air enters and leaves the gas exchange structures by the same route, there is no anatomical way that blood could flow countercurrent to the air flow. To offset these inefficiencies of tidal breathing, mammalian lungs have some design features that maximize the rate of gas exchange: an enormous surface area and a very short path length for diffusion. Mammalian lungs serve the respiratory needs of mammals well. We will use as our example the respiratory system of humans.

Gas Exchange in Human Lungs

Air enters the lungs through the oral cavity or nasal passage, which join together in the *pharynx* (Figure 48.10*a*). Below the pharynx, the esophagus conducts food to the stomach, and a single trachea leads to the lungs. At the beginning of this airway is the *larynx*, or voice box, which houses the vocal cords. The larynx is the "Adam's apple" that you can see or feel on the front of your neck. The trachea is about 2 cm in diameter. Its thin walls are prevented from collapsing by C-shaped bands of cartilage that support them as air pressure changes during the breathing cycle. If you run your fingers down the front of your neck just below your larynx, you can feel a couple of these bands of cartilage.

The trachea branches into two smaller *bronchi*, one leading to each lung. The bronchi branch repeatedly to generate a treelike structure of progressively smaller airways extending to all regions of the lungs. After 4 branchings, the cartilage supports disappear, marking the transition to **bronchioles**. After about 16 branchings, the bronchioles are smaller than the diameter of a pencil lead, and tiny, thin-walled air sacs called **alveoli** begin to appear. After about 6 more branchings, the airways end in clusters of alveoli (Figure 48.10*b*). Because the airways only conduct air to and from the alveoli and do not themselves conduct gas exchange, their volume is anatomical dead space.

The alveoli are the sites of gas exchange. The total number of alveoli in human lungs is about 300 million. Even though each alveolus is very small, their combined surface area for diffusion of respiratory gases is about 70 m^2— about one-fourth the size of a basketball court. Each alveolus is made of very thin cells. Between and surrounding the alveoli are networks of capillaries, whose walls are also made up of exceedingly thin cells. Where capillary meets alveolus, very little tissue separates them (Figure 48.10*c*), so the length of the diffusion path between air and blood is less than 2 μm.

Respiratory tract secretions aid ventilation

Mammalian lungs have two other important adaptations that do not directly influence their gas exchange properties, but do affect the process of ventilation: the production of mucus and the production of surfactant.

Many cells lining the airways produce a sticky mucus that captures bits of dirt and microorganisms that are inhaled. Other cells lining the airways have cilia whose beating continually sweeps the mucus, with its trapped debris, up toward the pharynx, where it can be swallowed or spit out. This phenomenon, called the *mucus escalator*, can be adversely affected by inhaled pollutants. Smoking one cigarette can immobilize the cilia of the airways for hours. A smoker's cough results from the need to clear the obstructing mucus from the airways when the mucus escalator is out of order. The genetic disease *cystic fibrosis* causes respiratory problems by affecting the respiratory mucus (see Figure 17.3). Due to a faulty chloride channel, the mucus that is produced is dehydrated, thick, and sticky. This mucus is difficult to clear, so debris and bacteria remain in the airways, resulting in blockage and infections.

A **surfactant** is a chemical substance that reduces the surface tension of a liquid by interfering with the cohesive forces that create it (see Chapter 2). Surface tension gives the surface of a liquid the properties of an elastic membrane. The thin film of fluid covering the air-facing surfaces of the alveoli has surface tension, which contributes to the elasticity of the lungs. This elasticity must be overcome to inflate the lungs.

Surface tension normally is reduced by certain cells in the alveoli, which are stimulated to produce surfactant molecules when they are stretched. If a baby is born more than a month prematurely, however, these cells may not yet be producing surfactant. Such a premature baby has great difficulty breathing because an enormous effort is required to stretch the alve-

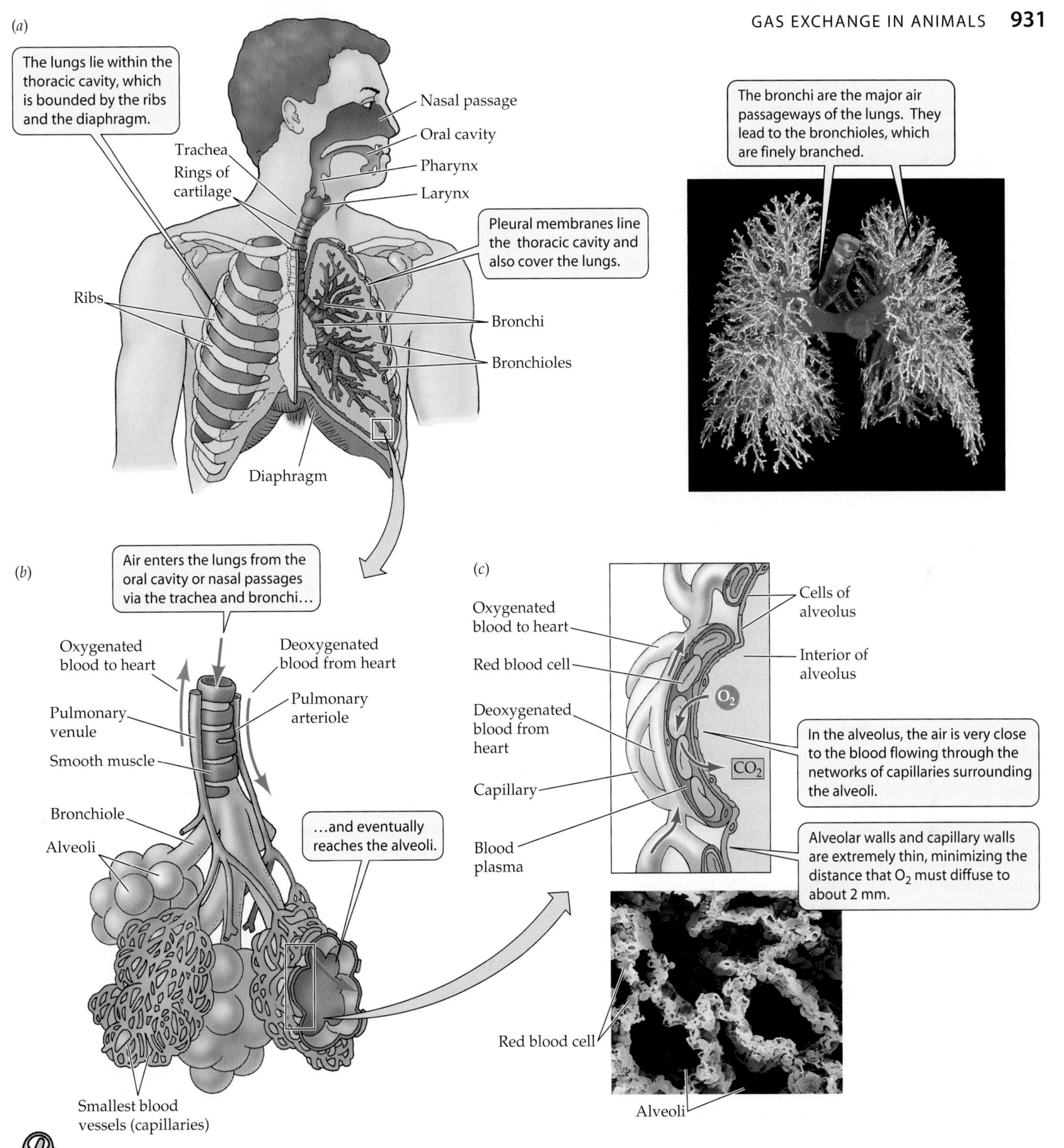

48.10 The Human Respiratory System The diagrams trace the hierarchy of human respiratory structures, from the lungs to the minuscule alveoli.

oli. A baby with this condition, known as *respiratory distress syndrome*, may die from exhaustion and suffocation. Common treatments have been to put the baby on a respirator to assist its breathing and to give it hormones to speed its lung development. A new approach, however, is to apply surfactant to the lungs via an aerosol.

Lungs are ventilated by pressure changes in the thoracic cavity

As Figure 48.10*a* shows, human lungs are suspended in a right and a left **thoracic cavity**. Each thoracic cavity is a closed compartment bounded on the top by the shoulder girdle, on the side by the rib cage, at the midline by tissues surrounding the heart and digestive tract, and on the bottom by a domed sheet of muscle, the **diaphragm**. Each thoracic cavity is lined by a continuous sheet of **pleural membrane**, which doubles back on itself to cover the lungs. Because the pleural membranes also constitute closed compartments within the thoracic cavity, they are called **pleural cavities**. Within the pleural cavity is a thin film of fluid that lubricates the inner surfaces of the pleural membranes so they can slip and slide against each other during breathing movements, but surface tension makes it difficult to pull the pleural membranes apart. Think of two wet panes of glass, or two wet microscope slides. You can easily slide them past each other, but it is difficult to separate them. In addition to being "stuck" to each other by surface tension, the pleural membranes are attached to the wall of the thoracic cavity and to the surface of the lung.

Breathing involves changes in the volume of the thoracic cavity. Because the pleural membranes are attached to the walls of the thoracic cavity, and because the pleural cavity is a closed compartment, any increase in the thoracic cavity volume creates a subatmospheric pressure (which we will refer to as a negative pressure) inside the pleural cavity. Even between breaths, there is normally a slight negative pressure in the pleural cavity because the rib cage is pulling outward and the elasticity of the lung tissue is pulling inward. This slight suction keeps the alveoli partly inflated. If the thoracic cavity is punctured—by a knife wound, for example—air leaks into the pleural cavity, and the pressure from this air causes the lung to collapse. If the wound is not sealed, breathing movements pull air into the pleural cavity rather than into the lung, and ventilation of the alveoli in that lung ceases.

Inhalation is initiated by contraction of the muscular diaphragm. As the diaphragm contracts, it expands the thoracic cavity, pulls on the pleural membranes, and increases the negative pressure in the pleural cavity. The closed pleural cavity cannot expand, so it pulls on the lungs. Because the lungs are not a closed cavity, but have an airway to the atmosphere, they can expand in volume, and air rushes in through the trachea from the outside. Exhalation begins when the contraction of the diaphragm ceases. The diaphragm relaxes and moves up, and the elastic recoil of the lung tissues pushes air out through the airways. When a person is at rest, inhalation is an active process and exhalation is a passive process (Figure 48.11).

The diaphragm is not the only muscle that can change the volume of the thoracic cavity. Between the ribs are two sets of **intercostal muscles**. The *external intercostal muscles* expand the thoracic cavity by lifting the ribs up and outward. The *internal intercostal muscles* decrease the volume of the thoracic cavity by pulling the ribs down and inward. When heavy demands are placed on the gas exchange system, such as during strenuous exercise, the external intercostal muscles increase the volume of air inhaled, making use of the inspiratory reserve volume, and the internal intercostal muscles increase the amount of air exhaled, making use of the expiratory reserve volume. The abdominal muscles can also aid in breathing. When they contract, they cause the abdominal contents to push up on the diaphragm and contribute to the expiratory reserve volume.

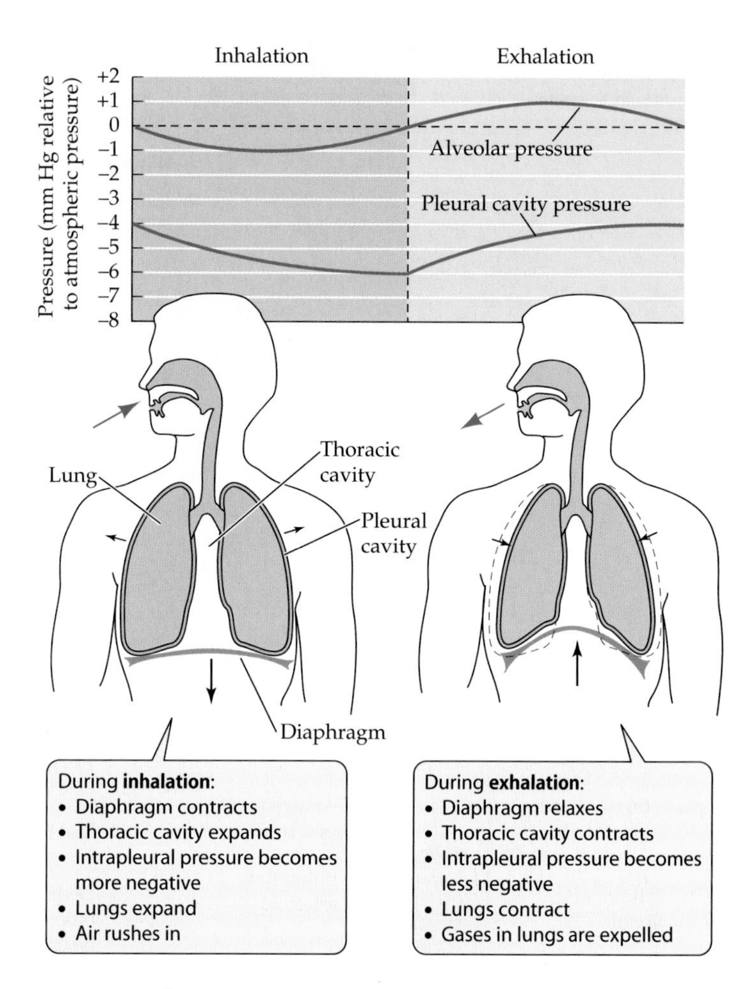

48.11 Into the Lungs and Out Again Inhalation is an active process spurred by the contraction of the diaphragm. Exhalation generally is a passive process as the diaphragm relaxes.

Remember that ventilation and perfusion work together to maximize the partial pressure gradients across the gas exchange surface. Ventilation delivers O_2 to the environmental side of the exchange surface, where it diffuses into the body and is swept away by perfusion. The reverse is true for the exchange of CO_2: Perfusion delivers it to the exchange surface, where it diffuses out and is swept away by ventilation.

Blood Transport of Respiratory Gases

Perfusion of the lungs is one of the functions of the circulatory system. The circulatory system uses a pump (the heart) and a network of vessels to transport extracellular fluids and associated cells (blood) around the body. Circulatory systems are the subject of the next chapter, so here we will discuss only one aspect of perfusion: how the respiratory gases are transported in the blood.

The liquid part of the blood, the *blood plasma*, carries some O_2 in solution, but its ability to transport O_2 is quite limited. The blood plasma of a human, for example, can contain in solution only about 0.3 ml of O_2 per 100 ml of plasma, which is inadequate to support even basal metabolism. To increase its O_2 transport capacity, the blood of most animals, vertebrate and invertebrate, also contains molecules that can bind reversibly to O_2 depending on its partial pressure. These molecules pick up or bind O_2 where its partial pressure is high and release it where its partial pressure is lower. There are many O_2 transport molecules in the animal kingdom, but in vertebrates this role is played by hemoglobin contained in red blood cells. Hemoglobin increases the capacity of blood to transport O_2 by about 60-fold.

Hemoglobin combines reversibly with oxygen

Red blood cells contain enormous numbers of hemoglobin molecules. Hemoglobin is a protein consisting of four polypeptide subunits (see Figure 3.8). Each of these polypeptides surrounds a *heme group*—an iron-containing ring structure that can reversibly bind a molecule of O_2. Thus, each molecule of hemoglobin can bind to four molecules of O_2.

As O_2 diffuses into the red blood cells, it binds to hemoglobin. Once O_2 is bound, it cannot diffuse back across the red cell plasma membrane. By binding O_2 molecules as they enter the red blood cells, hemoglobin maximizes the partial pressure gradient driving the diffusion of O_2 into the cells. In addition, it enables the red blood cells to carry a large amount of O_2 to the tissues of the body.

The ability of hemoglobin to pick up or release O_2 depends on the P_{O_2} of its environment. When the P_{O_2} of the blood plasma is high, as it usually is in the lung capillaries, each molecule of hemoglobin can carry its maximum load of four molecules of O_2. As the blood circulates through the rest of the body, it encounters lower P_{O_2} values. At these lower P_{O_2} values, the hemoglobin releases some of the O_2 it is carrying (Figure 48.12).

The relation between P_{O_2} and the amount of O_2 bound to hemoglobin is not linear, but S-shaped (sigmoidal). The sigmoidal hemoglobin–O_2 binding curve in Figure 48.12 reflects interactions between the four subunits of the hemoglobin molecule. At low P_{O_2} values, only one subunit will bind an O_2 molecule. When it does so, the shape of that subunit changes, causing an alteration in the quaternary structure of the whole hemoglobin molecule. That structural change makes it easier for the other subunits to bind a molecule of O_2; that is, their O_2 *affinity* is increased. Therefore, a smaller increase in P_{O_2} is necessary to get most of the hemoglobin molecules to bind two O_2 molecules (that is, to become 50% saturated) than was necessary to get them to bind one O_2 molecule (to become 25% saturated). This influence of the binding of O_2 by one subunit on the O_2 affinity of the other subunits is called **positive cooperativity**.

Once the third molecule of O_2 is bound, the relationship seems to change, as a larger increase in P_{O_2} is required for the hemoglobin to reach 100 percent saturation. This upper bend of the sigmoid curve is due to a probability phenomenon: The closer we get to having all subunits occupied, the less likely it is that any particular O_2 molecule will find a place to bind. Therefore, it takes a relatively greater P_{O_2} to achieve 100 percent saturation.

This is a good place to mention the danger posed by carbon monoxide (CO), which can come from a faulty furnace or from burning a fuel such as charcoal or kerosene without adequate ventilation. CO binds to hemoglobin with a higher affinity (240 times higher!) than O_2. Thus, CO prevents hemoglobin from transporting and releasing O_2 to the tissues of the body. The victim loses consciousness and can die because the brain lacks O_2.

The O_2-binding properties of hemoglobin help get O_2 to the tissues that need it most. In the lungs, where the P_{O_2} is about 100 mm Hg, hemoglobin is 100 percent saturated. The P_{O_2} in blood returning to the heart from the body is usually about 40 mm Hg. You can see from Figure 48.12 that at this P_{O_2}, the hemoglobin is still about 75 percent saturated. This means that as the blood circulates around the body, only about one in four of the O_2 molecules it carries is released to the tissues. This system seems inefficient, but it is really quite

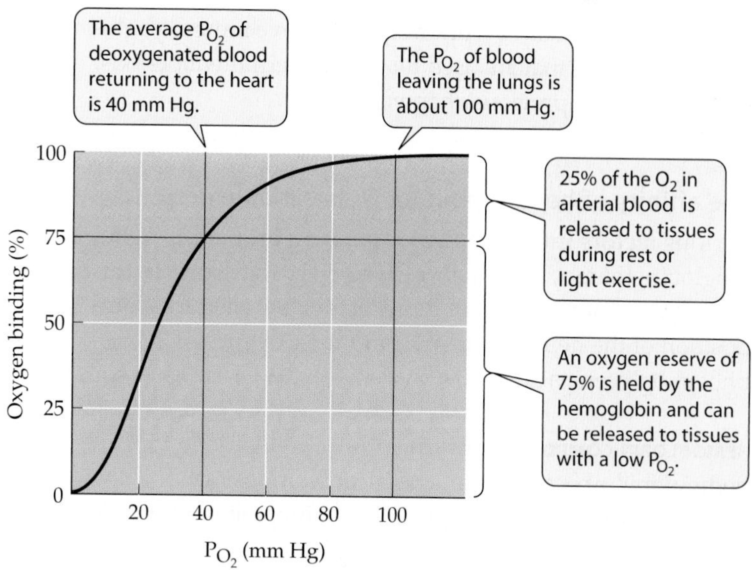
The average P_{O_2} of deoxygenated blood returning to the heart is 40 mm Hg.
The P_{O_2} of blood leaving the lungs is about 100 mm Hg.
25% of the O_2 in arterial blood is released to tissues during rest or light exercise.
An oxygen reserve of 75% is held by the hemoglobin and can be released to tissues with a low P_{O_2}.
Oxygen binding (%)
100
75
50
25
0
20
40
60
80
100
P_{O_2} (mm Hg)

48.12 The Binding of O_2 to Hemoglobin Depends on P_{O_2} Hemoglobin in blood leaving the lungs is 100 percent saturated (four molecules of O_2 are bound to each hemoglobin molecule). Most hemoglobin molecules will drop only one of their four O_2 molecules as they circulate through the body, and are still 75 percent saturated when the blood returns to the lungs. The steep portion of this oxygen-binding curve comes into play when tissue P_{O_2} falls below the normal 40 mm Hg, at which point the hemoglobin will "unload" its O_2 reserves.

adaptive, because the hemoglobin keeps 75 percent of its O_2 in reserve to meet peak demands.

When a tissue becomes starved of O_2 and its local P_{O_2} falls below 40 mm Hg, the hemoglobin flowing through that tissue is on the steep portion of its sigmoid binding curve. That means that relatively small decreases in P_{O_2} below 40 mm Hg will result in the release of lots of O_2 to the tissue. Thus the positive cooperativity of O_2 binding by hemoglobin is very effective in making O_2 available to the tissues precisely when and where it is needed most.

Myoglobin holds an O_2 reserve

Muscle cells have their own oxygen-binding molecule, **myoglobin**. Myoglobin consists of just one polypeptide chain associated with an iron-containing ring structure that can bind one molecule of O_2. Myoglobin has a higher affinity for O_2 than hemoglobin does, so it picks up and holds O_2 at P_{O_2} values at which hemoglobin is releasing its bound O_2 (Figure 48.13).

Myoglobin provides a reserve of O_2 for the muscle cells for times when metabolic demands are high and blood flow is interrupted. Interruption of blood flow in muscles is common because contracting muscles constrict blood vessels. When tissue P_{O_2} values are low and hemoglobin can no longer supply more O_2, myoglobin releases its bound O_2. Diving mammals such as seals have high concentrations of myoglobin in their muscles, which is one reason they can stay under water for so long. (We will learn more about adaptations for diving in the next chapter.) Even in nondiving animals, muscles called on for extended periods of work frequently have more myoglobin than muscles that are used for short, intermittent periods, as we saw in the previous chapter.

The affinity of hemoglobin for O_2 is variable

Various factors influence the O_2-binding properties of hemoglobin, thereby influencing O_2 delivery to tissues. In this section we examine three of those factors: the chemical composition of the hemoglobin, pH, and the presence of 2,3 bisphosphoglyceric acid (BPG).

HEMOGLOBIN COMPOSITION. There is more than one type of hemoglobin, because the chemical composition of the polypeptide chains that form the hemoglobin molecule varies. The normal hemoglobin of adult humans has two each of two kinds of polypeptide chains—two α-globin chains and two β-globin chains—and the oxygen-binding characteristics shown in Figure 48.12.

Before birth, the human fetus has a different form of hemoglobin, consisting of two α-globin and two γ-globin chains. The functional difference between these two types of hemoglobin is that the fetal hemoglobin has a higher affinity for O_2. Therefore, the hemoglobin–O_2 binding curve of fetal hemoglobin is shifted to the left in comparison to the curve for adult hemoglobin (see Figure 48.13). You can see from these curves that if both types of hemoglobin are at the same P_{O_2}, the fetal hemoglobin will pick up O_2 released by its mother's adult hemoglobin. This difference in O_2 affinities facilitates the transfer of O_2 from the mother's blood to the blood of the fetus in the placenta.

Llamas and vicuñas are native to the Andes Mountains of South America. In the natural habitat of these mammals, more than 5,000 m above sea level, the P_{O_2} is below 85 mm Hg, and the P_{O_2} in their lungs is about 50 mm Hg. Thus, the hemoglobins of these animals must pick up O_2 in an environment that has a low P_{O_2}. The hemoglobins of llamas and vicuñas have oxygen-binding curves to the left of the curves of hemoglobins of most other mammals—in other words,

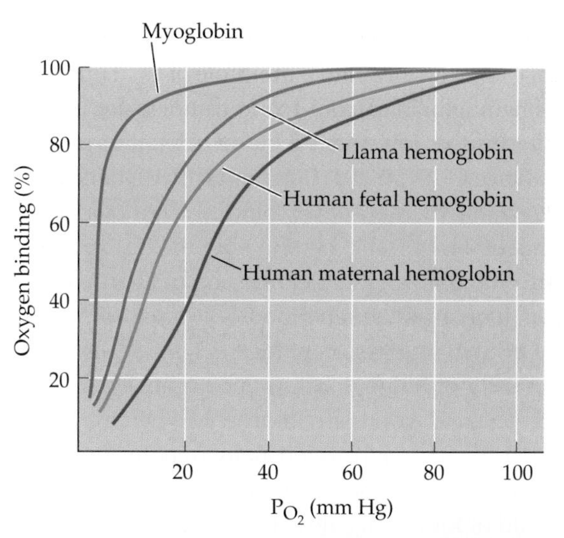

Llama guanaco

48.13 Oxygen-Binding Adaptations The different hemoglobins and myoglobin have different oxygen-binding properties adapted to different circumstances. The hemoglobin of llamas, for example, is adapted for binding O_2 at high altitudes, where P_{O_2} is low.

their hemoglobin can become saturated with O_2 at lower P_{O_2} values than those of other mammals.

PH. The oxygen-binding properties of hemoglobin are also influenced by physiological conditions. The influence of pH on the function of hemoglobin is known as the **Bohr effect**. As blood passes through metabolically active tissue such as exercising muscle, it picks up acidic metabolites such as lactic acid, fatty acids, and CO_2, and blood pH falls. The excess H^+ ions bind preferentially to deoxygenated hemoglobin and decrease the affinity of that hemoglobin for O_2. As a result, the oxygen-binding curve of hemoglobin shifts to the right. This shift means that the hemoglobin will release more O_2 in tissues where pH is low—another way that O_2 is supplied where and when it is most needed.

2,3 BISPHOSPHOGLYCERIC ACID. BPG is a metabolite of glycolysis (see Figure 7.6). Mammalian red blood cells have a high concentration of BPG, which serves as an important regulator of hemoglobin function. BPG, like excess H^+ ions, reversibly combines with deoxygenated hemoglobin and lowers its affinity for O_2. The result is that at any P_{O_2}, hemoglobin releases more of its bound O_2 than it otherwise would. In other words, BPG shifts the oxygen-binding curve of mammalian hemoglobin to the right. When humans go to high altitudes, or when they cease being sedentary and begin to exercise, the level of BPG in their red blood cells goes up and makes it easier for the hemoglobin to deliver more O_2 to the tissues. The reason that fetal hemoglobin has a left-shifted hemoglobin–O_2 binding curve is that the γ-globin chains of fetal hemoglobin have a lower affinity for BPG than do the β-globin chains of adult hemoglobin.

CO_2 is transported as bicarbonate ions in the blood

Delivering O_2 to the tissues is only half of the respiratory function of the blood. The blood also must take carbon dioxide, a metabolic waste product, away from the tissues (Figure 48.14). CO_2 is highly soluble and readily diffuses through cell membranes, moving from its site of production in the tissues into the blood, where the partial pressure of carbon dioxide (P_{CO_2}) is lower. However, very little dissolved CO_2 is transported by the blood. Most CO_2 produced by the tissues is transported to the lungs in the form of **bicarbonate ions**, HCO_3^-. CO_2 is converted to HCO_3^-, transported to the lungs, and then converted back to CO_2 in several steps.

When CO_2 dissolves in water, some of it slowly reacts with the water molecules to form carbonic acid (H_2CO_3), some of which then dissociates into a proton (H^+) and a bicarbonate ion (HCO_3^-). This reversible reaction is expressed as follows:

$$CO_2 + H_2O \rightleftharpoons H_2CO_3 \rightleftharpoons H^+ + HCO_3^-$$

In the blood plasma, the reaction between CO_2 and H_2O proceeds slowly. But it is a different story in the endothelial cells of the capillaries and in the red blood cells, where the enzyme *carbonic anhydrase* speeds up the conversion of CO_2 to H_2CO_3. The newly formed carbonic acid dissociates, and the resulting bicarbonate ions enter the plasma in exchange for Cl^- (see Figure 48.14). By converting CO_2 to H_2CO_3, carbonic anhydrase reduces the P_{CO_2} in these cells and in the plasma, facilitating the diffusion of CO_2 from tissue cells to endothelial

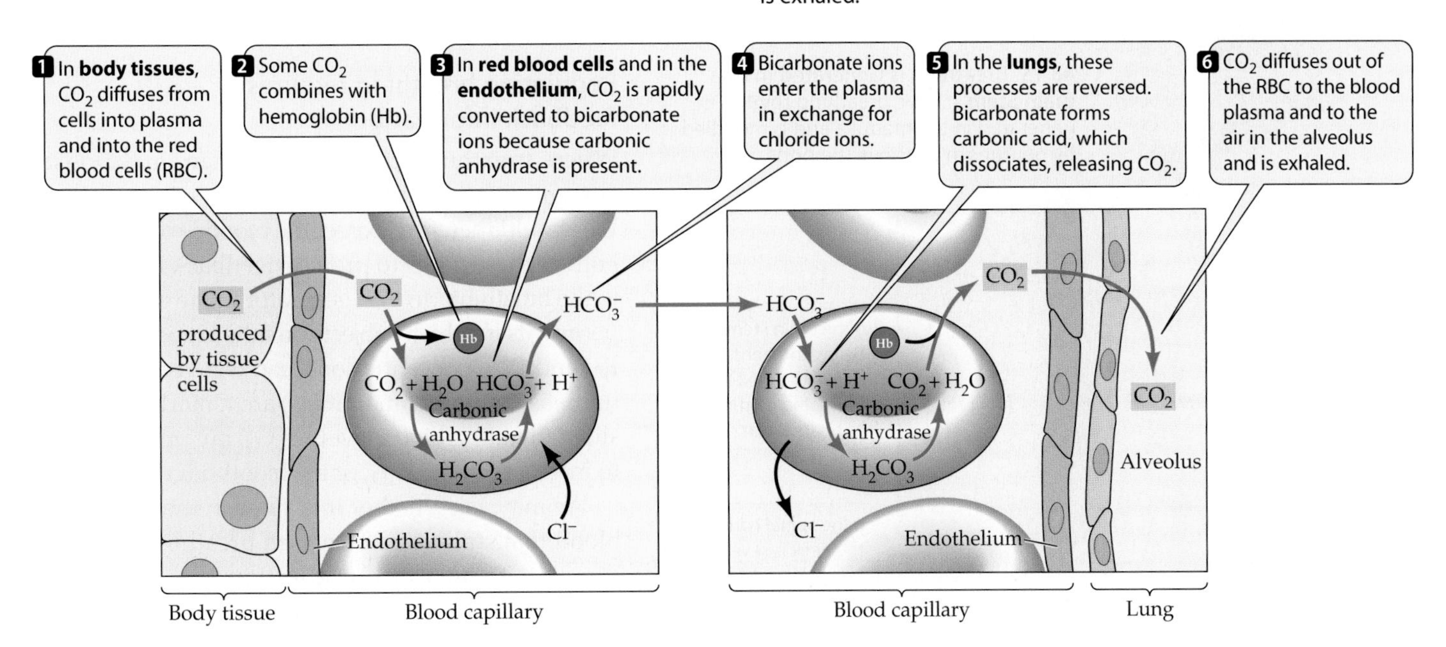

1 In body tissues, CO2 diffuses from cells into plasma and into the red blood cells (RBC).
2 Some CO2 combines with hemoglobin (Hb).
3 In red blood cells and in the endothelium, CO2 is rapidly converted to bicarbonate ions because carbonic anhydrase is present.
4 Bicarbonate ions enter the plasma in exchange for chloride ions.
5 In the lungs, these processes are reversed. Bicarbonate forms carbonic acid, which dissociates, releasing CO2.
6 CO2 diffuses out of the RBC to the blood plasma and to the air in the alveolus and is exhaled.
CO2 produced by tissue cells
CO2
Hb
CO2 + H2O HCO3- + H+
Carbonic anhydrase
H2CO3
HCO3-
Cl-
Endothelium
HCO3-
HCO3- + H+ CO2 + H2O
Carbonic anhydrase
H2CO3
Cl-
Endothelium
CO2
CO2
Alveolus
Body tissue
Blood capillary
Blood capillary
Lung

48.14 Carbon Dioxide Is Transported as Bicarbonate Ions Carbonic anhydrase in capillary endothelial cells and in red blood cells facilitates conversion of CO_2 produced by tissues into bicarbonate ions carried by the plasma. In lungs, the process is reversed as CO_2 is exhaled.

cells, plasma, and red blood cells. Some CO_2 is also carried in chemical combination with hemoglobin.

In the lungs, the reactions involving CO_2 and bicarbonate ions are reversed. Remember that an enzyme such as carbonic anhydrase only speeds up a reversible reaction; it does not determine its direction. The direction is determined by concentrations of reactants and products (see Chapter 6). Ventilation keeps the CO_2 concentration in the alveoli low, so CO_2 diffuses from the blood plasma into the alveoli, lowering the CO_2 concentration in the blood, which favors the conversion of HCO_3^- into CO_2.

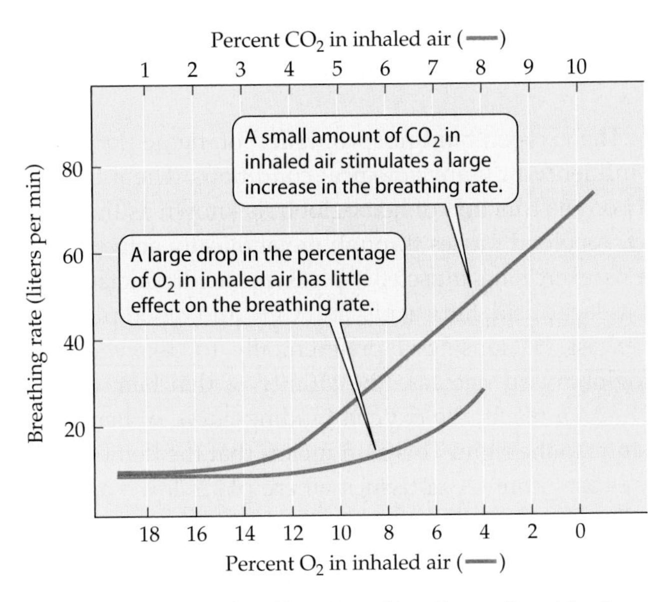

48.16 Carbon Dioxide Affects Breathing Rate Breathing is more sensitive to increased carbon dioxide content in inhaled air than to it's decreased oxygen content.

Regulation of Breathing

We must breathe every minute of our lives, but we don't usually worry about our need to breathe, or even think about it very often. Breathing is an autonomic function of the nervous system. The breathing pattern easily adjusts itself around other activities (such as speech and eating), and breathing rates change to match the metabolic demands of our bodies. In this section we examine how the regular breathing cycle is generated and controlled.

Breathing is controlled in the brain stem

The autonomic nervous system maintains breathing and modifies its depth and frequency to meet the demands of the body for O_2 supply and CO_2 elimination. Breathing ceases if the spinal cord is severed in the neck region, showing that the breathing pattern is generated in the brain. If the brain stem is cut just above the medulla, the segment of the brain stem just above the spinal cord, an irregular breathing pattern remains (Figure 48.15).

Groups of neurons within the medulla increase their firing rates just before an inhalation begins. As more and more of these neurons fire—and fire faster and faster—the diaphragm contracts. Suddenly the neurons stop firing, the diaphragm relaxes, and exhalation begins. Exhalation is usually a passive process that depends on the elastic recoil of the lung tissues. When breathing demand is high, however, as during strenuous exercise, motor neurons for the intercostal muscles are recruited, which increases both the inhalation and the exhalation volumes. Brain areas above the medulla modify breathing to accommodate speech, ingestion of food, coughing, and emotional states.

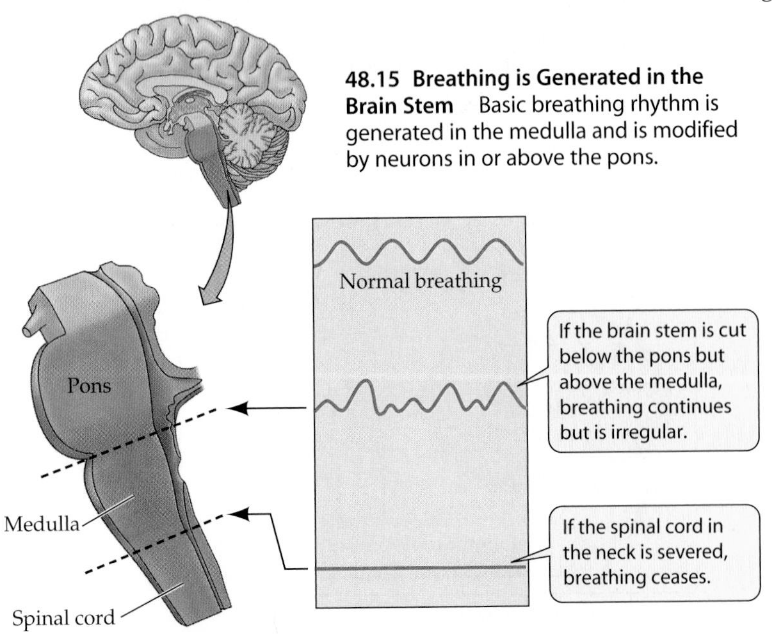

48.15 Breathing is Generated in the Brain Stem Basic breathing rhythm is generated in the medulla and is modified by neurons in or above the pons.

Regulating breathing requires feedback information

When breathing changes or when metabolism changes, it alters the P_{O_2} and the P_{CO_2} in the blood. We should therefore expect the blood levels of one or both of these gases to provide feedback information to the breathing rhythm generator in the medulla. Experiments in which subjects breathe air with different P_{O_2} and P_{CO_2} concentrations lead us to conclude that humans (and other mammals) are remarkably insensitive to falling blood levels of O_2, but very sensitive to increases in the P_{CO_2} of the blood (Figure 48.16).

We may ask whether it is a rise in the P_{CO_2} of the blood that stimulates increased breathing when we exercise. To answer this question, researchers ran dogs on treadmills at different speeds. As the speed of the treadmill increased, the respiratory gas exchange rate of the dogs increased but the P_{CO_2} of the dogs' blood remained constant. Before concluding that blood P_{CO_2} is not the metabolic feedback information for breathing rate, however, the researchers changed their experiment. Instead of increasing the speed of the treadmill, they gradually increased its slope so that the dogs were running at the same speed, but were working harder because they were running uphill.

In this experiment, the P_{CO_2} of the blood increased as the slope of the treadmill increased and as the respiratory gas exchange rate increased. The researchers concluded that the P_{CO_2} of the blood is the primary metabolic feedback information for breathing. However, when an animal starts to run or changes its running speed, additional feedback information from receptors in muscles and joints changes the sensitivity of the CO_2 sensors—an example of feedforward information.

Where are partial pressures of gases in the blood sensed? The major site of CO_2 sensitivity is an area on the ventral surface of the medulla, not far from the groups of neurons that generate the breathing rhythm. Primary sensitivity to P_{O_2} in the blood resides in nodes of tissue on the large blood vessels

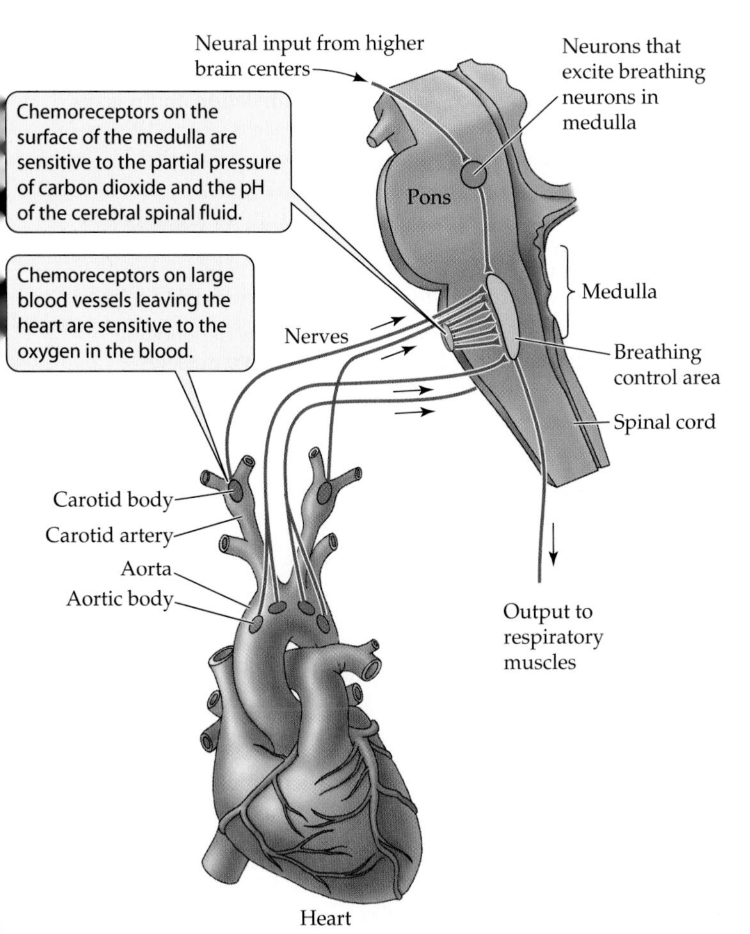
Neural input from higher brain centers
Neurons that excite breathing neurons in medulla
Chemoreceptors on the surface of the medulla are sensitive to the partial pressure of carbon dioxide and the pH of the cerebral spinal fluid.
Pons
Chemoreceptors on large blood vessels leaving the heart are sensitive to the oxygen in the blood.
Medulla
Nerves
Breathing control area
Spinal cord
Carotid body
Carotid artery
Aorta
Aortic body
Output to respiratory muscles
Heart

48.17 Feedback Information Controls Breathing The body uses feedback information from chemosensors in the heart and the brain to match breathing rate to metabolic demand.

leaving the heart: the aorta and the carotid arteries (Figure 48.17). These *carotid* and *aortic bodies* receive enormous supplies of blood relative to their small size, and they contain chemoreceptors. If the blood supply to these structures decreases, or if the P_{O_2} of the blood falls dramatically, the chemoreceptors are activated and send nerve impulses to the breathing control center. Although we are not very sensitive to changes in blood P_{O_2}, the carotid and aortic bodies can stimulate increases in breathing during exposure to very high altitudes or when blood volume or blood pressure is very low.

Chapter Summary

Physical Processes of Respiratory Gas Exchange

▶ Most cells require a constant supply of O_2 and continuous removal of CO_2. These respiratory gases are exchanged between the body fluids of an animal and its environment by diffusion.

▶ In aquatic animals, gas exchange is limited by the low diffusion rate and low amount of O_2 in water. Aquatic animals face a double bind in that the amount of O_2 in water decreases, but their metabolism and the amount of work required to move water over their gas exchange surfaces increase, as water temperature rises. **Review Figure 48.2**

▶ In air, the partial pressure of oxygen decreases with altitude.

▶ Fick's law of diffusion shows how various physical factors influence the rate of diffusion of gases. Adaptations to maximize respiratory gas exchange influence one or more components of Fick's law.

Adaptations for Respiratory Gas Exchange

▶ Adaptations to maximize gas exchange include increasing the surface areas for gas exchange, maximizing partial pressure gradients across those exchange surfaces by decreasing their thickness, ventilating the outer surface with the respiratory medium, and perfusing the inner surface with blood. **Review Figure 48.3**

▶ Insects distribute air throughout their bodies in a system of tracheae, tracheoles, and air capillaries. **Review Figures 48.4**

▶ Fish have large gas exchange surface areas that are ventilated continuously and unidirectionally with water. Countercurrent blood flow helps increase the efficiency of gas exchange. **Review Figures 48.5, 48.6**

▶ The gas exchange system of birds includes air sacs that communicate with the lungs, but are not used for gas exchange. Air flows unidirectionally through bird lungs in parabronchi. Gases are exchanged in air capillaries that run between parabronchi. **Review Figure 48.7**

▶ Each breath of air remains in the bird respiratory system for two breathing cycles. The air sacs work as bellows to supply the air capillaries with a continuous, unidirectional flow of fresh air. **Review Figure 48.8.** **See Web/CD Tutorial 48.1**

▶ Breathing in vertebrates other than birds is tidal and is therefore less efficient than gas exchange in fish or birds. Even though the volume of air exchanged with each breath can vary considerably, the inhaled air is always mixed with stale air. **Review Figure 48.9**

Gas Exchange in Human Lungs

▶ In mammalian lungs, the gas exchange surface area provided by the millions of alveoli is enormous, and the diffusion path length between the air and perfusing blood is very short. **Review Figure 48.10.** **See Web/CD Activity 48.1**

▶ Surface tension in the alveoli would make inflation of the lungs difficult if the alveoli did not produce surfactant.

▶ Inhalation occurs when contractions of the diaphragm create subatmospheric pressure in the thoracic cavity. Relaxation of the diaphragm increases pressure in the thoracic cavity and causes exhalation. **Review Figure 48.11.** **See Web/CD Tutorial 48.2**

▶ During periods of heavy metabolic demands, such as strenuous exercise, the intercostal muscles, located between the ribs, increase the volume of air inhaled and exhaled.

Blood Transport of Respiratory Gases

▶ Oxygen is reversibly bound to hemoglobin in red blood cells. Each molecule of hemoglobin can carry a maximum of four molecules of O_2. Because of positive cooperativity, the affinity of hemoglobin for O_2 depends on the P_{O_2} to which the hemoglobin is exposed. Therefore, hemoglobin picks up O_2 as it flows through respiratory exchange structures and gives up O_2 in metabolically active tissues. **Review Figure 48.12**

▶ Myoglobin has a high affinity for O_2 and serves as an O_2 reserve in muscle.

▶ There is more than one type of hemoglobin. Fetal hemoglobin has a higher affinity for O_2 than does maternal hemoglobin, allowing fetal blood to pick up O_2 from the maternal blood in the placenta. **Review Figure 48.13**

▶ The affinity of hemoglobin for O_2 is decreased by the presence of hydrogen ions or 2,3 bisphosphoglyceric acid.

▶ Carbon dioxide is transported in the blood principally as bicarbonate ions. **Review Figure 48.14**

See Web/CD Activity 48.2

Regulation of Breathing

▶ The breathing rhythm is an autonomic function generated by neurons in the medulla and modulated by higher brain centers. **Review Figure 48.15**

▶ The most important feedback stimulus for breathing is the level of CO_2 in the blood. **Review Figure 48.16**

▶ The breathing rhythm is sensitive to feedback from chemoreceptors on the ventral surface of the medulla and in the carotid and aortic bodies on the large vessels leaving the heart. **Review Figure 48.17**

See Web/CD Activity 48.3 for a concept review of this chapter.

Self-Quiz

1. Which of the following statements is *not* true?
   a. Respiratory gases are exchanged only by diffusion.
   b. Oxygen has a lower rate of diffusion in water than in air.
   c. The O_2 content of water falls as the temperature of water rises.
   d. The amount of O_2 in the atmosphere decreases with increasing altitude.
   e. Birds have evolved active transport mechanisms to augment their respiratory gas exchange.
2. Which statement about the gas exchange system of birds is *not* true?
   a. Respiratory gases are not exchanged in the air sacs.
   b. It can achieve more complete exchange of O_2 from air to blood than the human gas exchange system can.
   c. Air passes through birds' lungs in only one direction.
   d. The gas exchange surfaces in bird lungs are the alveoli.
   e. A breath of air remains in the system for two breathing cycles.
3. Which statement about gas exchange in fish is true?
   a. Blood flows over the gas exchange surfaces in a direction opposite to the flow of water.
   b. Gases are exchanged across the gill filaments.
   c. Ventilation of the gills is tidal in fast swimming fishes.
   d. Less work is needed to ventilate gills in warm water than in cold water.
   e. The path length for diffusion of respiratory gases is determined by the length of the gill filaments.
4. In the human gas exchange system,
   a. the lungs and airways are completely collapsed after a forceful exhalation.
   b. the average P_{O_2} concentration of air inside the lungs is always lower than that in the air outside the lungs.
   c. the P_{O_2} of the blood leaving the lungs is greater than the P_{O_2} of the exhaled air.
   d. the amount of air that is moved per breath during normal, at-rest breathing is termed the total lung capacity.
   e. oxygen and carbon dioxide are actively transported across the alveolar and capillary membranes.
5. Which statement about the human gas exchange system is *not* true?
   a. During inhalation, a subatmospheric pressure exists in the space between the lung and the thoracic wall.
   b. Smoking one cigarette can immobilize the cilia lining the airways for hours.
   c. The respiratory control center in the medulla responds more strongly to changes in arterial O_2 concentration than to changes in arterial CO_2 concentration.
   d. Without surfactant, the work of breathing is greatly increased.
   e. The diaphragm contracts during inhalation and relaxes during exhalation.
6. The hemoglobin of a human fetus
   a. is the same as that of an adult.
   b. has a higher affinity for O_2 than adult hemoglobin has.
   c. has only two protein subunits instead of four.
   d. is supplied by the mother's red blood cells.
   e. has a lower affinity for O_2 than adult hemoglobin has.
7. The amount of O_2 carried by hemoglobin depends on the P_{O_2} in the blood. Hemoglobin in active muscles
   a. becomes saturated with O_2.
   b. takes up only a small amount of O_2.
   c. readily unloads O_2.
   d. tends to decrease the P_{O_2} in the muscle tissues.
   e. is denatured.
8. Most CO_2 in the blood is carried
   a. in the cytoplasm of red blood cells.
   b. dissolved in the plasma.
   c. in the plasma as bicarbonate ions.
   d. bound to plasma proteins.
   e. in red blood cells bound to hemoglobin.
9. Myoglobin
   a. binds O_2 at P_{O_2} values at which hemoglobin is releasing its bound O_2.
   b. has a lower affinity for O_2 than hemoglobin does.
   c. consists of four polypeptide chains, just as hemoglobin does.
   d. provides an immediate source of O_2 for muscle cells at the onset of activity.
   e. can bind four O_2 molecules at once.

10. When the level of CO_2 in the bloodstream *increases*,
    a. the rate of respiration decreases.
    b. the pH of the blood rises.
    c. the respiratory centers become dormant.
    d. the rate of respiration increases.
    e. the blood becomes more alkaline.

For Discussion

1. A species of fish that lives in Antarctica has no hemoglobin. What anatomical and behavioral characteristics would you expect to find in this fish, and why is its distribution limited to the waters of Antarctica?
2. Blood banks store whole blood for a much shorter period than they store blood plasma. The reason is that when blood that has been stored for too long is infused into a patient, it can actually decrease the O_2 availability to the patient's tissues. Why is this so? Explain in terms of the different physiological functions of 2,3 bisphosphoglyceric acid.
3. In the early 1800s, three French scientists went up in a hot air balloon to observe the effects of low oxygen levels on the human body. They took measurements on each other, but as the balloon went higher and higher, their writing became increasingly illegible and nonsensical. They recorded no realization of danger, but in fact continued to throw ballast until all went unconscious. The balloon descended on its own, but only one scientist survived. From your knowledge of respiratory gas exchange and regulation of respiration, how would you explain this tragic episode?
4. In the disease emphysema, the fine structures of alveoli break down, resulting in the formation of larger air cavities in the lungs. Also, the tissue of the lungs becomes less elastic. Explain at least two reasons why patients with emphysema have a low tolerance for exercise.
5. A condition called "the bends" occurs in scuba divers who come too quickly to the surface after spending an extended period in deep water, where they have been breathing pressurized air. The cause of the bends is tiny bubbles of nitrogen coming out of solution in the blood plasma. Seals spend much more time under water and at deeper depths than scuba divers, yet they do not suffer the bends. Why?

49 Circulatory Systems

When you work or exercise, your heart rate and the amount of blood your heart pumps each minute rise as much as three or four times above their resting levels to meet your increasing metabolic demands. It makes sense that working muscle requires more blood, and therefore that the heart has to work harder during exercise. This rationale seems to fall apart, however, when we examine the heart rate of a marine mammal such as an elephant seal while it is at sea.

Elephant seals spend weeks in the open ocean making repeated dives to obtain food. While at sea, a seal may spend only 10 percent of its time at the surface. The rest of the time, it is under water and holding its breath. Can you imagine being able to breathe for only 6 minutes each hour? The average duration of a seal's dive is about 25 minutes, but dives of over 2 hours have been recorded.

Even when the seals are actively pursuing prey under water, the responses of their hearts are very different from the response of your heart when you engage in exercise. The seal's heart slows dramatically, from about 110 beats per minute when it is at the surface to an average of about 30 beats per minute during a dive. At times during the dive, its heart rate may fall to 3 or 4 beats per minute. With the fall in heart rate, the total output of blood falls proportionally. Blood flow to the swimming muscles falls practically to zero. Blood flow to the heart is less than a third what it was before the dive. Only blood flow to its nervous system is maintained at pre-dive levels. Because the circulatory system of the seal responds differently during exercise than yours does, the seal is able to conserve its oxygen supplies for critical functions and remain under water for long periods.

In this chapter, you will learn about the adaptations of circulatory systems that enable them to match blood supply with demand. We will begin the chapter by contrasting the open and closed circulatory systems of invertebrates. Then we will discuss the evolution, structure, and function of vertebrate circulatory systems. Taking the human circulatory system as a model, we will explore the mechanics of the beating heart and the characteristics of the vascular system: the arteries, capillaries, and veins. Another component of the circulatory system is the blood, and we will describe the features of this

Champion Diver This young northern elephant seal (*Mirounga angustirostris*) will spend its days making repeated dives at sea. Visits to the surface average only 6 minutes out of each hour.

fluid tissue and the dynamics of fluid exchange between the blood and the tissue fluids. The chapter ends with a discussion of the hormonal and neuronal regulation of the mammalian circulatory system, in which we will return to the diving seal to understand more about its unusual circulatory adaptations for life at sea.

Circulatory Systems: Pumps, Vessels, and Blood

A **circulatory system** consists of a muscular pump (**heart**), a fluid (**blood**) that can transport materials, and a series of conduits (**blood vessels**) through which the fluid can be pumped around the body. Heart, blood, and vessels are also known collectively as a **cardiovascular system** (from the Greek *kardia*, "heart," and the Latin *vasculum*, "small vessel"). In this section, we will compare the circulatory systems of different groups of animals.

Some simple aquatic animals do not have circulatory systems

A circulatory system is unnecessary if the cells of an organism are close enough to the external environment that nutrients, respiratory gases, and wastes can diffuse between the cells and the environment. Small aquatic invertebrates have structures and body shapes that permit direct exchanges between cells and environment. Many of these animals have flattened body shapes that maximize the amount of surface area that is in contact with the external environment (see Figure 48.1*a*). The cells of some other aquatic invertebrates are served by highly branched central cavities called *gastrovascular systems* that exchange their contents with the external environment. All the cells of a sponge, for example, are in contact with, or very close to, the water that surrounds the animal and circulates through its central cavity (see Figure 48.1*b*).

Large surface areas and branched internal cavities cannot satisfy the needs of larger animals with many layers of cells. The cells of such animals are surrounded by an internal environment of extracellular fluids, which we will refer to in this chapter as *tissue fluids*. Circulatory systems carry materials to and from all regions of the body to maintain the optimum composition of the tissue fluids, which in turn serve the needs of the cells. Circulatory systems may or may not keep the circulating fluid (blood) separate from the tissue fluid; such circulatory systems are considered closed or open, respectively.

Open circulatory systems move tissue fluid

The simplest circulatory systems squeeze tissue fluid through intercellular spaces as the animal moves. In these **open circulatory systems**, there is no distinction between tissue fluid and blood. Usually a heart assists the distribution of the fluid. The contractions of the heart propel the tissue fluid through vessels leading to different regions of the body, but the fluid leaves those vessels to trickle through the tissues and eventually return to the heart. Open circulatory systems are found in arthropods and mollusks as well as in some other invertebrate groups. In the generalized arthropod shown in Figure 49.1*a*, the fluid returns to the heart through valved openings called *ostia*. In the mollusk in Figure 49.1*b*, open vessels aid in the return of tissue fluid to the heart.

Closed circulatory systems circulate blood through tissues

In a **closed circulatory system**, a system of vessels keeps circulating blood separate from the tissue fluid. Blood is pumped through this *vascular system* by one or more muscular hearts, and some components of the blood never leave the vessels. Closed circulatory systems characterize vertebrates, annelids, and some other invertebrate groups.

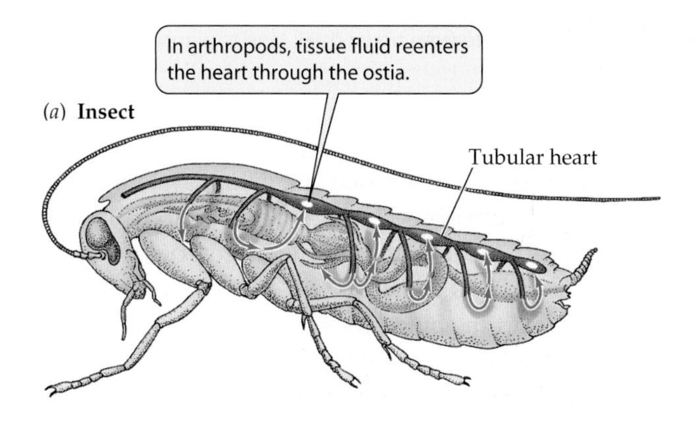
In arthropods, tissue fluid reenters the heart through the ostia.
(*a*) Insect
Tubular heart

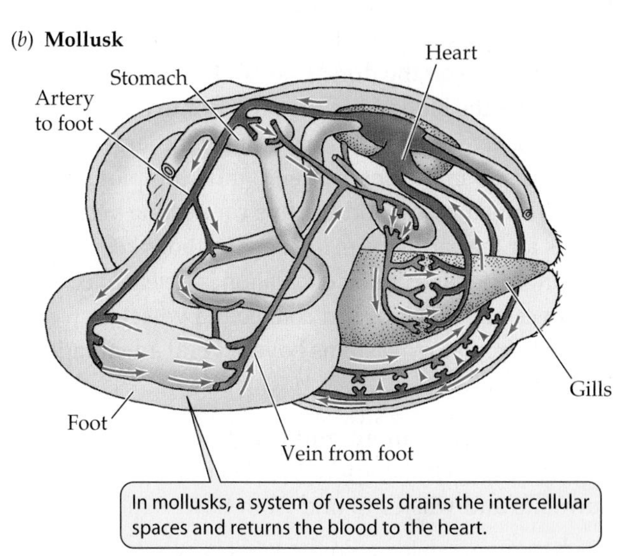
(*b*) Mollusk
Heart
Stomach
Artery to foot
Gills
Foot
Vein from foot
In mollusks, a system of vessels drains the intercellular spaces and returns the blood to the heart.

49.1 Open Circulatory Systems In both arthropods (*a*) and mollusks (*b*), blood is pumped by a tubular heart and directed to different regions of the body through vessels that open into intercellular spaces.

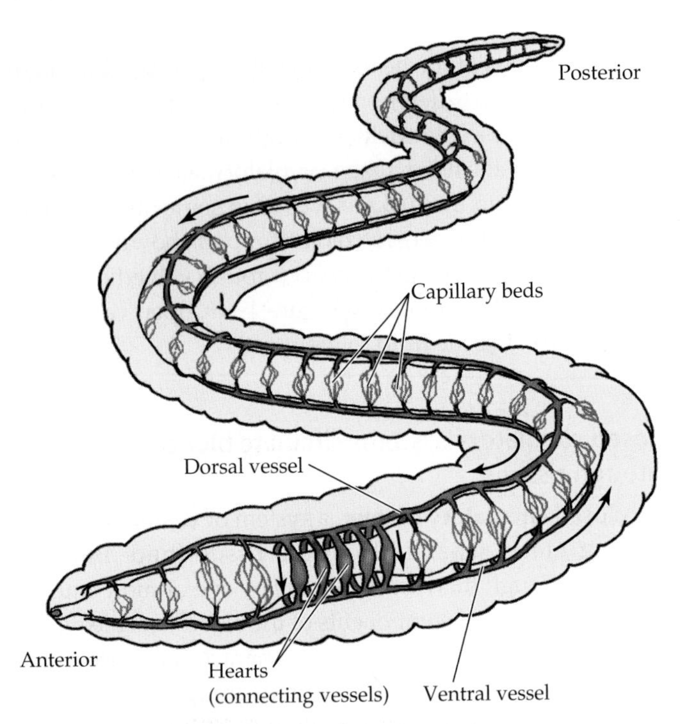

49.2 A Closed Circulatory System An earthworm exemplifies a closed circulatory system, in which blood is confined to the blood vessels, is kept separate from tissue fluids, and is pumped by one or more muscular hearts.

A simple example of a closed circulatory system is that of the common earthworm (Figure 49.2). One large blood vessel on the ventral side of the earthworm carries blood from its anterior end to its posterior end. Smaller vessels branch off and transport the blood to even smaller vessels serving the tissues in each segment of the worm's body. In the smallest vessels, respiratory gases, nutrients, and metabolic wastes diffuse between the blood and the tissue fluid. The blood then flows from these vessels into larger vessels that lead into one large vessel on the dorsal side of the worm. The dorsal vessel carries the blood from the posterior to the anterior end of the body. Five pairs of vessels connect the large dorsal and ventral vessels in the anterior end, thus completing the circuit. The dorsal vessel and the five connecting vessels serve as hearts for the earthworm; their contractions keep the blood circulating. The direction of circulation is determined by one-way valves in the dorsal and connecting vessels.

Closed circulatory systems have several advantages over open systems:

- Blood can flow more rapidly through vessels than through intercellular spaces, and can therefore transport nutrients and wastes to and from tissues more rapidly.
- By changing resistance in the vessels, closed systems can be more selective in directing blood to specific tissues.
- Specialized cells and large molecules that aid in the transport of hormones and nutrients can be kept within the vessels, but can drop their cargo in the tissues where it is needed.

Overall, closed circulatory systems can support higher levels of metabolic activity than open systems can, especially in larger animals. How, then, do highly active insect species achieve high levels of metabolic output with their open circulatory systems? One way is by not depending on their circulatory systems for respiratory gas exchange (see Figure 48.4).

Vertebrate Circulatory Systems

Vertebrates have closed circulatory systems and hearts with two or more chambers. Valves between the chambers, and between the chambers and the vessels, prevent the backflow of blood when the heart contracts.

As we explore the features of the circulatory systems of the different classes of vertebrates, a general evolutionary theme will become apparent: There is a progressively more complete separation of the blood that circulates to the gas exchange organs from the blood that circulates to the rest of the body. In fish, blood is pumped from the heart to the gills and then to the tissues of the body and back to the heart. In birds and mammals, at the other end of the evolutionary scale, blood is pumped from the heart to the lungs and back to the heart in a **pulmonary circuit**, and then from the heart to the rest of the body and back to the heart in a **systemic circuit**. We will trace the evolution of the separation of the circulation into two circuits.

The closed vascular system of vertebrates begins with vessels called **arteries** that carry blood away from the heart. Arteries give rise to smaller vessels called **arterioles**, which feed blood into capillary beds. **Capillaries** are the tiny, thin-walled vessels where materials are exchanged between the blood and the tissue fluid. Small vessels called **venules** drain capillary beds. The venules join together to form larger vessels called **veins**, which deliver blood back to the heart.

We can trace the evolutionary history of vertebrate circulatory systems by examining the circulatory systems of fish, lungfishes, amphibians, reptiles, crocodilians, and mammals.

Fish have two-chambered hearts

The fish heart has two chambers. An **atrium** receives blood from the body and pumps it into a more muscular chamber, the **ventricle**. The ventricle pumps the blood to the gills, where gases are exchanged. Blood leaving the gills collects in a large dorsal artery, the **aorta**, which distributes blood to smaller arteries and arterioles leading to all the organs and tissues of the body. In the tissues, blood flows through beds of tiny capillaries, collects in venules and veins, and eventually returns to the atrium of the heart.

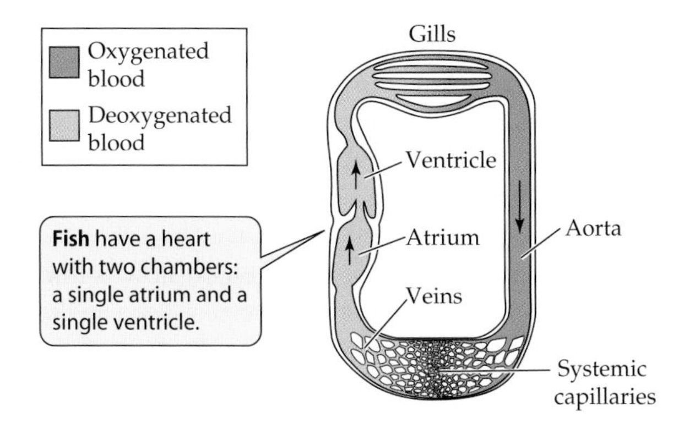
Oxygenated blood
Deoxygenated blood
Gills
Ventricle
Atrium
Aorta
Veins
Systemic capillaries
Fish have a heart with two chambers: a single atrium and a single ventricle.

Most of the pressure imparted to the blood by the contraction of the ventricle is dissipated by the high resistance of the narrow spaces through which blood flows in the gill lamellae. As a result, blood leaving the gills and entering the aorta is under low pressure, limiting the maximum capacity of the fish circulatory system to supply the tissues with oxygen and nutrients. This limitation on arterial blood pressure does not seem to hamper the performance of many rapidly swimming species, such as tuna and marlin, however.

The evolutionary transition from breathing water to breathing air had important consequences for the vertebrate circulatory system. An example of how the system changed to serve a primitive lung can be seen the African lungfish.

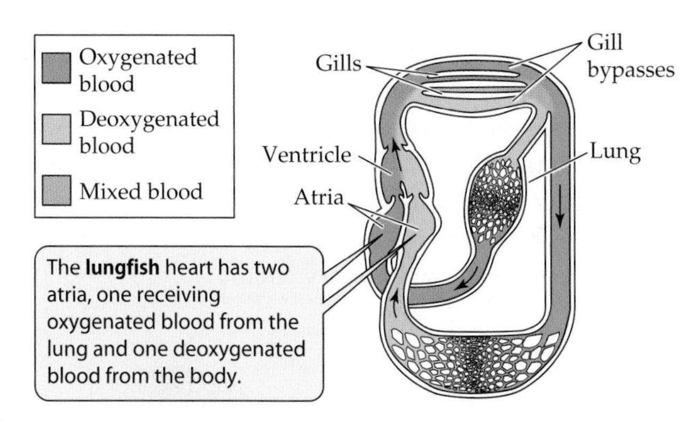
Oxygenated blood
Deoxygenated blood
Mixed blood
Gills
Gill bypasses
Ventricle
Lung
Atria
The lungfish heart has two atria, one receiving oxygenated blood from the lung and one deoxygenated blood from the body.

Lungfish are periodically exposed to water with low oxygen content or to situations in which their aquatic environment dries up. Their adaptation for dealing with these conditions is an outpocketing of the gut that serves as a lung. The lung contains many thin-walled blood vessels, so blood flowing through those vessels can pick up oxygen from air gulped into the lung.

How does the circulatory system take advantage of this new organ? The posterior pair of gill arteries has been modified to carry blood to the lung, and a new vessel carries oxygenated blood from the lung back to the heart. In addition, two anterior gill arches have lost their gills, and their blood vessels deliver blood from the heart directly to the dorsal aorta. Because a few of the gill arches retain gills, the African lungfish can breathe either air or water.

The lungfish heart has adaptations that partially separate the flow of its blood into pulmonary and systemic circuits. Unlike other fishes, the lungfish has a partly divided atrium; the left side receives oxygenated blood from the lungs, and the right side receives deoxygenated blood from the other tissues. These two bloodstreams stay mostly separate as they flow through the ventricle and the large vessel leading to the gill arches. As a result, oxygenated blood mostly goes to the anterior gill arteries leading to the dorsal aorta, and the deoxygenated blood mostly goes to the posterior arches with functional gills and to the lung.

We can conclude that the lung of the lungfish evolved as a means of supplementing oxygen uptake from the gills, but the associated modifications of the vascular system set the stage for the evolution of separate pulmonary and systemic circulations.

Amphibians have three-chambered hearts

Pulmonary and systemic circulation are partly separated in adult amphibians. A single ventricle pumps blood to the lungs and to the rest of the body. Two atria receive blood returning to the heart. One receives oxygenated blood from the lungs, and the other receives deoxygenated blood from the body.

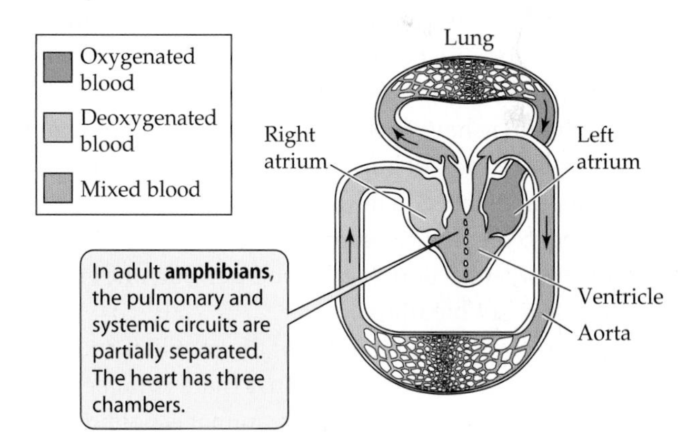
Oxygenated blood
Deoxygenated blood
Mixed blood
Lung
Right atrium
Left atrium
Ventricle
Aorta
In adult amphibians, the pulmonary and systemic circuits are partially separated. The heart has three chambers.

Because both atria deliver blood to the same ventricle, the oxygenated and deoxygenated blood could mix, so that blood going to the tissues would not carry a full load of oxygen. Mixing is limited, however, because anatomical features of the ventricle direct the flow of deoxygenated blood from the right atrium to the pulmonary circuit and the flow of oxygenated blood from the left atrium to the aorta.

The advantage of this partial separation of pulmonary and systemic circulation is that the high flow resistance of the gas exchange organ no longer lies between the heart and the tissues. Therefore, the amphibian heart delivers blood to the aorta, and hence to the body, at a higher pressure than the fish heart does.

Reptiles have exquisite control of pulmonary and systemic circulation

Turtles, snakes, and lizards are commonly said to have three-chambered hearts, while crocodilians (crocodiles and alligators) are said to have four-chambered hearts. But this statement is an oversimplification. The hearts of all these animals have two separate atria and a ventricle that is divided in a complex way so that mixing of oxygenated and deoxygenated blood is minimized.

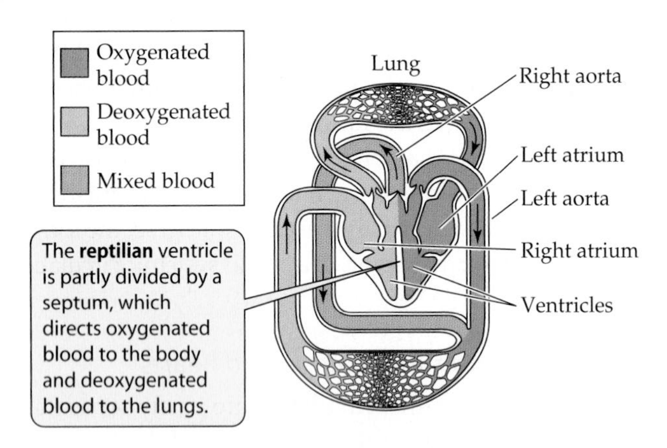
Oxygenated blood
Deoxygenated blood
Mixed blood
Lung
Right aorta
Left atrium
Left aorta
Right atrium
Ventricles
The **reptilian** ventricle is partly divided by a septum, which directs oxygenated blood to the body and deoxygenated blood to the lungs.

The most important and unusual feature of reptilian and crocodilian hearts is their ability to alter the proportion of blood going to the lungs and to the rest of the body. Consider the behavior, ecology, and physiology of these animals. Despite the common image of turtles as being slow and plodding, reptiles and crocodilians can be fast, active, powerful animals. They can also be inactive for long periods of time, during which they have metabolic rates much lower than the resting metabolic rates of birds and mammals. The enormous range of metabolic needs in these animals means that they do not have to breathe continuously. Some species are also accomplished divers and spend long periods under water, where they cannot breathe.

To understand the adaptations of the reptilian and crocodilian hearts, you have to realize that there is no benefit in sending blood to the lungs when an animal is not breathing. The hearts of these animals circulate blood through their lungs and then to the rest of their bodies when they are breathing, but when they are not breathing, they can bypass the pulmonary circuit and pump all the blood around the body. How do they accomplish this switching?

Reptiles have two aortas instead of one. The right aorta can receive blood from either the right side or the left side of the ventricle. The two sides of the ventricle are partially divided by a *septum*. When the animal is breathing air, two factors cause blood from the right side of the ventricle to go preferentially into the pulmonary circuit rather than into the systemic circuit. First, the resistance in the pulmonary circuit is lower than that in the systemic circuit. Second, there is a slight asynchrony in the timing of ventricular contraction, so the blood in the right side of the ventricle tends to be ejected slightly before the blood in the left side. As the ventricle contracts, the deoxygenated blood in the right side of the ventricle moves first into the pulmonary circuit. When the oxygenated blood in the left side of the ventricle starts to move, it encounters resistance in the pulmonary circuit, which is already filled with the deoxygenated blood from the right side. Therefore, the blood from the left side tends to flow into the two aortas.

When the reptile stops breathing, blood flow is rerouted by constriction of vessels in the lung. As resistance in the pulmonary circuit increases, the blood from the right side of the ventricle tends to be directed into one of the aortas. As a result, blood from both sides of the ventricle flows through the aortas to the systemic circuit.

The ability of snakes, lizards, and turtles to redirect blood flow from the pulmonary circuit to the systemic circuit depends on the incomplete division of their ventricles. Crocodilians have true four-chambered hearts with completely divided ventricles. Yet the crocodilians have not lost the ability to shunt blood from the pulmonary circuit when they are not breathing. The crocodilians have one aorta originating in the left ventricle and one aorta originating in the right ventricle. However, a short channel connects these two aortas just after they leave the heart, making it possible for blood to flow from one aorta to the other.

Because the crocodilians' ventricles are separate, they can generate different pressures when they contract. When the animal is breathing, the pressure in the left ventricle and the left aorta is higher than the pressure in the right ventricle. This higher pressure is communicated through the connecting channel to the right aorta and prevents right-ventricle blood from entering that aorta. As a result, both aortas carry blood from the left ventricle, and the blood from the right ventricle flows to the pulmonary circuit.

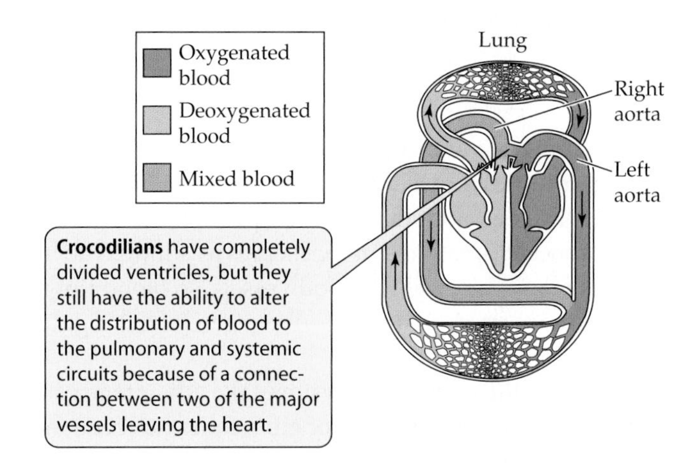
Oxygenated blood
Deoxygenated blood
Mixed blood
Lung
Right aorta
Left aorta
Crocodilians have completely divided ventricles, but they still have the ability to alter the distribution of blood to the pulmonary and systemic circuits because of a connection between two of the major vessels leaving the heart.

When a crocodilian is not breathing, constriction of vessels in the lung increases the resistance in the pulmonary circuit. As a result, pressure builds up in the right ventricle to a level that exceeds the pressure in the right aorta. Under these conditions, blood from both ventricles flows through the two aortas and the systemic circuit, and little blood flows into the pulmonary circuit.

Birds and mammals have fully separated pulmonary and systemic circuits

The four-chambered hearts of birds and mammals completely separate their pulmonary and systemic circuits. Separate circuits have several advantages:

- Oxygenated and deoxygenated blood cannot mix; therefore, the systemic circuit is always receiving blood with the highest oxygen content.
- Respiratory gas exchange is maximized because the blood with the lowest oxygen content and highest CO_2 content is sent to the lungs.
- Separate systemic and pulmonary circuits can operate at different pressures.

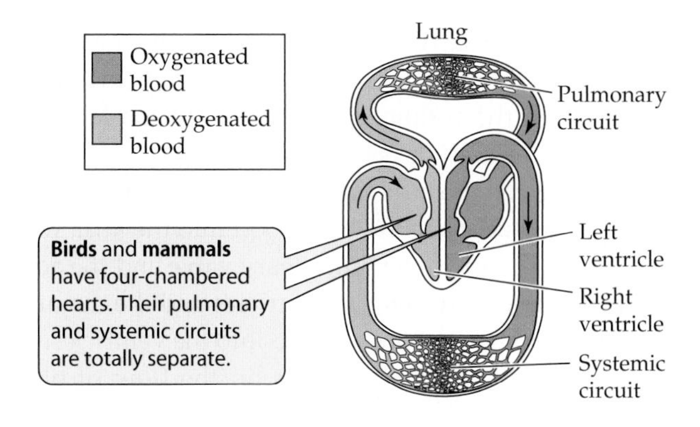
Oxygenated blood
Deoxygenated blood
Lung
Pulmonary circuit
Left ventricle
Right ventricle
Systemic circuit
Birds and **mammals** have four-chambered hearts. Their pulmonary and systemic circuits are totally separate.

The tissues of birds and mammals have high nutrient demands and thus a very high density of the smallest vessels, the capillaries. Many small vessels present high resistance to the flow of blood. Therefore, high pressure is required in the systemic circuits of birds and mammals. Their pulmonary circuits have fewer capillaries, and thus lower resistance, than their systemic circuits, so the pulmonary circuits of birds and mammals can function at lower pressures.

The Human Heart: Two Pumps in One

Like all other mammalian hearts, the human heart has four chambers: two atria and two ventricles (Figure 49.3). The atrium and ventricle on the right side of your body are called

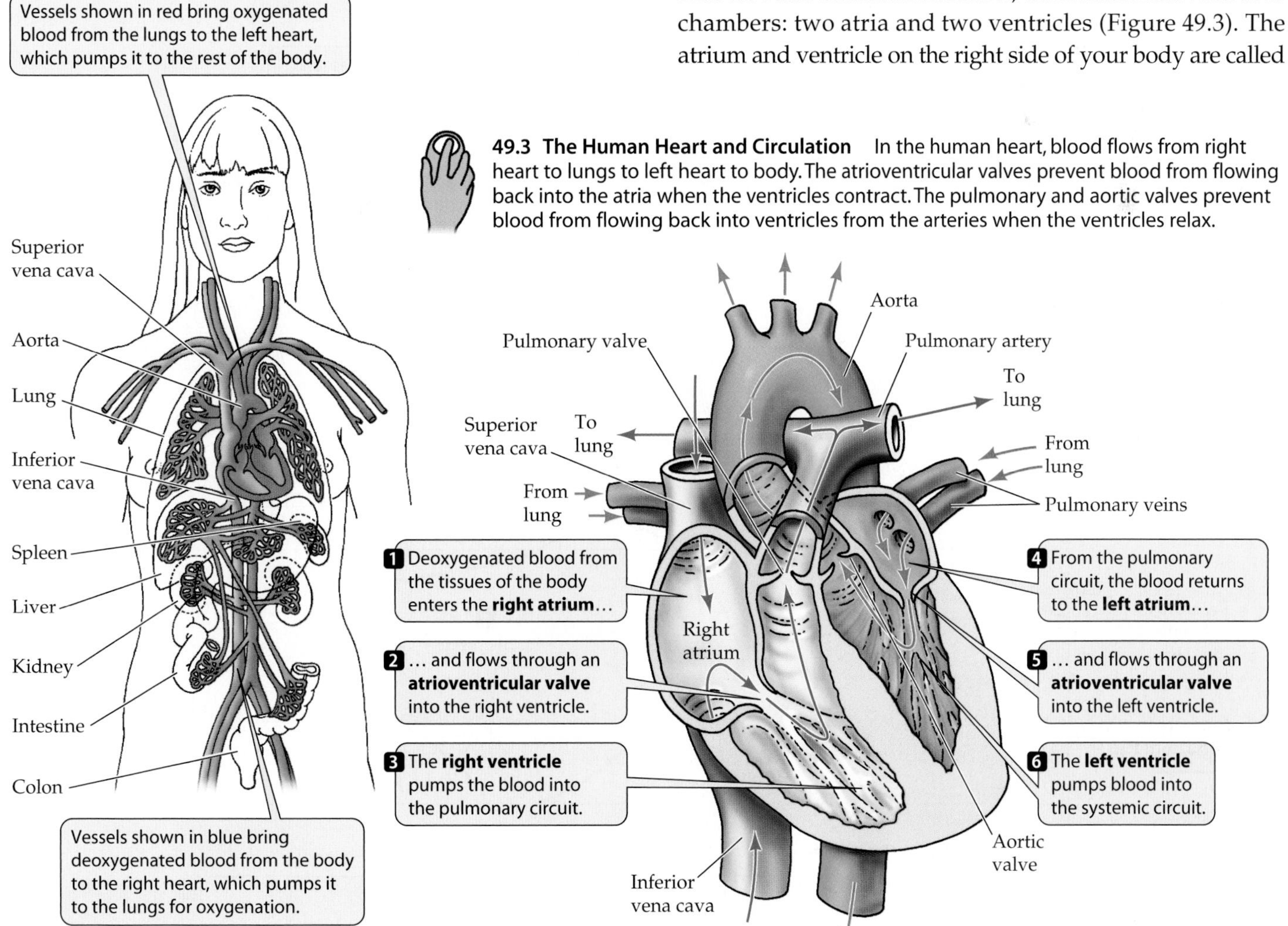
Vessels shown in red bring oxygenated blood from the lungs to the left heart, which pumps it to the rest of the body.
Superior vena cava
Aorta
Lung
Inferior vena cava
Spleen
Liver
Kidney
Intestine
Colon
Vessels shown in blue bring deoxygenated blood from the body to the right heart, which pumps it to the lungs for oxygenation.
Aorta
Pulmonary valve
Pulmonary artery
To lung
Superior vena cava
To lung
From lung
From lung
Pulmonary veins
1 Deoxygenated blood from the tissues of the body enters the **right atrium**…
Right atrium
2 … and flows through an **atrioventricular valve** into the right ventricle.
3 The **right ventricle** pumps the blood into the pulmonary circuit.
4 From the pulmonary circuit, the blood returns to the **left atrium**…
5 … and flows through an **atrioventricular valve** into the left ventricle.
6 The **left ventricle** pumps blood into the systemic circuit.
Aortic valve
Inferior vena cava

49.3 The Human Heart and Circulation In the human heart, blood flows from right heart to lungs to left heart to body. The atrioventricular valves prevent blood from flowing back into the atria when the ventricles contract. The pulmonary and aortic valves prevent blood from flowing back into ventricles from the arteries when the ventricles relax.

the right heart. The atrium and ventricle on the left side of your body are called the left heart. The right heart pumps blood through the pulmonary circuit, and the left heart pumps blood through the systemic circuit.

Valves between the atria and ventricles, the **atrioventricular valves**, prevent backflow of blood into the atria when the ventricles contract. The **pulmonary valve** and the **aortic valve**, positioned between the ventricles and the major arteries, prevent the backflow of blood into the ventricles.

In this section, we'll first focus on the flow of blood through the heart and through the body. Then we'll examine the unique electrical properties of cardiac muscle that result in the heartbeat, and we'll see how the heart's electrical activity can be recorded in an EKG (electrocardiogram).

Blood flows from right heart to lungs to left heart to body

Let's follow the circulation of the blood through the heart, starting in the right heart. The right atrium receives deoxygenated blood from the **superior** (upper) **vena cava** and the **inferior** (lower) **vena cava** (see Figure 49.3), large veins that collect blood from the upper and lower body, respectively. The veins of the heart itself also drain into the right atrium. From the right atrium, the blood flows into the right ventricle. Most of the filling of the ventricle results from passive flow while the heart is relaxed between beats. Just at the end of this period of ventricular filling, the atrium contracts and adds a little more blood to the ventricular volume. The right ventricle then contracts, pumping the blood into the **pulmonary artery**, which transports it to the lungs.

After gas exchange occurs in the lungs, the **pulmonary veins** return the oxygenated blood from the lungs to the left atrium, from which the blood enters the left ventricle. As with the right side of the heart, most left ventricular filling is passive, but the ventricle is topped off by contraction of the atrium just at the end of the period of passive filling.

The walls of the left ventricle are powerful muscles that contract around the blood with a wringing motion starting from the bottom. When pressure in the left ventricle is high

49.4 The Cardiac Cycle The rhythmic contraction (systole) and relaxation (diastole) of the ventricles is called the cardiac cycle. The graphical representation below shows pressure and volume changes during the cardiac cycle for the left ventricle only.

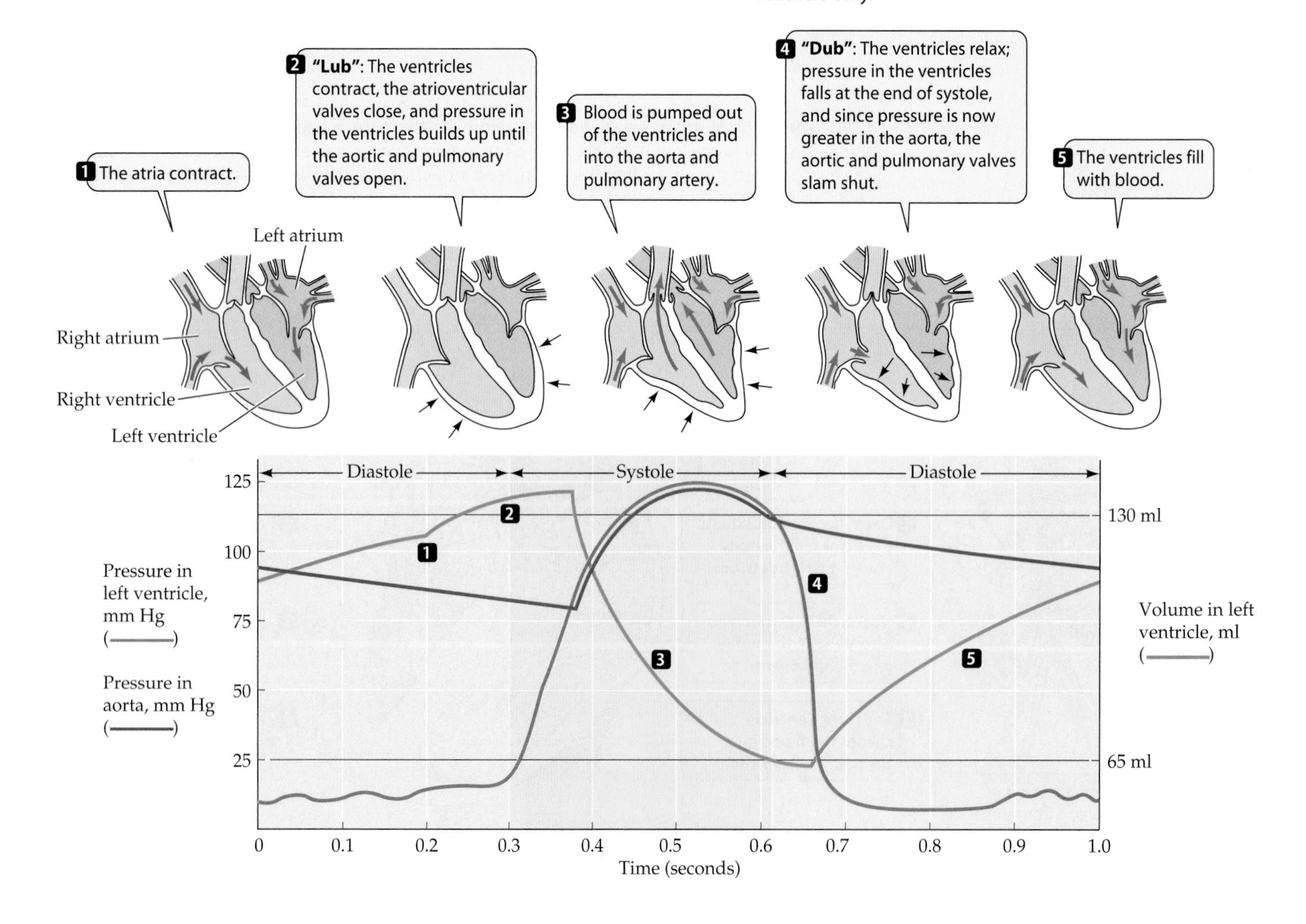

enough to push open the aortic valve, the blood rushes into the aorta to begin its circulation throughout the body. In Figure 49.3, observe that the left ventricle is more massive than the right ventricle. The left ventricle has to propel the blood through many more kilometers of blood vessels than does the right ventricle and must therefore push against more resistance, even though both pump the same volume of blood.

Both sides of the heart contract at the same time. The contraction of the two atria, followed by the contraction of the two ventricles and then relaxation, is called the **cardiac cycle**. Contraction of the ventricles is called ventricular **systole**, and relaxation of the ventricles called ventricular **diastole** (Figure 49.4). Just at the end of diastole, the atria contract and top off the volume of blood in the ventricles. The sounds of the cardiac cycle, the "lub-dub" heard through a stethoscope placed on the chest, are created by the slamming shut of the heart valves. The closing and opening of these valves are simple mechanical events resulting from pressure differences on the two sides of the valves. As the ventricles begin to contract, the pressure in the ventricles rises above the pressure in the atria, blood starts flowing back into the atria, and the atrioventricular valves close ("lub"). When the ventricles begin to relax, the high pressure in the aorta and pulmonary artery causes blood to start to flow back into the ventricles, and this flow of blood closes the aortic and pulmonary valves ("dub"). Defective valves produce turbulent blood flow and produce the sounds known as *heart murmurs*. For example, if an atrioventricular valve is defective, blood will flow back into the atrium with a "whoosh" sound following the "lub."

The cardiac cycle can be felt in the pulsation of arteries such as the one that supplies blood to your hand. You can feel your pulse by placing two fingers from one hand lightly over the wrist of the other hand just below the thumb. During systole, blood surges through the arteries of your arm and hand, and you can feel the surge as a pulsing of the artery in your wrist.

Blood pressure changes associated with the cardiac cycle can be measured in the large artery in your arm by using an inflatable pressure cuff and a pressure gauge, together called a *sphygmomanometer*, and a stethoscope (Figure 49.5). This method measures the minimum pressure necessary to compress an artery so that blood does not flow through it at all (the systolic value) and the minimum pressure that permits intermittent flow through the artery (the diastolic value). In a conventional blood pressure reading, the systolic value is placed over the diastolic value. Normal values for a young adult might be 120 mm of mercury (Hg) during systole and 80 mm Hg during diastole, or 120/80.

The heartbeat originates in the cardiac muscle

Cardiac muscle, as we saw in Chapter 47, has some unique properties that allow it to function as an effective pump. First,

49.5 Measuring Blood Pressure Blood pressure in the major artery of the arm can be measured with a device called a sphygmomanometer, which combines an inflatable cuff and a pressure gauge. A stethoscope is also used to detect sounds created by the blood vessels in response to changes in pressure during the cardiac cycle.

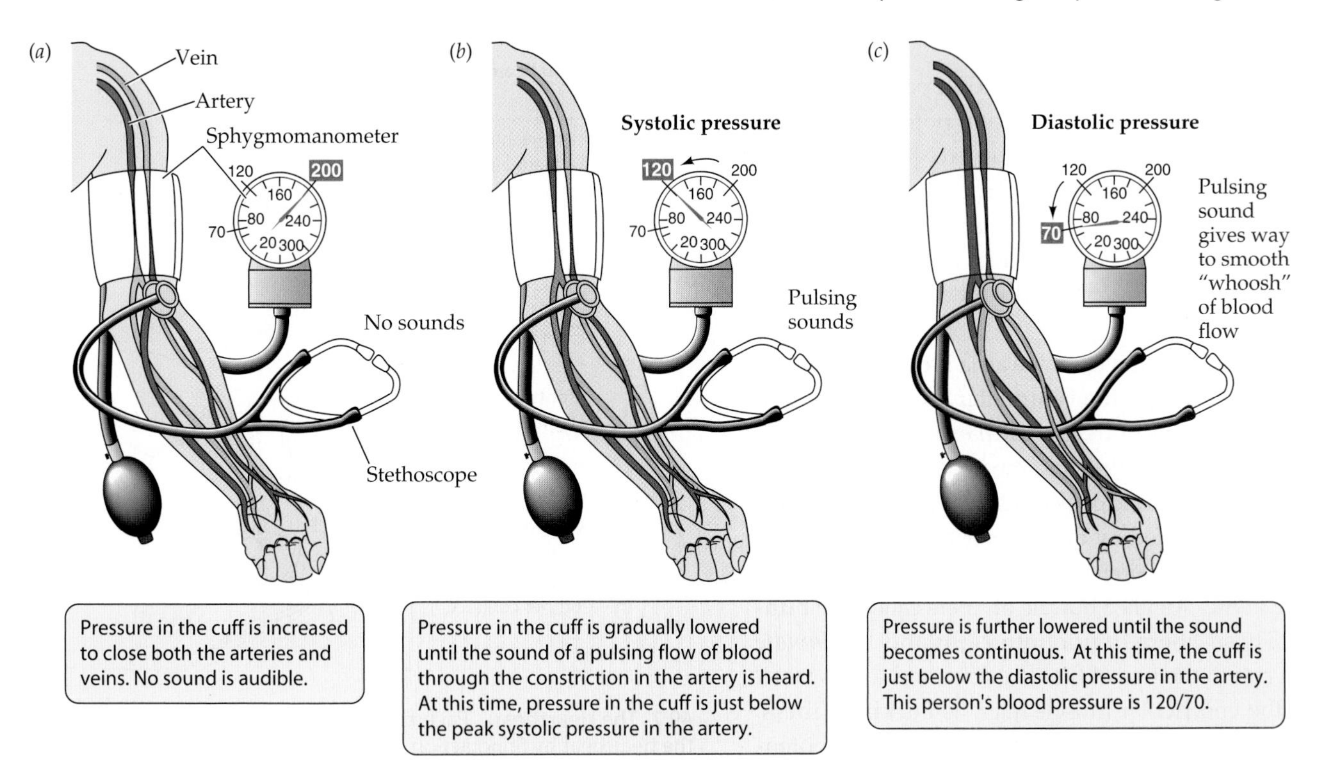

cardiac muscle cells are in electrical contact with one another through gap junctions, which enable action potentials to spread rapidly from cell to cell. Because a spreading action potential stimulates contraction, large groups of cardiac muscle cells contract in unison. This coordinated contraction is essential for pumping blood effectively.

Second, some cardiac muscle cells are *pacemaker cells.* These cells have the ability to initiate action potentials without stimulation from the nervous system. When they fire action potentials, they stimulate neighboring cells to contract. The primary pacemaker of the heart is a nodule of modified cardiac muscle cells, the **sinoatrial node**, located at the junction of the superior vena cava and right atrium. The resting membrane potentials of these cells are not stable, but gradually becomes less negative (more positive) until they reach the threshold for initiating an action potential (Figure 49.6). The action potentials of pacemaker cells result from the opening of voltage-gated calcium channels, so they look different from the sodium action potentials graphed in Figure 44.10: They are slower and last longer.

The nervous system controls the heartbeat (speeds it up or slows it down) by influencing the rate at which pacemaker cells gradually depolarize between action potentials. Acetylcholine released by parasympathetic neurons onto the pacemaker cells slows their rate of depolarization and thereby slows the heart rate. Norepinephrine released by sympathetic neurons onto the pacemaker cells increases their rate of depolarization and thereby speeds the heart rate (see Figure 49.6).

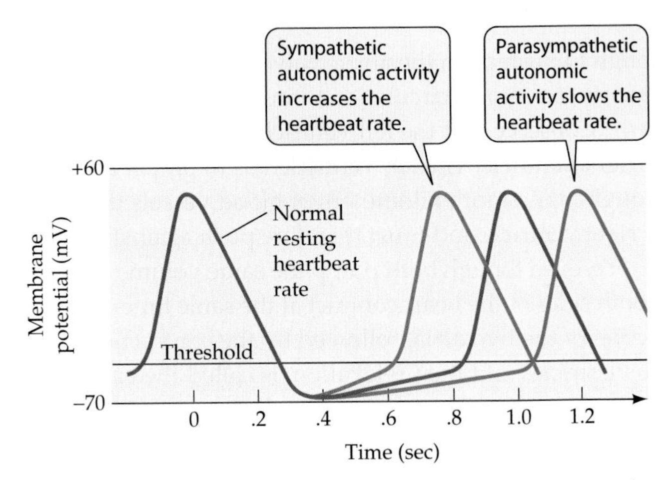

49.6 The Autonomic Nervous System Controls Heart Rate The plasma membranes of pacemaker cells spontaneously depolarize to threshold and fire action potentials. Signals from the two divisions of the autonomic nervous system raise and lower the heart rate by altering the rate at which the cells depolarize. Their effects can be seen in the altered slope of the membrane potential curve prior to reaching threshold.

A conduction system coordinates the contraction of heart muscle

A normal heartbeat begins with an action potential in the sinoatrial node (Figure 49.7). This action potential spreads rapidly throughout the electrically coupled cells of the atria, causing them to contract in unison. Since there are no gap junctions between the cells of the atria and those of the ventricles, however, the action potential does not spread directly to the ventricles, and the ventricles do not contract in unison with the atria.

The action potential initiated in the atria is regenerated and conducted to and through the ventricular muscle mass by a system of modified, noncontractile cardiac muscle cells. Situated at the junction of the atria and the ventricles is a nodule of modified cardiac muscle cells called the **atrioventricular node**, which is stimulated by the depolarization of the atria. With a slight delay, it generates action potentials that are conducted to the ventricles via a bundle of fibers called the **bundle of His**. These fibers divide into right and left *bundle branches* that run to the tips of the ventricles and then spread throughout the ventricular muscle mass as **Purkinje fibers**. These conducting fibers ensure that the cardiac action poten-

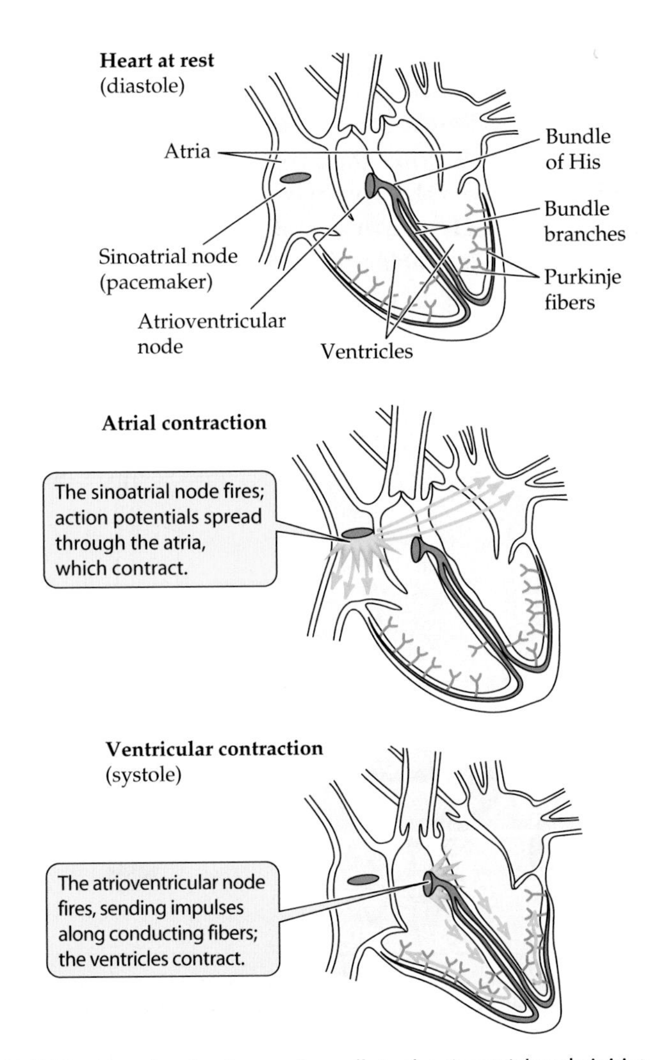

49.7 The Heartbeat Pacemaker cells in the sinoatrial node initiate the heartbeat by firing action potentials.

tial spreads rapidly and evenly throughout the ventricular muscle mass, starting at the very bottom of the ventricles. The short delay in the spread of the action potential imposed by the atrioventricular node ensures that the atria contract before the ventricles do, so that the blood passes progressively from the atria to the ventricles to the arteries.

Electrical properties of ventricular muscles sustain heart contraction

Contractions of ventricular muscle fibers last for about 300 milliseconds—much longer than those of skeletal muscle fibers. The reason is found in the electrical properties of these cells. Like neuronal and skeletal muscle action potentials, ventricular muscle cell action potentials are initiated by the opening of voltage-gated sodium channels. Unlike neurons and skeletal muscle fibers, however, ventricular muscle cells remain depolarized for a long time. This plateau of the action potential is due to sustained opening of voltage-gated calcium channels (Figure 49.8). The contraction of cardiac muscle, like other muscle, is stimulated when Ca^{2+} is available to bind with troponin (see Figure 47.6). As long as their Ca^{2+} channels remain open, the ventricular muscle cells continue to contract.

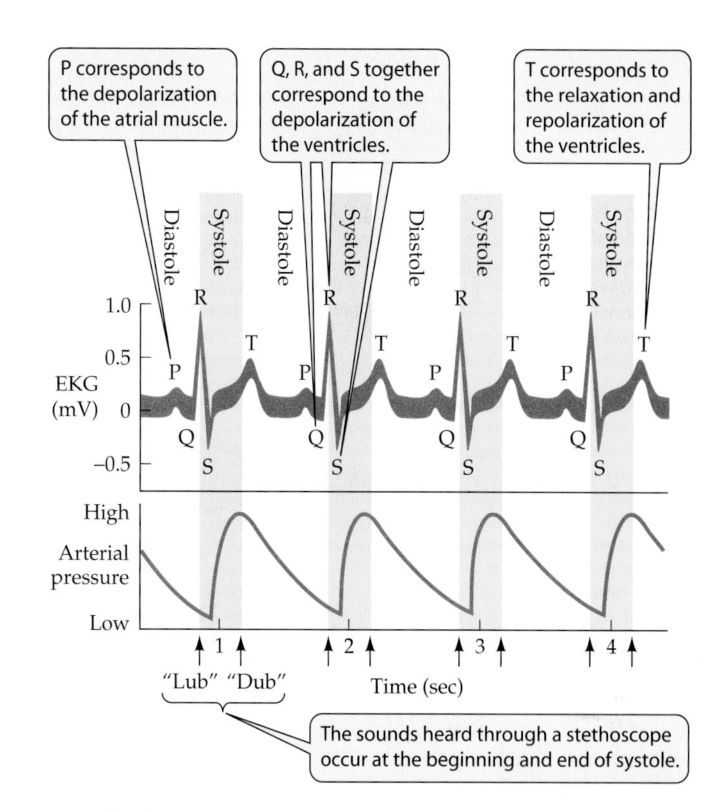

49.9 The Electrocardiogram An EKG can be used to monitor heart function. Variations from the normal pattern shown here can be used to diagnose heart problems.

The EKG records the electrical activity of the heart

Electrical events in the cardiac muscle during the cardiac cycle can be recorded by electrodes placed on the surface of the body. Such a recording is called an **electrocardiogram**, or **EKG** ("EKG" because the Greek word for heart is *kardia*, but "ECG" is also used). The EKG is an important tool for diagnosing heart problems.

The action potentials that sweep through the muscles of the atria and the ventricles before they contract are such massive, localized electrical events that they cause electric currents to flow outward from the heart to all parts of the body. Surface electrodes placed at different locations on the body detect those electric currents at different times, and therefore register a voltage difference. The appearance of the EKG depends on the placement of the electrodes. Electrodes placed on the right wrist and left ankle produced the normal EKG shown in Figure 49.9. The wave patterns of the EKG are designated P, Q, R, S, and T, each letter representing a particular event in the cardiac muscle, as shown in the figure.

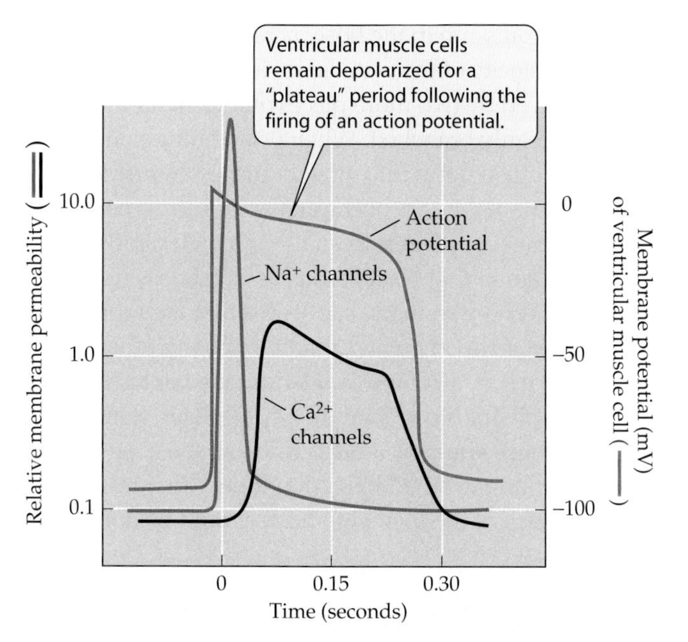

49.8 The Action Potential of Ventricular Muscle Fibers The rising phase of the action potential of ventricular muscle fibers (graphed in red) is due to the opening of voltage-gated Na^+ channels (blue). However, the membrane remains in a depolarized state for a prolonged time because of the opening of voltage-gated Ca^{2+} channels (black).

The Vascular System: Arteries, Capillaries, and Veins

Blood circulates throughout the body in a system of blood vessels: arteries, capillaries, and veins. In this section, we will see how the structure of each of these vessel types supports its functions. We will also consider another set of vessels, the lymphatic vessels, which return tissue fluid to the blood.

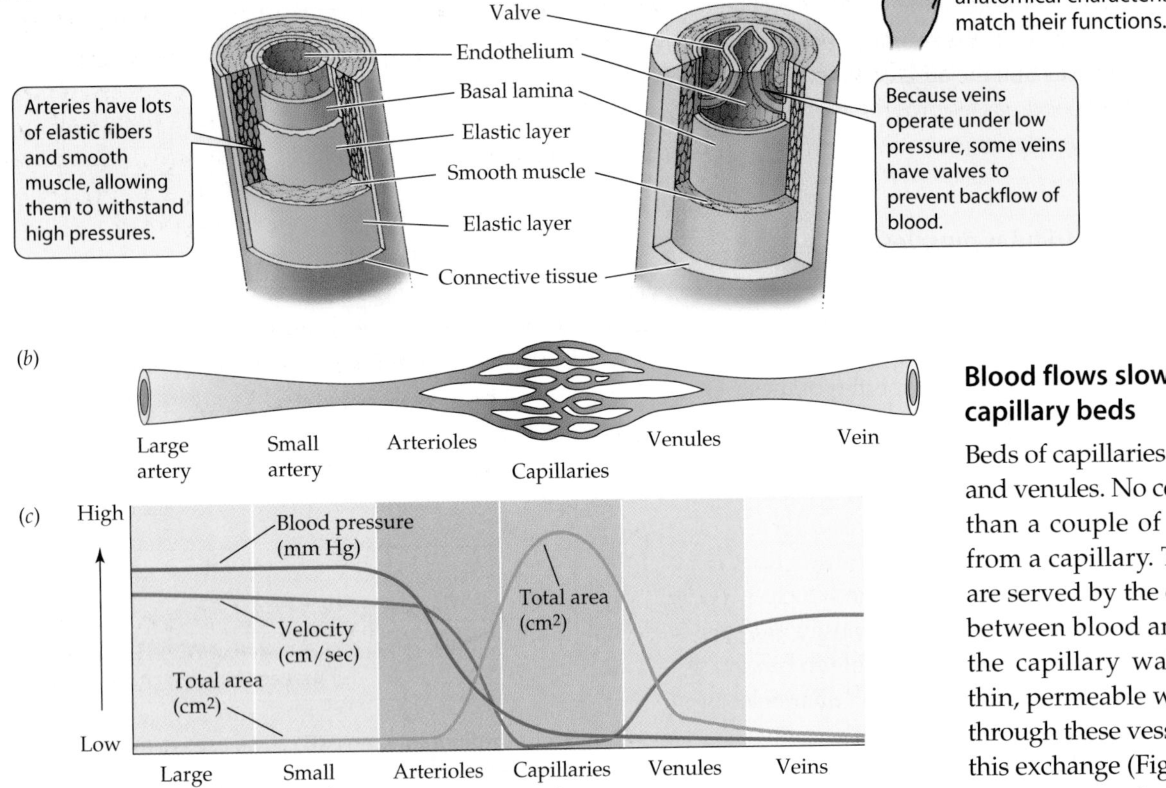

49.10 Anatomy of Blood Vessels The anatomical characteristics of blood vessels match their functions.

Arteries and arterioles have abundant elastic and muscle fibers

The walls of the large arteries have many collagen and elastic fibers, which enable them to withstand the high pressures of blood flowing rapidly from the heart (Figure 49.10). The elastic fibers have another important function as well: During systole, they are stretched, and thereby store some of the energy imparted to the blood by the heart. During diastole, they return this energy to the blood by elastic recoil, squeezing it and pushing it forward. As a result, even though the flow of blood through the arteries pulsates with the beating of the heart, it is smoother than it would be through a system of rigid pipes.

Smooth muscle cells in the walls of the arteries and arterioles allow those vessels to be constricted or dilated. When the diameter of the vessels changes, their resistance to blood flow changes as well, and the amount of blood flowing through them changes as a result. By influencing the contraction and relaxation of the smooth muscle in the vessel walls, neuronal and hormonal mechanisms can control the resistance of the vessels and therefore the distribution of blood to the different tissues of the body. (We'll see how these mechanisms work later in this chapter.) The arteries and arterioles are referred to as the *resistance vessels* because their resistance can vary.

Blood flows slowly through capillary beds

Beds of capillaries lie between arterioles and venules. No cell of the body is more than a couple of cell diameters away from a capillary. The needs of the cells are served by the exchange of materials between blood and tissue fluid across the capillary walls. Capillaries have thin, permeable walls, and blood flows through these vessels slowly, facilitating this exchange (Figure 49.11).

To anyone who has played with a garden hose, it may seem strange that blood flows through the large arteries rapidly at high pressures, but when it reaches the small capillaries, the pressure and rate of flow decrease. When you restrict the diameter of a garden hose by placing your thumb over the opening, the pressure in the hose increases, which in turn increases the velocity of the water spraying out of the hose. This puzzle is solved by two more pieces of information. First, arterioles are highly branched. When flow through one branch is restricted, blood flows into other branches, so pressure does not build up quickly. Second, each arteriole gives rise to a large number of capillaries. Even though each capillary has a diameter so small that red blood cells must pass through in single file, there are so many capillaries that their total cross-sectional area is much greater than that of any other class of vessels. As a result, all of the capillaries together have a much greater capacity for blood than do the arterioles. Returning to our garden hose analogy, if we connected the hose to many junctions leading to small irrigation tubes, the pressure and the flow in each of the irrigation tubes would be quite low.

Materials are exchanged in capillary beds by filtration, osmosis, and diffusion

The walls of capillaries are made of a single layer of thin endothelial cells. In most tissues of the body other than the brain, capillaries have tiny holes called *fenestrations* (Latin,

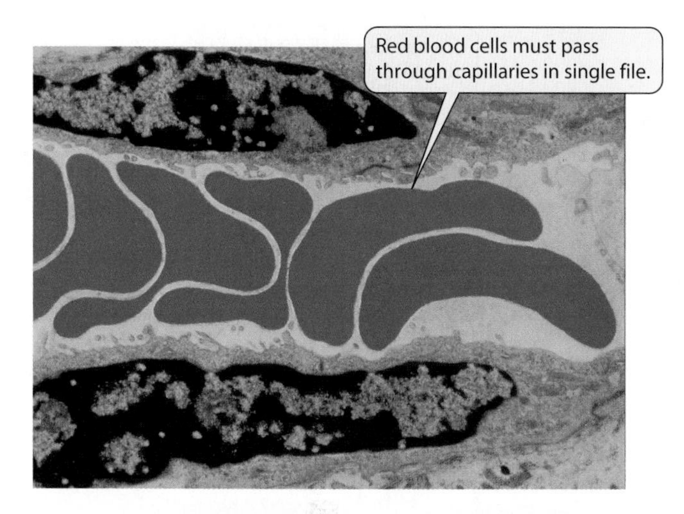

49.11 A Narrow Lane Capillaries have a very small diameter, and blood flows through them slowly.

"windows"). Capillaries are permeable to water, to some ions, and to some small molecules, but not to large molecules such as proteins. Blood pressure therefore squeezes water and some small solutes out of the capillaries and into the surrounding intercellular spaces. Why don't water and small-molecular-weight solutes collect in the intercellular spaces? How is the blood volume maintained if fluid is continuously leaking out of the capillaries?

An answer to this question was put forth more than a hundred years ago by the physiologist E. H. Starling. Starling suggested that water balance in capillary beds is a result of two opposing forces, which have come to be known as **Starling's forces**. One force is blood pressure, which squeezes water and small solutes out of the capillaries, and the other is osmotic pressure created by the large protein molecules that cannot leave the capillaries. Starling called this second force *colloidal osmotic pressure*. He hypothesized that blood pressure is high at the arterial end of a capillary bed and drops steadily as blood flows to the venous end (Figure 49.12). The colloidal osmotic pressure, however, is constant along the capillary. As long as the blood pressure is above the osmotic pressure, water leaves the capillary, but when blood pressure falls below the osmotic pressure, water returns to the capillary. The actual numbers for a normal capillary bed in a resting person suggest that there would be a *slight* net loss of water to the intercellular spaces.

Several observations supported Starling's model. In people with severe liver disease or protein starvation, there is a fall in blood protein concentration that leads to an accumulation of water in the extracellular spaces, which results in tissue swelling, or **edema**. Edema is also a characteristic of the inflammation reaction that accompanies tissue damage or allergic responses (see Figure 18.4). *Histamine*, a mediator of inflammation that is released by certain white blood cells, increases the permeability of capillaries and also relaxes the smooth muscles of the arterioles, resulting in higher blood pressure in the capillaries. The hypothesis modeled in Figure 49.12 predicts that edema should occur in all of these cases.

Only recently have Starling's forces come into question as the complete explanation for fluid exchange in capillary beds. There are situations that are not explained by Starling's hypothesis. During strenuous exercise, the blood pressure in the arterioles serving the muscles rises substantially, yet edema does not occur. In birds, the blood pressure in arterioles is much higher than in mammals, and the colloidal osmotic pressure is lower. If edema is not a chronic problem in exercising muscles and in birds, what is missing from Starling's model?

Recent research suggests that bicarbonate ions (HCO_3^-) in the blood plasma are an important contributor to the osmotic attraction of water back into the capillary. As the blood flows through the capillary, CO_2 diffuses into the plasma and is converted into bicarbonate ions (see Figure 48.14); therefore, there is a substantial rise in HCO_3^- concentration as blood

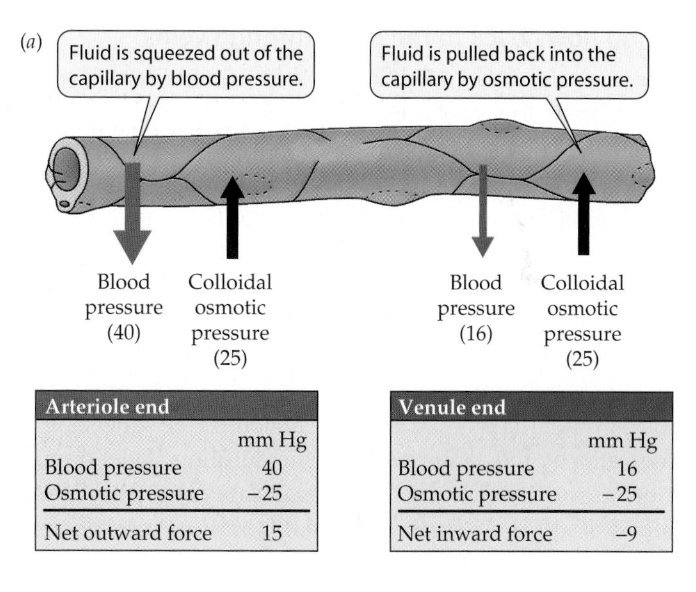

Arteriole end	mm Hg
Blood pressure	40
Osmotic pressure	–25
Net outward force	15

Venule end	mm Hg
Blood pressure	16
Osmotic pressure	–25
Net inward force	–9

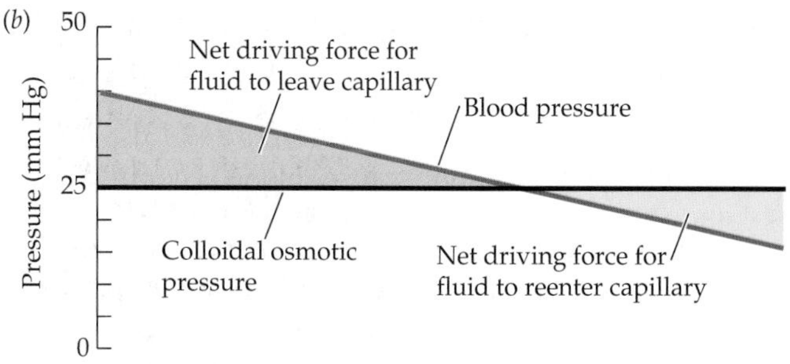

49.12 Starling's Forces Starling's model explains how blood volume is maintained in the capillary beds. (*a*) When blood pressure is greater than the colloidal osmotic pressure, water leaves the capillary; when blood pressure falls below this osmotic pressure, water returns to the capillary. (*b*) The balance of these two forces changes over the capillary bed as blood pressure falls.

flows through the capillary. When a person is at rest, the increasing HCO_3^- concentration can cause the osmotic pressure of the blood at the venous end to be 30 mm Hg higher than at the arterial end, and during strenuous exercise this difference can be hundreds of mm Hg. Thus it appears that CO_2 and HCO_3^- are the major factors that pull water back into the capillaries, not colloidal osmotic pressure.

Lipid-soluble substances and many small solute molecules can easily pass through capillary walls from an area of higher concentration to one of lower concentration. The capillaries in different tissues, however, are differentially selective as to the sizes of molecules that can pass through them. All capillaries are permeable to O_2, CO_2, glucose, lactate, and small ions such as Na^+ and Cl^-. The capillaries of the brain do not have fenestrations, and therefore not much else can pass through them other than lipid-soluble substances, such as alcohol. This high selectivity of brain capillaries is known as the *blood–brain barrier*. Much less selective capillaries are found in the digestive tract, where nutrients are absorbed, and in the kidneys, where wastes are filtered.

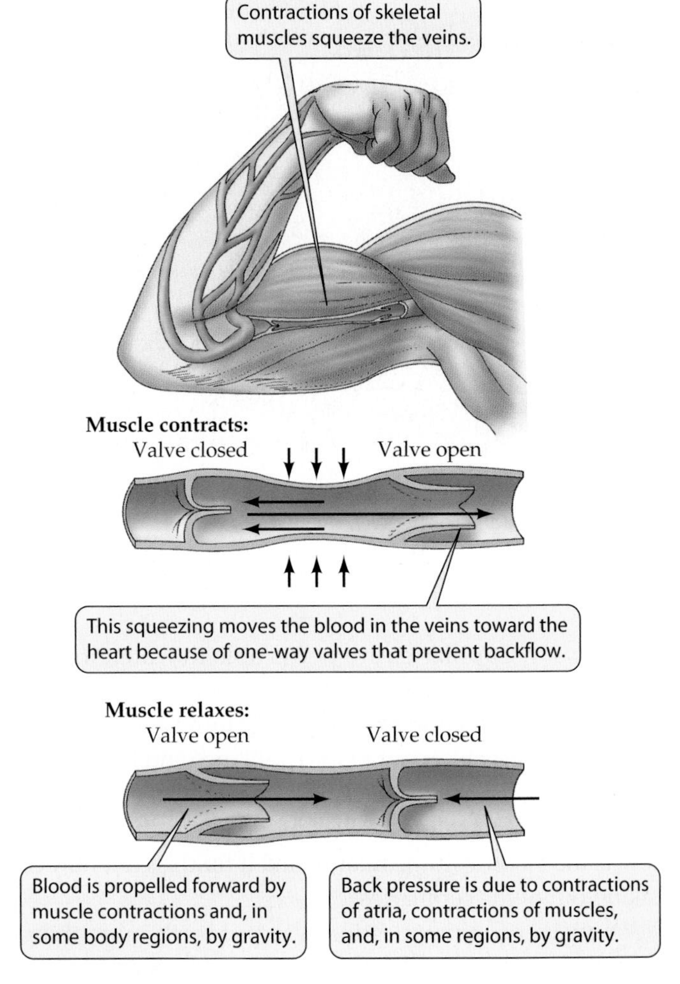

49.13 One-Way Flow Veins have valves that prevent blood from flowing backward.

Blood flows back to the heart through veins

The pressure of the blood flowing from capillaries to venules is extremely low, and is insufficient to propel blood back to the heart. The walls of veins are more expandable than the walls of arteries, and blood tends to accumulate in veins. As much as 80 percent of the total blood volume may be in the veins of a resting individual. Because of their high capacity to store blood, veins are called *capacitance vessels*.

Blood flow through veins that are above the level of the heart is assisted by gravity. Below the level of the heart, however, venous return is against gravity. The most important force propelling blood from these regions is the squeezing of the veins by the contractions of surrounding skeletal muscles. As muscles contract, the vessels are compressed, and blood is squeezed through them. Blood flow may be temporarily obstructed during a prolonged muscle contraction, but when muscles relax, blood is free to move again. One-way valves within the veins of the extremities prevent backflow of blood. Thus, whenever a vein is squeezed, blood is propelled forward toward the heart (Figure 49.13).

In a resting person, gravity causes blood accumulation in the veins of the lower body and exerts back pressure on the capillary beds. This back pressure shifts the balance between blood pressure and osmotic pressure so that there is an increased loss of fluid to the intercellular spaces. That is why you have trouble putting your shoes back on after you sit for a long time with your shoes off, such as on an airline flight. In persons with very expandable veins, the veins may become so stretched that the valves can no longer prevent backflow. This condition produces *varicose* (swollen) *veins*. Draining of these veins is highly desirable and can be aided by wearing support hose and periodically elevating the legs above the level of the heart.

When an animal walks or runs, its legs act as auxiliary vascular pumps, returning blood to the heart from the veins of the lower body. As a greater volume of blood is returned to the heart, the heart contracts more forcefully, and its pumping action is enhanced. This strengthening of the heartbeat is due to a property of cardiac muscle cells described by the *Frank–Starling law*: If the cells are stretched, as they are when the volume of returning blood increases, they contract more forcefully. The actions of breathing also help return venous blood to the heart. The ventilatory muscles create negative pressure that pulls air into the lungs (see Figure 48.11), and this negative pressure also pulls blood toward the chest, increasing venous return to the right atrium. In addition, some of the largest veins closest to the heart contain smooth muscle that contracts at the onset of exercise. This contraction can rapidly increase venous return and stimulate the heart in accord with the Frank–Starling law, thus increasing cardiac output.

Lymphatic vessels return tissue fluid to the blood

The tissue fluid that accumulates outside the capillaries contains water and small molecules, but no red blood cells, and less protein than there is in blood. A separate system of vessels—the **lymphatic system**—returns tissue fluid to the blood.

After entering the lymphatic vessels, the tissue fluid is called **lymph**. Fine lymphatic capillaries merge progressively into larger and larger vessels and end in two lymphatic vessels—the **thoracic ducts**—that empty into large veins at the base of the neck (see Figure 18.1). The left thoracic duct carries most of the lymph from the lower part of the body and is much larger than the right thoracic duct. Lymph, like blood, is propelled toward the heart by skeletal muscle contractions and breathing movements, and lymphatic vessels, like veins, have one-way valves that keep the lymph flowing toward the thoracic duct.

Mammals and birds have **lymph nodes** along the major lymphatic vessels. Lymph nodes are a major site of lymphocyte production and of the phagocytic action that removes microorganisms and other foreign materials from the circulation. The lymph nodes also act as filters. Particles become trapped there and are digested by phagocytes in the nodes. When you get an infection, the lymph nodes closest to the infection become swollen and sore. The swelling is due to the accumulation of immune system cells that have been activated to fight the infection.

Will you die of cardiovascular disease?

Cardiovascular disease is responsible for about half of all deaths each year in the United States and Europe; it is by far the largest single killer in those countries. The immediate cause of most of these deaths is heart attack or stroke, but those events are frequently the end result of a disease called **atherosclerosis** ("hardening of the arteries") that begins many years before symptoms are detected. Hence atherosclerosis is called the "silent killer."

Healthy arteries have a smooth internal lining of endothelial cells (Figure 49.14*a*). This lining can be damaged by chronic high blood pressure, smoking, a high-fat diet, or microorganisms. Deposits called **plaque** begin to form at sites of endothelial damage. First, the damaged endothelial cells attract certain white blood cells to the site. These cells are then joined by smooth muscle cells migrating from the deeper layers of the arterial wall. Lipids, especially cholesterol, are deposited in these cells, so that the developing plaque becomes fatty. Fibrous connective tissue made by the invading smooth muscle cells in the plaque, along with deposits of calcium, makes the artery wall less elastic—hence, "hardening of the arteries." The growing plaque deposit narrows the artery and causes turbulence in the blood flowing over it. Local conditions can cause blood platelets (discussed later in this chapter) to stick to the plaque and initiate the formation of an intravascular blood clot, called a **thrombus**, which can quickly block the artery (Figure 49.14*b*).

The blood supply to the heart itself flows through the **coronary arteries**, which are highly susceptible to atherosclerosis. As these arteries narrow, blood flow to the heart muscles decreases. Chest pain and shortness of breath during mild exertion are symptoms of this condition. A person with atherosclerosis is at high risk of forming a thrombus in a coronary artery. This condition, called **coronary thrombosis**, can totally block the vessel, causing a *heart attack*, or **myocardial infarction** (**MI**).

A piece of a thrombus that breaks loose, called an **embolus**, is likely to travel to and become lodged in a vessel of

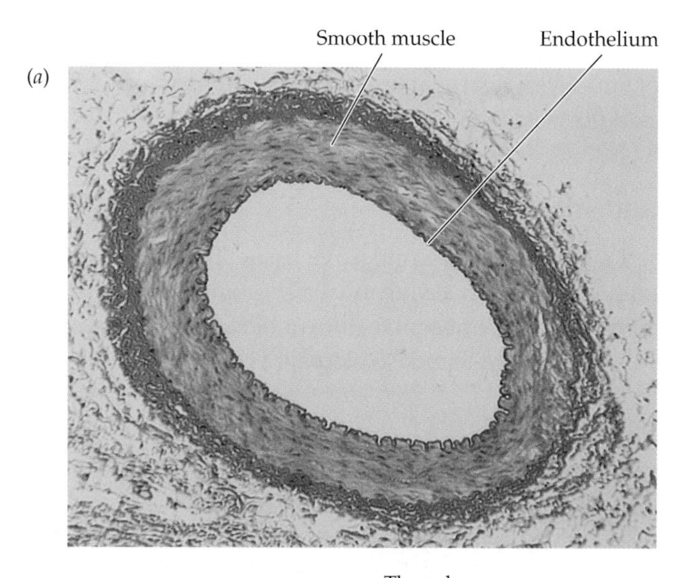

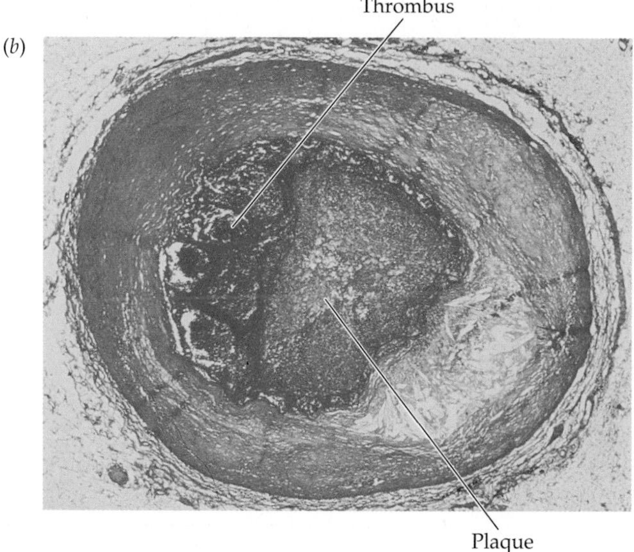

49.14 Atherosclerotic Plaque (*a*) A healthy, clear artery. (*b*) An atherosclerotic artery, clogged with plaque and a thrombus.

smaller diameter, blocking its flow (an **embolism**). Arteries already narrowed by plaque formation are likely places for an embolism. An embolism in an artery in the brain causes the cells fed by that artery to die. This event is called a **stroke**. The specific damage resulting from a stroke, such as memory loss, speech impairment, or paralysis, depends on the location of the blocked artery.

The most effective approach to cardiovascular disease is prevention by healthy lifestyle choices, not treatment. Probably the most important determinants of whether or not you will get atherosclerosis are your genetic predisposition and your age. Environmental risk factors also play a large role, however. If you do have a genetic predisposition to atherosclerosis, it is even more important to minimize environmental risk factors. These factors include high-fat and high-cholesterol diets, smoking, and a sedentary lifestyle. Certain untreated medical conditions, such as *hypertension* (high blood pressure), obesity, and diabetes, are also risk factors for atherosclerosis. Changes in diet and behavior can prevent and reverse early atherosclerosis and help fend off the silent killer.

Blood: A Fluid Tissue

Blood is classified as a connective tissue: It consists of cells suspended in an extracellular matrix of complex, yet specific, composition. The unusual feature of blood is that the extracellular matrix is a liquid, so blood is a fluid tissue.

The cells of the blood can be separated from the fluid matrix, called **plasma**, by centrifugation (Figure 49.15). If a sample of blood is spun in a centrifuge, all the cells move to the bottom of the tube, leaving the clear, straw-colored plasma on top. The *packed-cell volume*, or *hematocrit*, is the percentage of the blood volume made up by cells. Normal hematocrit is about 38 percent for women and 46 percent for men, but these values can vary considerably. They are usually higher, for example, in people who live and work at high altitudes because the low oxygen concentrations at high altitudes stimulate the production of more red blood cells.

In this section, we will consider two classes of cellular elements in blood: the red blood cells and the platelets, which are pinched-off fragments of cells. We discussed the other important class of blood cells—white blood cells, or leukocytes—in Chapter 18.

Red blood cells transport respiratory gases

Most of the cells in the blood are **erythrocytes**, or red blood cells. Mature red blood cells are biconcave, flexible discs packed with hemoglobin. Their function is to transport respiratory gases. Their shape gives them a large surface area for gas exchange, and their flexibility enables them to squeeze through narrow capillaries. There are 5 to 6 million red blood cells per cubic milliliter of blood (about one drop).

Red blood cells, as well as all the other cellular components of blood, are generated by stem cells in the bone mar-

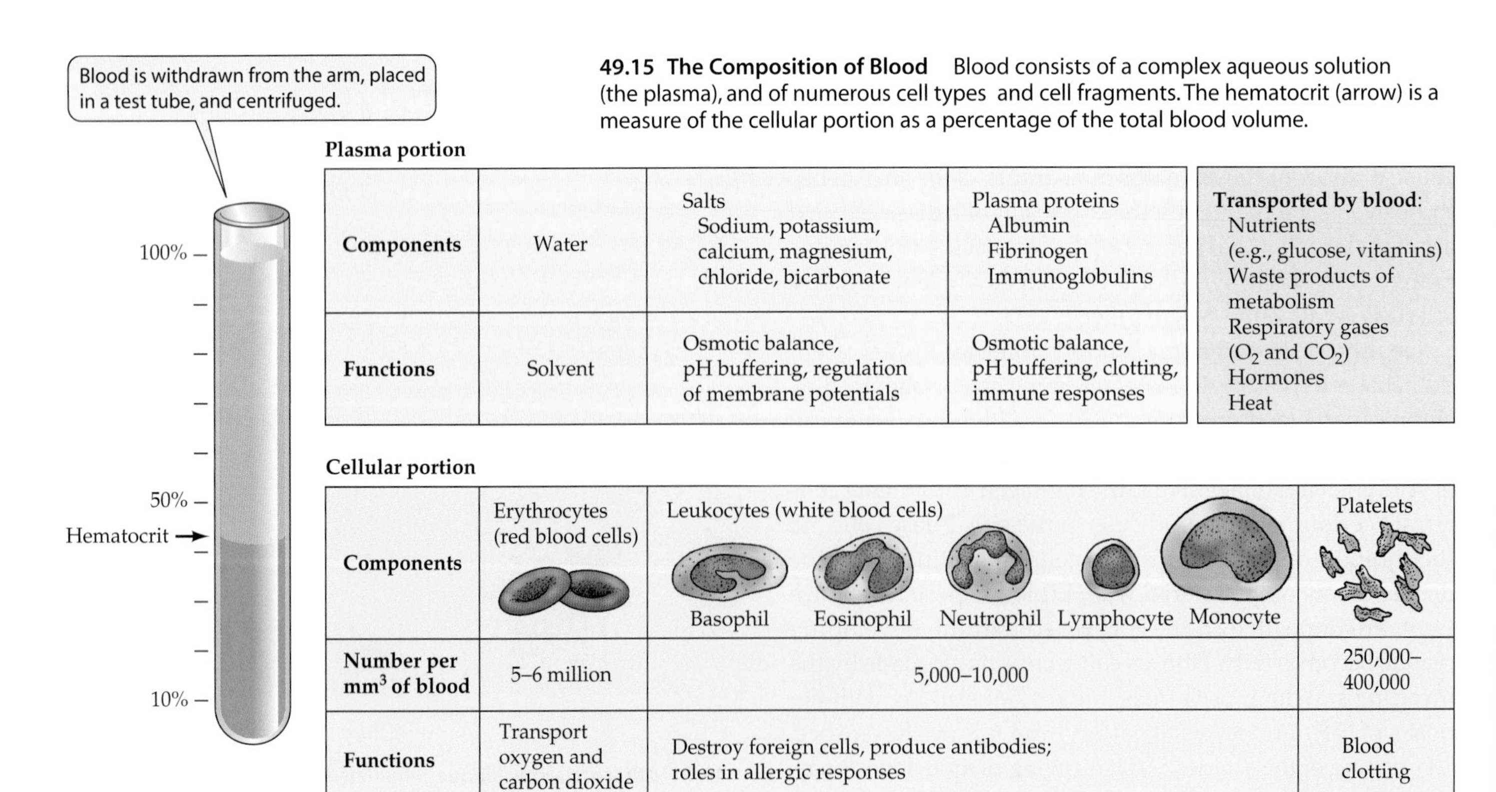

Plasma portion

Components	Water	Salts Sodium, potassium, calcium, magnesium, chloride, bicarbonate	Plasma proteins Albumin Fibrinogen Immunoglobulins
Functions	Solvent	Osmotic balance, pH buffering, regulation of membrane potentials	Osmotic balance, pH buffering, clotting, immune responses

Cellular portion

Components	Erythrocytes (red blood cells)	Leukocytes (white blood cells) Basophil, Eosinophil, Neutrophil, Lymphocyte, Monocyte	Platelets
Number per mm^3 of blood	5–6 million	5,000–10,000	250,000–400,000
Functions	Transport oxygen and carbon dioxide	Destroy foreign cells, produce antibodies; roles in allergic responses	Blood clotting

49.15 The Composition of Blood Blood consists of a complex aqueous solution (the plasma), and of numerous cell types and cell fragments. The hematocrit (arrow) is a measure of the cellular portion as a percentage of the total blood volume.

row, particularly in the ribs, breastbone, pelvis, and vertebrae. Red blood cell production is controlled by a hormone, **erythropoietin**, which is released by cells in the kidney in response to insufficient oxygen (*hypoxia*) in the tissues. Many tissues respond to hypoxia by expressing a transcription factor called *hypoxia-inducible factor 1* (*HIF-1*). When it reaches the kidney, HIF-1 activates the gene encoding erythropoietin.

Under normal conditions, your bone marrow produces about 2 million red blood cells every second. Developing, immature red blood cells divide many times while still in the bone marrow, and during this time they produce hemoglobin. When the hemoglobin content of a red blood cell approaches about 30 percent, its nucleus, endoplasmic reticulum, Golgi apparatus, and mitochondria begin to break down. This process is almost complete when the new red blood cell squeezes between the endothelial cells of blood vessels in the bone marrow and enters the circulation.

Each red blood cell circulates for about 120 days. As it gets older, its membrane becomes less flexible and more fragile. Therefore, old red blood cells can rupture as they bend to fit through narrow capillaries. One place where they are really squeezed is in the **spleen**, an organ that sits near the stomach in the upper left side of the abdominal cavity. The spleen has many venous cavities, or sinuses, that serve as a reservoir for red blood cells, but to get into the sinuses, the red blood cells must squeeze between spleen cells. When old red blood cells are ruptured by this squeezing, their remnants are taken up and degraded by macrophages.

Platelets are essential for blood clotting

Besides producing erythrocytes and leukocytes, the stem cells in the bone marrow produce cells called **megakaryocytes**. Megakaryocytes are large cells that remain in the bone marrow and continually break off cell fragments called **platelets**. A platelet is just a tiny fragment of a cell without cell organelles, but it is packed with enzymes and chemicals necessary for its function: sealing leaks in blood vessels and initiating blood clotting (Figure 49.16).

Damage to a blood vessel exposes collagen fibers. When a platelet encounters collagen fibers, it is activated. It swells, becomes irregularly shaped and sticky, and releases chemicals called **clotting factors**, which activate other platelets and initiate the clotting of blood. The sticky platelets also form a patch at the damaged site.

The clotting of blood requires many steps and many clotting factors. The absence of any one of these factors can impair clotting and cause excessive bleeding. Because the liver produces most of the clotting factors, liver diseases such as hepatitis and cirrhosis can result in excessive bleeding. The sex-linked genetic disorder hemophilia (see Chapters 10 and 17) is an example of a genetic inability to produce one of the clotting factors.

Blood clotting factors participate in a cascade of chemical reactions that activate other substances circulating in the blood. The cascade begins with cell damage and platelet activation and leads to the conversion of an inactive circulating enzyme, **prothrombin**, to its active form, **thrombin**. Thrombin causes molecules of a plasma protein called **fibrinogen** to polymer-

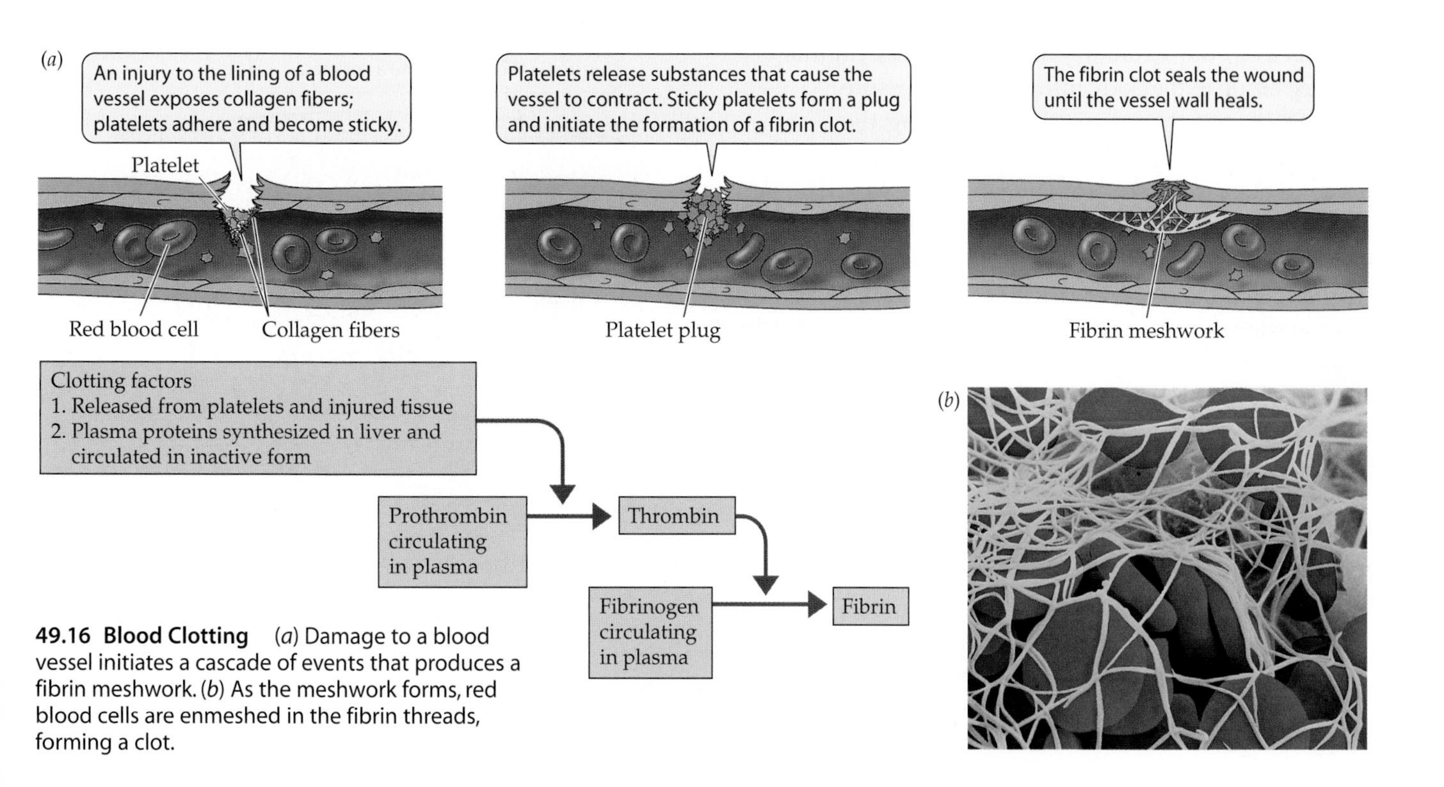

49.16 Blood Clotting (*a*) Damage to a blood vessel initiates a cascade of events that produces a fibrin meshwork. (*b*) As the meshwork forms, red blood cells are enmeshed in the fibrin threads, forming a clot.

ize and form threads of **fibrin**. The fibrin threads form the meshwork that clots the blood, seals the vessel, and provides a scaffold for the formation of scar tissue (see Figure 49.16).

Plasma is a complex solution

Plasma, the clear, straw-colored liquid portion of the blood, contains dissolved gases, ions, nutrient molecules, proteins, and other molecules, such as hormones. Most of the ions are Na^+ and Cl^- (hence the salty taste of blood), but many other ions are also present. The nutrient molecules in plasma include glucose, amino acids, lipids, cholesterol, and lactic acid. The circulating proteins in plasma include the blood clotting factors that we have just mentioned. Plasma is very similar to tissue fluid in composition, and most of its components move readily between these two fluid compartments of the body. The main difference between the two fluid compartments is the higher concentration of proteins in the plasma.

Control and Regulation of Circulation

The circulatory system is controlled and regulated by neuronal and hormonal mechanisms at both the local and systemic levels. Every tissue in the body requires an adequate supply of blood that is saturated with oxygen, carries essential nutrients, and is relatively free of waste products. But the nervous system cannot monitor and control every capillary bed in the body. Instead, each tissue regulates its own blood flow through *autoregulatory mechanisms* that cause the arterioles supplying the tissue to constrict or dilate.

The collective autoregulatory actions of every capillary bed in every tissue in the body influence the pressure and composition of the arterial blood leaving the heart. If many arterioles suddenly dilate, for example, allowing blood to flow through many more capillary beds, arterial blood pressure falls. If all these newly filled capillary beds contribute metabolic waste products to the blood at one time, the concentration of wastes in the blood returning to the heart increases. The nervous and endocrine systems respond to such changes by changing breathing rate, heart rate, and blood distribution to match the metabolic needs of the body.

Autoregulation matches local blood flow to local need

The autoregulatory mechanisms that adjust the flow of blood to a tissue are local mechanisms, but they can be influenced by the nervous system and by certain hormones.

The amount of blood that flows through a capillary bed is controlled by the smooth muscle of the arteries and arterioles feeding that bed. The flow of blood in a typical capillary bed is diagrammed in Figure 49.17. Blood flows into the bed from an arteriole. Smooth muscle "cuffs," or **precapillary sphincters**, on the arteriole can shut off the supply of blood to the capillary bed. When the precapillary sphincters are relaxed and the arteriole is open, the arterial blood pressure pushes blood into the capillaries.

Autoregulation depends on the sensitivity of the smooth muscle to its local chemical environment. Low O_2 concentrations and high CO_2 concentrations cause the smooth muscle to relax, thus increasing the supply of blood, which brings in more O_2 and carries away CO_2—a response known as *hyperemia*, which means "excess blood." Increases in other by-products of metabolism, such as lactic acid, hydrogen ions, potassium, and adenosine (all of which increase in exercising muscle), promote hyperemia through the same mechanism. Hence, activities that increase the metabolism of a tissue also induce hyperemia in that tissue.

Arterial pressure is controlled and regulated by hormonal and neuronal mechanisms

Control and regulation of the cardiovascular system begins with the local autoregulatory mechanisms we have just described. As more blood flows into the tissues, the central blood pressure falls, and the composition of the blood re-

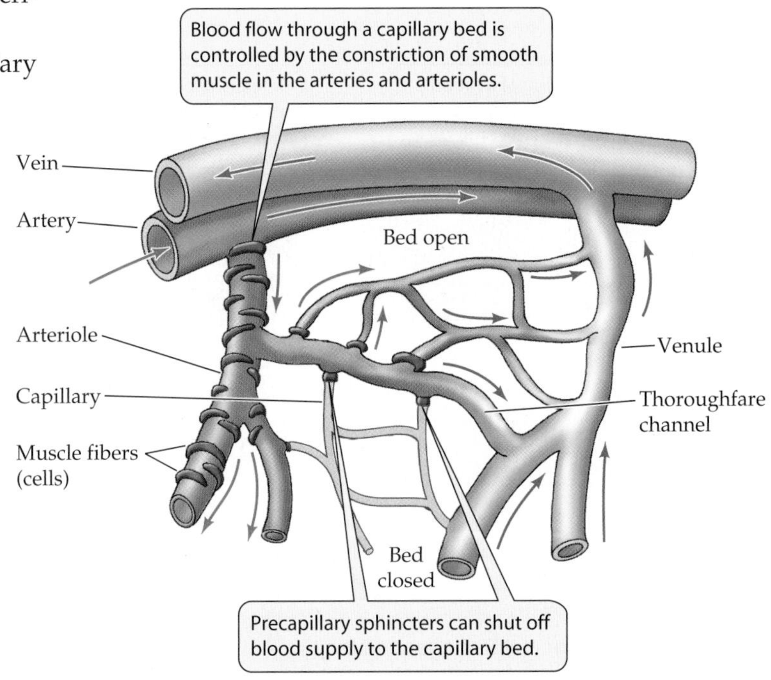

49.17 Local Control of Blood Flow Low O_2 concentrations or high levels of metabolic by-products cause the smooth muscle of the arteries and arterioles to relax, thus increasing the supply of blood to the capillary bed.

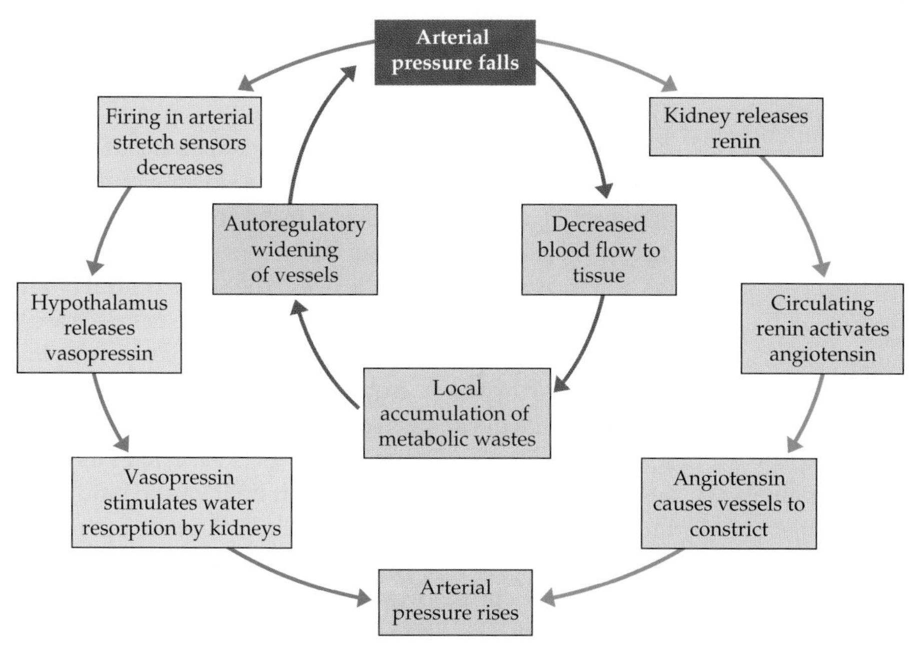

49.18 Control of Blood Pressure through Vascular Resistance A drop in arterial pressure reduces blood flow to tissues, resulting in local accumulation of metabolic wastes. This change in the extracellular environment stimulates autoregulatory opening of the arteries and would lead to a further decrease in central blood pressure if this were not prevented by the negative feedback mechanisms shown in this diagram, which work by promoting the constriction of arteries in less essential tissues.

turning to the heart reflects the exchanges that are occurring in the tissues. Changes in central blood pressure and blood composition are sensed, and both endocrine and central nervous system responses are activated to return blood pressure and composition to normal. Thus circulatory functions are matched to the regional and overall needs of the body.

Most arteries and arterioles are innervated by the autonomic nervous system, particularly the sympathetic division. Most sympathetic neurons release norepinephrine, which causes the smooth muscle cells to contract, thus constricting the vessels and reducing blood flow. An exception is found in skeletal muscle, in which specialized sympathetic neurons release acetylcholine, causing the smooth muscle of the arterioles to relax and the vessels to dilate, increasing blood to flow to the muscle.

Hormones also can cause arterioles to constrict. *Epinephrine*, which has actions similar to those of norepinephrine, is released from the adrenal medulla during massive sympathetic activation—the fight-or-flight response. *Angiotensin*, produced when blood pressure in the kidneys falls, causes arterioles to constrict. *Vasopressin*, released by the posterior pituitary when blood pressure falls, has similar effects (Figure 49.18). These hormones influence arterioles located for the most part in peripheral tissues (extremities) or in tissues whose functions need not be maintained continuously (such as the gut). By reducing blood flow in those arterioles, these hormones increase central blood pressure and blood flow to essential organs such as the heart, brain, and kidneys.

The autonomic nervous system activity that controls heart rate and constriction of blood vessels originates in cardiovascular centers in the medulla. Many inputs converge on this central integrative network and influence the commands it issues via parasympathetic and sympathetic nerves (Figure 49.19). Of special importance is incoming information about changes in blood pressure and composition from stretch re-

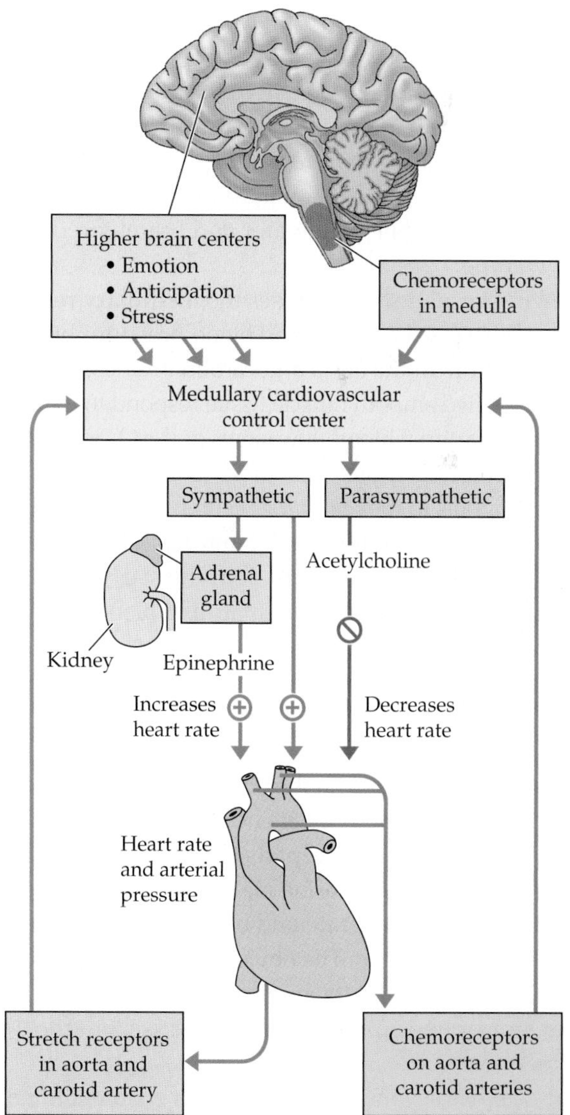

49.19 Regulating Blood Pressure The autonomic nervous system controls heart rate in response to information about blood pressure and blood composition that is integrated by regulatory centers in the medulla.

ceptors and chemoreceptors in the walls of the large arteries leading to the brain: the aorta and the carotid arteries.

Increased activity in the stretch receptors of the large arteries indicates rising blood pressure and inhibits sympathetic nervous system signaling to arteries and arterioles while increasing parasympathetic signaling to the heart's pacemaker. As a result, the heart slows, and arterioles in peripheral tissues dilate. If pressure in the large arteries falls, the activity of the stretch receptors decreases, stimulating sympathetic output to the arteries and arterioles while reducing parasympathetic output to the heart's pacemaker. As a result, the heart beats faster, and the arterioles in peripheral tissues constrict. When arterial pressure falls, the change in stretch receptor activity also causes the hypothalamus to release vasopressin, which, in addition to constricting arterioles, helps stabilize blood pressure by stimulating greater water resorption by the kidneys. We'll learn more about this mechanism in Chapter 51.

The heart itself is also sensitive to stretching. When the atria are receiving too much venous return, they release a hormone called *atrial natriuretic factor*. This hormone, which we will discuss in more detail in Chapter 51, stimulates the kidney to excrete Na^+ and water and thereby reduces the blood volume.

Other information that causes the medullary regulatory system to increase heart rate and blood pressure comes from chemoreceptors in the aorta and carotid arteries. These nodules of modified smooth muscle tissue respond to inadequate O_2 supply. If arterial blood flow slows or the O_2 content of the arterial blood falls drastically, these receptors are activated and send signals to the regulatory center. The regulatory center also receives input from other brain areas. Emotions or the anticipation of intense activity, as at the start of a race, can cause the center to increase heart rate and blood pressure.

Cardiovascular control in diving mammals conserves oxygen

We began this chapter with the observation that when a seal begins underwater activity, its heart rate slows and blood flow to all of its tissues except its brain drops dramatically. This "diving reflex" of marine mammals is in stark contrast to the increase in heart rate and blood flow we experience when we begin exercise. The obvious difference between the situation of the seal and the human is that the human has access to atmospheric oxygen during exercise, but the diving seal does not.

The seal has several adaptations that enable it to remain under water for a long time. The seal's oxygen storage capacity is about twice that of human due to the seal's greater blood volume, the greater oxygen-carrying capacity of its blood, and the greater amount of myoglobin in its muscles. These adaptations are not sufficient, however, to explain dives of half an hour or more. The seal's most important adaptation for diving is the *diving reflex*, which is a slowing of the heart (Figure 49.20) and a constriction of the major blood vessels going to all tissues except certain critical ones, such as the nervous system, the heart, and the eyes. The seal's central blood pressure remains high, but blood flow to its tissues decreases. This reduced blood flow has two effects: One is to switch the tissues to glycolytic (anaerobic) metabolism, and the other is to suppress the metabolism of the tissue.

While diving, the seal accumulates lactic acid in its muscles, which constitutes an "oxygen debt" to be paid back through elevated metabolism after the dive ends. But the total metabolic "debt" is much less than the metabolism that would have occurred over the same period of time had the seal not dived. The diving reflex causes the seal to be *hypometabolic* (to have a metabolic rate below its basal rate) during the dive. Hypometabolism, increased oxygen stores, and a high capacity for anaerobic metabolism make it possible for the seal to perform its amazing diving feats.

The seal's diving reflex may seem like a unique adaptation, but it provides yet another example of how natural selection shapes biological traits that are widely shared among related species. Humans also have a diving reflex. It is controlled by the vagus nerve and the parasympathetic nervous system. When our faces are submerged, we experience a mild slowing of our heart rate. This reflex probably serves as a protective response during the birth process, when pressure on the umbilical cord can deprive the fetus of maternal oxygen before breathing can begin. There are many cases, how-

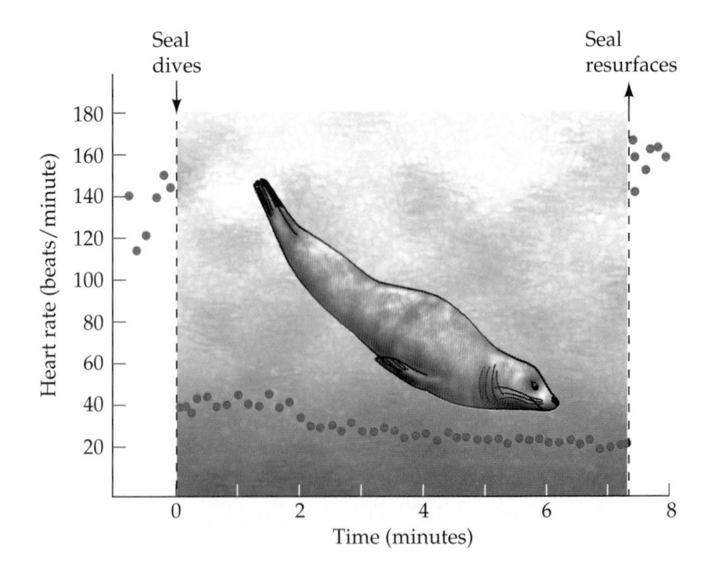

49.20 The Diving Reflex When a marine mammal dives, its heart rate slows and the arteries to most of its organs constrict, so that almost all blood flow and available oxygen goes to the animal's heart and brain. These adaptations enable some seals to remain underwater for up to an hour.

ever, in which drowning victims have been submerged in cold water for rather long periods of time, yet have survived with no brain damage. The human diving reflex, along with the rapid cooling of the brain as body heat is lost to the water, is the probable explanation for these remarkable cases of survival.

Chapter Summary

Circulatory Systems: Pumps, Vessels, and Blood

▶ The metabolic needs of the cells of small aquatic animals are met by direct exchange of materials with the external medium. The metabolic needs of the cells of larger animals are met by a circulatory system that transports nutrients, respiratory gases, and metabolic wastes throughout the body.

▶ In open circulatory systems, the blood or tissue fluid leaves vessels and percolates through tissues. **Review Figure 49.1**

▶ In closed circulatory systems, the blood is contained in a system of vessels. **Review Figure 49.2**

Vertebrate Circulatory Systems

▶ The circulatory systems of vertebrates consist of a heart and a closed system of vessels containing blood that is separate from the tissue fluid. Arteries and arterioles carry blood from the heart; capillaries are the site of exchange between blood and tissue fluid; venules and veins carry blood back to the heart.

▶ The vertebrate heart evolved from two chambers in fishes to three in amphibians and reptiles and four in crocodilians, mammals, and birds. This evolutionary progression has led to an increasing separation of blood that flows to the gas exchange organs and blood that flows to the rest of the body. **Review Pages 943–945. See Web/CD Activity 49.1**

▶ In birds and mammals, blood circulates through two circuits: the pulmonary circuit and the systemic circuit.

The Human Heart: Two Pumps in One

▶ The human heart has four chambers. Valves in the heart prevent the backflow of blood. **Review Figure 49.3. See Web/CD Activity 49.2**

▶ The cardiac cycle has two phases: systole, in which the ventricles contract; and diastole, in which the ventricles relax. The sequential heart sounds ("lub-dub") are made by the closing of the heart valves. **Review Figure 49.4. See Web/CD Tutorial 49.1**

▶ Blood pressure can be measured using a sphygmomanometer and a stethoscope. **Review Figure 49.5**

▶ The autonomic nervous system controls heart rate: Sympathetic activity increases heart rate, and parasympathetic activity decreases it. These actions are due to the effects of norepinephrine and acetylcholine on the rate of depolarization of the plasma membranes of pacemaker cells. **Review Figure 49.6**

▶ The sinoatrial node controls the cardiac cycle by initiating a wave of depolarization in the atria, which is conducted to the ventricles through a system consisting of the atrioventricular node, the bundle of His, and the Purkinje fibers. **Review Figure 49.7**

▶ The sustained contraction of ventricular muscle cells is due to long action potentials that are generated by voltage-gated calcium channels. **Review Figure 49.8**

▶ An EKG records electrical events associated with the contraction and relaxation of the cardiac muscles. **Review Figure 49.9**

The Vascular System: Arteries, Capillaries, and Veins

▶ Arteries and arterioles have many elastic fibers that enable them to withstand high pressures. Abundant smooth muscle cells allow these vessels to change their diameter, altering their resistance and thus blood flow. **Review Figure 49.10. See Web/CD Activity 49.3**

▶ Capillary beds are the site of exchange of materials between blood and tissue fluid.

▶ The Starling hypothesis offers an explanation for the exchange of fluids between blood and tissues that is based on the balance between blood pressure and osmotic pressure in the capillaries. **Review Figure 49.12**

▶ The ability of a specific molecule to cross a capillary wall depends on the architecture of the capillary, the type of substance, and the concentration gradient between the blood and the tissue fluid.

▶ Veins have a high capacity for storing blood. Aided by gravity, by contractions of skeletal muscle, and by the actions of breathing, they return blood to the heart. **Review Figure 49.13**

▶ The lymphatic system returns the tissue fluid to the blood.

▶ Cardiovascular disease is responsible for about half of all deaths in the United States and Europe.

Blood: A Fluid Tissue

▶ Blood can be divided into a plasma portion (water, salts, and proteins) and a cellular portion (red blood cells, white blood cells, and platelets). All of the cellular components are produced from stem cells in the bone marrow. **Review Figure 49.15**

▶ Red blood cells transport respiratory gases. Their production in the bone marrow is stimulated by erythropoietin, which is produced in response to hypoxia in the tissues.

▶ Platelets, along with circulating proteins, are involved in blood clotting. **Review Figure 49.16**

▶ Plasma is a complex solution that contains gases, ions, nutrient molecules, proteins, and other molecules.

Control and Regulation of Circulation

▶ Blood flow through capillary beds is controlled by local autoregulatory mechanisms, hormones, and the autonomic nervous system. **Review Figure 49.17**

▶ Blood pressure is controlled in part by the hormones vasopressin and angiotensin, which stimulate contraction of blood vessels. **Review Figure 49.18**

▶ Heart rate is controlled by the autonomic nervous system, which responds to information about blood pressure and blood composition that is integrated by regulatory centers in the medulla. **Review Figure 49.19**

▶ Diving mammals conserve blood oxygen stores by slowing the heart rate during dives. **Review Figure 49.20**

Self-Quiz

1. An open circulatory system is characterized by
   a. the absence of a heart.
   b. the absence of blood vessels.
   c. blood with a composition different from that of tissue fluid.
   d. the absence of capillaries.
   e. a higher-pressure circuit through gills than to other organs.

2. Which statement about vertebrate circulatory systems is *not* true?
   a. In fish, oxygenated blood from the gills returns to the heart through the left atrium.
   b. In mammals, deoxygenated blood leaves the heart through the pulmonary artery.
   c. In amphibians, deoxygenated blood enters the heart through the right atrium.
   d. In reptiles, the blood in the pulmonary artery has a lower oxygen content than the blood in the aorta.
   e. In birds, the pressure in the aorta is higher than the pressure in the pulmonary artery.
3. Which statement about the human heart is *true*?
   a. The walls of the right ventricle are thicker than the walls of the left ventricle.
   b. Blood flowing through atrioventricular valves is always deoxygenated blood.
   c. The second heart sound is due to the closing of the aortic valve.
   d. Blood returns to the heart from the lungs in the vena cava.
   e. During systole, the aortic valve is open and the pulmonary valve is closed.
4. The pacemaker actions of cardiac muscle
   a. are due to opposing actions of norepinephrine and acetylcholine.
   b. are localized in the bundle of His.
   c. depend on the gap junctions between the cells that make up the atria and those that make up the ventricles.
   d. are due to spontaneous depolarization of the plasma membranes of some cardiac muscle cells.
   e. result from hyperpolarization of cells in the sinoatrial node.
5. Blood flow through capillaries is slow because
   a. lots of blood volume is lost from the capillaries.
   b. the pressure in venules is high.
   c. the total cross-sectional area of capillaries is larger than that of arterioles.
   d. the osmotic pressure in capillaries is very high.
   e. red blood cells are bigger than capillaries and must squeeze through.
6. How are lymphatic vessels like veins?
   a. Both have nodes where they join together into larger common vessels.
   b. Both carry blood under low pressure.
   c. Both are capacitance vessels.
   d. Both have valves.
   e. Both carry fluids rich in plasma proteins.
7. The production of red blood cells
   a. ceases if the hematocrit falls below normal.
   b. is stimulated by erythropoietin.
   c. is about equal to the production of white blood cells.
   d. is inhibited by prothrombin.
   e. occurs in bone marrow before birth and in lymph nodes after birth.
8. Which of the following does *not* increase blood flow through a capillary bed?
   a. High concentration of CO_2
   b. High concentration of lactate and hydrogen ions
   c. Histamine
   d. Vasopressin
   e. Increase in arterial pressure
9. Blood clotting
   a. is impaired in patients with hemophilia because they don't produce platelets.
   b. is initiated when platelets release fibrinogen.
   c. involves a cascade of factors produced in the liver.
   d. is initiated by leukocytes forming a meshwork.
   e. requires conversion of angiotensinogen to angiotensin.
10. Autoregulation of blood flow to a tissue is due to
   a. sympathetic innervation.
   b. the release of vasopressin by the hypothalamus.
   c. increased activity of stretch receptors.
   d. chemoreceptors in the aorta and the carotid arteries.
   e. the effect of the local chemical environment on arterioles.

For Discussion

1. At the beginning of a race, cardiac output increases immediately before there is any change in blood O_2 or CO_2 concentrations. Explain two factors that contribute to this effect. Include the Frank–Starling law in your answer.
2. Explain how the hearts of crocodilians have the advantages of mammalian hearts during exercise but the efficiency of reptilian hearts during rest.
3. A sudden and massive loss of blood results in a decrease in blood pressure. Describe several mechanisms that help return blood pressure to normal.
4. You can describe the cycle of events in a ventricle of the heart by a graph that plots the pressure in the ventricle on the y axis and the volume of blood in the ventricle on the x axis. What would such a graph look like? Where would the heart sounds be on this graph? How would the graph differ for the left and the right ventricles?
5. If the major arteries become clogged with plaque and become less elastic because of atherosclerosis, the left ventricle must work harder and harder to pump an adequate supply of blood to the body. As a result, the left ventricle can become weakened and begin to fail, even though the right ventricle is healthy. A heart attack primarily affecting the left ventricle can have the same effect. This condition is known as congestive heart failure, and it commonly leads to fatal pulmonary edema. Explain how left ventricular failure can result in pulmonary edema, and why is it said that this condition creates a vicious circle that makes itself worse rapidly.

50 Nutrition, Digestion, and Absorption

For thousands of years, the Pima of southwestern North America were hunters and gatherers who supplemented their diet with subsistence agriculture. Their environment was arid, so they developed sophisticated irrigation systems; even so, they frequently encountered drought and subsequent starvation. Today many individuals of the ethnic Pima population are clinically obese and plagued with obesity-related illnesses such as diabetes, high blood pressure, and heart disease. The incidence of diabetes among these native American people rose from 45 percent of adults in 1965 to 80 percent in 1999. What could cause such a radical health change in an entire population? There are at least two factors involved: genetics and lifestyle.

Geneticists hypothesize that recurring episodes of starvation produce strong selective pressure for "efficiency genes"—alleles of those genes involved in digestive, absorptive, and energy storage functions that result in greater efficiency in converting food into energy and energy reserves, such as fat. Efficiency genes would give individuals a strong selective advantage when food is scarce. An example of an efficiency gene phenotype is seen among the Pima. As we will see later in this chapter, the hormone insulin facilitates the conversion of dietary sugar into fat tissue. For many Pima, consuming a standard amount of glucose causes their insulin levels to rise three times higher than those of Americans of European ancestry.

The other factor in the Pima obesity epidemic is an abrupt change in their traditional lifestyle. When food is plentiful and has a high calorie content, efficiency genes contribute to obesity by maximizing fat storage. Today the Pima eat a modern Western diet, including high-calorie fast foods. In general, they also engage in less physical work than their ancestors did. High-calorie diet and sedentary lifestyle are factors that affect not just the Pima, but are contributing to the overall increase in obesity throughout the U.S. population. Researchers are studying the Pima to learn more about the genetics of obesity and related diseases.

In this chapter we will review the nutrients that organisms require for energy, as molecular building blocks, and for

Efficiency Genes The Pima are an example of a population that has probably undergone selection in the past for genes that improve the efficiency of managing the energy obtained from food. With modern diets and lifestyles, these "efficiency genes" can contribute to obesity.

specific biochemical functions. We will examine diverse adaptations for acquiring, ingesting, and digesting food and absorbing nutrients. Then we will learn how the body regulates its traffic in metabolic fuels, and we will return to the question of control of body mass. Finally, we will raise the issue of natural and artificial toxins in food.

Nutrient Requirements

Animals must eat other organisms to stay alive. Since they derive their nutrition from other organisms, they are called **heterotrophs**. In contrast, **autotrophs** (most plants, some bacteria, and some protists) trap solar energy through photosynthesis and use that energy to synthesize all of their components. Directly or indirectly, heterotrophs take advantage of—indeed, depend on—the organic synthesis carried out by autotrophs. Heterotrophs have evolved an enormous diversity of adaptations to exploit, directly or indirectly, the resources made available through the actions of autotrophs (Figure 50.1). In this section we will discover how animals use those resources as energy sources and as building blocks for complex molecules.

Energy can be measured in calories

In Chapters 6 and 7, we learned that energy in the chemical bonds of food molecules is transferred to the high-energy phosphate bonds of ATP. ATP provides animals with energy for cellular work. Each conversion of energy from food molecules to ATP and from ATP to cellular work is inefficient, however; in fact, most of the energy that was in the food is lost as heat. Even the energy the animal uses is eventually reduced to heat, as molecules that were synthesized are broken down and the energy of movement is dissipated by friction. Therefore, we can talk about the energy requirements of animals and the energy content of food in terms of a measure of heat energy: the calorie.

A **calorie** is the amount of heat necessary to raise the temperature of 1 gram of water 1°C. Since this value is a tiny amount of energy compared with the energy requirements of many animals, physiologists commonly use the **kilocalorie** (kcal) as a unit of measure (1 kcal = 1,000 calories). Nutritionists also use the kilocalorie as a standard unit of energy, but they traditionally refer to it as the **Calorie** (**Cal**), which is always capitalized to distinguish it from the single calorie. (Scientists are gradually abandoning the calorie as an energy unit as they switch to the International System of Units. In this system, the basic unit of energy is the joule: 1 joule = 0.239 calories.)

The *metabolic rate* of an animal (see Chapter 41) is a measure of the overall energy needs that must be met by the animal's ingestion and digestion of food. The basal metabolic rate of a human is about 1,300–1,500 kcal/day for an adult female and 1,600–1,800 kcal/day for an adult male. Physical activity adds to this basal energy requirement. For a person doing sedentary work, about 30 percent of total energy consumption is due to skeletal muscle activity, and for a person doing heavy physical labor, 80 percent or more of total caloric expenditure is due to skeletal muscle activity. The components of food that provide energy are fats, carbohydrates,

50.1 Heterotrophs Get Energy from Autotrophs (*a*) Herbivores get their energy directly from autotrophs. Large herbivores such as the giant panda must consume huge amounts of plant matter (this particular animal eats only bamboo plants) to fulfill their nutritional needs. (*b*) Polar bears are carnivores, and ferocious predators. A carnivore's energy is indirectly obtained from autotrophs, since the energy stored in a prey animal was originally obtained from autotrophs.

(*a*) *Ailuropoda melanoleuca*

(*b*) *Ursus maritimus*

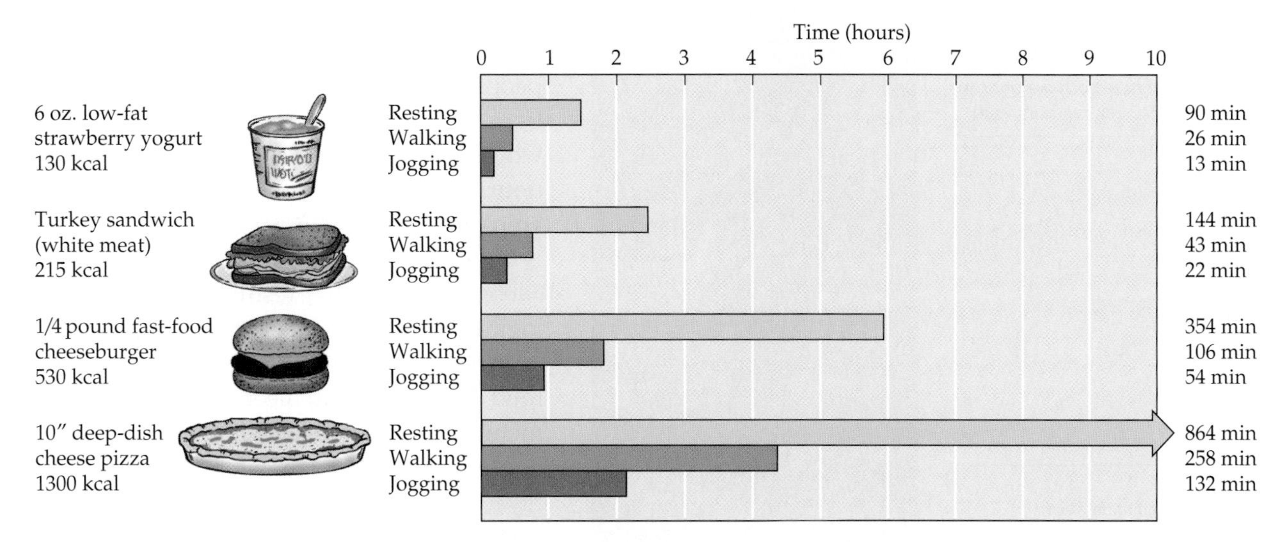

50.2 Food Energy and How We Use It The energy contained in several common food items is shown at the left. The graphs indicate about how long it would take a person with a basal metabolic rate of about 1,800 kcal/day to utilize the equivalent amount of energy while involved in various activities.

and proteins. Fats yield 9.5 kcal/gram, carbohydrates 4.2 kcal/gram, and proteins about 4.1 kcal/gram. Some equivalencies of food, energy, and energy consumption are shown in Figure 50.2.

Energy budgets reveal how animals use their resources

It is possible, of course, to quantify the caloric value of any food an animal eats. It is also possible to quantify the caloric cost of anything an animal does. By comparing calories consumed with calories expended, it is possible to construct *energy budgets* for any set of circumstances. Energy budgets allow ecologists and evolutionists to apply a cost–benefit analysis to any behavior.

Consider, for example, territorial aggression: Under what circumstances does it "pay" to fight over a food resource? Such a study was done on African sunbirds, which feed on the nectar of a particular flower. These birds defend feeding territories in some habitats, but not in others. Investigators hypothesized that the birds could "afford" to be territorial only if the food resource was rich enough to support the metabolic cost of aggressive behavior in addition to other essential activities. Accordingly, they determined how much nectar a flower produces, and how this amount differs with habitat type. They then determined how many calories the birds spent when they were resting, when they were foraging, and when they were aggressively defending a patch of flowers.

According to the investigators' calculations, if a bird had a choice between a patch of flowers that produced nectar at a rate of 1 μl/day/blossom or a patch that produced nectar at 2 μl/day/blossom, by selecting the richer patch, it could meet its daily caloric requirement with 4 hours of foraging instead of 8 hours of foraging. Since the caloric expenditure during foraging is 600 calories/hour greater than the caloric expenditure when resting, the bird could save 2,400 calories (4 hours × 600 calories/hour) by feeding on the more productive flowers. However, territorial defense costs 2,000 calories more per hour than foraging. Therefore, if a bird had to spend much more than an hour a day chasing intruders away from its rich flower patch, it would be better off being nonaggressive and feeding on the less productive flowers. Direct observation and measurement showed that the actual behavior of the birds agreed with these predictions based on energy budget calculations.

Sources of energy can be stored in the body

Although the cells of the body use energy continuously, most animals do not eat continuously. Therefore, animals must store fuel molecules that can be released as needed between meals.

Carbohydrates are stored in liver and muscle cells as *glycogen*, but the total glycogen store represents only about a day's basal energy requirements (1,500–2,000 Cal). Fat is the most important form of stored energy in the bodies of animals. Not only does fat have more energy per gram than glycogen, but it can be stored with little associated water, making it more compact. Migrating birds store energy as fat to fuel their long flights; if they had to store the same amount of energy as glycogen, they would be too heavy to fly! Proteins are not used as energy storage compounds, although body protein can be metabolized as an energy source of last resort.

If an animal takes in too little food to meet its energy requirements, it is **undernourished**, and must make up the shortfall by metabolizing some of the molecules of its own body. This consumption of self begins with the energy storage compounds glycogen and fat. Protein loss is minimized

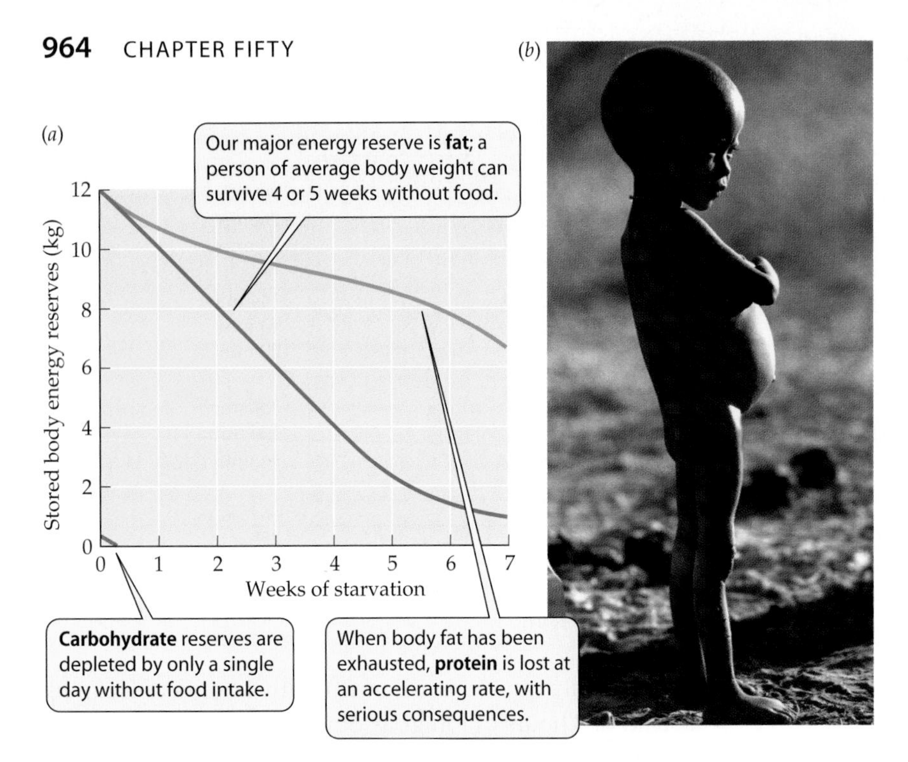
(a)
(b)
Our major energy reserve is **fat**; a person of average body weight can survive 4 or 5 weeks without food.
Stored body energy reserves (kg)
12
10
8
6
4
2
0
0 1 2 3 4 5 6 7
Weeks of starvation
Carbohydrate reserves are depleted by only a single day without food intake.
When body fat has been exhausted, **protein** is lost at an accelerating rate, with serious consequences.

50.3 The Course of Starvation (*a*) In a person subjected to undernutrition, the energy reserves of the body are eventually depleted. (*b*) The swollen (due to edema) abdomen, face, hands, and feet of this young Somalian girl, as well as her spindly limbs, are symptoms of kwashiorkor. This syndrome results from the body breaking down blood proteins and muscle tissue to obtain needed amino acids.

for as long as possible, but eventually a starving animal begins to break down its own proteins for fuel (Figure 50.3*a*). The syndrome that results is called *kwashiorkor* (Figure 50.3*b*). Blood proteins are among the first to be used, resulting in loss of fluid to the intercellular spaces (edema; see Chapter 49). Additional consequences of protein deficiency are breakdown of the immune system and degeneration of the liver. Muscles waste away, and eventually even brain protein is lost, leading to mental retardation. If starvation continues, the breakdown of body proteins eventually leads to death.

Undernourishment is rampant among people in nonindustrialized and war-torn nations, and a billion people—one-sixth of the world's population—are chronically undernourished. Ironically, one cause of life-threatening undernourishment in Western nations is a self-imposed starvation syndrome called *anorexia nervosa* that results from a psychological aversion to real or imagined body fat.

When an animal consistently takes in *more* food than it needs to meet its energy requirements, it is **overnourished**. The excess nutrients are stored as increased body mass. First, glycogen reserves build up; then additional dietary carbohydrates, fats, and proteins are converted to body fat. In some species, such as hibernators, seasonal overnutrition is an important adaptation for surviving periods when food is not available. In humans, however, overnutrition can be a serious health hazard, increasing the risk of high blood pressure, heart attack, diabetes, and other disorders.

Food provides carbon skeletons for biosynthesis

Every animal requires certain basic organic molecules that it cannot synthesize for itself, but needs as building blocks for its own complex organic molecules (see Chapter 7). An example of such a required *carbon skeleton* is the acetyl group (Figure 50.4). Animals cannot make acetyl groups from carbon, oxygen, and hydrogen molecules; they must obtain acetyl groups by metabolizing carbohydrates, fats, or proteins obtained by eating other organisms.

Acetyl groups can be derived from the metabolism of almost any food; they are unlikely ever to be in short supply

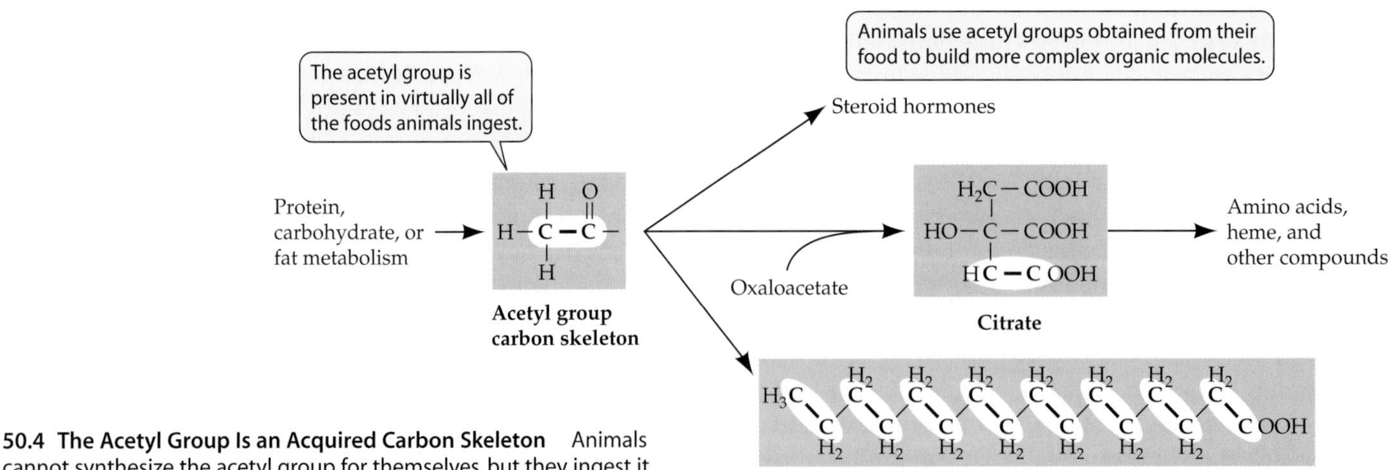
The acetyl group is present in virtually all of the foods animals ingest.
Animals use acetyl groups obtained from their food to build more complex organic molecules.
Protein, carbohydrate, or fat metabolism
Acetyl group carbon skeleton
Steroid hormones
Oxaloacetate
Citrate
Amino acids, heme, and other compounds
Palmitic acid (and other fatty acids)

50.4 The Acetyl Group Is an Acquired Carbon Skeleton Animals cannot synthesize the acetyl group for themselves, but they ingest it in their food and use it to synthesize a wide variety of molecules.

for an adequately nourished animal. Other carbon skeletons, however, are derived from more limited sources, and an animal can suffer a deficiency of these materials even if its caloric intake is adequate.

Amino acids, the building blocks of proteins, are a good example of carbon skeletons that can be in short supply. Humans obtain amino acids by breaking down proteins from food. Another source of amino acids is the breakdown of existing body proteins, which are in constant turnover as the tissues of the body undergo normal remodeling and renewal. From all these amino acids, the body synthesizes its own protein molecules as specified by its DNA.

Animals can synthesize some of their own amino acids by taking carbon skeletons synthesized from acetyl or other groups and transferring to them amino groups (—NH_2) derived from other amino acids. But most animals cannot synthesize all the amino acids they need. Each species has certain **essential amino acids** that it must obtain from food. Different species have different essential amino acids, and in general, herbivores have fewer essential amino acids than carnivores. If an animal does not take in enough of one of its essential amino acids, its protein synthesis is impaired.

There are eight essential amino acids that humans must obtain from their food: isoleucine, leucine, lysine, methionine, phenylalanine, threonine, tryptophan, and valine. All eight are available in milk, eggs, meat, and soybean products, but most plant foods do not contain all eight. A strict vegetarian diet, therefore, poses a risk of protein malnutrition. A *complementary* dietary mixture of plant foods, however, supplies all eight essential amino acids (Figure 50.5). In general, grains are complemented by legumes or by milk products; legumes are complemented by grains, seeds, and nuts. Long before the chemical basis for this complementarity was understood, societies with little access to meat developed complementary diets. Many Central and South American peoples traditionally eat beans with corn, and the native peoples of North America complemented their beans with squash.

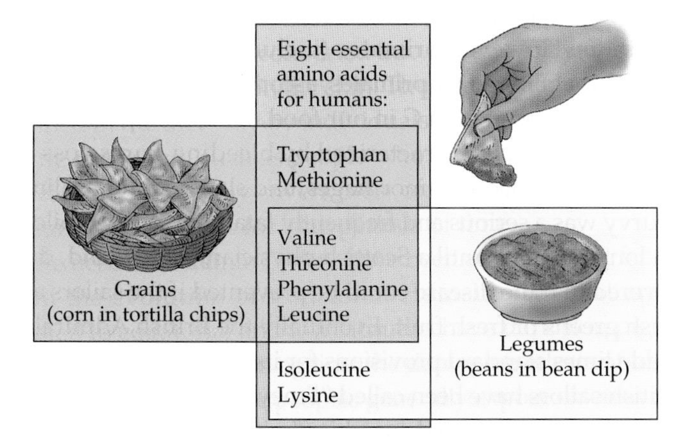

50.5 A Strategy for Vegetarians By combining cereal grains and legumes, a vegetarian can obtain all eight essential amino acids.

Why are dietary proteins completely digested to their constituent amino acids before being used by the body? Wouldn't it be more energy-efficient to reuse some dietary proteins directly? There are several reasons why ingested proteins are not used "as is":

- Macromolecules such as proteins are not readily absorbed by the cells of the gut, but their constituent monomers (such as amino acids) are readily absorbed.
- Protein structure and function are highly species-specific. A protein that functions optimally in one species might not function well in another.
- Foreign proteins entering the body directly from the gut would be recognized as invaders and would be attacked by the immune system.

Similarly, humans can synthesize almost all the lipids required by the body using acetyl groups obtained from food (see Figure 50.4), but we must have a dietary source of certain **essential fatty acids**—notably, linoleic acid—that we cannot synthesize. Linoleic acid is an unsaturated fatty acid needed by mammals to synthesize other unsaturated fatty acids, such as arachidonic acid, which is a component of several signaling molecules, including prostaglandins. A deficiency of linoleic acid can lead to problems such as infertility and impaired lactation. Essential fatty acids are also necessary components of membrane phospholipids.

Animals need mineral elements for a variety of functions

The principal mineral elements that animals require are listed in Table 50.1. Elements required in large amounts are called **macronutrients**; elements required in only tiny amounts are called **micronutrients**. Some micronutrients are required in such minute amounts that deficiencies are never observed, but they are nevertheless essential elements.

Calcium is an example of a macronutrient. It is the fifth most abundant element in the body; a 70-kg person contains about 1.2 kg of calcium. Calcium phosphate is the principal structural material in bones and teeth. Muscle contraction, neuronal function, and many other intracellular functions in animals require calcium ions. The turnover of calcium in the extracellular fluid is high, as bones are constantly being remodeled and calcium is constantly entering and leaving cells. Calcium is lost from the body in urine, sweat, and feces, so it must be replaced regularly. Humans require about 800 to 1,000 mg of calcium per day in the diet.

Iron is an example of a micronutrient. Iron is found everywhere in the body because it is the oxygen-binding atom in hemoglobin and myoglobin and is a component of enzymes

veloped beriberi-like symptoms when fed only polished rice. In 1912, Casimir Funk cured pigeons of beriberi by feeding them the discarded hulls.

At the time of Funk's discovery, all diseases were thought to be either caused by microorganisms or inherited. Funk suggested the radical idea that beriberi and some other diseases are dietary in origin and result from deficiencies in specific substances. Funk coined the term "vitamines" because he mistakenly thought that all these substances were amines (compounds with amino groups) vital for life. In 1926, thiamin (vitamin B_1)—the substance lost in the rice milling process—was the first vitamin to be isolated in pure form.

Deficiency diseases can also result from an inability to absorb or process an essential nutrient even if it is present in the diet. Vitamin B_{12} (cobalamin), for example, is present in all foods of animal origin. Since plants neither use nor produce vitamin B_{12}, a strictly vegetarian diet (not supplemented by vitamin pills) can lead to a B_{12} deficiency disease called *pernicious anemia*, characterized by a failure of red blood cells to mature. The most common cause of pernicious anemia, however, is not a lack of vitamin B_{12} in the diet, but an inability to absorb it. Normally, cells in the stomach lining secrete a peptide called *intrinsic factor*, which binds to vitamin B_{12} and makes it possible for it to be absorbed in the small intestine. Conditions that damage the stomach lining can therefore cause pernicious anemia.

Inadequate mineral nutrition can also lead to deficiency diseases. Iodine, for example, is a constituent of the hormone thyroxine, which is produced in the thyroid gland. If there is insufficient iodine in the diet, the thyroid gland grows larger in an attempt to compensate for inadequate production of thyroxine. The swelling of the neck that results is called a *goiter*. The introduction of iodized table salt has greatly reduced the incidence of goiter in the United States.

Adaptations for Feeding

Heterotrophic organisms can be classified by how they acquire their nutrition. **Saprobes** (also called *saprotrophs* or *decomposers*) are organisms—mostly protists and fungi—that absorb nutrients from dead organic matter. **Detritivores**, such as earthworms and crabs, actively feed on dead organic material. Animals that feed on living organisms are **predators**: **Herbivores** prey on plants, **carnivores** prey on animals, and **omnivores** prey on both. **Filter feeders**, such as clams and blue whales, prey on small organisms by filtering them from the aquatic environment. **Fluid feeders** include mosquitoes, aphids, and leeches, as well as birds such as hummingbirds that feed on plant nectar. The anatomical adaptations that enable a species to exploit a particular source of nutrition are usually quite obvious, but physiological and biochemical adaptations can be just as important.

The food of herbivores is often low in energy and hard to digest

Most vegetation is coarse and difficult to break down physically, but herbivores must process large amounts of it, since its energy content is low. Most herbivores spend a great deal of their time feeding. Many have striking adaptations for feeding, such as the trunk (a flexible, gripping nose) of the elephant or the long neck of the giraffe. Many types of grinding, rasping, cutting, and shredding mouthparts have evolved in invertebrates for ingesting plant material, and the teeth of herbivorous vertebrates have been shaped by selection to process coarse plant matter. The digestive processes of herbivores can also be quite specialized, as we will see below.

Carnivores must detect, capture, and kill prey

The predatory behaviors of many carnivores are legendary. One need only call to mind the hunting skills of hawks, wolves, or tigers. Carnivores have evolved stealth, speed, power, large jaws, sharp teeth, and strong gripping appendages. Carnivores also have evolved remarkable means of detecting prey. Bats use echolocation, pit vipers sense infrared radiation from the warm bodies of their prey, and certain fishes detect electric fields created in the water by their prey (see Chapter 45).

Adaptations for killing and ingesting prey are diverse and highly specialized. These adaptations can be especially important when the prey are capable of inflicting damage on the predator. A snake may strike with poisonous fangs, using its venom to immobilize its prey before ingesting it. To swallow large prey, a snake disengages its lower jaw from its joint with the skull (Figure 50.6). The tentacles of jellyfishes, the long, sticky tongues of chameleons, and the webs of spiders are other fascinating examples of adaptations for capturing and immobilizing prey. Some predators digest their prey externally. For example, a spider may inject its insect prey with digestive enzymes and then suck out the liquefied contents, leaving behind the empty exoskeletons frequently seen in old spider webs.

Vertebrate species have distinctive teeth

Teeth are adapted for the acquisition and initial processing of specific types of foods. Because they are one of the hardest structures of the body, an animal's teeth remain in the environment long after it dies. Paleontologists use teeth to identify animals that lived in the distant past and to deduce what their feeding behavior might have been.

All mammalian teeth have the same general structure, consisting of three layers (Figure 50.7*a*). An extremely hard material called **enamel**, composed principally of calcium phosphate, covers the crown of the tooth. Both the crown and

50.6 An Adaptation for Carnivory Snakes such as this Texas rat snake (*Elaphe obsoleta*) can ingest large prey (in this case, a lizard) by dislocating their jaws.

the root contain a layer of bony material called **dentine**, inside of which is a **pulp cavity** containing blood vessels, nerves, and the cells that produce the dentine.

The shapes and organization of mammalian teeth, however, can be very different, since they are adapted to specific diets (Figure 50.7*b*). In general, *incisors* are used for cutting, chopping, or gnawing; *canines* are used for stabbing, ripping, and shredding; and *molars* and *premolars* (the cheek teeth) are used for shearing, crushing, and grinding. The highly varied diet of humans is reflected by our multipurpose set of teeth, as is common among omnivores.

Digestion

Most animals digest their food extracellularly. That is, animals take food into a body cavity that is continuous with the outside environment, into which they secrete digestive enzymes. The enzymes act on the food, reducing it to nutrient molecules that can be absorbed by the cells lining the cavity.

The simplest digestive systems are *gastrovascular cavities*, which connect to the outside world through a single opening. Cnidarians, for example, capture prey using their stinging nematocysts and cram it into their gastrovascular cavity with their tentacles (see Figure 32.7). Enzymes in the gastrovascular cavity partly digest the prey. Cells lining the cavity take in small food particles by endocytosis. The vesicles created by endocytosis then fuse with lysosomes containing digestive enzymes, and intracellular digestion completes the breakdown of the food. Nutrients are released to the cytoplasm as the vesicle breaks down.

Tubular guts have an opening at each end

The guts of most animals are tubular: A *mouth* takes in food; molecules are digested and absorbed throughout the length of the gut; and solid digestive wastes are excreted through

50.7 Mammalian Teeth (*a*) A mammalian tooth has three layers: enamel, dentine, and pulp cavity. (*b*) The teeth of different mammalian species are specialized for different diets.

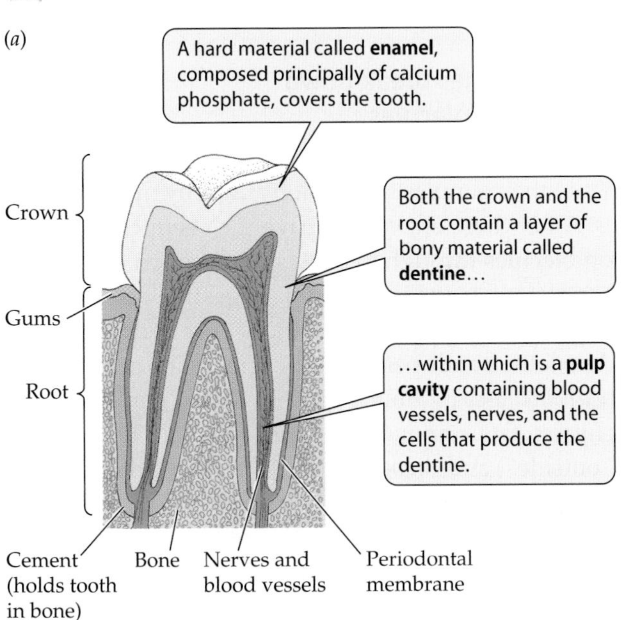

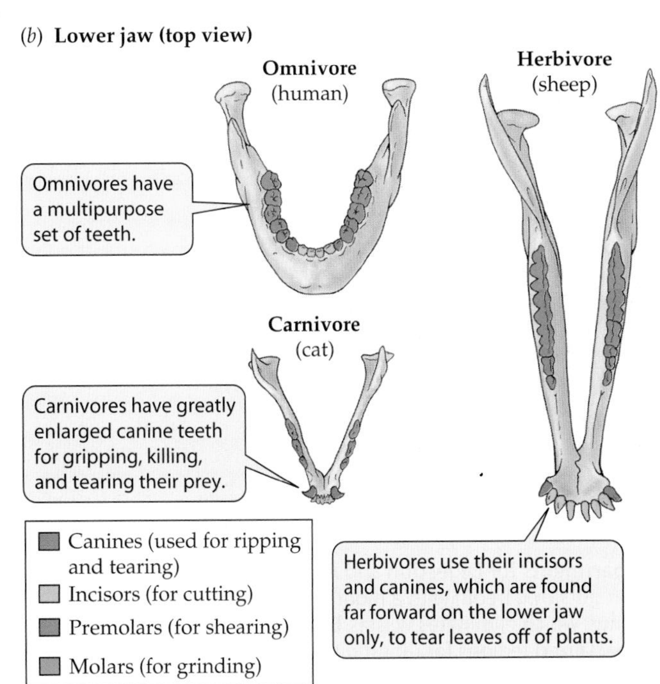

an *anus*. Different regions in the tubular gut are specialized for particular functions (Figure 50.8). These functions must be coordinated so that they occur in the proper sequence and at appropriate rates to maximize the efficiency of digestion and absorption of nutrients.

At the anterior end of the gut are the mouth (the opening itself) and the mouth cavity. Food may be broken up by teeth (in some vertebrates), by the radula (in snails), or by mandibles (in insects). In earthworms and most birds, a muscular portion of the gut called the *gizzard* grinds the food together with small stones. Some animals, such as snakes, simply ingest large chunks of food, with little or no fragmentation (see Figure 50.6).

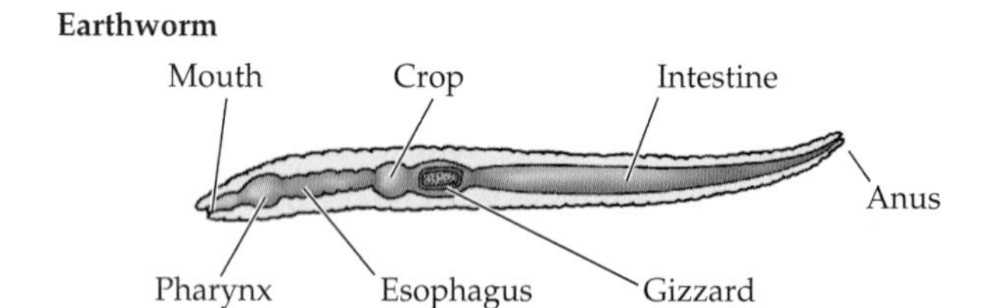

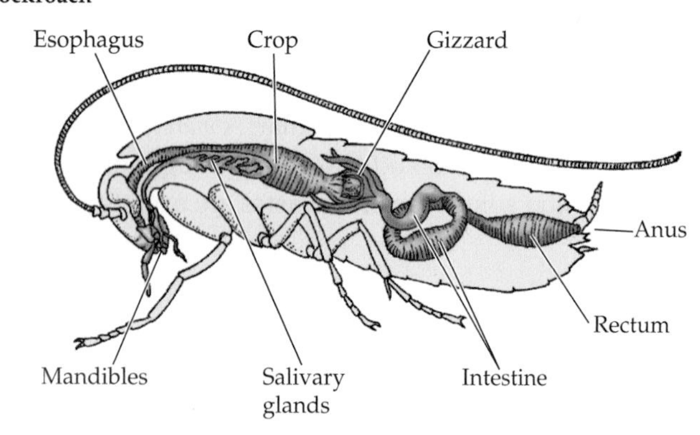

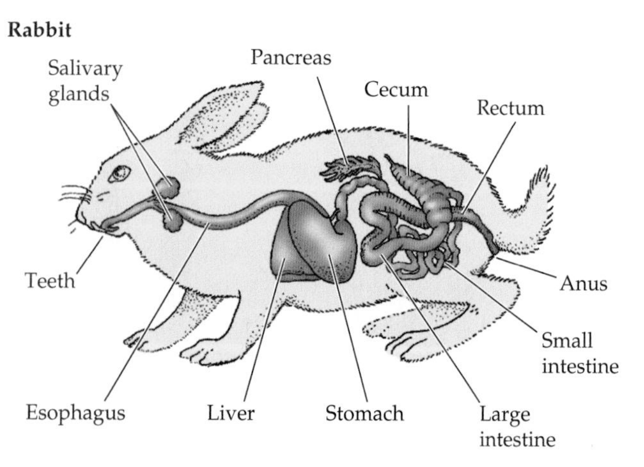

50.8 Compartments for Digestion and Absorption Most animals have tubular guts that begin with a mouth, which takes in food, and end in an anus, which eliminates wastes. Between these two structures are specialized regions for digestion and nutrient absorption; the structures in these regions vary from species to species.

Stomachs and **crops** are storage chambers that enable animals to ingest relatively large amounts of food and digest it later. In these storage chambers, food may be further fragmented and mixed, but digestion may or may not occur there, depending on the species. In any case, food delivered into the next section of the gut, the **midgut** or **intestine**, is well minced and well mixed.

Most nutrients, water, and ions are absorbed in the midgut. To digest food materials, specialized glands secrete some digestive enzymes into the midgut, and the gut wall itself secretes other enzymes. The **hindgut** recovers water and ions and stores undigested wastes, or **feces**, so that they can be released to the environment at an appropriate time or place. A muscular **rectum** near the anus assists in the expulsion of feces.

Within the hindguts of many species are colonies of endosymbiotic bacteria. These bacteria obtain their own nutrition from the food passing through the host's gut while contributing to the digestive processes of the host. Members of the leech genus *Hirudo*, for example, produce no enzymes that can digest the proteins in the blood they suck from vertebrates. A colony of gut bacteria produces the enzymes necessary to break down those proteins into amino acids, which are subsequently used by both the leech and the bacteria. Some animals, including humans, rely on microorganisms in their guts to supply them with vitamins.

In many animals, the parts of the gut that absorb nutrients have evolved extensive surface areas (Figure 50.9*a, b*). In vertebrates, the wall of the gut is richly folded, with the individual folds bearing legions of tiny fingerlike projections called **villi** (Figure 50.9*c*). The cells that line the surfaces of the villi, in turn, have microscopic projections, called **microvilli**. The microvilli give the gut an enormous internal surface area for the absorption of nutrients.

Digestive enzymes break down complex food molecules

Protein, carbohydrate, and fat macromolecules are broken down into their simplest monomeric units by hydrolytic enzymes. All of these enzymes cleave the chemical bonds of macromolecules through hydrolysis, a reaction that adds a water molecule (see Figure 3.3*b*). Digestive enzymes are classified according to the substances they hydrolyze: *Proteases* break the bonds between adjacent amino acids in proteins; *carbohydrases* hydrolyze carbohydrates; *peptidases*, peptides; *lipases*, fats; and *nucleases*, nucleic acids. The prefixes *exo-* ("outside") and *endo-* ("within") indicate where the enzyme cleaves the molecule: An *endoprotease* hydrolyzes a protein at an internal site along the polypeptide chain, and an *exoprotease* snips away amino acids at the ends of the molecule.

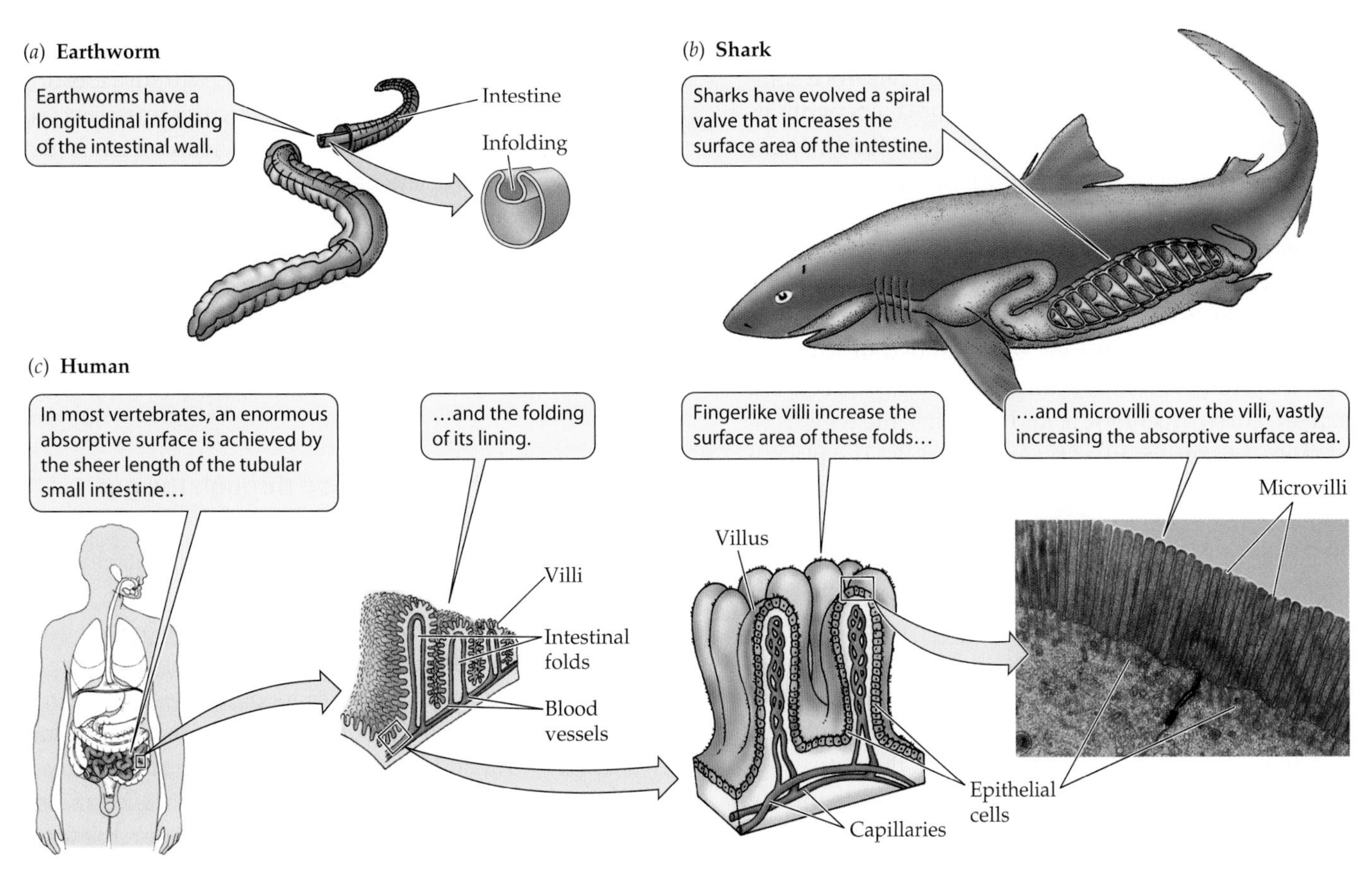

50.9 Greater Intestinal Surface Area Means More Nutrient Absorption The guts of most animals have evolved to maximize their surface area.

How can an organism produce enzymes that hydrolyze biological macromolecules without digesting itself? Most digestive enzymes are produced in an inactive form, known as a **zymogen**, so that they cannot act on the cells that produce them. When secreted into the gut, zymogens are generally activated by another enzyme, as we will explain below. The lining of the gut is not digested because it is protected by a covering of mucus.

Structure and Function of the Vertebrate Gut

The digestive tract of vertebrates is a tubular gut that runs from mouth to anus (Figure 50.10). The vertebrate gut can be divided into several compartments that are specialized for different digestive and absorptive functions. In addition, several accessory structures produce compounds that contribute to the digestive process and release them into the gut.

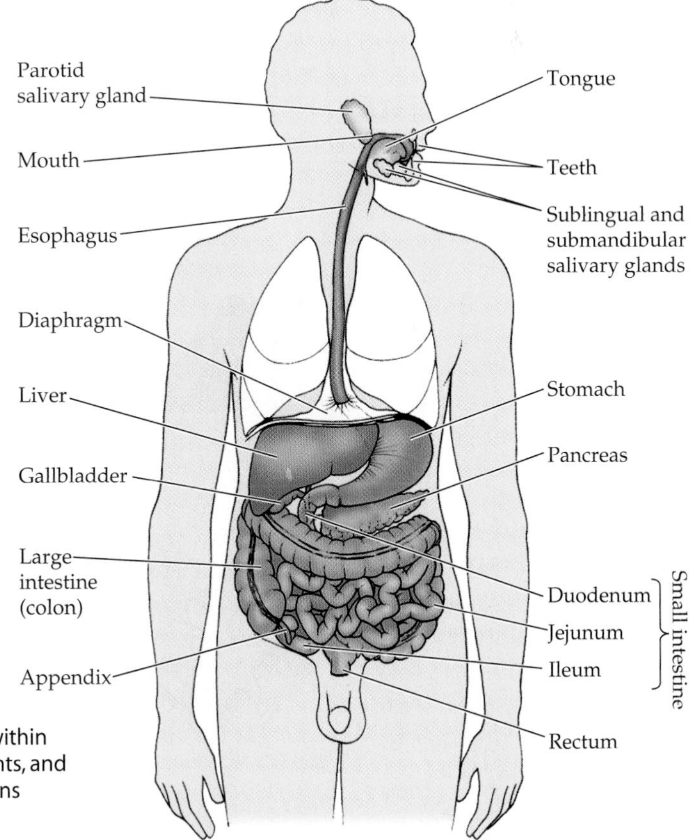

50.10 The Human Digestive System Different compartments within the long tubular gut specialize in digesting food, absorbing nutrients, and storing and expelling wastes. Accessory organs contribute secretions containing enzymes and other molecules.

The vertebrate gut has four tissue layers

The cellular architecture of the vertebrate gut follows a common plan throughout its length: Four major layers of different cell types form the wall of the gut (Figure 50.11). These layers differ somewhat from compartment to compartment, but they are always present.

Starting in the cavity, or *lumen*, of the gut, the first tissue layer is the **mucosa**. Mucosal cells have secretory and absorptive functions. Some secrete mucus, which lubricates and protects the walls of the gut; others secrete digestive enzymes, and still others secrete hormones. Mucosal cells in the stomach secrete hydrochloric acid (HCl). In some regions of the gut, nutrients are absorbed by mucosal cells. The plasma membranes of these absorptive cells have many folds that increase their surface area (see Figure 50.9*c*)

At the base of the mucosa are some smooth muscle cells, and just outside the mucosa is the second tissue layer, the **submucosa**. Here we find the blood and lymph vessels that carry absorbed nutrients to the rest of the body. The submucosa also contains a network of nerves; the neurons in this network are both sensory (responsible for stomach aches) and regulatory (controlling the various secretory functions of the gut).

External to the submucosa are two layers of smooth muscle tissue responsible for the movements of the gut. Innermost is the *circular muscle layer*, with its cells oriented around the gut. Outermost is the *longitudinal muscle layer*, with its cells oriented along the length of the gut. The circular muscles constrict the gut, and the longitudinal muscles shorten it. Between the two layers of smooth muscle is another network of nerves, which controls and coordinates the movements of the gut. The coordinated activity of the two smooth muscle layers moves the gut contents continuously toward the rectum.

Surrounding the gut is a coat of fibrous tissue called the **serosa**. Like other abdominal organs, the gut is also covered and supported by a tissue called the *peritoneum*.

50.11 Tissue Layers of the Vertebrate Gut In all compartments of the gut, the organization of the tissue layers is the same, but specialized adaptations of specific tissues characterize different regions.

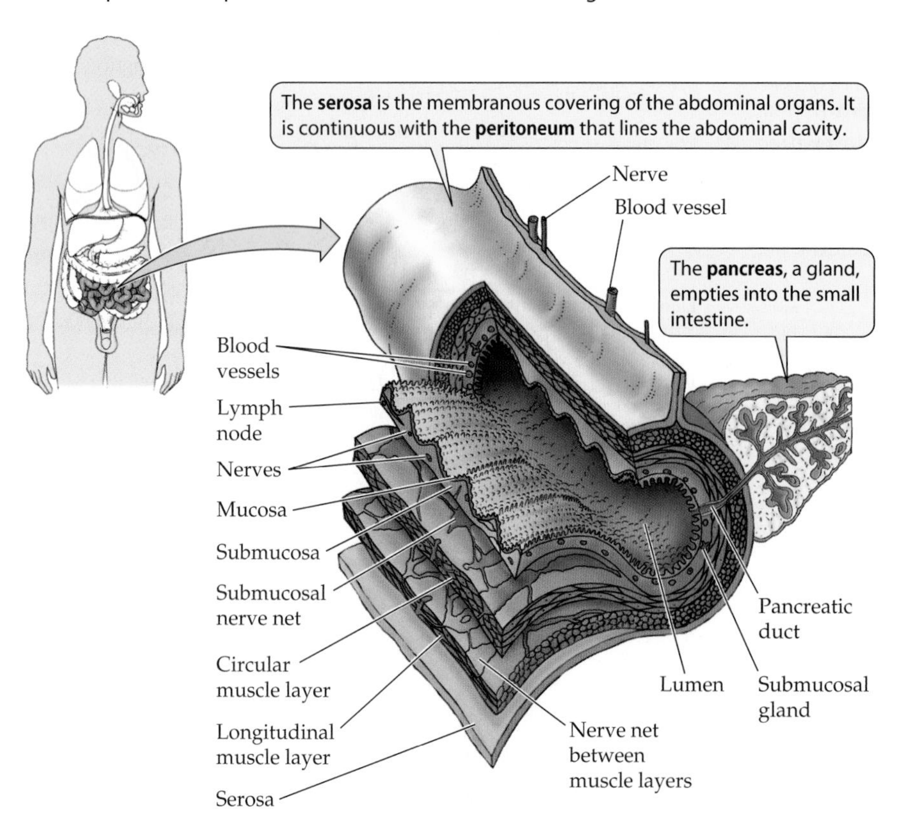

Mechanical activity moves food through the gut and aids digestion

In most vertebrates, including humans, food entering the mouth is chewed and mixed with saliva. A muscular *tongue* then pushes a chunk, or *bolus*, of chewed food toward the back of the mouth cavity. By making contact with the *soft palate* at the back of the throat, the bolus of food initiates *swallowing*, which is a complex series of autonomic reflexes. If you stand in front of a mirror and gently touch this tissue at the back of your throat with a cotton swab, you will experience an uncontrollable urge to swallow. Swallowing propels the food through the pharynx (where the mouth cavity and the nasal passages join) and into the **esophagus** (the food tube). To prevent the food from entering the trachea (windpipe), the *larynx* (voice box) closes, and a flap of tissue called the *epiglottis* covers the entrance to the trachea (Figure 50.12).

Once the food is in the esophagus, a wave of smooth muscle contraction, called **peristalsis**, takes over and pushes it toward the stomach. The smooth muscle layers of the gut contract in response to being stretched. Swallowing a bolus of food stretches the upper end of the esophagus, and this stretching initiates a wave of contraction that moves progressively down the gut from the pharynx toward the anus.

The movement of food from the stomach into the esophagus is normally prevented by the *lower esophageal sphincter*, a thick ring of circular smooth muscle at the junction of the esophagus and the stomach. This sphincter is normally constricted, but waves of peristalsis cause it to relax enough to let food pass from the esophagus into the stomach. Sphincter muscles are found elsewhere in the digestive tract as well. The *pyloric sphincter* governs the passage of stomach contents into the intestine. Another important sphincter surrounds the anus.

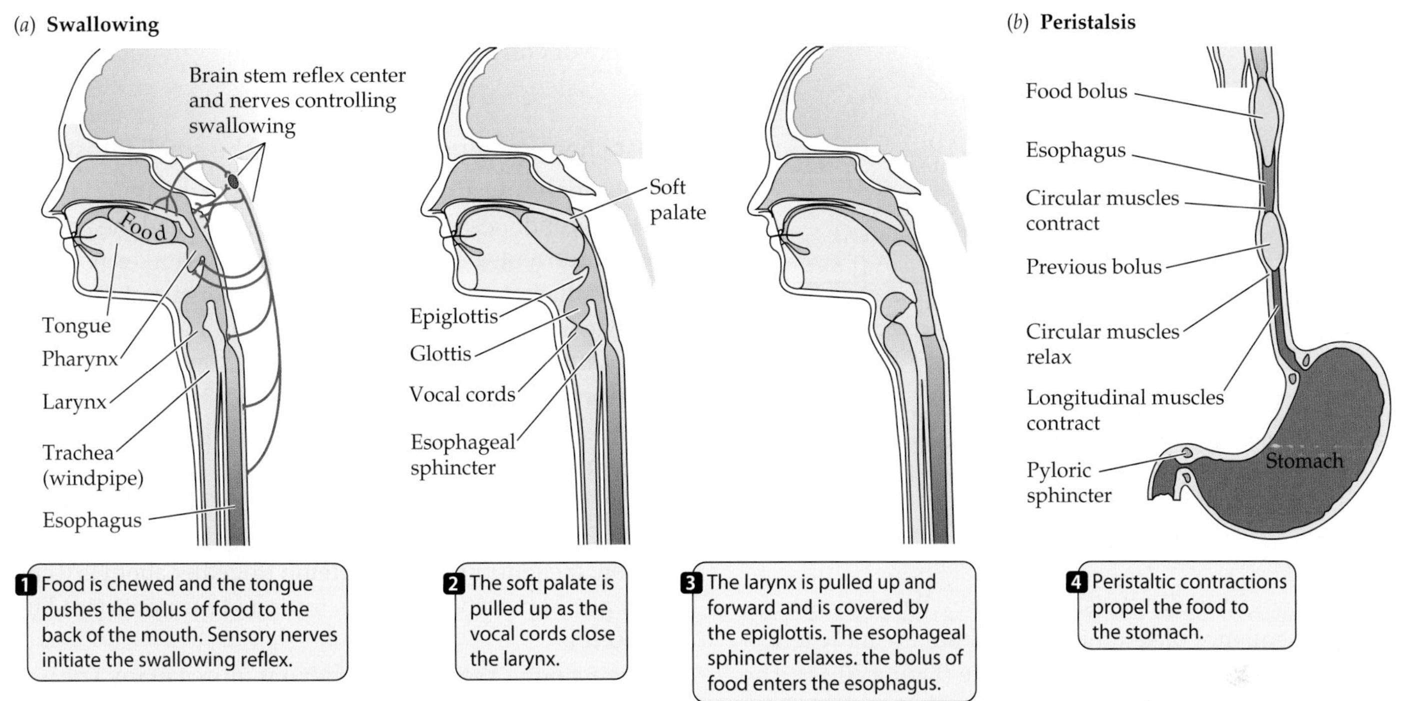

50.12 Swallowing and Peristalsis Food pushed to the back of the mouth triggers the swallowing reflex. Once food enters the esophagus, peristalsis propels it through the gut.

Chemical digestion begins in the mouth and the stomach

The enzyme *amylase* is secreted by the salivary glands and mixed with food as it is chewed. Amylase hydrolyzes the bonds between the glucose monomers that make up starch molecules. The action of amylase is what makes a chewed piece of bread or cracker taste slightly sweet if you hold it in your mouth long enough.

Most vertebrates can rapidly consume a large volume of food, but digesting that food is a slower process. The stomach stores the consumed food until it can be digested. The secretions of the stomach kill microorganisms that are taken in with the food and begin the digestion of proteins.

The major enzyme produced by the stomach is an endopeptidase called **pepsin**. Pepsin is secreted as a zymogen called **pepsinogen** by cells in the *gastric glands* that are deep folds in the stomach lining (Figure 50.13). Other cells in the gastric glands produce hydrochloric acid, and still others near the openings of the gastric glands and throughout the stomach mucosa secrete mucus.

Hydrochloric acid (HCl) maintains the stomach fluid (the *gastric juice*) at a pH between 1 and 3. This extremely low pH activates the conversion of pepsinogen to pepsin, and provides the right pH for pepsin's enzymatic action.

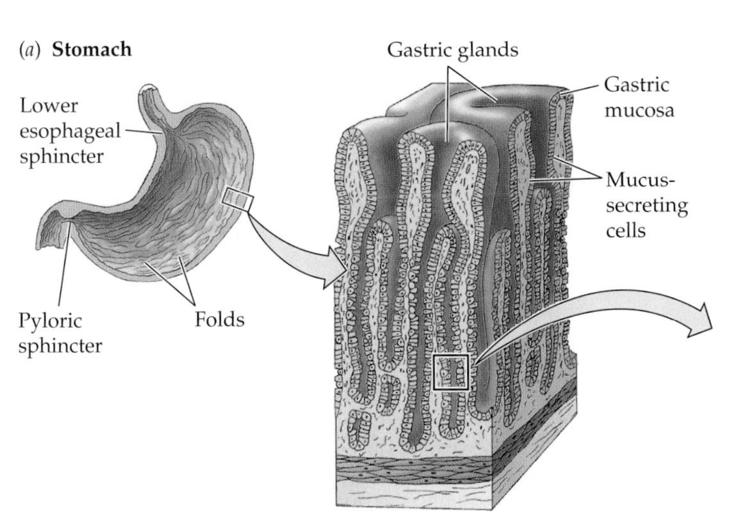

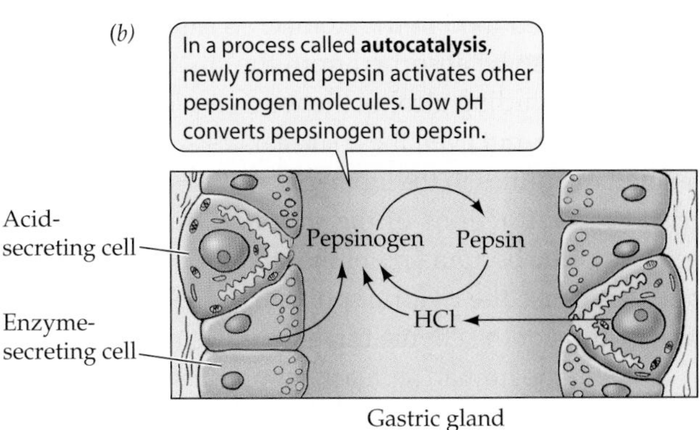

50.13 The Stomach (*a*) The human stomach stores and breaks down ingested food. (*b*) Cells in the gastric glands secrete hydrochloric acid and pepsin. Both the gastric glands and the gastric mucosa secrete mucus that protects the stomach.

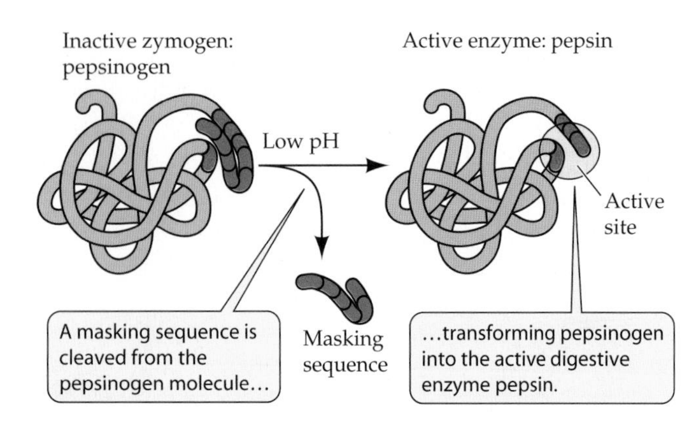

50.14 Activating a Zymogen Low pH in the stomach stimulates cleavage of a masking sequence of amino acids, transforming the zymogen pepsinogen into the active digestive enzyme pepsin. Pepsin itself also activates pepsinogen, through autocatalysis.

The pepsinogen molecule is activated by the cleavage of a masking sequence of 44 amino acids from the its N-terminal end, exposing the pepsin active site (Figure 50.14). The newly formed pepsin activates other pepsinogen molecules in a positive feedback process called *autocatalysis*. Hydrochloric acid also helps dissolve the intercellular substances holding the ingested tissues together. The breakdown of these tissues exposes more food surface area to the action of pepsin and, eventually, other digestive enzymes in the small intestine.

Mucus secreted by the stomach mucosa coats the walls of the stomach and protects them from being eroded and digested by HCl and pepsin. Sometimes, however, the walls of the stomach are exposed, and the resulting damage is called an *ulcer*. It was once thought that ulcers were caused by stress and oversecretion of digestive juices. However, it turns out that the basis for most ulcers is an infectious bacterium called *Helicobacter pylori*, which has the remarkable ability to live in the highly acidic environment of the stomach. Damage to the stomach lining initiated by the bacterial infection is made worse by HCl and pepsin.

Contractions of the smooth muscles in the walls of the stomach churn its contents, thoroughly mixing them with the stomach secretions. The acidic, fluid mixture of gastric juice and partly digested food in the stomach is called **chyme**. A few substances can be absorbed from the chyme across the stomach wall, including alcohol (hence its rapid effects), aspirin, and caffeine, but even these substances are absorbed in rather small quantities in the stomach.

Peristaltic contractions of the stomach walls push the chyme toward the bottom end of the stomach. These waves of peristalsis cause the pyloric sphincter to relax briefly so that little squirts of the chyme can enter the small intestine. In this manner, the human stomach empties itself gradually over a period of approximately 4 hours. This slow introduction of food into the small intestine enables the small intestine to work on a little material at a time.

Most chemical digestion occurs in the small intestine

In the **small intestine**, the digestion of carbohydrates and proteins continues, and the digestion of fats and the absorption of nutrients begin. The small intestine takes its name from its diameter; it is in fact very large organ, about 3 meters long in an adult. Because of its length, and because of the folds, villi, and microvilli of its lining, its inner surface area is enormous: about 550 m^2, or roughly the size of a tennis court. Across this surface, the small intestine absorbs all the nutrient molecules derived from food.

The small intestine has three sections. The initial section (about 25 cm), called the **duodenum**—is the site of most digestion; the **jejunum** and the **ileum** (together about 270 cm) carry out 90 percent of the absorption of nutrients (see Figure 50.10).

Digestion in the small intestine requires many specialized enzymes, as well as several other secretions. Two accessory organs that are not part of the digestive tract—the liver and the pancreas—provide many of these enzymes and secretions.

LIVER. The liver synthesizes **bile** from cholesterol. Bile secreted from the liver flows through the *hepatic duct* to the

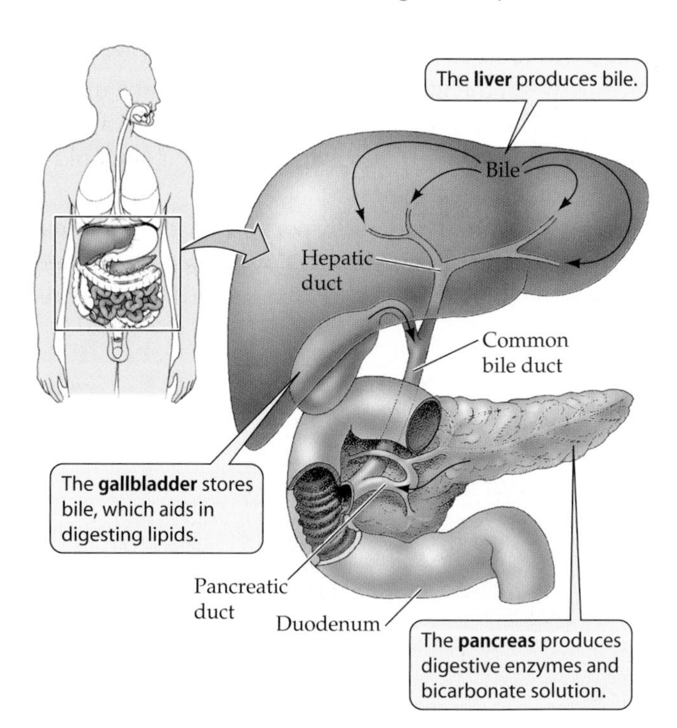

The liver produces bile.
Bile
Hepatic duct
Common bile duct
The gallbladder stores bile, which aids in digesting lipids.
Pancreatic duct
Duodenum
The pancreas produces digestive enzymes and bicarbonate solution.

50.15 The Ducts of the Gallbladder and Pancreas Bile produced in the liver leaves the liver via the hepatic duct. Branching off this duct is the gallbladder, which stores bile. Below the gallbladder, the hepatic duct is called the common bile duct and is joined by the pancreatic duct before entering the duodenum.

duodenum and through a side branch of the hepatic duct to the **gallbladder** (Figure 50.15), where it is stored until it is needed to assist in fat digestion. When fat enters the duodenum, a hormonal signal causes the walls of the gallbladder to contract rhythmically, squeezing bile out of the gallbladder and into the hepatic duct. Below the branch to the gallbladder, the hepatic duct is called the *common bile duct*. Bile from the gallbladder flows down the common bile duct to the duodenum.

To understand the role of bile in fat digestion, think of an oil and vinegar salad dressing. The oil, which is hydrophobic, tends to aggregate in large globules. For that reason, many salad dressings include an *emulsifier*—something that prevents oil droplets from aggregating. Mayonnaise, for example, is oil and vinegar with egg yolk added as an emulsifier. Bile emulsifies fats in the chyme, and thereby greatly enlarges the surface area of the fats exposed to the *lipases*—the enzymes that digest fats. One end of each bile molecule is soluble in fat (it is lipophilic, or hydrophobic); the other end is soluble in water (it is hydrophilic, or lipophobic). Bile molecules bury their lipophilic ends in fat droplets, leaving their lipophobic ends sticking out. As a result, they prevent the fat droplets from sticking together. The very small fat particles that result are called *micelles* (Figure 50.16*a*).

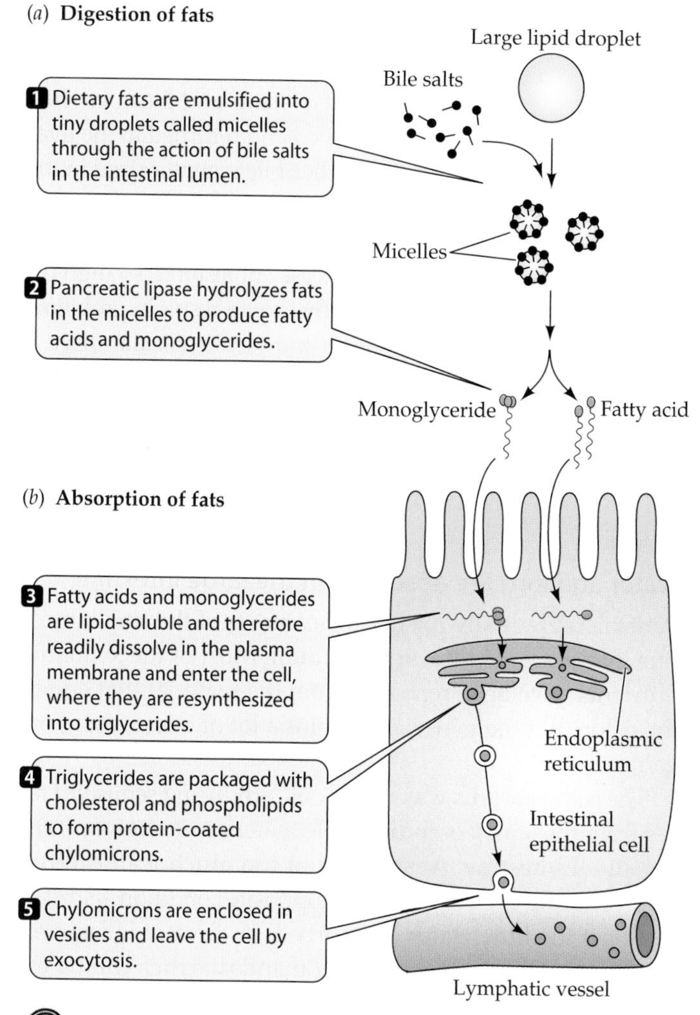

(a) Digestion of fats
Large lipid droplet
Bile salts
1 Dietary fats are emulsified into tiny droplets called micelles through the action of bile salts in the intestinal lumen.
Micelles
2 Pancreatic lipase hydrolyzes fats in the micelles to produce fatty acids and monoglycerides.
Monoglyceride
Fatty acid
(b) Absorption of fats
3 Fatty acids and monoglycerides are lipid-soluble and therefore readily dissolve in the plasma membrane and enter the cell, where they are resynthesized into triglycerides.
Endoplasmic reticulum
4 Triglycerides are packaged with cholesterol and phospholipids to form protein-coated chylomicrons.
Intestinal epithelial cell
5 Chylomicrons are enclosed in vesicles and leave the cell by exocytosis.
Lymphatic vessel

50.16 The Digestion and Absorption of Fats (*a*) Dietary fats are broken up by bile into small micelles that present a large surface area to lipase action. (*b*) The products of fat digestion are absorbed by intestinal mucosal cells, where they are resynthesized into triglycerides and exported to lymphatic vessels.

PANCREAS. The **pancreas** is a large gland that lies just beneath the stomach (see Figures 50.10 and 50.15). It functions as both an endocrine gland (secreting hormones to the blood and tissue fluid; see Chapter 42) and an exocrine gland (secreting other substances through ducts to the outside of the body). Its exocrine products are delivered to the gut through the pancreatic duct, which joins the common bile duct.

The exocrine tissues of the pancreas produce a host of digestive enzymes, including lipases, amylases, and proteases (Table 50.3). As in the stomach, some of these enzymes—most notably the proteases—are released as zymogens; otherwise, they would digest the pancreas and its ducts before they ever reached the duodenum. Once in the duodenum, one of these zymogens, **trypsinogen**, is activated by an enzyme called *enterokinase*, which is produced by cells lining the duodenum. This process is similar to the activation of pepsinogen by low pH in the stomach: Active **trypsin** can cleave other trypsinogen molecules to release even more active trypsin. Similarly, trypsin activates other zymogens secreted by the pancreas.

The mixture of zymogens produced by the pancreas can be very dangerous if the pancreatic duct is blocked or if the pancreas is injured by an infection or a severe blow to the abdomen. A few trypsinogen molecules spontaneously converting to trypsin can initiate a chain reaction of enzyme activity that digests the pancreas in a short time, destroying both its endocrine and exocrine functions.

The pancreas also produces a secretion rich in bicarbonate ions (HCO_3^-). Bicarbonate ions are basic and neutralize the acidic pH of the chyme that enters the duodenum from the stomach. Intestinal enzymes function best at a neutral or slightly alkaline pH.

Nutrients are absorbed in the small intestine

The final digestion of proteins and carbohydrates that produces absorbable products takes place among the microvilli. The mucosal cells with microvilli produce peptidases, which cleave polypeptides into tripeptides, dipeptides, and individual amino acids that the cells can absorb. These cells also produce the enzymes maltase, lactase, and sucrase, which cleave the common disaccharides into their constituent, absorbable monosaccharides: glucose, galactose, and fructose.

50.3 Sources and Functions of the Major Digestive Enzymes of Humans

ENZYME	SOURCE	ACTION
Salivary amylase	Salivary glands	Starch → Maltose
Pepsin	Stomach	Proteins → Peptides; autocatalysis
Pancreatic amylase	Pancreas	Starch → Maltose
Lipase	Pancreas	Fats → Fatty acids and glycerol
Nuclease	Pancreas	Nucleic acids → Nucleotides
Trypsin	Pancreas	Proteins → Peptides; activation of zymogens
Chymotrypsin	Pancreas	Proteins → Peptides
Carboxypeptidase	Pancreas	Peptides → Peptides and amino acids
Aminopeptidase	Small intestine	Peptides → Peptides and amino acids
Dipeptidase	Small intestine	Dipeptides → Amino acids
Enterokinase	Small intestine	Trypsinogen → Trypsin
Nuclease	Small intestine	Nucleic acids → Nucleotides
Maltase	Small intestine	Maltose → Glucose
Lactase	Small intestine	Lactose → Galactose and glucose
Sucrase	Small intestine	Sucrose → Fructose and glucose

Many humans stop producing the enzyme lactase around the age of 4 years and thereafter have difficulty digesting lactose, which is the sugar in milk. Lactose is a disaccharide and cannot be absorbed without being cleaved into its constituent monosaccharides, glucose and galactose. If a substantial amount of lactose remains unabsorbed and passes into the large intestine, its metabolism by bacteria in the large intestine causes abdominal cramps, gas, and diarrhea.

THE ROLE OF SODIUM. The mechanisms by which cells lining the intestine absorb nutrient molecules and inorganic ions are diverse and not completely understood. Many inorganic ions are actively transported by these cells. Transport proteins exist for sodium, calcium, and iron. Transporter proteins also exist for certain classes of amino acids and for glucose and galactose, but their activity is much reduced if active sodium transport is blocked.

Sodium diffuses from the gut contents into the mucosal cells and is then actively transported from the mucosal cells into the submucosa. To diffuse into a mucosal cell, a sodium ion binds to a symport in the mucosal cell plasma membrane. Symport also binds a nutrient molecule, such as glucose or an amino acid. The diffusion of the sodium ion, driven by a concentration gradient, thus drives the absorption of the nutrient molecule. This mechanism is called *sodium cotransport*.

FAT ABSORPTION. The absorption of the products of fat digestion does not involve transporter proteins (see Figure 50.15*b*). Pancreatic lipases break down fats into diglycerides, monoglycerides, and fatty acids, all of which are lipid-soluble and thus able to pass through the plasma membranes of microvilli and diffuse into the intestinal mucosal cells. Once in the cells, the fatty acids and monoglycerides are resynthesized into triglycerides, combined with cholesterol and phospholipids, and coated with protein to form water-soluble **chylomicrons**, which are little particles of fat. Rather than entering the blood directly, the chylomicrons pass into the lymphatic vessels in the submucosa. They then flow through the lymphatic system and enter the bloodstream through the thoracic ducts at the base of the neck. After a meal rich in fats, the chylomicrons can be so abundant in the blood that they give it a milky appearance.

The bile that emulsifies the fats is not absorbed along with the monoglycerides and the fatty acids, but shuttles back and forth between the gut contents and the microvilli. In the ileum, bile is actively resorbed and returned to the liver via the bloodstream. As noted earlier, bile is synthesized in the liver from cholesterol. Cholesterol comes from food, but it is also synthesized by liver cells and intestinal cells. The body has no way of breaking down excess cholesterol, so high dietary intake or high levels of synthesis can result in problems such as arterial plaque formation and cardiovascular disease (see Figure 49.14). One major way in which cholesterol leaves the body is through the elimination of unresorbed bile in the feces. Certain kinds of fiber bind bile, decreasing its resorption in the ileum and thus helping to lower cholesterol levels.

Water and ions are absorbed in the large intestine

Peristalsis gradually pushes the contents of the small intestine into the large intestine, or **colon**. Most of the available nutrients have been removed from the material that enters the colon, but the material contains a lot of water and inorganic ions.

The colon absorbs water and ions, producing semisolid feces from the slurry of indigestible materials it receives from the small intestine. Absorption of too much water in the colon can cause *constipation*. The opposite condition, *diarrhea*, results if too little water is absorbed, or if water is secreted into the colon. (Both constipation and diarrhea can be induced by toxins from certain microorganisms.) Feces are stored in the last segment of the colon until they are excreted.

Immense populations of bacteria live within the human colon. One of the resident species is *Escherichia coli*, a bacterium popular among researchers in biochemistry, genetics, and molecular biology. *E. coli* lives on matter that is indigestible to humans and produces some products—such as vitamin K and biotin—that are useful to its host. Excessive or prolonged intake of antibiotics can lead to vitamin deficiency because the antibiotics kill the normal intestinal bacteria at the same time they are killing the disease-causing organisms for which they are intended.

Intestinal bacteria produce gases such as methane and hydrogen sulfide as by-products of their largely anaerobic metabolism. Humans expel gas after eating beans because the beans contain certain carbohydrates that the bacteria can break down, but their human hosts cannot. A large percentage of the mass of feces consists of the cell walls of dead intestinal bacteria.

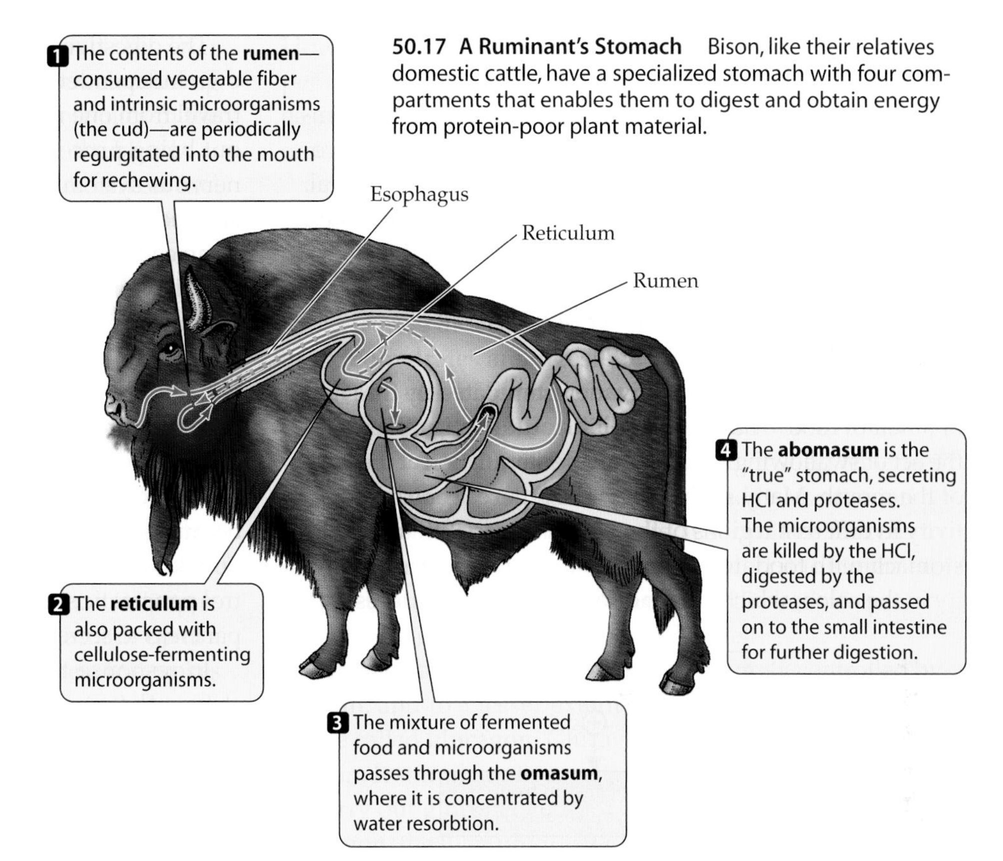

1 The contents of the **rumen**—consumed vegetable fiber and intrinsic microorganisms (the cud)—are periodically regurgitated into the mouth for rechewing.
Esophagus
Reticulum
Rumen
2 The **reticulum** is also packed with cellulose-fermenting microorganisms.
3 The mixture of fermented food and microorganisms passes through the **omasum**, where it is concentrated by water resorbtion.
4 The **abomasum** is the "true" stomach, secreting HCl and proteases. The microorganisms are killed by the HCl, digested by the proteases, and passed on to the small intestine for further digestion.

50.17 A Ruminant's Stomach Bison, like their relatives domestic cattle, have a specialized stomach with four compartments that enables them to digest and obtain energy from protein-poor plant material.

Herbivores have special adaptations for digesting cellulose

Cellulose is the principal organic compound in the diets of herbivores. Most herbivores, however, cannot produce *cellulases*, the enzymes that hydrolyze cellulose. Exceptions include silverfish (insects well known for eating books and stored papers), earthworms, and shipworms. Other herbivores, from termites to cattle, rely on microorganisms living in their digestive tracts to digest cellulose for them.

The digestive tracts of **ruminants** (cud chewers) such as cattle, goats, and sheep are specialized to maximize the benefits of their endosymbiotic microorganisms. In place of the usual mammalian stomach, ruminants have a large, four-chambered organ (Figure 50.17). The first two chambers, the *rumen* and the *reticulum*, are packed with anaerobic microorganisms that break down cellulose by fermentation. The ruminant periodically regurgitates the contents of the rumen (the *cud*) into the mouth for rechewing. When the more thoroughly ground-up vegetable fibers are swallowed again, they present more surface area to the microorganisms for their digestive actions.

The microorganisms in the rumen and reticulum metabolize cellulose and other nutrients to simple fatty acids, which become nutrients for their host. In addition, the microorganisms themselves provide an important source of protein for the host. A cow can derive more than 100 grams of protein per day from digestion of its endosymbiotic microorganisms.

The food leaving the rumen carries with it enormous numbers of cellulose-fermenting microorganisms. This mixture passes through the *omasum*, where it is concentrated by water absorption. It then enters the true stomach, the *abomasum*, which secretes hydrochloric acid and proteases. The microorganisms are killed by the acid, digested by the proteases, and passed on to the small intestine for further digestion and absorption. The rate of multiplication of microorganisms in the rumen is great enough to offset their loss, so a well-balanced, mutually beneficial relationship is maintained.

Some mammalian herbivores other than ruminants have microbial fermentation chambers in a branch off the large intestine, called the **cecum**. Rabbits and hares are good examples (see Figure 50.8). Since the cecum empties into the large intestine, the absorption of the nutrients produced by the microorganisms is inefficient and incomplete. Therefore, some of these animals re-ingest some of their own feces, a behavior known as *coprophagy*. Coprophagous species usually produce two kinds of feces, one consisting of pure waste (which they discard), and one consisting mostly of cecal material, which they reingest directly from the anus. As this cecal material passes through the stomach and small intestine, the nutrients it contains are digested and absorbed. In humans, the cecum has become the vestigial *appendix*, which serves no digestive function.

of the cells of the body preferentially use fatty acids as their metabolic fuel. One tissue that does not switch fuel sources during the postabsorptive period is the nervous system.

The cells of the nervous system require a constant supply of glucose, and they do not require the action of insulin to enable them to take up glucose from the blood. The nervous system can use other fuels only to a very limited extent. The overall dependence of neuronal tissue on glucose is the reason it is so important for other cells of the body to shift to fat metabolism during the postabsorptive period. This shift preserves the available glucose and glycogen stores for the nervous system for as long as possible.

The traffic of fuel molecules during the absorptive and postabsorptive periods is summarized in Figure 50.20, which indicates the steps controlled by insulin and glucagon.

The Regulation of Food Intake

Obesity is a major health issue in the United States. People spend billions of dollars every year on schemes to lose weight, but the problem still increases. A simple rule—take in fewer calories than your body burns, but maintain a balanced diet—should solve the problem, but it doesn't. Why? As we noted at the beginning of this chapter, social and lifestyle factors play a major role in obesity, but these factors play out against a genetic and regulatory background.

50.20 Fuel Molecule Traffic during the Absorptive and Postabsorptive Periods Insulin promotes glucose uptake by liver, muscle, and fat cells during the absorptive period. During the postabsorptive period, the lack of insulin blocks glucose uptake by these same tissues and promotes fat and glycogen breakdown to supply metabolic fuel.

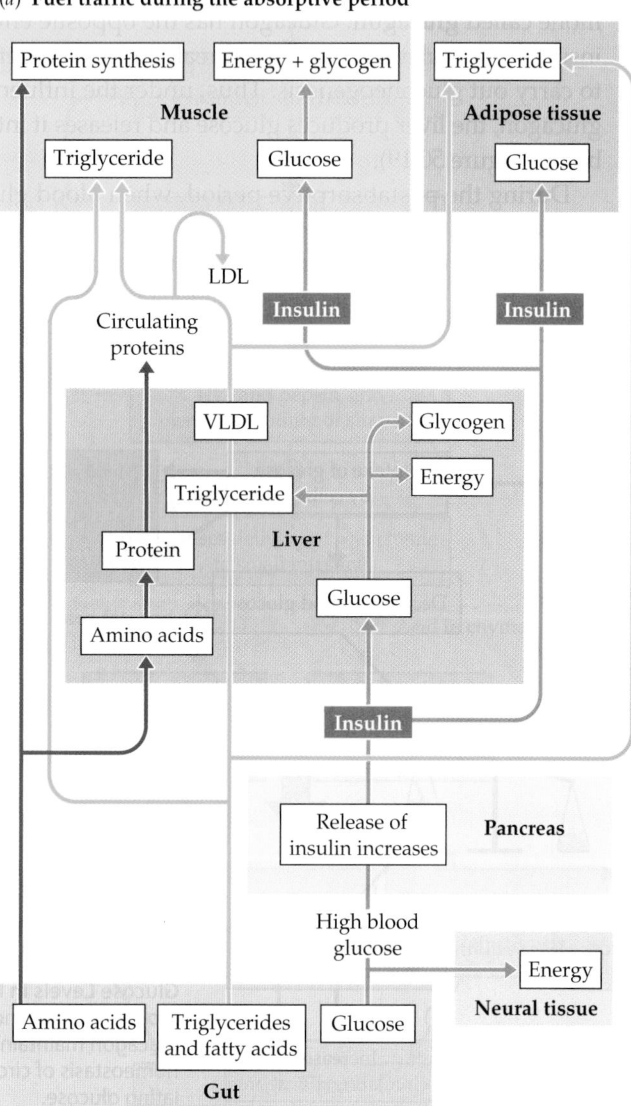

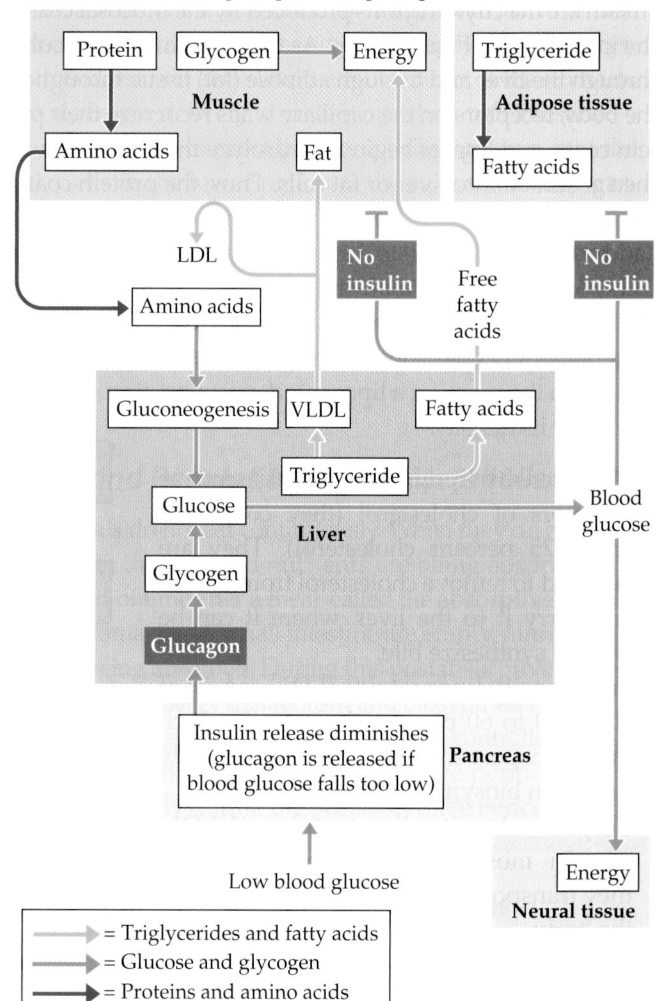

50.21 A Single-Gene Mutation Leads to Obesity in Mice Leptin serves as a negative feedback signal to the brain to limit food intake. The fat cells of the *ob/ob* mouse on the left do not produce leptin. The wild-type mouse on the right does produce leptin and did not become obese when kept under the same conditions as the *ob/ob* mouse.

The amount of food an animal eats is governed by its sensations of hunger and satiety. These sensations are influenced by a region of the brain called the hypothalamus. If a region in the middle of the hypothalamus of rats, called the *ventromedial hypothalamus*, is damaged, the animals will increase their food intake and become obese. If a different region of the hypothalamus, called the *lateral hypothalamus*, is damaged, rats will decrease their food intake and become thin. In both cases, the rats eventually reach a new equilibrium body weight, which they maintain. Thus, regulation is maintained, but the set point has been changed. Other brain regions have also been implicated in control of hunger and satiety.

In Chapter 41, we learned that regulation involves feedback information and a means of comparing that information with a set point. There is some evidence that cells in the hypothalamus and in the liver are sensitive to the levels of glucose and insulin in the blood, with high levels stimulating satiety and low levels stimulating hunger. There is even stronger evidence, however, that signals from fat metabolism influence hunger and satiety.

A single-gene mutation in mice, when present in the homozygous condition, results in mice that eat enormous amounts of food and become obese (Figure 50.21). Using genetics terms, these mice are called *ob/ob* mice, due to their double dose of the recessive "obese" allele. The wild-type *ob* allele codes for a protein that has been named **leptin** (from the Greek *leptos*, "thin"). When leptin was injected into *ob/ob* mice, they ate less and lost body fat. Leptin is produced by fat cells and circulates in the blood. Receptors for leptin are found in the regions of the hypothalamus that are involved in control of hunger and satiety. It seems that leptin provides feedback information about the status of the body fat reserves to the brain.

Could leptin be used to reduce human obesity? A very few obese people do not produce the hormone leptin, and injections of leptin can curb their appetites and enable them to lose body mass. Most obese people, however, have higher than normal circulating levels of leptin. It is likely that they have leptin receptors with reduced sensitivity. Understanding the actions of leptin in normal and obese individuals might provide a partial answer as to why some individuals find it easier than others to avoid excessive food intake and increases in body fat.

Additional feedback signals are most certainly involved in the regulation of food intake. The most recent one to be discovered is a hormone called *ghrelin*, which is produced and secreted by cells in the stomach. Normally, ghrelin levels rise before a meal and fall after a meal. Fasting causes an increase in ghrelin levels. Ghrelin binding to its receptors in the hypothalamus stimulates appetite. Ghrelin also stimulates cells in the pituitary gland to release growth hormone.

Toxic Compounds in Food

Plant and animal tissues contain nutrients, but as we have seen in other chapters of this book, they can also contain toxic compounds. Some plants produce toxic secondary metabolites as defenses against herbivores. One example is the nicotine in tobacco. Animals may use toxins for capturing prey as well as for self-defense (Figure 50.22). Ingesting many plant and animal tissues, therefore, can be dangerous. Human activities add millions of tons of synthetic toxic com-

50.22 Naturally Toxic The Japanese puffer fish is a delicacy in sushi restaurants, but only highly skilled chefs can prepare the fish so that its natural and highly potent neurotoxins do not endanger diners.

pounds to our environment every year, and many of these compounds enter the air we breathe and the water we drink, as well as the food we eat. A whole new field, called *environmental toxicology*, has developed to address the problems of poisons in the environment.

Some toxins are retained and concentrated in organisms

The physical and chemical properties of a toxic compound affect its retention within a biological system. If a compound can dissolve in water, it may be quickly metabolized (and thus detoxified) because it is easily accessible to the wide variety of enzymes that can break down complex molecules in food. In addition to being broken down or metabolized, many water-soluble compounds can be filtered out of the blood by the kidneys, and therefore do not accumulate in the body. However, some potentially dangerous water-soluble compounds can be incorporated into the body and disrupt normal functions. An example is lead, which can replace iron in blood and calcium in bone.

Lipid-soluble compounds are usually metabolized more slowly than water-soluble compounds, and they are often stored in the body for a long time because they dissolve in adipose tissues. Lipid-soluble compounds can accumulate in the body and reach very high concentrations. Some lipid-soluble toxins, including many pesticides, can *bioaccumulate* in the environment; that is, they can become more and more concentrated in predators that eat contaminated prey. The pesticide load is passed up the food chain from prey to predator, growing increasingly concentrated in the tissues of each consumer in turn. In the top predator, the pesticide may be concentrated thousands or millions of times. Long-lived predators, such as eagles and bears, are particularly at risk for heavy pesticide burdens because they have many years to accumulate them. Bioaccumulated toxins may be responsible for the high rates of cancer and infertility found in some wildlife populations.

The body cannot metabolize many synthetic toxins

How does the animal body handle synthetic toxins? In many cases, the systems that metabolize natural chemicals can also metabolize synthetic toxins, breaking them apart and eliminating them through the urine. Liver enzymes called *cytochrome P450s* are responsible for much of this detoxification. P450s are less specific in their abilities to bind substrates than are most enzymes; thus, each P450 can catalyze reactions with a wide range of compounds, and there are many P450s. Few natural compounds can escape the P450s, even when the body encounters them for the first time.

Some synthetic chemicals, however, fall outside the range of structures that P450s and other enzymes can metabolize. If a synthetic chemical that cannot be metabolized is structurally similar to a hormone, that synthetic chemical may activate the hormonal signaling pathway within target cells. Whereas the natural hormonal signal can be turned off, the synthetic signal cannot be, and control of function is lost.

One example of a class of toxic synthetic chemicals that has bioaccumulated in animals, including humans, is the polychlorinated biphenyls (PCBs). PCBs were produced extensively for use as an insulating fluid in electrical transformers from the 1930s until recently. They are chemically stable, lipophilic, and are now found throughout the environment. They have been shown to bioaccumulate, reaching dangerously high levels in fish from contaminated waters such as the Great Lakes. In communities around the Great Lakes, studies have indicated cognitive impairment in children of mothers with a high body burden of PCBs, probably from eating fish caught in the Great Lakes. The risks of PCBs are now clear, but it is usually difficult to make a causal connection between a toxin in the environment and specific health effects in a population. Environmental toxicologists must be able to study large populations, use powerful statistics, and do controlled laboratory studies to obtain evidence that will support policy changes to stop and reverse the effects of synthetic environmental toxins.

Chapter Summary

Nutrient Requirements

▶ Animals are heterotrophs that derive their energy and molecular building blocks, directly or indirectly, from autotrophs.

▶ Carbohydrates, fats, and proteins in food supply animals with metabolic energy. A measure of the energy content of food is the calorie. Excess caloric intake is stored as glycogen and fat. **Review Figure 50.2**

▶ An animal with insufficient caloric intake is undernourished and must metabolize its stored glycogen and fat, and finally its own proteins, for energy. In humans, overnutrition can also be a serious health hazard. **Review Figure 50.3**

▶ For many animals, food provides essential carbon skeletons that they cannot synthesize themselves. **Review Figure 50.4**

▶ Humans require eight essential amino acids in the diet. All are available in milk, eggs, or meat, but not in all vegetables. Thus, vegetarians must eat a mix of complementary foods. **Review Figure 50.5**

▶ Different animals need mineral elements in different amounts. Macronutrients, such as calcium, are needed in large quantities. Micronutrients, such as iron, are needed in small amounts. **Review Table 50.1.** See Web/CD Activity 50.1

▶ Vitamins are organic molecules that must be obtained in food. **Review Table 50.2.** See Web/CD Activity 50.2

▶ Malnutrition results when any essential nutrient is lacking from the diet. A chronic state of malnutrition causes a deficiency disease.

Adaptations for Feeding

▶ Animals can be characterized by how they acquire nutrients: Saprotrophs and detritivores depend on dead organic matter, filter feeders strain the aquatic environment for small food items, herbivores eat plants, and carnivores eat animals.

▶ Behavioral and anatomical adaptations reflect feeding strategies. In vertebrates, teeth have evolved to match the diet. **Review Figure 50.7.** **See Web/CD Activity 50.3**

Digestion

▶ Digestion involves the breakdown of complex food molecules into monomers that can be absorbed and utilized by cells. In most animals, digestion is extracellular and external to the body, taking place in a tubular gut that has different regions specialized for different digestive functions. **Review Figure 50.8**

▶ Absorptive areas of the gut are characterized by a large surface area. **Review Figure 50.9**

▶ Hydrolytic enzymes break down proteins, carbohydrates, and fats into their monomeric units. To prevent the organism itself from being digested, these enzymes are released as inactive zymogens, which become activated when secreted into the gut.

Structure and Function of the Vertebrate Gut

▶ The vertebrate gut can be divided into several compartments with different functions. **Review Figure 50.10.** **See Web/CD Activity 50.4**

▶ The cells and tissues of the vertebrate gut are organized in the same way throughout its length. The innermost tissue layer, the mucosa, is the secretory and absorptive surface. The submucosa contains secretory cells and glands, blood and lymph vessels and nerves. External to the submucosa are two smooth muscle layers (circular and longitudinal) that move food through the gut. Between the two muscle layers is a nerve network that controls the movements of the gut. **Review Figure 50.11**

▶ Swallowing is a reflex that pushes food into the esophagus. Peristalsis moves food from the beginning of the esophagus through the entire length of the gut. Sphincters block the gut at certain locations, but they relax as a wave of peristalsis approaches. **Review Figure 50.12**

▶ Enzymatic digestion begins in the mouth, where amylase is secreted with the saliva. Protein digestion begins in the stomach, where pepsin and HCl are secreted by the stomach mucosa. The mucosa also secretes mucus, which protects the tissues of the gut. **Review Figures 50.13, 50.14**

▶ In the duodenum, pancreatic enzymes carry out most of the digestion of food. Bile from the liver and gallbladder assists in the digestion of fats by breaking them into micelles. Bicarbonate ions from the pancreas neutralize the pH of the chyme entering from the stomach to produce an environment conducive to the actions of pancreatic enzymes. **Review Figure 50.15, Table 50.3**

▶ Final enzymatic cleavage of polypeptides and disaccharides occurs among the microvilli of the intestinal mucosa. Amino acids, monosaccharides, and many inorganic ions are absorbed by the microvilli. In many cases, specific transporter proteins in the plasma membranes of the mucosal cells transport nutrients into the cells. Sodium cotransport is a common mechanism for actively absorbing nutrient molecules and ions.

▶ Fats are broken down by lipases and absorbed mostly as monoglycerides and fatty acids. These products pass through the membranes of mucosal cells and are then resynthesized into triglycerides within the cells. The triglycerides are combined with cholesterol and phospholipids and coated with protein to form chylomicrons, which pass out of the mucosal cells and into lymphatic vessels in the submucosa. **Review Figure 50.16*b*.** **See Web/CD Tutorial 50.1**

▶ Water and ions are absorbed in the large intestine as waste matter is consolidated into feces, which are periodically excreted.

▶ In herbivores such as ruminants and rabbits, some compartments of the gut have large populations of microorganisms that aid in digesting materials that otherwise would be indigestible to their host. **Review Figure 50.17**

Control and Regulation of Digestion

▶ Autonomic reflexes coordinate activity in different regions of the digestive tract, which has an intrinsic nervous system that can act independently of the CNS.

▶ The actions of the stomach and small intestine are largely controlled by the hormones gastrin, secretin, and cholecystokinin. **Review Figure 50.18**

Control and Regulation of Fuel Metabolism

▶ The liver plays a central role in directing the traffic of fuel molecules. During the absorptive period, the liver takes up and stores fats and carbohydrates, converting monosaccharides to glycogen or fats. The liver also takes up amino acids and uses them to produce blood plasma proteins.

▶ Fat and cholesterol are shipped out of the liver as low-density lipoproteins. High-density lipoproteins act as acceptors of cholesterol and are believed to bring fat and cholesterol back to the liver.

▶ Fuel metabolism during the absorptive period is controlled largely by insulin, which promotes glucose uptake and utilization by most cells of the body, as well as glycogen and fat synthesis. During the postabsorptive period, the lack of insulin blocks the uptake and utilization of glucose by most cells of the body except neurons. If blood glucose levels fall, glucagon is secreted, stimulating the liver to break down glycogen and release glucose to the blood. **Review Figures 50.17, 50.19.** **See Web/CD Tutorial 50.2**

The Regulation of Food Intake

▶ Food intake is governed by sensations of hunger and satiety, which are determined by brain mechanisms.

▶ Leptin is a hormone produced by fat cells that inhibits food intake, apparently by providing feedback information about fat reserves to the brain.

Toxic Compounds in Food

▶ Natural plant and animal foods can contain toxic compounds in addition to nutrients. Human activities such as the use of pesticides and the release of pollutants into the environment have made the problem of toxins in food even worse.

▶ An organism can accumulate toxic compounds in its body, especially if those compounds are lipid-soluble or take the structural place of a natural molecule.

▶ Toxins such as PCBs that accumulate in the bodies of prey are transferred to and further concentrated in the bodies of their predators. This bioaccumulation produces high concentrations of toxins in animals high up the food chain.

Self-Quiz

1. Most of the metabolic energy that a bird requires for a long-distance migratory flight is stored as
   a. glycogen.
   b. fat.
   c. protein.
   d. carbohydrates.
   e. ATP.

2. Which statement about essential amino acids is true?
   a. They are not found in vegetarian diets.
   b. They are stored by the body for the times when they are needed.
   c. Without them, one is undernourished.
   d. All animals require the same ones.
   e. Humans can acquire all of theirs by eating milk, eggs, and meat.
3. Which statement about vitamins is true?
   a. They are essential inorganic nutrients.
   b. They are required in larger amounts than are essential amino acids.
   c. Many serve as coenzymes.
   d. Vitamin D can be acquired only by eating meat or dairy foods.
   e. When vitamin C is eaten in large quantities, the excess is stored in fat for later use.
4. The digestive enzymes of the small intestine
   a. do not function best at a low pH.
   b. are produced and released in response to circulating secretin.
   c. are produced and released under neuronal control.
   d. are all secreted by the pancreas.
   e. are all activated by an acidic environment.
5. Which statement about nutrient absorption by the intestinal mucosal cells is true?
   a. Carbohydrates are absorbed as disaccharides.
   b. Fats are absorbed as fatty acids and monoglycerides.
   c. Amino acids move across the plasma membrane only by diffusion.
   d. Bile transports fats across the plasma membrane.
   e. Most nutrients are absorbed in the duodenum.
6. Chylomicrons are like the tiny micelles of dietary fat in the lumen of the small intestine in that both
   a. are coated with bile.
   b. are lipid-soluble.
   c. travel through the lymphatic system.
   d. contain triglycerides.
   e. are coated with lipoproteins.
7. Microbial fermentation in the gut of a cow
   a. produces fatty acids as a major nutrient for the cow.
   b. occurs in specialized regions of the small intestine.
   c. occurs in the cecum, from which food is regurgitated, chewed again, and swallowed into the true stomach.
   d. produces methane as a major nutrient.
   e. is possible because the stomach wall does not secrete hydrochloric acid.
8. Which of the following is stimulated by cholecystokinin?
   a. Stomach motility
   b. Release of bile
   c. Secretion of hydrochloric acid
   d. Secretion of bicarbonate ions
   e. Secretion of mucus
9. During the absorptive period,
   a. breakdown of glycogen supplies glucose to the blood.
   b. glucagon secretion is high.
   c. the number of circulating lipoproteins is low.
   d. glucose is the major metabolic fuel.
   e. the synthesis of fats and glycogen in muscle is inhibited.
10. During the postabsorptive period,
   a. glucose is the major metabolic fuel.
   b. glucagon stimulates the liver to produce glycogen.
   c. insulin facilitates the uptake of glucose by brain cells.
   d. fatty acids constitute the major metabolic fuel.
   e. liver functions slow down because of low insulin levels.

For Discussion

1. Several currently popular diet books recommend high fat and protein intake and low carbohydrate intake as a means of losing body mass. What could the rationale of a high-fat and high-protein diet be, and what health issues should be considered when someone considers going on such a diet?
2. Carnivores generally have more dietary vitamin requirements than herbivores do. Why?
3. It is said that the most important hormonal control of fuel metabolism in the postabsorptive period is the lack of insulin. Explain.
4. Why is obstruction of the common bile duct so serious? Consider in your answer the multiple functions of the pancreas and the way in which digestive enzymes are processed.
5. Trace the history of a fatty acid molecule from a piece of buttered toast to a plaque on a coronary artery. What possible forms and structures could it have passed through in the body? Describe a direct and an indirect route it could have taken.

51 Salt and Water Balance and Nitrogen Excretion

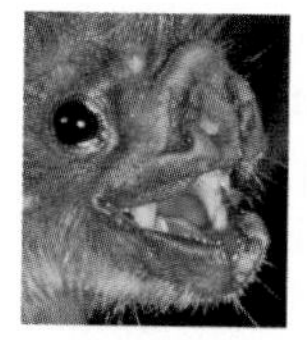

Blood, sweat, and tears taste salty because they reflect the composition of the tissue fluid that bathes the cells of the body. The volume and the composition of the tissue fluid must remain within certain limits, and it must be kept relatively free of wastes. Maintaining homeostasis of the tissue fluid can be challenging. Consider the problems of vampires—not the horror film kind, but the bat kind.

Vampire bats are small, tropical mammals that feed on the blood of other mammals, such as cattle. The bat lands on an unsuspecting (usually sleeping) victim, bites into a vein, and drinks blood—a high-protein, liquid food. The bat has only a short time to feed before the victim wakes and shakes it off. To maximize the volume of blood it can ingest, it eliminates water from its food as fast as it can by producing a lot of very dilute urine. The warm trickle down the neck of the victim is not blood!

Once feeding ends, however, this high rate of water loss cannot continue. Now the vampire bat is digesting protein and must excrete large amounts of nitrogenous breakdown products while conserving its body water. Within minutes, the excretory system of the vampire bat switches from producing lots of very dilute urine to producing a tiny amount of highly concentrated urine. The vampire bat rapidly switches from an excretory physiology typical of a mammal living in an environment with abundant fresh water (copious amounts of dilute urine) to an excretory physiology typical of a mammal living in an arid desert (small amounts of concentrated urine).

In this chapter, we will discover how animals accomplish the various feats of salt and water balance and excretion of wastes that adapt them to many different environments. We will begin by discussing the challenges presented by those different environments Then we will use some invertebrate examples to illustrate the basic mechanisms used in the excretory systems of all animals. Turning to vertebrates, we will see that the basic anatomical unit of the excretory system is the nephron. The earliest nephron probably evolved as a structure that enabled vertebrate ancestors living in fresh water to excrete water while conserving salts. As vertebrates evolved to inhabit marine and terrestrial habitats, their excretory systems had to conserve water. We will see how the nephron evolved in mammals to perform this function. Finally, we will present the mechanisms that control and regulate salt and water balance in mammals, giving the vampire bat and other species their remarkable abilities to exploit unusual diets and extreme environments.

Blood as a Fast Food The vampire bat (*Desmodus rotundus*) is able to adjust its excretory physiology from water-excreting to water-conserving, depending on whether it is ingesting or digesting its blood meal.

Tissue Fluid and Water Balance

Life evolved in the seas, and seawater is the extracellular environment for the cells of the simplest marine animals. More complex marine animals have an internal environment consisting of *tissue fluid* (extracellular fluid), which is isolated from seawater but is similar to it in composition and solute concentration. Most marine and all freshwater vertebrates and terrestrial animals maintain tissue fluids whose concentration and composition differ considerably from that of seawater. The concentration of the tissue fluid determines the water balance of the cells, and the composition of the tissue fluid influences the health and functions of the cells it bathes, as we saw in Chapter 41. Recall, for example, the importance of ionic concentration gradients between the tissue fluid and the cytoplasm of nerve and muscle cells (see Chapter 44).

To understand what is meant by *water balance*, remember that cell plasma membranes are permeable to water and that the movement of water across membranes depends on differences in solute concentration. (You may wish to review the discussion of osmosis in Chapter 5.) Recall the principle you learned in Chapter 36, in the discussion of the movement of water in plants: The greater the solute concentration of a solution, the greater the tendency of water to move into it from another solution of lower solute concentration. Likewise, if the solute concentration of the tissue fluid surrounding animal cells is less than that of the cytoplasm, water moves into the cells, causing them to swell and possibly burst. If the solute concentration of the tissue fluid is greater than that of the cytoplasm, the cells lose water and shrink. The solute concentration of the tissue fluid determines both the volume and the solute concentration of the cells themselves.

Animal physiologists use the term **osmolarity** in discussing osmotic phenomena. The osmolarity of a solution is the moles of osmotically active particles per liter of solvent. Thus, a 1 molar solution of glucose is also a 1 osmolar solution, but a 1 molar solution of sodium chloride (NaCl) is a 2 osmolar solution, because each NaCl molecule dissociates into two osmotically active particles.

To achieve water balance, animals must be able to maintain the osmolarity of their tissue fluid within an appropriate range. In addition, they must maintain an appropriate solute composition of the tissue fluid, saving some substances and eliminating others. To accommodate these needs, most animals have excretory organs.

Excretory organs control the osmolarity of tissue fluid by filtration, secretion, and resorption

Excretory organs control the osmolarity and the volume of blood and tissue fluid by excreting solutes that are present in excess (such as NaCl when we eat lots of salty food) and conserving solutes that are valuable or in short supply (such as glucose and amino acids). In terrestrial organisms, these excretory organs also eliminate the waste products of nitrogen metabolism. The output of the excretory organs is called **urine**.

The same basic principles apply to the excretory organs of many species. In some way, the excretory organ filters tissue fluid to produce a filtrate that contains no cells or large molecules, such as proteins. The composition of this filtrate is then modified to produce urine. The filtration process is usually carried out on blood plasma driven across capillary walls in the excretory organ by blood pressure. Water and small molecular weight solutes cross the capillary wall, but large molecular weight solutes and cells remain in the blood. The filtrate (water and small molecules) then enters a system of tubules. The cells of the tubules change the composition of the filtrate by active secretion and resorption of specific solute molecules. We will see that these three mechanisms—filtration, secretion, and resorption—are used in the excretory systems of a wide variety of animals.

The functions of the excretory organs of a species, and therefore the composition of its urine, depend on the environment in which it lives. We will examine the different excretory systems that maintain salt and water balance and eliminate nitrogen in marine, freshwater, and terrestrial habitats. As we shall see, in spite of the evolutionary diversity of the anatomical and physiological details, all these systems obey a common rule: *There is no active transport of water*. Water must be moved either by pressure or by a difference in osmolarity.

Environments and animals can be classified in terms of salts and water

The salt concentration, or osmolarity, of ocean water is about 1,070 milliosmoles/liter (mosm/l), and fresh water is generally between 1 and 10 mosm/l. Aquatic environments grade continuously from fresh to extremely salty. Consider a place where a river enters the sea through a bay or a marsh. Aquatic environments within that bay or marsh range in osmolarity from that of the fresh water of the river to that of the open sea. Evaporating tide pools can reach an even greater osmolarity than seawater. Animals live in all these environments. Some aquatic species are **osmoconformers**; they allow the osmolarity of their tissue fluid to equilibrate with their environment. In contrast, **osmoregulators** maintain the osmolarity of their tissue fluid at a constant level as the environment changes.

OSMOCONFORMERS. Over a wide range of environmental osmolarities, marine invertebrates simply equilibrate the osmolarity of their tissue fluid with that of the environment. There are limits to osmoconformity, however. No animal could survive if its tissue fluid had the osmolarity of fresh water; nor could animals survive with internal osmolarities as high as those that may be reached in an evaporating tide pool. Such solute concentrations cause proteins to denature.

OSMOREGULATORS. All animals have solutes in their tissue fluids. Therefore, in fresh water, osmosis causes water to invade their bodies, so osmoregulation is essential. *To osmoregulate in fresh water, animals must excrete water and conserve solutes;* hence they produce large amounts of dilute urine.

In salt water, the opposite problem exists: For animals that maintain the osmolarity of their tissue fluids below that of the sea, osmosis causes a loss of water. *To osmoregulate in salt water, animals must conserve water and excrete salts;* thus they tend to produce small amounts of urine, and they have means of excreting salts.

Even animals that osmoconform over a wide range of environmental osmolarities must osmoregulate in extreme environments. The brine shrimp *Artemia* (Figure 51.1) can live in environments of almost any salinity. *Artemia* are found in huge numbers in the most saline environments known, such as Great Salt Lake in Utah and coastal evaporation ponds where salt is concentrated for commercial purposes (see Figure 27.20). The osmolarity of such water reaches 2,500 mosm/l. At these high environmental osmolarities, *Artemia* is capable of maintaining its tissue fluid osmolarity considerably below that of the environment, and therefore acts as a *hypoosmotic regulator*. Very few organisms can survive in the crystallizing brine in which *Artemia* thrives. The main mechanism this small crustacean uses for hypoosmoregulation is the active transport of Cl^- from its tissue fluid out across its gill membranes to the environment. Na^+ ions follow.

Artemia cannot survive in fresh water, but it can live in dilute seawater, in which it maintains the osmolarity of its tissue fluid above that of the environment. Under these conditions, *Artemia* behaves as a *hyperosmotic regulator;* that is, it maintains the osmolarity of its tissue fluid above the osmolarity of the environment. It achieves this by reversing the direction of Cl^- ion transport across its gill membranes.

IONIC CONFORMERS AND REGULATORS. Osmoconformers can also be **ionic conformers**, allowing the ionic composition, as well as the osmolarity, of their tissue fluid to match that of the environment. Most osmoconformers, however, are **ionic regulators** to some degree: They employ active transport mechanisms to maintain specific ions in their tissue fluid at concentrations different from those in the environment.

The terrestrial environment presents problems of osmoregulation and ionic regulation that are entirely different from those faced by aquatic organisms. Because the terrestrial environment is extremely desiccating (drying), most terrestrial animals must conserve water. (Exceptions are animals such as muskrats and beavers that spend most of their time in fresh water.)

Terrestrial animals obtain their salts from food. But plants generally have low concentrations of sodium, so most herbivores must conserve sodium ions or obtain them elsewhere. Some terrestrial herbivores travel long distances to naturally occurring salt licks. By contrast, birds that feed on marine animals must excrete the excess of sodium they ingest with their food. Their *nasal salt glands* excrete a concentrated solution of NaCl via a duct that empties into the nasal cavity. Birds, such as penguins and seagulls, that have nasal salt glands can be seen frequently sneezing or shaking their heads to get rid of the very salty droplets excreted from their nasal salt glands (Figure 51.2).

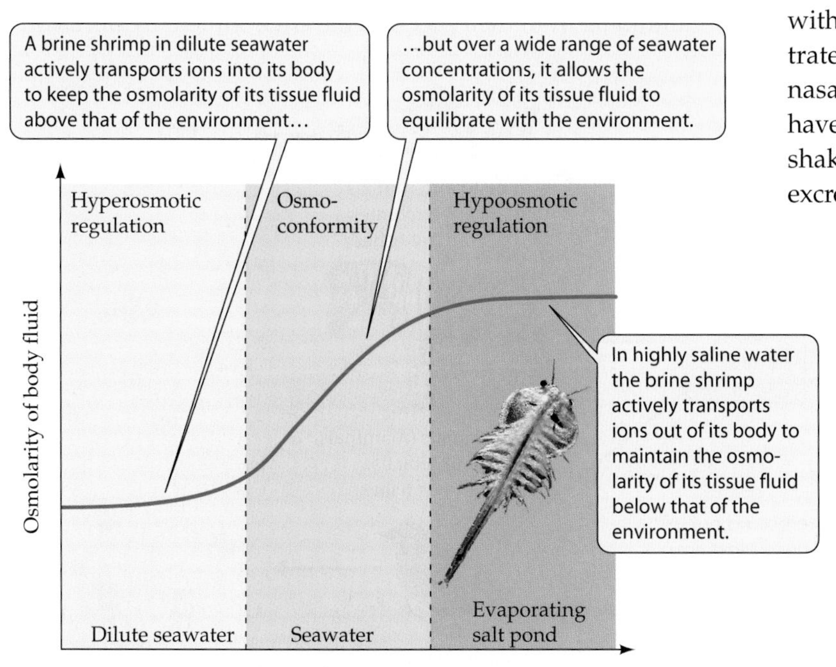
A brine shrimp in dilute seawater actively transports ions into its body to keep the osmolarity of its tissue fluid above that of the environment...
...but over a wide range of seawater concentrations, it allows the osmolarity of its tissue fluid to equilibrate with the environment.
Hyperosmotic regulation
Osmo-conformity
Hypoosmotic regulation
In highly saline water the brine shrimp actively transports ions out of its body to maintain the osmolarity of its tissue fluid below that of the environment.
Osmolarity of body fluid
Dilute seawater
Seawater
Evaporating salt pond
Osmolarity of environment

51.1 Environments Can Vary Greatly in Salt Concentration Animals such as the brine shrimp that live at the extremes of environmental osmolarities display flexible osmoregulatory abilities. They become hyperosmotic regulators in very dilute water and hypoosmotic regulators in very saline water.

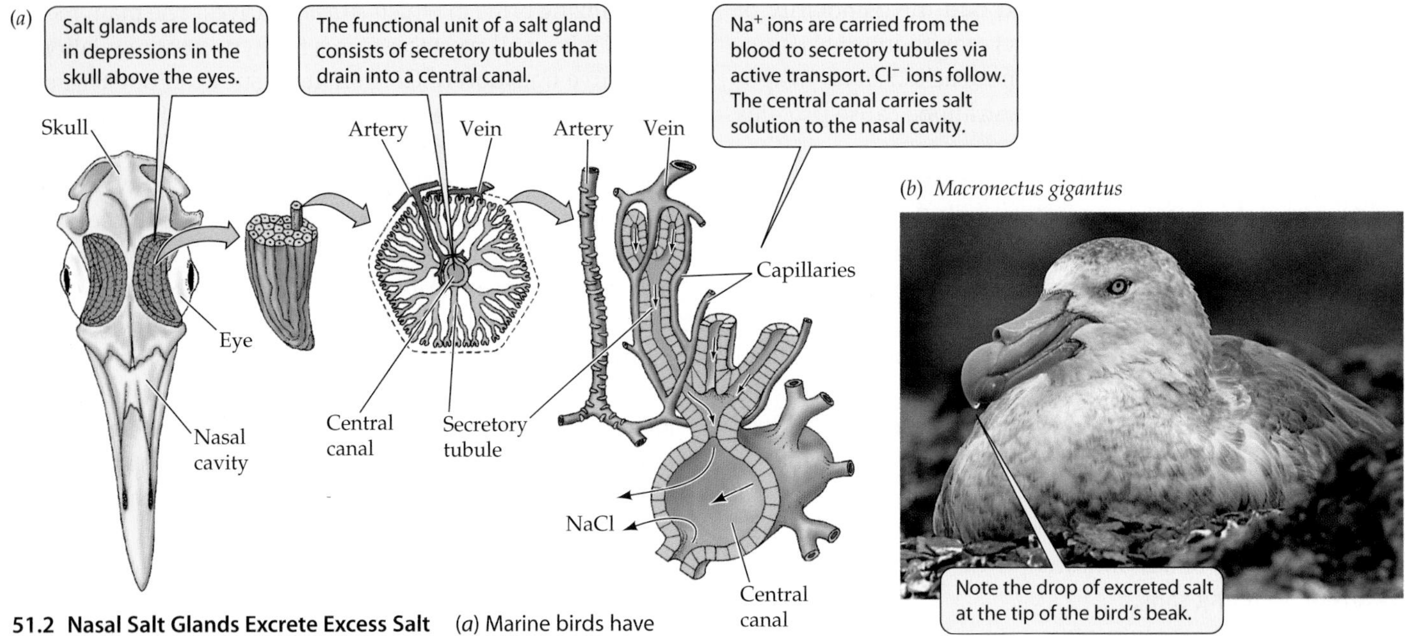

51.2 Nasal Salt Glands Excrete Excess Salt (*a*) Marine birds have nasal salt glands adapted to excrete the excess salt from the seawater they consume with their food. (*b*) This giant petrel has returned from a feeding trip at sea and is excreting salt through its nasal salt gland.

Excreting Nitrogenous Wastes

In addition to maintaining salt and water balance, animals must eliminate the waste products of metabolism from their tissue fluid. The end products of the metabolism of carbohydrates and fats are water and carbon dioxide, which are not difficult to eliminate. Proteins and nucleic acids, however, contain nitrogen, so their metabolism produces nitrogenous wastes in addition to water and carbon dioxide. The most common nitrogenous waste is **ammonia** (NH_3), which is highly toxic to animals. Ammonia must be excreted continuously to prevent its accumulation, or it must be detoxified by conversion into other molecules before it is excreted. Those molecules are principally **urea** and **uric acid** (Figure 51.3).

Aquatic animals excrete ammonia

Continuous excretion of ammonia is relatively simple for aquatic animals. Ammonia is highly soluble in water and diffuses rapidly. Animals that breathe water continuously lose ammonia from their blood to the environment by diffusion

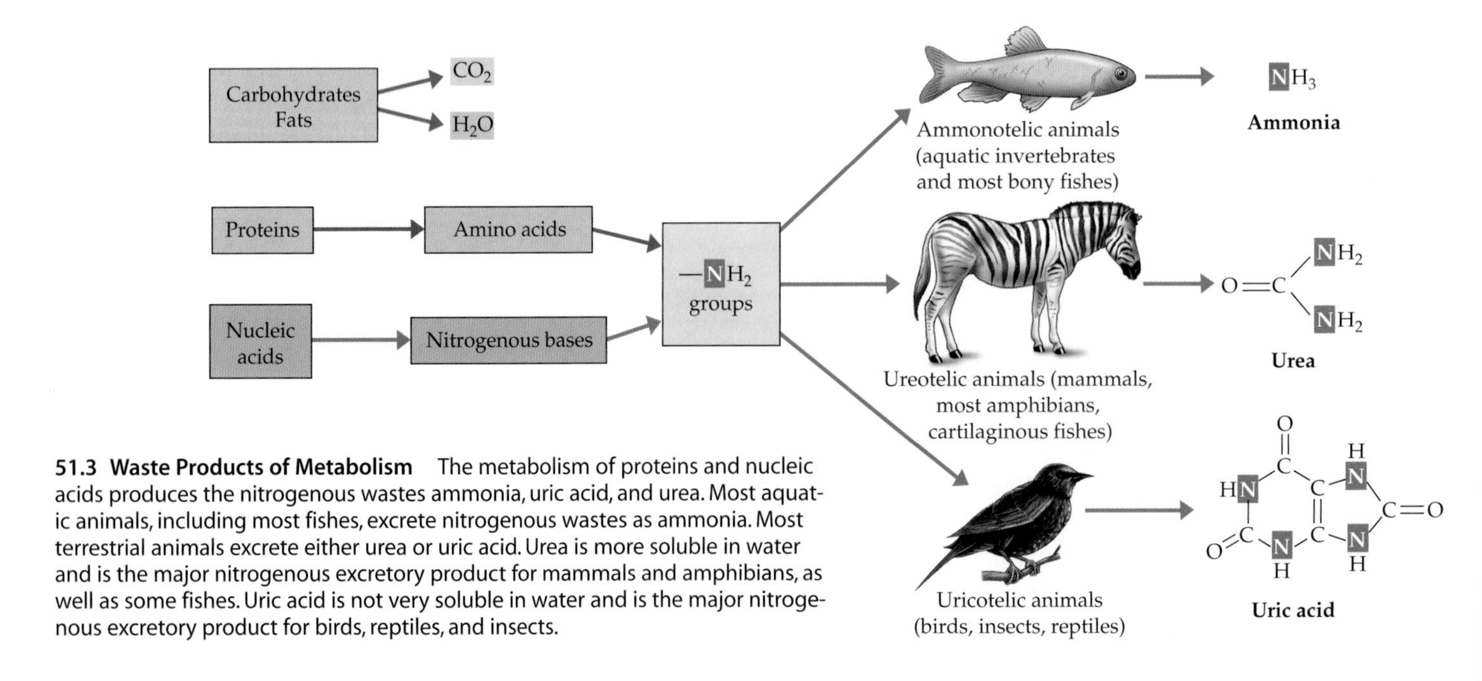

51.3 Waste Products of Metabolism The metabolism of proteins and nucleic acids produces the nitrogenous wastes ammonia, uric acid, and urea. Most aquatic animals, including most fishes, excrete nitrogenous wastes as ammonia. Most terrestrial animals excrete either urea or uric acid. Urea is more soluble in water and is the major nitrogenous excretory product for mammals and amphibians, as well as some fishes. Uric acid is not very soluble in water and is the major nitrogenous excretory product for birds, reptiles, and insects.

across their gill membranes. Animals that excrete ammonia, such as aquatic invertebrates and bony fishes, are said to be **ammonotelic**.

Many terrestrial animals and some fishes excrete urea

Ammonia is a dangerous metabolite for terrestrial animals that have limited access to water. In mammals, ammonia is lethal when it reaches a concentration of only 5 mg/100 ml of blood. Therefore, terrestrial (and some aquatic) animals convert ammonia into either urea or uric acid. **Ureotelic** animals, such as mammals, amphibians, and cartilaginous fishes (sharks and rays), excrete urea as their principal nitrogenous waste product.

Urea is quite soluble in water, but excretion of urea solutions at low concentrations could result in a large loss of water that many terrestrial animals can ill afford. As we will see later in this chapter, mammals have evolved excretory systems that conserve water by producing concentrated urea solutions. The cartilaginous fishes are another story. These marine species keep their body fluids almost isotonic to the marine environment by retaining high concentrations of urea.

Some terrestrial animals excrete uric acid

Animals that conserve water by excreting nitrogenous wastes as uric acid are said to be **uricotelic**. Insects, reptiles, birds, and some amphibians are uricotelic. Uric acid is very insoluble in water and is excreted as a semisolid (for example, it is the whitish material in bird droppings). Therefore, a uricotelic animal loses very little water as it disposes of its nitrogenous wastes.

Most species produce more than one nitrogenous waste

Humans are ureotelic, yet we also excrete uric acid and ammonia. The uric acid in human urine comes largely from the metabolism of nucleic acids and caffeine. In the condition known as *gout*, uric acid levels in the tissue fluid increase, and uric acid precipitates out of solution in the joints and elsewhere, causing swelling and pain. The excretion of ammonia is an important mechanism for regulating the pH of the tissue fluid.

In some species, different developmental stages live in different habitats and have different forms of nitrogen excretion. The tadpoles of frogs and toads, for example, excrete ammonia across their gill membranes, but when they develop into adult frogs or toads, they generally excrete urea. Some adult frogs and toads that live in arid habitats excrete uric acid. These examples show the considerable evolutionary flexibility in the excretion of nitrogenous wastes.

The Diverse Excretory Systems of Invertebrates

Freshwater and terrestrial invertebrates have a wide variety of adaptations for maintaining salt and water balance and excreting nitrogen. In this section, we will explore three examples of invertebrate excretory systems: protonephridia, metanephridia, and Malpighian tubules.

The protonephridia of flatworms excrete water and conserve salts

Many flatworms, such as *Planaria*, live in fresh water. These animals excrete water through an elaborate network of tubules running throughout their bodies. The tubules end in *flame cells*, so called because each cell has a tuft of cilia projecting into the tubule. The beating of the cilia give the appearance of a flickering flame (Figure 51.4). A flame cell and a tubule together form a **protonephridium** (plural, protonephridia; from the Greek *proto*, "before," and *nephros*, "kidney").

Tissue fluid enters the tubules (how it does so is not entirely clear), and the beating of the cilia causes this fluid to flow through the tubules toward the animal's excretory pore. As it flows, the cells of the tubules modify the composition of the fluid. As the modified tubule fluid (urine) leaves the flatworm's body, it is less concentrated than the animal's tissue fluid, so ions are conserved and water is excreted by the protonephridium.

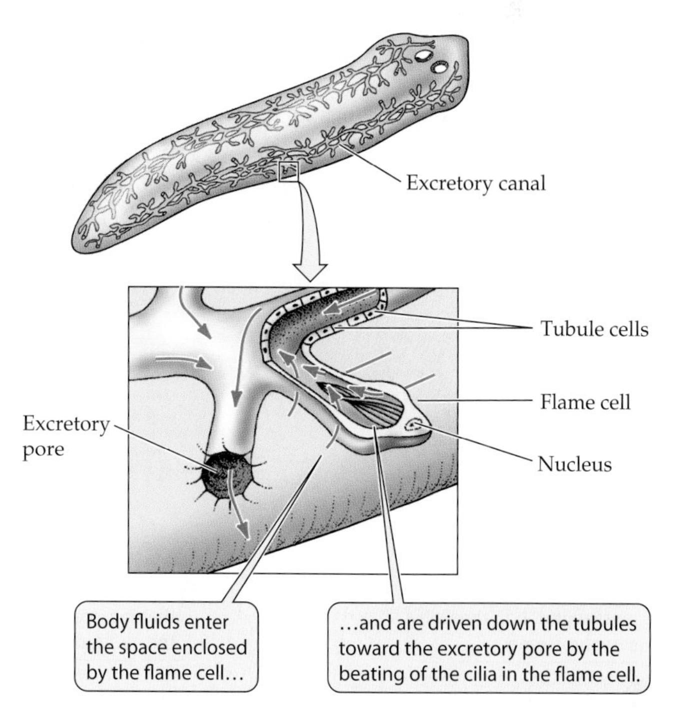
Excretory canal
Tubule cells
Flame cell
Excretory pore
Nucleus
Body fluids enter the space enclosed by the flame cell...
...and are driven down the tubules toward the excretory pore by the beating of the cilia in the flame cell.

51.4 Protonephridia in Flatworms The protonephridia of the flatworm *Planaria* consist of tubules ending in flame cells. The tubule cells modify the composition of the fluid passing through them.

The metanephridia of annelids process coelomic fluid

Filtration of body fluids and modification of urine by tubules are highly developed processes in annelid worms, such as the earthworm. Recall that annelids are segmented and have a fluid-filled body cavity, called a *coelom*, in each segment (see Figure 32.22). Annelids have a closed circulatory system through which blood is pumped under pressure. The pressure causes the blood to be filtered across the thin, permeable capillary walls into the coelom. The cells and large protein molecules of the blood stay behind in the capillaries, while water and small molecules leave them and enter the coelom. In addition, some waste products, such as ammonia, diffuse directly from the tissues into the coelom. But where does this coelomic fluid go?

Each segment of the earthworm contains a pair of **metanephridia** (singular, metanephridium; from the Greek *meta*, "akin to," and *nephros*, "kidney"). Each metanephridium begins in one segment as a ciliated, funnel-like opening in the coelom, called a *nephrostome*, which leads into a tubule in the next segment. The tubule ends in a pore, called the *nephridiopore*, that opens to the outside of the animal (Figure 51.5). Coelomic fluid enters the metanephridia through the nephrostomes. As the fluid passes through the tubules, their cells actively resorb certain molecules from it and actively secrete other molecules into it. What leaves the animal through the nephridiopores is a hypotonic (dilute) urine containing nitrogenous wastes, among other solutes.

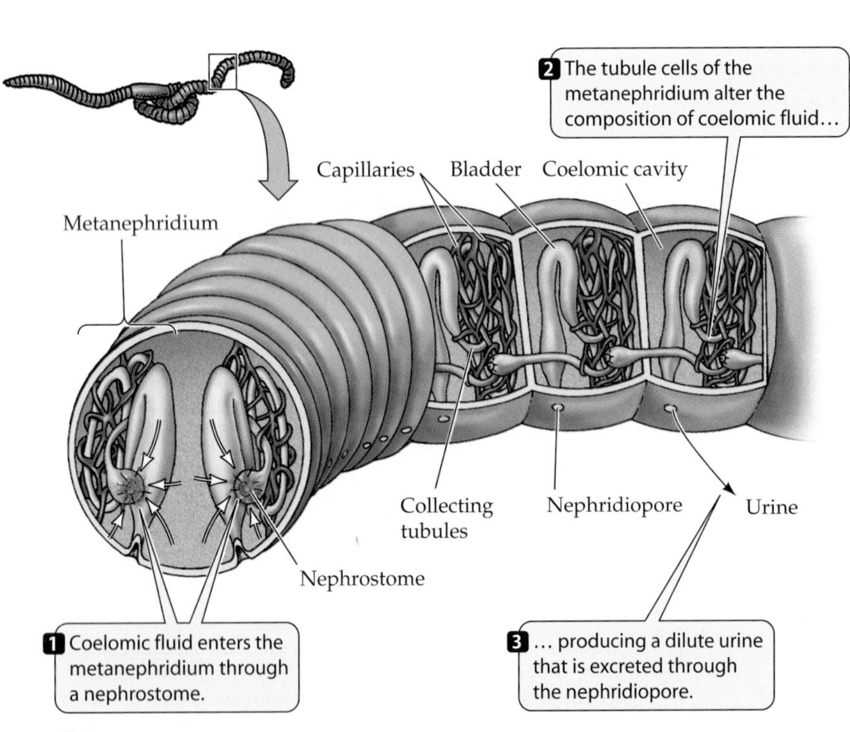

2 The tubule cells of the metanephridium alter the composition of coelomic fluid...
Capillaries
Bladder
Coelomic cavity
Metanephridium
Collecting tubules
Nephridiopore
Urine
Nephrostome
1 Coelomic fluid enters the metanephridium through a nephrostome.
3 ... producing a dilute urine that is excreted through the nephridiopore.

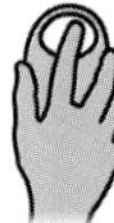

51.5 Metanephridia in Earthworms The metanephridia of annelids are arranged segmentally. The cross section (at the left) shows a pair of metanephridia. Longitudinal sections (at the right) show only one metanephridium of the two in each segment.

The Malpighian tubules of insects depend on active transport

Insects can excrete nitrogenous wastes with very little loss of water. Therefore, some insect species can live in the driest habitats on Earth. The insect excretory system consists of **Malpighian tubules**. An individual insect has from two to more than a hundred of these tubules attached to the gut between the midgut and hindgut. The tubules are closed at the other end, and they project throughout the insect's body cavity (Figure 51.6). Insects have open circulatory systems and therefore cannot use filtration to produce an excretory fluid.

The cells of the Malpighian tubules actively transport uric acid, potassium ions, and sodium ions from the tissue fluid into the tubules. As these solutes are secreted into the tubules, water follows because of the difference in osmolarity. The walls of the Malpighian tubules have muscle cells that contract to help move the contents of the tubules toward the hindgut.

The tubule fluid changes in composition while it is in the hindgut. The contents of the hindgut are more acidic than the tubule fluid; as a result, uric acid becomes less soluble and precipitates out of solution as it approaches and enters the rectum. The epithelial cells of the hindgut and rectum actively transport sodium and potassium ions from the gut contents back into the tissue fluid. Because the uric acid molecules have precipitated out of solution, water is free to follow the resorbed salts back into the tissue fluid through osmosis. Remaining in the rectum are crystals of uric acid mixed with undigested food; this dry matter is what the insect eliminates. The Malpighian tubule system is a highly effective mechanism for excreting nitrogenous wastes and some salts without giving up a significant fraction of the animal's precious water supply.

Vertebrate Excretory Systems

The major excretory organ of vertebrates is the **kidney**. The functional unit of the kidney is the **nephron**. Each human kidney has about a million nephrons. All vertebrate kidneys contain nephrons, yet the kidneys of different species can contribute to water and salt balance in opposite ways.

Both marine and terrestrial vertebrates must conserve water

Some paleontologists believe that the immediate ancestors of the vertebrates lived in fresh water. If that is true, the nephron would have evolved as a structure to excrete excess water. Indeed, the

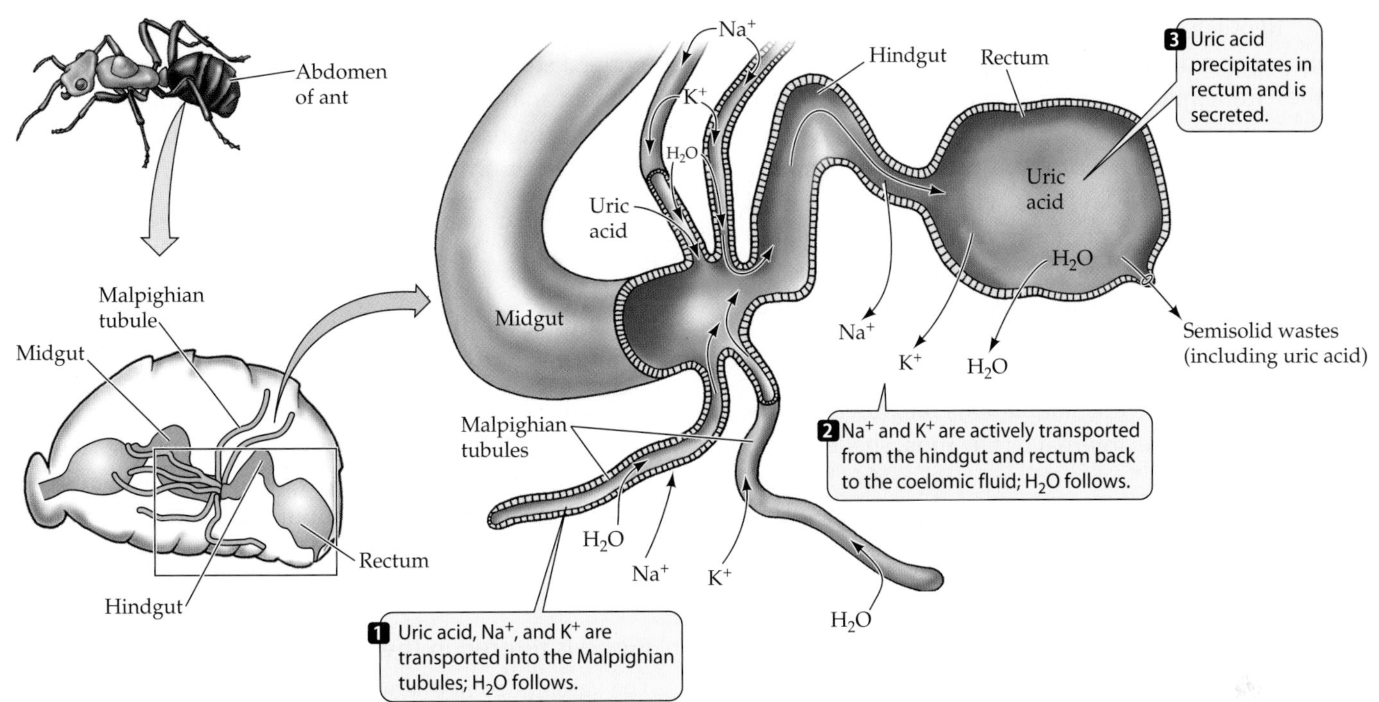

51.6 Malpighian Tubules in Insects The blind, thin-walled Malpighian tubules are attached to the junction of the insect's midgut and hindgut and project into the spaces containing tissue fluid.

nephron can filter large quantities of blood and conserve valuable solutes such as glucose, amino acids, and certain ions while excreting water and nitrogenous wastes. How, then, have vertebrates adapted to environments where water must be conserved and salts excreted? The answer to this question differs among vertebrate groups. Even among the marine fishes, the adaptations of the bony fishes are different from those of the cartilaginous fishes.

MARINE BONY FISHES. Marine bony fishes cannot produce urine that is more concentrated than their tissue fluid, but unlike most marine animals, they osmoregulate their tissue fluid at only one-fourth to one-third the osmolarity of seawater. They prevent excessive loss of water by producing very little urine.

Marine bony fishes take in seawater with their food, which results in large salt loads. The fish handle these salt loads by simply not absorbing some ions (such as Mg^{2+} or SO_4^{2-}) from their guts and by actively excreting others (such as Cl^-) from the gill membranes and from the renal tubules. Nitrogenous wastes are lost as ammonia from the gill membranes.

CARTILAGINOUS FISHES. Cartilaginous fishes are osmoconformers, but not ionic conformers. Unlike marine bony fishes, cartilaginous fishes convert nitrogenous wastes to urea and another compound called trimethylamine oxide, and they retain large amounts of these compounds in their tissue fluids. As a result, their tissue fluids have an osmolarity close to that of seawater, so they do not lose body water to the environment by osmosis. These species have adapted to a concentration of urea in the body fluids that would be toxic to other vertebrates.

Sharks and rays still have the problem of excreting the large amounts of salts they take in with their food. One adaptation to solve this problem is a NaCl-secreting *rectal gland.*

AMPHIBIANS. Most amphibians live in or near fresh water and stay in humid habitats when they venture from the water. Like freshwater fishes, most amphibian species produce large amounts of dilute urine and conserve salts. Some amphibians, however, have adapted to habitats that require water conservation.

Amphibians living in very dry terrestrial environments have reduced the permeability of their skin to water. Some secrete a waxy substance that they spread over the skin to waterproof it. Several species of frogs that live in arid regions of Australia burrow deep into the ground and remain there during long dry periods. They enter *estivation,* a state of very low metabolic activity and therefore low water turnover. When it rains, the frogs come out of estivation, feed, and reproduce. Their most interesting adaptation is an enormous urinary bladder. Before entering estivation, they fill the bladder with dilute urine, which can amount to one-third of their body weight. This dilute urine serves as a water reservoir that is gradually resorbed into the blood during the long period of estivation. Australian aboriginal peoples dig up estivating frogs as an emergency source of drinking water.

REPTILES. Reptiles occupy habitats ranging from aquatic to extremely hot and dry. Three major adaptations have freed the reptiles from maintaining the close association with water that is necessary for most amphibians. First, reptiles do not need fresh water to reproduce, because they employ internal fertilization and lay eggs with shells that retard evaporative water loss. Second, they have scaly, dry skin that retards evaporative water loss. Third, they excrete nitrogenous wastes as uric acid solids, losing little water in the process.

BIRDS. Birds have the same adaptations for water conservation that reptiles have: internal fertilization, shelled eggs, skin that retards water loss, and uric acid as the nitrogenous waste product. In addition, birds can produce urine that is more concentrated than their tissue fluids.

The nephron is the functional unit of the kidney

To understand how the kidney can fulfill different functions in different animal groups, we need to understand the structure of a nephron and how its parts work together. Vertebrate nephrons generally have three main constituents:

- A dense ball of capillaries called the **glomerulus** filters a portion of the blood plasma.
- **Renal tubules** receive and modify the glomerular filtrate.
- **Peritubular capillaries** bring substances to and take substances away from the renal tubules.

The glomerulus and peritubular capillaries are vascular structures closely associated with the renal tubule (Figure 51.7) The different ways in which these three elements can work together influence the amount and composition of the urine.

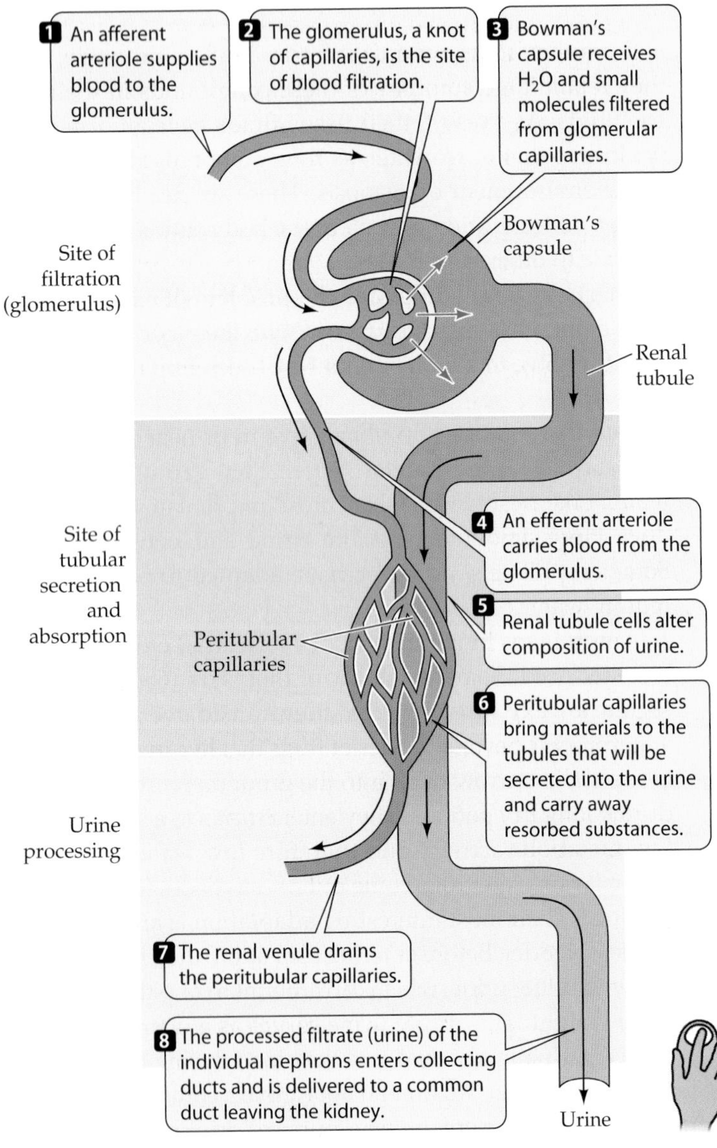

Blood is filtered in the glomerulus

The vascular component of the nephron is unusual in that the two capillary beds—the glomerulus and the peritubular capillaries—lie in series between the arteriole that supplies them and the venule that drains them. The glomerulus (plural, glomeruli) is a dense knot of very permeable vessels (Figure 51.8*a*). Blood enters the glomerulus through an **afferent arteriole** and exits through an **efferent arteriole**. The efferent arteriole gives rise to the peritubular capillaries, which surround the tubular component of the nephron.

The tubule component of the nephron—the renal tubule—begins with **Bowman's capsule**, which encloses the glomerulus. The glomerulus appears to be pushed into Bowman's capsule much like a fist pushed into an inflated balloon. The cells of the capsule that come into direct contact with the glomerular capillaries are called **podocytes** (Figure 51.8*b*). These highly specialized cells have numerous armlike extensions, each with hundreds of fine, fingerlike projections. The podocytes wrap around the capillaries so that their fingerlike projections interdigitate and cover the capillaries completely.

The glomerulus filters the blood to produce a fluid (the renal filtrate) that lacks cells and large molecules. The walls of the capillaries, the basal lamina of the capillary endothelium, and the podocytes of Bowman's capsule all participate in filtration. As we saw in Chapter 49, fenestrations between the endothelial cells of the capillaries allow water and small molecules to leave them, but are too

1 An afferent arteriole supplies blood to the glomerulus.
2 The glomerulus, a knot of capillaries, is the site of blood filtration.
3 Bowman's capsule receives H_2O and small molecules filtered from glomerular capillaries.
Bowman's capsule
Site of filtration (glomerulus)
Renal tubule
Site of tubular secretion and absorption
4 An efferent arteriole carries blood from the glomerulus.
5 Renal tubule cells alter composition of urine.
Peritubular capillaries
6 Peritubular capillaries bring materials to the tubules that will be secreted into the urine and carry away resorbed substances.
Urine processing
7 The renal venule drains the peritubular capillaries.
8 The processed filtrate (urine) of the individual nephrons enters collecting ducts and is delivered to a common duct leaving the kidney.
Urine

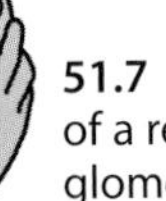

51.7 The Vertebrate Nephron The vertebrate nephron consists of a renal tubule closely associated with two capillary beds, the glomerulus and the peritubular capillaries.

(a)

(b)

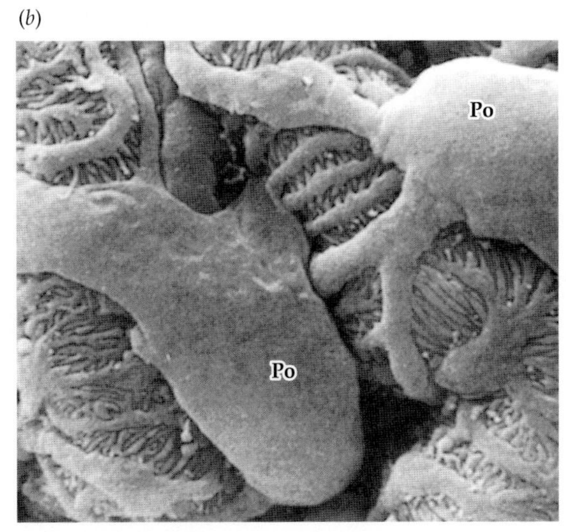

(c)

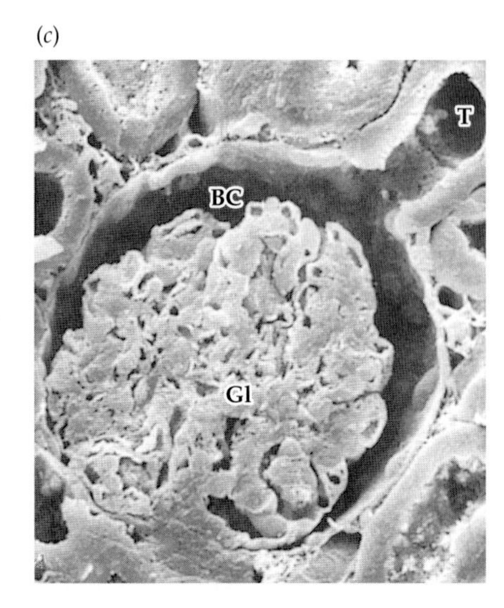

51.8 A Tour of the Nephron These scanning electron micrographs (SEMs) show the anatomical bases for kidney function. (*a*) The blood vessels of the kidney, shown without the tubule components. Each glomerulus (Gl) has an afferent (AAr) and an efferent (EAr) arteriole. The peritubular capillaries (Pt) are looser networks that surround the tubules of the nephron. (*b*) The podocytes (Po) are in direct contact with the capillaries of the glomerulus. Each podocyte has hundreds of tiny fingerlike projections that create filtration slits between them. Anything passing from the capillaries into the tubule of the nephron must pass through these slits. (*c*) A cross section of a glomerulus surrounded by the tubule cells that form Bowman's capsule (BC), which collects the filtrate and funnels it into the tubule (T) of the nephron.

small to permit red blood cells to pass through. The meshwork of the basal lamina is even finer than the fenestrations, and it prevents large molecules from leaving the capillaries. Also smaller than the fenestrations are the narrow slits between the fingerlike projections of the podocytes. As a result of these anatomical adaptations, water and small molecules enter the renal tubule of the nephron (Figure 51.8*c*), but cells and proteins remain in the capillaries.

The force that drives filtration in the glomerulus is the pressure of the arterial blood. As in every other capillary bed, the pressure of the blood entering the permeable capillaries causes the filtration of water and small molecules. The glomerular filtration rate is high because glomerular capillary blood pressure is unusually high, and because the capillaries of the glomerulus, along with their covering of podocytes, are more permeable than other capillary beds in the body.

The renal tubules convert glomerular filtrate to urine

The composition of the filtrate that first enters the nephron is similar to that of the blood plasma. This filtrate contains glucose, amino acids, ions, and nitrogenous wastes in the same concentrations as in the blood plasma, but it lacks the plasma proteins. As this fluid passes down the renal tubule, its composition changes as the cells of the tubule actively resorb certain molecules from the tubule fluid and secrete other molecules into it. When the tubule fluid leaves the kidney as urine, its composition is very different from that of the original filtrate.

The function of the renal tubules is to control the composition of the urine by actively secreting and resorbing specific molecules. The blood in the peritubular capillaries brings to the renal tubules molecules that the tubule cells will secrete into the urine. It also carries away any molecules that have been resorbed from the urine.

The Mammalian Excretory System

The adaptations of mammals and birds that allowed them to produce urine hyperosmotic to their tissue fluid were an important step in vertebrate evolution. These adaptations enabled the excretory system to conserve water while still excreting excess salts and nitrogenous wastes. Mammals and birds have high body temperatures and high metabolic rates, and therefore have the potential for a high rate of water loss. Being able to minimize water loss from their excretory systems made it possible for these highly active species to occupy arid habitats.

Kidneys produce urine, which the bladder stores

We will use the excretory system of humans as our example of the mammalian excretory system. Humans have two kidneys just under the dorsal wall of the abdominal cavity in the mid-back region (Figure 51.9*a*). Each kidney filters blood, processes the filtrate into urine, and releases that urine into

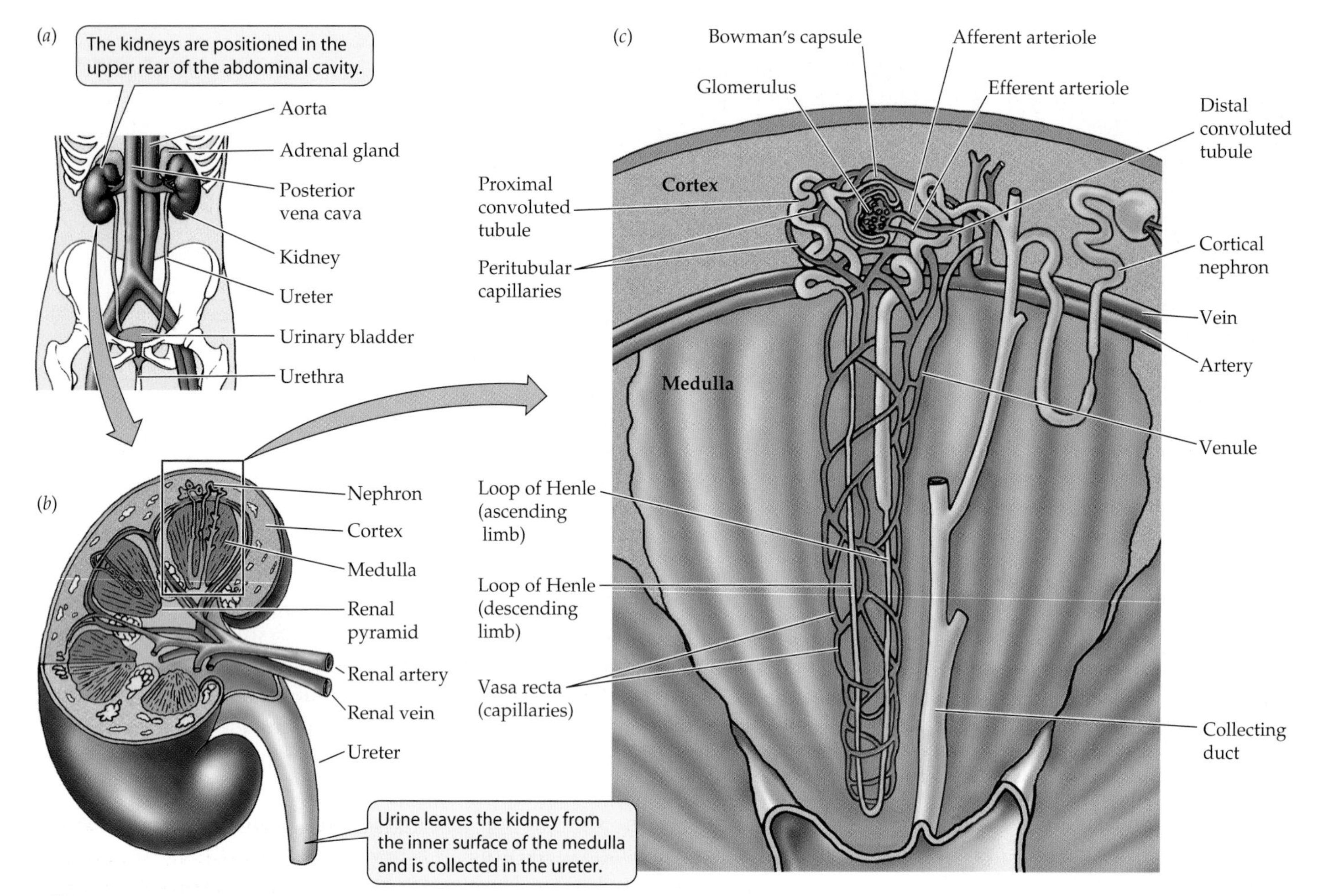

51.9 The Human Excretory System (*a*) The human kidneys are positioned in the upper dorsal region of the abdominal cavity. (*b*) The human kidney has a regular internal tissue structure that is the basis for its function. (*c*) The glomeruli and the proximal and distal convoluted tubules are located in the cortex of the kidney, but the loops of Henle run in parallel as straight sections down into the renal medulla and back up to the cortex. Collecting ducts run from the cortex to the inner surface of the medulla, where they open into the ureter.

a duct called the **ureter**. The ureter of each kidney leads to the **urinary bladder**, where the urine is stored until it is excreted through the urethra. The **urethra** is a short tube that opens to the outside at the end of the penis in males or just anterior to the vaginal opening in females.

Two sphincter muscles surrounding the base of the urethra control the timing of urination. One of these sphincters is a smooth muscle and is controlled by the autonomic nervous system. As the bladder fills, stretch receptors in the walls of the bladder trigger a spinal reflex that relaxes this sphincter. This reflex is the only control of urination in infants. The other sphincter is a skeletal muscle and is controlled by the voluntary nervous system. When the bladder is *very* full, only deliberate conscious effort prevents urination.

Nephrons have a regular arrangement in the kidney

The kidney is shaped like a kidney bean; when cut down its long axis and split open as a bean splits open, its important anatomical features are revealed (Figure 51.9*b*). The ureter and the **renal artery** and **renal vein** enter the kidney on its concave (punched-in) side. The ureter extends into the kidney in several branches, the ends of which envelop kidney tissues called **renal pyramids**. The renal pyramids make up the internal core, or **medulla**, of the kidney. The medulla is surrounded by tissue with a different appearance, called the **cortex**. The renal artery and vein give rise to many arterioles and venules in the region between the cortex and the medulla.

Each human kidney contains about a million nephrons, and their organization within the kidney is very regular. All of the glomeruli are located in the cortex. The initial segment of a renal tubule is called the **proximal convoluted tubule**—"proximal" because it is the first segment of renal tubule to receive the filtrate from the glomerulus, and "convoluted" because it is twisted (Figure 51.9*c*). All the proximal convoluted tubules are also located in the cortex.

At a certain point, the renal tubule takes a dive directly down into the medulla. The portion of the tubule in the medulla is the **loop of Henle**. It is called a loop because it runs straight down into the medulla, makes a hairpin turn, and rises straight back to the cortex. Not all nephrons have long loops of Henle; so-called cortical nephrons do not. Where the ascending limb of the loop of Henle reaches the cortex, it becomes the **distal convoluted tubule**—"distal" because it receives the filtrate from the glomerulus after the other renal tubule segments. The distal convoluted tubules of many nephrons join a common **collecting duct** in the cortex. The collecting ducts then run in parallel with the loops of Henle down through the renal pyramid. The collecting ducts, however, empty into the funnel-shaped *pelvis*, which narrows down to leave the kidney as the ureter.

Blood vessels also have a regular arrangement in the kidney

The organization of the blood vessels of the kidney closely parallels the organization of the nephrons (see Figure 51.9*c*). Smaller and smaller arteries branch from the renal artery and radiate into the cortex. Finally an afferent arteriole carries blood to each glomerulus. Draining each glomerulus is an efferent arteriole that gives rise to the peritubular capillaries, most of which surround the cortical portions of the tubules.

A few peritubular capillaries run into the medulla in parallel with the loops of Henle and the collecting ducts. These capillaries form the **vasa recta**. All the peritubular capillaries from a nephron join back together into a venule that joins with venules from other nephrons and eventually leads to the renal vein, which takes blood from the kidney.

The volume of glomerular filtration is greater than the volume of urine

Most of the water and solutes filtered out of the glomerulus are resorbed and do not appear in the urine. We can reach this conclusion by comparing the rate of filtration by the glomeruli with the rate of urine production. The kidneys receive about 1 liter of blood per minute, or about 1,500 liters of blood per day. How much of this huge volume is filtered out of the glomeruli? The answer is about 12 percent. This is still a large volume—180 liters per day! Since we normally urinate 2–3 liters per day, about 98 percent of the fluid volume that is filtered out of the glomerulus is returned to the blood. Where and how is this enormous fluid volume resorbed?

Most filtrate is resorbed by the proximal convoluted tubule

The proximal convoluted tubule is responsible for most of the resorption of water and solutes from the glomerular filtrate. The cells of this section of the renal tubule are cuboidal, and their surfaces facing into the tubule have many microvilli, which increase their surface area for resorption. These cells have lots of mitochondria—an indication that they are metabolically active. They actively transport Na^+ (with Cl^- following passively) and other solutes, such as glucose and amino acids, out of the tubule fluid. Almost all glucose and amino acid molecules that are filtered from the blood are actively resorbed by these cells and transported back into the tissue fluid. The active transport of solutes into the tissue fluid causes water to follow osmotically. The water and solutes moved into the tissue fluid are taken up by the peritubular capillaries and returned to the venous blood leaving the kidney.

Despite the large volume of water and solutes resorbed by the proximal convoluted tubule, the overall osmolarity of the renal filtrate does not change. The fluid that enters the loop of Henle has the same osmolarity as the blood plasma, although its composition is different. How, then, does the kidney produce urine that is hypertonic to the blood plasma?

The loop of Henle creates a concentration gradient in the surrounding tissue

Humans can produce urine that is four times more concentrated than their blood plasma. The vampire bat we encountered at the beginning of this chapter can produce urine that is twenty times more concentrated than its blood plasma, and some desert-dwelling animals, as we will see, can produce even greater concentrations. This concentrating ability of the mammalian kidney is due to the loops of Henle, which function as a *countercurrent multiplier system*. The term "countercurrent" refers to the fact that the tubule fluid in the descending limb of the loop flows in the opposite direction from that in the ascending limb. "Multiplier" refers to the ability of this system to create a solute concentration gradient in the renal medulla. The loops of Henle do not themselves produce concentrated urine; rather, they increase the osmolarity of the tissue fluid in the medulla in such a way that the tissue fluid concentration increases from the top to the bottom of the medulla.

The segments of the loop of Henle differ anatomically and functionally. Cells of the descending limb and the initial cells of the ascending limb are flat, with no microvilli and few mitochondria. They are not specialized for transport. Partway up the ascending limb, the cells become specialized for active transport. These cells are cuboidal and have lots of mitochondria. Accordingly, the loop of Henle is divided into the *thin descending limb*, the *thin ascending limb*, and the *thick ascending limb.*

To understand the countercurrent multiplier mechanism, it is easiest to move backward through the renal tubule, starting with the thick ascending limb (Figure 51.10). The thick ascending limb actively resorbs Cl^- (with Na^+ following pas-

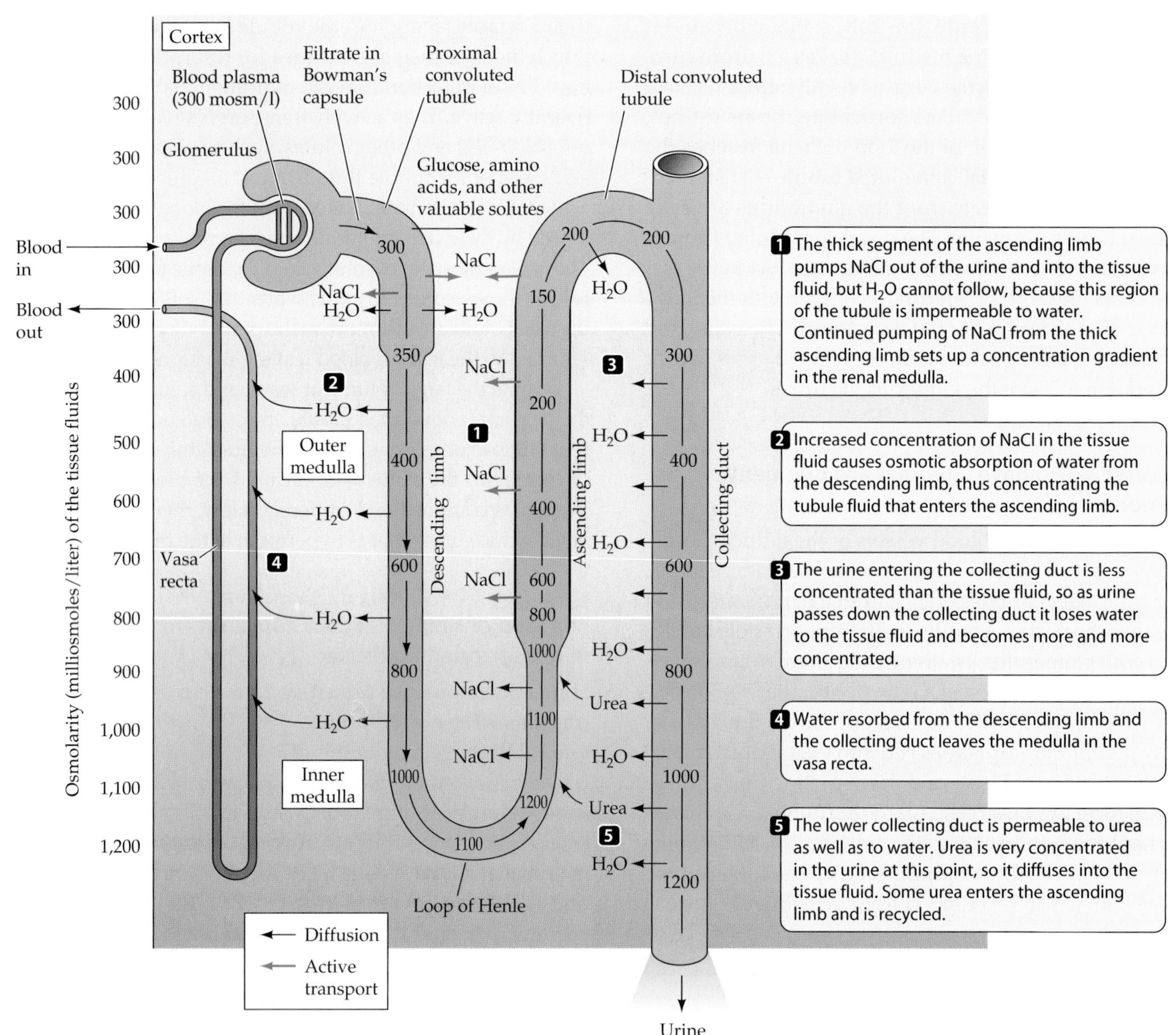

51.10 Concentrating the Urine A countercurrent multiplier mechanism enables the kidney to produce urine that is far more concentrated than mammalian blood plasma.

sively) from the tubule fluid and moves it into the surrounding tissue fluid. The thick ascending limb is not permeable to water, so the resorption of Na^+ and Cl^- raises the concentration of those solutes in the surrounding tissue fluid.

The thin descending limb, in contrast, is rather permeable to water, but not very permeable to Na^+ and Cl^-. Since the surrounding tissue fluid has been made more concentrated by the Na^+ and Cl^- resorbed from the neighboring thick ascending limb, water is withdrawn osmotically from the tubule fluid in the descending limb. Therefore, the fluid in the descending limb becomes more and more concentrated as it flows toward the hairpin turn at the bottom of the renal medulla.

The thin ascending limb, like the thick ascending limb, is not permeable to water. It is, however, permeable to Na^+ and Cl^-. As the concentrated tubule fluid flows up the thin ascending limb, it is more concentrated than the surrounding tissue fluid, so Na^+ and Cl^- diffuse out of it. When the tubule fluid reaches the thick ascending limb, active transport continues to move Na^+ and Cl^- from the tubule fluid to the tissue fluid.

As a result of the processes described above, the tubule fluid reaching the distal convoluted tubule is less concentrated than the blood plasma, and the solutes that have been left behind in the renal medulla have created a concentration gradient in the surrounding tissue fluid. The tissue fluid of the renal medulla becomes more and more concentrated as we move from the border with the cortex down to the tips of the renal pyramids.

You may wonder why the blood flow through the medulla does not wash out the concentration gradient established by the loops of Henle. This is the significance of the parallel arrangement of the descending and ascending peritubular capillaries—the vasa recta—in the medulla. Countercurrent exchange between the descending and ascending vessels facilitates preservation of the concentration gradient.

Water resorption begins in the distal convoluted tubule

The first portion of the distal convoluted tubule has properties similar to the thick ascending limb of the loop of Henle. Na^+ and Cl^- are transported out of the tubule fluid, and water cannot follow. As a result, the tubule fluid becomes even more dilute. The later sections of the distal convoluted tubule, however, can be permeable to water, and water can be osmotically drawn into the surrounding tissue fluid. As the tubule fluid flows from the distal tubule to the collecting duct, it can be below or equal to the osmolarity of the blood plasma.

Urine is concentrated in the collecting duct

The tubule fluid entering the collecting duct is at the same solute *concentration* as the blood plasma, but its *composition* is considerably different from that of the plasma. The major solute in the tubule is now urea, since salts were resorbed earlier in the nephron. As the tubule fluid flows down the collecting duct, it loses water osmotically to the surrounding tissue fluid.

The concentration gradient established in the renal medulla by the loops of Henle creates the osmotic potential that withdraws water from the collecting ducts. The collecting ducts begin in the renal cortex and run through the renal medulla before emptying into the ureter at the tips of the renal pyramids. As the solute concentration of the surrounding tissue fluid increases, more and more water can be absorbed from the urine in the collecting duct. By the time it reaches the ureter, the urine can become greatly concentrated, with urea as a major solute.

As water is withdrawn from the collecting duct, some urea also leaks out into the medullary tissue, adding to its osmotic potential. This urea diffuses back into the loop of Henle and is returned to the collecting duct. The recycling of urea in the renal medulla contributes significantly to the ability of the kidney to concentrate the urine in the collecting duct.

Overall, the ability of a mammal to concentrate its urine is determined by the maximum concentration gradient it can establish in its renal medulla. An important adaptation for increasing the concentration gradient is to increase the lengths of the loops of Henle. The desert gerbil, for example, has such extremely long loops of Henle that its renal pyramid (each of its kidneys has only one, in contrast to ours) extends far out of the concave surface of the kidney (Figure 51.11). These animals are so effective in conserving water that they can survive on the water released by the metabolism of their food; they do not need to drink!

51.11 The Ability to Concentrate The ability of the mammalian kidney to concentrate urine depends on the lengths of its loops of Henle relative to the overall size of the kidney. The kidney shown here, from a desert gerbil, has a single renal pyramid with loops of Henle so long that they protrude out of the medulla and into the ureter.

The kidneys help regulate acid–base balance

Besides regulating salt and water balance and excreting nitrogenous wastes, the kidneys have another important role: They regulate the hydrogen ion concentration (the pH) of the blood. Blood pH is a critical variable because it influences the structure, and therefore the function, of proteins.

One way to minimize pH changes in a chemical solution is to add a *buffer*—a substance that can either absorb or release hydrogen ions (see Chapter 2). The major buffers in the blood are bicarbonate ions (HCO_3^-; see Figure 48.14) that are formed from the disassociation of carbonic acid, which in turn is formed by the hydration of CO_2 according to the following equilibrium reaction:

$$CO_2 + H_2O \rightleftharpoons H_2CO_3 \rightleftharpoons H^+ + HCO_3^-$$

You can see that if excess H^+ ions are added to this reaction mixture, the reaction will move to the left and absorb the excess H^+. On the other hand, if H^+ ions are removed from the reaction mixture, the reaction will move to the right and supply more H^+ ions.

The HCO_3^- buffer system is important for controlling the pH of the blood because the reaction can be pushed to the right and pulled to the left physiologically. The lungs control the levels of CO_2 in the blood, thus altering the acid portion of the reaction. The kidneys control the base portion of the

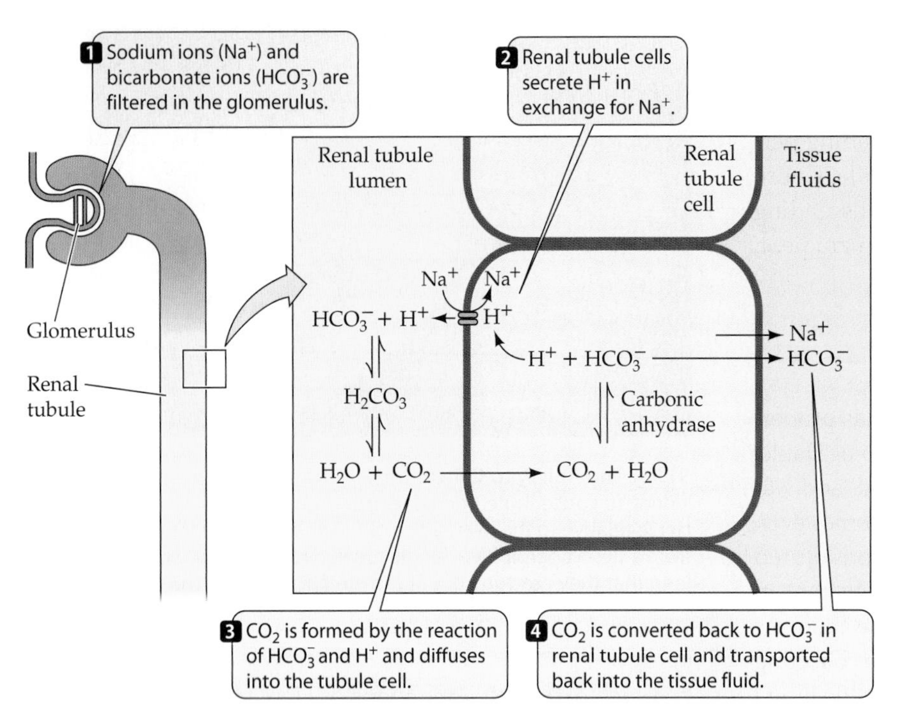

51.12 The Kidney Excretes Acids and Conserves Bases Bicarbonate ions are filtered out of the glomerulus, and renal tubule cells secrete hydrogen ions into the tubule fluid. In the renal tubule, the filtered bicarbonate buffers the secreted hydrogen ions and keeps the urine from becoming too acidic. The CO_2 formed by the reaction of bicarbonate and hydrogen ions is converted back to bicarbonate by the renal tubule cells and transported back into the tissue fluid.

reaction by altering the levels of H^+ and HCO_3^- ions in the blood. The renal tubules secrete H^+ into the tubule fluid and resorb HCO_3^- (Figure 51.12). The kidney has other buffering systems as well, and together they greatly enhance the ability of the kidney to eliminate acid from the blood.

Regulation of Kidney Functions

Several regulatory mechanisms act on the kidneys to maintain blood osmolarity and blood pressure. We will discuss these mechanisms separately, but keep in mind that they are always working together.

The kidneys act to maintain the glomerular filtration rate

If the kidneys stop filtering blood, they cannot accomplish any of their functions. The maintenance of a constant **glomerular filtration rate** (**GFR**) depends on an adequate blood supply to the kidneys at an adequate blood pressure. Therefore, the kidneys have mechanisms to maintain their blood supply and blood pressure regardless of what is happening elsewhere in the body. Because these adaptations of the kidney support the maintenance of kidney function, they are called *autoregulatory* mechanisms. The kidney's autoregulatory adjustments compensate for decreases in cardiac output or decreases in blood pressure so that the GFR remains constant (Figure 51.13).

One autoregulatory mechanism is the dilation (expansion) of the afferent renal arterioles when blood pressure falls. This dilation decreases the resistance in the arterioles and helps maintain blood pressure in the glomerular capillaries. If arteriole dilation does not keep the GFR from falling, the kidney releases an enzyme, **renin**, into the blood. Renin acts on a circulating protein to convert it into an active hormone called **angiotensin**.

Angiotensin has several effects that help restore the GFR to normal. First, angiotensin constricts the efferent renal arterioles, which elevates blood pressure in the glomerular capillaries. Second, it constricts peripheral blood vessels all over the body—an action that elevates central blood pressure. Third, it stimulates the adrenal cortex to release the hormone **aldosterone**. Aldosterone stimulates sodium resorption by the kidney, thereby making its resorption of water more effective. Enhanced water resorption helps maintain blood volume and therefore central blood pressure. Finally, angiotensin acts on the brain to stimulate thirst. Increased water intake in response to thirst increases blood volume and blood pressure.

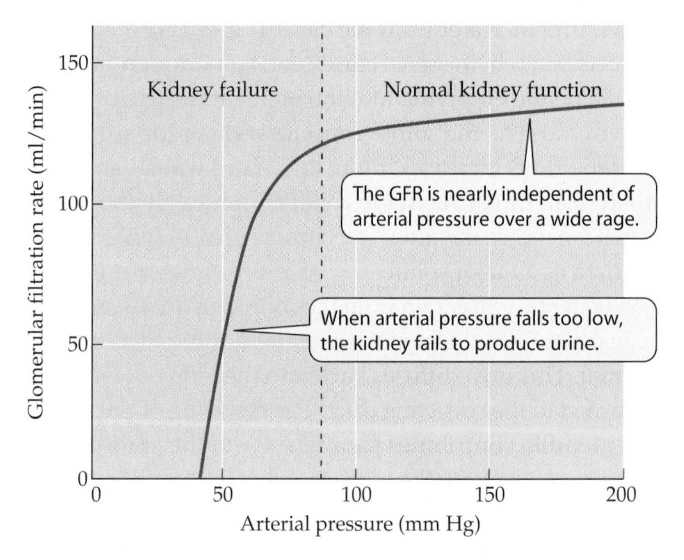

51.13 Maintaining the Glomerular Filtration Rate Glomerular filtration is driven by arterial pressure, but autoregulatory mechanisms prevent rises and falls in glomerular filtration rate (GFR) over a wide range of pressures.

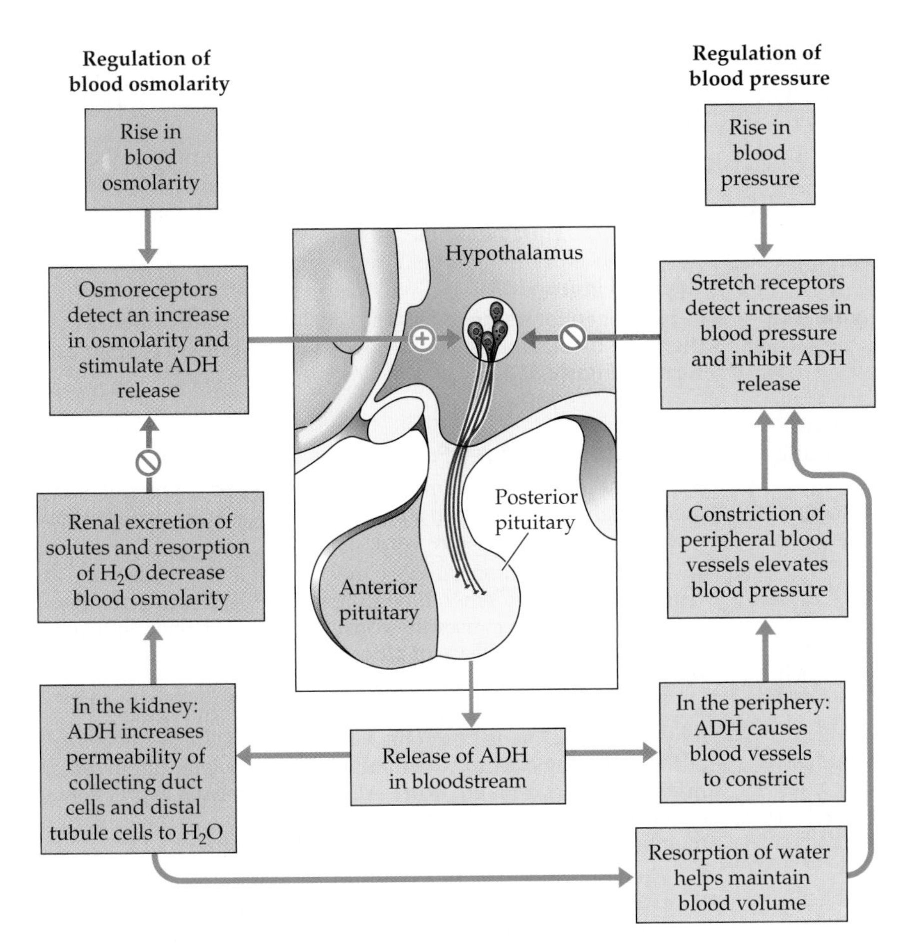
Regulation of blood osmolarity
Rise in blood osmolarity
Osmoreceptors detect an increase in osmolarity and stimulate ADH release
Renal excretion of solutes and resorption of H_2O decrease blood osmolarity
In the kidney: ADH increases permeability of collecting duct cells and distal tubule cells to H_2O
Hypothalamus
Posterior pituitary
Anterior pituitary
Release of ADH in bloodstream
Regulation of blood pressure
Rise in blood pressure
Stretch receptors detect increases in blood pressure and inhibit ADH release
Constriction of peripheral blood vessels elevates blood pressure
In the periphery: ADH causes blood vessels to constrict
Resorption of water helps maintain blood volume

51.14 Antidiuretic Hormone Increases Blood Pressure and Promotes Water Resorption ADH is produced by neurons in the hypothalamus and released from their axons in the posterior pituitary. The release of ADH is stimulated by hypothalamic osmoreceptors and inhibited by stretch receptors in the great arteries.

ADH controls the permeability of the collecting ducts by stimulating the production and activity of membrane proteins that form water channels. These proteins, called **aquaporins**, are found in many tissues that are permeable to water—for example, the capillary endothelium, red blood cells, and the proximal convoluted tubules of the kidney. Differences among tissues in water permeability can be related to the presence or absence of aquaporins. Aquaporins are expressed in the descending limb of the loop of Henle, for example, but not in the ascending limb. One particular aquaporin that is found in distal convoluted tubule and collecting duct cells is controlled by ADH on both a long-term and a short-term basis. Over the long term, ADH levels influence the expression of the gene for this aquaporin; over the short term, ADH controls the insertion of the aquaporin into the cells' plasma membranes.

ADH also helps regulate blood osmolarity by controlling water resorption. Sensory cells in the hypothalamus monitor the osmolarity of the blood. If blood osmolarity increases, these *osmoreceptors* stimulate increased release of ADH to enhance water resorption from the kidneys. The osmoreceptors also stimulate thirst. The resulting water retention and water intake dilutes the blood as it expands blood volume.

Blood pressure and osmolarity are regulated by ADH

When you lose blood volume, your blood pressure tends to fall. Besides activating the kidney autoregulatory mechanisms described in the previous section, a drop in blood pressure decreases the activity of the stretch receptors in the walls of the aorta and the carotid arteries (see Figure 49.17). These stretch receptors provide information to cells in the hypothalamus that produce **antidiuretic hormone** (**ADH**, also called *vasopressin*) and send it down their axons to the posterior pituitary gland. As stretch receptor activity decreases, the production and release of this hormone increases (Figure 51.14).

ADH acts on the distal convoluted tubules and collecting ducts of the kidney to increase their permeability to water. When the circulating level of ADH is high, the distal tubules and the collecting ducts are very permeable to water, more water is resorbed from the urine, and only small quantities of concentrated urine are produced, thus conserving blood volume and maintaining blood pressure. When ADH levels are low, water is not resorbed from the collecting ducts, and lots of dilute urine is produced.

The heart produces a hormone that influences kidney function

When systemic venous return to the heart increases and the atria of the heart become more stretched, the atrial muscle fibers release a peptide hormone called **atrial natriuretic peptide** (**ANP**). This peptide hormone enters the circulation, and when it reaches the kidney, it decreases the resorption of sodium. The result is an increased loss of sodium and water, which has the effect of lowering blood volume and blood pressure. In pathological situations in which a weakened heart cannot pump enough blood to keep up with the venous return (*congestive heart failure*), the atria become stretched more than normal, and the resulting high levels of ANP in the blood can be a useful diagnostic measure.

Chapter Summary

Tissue Fluid and Water Balance

▶ Most adaptations for maintaining salt and water balance and for excreting nitrogen wastes employ the same basic mechanisms: filtration of tissue fluid and active secretion and resorption of specific molecules.

▶ The problems of salt and water balance and nitrogen excretion that animals face depend on their environments, but in all animal excretory systems, there is no active transport of water.

▶ Marine animals can be osmoconformers or osmoregulators. Freshwater animals must be osmoregulators and must continually excrete water and conserve salts. Most animals are ionic regulators to some degree. **Review Figure 51.1**

▶ On land, water conservation is essential, and diet determines whether salts must be conserved or excreted. Marine birds excrete excess salt through nasal salt glands. **Review Figure 51.2**

Excreting Nitrogenous Wastes

▶ Aquatic animals can eliminate nitrogenous wastes such as ammonia by diffusion across their gill membranes. Terrestrial animals must detoxify ammonia by converting it to urea or uric acid before excretion. **Review Figure 51.3**

▶ Depending on the form in which they excrete their nitrogenous wastes, animals are classified as ammonotelic, ureotelic, or uricotelic.

The Diverse Excretory Systems of Invertebrates

▶ The protonephridia of flatworms consist of flame cells and excretory tubules. Tissue fluid is filtered into the tubules, which process the filtrate to produce a dilute urine. **Review Figure 51.4**

▶ In annelid worms, blood pressure causes filtration of the blood across capillary walls. The filtrate enters the coelomic cavity, where it is taken up by metanephridia. As the filtrate passes through the metanephridia to the outside, its composition is changed by active transport mechanisms. **Review Figure 51.5.** See Web/CD Activity 51.1

▶ The Malpighian tubules of insects receive ions and nitrogenous wastes by active transport across the tubule cells. Water follows by osmosis. Ions and water are resorbed from the rectum, so the insect excretes semisolid wastes. **Review Figure 51.6**

Vertebrate Excretory Systems

▶ The nephron, the functional unit of the vertebrate kidney, consists of a glomerulus, in which blood is filtered across the walls of a knot of capillaries, a renal tubule, which processes the filtrate into urine to be excreted, and a system of peritubular capillaries, which surround the tubule. **Review Figure 51.7.** See Web/CD Activity 51.2

▶ The adaptations of marine and terrestrial animals for conserving water are diverse. Marine bony fishes have few glomeruli and produce little urine. Cartilaginous fishes retain urea so that the osmolarity of their body fluids remains close to that of seawater. Amphibians remain close to water or have waxy skin coverings. Reptiles have scaly skin, lay shelled eggs, and excrete nitrogenous wastes as uric acid.

▶ Birds share the adaptations of reptiles; in addition, they can produce urine more concentrated than their tissue fluid. Only birds and mammals can produce such hypertonic urine.

The Mammalian Excretory System

▶ The concentrating ability of the mammalian kidney depends on its anatomy. **Review Figure 51.9*a,b***

▶ The glomeruli and the proximal and distal convoluted tubules are located in the cortex of the kidney. Certain molecules are actively resorbed from the glomerular filtrate by the tubule cells, and other molecules are actively secreted. Straight sections of renal tubules called loops of Henle and collecting ducts are arranged in parallel in the medulla of the kidney. **Review Figure 51.9*c*.** See Web/CD Activity 51.3

▶ Salts and water are resorbed in the proximal convoluted tubule without the renal filtrate becoming more concentrated, although its composition changes.

▶ The loops of Henle create a concentration gradient in the tissue fluid of the renal medulla by a countercurrent multiplier mechanism. Urine flowing down the collecting ducts to the ureter is concentrated by the osmotic resorption of water caused by the concentration gradient in the surrounding tissue fluid. **Review Figure 51.10**

▶ Hydrogen ions secreted by the renal tubules are buffered in the urine by bicarbonate and other chemical buffering systems. **Review Figure 51.12**

See Web/CD Tutorial 51.1

Regulation of Kidney Functions

▶ Kidney function in mammals is controlled by autoregulatory mechanisms that maintain a constant high glomerular filtration rate even if blood pressure varies.

▶ An important autoregulatory mechanism is the release of renin by the kidney when blood pressure falls. Renin activates angiotensin, which causes the constriction of peripheral blood vessels, causes the release of aldosterone (which enhances water resorption), and stimulates thirst.

▶ Changes in blood pressure and osmolarity influence the release of antidiuretic hormone, which controls the permeability of the collecting duct to water and therefore the amount of water that is resorbed from the urine. ADH stimulates the expression of proteins called aquaporins that serve as water channels in the membranes of collecting duct cells. **Review Figure 51.13**

▶ When the volume of blood returning to the heart increases and stretches the atrial walls, they release atrial natriuretic peptide (ANP), which causes increased excretion of salt and water.

Self-Quiz

1. Which statement about osmoregulators is true?
   a. Most marine invertebrates are osmoregulators.
   b. All freshwater invertebrates are hyperosmotic regulators.
   c. Cartilaginous fishes are hypoosmotic regulators.
   d. Bony marine fishes are hyperosmotic regulators.
   e. Mammals are hypoosmotic regulators.
2. The excretion of nitrogenous wastes
   a. by humans can be in the form of urea and uric acid.
   b. by mammals is never in the form of uric acid.
   c. by marine fishes is in the form of urea.
   d. does not contribute to the osmolarity of the urine.
   e. requires more water if the waste product is the rather insoluble uric acid.
3. How are earthworm metanephridia like mammalian nephrons?
   a. Both process coelomic fluid.
   b. Both take in fluid through a ciliated opening.
   c. Both produce hypertonic urine.
   d. Both employ tubular secretion and resorption to control urine composition.
   e. Both deliver urine to a urinary bladder.
4. What is the role of renal podocytes?
   a. They control the glomerular filtration rate by changing the resistance of renal arterioles.
   b. They resorb most of the glucose that is filtered from the plasma.
   c. They prevent red blood cells and large molecules from entering the renal tubules.

d. They provide a large surface area for tubular secretion and resorption.
e. They release renin when the glomerular filtration rate falls.

5. Which of the following are *not* found in a renal pyramid?
   a. Collecting ducts
   b. Vasa recta
   c. Peritubular capillaries
   d. Convoluted tubules
   e. Loops of Henle
6. Which part of the nephron is responsible for most of the difference in mammals between the glomerular filtration rate and the urine production rate?
   a. The glomerulus
   b. The proximal convoluted tubule
   c. The loop of Henle
   d. The distal convoluted tubule
   e. The collecting duct
7. For mammals of the same size, what feature of their excretory systems would give them the greatest ability to produce a hypertonic urine?
   a. Higher glomerular filtration rate
   b. Longer convoluted tubules
   c. Increased number of nephrons
   d. More permeable collecting ducts
   e. Longer loops of Henle
8. Which of the following would *not* be a response stimulated by a large drop in blood pressure?
   a. Constriction of afferent renal arterioles
   b. Increased release of renin
   c. Increased release of antidiuretic hormone
   d. Increased thirst
   e. Constriction of efferent renal arterioles
9. Which statement about angiotensin is true?
   a. It is secreted by the kidney when the glomerular filtration rate falls.
   b. It is released by the posterior pituitary when blood pressure falls.
   c. It stimulates thirst.
   d. It increases the permeability of the collecting ducts to water.
   e. It decreases glomerular filtration rate when blood pressure rises.
10. Birds that feed on marine animals ingest a lot of salt, but they excrete most of it by means of
   a. Malpighian tubules.
   b. rectal salt glands.
   c. gill membranes.
   d. hypertonic urine.
   e. nasal salt glands.

For Discussion

1. Why is it said that the oceans are a physiological desert? For what animals would this apply?
2. Persons with uncontrolled diabetes mellitus can have very high levels of glucose in their blood. Why do such individuals have a high level of urine production?
3. Inulin is a molecule that is filtered out of the glomerulus, but is not secreted or resorbed by the renal tubules. If you injected inulin into an animal and after a brief time measured the concentration of inulin in its blood and urine, how could you determine the animal's glomerular filtration rate? Assume that the rate of urine production is 1 ml per minute.
4. After you did the inulin experiment to measure glomerular filtration rate, how could you use that information to determine whether another substance is secreted or resorbed by the renal tubules? Assume you can measure the concentration of that substance in the blood and in the urine. Urine production is still 1 ml per minute.
5. Explain what happens with respect to control and regulation of your salt and water balance when you go to a movie and eat a lot of very salty popcorn.

52 Animal Behavior

A troop of Japanese macaques living on an island was being studied by scientists, who fed the monkeys by throwing pieces of sweet potatoes onto the beach from a passing boat. The monkeys tried to brush the sand off the sweet potatoes, but they were still gritty. One day a young female monkey began taking her sweet potatoes to the water and washing them. Soon her siblings and other juveniles in her play group imitated her new behavior. Next their mothers began washing their potatoes. No adult males imitated this behavior, but young males learned the behavior from their mothers and their siblings.

The scientists were fascinated by the way the creative, insightful behavior of one juvenile female spread through the population, so they presented the monkeys with a new challenge: They threw wheat onto the beach. Picking grains of wheat out of the sand was tedious and difficult. The same juvenile female came up with a solution: She carried handfuls of sand and grain to the water and threw them in. The sand sank but the grain floated, enabling her to skim it off the surface and eat it. This behavior spread through the population in the same way potato washing did—first to other juveniles, then to mothers, and then from mothers to both their male and female offspring.

The macaques now routinely wash their food. They play in the water, which they did not do before, and they have added some marine items to their diet. Clearly, this population of monkeys has invented new behaviors that have spread by imitative learning and have become traditions in the population. One could say that they have acquired a *culture:* a set of behaviors shared by members of the population and transmitted by learned traditions.

The reason this study of macaques is so interesting is that it erodes what once seemed to be a clear distinction between human behavior and the behavior of other animals. The behavior of most animals is largely determined by inheritance, with learning playing a relatively minor role. In contrast, most human behaviors are acquired through cultural traditions and learning. However, genetic components contributing to human personality traits and predispositions are also being demonstrated. These observations,

Learned Behaviors Shared by a Population Become a Culture In the space of only a few generations, a population of Japanese macaques (*Macaca fuscata*) learned and transmitted a set of behaviors that included washing food, playing in the water, and eating marine food items—a new "culture" of water-related behaviors.

along with the fact that other primates can invent novel behaviors and pass them on culturally, show that there is no absolute dividing line between human and animal behavior.

We will begin this chapter with descriptions of some classic studies of behavior that is largely shaped by inheritance, but to varying degrees is modified by experience. The rest of the chapter will be devoted to discussions of the physiological mechanisms underlying behavior, the development of behavior, and the modifiability of behavior. Throughout the chapter, use what you read to raise your own questions about human behavior, to which we will return at the end of the chapter.

Argiope aurantia

52.1 Spider Web Designs Are Species-Specific Each spider performs a stereotypic sequence of movements typical of its species that results in a species-specific web design.

What, How, and Why Questions

Most animal species can be identified by certain behaviors. Such behaviors have been shaped by natural selection and are species-specific. On the other hand, behavioral flexibility can be extremely valuable to an animal that has to respond to changing conditions and complex situations, as in social interactions. Therefore, to varying degrees, behavior is modifiable by learning.

In studying any behavior, we can ask *what*, *how*, and *why* questions. *What* questions focus on describing the details of a behavior, including its *proximate* cause—in other words, what stimuli cause the animal to express the behavior. *How* questions are about the mechanisms of behavior—the underlying neuronal, hormonal, and anatomical mechanisms that we have been studying throughout Part Seven of this book. *How* questions can also focus on the means by which an animal acquires a behavior—especially on the relative roles of genetically determined mechanisms and experience. Most behaviors result from complex interactions between inherited anatomical and physiological mechanisms and the ability to modify behavior through learning.

Why questions have to do with the *ultimate* causes of behavior—the selective pressures that shaped its evolution. In this chapter, we will frequently discuss the adaptive nature of behavior, but the evolution of behavior will be the major focus of the Chapter 53.

Behavior Shaped by Inheritance

Much of the behavior of many animals is highly *stereotypic* (it is performed in the same way every time) and *species-specific* (there is little variation in the way different individuals of the same species perform it). We can identify species of spiders, for example, by their web designs (Figure 52.1). Web spinning requires thousands of movements performed in just the right sequence, and for a given species, most of that sequence is performed the same way every time. Different spider species spin webs of different designs, using different sequences of movements.

Web spinning by spiders is also an example of a complex behavior that requires no learning or prior experience. In many species of spiders, when juvenile spiders hatch, their mother is already dead, and they disperse immediately (remember *Charlotte's Web* by E. B. White?). They have no experience of their mother's web. Yet when they construct their own webs, they do it perfectly, without the benefit of experience or a model to copy. In fact, their web-spinning behavior is resistant to modification by learning. When confronted experimentally with obstacles to web construction, young spiders appear incapable of learning how to modify the design of their webs.

Many classic studies of stereotypic and species-specific behavior were performed by scientists who studied the behavior of animals in nature. These scientists became known as **ethologists**. A parallel field of animal behavior studies, called **comparative psychology**, focused on learning by animals—mostly rats and pigeons—in laboratory situations. In contrast, the early ethologists asked to what extent behaviors are determined by genetic inheritance and to what extent they are modifiable by experience.

Experiments can determine whether a behavior is inherited

Two experimental approaches have been used extensively by ethologists to determine whether behavior is inherited. In a *deprivation experiment*, an animal is reared so that it is deprived of all experience relevant to the behavior under study. If it still exhibits the behavior, the behavior is presumed to be inherited. In one such experiment, a tree squirrel was reared in isolation, on a liquid diet, and in a cage without soil or other particulate matter. When the young squirrel was given a nut, it put the nut in its mouth and ran around the cage. Eventually it made stereotypic digging movements in the corner of its cage, placed the nut in the imaginary hole, went through the motions of refilling the hole, and ended by tamping the nonexistent soil with its nose. The squirrel had never handled a food object and had never experienced soil, yet the stereotypic behavior of a squirrel burying a nut was fully expressed.

In a *hybridization experiment*, closely related species are interbred and the behavior of their offspring is observed. Many closely related species show distinct differences in certain kinds of behavior. When such species can be interbred, it is possible to see whether their offspring have inherited elements of the behavior of one or both parents.

Konrad Lorenz, a pioneer in the field of ethology, conducted hybridization experiments on ducks to investigate the hereditary basis of their elaborate courtship displays. Dabbling duck species such as mallards, teals, pintails, and gadwalls are closely related to one another and can interbreed, but because of the specificity of their courtship displays, they rarely do so in nature. Each male duck performs a carefully choreographed water ballet that is typical of his species (Figure 52.2). A female is not likely to accept his advances unless the entire display is successfully and correctly completed.

When Lorenz crossbred dabbling duck species, he found that the hybrid offspring expressed some elements of the courtship displays of each parent species, but expressed them in new combinations. Of particular interest was his observation that the hybrids sometimes showed display components that were not in the repertoire of either parent species, but were characteristic of the displays of other species. Lorenz's hybridization studies clearly demonstrated that the stereotypic motor patterns of the courtship displays were inherited. The observation that females were not interested in males performing hybrid displays was evidence that natural selection was shaping these genetically determined behaviors.

Simple stimuli can trigger behavior

If a behavior is not expressed during a deprivation experiment, it may nonetheless have genetic determinants. The right conditions may not have been available to stimulate the behavior during the experiment. The squirrel described above, for example, had to be given a nut for its digging and burying behaviors to be triggered. Specific stimuli are required to elicit the expression of many inherited behaviors. Konrad Lorenz and Niko Tinbergen conducted classic studies of the nature of the stimuli that elicit such behaviors and called such stimuli **releasers**.

Releasers are usually a specific component of the sensory information available to an animal. Adult male European robins, for example, have red feathers on their breasts, which serve as releasers of aggressive behavior in other males. During the breeding season, the sight of an adult male robin stimulates another male robin to sing, perform aggressive displays, and attack the intruder if he does not heed these warnings and retreat. An immature male robin, whose feathers are all brown, does not elicit this aggressive behavior. A tuft of red feathers on a stick, however, is a sufficient releaser for male aggressive behavior, even though it looks nothing like a real robin.

Tinbergen and A. C. Perdeck carefully examined the releasers involved in the interactions between herring gulls and their chicks during feeding. An adult herring gull has a red dot at the end of its bill (Figure 52.3). When the gull returns to its nest with food in its stomach, its chicks beg for food by pecking at the red dot, thereby stimulating the adult to regurgitate the food for the chicks to eat. Tinbergen and Perdeck hypothesized that the red dot was a releaser for the chicks' begging behavior.

To test their hypothesis, Tinbergen and Perdeck made paper cutout models of gull heads and bills, varying the colors and the shapes. Then they rated each model according to how many begging pecks it received from naive, newly hatched gull chicks (see Figure 52.3). The shape or color of the model head made no difference. In fact, a head was not even necessary; the chicks responded just as well to models of bills alone—as long as they included the red dot. The

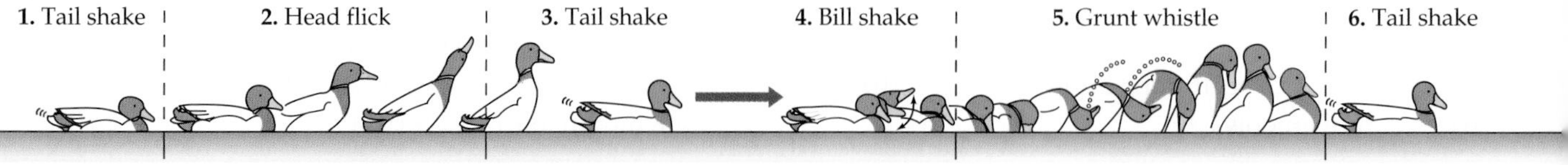

52.2 The Mallard Courtship Ballet The courtship display of the male mallard duck contains ten elements. The displays of closely related duck species contain some of the same ten elements, but have other elements not displayed by mallards. The elements of the courtship display and their sequence are species-specific and act to prevent hybridization.

1. Tail shake | **2.** Head flick | **3.** Tail shake | **4.** Bill shake | **5.** Grunt whistle | **6.** Tail shake

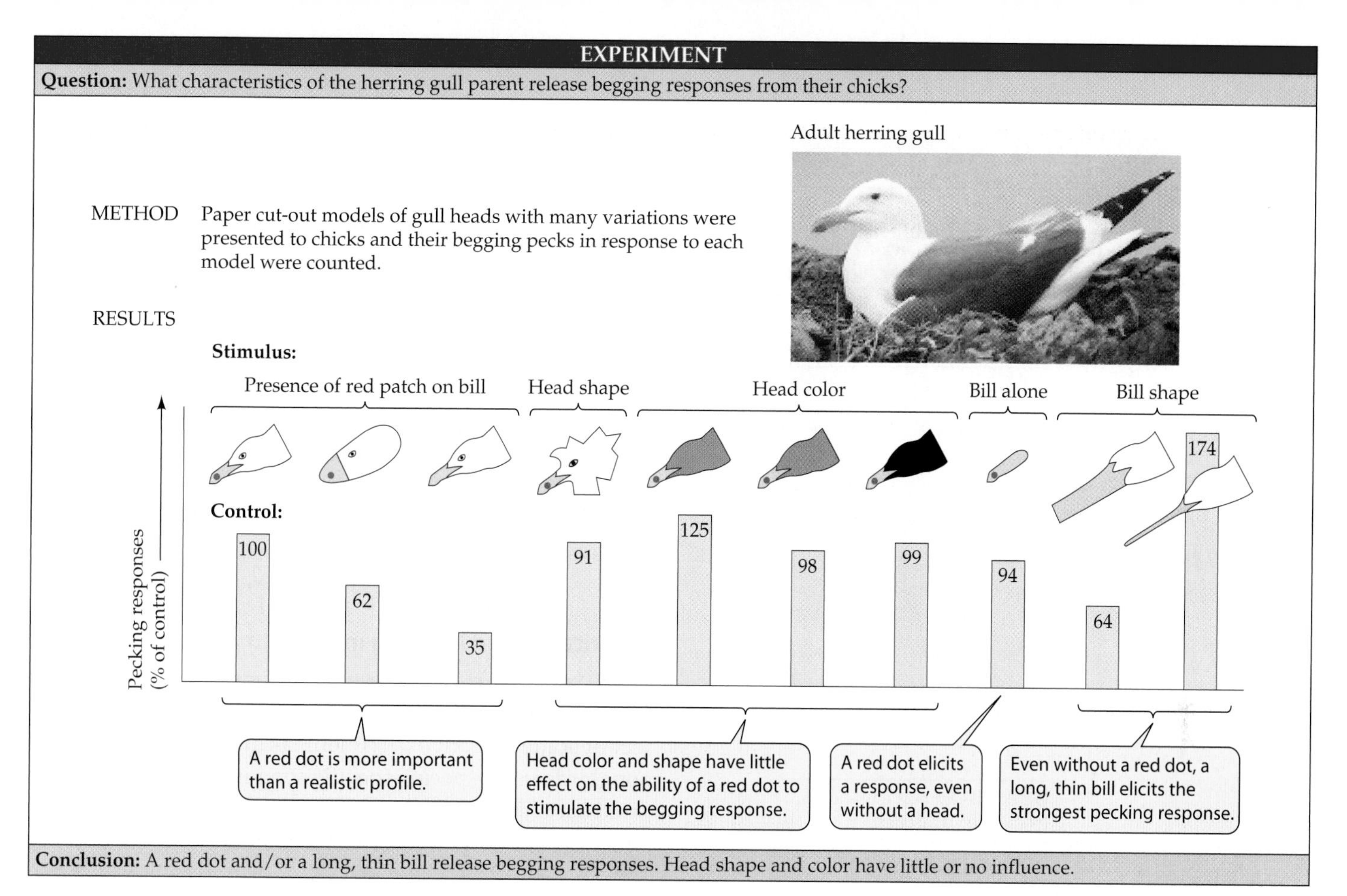

EXPERIMENT
Question: What characteristics of the herring gull parent release begging responses from their chicks?
Adult herring gull
METHOD
Paper cut-out models of gull heads with many variations were presented to chicks and their begging pecks in response to each model were counted.
RESULTS
Stimulus:
Presence of red patch on bill
Head shape
Head color
Bill alone
Bill shape
Control:
Pecking responses (% of control)
100
62
35
91
125
98
99
94
64
174
A red dot is more important than a realistic profile.
Head color and shape have little effect on the ability of a red dot to stimulate the begging response.
A red dot elicits a response, even without a head.
Even without a red dot, a long, thin bill elicits the strongest pecking response.
Conclusion: A red dot and/or a long, thin bill release begging responses. Head shape and color have little or no influence.

52.3 Releasing the Begging Response A series of experiments rated the begging responses of herring gull chicks to artificial models of gull heads and bills to discover which features of the parent—head shape, bill shape, or red dot—were releasers of this behavior.

shape of the bill was important, however; a model with a long, thin bill elicited stronger responses than a realistic model. Clearly the chicks had inherited the ability to recognize a simple set of stimuli and respond to them with their also inherited begging behavior. To the ethologists, this finding represented an example of a behavior that was genetically determined rather than learned.

Learning also shapes behavior

Lorenz, Tinbergen, and Karl von Frisch (whose work on honeybees you will encounter later in this chapter) shared a Nobel prize in Physiology and Medicine in 1973 for their contributions to our understanding of animal behavior. New generations of behavioral biologists, however, have moved beyond the ethologists' focus on inherited behavior to show that most behavior involves an interaction between inheritance and learning.

The begging behavior of gull chicks is an example of such an interaction. Although newly hatched chicks respond maximally to simplistic artificial releasers, they gradually learn to discriminate between models and real gull heads, and they eventually beg only from their own parents. Thus, the inherited ability to recognize a simple releaser is subsequently refined by learning.

The early ethologists did not ignore learning or deny that it took place; in fact, they pioneered the study of learning. Tinbergen performed an early study of *spatial learning*, by which an animal learns to recognize features in its environment. In a classic experiment, he placed objects such as pine cones near the entrance of a nest dug by a female digger wasp. After the wasp left her nest, he moved the objects a short distance away. Upon returning, the wasp oriented to the moved objects and could not find her nest entrance (Figure 52.4). She had learned to recognize objects in the environment to use as orientation cues.

7. Head up, tail up | **8.** Turn toward female | **9.** Nod swimming | **10.** Turn the back of the head

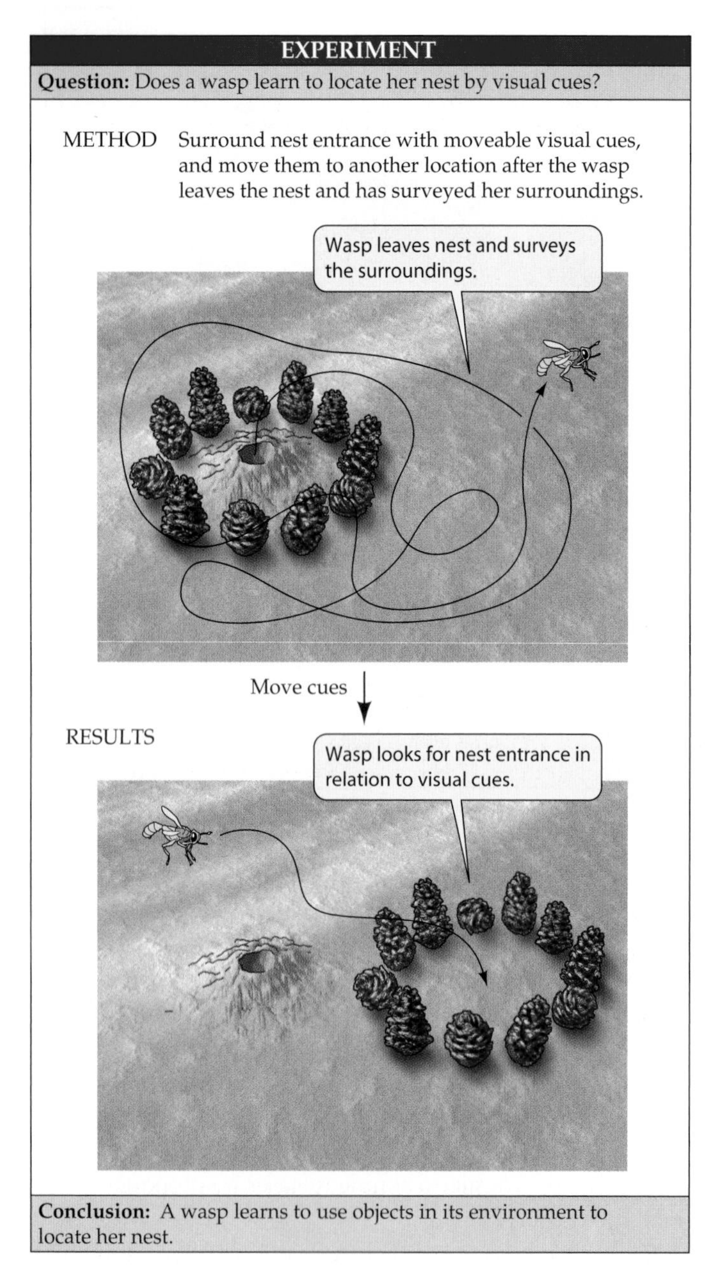

EXPERIMENT
Question: Does a wasp learn to locate her nest by visual cues?
METHOD Surround nest entrance with moveable visual cues, and move them to another location after the wasp leaves the nest and has surveyed her surroundings.
Wasp leaves nest and surveys the surroundings.
Move cues
RESULTS
Wasp looks for nest entrance in relation to visual cues.
Conclusion: A wasp learns to use objects in its environment to locate her nest.

52.4 Spatial Learning Tinbergen's classic experiment showed that a female digger wasp learns the positions of objects in her environment.

Imprinting is the learning of a complex releaser

Releasers are generally simple subsets of the information available to an animal because there are limits to what can be programmed genetically. A type of learning called **imprinting** makes it possible to learn, during a limited **critical period**, a complex set of stimuli that can later serve as a releaser. The classic example is the imprinting of offspring on their parents and parents on their offspring to ensure individual recognition even in a crowded situation such as a colony or a herd.

Lorenz hatched goose eggs in an incubator, and because he was the first thing the goslings saw when they hatched, they imprinted on him. The goslings followed him everywhere, interacting with him as if he were their parent (Figure 52.5). When the experiment was repeated by his assistants, each wearing boots with a different design, the goslings imprinted on the boots, and would follow only a person wearing the boots they first saw when they hatched.

The critical period for imprinting is determined by a developmental or hormonal state and can be quite brief. If a mother goat, for example, does not nuzzle and lick her newborn within 5 to 10 minutes after its birth, she will not recognize it as her own later. In this case, imprinting depends on olfactory cues, and the critical period is determined by the high levels of the hormone oxytocin in the mother's circulatory system at the time of birth.

Inheritance and learning interact to produce bird song

Many behavior patterns result from intricate interactions of inheritance and learning. One example that has been the subject of some elegant experiments is bird song. Adult male songbirds use a species-specific song in territorial displays and courtship. For most species, such as the white-crowned sparrow, learning is an essential step in the acquisition of song.

If the eggs of white-crowned sparrows are hatched in an incubator and the young male birds are reared in isolation, their adult songs will be unusual assemblages of sounds, not the typical species-specific song. White-crowned sparrows cannot express their species-specific song without imprinting on that song as nestlings (Figure 52.6, experiment 1). But even though the male white-crowned sparrow must hear the song of his own species as a nestling to sing it as an adult, he does not sing it as a juvenile. Instead, the auditory imprinting he experiences as a nestling forms a song memory in his nervous system.

52.5 Imprinting Enables an Animal to Learn a Complex Releaser Because ethologist Konrad Lorenz was the first thing these goslings saw when they hatched, they imprinted on him, and interacted with him thereafter as if he were their parent.

EXPERIMENT 1

Question: Is learning essential for song acquisition in white-crowned sparrows?

METHOD Raise young sparrows in the presence of an adult male sparrow singing. Record song of these control birds when they mature and plot as a sonogram.

RESULTS

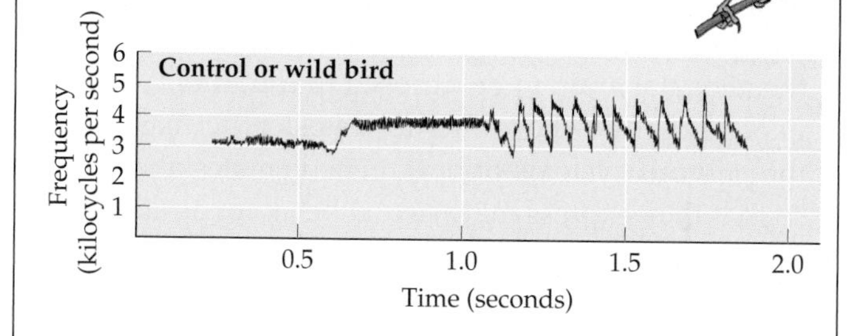

METHOD Hatch eggs in an incubator and rear the birds in isolation. Record and plot their song. Compare to the control birds' song.

RESULTS

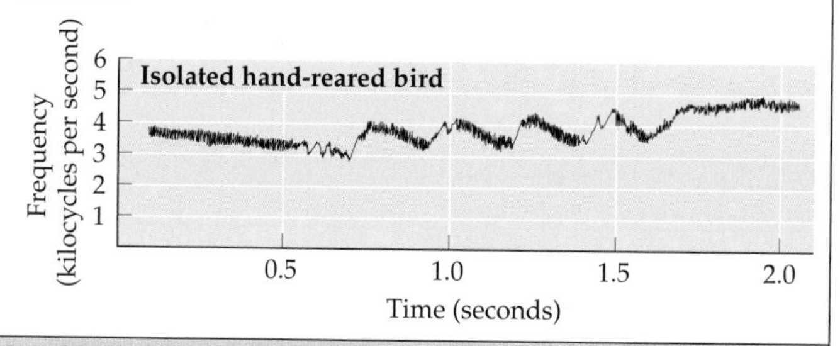

Conclusion: White-crowned sparrows that do not hear adult song as nestlings do not express their species-specific song when they mature. Therefore nestling birds form a song memory.

EXPERIMENT 2

Question: Do maturing white-crowned sparrows require auditory feedback to learn to express their species-specific song?

METHOD Deafen a subadult bird that has heard the song of an adult male when he was a nestling.

RESULTS

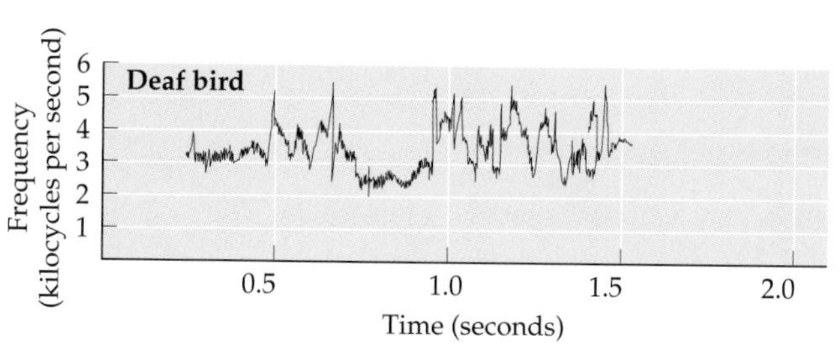

Conclusion: Even if the bird has formed a song memory, he needs auditory feedback to learn to match it.

52.6 Two Critical Periods for Song Learning To sing his species-specific song as an adult, a male white-crowned sparrow must acquire a song memory by hearing the song as a nestling, and must be able to hear himself as he attempts to match his singing to that song memory.

As the young male sparrow approaches sexual maturity the following spring, he tries to sing, and eventually he matches his imprinted song memory through trial and error. If a bird that has heard his species-specific song as a nestling is deafened before he begins to express his song, he will not develop his species-specific song (Figure 52.6, experiment 2). The bird must be able to hear himself to match his song memory. If he is deafened *after* he expresses his correct species-specific song, however, he will continue to sing like a normal bird. Two periods of learning are essential: the first in the nestling stage, the second as the bird approaches sexual maturity.

Genetically determined behavior is adaptive under certain conditions

The ability to learn and to modify behavior as a result of experience is often highly adaptive. Why, then, are so many behavior patterns in so many species genetically determined? We have already touched on one answer to this question: If role models and opportunities to learn are not available—as in species with nonoverlapping generations, such as spiders—then there is no alternative to inherited behavior.

Inherited behavior is also adaptive when mistakes are costly or dangerous. Mating with a member of the wrong species is a costly mistake; thus the function of much courtship behavior, such as that of dabbling ducks, is to guarantee species recognition. In an environment in which incorrect as well as correct models exist, learning the wrong pattern of courtship behavior would be possible.

Inheritance of behavior patterns used to avoid predators or capture dangerous prey is also adaptive. These situations allow no room for mistakes: If the behavior is not performed promptly and accurately the first time, there may not be a second chance.

Hormones and Behavior

All behavior depends on the nervous system for initiation, coordination, and execution. Frequently, however, it is the endocrine system, through its controlling influences on the development and the physiological state of the animal, that determines when a particular behavior is performed, and even when certain behaviors can be learned. In this section we will present two complex cases in which hormones control the development, learning, and expression of behavior: the development and expression of sexual behavior in rats and the maturation of the brain regions required for song learning and expression in birds.

Sex steroids determine the development and expression of sexual behavior in rats

Differences in the behavior of males and females of a species are clear examples of genes influencing the development and expression of behavior. Such sex differences in behavior are the result of actions of the sex steroids on the brain.

Rats, like most other animals, have stereotypic sexual behaviors. A female rat in *estrus* (a period of sexual receptivity) responds to tactile stimulation of her hindquarters by assuming a mating posture called *lordosis*. A male rat encountering a female in estrus engages in stereotypic copulatory behavior. The roles of genes and sex steroids in the development and expression of lordosis and male copulatory behavior have been investigated through experiments that manipulated the exposure of the newborn and adult rat brain to sex steroids. Experiments such as those shown in Figure 52.7 led to three conclusions:

- Development of male sexual behavior requires the brain of the newborn rat to be exposed to testosterone, but development of female sexual behavior does not require the neonatal brain to be exposed to estradiol.
- Exposure of newborn rats to testosterone masculinizes the nervous systems of both genetic males and females so that they express male sexual behavior as adults.
- Sex steroids are necessary for adult rats to express sexual behavior. Moreover, the male sex steroid, testosterone, has an effect only in males, and the female sex steroid, estradiol, has an effect only in females.

Thus, the sex steroids that are present during development determine which pattern of sexual behavior develops, and

52.7 Hormonal Control of Sexual Behavior Experimental hormonal treatments of rats that have been spayed (had their ovaries or testes removed) demonstrate that the sex steroids present during development determine what sexual behavior patterns develop, and the sex steroids present in the adult control the expression of those patterns.

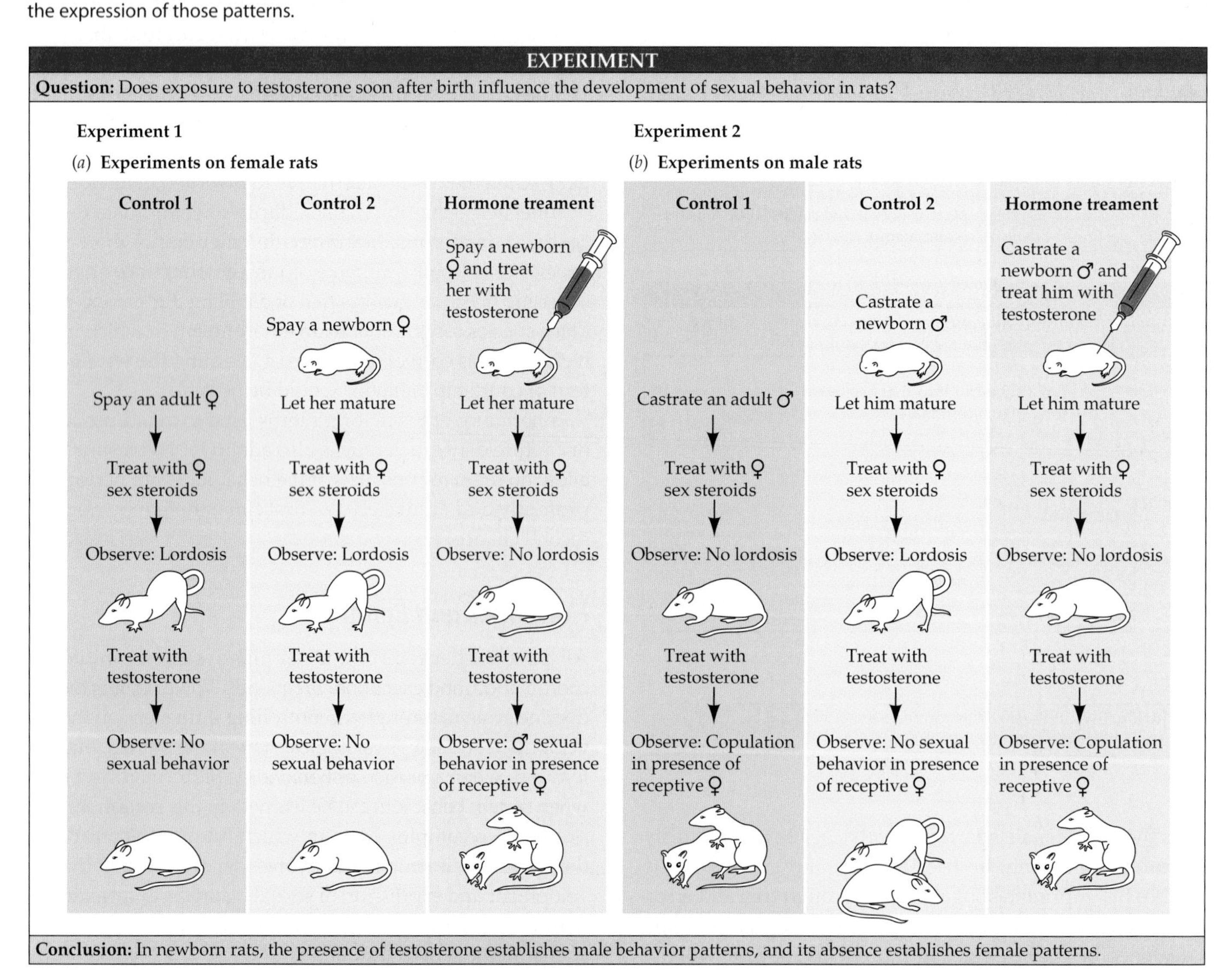

the sex steroids that are present in adulthood determine whether that pattern is expressed.

Testosterone affects the development of the brain regions responsible for song in birds

We have already seen that learning is essential for the acquisition of bird song. Both male and female songbirds hear their species-specific song as nestlings, but only the males of most species sing as adults. Male birds use song to claim and advertise territory, compete with other males, and declare dominance. They also use song to attract females, which suggests that the females know the song of their species even if they do not sing it. Do sex steroids control the learning and expression of song in male and female songbirds?

After leaving the nest where they heard their father's song, young songbirds from temperate and Arctic habitats migrate and associate with other species in mixed-species flocks. During this time they do not sing, and they do not hear their species-specific song again until the following spring. As that spring approaches and the days become longer, the young male's testes begin to grow and mature. As his testosterone level rises, he begins to try to sing, matching his own vocalizations to his imprinted memory of his father's song. Even if he is isolated at this time from all other males of his species, his song will gradually improve until it is a proper rendition of his species-specific song. At that point, the song is *crystallized*—the bird expresses it in similar form every spring thereafter.

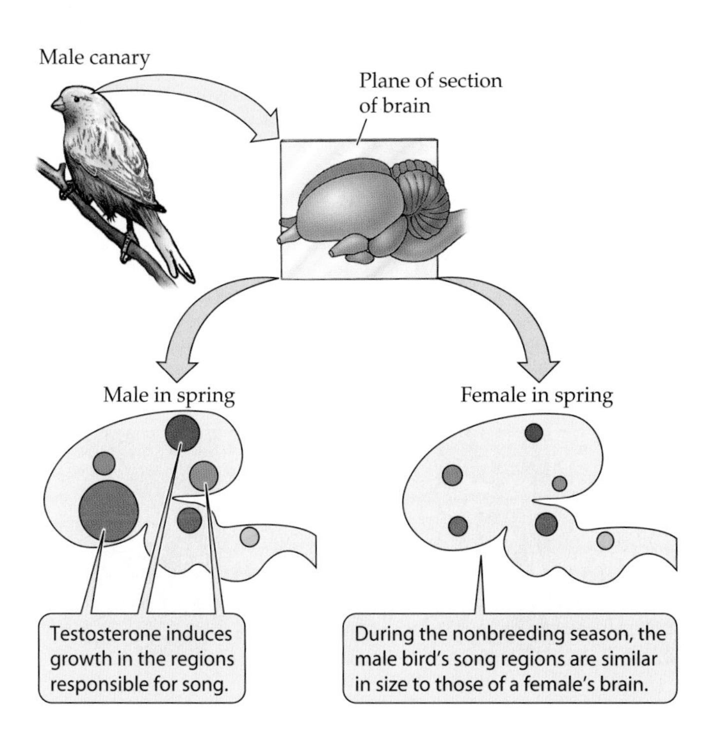
Male canary
Plane of section of brain
Male in spring
Female in spring
Testosterone induces growth in the regions responsible for song.
During the nonbreeding season, the male bird's song regions are similar in size to those of a female's brain.

52.8 Effects of Testosterone on Bird Brains In spring, rising testosterone levels in the male cause the song regions of the brain to develop. The size of each circle is proportional to the volume of the brain occupied by that region.

Why don't the females of most songbird species sing? Can't they learn the patterns of their species-specific song? Do they lack the muscular or nervous system capabilities necessary to sing? Or do they simply lack the hormonal stimulus for developing the behavior? To answer these questions, investigators injected female songbirds with testosterone in the spring. In response to these injections, the females developed their species-specific song and sang just as the males did. Apparently females form a memory of their species-specific song when they are nestlings and have the capability to express it, but they normally lack the hormonal stimulation.

What does testosterone do to the brain of the songbird? A remarkable discovery revealed that each spring, an increase in testosterone causes certain parts of the male's brain necessary for learning and expressing song to grow larger (Figure 52.8). Individual neurons in those regions of the brain increase in size and grow longer extensions, and the numbers of neurons in those regions increase. Such research on the neurobiology of bird song has revealed that hormones can control behavior by influencing brain structure as well as brain function, both developmentally and seasonally.

The Genetics of Behavior

To say that behavior is inherited does not mean that specific genes code for specific behaviors. Genes code for proteins, and there are many complex steps between the expression of a gene as a protein product and the expression of a behavior. In no case are all the steps between a gene and its influence on a behavior known. Nevertheless, it is clear that behavior has genetic determinants. In this section we will look at three approaches to investigating how genes affect behavior: hybridization, artificial selection and crossing of the selected strains, and molecular analysis of genes and gene products.

Hybridization experiments show whether a behavior is genetically determined

The effects of hybridization on the courtship displays of dabbling duck species were the subject of a classic ethological experiment, as we saw above. A more recent set of hybridization experiments was performed on crickets. The songs of crickets, like bird songs, are species-specific, and as in birds, only male crickets "sing." They do so by rubbing one wing against another wing that has a serrated edge. The sounds they produce can be recorded and analyzed quantitatively.

When two species of crickets were crossed, their offspring (the F_1 generation) expressed songs that had features of the songs of both parental species. Backcrosses of F_1 individuals with the parental species produced individuals that had songs closer to that of the parental species used in the backcross. Clearly the genetic background determined the song pattern. What was amazing, however, was the demonstration that female preferences for male songs were under similar genetic control. Given a choice, females from each parental species preferred the calls of males from their own species, but hybrid females preferred the calls of hybrid males.

These genetic differences between the two parental cricket species and the hybrids were reflected in the properties of their nervous systems. When specific neurons in the crickets' brains were stimulated, songs that reflected the genotypes of the crickets were expressed.

Artificial selection and crossbreeding experiments reveal the genetic complexity of behavior

Domesticated animals provide abundant evidence that artificial selection of mating pairs on the basis of their behavior can result in strains with distinct behavioral as well as anatomical characteristics. Among dogs, consider retrievers, pointers, and shepherds. Each has a particular behavioral tendency that can be honed to a fine degree by training. However, dogs and other large animals are not the best subjects for genetic studies. Most artificial selection experiments in behavioral genetics have been done on more convenient laboratory animals with short life cycles and large numbers of offspring, such as the fruit fly (*Drosophila*).

Artificial selection has been successful in shaping a variety of behavior patterns in fruit flies, especially aspects of their courtship and mating behavior. Crossing of these artificially selected strains reveals that most of these behavioral differences are due to multiple genes that probably influence the behavior indirectly by altering general properties of the nervous system.

Few behavioral genetic studies reveal simple Mendelian segregation of behavioral traits. An exception is nest-cleaning behavior in honeybees. One genetic strain of honeybees practices nest-cleaning, or *hygienic*, behavior, which makes them resistant to a bacterium that infects and kills the larvae of honeybees. When a larva dies, workers of this strain uncap its brood cell and remove the carcass from the hive. Another strain of honeybees does not show this hygienic behavior and therefore is more susceptible to the spread of the disease (Figure 52.9).

When these two strains of honeybees were crossed, all members of the F_1 generation were nonhygienic, indicating that the behavior is controlled by recessive alleles. Backcrossing the F_1 with the hygienic parental strain produced the typical 3:1 ratio expected for a two-gene trait in the F_2 generation (see Chapter 10). The behavior of the nonhygienic F_2 individuals was also interesting. One-third of them showed no hygienic behavior at all; one-third uncapped the cells of dead larvae, but did not remove them; and one-third did not uncap cells, but did remove carcasses if the cells were open.

Even though these results suggest a gene for uncapping and a gene for removal, these behavior patterns are complex. They involve sensory mechanisms, orientation movements, and motor patterns, each of which depends on multiple properties of many cells. The genetic deficits of nonhygienic bees could influence very small, specific, yet critical properties of some cells. If even a single critical prop-

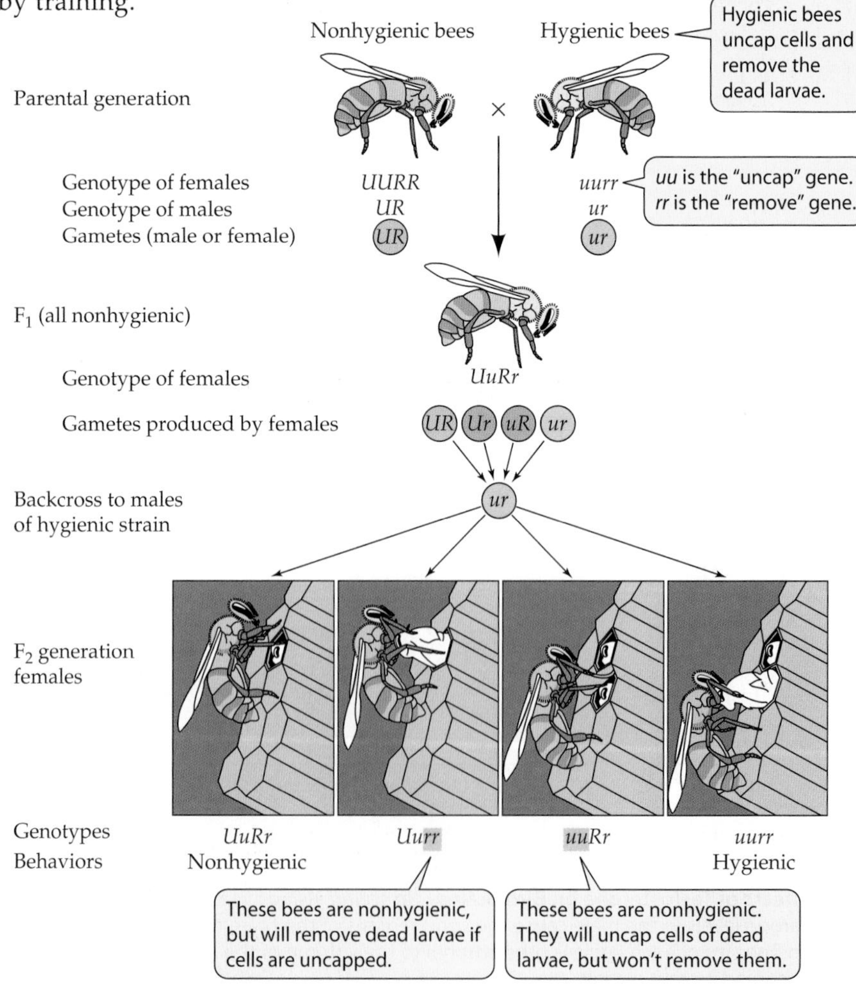

52.9 Genes and Hygienic Behavior in Honeybees Some honeybee strains remove the carcasses of dead larvae from their nests. This behavior seems to have two components: uncapping the larval cell (*u*) and removing the carcass (*r*), each of which is under the control of a recessive gene.

erty, such as a crucial synapse or a particular sensory receptor, were lacking, the whole behavior pattern would fail to be expressed. Thus there is no specific gene that codes for the entire behavior.

Molecular genetic techniques reveal specific genes that influence behavior

Molecular geneticists are investigating specific genes that influence behavior. Male courtship behavior in fruit flies is a subject of many such studies. This behavior is stereotypic, species-specific, and requires no learning. A male recognizes a potential mate, follows her, taps her body with his forelegs, extends and vibrates one wing, and licks the female's genitals. If the female is receptive, the male copulates with her (Figure 52.10*a*). Research in molecular genetics has now shown that most of this male courtship behavior is controlled by a single gene.

In fruit flies with two X chromosomes (females), a gene called *sex-lethal* (*sxl*) is expressed. This gene is at the top of a genetic hierarchy that determines all aspects of sexual differentiation and behavior in fruit flies (Figure 52.10*b*). The Sxl protein causes another gene, called *transformer* (*tra*), to produce the female-specific Tra protein. Fruit flies without the *tra* gene develop into males anatomically and behaviorally, regardless of how many X chromosomes they have. But it is still another gene in the sex determination hierarchy that is responsible for male behavior.

In the absence of the Tra protein, two additional genes, called *doublesex* (*dsx*) and *fruitless* (*fru*), are expressed. The *dsx* gene controls the anatomical differentiation of males, and *fru* causes the formation of a nervous system that expresses male courtship behavior. Mutations of the *fru* gene do not affect male body form, but they disrupt male courtship behavior. We don't know all of the actions that the male-specific Fru protein has in the development of the fruit fly nervous system, but this is about as close as we can get at present to identifying a gene that controls a complex behavior.

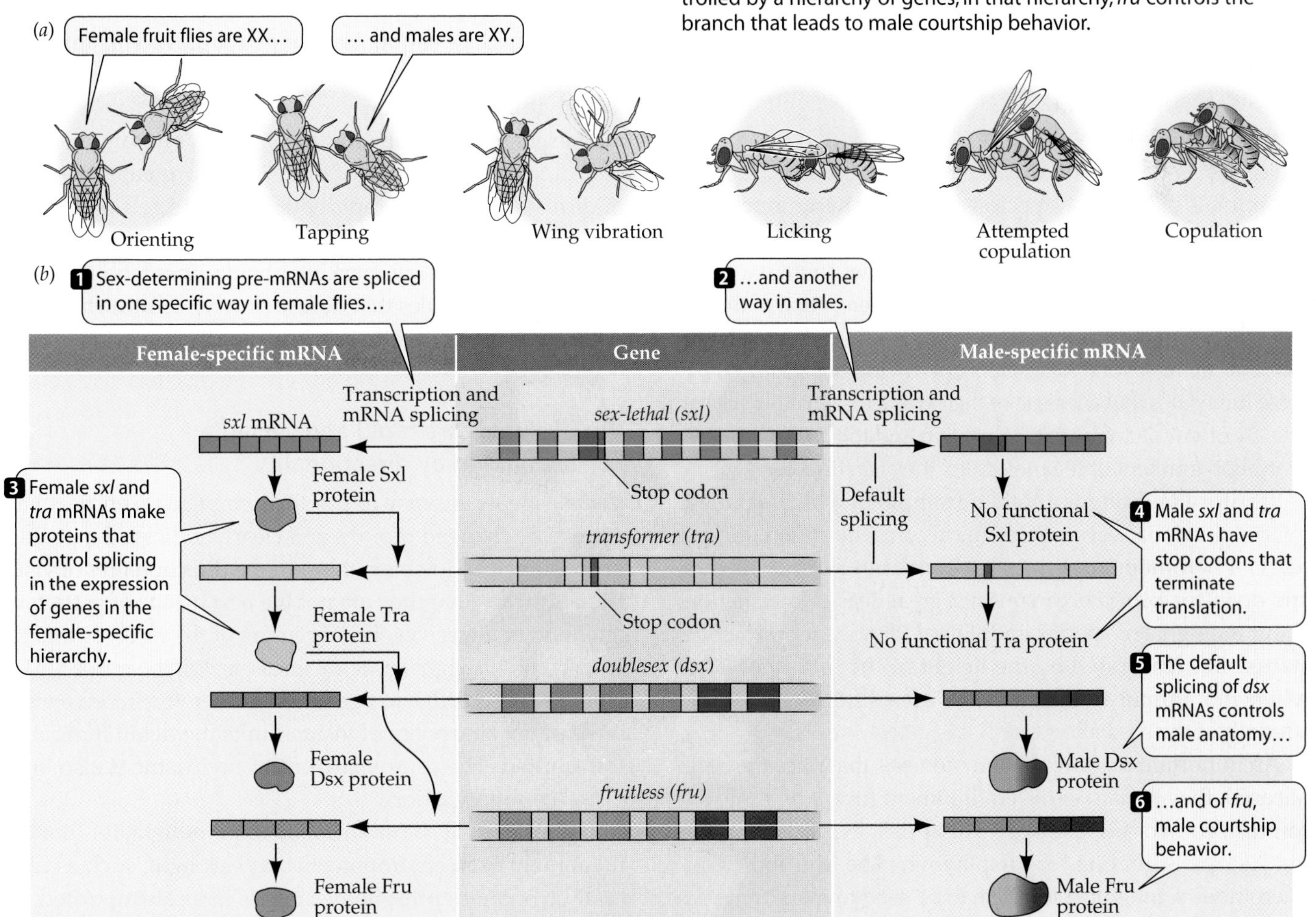

52.10 The *fruitless* Gene Controls Male Courtship Behavior in Fruit Flies (*a*) Male fruit flies display stereotypic, species-specific courtship behavior. (*b*) Sexual differentiation in *Drosophila* is controlled by a hierarchy of genes; in that hierarchy, *fru* controls the branch that leads to male courtship behavior.

Communication

Communication is behavior that influences the actions of other individuals. It consists of **displays** or **signals** that can be perceived by other individuals and which convey information to them. Natural selection shapes displays or signals into systems of communication if the transmission of information benefits both the sender and the receiver. Thus, the ultimate cause of communication is the selective advantage it gives to individuals that engage in it. The courtship displays of a male, for example, benefit the male if they attract females. The displays also benefit the female if they allow her to assess whether the male is of the right species and whether he is strong, vigorous, and has other attributes that will make him a good father. A common mutual benefit of communication is the reduction of uncertainty about the status or intentions of the signaler. Even in aggressive interactions, reducing uncertainty helps both sender and receiver to avoid physical harm.

Studies of communication can be complex because they must take into account the sender, the receiver, and the environment. The displays or signals that an animal can generate depend on its physiology and anatomy. Likewise, an animal's ability to perceive displays or signals depends on its sensory physiology and on the environment through which the display or signal must be transmitted.

Chemical signals are durable

Molecules used for chemical communication between individual animals of the same species are called **pheromones**. Because of the diversity of their molecular structures, pheromones can communicate very specific messages that contain a great deal of information. The mate attraction pheromone of the female silkworm moth is a good example (see Figure 45.3). Male moths as far as several kilometers downwind are informed by these molecules that a female of their species is sexually receptive. By orienting to the wind direction and following the concentration gradient of the molecules, they can find her.

Territory marking is another example in which detailed information is conveyed by chemical communication (Figure 52.11). Pheromone messages left by mammals such as cats and dogs, for example, can reveal a great deal of information about the signaler: species, individual identity, reproductive status, size (indicated by the height of the message), and when the animal was last in the area (indicated by the strength of the scent).

An important feature of pheromones is that once they are released, they remain in the environment for a long time. By contrast, vocal or visual displays disappear as soon as the animal stops signaling or displaying. The durability of pheromone signals enables them to be used to mark trails, as ants do, to mark territory, or to indicate directionality, as in

Panthera tigris

52.11 Many Animals Communicate with Pheromones To mark her territory, this female tiger is spraying urine, which contains pheromonal secretions from a scent gland in her hindquarters, onto a tree. Other tigers passing the spot will know that the area is "claimed," and they will know something about the animal who claimed it.

the case of the moth sex attractant. However, it also means that the message cannot be changed rapidly. This inflexibility makes pheromonal communication unsuitable for a rapid exchange of information.

The chemical nature and the size of a pheromone molecule determine its speed of diffusion. The greater the speed of diffusion, the more rapidly the message gets out and the farther it will travel, but the sooner it will disappear. Trail-marking and territory-marking pheromones tend to be relatively large molecules that diffuse slowly; sex attractants tend to be small molecules that diffuse rapidly.

Visual signals are rapid and versatile but are limited by directionality

Visual signals are easy to produce, come in an endless variety, can be changed rapidly, and clearly indicate the position of the signaler. However, the extreme directionality of visual signals means that they are not the best means of getting the attention of a receiver. The receptors of the receiver must be focused on the signaler, or the message will be missed. Most animals are sensitive to light and can therefore receive visual signals, but sharpness of vision limits the detail that can be transmitted. The complexity of the environment also limits visual communication.

Because visual communication requires light, it is not useful at night or in environments that lack light, such as caves and the ocean depths. Some species have surmounted this constraint by evolving their own light-emitting mechanisms.

Fireflies, for example, use a enzymatic mechanism to create flashes of light. By emitting flashes in species-specific patterns, fireflies can advertise for mates at night.

Fireflies also illustrate how some species can exploit the communication systems of other species. There are predatory species of fireflies that mimic the mating flashes of other species. When an eager suitor approaches the signaling individual, it is eaten. Thus, deception can be part of animal communication systems, just as it is part of human use of language.

Auditory signals communicate well over a distance

Compared with visual communication, auditory communication has advantages and disadvantages. Sound can be used at night and in dark environments. It can go around objects that would interfere with visual signals, so it can be used in complex environments such as forests. It is better than visual signals at getting the attention of a receiver because the receiver does not have to be focused on the signaler for the message to be received.

Like visual signals, sound can provide directional information, as we saw in Chapter 44, as long as the receiver has at least two receptors spaced somewhat apart. By maximizing or minimizing certain features of the sounds they emit, animals can make their location easier or more difficult to determine.

Sound is useful for communicating over long distances. Even though the intensity of sound decreases with distance from the source, loud sounds can be used to communicate over distances much greater than those possible with visual signals. An extreme example is the communication of whales. Some whales, such as the humpback, have complex songs. When these sounds are produced at a certain depth (around 1,000 m), they can be heard hundreds of kilometers away. In this way, humpback whales can locate each other over vast areas of ocean.

Auditory signals cannot convey complex information as rapidly as visual signals can, as is implied by the expression "A picture is worth a thousand words." When individuals are in visual contact, an enormous amount of information is exchanged instantaneously (for example, species, sex, individual identity, reproductive status, level of motivation, dominance, vigor, alliances with other individuals, and so on). Coding that amount of information, with all of its subtleties, as auditory signals would take considerable time, thus increasing the possibility that the communicators could be located by predators.

Tactile signals can communicate complex messages

Communication by touch is common, although not always obvious. Animals in close contact use tactile interactions extensively, especially under conditions that do not favor visual communication. When eusocial insects such as ants, termites, or bees meet, they contact each other with their antennae and front legs.

One of the best-studied uses of tactile communication, beginning with the work of Karl von Frisch, is the dance of honeybees. When a forager bee finds food, she returns to the hive and communicates her discovery to her hivemates by dancing in the dark on the vertical surface of the honeycomb. The dance is monitored by other bees, who follow and touch the dancer to interpret the message.

If the food source is more than ~80 meters away from the hive, the bee performs a *waggle dance* (Figure 52.12), which conveys information about both the distance and the direction of the food source. The bee repeatedly traces out a figure-eight pattern as she runs on the vertical surface. She alternates half-circles to the left and right with vigorous wagging of her abdomen in the short, straight run between turns. The angle of the straight run indicates the direction of the food source relative to the direction of the sun. The speed

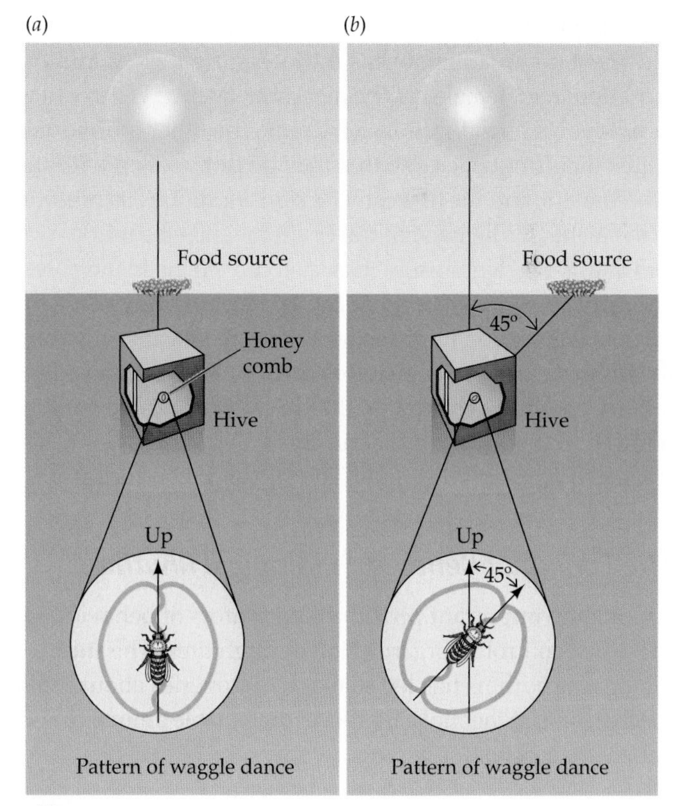

52.12 The Waggle Dance of the Honeybee *(a)* A honeybee runs straight up on the vertical surface of the honeycomb in a dark hive while waggling her abdomen to tell her hivemates that there is a food source in the direction of the sun. The intensity of the waggle indicates exactly how far the food source is. If the food source were in the opposite direction from the sun, she would orient her waggle runs straight down. *(b)* When her waggle runs are at an angle from the vertical, the other bees know that the same angle separates the direction of the food source from the direction of the sun.

of the dancing indicates the distance to the food source: The farther away it is, the slower the waggle run.

If the food she has found is less than 80 meters from the hive, the forager performs a *round dance*, running rapidly in a circle and reversing her direction after each circumference. The odor on her body and the round dance combine tactile and chemical cues: The odor indicates the flower to be looked for, and the dance communicates the fact that the food source is within 100 meters of the hive.

Electric signals can also communicate messages

Some species of fish have evolved the ability to generate electric fields in the water around them by emitting a series of electric pulses (see Chapter 45). These trains of electric pulses can be used for sensing objects in the immediate surroundings, and they can also be used for communication.

An electrode connected to an amplifier and a speaker can be used to "listen" to the signals generated by glass knife fish in a tank. Each individual fish emits a pulse at a different frequency, and the frequency each fish uses relates to its status in the population. Males emit lower frequencies than females. The most dominant male has the lowest frequency, and the most dominant female has the highest frequency. When a new individual is introduced into the tank, the other individuals adjust their frequencies so that they do not overlap with the newcomer's, and the newcomer's signal indicates its position in the hierarchy. In their natural environment—the murky waters of tropical rainforests—these fish can tell the identity, sex, and social position of another fish by its electric signals.

Communication has been a fruitful area for investigating the ultimate causes of behavior and how the resulting adaptations have been shaped by the environment. Next we will return to some studies of proximate causes of behavior to see some examples of how "how" questions can be addressed.

The Timing of Behavior: Biological Rhythms

Among the important proximate regulators of behavior are those that control its organization through time. The study of biological rhythms has led to major discoveries about brain mechanisms at the molecular level that enable animals to organize their behavior in time.

Circadian rhythms control the daily cycle of behavior

Our planet turns on its axis once every 24 hours, creating a cycle of environmental conditions that has existed throughout the evolution of life. Daily biological cycles are characteristic of almost all organisms. What is surprising, however, is that this daily rhythmicity does not depend on the 24-hour cycle of light and dark.

If animals are kept in constant darkness, at a constant temperature with food and water available all the time, they still demonstrate daily cycles of activities such as sleeping, eating, drinking, and just about anything else that can be measured. The persistence of these daily cycles in the absence of environmental time cues suggests that animals have an endogenous (internal) clock. Without time cues from the environment, however, these daily cycles are not exactly 24 hours long. They are therefore called **circadian rhythms** (from the Latin *circa*, "about," and *dies*, "day").

To discuss biological rhythms, we must review some terminology that was introduced in Chapter 39. A biological rhythm can be thought of as a series of cycles, and the length of one of those cycles is the *period* of the rhythm (see Figure 39.13). Any point on the cycle is a *phase* of that cycle: Hence, when two rhythms completely match, they are *in phase*, and if a rhythm is shifted (as in the resetting of a clock), it is *phase-advanced* or *phase-delayed*. Since the period of a circadian rhythm is not exactly 24 hours, it must be phase-advanced or phase-delayed every day to remain in phase with the daily cycle of the environment. In other words, the rhythm has to be *entrained* to the cycle of light and dark in the environment.

ENTRAINMENT. The resetting of the circadian rhythm by environmental cues is called **entrainment**. An animal kept in constant conditions will not be entrained to the 24-hour cycle of the environment, and its circadian clock will run according to its natural period—it will be **free-running**. If its period is less than 24 hours, the animal will begin its activity a little earlier each day (see the middle panel of Figure 52.13). The free-running circadian rhythm is under genetic control. Different species may have different average periods, and within a species, mutations can lead to different period lengths.

Animals with free-running circadian rhythms can be used in experiments to identify and investigate the characteristics of stimuli that phase-shift or entrain the circadian clock. Under natural conditions, environmental time cues, such as the onset of light or dark, entrain the free-running rhythm to the 24-hour cycle of the real world. In the laboratory, it is possible to entrain the circadian rhythms of free-running animals with short pulses of light or dark administered every 24 hours (bottom panel of Figure 52.13).

When you fly across several time zones, your circadian rhythm is out of phase with the real world at your destination; the result is jet lag. Gradually your endogenous clock synchronizes itself with the real world as it is reentrained by environmental cues. But because your endogenous rhythm cannot be shifted by more than 30 to 60 minutes each day, it takes several days to reentrain your clock to real time in your new location. This period of reentrainment is the time during which you experience jet lag, because your endogenous

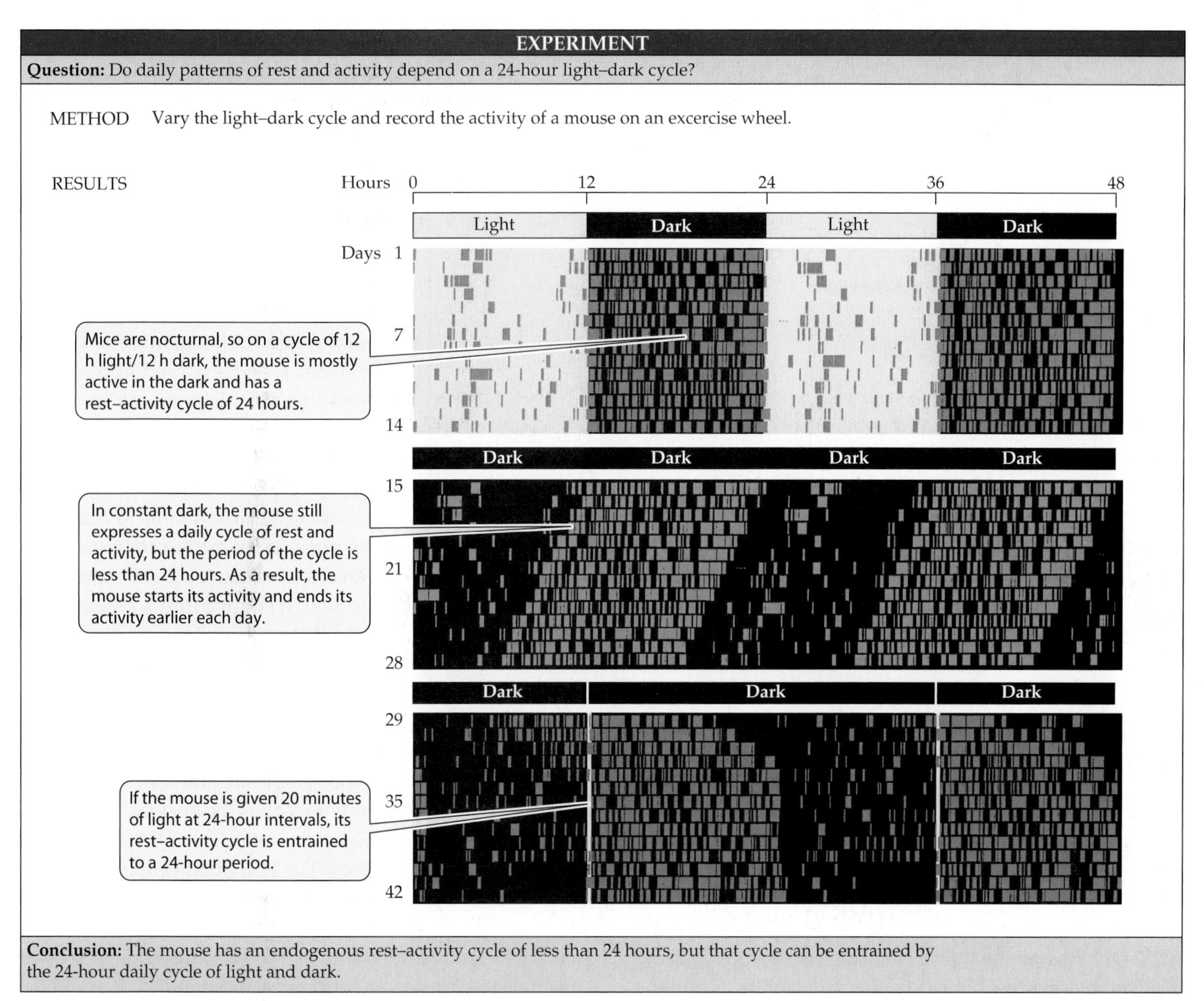
EXPERIMENT
Question: Do daily patterns of rest and activity depend on a 24-hour light–dark cycle?
METHOD Vary the light–dark cycle and record the activity of a mouse on an excercise wheel.
RESULTS
Hours 0 12 24 36 48
Light Dark Light Dark
Days 1 7 14
Mice are nocturnal, so on a cycle of 12 h light/12 h dark, the mouse is mostly active in the dark and has a rest–activity cycle of 24 hours.
Dark Dark Dark Dark
15 21 28
In constant dark, the mouse still expresses a daily cycle of rest and activity, but the period of the cycle is less than 24 hours. As a result, the mouse starts its activity and ends its activity earlier each day.
Dark Dark Dark
29 35 42
If the mouse is given 20 minutes of light at 24-hour intervals, its rest–activity cycle is entrained to a 24-hour period.
Conclusion: The mouse has an endogenous rest–activity cycle of less than 24 hours, but that cycle can be entrained by the 24-hour daily cycle of light and dark.

52.13 Circadian Rhythms The marks indicate times when a mouse is running on an activity wheel. Two days of activity are recorded on each horizontal line, such that the data for each day are plotted twice, once on the right half of a line and again on the left half of the next line below; this double plotting makes patterns easier to see. The schedule of light and dark exposure is indicated by the solid bars running across the top of the figure. First the mouse experiences 12 hours of light and 12 hours of dark every day (top panel), then it is placed in constant darkness (middle panel), and finally it is given a 20-minute exposure to light each day (bottom panel). In constant darkness, the circadian rhythm is free-running, but a 20-minute flash of light at 24-hour intervals can entrain it.

rhythm is waking you up, making you sleepy, initiating activities in your digestive tract, and stimulating many other physiological functions at inappropriate times of the day.

THE CIRCADIAN CLOCK. Where is the clock that controls the circadian rhythm? In mammals, the master circadian clock is located in two tiny groups of cells just above the *optic chiasm*, the place where the two optic nerves cross. These structures are called the **suprachiasmatic nuclei (SCN)**. If a mammal's SCN are destroyed, it loses circadian rhythmicity. Under constant conditions, it is equally likely to be active or asleep at any time of day (Figure 52.14).

Remarkable experiments have shown that circadian rhythms of rest and activity can be restored in an animal whose SCN have been destroyed if it receives a transplant of those nuclei from another animal. Because the restored rhythm has the period of the donor animal, the transplanted tissue clearly controls the recipient's behavior.

Circadian rhythms are found in every animal group, as well as in protists, plants, and fungi, but only vertebrates have SCN. Thus, natural selection has produced a variety of circadian clocks. In most non-mammals, the master clock contains photoreceptors that directly sense changes in light and synchronize behavior with environmental cycles of light

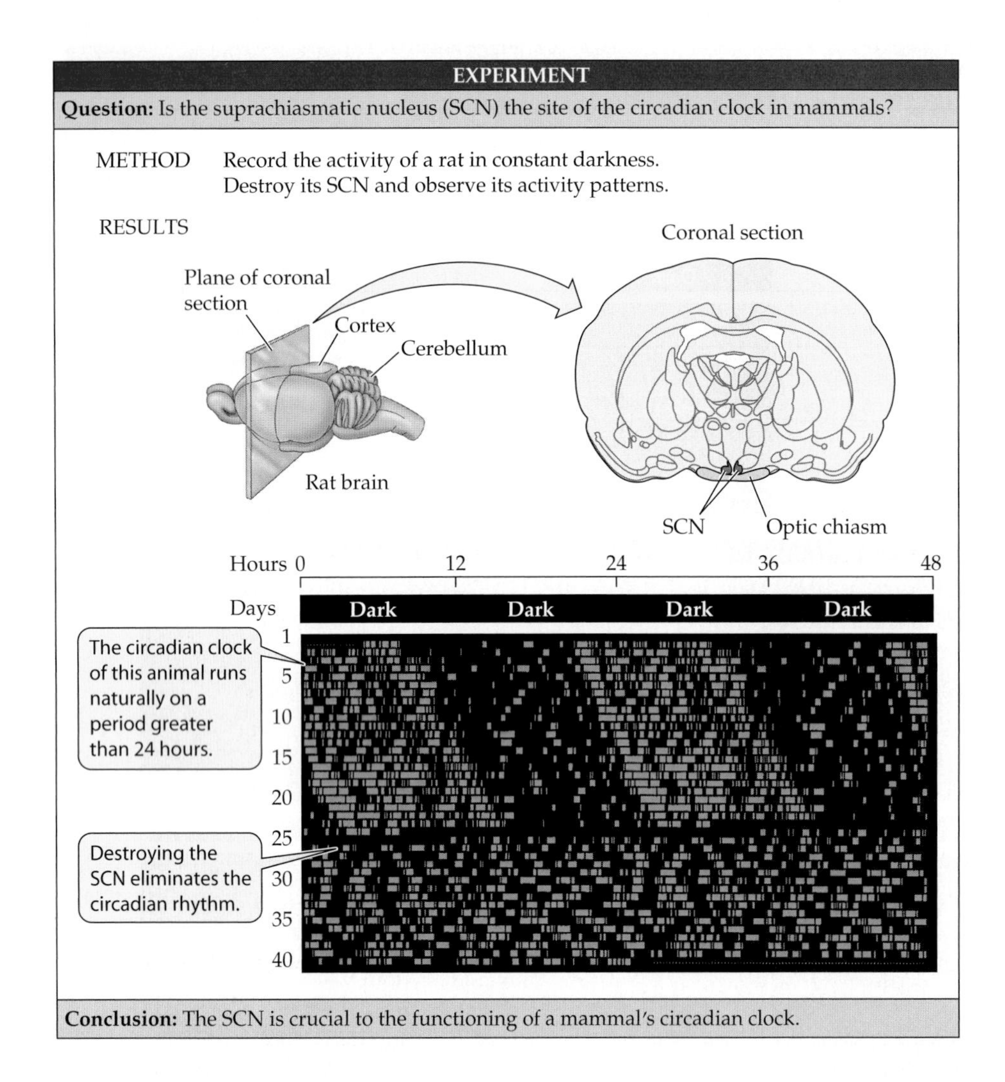

EXPERIMENT
Question: Is the suprachiasmatic nucleus (SCN) the site of the circadian clock in mammals?
METHOD
Record the activity of a rat in constant darkness.
Destroy its SCN and observe its activity patterns.
RESULTS
Plane of coronal section
Cortex
Cerebellum
Rat brain
Coronal section
SCN
Optic chiasm
Hours 0 12 24 36 48
Days
Dark
Dark
Dark
Dark
1
5
10
15
20
25
30
35
40
The circadian clock of this animal runs naturally on a period greater than 24 hours.
Destroying the SCN eliminates the circadian rhythm.
Conclusion: The SCN is crucial to the functioning of a mammal's circadian clock.

52.14 Where the Clock Is The circadian clock of mammals is in the suprachiasmatic nuclei (SCN) of the brain. If its suprachiasmatic nuclei are destroyed, a mammal loses its circadian rhythm.

and dark. In the mollusk *Bulla*, for example, the cells driving circadian rhythms are in the eyes. Birds do have SCN, but the master clock of at least some species resides in the *pineal gland*, a mass of tissue between the cerebral hemispheres that produces the hormone melatonin. If the pineal gland of a bird is removed, the bird loses its circadian rhythm. In protists and fungi, circadian rhythmicity is a property of individual cells, and the individual cells of many multicellular animals can generate circadian rhythms. What are the molecular mechanisms of these circadian clocks?

CLOCK GENES. Enormous progress has been made in recent years toward discovering the molecular and biochemical basis of circadian rhythms. In all organisms that have been studied, self-regulating positive and negative feedback loops of DNA transcription and mRNA translation interact to generate circadian rhythms. The main set of genes involved in these loops is homologous over a wide range of organisms, from bread molds to humans. This surprising discovery means that circadian clocks probably arose very early in the history of life on Earth.

Many of the genetic mechanisms of mammalian circadian rhythms have been revealed by studying gene mutations that alter circadian rhythms. The story began with a gene called *period* (*per*) that was discovered in fruit flies. Mutations of this gene cause flies to have either short or long free-running circadian periods. Mutations of another circadian gene, *timeless* (*tim*), cause a loss of circadian rhythms in fruit flies. The presence of mRNA for *per* and *tim* shows a circadian cycle of expression, and with a slight delay, so do the PER and TIM proteins. Thus, the transcription and translation of these two genes have a circadian rhythm (Figure 52.15). But what controls the rhythm?

The PER and TIM proteins dimerize in the cytoplasm, and the resulting heterodimer enters the nucleus, where it inhibits transcription of the *per* and *tim* genes. The PER/TIM heterodimer does this by blocking a heterodimer of two other clock genes, *clock* (*clk*) and *cycle* (*cyc*), that promotes *per* and *tim* transcription. This molecular negative feedback loop is just one of several such loops in the circadian clock of a fruit fly, but it is the basic model for a clock mechanism: genes whose products shut down their own expression with a delay.

The circuit design of the mammalian clock mechanism is similar to that of the fruit fly's, but it involves different genes. Some of these genes are homologous with the clock genes found in fruit flies, but their molecular interactions and functions differ in mammals.

Circannual rhythms control seasonal behaviors

In addition to turning on its axis every 24 hours, our planet revolves around the sun once every 365 days. Because Earth is tilted on its axis, its revolution around the sun results in seasonal changes in day length at all locations except the equator. These changes secondarily create seasonal changes in temperature, rainfall, and other variables, as we will see in Chapter 56. Because the behavior of animals must adapt to these seasonal changes, animals must be able to anticipate

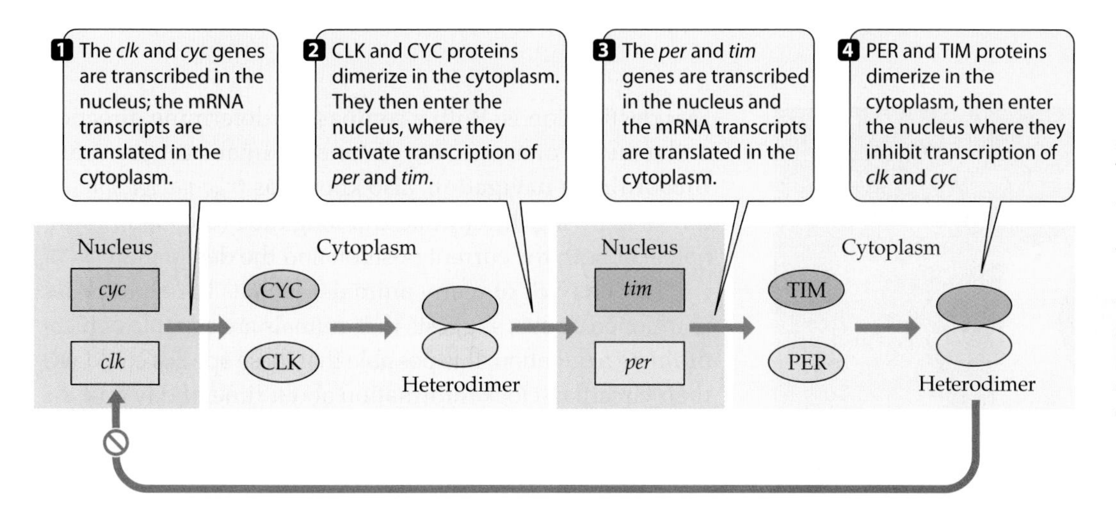
1 The *clk* and *cyc* genes are transcribed in the nucleus; the mRNA transcripts are translated in the cytoplasm.
2 CLK and CYC proteins dimerize in the cytoplasm. They then enter the nucleus, where they activate transcription of *per* and *tim*.
3 The *per* and *tim* genes are transcribed in the nucleus and the mRNA transcripts are translated in the cytoplasm.
4 PER and TIM proteins dimerize in the cytoplasm, then enter the nucleus where they inhibit transcription of *clk* and *cyc*
Nucleus
Cytoplasm
Nucleus
Cytoplasm
cyc
clk
CYC
CLK
Heterodimer
tim
per
TIM
PER
Heterodimer

52.15 Circadian Rhythms Are Generated by a Molecular Clock The *period* and *timeless* genes and the *clock* and *cycle* genes are homologous to "clock genes" found in a wide range of organisms. Simple feedback loops in which transcription of these genes is controlled by their own protein products result in cycles of gene expression with a period of roughly 24 hours.

the seasons and adjust their behavior accordingly. Most animals, for example, reproduce more successfully if they do not produce young in the winter.

For many species, change in day length, or *photoperiod*, is a reliable indicator of seasonal changes to come. If day length has a direct effect on the physiology and behavior of a species, that species is said to be *photoperiodic*. If male deer, for example, are held in captivity and subjected to two cycles of change in day length in one year, they will grow and shed their antlers twice during that year.

For some animals, change in day length is not a reliable seasonal cue. Hibernators spend long months in dark burrows underground, away from any indicators of day length, but have to be physiologically prepared to breed almost as soon as they emerge in the spring. A bird overwintering in the Tropics cannot use changes in photoperiod as a cue to time its migration north to the breeding grounds. Hibernators and equatorial migrants have endogenous annual rhythms, called **circannual rhythms**. In other words, their nervous systems have a built-in calendar. Just as circadian rhythms are not exactly 24 hours long, circannual rhythms are not exactly 365 days long, but usually shorter. The brain mechanisms of circannual rhythms are unknown.

Finding Their Way: Orientation and Navigation

Within its local environment, an animal can organize its behavior spatially by orienting to landmarks, as honeybees and digger wasps do. But what if the destination is a considerable distance away? In this section we examine several modes of long-distance navigation.

Piloting animals orient themselves by means of landmarks

In most cases an animal finds its way using simple means: It knows and remembers the structure of the environment through which it moves. Navigating by means of landmarks is called **piloting**. Gray whales, for example, migrate seasonally between the Bering Sea and the coastal lagoons of Mexico. They find their way by following the west coast of North America. Coastlines, mountain chains, rivers, water currents, and wind patterns can all serve as piloting cues. But some remarkable cases of long-distance orientation and movement cannot be explained by piloting.

Homing animals can return repeatedly to a specific location

The ability of an animal to return to a nest site, burrow, or other specific location is called **homing**. In most cases, homing is merely piloting in a known environment, but some animals are capable of much more sophisticated homing.

Marine birds provide many dramatic examples of homing over great distances in an environment where landmarks are rare. Many marine birds fly over hundreds of miles of featureless ocean on their daily feeding trips and then return directly to a nest site on a tiny island. Albatrosses display remarkable feats of homing. When a young albatross first leaves its parents' nest on an oceanic island, it flies widely over the southern oceans for 8 or 9 years before it reaches reproductive maturity. At that time, it flies back to the island where it was raised to select a mate and build a nest (Figure 52.16). After their first mating season, the pair separate, and each bird resumes its solitary wanderings. The next year they return to the same nest site at the same time, reestablish their pair bond, and breed. Thereafter they return to the nest to breed every other year, spending many months in between at sea.

Homing pigeons can be transported to remote sites where they have never been, and when they are released, fly home. Data on departure directions, known flying speeds, and distances traveled show that homing pigeons fly fairly directly from the point of release to home. They do not randomly search until they encounter familiar territory.

Scientists have used homing pigeons to investigate the mechanisms of navigation. One series of experiments tested the hypothesis that the pigeons depend on visual cues. Pigeons were fitted with frosted contact lenses so that they

52.16 Coming Home A pair of black-browed albatrosses engage in courtship display over their partially completed mud nest. Many albatrosses return to the site of their own birth to find a mate and breed.

could see nothing but the degree of light and dark. These pigeons still homed and fluttered down to the ground in the vicinity of their loft. Thus, they were able to navigate without visual images of the landscape.

Migrating animals travel great distances with remarkable accuracy

For as long as humans have inhabited temperate and subpolar latitudes, they must have been aware that whole populations of animals, especially birds, disappear and reappear seasonally—that is, they **migrate**. Not until the early nineteenth century, however, were patterns of migration established by marking individual birds with identification bands around their legs. Being able to identify individual birds in a population made it possible to demonstrate that the same birds and their offspring returned to the same breeding grounds year after year, and that these same birds were found during the nonbreeding season at locations hundreds or even thousands of kilometers from their breeding grounds.

Because many homing and migrating species are able to take direct routes to their destinations through environments they have never experienced, they must have mechanisms of navigation other than piloting. Humans use two systems of navigation: distance-and-direction navigation and bicoordinate navigation. **Distance-and-direction navigation** requires knowing the direction to the destination and how far away that destination is. With a compass to determine direction and a means of measuring distance, humans can navigate. **Bicoordinate navigation**, also known as *true navigation*, requires knowing the latitude and longitude (the map coordinates) of both the current position and the destination.

The behavior of many animals, such as the albatrosses mentioned above, suggests that animals are capable of bicoordinate navigation. It is possible that these species could use their circadian clock information about time of day and the position of the sun to determine their coordinates—much as sailors did in the days before global positioning satellites. However, there is no strong scientific evidence for such mechanisms to date—though, of course, it is not easy to do experiments on world-traveling animals such as albatrosses. The best evidence for mechanisms of animal navigation comes from studies of distance-and-direction navigation.

Researchers conducted an experiment with European starlings to determine their method of navigation. These birds migrate between their breeding grounds in the Netherlands and northern Germany and their wintering grounds to the west and southwest, in southern England and northern France (Figure 52.17). The researchers captured birds on their breeding grounds, marked them, transported them to Switzerland—south of their breeding grounds—and released them. The researchers expected that if the starlings were using distance-and-direction navigation, the marked birds would be recovered in southern France and Spain, to the southwest of where they were released. Naive juvenile starlings did use distance-and-direction navigation and ended up in Spain, but experienced adult birds were less disrupted by their geographic displacement.

How do animals determine distance and direction? In many instances, determining distance is not a problem as long as the animal recognizes its destination. Homing animals recognize landmarks and can pilot once they reach familiar areas. Evidence suggests that circannual rhythms play a role in determining migration distances for some species. Birds kept in captivity display increased and oriented activity at the time of year when they would normally migrate. Such *migratory restlessness* has a definite duration, which corresponds to the usual duration of migration for the species. Because distance is determined by how long an animal moves in a given direction, the duration of migratory restlessness could set the distance for its migration.

Two obvious means of determining direction are the sun and the stars. During the day, the sun can serve as a compass, as long as the time of day is known. In the Northern Hemisphere, the sun rises in the east, sets in the west, and points south at noon. As we have seen, animals can tell the time of day by means of their circadian clocks. Clock-shifting experiments have demonstrated that animals use their circadian clocks to determine direction from the position of the sun.

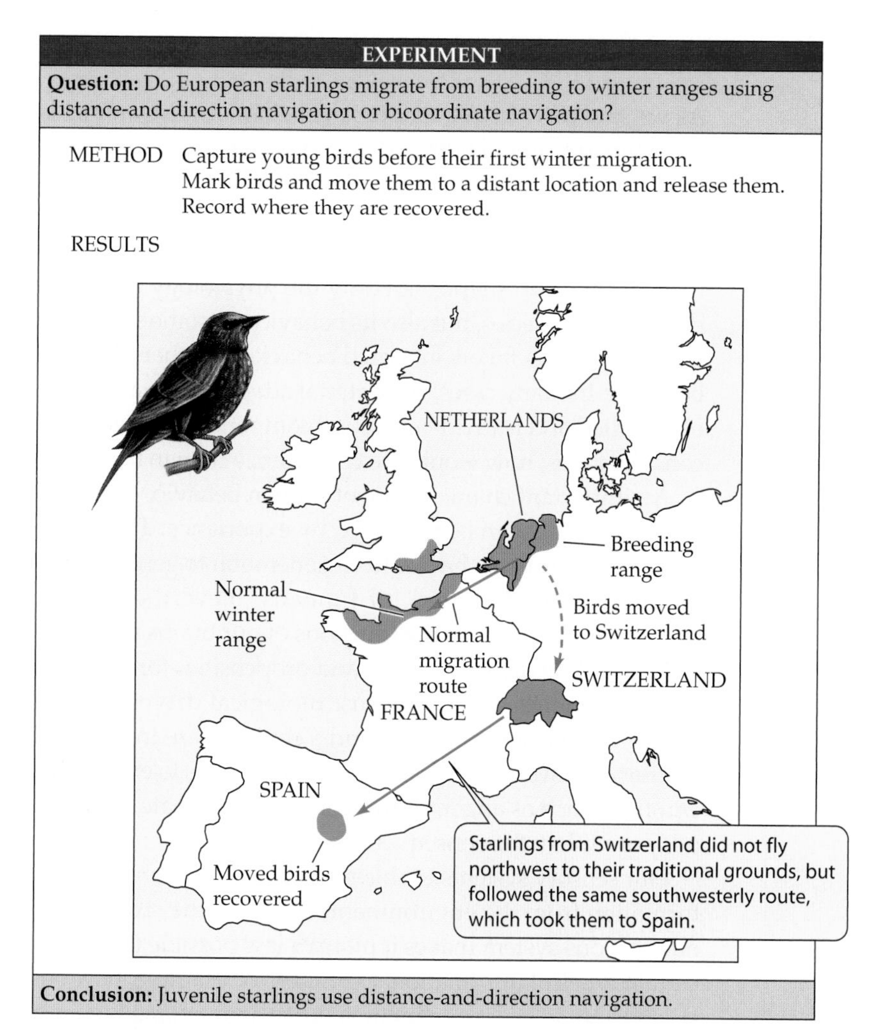

EXPERIMENT
Question: Do European starlings migrate from breeding to winter ranges using distance-and-direction navigation or bicoordinate navigation?
METHOD Capture young birds before their first winter migration.
Mark birds and move them to a distant location and release them.
Record where they are recovered.
RESULTS
NETHERLANDS
Breeding range
Normal winter range
Normal migration route
Birds moved to Switzerland
SWITZERLAND
FRANCE
SPAIN
Moved birds recovered
Starlings from Switzerland did not fly northwest to their traditional grounds, but followed the same southwesterly route, which took them to Spain.
Conclusion: Juvenile starlings use distance-and-direction navigation.

52.17 Distance-and-Direction Navigation European starlings normally make a short winter migration in a southwesterly direction, from the Netherlands to coastal France and southern England (red arrow). Juvenile starlings that were moved to a site in Switzerland did not fly northwest to their traditional wintering grounds, but flew in the same southwesterly direction they would normally take (blue arrow), which took them to Spain.

Researchers placed pigeons in a circular cage that enabled them to see the sun and sky, but no other visual cues (Figure 52.18). Food bins were arranged around the sides of the cage, and the birds were trained to expect food in the bin at one particular direction—south, for example. After training, no matter what time they were fed, and even if the cage was rotated between feedings, the birds always went to the bin at the southern end of the cage for food, even if that bin contained no food.

Next, the birds were placed in a room with a controlled light cycle, and their circadian rhythms were phase-shifted by turning the lights on at midnight and off at noon. After about 2 weeks, the birds' circadian clocks had been phase-advanced by 6 hours. Then the birds were returned to the circular cage under natural light conditions, with sunrise at 6:00 A.M. Because of the shift in their circadian rhythms, their endogenous clocks were indicating noon at the time the sun came up.

If food was always in the south bin, and it was sunrise, the birds should have looked for food 90 degrees to the right of the direction of the sun. But because their circadian clocks were telling them it was noon, they looked for food in the direction of the sun—in the east bin. The 6-hour phase shift in their circadian clocks resulted in a 90-degree error in their orientation. These kinds of experiments on many species have shown that animals can orient by means of a *time-compensated solar compass.*

Many animals are normally active at night; in addition, many day-active bird species migrate at night and thus cannot use the sun to determine direction. The stars offer two sources of information about direction: moving constellations and a fixed point. The positions of constellations change because Earth is rotating. With a star map and a clock, direction can be determined from any constellation. But one point that does not change position during the night is the point directly over the axis on which Earth turns. In the Northern Hemisphere, a star called Polaris, or the North Star, lies in that position and always indicates north.

Stephen Emlen at Cornell University investigated whether birds use these sources of directional information from the stars. He raised young birds in a planetarium, in which star patterns are projected on the ceiling of a large, domed room. The star patterns in the planetarium could be slowly rotated to simulate the rotation of Earth. If the star patterns were rotated each night as the young birds matured, they were able to orient in the planetarium. but birds raised in the planetarium under a nonmoving sky could not. These experiments showed that birds can learn to use star patterns for orientation if the sky rotates.

Animals cannot use sun and star compasses when the sky is overcast, yet they still home and migrate under such conditions. There appears to be considerable redundancy in animals' abilities to sense direction. Pigeons are able to home as well on overcast days as on clear days, but this ability is severely impaired if small magnets are attached to their heads—evidence that the birds use a magnetic sense, although the neurophysiology of this sense is largely unknown. Another possible cue is the plane of polarization of light, which can give directional information even under heavy cloud cover. Very low frequencies of sound can provide information about coastlines and mountain chains. Weather patterns can also provide considerable directional information.

52.18 The Time-Compensated Solar Compass Pigeons whose circadian rhythms were phase-shifted forward by 6 hours oriented as though the dawn sun was at its noon position. These results show that birds are capable of using their circadian clocks to determine direction from the position of the sun.

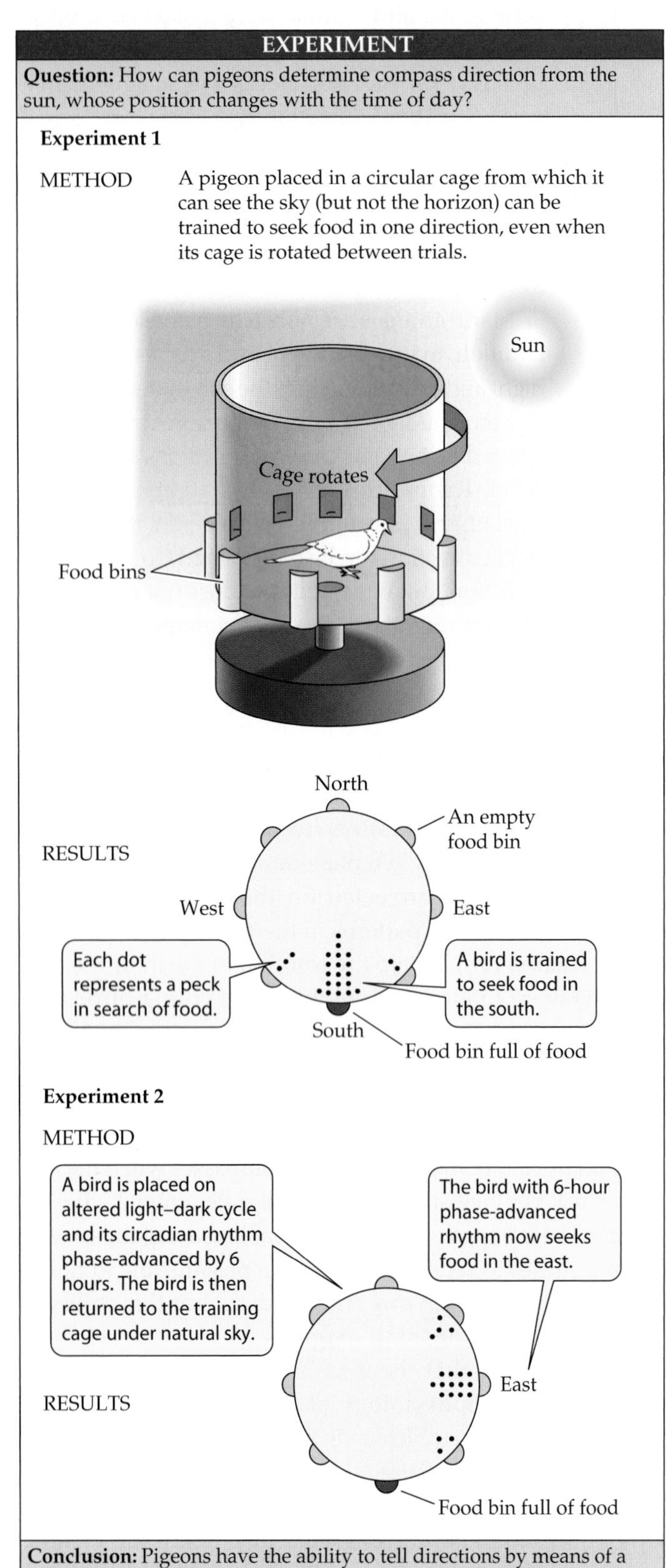

EXPERIMENT
Question: How can pigeons determine compass direction from the sun, whose position changes with the time of day?
Experiment 1
METHOD
A pigeon placed in a circular cage from which it can see the sky (but not the horizon) can be trained to seek food in one direction, even when its cage is rotated between trials.
Sun
Cage rotates
Food bins
RESULTS
North
An empty food bin
West
East
Each dot represents a peck in search of food.
A bird is trained to seek food in the south.
South
Food bin full of food
Experiment 2
METHOD
A bird is placed on altered light-dark cycle and its circadian rhythm phase-advanced by 6 hours. The bird is then returned to the training cage under natural sky.
The bird with 6-hour phase-advanced rhythm now seeks food in the east.
RESULTS
East
Food bin full of food
Conclusion: Pigeons have the ability to tell directions by means of a time-compensated solar compass.

Human Behavior

As we saw early in this chapter, the behavior of an animal is a mixture of components that are inherited and components that can be molded by learning. However, even some aspects of learned behavior patterns—such as what can be learned and when it can be learned—have genetic determinants. Thus natural selection shapes not only the physiology and morphology of a species, but also its behavior. In some situations natural selection favors inherited behavior; in others, learned behavior. In many cases, the optimal adaptation is a mixture of inherited and learned behavioral components. Given these considerations, how would we characterize human behavior?

An important characteristic of human behavior is the extent to which it can be modified by experience. The transmission of learned behavior from generation to generation—**culture**—is the hallmark of humans. Nevertheless, the structures and many of the functions of our brains are inherited, including drives, limits to and propensities for learning, and even some motor patterns. Biological drives such as hunger, thirst, sexual desire, and sleepiness are inherent in our nervous systems. Is it reasonable, therefore, to expect that emotions such as anger, aggression, fear, love, hate, and jealousy are solely the consequences of learning?

Our sensory systems enable us to use certain subsets of information from the environment; similarly, the structure of our nervous system makes it more or less possible to process certain types of information. Consider, for example, how basic and simple it is for an infant to learn spoken language, yet how many years that same child must struggle to master reading and writing. Verbal communication is deeply rooted in our evolutionary past, whereas reading and writing are relatively recent products of human culture.

Some motor patterns seem to be programmed into our nervous system. Studies of diverse human cultures from around the world reveal basic similarities in facial expressions and body language among human populations that have had little or no contact with one another. Infants born blind still smile, frown, and show other facial expressions at appropriate times, even though they have never observed such expressions in others.

Acknowledging that aspects of our behavior have been shaped by evolution in no way detracts from the value we place on our ability to learn and the importance of cultural transmission of information to our species. Even so, we are recognizing that culture, in its simplest form, is not uniquely human. In the introduction to this chapter we saw what has been characterized as culture in Japanese macaques. Individuals invented new behaviors, and those new behaviors were transmitted by imitative learning through the population.

In a recent study, scientists who had spent years studying chimpanzee behavior in seven widely separated areas of

Africa compared their findings on chimpanzee behavior. They were able to identify 39 behaviors, ranging from tool use to courtship behavior, that were common in some populations but absent in others. Moreover, the variation in these behaviors was much greater between populations than within populations, and each population had a distinct repertoire of these behaviors. Just as human societies are characterized by different assemblages of culturally transmitted customs or customary behaviors, so are these chimpanzee populations.

It is increasingly difficult to draw a line between human behavior and animal behavior, especially that of our closest primate relatives. But why should we expect such a line to exist? We do not expect such a lack of continuity in molecular, biochemical, physiological, or anatomical characteristics. Similarly, human and animal behavior are points on a continuum. Our challenge is to understand their common mechanisms and the reasons for their quantitative differences.

Chapter Summary

What, How, and Why Questions

▶ Studies of animal behavior seek to describe behavior, understand its mechanisms, and understand its evolution.

Behavior Shaped by Inheritance

▶ Many behaviors of many species are stereotypic and species-specific, and are thus largely determined by inheritance. They do not require experience and are minimally modifiable by learning.

▶ Deprivation experiments deprive an animal of opportunities to learn a behavior and can therefore reveal that a behavior is inherited.

▶ Hybridization experiments can also reveal genetic influences on behavior. **Review Figure 52.2**

▶ Some behaviors are triggered by simple stimuli called releasers. **Review Figure 52.3**

▶ Spatial learning enables an animal to learn and use information about its physical environment. **Review Figure 52.4**

▶ Imprinting enables an animal to learn the features of a complex releaser, such as the identity of its parents.

▶ The acquisition of bird song is an example in which inheritance and learning interact, enabling an animal to learn a behavior if exposed to the correct stimuli during certain critical periods. **Review Figure 52.6**

▶ Genetically programmed behavior is highly adaptive for species, such as those with nonoverlapping generations, that have little opportunity to learn, for species that might learn the wrong behavior, and in situations in which mistakes are costly or dangerous.

Hormones and Behavior

▶ In rats, the sex steroids present during development determine what sexual behavior patterns develop, and the sex steroids present in the adult control the expression of those patterns. **Review Figure 52.7. See Web/CD Tutorial 52.1**

▶ In birds, testosterone determines a bird's ability to sing by causing the brain regions responsible for song to develop. **Review Figure 52.8**

The Genetics of Behavior

▶ There are many complex steps between the expression of a gene as a protein product and the expression of a behavior. Several types of experiments help reveal how genes affect behavior.

▶ Artificial selection and crossbreeding can produce individuals with particular behavioral traits that are inherited. **Review Figure 52.9**

▶ The techniques of molecular genetics can reveal the functions of specific genes that influence behavior. **Review Figure 52.10**

Communication

▶ Communication consists of displays or signals that can be perceived by other individuals and which influence their behavior. Natural selection favors communication systems when both sender and receiver benefit from the exchange of information.

▶ Many animals communicate by emitting pheromones into the environment and by sensing the pheromones of other animals. Pheromonal messages can last a long time, but they cannot be changed quickly.

▶ Visual communication is easy, versatile, and rapid, but it is limited by its directionality, by the visual acuity of the receiver, and by environmental conditions such as darkness.

▶ Auditory signals can be used at night, can go around objects that would interfere with visual communication, can easily get the receiver's attention, can provide directional information, and can travel long distances.

▶ Tactile signals can communicate complex messages, as the dance of the honeybee demonstrates. **Review Figure 52.12. See Web/CD Activity 52.1**

▶ The electric signals generated by some fishes can be used for communication.

The Timing of Behavior: Biological Rhythms

▶ Animal behavior is expressed in daily cycles called circadian rhythms. A circadian rhythm is an endogenous rhythm with a period not equal to 24 hours. To remain in phase with the 24-hour daily cycle of the environment, a circadian rhythm must be phase-shifted every day. Phase-shifting cues, such as the onset of light and dark, entrain circadian rhythms to the natural 24-hour period. **Review Figure 52.13. See Web/CD Tutorial 52.2**

▶ In mammals, the clock that controls the circadian rhythm is located in the suprachiasmatic nuclei of the brain. In other animals, different structures function as the circadian clock. **Review Figure 52.14**

▶ Genes with self-regulating feedback loops of transcription and translation have been identified as the cellular clock mechanism in a variety of species. **Review Figure 52.15**

▶ Circannual rhythms ensure that animals, such as hibernators and equatorial migrants, that cannot rely on changes in day length as seasonal cues perform the appropriate behaviors at the appropriate times of year.

Finding Their Way: Orientation and Navigation

▶ Piloting animals find their way by orienting to landmarks.

▶ Animals that navigate by distance and direction determine distance in part by recognizing landmarks in the vicinity of their destination and in part by biological rhythms that determine how far they travel. **Review Figure 52.17**

▶ Sources of directional information include a time-compensated solar compass and an ability to locate a fixed point in the night sky. **Review Figure 52.18. See Web/CD Tutorial 52.3**

Human Behavior

▶ Human behavior, like that of all other animals, consists of genetically determined and learned components. What distinguishes humans is the extent to which we can modify our behavior on the basis of experience and pass those modifications on to others.

Self-Quiz

1. Birds that migrate at night
   a. inherit a star map.
   b. determine direction by knowing the time and the position in the sky of a constellation.
   c. orient to a fixed point in the sky.
   d. imprint on one or more key constellations.
   e. determine distance, but not direction, from the stars.
2. If a bird is trained to seek food on the western side of a cage open to the sky, the bird's circadian rhythm is then phase-delayed by 6 hours, and after phase-shifting the bird is returned to the open cage at noon real time, it will seek food in the
   a. north.
   b. south.
   c. east.
   d. west.
3. If an animal sees light earlier than expected on the basis of a circadian rhythm,
   a. it could cause symptoms of jet lag.
   b. it could phase-advance the circadian rhythm.
   c. the animal could be east of home.
   d. it could entrain the animal's circadian rhythm.
   e. all of the above
4. If you do not see courtship behavior in a deprivation experiment investigating the proximate causes of sexual behavior, you can conclude that
   a. the animal is not sexually mature.
   b. the animal has low sexual drive.
   c. it is the wrong time of year.
   d. the appropriate releaser is not present.
   e. none of the above
5. The most likely explanation for the observation that humans from entirely different societies smile when they greet a friend is that
   a. they share a common culture.
   b. they have imprinted on smiling faces when they were infants.
   c. they have learned that smiling does not stimulate aggression.
   d. smiling is an inherited behavior pattern.
   e. smiling is a learned behavior.
6. To be able to pilot, an animal must
   a. have a time-compensated solar compass.
   b. orient to a fixed point in the night sky.
   c. know the distance between two points.
   d. know landmarks.
   e. know its longitude and latitude.
7. Which of the following is true about the building of a web by a spider?
   a. Spiders use a different design depending on the environment.
   b. A young spider learns to build a web by copying the web of its mother.
   c. A young spider imprints on its mother's web, and when it is sexually mature, it replicates that design.
   d. The motor patterns for web building are largely inherited.
   e. Female spiders select mates on the basis of the quality of their webs.
8. Which of the following statements about communication is *true*?
   a. Complex information can be conveyed most rapidly by pheromones.
   b. Visual signaling is advantageous in complex environments.
   c. A disadvantage of auditory communication is that it always reveals the location of the signaller.
   d. An advantage of pheromones is that the message can persist through time.
   e. The dance of bees is an example of using visual signaling to communicate.
9. Which statement about releasers is *true*?
   a. The appropriate releaser always triggers a response.
   b. Releasers are simple subsets of the sensory signals available to the animal.
   c. Releasers are learned through imprinting.
   d. Releasers trigger learned behavior patterns.
   e. An animal responds to a releaser only when it is sexually mature.
10. Which statement about the genetics of behavior is *true*?
    a. Approximately 20 genes control the courtship displays of male dabbling ducks.
    b. The loss of a single gene can eliminate male sexual behavior in fruit flies.
    c. Genes for retrieving, pointing, and herding have been identified in dogs.
    d. Inherited behaviors are highly modifiable because learning can influence gene expression.
    e. Hygienic behavior in bees has been shown to be controlled by two dominant genes.

For Discussion

1. An oystercatcher is a bird that normally lays a clutch of two eggs. If you place an artificial nest with either three artificial but normal-sized eggs or one very large artificial egg near the oystercatcher's nest, the oystercatcher will abandon its own two eggs and attempt to incubate the artificial eggs. How can you explain this behavior?
2. Cowbirds are nest parasites. A female cowbird lays her eggs in the nest of another bird species, which then incubates the eggs and raises the young. What do you think would characterize the acquisition of song in cowbirds? In a given area, cowbirds tend to parasitize the nests of particular bird species. How do you think female cowbirds learn this behavior? How would you test your hypothesis?
3. The short-tailed shearwater is a bird that winters in Antarctica and summers in the Arctic. What problems would this species have in using either the sun or the stars for navigation? What is the most likely means it uses to find its way to its summer and its winter feeding grounds?
4. Male dogs lift a hind leg when they urinate; female dogs squat. If a male puppy receives an injection of estrogen when it is a newborn, it will never lift its leg to urinate for the rest of its life; it will always squat. How might this result be explained?
5. If you were able to be the first person to visit a human population that had never been in contact with another culture, how could you use that opportunity to explore whether there were any human behaviors that were genetically determined?

LIFE ESSAY

What are the ethical issues surrounding medical treatment?

- by Nancy S. Jecker -

The desire to stave off death as long as possible is deeply rooted. We herald advances in medicine as "successes" because they enable us to meet a deep desire to live, and to postpone death as long as possible. Yet despite the enormous benefit of prolonging lives by medical means, in some cases prolonging life does not help the patient.

Patients sometimes refuse lifesaving interventions because they do not wish to be kept alive. Consider, for example, the case of a 22-year-old man with a gunshot wound to the neck who cannot move anything but his face and must blow into a straw to call a nurse to change his diaper. The patient refused a lifesaving blood transfusion for anemia. Although the transfusion would have cured the young man's anemia, it would have done nothing to ameliorate his underlying medical condition.

When patients are not competent to make decisions on their own behalf, family members sometimes make such decisions. Like patients, family members sometimes choose to forego lifesaving interventions. For example, the family of a severely retarded 67-year-old man who was diagnosed with leukemia determined that chemotherapy and blood transfusions should be withheld due to the trauma of treatment.The patient had an I.Q. of 10 and a mental age of approximately 2 years and 8 months. Unable to communicate verbally, he responded only to gestures or physical contact and used grunts and gestures to express himself. The family believed that side effects of chemotherapy—severe nausea, bladder irritation, numbness and tingling of the extremities, and loss of hair—would cause undue suffering because the patient would be unable to understand why they were occurring.

In other instances the medical team caring for a patient may recommend that lifesaving treatment be withheld or withdrawn because it no longer provides a meaningful benefit to the patient. This was the case when an 85-year-old woman tripped on a rug, tumbled to the floor and broke her hip. During hospitalization, she developed respiratory failure requiring tracheal intubation (feeding tube placement) and placement on a mechanical respirator. Subsequently, she suffered a cardiac arrest, requiring emergency resuscitation. Following this episode, the patient never regained consciousness, and was eventually determined to be in a permanently unconscious state known as "persistent vegetative state." After months of meeting the family's request to continue lifesaving medical treatment, the medical team recommended withdrawing life-sustaining treatment and letting the patient die.

In some cases government decisions determine who receives lifesaving treatment. This can occur when state or federal governments restrict public funding for medical services. In Oregon, for example, a law was passed in 1988 that the state would no longer use public (Medicaid) funds to pay for heart, liver, pancreas, or bone marrow transplants for the poor. The state decided instead to use the $1.1 million it spent annually on organ transplantation toward a program of prenatal care.

While is it widely held that under certain conditions lifesaving medical treatment can be ethically withheld or withdrawn, people may disagree about the justification for withholding or withdrawing treatment in particular cases. This leads to reflection on the ethical values underlying such decisions. These values are often expressed in the form of health-care principles, which require health providers to respect patient autonomy, benefit the patient, avoid harming the patient, and justly allocate scarce medical resources. A standard approach to justifying ethical judgments in these particular cases is to show that one's position is supported by ethical principles. The principles, in turn, may be justified by showing that they follow from an ethical theory that is itself widely held or from a common morality that we all can accept.

Whether in our professional lives, as members of a health care team, or in our personal lives, as patients or decision makers for family members, we each will face ethical decisions about medical treatment. If we engage ourselves fully in these choices, they are sure to challenge some of our deepest beliefs about the meaning and value of human life.

Nancy S. Jecker is Professor of Medical Ethics at the University of Washington School of Medicine, Department of Medical History and Ethics. She is also Adjunct Professor at the University of Washington School of Law and Department of Philosophy. Dr. Jecker has written or edited three books and over 90 articles and chapters on ethics and health care. Her articles have appeared in *The Journal of the American Medical Association, The Hastings Center Report, Annals of Internal Medicine, The Journal of Medicine and Philosophy*, and other publications.

Discussion Questions

1. In the first example given, should health care professionals honor the patient's refusal of the blood transfusion? Why or why not?
2. Should the 67-year-old man receive chemotherapy to treat leukemia? How would you defend your position with ethical argument?
3. What are some of the ethical issues raised by the Oregon plan to eliminate Medicaid funding for certain organ transplantation procedures?
4. Should the feeding tube and respirator keeping the 85-year-old woman alive be withdrawn? Why or why not?

Web Links

National Library of Medicine
www.nlm.nih.gov

Center for Bioethics, University of Pennsylvania
www.med.upenn.edu/bioethics/

Bioethics Education Project, University of Washington School of Medicine
http://eduserv.hscer.washington.edu/bioethics/

53 Behavioral Ecology

Spices have played a major role in human history. The Gothic leader Alaric, who laid siege to Rome nearly 2,500 years ago, demanded (in addition to large quantities of precious metals) 1,364 kilograms of pepper as a ransom. The voyages of Marco Polo, Ferdinand Magellan, and Christopher Columbus were underwritten by kings and undertaken at great risk to sailors to find new and faster routes to spice-growing countries.

Why do humans crave spices? We know that spices enhance the flavors and colors of foods. However, that simple answer quickly suggests other questions. Why do we find foods more appealing when they contain pungent plant products? Why do people use dozens of different kinds of spices? Why are the foods of some cultures spicier than others? These questions are typical of the kinds of questions biologists ask about how and why animals make choices about what kinds of foods to eat. Such questions are the concern of behavioral ecology, a field that merges two areas of study within the life sciences.

Ecology is the science that deals with all kinds of biological interactions in the living world. Interactions among individuals of the same species may give rise to complex social behaviors and elaborate social systems. Biological interactions also include those between individuals of different species and between organisms and their physical environment. These interactions, in turn, influence the structure of communities (the organisms living together in the same area), ecosystems (all organisms in an area and their physical environment), and the biosphere (see Figure 1.6).

In this first chapter of Part Eight, we will look at the field of behavioral ecology. We will discuss how organisms respond to changes in the environment, decide where to carry out their activities, select the resources they need (food, water, shelter, nest sites), respond to predators and competitors, and associate with other members of their own species. Individual behavioral choices are the foundation of much of ecology because changes in the densities and distributions of populations are the cumulative results of the decisions of many individuals.

Our use of the words "decision" and "decide" here does not imply that the behavioral choices animals make are conscious. Rather, we mean that behavioral choices have been molded by natural selection such that individuals act "as if" they knew how their choices would influence their survival and reproductive success.

The term **environment**, as used by ecologists, includes both **abiotic** (physical and chemical) factors, such as water, nutrients,

The Quest for Spice Over the centuries humans have traveled far and endured great risks in order to provide themselves with pungent spices from tropical plants. What is it about spices that produces such a profound effect on our behavior?

light, temperature, and wind, and **biotic** factors, which includes all other organisms living in an area. Interactions between organisms and their environments are two-way processes: Organisms both influence and are influenced by their environments. Indeed, dealing with environmental changes caused by our own species is one of the major challenges facing organisms in the modern world. For this reason, ecologists are often asked to help analyze causes of environmental problems and to assist in finding solutions for them. However, it is important not to confuse the science of ecology with "environmentalism," or with the term "ecology," as it is often used in popular writing, to describe nature as some kind of superorganism.

Responding to Environmental Variation

Any organism that reaches old age has made decisions throughout its life history as it grew to maturity, reproduced, and suffered the effects of aging. Animals choose where to settle, how long to stay there, and when, if ever, to leave. They also select places for specific activities, such as resting and nesting, and they choose which things to eat from among the rich array of potential food sources in their immediate environments. Most animals also choose with whom to associate and for what purposes. And they make these choices in an environment that is continually changing. Plants, because they lack nervous systems and (except as seeds or spores) generally can move only by growing, make fewer choices than animals, but the same principles apply to them.

Environmental changes to which organisms must respond happen at many different time scales. Some changes, such as the approach of a fire, storm, or predator, require immediate responses; other changes allow time for a more gradual response. Some plants detach their leaves when storm winds reach a critical velocity and regrow new leaves afterward (Figure 53.1). Many plants reduce water loss and overheating by shifting the position of their leaves during the day so that they intercept sunlight early and late in the day, but do not overheat at midday. Lizards bask in sunshine in the morning to raise their body temperature but move into the shade when it gets too hot (see Figure 41.8).

All organisms have the ability to change their locations, either actively or passively, at some time during their lives; that is, few individuals die exactly where they were born. Individuals may leave the site of their birth to find a place where they can reproduce. Others may seek new locations when local conditions deteriorate. If repeated seasonal changes alter an environment, organisms may evolve life cycles that appear to anticipate those changes. Migration is one response to cyclical environmental changes. Most insectivorous birds, for example, leave high (temperate) latitudes in autumn for more favorable wintering grounds at low latitudes. Grazing mammals migrate away from seasonally dry areas, following the rains that produce lush grass (Figure 53.2). Other animals enter a resting state (hibernation; see Figure 41.19) before adverse conditions materialize and remain in that state until environmental signals indicate that conditions have improved.

Animals choose where to live

Selecting a place in which to live is one of the most important decisions an individual makes. The environment in which an organism lives is called its **habitat**. Once a habitat is chosen, an animal seeks its food, resting places, nest sites, and escape routes within that habitat. Choice of habitat may strongly influence survival and reproductive success, but some of the ways in which organisms make their choices are surprisingly simple. The cues most organisms use to select suitable habitats have a common feature: They are good predictors of general conditions suitable for future survival and reproduction.

A young red abalone, a kind of gastropod mollusk, begins its life as an egg that is fertilized in the open ocean. About

53.1 Plants Can Respond to Environmental Changes These palm trees in Quintana Roo on the Yucatan Peninsula dropped their leaves during a hurricane, which saved their trunks from being blown out of the ground. New leaves will now grow from the top of the trunk.

53.2 Migration Is a Response to Predictable Seasonal Changes East African wildebeest live in large herds that follow the rains to places with fresh grass.

Connochaetes taurinus

14 hours after fertilization, the egg hatches, but the swimming larva has enough yolk to continue developing for another 7 days without eating. Then the larva stops developing, swims to the seafloor, chooses a place in which to settle, and metamorphoses.

Red abalone larvae settle only on coralline algae, upon which they feed. They recognize coralline algae by a chemical these algae produce. In the laboratory, abalone larvae will settle on any surface on which this molecule has been placed, but in nature only coralline algae produce it. By using this simple cue, the larvae always settle on a surface that is suitable for their future development.

Many animals use the presence and success of already settled individuals as an indication that the habitat may be good. After collared flycatchers arrive on their breeding grounds in spring, they regularly peer into the nests of other individuals. Seeing this behavior, researchers hypothesized that the flycatchers were assessing the quality of the habitat by seeing how well their neighbors were doing. To test this hypothesis, the researchers created some areas with super-sized broods by taking young birds from some nests and adding them to nests in another area. The next year, flycatchers preferentially settled in those areas where broods had been artificially enlarged (Figure 53.3).

An animal may leave an area, either temporarily or permanently, if its population has grown too large to be supported by the local resources or if the environment has deteriorated. When a colony of the ant *Lepidothorax albipennis* has grown too large for its nest site (or the nest site has been damaged), recruiter ants—all of which are female workers (which we will describe later in this chapter)—look for potential new nest sites. When an ant finds a suitable site, she returns to the nest and releases a pheromone that attracts another recruiter. The two run together back to the site. If the other recruiter likes the site better than others she has visited, she returns to the nest and recruits another worker. The behavior of recruiters is governed by only two rules. First, if a site is not very attractive, a recruiter delays her return to it. As a result, ants gather at mediocre sites more slowly than at better sites. Second, once a threshold number of workers have been recruited to a particular site, the recruiters change their behavior and begin carrying eggs and larvae from the colony to the new site. In this way the colony reaches agreement on the best site, which may not be the first one discovered.

Defending a territory may improve fitness

In many cases, an animal can improve its survival and reproductive success by establishing exclusive use of the resources of part of its habitat. The most common way of

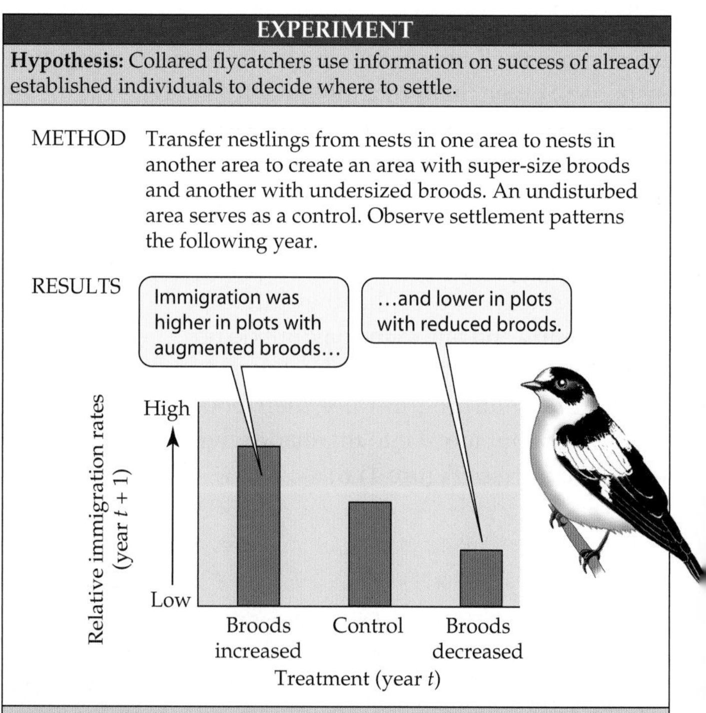

EXPERIMENT
Hypothesis: Collared flycatchers use information on success of already established individuals to decide where to settle.
METHOD
Transfer nestlings from nests in one area to nests in another area to create an area with super-size broods and another with undersized broods. An undisturbed area serves as a control. Observe settlement patterns the following year.
RESULTS
Immigration was higher in plots with augmented broods…
…and lower in plots with reduced broods.
Relative immigration rates (year t + 1)
High
Low
Broods increased
Control
Broods decreased
Treatment (year t)
Conclusion: Collared flycatchers assess brood sizes in the current year and use that information when making habitat choices the next year.

53.3 Flycatchers Use Neighbors' Success to Assess Habitat Quality Collared flycatchers (*Ficedula albicollis*) settled at higher densities in areas where experimenters had artificially enlarged the broods of other flycatchers.

doing so is to establish a **territory** from which the resident excludes *conspecifics* (other individuals of the same species)—and sometimes individuals of other species as well—by advertising that it owns the area and, if necessary, chasing others away. But advertising and chasing take time and energy that could have been used for other beneficial purposes, such as finding food and watching out for predators.

To understand the evolution of these kinds of behavior, ecologists often use a *cost–benefit approach*. This approach assumes that an animal has only a limited amount of time and energy to devote to its activities. Animals seldom perform behaviors whose total costs are greater than the sum of their benefits—the improvements in survival and reproductive success that the animal achieves by performing the behavior. A cost–benefit approach provides a framework that behavioral ecologists can use to design experiments and make observations that enable them to understand why behavior patterns evolve as they do.

The total cost of any particular behavior has three components:

- **Energetic cost** is the difference between the energy the animal would have expended had it rested and the energy expended in performing the behavior.
- **Risk cost** is the increased chance of being injured or killed as a result of performing the behavior, compared with resting.
- **Opportunity cost** is the sum of the benefits the animal forfeits by not being able to perform other behaviors during the same time interval. An animal that devotes all of its time to foraging, for example, cannot achieve high reproductive success!

An experiment estimated the costs incurred by male lizards when defending a territory. Male Yarrow's spiny lizards defend territories, from which they exclude conspecific males. They normally do so most vigorously during September and October, when females are most receptive to mating (in this case, potential mates living in the territory are a resource for the males). To assess the costs of territorial behavior, experimenters inserted small capsules containing testosterone, a hormone that the lizards normally produce in the fall and which induces territorial behavior, beneath the skin of some males. They performed the experiment in June and July, a time of year when the lizards are normally only weakly territorial. Control males were also captured and released, but they received no testosterone implants.

Males with implanted testosterone capsules patrolled their territories more, performed more advertising displays, and expended about one-third more energy (energetic cost) than control males. As a result, they had less time to feed (opportunity cost), captured fewer insects, stored less energy, and died at a faster rate (risk cost) (Figure 53.4). This experiment demonstrated that the costs of active territorial defense are high. In June and July, when females are less receptive, these costs probably outweigh the benefits. Probably that is why the lizards normally reduce their territorial behavior at that time of year.

Animals choose what foods to eat

After choosing a habitat, individuals use the resources of that habitat, including food. Because food is so important, we consider it here in some detail. When an animal *forages* (looks for food), how much time should it spend searching an area

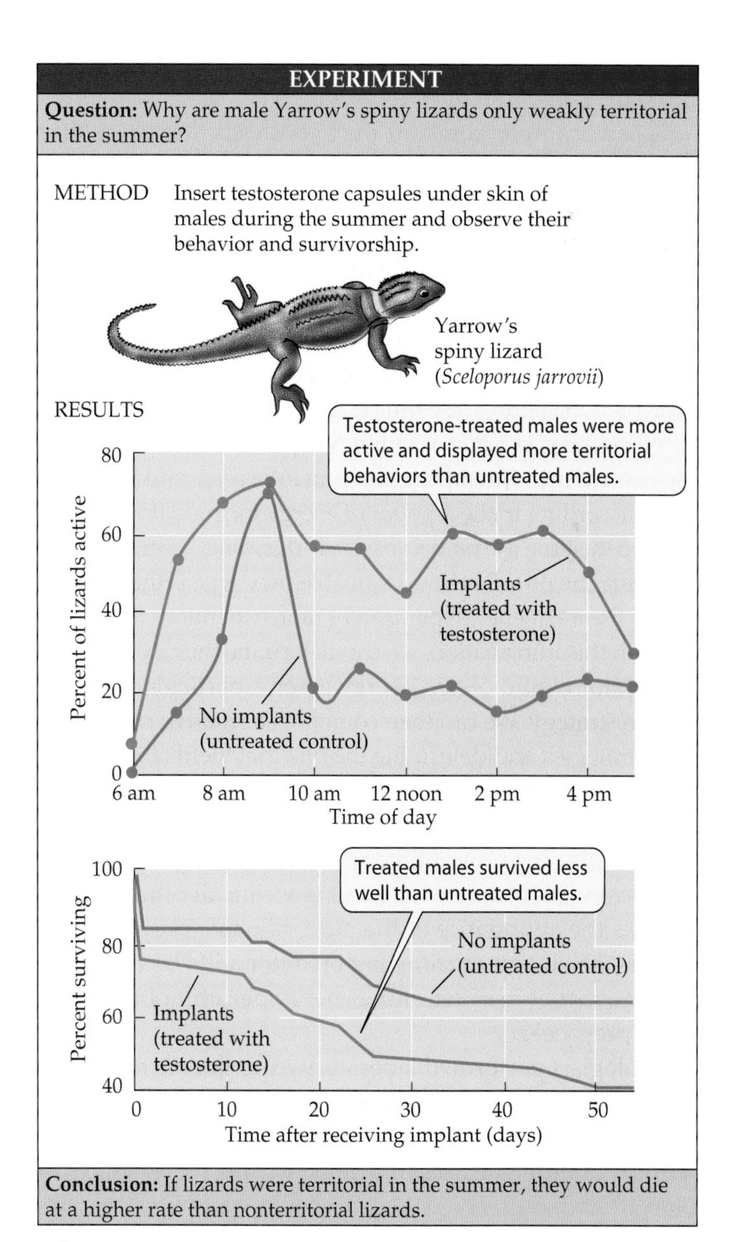

EXPERIMENT
Question: Why are male Yarrow's spiny lizards only weakly territorial in the summer?
METHOD Insert testosterone capsules under skin of males during the summer and observe their behavior and survivorship.
Yarrow's spiny lizard (Sceloporus jarrovii)
RESULTS
Testosterone-treated males were more active and displayed more territorial behaviors than untreated males.
Percent of lizards active
80
60
40
20
0
Implants (treated with testosterone)
No implants (untreated control)
6 am
8 am
10 am
12 noon
2 pm
4 pm
Time of day
Treated males survived less well than untreated males.
Percent surviving
100
80
60
40
No implants (untreated control)
Implants (treated with testosterone)
0
10
20
30
40
50
Time after receiving implant (days)
Conclusion: If lizards were territorial in the summer, they would die at a higher rate than nonterritorial lizards.

53.4 The Costs of Defending a Territory By using hormone implants to increase territorial behavior, experimenters measured the costs to male lizards of defending a territory during the summer months.

before moving to another site? When many different types of prey are available, which ones should a predator take, and which ones should it ignore? **Foraging theory** helps us answer these questions.

To predict how a foraging animal should behave, a scientist first specifies the objective of the behavior and then attempts to determine the behavioral choices that would best achieve that objective. This approach to foraging theory is known as **optimality modeling**. Its underlying assumption is that natural selection molds the behavior of animals so that, generally, they make the best choices available to them. A number of hypotheses can be proposed because a forager may have a number of objectives: It may attempt to maximize the rate at which it obtains energy (calories), vitamins, or minerals, to avoid toxins, or to reduce its risk of being captured by a predator while it is foraging.

As an example, consider the hypothesis that a predator should choose among available prey in order to maximize the rate at which it obtains energy. This is a plausible hypothesis because the more rapidly a predator captures food, the more time and energy it will have for other activities, such as reproduction or avoiding its own predators. To determine how a predator should choose prey if its objective is to maximize its energy intake rate, we characterize each type of available prey by two features: the time it takes the predator to pursue, capture, and consume an individual prey, and the amount of energy an individual prey contains. We then rank the prey types according to the amount of energy the predator gets relative to the time the predator spends pursuing, capturing, and handling the prey. The most valuable prey type is the one that yields the most energy per unit of time expended.

With this information, we can determine the rate at which a predator would obtain energy given a particular prey selection strategy. We can then compare alternative prey selection strategies and determine the one that yields the highest rate of energy intake. Such calculations show that, if the most valuable prey type is abundant enough, a predator gains the most energy per unit of time spent foraging by taking only the most valuable prey type and ignoring all others. However, as the abundance of the most valuable prey type decreases, an energy-maximizing predator adds less valuable prey to its diet in order of the energy per unit of time that those prey yield.

Ecologists performed laboratory experiments to test the energy maximization hypothesis. In preparation for their experiments, the scientists measured the energy content of water fleas of different sizes (the different prey types), the time bluegill sunfish (the predators) needed to capture and eat different prey types, the energy they spent pursuing and capturing prey, and rates at which they encountered prey under different prey densities. The scientists then stocked experimental environments with different densities and proportions of large, medium, and small water fleas. They made two predictions from the hypothesis: (1) that in an environment stocked with low densities of all three sizes of prey, the fish would eat every water flea they encountered, and (2) that in an environment with abundant large water fleas, the fish would ignore smaller water fleas. The proportions of large, medium, and small water fleas taken by the fish under different conditions were close to those predicted by the hypothesis (Figure 53.5).

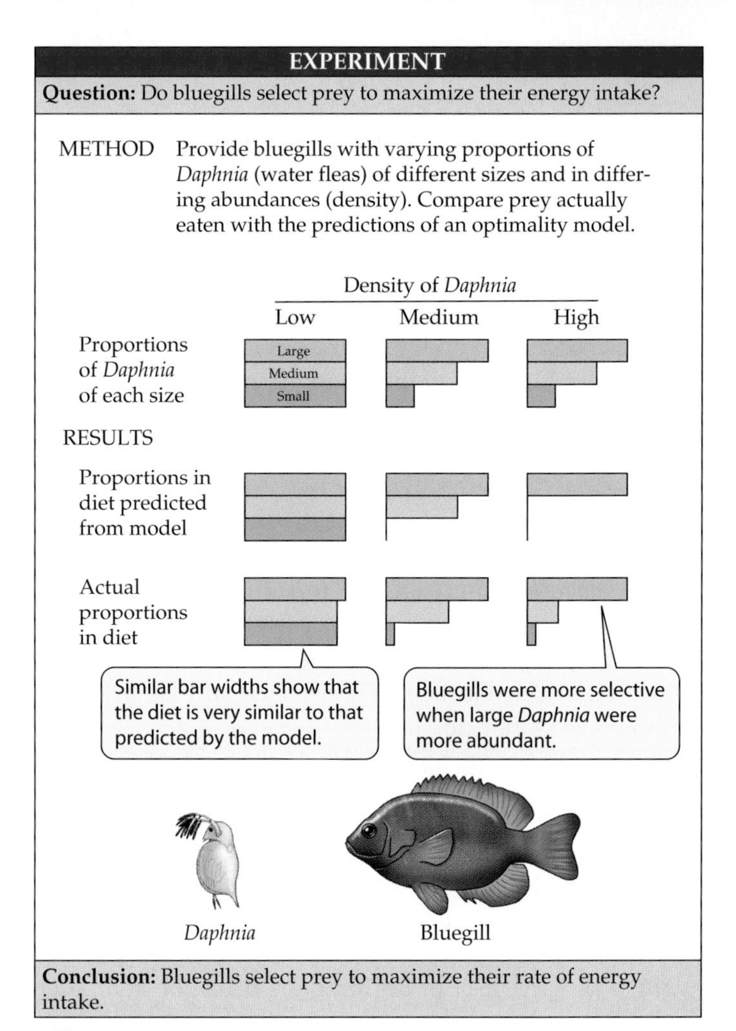

EXPERIMENT
Question: Do bluegills select prey to maximize their energy intake?
METHOD Provide bluegills with varying proportions of Daphnia (water fleas) of different sizes and in differing abundances (density). Compare prey actually eaten with the predictions of an optimality model.
Density of Daphnia
Low
Medium
High
Proportions of Daphnia of each size
Large
Medium
Small
RESULTS
Proportions in diet predicted from model
Actual proportions in diet
Similar bar widths show that the diet is very similar to that predicted by the model.
Bluegills were more selective when large Daphnia were more abundant.
Daphnia
Bluegill
Conclusion: Bluegills select prey to maximize their rate of energy intake.

53.5 Bluegills Are Energy Maximizers The prey choices of bluegill sunfish (*Lepomis macrochirus*) were very similar to those predicted by an energy-maximizing optimality model.

In the bluegill example, only the energy content of prey mattered, but minerals are also important to foragers. Many species of mammals and birds, all of which are primarily or exclusively herbivores or seed eaters, get mineral nutrients by eating soil at particular sites where mineral-rich soil is exposed (Figure 53.6). Humans, especially pregnant women in many traditional societies, consume soil, either by itself or mixed with otherwise toxic or bitter foods, such as acorns and wild potatoes. About 500 tons per year of mineral-rich clay are extracted from the ground or from termite mounds in Nigeria and exported for sale in markets throughout West Africa.

Ara chloroptera and *A. macao*

53.6 Mineral Seekers These macaws are obtaining mineral nutrients from the clay at this mineral lick in the Amazon jungle of Peru.

Like minerals, the spices humans add to our food contain little energy. They must provide some other benefit for us to value them so highly. One hypothesis proposed to explain why we find spicy foods tasty is the antimicrobial hypothesis. This hypothesis is based on the fact that spices are chemicals known to protect the plants that produce them against bacteria and fungi. It is reasonable to assume that they might also protect food from attacks by bacteria and fungi, and thus it might be adaptive for people to use spices in cooking to protect themselves from contaminated food.

The prediction that spices used in cooking should have antimicrobial activity has been tested experimentally in the laboratory by challenging food-borne bacteria and fungi with chemicals found in spices. In one such experiment, scientists prepared alcohol extracts of spices and added them to cultures of food-borne bacteria. Most of the commonly used spices were found to inhibit the growth of more than one kind of food-borne bacteria (Figure 53.7). These findings support the antimicrobial hypothesis.

Although these tests support the antimicrobial hypothesis, they do not exclude the possibility that an alternative hypothesis might also explain the human taste for spices. One such alternative hypothesis is that we enjoy spices because they disguise the smell and taste of spoiled foods. This hypothesis cannot be tested easily. However, some data suggest that toxins from food-borne bacteria kill thousands of people every year and debilitate millions more. Therefore, eating spoiled food by covering up its bad flavors is a dangerous thing to do, even for a starving person. Natural selection is not likely to have favored people who ate rancid food because the flavors that signaled danger were disguised.

Choice of associates influences fitness

Most animals do not lead solitary lives. They associate with other individuals for a variety of reasons. Consider, for example, one important decision made by individuals of sexually reproducing species: the choice of mating partners. These

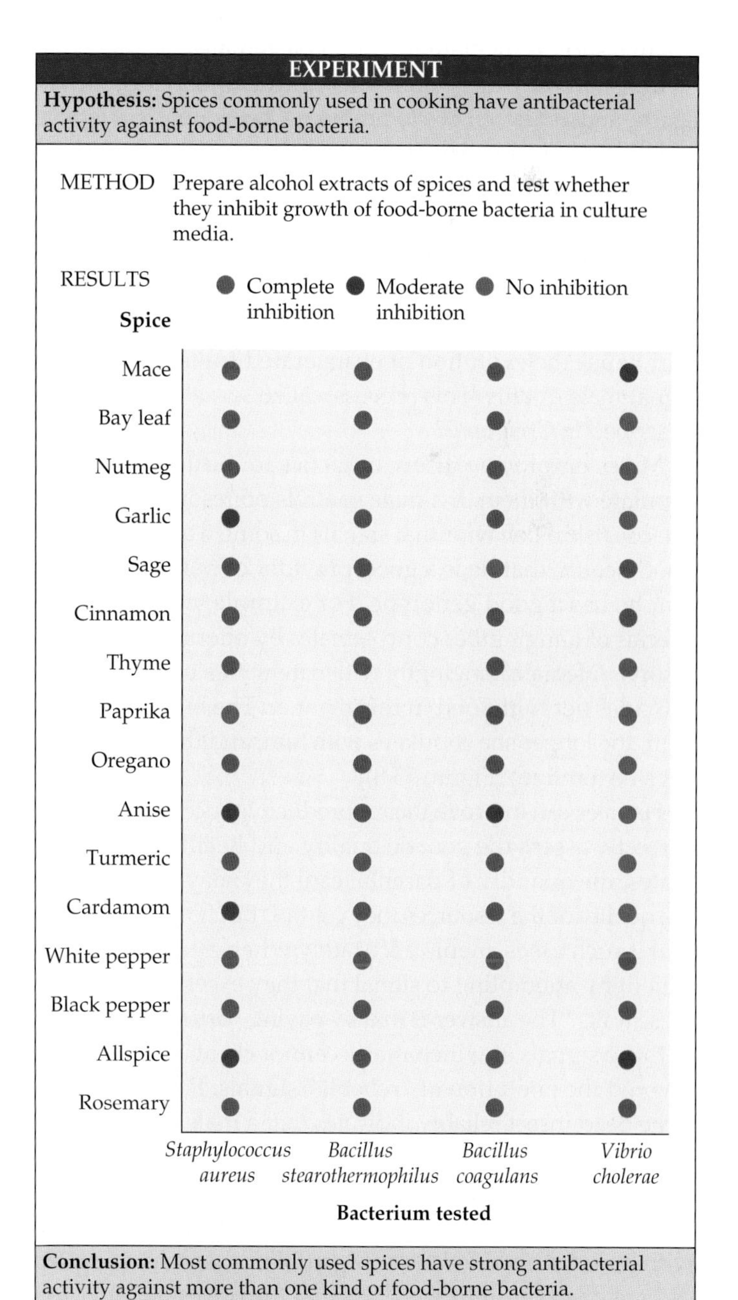

53.7 Most Spices Have Antimicrobial Activity Laboratory tests show that most commonly used spices have moderate to strong antimicrobial activity.

choices may be based on the inherent qualities of a potential mate, on the resources it controls (food, nest sites, escape places), or on a combination of the two. How do individuals choose mates?

The reproductive behaviors of males and females are often very different. Males usually initiate courtship, and they often fight for opportunities to mate with females. Females seldom fight over males, and they often reject courting males. Why are these sexual roles so different?

The answer lies in part in the costs of producing sperm and eggs. Because sperm are small and cheap to produce, one male produces enough to father a very large number of offspring—usually many more than the number of eggs a female can produce or the number of young she can nourish. Therefore, males of most species can increase their reproductive success by mating with many females.

Eggs, on the other hand, are typically much larger than sperm and are expensive to produce. Consequently, a female is unlikely to increase her reproductive output very much by increasing the number of males with which she mates. The reproductive success of a female depends primarily upon the quality of the genes she receives from her mate, the resources he controls, or the amount of assistance he provides in the care of her offspring. By their choices among males, females may cause the evolution of exaggerated traits that reliably signal male quality. This process, called *sexual selection*, was described in Chapter 23.

Males employ a variety of tactics to induce females to copulate with them. If a male controls no resources, he may use courtship behavior that signals in some way that he is in good health, that he is a good provider of parental care, or that he has a good genotype. For example, males of some species of hangingflies court females by offering them dead insects. A female hangingfly will mate with a male only if he provides her with food in this manner. The bigger the food item, the longer she copulates with him, and the more of her eggs he fertilizes (Figure 53.8).

Females can improve their reproductive success if they can correctly assess the genetic quality and health of potential mates, the quantity of parental care they may provide, and the quality of the resources they control. But how can females make such assessments accurately when all males would benefit by attempting to signal that they excel in all three of these traits? The answer is that by paying particular attention to those signals at which males cannot cheat, females have favored the evolution of "reliable" signals. Possession of a large dead insect reliably indicates that a male hangingfly is a good forager.

Biologists were slow to discover some of the signals used by animals in mate choice because they could not be detected with humans' unaided sense organs. For example, scientists did not discover until the 1970s that ultraviolet (UV) vision is widespread among birds, because humans cannot see UV light. Experiments with the bluethroat, a small bird that breeds in northern Europe and Asia, showed that females respond to UV light reflected by the bright blue throat patches of males. They prefer normal males rather than males whose throat patches have been dulled by applying a sunscreen chemical that absorbs UV wavelengths (Figure 53.9). Why should females use UV reflectance to assess potential mates? Laboratory research has shown that a bird's physical condition can influence the intensity with which its plumage reflects UV. Thus, UV reflectance is a reliable indicator of a male's health.

53.8 A Male Wins His Mate The male hangingfly on the left has just presented a moth to his mate, thus demonstrating his foraging skills. She feeds on the moth while they copulate. The bigger the moth, the longer they copulate, and the more eggs he fertilizes.

The throat feathers of a male bluethroat reflect ultraviolet light.
Luscinia svecica

53.9 Ultraviolet-Reflecting Plumage Affects Female Choice Female bluethroats are attracted to males whose throat feathers have high UV reflectance, which signals a healthy, high-quality male.

The Evolution of Animal Societies

Social behavior evolves when cooperation among conspecifics produces, on average, higher rates of survival and reproduction than solitary individuals can achieve. Associations for reproduction may consist of little more than a coming together of eggs and sperm, but individuals of many species associate for longer times to provide care for their offspring. Associating with conspecifics may also improve survival in ways unrelated to reproduction, such as by reducing the risk of being captured by a predator.

The social systems of many animals are very simple: Males court females, the fertilized females disperse and lay eggs, and the eggs and larvae grow to maturity untended. Other systems—such as the elaborate colonies of ants, bees, and wasps or the social groups of lions and primates—are very complex. How did these complex animal societies evolve?

Although today's social systems are the result of long periods of evolution, behavior leaves few traces in the fossil record. Biologists must infer possible routes of the evolution of social systems by studying current patterns of social organization. Fortunately, many degrees of social system complexity exist among living species; the simpler systems suggest stages through which the more complex ones may have passed.

We will describe only a few animal social systems, but as we look at these examples, we will keep in mind three important concepts:

- Social systems are best understood not by asking how they benefit the species as a whole, but by asking how the individuals that join together benefit by the association.
- Social systems are dynamic; individuals constantly communicate with one another and adjust their relationships.
- The costs and benefits experienced by individuals in a social system differ according to their age, sex, physiological condition, and status.

Group living confers benefits but also imposes costs

Living in groups may confer many types of benefits. It may improve hunting success or expand the range of prey that can be captured. For example, by hunting in groups, our ancestors were able to kill large mammals they could not have subdued as individual hunters. These social humans could also defend their prey and themselves from other carnivores, and could tell one another about, for example, the locations of food and predators.

Many small birds forage in flocks. To test whether flocking provides protection against predators, an investigator released a trained goshawk near wood pigeons in England. The hawk was most successful when it attacked solitary pigeons. Its success in capturing a pigeon decreased as the number of pigeons in the flock increased (Figure 53.10). The larger the flock of pigeons, the sooner some individual in the flock spotted the hawk and flew away. This escape behavior stimulated other individuals in the flock to take flight as well. But foraging in a flock also imposes a cost: The pigeons in a flock interfere with one another's ability to find seeds.

Social behavior has many costs as well as benefits. In some social species, individuals inhibit one another's reproduction or injure one another's offspring. An almost universal cost associated with group living is higher exposure to diseases and parasites. Long before the causes of diseases were known, people knew that association with sick persons increased their chances of getting sick. Quarantine has been used to combat the spread of illness for as long as we have

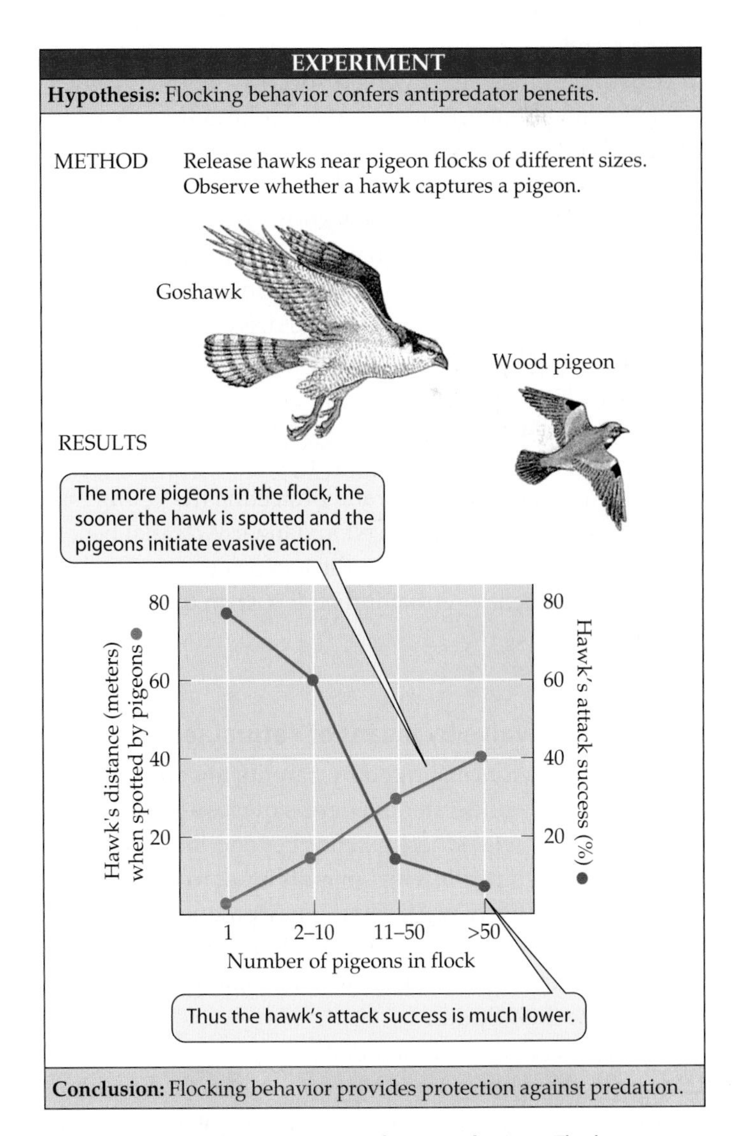

EXPERIMENT
Hypothesis: Flocking behavior confers antipredator benefits.
METHOD
Release hawks near pigeon flocks of different sizes.
Observe whether a hawk captures a pigeon.
Goshawk
Wood pigeon
RESULTS
The more pigeons in the flock, the sooner the hawk is spotted and the pigeons initiate evasive action.
80
60
40
20
Hawk's distance (meters) when spotted by pigeons
Hawk's attack success (%)
1
2–10
11–50
>50
Number of pigeons in flock
Thus the hawk's attack success is much lower.
Conclusion: Flocking behavior provides protection against predation.

53.10 Groups Provide Protection from Predators The larger a flock of pigeons, the greater the distance at which they detect an approaching hawk, and the less likely the hawk is to succeed in capturing a pigeon.

written records. The diseases of wild animals are not as well known, but they too are spread mostly by close contact.

In some species, parents care for their offspring

The most widespread form of social system is the family, an association of one or more adults and their dependent offspring. If parental care lasts a long time, or if the breeding season is longer than the time it takes for offspring to mature, adults may still be caring for younger offspring when older offspring reach parenting age. These older offspring may help their parents care for their younger siblings. Among birds, many communal breeding systems probably evolved by this route. Florida scrub jays live all year on territories, each of which contains a breeding pair and up to six helpers that bring food to the nest. Nearly all helpers are offspring from the previous breeding season that remain with their parents.

Most mammals also evolved social systems via an extended family. In simple mammalian social systems, solitary females or male–female pairs care for their young. As the period of parental care increases, older offspring are still present when the next generation is born, and they often help rear their younger siblings. In most social mammal species, female offspring remain in the group in which they were born, but males tend to leave, or are driven out, and must seek other social groups. Therefore, among mammals, most helpers are females.

Raising a family involves tremendous costs for parents and helpers. Animals who provide food for their young may sacrifice food for themselves, and protecting the young may involve the animal putting itself in danger. Acts that benefit another individual at a cost to the performer are **altruistic acts**. How can behavior that inflicts a cost on the performer evolve?

Altruism can evolve by means of natural selection

Altruistic behaviors exhibited by parents toward their offspring are easily understood in terms of close genetic relatedness. Genetic relatedness extends beyond the parent–offspring relationship, allowing an individual to influence its fitness in two different ways. First, it may produce its own offspring, contributing to its own **individual fitness**. Second, it may help relatives (who bear some of the same genes) in ways that increase their fitness.

Because relatives are descended from a common ancestor, they are likely to bear some of the same alleles. In diploid organisms, two offspring of the same parents share on average 50 percent of the same alleles; an individual is likely to share 25 percent of its alleles with its sibling's offspring. Therefore, by helping its relatives, an individual can increase the representation of some of its own alleles in the population. This process is called **kin selection**. Together, individual fitness and fitness gained through helping non-descendent kin determine the **inclusive fitness** of an individual. Occasional altruistic acts may eventually evolve into altruistic behavior patterns if the benefits of increasing the reproductive success of relatives exceed the costs of decreasing the altruist's own reproductive success.

Many social groups consist of some individuals that are close relatives and others that are unrelated or distantly related. Individuals of some species recognize their relatives and adjust their behavior accordingly. White-fronted bee-eaters are African birds that nest colonially. Most breeding pairs are assisted by nonbreeding adults that help incubate eggs and feed nestlings. Nearly all of these helpers assist close relatives (Figure 53.11). When helpers have a choice of two nests at which to help, about 95 percent of the time they choose the nest with the young more closely related to them.

Several other pieces of evidence suggest that the helping behavior of white-fronted bee-eaters evolved through kin selection. First, both males and females help to care for nestlings, but males help more often than females. Males remain in the social group in which they were born, but females join other social groups when they mature. Therefore, females typically live in social groups composed primarily of nonrelatives.

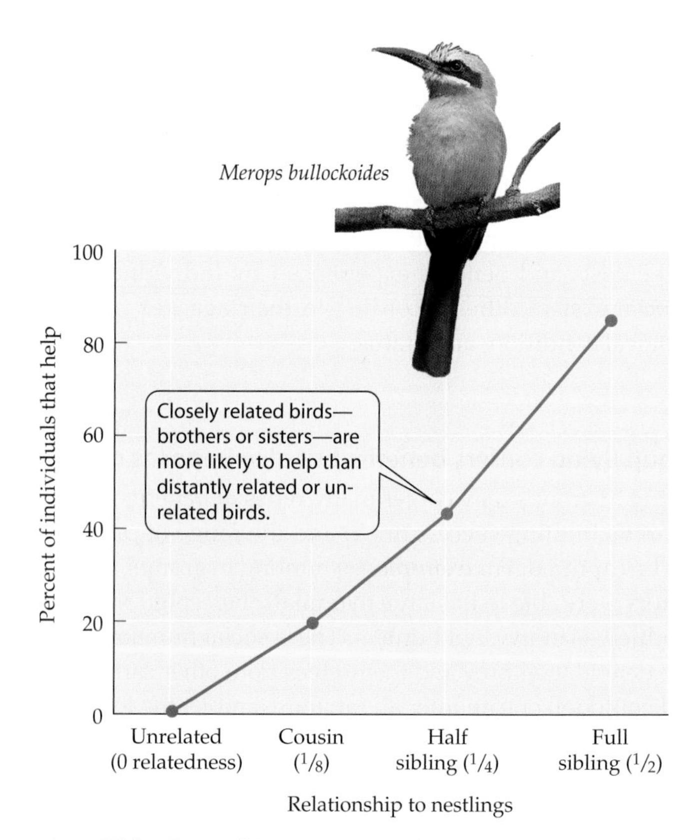

53.11 White-Fronted Bee-Eaters are Altruists Bee-eaters that help to care for nestlings preferentially help close relatives.

Second, individual bee-eaters do not appear to gain anything in addition to inclusive fitness by helping—helpers do not gain experience that improves their performances when they become breeders. Finally, nests with helpers produce more fledglings than do nests without helpers, showing that helpers do increase the number of fledglings produced by their close relatives. Notice that all these patterns are consistent with the principle that bee-eaters behave in ways that improve their individual fitness, not in ways that benefit the species.

Eusociality is extreme social behavior

Species whose social groups include sterile individuals are said to be **eusocial**. This extreme form of social behavior has evolved in termites and many hymenopterans (ants, bees, and wasps). In these species, most females are *workers* that forage for the colony and/or defend it against predators, but do not reproduce. Workers may include soldiers with large, specialized defensive weapons (Figure 53.12), which may be killed while defending the colony. Only a few females, known as *queens*, are fertile, and they produce all the offspring of the colony.

Both genetic and environmental factors facilitate the evolution of eusociality. The British evolutionist W. D. Hamilton first suggested that eusociality evolved among the Hymenoptera because its members have an unusual sex determination system in which males are haploid but females are diploid. Among the Hymenoptera, a fertilized (diploid) egg hatches into a female; an unfertilized (haploid) egg hatches into a male.

If a female copulates with only one male, all the sperm she receives are identical because a haploid male has only one set of chromosomes, all of which are transmitted to every sperm cell. Therefore, a female's daughters share all of their father's genes. They also share, on average, half of the genes they receive from their mother. As a result, they share 75 percent of their alleles on average, rather than the 50 percent they would share if both parents were diploid. Since workers are more genetically similar to their sisters than they would be to their own offspring, they can increase their fitness more by caring for their sisters than by producing and caring for their own offspring.

Eusociality may also be favored if establishment of new colonies is difficult and dangerous. Nearly all eusocial animals construct elaborate nests or burrow systems within which their offspring are reared. Naked mole-rats—the most eusocial mammals—live in underground colonies containing 70 to 80 individuals. The colony's tunnel systems are maintained by sterile workers. Breeding is restricted to a single queen and several kings that live in a nest chamber in the center of the colony. Individuals attempting to found new colonies are at high risk of being captured by predators, and most founding events fail. Thus, high predation rates, which favor cooperation among founding individuals, may facilitate the evolution of eusociality.

Inbreeding—the mating of individuals who are genetically related—can generate increased genetic relatedness within a group. Even if two parents are unrelated, but each is the product of generations of intense inbreeding, all of their offspring may be genetically nearly identical. Such offspring would increase their fitness by helping to rear siblings. Genetic similarity generated by inbreeding could explain the evolution of eusociality among the many hymenopteran species in which queens mate with many males and among termites and naked mole-rats, in which both sexes are diploid.

Eciton burchelli

53.12 Sterile Workers are Extreme Altruists Eusocial insect species contain classes of sterile worker individuals. These soldier army ants from Panama protect their colonies with their large, powerful jaws.

Behavioral Ecology, Population Dynamics, and Community Structure

The ways in which organisms make decisions about habitats, food, and associates have many important implications for the structure and functioning of ecological systems. We will describe two of those implications here. First, animals with complex social organization often achieve remarkably high abundances. Second, the ways in which animals select habitats and food, combined with their interactions with individuals of other species, may influence the range of habitat and foods a species uses in nature.

Social animals may achieve great abundances

The abundances achieved by some social animals are impressive. For example, up to 94 percent of the individuals and 86 percent of the biomass of arthropods in the *canopies* (tree-

53.13 Termite Mounds Are Large and Complex These immense Australian termite mounds are constructed over many years by millions of worker termites. Elaborate nests or burrows, which are very costly to construct and maintain, characterize nearly all eusocial animals.

tops) of tropical rainforests are social ants. Termites (also social insects) are the primary consumers of plant tissues in the savannas of Africa. They live in and build large mounds, within which many other species of animals live. Termites may extract nutrients from the soil at depths as great as 80 meters. In parts of Australia, termite density may reach 1,000 colonies per hectare (Figure 53.13).

Ants and termites have achieved these remarkable abundances in part because their social organization allows them to exploit the services of other organisms in harvesting vital resources. The most abundant and highly productive ants and termites actively cultivate fungi that break down difficult-to-digest plant tissues, including wood. Some ants tend aphids and other insects that tap phloem fluids, protecting the phloem-suckers from predators. Because phloem is rich in carbohydrates but poor in proteins, phloem-suckers ingest more carbohydrates than they can use. They eject the excess in the form of sugar-rich anal drops (see Figure 36.13), which the ants eat. Because the ants can easily obtain enough carbohydrates in this manner, they need to get only proteins in other ways, such as by eating other insects. Moreover, with their high, sugar-based metabolic rates, the ants can expend the energy needed to drive other predatory insects away from their food sources. In this way, ants strongly influence the community of insects in tropical forest canopies.

Social living also enables organisms to find and use temporally and spatially patchy foods. The wildebeest, which travels in large herds, is the most abundant large mammal in Africa (see Figure 53.2). More than a million individuals are found in the herd that migrates between the Masai Mara in Kenya and the Serengeti in Tanzania to feed on the rapidly growing grass that follows seasonal rains in each area.

Even more striking is the abundance achieved by our own species (Figure 53.14). Social living enabled members of human groups to specialize in different activities. Among the benefits of specialization were domestication of plants and animals and cultivation of land. These innovations enabled our ancestors to increase the resources at their disposal dramatically. Those increases, in turn, stimulated rapid population growth up to the limit determined by the agricultural productivity that was possible with human- and animal-powered tools. Agricultural machines and artificial fertilizers, made possible by the tapping of fossil fuels, greatly increased agricultural productivity and removed that earlier limit. In addition, the development of modern medicine reduced the mortality rate in human populations. Medicine and better hygiene have also allowed people to live in large numbers in areas where diseases formerly kept numbers very low. However, these successes have been accompanied by many problems, some of which we will discuss in subsequent chapters.

Interspecific interactions influence animal distributions

As we have seen, animals assess habitat quality and settle preferentially in better places. They also select the food items that give them the best return for the time and energy they expend in getting them. The optimality modeling approach used to develop and test hypotheses about how such choices are made has yielded an important general "rule of thumb" of behavioral ecology: As much as possible, organisms concentrate on doing what they do best and avoid doing what they do poorly.

However, interspecific interactions may prevent animals from living in those environments in which they would do best. Individuals of a behaviorally dominant species may be able to exclude individuals of a subordinate species from its preferred foraging areas. How such behavioral dominance influences use of foraging areas can be illustrated by observing hummingbird behavior.

Hummingbirds extract nectar from flowers and often defend patches of flowers from other hummingbirds. In an ex-

53.14 Social Organization Allows Humans to Live at High Densities Human cities such as Benidorm on Spain's Costa Blanca are examples of how social organization allows our species to achieve and sustain extreme population densities.

periment done in southeastern Arizona, investigators created artificial "flower patches" by setting up an array of feeders. Some feeders contained artificial nectar that was rich in sucrose; others contained a more dilute solution that was a poorer source of sucrose. Hummingbirds quickly learned which were the high-quality feeders because the rich ones had blue bee guards; the poor ones had yellow bee guards.

Males of three hummingbird species visited the feeders. Interactions were strongest between two of them: Male blue-throated hummingbirds, which weigh about 8.3 g on average, behaviorally dominated the smaller male black-chinned hummingbirds, which weighed only about 3.2 g. When no male blue-throats were present, black-chinned males fed almost exclusively at the rich feeders; but when male blue-throats were present, black-chinned males fed at poor feeders as often as they fed at rich feeders. Even though the nectar at the poor feeders was more dilute, the black-chins achieved about the same rate of energy from them as from the rich feeders because they were able to feed longer at the poor feeders without being chased away by the larger blue-throat males.

These kinds of observations show us that the ways in which animals choose what to eat, where to seek food, and with whom to associate influence the sizes and distributions of populations of many species and how they interact in nature. These aspects of populations will form the focus of the next chapter, on population ecology.

Chapter Summary

▶ Behavioral ecology is the study of how animals decide where to carry out different activities, select the resources they need, respond to predators and competitors, and interact with other members of their own species.

▶ An organism's environment includes both abiotic and biotic components.

Responding to Environmental Variation

▶ Organisms respond adaptively to environmental changes.

▶ The cues animals use to select habitats may be simple, but they must be good predictors of conditions suitable for future survival and reproduction.

▶ The success of already settled individuals may provide evidence of habitat quality. **Review Figure 53.3**

▶ Cost–benefit analyses of behavior are based on the principle that animals have only limited amounts of time and energy to devote to their activities.

▶ Behaviors such as defending a territory may have three kinds of costs: energetic cost, opportunity cost, and risk cost. **Review Figure 53.4. See Web/CD Tutorial 53.1**

▶ Foraging theory was developed to understand how animals select foods from those present in the environment. **Review Figure 53.5. See Web/CD Tutorial 53.2**

▶ The human taste for spices may have evolved because spices have antimicrobial activity. **Review Figures 53.7**

▶ Because males produce enough sperm to fertilize many more eggs than a single female can produce, males typically increase their reproductive success by mating with many females. The reproductive success of females, on the other hand, is typically limited by the cost of producing eggs. As a result, males usually initiate courtship and often fight for opportunities to mate with females. Females seldom fight over males and often reject courting males.

▶ Courting males perform behaviors that signal their desirability as mating partners By paying particular attention to those signals at which males cannot cheat, females have favored the evolution of "reliable" signals.

The Evolution of Animal Societies

▶ Social systems are best understood not by asking how they benefit the species as a whole, but by asking how the individuals that join together benefit by the association.

▶ Social systems are dynamic; individuals constantly communicate with one another and adjust their relationships.

▶ Living in a group may provide protection against predators. **Review Figure 53.10**

▶ The origin of most animal societies is the family, an association of one or more adults and their dependent offspring.

▶ Altruism among closely related individuals can evolve by means of kin selection because individuals who help close relatives can improve their inclusive fitness. **Review Figure 53.11**

▶ Eusocial systems with sterile individuals have evolved among hymenopterans (ants, bees, and wasps), termites, and one mammal, the naked mole-rat.

▶ The more closely related the individuals in a colony are to one another, and the greater the difficulty of establishing independent colonies, the more likely eusociality is to evolve.

Behavioral Ecology, Population Dynamics, and Community Structure

▶ Social animals may achieve great abundances.

▶ Interspecific interactions, as well as habitat and food choices, influence animal distributions.

See Web/CD Activity 53.1 for a concept review of this chapter.

Self-Quiz

1. Which of the following is *not* a component of the cost of performing a behavior?
   a. Its energetic cost
   b. The risk of being injured
   c. Its opportunity cost
   d. The risk of being attacked by a predator
   e. Its information cost
2. An almost universal cost associated with group living is
   a. increased risk of predation.
   b. interference with foraging.
   c. higher exposure to diseases and parasites.
   d. poorer access to mates.
   e. poorer access to sleeping sites.
3. Which is *not* an important assumption of foraging behavior theory?
   a. Efficient foragers spend less time fulfilling their energy needs than inefficient ones.
   b. Superior foragers will generally produce more surviving offspring.
   c. A successful predator will choose its prey in such a way as to maximize its energy intake.
   d. An efficient predator will always choose the most abundant prey.
   e. The ability of a predator to discriminate among prey items has a genetic basis.
4. The basic components of an optimality model of behavior are
   a. the type of behavior and its neural control mechanisms.
   b. the objective of the behavior and the choices that would best achieve it.
   c. the objective of the behavior and its neural control mechanisms.
   d. the goal of the behavior and the constraints imposed by the animal's structure.
   e. the objective to be maximized and the currency used to measure it.
5. The choice of a mating partner may be based on
   a. the inherent qualities of a potential mate.
   b. the resources held by a potential mate.
   c. both the inherent qualities of a potential mate and the resources it holds.
   d. the success of individuals of the opposite sex in courtship.
   e. All of the above
6. Altruistic behavior
   a. confers a benefit on the performer by inflicting some cost on some other individual.
   b. confers a benefit both on the performer and on some other individual.
   c. inflicts a cost both on the performer and on some other individual.
   d. confers a benefit on another individual at some cost to the performer.
   e. confers a cost on the performer without benefiting any other individual.
7. Kin selection is
   a. mating between relatives.
   b. the adoption of young by an unrelated adult.
   c. the ability to recognize one's relatives in a social group.
   d. a behavior that increases the survivorship of an individual's relatives.
   e. only found among social mammals.
8. Species whose social groups include sterile individuals are said to be
   a. eusocial.
   b. semisocial.
   c. oligosocial.
   d. sterisocial.
   e. supersocial.

For Discussion

1. Most hawks are solitary hunters. Swallows often hunt in groups. What are some plausible explanations for this difference? How could you test your ideas?
2. Among birds, males of species that mate with many females and perform communal courtship displays are usually much larger and more brightly colored than females, whereas among species that form monogamous pairs, males are usually similar in size to females, whether or not they are more brightly colored. What hypotheses can be advanced to explain this difference?
3. Many animals defend space, but the sizes of the territories they defend and the resources these areas provide vary enormously. Why don't all animals defend the same type and size of territory?
4. When frogs mate, a male clasps a female behind her front legs and stays with her until she lays her eggs, at which time he fertilizes them. In most species of frogs, the male remains clasped to the female for a short time, usually no longer than a few hours. However, in some species, pairs may remain together for up to several weeks. In view of the fact that a male cannot court or mate with any other female while clasping one, and that a female lays only a single clutch of eggs, why is it advantageous for males to behave this way? What can you guess about the breeding ecology of frogs that remain clasped for long periods? Why should females permit males to clasp them for so long? (Females do not struggle!)
5. Among vertebrates, helpers are individuals capable of reproducing, and most of them later breed on their own. Among eusocial insects, sterile castes have evolved repeatedly. What differences between vertebrates and insects might explain the failure of sterile castes to evolve in the former?
6. The use of DNA fingerprinting technology (see Chapter 16) has shown that in many species, social partners and genetic partners differ. Under what conditions do individuals benefit from copulating with individuals other than their social mates? Do males and females benefit equally from this behavior?

54 Population Ecology

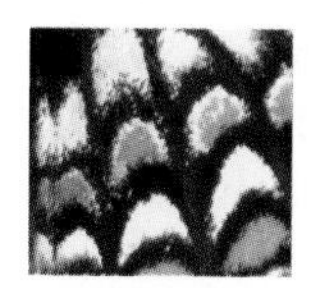

The bay checkerspot butterfly lives in the hills south of San Francisco, California. If you were to travel through this region in search of butterflies, you would find checkerspots in some areas, but not in others. If you looked for butterflies in the region over a number of years, you would find more butterflies in some years than in others. In some places where you found checkerspots one year, you would go back the following year and find none, and in other places where there were previously no checkerspots, you would find them. What is the reason for these variations?

Ecologists who study populations attempt to answer this and a number of other questions: Why do the numbers of individuals of a species in a certain area fluctuate? Why do the geographic ranges of species vary so much? Why is a species abundant in some parts of its geographic range and rare in others, and why does its range change over time?

To understand how and why population sizes fluctuate, ecologists count individuals in different places and try to understand the relative importance of the processes that determine the number of individuals of a species in any particular location. These processes are influenced by individuals of the same species and by individuals of other species living in the same environment, as well as by abiotic environmental factors.

In this chapter we will discuss how and why the sizes of populations of species vary over space and time, and show how this knowledge is used to predict and manage the growth of populations of special interest to people. To set the stage for answering questions about populations, we first describe the kinds of information ecologists gather about the populations they study and how they use that information to answer the questions we just posed. Then we describe how populations of different species interact and how those interactions influence numbers of individuals and where they live. In the next chapter we will describe how interactions among populations and between populations and the physical environment influence the structure of ecological communities.

A Case Study in Subpopulations Patchy subpopulations of the rare bay checkerspot butterfly, *Euphydryas editha bayensis*, provide a well-studied example of population dynamics. This individual was photographed in Morgan Hill, a large patch of suitable habitat for this species in the San Francisco Bay area (see Figure 54.13).

Populations in Space and Time

The individuals of a species within a given area constitute a **population**. At any given moment, an individual organism occupies only one spot in space, and is of a particular age and size. The members of a population, however, are distributed over space, and they differ in age and size. The distribution of the ages of individuals in a population and the way those individuals are distributed over the environment describe **population structure**. Ecologists study population structure because the spatial distributions of individuals and their ages influence the stability of populations and affect interactions among species.

The number of individuals of a species per unit of area (or volume) is its **population density**. Ecologists are interested in population densities because dense populations often exert strong influences on their members and on populations of other species. Scientists working in agriculture, conservation, and medicine typically try to maintain high population densities of some species (crop plants, game animals, aesthetically attractive species, threatened or endangered species) and reduce the densities of others (agricultural pests, disease organisms). To manage populations, we need to know what factors cause populations to grow or decrease in size and how these factors work.

Because organisms and their environments differ, population densities may be measured in more than one way. Ecologists usually measure the densities of organisms in terrestrial environments as the number of individuals per unit of area. However, number per unit of volume is generally a more useful measure for organisms living in the water column. For species whose members differ markedly in size, as do most plants and some animals (such as mollusks, fishes, and reptiles), the percentage of ground covered or the total mass of individuals may be more useful measures of density than the number of individuals.

The structure of a population changes continually because **demographic events**—births, deaths, immigration (movement of individuals into the area), and emigration (movement of individuals out of the area)—are common occurrences. Knowledge of when individuals are born and when they die provides a surprising amount of information about a population. Let's examine how ecologists measure birth and death rates and use that information to understand **population dynamics**—the change in population density through time and space. The study of birth, death, and movement rates that give rise to population dynamics is known as **demography**.

Births, deaths, and movements drive population dynamics

Ecologists measure the *rates* (number per unit of time) at which births, deaths, and movements take place in a population, and they study how these rates are influenced by environmental factors, life histories, and population densities.

The number of individuals in a population at any given time is equal to the number present at some time in the past, plus the number born between then and now, minus the number that died, plus the number that immigrated, minus the number that emigrated. That is, the number of individuals at a given time, N_1, is given by the equation

$$N_1 = N_0 + B - D + I - E$$

where N_1 is the number of individuals at time 1; N_0 is the number of individuals at time 0; B is the number of individuals born, D the number that died, I the number that immigrated, and E the number that emigrated between time 0 and time 1. If we measure these rates over many time intervals, we can determine how a population's density changes over time.

Life tables summarize patterns of births and deaths

Life tables provide summaries of births and deaths in a population. Life tables were developed by the Romans nearly 2,000 years ago to determine how much money needed to be set aside to compensate families of soldiers that might be killed in battle. Today, life insurance companies use life tables to determine how much to charge people for insurance policies. Biologists use life tables to predict future trends in populations.

We can construct a life table by determining for a group of individuals born at the same time (called a *cohort*) the number that are still alive at later dates (*survivorship*). Some life tables also include the number of offspring produced by the cohort during each time interval. An example of a life table based on an intensive study of the cactus finch carried out on Isla Daphne in the Galápagos archipelago, is shown in Table 54.1.

The data in Table 54.1 come from 210 birds that hatched in 1978 and were followed until 1991, when only 3 individuals were still alive. The table shows that the mortality rate for these birds was high during the first year of life. It then dropped dramatically for several years, followed by a general increase in later years. Mortality rates fluctuated among years because the survival of these birds depends on seed production, which strongly correlates with rainfall. The Galápagos archipelago experiences both drought years and years of heavy rain. During drought years, plants produce few seeds, birds do not nest, and adult survival is poor. In years when rainfall is heavy, seed production is high, most birds breed several times, and adult survival is high. The survival rates in the table reflect these rainfall fluctuations. Variation in seed production resulting from the alternation of wet and dry years is a major reason why the cactus finch population fluctuates so greatly.

54.1 Life Table of the 1978 Cohort of the Cactus Finch (*Geospiza scandens*) on Isla Daphne

AGE IN YEARS (*X*)	NUMBER ALIVE	SURVIVORSHIP[a]	SURVIVAL RATE[b]	MORTALITY RATE[c]
0	210	1.000	0.434	0.566
1	91	0.434	0.857	0.143
2	78	0.371	0.898	0.102
3	70	0.333	0.928	0.072
4	65	0.309	0.955	0.045
5	62	0.295	0.678	0.322
6	42	0.200	0.548	0.452
7	23	0.109	0.652	0.348
8	15	0.071	0.933	0.067
9	14	0.067	0.786	0.214
10	11	0.052	0.909	0.091
11	10	0.048	0.400	0.600
12	4	0.019	0.750	0.250
13	3	0.014	0.996	

[a]Survivorship = the proportion of newborns who survive to age x.
[b]Survival rate = the proportion of individuals of age x who survive to age $x + 1$.
[c]Mortality rate = the proportion of individuals of age x who die before the age of $x + 1$.

Ecologists often use graphs to highlight the most important changes in populations. Graphs of survivorship in relation to age show when individuals survive well and when they do not. Survivorship curves in many populations fall into one of three patterns. In some populations, most individuals survive for most of their potential life span, then die at about the same age. For example, because of intensive parental care and the availability of medical services, the survivorship of humans in the United States is high for many decades, but then declines rapidly in older individuals (Figure 54.1*a*). In a second pattern, which is characteristic of many songbirds, the probability of surviving is about the same over most of the life span once individuals are a few months old (Figure 54.1*b*). A third widespread survivorship pattern is found among organisms that produce a large number of offspring, each of which receives little energy or parental care. In these species, high death rates of young individuals are followed by high survival rate during the middle part of the life span. *Spergula vernalis*, an annual plant that grows on sand dunes in Poland, illustrates this pattern (Figure 54.1*c*).

The age distribution of individuals in a population reveals much about the recent history of births and deaths in the population. The timing of births and deaths can influence age distributions for many years in populations of long-lived species. The human population of the United States is a good example. Between 1947 and 1964, the United States experienced what is known as the post-World War II baby boom. During these years, average family size grew from 2.5 to 3.8 children; an unprecedented 4.3 million babies were born in 1957. Birth rates declined during the 1960s, but Americans born during the baby boom still constitute the dominant age class in the first part of the twenty-first century (Figure 54.2). "Baby boomers" became parents in the 1980s, producing another bulge in the age distribution—a "baby boom echo"—but they had, on average, fewer children than their parents did, so the bulge is not as large.

By summarizing information on when individuals are born and die, life tables help us understand why population densities change over time. Life table data can also be used to determine how heavily a population can be harvested and which age groups should be the focus of our efforts to save rare species. We will discuss the management of populations later in this chapter, but first we'll see how interactions among species influence the dynamics of particular populations.

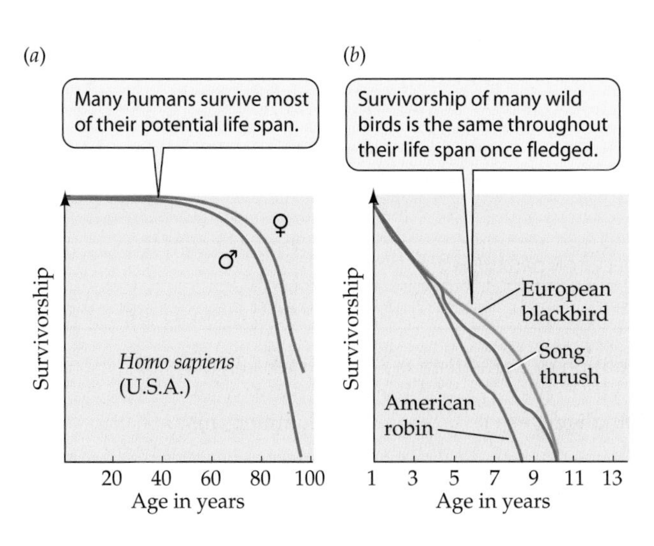

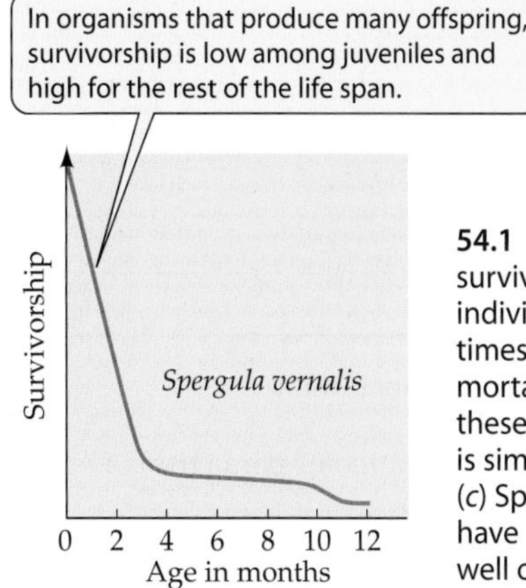

54.1 Survivorship Curves Three common survivorship curves show the number of individuals in a cohort still alive at different times over the life span. (*a*) For this curve, mortality is highest at advanced ages. (*b*) For these species, the probability of survivorship is similar throughout much of the life span. (*c*) Species with this survivorship pattern have high mortality at early ages, but survive well once past a critical point.

54.2 Age Distributions Change over Time The graphs shows age distributions for the human population of the United States from 1960 to 2020. The high birth rates during the "baby boom" have influenced the structure of the population over many decades.

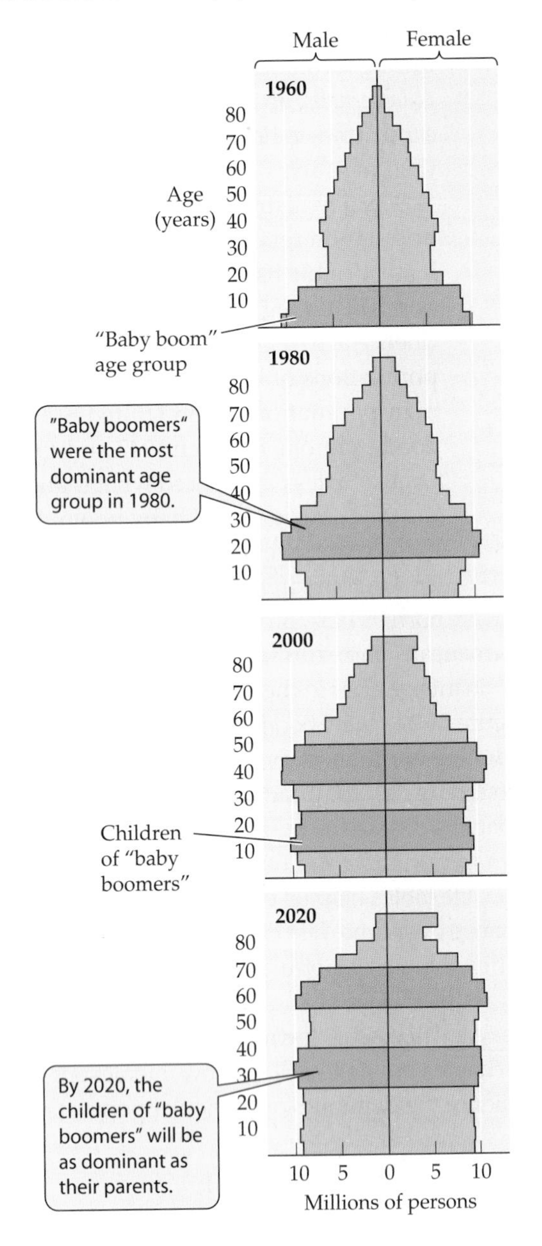

Types of Ecological Interactions

So far, we have considered only survival and reproductive rates of single species. Before we can answer the questions we have posed about populations, we also need to look at the ways in which populations of different species interact with one another (Table 54.2) These interactions fall into five general categories:

- A **mutualism** is an interaction in which both participants benefit (+/+ interaction).
- A **commensalism** is an interaction in which one participant benefits but the other is unaffected (+/0 interaction).
- An **amensalism** is an interaction in which one participant is harmed but the other is unaffected (0/– interaction).
- A **predator–prey** or **parasite–host interaction** is one in which one participant is harmed, but the other benefits (+/– interaction).
- If two organisms use the same resources and those resources are insufficient to supply their combined needs, the organisms are called competitors, and their interactions constitute **competition** (–/– interactions).

Mutualistic interactions exist between plants and microorganisms, protists and fungi, plants and insects, among animals, and among plants. Most plants have beneficial associations with soil-inhabiting fungi, called mycorrhizae, which enhance the plant's ability to extract minerals from the soil (see Figure 31.16). Some plants have mutualistic relationships with nitrogen-fixing bacteria of the genus *Rhizobium* (see Figure 37.7).

Animals have important mutualistic interactions with protists, plants, and other animals. Corals and some tunicates gain most of their energy from photosynthetic protists living within their tissues. In exchange, they provide the protists with nutrients from the small animals they capture. Termites have nitrogen-fixing protists in their guts that help them digest the cellulose in the wood they eat. The termites provide the protists with a suitable environment in which to live and an abundant supply of cellulose.

54.2 Types of Ecological Interactions

		EFFECT ON ORGANISM 2		
		HARM	BENEFIT	NO EFFECT
EFFECT ON ORGANISM 1	HARM	Competition (–/–)	Predation or parasitism (–/+)	Amensalism (–/0)
	BENEFIT	Predation or parasitism (+/–)	Mutualism (+/+)	Commensalism (+/0)
	NO EFFECT	Amensalism (0/–)	Commensalism (0/+)	—

54.3 Commensalism Benefits One Partner Cattle egrets (*Bubulcus ibis*) capture more insects with less effort when they forage around large grazing mammals such as this Cape buffalo (*Syncerus caffer*). The buffaloes are neither harmed nor helped by the egrets.

An example of a commensalism is the relationship between cattle egrets and grazing mammals. Cattle egrets are found throughout the tropics and subtropics. They typically forage on the ground around cattle or other large mammals, concentrating their attention near the mammals' heads and feet, where they capture insects flushed by their hooves and mouths (Figure 54.3). Cattle egrets foraging close to grazing mammals capture more food for less effort than egrets foraging away from grazing mammals. The benefit to the egrets is clear; the mammals neither gain nor lose.

Amensalisms are widespread and important interactions. Mammals, for example, may congregate around water holes, trampling and killing many plants. The mammals benefit by drinking water, but not by trampling and killing the plants. Leaves and branches falling from trees often damage smaller plants beneath them. The trees drop their old structures regardless of whether or not they damage other plants.

Predation and competition have particularly important influences on population dynamics. For that reason, we will illustrate several examples of these interactions later in this chapter. All five types of interactions, combined with the effects of the physical environment, determine the range of environmental conditions under which a species can persist. If there were no competitors, predators, or pathogens in its environment, a species would be able to persist under a broader array of physical conditions than it can in the presence of other species that negatively affect it. On the other hand, the presence of beneficial species may increase the range of physical conditions in which a species can persist.

With this background information on population structure and dynamics and interactions among populations, we can now turn to the questions that we posed at the beginning of this chapter. We will begin with abundance and rarity.

Factors Influencing Population Densities

You have probably observed that in a particular area, some species are much more abundant than others. Some locally rare species may be abundant somewhere else; other species may exist at low population densities everywhere. Some species that are rare at a given time may be abundant at some later time, or vice versa. Four factors—resource abundance, the size of individuals, the length of time a species has lived in an area, and social organization—exert strong influences on population density.

- *Species that use abundant resources often reach higher population densities than species that use scarce resources.* Thus, on average, animals that eat plants are typically more common than animals that eat other animals. We will explore this pattern in greater detail in Chapter 55.
- *Species with small individuals generally reach higher population densities than species with large individuals.* In general, population density decreases as body size increases, because small individuals require less energy to survive than large individuals.

This relationship can be demonstrated by a logarithmic plot of population density against body size for a variety of mammals worldwide (Figure 54.4). Although there is a strong relationship between population density and body size, the great scatter of points on the graph shows that some small

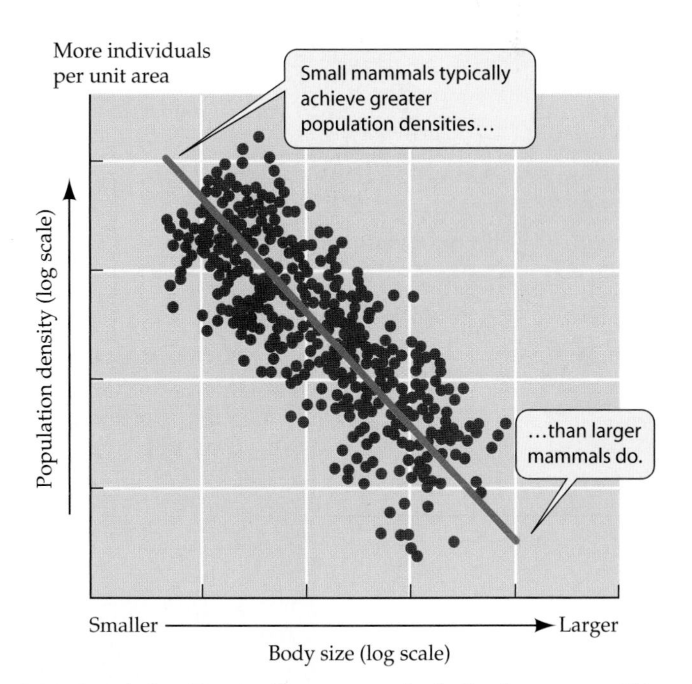

54.4 Population Density Decreases as Body Size Increases This trend is illustrated by a logarithmic plot (that is, each tick mark is 10 times greater than the one before it) of population density against body size for mammals of different sizes; each dot represents a different species, and the resulting slope (straight line) is determined algebraically.

species may use scarce resources and some large species may use abundant resources.

- *Some newly introduced species reach high population densities.* Species that have recently escaped control by the factors that normally prevent them from becoming more abundant may achieve temporarily high population densities. Species that are introduced into a new region, where their normal predators and diseases are absent, sometimes reach population densities much higher than ever found in their native ranges.

The zebra mussel, *Dreissena polymorpha*, whose larvae were carried from Europe in the ballast water of commercial cargo ships, became established in the Great Lakes in about 1985. Zebra mussels spread rapidly, and today they occupy much of the Great Lakes and the Mississippi River drainage (Figure 54.5). In some places these mussels have reached densities as high as 400,000 individuals per square meter; such densities are never found in Europe. Densities of zebra mussels in North America are likely to decrease in the future as local predators and diseases begin to attack them.

- *Complex social organization may facilitate high densities.* As we saw in Chapter 53, some highly social species, including ants, termites, and humans, can achieve remarkably high population densities.

The factors that influence population density may strengthen or weaken over time. When this happens, population densities change. Let's look now at how and why population densities fluctuate.

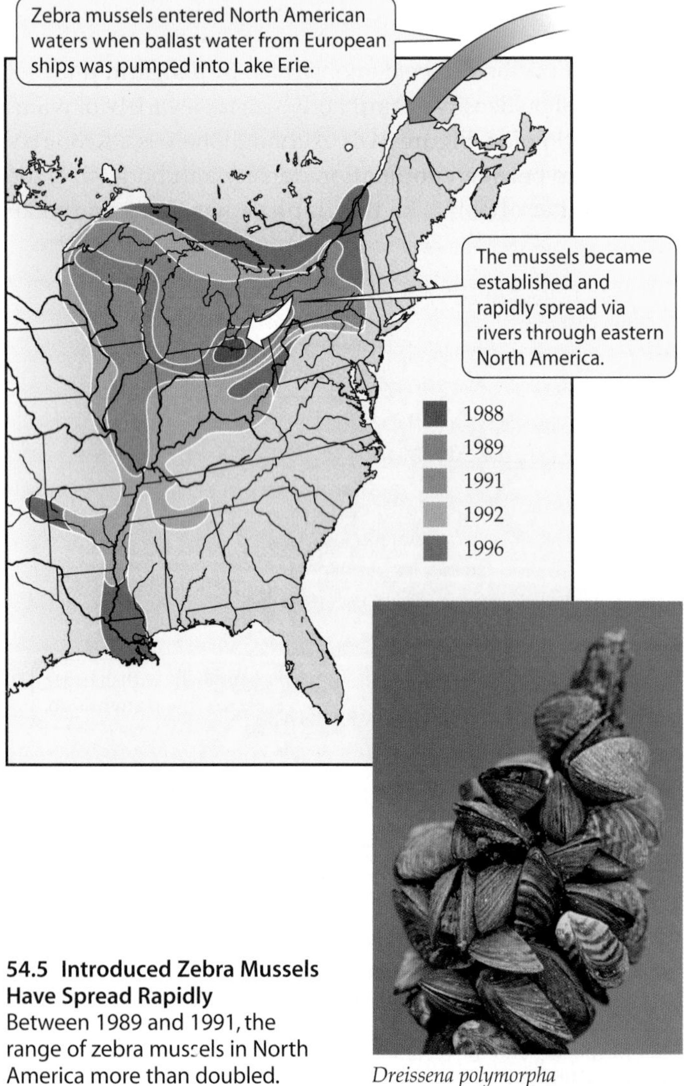

Zebra mussels entered North American waters when ballast water from European ships was pumped into Lake Erie.
The mussels became established and rapidly spread via rivers through eastern North America.
1988
1989
1991
1992
1996

54.5 Introduced Zebra Mussels Have Spread Rapidly
Between 1989 and 1991, the range of zebra mussels in North America more than doubled.

Dreissena polymorpha

Fluctuations in Population Densities

Although some populations fluctuate markedly in density, even the most dramatic fluctuations are much less than those that are theoretically possible. To visualize those possibilities, consider a single bacterium selected at random from the surface of this book. If all its descendants were able to grow and reproduce in an unlimited environment, explosive population growth would result. In a month, this bacterial colony would weigh more than the visible universe and would be expanding outward at the speed of light. Similarly, a single pair of Atlantic cod and their descendants, reproducing at the maximum rate of which they are capable, would fill the Atlantic Ocean basin in 6 years if none of them died. Obviously, such dramatic population growth does not occur in nature. What prevents it from happening?

All populations have the potential for exponential growth

Bacteria and cod illustrate the fact that all populations have the potential for explosive growth. As the number of individuals in a population increases, the number of new individuals added per unit of time accelerates, even if the rate of increase expressed on a per individual basis—called the *per capita growth rate*—remains constant. If births and deaths occur continuously and at constant rates, a graph of the population size over time forms a continuous, J-shaped curve (Figure 54.6). This form of explosive increase is called **exponential growth**. It can be expressed mathematically in the following way:

Rate of increase in number of individuals

$$= \begin{pmatrix} \text{Average per capita birth rate} \\ -\text{ Average per capita death rate} \end{pmatrix}$$

$$\times \text{ Number of individuals}$$

or, more concisely,

$$= \frac{\Delta N}{\Delta t} = (b - d)N$$

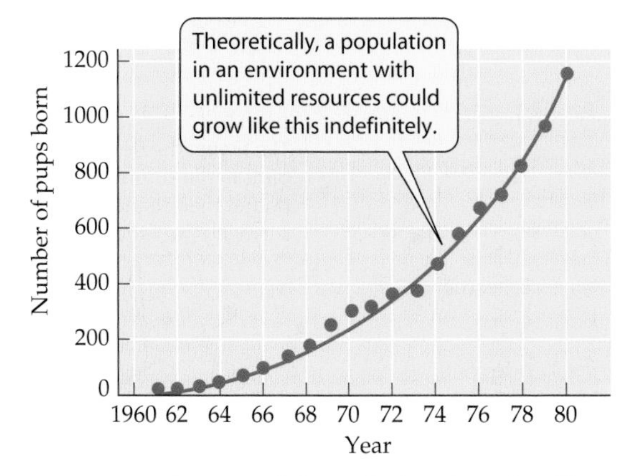

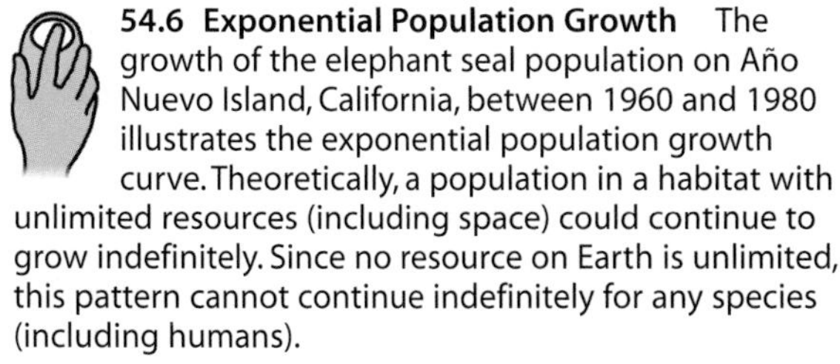

54.6 Exponential Population Growth The growth of the elephant seal population on Año Nuevo Island, California, between 1960 and 1980 illustrates the exponential population growth curve. Theoretically, a population in a habitat with unlimited resources (including space) could continue to grow indefinitely. Since no resource on Earth is unlimited, this pattern cannot continue indefinitely for any species (including humans).

Mirounga angustirostris

where $\Delta N/\Delta t$ is the rate of change in the size of the population (ΔN = change in number of individuals; Δt = change in time).

The difference between the average per capita birth rate in a population (b) and its average per capita death rate (d) is the *net reproductive rate* (r). (In these equations, b includes both births and immigrations, and d includes both deaths and emigrations.) When conditions are optimal for the population, the net reproductive rate has its highest value, called r_{max}, or the *intrinsic rate of increase*; r_{max} has a characteristic value for each species. Therefore, the rate of growth of a population under optimal conditions is

$$\frac{\Delta N}{\Delta t} = r_{max} N$$

For very short time periods, some populations may grow at rates close to the intrinsic rate of increase. For example, northern elephant seals were hunted nearly to extinction in the late nineteenth century. In 1890, only about 20 animals remained, confined to Isla Guadalupe off the northwestern coast of Mexico. Once the hunting was stopped, the population was protected from its major predator, and ample elephant seal habitat remained available, so the population began to increase rapidly. Elephant seals recolonized Año Nuevo Island near Santa Cruz, California, in 1960. In the 20 years after colonization, the population breeding on the island expanded exponentially (see Figure 54.6).

Population growth is influenced by environmental limits

No real population can maintain exponential growth for very long. As a population increases in size, environmental limits cause birth rates to drop and death rates to rise. In fact, over long time periods, the densities of most populations fluctuate around a relatively constant number. The simplest way to picture the limits imposed by the environment is to assume that an environment can support no more than a certain number of individuals of any particular species per unit of area. This number, called the **environmental carrying capacity** (K), is determined by the availability of resources—food, nest sites, shelter—as well as by disease, predators, and, in some cases, social interactions.

Because of environmental limits, the growth of a population typically slows down as its density approaches the environmental carrying capacity. A graph of the population size over time results in an S-shaped curve (Figure 54.7). This pattern is called **logistic growth**. The simplest way to generate an S-shaped growth curve is to add to the equation for exponential growth a term, $(K - N)/K$, that slows the population's growth as it approaches the carrying capacity. This

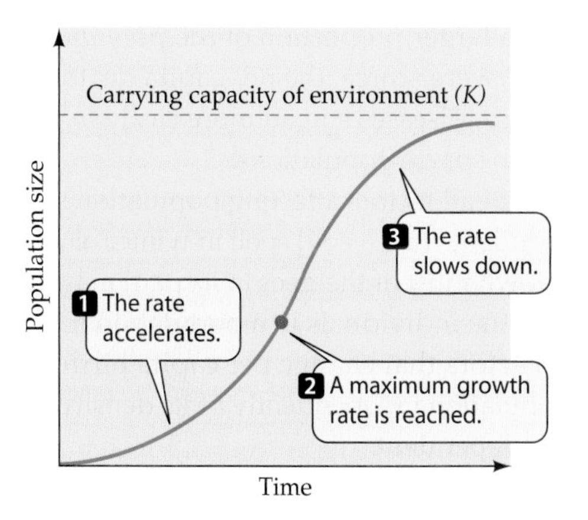

54.7 Logistic Population Growth Typically, a population in an environment with limited resources stops growing exponentially long before it reaches the environmental carrying capacity.

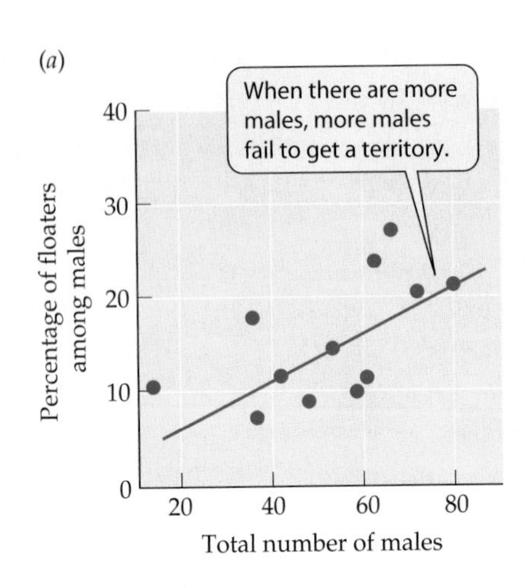

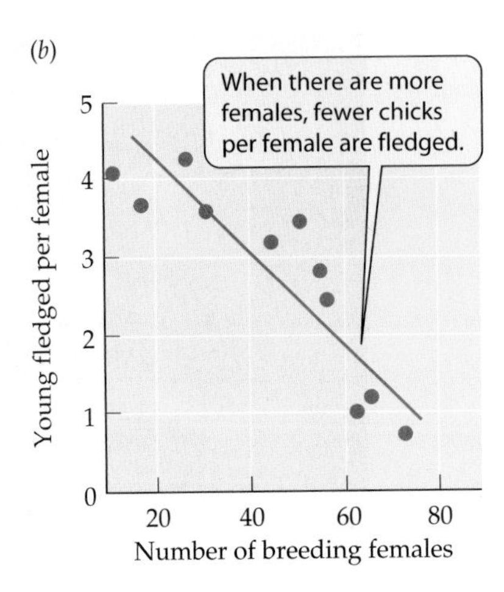

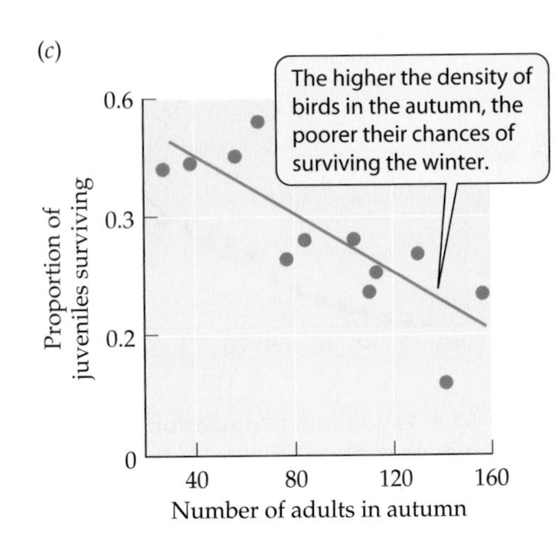

54.8 Regulation of an Island Population of Song Sparrows The size of the population of song sparrows (*Melospiza melodia*) on Mandarte Island, British Columbia, is determined in part by the severity of winter weather. In addition, the population is regulated by (*a*) the territorial behavior of males, (*b*) the reproductive success of females, and (*c*) the survival of juveniles in relation to population density.

term implies that each individual added to the population depresses population growth by an equal amount:

$$\frac{\Delta N}{\Delta t} = r\left(\frac{K-N}{K}\right)N$$

Population growth stops when $N = K$ because then $(K - N) = 0$, so $(K - N)/K = 0$, and thus $\Delta N/\Delta t = 0$.

Population densities influence birth and death rates

Because each additional individual typically makes things worse for other members of the population in a limited environment, per capita birth and death rates usually change in response to population density; that is, they are **density-dependent**. Birth and death rates may be density-dependent for several reasons. First, as a species increases in abundance, it may deplete its food supply, reducing the amount of food available to each individual. Poorer nutrition may increase death rates and decrease birth rates. Second, predators may be attracted to areas with high densities of their prey. If predators capture a larger proportion of the prey than they did when the prey were scarce, the per capita death rate of the prey rises. Third, diseases spread more easily in dense populations than in sparse populations.

However, not all factors affecting population size act in a density-dependent way. A cold spell in winter or a hurricane that blows down most of the trees in its path may kill a large proportion of the individuals in a population regardless of its density. Factors that change per capita birth and death rates in a population independently of its density are said to be **density-independent**.

Fluctuations in the density of a population are determined by all the density-dependent and density-independent factors acting on it. The combined action of these factors is shown by the dynamics of a population of song sparrows on Mandarte Island, off the coast of British Columbia, Canada. Over a period of 12 years, the number of song sparrows fluctuated between 4 and 72 breeding females and between 9 and 100 territorial males. Death rates are high during particularly cold, snowy winters, regardless of the density of the population. Several density-dependent factors also contribute to fluctuations in the density of the population. The number of breeding males, for example, is limited by territorial behavior: The larger the number of males, the larger the number that fail to gain territories and must live as "floaters" with little chance of reproducing (Figure 54.8*a*). Also, the larger the number of breeding females, the fewer offspring each female fledges (raises to the age when it can leave the nest) (Figure 54.8*b*). And the more birds alive in autumn, the poorer the chances of juveniles born in that year surviving the winter (Figure 54.8*c*). Thus, the number of males and females breeding each year is influenced by both density-independent and density-dependent factors.

Population Fluctuations

The cactus finch, which we met in our earlier discussion of life tables, is a small, short-lived seed-eating bird that lives only in the Galápagos archipelago. The south polar skua is a long-lived carnivorous seabird with a broad geographic range in the southern oceans. Over several decades, the number of cactus finches fluctuated widely, as we have already noted. The number of skuas fluctuated very little over an even longer time period (Figure 54.9). Why did the population of skuas fluctuate so much less than the population of cactus finches?

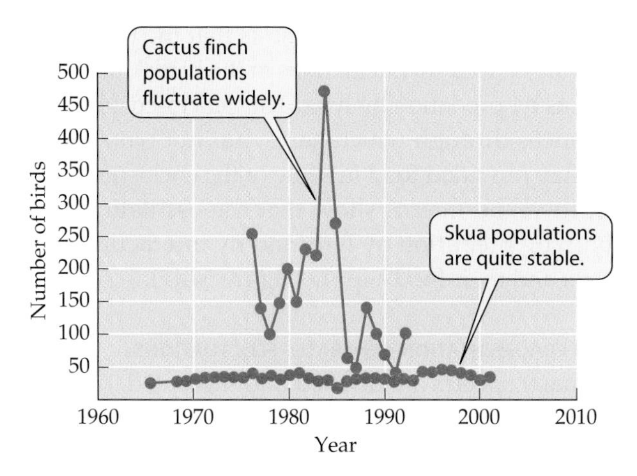

54.9 Population Sizes May Be Stable or Highly Variable Populations of small, short-lived species, such as that of cactus finches (*Geospiza scandens*) on Isla Daphne in the Galápagos Archipelago, tend to fluctuate much more than do populations of larger, longer-lived species, such as that of south polar skuas (*Catharacta maccormicki*).

All populations fluctuate less than the theoretical maximum, but the sizes of some populations fluctuate remarkably little. The comparison between south polar skuas and cactus finches illustrates one cause of such differences: Species with long-lived individuals that have low reproductive rates, such as south polar skuas, typically have more stable populations than species with short-lived individuals that have high reproductive rates, such as cactus finches. Small, short-lived individuals are generally more vulnerable to environmental changes than long-lived individuals. That is why insect population densities tend to fluctuate much more than those of birds and mammals, and population densities of annual plants fluctuate much more than those of trees.

EPISODIC REPRODUCTION GENERATES FLUCTUATIONS. For most species, some years are better for reproducing than other years. In Lake Erie, 1944 was such an excellent year for reproduction of whitefish that individuals born in that year dominated whitefish catches in the lake for several years (Figure 54.10*a*). Similarly, most of the individuals found in a population of black cherry trees in a Wisconsin forest in 1971 became established between 1931 and 1941 (Figure 54.10*b*). Population densities increase following years of good reproductive success, but they decrease following years of poor reproduction.

RESOURCE FLUCTUATIONS GENERATE CONSUMER FLUCTUATIONS. Densities of populations of species that depend on a single or just a few resources are likely to fluctuate more than those of species that use a greater variety of resources. As we have seen, cactus finch populations fluctuate with the annual production of the seeds they eat, which varies greatly. Similarly, several species of birds and mammals that live in northern coniferous forests depend on seeds in conifer cones. Most trees in northern coniferous forests reproduce synchronously and episodically; consequently, over large areas, there are years of massive seed production and years of little or no seed production. Some birds (such as crossbills) wander over large areas, looking for places where

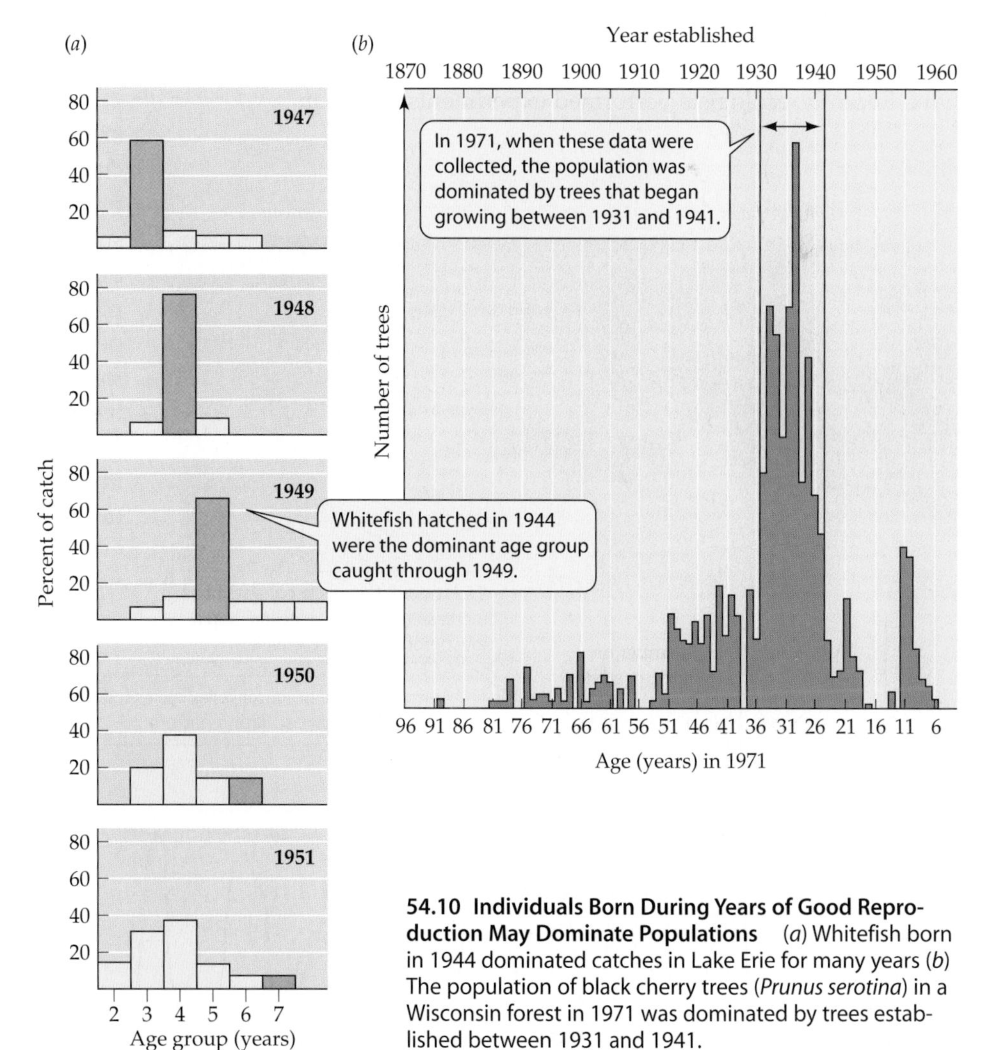

54.10 Individuals Born During Years of Good Reproduction May Dominate Populations (*a*) Whitefish born in 1944 dominated catches in Lake Erie for many years (*b*) The population of black cherry trees (*Prunus serotina*) in a Wisconsin forest in 1971 was dominated by trees established between 1931 and 1941.

cones have been produced. Other birds (such as jays and nutcrackers) and some mammals (squirrels) store cones during years of high production, but they often suffer high mortality rates during years when the trees in their area produce few or no cones.

POPULATION INTERACTIONS GENERATE FLUCTUATIONS Predators often cause fluctuations in the densities of their prey because predator population growth nearly always lags behind growth in populations of their prey. The predator population grows and eats most of its prey population, followed by a crash in the predator population, which no longer has enough food. Population oscillations among small mammals and their predators living at high latitudes, where many predators depend on only one or a few prey species, are the best-known examples of fluctuations in population densities driven by predator–prey interactions. Populations of Arctic lemmings and their chief predators—snowy owls, jaegers, and Arctic foxes—oscillate with a 3- to 4-year periodicity. Populations of Canadian lynx and their principal prey, snowshoe hares, oscillate on a 9- to 11-year cycle (Figure 54.11).

For many years, ecologists thought that hare–lynx oscillations were caused only by interactions between hares and lynx. Recently, ecologists performed experiments in Yukon Territory, Canada, to test the hypothesis that the lynx–hare oscillations are caused by fluctuations in the hares' food supply as well as by predation by lynx. They enclosed some areas with fences through which hares, but not lynx, could pass, and they provided food in some of the enclosures. The results of the experiments show that the oscillations are driven both by predation by lynx and by interactions between hares and their food supply (Figure 54.12).

POPULATION FRAGMENTATION GENERATES FLUCTUATIONS. Populations of many species are divided into separated, discrete subpopulations living in distinct habitat patches, among which some exchange of individuals occurs. Each subpopulation has a probability of "birth" (colonization) and "death" (extinction). Within each subpopulation, growth occurs in the ways we have just described, but because the subpopulations are much smaller than the population as a whole, local disturbances and random fluctuations in numbers of individuals are more likely to cause the extinction of subpopulations than the extinction of an entire population. However, if individuals move frequently between subpopulations, immigrants may prevent declining subpopulations from becoming extinct. This process is known as the **rescue effect**.

EXAMPLES OF SUBPOPULATION DYNAMICS. The bay checkerspot butterfly (*Euphydryas editha bayensis*) provides a good illustration of the dynamics of subpopulations. The caterpillars (larvae) of this butterfly feed on only a few species of annual plants, which are restricted to outcrops of serpentine rock on hills south of San Francisco, California. The bay checkerspot has been studied for many years by Stanford University biologists. During drought years, most host plants die early in spring, before the caterpillars have developed far enough to be able to enter their summer rest-

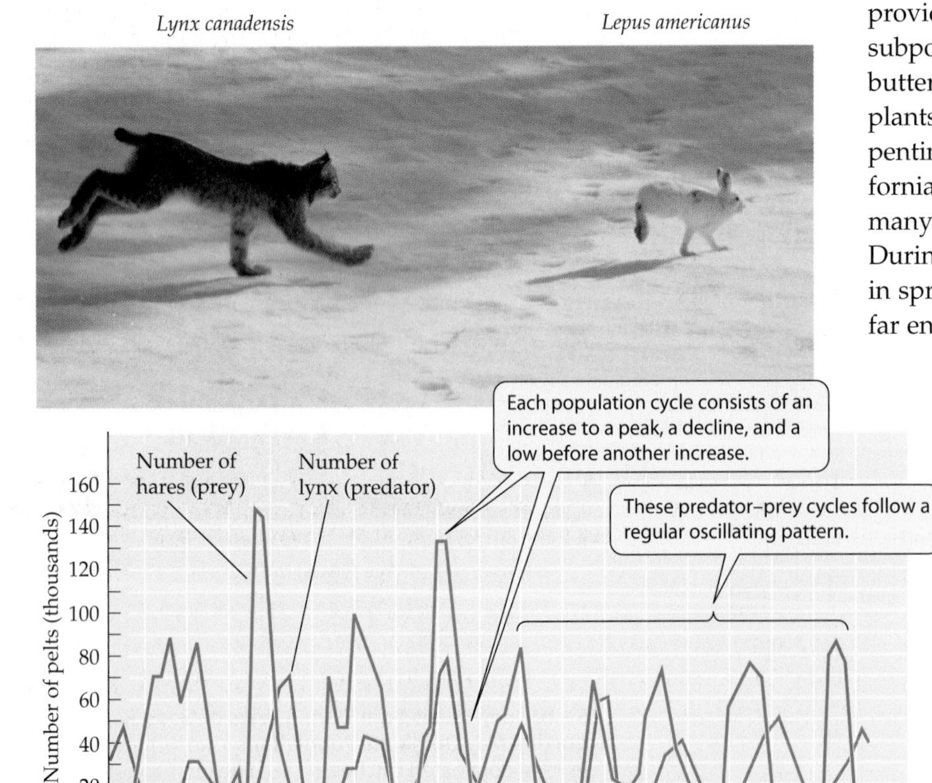

54.11 Hare and Lynx Populations Cycle in Nature The 9–11-year population cycle of the snowshoe hare and its major predator, the Canadian lynx, was revealed in records of the number of pelts that were sold to the Hudson's Bay Company by fur trappers.

EXPERIMENT

Hypothesis: Population cycles of hares are influenced by both food supply and predators.

METHOD Select 9 1-km^2 blocks of undisturbed coniferous forest. In two of the blocks give the hares supplemental food year-round. Erect an electric fence around two other blocks, with mesh large enough to allow hares, but not lynxes, to pass. Provide extra food in one of these blocks. In two other blocks add fertilizer to increase food quality. Use three other blocks as unmanipulated controls.

RESULTS

Ratio of hare density to controls

Food added — Adding food tripled hare density.

Predators excluded — Excluding predators doubled hare density.

Fertilizer added — Fertilizing vegetation to increase its food quality had no significant effect.

Food added and predators excluded — Both adding food and excluding predators increased hare density dramatically.

Control

One hare population cycle (11 years)

Conclusion: Population cycles of the snowshoe hare are influenced by their food supply as well as by interactions with their predators.

54.12 Prey Population Cycles May Have Multiple Causes Experiments showed that both food supply and predation (but not food quality) affect the population densities of snowshoe hares.

ing stage. At least three butterfly subpopulations became extinct during a severe drought in 1975–1977. The largest patch of suitable butterfly habitat, Morgan Hill, typically supports thousands of butterflies (Figure 54.13). It probably served as a source of individuals that dispersed to and recolonized small patches where the butterflies had become extinct.

In another study, ecologists manipulated the habitat of tiny arthropods (springtails—tiny insects without wings—and mites) to investigate the subpopulation dynamics of these animals. In one experiment, they created isolated patches of the animals' habitat—mosses growing on rocks—by clearing moss from parts of the rock surface (Figure 54.14, Experiment 1). The number of species present in these patches declined about 40 percent within a year, with more rare species than common species disappearing from the patches. The experiment illustrated that small, isolated populations are more likely to become extinct than large populations are.

In a second experiment, the investigators created similar patches, but these patches were connected by narrow corridors of moss that were either intact or disrupted by a barrier only 10 mm wide (Figure 54.14, Experiment 2). Moss patches connected by unbroken corridors contained more species of arthropods a year later than patches whose corridors were discontinuous. Thus, a gap of only 10 mm was sufficient to reduce the rescue effect.

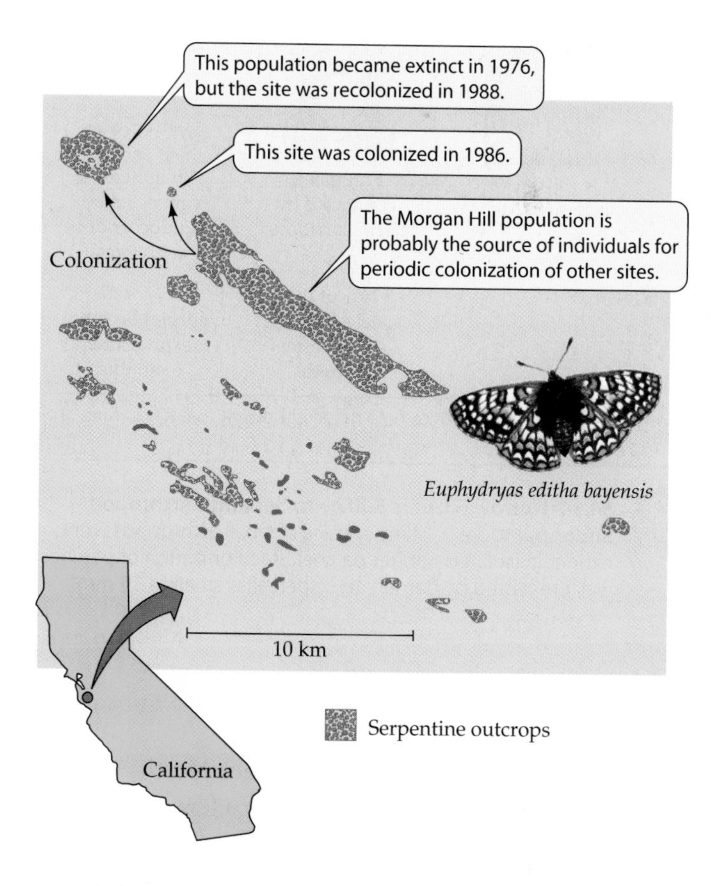

54.13 Subpopulation Dynamics The bay checkerspot butterfly population is divided into a number of subpopulations confined to patches of habitat (serpentine rock) that contain the food plants of its larvae. Extinction of these subpopulations is common.

EXPERIMENT

Hypothesis: Even small barriers to dispersal may reduce the number of species in a habitat patch.

METHOD Moss growing on rocks was trimmed to form distinct habitat patches. The number of small organisms (mostly arthropods) living in the patches was observed over time.

Control | Experiment 1 patches | Experiment 2 patch | Moss patches

Experiment 1

50 cm | 50 cm | Control patch | Fragments, each 20 cm^2

RESULTS In fragments, 40% of species became extinct after 1 year.

Experiment 2

50 cm | 50 cm | Control patch | Fragments connected by 7-cm corridors | 10-mm gaps | Fragments connected by pseudocorridors with gaps

RESULTS 14% of the species became extinct after 6 months. 41% of the species became extinct after 6 months.

Conclusion: Even small barriers to dispersal raised extinction rates in a fragmented habitat.

54.14 Narrow Barriers Suffice to Separate Arthropod Subpopulations Many species of small arthropods went extinct in isolated habitat patches. Recolonization of patches was prevented by barriers to dispersal as small as 10 mm.

Variations in Species' Ranges

Douglas firs are widespread in western North America, whereas giant sequoias are restricted to a few groves in the southern Sierra Nevada of California. The desert pupfish is restricted to a single spring in Death Valley, California, whereas smallmouth bass live in most of the rivers and lakes in eastern North America. Why do the geographic ranges of species vary so much? The factors contributing to this variation include speciation processes, dispersal abilities, and interactions with other species. As you might suspect, not all of the factors that influence geographic ranges are important for all species.

SPECIATION PROCESSES INFLUENCE RANGE SIZES. As we saw in Chapter 24, there are several ways in which a new species can originate. A species that arises by polyploidy inevitably begins with a very small range. Because many polyploid plant species have formed only recently and have not spread much beyond the site of their origin, many of these plants have small ranges. Similarly, species that arise through founder events begin their history with small ranges. In contrast, most species that arise via allopatric speciation begin with large ranges. Finally, as a species declines toward extinction, as may be happening to giant sequoias (Figure 54.15), its range shrinks until it vanishes when the last individual dies.

DISPERSAL ABILITIES RESTRICT GEOGRAPHIC RANGES. As we also saw in Chapter 24, the dispersal abilities of different species vary greatly. The experiments with small arthropods living in mosses on rocks show that even narrow barriers may prevent some species from reaching and colonizing an area. The solitary spring that is home to the desert pupfish is isolated from other bodies of fresh water, so the fish cannot disperse. Thus, the absence of many species from an area is simply due to a failure to get there. Zebra mussels, for

Sequoiadendron giganteum

54.15 The Last Refuge The range of giant sequoias has progressively shrunk to a few remaining groves of trees scattered in the southern Sierra Nevada mountains of California.

example, were not found in North America before 1985 because they were unable to disperse across the Atlantic Ocean from Europe. Lack of suitable habitat was not the reason for their absence, as demonstrated by their dramatic population growth in North America once they were transported there by human activities. Once they reached North America, they were able to disperse rapidly because the larvae are free-swimming and the adults can attach to moving objects, such as boat hulls.

PREDATORS MAY RESTRICT SPECIES' RANGES. Predators may eliminate their prey in some places, but not in others. For example, chorus frogs (*Pseudacris triseriata*) are found in only some of the ponds on islands in Lake Superior. Three major predators—the larvae of a salamander, the nymphs of a large dragonfly, and dytiscid beetles—eat chorus frog tadpoles. An ecologist noticed that the tadpoles were common in ponds with beetles, but rare in ponds with salamander larvae and dragonfly nymphs. In laboratory experiments, he established that the salamander larvae could eat only small tadpoles, but that dragonfly nymphs could eat tadpoles of all sizes. Therefore, he hypothesized that dragonfly nymphs were responsible for eliminating chorus frogs from many ponds. He tested his hypothesis by manipulating densities of predators and prey in ponds. The results showed that dragonfly nymphs can eliminate chorus frogs from ponds that would otherwise be suitable for them (Figure 54.16).

COMPETITION MAY RESTRICT SPECIES' RANGES. How competitive interactions may restrict the ranges of species is illustrated by interactions between two species of barnacles, *Balanus balanoides* and *Chthamalus stellatus*, on rocky North Atlantic seashores. These barnacles have planktonic larvae, which settle between high and low tide levels on the shoreline and become sessile adults. Adult *Chthamalus* generally live higher in the intertidal zone than do adult *Balanus*, and there is little overlap between the two species. What keeps their ranges so distinct?

By experimentally removing one or the other species, researchers have shown that the vertical ranges of adults of both species are greater in the absence of the other species. *Chthamalus* larvae normally settle in large numbers in the *Balanus* zone. If *Balanus* are absent, young *Chthamalus* survive and grow well in the *Balanus* zone, but if *Balanus* are present, they smother, crush, or undercut the *Chthamalus*. *Balanus* larvae also settle in the *Chthamalus* zone, but the young *Balanus* grow slowly there because they lose water rapidly when exposed to air, so *Chthamalus* outcompete *Balanus* in that zone. The result of the competitive interaction between the two species is intertidal zonation, with *Chthamalus* growing above *Balanus* (Figure 54.17).

54.16 Predators Exclude Prey from Some Habitats The speed with which dragonfly nymphs can eliminate tadpoles of the chorus frog from a pond is illustrated by the results of experiments in which populations of predators and prey were manipulated.

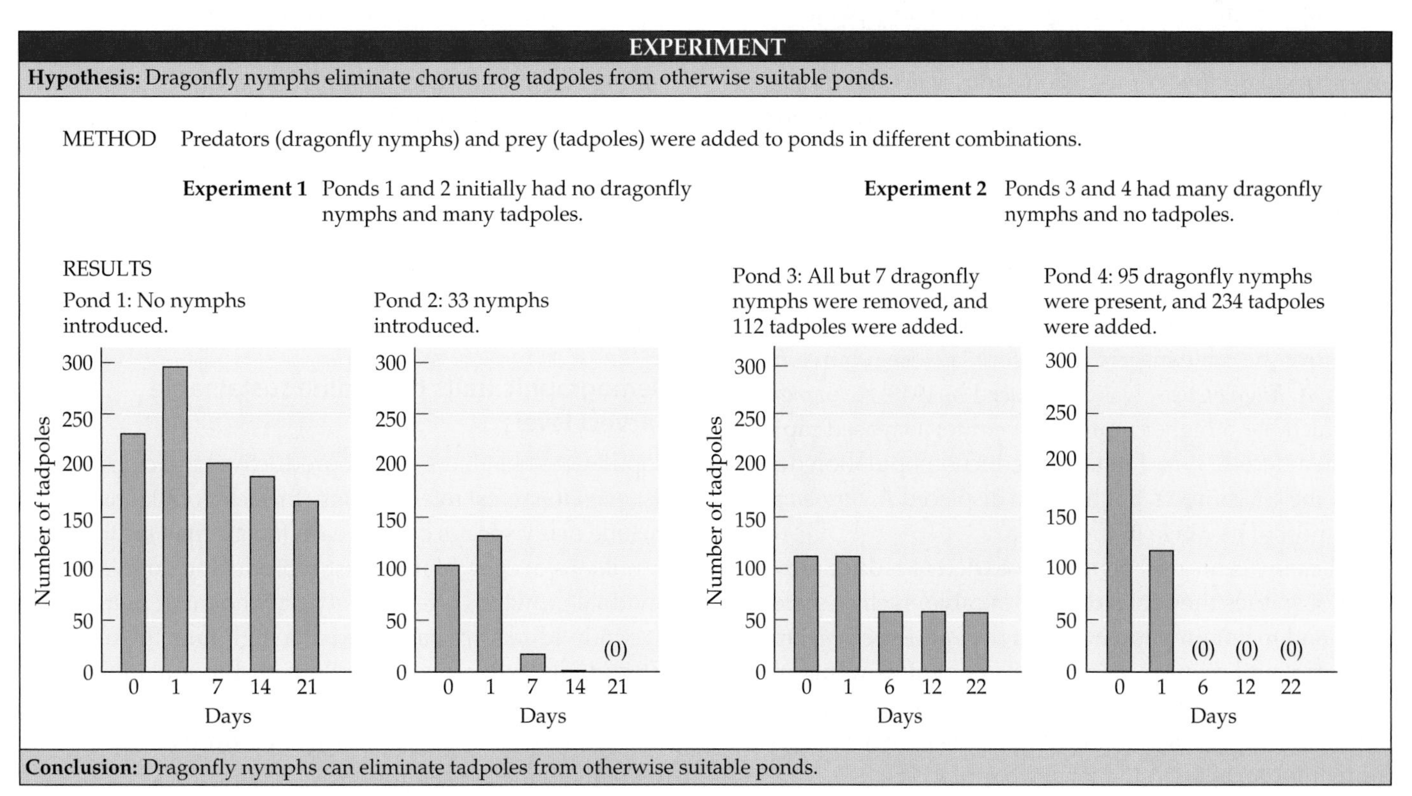

EXPERIMENT
Hypothesis: Dragonfly nymphs eliminate chorus frog tadpoles from otherwise suitable ponds.
METHOD Predators (dragonfly nymphs) and prey (tadpoles) were added to ponds in different combinations.
Experiment 1 Ponds 1 and 2 initially had no dragonfly nymphs and many tadpoles.
Experiment 2 Ponds 3 and 4 had many dragonfly nymphs and no tadpoles.
RESULTS
Pond 1: No nymphs introduced.
Pond 2: 33 nymphs introduced.
Pond 3: All but 7 dragonfly nymphs were removed, and 112 tadpoles were added.
Pond 4: 95 dragonfly nymphs were present, and 234 tadpoles were added.
Number of tadpoles
0
50
100
150
200
250
300
Days
0 1 7 14 21
0 1 6 12 22
(0)
Conclusion: Dragonfly nymphs can eliminate tadpoles from otherwise suitable ponds.

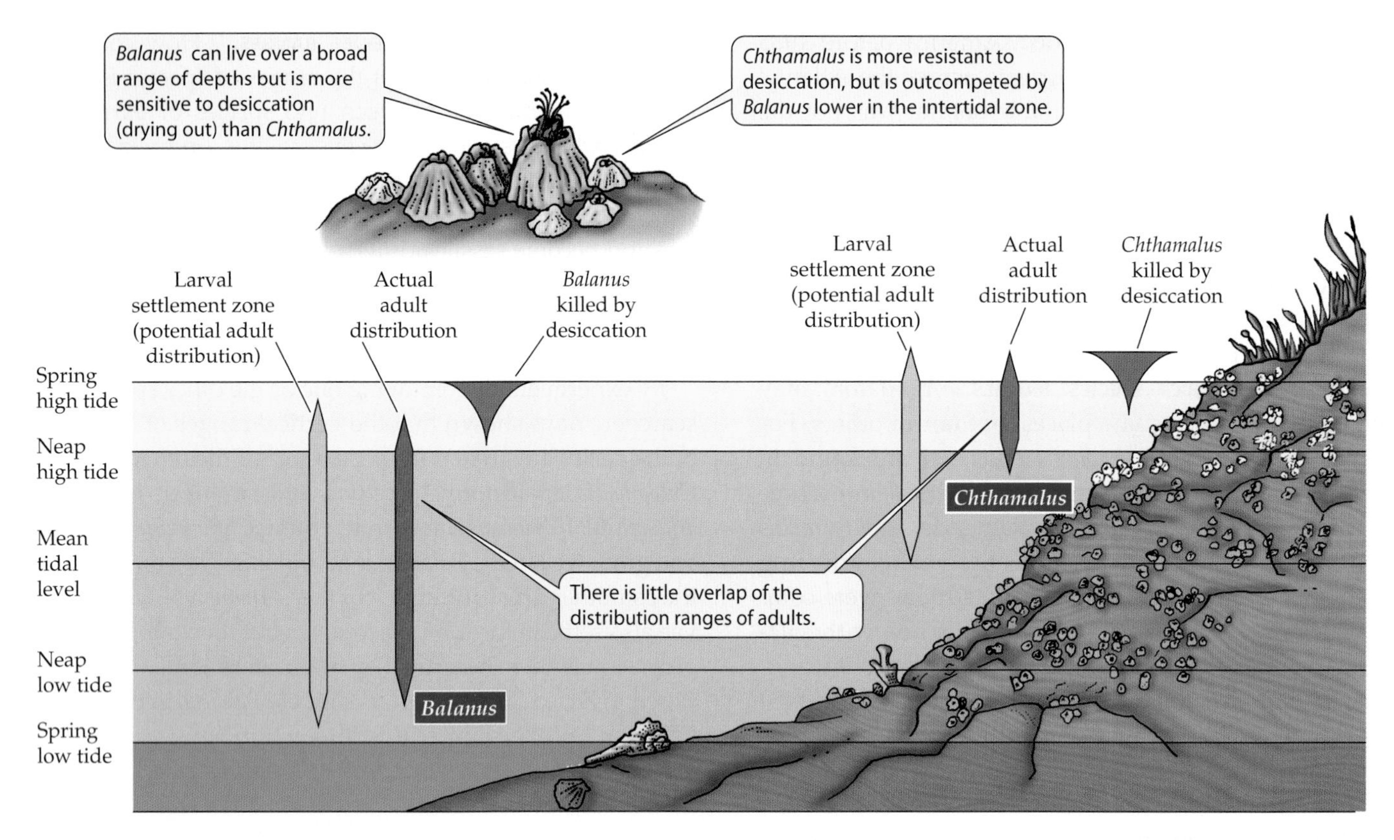

54.17 Competition Restricts the Intertidal Ranges of Barnacles Interspecific competition between *Balanus* and *Chthamalus* makes the zone each species occupies smaller than the zone it could occupy in the absence of the other species. The width of the red and gold bars is proportional to the density of the populations.

Sessile animals such as barnacles and many plants compete for space, but most mobile animals compete for food. As an example of how competition can restrict the ranges of such species, consider the distribution of two species of wasps in California. These wasps lay their eggs on scale insects, and the larvae that hatch from those eggs burrow into, eat, and kill the scale insects. Both wasps were introduced to control outbreaks of scale insects that were damaging citrus orchards. The Mediterranean wasp *Aphytis chrysomphali* was introduced to southern California around 1900, but it failed to control the scale insects. Therefore, a close relative from China, *A. lingnanensis*, was introduced in 1948. *A. lingnanensis*, which has a higher reproductive rate, increased rapidly. Within a decade it had not only reduced population densities of the scale insects, but had also displaced *A. chrysomphali* from most of its range in California.

For many centuries, people have tried to reduce populations of species they consider undesirable, such as scale insects, and maintain populations of desirable species. Efforts to control and manage populations of organisms are more likely to be successful if they are based on knowledge of how populations grow and are regulated. Let's see how such information can be used to manage populations.

Managing Populations

A general principle of population dynamics is that both the total number of births and the growth rates of individuals tend to be highest when a population is well below its carrying capacity (see Figure 54.8). Therefore, if we wish to maximize the number of individuals that can be harvested from a population, we should manage the population so that it is far enough below carrying capacity to have high birth and growth rates. Hunting seasons for game birds and mammals are established with this objective in mind.

Demographic traits determine sustainable harvest levels

Populations that have high reproductive capacities can persist even if harvest rates are high. In such populations (which include many species of fish), each female may lay thousands or millions of eggs. In these fast-reproducing populations, individual growth is often density-dependent. If prereproductive individuals are harvested at a high rate, the remaining individuals may grow faster. Some fish populations can be harvested heavily because only a modest number of females must survive to reproductive age to produce the eggs needed to maintain the population.

Fish can, of course, be overharvested. Many fish populations have been greatly reduced because so many individuals were harvested that too few reproductive adults survived to maintain the population. The Georges Bank off the coast of New England—a source of cod, halibut, and other prime food fishes—was exploited so heavily during the twentieth century that many fish stocks were reduced to levels insufficient to support a commercial fishery. The fishery has remained closed into the twenty-first century.

The whaling industry has also engaged in excessive harvests. The blue whale, Earth's largest animal, was the first whale species to be hunted nearly to extinction. The industry then turned to smaller species of whales that were still numerous enough to support commercially viable whaling operations (Figure 54.18).

Management of whale populations is difficult for two reasons. First, unlike fish, whales reproduce at very low rates. They have long prereproductive periods before they mature, produce only one offspring at a time, and have long intervals between births. Thus, many adult whales are needed to produce even a small number of offspring. Second, because whales are distributed widely throughout Earth's oceans, they are an international resource whose conservation and wise management depends upon cooperative action by all whaling nations. This goal continues to be difficult to achieve.

Demographic information is used to control populations

The same management principles apply if we wish to reduce the size of populations of undesirable species and keep them at low densities. At densities well below carrying capacity, populations typically have high birth rates, and can therefore withstand higher death rates than they can when they are closer to carrying capacity.

When population dynamics are influenced primarily by factors that operate in a density-dependent manner, killing part of a population typically reduces it to a density at which it reproduces at a higher rate. A more effective approach to reducing such a population is to remove its resources, thereby lowering the carrying capacity of its environment. We can rid our dumps and cities of rats more easily by making garbage unavailable (reducing the carrying capacity of the rats' environment) than by poisoning rats (which only increases their reproductive rate). However, this option may not exist in agriculture, in which a high density of the crop is the management objective.

Similarly, if we wish to preserve a rare species, the most important step usually is to provide it with suitable habitat. If habitat is available, the species will usually reproduce at rates sufficient to maintain its population. If the habitat is insufficient, preserving the species usually requires expensive and continuing intervention, such as providing extra food.

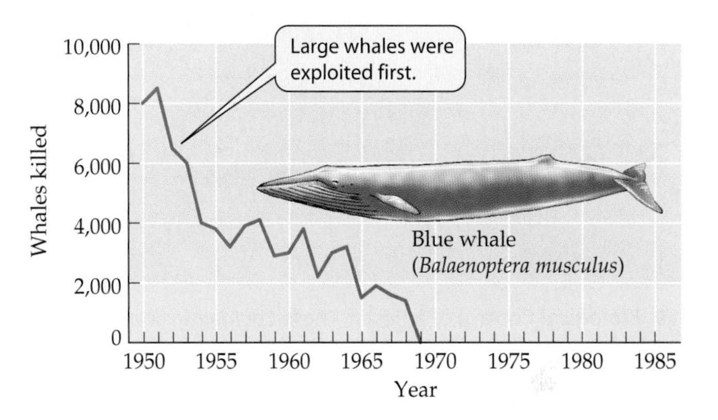

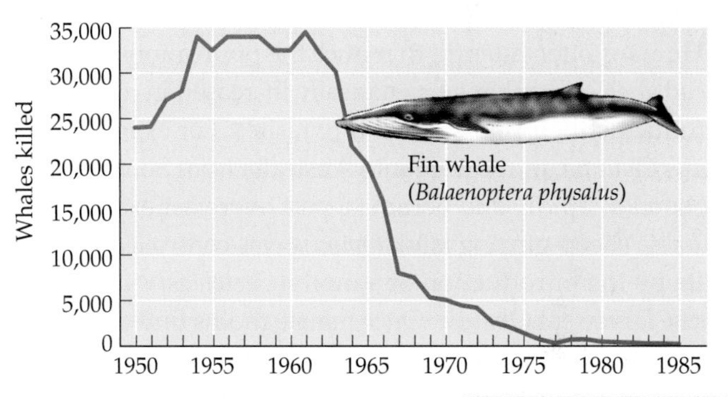

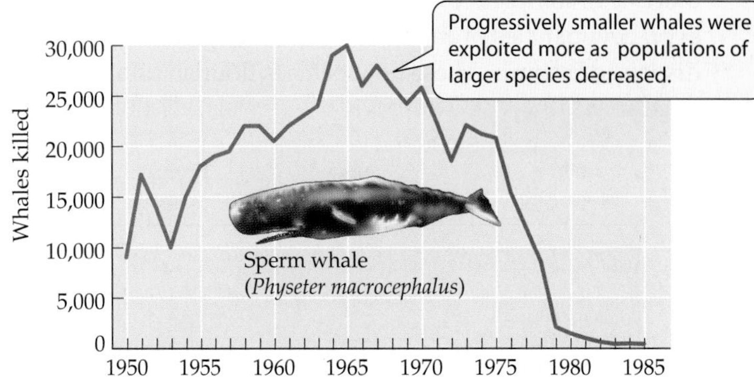

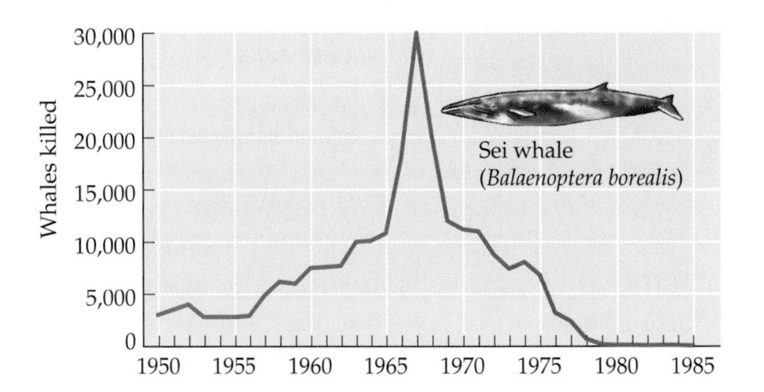

54.18 Overexploitation of Whales These graphs show the numbers of whales of four species killed each year from 1950 to 1985. All four species were driven to very low population levels by sustained hunting.

54.19 Biological Control of a Pest These *Cactoblastis* caterpillars are consuming an *Opuntia* cactus in Australia.

Humans often attempt to reduce the populations of introduced species that have dramatically increased in density by introducing their predators and parasites. For example, the cactus *Opuntia*, introduced into Australia from South America, spread rapidly and became a pest over vast expanses of valuable sheep-grazing land. *Opuntia* was controlled in Australia by the introduction of a moth (*Cactoblastis cactorum*) whose larvae eat *Opuntia*. Once female moths find a patch of cactus and lay their eggs on it, the larvae that hatch from those eggs completely destroy the patch (Figure 54.19). However, new patches of cactus arise in other places from seeds dispersed by birds. These new patches flourish until they are found and destroyed by *Cactoblastis*. Over a large region, the numbers of both *Opuntia* and *Cactoblastis* are now fairly constant and low, but in local areas, there are extreme oscillations caused by the extermination of first the cactus and then the moth. Today, both *Opuntia* and *Cactoblastis* in Australia are distributed as scattered subpopulations among which individuals occasionally disperse.

Can we manage our own population?

Managing our own population has become a matter of great concern because the size of the human population is responsible for most of the environmental problems we are facing today, from pollution to extinctions of other species. For thousands of years, Earth's carrying capacity for human populations was set at a low level by food and water supplies and disease. We saw in Chapter 53 how human social behavior and specialization has allowed us to develop technologies for increasing our resources and combating diseases. The domestication of plants and animals, improved crops and farm yields, mining and use of fossil fuels, and the development of modern medicine have all contributed to the staggering increase in Earth's human population (Figure 54.20).

What is Earth's present carrying capacity for people? Today's carrying capacity is set in part by Earth's ability to absorb the by-products, especially carbon dioxide, of our enormous consumption of fossil fuel energy; by water availability (in many areas); and by whether we are willing to cause the extinction of millions of other species to accommodate our increasing use of Earth's resources. We will explore some of the consequences of high human population densities and high per capita use of resources for the survival of other species in Chapter 57.

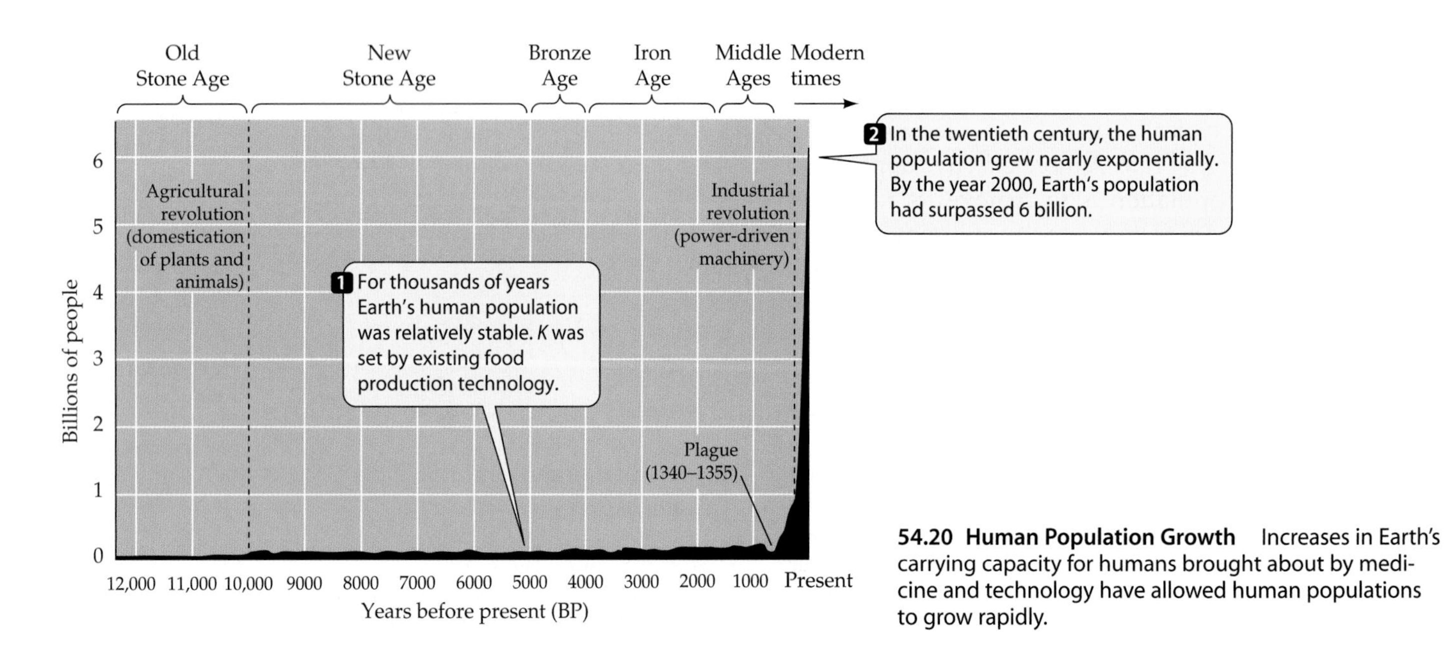

54.20 Human Population Growth Increases in Earth's carrying capacity for humans brought about by medicine and technology have allowed human populations to grow rapidly.

Regional and Global Processes Influence Local Population Dynamics

Between 1950 and 1980, annual counts of breeding birds were conducted in Eastern Wood, in southeastern England. During that time, populations of some species increased while others decreased (Figure 54.21). For example, the population of wood pigeons more than doubled, but the population of garden warblers decreased to zero in 1971; no more than two pairs have bred in the wood since then. The population of blue tits increased from just a few pairs to an average of more than 15 pairs. Why did these populations change so differently?

No matter how intensively ecologists might have studied the birds of Eastern Wood, they could not have answered that question, because populations of two of these three species were strongly influenced by events remote from Eastern Wood. Wood pigeons increased greatly over most of southern England during this 30-year period because of the widespread adoption of oilseed rape as an agricultural crop. Rape fields provide wood pigeons with abundant winter food. Garden warblers decreased because their overwinter survival was poor due to a severe drought on their wintering grounds in West Africa.

The population of blue tits was influenced primarily by changes within Eastern Wood itself. Until the early 1950s, trees in Eastern Wood were periodically felled and sold for timber. After the cutting stopped, more holes, in which blue tits nest, became available in mature and dead trees.

Local population dynamics are often influenced both by local interactions and by more distant or remote events, perhaps even on different continents. How processes occurring at varying spatial and temporal scales influence the structure of ecological communities is the focus of the next chapter.

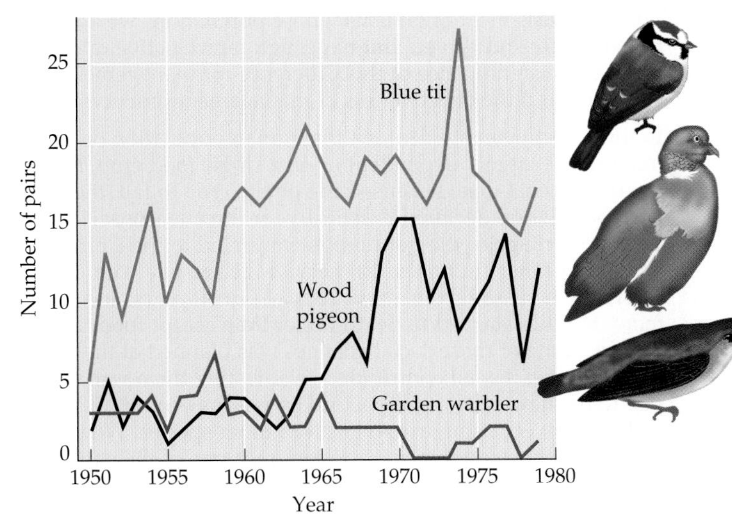

54.21 Populations May Be Influenced by Remote Events Populations of some birds increased in Eastern Wood, England, while others decreased. The wood pigeon (*Columba palumbus*) and garden warbler (*Sylvia borin*) population shifts were strongly influenced by different events that took place far from Eastern Wood. Only the blue tit (*Parus caeruleus*) population was affected most strongly by events within Eastern Wood itself.

Chapter Summary

Populations in Space and Time

▶ A population consists of all the individuals of a species within a given area.

▶ The number of individuals of a species per unit of area (or volume) is its population density. Dense populations often exert strong influences on populations of other species.

▶ Life tables summarize information about births and deaths in populations. **Review Table 54.1**

▶ Graphs of survivorship in relation to age show when individuals survive well and when they do not. **Review Figure 54.1**

▶ The age distribution of individuals in a population reveals much about the recent history of births and deaths in the population. The timing of births and deaths may influence age distributions for many years. **Review Figure 54.2**

Types of Ecological Interactions

▶ Individuals of two populations may interact in ways that may benefit or harm either or both participants. **Review Table 54.2. See Web/CD Activity 54.1**

Factors Influencing Population Densities

▶ Species with small individuals typically achieve higher population densities than species with large individuals. **Review Figure 54.4**

▶ Introduced species sometimes achieve great population densities. **Review Figure 54.5**

Fluctuations in Population Densities

▶ All populations have the potential to grow exponentially under optimal conditions. **Review Figure 54.6. See Web/CD Tutorial 54.1**

▶ No population can maintain exponential growth for very long because environmental limits cause birth rates to drop and death rates to rise.

▶ The number of individuals of a particular species that an environment can support—called the carrying capacity (*K*)—is determined by the availability of resources and by disease and predators.

▶ A population in a limited environment shows a logistic growth pattern, in which growth rates decrease as the carrying capacity is approached. **Review Figure 54.7. See Web/CD Activity 54.2 and Tutorial 54.2**

▶ The density of a population is influenced by the combined effects of all density-dependent and density-independent factors affecting it. **Review Figure 54.8**

Population Fluctuations

▶ Populations do not fluctuate as much as theoretically possible, but some fluctuate much more than others. The amount of fluctuation is influenced by body size, reproductive rate, and range size. **Review Figure 54.9**

▶ Population fluctuations may be strongly influenced by years of good reproduction. **Review Figure 54.10**

▶ Predator–prey interactions may generate population cycles. **Review Figures 54.11, 54.12**

▶ Populations of many species exist as small, fragmented subpopulations. Extinction of subpopulations is common, but indi-

viduals from other fragments may recolonize them. **Review Figures 54.13, 54.14.** See Web/CD Tutorial 54.4
See Web/CD Tutorial 54.3

Variation in Species' Ranges

▶ Some species are restricted to very small areas, whereas others are widely distributed on Earth.

▶ Species' ranges are influenced by speciation processes, dispersal abilities, predators, and competition. **Review Figures 54.16, 54.17**

Managing Populations

▶ Humans can use the principles of population dynamics to control and manage populations of desirable and undesirable species. Nevertheless, humans have overexploited many populations. **Review Figure 54.18**

Regional and Global Processes Influence Local Population Dynamics

▶ Population densities may be influenced by both local conditions and remote events. **Review Figure 54.20**

Self-Quiz

1. The number of individuals of a species per unit of area is known as its
   a. population size.
   b. population density.
   c. population structure.
   d. subpopulation.
   e. biomass.
2. The age distribution of a population is determined by
   a. the timing of births.
   b. the timing of deaths.
   c. the timing of both births and deaths.
   d. the rate at which the population is growing.
   e. All of the above
3. Which of the following is *not* a demographic event?
   a. Growth
   b. Birth
   c. Death
   d. Immigration
   e. Emigration
4. A group of individuals born at the same time is known as a
   a. deme.
   b. subpopulation.
   c. Mendelian population.
   d. cohort.
   e. taxon.
5. Two organisms that use the same resources when those resources are in short supply are said to be
   a. predators.
   b. competitors.
   c. mutualists.
   d. commensalists.
   e. amensalists.
6. Damage caused to shrubs by branches falling from overhead trees is an example of
   a. interference competition.
   b. partial predation.
   c. amensalism.
   d. commensalism.
   e. diffuse coevolution.
7. A population grows at a rate closest to its intrinsic rate of increase when
   a. its birth rates are the highest.
   b. its death rates are the lowest.
   c. environmental conditions are optimal.
   d. it is close to the environmental carrying capacity.
   e. it is well below the environmental carrying capacity.
8. Immigrants that prevent a subpopulation from becoming extinct result in a
   a. colonization effect.
   b. rescue effect.
   c. metapopulation effect.
   d. genetic drift effect.
   e. salvage effect.
9. Density-dependent population regulation is strongest when
   a. only birth rates change in response to density.
   b. only death rates change in response to density.
   c. diseases spread in populations at all densities.
   d. both birth and death rates change in response to density.
   e. population densities fluctuate very little.
10. The best way to reduce the population of an undesirable species in the long term is to
    a. reduce the carrying capacity of the environment for the species.
    b. selectively kill reproducing adults.
    c. selectively kill pre-reproductive individuals.
    d. attempt to kill individuals of all ages.
    e. sterilize individuals.

For Discussion

1. Why are big, fierce animals rare?
2. Why do predator–prey interactions often generate cycles or great fluctuations in population densities? Would you expect lynx populations to fluctuate as much as they do if lynx had a variety of abundant prey species available to them?
3. Most organisms whose populations we wish to manage for higher densities are long-lived and have low reproductive rates, whereas most organisms whose populations we attempt to reduce are short-lived, but have high reproductive rates. What is the significance of this difference for management strategies and the effectiveness of management practices?
4. In the mid-nineteenth century, the human population of Ireland was largely dependent upon a single food crop, the potato. When a disease caused the potato crop to fail, the Irish population declined drastically for three reasons: (1) a large percentage of the population emigrated to the United States and other countries; (2) the average age of a woman at marriage increased from about 20 to about 30 years; and (3) many families starved to death rather than accept food from Britain. None of these social changes was planned at the national level, yet all contributed to adjusting the population size to the new carrying capacity. Discuss the ecological principles involved, using examples from other species. What would you have done had you been in charge of the national population policy for Ireland at that time?
5. Because some species introduced to control a pest have become pests themselves, some scientists argue that species introductions should not be used under any circumstances to control pests. Others argue that, provided they are properly researched and controlled, we should continue to use introductions as part of our set of tools for managing pest populations. Which view do you support? Why?

55 Communities and Ecosystems

It is interesting to contemplate an entangled bank, clothed with many plants of many kinds, with birds singing on the bushes, with various insects flitting about, and with worms crawling through the damp earth, and to reflect that these elaborately constructed forms, so different from each other, and dependent on each other in so complex a manner, have all been produced by the laws acting around us.

—CHARLES DARWIN, 1859

Charles Darwin is remembered mostly for his contributions to evolutionary theory, but this quote from *The Origin of Species* shows that he was also a pioneering ecologist who understood the nature and complexity of the interactions among the species of organisms that live in a particular place.

The species that live and interact in an area constitute an **ecological community**. Darwin's "entangled bank" was an ecological community that had obvious boundaries defined by adjacent crops, pastures, and gardens. But the organisms living in the bank were not confined within those boundaries. Some of the seeds that landed in the bank and grew into trees and shrubs came from parent plants living far away. The insects and birds Darwin observed must have flown into and out of the bank from a large area. To understand which species live in such a bank and how they interact, he would have needed to know about such movements, just as we need to know about drought in Africa in order to understand fluctuations in the population of garden warblers in Eastern Wood, as we saw at the end of Chapter 54.

Darwin's entangled bank had also changed over time. Glaciers had covered the area 10,000 years earlier. The plant species Darwin observed colonized Britain at different times over the several thousand years since the glaciers melted. Ecological communities are not assemblages of organisms that move together as units when environmental conditions change. Rather,

"Clothed with Many Plants of Many Kinds" Charles Darwin's eloquent description of an "entangled bank" reveals his understanding of the nature and complexity of interactions among species living in the same community.

each species has unique interactions with its biotic and abiotic environments.

The species that form an ecological community, together with the physical environment, constitute an **ecosystem**. To understand the processes that influence ecosystems and the patterns they produce, we must study both the interactions of organisms with one another and their interactions with the physical environment; that is, we must study both communities and ecosystems.

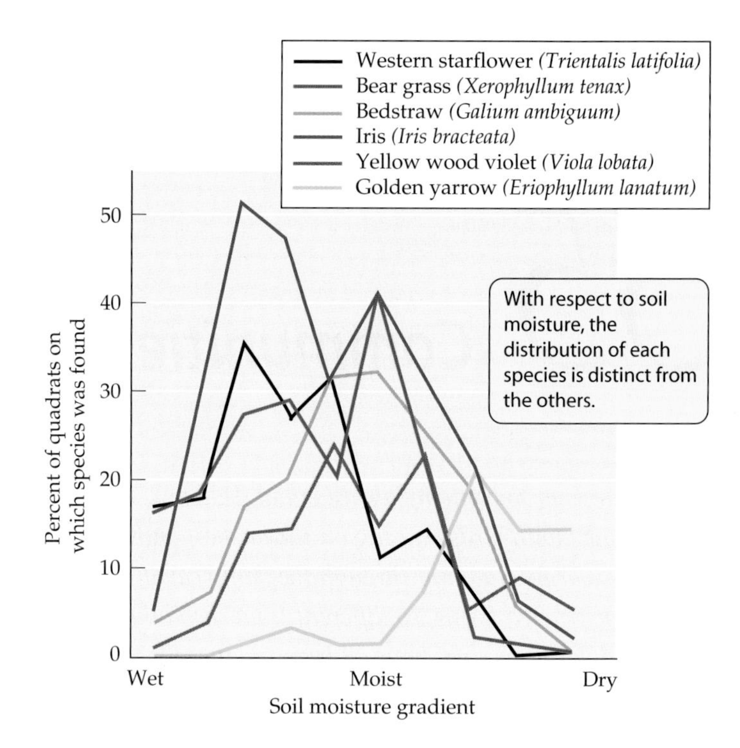

55.1 Plant Distributions along an Environmental Gradient The abundances of different plant species change gradually and individually along a soil moisture gradient in Oregon's Siskiyou Mountains. Each species performs best at a different optimum soil moisture. (*Quadrats* are sample plots of ground designated and marked for an ecological study.)

Communities: Loose Assemblages of Species

Ecological communities contain many species that interact with one another via the processes we discussed in Chapter 54: competition, predation, mutualism, commensalism, and amensalism. The importance of these interactions changes as a result of changes in the physical environment, gains and losses of species, and changes in the population densities of species. These interactions also change over longer times as the interacting species evolve.

Early in the twentieth century, there was a major debate between two leading North American plant ecologists over the nature of communities. In 1926, Henry Gleason argued that plant communities were loose assemblages of species, each of which was individualistically distributed according to its unique interactions with the physical environment. In a paper published in 1936, Frederick Clements argued that plant communities were tightly integrated "superorganisms." He believed that there were places where one group of species dropped out and was replaced by a very different group.

The debate was resolved by detailed studies of the distributions of plants. Especially influential were analyses of the vegetation of the Siskiyou Mountains of Oregon. Those studies showed that different combinations of plant species were found at different locations. Species entered and dropped out of communities independently over environmental gradients (Figure 55.1). These and other results generally supported Gleason's view of the nature of communities. However, where environmental conditions change abruptly, as they do at the edges of lakes and streams, the ranges of many species may terminate at the same place.

Ecologists try to understand ecological communities and ecosystems by asking several general questions: What patterns exist in ecological communities and ecosystems? How does the physical environment influence those patterns? What are the relative roles of historical accident and current interactions in determining those patterns? How does evolution, acting on members of the community, influence the assemblage of species that live together?

Given that each community contains a multitude of species, each interacting with many other species, this goal might appear to be unattainable. Fortunately, we do not need to know all of the details to make considerable progress, because a few interactions often determine most of the features of an ecosystem. Let's begin by looking at such interactions in one community.

A few interactions may determine the features of a community

Although hundreds of species live in oak forest communities in the eastern United States, only a few species dominate the ecological interactions in those communities: oak trees (and their leaves and acorns) white-footed mice (*Peromyscus maniculatus*), gypsy moths (*Lymantria dispar*), and white-tailed deer (*Odocoileus virginianus*). Oak trees produce large crops of acorns every 2–5 years, but produce few acorns during intervening years. Acorns are a critical food for white-footed mice and deer. During years of heavy acorn production, white-footed mice survive well during winter and have great reproductive success. During years of poor acorn crops, the mice survive poorly.

Gypsy moth larvae eat oak leaves. When they have completed their development, the larvae crawl to the tree trunks, where they pupate. While the larvae inside the pupae are transforming themselves into moths, they may be eaten by white-footed mice, which search for them on the tree trunks. Every 6 or 7 years, gypsy moths become extremely abundant.

The population cycles of gypsy moths and white-footed mice are of interest to us because these species greatly affect humans and their resources. In years when gypsy moth caterpillars are extremely abundant, they defoliate millions of hectares of oak forest, damaging and killing many trees. Also, white-footed mice, along with white-tailed deer, are the primary hosts of the black-legged tick (*Ixodes scapularis*), the vector of the spirochete bacterium (*Borrelia burgdorferi*) that causes Lyme disease in humans.

Ecologists wanted to test the hypothesis that white-footed mice generate fluctuations in gypsy moth populations by eating their pupae. They performed an experiment during a year of low gypsy moth but high white-footed mouse population densities. They removed mice from three experimental areas by trapping them, but did nothing to the mice in three other areas, which served as controls. On the control plots, which had high mouse densities, they searched tree trunks, but found no moth pupae. To determine that predation was responsible for the absence of pupae, they attached previously collected freeze-dried pupae to tree trunks. Within 2–4 days, all of the pupae in the control plots had been eaten, most of them by white-footed mice. In the experimental plots, from which white-footed mice had been removed, 22 percent of the freeze-dried pupae remained uneaten for at least 13 days. That would have been long enough for the pupae to complete their metamorphosis into adult moths.

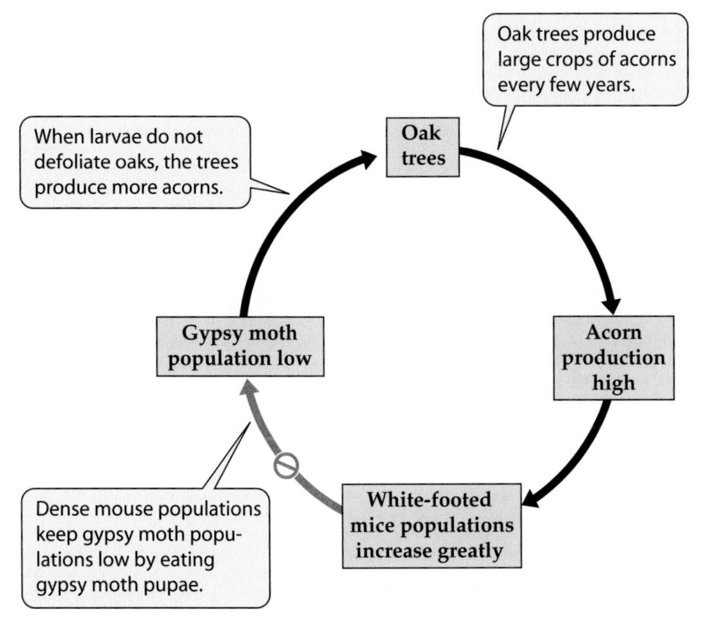

55.2 Interactions within Communities Control Populations Experiments in oak forests in eastern North America have shown that acorn abundance affects the abundance of the white-footed mouse, and that the mice in turn are a major control on the gypsy moth population.

If white-footed mice control gypsy moth populations, what controls mouse populations? In another experiment, the ecologists added more than 811,000 acorns to experimental plots during a year when oak trees were producing very few acorns. Mouse populations became much more dense on the plots with added acorns than on the control plots.

These experiments demonstrated that when white-footed mice are abundant, they can prevent gypsy moths from completing their life cycle. The experiments also demonstrated that white-footed mouse population densities are determined primarily by acorn density (Figure 55.2). Without performing those experiments, investigators could only speculate about the interactions of the species in oak forest communities and their importance.

With this background, we will now discuss the major processes that influence communities and ecosystems and the patterns they generate.

Process and Pattern in Communities and Ecosystems

We begin our discussion of processes with solar energy input and precipitation because nearly all ecological processes depend, either directly or indirectly, on the amount and seasonal pattern of solar energy input and supply of water. Next we'll discuss what ecologists have learned by studying interactions among species and how the influences of various factors on community patterns change over space and time. As you read, be aware that the structure of human language forces us to discuss these factors one at a time, but several of them typically operate simultaneously.

Solar energy and precipitation drive ecosystem processes

All organisms depend on inputs of energy (in the form of sunlight or high-energy molecules), water, and nutrients for their metabolism and growth. With the exception of a few ecosystems (some caves, deep-sea hydrothermal vent systems) in which solar energy is not the main energy source, all energy utilized by organisms comes (or once came) from the sun. Even the fossil fuels—coal, oil, and natural gas—upon which the economy of modern human civilization is based are reserves of captured solar energy locked up in the remains of organisms that lived millions of years ago.

Solar energy enters ecosystems by way of plants and other photosynthetic organisms. Only about 5 percent of the solar energy that arrives on Earth is captured by photosynthesis. The remaining energy is either radiated back into the atmosphere as heat or consumed by the evaporation of water from plants and other surfaces. **Gross primary productivity** is the rate at which energy is incorporated into the bodies of pho-

55.3 Energy Flow through an Ecosystem In this diagram, the width of each channel is roughly proportional to the amount of energy flowing through it. The arrows indicate directions of energy flow.

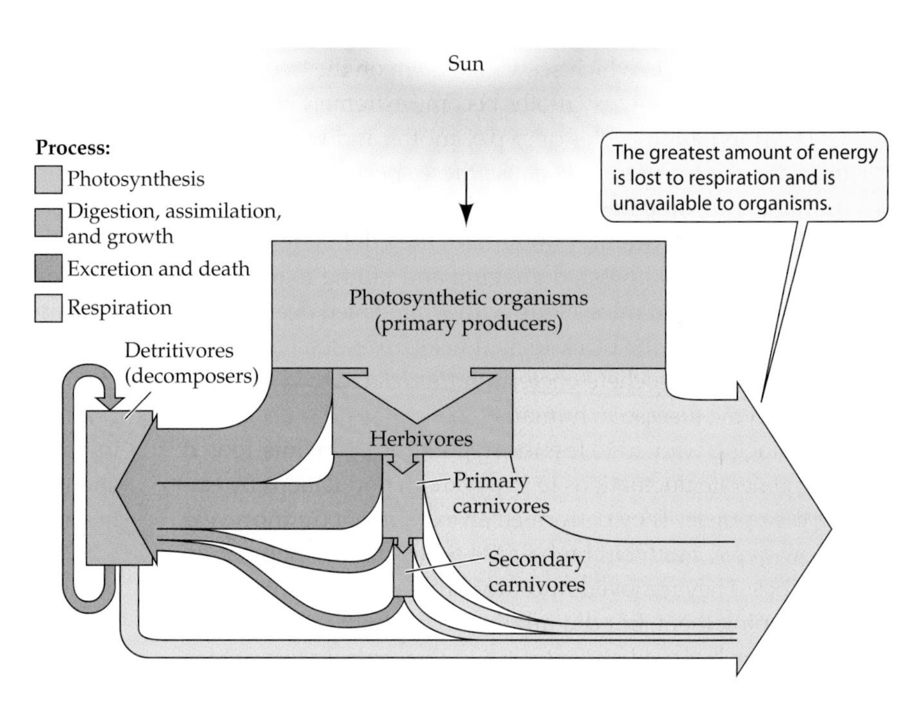

tosynthetic organisms. The accumulated energy is called **primary production**; that is, productivity is a rate, and production is a product. Plants use some of this energy for their own metabolism; the rest is stored in their bodies or used for their growth and reproduction. The energy available to organisms that eat plants, called **net primary production**, is gross primary production minus the energy expended by the plants on their respiration. Only the energy content of an organism's net production—its growth plus reproduction—is available to other organisms that consume it (Figure 55.3).

The distribution of the total amount of energy that plants assimilate by means of photosynthesis reflects the distribution of land masses, temperature, and moisture on Earth (Figure 55.4). Close to the equator at sea level, temperatures are high throughout the year, and the water supply typically is adequate for plant growth much of the time. In these climates, productive forests thrive. In lower-latitude and mid-

55.4 Primary Production in Different Ecosystems The primary production of Earth's ecosystems can be measured (*a*) by the geographic extent of the different ecosystems; (*b*) by net annual primary production; and (*c*) by the percentage of Earth's total primary production contributed by each ecosystem.

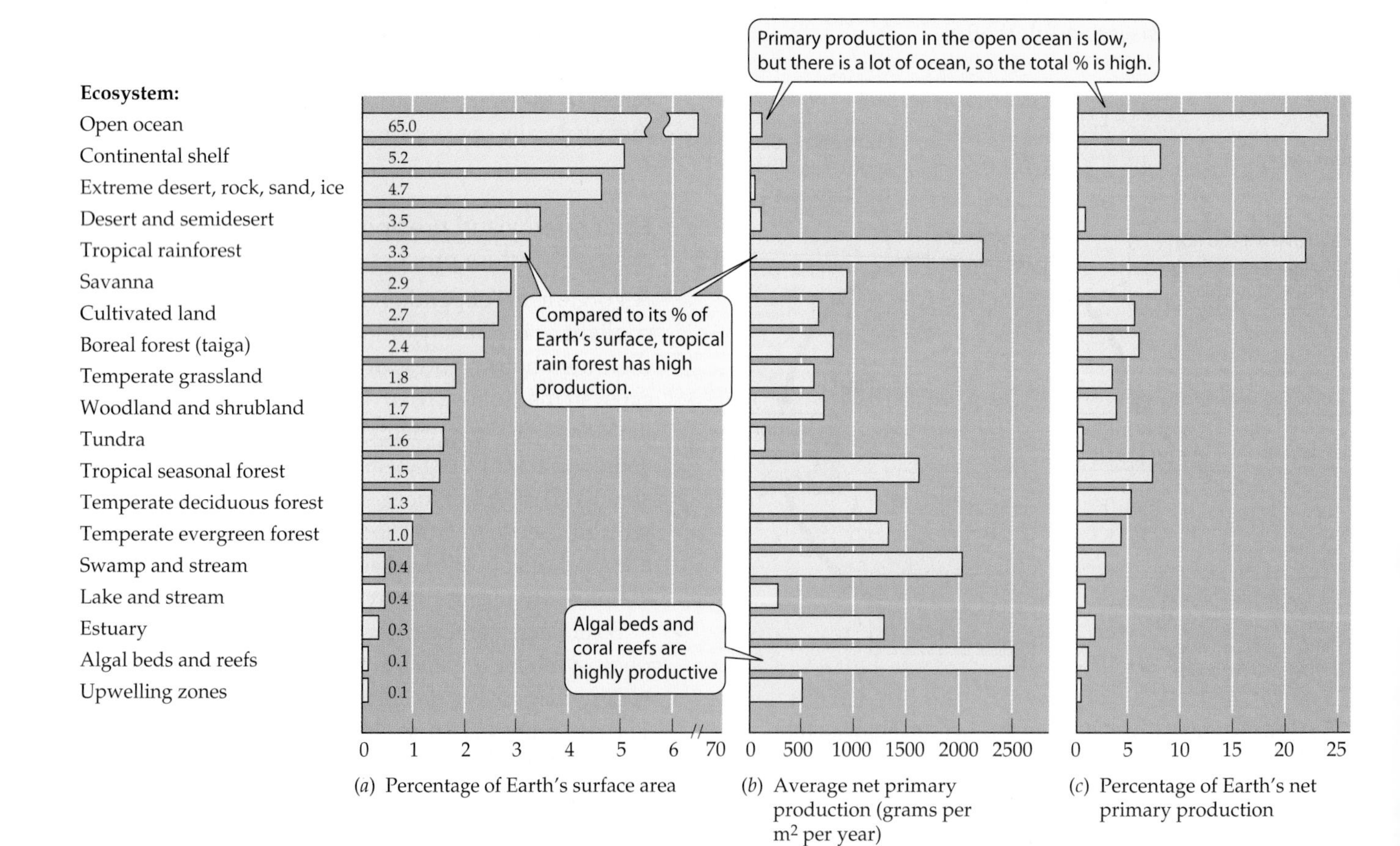

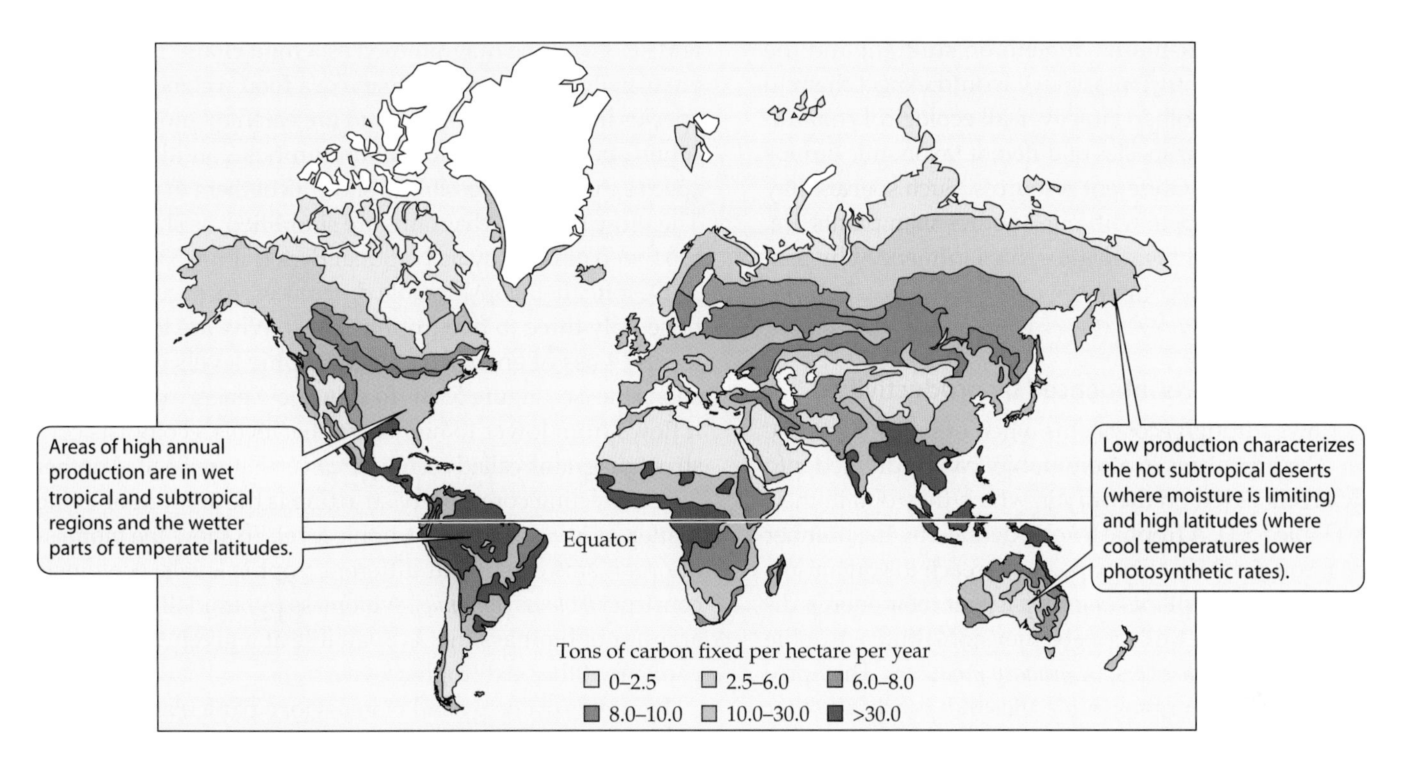
Areas of high annual production are in wet tropical and subtropical regions and the wetter parts of temperate latitudes.
Equator
Low production characterizes the hot subtropical deserts (where moisture is limiting) and high latitudes (where cool temperatures lower photosynthetic rates).
Tons of carbon fixed per hectare per year
0–2.5
2.5–6.0
6.0–8.0
8.0–10.0
10.0–30.0
>30.0

55.5 Net Primary Production of Terrestrial Ecosystems Variations in temperature and water availability over Earth's land surface influence the annual production of ecosystems.

latitude deserts, where plant growth is limited by lack of moisture, annual primary production is low. At still higher latitudes, even though moisture is generally available, annual primary production is also low because it is cold much of the year (Figure 55.5). Production in aquatic systems is limited by light, which decreases rapidly with depth; by nutrients, which sink and must be replaced by upwelling of water; and by temperature (see Chapter 58). Primary productivity strongly influences two other important features of ecological communities: species richness and food web structure.

Species richness is influenced by primary productivity

The number of species living in a community (its **species richness**) is correlated with gross primary productivity, but the relationship between these two factors is complex. Ecologists first observed that species richness often increases with productivity up to a point, but then decreases (Figure 55.6). The increase occurs because the number of individuals an area can support increases with productivity, and with larger population sizes, species extinction rates are lower. But why should species richness decrease when productivity is still higher?

One hypothesis proposed to explain this decrease postulates that interspecific competition becomes more intense when productivity is higher, resulting in **competitive exclusion** of some species. This hypothesis is supported by the results of a long-term experiment at the Rothamstead Experiment Station in England in which fertilizer has been added to some plots of land to increase productivity. Fertilized and unfertilized plots have been monitored continuously at Rothamstead since 1856. Over this time period, the number of plant species in unfertilized plots has remained roughly constant, whereas species richness has declined in the fertilized plots.

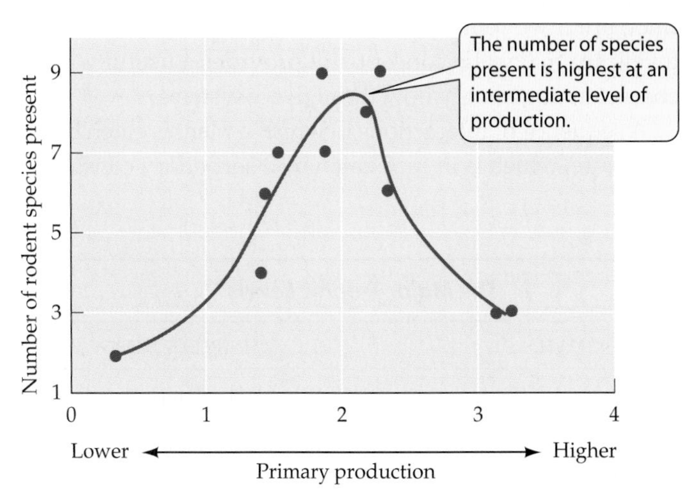
The number of species present is highest at an intermediate level of production.
Number of rodent species present
9
7
5
3
1
0
1
2
3
4
Lower
Higher
Primary production

55.6 Local Species Richness Peaks at Intermediate Productivity The number of species of rodents that live on gravel and rocky plains in the Gobi desert peaks at intermediate levels of primary production. Ecologists believe that when productivity is highest, interspecific competition becomes stronger, resulting in the exclusion of some rodent species by others.

All photosynthetic plants depend on sunlight and the same set of mineral nutrients, so we would expect them to compete with one another. However, all ecological communities are composed of species of different taxonomic groups that depend on many different resources. Such species may be less likely to compete with one another than plants are. How is the pattern of their relationships influenced by productivity?

Food web structure is influenced by productivity

Energy flows through ecosystems when organisms eat one another. The organisms in a community can be divided into **trophic levels** based on the way in which they obtain their energy (Table 55.1). A trophic level is defined by the number of steps through which energy passes to reach the organisms in it. Photosynthetic plants (*autotrophs*) get their energy directly from sunlight. Collectively, they constitute a trophic level called *photosynthesizers*, or *primary producers*. They produce the energy-rich organic molecules that nearly all other organisms consume.

In most ecological communities, all nonphotosynthetic organisms (*heterotrophs*) consume, either directly or indirectly, the energy-rich organic molecules produced by primary producers. Organisms that eat plants constitute a trophic level called *herbivores* or *primary consumers*. Organisms that eat herbivores are called *secondary consumers*. Those that eat secondary consumers are called *tertiary consumers*, and so on. Organisms that eat the dead bodies of organisms or their waste products are called *detritivores* or *decomposers*. Organisms that obtain their food from both primary producers and another trophic level are called *omnivores*. Because many species are omnivores, trophic levels are often not clearly distinct, but if we remember that boundaries between trophic levels are fuzzy, the concept still provides a useful way of characterizing energy flow through ecosystems.

A sequence of interactions in which a plant is eaten by an herbivore, which is in turn eaten by a secondary consumer, and so on, can be diagrammed as a **food chain**. Food chains are usually interconnected to make a **food web** because most species in a community eat and are eaten by more than one other species. Ecological communities contain so many species that it is impossible to show all of them in a food web. Therefore, all such diagrams are simplified, as shown by the food web of Isle Royale National Park (Figure 55.7).

Despite their considerable differences, most communities have only three to five trophic levels. Why are there so few levels? Loss of energy between trophic levels is partly responsible (see Figure 55.3). To show how energy decreases as it flows from lower to higher trophic levels, ecologists construct diagrams called *energy pyramids* (Figure 55.8). Another factor affecting community structure is the amount of living matter, or *biomass*, at each trophic level. To show the biomass of organisms existing at different trophic levels, ecologists construct *biomass pyramids*. A biomass pyramid illustrates the amount of biomass that is available at a given time for organisms at the next trophic level.

Pyramids of energy and biomass for a particular ecosystem usually have similar shapes, but sometimes they do not. Their shapes depend on the dominant organisms and how they allocate their energy. In most terrestrial ecosystems, the dominant photosynthetic plants are large. They store energy for long periods, some of it in difficult-to-digest forms (such as cellulose and lignin). Therefore, the primary producer level in these systems contains a large biomass. However, grasslands and forests have strikingly different patterns of energy flow. Trees store a great deal of their energy as wood, which is composed of difficult-to-digest material. Wood is rarely eaten unless the tree is diseased or otherwise weakened. In contrast, grassland plants produce few hard-to-digest woody tissues. Mammals may consume 30–40 percent of the annual aboveground net primary production of grasslands, and insects may consume an additional 5–15 percent. Soil organisms, primarily nematodes, may consume 6–40 percent of the belowground production (Figure 55.8*a*,*b*). Thus, the herbivore level has a relatively larger biomass in grasslands than in forests.

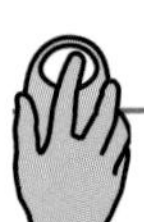

55.1 The Major Trophic Levels

TROPHIC LEVEL	SOURCE OF ENERGY	EXAMPLES
Photosynthesizers (primary producers)	Solar energy	Green plants, photosynthetic bacteria and protists
Herbivores	Tissues of primary producers	Termites, grasshoppers, gypsy moth larvae, anchovies, deer, geese, white-footed mice
Primary carnivores	Herbivores	Spiders, warblers, wolves, copepods
Secondary carnivores	Primary carnivores	Tuna, falcons, killer whales
Omnivores	Several trophic levels	Humans, opossums, crabs, robins
Detritivores (decomposers)	Dead bodies and waste products of other organisms	Fungi, many bacteria, vultures, earthworms

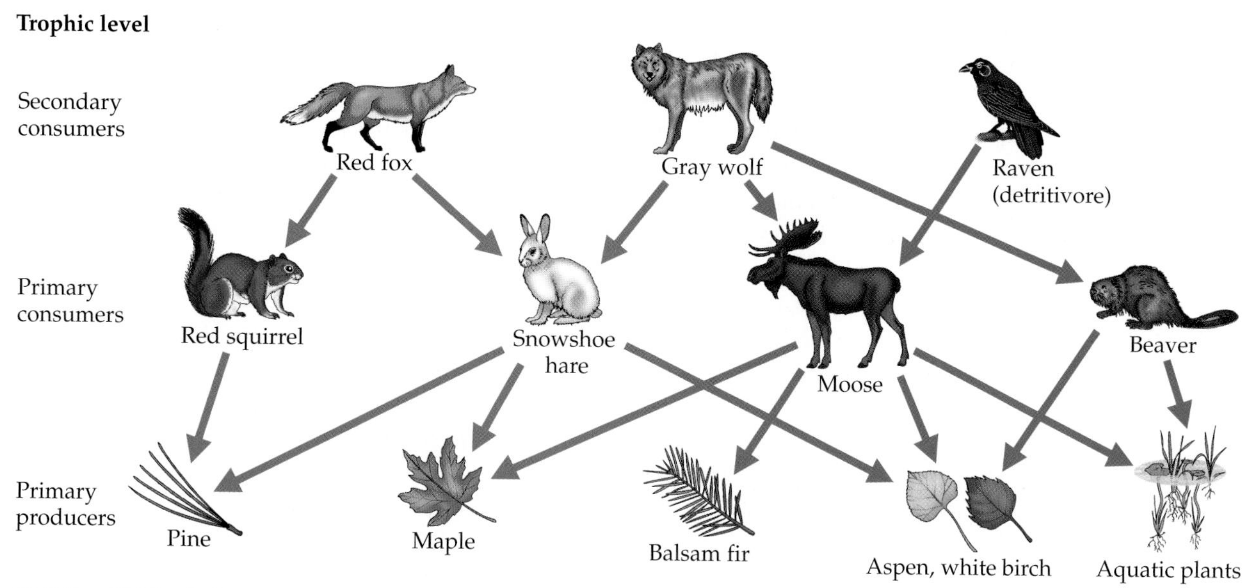

55.7 Food Web of Isle Royale National Park This food web includes only large vertebrates and the plants on which they depend. Even with these restrictions, the web is complex. The arrows show who eats whom.

Energy flow	Trophic level
Energy flow (calories/m²/day)	**Biomass** (grams/m²)

(*a*) **Grassland**

Most of the biomass in a grassland is found in the green plants, and most of the energy flows through them.

(*b*) **Forest**

In forests, the majority of biomass is tied up in wood and is not available to most herbivores.

(*c*) **Open ocean**

A marine community produces an inverted pyramid of biomass. The producers are unicellular algae, which divide so rapidly that a small biomass can support a much larger biomass of herbivores.

Trophic levels

- Carnivores
- Herbivores
- Producers (photosynthesizers)

55.8 Pyramids of Biomass and Energy Energy pyramids (left column) allow ecologists to compare patterns of energy flow through trophic levels in different ecosystems. Biomass pyramids (right column) allow them to compare the amount of material present in living organisms at different trophic levels.

In most aquatic ecosystems, the dominant photosynthesizers are bacteria and protists. Those unicellular organisms have such high rates of cell division that a small biomass of photosynthesizers can feed a much larger biomass of herbivores, which grow and reproduce much more slowly. This pattern can produce an inverted biomass pyramid, even though the energy pyramid for the same ecosystem has the typical shape (Figure 55.8*c*).

Much of the energy ingested by organisms is converted to biomass that is eventually consumed by detritivores, such as bacteria, fungi, worms, mites, and insects. These organisms transform the dead remains and waste products of organisms into free mineral nutrients that can again be taken up by plants. If there were no detritivores, most nutrients would eventually be tied up in dead bodies, where they would be unavailable to plants. Continued ecosystem productivity depends on the rapid decomposition of detritus.

Species richness and productivity influence ecosystem stability

We have seen that, up to a point, high primary productivity favors increased species richness, but is the reverse true? That is, does species richness also influence ecosystem productivity? Ecologists hypothesized that species richness might enhance ecosystem productivity because no two species in a community have the same relationship with the environment. Therefore, a richer mixture of species should result in a more complete use of the available resources. In addition, if the environment changes, a species-rich ecosystem is more likely to contain species that are already adapted to the new conditions than is a species-poor ecosystem. Therefore, ecologists hypothesized that a species-rich ecosystem should also be more stable—that is, over time it should change less in both productivity and species composition than a species-poor ecosystem.

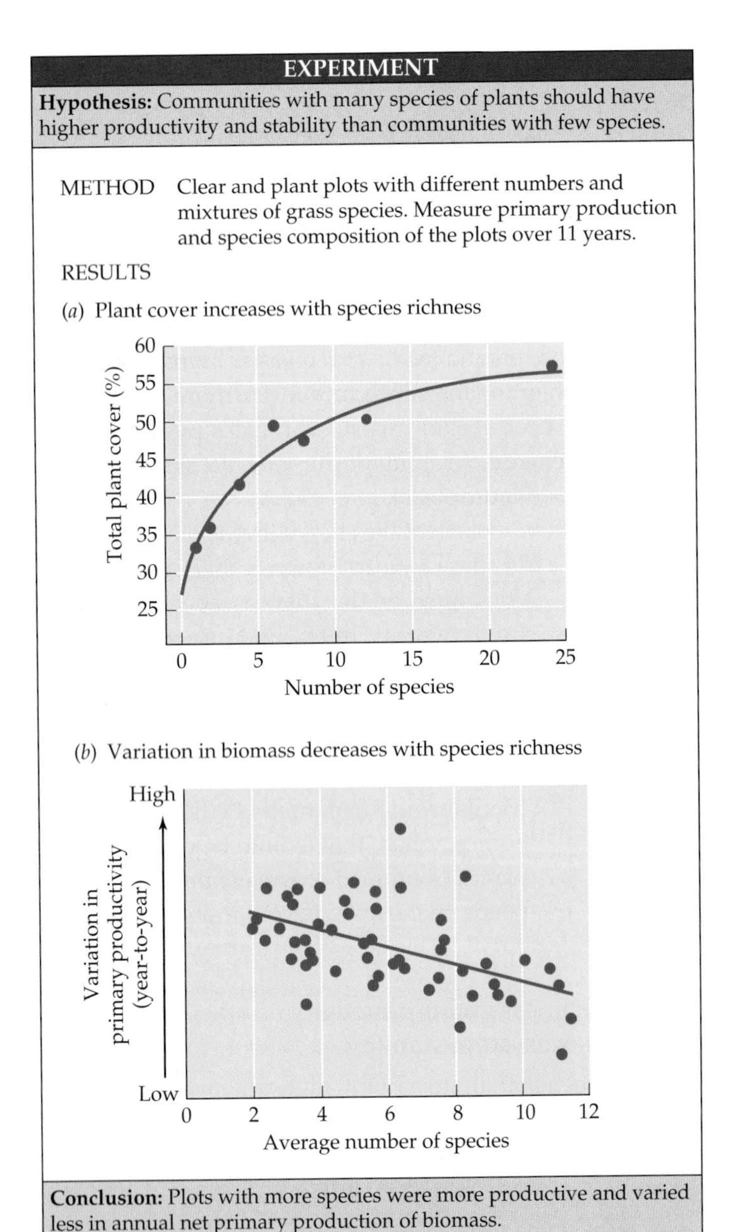

EXPERIMENT
Hypothesis: Communities with many species of plants should have higher productivity and stability than communities with few species.
METHOD Clear and plant plots with different numbers and mixtures of grass species. Measure primary production and species composition of the plots over 11 years.
RESULTS
(a) Plant cover increases with species richness
Total plant cover (%)
60
55
50
45
40
35
30
25
0
5
10
15
20
25
Number of species
(b) Variation in biomass decreases with species richness
Variation in primary productivity (year-to-year)
High
Low
0
2
4
6
8
10
12
Average number of species
Conclusion: Plots with more species were more productive and varied less in annual net primary production of biomass.

55.9 Species Richness Enhances Community Productivity and Stability Net primary productivity was greater, and variation in net primary productivity from one year to the next was less, in species-rich than in species-poor plots of grasses.

To test this hypothesis, ecologists cleared several outdoor plots, in which they planted grasses in a variety of mixtures of different species richness, from a few to 25 species. At the end of each growing season, they measured grass biomass (a measure of net primary production) and the population densities of all the grasses in the plots. These measurements were made over a period of 11 years, which included a serious drought. The plots with more species were more productive and their productivity varied less from one year to another, supporting the hypothesis (Figure 55.9). However, the population densities of individual species in the plots were not stable over the years (regardless of a plot's species richness), because different species performed better during drought years and during wet years.

Although species richness and productivity are often positively correlated, as they were in this experiment, such a correlation could result if only one or a few species exerted very strong influences on the flow of matter and energy through an ecosystem. Ecologists have made an effort to identify and study such species.

Individual species may influence community processes

Species whose influences on ecosystems are greater than would be expected on the basis of their abundance are called **keystone species**. Keystone species may influence both the species richness of communities and the flow of energy and materials through ecosystems. Beavers, for example, are keystone species. They create meadows and ponds—habitats for other species—by cutting down trees and building dams.

Large grazing and browsing mammals, such as bison (Figure 55.10*a*), also are keystone species. They often change the structure and composition of the vegetation dramatically. To determine the influence of bison on prairie vegetation, ecologists established a herd on the Konza Prairie Research Natural Area in northeastern Kansas, the largest tract of unplowed tallgrass prairie in North America. The bison were allowed to graze in some areas, but excluded from others. The bison herd numbered about 200 animals in 2003. The prairie was regularly burned in spring to mimic the fires of prehistoric times. Bison primarily eat grasses; they eat few of the broad-leaved plants (called *forbs*) that grow among the grasses. Bison also prefer to graze on recently burned areas.

In the Konza Prairie ecosystem, the areas from which bison have been excluded are now dominated by tall grasses and have few plant species. In contrast, areas that have been grazed by bison have many more species of forbs because the bison create spaces for forbs by preferentially grazing on grasses (Figure 55.10*b*). Furthermore, urea in bison urine is hydrolyzed to ammonium within a few days, and the nitrogen in the ammonium is immediately available for plants. In contrast, decomposing leaf litter releases nitrogen much more

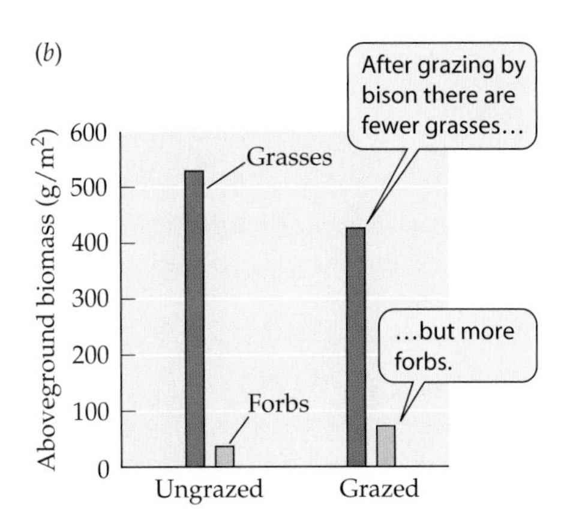

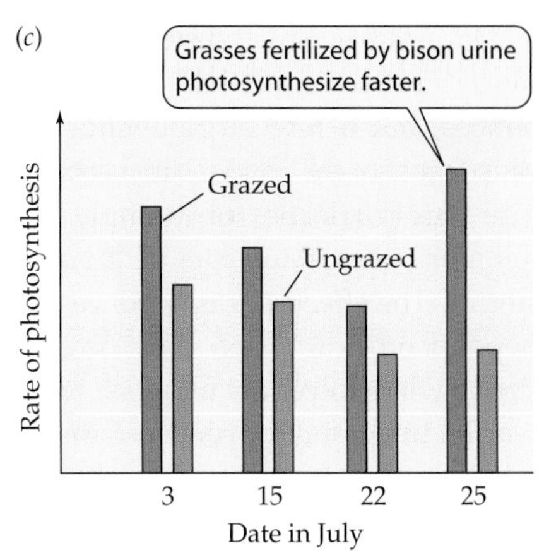

Bison bison

55.10 Grazing Increases Plant Species Richness and Productivity (*a*) Bison are keystone species in the tallgrass prairie ecosystem. By grazing preferentially on grasses, they increase both the density of forbs (*b*) and overall plant productivity (*c*).

slowly. Therefore, plants in patches grazed by bison have higher leaf nitrogen levels and photosynthesize faster than plants growing in ungrazed patches (Figure 55.10*c*). From these observations, the investigators concluded that bison are keystone herbivores that influence species composition, nutrient cycling, and energy flow by their grazing.

Another keystone species is the sea star *Pisaster ochraceous*, which lives in rocky intertidal ecosystems on the Pacific coast of North America. Its preferred prey is the mussel *Mytilus californianus*. In the absence of sea stars, these mussels crowd out other competitors in a broad belt of the intertidal zone. By consuming mussels, *Pisaster* creates bare spaces that are taken over by a variety of other species (Figure 55.11).

The influence of *Pisaster* on species richness was demonstrated by experimentally removing sea stars from selected parts of the intertidal zone repeatedly over a 5-year period. Two major changes occurred in the areas from which sea stars were removed. First, the lower edge of the mussel bed extended lower down into the intertidal zone, showing that sea stars are able to eliminate mussels completely where they are covered with water most of the time. Second, and more dramatically, 28 species of animals and algae disappeared from the sea star removal zone. Eventually only *Mytilus*, the dominant competitor, occupied the entire substratum. By altering competitive relationships, predation by *Pisaster* largely determines which species live in these rocky intertidal ecosystems.

Disturbance and Community Structure

By their activities, keystone species generate disturbances. A *disturbance* is an event that changes the survival rate of one or more species in an ecological community. Logs carried by waves may crush algae and animals attached to rocks in an intertidal community or a windstom may blow down a tree, crushing shrubs and herbs. The effects of such disturbances are typically limited to a small area. Other kinds of disturbances, such as hurricanes and volcanic eruptions, may af-

55.11 Sea Stars are Keystone Predators Ochre sea stars (*Pisaster ochraceus*) have harvested all the mussels from the lower parts of these rocks on the Olympic Peninsula of Washington. By consuming mussels, *Pisaster* creates bare spaces that are taken over by a variety of other species, thus exerting a keystone influence on the intertidal ecosystems of the Pacific Northwest.

fect large areas. Small disturbances are much more common than large disturbances, but a few large events may cause most of the changes in a community. One hurricane, for example, may fell more trees than years of "normal" storms. The effects of disturbances also depend on how often they occur. If strong windstorms are frequent, for example, trees may never have the opportunity to grow tall.

A particular type of disturbance can have a variety of effects. For example, in 1988, massive fires burned one-third of Yellowstone National Park, but they created a mosaic that included unburned patches, areas where only herbs and shrubs were burned, and areas where all trees were consumed (Figure 55.12).

55.12 Fires Create Mosaics of Burned and Unburned Patches This view of Mount Washburn in Yellowstone National Park was taken 10 years after the massive forest fires of 1988.

Although the consequences of various kinds of disturbances are highly variable, their results conform to a general pattern: Communities with very high levels of disturbance and those with very low levels of disturbance have fewer species than communities subjected to intermediate levels of disturbance. The discovery of this general pattern generated the **intermediate disturbance hypothesis**. This hypothesis explains the low species richness in areas with high disturbance levels by suggesting that only species with great dispersal abilities and rapid reproductive rates can persist in such areas. Conversely, the hypothesis explains the decline in species richness where disturbance levels are low by suggesting that competitively dominant species displace other species, as mussels did when sea stars were removed.

The intermediate disturbance hypothesis was tested using boulders on intertidal beaches in California. An ecologist observed that boulders of intermediate size had more species of algae and barnacles on them than either larger or smaller boulders did. He hypothesized that this pattern existed because waves move small boulders more easily, more often, and farther than large boulders. When a wave moves a boulder, its motion destroys organisms living on the boulder's surface. To test his hypothesis, the ecologist altered disturbance levels by gluing small boulders to the substratum. These secured small boulders accumulated species more rapidly than unsecured small boulders, supporting the hypothesis (Figure 55.13). The experiment also showed that species not normally found on small boulders can survive there if the boulders are not moved often.

This experiment also demonstrated that the number of species in a community changes over time following a disturbance. A change in community composition following a disturbance is called **ecological succession**. The changes on

EXPERIMENT

Hypothesis: Medium-sized boulders have more species growing on them than either smaller or larger boulders because they are subjected to intermediate levels of disturbance.

METHOD Sterilize a number of small boulders. Secure some of them to the natural substratum with glue. Leave other boulders unsecured to serve as controls. Observe accumulation of species on the boulders over time.

RESULTS Secured small boulders accumulate many more species than unsecured small boulders.

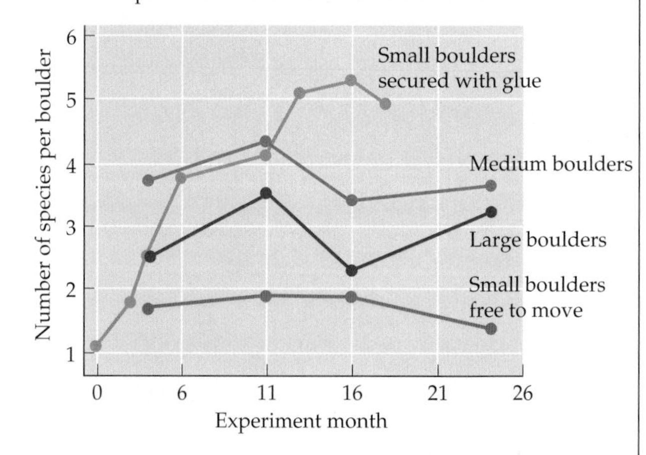

Conclusion: Small boulders have fewer species not because they are unsuitable for local species, but because the high rates at which they are moved by waves prevent many species from surviving on them.

55.13 Species Richness Is Greatest at Intermediate Levels of Disturbance By gluing boulders to the substratum, an ecologist showed that small boulders can support more species if they are not disturbed at high rates.

the intertidal boulders were rapid: The secured small boulders had more species than the unsecured boulders within 16 months. Most ecological successions progress much more slowly.

Ecologists divide successions into two major types: primary succession and secondary succession. A **primary succession** begins on sites that lack living organisms. A **secondary succession** begins on sites where some organisms have survived the most recent disturbance. The patterns and causes of ecological succession are varied, but the species that colonize a site soon after the disturbance often alter environmental conditions so that they become favorable for other species.

A good example of a primary succession is the changes in the plant community that followed the retreat of a glacier in Glacier Bay, Alaska, over the last 200 years. The melting and retreating glacier left a series of *moraines*—gravel deposits formed where the glacial front was stationary for a number of years. No human observer was present to measure changes over the 200-year period, but ecologists have inferred the temporal pattern of succession by measuring plant communities on moraines of different ages. The youngest moraines, close to the current glacial front, are populated with bacteria, fungi, and photosynthetic microorganisms. Slightly older moraines farther from the glacial front have lichens, mosses, and a few species of shallow-rooted herbs. Still farther from the glacial front, successively older moraines have shrubby willows, alders, and spruces.

By comparing moraines of different ages, ecologists deduced the pattern of plant succession and changes in soil nitrogen content at Glacier Bay (Figure 55.14). Succession is caused in part by changes in the soil brought about by the plants themselves. Nitrogen is virtually absent from glacial moraines, so the only plants that can grow on recently exposed moraines at Glacier Bay are a herbaceous plant (*Dryas*) and alder trees (*Alnus*), both of which have nitrogen-fixing bacteria in nodules on their roots. Nitrogen fixation by *Dryas* and alders improves the soil so that spruces can grow. Spruces then outcompete and displace the alders and *Dryas*. If the local climate does not change dramatically, a forest community dominated by spruces is likely to persist for many centuries on old moraines at Glacier Bay.

A secondary succession may begin with the dead parts of organisms. The succession of fungal species that decompose pine needles in litter beneath Scots pines (*Pinus sylvestris*) is shown in Figure 55.15. New needles continuously fall from the pines, so the surface layer of litter is young and deeper layers are progressively older. Decomposition begins when the first group of fungi starts consuming the needles soon after they fall. Each group of fungi derives its energy by decomposing certain compounds, converting them to other compounds that are used by the next group of species. This process continues over about 7 years, by which time the last group of fungi—basidiomycetes—has decomposed the last remaining compounds.

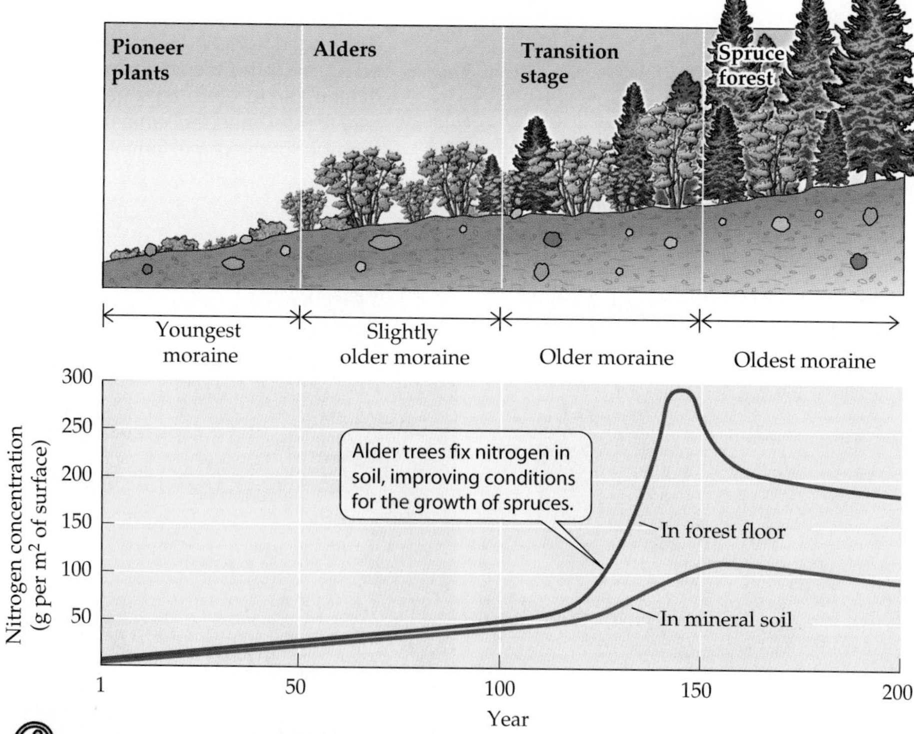

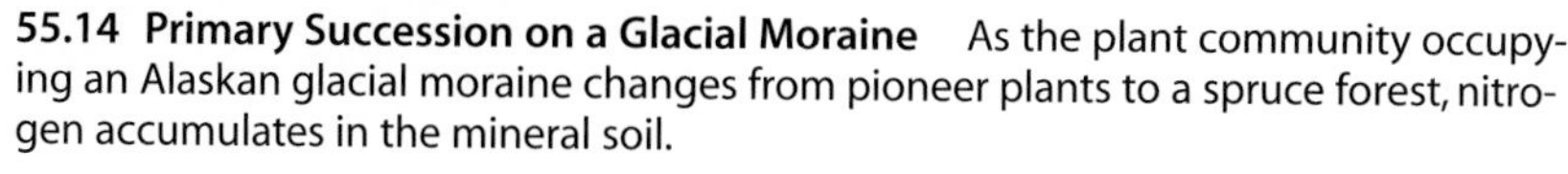

55.14 Primary Succession on a Glacial Moraine As the plant community occupying an Alaskan glacial moraine changes from pioneer plants to a spruce forest, nitrogen accumulates in the mineral soil.

Dispersal, Extinction, and Community Structure

When we discussed population ecology in Chapter 54, we noted that local subpopulations often become extinct, but are reestablished by immigrants from other subpopulations. We know that immigration and emigration influence the structure of communities because species deliberately or inadvertently introduced by humans, such as gypsy moths and zebra mussels, may come to dominate the communities they invade. The rate of introduction of new species and the extinction of existing species has been greatly increased by human activities over the past few centuries, but throughout the history of life on Earth, species have colonized new areas and gone extinct. The composition of ecological communities is influenced not only by the relatively

55.15 Secondary Succession on Pine Needles As indicated by the widths of the bars, the abundances of ten types of fungi in pine needle litter change with time, which increases with depth within the layer.

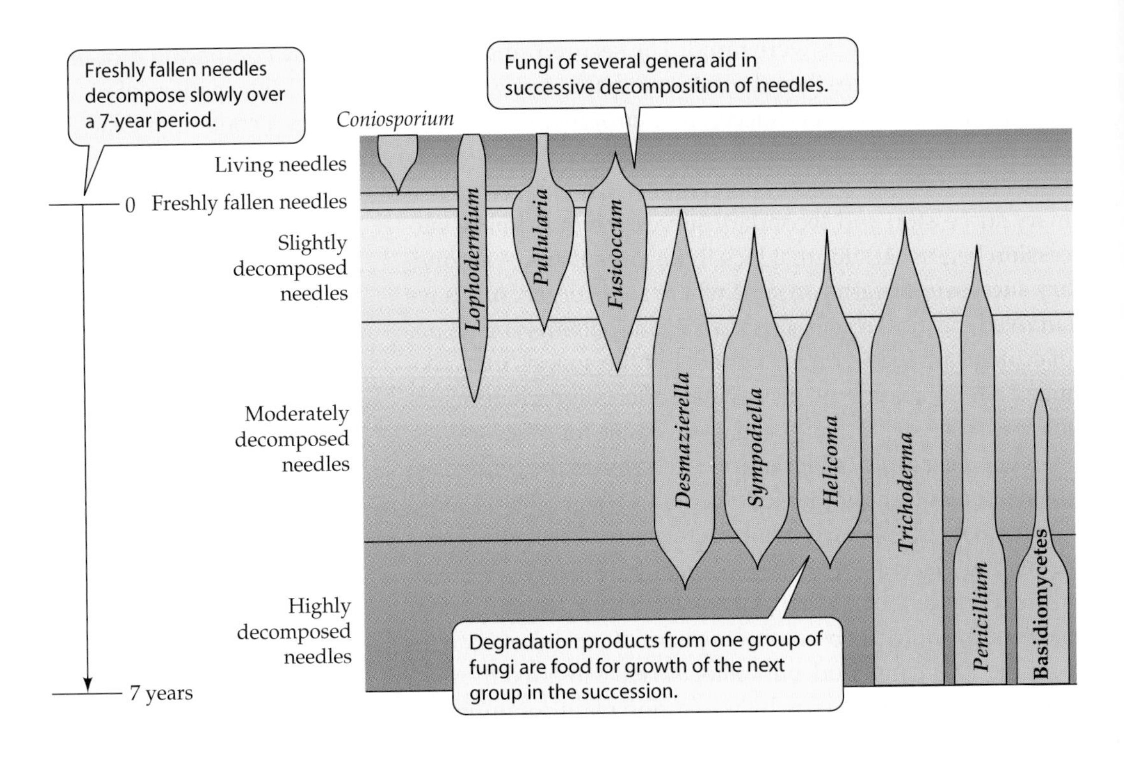

Freshly fallen needles decompose slowly over a 7-year period.
Fungi of several genera aid in successive decomposition of needles.
Coniosporium
Living needles
0 Freshly fallen needles
Slightly decomposed needles
Moderately decomposed needles
Highly decomposed needles
7 years
Lophodermium
Pullularia
Fusicoccum
Desmazierella
Sympodiella
Helicoma
Trichoderma
Penicillium
Basidiomycetes
Degradation products from one group of fungi are food for growth of the next group in the succession.

(*a*)

Dasypus novemcinctus

Erethizon dorsatum

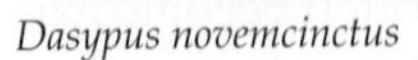

55.16 North and South America Exchanged Mammals (*a*) The nine-banded armadillo and the porcupine are among only a handful of species that colonized North American from South America. (*b*) Some species that exist today only in South America are descended from ancestors who migrated from North America, including the Patagonian fox (left), the Chacoan peccary (center), and the llama (right), a member of the camel family.

(*b*)

Dusicyon griseus

Catagonus wagneri

Lama glama

short-term interactions we have already discussed, but also by these long-term, large-scale ecological and evolutionary processes.

A massive colonization of new areas by mammals occurred when the Central American land bridge formed about 4 million years ago. This land bridge connected North and South America for the first time in about 65 million years. While the two continents were separated, their mammals evolved independently of one another because terrestrial mammals (with the exception of bats) are poor dispersers across water barriers. South America evolved a distinctive mammalian fauna dominated by marsupials, primates, edentates (armadillos, sloths, ground sloths, and anteaters), and caviomorph rodents (porcupines, capybaras, pacas, agoutis, guinea pigs, chinchillas, and others) (Figure 55.16*a*).

Many species of mammals dispersed across the newly established land bridge. Only a few South American species—the porcupine, nine-banded armadillo, and Virginia opossum—became established north of the tropical forests of Central America, but many North America mammals—rabbits, mice, foxes, bears, raccoons, weasels, cats, horses, tapirs, peccaries, camels, and deer—successfully colonized South America. The North American invasion apparently caused the extinction of several kinds of large marsupial carnivores and the large herbivores that were their prey. Subsequently, the northern invaders formed new species that today exist only in South America (Figure 55.16*b*).

The exchange of mammals between North and South America is but one example of the profound influence of immigration, extinction, and subsequent evolution on the patterns of distribution of life on Earth. We will explore these patterns and their causes in detail in Chapter 56.

Chapter Summary

▶ Ecological communities are assemblages of species, each of which interacts in unique ways with its environment. In most cases, species drop out of and are added to communities gradually across environmental gradients. **Review Figure 55.1**

Communities: Loose Assemblages of Species

▶ Experiments can tell us which interactions among species exert the strongest effects on community structure. **Review Figure 55.2**

Process and Pattern in Communities and Ecosystems

▶ Most of the energy incorporated by an organism is used in its respiration. Only a small proportion can be captured by other organisms that consume it. **Review Figure 55.3. See Web/CD Activity 55.1**

▶ Primary production is determined by temperature and precipitation. Therefore, it varies over Earth's surface. **Review Figures 55.4, 55.5**

▶ Species richness increases with primary production up to a point, after which it declines. **Review Figure 55.6**

▶ A trophic level consists of those organisms whose major food source has passed through the same number of steps. **See Web/CD Activity 55.2**

▶ Food webs are diagrams of who eats whom in ecological communities. Most food webs have only three to five trophic levels. **Review Figure 55.7**

▶ Energy pyramids show the flow of energy through trophic levels. Biomass pyramids show the amount of living matter at each trophic level. **Review Figure 55.8**

▶ Communities with more species are generally more productive and more stable than communities with fewer species. **Review Figure 55.9**

▶ Keystone species influence community structure and dynamics out of proportion to their abundances. **Review Figure 55.10**

Disturbance and Community Structure

▶ All ecological communities are subjected to a variety of disturbances. Typically, small disturbances are much more common than large ones.

▶ Communities subjected to moderate levels of disturbance typically have more species than communities subjected to lower or higher levels of disturbance. **Review Figure 55.13**

▶ After a disturbance, the structure and composition of an ecological community changes as organisms modify the physical environment and interact with one another. **Review Figures 55.14, 55.15. See Web/CD Tutorial 55.1**

Dispersal, Extinction, and Community Structure

▶ The composition of ecological communities is influenced by ecological and evolutionary events taking place over long time periods and large spatial scales.

Self-Quiz

1. An ecological community is
   a. all the species of organisms that live and interact with one another in an area.
   b. all the species that live and interact in an area together with the abiotic environment.
   c. all the species in an area that belong to a particular trophic level.
   d. all the species that are members of a local food web.
   e. all of the above
2. What is the difference between primary productivity and primary production?
   a. Primary productivity is always greater than primary production.
   b. Primary productivity is always less that primary production.
   c. Primary productivity is a rate, whereas primary production is a product.
   d. Primary productivity is a product, but primary production is a rate.
   e. There is no real difference between primary productivity and primary production.
3. The total amount of energy that plants assimilate by photosynthesis is called
   a. gross primary production.
   b. net primary production.

   c. biomass.
   d. a pyramid of energy.
   e. succession.

4. The amount of energy reaching a higher trophic level is determined by
   a. net primary production.
   b. net primary production and the efficiencies with which food energy is converted to biomass.
   c. gross primary production.
   d. gross primary production and the efficiencies with which food energy is converted to biomass.
   e. gross primary production and net primary production.

5. The pyramids of energy and biomass of forests and grasslands differ because
   a. forests are more productive than grasslands.
   b. forests are less productive than grasslands.
   c. large mammals avoid living in forests.
   d. trees store much energy in difficult-to-digest wood, whereas grassland plants produce few difficult-to-digest tissues.
   e. grasses grow faster than trees.

6. Keystone species
   a. influence the structure of the communities in which they live more than expected on the basis of their abundance.
   b. strongly influence the species composition of communities.
   c. may speed up the rate of nutrient cycling.
   d. may be herbivores or carnivores.
   e. all of the above

7. What is the general relationship between species richness and disturbance?
   a. Species richness peaks at low levels of disturbance.
   b. Species richness peaks at high levels of disturbance.
   c. Species richness peaks at intermediate levels of disturbance.
   d. Species richness is less at intermediate levels of disturbance.
   e. There is no general relationship between species richness and level of disturbance.

8. Ecological succession is
   a. the changes in species over time.
   b. the gradual process by which the species composition of a community changes.
   c. the changes in a forest as the trees grow larger.
   d. the process by which a species becomes abundant.
   e. the buildup of soil nutrients.

9. Primary succession begins
   a. soon after a disturbance ends.
   b. at varying times after a disturbance ends.
   c. at sites where some species survived the disturbance.
   d. at sites were no species survived the disturbance.
   e. at sites where only primary producers survived the disturbance.

10. The South American mammals that became established in North America after crossing the Central American land bridge include
   a. porcupines, armadillos, and caviomorph rodents.
   b. porcupines, caviomorph rodents, and Virginia opossums.
   c. Virginia opossums, porcupines, and anteaters.
   d. Virginia opossums, porcupines, and armadillos.
   e. armadillos, anteaters, and caviomorph rodents.

For Discussion

1. Some evidence suggests that interspecific competition may be responsible for the decrease in species richness at high levels of productivity. What other hypotheses might explain this puzzling relationship? How would you test them?
2. The increased productivity and stability of species-rich communities could be explained by ecological differences among the species or by the fact that the more species in a community, the greater the chance that it will contain an unusually productive species. How could you distinguish between these competing hypotheses?
3. If species-rich communities are more productive than species-poor communities, how can modern agriculture, which is based almost entirely on cultivating a single species on a plot, be so productive?
4. We illustrated succession with two examples from forests. How might ecological succession differ in grasslands? In deserts? In the rocky intertidal zone?
5. Many conservationists believe that our greatest efforts should be expended to save undisturbed environments. Many users of natural resources, on the other hand, argue that disturbing environments to extract resources will actually improve species preservation. Is the latter view an appropriate invocation of the intermediate disturbance hypothesis?

56 Biogeography

[I]t is... those [species] which range widely over the world, are the most diffused within their own country, and are the most numerous in individuals, which oftenest produce well-marked varieties, or as I consider them, incipient species.

—Charles Darwin, 1859

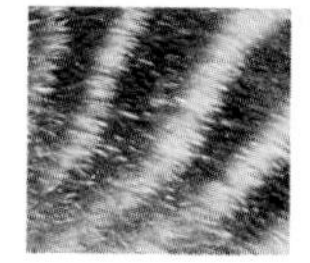

In this passage from the second chapter of *The Origin of Species*, Charles Darwin was reporting the results of his tabulations of several well-studied regional floras. Darwin never published those data, but his suggestion that species that are widespread tend to be both more abundant and more variable than species with narrower ranges has been supported by recent evidence. Thus, in addition to his contributions to evolutionary theory and ecology, Darwin anticipated modern advances in the field of biogeography.

Widespread species are often abundant locally, but no species is found everywhere. The study of the distribution of organisms over Earth's surface began when eighteenth-century European explorers, settlers, and travelers started to take note of the vast differences among the biota on the different continents and attempted to understand them.

When the first Europeans arrived in Australia, they saw plants and animals that differed in perplexing ways from the ones they knew at home. Among these oddities were flowers pollinated by brush-tongued parrots and mammals that hopped around on their hind legs, carrying their offspring in pouches. In contrast, the first Europeans to visit North America felt at home because the plants (such as oaks, elms, and pines) and animals (such as deer, rabbits, foxes, thrushes, and crows) of North America were similar to those of Europe. Why was North America's biota so similar to Europe's while Australia's was so bizarrely different?

Biogeography is the science that documents and attempts to explain the patterns of distribution of populations, species, and ecological communities across Earth. In this chapter, we will show how biogeographers identify the processes that influence the distributions of species, both those that operated in the remote past and those that are operating today. We will also review Earth's major biogeographic regions. Finally, we will look at the factors that influence the number of species that live together.

An Australian Endemic The numbat (*Myrmecobius fasciatus*) is an Australian marsupial mammal that uses its highly specialized tongue to feed on ants and termites. Unrelated and different-appearing animals on other continents also feed on these insects, but the numbat, like many other mammals, is unique to Australia.

Earth's Biogeographic Regions

Explaining species' distributions might seem to be a simple matter, because the question of why a species is or is not found in a certain location has only a few possible answers:

- If a species occupies a particular area, either it evolved there, or it evolved elsewhere and dispersed to the area.
- If a species is not found in a particular area, either it evolved elsewhere and never dispersed to the area, or it was once present in the area but no longer lives there.

Determining which of these possible answers is correct requires information about the evolutionary histories of species, which comes from fossils and from knowledge of their phylogenetic relationships. It also requires information about changes in Earth's geography (such as continental drift, glacial advances and retreats, sea level changes, and mountain building) that occurred as the organisms were evolving. Such geological information can tell us whether organisms evolved where they are currently found or dispersed and colonized new areas from a distant area of origin.

The biotas of the continents differ enough to allow the division of Earth into several major **biogeographic regions**. Biogeographic regions are based on the taxonomic similarities of the organisms living in them. The boundaries of biogeographic regions are set where species compositions change dramatically over short distances (Figure 56.1). The biotas of the biogeographic regions differ because oceans, mountains, deserts, and other barriers restrict the dispersal of organisms. Although there has been dispersal of organisms between adjacent biogeographic regions (as happened between North and South America, as we described at the end of Chapter 55), such interchanges have not been frequent enough to eliminate the striking differences that have resulted from speciation and extinction within each region.

A species found only within a certain region is said to be **endemic** to that region. Remote islands typically have distinctive endemic biotas because water barriers greatly restrict immigration. If the islands are large enough or form part of an archipelago, allopatric speciation often produces unique species and communities, as we saw in Chapter 24. For example, nearly all the tracheophytes and vertebrates of Madagascar, a large island off the eastern coast of Africa, are endemic to that island (Figure 56.2). Madagascar by itself could be called a biogeographic region, but because dozens of is-

56.1 Major Biogeographic Regions The biotas of Earth's major biogeographic regions differ strikingly from one another.

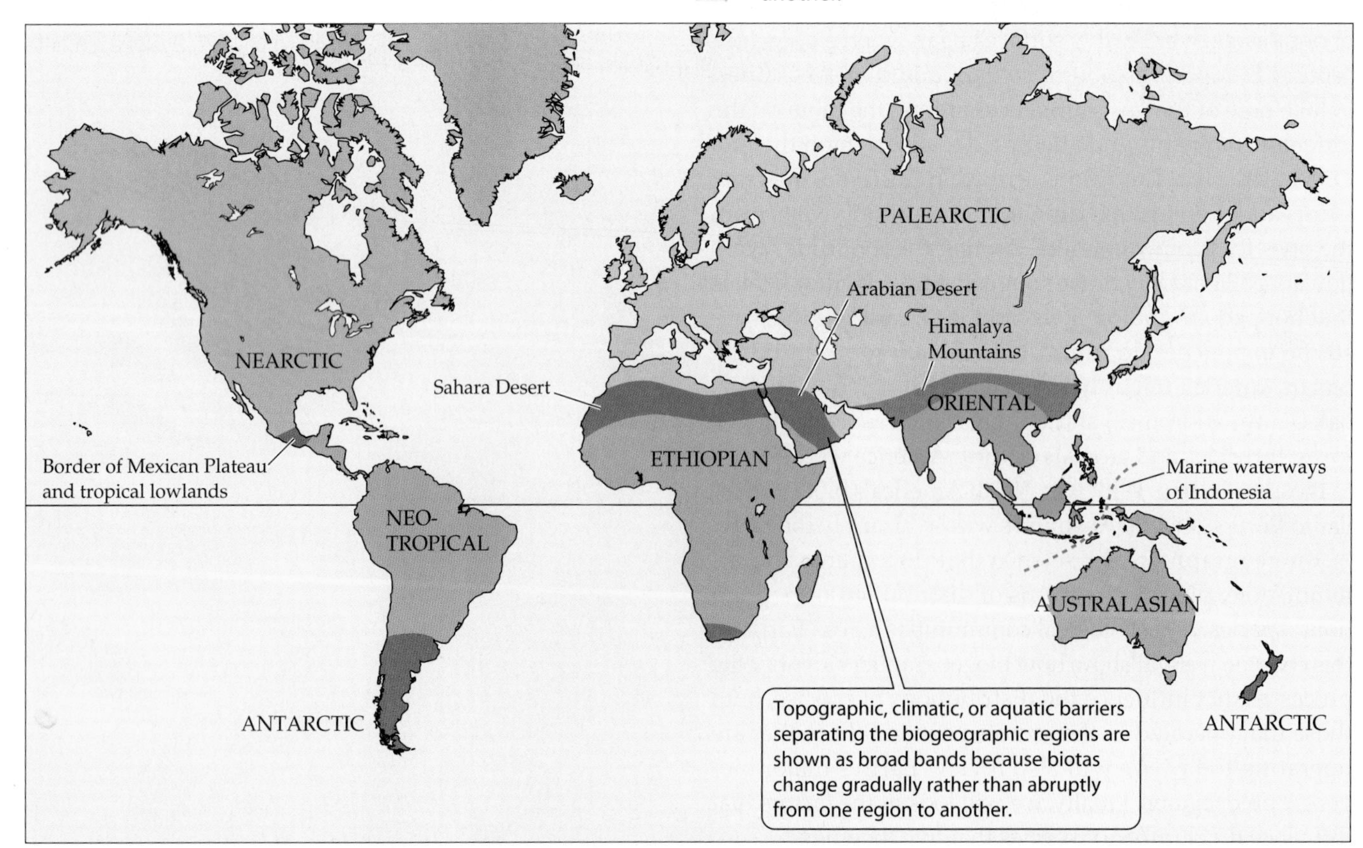

Chamaeleo pantheri (panther chameleon)

Hemicentetes semispinosus (yellow-streaked tenrec)

Cryptoprocta ferox (fossa)

Lemur catta (ring-tailed lemur)
climbing *Alluaudia procera* (Madagascar ocotillo)

Adansonia grandidieri (giant baobab tree)

Pachypodium rosulatum (elephant's foot)

56.2 Madagascar Abounds with Endemic Species The majority of tracheophyte and vertebrate species found on the island of Madagascar are found nowhere else on Earth.

lands would qualify as biogeographic regions on the basis of the distinctness of their biotas, islands are treated differently from continents.

Most species are confined to a single biogeographic region. The most widespread species today is probably *Homo sapiens*, but a few other species—for example, the great egret, the osprey, the peregrine falcon, and the barn owl—are found on all continents except Antarctica.

Next we will discuss the influence of speciation, extinction, and other historical processes on biogeographic patterns, and then consider the influence of processes operating today.

History and Biogeography

Before 1850, as we saw in Chapter 22, most people, including biogeographers, believed in a relatively unchanging Earth that was too young for long-term processes to account for the diversity and distribution of life. Linnaeus (1758), for example, believed that all organisms had been created in one place, which he called Paradise, from which they later dispersed. Indeed, because most people believed that the continents were fixed in their positions, the only way to account for the current distributions of organisms was to invoke massive dispersal.

The notion that the continents might have moved was not seriously considered until 1912, when Alfred Wegener, a German meteorologist, argued that the continents had drifted over time. Wegener based his theory on several observations:

- The shapes of continents (the outlines of Africa and South America seem to fit together like pieces of a puzzle)
- The alignment of mountain chains, rock strata, coal beds, and glacial deposits on different continents
- The distributions of closely related species that were shared between Africa and South America, which were difficult to explain if the continents had always been where they are now

When Wegener proposed that the continents had moved, few scientists took him seriously. There were no known mechanisms to move continents, and no convincing geological evidence of such movements existed. As we learned in Chapter 22, geological evidence and plausible mechanisms were eventually discovered. The broad pattern of continental movement, which continues today, is now clear.

About 280 million years ago, the continents were united to form a single land mass, called Pangaea (see Figure 22.13). The continents then began to separate from one another, but when the continents were still very close to one another (about 245 mya), many groups of terrestrial and freshwater organisms, such as insects, freshwater fishes, frogs, and tracheophytes, had already evolved. The ancestors of some organisms that live on widely separated continents today were probably present on those land masses when they were part of Pangaea.

By 100 mya, continental drift had separated Pangaea into northern (Laurasia) and southern (Gondwana) land masses, and the southern continents were moving away from each other (see Figure 22.15). Over time, India separated from Africa and slammed into southern Asia, Australia moved closer to Southeast Asia, and South America, which had drifted as an island for 60 million years, came into contact with North America. Throughout the history of life, continental drift has both separated and combined biotas, thus greatly influencing the distribution and evolution of species.

Biogeographers convert phylogenies to "area phylogenies"

As the age of Earth, the geological processes that shaped it, and the mechanics of evolution became better understood, biogeographers were able to ask questions such as, Where and when did evolutionary lineages originate? How did they spread? What do the present-day distributions of organisms tell us about their past histories?

A technique that was developed to help answer these questions was the creation of *area phylogenies*. To generate an area phylogeny, biogeographers alter a taxonomic phylogeny by replacing the names of the taxa with the names of the places where those taxa live or lived. For example, an area phylogeny suggests that horses speciated as they moved from Asia to Africa, whereas the speciation of zebras took place entirely in Africa (Figure 56.3).

Biogeographers use several approaches to infer the approximate times of separation of taxa within a lineage. First, if a molecular clock has been ticking at a relatively constant rate, the amount of difference in the molecules of species should be strongly correlated with the length of time their lineages have been evolving independently, as we saw in Chapter 26. Second, fossils can help to show how long a taxon has been present in an area and whether its members formerly lived in areas where they are no longer found. A third valuable source of information is the distribution of living species. Much more information can be gathered on current distributions than will ever be available from fossils. Similarities in the distributions of many lineages of organisms provide clues about past events that affected them.

Vicariant events and dispersal both influence distributions

The appearance of a barrier that splits the range of a species is called a **vicariant event**. A vicariant event divides the population of a species even though no individuals have dis-

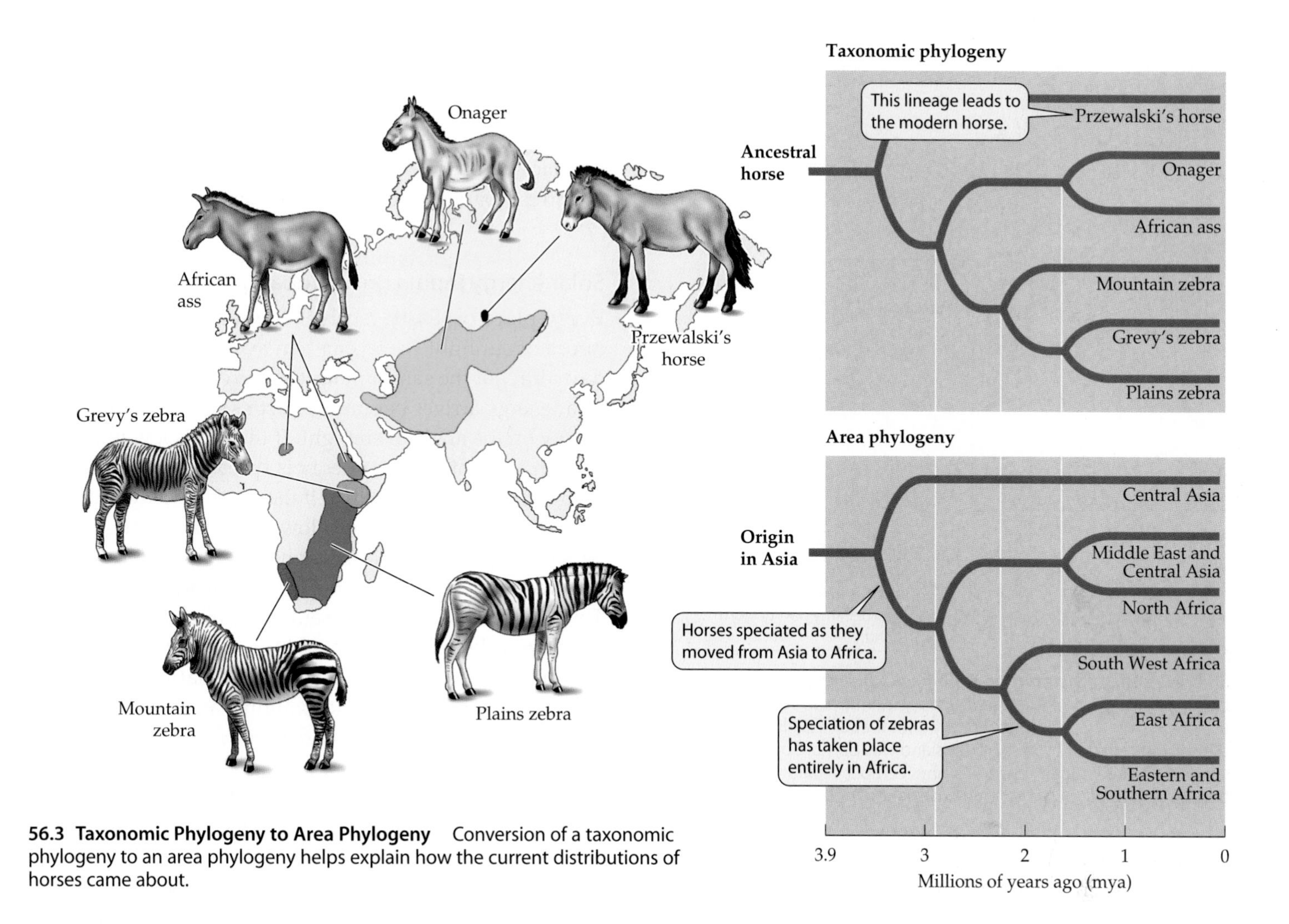

56.3 Taxonomic Phylogeny to Area Phylogeny Conversion of a taxonomic phylogeny to an area phylogeny helps explain how the current distributions of horses came about.

persed to new areas. If, however, members of a species cross an already existing barrier and establish a new population, the species' disjunct range is a result of **dispersal**.

By studying a single evolutionary lineage, a biogeographer may discover evidence suggesting that the distribution of an ancestral species was influenced by a vicariant event such as a change in sea level or mountain building (see Figure 24.4). If that inference is correct, then species in other lineages are likely to have been influenced by the same event; that is, a number of lineages may have similar distribution patterns. Differences in distribution patterns among lineages may indicate that the lineages responded differently to the same vicariant events, that the lineages separated at different times, or that the lineages have had very different dispersal histories. By analyzing such similarities and differences, biogeographers can discover the relative roles of vicariant events and dispersal in determining today's distribution patterns.

The longer an area has been isolated from other areas by a vicariant event, such as continental drift, the more endemic taxa it is likely to have, because there has been more time for evolutionary divergence to take place. Australia, which has been separated the longest from the other continents (about 65 million years), has the most distinctive biota of any continent. South America has the next most distinctive biota, having been isolated from other continents for nearly 60 million years. North America and Eurasia, which were joined together for much of Earth's history, have very similar biotas. That is why the early European travelers felt more at home in North America than in Australia.

When several hypotheses can explain a pattern, scientists typically prefer the most *parsimonious* one—the one that requires the smallest number of unobserved events to account for it. We saw how the parsimony principle is used in the reconstruction of phylogenies in Chapter 25. To see how it is applied to biogeography, consider the distribution of the New Zealand flightless weevil *Lyperobius huttoni*, a species that is found in the mountains of South Island and on sea cliffs at the extreme southwestern corner of North Island (Figure 56.4). If you knew only its current distribution and the current positions of the two islands, you might surmise that, even though this weevil cannot fly, it had somehow managed

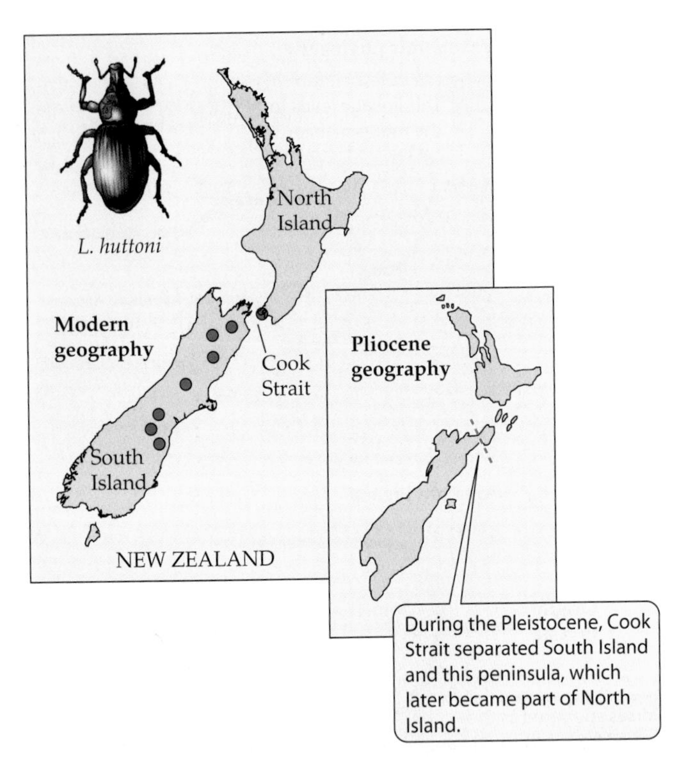

56.4 A Vicariant Distribution Explained Blue circles indicate the current distribution of the weevil *Lyperobius huttoni*. A comparison of the present New Zealand geography with that of the Pliocene, when the southern part of today's North island was part of South Island, suggests that a vicariant event—a physical split separating populations—is the most parsimonious explanation for this distribution.

to cross Cook Strait, the 25-kilometer body of water that separates the two islands.

However, more than 60 other animal and plant species, including other species of flightless insects, live on both sides of Cook Strait. Although organisms do cross marine and terrestrial barriers, it is unlikely that all of these species made the same ocean crossing. In fact, we do not need to make that assumption. Geological evidence indicates that the present-day southwestern tip of North Island was formerly united with South Island. Therefore, none of the 60 species need have made a water crossing. A single vicariant event, the separation of the northern tip of South Island from the remainder of the island by the newly formed Cook Strait, could have split all of the distributions.

As we have just seen, the distributions of species today have been determined, in part, by history. However, because all organisms must be able to survive in today's environmental conditions, Earth's climates also exert a powerful influence on the distributions, abundances, and evolution of species.

Ecology and Biogeography

The **climate** of a region is the average of the atmospheric conditions (temperature, precipitation, and wind velocity) found there over time. Climates vary greatly from place to place on Earth, primarily because different places receive different amounts of solar energy. We will first examine how these differences in solar energy input determine atmospheric and oceanic circulation. Then we will show how climates influence the geographic distributions of organisms.

Solar energy inputs drive global climates

Every place on Earth receives the same total number of hours of sunlight each year—an average of 12 hours per day—but not the same amount of *energy*. The rate at which solar energy arrives per unit of Earth's surface depends primarily on the angle of sunlight. If the sun is low in the sky, a given amount of solar energy is spread over a larger area (and is thus less intense) than if the sun is directly overhead. In addition, when the sun is low in the sky, sunlight must pass through more of Earth's atmosphere, so more of its energy is absorbed and reflected before it reaches the ground. Thus, at higher latitudes (closer to the poles), there is greater variation in both day length and the angle of arriving solar energy over the course of a year than at latitudes closer to the equator. On average, mean annual air temperature decreases about 0.4°C for every degree of latitude (about 110 kilometers) at sea level.

Air temperature also decreases with elevation. As a parcel of air rises, it expands, its molecules move farther apart, its pressure and temperature drop, and it releases moisture. When a parcel of air descends, it is compressed, its pressure rises, its temperature increases, and it takes up moisture.

Earth's climates are strongly influenced by global air circulation patterns, which result from the global variation in solar energy input that we have just described and from the spinning of Earth on its axis. Air rises when heated by the sun. Warm air rises in the Tropics, which receive the greatest solar energy input. This rising air is replaced by air that flows toward the equator from the north and south. The coming together of these air masses produces the *intertropical convergence zone*. Typically, heavy rains fall in a region when it is close to the intertropical convergence zone as the rising air releases its moisture. The intertropical convergence zone shifts latitudinally with the seasons, following the shift in the zone of greatest solar energy input. This shift results in a characteristic latitudinal pattern of distribution of rainy and dry seasons in tropical and subtropical regions (Figure 56.5).

The air that moves into the intertropical convergence zone to replace the rising air is replaced, in turn, by air from aloft that descends at roughly 30° north and south latitudes after having traveled away from the equator at great heights. This air, which cooled and lost its moisture while rising at the equator, now descends, warms, and takes up rather than releasing moisture. Many of Earth's deserts, such as the Sahara

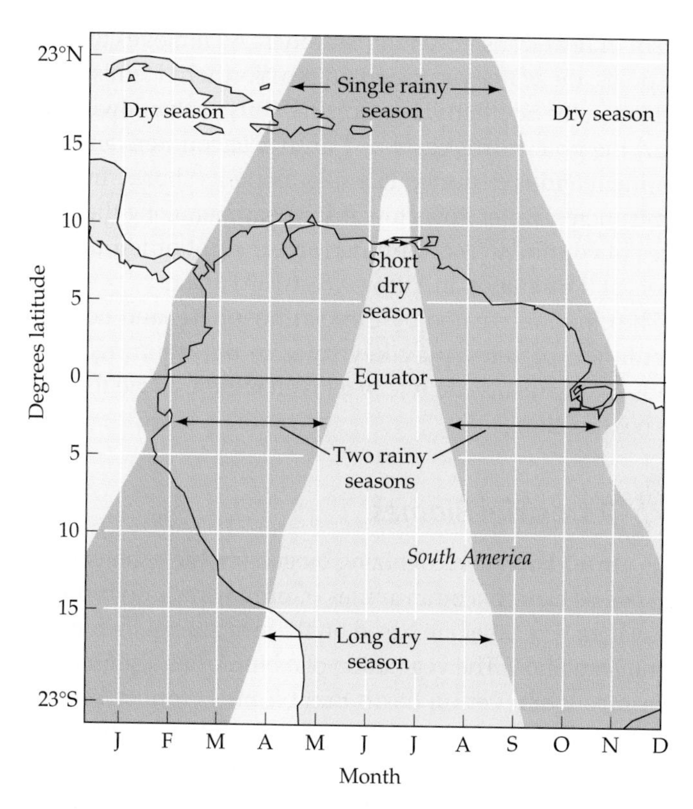

56.5 Rainy and Dry Seasons Change with Latitude In the Tropics and Subtropics, rainy and dry months are highly predictable based on the region's latitude.

and the Australian deserts, are located at these latitudes where dry air descends.

At about 60° north and south latitudes, air rises again and moves either toward or away from the equator. At the poles, where there is little input of solar energy, air descends. The black arrows around the edge of Figure 56.6 show these vertical air circulation patterns. These movements of air masses are responsible, in part, for global wind patterns.

The spinning of Earth on its axis also influences surface winds because Earth's velocity is rapid at the equator, where its diameter is greatest, but relatively slow close to the poles. An air mass at a particular latitude has the same velocity as Earth has at that latitude. As an air mass moves toward the equator, it confronts an increasingly faster spin, and its rotational movement is slower than that of Earth beneath it. Conversely, as an air mass moves poleward, it confronts an increasingly slower spin, and it speeds up relative to Earth beneath it. Therefore, air masses moving latitudinally are deflected to the right in the Northern Hemisphere and to the left in the Southern Hemisphere. Winds blowing toward the equator from the north and south veer to become the northeast and southeast trade winds, respectively. Winds blowing away from the equator also veer and become the westerly winds that prevail at mid-latitudes. The blue arrows in Figure 56.6 show the average directions of these surface winds.

When wind patterns bring air into contact with a mountain range, the air rises to pass over the mountains, cooling as it does so. Because cool air cannot hold as much moisture as warm air, clouds frequently form and release moisture as rain or snow. On the leeward side of the mountain range, the now-dry air descends, warms, and once again picks up moisture. This pattern often results in a dry area called a *rain shadow* on the leeward sides of moutain ranges (Figure 56.7).

Global oceanic circulation is driven by wind patterns

The global pattern of wind circulation drives the circulation of ocean water. Ocean water generally moves in the direction of the prevailing winds (Figure 56.8). Winds blowing toward the equator from the northeast and southeast cause water to converge at the equator and move westward until it encounters a continental land mass. At that point the water splits, some of it moving north and some of it moving south along continental shores. This poleward movement of ocean water that has been warmed in the Tropics is a major mech-

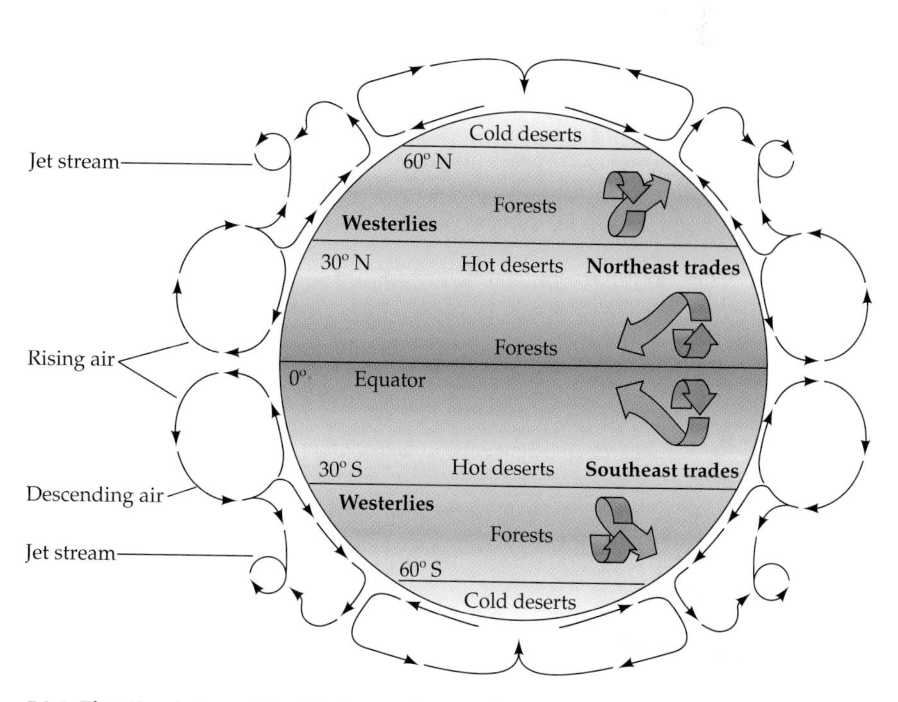

56.6 The Circulation of Earth's Atmosphere If we could stand outside Earth and observe its air movements, we would see vertical air circulation patterns similar to those indicated by the black arrows and surface winds similar to those shown by the blue arrows. Both the vertical and horizontal circulation patterns shift to the north during the northern summer and to the south during the northern winter.

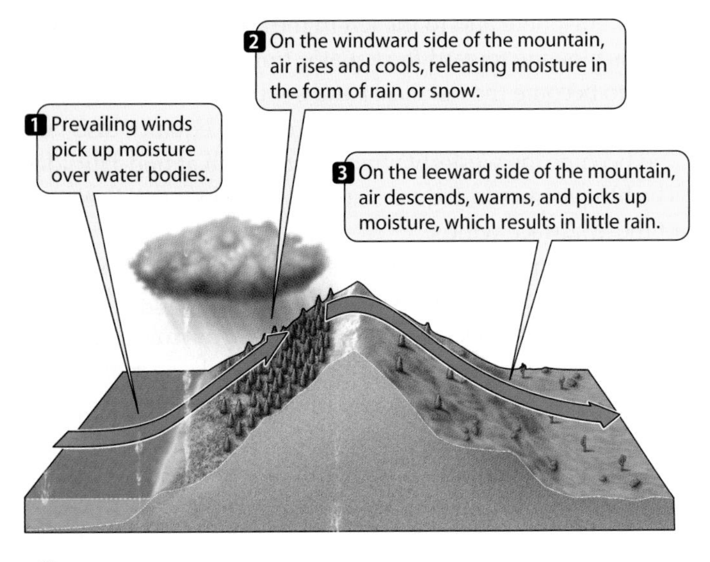

56.7 A Rain Shadow Average annual rainfall tends to be lower on the leeward side of a mountain range than on the windward side.

56.8 Global Oceanic Circulation To see that ocean currents are driven primarily by winds, compare the surface currents shown here with the prevailing surface winds shown in Figure 56.6. Deep ocean currents differ strikingly from the surface ones shown here.

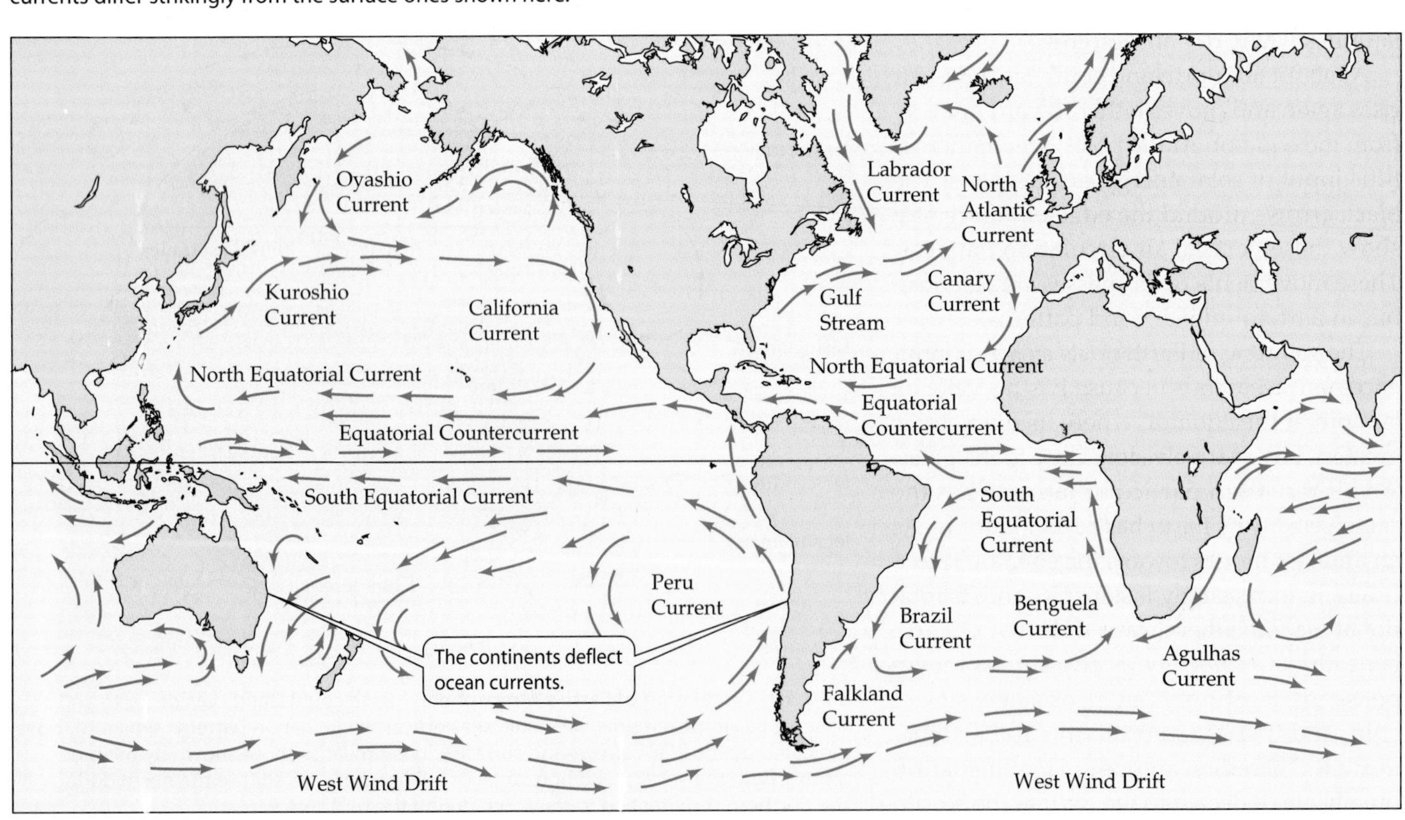

anism of heat transfer to high latitudes. As it moves toward the poles, the water veers right in the Northern Hemisphere and left in the Southern Hemisphere. Thus, water flowing toward the poles turns eastward until it encounters another continent and is deflected laterally along its shores. In both hemispheres, water flows toward the equator along the west sides of continents, continuing to veer right or left until it meets at the equator and flows westward again.

The climates created by these atmospheric and oceanic circulation patterns play key roles in determining what kinds of organisms can live in a given region, as we'll see in the next section.

Terrestrial Biomes

In addition to recognizing biogeographic regions, ecologists also classify communities of organisms into **biomes**, ecosystem types that are based on the structure of their dominant vegetation. The vegetation of a biome has a similar appearance wherever on Earth that biome is found, but the plant species in these communities, despite their physical similarities, may not be evolutionarily closely related. Biomes are named for and identified by their characteristic vegetation, sometimes supplemented by a description of their location or climate, but each biome contains many species in all other taxonomic groups.

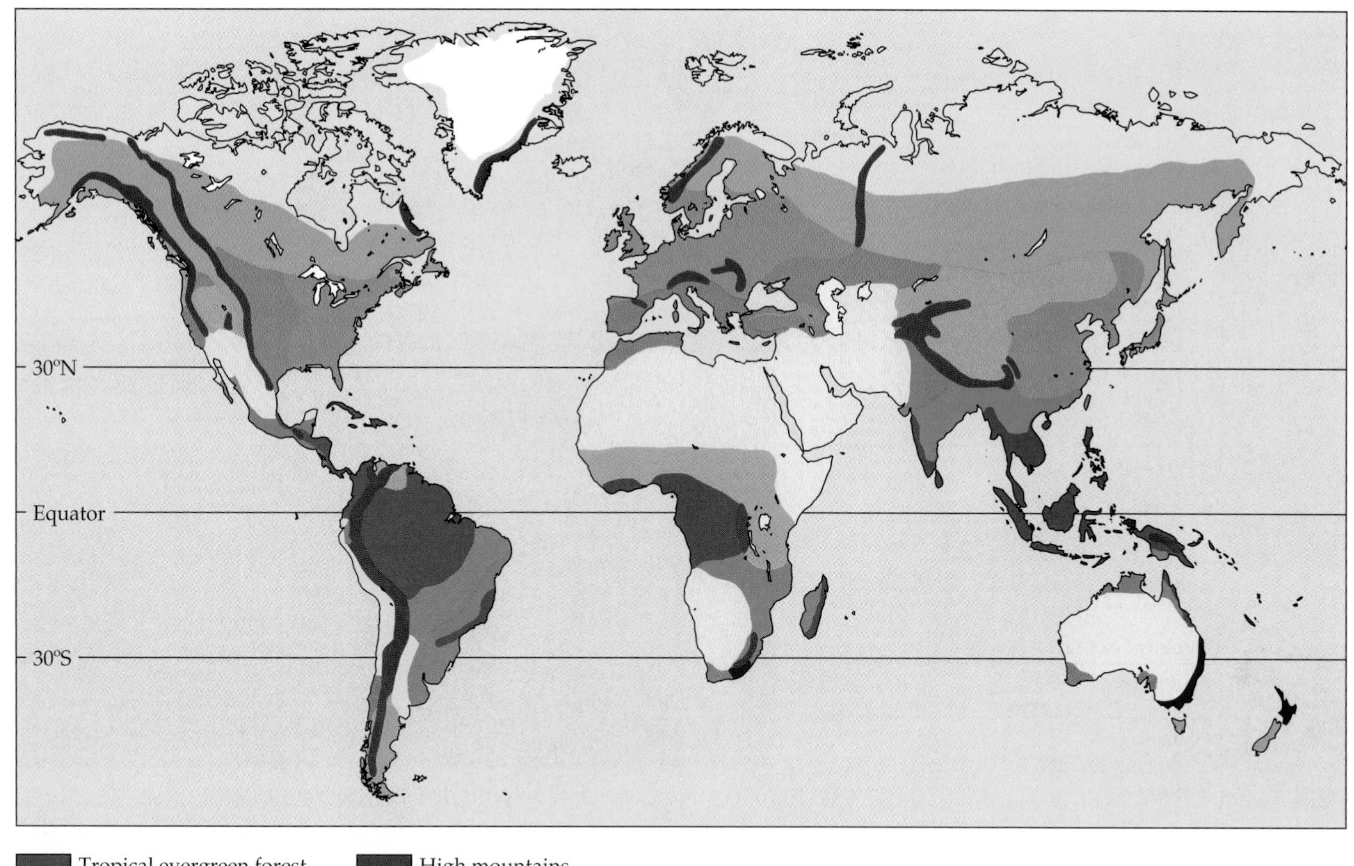

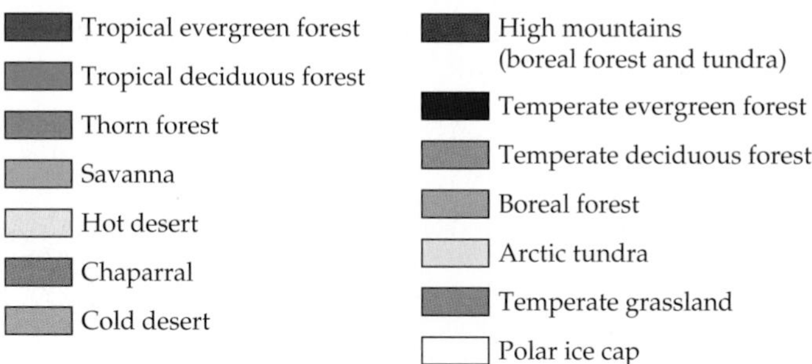

56.9 Biomes Have Distinct Geographic Distributions The distribution of biomes is strongly influenced by patterns of temperature and rainfall.

The distribution of biomes on Earth is strongly influenced by annual patterns of temperature and rainfall (Figure 56.9). In some biomes, such as temperate deciduous forest, precipitation is relatively constant throughout the year, but temperature varies strikingly between summer and winter. In other biomes, both temperature and precipitation change seasonally. In still other biomes, such as tropical rainforest, temperatures are nearly constant, but rainfall varies seasonally. In the Tropics, where seasonal temperature fluctuations are small, annual cycles are dominated by wet and dry seasons (see Figure 56.5).

It is easiest to grasp the similarities and differences among terrestrial biomes by means of a combination of photographs and graphs of temperature, precipitation, and biological activity, supplemented by a few words that describe the species richness and other attributes of those biomes. We use this method in the following pages to describe the major terrestrial biomes of the world.

- Each biome is represented by a map showing its locations and two photographs that illustrate either the biome at different times of year or representatives of the biome in different places on Earth.
- One set of graphs plots seasonal patterns of temperature and precipitation at a site in the biome.
- Other graphs show how active different kinds of organisms are during the year. (For high-latitude biomes, patterns in the Southern Hemisphere are six months out of phase with those shown, which represent the Northern Hemisphere.) Levels of biological activity, shown by the width of the horizontal bars, change either because resident organisms become more or less active (produce leaves, come out of hibernation, hatch, or reproduce) or because organisms migrate into and out of the biome at different times of the year.
- A small box describes the growth forms of the plants that dominate the vegetation in the biome and its patterns of species richness.

These descriptions are very general and cannot describe the variation that exists within each biome.

TUNDRA

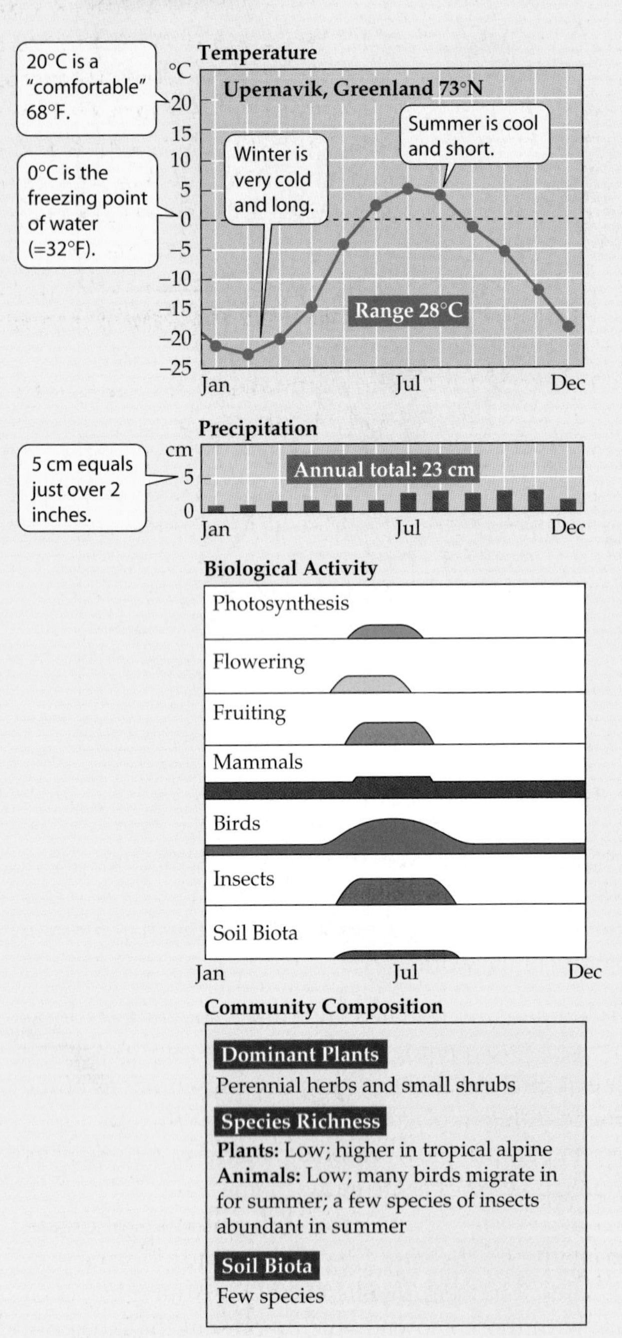

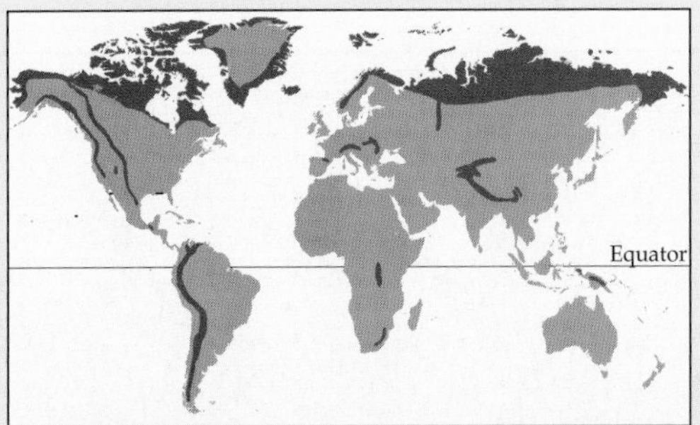

Arctic tundra, Northwest Territories, Canada

Tropical alpine tundra, Teleki Valley, Mt. Kenya

Tundra is found at high latitudes and in high mountains

The **tundra** biome is found in the Arctic and high in mountains at all latitudes. Arctic tundra vegetation, which consists of short perennial plants, is underlain by *permafrost*—soil whose water is permanently frozen. The top few centimeters of soil thaw during the short summers, when the sun shines 24 hours a day. Even though there is little precipitation, lowland Arctic tundra is very wet because water cannot drain down through the permafrost. Plants grow for only a few months each year. Most Arctic tundra animals either migrate into the area only for the summer or are dormant for most of the year.

Tropical alpine tundra is not underlain by permafrost, so photosynthesis and most other biological activities continue (albeit slowly) throughout the entire year. As the photo of alpine vegetation on Mt. Kenya shows, more plant growth forms are present in tropical alpine than in arctic tundra vegetation.

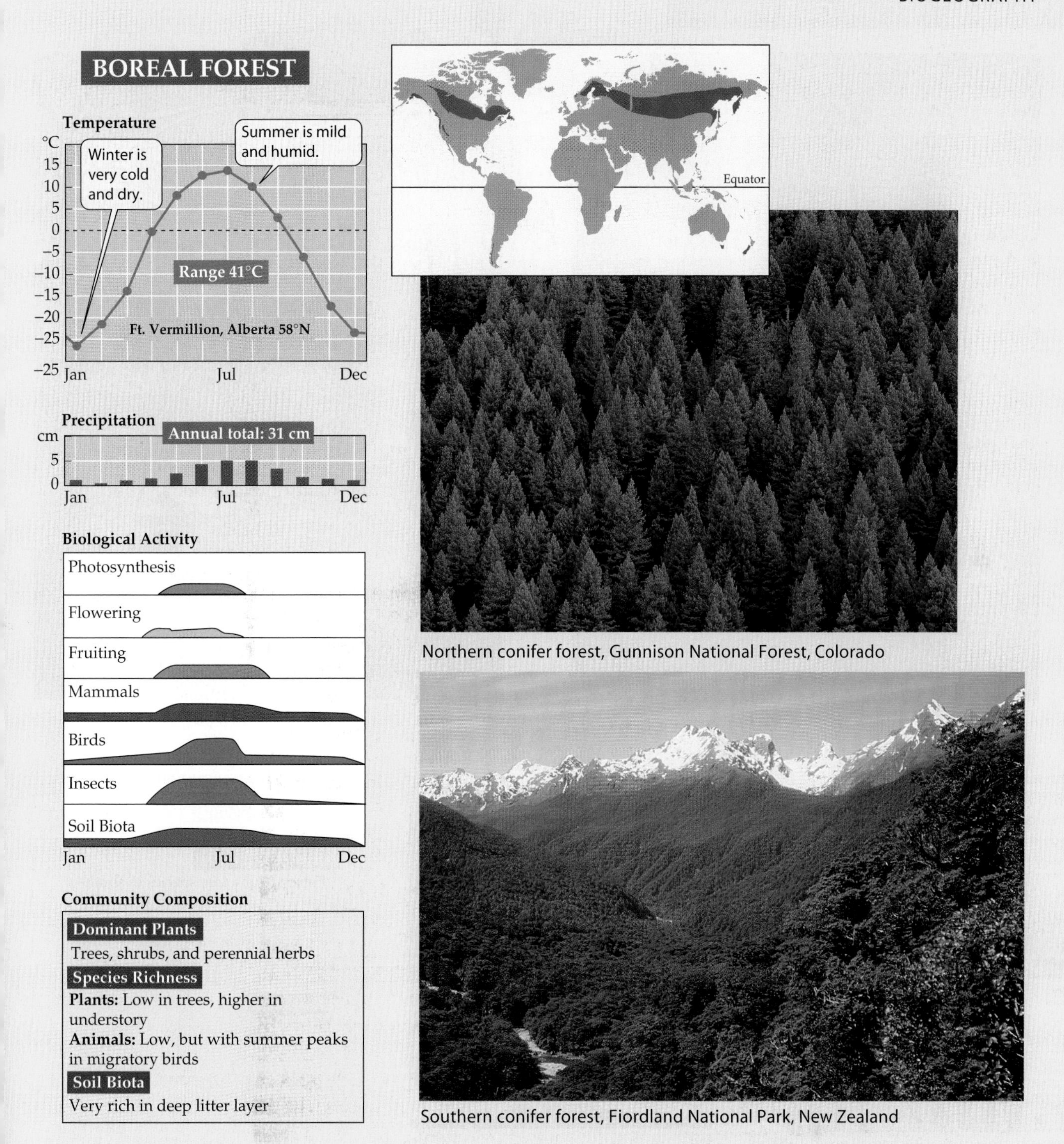

Northern conifer forest, Gunnison National Forest, Colorado

Southern conifer forest, Fiordland National Park, New Zealand

Most boreal forests are dominated by evergreen trees

Boreal forest is found equatorward from tundra and at lower elevations on temperate-zone mountains. Boreal forest winters are long and very cold; summers are short (although often warm). The shortness of the summers favors trees with evergreen leaves because these trees are ready to photosynthesize as soon as temperatures warm in spring.

The boreal forests of the Northern Hemisphere are dominated by evergreen coniferous gymnosperms. In the Southern Hemisphere the dominant trees are southern beeches (*Nothofagus*), some of which are evergreen. Evergreen forests also grow along the west coasts of continents at middle to high latitudes in both hemispheres, where winters are mild but very wet and summers are cool and dry. These forests are home to Earth's tallest trees.

Boreal forests have only a few tree species. The dominant animals (e.g., moose, hares) eat leaves. The seeds in the cones of conifers support a fauna of rodents and birds.

TEMPERATE DECIDUOUS FOREST

A Rhode Island forest in summer and...

...in winter

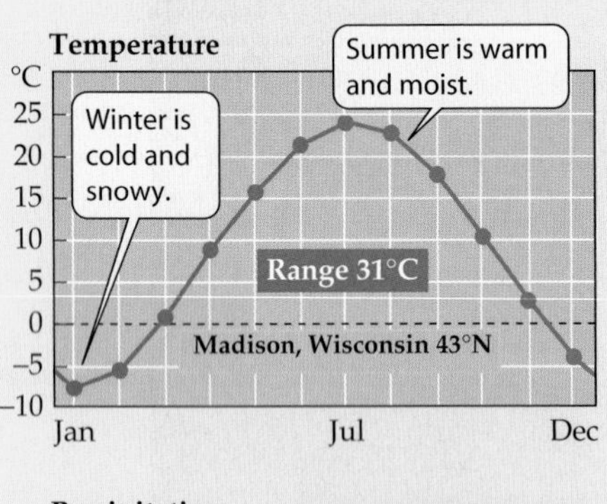

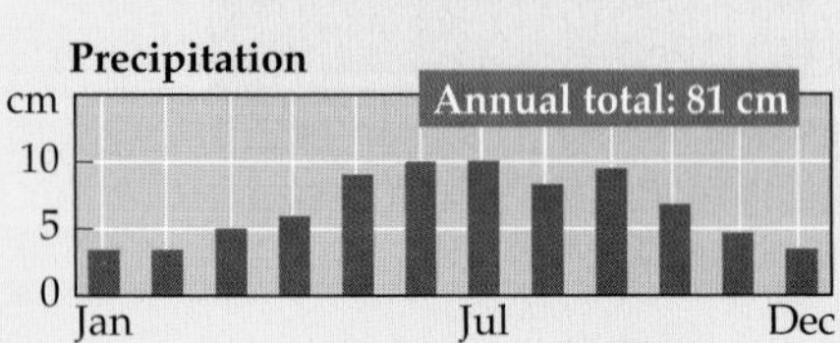

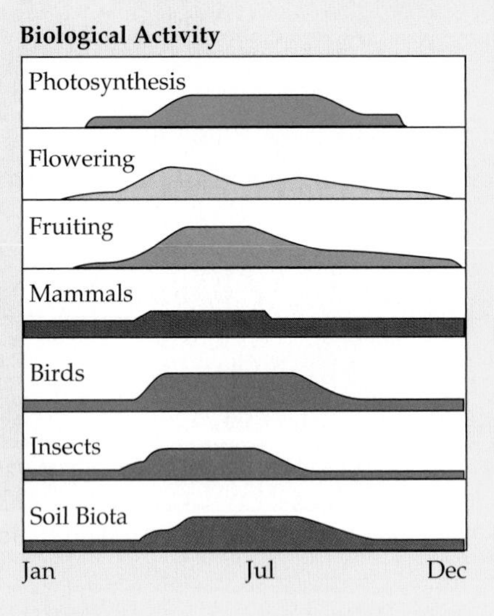

Community Composition

Dominant Plants
Trees and shrubs

Species Richness
Plants: Many tree species in southeastern U.S. and eastern Asia, rich shrub layer
Animals: Rich; many migrant birds, richest amphibian communities on Earth, rich summer insect fauna

Soil Biota
Rich

Temperate deciduous forests change with the seasons

The **temperate deciduous forest** biome is found in eastern North America, eastern Asia, and Europe. Temperatures in these regions fluctuate dramatically between summer and winter. Precipitation is relatively evenly distributed throughout the year.

Deciduous trees, which dominate these forests, lose their leaves during the cold winters and produce leaves that photosynthesize rapidly during the warm, moist summers. Many more tree species live here than in boreal forests. The temperate forests richest in species are in the southern Appalachian Mountains of the United States and in eastern China and Japan—areas that were not covered by glaciers during the Pleistocene. Many genera of plants and animals are shared among the three geographically separate deciduous forest biomes.

TEMPERATE GRASSLANDS

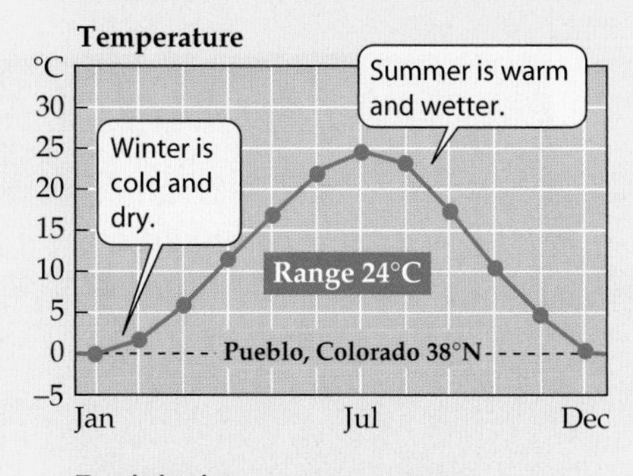

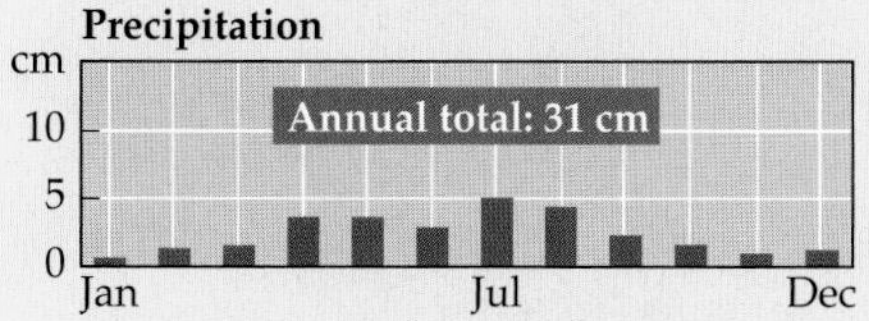

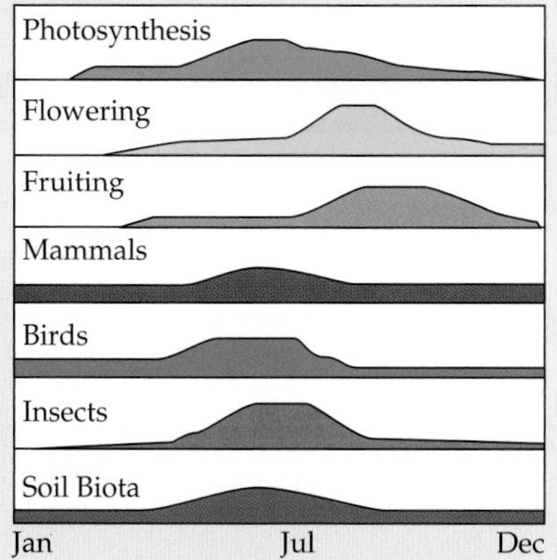

Community Composition

Dominant Plants

Perennial grasses and forbs

Species Richness

Plants: Fairly high

Animals: Relatively few birds because of simple structure; mammals fairly rich

Soil Biota

Rich

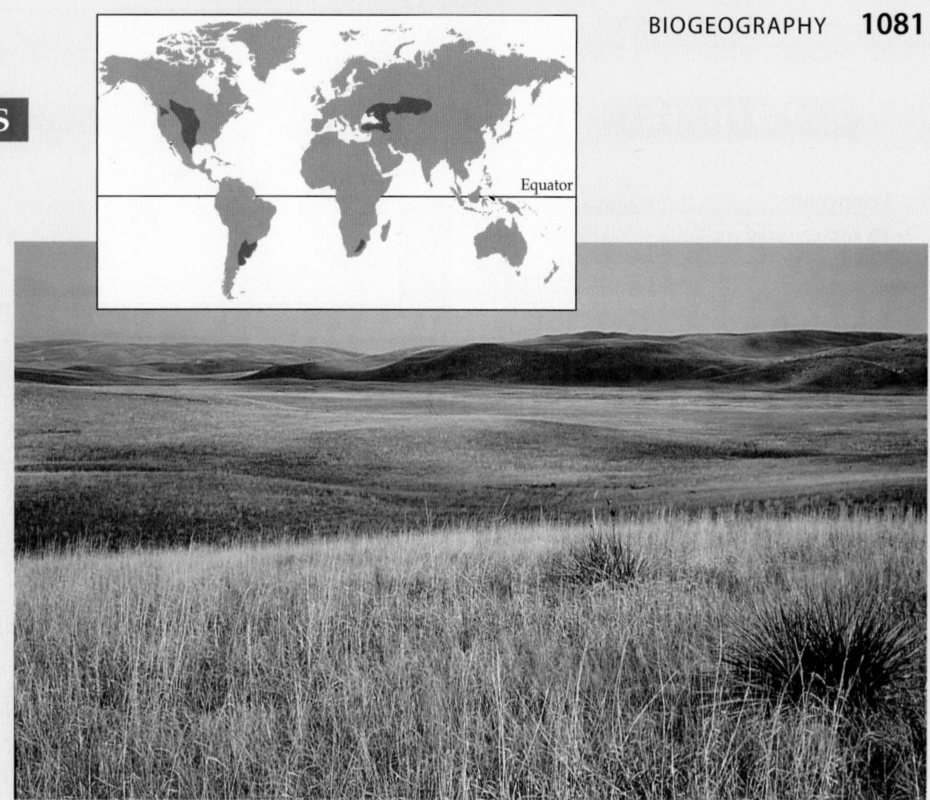

Nebraska prairie in spring

The Veldt, Natal, South Africa

Temperate grasslands are widespread

The **temperate grassland** biome is found in many parts of the world, all of which are relatively dry for much of the year. Most grasslands, such as the pampas of Argentina, the veldt of South Africa, and the Great Plains of North America, have hot summers and relatively cold winters. Most of this biome has been converted to agriculture. In some grasslands, most of the precipitation falls in winter (California grasslands); in others, the majority falls in summer (Great Plains, Russian steppe).

Grassland vegetation is structurally simple, but it is rich in species of perennial grasses, sedges, and forbs. Grasslands are often riots of color when forbs are in bloom. Grassland plants are adapted to grazing and fire. They store much of their energy underground and quickly resprout after they are burned or grazed.

COLD DESERT

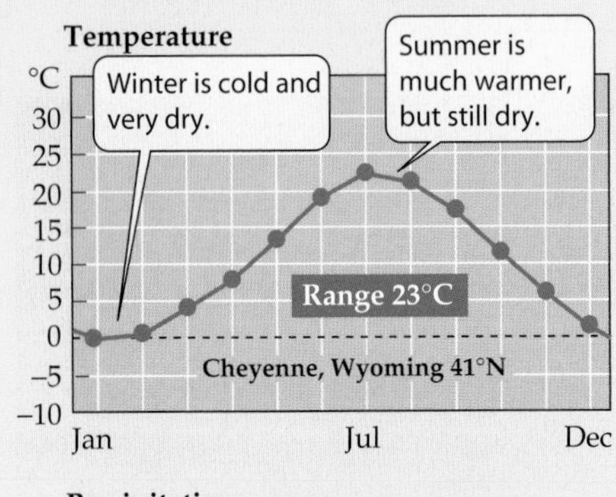

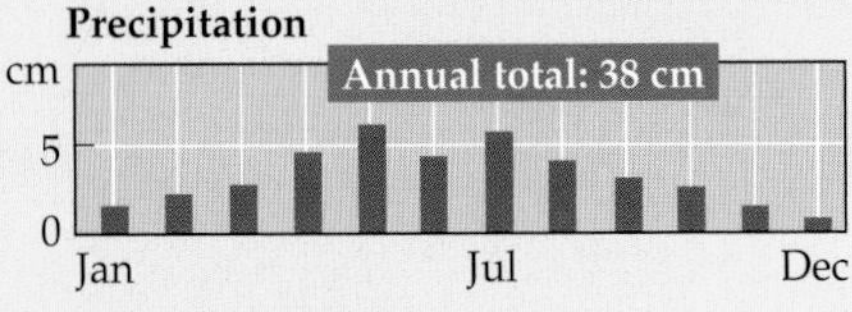

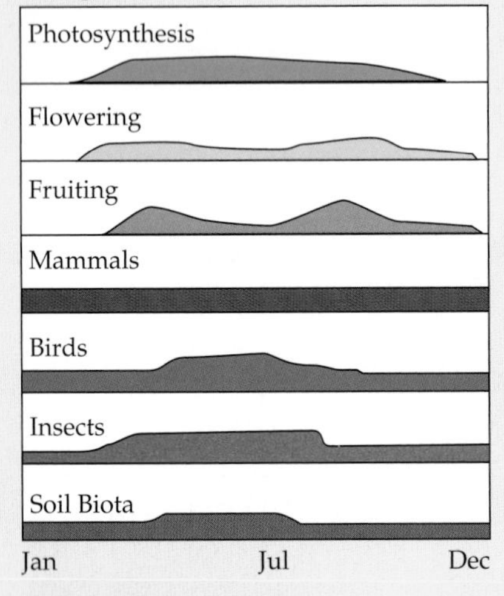

Community Composition

Dominant Plants
Low stature shrubs and herbaceous plants

Species Richness
Plants: Few species
Animals: Rich in seed-eating birds, ants, and rodents; low in all other taxa

Soil Biota
Poor in species

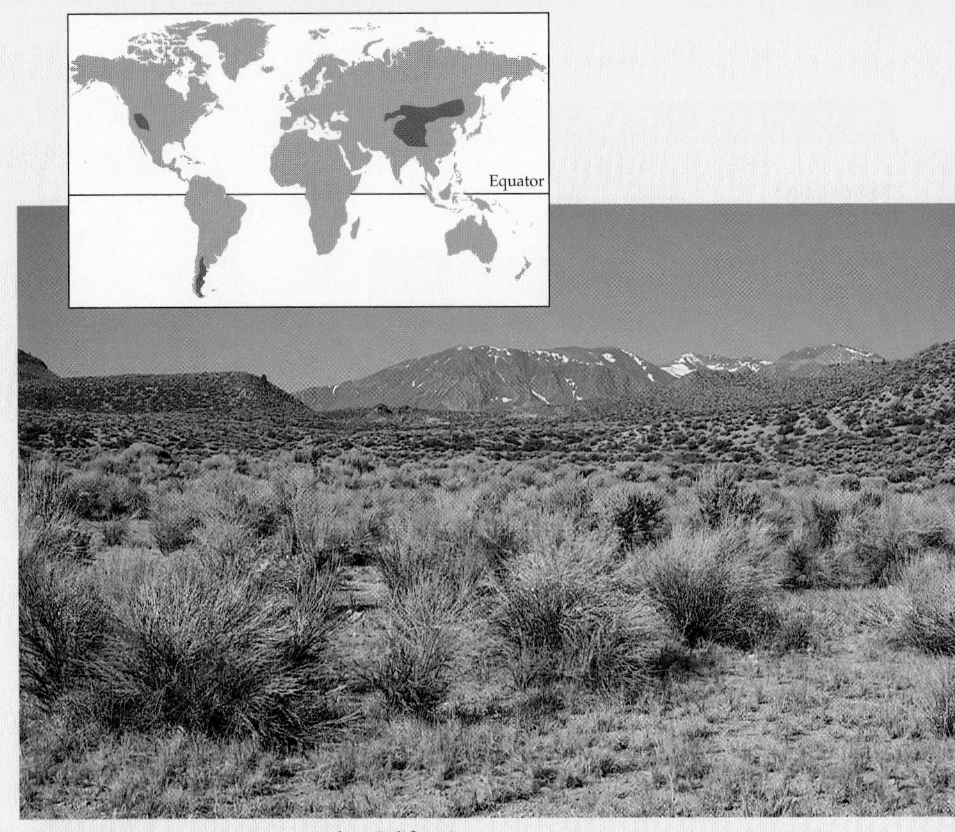

Sagebrush steppe near Mono Lake, California

Los Glaciares National Park, Argentina

Cold deserts are high and dry

The **cold desert** biome is found in dry regions at middle to high latitudes, especially in the interiors of large continents in the rain shadows of mountain ranges. Seasonal changes in temperature are great.

Cold deserts are dominated by a few species of low-growing shrubs. The surface layers of the soil are recharged with moisture in winter, and plant growth is concentrated in spring. Because soils dry rapidly in spring, annual primary production is low.

Cold deserts are relatively poor in species of most taxonomic groups, but the plants of this biome tend to produce large numbers of seeds, supporting a rich fauna of seed-eating birds, ants, and rodents.

HOT DESERT

Anzo Borrego Desert, California

Simpson Desert, following rain, Australia

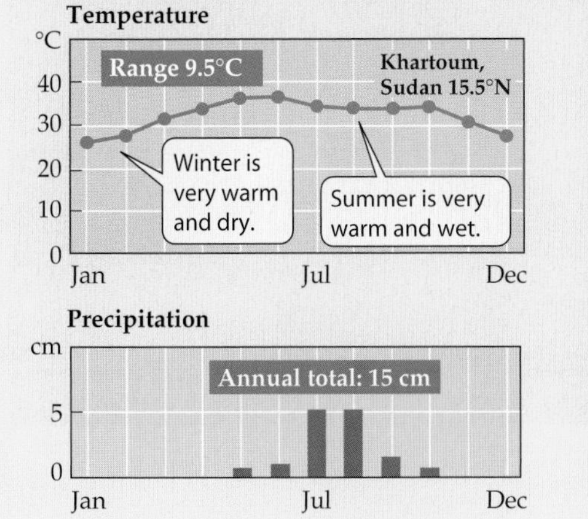

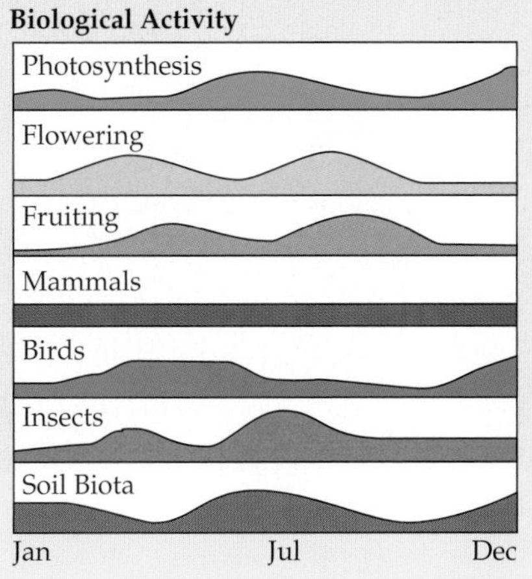

Community Composition

Dominant Plants
Many different growth forms

Species Richness
Plants: Fairly high; many annuals
Animals: Very rich in rodents; richest bee communities on Earth; very rich in reptiles and butterflies

Soil Biota
Poor in species

Hot deserts form around 30° latitude

The **hot desert** biome is found in two belts, centered around 30° north and 30° south latitudes, where air descends, warms, and picks up moisture. Hot deserts receive most of their rainfall in summer, but they also receive winter rains from storms that form over the mid-latitude oceans. The driest large regions, where summer and winter rains rarely penetrate, are in the center of Australia and the middle of the Sahara Desert of Africa.

Except in these driest regions, hot deserts have richer and structurally more diverse vegetation than cold deserts. Succulent plants that store large quantities of water in their expandable stems are conspicuous in some hot deserts. Annual plants germinate in abundance and grow when rain falls. Pollination and dispersal of fruits by animals are common. Rodents, termites, and ants are often remarkably abundant, and lizards and snakes typically are rich in species and abundant.

CHAPARRAL

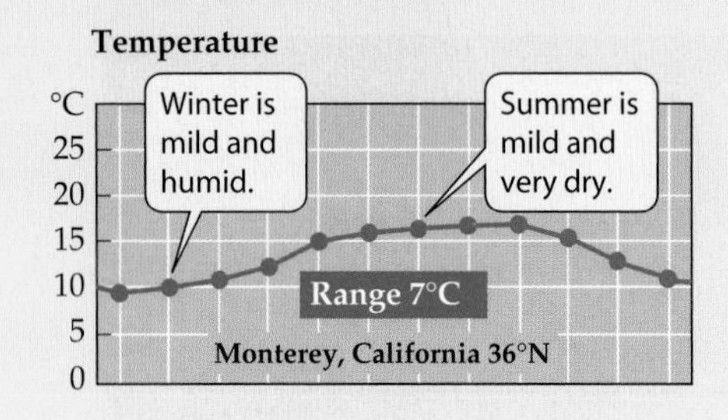

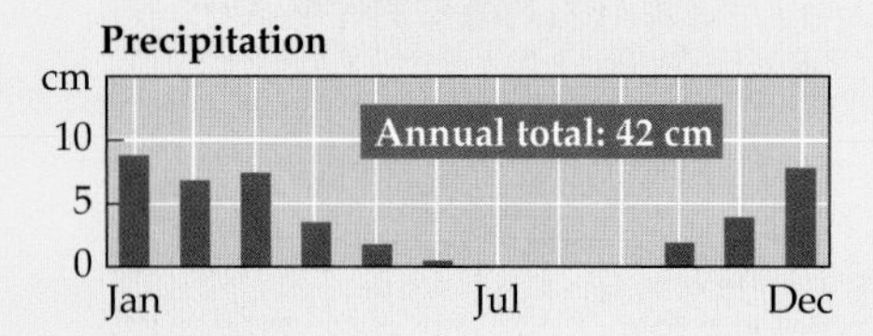

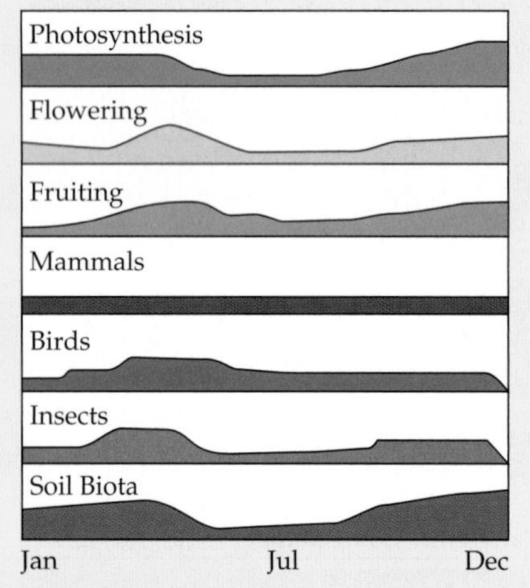

Community Composition

Dominant Plants

Low stature shrubs and herbaceous plants

Species Richness

Plants: Extremely high in South Africa and Australia

Animals: Rich in rodents and reptiles; very rich in insects, especially bees

Soil Biota

Moderately rich

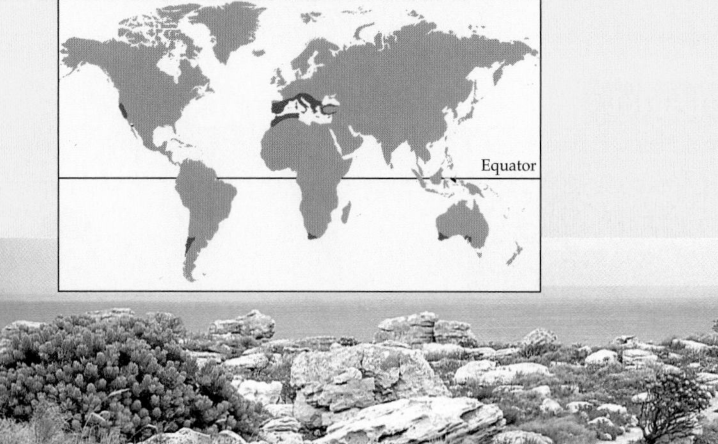

Fynbos vegetation, Cape of Good Hope, South Africa

Mendocino, California

The chaparral climate is dry and pleasant

The **chaparral** biome is found on the west sides of continents at moderate latitudes (around 30°), where cool ocean waters flow offshore. Winters in this biome are cool and wet; summers are warm and dry. Such climates are found in the Mediterranean region of Europe, coastal California, central Chile, extreme southern Africa, and southwestern Australia.

The dominant plants of chaparral vegetation are low-growing shrubs and trees with tough, evergreen leaves. The shrubs carry out most of their growth and photosynthesis in early spring, when insects are active and birds breed. Annual plants are abundant and produce copious seeds that fall onto the soil. This biome thus supports large populations of small rodents, most of which store seeds in underground burrows. Chaparral vegetation is naturally adapted to survive periodic fires. Many shrubs of Northern Hemisphere chaparral produce bird-dispersed fruits that ripen in the late fall, when large numbers of migrant birds arrive from the north.

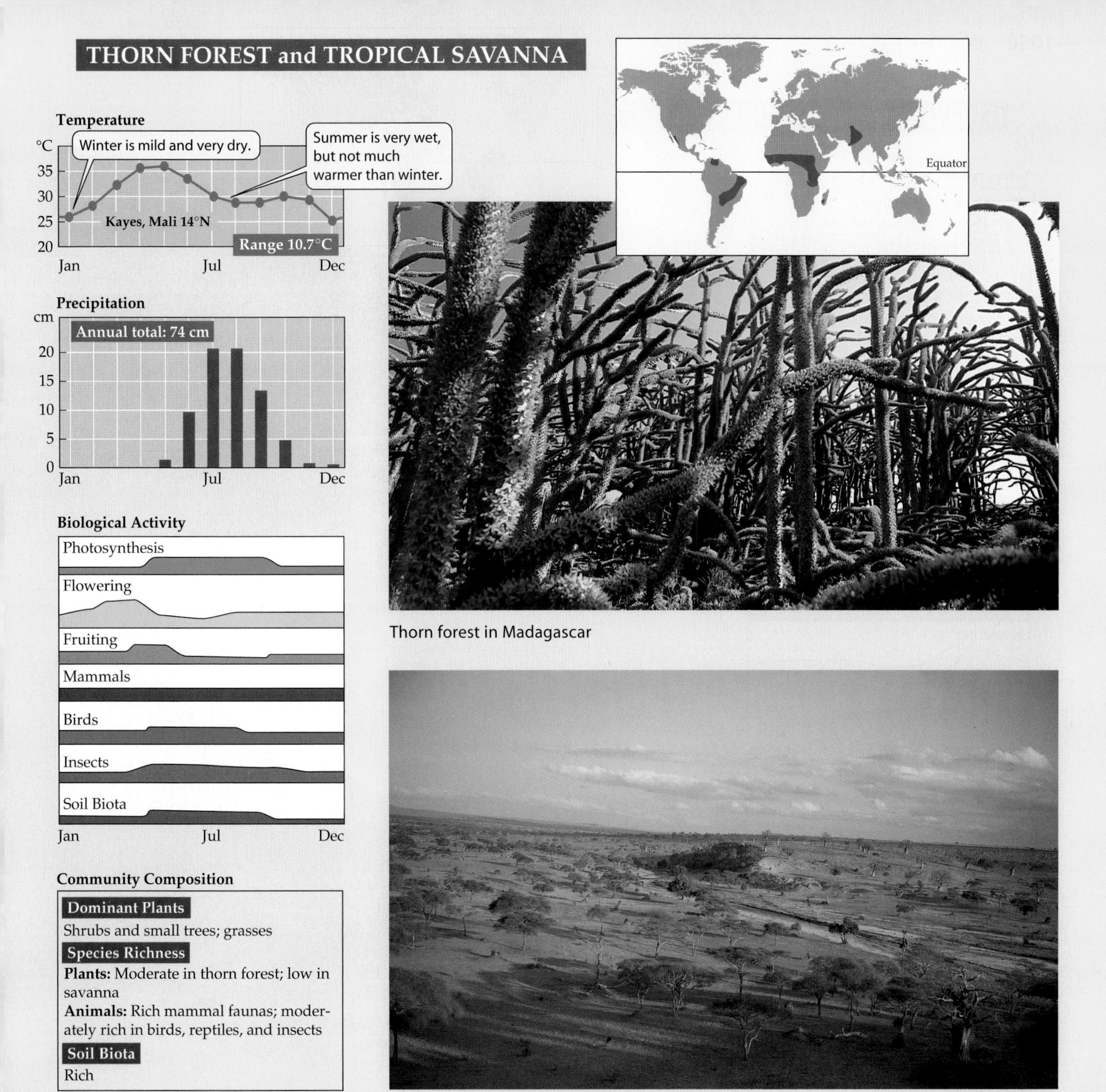

Thorn forest in Madagascar

Savanna in Tanzania

Thorn forests and savannas have similar climates

Thorn forests are found on the equatorial sides of hot deserts. The climate is semiarid; little or no rain falls during winter, but rainfall may be heavy during summer. Thorn forests contain many plants similar to those found in hot deserts. The dominant plants are spiny shrubs and small trees, many of which drop their leaves during the long dry winter. Members of the genus *Acacia* are common in thorn forests worldwide.

The dry tropical and subtropical regions of Africa, South America, and Australia have extensive areas of **savannas**—expanses of grasses and grasslike plants with scattered trees. The largest savannas are found in central and eastern Africa, where the biome supports huge numbers of grazing and browsing mammals and many large carnivores that prey on them. The grazers and browsers maintain the savannas. If savanna vegetation is not grazed, browsed, or burned, it typically reverts to dense thorn forest.

TROPICAL DECIDUOUS FOREST

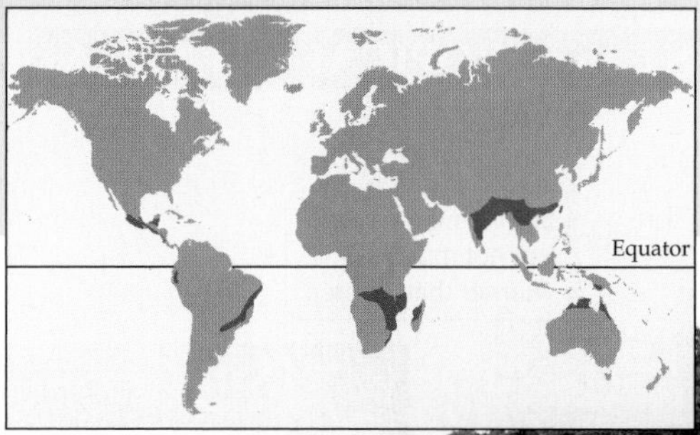
Equator

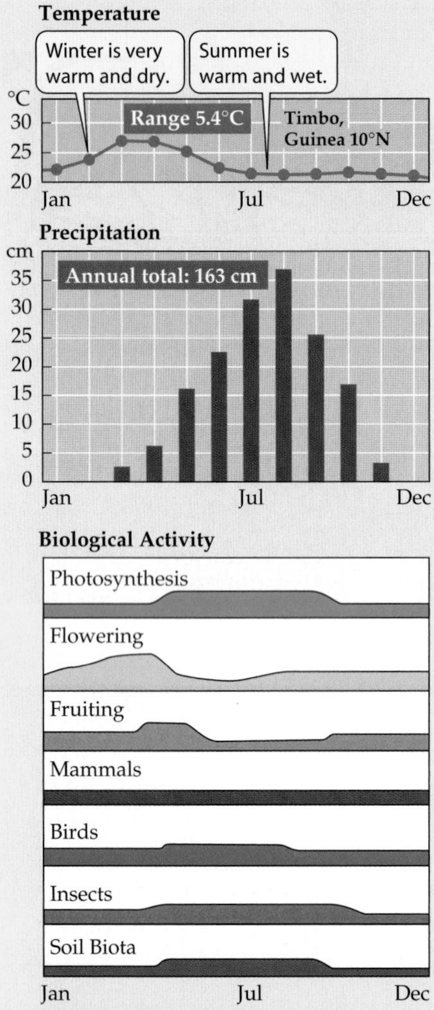
Temperature
Winter is very warm and dry.
Summer is warm and wet.
°C
30
25
20
Range 5.4°C
Timbo, Guinea 10°N
Jan
Jul
Dec
Precipitation
cm
35
30
25
20
15
10
5
0
Annual total: 163 cm
Jan
Jul
Dec
Biological Activity
Photosynthesis
Flowering
Fruiting
Mammals
Birds
Insects
Soil Biota
Jan
Jul
Dec

Community Composition

Dominant Plants
Deciduous trees

Species Richness
Plants: Moderately rich in tree species
Animals: Rich mammal, bird, reptile, and amphibian communities; rich in insects

Soil Biota
Rich, but poorly known

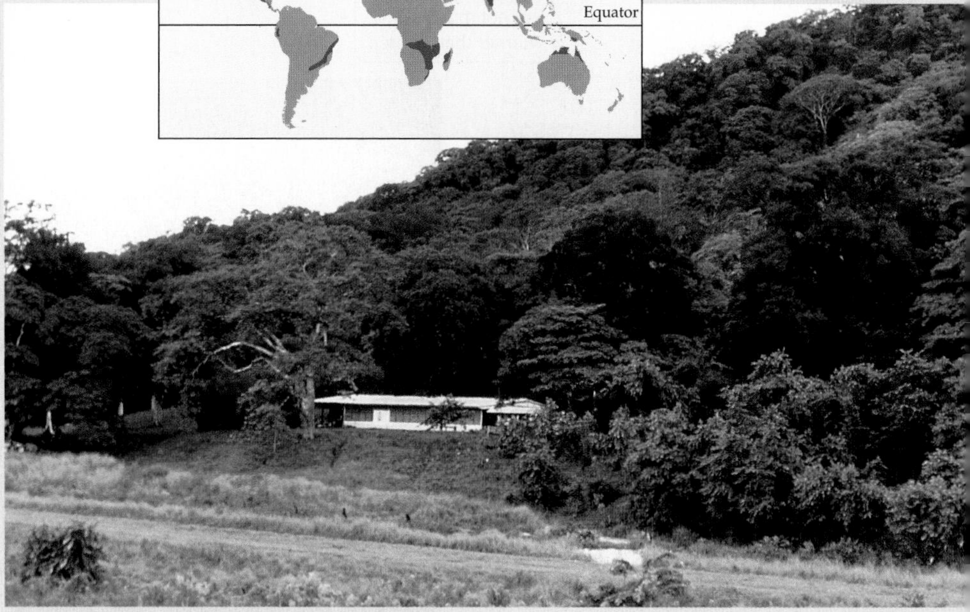

Palo Verde National Park, Costa Rica, in the rainy season…

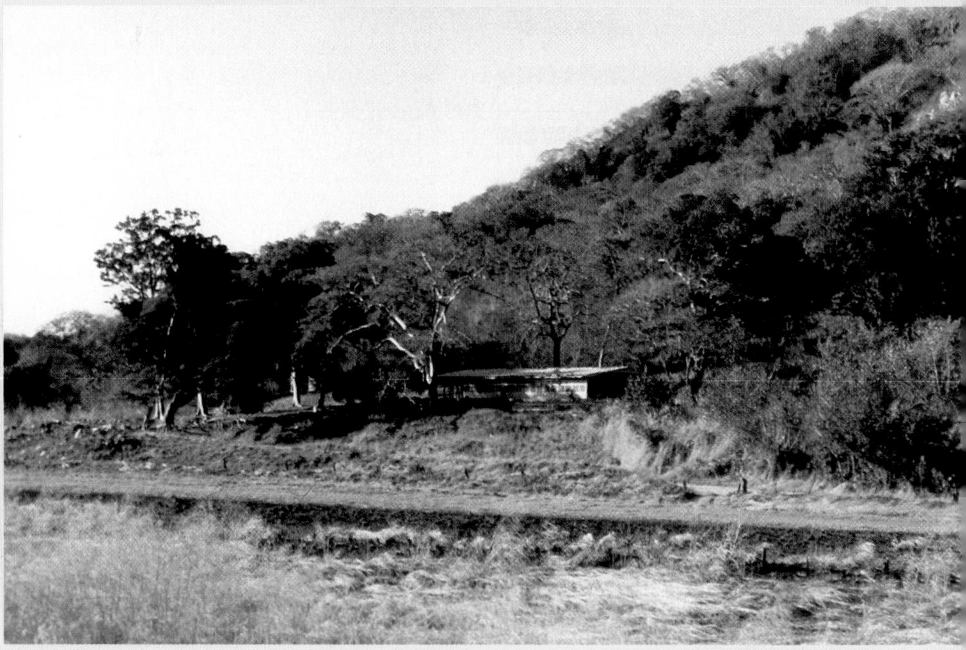

…and in the dry season

Tropical deciduous forests occur in hot lowlands

As the length of the rainy season increases toward the equator, **tropical deciduous forests** replace thorn forests. These forests have taller trees and fewer succulent plants than thorn forests, and they are much richer in plant and animal species. Most of the trees, except for those growing along rivers, lose their leaves during the long, hot dry season. Many of them flower while they are leafless, and most species are pollinated by animals. During the hot rainy season, biological activity is intense.

The soils of the tropical deciduous biome are some of the best soils in the tropics for agriculture, because they are less leached of nutrients than the soils of wetter areas. As a result, most tropical deciduous forests have been cleared for agriculture and cattle grazing. Restoration efforts are underway on several continents.

TROPICAL EVERGREEN FOREST

The exterior of lowland wet forest…

…and its interior, Cocha Cashu, Peru

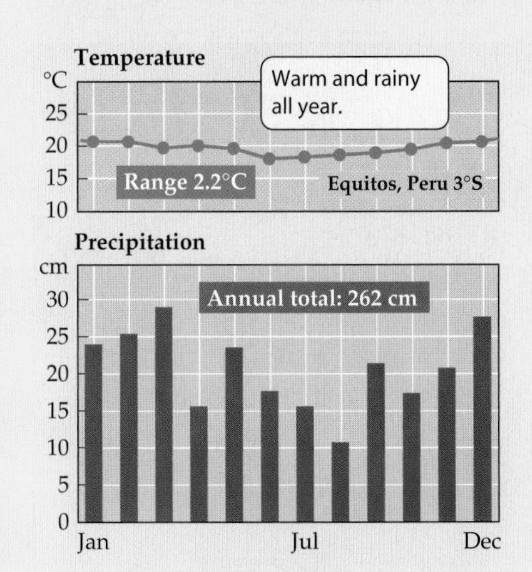

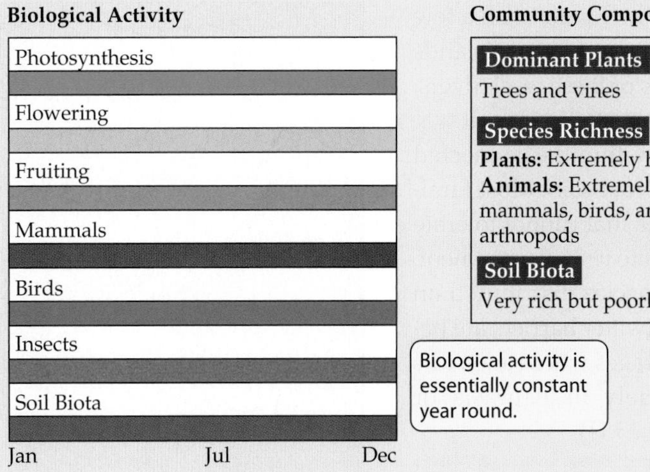

Biological activity is essentially constant year round.

Community Composition

Dominant Plants
Trees and vines

Species Richness
Plants: Extremely high
Animals: Extremely high in mammals, birds, amphibians, and arthropods

Soil Biota
Very rich but poorly known

Tropical evergreen forests are species-rich

Tropical evergreen forests are found in equatorial regions where total rainfall exceeds 250 cm annually and the dry season lasts no longer than 2 or 3 months. They are the richest of all biomes in number of species of both plants and animals, with up to 500 species of trees per km^2. Along with their immense species richness, tropical evergreen forests have the highest overall productivity of all ecological communities. However, most mineral nutrients are tied up in the vegetation. The soils usually cannot support agriculture without massive applications of fertilizers.

On the slopes of tropical mountains, trees are shorter than lowland tropical trees. Their leaves are smaller, and there are more *epiphytes* (plants that grow on other plants, deriving their nutrients and moisture from air and water rather than soil).

Aquatic Biogeography

Three-fourths of Earth's surface is covered by water, most of it in the oceans. Earth's oceans form one large, interconnected water mass with only partial barriers to dispersal. Fresh waters, in contrast, are divided into river basins and thousands of relatively isolated lakes. For organisms that cannot survive out of water, terrestrial habitats between bodies of water are barriers to dispersal. However, some aquatic species have flying adults that can disperse widely among water bodies. Others have windborne, desiccation-resistant spores and seeds. Still others are small enough to be transported by mud on the feet of birds.

Freshwater environments have little water but many species

Although only about 2.5 percent of Earth's water is found in ponds, lakes, and streams, about 10 percent of all aquatic species live in freshwater habitats. Many freshwater taxa that are capable of dispersing across terrestrial barriers are found over several continents. Prominent among freshwater taxa are the more than 25,000 species of insects that have at least one aquatic stage in their life cycle. Typically, eggs and larvae are aquatic; the adults have wings. Some of these insects, such as dragonflies, are powerful flyers, but mayflies and some other species are weak flyers, desiccate rapidly in air, and live no longer than a few days. As you would expect, oceanic islands have few, if any, species of these weak flyers.

Similarly, fishes unable to live in salt water can disperse only within the connected rivers and lakes of a river basin. Most families of freshwater fishes that cannot tolerate salt water are restricted to a single continent. Those families with species distributed on both sides of major saltwater barriers are believed to be ancient lineages whose ancestors were distributed widely in Laurasia or Gondwana (see Figure 25.11).

Water temperature defines marine biogeographic regions

As we saw in Figure 56.8, ocean water moves in great circular patterns—clockwise in the Northern Hemisphere and counterclockwise in the Southern Hemisphere. Even organisms with limited swimming abilities can move long distances simply by floating with ocean currents. Nevertheless, most marine organisms have restricted ranges. Why is this true?

The oceans may be connected, but water temperatures, salinities, and food supplies all may change spatially. Living successfully in different regions of the ocean requires different physiological tolerances and morphological attributes. Ocean temperatures, for example, can be barriers to dispersal because many marine organisms function well in only a relatively narrow range of temperatures. The main biogeographic divisions of the ocean coincide with regions where the surface water temperatures and salinities change relatively abruptly as a result of horizontal and vertical ocean currents (Figure 56.10). These temperature changes, in combination with seasonal changes in the amount of daylight, determine the seasons of maximum primary production. Species of marine algae photosynthesize either in summer or in winter, but not during both seasons.

Deep ocean waters prevent the dispersal of marine organisms that live only in shallow water. The distance that eggs and larvae of many marine organisms can be carried by ocean currents is determined in large part by the time it takes for larvae to metamorphose into sedentary adults. Relatively few species have eggs and larvae that survive long enough to dis-

56.10 Oceanic Biogeographic Regions are Determined by Ocean Currents The arrows represent ocean currents. Different biogeographic regions, in which photosynthesis is maximized at different seasons, are indicated by different colors.

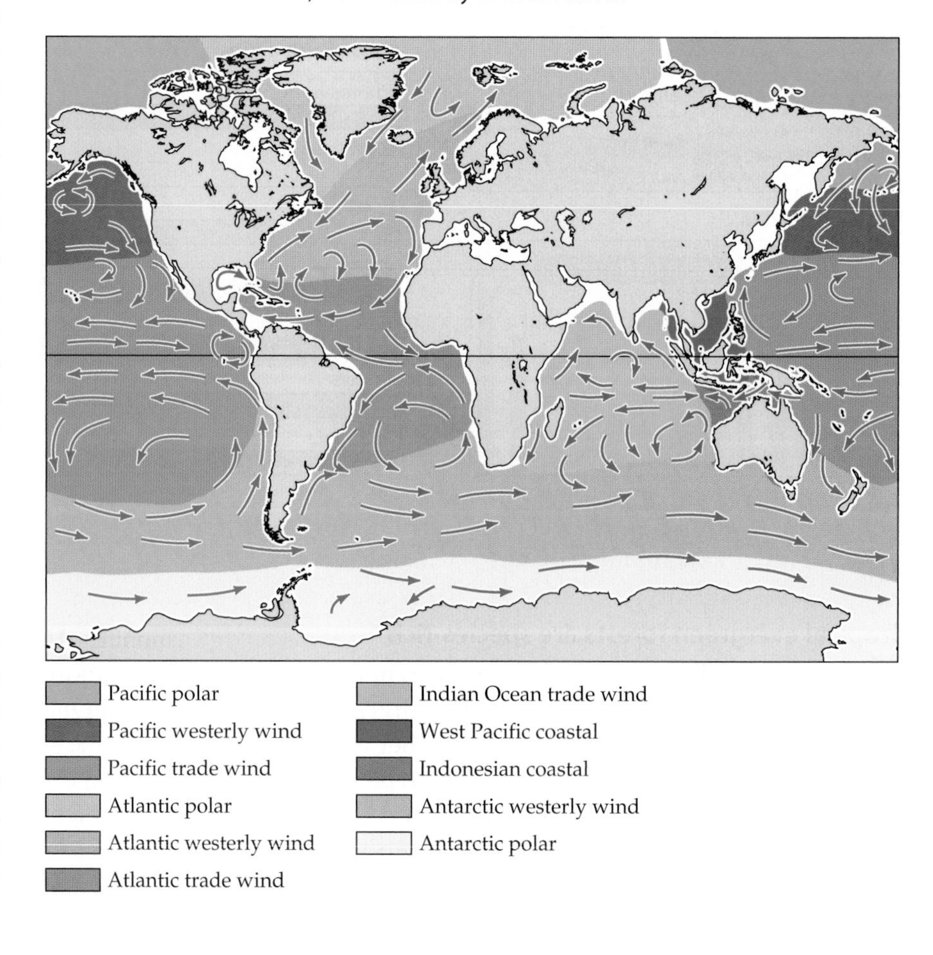

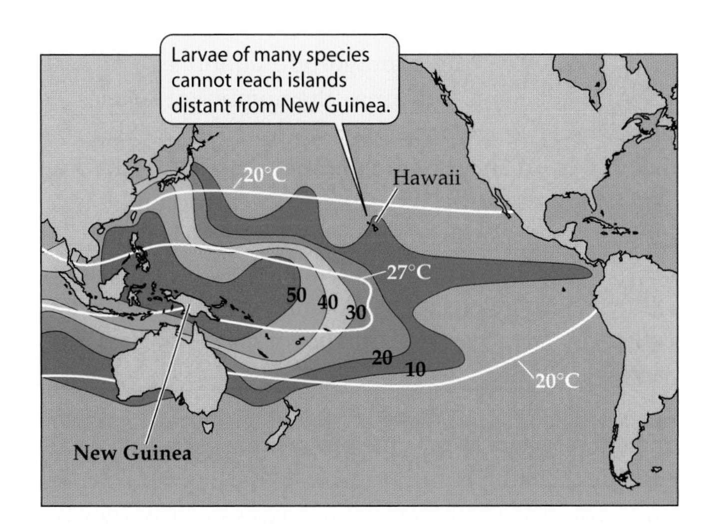

56.11 Generic Richness of Reef-Building Corals Declines with Distance from Indonesia The colored zones represent areas with equal numbers of coral genera. The 20° and 27° mean annual temperature isotherms are also shown.

Number of species (log scale)
5.5
5.0
4.5
4.0
3.5
3.0
2.5
Earth
Africa (sub-Sahara)
Canada and U.S.
Europe
Madagascar
Australia
Increasing area in Britain
New Zealand
10^0 10^1 10^2 10^3 10^4 10^5 10^6 10^7 10^8
Area (square miles)

56.12 Species Richness Increases with Area Sampled Plotted here are the number of species of vascular plants in eight increasingly large samples in Britain; in several larger areas, and, finally, on the entire Earth. Recall that in a logarithmic scale, each increment of measurement is 10 times larger than the preceding one.

perse across wide barriers of deep water. As a result, the richness of shallow-water species in the intertidal and subtidal zones of isolated islands in the Pacific Ocean decreases with distance from the larger islands of Indonesia (Figure 56.11).

Regional Patterns of Species Richness

As we saw in Chapter 55, local species richness is often positively correlated with both productivity and disturbance level. Other patterns of species richness appear at larger spatial scales. As we increase the area we are sampling, the number of species we record increases slowly (Figure 56.12). However, if our sampling area crosses a biogeographic boundary, the rate at which we add new species suddenly increases. At that point, we have added to our sample another biogeographic region with a different evolutionary history and a different biota.

One of the first geographic patterns of species richness observed by biologists was that more species are found in low-latitude than in high-latitude regions. Figure 56.13 shows this latitudinal gradient in species richness for mammals in North and Central America. Similar patterns exist for birds, frogs, and trees and for many marine taxa. The figure also shows that more species are found in mountainous regions than in relatively flat areas because more vegetation types and climates exist within these topographically complex areas.

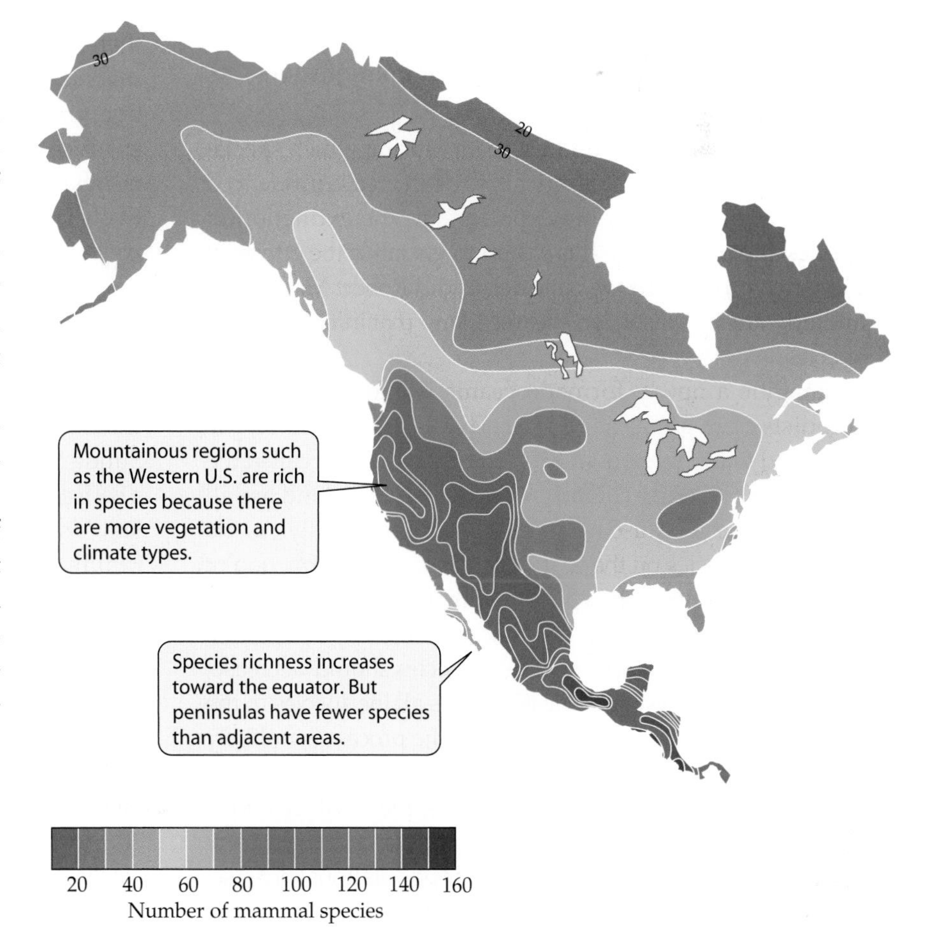

56.13 The Latitudinal Gradient of Species Richness of North American Mammals The colored zones represent regions with equal numbers of species. An increase in species richness toward the equator typifies many other taxa, such as birds, amphibians, and trees.

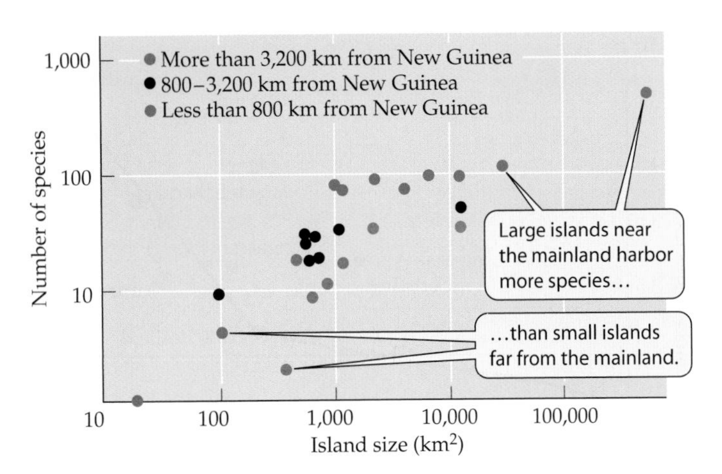

56.14 Small, Distant Islands Have Fewer Bird Species The dots show the numbers of land and freshwater bird species on islands of different sizes in the Moluccas, Melanesia, Micronesia, and Polynesia. These islands have been divided into three groups according to their distance from the "mainland," which in this case is New Guinea.

Species richness on islands and peninsulas is always less than that in an equivalent area on the nearest mainland. On islands, species richness is positively correlated with island size, but inversely correlated with distance from the mainland (Figure 56.14). An influential model relates this pattern to the island's history of immigrations and extinctions.

Species richness is related to rates of immigration and extinction

Over periods of a few hundred years (during which speciation is unlikely), the species richness of an area is influenced by the immigration of new species and the extinction of species already present. It is easiest to visualize the effects of these two processes if we consider, as did Robert MacArthur and Edward O. Wilson, an oceanic island that initially has no species.

Imagine a newly formed oceanic island that receives colonists from a mainland area. The list of species on the mainland that might possibly colonize the island is called the *species pool*. The first colonists to arrive on the island are all "new" species because no species live there initially. As the number of species on the island increases, a larger fraction of colonizing individuals will be members of species already present. Therefore, even if the same number of species arrive as before, the rate of arrival of new species should decrease, until it reaches zero when the island has all the species in the species pool. As we will see, however, the process is unlikely to proceed that far.

Now consider extinction rates. At first there will be only a few species on the island, and their populations may grow large. As more species arrive and their populations increase, the resources of the island will be divided among more species. Therefore, the average population size of each species will become smaller as the number of species increases. The smaller a population, the more likely it is to become extinct. In addition, the number of species that can possibly become extinct increases as species accumulate on the island. Furthermore, new arrivals on the island may include pathogens and predators that increase the probability of extinction for other species. For all these reasons, the rate of extinction increases as the number of species on the island increases.

Because the rate of arrival of new species decreases and the extinction rate increases as the number of species increases, eventually the number of species on the island should reach an equilibrium at which the rates of arrival and extinction are equal (Figure 56.15*a*). If there are more species than the equilibrium number, extinctions should exceed arrivals, and species richness should decline. If there are fewer species than the equilibrium number, arrivals should exceed extinctions, and species richness should increase. The equilibrium is dynamic because if either rate fluctuates, as they generally do, the equilibrium number of species shifts up and down.

MacArthur and Wilson's model can also be used to predict how species richness should differ among islands of different sizes and different distances from the mainland. We expect extinction rates to be higher on small islands than on large islands because species' populations are, on average, smaller there. Similarly, we expect fewer immigrants to reach islands that are more distant from the mainland. Figure 56.15*b* gives hypothetical relative species richnesses for islands of different sizes and distances from the mainland. As you can see, the number of species should be highest for islands that are relatively large and relatively close to the mainland.

The MacArthur-Wilson model has been tested

Major disturbances, which serve as "natural experiments," sometimes permit colonization and extinction rates to be estimated directly. In August 1883, Krakatau, an island in the Sunda Strait between Sumatra and Java, was devastated by a series of volcanic eruptions that destroyed all life on the island's surface. After the lava cooled, Krakatau was colonized rapidly by plants and animals from Sumatra to the east and Java to the west. By 1933, the island was again covered with a tropical evergreen forest, and 271 species of plants and 27 species of resident land birds were found there.

During the 1920s, when a forest canopy was developing, there were high rates of colonization by both birds and plants (Table 56.1). Birds probably brought the seeds of many plants because, between 1908 and 1934, both the percentage (from 20% to 25%) and the absolute number (from 21 to 54) of plant species with bird-dispersed seeds increased. Today the numbers of species of plants and birds are not increasing as fast

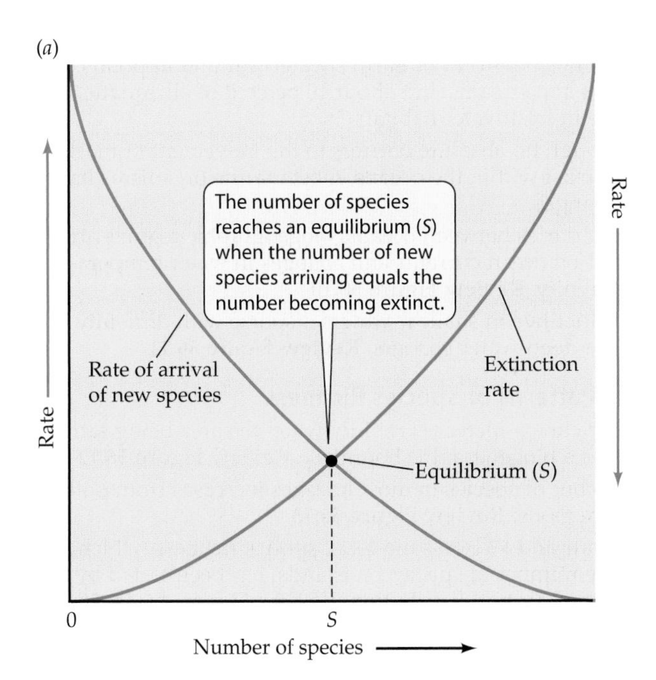

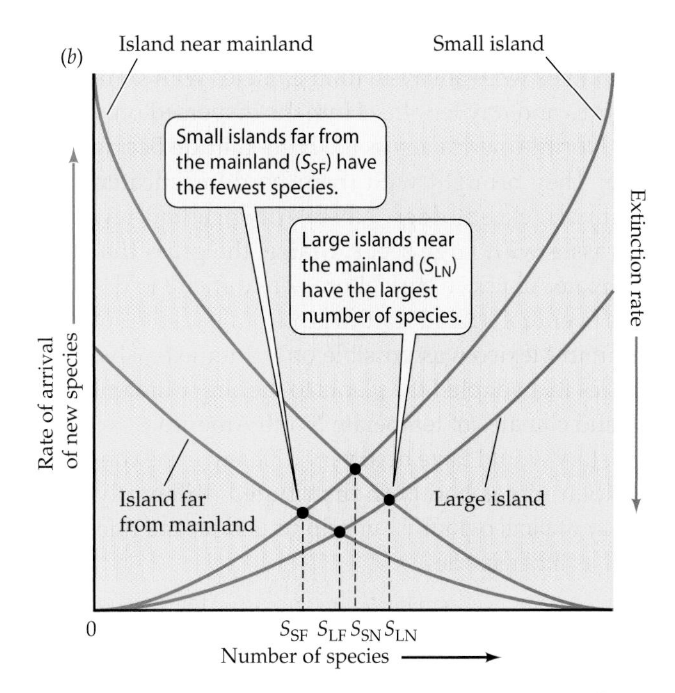

56.15 MacArthur and Wilson's Model of Species Richness on Islands (*a*) The rate of arrival of new species and the rate of extinction of species already present determine the equilibrium number of species on an island. (*b*) These rates are affected by the size of the island and its distance from the mainland.

as they did during the 1920s, but colonizations and extinctions continue, as predicted by the model.

Biogeography and Human History

The distributions of land masses and species on Earth have had a strong influence on human history. Humans first evolved in Africa, but eventually dispersed throughout the world. In recent times, human cultures from Eurasia came to dominate other cultures. Biogeography gave these cultures a number of advantages.

56.1 *Number of Species of Resident Land Birds on Krakatau*

PERIOD	NUMBER OF SPECIES	EXTINCTIONS	COLONIZATIONS
1908	13		
1908–1919		2	17
1919–1921	28		
1921–1933		3	4
1933–1934	29		
1934–1951		3	7
1951	33		
1952–1984		4	7
1984–1996	36		

Eurasia happened to have a large number of species of plants and animals that were suitable for domestication. Eurasia was home to 39 species of large-seeded grasses, many more than were found in Africa or the Americas. It also had 72 species of large mammals, compared with 51 in sub-Saharan Africa and 24 in the Americas. Thirteen of these species, including pigs, horses, cattle, sheep, goats, and camels, were domesticated in Eurasia. None were domesticated in Africa, and only one, the llama, in the Americas.

To be amenable to domestication, large mammals had to have three important social characteristics: They had to live in herds, have well-developed male dominance hierarchies, and not defend territories. These traits enabled humans to tame the animals, exert behavioral dominance over them, and keep them in herds. All of the large mammals of Africa lacked one or more of these traits.

Besides providing people with food, the domestication of large mammals had other important influences on human history. Many human diseases, such as smallpox and measles, were acquired from domesticated mammals. Eurasian people acquired immunity to these diseases, but people on other continents did not. Thus, when Europeans colonized the New World, they brought with them diseases that devastated the indigenous people. Those unfortunate people transmitted no fatal diseases to the Europeans. In addition, the Europeans had horses, the only domesticated mammals capable of carrying a person at high speeds. Throughout human history, cultures with horses have defeated and dominated cultures without them.

In Eurasia, most mountain ranges are oriented in an east-west direction. Therefore, dispersal of people and their domesticated plants and animals was relatively easy, and dispersing individuals were always within climates with similar temperatures and day lengths. Humans dispersed only recently into North America across the high-latitude Bering Land Bridge. They brought with them no domesticated plants or animals, except dogs. North America had few species of grasses with large seeds. Maize, the grass that came to dominate American agriculture, was difficult to domesticate. Its eventual spread northward from its center of domestication in Mexico was possible only after extensive genetic changes that adapted the plants to the very different day lengths and climates of temperate North America.

Human history would have been very different if the continents and their biotas had been distributed differently. Thus, the study of biogeography can help us understand ourselves as well as other species.

Chapter Summary

▶ Biogeography is the science that attempts to describe and explain patterns in the distribution of life on Earth.

Earth's Biogeographic Regions

▶ If a species occupies a particular area, either it evolved there, or it evolved elsewhere and dispersed to that area.

▶ If a species is not found in a particular area, either it evolved elsewhere and never dispersed to that area, or it was once present in that area but no longer lives there.

▶ Biogeographers divide Earth into several major biogeographic regions. **Review Figure 56.1.** See Web/CD Activity 56.1

▶ Remote islands contain many endemic species.

History and Biogeography

▶ Continental drift has influenced the distributions of organisms throughout Earth's history.

▶ Biogeographers often analyze species distributions by converting phylogenies into area phylogenies. **Review Figure 56.3**

▶ Distribution patterns can result from vicariant events or from dispersal.

▶ The principle of parsimony is used to explain distribution patterns. **Review Figure 56.4**

Ecology and Biogeography

▶ Global atmospheric circulation is driven by solar energy input. **Review Figure 56.6**

▶ Wind circulation patterns influence the amount and seasonal nature of rainfall. **Review Figures 56.5, 56.7.** See Web/CD Tutorial 56.1

▶ Global oceanic circulation is driven by winds. **Review Figure 56.8**

Terrestrial Biomes

▶ Terrestrial biomes are major ecosystem types that differ from one another in the structure of their dominant vegetation.

▶ The distribution of biomes on Earth is strongly influenced by annual patterns of temperature and rainfall. **Review Figure 56.9**

▶ The major terrestrial biomes are tundra, boreal forest, temperate deciduous forest, temperate grassland, cold desert, hot desert, chaparral, thorn forest and savanna, tropical deciduous forest, and tropical evergreen forest. **Review Pages 1076–1085.** See Web/CD Tutorial 56.2

Aquatic Biogeography

▶ Only about 2.5 percent of Earth's water is found in ponds, lakes, rivers and streams, but about 10 percent of all aquatic species live in freshwater habitats.

▶ Even though no absolute barriers to the movement of marine organisms exist within the oceans, most marine organisms have restricted ranges.

▶ The boundaries between oceanic biogeographic regions are determined by ocean currents and changes in water temperature and salinity. **Review Figure 56.10**

▶ Species that live in shallow waters disperse with difficulty across wide deep-water barriers. **Review Figure 56.11**

Regional Patterns of Species Richness

▶ Species richness increases rapidly when the area being sampled crosses a biogeographic boundary. **Review Figure 56.12**

▶ The number of species in most lineages increases from polar to tropical regions. **Review Figure 56.13**

▶ MacArthur and Wilson's model of species richness, which predicts the number of species on islands, has been tested by examining patterns of distribution. **Review Figures 56.14, 56.15, Table 56.1**

See Web/CD Tutorial 56.3

Biogeography and Human History

▶ The distributions of plants, animals, and continents have exerted powerful influences on human history.

Self-Quiz

1. Biogeography as a science began when
   a. eighteenth-century travelers first noted intercontinental differences in the distributions of organisms.
   b. Europeans went to the Middle East during the Crusades.
   c. phylogenetic methods were developed.
   d. the fact of continental drift was accepted.
   e. Charles Darwin proposed the theory of natural selection.
2. Vicariant events
   a. are infrequent in nature.
   b. were common in the past but are rare today.
   c. separate species ranges in the absence of dispersal.
   d. were rare in the past but are common today.
   e. caused most of today's disjunct distributions.
3. Marine biogeographic regions exist even though the oceans are all connected because
   a. primary production is low in the oceans.
   b. ocean currents keeps organisms close to where they were born.
   c. most families and higher taxa of marine organisms evolved before the oceans were separated by continental drift.
   d. water temperatures and salinities often change abruptly where ocean currents meet.
   e. oceanic circulation is too slow to carry marine organisms from one ocean to another.
4. A parsimonious interpretation of a distribution pattern is one that
   a. requires the smallest number of undocumented vicariant events.
   b. requires the smallest number of undocumented dispersal events.
   c. requires the smallest total number of undocumented vicariant plus dispersal events.
   d. accords with the phylogeny of a lineage.
   e. accounts for centers of endemism.

5. The only major biogeographic region that today is completely isolated by water from other regions is
   a. Greenland.
   b. Africa.
   c. South America.
   d. Australasia.
   e. North America.
6. In MacArthur and Wilson's model, equilibrium species richness is reached when
   a. immigration rates of new species and extinction rates of species are equal.
   b. immigration rates of all species and extinction rates of species are equal.
   c. the rate of vicariant events equals the rate of dispersal.
   d. the rate of island formation equals the rate of island loss.
   e. No equilibrium number of species exists in that model.
7. Chaparral vegetation is dominated by
   a. deciduous trees.
   b. evergreen trees.
   c. deciduous shrubs.
   d. evergreen shrubs.
   e. grasses.
8. Which of the following is *not* true of tropical evergreen forests?
   a. They have large numbers of species of trees.
   b. Most plant species are animal-pollinated.
   c. Most plant species have animal-dispersed fruits.
   d. Biological energy flow is very high.
   e. High productivity depends on a rich supply of soil nutrients.
9. Cold deserts
   a. are dominated by a few species of low-growing shrubs.
   b. are dominated by a rich flora of low-growing shrubs.
   c. have few species of woody plants, but of varied growth forms.
   d. have many species of woody plants of varied growth forms.
   e. are dominated by a few species of tall shrubs.
10. Biogeography exerted a strong influence on human history because
   a. humans first evolved in Africa.
   b. Eurasia had more species of plants and animals that were easily domesticated.
   c. Old World mountain ranges are oriented in an east-west direction.
   d. horses were found only in Eurasia.
   e. All of the above

For Discussion

1. Horses evolved in North America, but subsequently became extinct there. They survived to modern times only in Africa and Asia. In the absence of a fossil record, we would probably infer that horses originated in the Old World. Today, the Hawaiian Islands have by far the greatest number of species of fruit flies (*Drosophila*). Would you conclude that the genus *Drosophila* originally evolved in Hawaii and spread to other regions? Under what circumstances do you think it is safe to conclude that a group of organisms evolved close to where the greatest number of species live today?
2. Processes in nature do not always conform to the parsimony principle. Why, then, do biogeographers often use the parsimony principle to infer geographic histories of species and lineages?
3. A well-known legend states that Saint Patrick drove the snakes out of Ireland. Give some alternative explanations, based on sound biogeographic principles, for the absence of indigenous snakes in that country.
4. Most of the world's flightless birds are either nocturnal and secretive (such as the kiwi of New Zealand) or large, swift, and powerful (such as the ostrich of Africa). The exceptions are found primarily on islands. Many of these island species have become extinct with the arrival of humans and their domestic animals. What special conditions on islands might permit the survival of flightless birds? Why has human colonization so often resulted in the extinction of such birds? The power of flight has been lost secondarily in representatives of many groups of birds and insects; what are some possible evolutionary advantages of flightlessness that might offset its obvious disadvantages?
5. MacArthur and Wilson's model of species richness on islands incorporates almost nothing about the biology of the species. What traits of species should be incorporated into more realistic models of rates of colonization and extinction of species on islands?
6. A legislator introduces a controversial bill into the U.S. Congress that would ban all introductions of exotic species to the Hawaiian Islands. Would you vote in favor of this bill if you were in Congress? Why or why not?

57 Conservation Biology

In 1998, botanists of the Massachusetts Natural Heritage and Endangered Species Program published a booklet called *A Guide to Invasive Plants in Massachusetts*. The purpose of the booklet was to warn citizens about certain non-native plants that were becoming established and interfering with the state's natural ecosystems. But many nursery owners were upset; some of these invasive exotic plants were big money-makers for the horticultural industry. Horticulturalists lobbied the state government and succeeded in getting the booklet withdrawn from publication.

Fortunately, many people in the horticultural world now recognize that even though most introduced plants do not become invasive, some plant species have become serious pests on several continents. Indeed, colonization of new areas by introduced plants, animals, and microorganisms that become abundant in their new ranges is second only to habitat loss as a threat to Earth's biodiversity.

Humans have caused extinctions for thousands of years. When people first crossed the Bering Land Bridge and arrived in North America about 20,000 years ago, they encountered a rich fauna of large mammals. Most of those species were exterminated—probably by overhunting—within a few thousand years. A similar extermination of large animals followed the human colonization of Australia, about 40,000 years ago. At that time, Australia had 13 genera of marsupials larger than 50 kg, a genus of gigantic lizards, and a genus of heavy, flightless birds. All the species in 13 of those 15 genera had become extinct by 18,000 years ago. When Polynesian people settled in Hawaii about 2,000 years ago, they exterminated, probably by overhunting, at least 39 species of endemic land birds. Among them were 7 species of geese, 2 species of flightless ibises, a sea eagle, a small hawk, 7 flightless rails, 3 species of owls, 2 large crows, a honeyeater, and at least 15 species of finches.

The pace of human-caused extinction of species is accelerating rapidly. Most of the human activities that are currently causing extinctions are not new—but today there are many more humans living on Earth, doing more things that endanger species. Current extinction rates have raised serious concerns about the future of biological diversity on Earth. These concerns led to the rapid development during the 1980s of the applied discipline of **conservation biology**: the scientific study of how to preserve the diversity of life. Conservation biologists study the factors that threaten species with extinction, and they develop methods to help preserve genes, species, communities, and ecosystems. The

A Successful Invasion Introduced into the northeastern United States from Europe during the 1800s, *Lythrum salicaria*—purple loosestrife—was sold as an ornamental plant and for medicinal uses. Loosestrife establishes itself readily in natural wetlands, such as this riverbank in Massachusetts, where it outcompetes native species and changes the habitat of waterfowl and other animals.

science of conservation biology draws heavily on concepts and knowledge from population genetics, evolution, ecology, biogeography, wildlife management, economics, and sociology. In turn, the needs of conservation are stimulating new research in those fields.

In this chapter, we will see how conservation biologists estimate rates of species extinction and determine the causes of extinctions. We will learn how science is used to reduce extinction rates and help populations recover. But why should we care about species extinctions?

Why Care about Species Extinctions?

Extinction is forever. If we purposely or inadvertently exterminate a species, we have irreversibly destroyed a resource of unknown value. But people value biodiversity for many reasons:

- Humans depend on other species for food, fiber, and medicine. More than half the medical prescriptions written in the United States contain a natural plant or animal product.
- Humans derive enormous aesthetic pleasure from interacting with other organisms. Many people would consider a world with far fewer species to be a less desirable place in which to live.
- Living in ways that cause the extinction of other species raises serious ethical issues. These issues are receiving increased attention from philosophers, ethicists, and religious leaders.
- Extinctions deprive us of opportunities to study and understand ecological relationships among organisms. The more species are lost, the more difficult it will be to understand the structure and functioning of ecological communities and ecosystems.
- Species are necessary for the functioning of the ecosystems of which they are a part and the many benefits those ecosystems provide to humanity.

Among the benefits provided by ecosystems are generation and maintenance of fertile soils, prevention of soil erosion, detoxification and recycling of waste products, regulation of the hydrological cycle and the composition of the atmosphere, control of agricultural pests, and pollination of plants.

The benefits provided to humans by functioning ecosystems are very hard to calculate, but their value can be estimated. The benefits provided by the native vegetation of the Western Cape Province, South Africa, were estimated by a group of economists, ecologists, and land managers. The native vegetation of the highlands of this area is a species-rich community of shrubs, known as *fynbos* (pronounced "fainbos"). These shrubs can survive regular summer droughts, nutrient-poor soils, and the fires that periodically sweep through the highlands (Figure 57.1*a*). The fynbos-clad highlands provide about two-thirds of the Western Cape's water requirements. In addition, some species of the endemic flora

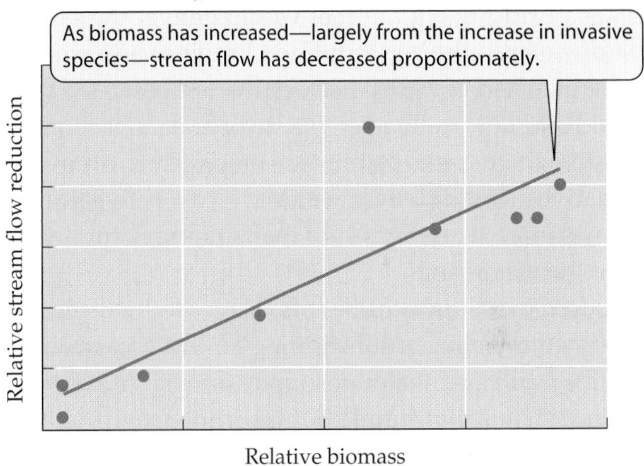

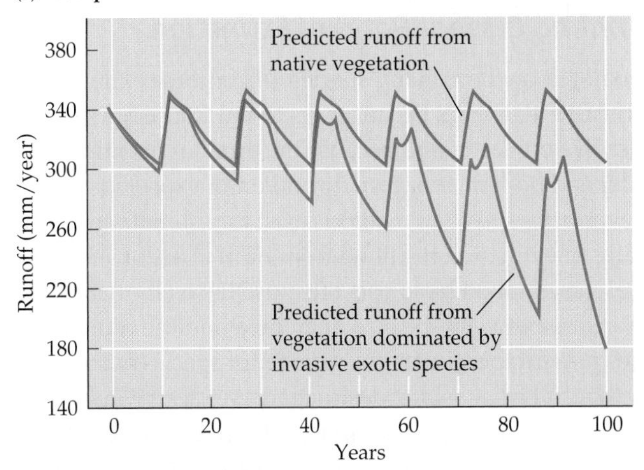

57.1 Invasive Species Disrupt Ecosystem Function (*a*) The unique fynbos ecosystems of South Africa provide much of the area's water. (*b*) Stream flow from fynbos watersheds is inversely proportional to plant biomass. (*c*) A computer simulation of stream flows from watersheds that have and have not been invaded by exotic trees.

are harvested for cut and dried flowers and thatching grass. The combined value of these harvests in 1993 was about $19 million. Some of the income from tourism in the region comes from people who want to see the fynbos. About 400,000 people visit the Cape of Good Hope Nature Reserve each year, primarily to see the many endemic plants.

During recent decades, a number of plants introduced into South Africa from other continents have invaded the fynbos. Because they are taller and grow faster than the native plants, these exotics increase the intensity and severity of fires. By transpiring larger quantities of water, they decrease stream flows to less than half the amount flowing from mountains covered with native plants, reducing the water supply (Figure 57.1*b*). Removing the exotic plants by felling and digging out invasive trees and shrubs and managing fire is estimated to cost between $140 and $830 per hectare, depending on the densities of invasive plants. Annual follow-up operations cost about $8 per hectare.

When natural ecosystems are lost, the services they provided must be replaced, often at a much higher cost. A sewage purification plant that would deliver the same volume of water to the Western Cape Province as a well-managed watershed of 10,000 hectares would cost $135 million to build and $2.6 million per year to operate. Desalination of seawater would cost four times as much. Thus, the available alternatives would deliver water at a cost between 1.8 and 6.7 times more than the cost of maintaining natural vegetation in the watershed.

Modern industrial societies often favor technologically sophisticated methods of substituting for lost ecosystem services. The study of water resources in the Western Cape Province shows that simple but labor-intensive methods—cutting and burning—may, in some cases, be cheaper.

Estimating Current Rates of Extinction

We do not know how many species will become extinct during the next 100 years because we do not know how many species live on Earth, and because the number of extinctions will depend both on what we do and on unexpected events.

Nevertheless, several methods exist for estimating probable rates of extinction resulting from human actions. For example, conservation biologists often use the well-established relationship between the size of an area and the number of species present to estimate the number of species extinctions likely to result from habitat destruction. We saw in Chapter 56 that the number of species on an island increases with the size of the island. This **species–area relationship** can be applied to habitat patches on the mainland as well. Biologists have measured the rate at which species richness tends to decrease with decreasing patch size. Their findings suggest that, on average, a 90 percent loss of habitat will result in the loss of half of the species living in that habitat. The current rate of loss of tropical evergreen forests—Earth's richest biome—is about 2 percent of the remaining forest each year. If this rate of loss continues, about 1 million species that live in tropical evergreen forests will become extinct during the coming century.

To estimate the risk that a population will become extinct, conservation biologists develop models that incorporate information about a population's size, its genetic variation, and the morphology, physiology, and behavior of its members. Species in imminent danger of extinction over all or a significant part of their range are labeled *endangered species*. *Threatened species* are those that are likely to become endangered in the near future. Although rarity in and of itself is not always a cause for concern, species whose populations are shrinking rapidly usually are at risk. Species with only a few individuals confined to a small range are likely to be eliminated by local disturbances such as fires, unusual weather, disease, and predators.

In an example of such a population study, an ecologist constructed a quantitative model of the dynamics of the grizzly bear population in Yellowstone National Park, using detailed data collected over a 12-year period. The model kept track of individual bears and incorporated the effects of chance events, such as fires. The output of the model suggested that for the grizzly bear population to have a 95 percent chance of persisting for a century, there must be enough habitat to support 70–90 bears. To achieve a higher probability of survival, or the same probability of survival for 200 years, more bear habitat would be needed.

Preserving Biodiversity

The human activities that threaten species include habitat destruction, the introduction of invasive species, overexploitation, disease, alteration of disturbance patterns, and climate change. Conservation biologists determine how these activities are affecting species and use that information to devise actions to preserve species that are endangered or threatened.

Habitat loss is studied by observation and experimentation

Habitat loss is the most important cause of endangerment of species in the United States, especially species that live in fresh waters (Figure 57.2). As habitats are progressively destroyed by human activities, the remaining habitat patches become smaller and more isolated. In other words, the habitat becomes **fragmented**. Small habitat patches are qualitatively different from larger patches of the same habitat in ways that affect the survival of species. Small patches cannot maintain populations of species that require large areas, and they can support only small populations of many of the species that can survive in them.

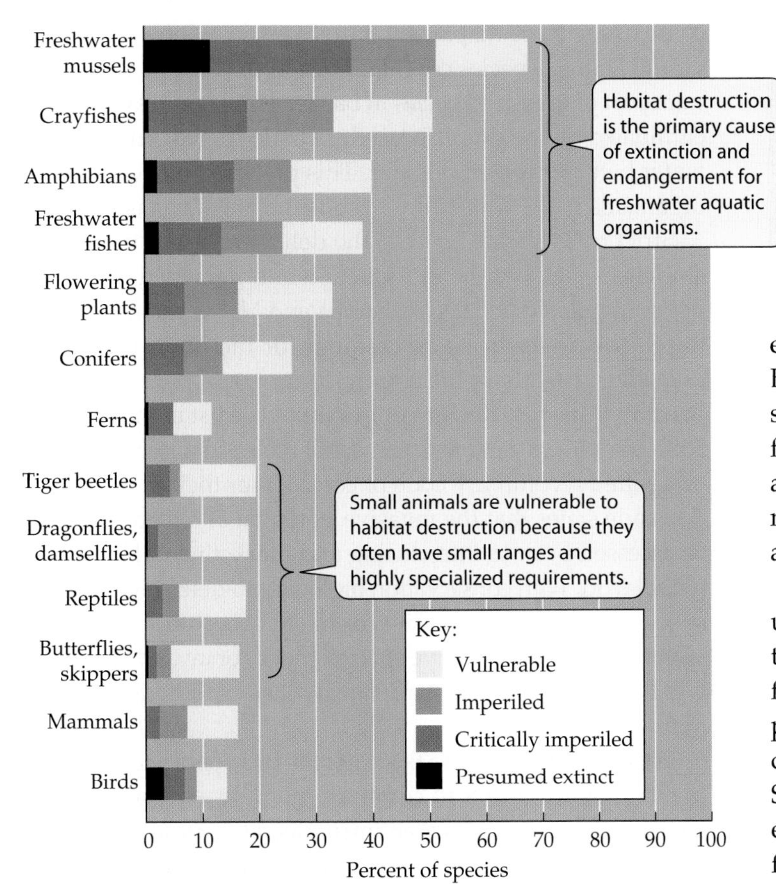

57.2 Proportions of U.S. Species Extinct or In Peril The groups of species that are most endangered—mussels, crayfishes, amphibians, and fishes—live in freshwater habitats, which have been extensively destroyed and polluted.

In addition, the fraction of a patch that is influenced by effects originating outside the habitat—**edge effects**—increases rapidly as patch size decreases (Figure 57.3). Close to the edges of forest patches, for example, winds are stronger, temperatures are higher, humidity is lower, and light levels are higher than they are farther inside the forest. Species from surrounding habitats often colonize the edges of patches to compete with or prey upon the species living there.

Usually we do not know which organisms lived in an area before its habitats became fragmented. To address this problem, a major research project in a tropical evergreen forest near Manaus, Brazil, was launched before logging took place. Landowners agreed to preserve forest patches of certain sizes and configurations (Figure 57.4). Biologists counted species in those patches while they were still part of the continuous forest. Soon after the surrounding forest was cut and converted to pasture, species began to disappear from isolated patches. The first species to be eliminated were monkeys that travel over large areas. Army ants and the birds that follow army ant swarms also disappeared.

Species that become extinct in small habitat fragments are unlikely to become reestablished because the more isolated the patches are, the less likely dispersing individuals are to find them. However, as we saw in Chapter 54, a species may persist in a small patch if it is connected to other patches by corridors of habitat through which individuals can disperse. Some of the pastures that surrounded the experimental forest fragments in Brazil have been abandoned, and a young forest is growing on them. Within 7–9 years of abandonment, some ant-following birds recolonized forest fragments connected to larger forest patches by young forests. Other species of birds that forage in the forest canopy also reestablished themselves. The young forest is not a suitable permanent habitat for most of these species, but it is an environment through which individuals can disperse to find new places where they can live.

Introduced predators, competitors, and pathogens have eliminated many species

Some species that have been introduced to regions outside their original range have become *invasive*—that is, they have spread widely and become unduly abundant, at a cost to the native species of the region. Invasive species are a major component of human-caused environmental change. Deliberately or accidentally, people move many species of organisms from one continent to another. Hundreds of species of plants have been introduced to new areas as ornamentals, as we saw at the beginning of this chap-

This area is influenced by edge effects.
This area is not influenced by edge effects.
Habitat patch
Because the width of the edges is relatively constant, as the total area becomes smaller, the edge becomes proportionately larger.
30.55% 43.75% 64% 88.8%
Percentage of patch influenced by edge effects

57.3 Edge Effects The smaller a patch of habitat, the greater the proportion of that patch that is influenced by conditions in the surrounding environment.

1 Isolated patches lose species much more quickly...
2 ...than patches connected to the main forest.
3 Even larger patches are strongly influenced by edge effects.

57.4 Brazilian Forest Fragments Studied for Species Loss Isolated patches lost species much more quickly than patches connected to the main forest. Even the larger patches, such as the one in the foreground, were too small to maintain populations of some species.

ter. Weed seeds have been carried around the world accidentally in sacks of crop seeds. Europeans deliberately introduced rabbits and foxes to Australia for sport hunting. Nearly half of the small to medium-sized marsupials and rodents of Australia have been exterminated during the last 100 years by a combination of competition with introduced rabbits and predation by introduced domestic cats and foxes.

Some pathogens have proliferated quickly following their introduction to new continents. Exotic disease-causing organisms have decimated populations of several eastern North American forest trees. The chestnut blight, caused by a European fungus, virtually eliminated the American chestnut, formerly an abundant tree in Appalachian Mountain forests. Nearly all American elms over large areas of the East and Midwest have been killed by Dutch elm disease, caused by the fungus *Ceratocystis ulmi*, which reached North America in 1930. Ecologists suspect that intercontinental movement of disease organisms caused extinctions in the past, but evidence of such disease outbreaks is not usually preserved in the fossil record.

The best way to reduce the damage caused by invasive species is to prevent their establishment in the first place. For example, the shipping industry often spreads invasive species (bacteria, dinoflagellates, invertebrates, and fish) in ballast water, which is pumped into a ship at one port and discharged at another. (That is how zebra mussels were introduced into North America from Europe, as we saw in Chapter 54.) San Francisco Bay is now home to at least 234 exotic species, most of which arrived in ballast water, and some of them are displacing native species. Controlling invasive aquatic species costs millions of dollars per year, but transport of invasive species in ballast water could largely be eliminated by the simple procedure of deoxygenating ballast water before it is pumped out. This practice both kills most organisms in the water and extends the life of ballast tanks.

Strict rules already govern the deliberate introduction of animal species, but the introduction of ornamental plants is poorly regulated. In 1998, Australia and New Zealand began to require a weed risk assessment for the importation of plants not already in the country or not on a "clean list" of permitted species. Regulations do not yet exist in the United States, but in 2002 some members of the horticultural industry crafted a voluntary code of conduct for their profession. The code states that the invasive potential of a plant should be assessed prior to introducing and marketing it. Horticulturists work with conservation biologists to determine which species are currently invasive, or likely to become so, and to identify suitable alternative species. Stocks of invasive species will be phased out, and gardeners will be encouraged to use noninvasive plants.

But how can we assess the potential of a species to become invasive? One way is to compare the traits of species that have become invasive when introduced to a new area with those of other species that have not. Such comparisons show that a plant species is more likely to become invasive if it has a short generation time, small seeds, is dispersed by vertebrates, has a large range in its native continent, depends on nonspecific mutualists (root symbionts, pollinators, and seed dispersers), and is not evolutionarily closely related to plants in the area to which it is introduced. The best predictor, however, is whether the species is already known to be invasive elsewhere.

Using the traits that characterize most invasive species, conservation biologists have developed a decision tree to be used to determine whether an exotic species should be introduced into North America (Figure 57.5). Using such a decision tree cannot eliminate the introduction of all potentially invasive species, but if used conscientiously, its application can greatly reduce the risk.

Overexploitation has driven many species to extinction

Until recently, humans caused extinctions primarily by overhunting. Overexploitation of other species continues today. Elephants and rhinoceroses are threatened in Africa because poachers kill them for their tusks and horns, which are used for ornaments and knife handles, and because some men believe that powdered rhinoceros horn enhances their sexual potency. Massive international trade in pets, ornamental plants, and tropical forest hardwoods has decimated many species of orchids, tropical fishes, corals, parrots, and reptiles.

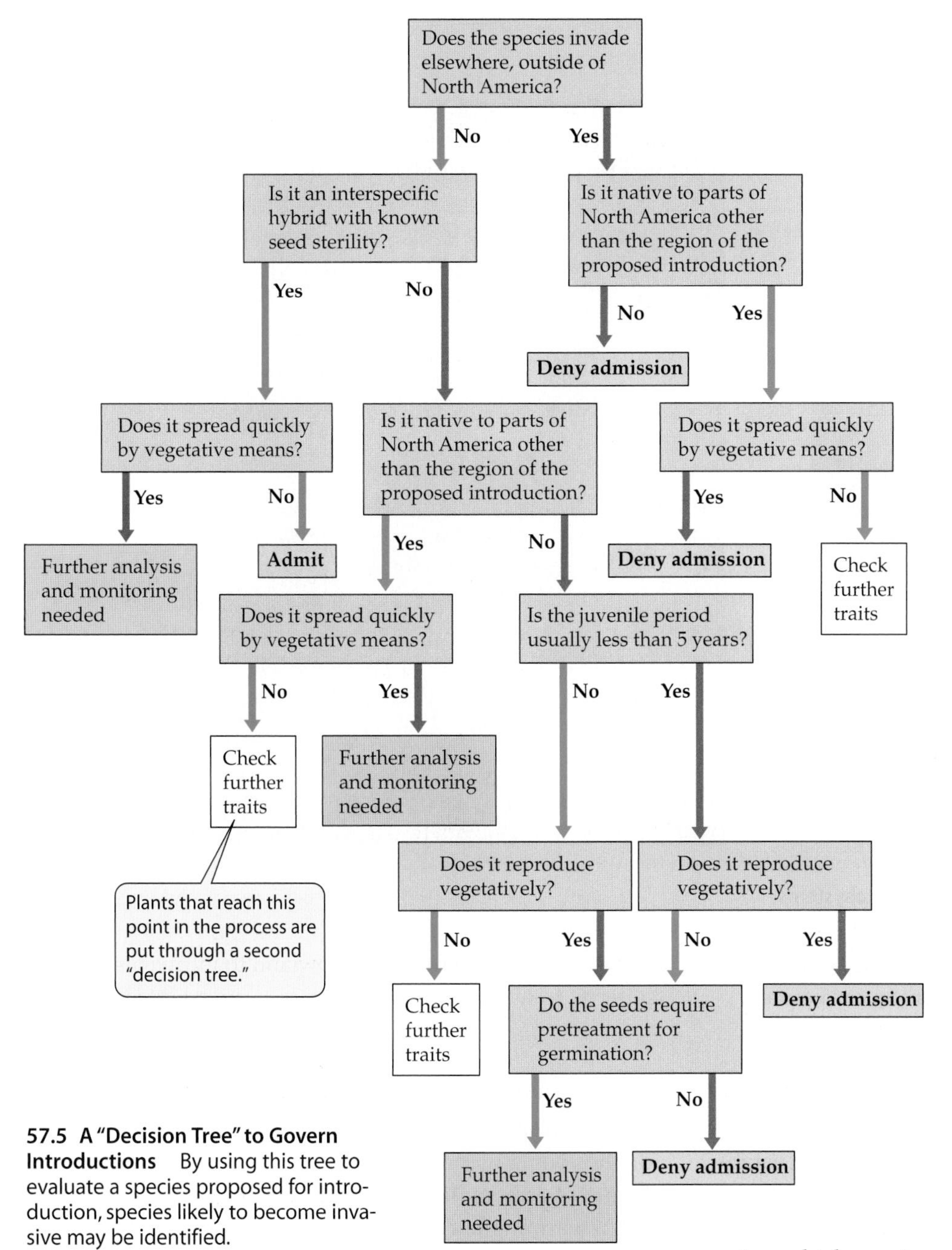

57.5 A "Decision Tree" to Govern Introductions By using this tree to evaluate a species proposed for introduction, species likely to become invasive may be identified.

Several programs have been initiated to help us continue to use species in ways that do not threaten their survival.

CERTIFICATION PROGRAMS. Many purchasers of wood products would like to buy only products that have been harvested in ways that protect biodiversity and ecosystem productivity. To enable them to exercise that choice, the Forest Stewardship Council (FSC) was established in 1993 by a consortium of environmental organizations and members of the forest product industry. FSC establishes criteria that a forest product company must meet for its products to be certified. Certification companies determine whether a forestry operation meets the criteria and ensure that there is a chain of custody that tracks certified products on their way to market. More than 400 companies in 18 countries have committed to purchasing certified wood products. By May 2003, more than 88 million acres of managed forests worldwide had been certified by FSC; 18.4 million of these acres were in North America.

To serve the same function for marine products, the Marine Stewardship Council was formed through an alliance between the World Wildlife Fund and Unilever, one of the largest marketers of frozen seafood. The first marine certified product, Australian rock lobster, came to market in 2000. Alaskan salmon has also been certified; other major fisheries are in the process of becoming certified. This action, combined with the elimination of government subsidies and perverse incentives, can help reduce the current overexploitation of many marine fish stocks.

ENDING TRADE IN ENDANGERED SPECIES. Species that are truly endangered typically cannot withstand any rate of harvest. The mechanism for prohibiting exploitation of these species is the Convention on International Trade in Endangered Species (CITES). Most nations of the world are members of CITES. National representatives meet every two years to review the status of species currently under protection, to determine which species may no longer need protection, and to add new species to its lists. CITES rules currently prohibit international trade in items such as whale meat, rhinoceros horn, and many species of parrots and orchids.

Some species depend on particular disturbance patterns

In Chapter 56, we saw that local species richness is sometime greatest at intermediate levels of disturbance. Many species depend on particular patterns of disturbance to persist. Some plant species, for example, germinate only after a fire; others depend on flooding to open sites where they can become established. Humans often try to reduce the frequency and intensity of such disturbances for their own purposes. Conservation biologists work to assess whether reestablishment of historic disturbance patterns may help preserve biodiversity.

(*a*)

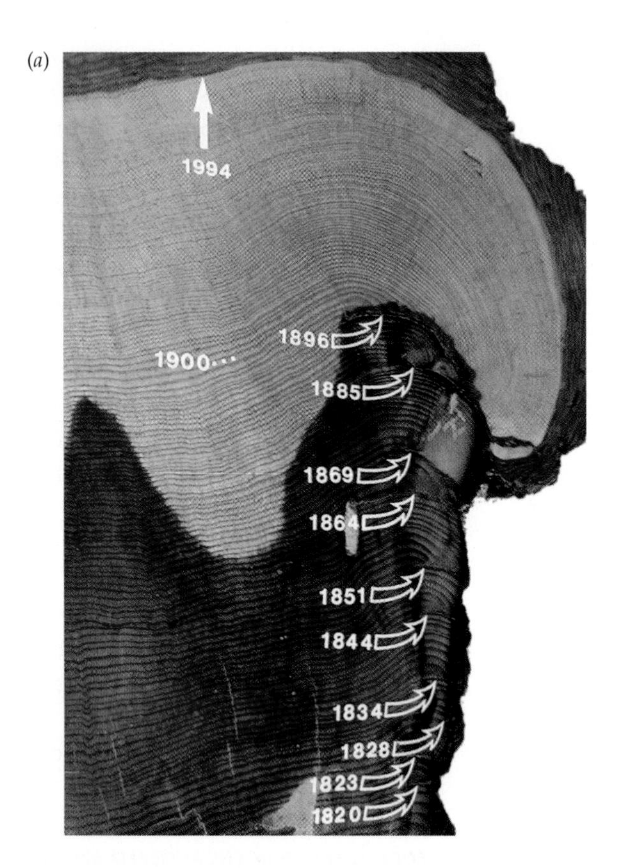

(*b*)

57.6 The Frequency and Intensity of Fires Affect Ecosystems
(*a*) As revealed by scars (arrows) in tree growth rings, low-intensity ground fires were frequent in the pine forests of the southwestern United States prior to fire suppression. (*b*) Fire suppression results in the buildup of large quantities of fuel, so that subsequent fires are likely to spread to the canopy and kill most trees.

Many species require periodic fires for successful establishment and survival, but for many years the official policy in the United States, symbolized by Smokey Bear, was to suppress all forest fires. It is now generally regarded as appropriate to use controlled burning as a management tool, particularly in Western North America. But to determine how to do so, we need to know the historical pattern of fires in an area.

Scars in the annual growth rings of trees preserve evidence of past fires that did not kill them. Therefore, tree-ring researchers can determine when fires occurred, how severe they were, and when fire patterns changed. Annual growth rings on ponderosa pines show that low-intensity ground fires were common near Los Alamos, New Mexico, until about 1900 (Figure 57.6*a*). After that date, cattle and sheep grazing in pine forests and fire suppression greatly reduced the frequency of low-intensity fires. Without these fires, dead branches and needles accumulated in the forest. The buildup of these fuels meant that when fires inevitably did occur, they were much more likely to become intense, tree-consuming canopy fires (Figure 57.6*b*). Today, ground fires are deliberately started in many areas to keep fuel loads low and to mimic historic fire patterns, to which many native species are adapted.

Rapid climate change may cause species extinctions

Scientists from many fields believe that Earth's climate is rapidly becoming warmer as a result of human-caused changes in Earth's atmosphere. We will examine the causes of this global warming in Chapter 58. Conservation biologists cannot alter rates of global warming, but their research can help us to predict how the resulting climate changes will affect organisms and find ways of mitigating those effects. Such research activities include analyses of past climatic events and studies of sites currently undergoing rapid climate change.

Atmospheric scientists predict that temperatures in North America will increase 2° to 5°C by the end of the twenty-first century. If the climate warms by only 1°C, the average temperature currently found at any particular location in North America will be found 150 km to the north. If the climate warmed 2°–5°C, species would need to shift their ranges as much as 500 to 800 km in a single century. Some habitats, such as alpine tundra, could be eliminated as forests expand up mountains.

Knowledge of how organisms responded to past climate changes can help us predict the effects of the current warming trend. Biologists are studying how rapidly species ranges

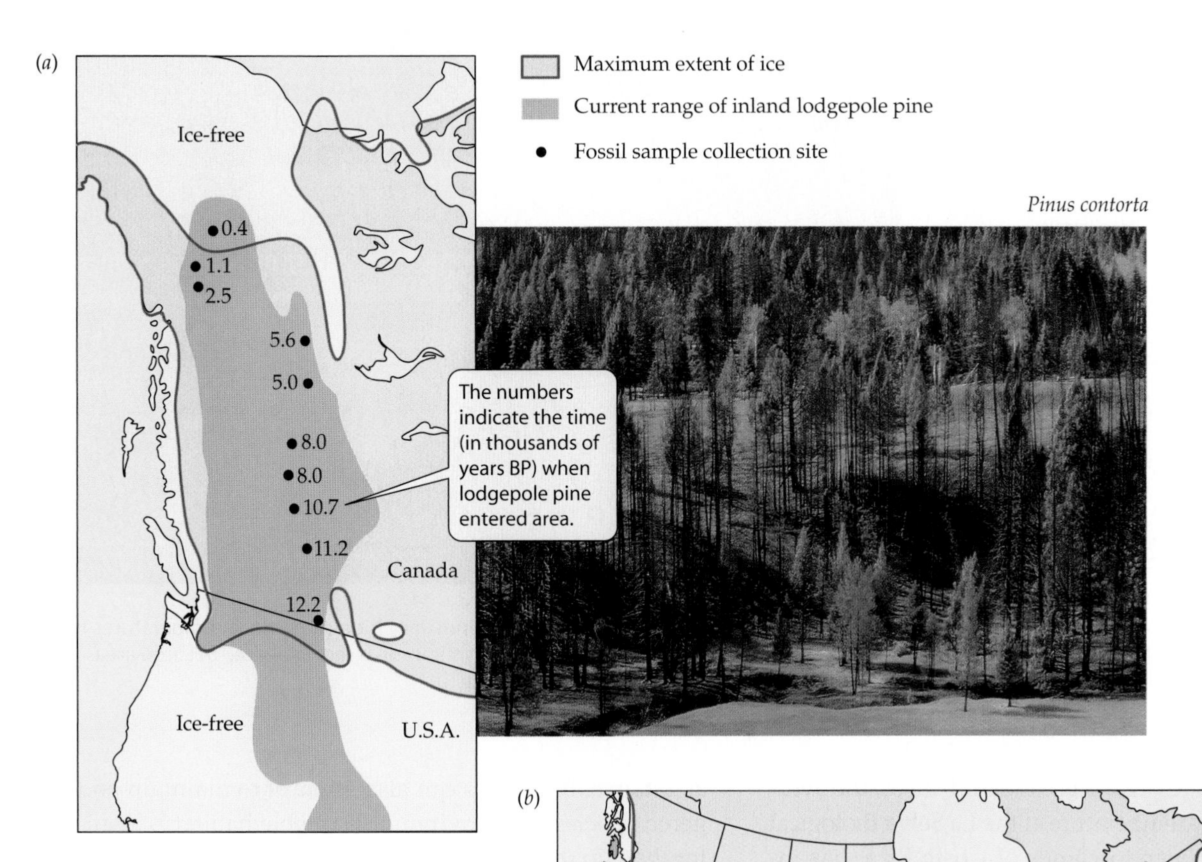

57.7 Some Species Shift Their Ranges in Response to Climate Change (*a*) The range of lodgepole pines in North America expanded north nearly as fast as glaciers retreated. (*b*) Some native earthworm species disperse so slowly they have hardly moved into glaciated regions.

shifted during the last 10,000 years of postglacial warming, which species were and were not able to keep pace with climate change by shifting their ranges, and how past ecological communities differed from those of today. Some organisms with good dispersal abilities, such as birds, can shift their ranges as rapidly as the climate changes, provided that appropriate habitat exists in new areas. However, the ranges of many species with sedentary adults are likely to shift slowly. As the glaciers retreated in North America, the ranges of some coniferous trees expanded northward, so that today they grow as far north as the current climate permits (Figure 57.7*a*). Some species of earthworms, on the other hand, spread very slowly into the areas that had been covered by ice (Figure 5.7*b*). Introduced European earthworms survive well in parts of Canada north of the ranges of native earthworms, indicating that slow dispersal, not lack of suitable habitat, is responsible for the range limitations of this group.

If Earth's surface warms as predicted, climatic zones will not simply shift northward. In addition to such shifts, entirely new climates will develop, and some existing climates will disappear. New climates are certain to develop at low elevations in the Tropics because a warming of even 2°C would result in climates near sea level that are warmer than those found anywhere in the humid Tropics today. Adaptation to those climates may prove difficult for many tropical organisms. Although there has been little recent climate warming in tropical regions, nights are now slightly warmer than they

(a)

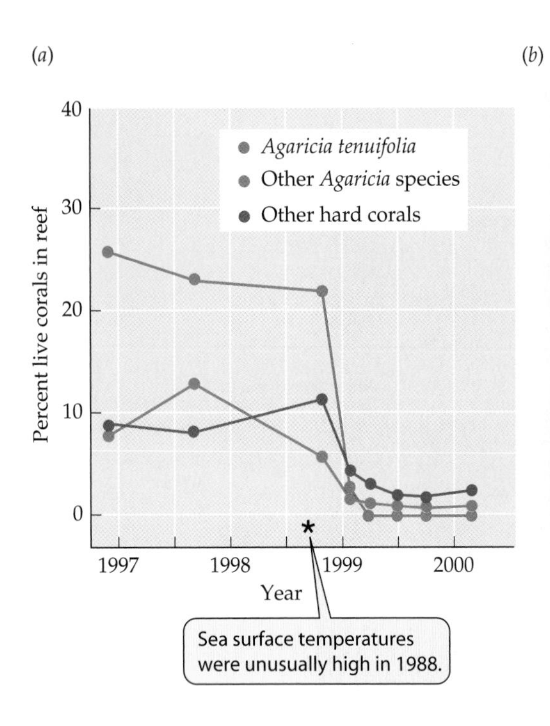

(b)

57.8 Global Warming Affects Corals (*a*) Unusually high sea surface temperatures in 1988 caused massive bleaching and death of corals on a reef in Belize. (*b*) Large areas of coral reefs in Florida Bay have been bleached.

were only a few decades ago. Since the mid-1980s, the average minimum nightly temperature at the La Selva Biological Station, in the Caribbean lowlands of Costa Rica, has increased from about 20°C to 22°C. During the warmer nights, trees use more of their energy reserves. The result has been a reduction of about 20 percent in the average growth rates of trees of six different species.

In 1988, the highest sea surface temperatures ever recorded caused corals to lose their symbiotic dinoflagellates (a phenomenon called *bleaching*) and increased their mortality worldwide (Figure 57.8). If warming of the oceans continues as predicted, about 40 percent of coral reefs worldwide are likely to be killed by 2010. To identify possible ways to help preserve coral reefs, biologists are measuring conditions in places where corals have escaped bleaching. They have found that reefs adjacent to cool, upwelling waters and reefs with cloudy waters, both of which have relatively low temperatures, are generally healthy. These reefs are receiving special protection because corals are likely to continue to survive well there. Corals from those reefs could be sources of colonists for reestablishing reefs where the corals have died if cooler ocean temperatures return in the future.

Habitat Restoration and Species Recovery

If the cause of a species' endangerment is the loss or modification of its habitat , conservation biologists can attempt to find ways of restoring that habitat. A field called **restoration ecology** has developed to study methods of restoring natural habitats. Such methods are needed because many ecosystems will not recover, or will do so only very slowly, without assistance. Biologists can also attempt to maintain endangered species in captivity until suitable habitat is available for them in the wild.

Restoring ecosystem processes is difficult

Conservation biologists have only a limited ability to restore natural ecosystems. In the United States, the false belief that humans can create functioning ecosystems has resulted in policies that make it easy to get permits for developments that destroy habitats. Developers need only state that they will create habitats to substitute for the ones they are destroying. However, even the most experienced wetland ecologists have great difficulty creating new wetlands that support the species that live in those being destroyed.

In southern California, where 90 percent of the coastal wetlands have been destroyed, wetland restoration is a high priority. Because species have been lost from degraded coastal wetlands, restoration requires species introductions, but which species should be introduced? In early attempts at restoration, only one or two common, easily grown wetland species were planted. Many wetland-associated species failed to recolonize these "rehabilitated" wetlands. To understand why, biologists established a large field experiment at the Tijuana Estuary to examine the effects of plant species richness on several factors that might affect the success of wetland restoration. They found that experimental plots planted with species-rich mixtures developed a complex vegetation structure, which is important to insects and birds. The species-rich plots also accumulated nitrogen faster than species-poor experimental communities (Figure 57.9).

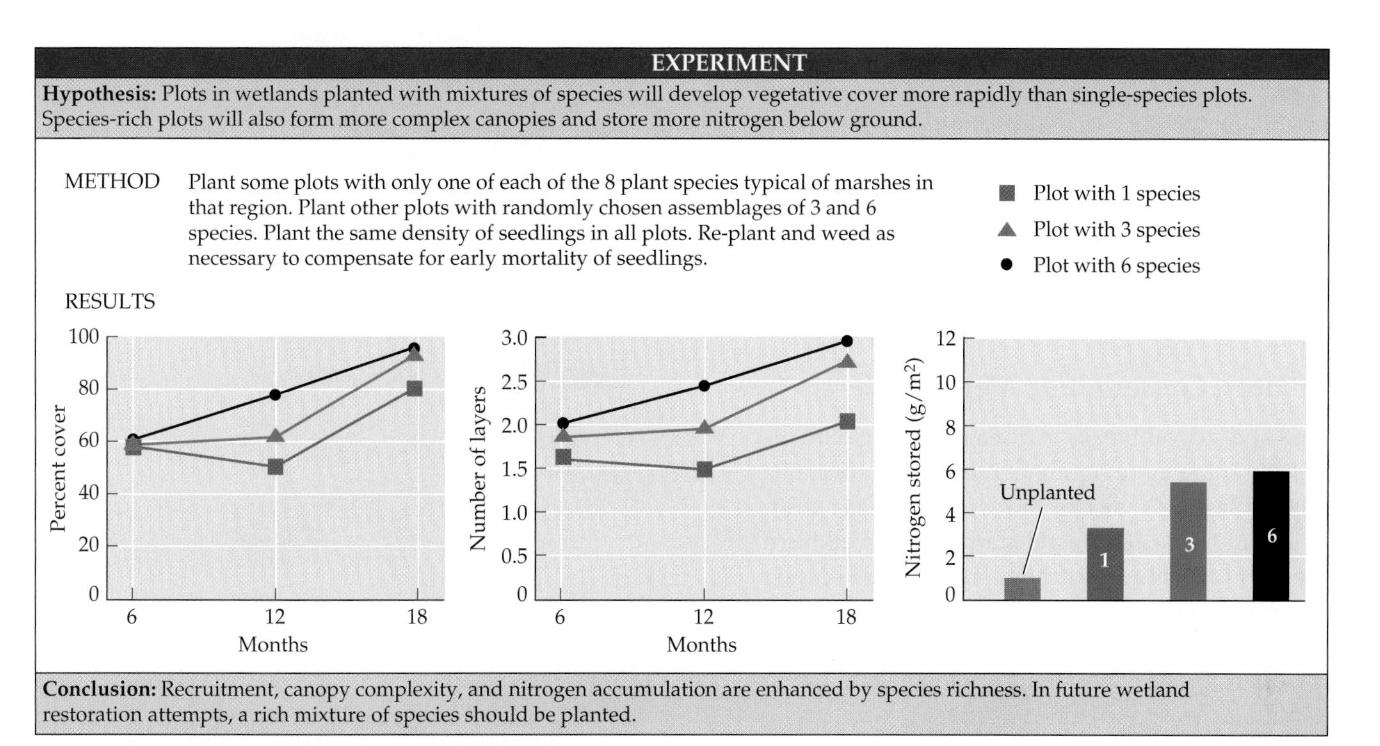

EXPERIMENT
Hypothesis: Plots in wetlands planted with mixtures of species will develop vegetative cover more rapidly than single-species plots. Species-rich plots will also form more complex canopies and store more nitrogen below ground.
METHOD Plant some plots with only one of each of the 8 plant species typical of marshes in that region. Plant other plots with randomly chosen assemblages of 3 and 6 species. Plant the same density of seedlings in all plots. Re-plant and weed as necessary to compensate for early mortality of seedlings.
Plot with 1 species
Plot with 3 species
Plot with 6 species
RESULTS
Percent cover
Months
Number of layers
Months
Nitrogen stored (g/m²)
Unplanted
1
3
6
Conclusion: Recruitment, canopy complexity, and nitrogen accumulation are enhanced by species richness. In future wetland restoration attempts, a rich mixture of species should be planted.

57.9 Species Richness Enhances Wetland Restoration Both vegetation complexity and nitrogen accumulation are greater in species-rich than in species-poor experimental plots.

Captive propagation can prevent some species from becoming extinct

Sometimes an endangered species can be maintained in captivity while the external threats to its existence are reduced or removed. However, captive propagation is only a temporary measure that buys conservation biologists time to deal with those threats. Existing zoos, aquariums, and botanical gardens do not have enough space to maintain adequate populations of more than a small fraction of Earth's rare and endangered species. Nonetheless, captive propagation can play an important role by maintaining species during critical periods and by providing a source of individuals for reintroduction into the wild. Captive propagation projects in zoos also have raised public awareness of threatened and endangered species.

Captive propagation is helping to save the California condor, North America's largest bird (Figure 57.10). Two hundred years ago, condors ranged from southern British Columbia to northern Mexico, but by 1978, the wild population was plunging toward extinction—only 25 to 30 birds remained in southern California. Many birds were poisoned by ingesting carcasses containing lead shot.

To save the condor from certain extinction, biologists initiated a captive propagation program in 1983. The first chick conceived in captivity hatched in 1988. By 1993, nine captive pairs were producing chicks, and the captive population had increased to more than 60 birds. The captive population was large enough that six captive-bred birds could be released in the mountains north of Los Angeles in 1992. These birds are provided with lead-free food in remote areas, and they are using the same roosting sites, bathing pools, and mountain

Gymnogyps californianus

57.10 Soaring High Once More Captive propagation has enabled California condor populations to be reestablished. Captive-reared birds have successfully survived after being released into the wild in California and Arizona.

ridges as did their predecessors. Captive-reared birds also were released late in 1996 in northern Arizona. It is still too early to pronounce the program a success, but as of February 2003, there were 81 wild condors in California and Arizona. Lead poisoning is still a problem, but an effort to encourage hunters to use non-lead ammunition is under way. Without captive propagation, the California condor would probably be extinct today.

Healing Biotas: Conservation Medicine

On both land and sea, outbreaks of diseases among wild organisms are becoming more common threats to biodiversity. The Caribbean Basin is a disease hot spot. *Diadema antillarum*, a dominant sea urchin, and staghorn and elkhorn corals have been virtually eradicated, and disease among corals is increasing rapidly. Outbreaks of several diseases have affected large areas of corals in the Indo-Pacific region. Mortality rates of marine mammals are increasing in the North Atlantic.

The impressive endemic bird fauna of the Hawaiian Islands has been decimated by habitat destruction, overhunting, and introduced predators and diseases (Figure 57.11). For example, wild pigs introduced by the native Polynesians damage the ground cover and soils of Hawaiian forests. A side effect of this habitat destruction is that indentations left by the pigs' foraging fill with water and are breeding grounds for mosquitoes that carry avian malaria. Below 1,000 meters elevation, nearly all endemic Hawaiian bird species have been eliminated by this disease, which was introduced to the islands with exotic birds. The native birds, never having been exposed to malaria, were highly susceptible. Species that inhabit altitudes above the current range of mosquitoes have fared better, but the insects' range may be expanding upward as the climate warms.

Another disease of birds, the mosquito-borne West Nile virus, has exploded across the United States, where it has killed more than 250,000 birds. The virus primarily infects birds, but can be transmitted to humans. First detected in New York in autumn 1999, within 4 years the virus was found in 43 of the contiguous 48 states and 6 of Canada's 10 provinces. By November 2003, there had been a reported 11,516 human cases in the U.S., with 439 deaths. How West Nile virus spread so rapidly is not understood. To find out, biologists are studying where its mosquito vectors feed, how long they survive, and where they hibernate.

A new field of **conservation medicine** is developing to help identify the causes of such increases in wildlife diseases and to devise effective solutions. Molecular techniques are being used to identify species, strains, and life cycle stages of microbial pathogens. Life histories of disease vectors are being studied to discover the vulnerable stages where interventions are most likely to prevent transmission of the pathogen and limit its effects.

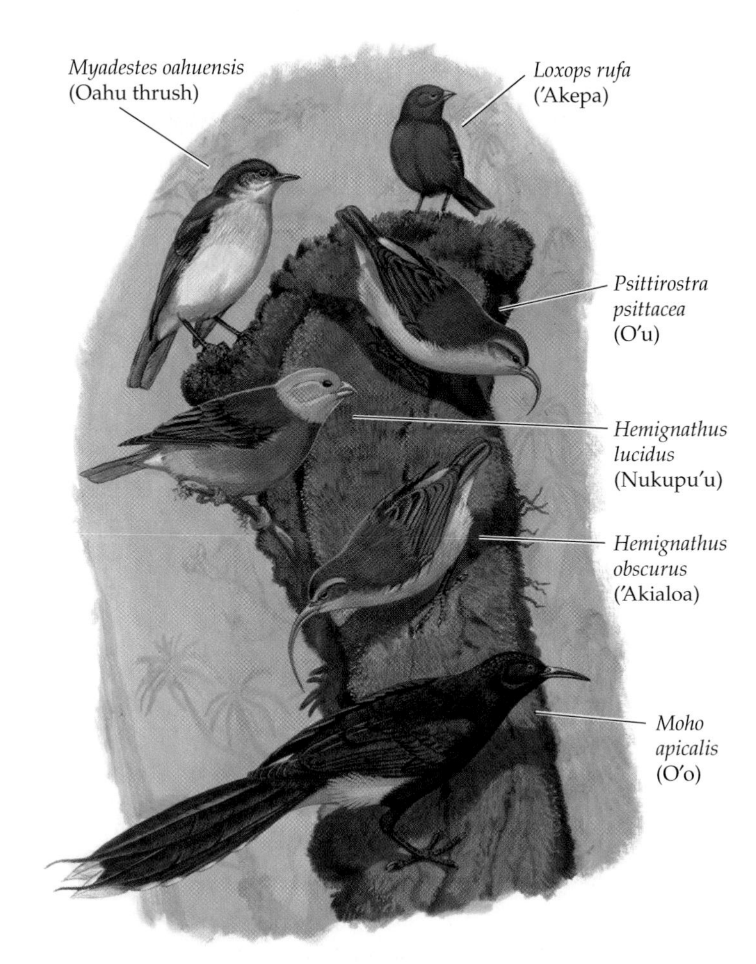

57.11 Extinct Hawaiian Honeycreepers Shown here are just six of the many Hawaiian bird species that have disappeared over the past 150 years. The O'o was among the birds native Polynesians hunted for their feathers, hundreds of thousands of which were used in ceremonial capes for the chiefs. Since 1900, many honeycreeper species have become extinct largely due to avian malaria, an introduced disease to which most endemic birds have no resistance.

Setting Limits: The Legacy of Samuel Plimsoll

During the nineteenth century, many British merchant ships sailed Earth's oceans. At that time, there were no undersea telegraph cables or shipboard radios. Once a ship left a harbor, it was out of contact with the rest of the world; in the case of a shipwreck, rescue was impossible. Owners could maximize their profits by overloading their ships, even though this caused some of them to be unseaworthy and sink. Samuel Plimsoll, a member of England's Parliament, became concerned about the rate of loss of British ocean-going vessels and sailors. He convinced Parliament to require that a "load line" be painted on the hull of every large vessel. The position of the line was calculated using factors such as the structural strength of the vessel and the shape of its hull. If the load line was under water, the ship was not permitted to leave the harbor. The "Plimsoll line," as it has come to be known, dramatically reduced the rate of loss of British ships and sailors at sea.

The increasing loss of Earth's species suggests that the load of human activities has pushed the hull of Noah's Ark below the Plimsoll line. But where and how should society draw that line? The decision should be based on scientific information, but just as in Samuel Plimsoll's time, science cannot determine an "acceptable rate of loss." Moreover, we must be concerned not only with species extinctions and ecosystem functioning, but with the overall functioning of the biosphere as well. To help you think more about how society should decide where to draw its Plimsoll line, we turn in the next and final chapter of this book to the functioning of the entire Earth system and how human activities are changing its processes at a global scale.

Chapter Summary

▶ Humans have caused extinctions of species for thousands of years, but the rate of human-caused extinctions is rising rapidly today.

Why Care about Species Extinctions?

▶ Species provide the food, fiber, medicines, and aesthetic opportunities upon which the quality of human life depends.

▶ The extinction of species as a result of human activities raises serious ethical issues.

▶ Extinctions deprive us of opportunities to understand ecological relationships among organisms.

▶ Ecosystems provide valuable services that can be replaced only by expensive and continuing human effort. **Review Figure 57.1**

Estimating Current Rates of Extinction

▶ Estimates of current rates of extinction are based primarily on species–area relationships and population models.

Preserving Biodiversity

▶ Habitat destruction is the most important cause of species extinction today. **Review Figure 57.2**

▶ A greater proportion of small than large habitat patches is affected by external influences. **Review Figures 57.3, 57.4**

▶ Invasive species are major causes of extinction. Biologists use information on species that have become invasive to identify species likely to become invasive if introduced. **Review Figure 57.5**

▶ Certification programs enable consumers to purchase materials produced in ways that do not harm biodiversity.

▶ Overexploitation, which historically resulted in most human-caused extinctions, is still an important cause of extinctions today.

▶ Information on how species are affected by disturbances helps conservation biologists decide where to reestablish historic disturbance patterns.

▶ Species have responded at different rates to past climate changes. **Review Figure 57.7**

Habitat Restoration and Species Recovery

▶ Restoration of habitats is often necessary to preserve species. Restoration of some ecosystem types, especially wetlands, is difficult. **Review Figure 57.9**

▶ Captive propagation plays a useful but limited role in conservation.

Healing Biotas: Conservation Medicine

▶ Disease outbreaks among wild species are increasing. Some of these diseases can be transmitted to humans. The new field of conservation medicine is helping to identify the causes of increases in diseases and to devise effective solutions.

Setting Limits: The Legacy of Samuel Plimsoll

▶ Like an overloaded merchant ship, the "Noah's Ark" of Earth's biodiversity may be in danger of sinking from an overload of stresses and extinctions attributable to human activities.

See Web/CD Activity 57.1 for a concept review of this chapter.

Self-Quiz

1. Which of the following is *not* currently a major cause of species extinctions?
   a. Habitat destruction
   b. Rising sea levels
   c. Overexploitation
   d. Introduction of predators
   e. Introduction of diseases
2. The most important cause of endangerment of species in the United States currently is
   a. pollution.
   b. exotic species.
   c. overexploitation.
   d. habitat loss.
   e. loss of mutualists.
3. People care about species extinctions because
   a. more than half of the medical prescriptions written in the United States contain a natural plant or animal product.
   b. people derive aesthetic pleasure from interacting with other organisms.
   c. causing species extinctions raises serious ethical issues.
   d. biodiversity helps maintain ecosystem services.
   e. All of the above
4. As a habitat patch gets smaller, it
   a. cannot support populations of species that require large areas.
   b. supports only small populations of many species.
   c. is influenced to an increasing degree by edge effects.
   d. is invaded by species from surrounding habitats.
   e. All of the above
5. A plant species is most likely to become invasive when introduced to a new area if it
   a. grows tall.
   b. has become invasive in other places where it has been introduced.
   c. is closely related to species living in the area into which it is introduced.
   d. has specialized dispersers of its seeds.
   e. has a long life span.

6. Conservation biologists are concerned about global warming because
   a. the rate of change in climate is projected to be faster than the rate at which many species can shift their ranges.
   b. it is already too hot in the Tropics.
   c. climates have been so stable for thousands of years that many species lack the ability to tolerate variable temperatures.
   d. climate change will be especially harmful to rare species.
   e. None of the above
7. Scientists can determine the historical frequency of fires in an area by
   a. examining charcoal in sites of ancient villages.
   b. measuring carbon in soils.
   c. radioactively dating fallen tree trunks.
   d. examining fire scars in growth rings of living trees.
   e. determining the age structure of forests.
8. Captive propagation is a useful conservation tool, provided that
   a. there is space in zoos, aquariums, and botanical gardens for breeding a few individuals.
   b. the genetic pedigree of all individuals is known.
   c. the threats that endangered the species are being alleviated so that captive-reared individuals can later be released back into the wild.
   d. there are sufficient caretakers.
   e. Captive propagation should not be used because it directs attention away from the need to protect the species in their natural habitats.
9. Restoration ecology is an important field because
   a. many areas have been highly degraded.
   b. many areas are vulnerable to global climate change.
   c. many species suffer from demographic stochasticity.
   d. many species are genetically impoverished.
   e. fire is a threat to many areas.
10. The new discipline of conservation medicine has developed because
    a. the frequency of diseases has increased among marine organisms.
    b. the frequency of diseases has increased among terrestrial organisms.
    c. the frequency of diseases has increased among both marine and terrestrial organisms.
    d. scientists can better control diseases today than they previously could.
    e. diseases can be readily diagnosed today.

For Discussion

1. Most species driven to extinction by humans in the past were large vertebrates. Do you expect this pattern to persist into the future? If not, why not?
2. Conservation biologists have debated extensively which is better: many small nature reserves or a few large ones. What ecological processes should be evaluated in making judgments about the size and location of reserves? To what extent should we be concerned with preserving the largest number of species rather than those species judged to be of unusual importance for scientific, aesthetic, or commercial reasons?
3. During World War I, French doctors adopted a "triage" system for dealing with wounded soldiers. The wounded were divided into three categories: those almost certain to die no matter what was done to help them, those likely to recover even if not assisted, and those whose probability of survival was greatly increased if they were given immediate medical attention. Limited medical resources were directed primarily at the third category. What are some implications of adopting a similar attitude toward species preservation?
4. Utilitarian arguments dominate discussions about the importance of preserving the biological richness of the planet. In your opinion, what role should ethical and moral arguments play?
5. The desert bighorn sheep of the southwestern United States is endangered. Its major predator, the puma, is also threatened in the region. Under what conditions, if any, would it be appropriate to suppress the population of one rare species to assist another rare species?

58 Earth System Science

An atom of phosphorus "X" had marked time in the limestone ledge since the Paleozoic seas covered the land. Time, to an atom locked in a rock, does not pass. The break came when a bur-oak root nosed down a crack and began prying and sucking. In the flash of a century the rock decayed, and X was pulled out and up into the world of living things. He helped build a flower, which became an acorn, which fattened a deer, which fed an Indian all in a single year.

—*Aldo Leopold*, A SAND COUNTY ALMANAC

In this account, Leopold, a pioneer wildlife manager and promoter of environmental stewardship, vividly portrays the flow of atoms between the physical and biological environment and among different organisms. A few of Earth's atoms have changed by radioactive decay, a few have escaped to space, and meteors and meteorites have delivered a few new atoms to Earth. Nevertheless, virtually all of the atoms in our own bodies, and in all other living organisms, have been present on Earth since its formation about 4.5 billion years ago, cycling among its various components.

We began Part One of this book with an introduction to the non-living atoms and molecules that are the building blocks of Earth and of the life found here. It is perhaps fitting that the final chapter should return to these atoms, including them now as part of a larger story. For scientists trained to read them, the atmosphere, rocks, soil, and living organisms harbor clues that tell us about their histories, what was happening on Earth when they were formed, and how they were subsequently transformed. Knowledge of Earth's history helps us understand the changes taking place on Earth today.

Earth system science has emerged as a new field of inquiry that focuses on Earth as a whole. A *system* is a group of entities that interact to yield some product. For example, an individual animal is made up of systems of organs that work together to perform some function, such as digestion. Earth's system is composed of cycles of materials, inputs of solar energy, and interactions between living organisms and the physical environment. Interactions among these components determine how Earth as a planet functions.

"Time, to an Atom Locked in a Rock, Does Not Pass." The "pillar peaks" of Zhangjiajie National Forest Park in China's Hunan province stand like sentinels in the mist. The atoms in these rocks move through biogeochemical cycles as inorganic atoms from the rock-based soil are taken up by plants.

In this chapter we will describe the major cycles of materials among the compartments of Earth's system. We will show how life has modified Earth's features throughout evolutionary history. We will also show how human alterations of the great biogeochemical cycles continue to modify Earth's system today.

Earth's System Has Four Compartments

Earth is an essentially closed system with respect to atomic matter, but it is an open system with respect to energy. The sun delivers a nearly constant amount of energy to Earth every day, and has done so for billions of years. Energy from the sun, combined with the energy from radioactive decay that melts the magma in Earth's interior, drives the processes that move materials around the planet. Many of these processes are cyclic (Figure 58.1). Almost all the rocks that compose the continents have been processed at least once through a chemical and physical cycle involving weathering, formation of sediments, and movement into Earth's interior through continental drift. Deep within Earth, these rocks are subjected to great heat and pressure to form new rocks. The water in the oceans has been evaporated, condensed, precipitated, and returned to the oceans via rivers and groundwater flow many thousands of times. The carbon (C), nitrogen (N), oxygen (O), and sulfur (S) in atmospheric gases have been cycled repeatedly through living organisms.

We take for granted many features of Earth, but Earth is actually a very unusual planet. Its unusual properties include the presence of life, an ocean, a moderate surface temperature, continental drift, and a large moon. The moon has had a profound effect on Earth's history. First, it stabilized the tilt of Earth on its axis. The degree of tilt strongly influences climate; for example, if the tilt were 50° or more, equatorial regions would receive less solar energy over the year than would the polar regions! The moon also plays a major role in producing ocean tides, as well as slowing Earth's rotation.

Planets without life are in perpetual thermodynamic equilibrium, but Earth is far from thermodynamic equilibrium. Oxygen gas (O_2), nitrogen gas (N_2), and water vapor (H_2O) are the main molecules of Earth's atmosphere, but without living organisms, these gases would produce nitric acid (HNO_3), which would subsequently dissolve in the ocean and remain there. It is because of the activities of living organisms that this does not happen.

As we saw in Chapter 55, energy flows through ecosystems from producers to consumers. At each transformation, much of this energy is used to power metabolism and is dissipated as heat, a form that cannot be used by other organisms to power their metabolism. Chemical elements, on the other hand, are not altered when they are transferred among organisms. The availability of the chemical elements of which living organisms are composed—carbon, nitrogen, phosphorus, calcium, sodium, sulfur, hydrogen, oxygen, and a few others—is strongly influenced by how organisms get them, how long they retain them, and what they do with them while they have them.

To understand the cycling of elements, it is convenient to divide the physical environment into four interacting com-

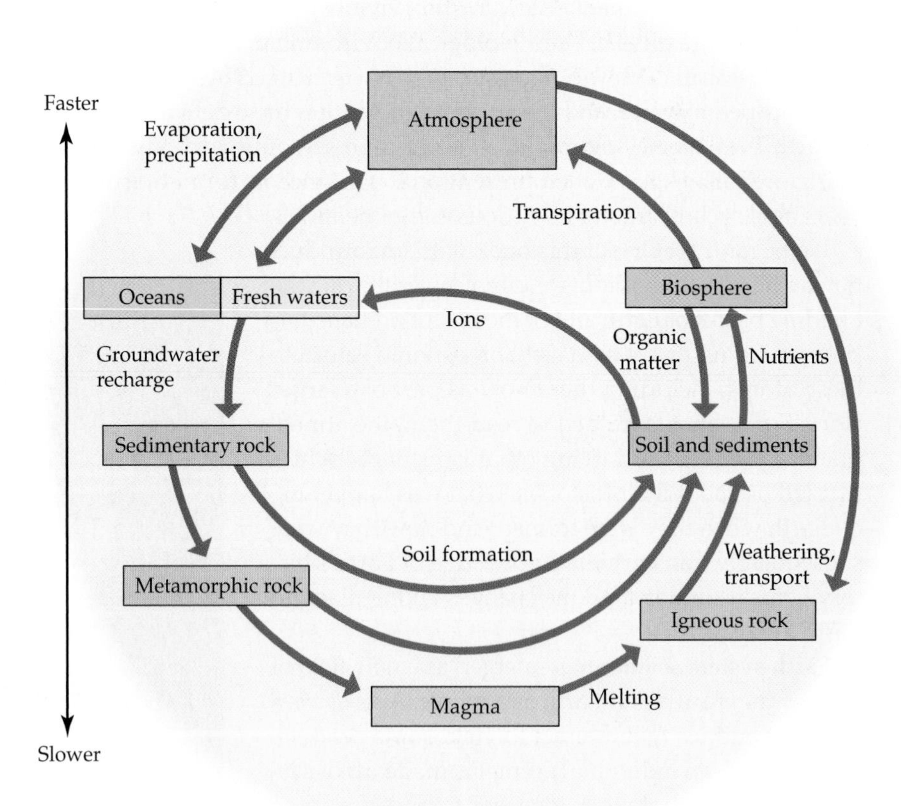

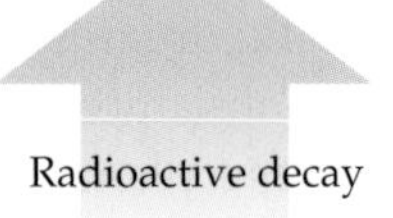

58.1 A Generalized Biogeochemical Cycle The arrows indicate fluxes of material between various compartments. Fluxes between rocks and soils are slower than fluxes between organisms, oceans, fresh waters, and the atmosphere.

partments: oceans, fresh waters, atmosphere, and land. These four compartments, and the types of organisms living in them, are very different. Therefore, the quantities of different elements in each compartment, what happens to them, and the rates at which they enter and leave the compartments differ strikingly. After we have described the four compartments, we will discuss how elements cycle among them.

Oceans receive materials from the land and atmosphere

Oceans receive materials from land primarily in runoff from rivers. Over time scales of hundreds to thousands of years, most materials that cycle through the four compartments end up in the oceans. The oceans are enormous, and they exchange materials with the atmosphere only at their surface, so they respond very slowly to inputs from other compartments.

Except near land on continental shelves, ocean waters mix slowly. Most elements that enter the oceans from other compartments gradually sink to the seafloor. They may remain there for millions of years until the seafloor sediments are elevated above sea level by uplifts of Earth's crust. Therefore, concentrations of mineral nutrients are very low in most ocean waters. Some elements are brought back to the surface near the coasts of continents where offshore winds push surface waters away from shore, causing cold bottom water to rise to the surface (Figure 58.2). Waters in these *zones of upwelling* are rich in mineral nutrients, and most of the world's great fisheries are concentrated there.

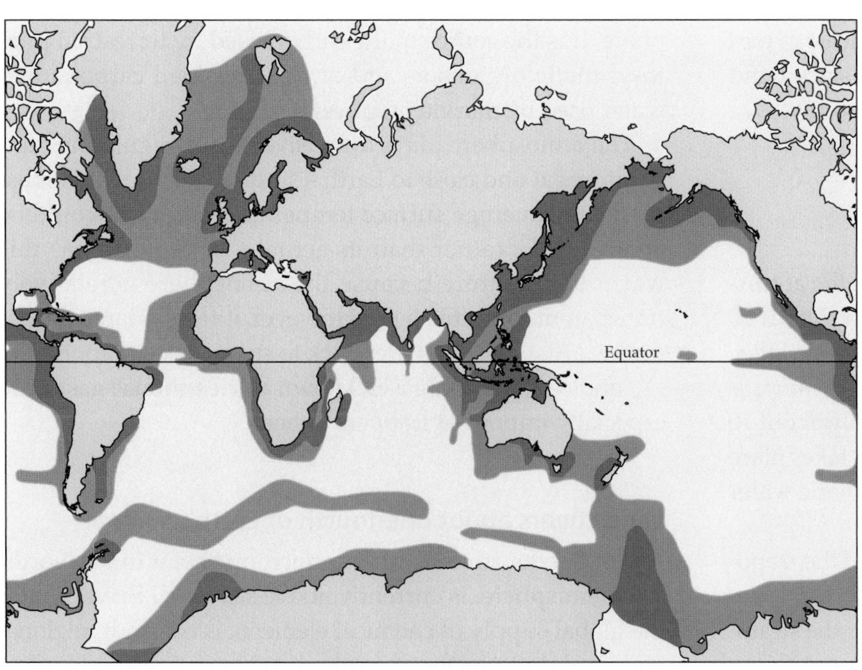

58.2 Primary Production Is High in Zones of Upwelling Primary production in the oceans is highest near continents where surface waters, driven by prevailing winds, move offshore and are replaced by cool, nutrient-rich water upwelling from below.

Earth's water moves rapidly through lakes and rivers

Only a small fraction of Earth's water resides at any one time in lakes and rivers. However, water moves rapidly through them. Some mineral nutrients enter the freshwater compartment in rainfall, but most are released by the weathering of rocks and are carried to lakes and rivers by surface flow or the by movement of *groundwater* (water in soil and rocks).

After entering rivers, mineral nutrients are usually carried rapidly to lakes or to the oceans. In lakes, they are taken up by organisms and incorporated into their cells. These organisms eventually die and sink to the bottom, taking the nutrients with them. Decomposition of their tissues consumes the O_2 in the bottom water. The surface waters of lakes thus quickly become depleted of nutrients, while deeper waters become depleted of O_2. This process is countered, however, by vertical movements of water called *turnover*. Turnover brings nutrients and dissolved CO_2 to the surface and O_2 to deeper water. Wind is an important agent of turnover in shallow lakes, but in deeper lakes it usually mixes only surface waters. Deep lakes in temperate climates have an annual turnover cycle that is driven by temperature (Figure 58.3).

Lakes in temperate regions turn over because water is most dense at 4°C; above and below that temperature, it expands. The spring sun warms the surface layer of a lake, with the depth of the warm layer gradually increasing as spring and summer progress. However, there is still a well-defined thermocline where the temperature drops abruptly to about 4°C. Only if the lake is shallow enough to warm right to the bottom does the temperature of the deepest water rise above 4°C.

In autumn, as the surface of the lake cools, the cooler surface water—which is denser than the warmer water below it—sinks and is replaced by warmer water from below. This process continues until the entire lake has reached 4°C. At this point, the density of the water is uniform throughout the lake, and even modest winds readily mix the entire water column. As colder weather then cools the surface water below 4°C, that water becomes less dense than the 4°C water below it. Therefore, it floats at the top. Another turnover occurs in spring, when the surface layers above the thermocline warm to 4°C and the water column, again being of uniform density throughout, is easily mixed by wind.

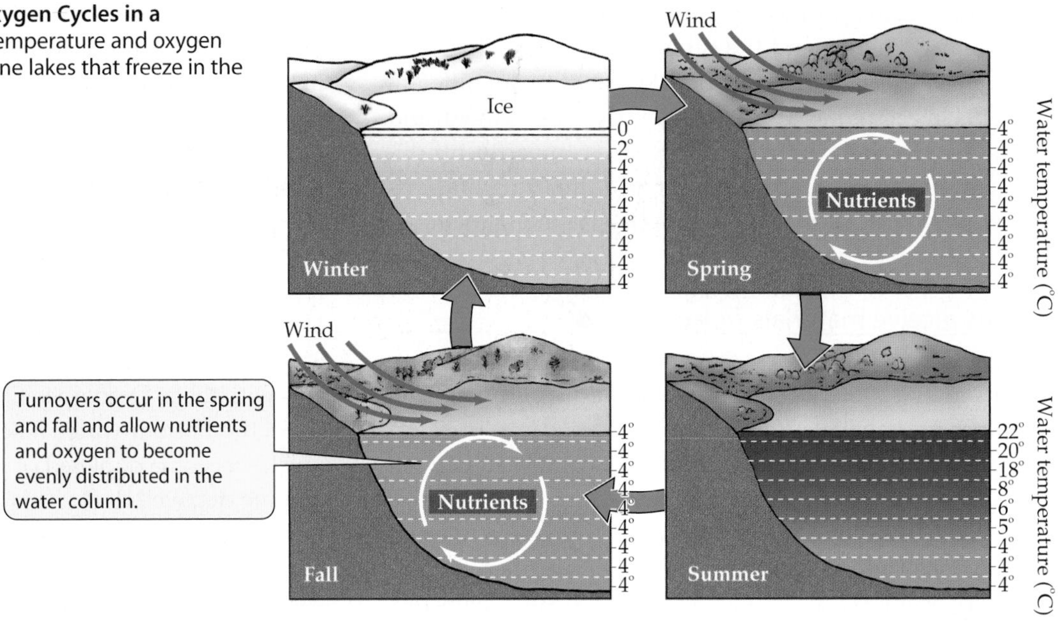

58.3 Annual Temperature and Oxygen Cycles in a Temperate Lake These vertical temperature and oxygen profiles are typical of temperate-zone lakes that freeze in the winter.

Turnovers occur in the spring and fall and allow nutrients and oxygen to become evenly distributed in the water column.

Arctic lakes turn over only once each year. Deep tropical and subtropical lakes may be permanently stratified because they never become cool enough to have uniformly dense water. Their bottom waters lack oxygen because decomposition quickly depletes any oxygen that reaches them. However, many tropical lakes are overturned at least periodically by strong winds so that their deeper waters are occasionally oxygenated. When this does not happen, lake productivity can be drastically affected. Fishermen of Lake Tanganyika, an extremely large and deep tropical lake in East Africa, have seen their catches decline by as much as 50 percent over the past 30 years. This decline has been attributed at least in part to warmer water temperatures and a decade of slack wind conditions.

The atmosphere regulates temperatures close to Earth's surface

The third major compartment of Earth's system, the **atmosphere**, is a thin layer of gases surrounding Earth. About 80 percent of the mass of the atmosphere lies in its lowest layer, the *troposphere*, which extends upward from Earth's surface about 17 km in the Tropics and Subtropics, but only about 10 km at higher latitudes. Most global air circulation takes place within the troposphere, and virtually all atmospheric water vapor is found there.

The *stratosphere*, which extends from the top of the troposphere up to about 50 kilometers above Earth's surface, contains very little water vapor. Most materials enter the stratosphere from the troposphere near the equator, where air rises to high altitudes, as we saw in Chapter 56. These materials tend to remain in the stratosphere for a relatively long time. Ozone (O_3) in the stratosphere absorbs most of the biologically damaging short-wavelength ultraviolet radiation that enters the atmosphere. (Ozone released into the troposphere by human activities, however, may contribute to global warming, as we will see below.)

The atmosphere is 78.08 percent N_2, 20.95 percent O_2, 0.93 percent argon, and 0.03 percent carbon dioxide (CO_2). It also contains traces of hydrogen gas, neon, helium, krypton, xenon, ozone, and methane. The atmosphere contains Earth's biggest pool of nitrogen as well as large supplies of O_2. Although CO_2 constitutes a very small fraction of the atmosphere, it is the source of the carbon used by terrestrial photosynthetic organisms and of the dissolved carbonate in water used by marine producers.

The atmosphere plays a decisive role in regulating temperatures at and close to Earth's surface. Without an atmosphere, the average surface temperature of Earth would be about –18°C, rather than its actual +17°C. Earth has this warm temperature because the atmosphere is relatively transparent to visible light. However, it traps a large part of the heat that Earth radiates back to space. Water vapor, CO_2, O_3, and certain other gases, known as **greenhouse gases**, are especially important trappers of heat.

Land covers about one-fourth of Earth's surface

About one-fourth of Earth's surface, most of it in the Northern Hemisphere, is currently above sea level. Even though the global supply of chemical elements is constant, regional and local deficiencies of particular elements strongly affect ecosystem processes on land, where elements move slowly and usually over only short distances.

The land compartment is connected to the atmospheric compartment by organisms that take chemical elements from and release them to the air. Chemical elements in soils are carried in solution into groundwater and eventually into the oceans, where they are unavailable to organisms until an episode of uplifting raises marine sediments and a new cycle of erosion and weathering begins. The type of soil that exists in an area depends on the underlying rock, as well as on climate, topography, the organisms living there, and the length of time that soil-forming processes have been acting. Very old soils are much less fertile than most young soils because nutrients leach out of them over time.

Although land covers such a small proportion of Earth's surface, human life depends intimately on soil fertility and the productivity of terrestrial ecosystems. And the land is the Earth system compartment that has been most strongly affected by human activities.

Biogeochemical Cycles, Water, and Fire

The chemical elements organisms need in large quantities—carbon, hydrogen, oxygen, nitrogen, phosphorus, and sulfur—cycle through organisms to the physical environment and back again. The pattern of movement of a chemical element through organisms and the four compartments of the physical environment is called its **biogeochemical cycle**. Each chemical element used by organisms has a distinctive biogeochemical cycle whose properties depend on the physical and chemical nature of the element and how organisms use it. All chemical elements cycle quickly through organisms because no individual, even of the longest-lived species, lives very long in geological terms.

Before we describe the cycles of these elements, however, we must discuss water and fire. As we have just seen, the movement of water transfers many elements between the atmosphere, land, fresh waters, and oceans. Fire is a powerful agent that speeds the cycling of chemical elements.

Water transfers materials from one compartment to another

The cycling of water through the oceans, atmosphere, fresh waters, and land is known as the **hydrological cycle**. The hydrological cycle operates because more water is evaporated from the surface of the oceans than is returned to them as precipitation (rain or snow). The excess evaporated water is carried by winds over the land, where it falls as precipitation.

Water also evaporates from soils, from lakes and rivers, and from the leaves of plants (transpiration), but the total amount evaporated from those surfaces is less than the amount that falls on them as precipitation. The excess terrestrial precipitation eventually returns to the oceans via rivers, coastal runoff, and groundwater flows (Figure 58.4).

Earth's 16 largest rivers account for more than one-third of total water discharge; more than half of the discharge comes from the three largest rivers (Amazon, Congo, and Yangtze). Despite their relatively small volume, rivers play a disproportionate role in the hydrological cycle because the average residence time of water in lakes and rivers is only 4.3 years, compared with 2,640 years in the oceans. The average turnover time of water in living things is much shorter—about 5.6 days.

By building dams, canals, and reservoirs and by diverting huge quantities of water to irrigated fields, humans have had major effects on the temporal and spatial distribution of fresh water on Earth. The most important consequence of human activities is that more water now evaporates from land and less flows to the oceans than before the Industrial Revolution. In addition, freshwater flow patterns are being seriously altered. For example, dams on the Columbia River in Washington State are

59
Evaporation
95
Precipitation
36
Transport over land
283
Precipitation
319
Evaporation
36
Runoff
210,000
(in sedimentary rocks near surface)
1,380,000
(in oceans)
The greatest exchanges of water take place at the ocean surface.
Although rocks contain large quantities of water, this "locked-in" water plays a very small role in the hydrological cycle.

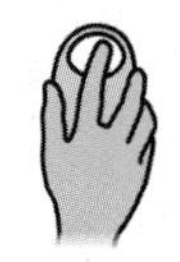

58.4 The Global Hydrological Cycle The numbers show the relative amounts of water (expressed as units of 10^{18} g) held in or exchanged annually by ecosystem compartments. The widths of the arrows are proportional to the sizes of the fluxes.

(a)

(b)

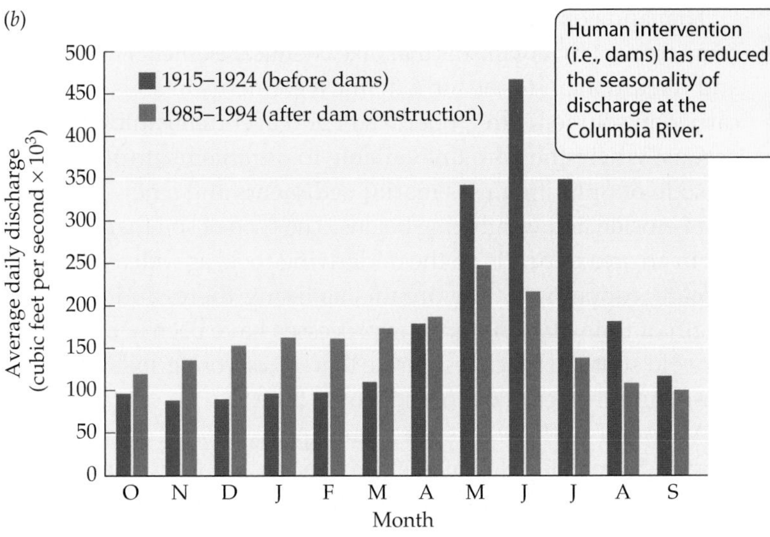

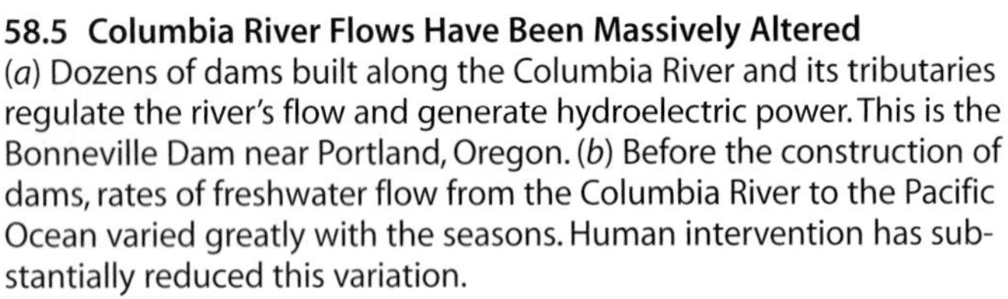

58.5 Columbia River Flows Have Been Massively Altered
(*a*) Dozens of dams built along the Columbia River and its tributaries regulate the river's flow and generate hydroelectric power. This is the Bonneville Dam near Portland, Oregon. (*b*) Before the construction of dams, rates of freshwater flow from the Columbia River to the Pacific Ocean varied greatly with the seasons. Human intervention has substantially reduced this variation.

managed to reduce the variation in water flow rates during the year (Figure 58.5). An unintended result of this practice is reduced autumn and winter surface salinity from the mouth of the river north along the coast to the Aleutian Islands, with negative consequences for the growth and survival of salmon in that part of the ocean.

In some places, groundwater is being seriously depleted because humans are using it more rapidly than it can be replaced, primarily by pumping it for irrigation. In some areas of the High Plains of the central United States, more than half of the groundwater has been removed. On the North China Plain, depletion of shallow aquifers is forcing people to sink wells more than 1,000 meters deep to reach water. Much of the groundwater we are using today was deposited during the last Ice Age, when regional precipitation was much greater than it is today. Flows of water to the oceans have been increased by the pumping of groundwater, thereby contributing to sea level rises during the past century.

The World Resources Institute's *Pilot Analysis of Global Ecosystems*, published in 2000, predicts that if current water consumption patterns continue, by 2025, at least 3.5 billion people (48% of the world population) will live in areas with inadequate water supplies, primarily in Asia. Fortunately, current water consumption patterns need not persist. Per capita consumption in the United States and Europe is dropping because of increasing use of water-efficient home appliances, increasing prices charged for water, and development of regulations that restrict water use. If programs to improve water use efficiency are implemented vigorously, global water use could be even less in 2025, despite continued population growth.

Fire is a major mover of elements

Fire is an important disturber of ecosystems worldwide. Every year, 200–400 hectares of savannas, 5–15 million hectares of boreal forests, and lesser amounts of other biomes burn. Lightning ignites some fires, but most fires are started by humans as a way of managing vegetation. Fires consume the energy of, and release chemical elements from, the vegetation they burn. Some nutrients, such as nitrogen, sulfur, and selenium, are easily vaporized by fire. They are discharged to the atmosphere in smoke or carried to groundwater by rain falling on the burned ground.

Fires also release large amounts of carbon into the atmosphere. The global annual flux of carbon to the atmosphere from savanna and forest fires is estimated to be in the range of 1.7 to 4.1 petagrams (one petagram = 10^{15} grams). Biomass burning is responsible for about 40 percent of Earth's annual production of CO_2. It is also a significant contributor to the production of other greenhouse gases, such as carbon monoxide (CO) (32%), methane (CH_4) (10%), and tropospheric O_3 (38%).

Because humans deliberately start most fires, we can take steps to reduce their frequency and intensity. As we saw in Chapter 57, periodic burning of fire-prone vegetation can prevent the buildup of large fuel loads and reduce the frequency of high-intensity canopy fires, which discharge great quantities of materials to the atmosphere. Improvements in the productivity of tropical agriculture can reduce the rate at which forests are cleared and burned to create more agricultural land. Expanded use of solar cookers can reduce the need to cut trees and burn the wood to cook meals.

58.6 The Global Carbon Cycle The numbers show the quantities of carbon (expressed as units of 10^{15} g) held in or exchanged annually by ecosystem compartments. The widths of the arrows are proportional to the sizes of the fluxes.

Our discussion of water and fire shows that the biogeochemical cycles of different chemical elements are connected to one another. However, it is easier to understand the interactions of these cycles if we discuss each one separately.

The Carbon Cycle

Organisms are triumphs of organic chemistry; to survive, they must have access to carbon atoms. Nearly all the carbon in organisms comes from CO_2 in the atmosphere or dissolved carbonate ions (HCO_3^-) in water. Carbon is incorporated into organic molecules by photosynthesis in the cells of autotrophs. All heterotrophic organisms get their carbon by consuming autotrophs or other heterotrophs, their remains, or their waste products. Biological processes move carbon between the atmospheric and terrestrial compartments, removing it from the atmosphere during photosynthesis and returning it to the atmosphere during respiration (Figure 58.6). Most of Earth's carbon is stored in the oceans; on land, most carbon that is available to organisms is stored in soils.

At times in the remote past, great quantities of carbon were removed from the global carbon cycle when organisms died in large numbers and were buried in sediments lacking O_2. In such anaerobic environments, detritivores do not reduce organic carbon to CO_2. Instead, organic molecules accumulate and eventually are transformed into deposits of oil, natural gas, coal, or peat. Humans have discovered and used these deposits, known as **fossil fuels**, at ever-increasing rates during the past 150 years. As a result, CO_2, the final product of the burning of these fuels, is being released into the atmosphere faster today than it is dissolving in surface waters of the oceans or being incorporated into terrestrial biomass.

Based on a variety of calculations, atmospheric scientists believe that 150 years ago, before the Industrial Revolution, the concentration of atmospheric CO_2 was probably about 265 parts per million. Today it is 350 parts per million (Figure 58.7). This difference represents a rate of increase more than 10 times faster than at any other time for millions of years. How will global climate and Earth's ecosystems change in response to this rapid CO_2 enrichment? What are the main factors driving these changes in the carbon cycle, and what processes may tend to stabilize them?

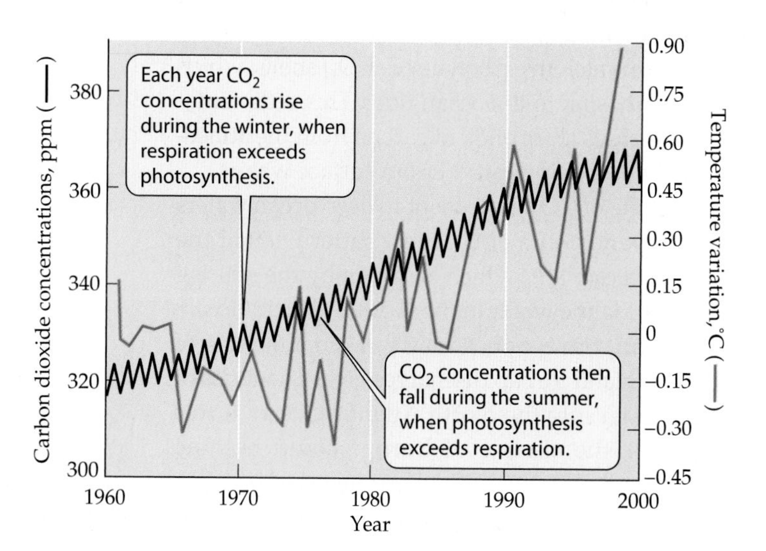

58.7 Atmospheric Carbon Dioxide Concentrations Are Increasing These carbon dioxide concentrations, expressed as parts per million by volume of dry air, were recorded on top of Mauna Loa, Hawaii.

Carbon concentrations in the atmosphere influence Earth's climate

The buildup of atmospheric CO_2 that has resulted from the burning of fossil fuels is warming Earth. The concentration of CO_2 in air trapped in the Antarctic and Greenland ice caps during the last Ice Age—between 15,000 and 30,000 years ago—was as low as 200 parts per million. During a warm interval 5,000 years ago, atmospheric CO_2 may have been slightly higher than it is today. This long-term record shows that Earth was warmer when CO_2 levels were higher and cooler when they were lower.

Complex computer models of Earth's system indicate that a doubling of the atmospheric CO_2 concentration would shift climates toward the poles, would probably cause droughts in the central regions of continents, and would increase precipitation in coastal areas. Global warming could result in the melting of the Greenland and Antarctic ice caps, and would warm the oceans. If so, the oceans would expand, raising sea levels and flooding coastal cities and agricultural lands.

Both physical and biological processes control the carbon cycle

Over decades to centuries, the oceans, which have 50 times the amount of dissolved inorganic carbon as the atmosphere, determine atmospheric CO_2 concentrations. The rate at which CO_2 moves from the atmosphere to the oceans depends, in part, on photosynthesis by plankton in the surface waters. These organisms remove carbon from the water, thereby increasing its absorption of carbon from the atmosphere. In addition, many marine organisms form calcium carbonate ($CaCO_3$) shells, which eventually sink to the ocean floor. This sedimentation increases absorption of carbon from the atmosphere by removing carbon from surface waters.

The rate of CO_2 movement to deep ocean waters also depends on a circulation pattern called the *ocean conveyor belt*, which is driven by the sinking of dense, saline water in the North Atlantic Ocean. The ocean conveyor belt may weaken if melting of the Greenland ice cap discharges great quantities of fresh water into the North Atlantic Ocean. If this happens, the climate of Europe could become colder while climates elsewhere are warming.

Each year, photosynthesis by terrestrial vegetation, principally in forests and savannas, absorbs about 60 billion metric tons of carbon. About the same amount of carbon is released by respiration, about half by the plants themselves and half by microbes decomposing organic matter produced by the plants. Currently photosynthetic consumption of CO_2 appears to exceed respiratory production of CO_2 by 2 billion metric tons of carbon per year. Thus, Earth's forests are storing carbon that would otherwise be increasing atmospheric CO_2 concentrations.

Ecologists are conducting experiments to determine the effects of higher atmospheric concentrations of CO_2 on rates

EXPERIMENT

Hypothesis: Higher atmospheric concentrations of CO_2 will result in increased carbon uptake and increased storage of carbon in below-ground carbon pools.

METHOD Establish circular, open-top chambers. Blow CO_2-enriched air into experimental chambers to achieve a concentration of 720 ppm. Blow natural air into other chambers that serve as controls. Measure shoot and surface litter mass in each chamber. Remove soil cores from each chamber and measure carbon content of roots and detritus.

RESULTS

(*a*) Sandstone grassland ecosystem

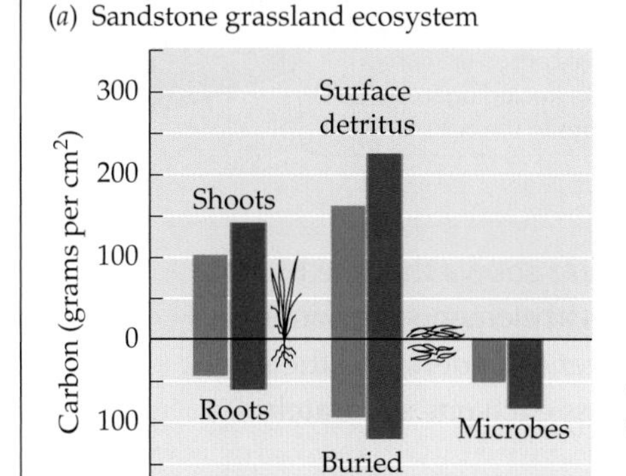

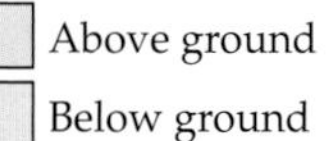

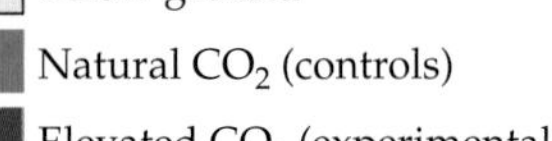

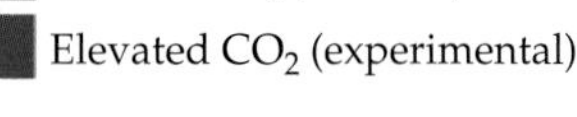

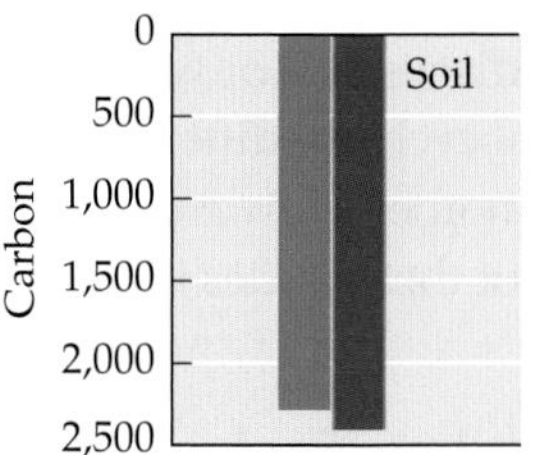

(*b*) Serpentine grassland ecosystem

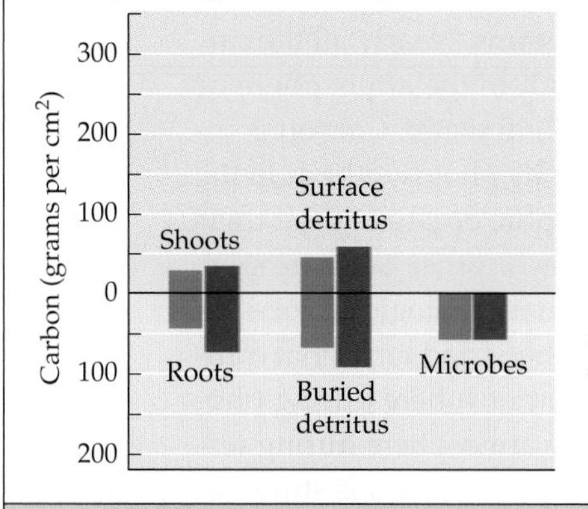

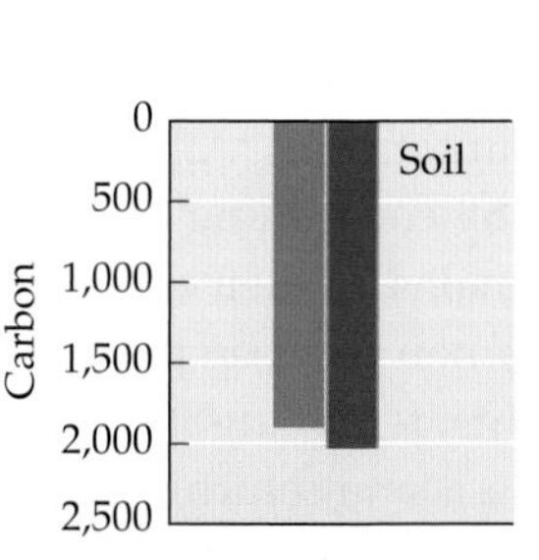

Conclusion: Experimental doubling of CO_2 did increase carbon uptake, but, contrary to predictions, carbon was partitioned to rapidly cycling carbon pools below ground. As a result, carbon storage did not increase. Thus, increased productivity may not lead to increased long-term storage of carbon.

58.8 Will Increased CO_2 Levels Increase Carbon Storage? The results of these experiments, which took place over 3 years on two different grassland ecosystems, suggest that in these ecosystems elevated levels of atmospheric CO_2 result in higher rates of photosynthesis (carbon uptake), but the amount of carbon stored in above-ground plant parts does not increase significantly.

of photosynthesis and carbon storage in ecosystems (Figure 58.8). In the experiment shown, photosynthetic rates increased but long-term carbon storage did not. However, this lack of increased carbon storage in grassland ecosystems does not mean that carbon storage in forests—where carbon is stored in wood—will not increase.

Humans must try to influence the carbon cycle

The adverse consequences of increased atmospheric concentrations of CO_2 are likely to be severe. Therefore, scientists are exploring various methods to reduce the rate of release of CO_2 to the atmosphere and to increase the rate of its removal from the atmosphere and its transfer to long-term storage. No single, simple action can solve the problem, but many steps can be taken that would help to reduce atmospheric concentrations of CO_2.

For example, we can increase fuel efficiency of motor vehicles and airplanes, the efficiency of applicances, and improve mass public transportation. We could also impose carbon taxes that would encourage more efficient use of fossil fuels.

The Nitrogen Cycle

Nitrogen gas (N_2) makes up 78 percent of the atmosphere, but most organisms cannot use nitrogen in its gaseous form. Only a few species of microorganisms can convert atmospheric N_2 into forms that are usable by plants, a process called *nitrogen fixation*. *Denitrification*, the principal process that removes nitrogen from the biosphere and returns it to the atmosphere, is also carried out by microorganisms. These movements of nitrogen among organisms and between organisms and the atmosphere were detailed in Chapter 37 (see Figure 37.8). They account for about 95 percent of all nitrogen fluxes on Earth (Figure 58.9).

Biologically usable nitrogen is often in short supply in ecosystems. That is why nearly all fertilizers contain compounds of nitrogen. Populations of nitrogen-fixing organisms do not increase to such an extent that nitrogen is no longer limiting because nitrogen tends to be lost rapidly from ecosystems by leaching, vaporization of ammonia, and denitrification. Also, fixing nitrogen requires large amounts of energy. As a result, nitrogen-fixing organisms often lose out in competition with non-fixers when nitrogen becomes more readily available.

As a result of extensive use of fertilizers on agricultural crops and the burning of fossil fuels (which generates nitric oxide), the total nitrogen fixation by humans today is about equal to global natural nitrogen fixation. The human-caused flux is expected to continue to increase during the coming decades. A variety of adverse effects are associated with these large perturbations of the nitrogen cycle. They include the contamination of groundwater by nitrate (NO_3^-) from agricultural runoff, increases in atmospheric nitrous oxide (N_2O) and tropospheric O_3 (both greenhouse gases), and smog production.

When more nitrogen is applied to croplands than is taken up by plants, the excess nitrogen moves downward into groundwater and, eventually, into rivers, lakes, and oceans. The addition of nutrients to these bodies of water, known as **eutrophication**, can have a number of negative effects on aquatic ecosystems, as we'll see in our discussion of the phosphorus cycle. The "dead zone" in the Gulf of Mexico that has formed around the mouth of the Mississippi River is a result of flows of nitrogen-enriched water from agricultural fields in the Upper Midwest (Figure 58.10). Outbreaks of the toxic dinoflagellate *Pfiesteria* in estuaries on the Atlantic coast of North America are another example of the adverse consequences of nitrogen enrichment of ocean waters.

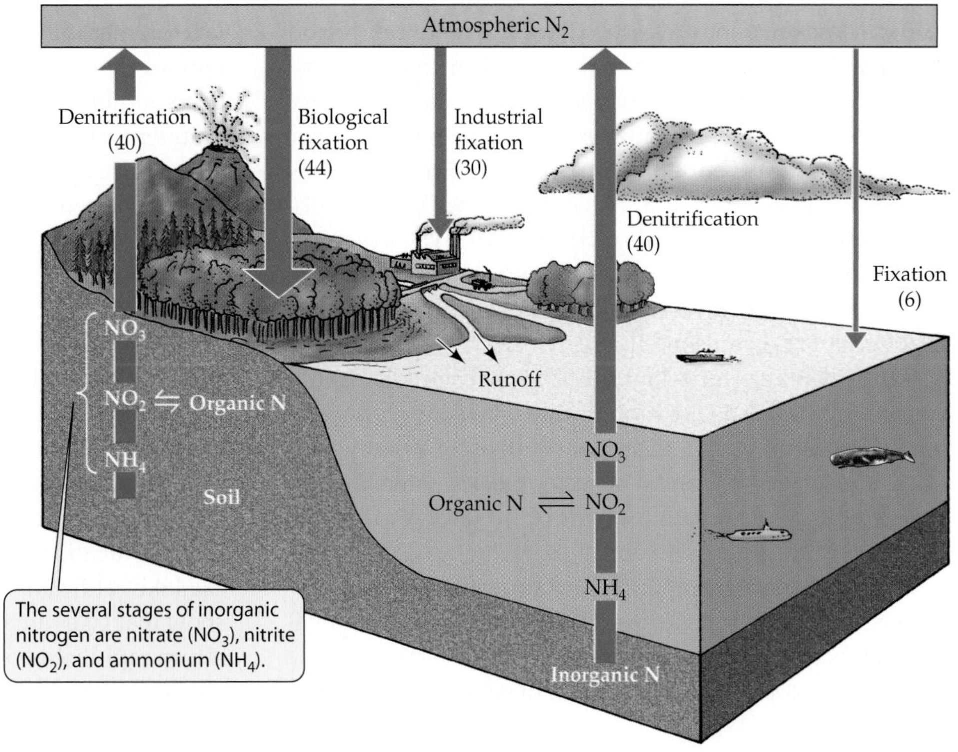

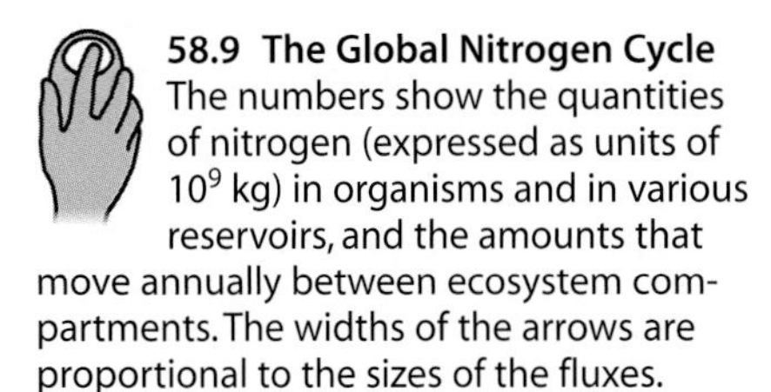

58.9 The Global Nitrogen Cycle The numbers show the quantities of nitrogen (expressed as units of 10^9 kg) in organisms and in various reservoirs, and the amounts that move annually between ecosystem compartments. The widths of the arrows are proportional to the sizes of the fluxes.

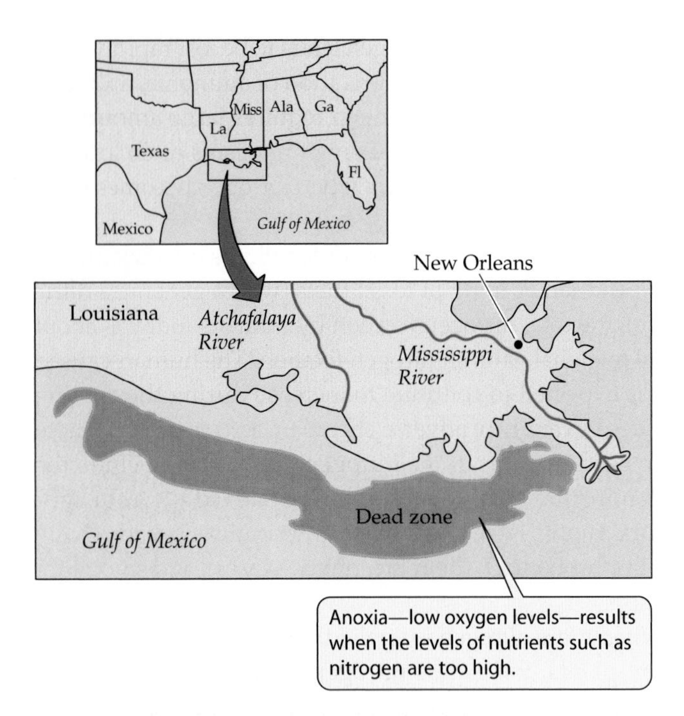

58.10 A "Dead Zone" at the Mouth of the Mississippi River Rising amounts of nitrogen and phosphorus in the runoff from agricultural lands in the midwest United States reach the Gulf of Mexico. Resulting algal growth from this nutrient enrichment depletes the oxygen in the water, creating a deoxygenated "dead zone" in which most aquatic organisms cannot survive.

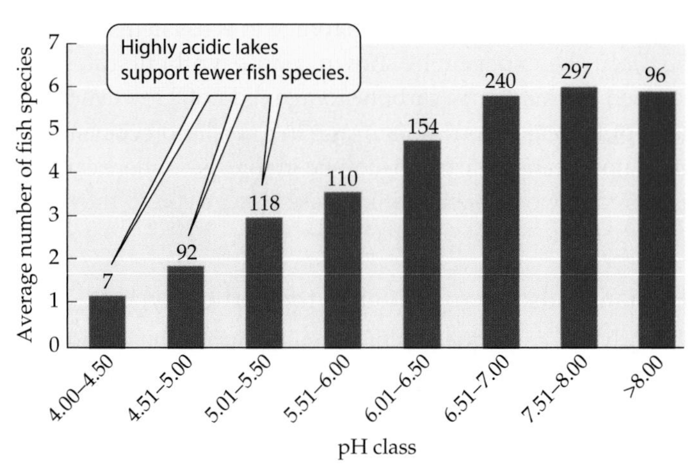

58.11 Acidification of Lakes Exterminates Fish Species The number of fish species found in lakes in the Adirondack region of New York State is inversely correlated with pH. (Recall from Chapter 2 that lower numbers indicate greater acidity, and a pH of 7 is neutral. The numbers above the bars indicate the number of lakes in each pH category.)

The Sulfur Cycle

Emissions of the gases sulfur dioxide (SO_2) and hydrogen sulfide (H_2S) from volcanoes and fumaroles (vents for hot gases) are the only significant natural nonbiological fluxes of sulfur. These emissions release, on average, between 10 and 20 percent of the total natural flux of sulfur to the atmosphere, but they vary greatly in time and space. Large volcanic eruptions spread great quantities of sulfur over broad areas, but they are rare events. Terrestrial and marine organisms also emit compounds of sulfur. Certain marine algae produce large amounts of dimethyl sulfide (CH_3SCH_3), which accounts for about half the biotic component of the sulfur cycle. Sulfur is apparently always abundant enough to meet the needs of living organisms.

Sulfur plays an important role in global climate. Even if air is moist, clouds do not form readily unless there are small particles around which water can condense. Dimethyl sulfide is the major component of such particles. Therefore, increases or decreases in atmospheric sulfur levels can change cloud cover and hence climate.

Humans have altered the sulfur cycle, as well as the nitrogen cycle, by the burning of fossil fuels. An important regional effect of these alterations is **acid precipitation**—rain or snow whose pH is lowered by the presence of sulfuric acid (H_2SO_4) and nitric acid (HNO_3), derived in large part from the burning of fossil fuels. These acids enter the atmosphere and may travel hundreds of kilometers before they settle to Earth in precipitation or as dry particles.

Acid precipitation now characterizes all major industrial countries and is particularly widespread in eastern North America and Europe. The normal pH of precipitation in New England is about 5.6, but precipitation there now averages about pH 4.4, and there are occasional storms with a precipitation pH as low as 3.0. Precipitation with a pH of about 3.5 or lower causes direct damage to the leaves of plants and reduces photosynthetic rates. Acidification of lakes in the Adirondack region of New York has reduced fish species richness by causing the extinction of acid-sensitive species (Figure 58.11). Fortunately, as a result of the establishment of a flexible regulatory system under the 1990 Clean Air Act Amendments, precipitation in much of the eastern United States is less acid today than it was 18 years ago, primarily because of reductions in sulfur emissions (Figure 58.12).

Ecologists in Canada studied the effects of acid precipitation on small lakes by adding enough H_2SO4 to two lakes to reduce their pH from about 6.6 to 5.2. In both lakes, nitrifying bacteria failed to adapt to these moderately acidic conditions. As a result, the nitrogen cycle was blocked, and ammonium accumulated in the water. When the ecologists stopped adding acid to one of the lakes, its pH increased to 5.4, and nitrification resumed. After about 1 year, the pH of the lake returned to its original value. These experiments show that lakes are very sensitive to acidification, but pH can return rapidly to normal values because water in lakes is exchanged at a rapid rate.

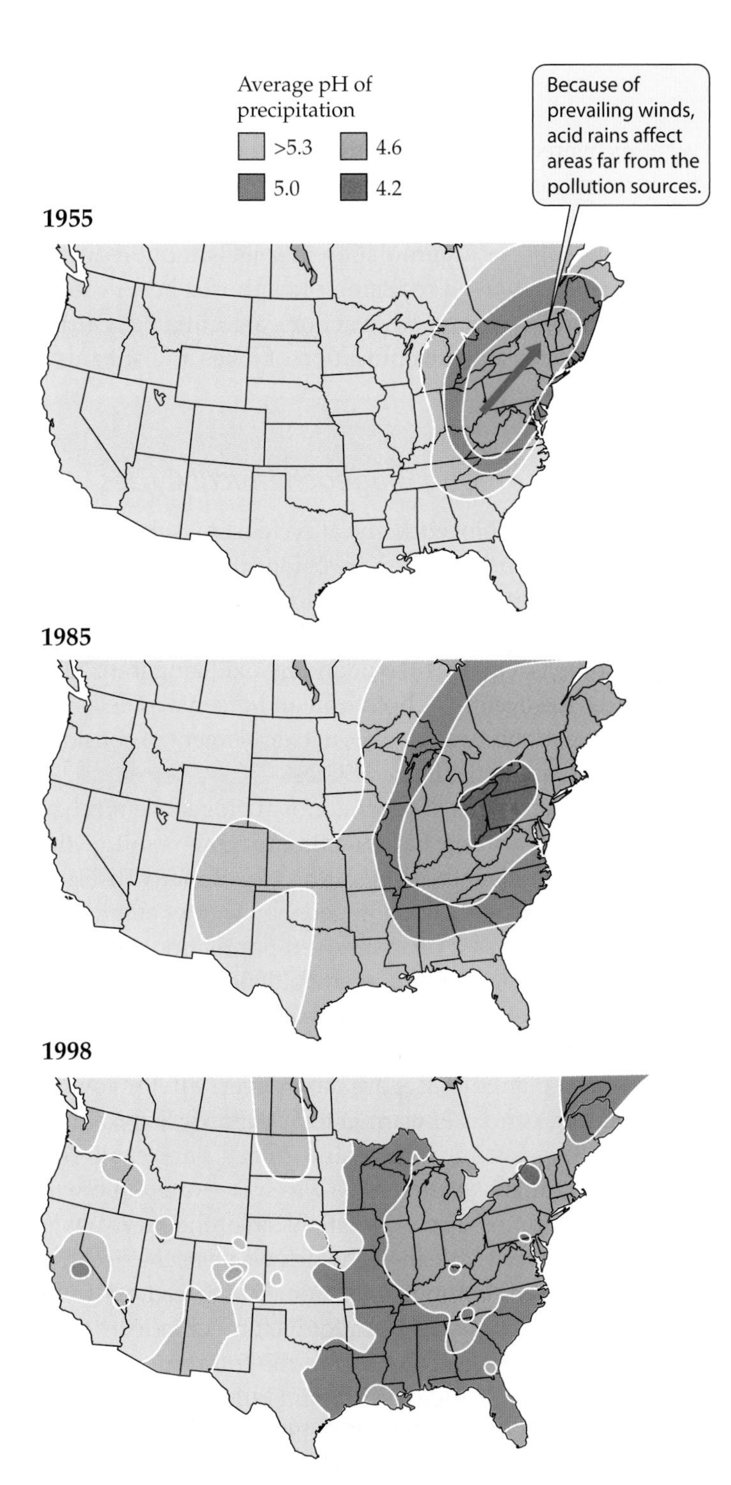
Average pH of precipitation
>5.3
4.6
5.0
4.2
Because of prevailing winds, acid rains affect areas far from the pollution sources.
1955
1985
1998

58.12 Acid Precipitation Is Decreasing in the Eastern United States Thanks to emission controls, precipitation in many parts of the eastern United States is less acid than it was three decades ago.

The Phosphorus Cycle

Phosphorus accounts for only about 0.1 percent of Earth's crust, but it is an essential nutrient for all life forms. It is a key component of DNA and ATP. Unlike the other elements discussed in this section, phosphorus lacks a gaseous phase. Some phosphorus is transported on dust particles, but in general the atmospheric compartment plays a very minor role in the phosphorus cycle. The global phosphorus cycle takes millions of years to complete because the processes of rock formation on the ocean bottom, subsequent uplifting, and weathering of rock into soil all act slowly (Figure 58.13). However, an atom of phosphorus may cycle rapidly among organisms, as did *X* in Aldo Leopold's account at the beginning of this chapter.

Human activity has radically accelerated some parts of the phosphorus cycle. About 90 percent of the phosphorus that is mined is used to produce fertilizers and animal feeds. One consequence of our massive use of phosphorus for fertilizer is that between 10.5 and 15.5 million metric tons of phosphorus are accumulating in soils each year, primarily on agricultural lands (Figure 58.14). Increasing concentrations of phosphorus in agricultural soils is certain to lead to increased flows of phosphorus to streams and lakes.

Phosphorus is often a limiting nutrient in soils and lakes, which is why adding phosphorus to farmland and lakes increases their biological productivity. Most of this extra phosphorus enters lakes in the form of phosphates derived from fertilizers and household detergents. The resulting eutrophication allows algae and bacteria to multiply, forming blooms that turn the water green. The decomposition of dead cells produced by this increased biological activity consumes all the O_2 in the lake, and anaerobic organisms come to dominate the bottom sediments. These anaerobic organisms do not break down carbon compounds all the way to CO_2. Their metabolic end products build up; many of these products have unpleasant odors.

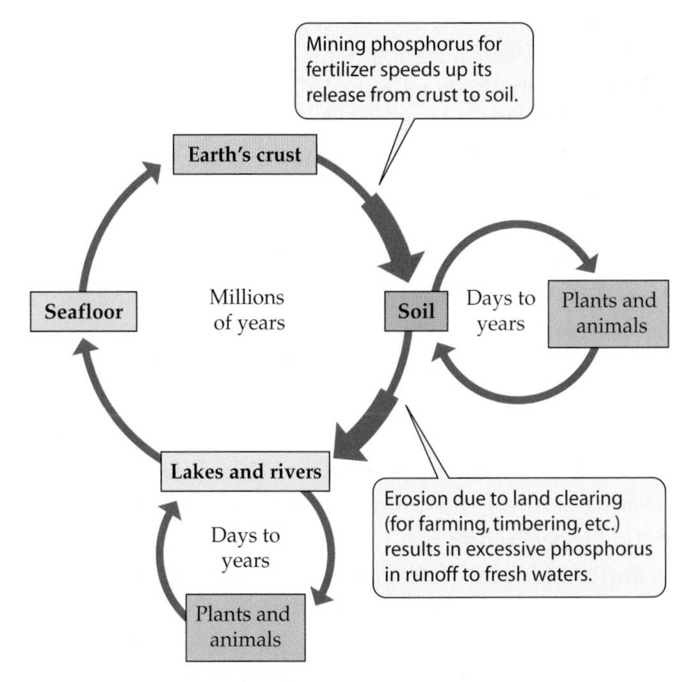
Mining phosphorus for fertilizer speeds up its release from crust to soil.
Earth's crust
Seafloor
Millions of years
Soil
Days to years
Plants and animals
Lakes and rivers
Days to years
Plants and animals
Erosion due to land clearing (for farming, timbering, etc.) results in excessive phosphorus in runoff to fresh waters.

58.13 The Phosphorus Cycle The width of the arrows is proportional to flux rates. The two great increases are due to human activities.

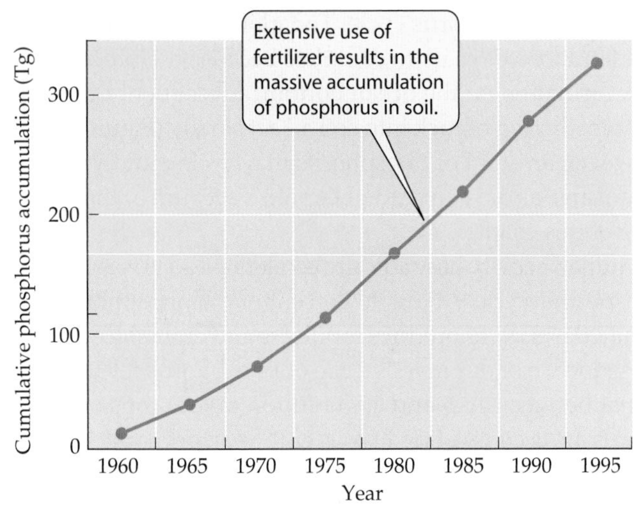

58.14 Phosphorus Is Accumulating in Agricultural Soils Between 10.5 and 15.5 million metric tons of phosphorus are being added to agricultural soils each year. This graph shows the cumulative effects of this excess over a period of 40 years. (A teragram, Tg, equals 10^{12} grams.)

Lake Erie, one of the Great Lakes on the border between the United States and Canada, is a eutrophic lake today. Two hundred years ago, it had only moderate rates of photosynthesis and clear, oxygenated water. More than 15 million people live in the Lake Erie drainage basin. Nearby cities pour more than 250 billion liters of domestic and industrial wastes into the lake annually. The entire basin is intensely farmed and heavily fertilized.

In the early part of the twentieth century, nutrient concentrations in the lake increased greatly, and algae proliferated. At the water filtration plant in Cleveland, Ohio, algae increased from 81 per milliliter in 1929 to 2,423 per milliliter in 1962. Algal blooms and populations of bacteria also increased. The numbers of the colon bacterium *Escherichia coli* increased enough to cause the closing of many of the lake's beaches as health hazards.

Since 1972, the United States and Canada have invested more than 9 billion dollars to improve municipal waste facilities and reduce discharges of phosphorus into Lake Erie. As a result, the amount of phosphate added to Lake Erie has decreased more than 80 percent from the maximum level, and phosphorus concentrations in the lake have declined substantially. The deeper waters of Lake Erie still become poor in O_2 during the summer months, but the rate of O_2 depletion is declining.

Fortunately, the potential for recovery and recycling of phosphorus is high. The amount of phosphorus contained in sewage and animal wastes could supply much of the needs of the detergent and fertilizer industries. In the United Kingdom, for example, if only 50 percent of the phosphorus in 25 percent of the sewage and 15 percent of the animal wastes were recovered and recycled, recycling could supply half of the country's industrial demand for phosphorus. Careful application of fertilizers on agricultural lands can reduce the rate of phosphorus accumulation in soils without reducing crop yields. However, reducing phosphorus in soils takes many decades after remedial actions are initiated. During that time, increased eutrophication of lakes and streams is certain to happen.

Interactions among Biogeochemical Cycles

Our discussion of biogeochemical cycles shows that they interact strongly, and that they are being substantially altered by human activities. The acidity of precipitation, for example, is a result of the combined influences of SO_2, NO_3^-, and NH_3. Nitrate is a powerful oxidant. By oxidizing iron, NO_3^- influences the cycling of both iron and arsenic in lakes, increasing the mobility of arsenic, a toxic element of considerable importance in the United States.

Human alterations of several global biogeochemical cycles are also warming Earth's climate. One result of this global warming is increasing outbreaks of many diseases. Winter cold typically kills pathogens, sometimes eliminating as much as 99 percent of a pathogen population. If climate warming reduces this population bottleneck, some diseases will become more common. For example, dengue fever is now spreading to higher latitudes where it was formerly absent. Several plant diseases are more severe after mild winters or during periods of warmer temperatures. For example, during a 39-year study in Maine, beech bark cankering caused by a fungus (*Nectria* spp.) was worse after mild winters or dry autumns. These conditions favor the survival and spread of the beech scale insect, which weakens the trees and predisposes them to fungal infection. The spread of this infection poses a serious problem for the timber industry.

The 1990 Assessment Report of the Intergovernmental Panel on Climate Change (IPCC) paid little attention to risks to human health. In contrast, IPCC's 1996 Assessment Report gave detailed consideration to the potential effects of climate change on human health. Our increasingly interconnected global society enables disease organisms to travel rapidly around the world, as in the case of the SARS virus, which was carried by infected people from China to Canada within a few days. The combination of human mobility and climate warming poses serious challenges to human health throughout the world.

Visions of the Future

As we saw in Chapter 54, the size of Earth's human population is a factor in many environmental problems, and it is increasing rapidly (see Figure 54.20). Anyone 40 years old or

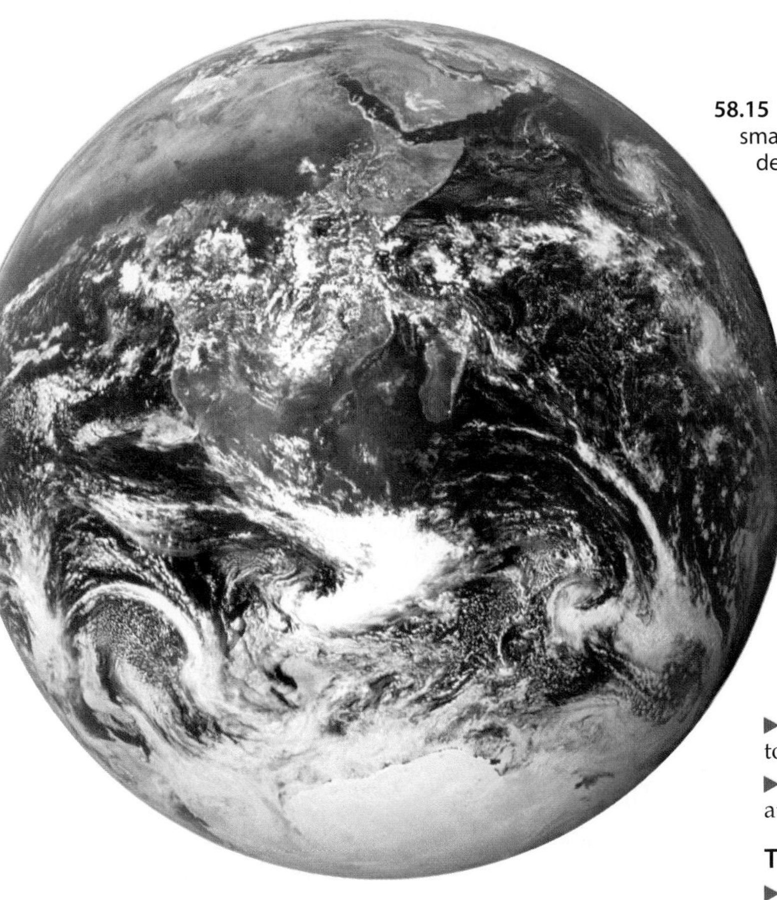

58.15 Earth from Space The view from space reminds us that Earth is a small planet. The future of humanity, as well as that of other species, will depend on how we function as stewards of Earth and its resources.

older has lived through a doubling of the human population. But around 1965, the population growth rate peaked and began to decline. The most recent decline before that time, in the fourteenth century, was caused by increased death rates from plagues, wars, and famines. In contrast, the current decline is being caused by voluntary reductions in fertility. Fertility rates have been declining in rich countries for more than a century, but they began to fall in poor countries as well in 1965. Nobody had predicted this change, but the fact that it was voluntary gives us hope that humanity may be able, at least in part, to choose the kind of world in which its children and grandchildren will live.

If humans are to live on Earth in a sustainable manner that provides opportunities for the development of human creativity and rewarding lifestyles, major changes will need to be made in how we use environmental resources. Because we have limited abilities to regenerate and restore degraded natural ecosystems, we will need to develop an increased appreciation of the roles of living organisms in Earth's system. We must not load our ecosystems so heavily that their Plimsoll lines sink below the water line.

Chapter Summary

Earth's System Has Four Compartments

▶ The elements on which life depends cycle among the four compartments of Earth's physical environment: oceans, fresh waters, atmosphere, and land. **Review Figure 58.1**

▶ Primary production in oceans is highest in zones of upwelling adjacent to continents, where nutrient-rich waters rise to the surface. **Review Figure 58.2**

▶ Gases in the atmosphere are important regulators of temperatures on Earth.

▶ Deep lakes in the temperate zone have an annual turnover cycle that is driven by temperature. **Review Figure 58.3**

Biogeochemical Cycles, Water, and Fire

▶ The hydrological cycle is driven by evaporation of water, most of it from ocean surfaces. **Review Figure 58.4. See Web/CD Tutorial 58.1**

▶ Human activities have altered the flux of water from the land to the oceans. **Review Figure 58.5**

▶ Biomass burning contributes great quantities of carbon to the atmosphere.

The Carbon Cycle

▶ Atmospheric carbon dioxide is the immediate source of carbon for terrestrial organisms, but only a small part of Earth's carbon is in the atmosphere. **Review Figure 58.6. See Web/CD Tutorial 58.2**

▶ Increasing concentrations of CO_2 in the atmosphere are changing climates and influencing ecological processes. **Review Figure 58.7, 58.8**

▶ The ocean conveyor belt carries great quantities of carbon to the deep ocean. **See Web/CD Tutorial 58.3**

The Nitrogen Cycle

▶ Nitrogen makes up 78 percent of Earth's atmosphere, but nitrogen can be converted into biologically useful forms only by a few species of microorganisms. **Review Figure 58.9. See Web/CD Tutorial 58.4**

▶ Runoff of nitrogen from agricultural lands causes eutrophication in aquatic ecosystems. **Review Figure 58.10**

The Sulfur Cycle

▶ Acid precipitation results from the combined effects of human alterations in the nitrogen and sulfur cycles. **Review Figures 58.11, 58.12**

The Phosphorus Cycle

▶ The phosphorus cycle differs from the cycles of carbon, nitrogen, and sulfur in that it lacks a gaseous phase. **Review Figure 58.13**

▶ As a result of high fertilization rates, large amounts of phosphorus are accumulating in agricultural soils. **Review Figure 58.14**

▶ Eutrophication resulting from human inputs of phosphorus has damaged ecosystems in many lakes.

Interactions among Biogeochemical Cycles

▶ Human alterations of global biogeochemical cycles are changing Earth's climate. Outbreaks of diseases are one effect of this global warming.

Visions of the Future

▶ Actions people might undertake today could greatly influence the future quality of human life and the welfare of other species that share our small planet with us.

Self-Quiz

1. Earth is in chemical disequilibrium because
   a. Earth has a moon.
   b. organisms dissipate energy as heat.
   c. most continents are in the Northern Hemisphere.
   d. Earth has living organisms that maintain the O_2, N_2, and H_2O of the atmosphere.
   e. Earth is tilted on its axis.
2. Zones of marine upwelling are important because
   a. they help scientists measure the chemistry of deep ocean water.
   b. they bring to the surface organisms that are difficult to observe elsewhere.
   c. ships can sail faster in these zones.
   d. they increase marine productivity by bringing nutrients back to surface ocean waters.
   e. they bring oxygenated water to the surface.
3. Which of the following is *not* true of the troposphere?
   a. It contains nearly all of the atmospheric water vapor.
   b. Materials enter it primarily at the intertropical convergence zone.
   c. It is about 17 km deep in the Tropics.
   d. Most global air circulation takes place there.
   e. It contains about 80 percent of the mass of the atmosphere.
4. The hydrological cycle is driven by the
   a. flow of water into the oceans via rivers.
   b. evaporation (transpiration) of water from the leaves of plants.
   c. evaporation of water from the surface of the oceans.
   d. precipitation falling on the land.
   e. fact that less water falls on the ocean as precipitation than evaporates from its surface.
5. Carbon dioxide is called a greenhouse gas because
   a. it is used in greenhouses to increase plant growth.
   b. it is transparent to heat, but traps sunlight.
   c. it is transparent to sunlight, but traps heat.
   d. it is transparent to both sunlight and heat.
   e. it traps both sunlight and heat.
6. The ocean conveyor belt carries water to the deep ocean in the
   a. North Pacific Ocean
   b. South Pacific Ocean
   c. North Atlantic Ocean
   d. South Atlantic Ocean
   e. Indian Ocean
7. The cycle of phosphorus differs from the cycles of carbon and nitrogen in that
   a. it lacks a gaseous phase.
   b. it lacks a liquid phase.
   c. only phosphorus is cycled through marine organisms.
   d. living organisms do not need phosphorus.
   e. The phosphorus cycle does not differ importantly from the carbon and nitrogen cycles.
8. The sulfur cycle influences the global climate because
   a. sulfur compounds are important greenhouse gases.
   b. sulfur compounds help transfer carbon from the atmosphere to the oceans.
   c. sulfur compounds in the atmosphere are particles around which water condenses to form clouds.
   d. sulfur compounds contribute to acid precipitation.
   e. The sulfur cycle does not influence the global climate.
9. Acid precipitation results from human modifications of
   a. the carbon and nitrogen cycles.
   b. the carbon and sulfur cycles.
   c. the carbon and phosphorus cycles.
   d. the nitrogen and sulfur cycles.
   e. the nitrogen and phosphorus cycles.
10. Acid precipitation may change lakes by
    a. making phosphorus more available to plants, thereby increasing productivity.
    b. making nitrogen more available to plants, thereby increasing productivity.
    c. increasing populations of nitrogen-fixing microorganisms.
    d. accelerating the carbon cycle within lakes.
    e. causing the local extinction of species that cannot tolerate the lower pH.

For Discussion

1. A powerful hurricane strikes the East Coast of the United States. Some people claim that this disaster was due to warming of the oceans caused by greenhouse gases in the atmosphere. Others assert that global warming is not responsible because hurricanes have occurred for many centuries. How would you evaluate these conflicting claims?
2. The waters of Lake Washington, adjacent to the city of Seattle, rapidly returned to their pre-industrial condition when sewage was diverted from the lake to Puget Sound, an arm of the Pacific Ocean. Would all lakes being polluted with sewage clean themselves up as quickly as Lake Washington if pollutant inputs were stopped? What characteristics of a lake are most important to its rate of recovery following reduction of pollutant inputs?
3. Tropical forests currently are being cut at a very rapid rate. Does this necessarily mean that deforestation is a major source of carbon dioxide to the atmosphere? If not, why not?
4. What types of experiments would you conduct to assess the likely consequences of fertilization of the oceans with iron to increase rates of photosynthesis? At what spatial and temporal scales should they be conducted?
5. A government official authorizes the construction of a large coal-burning power plant in a former wilderness area. Its smokestacks discharge great quantities of combustion wastes. List and describe all likely effects on Earth's system at local, regional, and global levels. Now suppose the wastes were thoroughly scrubbed from the stack gases. Which of the effects you have just outlined would still happen?
6. Many nations recently signed the Kyoto Accord, which obligates them to reduce their emissions of carbon dioxide to the atmosphere. The United States refused to sign the treaty, claiming it was against the country's interests. The United States has been severely criticized for abandoning the Kyoto process. Is such criticism warranted? Justify your answer.

LIFE ESSAY

Toward economic principles for sustainable ecosystems management

- by William E. Rees -

Most economic analysis deals only with those things of direct use to humans that can be readily measured in monetary terms. All other values, including most of those associated with nature, are deemed *externalities* (outside the compass of the market) by economists.

Environmental economists have responded to this limitation by trying to "price" certain features of ecosystems that have not historically been perceived to have monetary value. This effort at least recognizes that nature makes substantial contributions to the economy. The question is, Do these evaluations enable us to incorporate all the values associated with ecosystems into the prices people pay for commodities derived from "natural capital"? Regrettably, the answer is no.

Many of the values associated with ecosystems are difficult to quantify, let alone price. In particular, consider those ecosystem components and processes whose precise contribution to life support and the economy is not known, and may not be even be suspected until they are lost or destroyed. An economics that does not reflect this reality would be dangerously misleading in the decision process.

The emerging discipline of ecological economics attempts to address this problem, recognizing that the human economy is an open, growing, *dependent* subsystem of the materially closed, nongrowing ecosphere. Ecological economics sees economic activity as a tranformation process subject to biophysical laws, particularly the second law of thermodynamics. In this light, both ecosystems and the economy are revealed as structures that grow and develop by extracting available energy and material from their host systems and exporting their wastes (entropy) back into their hosts. In short, they maintain their internal order at the expense of increasing disorder elsewhere in the hierarchy.

The problem is that, while the ecosphere *evolves* by dissipating solar energy, the economy *grows* by dissipating the ecosphere. In other words, thermodynamically, the economy is positioned to consume and pollute the ecosphere from the inside out—a second law explanation for the degradation of the ecosphere.

Moreover, recognizing that a certain minimal amount of nature must be conserved to support the human enterprise, ecological economists have advanced the "constant capital stocks criterion" for sustainability. In its strongest form, this criterion states that *each generation must inherit from the previous generation at least an equivalent adequate per capita stock of both human-made and natural capital.*

But how can we determine what constitutes an "adequate" supply of nature? The behavior of ecosystems under stress is often chaotic and characterized by multiple possible equilibrium states. An overexploited ecosystem may approach some critical threshold with little warning that the system is about to flip into a new equilibrium—and it is possible that the new state may not be compatible with human needs.

This is not mere conjecture. Studies show that industrial fishing has reduced the predatory fish biomass of the oceans to about 10 percent of preindustrial levels in just 50 years. Many depleted fish stocks, such as Canada's northern cod (which once supported one of the richest fisheries in the world) show no sign of recovery after a decade or more of suspended fishing effort. Taming humanity's expansionist tendencies must be the primary goal of sustainability economics. Humans cannot harvest more than ecosystems can produce sustainably without being pushed beyond some unforeseen, dangerous threshold.

In summary, sustainability economics must be a risk-averse economics designed to keep human society at safe distance from treacherous boundaries. But mere economics is not enough. Sustainable ecosystems management must factor nonmarket values into the decision-making equation. People living in today's techno-industrial world would benefit greatly from a renewed sense of *biophilia*, a feeling of oneness with nature. The simple fact is that humans do not conserve what they do not love.

William E. Rees received his Ph.D. in bioecology from the University of Toronto. He has taught at the University of British Columbia since 1969 and is currently a professor in the University's School of Community and Regional Planning. His recent work is in ecological economics, urban ecology, and human carrying capacity. He is co-author of *Our Ecological Footprint: Reducing Human Impact on Earth*, published in 1996.

Discussion Questions

1. Environmental economists argue that a total economic valuation of nature may show that an ecosystem or species has sufficient dollar value in its natural state to protect it against any "development" that would produce less value. Therefore, economic valuation achieves the conservationist's purpose without having to resort to moral, ethical, spiritual, or other arguments. A counter argument is that a proper consideration of moral and spiritual values and ecological uncertainty would enable us to come to the right decision without having to resort to fundamentally flawed economic analyses. Discuss the pros and cons of these arguments.
2. It can be argued that, like other species, *Homo sapiens* has an innate tendency to expand into all the ecological space available to it regardless of resultant habitat destruction and displacement of other species. Economic growth as promoted by conventional economics is therefore merely a reflection of biological reality, and since the dynamics are "natural" there's not much we can do about it. Eventually nature will have her revenge. Discuss.

Web Links

International Society for Ecological Economics www.ecologicaleconomics.org/

United States Society for Ecological Economics www.ussee.org/

Post-Autistic Economics Network www.paecon.net/

The following site is a rich source of classic papers and data on ecological decay and ecological economics: www.dieoff.com

Appendix: Some Measurements Used in Biology

QUANTITY	NAME OF UNIT	SYMBOL	DEFINITION
Length	meter (*also* metre)	m	A base unit. 1 m = 100 cm (39.37 inches)
	kilometer	km	1 km = 1000 m = 10^3 m (0.62 miles)
	centimeter	cm	1 cm = $\frac{1}{100}$ m = 10^{-2} m
	millimeter	mm	1 mm = $\frac{1}{1000}$ m = 10^{-3} m
	micrometer	μm	1 μm = $\frac{1}{1000}$ mm = 10^{-6} m
	nanometer	nm	1 nm = $\frac{1}{1000}$ μm = 10^{-9} m
Area	square meter	m^2	Area encompassed by a square, each side of which is 1 m in length
	hectare	ha	1 ha = 10,000 m^2 = 10^4 m^2 (2.47 acres)
	square centimeter	cm^2	1 cm^2 = $\frac{1}{10,000}$ m^2 = 10^{-4} m^2
Volume	liter (*also* litre)	l	1 l = $\frac{1}{1000}$ m^3 = 10^{-3} m^3 (1.057 qts)
	milliliter	ml	1 ml = $\frac{1}{1000}$ l = 10^{-3} l = 1 cm^3 = 1 cc
	microliter	μl	1 μl = $\frac{1}{1000}$ ml = 10^{-3} ml = 10^{-6} l
Mass	kilogram	kg	A base unit. 1 kg = 1000 g = 2.20 lbs
	gram	g	1 g = $\frac{1}{1000}$ kg = 10^{-3} kg
	milligram	mg	1 mg = $\frac{1}{1000}$ g = 10^{-3} g = 10^{-6} kg
Time	second	s	A base unit. 1 s = $\frac{1}{60}$ min
	minute	min	1 min = 60 s
	hour	h	1 h = 60 min = 3,600 s
	day	d	1 d = 24 h = 86,400 s
Temperature	kelvin	K	A base unit. 0 K = −273.15°C = absolute zero
	degree Celsius	°C	0°C = 273.15 K = melting point of ice
Heat, work	calorie	cal	1 cal = heat necessary to raise 1 gram of pure water from 14.5°C to 15.5°C = 4.184 J
	kilocalorie	kcal	1 kcal = 1000 cal = 10^3 cal = (in nutrition) 1 Calorie
	joule	J	1 J = 0.2389 cal (The joule is now the accepted unit of heat in most sciences.)
Electric potential	volt	V	A unit of potential difference or electromotive force
	millivolt	mV	1 mV = $\frac{1}{1000}$ V = 10^{-3} V

Glossary

Abdomen (ab' duh mun) [L. *abdomin*: belly] In arthropods, the posterior segments of the body; in mammals, the part of the body containing the intestines and most other internal organs, posterior to the thorax.

Abiotic (a' bye ah tick) [Gk. *a*: not + *bios*: life] Nonliving.

Abscisic acid (ab sighs' ik) A plant growth substance having growth-inhibiting action. Causes stomata to close.

Abscission (ab sizh' un) [L. *abscissio*: break off] The process by which leaves, petals, and fruits separate from a plant.

Absolute temperature scale Also known as the Kelvin scale. A temperature scale in which zero is the state of no molecular motion, or "absolute zero" (–273° on the Celsius scale).

Absorption (1) Of light: complete retention, without reflection or transmission. (2) Of liquids: soaking up (taking in through pores or cracks).

Absorption spectrum A graph of light absorption versus wavelength of light; shows how much light is absorbed at each wavelength.

Abyssal zone (uh biss' ul) [Gr. *abyssos*: bottomless] The deep ocean, below the point that light can penetrate.

Accessory pigments Pigments that absorb light and transfer energy to chlorophylls for photosynthesis.

Acetylcholine A neurotransmitter substance that carries information across vertebrate neuromuscular junctions and some other synapses.

Acetylcholinesterase An enzyme that breaks down acetylcholine.

Acetyl coenzyme A (acetyl CoA) Compound that reacts with oxaloacetate to produce citrate at the beginning of the citric acid cycle; a key metabolic intermediate in the formation of many compounds.

Acid [L. *acidus*: sharp, sour] A substance that can release a proton in solution. (Contrast with base.)

Acid precipitation Precipitation that has a lower pH than normal as a result of acid-forming precursors introduced into the atmosphere by human activities.

Acidic Having a pH of less than 7.0 (a hydrogen ion concentration greater than 10^{-7} molar).

Acoelomate Lacking a coelom.

Acquired Immune Deficiency Syndrome See AIDS.

Acrosome (a' krow soam) [Gr. *akros*: highest + *soma*: body] The structure at the forward tip of an animal sperm which is the first to fuse with the egg membrane and enter the egg cell.

ACTH (adrenocorticotropin) A pituitary hormone that stimulates the adrenal cortex.

Actin [Gr. *aktis*: ray] One of the two major proteins of muscle; it makes up the thin filaments. Forms the microfilaments found in most eukaryotic cells.

Action potential An impulse in a neuron taking the form of a wave of depolarization or hyperpolarization imposed on a polarized cell surface.

Action spectrum A graph of a biological process versus light wavelength; shows which wavelengths are involved in the process.

Activating enzymes Also called aminoacyl-tRNA synthetases; these enzymes catalyze the addition of amino acids to their appropriate tRNAs.

Activation energy (E_a) The energy barrier that blocks the tendency for a set of chemical substances to react.

Active site The region on the surface of an enzyme where the substrate binds, and where catalysis occurs.

Active transport The energy-dependent transport of a substance across a biological membrane against a concentration gradient—that is, from a region of low concentration (of that substance) to a region of high concentration. (See primary active transport, secondary active transport; contrast with facilitated diffusion.)

Adaptation (a dap tay' shun) In evolutionary biology, a particular structure, physiological process, or behavior that makes an organism better able to survive and reproduce. Also, the evolutionary process that leads to the development or persistence of such a trait.

Adenine (A) (a' den een) A nitrogen-containing base found in nucleic acids, ATP, NAD, and other compounds.

Adenosine triphosphate See ATP.

Adenylate cyclase Enzyme catalyzing the formation of cyclic AMP (cAMP) from ATP.

Adrenal (a dree' nal) [L. *ad*: toward + *renes*: kidneys] An endocrine gland located near the kidneys of vertebrates, consisting of two glandular parts, the cortex and medulla.

Adrenaline See epinephrine.

Adrenocorticotropin See ACTH.

Adsorption Binding of a gas or a solute to the surface of a solid.

Aerobic (air oh' bic) [Gr. *aer*: air + *bios:* life] In the presence of oxygen; requiring oxygen.

Afferent (af' ur unt) [L. *ad*: toward + *ferre*: to carry] Carrying to, as in a neuron that carries impulses to the central nervous system, or a blood vessel that carries blood to a structure. (Contrast with efferent.)

AIDS (acquired immune deficiency syndrome) Condition caused by a virus (HIV) in which the body's helper T lymphocytes are reduced, leaving the victim subject to opportunistic diseases.

Aldehyde (al' duh hide) A compound with a —CHO functional group. Many sugars are aldehydes. (Contrast with ketone.)

Aldosterone (al dohs' ter own) A steroid hormone produced in the adrenal cortex of mammals. Promotes secretion of potassium and reabsorption of sodium in the kidney.

Alga (al' gah) (plural: algae) [L.: seaweed] Any one of a wide diversity of mostly photosynthetic protests.

Allantois (al lan' to is) A sac-like extraembryonic membrane that contains nitrogen waste from embryo.

Allele (a leel') [Gr. *allos*: other] The alternate forms of a genetic character found at a given locus on a chromosome.

Allele frequency The relative proportion of a particular allele in a specific population.

Allergy [Ger. *allergie*: altered reaction] An overreaction to amounts of an antigen that do not affect most people; often involves IgE antibodies.

Allometric growth A pattern of growth in which some parts of the body of an organism grow faster than others, resulting in a change in body proportions as the organism grows.

Allopatric speciation (al' lo pat' rick) [Gr. *allos*: other + *patria*: homeland] Also called geographical speciation, this is the formation of two species from one when reproductive isolation occurs because of the interposition of (or crossing of) a physical geographic barrier such as a river. (Contrast with parapatric speciation, sympatric speciation.)

Allostery (al' lo steer y) [Gr. *allos*: other + *stereos*: structure] Regulation of the activity of a protein by the binding of an effector molecule at a site other than the active site.

Alpha (α) helix Type of protein secondary structure; a right-handed spiral.

Alternation of generations The succession of multicellular haploid and diploid phases in some sexually reproducing organisms, notably plants.

Altruism Behavior that harms the individual who performs it but benefits other individuals.

Alveolus (al ve' o lus) (plural: alveoli) [L. *alveus*: cavity] A small, baglike cavity, especially the blind sacs of the lung.

Ambient That which surrounds; the immediate environment.

Amensalism (a men' sul ism) Interaction in which one animal is harmed and the other is unaffected. (Contrast with commensalism, mutualism.)

Amine An organic compound with an amino group (see Amino acid).

Amino acid An organic compound containing both NH_2 and COOH groups. Proteins are polymers of amino acids.

Ammonotelic (am moan' o teel' ic) [Gr. *telos*: end] Describes an organism in which the final product of breakdown of nitrogen-containing compounds (primarily proteins) is ammonia. (Contrast with ureotelic, uricotelic.)

Amnion (am′ nee on) The fluid-filled sac in which the embryos of reptiles, birds, and mammals develop.

Amniote Any of the vertebrate animals whose embryos are enclosed in an amnion: reptiles, birds, and mammals.

Amniote egg A shelled egg surrounding four extraembryonic membranes and embryo-nourishing yolk. This adaptation allowed animals to colonize the terrestrial environment.

Amphipathic (am′ fi path′ ic) [Gr. *amphi*: both + *pathos*: emotion] Of a molecule, having both hydrophilic and hydrophobic regions.

Amylase (am′ ill ase) Any of a group of enzymes that digest starch.

Anabolism (an ab′ uh liz′ em) [Gr. *ana*: upward + *ballein*: to throw] Synthetic reactions of metabolism, in which complex molecules are formed from simpler ones. (Contrast with catabolism.)

Anaerobic (an ur row′ bic) [Gr. *an*: not + *aer*: air + *bios*: life] Occurring without the use of molecular oxygen, O_2.

Anagenesis Evolutionary change in a single lineage over time.

Analogy (a nal′ o jee) [Gr. *analogia*: resembling] A resemblance in function, and often appearance as well, between two structures that is due to convergent evolution rather than to common ancestry. (Contrast with homology.)

Anaphase (an′ a phase) [Gr. *ana*: upward progress] The stage in nuclear division at which the first separation of sister chromatids (or, in the first meiotic division, of paired homologues) occurs.

Anaphylactic shock A precipitous drop in blood pressure caused by loss of fluid from capillaries because of an increase in their permeability stimulated by an allergic reaction.

Ancestral trait Trait shared by a group of organisms as a result of descent from a common ancestor.

Androgens (an′ dro jens) The male sex steroids.

Aneuploidy (an′ you ploy dee) A condition in which one or more chromosomes or pieces of chromosomes are either lacking or present in excess.

Angiosperm (an′ jee oh spurm) [Gr. *angion*: vessel + *sperma*: seed] One of the flowering plants; literally, one whose seed is carried in a "vessel" (i.e., the fruit).

Angiotensin (an′ jee oh ten′ sin) A peptide hormone that raises blood pressure by causing peripheral vessels to constrict. Also maintains glomerular filtration by constricting efferent vessels and stimulates thirst and the release of aldosterone.

Animal [L. *animus*: breath, soul] A member of the kingdom Animalia. In general, a multicellular eukaryote that obtains its food by ingestion.

Animal hemisphere The metabolically active upper portion of some animal eggs, zygotes, and embryos; does not contain the dense nutrient yolk. (Contrast with vegetal hemisphere.)

Anion (an′ eye on) [Gk. *ana*: upward progress] A negatively charged ion. (Contrast with cation.)

Anisogamy (an eye sog′ a mee) [Gr. *aniso*: unequal + *gamos*: marriage] The existence of two dissimilar gametes (egg and sperm).

Annual Referring to a plant whose life cycle is completed in one growing season. (Contrast with biennial, perennial.)

Antenna system In photosynthesis, a group of different molecules that cooperate to absorb light energy and transfer it to a reaction center.

Anterior pituitary The portion of the vertebrate pituitary gland that derives from gut epithelium and produces tropic hormones.

Anther (an′ thur) [Gr. *anthos*: flower] A pollen-bearing portion of the stamen of a flower.

Antheridium (an′ thur id′ ee um) (plural: antheridia) [Gr. *antheros*: blooming] The multicellular structure that produces the sperm in bryophytes and ferns.

Antibody One of the millions of proteins produced by the immune system that specifically binds to a foreign substance and initiates its removal from the body.

Anticodon The three nucleotides in transfer RNA that pair with a complementary triplet (a codon) in messenger RNA.

Antidiuretic hormone A hormone that controls water reabsorption in the mammalian kidney. Also called vasopressin.

Antigen (an′ ti jun) Any substance that stimulates the production of an antibody or antibodies in the body of a vertebrate.

Antigenic determinant A specific region of an antigen, which is recognized by and binds to a specific antibody.

Antiparallel Pertaining to molecular orientation in which a molecule or parts of a molecule have opposing directions.

Antipodal cell At one end of the megagametophyte, one of the three cells which eventually degenerate.

Antiport A membrane transport process that carries one substance in one direction and another in the opposite direction. (Contrast with symport.)

Antisense nucleic acid A single-stranded RNA or DNA complementary to and thus targeted against the mRNA transcribed from a harmful gene such as an oncogene.

Anus (a′ nus) Opening through which digestive wastes are expelled, located at the posterior end of the gut.

Aorta (a or′ tah) [Gr. *aorte:* aorta] The main trunk of the arteries leading to the systemic (as opposed to the pulmonary) circulation.

Apex (a′ pecks) The tip or highest point of a structure, as the apex of a growing stem or root.

Apical (a′ pi kul) Pertaining to the apex, or tip, usually in reference to plants.

Apical dominance Inhibition by the apical bud of the growth of axillary buds.

Apical meristem The meristem at the tip of a shoot or root; responsible for the plant's primary growth.

Apomixis (ap oh mix′ is) [Gr. *apo*: away from + *mixis*: sexual intercourse] The asexual production of seeds.

Apoplast (ap′ oh plast) in plants, the continuous meshwork of cell walls and extracellular spaces through which material can pass without crossing a plasma membrane. (Contrast with symplast.)

Apoptosis (ay pu toh′ sis) A series of genetically programmed events leading to cell death.

Aquaporin A transport protein in plant and animal cells through which water passes in osmosis.

Aquatic (a kwa′ tic) [L. *aqua*: water] Living in water. (Compare with marine, terrestrial.)

Aqueous (a′ kwee us) [L. *aqua*: water] Pertaining to water or a watery solution.

Archegonium (ar′ ke go′ nee um) [Gr. *archegonos*: first, foremost] The multicellular structure that produces eggs in bryophytes, ferns, and gymnosperms.

Archenteron (ark en′ ter on) [Gr. *archos*: first + *enteron*: bowel] The earliest primordial animal digestive tract.

Arteriosclerosis See atherosclerosis.

Artery A muscular blood vessel carrying oxygenated blood away from the heart to other parts of the body. (Contrast with vein.)

Ascus (ass′ cuss) [Gr. *askos*: bladder] In ascomycete fungi (sac fungi), the club-shaped sporangium within which spores (ascospores) are produced by meiosis.

Asexual Without sex.

Assortative mating A breeding system in which mates are selected on the basis of a particular trait or group of traits.

Atherosclerosis (ath′ er oh sklair oh′ sis) [Gk. *athero*: gruel, porridge + *skleros*: hard] A disease of the lining of the arteries characterized by fatty, cholesterol-rich deposits in the walls of the arteries. When fibroblasts infiltrate these deposits and calcium precipitates in them, the disease become arteriosclerosis, or "hardening of the arteries."

Atmosphere The gaseous mass surrounding our planet. Also a unit of pressure, equal to the normal pressure of air at sea level.

Atom [Gr. *atomos*: indivisible] The smallest unit of a chemical element. Consists of a nucleus and one or more electrons.

Atomic mass The average mass of an atom of an element; the average depends on the relative amounts of different isotopes of the element on Earth. Also called atomic weight.

Atomic number The number of protons in the nucleus of an atom; also equals the number of electrons around the neutral atom. Determines the chemical properties of the atom.

ATP (adenosine triphosphate) An energy-storage compound containing adenine, ribose, and three phosphate groups. When it is formed from ADP, useful energy is stored; when it is broken down (to ADP or AMP), energy is released to drive endergonic reactions.

ATP synthase An integral membrane protein that couples the transport of proteins with the formation of ATP.

Atrium (a′ tree um) [L. *atrium*: central hall] An internal chamber. In the hearts of vertebrates, the thin-walled chamber(s) entered by blood on its way to the ventricle(s). Also, the outer ear.

Autoimmune disease A disorder in which the immune system attacks the animal's own antigens.

Autonomic nervous system The system that controls such involuntary functions as those of guts and glands.

Autosome Any chromosome (in a eukaryote) other than a sex chromosome.

Autotroph (au' tow trow' fik) [Gr. *autos:* self + *trophe*: food] An organism that is capable of living exclusively on inorganic materials, water, and some energy source such as sunlight or chemically reduced matter. (Contrast with heterotroph.)

Auxin (awk' sin) [Gr. *auxein*: to grow] In plants, a substance (the most common being indoleacetic acid) that regulates growth and various aspects of development.

Auxotroph (awks' o trofe) [Gr. *auxein*: to grow + *trophe*: food] A mutant form of an organism that requires a nutrient or nutrients not usually required by the wild type. (Contrast with prototroph.)

Axon [Gr. *axon*: axle] The part of a neuron that conducts action potentials away from the cell body.

Axon hillock The junction between an axon and its cell body, where action potentials are generated.

Axon terminals The endings of an axon; they form synapses and release neurotransmitter.

Axoneme (ax' oh neem) The complex of microtubules and their crossbridges that forms the motile apparatus of a cilium.

Bacillus (bah sil' us) [L: little rod] Any of various rod-shaped bacteria.

Bacteria (bak teer' ee ah) (singular: bacterium) [Gr. *bakterion*: little rod] Prokaryote in the Domain Bacteria. The chromosomes of bacteria are not contained in nuclear envelopes.

Bacteriophage (bak teer' ee o fayj) [Gr. *bakterion*: little rod + *phagein*: to eat] One of a group of viruses that infect bacteria and ultimately cause their disintegration.

Bacteroids Nitrogen-fixing organelles that develop from endosymbiotic bacteria.

Balanced polymorphism [Gr. *polymorphos*: many forms] The maintenance of more than one form, or the maintenance at a given locus of more than one allele, at frequencies of greater than 1 percent in a population. Often results when heterozygotes are more fit than either homozygote.

Bark All tissues outside the vascular cambium of a plant.

Baroreceptor [Gr. *baros*: weight] A pressure-sensing cell or organ.

Barr body In mammals, an inactivated X chromosome.

Basal Pertaining to one end—the base—of an axis.

Basal body Centriole found at the base of a eukaryotic flagellum or cilium.

Basal metabolic rate (BMR) The minimum rate of energy turnover in an awake (but resting) bird or mammal that is not expending energy for thermoregulation.

Base (1) A substance which can accept a hydrogen ion in solution. (Contrast with acid.) (2) In nucleic acids, the purine or pyrimidine that is attached to each sugar in the backbone.

Base pairing See complementary base pairing.

Basic Having a pH greater than 7.0 (i.e., having a hydrogen ion concentration lower than 10^{-7} molar).

Basidium (bass id' ee yum) In basidiomycete fungi, the characteristic sporangium in which four spores are formed by meiosis and then borne externally before being shed.

Basophils One type of phagocytic white blood cell that releases histamine and may promote T cell development.

B cell A type of lymphocyte involved in the humoral immune response of vertebrates. Upon recognizing an antigenic determinant, a B cell develops into a plasma cell, which secretes an antibody. (Contrast with T cell.)

Benefit An improvement in survival and reproductive success resulting from performing a behavior or having a trait. (Contrast with cost.)

Benign (be nine') A tumor that grows to a certain size and then stops, uaually with a fibrous capsule surrounding the mass of cells. Benign tumors do not spread (metastasize) to other organs.

Benthic zone [Gr. *benthos*: bottom] The bottom of the ocean.

Beta (β) pleated sheet Type of protein secondary structure; results from hydrogen bonding between polypeptide regions running antiparallel to each other.

Biennial Referring to a plant whose life cycle includes vegetative growth in the first year and flowering and senescence in the second year. (Contrast with annual, perennial.)

Bilateral symmetry The condition in which only the right and left sides of an organism, divided exactly down the back, are mirror images of each other. (Contrast with biradial symmetry.)

Bile A secretion of the liver delivered to the small intestine via the common bile duct. In the intestine, bile emulsifies fats.

Binocular cells Neurons in the visual cortex that respond to input from both retinas; involved in depth perception.

Binomial (bye nome' ee al) Consisting of two elements; for example, the binomial nomenclature of biology in which each species has two names (the genus name followed by the species name).

Biodiversity crisis The current high rate of loss of species, caused primarily by human activities.

Biogeochemical cycles Movement of elements through living organisms and the physical environment.

Biogeographic region A continental-scale part of Earth that has a biota distinct from that of other such regions.

Bioluminescence The production of light by biochemical processes in an organism.

Biomass The total weight of all the living organisms, or some designated group of living organisms, in a given area.

Biome (bye' ome) A major division of the ecological communities of Earth; characterized by distinctive vegetation.

Biota (bye oh' tah) All of the organisms, including animals, plants, fungi, and microorganisms, found in a given area.

Biradial symmetry Radial symmetry modified so that only two planes can divide the animal into similar halves.

Blastocoel (blass' toe seal) [Gr. *blastos*: sprout + *koilos*: hollow] The central, hollow cavity of a blastula.

Blastocyst (blass' toe cist) An early embryo formed by the first divisions of the fertilized egg (zygote). In mammals, a hollow ball of cells.

Blastodisc (blass' toe disk) A disk of cells forming on the surface of a large yolk mass, comparable to a blastula, but occurring in animals such as birds and reptiles, in which the massive yolk restricts cleavage to one side of the egg only.

Blastomere A cell produced by the division of a fertilized egg.

Blastopore The opening from the archenteron to the exterior of a gastrula.

Blastula (blass' chu luh) An early stage in animal embryology; in many species, a hollow sphere of cells surrounding a central cavity, the blastocoel. (Contrast with blastodisc.)

Blood–brain barrier A property of the blood vessels of the brain that prevents most chemicals from diffusing from the blood into the brain.

Blue light receptor Molecule in plants that absorbs blue light (400–500 nm). Mediates many plant responses including phototropism, stomatal movements, and expression of some genes.

Body cavity Membrane-lined, fluid-filled compartment that lies between the cell layers of many animals.

Body plan A basic structural design that includes an entire animal, its organ systems, and the integrated functioning of its parts.

Bottleneck Refers to stressful periods that only a few individuals of a once-large population survive, resulting in substantially reduced genetic variation in the population.

Bowman's capsule An elaboration of kidney tubule cells that surrounds a know of capillaries (the glomerulus). Blood is filtered across the walls of these capillaries and the filtrate is collected into Bowman's capsule.

Brain stem The portion of the vertebrate brain between the spinal cord and the forebrain.

Brassinosteroids Plant steroid hormones that mediate light effects promoting the elongation of stems and pollen tubes.

Bronchus (plural: bronchi) The major airway(s) branching off the trachea into the vertebrate lung.

Brown fat Fat tissue in mammals that is specialized to produce heat. It has many mitochondria and capillaries, and a protein that uncouples oxidative phosphorylation.

Browser An animal that feeds on the tissues of woody plants.

Bryophyte (bry' uh fite) [Gk. *bryon*: moss; *phyton*: plant] A moss. This term was once frequently used to refer to all nontracheophyte plants.

Budding Asexual reproduction in which a more or less complete new organism simply grows from the body of the parent organism and eventually detaches itself.

Buffer A substance that can transiently accept or release hydrogen ions and thereby resist changes in pH.

C_3 photosynthesis The form of photosynthesis in which 3-phosphoglycerate is the first stable product, and ribulose bisphosphate is the CO_2 receptor.

C_4 photosynthesis The form of photosynthesis in which oxaloacetate is the first stable product, and phosphoenolpyruvate is the CO_2 acceptor. C_4 plants also perform the reactions of C_3 photosynthesis.

Calcitonin A hormone produced by the thyroid gland; it lowers blood calcium and promotes bone formation. (Compare with parathyroid hormone.)

Calmodulin (cal mod′ joo lin) A calcium-binding protein found in all animal and plant cells; mediates many calcium-regulated processes.

calorie [L. *calor*: heat] The amount of heat required to raise the temperature of one gram of water by one degree Celsius (1°C) from 14.5°C to 15.5°C. Calorie spelled with a capital C refers to the kilocalorie (1 kcal = 1,000 cal).

Calvin–Benson cycle The stage of photosynthesis in which CO_2 reacts with RuBP to form 3PG, 3PG is reduced to a sugar, and RuBP is regenerated, while other products are released to the rest of the plant.

Calyx (kay′ licks) [Gr. *kalyx*: cup] All of the sepals of a flower, collectively.

CAM See Crassulacean acid metabolism.

Cambium (kam′ bee um) [L. *cambiare*: to exchange] A meristem that gives rise to radial rows of cells in stem and root, increasing them in girth; commonly applied to the vascular cambium which produces wood and phloem, and the cork cambium, which produces bark.

cAMP (cyclic AMP) A compound formed from ATP that mediates the effects of numerous animal hormones.

Canopy The leaf-bearing part of a tree. Collectively, the aggregate of the leaves and branches of the larger woody plants of an ecological community.

Capillaries [L. *capillaris*: hair] Very small tubes, especially the smallest blood-carrying vessels of animals between the termination of the arteries and the beginnings of the veins.

Carbohydrates Organic compounds containing carbon, hydrogen, and oxygen in the ratio 1:2:1 (i.e., with the general formula $C_nH_{2n}O_n$). Common examples are sugars, starch, and cellulose.

Carboxylic acid (kar box sill′ ik) An organic acid containing the carboxyl group, —COOH, which dissociates to the carboxylate ion, $—COO^-$.

Carcinogen (car sin′ oh jen) A substance that causes cancer.

Cardiac (kar′ dee ak) [Gr. *kardia*: heart] Pertaining to the heart and its functions.

Carnivore [L. *carn*: flesh + *vovare*: to devour] An organism that eats animal tissues. (Contrast with detritivore, herbivore, omnivore.)

Carotenoid (ka rah′ tuh noid) A yellow, orange, or red lipid pigment commonly found as an accessory pigment in photosynthesis; also found in fungi.

Carpel (kar′ pel) [Gr. *karpos*: fruit] The organ of the flower that contains one or more ovules.

Carrier (1) In facilitated diffusion, a membrane protein that binds a specific molecule and transports it through the membrane. (2) In respiratory and photosynthetic electron transport, a participating substance such as NAD that exists in both oxidized and reduced forms. (3) In genetics, a person heterozygous for a recessive trait.

Carrying capacity. In ecology, the maximum number of individuals of a particular species that can be supported on a sustained basis in a suitable habitat.

Cartilage In vertebrates, a tough connective tissue found in joints, the outer ear, and elsewhere. Forms the entire skeleton in some animal groups.

Casparian strip A band of cell wall containing suberin and lignin, found in the endodermis. Restricts the movement of water across the endodermis.

Catabolism [Gr. *kata*: to break down + *ballein*: to throw] Degradational reactions of metabolism, in which complex molecules are broken down. (Contrast with anabolism.)

Catabolite repression In the presence of abundant glucose, the diminished synthesis of catabolic enzymes for other energy sources.

Catalyst (cat′ a list) [Gr. *kata*: to break down] A chemical substance that accelerates a reaction without itself being consumed in the overall course of the reaction. Catalysts lower the activation energy of a reaction. Enzymes are biological catalysts.

Cation (cat′ eye on) An ion with one or more positive charges. (Contrast with anion.)

Caudal [L. *cauda*: tail] Pertaining to the tail, or to the posterior part of the body.

cDNA See complementary DNA.

Cecum (see′ cum) [L. *caecus*: blind] A blind branch off the large intestine. In many nonruminant mammals, the cecum contains a colony of microorganisms that contribute to the digestion of food.

Cell adhesion molecules Molecules on animal cell surfaces that affect the selective association of cells during development of the embryo.

Cell cycle The stages through which a cell passes between one division and the next. Includes all stages of interphase and mitosis.

Cell division The reproduction of a cell to produce two new cells. In eukaryotes, this process involves nuclear division (mitosis) and cytoplasmic division (cytokinesis).

Cell junctions Specialized structures associated with the plasma membranes of epithelial cells. Some contribute to cell adhesion, others to intercellular communication.

Cell plate Following mitosis in plant cells, the initial wall-like structure that forms separating the nuclei from the surrounding cytoplasm. Later becomes the cell wall.

Cell theory The theory, well established, that all organisms consist of cells, and that all cells come from preexisting cells.

Cell wall A relatively rigid structure that encloses cells of plants, fungi, many protists, and most prokaryotes. Gives these cells their shape and limits their expansion in hypotonic media.

Cellular immune response Action of the immune system based on the activities of T cells. Directed against parasites, fungi, intracellular viruses, and foreign tissues (grafts). (Contrast with humoral immune system.)

Cellular respiration See respiration.

Cellulose (sell′ you lowss) A straight-chain polymer of glucose molecules, used by plants as a structural supporting material.

Central dogma The statement that information flows from DNA to RNA to polypeptide (in retroviruses, there is also information flow from RNA to cDNA).

Central nervous system That part of the nervous system which is condensed and centrally located, e.g., the brain and spinal cord of vertebrates; the chain of cerebral, thoracic and abdominal ganglia of arthropods.

Centrifuge [L. *centrum*: center + *fugere*: to flee] A laboratory device in which a sample is spun around a central axis at high speed. Used to separate suspended materials of different densities.

Centriole (sen′ tree ole) A paired organelle that helps organize the microtubules in animal and protist cells during nuclear division.

Centromere (sen′ tro meer) [Gr. *centron*: center + *meros*: part] The region where sister chromatids join.

Centrosome (sen′ tro soam) The major microtubule organizing center of an animal cell.

Cephalization (sef ah luh zay′ shun) [Gr. *kephale*: head] The evolutionary trend toward increasing concentration of brain and sensory organs at the anterior end of the animal.

Cerebellum (sair uh bell′ um) [L.: diminutive of *cerebrum*, brain] The brain region that controls muscular coordination; located at the anterior end of the hindbrain.

Cerebral cortex The thin layer of gray matter (neuronal cell bodies) that overlays the cerebrum.

Cerebrum (su ree′ brum) [L. *cerebrum*: brain] The dorsal anterior portion of the forebrain, making up the largest part of the brain of mammals. In mammals, the chief coordination center of the nervous system; consists of two cerebral hemispheres.

Cervix (sir′ vix) [L. *cervix*: neck] The opening of the uterus into the vagina.

cGMP (cyclic guanosine monophosphate) An intracellular messenger that is part of signal transmission pathways involving G proteins. (See G protein.)

Channel protein A membrane protein that forms an aqueous passageway though which specific solutes may pass.

Chaperonins A group of proteins that limit inappropriate interactions between cellular proteins under denaturing conditions such as high temperature.

Chaperone protein A protein that assists a newly forming protein in adopting its appropriate tertiary structure.

Chemical bond An attractive force stably linking two atoms.

Chemical reaction The change in the composition or distribution of atoms of a substance with consequent alterations in properties.

Chemiosmosis The formation of ATP in mitochondria and chloroplasts, resulting from a pumping of protons across a membrane (against a gradient of electrical charge and of pH), followed by the return of the protons through a protein channel with ATPase activity.

Chemoautotroph See Chemolithotroph.

Chemoheterotroph An organism that must obtain both carbon and energy from organic substances. (Contrast with chemolithotroph, photoautotroph, photoheterotroph.)

Chemolithotroph [Gk. *lithos*: stone, rock] An organism that uses carbon dioxide as a carbon source and obtains energy by oxidizing inorganic substances from its environment. (Contrast with chemoheterotroph, photoautotroph, photoheterotroph.)

Chemoreceptor A cell or tissue that senses specific substances in its environment.

Chemosynthesis Synthesis of food substances, using the oxidation of reduced materials from the environment as a source of energy.

Chiasma (kie az' muh) (plural: chiasmata) [Gr. *chiasmata*: cross] An X-shaped connection between paired homologous chromosomes in prophase I of meiosis. A chiasma is the visible manifestation of crossing over between homologous chromosomes.

Chitin (kye' tin) [Gr. *kiton*: tunic] The characteristic tough but flexible organic component of the exoskeleton of arthropods, consisting of a complex, nitrogen-containing polysaccharide. Also found in cell walls of fungi.

Chlorophyll (klor' o fill) [Gr. *kloros*: green + *phyllon*: leaf] Any of a few green pigments associated with chloroplasts or with certain bacterial membranes; responsible for trapping light energy for photosynthesis.

Chloroplast [Gr. *kloros*: green + *plast*: a particle] An organelle bounded by a double membrane containing the enzymes and pigments that perform photosynthesis. Chloroplasts occur only in eukaryotes.

Choanocyte (ko' an uh site) The collared, flagellated feeding cells of sponges.

Cholecystokinin (ko' luh sis tuh kai' nin) A hormone produced and released by the lining of the duodenum when it is stimulated by undigested fats and proteins. It stimulates the gallbladder to release bile and slows stomach activity.

Chorion (kor' ee on) [Gr. *khorion*: afterbirth] The outermost of the membranes protecting mammal, bird, and reptile embryos; in mammals it forms part of the placenta.

Chromatid (kro' ma tid) Each of a pair of new sister chromosomes from the time at which the molecular duplication occurs until the time at which the centromeres separate at the anaphase of nuclear division.

Chromatin The nucleic acid–protein complex found in eukaryotic chromosomes.

Chromatophore (krow mat' o for) [Gr. *kroma*: color + *phoreus*: carrier] A pigment-bearing cell that expands or contracts to change the color of the organism.

Chromosomal mutation Loss of or changes in position/direction of a DNA segment on a chromosome.

Chromosome (krome' o sowm) [Gr. *kroma*: color + *soma*: body] In bacteria and viruses, the DNA molecule that contains most or all of the genetic information of the cell or virus. In eukaryotes, a structure composed of DNA and proteins that bears part of the genetic information of the cell.

Chylomicron (ky low my' cron) Particles of lipid coated with protein, produced in the gut from dietary fats and secreted into the extracellular fluids.

Chyme (kime) [Gr. kymus, juice] Created in the stomach; a mixture of ingested food with the digestive juices secreted by the salivary glands and the stomach lining.

Cilium (sil' ee um) (plural: cilia) [L.: eyelash] Hairlike organelle used for locomotion by many unicellular organisms and for moving water and mucus by many multicellular organisms. Generally shorter than a flagellum.

Circadian rhythm (sir kade' ee an) [L. *circa*: approximately + *dies*: day] A rhythm in behavior, growth, or some other activity that recurs about every 24 hours under constant conditions.

Circannual rhythm [L. *circa*: approximately + *annus*: year) A rhythm of behavior, growth, or some other activity that recurs on a yearly basis.

Citric acid cycle A set of chemical reactions in cellular respiration, in which acetyl CoA is oxidized to carbon dioxide, and hydrogen atoms are stored as NADH and $FADH_2$. Also called the Krebs cycle.

Clade (Gk. *klados*: branch] In taxonomy, a monophyletic group made up of an ancestor and all of its descendants.

Class I MHC molecules These cell surface proteins participate in the cellular immune response directed against virus-infected cells.

Class II MHC molecules These cell surface proteins participate in the cell-cell interactions (of helper T cells, macrophages, and B cells) of the humoral immune response.

Class switching The process whereby a plasma cell changes the class of immunoglobulin that it synthesizes by changing the DNA region coding for the C segment.

Clathrin A fibrous protein on the inner surfaces of animal cell membranes that strengthens coated vesicles and thus participates in receptor-mediated endocytosis.

Cleavage First divisions of the fertilized egg of an animal.

Cline A gradual change in the traits of a species over a geographical gradient.

Cloaca (klo ay' kuh) [L. *cloaca*: sewer] In some invertebrates, the posterior part of the gut; in many vertebrates, a cavity receiving material from the digestive, reproductive, and excretory systems.

Clonal anergy Prevention of the synthesis of antibodies against the body's own antigens. When a T cell binds to a self-antigen, it does not receive signals from an antigen-presenting cell; thus the T cell dies (becomes anergic) rather than yielding a clone of active cells.

Clonal deletion The inactivation or destruction of lymphocyte clones that would produce immune reactions against the animal's own body.

Clonal selection The mechanism by which exposure to antigen results in the activation of selected T- or B-cell clones, resulting in an immune response.

Clone [Gr. *klon*: twig, shoot] Genetically identical cells or organisms produced from a common ancestor by asexual means.

Cnidocytes (nye' duh sites) The feeding cells of cnidarians, within which nematocysts are housed.

Coacervate (ko as' er vate) [L. *coacervare*: to heap up] An aggregate of colloidal particles in suspension.

Coated vesicle Cytoplasmic vesicle containing distinctive proteins, including clathrin.

Coccus (kock' us) [Gr. *kokkos*: berry, pit] Any of various spherical or spheroidal bacteria.

Cochlea (kock' lee uh) [Gr. *kokhlos* snail] A spiral tube in the inner ear of vertebrates; it contains the sensory cells involved in hearing.

Codominance A condition in which two alleles at a locus produce different phenotypic effects and both effects appear in heterozygotes.

Codon Three nucleotides in messenger RNA that direct the placement of a particular amino acid into a polypeptide chain. (Contrast with anticodon.)

Coelom (see' lum) [Gr. *koiloma*: cavity] The body cavity of certain animals; the coelom is lined with cells of mesodermal origin.

Coelomate Having a coelom.

Coenocyte (seen' a sight) [Gr. *koinos*: common + *kytos*: container] A "cell" enclosed by a single plasma membrane but containing many nuclei.

Coenzyme A nonprotein organic molecule that plays a role in catalysis by an enzyme.

Cofactor An inorganic ion that is weakly bound to an enzyme and required for its activity.

Cohesin Proteins involved in binding chromatids together.

Coevolution Concurrent evolution of two or more species that are mutually affecting each other's evolution.

Cohort (co' hort) [L. *cohors*: company of soldiers] A group of similar-age organisms, considered as it passes through time.

Cold hardening Increased capacity of some plant species to withstand cold spells by their repeated exposure to cool but not damaging temperatures.

Coleoptile A sheath that surrounds and protects the apical meristem and young primary leaves of a seedling as they move through the soil.

Collagen [Gr. *kolla*: glue] A fibrous protein found extensively in bone and connective tissue.

Collecting duct In vertebrates, a tubule that receives urine produced in the nephrons of the kidney and delivers that fluid to the ureter for excretion.

Collenchyma (cull eng' kyma) [Gr. *kolla*: glue + *enchyma*: infusion] A type of plant cell, living at functional maturity, which lends flexible support by virtue of primary cell walls thickened at the corners. (Contrast with parenchyma, sclerenchyma.)

Colon [Gr. *kolon*: large intestine] The large intestine.

Common bile duct A single duct that delivers bile from the gallbladder and secretions from the pancreas into the small intestine.

Communication A signal from one organism (or cell) that alters the functioning or behavior of another organism (or cell).

Community Any ecologically integrated group of species of microorganisms, plants, and animals inhabiting a given area.

Companion cell Specialized cell found adjacent to a sieve tube member in flowering plants.

Comparative genomics Computer-aided comparison of DNA sequences between different organisms to reveal genes with related functions.

Comparative method An approach to studying evolution and ecology in which hypotheses are tested by measuring the distribution of states among a large number of species.

Compensation point The light intensity at which the rates of photosynthesis and of cellular respiration are equal.

Competition In ecology, use of the same resource by two or more species, when the resource is present in insufficient supply for the combined needs of the species.

Competitive exclusion A result of competition between species for a limiting resource in which one species completely eliminates the other.

Competitive inhibitor A nonsubstrate that binds to the active site of an enzyme and thereby inhibits binding of substrate and reaction from part of the environment.

Complement system A group of eleven proteins that play a role in some reactions of the immune system. The complement proteins are not immunoglobulins.

Complementary base pairing The AT (or AU), TA (or UA), CG, and GC pairing of bases in double-stranded DNA, in transcription, and between tRNA and mRNA.

Complementary DNA (cDNA) DNA formed by reverse transcriptase acting with an RNA template; essential intermediate in the reproduction of retroviruses; used as a tool in recombinant DNA technology; lacks introns.

Complete metamorphosis A change of state during the life cycle of an organism in which the body is almost completely rebuilt to produce an individual with a very different body form. Characteristic of insects such as butterflies, moths, beetles, ants, wasps, and flies.

Compound (1) A substance made up of atoms of more than one element. (2) A structure made up of many units, as the compound eyes of arthropods.

Condensation reaction A reaction in which two molecules become connected by a covalent bond and a molecule of water is released. ($AH + BOH \rightarrow AB + H_2O$.)

Conditional mutations Mutations that show characteristic phenotype only under certain environmental conditions such as temperature.

Conformation. The three-dimensional shape of a protein or other macromolecule.

Cones (1) In the vertebrate retina: photoreceptors responsible for color vision. (2) In gymnosperms: reproductive structures consisting of many sporophylls packed relatively tightly.

Conidium (ko nid' ee um) [Gr. *konis*: dust] An asexual fungus spore borne singly or in chains either apically or laterally on a hypha.

Conifer (kahn' e fer) [Gr. *konos*: cone + *phero*: carry] One of the cone-bearing gymnosperms, mostly trees, such as pines and firs.

Conjugation (kon ju gay' shun) [L. *conjugare*: yoke together] The close approximation of two cells during which they exchange genetic material, as in *Paramecium* and other ciliates, or during which DNA passes from one to the other through a tube, as in bacteria.

Connective tissue An animal tissue that connects or surrounds other tissues; its cells are embedded in a collagen-containing matrix.

Connexon In a gap junction, a protein channel linking adjacent animal cells.

Consensus sequences Short stretches of DNA that appear, with little variation, in many different genes.

Constant region For a particular class of immunoglobulin molecules, the region with identical amino acid composition.

Constitutive enzyme An enzyme that is present in approximately constant amounts in a system, whether its substrates are present or absent. (Contrast with inducible enzyme.)

Consumer An organism that eats the tissues of some other organism.

Continental drift The gradual movements of the world's continents that has occurred over billions of years.

Convergent evolution The evolution of similar features independently in unrelated taxa from different ancestral structures.

Copulation Reproductive behavior that results in a male depositing sperm in the reproductive tract of a female.

Corepressor A low-molecular-weight compound that unites with a protein (the repressor) to prevent transcription in a repressible operon.

Cork A waterproofing tissue in plants, with suberin-containing cell walls. Produced by a cork cambium.

Corolla (ko role' lah) [L. *corolla*: a small crown] All of the petals of a flower, collectively.

Coronary (kor' oh nair ee) (L. *corona*: crown) Referring to the blood vessels of the heart.

Corpus luteum (kor' pus loo' tee um) [L.: yellow body] A structure formed from a follicle after ovulation; it produces hormones important to the maintenance of pregnancy.

Cortex [L. *cortex*: bark, rind] (1) In plants, the tissue between the epidermis and the vascular tissue of a stem or root. (2) In animals, the outer tissue of certain organs, such as the adrenal cortex and cerebral cortex.

Corticosteroids Steroid hormones produced and released by the cortex of the adrenal gland.

Cotyledon (kot' ul lee' dun) [Gr. *kotyledon*: hollow space] A "seed leaf." An embryonic organ that stores and digests reserve materials; may expand when seed germinates.

Countercurrent exchange An adaptation that promotes maximum exchange of heat or any diffusible substance between two fluids by the fluids flow in opposite directions through parallel tubes close together

Covalent bond A chemical bond that arises from the sharing of electrons between two atoms. Usually a strong bond.

Crassulacean acid metabolism (CAM) A metabolic pathway enabling the plants that possess it to store carbon dioxide at night and then perform photosynthesis during the day with stomata closed.

Crista (plural: cristae) A small, shelflike projection of the inner membrane of a mitochondrion; the site of oxidative phosphorylation.

Critical night length In the photoperiodic flowering response of short-day plants, the length of night above which flowering occurs and below which the plant remains vegetative. (The reverse applies in the case of long-day plants.)

Critical period The age during which some particular type of learning must take place or during which it occurs much more easily than at other times. Typical of song learning among birds.

Cross section A section taken perpendicular to the longest axis of a structure. Also called a transverse section.

Crossing over The mechanism by which linked markers undergo recombination. In general, the term refers to the reciprocal exchange of corresponding segments between two homologous chromatids.

Cryptic [Gr. *kryptos*: hidden] The resemblance of an animal to some part of its environment, which helps it to escape detection by predators.

Cryptochromes [Gr. *kryptos*: hidden + *kroma*: color] Photoreceptors mediating some blue-light effects in plants and animals.

Culture (1) A laboratory association of organisms under controlled conditions. (2) The collection of knowledge, tools, values, and rules that characterize a human society.

Cuticle A waxy layer on the outer surface of a plant or an insect, tending to retard water loss.

Cyanobacteria (sigh an' o bacteria) [Gr. *kuanos*: blue] A lineage of photosynthetic bacteria, formerly referred to as blue-green algae; they use chlorophyll *a* in photosynthesis.

Cyclic AMP See cAMP.

Cyclic electron transport In photosynthetic light reactions, the flow of electrons that produces ATP but no NADPH or O_2.

Cyclins Proteins that activate cyclin-dependent kinases, bringing about transitions in the cell cycle.

Cyclin-dependent kinase (cdk) A *kinase* catalyzes the addition of phosphate groups from ATP to target molecules. Cdk's target proteins are involved in transitions in the cell cycle and are active only when complexed to additional protein subunits, cyclins.

Cytochromes (sy′ toe chromes) [Gr. *kytos*: container + *chroma*: color] Iron-containing red proteins, components of the electron-transfer chains in photophosphorylation and respiration.

Cytokinesis (sy′ toe kine ee′ sis) [Gr. *kytos*: container + *kinein*: to move] The division of the cytoplasm of a dividing cell. (Compare with mitosis.)

Cytokinin (sy′ toe kine′ in) A member of a class of plant growth substances playing roles in senescence, cell division, and other phenomena.

Cytoplasm The contents of the cell, excluding the nucleus.

Cytoplasmic determinants In animal development, gene products whose spatial distribution may determine such things as embryonic axes.

Cytosine (C) (site′ oh seen) A nitrogen-containing base found in DNA and RNA.

Cytoskeleton The network of microtubules and microfilaments that gives a eukaryotic cell its shape and its capacity to arrange its organelles and to move.

Cytosol The fluid portion of the cytoplasm, excluding organelles and other solids.

Cytotoxic T cells (T_C) Cells of the cellular immune system that recognize and directly eliminate virus-infected cells. (Compare with helper T cells.)

DAG See Diacylglycerol.

Deciduous [L. *deciduus*: falling off] Refers to a woody plant that sheds it leaves but does not die.

Decomposer See detritivore.

Degeneracy The situation in which a single amino acid may be represented by any of two or more different codons in messenger RNA. Most of the amino acids can be represented by more than one codon.

Deletion A mutation resulting from the loss of a continuous segment of a gene or chromosome. Such mutations never revert to wild type. (Contrast with duplication, point mutation.)

Deme (deem) [Gr. *demos:* the people] Any local population of individuals belonging to the same species that interbreed with one another.

Demographic processes The events—such as births, deaths, immigration, and emigration—that determine the number of individuals in a population.

Demographic stochasticity Random variations in the factors influencing the size, density, and distribution of a population.

Denaturation Loss of activity of an enzyme or nucleic acid molecule as a result of structural changes induced by heat or other means.

Dendrite [Gr. *dendron*: tree] A fiber of a neuron which often cannot carry action potentials. Usually much branched and relatively short compared with the axon, and commonly carries information to the cell body of the neuron.

Denitrification Metabolic activity by which inorganic nitrogen-containing ions are reduced to form nitrogen gas and other products; carried on by certain soil bacteria.

Density dependence Change in the severity of action of agents affecting birth and death rates within populations that are directly or inversely related to population density.

Density independence The state where the severity of action of agents affecting birth and death rates within a population does not change with the density of the population.

Deoxyribonucleic acid See DNA.

Depolarization A change in the electric potential across a membrane from a condition in which the inside of the cell is more negative than the outside to a condition in which the inside is less negative, or even positive, with reference to the outside of the cell. (Contrast with hyperpolarization.)

Derived trait A trait found among members of a lineage that was not present in the ancestors of that lineage.

Dermal tissue system The outer covering of a plant, consisting of epidermis in the young plant and periderm in a plant with extensive secondary growth. (Contrast with ground tissue system and vascular tissue system.)

Desmosome (dez′ mo sowm) [Gr. *desmos*: bond + *soma*: body] An adhering junction between animal cells.

Determination Process whereby an embryonic cell or group of cells becomes fixed into a predictable developmental pathway.

Detritivore (di try′ ti vore) [L. *detritus*: worn away + *vorare*: to devour] An organism that obtains its energy from the dead bodies and/or waste products of other organisms.

Deuterostome A major evolutionary lineage in animals, characterized by radial cleavage, enterocoelous development, and other traits. (Compare with protostome.)

Development Progressive change, as in structure or metabolism; in most kinds of organisms, development continues throughout the life of the organism.

Diacylglycerol (DAG) In hormone action, the second messenger produced by hydrolytic removal of the head group of certain phospholipids.

Diaphragm (dye′ uh fram) [Gr. *diaphrassein*: barricade] A sheet of muscle that separates the thoracic and abdominal cavities in mammals; responsible for breathing. (2) A method of birth control in which a sheet of rubber is fitted over the woman's cervix, blocking the entry of sperm.

Diastole (dye ass′ toll ee) [Gr. : dilation] The portion of the cardiac cycle when the heart muscle relaxes. (Contrast with systole.)

Dicot (short for dicotyledon) [Gr. *di*: two + *kotyledon*: a hollow space] This term, not used in this book, formerly referred to all angiosperms other than the monocots. (See eudicot, monocot.)

Differentiation Process whereby originally similar cells follow different developmental pathways. The actual expression of determination.

Diffusion Random movement of molecules or other particles, resulting in even distribution of the particles when no barriers are present.

Digestion Enzyme-catalyzed process by which large, usually insoluble, molecules (foods) are hydrolyzed to form smaller molecules of soluble substances.

Dihybrid cross A mating in which the parents differ with respect to the alleles of two loci of interest.

Dikaryon (di care′ ee ahn) [Gr. *di*: two + *karyon*: kernel] A cell or organism carrying two genetically distinguishable nuclei. Common in fungi.

Dioecious (die eesh′ us) [Gr.: *di*: two + *oikos*: house] Refers to organisms in which the two sexes are "housed" in two different individuals, so that eggs and sperm are not produced in the same individuals. Examples: humans, fruit flies, date palms. (Contrast with monoecious.)

Diploblastic Having two cell layers. (Contrast with triploblastic.)

Diploid (dip′ loid) [Gr. *diplos*: double] Having a chromosome complement consisting of two copies (homologues) of each chromosome. Designated $2n$.

Directional selection Selection in which phenotypes at one extreme of the population distribution are favored. (Contrast with disruptive selection, stabilizing selection.)

Disaccharide A carbohydrate made up of two monosaccharides (simple sugars).

Displacement activity Apparently irrelevant behavior performed by an animal under conflict situations, especially when tendencies to attack and escape are closely balanced.

Display A behavior that has evolved to influence the actions of other individuals.

Disruptive selection Selection in which phenotypes at both extremes of the population distribution are favored. (Contrast with directional selection; stabilizing selection.)

Distal Away from the point of attachment or other reference point. (Contrast with proximal.)

Disturbance A short-term event that disrupts populations, communities, or ecosystems by changing the environment.

Disulfide bridge The covalent bond between twosulfur atoms (–S—S–) linking to molecules or remote parts of the same molecule.

Diverticulum (di ver tik′ u lum) [L. *divertere*: turn away] A small cavity or tube that connects to a major cavity or tube.

Division A term used by some microbiologists and formerly by botanists, corresponding to the term phylum.

DNA (deoxyribonucleic acid) The fundamental hereditary material of all living organisms. In eukaryotes, stored primarily in the cell nucleus. A nucleic acid using deoxyribose rather than ribose.

DNA chip A small glass or plastic square onto which thousands of single-stranded DNA sequences are fixed. Hybridization of cell-derived RNA or DNA to the target sequences can be performed. (See DNA hybridization.)

DNA fingerprint An individual's unique DNA fragments produced by action of restriction endonucleases and separated by electrophoresis.

DNA helicase. An enzyme that functions during DNA replication to unwind the double helix.

DNA hybridization A process by which DNAs from two species are mixed and heated so that interspecific double helixes are formed.

DNA ligase Enzyme that unites Okazaki fragments of the lagging strand during DNA replication; also mends breaks in DNA strands. It connects pieces of a DNA strand and is used in recombinant DNA technology.

DNA methylation Addition of methyl groups to DNA; plays role in regulation of gene expression; protects a bacterium's DNA against its restriction endonucleases.

DNA polymerase Any of a group of enzymes that catalyze the formation of DNA strands from a DNA template.

DNA sequencing Determining the precise sequence of nucleotides in DNA.

DNA topoisomerase Enzymes that introduce positive or negative supercoils into the double-stranded DNA of continuous (circular) chromosomes.

Domain The largest unit in the current taxonomic nomenclature. Members of the three domains (Bacteria, Archaea, and Eukarya) are believed to have been evolving independently of each other for at least a billion years.

Dominance In genetics, the ability of one allelic form of a gene to determine the phenotype of a heterozygous individual, in which the homologous chromosomes carries both it and a different (recessive) allele. (Contrast with recessive.)

Dormancy A condition in which normal activity is suspended, as in some seeds and buds.

Dorsal [L. *dorsum*: back] Pertaining to the back or upper surface. (Contrast with ventral.)

Dorsal lip In amphibian embryo, the dorsal part of the blastpore which directs the development of nearby regions.

Double fertilization Process virtually unique to angiosperms in which one sperm nucleus combines with the egg to produce a zygote, and the other sperm nucleus combines with the two polar nuclei to produce the first cell of the triploid endosperm.

Double helix In DNA, the natural, right-handed coil configuration of two complementary, antiparallel strands.

Duodenum (do' uh dee' num) The beginning portion of the vertebrate small intestine. (Compare with ileum, jejunum.)

Dynein [Gr. *dynamis*: power] A protein that plays a part in the movement of eukaryotic flagella and cilia by means of conformational changes.

Ecdysone (eck die' sone) [Gr. *ek*: out of + *dyo*: to clothe] In insects, a hormone that induces molting.

Ecological community The species living together at a particular site.

Ecological niche (nitch) [L. *nidus*: nest] The functioning of a species in relation to other species and its physical environment.

Ecological succession The sequential replacement of one assemblage of populations by another in a habitat following some disturbance.

Ecology [Gr. *oikos*: house + *logos*: study] The scientific study of the interaction of organisms with their living and nonliving (abiotic) environment.

Ecosystem (eek' oh sis tum) The organisms of a particular habitat, such as a pond or forest, together with the physical environment in which they live.

Ectoderm [Gr. *ektos*: outside + *derma*: skin] The outermost of the three embryonic tissue layers first delineated during gastrulation. Gives rise to the skin, sense organs, nervous system, etc.

Ectotherm [Gr. *ektos*: outside + *thermos*: heat] An animal unable to control its body temperature. (Contrast with endotherm.)

Edema (i dee' mah) [Gr. *oidema*: swelling] Tissue swelling caused by the accumulation of fluid.

Edge effect The changes in ecological processes in a community caused by physical and biological factors originating in an adjacent community.

Effector Any organ, cell, or organelle that moves the organism through the environment or else alters the environment; for example, muscle, exocrine glands, chromatophores.

Effector phase Stage of the immune response, when cytotoxic T cells attack virus-infected cells, and helper T cells assist B cells to differentiate into plasma cells.

Efferent [L. *ex*: out + *ferre*: to bear] In physiology, conducting outward or away from an organ or structure. (Contrast with afferent.)

Egg In all sexually reproducing organisms, the female gamete; in birds, reptiles, and some other vertebrates, a structure witin which early embryonic development occurs. (Compare with amniote egg.)

Elasticity The property of returning quickly to a former state after a disturbance.

Electrocardiogram (EKG) A graphic recording of electrical potentials from the heart.

Electroencephalogram (EEG) A graphic recording of electrical potentials from the brain.

Electromyogram (EMG) A graphic recording of electrical potentials from muscle.

Electron transport. The passage of electrons through a series of proteins with a release of energy which may be captured in a concentration gradient or chemical form such as NADH or ATP.

Electronegativity The tendency of an atom to attract electrons when it occurs as part of a compound.

Electrostatic Pertaining to the attraction and repulsion of negative and positive charges on atoms due to the number and distribution of electrons.

Electrophoresis (e lek' tro fo ree' sis) [L. *electrum*: amber + Gr. *phorein*: to bear] A separation technique in which substances are separated from one another on the basis of their electric charges and molecular weights.

Elongation Growth of a plant axis or cell primarily in the longitudinal direction.

Embolus (em' buh lus) [Gr. *embolos*: inserted object; stopper] A circulating blood clot. Blockage of a blood vessel by an embolus or by a bubble of gas is referred to as an embolism. (Contrast with thrombus.)

Embryo [Gr. *en*: within + *bryein*: to grow] A young animal, or young plant sporophyte, while it is still contained within a protective structure such as a seed, egg, or uterus.

Embryo sac In angiosperms, the female gametophyte. Found within the ovule, it consists of eight or fewer cells, membrane bounded, but without cellulose walls between them.

Emergent property A property of a complex system that is not exhibited by its individual component parts.

Emigration The deliberate and usually oriented departure of an organism from the habitat in which it has been living.

3′ End (3 prime) The end of a DNA or RNA strand that has a free hydroxyl group at the 3′ carbon of the sugar (deoxyribose or ribose).

5′ End (5 prime) The end of a DNA or RNA strand that has a free phosphate group at the 5′ carbon of the sugar (deoxyribose or ribose).

Endemic (en dem' ik) [Gr. *endemos*: native, dwelling in] Confined to a particular region, thus often having a comparatively restricted distribution.

Endergonic reaction A chemical reaction that requires the input of energy in order to proceed. (Contrast with exergonic reaction.)

Endocrine gland (en' doh krin) [Gr. *endo*: within + *krinein*: to separate] Any gland, such as the adrenal or pituitary gland of vertebrates, that secretes certain substances, especially hormones, into the body through the blood.

Endocytosis A process by which liquids or solid particles are taken up by a cell through invagination of the plasma membrane. (Contrast with exocytosis.)

Endoderm [Gr. *endo*: within + *derma*: skin] The innermost of the three embryonic tissue layers delineated during gastrulation. Gives rise to the digestive and respiratory tracts and structures associated with them.

Endodermis In plants, a specialized cell layer marking the inside of the cortex in roots and some stems. Frequently a barrier to free diffusion of solutes.

Endomembrane system Endoplasmic reticulum plus Golgi apparatus; also lysosomes, when present. A system of membranes that exchange material with one another.

Endoplasmic reticulum (ER) [Gr. *endo*: within + L. *plasma*: form + L. *reticulum*: net] A system of membranous tubes and flattened sacs found in the cytoplasm of eukaryotes. Exists in two forms: rough ER, studded with ribosomes; and smooth ER, lacking ribosomes.

Endorphins Naturally occurring, opiate-like substances in the mammalian brain.

Endoskeleton [Gr. *endo*: within + *skleros*: hard] An internal skeleton covered by other, soft body tissues. (Contrast with exoskeleton.)

Endosperm [Gr. *endo*: within + *sperma*: seed] A specialized triploid seed tissue found only in angiosperms; contains stored nutrients for the developing embryo.

Endosymbiosis [Gr. *endo*: within + *sym*: together + *bios*: life] Two species living together, with one living inside the body (or even the cells) of the other.

Endosymbiotic theory The theory that the eukaryotic cell evolved from a prokaryote that contained other endosymbiotic prokaryotes.

Endotherm [Gr. *endo*: within + *thermos*: heat] An animal that can control its body temperature by the expenditure of its own metabolic energy. (Contrast with ectotherm.)

End product inhibition A control capacity of some metabolic pathways in which the final product produced inhibits an early enzyme in the pathway.

Energetic cost The difference between the energy an animal expends in performing a behavior and the energy it would have expended had it rested.

Energy The capacity to do work or move matter against an opposing force. The capacity to accomplish change.

Enhancer In eukaryotes, a DNA sequence, lying on either side of the gene it regulates, that stimulates a specific promoter.

Enterocoelous development A pattern of development in which the coelum is formed by an outpocketing of the embryonic gut (enteron).

Enterokinase (ent uh row kine' ase) An enzyme secreted by the mucosa of the duodenum. It activates the zymogen trypsinogen to create the active digestive enzyme trypsin.

Entrainment With respect to circadian rhythms, the process whereby the period is adjusted to match the 24-hour environmental cycle.

Entropy (en' tro pee) [Gr. *tropein*: to change] A measure of the degree of disorder in any system. Spontaneous reactions in a closed system are always accompanied by an increase in disorder and entropy.

Environment Whatever surrounds and interacts with a population, an organism or cell. May be external or internal.

Enzyme (en' zime) [Gr. *en*: in + *zyme*: yeast] A protein, on the surface of which are chemical groups so arranged as to make the enzyme a catalyst for a chemical reaction.

Eosinophils Phagocytic white blood cells that attack multicellular parasites once they have been coated with antibodies.

Epi- [Gr.: upon, over] A prefix used to designate a structure located on top of another; for example: epidermis, epiphyte.

Epiblast [Gr. *epi*: upon, over] The upper or overlying portion of the avian blastula which is joined to the hypoblast at the margins of the blastodisc.

Epicotyl (epp' i kot' il) [Gr. *epi*: over + *kotyle*: something hollow] That part of a plant embryo or seedling that is above the cotyledons.

Epidermis [Gr. *epi*: over + *derma*: skin] In plants and animals, the outermost cell layers. (Only one cell layer thick in plants.)

Epididymis (epuh did' uh mus) [Gr. *epi*: over + *didymos*: testicle] Coiled tubules in the testes that store sperm and conduct sperm from the seiminiferous tubules to the vas deferens.

Epinephrine (ep i nef' rin) [Gr. *epi*: over + *nephros*: kidney] The "fight or flight" hormone produced by the medulla of the adrenal gland; it also functions as a neurotransmitter. (Also known as adrenaline.)

Epiphyte (ep' e fyte) [Gr. *epi*: over + *phyton*: plant] A specialized plant that grows on the surface of other plants but does not parasitize them.

Epistasis Interaction between genes in which the presence of a particular allele of one gene determines whether another gene will be expressed.

Epithelium In animals, a layer of cells covering or lining an external surface or a cavity.

Equatorial plate In a cell undergoing mitosis, the region in the middle of a cell where the centromeres will align during metaphase.

Equilibrium Any state of balanced opposing forces and no net change.

ER See Endoplasmic reticulum.

Erythrocyte (ur rith' row site) [Gr. *erythros*: red + *kytos*: container] A red blood cell.

Esophagus (i soff' i gus) [Gr. *oisophagos*: gullet] That part of the gut between the pharynx and the stomach.

Ester linkage A condensation (water-releasing) reaction in which the carboxyl group of a fatty acid reacts with the hydroxyl group of an alcohol. Lipids are formed in this way.

Estivation (ess tuh vay' shun) [L. *aestivalis*: summer] A state of dormancy and hypometabolism that occurs during the summer; usually a means of surviving drought and/or intense heat. Contrast with hibernation.

Estrogen Any of several steroid sex hormones; produced chiefly by the ovaries in mammals.

Estrus (es' truss) [L. *oestrus*: frenzy] The period of heat, or maximum sexual receptivity, in some female mammals. Ordinarily, the estrus is also the time of release of eggs in the female.

Ethylene One of the plant hormones, the gas $H_2C{=}CH_2$.

Euchromatin Chromatin that is diffuse and nonstaining during interphase; may be transcribed. (Contrast with heterochromatin.)

Eudicots (yew die' kots) [Gr. *eu*: true + *di*: two + *kotyledon*: a cup-shaped hollow] The most diverse and abundant lineage of angiosperms. Eudicot embryos have two cotyledons, and eudicot flowers usually have parts (sepals, petals, etc.) in fours and fives.

Eukaryotes (yew car' ree oats) [Gr. *eu*: true + *karyon*: kernel or nucleus] Organisms whose cells contain their genetic material inside a nucleus. Includes all life other than the viruses, archaea, and bacteria.

Eusocial Term applied to insects, such as termites, ants, and many bees and wasps, in which individuals cooperate in the care of offspring, there are sterile castes, and generations overlap.

Eutrophication (yoo trofe' ik ay' shun) [Gr. *eu*: truly + *trephein*: to flourish] The addition of nutrient materials to a body of water, resulting in changes in ecological processes and species composition therein.

Evolution Any gradual change. Organic evolution, often referred to as evolution, is any genetic and resulting phenotypic change in organisms from generation to generation.

Evolutionary agent Any factor that influences the direction and rate of evolutionary changes.

Evolutionarily conserved Refers to traits that have evolved very slowly and are similar or even identical in individuals of many different phyla.

Evolutionary radiation The proliferation of species within a single evolutionary lineage.

Evolutionary reversal The reappearance of the ancestral state of a trait in a lineage in which that trait had acquired a derived state.

Excision repair The removal and damaged DNA and its replacement by the appropriate nucleotides.

Excitatory postsynaptic potential (EPSP) A change in the resting potential of a postsynaptic membrane in a positive (depolarizing) direction. (Contrast with inhibitory postsynaptic potential.)

Excretion Release of metabolic wastes by an organism.

Exergonic reaction A reaction in which free energy is released. (Contrast with endergonic reaction.)

Exocrine gland (eks' oh krin) [Gr. *exo*: outside + *krinein*: to separate] Any gland, such as a salivary gland, that secretes to the outside of the body or into the gut. (Contrast with endocrine gland.)

Exocytosis A process by which a vesicle within a cell fuses with the plasma membrane and releases its contents to the outside. (Contrast with endocytosis.)

Exon A portion of a DNA molecule, in eukaryotes, that codes for part of a polypeptide. (Contrast with intron.)

Exoskeleton (eks' oh skel' e ton) [Gr. *exos*: outside + *skleros*: hard] A hard covering on the outside of the body to which muscles are attached. (Contrast with endoskeleton.)

Exotoxins Highly toxic proteins released by living, multiplying bacteria.

Experiment The basis of the scientific method, in which particular factors are manipulated while other factors are held constant so that the potential influences of the manipulated factors can be determined.

Exponential growth Growth, especially in the number of organisms in a population, which is a geometric function of the size of the growing entity: the larger the entity, the faster it grows. (Contrast with logistic growth.)

Expression vector A DNA vector, such as a plasmid, that carries a DNA sequence that includes the adjacent sequences for its expression into mRNA and protein in a host cell.

Expressivity The degree to which a genotype is expressed in the phenotype; may be affected by the environment.

Extensor A muscle the extends an appendage.

Extinction The termination of a lineage of organisms.

Extrinsic protein A membrane protein found only on the surface of the membrane. (Contrast with intrinsic protein.)

Extracellular matrix. In animal tissues, a material of heterogeneous composition surrounding cells and performing many functions including adhesion of cells.

Extraembryonic membranes. The four membranes that support the developing embryo of reptiles, birds, and mammals but are not part of the embryo (amnion, allantois, chorion, and yolk sac)

F_1 (first filial generation) The immediate progeny of a parental (P) mating.

F_2 (second filial generation) The immediate progeny of a mating between members of the F_1 generation.

Facilitated diffusion Passive movement through a membrane involving a specific carrier protein; does not proceed against a concentration gradient. (Contrast with active transport, diffusion.)

Facultative anaerobes Prokaryotes that can shift their metabolism between anaerobic and aerobic operations depending on the presence or absence of O_2.

FAD See Flavin adenine dinucleotide.

Fat A triglyceride that is solid at room temperature. (Contrast with oil.)

Fate map. A map of the blastula showing which blastomers will contribute to specific tissues and organs in the mature body.

Fatty acid A molecule with a long hydrocarbon tail and a carboxyl group at the other end. Found in many lipids.

Fauna (faw′ nah) All of the animals found in a given area. (Contrast with flora.)

Feces [L. *faeces*: dregs] Waste excreted from the digestive system.

Feedback control Control of a particular process induced, directly or indirectly, by the presence or absence of a product of that process.

Fermentation (fur men tay′ shun) [L. *fermentum*: yeast] The anaerobic degradation of a substance such as glucose to smaller molecules with the extraction of energy.

Fertilization Union of gametes. Also known as syngamy.

Fertilization membrane A membrane surrounding an animal egg which becomes rapidly raised above the egg surface within seconds after fertilization, serving to prevent entry of a second sperm.

Fetus The latter stages of an embryo that is still contained in an egg or uterus; in humans, the unborn young from the eighth week of pregnancy to the moment of birth.

Fiber An elongated, tapering cell of flowering plants, usually with a thick cell wall. Serves a support function.

Fibrin A protein that polymerizes to form long threads that provide structure to a blood clot.

Filter feeder An organism that feeds upon much smaller organisms, that are suspended in water or air, by means of a straining device.

Filtration In the excretory physiology of some animals, the process by which the initial urine is formed; water and most solutes are transferred into the excretory tract, while proteins are retained in the blood or hemolymph.

First law of thermodynamics Energy can be neither created nor destroyed.

Fission Reproduction of a prokaryote by division of a cell into two comparable progeny cells.

Fitness The contribution of a genotype or phenotype to the genetic composition of subsequent generations, relative to the contribution of other genotypes or phenotypes. (See inclusive fitness.)

Fixed action pattern A behavior that is genetically programmed.

Flagellum (fla jell′ um) (plural: flagella) [L. *flagellum*: whip] Long, whiplike appendage that propels cells. Prokaryotic flagella differ sharply from those found in eukaryotes.

Flavin adenine dinucleotide (FAD) A coenzyme involved in redox reactions and containing the vitamin riboflavin (B_2).

Flexor A muscle that flexes an appendage.

Flora (flore′ ah) All of the plants found in a given area. (Contrast with fauna.)

Floral meristem Meristem that forms the sexual parts of flowering plants (sepals, petals, stamens, and carpels).

Florigen A plant hormone (not yet isolated) involved in the conversion of a vegetative shoot apex to a flower.

Flower The total reproductive structure of an angiosperm; its basic parts include the calyx, corolla, stamens, and carpels.

Fluid mosaic model A molecular model for the structure of biological membranes consisting of a fluid phospholipid bilayer in which suspended proteins are free to move in the plane of the bilayer.

Fluorescence The emission of a photon of visible light by an excited atom or molecule.

Follicle [L. *folliculus*: little bag] In female mammals, an immature egg surrounded by nutritive cells.

Follicle-stimulating hormone A gonadotropic hormone produced by the anterior pituitary.

Food chain A portion of a food web, most commonly a simple sequence of prey species and the predators that consume them.

Food vacuole Membrane enclosed structure formed by phagocytosis in which engulfed food particles are digested by the action of lysosomal enzymes.

Food web The complete set of food links between species in a community; a diagram indicating which ones are the eaters and which are eaten.

Forb Any broad-leaved herbaceous plant. Especially applied to such plants growing in grasslands.

Fossil Any recognizable structure originating from an organism, or any impression from such a structure, that has been preserved over geological time.

Fossil fuel A fuel (particularly petroleum products) composed of the remains of organisms that lived in the remote past.

Founder effect Random changes in allele frequencies resulting from establishment of a population by a very small number of individuals.

Fovea [L. *fovea*; a small pit] The area, in the vertebrate retina, of most distinct vision.

Frame-shift mutation A mutation resulting from the addition or deletion of one or two consecutive base pairs in the DNA sequence of a gene, resulting in misreading mRNA during translation and production of a nonfunctional protein. (Contrast with missense mutation, nonsense mutation, synonymous mutation.)

Free energy That energy which is available for doing useful work, after allowance has been made for the increase or decrease of disorder.

Frequency-dependent selection Selection that changes in intensity with the proportion of individuals in a population having the trait.

Fruit In angiosperms, a ripened and mature ovary (or group of ovaries) containing the seeds. Sometimes applied to reproductive structures of other groups of plants.

Fruiting body A structure that bears spores.

Functional genomics The assignment of functional roles to genes first identified by sequencing entire genomes.

Functional group A characteristic combination of atoms that contribute specific properties when attached to larger molecules.

Functional mRNA Eukaryotic mRNA that has been modified after transcription by the removal of introns and the addition of a 5′ cap and a 3′ poly(A) tail.

Fungus A member of the kingdom Fungi, a (usually) multicellular eukaryote with absorptive nutrition. (Yeasts are unicellular fungi.)

G cap A chemically modified GTP added to the 5′ end of mRNA; facilitates binding of mRNA to ribosome and prevents mRNA breakdown.

G_1 phase In the cell cycle, the gap between the end of mitosis and the onset of the S phase.

G_2 phase In the cell cycle, the gap between the S (synthesis) phase and the onset of mitosis.

G protein A membrane protein involved in signal transduction; characterized by binding GDP or GTP.

Gametangium (gam uh tan′ gee um) [Gr. *gamos*: marriage + *angeion*: vessel] Any plant or fungal structure within which a gamete is formed.

Gamete (gam′ eet) [Gr. *gamete/gametes*: wife, husband] The mature sexual reproductive cell: the egg or the sperm.

Gametogenesis (ga meet′ oh jen′ e sis) [Gr. *gamete/gametes*: wife, husband + *genesis*: source] The specialized series of cellular divisions that leads to the production of sex cells (gametes). (Contrast with oogenesis and spermatogenesis.)

Gametophyte (ga meet′ oh fyte) In plants and photosynthetic protists with alternation of generations, the multicellular haploid phase that produces the gametes. (Contrast with sporophyte.)

Ganglion (gang′ glee un) [Gr.: tumor] A group or concentration of neuron cell bodies.

Gap genes During insect development, the first step of segmentation genes to act organizing the anterior-posterior axis.

Gap junction A 2.7-nanometer gap between plasma membranes of two animal cells, spanned by protein channels. Gap junctions allow chemical substances or electrical signals to pass from cell to cell.

Gas exchange In animals, the process of taking up oxygen from the environment and releasing carbon dioxide to the environment.

Gastrovascular cavity Serving for both digestion (gastro) and circulation (vascular); in particular, the central cavity of the body of jellyfish and other cnidarians.

Gastrula (gas' true luh) [Gr. *gaster*: stomach] An embryo forming the characteristic three cell layers (ectoderm, endoderm, and mesoderm) which will give rise to all of the major tissue systems of the adult animal.

Gastrulation Development of a blastula into a gastrula.

Gated channel A membrane protein that opens and closes in response to binding of specific molecules or to changes in membrane potential. When open, it allows specific ions to move across the membrane.

Gene [Gr. *genes*: to produce] A unit of heredity. Used here as the unit of genetic function which carries the information for a single polypeptide or RNA.

Gene amplification Creation of multiple copies of a particular gene, allowing the production of large amounts of the RNA transcript (as in rRNA synthesis in oocytes).

Gene cloning Formation of a clone of bacteria or yeast cells containing a particular foreign gene.

Gene family A set of identical, or once-identical, genes, derived from a single parent gene; need not be on the same chromosomes; classic example is the globin family in vertebrates.

Gene flow The exchange of genes between different species (an extreme case referred to as hybridization) or between different populations of the same species caused by migration following breeding.

Gene frequency See Allele frequency.

Gene library All of the cloned DNA fragments generated by action of a restriction endonuclease on a genome or chromosome.

Gene pool All of the alleles of all of the genes in a population.

Gene therapy Treatment of a genetic disease by providing patients with cells containing functioning alleles of the genes that are nonfunctional in their bodies.

Generative cell In a pollen tube, a haploid nucleus that undergoes mitosis to produce the two sperm nuclei that participate in double fertilization. (Contrast with tube cell.)

Genetics The study of the structure, functioning, and inheritance of genes, the units of hereditary information.

Genetic drift Changes in gene frequencies from generation to generation in a small population as a result of random (chance) processes.

Genetic screening The application of medical tests to determine whether an individual carries a specific allele.

Genetic stochasticity Random variation in the frequencies of alleles and genotypes in a population over time. (Compare with demographic stochasticity.)

Genome (jee' nome) All the genes in a complete haploid set of chromosomes. (Compare with proteome.)

Genomics The study of entire sets of genes and their interactions.

Genomic imprinting When a given gene's phenotype is determined by whether that gene is inherited from the male or the female parent.

Genotype (jean' oh type) [Gr. *gen*: to produce + *typos*: impression] An exact description of the genetic constitution of an individual, either with respect to a single trait or with respect to a larger set of traits. (Contrast with phenotype.)

Genus (jean' us) (plural: genera) [Gr. *genos*: stock, kind] A group of related, similar species.

Geotropism See gravitropism.

Germ cell [L. *germen*: to beget] A reproductive cell or gamete of a multicellular organism. Contrast with somatic cell.

Germ layers The three embryonic tissue layers formed during gastrulation (ectoderm, mesoderm, endoderm).

Germination The sprouting of a seed or spore.

Gestation (jes tay' shun) [L. *gestare*: to bear] The period during which the embryo of a mammal develops within the uterus. Also known as pregnancy.

Gibberellin (jib er el' lin) A class of plant growth substances playing roles in stem elongation, seed germination, flowering of certain plants, etc. Named for the fungus *Gibberella*.

Gill An organ for gas exchange in aquatic organisms.

Gill arch A skeletal structure that supports gill filaments and the blood vessels that supply them.

Gizzard (giz' erd) [L. *gigeria*: cooked chicken parts] A muscular port of the stomach of birds that grinds up food, sometimes with the aid of fragments of stone.

Gland An organ or group of cells that produces and secretes one or more substances.

Glans penis Sexually sensitive tissue at the tip of the penis.

Glia (glee' uh) [Gr. *glia*: glue] Cells, found only in the nervous system, that do not conduct action potentials.

Glomerulus (glo mare' yew lus) [L. *glomus*: ball] Sites in the kidney where blood filtration takes place. Each glomerulus consists of a knot of capillaries served by afferent and efferent arterioles.

Glucocorticoids Steroid hormones produced by the adrenal cortex. Secreted in response to ACTH, they inhibit glucose uptake by many tissues in addition to mediating other stress responses.

Glucagon A hormone produced and released by cells in the islets of Langerhans of the pancreas. It stimulates the breakdown of glycogen in liver cells.

Gluconeogenesis The biochemical synthesis of glucose from other substances, such as amino acids, lactate, and glycerol.

Glycerol (gliss' er ole) A three-carbon alcohol with three hydroxyl groups; a component of phospholipids and triglycerides.

Glycogen (gly' ko jen) An energy storage polysaccharide found in animals and fungi; a branched-chain polymer of glucose, similar to starch.

Glycolysis (gly kol' li sis) [Gr. *gleukos*: sugar + *lysis*: break apart] The enzymatic breakdown of glucose to pyruvic acid. One of the evolutionarily oldest of the cellular energy-yielding mechanisms.

Glycosidic linkage The bond between sugar molecules through an intervening oxygen atom (–O–).

Glyoxysome (gly ox' ee soam) An organelle found in plants, in which stored lipids are converted to carbohydrates.

Golgi apparatus (goal' jee) A system of concentrically folded membranes found in the cytoplasm of eukaryotic cells; functions in secretion from cell by exocytosis.

Gonad (go' nad) [Gr. *gone*: seed] An organ that produces sex cells in animals: either an ovary (female gonad) or testis (male gonad).

Gonadotropin A hormone that stimulates the gonads.

Gondwana The large southern land mass that existed from the Cambrian (540 mya) to the Jurassic (138 mya). Present-day remnants are South America, Africa, India, Australia, and Antarctica.

Graft A bud or stem segment from one plant artificially and viably attached to another plant. A form of asexual reproduction.

Gram stain A differential purple stain useful in characterizing bacteria.

Granum (plural: grana) Within a chloroplast, a stack of thylakoids.

Gravitropism A directed plant growth response to gravity.

Grazer An animal that eats the vegetative tissues of herbaceous plants.

Green gland An excretory organ of crustaceans.

Greenhouse effect The heating of Earth's atmosphere by gases that are transparent to sunlight but opaque to heat.

Greenhouse gases Gases that contribute climate warming (water vapor, carbon dioxide, and methane) because they are transparent to light but opaque to heat.

Gross primary production The total energy captured by plants growing in a particular area.

Ground meristem That part of an apical meristem that gives rise to the ground tissue system of the primary plant body.

Ground tissue system Those parts of the plant body not included in the dermal or vascular tissue systems. Ground tissues function in storage, photosynthesis, and support.

Group transfer The exchange of atoms between molecules.

Growth Irreversible increase in volume (probably the most accurate definition, but at best a dangerous oversimplification).

Growth factors A group of proteins that circulate in the blood and trigger the normal growth of cells. Each growth factor acts only on certain target cells.

Guanine (G) (gwan' een) A nitrogen-containing base found in DNA, RNA, and GTP.

Guard cells In plants, specialized, paired epidermal cells that surround and control the opening of a stoma (pore). See stoma.

Gut An animal's digestive tract.

Guttation The extrusion of liquid water through openings in leaves, caused by root pressure.

Gymnosperm (jim′ no sperm) [Gr. *gymnos*: naked + *sperma*: seed] A plant, such as a pine or other conifer, whose seeds do not develop within an ovary (hence, the seeds are "naked").

Gyrus The raised or ridged portion of the convoluted surface of the brain. (Contrast to sulcus.)

Habit The form or pattern of growth characteristic of an organism.

Habitat The environment in which an organism lives.

Habituation (ha bich′ oo ay shun) The simplest form of learning, in which an animal presented with a stimulus without reward or punishment eventually ceases to respond.

Hair cell A type of mechanoreceptor in animals. Detects sound waves and other forms of motion in air or water.

Half-life The time required for half of a sample of a radioactive isotope to decay to its stable, nonradioactive form.

Halophyte (hal′ oh fyte) [Gr. *halos*: salt + *phyton*: plant] A plant that grows in a saline (salty) environment.

Haploid (hap′ loid) [Gr. *haploeides*: single] Having a chromosome complement consisting of just one copy of each chromosome; designated 1*n* or *n*. (Contrast with diploid.)

Hardy–Weinberg equililbrium The allele frequency at a given locus in a sexually reproducing population that is not being acted on by agents of evolution.

Haustorium (haw stor′ ee um) [L. *haustus*: draw up] A specialized hypha or other structure by which fungi and some parasitic plants draw food from a host plant.

Haversian systems Units of organization in compact bone that reflect the action of intercommunicating osteoblasts.

Heat-shock proteins Chaperone proteins expressed in cells exposed to high temperatures or other forms of environmental stress.

Helper T cells (T_H) T cells that participate in the activation of B cells and of other T cells; targets of the HIV-I virus, the agent of AIDS. (Contrast with cytotoxic T cells.)

Hematocrit (heme at′ o krit) [Gr. *heaema*: blood + *krites*: judge] The proportion of 100 cc of blood that consists of red blood cells.

Hemizygous (hem′ ee zie′ gus) [Gr. *hemi*: half + *zygotos*: joined] In a diploid organism, having only one allele for a given trait, typically the case for X-linked genes in male mammals and Z-linked genes in female birds. (Contrast with homozygous, heterozygous.)

Hemoglobin (hee′ mo glow bin) [Gr. *heaema*: blood + L. *globus*: globe] Oxygen-transporting protein found in the red blood cells of vertebrates (and found in some invertebrates).

Hensen's node In avian embryos, a structure at the anterior end of the primitive groove; determines the fates of cells passing over it during gastrulation.

Hepatic (heh pat′ ik) [Gr. *hepar*: liver] Pertaining to the liver.

Hepatic duct Duct that conveys bile from the liver to the gallbladder.

Herbivore (ur′ bi vore) [L. *herba*: plant + *vorare*: to devour] An animal that eats plant tissues. (Contrast with carnivore, detritivore, omnivore.)

Heritable Able to be inherited; in biology refers to genetically influenced traits.

Hermaphroditism (her maf′ row dite ism) [Gr. Hermes (messenger god) + Aphrodite (goddess of love)] The coexistence of both female and male sex organs in the same organism.

Hertz (abbreviated Hz) Cycles per second.

Hetero- [Gr.: *heteros*: other, different] A prefix specifying that two or more different conditions are involved; for example, heterotroph, heterozygous.

Heterochromatin Chromatin that retains its coiling during interphase; generally not transcribed. (Contrast with euchromatin.)

Heterocyst A large, thick-walled cell in the filaments of certain cyanobacteria; performs nitrogen fixation.

Heterogeneous nuclear RNA (hnRNA) The product of transcription of a eukaryotic gene, including transcripts of introns.

Heterokaryon In fungi, hypha containing two genetically different nuclei.

Heteromorphic (het′ er oh more′ fik) [Gr. *heteros*: different + *morphe*: form] having a different form or appearance, as two heteromorphic life stages of a plant. (Contrast with isomorphic.)

Heterosporous (het′ er os′ por us) Producing two types of spores, one of which gives rise to a female megaspore and the other to a male microspore. (Contrast with homosporous.)

Heterosis Situation in which heterozygous genotypes are superior to homozygous genotypes with respect to growth, survival, or fertility. Also called hybrid vigor.

Heterotherm An animal that regulates its body temperature at a constant level at some times but not others, such as a hibernator.

Heterotroph (het′ er oh trof) [Gr. *heteros*: different + *trophe*: food] An organism that requires preformed organic molecules as food. (Contrast with autotroph.)

Heterozygous (het′ er oh zie′ gus) [Gr. *heteros*: different + *zygotos*: joined] Of a diploid organism having different alleles of a given gene on the pair of homologues carrying that gene. (Contrast with homozygous.)

Hibernation [L. *hibernum*: winter] The state of inactivity of some animals during winter; marked by a drop in body temperature and metabolic rate.

Hierarchical sequencing An approach to DNA sequencing in which markers are mapped and DNA sequences are aligned by matching overlapping sites of known sequence.

Highly repetitive DNA Short DNA sequences present in millions of copies in the genome, next to each other (in tandem). In reassociation experiments, denatured highly repetitive DNA reanneals very quickly.

Hippocampus A part of the forebrain that takes part in long-term memory formation.

Histamine (hiss′ tah meen) A substance released by damaged tissue, or by mast cells in response to allergens. Histamine increases vascular permeability, leading to edema (swelling).

Histology The study of tissues.

Histone Any one of a group of basic proteins forming the core of a nucleosome, the structural unit of a eukaryotic chromosome. (Compare with nucleosome.)

Hierarchical sequencing An approach to DNA sequencing in which markers are mapped and DNA sequences are aligned by matching overlapping sites of known sequence.

hnRNA See heterogeneous nuclear RNA.

Homeobox A 180-base-pair segment of DNA found in certain genes (called Hox genes), perhaps regulating the expression of other genes and thus controlling large-scale developmental processes.

Homeostasis (home′ ee o sta′ sis) [Gr. *homos*: same + *stasis*: position] The maintenance of a steady state, such as a constant temperature or a stable social structure, by means of physiological or behavioral feedback responses.

Homeotherm (home′ ee o therm) [Gr. *homos*: same + *thermos*: heat] An animal that maintains a constant body temperature by its own internal heating and cooling mechanisms. (Contrast with heterotherm, poikilotherm.)

Homeotic genes (home ee ot′ ic) Genes that determine the developmental fate of entire segments of an animal.

Homeotic mutations Mutations in homeotic genes that drastically alter the characteristics of a particular body segment, giving it the characteristics of other segments (as when wings grow from a *Drosophila* thoracic segment that should have produced legs).

Homo- [Gr. *homos*: same] Prefix indicating two or more similar conditions, structures, or processes. Contrast to hetero-.

Homolog (home′ o log′) [Gr. *homos*: same + *logos*: word] One of a pair (or larger set, of chromosomes having the same overall genetic composition and sequence. In diploid organisms, each chromosome inherited from one parent is matched by an identical (except for mutational changes) chromosome—its homolog—from the other parent.

Homology (ho mol′ o jee) [Gr. *homologia*: of one mind; agreement] A similarity between two structures that is due to inheritance from a common ancestor. The structures are said to be homologous. (Contrast with analogy.)

Homoplasy (home′ uh play zee) [Gr. *homos*: same + *plastikos*: shape, mold] The presence in several species of a trait not present in their most common ancestor. Can result from convergent evolution, reverse evolution, or parallel evolution.

Homosporous Producing a single type of spore that gives rise to a single type of gametophyte, bearing both female and male reproductive organs. (Contrast with heterosporous.)

Homozygous (home′ oh zie′ gus) [Gr. *homos*: same + *zygotos*: joined] In a diploid organism, having identical alleles of a given gene on both homologous chromosomes. An individual may be a homozygote with respect to one gene and a heterozygote with respect to another. (Contrast with heterozygous.)

Hormone (hore' mone) [Gr. *hormon*: to excite, stimulate] A substance produced in minute amount at one site in a multicellular organism and transported to another site where it acts on target cells.

Host An organism that harbors a parasite or symbiont and provides it with nourishment.

Hox genes See homeobox.

Humoral immune response The part of the immune system mediated by B cells that produce circulating antibodies active against extracellular bacterial and viral infections.

Humus (hew' muss) The partly decomposed remains of plants and animals on the surface of a soil.

Hyaluronidase ((high' uh loo ron' uh dase) An enzyme that digests proteoglycans. In sperm cells, it digests the coatings surrounding an egg so the sperm can enter.

Hybrid (high' brid) [L. *hybrida*: mongrel] The offspring of genetically dissimilar parents. In molecular biology, a double helix formed of nucleic acids from different sources.

Hybrid vigor See heterosis.

Hybridoma A cell produced by the fusion of an antibody-producing cell with a myeloma cell; it produces monoclonal antibodies.

Hybrid zone A narrow zone where two populations interbreed, producing hybrid individuals.

Hydrocarbon A compound containing only carbon and hydrogen atoms.

Hydrogen bond A weak electrostatic bond which arises from the attraction between the slight positive charge on a hydrogen atom and a slight negative charge on a nearby oxygen or nitrogen atom.

Hydrological cycle The movement of water from the oceans to the atmosphere, to the soil, and back to the oceans.

Hydrolysis (high drol' uh sis) [Gr. *hydro*: water + *lysis*: break apart] A chemical reaction that breaks a bond by inserting the components of water: AB + $H_2O \rightarrow$ AH + BOH.

Hydrophilic (high dro fill' ik) [Gr. *hydro*: water + *philia*: love] Having an affinity for water. (Contrast with hydrophobic.)

Hydrophobic (high dro foe' bik) [Gr. *hydro*: water + *phobia*: fear] Having no affinity for water. Uncharged and nonpolar groups of atoms are hydrophobic, for example fats and side chain of the amino acid phenylalanine. (Contrast with hydrophilic.)

Hydrostatic pressure Pressure generated by compression of liquid in a confined space. Generated in plants, fungi, and some protists with cell walls by the osmotic uptake of water. Generated in animals with closed circulatory systems by the beating of a heart.

Hydrostatic skeleton The incompressible internal liquids of some animals that transfer forces from one part of the body to another when acted upon by the surrounding muscles.

Hydroxyl group The —OH group found on alcohols and sugars.

Hyper- [Gk. *hyper*: above, over] Prefix indicating above, higher, more.

Hyperpolarization A change in the resting potential of a membrane so the inside of a cell becomes more electronegative. (Contrast with depolarization.)

Hypersensitive response A defensive response of plants to microbial infection; it results in a "dead spot."

Hypertension High blood pressure.

Hypertonic Having a greater solute concentration. Said of one solution compared to another. (Contrast with hypotonic, isotonic.)

Hypha (high' fuh) (plural: hyphae) [Gr. *hyphe*: web] In the fungi and oomycetes, any single filament.

Hypo- [Gk. *hypo*: beneath, under] Prefix indicating underneath, below, less.

Hypoblast The lower tissue portion of the avian blastula which is joined to the epiblast at the margins of the blastodisc.

Hypocotyl [Gk. *hypo*: beneath + *kotyledon*: hollow space] That part of the embryonic or seedling plant shoot that is below the cotyledons.

Hypothalamus The part of the brain lying below the thalamus; it coordinates water balance, reproduction, temperature regulation, and metabolism.

Hypothesis A tentative answer to a question, from which testable predictions can be generated. (Contrast with theory.)

Hypothesis-prediction method A method of science in which hypotheses are generated, predictions are made from them, and experiments and observations are performed to test the predictions.

Hypotonic Having a lesser solute concentration. Said of one solution in comparing it to another. (Contrast with hypotonic, isotonic.)

Imaginal disc [L. *imagos*: image, form] In insect larvae, groups of cells that develop into specific adult organs.

Imbibition Water uptake by a seed; first step in germination.

Immune system [L. *immunis*: exempt from] A system in vertebrates that recognizes and attempts to eliminate or neutralize foreign substances (e.g., bacteria, viruses, pollutants).

Immunization The deliberate introduction of antigen to bring about an immune response.

Immunoglobulins A class of proteins, with a characteristic structure, active as receptors and effectors in the immune system.

Immunological memory The capacity to more rapidly and massively respond to a second exposure to an antigen than occurred on first exposure.

Immunological tolerance A mechanism by which an animal does not mount an immune response to the antigenic determinants of its own macromolecules.

Imprinting (1) In genetics, the differential modification of a gene depending on whether it is present in a male or a female. (2) In animal behavior, a rapid form of learning in which an animal comes to make a particular response, which is maintained for life, to some object or other organism.

Inclusive fitness The sum of an individual's genetic contribution to subsequent generations both via production of its own offspring and via its influence on the survival of relatives who are not direct descendants.

Incomplete dominance Condition in which the heterozygous phenotype is intermediate between the two homozygous phenotypes.

Incomplete metamorphosis Insect development in which changes between instars are gradual.

Incus (in' kus) [L. *incus*: anvil] The middle of the three bones that conduct movements of the eardrum to the oval window of the inner ear. (See malleus, stapes.)

Independent assortment During meiosis, the random separation of genes carried on nonhomologous chromosomes. Articulated by Mendel as his second law.

Individual fitness That component of inclusive fitness resulting from an organism producing its own offspring. (Contrast with kin selection component.)

Indoleacetic acid See auxin.

Induced fit A change in enzyme conformation upon binding to substrate with an increase in the rate of catalysis.

Induced mutation A mutation resulting from treatment with a chemical or other agent.

Inducer (1) In enzyme systems, a small molecule which, when added to a growth medium, causes a large increase in the level of some enzyme. (2) In embryology, a substance that causes a group of target cells to differentiate in a particular way.

Inducible enzyme An enzyme that is present in much larger amounts when a particular compound (the inducer) has been added to the system. (Contrast with constitutive enzyme.)

Inflammation A nonspecific defense against pathogens; characterized by redness, swelling, pain, and increased temperature.

Inflorescence A structure composed of several flowers.

Inflorescence meristem A meristem that produces floral meristems as well as other small leafy structures (bracts).

Inhibitor A substance that binds to the surface of an enzyme and interferes with its action on its substrates.

Inhibitory postsynaptic potential A change in the resting potential of a postsynaptic membrane in the hyperpolarizing (negative) direction.

Initial cells In plant meristems, undifferentiated cells that retain the capacity to divide producing both undifferentiated cells (initials) and cells committed to differentiation. (Compare with stem cells.)

Initiation complex Combination of a ribosomal light subunit, an mRNA molecule, and the tRNA charged with the first amino acid coded for by the mRNA; formed at the onset of translation.

Initiation factors Proteins that assist in forming the translation initiation complex at the ribosome.

Inner cell mass Derived from the mammalian blastula (bastocyst), the inner cell mass will give rise to the yolk sac (via hypoblast) and embryo (via epiblast).

Inositol triphosphate (IP_3) An intracellular second messenger derived from membrane phospholipids.

Instar (in' star) An immature stage of an insect between molts.

Insulin (in' su lin) [L. *insula*: island] A hormone synthesized in islet cells of the pancreas that promotes the conversion of glucose into the storage material, glycogen.

Integral membrane protein A membrane protein embedded in the bilayer of the membrane. (Contrast with peripheral membrane protein.)

Integrase An enzyme that integrates retroviral cDNA into the genome of the host cell.

Integrated pest management Control of pests by the use of natural predators and parasites in conjunction with sparing use of chemicals; an attempt to limit environmental damage.

Integument [L. *integumentum*: covering] A protective surface structure. In gymnosperms and angiosperms, a layer of tissue around the ovule which will become the seed coat.

Intercalary meristem A meristematic region in plants which occurs not apically, but between two regions of mature tissue. Intercalary meristems occur in the nodes of grass stems, for example.

Intercostal muscles Muscles between the ribs that can augment breathing movements by elevating and suppressing the rib cage.

Interferon A glycoprotein produced by virus-infected animal cells; increases the resistance of neighboring cells to the virus.

Interleukins Regulatory proteins, produced by macrophages and lymphocytes, that act upon other lymphocytes and direct their development.

Intermediate filaments Cytoskeletal component with diameters between the larger microtubules and smaller microfilaments.

Internode The region between two nodes of a plant stem.

Interphase The period between successive nuclear divisions during which the chromosomes are diffuse and the nuclear envelope is intact. It is during this period that the cell is most active in transcribing and translating genetic information.

Interspecific competition Competition between members of two or more species. (Contrast with intraspecific competition.)

Intertropical convergence zone The tropical region where the air rises most strongly; moves north and south with the passage of the sun overhead.

Intraspecific competition Competition among members of the same species. (Contrast with interspecific competition.)

Intrinsic protein A membrane protein that is embedded in the phospholipid bilayer of the membrane. (Contrast with extrinsic protein.)

Intrinsic rate of increase The rate at which a population can grow when its density is low and environmental conditions are highly favorable.

Intron A portion of a DNA molecule that, because of RNA splicing, is not involved in coding for part of a polypeptide molecule. (Contrast with exon.)

Invagination An infolding of cells during animal embryonic development.

Inversion A rare 180° reversal of the order of genes within a segment of a chromosome.

Invertebrate Any animal that is not a vertebrate, that is, whose nerve cord is not enclosed in a backbone of bony segments.

In vitro [L.: in glass] In a test tube, rather than in a living organism. (Contrast with in vivo.)

In vivo [L.: in the living state] In a living organism. Many processes that occur in vivo can be reproduced in vitro with the right selection of cellular components. (Contrast with in vitro.)

Ion (eye' on) [Gr.: *ion*: wanderer] An atom or group of atoms with electrons added or removed, giving it a negative or positive electrical charge.

Ion channel A membrane protein that can let ions diffuse across the membrane. The channel can be ion-selective, and it can be voltage-gated or ligand-gated.

Ionic bond An electrostatic attraction between positively and negatively charged ions. Usually a strong bond.

Iris (eye' ris) [Gr. *iris*: rainbow] The round, pigmented membrane that surrounds the pupil of the eye and adjusts its aperture to regulate the amount of light entering the eye.

Irruption A rapid increase in the density of a population. Often followed by massive emigration.

Islets of Langerhans Clusters of hormone-producing cells in the pancreas.

Iso- [Gr. *isos*: equal] Prefix used two separate entities that share some element of identity.

Isogamous Describes male and female gametes that are morphologically identical.

Isolating mechanism Geographical, physiological, ecological, or behavioral mechanisms that lead to a reduction in the frequency of successful matings between individuals in separate populations of a species. Can lead to the eventual evolution of separate species.

Isomers Molecules consisting of the same numbers and kinds of atoms, but differing in the bonding patterns by which the atoms are held together.

Isomorphic (eye so more' fik) [Gr. *isos*: equal + *morphe*: form] Having the same form or appearance, as when the haploid and diploid life stages of an organism appear identical. (Contrast with heteromorphic.)

Isotonic Having the same solute concentration; said of two solutions. (Contrast with hypertonic, hypotonic.)

Isotope (eye' so tope) [Gr. *isos*: equal + *topos*: place] Isotopes of a given chemical element have the same number of protons in their nuclei (and thus are in the same position on the periodic table), but differ in the number of neutrons

Isozymes Forms of an enzyme that have somewhat different amino acid sequences but catalyze the same reaction.

Jasmonates Plant hormones that trigger defenses against pathogens and herbivores.

Jejunum (jih jew' num) The middle division of the small intestine, where most absorption of nutrients occurs. (See duodenum, ileum.)

Joule (jool, or jowl) A unit of energy, equal to 0.24 calories.

Juvenile hormone In insects, a hormone maintaining larval growth and preventing maturation or pupation.

Karyotype The number, forms, and types of chromosomes in a cell.

Kelvin temperature scale See absolute temperature scale.

Keratin (ker' a tin) [Gr. keras: horn] A protein which contains sulfur and is part of such hard tissues as horn, nail, and the outermost cells of the skin.

Ketone (key' tone) A compound with a C=O group attached to two other groups, neither of which is an H atom. Many sugars are ketones. (Contrast with aldehyde.)

Keystone species A species that exerts a major influence on the composition and dynamics of the community in which it lives.

Kidneys A pair of excretory organs in vertebrates.

Kin selection The component of inclusive fitness resulting from helping the survival of relatives containing the same alleles by descent from a common ancestor.

Kinase (kye' nase) An enzyme that transfers a phosphate group from ATP to another molecule. Protein kinases transfer phosphate from ATP to specific proteins, playing important roles in cell regulation.

Kinesin Motor protein having the capacity to attach to organelles or vesicles and move them along microtubules of the cytoskeleton.

Kinetic energy The energy associated with movement.

Kinetochore (kin net' oh core) [Gr. *kinetos*: moving] Specialized structure on a centromere to which microtubules attach.

Koch's posulates Four rules for establishing that a particular microorganism causes a particular disease.

Krebs cycle See citric acid cycle.

Lactic acid fermentation Fermentation whose end product is lactic acid (lactate).

Lagging strand In DNA replication, the daughter strand that is synthesized in discontinuous stretches. (See Okazaki fragments.)

Lamella (la mell' ah) (L. *lamina*: thin sheet] Layer.

Larva (plural: larvae) [L. *lares*: guiding spirits] An immature stage of any invertebrate animal that differs dramatically in appearance from the adult.

Larynx (lar' inks) [Gk. *larynx*: voice box] A structure between the pharynx and the trachea that includes the vocal cords.

Lateral Pertaining to the side.

Lateral bud Located above the point of attachment of leaf to stem, an axillary meristem, short stem, immature leaves, and covering scales.

Lateral gene transfer The transfer of genes from one prokaryotic species to another.

Lateral meristems The vascular cambium and cork cambium, which give rise to secondary tissue in plants.

Laticifers (luh tiss' uh furs) In some plants, elongated cells containing secondary plant products such as latex.

Leader sequence A sequence of amino acids at the amino-terminal end of a newly synthesized protein; determines where the protein will be placed in the cell.

Leading strand In DNA replication, the daughter strand that is synthesized continuously. (Contrast with lagging strand.)

Lenticel (len' ti sill) Spongy region in a plant's periderm, allowing gas exchange.

Leukocyte (loo' ko sight) [Gr. *leukos*: clear + *kytos*: container] A white blood cell.

Lichen (lie' kun) An organism resulting from the symbiotic association of a true fungus and either a cyanobacterium or a unicellular alga.

Life cycle The entire span of the life of an organism from the moment of fertilization (or asexual generation) to the time it reproduces in turn.

Life history The stages an individual goes through during its life.

Life table A table showing, for a group of equal-aged individuals, the proportion still alive at different times in the future and the number of offspring they produce during each time interval.

Ligament A band of connective tissue linking two bones in a joint.

Ligand (lig' and) Any molecule that binds to a receptor site of another (usually larger) molecule.

Lignin The principal noncarbohydrate component of wood, a polymer that binds together cellulose fibrils in some plant cell walls.

Limbic system A group of primitive vertebrate forebrain nuclei that form a network and are involved in emotions, drives, instinctive behaviors, learning, and memory.

Limiting resource The required resource whose supply most strongly influences the size of a population.

Linkage Association between genetic markers on the same chromosome such that they do not show random assortment and seldom recombine; the closer the markers, the lower the frequency of recombination.

Lipase (lip' ase; lye' pase) An enzyme that digests fats.

Lipids (lip' ids) [Gr. *lipos*: fat] Substances in a cell which are easily extracted by organic solvents; fats, oils, waxes, steroids, and other large organic molecules, including those which, with proteins, make up the cell membranes. (Compare with phospholipids.)

Littoral zone The coastal zone from the upper limits of tidal action down to the depths where the water is thoroughly stirred by wave action.

Liver A large digestive gland. In vertebrates, it secretes bile and is involved in the formation of blood.

Lobes Regions of the human cerebral hemispheres; includes the temporal, frontal, parietal, and occipital lobes.

Locus In genetics, a specific location on a chromosome. May be considered to be synonymous with *gene*.

Logistic growth Growth, especially in the size of an organism or in the number of organisms that constitute a population, which slows steadily as the entity approaches its maximum size. (Contrast with exponential growth.)

Long-day plants. A plant that requires long day to flower.

Loop of Henle (hen' lee) Long, hairpin loop of the mammalian renal tubule that runs from the cortex down into the medulla, and back to the cortex. Creates a concentration gradient in the interstitial fluids in the medulla.

Lophophore A U-shaped fold of the body wall with hollow, ciliated tentacles that encircles the mouth of animals in several different phyla. Used for filtering prey from the surrounding water.

Lordosis (lor doe' sis) [Gk. *lordosis*: curving forward] A posture assumed by females of some mammalian species (especially rodents) to signal sexual receptivity.

Lumen (loo' men) [L. *lumen*: light] The cavity inside any tubular organ or structure, such as the gut or a kidney tubule.

Luteinizing hormone A gonadotropin produced by the anterior pituitary. It stimulates the gonads to produce sex hormones.

Lymph [L. *lympha*: liquid] A clear, watery fluid that is formed as a filtrate of blood; it contains white blood cells; it collects in a series of special vessels and is returned to the bloodstream.

Lymph nodes Specialized tissue regions that act as filters for cells, bacteria and foreign matter.

Lymphocyte A major class of white blood cells. Includes T cells, B cells, and other cell types important in the immune response.

Lymphoid tissue Tissues of the immune defense system dispersed throughout the body and consisting of: thymus, spleen, bone marrow, lymph nodes, blood, and lymph.

Lysis (lie' sis) [Gr. *lysis*: break apart] Bursting of a cell.

Lysogenic cycle A form of viral replication in which the virus becomes incorporated into the bacterial chromosome and the host cell is not killed. (Contrast with lytic cycle.)

Lysosome (lie' so soam) [Gr. *lysis*: break away + *soma*: body] A membrane-enclosed organelle found in eukaryotic cells (other than plants). Lysosomes contain a mixture of enzymes that can digest most of the macromolecules found in the rest of the cell.

Lysozyme (lie' so zyme) An enzyme in saliva, tears, and nasal secretions that attacks bacterial cell walls, as one of the body's nonspecific defense mechanisms.

Lytic cycle A form of viral reproduction that lyses the host bacterium releasing the new viruses. (Contrast with lysogenic cycle.)

M phase The portion of the cell cycle in which mitosis takes place.

Macroevolution [Gr. *makros*: large, long] Evolutionary changes occurring over long time spans and usually involving changes in many traits. (Contrast with microevolution.)

Macromolecule A giant polymeric molecule. The macromolecules are proteins, polysaccharides, and nucleic acids.

Macronutrient A mineral element required by plant tissues in concentrations of at least 1 milligram per gram of their dry matter.

Macrophage (mac' roh faj) A type of white blood cell that endocytoses bacteria and other cells.

Major histocompatibility complex (MHC) A complex of linked genes, with multiple alleles, that control a number of cell surface antigens that identify self and can lead to graft rejection.

Malleus (mal' ee us) [L. *malleus*: hammer] The first of the three bones that conduct movements of the eardrum to the oval window of the inner ear. (See incus, stapes.)

Malpighian tubule (mal pee' gy un) A type of protonephridium found in insects.

Mammal [L. *mamma*: breast, teat] Any animal of the class Mammalia. Mammals are characterized by the production of milk by the female mammary glands and the possession of hair for body covering.

Mantle A sheet of specialized tissues that covers most of the viscera of mollusks; provides protection to internal organs and secretes the shell.

Mapping In genetics, determining the order of genes on a chromosome and the distances between them.

Marine [L. *mare*: sea, ocean] Pertaining to or living in the ocean. (Contrast with aquatic, terrestrial.)

Marker A gene of identifiable phenotype that indicates the presence on another gene, DNA segment, or chromosome fragment.

Marsupial (mar soo' pee al) A mammal belonging to the subclass Metatheria, such as opossums and kangaroos. Most have a pouch (marsupium) that contains the milk glands and serves as a receptacle for the young.

Mass extinctions Geological periods during which rates of extinction were much higher than during intervening times.

Mass number The sum of the number of protons and neutrons in an atom's nucleus.

Mast cells Typically found in connective tissue, mast cells can be provoked by antigens or inflammation to release histamine.

Maternal effect genes These genes code for morphogens that determine the polarity of the egg and larva in the fruit fly *Drosophila melanogaster*.

Maternal inheritance Inheritance in which the mother's phenotype is exclusively expressed. Mitochondria and chloroplasts are maternally inherited via egg cytoplasm. Also known as cytoplasmic inheritance.

Mating types A mating system in which the sexes are morphologically identical but carry different alleles and will mate.

Mechanoreceptor A cell that is sensitive to physical movement and generates action potentials in response.

Medulla (meh dull' luh) (1) The inner, core region of an organ, as in the adrenal medulla (adrenal gland) or the renal medulla (kidneys). (2) The portion of the brain stem that connects to the spinal cord.

Megagametophyte A female gametophyte that produces eggs only.

Megaspore [Gr. *megas*: large + *spora*: to sow] In plants, a haploid spore that produces a female gametophyte.

Meiosis (my oh' sis) [Gr. *meiosis*: diminution] Division of a diploid nucleus to produce four haploid daughter cells. The process consists of two successive nuclear divisions with only one cycle of chromosome replication.

Membrane potential The difference in electrical charge between the inside and the outside of a cell, caused by a difference in the distribution of ions.

Memory cells Long-lived lymphocytes produced by exposure to antigen. They persist in the body and are able to mount a rapid response to subsequent exposures to the antigen.

Mendelian population A local population of individuals belonging to the same species and exchanging genes with one another.

Mendel's first law See Segregation.

Mendel's second law See Independent assortment.

Menstrual cycle The monthly sloughing off of the uterine lining if fertilization does not occur in the female. Occurs between puberty and menopause.

Meristem [Gr. *meristos*: divided] Plant tissue made up of undifferentiated actively dividing cells.

Mesenchyme (mez' en kyme) [Gr. *mesos*: middle + *enchyma*: infusion] Embryonic or unspecialized cells derived from the mesoderm.

Mesoderm [Gr. *mesos*: middle + *derma*: skin] The middle of the three embryonic tissue layers first delineated during gastrulation. Gives rise to skeleton, circulatory system, muscles, excretory system, and most of the reproductive system.

Mesophyll (mez' uh fill) [Gr. *mesos*: middle + *phyllon*: leaf] Chloroplast-containing, photosynthetic cells in the interior of leaves.

Mesosome (mez' uh soam') [Gr. *mesos*: middle + *soma*: body] A localized infolding of the plasma membrane of a bacterium.

Messenger RNA (mRNA) A transcript of one of the strands of DNA; carries information (as a sequence of codons) for the synthesis of one or more proteins.

Meta- [Gr.: between, along with, beyond] A prefix used in biology to denote a change or a shift to a new form or level; for example, as used in metamorphosis.

Metabolism (meh tab' a lizm) [Gr. *metabole*: to change] The sum total of the chemical reactions that occur in an organism, or some subset of that total (as in respiratory metabolism).

Metabolic compensation Changes in metabolic properties of an organism that render it less sensitive to temperature changes.

Metabolic pathway A series of enzyme-catalyzed reactions so arranged that the product of one reaction is the substrate of the next.

Metamorphosis (met' a mor' fo sis) [Gr. *meta*: between + *morphe*: form, shape] A change occurring between one developmental stage and another, as for example from a tadpole to a frog. (See complete metamorphosis, incomplete metamorphosis.)

Metaphase (met' a phase) The stage in nuclear division at which the centromeres of the highly supercoiled chromosomes are all lying on a plane (the metaphase plane or plate) perpendicular to a line connecting the division poles.

Metapopulation A population divided into subpopulations, among which there are occasional exchanges of individuals.

Metastasis (meh tass' tuh sis) The spread of cancer cells from their original site to other parts of the body.

Methanogen Any member of a group of archaea that release methane as a metabolic product. This group is considered to be an extremely ancient one.

Methylation The addition of a methyl group (—CH_3) to a molecule. Extensive methylation of cytosine in DNA is correlated with reduced transcription.

MHC See Major histocompatibility complex.

Microbiology [Gr. *mikros*: small + *bios*: life + *logos*: discourse] The scientific study of microscopic organisms, particularly bacteria, protists, and viruses.

Microevolution The small evolutionary changes typically occurring over short time spans; generally involving a small number of traits and minor genetic changes. (Contrast with macroevolution.)

Microfilament Minute fibrous structure generally composed of actin found in the cytoplasm of eukaryotic cells. They play a role in the motion of cells.

Microgametophyte A male gametophyte that produces sperm only.

Micronutrient A mineral element required by plant tissues in concentrations of less than 100 micrograms per gram of their dry matter.

Micropyle (mike' roh pile) [Gr. *mikros*: small + *pylon*: gate] Opening in the integument(s) of a seed plant ovule through which pollen grows to reach the female gametophyte within.

Microspore [Gr. *mikros*: small + *spora*: to sow] In plants, a haploid spore that produces a male gametophyte.

Microtubules Minute tubular structures found in centrioles, spindle apparatus, cilia, flagella, and cytoskeleton of eukaryotic cells. These tubules play roles in the motion and maintenance of shape of eukaryotic cells.

Microvilli (singular: microvillus) The projections of epithelial cells, such as the cells lining the small intestine, that increase their surface area.

Middle lamella A layer of polysaccharides that separates plant cells; a shared middle lamella lies outside the primary walls of the two cells.

Migration The regular, seasonal movements of animals.

Mineral An inorganic substance other than water.

Mineral nutrients Inorganic ions required by organisms for normal growth and reproduction.

Mismatch repair When a single base in DNA is changed into a different base, or the wrong base inserted during DNA replication, there is a mismatch in base pairing with the base on the opposite strand. A repair system removes the incorrect base and inserts the proper one for pairing with the opposite strand.

Missense mutation A nonsynonymous mutation, or one that changes a codon for one amino acid to a codon for a different amino acid. (Contrast with frame-shift mutation, nonsense mutation, synonymous mutation.)

Mitochondrial matrix The fluid interior of the mitochondrion, enclosed by the inner mitochondrial membrane.

Mitochondrion (my' toe kon' dree un) [Gr. *mitos*: thread + *chondros*: grain] An organelle in eukaryotic cells that contains the enzymes of the citric acid cycle, the respiratory chain, and oxidative phosphorylation.

Mitosis (my toe' sis) [Gr. *mitos*: thread] Nuclear division in eukaryotes leading to the formation of two daughter nuclei each with a chromosome complement identical to that of the original nucleus.

Mitotic center Cellular region that organizes the microtubules for mitosis. In animals a centrosome serves as the mitotic center.

Moderately repetitive DNA DNA sequences that appear hundreds to thousands of times in the genome. They include the DNA sequences coding for rRNAs and tRNAs, as well as the DNA at telomeres.

Modular organism An organism which grows by producing additional units of body construction (modules) that are very similar to the units of which it is already composed.

Mole A quantity of a compound whose weight in grams is numerically equal to its molecular weight expressed in atomic mass units. Avogadro's number of molecules: 6.023×10^{23} molecules.

Molecular clock The theory that macromolecules diverge from one another over evolutionary time at a constant rate; this rate may provide insight into the phylogenetic relationships among organisms.

Molecular weight The sum of the atomic weights of the atoms in a molecule.

Molecule A particle made up of two or more atoms joined by covalent bonds or ionic attractions.

Molting The process of shedding part or all of an outer covering, as the shedding of feathers by birds or of the entire exoskeleton by arthropods.

Monoclonal antibody Antibody produced in the laboratory from a clone of hybridoma cells, each of which produces the same specific antibody.

Monocot [Gr. *mono*: one + *kotyledon*: a cup-shaped hollow] Any member of the angiosperm lineage in which the embryo produces a single cotyledon (seed leaf). Leaves of most monocots have their major veins arranged parallel to each other.

Monocytes White blood cells that produce macrophages.

Monoecious (mo nee' shus) [Gr. *mono*: one + *oikos*: house] Describes organisms in which both sexes are "housed" in a single individual that produces both eggs and sperm. (In some plants, these are found in different flowers within the same plant.)

Examples: corn, peas, earthworms, hydras. (Contrast with dioecious, perfect flower.)

Monohybrid cross A mating in which the parents differ with respect to the alleles of only one locus of interest.

Monomer [Gr. *mono*: one + *meros:* unit] A small molecule, two or more of which can be combined to form oligomers (consisting of a few monomers) or polymers (consisting of many monomers).

Monophyletic (mon′ oh fih leht′ ik) [Gk. *mono*: one + *phylon*: tribe] Descended from a single ancestral stock.

Monosaccharide A simple sugar. Oligosaccharides and polysaccharides are made up of monosaccharides.

Monosynaptic reflex A neural reflex that begins in a sensory neuron and makes a single synapse before activating a motor neuron.

Morphogen A diffusible substances whose concentration gradients determine patterns of development in animals and plants.

Morphogenesis (more′ fo jen′ e sis) [Gr. *morphe*: form + *genesis*: origin] The development of form; the overall consequence of determination, differentiation, and growth.

Morphology (more fol′ o jee) [Gr. *morphe*: form + *logos*: study, discourse] The scientific study of organic form, including both its development and function.

Mosaic development Pattern of animal embryonic development in which each blastomere contributes a specific part of the adult body. (Contrast with regulative development.)

Motor end plate The modified area on a muscle cell membrane where a synapse is formed with a motor neuron.

Motor neuron A neuron carrying information from the central nervous system to an effector such as a muscle fiber.

Motor proteins Specialized proteins that use energy to change shape and move cells or structures within cells. See dynein, kinesin.

Motor unit A motor neuron and the set of muscle fibers it controls.

mRNA See messenger RNA.

Mucosa (mew koh′ sah) An epithelial membrane containing cells that secrete mucus. The inner cell layers of the digestive and respiratory tracts.

Muscle Contractile tissue containing actin and myosin organized into polymeric chains called microfilaments. Muscle fiber A single muscle cell. In the case of striated muscle, a syncitial, multinucleate cell.

Muscle spindle Modified muscle fibers encased in a connective sheat and functioning as stretch receptors.

Mutagen (mute′ ah jen) [L. *mutare*: change + Gr. *genesis*: source] Any agent (e.g., chemicals, radiation) that increases the mutation rate.

Mutation A detectable, heritable change in the genetic material not caused by recombination.

Mutation pressure Evolution (change in gene proportions) by different mutation rates alone (i.e., without the influence of natural selection).

Mutualism The type of symbiosis, such as that exhibited by fungi and algae or cyanobacteria in forming lichens, in which both species profit from the association.

Mycelium (my seel′ ee yum) [Gr. *mykes*: fungus] In the fungi, a mass of hyphae.

Mycorrhiza (my′ ko rye′ za) [Gr. *mykes*: fungus + *rhiza*: root] An association of the root of a plant with the mycelium of a fungus.

Myelin (my′ a lin) A material forming a sheath around some axons. Formed by Schwann cells that wrap themselves about the axon, myelin insulates the axon electrically and increases the rate of transmission of a nervous impulse.

Myofibril (my′ oh fy′ bril) [Gr. *mys*: muscle + L. *fibrilla*: small fiber] A polymeric unit of actin or myosin in a muscle.

Myogenic (my oh jen′ ik) [Gr. *mys*: muscle + *genesis*: source] Originating in muscle.

Myoglobin (my′ oh globe′ in) [Gr. *mys*: muscle + L. *globus*: sphere] An oxygen-binding molecule found in muscle. Consists of a heme unit and a single globiin chain, and carrys less oxygen than hemoglobin.

Myosin One of the two major proteins of muscle, it makes up the thick filaments. (See actin.)

NAD (nicotinamide adenine dinucleotide) A compound found in all living cells, existing in two interconvertible forms: the oxidizing agent NAD^+ and the reducing agent $NADH + H^+$.

NADP (nicotinamide adenine dinucleotide phosphate) A compound similar to NAD, but possessing another phosphate group; plays similar roles but is used by different enzymes.

Natural killer cells A nonspecific defensive cell (lymphocyte) that attacks tumor cells and virus infected cells.

Natural selection The differential contribution of offspring to the next generation by various genetic types belonging to the same population. The mechanism of evolution proposed by Charles Darwin.

Necrosis (nec roh′ sis) [Gk. *nekros*: death] Tissue damage resulting from cell death.

Negative control The situation in which a regulatory macromolecule (generally a repressor) functions to turn off transcription. In the absence of a regulatory macromolecule, the structural genes are turned on.

Nematocyst (ne mat′ o sist) [Gr. *nema*: thread + *kystis*: cell] An elaborate, threadlike structure produced by cells of jellyfish and other cnidarians, used chiefly to paralyze and capture prey.

Nephridium (nef rid′ ee um) [Gr. *nephros*: kidney] An organ which is involved in excretion, and often in water balance, involving a tube that opens to the exterior at one end.

Nephron (nef′ ron) [Gr. *nephros*: kidney] The functional unit of the kidney, consisting of a structure for receiving a filtrate of blood, and a tubule that absorbs selected parts of the filtrate back into the bloodstream.

Nephrostome (nef′ ro stome) [Gr. *nephros*: kidney + *stoma*: opening] An opening in a nephridium through which body fluids can enter.

Nerve A structure consisting of many neuronal axons and connective tissue.

Net primary production Total photosynthesis minus respiration by plants.

Neural plate A thickened strip of ectoderm along the dorsal side of the early vertebrate embryo; gives rise to the central nervous system.

Neural tube An early stage in the development of the vertebrate nervous system consisting of a hollow tube created by two opposing folds of the dorsal ectoderm along the anterior–posterior body axis.

Neuromuscular junction The region where a motor neuron contacts a muscle fiber, creating a synapse.

Neuron (noor′ on) [Gr. *neuron*: nerve] A nervous system cell that can generate and conduct action potentials along an axon to a synapse with another cell.

Neurotransmitter A substance produced in and released by one a neuron (the presynaptic cell) that diffuses across a synapse and excites or inhibits another cell (the postsynaptic cell).

Neurula (nure′ you la) Embryonic stage during the dorsal nerve cord forms from two ectodermal ridges.

Neutral allele An allele that does not alter the functioning of the proteins for which it codes.

Neutral theory A view of molecular evolution that postulates that most mutations do not affect the amino acid being coded for, and that such mutations accumulate in a population at rates driven by genetic drift and mutation rates.

Neutron (new′ tron) One of the three most fundamental particles of matter, with mass approximately 1 amu and no electrical charge.

Neutrophils Abundant, short-lived phagocytic leukocytes that attack antibody-coated antigens.

Niche See ecological niche.

Nitrate reduction The process by which nitrate (NO_3^-) is reduced to ammonia (NH_3).

Nitric oxide (NO) An unstable molecule (a gas) that serves as a second messenger causing smooth muscle to relax. In the nervous system it operates as a neurotransmitter.

Nitrification The oxidation of ammonia to nitrite and nitrate ions, performed by certain soil bacteria.

Nitrogenase In nitrogen-fixing organisms, an enzyme complex that mediates the stepwise reduction of atmospheric N_2 to ammonia.

Nitrogen fixation Conversion of nitrogen gas to ammonia, which makes nitrogen available to living things. Carried out by certain prokaryotes, some of them free-living and others living within plant roots.

Node [L. *nodus*: knob, knot] In plants, a (sometimes enlarged) point on a stem where a leaf is or was attached.

Node of Ranvier A gap in the myelin sheath covering an axon; the point where the axonal membrane can fire action potentials.

Noncompetitive inhibitor An inhibitor that binds the enzyme at a site other than the active site. (Contrast with competitive inhibitor.)

Nondisjunction Failure of sister chromatids to separate in meiosis II or mitosis, or failure of homologous chromosomes to separate in meiosis I. Results in aneuploidy.

Nonpolar molecule A molecule whose electric charge is evenly balanced from one end of the molecule to the other.

Nonsense mutation Mutations that prematurely terminate a polypeptide by changing a codon for an amino acid to one of the codons (UAG, UAA, or UGA) that signal termination of translation. (Contrast with frame-shift mutation, missense mutation, synonymous mutation.)

Nonspecific defenses Immunologic responses directed against any invading agent without reacting to apecific antigens.

Nonsynonymous mutation A nucleotide substitution that that changes the amino acid specified (i.e., AGC → AGA, or serine → arginine). (Contrast with synonymous mutation.)

Nonsynonymous substitution The situation when a nonsynonymous mutation becomes dominant in a population. (Contrast with synonymous substitution.)

Nontracheophytes Those plants lacking well-developed vascular tissue; the liverworts, hornworts, and mosses. (Contrast with tracheophytes.)

Norepinephrine A neurotransmitter found in the central nervous system and also at the postganglionic nerve endings of the sympathetic nervous system. Also called noradrenaline.

Notochord (no' tow kord) [Gr. *notos*: back + *chorde*: string] A flexible rod of gelatinous material serving as a support in the embryos of all chordates and in the adults of tunicates and lancelets.

Nuclear envelope The surface, consisting of two layers of membrane, that encloses the nucleus of eukaryotic cells.

Nuclear pore complex Protein structure situated in nuclear pores through which RNA and proteins enter and leave the nucleus.

Nucleic acid (new klay' ik) A long-chain alternating polymer of deoxyribose or ribose and phosphate groups, with nitrogenous bases—adenine, thymine, uracil, guanine, or cytosine (A, T, U, G, or C)—as side chains. DNA and RNA are nucleic acids.

Nucleoid (new' klee oid) The region that harbors the chromosomes of a prokaryotic cell. Unlike the eukaryotic nucleus, it is not bounded by a membrane.

Nucleolar organizer (new klee' o lar) A region on a chromosome that is associated with the formation of a new nucleolus following nuclear division. The site of the genes that code for ribosomal RNA.

Nucleolus (new klee' oh lus) A small, generally spherical body found within the nucleus of eukaryotic cells. The site of synthesis of ribosomal RNA.

Nucleoplasm (new' klee o plazm) The fluid material within the nuclear envelope of a cell, as opposed to the chromosomes, nucleoli, and other particulate constituents.

Nucleosome A portion of a eukaryotic chromosome, consisting of part of the DNA molecule wrapped around a group of histone molecules, and held together by another type of histone molecule. The chromosome is made up of many nucleosomes.

Nucleotide The basic chemical unit in a nucleic acid. A nucleotide in RNA consists of one of four nitrogenous bases linked to ribose, which in turn is linked to phosphate. In DNA, deoxyribose is present instead of ribose.

Nucleoside A nucleotide without the phosphate group.

Nucleus (new' klee us) [L. *nux*: kernel or nut] In cells, the centrally located compartment of eukaryotic cells that is bounded by a double membrane and contains the chromosomes.

Null hypothesis The assertion that an effect proposed by its companion hypothesis does not in fact exist.

Nutrient A food substance; or, in the case of mineral nutrients, an inorganic element required for completion of the life cycle of an organism.

Obligate anaerobe An anaerobic prokaryote that cannot survive exposure to O_2.

Oil A triglyceride that is liquid at room temperature. (Contrast with fat.)

Okazaki fragments Newly formed DNA making up the lagging strand in DNA replication. DNA ligase links Okazaki fragments together to give a continuous strand.

Olfactory [L. *olfacere*: to smell] Having to do with the sense of smell.

Oligomer [Gr.: *oligo*: a few + *meros*: units] A compound molecule of intermediate size, made up of two to a few monomers. (Contrast with monomer, polymer.)

Oligosaccharins Plant hormones, derived from the plant cell wall, that trigger defenses against pathogens.

Ommatidium [Gr. *omma*: eye] One of the units which, collected into groups of up to 20,000, make up the compound eye of arthropods.

Omnivore [L. *omnis*: everything + *vorare*: to devour] An organism that eats both animal and plant material. (Contrast with carnivore, detritivore, herbivore.)

Oncogene [Gr. *onkos*: mass, tumor + *genes*: born] Genes that greatly stimulate cell division, giving rise to tumors.

Oocyte (oh' eh site) [Gr. *oon*: egg + *kytos*: container] The cell that gives rise to eggs in animals.

Oogenesis (oh' eh jen e sis) [Gr. *oon*: egg + *genesis*: source] Female gametogenesis, leading to production of the egg.

Oogonium (oh' eh go' nee um) In some algae and fungi, a cell in which an egg is produced.

Operator The region of an operon that acts as the binding site for the repressor.

Operon A genetic unit of transcription, typically consisting of several structural genes that are transcribed together; the operon contains at least two control regions: the promoter and the operator.

Opportunity cost The sum of the benefits an animal forfeits by not being able to perform some other behavior during the time when it is performing a given behavior.

Opsin (op' sin) [Gr. *opsis*: sight] The protein portion of the visual pigment rhodopsin. (See rhodopsin.)

Optic chiasm [Gr. *chiasma*: cross] Structure on the lower surface of the vertebrate brain where the two optic nerves come together.

Optical isomers Two isomers that are mirror images of one another.

Organ [Gk. *organon*: tool] A body part, such as the heart, liver, brain, root, or leaf. Organs are composed of different tissues integrated to perform a distinct function. Organs are in turn often integrated into systems, such as the digestive or reproductive system.

Organ identity genes Plant genes that specify the various parts of the flower. See homeotic genes.

Organ of Corti Structure in the inner ear that transforms mechanical forces produced from pressure waves ("sound waves") into action potentials that are sensed as sound.

Organelles (or gan els') Organized structures that are found in or on cells. Examples: ribosomes, nuclei, mitochrondria, chloroplasts, cilia, and contractile vacuoles.

Organic Pertaining to any aspect of living matter, e.g., to its evolution, structure, or chemistry. The term is also applied to any chemical compound that contains carbon.

Organism Any living being.

Organizer Region of an early embryo that directs the development of nearby regions. In amphibian early gastrulas, the dorsal lip of the blastopore is the organizer.

Organogenesis The formation of organs and organ systems during development.

Origin of replication A DNA sequence at which helicase unwinds the DNA double helix and DNA polymerase binds to initiate DNA replication.

Osmolarity The concentration of osmotically active particles in a solution.

Osmoregulation Regulation of the chemical composition of the body fluids of an organism.

Osmoreceptor A neuron that converts changes in the osmotic potential of interstial fluids into action potentials.

Osmosis (oz mo' sis) [Gr. *osmos*: to push] The movement of water across a differentially permeable membrane, from one region to another region where the water potential is more negative.

Ossicle (oss' ick ul) [L. *os*: bone] The calcified construction unit of echinoderm skeletons.

Osteoblasts (oss' tee oh blast) [Gk. *osteon*: bone + *blastos*: sprout] Cells that lay down the protein matrix of bone.

Osteoclasts (oss' tee oh clast) [Gk. *osteon*: bone + *klastos*: broken] Cells that dissolve bone.

Otolith (oh' tuh lith) [Gk. *otikos*: ear + *lithos*: stone[Structures in the vertebrate vestibular apparatus that mechanically stimulate hair cells when the head moves or changes position.

Oval window The flexible membrane that, when moved by the bones of the middle ear, produces pressure waves in the inner ear

Ovary (oh' var ee) [L. *ovum*: egg] Any female organ, in plants or animals, that produces an egg.

Oviduct [L. *ovum*: egg + *ducere*: to lead] In mammals, the tube serving to transport eggs to the uterus or to outside of the body.

Oviparous (oh vip' uh rus) Reproduction in which eggs are released by the female and development is external to the mother's body. (Contrast with viviparous.)

Ovulation The release of an egg from an ovary.

Ovule (oh' vule) In plants, a structure that contains a gametophyte and, within the gametophyte, an egg; when it matures, an ovule becomes a seed.

Ovum (oh' vum) [L. *ovum*: egg] The egg, the female sex cell.

Oxidation (ox i day' shun) Relative loss of electrons in a chemical reaction; either outright removal to form an ion, or the sharing of electrons with substances having a greater affinity for them, such as oxygen. Most oxidation, including biological ones, are associated with the liberation of energy. (Contrast with reduction.)

Oxidative phosphorylation ATP formation in the mitochondrion, associated with flow of electrons through the respiratory chain.

Oxidizing agent A substance that can accept electrons from another. The oxidizing agent becomes reduced; its partner becomes oxidized.

P generation Parental generation. The individuals that mate in a genetic cross. Their immediate offspring are the F_1 generation.

Pacemaker That part of the heart which undergoes most rapid spontaneous contraction, thus setting the pace for the beat of the entire heart. In mammals, the sinoatrial (SA) node. Also, an artificial device, implanted in the heart, that initiates rhythmic contraction of the organ.

Pacinian corpuscle A modified nerve ending that senses touch and vibration.

Pair rule genes Segmentation genes that divide the *Drosophila* larva into two segments each.

Paleontology (pale' ee on tol' oh jee) [Gr. *palaios*: ancient + *logos*: discourse] The scientific study of fossils and all aspects of extinct life.

Pancreas (pan' cree us) A gland located near the stomach of vertebrates that secretes digestive enzymes into the small intestine and releases insulin into the bloodstream.

Pangaea (pan jee' uh) [Gk. *pan*: all, every] The single land mass formed when all the continents came together in the Permian period.

Para- [Gk. *para*: akin to, beside] Prefix indicating association in being along side or accessory to.

Parabronchi Passages in the lungs of birds through which air flows.

Paracrine A hormone that acts locally, near the site of its secretion. (Compare with endocrine gland.)

Parallel evolution Evolutionary patterns that exist in more than one lineage. Often the result of underlying developmental processes.

Parapatric speciation [Gr. *para*: along side + *patria*: homeland] Reproductive isolation between subpopulations arising from some non-geographic but physical condition, such as soil nutrient content. (Contrast with allopatric speciation, sympatric speciation.)

Paraphyletic taxon A taxon that includes some, but not all, of the descendants of a single ancestor.

Parasite An organism that attacks and consumes parts of an organism much larger than itself. Parasites sometimes, but not always, kill the host.

Parasympathetic nervous system A portion of the autonomic (involuntary) nervous system. (Contrast with sympathetic nervous system.)

Parathyroids Four glands on the posterior surface of the thyroid that produce and release parathormone.

Parathyroid hormone Hormone secreted by the parathyroid glands. Stimulates osteoclast activity and raises blood calcium levels.

Parenchyma (pair eng' kyma) A plant tissue composed of relatively unspecialized cells without secondary walls.

Parsimony The principle of preferring the simplest among a set of plausible explanations of any phenomenon.

Parthenocarpy Formation of fruit from a flower without fertilization.

Parthenogenesis (par' then oh jen' e sis) [Gr. parthenos: virgin + genesis: source] The production of an organism from an unfertilized egg.

Partial pressure The portion of the barometric pressure of a mixture of gases that is due to one component of that mixture. For example, the partial pressure of oxygen at sea level is 20.9% of barometric pressure.

Passive transport Diffusion across a membrane; may or may not require a channel or carrier protein. Contrast to active transport.

Patch clamping A technique for isolating a tiny patch of membrane to allow the study of ion movement through a particular channel.

Pathogen (path' o jen) [Gr. *pathos*: suffering + *genesis:* source] An organism that causes disease.

Pathway Any sequence of enzyme-catalyzed, chemical reactions in which the product(s) of a reaction is/are the substrate(s) for another reaction.

Pattern formation In animal embryonic development, the organization of differentiated tissues into specific structures such as wings.

Pedigree The pattern of transmission of a genetic trait within a family.

Penetrance Of a genotype, the proportion of individuals with that genotype who show the expected phenotype.

PEP carboxylase The enzyme that combines carbon dioxide with PEP to form a 4-carbon dicarboxylic acid at the start of C4 photosynthesis or of crassulacean acid metabolism (CAM).

Pepsin [Gr. *pepsis*: digestion] An enzyme, in gastric juice, that digests protein.

Peptide linkage The bond between amino acid residues in a protein. Formed between a carboxyl group and amino group ($CO—NH^-$) with the loss of water molecules.

Peptidoglycan The cell wall material of many prokaryotes, consisting of a single enormous molecule that surrounds the entire cell.

Perennial (per ren' ee al) [L. *per*: throughout + *annus*: year] Refers to a plant that survives from year to year. (Contrast with annual, biennial.)

Perfect flower A flower with both stamens and carpels, therefore hermaphroditic.

Pericycle [Gr. *peri*: around + *kyklos*: ring or circle] In plant roots, tissue just within the endodermis, but outside of the root vascular tissue. Meristematic activity of pericycle cells produces lateral root primordia.

Periderm The outer tissue of the secondary plant body, consisting primarily of cork.

Period (1) A minor category in the geological time scale. (2) The duration of a single cycle in a cyclical event, such as a circadian rhythm.

Peripheral membrane protein Membrane protein not embedded in the bilayer. Contrast to integral membrane protein.

Peripheral nervous system Neurons that transmit information to and from the central nervous system and whose cell bodies reside outside the brain or spinal cord.

Peristalsis (pair' i stall' sis) [Gr. *peri*: around + *stellein*: place] Wavelike muscular contractions proceeding along a tubular organ, propelling the contents along the tube.

Peritoneum The mesodermal lining of the body cavity among coelomate animals.

Permease A membrane protein that specifically transports a compound or family of compounds across the membrane.

Peroxisome An organelle that houses reactions in which toxic peroxides are formed. The peroxisome isolates these peroxides from the rest of the cell.

Petal [Gk. *petalon*: spread out] In an angiosperm flower, a sterile modified leaf, nonphotosynthetic, frequently brightly colored, and often serving to attract pollinating insects.

Petiole (pet' ee ole) [L. *petiolus*: small foot] The stalk of a leaf.

pH The negative logarithm of the hydrogen ion concentration; a measure of the acidity of a solution. A solution with pH = 7 is said to be neutral; pH values higher than 7 characterize basic solutions, while acidic solutions have pH values less than 7.

Phage (fayj) Short for bacteriophage. A virus that infects bacteria.

Phagocyte [Gk. *phagein*: to eat + *kystos*: sac] A white blood cell that ingests microorganisms by endocytosis.

Pharynx [Gr. *pharynx*: throat] The part of the gut between the mouth and the esophagus.

Phenotype (fee' no type) [Gr. *phanein*: to show] The observable properties of an individual resulting from both genetic and environmental factors. (Contrast with genotype.)

Phenotypic plasticity Refers to the fact that the phenotype of a developing organism is determined by a complex series of processes that are affected by both its genotype and its environment.

Pheromone (feer' o mone) [Gr. *pheros*: carry + *hormon*: excite, arouse] A chemical substance used in communication between organisms of the same species.

Phloem (flo' um) [Gr. *phloos*: bark] In vascular plants, the tissue that transports sugars and other solutes from sources to sinks. It consists of sieve cells or sieve tubes, fibers, and other specialized cells.

Phosphate group The functional group —OPO_3H_2. The transfer of energy from one compound to another is often accomplished by the transfer of a phosphate group.

Phosphodiester linkage The connection in a nucleic acid strand, formed by linking two nucleotides.

Phospholipids Lipids containing a phosphate group; important constituents of cellular membranes. (See lipids.)

Phosphorylation The addition of a phosphate group.

Photoautotroph An organism that obtains energy from light and carbon from carbon dioxide. (Contrast with chemolithotroph, chemoheterotroph, photoheterotroph.)

Photoheterotroph An organism that obtains energy from light but must obtain its carbon from organic compounds. (Contrast with chemolithotroph, chemoheterotroph, photoautotroph.)

Photon (foe' ton) [Gr. *photos*: light] A quantum of visible radiation; a "packet" of light energy.

Photoperiod (foe' tow peer' ee ud) The duration of a period of light, such as the length of time in a 24-hour cycle in which daylight is present.

Photoreceptor (1) A pigment that triggers a physiological response when it absorbs a photon. (2) A cell that senses and responds to light energy.

Photorespiration Light-driven uptake of oxygen and release of carbon dioxide, the carbon being derived from the early reactions of photosynthesis.

Photosynthesis (foe tow sin' the sis) [literally, "synthesis from light"] Metabolic processes, carried out by green plants, by which visible light is trapped and the energy used to synthesize compounds such as ATP and glucose.

Phototropism [Gr. *photos*: light + *trope*: turning] A directed plant growth response to light.

Phylogenetic tree Graphic representation of lines of descent among organisms.

Phylogeny (fy loj' e nee) [Gr. *phylon*: tribe, race + *genesis*: source] The evolutionary history of a particular group of organisms; also, the diagram of the "family tree" that shows genetic linkages between ancestors and descendants.

Phylum (plural: phyla) In taxonomy, a high-level category just beneath kingdom and above the class; a group of related, similar classes.

Physiology (fiz' ee ol' o jee) [Gr. *physis*: natural form + *logos*: discourse, study] The scientific study of the functions of living organisms and the individual organs, tissues, and cells of which they are composed.

Phytoalexins Substances toxic to pathogens, produced by plants in response to fungal or bacterial infection.

Phytochrome (fy' tow krome) [Gr. *phyton*: plant + *chroma*: color] A plant pigment regulating a large number of developmental and other phenomena in plants.

Pigment A substance that absorbs visible light.

Pilus (pill' us) [L. *pilus*: hair] A surface appendage by which some bacteria adhere to one another during conjugation.

Pistil [L. *pistillum*: pestle] The structure of an angiosperm flower within which the ovules are borne. May consist of a single carpel, or of several carpels fused into a single structure. Usually differentiated into ovary, style, and stigma.

Pith In plants, relatively unspecialized tissue found within a cylinder of vascular tissue.

Pituitary A small gland attached to the base of the brain in vertebrates. Its hormones control the activities of other glands. Also known as the hypophysis.

Pits Recessed cavities in the cell walls of a plant vascular element where only the primary wall is present. facilitating the movement of sap between cells.

Placenta (pla sen' ta) [Gr. *plax*: flat surface] The organ found in most mammals that provides for the nourishment of the fetus and elimination of the fetal waste products.

Placental (pla sen' tal) Pertaining to mammals of the subclass Eutheria, a group characterized by the presence of a placenta; contains the majority of living species of mammals.

Plant A member of the kingdom Plantae. Multicellular, gaining its nutrition by photosynthesis.

Planula (plan' yew la) [L. *planum*: flat] The free-swimming, ciliated larva of the cnidarians.

Plaque (plack) [Fr.: a metal plate or coin] (1) A circular clearing in a turbid layer (lawn) of bacteria growing on the surface of a nutrient agar gel. (2) An accumulation of prokaryotic organisms on tooth enamel. Acids produced by these microorganisms can cause tooth decay. (3) A region of arterial wall invaded by fibroblasts and fatty deposits (see atherosclerosis).

Plasma (plaz' muh) [Gr. *plassein*: to mold] The liquid portion of blood, in which blood cells and other particulates are suspended.

Plasma cell An antibody-secreting cell that developed from a B cell. The effector cell of the humoral immune system.

Plasma membrane The membrane that surrounds the cell, regulating the entry and exit of molecules and ions. Every cell has a plasma membrane.

Plasmid A DNA molecule distinct from the chromosome(s); that is, an extrachromosomal element. May replicate independently of the chromosome.

Plasmodesma (plural: plasmodesmata) [Gr. *plassein*: to mold + *desmos*: band] A cytoplasmic strand connecting two adjacent plant cells.

Plasmolysis (plaz mol' i sis) Shrinking of the cytoplasm and plasma membrane away from the cell wall, resulting from the osmotic outflow of water. Occurs only in cells with rigid cell walls.

Plastid Organelle in plants that serves for food manufacture (by photosynthesis) or food storage; bounded by a double membrane.

Plastoquinone A mobile electron carrier within the thylakoid membrane of the chloroplast linking photosystems I and II of photosynthesis.

Platelet A membrane-bounded body without a nucleus, arising as a fragment of a cell in the bone marrow of mammals. Important to blood-clotting action.

Pleiotropy (plee' a tro pee) [Gr. *pleion*: more] The determination of more than one character by a single gene.

Pleural membrane [Gk. *pleuras*: rib, side] The membrane lining the outside of the lungs and the walls of the thoracic cavity. Inflammation of these membranes is a condition known as pleurisy.

Podocytes Cells of Bowman's capsule of the nephron that cover the capillaries of the glomerulus, forming filtration slits.

Poikilotherm (poy' kill o therm) [Gr. *poikilos*: varied + *thermos*: heat] An animal whose body temperature tends to vary with the surrounding environment. (Contrast with homeotherm, heterotherm.)

Point mutation A mutation that results from a small, localized alteration in the chemical structure of a gene; can revert to wild type. (Contrast with deletion.)

Polar body A nonfunctional nucleus produced by meiosis, accompanied by very little cytoplasm. The meiosis which produces the mammalian egg produces in addition three polar bodies.

Polar molecule A molecule in which the electric charge is not distributed evenly in the covalent bonds.

Polar nuclei In flowering plants, the two nuclei in the central cell of the megagametophyte; following fertilization they give rise to the endosperm.

Polarity In development, the difference between one end and the other. In chemistry, the property that makes a polar molecule.

Pollen [L. *pollin*: fine, powdery flour] In seed plants, the microscopic grains containing the male gametophyte (microgametophyte) and gamete (microspore).

Pollination The process of transferring pollen from the anther to the receptive surface (stigma) of the pistil in plants.

Poly- [Gr. *poly*: many] A prefix denoting multiple entities.

Poly(A) tail A long sequence of adenine nucleotides (50–250) added after transcription to the 3' end of most eukaryotic mRNAs.

Polygenes Multiple loci whose alleles increase or decrease a continuously variable phenotypic trait.

Polymer [Gr. *poly*: many + *meros*: unit] A large molecule made up of similar or identical subunits called monomers. (Contrast with monomer, oligomer.)

Polymerase chain reaction (PCR) An enzymatic technique for the rapid production of millions of copies of a particular stretch of DNA.

Polymerization reactions Chemical reactions that generate polymers by linking monomers.

Polymorphism (pol' lee mor' fiz um) [Gr. *poly*: many + *morphe*: form, shape] In genetics, the coexistence in the same population of two distinct hereditary types based on different alleles.

Polyp The sessile asexual stage in the life cycle of most cnidarians.

Polypeptide A large molecule made up of many amino acids joined by peptide linkages. Large polypeptides are called proteins.

Polyphyletic group A group containing taxa, not all of which share the most recent common ancestor.

Polyploid (pol′ lee ploid) A cell or an organism in which the number of complete sets of chromosomes is greater than two.

Polysaccharide A macromolecule composed of many monosaccharides (simple sugars). Common examples are cellulose and starch.

Polysome (polyribosome) A complex consisting of a threadlike molecule of messenger RNA and several (or many) ribosomes. The ribosomes move along the mRNA, synthesizing polypeptide chains as they proceed.

Polytene (pol′ lee teen) [Gr. *poly*: many + *taenia*: ribbon] An adjective describing giant interphase chromosomes, such as those found in the salivary glands of fly larvae. The characteristic pattern of bands and bulges seen on these chromosomes provided a method for preparing detailed chromosome maps of several organisms.

Pons [L. *pons*: bridge] Region of the brain stem anterior to the medulla.

Population Any group of organisms coexisting at the same time and in the same place and capable of interbreeding with one another.

Population bottleneck See bottleneck.

Population density The number of individuals (or modules) of a population in a unit of area or volume.

Population genetics The study of genetic variation and its causes within populations.

Population structure The proportions of individuals in a population belonging to different age classes (age structure). Also, the distribution of the population in space.

Portal vein [L. *portal*: gate] A vein connecting two capillary beds, as in the hepatic portal system.

Positional cloning A technique for isolating a gene associated with a disease on the basis of its approximate chromosomal location.

Positional information Signals by which genes regulate cell functions to locate cells in a tissue during development.

Positive control The situation in which a regulatory macromolecule is needed to turn transcription of structural genes on. In its absence, transcription will not occur.

Positive cooperativity Occurs when a molecule can bind several ligands and each one that binds alters the conformation of the molecule so that it can bind the next ligand more easily. The binding of four molecules of O_2 by hemoglobin is an example of positive cooperativity.

Post [L. *postere*: behind, following after] Prefix denoting something that comes after.

Postabsorptive period When there is no food in the gut and no nutrients are being absorbed.

Postsynaptic cell The cell whose membranes receive neurotransmitter after its release by another cell (the presynaptic cell) at a synapse.

Postzygotic reproductive barrier Any mechanism that prevents the hybrid gametes of two different species from developing into viable reproductive adults. (Contrast with prezygotic reproductive barrier.)

Pre-mRNA (precursor mRNA) Initial gene transcript before it is modified to produce functional mRNA. Also known as the primary transcript.

Predator An organism that kills and eats other organisms.

Pressure flow model An effective model for phloem transport in angiosperms. It holds that sieve element transport is driven by an osmotically driven pressure gradient between source and sink.

Pressure potential The hydrostatic pressure of an enclosed solution in excess of the surrounding atmospheric pressure.

Presynaptic excitation/inhibition Occurs when a neuron modifies activity at a synapse by releasing a neurotransmitter onto the presynaptic nerve terminal.

Prezygotic reproductive barrier Any barrier to gene exchange that operates before mating. (Contrast with postzygotic reproductive barrier.)

Prey [L. *praeda*: booty] An organism consumed as an energy source.

Primary active transport Form of active transport in which ATP is hydrolyzed, yielding the energy required to transport ions against their concentration gradients. (Contrast with secondary active transport.)

Primary embryonic organizer. See Organizer.

Primary growth In plants, growth produced by the apical meristems. (Contrast with secondary growth.)

Primary producer A photosynthetic or chemosynthetic organism that synthesizes complex organic molecules from simple inorganic ones.

Primary succession Succession that begins in an area initially devoid of life, such as on recently exposed glacial till or lava flows.

Primary structure The specific sequence of amino acids in a protein.

Primary wall Cellulose-rich cell wall layers laid down by a growing plant cell.

Primate (pry′ mate) A member of the order Primates: a prosimian, monkey, ape, or human.

Primer A short, single-stranded segment of DNA that is the necessary starting material for the synthesis of a new DNA strand, which is synthesized from the 3′ end of the primer.

Primitive streak A line running axially along the blastodisc, the site of inward cell migration during formation of the three-layered embryo. Formed in the embryos of birds and fish.

Primordium [L. *primordium*: origin] The most rudimentary stage of an organ or other part.

Prion An infectious protein that can proliferate by converting other proteins.

Pro- [L.: first, before, favoring] A prefix often used in biology to denote a developmental stage that comes first or an evolutionary form that appeared earlier than another. For example, prokaryote, prophase.

Probe A segment of single stranded nucleic acid used to identify DNA molecules containing the complementary sequence.

Procambium Primary meristem that produces the vascular tissue.

Progesterone [L. *pro*: favoring + *gestare*: to bear] A vertebrate female sex hormone that maintains pregnancy.

Prokaryotes (pro kar′ ry otes) [L. *pro*: before + Gk. *karyon*: kernel, nucleus] Organisms whose genetic material is not contained within a nucleus: the bacteria and archaea. Considered an earlier stage in the evolution of life than the eukaryotes.

Prometaphase The phase of nuclear division that begins with the disintegration of the nuclear envelope.

Promoter The region of an operon that acts as the initial binding site for RNA polymerase.

Proofreading The correction of an error in DNA replication just after an incorrectly paired base is added to the growing polynucleotide chain.

Prophage (pro′ fayj) The noninfectious units that are linked with the chromosomes of the host bacteria and multiply with them but do not cause dissolution of the cell. Prophage can later enter into the lytic phase to complete the virus life cycle.

Prophase (pro′ phase) The first stage of nuclear division, during which chromosomes condense from diffuse, threadlike material to discrete, compact bodies.

Prostaglandin Any one of a group of specialized lipids with hormone-like functions. It is not clear that they act at any considerable distance from the site of their production.

Prosthetic group Any nonprotein portion of an enzyme.

Proteasome In the eukaryotic cytoplasm, a huge protein structure that binds to and digests cellular proteins that have been tagged by ubiquitin.

Protein (pro′ teen) [Gr. *protos*: first] One of the most fundamental building substances of living organisms. A long-chain polymer of amino acids with twenty different common side chains. Occurs with its polymer chain extended in fibrous proteins, or coiled into a compact macromolecule in enzymes and other globular proteins.

Proteolysis [protein + Gk. *lysis*: break apart] An enzymatic digestion of a protein or polypeptide.

Proteome The total of the different proteins that can be made by an organism. Because of alternate splicing of pre-mRNA, the number of proteins that can be made is usually much larger than the number of protein-coding genes present in the organism's genome.

Protobiont [Gr. *protos*: first, before + *bios*: life] Aggregates of abiotically produced molecules that cannot reproduce but do maintain internal chemical environments that differ from their surroundings.

Protoderm Primary meristem that gives rise to the plant epidermis.

Proton (pro′ ton) [Gr. *protos*: first, before] (1) A subatomic particle with a single positive charge. The number of protons in the nucleus of an atom determine its element. (2) A hydrogen ion, H^+.

Proton pump An active transport system that uses ATP energy to move hydrogen ions across a membrane generating an electric potential (voltage).

Proton motive force A force generated across a membrane expressed in millivolts having two components: a chemical potential (difference in proton concentration) plus an electrical potential due to the electrostatic charge on the proton.

Proto-oncogenes The normal alleles of genes possessing oncogenes (cancer-causing genes) as mutant alleles. Proto-oncogenes encode growth factors and receptor proteins.

Protostome [Gr. *protos*: first + *stoma*: mouth] One of the major lineages of animal evolution. Characterized by spiral, determinate cleavage of the egg, and by schizocoelous development. (Compare with deuterostome.)

Prototroph (pro' tow trofe') [Gr. *protos*: first + *trophein*: to nourish] The nutritional wild type, or reference form, of an organism. Any deviant form that requires growth nutrients not required by the prototrophic form is said to be a nutritional mutant, or auxotroph.

Protozoa (pro to zoe' ah) [Gk. *protos*: first, before + *zoon*: animal) A term formerly used for a single polyphyletic phylum of single-celled eukaryotic organisms including the flagellates, amoebas, and ciliates. This book does not use the term.

Provirus Viral DNA inserted into a bacterial host genome. (See Lysogenic cycle.)

Proximal Near the point of attachment or other reference point. (Contrast with distal.)

Pseudocoelom [Gr. *pseudes*: false] A body cavity not surrounded by a peritoneum. Characteristic of nematodes and rotifers.

Pseudogene [Gr. *pseudes*: false] A DNA segment that is homologous to a functional gene but contains a nucleotide change that prevents its expression.

Pseudopod (soo' do pod) [Gr. *pseudes*: false + *podos*: foot] A temporary, soft extension of the cell body that is used in location, attachment to surfaces, or engulfing particles.

Pulmonary [L. *pulmo*: lung] Pertaining to the lungs.

Punctuated equilibrium An evolutionary pattern in which periods of rapid change are separated by longer periods of little or no change.

Pupa (pew' pa) [L. *pupa*: doll, puppet] In certain insects (the Holometabola), the encased developmental stage between the larva and the adult.

Pupil The opening in the vertebrate eye through which light passes.

Purine (pure' een) One of the types of nitrogenous bases. The purines adenine and guanine are found in nucleic acids. (Contrast with pyrimidine.)

Purkinje fibers Specialized heart muscle cells that conduct excitation throughout the ventricular muscle.

Pyramid of biomass Graphical representation of the total body masses at different trophic levels in an ecosystem.

Pyramid of energy Graphical representation of the total energy contents at different trophic levels in an ecosystem.

Pyrimidine (per im' a deen) A type of nitrogenous base. The pyrimidines cytosine, thymine, and uracil are found in nucleic acids.

Pyruvate A three-carbon acid; the end product of glycolysis and the raw material for the citric acid cycle.

Q_{10} A value that compares the rate of a biochemical process or reaction over a 10°C range of temperature. A process that is not temperature-sensitive has a Q_{10} of 1; values of 2 or 3 mean the reaction speeds up as temperature increases.

Quantum (kwon' tum) [L. *quantus*: how great] An indivisible unit of energy.

Quaternary structure The specific three dimensional arrangement of protein subunits.

Quiescent center In root meristem, central region where cells do not divide or divide very slowly.

R factor (resistance factor) A plasmid that contains one or more genes that encode resistance to antibiotics.

R gene Resistance gene that functions in plant defenses against bacteria, fungi, and nematodes.

R group The distinguishing group of atoms of a particular amino acid.

Radial symmetry The condition in which two halves of a body are mirror images of each other regardless of the angle of the cut, providing the cut is made along the center line. Thus, a cylinder cut lengthwise down its center displays this form of symmetry. (Contrast with biradial symmetry.)

Radioisotope A radioactive isotope of an element. Examples are carbon-14 (^{14}C) and hydrogen-3, or tritium (^{3}H).

Radiometry The use of the regular, known rates of decay of radioisotopes of elements to determine dates of events in the distant past.

Reactant A chemical substance that enters into a chemical reaction with another substance.

Reaction A chemical change in which changes take place in the kind, number, or position of atoms making up a substance.

Reaction center A group of electron transfer proteins that receive energy from light-absorbing pigments and convert it to chemical energy by redox reactions.

Receptacle The end of a plant stem to which the parts of the flower are attached.

Receptor A site or protein on the outer surface of the plasma membrane or in the cytoplasm to which a specific ligand from another cell binds.

Receptor-mediated endocytosis Endocytosis initiated by macromolecular binding to a specific membrane receptor.

Receptor potential The change in the resting potential of a sensory cell when it is stimulated.

Recessive In genetics, an allele that does not determine phenotype in the presence of a dominant allele. Contrast with dominance.

Reciprocal crosses A pair of crosses, in one of which a female of genotype A mates with a male of genotype B and in the other of which a female of genotype B mates with a male of genotype A.

Recognition site See restriction site.

Recombinant An individual, meiotic product, or single chromosome in which genetic materials originally present in two individuals end up in the same haploid complement of genes. The reshuffling of genes can be either by independent segregation, or by crossing over between homologous chromosomes.

Recombinant DNA DNA generated in vitro, from more than one source.

Recombinant DNA technology The application of restriction endonucleases, plasmids, and transformation to alter and assemble recombinant DNA, with the goal of producing specific proteins.

Rectum The terminal portion of the gut, ending at the anus.

Redox reaction A chemical reaction in which one reactant becomes oxidized and the other becomes reduced.

Reducing agent A substance that can donate electrons to another substance. The reducing agent becomes oxidized, and its partner becomes reduced.

Reduction Gain of electrons by a chemical reactant; any reduction is accompanied by an oxidation. (Contrast with oxidation.)

Reflex An automatic action, involving only a few neurons (in vertebrates, often in the spinal cord), in which a motor response swiftly follows a sensory stimulus.

Refractory period Of a neuron, the time interval after an action potential, during which another action potential cannot be elicited.

Regulative development A pattern of animal embryonic development in which the fates of the first blastomeres are not absolutely fixed. (Contrast with mosaic development.)

Regulatory gene A gene that codes for a protein that controls the transcription of another gene(s).

Releaser A sensory stimulus that triggers a fixed action pattern.

Releasing hormone One of several hypothalamic hormones that stimulates the secretion of anterior pituitary hormone.

REM sleep A sleep state characterized by dreaming, skeletal muscle relaxation, and rapid eye movements.

Renal [L. *renes*: kidneys] Relating to the kidneys.

Replication Pertaining to the duplication of genetic material.

Replication complex The close association of several proteins operating in the replication of DNA.

Replication fork A point at which a DNA molecule is replicating. The fork forms by the unwinding of the parent molecule.

Reporter gene Marker genes included in recombinant DNA to indicate the presence of the recombinant DNA in a host cell.

Repressible enzyme An enzyme whose synthesis can be decreased or prevented by the presence of a particular compound. A repressible operon often controls the synthesis of such an enzyme.

Repressor A protein coded by the regulatory gene. The repressor can bind to a specific operator and prevent transcription of the operon.

Reproductive isolating mechanism Any trait that prevents individuals from two different populations from producing fertile hybrids.

Reproductive isolation The condition in which a population is not exchanging genes with other populations of the same species.

Rescue effect The process by which a few individuals moving among declining subpopulations of a species and reproducing may prevent their extinction.

Resolving power Of an optical device such as a microscope, the smallest distance between two lines that allows the lines to be seen as separate from one another.

Resource Something in the environment required by an organism for its maintenance and growth that is consumed in the process of being used.

Respiration (res pi ra' shun) [L. *spirare*: to breathe] (1) Cellular respiration; the catabolic pathways by which electrons are removed from various molecules and passed through intermediate electron carriers to O_2, generating H_2O and releasing energy. (2) Breathing.

Respiratory chain The terminal reactions of cellular respiration, in which electrons are passed from NAD or FAD, through a series of intermediate carriers, to molecular oxygen, with the concomitant production of ATP.

Resting potential The membrane potential of a living cell at rest. In cells at rest, the interior is negative to the exterior. (Contrast with action potential, electrotonic potential.)

Restoration ecology The science and practice of restoring damaged or degraded ecosystems.

Restriction endonuclease Any one of several enzymes, produced by bacteria, that break foreign DNA molecules at very specific sites. Some produce "sticky ends." Extensively used in recombinant DNA technology.

Restriction fragment length polymorphism See RFLP.

Restriction site A specific DNA base sequence recognized and acted on by a restriction endonuclease cutting the DNA.

Reticular system A central region of the vertebrate brain stem that includes complex fiber tracts conveying neural signals between the forebrain and the spinal cord, with collateral fibers to a variety of nuclei that are involved in autonomic functions, including arousal from sleep.

Retina (rett' in uh) [L. *rete*: net] The light-sensitive layer of cells in the vertebrate or cephalopod eye.

Retinal The light-absorbing portion of visual pigment molecules. Derived from β-carotene.

Retrovirus An RNA virus that contains reverse transcriptase. Its RNA serves as a template for cDNA production, and the cDNA is integrated into a chromosome of the mammalian host cell.

Reverse transcriptase An enzyme that catalyzes the production of DNA (cDNA), using RNA as a template; essential to the reproduction of retroviruses.

RFLP (Restriction fragment length polymorphism) Coexistence of two or more patterns of restriction fragments (patterns produced by restriction enzymes), as revealed by a probe. The polymorphism reflects a difference in DNA sequence on homologous chromosomes.

Rhizoids (rye' zoids) [Gr. *rhiza*: root] Hairlike extensions of cells in mosses, liverworts, and a few vascular plants that serve the same function as roots and root hairs in vascular plants. The term is also applied to branched, rootlike extensions of some fungi and algae.

Rhizome (rye' zome) A special underground stem (as opposed to root) that runs horizontally beneath the ground.

Rhodopsin A photopigment used in the visual process of transducing photons of light into changes in the membrane potential of photoreceptor cells.

Ribonucleic acid See RNA.

Ribosomal RNA (rRNA) Several species of RNA that are incorporated into the ribosome. Involved in peptide bond formation.

Ribosome A small organelle that is the site of protein synthesis.

Ribozyme An RNA molecule with catalytic activity.

Risk cost The increased chance of being injured or killed as a result of performing a behavior, compared to resting.

RNA (ribonucleic acid) An often single stranded nucleic acid whose nucleotides use ribose rather than deoxyribose and in which the base uracil replaces thymine found in DNA. Serves as genome from some viruses. (See rRNA, tRNA, mRMA, and ribozyme.)

RNA editing The alteration of bases on mRNA prior to its translation.

RNA polymerase An enzyme that catalyzes the formation of RNA from a DNA template.

RNA primase A replication complex enzyme that makes the primer strand of DNA needed to initiate DNA replication.

RNA splicing The last stage of RNA processing in eukaryotes, in which the transcripts of introns are excised through the action of small nuclear ribonucleoprotein particles (snRNP).

Rods Light-sensitive cells (photoreceptors) in the retina. (Contrast with cones.)

Root The organ responsible for anchoring the plant in the soil, absorbing water and minerals, and producing certain hormones. Some roots are storage organs.

Root cap A thimble-shaped mass of cells, produced by the root apical meristem, that protects the meristem; the organ that perceives the gravitational stimulus in root gravitropism.

Root hair A long, thin process from a root epidermal cell that absorbs water and minerals from the soil solution.

Rough ER That portion of the endoplasmic reticulum whose outer surface has attached ribosomes. Compare with smooth ER.

rRNA See ribosomal RNA.

Rubisco (Ribulose bisphosphate carboxylase/oxygenase) Acronym for the enzyme that combines carbon dioxide or oxygen with ribulose bisphosphate to catalyze the first step of the Calvin-Benson cycle.

Rumen (rew' mun) The first division of the ruminant stomach. It stores and initiates bacterial fermentation of food. Food is regurgitated from the rumen for further chewing.

Ruminant An herbivorous, cud-chewing mammal such as a cow, sheep, or deer, having a stomach consisting of four compartments.

S phase In the cell cycle, the stage of interphase during which DNA is replicated. (Contrast with G_1 phase, G_2 phase, M phase.)

Saprobe [Gr. *sapros*: rotten] An organism (usually a bacterium or fungus) that obtains its carbon and energy directly from dead organic matter.

Sarcomere (sark' o meer) [Gr. *sark*: flesh + *meros*: unit] The contractile unit of a skeletal muscle.

Saturated fatty acid A fatty acid usually containing from 12 to 18 carbon atoms and no double bonds.

Schizocoelous development [Gk. *schizo*: split + *koiloma*: cavity] Formation of a coelom during embryological development by a splitting of mesodermal masses.

Schwann cell A glial cell that wraps around part of the axon of a peripheral neuron, creating a myelin sheath.

Sclereid [Gr. *skleros*: hard] A type of sclerenchyma cell, commonly found in nutshells, that is not elongated.

Sclerenchyma (skler eng' kyma) [Gr. *skleros*: hard + *kymus*: juice] A plant tissue composed of cells with heavily thickened cell walls, dead at functional maturity. The principal types of sclerenchyma cells are fibers and sclereids.

Second law of thermodynamics States that in any real (irreversible) process, there is a decrease in free energy and an increase in entropy.

Second messenger A compound, such as cAMP, that is released within a target cell after a hormone (first messenger) has bound to a surface receptor on a cell; the second messenger triggers further reactions within the cell.

Secondary active transport Form of active transport which does not use ATP as an energy source; rather, transport is coupled to ion diffusion down a concentration gradient established by primary active transport.

Secondary growth In plants, growth produced by vascular and cork cambia, contributing to an increase in girth. (Contrast with primary growth.)

Secondary metabolite A compound synthesized by a plant that is not needed for basic cellular metabolism. Typically has an antiherbivore or antiparasite function.

Secondary structure Of a protein, localized regularities of structure, such as the α helix and the β pleated sheet.

Secondary succession Ecological succession after a disturbance that does not eliminate all the organisms that originally lived on the site.

Secondary wall Wall layers laid down by a plant cell that has ceased growing; often impregnated with lignin or suberin.

Secretin (si kreet' in) A peptide hormone secreted by the upper region of the small intestine when acidic chyme is present. Stimulates the pancreatic duct to secrete bicarbonate ions.

Section A thin slice, usually for microscopy, as a tangential section or a transverse section.

Seed A fertilized, ripened ovule of a gymnosperm or angiosperm. Consists of the embryo, nutritive tissue, and a seed coat.

Seed plant Plants in which the embryo is protected and nourished within a seed; the gymnosperms and angiosperms.

Seedling A young plant that has grown from a seed (rather than by grafting or by other means.)

Segmentation genes In insect larvae, genes that determine the number and polarity of larval segments.

Segment polarity genes Genes that determine the boundaries and front-to-back organization of the segments in the *Drosophila* larva.

Segregation In genetics, the separation of alleles, or of homologous chromosomes, from one another during meiosis so that each of the haploid daughter nuclei produced by meiosis contains one or the other member of the pair found in the diploid mother cell, but never both. This principle was articulated by Mendel as his "first law."

Selective permeability Allowing certain substances to pass through while other substances are excluded; a characteristic of membranes.

Self incompatability In plants, the rejection of their own pollen; promotes genetic variation and limits inbreeding.

Selfish act A behavioral act that benefits its performer but harms the recipients.

Semen (see' men) [L. *semin*: seed] The thick, whitish liquid produced by the male reproductive organ in mammals, containing the sperm.

Semiconservative replication The common way in which DNA is synthesized. Each of the two partner strands in a double helix acts as a template for a new partner strand. Hence, after replication, each double helix consists of one old and one new strand.

Seminiferous tubules The tubules within the testes within which sperm production occurs.

Senescence [L. *senescere*: to grow old] Aging; deteriorative changes with aging; the increased probability of dying with increasing age.

Sensory neuron A neuron leading from a sensory cell to the central nervous system. (Contrast with motor neuron.)

Sepal (see' pul) [L. *sepalum*: covering] One of the outermost structures of the flower, usually protective in function and enclosing the rest of the flower in the bud stage.

Septum [L. *saeptum*: partition, fence] A membrane or wall between two cavities.

Sertoli cells Cells in the seminiferous tubules that nuture the developing sperm.

Sessile (sess' ul) [L. *sedere*: to sit] Permanently attached; not moving.

Set point In a regulatory system, the threshold sensitivity to the feedback stimulus.

Sex chromosome In organisms with a chromosomal mechanism of sex determination, one of the chromosomes involved in sex determination.

Sex linkage The pattern of inheritance characteristic of genes located on the sex chromosomes of organisms having a chromosomal mechanism for sex determination.

Sexual reproduction Reproduction involving union of gametes.

Sexual selection Selection by one sex of characteristics in individuals of the opposite sex. Also, the favoring of characteristics in one sex as a result of competition among individuals of that sex for mates.

Shared derived trait A trait that arose in the ancestor of a phylogenetic group and is present (sometimes in modified form) in all of its members, thus helping define that group. Also called a synapomorphy.

Shoot system The aerial parts of a vascular plant, consisting of the leaves, stem(s), and flowers.

Short-day plant (SDP) A plant that requires short days (or long nights) in order to flower.

Sieve tube A column of specialized cells found in the phloem, specialized to conduct organic matter from sources (such as photosynthesizing leaves) to sinks (such as roots). Found principally in flowering plants.

Sieve tube element A single cell of a sieve tube, containing cytoplasm but relatively few organelles, with highly specialized perforated end walls leading to elements above and below.

Signal A chemical (neurotransmitter or hormone) or light message emitted from one cell/cells or organism(s) and received by others to cause some change in function or behavior.

Signal recognition particle (SRP) A complex of RNA and protein that recognizes both the signal sequence on a growing polypeptide and receptor protein on the surface of the ER.

Signal sequence The sequence of a protein that directs the protein through a particular cellular membrane.

Signal transduction pathway The series of biochemical steps whereby a stimulus to a cell (such as a hormone or neurotransmitter binding to a receptor) is translated into a response of the cell.

Silencer sequence A sequence of eukaryotic DNA that binds proteins that inhibit the transcription of an associated gene.

Silent mutation A change in gene sequence that, due to the redundancy of the genetic code, has no effect on the amino acid produced, and thus no effect on the protein phenotype. See synonymous mutation.

Similarity matrix A matrix to compare the structures of two molecules constructed by adding the number of their amino acids that are identical or different.

Sinoatrial node (sigh' no ay' tree al) [L. *sinus*: curve + *atrium*: hall, chamber] The pacemaker of the mammalian heart.

Sink In plants, any organ that imports the products of photosynthesis, such as roots, developing fruits, immature leaves. Contrast with source.

Sinus (sigh' nus) [L. *sinus*: curve, hollow] A cavity in a bone, a tissue space, or an enlargement in a blood vessel.

Sister chromatid In the eukaryotic cell, a chromatid resulting from chromosome replication during interphase.

Sister group Two phylogenetic groups that are each other's closes relative.

Skeletal muscle See striated muscle.

Sliding filament theory A proposed mechanism of muscle contraction based on formation and breaking of crossbridges between actin and myosin filaments, causing them to slide together.

Small intestine The portion of the gut between the stomach and the colon, consisting of the duodenum, the jejunum, and the ileum.

Small nuclear ribonucleoprotein particle (snRNP) A complex of an enzyme and a small nuclear RNA molecule, functioning in RNA splicing.

Smooth muscle One of three types of muscle tissue. Usually consists of sheets of mononucleated cells innervated by the autonomic nervous system.

Sodium–potassium pump The complex protein in plasma membranes that is responsible for primary active transport; it pumps sodium ions out of the cell and potassium ions into the cell, both against their concentration gradients.

Solute A substance that is dissolved in a liquid (solvent).

Solute potential A property of any solution, resulting from its solute contents; it may be zero or have a negative value.

Solution A liquid (solvent) and its dissolved solutes.

Solvent A liquid that has dissolved or can dissolve one or more solutes.

Somatic [Gr. *soma*: body] Pertaining to the body. Somatic cells are cells of the body (as opposed to germ cells).

Somite (so' might) One of the segments into which an embryo becomes divided longitudinally, leading to the eventual segmentation of the animal as illustrated by the spinal column, ribs, and associated muscles.

Source In plants, an organ exporting photosynthetic products in excess of its own needs. For example, a mature leaf or storage organ. Contrast with sink.

Spatial summation In the production or inhibition of action potentials in a postsynaptic neuron, the interaction of depolarizations and hyperpolarizations produced by several terminal boutons.

Spawning The direct release of sex cells into the water.

Speciation (spee' shee ay' shun) The process of splitting one population into two populations that are reproductively isolated from one another.

Species (spee' shees) [L. *specie*: kind] The basic lower unit of classification, consisting of a population or series of populations of closely related and similar organisms. The more narrowly defined "biological species" consists of individuals capable of interbreeding freely with each other but not with members of other species.

Species diversity A weighted representation of the species of organisms living in a region; large and common species are given greater weight than are small and rare ones. (Contrast with species richness.)

Species richness The total number of species living in a region. (Contrast with species diversity.)

Specific defenses Defensive reactions of the immune system that are based on antibody reaction with a specific antigen.

Specific heat The amount of energy that must be absorbed by a gram of a substance to raise its temperature by one degree centigrade. By convention, water is assigned a specific heat of one.

Sperm [Gr. *sperma*: seed] A male gamete (reproductive cell).

Spermatocyte (spur mat' oh site) [Gr. *sperma*: seed + *kytos*: container] The cell that gives rise to the sperm in animals.

Spermatogenesis (spur mat' oh jen' e sis) [Gr. *sperma*: seed + *genesis*: source] Male gametogenesis, leading to the production of sperm.

Spermatogonia Undifferentiated germ cells that give rise to primary spermatocytes and hence to sperm.

Sphincter (sfink' ter) [Gr. *sphinkter*: something that binds tightly] A ring of muscle that can close an orifice, for example at the anus.

Spindle apparatus An array of microtubules stretching from pole to pole of a dividing nucleus and playing a role in the movement of chromosomes at nuclear division. Named for its shape.

Spiracle (spy' rih kel) [L. *spirare*: to breathe] An opening of the treacheal respiratory system of terrestrial arthorpods.

Spliceosome An RNA–protein complex that splices out introns from eukaryotic pre-mRNAs.

Splicing The removal of introns and connecting of exons in eukaryotic pre-mRNAs.

Spontaneous generation The idea that life is generated continually from nonliving matter.
Spontaneous reaction A chemical reaction which will proceed on its own, without any outside influence. A spontaneous reaction need not be rapid.

Sporangium (spor an' gee um) [Gr. *spora*: seed + *angeion*: vessel or reservoir] In plants and fungi, any specialized stucture within which one or more spores are formed.

Spore [Gr. *spora*: seed] Any asexual reproductive cell capable of developing into an adult organism without gametic fusion. In plants, haploid spores develop into gametophytes, diploid spores into sporophytes. In prokaryotes, a resistant cell capable of surviving unfavorable periods.

Sporocyte Specialized cells of the diploid sporophyte that will divide by meiosis to produce four haploid spores. Germination of these spores produces the haploid gametophyte.

Sporophyte (spor' o fyte) [Gr. *spora*: seed + *phyton*: plant] In plants and protists with alternation of generations, the diploid phase that produces the spores. (Contrast with gametophyte.)

Stabilizing selection Selection against the extreme phenotypes in a population, so that the intermediate types are favored. (Contrast with disruptive selection.)

Stamen (stay' men) [L. *stamen*: thread] A male (pollen-producing) unit of a flower, usually composed of an anther, which bears the pollen, and a filament, which is a stalk supporting the anther.

Starch [O.E. *stearc*: stiff] A polymer of glucose; used by plants to store energy.

Start codon The mRNA triplet (AUG) that acts as a signal for the beginning of translation at the ribosome. (Compare with stop codons.)

Stasis [Gk. *stasis*: to stop, stand still] Period during which little or no evolutionary change takes place within a lineage or groups of lineages.

Statocyst (stat' oh sist) [Gk. *statos*: stationary + *kystos*: cell] An organ of equilibrium in some invertebrates.

Statolith (stat' oh lith) [Gk. *statos*: stationary + *lithos*: stone] A solid object that responds to gravity or movement and stimulates the mechanoreceptors of a statocyst.

Stele (steel) [Gr. *stylos*: pillar] The central cylinder of vascular tissue in a plant stem.

Stem Plant structure that holds leaves and/or flowers; it is the site for transporting and distributing material throughout the plant.

Stem cells In animals, undifferentiated cells that are capable of extensive proliferation. A stem cell generates more stem cells and a large clone of differentiated progeny cells. Compare with initial cells.

Steroid Any of numerous lipids based on a 17-carbon atom ring system.

Sticky ends On a piece of two-stranded DNA, short, complementary, one-stranded regions produced by the action of a restriction endonuclease. Sticky ends allow the joining of segments of DNA from different sources.

Stigma [L. *stigma*: mark, brand] The part of the pistil at the apex of the style that is receptive to pollen, and on which pollen germinates.

Stimulus [L. *stimulare*: to goad] Something causing a response; something in the environment detected by a receptor.

Stolon [L. *stolon*: branch, sucker] A horizontal stem that forms roots at intervals.

Stoma (plural: stomata) [Gr. *stoma*: mouth, opening] Small opening in the plant epidermis that permits gas exchange; bounded by a pair of guard cells whose osmotic status regulates the size of the opening.

Stop codons The mRNA codons that signal the end of protein translation at the ribosome: UAG, UGA, UAA.

Stratosphere The upper part of Earth's atmosphere, above the troposphere; extends from approximately 18 kilometers upward to approximately 50 kilometers above the surface.

Stratum (plural strata) [L. *stratos*: layer] A layer or sedimentary rock laid down at a particular time in a past.

Striated muscle Contractile tissue characterized by multinucleated cells containing highly ordered arrangements of actin and myosin microfilaments. Also known as skeletal muscle.

Stroma The fluid contents of an organelle such as a chloroplast.

Stromatolites Composite, flat-to-domed structures composed of successive mineral layers produced by the action of cyanobacteria in water; ancient ones provide evidence for early life on the earth.

Structural gene A gene that encodes the primary structure of a protein.

Style [Gr. *stylos*: pillar or column] In flowering plants, a column of tissue extending from the tip of the ovary, and bearing the stigma or receptive surface for pollen at its apex.

Sub- [L. *sub*: under] A prefix often used to designate a structure that lies beneath another or is less than another. For example, subcutaneous (beneath the skin); subspecies.

Suberin A waxlike lipid that acts as a barrier to water and solute movement across the Casparian strip of the endodermis. Suberin is the waterproofing element in the cell walls of cork.

Submucosa (sub mew koe' sah) The tissue layer just under the epithelial lining of the lumen of the digestive tract.

Substrate (sub' strayte) The molecule or molecules on which an enzyme exerts catalytic action.

Substratum The base material on which a sessile organism lives.

Succession In ecology, the gradual, sequential series of changes in species composition of a community following a disturbance.

Sulcus [L. *sulcare*: to plow] The valleys or creases between the raised portions of the convoluted surface of the brain. (Contrast with gyrus.)

Summation The ability of a neuron to fire action potentials in response to numerous subthreshold postsynaptic potentials arriving simultaneously at differentiated places on the cell, or arriving at the same site in rapid succession.

Surface area-to-volume ratio For any cell, organism, or geometrical solid, the ratio of surface area to volume; this is an important factor in setting an upper limit on the size a cell or organism can attain.

Surfactant A substance that decreases the surface tension of a liquid. Lung surfactant, secreted by cells of the alveoli, is mostly phospholipid and decreases the amount of work necessary to inflate the lungs.

Suspensor In the embryos of seed plants, the stalk of cells that pushes the embryo into the endosperm and is a source of nutrient transport to the embryo.

Symbiosis (sim' bee oh' sis) [Gr. *sym*: together + *bios*: living] The living together of two or more species in a prolonged and intimate ecological relationship. (Compare with parasitism and mutualism.)

Symmetry Describes an attribute of an animal body in which at least one plane can divide the body into similar, mirror-image halves. (See bilateral symmetry, biradial symmetry, radial symmetry.)

Sympathetic nervous system A division of the autonomic (involuntary) nervous system. (Contrast with parasympathetic nervous system.)

Sympatric speciation (sim pat' rik) [Gr. *sym*: same + *patria*: homeland] Speciation due to reproductive isolation without any physical separation of the subpopulation. (Contrast with allopatric speciation, parapatric speciation.)

Symplast The continuous meshwork of the interiors of living cells in the plant body, resulting from the presence of plasmodesmata. (Contrast with apoplast.)

Symport A membrane transport process that carries two substances in the same direction across the membrane. (Contrast with antiport.)

Synapse (sin' aps) [Gr. *syn*: together + *haptein*: to fasten] The narrow gap between the terminal bouton of one neutron and the dendrite or cell body of another.

Synapsis (sin ap′ sis) The highly specific parallel alignment (pairing) of homologous chromosomes during the first division of meiosis.

Synaptic vesicle A membrane-bounded vesicle, containing neurotransmitter, which is produced in and discharged by the presynaptic neuron.

Synapomorphy See shared derived trait.

Synergids [Gk. *syn*: together + *ergos*: performing work] In flowering plants, the two cells accompanying the egg cell at one end of the megmagametophyte.

Syngamy (sing′ guh mee) [Gr. *syn*: together + *gamos*: marriage] Union of gametes. Also known as fertilization.

Synonymous mutation A mutation that substitutes one nucleotide for another but does not change the amino acid specified (i.e., UUA → UUG, both specifying leucine). (Compare with frame-shift mutation, missense mutation, nonsense mutation.)

Synonymous substitution The situation when a synonymous mutation becomes widespread in a population. Typically not influenced by natural selection, these substitutions can accumulate in a population. (Contrast with nonsynonymous substitution.)

Systematics The scientific study of the diversity of organisms, and of their relationships. Includes both taxonomy (classification) and phylogeny (evolutionary relationships).

Systemic circulation The part of the circulatory system serving those parts of the body other than the lungs or gills.

Systemic acquired resistance A general resistance to many plant pathogens following infection by a single agent.

Systemin The only polypeptide plant hormone; participates in response to tissue damage.

Systole (sis′ tuh lee) [Gr. *systole*: contraction] Contraction of a chamber of the heart, driving blood forward in the circulatory system.

T cell A type of lymphocyte, involved in the cellular immune response. The final stages of its development occur in the thymus gland. (Contrast with B cell; see also cytotoxic T cell, helper T cell, suppressor T cell.)

T cell receptor A protein on the surface of a T cell that recognizes the antigenic determinant for which the cell is specific.

T tubules A system of tubules that runs throughout the cytoplasm of muscle fibers, through which action potentials spread.

Target cell A cell with the appropriate receptors to bind and respond to a particular hormone or other chemical mediator.

Taste bud A structure in the epithelium of the tongue that includes a cluster of chemoreceptors innervated by sensory neurons.

TATA box An eight-base-pair sequence, found about 25 base pairs before the starting point for transcription in many eukaryotic promoters, that binds a transcription factor and thus helps initiate transcription.

Taxis (tak′ sis) [Gr. *taxis*: arrange, put in order] The movement of an organism or its part directly toward or away from the stimulus. For example, positive phototaxis is movement toward a light source, negative geotaxis is movement away from gravity).

Taxon A unit such as genus, family, class, or order in a taxonomic system.

Taxonomy (taks on′ oh me) [Gr. *taxis*: arrange, put in order] The science of classifying organisms.

Telomeres (tee′ lo merz) [Gr. *telos*: end + *meros*: units, segments] Repeated DNA sequences at the ends of eukaryotic chromosomes.

Telophase (tee′ lo phase) [Gr. *telos*: end] The final phase of mitosis or meiosis during which chromosomes became diffuse, nuclear envelopes reform, and nucleoli begin to reappear in the daughter nuclei.

Template In biochemistry, a molecule or surface upon which another molecule is synthesized in complementary fashion, as in the replication of DNA. In the brain, a pattern that responds to a normal input but not to incorrect inputs.

Template strand In a stretch of double-stranded DNA, the strand that is transcribed.

Temporal summation [L. *tempus*: time; *summus*: highest amount] In the production or inhibition of action potentials in a postsynaptic neuron, the interaction of depolarizations or hyperpolarizations produced by rapidly repeated stimulation of a single point.

Tendon A collagen-containing band of tissue that connects a muscle with a bone.

Termination The end of protein synthesis triggered by a stop codon which binds release factor that causes the polypeptide to release from the ribosome.

Terrestrial (ter res′ tree al) [L. *terra*: earth] Pertaining to the land. (Contrast with aquatic, marine.)

Territory A fixed area from which an animal or group of animals excludes other members of the same (and sometimes other) species by aggressive behavior or display.

Tertiary structure In reference to a protein, the relative locations in three-dimensional space of all the atoms in the molecule. The overall shape of a protein. (Contrast with primary, secondary, and quaternary structures.)

Test cross A cross of a dominant-phenotype individual (which may be either heterozygous or homozygous) with a homozygous-recessive individual.

Testis (tes′ tis) (plural: testes) [L. *testis*: witness] The male gonad; the organ that produces the male sex cells.

Testosterone (tes toss′ tuhr own) A male sex steroid hormone.

Tetanus [Gr. *tetanos*: stretched] (1) A state of sustained maximal muscular contraction caused by rapidly repeated stimulation. (2) In medicine, an often fatal disease ("lockjaw") caused by the bacterium *Clostridium tetani.*

Tetrad [Gk. *tettares*: four] During prophase I of meiosis, the association of a pair of homologous chromosomes or four chromatids.

Thalamus [Gk. *thalamos*: chamber] A region of the vertebrate forebrain; involved in integration of sensory input.

Thallus (thal′ us) [Gr. *thallos*: sprout] Any algal body which is not differentiated into root, stem, and leaf.

Theory [Gk. *theoria*: analysis of facts] A far-reaching explanation of observed facts that is supported by such a wide body of evidence, with no significant contradictory evidence, that it is scientifically accepted as a factual framework. Examples are Newton's theory of gravity and Darwin's theory of evolution. (Contrast with hypothesis.)

Thermoneutral zone The range of temperatures over which an endotherm does not have to expend extra energy to thermoregulate.

Thermoreceptor A cell or structure that responds to changes in temperature.

Thoracic cavity [Gk. *thorax*: breastplate] The portion of the mammalian body cavity bounded by the ribs, shoulders, and diaphragm. Contains the heart and the lungs.

Thorax [Gk. *thorax*: breastplate] In an insect, the middle region of the body, between the head and abdomen. In mammals, the part of the body between the neck and the diaphragm.

Thrombin An enzyme that converts fibrinogen to fibrin, thus triggering the formation of blood clots.

Thrombus (throm′ bus) [Gk. *thrombos*: clot] A blood clot that forms within a blood vessel and remains attached to the wall of the vessel. (Contrast with embolus.)

Thylakoid (thigh la koid) [Gk. *thylakos*: sack or pouch] A flattened sac within a chloroplast. Thylakoid membranes contain all of the chlorophyll in a plant, in addition to the electron carriers of photophosphorylation. Thylakoids stack to form grana.

Thymine (T) A nitrogen-containing base found in DNA.

Thymus [Gr. *thymos*: warty] A ductless, glandular portion of the lymphoid system, involved in development of the immune system of vertebrates.

Thyroid [Gr. *thyreos*: door-shaped] A two-lobed gland in vertebrates. Produces the hormone thyroxin.

Tight junction A junction between epithelial cells, in which there is no gap whatever between the adjacent cells. Materials may pass through a tight junction only by entering the epithelial cells themselves.

Tissue A group of similar cells organized into a functional unit; usually integrated with other tissues to form part of an organ.

Tonus (toe′ nuss) [L. *tonus*: tension] A low level of muscular tension that is maintained even when the body is at rest.

Topsoil The uppermost soil layer; contains most of the organic matter of soil, but may be depleted of most mineral nutrients.

Totipotency [L. *toto*: whole, entire + *potens*: powerful] In a cell, the condition of possessing all the genetic information and other capacities necessary to form an entire individual.

Toxic [L. *toxicum*: poison] Injurious to the tissues of the host organism.

Trachea (tray′ kee ah) [Gr. *trakhoia*: tube] A tube that carries air to the bronchi of the lungs of vertebrates, or to the cells of arthropods.

Tracheary element Refers to either or both types of xylem cells: tracheids and vessel elements.

Tracheid (tray' kee id) A distinctive conducting and supporting cell found in the xylem of nearly all vascular plants, characterized by tapering ends and walls that are pitted but not perforated.

Tracheophytes [Gr. *trakhoia*: tube + *phyton*: plant] Those plants with xylem and phloem, including psilophytes, club mosses, horsetails, ferns, gymnosperms, and angiosperms. (Contrast with non-trachoephytes.)

Trait One form of a character: Eye color is a character; brown eyes and blue eyes are traits.

Transcription The synthesis of RNA, using one strand of DNA as the template.

Transcription factors Proteins that assemble on a eukaryotic chromosome, allowing RNA polymerase II to perform transcription.

Transduction (1) Transfer of genes from one bacterium to another, with a bacterial virus acting as the carrier of the genes. (2) In sensory cells, the transformation of a stimulus (e.g., light energy, sound pressure waves, chemical or electrical stimulants) into action potentials.

Transfection Uptake, incorporation, and expression of recombinant DNA.

Transfer cell A modified parenchyma cell that transports solutes from its cytoplasm into its cell wall, thus moving the solutes from the symplast into the apoplast.

Transfer RNA (tRNA) A family of double stranded RNA molecules. Each kind of tRNA carries a specific amino acid and anticodon that will pair with the complementary codon in mRNA during translation.

Transformation Mechanism for transfer of genetic information in bacteria in which pure DNA extracted from bacteria of one genotype is taken in through the cell surface of bacteria of a different genotype and incorporated into the chromosome of the recipient cell.

Transgenic organism An organism containing recombinant DNA incorporated into its genetic material.

Translation The synthesis of a protein (polypeptide). This occurs on ribosomes, using the information encoded in messenger RNA.

Translocation (1) In genetics, a rare mutational event that moves a portion of a chromosome to a new location, generally on a nonhomologous chromosome. (2) In vascular plants, movement of solutes in the phloem.

Transpiration [L. *spirare*: to breathe] The evaporation of water from plant leaves and stem, driven by heat from the sun, and providing the motive force to raise water (plus ions) from the roots.

Transposable element A segment of DNA that can move to, or give rise to copies at, another locus on the same or a different chromosome.

Transposon A mobile DNA segment that can insert into a chromosome and cause genetic change.

Triglyceride A simple lipid in which three fatty acids are combined with one molecule of glycerol.

Triplet See codon.

Triplet repeat Occurrence of repeated triplet of bases in a gene, often leading to genetic disease, as does excessive repetition of CGG in the gene responsible for fragile-X syndrome.

Triploblastic Having three cell layers. (Contrast with diploblastic.)

Trisomic Containing three rather than two members of a chromosome pair.

tRNA See transfer RNA.

Trophoblast At the 32-cell stage of mammalian development, the outer group of cells that will become part of the placenta. See also Inner cell mass.

Trochophore (troke' o fore) [Gr. *trochos*: wheel + *phoreus*: bearer] The free-swimming larva of some annelids and mollusks. Distinguished by a wheel-like band of cilia around the middle, the trochophore suggests an evolutionary relationship between these two groups.

Trophic level A group of organisms united by obtaining their energy from the same part of the food web of a biological community.

Tropic hormones Hormones of the anterior pituitary that control the secretion of hormones by other endocrine glands.

Tropism [Gr. *tropos*: to turn] In plants, growth toward or away from a stimulus such as light (phototropism) or gravity (gravitropism).

Tropomyosis [troe poe my' oh sin] A protein that, along with actin, constitutes the thin filaments of myofibrils. It controls the interactions of actin and myosin necessary for muscle contraction.

roposphere The lowest atmospheric zone, reaching upward from the Earth's surface approximately 17 km in the tropics and subtropics but only to about 10 km at higher latitudes. The zone in which virtually all the water vapor in the atmosphere is located.

Trypsin A protein-digesting enzyme. Secreted by the pancreas in its inactive form (trypsinogen), it becomes active in the duodenum of the small intestine.

T-tubules A set of transverse tubes that penetrates skeletal muscle fibers and terminates in the sarcoplasmic reticulum. The T-system transmits impulses to the sacs, which then release Ca^{2+} to initiate muscle contraction.

Tube cell The larger of the two cells in a pollen grain; responsible for growth of the pollen tube. See Generative cell.

Tubulin A protein that polymerizes to form microtubules.

Tumor [L. *tumor*: a swollen mass] A disorganized mass of cells, often growing out of control. Malignant tumors spread to other parts of the body.

Tumor suppressor genes Genes which, when homozygous mutant, result in cancer. Such genes code for protein products that inhibit cell proliferation.

Turgor pressure [L. *turgidus*: swollen] See Hydrostatic pressure.

Twitch A single unit of muscle contraction.

Tympanic membrane [Gr. *tympanum*: drum] The eardrum.

Ubiquinone A mobile electron carrier of the mitochondrial respiratory chain. Similar to plastoquinone found in chloroplasts.

Ubiquitin A small protein that is covalently linked to other cellular proteins identified for breakdown by the proteosome.

Umbilical cord Tissue made up of embryonic membranes and blood vessels that connects the embryo to the placenta in eutherian mammals.

Understory The aggregate of smaller plants growing beneath the canopy of dominant plants in a forest.

Unicellular (yoon' e sell' yer ler) [L. *unus*: one + *cella*: chamber] Consisting of a single cell; as for example a unicellular organism. (Contrast with multicellular.)

Uniport A membrane transport process that carries a single substance. (Contrast with antiport, symport.)

Unsaturated hydrocarbon A compound containing only carbon and hydrogen atoms, with one or more pairs of carbon atoms that are connected by double bonds.

Upwelling The upward movement of nutrient-rich, cooler water from deeper layers of the ocean.

Uracil (U) A pyrimidine base found in nucleotides of RNA.

Urea A compound serving as the main excreted form of nitrogen by many animals, including mammals.

Ureotelic Describes an organism in which the final product of the breakdown of nitrogen-containing compounds (primarily proteins) is urea. (Contrast with ammonotelic, uricotelic.)

Ureter (your' uh tur) A long duct leading from the vertebrate kidney to the urinary bladder or the cloaca.

Urethra (you ree' thra) In most mammals, the canal through which urine is discharged from the bladder and which serves as the genital duct in males.

Uric acid A compound that serves as the main excreted form of nitrogen in some animals, particularly those which must conserve water, such as birds, insects, and reptiles.

Uricotelic Describes an organism in which the final product of the breakdown of nitrogen-containing compounds (primarily proteins) is uric acid. (Contrast with ammonotelic, ureotelic.)

Urine (you' rin) [Gk. *ouron*: urine] In vertebrates, the fluid waste product containing the toxic nitrogenous by-products of protein and amino acid metabolism.

Uterus (yoo' ter us) [L. *utero*: womb] The uterus or womb is a specialized portion of the female reproductive tract in certain mammals. It receives the fertilized egg and nurtures the embryo in its early development.

Vaccination Injection of virus or bacteria or their proteins into the body, to induce immunization. The injected material is usually attenuated (weakened) before injection.

Vacuole (vac' yew ole) [Fr.: small vacuum] A liquid-filled, membrane-enclosed compartment in cytoplasm; may function as digestive chambers, storage chambers, waste bins.

Vagina (vuh jine' uh) [L.: sheath] In female mammals, the passage leading from the external genital orifice to the uterus; receives the copulatory organ of the male in mating.

van der Waals forces Weak attractions between atoms resulting from the interaction of the electrons of one atom with the nucleus of another. This type of attraction is about one-fourth as strong as a hydrogen bond.

Variable regions The part of an immunoglobulin molecule or T-cell receptor that includes the antigen-binding site.

Vascular (vas' kew lar) [L. *vasculum*: a small vessel] Pertaining to organs and tissues that conduct fluid, such as blood vessels in animals and phloem and xylem in plants.

Vascular bundle In vascular plants, a strand of vascular tissue, including conducting cells of xylem and phloem as well as thick-walled fibers.

Vascular cambium A lateral meristem giving rise to secondary xylem and phloem.

Vascular rays In vascular plants, radially oriented sheets of cells produced by the vascular cambium, carrying materials laterally between the wood and the phloem.

Vascular system The conductive system of the plant, consisting primarily of xylem and phloem.

Vasopressin See antidiuretic hormone.

Vector (1) An agent, such as an insect, that carries a pathogen affecting another species. (2) A plasmid or virus that carries an inserted piece of DNA into a bacterium for cloning purposes in recombinant DNA technology.

Vegetal hemisphere The lower portion of some animal eggs, zygotes, and embryos, in which the dense nutrient yolk settles. The vegetal pole refers to the very bottom of the egg or embyro. (Contrast with animal hemisphere.)

Vegetative Nonreproductive, or nonflowering, or asexual.

Vegetative reproduction Asexual reproduction.

Vein [L. *vena*: channel] A blood vessel that returns blood to the heart. (Contrast with artery.)

Ventral [L. *venter*: belly, womb] Toward or pertaining to the belly or lower side. (Contrast with dorsal.)

Ventricle A muscular heart chamber that pumps blood through the body.

Vernalization [L. *vernalis*: spring] Events occurring during a required chilling period, leading eventually to flowering.

Vertebral column The jointed, dorsal column that is the primary support structure of vertebrates.

Vertebrate An animal whose nerve cord is enclosed in a backbone of bony segments, called vertebrae. The principal groups of vertebrate animals are the fishes, amphibians, reptiles, birds, and mammals.

Vesicle A membrane enclosed compartment within the cytoplasm.

Vessel elements In plants, nonliving water conducting cells with perforated end walls. Compare with tracheids.

Vestibular apparatus (ves tib' yew lar) [L. *vestibulum*: an enclosed passage] Structures associated with the vertebrate ear; these structures sense changes in position or momentum of the head, affecing balance and motor skills.

Vestigial (ves tij' ee al) [L. *vestigium*: footprint, track] The remains of body structures that are no longer of adaptive value to the organism and therefore are not maintained by selection.

Vicariant distribution A population distribution resulting from the disruption of a formerly continuous range by a vicariant event.

Vicariant event (vye care' ee unce) [L. *vicus*: change] The splitting of the range of a taxon by the imposition of some barrier to interchange among its members.

Villus (vil' lus) (plural: villi) [L. *villus*: shaggy hair or beard] A hairlike projection from a membrane; for example, from many gut walls.

Virion (veer' e on) The virus particle, the minimum unit capable of infecting a cell.

Viroid (vye' roid) An infectious agent consisting of a single-stranded RNA molecule with no protein coat; produces diseases in plants.

Virulent [L. *virus*: poison, slimy liquid] Causing or capable of causing disease and death.

Virus Any of a group of ultramicroscopic infectious particles constructed of nucleic acid and protein (and, sometimes, lipid) that can reproduce only in living cells.

Vitamins [L. *vita*: life] Organic compounds that an organism cannot synthesize, but nevertheless requires in small quantity for normal growth and metabolism.

Viviparous (vye vip' uh rus) [L. *vivus*: alive] Reproduction in which fertilization of the egg and development of the embryo occur inside the mother's body. (Contrast with oviparous.)

VNTRs (variable number of tandem repeats) In the human genome, short DNA sequences that are repeated a characteristic number of times in related individuals. Can be used to make a DNA fingerprint.

Waggle dance The running movement of a working honey bee on the hive, during which the worker traces out a repeated figure eight. The dance contains elements that transmit to other bees the location of the food.

Water potential In osmosis, the tendency for a system (a cell or solution) to take up water from pure water, through a differentially permeable membrane. Water flows toward the system with a more negative water potential. (Contrast with osmotic potential, turgor pressure.)

Water vascular system The array of canals and tubelike appendages that serves as the circulatory system, locomotory system, and food-capturing system of many echinoderms; is in direct connection with the surrounding sea water.

Wavelength The distance between successive peaks of a wave train, such as electromagnetic radiation.

Wild type Geneticists' term for standard or reference type. Deviants from this standard, even if the deviants are found in the wild, are said to be mutant.

Wood Secondary xylem tissue.

Xanthophyll (zan' tho fill) [Gr. *xanthos*: yellowish-brown + *phyllon*: leaf] A yellow or orange pigment commonly found as an accessory pigment in photosynthesis, but found elsewhere as well. An oxygen-containing carotenoid.

X-linked A character that is coded for by a gene on the X chromosome; a sex-linked trait.

Xerophyte (zee' row fyte) [Gr. *xerox*: dry + *phyton*: plant] A plant adapted to an environment with a limited water supply.

Xylem (zy' lum) [Gr. *xylon*: wood] In vascular plants, the tissue that conducts water and minerals; xylem consists, in various plants, of tracheids, vessel elements, fibers, and other highly specialized cells.

Yeast artificial chromosome (YAC) A laboratory-made DNA molecule containing sequences of yeast chromosomes (origin of replication, telomeres, centromere, and selectable markers) so that it can be used as a vector in yeast.

Yolk [M.E. *yolke*: yellow] The stored food material in animal eggs, usually rich in protein and lipid.

Yolk sac In embryonic development of reptiles, birds, and mammals, the extraembryonic membrane that forms from the endoderm of the hypoblast; it encloses and digests the yolk.

Z-DNA A form of DNA in which the molecule spirals to the left rather than to the right.

Zeaxanthin A carotenoid pigment that is a blue-light receptor.

Zoospore (zoe' o spore) [Gr. *zoon*: animal + *spora*: seed] In algae and fungi, any swimming spore. May be diploid or haploid.

Zygote (zye' gote) [Gr. *zygotos*: yoked] The cell created by the union of two gametes, in which the gamete nuclei are also fused. The earliest stage of the diploid generation.

Zymogen An inactive precursor of a digestive enzyme secreted into the lumen of the gut, where a protease cleaves it to form the active enzyme.

Answers to Self-Quizzes

Chapter 2

1. b	6. a
2. d	7. d
3. c	8. a
4. c	9. e
5. d	10. b

Chapter 3

1. e	6. a
2. e	7. c
3. c	8. e
4. d	9. a
5. b	10. d

Chapter 4

1. b	6. e
2. d	7. a
3. c	8. d
4. e	9. b
5. a	10. d

Chapter 5

1. e	6. c
2. c	7. c
3. a	8. b
4. d	9. e
5. c	10. c

Chapter 6

1. c	6. e
2. e	7. d
3. b	8. b
4. c	9. a
5. c	10. e

Chapter 7

1. d	6. d
2. d	7. a
3. e	8. b
4. e	9. a
5. c	10. e

Chapter 8

1. c	6. c
2. b	7. c
3. d	8. d
4. b	9. d
5. e	10. b

Chapter 9

1. d	6. d
2. c	7. e
3. d	8. d
4. d	9. c
5. c	10. c

Chapter 10*

1. e	6. d
2. a	7. b
3. d	8. b
4. d	9. b
5. d	10. e

Chapter 11

1. c	6. b
2. a	7. d
3. c	8. d
4. b	9. c
5. e	10. c

Chapter 12

1. c	6. d
2. d	7. b
3. e	8. d
4. b	9. d
5. a	10. a

Chapter 13

1. b	6. d
2. e	7. d
3. a	8. c
4. c	9. b
5. c	10. d

Chapter 14

1. c	6. c
2. c	7. c
3. a	8. b
4. a	9. e
5. c	10. d

Chapter 15

1. d	6. a
2. d	7. e
3. c	8. b
4. c	9. c
5. d	10. a

Chapter 16

1. b	6. b
2. a	7. c
3. a	8. a
4. c	9. e
5. e	10. e

*Answers to Chapter 10 Genetics Problems

1. Each of the eight boxes in the Punnett squares should contain the genotype *Tt*, regardless of which parent was tall and which dwarf.

2. Yellow parent = s^Ys^b; offspring 3 yellow (s^Y–): 1 black (s^bs^b). Black parent = s^bs^b; offspring all black (s^bs^b). Orange parent = s^Os^b; offspring 3 orange (s^O–): 1 black (s^bs^b). Both s^O and s^Y are dominant to s^b.

3. See Figure 10.4, page 192.

4. The trait is autosomal. Mother *dp dp*, father *Dp dp*. If the trait were sex-linked, all daughters would be wild-type and sons would be *dumpy*.

5. All females wild-type; all males spotted.

6. F_1 all wild-type, *PpSwsw*; F_2 9:3:3:1 in phenotypes. See Figure 10.7, page 194, for analogous genotypes.

7*a*. Ratio of phenotypes in F_2 is 3:1 (double dominant to double recessive).

7*b*. The F_1 are *Pby* pB^Y; they produce just two kinds of gametes (*Pby* and *pBy*). Combine them carefully and see the 1:2:1 phenotypic ratio fall out in the F_2.

7*c*. Pink-blistery.

7*d*. See Figures 9.14 and 9.16 (pages 178–180). Crossing over took place in the F_1 generation.

8. The genotypes are:
PpSwsw
Ppswsw
ppSwsw
ppswsw
Ratio: 1:1:1:1

The phenotypes are:
wild eye, long wing — pink eye, long wing
wild eye, short wing — pink eye, short wing

Ratio: 1:1:1:1

9*a*. 1 black:2 blue:1 splashed white

9*b*. Always cross black with splashed white.

10*a*. $w^+ > w^e > w$

10*b*. Parents w^ew and w^+Y. Progeny $w{+}w^e$, $w{+}w$, w^eY, and wY.

11. All will have normal vision because they inherit dad's wild-type X chromosome, but half of them will be carriers.

12. Agouti parent *AaBb*. Albino offspring *aaBb* and *aabb*; black offspring *Aabb*; agouti offspring *AaBb*.

13. Because the gene is carried on mitochondrial DNA, it is passed through the mother only. Thus if the woman does not have the disease but her husband does, their child will not be affected. On the other hand, if the woman has the disease but her husband does not, their child *will* have the disease.

Chapter 17
1. a 6. b
2. c 7. e
3. b 8. d
4. b 9. c
5. d 10. b

Chapter 18
1. a 6. a
2. b 7. d
3. a 8. d
4. e 9. a
5. c 10. d

Chapter 19
1. c 6. c
2. a 7. d
3. a 8. b
4. b 9. a
5. b 10. b

Chapter 20
1. a 6. c
2. c 7. b
3. e 8. d
4. c 9. b
5. d 10. a

Chapter 21
1. e 6. b
2. d 7. a
3. a 8. c
4. b 9. e
5. c 10. b

Chapter 22
1. d 6. a
2. b 7. c
3. e 8. b
4. c 9. c
5. a 10. e

Chapter 23
1. b 7. d
2. d 8. b
3. d 9. e
4. c 10. b
5. d 11. c
6. e

Chapter 24
1. c 7. d
2. e 8. a
3. d 9. a
4. c 10. c
5. a 11. e
6. b

Chapter 25
1. c 7. a
2. a 8. d
3. d 9. b
4. e 10. d
5. b 11. e
6. c

Chapter 26
1. d 7. a
2. b 8. a
3. e 9. a
4. c 10. a
5. a 11. c
6. e 12. b

Chapter 27
1. e 6. b
2. e 7. d
3. b 8. a
4. c 9. c
5. e 10. b

Chapter 28
1. a 6. d
2. e 7. c
3. c 8. b
4. d 9. b
5. a 10. d

Chapter 29
1. d 6. e
2. c 7. c
3. e 8. b
4. b 9. b
5. b 10. d

Chapter 30
1. d 6. c
2. c 7. a
3. d 8. e
4. a 9. c
5. d 10. a

Chapter 31
1. b 6. a
2. d 7. e
3. e 8. a
4. c 9. c
5. d 10. c

Chapter 32
1. c 6. d
2. d 7. d
3. b 8. b
4. e 9. e
5. b 10. c

Chapter 33
1. d 6. c
2. c 7. d
3. a 8. d
4. e 9. e
5. b 10. e

Chapter 34
1. b 7. e
2. a 8. a
3. c 9. e
4. c 10. c
5. b 11. b
6. d

Chapter 35
1. d 6. b
2. b 7. b
3. e 8. c
4. e 9. a
5. a 10. d

Chapter 36
1. c 6. d
2. d 7. d
3. b 8. e
4. b 9. e
5. b 10. a

Chapter 37
1. d 6. c
2. d 7. e
3. c 8. a
4. a 9. d
5. a 10. e

Chapter 38
1. a 6. c
2. e 7. e
3. c 8. c
4. d 9. a
5. b 10. b

Chapter 39
1. d 6. e
2. b 7. a
3. e 8. b
4. b 9. c
5. d 10. d

Chapter 40
1. e 6. a
2. b 7. b
3. c 8. c
4. c 9. d
5. d 10. a

Chapter 41
1. c 6. e
2. a 7. a
3. d 8. e
4. b 9. e
5. b 10. c

Chapter 42
1. b 6. b
2. a 7. d
3. b 8. d
4. e 9. c
5. e 10. c

Chapter 43
1. e 6. d
2. e 7. d
3. a 8. c
4. d 9. d
5. d 10. a

Chapter 44
1. d 6. e
2. a 7. e
3. d 8. c
4. c 9. d
5. c 10. d

Chapter 45
1. d 6. e
2. d 7. b
3. a 8. c
4. b 9. c
5. e 10. d

Chapter 46
1. c 6. c
2. a 7. a
3. e 8. c
4. d 9. a
5. d 10. a

Chapter 47
1. e 6. d
2. a 7. e
3. b 8. a
4. c 9. a
5. b 10. a

Chapter 48
1. e 6. b
2. d 7. c
3. a 8. c
4. b 9. a
5. c 10. d

Chapter 49
1. d 6. d
2. a 7. b
3. c 8. d
4. d 9. c
5. c 10. e

Chapter 50
1. b 6. d
2. e 7. a
3. c 8. b
4. a 9. d
5. b 10. d

Chapter 51

1. b
2. a
3. d
4. c
5. d
6. b
7. e
8. a
9. c
10. e

Chapter 52

1. c
2. a
3. c
4. e
5. d
6. d
7. d
8. d
9. b
10. b

Chapter 53

1. e
2. c
3. d
4. b
5. c
6. d
7. d
8. a

Chapter 54

1. b
2. c
3. a
4. d
5. b
6. c
7. c
8. b
9. d
10. a

Chapter 55

1. a
2. c
3. a
4. b
5. d
6. e
7. c
8. a
9. d
10. d

Chapter 56

1. a
2. c.
3. d
4. c
5. d
6. a
7. d
8. e
9. a
10. e

Chapter 57

1. b
2. d
3. e
4. e
5. b
6. a
7. d
8. c
9. a
10. c

Chapter 58

1. d
2. d
3. e
4. e
5. c
6. c
7. a
8. c
9. d
10. e

Illustration Credits

Cover, Title Page, and Frontispiece
Steve Bloom/stevebloom.com.

Authors' Photograph
Christopher Small

Table of Contents Photographs
p. xviii *E. coli*: Dennis Kunkel Microscopy, Inc.
p. xx *Golgi*: Dennis Kunkel Microscopy, Inc.
p. xxi *chromosomes*: Andrew Syred/SPL/Photo Researchers, Inc.
p. xxiii *virus*: Dennis Kunkel Microscopy, Inc.
p. xxv *mouse embryo*: Dr. Fred Hossler/Visuals Unlimited.
p. xxvi *tortoise*: DigitalVision/PictureQuest.
p. xxvii *chimpanzee*: DigitalVision/PictureQuest.
p. xxviii *diatoms*: Andrew Syred/SPL/Photo Researchers, Inc.
p. xxix *fungus*: Photodisc Blue/Getty Images.
p. xxx *water lily*: Judith Worley/Painet, Inc.
p. xxxi *cycad*: IT Stock Free/PictureQuest.
p. xxxii *leaf*: Courtesy of Thomas Eisner, Cornell U.
p. xxxiii macaque: DigitalVision/PictureQuest.
p. xxxiv *penguin*: DigitalVision/PictureQuest.
p. xxxv *neuron*: Dennis Kunkel/Visuals Unlimited.
p. xxxvii *locusts*: IT Stock Free/PictureQuest.
p. xxxviii *elephants*: DigitalVision/PictureQuest.

Chapter 1 *Opener*: Steven Holt/stockpix.com. 1.1 *larva*: Valorie Hodgson/Visuals Unlimited. 1.1 *pupa*: Dick Poe/Visuals Unlimited. 1.1 *butterfly*: Bill Beatty/Visuals Unlimited. 1.4, 1.5: Dennis Kunkel Microscopy, Inc. 1.6 *organism*: Sami Sarkis/Painet Inc. 1.6 *population*: LogicStock/Painet Inc. 1.6 *community*: Georgie Holland/AGE Fotostock. 1.6 *biosphere*: Courtesy of NASA. 1.7*a eagle, curlew*: Stockbyte/PictureQuest. 1.7*a spoonbill*: E. J. Peiker/Painet Inc. 1.7*b coconut*: Inga Spence/Tom Stack & Assoc. 1.7*b blackberries*: Colin Varndell/Nature Picture Library. 1.7*b milkweed*: Bill Beatty/Painet Inc. 1.9: Data from S. D. Broomhall et al., 2000. *Conservation Biology* 14: 420–428. 1.10: Data from C. Davidson et al., 2002. *Conservation Biology* 16: 1588–1601.

Chapter 2 *Opener*: Courtesy of R. P. Irwin III & G. A. Franz, Smithsonian Institution. 2.5: SIU/Visuals Unlimited. 2.16: Robert Barber/Painet Inc. 2.17: P. Armstrong/Visuals Unlimited.

Chapter 3 *Opener*: Courtesy of NASA. 3.7: Data from PDB 1IVM, T. Obita, T. Ueda, & T. Imoto. 2003. *Cell. Mol. Life Sci.* 60: 176. 3.8*a*: Data from PDB 2HHB, G. Fermi et al. 1984. *J. Mol. Biol.* 175: 159. 3.10: Data from PDB 1HSG, Z. Chen et al. 1994. *J. Biol. Chem.* 269: 26344. 3.16*c left*: Biophoto Associates/Photo Researchers, Inc. 3.16*c middle*: Ken Wagner/Visuals Unlimited. 3.16*c right*: CNRI/SPL/Photo Researchers, Inc. 3.17 *cartilage*: Robert Brons/Biological Photo Service. 3.17 *beetle*: Peter J. Bryant/Biological Photo Service. 3.27: Structure prepared by Jason Kahn with InsightII (Accelrys); data from S. Arnott & D. W. Hukins. 1972. *Biochem. Biophys. Res. Commun.* 47(6):1504.

Chapter 4 *Opener*: Stanley M. Awramik/Biological Photo Service. *inset*: Dennis Kunkel Microscopy, Inc. 4.1: Sidney Fox/Science VU/Visuals Unlimited. 4.2: After N. Campbell, 1990. *Biology*, 2nd Ed., Benjamin Cummings Publishing Co. 4.4 *upper row*: David M. Phillips/Visuals Unlimited. 4.4 *middle row, left*: Conly L. Rieder/Biological Photo Service. 4.4 *middle row, center*: Courtesy of David Albertini, Tufts U. School of Medicine. 4.4 *middle row, right*: M. Abbey/Photo Researchers, Inc. 4.4 *bottom row, left*: D. P. Evenson/Biological Photo Service. 4.4 *bottom row, center*: Gerald Schatten/Biological Photo Service. 4.4 *bottom row, right*: Courtesy of Dr. Marisa Otegui, U. Colorado at Boulder. 4.5: J. J. Cardamone Jr. & B. K. Pugashetti/Biological Photo Service. 4.6*a*: J. J. Cardamone Jr./Biological Photo Service. 4.6*b*: Courtesy of S. Abraham & E. H. Beachey, VA Medical Center, Memphis, TN. 4.7 *nucleolus*: Richard Rodewald/Biological Photo Service. 4.7 *mitochondrion*: K. Porter, D. Fawcett/Visuals Unlimited. 4.7 *cytoskeleton*: D. Fawcett & J. Heuser/Photo Researchers, Inc. 4.7 *rough ER*: D. Fawcett/Science Source/Photo Researchers, Inc. 4.7 *plasma membrane*: Courtesy of J. David Robertson, Duke U. Medical Center. 4.7 *centrioles*: Barry F. King/Biological Photo Service. 4.7 *golgi apparatus*: Courtesy of L. Andrew Staehelin, U. Colorado. 4.7 *ribosome*: From Boublik et al., 1990, *The Ribosome*, p. 177. Courtesy of American Society for Microbiology. 4.7 *smooth ER*: D. Fawcett & D. Friend/Science Source/Photo Researchers, Inc. 4.7 *peroxisome*: E. H. Newcomb & S. E. Frederick/Biological Photo Service. 4.7 *cell wall*: Biophoto Associates/Photo Researchers, Inc. 4.7 *chloroplast*: W. P. Wergin & E. H. Newcomb/Biological Photo Service. 4.9 *left*: From Aebi, U. et al. 1986. *Nature* 323: 560. Macmillan Publishers Ltd. 4.9 *upper right*: Richard Rodewald/Biological Photo Service. 4.9 *lower right*: Courtesy of Dr. Ron Milligan, Scripps Research Institute. 4.10*a*: Barry King, U. California, Davis/Biological Photo Service. 4.10*b*: Biophoto Associates/Science Source/Photo Researchers, Inc. 4.11: D. Fawcett/Visuals Unlimited. 4.14: K. Porter & D. Fawcett/Visuals Unlimited. 4.15: W. P. Wergin & E. H. Newcomb/Biological Photo Service. 4.16*a*: John Durham/SPL/Photo Researchers, Inc. 4.16*b*: Paul W. Johnson/Biological Photo Service. 4.16*c*: Larry Jon Friesen. 4.17*a*: Geoff Bryant/Photo Researchers, Inc. 4.17*a inset*: Richard Green/Photo Researchers, Inc. 4.17*b*: G. Büttner/Naturbild/OKAPIA/Photo Researchers, Inc. 4.17*b inset*: Courtesy of R. R. Dute. 4.19: E. H. Newcomb & S. E. Frederick/Biological Photo Service. 4.20: Courtesy of M. C. Ledbetter, Brookhaven National Laboratory. 4.21: Courtesy of Vic Small, Austrian Academy of Sciences, Salzburg, Austria. 4.22: Courtesy of N. Hirokawa. 4.23: W. L. Dentler/Biological Photo Service. 4.24: B. J. Schnapp et al., 1985. *Cell* 40: 455; courtesy of B. J. Schnapp, R. D. Vale, M. P. Sheetz, & T. S. Reese. 4.25: Biophoto Associates/Photo Researchers, Inc. 4.26 *left*: Courtesy of David Sadava. 4.26 *upper right*: From J. A. Buckwalter & L. Rosenberg, 1983. *Coll. Rel. Res.* 3: 489; courtesy of L. Rosenberg. 4.26 *lower right*: J. Gross, Biozentrum/SPL/Photo Researchers, Inc.

Chapter 5 *Opener*: Courtesy of Stewart Cohen Pictures, Dallas, Texas. 5.2: After L. Stryer, 1981. *Biochemistry*, 2nd Ed., W. H. Freeman. 5.3: Courtesy of L. Andrew Staehelin, U. Colorado. 5.6*a*: Courtesy of D. S. Friend, U. California, San Francisco. 5.6*b*: Courtesy of Darcy E. Kelly, U. Washington. 5.6.*c*: Courtesy of C. Peracchia. 5.16: From M. M. Perry, 1979. *J. Cell Sci.* 39: 26.

Chapter 6 *Opener*: Data from PDB 1CX2, R. Kurumbail et al. 1996. *Nature* 384: 644. 6.5*b*: Darwin Dale/Photo Researchers, Inc. 6.12*a*: Data from PDB 1AL6, B. Schwartz et al. 1997. 6.12*c*: Data from PDB 1AB9, N. H. Yennawar et al. 1994. *Biochemistry* 33: 7326. 6.13: Data from PDB 1BB6, V. B. Vollan et al. 1999. *Acta Crystallogr. D. Biol. Crystallogr.* 55: 60. 6.15: Data from PDB 1A7K, H. Kim & W. G. Hol. 1998. *J. Mol. Biol.* 278: 5.

Chapter 7 *Opener*: Gianni Dagli Orti/Corbis. 7.12: Courtesy of Ephraim Racker.

Chapter 8 *Opener*: Inga Spence/Visuals Unlimited. 8.1: Andrew Syred/SPL/Photo Researchers, Inc. 8.3: Dr. Kenneth R. Miller/SPL/Photo Researchers, Inc. 8.12: Courtesy of Lawrence Berkeley National Laboratory. 8.15: E. H. Newcomb & S. E. Frederick/Biological Photo Service. 8.17: E. H. Newcomb & S. E. Frederick/Biological Photo Service.

Chapter 9 *Opener*: Nancy Kedersha/SPL/Photo Researchers, Inc. 9.1*a,c*: John D. Cunningham/Visuals, Unlimited. 9.1*b*: David M. Phillips/Visuals Unlimited. 9.2*b*: J. J. Cardamone Jr./Biological Photo Service. 9.5: Courtesy of G. F. Bahr. 9.6 *inset*: Biophoto Associates/Science Source/Photo Researchers,

Inc. 9.7*b*: Conly L. Rieder/Biological Photo Service. 9.8: Andrew S. Bajer, U. Oregon. 9.10*a*: T. E. Schroeder/Biological Photo Service. 9.10*b*: B. A. Palevitz & E. H. Newcomb/Biological Photo Service. 9.11: Garry T. Cole/Biological Photo Service. 9.12 *left*: Andrew Syred/SPL/Photo Researchers, Inc. 9.12 *center*: E. Webber/Visuals Unlimited. 9.12 *right*: Bill Kamin/Visuals Unlimited. 9.13: Courtesy of Dr. Thomas Ried & Dr. Evelin Schröck, NIH. 9.14: C. A. Hasenkampf/Biological Photo Service. 9.15: Klaus W. Wolf, U. West Indies. 9.19: Gopal Murti/Photo Researchers, Inc.

Chapter 10 *Opener*: David H. Wells/Corbis. 10.2: Wally Eberhart/Visuals Unlimited. 10.12: Courtesy the American Netherland Dwarf Rabbit Club. 10.15: NCI/Photo Researchers, Inc. 10.16: Courtesy of Pioneer Hi-Bred International, Inc. 10.24: Science VU/Visuals Unlimited. *Bay scallops*: Barbara J. Miller/Biological Photo Service.

Chapter 11 *Opener*: Universal City Studios/Shooting Star. 11.2: Biozentrum, Universtiy of Basel/SPL/Photo Researchers, Inc. 11.4: Courtesy of Prof. M. H. F. Wilkins, Dept. of Biophysics, King's College, U. London. 11.6*a*: A. Barrington Brown/Photo Researchers, Inc. 11.6*b*: Structure prepared by Jason Kahn with InsightII (Accelrys); data from S. Arnott & D. W. Hukins. 1972. *Biochem. Biophys. Res. Commun.* 47(6): 1504. 11.18: Dr. Peter Lansdorp/Visuals Unlimited.

Chapter 12 *Opener*: Alan L. Detrick/Photo Researchers, Inc. 12.7: Data from PDB 1EHZ, H. Shi & P. B. Moore. 2000. *RNA* 6: 1091. 12.13*b*: Courtesy of J. E. Edström and *EMBO J.* 12.17: Stanley Flegler/Visuals Unlimited.

Chapter 13 *Opener*: Dennis Kunkel Microscopy, Inc. 13.1*a*: Dennis Kunkel Microscopy, Inc. 13.1*b*: E.O.S./Gelderblom/Photo Researchers, Inc. 13.1*c*: Dennis Kunkel Microscopy, Inc. 13.8: Courtesy of L. Caro & R. Curtiss. 13.21: Based on an illustration by Anthony R. Kerlavage, Institute for Genomic Research. *Science* 269: 449 (1995).

Chapter 14 *Opener*: Inga Spence/Visuals Unlimited. 14.6: Tiemeier et al. 1978. *Cell* 14: 237. 14.17: Courtesy of Murray L. Barr, U. Western Ontario. 14.19: Courtesy of O. L. Miller, Jr.

Chapter 15 *Opener*: Ryan McVay, Photodisc Green/Getty Images. 15.2 *inset*: Biophoto Associates/Photo Researchers, Inc. 15.3: From de Vos et al., 1992. *Science* 255: 306. 15.12: Stephen A. Stricker, courtesy of Molecular Probes, Inc.

Chapter 16 *Opener*: Keith V. Wood/Science VU/Visuals Unlimited. 16.2: Philippe Plailly/Photo Researchers, Inc. 16.15 *left*: Courtesy of Ingo Potrykus, Swiss Federal Institute of Technology. 16.15 *right*: Joan Gemme & David McIntyre. 16.16: Courtesy of Eduardo Blumwald. 16.18: Bettmann/Corbis.

Chapter 17 *Opener*: Data from PDB 1IEP, B. Nagar et al. 2002. *Cancer Res.* 62: 4236. 17.5: C. Harrison et al., 1983. *J. Med. Genet.* 20: 280. 17.10: Courtesy of Harvey Levy & Cecelia Walraven, New England Newborn Screening Program. 17.13: Dennis Kunkel Microscopy, Inc. 17.18*b*: David M. Martin, M.D./SPL/Photo Researchers, Inc. 17.24: From P. H. O'Farrell. 1975. High resolution two-dimensional electrophoresis of proteins. *J. Biol. Chem.* 250: 4007–4021. Courtesy of Patrick H. O'Farrell.

Chapter 18 *Opener*: Francis G. Mayer/Corbis. 18.3: Dennis Kunkel Microscopy, Inc. 18.9: Dr. Gopal Murti/SPL/Photo Researchers, Inc. 18.14: Dr. Andrejs Liepins/SPL/Photo Researchers, Inc. 18.16: David Phillips/Science Source/Photo Researchers, Inc.

Chapter 19 *Opener*: Courtesy of Advanced Cell Technology, Worcester, Mass. USA. 19.4: Roddy Field, the Roslin Institute. 19.5: Courtesy of T. Wakayama & R. Yanagimachi. 19.10: J. E. Sulston & H. R. Horvitz, 1977. *Dev. Bio.* 56: 100. 19.12*b*: Courtesy of J. Bowman. 19.13 *left*: Courtesy of J. Bowman. 19.13 *right*: Courtesy of Detlef Weigel. 19.14: Courtesy of W. Driever & C. Nüsslein-Vollhard. 19.16: Courtesy of F. R. Turner, Indiana U.

Chapter 20 *Opener*: Dave B. Fleetham/Tom Stack & Assoc. 20.1: D. M. Phillips/Science Source/Photo Researchers, Inc. 20.2: Courtesy of Richard Elinson, U. Toronto. 20.5: Courtesy of J. G. Mulnard. 20.21*a*: Dr. G. Moscoso/SPL/Photo Researchers, Inc. 20.21*b*: Tissuepix/SPL/Photo Researchers, Inc.

Chapter 21 *Opener*: LogicStock/Painet Inc. 21.1 *upper left*: Colin Milkins/Oxford Scientific Films. 21.1 *upper right*: Jan Hinsch/SPL/Photo Researchers, Inc. 21.1 *lower left*: Oxford Scientific Films. 21.1 *lower right*: Robert Brons/Biological Photo Service. 21.2: Courtesy of W. J. Gehring & G. Halder. From Halder et al., 1995. *Science* 267: 1788. 21.4*a*: Courtesy of E. B. Lewis. 21.4*b*: Courtesy of H. Le Mouellic, Y. Lallemand, & P. Brûlet. From Le Mouellic et al. 1992. *Cell* 69: 251. 21.5 *cladogram*: After R. Galant and S. Carroll, 2002. *Nature* 415: 910. 21.5 *beetle*: Stockbyte/PictureQuest. 21.5 *centipede*: Burke/Triolo/Brand X Pictures/PictureQuest. 21.6: Courtesy of J. Hurle & E. Laufer. 21.7: Courtesy of J. Hurle. 21.9: Courtesy of S. Carroll & P. Brakefield. 21.10: Erick Greene. 21.11: Courtesy of A. A. Agrawal. 21.12: Nigel Cattlin, Holt Studios International/Photo Researchers, Inc.

Chapter 22 *Opener*: PhotoLink/Photodisc/PictureQuest. 22.1: Robert Fried/Tom Stack & Assoc. 22.3*a*: Tom & Therisa Stack/Tom Stack & Assoc. 22.3*b*: Stanley M. Awramik/Biological Photo Service. 22.6: François Gohier/Photo Researchers, Inc. 22.7: Jeff J. Daly/Visuals Unlimited. 22.8 *left*: Ken Lucas/Biological Photo Service. 22.8 *right*: Stanley M. Awramik/Biological Photo Service. 22.9: Chip Clark. 22.10: Hans Steur/Visuals Unlimited. 22.11: Tom McHugh/Field Museum, Chicago/Photo Researchers, Inc. 22.12: Chase Studios, Cedarcreek, MO. 22.14: Chris Butler/SPL/Photo Researchers, Inc. 22.16: K. Simons & David Dilcher. 22.18*a*: Kjell Sandved/Visuals Unlimited. 22.18*b*: Hans Reinhard/Okapia/Photo Researchers, Inc. 22.21: Calvin Larsen/Photo Researchers, Inc.

Chapter 23 *Snake*: Joseph T. Collins/Photo Researchers, Inc. *Newt*: Robert Clay/Visuals Unlimited. 23.1*a,b*: SPL/Photo Researchers, Inc. 23.2: Levi, W. 1965. *Encyclopedia of Pigeon Breeds.* T. F. H. Publications, Jersey City, NJ. (*a,b*: photos by R. L. Kienlen, courtesy of Ralston Purina Company; *c,d*: photos by Stauber.). 23.9: S. Maslowski/Visuals Unlimited. 23.11: Judith Worley/Painet Inc. 23.17*b*: Tony Tilford/Oxford Scientific Films. 23.22*a*: Marilyn Kazmers/Dembinsky Photo Assoc. 23.22*b*: Paul Osmond/Painet Inc. 23.23: Anup Shah/Dembinsky Photo Assoc.

Chapter 24 *Opener*: Rob Simpson/Painet Inc. 24.2*a left*: Gary Meszaros/Dembinsky Photo Assoc. 24.2*a right*: Lior Rubin/Peter Arnold, Inc. 24.2*b*: Fi Rust/Painet Inc. 24.9*a*: Virginia P. Weinland/Photo Researchers, Inc. 24.9*b*: José Manuel Sánchez de Lorenzo Cáceres. 24.11*a*: D. Cavagnaro/DRK Photo. 24.11*b*: Charles Webber/California Academy of Sciences. 24.13*a*: Stephen Dalton/NHPA. 24.13*b*: Daniel Heuclin/NHPA. 24.14 *upper, lower*: Peter J. Bryant/Biological Photo Service. 24.14 *center*: Courtesy of Kenneth Y. Kaneshiro, U. Hawaii. 24.16 *Madia*: Peter K. Ziminsky/Visuals Unlimited. 24.16 *Argyroxiphium*: Elizabeth N. Orians. 24.16 *Dubautia*: Noble Proctor/Photo Researchers, Inc. 24.16 *Wilkesia*: Gerald D. Carr.

Chapter 25 *Opener*: Lorne Resnick/AGE Fotostock. 25.3 *left*: Adam Jones/Dembinsky Photo Assoc. 25.3 *center*: Brian Parker/Tom Stack & Assoc. 25.3 *right*: Joe McDonald/DRK Photo. 25.11*a* : Mark Smith/Photo Researchers, Inc. 25.11*b*: After E. Verheyen et al., 2003. *Science* 300: 328.

Chapter 26 *Opener*: Courtesy of James Gathany/CDC. 26.4: Courtesy of Richard Alexander, U. Pennsylvania. 26.6*a*: Belinda Wright/DRK Photo. 26.6*b*: M. Graybill/J. Hodder/Biological Photo Service.

Chapter 27 *Opener*: Thomas Dressler/DRK Photo. 27.1: Kari Lounatmaa/Photo Researchers, Inc. 27.3*a*: D. M. Phillips/Photo Researchers, Inc. 27.3*b*: R. Kessel-G. Shih/Visuals Unlimited. 27.3*c*: Courtesy of Janice Carr/NCID/CDC. 27.4*a*: J. A. Breznak & H. S. Pankratz/Biological Photo Service. 27.4*b*: J. Robert Waaland/Biological Photo Service. 27.5: USDA/Visuals Unlimited. 27.6*a left*: D. M. Phillips/Visuals Unlimited. 27.6*a right*: Courtesy of Peter Hirsch & Stuart Pankratz. 27.6*b left*: Courtesy of the CDC. 27.6*b right*: Courtesy of Peter Hirsch & Stuart Pankratz. 27.10: P. Gates/Biological Photo Service. 27.11*a*: Paul W. Johnson/Biological Photo Service. 27.11*b*: H. S. Pankratz/Biological Photo Service. 27.11*c*: Bill Kamin/Visuals Unlimited. 27.12:

Courtesy of David Cox/CDC. 27.13: Randall C. Cutlip. 27.14: T. J. Beveridge/Biological Photo Service. 27.15: Dr. Gary Gaugler/Visuals Unlimited. 27.16: D. M. Phillips/Visuals Unlimited. 27.17: M. Gabridge/Visuals Unlimited. 27.19: Krafft/Hoa-qui/Photo Researchers, Inc. 27.20: Martin G. Miller/Visuals Unlimited.

Chapter 28 *Opener*: London School of Hygiene/SPL/Photo Researchers, Inc. 28.1*a*: Dennis Kunkel Microscopy, Inc. 28.1*b*: J. Paulin/Visuals Unlimited. 28.1*c*: Randy Morse/Tom Stack & Assoc. 28.4: Mike Abbey/Visuals Unlimited. 28.7*a*: Christian Gautier/Jacana/Photo Researchers, Inc. 28.7*b*: David Patterson/SPL/Photo Researchers, Inc. 28.7*c*: James Solliday/Biological Photo Service. 28.8: David Patterson, Linda Amaral Zettler, Mike Peglar, & Tom Nerad/micro*scope. 28.12: Oliver Meckes/Photo Researchers, Inc. 28.13: Sanford Berry/Visuals Unlimited. 28.15*a*: Mike Abbey/Visuals Unlimited. 28.15*b*: Dennis Kunkel Microscopy, Inc. 28.15*c*: Paul W. Johnson/Biological Photo Service. 28.15*d*: M. Abbey/Photo Researchers, Inc. 28.18*a*: Manfred Kage/Peter Arnold, Inc. 28.18*b*: Biophoto Associates/Photo Researchers, Inc. 28.20*a*: Joyce Photographics/Photo Researchers, Inc. 28.20*b*: J. Robert Waaland/Biological Photo Service. 28.21*a*: Jeff Foott/Tom Stack & Assoc. 28.21*b*: J. N. A. Lott/Biological Photo Service. 28.23: James W. Richardson/Visuals Unlimited. 28.24*a*: Milton Rand/Tom Stack & Assoc. 28.24*b*: J. N. A. Lott/Biological Photo Service. 28.25*a*: Carolina Biological/Visuals Unlimited. 28.25*b*: Andrew J. Martinez/Photo Researchers, Inc. 28.28*a*: William Bourland/micro*scope. 28.28*b*: David Patterson & Aimlee Laderman/micro*scope. 28.30*a*: Andrew Syred/SPL/Photo Researchers, Inc. 28.30*b*: Eric Grave/SPL/Photo Researchers, Inc. 28.31*a*: Barbara J. Miller/Biological Photo Service. 28.31*b*: Carolina Biological/Visuals Unlimited. 28.32: Courtesy of R. Blanton & M. Grimson.

Chapter 29 *Opener*: Dr. Ray Clark & Mervyn De Calcina/SPL/Photo Researchers, Inc. 29.3*a*: Ron Dengler/Visuals Unlimited. 29.3*b*: Larry Mellichamp/Visuals Unlimited. 29.6: J. Robert Waaland/Biological Photo Service. 29.7*a*: Rod Planck/Dembinsky Photo Assoc. 29.7*b*: William Harlow/Photo Researchers, Inc. 29.7*c*: Science VU/Visuals Unlimited. 29.8: Daniel Vega/AGE Fotostock. 29.9*a*: Brian Enting/Photo Researchers, Inc. 29.9*b*: Courtesy of J. H. Troughton. 29.11: University of Michigan Exhibit Museum. 29.15*a*: Ed Reschke/Peter Arnold, Inc. 29.15*b*: Carolina Biological/Visuals Unlimited. 29.16*a*: J. N. A. Lott/Biological Photo Service. 29.16*b*: David Sieren/Visuals Unlimited. 29.17: W. Ormerod/Visuals Unlimited. 29.18*a*: Rod Planck/Dembinsky Photo Assoc. 29.18*b*: Nuridsany et Perennou/Photo Researchers, Inc. 29.18*c*: Dick Keen/Visuals Unlimited. 29.19: L. West/Photo Researchers, Inc.

Chapter 30 *Opener*: Warren Faidley/DRK Photo. 30.3: Phil Gates/Biological Photo Service. 30.4*a*: Roland Seitre/Peter Arnold, Inc. 30.4*b*: Bernd Wittich/Visuals Unlimited. 30.4*c*: M. Graybill & J. Hodder/Biological Photo Service. 30.4*d*: N. H. Cheatham/DRK Photo. 30.5*a left*: Michael P. Gadomski/Photo Researchers, Inc. 30.5*a right*: Stan W. Elems/Visuals Unlimited. 30.5*b left*: Gerald & Buff Corsi/Visuals Unlimited. 30.5*b right*: John D. Cunningham/Visuals Unlimited. 30.8*a*: Dick Poe/Visuals Unlimited. 30.8*b*: Richard Shiell. 30.8*c*: Richard Shiell/Dembinsky Photo Assoc. 30.9*a*: Sinclair Stammers/SPL/Photo Researchers, Inc. 30.9*b*: Rod Planck/Photo Researchers, Inc. 30.12*a*: Inga Spence/Tom Stack & Assoc. 30.12*b*: Holt Studios/Photo Researchers, Inc. 30.12*c*: Catherine M. Pringle/Biological Photo Service. 30.12*d*: Henry Beeker/AGE Fotostock. 30.14*a*: Courtesy of Stephen McCabe, U. California, Santa Cruz, & UCSC Arboretum. 30.14*b*: John Gerlach/Dembinsky Photo Assoc. 30.14*c*: Rob & Ann Simpson/Visuals Unlimited. 30.14*d*: R. C. Carpenter/Photo Researchers, Inc. 30.14*e*: Geoff Bryant/Photo Researchers, Inc. 30.14*f*: Chris Sharp/Photo Researchers, Inc. 30.15*a*: Ken Lucas/Visuals Unlimited. 30.15*b*: Ed Reschke/Peter Arnold, Inc. 30.15*c*: Adam Jones/Dembinsky Photo Assoc. 30.16*a*: Richard Shiell. 30.16*b*: Adam Jones/Dembinsky Photo Assoc. 30.16*c*: Alan & Linda Detrick/Photo Researchers, Inc.

Chapter 31 *Opener*: Dr. Gary Gaugler/Visuals Unlimited. 31.1*a*: Inga Spence/Tom Stack & Assoc. 31.1*b*: L. E. Gilbert/Biological Photo Service. 31.2: D. M. Phillips/Visuals Unlimited. 31.4: G. T. Cole/Biological Photo Service. 31.5: N. Allin & G. L. Barron/Biological Photo Service. 31.7: J. Robert Waaland/Biological Photo Service. 31.8: Biophoto Associates/Photo Researchers, Inc. 31.9: M. F. Brown/Visuals Unlimited. 31.10: John D. Cunningham/Visuals Unlimited. 31.11*a*: Richard Shiell/Dembinsky Photo Assoc. 31.11*b*: Matt Meadows/Peter Arnold, Inc. 31.12: Andrew Syred/SPL/Photo Researchers, Inc. 31.14*a*: Angelina Lax/Photo Researchers, Inc. 31.14*b*: Manfred Danegger/Photo Researchers, Inc. 31.14*c*: Stan Flegler/Visuals Unlimited. 31.15 *inset*: Biophoto Associates/Photo Researchers, Inc. 31.16*a*: R. L. Peterson/Biological Photo Service. 31.16*b*: Merton F. Brown/Visuals Unlimited. 31.17*a*: Ed Reschke/Peter Arnold, Inc. 31.17*b*: Gary Meszaros/Dembinsky Photo Assoc. 31.18*a*: J. N. A. Lott/Biological Photo Service.

Chapter 32 *Opener*: Doug Scott/AGE Fotopix. 32.5*a*: D. Fawcett/Visuals Unlimited. 32.5*b*: Paul Osmond/Painet Inc. 32.5*c*: Courtesy of Jean Vacelet. 32.6*a*: Larry Jon Friesen. 32.6*b*: Jett Britnell/DRK Photo. 32.6*c*: David J. Wrobel/Visuals Unlimited. 32.6*d*: Larry Jon Friesen. 32.7, 32.8: Adapted from F. M. Bayerand & H. B. Owre. 1968. *The Free-Living Lower Invertebrates*, Macmillan Publishing Co. 32.9*a*: David Hall/Photo Researchers, Inc. 32.9*b*: Ed Robinson/Tom Stack & Assoc. 32.10, 32.11: Adapted from F. M. Bayerand & H. B. Owre. 1968. *The Free-Living Lower Invertebrates*, Macmillan Publishing Co. 32.12*b*: Kathie Atkinson/Oxford Scientific Films. 32.13: M. W. Martin, from *Science* 288: 841. 32.15*a*: Denise Tackett/DRK Photo. 32.17*b*, 32.19*a*: Robert Brons/Biological Photo Service. 32.20: David J. Wrobel/Biological Photo Service. 32.21*b*: Oxford Scientific Films. 32.24*a*: Brian Parker/Tom Stack & Assoc. 32.24*b*: Roger K. Burnard/Biological Photo Service. 32.24*c*: Stanley Breeden/DRK Photo. 32.24*d*: Courtesy of R. R. Hessler, Scripps Institute of Oceanography. 32.26*a*: Jeff Foott/Nature Picture Library. 32.26*b*: Dave Fleetham/Tom Stack & Assoc. 32.26*c,d,e*: Larry Jon Friesen. 32.26*f*: Fred McConnaughey/Photo Researchers, Inc.

Chapter 33 *Opener*: Sharon Cummings/Dembinsky Photo Associates. 33.2: Courtesy of Jen Grenier & Sean Carroll, University of Wisconsin. 33.4: R. Calentine/Visuals Unlimited. 33.5*b,c*: James Solliday/Biological Photo Service. 33.6*a*: Michael Fogden/DRK Photo. 33.6*b*: Diane R. Nelson/Visuals Unlimited. 33.7: Ken Lucas/Visuals Unlimited. 33.9*a,b*: Larry Jon Friesen. 33.9*c*: Tom Branch/Photo Researchers, Inc. 33.9*d*: A. Flowers & L. Newman/Photo Researchers, Inc. 33.12: Oxford Scientific Films. 33.13*a*: Sinclair Stammers/Nature Picture Library. 33.13*b*: Larry Jon Friesen. 33.13*c*: David Maitland/Masterfile. 33.13*d*: Larry Jon Friesen. 33.13*e*: Colin Milkins/Oxford Scientific Films. 33.13*f*: Peter J. Bryant/Biological Photo Service. 33.13*g*: Courtesy of Scott Bauer/USDA ARS. 33.13*h*: L. West/Photo Researchers, Inc. 33.15*a*: Marty Cordano/DRK Photo. 33.15*b*: William Leonard/DRK Photo. 33.16*a*: Joel Simon. 33.16*b*: Fred Bruemmer/DRK Photo. 33.17*a,b*: Larry Jon Friesen. 33.12*c*: W. M. Beatty/Visuals Unlimited. 33.12*d*: Photo by Eric Erbe; colorization by Chris Pooley/USDA ARS.

Chapter 34 *Opener*: Michael Fogden/DRK Photo. 34.1: From Bengtson, S. 2000. Teasing fossils out of shales with cameras and computers. *Palaeontologia Electronica* 3(1). 34.4*a*: Hal Beral/Visuals Unlimited. 34.4*b*: Randy Morse/Tom Stack & Assoc. 34.4*c*: Mark J. Thomas/Dembinsky Photo Assoc. 34.4*d*: Randy Morse/Tom Stack & Assoc. 34.4*e*: Larry Jon Friesen. 34.5*a*: C. R. Wyttenbach/Biological Photo Service. 34.6*a*: Denise Tackett/DRK Photo. 34.6*b*: David Wrobel/Visuals Unlimited. 34.7*b*: Robert Brons/Biological Photo Service. 34.10*a*: Brian Parker/Tom Stack & Assoc. 34.10*b*: Gary Milburn/Tom Stack & Assoc. 34.12*a*: Dave Fleetham/Tom Stack & Assoc. 34.12*b*: Marty Snyderman/Masterfile. 34.12*c*: Dave Fleetham/Tom Stack & Assoc. 34.13*a*: Tobias Bernhard/Oxford Scientific Films. 34.13*b*: Fred Bavendam/Minden Pictures. 34.13*c*: Dave Fleetham/Visuals Unlimited. 34.13*d*: Dr. Paul A. Zahl/Photo Researchers, Inc. 34.14: Tom McHugh, Steinhart Aquarium/Photo Researchers, Inc. 34.15*a*: Ken Lucas/Biological Photo Service. 34.15*b*: Gary Meszaros/Dembinsky Photo Assoc. 34.15*c*: Nick Garbutt/Indri Images. 34.19*a*: Dave B. Fleetham/Tom Stack & Assoc. 34.19*b*: C. Alan Morgan/Peter Arnold, Inc. 34.19*c*: Gerry Ellis, DigitalVision/PictureQuest. 34.19*d*: Michael Fogden/DRK Photo. 34.19*e*: Mark J. Thomas/Dembinsky Photo Assoc. 34.20*a*: From Xu, X., et al., 2003.

Nature 421: 335. Macmillan Publishers Ltd. 34.20*b*: Tom & Therisa Stack/Tom Stack & Assoc. 34.20*c*: Fossil from the Natural History Museum of Basel, photographed by Severino Dahint. 34.21*a*: Joe McDonald/Tom Stack & Assoc. 34.21*b*: Skip Moody/Dembinsky Photo Assoc. 34.21*c*: John Shaw/Tom Stack & Assoc. 34.21*d*: Fred Bruemmer/DRK Photo. 34.22*a*: Ed Kanze/Dembinsky Photo Assoc. 34.22*b*: Dave Watts/Tom Stack & Assoc. 34.23*a*: Art Wolfe. 34.23*b*: Jany Sauvanet/Photo Researchers, Inc. 34.23*c*: Hans & Judy Beste/Animals Animals. 34.24*a*: Rod Planck/Dembinsky Photo Assoc. 34.24*b*: Joe McDonald/Tom Stack & Assoc. 34.24*c*: Michael S. Nolan/Tom Stack & Assoc. 34.24*d*: Erwin & Peggy Bauer/Tom Stack & Assoc. 34.26*a*: Martin Harvey/DRK Photo. 34.26*b*: Stanley Breeden/DRK Photo. 34.27*a*: Steve Kaufman/DRK Photo. 34.27*b*: John Bracegirdle/Masterfile. 34.28*a*: Stan Osolinsky/Dembinsky Photo Assoc. 34.28*b*: Anup Shah/Dembinsky Photo Assoc. 34.28*c*: Art Wolfe. 34.28*d*: Anup Shah/Dembinsky Photo Assoc. 34.30: Dembinsky Photo Assoc.

Chapter 35 *Opener*: D. Cavagnaro/Visuals Unlimited. 35.3*a*: Antonia Reeve/SPL/Photo Researchers, Inc. 35.3*b*: R. Calentine/Visuals Unlimited. 35.4*a*: Joyce Photographics/Photo Researchers, Inc. 35.4*b*: Renee Lynn/Photo Researchers, Inc. 35.4*c*: C. K. Lorenz/Photo Researchers, Inc. 35.7: Biophoto Associates/Photo Researchers, Inc. 35.9*a*: Biodisc/Visuals Unlimited. 35.9*b*: P. Gates /Biological Photo Service. 35.9*c*: Biophoto Associates/Photo Researchers, Inc. 35.9*d*: Jack M. Bostrack/Visuals Unlimited. 35.9*e*: John D. Cunningham/Visuals Unlimited. 35.9*f*: J. Robert Waaland/Biological Photo Service. 35.11, 35.13: J. Robert Waaland/Biological Photo Service. 35.16*a*: Jim Solliday/Biological Photo Service. 35.16*b*: Microfield Scientific LTD/Photo Researchers, Inc. 35.16*c*: Ray F. Evert, U. Wisconsin, Madison. 35.16*d*: John D. Cunningham/Visuals Unlimited. 35.18*a,b left*: Carolina Biological/Visuals Unlimited. 35.18*a,b right*: J. Robert Waaland/Biological Photo Service. 35.20: John N. A. Lott/Biological Photo Service. 35.21: Jim Solliday/Biological Photo Service. 35.22: P. Gates/Biological Photo Service. 35.23*b*: Courtesy of Thomas Eisner, Cornell U. 35.23*c*: C. G. Van Dyke/Visuals Unlimited.

Chapter 36 *Opener*: Bettman/Corbis. 36.6: From B. Bentwood & J. Cronshaw. 1978. *Planta* 140: 111. 36.7: Ed Reschke/Peter Arnold, Inc. 36.10: After M. A. Zwieniecki et al., 2001. *Science* 291: 1059–1061. 36.11*a*: D. M. Phillips/Visuals Unlimited. 36.13: M. H. Zimmermann.

Chapter 37 *Opener*: Courtesy of Emerson D. Nafziger. 37.3: Kathleen Blanchard/Visuals Unlimited. 37.5: Hugh Spencer/Photo Researchers, Inc. 37.7: E. H. Newcomb & S. R. Tandon/Biological Photo Service. 37.9*a*: J. H. Robinson/Photo Researchers, Inc. 37.9*b*: Milton Rand/Tom Stack & Assoc. 37.10: Gilbert S. Grant/Photo Researchers, Inc.

Chapter 38 *Opener*: Jeremy Woodhouse/DRK Photo. 38.2: Tom J. Ulrich/Visuals Unlimited. 38.3: John Eastcott, Yva Momatiuk/DRK Photo. 38.5: Courtesy of J. A. D. Zeevaart, Michigan State U. 38.11: Ed Reschke/Peter Arnold, Inc. 38.13: Biophoto Associates/Photo Researchers, Inc. 38.16: T. A. Wiewandt/DRK Photo. 38.19: Courtesy of Dr. Eva Huala, Carnegie Institution of Washington.

Chapter 39 *Opener*: RMF/Visuals Unlimited. 39.2: Stephen Dalton/Photo Researchers, Inc. 39.4: From Bowman, J. (ed.), 1994. *Arabiopsis: An Atlas of Morphology and Development.* Springer-Verlag, New York. Photo by S. Craig & A. Chaudhury, Plate 6.2. 39.8*a*: C. P. George/Visuals Unlimited. 39.8*b*: Tess & David Young/Tom Stack & Assoc. 39.14: After M. J. Yanovsky and S. A. Kay, 2002. *Nature* 419: 308–312. 39.16*a*: Nigel Cattlin, Holt Studios International/Photo Researchers, Inc. 39.16*b*: Jerome Wexler/Photo Researchers, Inc.

Chapter 40 *Opener*: Peter J. Bryant/Biological Photo Service. 40.2: D. Cavagnaro/Visuals Unlimited. 40.6: Courtesy of Thomas Eisner, Cornell U. 40.7: Adam Jones/Dembinsky Photo Assoc. 40.8: J. N. A. Lott/Biological Photo Service. 40.9: Richard Shiell. 40.10: Janine Pestel/Visuals Unlimited. 40.11: Chip Isenhart/Tom Stack & Assoc. 40.12: J. N. A. Lott/Biological Photo Service. 40.13: Robert & Linda Mitchell. 40.15: Budd Titlow/Visuals Unlimited.

Chapter 41 *Opener*: AP/Wide World Photos. 41.3*a*: CNRI/SPL/Photo Researchers, Inc. 41.3*b*: G. W. Willis/Visuals Unlimited. 41.9*a*: B. & C. Alexander/Photo Researchers, Inc. 41.9*b*: Ann & Steve Toon/NHPA. 41.11: Auscape (Parer-Cook)/Peter Arnold, Inc. 41.15: G. W. Willis/Biological Photo Service. 41.16*a*: Peter Chadwick/SPL/Photo Researchers, Inc. 41.16*b*: Jim Roetzel/Dembinsky Photo Assoc.

Chapter 42 *Opener*: Courtesy of R. D. Fernald, Stanford U. 42.6: Schwartzwald Lawrence/Corbis Sygma. 42.13*b*: Courtesy of Gerhard Heldmaier, Philipps University.

Chapter 43 *Opener*: Nik Wheeler. 43.1*a*: Biophoto Associates/Photo Researchers, Inc. 43.1*b*: Brian Parker/Tom Stack & Assoc. 43.2*a*: Patricia J. Wynne. 43.5: CNRI/SPL/Photo Researchers, Inc. 43.7*a*: Mitsuaki Iwago/Minden Pictures. 43.7*b*: Johnny Johnson/DRK Photo. 43.12: P. Bagavandoss/Photo Researchers, Inc. 43.17: Courtesy of The Institute for Reproductive Medicine and Science of Saint Barnabas, New Jersey.

Chapter 44 *Opener*: Dietmar Nill/Nature Picture Library. 44.3*b*: C. Raines/Visuals Unlimited.

Chapter 45 *Opener*: Courtesy of Grace Sours, ATF. 45.3 *left*: Courtesy of R. A. Steinbrecht. 45.3 *right*: G. I. Bernard/Animals Animals. 45.5, 45.10: P. Motta/Photo Researchers, Inc. 45.15*a*: Dennis Kunkel Microscopy, Inc. 45.18: Omikron/Science Source/Photo Researchers, Inc.

Chapter 46 *Opener*: From van Praag et al. 2002. *Nature* 415: 1030. Macmillan Publishers Ltd. 46.8: From Harlow, J. M., 1869. *Recovery from the passage of an iron bar through the head.* Boston: David Clapp & Son. 46.13*a*: David Joel Photography, Inc. 46.16: Wellcome Dept. of Cognitive Neurology/SPL/Photo Researchers, Inc.

Chapter 47 *Opener*: Mark Andersen/RubberBall Productions/PictureQuest. 47.1 *upper*: Innerspace Imaging/SPL/Photo Researchers, Inc. 47.1 *center*: SPL/Photo Researchers, Inc. 47.1 *lower*: Eric Grave/SPL/Photo Researchers, Inc. 47.3: Frank A. Pepe/Biological Photo Service. 47.8*a*: Courtesy of Jesper L. Andersen. 47.15: Robert Brons/Biological Photo Service. 47.19*b upper*: Ken Lucas/Visuals Unlimited. 47.19*b lower*: Fred McConnaughey/Photo Researchers, Inc.

Chapter 48 *Opener*: John Warden/Alaskan Express/PictureQuest. 48.1*a*: Larry Jon Friesen. 48.1*b*: Robert Brons/Biological Photo Service. 48.1*c*: Tom McHugh/Photo Researchers, Inc. 48.4*b*: Skip Moody/Dembinsky Photo Assoc. 48.4*c*: Courtesy of Thomas Eisner, Cornell U. 48.10*a*: SPL/Photo Researchers, Inc. 48.10*c*: P. Motta/Photo Researchers, Inc. 48.13: Fred Bruemmer/DRK Photo.

Chapter 49 *Opener*: Doc White/Nature Picture Library. 49.11: Dennis Kunkel Microscopy, Inc. 49.14*a*: Chuck Brown/Science Source/Photo Researchers, Inc. 49.14*b*: Biophoto Associates/Science Source/Photo Researchers, Inc. 49.15: After N. Campbell, 1990. *Biology*, 2nd Ed., Benjamin Cummings Publishing Co. 49.16*b*: CNRI/Photo Researchers, Inc.

Chapter 50 *Opener*: Marilyn "Angel" Wynn/Nativestock.com. 50.1*a*: Gerry Ellis, DigitalVision/PictureQuest. 50.1*b*: Tom Walker/Visuals Unlimited. 50.3: AP/Wide World Photos. 50.6: David Roberts/Nature's Images/Photo Researchers, Inc. 50.9: Dennis Kunkel Microscopy, Inc. 50.21: Jackson/Visuals Unlimited. 50.22: Katsutoshi Ito/Nature Productions.

Chapter 51 *Opener*: Michael Fogden/DRK Photo. 51.1*a*: Brian Kenney. 51.2*b*: Rod Planck/Photo Researchers, Inc. 51.8: From R. G. Kessel & R. H. Kardon. 1979. *Tissues and Organs.* W. H. Freeman, San Francisco. 51.11: Courtesy of Lise Bankir, INSERM Unit, Hôpital Necker, Paris.

Chapter 52 *Opener*: Frans de Waal, Emory U. 52.1: Bill Beatty/Visuals Unlimited. 52.3: Courtesy of Marc Chappell, U. California, Riverside. 52.5: Nina Leen/TimePix. 52.11: François Savigny/Animals Animals. 52.16: Fritz Pölking/Dembinsky Photo Assoc.

Chapter 53 *Opener*: Arturo M. Enríquez/Painet, Inc. 53.1*a*: Francis Caldwell/Painet, Inc. 53.2: Anup Shah/DRK Photo. 53.6: Pete Oxford/DRK Photo. 53.8: Courtesy of John Alcock, Arizona State U. 53.9: Courtesy of Arild Johnsen, University of Oslo. 53.11: Art Wolfe. 53.12: David Houston/Bruce Coleman, Inc. 53.13: Courtesy of John Alcock, Arizona State U. 53.14: José Fuste Raga/AGE Fotostock.

Chapter 54 *Opener*: T. W. Davies/California Academy of Sciences. 54.3: Stephen G. Maka/DRK Photo. 54.4: After P. A. Marquet, 2000. *Science* 289: 1487–1488. 54.5: Ed Reschke/Peter Arnold, Inc. 54.6: François Gohier/Photo Researchers, Inc. 54.9: From T. Coulson et al., 2002. *Science* 295: 2023–2024. 54.11: Tom Brakefield/DRK Photo. 54.13: T. W. Davies/California Academy of Sciences. 54.15: Adam Jones/Photo Researchers, Inc. 54.19: Courtesy of John R. Hosking, NSW Agriculture, Australia.

Chapter 55 *Opener*: Bob Gibbons/Oxford Scientific Films. 55.1: Data from R. H. Whittaker, 1960. *Ecological Monographs* 30: 279–338. 55.10: Vitit Kantabutra/Painet, Inc. 55.11: Jim Zipp/Photo Researchers, Inc. 55.12: Tom Bean/DRK Photo. 55.15: After M. Begon, J. Harper, & C. Townsend. 1986. *Ecology*. Blackwell Scientific Publications. 55.16*a left*: Bill Lea/Dembinsky Photo Associates. 55.16*a right*: Stephen J. Krasemann/Photo Researchers, Inc. 55.16*b left*: M. P. Kahl/DRK Photo. 55.16*b center*: Kenneth W. Fink/Photo Researchers, Inc. 55.16*b right*: Jeff Foott/DRK Photo.

Chapter 56 *Opener*: Peter Mead/Tom Stack & Assoc. 56.2 *chameleon*: Joe McDonald/Tom Stack & Assoc. 56.2 *tenrec*: Nigel J. Dennis/Photo Researchers, Inc. 56.2 *fossa*: Pete Oxford/Nature Picture Library. 56.2 *lemur*: Wendy Dennis/Dembinsky Photo Associates. 56.2 *baobab*: Pete Oxford/Nature Picture Library. 56.2 *Pachypodium*: Michael Leach/Oxford Scientific Films. *Tundra, upper*: Tom & Pat Leeson/DRK Photo. *Tundra, lower*: Elizabeth N. Orians. *Boreal, upper*: Carr Clifton/Minden Pictures. *Boreal, lower*: James P. Rowan/DRK Photo. *Temperate deciduous*: Paul W. Johnson/Biological Photo Service. *Temperate grasslands, upper*: Robert & Jean Pollock/Biological Photo Service. *Temperate grasslands, lower*: Elizabeth N. Orians. *Cold desert, upper*: Edward Ely/Biological Photo Service. *Cold desert, lower*: Art Wolfe. *Hot desert, left*: Terry Donnelly/Tom Stack & Assoc. *Hot desert, right*: Dave Watts/Tom Stack & Assoc. *Chaparral*: Elizabeth N. Orians. *Thorn forest*: Nick Garbutt/Indri Images. *Savanna*: Tim Davis/Photo Researchers, Inc. *Tropical deciduous*: Courtesy of Donald L. Stone. *Tropical evergreen*: Elizabeth N. Orians.

Chapter 57 *Opener*: Andrew J. Martinez/Photo Researchers, Inc. 57.1*a*: Elizabeth N. Orians. 57.4: Richard Bierregaard, Courtesy of the Smithsonian Institution, Office of Environmental Awareness. 57.5: Adapted from S. H. Reichard and C. W. Hamilton, 1997. *Conservation Biology* 11: 193–203. 57.6*a*: Courtesy of Christopher Baisan and the Laboratory of Tree-Ring Research, U. Arizona, Tucson. 57.6*b*: Karen Wattenmaker/Painet, Inc. 57.7 *photo*: Stan Osolinsky/Oxford Scientific Films. 57.7*b*: After *BioScience*, 2002. 52: 809. 57.8*a*: R. B. Aronson et al., 2000. *Nature* 405: 36. 57.8*b*: Paul Osmond/Painet, Inc. 57.10: Tom Vezo/Nature Picture Library. 57.11: Julian Pender Hume/NHMPL.

Chapter 58 *Opener*: John Wong/RubberBall Productions/PictureQuest. 58.1: Adapted from M. C. Jacobson et al., 2000. Introduction to *Earth System Science: From Biogeochemical Cycles to Global Climate Change*, Academic Press. 58.5*a*: Corbis Images/PictureQuest. 58.8: Adapted from B. A. Hungate et al., 1997. *Nature* 388: 576–578. 58.15: Courtesy of NASA.

Index

Numbers in **boldface** indicate a definition of the term; numbers in *italic* indicate there is information in a figure, figure caption, or table.

About the Book

Editor: Andrew D. Sinauer
Project Editor: Carol J. Wigg
Developmental Editor: James Funston
Review coordinator: Susan McGlew
Copy Editor: Norma Roche
Production Manager: Christopher Small
Book Layout and Production: Janice Holabird, Jefferson Johnson, and Joan Gemme
Art Editing and Illustration Program: Elizabeth Morales
Book and Cover Design: Jefferson Johnson
Photo Research: David McIntyre
Index: Acorn Indexing
Color Separations: Burt Russell Litho
Book and Cover Manufacture: Courier Companies, Inc.

ICONS *(Continued from inside front cover)*

E indicates an Experiment Tutorial. * indicates an activity that applies to an entire chapter.